Fourier Methods in Imaging

Wiley-IS&T Series in Imaging Science and Technology

Series Editor:
Michael A. Kriss

Consultant Editors:
Anthony C. Lowe
Lindsay W. MacDonald
Yoichi Miyake

Reproduction of Colour (6th Edition)
R. W. G. Hunt

Colour Appearance Models (2nd Edition)
Mark D. Fairchild

Colorimetry: Fundamentals and Applications
Noburu Ohta and Alan R. Robertson

Color Constancy
Marc Ebner

Color Gamut Mapping
Ján Morovič

Panoramic Imaging: Sensor-Line Cameras and Laser Range-Finders
Fay Huang, Reinhard Klette and Karsten Scheibe

Digital Color Management (2nd Edition)
Edward J. Giorgianni and Thomas E. Madden

The JPEG 2000 Suite
Peter Schelkens, Athanassios Skodras and Touradj Ebrahimi (Eds.)

Color Management: Understanding and Using ICC Profiles
Phil Green (Ed.)

Fourier Methods in Imaging
Roger L. Easton, Jr.

Published in Association with the Society for
Imaging Science and Technology

IS&T
imaging.org

Fourier Methods in Imaging

Roger L. Easton, Jr.

Chester F. Carlson Center for Imaging Science
Rochester Institute of Technology
Rochester NY, USA

WILEY

A John Wiley and Sons, Ltd, Publication

This edition first published 2010

Registered office
John Wiley & Sons Ltd, The Atrium, Southern Gate, Chichester, West Sussex, PO19 8SQ, United Kingdom

For details of our global editorial offices, for customer services and for information about how to apply for permission to reuse the copyright material in this book please see our website at www.wiley.com.

Library of Congress Cataloging-in-Publication Data

Easton, Roger L. Jr.
Fourier methods in imaging / Roger L. Easton, Jr.
p. cm.
Includes bibliographical references and index.
ISBN 978-0-470-68983-7 (cloth)
1. Image processing–Mathematics. 2. Fourier analysis. I. Title.
TA1637.E23 2010
621.36'701515723–dc22

2010000343

A catalogue record for this book is available from the British Library.

ISBN: 9780470689837

Set in 9/11pt Times by Sunrise Setting Ltd, Torquay, UK.
Printed in Singapore by Markono Print Media Pte Ltd.

To my parents and my students

Contents

Series Editor's Preface xix

Preface xxiii

1 Introduction 1
- 1.1 Signals, Operators, and Imaging Systems . . . 1
  - 1.1.1 The Imaging Chain . . . 1
- 1.2 The Three Imaging Tasks . . . 3
- 1.3 Examples of Optical Imaging . . . 4
  - 1.3.1 Ray Optics . . . 4
  - 1.3.2 Wave Optics . . . 5
  - 1.3.3 System Evaluation of Hubble Space Telescope . . . 7
  - 1.3.4 Imaging by Ground-Based Telescopes . . . 8
- 1.4 Imaging Tasks in Medical Imaging . . . 8
  - 1.4.1 Gamma-Ray Imaging . . . 9
  - 1.4.2 Radiography . . . 11
  - 1.4.3 Computed Tomographic Radiography . . . 13

2 Operators and Functions 15
- 2.1 Classes of Imaging Operators . . . 15
  - 2.1.1 Linearity . . . 15
  - 2.1.2 Shift Invariance . . . 16
- 2.2 Continuous and Discrete Functions . . . 16
  - 2.2.1 Functions . . . 16
  - 2.2.2 Functions with Continuous and Discrete Domains . . . 17
  - 2.2.3 Continuous and Discrete Ranges . . . 19
  - 2.2.4 Discrete Domain and Range – "Digitized" Functions . . . 20
  - 2.2.5 Periodic, Aperiodic, and Harmonic Functions . . . 21
  - 2.2.6 Symmetry Properties of Functions . . . 26
  - Problems . . . 27

3 Vectors with Real-Valued Components 29
- 3.1 Scalar Products . . . 29
  - 3.1.1 Scalar Product of Distinct Vectors . . . 32
  - 3.1.2 Projection of One Vector onto Another . . . 34
- 3.2 Matrices . . . 34
  - 3.2.1 Simultaneous Evaluation of Multiple Scalar Products . . . 34
  - 3.2.2 Matrix–Matrix Multiplication . . . 36
  - 3.2.3 Square and Diagonal Matrices, Identity Matrix . . . 37

3.2.4 Matrix Transposes 38
3.2.5 Matrix Inverses 39
3.3 Vector Spaces 41
3.3.1 Basis Vectors 43
3.3.2 Vector Subspaces Associated with a System 44
Problems 48

4 Complex Numbers and Functions **51**
4.1 Arithmetic of Complex Numbers 52
4.1.1 Equality of Two Complex Numbers 52
4.1.2 Sum and Difference of Two Complex Numbers 52
4.1.3 Product of Two Complex Numbers 53
4.1.4 Reciprocal of a Complex Number 53
4.1.5 Ratio of Two Complex Numbers 53
4.2 Graphical Representation of Complex Numbers 53
4.3 Complex Functions 56
4.4 Generalized Spatial Frequency – Negative Frequencies 62
4.5 Argand Diagrams of Complex-Valued Functions 62
Problems 63

5 Complex-Valued Matrices and Systems **65**
5.1 Vectors with Complex-Valued Components 65
5.1.1 Inner Product 66
5.1.2 Products of Complex-Valued Matrices and Vectors 67
5.2 Matrix Analogues of Shift-Invariant Systems 67
5.2.1 Eigenvectors and Eigenvalues 70
5.2.2 Projections onto Eigenvectors 72
5.2.3 Diagonalization of a Circulant Matrix 75
5.2.4 Matrix Operators for Shift-Invariant Systems 80
5.2.5 Alternative Ordering of Eigenvectors 84
5.3 Matrix Formulation of Imaging Tasks 84
5.3.1 Inverse Imaging Problem 84
5.3.2 Solution of Inverse Problems via Diagonalization 86
5.3.3 Matrix–Vector Formulation of System Analysis 87
5.4 Continuous Analogues of Vector Operations 88
5.4.1 Inner Product of Continuous Functions 88
5.4.2 Complete Sets of Basis Functions 91
5.4.3 Orthonormal Basis Functions 92
5.4.4 Continuous Analogue of DFT 93
5.4.5 Eigenfunctions of Continuous Operators 93
Problems 94

6 1-D Special Functions **97**
6.1 Definitions of 1-D Special Functions 98
6.1.1 Constant Function 99
6.1.2 Rectangle Function 99
6.1.3 Triangle Function 101
6.1.4 Signum Function 101
6.1.5 Step Function 102
6.1.6 Exponential Function 104
6.1.7 Sinusoid 105
6.1.8 *SINC* Function 109

6.1.9 $SINC^2$ Function . . . 111
6.1.10 Gamma Function . . . 112
6.1.11 Quadratic-Phase Sinusoid – "Chirp" Function . . . 115
6.1.12 Gaussian Function . . . 117
6.1.13 "SuperGaussian" Function . . . 119
6.1.14 Bessel Functions . . . 121
6.1.15 Lorentzian Function . . . 124
6.1.16 Thresholded Functions . . . 125
6.2 1-D Dirac Delta Function . . . 126
6.2.1 1-D Dirac Delta Function Raised to a Power . . . 131
6.2.2 Sifting Property of 1-D Dirac Delta Function . . . 132
6.2.3 Symmetric (Even) Pair of 1-D Dirac Delta Functions . . . 133
6.2.4 Antisymmetric (Odd) Pair of 1-D Dirac Delta Functions . . . 134
6.2.5 *COMB* Function . . . 135
6.2.6 Derivatives of 1-D Dirac Delta Function . . . 137
6.2.7 Dirac Delta Function with Functional Argument . . . 139
6.3 1-D Complex-Valued Special Functions . . . 142
6.3.1 Complex Linear-Phase Sinusoid . . . 143
6.3.2 Complex Quadratic-Phase Exponential Function . . . 143
6.3.3 "Superchirp" Function . . . 145
6.3.4 Complex-Valued Lorentzian Function . . . 147
6.3.5 Logarithm of the Complex Amplitude . . . 149
6.4 1-D Stochastic Functions – Noise . . . 149
6.4.1 Moments of Probability Distributions . . . 151
6.4.2 Discrete Probability Laws . . . 152
6.4.3 Continuous Probability Distributions . . . 156
6.4.4 Signal-to-Noise Ratio . . . 160
6.4.5 Example: Variance of a Sinusoid . . . 161
6.4.6 Example: Variance of a Square Wave . . . 161
6.4.7 Approximations to *SNR* . . . 161
6.5 Appendix A: Area of $SINC[x]$ and $SINC^2[x]$ . . . 162
6.6 Appendix B: Series Solutions for Bessel Functions $J_0[x]$ and $J_1[x]$ . . . 166
Problems . . . 169

7 2-D Special Functions 171
7.1 2-D Separable Functions . . . 171
7.1.1 Rotations of 2-D Separable Functions . . . 172
7.1.2 Rotated Coordinates as Scalar Products . . . 172
7.2 Definitions of 2-D Special Functions . . . 174
7.2.1 2-D Constant Function . . . 174
7.2.2 Rectangle Function . . . 175
7.2.3 Triangle Function . . . 176
7.2.4 2-D Signum and STEP Functions . . . 176
7.2.5 2-D *SINC* Function . . . 178
7.2.6 $SINC^2$ Function . . . 178
7.2.7 2-D Gaussian Function . . . 180
7.2.8 2-D Sinusoid . . . 180
7.3 2-D Dirac Delta Function and its Relatives . . . 182
7.3.1 2-D Dirac Delta Function in Cartesian Coordinates . . . 183
7.3.2 2-D Dirac Delta Function in Polar Coordinates . . . 184
7.3.3 2-D Separable *COMB* Function . . . 186

7.3.4 2-D Line Delta Function . . . 187
7.3.5 2-D "Cross" Function . . . 194
7.3.6 "Corral" Function . . . 195
7.4 2-D Functions with Circular Symmetry . . . 195
7.4.1 Cylinder (Circle) Function . . . 196
7.4.2 Circularly Symmetric Gaussian Function . . . 197
7.4.3 Circularly Symmetric Bessel Function of Zero Order . . . 197
7.4.4 Besinc or Sombrero Function . . . 200
7.4.5 Circular Triangle Function . . . 201
7.4.6 Ring Delta Function . . . 202
7.5 Complex-Valued 2-D Functions . . . 204
7.5.1 Complex 2-D Sinusoid . . . 204
7.5.2 Complex Quadratic-Phase Sinusoid . . . 205
7.6 Special Functions of Three (or More) Variables . . . 205
Problems . . . 206

8 Linear Operators **207**
8.1 Linear Operators . . . 208
8.2 Shift-Invariant Operators . . . 213
8.3 Linear Shift-Invariant (LSI) Operators . . . 216
8.3.1 Linear Shift-Variant Operators . . . 221
8.4 Calculating Convolutions . . . 222
8.4.1 Examples of Convolutions . . . 223
8.5 Properties of Convolutions . . . 223
8.5.1 Region of Support of Convolutions . . . 225
8.5.2 Area of a Convolution . . . 225
8.5.3 Convolution of Scaled Functions . . . 226
8.6 Autocorrelation . . . 226
8.6.1 Autocorrelation of Stochastic Functions . . . 228
8.6.2 Autocovariance of Stochastic Functions . . . 229
8.7 Crosscorrelation . . . 229
8.8 2-D LSI Operations . . . 232
8.8.1 Line-Spread and Edge-Spread Functions . . . 233
8.9 Crosscorrelations of 2-D Functions . . . 234
8.10 Autocorrelations of 2-D Functions . . . 235
8.10.1 Autocorrelation of the Cylinder Function . . . 236
Problems . . . 236

9 Fourier Transforms of 1-D Functions **239**
9.1 Transforms of Continuous-Domain Functions . . . 239
9.1.1 Example 1: Input and Reference Functions are Even Sinusoids . . . 242
9.1.2 Example 2: Even Sinusoid Input, Odd Sinusoid Reference . . . 245
9.1.3 Example 3: Odd Sinusoid Input, Even Sinusoid Reference . . . 246
9.1.4 Example 4: Odd Sinusoid Input and Reference . . . 247
9.2 Linear Combinations of Reference Functions . . . 250
9.2.1 Hartley Transform . . . 251
9.2.2 Examples of the Hartley Transform . . . 251
9.2.3 Inverse of the Hartley Transform . . . 252
9.3 Complex-Valued Reference Functions . . . 254
9.4 Transforms of Complex-Valued Functions . . . 256
9.5 Fourier Analysis of Dirac Delta Functions . . . 259
9.6 Inverse Fourier Transform . . . 261

9.7 Fourier Transforms of 1-D Special Functions . . . 263
9.7.1 Fourier Transform of $\delta[x]$ . . . 264
9.7.2 Fourier Transform of Rectangle . . . 264
9.7.3 Fourier Transforms of Sinusoids . . . 266
9.7.4 Fourier Transform of Signum and Step . . . 268
9.7.5 Fourier Transform of Exponential . . . 270
9.7.6 Fourier Transform of Gaussian . . . 275
9.7.7 Fourier Transforms of Chirp Functions . . . 276
9.7.8 Fourier Transform of *COMB* Function . . . 279
9.8 Theorems of the Fourier Transform . . . 280
9.8.1 Multiplication by Constant . . . 281
9.8.2 Addition Theorem (Linearity) . . . 281
9.8.3 Fourier Transform of a Fourier Transform . . . 281
9.8.4 Central-Ordinate Theorem . . . 284
9.8.5 Scaling Theorem . . . 284
9.8.6 Shift Theorem . . . 287
9.8.7 Filter Theorem . . . 289
9.8.8 Modulation Theorem . . . 295
9.8.9 Derivative Theorem . . . 297
9.8.10 Fourier Transform of Complex Conjugate . . . 298
9.8.11 Fourier Transform of Crosscorrelation . . . 299
9.8.12 Fourier Transform of Autocorrelation . . . 302
9.8.13 Rayleigh's Theorem . . . 302
9.8.14 Parseval's Theorem . . . 304
9.8.15 Fourier Transform of Periodic Function . . . 306
9.8.16 Spectrum of Sampled Function . . . 307
9.8.17 Spectrum of Discrete Periodic Function . . . 308
9.8.18 Spectra of Stochastic Signals . . . 308
9.8.19 Effect of Nonlinear Operations of Spectra . . . 310
9.9 Appendix: Spectrum of Gaussian via Path Integral . . . 320
Problems . . . 321

10 Multidimensional Fourier Transforms **325**
10.1 2-D Fourier Transforms . . . 325
10.1.1 2-D Fourier Synthesis . . . 326
10.2 Spectra of Separable 2-D Functions . . . 327
10.2.1 Fourier Transforms of Separable Functions . . . 328
10.2.2 Fourier Transform of $\delta[x, y]$ . . . 328
10.2.3 Fourier Transform of $\delta[x - x_0, y - y_0]$ . . . 330
10.2.4 Fourier Transform of $RECT[x, y]$ . . . 332
10.2.5 Fourier Transform of $TRI[x, y]$ . . . 332
10.2.6 Fourier Transform of $GAUS[x, y]$ . . . 332
10.2.7 Fourier Transform of $STEP[x] \cdot STEP[y]$ . . . 334
10.2.8 Theorems of Spectra of Separable Functions . . . 334
10.2.9 Superpositions of 2-D Separable Functions . . . 335
10.3 Theorems of 2-D Fourier Transforms . . . 335
10.3.1 2-D "Transform-of-a-Transform" Theorem . . . 336
10.3.2 2-D Scaling Theorem . . . 336
10.3.3 2-D Shift Theorem . . . 336
10.3.4 2-D Filter Theorem . . . 337
10.3.5 2-D Derivative Theorem . . . 338

10.3.6 Spectra of Rotated 2-D Functions . . . 340
10.3.7 Transforms of 2-D Line Delta and Cross Functions . . . 341
Problems . . . 345

11 Spectra of Circular Functions **347**
11.1 The Hankel Transform . . . 347
11.1.1 Hankel Transform of Dirac Delta Function . . . 351
11.2 Inverse Hankel Transform . . . 353
11.3 Theorems of Hankel Transforms . . . 354
11.3.1 Scaling Theorem . . . 354
11.3.2 Shift Theorem . . . 354
11.3.3 Central-Ordinate Theorem . . . 354
11.3.4 Filter and Crosscorrelation Theorems . . . 355
11.3.5 "Transform-of-a-Transform" Theorem . . . 355
11.3.6 Derivative Theorem . . . 355
11.3.7 Laplacian of Circularly Symmetric Function . . . 356
11.4 Hankel Transforms of Special Functions . . . 356
11.4.1 Hankel Transform of $J_0(2\pi r\rho_0)$ . . . 356
11.4.2 Hankel Transform of $CYL(r)$ . . . 358
11.4.3 Hankel Transform of r^{-1} . . . 360
11.4.4 Hankel Transforms from 2-D Fourier Transforms . . . 361
11.4.5 Hankel Transform of $r^2\ GAUS(r)$ . . . 363
11.4.6 Hankel Transform of $CTRI(r)$ . . . 364
11.5 Appendix: Derivations of Equations (11.12) and (11.14) . . . 365
Problems . . . 369

12 The Radon Transform **371**
12.1 Line-Integral Projections onto Radial Axes . . . 371
12.1.1 Radon Transform of Dirac Delta Function . . . 377
12.1.2 Radon Transform of Arbitrary Function . . . 379
12.2 Radon Transforms of Special Functions . . . 380
12.2.1 Cylinder Function $CYL(r)$ . . . 380
12.2.2 Ring Delta Function $\delta(r - r_0)$ . . . 382
12.2.3 Rectangle Function $RECT[x, y]$ . . . 384
12.2.4 Corral Function $COR[x, y]$ . . . 385
12.3 Theorems of the Radon Transform . . . 387
12.3.1 Radon Transform of a Superposition . . . 387
12.3.2 Radon Transform of Scaled Function . . . 388
12.3.3 Radon Transform of Translated Function . . . 389
12.3.4 Central-Slice Theorem . . . 389
12.3.5 Filter Theorem of the Radon Transform . . . 390
12.4 Inverse Radon Transform . . . 391
12.4.1 Recovery of Dirac Delta Function from Projections . . . 392
12.4.2 Summation of Projections over Azimuths . . . 398
12.5 Central-Slice Transform . . . 402
12.5.1 Radial "Slices" of $f[x, y]$ . . . 402
12.5.2 Central-Slice Transforms of Special Functions . . . 403
12.5.3 Inverse Central-Slice Transform . . . 409
12.6 Three Transforms of Four Functions . . . 410
12.7 Fourier and Radon Transforms of Images . . . 419
Problems . . . 420

13 Approximations to Fourier Transforms **421**
13.1 Moment Theorem . . . 421
13.1.1 First Moment – Centroid . . . 424
13.1.2 Second Moment – Moment of Inertia . . . 424
13.1.3 Central Moments – Variance . . . 425
13.1.4 Evaluation of 1-D Spectra from Moments . . . 427
13.1.5 Spectra of 1-D Superchirps via Moments . . . 431
13.1.6 2-D Moment Theorem . . . 433
13.1.7 Moments of Circularly Symmetric Functions . . . 435
13.2 1-D Spectra via Method of Stationary Phase . . . 436
13.2.1 Examples of Spectra via Stationary Phase . . . 440
13.3 Central-Limit Theorem . . . 452
13.4 Width Metrics and Uncertainty Relations . . . 454
13.4.1 Equivalent Width . . . 454
13.4.2 Uncertainty Relation for Equivalent Width . . . 455
13.4.3 Variance as a Measure of Width . . . 455
Problems . . . 457

14 Discrete Systems, Sampling, and Quantization **459**
14.1 Ideal Sampling . . . 460
14.1.1 Ideal Sampling of 2-D Functions . . . 461
14.1.2 Is Sampling a Linear Operation? . . . 462
14.1.3 Is the Sampling Operation Shift Invariant? . . . 462
14.1.4 Aliasing Artifacts . . . 465
14.1.5 Operations Similar to Ideal Sampling . . . 467
14.2 Ideal Sampling of Special Functions . . . 467
14.2.1 Ideal Sampling of $\delta[x]$ and $COMB[x]$ . . . 470
14.3 Interpolation of Sampled Functions . . . 472
14.3.1 Examples of Interpolation . . . 478
14.4 Whittaker–Shannon Sampling Theorem . . . 479
14.5 Aliasing and Interpolation . . . 480
14.5.1 Frequency Recovered from Aliased Samples . . . 480
14.5.2 "Unwrapping" the Phase of Sampled Functions . . . 482
14.6 "Prefiltering" to Prevent Aliasing . . . 483
14.6.1 Prefiltered Images Recovered from Samples . . . 484
14.6.2 Sampling and Reconstruction of Audio Signals . . . 485
14.7 Realistic Sampling . . . 486
14.8 Realistic Interpolation . . . 491
14.8.1 Ideal Interpolator for Compact Functions . . . 491
14.8.2 Finite-Support Interpolators in Space Domain . . . 491
14.8.3 Realistic Frequency-Domain Interpolators . . . 495
14.9 Quantization . . . 500
14.9.1 Quantization "Noise" . . . 503
14.9.2 *SNR* of Quantization . . . 505
14.9.3 Quantizers with Memory – "Error Diffusion" . . . 507
14.10 Discrete Convolution . . . 507
Problems . . . 509

15 Discrete Fourier Transforms **511**
15.1 Inverse of the Infinite-Support DFT . . . 513
15.2 DFT over Finite Interval . . . 514
15.2.1 Finite DFT of $f[x] = 1[x]$ . . . 522

15.2.2 Scale Factor in DFT 524
15.2.3 Finite DFT of Discrete Dirac Delta Function 526
15.2.4 Summary of Finite DFT 526
15.3 Fourier Series Derived from Fourier Transform 527
15.4 Efficient Evaluation of the Finite DFT 529
15.4.1 DFT of Two Samples – The "Butterfly" 530
15.4.2 DFT of Three Samples 531
15.4.3 DFT of Four Samples 532
15.4.4 DFT of Six Samples 532
15.4.5 DFT of Eight Samples 533
15.4.6 Complex Matrix for Computing 1-D DFT 534
15.5 Practical Considerations for DFT and FFT 534
15.5.1 Computational Intensity 534
15.5.2 "Centered" versus "Uncentered" Arrays 536
15.5.3 Units of Measure in the Two Domains 538
15.5.4 Ensuring Periodicity of Arrays – Data "Windows" 539
15.5.5 A Garden of 1-D FFT Windows 545
15.5.6 Undersampling and Aliasing 551
15.5.7 Phase 554
15.5.8 Zero Padding 554
15.5.9 Discrete Convolution and the Filter Theorem 555
15.5.10 Discrete Transforms of Quantized Functions 559
15.5.11 Parseval's Theorem for DFT 560
15.5.12 Scaling Theorem for Sampled Functions 562
15.6 FFTs of 2-D Arrays 563
15.6.1 Interpretation of 2-D FFTs 564
15.6.2 2-D Hann Window 567
15.7 Discrete Cosine Transform 567
Problems 571

16 Magnitude Filtering 573
16.1 Classes of Filters 574
16.1.1 Magnitude Filters 574
16.1.2 Phase ("Allpass") Filters 575
16.2 Eigenfunctions of Convolution 576
16.3 Power Transmission of Filters 577
16.4 Lowpass Filters 579
16.4.1 1-D Test Object 581
16.4.2 Ideal 1-D Lowpass Filter 581
16.4.3 1-D Uniform Averager 581
16.4.4 2-D Lowpass Filters 583
16.5 Highpass Filters 585
16.5.1 Ideal 1-D Highpass Filter 585
16.5.2 1-D Differentiators 586
16.5.3 2-D Differentiators 587
16.5.4 High-Frequency Boost Filters – Image Sharpeners 588
16.6 Bandpass Filters 589
16.7 Fourier Transform as a Bandpass Filter 594
16.8 Bandboost and Bandstop Filters 596
16.9 Wavelet Transform 599
16.9.1 Tiling of Frequency Domain with Orthogonal Wavelets 600

16.9.2 Example of Wavelet Decomposition . . . 602
Problems . . . 602

17 Allpass (Phase) Filters 603
17.1 Power-Series Expansion for Allpass Filters . . . 604
17.2 Constant-Phase Allpass Filter . . . 605
17.3 Linear-Phase Allpass Filter . . . 606
17.4 Quadratic-Phase Filter . . . 608
17.4.1 Impulse Response and Transfer Function . . . 608
17.4.2 Scaling of Quadratic-Phase Transfer Function . . . 612
17.4.3 Limiting Behavior of the Quadratic-Phase Allpass Filter . . . 615
17.4.4 Impulse Response of Allpass Filters of Order 0, 1, 2 . . . 615
17.5 Allpass Filters with Higher-Order Phase . . . 615
17.5.1 Odd-Order Allpass Filters with $n \geq 3$ . . . 618
17.5.2 Even-Order Allpass Filters with $n \geq 4$ . . . 619
17.6 Allpass Random-Phase Filter . . . 619
17.6.1 Information Recovery after Random-Phase Filtering . . . 626
17.7 Relative Importance of Magnitude and Phase . . . 626
17.8 Imaging of Phase Objects . . . 628
17.9 Chirp Fourier Transform . . . 632
17.9.1 1-D “M–C–M” Chirp Fourier Transform . . . 632
17.9.2 1-D “C–M–C” Chirp Fourier Transform . . . 634
17.9.3 M–C–M and C–M–C with Opposite-Sign Chirps . . . 637
17.9.4 2-D Chirp Fourier Transform . . . 638
17.9.5 Optical Correlator . . . 638
17.9.6 Optical Chirp Fourier Transformer . . . 641
Problems . . . 645

18 Magnitude–Phase Filters 647
18.1 Transfer Functions of Three Operations . . . 648
18.1.1 Identity Operator . . . 648
18.1.2 Differentiation . . . 648
18.1.3 Integration . . . 650
18.2 Fourier Transform of Ramp Function . . . 653
18.3 Causal Filters . . . 654
18.4 Damped Harmonic Oscillator . . . 658
18.5 Mixed Filters with Linear or Random Phase . . . 661
18.6 Mixed Filter with Quadratic Phase . . . 661
Problems . . . 666

19 Applications of Linear Filters 667
19.1 Linear Filters for the Imaging Tasks . . . 667
19.2 Deconvolution – “Inverse Filtering” . . . 669
19.2.1 Conditions for Exact Recovery via Inverse Filtering . . . 671
19.2.2 Inverse Filter for Uniform Averager . . . 672
19.2.3 Inverse Filter for Ideal Lowpass Filter . . . 675
19.2.4 Inverse Filter for Decaying Exponential . . . 678
19.3 Optimum Estimators for Signals in Noise . . . 679
19.3.1 Wiener Filter . . . 680
19.3.2 Wiener Filter Example . . . 688
19.3.3 Wiener–Helstrom Filter . . . 689

19.3.4 Wiener–Helstrom Filter Example 693
19.3.5 Constrained Least-Squares Filter 695
19.4 Detection of Known Signals – Matched Filter 696
19.4.1 Inputs for Matched Filters 701
19.5 Analogies of Inverse and Matched Filters 703
19.5.1 Wiener and Wiener–Helstrom "Matched" Filter 706
19.6 Approximations to Reciprocal Filters 708
19.6.1 Small-Order Approximations of Reciprocal Filters 711
19.6.2 Examples of Approximate Reciprocal Filters 713
19.7 Inverse Filtering of Shift-Variant Blur 719
Problems 720

20 Filtering in Discrete Systems **723**
20.1 Translation, Leakage, and Interpolation 724
20.1.1 1-D Translation 724
20.1.2 2-D Translation 726
20.2 Averaging Operators – Lowpass Filters 728
20.2.1 1-D Averagers 728
20.2.2 2-D Averagers 730
20.3 Differencing Operators – Highpass Filters 731
20.3.1 1-D Derivative 731
20.3.2 2-D Derivative Operators 732
20.3.3 1-D Antisymmetric Differentiation Kernel 734
20.3.4 Second Derivative 734
20.3.5 2-D Second Derivative 736
20.3.6 Laplacian 737
20.4 Discrete Sharpening Operators 740
20.4.1 1-D Sharpeners 740
20.4.2 2-D Sharpening Operators 742
20.5 2-D Gradient 743
20.6 Pattern Matching 744
20.6.1 Normalization of Contrast of Detected Features 747
20.6.2 Amplified Discrete Matched Filters 748
20.7 Approximate Discrete Reciprocal Filters 749
20.7.1 Derivative 749
Problems 751

21 Optical Imaging in Monochromatic Light **753**
21.1 Imaging Systems Based on Ray Optics Model 754
21.1.1 Seemingly "Plausible" Models of Light in Imaging 754
21.1.2 Imaging Systems Based on Ray "Selection" by Absorption 758
21.1.3 Imaging System that Selects and Reflects Rays 760
21.1.4 Imaging Systems Based on Refracting Rays 761
21.1.5 Model of Imaging Systems 761
21.2 Mathematical Model of Light Propagation 762
21.2.1 Wave Description of Light 762
21.2.2 Irradiance 765
21.2.3 Propagation of Light 765
21.2.4 Examples of Fresnel Diffraction 772
21.3 Fraunhofer Diffraction 783
21.3.1 Examples of Fraunhofer Diffraction 785

21.4 Imaging System based on Fraunhofer Diffraction . . . 790
21.5 Transmissive Optical Elements . . . 792
21.5.1 Optical Elements with Constant or Linear Phase . . . 793
21.5.2 Lenses with Spherical Surfaces . . . 794
21.6 Monochromatic Optical Systems . . . 796
21.6.1 Single Positive Lens with $z_1 \gg 0$ . . . 796
21.6.2 Single-Lens System, Fresnel Description of Both Propagations . . . 799
21.6.3 Amplitude Distribution at Image Point . . . 803
21.6.4 Shift-Invariant Description of Optical Imaging . . . 806
21.6.5 Examples of Single-Lens Imaging Systems . . . 807
21.7 Shift-Variant Imaging Systems . . . 811
21.7.1 Response of System at "Nonimage" Point . . . 811
21.7.2 Chirp Fourier Transform and Fraunhofer Diffraction . . . 816
Problems . . . 819

22 Incoherent Optical Imaging Systems 823
22.1 Coherence . . . 823
22.1.1 Optical Interference . . . 823
22.1.2 Spatial Coherence . . . 828
22.2 Polychromatic Source – Temporal Coherence . . . 838
22.2.1 Coherence Volume . . . 842
22.3 Imaging in Incoherent Light . . . 842
22.4 System Function in Incoherent Light . . . 845
22.4.1 Incoherent MTF . . . 846
22.4.2 Comparison of Coherent and Incoherent Imaging . . . 847
Problems . . . 853

23 Holography 855
23.1 Fraunhofer Holography . . . 856
23.1.1 Two Points: Object and Reference . . . 856
23.1.2 Multiple Object Points . . . 862
23.1.3 Fraunhofer Hologram of Extended Object . . . 864
23.1.4 Nonlinear Fraunhofer Hologram of Extended Object . . . 866
23.2 Holography in Fresnel Diffraction Region . . . 867
23.2.1 Object and Reference Sources in Same Plane . . . 868
23.2.2 Reconstruction of Virtual Image from Hologram with Compact Support . . . 872
23.2.3 Reconstruction of Real Image: $z_2 > 0$ . . . 872
23.2.4 Object and Reference Sources in Different Planes . . . 873
23.2.5 Reconstruction of Point Object . . . 878
23.2.6 Extended Object and Planar Reference Wave . . . 882
23.2.7 Interpretation of Fresnel Hologram as Lens . . . 883
23.2.8 Reconstruction of Real Image of 3-D Extended Object . . . 885
23.3 Computer-Generated Holography . . . 885
23.3.1 CGH in the Fraunhofer Diffraction Region . . . 886
23.3.2 Examples of Cell CGHs . . . 890
23.3.3 2-D Lohmann Holograms . . . 894
23.3.4 Error-Diffused Quantization . . . 895
23.4 Matched Filtering with Cell-Type CGH . . . 898
23.5 Synthetic-Aperture Radar (SAR) . . . 900
23.5.1 Range Resolution . . . 904
23.5.2 Azimuthal Resolution . . . 906

23.5.3 SAR System Architecture 907
Problems 914

References **917**

Index **921**

Series Editor's Preface

Science, like life, is full of unintended consequences and unanticipated benefits. For example, by 1893 M. J. Hadamard had developed a set of functions represented as matrices that could "break down" the nature of a "signal" into components taken from a specific set of "basis" functions, which had complicated waveforms but simple binary values of "1" or "-1". The amplitudes of these functions defined the signal in terms of the basis functions. By the 1930s this set of functions had been codified into the Walsh–Hadamard transform, which was formed from a complete and orthogonal set of basis functions (in one or two dimensions). While the Walsh–Hadamard transform was of great mathematical interest, it did not have a lot of practical value until the age of the computer and its first major impact may have been when it was used to convert raw image data from deep-space probes into a series of bits that represented the "1s" and "-1s" of the basis functions. These binary signals were ideal for transmission from deep space and the image was easily reconstructed on Earth using the inverse transform. A second example of the unanticipated benefits is more relevant to this the 10th offering of the Wiley/IS&T Series on Imaging Science and Technology: *Fourier Methods in Imaging* by Dr. Roger L. Easton, Jr. In the early 1800s many of the world greatest physicists and chemists were focusing on the nature of heat, heat conduction, and steam engines and were creating the foundations of classical thermodynamics. One of theses scientists was Joseph Fourier. Fourier focused on solving the most basic nature of how heat (and temperature) moved through solids and this resulted in his work entitled *The Analytical Theory of Heat.* His solutions resulted in unique series and integrals using sine and cosine functions to provide the final solution to the heat conduction, over time and space, for a given system (with a well-defined geometry). This expansion into harmonic functions came to be known as the Fourier series, Fourier integral, or more simply the Fourier transform. Fourier analysis is used today in all fields of science and engineering.

Fourier also served as Secretary of the Institut d'Egypte in 1798–1801 during Napoleon's expedition to that country. The most important artifact found during this mission was the Rosetta Stone, which included copies of the same text in Greek, demotic, and Egyptian hieroglyphics. Several years later, Fourier encouraged the young Jean-François Champollion to work on translating the writings, and Champollion discovered the secret in 1822. Thomas Young, another famous scientist with significant contributions in optics, had also searched for the solution to this enigma. This aspect of Fourier's career and that of Dr. Easton will show an interesting similarity as noted below.

How does Fourier analysis impact on imaging? Consider the following two cases. When one "looks" at a recorded scene, one sees a continuous two-dimensional array of light values, which is interpreted by means of the visual system as an image. This is the natural spatial domain representation of the image and can be used to understand and alter the image as one wishes. However, there is an alternate mathematical representation of this image that can also be used to understand and alter the image. This alternate representation is the spatial frequency domain or the Fourier transform of the image. Consider the following set of operations. Find the average value of the image. Subtract the average value of the image from each point in the image resulting in a new image that now has both negative and positive values (light has no negative values, so this is just a mathematical abstraction of the image). Now construct a large (really infinite) set of two-dimensional patterns that are made up of sine and cosine functions. They will also have negative and positive values. Take each of these patterns and multiply them point by point with the average adjusted image and then sum all the values. This sum is then the coefficient (amplitude) for the given pattern used; this can be thought of as the projection of each pattern onto the average adjusted image. Repeat for all the patterns (basis functions). The resulting set of coefficients, along with the average value, now represent the alternate mathematical representation of the image, its Fourier transform. One can reconstruct the image by taking all the basis functions and multiplying by the appropriate coefficient, summing point by point, and adding the average value; this is equivalent to the inverse Fourier transform. Once the image has been transformed to the spatial frequency domain, one can operate on the coefficients to alter the nature of the image and then follow the above process to reconstruct the altered image. How all this can be done mathematically is presented in this excellent and precise text. One other example of the Fourier transform is useful to establish a more complete frame of reference for this text. Using basic wave optic reconstruction it is possible to show that the focal plane of a lens (not the image plane) contains the Fourier transform of the image. Thus one can perform operations on the image (in real time) by placing active devices (that alter the image and re-emit light) or passive devices (that just attenuate the light) in the focal plane of the lens and then using an identical lens to perform the inverse Fourier transform to get the altered image. Hence we see how Fourier's quest to understand how heat and temperature flow through solids has led, unintentionally, to such a vast and rich branch of imaging science.

Fourier Methods in Imaging provides the reader with a complete and coherent view of operating on images in the spatial frequency domain and how these operations relate to methods in the spatial domain, but may be easier to implement or more flexible in achieving a given goal. The first part of the text provides a clear review and exposition of the mathematical nature of linear system analysis for images and carefully considers the impact of sampling (moving from the continuous domain to a discrete domain) that is found in most digital imaging systems. Dr. Easton provides a host of often-used functions (*SINC* functions, triangle functions, etc.) in one-dimensional and two-dimensional cases, each of which are encountered in many practical image processing applications. He also provides a clear exposition of Hankel transforms (Fourier transforms in circular coordinates) and the Radon transform that forms the basis of many medical imaging systems like X-ray tomography. Dr. Easton then provides a comprehensive review of discrete transforms (used on all computers and in digital signal processing systems embedded in digital cameras and other digital imaging devices). These discrete transforms are the equivalent of the more general continuous transforms, but used on sampled images. Once the mathematical methods have been clearly explained, Dr. Easton uses the mathematics to implement a series of filtering applications in the spatial frequency domain, which are equivalent to more complex and harder to implement operations in the spatial domain. The topics of operating on coherent and non-coherent light and holography round out the text; these operations are on the actual image rather than a digitally encoded image like that from a camera or scanner.

Fourier Methods in Imaging represents an outstanding and practical review of operating in the spatial frequency domain for both "live optical" images and captured digital images. Every scientist or engineer working in modern imaging systems will find this to be an indispensable reference, one that sits on his or her desk and will be used time and time again.

Dr. Easton received his Ph.D. in Optical Science from the University of Arizona's Optical Sciences Center in 1986. He joined the Carlson Center for Imaging Science at The Rochester Institute of Technology (RIT) upon graduation and has been an integral part of the Center as both an outstanding instructor and researcher. He received the Professor Raymond C. Bowman Award for undergraduate teaching in Imaging Science from the Society for Imaging Science and Technology in 1997. Over his years at RIT he has concentrated on the mathematical treatment of linear imaging systems and on the experimental image processing techniques of "real" and captured images. In addition to his basic research in image science, Dr. Easton has been part of a team of scientific and historical scholars who have focused on the preservation and reconstruction of ancient manuscripts, including the Dead Sea Scrolls and the Archimedes Palimpsest. In this small way, he shares one of Fourier's own interests in the meaning of historical artifacts. This work resulted in his winning of the Archie Mahan Prize from the Optical Society of America, for the article "Imaging and the Dead Sea Scrolls" in 1988, and the

2003 Imaging Solution of the Year Award, by *Advanced Imaging Magazine*, for "Multispectral Imaging of the Archimedes Palimpsest", January 2003.

On a personal note, I had the pleasure of working with Dr. Easton while I was with the Eastman Kodak Research Laboratories and at the University of Rochester's Center for Electronic Imaging Systems, a joint effort with RIT, from 1986 through 1999. Dr. Easton is a truly dedicated teacher with proven experience in finding experimental solutions to complex imaging problems. As such, his offering of *Fourier Methods in Imaging* reflects his deep understanding of imaging problems, applications, and solutions. I will be proud to have this text on my desk and I highly recommend it to all working in the field of imaging science and technology.

MICHAEL A. KRISS
Formerly of the Eastman Kodak Research Laboratories and the University of Rochester

Preface

This book is intended to introduce the mathematical tools that can be applied to model and predict the action of imaging systems under some simplifying assumptions. A discussion of the mathematics used to model imaging systems encompasses such breadth of material that any single book that aspired to consider all aspects of the subject would be a massive tome. It should be made clear at the outset that this book intends no such pretense. Rather, its primary goal is to help readers develop an intuitive grasp of the most common mathematical methods that are useful for describing the action of general linear systems on signals of one or more spatial dimensions. In other words, the goal is to "develop images" of the mathematics. To assist in this development, many graphical and pictorial examples will be used for emphasis and to facilitate development of the readers' intuition.

A second goal of this book is to develop a consistent mathematical formalism for characterizing imaging systems. This effort requires derivation of equations used to describe both the action of the imaging system and its effect on the *quality* of the output image. Success in meeting the first objective of developing intuition should facilitate the achievement of the second goal. In the course of this discussion, we will derive representations of images that are defined over both continuous and discrete domains and for continuous and discrete ranges. These same representations may be used to describe imaging *systems* as well. Representations that are defined over a continuous domain are convenient for describing realistic objects, imaging systems, and the resulting images. Representations in discrete coordinates (i.e., using *sampled* functions) are essential for modeling general objects, images, and systems in a computer. Discrete images and systems are conveniently represented as vectors and matrices.

Authors of technical books at this level must always be cognizant of the different levels of preparation by the readers for the subject. Though the treatment may target readers with a "median" background, there always will be some spread about that median.

The contents of the book can be roughly grouped into five parts. Some chapters (and even some sections) may be skipped by many readers depending on their particular needs. The particular order of consideration of topics was chosen with some care to ensure a sequential discussion. That said, the choice also reflects the particular biases of the author to some extent.

After the introduction, the first part of the book (Chapters 2–5) attempts to address the inevitable variation in reader experience and preparation. It introduces the basic mathematical concepts of linear algebra for vectors and functions that are necessary for understanding the subsequent discussions. These include complex-valued functions, vector spaces, and inner products of vectors and functions and are intended to provide broad and less than rigorous reviews for readers who have already encountered mathematical discussions of these topics in previous studies, such as in quantum mechanics. This discussion is similar in tone to treatments of these subjects presented by *Image Reconstruction in Radiology*, by Anthony J. Parker (1990), and many readers will likely be able to skim (or even skip) some or all of it. Readers desiring or requiring a deeper discussion should consult some of the standard texts, such as *Linear Algebra and its Applications*, by Gilbert Strang (2005), and *Advanced Mathematical Models for Engineering and Science Students*, by Geoffrey Stephenson and Paul M. Radmore (1990).

The second part (Chapters 6–13) defines a set of "special" functions and describes the mathematical operations and transformations of continuous functions that are useful for describing imaging systems. The Fourier transforms of 1-D and 2-D functions are considered in detail, and the Radon transform is introduced. The last chapter in this part considers approximations of the Fourier transform and figures of merit that are useful metrics of the representations in the two domains. Note that other sources exist for discussions of the special functions, including *Linear Systems, Fourier Transforms, and Optics*, by Jack Gaskill (1978), and *The Fourier Transform and its Applications*, by Ronald N. Bracewell (1986).

The third part spans Chapters 14 and 15, and considers the Fourier transform of discrete functions. The importance of this discussion cannot be overstated, as many (if not most) applications require operations with sampled functions. The fourth part (Chapters 16–20) considers the description of imaging systems as linear "filters", and applies the mathematical tools to solve specific imaging tasks. In particular, Chapter 20 considers the application of linear filters to discrete functions. The fifth part considers in the remaining chapters the application of linear systems to model optical imaging systems, including holography.

The selection of parts depends on the needs of the readers. Many may find the first part to be a review of concepts that were considered in other venues. For these readers, a logical progression would include skimming the first part, more careful study of the second and third parts, and then selection of the appropriate topics from the fourth and fifth parts.

Two software programs used to create the examples in this book are available online for free. The original DOS program, "signals", creates and processes 1-D functions. It was originally written for classroom demonstrations. The program may be downloaded from
http://www.cis.rit.edu/resources/software/index.html.
This program runs directly on a Windows PC or may also be used in Linux and the Macintosh OS in the DOSBox environment (http://www.dosbox.com). As part of her senior research project in 2009, Juliet Bernstein wrote the second program "SignalShow" in Java. It creates and processes both 1-D and 2-D functions. It is available from the website http://www.signalshow.com.

I must thank many individuals who have participated in the writing of this book. Many students have provided inspiration and impetus to the process. Of particular note, I acknowledge John Knapp, Derek Walvoord, Ranjit Bhaskar, Ted Tantalo, David Wilbur, Anthony Calabria, Gary Hoffmann, Sharon Cady, Sally Robinson, Scott Brown, Kate Johnson, Alec Greenfield, Alvin Spivey, Noah Block, and Katie Hoheusle. Harry Barrett, Kyle Myers, Fenella France, and Jack Gaskill inspired by their examples. Special thanks to Juliet Bernstein for creating the computer-generated holograms.

Several colleagues also contributed in positive ways to the preparation of this text, including Keith Knox, Zoran Ninkov, Elliott Horch, William Cirillo, Ed Przybylowicz, Rodney Shaw, Jeff Pelz, Jon Arney, William A. Christens-Barry, Mike Toth, P. R. Mukund, Ajay Pasupuleti, and Reiner Eschbach. A few "colleagues", especially some other faculty, who made negative contributions will not be mentioned by name.

I would also like to thank some others who provided personal inspiration over the last 30+ years of my professional life. Among those who inspired the work are Harry Barrett, William Noel, Sue Chan, Andrea Zizzi, Fenella France, Catherine Carlson, and Judith Knight, as well my parents Roger and Barbara Easton.

ROGER L. EASTON, JR.
Rochester, New York
September 2009

1

Introduction

1.1 Signals, Operators, and Imaging Systems

As a simple definition, we may consider an imaging system to map the distribution of the input "object" to a "similar" distribution at the output "image" (where the meaning of "similar" is to be determined). Often the input and output amplitudes are represented in different units. For example, the input is often electromagnetic radiation with units of, say, watts per unit area, while the output may be a transparent negative emulsion measured in dimensionless units of "density" or "transmittance". In other words, the system often changes the form of the energy; it is a "transducer". The goal of this book is to mathematically describe the properties of imaging systems, and it is often convenient to use the model of a system as a chain of links.

1.1.1 The Imaging Chain

An imaging system is often modeled as a "chain" of links that transfer information in the form of energy from an input (the *object*) to the output (the *image*) of the system. Many schemes of links in an imaging system are plausible, depending on details, but one eight-link model is appropriate in many imaging systems:

1. The source of energy (usually in the form of electromagnetic radiation).
2. The object to be imaged, which interacts with the energy from the source by reflection, refraction, absorption, scattering, and/or other mechanism.
3. Propagation of the energy to the imaging system.
4. Energy collection (often using an optical system composed of lenses and/or mirrors).
5. Sensing or detection by a transducer (converts incident energy to a measurable form, e.g., photons to electrons).
6. Image processing, including data compression (if any).
7. Storage and/or transmission (if any).
8. Display.

The source and object often are one and the same, e.g., radiating objects such as stars. Sometimes it is useful to model the imaging system with a second stage of energy propagation after collection, as when evaluating optical imaging systems composed of a single thin lens.

Fourier Methods in Imaging Roger L. Easton, Jr.

In this book, we consider a simplified picture of the imaging chain that combines the source, object, and energy propagation into one entity that we will call the "input function" (or just the "input"); it usually is specified by a single-valued physical quantity f that varies over continuous coordinates in space, and perhaps in time t and color, specified by wavelength λ (or equivalently by temporal frequency ν):

$$\textit{Input to imaging system} = f[x, y, z, t, \lambda] \textit{ or } f[x, y, z, t, \nu] \tag{1.1a}$$

Though these quantities have explicit spectral and temporal coordinates in addition to the spatial dimensions, the signals considered in this book most often will be functions of one or two spatial coordinates only and are specified by $f[x]$ or $f[x, y]$. In other words, the input and output signals will be constant in time and wavelength. The number of coordinates necessary to specify the function is the *dimensionality*, and the set of all possible such coordinates defines its *domain*. We will use the shorthand notation of "1-D" and "2-D" for one- and two-dimensional signals, respectively; 2-D signals such as $f[x, y]$ are of greatest interest in imaging applications, but the study of 1-D signals will be considered in depth in this book as well. This is because 1-D systems often are easy to visualize and the results may be directly transferable to problems with higher dimensionality.

The output of the imaging system also will usually be specified by a single-valued physical quantity that will be denoted by g. Though the domain of the image g may be identical to that of the object f, it is more common to require different coordinates, and these will be denoted, when necessary, by primed coordinates. In many cases, the output will be a function of two spatial coordinates with no dependence on wavelength:

$$\textit{Output of imaging system} = g[x, y, z, t, \lambda] \rightarrow g[x, y] \tag{1.1b}$$

In imaging applications, the numerical value assigned to the dependent variable of the input or output signals in Equation (1.1) represents a measurable physical quantity. Based on the familiarity of optical imaging, a common descriptive name for f is the "brightness" of the scene, though this terminology is not used in some subdisciplines of optics, such as radiometry. Regardless of the nomenclature, the appropriate quantity (i.e., the units of f) depends on the particular imaging system. In optical imaging, the relevant quantity is the *irradiance* at each coordinate: the time average of the square of the magnitude of the electric field. In X-ray or gamma-ray imaging, the measured quantity is the number of quanta incident on the sensor as a function of spatial location. In acoustic imaging (sonar or ultrasound), the acoustic power radiated, transmitted, or reflected by an object is the quantity of record.

Now consider some important examples in a bit more detail. For example, optical images may be generated by light that is *coherent* or *incoherent*. At this point, we can think of coherent light as composed of a single wavelength λ and incoherent light as composed of many wavelengths, such as *natural light*. The relevant quantity in optical imaging with coherent illumination is the complex-valued amplitude of the electric field (including both its magnitude and the phase). The appropriate measured quantity in incoherent (natural) light is the time average of the square of the magnitude of the complex-valued amplitude of the electromagnetic field; this is the *irradiance* and may be denoted by $\langle |f|^2 \rangle$. In still other applications, the physical quantity represented by f may have a very different form. Though the signal and the system may have different forms, most of the principles discussed in this book will be applicable to some extent in all imaging situations.

The description of the imaging system requires a mathematical model of its action upon the input function f to generate the output g. This action will be denoted by an operator represented as an upper-case script character, such as $\mathcal{O}\{f[x, y, \ldots]\}$. The operator symbolizes the mathematical rule that assigns a particular output amplitude g to every location in its domain. In many cases, it will be possible to describe the action of the system as the combination of a specific function associated with the imaging system (the "system function") and a particular mathematical operation (e.g., multiplication or integration).

It is obvious that the output image generally is affected both by the mathematical form of the specific input object and by the characteristics of the system. The functional expression for a common type of simple image would be:

$$\mathcal{O}\{f[x, y, z, t, \lambda]\} = g[x', y'] \tag{1.2}$$

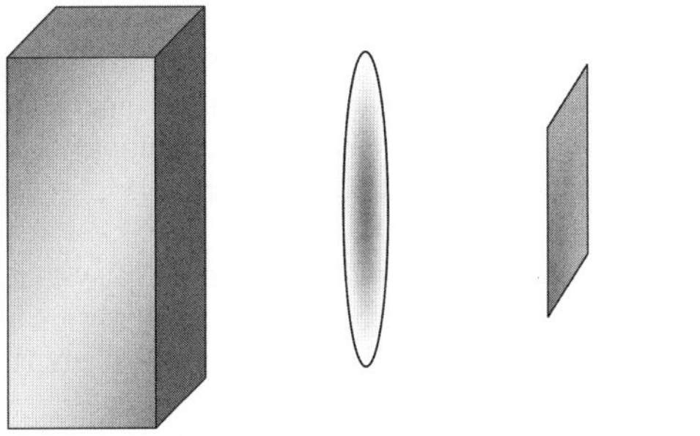

Figure 1.1 Schematic of an imaging system that acts on a time-varying input with three spatial dimensions and color, $f[x, y, z, t, \lambda]$ to produce a 2-D monochrome (gray-scale) image $g[x', y']$.

The schematic of the imaging process is shown in Figure 1.1. The spatial domain of the output image is often different from that of the input object, hence the use of primed characters in Equation (1.2). In realistic situations, the amplitude g also is affected by other parameters, such as the time, the exposure time Δt, and the spectral response of the sensor. In Equation (1.2), the effects of these additional parameters could be considered to be implicit in the system operator $\mathcal{O}$.

1.2 The Three Imaging Tasks

In many imaging applications, input objects and output images are functions of spatial dimensions only. Examples of mathematical relations for 1-D and 2-D systems are:

$$\mathcal{O}\{f[x]\} = g[x'] \tag{1.3a}$$

$$\mathcal{O}\{f[x, y]\} = g[x', y'] \tag{1.3b}$$

Simply put, the imaging chain relates three "entities": the input object, the action of the imaging system, and the output image. These three entities are denoted by the symbols f, $\mathcal{O}\{\ \}$, and g, respectively. A simple description of an imaging "task" is the process of specifying one of the three entities from knowledge of the other two. Three cases are evident:

1. The *forward* or *direct* problem: to find the mathematical expression for the image $g[x', \ldots]$ given complete knowledge of the input object $f[x, \ldots]$ and the system $\mathcal{O}$.
2. The *inverse problem*: to evaluate the input $f[x, \ldots]$ from the measured image $g[x', \ldots]$ and the system $\mathcal{O}$.
3. The *system analysis* problem: to determine the action of the operator $\mathcal{O}$ from the input $f[x, \ldots]$ and the image $g[x', \ldots]$ (the solution is often very similar in form to that of the inverse problem).

The solution of the direct task is often rather easy, while the others may be difficult or even mathematically impossible. Other and more complicated variants of these imaging problems are common, including the cases where knowledge of the entities (f, g, and/or $\mathcal{O}$) may be incomplete or contaminated by random noise. Some of the variants of the imaging problem will be considered in later chapters.

The additional complexity of the more general imaging system model is perhaps evident just from observation of the form of the general 1-D imaging relation in Equation (1.3a); the operator $\mathcal{O}$ must be a function of x and x' because it relates the object to the image. The most general form of $\mathcal{O}$

in Equation (1.3a) may modify the "brightness" f and/or the "location" x of all or part of the input signal by rearrangement, amplification, attenuation, or removal in an arbitrary fashion. For example, the image amplitude g at a specific location could be derived from the input amplitude at the corresponding location, from that at a different location, or from amplitudes at multiple locations in the input $f[x, \ldots]$. The functional form of the relationship between $f[x, \ldots]$ and $g[x', \ldots]$ may be linear or nonlinear, deterministic or random.

Though it is desirable to mathematically represent the action of system operators so that they are both concise and generally applicable, these two characteristics usually are mutually exclusive. In other words, a general system operator appropriate for the imaging task likely is impossible to specify in a concise mathematical notation.

Perhaps these examples give the readers some flavor of the difficulties to be attacked when specifying the action of an imaging system that is more general than the usual simple cases.

1.3 Examples of Optical Imaging

At this point we introduce a few simplified examples of optical and medical imaging systems to illustrate the imaging "tasks" and the mathematical concepts introduced in this book.

1.3.1 Ray Optics

Solution of a particular imaging task demands that the available "entities" f, g, and $\mathcal{O}$ be represented or modeled as mathematical expressions, which are then manipulated to derive an expression for the unknown entity. To illustrate the concept, consider the particularly simple, yet still very useful, mathematical model of optical imaging from introductory optics. A point source of energy emits geometrical "rays" of light that propagate in straight lines to infinity in all directions. The imaging "system" is an optical "element" that interacts with any ray it intercepts. The interaction mechanism is a physical process (usually refraction or reflection) that "diverts" the ray from its original direction. In this example, the optical element is a single thin lens located at a distance z_1 from the source. If the diameter of the lens is infinite, then all rays that move at all from left to right will intercept the lens and be diverted. Such a system may be described by the single parameter, the "focal length", which determines the "power" of the system to redirect rays. We will denote the focal length by the bold-faced roman "**f**" to distinguish it from the italic character "f" that will be used to represent the input amplitude. In the example of Figure 1.2, **f** is positive and the system redirects the light rays that emerge from the same object point so that they converge to an image point located at some distance z_2 from the lens. The mathematical descriptions relevant to the input "object", the imaging system, and output image are respectively the distance z_1 from the object to the lens, the focal length **f**, and the distance z_2 from the lens to the image point. The relationship of these three distances is the mathematical model of the imaging system, which is most commonly presented in the form:

$$\frac{1}{z_1} + \frac{1}{z_2} = \frac{1}{\mathbf{f}} \tag{1.4a}$$

This simple model of the optical system is "perfect" in the sense that all light rays emerging from a single source point of infinitesimal size at the object are assumed to converge to a single infinitesimal area about the location: the point "image". The relation of Equation (1.4a) may be rearranged into forms for each of the three tasks by placing the known parameters on one side of the equation and the unknown value on the other. The relations for the three tasks are trivial to derive from Equation (1.4a):

1. Direct task: given the object distance z_1 and the parameter **f** of the imaging system, find the output image distance z_2. The mathematical expression to be solved is a simple rearrangement of Equation (1.4a) with the two known quantities on the left-hand side and the unknown on the

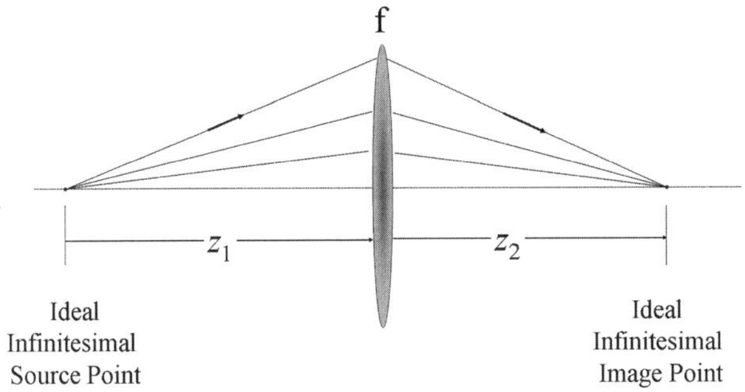

Figure 1.2 Optical imaging system in the ray optics model with no aberrations. All rays from a single source point converge to a single *image* point.

right:

$$\left(\frac{1}{\mathbf{f}} - \frac{1}{z_1}\right)^{-1} = z_2 \tag{1.4b}$$

2. Inverse task: given the output image point located at z_2 and the description of the system in the form of **f**, find the input object point via:

$$\left(\frac{1}{\mathbf{f}} - \frac{1}{z_2}\right)^{-1} = z_1 \tag{1.4c}$$

3. Analysis task: given the input object and output image points, determine the specification of the imaging system:

$$\left(\frac{1}{z_1} + \frac{1}{z_2}\right)^{-1} = \mathbf{f} \tag{1.4d}$$

In this particularly simple model, the mathematical relations to be solved in the direct and inverse tasks differ only in which parameter appears on the right-hand side, which means that the direct and inverse imaging tasks are equally "difficult" (or rather, equally trivial) to solve.

For a "fixed system" with focal length **f**, the imaging equations in Equation (1.4) "pair up" object and image planes located at distances z_1 and z_2. In other words, the imaging system constructs a "mapping" of source planes to image planes. However, we rarely seem to think about the action of the lens on objects with depth, or even on planar objects located at some distance z_1 that does not satisfy the imaging equation for a fixed focal length **f** and image distance z_2. In other words, we rarely even consider "out-of-focus" images in this simple model, though it is easy to imagine situations where we might want to calculate the appearance of such an image. It is possible to evaluate the "quality" of the image created by a nonideal imaging system using this simple ray optics model, but there are significant benefits from the more sophisticated model of light as a wave that is introduced next.

1.3.2 Wave Optics

Models of light that are more sophisticated than simple "rays" require a different mathematical description of how the "brightness" (the amplitude) generated by the object propagates through space. The resulting system operator $\mathcal{O}$ is significantly more complicated, and thus so are the corresponding equations that relate the input object and output image. A more complete model of optical imaging considers the physical observation that light "rays" are subject to optical "diffraction" that makes the energy deviate from straight-line propagation. A simple extension of the example already considered produces a finite-sized "patch" of light instead of an infinitesimal point, as suggested by Figure 1.3.

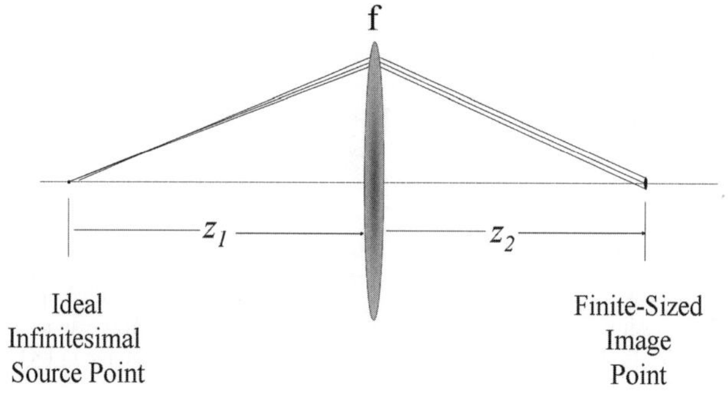

Figure 1.3 Ray model of optical imaging that includes *diffraction*, so that rays from a single source point "spread" while propagating to and from the lens. The image is not a single image "point", but rather a "blurry" spot.

It is more convenient to describe the light as a "wave" instead of as a "ray" in models of imaging systems that include diffraction. Each infinitesimal source point in this model emits spherical *wavefronts* of electromagnetic radiation that propagate outward from the source at the velocity of light. The radiation at all points on a particular such wave surface was emitted at the same instant of time. In other words, the vectors perpendicular to the local wave surface are the *rays*. One benefit of the wave model is due to the fact that a mathematical function exists that describes the spherical wave everywhere in space, thus making possible a "large-scale" or "global" picture of the radiation.

An optical element of the system (typically a lens or mirror) intercepts a portion of each spherical wave and acts to change its radius of curvature, perhaps even "reversing" the curvature so that successive propagating wave surfaces then *converge* toward an "image" of the point before diverging again (Figure 1.4). This model suggests another interpretation of the action of the optical system as an attempt to "replicate" the infinitesimal energy distribution of the point source. The fidelity of the reproduction is determined by the size of the image; a more faithful image exhibits a more compact distribution of energy. In real life, the size of the image produced by a "flawless" optical system decreases and the fidelity improves if the system intercepts are a larger area of the outgoing wave. In other words, the size of the image of the point object decreases as the area of the optic increases (assuming no defects in the optical system known as "aberrations"). We note at this point that it is physically impossible to replicate the infinitesimal area of the original source; even a "perfect" image has a finite area. The difference is ascribed to the phenomenon of optical "diffraction" and the finite-sized image of an infinitesimal point source is a "diffraction spot".

Images produced by multiple sources may be calculated in this model if the output of the system is the sum of the individual outputs from the point-source inputs; this is the first introduction to the concept of a *linear* system. In such a case, the image of two equally bright and closely spaced point sources in the wave optics model is the sum of two finite-sized "diffraction patterns" that may overlap. As the distance between the sources is decreased, the ability of the observer to distinguish the individual sources from the image will become more difficult. This leads to the concept of "spatial resolution" of an optical imaging system, as shown in Figure 1.4. The objects are pairs of point sources separated by different angles. The first pair of images is clearly distinguishable, while the overlap of the two images in the second example makes it more difficult to discern the true nature of the original object.

The wave model of light also allows estimation of the "pattern of energy" generated by the spherical wave at locations other than the image point, i.e., the distribution of light at an "out-of-focus" image may be calculated (Figure 1.5). This provides a means to establish the "appearance" or "quality" of the image at locations that are "in" or "out of" focus.

Optical imaging models that include diffraction are considered in Chapters 21–23.

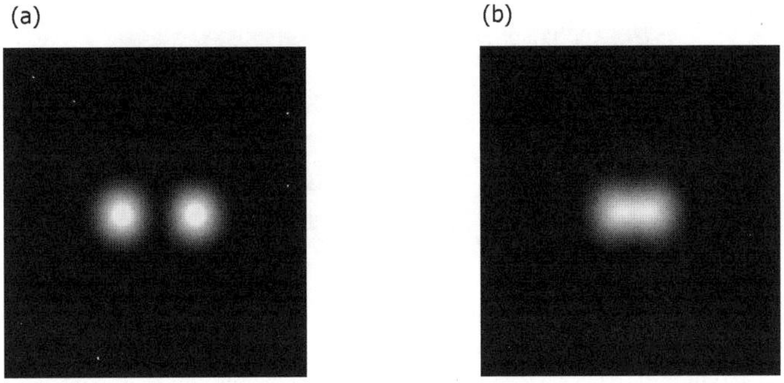

Figure 1.4 The effect of diffraction on the ability to resolve objects and fine structure in the image. The original objects are pairs of point objects (e.g., double stars) at two different separations viewed through an aberration-free system. In (a), the object clearly consists of two disjoint stars, but the result is less convincing in (b).

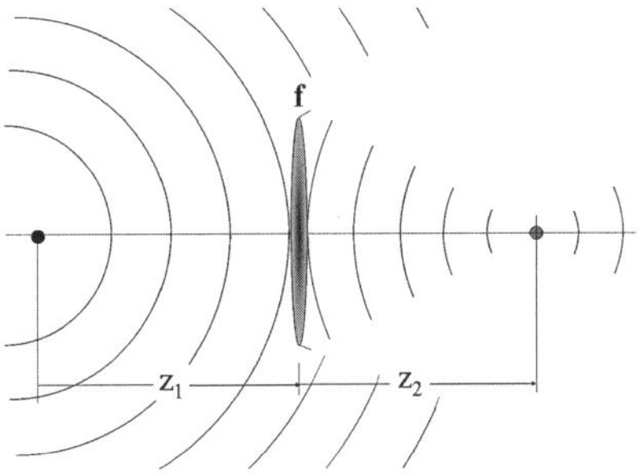

Figure 1.5 Schematic of the wave model of light propagation. Several spherical waves are shown that were emitted by the infinitesimal point source at times separated by equal intervals. The lens intercepts a portion of each wave and changes the curvatures of the wavefronts to make them converge. The size of the resulting image point is finite, not infinitesimal. The waves continue to propagate and expand away from the image.

1.3.3 System Evaluation of Hubble Space Telescope

One of the more impressive achievements of imaging science was the successful diagnosis of the optical system in the Hubble Space Telescope (HST) from images taken while in orbit. The HST was deployed from the Space Shuttle in 1990 after more than a decade of design and construction. Readers are probably aware that the primary mirror was improperly figured due to faulty testing and the quality of resulting images was much poorer than specified. Before the design and construction of corrective optics, the action of the optical system had to be characterized "at a distance" (to put it very mildly!). In the terms of our imaging-system description, the system operator $\mathcal{O}\{\ \}$ was only partly known when HST was deployed. By combining information from output images $g[x, y]$ of known object functions $f[x, y]$ with educated guesses of the cause of the optical faults, the action of the existing system was determined to sufficient accuracy to enable engineers to design the optical compensator dubbed COSTAR (the *Corrective Optics Space Telescope Axial Replacement*). COSTAR was retrofitted into HST during a Shuttle servicing mission in 1993. The results of this team effort to fix the problem were nothing short of spectacular; the quality of the corrected images met the original specifications. Two

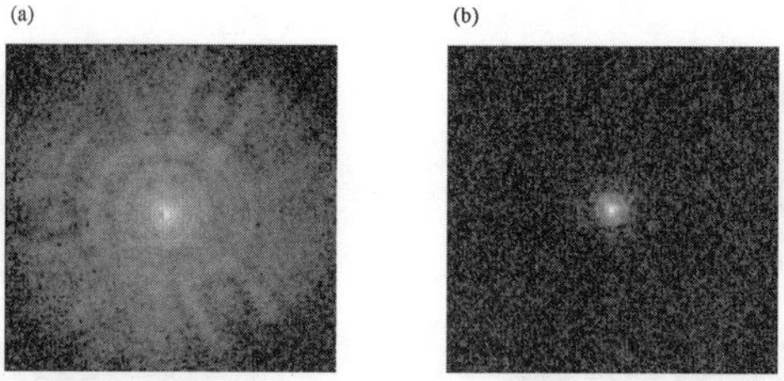

Figure 1.6 Images of a star from HST before and after adding the COSTAR optical corrector: (a) image before correction that exhibits "blur" due to the spherical aberration of the primary mirror; (b) image obtained after correction with COSTAR, where the light is concentrated in a much smaller spot.

images that compare the quality before and after the retrofit are shown in Figure 1.6, and other examples of the results before and after correction may be accessed at the HST website, hubble.stsci.edu.

1.3.4 Imaging by Ground-Based Telescopes

Now consider a practical example of an imaging system – the imaging of stellar objects through the atmosphere by ground-based telescopes. Stars seen from Earth are excellent approximations of point sources, so that their wavefronts are effectively spherical waves with $z_1 \cong \infty$. In other words, the wavefronts are effectively planar before encountering the atmosphere. At visible wavelengths and at standard temperature and pressure, the atmosphere is a transmissive medium that refracts light according to Snell's well-known law with index of refraction $n \cong 1.0002$, which varies with air density and thus with temperature. Local temporal variations in the temperature (and thus the density) of air produce local temporal variations in the refractive index, and thus change the angle of refraction. Since the air temperature varies locally and with depth in the atmosphere, the plane waves emitted by star(s) are refracted in a random pattern. The deformations translate and "defocus" the images over time intervals of the order of hundredths of seconds. The resulting image is a jumble of energy since light from a star is recorded at different locations in the image plane while light from nearby parts of the sky may be recorded at the same detector site. The clusters of recorded energy are commonly called *speckles*. The resolution of the recorded images is limited by these atmospheric aberrations, often to a very large degree. Fortunately, mathematical tools have been developed to utilize these speckle images to recover useful information out to the diffraction limit of the telescope. An early imaging tool for this purpose that was developed by Labeyrie in 1972 processed multiple short-exposure images taken through the atmosphere to provide high-resolution information that would normally be averaged to invisibility. The technique was modified by Knox and Thompson in 1974. A simulated example of Labeyrie's algorithm for "stellar speckle interferometry" is shown in Figure 1.7.

1.4 Imaging Tasks in Medical Imaging

Medical diagnostic imaging is another classic example of the inverse problem where the need is to determine the unknown "input" from the measured output and knowledge of the system. One essential difference between medical and astronomical imaging is due to the fact that the electromagnetic radiation used in the former is very much more "energetic"; the radiation is envisioned to be in the form of "photons" with very large energies that therefore propagate in straight lines in the manner of perfect geometrical rays. We now consider simplified versions of a few imaging systems with very different properties that are used in medical diagnostic applications.

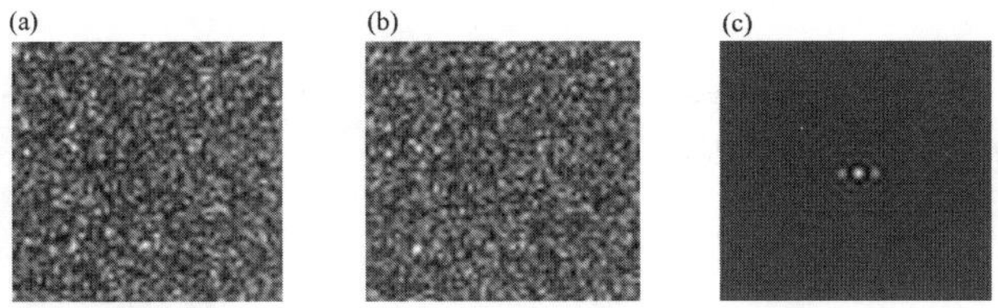

Figure 1.7 Simulation of Labeyrie's algorithm for stellar speckle interferometry: (a) image obtained from a single star through a turbulent atmosphere, showing "speckles" due to local variations in atmospheric refraction; (b) simulated image of a double star taken through the same atmosphere, showing "paired" speckles; (c) processed image from several instances of (b), which provides evidence of the existence and separation of the double star.

1.4.1 Gamma-Ray Imaging

In this first example, the individual photons, generated by nuclear decay in radioactive materials, are called *gamma rays*. Gamma-ray imaging is the basis for diagnostic nuclear medicine, where chemicals selectively absorbed by specific organs are tagged with radioactive atoms. For example, it is possible to attach radioactive technetium to iodine atoms that are injected into the body. The iodine is then selectively absorbed by the thyroid gland. It is medically useful to image gamma rays emitted by radioactive decay of the technetium to visualize the pathology of the thyroid gland. The object function may be written as $f[x, y, z, \lambda, t, \theta, \phi]$, where f is the number of photons emitted from the location in unit time in the spherical direction defined by the azimuth angle θ and latitude ϕ.

The kinetic energy of the gamma rays emitted by the technetium is approximately 0.14 MeV, which translates to an energy $E \cong 2.24 \times 10^{-7}$ ergs per photon. Imaging these gamma-ray photons presents a problem that does not exist in the imaging of visible light because the corresponding wavelength of light for the gamma ray is $\lambda_\gamma = hc/E \cong 9 \times 10^{-12}$ m, which is very much shorter than visible wavelengths with $\lambda \cong 5 \times 10^{-7}$ m. These gamma-ray photons are sufficiently energetic to pass through optical lenses and mirrors without effect (unless incident at very shallow ("grazing") angles of incidence, where they may be deviated by a small change in angle). Since the usual mechanisms for redirecting light do not work for gamma-ray photons, a different physical interaction must be used between energy and matter, such as absorption by dense materials. The "pinhole gamma camera" shown in Figure 1.8 is an example of such a system; it consists of an absorbing "plate" (typically made of lead) with a small hole of diameter d. The plate "selects" those photons from the source that travel along very specific paths through the hole to the sensor.

As a first example, consider a planar object that emits f photons with wavelength λ_γ per unit time from the location $[x, y]$ and that is located at a distance z_1 from the absorbing plate. The photons emitted by the object that pass through the hole continue to propagate to a planar photosensitive detector located at a distance z_2 from the absorber. The image is created by counting the photons with wavelength λ_γ over some exposure time to produce an "image" $g[x, y]$; the system equation has the form:

$$\mathcal{O}\{f[x, y, \lambda, t, \theta, \phi]\} = g[x, y] \tag{1.5}$$

The diameter of the pinhole and the distances z_1 and z_2 are parameters of the system operator $\mathcal{O}$. The goal of this modality of diagnostic imaging is to derive the spatial form of the planar object function f from the planar image g to solve the inverse problem.

The "quality" of the image is determined by whether the measured number of photons at each position of the detector is proportional to the number of photons emitted by the corresponding position of the object. Two parameters directly affect this measurement: the diameter d of the pinhole and the

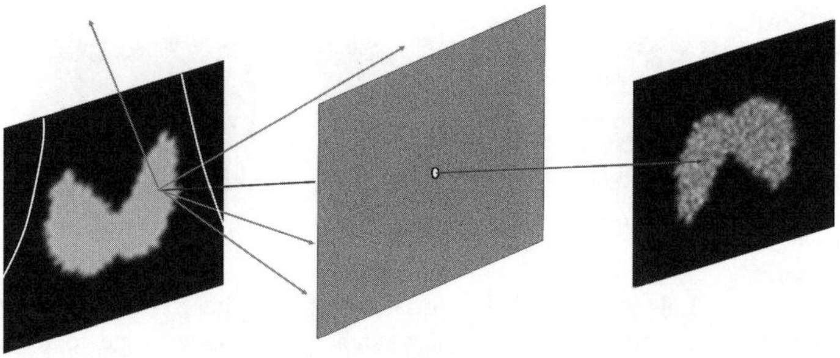

Figure 1.8 Schematic of gamma-ray pinhole imaging. The planar object $f[x, y, z, t, \lambda]$ emits energetic gamma-ray photons in all directions. An occasional emitted photon travels along a path that passes through the pinhole and is absorbed by the sensor, thus forming the image $g[x, y]$.

number of photons counted at each position in the detector, which is determined by the number of photons emitted by the object.

The action of the pinhole in the imaging system is to constrain the path of photons that reach the detector. If d is very small, the position of emission within the planar object may be determined very accurately and we might expect the resulting image to be a "faithful" replica of the original object. As d is increased, photons emitted from the same location in the object can expose different locations on the sensor and photons from different locations on the object expose the same location on the sensor. This ambiguity in location of sources degrades the image; it becomes "blurry", which means that the "spatial resolution" is somehow proportional to d^{-1}. It also is easy to see that the number of photons that reach the detector also is determined by the area of the pinhole; the number is small if $d \gtrapprox 0$ and larger if $d \gg 0$. For reasons not discussed here, the measured number of photons becomes more certain if more photons are counted. The resulting reduced variation in the number of counted photons means that the actual number of photons emitted by different points on the object may be more accurately estimated. In other words, the image becomes less "noisy" by counting more photons because the "brightness resolution" is improved.

The discussions of the last few paragraphs show that image quality in the pinhole gamma camera is determined by two countervailing requirements due to the hole diameter d; a smaller hole improves the spatial resolution but increases the statistical noise in the image. Simulations of images that illustrate these principles are shown in Figure 1.9.

Note also that the image is "inverted" by the gamma camera and that the relative sizes of the distances z_1 and z_2 determine the "magnification" of the image. If $z_1 > z_2$, then a particular area of the object is imaged onto a smaller area of the detector and the image is "minified" (smaller than the original). Obviously the image is "magnified" if $z_2 > z_1$.

At this point, we can practice being imaging scientists by using these observations to redesign the pinhole camera to compensate for its limitations. We might first want to decrease the statistical noise in the image by increasing the number of counted photons. One means to do this is to increase the dose of radioactivity to the patient so that more photons are emitted by the thyroid. Since this strategy creates additional problems for the patient, we seek other means. First, recognize that gamma-ray photons should be emitted from the technetium in any direction with equal likelihood. This allows us to record more of the emitted photons by adding more pinholes and more detectors. The resulting system creates the same "on-axis" image as before, but also additional "off-axis" images that contain similar information. The differences in geometry of the 3-D object and the sensors produce distortions in the images, but these may be corrected and the images combined to synthesize a single image formed from

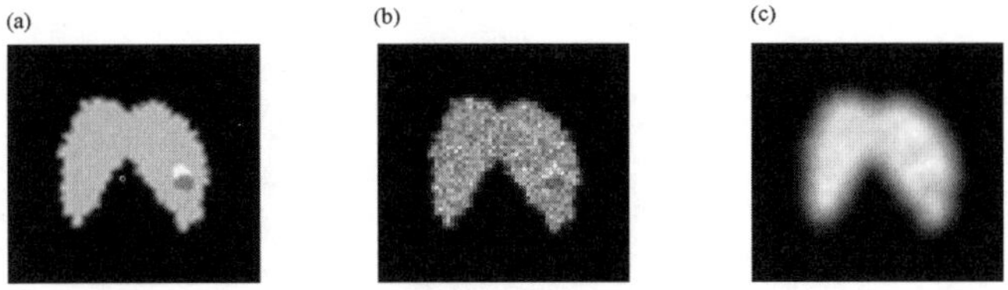

Figure 1.9 Simulation of the effect of pinhole size on spatial resolution. (a) Original object $f[x, y]$ is a simulated thyroid with a white "hot" spot (that emits more gamma rays) and an adjacent dark gray "cold" spot (the object has been rotated by 180° so that its orientation matches the images); (b) simulated image obtained using a small pinhole to produce good spatial resolution but with statistical noise due to the small number of counted photons; (c) simulated image with larger pinhole that reduces the statistical noise but "blurs" the image due to overlapping of different source points at the same location on the sensor.

a larger number of recorded photons, and thus with less statistical variation. For example, consider a system constructed from pinholes that are positioned sufficiently close together to produce overlapping images, meaning that photons from different locations on the object may expose the same location on the detector. Obviously, it is necessary to "unmix" the overlapping images to produce an image with improved quality. An example is shown in Figure 1.10. The technique for "unmixing" the overlapped images will be considered in the discussion of image filtering later in this book. This concept may be further extended by drilling more and more "pinholes" in the lead absorber. The pinholes may even merge together to form regions of "open space". In the example shown in Figure 1.11, 50% of the lead has been removed, thus transmitting many more photons to the detector. This pattern of detected photons is processed by a mathematical algorithm based on the pattern of pinhole apertures to "reconstruct" an approximation of the original object with less statistical variation. The process of collection and reconstruction is called *coded aperture imaging*.

1.4.2 Radiography

The next example of a medical imaging system is conventional radiography, which is what most people mean when they say "My dentist took an X-ray today". The X-radiation consists of photons with energies of the same order as gamma rays, but that are emitted from a source distinct from the object that may be characterized by its linear dimension d (area $\propto d^2$). The photons are selectively removed from the beam by structures within the 3-D object that absorb and/or scatter the incident radiation. The mathematical description of the 3-D object is a function f that measures its "ability" to remove X-ray photons from the beam, which may be called the X-ray *attenuation coefficient.* Also note that the dependence on exposure time t is ignored in the usual case of an object that does not change over the time scale of the image measurement. Just as was the case for gamma-ray imaging, the "spatial resolution" of the image is determined by the certainty that a recorded photon traveled a specific path from the source through the object, and thus the number reflects a physical feature of the object. The limiting factor in radiography is the diameter d of the X-ray source, which means that the uncertainty in the path traveled by a photon from a small source is small, while that uncertainty in the path is larger if d is larger.

In our simple example, we may consider that the photons that pass through the object are imaged using the same kind of sensor as in gamma-ray imaging. The form of the imaging equation is:

$$\mathcal{O}\{f[x, y, z]\} = g[x, y] \tag{1.6}$$

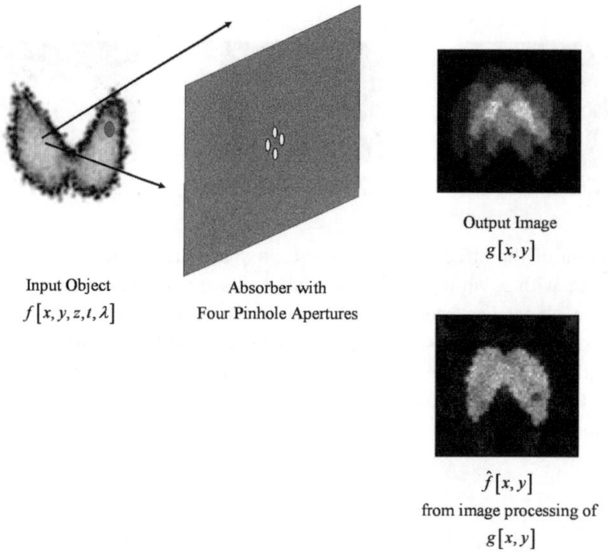

Figure 1.10 Simulation of output image $g[x, y]$ obtained of the object $f[x, y]$ through four pinholes. The overlapping gamma-ray images are then digitally processed in a computer to produce the estimate $\hat{f}[x, y]$ that more closely resembles $f[x, y]$.

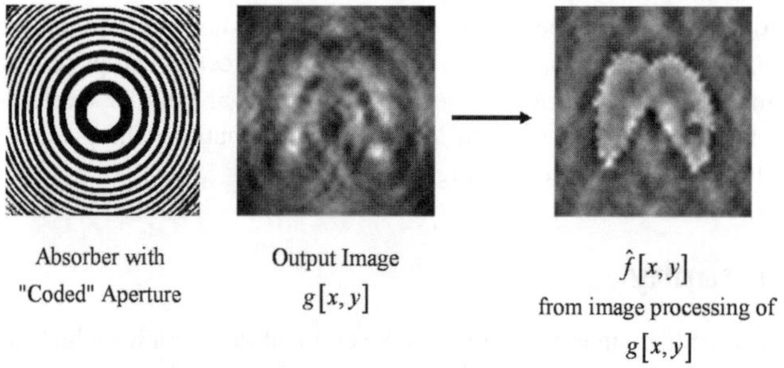

Figure 1.11 Simulated imaging through a continuum of pinholes that forms the "coded aperture": (a) aperture; (b) image of photons generated by the simulated thyroid through this aperture; (c) output image after processing based on knowledge to merge the continuum of raw images.

where the dependence on wavelength λ and time t has been deleted for clarity. Again, the parameters of the imaging system, including the source diameter d and the distances z_1 and z_2, are implicit in the system operator $\mathcal{O}\{\ \}$. The task is to determine the 3-D input function $f[x, y, z]$ from the 2-D output $g[x, y]$. The signal recorded by the sensor at a particular location is due to the integrated attenuation of the X-ray beam along the associated path through the object. This means that information about the third spatial dimension (the position in "depth") is lost by the recording process. For example, a dark spot on an X-ray image indicates only that the total attenuation of X-rays along that particular path is large, but it does not, by itself, tell us how the X-ray absorption was distributed along the path; it may have been concentrated in single region, in more than one location, or uniformly distributed along the path. This may be conveniently demonstrated on a 2-D function with "width" along the x-axis and

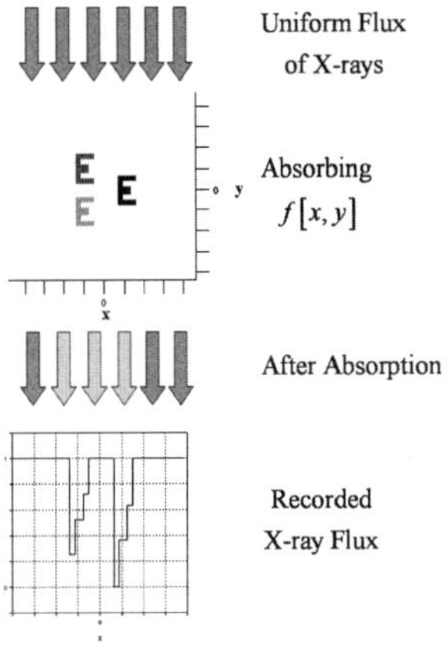

Figure 1.12 Simulation of imaging by a radiographic system. The object is a 2-D function $f[x, y]$ that describes the X-ray attenuation of a body. In this example, "white" represents a region that is completely transparent to X-rays and black represents a perfect absorber. The 1-D image $g[x]$ at the bottom is the line integral of the X-ray absorption by the object. The "depth" information of the structure is lost.

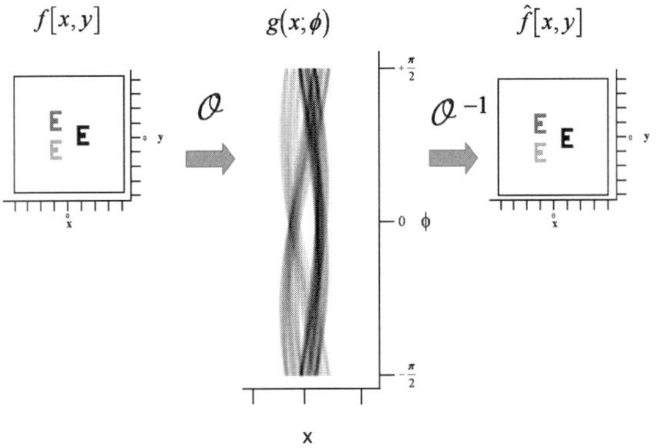

Figure 1.13 Simulation of X-ray CT system. The X-ray transmission of the object is measured at each of a set of angles ϕ to compute the 2-D image $g(x; \phi)$. An estimate $\widehat{f}[x, y]$ of the original object is computed from $g(x; \phi)$.

"depth" along z, i.e., $f[x, z]$. The X-ray absorption integrated along the z-direction yields a 1-D image $g[x]$, as shown in Figure 1.12.

This fundamental limitation of the radiographic system gives us a second opportunity to act as imaging scientists by modifying the system to recover the information about the absorber distribution in "depth" within the object. The machine that first accomplished this goal, the CT scanner, won a Nobel prize for its inventors.

1.4.3 Computed Tomographic Radiography

The modification to the radiographic system to retrieve depth information is somewhat similar to that used in gamma-ray imaging in the sense that more than one image of the same object is collected and processed. Because the source is distinct from the object in radiography, the images are not obtained simultaneously, but rather the source location must be changed between images. Consider the situation if two radiographs of the same object are made so that the X-rays pass through the object at two angles, say ϕ_1 and ϕ_2, relative to some coordinate system. In other words, we measure the X-ray attenuation of the object at two different locations about the object, as shown in Figure 1.13. The system operator now has the form:

$$\mathcal{O}\{f[x, y, z]\} = g[x, y, \phi_n] \tag{1.7}$$

In short, we have constructed a system that generates a third *angular* dimension of image data, though in this case the third dimension is sampled at only two locations and differs from the desired spatial coordinate z. Again, we need to solve the inverse problem by evaluating $f[x, y, z]$. As the next step, we can "fill in" the spaces between the azimuthal samples by gathering data at more angles ϕ. Finally, it is necessary to determine the appropriate mathematical operation that will "reconstruct" the X-ray attenuation at each location in the original object $f[x, y, z]$ from the measurements $g(x, y, \phi)$. The fact that we now have three dimensions of data should indicate that this gives us a "fighting chance" to solve the problem, but the mathematical derivation requires tools yet to be developed. We will derive one means to solve the inverse problem for this system in Chapter 12.

2

Operators and Functions

2.1 Classes of Imaging Operators

If the first chapter achieved its goal, it has whetted our collective appetites about the kinds of imaging systems that may be analyzed by the mathematical principles to be studied. We begin our discussion by considering the general system operator $\mathcal{O}$, and by stating immediately that many, if not most, imaging tasks are not solvable. It usually is necessary to restrict the possible actions of the system, and often rather severely, for an imaging task to be solved, even if we have freedom to modify the system. In these cases, the solution for a restricted system may be applicable only to a very narrow range of problems, but we adopt the philosophy that a "solution of limited applicability is better than no solution at all" and will plunge ahead.

As it happens, two restrictions on the action of the system operator $\mathcal{O}$ are particularly useful: these are (1) linearity and (2) shift invariance, often called *space invariance* or *time invariance* in systems with those dependencies. The descriptions of the system operator are considered in detail in Chapter 8, but the constraints are introduced and discussed briefly now because of their relevance in the description of matrix operators. The concept of "linearity" restricts the potential action of the system on the amplitude of the object, while "shift invariance" limits the ability of the imaging system to transfer energy from locations within the input function to locations in the output. It is important to recognize from the start that these restrictions are *idealizations*; no realistic system is truly linear or truly shift invariant. However, it is often true that any deviations from ideal behavior may be considered independently to derive appropriate corrections to the ideal result.

2.1.1 Linearity

The linearity criterion refers to the action of a system on a weighted sum (or *superposition*) of amplitudes of input functions In the context of this chapter, if the output of the 1-D operator $\mathcal{O}$ for each component function is $\mathcal{O}\{f_n[x]\} = g_n[x]$, then a sufficiently general criterion for linearity of the system is:

$$\begin{aligned}\mathcal{O}\{\alpha_1 f_1[x] + \alpha_2 f_2[x]\} &= \alpha_1 \mathcal{O}\{f_1[x]\} + \alpha_2 \mathcal{O}\{f_2[x]\} \\ &= \alpha_1 g_1[x] + \alpha_2 g_2[x] \end{aligned} \tag{2.1a}$$

where the α_n are weighting constants that generally are complex valued. In words, a linear operator acting on the weighted sum of input functions yields the identically weighted sum of the individual outputs. The corresponding statement for 2-D functions evidently is:

Fourier Methods in Imaging Roger L. Easton, Jr.

if $\mathcal{O}\{f[x, y]\} = g[x, y]$, *then*

$$\mathcal{O}\{\alpha_1 f_1[x, y] + \alpha_2 f_2[x, y]\} = \alpha_1 \mathcal{O}\{f_1[x, y]\} + \alpha_2 \mathcal{O}\{f_2[x, y]\}$$
$$= \alpha_1 g_1[x, y] + \alpha_2 g_2[x, y] \quad \textit{for all } x_0, y_0 \tag{2.1b}$$

2.1.2 Shift Invariance

Shift invariance is satisfied if the action of the operator is independent of the absolute position of the input object within its domain. In other words, the mathematical statement of the criterion to be met by a 1-D shift-invariant operator $\mathcal{O}$ is unaffected by translation of the function $f[x]$, which is in turn denoted by adding a constant to the argument x; this has the effect of "shifting" (translating) the origin of coordinates. For example, a translated function $f[x]$ with the former origin of coordinates relocated to x_0 is $f[x - x_0]$. Though this concept will be considered in much more detail later in this chapter, we can state that a shift-invariant 1-D operator $\mathcal{O}$ must satisfy the condition:

$$\textit{if } \mathcal{O}\{f[x]\} = g[x], \ \textit{then } \mathcal{O}\{f[x - x_0]\} = g[x - x_0] \quad \textit{for all } x_0 \tag{2.2a}$$

The corresponding statement for 2-D functions evidently is:

if $\mathcal{O}\{f[x, y]\} = g[x, y]$,

$$\textit{then } \mathcal{O}\{f[x - x_0, y - y_0]\} = g[x - x_0, y - y_0] \quad \textit{for all } x_0, \ y_0 \tag{2.2b}$$

In other words, shift invariance describes a quality of the action of an operator upon a *location* $[x, y, \ldots]$ within the source function.

The combined restrictions of linearity and shift invariance allow the system operator $\mathcal{O}$ to be expressed in a very concise form, as will be considered in Chapter 8. In the situations where both constraints are valid, the analysis of the system (imaging task #3) is simplified to a significant degree.

2.2 Continuous and Discrete Functions

The bulk of this book is devoted to the description and analysis of objects, images, and systems as mathematical functions. For this reason, we now review some basic and relevant concepts of functions (real and complex, continuous and discrete). Much of this discussion will be superfluous for many readers, but is included for completeness.

2.2.1 Functions

The basic concept of a mathematical *function* was mentioned in Chapter 1 as a rule that assigns a numerical value (the *dependent variable*, which we will often call the *amplitude*) to another numerical value (the *independent variable* or input *coordinate*). Functions are used to describe all aspects of the imaging process: the input scene f, the imaging system $\mathcal{O}$, and the output image g. As mentioned in Section 1.1, the input to a general imaging system usually varies over time, wavelength, and three spatial coordinates, so the appropriate general expression is $f[x, y, z, t, \lambda]$. We have already mentioned that the number of independent variables required to specify a unique location in that domain is the *dimensionality* of the function and that the set of possible values of the independent variables defines the *domain* of the function. Hence, $f[x, y, z, t, \lambda]$ is a five-dimensional function with a domain that includes the infinite spatial volume, all possible times, and all wavelengths of radiation. The set of possible values of the dependent variable f defines the *range* of the function.

A function with multiple spatial dimensions may be expressed in various equivalent coordinate systems, such as the 2-D Cartesian and polar coordinates or the Cartesian, cylindrical, and spherical representations that may be used for 3-D functions. The choice of coordinate system depends on the conditions of the specific imaging problem. In an attempt to clarify the representations, expressions in Cartesian coordinates will be indicated by enclosing the coordinates in brackets, such as the domains $[x, y]$ and $[x, y, z]$. Representations defined over domains that include at least one angle will be enclosed in parentheses; the common 2-D example is the polar representation over the domain (r, θ). This convention occasionally is useful to specify the nature of more complicated coordinate representations, but may be confused with the common notation for "open-ended" and "closed" intervals of real-valued coordinates, which also is used in this chapter to specify the domain and range of a function. In that context, combinations of brackets and parentheses may be used to specify the limits to the domain. Bracketed endpoints, e.g., $[-1, +1]$, specify a domain that includes the specified endpoints, while parentheses, such as $(-1, +1)$, are used when the interval does not include the endpoints. The infinite domain will be denoted by $(-\infty, +\infty)$. The potential for confusion arises when a bracket and a parenthesis are used together to specify an interval that includes one endpoint (the "closed" coordinate) but not the other (the "open" coordinate). For example, the notation $[0, +1)$ indicates an interval of unit length on the real line that includes the origin but not $x = +1$, and thus is equivalent to the statement that $0 \leq x < +1$. Whether the notation refers to a type of representation (Cartesian or polar) or a type of domain should be clear from the context.

We already have established classes for system operators based on their action, using the categories of "linear" and "shift invariant". To simplify the discussion of functions, it is often useful to establish classes based on common characteristics of the domain and/or the range of the members of the class. As examples, consider these variations in the form of a function: the domain may be specified either by continuous coordinates or by a discrete set of "samples"; the coordinates in the domain may be real, imaginary, or complex valued; the range of the dependent variable f may be real, imaginary, or complex valued; the dependent variable f may repeat at regular intervals to create a *periodic* function, etc. Another useful descriptor of the function $f[x, y, \ldots]$ is the set of its "zeros", the locations in the domain where the amplitude f is zero. Yet another useful descriptor is the "shape" of the function over its domain, which may be specified in several ways. One useful method for describing "shape" that is often useful is its similarity to a numerical power of the coordinate x. For example, the amplitude f of a *linear* 1-D function is proportional to the coordinate ($f[x] = \alpha x \propto x^{+1}$). Note that the description of a linear function differs significantly from that of a linear operator, where the output of the sum of individual inputs is identical to the sum of the individual outputs. The second power $f[x] = \beta x^2 \propto x^2$ defines a parabola and is a *quadratic* function. As we will see, it is often useful to express a 1-D function $f[x]$ in terms of its unique set of power-law constituent functions.

All of these classifications for functions (and others yet to be introduced) will be useful in the course of this discussion. We begin by considering the classes based on the continuity of the domain and range.

2.2.2 Functions with Continuous and Discrete Domains

The concept of a real continuous function of real coordinates likely is so familiar to readers that an extended discussion is unnecessary. However, to ensure that this discussion is at least somewhat independent of other sources, some examples of 1-D functions of continuous coordinates are considered to clarify the concepts; these examples are shown in Figure 2.1. The first of these is the simple linear function:

$$f_1[x] \equiv y = x \tag{2.3a}$$

Both the domain and range of f_1 span all real numbers in the open intervals $(-\infty, +\infty)$. We say that the range of the function is "bipolar" because it includes positive and negative values. It has one zero, i.e., $f_1[x] = 0$ at one coordinate $x = 0$.

The second function:

$$f_2[x] = x^2 \tag{2.3b}$$

obviously is quadratic and therefore not linear. Both the domain and range of $f_2[x]$ are continuous: the domain is $(-\infty, +\infty)$, while the range consists of all nonnegative real numbers $[0, +\infty)$. The single zero of $f_2[x]$ is located at $x = 0$.

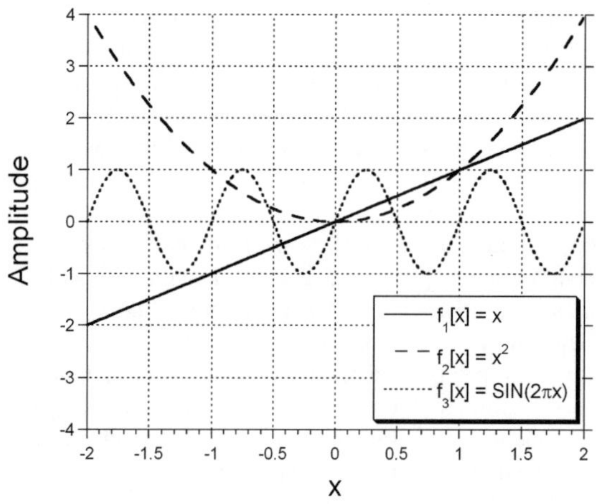

Figure 2.1 Graphs of the functions in Equation (2.3) with continuous domains and different ranges. Though shown with different line styles, all represent continuous functions.

The third function is a 1-D sinusoid with period X_3:

$$f_3[x] = \sin\left[\frac{2\pi x}{X_3}\right] \tag{2.3c}$$

which is shown in Figure 2.1 for $X_3 = 1$. Again both the domain and range are continuous: the domain is the set of real numbers $(-\infty, +\infty)$, while the range is the real-valued closed finite interval $[-1, +1]$. Note that the amplitude f_3 repeats for coordinates x separated by intervals of width X_3, which means that $f_3[x]$ satisfies the requirement for a *periodic* function, to be discussed in detail later in this chapter. The shape of this function $f_3[x]$ is not so easily expressed as a power of the coordinate x, but may be decomposed into an infinite number of such functions. It has an infinite number of isolated zeros that are uniformly spaced at intervals of width $\Delta x = X_3/2$.

The amplitude of $f_3[x]$ is nonzero at all but a discrete set of uniformly spaced isolated coordinates. A second class of functions has, in a sense, a "reciprocal" characteristic in that the amplitude is specified only at a discrete set of uniformly spaced coordinates. An example is:

$$f_4[x] = \cos[2\pi n \cdot \Delta x], \quad n = 0, \pm 1, \pm 2, \ldots \tag{2.4}$$

where Δx is some increment in the space domain. The function is evaluated only at values of the independent variable that satisfy $x = n \cdot \Delta x$. We speak of a *discrete* function in this case, which may be constructed from continuous functions by the common process of *sampling*. The mathematical description of sampling and its effects are very important in contemporary imaging and are considered in detail in Chapter 14. An example of a sampled image is shown in Figure 2.2.

The amplitude of the example functions in Equation (2.3) vary "smoothly" with x over their continuous domains. In other words, there are no "transition" coordinates where the derivative is not defined. Clearly this statement is not valid for the tangent function:

$$f_5[x] = \tan\left[\frac{2\pi x}{X}\right] \tag{2.5}$$

As shown in Figure 2.3 for $X = 2$, its amplitude "jumps" from $f_5 \to +\infty$ at $x = +(X/4) - \epsilon$ to $f_5 \to -\infty$ at $x = +(X/4) + \epsilon$, where ϵ is a very small positive real number (denoted by "$\epsilon \to 0$"). The amplitude of $f_5[x]$ is not defined at $x = X/4$ and its derivative is not finite there.

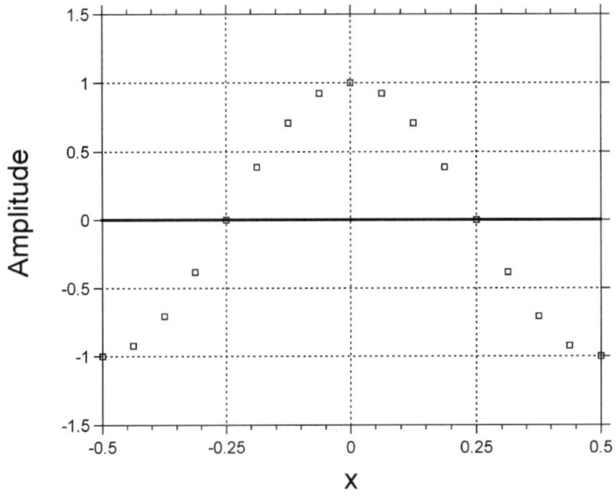

Figure 2.2 Discrete ("sampled") function $f[x] = \cos[2\pi n \cdot \frac{1}{16}]$, which has a discrete domain and a continuous range on the interval $[+1, -1]$.

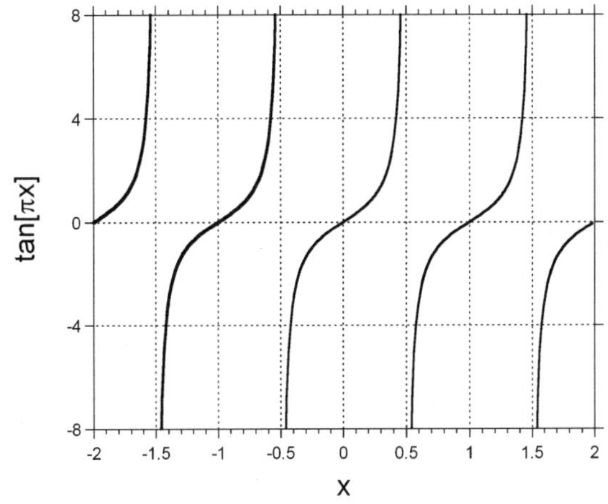

Figure 2.3 "Discontinuous" function $\tan[2\pi x/2]$, which has continuous domain and range over $(-\infty, +\infty)$.

2.2.3 Continuous and Discrete Ranges

The range of a real-valued function $f[x]$ already has been defined as the set of all allowed values of the dependent variable f. As was true for the domains, different flavors of range exist for different functions. For example, the range may have an infinite or finite extent, and all values between these extreme values may be allowed or only some discrete set. The range is continuous in the former case where all real numbers are allowed, though this description may be confused with the use of the descriptor "continuous" for a function with finite derivative at all coordinates. The trigonometric tangent function $f_5[x]$ in Equation (2.5) is "discontinuous" in the second sense due to its transitions

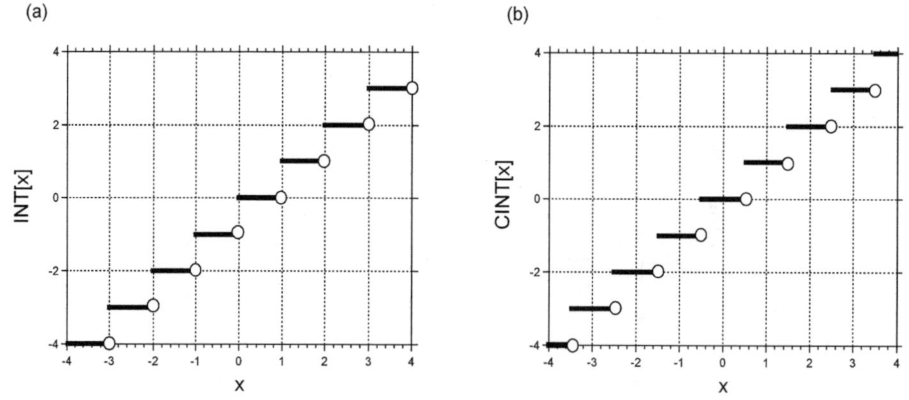

Figure 2.4 (a) "Greatest integer" function (truncation of amplitude) compared to (b) "closest integer" function *CINT*$[x]$ that "rounds" the amplitude to the nearest integer. The circles at the right end of each line indicate the "open end" of the interval in the integer mapping.

at periodic isolated values of X. However, because the range includes all real numbers in the interval $(-\infty, +\infty)$, $f_5[x]$ is "continuous" in the first sense of the term.

To introduce the concept of a function with a *discrete* range, consider the operator that evaluates the integer part of a real-valued independent variable x. In other words, the operation discards ("truncates") any decimal part of the amplitude. The process is often denoted by the name *INT*[ ] for "largest integer":

$$f_6[x] = INT[x] \tag{2.6}$$

which is shown in Figure 2.4. The domain of *INT*$[x]$ is the entire real line $(-\infty, +\infty)$, while the range includes the discrete (though infinite) set of integers $(0, \pm 1, \pm 2, \ldots)$. A common variant, the "closest integer" function *CINT*$[x]$, rounds the amplitude to the nearest integer, and may be evaluated via the *INT* function:

$$f_7[x] = CINT[x] \equiv INT\left[x + \frac{1}{2}\right] \tag{2.7}$$

INT[ ] and *CINT*[ ] are compared in Figure 2.4.

A truncation or rounding operator may be applied to the amplitude of any function $f[x]$ to convert from a continuous to a discrete range, e.g.,

$$g[x] = CINT\{f[x]\} \Longrightarrow g = CINT[f] \tag{2.8}$$

The domain of the "output" function $g[x]$ is continuous and its range is discrete. An example of the action of this operator on an input function $f[x]$ is shown in Figure 2.5. The process in Equation (2.8) is often called *quantization*, which is the second fundamental operation required in any imaging application that involves digital computation. Because upper-case script characters (such as $\mathcal{O}$) are used to denote system operators, we adopt the specific notation $\mathcal{Q}$ for quantization processes that generate functions with a discrete range from measurements of the continuous amplitude of the input function.

2.2.4 Discrete Domain and Range – "Digitized" Functions

A large fraction of imaging scientists spend much of their time evaluating and processing images in digital computers, which are capable of operating only on functions whose domain and range are discrete. These functions are evaluated only at a discrete set of coordinates *and* the ranges are restricted to a discrete set of values. The process is called *digitization* and such entities often are called *digital*

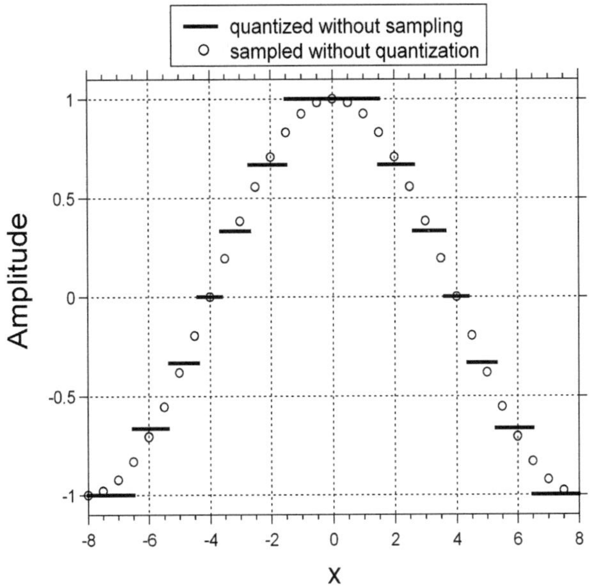

Figure 2.5 The 1-D function $f[x] = \cos[2\pi x/18]$, which has continuous domain and range, and the results from quantization alone (at eight levels) and from sampling alone (at 16 locations).

functions. An example of a digital rendering of a cosine function is compared to its sampled version in Figure 2.6.

2.2.5 Periodic, Aperiodic, and Harmonic Functions

The concept of a *periodic* function is another that is so familiar that in-depth discussion is unnecessary for most readers, so this short review again is included mainly for completeness. The criterion for a 1-D function $f[x]$ to be periodic may be expressed in a concise form:

$$f[x_0] = f[x_0 + nX_0] \tag{2.9}$$

where n is any integer and X_0 (the period of $f[x]$) is the smallest possible increment of the independent variable x such that Equation (2.9) is satisfied. In words, a nonnull function $f[x]$ is periodic if its amplitude is identical at all coordinates separated by integer multiples of some distance X_0. Note that the amplitude of any such function within a period may have a very irregular form; the criterion for periodicity merely requires that the amplitude repeat at regular intervals. An example is shown in Figure 2.7. The term *aperiodic function* evidently refers to any function that does not satisfy the criterion for periodicity, but is often applied to functions that are approximately periodic.

Functions of dimension two (or larger) may be periodic over all coordinates or over just a subset. For example, a 2-D function that is periodic over x but not over y may still be considered to be a periodic function in some contexts.

Harmonic functions are a specific subclass of periodic functions produced by oscillatory motions common in virtually all disciplines of science, but particularly in acoustics and in electromagnetism (and therefore in optics). A harmonic function is a sinusoid with a single wavelength (or frequency):

$$f[x] = A_0 \cos[\Phi[x]] = A_0 \cos\left[\frac{2\pi x}{X_0} + \phi_0\right] \tag{2.10}$$

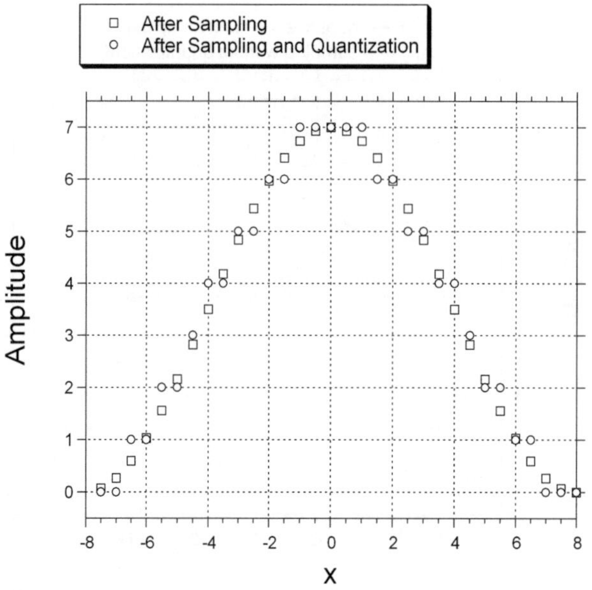

Figure 2.6 Results from the cascade of sampling and subsequent quantization to "digitize" a nonnegative sinusoid. The continuous values after sampling are shown as squares and the discrete values after subsequent quantization as circles.

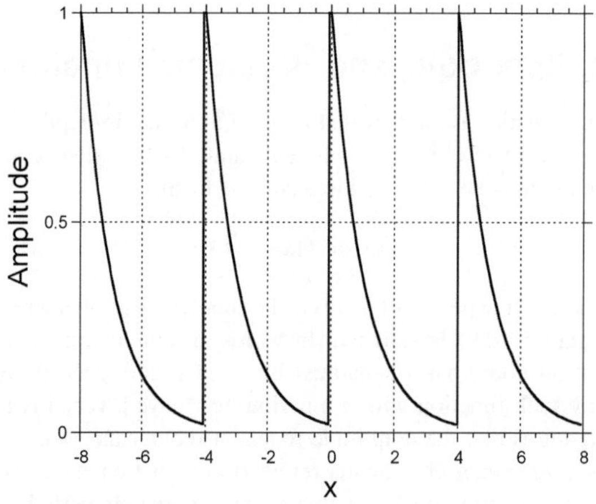

Figure 2.7 Function that obeys condition for periodicity: $f[x] = f[x + 4n]$ for all integers n.

which is plotted in Figure 2.8, where $\Phi[x]$ is the *phase angle* measured in radians; the phase angle of a harmonic function must be a linear function of the coordinate x. The additive constant ϕ_0 is the phase angle at $x = 0$, and therefore is often called the *initial phase*, though often (and confusingly) called just the *phase*. The concept of the phase function $\Phi[x]$ may be generalized to nonsinusoidal and

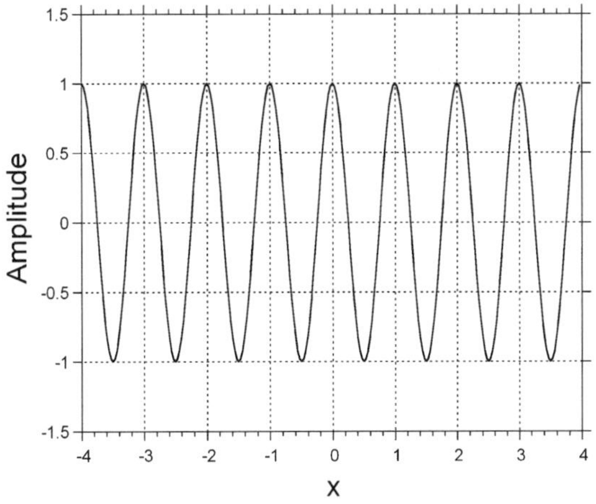

Figure 2.8 Harmonic function composed of a single sinusoid $f[x] = \cos[2\pi x/X_0]$ for $X_0 = 1$.

even to nonperiodic functions, and its interpretation is exceedingly important in many types of physical problems.

Of course, the argument of the harmonic function need not be a spatial coordinate; harmonic functions of time and of both space and time are common. Consider the common sinusoidal function whose phase is a linear function of both space and time:

$$f[x,t] = A_0 \cos[\Phi[x,t]] = A_0 \cos\left[\frac{2\pi x}{X_0} - \frac{2\pi t}{T_0} + \phi_0\right] \tag{2.11}$$

The amplitude is A_0 where the entire phase function $\Phi[x, t]$ evaluates to an integer multiple of 2π radians. Any location $[x_1, t_1]$ in the 2-D space–time domain for which the phase evaluates to zero specifies a particular maximum of the sinusoid. As the value of the time variable t increases to t_2, the spatial position x of this particular maximum also must change to x_2 to maintain $\Phi = 0$, i.e., $\Phi[x_1, t_1] = \Phi[x_2, t_2] = 0$. In the example of Equation (2.11), the location x of the point with zero phase must increase as t increases to maintain the value of Φ. In other words, this "point of constant phase" moves toward $x = +\infty$ with increasing t. In physics, sinusoids of the form of Equation (2.11) are called *traveling waves*.

Much of the descriptive jargon to be used in this book originated in the early use of harmonic functions to describe physical problems. For example, an amplitude of the form of Equation (2.11) may be used to describe an acoustic wave propagating along the x-axis. As it happens, the rate of energy transfer of the acoustic wave $f[x, t]$ is proportional to $(f[x, t])^2$, which is called the *power* of the wave.

The period X_0 of the harmonic sinusoid in Equation (2.11) is the interval of x over which the phase changes by 2π radians if t is kept fixed. It therefore is the distance between adjacent maxima and clearly must have units of length (e.g., mm). However, it is often convenient to recast the expression for the phase in terms of the reciprocal of the period $\xi_0 \equiv X_0^{-1}$, which specifies the number of periods ("cycles") of the sinusoid in one unit of the independent variable x (again perhaps measured in mm). Because it specifies a "rate" of sinusoidal oscillation, ξ_0 is called the *spatial frequency*. The equivalent expression for the sinusoid in Equation (2.10) written in terms of ξ_0 is:

$$f[x] = A_0 \cos[\Phi[x]] = A_0 \cos[2\pi\xi_0 x + \phi_0] \tag{2.12}$$

It is very common to measure ξ_0 in units of "cycles per mm".

An alternate measure of spatial frequency that is often used is based upon the number of radians of phase that exist within one unit of length. This *angular spatial frequency* will be denoted by k_0, which also is often called the *wavenumber* of a spatial wave (another common notation for the angular spatial frequency is σ_0). The equivalent expression to Equation (2.10) and Equation (2.12) is:

$$f[x] = A_0 \cos[\Phi[x]] = A_0 \cos[k_0 x + \phi_0] \tag{2.13}$$

where $k_0 = 2\pi\xi_0$.

If the phase function $\Phi[x]$ is known for a spatial sinusoid such as Equation (2.10), then the angular spatial frequency is most conveniently defined as the spatial derivative of the phase:

$$\textit{Angular spatial frequency } k_0 \equiv \frac{\partial\Phi[x]}{\partial x} \quad [\textit{radians per unit length}] \tag{2.14a}$$

which means that the angular spatial frequency will be constant over the entire domain if $\Phi[x]$ is a linear function of x. The *spatial frequency* is obtained by dividing by the factor of 2π radians per cycle:

$$\textit{Spatial frequency } \xi_0 = \frac{1}{2\pi}\frac{\partial\Phi[x]}{\partial x} \quad [\textit{cycles per unit length}] \tag{2.14b}$$

If the independent variable of the harmonic function has dimensions of time (e.g., seconds), then the parameter corresponding to the spatial frequency ξ_0 is the associated *temporal frequency*, often indicated by ν_0 and having units of "cycles per time interval" (e.g., *cycles per second*, now called *hertz* and abbreviated Hz). The units of the corresponding definition for the angular temporal frequency ω_0 are "radians per unit time", so that the conversion is $\omega_0 = 2\pi\nu_0$. The expressions for the temporal frequencies in terms of the phase function obviously are:

$$\textit{Angular temporal frequency } \omega_0 = \frac{\partial\Phi[t]}{\partial t} \quad [\textit{radians per unit time}] \tag{2.14c}$$

$$\textit{Temporal frequency } \nu_0 = \frac{1}{2\pi}\frac{\partial\Phi[t]}{\partial t} \quad [\textit{cycles per unit time}] \tag{2.14d}$$

The sinusoid in Equation (2.11) whose phase is a linear function of space and time may be rewritten in terms of these frequencies:

$$f[x, t] = A_0 \cos[\Phi[x, t]] = A_0 \cos[2\pi\xi_0 x_0 - 2\pi\nu_0 t_0 + \phi_0] \tag{2.15}$$

One significant consequence of the definition of the various flavors of frequency as derivatives of the harmonic phase is the evident possibility of *negative* harmonic frequencies, which result if the phase angle *decreases* as the corresponding coordinate (x or t) increases.

Note that any function $f[x]$ known to be a sinusoid with one of the forms presented in Equations (2.10)–(2.13) may be specified completely at any coordinate from the knowledge of three parameters: the amplitude A_0, the period X_0 (or spatial frequency ξ_0 or angular spatial frequency k_0), and the phase angle at any known coordinate (usually the initial phase ϕ_0). This equivalence is the basis for the alternate representation of a function obtained via the Fourier transform. The independent variable x and its corresponding frequency ξ_0 (or the time t and temporal frequency ν_0) are called *conjugate* variables.

The consequences of this discussion may be encapsulated into a concise definition of a harmonic function; it is a sinusoid whose frequency (spatial or temporal) does not vary with coordinate, and thus that the phase must be a linear function of coordinate. However, it is easy (and useful) to define sinusoids whose phase is a nonlinear function of coordinate, and thus whose frequency varies in space and/or time. An example of such a function that is very useful in imaging applications has a quadratic phase (Figure 2.9):

$$f[x] = A_0 \cos\left[\pi\left(\frac{x}{\alpha_0}\right)^2 + \phi_0\right] \tag{2.16}$$

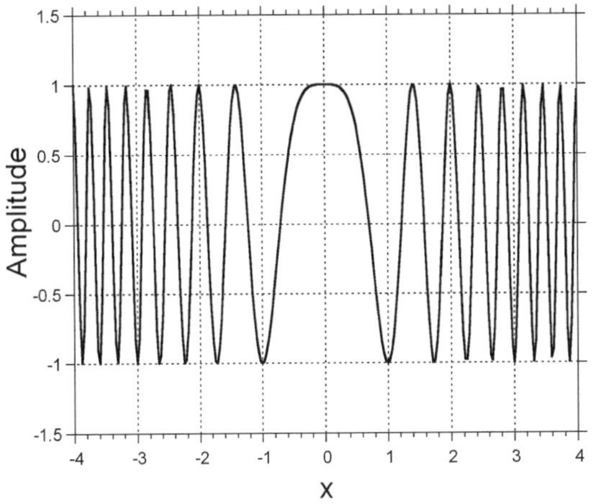

Figure 2.9 Sinusoidal function with quadratic phase, $f[x] = \cos[\pi x^2]$.

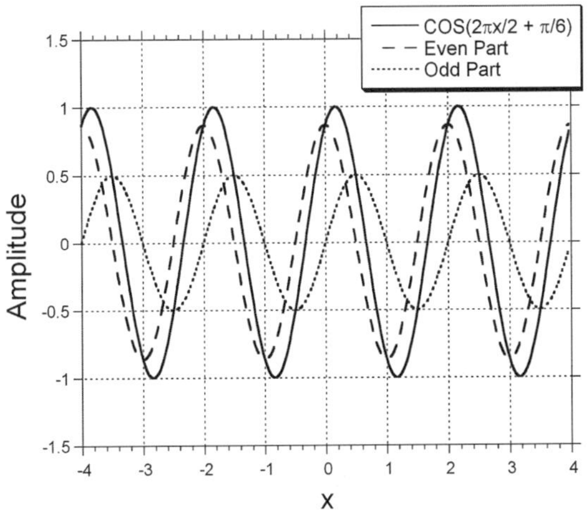

Figure 2.10 Sinusoidal function $f[x] = \cos[2\pi x/2 + \pi/6]$ decomposed into its even and odd parts: $f_e[x] = (\sqrt{3}/2)\cos[\pi x]$ and $f_o[x] = \frac{1}{2}\sin[\pi x]$.

Simple substitution of the quadratic phase into the definition of spatial frequency in Equation (2.14) demonstrates that the frequency is proportional to the coordinate:

$$\xi = \frac{1}{2\pi} \cdot \frac{\partial(\Phi[x])}{\partial x} = \frac{1}{2\pi} \cdot \left(2\pi \frac{x}{\alpha_0^2}\right) = \frac{x}{\alpha_0^2} \propto x \tag{2.17}$$

The properties of 1-D and 2-D quadratic-phase sinusoidal functions are considered in more detail in Chapter 6 and Chapter 7, and applications will be developed in Chapter 17 and Chapter 21.

2.2.6 Symmetry Properties of Functions

Another useful scheme for classifying functions and operators is based on the principle of algebra that allows a function $f[x]$ to be decomposed into unique *even* and *odd* parts $f_e[x]$ and $f_o[x]$ that satisfy the respective criteria:

$$f_e[x] = f_e[-x] \tag{2.18a}$$

$$f_o[x] = -f_o[-x] \tag{2.18b}$$

The even part of $f[x]$ is *symmetric* with respect to the origin, while the odd part is *antisymmetric*. It is easy to demonstrate that the even and odd parts of $f[x]$ are evaluated from $f[x]$ via:

$$f_e[x] \equiv \frac{1}{2}(f[x] + f[-x]) \tag{2.19a}$$

$$f_o[x] \equiv \frac{1}{2}(f[x] - f[-x]) \tag{2.19b}$$

and the sum of these two demonstrates that:

$$f[x] = f_e[x] + f_o[x] \tag{2.20}$$

It is evident that $\cos[2\pi\xi_0 x]$ is an even harmonic function and $\sin[2\pi\xi_0 x]$ is its odd counterpart. The general harmonic function with spatial frequency ξ_0 and initial phase ϕ_0 may be decomposed into its constituent even and odd parts by applying a (probably already familiar) trigonometric identity that will be derived in Chapter 4:

$$\cos[\alpha \pm \beta] = \cos[\alpha]\cos[\beta] \mp \sin[\alpha]\sin[\beta] \tag{2.21}$$

This expression yields the appropriate weightings of the even and odd parts necessary to generate the general harmonic function:

$$f[x] = \cos[2\pi\xi_0 x + \phi_0] = \cos[\phi_0]\cos[2\pi\xi_0 x] + (-\sin[\phi_0])\sin[2\pi\xi_0 x] \tag{2.22a}$$

$$f_e[x] = \cos[\phi_0]\cos[2\pi\xi_0 x] \tag{2.22b}$$

$$f_o[x] = (-\sin[\phi_0])\sin[2\pi\xi_0 x] \tag{2.22c}$$

In words, the even part and odd parts of the general harmonic function are themselves harmonic functions with the same spatial frequency ξ_0. The amplitudes of the even and odd parts are "weighted" by the cosine and sine of the initial phase ϕ_0, respectively (Figure 2.10). The converse of this statement also is true: that the sum of two sinusoids with the same frequency and arbitrary amplitudes and phases yields a sinusoid with that same frequency. For a 2-D function $f[x, y]$, the relationships corresponding to those in Equation (2.19) are:

$$f[x, y] = f_e[x, y] + f_o[x, y] \tag{2.23a}$$

$$f_e[x, y] \equiv \frac{1}{2}(f[x, y] + f[-x, -y]) \tag{2.23b}$$

$$f_o[x, y] \equiv \frac{1}{2}(f[x, y] - f[-x, -y]) \tag{2.23c}$$

Most (if not all) readers are familiar with the concept of the polar representation of a 2-D function's coordinates, such as $f_r(r, \theta)$, where r is the radial coordinate $(0 \leq r < \infty)$ and θ is the azimuthal angle that spans 2π radians. The even and odd parts of a function in polar coordinates must satisfy the relationships:

$$f_e(r, \theta) = f_e(r, \theta \pm n\pi) \tag{2.24a}$$

$$f_o(r, \theta) = -f_o(r, \theta \pm n\pi) \tag{2.24b}$$

where n is any integer.

PROBLEMS

2.1 Use the definitions of Equation (2.18) to derive expressions for the even and odd functions in Equation (2.19).

2.2 Consider two spatial sinusoids with the same spatial frequency ξ_0 but arbitrary amplitudes A_1 and A_2 and arbitrary phases ϕ_1 and ϕ_2:

(a) Prove that the sum of these two sinusoids is a sinusoid with that same frequency ξ_0.

(b) Find the expression that relates the amplitude A and phase ϕ of the summation sinusoid.

2.3 For a sinusoidal functions whose phase is a power of the coordinate:

$$f[x] = \cos\left[\pi\left(\frac{x}{\alpha}\right)^n + \phi_0\right]$$

(a) Graph the function for $\alpha = 1$, $\phi_0 = \pi/4$, and $n = 1$, 3, and 4.

(b) Find the equation for the spatial frequency of $f[x]$.

(c) Determine the dimensions of the parameter α.

3

Vectors with Real-Valued Components

3.1 Scalar Products

At this point, the path of this discussion makes a significant change in direction to consider the concepts of vector analysis. This study is useful for two reasons. First, many of the mathematical concepts used to analyze imaging systems may be usefully expressed in terms of scalars, vectors, and matrices. Second, the mathematical tools of discrete vectors and matrices may be generalized to functions of continuous variables. For example, the concepts of the scalar product of two vectors and the *projection* of one vector onto another may be generalized easily and are essential in the analysis of imaging systems. The discussion of this chapter follows that of many texts in linear algebra, such as that of Strang (2005). Similar material is presented in Parker (1990).

Any physical or mathematical quantity whose amplitude may be decomposed into "directional" components, i.e., components directed along different axes, is often represented conveniently as a vector. In this discussion, vectors are denoted by bold-faced underscored lower-case letters, e.g., $\underline{\mathbf{x}}$. For example, for a vector with two real-valued components the 3-D vector $\underline{\mathbf{x}}$ is specified by a vertical column of the three ordered numerical components:

$$\underline{\mathbf{x}} \equiv \begin{bmatrix} x_1 \\ x_2 \\ x_3 \end{bmatrix} \tag{3.1}$$

The usual notation for a vector with N elements is a column of N individual numerical scalars, where the parameter N is the *dimensionality* of the vector. If $N = 2$ or $N = 3$, the vector is usually interpreted as specifying a location in the 2-D plane or 3-D volume. The interpretation is extended to use vectors with four or more components to specify Cartesian coordinates of a location in a nonphysical space. Though such a vector is more difficult to visualize in these terms, it is easy to visualize if we think of it as a 1-D array of values. Consider the *transpose* of the column vector $\underline{\mathbf{x}}$, which is defined to be the same set of scalar components arrayed as a horizontal row, and is denoted in this discussion by a superscript T (another common notation uses an overscored tilde):

$$\underline{\mathbf{x}}^T = \begin{bmatrix} \boxed{x_1} & \boxed{x_2} & \boxed{x_3} & \boxed{\cdots} \end{bmatrix} = \tilde{\underline{\mathbf{x}}} \tag{3.2}$$

Fourier Methods in Imaging Roger L. Easton, Jr.

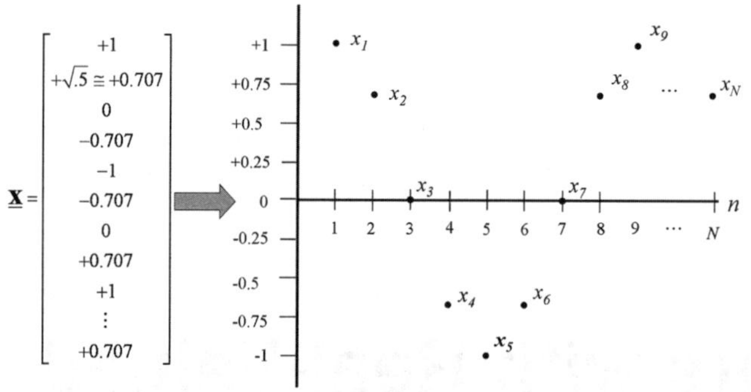

Figure 3.1 N-dimensional column vector $\underline{\mathbf{x}}$ and its graphical interpretation as a 1-D array of scalar components.

This notation naturally suggests a graphical rendering as samples of a 1-D function, where the x-coordinate of the function is the index of the coordinate and the y-axis is the amplitude, as shown in Figure 3.1. This interpretation suggests a mild notational discrepancy about *dimensions* because we are rendering an N-dimensional vector with a 1-D function. The specific meaning should be clear from the context.

Both real- and complex-valued scalars will be denoted by the same symbol x_n. If the components x_n are real, then the typical interpretation of a vector $\underline{\mathbf{x}}$ in Equation (3.1) is that it specifies a location in 3-D Cartesian space. The individual scalar components x_1, x_2, and x_3 are equivalent to the distances along the three axial directions (commonly labeled x, y, and z, respectively, in the space domain). In many common situations, the components of the vector $\underline{\mathbf{x}}$ have dimensions of length, but other representations are possible and useful. For example, we will often use a convenient representation of a sinusoid in the x–y plane that is specified by a spatial frequency vector whose components have the dimensions of cycles per mm.

To minimize any confusion resulting from the use of the symbol "x" to represent both a vector and a particular component of a vector, a normal-faced "x_i" with a subscript will be used to indicate the ith component of the vector $\underline{\mathbf{x}}$, while the bold-faced subscripted symbol "$\underline{\mathbf{x}}_i$" denotes the ith member of a set of vectors. Other notations also will be employed during certain aspects of the discussion, but these cases will be explicitly noted.

Algebraic operations of vectors are essential in this discussion. For example, the sum of two N-D vectors $\underline{\mathbf{x}}$ and $\underline{\mathbf{y}}$ is generated by summing the pairs of corresponding components:

$$\underline{\mathbf{x}} + \underline{\mathbf{y}} = \begin{bmatrix} x_1 \\ x_2 \\ \vdots \\ x_N \end{bmatrix} + \begin{bmatrix} y_1 \\ y_2 \\ \vdots \\ y_N \end{bmatrix} = \begin{bmatrix} x_1 + y_1 \\ x_2 + y_2 \\ \vdots \\ x_N + y_N \end{bmatrix} \tag{3.3}$$

The notations "x" and "y" used here merely distinguish between the two vectors and their components; they are *not* references to the x- and y-coordinates of 2-D or 3-D space. Note that this definition implies that the corresponding components of the two vectors must have the same dimension for their sum to exist.

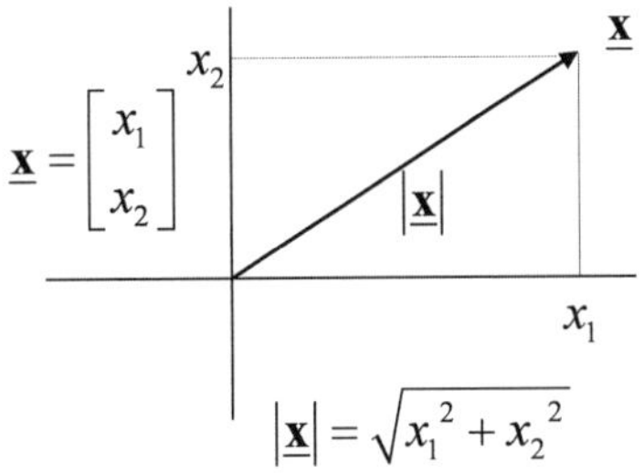

Figure 3.2 Length or "norm" of 2-D vector with real-valued components.

The definition of the difference of two vectors is evident from Equation (3.3):

$$\underline{\mathbf{x}} - \underline{\mathbf{y}} = \begin{bmatrix} x_1 \\ x_2 \\ \vdots \\ x_N \end{bmatrix} - \begin{bmatrix} y_1 \\ y_2 \\ \vdots \\ y_N \end{bmatrix} = \begin{bmatrix} x_1 - y_1 \\ x_2 - y_2 \\ \vdots \\ x_N - y_N \end{bmatrix} \tag{3.4}$$

Obviously, if the number of dimensions N of the vector is 1, 2, or 3, then the corresponding vector $\underline{\mathbf{x}}$ specifies a location on a line, on a plane, or within a volume, respectively. This interpretation of a vector as the location of a point in space is so pervasive and intuitive that it may obscure other useful and perhaps more general interpretations of vectors and vector components. For example, we can use the vector notation to represent a 2-D sampled object. Such an object is formed from an $N \times N$ array of samples or by "stacking" the N columns to create a 1-D vector with N^2 components. This stacking process is known as *lexicographic* ordering of the matrix. Such a representation is often used when constructing computer algorithms for processing digital images, but will not be considered further here.

By analogy with the usual interpretation of a vector in Cartesian space, the length of a vector with real-valued components is a real-valued scalar computed from the 2-D or 3-D "Pythagorean" sum of the components:

$$\sum_{n=1}^{N} (x_n)^2 \equiv |\underline{\mathbf{x}}|^2 \tag{3.5}$$

The result is the *squared magnitude* of the vector. The vector's length, or *norm*, is the square root of Equation (3.5), as shown in Figure 3.2, and thus also is real valued:

$$|\underline{\mathbf{x}}| = \sqrt{\sum_{n=1}^{N} (x_n)^2} \tag{3.6}$$

This definition implies that the norm of a vector must be nonnegative ($|\underline{\mathbf{x}}| \geq 0$) and that it is zero only if all scalar components of the vector are zero.

Vectors with unit length will be essential in the discussion of transformations into alternate representations. Such a *unit vector* is often indicated by a circumflex ($\hat{\underline{\mathbf{x}}}$). The unit vector pointing in the direction of any vector $\underline{\mathbf{x}}$ may be generated by dividing each component of $\underline{\mathbf{x}}$ by the scalar length

$|\underline{\mathbf{x}}|$ of the vector:

$$\hat{\underline{\mathbf{x}}} = \frac{\underline{\mathbf{x}}}{|\underline{\mathbf{x}}|} = \begin{bmatrix} \left(\frac{x_1}{|\underline{\mathbf{x}}|}\right) \\ \left(\frac{x_2}{|\underline{\mathbf{x}}|}\right) \\ \vdots \\ \left(\frac{x_N}{|\underline{\mathbf{x}}|}\right) \end{bmatrix} \tag{3.7}$$

The squared-magnitude operation of Equation (3.5) is the first example of the vector *scalar product* (also called the *dot product*), which defines a "product" of two vectors of the same dimension that generates a scalar. Following common mathematical notation, the scalar product operation will be denoted by a "dot" (•) between the symbols for the vectors. The process also may be written as the transpose of $\underline{\mathbf{x}}$ multiplied from the right by $\underline{\mathbf{x}}$. Therefore, the scalar product of a vector $\underline{\mathbf{x}}$ with itself may be written in equivalent ways:

$$|\underline{\mathbf{x}}|^2 = (\underline{\mathbf{x}} \bullet \underline{\mathbf{x}}) \equiv \underline{\mathbf{x}}^T \underline{\mathbf{x}} = \sum_{n=1}^{N} x_n^2 \tag{3.8}$$

3.1.1 Scalar Product of Distinct Vectors

It is easy to generalize Equation (3.8) to define the scalar product of distinct vectors $\underline{\mathbf{a}}$ and $\underline{\mathbf{x}}$ that have real-valued components **and** that have the same dimension N:

$$\underline{\mathbf{a}} \bullet \underline{\mathbf{x}} \equiv \underline{\mathbf{a}}^T \underline{\mathbf{x}} = \begin{bmatrix} \boxed{a_1} & \boxed{a_2} & \boxed{\cdots} & \boxed{a_N} \end{bmatrix} \begin{bmatrix} \boxed{x_1} \\ \boxed{x_2} \\ \boxed{\vdots} \\ \boxed{x_N} \end{bmatrix}$$

$$= a_1 x_1 + a_2 x_2 + \cdots + a_N x_N = \sum_{n=1}^{N} a_n x_n \tag{3.9}$$

In words, the scalar product of two vectors is obtained by multiplying pairs of vector components with the same indices and summing these products. Note that the scalar product of two distinct vectors may be positive, negative, or zero, whereas the scalar product of a vector with itself produces the nonnegative squared magnitude in Equation (3.8). It is evident that the scalar products of vectors with real-valued components in either order are identical:

$$\underline{\mathbf{a}} \bullet \underline{\mathbf{x}} = \underline{\mathbf{x}} \bullet \underline{\mathbf{a}} \tag{3.10}$$

Any process that performs an action between two entities for which the order is immaterial is *commutative*. The simple concept of the scalar product is the basis (future pun intended) for some very powerful tools for describing vectors and, after appropriate generalization, for functions of continuous variables. The features of the various forms of scalar product are the subject of much of the remainder of this chapter.

The scalar product of an arbitrary "input" vector $\underline{\mathbf{x}}$ with a "reference" vector $\underline{\mathbf{a}}$ has the form of an operator acting on $\underline{\mathbf{x}}$ to produce a scalar g. The appropriate process was defined in Equation (3.9):

$$\mathcal{O}\{\underline{\mathbf{x}}\} = \underline{\mathbf{a}} \bullet \underline{\mathbf{x}} = \sum_{n=1}^{N} a_n\, x_n = g \tag{3.11}$$

It is apparent that a real-valued multiplicative scale factor k applied to each component of the real-valued input vector $\underline{\mathbf{x}}$ results in the same scaling of the output scalar:

$$\mathcal{O}\{k\,\underline{\mathbf{x}}\} = \sum_{n=1}^{N} a_n (k \cdot x_n) = k \sum_{n=1}^{N} a_n\, x_n = k \cdot g \tag{3.12}$$

which demonstrates that the scalar product "operator" of Equation (3.11) satisfies the linearity condition of Equation (2.1).

The geometrical interpretation of a 2-D vector as the endpoint of a line drawn from the origin on the 2-D plane leads to an alternate expression for the scalar product of two vectors. It is convenient to use 2-D vectors denoted by $\underline{\mathbf{f}}_n$ with Cartesian components $[x_n, y_n]$, or represented in polar coordinates by the length $|\underline{\mathbf{f}}_n|$ and the azimuth angles θ_n. The geometric picture of the vector establishes the relationship between the polar and Cartesian representations to be:

$$\underline{\mathbf{f}}_n = [x_n, y_n] = [|\underline{\mathbf{f}}_n| \cos[\theta_n], |\underline{\mathbf{f}}_n| \sin[\theta_n]] \tag{3.13}$$

where, in this case, x_n and y_n represent x- and y-coordinates of the vector $\mathbf{f}_n$. The scalar product of two such vectors $\underline{\mathbf{f}}_1$ and $\underline{\mathbf{f}}_2$ is obtained by applying the definition of Equation (3.9), and may be cast into a different form by using the still-unproven trigonometric identity from Equation (2.21):

$$\begin{aligned}
\underline{\mathbf{f}}_1 \bullet \underline{\mathbf{f}}_2 &= x_1 x_2 + y_1 y_2 \\
&= (|\underline{\mathbf{f}}_1| \cos[\theta_1])(|\underline{\mathbf{f}}_2| \cos[\theta_2]) + (|\underline{\mathbf{f}}_1| \sin[\theta_1])(|\underline{\mathbf{f}}_2| \cos[\theta_2]) \\
&= |\underline{\mathbf{f}}_1||\underline{\mathbf{f}}_2|(\cos[\theta_1]\cos[\theta_2] + \sin[\theta_1]\sin[\theta_2]) \\
&= |\underline{\mathbf{f}}_1|\,|\underline{\mathbf{f}}_2| \cos[\theta_1 - \theta_2] = |\underline{\mathbf{f}}_1|\,|\underline{\mathbf{f}}_2| \cos|\theta_1 - \theta_2|
\end{aligned} \tag{3.14}$$

where the symmetry of the cosine function has been used in the last step. In words, the scalar product of two 2-D vectors is equal to the product of the lengths of the vectors and the cosine of the included angle $|\theta_1 - \theta_2|$. Knowledgeable readers will be aware that this result has been obtained by circular reasoning: we are defining the scalar product form in Equation (3.14) by using the Cartesian components of polar vectors in Equation (3.13), which were themselves determined by scalar products with the Cartesian basis vectors. This quandary is due in part to the familiarity of these concepts. Rather than resolve the issue from first principles, we will instead "sweep it under the rug" while continuing to use our existing intuition as a springboard to generalize these concepts to other applications. For example, it is easy now to generalize the scalar product in Equation (3.14) to real-valued vectors $\underline{\mathbf{a}}$ and $\underline{\mathbf{x}}$ with arbitrary dimension N:

$$\underline{\mathbf{a}} \bullet \underline{\mathbf{x}} = |\underline{\mathbf{a}}||\underline{\mathbf{x}}| \cos[\theta_a - \theta_x] = |\underline{\mathbf{a}}||\underline{\mathbf{x}}| \cos[\theta] \tag{3.15}$$

where θ is a simplified notation for the "included" angle between the two N-D vectors.

Equation (3.15) may be used to derive the *Schwarz inequality* for vectors by recognizing that $\cos[\theta] \leq 1$:

$$\underline{\mathbf{a}} \bullet \underline{\mathbf{x}} \leq |\underline{\mathbf{a}}||\underline{\mathbf{x}}| \tag{3.16}$$

The equality is satisfied only for vectors $\underline{\mathbf{a}}$ and $\underline{\mathbf{x}}$ that "point" in the same direction, which means that all ratios of corresponding components of $\underline{\mathbf{a}}$ and $\underline{\mathbf{x}}$ are identical, so that the included angle $\theta = 0$ radians; in other words, the vectors are scaled replicas. Note both the similarity and difference between the Schwarz inequality and *triangle inequality* for vectors, which has the form:

$$|\underline{\mathbf{a}} + \underline{\mathbf{x}}| \leq |\underline{\mathbf{a}}| + |\underline{\mathbf{x}}| \tag{3.17}$$

In words, the Schwarz inequality says that the scalar product of two vectors can be no larger than the product of their lengths, while the triangle inequality establishes that one side of a triangle can be no longer than the sum of the other two sides. Both relations are illustrated in Figure 3.3. Equation (3.15) may be combined with the definition of the unit vector in Equation (3.7) to obtain an expression for the

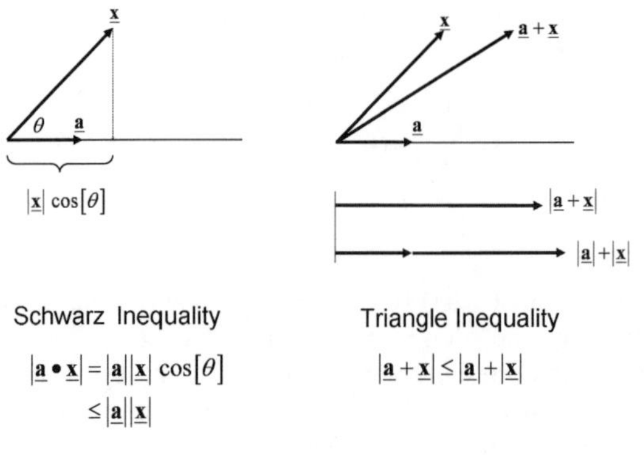

Figure 3.3 Graphical comparison of the Schwarz and triangle inequalities for the same pair of 2-D vectors $\underline{\mathbf{x}}$ and $\underline{\mathbf{a}}$.

included angle between two unit vectors:

$$\frac{\underline{\mathbf{a}}}{|\underline{\mathbf{a}}|} \bullet \frac{\underline{\mathbf{x}}}{|\underline{\mathbf{x}}|} = \underline{\hat{\mathbf{a}}} \bullet \underline{\hat{\mathbf{x}}} = \cos[\theta] \leq 1 \tag{3.18}$$

3.1.2 Projection of One Vector onto Another

The scalar product of an arbitrary vector $\underline{\mathbf{x}}$ and a unit "reference" vector $\hat{\mathbf{a}}$ evidently is:

$$\underline{\hat{\mathbf{a}}} \bullet \underline{\mathbf{x}} = \frac{\underline{\mathbf{a}}}{|\underline{\mathbf{a}}|} \bullet \underline{\mathbf{x}} = |\underline{\hat{\mathbf{a}}}||\underline{\mathbf{x}}| \cos[\theta] = |\underline{\mathbf{x}}| \cos[\theta] \tag{3.19}$$

The geometric picture of vectors leads to the interpretation that Equation (3.19) determines the length of that component of the "input vector" $\underline{\mathbf{x}}$ along the same direction as the unit "reference vector" $\underline{\hat{\mathbf{a}}}$. In other words, this scalar product is a measure of the geometric *projection* of $\underline{\mathbf{x}}$ onto $\underline{\hat{\mathbf{a}}}$, as shown in Figure 3.4. If the reference vector is not normalized to unit length, then the linearity of the scalar product allows the result to be scaled by $|\underline{\mathbf{a}}|^{-1}$ to obtain the correct-length projection of $\underline{\mathbf{x}}$ in the direction of $\underline{\mathbf{a}}$:

$$\textit{Projection of } \underline{\mathbf{x}} \textit{ onto } \underline{\mathbf{a}} = \frac{\underline{\mathbf{a}} \bullet \underline{\mathbf{x}}}{|\underline{\mathbf{a}}|} \tag{3.20}$$

That the scalar product of any two vectors with $\theta = \pi/2$ radians is zero is evident from its dependence on $\cos[\theta]$ in Equation (3.19); such vectors are said to be perpendicular or *orthogonal*. Two vectors that have unit length and are orthogonal are said to be *orthonormal*.

Equation (3.19) also establishes that the scalar product of two vectors will be negative when θ lies within the semi-closed interval $(+\pi/2, +\pi]$, which means that the projection of $\underline{\mathbf{x}}$ onto $\underline{\mathbf{a}}$ points in the direction opposite to that of $\underline{\mathbf{a}}$.

3.2 Matrices

3.2.1 Simultaneous Evaluation of Multiple Scalar Products

The concept of the scalar product of two vectors in Equation (3.9) may be extended to compute an ensemble of projections of a single N-D "input" vector $\underline{\mathbf{x}}$ onto each of a set of M N-D unit-length "reference" vectors $\underline{\hat{\mathbf{a}}}_m$. For the moment, all components of all vectors still are assumed to be real

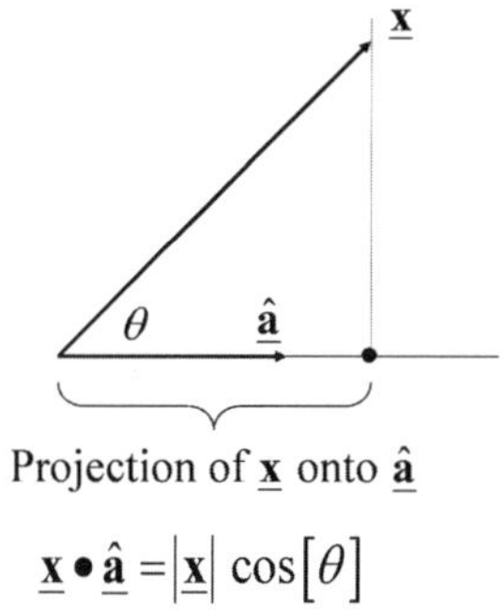

Figure 3.4 Graphical interpretation of the projection of the 2-D vector $\underline{\mathbf{x}}$ with real-valued components onto the unit vector $\hat{\underline{\mathbf{a}}}$ to produce $|\underline{\mathbf{x}}|\cos[\theta]$.

valued; the extension to complex-valued vectors will be discussed later. The ensemble of scalar products yields a set of M scalars that are distinguished by the index m assigned to each reference vector $\hat{\underline{\mathbf{a}}}_m$:

$$\hat{\underline{\mathbf{a}}}_m \bullet \underline{\mathbf{x}} = (\hat{\underline{\mathbf{a}}}_m)^T \underline{\mathbf{x}} = \sum_{n=0}^{N-1} (\hat{a}_m)_n \, x_n \equiv b_m \tag{3.21}$$

Each of the M output scalars $\{b_m\}$ represents the *projection* of the N-D vector $\underline{\mathbf{x}}$ onto a specific N-D "reference" vector $\hat{\underline{\mathbf{a}}}_m$, where the index m specifies the particular reference vector. These M scalars may be "stacked" to create an M-D column vector whose components are the projections of the input vector $\underline{\mathbf{x}}$ onto each of the set of unit "reference" vectors $\hat{\underline{\mathbf{a}}}_m$. The entire projection process may be implemented in one step by "stacking" the M N-D unit-length row vectors $(\hat{\underline{\mathbf{a}}}_m)^T$ in Equation (3.21) to create an M-row by N-column array of numbers that is called a *matrix*. The components $(\hat{a}_m)_n$ of the matrix may be designated by deleting the parentheses to create the "shorthand" notation $(\hat{a}_m)_n = \hat{a}_{mn}$, where m and n respectively denote the row and column of the component within the array:

$$\begin{bmatrix} (\hat{\underline{\mathbf{a}}}_1)^T \\ (\hat{\underline{\mathbf{a}}}_2)^T \\ \vdots \\ (\hat{\underline{\mathbf{a}}}_M)^T \end{bmatrix} = \begin{bmatrix} [\ (\hat{\underline{\mathbf{a}}}_1)_1 & (\hat{\underline{\mathbf{a}}}_1)_2 & \cdots & (\hat{\underline{\mathbf{a}}}_1)_N\ ] \\ [\ (\hat{\underline{\mathbf{a}}}_2)_1 & (\hat{\underline{\mathbf{a}}}_2)_2 & \cdots & (\hat{\underline{\mathbf{a}}}_2)_N\ ] \\ \vdots & \vdots & \ddots & \vdots \\ [\ (\hat{\underline{\mathbf{a}}}_M)_1 & (\hat{\underline{\mathbf{a}}}_M)_2 & \cdots & (\hat{\underline{\mathbf{a}}}_M)_N\ ] \end{bmatrix}$$

$$= \begin{bmatrix} \hat{a}_{11} & \hat{a}_{12} & \cdots & \hat{a}_{1N} \\ \hat{a}_{21} & \hat{a}_{22} & \cdots & \hat{a}_{2N} \\ \vdots & \vdots & \ddots & \vdots \\ \hat{a}_{M1} & \hat{a}_{M2} & \cdots & \hat{a}_{MN} \end{bmatrix} \tag{3.22}$$

Clearly scalar products may be evaluated for row vectors $\hat{\underline{\mathbf{a}}}_m$ with any norm (length), so long as they all have the same number of dimensions. However, the resulting scalar components b_m must be scaled by the length of the corresponding row vector before they may be interpreted as projections.

The matrix of components a_{mn} is denoted by the upper-case bold-faced character (e.g., $\underline{\mathbf{A}}$ in this case). Common concise and equivalent notational forms for matrix–vector multiplication are:

$$\underline{\mathbf{A}} \bullet \underline{\mathbf{x}} = \underline{\mathbf{A}}\,\underline{\mathbf{x}} = \{b_m\} = \underline{\mathbf{b}} \tag{3.23}$$

Clearly, the dimensionality of the input vector $\underline{\mathbf{x}}$ must be identical to that of all row vectors $\underline{\mathbf{a}}_m$ in $\underline{\mathbf{A}}$ for the scalar products to be defined, and the number M of row vectors in $\underline{\mathbf{A}}$ and the dimensionality of the output vector $\underline{\mathbf{b}}$ also are equal.

The representation of the scalar products of one "input" vector $\underline{\mathbf{x}}$ with multiple "reference" vectors $\underline{\mathbf{a}}_m$ is:

$$\underline{\mathbf{A}}\,\underline{\mathbf{x}} = \begin{bmatrix} a_{11} & a_{12} & \cdots & a_{1N} \\ a_{21} & a_{22} & \cdots & a_{2N} \\ \vdots & \vdots & \ddots & \vdots \\ a_{M1} & a_{M2} & \cdots & a_{MN} \end{bmatrix} \begin{bmatrix} (\underline{\mathbf{x}})_1 \\ (\underline{\mathbf{x}})_2 \\ \vdots \\ (\underline{\mathbf{x}})_N \end{bmatrix}$$

$$= \begin{bmatrix} a_{11} & a_{12} & \cdots & a_{1N} \\ a_{21} & a_{22} & \cdots & a_{2N} \\ \vdots & \vdots & \ddots & \vdots \\ a_{M1} & a_{M2} & \cdots & a_{MN} \end{bmatrix} \begin{bmatrix} x_1 \\ x_2 \\ \vdots \\ x_N \end{bmatrix} = \begin{bmatrix} \sum_{n=1}^{N} a_{1n}x_n \\ \sum_{n=1}^{N} a_{2n}x_n \\ \vdots \\ \sum_{n=1}^{N} a_{Nn}x_n \end{bmatrix} = \begin{bmatrix} b_1 \\ b_2 \\ \vdots \\ b_N \end{bmatrix} \equiv \underline{\mathbf{b}} \tag{3.24}$$

This notation suggests an interpretation of the concept of an imaging system; it is a set of M N-D "reference" vectors $\{\underline{\mathbf{a}}_m\}$ that act on the "input" vector $\underline{\mathbf{x}}$ via the scalar product to produce an M-D "output" vector $\underline{\mathbf{b}}$. Thus the matrix $\underline{\mathbf{A}}$ may be considered to be an imaging "system" and its action indicated by the script character $\mathcal{O}$ acting on the input vector:

$$\mathcal{O}\{\underline{\mathbf{x}}\} = \underline{\mathbf{A}\mathbf{x}} = \underline{\mathbf{b}} \tag{3.25}$$

The process of computing the scalar product of an input vector $\underline{\mathbf{x}}$ with a "reference" vector $\underline{\mathbf{a}}$ satisfies the necessary criterion in Equation (2.1) for a linear operator. This formulation (or notation) of a discrete linear operation as a matrix–vector product has several applications in imaging. The matrix $\underline{\mathbf{A}}$ need not be square in the general imaging problem; this more general case will be considered in more detail shortly.

We can consider the discussion of the imaging "tasks" in Section 1.1 in the light of this interpretation of matrix–vector multiplication as an imaging operator. The "direct" problem requires the solution for $\underline{\mathbf{b}}$ given complete knowledge of $\underline{\mathbf{A}}$ and $\underline{\mathbf{x}}$. The inverse problem assumes that $\underline{\mathbf{A}}$ and $\underline{\mathbf{b}}$ are known and that $\underline{\mathbf{x}}$ must be found. Finally, in system analysis the input and output vectors $\underline{\mathbf{x}}$ and $\underline{\mathbf{b}}$ are known and the matrix $\underline{\mathbf{A}}$ must be found. We will investigate this interpretation in more detail in a subsequent chapter.

3.2.2 Matrix–Matrix Multiplication

The process of scalar multiplication of two vectors may be generalized still further to compute an ensemble of scalar products evaluated from each of M N-D row vectors $\underline{\mathbf{a}}_m$ of the matrix $\underline{\mathbf{A}}$ and each of P N-D column vectors $\underline{\mathbf{x}}_p$ $(1 \le p \le P)$. The result is a set of P M-D output column vectors $\underline{\mathbf{b}}_p$. It is clear that the array of P input column vectors $\underline{\mathbf{x}}_p$ also may be represented as a matrix, with M rows and N columns, while the ensemble of P M-D output column vectors $\underline{\mathbf{b}}_p$ may be combined to form a matrix $\underline{\mathbf{B}}$ with M rows and P columns via $M \times P$ scalar products. The schematic of the process is shown in Equation (3.26):

$$\underline{\mathbf{A}}\,\underline{\mathbf{X}} = \begin{bmatrix} \boxed{a_{11} \quad a_{12} \quad \cdots \quad a_{1N}} \\ a_{21} \quad a_{22} \quad \cdots \quad a_{2N} \\ \vdots \quad\; \vdots \quad\; \ddots \quad\; \vdots \\ a_{M1} \quad a_{M2} \quad \cdots \quad a_{MN} \end{bmatrix} \begin{bmatrix} (x_1)_1 & (x_2)_1 & \cdots & (x_P)_1 \\ (x_1)_2 & (x_2)_2 & \cdots & (x_P)_2 \\ \vdots & \vdots & \ddots & \vdots \\ (x_1)_N & (x_2)_N & \cdots & (x_P)_N \end{bmatrix}$$

$$= \begin{bmatrix} (b_1)_1 & \boxed{(b_1)_2} & \cdots & (b_1)_P \\ (b_2)_1 & (b_2)_2 & \cdots & (b_2)_P \\ \vdots & \vdots & \ddots & \vdots \\ (b_M)_1 & (b_M)_2 & \cdots & (b_M)_P \end{bmatrix} = \begin{bmatrix} b_{11} & b_{12} & \cdots & b_{1P} \\ b_{21} & b_{22} & \cdots & b_{2P} \\ \vdots & \vdots & \ddots & \vdots \\ b_{M1} & b_{M2} & \cdots & b_{MP} \end{bmatrix} = \underline{\mathbf{B}} \tag{3.26}$$

The boxed elements illustrate the scalar product of the first "reference" vector $\underline{\mathbf{a}}_1$ with the second "input" vector $\underline{\mathbf{x}}_2$ to produce the element b_{12} in the "output" matrix $\underline{\mathbf{B}}$. The general output b_{mp} may be written as the summation over the elements of $\underline{\mathbf{A}}$ and $\underline{\mathbf{X}}$:

$$b_{mp} = (\underline{\mathbf{A}}\,\underline{\mathbf{X}})_{mp} = \sum_{k=1}^{N} a_{mk}\, x_{kp} \tag{3.27}$$

Since the number of columns in the first matrix $\underline{\mathbf{A}}$ must equal the number of rows in the second matrix $\underline{\mathbf{X}}$, the operation of matrix–matrix multiplication cannot generally be applied in the opposite order, i.e., $\underline{\mathbf{X}}\,\underline{\mathbf{A}}$ is generally not defined. However, the product of two matrices may be evaluated in either order *if* both $\underline{\mathbf{A}}$ and $\underline{\mathbf{X}}$ are $N \times N$ square, though the numerical results from the two computations generally differ:

$$(\underline{\mathbf{A}}\,\underline{\mathbf{X}})_{mp} \neq (\underline{\mathbf{X}}\,\underline{\mathbf{A}})_{mp} \tag{3.28}$$

In other words, matrix–matrix multiplication generally does not *commute*.

It is easy to use Equation (3.27) to demonstrate that matrix multiplication is associative, so that the product of three matrices may be computed in any combination as long as the order of the matrices is preserved:

$$\underline{\mathbf{A}}\,\underline{\mathbf{B}}\,\underline{\mathbf{C}} = (\underline{\mathbf{A}}\,\underline{\mathbf{B}})\,\underline{\mathbf{C}} = \underline{\mathbf{A}}\,(\underline{\mathbf{B}}\,\underline{\mathbf{C}}) \tag{3.29}$$

The interpretation of matrix–matrix multiplication as the action of the imaging "system" $\underline{\mathbf{A}}$ on a set of input column vectors in the matrix $\underline{\mathbf{X}}$ to produce a set of output column vectors in the matrix $\underline{\mathbf{B}}$ is useful when considering the representation of vectors and matrices in terms of different sets of basis vectors, which we consider in the next sections.

3.2.3 Square and Diagonal Matrices, Identity Matrix

Now consider an "imaging" system where all input "object" vectors $(\underline{\mathbf{x}})_n$ and output "image" vectors $(\underline{\mathbf{b}})_m$ have the same dimension N, so that the corresponding matrices $\underline{\mathbf{A}}$, $\underline{\mathbf{X}}$, and $\underline{\mathbf{B}}$ are all $N \times N$ and all of the various matrix products are defined and are $N \times N$, e.g., $\underline{\mathbf{A}}\,\underline{\mathbf{X}}$ and $\underline{\mathbf{X}}\,\underline{\mathbf{B}}$.

A particularly important subset of the square matrices are the *diagonal* matrices, for which all "off-diagonal" elements are null (i.e., $a_{mn} = 0$ for all $m \neq n$). The only elements of a diagonal matrix that may be nonzero are those for which $m = n$, such as a_{nn}. It is easy to see that the matrix product of two diagonal matrices in either order produces the same matrix that also is diagonal:

$$\begin{aligned} \underline{\mathbf{A}}\,\underline{\mathbf{X}} &= \begin{bmatrix} a_{11} & 0 & \cdots & 0 \\ 0 & a_{22} & \cdots & 0 \\ \vdots & \vdots & \ddots & \vdots \\ 0 & 0 & \cdots & a_{nn} \end{bmatrix} \begin{bmatrix} x_{11} & 0 & \cdots & 0 \\ 0 & x_{22} & \cdots & 0 \\ \vdots & \vdots & \ddots & \vdots \\ 0 & 0 & \cdots & x_{nn} \end{bmatrix} \\ &= \begin{bmatrix} x_{11} & 0 & \cdots & 0 \\ 0 & x_{22} & \cdots & 0 \\ \vdots & \vdots & \ddots & \vdots \\ 0 & 0 & \cdots & x_{nn} \end{bmatrix} \begin{bmatrix} a_{11} & 0 & \cdots & 0 \\ 0 & a_{22} & \cdots & 0 \\ \vdots & \vdots & \ddots & \vdots \\ 0 & 0 & \cdots & a_{nn} \end{bmatrix} \\ &= \underline{\mathbf{X}}\,\underline{\mathbf{A}} \textit{ if both } \underline{\mathbf{A}} \textit{ and } \underline{\mathbf{X}} \textit{ are diagonal} \end{aligned} \tag{3.30}$$

This observation shows that multiplication of diagonal matrices is commutative, even though multiplication of general matrices is not. The diagonal matrix with all diagonal elements equal to one is the *identity* matrix and will be denoted by its own symbol $\underline{\mathbf{I}}$. Its name describes the property that the product of any $N \times N$ matrix $\underline{\mathbf{A}}$ with the $N \times N$ identity matrix $\underline{\mathbf{I}}$ from either side is $\underline{\mathbf{A}}$:

$$\underline{\mathbf{A}}\,\underline{\mathbf{I}} = \underline{\mathbf{I}}\,\underline{\mathbf{A}} = \underline{\mathbf{A}} \tag{3.31}$$

In other words, the $N \times N$ diagonal identity matrix commutes with any $N \times N$ matrix, whether diagonal or not.

Two scalar metrics derived from square matrices provide useful information about some characteristics of operators based on those matrices. The simpler of these is the *trace* of the matrix, which is just the sum of the diagonal elements:

$$TR[\underline{\mathbf{A}}] = \sum_{n=1}^{N} a_{nn} \tag{3.32}$$

Note that the trace is defined for any square matrix, diagonal or not. The trace of a diagonal matrix obviously equals the sum of the matrix elements, and thus the trace of the $N \times N$ identity matrix is N.

The second useful scalar metric of a matrix is the *determinant*, which is a specific weighted sum of all elements of the matrix, where the possible values of the weights are ± 1. The sum that defines the determinant may be represented in many forms. Several notations are common for the determinant, such as $|\underline{\mathbf{A}}|$, but we will use $\det[\underline{\mathbf{A}}]$, which gives the flavor of an operator applied to the matrix. Readers who seek a more complete discussion of determinants should consult other sources on matrices or linear algebra (e.g., Strang, 2005). This treatment will use the simple formula for the determinant of a 2×2 matrix to describe the applications:

$$\underline{\mathbf{A}} = \begin{bmatrix} a_{11} & a_{12} \\ a_{21} & a_{22} \end{bmatrix} \Longrightarrow \det[\underline{\mathbf{A}}] \equiv a_{11}a_{22} - a_{21}a_{12} \tag{3.33}$$

In words, the determinant of a 2×2 matrix is the difference between the products of the diagonal elements and of the off-diagonal elements. The determinant of a diagonal 2×2 matrix evidently is the product of the diagonal elements $a_{11}a_{22}$.

The following relevant properties of determinants of $N \times N$ matrices are stated without proof, but confirmation for the 2×2 case is quite easy:

$$\det[\underline{\mathbf{I}}] = 1 \tag{3.34a}$$

$$\textit{if all elements in any row or column of } \underline{\mathbf{A}} \textit{ are zero, then } \det[\underline{\mathbf{A}}] = 0 \tag{3.34b}$$

$$\textit{if any two rows or columns of } \underline{\mathbf{A}} \textit{ are equal, then } \det[\underline{\mathbf{A}}] = 0 \tag{3.34c}$$

$$\det[\underline{\mathbf{A}}] \textit{ is unchanged if a multiple of a } \begin{Bmatrix} \textit{row} \\ \textit{column} \end{Bmatrix} \textit{ is added to any } \begin{Bmatrix} \textit{row} \\ \textit{column} \end{Bmatrix} \tag{3.34d}$$

Expressions for evaluating determinants of arbitrary square matrices are available in all mathematical software packages that deal with matrices and in any book on linear algebra.

3.2.4 Matrix Transposes

Recall that the transpose of a column vector was defined in Equation (3.4) as the row vector created from the same elements in the same order. The extension of the definition of the transpose to matrices is obvious. The *transpose* of the matrix $\underline{\mathbf{A}}$ with M rows and N columns is generated from the transpose of each column vector to create an N-row by M-column matrix denoted $\underline{\mathbf{A}}^T$. Equivalently, $\underline{\mathbf{A}}^T$ may be constructed by transposing its row vectors into column vectors. This second interpretation is more appropriate in our context as $\underline{\mathbf{A}}$ was assumed in the beginning to be constructed from the set of M N-D row vectors $\{\underline{\mathbf{A}}_m^T\}$. The third, and probably simplest, view is that a matrix transpose is obtained after reordering the elements via an exchange of their indices:

$$\underline{\mathbf{A}} = \{a_{mn}\} \Longrightarrow \underline{\mathbf{A}}^T = \{a_{nm}\} \tag{3.35}$$

The interpretation of matrix–matrix multiplication as multiple scalar products in Equation (3.26) may be used to infer a useful relation involving the product of the transposes of the matrices $\underline{\mathbf{A}}$ and $\underline{\mathbf{X}}$. Consider again the product of an M-row by N-column matrix $\underline{\mathbf{A}}$ and an N-row by P-column matrix $\underline{\mathbf{X}}$ that yields the M-row by P-column matrix $\underline{\mathbf{B}}$ in Equation (3.27). The scalar product of the first row vector of $\underline{\mathbf{A}}$ and the second column vector of $\underline{\mathbf{X}}$ yields a scalar that we associate with element b_{12} of the output matrix $\underline{\mathbf{B}}$:

$$\underline{\mathbf{a}}_1 \bullet \underline{\mathbf{x}}_2 = \underline{\mathbf{a}}_1^T \, \underline{\mathbf{x}}_2 = b_{12} \tag{3.36}$$

Because the scalar product of two vectors commutes, the same numerical value is obtained after exchanging the order of the vectors:

$$\underline{\mathbf{x}}_2 \bullet \underline{\mathbf{a}}_1 = \underline{\mathbf{x}}_2^T \, \underline{\mathbf{a}}_1 = b_{12} \tag{3.37}$$

The schematic layout of the calculation is:

$$\begin{aligned}
\underline{\mathbf{B}}^T &= \begin{bmatrix} b_{11} & b_{21} & \cdots & b_{P1} \\ \boxed{b_{12}} & b_{22} & \cdots & b_{P2} \\ \vdots & \vdots & \ddots & \vdots \\ b_{1M} & b_{2M} & \cdots & b_{PM} \end{bmatrix} \\
&= \begin{bmatrix} x_{11} & x_{12} & \cdots & x_{1P} \\ x_{21} & x_{22} & \cdots & x_{2P} \\ \vdots & \vdots & \ddots & \vdots \\ x_{N1} & x_{N2} & \cdots & x_{NP} \end{bmatrix} \begin{bmatrix} a_{11} & a_{21} & \cdots & a_{N1} \\ a_{12} & a_{22} & \cdots & a_{N2} \\ \vdots & \vdots & \ddots & \vdots \\ a_{1M} & a_{2M} & \cdots & a_{NM} \end{bmatrix} \\
&= \underline{\mathbf{X}}^T \, \underline{\mathbf{A}}^T
\end{aligned} \tag{3.38}$$

where the scalar product of the boxed second row vector of $\underline{\mathbf{X}}$ with the first column vector of $\underline{\mathbf{A}}$ produces the scalar b_{12} of the matrix $\underline{\mathbf{B}}$. Similarly, the scalar product of the third row vector of $\underline{\mathbf{A}}$ and the second column vector of $\underline{\mathbf{X}}$ yields b_{32}, which also may be obtained as the scalar product of the second row vector of $\underline{\mathbf{X}}^T$ with the third column vector of $\underline{\mathbf{A}}^T$:

$$\underline{\mathbf{a}}_3 \bullet \underline{\mathbf{x}}_2 = \underline{\mathbf{a}}_3^T \, \underline{\mathbf{x}}_2 = \underline{\mathbf{x}}_2 \bullet \underline{\mathbf{a}}_3 = \underline{\mathbf{x}}_2^T \, \underline{\mathbf{a}}_3 = b_{32} \tag{3.39}$$

From the schematic in Equation (3.38), it is evident that the ensemble of scalar products may be expressed equivalently in terms of the transposes of the matrices:

$$\underline{\mathbf{A}} \, \underline{\mathbf{X}} = \underline{\mathbf{B}} \Longrightarrow \underline{\mathbf{X}}^T \, \underline{\mathbf{A}}^T = \underline{\mathbf{B}}^T \tag{3.40}$$

Particularly note the exchange in the order of the matrix transposes. This result will be useful when discussing transformations of vectors to different representations later in this chapter.

3.2.5 Matrix Inverses

Given a square $N \times N$ matrix $\underline{\mathbf{A}}$, which we will use as a mathematical model of an imaging system, we now apply the interpretation of matrix–matrix multiplication to investigate the properties of the particular square $N \times N$ matrix $\underline{\mathbf{C}}$ that satisfies the condition:

$$\underline{\mathbf{A}} \, \underline{\mathbf{C}} = \underline{\mathbf{I}} \tag{3.41}$$

where $\underline{\mathbf{I}}$ is the $N \times N$ identity matrix. Equation (3.41) requires that the projection of the nth row vector of $\underline{\mathbf{A}}$ onto the nth column vector of $\underline{\mathbf{C}}$ evaluate to unity for all n, while the projections of the nth row onto all other columns of $\underline{\mathbf{C}}$ must be null. This is equivalent to the statement that the nth row vector of $\underline{\mathbf{A}}$ must be *orthonormal* to the column vectors of $\underline{\mathbf{C}}$. The specific and unique matrix $\underline{\mathbf{C}}$ that satisfies Equation (3.41) is called the *matrix inverse* of $\underline{\mathbf{A}}$ and is denoted $\underline{\mathbf{A}}^{-1}$:

$$\underline{\mathbf{A}}\,\underline{\mathbf{A}}^{-1} = \underline{\mathbf{I}} \tag{3.42}$$

The notation $\underline{\mathbf{A}}^{-1}$ must not be confused with the meaningless expression $1/\underline{\mathbf{A}}$.

Now consider the product of three matrices computed two different ways via the associativity property of matrix multiplication from Equation (3.29):

$$\begin{aligned}\underline{\mathbf{A}}\,\underline{\mathbf{A}}^{-1}\,\underline{\mathbf{A}} &= (\underline{\mathbf{A}}\,\underline{\mathbf{A}}^{-1})\,\underline{\mathbf{A}} = \underline{\mathbf{I}}\,\underline{\mathbf{A}} = \underline{\mathbf{A}}\\ \underline{\mathbf{A}}\,\underline{\mathbf{A}}^{-1}\,\underline{\mathbf{A}} &= \underline{\mathbf{A}}\,(\underline{\mathbf{A}}^{-1}\,\underline{\mathbf{A}}) = \underline{\mathbf{A}} = \underline{\mathbf{A}}\,\underline{\mathbf{I}}\end{aligned} \tag{3.43}$$

This demonstrates that both $\underline{\mathbf{A}}^{-1}\,\underline{\mathbf{A}}$ and $\underline{\mathbf{A}}\,\underline{\mathbf{A}}^{-1}$ must evaluate to the identity matrix $\underline{\mathbf{I}}$, thus ensuring that $\underline{\mathbf{A}}$ commutes with its inverse $\underline{\mathbf{A}}^{-1}$:

$$\underline{\mathbf{A}}\,\underline{\mathbf{A}}^{-1} = \underline{\mathbf{A}}^{-1}\,\underline{\mathbf{A}} = \underline{\mathbf{I}} \tag{3.44}$$

Even in cases where the inverse of a matrix operator $\underline{\mathbf{A}}^{-1}$ is known to exist, its computation is often very tedious. In imaging applications where the dimensionality N is often quite large, computation of the inverse may be impractical. Algorithms for computing matrix inverses are considered in many books on linear algebra and applied mathematics. In this book, the discussion of matrix inverses is restricted to two specific types of matrices that are particularly relevant in imaging.

Consider first the (very) special case of the inverse of a diagonal matrix $\underline{\mathbf{A}}$ when all diagonal elements are nonzero. The inverse of the diagonal matrix also is diagonal, and each diagonal element is the reciprocal of the corresponding element a_{mm}:

$$\begin{bmatrix} a_{11} & 0 & \cdots & 0\\ 0 & a_{22} & \cdots & 0\\ \vdots & \vdots & \ddots & 0\\ 0 & 0 & 0 & a_{NN}\end{bmatrix}\begin{bmatrix} (a_{11})^{-1} & 0 & \cdots & 0\\ 0 & (a_{22})^{-1} & \cdots & 0\\ \vdots & \vdots & \ddots & 0\\ 0 & 0 & 0 & (a_{NN})^{-1}\end{bmatrix} = \begin{bmatrix} 1 & 0 & \cdots & 0\\ 0 & 1 & \cdots & 0\\ \vdots & \vdots & \ddots & 0\\ 0 & 0 & 0 & 1\end{bmatrix} \tag{3.45}$$

This simplicity of the inverse of a diagonal matrix will provide the motivation to recast square matrices into "equivalent diagonal forms", as will be discussed in Chapter 5.

It is evident from Equation (3.45) that the inverse of a diagonal matrix exists only if all diagonal elements are nonzero. Recall that the determinant of a diagonal matrix is the product of the diagonal elements, which means that the determinant of a diagonal matrix with any null diagonal elements must be zero. Therefore, at least for diagonal matrices, $\det[\underline{\mathbf{A}}] = 0$ implies that $\underline{\mathbf{A}}^{-1}$ does not exist. Though we will not do so, it is possible to prove that the inverse does not exist for any matrix with a null determinant.

Next, consider the (even more) special example where the inverse of a square matrix exists and is in fact identical to the transpose of the matrix:

$$\underline{\mathbf{A}}\,\underline{\mathbf{A}}^{-1} = \underline{\mathbf{A}}\,\underline{\mathbf{A}}^{T} = \underline{\mathbf{I}} \Longrightarrow \underline{\mathbf{A}}^{-1} = \underline{\mathbf{A}}^{T} \tag{3.46}$$

From the interpretation of matrix–matrix multiplication as multiple scalar products, it is evident that this result is true only in cases for which the rows (and columns) of $\underline{\mathbf{A}}$ are mutually orthogonal and the norm of each row (or column) vector is unity. Such a special matrix is called *orthogonal*, though it would be more appropriate to say that it is *orthonormal*. From this observation, it is evident that the inverse of a matrix $\underline{\mathbf{A}}$ whose rows (or columns) are orthonormal is identical to $\underline{\mathbf{A}}^{T}$.

3.3 Vector Spaces

The concept of a vector $\underline{\mathbf{x}}$ with N real-valued components leads naturally to the idea of the set of all possible such vectors. A set of vectors will be denoted by enclosing the symbol for one member vector within braces, e.g., $\{\underline{\mathbf{x}}\}$. Membership of a specific vector $\underline{\mathbf{x}}_1$ in the set $\{\underline{\mathbf{x}}\}$ is signified by the notation $\underline{\mathbf{x}}_1 \in \{\underline{\mathbf{x}}\}$.

A *vector space* is defined as a set of vectors combined with the definition of two operations that may be applied to all of these vectors: addition of pairs of vectors and multiplication of individual vectors by scalars. To be a vector space, the set of vectors and the two operations must satisfy 10 criteria:

1. The sum of *any* two vectors in the set yields a third vector that also is a member of the set. This quality is known as *closure*:
$$\underline{\mathbf{x}}_1 + \underline{\mathbf{x}}_2 = \underline{\mathbf{x}}_3 \in \{\underline{\mathbf{x}}\} \tag{3.47a}$$
2. Addition of vectors is associative:
$$\underline{\mathbf{x}}_1 + \underline{\mathbf{x}}_2 + \underline{\mathbf{x}}_3 = (\underline{\mathbf{x}}_1 + \underline{\mathbf{x}}_2) + \underline{\mathbf{x}}_3 = \underline{\mathbf{x}}_1 + (\underline{\mathbf{x}}_2 + \underline{\mathbf{x}}_3) \tag{3.47b}$$
3. Addition of vectors is commutative:
$$\underline{\mathbf{x}}_1 + \underline{\mathbf{x}}_2 = \underline{\mathbf{x}}_2 + \underline{\mathbf{x}}_1 \tag{3.47c}$$
4. A vector denoted $\underline{\mathbf{0}}$ exists in $\{\underline{\mathbf{x}}\}$ (i.e., $\underline{\mathbf{0}} \in \{\underline{\mathbf{x}}\}$) that satisfies the relation:
$$\underline{\mathbf{x}} + \underline{\mathbf{0}} = \underline{\mathbf{x}} \tag{3.47d}$$
for all vectors $\underline{\mathbf{x}}$ in the set. The vector $\underline{\mathbf{0}}$ is called the *zero* or *null* vector. All components of the null vector are identically zero.
5. The components of any vector in the set may be multiplied by a scalar a to obtain another vector that also is in the set:
$$a(\underline{\mathbf{x}}_1) \in \{\underline{\mathbf{x}}\} \tag{3.47e}$$
6. A corollary to the previous condition demonstrates that for each vector $\underline{\mathbf{x}}$ in the set, the vector $-1 \cdot \underline{\mathbf{x}} \in \{\underline{\mathbf{x}}\}$, and this pair satisfies the condition that their sum yields the null vector which also is in the set (condition 4):
$$\underline{\mathbf{x}} + (-\underline{\mathbf{x}}) = \underline{\mathbf{0}} \tag{3.47f}$$
7. Multiplication of vectors by scalars is associative:
$$a_1 a_2 \underline{\mathbf{x}} = a_1 (a_2 \underline{\mathbf{x}}) \tag{3.47g}$$
8. Multiplication of a vector by a sum of scalars is defined, is distributive, and yields a vector in the set:
$$(a_1 + a_2)\underline{\mathbf{x}} = a_1 \underline{\mathbf{x}} + a_2 \underline{\mathbf{x}} \in \{\underline{\mathbf{x}}\} \tag{3.47h}$$
9. Multiplication of a sum of vectors by a scalar is defined, is distributive, and yields a vector in the set:
$$a(\underline{\mathbf{x}}_1 + \underline{\mathbf{x}}_2) = a\underline{\mathbf{x}}_1 + a\underline{\mathbf{x}}_2 \in \{\underline{\mathbf{x}}\} \tag{3.47i}$$
10. Multiplication of all components of a vector by the unit scalar yields the original vector (identity operation):
$$1\underline{\mathbf{x}} = \underline{\mathbf{x}} \tag{3.47j}$$

A *vector subspace* of the vector space $\{\underline{\mathbf{x}}_1\}$ is a set of vectors $\{\underline{\mathbf{x}}_2\}$ that satisfies the 10 conditions in Equation (3.47) and such that all members of the set $\{\underline{\mathbf{x}}_2\}$ also are contained in the set $\{\underline{\mathbf{x}}_1\}$. The first set of vectors may include more members than the second set, but it cannot have fewer members. If the set $\{\underline{\mathbf{x}}_1\}$ and its subset $\{\underline{\mathbf{x}}_2\}$ have the same number of members, the two sets have the same members. From the list of required properties, it is easy to see that a set containing only the null vector $\underline{\mathbf{0}}$ satisfies all requirements for a vector space, and therefore the set consisting of the single vector $\{\underline{\mathbf{0}}\}$ is a subspace of any vector space. In Section 3.2, the set of all possible image and object vectors associated with a system matrix operator $\underline{\mathbf{A}}$ will be grouped into subspaces based on the form of $\underline{\mathbf{A}}$. First, we define some other terms that are used in the study of vector spaces.

In problems involving imaging systems, four of the properties required for a set of vectors to be a vector space arguably are most important. This is because these statements are often particularly easy to test. These four properties are: closure under vector addition in Equation (3.47a), the presence in the set of the zero vector in Equation (3.47d) and the negative vector in Equation (3.47f), and distributive multiplication by scalars in Equation (3.47h). The first and last of these four requirements ensure that any linear combination (weighted sum) of N vectors $\underline{\mathbf{x}}_n$ selected from the vector space yields a vector $\underline{\mathbf{a}}$ that also is a member of the vector space. For example, the vector $\underline{\mathbf{a}}$ that is created from a weighted sum of the N vectors $\underline{\mathbf{x}}_n$ must be a member of the vector space:

$$\sum_{n=1}^{N} \beta_n \underline{\mathbf{x}}_n = \underline{\mathbf{a}} \tag{3.48}$$

where the weights β_n are scalars. Though this expression resembles the scalar product of two vectors in Equation (3.9), it is important to note that they are very different. In Equation (3.9), the components of two vectors are multiplied and the products are summed to create a scalar. In Equation (3.48), the scalars β_n are applied to each of a set of vectors $\underline{\mathbf{x}}_n$and summed to generate a new vector $\underline{\mathbf{x}}_n$.

Of particular interest are those linear combinations of N vectors in Equation (3.48) that yield the null vector $\underline{\mathbf{a}} = \underline{\mathbf{0}}$. Obviously, one way to construct the null vector is to set all scalars $\beta_n = 0$. If this is the *only* way to generate the zero vector from a weighted sum of a set of N vectors $\underline{\mathbf{x}}_n$, then these N vectors $\underline{\mathbf{x}}_n$ are said to be linearly *independent*. Among other things, the property of linear independence ensures that no one of the N vectors $\underline{\mathbf{x}}_n$ may be expressed as a weighted sum of the other $N-1$ vectors. If the sum in Equation (3.48) yields the null vector $\underline{\mathbf{a}} = \underline{\mathbf{0}}$ when one or more of the weights $\beta_n \neq 0$, then the set of vectors is *linearly dependent*. This means that at least one vector in a set of N linearly dependent vectors can be specified as a linear combination of the other $N-1$ members in the set; no member vector of a linearly independent set can be so specified. The maximum number of linearly independent vectors in a set that satisfies the 10 requirements for a vector space in Equation (3.47) is the *dimension* of that vector space, and specifies the minimum number of vectors that may be scaled and summed to generate any vector in the vector space $\{\underline{\mathbf{x}}\}$.

The set of weights $\{\beta_n\}$ in Equation (3.48) specifies how much of each of the N vectors $\underline{\mathbf{x}}_n$ is needed to produce the output vector $\underline{\mathbf{a}}$, and thus the set of weights may be interpreted as a *representation* of $\underline{\mathbf{a}}$ in terms of the N component vectors $\underline{\mathbf{x}}_n$. In other words, the set of weights β_n describes the vector $\underline{\mathbf{a}}$ in terms of that particular set of N vectors $\underline{\mathbf{x}}_n$. If every vector in a vector space may be expressed as a weighted sum of a specific linearly independent set of the N vectors $\{\underline{\mathbf{x}}_n\}$, then those N vectors $\{\underline{\mathbf{x}}_n\}$ constitute a *basis* for the vector space. The set of projections of any arbitrary vector $\underline{\mathbf{a}}$ onto the set of basis vectors completely describes $\underline{\mathbf{a}}$. Other synonyms for such a set of N reference vectors $\{\underline{\mathbf{x}}_n\}$ are that it is *complete* or that the set *spans* the vector space. The number N of vectors in the basis is the dimension of the vector space.

The usual representation of a vector in a 3-D volume specifies the projections of that vector onto three linearly independent unit vectors directed along three orthogonal axes. The maximum number of linearly independent vectors in a volume is three, which is the dimension of the space. Because any other vector in the space can be specified as a linear combination of three linearly independent vectors, then any set of four such vectors is linearly dependent. Any set of three basis vectors may be rotated in any direction without losing their property of linear independence, and thus define an equivalent basis set. The representations of a vector in terms of these two bases will (of course) be different, but transformations between the representations are straightforward to derive.

Any set of N linearly independent N-D vectors may serve as a basis for an N-D vector space. This motivates the search for a method to determine whether the N vectors in a particular set $\{\underline{\mathbf{x}}_n\}$ are linearly independent. Probably the most convenient way is to construct an $N \times N$ square matrix $\underline{\mathbf{X}}$ from the N column vectors and determine the appropriate properties of the matrix. The vectors are linearly independent if no column vector in this matrix $\underline{\mathbf{X}}$ may be expressed as a weighted sum of the other $N - 1$ column vectors. The properties of determinants expressed in Equation (3.33) allow a column vector to be replaced by the sum of that column and a linear combination of the other columns without affecting the value of the determinant. If the N vectors are linearly dependent, then Equation (3.34d) ensures that any column of $\underline{\mathbf{X}}$ may be replaced by the difference between the original column vector and the equivalent weighted sum of the other vectors; in other words, the replacement column vector is null. Since the new matrix thus created (call it $\underline{\mathbf{Y}}$) contains a column of zeros, Equation (3.34c) establishes that $\det[\underline{\mathbf{Y}}] = 0$. The invariance of the determinant under the replacement of the column vector by a linear combination ensures that $\det[\underline{\mathbf{X}}]$ must be 0 as well. This result demonstrates another property of square matrices:

$$\det[\underline{\mathbf{A}}] \neq 0 \textit{ if the column vectors of } \underline{\mathbf{A}} \textit{ are linearly independent} \tag{3.49}$$

This result will be used when discussing the existence of the matrix inverse.

3.3.1 Basis Vectors

In the "usual" 3-D Cartesian coordinate system, arbitrary vectors are described in terms of their scalar projections defined by Equation (3.20) onto the particular set of three unit vectors that are aligned with the Cartesian coordinate axes. These three unit vectors often are designated by:

$$\hat{\underline{\mathbf{x}}}_1 = \begin{bmatrix} 1 \\ 0 \\ 0 \end{bmatrix}, \quad \hat{\underline{\mathbf{x}}}_2 = \begin{bmatrix} 0 \\ 1 \\ 0 \end{bmatrix}, \quad \hat{\underline{\mathbf{x}}}_3 = \begin{bmatrix} 0 \\ 0 \\ 1 \end{bmatrix} \tag{3.50}$$

Of course, these three unit vectors are mutually orthogonal, as may be shown by substituting all possible pairs into Equation (3.18). Therefore the included angle of all pairs of distinct vectors must be $\pi/2$ radians and the three vectors in this set are mutually orthonormal:

$$\hat{\underline{\mathbf{x}}}_n \bullet \hat{\underline{\mathbf{x}}}_m = \begin{Bmatrix} 1 & \textit{if} & n = m \\ 0 & \textit{if} & n \neq m \end{Bmatrix} \equiv \delta_{nm} \tag{3.51}$$

The notation δ_{nm} stands for the so-called *Kronecker delta*, which takes on the values 1 or 0 depending on whether the two integer indices n and m are equal or not. In the subsequent discussion, the three unit vectors in Equation (3.50) will be labeled by $\hat{\underline{\mathbf{x}}}_n$ $(n = 1, 2, 3)$. The set constitutes a basis for all 3-D vectors with real-valued components.

In the more general N-D case, a Cartesian basis consists of a set of N unit vectors directed along mutually orthogonal axes. The vectors in this set may be represented in concise form as $\{\hat{\underline{\mathbf{x}}}_n\}$, where the subscript $n \leq N$ specifies the particular Cartesian axis along which the vector "points". From Equation (3.19), the projection of an arbitrary N-D vector $\underline{\mathbf{a}}$ onto the ith member of the basis set $\{\hat{\underline{\mathbf{x}}}_n\}$ is a real number β:

$$\underline{\mathbf{a}} \bullet \hat{\underline{\mathbf{x}}}_i = \sum_{n=1}^{N} a_n (\hat{x}_i)_n = |\underline{\mathbf{a}}| \cos[\theta_i] \equiv \beta_i \tag{3.52}$$

where θ_i is the included angle between $\underline{\mathbf{a}}$ and the basis vector $\hat{\underline{\mathbf{x}}}_i$. By projecting the vector $\underline{\mathbf{a}}$ onto each of the normalized basis vectors, a set of N real numbers β_i is produced which are the individual projected lengths of the components of $\underline{\mathbf{a}}$ along the basis vectors. In other words, the set of N real-valued scalars $\{\beta_n\}$ describes the set of "weights" that must be applied to the individual basis vectors so that their sum

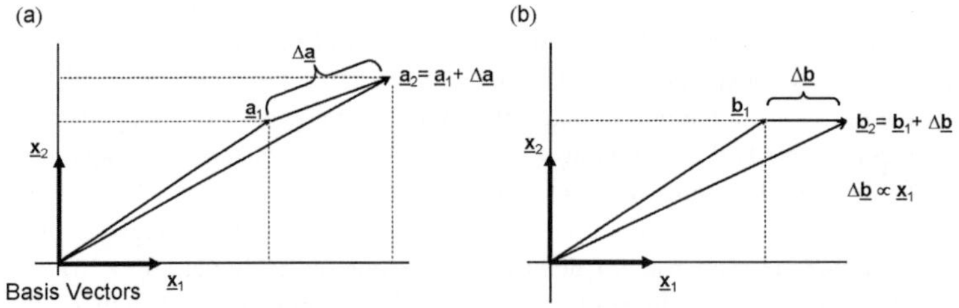

Figure 3.5 Addition of vector increments: (a) arbitrary incremental vector; (b) incremental vector along the first basis vector, which changes the projection along that basis vector only.

is identical to the arbitrary vector $\underline{\mathbf{a}}$:

$$\underline{\mathbf{a}} = \sum_{n=1}^{N} \beta_n \, \hat{\underline{\mathbf{x}}}_n \tag{3.53}$$

The set of N weight coefficients $\{\beta_n\}$ in the linear combination of basis vectors $\underline{\mathbf{x}}_n$ in Equation (3.52) comprises the unique *representation* of the vector $\underline{\mathbf{a}}$ in terms of that particular set of basis vectors $\hat{\underline{\mathbf{x}}}_n$. The process will be illustrated by expressing the arbitrary 3-D vector $\underline{\mathbf{a}}$ in terms of the common 3-D Cartesian basis in Equation (3.50). The representation of the vector in terms of the unknown coefficients β_n is:

$$\underline{\mathbf{a}} = \begin{bmatrix} a_1 \\ a_2 \\ a_3 \end{bmatrix} = \sum_{n=1}^{3} \beta_n \hat{\underline{\mathbf{x}}}_n = \beta_1 \begin{bmatrix} 1 \\ 0 \\ 0 \end{bmatrix} + \beta_2 \begin{bmatrix} 0 \\ 1 \\ 0 \end{bmatrix} + \beta_3 \begin{bmatrix} 0 \\ 0 \\ 1 \end{bmatrix} \tag{3.54}$$

The projection β_1 onto the first basis vector is obtained by applying Equation (3.52):

$$\beta_1 = \underline{\mathbf{a}} \bullet \hat{\underline{\mathbf{x}}}_1 = \begin{bmatrix} a_1 \\ a_2 \\ a_3 \end{bmatrix} \bullet \begin{bmatrix} 1 \\ 0 \\ 0 \end{bmatrix} = a_1 \tag{3.55}$$

The projections onto the other basis vectors are obtained in the same simple manner: $\beta_2 = a_2$ and $\beta_3 = a_3$. Though a trivial example, this result does illustrate the particular convenience of representations over an orthonormal basis because any one projection of an arbitrary vector specified in Equation (3.52) is not affected by changes in the vector along the directions of the other basis vectors; there is no "crosstalk" among the projections (Figure 3.5). Obviously, the projection of a specific vector $\underline{\mathbf{a}}$ onto different basis sets results in different sets of weight coefficients $\{\beta_n\}$ and thus different (though equivalent) representations for the vector $\underline{\mathbf{a}}$. It is apparent from the geometric interpretation of 2-D and 3-D spaces that an infinite number of possible sets of orthonormal basis vectors exist.

3.3.2 Vector Subspaces Associated with a System

At this point, we divert the discussion a bit to consider four particular sets of vectors that are associated with a matrix operator $\underline{\mathbf{A}}$ that may be the mathematical model of an imaging system. In this discussion, the matrix operator $\underline{\mathbf{A}}$ need not be square. In Equation (3.22), the M-row by N-column matrix $\underline{\mathbf{A}}$ was formed by stacking the M N-D row vectors $(\underline{\mathbf{A}})^T$.

3.3.2.1 Row Subspace

The first set of vectors associated with $\underline{\mathbf{A}}$ is constructed by computing all possible linear combinations of the N-D row vectors, and thus the set of vectors is N dimensional. The set of such vectors defines the N-D *row-vector subspace*, also called just the *row subspace*, of the matrix $\underline{\mathbf{A}}$:

$$\sum_{m=0}^{M-1} \beta_m \underline{\mathbf{a}}_m \equiv \{\underline{\mathbf{x}}_r\} \tag{3.56}$$

where the numbers β_m are the arbitrary scalar weighting coefficients applied to the N-D row vectors $\underline{\mathbf{a}}_m$. The subscript "r" within $\{\underline{\mathbf{x}}_r\}$ designates that the vectors are members of the row subspace. The discussion in Section 3.2 demonstrated that the specific row vectors of $\underline{\mathbf{A}}$ span the row subspace, i.e., that any vector in the row subspace may be represented as a weighted sum of the row vectors. Therefore, any linearly independent set of the M N-D row vectors is a suitable basis for the row subspace. Obviously, the N-D null vector is a member of the row subspace because it is generated by Equation (3.56) when all coefficients β_m are zero.

3.3.2.2 Null Subspace

A second vector subspace associated with $\underline{\mathbf{A}}$ is composed of all N-D input vectors $\underline{\mathbf{x}}_n$ which project onto the rows of $\underline{\mathbf{A}}$ to generate the M-D null output vector:

$$\underline{\mathbf{A}}\, \underline{\mathbf{x}}_n = \underline{\mathbf{0}} \tag{3.57}$$

The set of all such N-D vectors $\{\underline{\mathbf{x}}_n\}$ constitutes the *null vector subspace* (or *null subspace*) of $\underline{\mathbf{A}}$. Evidently, the N-D null vector $\underline{\mathbf{0}}$ must belong to the null subspace. After combining the picture of matrix–vector multiplication as a collection of scalar products with the definition that two vectors are orthogonal when their scalar product is zero, it is evident that all vectors in the row subspace of $\underline{\mathbf{A}}$ must be orthogonal to all vectors in the null subspace of $\underline{\mathbf{A}}$:

$$\underline{\mathbf{x}}_r \bullet \underline{\mathbf{x}}_n = 0 \tag{3.58}$$

Note that the N-D null vector $\underline{\mathbf{0}}$ "belongs" to both the row and null subspaces of $\underline{\mathbf{A}}$, and is the only such common member. From the condition of Equation (3.57), the null space of the matrix may be interpreted as the set of input vectors that are "blocked" by the action of the imaging system so that the resulting output vector is identical to the output from a null input "stimulus". This interpretation may be extended to describe the vectors in the row subspace as those inputs that "pass through" the system to generate a nonnull output.

If the null subspace of the matrix operator $\underline{\mathbf{A}}$ contains any nonnull vectors, then distinct input vectors may be constructed that generate the same output vector $\underline{\mathbf{b}}$ under the action of $\underline{\mathbf{A}}$; the distinct input vectors are constructed by adding any scaled vector in the null subspace to the original input. Therefore, these multiple input vectors may not be distinguished from their identical outputs, even if the system matrix $\underline{\mathbf{A}}$ is known. To illustrate this result, consider a null subspace of a particular matrix operator $\underline{\mathbf{A}}$ that includes a single nonnull vector $\underline{\mathbf{x}}_1$:

$$\underline{\mathbf{A}}\, \underline{\mathbf{x}}_1 = \underline{\mathbf{0}} \tag{3.59}$$

A different nonnull vector $\underline{\mathbf{x}}_0$ is assumed to yield the nonnull vector $\underline{\mathbf{b}}_0$ under the action of the matrix:

$$\underline{\mathbf{A}}\, \underline{\mathbf{x}}_0 = \underline{\mathbf{b}}_0 \tag{3.60}$$

The linearity property of matrix–vector multiplication ensures that the operation applied to the sum vector yields the sum of the individual outputs:

$$\underline{\mathbf{A}}(\underline{\mathbf{x}}_0 + \underline{\mathbf{x}}_1) = \underline{\mathbf{A}}\, \underline{\mathbf{x}}_0 + \underline{\mathbf{A}}\, \underline{\mathbf{x}}_1 = \underline{\mathbf{b}}_0 + \underline{\mathbf{0}} \tag{3.61}$$

In words, the action of the system $\underline{\mathbf{A}}$ on two distinct input vectors $\underline{\mathbf{x}}_0$ and $(\underline{\mathbf{x}}_0 + \underline{\mathbf{x}}_1)$ produces the same output "image" $\underline{\mathbf{b}}_0$. Therefore, which of the two input vectors was applied to $\underline{\mathbf{A}}$ cannot be determined from knowledge of $\underline{\mathbf{A}}$ and $\underline{\mathbf{b}}_0$. Such a matrix operator $\underline{\mathbf{A}}$ is *singular*; its inverse does not exist and thus the inverse imaging problem cannot be solved.

3.3.2.3 Column Subspace

In a fashion similar to the derivation of the row subspace, a third vector subspace associated with a system matrix $\underline{\mathbf{A}}$ is composed of all linear combinations of its N M-D columns; the name for this set of M-D vectors obviously is the *column subspace* of $\underline{\mathbf{A}}$. If the symbol $\underline{\mathbf{h}}_n$ is used to represent the nth M-D column of $\underline{\mathbf{A}}$, the vectors in the column subspace may be written as their linear combinations:

$$\sum_{n=0}^{N-1} \gamma_n \underline{\mathbf{h}}_n \equiv \{\underline{\mathbf{b}}_c\} \tag{3.62}$$

where again the coefficients γ_m are arbitrary real numbers and the subscript "c" designates that the output vector belongs to the column subspace. The set $\{\underline{\mathbf{b}}_c\}$ contains all possible M-D output vectors that result from the action of the system on all possible N-D input vectors. Obviously, the M-D *null* vector is a member of the column space, because it is generated by fixing all weights $\gamma_n = 0$.

Though not proven here, it is not difficult to show that the numbers of linearly independent vectors in the column subspace and in the row subspace are identical. In other words, the dimensionalities of the row subspace and the column subspace are equal, which is equivalent to saying that a basis for $\{\underline{\mathbf{x}}_r\}$ and a basis for $\{\underline{\mathbf{b}}_c\}$ contain the same number of vectors. This number defines the *rank* of the matrix $\underline{\mathbf{A}}$. Note that the rank of $\underline{\mathbf{A}}$ is less than the dimensionality of $\underline{\mathbf{A}}$ if the null space of $\underline{\mathbf{A}}$ includes any nonnull vectors. In fact, the rank of the $N \times N$ matrix $\underline{\mathbf{A}}$ is the difference between the dimensionality N of $\underline{\mathbf{A}}$ and that of the null subspace $\{\underline{\mathbf{x}}_n\}$. Put another way, the sum of the dimensionalities of the row and null subspaces must equal N, the dimensionality of the input vectors.

3.3.2.4 Left Null Subspace

A fourth subspace may be constructed by analogy with the null subspace by finding all M-D vectors that are orthogonal to the members of the column subspace. For a reason that is probably not obvious, this fourth vector subspace is called the *left null subspace* of $\underline{\mathbf{A}}$. The nonnull M-D vectors in this subspace cannot be created by the action of the system on the set of all possible input vectors. The M-D null vector is the only common member of the column and left null subspaces. The sum of the dimensionalities of the column and left null subspaces must equal M, the dimensionality of the output vectors.

To illustrate an example of the four vector subspaces associated with a matrix operator, consider the matrix $\underline{\mathbf{A}}$ consisting of three rows and two columns:

$$\underline{\mathbf{A}} = \begin{bmatrix} 1 & 0 \\ 2 & 4 \\ 4 & 4 \end{bmatrix} \tag{3.63}$$

The input vectors $\{\underline{\mathbf{x}}\}$ and output vectors $\{\underline{\mathbf{b}}\}$ associated with $\underline{\mathbf{A}}$ have two and three components, respectively. The row subspace consists of all 2-D vectors that are linear combinations of the rows, i.e., it is the set of vectors that can be expressed in the form:

$$\underline{\mathbf{x}}_r = \beta_1 \begin{bmatrix} 1 \\ 0 \end{bmatrix} + \beta_2 \begin{bmatrix} 2 \\ 4 \end{bmatrix} + \beta_3 \begin{bmatrix} 4 \\ 4 \end{bmatrix} \tag{3.64}$$

Recall that at most two 2-D vectors may be linearly independent; in this case, any of the vectors may be represented as a weighted sum of the other two. For example, the third row vector is this specific combination of the first two:

$$\begin{bmatrix} 4 \\ 4 \end{bmatrix} = \begin{bmatrix} 2 \\ 4 \end{bmatrix} + 2 \begin{bmatrix} 1 \\ 0 \end{bmatrix} \tag{3.65}$$

Obviously, the first row vector may be constructed from a different specific combination of the second and third, etc. The first two 2-D row vectors are linearly independent because they are not scaled replicas of each other, and therefore they constitute a basis for the 2-D row subspace. In other words, all vectors in the row subspace may be expressed as a linear combination of the form:

$$\underline{\mathbf{x}}_r = \beta_1 \begin{bmatrix} 1 \\ 0 \end{bmatrix} + \beta_2 \begin{bmatrix} 2 \\ 4 \end{bmatrix} \tag{3.66}$$

where β_1 and β_2 are the appropriate weights for constructing a particular vector. Because the number of vectors in a basis for the row subspace is equal to the dimensionality of the vectors, then the row subspace includes all 2-D vectors.

The null subspace of $\underline{\mathbf{A}}$ is the set of 2-D vectors that are orthogonal to vectors in the row subspace. Because the entire 2-D plane is spanned by the row-space basis vectors, the null subspace has dimension zero and consists of only the 2-D null vector

$$\underline{\mathbf{0}} \equiv \begin{bmatrix} 0 \\ 0 \end{bmatrix}$$

The column subspace is spanned by the two 3-D column vectors of $\underline{\mathbf{A}}$:

$$\underline{\mathbf{h}}_1^T = \begin{bmatrix} 1 \\ 2 \\ 4 \end{bmatrix} \quad \textit{and} \quad \underline{\mathbf{h}}_2^T = \begin{bmatrix} 0 \\ 4 \\ 4 \end{bmatrix}$$

These two vectors are not scaled replicas and therefore must be linearly independent. Though the column vectors have three components, the fact that the basis for the subspace includes only two 3-D vectors means that the column space has dimension two; it is the 2-D plane defined by the two 3-D basis vectors $\underline{\mathbf{h}}_1$ and $\underline{\mathbf{h}}_2$.

The left null subspace consists of all 3-D vectors that are simultaneously orthogonal to $\underline{\mathbf{h}}_1$ and $\underline{\mathbf{h}}_2$. By computing the scalar product of an arbitrary 3-D vector $\underline{\mathbf{b}}_n$ with $\underline{\mathbf{h}}_1$ and $\underline{\mathbf{h}}_2$, a relation for the components of the vectors in the left null subspace may be found. Since the vectors have three components and the dimensionality of the column subspace is two, then nonnull vectors must belong to the left null subspace. In this example, a suitable basis of the left null subspace is the vector

$$\underline{\mathbf{b}}_n = \begin{bmatrix} 2 \\ 1 \\ -1 \end{bmatrix}$$

which may be easily shown to be orthogonal to both basis vectors for $\underline{\mathbf{b}}_c$ by applying the criterion in Equation (3.20).

The description of the subspaces of $\underline{\mathbf{A}}$ may be used to specify the requirement for the existence of the inverse matrix $\underline{\mathbf{A}}^{-1}$corresponding to a square matrix operator $\underline{\mathbf{A}}$. The criterion may be expressed in several equivalent ways:

1. the null subspace of $\underline{\mathbf{A}}$ consists of $\underline{\mathbf{0}}$ only; or
2. the determinant $|\underline{\mathbf{a}}| \neq 0$; or
3. the row vectors of $\underline{\mathbf{A}}$ span the vector space on which $\underline{\mathbf{A}}$ acts; or
4. the column vectors of $\underline{\mathbf{A}}$ span the vector space on which $\underline{\mathbf{A}}$ acts; or
5. the dimensions of the row and the column subspaces are identical to $RANK[\underline{\mathbf{A}}]$.

The first of these has been considered already in the discussion of the null subspace. Derivations of the others are considered in most texts on linear algebra.

PROBLEMS

3.1 The three vectors $\underline{\mathbf{x}}$, $\underline{\mathbf{y}}$, $\underline{\mathbf{z}}$ have real-valued components and form the sides of a triangle. Prove the "law of cosines" for these vectors, i.e., that:

$$|\underline{\mathbf{z}}|^2 = |\underline{\mathbf{x}}|^2 + |\underline{\mathbf{y}}|^2 - 2\underline{\mathbf{x}} \bullet \underline{\mathbf{y}} = |\underline{\mathbf{x}}|^2 + |\underline{\mathbf{y}}|^2 - 2|\underline{\mathbf{x}}||\underline{\mathbf{y}}| \cos[\theta]$$

where θ is the angle between the vectors $\underline{\mathbf{x}}$ and $\underline{\mathbf{y}}$. (HINT: use the scalar product.)

3.2 Find the lengths of the 3-D vectors:

(a) $\underline{\mathbf{v}}_a = \begin{bmatrix} +1 \\ -2 \\ +1 \end{bmatrix}$

(b) $\underline{\mathbf{v}}_b = \begin{bmatrix} -1 \\ +2 \\ -2 \end{bmatrix}$

(c) $\underline{\mathbf{v}}_c = \begin{bmatrix} 0 \\ 2 \\ 2 \end{bmatrix}$

3.3 In each case, find the scalar product $\underline{\mathbf{v}}_1 \bullet \underline{\mathbf{v}}_2$:

(a) $\underline{\mathbf{v}}_1 = \begin{bmatrix} 1 \\ 1 \\ 1 \end{bmatrix}; \quad \underline{\mathbf{v}}_2 = \begin{bmatrix} 1 \\ 1 \\ 1 \end{bmatrix}$

(b) $\underline{\mathbf{v}}_1 = \begin{bmatrix} 1 \\ 0 \\ 0 \end{bmatrix}; \quad \underline{\mathbf{v}}_2 = \begin{bmatrix} 0 \\ 1 \\ 1 \end{bmatrix}$

(c) $\underline{\mathbf{v}}_1 = \begin{bmatrix} \frac{1}{\sqrt{3}} \\ \frac{1}{\sqrt{3}} \\ \frac{1}{\sqrt{3}} \end{bmatrix}; \quad \underline{\mathbf{v}}_2 = \begin{bmatrix} \frac{1}{\sqrt{3}} \\ \frac{1}{\sqrt{3}} \\ \frac{1}{\sqrt{3}} \end{bmatrix}$

3.4 Find the *projection* of the first vector onto the second *and* the projection of the second vector onto the first.

(a) $\underline{\mathbf{v}}_1 = \begin{bmatrix} 1 \\ 2 \\ -1 \end{bmatrix}; \quad \underline{\mathbf{v}}_2 = \begin{bmatrix} 0 \\ 1 \\ 1 \end{bmatrix}$

(b) $\underline{\mathbf{v}}_1 = \begin{bmatrix} 1 \\ 2 \\ -1 \end{bmatrix}; \quad \underline{\mathbf{v}}_2 = \begin{bmatrix} 0 \\ 2 \\ 2 \end{bmatrix}$

3.5 Find the vectors obtained by rotating the two 2-D basis vectors $\begin{bmatrix} 1 \\ 0 \end{bmatrix}, \begin{bmatrix} 0 \\ 1 \end{bmatrix}$ by angles of $\pi/2$ and $-\pi/6$ radians.

3.6 Consider an arbitrary 2-D real-valued vector $\underline{\mathbf{a}}$ that has the representation $\begin{bmatrix} a_1 \\ a_2 \end{bmatrix}$ in the usual Cartesian basis, i.e.,

$$\underline{\mathbf{a}} = a_1 \begin{bmatrix} 1 \\ 0 \end{bmatrix} + a_2 \begin{bmatrix} 0 \\ 1 \end{bmatrix}$$

Derive the representation for the same vector in a basis that has been rotated by an angle of $2\pi/3$ radians.

3.7 Find the vector subspaces (row, column, null, and left null) associated with these matrices:

(a) $\begin{bmatrix} 1 & 0 & 0 \\ 0 & 1 & 0 \\ 0 & 0 & 0 \end{bmatrix}$

(b) $\begin{bmatrix} 1 & 2 & 0 \\ -2 & 1 & 0 \\ 0 & 0 & 1 \end{bmatrix}$

3.8 A 2×2 sampled "object" with real-valued gray values $\begin{bmatrix} a & c \\ b & d \end{bmatrix}$ may be represented as a 4-D vector by "stacking" the columns in "lexicographic order" to obtain

$$\underline{\mathbf{x}} \equiv \begin{bmatrix} a \\ b \\ c \\ d \end{bmatrix}$$

Find the matrix operators $\underline{\mathbf{A}}_n$ that, when applied to $\underline{\mathbf{x}}$, produce output "images" $\underline{\mathbf{b}}_n$ with the following qualities:

(a) $\underline{\mathbf{b}}_1$ corresponds to the original image with the rows and columns exchanged, i.e., the output 2×2 image is $\begin{bmatrix} a & b \\ c & d \end{bmatrix}$.

(b) The elements of the output image are the sums of the rows and of the columns of $\underline{\mathbf{x}}$, i.e.,

$$\underline{\mathbf{b}}_2 = \begin{bmatrix} a+c \\ b+d \\ a+b \\ c+d \end{bmatrix}$$

3.9 Find the inverses $\underline{\mathbf{A}}_n^{-1}$ of the two matrix operators derived above if they exist. If $\underline{\mathbf{A}}_n^{-1}$ does not exist, then explain why.

3.10 For each of the following matrices:

$$\underline{\mathbf{A}}_1 = \begin{bmatrix} 1 & -1 & 0 \\ 0 & 1 & -1 \\ -1 & 0 & 1 \end{bmatrix}$$

$$\underline{\mathbf{A}}_2 = \frac{1}{2}\begin{bmatrix} 1 & 1 & 0 \\ 0 & 1 & 1 \\ 1 & 0 & 1 \end{bmatrix}$$

$$\underline{\mathbf{A}}_3 = \frac{1}{2}\begin{bmatrix} 1 & 1 & 0 & 0 \\ 0 & 1 & 1 & 0 \\ 0 & 0 & 1 & 1 \\ 1 & 0 & 0 & 1 \end{bmatrix}$$

$$\underline{\mathbf{A}}_4 = \frac{1}{3}\begin{bmatrix} 1 & 1 & 1 & 0 \\ 0 & 1 & 1 & 1 \\ 1 & 0 & 1 & 1 \\ 1 & 1 & 0 & 1 \end{bmatrix}$$

(a) Find the diagonal forms $\mathbf{\Lambda}_n$ for each of the system matrices presented below (note the leading multiplicative factors). You may use a computing "toolbox" or do the computation by hand (it is not difficult).

(b) Determine whether each system is invertible.

(c) For those systems that are invertible, find expressions for the inverse matrices in the "standard" basis, i.e.,

$$\underline{\mathbf{x}}_1 = \begin{bmatrix} 1 \\ 0 \\ 0 \\ \vdots \\ 0 \end{bmatrix}, \underline{\mathbf{x}}_2 = \begin{bmatrix} 0 \\ 1 \\ 0 \\ \vdots \\ 0 \end{bmatrix}, \ldots, \underline{\mathbf{x}}_N = \begin{bmatrix} 0 \\ 0 \\ 0 \\ \vdots \\ 1 \end{bmatrix}$$

(d) For the systems that are not invertible, find the matrix "pseudoinverses" in the "standard" basis.

(e) In words, describe the "action" of these systems on discrete input vectors of the appropriate dimension.

3.11 Consider the vectors $\mathbf{e}_1 = [-2, 1]$, $\mathbf{e}_2 = [2, 4]$.

(a) Are they linearly independent?

(b) What space do they span?

(c) Are they a basis for 2-D real vectors?

(d) Using that set, expand the vector $\mathbf{u} = [6, 2]$.

4

Complex Numbers and Functions

At this point, the course of the discussion changes direction once again. We divert from the consideration of vectors to review the concepts of complex numbers and complex algebra prior to considering the properties of vectors with complex-valued components.

Complex-valued quantities are convenient (if not essential) for describing the behavior of systems used to generate images of signals that are sinusoidal functions of space and/or time. The representation of sinusoidal waves is simplified by using notation based on the magnitude and phase angle of the signal; the concise notation is convenient even though the represented quantities may be real valued. For example, the electric field amplitude (voltage) of a traveling sinusoidal electromagnetic wave is a vector with real-valued amplitude that varies over both temporal and spatial coordinates. The variations in time and space in turn are related through the dispersion equation that relates the frequency and velocity of the wave. In such a case, the representation of the wave in terms of complex-valued components is often the most convenient description of the electric field.

The discussion of this section also will describe vectors constructed from complex-valued components. This extension of the vector concept will prove to be very useful when interpreting the Fourier transform.

Complex numbers are a generalization of *imaginary numbers* and often are denoted by the letter "z". The imaginary numbers are an outgrowth of the concept of $\sqrt{-1}$, which has no real-valued solution. This numerical value was assigned the symbol i by Leonhard Euler in 1777 (Nahin, 1998), so that:

$$\sqrt{-1} \equiv i \implies i^2 = -1 \tag{4.1}$$

Electrical engineers use the letter "j" for $\sqrt{-1}$, which ostensibly was selected to avoid confusion with the use of "I" for current in Ohm's law (how they avoid confusion with the common use of "J" to represent current density is not mentioned). At any rate, we will retain the historic notation that is common in mathematics and physics. An arbitrary imaginary number is obtained by scaling i by a real-valued weight b, and the general complex number z is a composite number formed from the sum of real and imaginary components:

$$z \equiv a + ib \tag{4.2}$$

where both a and b are real numbers. The quantity a is the *real part* of z (denoted $a = \Re\{z\}$), while b is called the *imaginary part* of z ($b = \Im\{z\}$, though note carefully that b is still a real number). The *complex conjugate* of $z = a + i\,b$ is obtained by multiplying the imaginary part of z by -1 and will be

Fourier Methods in Imaging Roger L. Easton, Jr.

denoted by z^*:

$$z^* \equiv a - ib \tag{4.3}$$

Note that this concept of the "conjugate" of a complex number is very different from that of "conjugate coordinates", which were introduced in Chapter 2.

The real and imaginary parts of z may be expressed in terms of z and its complex conjugate z^* via two relations that are easily confirmed:

$$\Re\{z\} = \frac{1}{2}(z + z^*) \tag{4.4}$$

$$\Im\{z\} = \frac{1}{2i}(z - z^*) \tag{4.5}$$

Note the similarity between these relations and those for the even and odd parts of a function $f[x]$ in Equation (2.18).

4.1 Arithmetic of Complex Numbers

If complex numbers are to be used to represent objects, images, and imaging systems, rules must be defined to manipulate them to model the action of the imaging system. The simplest rules describe the sum and difference of two complex numbers: $z_1 = a_1 + ib_1$ and $z_2 = a_2 + ib_2$. These rules are based on customary arithmetic after realizing that the real and imaginary parts of complex numbers are "independent" (or "orthogonal").

4.1.1 Equality of Two Complex Numbers

For example, two complex numbers z_1 and z_2 are equal if their real parts and their imaginary parts are equal:

$$z_1 = z_2 \textit{ if and only if } a_1 = a_2 \textit{ and } b_1 = b_2 \tag{4.6}$$

4.1.2 Sum and Difference of Two Complex Numbers

The sum or difference of the complex numbers is obtained by adding or subtracting their real and imaginary parts separately:

$$z_1 \pm z_2 = (a_1 + i\,b_1) \pm (a_2 + i\,b_2) = (a_1 \pm a_2) + i(b_1 \pm b_2) \tag{4.7}$$

$$\Longrightarrow \Re\{z_1 \pm z_2\} = a_1 \pm a_2 = \Re\{z_1\} \pm \Re\{z_2\} \tag{4.7a}$$

$$\Longrightarrow \Im\{z_1 \pm z_2\} = b_1 \pm b_2 = \Im\{z_1\} \pm \Im\{z_2\} \tag{4.7b}$$

The similarity of this definition to that for the sum and difference of two vectors in Equations (3.2) and (3.3) immediately suggests that complex numbers and 2-D vectors with real-valued components have analogous properties. This interpretation will be discussed in detail in Section 4.2.

4.1.3 Product of Two Complex Numbers

The product of the two complex numbers is obtained by following the rules of arithmetic multiplication while retaining the factors of i and applying the definition that $i^2 = -1$:

$$\begin{aligned} z_1 \cdot z_2 &= (a_1 + i\, b_1) \cdot (a_2 + i\, b_2) = a_1 a_2 + a_1(i\, b_2) + a_2(i\, b_1) + (i\, b_1)(i\, b_2) \\ &= (a_1 a_2 + (i)^2 b_1 b_2) + i(a_1 b_2 + a_2 b_1) \\ &= (a_1 a_2 - b_1 b_2) + i(a_1 b_2 + a_2 b_1) \end{aligned} \tag{4.8}$$

$$\Longrightarrow \Re\{z_1 z_2\} = a_1 a_2 - b_1 b_2 \tag{4.8a}$$

$$\Longrightarrow \Im\{z_1 z_2\} = a_1 b_2 + a_2 b_1 \tag{4.8b}$$

4.1.4 Reciprocal of a Complex Number

If $z_1 \neq 0$ (meaning that $\Re\{z_1\} \neq 0$ and/or that $\Im\{z_1\} \neq 0$), then the reciprocal of the complex number (denoted z_1^{-1}) may be evaluated as real and imaginary parts by multiplying z_1 by the unit constant in the form of z_1^* divided by itself. This is allowed since $z_1^* = a_1 - ib_1 \neq 0$:

$$\frac{1}{z_1} = \frac{1}{z_1} \cdot \frac{z_1^*}{z_1^*} = \frac{z_1^*}{|z_1|^2} = \frac{a_1 - ib_1}{a_1^2 + b_1^2} \quad \textit{if } z_1 \neq 0 \tag{4.9}$$

4.1.5 Ratio of Two Complex Numbers

The definition for the ratio of two complex numbers is obtained by combining the definition of the product of two complex numbers in Equation (4.8) with that for the reciprocal in Equation (4.9):

$$\begin{aligned} \frac{z_1}{z_2} &= \frac{z_1}{z_2} \cdot \frac{z_2^*}{z_2^*} = \frac{a_1 + i\, b_1}{a_2 + i\, b_2} \cdot \frac{a_2 - ib_2}{a_2 - ib_2} \\ &= \frac{(a_1 a_2 + b_1 b_2) + i(a_2 b_1 - a_1 b_2)}{a_2^2 + b_2^2} \end{aligned} \tag{4.10}$$

$$\Re\left\{\frac{z_1}{z_2}\right\} = \frac{(a_1 a_2 + b_1 b_2)}{a_2^2 + b_2^2} \tag{4.10a}$$

$$\Im\left\{\frac{z_1}{z_2}\right\} = \frac{a_2 b_1 - a_1 b_2}{a_2^2 + b_2^2} \tag{4.10b}$$

Special care must be exercised when applying some familiar rules of algebra when imaginary or complex numbers are used. Nahin (1998) points out some examples of such relationships that fail, such as:

$$\sqrt{ab} = \sqrt{a} \times \sqrt{b} \tag{4.11}$$

which yields an incorrect result when both a and b are negative.

4.2 Graphical Representation of Complex Numbers

Because the expression for the sum of two complex numbers in Equation (4.7) is identical in form to that for the sum of two 2-D vectors in Equation (3.2), it seems reasonable to suggest that the arithmetic of complex numbers is analogous to that of 2-D vectors with real-valued components. If this is so, then the complex number $z = a + ib$ is equivalent to the ordered pair of real numbers $[a, b]$. In other words, the domain of the individual complex numbers is equivalent to the 2-D domain of real numbers. This observation suggests that the set of individual complex numbers (a "one-dimensional" set) does not exhibit the property of *ordered size* that exists for the 1-D array of real numbers. This concept is perhaps best illustrated by example. Consider two real numbers a and b. If both $a > 0$ and $b > 0$, then $ab > 0$. This relationship establishes a metric for the relative sizes of the real numbers. However, the corresponding relationship does not exist for the set of "1-D" complex numbers $a + ib$ because the statement $i > 0$ is ambiguous at best and meaningless at worst. Complex numbers may be ordered in size only by using a true 1-D metric. For example, our observation of the equivalence of vectors and complex numbers may be applied to define analogies for the "length" and "azimuth angle" of a complex number. The "length" of the complex number $z = a + ib$ is equivalent to the length of the equivalent 2-D vector $[a, b]$. Mathematicians typically call this quantity the *modulus* or *absolute value* of the complex number, which is the source of its common notation $|z|$. We will adopt terminology more commonly used in the physical sciences by using the term *magnitude* to refer to this measure of a complex number. This choice is dictated in part by common usage in optics, but it also serves to distance the name from the similar term of *modulation*, which has two different meanings of its own. The Pythagorean theorem specifies that the equation for the magnitude of a complex number is:

$$|\underline{\mathbf{r}}| = \sqrt{a^2 + b^2} \tag{4.12}$$

It also may be expressed conveniently via the complex conjugate:

$$|z| = \sqrt{z\, z^*} = \sqrt{(a + ib)(a - ib)} = \sqrt{a^2 + b^2} \tag{4.13}$$

In words, the magnitude of z is an appropriate metric of ordered size for the complex numbers.

The analogy between a complex number and an ordered pair also ensures that the former may be depicted graphically by assigning the imaginary part to the y-axis coordinate in a 2-D plane. In other words, the real-valued y-axis coordinate is multiplied by $i = \sqrt{-1}$. Such a plot has been called the *Argand* diagram in recognition of its invention by Jean-Robert Argand in 1806. Examples of Argand diagrams are shown in Figure 4.1, including the locations of a particular complex number z, its complex conjugate z^* and its reciprocal z^{-1}. The vector equivalence of the complex number suggests the equivalent representation in polar coordinates using the magnitude $|z|$ and polar angle ϕ of the vector: $z = (|z|, \phi)$ (note the use of parentheses to enclose polar coordinates). Again, we adopt the notation of physical science by naming ϕ the "azimuth angle" (also called the "phase angle") of the complex number. This is the angle whose tangent is the ratio of the imaginary and real parts:

$$\phi = \tan^{-1}\left[\frac{b}{a}\right] = \tan^{-1}\left[\frac{\Im\{z\}}{\Re\{z\}}\right] \tag{4.14}$$

Mathematicians sometimes assign the terms *amplitude* or *argument* to ϕ; the reason why we adopt alternate terminology should be obvious! The polar representation of the complex number is often called a *phasor* and consequently the Argand diagram is often called a *phasor diagram*. Note that this definition of phase is more general than was used for real-valued sinusoidal functions in Equation (2.10). From this equation, the phase angle of any complex number is zero radians when the real part is nonnegative and the imaginary part is zero. If the real and imaginary parts are both zero, then the corresponding 2-D real vector has zero length and therefore an indeterminate direction; the phase angle in this case usually is defined arbitrarily to be zero radians. This definition of phase based on the graphical representation specifies a numerical value for the phase for the complex number. The concept of phase will be applied to complex-valued functions in Section 4.3.

There is a subtle problem with the definition of phase in Equation (4.14). Though the range of valid phase angles is $(-\infty, +\infty)$, the inverse tangent function is multiple valued over any contiguous range exceeding 2π radians. For this reason, a complex number with phase angle ϕ_0 is indistinguishable on the Argand diagram from a complex number with the same magnitude and a phase angle of $\phi_0 + 2\pi n$, where n is any integer. By convention, one interval of 2π radians is selected as the *principal*

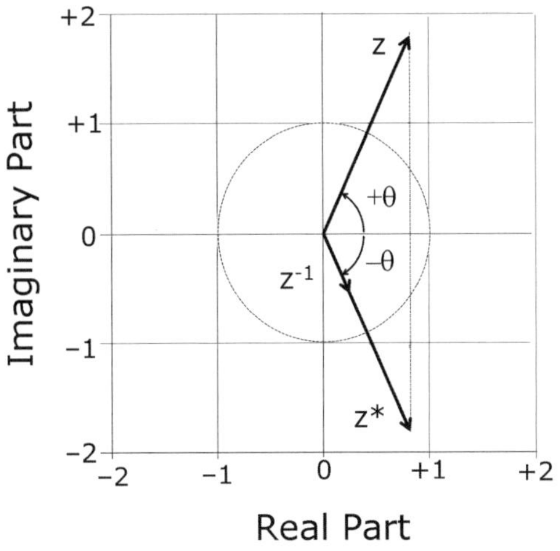

Figure 4.1 Argand diagrams of z, z^{-1}, and z^*, showing that the phase angles of z^{-1} and z^* are equal and the negative of the phase of z.

value of the phase that describes the possible values of ϕ in Equation (4.14). Two semi-closed intervals are commonly used: $[-\pi, +\pi)$ and $[0, 2\pi)$. The first choice is used in this discussion because it is (effectively) symmetric with respect to the origin.

An additional related problem often appears when phase angles are evaluated by computer. The inverse tangent function in Equation (4.14) available in some computer languages has only a single argument, which for complex numbers is the ratio of the imaginary and real parts. Because the ratios of two positive and of two negative amplitudes are both positive, the inverse tangent function cannot distinguish between the two cases; the range of possible values of the phase angle thus computed lies within the interval $[-\pi/2, +\pi/2)$. Additional calculations must be performed based on the algebraic signs of the real and imaginary parts to select the appropriate quadrant and assign the correct angle. Some computer language compilers do include a two-argument inverse tangent function whose principal value spans the full interval of 2π radians, though the user still needs to be aware of the domain used.

Again by analogy with the arithmetic laws for vectors, the real and imaginary parts of a complex number may be evaluated from its magnitude and angle via the relations (Figure 4.1):

$$\Re\{z\} = \Re\{a + i\,b\} = a = r\cos[\phi] \tag{4.15}$$

$$\Im\{z\} = \Im\{a + i\,b\} = b = r\sin[\phi] \tag{4.16}$$

$$z = \Re\{z\} + i\Im\{z\} = r\cos[\phi] + r(i\sin[\phi]) = r(\cos[\phi] + i\sin[\phi]) \tag{4.17}$$

To recast this last expression into yet another useful form, the well-known Taylor-series representations for cosine, sine, and e^u are substituted:

$$\cos[\phi] = \frac{\phi^0}{0!} - \frac{\phi^2}{2!} + \frac{\phi^4}{4!} - \cdots = \sum_{n=0}^{+\infty}(-1)^n\frac{\phi^{2n}}{(2n)!} \tag{4.18a}$$

$$\sin[\phi] = \frac{\phi^1}{1!} - \frac{\phi^3}{3!} + \frac{\phi^5}{5!} - \cdots = \sum_{m=0}^{+\infty}(-1)^m\frac{\phi^{2m+1}}{(2m+1)!} \tag{4.18b}$$

$$e^u = \frac{u^0}{0!} + \frac{u^1}{1!} + \frac{u^2}{2!} + \cdots = \sum_{n=0}^{+\infty}\frac{u^n}{n!} \tag{4.18c}$$

Note that the series for the symmetric (even) cosine includes only even powers of ϕ, while that of the antisymmetric (odd) sine wave includes only the odd powers.

After a bit of manipulation to substitute i^2 for -1, i^3 for $-i$, i^4 for $+1$, etc., a relation that is essential in complex analysis is obtained:

$$\begin{aligned}\cos[\phi] + i\,\sin[\phi] &= \left(\frac{\phi^0}{0!} - \frac{\phi^2}{2!} + \frac{\phi^4}{4!} - \cdots\right) + i\left(\frac{\phi^1}{1!} - \frac{\phi^3}{3!} + \frac{\phi^5}{5!} - \cdots\right) \\ &= \frac{\phi^0}{0!} + i\frac{\phi^1}{1!} + i^2\frac{\phi^2}{2!} + i^3\frac{\phi^3}{3!} + i^4\frac{\phi^4}{4!} + \cdots \\ &= \sum_{n=0}^{\infty} \frac{(i\phi)^n}{n!} = e^{+i\phi} \end{aligned} \tag{4.19}$$

This result is the well-known and vital *Euler relation*, named to commemorate the man who originally derived it in 1748.

Once again, it is evident that the reasoning just used to obtain the Euler relation is circular, as it used two particular Taylor-series representations to derive Equation (4.19) for complex numbers, which are often obtained in the first place by complex analysis. It is not strictly appropriate to use the series to derive the Euler relation. Readers requiring more detail on these series expansions should consult any of the popular texts on complex analysis.

In summary, the complex number z may be represented in several equivalent ways:

$$z = \Re\{z\} + i\Im\{z\} = |z|\,e^{i\phi} = |z|(\cos[\phi] + i\,\sin[\phi]) \tag{4.20}$$

The Euler relation may be used to derive other relationships for the product, reciprocal, and ratio of complex numbers based on Equations (4.8) to (4.10). The product of z_1 and z_2 may be cast in the form:

$$z_1 z_2 = (|z_1|\,e^{i\Phi\{z_1\}})(|z_2|\,e^{i\Phi\{z_2\}}) = (|z_1||z_2|)\,e^{i(\Phi\{z_1\}+\Phi\{z_2\})} \tag{4.21}$$

so the magnitudes multiply and the phases add when computing the product of two complex numbers. In similar fashion, the reciprocal of a complex number is:

$$\frac{1}{z_2} = \frac{1}{|z_2|\,e^{+i\Phi\{z_2\}}} = \frac{1}{|z_2|}\,e^{-i\Phi\{z_2\}} \tag{4.22}$$

The phase of the reciprocal is the negative of the phase of the complex number. From these two relations, it is simple to demonstrate that the ratio of two complex numbers may be represented in magnitude/phase as:

$$\frac{z_1}{z_2} = \frac{|z_1|}{|z_2|}\,e^{i(\Phi\{z_1\}-\Phi\{z_2\})} \tag{4.23}$$

In words, the magnitude of the ratio of two complex numbers is the ratio of the magnitudes, while the phase of the ratio is the difference of the phases.

4.3 Complex Functions

The "complex functions" that are most common in imaging applications have a real-valued domain and a complex-valued range. This is a more restrictive definition than that used in mathematical analysis where both the domain and range of the function are complex valued. As an example of this more general case, consider the notation:

$$w[z] = w[x + iy] \tag{4.24}$$

where x and y are real valued so that the independent variable z is complex. The dependent variable w in this function also is assumed to be a complex number, and so may be expressed equivalently as real

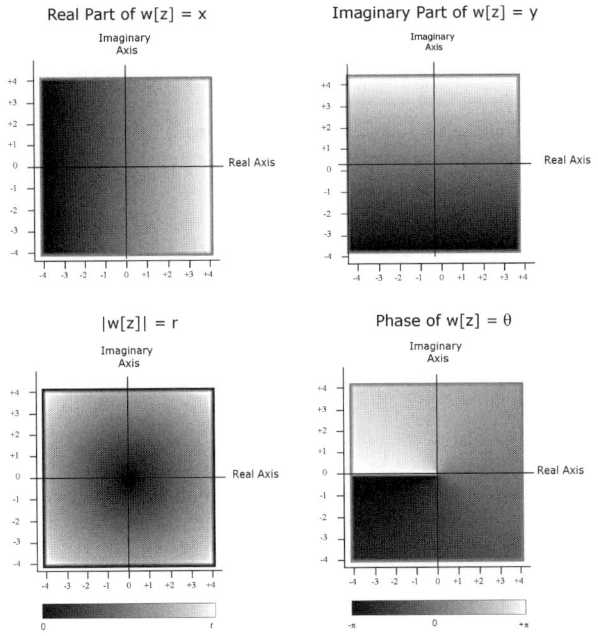

Figure 4.2 The complex function $w[z] = z = x + iy$. The magnitude is $|x + iy| = \sqrt{x^2 + y^2}$, which increases with radial distance from the origin, while the phase is $\tan^{-1}[y/x] = \phi$.

and imaginary parts or as magnitude and phase:

$$w[z] = \Re\{w[x + iy]\} + i\,\Im\{w[x + iy]\} = |w[z]|\,e^{i\Phi\{w[z]\}} \tag{4.25}$$

Both $|w[z]|$ and $\mathbf{\Phi}\{w[z]\}$ denote real-valued functions evaluated for each location z in the complex plane. Because both representations of $w[z]$ may be represented pairs of 2-D "images", the real and imaginary parts of w are depicted as "gray values" specified for each ordered pair $[x, y]$ that is equivalent to z. Alternatively, the pair of 2-D arrays of the nonnegative magnitude and initial phase are an equivalent representation of $w[z]$. The example $w[z] = z = x + iy$ is depicted in Figure 4.2.

The techniques for analyzing and manipulating general complex functions $w[z]$ are very useful for deriving and interpreting some aspects of the theory of linear systems. In particular, the process for integrating a complex function $w[z]$ along specific paths (contours) in the 1-D complex domain (equivalent to the 2-D real plane) is very useful when evaluating properties of some real-valued special functions such as *SINC*$[x]$ in Chapter 6. That said, many of these same results may be derived much more easily from readily proven properties of the Fourier transform. For this reason, a rigorous study of complex functions will not be considered in this book. Interested readers should consult the literature that deals with the theory of complex variables, such as Arfken (2000).

In the current discussion, we will be concerned only with the more restrictive definition of complex functions that have real-valued domains. These will be denoted by the same symbols that have been used for functions with real-valued ranges, such as $f[x]$, but f will be a complex number unless otherwise noted. Obviously, any such $f[x]$ may be decomposed into its unique real and imaginary parts $f_R[x]$ and $f_I[x]$, respectively, which in turn are real-valued functions of x:

$$f[x] = \Re\{f[x]\} + i\Im\{f[x]\} \equiv f_R[x] + if_I[x] \tag{4.26}$$

Note that $f_R[x]$ and $f_I[x]$ may be decomposed further into their respective even and odd parts by applying the relationships in Equation (2.18). As will be shown in Chapter 9, the even–odd decomposition of $f[x]$ provides useful information about its Fourier transform.

A representation of $f[x]$ that is equivalent to that in Equation (4.26) is obtained by separately graphing the magnitude $|f|$ and phase angle Φ_f as functions of x via Equations (4.13) and (4.14). The relationships of the two pairs are identical to those for individual complex numbers that were considered in Equations (4.13) to 4.16):

$$|f[x]| = \sqrt{(f_R[x])^2 + (f_I[x])^2} \tag{4.27a}$$

$$\Phi\{f[x]\} = \tan^{-1}\left[\frac{f_I[x]}{f_R[x]}\right] \tag{4.27b}$$

$$f_R[x] = |f[x]| \cos[\Phi\{f[x]\}] \tag{4.27c}$$

$$f_I[x] = |f[x]| \sin[\Phi\{f[x]\}] \tag{4.27d}$$

Note that the phase angle in Equation (4.27b) may be evaluated for *any* complex-valued function, whereas the definition of "phase" in Equation (2.10) applied only to sinusoids. For example, the phase of any real-valued or imaginary-valued function is obtained by recognizing that the imaginary part of the former and the real part of the latter are zero for all x. The complex amplitude f of any real-valued function always lies on the real axis of the complex domain, which means that its phase angle may take on only two possible values in the semi-closed interval of angles. If the range is assumed to be $[-\pi, +\pi)$, as in our adopted convention, the phase of a real-valued function $f[x]$ is 0 radians at all coordinates x where $f_R[x] \geq 0$ and $-\pi$ radians where $f_R[x] < 0$. The corresponding angles for the "one-sided" interval $[0, 2\pi)$ are 0 and $+\pi$ radians. The phase of any function with $f_R[x] = 0[x]$ has three possible values: $\mathbf{\Phi}\{f[x]\} = 0$ radians where $f_I[x] = 0$, $\mathbf{\Phi}\{f[x]\} = +\pi/2$ where $f_I[x] > 0$, and $\mathbf{\Phi}\{f[x]\} = -\pi/2$ where $f_I[x] < 0$; of course this is $+3\pi/2$ radians in the other convention.

A complex function $f[x]$ is said to be *Hermitian* if its real part is even and its imaginary part is odd. This concept will be of particular interest in the discussion of Fourier transforms. The complex conjugate of a Hermitian function has the interesting property that it is equal to the "reversed" function:

$$f^*[x] = f[-x] \Longrightarrow f^*[-x] = f[x] \textit{ if } f[x] \textit{ is Hermitian} \tag{4.28}$$

An oft-used descriptor of a complex-valued function $f[x]$ is its *power*, which is a generalization of the definition introduced in Section 2.2.5 for a real-valued sinusoid. The power is the 1-D real-valued function obtained by squaring the real-valued magnitude defined in Equation (4.13):

$$|f[x]|^2 = f[x] \cdot f^*[x] = (f_R[x])^2 + (f_I[x])^2 \tag{4.29}$$

and obviously is called the *squared magnitude* of $f[x]$.

Later in this book, when the principles of linear systems are applied to images, complex-valued functions of 2-D real-valued input coordinates such as $f[x, y]$ will be required. In the discussion of imaging systems that act on functions with continuous real-valued domains, images and systems will be represented by complex functions constructed from members of a group of real-valued *special functions* which will be defined in Chapter 6. However, one particularly important complex function will be introduced now, namely the *complex sinusoid* or *complex linear-phase exponential*:

$$f[x] = e^{+2\pi i\xi_0 x} = \exp[+2\pi i\xi_0 x] \tag{4.30}$$

where the two notations are equivalent. The real and imaginary parts of this function are obtained by applying the Euler relation in Equation (4.19) at each value of the coordinate x:

$$e^{+i\theta} = \cos[\theta] + i\sin[\theta]$$

$$\Longrightarrow e^{+2\pi i\xi_0 x} = \cos[2\pi\xi_0 x] + i\sin[2\pi\xi_0 x] \tag{4.31}$$

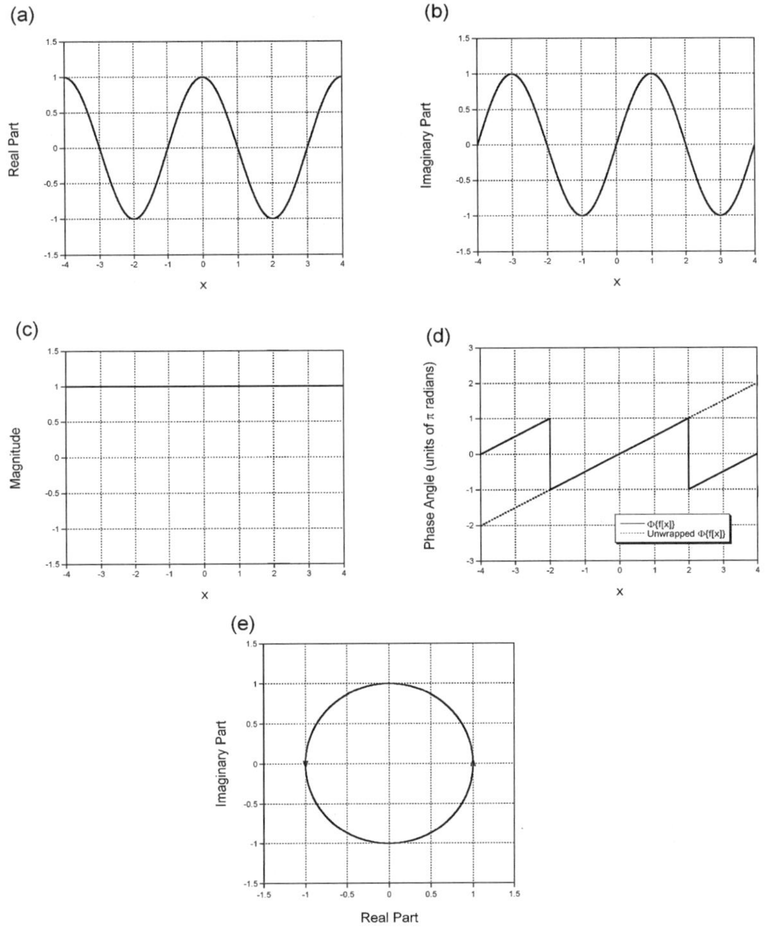

Figure 4.3 Representations of the complex sinusoidal function $f[x] = e^{+2\pi i\xi_0 x} = \cos[2\pi\xi_0 x] + i\sin[2\pi\xi_0 x]$ where $\xi_0 = 0.25$ as real part, imaginary part, magnitude, phase, and Argand diagram; the magnitude is the unit constant, while the phase is proportional to the coordinate.

Note that the real part and the imaginary part of $e^{+2\pi i\xi_0 x}$ are identical to its even and odd parts, respectively. The magnitude of the complex sinusoid is unity for all x, while its phase angle is a linear function of x:

$$|\, e^{+2\pi i\xi_0 x}\,| = \sqrt{\cos^2[2\pi\xi_0 x] + \sin^2[2\pi\xi_0 x]} = 1 \tag{4.32}$$

$$\begin{aligned}\Phi\{e^{+2\pi i\xi_0 x}\} &= \tan^{-1}\left[\frac{+\sin[+2\pi\xi_0 x]}{\cos[+2\pi\xi_0 x]}\right]\\ &= \tan^{-1}[\tan[+2\pi\xi_0 x]] = +2\pi\xi_0 x\end{aligned} \tag{4.33}$$

Graphs of the real/imaginary parts and magnitude/phase of the complex linear-phase exponential are shown in Figure 4.3.

The graph of the phase of the complex sinusoid illustrates the "subtle problem" mentioned in Chapter 2. The real and imaginary parts of $f[x]$ are smoothly varying functions of x, and Equation

(4.32) demonstrates that its phase is proportional to x. However, the inverse tangent function used in Equation (4.27b) is single valued and continuous only over the range of the so-called "principal value", assumed to be $[-\pi, +\pi)$ in our convention. In Figure 4.3d, the "solid line" is the phase computed in the principal interval; the calculated phase exhibits discontinuities from $+\pi$ to $-\pi$ at intervals of ξ_0^{-1}. Often this is not a problem, but some applications require that the phase discontinuities be removed. The "unwrapped" phase is computed from the principal values of the phase at neighboring coordinates and the assumption that the discontinuous transitions are artifacts of the phase calculation. Increments of $\pm 2\pi n$ radians are added to the phase to eliminate the discontinuities in the unwrapped phase shown as the "dotted" line in Figure 4.3d. In other words, the phase unwrapping algorithm assumes that the derivative of the phase is constrained to be finite. Note that this assumption is not always true, and that implementation of phase unwrapping may be difficult, particularly for functions of two or more dimensions (Ghiglia and Pritt, 1998).

The definition of the complex conjugate in Equation (4.3) may be applied to the complex sinusoid in Equation (4.31) to obtain another useful expression:

$$\begin{aligned}(e^{+2\pi i\xi_0 x})^* &= e^{+2\pi(-i)\xi_0 x} = e^{-2\pi i\xi_0 x} \\ &= (\cos[2\pi\xi_0 x] + i\sin[2\pi\xi_0 x])^* = \cos[2\pi\xi_0 x] - i\sin[2\pi\xi_0 x] \\ &= \cos[-2\pi\xi_0 x] + i\sin[-2\pi\xi_0 x]\end{aligned} \tag{4.34}$$

where the respective evenness and oddness of the cosine and sine functions have been used. The corresponding expressions for the magnitude and phase of the complex conjugate of the linear-phase complex exponential are:

$$|e^{-2\pi i\xi_0 x}| = \sqrt{\cos^2[-2\pi\xi_0 x] + \sin^2[-2\pi\xi_0 x]} = 1 \tag{4.35}$$

$$\begin{aligned}\boldsymbol{\Phi}\{e^{-2\pi i\xi_0 x}\} &= \tan^{-1}\left[\frac{\sin[-2\pi\xi_0 x]}{\cos[-2\pi\xi_0 x]}\right] \\ &= \tan^{-1}[\tan[-2\pi\xi_0 x]] = -2\pi\xi_0 x\end{aligned} \tag{4.36}$$

which demonstrate that the complex conjugate of the complex sinusoid also has unit magnitude and linear phase, as expected. This result leads to a more general recipe for constructing the complex conjugate than that in Equation (4.3). Rather than expressing the complex number in the form of real and imaginary parts, and then multiplying the latter by -1, the complex conjugate is obtained simply by negating all occurrences of i in either real/imaginary or magnitude/phase notation.

Other useful relations for linear-phase exponentials are derived readily from sums, differences, and products of angles by applying the Euler relation. For example, the definitions of the real and imaginary parts in terms of the complex number and its conjugate in Equation (4.4) and Equation (4.5) yield expected equivalent expressions for the cosine and sine:

$$\frac{1}{2}(e^{+2\pi i\xi_0 x} + e^{-2\pi i\xi_0 x}) = \cos[2\pi\xi_0 x] \tag{4.37}$$

$$\frac{1}{2i}(e^{+2\pi i\xi_0 x} - e^{-2\pi i\xi_0 x}) = \sin[2\pi\xi_0 x] \tag{4.38}$$

The Euler relation may be used to derive the outcome of a complex number expressed in magnitude/phase form raised to a real-valued power. This may be evaluated in two different but equivalent ways:

$$z^n = (|z|\,e^{i\theta})^n = |z|^n[e^{i\theta}]^n = |z|^n[\cos[\theta] + i\sin[\theta]]^n \tag{4.39a}$$

$$= |z|^n[e^{in\theta}] = |z|^n[\cos[n\theta] + i\sin[n\theta]] \tag{4.39b}$$

By equating the functions of θ, we obtain *De Moivre's theorem*:

$$(e^{i\theta})^n = e^{in\theta} = \cos[n\theta] + i\sin[n\theta] \tag{4.40}$$

which may be viewed as a generalization of Equation (4.20). In words, it states that the value of a complex number raised to a numerical power n is obtained by raising the magnitude to that power while multiplying the phase angle by the same number. For example, the square of a complex number is obtained by squaring the magnitude and doubling the phase angle. De Moivre's theorem is essential for finding complex-valued roots of equations, as will be required later in this discussion.

The Euler relation also may be used to derive expressions for the cosine and sine of the sum and difference of two angles, including the identity that has already been used several times. First, the same expression is evaluated in two different ways:

$$e^{+i\theta} \cdot e^{\pm i\phi} = e^{i(\theta \pm \phi)} = \cos[\theta \pm \phi] + i\sin[\theta \pm \phi] \tag{4.41a}$$

$$\begin{aligned} e^{+i\theta} \cdot e^{\pm i\phi} &= (\cos[\theta] + i\sin[\theta]) \cdot (\cos[\phi] \pm i\sin[\phi]) \\ &= \cos[\theta]\cos[\phi] \pm i^2 \sin[\theta]\sin[\phi] + i\sin[\theta]\cos[\phi] \pm i\cos[\theta]\sin[\phi] \\ &= \cos[\theta]\cos[\phi] \mp \sin[\theta]\sin[\phi] + i(\sin[\theta]\cos[\phi] \pm \cos[\theta]\sin[\phi]) \end{aligned} \tag{4.41b}$$

By equating the real and imaginary parts of the right-hand sides of these equations, two familiar and useful trigonometric identities are derived:

$$\cos[\theta \pm \phi] = \cos[\theta]\cos[\phi] \mp \sin[\theta]\sin[\phi] \tag{4.42a}$$

$$\sin[\theta \pm \phi] = \sin[\theta]\cos[\phi] \pm \cos[\theta]\sin[\phi] \tag{4.42b}$$

The first of these already has been used to derive the amplitudes of the even and odd parts of the sinusoid with arbitrary initial phase in Equation (2.22) and to derive the form of the scalar product of two vectors in terms of the included angle in Equation (3.15).

The real and imaginary parts of a complex number with known magnitude and phase obviously are readily derived by applying the Euler relation. For example, the numerical value of $e^{\pm i\pi}$ is:

$$e^{\pm 2\pi i \cdot \frac{1}{2}} = e^{\pm i\pi} = \cos[\pi] \pm i\sin[\pi] = -1 \pm 0i = -1 \tag{4.43}$$

which can be used to show that:

$$e^{i(+\theta \pm \pi)} = e^{+i\theta}\, e^{\pm i\pi} = e^{+i\theta} \cdot (-1) = -(\cos[\theta] + i\sin[\theta]) \tag{4.44}$$

This last result demonstrates that incrementing or decrementing the phase of a sinusoidal function by an odd multiple of π radians multiplies the amplitude by -1.

Any complex function known to have constant magnitude and linear phase may be specified completely by three real-valued quantities: the spatial frequency ξ_0, the maximum amplitude A_0, and the phase angle at the origin ϕ_0 (the *initial phase*). The equations that relate the maximum amplitude and initial phase of the complex function to the maximum amplitudes of the real/imaginary parts are:

$$(A_0)_R = A_0 \cos[\phi_0] \tag{4.45a}$$

$$(A_0)_I = A_0 \sin[\phi_0] \tag{4.45b}$$

$$A_0 = \sqrt{(A_0)_R^2 + (A_0)_I^2} \tag{4.45c}$$

$$\phi_0 = \tan^{-1}\left[\frac{(A_0)_I}{(A_0)_R}\right] \tag{4.45d}$$

Since the initial phase is defined via the inverse tangent, its principal value lies in the range $[-\pi, +\pi)$ in our convention. Therefore, sinusoids with the same frequency and amplitude are indistinguishable if their initial phases differ by any multiple of 2π.

4.4 Generalized Spatial Frequency – Negative Frequencies

In Equation (2.14), the spatial frequency of a real-valued sinusoid was shown to be proportional to the rate of change of the phase angle. The definition of the phase of a complex number in Equation (4.27b) may be used to extend the concept of spatial frequency to complex-valued functions other than sinusoids:

$$\boldsymbol{\Phi}\{f[x]\} \equiv \tan^{-1}\left[\frac{f_I[x]}{f_R[x]}\right] \tag{4.46}$$

The generalized concept of the spatial frequency of a complex-valued function is obtained by applying Equation (2.14b) to this phase:

$$\xi[x] = \frac{1}{2\pi}\frac{\partial \boldsymbol{\Phi}\{f[x]\}}{\partial x} \tag{4.47}$$

For example, the spatial frequency of the complex linear-phase exponential in Equation (4.47) is:

$$f[x] = 1[x]\, e^{+i\boldsymbol{\Phi}\{f[x]\}} = e^{+2\pi i\xi_0 x}$$

$$\Longrightarrow \xi[x] = \frac{1}{2\pi}\frac{\partial \Phi\{f[x]\}}{\partial x} = \xi_0 \tag{4.48}$$

In words, the spatial frequency of a complex function is a constant if the phase includes at most the sum of linear and constant functions of x. However, it is probably evident that the phase of a 1-D complex-valued function may include higher-order functions of x, and so the spatial frequency will vary over the coordinate x. The value of ξ evaluated at a specific coordinate x_0 is often called the *instantaneous* spatial frequency at that location. Because the spatial frequency in Equation (4.47) is defined by a derivative, it may take on any real value in the infinite range $(-\infty, +\infty)$. In other words, negative values of the spatial frequency are possible. The concept of a negative frequency may seem disconcerting at first glance; the meaning of a spatial oscillation with a frequency of -1 sinusoidal cycle per mm may not be obvious. However, Equation (4.47) establishes that negative spatial frequency is analogous to the readily accepted and visualized concept of negative velocity. Velocity is the rate of change of position with respect to time, $v = \partial x/\partial t$, which is negative in all parts of the domain where the coordinate x *decreases* with increasing time. In identical fashion, the spatial frequency ξ of a complex function $f[x]$ is negative in all regions of the domain where the phase angle $\boldsymbol{\Phi}\{f[x]\}$ *decreases* as x increases.

4.5 Argand Diagrams of Complex-Valued Functions

We have already seen that any function $f[x]$ with a real-valued domain and complex-valued range may be plotted as either of two equivalent pairs of 1-D real-valued functions of x: the real and imaginary parts, or the magnitude and phase. Such a function also may be graphed on an Argand diagram; for each coordinate x, the real part of the function is plotted along the horizontal axis and the imaginary part on the vertical axis. This location represents the "tip" of the phasor generated by the function at that value of x, and may be equivalently specified by its length (magnitude) and angle (initial phase). The ensemble of phasor tips evaluated for each value of x describes a curve on the Argand diagram that is an equivalent representation of $f[x]$. This representation of the complex function on the Argand diagram is analogous to the simultaneous display of two functions on an oscilloscope that is called a *Lissajous figure*. Consider the Argand diagram of the complex linear-phase sinusoid $f[x] = e^{+2\pi i\xi x}$ shown in Figure 4.3e; because $|f[x]| = 1[x]$, all phasors have unit length and thus all points of the function lie on the unit circle in the Argand diagram. Because the phase angle increases in proportion to x, the phasor tip traces out the unit circle in the counterclockwise direction at a constant rate. This provides a visual image of the concept of positive and negative spatial frequencies of a function as the direction of rotation of the phasor. The spatial frequency is positive in all regions of the domain where

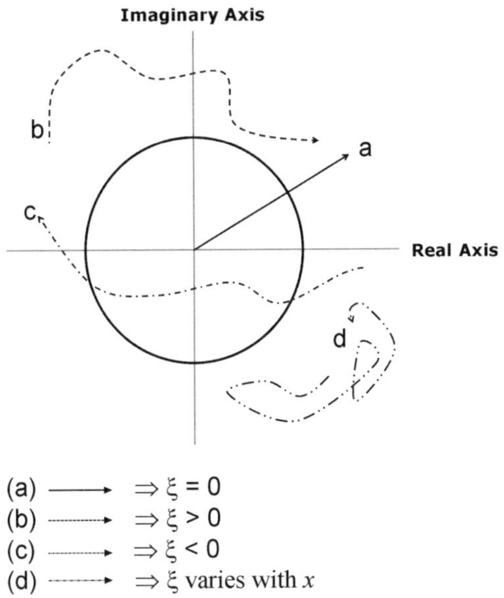

Figure 4.4 Argand diagrams (Lissajous figures) of complex-valued 1-D functions of x with spatial frequencies that are: (a) null; (b) positive; (c) negative; and (d) varies with x.

the phasor angle rotates about the origin in a counterclockwise direction (toward positive angles) as x increases; the spatial frequency is negative at all values of x for which the angular motion of the phasor tip is clockwise (negative angles) as x increases. Note that the spatial frequency is zero in those regions of x where the angle of the phasor tip does not change, i.e., where the phasor tip is stationary or moves directly toward or away from the origin. These cases are illustrated in Figure 4.4.

PROBLEMS

4.1 For $z_1 = 2 - i$ and $z_2 = 1 + 3\,i$, locate the following points in a single sketch of the complex plane:

(a) z_1
(b) z_1^*
(c) z_1^{-1}
(d) z_2

4.2 For $z_1 = 2 - i$ and $z_2 = 1 + 3\,i$, evaluate the operations and locate the resulting complex numbers on the complex plane.

(a) $z_1 + z_2$
(b) $z_1 - z_2$
(c) $z_1 z_1^*$
(d) $z_1 z_2$
(e) $z_1 z_2^*$
(f) $z_1^* z_2$
(g) $z_1^* z_2^*$
(h) z_2 / z_1

4.3 Calculate all roots of the following equations and express them as both real/imaginary parts and as magnitude/phase.

(a) $z^2 + i = 0$
(b) $z^3 - i = 0$
(c) $z^3 + 1 = i$
(d) $z^2 - i = 4$

4.4 For $w[z] = z^{-1}$, compute and plot the real and imaginary parts $a[x, y]$ and $b[x, y]$ where $z = x + iy$.

4.5 Use complex analysis to demonstrate that:

(a) $\cos[5\theta] = 16\cos^5[\theta] - 20\cos^3[\theta] + 5\cos[\theta]$
(b) $\sin[5\theta] = 16\sin^5[\theta] - 20\sin^3[\theta] + 5\sin[\theta]$

4.6 Show that the complex conjugate of the sum of two complex numbers is the sum of the complex conjugates: $(z_1 + z_2)^* = z_1^* + z_2^*$.

4.7 Show that the complex conjugate of the product of two complex numbers is the product of the complex conjugates: $(z_1\, z_2)^* = z_1^*\, z_2^*$.

4.8 Determine the two values of $(-i)^{\frac{1}{2}}$.

4.9 Find the projections of $f[x] = x, x^2, \ldots, x^n$ onto the function defined by $m[x] = e^{-x}$ for $x \geq 0$ and $m[x] = 0$ for $x < 0$. (Gamma function)

4.10 Use De Moivre's theorem in Equation (4.40) to calculate all roots of the following equations and express them as both real/imaginary parts and as magnitude/phase.

(a) $z^5 - i = 0$
(b) $z^3 + 1 = 0$
(c) $z^2 + i = 4$

4.11 Determine the requirement that must be satisfied for the three complex numbers z_1, z_2, and z_3 to lie on a straight line in the complex plane.

4.12 If $z_1 = 4 - 3i$ and $z_2 = -1 + 2i$, find the analytic and graphical solutions to:

(a) $|z_1 + z_2|$
(b) $|z_1 - z_2|$
(c) $z_1^* - z_2^*$
(d) $|2z_1^* - 3z_2^* - 2|$

4.13 Describe the set of complex numbers z that satisfy the constraint $z\, z^* = 4$.

(a) Use De Moivre's theorem to prove that $\sin[3\theta] = 3\sin[\theta] - 4\sin^3[\theta]$.
(b) Find the members of the set of complex numbers z_0 that satisfy $z_0^* = z_0^5$.

4.14 Show that

$$\left(\frac{ib - 1}{ib + 1}\right)^{ia} = e^{-2a\cot^{-1}[b]}$$

(HINT: use a graphical derivation for the numerator and denominator in terms of the cotangent.)

5

Complex-Valued Matrices and Systems

5.1 Vectors with Complex-Valued Components

The two general disjoint subjects of the discussion thus far – vectors and matrices in Chapter 3 and complex numbers and functions in Chapter 4 – will now be united to consider vectors and matrices formed from complex-valued components. Though of interest in their own right, the details of these vectors will be helpful during the discussion of an alternative representation of continuous functions (the Fourier transform) in Chapter 9.

Following Equation (3.1), we can define two-element vectors with complex components as an ordered pair of complex numbers:

$$\underline{\mathbf{x}} = \begin{bmatrix} z_1 \\ z_2 \end{bmatrix} = \begin{bmatrix} a_1 + i\, b_1 \\ a_2 + i\, b_2 \end{bmatrix} \tag{5.1}$$

where the a_n and b_n are the real-valued components of the complex number z_n. Though $\underline{\mathbf{x}}$ is a 2-D complex vector (because it is composed of two complex-valued components), its representation requires four real numbers. For this reason alone, complex-valued vectors are more difficult to visualize than real-valued vectors; this 2-D complex vector requires a 4-D real space to locate all components.

The concept of the "length", or "norm", of a vector with real-valued components defined in Equation (3.6) had to be real valued and nonnegative. The geometric picture of a 2-D complex-valued vector as a 4-D real-valued vector means that the length of the complex vector is still a real-valued and nonnegative quantity. To calculate the length, the concept of the "scalar product" in Equation (3.9) must be generalized based on the definition of the magnitude of a complex number in Equation (4.13). This modification will ensure that the sum of products of the complex-valued components with themselves will be real valued. For example, the complex analogue of the scalar product of $\underline{\mathbf{x}}$ in Equation (5.1) with itself may be defined as:

$$\begin{aligned} \underline{\mathbf{x}} \bullet \underline{\mathbf{x}} = |\underline{\mathbf{x}}|^2 &\Longrightarrow (\hat{\underline{\mathbf{x}}}^*)^T \underline{\mathbf{x}} = (a_1 - ib_1)(a_1 + ib_1) + (a_2 - i\, b_2)(a_2 + ib_2) \\ &= (a_1^2 + b_1^2) + (a_2^2 + b_2^2) = (a_1^2 + a_2^2) + (b_1^2 + b_2^2) \end{aligned} \tag{5.2}$$

Fourier Methods in Imaging Roger L. Easton, Jr.

which is guaranteed to be real valued because all of the a_n and b_n are real valued. This concept may be extended to M-D vectors with complex components:

$$\underline{\mathbf{x}} = \begin{bmatrix} z_1 \\ z_2 \\ \vdots \\ z_M \end{bmatrix} = \begin{bmatrix} a_1 + ib_1 \\ a_2 + ib_2 \\ \vdots \\ a_M + ib_M \end{bmatrix}$$

$$\Longrightarrow \underline{\mathbf{x}} \bullet \underline{\mathbf{x}} = \sum_{m=1}^{M} (a_m^2 + b_m^2) = \sum_{m=1}^{M} a_m^2 + \sum_{m=1}^{M} b_m^2 \tag{5.3}$$

In words, the squared magnitude of a vector with complex-valued components is the sum of the squares of the real parts and of the imaginary parts of each component. Note that the units of the scalar product must be the square of the units of the elements of the vector. For example, for a 3-D vector whose coordinates are measured in mm, the scalar product of the vector with itself will be measured in units of mm^2. The magnitude (or length or norm) of $\underline{\mathbf{x}}$ is the nonnegative square root of the result of Equation (5.2).

5.1.1 Inner Product

This generalized expression for the scalar product of a vector with itself may be applied to distinct N-D vectors $\underline{\mathbf{a}}$ and $\underline{\mathbf{x}}$ that have complex-valued components:

$$\underline{\mathbf{a}} \bullet \underline{\mathbf{x}} = \sum_{n=1}^{N} (a_n^*) x_n \tag{5.4}$$

Note that the scalar resulting from Equation (5.4) is complex valued, in general. It is easy to show that the generalized scalar product in Equation (5.4) does not commute under exchange of the vectors:

$$\underline{\mathbf{a}} \bullet \underline{\mathbf{x}} = (\underline{\mathbf{x}} \bullet \underline{\mathbf{a}})^* \neq \underline{\mathbf{x}} \bullet \underline{\mathbf{a}} \tag{5.5}$$

This is a significant departure from the behavior of the dot product of real-valued vectors in Equation (3.9).

Though the name *scalar product* for the operation in Equation (5.4) is retained by some authors when the component vectors have complex-valued components, we will use the also common name of *inner product* to distinguish the process of Equation (5.4) from the simpler scalar product of real-valued vectors in Equation (3.9). This name has the advantage of distinguishing this definition from the *outer product* that generates a matrix from two component vectors. The outer product is not considered further here, but is discussed in many texts on matrices (e.g., Golub and Van Loan, 1996). By the same argument that led to Equation (3.12), the inner product of an arbitrary vector $\underline{\mathbf{x}}$ and a "reference" vector $\underline{\mathbf{a}}$ is an operation that satisfies the linearity condition of Equation (2.1).

Note that an alternate definition of the magnitude of a complex vector in Equation (5.2) and of the inner product of distinct vectors in Equation (5.4) results by instead applying the complex conjugate to the second term:

$$(\underline{\mathbf{a}} \bullet \underline{\mathbf{x}})_{\text{alternate}} \equiv \underline{\mathbf{a}}^T \underline{\mathbf{x}}^* = \sum_{n=1}^{N} a_n (x_n)^* \tag{5.6}$$

This definition is completely self-consistent, but produces different results. This construction is commonly used in mathematical analysis (e.g., Rudin, 1976), while that of Equation (5.4) is more common in physics. We will use the first convention because our interpretation that the vector $\underline{\mathbf{x}}$ represents the input to the system makes it more reasonable for the system operator $\underline{\mathbf{a}}$ to change. This convention also leads to a pleasing interpretation of the Fourier transform operation (Chapter 9).

The definition of the inner product may be applied to generalize the Schwarz inequality in Equation (3.16) to vectors with complex-valued components:

$$|\underline{\mathbf{a}} \bullet \underline{\mathbf{x}}| \leq |\underline{\mathbf{a}}| \cdot |\underline{\mathbf{x}}| \tag{5.7}$$

This relationship establishes that the product of the lengths of two vectors with complex-valued components is at least as large as the magnitude of their inner product. Again, the equality condition is satisfied only in those cases where the two vectors "point" in the same direction, meaning that one vector is a scaled replica of the other.

Despite the differences in definitions between the scalar product of vectors with real-valued components and the inner product of vectors with complex components, we retain the interpretation that the inner product of a unit-norm complex vector (such as $\hat{\underline{\mathbf{a}}}$) with an arbitrary "input" vector $\underline{\mathbf{x}}$ yields the (now possibly complex-valued) projection of $\underline{\mathbf{x}}$ onto $\hat{\underline{\mathbf{a}}}$. As just described, the order of the vectors in the operation is important since the inner product operation does not commute.

Unfortunately, the analogous expression to Equation (3.18) for the angle between two vectors with complex components does not exist for real-valued angles. Because the inner product of two vectors generally is complex, this would require a complex-valued cosine and thus a complex-valued angle. Even though the concept of the angle between complex-valued vectors is not well defined, it is still common to say that two vectors with complex-valued components and null inner product are "orthogonal".

5.1.2 Products of Complex-Valued Matrices and Vectors

The process of matrix–vector multiplication defined for real-valued components in Equation (3.24) is applied without modification to vectors and matrices with complex-valued components. The complex-conjugation operation that is an integral part of the definition of the inner product is not explicitly included in the definition of complex matrix–vector multiplication. Therefore, multiplication of a complex-valued vector by a matrix with complex-valued elements computes the *scalar* product of the row vector of $\underline{\mathbf{A}}$ and the column vector $\underline{\mathbf{x}}$ rather than their *inner* product. Put another way, the product of a complex-valued column vector with a complex-valued row vector in the matrix $\underline{\mathbf{A}}$ is the projection of the input column vector onto the *complex conjugate* of the corresponding row vector of the matrix, rather than the projection onto the row vector itself. Therefore, to obtain the true projections b_m of a complex-valued input vector $\underline{\mathbf{x}}$ onto each of the set of complex-valued unit reference vectors $\{\hat{\underline{\mathbf{a}}}_m\}$ arrayed as a matrix $\underline{\mathbf{A}}$, the complex conjugation of the matrix components must be included explicitly in the matrix–vector product. The scalar product of a single row vector is the product of the complex conjugate of the system matrix with the input column vector:

$$\hat{\underline{\mathbf{a}}}_m \bullet \underline{\mathbf{x}} = \sum_{n=0}^{N-1} (\hat{a}_m^*)_n x_n = b_m = (\underline{\mathbf{A}}^* \, \underline{\mathbf{x}})_m \tag{5.8a}$$

$$\Longrightarrow \underline{\mathbf{A}}^* \, \underline{\mathbf{x}} = \underline{\mathbf{b}} \tag{5.8b}$$

This distinction will be important when we consider the interpretation of the discrete Fourier transform in Chapter 15.

5.2 Matrix Analogues of Shift-Invariant Systems

Any 1-D discrete imaging system described by a matrix–vector product is guaranteed to be linear because the appropriate scalar (or inner) product operation is linear. However, the additional imposition of shift invariance constrains the allowable forms of the matrix. Recall that the criterion for shift invariance in Equation (2.2a) requires that the only effect of a translation of the input function be

$$\text{(a)}\quad \begin{matrix}\vdots\\0\\0\\ \begin{bmatrix}x_0\\x_1\\x_2\\x_3\end{bmatrix}\\0\\0\\\vdots\end{matrix} \longrightarrow \begin{matrix}\vdots\\0\\0\\x_0\\ \begin{bmatrix}x_1\\x_2\\x_3\\0\end{bmatrix}\\0\\\vdots\end{matrix} \qquad \text{(b)}\quad \begin{matrix}\vdots\\x_2\\x_3\\ \begin{bmatrix}x_0\\x_1\\x_2\\x_3\end{bmatrix}\\x_0\\x_1\\\vdots\end{matrix} \longrightarrow \begin{matrix}\vdots\\x_2\\x_3\\x_0\\ \begin{bmatrix}x_1\\x_2\\x_3\\x_0\end{bmatrix}\\x_1\\\vdots\end{matrix}$$

Figure 5.1 Two concepts of translation of the components of a 4-D vector $\underline{\mathbf{x}}$: (a) the components translated from off the edge are assumed to be zero; (b) the components translated from off the edge are identical to the components displaced "off the edge".

an identical translation of the output. In the current interpretation of the input and output as vectors, the concept of translation is not well defined, though Figure 5.1 illustrates two possible choices: either the displaced components could "disappear off the edge" so that the translated vector includes null components, or the input vector could be assumed periodic so that translation of an N-D vector by N units returns the original vector.

Consider the specific case of an N-D input vector $\underline{\mathbf{x}}$ that includes a single nonzero component:

$$f[x] \rightarrow \underline{\mathbf{x}} \equiv \begin{bmatrix}1\\0\\0\\0\\\vdots\\0\end{bmatrix} \tag{5.9}$$

In either convention, the process of translating the components of this N-D vector by integer units yields the forms:

$$f[x-1] \rightarrow \begin{bmatrix}0\\1\\0\\0\\\vdots\\0\end{bmatrix}, \quad f[x-2] \rightarrow \begin{bmatrix}0\\0\\1\\0\\\vdots\\0\end{bmatrix}, \quad \ldots, \quad f[x-(N-1)] \rightarrow \begin{bmatrix}0\\0\\0\\0\\\vdots\\1\end{bmatrix} \tag{5.10}$$

If we extend this one more component to a translation of the N-D vector by N components, the result is the null vector if we use the first convention. The second convention yields the original vector:

$$f[x-N] = f[x] \rightarrow \begin{bmatrix}1\\0\\0\\0\\\vdots\\0\end{bmatrix} \tag{5.11}$$

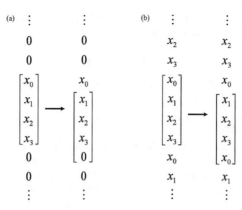

This definition is used in the subsequent discussion, in part because the norm of the input vector is always preserved. The corresponding condition of shift invariance for matrix–vector multiplication has the form:

$$\text{if } \underline{\mathbf{A}} \text{ is ``shift invariant'' and } \underline{\mathbf{A}} \begin{bmatrix} x_0 \\ x_1 \\ x_2 \\ x_3 \\ \vdots \\ x_{N-1} \end{bmatrix} = \begin{bmatrix} b_0 \\ b_1 \\ b_2 \\ b_3 \\ \vdots \\ b_{N-1} \end{bmatrix}$$

$$\text{then } \underline{\mathbf{A}} \begin{bmatrix} x_{N-1} \\ x_0 \\ x_1 \\ x_2 \\ \vdots \\ x_{N-2} \end{bmatrix} = \begin{bmatrix} b_{N-1} \\ b_0 \\ b_1 \\ b_2 \\ \vdots \\ b_{N-2} \end{bmatrix} \tag{5.12}$$

where the indices of the vector components are used to specify the original location within the untranslated vector.

For example, consider a simple matrix $\underline{\mathbf{A}}$ with $N = M = 3$ that satisfies the following relation when applied to a specific 3-D input vector:

$$\underline{\mathbf{A}} \begin{bmatrix} 1 \\ 0 \\ 0 \end{bmatrix} = \begin{bmatrix} \alpha \\ \beta \\ \gamma \end{bmatrix} \tag{5.13a}$$

It is easy to see that this equation specifies the first column of $\underline{\mathbf{A}}$. For $\underline{\mathbf{A}}$ to satisfy the criterion of "circular shift invariance" in Equation (5.11), then two more relationships are determined:

$$\underline{\mathbf{A}} \begin{bmatrix} 0 \\ 1 \\ 0 \end{bmatrix} = \begin{bmatrix} \gamma \\ \alpha \\ \beta \end{bmatrix} \quad \text{and} \quad \underline{\mathbf{A}} \begin{bmatrix} 0 \\ 0 \\ 1 \end{bmatrix} = \begin{bmatrix} \beta \\ \gamma \\ \alpha \end{bmatrix} \tag{5.13b}$$

It is easy to see that the form of the shift-invariant matrix operator that yields these combinations of input and output vectors must have the form where the diagonals are constant:

$$\underline{\mathbf{A}} = \begin{bmatrix} \alpha & \gamma & \beta \\ \beta & \alpha & \gamma \\ \gamma & \beta & \alpha \end{bmatrix} \tag{5.14}$$

Note that the process just performed is an example of the third class of imaging problem; we have "analyzed the system" to derive the operator $\underline{\mathbf{A}}$ from three pairs of input and output vectors. Also note that each row (or column) of this shift-invariant matrix $\underline{\mathbf{A}}$ in Equation (5.14) is a "circular translation" of the adjacent row (or column) by one component; adjacent rows (or columns) of $\underline{\mathbf{A}}$ are obtained by shifting the elements to the right (or down), and returning the last component to the position of the first. All linear matrix operators that are shift invariant in the sense of Equation (5.12) have this "circular" form and are called *circulant* matrices. Note that $N \times N$ circulant matrices are specified completely from knowledge of the N components in any row (or column). Circulant matrices have some special properties that will be exploited later in this chapter, and their extension to the continuous case defines the operation of *convolution*.

If the first convention for translation of the vector is used, the resulting shift-invariant matrix operator would have the form:

$$\underline{\mathbf{A}} = \begin{bmatrix} \alpha & 0 & 0 \\ \beta & \alpha & 0 \\ \gamma & \beta & \alpha \end{bmatrix} \tag{5.15}$$

where all the elements along the main diagonal and the subdiagonals are identical. Such a form is called a *Toeplitz* matrix. In its most general form, an $N \times N$ Toeplitz matrix has at most $2N - 1$ different elements. Note that a circulant matrix also satisfies the requirements for a Toeplitz matrix (Jain, 1988). From this point forward, we will assume the circulant form of the matrix operator, such as Equation (5.14).

5.2.1 Eigenvectors and Eigenvalues

The discussion of Chapter 3 introduced the idea that a vector is "represented" by its scalar product projections onto each member of a set of basis ("reference") vectors. The most convenient bases consist of mutually orthonormal vectors that establish a Cartesian coordinate system; the fact that the vectors in the basis are orthogonal ensures that any change in the input vector along the direction of a particular basis vector affects only the corresponding projection. The fact that the basis vectors are orthonormal simplifies the computation of the projections in Equation (3.20). An infinite number of orthonormal basis sets exists for any vector space; these differ by rigid rotations of the vector ensemble. At this point, we raise the question of whether a particular set of basis vectors might be "preferred" among all possible choices. If the only goal is to describe arbitrary vectors without performing any imaging operation, it is difficult to imagine a criterion that would make one particular representation more desirable than any other. However, recall from Chapter 3 that the action of an imaging system naturally divides the input and output vectors into pairs of orthogonal groups. As it happens, these vector subspaces associated with the system matrix $\underline{\mathbf{A}}$ may be used to construct particularly convenient bases for representing all input and output vectors when considering the specific imaging operator $\underline{\mathbf{A}}$. We will see that this representation is even more important in the case of an $N \times N$ shift-invariant system operator as defined in Equation (5.14), because all such systems share the same convenient set of basis vectors regardless of the specific numerical elements in the circulant matrix $\underline{\mathbf{A}}$.

Consider an $N \times N$ matrix operator $\underline{\mathbf{A}}$ so that the input and output vectors have the same dimensionality N. The specific form of $\underline{\mathbf{A}}$ is not important at this point. Without any obvious motivation, we seek to determine the form of any input vector(s) $\underline{\mathbf{x}}_p$ that exhibit a special property: that the action of $\underline{\mathbf{A}}$ on the N-D vector $\underline{\mathbf{x}}_p$ yields an N-D output vector $\underline{\mathbf{b}}_p$ that is a scaled replica of $\underline{\mathbf{x}}_p$. Put another way, the ratio of the corresponding components of $\underline{\mathbf{b}}_p$ and $\underline{\mathbf{x}}_p$ must all be identical: $b_1/x_1 = b_2/x_2 = \cdots = b_N/x_N$. The resulting form of the imaging equation in this case is:

$$\underline{\mathbf{A}}\,\underline{\mathbf{x}}_p = \underline{\mathbf{b}}_p \propto \underline{\mathbf{x}}_p \tag{5.16}$$

Note that more than one pair of vectors may satisfy this property and that different such pairs may be related by different scale factors. By defining the scale factor that relates a particular pair of input and output vectors $\underline{\mathbf{x}}_p$ and $\underline{\mathbf{b}}_p$ to be λ_p, the imaging equation may be rewritten:

$$\underline{\mathbf{A}}\,\underline{\mathbf{x}}_p = \lambda_p \cdot \underline{\mathbf{x}}_p = \lambda_p \cdot (\underline{\mathbf{I}}\,\underline{\mathbf{x}}_p) = (\underline{\mathbf{I}}\,\lambda_p)\,\underline{\mathbf{x}}_p \tag{5.17}$$

Any vector $\underline{\mathbf{x}}_p$ that satisfies this condition is an *eigenvector* or *principal vector* of the matrix $\underline{\mathbf{A}}$; the proportionality constant λ_p is the *eigenvalue* associated with that eigenvector $\underline{\mathbf{x}}_p$. For emphasis, we repeat that the eigenvectors and eigenvalues are determined by (and associated with) the system matrix $\underline{\mathbf{A}}$; different system operators generally have different sets of eigenvectors and eigenvalues. Also note that any eigenvector that is scaled by a complex-valued constant also is an eigenvector of $\underline{\mathbf{A}}$ with the same eigenvalue. For this reason, it is convenient to normalize all eigenvectors to unit length. Though not be proven here, it is possible to show that at most N distinct eigenvectors are associated with any $N \times N$ operator $\underline{\mathbf{A}}$.

Equation (5.17) determines that the eigenvalue specifies the proportion of the corresponding eigenvector that is transmitted from the input vector to the output vector. If it is possible to represent an input vector as a weighted sum of its projections onto the eigenvectors (i.e., if the eigenvectors constitute a basis), then the output vector may be derived by scaling the projections of the input vector onto the eigenvectors by the corresponding eigenvalues. The set of N eigenvalues associated with the $N \times N$ matrix operator $\underline{\mathbf{A}}$ is called the *spectrum* of the operator. This terminology suggests that there may be an analogy between the set of eigenvalues and the spectrum of white light in a rainbow. The "brightness" of each color in the white-light spectrum is a measure of the amplitude at that wavelength present in the original undispersed white light. In similar fashion, the set of eigenvalues describes "how much" of each eigenvector is transmitted by the action of $\underline{\mathbf{A}}$ to the output.

To illustrate the derivation of eigenvalues and associated eigenvectors of matrix operators, we will consider three 2×2 matrices $\underline{\mathbf{A}}_n$. The first matrix, $\underline{\mathbf{A}}_1$, is the 2×2 identity matrix:

$$\underline{\mathbf{A}}_1\,\underline{\mathbf{x}} = \begin{bmatrix} 1 & 0 \\ 0 & 1 \end{bmatrix} \begin{bmatrix} x_1 \\ x_2 \end{bmatrix} = \begin{bmatrix} x_1 \\ x_2 \end{bmatrix} = 1 \cdot \begin{bmatrix} x_1 \\ x_2 \end{bmatrix} = \lambda \begin{bmatrix} x_1 \\ x_2 \end{bmatrix} \tag{5.18}$$

which is a diagonal matrix that also is circulant and therefore Toeplitz. Because every output vector resulting from application of this operator $\underline{\mathbf{A}}_1$ is identical to the corresponding input vector, all 2-D vectors are eigenvectors of $\underline{\mathbf{A}}_1$ with identically unit eigenvalues. In this case, any pair of 2-D unit-length orthogonal vectors is a satisfactory basis for the 2-D vector space. The obvious choice for such a set is the unit vectors $\left[\begin{smallmatrix} 1 \\ 0 \end{smallmatrix}\right]$ and $\left[\begin{smallmatrix} 0 \\ 1 \end{smallmatrix}\right]$.

The second 2×2 matrix example is obtained by exchanging the rows (or the columns) of the 2×2 identity matrix. This matrix $\underline{\mathbf{A}}_2$ has "0" on the diagonal and "1" elsewhere; it is circulant and Toeplitz. The eigenvectors and eigenvalues of this matrix are found by simultaneously solving two scalar product equations.

$$\underline{\mathbf{A}}_2 \begin{bmatrix} x_1 \\ x_2 \end{bmatrix} = \begin{bmatrix} 0 & 1 \\ 1 & 0 \end{bmatrix} \begin{bmatrix} x_1 \\ x_2 \end{bmatrix} = \lambda \begin{bmatrix} x_1 \\ x_2 \end{bmatrix} \Longrightarrow x_1 = \lambda x_2 \textit{ and } x_2 = \lambda x_1 \tag{5.19}$$

where x_1 and x_2 represent the two components of the 2-D eigenvector associated with λ. The simultaneous solutions of the components in Equation (5.19) require that λ be either $+1$ or -1. For no good reason, we arbitrarily assign $+1$ to λ_1 and -1 to λ_2. The associated eigenvector for each eigenvalue is then derived and distinguished by the corresponding subscript. It is easy to see that the two components of the eigenvector $\underline{\mathbf{x}}_1$ associated with $\lambda_1 = +1$ *must* be equal: $x_1 = x_2$, e.g.,

$$\underline{\mathbf{x}}_1 = \begin{bmatrix} 1 \\ 1 \end{bmatrix}$$

The components of the eigenvector associated with $\lambda_2 \equiv -1$ must have equal magnitude and opposite sign: $x_1 = -x_2$, e.g.,

$$\underline{\mathbf{x}}_2 = \begin{bmatrix} 1 \\ -1 \end{bmatrix} \quad \textit{or} \quad \begin{bmatrix} -1 \\ 1 \end{bmatrix}$$

which are replicas scaled by the factor -1. The two normalized eigenvectors are found by scaling the eigenvector components via Equation (3.7):

$$\lambda_1 = +1 \Longrightarrow \hat{\underline{\mathbf{x}}}_1 = \frac{1}{\sqrt{2}} \begin{bmatrix} 1 \\ 1 \end{bmatrix} \tag{5.20a}$$

$$\lambda_2 = -1 \Longrightarrow \hat{\underline{\mathbf{x}}}_2 = \frac{1}{\sqrt{2}} \begin{bmatrix} 1 \\ -1 \end{bmatrix} \quad \textit{or} \quad \frac{1}{\sqrt{2}} \begin{bmatrix} -1 \\ 1 \end{bmatrix} \tag{5.20b}$$

Also for no obvious reason, we adopt the first choice given for the second eigenvector $\hat{\underline{\mathbf{x}}}_2$ so that the first component is positive. Note that the scalar product of the two real-valued unit-length eigenvectors is zero, which means that they also are orthonormal. These two normalized vectors also constitute a satisfactory orthonormal basis for any 2-D vector.

The third 2×2 matrix and its corresponding "eigen-equation" are:

$$\underline{\mathbf{A}}_3 \begin{bmatrix} x_1 \\ x_2 \end{bmatrix} = \begin{bmatrix} 0 & 1 \\ 0 & 0 \end{bmatrix} \begin{bmatrix} x_1 \\ x_2 \end{bmatrix} = \lambda \begin{bmatrix} x_1 \\ x_2 \end{bmatrix} \Longrightarrow x_2 = \lambda x_1 \textit{ and } 0 = \lambda x_2 \tag{5.21}$$

The simultaneous solution of this equation requires that $x_2 = 0$, while x_1 is arbitrary. There is a single eigenvalue, $\lambda = 0$. The associated normalized eigenvector is:

$$\begin{bmatrix} 0 & 1 \\ 0 & 0 \end{bmatrix} \begin{bmatrix} x_1 \\ y_1 \end{bmatrix} = 0 \cdot \begin{bmatrix} x_1 \\ y_1 \end{bmatrix} \Longrightarrow \begin{bmatrix} x_1 \\ y_1 \end{bmatrix} = \begin{bmatrix} 1 \\ 0 \end{bmatrix} = \begin{bmatrix} x_2 \\ y_2 \end{bmatrix} \tag{5.22}$$

In words, the action of this system scales the piece of the input vector that is proportional to this eigenvector by a null factor; the system "blocks" the eigenvector so that it cannot be part of any output vector. Since there is only a single eigenvector associated with $\underline{\mathbf{A}}_3$, it alone cannot serve as a basis for all 2-D vectors.

The system equation in Equation (5.17) that is valid for eigenvectors may be rewritten in a form that leads to a more systematic method for computing the eigenvalues. The N-D eigenvector is rewritten as the product of the $N \times N$ identity matrix and the input vector scaled by the eigenvalue:

$$(\underline{\mathbf{A}} - \underline{\mathbf{I}}\lambda_p)\,\underline{\mathbf{x}}_p \equiv \underline{\mathbf{C}}\,\underline{\mathbf{x}}_p = \underline{\mathbf{0}} \tag{5.23}$$

Evidently the null subspace of this newly defined matrix operator $\underline{\mathbf{C}}$ includes at least one nonnull vector $\underline{\mathbf{x}}_p$. The discussion following Equation (3.66) indicates that $\underline{\mathbf{C}}^{-1}$ does not exist and that the determinant of $\underline{\mathbf{C}}$ therefore must be zero:

$$\det[\underline{\mathbf{C}}] = \det[\underline{\mathbf{A}} - \underline{\mathbf{I}}\,\lambda_p] = 0 \tag{5.24}$$

Evaluation of this determinant yields an Nth-order equation in terms of the eigenvalues λ_p. The N roots of this so-called *secular equation* are the N eigenvalues of the original matrix $\underline{\mathbf{A}}$. In the example of the 2×2 matrix $\underline{\mathbf{A}}_3$ in Equation (5.21), the two eigenvalues were identically zero; this matrix has *degenerate* eigenvalues that are associated with a single eigenvector. The eigenvalues are introduced individually into the eigenvector equation of Equation (5.17) to find relationships among the N components of the eigenvectors, as was done to find the eigenvectors of the 2×2 operator in Equation (5.20).

5.2.2 Projections onto Eigenvectors

We now use the concepts of eigenvectors and eigenvalues to derive a very useful "equivalent" expression for the system matrix $\underline{\mathbf{A}}$, which may be considered to be the representation of $\underline{\mathbf{A}}$ in a more convenient set of basis vectors. This result applies in all cases where the system matrix $\underline{\mathbf{A}}$ is $N \times N$ square and has N distinct eigenvalues λ_n with N distinct associated eigenvectors $\underline{\mathbf{x}}_n$. The N eigenvector relationships are:

$$\begin{aligned} \underline{\mathbf{A}}\,\underline{\mathbf{x}}_1 &= \lambda_1\,\underline{\mathbf{x}}_1 \\ \underline{\mathbf{A}}\,\underline{\mathbf{x}}_2 &= \lambda_2\,\underline{\mathbf{x}}_2 \\ \underline{\mathbf{A}}\,\underline{\mathbf{x}}_3 &= \lambda_3\,\underline{\mathbf{x}}_3 \\ &\vdots \\ \underline{\mathbf{A}}\,\underline{\mathbf{x}}_N &= \lambda_N\,\underline{\mathbf{x}}_N \end{aligned} \tag{5.25}$$

A concise expression for this set of N "eigen-equations" is obtained by constructing a matrix from the N N-D column eigenvectors $\underline{\mathbf{x}}_n$ that is multiplied by the system matrix $\underline{\mathbf{A}}$ and a matrix of the N N-D column vectors $\lambda_n\,\underline{\mathbf{x}}_n$:

$$\underline{\mathbf{A}}\begin{bmatrix} \begin{bmatrix} \underline{\mathbf{x}}_1 \end{bmatrix} & \begin{bmatrix} \underline{\mathbf{x}}_2 \end{bmatrix} & \cdots & \begin{bmatrix} \underline{\mathbf{x}}_N \end{bmatrix} \end{bmatrix} = \begin{bmatrix} \begin{bmatrix} \lambda_1\,\underline{\mathbf{x}}_1 \end{bmatrix} & \begin{bmatrix} \lambda_2\,\underline{\mathbf{x}}_2 \end{bmatrix} & \cdots & \begin{bmatrix} \lambda_N\,\underline{\mathbf{x}}_N \end{bmatrix} \end{bmatrix} \tag{5.26}$$

This $N \times N$ matrix of the eigenvectors $\underline{\mathbf{x}}_n$ will be assigned an arbitrary name $\underline{\mathbf{D}}$, whose specific form depends on the order of the eigenvectors, and thus of the eigenvalues; in other words, there are several possible forms of $\underline{\mathbf{D}}$ that differ in the order of the columns. A particularly useful order of eigenvectors will be specified later in the discussion. Note also that the problem assumes that the N eigenvectors are distinct, which means that no vector is a scaled copy of any other.

The right-hand side of Equation (5.26) may be rewritten as the matrix product of $\underline{\mathbf{D}}$ and an $N \times N$ diagonal matrix $\underline{\mathbf{\Lambda}}$ that is formed by placing the N eigenvalues on the diagonal in the same order as the associated eigenvectors in $\underline{\mathbf{D}}$:

$$\left[\begin{bmatrix} \lambda_1 \, \underline{\mathbf{x}}_1 \end{bmatrix} \begin{bmatrix} \lambda_2 \, \underline{\mathbf{x}}_2 \end{bmatrix} \cdots \begin{bmatrix} \lambda_N \, \underline{\mathbf{x}}_N \end{bmatrix}\right]$$

$$= \left[\begin{bmatrix} \underline{\mathbf{x}}_1 \end{bmatrix} \begin{bmatrix} \underline{\mathbf{x}}_2 \end{bmatrix} \cdots \begin{bmatrix} \underline{\mathbf{x}}_N \end{bmatrix}\right] \begin{bmatrix} \lambda_1 & 0 & \cdots & 0 \\ 0 & \lambda_2 & \cdots & 0 \\ \vdots & \vdots & \ddots & 0 \\ 0 & 0 & 0 & \lambda_N \end{bmatrix} \equiv \underline{\mathbf{D}}\,\underline{\mathbf{\Lambda}} \tag{5.27}$$

Particularly note the order of matrices on the right-hand side; the diagonal matrix of eigenvalues is on the right. Readers unfamiliar with this expression should try a few examples to confirm its validity. The resulting concise expression for Equation (5.26) relates two products of $N \times N$ matrices:

$$\underline{\mathbf{A}}\,\underline{\mathbf{D}} = \underline{\mathbf{D}}\,\underline{\mathbf{\Lambda}} \tag{5.28}$$

Now, if the N columns of $\underline{\mathbf{D}}$ are linearly independent (a more stringent condition than being distinct), then $\det \underline{\mathbf{D}} \neq 0$ and the inverse $N \times N$ matrix $\underline{\mathbf{D}}^{-1}$ exists. Under these conditions, Equation (5.28) may be rewritten in two additional equivalent forms by multiplying both sides by $\underline{\mathbf{D}}^{-1}$. First, both sides of Equation (5.28) may be multiplied from the *left* by $\underline{\mathbf{D}}^{-1}$:

$$\begin{aligned} \underline{\mathbf{D}}^{-1}\,(\underline{\mathbf{A}}\,\underline{\mathbf{D}}) &= \underline{\mathbf{D}}^{-1}\,(\underline{\mathbf{D}}\,\underline{\mathbf{\Lambda}}) \\ &= (\underline{\mathbf{D}}^{-1}\,\underline{\mathbf{D}})\,\underline{\mathbf{\Lambda}} = \underline{\mathbf{\Lambda}} \end{aligned} \tag{5.29a}$$

where the definition of the inverse matrix in Equation (3.41) has been used. Alternatively, both sides could be multiplied by $\underline{\mathbf{D}}^{-1}$ from the *right* to obtain:

$$\begin{aligned} (\underline{\mathbf{A}}\,\underline{\mathbf{D}})\,\underline{\mathbf{D}}^{-1} &= (\underline{\mathbf{D}}\,\underline{\mathbf{\Lambda}})\,\underline{\mathbf{D}}^{-1} \\ &= \underline{\mathbf{A}}\,(\underline{\mathbf{D}}\,\underline{\mathbf{D}}^{-1}) = \underline{\mathbf{A}} \end{aligned} \tag{5.29b}$$

In words, these two expressions define forward and inverse *transformations* between the matrix operator $\underline{\mathbf{A}}$ and the diagonal form $\underline{\mathbf{\Lambda}}$. Matrices $\underline{\mathbf{A}}$ and $\underline{\mathbf{\Lambda}}$ that satisfy the conditions in Equation (5.29) for some transformation matrix $\underline{\mathbf{D}}$ are *similar matrices*, and the relationships in Equations (5.28a) and (5.29b) are *similarity transformations*. The matrix $\underline{\mathbf{D}}$ is often called the *diagonalizing operator* for $\underline{\mathbf{A}}$. Note that the transformations in Equation (5.29) exist for all cases where the matrix $\underline{\mathbf{D}}$ of the eigenvectors has been determined and is invertible.

If the secular equation in Equation (5.24) has any repeated roots, then only a single eigenvector may be determined for those roots, as was true for $\underline{\mathbf{A}}_3$ in Equation (5.21). Because the degenerate eigenvalues all have the same associated eigenvector, the matrix $\underline{\mathbf{D}}$ constructed from the eigenvectors would have as many identical column vectors as the number of degenerate eigenvalues. As asserted in Equation (3.34c), the determinant of any such $\underline{\mathbf{D}}$ must be zero. In these cases of degenerate eigenvalues, the inverse matrix $\underline{\mathbf{D}}$ does not exist and so the similarity transformation may not be constructed. However, there is a special case that is relevant to imaging where the diagonalizing matrix does exist even when the eigenvalues are degenerate: the diagonalizing matrix is predetermined by the form of $\underline{\mathbf{A}}$. As we will see, all $N \times N$ circulant matrices have the same set of associated eigenvectors, so that the diagonalizing transformations in Equation (5.29) may be applied even in those cases where the eigenvalues are degenerate or null.

The diagonal form $\underline{\mathbf{\Lambda}}$ may be used to recast the matrix–vector product form of the imaging equation in Equation (3.23) into a different and often more convenient form by substituting the expression for $\underline{\mathbf{A}}$ in Equation (5.29b):

$$\underline{\mathbf{A}}\,\underline{\mathbf{x}} = (\underline{\mathbf{D}}\,\underline{\mathbf{\Lambda}}\,\underline{\mathbf{D}}^{-1})\underline{\mathbf{x}} = \underline{\mathbf{b}} \tag{5.30}$$

These equivalent forms may be multiplied from the left by an additional factor of $\underline{\mathbf{D}}^{-1}$ to obtain yet another equivalent form for the imaging equation:

$$\underline{\mathbf{D}}^{-1}(\underline{\mathbf{D}}\,\underline{\mathbf{\Lambda}}\,\underline{\mathbf{D}}^{-1})\underline{\mathbf{x}} = (\underline{\mathbf{D}}^{-1}\,\underline{\mathbf{D}})\,\underline{\mathbf{\Lambda}}\,(\underline{\mathbf{D}}^{-1}\underline{\mathbf{x}}) = \underline{\mathbf{\Lambda}}\,(\underline{\mathbf{D}}^{-1}\,\underline{\mathbf{x}}) = (\underline{\mathbf{D}}^{-1}\,\underline{\mathbf{b}}) \tag{5.31}$$

The matrix–vector products enclosed in parentheses compute the projections of the input and output vectors onto the row vectors of $(\underline{\mathbf{D}}^*)^{-1}$:

$$\underline{\mathbf{D}}^{-1}\,\underline{\mathbf{x}} \equiv \underline{\mathbf{x}}' \tag{5.32a}$$

$$\underline{\mathbf{D}}^{-1}\,\underline{\mathbf{b}} \equiv \underline{\mathbf{b}}' \tag{5.32b}$$

In words, the individual components of $\underline{\mathbf{x}}'$ and $\underline{\mathbf{b}}'$ are the respective projections of the input and output vectors $\underline{\mathbf{x}}$ and $\underline{\mathbf{b}}$ onto each of a common set of basis vectors (which happen to be the eigenvectors of the system matrix $\underline{\mathbf{A}}$), instead of onto the usual set of Cartesian basis vectors aligned with the coordinate axes. The first component of $\underline{\mathbf{x}}'$ is the scalar projection of the input vector $\underline{\mathbf{x}}$ onto the first eigenvector $\underline{\mathbf{x}}_1$ of $\underline{\mathbf{A}}$, the second component of $\underline{\mathbf{x}}'$ is the projection of the input onto the second eigenvector, etc. The expression in Equation (5.31) is an alternate representation of the imaging system and may be written in the form of Equation (3.23):

$$\underline{\mathbf{\Lambda}}\,\underline{\mathbf{x}}' = \underline{\mathbf{b}}' \tag{5.33}$$

The advantages that accrue from this "new" representation for the imaging equation are based on the diagonal form of the "transformed system matrix" $\underline{\mathbf{\Lambda}}$ and that its diagonal elements are the eigenvalues of $\underline{\mathbf{A}}$. Any eigenvector of $\underline{\mathbf{A}}$ that is part of the input vector $\underline{\mathbf{x}}$ passes through the system and becomes part of the output vector $\underline{\mathbf{b}}$ after being scaled by the corresponding eigenvalue, which is immediately evident from its location on the diagonal of $\underline{\mathbf{\Lambda}}$. The eigenvector corresponding to any null eigenvalue of $\underline{\mathbf{A}}$ is "blocked" completely by the system, meaning that no part of that eigenvector exists in $\underline{\mathbf{b}}$. In addition, we see again that the inverse problem cannot be solved for matrices with any null eigenvalues because the "amount" of the corresponding eigenvector that was present in the input vector $\underline{\mathbf{x}}$ cannot be determined from $\underline{\mathbf{b}}$ and $\underline{\mathbf{A}}$. In short, a system matrix with one or more null eigenvalues cannot be inverted. This should be no surprise, as we already showed in Equation (3.44) that a diagonal matrix with any null diagonal components cannot be inverted, and we showed in Equation (5.30) that the original and diagonal forms of the matrix are equivalent.

The relations in Equation (5.32) may be recast by multiplying both sides of both equations from the left by $\underline{\mathbf{D}}$:

$$\underline{\mathbf{D}}\,(\underline{\mathbf{D}}^{-1}\,\underline{\mathbf{x}}) = (\underline{\mathbf{D}}\,\underline{\mathbf{D}}^{-1})\,\underline{\mathbf{x}} = \underline{\mathbf{x}} = \underline{\mathbf{D}}\,\underline{\mathbf{x}}' \tag{5.34a}$$

$$\underline{\mathbf{D}}\,(\underline{\mathbf{D}}^{-1}\,\underline{\mathbf{b}}) = (\underline{\mathbf{D}}\,\underline{\mathbf{D}}^{-1})\,\underline{\mathbf{b}} = \underline{\mathbf{b}} = \underline{\mathbf{D}}\,\underline{\mathbf{b}}' \tag{5.34b}$$

These define the transformations of the vectors expressed in the "eigenvector basis" back to the usual "Cartesian basis". In words, the components of the vectors in the original basis are obtained by projecting the vectors in the "eigenvector basis" onto the complex conjugates of the rows of $\underline{\mathbf{D}}$.

The ensemble of expressions in Equations (5.29), (5.32), and (5.34) allow the constituent vectors and system matrix in any imaging problem to be transformed back and forth between these representations as desired. The recipes to solve the imaging problem may be represented graphically, as shown in Figure 5.2.

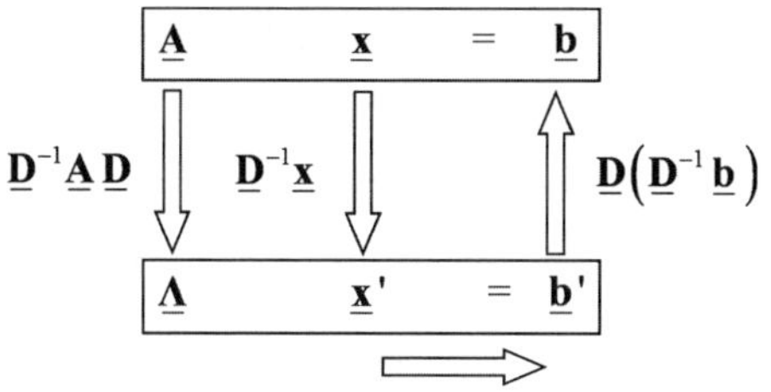

Figure 5.2 Schematic of matrix representation of imaging system. The matrix operator $\underline{\mathbf{A}}$ is transformed to the diagonal form $\underline{\mathbf{\Lambda}}$ via $\underline{\mathbf{D}}^{-1}\underline{\mathbf{A}}\underline{\mathbf{D}}$. The input vector is transformed to the diagonal form via $\underline{\mathbf{x}}' = \underline{\mathbf{D}}^{-1}\underline{\mathbf{x}}$. The output vector in the diagonal formed is transformed back to the original basis via $\underline{\mathbf{b}} = \underline{\mathbf{D}}\underline{\mathbf{b}}'$. The block diagram shows that this solution operates "around the block".

5.2.3 Diagonalization of a Circulant Matrix

To illustrate the utility of these transformations, we will derive the eigenvectors, eigenvalues, and diagonal form of a particular 3 × 3 circulant (shift-invariant) matrix:

$$\underline{\mathbf{A}} = \begin{bmatrix} 0 & 1 & 0 \\ 0 & 0 & 1 \\ 1 & 0 & 0 \end{bmatrix} \tag{5.35}$$

and use them to solve the direct imaging problem for a particular complex-valued input vector:

$$\underline{\mathbf{x}} = \begin{bmatrix} 0 \\ i \\ 0 \end{bmatrix} = \begin{bmatrix} 0 \\ e^{+i\pi/2} \\ 0 \end{bmatrix} \tag{5.36}$$

Clearly, the matrix–vector product may be found easily by direct multiplication using Equation (3.24):

$$\underline{\mathbf{A}}\,\underline{\mathbf{x}} = \begin{bmatrix} e^{+i\pi/2} \\ 0 \\ 0 \end{bmatrix} = \begin{bmatrix} i \\ 0 \\ 0 \end{bmatrix} \tag{5.37}$$

so the motivation for this exercise may not be clear at this point. The author will merely invoke the words of many American politicians and ask the reader to "Trust me".

First, find the eigenvectors of $\underline{\mathbf{A}}$. The eigenvector equation for the nth eigenvalue is:

$$\begin{bmatrix} 0 & 1 & 0 \\ 0 & 0 & 1 \\ 1 & 0 & 0 \end{bmatrix} \begin{bmatrix} (x_n)_1 \\ (x_n)_2 \\ (x_n)_3 \end{bmatrix} = \lambda_n \begin{bmatrix} (x_n)_1 \\ (x_n)_2 \\ (x_n)_3 \end{bmatrix} \tag{5.38}$$

where the numerical subscripts again refer to the location of the component within the eigenvector. The determinant of the matrix $\underline{\mathbf{C}} = \underline{\mathbf{A}} - \underline{\mathbf{I}}\lambda$ in Equation (5.23) yields a third-order secular equation in terms of the eigenvalues of $\underline{\mathbf{A}}$:

$$\det[\underline{\mathbf{A}} - \underline{\mathbf{I}}\lambda] = \det \begin{bmatrix} -\lambda & 1 & 0 \\ 0 & -\lambda & 1 \\ 1 & 0 & -\lambda \end{bmatrix} = \lambda^3 - 1 = 0 \tag{5.39}$$

The only real-valued root of this equation is $\lambda = +1$. Because there should be three roots of a cubic equation, a naive reader might conclude that the eigenvalues are triply degenerate (all three eigenvalues are unity). However, complex analysis comes to the rescue in the form of De Moivre's theorem in

Equation (4.40) to determine three distinct complex-valued eigenvalues:

$$\lambda_1 = (e^0)^{\frac{1}{3}} = 1 \tag{5.40a}$$

$$\lambda_2 = (e^{+2\pi i})^{\frac{1}{3}} = \exp\left[+\frac{2\pi i}{3}\right] = -\frac{1}{2} + i\frac{\sqrt{3}}{2} \tag{5.40b}$$

$$\lambda_3 = (e^{+4\pi i})^{\frac{1}{3}} = \exp\left[+\frac{4\pi i}{3}\right] = -\frac{1}{2} - i\frac{\sqrt{3}}{2} = \lambda_2^* \tag{5.40c}$$

The order of the eigenvalues is (at this point) completely arbitrary. To derive the corresponding eigenvectors, the three eigenvalues are substituted into the definition in Equation (5.17) one at a time. Each substitution leads to a set of three equations that must be satisfied simultaneously by the components of the corresponding 3-D eigenvector. The simplest result is obtained for the first eigenvalue $\lambda_1 = 1$, whose components must simultaneously satisfy the relations:

$$\begin{aligned}(x_1)_2 &= (x_1)_1\\ (x_1)_3 &= (x_1)_2\\ (x_1)_1 &= (x_1)_3\end{aligned} \tag{5.41}$$

From this point forward, we will drop the parentheses so that $(x_1)_1 \equiv x_{11}$. The simultaneous solution of these three equations requires that all components of the eigenvector be equal. After normalization to unit length, the eigenvector corresponding to $\lambda_1 = 1$ is:

$$\hat{\underline{\mathbf{x}}}_1 = \begin{bmatrix} \frac{1}{\sqrt{3}} \\ \frac{1}{\sqrt{3}} \\ \frac{1}{\sqrt{3}} \end{bmatrix} \equiv \frac{1}{\sqrt{3}} \begin{bmatrix} 1 \\ 1 \\ 1 \end{bmatrix} \tag{5.42}$$

where we use the definition of the operation of scalar multiplication of a vector to extract the common factor of $(\sqrt{3})^{-1}$. This eigenvector $\underline{\mathbf{x}}_1$ points along the diagonal in the first quadrant of the real-valued 3-D volume.

The three equations that relate the components of the second eigenvector are:

$$\begin{aligned}x_{22} &= \lambda_2\, x_{21} = e^{+2\pi i/3}\, x_{21}\\ x_{23} &= e^{+2\pi i/3}\, x_{22}\\ x_{21} &= e^{+2\pi i/3}\, x_{23} = e^{+4\pi i/3}\, x_{22} = e^{+6\pi i/3}\, x_{21}\end{aligned} \tag{5.43}$$

In words, each component is obtained by scaling the component with the next lower index (modulo 3) by the phase factor $e^{+2\pi i/3}$; to change from one component to the next (modulo 3), just increment the phase by $2\pi/3$ radians. A normalized "circular" solution exists for the components of this vector:

$$\hat{\underline{\mathbf{x}}}_2 = \frac{1}{\sqrt{3}} \begin{bmatrix} 1 \\ e^{+2\pi i/3} \\ e^{+4\pi i/3} \end{bmatrix} = \frac{1}{\sqrt{3}} \begin{bmatrix} 1 \\ e^{+2\pi i/3} \\ e^{-2\pi i/3} \end{bmatrix} \tag{5.44}$$

where the periodicity of $e^{+i\theta}$ over 2π radians has been used to obtain the last expression. Note that other forms for this eigenvector are possible, but (for no obvious reason) we have chosen that for which the first component is real valued. This second eigenvector is more difficult to visualize than $\underline{\mathbf{x}}_1$; two real-valued 3-D volumes are required to individually graph the real part and imaginary part. The real and imaginary parts of the second eigenvector are oriented along the directions in the 3-D plane:

$$\Re\{\hat{\underline{\mathbf{x}}}_2\} = \frac{1}{\sqrt{3}}\begin{bmatrix} 1 \\ -0.5 \\ -0.5 \end{bmatrix}, \quad \Im\{\hat{\underline{\mathbf{x}}}_2\} = \frac{1}{\sqrt{3}}\begin{bmatrix} 0 \\ +\frac{\sqrt{3}}{2} \\ -\frac{\sqrt{3}}{2} \end{bmatrix} \tag{5.45}$$

The same procedure applied to the third eigenvalue yields an eigenvector for which each component is scaled from the previous by the complex factor $e^{+4\pi i/3}$; the form of the normalized eigenvector is:

$$\hat{\underline{\mathbf{x}}}_3 = \frac{1}{\sqrt{3}}\begin{bmatrix} 1 \\ e^{+4\pi i/3} \\ e^{+8\pi i/3} \end{bmatrix} = \frac{1}{\sqrt{3}}\begin{bmatrix} 1 \\ e^{-2\pi i/3} \\ e^{+2\pi i/3} \end{bmatrix} = (\hat{\underline{\mathbf{x}}}_2)^* \tag{5.46}$$

Each subsequent component (modulo 3) of $\hat{\underline{\mathbf{x}}}_3$ is obtained by incrementing the phase of the previous component by $+4\pi/3$ radians. By considering all three eigenvectors at once, a pattern is observed. The (unnormalized) first component of each eigenvector is unity; each subsequent component is obtained by incrementing the phase of the previous component by 0, $+2\pi/3$, or $+4\pi/3$ radians, respectively. In other words, the elements of each eigenvector are samples of a complex sinusoid that oscillates at a specific rate. The samples of the first eigenvector are constant, and so represent samples of a sinusoid that oscillates 0 times. The elements of the second and third eigenvectors are samples of sinusoids that "go through" one and two cycles around the Argand diagram in three samples, respectively. This interpretation of the eigenvectors of this circulant matrix is illustrated in Figure 5.3 and will be useful in future examples.

The mutual inner products of these three vectors may be computed by applying Equation (5.4), which recalls that the components of the first vector must be conjugated to compute the inner product of vectors with complex-valued components:

$$\begin{array}{lll} \hat{\underline{\mathbf{x}}}_1 \bullet \hat{\underline{\mathbf{x}}}_1 = 1 & \hat{\underline{\mathbf{x}}}_1 \bullet \hat{\underline{\mathbf{x}}}_2 = 0 & \hat{\underline{\mathbf{x}}}_1 \bullet \hat{\underline{\mathbf{x}}}_3 = 0 \\ \hat{\underline{\mathbf{x}}}_2 \bullet \hat{\underline{\mathbf{x}}}_1 = 0 & \hat{\underline{\mathbf{x}}}_2 \bullet \hat{\underline{\mathbf{x}}}_2 = 1 & \hat{\underline{\mathbf{x}}}_2 \bullet \hat{\underline{\mathbf{x}}}_3 = 0 \\ \hat{\underline{\mathbf{x}}}_3 \bullet \hat{\underline{\mathbf{x}}}_1 = 0 & \hat{\underline{\mathbf{x}}}_3 \bullet \hat{\underline{\mathbf{x}}}_2 = 0 & \hat{\underline{\mathbf{x}}}_3 \bullet \hat{\underline{\mathbf{x}}}_3 = 1 \end{array} \tag{5.47}$$

Therefore the three eigenvectors of this matrix are orthonormal.

The diagonalizing matrix $\underline{\mathbf{D}}$ in Equation (5.27) is constructed from the three eigenvectors of $\underline{\mathbf{A}}$:

$$\begin{aligned} \underline{\mathbf{D}} &= \frac{1}{\sqrt{3}}\left[\left[\hat{\underline{\mathbf{x}}}_1\right]\left[\hat{\underline{\mathbf{x}}}_2\right]\left[\hat{\underline{\mathbf{x}}}_3\right]\right] \\ &= \frac{1}{\sqrt{3}}\begin{bmatrix} 1 & 1 & 1 \\ 1 & e^{+2\pi i/3} & e^{+4\pi i/3} \\ 1 & e^{+4\pi i/3} & e^{+8\pi i/3} \end{bmatrix} = \frac{1}{\sqrt{3}}\begin{bmatrix} 1 & 1 & 1 \\ 1 & e^{+2\pi i/3} & e^{-2\pi i/3} \\ 1 & e^{-2\pi i/3} & e^{+2\pi i/3} \end{bmatrix} \end{aligned} \tag{5.48}$$

where the periodicity of $e^{i\theta}$ again has been used to obtain the last expression and the order of the eigenvectors in $\underline{\mathbf{D}}$ matches the arbitrary order of the eigenvalues. Both numerical forms of $\underline{\mathbf{D}}$ given are useful in different circumstances; the first is more easily remembered as the first components of each row are unity, and each subsequent component of the columns is the "previous" component multiplied by the respective numerical constants 1, $e^{+2\pi i/3}$, and $e^{+4\pi i/3}$. Note that the rows and columns of the second expression for $\underline{\mathbf{D}}$ are identical so that the matrix is equal to its transpose. In mathematical jargon, matrices that are unchanged by transposition are *symmetrical*:

$$\underline{\mathbf{D}}^T = \underline{\mathbf{D}} \Longrightarrow \underline{\mathbf{D}} \textit{ is a "symmetric" matrix} \tag{5.49}$$

The different usages of the term *symmetry* for matrices in Equation (5.49) and for functions in Equation (2.18) are distinguished by context and rarely create confusion. Also note that a circulant matrix

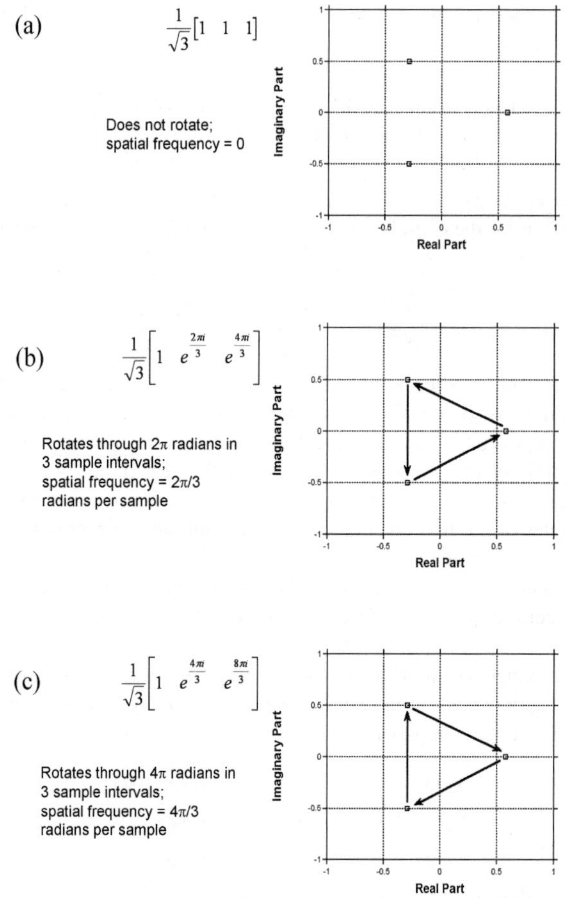

(a) $\frac{1}{\sqrt{3}}\begin{bmatrix} 1 & 1 & 1 \end{bmatrix}$

Does not rotate; spatial frequency = 0

(b) $\frac{1}{\sqrt{3}}\begin{bmatrix} 1 & e^{\frac{2\pi i}{3}} & e^{\frac{4\pi i}{3}} \end{bmatrix}$

Rotates through 2π radians in 3 sample intervals; spatial frequency = $2\pi/3$ radians per sample

(c) $\frac{1}{\sqrt{3}}\begin{bmatrix} 1 & e^{\frac{4\pi i}{3}} & e^{\frac{8\pi i}{3}} \end{bmatrix}$

Rotates through 4π radians in 3 sample intervals; spatial frequency = $4\pi/3$ radians per sample

Figure 5.3 Argand diagrams of eigenvectors that diagonalize a 3×3 circulant matrix: (a) elements of first eigenvector are identical; (b) elements of second eigenvector "rotate" counterclockwise about the origin of the Argand diagram one time, with spatial frequency of $2\pi/3$ radians per sample; (c) elements of third eigenvector rotate clockwise about the origin, with spatial frequency of $-2\pi/3$ radians per sample.

generally is not symmetric except in the 2×2 case. Two well-known and easily proven properties of square symmetric matrices are very useful in many areas of the sciences: the eigenvalues of a real-valued symmetric matrix must be real, and the eigenvectors of a real-valued symmetric matrix that correspond to distinct (nondegenerate) eigenvalues must be orthogonal. Proofs of these assertions are available in most texts on mathematical physics, but they are not considered further here because they do not appear often in the context of imaging.

The symmetry of $\underline{\mathbf{D}}$ may be combined with the observation that the rows of $\underline{\mathbf{D}}$ are orthonormal to derive the expression for the inverse matrix $\underline{\mathbf{D}}^{-1}$ fairly easily. We know from Equation (3.45) that the $N \times N$ inverse matrix $\underline{\mathbf{D}}^{-1}$ must satisfy:

$$\underline{\mathbf{D}}\,\underline{\mathbf{D}}^{-1} = \underline{\mathbf{D}}^{-1}\,\underline{\mathbf{D}} = \underline{\mathbf{I}} \tag{5.50}$$

where $\underline{\mathbf{I}}$ is the $N \times N$ identity matrix defined in Equation (3.30). The matrix product on the left-hand side requires that the nth row vector $\underline{\mathbf{a}}_n$ of $\underline{\mathbf{D}}$ be orthonormal to the mth column vector $\underline{\mathbf{x}}_m$ of $\underline{\mathbf{D}}^{-1}$, so that $\underline{\mathbf{a}}_n \bullet \underline{\mathbf{x}}_m = 1$ when $n = m$ and 0 otherwise. In the same manner, the second expression demands that the scalar product of the row vectors of $\underline{\mathbf{D}}^{-1}$ be orthonormal to the column vectors of $\underline{\mathbf{D}}$. We showed in Equation (5.47) that the column vectors of $\underline{\mathbf{D}}$ are mutually orthonormal, and the symmetry of $\underline{\mathbf{D}}$ ensures that the row vectors must be so as well. The task of constructing $\underline{\mathbf{D}}^{-1}$ reduces to the problem of ensuring that the *scalar products* of the rows of $\underline{\mathbf{D}}$ with the columns of $\underline{\mathbf{D}}^{-1}$ are identical to the mutual *inner products* of the vectors of $\underline{\mathbf{D}}$. Therefore, the rows of $\underline{\mathbf{D}}^{-1}$ must be the complex conjugates of the columns of $\underline{\mathbf{D}}$, and vice versa. In short, the inverse of this symmetric matrix $\underline{\mathbf{D}}$ with orthonormal columns is merely the complex conjugate of its transpose:

$$\underline{\mathbf{D}}^{-1} = (\underline{\mathbf{D}}^*)^T = (\underline{\mathbf{D}}^T)^* \tag{5.51}$$

This result is easily confirmed for the 3×3 matrix in Equation (5.48):

$$\underline{\mathbf{D}}\,\underline{\mathbf{D}}^{-1} = \frac{1}{\sqrt{3}}\begin{bmatrix} 1 & 1 & 1 \\ 1 & e^{+2\pi i/3} & e^{-2\pi i/3} \\ 1 & e^{-2\pi i/3} & e^{+2\pi i/3} \end{bmatrix} \cdot \frac{1}{\sqrt{3}}\begin{bmatrix} 1 & 1 & 1 \\ 1 & e^{-2\pi i/3} & e^{+2\pi i/3} \\ 1 & e^{+2\pi i/3} & e^{-2\pi i/3} \end{bmatrix} = \begin{bmatrix} 1 & 0 & 0 \\ 0 & 1 & 0 \\ 0 & 0 & 1 \end{bmatrix} \tag{5.52}$$

Following the concept for functions introduced in Equation (4.28), square matrices that satisfy the condition of Equation (5.51) are called *Hermitian*, and are very important in physics (particularly in quantum mechanics), though for different reasons than in imaging. By multiplying the system matrix $\underline{\mathbf{A}}$ by the transformation matrix $\underline{\mathbf{D}}$ and its inverse in Equation (5.29a), the diagonal form of $\underline{\mathbf{A}}$ is obtained:

$$\begin{aligned}\underline{\mathbf{\Lambda}} &= \frac{1}{\sqrt{3}}\begin{bmatrix} 1 & 1 & 1 \\ 1 & e^{-2\pi i/3} & e^{+2\pi i/3} \\ 1 & e^{+2\pi i/3} & e^{-2\pi i/3} \end{bmatrix} \cdot \begin{bmatrix} 0 & 1 & 0 \\ 0 & 0 & 1 \\ 1 & 0 & 0 \end{bmatrix} \cdot \frac{1}{\sqrt{3}}\begin{bmatrix} 1 & 1 & 1 \\ 1 & e^{+2\pi i/3} & e^{-2\pi i/3} \\ 1 & e^{-2\pi i/3} & e^{+2\pi i/3} \end{bmatrix} \\ &= \begin{bmatrix} 1 & 0 & 0 \\ 0 & e^{+2\pi i/3} & 0 \\ 0 & 0 & e^{+4\pi i/3} \end{bmatrix}\end{aligned} \tag{5.53}$$

which, as predicted, consists of the eigenvalues of $\underline{\mathbf{A}}$ on the diagonal. Note that the order of the eigenvalues along the diagonal of $\underline{\mathbf{\Lambda}}$ is the same as the order of the corresponding eigenvectors in $\underline{\mathbf{D}}$.

The transformation of the input vector $\underline{\mathbf{x}}$ from Equation (5.36) to the diagonal representation is the direct application of Equation (5.32a):

$$\underline{\mathbf{x}}' = \underline{\mathbf{D}}^{-1}\,\underline{\mathbf{x}} = \frac{1}{\sqrt{3}}\begin{bmatrix} 1 & 1 & 1 \\ 1 & e^{-2\pi i/3} & e^{+2\pi i/3} \\ 1 & e^{+2\pi i/3} & e^{-2\pi i/3} \end{bmatrix}\begin{bmatrix} 0 \\ e^{+i\pi/2} \\ 0 \end{bmatrix} = \frac{1}{\sqrt{3}}\begin{bmatrix} e^{+i\pi/2} \\ e^{-i\pi/6} \\ e^{-5\pi i/6} \end{bmatrix} \tag{5.54}$$

which expresses the input vector $\underline{\mathbf{x}}$ in terms of the basis composed of the three eigenvectors of $\underline{\mathbf{A}}$. In other words, the projections of $\underline{\mathbf{x}}$ onto the eigenvectors of $\underline{\mathbf{A}}$ are the components of $\underline{\mathbf{x}}'$:

$$\begin{aligned}&(e^{+i\pi/2})\underline{\mathbf{x}}_1 + (e^{-i\pi/6})\underline{\mathbf{x}}_2 + (e^{-5\pi i/6})\underline{\mathbf{x}}_3 \\ &= e^{+i\pi/2}\left(\frac{1}{\sqrt{3}}\right)^2\begin{bmatrix} 1 \\ 1 \\ 1 \end{bmatrix} + e^{-i\pi/6}\left(\frac{1}{\sqrt{3}}\right)^2\begin{bmatrix} 1 \\ e^{+2\pi i/3} \\ e^{+4\pi i/3} \end{bmatrix} + e^{-5\pi i/6}\left(\frac{1}{\sqrt{3}}\right)^2\begin{bmatrix} 1 \\ e^{+2\pi i/3} \\ e^{+4\pi i/3} \end{bmatrix} \\ &= \begin{bmatrix} 0 \\ e^{+i\pi/2} \\ 0 \end{bmatrix}\end{aligned} \tag{5.55}$$

As just noted, the elements of these eigenvectors are samples of complex sinusoids that oscillate 0, 1, and 2 times over three samples, respectively. The product of the diagonal matrix in Equation (5.53) and the transformed input vector $\underline{\mathbf{x}}'$ in Equation (5.55) yields the transformed output vector $\underline{\mathbf{b}}'$, which in this case is:

$$\underline{\underline{\mathbf{\Lambda}}}\,\underline{\mathbf{x}}' = \underline{\mathbf{b}}' = \frac{1}{\sqrt{3}}\begin{bmatrix} e^{+i\pi/2} \\ e^{+i\pi/2} \\ e^{+i\pi/2} \end{bmatrix} = \frac{1}{\sqrt{3}}\begin{bmatrix} i \\ i \\ i \end{bmatrix} \tag{5.56}$$

And finally, the output vector in the original basis is obtained by transforming back via multiplication by $\underline{\underline{\mathbf{D}}}$ as in Equation (5.34b):

$$\underline{\mathbf{b}} = \underline{\underline{\mathbf{D}}}\,\underline{\mathbf{b}}' = \frac{1}{\sqrt{3}}\begin{bmatrix} 1 & 1 & 1 \\ 1 & e^{+2\pi i/3} & e^{-2\pi i/3} \\ 1 & e^{-2\pi i/3} & e^{+2\pi i/3} \end{bmatrix}\left(\frac{1}{\sqrt{3}}\begin{bmatrix} i \\ i \\ i \end{bmatrix}\right) = \begin{bmatrix} i \\ 0 \\ 0 \end{bmatrix} \tag{5.57}$$

which agrees with the direct result obtained in Equation (5.37).

Readers are probably asking, "If the matrix–vector product is so easy to solve, then why bother with all this transformation stuff?" The answer to this question depends on the additional insight that is obtained by applying the eigenvectors and eigenvalues to inverse and system analysis problems. These ideas are discussed in the next section.

5.2.4 Matrix Operators for Shift-Invariant Systems

The example of matrix and vector transformation used in the last section was based on the particular 3×3 circulant matrix in Equation (5.35). Now consider the properties of this representation for the general 3×3 circulant matrix:

$$\underline{\underline{\mathbf{A}}} = \begin{bmatrix} \alpha & \beta & \gamma \\ \gamma & \alpha & \beta \\ \beta & \gamma & a \end{bmatrix} \tag{5.58}$$

where the elements α, β, and γ may be complex valued. The eigenvalues and eigenvectors of this matrix are obtained in a straightforward (though perhaps tedious) fashion using the principles already discussed. Rather than do that, we will take a shortcut based on the example just completed. Let us apply the general circulant 3×3 matrix operator to the eigenvectors of the specific circulant matrix in Equation (5.35). The first eigenvector $\underline{\mathbf{x}}_1$ was derived in Equation (5.42):

$$\begin{aligned} \underline{\underline{\mathbf{A}}}\,\underline{\hat{\mathbf{x}}}_1 &= \begin{bmatrix} \alpha & \beta & \gamma \\ \gamma & \alpha & \beta \\ \beta & \gamma & a \end{bmatrix}\left(\frac{1}{\sqrt{3}}\begin{bmatrix} 1 \\ 1 \\ 1 \end{bmatrix}\right) = \frac{1}{\sqrt{3}}\begin{bmatrix} \alpha+\beta+\gamma \\ \alpha+\beta+\gamma \\ \alpha+\beta+\gamma \end{bmatrix} \\ &= (\alpha+\beta+\gamma)\left(\frac{1}{\sqrt{3}}\begin{bmatrix} 1 \\ 1 \\ 1 \end{bmatrix}\right) \equiv \lambda_1\underline{\hat{\mathbf{x}}}_1 \end{aligned} \tag{5.59}$$

So we observe that the action of the general circulant matrix on the first eigenvector of the specific circulant matrix yields a scaled replica of that eigenvector. In other words, $\underline{\mathbf{x}}_1$ also is an eigenvector of the general circulant matrix, with eigenvalue $\lambda_1 = \alpha + \beta + \gamma$.

This property of the first eigenvector of the specific circulant matrix suggests that we try the same strategy for the remaining eigenvectors in Equations (5.44) and (5.46). The second gives the result:

$$\begin{aligned} \underline{\underline{\mathbf{A}}}\,\underline{\hat{\mathbf{x}}}_2 &= \begin{bmatrix} \alpha & \beta & \gamma \\ \gamma & \alpha & \beta \\ \beta & \gamma & a \end{bmatrix}\left(\frac{1}{\sqrt{3}}\begin{bmatrix} 1 \\ e^{+2\pi i/3} \\ e^{+4\pi i/3} \end{bmatrix}\right) = \frac{1}{\sqrt{3}}\begin{bmatrix} \alpha + \beta\, e^{+2\pi i/3} + \gamma\, e^{+4\pi i/3} \\ \gamma + \alpha\, e^{+2\pi i/3} + \beta\, e^{+4\pi i/3} \\ \beta + \gamma\, e^{+2\pi i/3} + \alpha\, e^{+4\pi i/3} \end{bmatrix} \\ &= (\alpha + \beta\, e^{+2\pi i/3} + \gamma\, e^{+4\pi i/3})\left(\frac{1}{\sqrt{3}}\begin{bmatrix} 1 \\ e^{+2\pi i/3} \\ e^{+4\pi i/3} \end{bmatrix}\right) = \lambda_2\underline{\hat{\mathbf{x}}}_2 \end{aligned} \tag{5.60}$$

where the periodicity of $e^{i\theta}$ has been used. Again we see that the second eigenvector of the specific circulant matrix turns out to be an eigenvector of the general circulant matrix $\underline{\mathbf{A}}$, and the eigenvalue of the general matrix is a linear combination of the components of $\underline{\mathbf{A}}$ with complex-valued weights.

The same process applied to $\hat{\underline{\mathbf{x}}}_3$ from Equation (5.46) yields:

$$\underline{\mathbf{A}}\,\hat{\underline{\mathbf{x}}}_3 = \begin{bmatrix} \alpha & \beta & \gamma \\ \gamma & \alpha & \beta \\ \beta & \gamma & a \end{bmatrix} \left(\frac{1}{\sqrt{3}} \begin{bmatrix} 1 \\ e^{+4\pi i/3} \\ e^{+8\pi i/3} \end{bmatrix} \right) = \frac{1}{\sqrt{3}} \begin{bmatrix} \alpha + \beta\, e^{+4\pi i/3} + \gamma\, e^{+8\pi i/3} \\ \gamma + \alpha\, e^{+4\pi i/3} + \beta\, e^{+8\pi i/3} \\ \beta + \gamma\, e^{+4\pi i/3} + \alpha\, e^{+8\pi i/3} \end{bmatrix}$$

$$= (\alpha + \beta\, e^{+4\pi i/3} + \gamma\, e^{+8\pi i/3}) \left(\frac{1}{\sqrt{3}} \begin{bmatrix} 1 \\ e^{+4\pi i/3} \\ e^{+8\pi i/3} \end{bmatrix} \right) = \lambda_3 \underline{\mathbf{x}}_3 \tag{5.61}$$

The observation that the normalized eigenvectors for the general 3×3 circulant matrix are identical to those derived for the specific case in Equations (5.42), (5.44), and (5.46) is very important. We will generalize this result without proof to state that *all* $N \times N$ circulant matrices have the same set of N N-D eigenvectors and thus the same diagonalization matrices $\underline{\mathbf{D}}$ and $\underline{\mathbf{D}}^{-1}$. Therefore, the diagonalization process in Equation (5.29a) may be "preprogrammed" to compute $\underline{\boldsymbol{\Lambda}}$ for any $N \times N$ shift-invariant matrix. It is more convenient to determine the eigenvalues of a circulant matrix via the "preprogrammed" diagonalization of Equation (5.29a) rather than by finding the roots of the secular equation. For example, the expressions in Equations (5.59), (5.60), and (5.61) for the eigenvalues of the general 3×3 circulant matrix may be used to immediately write down the diagonal form of the general 3×3 circulant matrix obtained when using the same order of eigenvectors as in Equation (5.48):

$$\underline{\boldsymbol{\Lambda}} = \begin{bmatrix} (\alpha+\beta+\gamma) & 0 & 0 \\ 0 & (\alpha + \beta\, e^{+2\pi i/3} + \gamma\, e^{-2\pi i/3}) & 0 \\ 0 & 0 & (\alpha + \beta\, e^{-2\pi i/3} + \gamma\, e^{+2\pi i/3}) \end{bmatrix} \tag{5.62}$$

This result immediately establishes the conditions on α, β, γ that ensure that the individual eigenvalues are nonzero, and hence the conditions for the 3×3 circulant matrix to be invertible. For example, one of the conditions is that $\alpha + \beta + \gamma \neq 0$.

The general 4×4 circulant matrix has the form:

$$\underline{\mathbf{A}} = \begin{bmatrix} \alpha & \beta & \gamma & \delta \\ \delta & \alpha & \beta & \gamma \\ \gamma & \delta & \alpha & \beta \\ \beta & \gamma & \delta & \alpha \end{bmatrix} \tag{5.63}$$

The eigenvalues and corresponding normalized eigenvectors again are found by solving the secular equation and establishing the relationships among the components of the vectors. They are:

$$\lambda_1 = \alpha + \beta + \gamma + \delta,$$

$$\hat{\underline{\mathbf{x}}}_1 = \frac{1}{\sqrt{4}} \begin{bmatrix} 1 \\ 1 \\ 1 \\ 1 \end{bmatrix} \tag{5.64a}$$

$$\lambda_2 = \alpha + \beta\, e^{+i\pi/2} + \gamma\, e^{+i\pi} + \delta\, e^{+3\pi i/2} = \alpha + i\beta - \gamma - i\delta$$

$$\hat{\underline{\mathbf{x}}}_2 = \frac{1}{\sqrt{4}} \begin{bmatrix} 1 \\ e^{+i\pi/2} \\ e^{+i\pi} \\ e^{+3\pi i/2} \end{bmatrix} = \frac{1}{2} \begin{bmatrix} 1 \\ i \\ -1 \\ -i \end{bmatrix} \tag{5.64b}$$

$$\lambda_3 = \alpha + \beta\, e^{+i\pi} + \gamma\, e^{+2\pi i} + \delta\, e^{+3\pi i} = \alpha - \beta + \gamma - \delta$$

$$\underline{\hat{\mathbf{x}}}_3 = \frac{1}{\sqrt{4}} \begin{bmatrix} 1 \\ e^{+i\pi} \\ e^{+2\pi i} \\ e^{+3\pi i} \end{bmatrix} = \frac{1}{2} \begin{bmatrix} +1 \\ -1 \\ +1 \\ -1 \end{bmatrix} \tag{5.64c}$$

$$\lambda_4 = \alpha + \beta\, e^{+3\pi i/2} + \gamma\, e^{+3\pi i} + \delta\, e^{+9\pi i/2} = \alpha - i\,\beta - \gamma + i\,\delta$$

$$\underline{\hat{\mathbf{x}}}_4 = \frac{1}{\sqrt{4}} \begin{bmatrix} 1 \\ e^{+3\pi i/2} \\ e^{+3\pi i} \\ e^{+9\pi i/2} \end{bmatrix} = \frac{1}{2} \begin{bmatrix} 1 \\ -i \\ -1 \\ +i \end{bmatrix} \tag{5.64d}$$

These results may be confirmed easily by direct substitution. Sharp-eyed readers may detect a pattern in these 4-D eigenvectors resembling that already noted in the 3-D eigenvectors. The elements of the first eigenvector of the 4×4 circulant matrix are identical, and may be interpreted as samples of a complex sinusoid that does not oscillate. The (unnormalized) first element of the second eigenvector is the unit constant, and each subsequent element is obtained by incrementing the phase by $\pi/2$ radians; these are samples of a complex sinusoid that goes through one cycle in four samples. The elements of the third and fourth eigenvectors are samples of sinusoids with two and three cycles in four samples, respectively. This pattern leads to easily remembered expressions for the $N \times N$ transformation matrix $\underline{\mathbf{D}}$ and its inverse $\underline{\mathbf{D}}^{-1}$ when using this order of eigenvectors. The element of $\underline{\mathbf{D}}$ in the nth row and the kth column is denoted by $\underline{\mathbf{D}}_{kn}$:

$$(\underline{\mathbf{D}})_{kn} = \frac{1}{\sqrt{N}} \exp\left[+2\pi i \frac{(k-1)(n-1)}{N}\right] \quad \textit{for } 1 \le k, n \le N \tag{5.65}$$

The elements of these eigenvectors are samples of complex sinusoids that respectively oscillate over 0, 1, 2, up to $N - 1$ cycles over the N samples of the vector. Note also that this matrix $\underline{\mathbf{D}}$ is symmetric since the row and column indices appear in identical forms. As before, the elements of $\underline{\mathbf{D}}^{-1}$ are just the complex conjugates of the transposed elements of $\underline{\mathbf{D}}$:

$$(\underline{\mathbf{D}}_{mn})^{-1} = (\underline{\mathbf{D}}^*)_{nm} = (\underline{\mathbf{D}}^*)_{mn} = \frac{1}{\sqrt{N}} \exp\left[-2\pi i \frac{(m-1)(n-1)}{N}\right] \quad \textit{for } 1 \le m, n \le N \tag{5.66}$$

The set of elements in either the nth row or column of $\underline{\mathbf{D}}$ is just equally spaced samples of the complex linear-phase exponential (sinusoid) that oscillates $(n - 1)$ times over the N samples. Therefore, the transformation of the vectors $\underline{\mathbf{x}}$ and $\underline{\mathbf{b}}$ to the eigenvector basis as stated in Equation (5.32) may now be interpreted in a very useful way. As we have seen several times, the product of $\underline{\mathbf{D}}^{-1}$ and $\underline{\mathbf{x}}$ yields the projections of $\underline{\mathbf{x}}$ onto the rows of $(\underline{\mathbf{D}}^{-1})^*$, which are the column vectors of $\underline{\mathbf{D}}$ in Equation (5.65). In short, Equation (5.32a) derives $\underline{\mathbf{x}}'$ by projecting $\underline{\mathbf{x}}$ onto a specific set of sampled complex sinusoids. Equation (5.29) defines the "inverse transformation" of the vector by projecting $\underline{\mathbf{x}}'$ onto the complex conjugates of these same sinusoids.

5.2.4.1 Discrete Fourier Transform

The discussion just completed motivated (even if it did not derive) the matrix form of the operation that transforms the values of a vector of any dimension N to the coordinate system for which the

N-dimensional circulant matrix operator $\underline{\mathbf{A}}$ is diagonal:

$$\underline{\mathbf{x}}' = \underline{\mathbf{D}}^{-1}\underline{\mathbf{x}} = \frac{1}{\sqrt{N}} \begin{bmatrix} 1 & 1 & 1 & \cdots & 1 \\ 1 & e^{-2\pi i/N} & e^{-4\pi i/N} & \cdots & e^{-2\pi i(N-1)/N} \\ 1 & e^{-4\pi i/N} & e^{-8\pi i/N} & \cdots & e^{-4\pi i(N-1)/N} \\ \vdots & \vdots & \vdots & \ddots & \vdots \\ 1 & e^{-2\pi i(N-1)/N} & e^{-4\pi i(N-1)/N} & \cdots & e^{-2\pi i(N-1)^2/N} \end{bmatrix} \begin{bmatrix} x_0 \\ x_1 \\ x_2 \\ \vdots \\ x_{N-1} \end{bmatrix} \tag{5.67}$$

The operation may be written in more concise "equation" form that evaluates the mth component of the transformed vector as a weighted sum of the N components of the input vector:

$$x'_k = \frac{1}{\sqrt{N}} \sum_{n=1}^{N} \exp\left[-2\pi i \frac{(k-1)(n-1)}{N}\right] \cdot x_n \tag{5.68}$$

where $1 \le k \le N$. A specific component of the transformed vector $\underline{\mathbf{x}}'$, say $(\underline{\mathbf{x}}')_k$, is the inner product of one row of the transformation matrix $\underline{\mathbf{D}}^{-1}$ with the "input" vector $\underline{\mathbf{x}}$. From the discussion of the interpretation of matrix–vector multiplication as an inner product, we can see that

$$\begin{aligned} (\underline{\mathbf{x}}')_k &= \frac{1}{\sqrt{N}} \left(\exp\left[-2\pi i \frac{(k-1)(n-1)}{N}\right] \right)^* \bullet \underline{\mathbf{x}} \\ &= \frac{1}{\sqrt{N}} \exp\left[+2\pi i \cdot (n-1) \cdot \left(\frac{k-1}{N}\right)\right] \bullet \underline{\mathbf{x}} \end{aligned} \tag{5.69}$$

In words, the kth component of $\underline{\mathbf{x}}'$ is the inner product of the input vector $\underline{\mathbf{x}}$ with a vector whose components are samples of the complex-valued sinusoid that oscillates with frequency $\xi_k = (k-1)/N$ (period $X_k = N/(k-1)$). In other words, the components of $\underline{\mathbf{x}}'$ are the projections of $\underline{\mathbf{x}}$ onto vectors that oscillate sinusoidally with different spatial frequencies. Clearly this can be generalized to a case where the output vector has a different number of components, e.g., $0 \le k < M$.

It will be convenient to substitute a "function" notation for the input and output vectors; we will rename the numerical value of the nth component of the input vector and of the kth component of the transformed vector to the amplitudes of input and output functions f at coordinate n and F at coordinate k, respectively:

$$x_n \to f[n] \quad (0 \le n < N) \tag{5.70a}$$

$$x'_k \to F[k] \quad (0 \le k < M) \tag{5.70b}$$

Note that $M = N$ in our case. The resulting expression for the transformation from $f[n]$ to $F[k]$ is:

$$F[k] = \frac{1}{\sqrt{N}} \sum_{n=1}^{N} f[n] \cdot \exp\left[-2\pi i \frac{(k-1)(n-1)}{N}\right] \tag{5.71}$$

As we will see later, it is often convenient to renumber the indices to the interval $0 \le n, k \le N-1$ to produce the final expression:

$$F[k] = \frac{1}{\sqrt{N}} \sum_{n=0}^{N-1} \left(f[n] \cdot \exp\left[-2\pi i \cdot n \cdot \frac{k}{N}\right] \right) \tag{5.72}$$

where k/N is the number of cycles in N samples of the complex sinusoid $\exp[-2\pi i \cdot n \cdot k/N]$.

The expression in Equation (5.72) is the *discrete Fourier transform* (DFT). The name implies the inference that this is the discrete version of some operation on continuous functions. In Chapters 9–12, we will consider the continuous transformation in detail, and return to the discrete transform in Chapters 14–15.

5.2.5 Alternative Ordering of Eigenvectors

5.2.5.1 Principal Components

The eigenvectors in the diagonalizing matrix $\underline{\mathbf{D}}$ in Equation (5.48) were arranged in order of increasing size of the phase increment between adjacent elements of the vector. This choice fixes the order of the eigenvalues in the diagonal form $\underline{\mathbf{\Lambda}}$ of the matrix. It is probably obvious that other orderings of the eigenvectors in $\underline{\mathbf{D}}$ and of the eigenvalues in $\underline{\Lambda}$ are feasible. For example, we may construct a similarity transformation that generates a diagonal form $\underline{\mathbf{\Lambda}}$ with the eigenvalues arranged in order of decreasing magnitude. The first column in the diagonalizing matrix $\underline{\mathbf{D}}$ is the eigenvector associated with this eigenvalue. The first element in the transformed input vector $\underline{\mathbf{x}}'$ will be the projection of the original input vector onto the eigenvector that is best transmitted by the imaging system. In a sense, this eigenvector is the most important in that imaging system and the ordering of eigenvectors is optimum, since the action of the imaging system may be approximated by using only the eigenvectors associated with the largest eigenvalues. The most important projections of the input vector in that imaging system $\underline{\mathbf{A}}$ are called the *principal components* of the input vector.

5.3 Matrix Formulation of Imaging Tasks

We have studied in detail some methods for solving the discrete version of the direct imaging problem. We now turn our attention to the corresponding cases for the other types of imaging problem in the next two sections.

5.3.1 Inverse Imaging Problem

The solution of the discrete inverse imaging problem requires that the N-element input vector $\underline{\mathbf{x}}$ be determined from the M-row by N-column system matrix $\underline{\mathbf{A}}$ and the M-element output vector $\underline{\mathbf{b}}$. This is exactly the problem of solving a system of M linear equations in N unknowns. If the number of rows M in the matrix $\underline{\mathbf{A}}$ (and of elements in the known vector $\underline{\mathbf{b}}$) is smaller than the number of elements N in the unknown vector $\underline{\mathbf{x}}$, then a solution is not possible because there are more projections to be determined from fewer scalar product equations. Such an inverse problem is *underdetermined*. If $M = N$, then the number of components of the unknown vector $\underline{\mathbf{x}}$ is equal to the number of components of the known vector $\underline{\mathbf{b}}$ and the system operator matrix $\underline{\mathbf{A}}$ is square. If $M > N$, then the system includes more linear equations than unknowns and the system is *overdetermined*. Schematic examples of the three cases are:

$$\begin{bmatrix} a_{11} & a_{12} & a_{13} \\ a_{21} & a_{22} & a_{23} \\ a_{31} & a_{32} & a_{33} \\ a_{41} & a_{42} & a_{43} \\ a_{51} & a_{52} & a_{53} \end{bmatrix} \begin{bmatrix} x_1 \\ x_2 \\ x_3 \end{bmatrix} = \begin{bmatrix} b_1 \\ b_2 \\ b_3 \\ b_4 \\ b_5 \end{bmatrix} \quad \textit{(overdetermined)} \tag{5.73a}$$

$$\begin{bmatrix} a_{11} & a_{12} & a_{13} & a_{14} & a_{15} \\ a_{21} & a_{22} & a_{23} & a_{24} & a_{25} \\ a_{31} & a_{32} & a_{33} & a_{34} & a_{35} \\ a_{41} & a_{42} & a_{43} & a_{44} & a_{45} \\ a_{51} & a_{52} & a_{53} & a_{54} & a_{55} \end{bmatrix} \begin{bmatrix} x_1 \\ x_2 \\ x_3 \\ x_4 \\ x_5 \end{bmatrix} = \begin{bmatrix} b_1 \\ b_2 \\ b_3 \\ b_4 \\ b_5 \end{bmatrix} \quad \textit{(critically determined)} \tag{5.73b}$$

$$\begin{bmatrix} a_{11} & a_{12} & a_{13} & a_{14} & a_{15} \\ a_{21} & a_{22} & a_{23} & a_{24} & a_{25} \\ a_{31} & a_{32} & a_{33} & a_{34} & a_{35} \end{bmatrix} \begin{bmatrix} x_1 \\ x_2 \\ x_3 \\ x_4 \\ x_5 \end{bmatrix} = \begin{bmatrix} b_1 \\ b_2 \\ b_3 \end{bmatrix} \quad \textit{(underdetermined)} \tag{5.73c}$$

In this case, solution of the inverse problem is equivalent to finding a matrix operator $\underline{\mathbf{B}}$ which satisfies the condition:

$$\underline{\mathbf{B}}\,\underline{\mathbf{b}} = \underline{\mathbf{x}} \tag{5.74}$$

The expression for the known vector $\underline{\mathbf{b}}$ in Equation (5.8) may be substituted to yield:

$$\underline{\mathbf{B}}(\underline{\mathbf{A}}\,\underline{\mathbf{x}}) = (\underline{\mathbf{B}}\,\underline{\mathbf{A}})\underline{\mathbf{x}} = \underline{\mathbf{x}} = \underline{\mathbf{I}}\,\underline{\mathbf{x}} \tag{5.75}$$

where the linearity property of the scalar product has been used. Obviously, $\underline{\mathbf{B}}$ must be the inverse matrix $\underline{\mathbf{A}}^{-1}$ to produce the appropriate solution:

$$\underline{\mathbf{A}}^{-1}\,\underline{\mathbf{b}} = \underline{\mathbf{x}} \tag{5.76}$$

The existence conditions for the inverse of a square matrix were considered to some extent in Section 3.2.5. They may be expressed in several equivalent ways. For example, the N scalar product equations implicit in this matrix expression must not be redundant, which means that the N row vectors of $\underline{\mathbf{A}}$ are linearly independent. A second equivalent statement is that the null subspace of $\underline{\mathbf{A}}$ must include only the null vector.

A different and very useful interpretation of the imaging equation is possible in those cases where the dimensions of the input and output vectors are identical and the inverse operator $\underline{\mathbf{A}}^{-1}$ exists. The operator $\underline{\mathbf{A}}$ then is an invertible mapping between the input object $\underline{\mathbf{x}}$ and the output image $\underline{\mathbf{b}}$ so that either vector may be derived from complete knowledge of the other and of the system operator. Put another way, in imaging systems for which $\underline{\mathbf{A}}^{-1}$ exists, $\underline{\mathbf{x}}$ and $\underline{\mathbf{b}}$ may be interpreted as equivalent representations of the same vector. By analogy with the interpretation of $\underline{\mathbf{D}}$ and $\underline{\mathbf{D}}^{-1}$, the system operator $\underline{\mathbf{A}}$ and its inverse $\underline{\mathbf{A}}^{-1}$ are the transformations between these equivalent representations $\underline{\mathbf{x}}$ and $\underline{\mathbf{b}}$. Discussions of the form and use of transformations among equivalent representations constitute the major part of this book; the most common such transformation is the Fourier transform, but a few others also will be introduced.

The third and last class of discrete inverse problem is the case where $M > N$, meaning that the number of elements in the known output vector $\underline{\mathbf{b}}$ exceeds the number of components of the unknown input object $\underline{\mathbf{x}}$. Such an inverse problem is overdetermined; there are more scalar product equations available than unknown components of the input vector $\underline{\mathbf{x}}$, which clearly means that at least some of these M equations supply redundant information about $\underline{\mathbf{x}}$; the equations are linearly dependent. A naive student might assume that such problems are solved by discarding rows of the matrix $\underline{\mathbf{A}}$ until an $N \times N$ square matrix is obtained; the inverse of this "new" matrix exists if the N remaining scalar product equations are linearly independent. It is probably clear that this strategy is not guaranteed to work; there is always the danger of throwing out equations that convey "essential" information about the system. A better solution is available that utilizes all available information, redundant or not. This allows a new matrix operator $\underline{\mathbf{C}}$ to be derived from all information available in the output vector $\underline{\mathbf{b}}$ and in the rectangular matrix $\underline{\mathbf{A}}$. This new matrix $\underline{\mathbf{C}}$ is analogous to the inverse matrix $\underline{\mathbf{A}}^{-1}$ because it acts on the M-D output vector $\underline{\mathbf{b}}$ to compute the N-D input vector $\underline{\mathbf{x}}$:

$$\underline{\mathbf{C}}\,\underline{\mathbf{b}} = \underline{\mathbf{C}}(\underline{\mathbf{A}}\,\underline{\mathbf{x}}) = (\underline{\mathbf{C}}\,\underline{\mathbf{A}})\underline{\mathbf{x}} \tag{5.77}$$

From the discussion surrounding Equation (3.25), it is apparent that the matrix $\underline{\mathbf{C}}$ must have N rows and M columns to be able to generate the N-D vector $\underline{\mathbf{x}}$ from the M-D vector $\underline{\mathbf{b}}$. Only one nontrivial matrix is available in the statement of the problem that may be used as a tool to construct $\underline{\mathbf{C}}$; this is the system matrix $\underline{\mathbf{A}}$, which has M rows and N columns. The transpose of $\underline{\mathbf{A}}$ has the correct number of rows and columns to operate on $\underline{\mathbf{b}}$. Therefore, we pursue an ad hoc "solution" to the inverse imaging problem by using the only information available in the only way it may be applied. First, multiply both sides of the system equation from the left by the transpose of the system matrix:

$$\underline{\mathbf{A}}^T\,\underline{\mathbf{A}}\,\underline{\mathbf{x}} = \underline{\mathbf{A}}^T\,\underline{\mathbf{b}} \tag{5.78}$$

The discussion of the matrix–matrix product following Equation (3.25) establishes that the product matrix $\underline{\mathbf{A}}^T\underline{\mathbf{A}}$ is $N \times N$ square, and thus may be invertible. The inverse exists if the N column vectors (or equivalently, the N row vectors) are linearly independent. If this condition is satisfied, then both sides of Equation (5.73) are multiplied from the left by this matrix:

$$\begin{aligned}(\underline{\mathbf{A}}^T\,\underline{\mathbf{A}})^{-1}(\underline{\mathbf{A}}^T\,\underline{\mathbf{A}})\underline{\mathbf{x}} &= \underline{\mathbf{x}}\\ \Longrightarrow \underline{\mathbf{x}} = (\underline{\mathbf{A}}^T\,\underline{\mathbf{A}})^{-1}\underline{\mathbf{A}}^T\underline{\mathbf{b}} &\equiv \underline{\mathbf{A}}^\dagger\underline{\mathbf{b}}\end{aligned} \tag{5.79}$$

The resulting N-row by M-column matrix denoted by $\underline{\mathbf{A}}^\dagger$ acts "like" the inverse matrix of the overdetermined system operator $\underline{\mathbf{A}}$. It is commonly called the Moore–Penrose *pseudoinverse* of $\underline{\mathbf{A}}$. If the data in the output image $\underline{\mathbf{b}}$ are *consistent*, meaning that all sets of N linearly independent scalar product equations of $\underline{\mathbf{A}}$ would produce the same solution for $\underline{\mathbf{x}}$, then the operator defined in Equation (5.79) generates the unique and correct input vector $\underline{\mathbf{x}}$ that produces $\underline{\mathbf{b}}$ under the action of $\underline{\mathbf{A}}$. For this reason, the Moore–Penrose pseudoinverse is very much analogous to the inverse matrix in the case of consistent data. The benefit of the Moore–Penrose pseudoinverse really accrues when the set of scalar product equations is not consistent, as would occur when some or all components of $\underline{\mathbf{b}}$ are contaminated by additive random numbers ("noise") for whatever reason. This situation is the rule rather than the exception in realistic imaging problems where noise is introduced by errors in the measurements of the output vector $\underline{\mathbf{b}}$. We say that the resulting data are *inconsistent*; different sets of N linearly independent scalar product equations taken from the M available would yield different input vectors $\underline{\mathbf{x}}$, if indeed $\underline{\mathbf{x}}$ could even be found. In such cases, no unique and correct solution for $\underline{\mathbf{x}}$ is possible. When applied to inconsistent data, the Moore–Penrose pseudoinverse in Equation (5.79) generates the estimate of $\underline{\mathbf{x}}$ that is "closest" to the unavailable true solution based on the metric of minimum squared error. In other and perhaps more familiar words, the pseudoinverse operator $\underline{\mathbf{A}}^\dagger$ performs a linear regression to produce the "best" estimate of the input vector in a least-squares sense.

5.3.2 Solution of Inverse Problems via Diagonalization

The discussion of matrix diagonalization in Section 5.2 may be used to derive yet another recipe for solving the inverse problem in those cases where the system matrix $\underline{\mathbf{A}}$ is square. This method is very useful both as a practical solution and because of the insight that may be gleaned from the process.

First, the square system matrix $\underline{\mathbf{A}}$ is diagonalized to obtain the equivalent diagonal matrix $\underline{\mathbf{\Lambda}}$ of eigenvalues via the similarity transformation in Equation (5.29a). This diagonal representation is trivial to invert via Equation (3.44) in any case where all eigenvalues of $\underline{\mathbf{A}}$ are nonzero:

$$(\underline{\mathbf{\Lambda}}^{-1})_{nn} = \frac{1}{\lambda_n} \tag{5.80}$$

If any eigenvalue is zero, the inverse diagonal matrix $\underline{\mathbf{\Lambda}}^{-1}$ does not exist. However, we still can construct a type of pseudoinverse of the diagonal matrix (call it $\underline{\mathbf{\Lambda}}^\dagger$) by inverting all nonnull eigenvalues and setting the undefined reciprocal of any null eigenvalues to zero:

$$\begin{aligned}(\underline{\mathbf{\Lambda}}^\dagger)_{nn} &= \begin{cases}\lambda_n^{-1} = \dfrac{1}{\lambda_n} & \textit{if } \lambda_n \neq 0\\ 0 & \textit{if } \lambda_n = 0\end{cases}\\ (\underline{\mathbf{\Lambda}}^\dagger)_{nm} &= 0 \quad \textit{if } n \neq m\end{aligned} \tag{5.81}$$

This definition of the pseudoinverse $\underline{\Lambda}^{\dagger}$ of a diagonal matrix may be justified by considering the action of the system matrix $\underline{\mathbf{A}}$ on any eigenvector whose corresponding eigenvalue is zero. From the eigenvector equation in Equation (5.17), it is evident that any such eigenvector is blocked by the system and cannot become part of the output vector; $\underline{\mathbf{b}}$ may include no projections from any blocked eigenvectors. Since the amplitudes of any such eigenvectors cannot be determined from the output vector $\underline{\mathbf{b}}$ without other a priori information, the pseudoinverse matrix operator makes no attempt to try.

The inverse operator $\underline{\mathbf{A}}^{-1}$ in the original representation (or the corresponding pseudoinverse $\underline{\mathbf{A}}^{\dagger}$ if the equations are inconsistent) may be found easily by applying the inverse matrix transformation in Equation (5.29b):

$$\underline{\mathbf{A}}^{-1} = \underline{\mathbf{D}}\, \underline{\Lambda}^{-1}\, \underline{\mathbf{D}}^{-1} \tag{5.82}$$

The new path for determining the inverse or the pseudoinverse of the square matrix $\underline{\mathbf{A}}$ may be summarized in these steps:

1. Find the eigenvectors and eigenvalues of $\underline{\mathbf{A}}$.
2. Construct the diagonalizing matrix $\underline{\mathbf{D}}$ from the eigenvectors.
3. Compute the diagonal form $\underline{\Lambda} = \underline{\mathbf{D}}^{-1}\, \underline{\mathbf{A}}\, \underline{\mathbf{D}}$ (if it exists).
4. Invert (or pseudoinvert) $\underline{\Lambda}$ to obtain $\underline{\Lambda}^{-1}$ (or $\underline{\Lambda}^{\dagger}$, as appropriate).
5. Transform the diagonal inverse (or pseudoinverse) matrix back to the original basis via the appropriate process in Equation (5.78).

This recipe is particularly convenient for circulant matrix operators because the diagonalizing matrix $\underline{\mathbf{D}}$ and its inverse are available from Equations (5.65) and (5.66) without solving the secular equation.

5.3.3 Matrix–Vector Formulation of System Analysis

The third imaging problem to be considered in terms of the matrix–vector imaging equation is the determination of the system operator $\underline{\mathbf{A}}$ from knowledge of the input and output vectors $\underline{\mathbf{x}}$ and $\underline{\mathbf{b}}$. In the matrix formulation, the difficulty of the general problem becomes clear quite quickly: the $N \times M$ unknown elements of the system matrix $\underline{\mathbf{A}}$ must be determined from the elements of the N-D vector $\underline{\mathbf{x}}$ and of the M-D vector $\underline{\mathbf{b}}$. There are M N-D scalar product equations but $N \cdot M$ unknowns in the general operator $\underline{\mathbf{A}}$. We restrict ourselves to the case of a square matrix ($N = M$), where we much derive a simultaneous solution of N scalar product equations each with N unknown parameters. The largest number of unknown coefficients of $\underline{\mathbf{A}}$ that may be solved from N equations is N, which means that $\underline{\mathbf{A}}$ may be determined only if it includes N (or fewer) independent coefficients out of the N^2 possible. Though this may seem to be a severe constraint on the possible form of $\underline{\mathbf{A}}$, we already have seen that useful imaging matrix operators of this form do exist: the circulant matrices that model discrete shift-invariant imaging systems. The algorithm for determining the form of the matrix will be illustrated by example.

Assume first that a single N-D output vector $\underline{\mathbf{b}}_1$ is measured by applying a single N-D vector $\underline{\mathbf{x}}_1$ as the input to the unknown operator $\underline{\mathbf{A}}$. Since the eigenvectors of a circulant matrix are predetermined and linearly independent, the diagonalizing matrices $\underline{\mathbf{D}}$ and $\underline{\mathbf{D}}^{-1}$ exist and are known a priori. The known input and output vector may be transformed to the representation in the eigenvector basis via the relations in Equation (5.32). In this representation, each component of the output vector is a scaled replica of the same component of the input vector, where the scaling factor is the corresponding unknown diagonal element (eigenvalue) of the system matrix:

$$\lambda_i (\underline{\mathbf{x}}_1')_n = (\underline{\mathbf{b}}_1')_n \tag{5.83}$$

where the subscript n $(0 \leq n \leq N-1)$ indicates the specific component of the transformed vectors that is being considered. The unknown nth eigenvalue is the ratio of the nth components of the two transformed vectors:

$$\lambda_n = \frac{(\underline{\mathbf{b}}'_1)_n}{(\underline{\mathbf{x}}'_1)_n} \tag{5.84}$$

The solution for λ_n obviously exists for all components of $\underline{\mathbf{x}}'_1$ that are not zero. If any component of $\underline{\mathbf{x}}'_1$ is null, then the corresponding component of the output vector $\underline{\mathbf{b}}'_1$ must be zero as well (assuming that no noise is present in the measurement of $\underline{\mathbf{b}}$). For example, assume that one component of $\underline{\mathbf{x}}'_1$ is zero, say $(\underline{\mathbf{x}}'_1)_k$. The process may be repeated for other input vectors $\underline{\mathbf{x}}_n$ until finding one for which the kth component is nonzero; the corresponding ratio in Equation (5.80) will yield the necessary eigenvalue λ_k. After all eigenvalues of $\underline{\mathbf{A}}$ are computed in this fashion, the matrix is reconstructed from its diagonal form $\underline{\mathbf{\Lambda}}$ via Equation (5.29b).

From this discussion, it may be apparent that a matrix operator $\underline{\mathbf{A}}$ that is known to be circulant but with unknown coefficients may be determined by applying each of the N known eigenvectors to $\underline{\mathbf{A}}$ one at a time. The ratio of each element of the transformed output vector to the corresponding nonzero component of each transformed input vector will be the corresponding eigenvalue of $\underline{\mathbf{A}}$. The diagonal form of $\underline{\mathbf{A}}$ may be constructed from the N eigenvalues, followed by transformation to the original representation via Equation (5.29b). This procedure will be considered in more detail when discussing continuous systems in Chapter 17.

5.4 Continuous Analogues of Vector Operations

At this point, we change the direction of the discussion in this chapter one last time to adapt some of the concepts of discrete vectors to apply them to functions with continuous domains. This is perhaps redundant, at least in part, because we introduced the interpretation of an N-dimensional vector as a 1-D array of values in Figure 3.1. The important concepts include the inner product, projection of vectors onto other vectors, and the eigenvectors of a matrix operator. We will find that the analogies between the discrete and continuous cases are exact. The concepts of the continuous case will be very useful when investigating the solutions of the three imaging problems. More detailed theoretical treatments are available, e.g., Davis (1963).

Discrete row vectors with different dimensionalities are shown in Figure 5.4. As the number of components $N \to \infty$, the amplitudes of the components of the N-D may be described as a function f of a continuous variable, say x. In short, we may think of a function $f[x]$ as an infinite-dimensional vector. We may use this interpretation to extend the concepts of the inner product and of vector projections to continuous functions.

5.4.1 Inner Product of Continuous Functions

Recall from Equation (3.9) that the scalar product of two vectors with real-valued components is the sum of the products of the pairs of corresponding components. The process was modified for vectors with complex-valued components to obtain Equation (5.4), which included the explicit complex conjugation of the components of the "reference vector" to ensure a real-valued norm. The formula for the inner product may be applied to vectors with any number of discrete components; the dimensionality of the vector space may be an arbitrarily large integer. Once the concept of an "infinite-dimensional" discrete vector is accepted, it requires only a small conceptual leap to apply the definition to functions with continuous real-valued domains and continuous complex-valued ranges. The analogue of the discrete inner product is a continuous summation (i.e., an integral) of the product of the two complex-valued functions over the continuous real-valued domain x. In general, the output of a continuous inner product is a complex-valued scalar g.

To clarify the analogy between inner products of vectors and of functions, consider first the scalar product of two real-valued functions $f[x]$ and $m[x]$. These correspond to the input and "reference" vectors $\underline{\mathbf{x}}$ and $\underline{\mathbf{a}}$, respectively, in Equation (3.9). The analogue to the scalar product for continuous

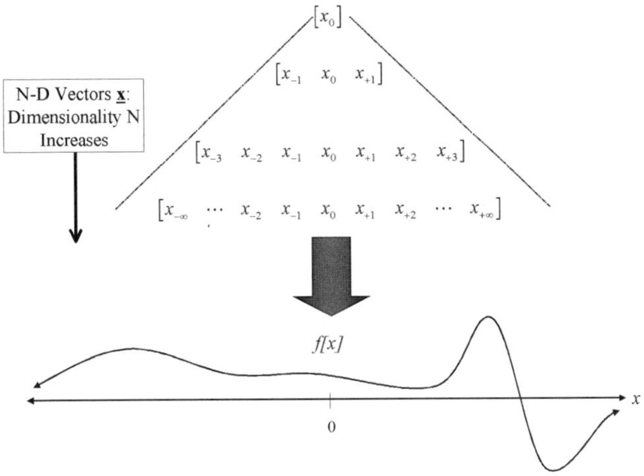

Figure 5.4 The analogy between a finite-dimensional vector $\underline{\mathbf{x}}$ and a continuous function $f[x]$. As the number of vector dimensions increases, the N-D vector $\underline{\mathbf{x}}$ may be visualized as the 1-D function $f[x]$.

functions evidently is the area of the "point-by-point" product of the two functions:

$$g = \int_{-\infty}^{+\infty} m[\alpha] f[\alpha]\, d\alpha \tag{5.85}$$

where α is a dummy variable of integration. Based on the vector interpretation, of the scalar product, the numerical value g is a measure of the "projection" of the input function $f[x]$ onto the "reference" function $m[x]$. The scalar product of two functions is defined only when both have the same domains, and thus the same dimensionality (1-D in this case).

The analogue to the inner product of two vectors applies in the case of two complex-valued continuous functions. The discussion surrounding Equation (5.4) establishes that the inner product of a function with itself should be real valued and nonnegative. Therefore, the definition must include the complex conjugate of one of the component functions:

$$g = \int_{-\infty}^{+\infty} f^*[\alpha] f[\alpha]\, d\alpha = \int_{-\infty}^{+\infty} |f[\alpha]|^2\, d\alpha \tag{5.86}$$

This real-valued quantity g is analogous to the squared magnitude of a vector in Equation (3.5) and may be called the *integrated power* of the continuous function $f[x]$. The continuous analogue to the inner product of distinct complex-valued vectors in Equation (5.4) is:

$$g = \int_{-\infty}^{+\infty} m^*[\alpha] f[\alpha]\, d\alpha \tag{5.87}$$

where $f[x]$ is the input function, $m[x]$ is the "reference", and the output scalar g is complex valued, in general. As was the case with vectors, Equation (5.87) applies to both real- and complex-valued functions. Little confusion should arise from the use of the term *inner product* for both real-valued and complex-valued functions; the cases are distinguished by context.

The scalar product of two real-valued vectors in Equation (3.11) is a measure of the component of the "input" vector $\underline{\mathbf{x}}$ "in the direction of" the reference vector $\underline{\mathbf{A}}$; this is conceptually identical to the projection of $\underline{\mathbf{x}}$ onto a unit "reference" vector $\hat{\underline{\mathbf{a}}}$. A corresponding interpretation is made for the inner product of two continuous functions in Equation (5.87). The scalar g is proportional to the "projection" of $f[x]$ onto $m[x]$, and so serves as a measure of "how much" of the reference function $m[x]$ is present within the input $f[x]$. The scalar g is equal to the projection when the reference function $m[x]$ has the analogous property to unit length. From Equation (5.86), this evidently requires that the projection of $m[x]$ onto itself be unity:

$$\int_{-\infty}^{+\infty} m^*[\alpha]m[\alpha]\,d\alpha = \int_{-\infty}^{+\infty} |m[\alpha]|^2\,d\alpha = 1 \tag{5.88}$$

In other words, the integrated power of the reference function must be unity for Equation (5.87) to measure the projection of $f[x]$ onto $m[x]$.

Equation (3.15) showed that the scalar product of two real-valued vectors of fixed length is maximized when the included angle $\theta = 0$ radians, i.e., when the vectors point in the same direction. Though the analogue to the concept of the included angle is not obvious for continuous functions, the result in Equation (3.16) leads us to surmise that the inner product of $f[x]$ and $m[x]$ satisfies a continuous analogue of the Schwarz inequality of Equation (5.7):

$$\frac{\left|\int_{-\infty}^{+\infty} m^*[\alpha]f[\alpha]\,d\alpha\right|}{\sqrt{\int_{-\infty}^{+\infty} |f[\alpha]|^2\,d\alpha}\sqrt{\int_{-\infty}^{+\infty} |m[\alpha]|^2\,d\alpha}} \leq 1 \tag{5.89}$$

The rigorous proof of this relationship is given in many sources, including Papoulis (1968). Note that the equality is satisfied only if $f[x] \propto m[x]$. This result will be useful during the discussion of correlation in Chapter 8, among other places. We use the analogy to vectors by saying that two functions are *orthogonal* when the projection of one onto the other is null. In such a case, the input function $f[x]$ "includes" no part of the reference function $m[x]$, and therefore that reference function is not required to synthesize $f[x]$.

To illustrate the concept of "projection" applied to real-valued functions, consider the important example of two same-frequency real-valued sinusoids with a relative phase of $\pi/2$, e.g., $m[x] = \cos[2\pi\xi x]$ and $f[x] = \sin[2\pi\xi x]$. The inner product of these two functions may be split into two additive terms by applying a trigonometric identity that may be obtained from Equation (4.42b):

$$\sin[\alpha]\cos[\beta] = \frac{1}{2}(\sin[\alpha+\beta] + \sin[\alpha-\beta]) \tag{5.90}$$

The inner product of the two real-valued sinusoids is:

$$\begin{aligned} g &= \int_{-\infty}^{+\infty} \sin[2\pi\xi\alpha]\cos[2\pi\xi\alpha]\,d\alpha \\ &= \frac{1}{2}\int_{-\infty}^{+\infty} (\sin[2\pi(\xi-\xi)\alpha] + \sin[2\pi(\xi+\xi)\alpha])\,d\alpha \\ &= \frac{1}{2}\int_{-\infty}^{+\infty} \sin[2\pi\cdot 0\cdot\alpha]\,d\alpha + \frac{1}{2}\int_{-\infty}^{+\infty} \sin[4\pi\xi\alpha]\,d\alpha \\ &= \frac{1}{2}\int_{-\infty}^{+\infty} 0\,d\alpha + \frac{1}{2}\int_{-\infty}^{+\infty} \sin[4\pi\xi\alpha]\,d\alpha \end{aligned} \tag{5.91a}$$

The first integral obviously is zero because the integrand is zero. The second term is the integral of an odd function over symmetric limits, and also evaluates to zero regardless of the value of the argument of the sine because the halves of the integrand over positive and negative x are antisymmetric (equal amplitudes and opposite signs). The end result therefore is:

$$\int_{-\infty}^{+\infty} \sin[2\pi\xi\alpha]\cos[2\pi\xi\alpha]\,d\alpha = 0 \tag{5.91b}$$

From the geometric interpretation of the inner product of vectors, this result indicates that cosine and sine functions with the same argument must be orthogonal. Orthogonal sinusoidal functions with the same spatial frequency are said to be *in quadrature*, as implied by the fact that the initial phases of the cosine and sine differ by $\pi/2$ radians, corresponding to one-quarter of a cycle. The projections of sinusoidal functions are very important in signal analysis and are considered in detail in Chapter 9.

5.4.2 Complete Sets of Basis Functions

A representation of a vector in terms of its projections onto each member of a complete set of basis vectors was considered in Chapter 3:

$$\underline{\mathbf{a}} = \sum_{n}^{N} b_n \, \underline{\hat{\mathbf{x}}}_n \tag{5.92}$$

If the basis vectors are mutually orthogonal, then the addition to $\underline{\mathbf{A}}$ of an increment of a single basis vector affects only the specific projection onto that basis vector; there is no crosstalk among projections. Recall also that there usually are many possible sets of basis vectors, and thus many different but equivalent *representations* of $\underline{\mathbf{A}}$. The equivalent representations are related by invertible transformations.

The projection of a complex-valued function $f[x]$ onto a complex-valued reference function $m[x]$ is defined by Equation (5.87):

$$g = \int_{-\infty}^{+\infty} m^*[\alpha] f[\alpha] \, d\alpha \tag{5.93}$$

We now surmise a generalization of this expression to create other representations of continuous functions $f[x]$ in a fashion analogous to Equation (5.68). In other words, we hope to project the input function $f[x]$ onto each of a set of continuous functions that includes $m[x]$ as one of its members. This ensemble of projections then may be used as an equivalent representation of $f[x]$ if some other conditions are met. By analogy with the set of basis vectors, we might denote the set of continuous reference functions, which also are called *basis* functions, by enclosing the function name in braces, e.g., $\{m[x]\}$. We also assume that the equivalent representations must have the same dimensionality for the transformation to be invertible, which means that the number of independent basis functions used to represent $f[x]$ must be identical to the dimensionality of $f[x]$.

Each of the infinite number of basis functions must be distinguished by a parameter. It will not be appropriate to use an integer parameter or subscript as we do for vectors, such as in $m_n[x]$, because the differences between adjacent members of the set will be more subtle. Instead, it is much more convenient to use a continuous parameter analogous to a spatial coordinate to specify a particular basis function. A suitable nomenclature for the set of reference functions might be $m[x; x']$, where the included semicolon distinguishes the spatial coordinate x of the reference function from the parameter x' associated with that particular function. In other words, a specific continuous 1-D function $m[x]$ is associated with each value of the parameter x'. Other notations also would be satisfactory, such as $m[x, x']$, though this might be confused with a 2-D function with independent variables x and x'.

By applying the former notation, the inner product of each reference function with the input function $f[x]$ will be a specific complex scalar distinguished by the parameter x':

$$\int_{-\infty}^{+\infty} m^*[\alpha; x'] f[\alpha] \, d\alpha \equiv g[x'] \tag{5.94}$$

where α has been used again as the dummy variable of integration. Note the similarity of form to the discrete matrix–vector product, where the ensemble of reference functions is a 2-D continuous function analogous to the matrix $\underline{\mathbf{A}}$, while the 1-D input function $f[x]$ plays the role of the input vector $\underline{\mathbf{x}}$. Based on Equation (2.1), the entire process acts like a linear operator applied to a 1-D continuous function $f[x]$ to generate a different 1-D function $g[x']$. If each row of $m[x, x']$ satisfies the continuous analogue of unit length in Equation (5.88), then the amplitude g at a specific coordinate x'_0 is the projection of

$f[x]$ onto a particular reference function $m[x, x'_0]$. Since the 1-D transformation of $f[x]$ requires a 2-D set of reference functions $m[x, x']$, the extension to a 2-D input function $f[x, y]$ requires a 4-D reference function that may be represented as $m[x, y; x', y']$. In any case, Equation (5.94) defines the transformation from $f[x]$ to $g[x']$, which is the representation in a new but equivalent coordinate system.

We have extended the idea of projection of one complex-valued vector onto another to construct the operator that projects a complex-valued function onto another via a linear operator in Equation (5.94). In Chapter 8 we will add the additional constraint of shift invariance to construct an analogous process to vector multiplication by a circulant matrix.

5.4.3 Orthonormal Basis Functions

Based on the analogy with the projection of a discrete vector $\underline{\mathbf{a}}$ onto a set of basis vectors $\underline{\mathbf{x}}_n$, the projection of a 1-D function onto a set of basis functions $m[x; x']$ is interpreted easily if the basis functions are mutually orthogonal. In this case, the inner product of any two distinct functions in the basis set is zero:

$$\int_{-\infty}^{+\infty} m^*[\alpha; x'_1]m[\alpha; x'_2]\, d\alpha = 0 \tag{5.95}$$

The two parameters x'_1 and x'_2 are used to indicate that the two basis functions are distinct.

In the discrete case, the scalar product of a vector with itself in Equation (3.5) yields the square of the length of the vector. The corresponding result for the continuous case is less clear; what is the meaning of the length or norm of a function? We hypothesize that it is the continuous version of Equation (3.5), which evidently is:

$$\int_{-\infty}^{+\infty} m^*[\alpha; x'_1]m[\alpha; x'_1]\, d\alpha = \int_{-\infty}^{+\infty} |m[\alpha; x'_1]|^2\, d\alpha \tag{5.96}$$

This leads rather obviously to a hypothesized expression for the norm of a general 1-D function $g[x]$. This quantity is often indicated by the symbol $||g[x]||$ and is defined as:

$$||g[x]|| \equiv \sqrt{\left(\int_{-\infty}^{+\infty} |g[\alpha]|^2\, d\alpha\right)} \tag{5.97}$$

which may be called the *root-mean-square (RMS)* norm of the 1-D function $g[x]$. This definition demands an illustrative example. Consider again the norm of the real-valued 1-D sinusoidal function $g[x] = A_0 \cos[2\pi\xi_0 x]$. From Equation (5.97), the norm of $g[x]$ is:

$$\begin{aligned} ||A_0 \cos[2\pi\xi_0 x]|| &= \sqrt{A_0^2}\sqrt{\int_{-\infty}^{+\infty} |\cos[2\pi\xi_0\alpha]|^2\, d\alpha} \\ &= |A_0|\sqrt{\int_{-\infty}^{+\infty} |\cos[2\pi\xi_0\alpha]|^2\, d\alpha} \end{aligned} \tag{5.98}$$

The integrand is nonnegative everywhere, which means that the norm is infinite for all spatial frequencies ξ_0. To avoid this indeterminate calculation, it is possible to define a norm over a finite domain such as:

$$||A_0 \cos[2\pi\xi_0 x]|| \equiv |A_0|\sqrt{\int_a^b |\cos[2\pi\xi_0\alpha]|^2\, d\alpha} \tag{5.99}$$

The limits of integration may be symmetric (so that $a = -b$) or not.

Though other classes of norms of a function may be defined, they generally are not applied to imaging problems and will not be considered here.

5.4.4 Continuous Analogue of DFT

The DFT in Equation (5.72) is the weighted sum of the components of the input vector, which we now call the samples of the input function $f[n]$:

$$F[k] = \sum_{n=0}^{N-1} f[n] \cdot \exp\left[-2\pi i \frac{kn}{N}\right] \tag{5.100}$$

If we identify k/N to be the spatial frequency of the sinusoid that is indexed by k that specifies the number of cycles of the complex sinusoid in N samples (we can call it ξ_k) and n to be the analogy to the space coordinate (call it x_n), then this equation is rewritten as:

$$F[\xi_k] = \sum_{n=0}^{N-1} f[x_n] \cdot \exp[-2\pi i \xi_k x_n] \tag{5.101}$$

The analogy of the discrete and continuous inner products leads directly to the generalization to continuous spatial coordinates and continuous spatial frequencies by substituting the integral over x for the sum over x_n:

$$\begin{aligned} F[\xi] &= \int_{-\infty}^{+\infty} (\exp[+2\pi i \xi x])^* \cdot f[x]\, dx \\ &= \int_{-\infty}^{+\infty} \cos[2\pi \xi x] \cdot f[x]\, dx - i \int_{-\infty}^{+\infty} \sin[2\pi \xi x] \cdot f[x]\, dx \end{aligned} \tag{5.102}$$

In words, this is the projection of the continuous function $f[x]$ onto each of the infinite set of orthogonal continuous basis functions $\exp[+2\pi i \xi x]$ that are distinguished by the continuous spatial frequency ξ. This transformation is the subject of Chapters 9–12.

5.4.5 Eigenfunctions of Continuous Operators

Recall from Section 5.2 that the action of a square matrix operator can affect only the length of an eigenvector, and that this change is multiplication by the corresponding eigenvalue. Also, it was demonstrated that the number of eigenvectors cannot exceed the dimensionality of the matrix. We saw in Section 5.2 that the eigenvectors of a circulant matrix form a complete set of orthogonal vectors, and thus may serve as a useful basis for the input and output vectors.

In an exactly analogous fashion, the only effect of a continuous operator $\mathcal{O}$ on an eigenfunction is to scale its amplitude by a factor equal to the eigenvalue. Therefore, a particular eigenfunction $f[x]$ of the continuous operator $\mathcal{O}$ must obey the condition:

$$\mathcal{O}\{f[x]\} = g[x] = \lambda \cdot f[x] \tag{5.103}$$

The domain of the input and output functions must be identical for this condition to be evaluated, and the eigenvalue λ generally is complex valued. As was true for the eigenvectors of circulant matrices, the eigenfunctions of linear and shift-invariant continuous operators include an infinite number of functions that obey the condition of orthogonality as defined by Equation (5.95). Therefore any function may be represented as a linear combination of the eigenfunctions specified by a weighted summation in the form of a superposition integral. The infinite set of weights that are applied to these eigenfunctions to generate a particular function $f[x]$ constitutes an equivalent *representation* of $f[x]$.

The rest of this book is devoted to investigations of the eigenfunctions of linear and shift-invariant operators. The properties are very useful when solving the imaging problems that were introduced in Chapter 1. Fortunately, the proofs of many of the useful properties of these continuous eigenfunctions are actually much simpler than for the corresponding discrete vectors.

PROBLEMS

5.1 Find the lengths of the following vectors:

(a) $\begin{bmatrix} 3 \\ 4 \end{bmatrix}$

(b) $\begin{bmatrix} -i \\ 1-i \end{bmatrix}$

(c) an N-D vector $\underline{\mathbf{x}}$ with all N components $x_n = 1 + i$.

5.2 For each of the following vectors, find an orthogonal unit vector:

(a) $\begin{bmatrix} -1+i \\ +1-i \end{bmatrix}$

(b) $\begin{bmatrix} 1+i \\ 2-i \end{bmatrix}$

(c) $\begin{bmatrix} 1 \\ i \end{bmatrix}$

(d) $\begin{bmatrix} e^{+i\pi/3} \\ e^{-i\pi/4} \end{bmatrix}$

5.3 In each case, calculate the "projection" of the vector $\underline{\mathbf{x}}$ onto the vector $\underline{\mathbf{a}}$:

(a) $\underline{\mathbf{x}} = \begin{bmatrix} 1 \\ 1 \\ 0 \end{bmatrix}, \underline{\mathbf{a}} = \begin{bmatrix} 0 \\ 1 \\ 1 \end{bmatrix}$

(b) $\underline{\mathbf{x}} = \begin{bmatrix} i \\ i \\ 0 \end{bmatrix}, \underline{\mathbf{a}} = \begin{bmatrix} 0 \\ i \\ i \end{bmatrix}$

(c) $\underline{\mathbf{x}} = \begin{bmatrix} 1+i \\ 1 \\ 1-i \end{bmatrix}, \underline{\mathbf{a}} = \begin{bmatrix} 1 \\ 1+i \\ 1-i \end{bmatrix}$

5.4 Consider these "imaging-system" matrices $\underline{\mathbf{A}}$ that act on 2-D vectors:

$$\underline{\mathbf{A}}_1 = \begin{bmatrix} 2i & 1 \\ 1 & 2i \end{bmatrix}$$

$$\underline{\mathbf{A}}_2 = \begin{bmatrix} 0 & -i \\ -i & 0 \end{bmatrix}$$

$$\underline{\mathbf{A}}_3 = \begin{bmatrix} 1 & 1 \\ 1 & 1 \end{bmatrix}$$

(a) In each case, find the "diagonalized" matrix $\underline{\mathbf{\Lambda}}_n$ corresponding to $\underline{\mathbf{A}}_n$.

(b) From the forms of $\underline{\mathbf{\Lambda}}_n$, determine which of the matrices $\underline{\mathbf{A}}_n$ are invertible.

(c) Find the inverses $\underline{\mathbf{A}}_n^{-1}$ that exist and the pseudoinverses $\underline{\mathbf{A}}_n^{\dagger}$ in those cases where $\underline{\mathbf{A}}_n^{-1}$ do not exist.

(d) Determine normalized basis vectors for the row subspace, null subspace, column subspace, and left null subspace.

5.5 Find the eigenvectors and corresponding eigenvalues of each matrix:

(a) $\underline{\mathbf{A}}_1 = \begin{bmatrix} 1 & 0 & 0 \\ 0 & 1 & 0 \\ 0 & 0 & 1 \end{bmatrix}$

(b) $\underline{\mathbf{A}}_2 = \begin{bmatrix} 1 & 0 & 0 \\ 0 & 1 & 0 \\ 0 & 0 & 0 \end{bmatrix}$

(c) $\underline{\mathbf{A}}_3 = \begin{bmatrix} +1 & -1 & 0 \\ 0 & +1 & -1 \\ -1 & 0 & +1 \end{bmatrix}$

5.6 The following 2 × 2 matrix operators represent "imaging systems":

$$\underline{\mathbf{A}}_1 = \begin{bmatrix} 2 & 0 \\ 0 & \frac{1}{3} \end{bmatrix}$$

$$\underline{\mathbf{A}}_2 = \begin{bmatrix} 1 & 1 \\ -1 & 1 \end{bmatrix}$$

$$\underline{\mathbf{A}}_3 = \begin{bmatrix} 1 & i \\ i & 1 \end{bmatrix}$$

$$\underline{\mathbf{A}}_4 = \begin{bmatrix} 2i & 0 \\ 0 & 0 \end{bmatrix}$$

For each, do the following:

(a) Describe in words the action of this system on an arbitrary 2-D input object vector $\underline{\mathbf{x}}$.

(b) Determine whether the system is invertible.

(c) If the answer to (b) is "yes", then find the inverse matrix.

5.7 For the 4 × 4 matrix $\underline{\mathbf{A}}$:

$$\underline{\mathbf{A}} = \begin{bmatrix} 0 & -i/2 & 0 & -i/2 \\ -i/2 & 0 & -i/2 & 0 \\ 0 & -i/2 & 0 & -i/2 \\ -i/2 & 0 & -i/2 & 0 \end{bmatrix}$$

(a) Find the eigenvalues λ_n of $\underline{\mathbf{A}}$; you may use any method.

(b) Find the diagonal form $\underline{\mathbf{\Lambda}}$.

(c) Explain why this matrix is or is not invertible.

6

1-D Special Functions

This chapter introduces the definitions and notation for several specific functions of one variable. As presented, the 1-D deterministic special functions are defined over a real-valued domain labeled by the independent variable x, which has dimensions of length. Obviously, functions of time, wavelength, or other dimension may be constructed by redefining the coordinates as appropriate. With one exception, these functions serve as building blocks for creating useful simple 1-D signals and 2-D images and to construct mathematical descriptors for the action of imaging systems. The exception is the gamma function, which is a convenient computational tool for solving many imaging-related problems. When used to construct input images or descriptors of imaging systems, the special functions are scaled in "width" and amplitude, translated as necessary, and may be summed and/or multiplied together. For example, useful input functions may be created by summing translated replicas of rectangles or by point-by-point multiplication of a sinusoid by a rectangle. The process of multiplying two functions together is called *modulation*, but note that this term also has another common meaning in imaging that will be described in Section 6.1.7.

Most functions defined in this chapter have real-valued amplitudes, and may be combined to create complex-valued functions by separately constructing the real and imaginary parts via Equations(4.4) and (4.5), or by creating functions for the magnitude and phase followed by conversion to real and imaginary parts via the relationships in Equations (4.15) and (4.16). A few functions with explicitly complex-valued amplitudes also are defined in this chapter due to their particular convenience. Some of the defined functions perhaps are obvious, but are included for completeness.

Most functions considered in this chapter are deterministic, meaning that the amplitude is specified exactly at each coordinate. Random (or "stochastic") functions are described briefly in Section 6.4. Only the statistics of the amplitudes of these functions may be specified; the amplitude at a particular coordinate is not determined or predictable. Random functions are necessary to create and describe "noisy" signals.

Of course, all authors bring their own special needs and prejudices to the task of defining sets of special functions; this author pleads guilty in advance to such charges (witness the discussion of the notation for $\sqrt{-1}$ in Chapter 4!). It therefore should not be surprising to readers that definitions will differ in some details. The particular advantages and disadvantages of each definition will be discussed, though not necessarily with dispassion. The notation and conventions specified by Gaskill (1978) are used in many cases, though several additional functions also are defined here.

The defining expressions for each deterministic function specify the amplitude at each location in the real-valued domain, which has one or more dimensions as appropriate. It is often convenient to classify functions based on their similar and distinctive "features". An example of such a feature is the

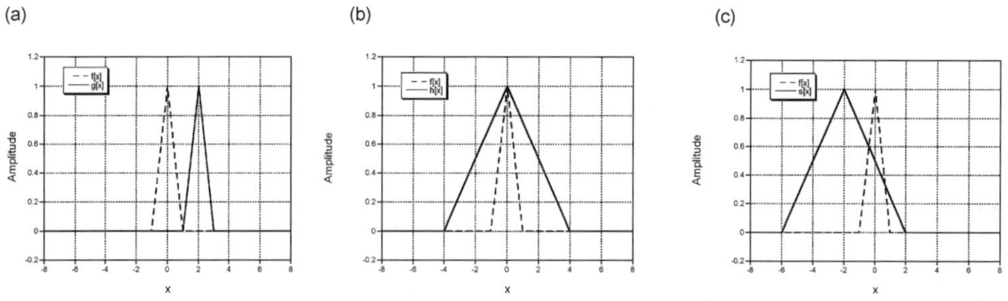

Figure 6.1 Varying the parameters in the argument of a special function $f[x]$ that has compact support: (a) $g[x] = f[x-2]$, showing the translation of the center of symmetry to $x_0 = +2$; (b) $h[x] = f[x/4]$, which scales the width by the factor of 4; (c) $s[x] = f[(x+2)/4]$, which both translates and scales.

set of "zeros" of the function, which are the locations within the domain where the amplitude is zero. Many functions exhibit no zeros or only a few isolated zeros within the infinite real domain. In these cases, it is useful to define the "support" of the function $f[x]$ to be the smallest region of the domain that includes all coordinates where $f \neq 0$. If this region has finite extent, then $f[x]$ is said to exhibit "compact" (or "finite") support. Some functions (such as $\cos[2\pi\xi_0 x]$) have "infinite support" because they have only isolated zeros or no zeros at all over the entire infinite domain. We will also see that functions may be represented using different sets of basis functions, and these different representations may have different regions of support in their respective domains.

Though less common in the imaging context, it also is possible to classify functions based on the locations of their "singularities" where the amplitude f is not defined. One type of singularity occurs at all coordinates x where $(f[x])^{-1}$ is zero.

The real-valued and deterministic 1-D functions introduced in Section 6.1 are defined over a 1-D domain. Section 6.2 describes the Dirac delta function, which is a unique 1-D function that has infinitesimal support and finite area. A few special functions with complex-valued amplitudes will be introduced in Section 6.3, while the 1-D stochastic functions are considered in Section 6.4. Multidimensional functions (functions with M-D domains where $M \geq 2$) are introduced beginning in Chapter 7.

6.1 Definitions of 1-D Special Functions

Many functions (e.g., *RECT*$[x]$, *TRI*$[x]$, *SGN*$[x]$) are defined by specifying the amplitude f at a coordinate by a "rule" or mathematical relation. These functions often are scaled and/or translated to create new functions. For example, the amplitude of the function at the origin of coordinates may be translated to $x = x_0$ by subtracting x_0 from the independent variable, producing the form $x - x_0$, as shown in Figure 6.1a. The "width" of these functions may be varied by scaling the independent variable x by a real-valued factor b to evaluate $f[x/b]$. For illustration, consider a function $f[x]$ with compact support such that $f[x] = 0$ for $|x| > 1$. The resulting scaled function $f[x/b] = 0$ for $|x| > b$, and so the scaled function is "wider" if $|b| > 1$ and "narrower" if $|b| < 1$. The process is illustrated in Figure 6.1b. If the original $f[x]$ has finite support, then b is the scale factor for both that support interval and of the area of the function. If the function has infinite support, then b is the scale factor of separation distances between specific features in the amplitude of the function (e.g., locations of zeros). Unless noted, the special functions with finite support are normalized so that their areas are identical to this scaling parameter b. Both scaling and translation may be applied to a function at one time by applying the general argument $(x - x_0)/b$, as shown in Figure 6.1c.

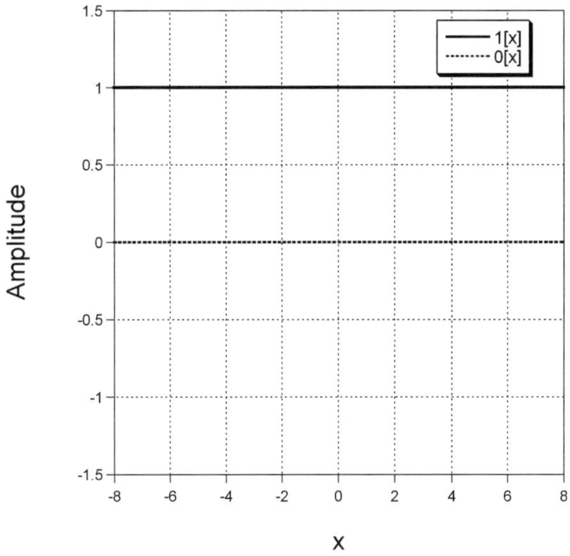

Figure 6.2 Constant functions $1[x]$ and $0[x]$.

A few of the functions are defined by a single mathematical expression that applies to the entire domain. Examples include the sinusoids, those functions based upon them, and the Gaussian function $\exp[-\pi x^2]$. These functions may be legitimately scaled by complex-valued width parameters to produce functions with complex-valued amplitudes.

6.1.1 Constant Function

Two 1-D constant-amplitude functions may seem unnecessarily trivial, but in fact will be useful as placeholders and as actual functions in several contexts. These are the unit constant:

$$1[x] \equiv 1 \quad \textit{for all } x \tag{6.1a}$$

and the null constant:

$$0[x] \equiv 0 \quad \textit{for all } x \tag{6.1b}$$

The unit constant is shown in Figure 6.2 and obviously has infinite support and infinite area. Both the support and the area of the null constant $0[x]$ are zero. A constant function with any desired complex-valued amplitude is generated trivially by multiplying $1[x]$ by that complex constant.

Obviously, any translation or scaling applied to the argument of any constant function has no effect on the amplitude at any coordinate:

$$1\left[\frac{x-x_0}{b}\right] = 1[x] = 1 \quad \textit{if } b \neq 0 \tag{6.2a}$$

$$0\left[\frac{x-x_0}{b}\right] = 0[x] = 0 \quad \textit{if} b \neq 0 \tag{6.2b}$$

6.1.2 Rectangle Function

The 1-D rectangle is useful in many imaging situations, such as to multiply (modulate) another function with wider support. For example, this operation might represent the effect of the finite field of view of

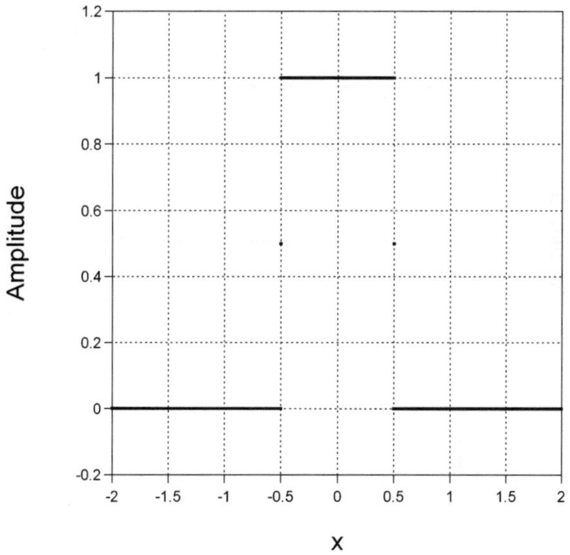

Figure 6.3 Rectangle function $RECT[x]$, which has unit support and unit area.

an imaging system on an object. As defined by Gaskill (1978), the rectangle function has unit amplitude within its finite support except at the endpoints, where the amplitude is the average of the neighboring amplitudes (Figure 6.3):

$$RECT[x] \equiv \begin{cases} 1 & \text{if } |x| < \dfrac{1}{2} \\ \dfrac{1}{2} & \text{if } |x| = \dfrac{1}{2} \\ 0 & \text{if } |x| > \dfrac{1}{2} \end{cases} \tag{6.3}$$

$RECT[x]$ clearly is symmetric (even) with respect to the origin of coordinates and has a discrete range. Take special note of the half-unit amplitudes defined at the "edges" of the support; Bracewell (1986a) asserts that these explicit edge amplitudes are "almost never important" because they contribute nothing to any integral of the *RECT* function modulated by a finite-support function. Though this statement is true and useful in some contexts, the endpoint amplitudes are *very* important when $RECT[x]$ is sampled to construct a discrete function, as will be considered in Chapter 14. For this reason, we retain the explicit endpoint amplitudes in the definition.

The more general form of the *RECT* includes parameters for the location of the center of symmetry x_0 and of the real-valued width b so that the argument is $(x - x_0)/b$.

$$RECT\left[\frac{x - x_0}{b}\right] \equiv \begin{cases} 1 & \text{if } |x - x_0| < \dfrac{|b|}{2} \text{ if } b \neq 0 \\ \dfrac{1}{2} & \text{if } |x - x_0| = \dfrac{|b|}{2} \implies x = x_0 \pm \dfrac{|b|}{2} \\ 0 & \text{if } |x - x_0| > \dfrac{|b|}{2} \end{cases} \tag{6.4}$$

Note that both the width of the region of support and the area of this function are equal to $|b|$. Because $RECT[x]$ is even, the algebraic sign of the scale parameter b does not affect the amplitude of the function, which means that $RECT[x - x_0] = RECT[x_0 - x]$.

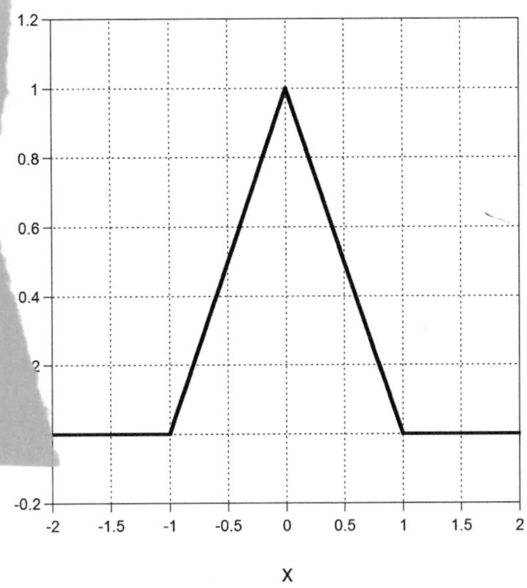

Figure 6.4 Triangle function *TRI*[*x*].

6.1.3 Triangle Function

Like *RECT*[*x*], the triangle function *TRI*[*x*] has unit amplitude at the origin and unit area, but its amplitude decreases from the maximum at a constant rate until reaching zero at $x = \pm 1$:

$$TRI[x] \equiv \begin{Bmatrix} 1-|x| & \textit{if } |x| < 1 \\ 0 & \textit{if } |x| \geq 1 \end{Bmatrix} = (1-|x|)\; RECT\left[\frac{x}{2}\right] \tag{6.5}$$

As shown in Figure 6.4, the support of *TRI*[*x*] is two units and the slopes are $+1$ for $-1 \leq x \leq 0$ and -1 for $0 \leq x \leq 1$. The more general expression for the triangle centered at $x = x_0$ with real-valued scale parameter b is:

$$\begin{aligned} TRI\left[\frac{x-x_0}{b}\right] &\equiv \begin{Bmatrix} 1-|(x-x_0)/b| & \textit{if } |x-x_0| < |b| \\ 0 & \textit{if } |x-x_0| \geq |b| \end{Bmatrix} \\ &= \left(1 - \frac{|x-x_0|}{b}\right) RECT\left[\frac{x-x_0}{2b}\right] \end{aligned} \tag{6.6}$$

The support of this scaled triangle is $2 \cdot |b|$ and its area is $|b|$. Like *RECT*[*x*], *TRI*[*x*] is often used to modulate other functions in imaging situations. For example, it is particularly useful as an apodizing (or "multiplicative weighting") function when modeling optical systems that are used in natural light.

6.1.4 Signum Function

The "signum" function (pronounced *sig′num)* assigns the numerical values 0 and ± 1 to the dependent variable based on the algebraic sign of the argument. Its name is intended to eliminate confusion of the homonyms "sign" and "sine". The definition explicitly includes a null value at the origin, which is the average of the adjacent amplitudes, just as in the definition of *RECT*[*x*]:

$$SGN[x] \equiv \begin{cases} 1 & \textit{if } x > 0 \\ 0 & \textit{if } x = 0 \\ -1 & \textit{if } x < 0 \end{cases} \tag{6.7}$$

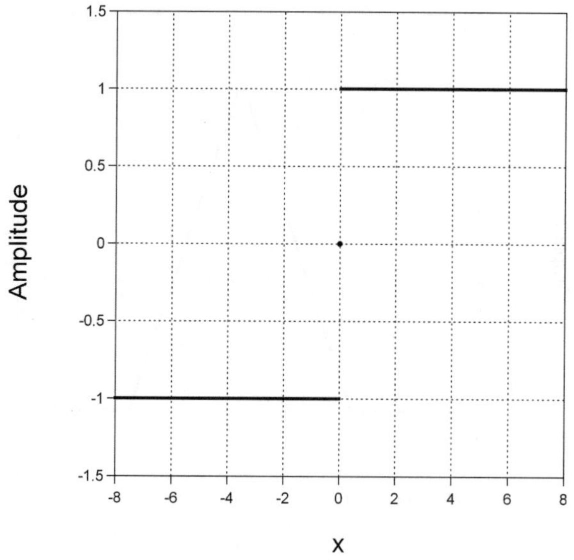

Figure 6.5 Signum function $SGN[x]$.

Again, Bracewell does not define $SGN[0] = 0$ for the same reason that he does not include the amplitudes at the endpoints of $RECT[x]$, but this amplitude is important when sampling $SGN[x]$. The function is illustrated in Figure 6.5.

As before, the abscissa where the amplitude transition occurs may be translated by inserting an additive constant factor into the argument:

$$SGN[x - x_0] \equiv \begin{cases} 1 & \text{if } x - x_0 > 0 \Longrightarrow x > x_0 \\ 0 & \text{if } x - x_0 = 0 \Longrightarrow x = x_0 \\ -1 & \text{if } x - x_0 < 0 \Longrightarrow x < x_0 \end{cases} \tag{6.8}$$

Because it has infinite support and a distinctive amplitude "feature" only at its center, the only effect of the scale factor b on the signum function is to reverse its form if $b < 0$:

$$SGN\left[\frac{x - x_0}{b}\right] = -SGN[x - x_0] \quad \text{if } b < 0 \tag{6.9}$$

6.1.5 Step Function

The *STEP* function resembles $SGN[x]$ in that its amplitude "switches" between its two extrema at the origin, but the amplitude for negative x is specified as 0, and the amplitude at the origin again is the average of the adjacent amplitudes:

$$STEP[x] \equiv \begin{cases} 1 & \text{if } x > 0 \\ \frac{1}{2} & \text{if } x = 0 \\ 0 & \text{if } x < 0 \end{cases} \tag{6.10}$$

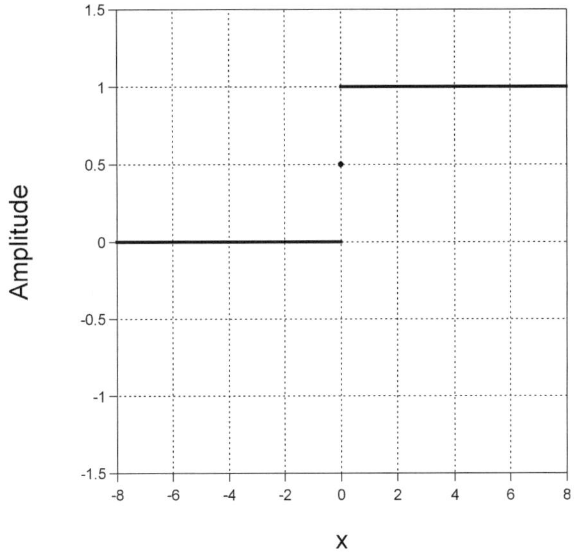

Figure 6.6 *STEP* function *STEP*$[x]$.

The more general *STEP* function is centered about the arbitrary coordinate x_0:

$$STEP[x - x_0] \equiv \begin{cases} 1 & if\ x - x_0 > 0 \Longrightarrow x > x_0 \\ \frac{1}{2} & if\ x - x_0 = 0 \Longrightarrow x = x_0 \\ 0 & if\ x - x_0 < 0 \Longrightarrow x < x_0 \end{cases} \tag{6.11}$$

The graph is shown in Figure 6.6. Again, Bracewell ignores the amplitude of *STEP*[0], and assigns the notation $H[x]$ for this "Heaviside unit step". We retain the name *STEP*$[x]$ following Gaskill to avoid confusion with the usage of H to represent the transfer function of a shift-invariant imaging system that will be introduced in Chapter 8.

The *STEP* function has semi-infinite support and is conveniently expressed in terms of its even and odd parts:

$$STEP[x] = \frac{1}{2}(1[x] + SGN[x]) \tag{6.12}$$

Just as was true for *SGN*$[x]$, the only impact of a scale factor b in the argument is to reverse the function if $b < 0$. The area of *STEP*$[x]$ is positive and infinite.

Recall that it is sometimes convenient to represent *RECT*$[x]$ as the difference of two *STEP* functions, as shown in Figure 6.7:

$$\begin{aligned} RECT[x] &= STEP\left[x - \left(-\frac{1}{2}\right)\right] - STEP\left[x - \left(+\frac{1}{2}\right)\right] \\ &= STEP\left[x + \frac{1}{2}\right] - STEP\left[x - \frac{1}{2}\right] \end{aligned} \tag{6.13a}$$

$$RECT\left[\frac{x}{b}\right] = STEP\left[x + \frac{b}{2}\right] - STEP\left[x - \frac{b}{2}\right] \tag{6.13b}$$

The *STEP* function in the time domain is often used to modulate other functions to produce a result that has null amplitude for $t < 0$. Such a modulated function may be used to describe a *causal* phenomenon.

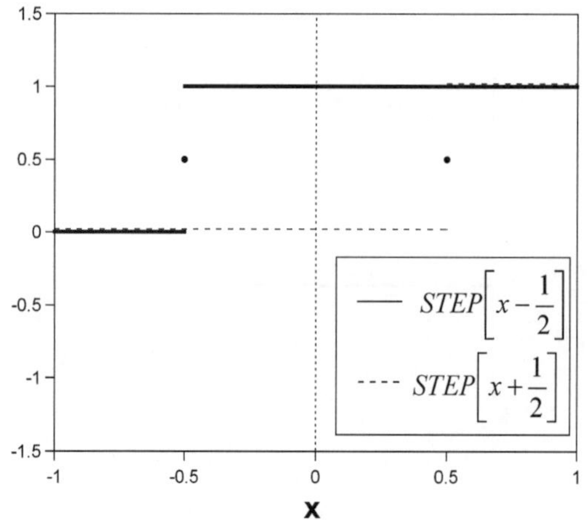

Figure 6.7 *RECT*[*x*] constructed from the difference of two translated *STEP* functions, $RECT[x] = STEP[x + \frac{1}{2}] - STEP[x - \frac{1}{2}]$.

6.1.6 Exponential Function

Other than perhaps the sinusoid, the 1-D function that arguably appears most often in physical and engineering problems is the exponential:

$$f_1[x] = e^{+x} \tag{6.14a}$$

which evaluates to zero at $x = -\infty$, to unity at the origin, and increases without limit for increasing positive x. Its reversed replica:

$$f_2[x] = e^{+(-x)} = e^{-x} \tag{6.14b}$$

has the same unit amplitude at the origin and both have infinite area:

$$\int_{-\infty}^{+\infty} e^{+x}\, dx = \int_{-\infty}^{+\infty} e^{-x}\, dx = +\infty \tag{6.15}$$

The second function whose amplitude "decays" with increasing x is of more interest to us and often appears in physical problems after modulation by the *STEP* function:

$$f_2[x] = e^{-x} \cdot STEP[x] \tag{6.16}$$

The resulting function (Figure 6.8b) has "semi-infinite" support and unit area:

$$\int_{-\infty}^{+\infty} e^{-x}\, STEP[x]\, dx = \int_{0}^{+\infty} e^{-x}\, dx = -\, e^{-x}\, \Big|_{x=0}^{x=+\infty} = +1 \tag{6.17}$$

The multiplication by *STEP*[*x*] ensures that $f[0] = \frac{1}{2}$ in Equation (6.16). Its amplitude at $x = +1$ is $e^{-1} \cong 0.367$. This "windowed" exponential function expressed in the time domain is useful when modeling dynamic systems that lose energy or amplitude due to losses such as friction.

Because it is the product of two functions that themselves have concise names, no special terminology is assigned to the function in Equation (6.16).

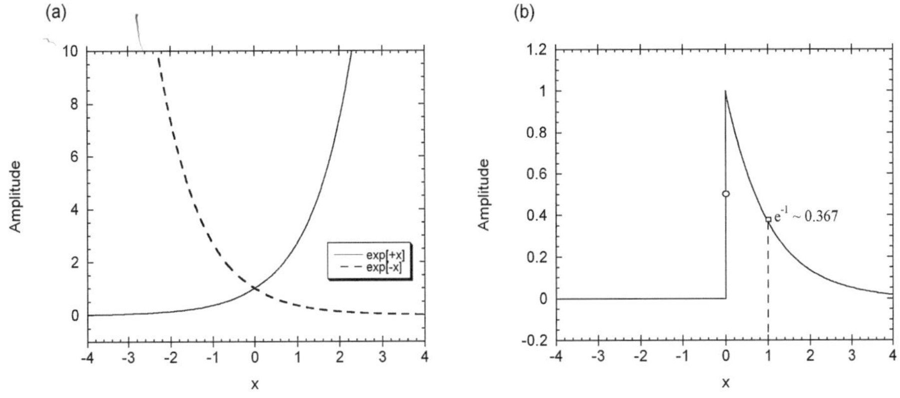

Figure 6.8 (a) Exponential function $\exp[+x]$ and its reversed replica $\exp[-x]$; both have infinite support and infinite area. (b) Product of decaying exponential function and STEP function $f[x] = \exp[-x] \cdot STEP[x]$, which has infinite support and unit area.

6.1.7 Sinusoid

The sinusoid arguably is the most pervasive function in the physical sciences. Its mathematical form may be derived in any of several ways, but one of the more common is as the solution to a specific second-order differential equation, the *wave equation*:

$$\frac{d^2}{dx^2} f[x] + \alpha^2 f[x] = 0 \tag{6.18}$$

This derivation is presented in many elementary physics texts. The sinusoid appears in classical mechanics as the projection of the endpoint of a vector that rotates about the origin at a uniform rate. The distance from the tip of the vector to the origin projected onto a line perpendicular to the distance of closest approach is a sinusoidal function of time (Figure 6.9). It is often more convenient to define the form of the general sinusoid in terms of the cosine function, which is symmetric with respect to the origin. As already introduced in Section 2.2, the equation for the general sinusoidal function in the space domain is:

$$A_0 \cos[\Phi[x]] = A_0 \cos\left[\frac{2\pi x}{X_0} + \phi_0\right] = A_0 \cos[2\pi\xi_0 x + \phi_0] \tag{6.19}$$

This also may be rewritten in terms of the complex exponential function via Equations (4.37) and (4.38) as:

$$A_0 \cos[2\pi\xi_0 x + \phi_0] = \frac{A_0}{2}(e^{+i(2\pi\xi_0 x+\phi_0)} + e^{-i(2\pi\xi_0 x+\phi_0)}) \tag{6.20}$$

The proportionality constant of the linear component of the phase in Equation (6.19) is $2\pi/X_0$, which is the number of radians traversed by the argument per unit distance, and thus is identical to the "angular spatial frequency" of the sinusoid that was introduced in Equation (2.14a). The reciprocal of the period X_0 (denoted by ξ_0) describes the number of cycles traversed by the sinusoid in unit distance. Also as considered in Equation (2.14b), ξ_0 is called the *spatial frequency*. Though the term *cycle* is often used to describe a single period of any periodic function, we constrain that usage to sinusoidal functions only; analogous terms may be defined for other periodic functions, such as "line pairs per mm" for square waves.

The other relevant parameter of Equation (6.19) is the amplitude A_0, which is measured in units appropriate to the problem. It is evident that the cosine may be computed for any real-valued argument,

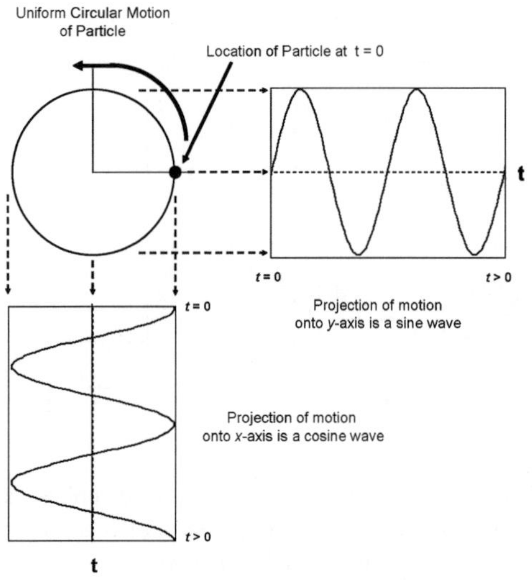

Figure 6.9 Projection of uniform circular motion onto orthogonal axes generates two sinusoidal functions in quadrature.

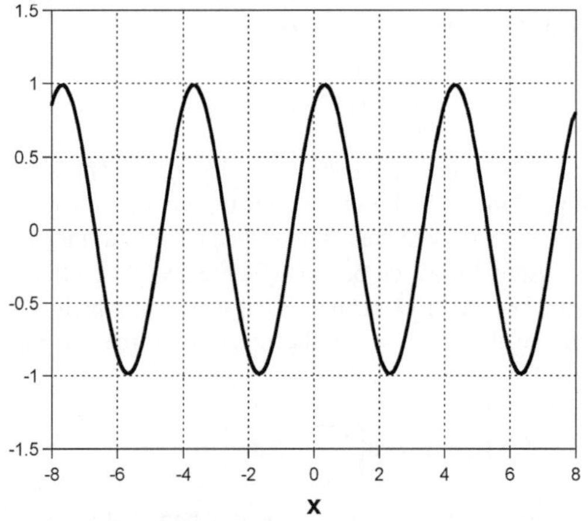

Figure 6.10 Sinusoidal function $\cos[2\pi\xi_0 x + \phi_0]$ with $\xi_0 = \frac{1}{4}$ cycle per unit length and $\phi_0 = -\pi/6$ radians.

including nonlinear functions of x, but the resulting function is not a harmonic sinusoid. Examples of such higher-order sinusoids are introduced later in this chapter.

If the initial phase ϕ_0 is an integer multiple of π radians, then the sinusoid in Equation (6.19) is the symmetric cosine. If $\phi_0 = \pm\pi/2$ radians, the resulting function is the odd sinusoid $\mp \sin[2\pi\xi_0 x]$. An example of the linear-phase sinusoid is shown in Figure 6.10. Note the significant potential for confusion due to the two uses of the term *phase* when dealing with sinusoids: the sinusoidal phase $\Phi[x]$ is the angle of rotation of the phasor about the origin, while the phase of a complex-valued function was defined in Equation (4.14) to be the inverse tangent of the ratio of the imaginary and real parts.

Equation (2.14b) may be applied to demonstrate that the spatial frequency is constant when the phase is at most a linear function of the independent variable x:

$$\xi = \frac{1}{2\pi}\frac{\partial \Phi[x]}{\partial x} = \frac{1}{2\pi}\frac{\partial}{\partial x}[2\pi\xi_0 x + \phi_0] = \left(\frac{1}{2\pi}\right)\cdot 2\pi\xi_0 = \xi_0 \tag{6.21}$$

Obviously, this condition may be used to define "harmonic" functions.

From Figure 6.10, it is evident that the positive and negative lobes of the sinusoid have equal areas of opposite sign; this leads to the observation that the area of a harmonic sinusoid is zero:

$$\int_{-\infty}^{+\infty} \cos[2\pi\xi_0 x + \phi_0]\,dx = 0 \quad \textit{if } \xi_0 \neq 0 \tag{6.22}$$

The validity of this result is often doubted by students newly exposed to the concept, based on the argument that the area will oscillate about zero as additional positive and negative lobes are included in the integral as the limits approach infinity. The area of the sinusoid also will be shown to be zero beyond doubt in Chapter 9 by applying an easily proven theorem of the Fourier transform.

Sinusoidal functions often are applied as modulations (multipliers) of other functions with compact support; this is particularly common in electrical engineering. For example, consider the product of an odd sinusoidal wave of the form in Equation (6.19) with a decaying exponential in Equation (6.16):

$$f[x] = \left(\cos\left[2\pi\xi_0 x - \frac{\pi}{2}\right]\right)\cdot(e^{-x}\cdot STEP[x]) = \sin[2\pi\xi_0 x]\cdot(e^{-x}\cdot STEP[x]) \tag{6.23}$$

The graph of this function is shown in Figure 6.11. The odd sinusoid is chosen as the modulating function because its amplitude is zero at the origin and increases "slowly" with increasing x; if $\cos[2\pi\xi_0 x]$ had been used instead, the amplitude of the product function would have exhibited a discontinuous transition at the origin. This function models the behavior of a damped harmonic oscillator in classical mechanics, and it and its relatives also appear frequently in other areas of physics. The behavior and uses of this function will be considered in some detail in Chapter 18.

The squared magnitude of a sinusoidal function also is called its *spatial power* or *intensity* and is guaranteed to be nonnegative. It may be recast into the sum of a constant part and a sinusoidal part by squaring the complex-exponential expression for the cosine from Equation (4.37):

$$\cos^2[2\pi\xi_0 x] = \left(\frac{e^{+2\pi i\xi_0 x} + e^{-2\pi i\xi_0 x}}{2}\right)^2 = \frac{1}{4}(2 + e^{4\pi i\xi_0 x} + e^{-4\pi i\xi_0 x})$$

$$= \frac{1}{2} + \frac{1}{2}\cos[4\pi\xi_0 x] = \frac{1}{2}(1 + \cos[2\pi(2\xi_0)x]) \tag{6.24}$$

In words, $\cos^2[2\pi\xi_0 x]$ is equivalent to the sum of a half-unit additive constant and a cosine function whose spatial frequency and amplitude are twice as large and half as large, respectively. The additive constant is called a *bias* applied to the doubled-frequency cosine, and which ensures that the amplitude is nonnegative in this case. The relative sizes of the maxima and minima of a nonnegative sinusoidal function may be combined to create a metric of the function that is used in optics and image processing. Assume that $f[x]$ is a sinusoid that has been biased so that its amplitude is nonnegative everywhere. The maximum and minimum amplitudes are called f_{max} and f_{min}, respectively. The numerical value

$$m_f \equiv \frac{f_{\text{max}} - f_{\text{min}}}{f_{\text{max}} + f_{\text{min}}} \tag{6.25}$$

is real valued and is called the *modulation* of the biased sinusoid. The range of allowed values of m_f for nonnegative sinusoids is $0 \le m_f \le 1$. Obviously, $m_f = 1$ for all sinusoids with $f_{\text{min}} = 0$, while $m_f < 1$ if $f_{\text{min}} > 0$. The modulation is zero if the amplitude of the sinusoid is zero, i.e., when $f_{\text{max}} = f_{\text{min}}$. From this definition, we see that the modulation of $\cos^2[2\pi\xi_0 x]$ in Equation (6.24) is unity; if

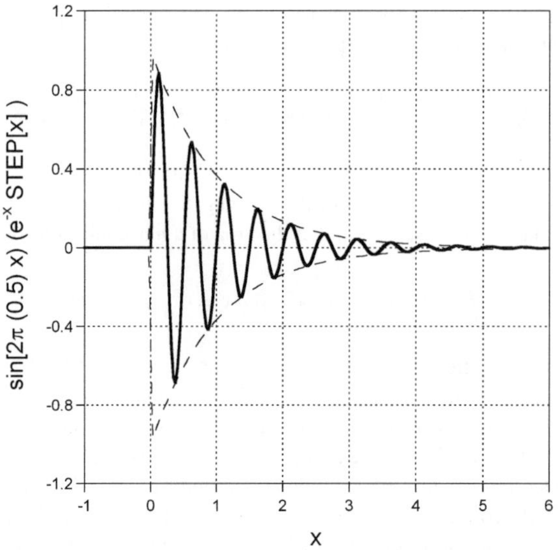

Figure 6.11 Modulation of a sinusoid by a decaying exponential: $f[x] = \cos[2\pi x/2] \cdot (\exp[-x] \cdot STEP[x])$.

an additional bias were added to Equation (6.24), the resulting modulation would be less than unity. Consider the general nonnegative sinusoid with an additive bias:

$$f[x] = A_0 + A_1 \cos[2\pi\xi_0 x + \phi_0] \tag{6.26}$$

where $A_0 \geq A_1$. The modulation of this function is obtained by simple substitution:

$$m_f = \frac{(A_0 + A_1) - (A_0 - A_1)}{(A_0 + A_1) + (A_0 - A_1)} = \frac{A_1}{A_0} \tag{6.27}$$

In words, the modulation of a nonnegative sinusoid is the ratio of the amplitude A_1 to the bias A_0, and provides a measure of the relative "brightnesses" of the maxima and minima. If m_f is large, then the difference between the maxima and minima is large compared to the bias and the sinusoidal pattern should be readily apparent to an observer. Examples of biased sinusoids with different modulations are presented in Figure 6.12.

That two different definitions of the term *modulation* have been introduced means that it will be quite possible to confuse them in any specific situation. In the following discussions, the appropriate meaning should be clear from the context. If only one function is considered and that function is sinusoidal, then modulation refers to the metric in Equation (6.25). To add further risk of confusion, other authors have used still other terms. Michelson called m_f the *visibility* of the sinusoid to relate to the ability of the human visual system to recognize the existence of sinusoidal variations in intensity. Still other authors have used the term *contrast*, though we reserve that name to refer to nonnegative square-wave signals rather than sinusoids. Use of "modulation" retains the relationship to the entrenched imaging terminology for the *modulation transfer function* (or MTF), which will be considered in detail later in the book. The MTF describes the effect of the system upon the modulation of a nonnegative sinusoidal input. The analogous relationship for square-wave signals will be called the *contrast transfer function* (CTF).

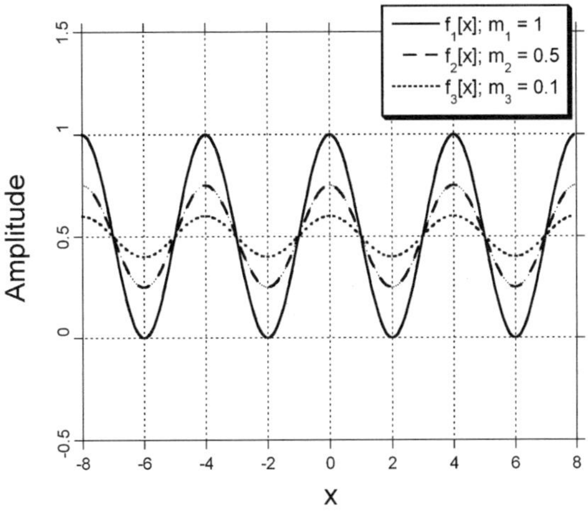

Figure 6.12 Nonnegative sinusoidal functions with the same bias but different modulations: $f_1[x] = 0.5 + 0.5 \cdot \cos[2\pi x/4]$, with $m_1 = 1$; $f_2[x] = 0.5 + 0.25 \cdot \cos[2\pi x/4]$ with $m_2 = 0.5$; $f_3[x] = 0.5 + 0.1 \cdot \cos[2\pi x/4]$ with $m_3 = 0.1$.

6.1.8 *SINC* Function

The *SINC* function is the product of the odd sinusoid with $\xi_0 = \frac{1}{2}$ cycles per unit and $\phi_0 = -\pi/2$ radians with the odd function $(\pi x)^{-1}$. The amplitude of the second function is undefined at the origin and approaches zero as $|x| \to +\infty$. The resulting function is:

$$SINC[x] \equiv \frac{\sin[\pi x]}{\pi x} = \frac{\cos[2\pi x/2 - \pi/2]}{\pi x} \tag{6.28}$$

Some authors define the *SINC* function without the factors of π, but we include them because the amplitude of the *SINC* conveniently vanishes at all nonzero integer values of x: $SINC[x] = 0$ for $x = \pm 1, \pm 2, \ldots$. Because both numerator and denominator are zero at the origin, the amplitude of $SINC[0]$ may be determined from the ratio of the derivatives via L'Hôpital's rule:

$$SINC[0] = \frac{\lim_{x\to 0}\{(d/dx)\sin[\pi x]\}}{\lim_{x\to 0}\{(d/dx)[\pi x]\}} = \frac{\pi \cos[0]}{\pi} = 1 \tag{6.29}$$

$SINC[x]$ is plotted in Figure 6.13 as a real function and as magnitude and phase. Between the zeros, the magnitude has a local extremum in the vicinity of (though not coincident with) the half-integer values of x. The first few local extrema of the magnitudes are located at approximately:

$$\begin{aligned} SINC[\pm 1.428] &\cong -0.2172 \\ SINC[\pm 2.459] &\cong +0.1284 \\ SINC[\pm 3.471] &\cong -0.091\,32 \end{aligned} \tag{6.30}$$

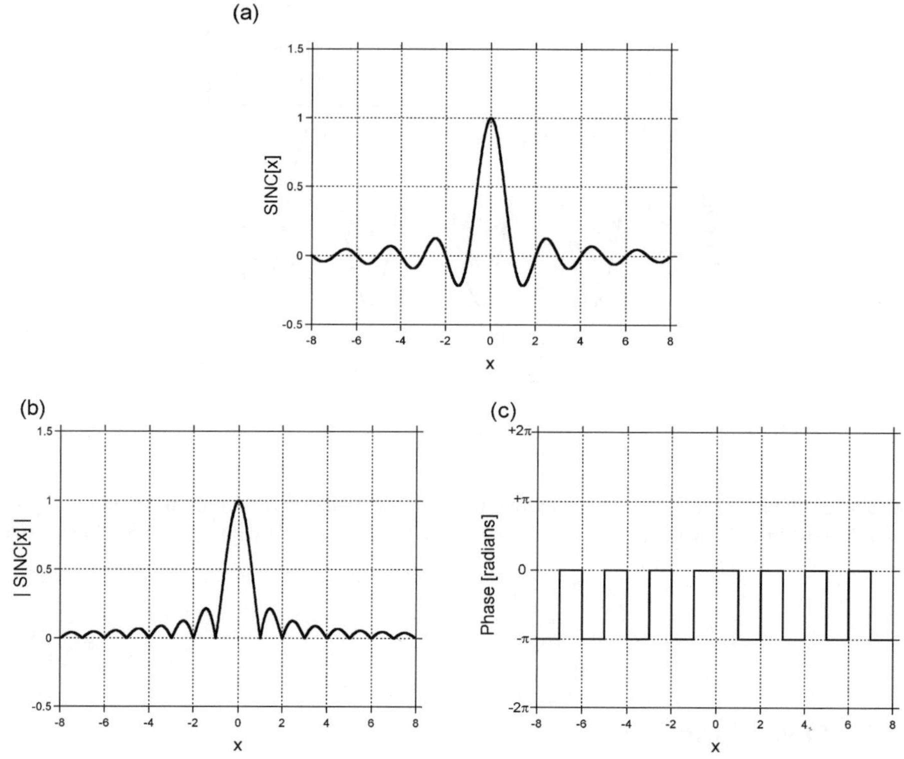

Figure 6.13 $SINC[x] \equiv \sin[\pi x]/\pi x$: (a) real-valued function; (b) magnitude $|SINC[x]|$, showing the "cusps" at the integer values of x other than 0; (c) phase, showing values of 0 and $-\pi$.

Note that the extrema are located near (but not *at*) half-integer values of the argument, where the amplitudes are:

$$\begin{aligned}
SINC\left[\frac{1}{2}\right] &= \frac{2}{\pi} \cong 0.6366 \\
SINC\left[\frac{3}{2}\right] &= -\frac{2}{3\pi} \cong -0.2122 \\
SINC\left[\frac{5}{2}\right] &= +\frac{2}{5\pi} \cong 0.1273 \\
SINC\left[\frac{7}{2}\right] &= -\frac{2}{7\pi} \cong -0.090\,95 \\
&\vdots \\
SINC\left[\frac{m}{2}\right] &= (-1)^m \frac{2}{(2m+1)\pi}
\end{aligned} \tag{6.31}$$

The phase of the real-valued *SINC* is 0 or $-\pi$ radians where the function is positive (or zero) and negative, respectively.

The *SINC* function also may be written as a series expansion based upon Equation (4.18b):

$$\frac{\sin[\pi x]}{\pi x} = \frac{1}{\pi x}\left(\frac{(\pi x)^1}{1!} - \frac{(\pi x)^3}{3!} + \frac{(\pi x)^5}{5!} - \cdots\right)$$

$$= \left(1 - \left(\frac{\pi^2}{6}\right)x^2 + \left(\frac{\pi^4}{120}\right)x^4 - \left(\frac{\pi^6}{5040}\right)x^6 + \cdots\right)$$

$$= \sum_{n=0}^{+\infty}(-1)^n \frac{(\pi x)^{2n}}{(2n+1)!} \tag{6.32}$$

This series also demonstrates that $SINC[0] = 1$. Note the denominators of the sequence of coefficients: we will have occasion to compare these to series expressions for other functions later in this chapter.

The area of $SINC[x]$ may be determined rigorously by evaluating the appropriate path integral in the complex plane (Silverman 1984, and Appendix 6A). However, a much simpler method uses the central-ordinate theorem of the Fourier transform, which will be proven in Chapter 9. At this point, we merely state without proof that the area of the unscaled *SINC* function is unity:

$$\int_{-\infty}^{+\infty} SINC[x]\, dx = 1 \tag{6.33}$$

The form of the *SINC* function after translation and scaling by real-valued factors obviously is:

$$SINC\left[\frac{x - x_0}{b}\right] = \frac{\sin[\pi[(x - x_0)/b]]}{[\pi[(x - x_0)/b]]} \tag{6.34}$$

and its amplitude vanishes at $x = n|b| + x_0$ $(n = \pm 1, \pm 2, \ldots)$. The area of the scaled *SINC* function is $|b|$ regardless of the translation x_0. Note that the support of $SINC[x/b]$ is infinite, and that its magnitude nowhere exceeds x^{-1}.

6.1.9 $SINC^2$ Function

This function is an obvious extension of Equation (6.28) and is shown in Figure 6.14. It also appears in many imaging contexts, including the description of the action of optical imaging systems constructed from optics with rectangular cross-sections and used in natural (incoherent) light. Obviously the zeros of both $SINC^2[x]$ and $SINC[x]$ occur at the same locations ($x = \pm n$ for $n \neq 0$), but note from its graph that $SINC^2[x]$ varies "smoothly" in the vicinity of its zeros (novices often visualize cusps in $SINC^2[x]$ similar to those of $|SINC[x]|$). Because $SINC^2[x]$ is nonnegative everywhere, the discussion following Equation (4.14) demonstrates that its phase angle is 0 radians for all x. Like $SINC[x]$, the area of $SINC^2[x]$ may be computed either by evaluating the appropriate integral in the complex plane or by applying the central-ordinate theorem from Chapter 9. The former method requires more familiarity with complex analysis than has been assumed herein, while the latter is not yet proven. Therefore, we state at this point, again without proof (yet), the possibly surprising result that the areas of $SINC^2[x]$ and $SINC[x]$ are identically unity:

$$\int_{-\infty}^{+\infty} SINC^2[x]\, dx = \int_{-\infty}^{+\infty} SINC[x]\, dx = 1 \tag{6.35}$$

The integrals are evaluated rigorously via Cauchy's theorem in the complex plane in Appendix 6.5. It is evident that the areas of $SINC^2[x/b]$ and $SINC[x/b]$ also must be identical:

$$\int_{-\infty}^{+\infty} SINC^2\left[\frac{x}{b}\right] dx = \int_{-\infty}^{+\infty} SINC\left[\frac{x}{b}\right] dx = |b| \tag{6.36}$$

The generalization to the shifted and scaled version $SINC^2[(x - x_0)/b]$ is obvious from Equation (6.34).

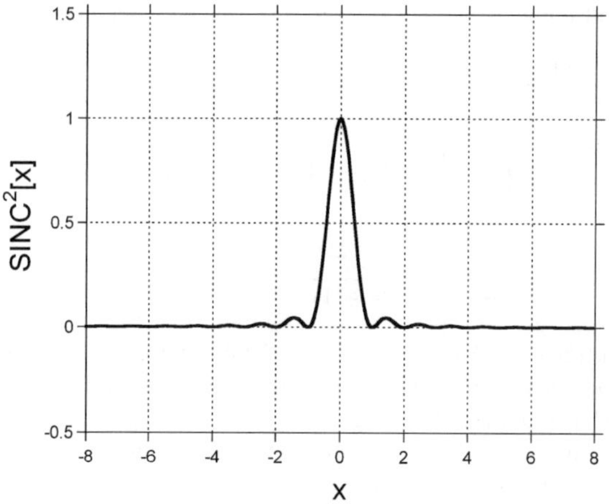

Figure 6.14 $SINC^2[x]$, showing its nonnegative character. Though not obvious, the areas of $SINC[x]$ and of $SINC^2[x]$ are identically unity.

6.1.10 Gamma Function

This section may be skipped without loss of continuity.

Unlike the other special functions already considered, the gamma function $\Gamma[x]$ does not represent the form of a useful signal or serve as the descriptive function of a useful imaging system. Rather, it is a computational tool for solving imaging problems that involve other special functions. For example, $\Gamma[x]$ provides an avenue for deriving the areas and Fourier transforms of the yet-to-be-considered quadratic-phase sinusoid and Gaussian functions. This section provides only the briefest introduction to some of the relevant properties of $\Gamma[x]$. Readers who require a more detailed discussion should consult a reference on complex analysis (e.g., Arfken, 2000).

The gamma function evaluated at a particular positive argument x_0 may be defined as the area of the decaying exponential $e^{-\alpha} \cdot STEP[\alpha]$ in Equation (6.16) modulated by α^{x-1}:

$$\Gamma[x] = \int_0^{+\infty} e^{-\alpha}\, \alpha^{x-1}\, d\alpha = \int_{-\infty}^{+\infty} (e^{-\alpha} \cdot STEP[\alpha])\alpha^{x-1}\, d\alpha \tag{6.37}$$

where $x > 0$. Note that the argument x appears in the integral only in the exponential term α^{x-1}. Therefore, the gamma function evaluated at $x = 1$ is the area of $e^{-\alpha} \cdot STEP[\alpha]$, which was shown in Equation (6.17) to be unity:

$$\Gamma[1] = \int_{-\infty}^{+\infty} e^{-\alpha} \cdot STEP[\alpha]\, d\alpha = 1 \tag{6.38}$$

For a fixed positive finite value of x, $e^{-\alpha}$ decreases with increasing α while α^{x-1} increases rapidly; these combined conflicting behaviors ensure that the area in Equation (6.37) remains finite for finite and positive values of x, but the area of the product function increases rapidly with x for $x > 1$.

The integral form of the gamma function may be recast by changing the variable of integration from α to $e^{-\alpha}$:

$$\Gamma[x] = \int_0^{+\infty} (\alpha^{x-1})\, d(-\, e^{-\alpha}) \tag{6.39}$$

New variables may be assigned to allow convenient integration by parts:

$$\mathrm{u} = \alpha^{x-1} \tag{6.40a}$$

$$d\mathrm{u} = (x-1)\alpha^{x-2}\, d\alpha \tag{6.40b}$$

$$\mathrm{v} = -\, e^{-\alpha} \Longrightarrow \mathrm{v} = -1 \textit{ for } \alpha = 0 \textit{ and } \mathrm{v} = 0 \textit{ for } \alpha = +\infty \tag{6.40c}$$

$$d\mathrm{v} = +\, e^{-\alpha}\, d\alpha \tag{6.40d}$$

These are substituted into the integration-by-parts formula to obtain a recursion relation for $\Gamma[x]$ that is valid for positive values of x:

$$\Gamma[x] = \int_{\alpha=0}^{\alpha=+\infty} \mathrm{u}\, d\mathrm{v} = \mathrm{uv}|_{\alpha=0}^{\alpha=+\infty} - \int_{\alpha=0}^{\alpha=+\infty} \mathrm{v}\, d\mathrm{u}$$

$$= (\alpha^{x-1}\, e^{-\alpha})|_{\alpha=0}^{\alpha=+\infty} + (x-1)\int_{\alpha=0}^{\alpha=+\infty} \alpha^{x-2}\, e^{-\alpha}\, d\alpha = 0 + (x-1)\Gamma[x-1]$$

$$\Longrightarrow \Gamma[x] = (x-1)\Gamma[x-1] \textit{ if } x > 0 \tag{6.41a}$$

This sequence of operations may be repeated to derive a recursion relation for the values of the gamma function separated by integer increments:

$$\begin{aligned}\Gamma[x] &= (x-1)\Gamma[x-1] \\ &= (x-1)((x-2)\Gamma[x-2]) \\ &= (x-1)(x-2)(x-3)\ldots(x-\mathit{INT}[x])\Gamma[x-\mathit{INT}[x]]\end{aligned} \tag{6.41b}$$

where $\mathit{INT}[x]$ is the integer part of the positive real number x; this relationship is valid because $x - \mathit{INT}[x] \geq 0$. In the cases where the argument x is an integer n, then this expression may be simplified to the well-known expression:

$$\Gamma[n] = (n-1)(n-2)(n-3)\ldots(1)\Gamma[1] = (n-1)! \tag{6.42}$$

where Equation (6.38) has been used. This result is the source of the often-used name of the *factorial* function for $\Gamma[x]$. It follows immediately from Equation (6.42) that $\Gamma[1] = \Gamma[2] = 1$ and that $0! = 1$. The integral form of the gamma function in Equation (6.37) may be interpreted as a generalization of the definition of the "factorial" to noninteger and/or negative arguments:

$$x! \equiv \Gamma[x+1] = \int_0^{+\infty} \alpha^x\, e^{-\alpha}\, d\alpha \tag{6.43}$$

Equation (6.41) may be used to extend the domain of $\Gamma[x]$ to negative arguments. For example, $\Gamma[-\frac{1}{2}] = -2 \cdot \Gamma[+\frac{1}{2}]$. Note that the amplitude of the gamma function is indeterminate when evaluated at any nonpositive integer argument. The gamma function in the vicinity of the origin is plotted in Figure 6.15 (Bender and Orszag, 1978).

Because $\Gamma[x]$ has no zeros, its reciprocal $(\Gamma[x])^{-1}$ must have no singularities and may be expressed as a Taylor series that is valid for all x. The first few terms of this series are (Smith, 1975):

$$(\Gamma[x])^{-1} \cong 1 + 0.577\,21\,x - 0.655\,87\,x^2 - 0.042\,00\,x^3 + 0.166\,54\,x^4 + \cdots \tag{6.44}$$

Though this series may be used to compute values of $\Gamma[x]$, other expansions exist that converge more efficiently to the correct value. The derivations of the Fourier transforms of the Gaussian and of the quadratic-phase sinusoid functions in Chapter 9 will require evaluation of the gamma function for half-integer arguments (e.g., $x = \frac{1}{2}, \frac{3}{2}$, etc.). These are obtained by applying the recursion relation in Equation (6.41) to $\Gamma[\frac{1}{2}]$, which may be evaluated by recasting the integral into the form of the

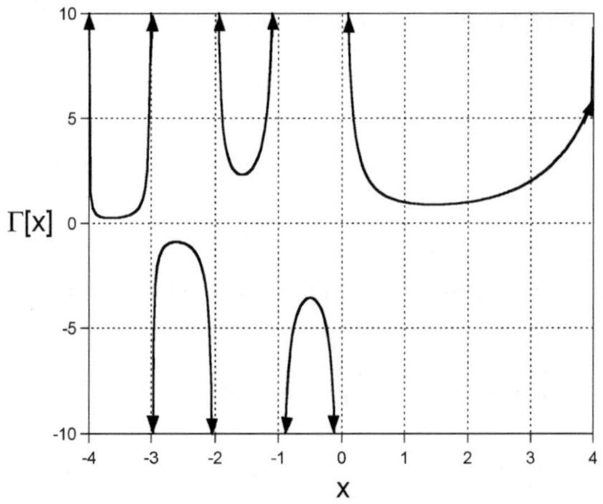

Figure 6.15 Graph of $\Gamma[x]$, showing that the amplitude increases without limit for $x \to +\infty$ and is undefined at all integer values of $x \leq 0$.

easily evaluated "error function", which is commonly used in statistics. Start by substituting $x = \frac{1}{2}$ into Equation (6.37):

$$\Gamma\left[\frac{1}{2}\right] = \int_0^{+\infty} \alpha^{(\frac{1}{2}-1)}\, e^{-\alpha}\, d\alpha = \int_0^{+\infty} \alpha^{-\frac{1}{2}}\, e^{-\alpha}\, d\alpha \tag{6.45}$$

Change the variable of integration to $\beta = \sqrt{\alpha}$ to obtain the definite integral I:

$$\begin{aligned} \Gamma\left[\frac{1}{2}\right] &= \int_{\beta=0}^{\beta=+\infty} \beta^{-1}\, e^{-\beta^2}\, 2\beta\, d\beta = 2\int_{\beta=0}^{\beta=+\infty} e^{-\beta^2}\, d\beta \\ &= \int_{\beta=-\infty}^{\beta=+\infty} e^{-\beta^2}\, d\beta \equiv I \end{aligned} \tag{6.46}$$

where the symmetry of the integrand has been used in the last step. An elegant evaluation of I is derived by constructing its square as the product of two integrals with independent variables, and then converting to polar coordinates:

$$\begin{aligned} I^2 &= \left(\int_{\beta=-\infty}^{\beta=+\infty} e^{-\beta^2}\, d\beta\right)\left(\int_{\gamma=-\infty}^{\gamma=+\infty} e^{-\gamma^2}\, d\gamma\right) \\ &= \int_{\beta=-\infty}^{\beta=+\infty}\int_{\gamma=-\infty}^{\gamma=+\infty} e^{-(\beta^2+\gamma^2)}\, d\beta\, d\gamma \\ &= \int_{\theta=-\pi}^{\theta=+\pi}\int_{\rho=0}^{\rho=+\infty} e^{-\rho^2}(\rho\, d\rho)\, d\theta \\ &= 2\pi\int_{\rho=0}^{\rho=+\infty} e^{-\rho^2}\, \rho\, d\rho \end{aligned} \tag{6.47}$$

The variable of integration is changed again, this time to $u = e^{-\rho^2}$, so that $du = -2\rho\, e^{-\rho^2}\, d\rho$:

$$I^2 = 2\pi\int_{u=1}^{u=0}\left(-\frac{1}{2}\right) du = 2\pi\left(\frac{1}{2}\right) = \pi \tag{6.48}$$

Therefore, the required result is:

$$\Gamma\left[\frac{1}{2}\right] = \sqrt{I^2} = \sqrt{\pi} \cong 1.7725 \tag{6.49}$$

The gamma function is evaluated for other half-integral values by applying the recursion relation in Equation (6.41):

$$\Gamma\left[\frac{5}{2}\right] = \frac{3}{2}\Gamma\left[\frac{3}{2}\right] = \frac{3\sqrt{\pi}}{4} \cong 1.3293 \tag{6.50a}$$

$$\Gamma\left[\frac{3}{2}\right] = \frac{1}{2}\Gamma\left[\frac{1}{2}\right] = \frac{\sqrt{\pi}}{2} \cong 0.8862 \tag{6.50b}$$

$$\Gamma\left[\frac{1}{2}\right] = -\frac{1}{2}\Gamma\left[-\frac{1}{2}\right] \Longrightarrow \Gamma\left[-\frac{1}{2}\right] = -2\Gamma\left[+\frac{1}{2}\right] = -2\sqrt{\pi} \cong -3.5449 \tag{6.50c}$$

and so forth. Again, note the growth in the amplitude of $\Gamma[x]$ for increasing values of $x > 1$.

The values of the gamma function of the positive "reciprocal integers" ($x = n^{-1}$, where $n = 1, 2, 3, \ldots$) may be evaluated by applying the series expansion in Equation (6.44); the first five examples are:

$$\Gamma\left[\frac{1}{1}\right] = 0! = 1 \tag{6.51a}$$

$$\Gamma\left[\frac{1}{2}\right] = \sqrt{\pi} \cong 1.7725 \tag{6.51b}$$

$$\Gamma\left[\frac{1}{3}\right] \cong 2.6789 \tag{6.51c}$$

$$\Gamma\left[\frac{1}{4}\right] \cong 3.6256 \tag{6.51d}$$

$$\Gamma\left[\frac{1}{5}\right] \cong 4.5908 \tag{6.51e}$$

Note that the amplitude of $\Gamma[1/n]$ increases as n increases. The amplitude of $\Gamma[1/n]$ in the limit of large n may be evaluated by applying Equation (6.41):

$$\lim_{n\to\infty}\left\{\frac{1}{n}\Gamma\left[\frac{1}{n}\right]\right\} = \lim_{n\to\infty}\left\{\Gamma\left[1+\frac{1}{n}\right]\right\} \Longrightarrow \Gamma[1] = 1 \Longrightarrow \lim_{n\to\infty}\left\{\Gamma\left[\frac{1}{n}\right]\right\} = n \tag{6.52}$$

These values are useful when considering the properties of the Fourier transforms of the "superGaussian" and "superchirp" functions, which will be introduced later in this chapter.

6.1.11 Quadratic-Phase Sinusoid – "Chirp" Function

The very name of "quadratic-phase sinusoid" indicates that this is a sinusoidal function of x^2. The requirement that the phase must be a dimensionless quantity (measured in radians) ensures that the argument must include other factors with aggregate dimensions of length^{-2}. One possible notation for the quadratic-phase sinusoid is:

$$f[x] = A\cos[\Phi[x]] = A\cos\left[\frac{\pi x^2}{\alpha^2} + \phi_0\right] \tag{6.53}$$

where the scale parameter α has dimensions of length. It is evident from this form of the phase that α specifies the particular coordinate where the phase differs from the phase ϕ_0 at the origin by π radians.

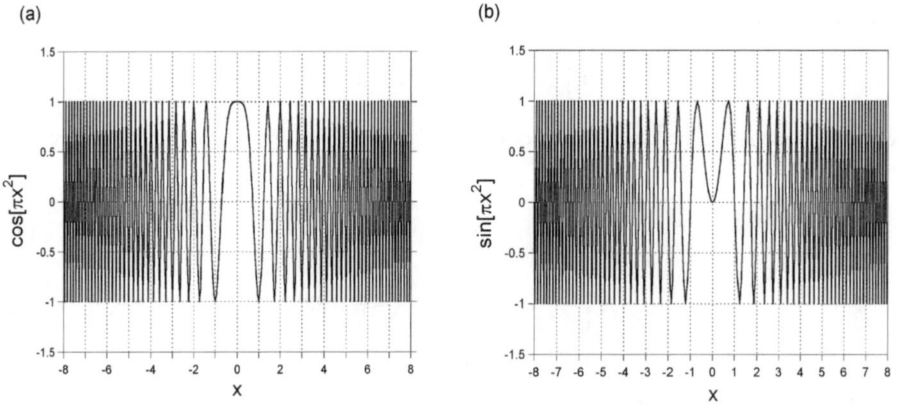

Figure 6.16 Quadratic-phase sinusoidal functions, showing symmetry and increasing spatial frequency as $|x| \to \infty$: (a) $\cos[\pi x^2]$; (b) $\cos[\pi x^2 - \pi/2] = \sin[\pi x^2]$.

Two realizations of the quadratic-phase sinusoid are illustrated: for $\phi_0 = 0$ radians in Figure 6.16a and for $\phi_0 = -\pi/2$ radians in Figure 6.16b. Note that the amplitude of $f[x]$ in Equation (6.53) is symmetric with respect to the origin regardless of the value of ϕ_0; this is obvious because the only term involving x in the argument of the sinusoid appears as x^2. The center of symmetry may be translated by adding a constant to the argument:

$$f[x] = A\cos\left[\pi\frac{(x-x_0)^2}{\alpha^2} + \phi_0\right] \tag{6.54}$$

The spatial frequency of the quadratic-phase sinusoid in Equation (6.53) is obtained by substitution into Equation (4.47):

$$\xi[x] = \frac{1}{2\pi}\frac{\partial\Phi[x]}{\partial x} = \frac{1}{2\pi}\frac{\partial}{\partial x}\left[\frac{\pi x^2}{\alpha^2} + \phi_0\right] = \frac{1}{2\pi}\left[\frac{2\pi x}{\alpha^2}\right] = \frac{x}{\alpha^2} \tag{6.55}$$

which has dimensions of reciprocal length, as appropriate for a spatial frequency. Because ξ is proportional to x, rather than being constant, the derivative of the phase in Equation (6.55) is often called the *instantaneous spatial frequency* of the quadratic-phase sinusoid. The linear dependence of ξ on x also leads to another often-used name of "linear frequency modulation" or "linear FM signal" for the quadratic-phase sinusoid. The change in frequency of temporal functions of the form of Equation (6.53) is often compared to the rising-frequency voicing of some birds, which in turn has led to the application of the term "chirp" to any quadratic-phase sinusoid. The parameter α is often called the *chirp rate*, as it describes the rate of change of the frequency of the sinusoid. Equation (6.53) shows that the magnitude of the phase difference of the chirp function measured between $x = 0$ and $x = \alpha$ is π radians. Examples demonstrating the effect of varying α are shown by comparing Figure 6.17 where $\alpha = 2$ to Figure 6.16 where $\alpha = 1$. Note that the phase of the chirp function changes more slowly if the chirp rate α is larger.

The area of the linear-phase sinusoid was shown to be zero in Equation (6.22), but the area of the chirp function is not. This may be demonstrated directly in a manner very similar to that used to evaluate $\Gamma[1/2]$ in Equation (6.46). A circularly symmetric 2-D chirp function is created and its area is evaluated in polar coordinates via the gamma function; see Problem 6.14. A simpler method to evaluate the area of Equation (6.54) uses the as-yet-unproven central-ordinate theorem of the Fourier transform in Chapter 9. As is probably obvious from the graphs, the area of a linear-phase sinusoid such as $\cos[\pi x]$ is zero precisely because the areas of adjacent positive and negative lobes cancel. In turn, this happens because the phase of the sinusoid is a linear function of the coordinate so that its rate of

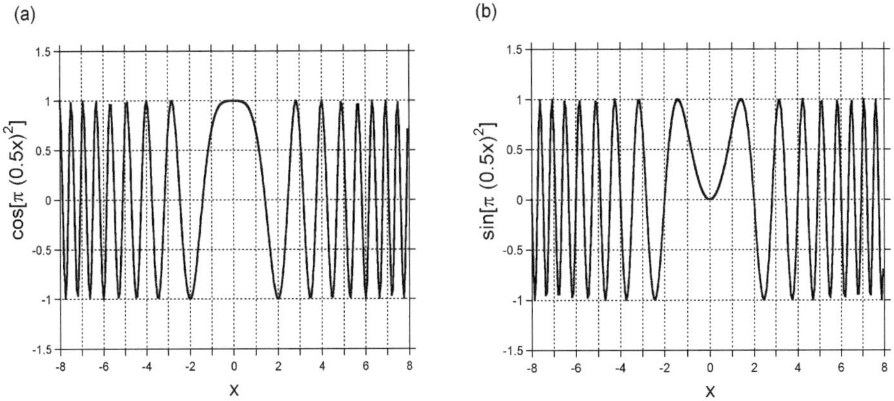

Figure 6.17 Effect of scale factor on quadratic-phase sinusoids: (a) $\cos[\pi(x/2)^2]$; (b) $\sin[\pi(x/2)^2]$. The functions have infinite support and contain all spatial frequencies, but the zero crossings have been moved outward.

change is constant. Because the phase of the chirp function (or of any sinusoidal function whose phase is a nonlinear function of x) changes with x at a variable rate, then adjacent positive and negative lobes must have different "widths" and thus different areas that do not cancel.

The rate of change of phase of the chirp function is seen from Equation (6.53) to be zero at the origin and increases with $|x|$. Any locations where the rate of change of phase is zero (and thus the spatial frequency also is zero) are "points of stationary phase" of the function. The major contributors to the area of the quadratic-phase sinusoid will be due to the lobes of the function centered about any such locations. For example, it is evident from the graph in Figure 6.16a that the area of the positive lobe of $\cos[\pi x^2]$ centered at point of stationary phase at $x = 0$ is larger than the combined areas of the adjacent negative lobes centered at $x = \pm\alpha \cdot \sqrt{1}$. Also, the areas of the positive lobes centered at $x = \pm\alpha \cdot \sqrt{2}$ are larger than the negative areas of lobes centered at $x = \pm a \cdot \sqrt{3}$. This pattern continues for the remaining lobes. This discussion leads to the observation that the area of $\cos[\pi x^2]$ must be positive. Similar arguments may be made for $\sin[\pi x^2]$, which is illustrated in Figure 6.16b. Without proof (as yet), we state that the areas of the chirp functions with $\phi_0 = 0$ (cosine chirp) and with $\phi_0 = -\pi/2$ (sine chirp) are identical and positive. For a chirp function with $\xi_0 = 1$ cycle per unit length and $\alpha = 1$ unit in Equation (6.53), these are:

$$\int_{-\infty}^{+\infty} \cos[\pi x^2]\, dx = \int_{-\infty}^{+\infty} \sin[\pi x^2]\, dx = \frac{1}{\sqrt{2}} \cong 0.7071 \qquad (6.56)$$

These results are closely related to the Fresnel integrals in optics, which will be considered in Chapter 21. The behavior of functions in the vicinity of points of stationary phase may be used to help compute the asymptotic behavior of Fourier transforms, as will be demonstrated in Chapter 13.

A complex-valued chirp function is considered in Section 6.3.2, where the real and imaginary parts are the cosine and sine chirps, respectively. Obviously, the concept of nonlinear phase can be extended to still higher-order functions of the coordinate x. These will be considered to some extent in Section 6.3.3.

6.1.12 Gaussian Function

The Gaussian is the familiar "bell curve" of probability theory and appears in many imaging contexts. The definition of the centered and unscaled Gaussian used in this book follows that of Gaskill:

$$GAUS[x] = e^{-\pi x^2} \qquad (6.57)$$

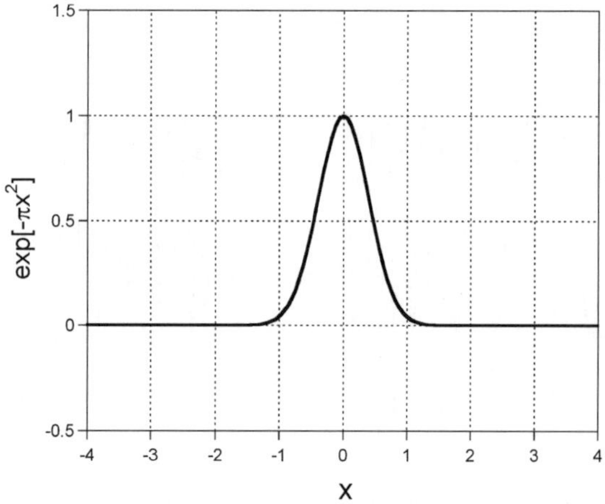

Figure 6.18 Gaussian function $\exp[-\pi x^2]$.

and its graph is shown in Figure 6.18. Even though the equation for $GAUS[x]$ is concise and easily remembered, the named function will be used on most occasions. Note the scale factor of π in the exponent, which is not universally used. From its peak amplitude of unity at the origin, the Gaussian of Equation (6.57) decays smoothly as $|x|$ increases, decreasing to $e^{-\pi} \cong 0.043$ at $x = \pm 1$ and approaching zero as $|x| \to \infty$. The Gaussian has infinite support.

The elegant method for computing the area of $\Gamma[1/2]$ in Equation (6.47) may be adapted to demonstrate that the area of $GAUS[x]$ is unity; we construct the circularly symmetric 2-D Gaussian $\exp[-\pi(x^2 + y^2)]$ and evaluate the area in polar coordinates. Other methods also are possible, including evaluation of the integral of $\exp[-\pi z^2]$ along the appropriate contour in the complex plane (Stephenson and Radmore, 1990). The simplest solution uses the central-ordinate theorem (to be proven in Chapter 9). Alternatively, the variable of integration may be changed to recast the area of the Gaussian into the form of the gamma function in Equation (6.37). This last method also will be used to evaluate the area of the so-called "superGaussian", of which the Gaussian function is a special case. All of these methods show that the area of the Gaussian is unity:

$$\int_{-\infty}^{+\infty} GAUS[x]\, dx = \int_{-\infty}^{+\infty} e^{-\pi x^2}\, dx = 1 \tag{6.58}$$

The more general form of the Gaussian may be constructed with real-valued scaling and translation factors:

$$GAUS\left[\frac{x - x_0}{b}\right] = e^{-\pi(x-x_0)^2/b^2} \tag{6.59}$$

In words, the coordinate of maximum amplitude has been translated to $x = x_0$ and the area of Equation (6.59) is $|b|$.

Readers already familiar with use of the bell curve in statistics may find it instructive to compare the definition of the Gaussian function in Equation (6.57) to that commonly used in probability theory. A random variable n is said to be normally distributed with mean $\langle n\rangle$ and standard deviation σ (variance σ^2) if it exhibits a probability density function $p[n]$ of the form:

$$p[n] = \frac{1}{\sqrt{2\pi\sigma^2}} \exp\left[-\frac{(n - \langle n\rangle)^2}{2\sigma^2}\right] \tag{6.60}$$

Comparison to Equation (6.58) after appropriate change of variables demonstrates that $p[n]$ in Equation (6.60) has unit area as required. The width parameter b in Equation (6.59) is proportional to the standard deviation σ in Equation (6.60):

$$b = (\sqrt{2\pi})\sigma \cong 2.5\sigma \Longrightarrow \sigma \cong \frac{b}{2.5} = 0.4b \tag{6.61a}$$

and the corresponding relationship between the variance and b^2 is:

$$b^2 = 2\pi\sigma^2 \cong 6.28\sigma^2 \Longrightarrow \sigma^2 \cong \frac{b^2}{6.28} \cong 0.16b^2 \tag{6.61b}$$

These relations will be useful when discussing the central-limit theorem in Chapter 13.

6.1.13 "SuperGaussian" Function

This section may be skipped without loss of continuity.

A more general function based on the Gaussian is useful in optics because it conveys a larger optical energy density within the same beam diameter than the corresponding Gaussian. This function, called the *superGaussian* by some authors, has the same form as the Gaussian in Equation (6.57) but with integer exponents other than 2. One possible notation retains the name of the function, but includes the power n as a parameter set off from the argument by a semicolon:

$$GAUS[x;\, n] \equiv e^{-\pi|x|^n} \tag{6.62}$$

where n is a positive integer. The absolute value of the coordinate in the definition ensures that $GAUS[x;\, n]$ remains finite and even for odd n. Note that the superGaussian for $n = 1$ is the sum of $e^{-\pi x} \cdot STEP[x]$ and a "reversed" replica $e^{-\pi|x|} \cdot STEP[-x]$.

Though $GAUS[x;\, n]$ is not one of the "standard" functions used by many authors, it will be useful in a few contexts in this book, including the discussion of functions used to interpolate sampled data in Chapter 14. Figure 6.19 illustrates the function for several values of n. The amplitudes of all superGaussian functions are unity at the origin and $e^{-\pi}$ at $x = \pm 1$, but they differ in rates of "decay". The amplitude decreases over longer distances near the origin and over shorter distances in the vicinity of $|x| \cong 1$ as n is allowed to increase. In the limit $n \to \infty$, the amplitudes of the function in the various regions are:

$$\textit{for } |x| < 1: \quad \lim_{n\to\infty} \{|x|^n\} = 0 \Longrightarrow \lim_{n\to\infty} \{e^{-\pi|x|^n}\} = 1 \tag{6.63a}$$

$$\textit{for } |x| > 1: \quad \lim_{n\to\infty} \{|x|^n\} = +\infty \Longrightarrow \lim_{n\to\infty} \{e^{-\pi|x|^n}\} = 0 \tag{6.63b}$$

$$\textit{for } |x| = 1: \quad \lim_{n\to\infty} \{|x|^n\} = 1 \Longrightarrow \lim_{n\to\infty} \{e^{-\pi|1|^n}\} = e^{-\pi} \cong 0.043\,21 \tag{6.63c}$$

Clearly the shape of the unscaled superGaussian for large values of n resembles the rectangle function with $b = 2$:

$$\lim_{n\to\infty} \{GAUS[x;\, n]\} \cong RECT\left[\frac{x}{2}\right] \tag{6.64}$$

Though the "endpoint" amplitudes of these two functions are not identical, these isolated values have no effect on the areas. The rectangle of unit width can be approximated by scaling the superGaussian:

$$RECT[x] \cong \lim_{n\to\infty} \{GAUS[2x;\, n]\} = \lim_{n\to\infty} \{e^{-\pi|2x|^n}\} \tag{6.65}$$

This indicates that superGaussians of finite order n may be used as approximations of the rectangle function.

The area of the superGaussian may be derived by applying the definition of the gamma function in Equation (6.37):

$$\Gamma[x] = \int_0^{+\infty} \alpha^{x-1}\, e^{-\alpha}\, d\alpha \tag{6.66}$$

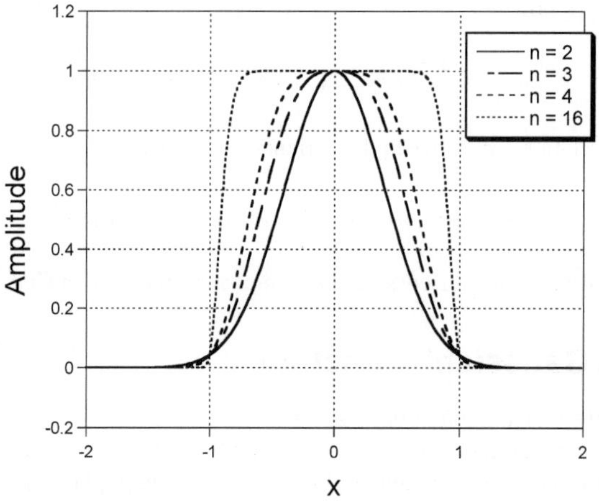

Figure 6.19 SuperGaussian functions $f[x] = \exp[-\pi|x|^n]$ for $n = 2$, 3, 4, and 16, showing that all curves evaluate to unity at $x = 0$ and to $e^{-\pi} \simeq 0.043\,21$ at $x = \pm 1$.

The variable of integration is changed from α to πu^n:

$$\begin{aligned}\Gamma[x] &= \int_0^{+\infty} (\pi u^n)^{x-1}\, e^{-\pi u^n}\, d(\pi u^n) \\ &= \int_0^{+\infty} \pi^{x-1}\, u^{nx-n}\, e^{-\pi u^n}\, n\pi u^{n-1}\, du \\ &= \int_0^{+\infty} n\pi^x u^{nx-n+n-1}\, e^{-\pi u^n}\, du \end{aligned} \tag{6.67}$$

Divide both sides by n to obtain:

$$\frac{1}{n}\,\pi^{-x}\,\Gamma[x] = \int_0^{+\infty} u^{nx-1}\, e^{-\pi u^n}\, du \tag{6.68}$$

The integral of $e^{-\pi u^n}$ over positive u is obtained by setting $x = n^{-1}$ to ensure that $u^{nx-1} = u^0 = 1$. The result may be expressed in terms of the gamma function for a "reciprocal-integer" argument that was tabulated in Equation (6.51):

$$\int_0^{+\infty} e^{-\pi u^n}\, du = \frac{1}{n}\pi^{-1/n}\Gamma\left[\frac{1}{n}\right] \tag{6.69}$$

where the variable of integration was renamed. The area of the symmetric function $e^{-\pi|x|^n}$ is twice the value of the semi-infinite integral:

$$\int_{-\infty}^{+\infty} e^{-\pi|x|^n}\, dx = \frac{2}{n}(\pi^{-1/n})\Gamma\left[\frac{1}{n}\right] = \frac{2}{\pi^{1/n}}\Gamma\left[\frac{n+1}{n}\right] \tag{6.70}$$

This result demonstrates that the area of the superGaussian of order n is proportional to the gamma function with argument $(n+1)/n$. The areas evaluated for $n = 1 - 4$ are:

$$n=1: \int_{-\infty}^{+\infty} e^{-\pi|x|}\, dx = 2\pi^{-1}\Gamma[1] = \left(\frac{2}{\pi}\right)0! = \frac{2}{\pi} \cong 0.6366 \tag{6.71a}$$

$$n=2: \int_{-\infty}^{+\infty} e^{-\pi|x|^2}\, dx = \frac{2}{2}\frac{1}{\sqrt{\pi}}\Gamma\left[\frac{1}{2}\right] = \frac{2}{\sqrt{\pi}}\Gamma\left[\frac{3}{2}\right] = 1 \tag{6.71b}$$

$$n=3: \int_{-\infty}^{+\infty} e^{-\pi|x|^3}\, dx = \frac{2}{3}\left(\frac{1}{\pi^{\frac{1}{3}}}\right)\Gamma\left[\frac{1}{3}\right] = \frac{2}{(\pi^{\frac{1}{3}})}\Gamma\left[\frac{4}{3}\right] \cong 1.2194 \tag{6.71c}$$

$$n=4: \int_{-\infty}^{+\infty} e^{-\pi|x|^4}\, dx = \frac{2}{4}\left(\frac{1}{\pi^{\frac{1}{4}}}\right)\Gamma\left[\frac{1}{4}\right] = \frac{2}{(\pi^{\frac{1}{4}})}\Gamma\left[\frac{5}{4}\right] \cong 1.3616 \tag{6.71d}$$

Equation (6.71b) confirms that the area of the "normal" Gaussian function is unity. The area increases with n, as was shown in Figure 6.19. The limiting value of the area as $n \to +\infty$ may be computed by applying Equation (6.52):

$$\lim_{n\to\infty}\left\{\int_{-\infty}^{+\infty} e^{-\pi|x|^n}\, dx\right\} = \lim_{n\to\infty}\left\{\frac{2\cdot}{\pi^{1/n}}\left(\frac{1}{n}\cdot\Gamma\left[\frac{1}{n}\right]\right)\right\} = \left(\frac{2}{\pi^0}\right)\cdot 1 = 2 \tag{6.72}$$

which agrees with the interpretation in Equation (6.64) that the limiting form of the superGaussian is $RECT[x/2]$.

6.1.14 Bessel Functions

This section may be skipped without loss of continuity.

Just as the linear-phase sinusoids are the solutions of the differential equation in Equation (6.18), other useful functions are the solutions of another 1-D differential equation:

$$x^2\frac{d^2}{dx^2}(Z_\nu[x]) + x\frac{d}{dx}(Z_\nu[x]) + (x^2 - \nu^2)Z_\nu[x] = 0 \tag{6.73}$$

where $x \geq 0$ and ν is a real number. The solutions $Z_\nu[x]$ are the *Bessel functions*, where the notation denotes particular solutions for each value of the order ν. Differential equations of this form appear often in physical problems involving planar circular symmetry or cylindrical coordinates, such as descriptions of imaging systems constructed from optics with circular cross-sections. A rigorous study of the origin and properties of the Bessel functions is beyond the scope of this book; interested readers should consult any of the standard books on mathematical physics for details, such as Arfken (2000) or Stephenson and Radmore (1990). Three types of solutions to Equation (6.73) are recognized. The "Bessel functions of the first kind" with integer and half-integer order ($\nu = n$ or $n/2$ for $n = 0, 1, 2, 3, \ldots$) are most relevant in imaging; these are labeled $J_n[x]$ and have finite amplitude for positive x. The "Bessel functions of the second kind" (also called the "Neumann functions") are denoted by $N_\nu[x]$ and have indeterminate amplitude at $x = 0$. The third type of solution to Bessel's differential function is the "Hankel function" (not to be confused with the "Hankel transform" that will be discussed later). The Hankel function is a complex-valued linear combination of the first two types using a scheme analogous to the Euler relation: $H_\nu[x] = J_\nu[x] \pm i\, N_\nu[x]$. Because the second and third solutions generally are not relevant to imaging problems, they are not considered here.

Several avenues are available to derive numerical values for the Bessel functions of interest, including the so-called *generating function*, contour integrals, and a series solution of the differential equation for integer indices ($\nu = n$). We mention only the last, and that only briefly. We assume that $J_n[x]$ has the form of a power series in x with unknown coefficients:

$$J_n[x] = \sum_{\ell=0}^{+\infty} a_\ell x^\ell \tag{6.74}$$

By inserting the series into Equation (6.73), evaluating the derivatives, equating terms with the same power of x, and juggling terms, two series solutions for $J_n[x]$ are obtained: one each for positive and negative values of n. We will be most interested in $J_0[x]$ and $J_1[x]$, so only the series for positive n is considered here. The power series solution for $J_0[x]$ is derived in Appendix 6.6 and the result is given in Equation (B6.7):

$$\begin{aligned} J_0[x] &= 1 - \frac{x^2}{2^2} + \frac{x^4}{2^2 \cdot 4^2} - \frac{x^6}{2^2 \cdot 4^2 \cdot 6^2} + \cdots \\ &= 1 - \frac{x^2}{4} + \frac{x^4}{64} - \frac{x^6}{2304} + \cdots \end{aligned} \tag{6.75}$$

The numerical coefficients of these first several terms should be compared to those for $SINC[x]$ in Equation (6.32). The algebraic signs of the coefficients in both series alternate. Also, the absolute values of the coefficients of $J_0[x]$ decrease more slowly with n than those of $SINC[x]$, which means that the extrema of the local amplitude of $J_0[x]$ decrease more slowly. This sequence of coefficients for $J_0[x]$ will reappear in the discussion of Fourier transforms of circularly symmetric functions in Chapter 11. Unlike the *SINC* function, the zeros of $J_0[x]$ are spaced at intervals that are only approximately uniform. For example, the three zeros nearest the origin occur at:

$$\begin{array}{ll} \textit{Location of zeros of } J_0[x] & \textit{Spacing } \Delta x \\ x_1 \cong 2.4048 \cong 0.7655\pi & \\ & x_2 - x_1 \cong 0.9916\pi \\ x_2 \cong 5.5201 \cong 1.7571\pi & \\ & x_3 - x_2 \cong 0.9974\pi \\ x_3 \cong 8.6537 \cong 2.7546\pi & \\ \vdots & \vdots \\ x_{M-1} & \\ & x_M - x_{M-1} \lessapprox \pi \\ x_M & \end{array} \tag{6.76}$$

It is evident from the series that the amplitude and slope of J_0 at $x = 0$ are unity and zero, respectively. Also note that the intervals between the first and second pair of adjacent zeros are approximately 0.9916π and 0.9974π, respectively; the interval between x_2 and x_3 is larger than that between x_1 and x_2. In fact, the incremental distance between successive pairs of zeros asymptotically approaches π as $x \to +\infty$. This result suggests that the asymptotic form of $J_0[x]$ is an oscillating function with period 2π. In fact, $J_0[x]$ resembles a cosine oscillation with period $(2\pi)^{-1}$ and initial phase $\phi_0 = -\pi/4$ modulated by a decaying function that happens to be proportional to $x^{-\frac{1}{2}}$:

$$\begin{aligned} \lim_{x \to +\infty} \{J_0[x]\} &= \sqrt{\frac{2}{\pi x}} \cos\left[x - \frac{\pi}{4}\right] \\ &= \sqrt{\frac{2}{\pi x}} \cos\left[2\pi\left(\frac{1}{2\pi}\right)x - \frac{\pi}{4}\right] \end{aligned} \tag{6.77}$$

The graphs of the first four integer-order Bessel functions are shown in Figure 6.20.

Another, and very convenient, representation of the 1-D zero-order Bessel function is as a 1-D radial profile (or "slice") of a circularly symmetric 2-D function that is generated by summing 2-D cosine waves with the same spatial period but directed along all azimuths. The derivation of this result will be considered during the discussion of 2-D circularly symmetric functions in Section 7.3.

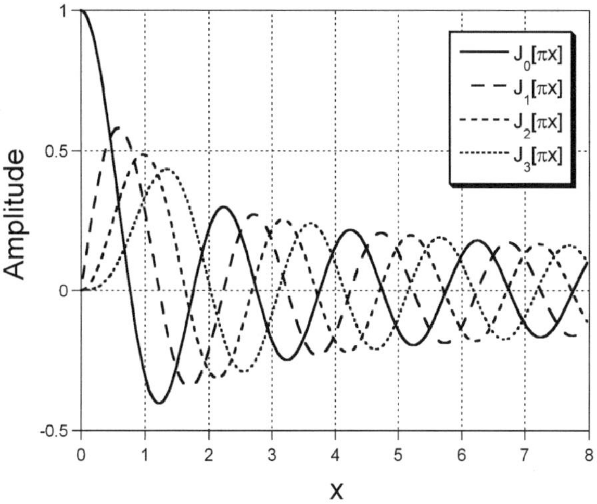

Figure 6.20 The Bessel function of the first kind $J_n[\pi x]$ for n = 0–3, showing the decay in amplitude with increasing x and the fact that the different orders are approximately in quadrature.

The series solution for $J_1[x]$ also is derived in the Appendix below. The result of Equation (B6.13) includes only odd powers of x, which ensures that $J_1[x]$ must be an odd function:

$$J_1[x] = \frac{x}{2} - \frac{x^3}{2^2 \cdot 4} + \frac{x^5}{2^2 \cdot 4^2 \cdot 6} - \frac{x^7}{2^2 \cdot 4^2 \cdot 6^2 \cdot 8} + \cdots$$
$$= \frac{x}{2} - \frac{x^3}{16} + \frac{x^5}{384} - \frac{x^7}{18\,432} + \cdots = \frac{x}{2}\left(1 - \frac{x^2}{8} + \frac{x^4}{192} - \frac{x^6}{9216} + \cdots\right) \tag{6.78}$$

The graph of $J_1[x]$ also is shown in Figure 6.20. As seen from the series, the amplitude and slope at the origin are equal to the coefficients $\alpha_0 = 0$ and $\alpha_1 = \frac{1}{2}$ respectively. As x increases from zero, the amplitude increases from zero and reaches a local maximum at $x \cong 1.8412 \cong 0.586\pi$. Also note that the absolute values of the numerical coefficients of $J_1[x]$ decrease more slowly than those of *SINC*$[x]$ in Equation (6.32), which again means that the absolute values of successive local maxima of $J_1[x]$ decrease more slowly than those of *SINC*$[x]$ as $x \to \infty$, which is the same behavior exhibited by $J_0[x]$.

The first three zeros of $J_1[x]$ are located at:

$$\begin{array}{ll} \textit{Location of zeros of } J_1[x] & \textit{Spacing } \Delta x \\ x_1 \cong 3.8317 \cong 1.219\pi & \\ & x_2 - x_1 \cong 1.0135\pi \\ x_2 \cong 7.0156 \cong 2.2331\pi & \\ & x_3 - x_2 \cong 1.0052\pi \\ x_3 \cong 10.1735 \cong 3.2383\pi & \\ \vdots & \vdots \\ x_{M-1} & \\ & x_M - x_{M-1} \gtrapprox \pi \\ x_M & \end{array} \tag{6.79}$$

The intervals between the first pairs of zeros are approximately 1.0135π and 1.0052π; the interval between adjacent zeros of $J_1[x]$ *decreases* for increasing x behavior that complements that of $J_0[x]$. The interval between zeros for both $J_0[x]$ and $J_1[x]$ asymptotically approach π as $x \to \infty$, though from different directions. The asymptotic form of $J_1[x]$ is in quadrature to $J_0[x]$, i.e., the phase difference of the oscillations of the two functions is $-\pi/2$ radians, and thus $J_1[x]$ is approximately equal to a sine wave with period $(2\pi)^{-1}$ and initial phase of $\phi_0 = -\pi/4$ modulated by $x^{-1/2}$:

$$\lim_{x\to+\infty} \{J_1[x]\} = \sqrt{\frac{2}{\pi x}} \cos\left[x - \frac{3\pi}{4}\right]$$
$$= \sqrt{\frac{2}{\pi x}} \cos\left[\left(x - \frac{\pi}{4}\right) - \frac{\pi}{2}\right] = \sqrt{\frac{2}{\pi x}} \sin\left[2\pi\left(\frac{1}{2\pi}\right)x - \frac{\pi}{4}\right] \tag{6.80}$$

Though we do not derive it, the series solution for the Bessel function that is valid for all positive integer values of n is (Stephenson and Radmore, 1990):

$$J_n[x] = \sum_{\ell=0}^{+\infty} \frac{(-1)^\ell}{\ell!(n+\ell)!}\left(\frac{x}{2}\right)^{n+2\ell} \tag{6.81}$$

The Bessel functions of the first kind with order $n > 2$ do not occur frequently in imaging applications, but we mention a few properties. All have null amplitude at the origin (i.e., $J_n[0] = 0$ for $n > 0$) and the next zero crossing occurs farther from the origin as n is increased. The interval between zeros is larger than π for small values of x and decreases to π in the limit of large x. The asymptotic behavior of $J_n[x]$ is out of phase to that of $J_{n-1}[x]$ by $-\pi/2$ radians, as illustrated by the cases of $n = 2$ and 3:

$$\lim_{x\to+\infty} \{J_2[x]\} = \sqrt{\frac{2}{\pi x}} \cos\left[x - \frac{5\pi}{4}\right] \tag{6.82a}$$

$$\lim_{x\to+\infty} \{J_3[x]\} = \sqrt{\frac{2}{\pi x}} \cos\left[x - \frac{7\pi}{4}\right] \tag{6.82b}$$

In short, $J_n[x]$ and $J_{n+1}[x]$ are in approximate quadrature as $x \to \infty$. Plots of $J_2[x]$ and $J_3[x]$ are included in Figure 6.20.

6.1.15 Lorentzian Function

This section may be skipped without loss of continuity.

The Lorentzian function is named after the Dutch physicist Hendrik Lorentz, who demonstrated the significance of this particular curve in his study of atomic radiation in the early 1900s. The Lorentzian curve is the theoretical shape of spectral lines created by atomic absorption or emission. Though different authors use definitions with different constants and scale factors, one appropriate formulation for the real-valued Lorentzian is:

$$LOR[x] = \frac{2}{1 + (2\pi x)^2} = \frac{2}{1 + 4\pi^2 x^2} \tag{6.83}$$

The amplitude of $LOR[x]$ evidently is proportional to the reciprocal of the sum of the unit constant and a quadratic function of the coordinate. The quadratic dependence on coordinate x ensures that the Lorentzian is an even function.

The multiplicative factor of 2 is included to ensure that $LOR[x]$ has unit area, as may be demonstrated by integrating Equation (6.83) after a trigonometric substitution:

$$\int_{-\infty}^{+\infty} \frac{2}{1 + (2\pi x)^2}\, dx = \int_{-\infty}^{+\infty} \frac{2}{1 + u^2} \frac{du}{2\pi} = \frac{1}{\pi} \tan^{-1}[u]\Big|_{u=-\infty}^{u=+\infty}$$
$$= \frac{1}{\pi}\left[\frac{\pi}{2} - \left(-\frac{\pi}{2}\right)\right] = 1 \tag{6.84}$$

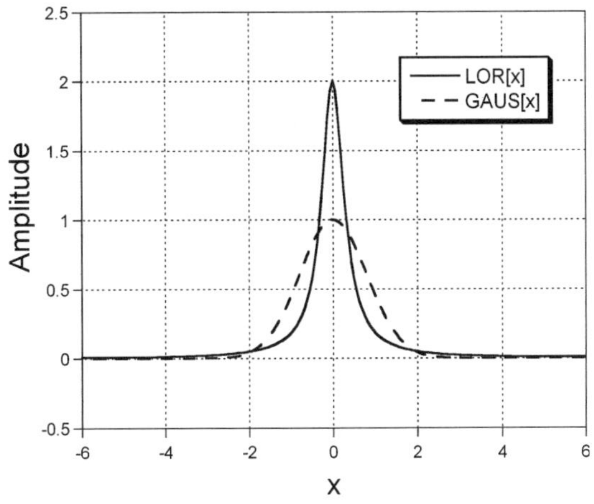

Figure 6.21 Comparison of the Lorentzian and Gaussian functions to unit area. *LOR*[x] decays more quickly than *GAUS*[x] for large values of $|x|$.

The amplitude obviously decays from $LOR[0] = 2$ at the origin through $LOR[\pm 1] \cong 0.0482$ and on to zero at $x = \pm\infty$. Note that $LOR[\pm 1] \cong GAUS[\pm 1]$. $LOR[x]$ decays to zero more rapidly than $GAUS[x]$ for $|x| < 1$, and more slowly for $|x| > 1$. The two functions are compared in Figure 6.21.

A variant of the Lorentzian function with complex amplitude will be defined in Section 6.3.3.

6.1.16 Thresholded Functions

The requirement that functions to be processed by digital computer have a discrete domain and range was described in Section 2.2.4. Functions with discrete ranges may be constructed by "mapping" the amplitude through a subsequent function that has a discrete range. The simplest such thresholding operation is generated by the SIGNUM function. A very common example of the resulting thresholded function is the square wave generated by mapping the amplitude of a sinusoidal input function:

$$SGN[b + \cos[2\pi\xi_0 x]] = \begin{cases} +1 & \textit{if } b + \cos[2\pi\xi_0 x] > 0 \\ 0 & \textit{if } b + \cos[2\pi\xi_0 x] = 0 \\ -1 & \textit{if } b + \cos[2\pi\xi_0 x] < 0 \end{cases} \tag{6.85}$$

as shown in Figure 6.22. The additive bias b may be varied to change the relative widths of the positive and negative "lobes" of the square wave; the widths are equal when the bias $b = 0$. The percentage of the width of the region with maximum amplitude to the period of the square wave is often called the *duty cycle*. Under this definition, addition of positive or negative biases results in a larger or smaller duty cycle, respectively. If $b = 0$, the duty cycle is 50%; if $b > 1$ or $b < -1$, the resulting duty cycles are 100% and 0%, respectively.

The thresholding process itself may be interpreted as the action of a nonlinear imaging system which generates a discrete three-state output from a continuous input, and is often interpreted as a form of lookup table. The allowed output amplitudes may be changed to 0, 1/2, and 1 by addition and multiplication of the output in a fashion analogous to that used to apply the *STEP* function in place of the SIGNUM function in Equation (6.7). The result may be expressed in terms of the SIGNUM

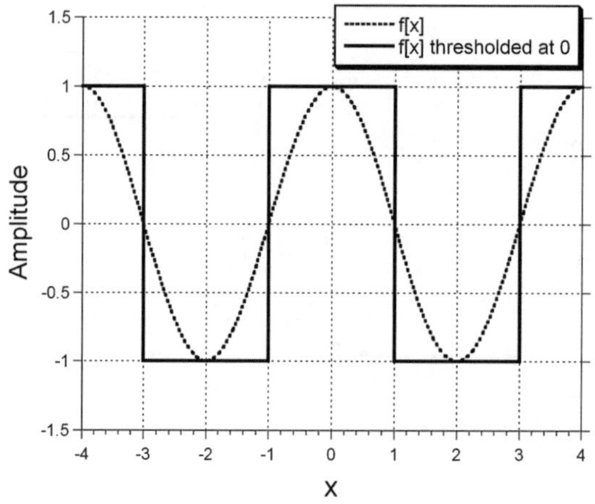

Figure 6.22 Thresholding of a sinusoidal function through a signum function lookup table produces a square wave.

function:

$$\frac{1}{2}(1 + SGN[b + \cos[2\pi\xi_0 x]]) = \begin{cases} 1 & \textit{if } b + \cos[2\pi\xi_0 x] > 0 \\ \frac{1}{2} & \textit{if } b + \cos[2\pi\xi_0 x] = 0 \\ 0 & \textit{if } b + \cos[2\pi\xi_0 x] < 0 \end{cases} \tag{6.86}$$

Note that the symmetry of the square wave in Equation (6.85) reflects that of the input sinusoid; the square wave is odd when the input is an odd sine wave. The spatial periods of the sine wave and resulting square wave also must be identical. From this result and the notation in Equation (6.19), we may be tempted to specify a spatial frequency ξ_0 of the square wave in "cycles per unit length". However, we will use a convention that allows the term *cycle* to be applied only to represent one period of a pure sinusoid. Because the square wave of the form in Equation (6.85) consists of a set of neighboring "bright" and "dark" bars, a different terminology will be used to describe the period of the thresholded sinusoid. We choose the term *line pair* for this period, and so the analogue to spatial frequency for a square wave is the measure of the number of line pairs per unit length. Many authors do not make so strict a distinction in terminology between the two cases, but we will do so in an attempt to maximize clarity.

The same thresholding process may be applied to other continuous special functions. One example that is useful in optics is the model of the 1-D "zone plate" illustrated in Figure 6.23, which is a biased and thresholded quadratic-phase sinusoid:

$$f[x] = \frac{1}{2}\left(1 + SGN\left[\cos\left[\frac{\pi x^2}{\alpha^2} + \phi_0\right]\right]\right) = STEP\left[\cos\left[\frac{\pi x^2}{\alpha^2} + \phi_0\right]\right] \tag{6.87}$$

6.2 1-D Dirac Delta Function

The name of the Dirac delta function $\delta[x]$ honors the physicist P.A.M. Dirac, who introduced the notation for this function during his work in quantum mechanics even though the function itself had been used by mathematicians for many years prior. Another common name is the *impulse* function. The Dirac delta function does not have a "proper" definition that assigns a definite amplitude to each coordinate x. In a strict sense, this means that $\delta[x]$ really is not a function at all. Rather, because its area

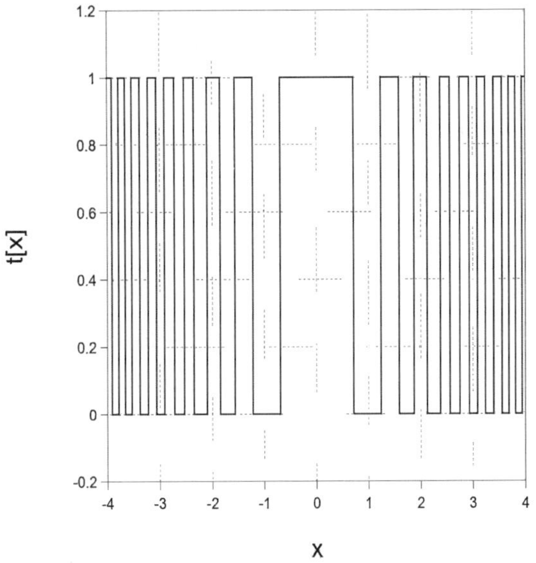

Figure 6.23 Example of thresholded chirp function that produces a Fresnel zone plate $f[x] = CINT\{\frac{1}{2}(1 + \cos[\pi x^2])\}$ showing that the distances between transitions decreases with increasing radial distance from the origin.

evaluated between two coordinates has a physical interpretation but its amplitude does not, the Dirac delta function is analogous to a probability density function. In other words, the notation of $\delta[x]$ really is meaningful only within the integrand of an integral. For this reason the Dirac delta function perhaps is more properly called a *delta distribution*.

Regardless of the nomenclature, the properties of the Dirac delta function make it an essential tool for solving problems involving many types of physical systems, including classical mechanics, quantum mechanics, and electrodynamics. The unrealistic but mathematically tractable model of the Dirac delta function often may be substituted for realistic physical entities to simplify calculations. One example is common in classical mechanics, where the motion of a spherical body of uniform mass density in a gravitational field may be predicted by modeling the body as a *point mass* located at the center of the sphere. The point mass has infinitesimal volume but finite mass and may be represented mathematically by a 3-D Dirac delta function.

The conditions to be satisfied by a 1-D Dirac delta function are infinitesimal support and finite area. The mathematical statements of these properties are respectively:

$$\delta[x - x_0] = 0 \quad for\ x \neq x_0 \tag{6.88a}$$

$$\int_{-\infty}^{+\infty} \delta[x - x_0]\, dx = 1 \tag{6.88b}$$

The second condition demonstrates that the dimension of $\delta[x]$ must be the reciprocal of the dimension of dx, or $(\text{length})^{-1}$.

Because its amplitude at that one coordinate x_0 is not defined, $\delta[x - x_0]$ cannot be depicted in a conventional graphical way. However, an "image" of its characteristics is conveyed by representing $\delta[x - x_0]$ as an arrow (or "spike") located at x_0 drawn so that the height of the arrowhead's tip above the x-axis is equal to the area of the function. In other words, the base of the arrow *always* rests on

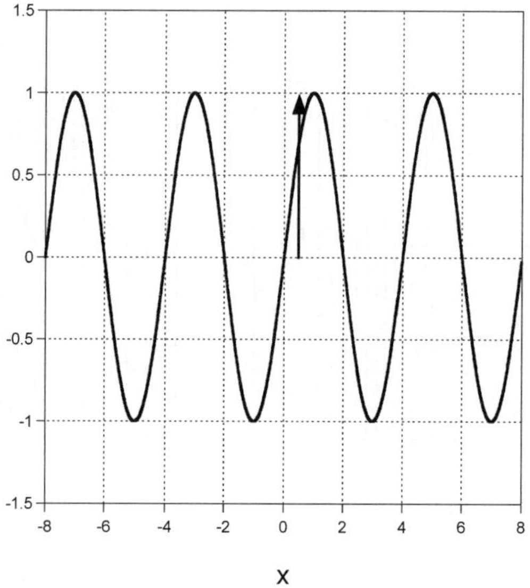

Figure 6.24 Graphical depiction of $f[x] = \delta[x] + \cos[2\pi x/4 - \pi/2]$. The "base" of the Dirac delta function always lies on the x-axis and the "height" of the tip of the arrow from the axis represents its area.

the x-axis, even in those cases where $\delta[x - x_0]$ has been added to a "proper" finite-amplitude function $f[x]$. An example of such a case is shown in Figure 6.24.

The properties of the Dirac delta function often are confusing to students at first. A fairly common error misconstrues the Dirac delta function and the function $f_0[x]$ defined by:

$$f_0[x] = 0 \quad \textit{if } x \neq 0 \tag{6.89a}$$

$$f_0[x] = 1 \quad \textit{if } x = 0 \tag{6.89b}$$

These conditions define a proper discontinuous function that has zero area. In fact, any number of such isolated points with finite amplitude have no effect on any integral that includes this function in the integrand. To illustrate, the following integrals give different results:

$$\int_{x_0-\frac{1}{2}}^{x_0+\frac{1}{2}} (f_0[x - x_0] + 1[x])\, dx = 1 \tag{6.90a}$$

$$\int_{x_0-\frac{1}{2}}^{x_0+\frac{1}{2}} (\delta[x - x_0] + 1[x])\, dx = 2 \tag{6.90b}$$

Beyond the constraints specified in Equation (6.88), the details of the functional form used to describe $\delta[x]$ virtually do not matter in many situations. In fact, $\delta[x]$ is often defined as the limit of any of several sequences of functions that differ in some details, such as whether the functions in the sequence are differentiable. Each representation is useful in an appropriate circumstance, and so we will devote some effort to describing several of the useful forms. The requirement of unit area is satisfied by several of the special functions already defined. Consider first that the scaled rectangle may be used to define a function that satisfies the requirement of Equation (6.88b):

$$\int_{-\infty}^{+\infty} \frac{1}{|b|} RECT\left[\frac{x}{b}\right] dx = 1 \tag{6.91}$$

In the limit $b \to 0$, this function also satisfies the first condition for the Dirac delta function in Equation (6.88a):

$$\lim_{b\to 0}\left\{\frac{1}{|b|}RECT\left[\frac{x}{b}\right]\right\} = 0 \quad \textit{if } x \neq 0 \tag{6.92}$$

The symmetry of $RECT[x/b]$ implies that the Dirac delta function must be symmetric with respect to the origin:

$$\delta[x] = \lim_{b\to 0}\left\{\frac{1}{|b|}RECT\left[\frac{x}{b}\right]\right\} = \lim_{b\to 0}\left\{\frac{1}{|-b|}RECT\left[\frac{x}{-b}\right]\right\} = \delta[-x] \tag{6.93}$$

which leads to the observation that the order of the arguments of a shifted Dirac delta function does not matter:

$$\delta[x - x_0] = \delta[-(x - x_0)] = \delta[x_0 - x] \tag{6.94}$$

Another useful representation of the Dirac delta function is obtained by scaling the triangle function:

$$\delta[x] = \lim_{b\to 0}\left\{\frac{1}{|b|}TRI\left[\frac{x}{b}\right]\right\} \tag{6.95}$$

which has three discontinuities located at $x = 0$ and $x = \pm b$. Similar limiting expressions for $\delta[x]$ are obtained from versions of the *SINC*, $SINC^2$, *GAUS*, and Lorentzian functions with appropriate scaling of the amplitude and width parameter, even though the support of each function is infinite when the width parameter b is finite. The amplitude at finite (but nonnull) coordinates becomes infinitesimal in the limit $b \to 0$:

$$\delta[x] = \lim_{b\to 0}\left\{\frac{1}{|b|}SINC\left[\frac{x}{b}\right]\right\} \tag{6.96a}$$

$$= \lim_{b\to 0}\left\{\frac{1}{|b|}SINC^2\left[\frac{x}{b}\right]\right\} \tag{6.96b}$$

$$= \lim_{b\to 0}\left\{\frac{1}{|b|}GAUS\left[\frac{x}{b}\right]\right\} \tag{6.96c}$$

$$= \lim_{b\to 0}\left\{\frac{1}{|b|}LOR\left[\frac{x}{b}\right]\right\} \tag{6.96d}$$

These representations of the Dirac delta function have no discontinuities and can be used to derive expressions for the derivatives of the Dirac delta function. The expression in terms of the *SINC* function in Equation (6.96a) also will be useful for representing an array of equally spaced Dirac delta functions, which will be defined in Section 6.2.5. The validity of representing $\delta[x]$ in terms of $SINC[x/b]$ may be demonstrated easily. First, the definition of the *SINC* function in Equation (6.28) is substituted into Equation (6.96a) to obtain a representation of the Dirac delta function:

$$f[x] = \lim_{b\to 0}\left\{\frac{1}{|b|}SINC\left[\frac{x}{b}\right]\right\} = \lim_{b\to 0}\left\{\frac{1}{|b|}\left(\frac{\sin[\pi x/b]}{[\pi x/b]}\right)\right\} = \lim_{b\to 0}\left\{\frac{\sin[\pi x/b]}{\pi x}\right\} \tag{6.97}$$

The reciprocal $N = b^{-1}$ may be substituted, which recasts the limit to infinity:

$$f[x] = \lim_{N\to\infty}\left\{\frac{\sin[N\pi x]}{\pi x}\right\} = \lim_{N\to\infty}\left\{\frac{\sin[2\pi(N/2)x]}{\pi x}\right\} \tag{6.98}$$

This evidently is the limiting behavior of the ratio of a sine wave with spatial frequency $\xi = N/2$ and the linear function πx. In the limit $N \to \infty$, the period of $\sin[N\pi x]$ approaches zero and the rapid oscillation ensures that the local amplitude of the $f[x]$ approaches zero for nonzero values of x. Both numerator and denominator have null amplitude at the origin, so L'Hôpital's rule must be applied to compute the ratio of the derivatives for $f[0]$. The slopes of the numerator and denominator are $N\pi$ and π, respectively. This means that the amplitude of the ratio evaluated at the origin is N, which becomes indeterminate in the limit:

$$f[x] = \frac{\lim_{N\to\infty}\{N\pi\}}{\lim_{N\to\infty}\{\pi\}} = \lim_{N\to\infty}\{N\} = \infty \tag{6.99}$$

This is a scaled *SINC*[x] with unit area. This discussion also established that the amplitude $f[x]=0$ for $x \neq 0$, so $f[x]$ satisfies the requirements of Equation (6.88) for the Dirac delta function.

The expression for the Dirac delta function in terms of the Lorentzian in Equation (6.96d) also may be recast into a new form:

$$\begin{aligned}\delta[x] &= \lim_{b\to 0}\left\{\frac{1}{|b|}LOR\left[\frac{x}{b}\right]\right\} = \lim_{b\to 0}\left\{\frac{1}{|b|}\frac{2}{1+(2\pi x/b)^2}\right\} \\ &= \lim_{b\to 0}\left\{\frac{(b/2\pi)/\pi}{(b/2\pi)^2+x^2}\right\} = \lim_{\epsilon\to 0}\left\{\frac{\epsilon/\pi}{\epsilon^2+x^2}\right\} \quad \textit{where } \epsilon \equiv \frac{b}{2\pi}\end{aligned} \tag{6.100}$$

This expression is convenient in some applications, such as in quantum mechanics (Merzbacher, 1997).

From the defining properties in Equation (6.88), it is evident that the amplitudes of the semi-infinite integral of $\delta[x]$ depend on the specific location of the upper limit:

$$\int_{-\infty}^{x} \delta[\alpha]\,d\alpha = \begin{cases}1 & \textit{if } x>0\\ 0 & \textit{if } x<0\end{cases} \tag{6.101}$$

The value of the integral evaluated at $x=0$ is not specified by this relation due to the discontinuity. A value may be assigned by using the Cauchy principal value, which is the mean value of the integral as $x \to 0$ from the negative and positive sides of the origin. The resulting expression is identical to the definition of *STEP*[x] in Equation (6.10):

$$\int_{-\infty}^{x} \delta[\alpha]\,d\alpha = \left\{\begin{array}{ll}1 & \textit{if } x>0\\ \frac{1}{2} & \textit{if } x=0\\ 0 & \textit{if } x<0\end{array}\right\} = STEP[x] \tag{6.102}$$

By differentiating both sides of this result and applying the fundamental theorem of calculus, we obtain another useful expression for the Dirac delta function:

$$\begin{aligned}\frac{d}{dx}STEP[x] &= \frac{d}{dx}\int_{-\infty}^{x}\delta[\alpha]\,d\alpha = \delta[x]-\delta[-\infty]\\ &= \delta[x]-0=\delta[x]\end{aligned} \tag{6.103a}$$

Therefore, the derivative of a step function centered at x_0 is a Dirac delta function located at x_0:

$$\frac{d}{dx}STEP[x-x_0] = \delta[x-x_0] \tag{6.103b}$$

Another and very important representation of the Dirac delta function is obtained by summing complex sinusoids with unit amplitudes and zero phase over all spatial frequencies:

$$\int_{-\infty}^{+\infty} e^{+2\pi i\xi x}\,d\xi = \int_{-\infty}^{+\infty}\cos[2\pi\xi x]\,d\xi + i\int_{-\infty}^{+\infty}\sin[2\pi\xi x]\,d\xi \tag{6.104}$$

To see that the expression satisfies the criteria required of the Dirac delta function in Equation (6.88), the integral may be evaluated over arbitrary finite and symmetric limits:

$$\begin{aligned}\int_{-B}^{+B} e^{+2\pi i\xi x}\,d\xi &= \frac{1}{2\pi i x}(e^{+2\pi\xi x})\Big|_{\xi=-B}^{\xi=B} = \frac{1}{\pi x}\frac{(e^{+2\pi i x B}-e^{-2\pi i x B})}{2i}\\ &= \frac{1}{\pi x}\sin[2\pi Bx] = 2B\,SINC[2Bx]\end{aligned} \tag{6.105}$$

In the limit $B \to +\infty$, this expression is equivalent to Equation (6.96a) and thus is a valid representation of the Dirac delta function. Note that the integral of the original complex Hermitian function over symmetric limits yields a real-valued result due to the cancellation of areas for positive and negative x in the antisymmetric imaginary part.

The representations of $\delta[x]$ as a summation of unit-amplitude cosines over all spatial frequencies may be visualized by considering the partial sum of sinusoidal components illustrated in Figure 6.25. Four cosine functions with periods of 4, 6, 8, and ∞ are added; obviously, the last cosine with infinite period is identical to the unit constant. Each of these constituent sinusoids has unit amplitude at coordinates that are even divisible by each period (e.g., at $x = 0, \pm 24, \pm 48, \ldots$). In the terminology often used in optics, we say that the four cosine functions "constructively interfere" at these locations. At other values of x, the amplitudes of the individual cosines scatter about zero so that their sum is less than the maximum amplitude. The sinusoids may sum to zero at other locations to produce "destructive interference". Addition of a fifth cosine with period $X_5 = 32$ increases the amplitude at the origin by one while changing the locations of constructive interference to $x = \pm 96, \pm 192, \ldots$. If cosine terms with shorter periods (higher spatial frequencies) are added, they will contribute negative amplitude closer to the origin and tend to "cancel" the positive amplitudes from the lower-frequency terms in that neighborhood. In the limit, summation of cosines with all possible positive and negative frequencies generates infinite amplitude at $x = 0$ where all component sinusoids have unit amplitude, and cancellation at $x \neq 0$ where the amplitudes are scattered over the range $[-1, +1]$. Functions resulting from summing additional sinusoids are shown in Figure 6.25b.

This representation of the Dirac delta function as a summation of a specific set of constituent sinusoids is our first example of Fourier synthesis of a function. In fact, this may be the most significant such result, as it demonstrates that a discontinuous function may be so represented, even though its amplitude jumps from zero to infinity and back again within an infinitesimal distance. Once this result is accepted, then it will not be surprising that similar syntheses may be performed that are valid for "better-behaved" functions.

By changing the variable of integration in Equation (6.104), it is easy to demonstrate that:

$$\delta[-x] = \int_{-\infty}^{+\infty} e^{-2\pi i\xi[-x]}\, d\xi = \int_{-\infty}^{+\infty} e^{+2\pi i\xi x}\, d\xi = \delta[x] \tag{6.106}$$

which confirms the symmetry of $\delta[x]$ from Equation (6.93).

The integral form of $\delta[x]$ also may be used to derive an equivalent expression for a Dirac delta function scaled by a "width parameter" d:

$$\begin{aligned} \delta\left[\frac{x}{d}\right] &= \int_{-\infty}^{+\infty} e^{-2\pi i\xi(x/d)}\, d\xi = \int_{-\infty}^{+\infty} e^{-2\pi i(\xi/d)x}\, d\xi \\ &= \int_{-\infty}^{+\infty} e^{-2\pi i\alpha x}\, |d|\, d\alpha \quad (\textit{for } \xi \equiv \alpha d \Longrightarrow d\xi = |d|\, d\alpha) \\ &= |d| \int_{-\infty}^{+\infty} e^{-2\pi i\alpha x}\, d\alpha = |d|\, \delta[x] \end{aligned} \tag{6.107}$$

This is the "scaling property" of the 1-D Dirac delta function. In words, scaling the "width" and scaling the "amplitude" of $\delta[x]$ by the factor b are equivalent operations.

6.2.1 1-D Dirac Delta Function Raised to a Power

Since the amplitude of the Dirac delta function is undefined at the central ordinate, the concept of a Dirac delta function raised to a power (e.g., $(\delta[x/d])^2$) often creates consternation. The first defining characteristic of $\delta[x]$ in Equation (6.88a) is easy to generalize since $0^n = 0$ for all $n > 0$, so our only task is to appropriately account for the behavior of Equation (6.88b) if the Dirac delta function is raised to a power. Since $\delta[x]$ is not a "function", there is no single correct answer, but we adopt the useful

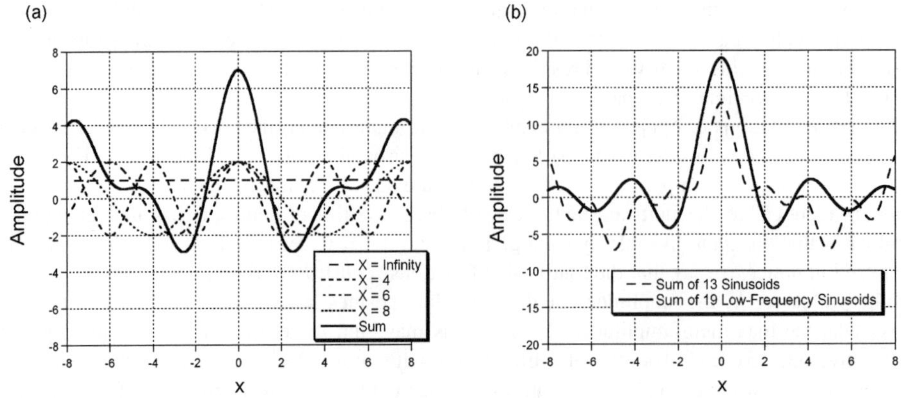

Figure 6.25 Approximations of $\delta[x]$ obtained by summing sinusoidal functions with small spatial frequencies. (a) Sinusoidal components with periods $X = \infty$, four units, six units, and eight units, and their sum, which has a larger amplitude and a narrower central lobe. It is a crude approximation of $\delta[x]$. (b) The sum of 17 and 19 sinusoidal components produces a taller and narrower *SINC*-like function that approaches $\delta[x]$ in the limit.

interpretation that the area of $(\delta[x/d_0])^n$ is the nth power of the area:

$$\int_{-\infty}^{+\infty} \left(\delta\left[\frac{x}{d_0}\right]\right)^n dx = \left(\int_{-\infty}^{+\infty} \delta\left[\frac{x}{d_0}\right] dx\right)^n = (d_0)^n \quad \text{for } n = 1, 2, \ldots \tag{6.108}$$

Because Equation (6.107) showed that a change in the "amplitude" of a scaled Dirac delta is equivalent to changing its area, and vice versa, there is some logic to the idea that raising the amplitude to a power is equivalent to increasing the area by that power. Under this interpretation, the integrand in Equation (6.108) may be rewritten as:

$$\left(\delta\left[\frac{x}{d_0}\right]\right)^n = (d_0)^n \cdot \delta[x] \quad \text{for } n = 1, 2, \ldots \tag{6.109}$$

Note that this interpretation implies that:

$$(\delta[x])^n = \delta[x] \quad \text{for } n = 1, 2, \ldots \tag{6.110}$$

Note again that the fact that $\delta[x]$ is not a "function" means that this result does not necessarily follow from the limiting expressions in terms of other functions in Equations (6.93), (6.95), and (6.96). For example, the limiting expression for $\delta[x]$ in terms of a triangle function in Equation (6.95) clearly does not satisfy Equation (6.110) because the area of $((1/b_0)\ \mathit{TRI}[x/b_0])^2$ is larger than the area of $(1/b_0)\ \mathit{TRI}[x/b_0]$ in the limit $b_0 \to 0$.

This interpretation of the power of a Dirac delta function will be useful in our discussion of optical imaging in Chapters 21 and 22.

6.2.2 Sifting Property of 1-D Dirac Delta Function

Perhaps the most significant property of the Dirac delta function is its ability to evaluate the amplitude of another function at any coordinate in its domain via an integral. In fact, some authors use the mathematical statement of this property as the definition of the Dirac delta function. The infinitesimal support of $\delta[x]$ allows this area of the product of $\delta[x]$ and an arbitrary (finite) function $f[x]$ to be evaluated by a method similar to the approximation of an integral as a summation of rectangular areas. Consider an arbitrary "well-behaved" function $f[x]$ for which all derivatives exist and are finite. The product of $f[x]$ and the representation of $\delta[x]$ as the limit of a sequence of scaled RECT functions is:

$$\int_{-\infty}^{+\infty} f[x]\delta[x]\,dx = \lim_{b\to 0}\left\{\frac{1}{|b|}\int_{-\infty}^{+\infty} RECT\left[\frac{x}{b}\right] f[x]\,dx\right\} \tag{6.111}$$

The area of the limiting rectangle is the width b multiplied by the amplitude evaluated at the origin:

$$\int_{-\infty}^{+\infty} f[x]\delta[x]\,dx = \lim_{b\to 0}\left\{\frac{1}{|b|}(|b|\,f[0])\right\} = f[0] \tag{6.112}$$

In words, the integral of the product of $f[x]$ and $\delta[x]$ has "sifted" out the specific amplitude $f[0]$ from $f[x]$; this is the reason for the name *sifting* property for this characteristic behavior of $\delta[x]$. It is straightforward to translate the Dirac delta function by adding a term of $-x_0$ to the argument of the *RECT* function in Equation (6.111); the analogous integral "sifts" out the amplitude of $f[x]$ at x_0:

$$\int_{-\infty}^{+\infty} f[x]\delta[x-x_0]\,dx = \int_{-\infty}^{+\infty} f[x]\delta[x_0-x]\,dx = f[x_0] \tag{6.113}$$

The form of this expression is identical to that in Equation (5.93) that defined the projection of $f[x]$ onto the "reference" function $m[x]$. Therefore the sifting property of the Dirac delta function may be interpreted as projecting $f[x]$ onto a translated Dirac delta function $\delta[x-x_0]$ to produce the amplitude $f[x_0]$.

The sifting property may be used to derive yet another useful result for a well-behaved $f[x]$:

$$\begin{aligned}\int_{-\infty}^{+\infty} f[x]\delta[x-x_0]\,dx &= f[x_0] = f[x_0]\cdot 1\\ &= f[x_0]\int_{-\infty}^{+\infty}\delta[x-x_0]\,dx = \int_{-\infty}^{+\infty} f[x_0]\,\delta[x-x_0]\,dx\end{aligned} \tag{6.114}$$

Because the two integrals over the same infinite limits evaluate to the same value, the integrands must be equivalent:

$$\boxed{f[x]\,\delta[x-x_0] = f[x_0]\,\delta[x-x_0]} \tag{6.115}$$

Substitution of $f[x]=x$ and $x_0=0$ produces the special case:

$$x\cdot\delta[x] = 0\cdot\delta[x] = 0 \tag{6.116}$$

which is illustrated in Figure 6.26. In words, the product of a Dirac delta function and a "well-behaved" function $f[x]$ is equivalent to a Dirac delta function that has been scaled by the amplitude of $f[x_0]$, where x_0 is the coordinate where the argument of the Dirac delta function is zero. This very useful relation will be applied often during the discussion of discrete linear systems in Chapter 14.

6.2.3 Symmetric (Even) Pair of 1-D Dirac Delta Functions

We follow the definitions of Bracewell and naming conventions of Gaskill for three additional special functions based on the Dirac delta function that are particularly useful for describing sampled systems and optical systems. These are the even and odd pairs of Dirac delta functions and the *COMB* function.

Unit-amplitude Dirac delta functions located at $x=\pm 1$ define the even (or symmetric) pair. The function is denoted by a "double-delta" symbol of the form:

$$\delta\delta[x] \equiv \delta[x+1] + \delta[x-1] \tag{6.117}$$

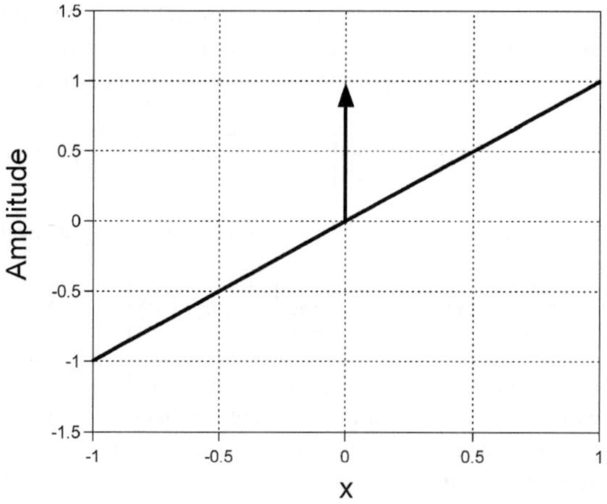

Figure 6.26 Demonstration of sifting property of $\delta[x]$ applied to the specific $f[x] = x$, producing $x \cdot \delta[x] = 0 \cdot \delta[x] = 0[x]$.

and is illustrated in Figure 6.27. The area of the pair is trivial to evaluate:

$$\int_{-\infty}^{+\infty} \delta\delta[x]\,dx = 2 \tag{6.118}$$

By applying the scaling property of the Dirac delta function in Equation (6.107), we obtain an expression for the scaled pair:

$$\begin{aligned}\delta\delta\left[\frac{x}{b}\right] &= \delta\left[\frac{x}{b}+1\right] + \delta\left[\frac{x}{b}-1\right] = \delta\left[\frac{x+b}{b}\right] + \delta\left[\frac{x-b}{b}\right] \\ &= |b|\,(\delta[x+b] + \delta[x-b])\end{aligned} \tag{6.119}$$

This is a pair of Dirac delta functions located at $x = \pm b$, each with area $|b|$. By dividing both sides by $|b|$, we obtain the expression for a pair of Dirac delta functions with unit area located at $\pm b$:

$$\frac{1}{|b|}\delta\delta\left[\frac{x}{b}\right] = \delta[x+b] + \delta[x-b] \tag{6.120}$$

6.2.4 Antisymmetric (Odd) Pair of 1-D Dirac Delta Functions

The odd pair of Dirac delta functions consists of Dirac deltas with areas ± 1 located at $x = \mp 1$, respectively. The notation for this function is:

$${}^{\delta}{}_{\delta}[x] \equiv \delta[x+1] - \delta[x-1] \tag{6.121}$$

In words, the odd pair is denoted by two "δ" symbols where their relative baselines indicate the algebraic sign of the amplitude: the left-hand symbol is above the baseline to indicate that the corresponding delta function located at $x = -1$ has positive amplitude, while the right-hand component ($x = +1$) has negative amplitude. The graphical representation of the odd pair is shown in Figure 6.28. The total area of the odd pair obviously is zero. The scaling property yields this expression for the scaled odd pair at $x = \pm b$:

$${}^{\delta}{}_{\delta}\left[\frac{x}{b}\right] = |b|(\delta[x+b] - \delta[x-b]) \tag{6.122}$$

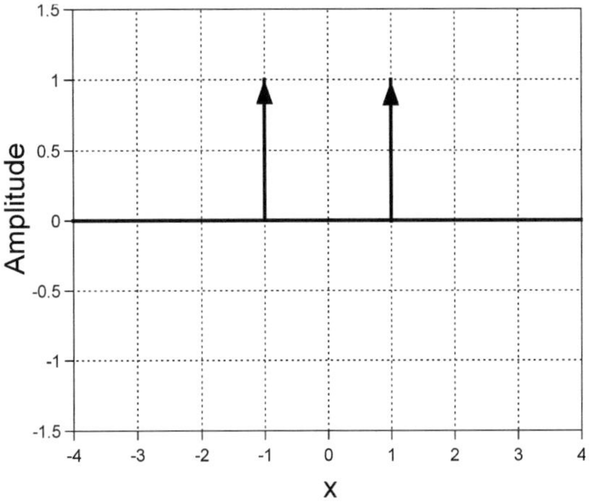

Figure 6.27 Even pair of Dirac delta functions: $\delta\delta[x] \equiv \delta[x+1] + \delta[x-1]$.

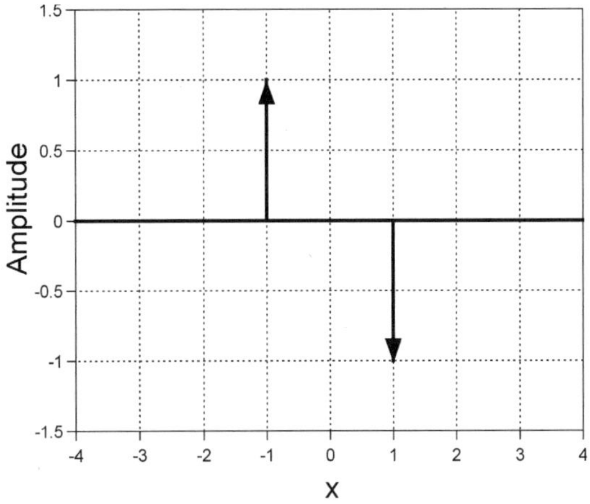

Figure 6.28 Odd pair of Dirac delta functions: ${}^{\delta}{}_{\delta}[x] \equiv \delta[x+1] - \delta[x-1]$.

which leads to the equivalent expression:

$$\frac{1}{|b|}{}^{\delta}{}_{\delta}\left[\frac{x}{b}\right] = \delta[x+b] - \delta[x-b] \tag{6.123}$$

6.2.5 *COMB* Function

A set of Dirac delta functions located at all integer values of x is a special function whose name was selected by Gaskill evidently because of its obvious resemblance of the function to the cosmetic implement (and of which he has fond – though distant – memories!). The graphical representation is

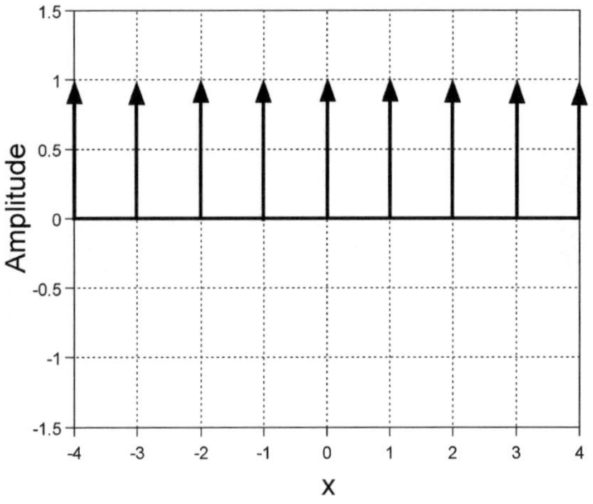

Figure 6.29 The *COMB* function: $COMB[x] \equiv \sum_{n=-\infty}^{+\infty} \delta[x-n]$.

shown in Figure 6.29. The mathematical expression that defines the *COMB* function is:

$$COMB[x] \equiv \sum_{n=-\infty}^{+\infty} \delta[x-n] \tag{6.124}$$

Note that the areas of $COMB[x]$ and of the unit constant $1[x]$ are infinite but identical because each unit-width segment encloses an area of unity. The location of the center of symmetry of the *COMB* function may be translated by adding a parameter $-x_0$ to the argument:

$$\sum_{n=-\infty}^{+\infty} \delta[x-x_0-n] = COMB[x-x_0] \tag{6.125}$$

The interval between the elements of the *COMB* may be scaled by applying the scaling property of the Dirac delta function from Equation (6.107):

$$\begin{aligned} COMB\left[\frac{x}{b}\right] &= \sum_{n=-\infty}^{+\infty} \delta\left[\frac{x}{b}-n\right] = \sum_{n=-\infty}^{+\infty} \delta\left[\frac{x-nb}{b}\right] \\ &= \sum_{n=-\infty}^{+\infty} |b|\,\delta[x-nb] = |b| \sum_{n=-\infty}^{+\infty} \delta[x-nb] \end{aligned} \tag{6.126}$$

This expression indicates that the *COMB* function composed of unit-area Dirac delta functions centered at the origin and separated by intervals b is:

$$\sum_{n=-\infty}^{+\infty} \delta[x-nb] = \frac{1}{|b|} COMB\left[\frac{x}{b}\right] \tag{6.127}$$

The *COMB* function is commonly used to modulate a function defined over a continuous domain to construct a function with a discrete domain. This process of "sampling" will be considered in detail in Chapter 14. However, we note at this point that the resulting product must be interpreted with care. For example, consider the modulation of a sinusoid by unit-amplitude Dirac delta functions separated by intervals of width Δx:

$$\cos[2\pi\xi_0 x+\phi_0]\left(\frac{1}{\Delta x}COMB\left[\frac{x}{\Delta x}\right]\right)=\cos[2\pi\xi_0 x+\phi_0]\left(\sum_{n=-\infty}^{+\infty}\delta[x-n\,\Delta x]\right)$$

$$=\sum_{n=-\infty}^{+\infty}\cos[2\pi\xi_0 x+\phi_0]\delta[x-n\,\Delta x]$$

$$=\sum_{n=-\infty}^{+\infty}\cos[2\pi\xi_0(n\Delta x)+\phi_0]\delta[x-n\,\Delta x] \tag{6.128}$$

where the last step follows from the sifting property in Equation (6.113). This result demonstrates that the product of this specific continuous function and a *COMB* function is a set of equispaced Dirac delta functions whose areas are weighted by the amplitude of the continuous function evaluated at those specific locations. An equivalent expression for the *COMB* function may be generated by extending the result in Equation (6.97), which expressed the Dirac delta function as a finite-area *SINC* function in the limit of zero width:

$$\delta[x]=\lim_{N\to\infty}\left\{\frac{\sin[N\pi x]}{\pi x}\right\}=\lim_{N\to\infty}\left\{\frac{\sin[2\pi(N/2)x]}{\pi x}\right\} \tag{6.129}$$

A *COMB* function is obtained by evaluating the ratio of two terms that have null amplitude and unit slope at all integer values of x. The sinusoid in the numerator already satisfies this requirement. A function that exhibits “almost” the desired behavior in the denominator is $\sin[2\pi x/2]=\sin[\pi x]$, which has null amplitude at all integer values of x. However, its slope is positive at all even integer coordinates x and negative at odd integer coordinates. Thus the ratio of slopes of the two functions will have the incorrect algebraic sign at the odd values of x. The correct ratio is obtained by restricting the values of N in the numerator function $\sin[N\pi x]$ to be odd so that its slope also is negative at integer values of x. At noninteger coordinates, $\sin[N\pi x]$ oscillates rapidly and $\sin[\pi x]$ oscillates slowly about zero. In the limit of large odd N, the rapid oscillation of $\sin[N\pi x]$ ensures that the local amplitude approaches zero. Thus an equivalent expression for the *COMB* function is:

$$COMB[x]=\lim_{\text{odd } N\to\infty}\left\{\frac{\sin[N\pi x]}{\sin[\pi x]}\right\} \tag{6.130}$$

Examples are presented in Figure 6.30.

6.2.6 Derivatives of 1-D Dirac Delta Function

A representation of the derivative of the Dirac delta function may be defined by using the scaled-*RECT* expression for $\delta[x]$ in Equation (6.92) and the representation of $RECT[x]$ in terms of $STEP[x]$ in Equation (6.13b):

$$\frac{d\delta[x]}{dx}\equiv\delta^{(1)}[x]=\lim_{b\to 0}\left\{\frac{d}{dx}\left(\frac{1}{|b|}RECT\left[\frac{x}{b}\right]\right)\right\}$$

$$=\lim_{b\to 0}\left\{\frac{1}{|b|}\frac{d}{dx}\left(STEP\left[x+\frac{b}{2}\right]-STEP\left[x-\frac{b}{2}\right]\right)\right\}$$

$$=\lim_{b\to 0}\left\{\frac{1}{|b|}\left(\delta\left[x+\frac{b}{2}\right]-\delta\left[x-\frac{b}{2}\right]\right)\right\}$$

$$=\lim_{b\to 0}\left\{\frac{1}{|b|}\left(\delta\left[\frac{((2x/b)+1)}{(2/b)}\right]-\delta\left[\frac{((2x/b)-1)}{(2/b)}\right]\right)\right\}$$

$$=\lim_{b\to 0}\left\{\frac{1}{|b|}\cdot\frac{2}{|b|}\delta_\delta\left[\frac{2x}{b}\right]\right\}=\lim_{b\to 0}\left\{\frac{2}{|b|^2}\delta_\delta\left[\frac{x}{(b/2)}\right]\right\} \tag{6.131}$$

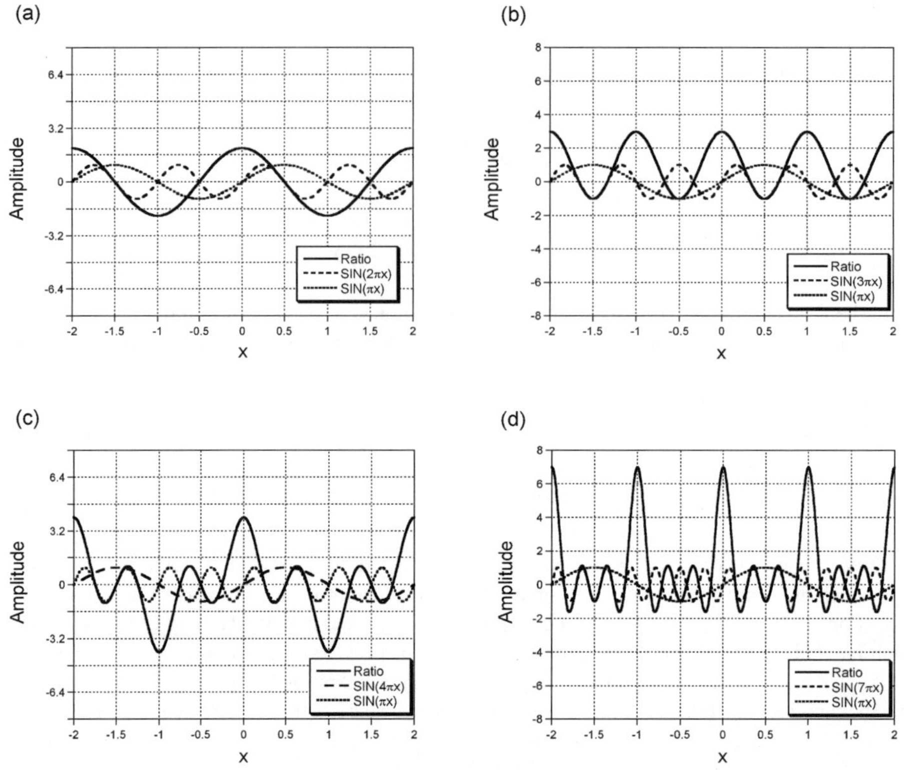

Figure 6.30 Approximations for *COMB*$[x]$ as ratios of sinusoids: (a) $\sin[2\pi x]/\sin[\pi x] = 2\cos[\pi x]$; (b) $\sin[3\pi x]/\sin[\pi x] = 2\cos[2\pi x] + 1$; (c) $\sin[4\pi x]/\sin[\pi x]$; (d) $\sin[7\pi x]/\sin[\pi x]$. To approximate *COMB*$[x]$, the numerator must be an odd multiple of πx.

Note that $\delta^{(1)}[x]$ is an odd function because $\delta[x]$ is even, and so the integral of $\delta^{(1)}[x]$ over symmetric limits must be zero. By differentiating the expression for $f[x]$ within the sifting property in Equation (6.113) over x, the sifting property for derivatives is obtained:

$$\begin{aligned}\frac{df}{dx} &= \frac{d}{dx}\left(\int_{-\infty}^{+\infty} f[\alpha]\delta[x-\alpha]\,d\alpha\right) = \int_{-\infty}^{+\infty} f[\alpha]\frac{d}{dx}(\delta[x-\alpha])\,d\alpha \\ &= \int_{-\infty}^{+\infty} f[\alpha]\delta^{(1)}[x-\alpha]\,d\alpha \end{aligned} \tag{6.132}$$

In words, this result demonstrates that the derivative of $f[x]$ may be generated by sifting it with the derivative of the Dirac delta function. Similarly, the nth derivative of $f[x]$ is obtained by sifting with the nth derivative of $\delta[x]$:

$$\begin{aligned}\int_{-\infty}^{+\infty} f[\alpha]\frac{d^n}{dx^n}(\delta[x-\alpha])\,d\alpha &= \frac{d^n}{dx^n}\int_{-\infty}^{+\infty} f[\alpha]\delta[x-\alpha]\,d\alpha \\ &= \frac{d^n f}{dx^n} \equiv f^{(n)}[x] \end{aligned} \tag{6.133}$$

Gaskill uses the graphical representation shown in Figure 6.31 for the derivative of the Dirac delta function. Both component Dirac delta functions in this so-called "doublet" rest on the horizontal axis at

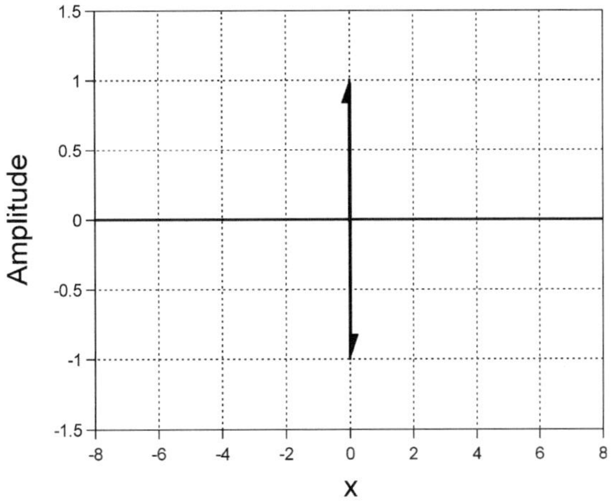

Figure 6.31 $\delta'[x]$ represented as "doublet" of Dirac delta functions.

the origin of coordinates. The left and right halves of the arrowhead are on the upward- and downward-pointing lines to indicate the analogous behavior of $\delta^{(1)}[x]$ and an odd pair of Dirac delta functions in Equation (6.121).

6.2.7 Dirac Delta Function with Functional Argument

Finally, consider the form of a Dirac delta function whose argument is itself a function of the continuous spatial coordinate x, such as $g[x]$. One obvious example is $g[x] = x/b$, which we showed to give the result in Equation (6.107). This result may be extended easily to the case where g is the general linear function, $g[x] = mx + b$, where m is the slope and b is the y-intercept of the line. The 1-D Dirac delta function may be rewritten and again evaluated by applying Equation (6.107):

$$\delta[mx + b] = \delta\left[\frac{x + (b/m)}{1/m}\right] = \left|\frac{1}{m}\right|\delta\left[x - \left(-\frac{b}{m}\right)\right] \tag{6.134}$$

where $-b/m$ is the x-intercept of the line. In words, the Dirac delta function whose argument is a linear function of x is located at the x-intercept and its area is the absolute value of the reciprocal of the slope. Examples with both positive and negative slopes are shown in Figure 6.32. Note that the Dirac delta function has a positive weight for both positive and negative slopes.

In the more general case, the functional form may be evaluated if $g[x]$ satisfies certain conditions to be described. Such expressions are sufficiently common in imaging (e.g., in the derivation of the inverse Radon transform in Chapter 12) to make its consideration useful. The discussion is straightforward and may help develop a better intuitive understanding of the Dirac delta function. From the first criterion for the Dirac delta function in Equation (6.88a), it is evident that $\delta[g[x]] = 0$ at all coordinates where the amplitude $g[x] \neq 0$. For example, if $g[x] = 2 + \cos[2\pi x]$, then $\delta[g[x]] = 0[x]$ because $g[x] > 0$ for all x. The second criterion of unit area in Equation (6.88b) for a functional argument is considered first in the case where $g[x]$ has a single zero located at x_0, so that $g[x_0] = 0$. If the derivatives of $g[x]$ exist and

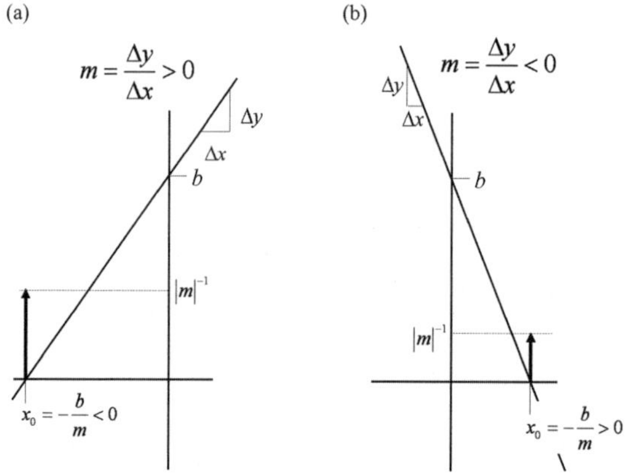

Figure 6.32 The 1-D Dirac delta function of a linear argument $\delta[mx + b]$: (a) if the slope m and y-intercept b are positive, then the argument of the Dirac delta function is zero at $x_0 = -b/m < 0$ and the scaling is $|m|^{-1} > 0$; (b) if $m < 0$ and $b > 0$, then the argument is zero at $x_0 = -b/m > 0$ and the weighting is still positive. If the y-intercept is negative, then the argument of the Dirac delta function is zero for positive and negative x_0, respectively.

are finite in the vicinity of x_0, then $g[x]$ may be expanded in a Taylor series about that coordinate:

$$\begin{aligned} g[x] &= \sum_{n=0}^{+\infty} a_n (x - x_0)^n = \sum_{n=0}^{\infty} \frac{1}{n!} \left(\frac{d^n g[x]}{dx^n} \right)\bigg|_{x=x_0} (x - x_0)^n \\ &= g[x_0] + (x - x_0) \left(\frac{dg}{dx} \right)\bigg|_{x=x_0} + \frac{(x - x_0)^2}{2} \left(\frac{d^2 g}{dx^2} \right)\bigg|_{x=x_0} + \cdots \\ &= 0 + (x - x_0) \left(\frac{dg}{dx} \right)\bigg|_{x=x_0} + \frac{(x - x_0)^2}{2} \left(\frac{d^2 g}{dx^2} \right)\bigg|_{x=x_0} + \cdots \end{aligned} \tag{6.135}$$

Because the amplitude of $\delta[g[x]]$ is zero except at $x = x_0$, the area of $\delta[g[x]]$ is concentrated in the vicinity of x_0, so it may be calculated by restricting the domain of the integral to $x_0 \pm \varepsilon$, where ε is an arbitrarily small positive real number:

$$\begin{aligned} \int_{-\infty}^{+\infty} \delta[g[x]]\, dx &\cong \int_{x_0-\epsilon}^{x_0+\epsilon} \delta[g[x]]\, dx \\ &= \int_{x_0-\epsilon}^{x_0+\epsilon} \delta\left[(x - x_0) \left(\frac{dg}{dx} \right)\bigg|_{x=x_0} + \frac{(x - x_0)^2}{2} \left(\frac{d^2 g}{dx^2} \right)\bigg|_{x=x_0} + \cdots \right] dx \end{aligned} \tag{6.136}$$

This restriction of the domain of x to the neighborhood of x_0 ensures that only the smallest-order term of the Taylor series with nonzero amplitude need be retained. For example, we recognize that $(x - x_0) \gg (x - x_0)^n$ for $n > 1$ if $x \cong x_0$, so if the first derivative of $g[x]$ is nonzero and finite at x_0, then Equation (6.136) may be approximated by a much simpler expression:

$$\begin{aligned}\int_{-\infty}^{+\infty}\delta[g[x]]\,dx &\cong \int_{x_0-\epsilon}^{x_0+\epsilon}\delta\left[(x-x_0)\left(\frac{dg}{dx}\right)\bigg|_{x=x_0}\right]dx = \int_{x_0-\epsilon}^{x_0+\epsilon}\frac{\delta[x-x_0]}{(|dg/dx|)|_{x=x_0}}\,dx\\ &= \left(\left|\frac{dg}{dx}\right|^{-1}\right)\bigg|_{x=x_0}\cdot\int_{x_0-\epsilon}^{x_0+\epsilon}\delta[x-x_0]\,dx\\ &= \left(\left|\frac{dg}{dx}\right|^{-1}\right)\bigg|_{x=x_0}\cdot 1\end{aligned} \tag{6.137}$$

where the scaling property of the Dirac delta function in Equation (6.107) and the unit area of $\delta[x-x_0]$ have been used. Therefore, the requirements for the Dirac delta function are satisfied if $\delta[g[x]]$ has the form:

$$\delta[g[x]] = \frac{\delta[x-x_0]}{(|dg/dx|)|_{x=x_0}} \tag{6.138}$$

which requires that $g[x_0]=0$ and $dg/dx|_{x=x_0}\neq 0$. If $g[x_0]=0$ and $dg/dx|_{x=x_0}\neq 0$ and finite, then all terms in the Taylor series of order 3 and higher are neglected in Equation (6.136) and Equation (6.107) may be used to obtain the expression:

$$\delta[g[x]] = \delta\left[\frac{(x-x_0)^2}{2}\left(\frac{d^2g}{dx^2}\right)\bigg|_{x=x_0}\right] = \frac{2\delta[(x-x_0)^2]}{|(d^2g/dx^2)|_{x=x_0}|} \tag{6.139}$$

However, we are now stuck; further simplification of this result requires evaluation of $\delta[x^2]$, which has the same qualities that were used to derive the expression in Equation (6.138): that $g[x_0]=g[0]=0$ and $(dg/dx)|_{x_0=0}=0$. The expression for the Dirac delta function of a functional argument in Equation (6.138) under these conditions is no more useful than its original form. In the more general case where $g[x_n]=0$ at the N locations x_n and $g'[x_n]\neq 0$ at each, the development leading to Equation (6.137) may be applied at each of the zeros to produce an expression for $\delta[g[x]]$ that is a summation over the zeros:

$$\delta[g[x]] = \sum_{n=1}^{N}\frac{\delta[x-x_n]}{|(dg/dx|_{x=x_n})|} \tag{6.140}$$

Consider some examples of the Dirac delta function of a functional argument. The first is the quadratic function $g_1[x]=x^2$ that was just mentioned and is shown in Figure 6.33. It has one zero at $x=0$, but its derivative also evaluates to zero there so that substitution into Equation (6.35) produces the worthless result $\delta[x^2]=\delta[x^2]$. We leave this situation as is, recalling the observation that some of the qualities of the improper Dirac delta function are not well defined and are the source of this difficulty. The second example function $g_2[x]=x^2-1$ has two zeros located at $x=\pm 1$ and the respective slopes at these points are ± 2. The conditions of Equation (6.140) are valid, and the equivalent expression is:

$$\begin{aligned}\delta[x^2-1] &= \frac{\delta[x+1]}{|-2|}+\frac{\delta[x-1]}{|+2|}\\ &= \frac{1}{2}(\delta[x+1]+\delta[x-1]) = \frac{1}{2}\delta\delta[x]\end{aligned} \tag{6.141}$$

where the definition of the even pair of Dirac delta functions in Equation (6.117) has been used. This is a useful identity for the even pair of Dirac delta functions in some contexts. As a third example, consider $g_3[x]=\sin[2\pi\xi_0 x]$, which has zeros at integer multiples of $1/(2\xi_0)$, at which points the slopes are

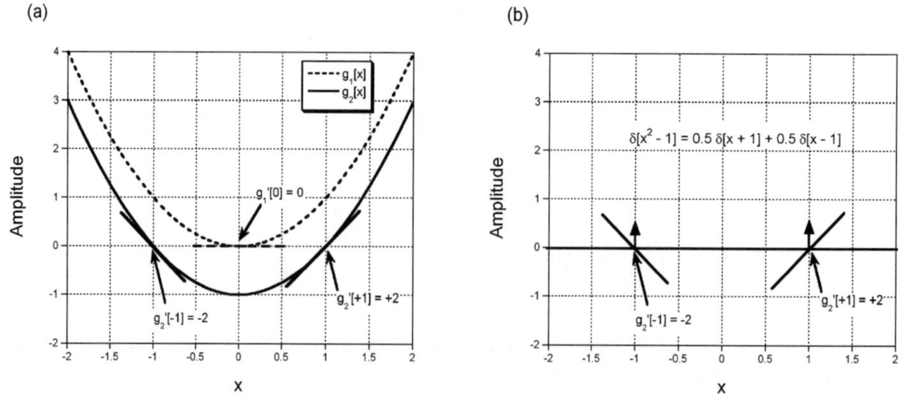

Figure 6.33 Dirac delta function of functional argument $\delta[g[x]]$: (a) $g_1[x] = x^2$ has one zero located at $x = 0$ and its slope $g_1'[0] = 0$, $g_2[x] = x^2 - 1$ has zeros at $x = \pm 1$ with slopes $g_2'[\pm 1] = \pm 2$; (b) the Dirac delta function $\delta[g_1[x]]$ is not defined, while $\delta[g_2[x]] = 0.5(\delta[x+1] + \delta[x-1])$.

$\pm 2\pi\xi_0$. The equivalent expression is obtained from Equation (6.138):

$$\begin{aligned}\delta[\sin[2\pi\xi_0 x]] &\\ &= \frac{\cdots + \delta[x + (2/2\xi_0)] + \delta[x + (1/2\xi_0)] + \delta[x] + \delta[x - (1/2\xi_0)] + \delta[x - (2/2\xi_0)] + \cdots}{|\pm 2\pi\xi_0|} \\ &= \frac{1}{2\pi|\xi_0|}\sum_{n=-\infty}^{+\infty}\delta\left[x - \frac{n}{2\xi_0}\right] = \frac{1}{2\pi\xi_0}\, COMB\left[\frac{2\xi_0 x - n}{2\xi_0}\right] \\ &= \frac{1}{\pi}\, COMB[2\xi_0 x]\end{aligned} \tag{6.142}$$

An equivalent expression for $COMB[x]$ may be written in terms of $\sin(\pi x)$ by setting $\xi_0 = \frac{1}{2}$:

$$COMB[x] = \pi\ \delta[\sin[\pi x]] = \pi\,\delta\left[\sin\left[2\pi\frac{x}{2}\right]\right] \tag{6.143}$$

6.3 1-D Complex-Valued Special Functions

A complex-valued 1-D function may be created by assigning examples of individual real-valued special functions considered in Section 6.1 as the real and imaginary parts. For example, consider the simple complex function whose real and imaginary parts are $RECT[x]$ and $RECT[x/2]$, respectively, as shown in Figure 6.34:

$$f[x] = f_R[x] + i\ f_I[x] = RECT\left[\frac{x}{2}\right] + i RECT\left[\frac{x}{4}\right] \tag{6.144}$$

Its magnitude and phase are easily calculated via the relations in Equation (4.27a):

$$|f[x]| \equiv \sqrt{(f_R[x])^2 + (f_I[x])^2} = (\sqrt{2} - 1)\ RECT\left[\frac{x}{2}\right] + RECT\left[\frac{x}{4}\right] \tag{6.145a}$$

$$\Phi\{f[x]\} \equiv \tan^{-1}\left[\frac{f_I[x]}{f_R[x]}\right] \tag{6.145b}$$

Particularly note the algebraic signs of the phase angles for this function.

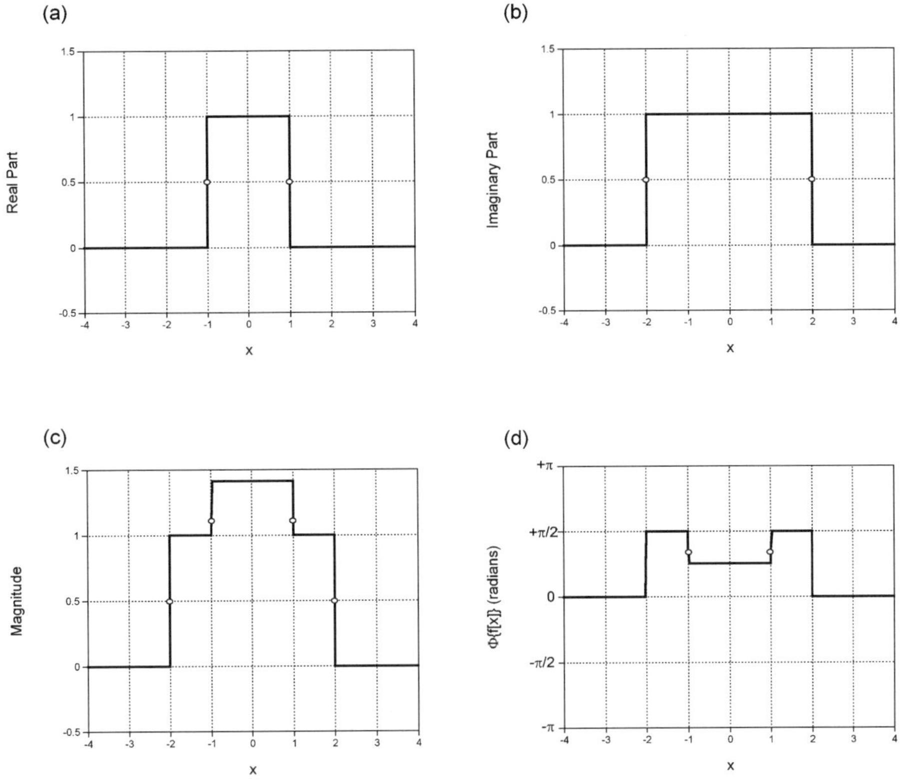

Figure 6.34 Complex-valued function $f[x] = RECT[x/2] + iRECT[x/4]$ constructed from real functions: (a) real part; (b) imaginary part; (c) magnitude; and (d) phase.

6.3.1 Complex Linear-Phase Sinusoid

Some important special functions are defined in complex-valued form. The most important is the complex linear-phase exponential introduced when deriving the Euler relation in Equation (4.19):

$$\begin{aligned} e^{\pm 2\pi i \xi_0 x} &= \cos[2\pi\xi_0 x] \pm i\,\sin[2\pi\xi_0 x] \\ &= \cos[2\pi\xi_0 x] \pm i\,\cos\left[2\pi\xi_0 x - \frac{\pi}{2}\right] \end{aligned} \tag{6.146}$$

The magnitude of the linear-phase exponential was shown to be unity in Equation (4.32) and the phase to be a linear function of x in Equation (4.33). The real/imaginary, magnitude/phase, and Argand representations are shown in Figure 6.35.

6.3.2 Complex Quadratic-Phase Exponential Function

The Euler relation may be used to define a complex function whose real and imaginary parts are quadratic-phase sinusoids in quadrature. This complex chirp has some very interesting and useful properties and is essential for describing optical imaging systems. Just as in the linear-phase case, the real and imaginary parts of the complex chirp are unit-amplitude quadratic-phase sinusoids originally

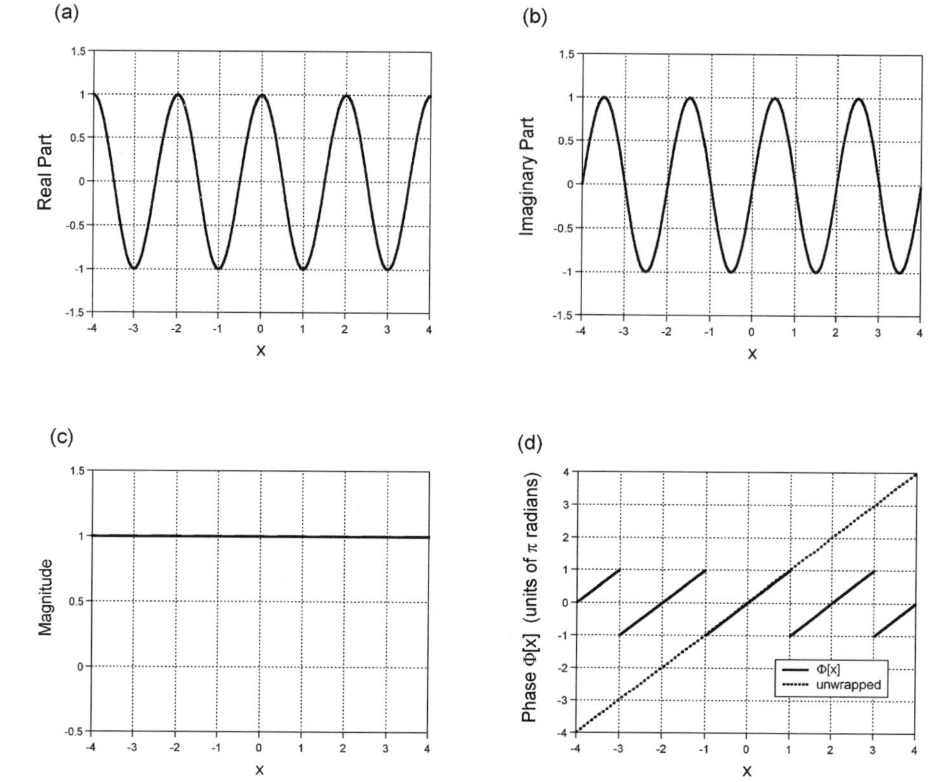

Figure 6.35 Complex sinusoid $f[x] = \exp[+2\pi i x/2]$:
(a) $\Re\{\exp[+2\pi i x/2]\} = \cos[\pi x]$;
(b) $\Im\{\exp[+2\pi i x/2]\} = \sin[\pi x]$;
(c) $|\exp[+2\pi i x/2]| = 1[x]$; and
(d) phase shown over the finite range $[-\pi, +\pi)$ and after "unwrapping" over the range $(-\infty, +\infty)$.

defined in Equation (6.53):

$$e^{\pm i\pi(x/\alpha)^2} = \cos\left[\frac{\pi x^2}{\alpha^2}\right] \pm i \sin\left[\frac{\pi x^2}{\alpha^2}\right] \tag{6.147}$$

The magnitude and phase from Equation (4.27) are the unit constant and a quadratic function of x, respectively. Note that an additive constant phase also may be applied to create $\exp[\pm i(\pi(x/\alpha)^2 + \phi_0)]$, but that the complex function still is symmetric due to the dependence of the argument on x^2. The spatial frequency of the complex chirp is proportional to the rate of change of phase, as was defined by Equation (4.47):

$$\xi[x] = \pm\frac{1}{2\pi}\frac{\partial\Phi}{\partial x} = \pm\frac{1}{2\pi}\frac{\partial(\pm\pi(x/\alpha)^2)}{\partial x} = \pm\frac{x}{\alpha^2} \tag{6.148}$$

The area of the complex chirp obviously is complex valued. The Gaussian and complex chirp functions are defined by a single functional form throughout their real-valued domains. Their "widths" may be scaled by complex-valued analogues of the factor b. For example:

$$\exp\left[\pm i\pi\left(\frac{x}{\sqrt{\mp i}}\right)^2\right] = \exp\left[\pm i\pi\left(\frac{x^2}{\mp i}\right)\right] = \exp[-\pi x^2] = GAUS[x] \tag{6.149a}$$

It also is possible to express the chirp function in the form of a Gaussian:

$$\exp[\pm i\pi x^2] = \exp[-(\mp i)\pi x^2] = \exp[-\pi(x\sqrt{\mp i})^2]$$
$$= GAUS[(\sqrt{\mp i})x] = GAUS\left[\frac{x}{\sqrt{\pm i}}\right] \tag{6.149b}$$

These expressions will be useful in several contexts later in the book.

6.3.3 "Superchirp" Function

By analogy with the definition of the superGaussian function of Equation (6.62), the complex chirp function in Equation (6.147) may be generalized to create a "superchirp" function that includes the exponent as a parameter:

$$e^{\pm i\pi x^n} = \cos[\pi x^n] \pm i\,\sin[\pi x^n] \tag{6.150}$$

Regardless of the integer value of the exponent n, the real part of the superchirp is always even because of the cosine function, while the imaginary part is even for even n and odd for odd n, so the complex-valued function is even when n is even and Hermitian when n is odd. Unlike the Gaussian, we will not introduce a special shorthand name for the superchirp as it will be used relatively rarely and the notation in Equation (6.150) already is rather concise. A comparison between the superchirp in Equation (6.150) and the superGaussian in Equation (6.62) is useful. Recall that it was necessary to define the superGaussian as a function of $|x|$ to ensure that its amplitude remained finite for $x < 0$ when n is odd. This constraint is not required for the superchirp because its magnitude is unity everywhere. However, it sometimes is useful to construct a symmetric superchirp for odd values of n by substituting $|x|$ in the argument:

$$f[x] = e^{\pm i\pi|x|^n} = \cos[\pi|x|^n] \pm i\,\sin[\pi|x|^n] \Longrightarrow f[-x] = f[x] \tag{6.151}$$

Similarly, it is possible to construct a Hermitian superchirp for even values of n by forcing the imaginary part to be odd via multiplication by $SGN[x]$:

$$f[x] = \cos[\pi x^n] \pm i\,SGN[x]\sin[\pi|x|^n] \Longrightarrow f^*[-x] = f[x] \tag{6.152}$$

The symmetric superchirp may be recast into the form of a superGaussian by following the recipe that relates the quadratic-phase sinusoid in Equation (6.147) and the normal Gaussian function in Equation (6.57). The appropriate power of i is included in the scale ("width") factor for the superGaussian. A symmetric superchirp may be expressed in terms of the corresponding nth-order superGaussian:

$$\exp[\pm i\pi|x|^n] = \exp[-\pi((\mp i)^{1/n}|x|)^n] = GAUS[((\mp i)^{1/n}|x|);\,n] \tag{6.153}$$

The real and imaginary parts of $e^{+i\pi x^n}$ are shown in Figure 6.36a for several values of n. Note that the amplitude of the superchirp at $x = 1$ is $e^{\pm i\pi} = -1 + i\cdot 0$ regardless of the value of n. In Equation (6.22), we demonstrated that the area of a linear-phase sinusoid is zero, while that of a quadratic-phase sinusoid such as $\cos[\pi x^2]$ was shown to be positive in Equation (6.56). This was interpreted in terms of the so-called "point of stationary phase" of the quadratic-phase sinusoid at $x = 0$. The higher-order superchirp functions exhibit a larger region where the phase is approximately stationary in the vicinity of the origin, so that the real part is approximately unity and the imaginary part is approximately zero. For this reason, the areas of the central lobes of the real and imaginary parts respectively increase and decrease with n. Just as for the quadratic-phase function, the main contribution to the area of the superchirp comes from the central lobes where the phase is approximately stationary. This is confirmed by deriving the area of the entire superchirp as a function of n by a simple variant of the derivation that began in Equation (6.67). The factor of $u = i\pi x^n$ is substituted in Equation (6.67), followed by the division of both sides by $n(i\pi)^\alpha$ in Equation (6.68). The resulting expression for the integral of $e^{\pm i\pi x^n}$

over "semi-infinite" limits is:

$$\int_0^{+\infty} e^{\pm i\pi x^n}\,dx = \frac{1}{n}\Gamma\left[\frac{1}{n}\right]\pi^{-1/n}\exp\left[\pm\frac{i\pi}{2n}\right] = \left(\frac{1}{\pi}\right)^{1/n}\exp\left[\pm\frac{i\pi}{2n}\right]\Gamma\left[1+\frac{1}{n}\right] \tag{6.154}$$

where the recursion relation for the gamma function in Equation (6.41) has been used. The area of the symmetric superchirp obviously is twice this value:

$$\int_{-\infty}^{+\infty} e^{\pm i\pi |x|^n}\,dx = 2\Gamma\left[1+\frac{1}{n}\right]\pi^{-1/n}\exp\left[\pm\frac{i\pi}{2n}\right] \tag{6.155}$$

Note that the area of the Hermitian superchirp (odd n) must be real valued because the area of the odd imaginary part is zero:

$$\begin{aligned}\int_{-\infty}^{+\infty} e^{\pm i\pi x^n}\,dx &= \Re\left\{2\cdot\pi^{-1/n}\Gamma\left[1+\frac{1}{n}\right]\exp\left[\pm\frac{i\pi}{2n}\right]\right\}\\ &= 2\Gamma\left[1+\frac{1}{n}\right]\pi^{-1/n}\cos\left[\frac{\pi}{2n}\right] \quad \textit{(for odd n)}\end{aligned} \tag{6.156}$$

The values of the gamma function for the reciprocal integers that are required to evaluate the areas of the superchirps for the first few values of n were presented in Equation (6.51). Clearly, substitution of $n = 2$ yields the standard form of the quadratic-phase complex exponential in Equation (6.146) with unit chirp rate ($\alpha = 1$). Equation (6.156) establishes that its area is:

$$\begin{aligned}\int_{-\infty}^{+\infty} e^{\pm i\pi x^2}\,dx &= e^{\pm i\pi/4}\,\pi^{-\frac{1}{2}}\frac{2}{|2|}\Gamma\left[\frac{1}{2}\right] = e^{\pm i\pi/4}\left(\frac{1}{\sqrt{\pi}}\right)\cdot 1\cdot\sqrt{\pi} = e^{\pm i\pi/4}\\ &= \left(\frac{1}{\sqrt{2}}\right)(1\pm i) \cong 0.707(1\pm i)\end{aligned} \tag{6.157a}$$

which confirms the hypothesis put forth in Equation (6.56) that the areas of the real-valued cosine and sine chirp functions are identical. The real-valued area of the Hermitian superchirp with $n = 3$ is obtained from Equation (6.156):

$$\begin{aligned}\int_{-\infty}^{+\infty} e^{\pm i\pi x^3}\,dx &= \Re\left\{2\,e^{\pm i\pi/6}\,\pi^{-\frac{1}{3}}\Gamma\left[\frac{4}{3}\right]\right\}\\ &\cong \cos\left[\frac{\pi}{6}\right]\frac{1}{\pi^{\frac{1}{3}}}\cdot 2\cdot 0.893 \cong 1.056\end{aligned} \tag{6.157b}$$

while the area of the corresponding symmetric chirp is complex valued:

$$\begin{aligned}\int_{-\infty}^{+\infty} e^{\pm i\pi |x|^3}\,dx &= 2\,e^{\pm i\pi/6}\,\pi^{-\frac{1}{3}}\Gamma\left[\frac{4}{3}\right]\\ &\cong 1.219\,e^{\mp i\pi/6} \cong 1.056 \pm i(0.609)\end{aligned} \tag{6.157c}$$

The area of the superchirp with $n = 4$ also is complex valued:

$$\begin{aligned}\int_{-\infty}^{+\infty} e^{\pm i\pi x^4}\,dx &= 2\,e^{+i\pi/8}\,\pi^{-\frac{1}{4}}\Gamma\left[\frac{5}{4}\right]\\ &= \frac{1}{2}\cdot 3.6256 \cong 1.2580 \pm i(0.5211)\end{aligned} \tag{6.157d}$$

Note that the real part has larger area than $e^{+i\pi x^2}$, while the area of the imaginary part is smaller. The area of the Hermitian chirp for $n = 5$ is real-valued:

$$\int_{-\infty}^{+\infty} e^{\pm i\pi x^5}\, dx = \Re\left\{2 \cdot e^{\pm i\pi/10}\, \pi^{-\frac{1}{5}} \Gamma\left[\frac{6}{5}\right]\right\}$$
$$= \cos\left[\frac{\pi}{10}\right] \cdot \pi^{-\frac{1}{5}} \cdot 2 \cdot 0.9182 \cong 1.3891 \tag{6.157e}$$

and larger than the area for $n = 3$ in Equation (6.157c). The real-valued area of the symmetric chirp for $n = 5$ is larger than for $n = 3$, while the imaginary part is smaller:

$$\int_{-\infty}^{+\infty} e^{\pm i\pi |x|^5}\, dx = 2\, e^{\pm i\pi/10}\, \pi^{-\frac{1}{5}} \Gamma\left[\frac{6}{5}\right]$$
$$\cong 1.2891 \pm i(0.4513) \tag{6.157f}$$

Some trends in these results are obvious. The real part of the area increases with n, while the magnitude of the areas of the imaginary part of the symmetric chirps decreases with increasing n. This is because the "bulk" of the total area is due to the "central lobes" of the functions, which are in quadrature. As n increases, a larger portion of the real lobe has an amplitude close to unity, while the amplitude of the corresponding portion of the imaginary lobe is close to zero. In the limit as $n \to +\infty$, the area of the superchirp becomes:

$$\lim_{n\to\infty}\left\{\int_{-\infty}^{+\infty} e^{-i\pi x^n}\, dx\right\} = \lim_{n\to\infty}\left\{e^{+i\pi/n}\, \pi^{-1/n} \frac{2}{n} \Gamma\left[\frac{1}{n}\right]\right\}$$
$$\Longrightarrow e^0 \cdot \pi^0 \cdot 2 \cdot \lim_{n\to\infty}\left\{\frac{1}{n}\Gamma\left[\frac{1}{n}\right]\right\} = 2 + 0i \tag{6.157g}$$

where Equation (6.52) has been used. Note that the area of this infinite-order superchirp is identical to that of the infinite-order superGaussian in Equation (6.72). The infinite-order superchirp exhibits constant unit amplitude out to the transition at $|x| = 1$. The function outside of this point oscillates at an "infinite rate" with unit magnitude; the resulting lobes have null area. These limiting cases are illustrated in Figure 6.36b:

n	$\int_{-\infty}^{+\infty} e^{\pm i\pi x^n}\, dx$	$\int_{-\infty}^{+\infty} e^{\pm i\pi \lvert x\rvert^n}\, dx$
2	$\frac{1}{\sqrt{2}}(1 \pm i) \cong 0.707(1 \pm i)$	$\frac{1}{\sqrt{2}}(1 \pm i) \cong 0.707(1 \pm i)$
3	$\cong 1.056$	$\cong 1.056 \pm i(0.609)$
4	$\cong 1.2580 \pm i(0.521)$	$\cong 1.2580 \pm i(0.521)$
5	$\cong 1.2891$	$\cong 1.2891 \pm i(0.451)$

(6.158)

The representation of the superchirp in terms of its constituent sinusoids will be considered in Chapter 13, and its properties and uses will be explored in greater detail in Chapter 17 when considering phase filters.

6.3.4 Complex-Valued Lorentzian Function

A complex-valued function whose real and imaginary parts are based on the real-valued Lorentzian of Equation (6.83) will be shown to be closely related to the decaying exponential function evaluated for $x > 0$ that was introduced in Equation (6.16). For a not-yet-obvious reason, we define the complex

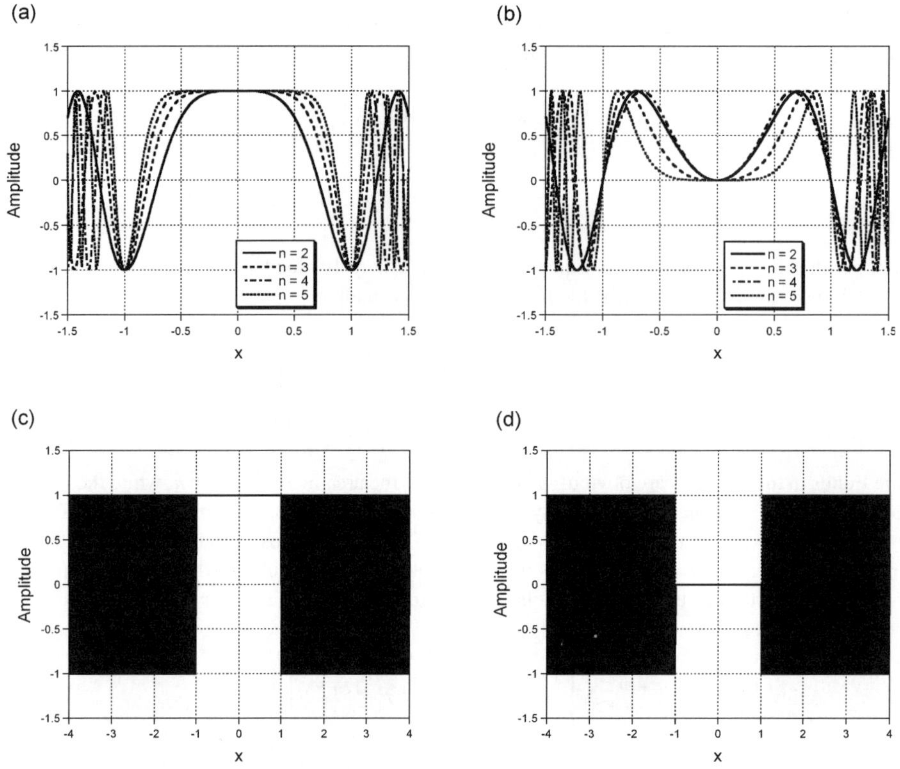

Figure 6.36 "Superchirp" functions $\exp[+i\pi x^n]$ for increasing values of n: (a) real part $\cos[\pi x^n]$; (b) imaginary part $\sin[\pi x^n]$; (c) limit of real part as $n \to \infty$ is approximately 1 for $|x| \leq 1$ and oscillates "rapidly" for $|x| > 1$, and thus approximates $RECT[x/2]$; (d) limit of the imaginary part oscillates and is approximated by $0[x]$.

Lorentzian to be the real-valued Lorentzian modulated by the complex-valued function $1 - 2\pi i x$:

$$f[x] = LOR[x] \cdot (1 - 2\pi i x) = \frac{2 \cdot (1 - 2\pi i x)}{1 + (2\pi x)^2}$$

$$= \frac{2}{1 + (2\pi x)^2} - i \frac{4\pi x}{1 + (2\pi x)^2} \equiv CLOR[x] \tag{6.159}$$

The even real part is the ratio of two symmetric functions while the odd imaginary part is the ratio of an odd and an even function. This means that the resulting function is Hermitian, as shown by the graph in Figure 6.37. Its area is the unit area of the real part. The magnitude of Equation (6.159) is easy to derive:

$$|CLOR[x]| = \left| 2 \cdot \frac{1 - 2\pi i x}{1 + (2\pi x)^2} \right| = 2\sqrt{\frac{1 + (2\pi x)^2}{(1 + (2\pi x)^2)^2}}$$

$$= \sqrt{\frac{4}{(1 + (2\pi x)^2)}} = \sqrt{2 \cdot LOR[x]} \tag{6.160}$$

which "decays" more slowly than the magnitude of the real-valued Lorentzian. The phase of Equation (6.159) is the arctangent of the ratio of the imaginary and real parts. The identical denominators of these parts cancel to leave the inverse tangent of $-2\pi x$ as the result:

$$\Phi\{CLOR[x]\} = \Phi\left\{\frac{2(1-2\pi i x)}{1+(2\pi x)^2}\right\} = \tan^{-1}\left[-\frac{4\pi x/[1+(2\pi x)^2]}{2/[1+(2\pi x)^2]}\right]$$
$$= \tan^{-1}[-2\pi x] = -\tan^{-1}[2\pi x] \qquad (6.161)$$

In words, the phase of the complex Lorentzian function follows the curve of the inverse tangent function. The Argand diagram has the form of a circle of unit radius centered at [1, 0].

6.3.5 Logarithm of the Complex Amplitude

Occasionally it is useful to compute the logarithm of the complex-valued amplitude of a signal or image. Though not so much a "function" as an "operator", it is sufficiently common to merit its consideration here. The operation is a straightforward extension of the real-valued logarithm for amplitudes expressed as magnitude/phase. Given a (generally complex-valued) function $f[x]$ expressed as magnitude/phase, the logarithm of the amplitude is:

$$LOG[f[x]] = LOG[|f[x]|\, e^{i\Phi\{f[x]\}}]$$
$$= LOG[|f[x]|] + LOG[e^{i\Phi\{f[x]\}}]$$
$$= LOG[|f[x]|] + i\Phi\{f[x]\} \qquad (6.162)$$

In words, the real and imaginary parts of the logarithm of a complex-valued amplitude are respectively the logarithm of the magnitude and the initial phase of the complex amplitude. In this context, the true "unwrapped" phase is required in the definition, so the ranges of both the real and imaginary parts are $(-\infty, +\infty)$. An example of the complex logarithm is shown in Figure 6.38. Note that the real part of the complex logarithm is not defined at any location where $|f[x]| = 0$; these are the singularities of $LOG[f[x]]$.

6.4 1-D Stochastic Functions – Noise

The term *noise* refers to any function whose amplitude at a location is not deterministic, but rather is derived from a random (or *stochastic*) process. The amplitude f at each coordinate x of a deterministic function is specified completely by its parameters (width, amplitude, etc.), but that of a stochastic "noise" function n at coordinate x is selected from the distribution that describes the probability of occurrence. The amplitude of any such function $n[x]$ is not "determined" by this probability law; rather, it only specifies statistical averages of the signal amplitude. The range of allowed amplitudes n of the resulting function may be discrete or continuous, depending on the physical process. An example of a discrete stochastic process is the number of individual photons that are measured by a detector in time interval Δt. The number of photons must be an integer, and the numbers counted during disjoint intervals of equal length typically exhibit a degree of "scatter" about some mean value. An example of a continuous stochastic process is the spatial distribution of photon arrivals that have experienced diffraction within an optical system; the location $[x_n, y_n]$ where the nth photon arrives at the image plane is different for each photon and the probability density is continuously distributed. The stochastic functions described in this section are real valued and are designated by the functional notation $n[x]$ instead of $f[x]$, where "n" is used to indicate its "noisy" character. Note that there is some potential for confusion because n also is used to specify an integer coordinate, e.g., $f[n]$, but the appropriate meaning is indicated by the context. Just as for deterministic functions, complex-valued noise may be constructed by adding two random variables at each coordinate after weighting one by i:

$$n[x] = n_1[x] + in_2[x] \qquad (6.163)$$

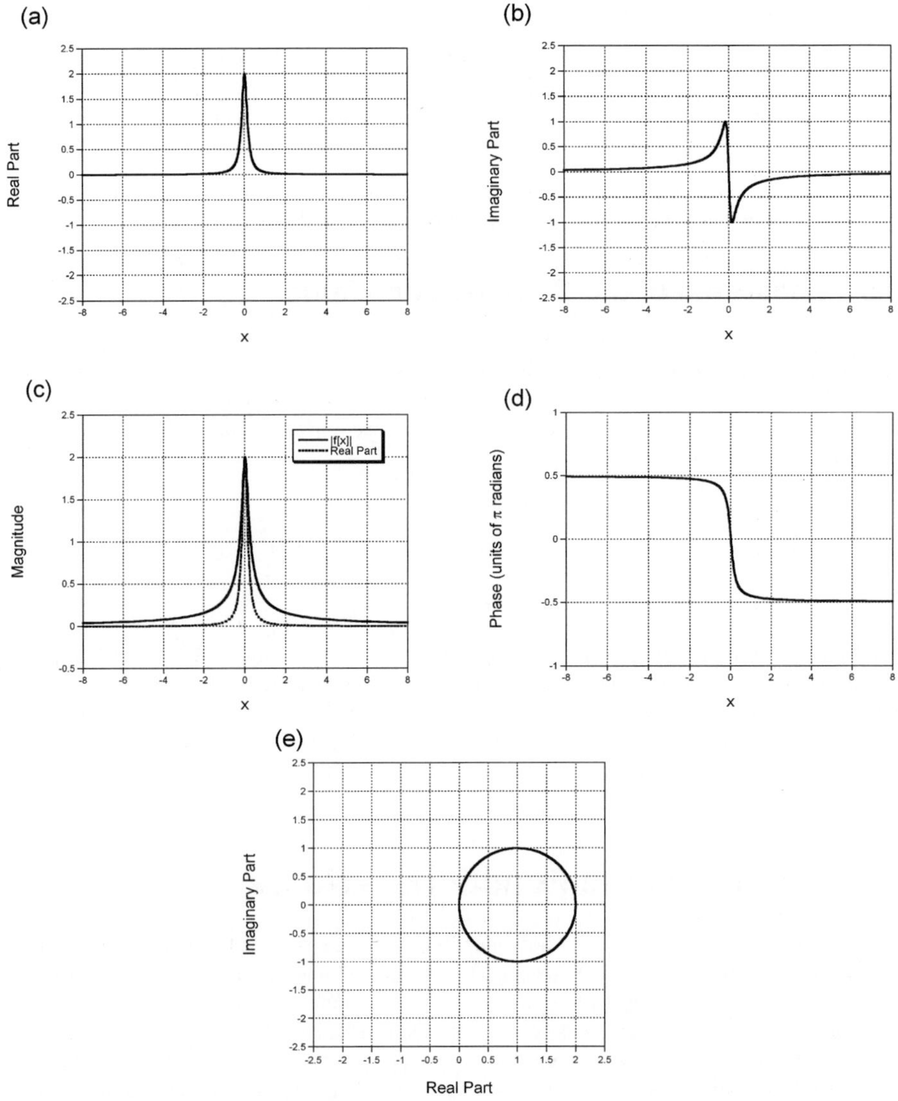

Figure 6.37 The complex Lorentzian function $2 \cdot (1 - 2\pi i x)/[1 + (2\pi x)^2]$: (a) $\Re\{CLOR[x]\} = 2/[1 + (2\pi x)^2]$; (b) $\Im\{CLOR[x]\} = -2\pi x/[1 + (2\pi x)^2]$; (c) $|CLOR[x]| = \sqrt{4/[1 + (2\pi x)^2]}$; (d) phase $\Phi\{CLOR[x]\} = -TAN^{-1}[2\pi x]$; (e) Argand diagram is a unit-radius circle centered at [1, 0].

The range and relative likelihoods of possible amplitudes n are specified by the probability distribution function. The fundamental differences between discrete and continuous probability distributions will require different notations for each: P_n denotes the probability of the integer value n for a discrete distribution, while $p[n]$ is the notation for the probability density of a real-valued amplitude n derived from a continuous distribution. The probability that the continuous variable n lies within a specific

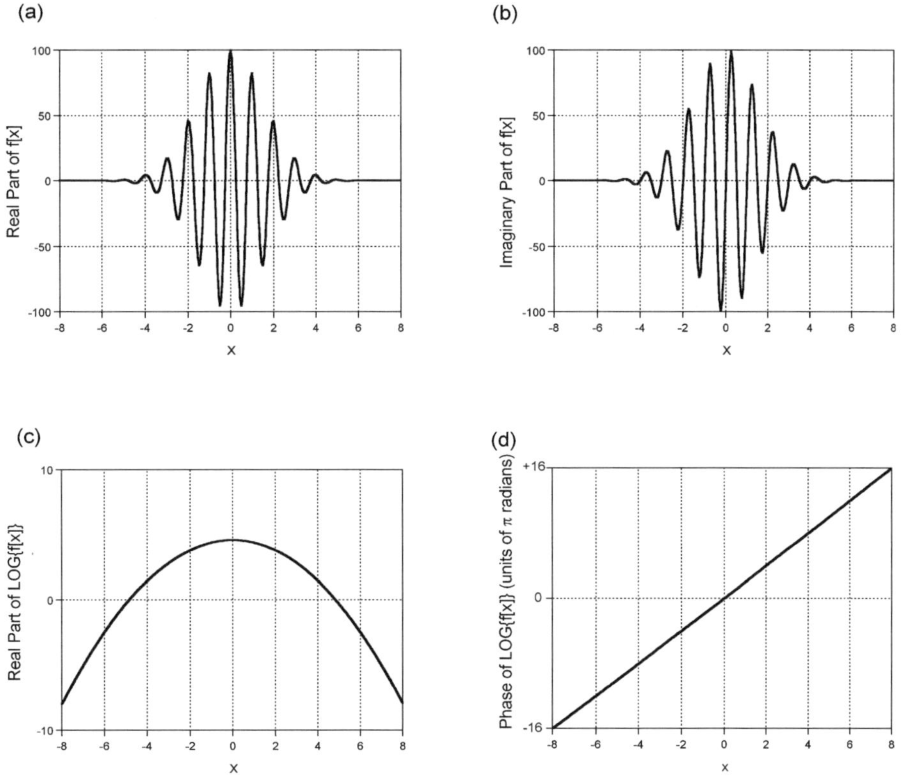

Figure 6.38 The complex logarithm of $f[x] = 100 \cdot e^{(+2\pi i x)}\, e^{-\pi (x/4)^2}$:
(a) $\Re\{f[x]\} = 100 \cdot \cos[2\pi x]\, e^{-\pi(x/4)^2}$;
(b) $\Im\{f[x]\} = 100 \cdot \sin[2\pi x]\, e^{-\pi(x/4)^2}$;
(c) $\Re\{\log_e[f[x]]\} = \log_e |f[x]| = \log_e |100\pi(x/4)^2|$; (d) $\Im\{\log_e(f[x])\} = \Phi\{f[x]\} = 2\pi x$.

interval is the integral of $p[n]$ over that interval, e.g.,

$$P_n(a \leq n \leq b) = \int_a^b p[n]\, dn \tag{6.164}$$

is the probability that the output lies in the range $a \leq n \leq b$. In experimental applications, $p[n]$ is often not known explicitly from the conditions of the system, but is often estimated from the histogram of a data set generated by the system. The histogram is constructed by determining how many samples from the probability distribution reside within different (and usually equally sized) intervals of the allowed range of n. Histograms of the probability distributions discussed in this section are presented below.

6.4.1 Moments of Probability Distributions

The "distribution" of probability about the origin of coordinates is determined by the moments of $p[n]$. The kth moment is the area of the product of $p[n]$ and a weight factor n^k:

$$\mathrm{m}_k\{p[n]\} \equiv \int_{-\infty}^{+\infty} n^k\, p[n]\, dn \tag{6.165}$$

This reduces to a summation if the probability law is discrete. The concepts of the inner product of two functions that were summarized by Equation (5.87) show that this relation may be interpreted as the projection of the monomial function n^k onto $p[n]$, or vice versa. The numerical result is the expected value of n^k and is often indicated by $\langle n^k \rangle$. The "zeroth moment" is evaluated by setting $k = 0$ in Equation (6.165). This is the projection of the unit constant onto $p[n]$, which must be the unit area of $p[n]$:

$$\mathrm{m}_0\{p[n]\} = \int_{-\infty}^{+\infty} n^0 p[n]\,dn = \int_{-\infty}^{+\infty} p[n]\,dn = \langle n^0 \rangle = 1 \tag{6.166}$$

In words, the area of the probability distribution function $p[x]$ is unity, which merely indicates that the amplitude of a sample of n must be one of the allowed values. The first moment is the projection of n^1 onto $p[n]$, and thus the calculation "amplifies" the contributions of large values of n. The result is the mean value of the amplitude n selected from the probability distribution $p[n]$ and is variously denoted by $\langle n \rangle$, $\hat{n}$, and μ, though we reserve the last notation for a variant of the moments to be considered very shortly. The first moment is:

$$\mathrm{m}_1\{p[n]\} = \int_{-\infty}^{+\infty} n^1 p[n]\,dn = \langle n \rangle \tag{6.167}$$

The first moment also is called the *centroid* of the probability distribution, which is the value of n that divides the area of $p[n]$ into equal parts. The central moments are variants of the moments; they are the expected values of $(n - \langle n \rangle)^k$ for that probability law and thus measure the weighted variation of $p[n]$ about the mean:

$$\mu_k\{p[n]\} \equiv \int_{-\infty}^{+\infty} (n - \langle n \rangle)^k \, p[n]\,dn \tag{6.168}$$

It is easy to show that the first central moment $\mu_1 = 0$. The second moment of a probability law is the projection of n^2 onto $p[n]$:

$$\mathrm{m}_2\{p[n]\} = \int_{-\infty}^{+\infty} n^2 \, p[n]\,dn \equiv \langle n^2 \rangle \tag{6.169}$$

The second central moment of $p[n]$ is the variance σ^2, which measures the "spread" of the noise probability about its mean value:

$$\mu_2\{p[n]\} \equiv \int_{-\infty}^{+\infty} (n - \langle n \rangle)^2 p[n]\,dn \equiv \sigma^2 \tag{6.170}$$

By expanding $(n - \langle n \rangle)^2$ and evaluating the three resulting integrals, it is easy to derive the useful relationship among the mean, variance, and the mean value of the square $\langle n^2 \rangle$:

$$\sigma^2 = \langle n^2 \rangle - \langle n \rangle^2 \tag{6.171}$$

In words, the variance of a probability law is the difference of the expected value of n^2 and the square of the expected value of n. A further discussion of the properties and applications of the moments of general functions will be considered in Chapter 13. At this point, we merely say that the moments and the Fourier transform of an arbitrary function are intimately related. Readers who are interested in a deeper discussion of noise are referred to some of the standard references (Frieden, 2002). However, it will be necessary to understand certain important properties of specific noise distributions when discussing optimum filtering in Chapter 19. Several specific noise distributions will be introduced here: the binomial and Poisson distributions are produced by summations of independent Bernoulli trials, while the uniform, normal, and Rayleigh distributions apply to continuous random variables.

6.4.2 Discrete Probability Laws

The discrete probability laws are used to model processes that have discrete, and often binary, outcomes. They are particularly useful when constructing models of such imaging processes as photon absorption

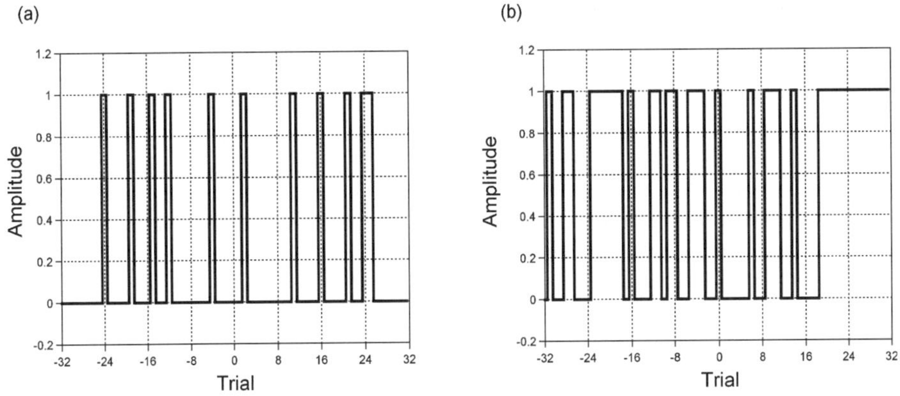

Figure 6.39 Examples of 64 Bernoulli trials: (a) $p = 0.2$; (b) $p = 0.5$.

by the sensor. The simplest type of discrete probability law applies to events that have only two possible outcomes: success or failure, true or false, on or off, head or tail. The individual implementation of such a binary event is the "Bernoulli trial", while collective realizations of many such events are described by the binomial and Poisson distributions.

6.4.2.1 Bernoulli Trials

As an example of a "binary" experiment, consider a flux of photons onto an "imperfect" absorbing material. Individual photons may be absorbed or not, and the testing of the absorption of a photon is a Bernoulli trial. We will indicate a "successful" absorption by 1 and a "failure" by 0. The statistics of a string of Bernoulli trials are specified by the probability of a "success" (outcome "1"), which will be denoted by "p". Obviously, the value of p must satisfy the constraint $0 \le p \le 1$. It also is evident that the probability of a "failure" (outcome "0") is the complement $1 - p$, which is often denoted by "q". These relative probabilities for a particular absorber may be determined either from a physical model of the interaction or from observed results of a large number of Bernoulli trials. "Images" of independent Bernoulli trials are shown in Figure 6.39 for different values of p. The most interesting aspects of Bernoulli trials arise when independent experiments are repeated.

6.4.2.2 Multiple Bernoulli Trials – Binomial Probability Distribution

Consider the range of possible results of N Bernoulli trials where the probability of outcome "1" is p and the probability of outcome "0" is $1 - p$. Clearly, the number of possible distinct outcomes is 2^N and the relative likelihood of these outcomes is determined by p. For example, the probability of the specific sequence 101010 ... 10 must be:

$$p(10101010\ldots 101010) = p \cdot q \cdot p \cdot q \cdot \cdots \cdot p \cdot q = p^{N/2} \cdot q^{N/2} \tag{6.172}$$

where the last step follows because this string of trials includes equal numbers of "successes" and "failures". Obviously, the probability of a different sequence where the first $N/2$ outcomes are "1" and the remaining $N/2$ outcomes are "0" must be identical:

$$p(1111111\ldots 0000000) = p \cdot p \cdot p \cdot p \cdot \cdots \cdot q \cdot q \cdot q \cdot q \cdot q = p^{N/2} \cdot q^{N/2} \tag{6.173}$$

because these two distinct outcomes have the same number of "successes" and "failures". Their histograms must also be identical. More generally, consider the outcome of N trials that contains exactly

n "successes". The probability of a specific such outcome is:

$$p_1 \cdot p_2 \cdot p_3 \cdot \dots \cdot p_n \cdot q_1 \cdot q_2 \cdot q_3 \cdot \dots \cdot q_{N-n} = p^n \cdot q^{N-n} \tag{6.174}$$

Identical probabilities will be associated to a number of other outcomes that will differ in the order of successes and failures. In most applications, this order of arrangement is not significant; the only parameter that matters is the total number of successes n. To determine the probability, we must compute the number of possible combinations of n successes in N trials. It is straightforward to show that this number is the so-called binomial coefficient:

$$\frac{N!}{(N-n)!n!} \equiv \binom{N}{n} \tag{6.175}$$

which is the number of combinations of n "successes" and $N - n$ "failures". The probability of n "successes" in N trials is:

$$P_n = \frac{N!}{n!\,(N-n)!} p^n (1-p)^{N-n} = \binom{N}{n} p^n (1-p)^{N-n} \tag{6.176}$$

For example, consider a coin flip with two outcomes, H and T. In four flips ($N = 4$), the number of combinations with $n = 2$ heads is $4!/2!2! = 6$. The realizations are *HHTT*, *HTHT*, *THHT*, *THTH*, *HTTH*, and *TTHH*. If H and T are equally likely so that $p = q = 0.5$, then the probability of two heads in four flips is $P_2 = 6(0.5)^2(0.5)^2 = 0.375$. The number of realizations of four heads in four flips is $4!/0!4! = 1$, so that the probability of four heads in four flips is $P_4 = 1(0.5)^4(0.5)^0 = 0.0625$. Of course, the binomial law also applies to cases where $p \neq q$, which is analogous to flipping an"unfair" coin. If the probability of a head were $p = 0.75$, then the probability that four flips would produce two heads would be $P_2 = 6(0.75)^2(0.25)^2 \cong 0.211$ and the probability of getting four heads is $P_4 = 1(0.75)^4(0.25)^0 \cong 0.316$, which is larger than P_2. The mean value and variance of the number of outcomes with individual probability p in a process consisting of N experiments is obtained by substituting Equation (6.176) into Equation (6.165):

$$\langle n \rangle = Np \tag{6.177}$$

$$\sigma^2 = Np(1-p) \tag{6.178}$$

A realization of a process with a binomial probability law is shown in Figure 6.40. Note that the "shape" of the histogram of the binomial distribution is approximately Gaussian. Though the Gaussian distribution is continuous and the binomial distribution refers to discrete variables, this observation suggests that samples that would be obtained from a large number of independent Bernoulli trials with probability p may be approximately generated by thresholding values generated by a Gaussian distribution.

6.4.2.3 Poisson Probability Law

The Poisson probability law is an approximation to the binomial law for large numbers of rarely occurring events, i.e., $N \gg 1$ and $p \cong 0$. The mean number of events is $\langle n \rangle = N \cdot p$, and is commonly denoted by λ. The form of the Poisson law is obtained by substituting into the binomial law of Equation (6.176) in the limit $N \to \infty$:

$$p_n = \lim_{N\to\infty} \left\{ \binom{N}{n} \left[\frac{\lambda}{N}\right]^n \left[1 - \frac{\lambda}{N}\right]^{N-n} \right\} \tag{6.179}$$

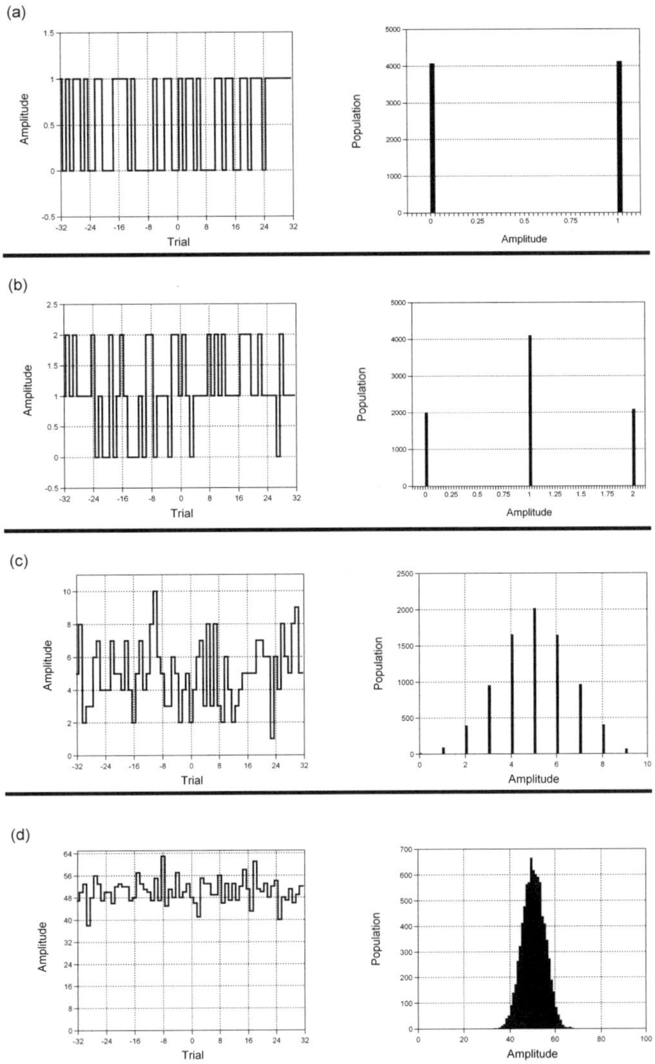

Figure 6.40 Realizations of N Bernoulli trials at 8192 samples with $p = 0.5$ and the resulting histograms: (a) $N = 1$ trial per sample, two (nearly) equally likely outcomes; (b) $N = 2$, for which three outcomes are possible (0, 1, 2); (c) $N = 10$; and (d) $N = 100$. The histogram approaches a Gaussian function for large N.

We may take the natural logarithm of both sides to obtain:

$$\begin{aligned}
\log_e[p_n] &= \lim_{N\to\infty}\left\{\log_e\left[\binom{N}{n}\left[\frac{\lambda}{N}\right]^n\left[1-\frac{\lambda}{N}\right]^{N-n}\right]\right\} \\
&= \lim_{N\to\infty}\left\{\log_e\left[\binom{N}{n}\left[\frac{\lambda}{N}\right]^n\right]\right\} + \lim_{N\to\infty}\left\{[N-n]\log_e\left[1-\frac{\lambda}{N}\right]\right\} \\
&= \lim_{N\to\infty}\left\{\log_e\left[\left(\frac{N!}{[N-n]!n!}\right)\left[\frac{\lambda}{N}\right]^n\right]\right\} + \lim_{N\to\infty}\left\{\frac{\log_e[1-\lambda/N]}{[N-n]^{-1}}\right\}
\end{aligned} \tag{6.180}$$

We may use the fact that n is a small number to evaluate the first term in the limit:

$$\lim_{N\to\infty}\left\{\log_e\left[\frac{N(N-1)(N-2)\ldots(N-n+1)}{n!}\left(\frac{\lambda}{N}\right)^n\right]\right\}\cong\log_e\left\{\frac{N^n}{n!}\left(\frac{\lambda}{N}\right)^n\right\} \tag{6.181}$$

The second term may be evaluated by recognizing that it is the ratio of two terms that both approach zero in the limit, so that L'Hôpital's rule may be applied:

$$\begin{aligned}\lim_{N\to\infty}\left\{\frac{\log_e(1-\lambda/N)}{(N-n)^{-1}}\right\} &= \lim_{N\to\infty}\left\{\frac{(d/dN)(\log_e(1-\lambda/N))}{(d/dN)(N-n)^{-1}}\right\}\\ &= \lim_{N\to\infty}\left\{\frac{(1-\lambda/N)^{-1}(\lambda/N^2)}{-(N-n)^{-2}}\right\}\\ &= \lim_{N\to\infty}\left\{-\lambda\left(\frac{N-n}{N}\right)^2\left(1-\frac{\lambda}{N}\right)^{-1}\right\}\end{aligned} \tag{6.182}$$

These are collected to give the expression for the logarithm of the binomial probability distribution in this limit, which then may easily be recast into the form:

$$\log_e[p_n]=\log_e\left[\frac{\lambda^n}{n!}\right]-\lambda\Longrightarrow p_n=\left[\frac{\lambda^n}{n!}\right]e^{-\lambda} \tag{6.183}$$

This derivation nicely demonstrates that the Poisson distribution really is the particular limiting case of the binomial distribution, which is in turn the probability for a sequence of Bernoulli trials. It is useful to consider the "goodness" of the Poisson approximation to the binomial law for different values of the probability p. From the several examples compared in Figure 6.41, we can see that the Poisson formula is a good approximation for $p \leq 0.1$. The mean value, variance, and third central moment of the Poisson distribution are identically λ. An example of an imaging process that yields a Poisson distribution is the number of arrivals in a fixed time interval of rare and independent photons (as from a faint star) if the arrivals occur with a uniform probability distribution over time. The number of arrivals in the specified time interval follows a Poisson distribution. The mean number of events for that interval size will occur most frequently, while more or fewer arrivals will occur less often.

6.4.3 Continuous Probability Distributions

6.4.3.1 Uniform Distribution

The uniform distribution generates what is perhaps the most intuitive type of noise; the amplitude n is equally likely to occur within any finite interval of equal size in the allowed range of amplitudes. In other words, the probability density function of a uniform distribution is a rectangle function:

$$p_{\text{Uniform}}[n]=\frac{1}{|b|}RECT\left[\frac{n-\langle n\rangle}{b}\right] \tag{6.184}$$

where b defines the width of allowed values of n and $\langle n\rangle$ is the mean value. The multiplicative scale factor b^{-1} ensures that the distribution function has unit area, i.e., that $\mathrm{m}_0 = 1$. Equation (6.184) may be substituted into Equation (6.171) to easily demonstrate that the variance of the uniform distribution is $\sigma^2 = b^2/12$, which means that the standard deviation $\sigma = b/\sqrt{12}$. Most digital computer languages include a random-number generator that selects amplitudes from a uniform distribution (at least, approximately) over the interval [0, 1). An image and the computed histogram of a 1-D "image" of random numbers taken from a uniform distribution are shown in Figure 6.42.

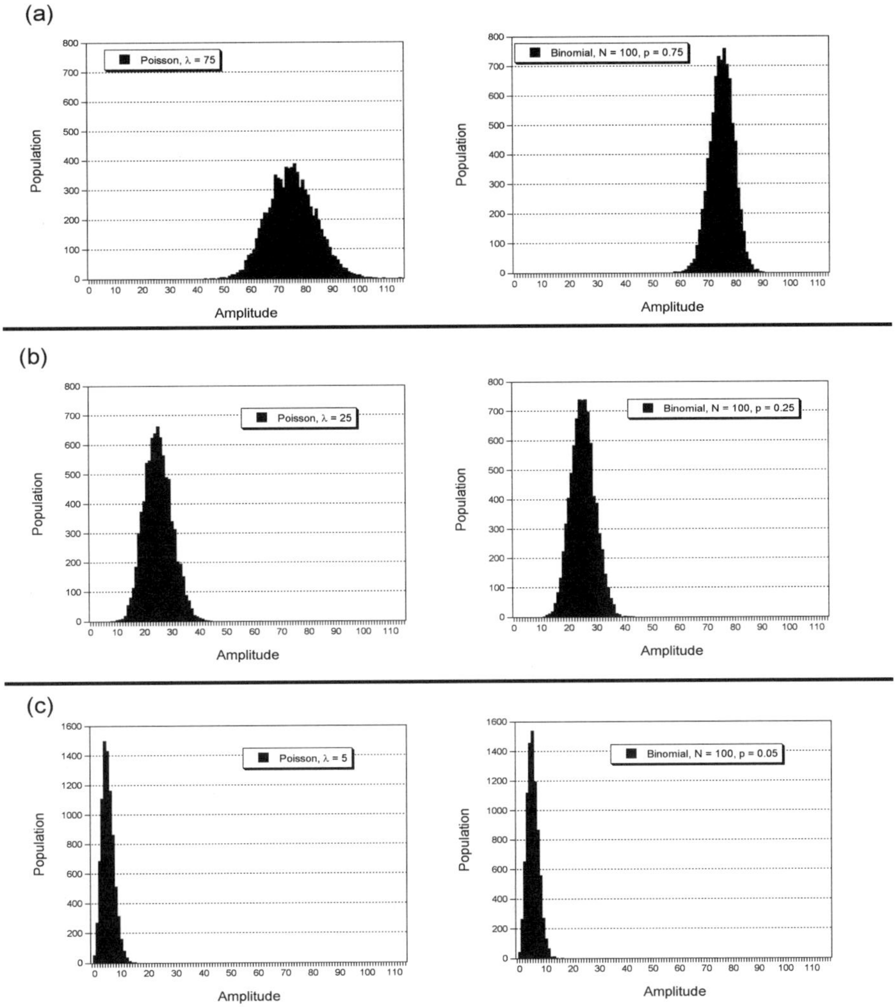

Figure 6.41 Comparison of binomial and Poisson random variables, showing that the Poisson distribution satisfies $\langle n \rangle = \sigma^2$. The binomial distribution is computed for $N = 100$: (a) binomial distribution with $p = 0.75$ produces $\langle n \rangle = 75.08$, $\sigma^2 = 18.76$, Poisson distribution with $\lambda = 75$ gives $\langle n \rangle = 75.05$, $\sigma^2 = 74.57$; (b) binomial distribution with $p = 0.25$ has $\langle n \rangle = 25.04$, $\sigma^2 = 18.91$, Poisson distribution with $\lambda = 25$ has $\langle n \rangle = 24.95$, $\sigma^2 = 24.80$; (c) binomial with $p = 0.05$ produces $\langle n \rangle = 4.97$, $\sigma^2 = 4.77$, Poisson distribution with $\lambda = 5$ results in $\langle n \rangle = 5.02$, $\sigma^2 = 5.08$.

6.4.3.2 Normal Distribution

The normal distribution is the familiar symmetric "bell curve" of probability that was mentioned during the discussion of Section 6.1.12, and is no doubt the most applicable of all the probability laws considered here. Its mean value $\langle n \rangle$ is identical to the most likely amplitude (peak of the probability density). The probability that an amplitude will differ from the mean progressively decreases as the value moves away from $\langle n \rangle$. The probability distribution is a Gaussian function of the form in Equation (6.57), where the width parameter b is proportional to the standard deviation σ of the probability

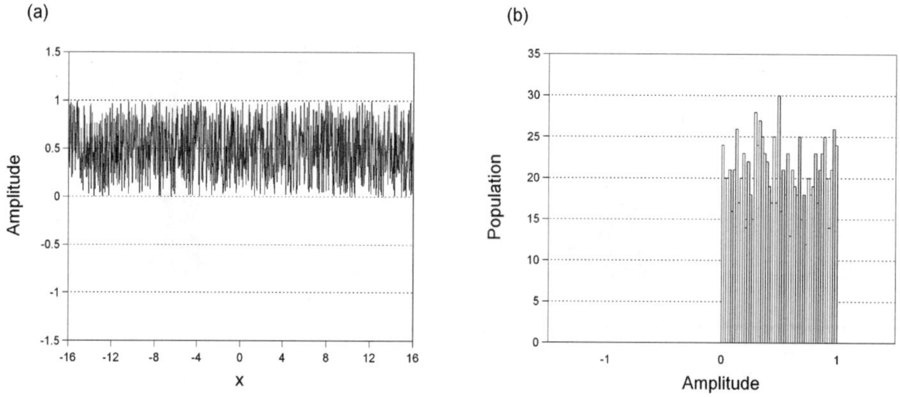

Figure 6.42 Samples of a uniformly distributed random variable in the interval [0, 1) with $\langle n \rangle = 0.5$ and $\sigma^2 = \frac{1}{12} \simeq 0.083$: (a) sample of $n[x]$; (b) histogram of $n[x]$, with measured $\langle n \rangle = 0.499$ and $\sigma^2 = 0.082$.

distribution. For given mean value $\langle n \rangle$ and variance σ^2 (standard deviation σ), the probability density function of the normal distribution is:

$$p_{\text{Normal}}[n] = \frac{1}{\sqrt{2\pi}\,\sigma} \exp\left[-\frac{(n - \langle n \rangle)^2}{2\sigma^2}\right] \tag{6.185}$$

The leading factor $(\sqrt{2\pi})^{-1}$ ensures that the area of the probability density function is unity. A zero-mean normal distribution (i.e., with $\langle n \rangle = 0$) is explicitly called a "Gaussian distribution". The image and measured histogram of a 1-D normal distribution are shown in Figure 6.43. Many stochastic functions may be approximated by the normal distribution due to the "central-limit theorem", which is described in Chapter 13, and which states that a cascade of stochastic processes derived from a (nearly) arbitrary set of probability density functions will generate a process with a normal distribution. Since many of the applications of stochastic variables in imaging involve cascaded systems, the central-limit theorem ensures that the probability law of the outputs is Gaussian to a good approximation.

6.4.3.3 Rayleigh Distribution

The Rayleigh distribution occurs in imaging applications that involve Fourier transforms of distributions of complex-valued random variables. Examples include the description of Fraunhofer diffraction from a random scatterer and computer-generated holography. The Rayleigh distribution may be generated from a complex-valued probability distribution where both the real and imaginary parts are random variables selected from the same Gaussian distribution. The magnitudes of complex amplitudes selected from this distribution exhibit a Rayleigh distribution. In other words, it is the statistical distribution of the magnitudes of the sum of a set of vectors whose x- and y-projections are independent and identically distributed Gaussian random variables. The correspondence between 2-D real vectors and the set of complex numbers establishes that the magnitude of the sum of a set of complex numbers (phasors) will follow a Rayleigh distribution if the real and imaginary parts are independent and identically distributed Gaussian random variables. The probability density function of a Rayleigh distribution is characterized by a single parameter a:

$$p_{\text{Rayleigh}}[n] = \frac{n}{a^2} \exp\left[-\left(\frac{n^2}{2a^2}\right)\right] \mathit{STEP}[n] \tag{6.186}$$

where the *STEP* function ensures that the allowed amplitudes n must be nonnegative, because they are the magnitudes of complex numbers. The mean $\langle n \rangle$, variance σ^2, and standard deviation σ of the

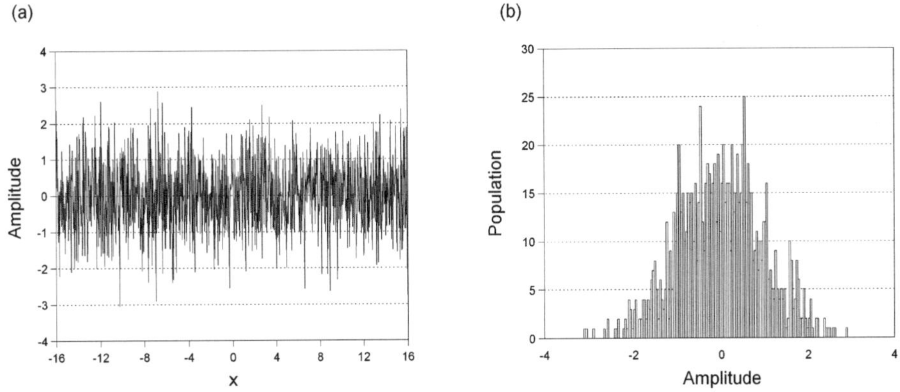

Figure 6.43 Samples of a random variable generated by a normal distribution with $\langle n \rangle = 0$, $\sigma^2 = 1$: (a) samples of n; (b) histogram of n with measured $\langle n \rangle = 0.08$, $\sigma^2 = 1.027$.

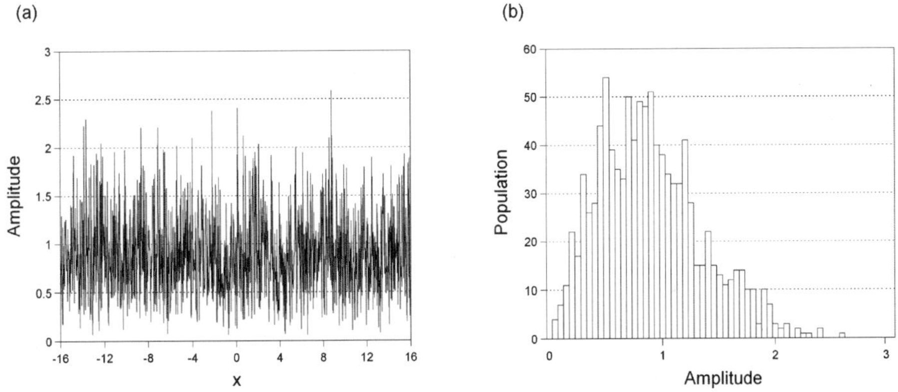

Figure 6.44 Rayleigh-distributed random variable $n[x]$ generated as the magnitude of a complex random variable whose real and imaginary parts are Gaussian distributed with zero mean and unit variance: (a) 1024 samples of n; (b) histogram of n. The mean and variance of the samples are $\langle n \rangle = 1.270$ and $\sigma^2 = 0.424$.

Rayleigh distribution must be functions of the parameter a, and may be shown to be:

$$\langle n \rangle = \sqrt{\frac{\pi}{2}}a \cong 1.25a \tag{6.187a}$$

$$\sigma^2 = \left(2 - \frac{\pi}{2}\right)a^2 \cong 0.429a^2 \implies \sigma \cong 0.655a \tag{6.187b}$$

An example of a Rayleigh distribution is shown in Figure 6.44.

6.4.4 Signal-to-Noise Ratio

A metric commonly applied to signals and images is the *signal-to-noise ratio* (*SNR*), which (as its name implies) is a measure of the uncertainty in the measurement of the amplitude of a continuous (not sampled or quantized) signal $f[x]$. The *SNR* is often easy to evaluate for an image, but even more often it does not convey a very accurate sense of the visual appearance of the image. For example, if two images of the same object have different *SNR*s, that with the larger may not "look" better to the eye. Even so, the concept merits exploration. Consider that a measured signal $g[x]$ is the sum of the deterministic signal $f[x]$ and some stochastic signal $n[x]$:

$$g[x] = f[x] + n[x] \tag{6.188}$$

The statistics (mean and variance) of the two component signals may be evaluated. For the deterministic input signal $f[x]$, these statistics may be approximated over finite domains:

$$\langle n \rangle_f \equiv \int_{-\infty}^{+\infty} f[x]\,dx \Longrightarrow \widehat{\langle n \rangle}_f = \frac{1}{X_0} \int_{-X_0/2}^{+X_0/2} f[x]\,dx \tag{6.189a}$$

$$\sigma_f^2 = \int_{-\infty}^{+\infty} [f[x] - \langle n \rangle_f]^2\,dx \Longrightarrow \widehat{\sigma_f^2} = \frac{1}{X_0} \int_{-X_0/2}^{+X_0/2} [f[x] - \widehat{\langle n \rangle}_f]^2\,dx \tag{6.189b}$$

The *signal-to-noise power ratio* is most rigorously defined as the dimensionless ratio of the variances of the signal and any additive noise:

$$SNR \equiv \frac{\sigma_f^2}{\sigma_n^2} \tag{6.190}$$

This means that there is a larger variation of the signal amplitude than of the noise amplitude in a signal with a large *SNR*. This definition of *SNR* as the ratio of variances often has a wide range – easily several orders of magnitude – so that the numerical values may become unwieldy. For this reason, its range generally is compressed by mapping to a logarithmic scale with dimensionless "units" of *bels*:

$$SNR = \log_{10}\left[\frac{\sigma_f^2}{\sigma_n^2}\right] = 2\log_{10}\left[\frac{\sigma_f}{\sigma_n}\right] \quad [bels] \tag{6.191}$$

It is yet more common to express this value in units of tenths of a bel so that the integer part of the number is more precise. The resulting metric is in terms of *decibels*:

$$SNR = 10\log_{10}\left[\frac{\sigma_f^2}{\sigma_n^2}\right] = 20\log_{10}\left[\frac{\sigma_f}{\sigma_n}\right] \quad [decibels] \tag{6.192}$$

Under this definition, if the signal variance is 10 times larger than the noise variance then $SNR = 10\,\text{dB}$; if the standard deviation is 10 times larger than that of the noise, then $SNR = 20\,\text{dB}$. The variances obviously depend on the statistics (the histograms) of the signal and noise. The variances depend only on the range of gray values and not on their "arrangement" (i.e., numerical "order" or "pictorial" appearance in the image). Since the noise is often determined by the measurement equipment, a single measurement of the noise variance is often used for many signal amplitudes. However, the signal variance must be measured each time.

6.4.5 Example: Variance of a Sinusoid

The variance of a sinusoid with amplitude A_0 is easily computed by direct integration:

$$\begin{aligned}
f[x] &= A_0 \cos\left[2\pi \frac{x}{X_0}\right] \\
\sigma_f^2 &= \frac{1}{X_0}\int_{-X_0/2}^{+X_0/2} (f[x] - \langle f[x]\rangle)^2 \, dx = \frac{1}{X_0}\int_{-X_0/2}^{+X_0/2} \left(A_0 \cos\left[2\pi \frac{x}{X_0}\right]\right)^2 dx \\
&= \frac{A_0^2}{X_0}\int_{-X_0/2}^{+X_0/2} \frac{1}{2}\left(1 + \cos\left[4\pi \frac{x}{X_0}\right]\right) dx = \frac{A_0^2}{2X_0}(X_0 + 0) \\
&= \sigma_f^2 = \frac{A_0^2}{2} \quad \textit{for sinusoid with amplitude } A_0
\end{aligned} \tag{6.193}$$

Note that the variance does not depend on the period (i.e., on the spatial frequency) or on the initial phase – it is a function only of the histogram of the values in a period and not of the "ordered" values. It also does not depend on any "bias" (additive constant) in the signal. The standard deviation of the sinusoid is just the square root of the variance:

$$\sigma_f = \frac{A_0}{\sqrt{2}} \cong 0.707 A_0 \quad \textit{for sinusoid with amplitude } A_0 \tag{6.194}$$

6.4.6 Example: Variance of a Square Wave

The variance of a square wave with the same amplitude also is easily evaluated by integration of the thresholded sinusoid:

$$\begin{aligned}
f[x] &= A_0 \, SGN\left[\cos\left[2\pi \frac{x}{X_0}\right]\right] \\
\sigma_f^2 &= \frac{1}{X_0}\int_{-X_0/2}^{+X_0/2} [f[x] - \langle f[x]\rangle]^2 \, dx = \frac{1}{X_0}\left(\int_{-X_0/4}^{+X_0/4} [-A_0]^2 \, dx + \int_{+X_0/4}^{+3X_0/4} [+A_0]^2 \, dx\right) \\
&= \frac{1}{X_0}\left(A_0^2 \frac{X_0}{2} + A_0^2 \frac{X_0}{2}\right) = A_0^2
\end{aligned} \tag{6.195}$$

$$\Longrightarrow \sigma_f = A_0 \quad \textit{for square wave with amplitude } A_0 \tag{6.196}$$

Note that the variance of the square wave is larger than that of the sine wave with the same amplitude, which makes intuitive sense because the amplitudes of the square wave generally are more "distant" from the mean than are those from a sinusoid.

6.4.7 Approximations to *SNR*

Since the variance depends on the statistics of the signal, it is common (though less rigorous) to approximate the variance by the square of the *dynamic range*, which is the "peak-to-peak signal amplitude" $f_{max} - f_{min}$. In the examples of the sinusoid and the square wave already considered, the approximations are:

$$\textit{Sinusoid with amplitude } A_0 \Longrightarrow \sigma_f^2 = \frac{A_0^2}{2}$$

$$\Longrightarrow (f_{\max} - f_{\min})^2 = (2A_0)^2 = 4A_0^2 = 8\sigma_f^2 \tag{6.197a}$$

$$\textit{Square wave with amplitude } A_0 \Longrightarrow \sigma_f^2 = A_0^2$$

$$\Longrightarrow (f_{\max} - f_{\min})^2 = (2A_0)^2 = 4A_0^2 = 4\sigma_f^2 \tag{6.197b}$$

For the example of Gaussian noise with variance $\sigma^2 = 1$ and mean μ, the dynamic range $f_{\max} - f_{\min}$ of the noise technically is infinite, but its extrema often may be approximated based on the observation that few amplitudes exist more than four standard deviations from the mean, so that $f_{\max} \cong \mu_f + 4\sigma_f$, $f_{\min} \cong \mu_f - 4\sigma_f$, leading to $f_{\max} - f_{\min} \cong 8\sigma_f$. The estimate of the variance of the signal is then $(f_{\max} - f_{\min})^2 \cong 64\sigma_f^2$, which is (obviously) 64 times larger than the actual variance. Because this estimate of the signal variance is too large, the estimates of the *SNR* thus obtained will be too optimistic.

Often, the signal and noise of images are measured by photoelectric detectors as differences in electrical potential in volts; the signal dynamic range is $V_f = V_{\max} - V_{\min}$. If the average noise voltage is V_n, then the *SNR* is:

$$SNR = 10 \log_{10}\left(\frac{V_f^2}{V_n^2}\right) = 20 \log_{10}\left(\frac{V_f}{V}\right) \quad [dB] \tag{6.198}$$

As an aside, we mention that the *signal amplitude* (or *level*) of analogue electrical signals is often described in terms of *dB* measured relative to some fixed reference. If the reference level is 1 V, then the signal level is measured in units of *dBV*:

$$Level = 10 \log_{10}[(V_f[\mathrm{V}])^2]\, dBV = 20 \log_{10}[V_f]\, dBV \tag{6.199}$$

The level is measured relative to 1 mV in units of *dBm*:

$$Level = 10 \log_{10}\left(\frac{(V_f)^2}{(10^{-3}V_f[\mathrm{V}])^2}\right) dBV = 10 \log_{10}(V_f^2)\, dBm \tag{6.200}$$

We will revisit this discussion of *SNR* during the discussion of quantization in Chapter 14.

6.5 Appendix A: Area of $SINC[x]$ and $SINC^2[x]$

The areas of $SINC[x]$ and $SINC^2[x]$ were stated without proof to be identically unity in Equation (6.33) and Equation (6.35). The demonstrations of these facts are quite simple after proving the so-called "central-ordinate theorem" of the Fourier transform, but that will not happen until Chapter 9. The areas are evaluated in this section by integrating the corresponding complex-valued functions along appropriate paths in the complex plane using Cauchy's theorem. The integral is recast by changing the variable of integration to $x = \pi\alpha$:

$$\int_{x=-\infty}^{+\infty} \frac{\sin[\pi\alpha]}{\pi\alpha}\, d\alpha = \int_{x=-\infty}^{+\infty} \frac{\sin[x]}{x}\left(\frac{1}{\pi}\, dx\right) = \frac{1}{\pi}\int_{x=-\infty}^{+\infty} \frac{\sin[x]}{x}\, dx \tag{A6.1}$$

We use the Euler relation to substitute for $\sin[x]$:

$$\frac{1}{\pi}\int_{-\infty}^{+\infty} \frac{\sin[x]}{x}\, dx = \Im\left\{\frac{1}{\pi}\int_{-\infty}^{+\infty} \frac{e^{+ix}}{x}\, dx\right\} \tag{A6.2}$$

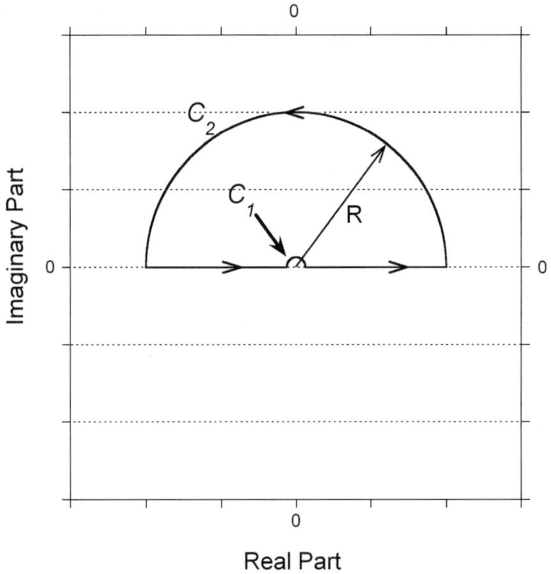

Figure A6.1 Integration path in the complex plane used to evaluate the areas of *SINC*$[x]$ and *SINC*$^2[x]$, in the limits $\rho \to 0$ and $R \to \infty$.

The corresponding integrand in the complex plane is obtained by substituting $z = x + iy$ and integrating along the x-axis:

$$\Im\left\{\frac{1}{\pi}\int_{-\infty}^{+\infty}\frac{e^{+ix}}{x}\,dx\right\} = \Im\left\{\frac{1}{\pi}\int_{z=-\infty+0i}^{+\infty+0i}\frac{e^{+iz}}{z}\,dz\right\} \tag{A6.3}$$

The integrand has a singularity ("pole") at the origin $z = 0$, though this singularity is actually "removable" because *SINC*$[x]$ is finite there. The 1-D integral is part of an integral along a closed path C shown in Figure A6.1 that does not enclose the origin. Cauchy's theorem states that the integral around this closed path must be zero because the path encloses no singularities. The resulting integral may be evaluated in four parts:

$$\oint_C \frac{e^{+iz}}{z}\,dz = \int_{z=-R+0i}^{-\rho+0i}\frac{e^{+iz}}{z}\,dz + \int_{C_1}\frac{e^{+iz}}{z}\,dz + \int_{z=\rho+0i}^{R+0i}\frac{e^{+iz}}{z}\,dz + \int_{C_2}\frac{e^{+iz}}{z}\,dz$$
$$= \left(\int_{-R}^{-\rho}\frac{e^{+ix}}{x}\,dx + \int_{\rho}^{R}\frac{e^{+ix}}{x}\,dx\right) + \left(\int_{C_1}\frac{e^{+iz}}{z}\,dz + \int_{C_2}\frac{e^{+iz}}{z}\,dz\right) = 0 \tag{A6.4}$$

where C_1 and C_2 are semicircles of radius ρ and R respectively in the upper half plane. The variable of integration in the two integrals along the x-axis has been changed back from z to x. We will evaluate these integrals in the limits $\rho \to 0$ and $R \to \infty$. The paths along curves C_1 and C_2 have the respective forms $z_1 = \rho\, e^{i\theta_1}$ $(+\pi \le \theta_1 \le 0)$ and $z_2 = R\, e^{i\theta_2}$ $(0 \le \theta \le +\pi)$. The corresponding length elements along these paths are $dz_1 = i\rho\, e^{i\theta_1}\, d\theta_1$ and $dz_2 = iR\, e^{i\theta_2}\, d\theta_2$.

Consider the integrand of the integral along path C_1 in the limit $\rho \to 0$. The numerator e^{iz} is rewritten in terms of z_1 to obtain:

$$\lim_{\rho\to 0}\left\{\int_{\theta_1=+\pi}^{\theta_1=0}\frac{e^{+i\rho\, e^{i\theta_1}}}{\rho\, e^{i\theta_1}} i\rho\, e^{i\theta_1}\, d\theta_1\right\} = i\lim_{\rho\to 0}\left\{\int_{\theta_1=+\pi}^{\theta_1=0} e^{+i\rho\, e^{i\theta_1}}\, d\theta_1\right\} \tag{A6.5}$$

By assuming the appropriate convergence properties so that the limit of the integral is the integral of the limiting expressions, we obtain:

$$i \lim_{\rho\to 0}\left\{\int_{\theta_1=+\pi}^{\theta_1=0} e^{+i\rho\, e^{i\theta_1}}\, d\theta_1\right\} = i\int_{\theta_1=+\pi}^{\theta_1=0} \lim_{\rho\to 0}\{e^{+i\rho\, e^{i\theta_1}}\}\, d\theta_1 \tag{A6.6}$$

The Euler relation is applied to rewrite the complex exponential:

$$\begin{aligned} i\int_{\theta_1=+\pi}^{\theta_1=0} \lim_{\rho\to 0}\{e^{+i\rho\, e^{i\theta_1}}\}\, d\theta_1 &= i\int_{\theta_1=+\pi}^{\theta_1=0} \lim_{\rho\to 0}\{e^{+i\rho(\cos[\theta_1]+i\sin[\theta_1])}\}\, d\theta_1 \\ &= i\int_{\theta_1=+\pi}^{\theta_1=0} \lim_{\rho\to 0}\{e^{+i\rho\cos[\theta_1]}\, e^{-\rho\sin[\theta_1]}\}\, d\theta_1 \\ &= i\int_{\theta_1=+\pi}^{\theta_1=0} d\theta_1 = i(0-\pi) = -i\pi \end{aligned} \tag{A6.7}$$

where $e^{-\rho\sin[\theta_1]} \to 1$ in the limit $\rho \to 0$.

The second integral over the large semicircle is rewritten in the same manner:

$$\lim_{R\to\infty}\left\{\int_{C_2} \frac{e^{+iR\, e^{i\theta_2}}}{R\, e^{i\theta_2}} iR\, e^{i\theta_2}\, d\theta_2\right\} = i\int_{\theta_2=0}^{+\pi} \lim_{R\to\infty}\{e^{+iR\cos[\theta_2]}\, e^{-R\sin[\theta]}\}\, d\theta_2 = 0 \tag{A6.8}$$

where $e^{-R\sin[\theta_1]} \to 0$ in the limit $R \to \infty$. Thus the sum of the four component integrals around the path is zero. The desired complex integral may be evaluated:

$$\begin{aligned} 0 &= \lim_{R\to\infty}\lim_{\rho\to 0}\left\{\int_{-R}^{-\rho} \frac{e^{+ix}}{x}\, dx + \int_{\rho}^{R} \frac{e^{+ix}}{x}\, dx\right\} - i\pi \\ &= \int_{x=-\infty}^{+\infty} \frac{e^{+ix}}{x}\, dx - i\pi \\ \Longrightarrow &\int_{x=-\infty}^{+\infty} \frac{e^{+ix}}{x}\, dx = 0 + i\pi \end{aligned} \tag{A6.9}$$

The area in Equation (6.33) is obtained by substitution into Equation (A6.1) and evaluating the imaginary part:

$$\begin{aligned} \int_{x=-\infty}^{+\infty} \frac{\sin[\pi\alpha]}{\pi\alpha}\, d\alpha &= \frac{1}{\pi}\Im\left\{\int_{x=-\infty}^{+\infty} \frac{e^{+ix}}{x}\, dx\right\} \\ &= \frac{1}{\pi}\Im\{i\pi\} = 1 \end{aligned} \tag{A6.10}$$

The area of $SINC^2[x]$ is evaluated in a similar fashion after some algebra:

$$\begin{aligned} \int_{\alpha=-\infty}^{+\infty} SINC^2[\alpha]\, d\alpha &= \int_{\alpha=-\infty}^{+\infty} \frac{\sin^2[\pi\alpha]}{(\pi\alpha)^2}\, d\alpha \\ &= \frac{1}{\pi}\int_{x=-\infty}^{+\infty} \frac{\sin^2[x]}{x^2}\, dx \end{aligned} \tag{A6.11}$$

The Euler relation is applied to obtain:

$$\begin{aligned} \int_{x=-\infty}^{+\infty} \frac{\sin^2[x]}{x^2}\, dx &= \int_{x=-\infty}^{+\infty} \frac{1-\cos[2x]}{2x^2}\, dx \\ &= \Re\left\{\int_{x=-\infty}^{+\infty} \frac{1-e^{+i(2x)}}{2x^2}\, dx\right\} \end{aligned} \tag{A6.12}$$

This is rewritten as a path integral on the complex plane:

$$\Re\left\{\int_{x=-\infty}^{+\infty} \frac{1-e^{+i(2x)}}{2x^2}\,dx\right\} = \int_{z=-\infty+0i}^{+\infty+0i} \frac{1-e^{+i(2z)}}{2z^2}\,dz \tag{A6.13}$$

The integral will be evaluated on the same closed path that was used for *SINC*$[x]$. The four component integrals are:

$$\begin{aligned}
\oint_C \frac{1-e^{+i(2z)}}{2z^2}\,dz &= \int_{z=-R+0i}^{-\rho+0i} \frac{1-e^{+i(2z)}}{2z^2}\,dz + \int_{C_1} \frac{1-e^{+i(2z)}}{2z^2}\,dz \\
&\quad + \int_{z=\rho+0i}^{R+0i} \frac{1-e^{+i(2z)}}{2z^2}\,dz + \int_{C_2} \frac{1-e^{+i(2z)}}{2z^2}\,dz \\
&= \left(\int_{x=-R}^{-\rho} \frac{1-e^{+i(2x)}}{2x^2}\,dx + \int_{x=+\rho}^{+R} \frac{1-e^{+i(2x)}}{2x^2}\,dx\right) \\
&\quad + \left(\int_{C_1} \frac{1-e^{+i(2z)}}{2z^2}\,dz + \int_{C_2} \frac{1-e^{+i(2z)}}{2z^2}\,dz\right)
\end{aligned} \tag{A6.14}$$

These will be evaluated in the limit $\rho \to 0$ and $R \to \infty$, as before. Consider the integral over path C_1 where $z = \rho\, e^{+i\theta_1}$:

$$\begin{aligned}
\int_{C_1} \frac{1-e^{+i(2z)}}{2z^2}\,dz &= \int_{\theta_1=\pi}^{0} \frac{1-e^{+i(2z)}}{(2\rho^2\, e^{+2i\theta_1})}(i\rho\, e^{+i\theta_1}\, d\theta_1) \\
&= \frac{i}{2}\int_{\theta_1=\pi}^{0} \frac{(1-e^{+i(2z)})\, e^{-i\theta_1}}{\rho}\, d\theta_1 \\
&= \frac{i}{2}\int_{\theta_1=\pi}^{0} \frac{(1-e^{+2i\rho\cos[\theta_1]}\, e^{-2\rho\sin[\theta_1]})\, e^{-i\theta_1}}{\rho}\, d\theta_1
\end{aligned} \tag{A6.15}$$

The exponential terms may be written as power series that may be truncated at the first-order term because $\rho \to 0$:

$$e^{+2i\rho\cos[\theta_1]} = \sum_{n=0}^{+\infty} \frac{(2i\rho\cos[\theta_1])^n}{n!} \cong 1 + i(2\rho\cos[\theta_1]) \quad \textit{if}\ \rho \to 0 \tag{A6.16a}$$

$$e^{-2\rho\sin[\theta_1]} = \sum_{n=0}^{+\infty} \frac{(-2\rho\sin[\theta_1])^n}{n!} \cong 1 - 2\rho\sin[\theta_1] \quad \textit{if}\ \rho \to 0 \tag{A6.16b}$$

Thus the limiting form of the numerator is:

$$\begin{aligned}
1 - e^{+i(2z)} &\cong 1 - (1 + i(2\rho\cos[\theta_1]))(1 - 2\rho\sin[\theta_1]) \\
&= 1 - (1 + i(2\rho\cos[\theta_1]) - 2\rho\sin[\theta_1] - 4i\rho^2\cos[\theta_1]\sin[\theta_1]) \\
&\cong -2\rho(i\cos[\theta_1] + \sin[\theta_1]) = -2i\rho(\cos[\theta_1] + i\sin[\theta_1]) = -2i\rho\, e^{+i\theta_1}
\end{aligned} \tag{A6.17}$$

where the second-order term is neglected, again because $\rho \to 0$. The resulting integral is:

$$\begin{aligned}
\int_{C_1} \frac{1-e^{+i(2z)}}{2z^2}\,dz &= \frac{i}{2}\int_{\theta_1=\pi}^{0} \frac{(-2i\rho\, e^{+i\theta_1})\, e^{-i\theta_1}}{\rho}\, d\theta_1 \\
&= \int_{\theta_1=\pi}^{0} d\theta_1 = -\pi
\end{aligned} \tag{A6.18}$$

The integral over the large semicircle evaluates to zero because of the effect of the term $e^{-2R\sin[\theta_2]}$ as $R \to \infty$:

$$\int_{C_2} \frac{1 - e^{+i(2z)}}{2z^2}\,dz = \lim_{R\to\infty}\left\{\int_{\theta_2=0}^{+\pi} \frac{(1 - e^{+2iR\cos[\theta_2]}\,e^{-2R\sin[\theta_2]})\,e^{-i\theta_2}}{\rho}\,d\theta_2\right\} \to 0 \tag{A6.19}$$

Cauchy's theorem again states that the integral around the closed path must evaluate to zero because no singularities are enclosed. The remaining terms are:

$$\int_{x=-\infty}^{+\infty} \frac{1 - e^{+i(2x)}}{2x^2}\,dx = \pi \tag{A6.20}$$

and the desired area is obtained by substitution into Equation (A6.12):

$$\int_{x=-\infty}^{+\infty} SINC^2[\alpha]\,d\alpha = \frac{1}{\pi}\Re\left\{\int_{x=-\infty}^{+\infty} \frac{1 - e^{+i(2x)}}{2x^2}\,dx\right\} = 1 \tag{A6.21}$$

Note that the fact that the areas of $SINC[x]$ and $SINC^2[x]$ are both equal to unity does not mean that the area of $SINC^n[x]$ is unity for integers $n > 2$. As we will see in Chapter 9, the central-ordinate and Rayleigh's theorems of the Fourier transform provide useful tools for evaluating the areas of the powers of $SINC[x]$. It will then be easy to show that the area of $SINC^n[x]$ decreases for increasing $n > 2$.

6.6 Appendix B: Series Solutions for Bessel Functions $J_0[x]$ and $J_1[x]$

Substitute $\nu = 0$ into the differential equation in Equation (6.73) to obtain the differential equation satisfied by the zero-order Bessel function $J_0[x]$:

$$x^2\frac{d^2}{dx^2}(J_0[x]) + x\frac{d}{dx}(J_0[x]) + x^2 J_0[x] = 0 \tag{B6.1}$$

The power-series solution for $J_0[x]$ has the form:

$$J_0[x] = \sum_{\ell=0}^{+\infty} a_\ell x^\ell \tag{B6.2}$$

After substitution in Equation (B6.1) and evaluating the derivatives of each term, we obtain a relationship that must be satisfied by a_ℓ for each power of x:

$$\begin{aligned}
&x^2\frac{d^2}{dx^2}\left(\sum_{\ell=0}^{+\infty} a_\ell x^\ell\right) + x\frac{d}{dx}\left(\sum_{\ell=0}^{+\infty} a_\ell\, x^\ell\right) + x^2\sum_{\ell=0}^{+\infty} a_\ell x^\ell\\
&= x^2\sum_{\ell=0}^{+\infty} a_\ell(\ell(\ell-1))x^{\ell-2} + \left(\sum_{\ell=0}^{+\infty} a_\ell \ell x^\ell\right) + \left(\sum_{\ell=0}^{+\infty} a_\ell x^{\ell+2}\right)\\
&= \sum_{\ell=0}^{+\infty} a_\ell(\ell(\ell-1))x^\ell + a_\ell \ell x^\ell + a_\ell x^{\ell+2} = 0
\end{aligned} \tag{B6.3}$$

The sum may be rewritten to collect the coefficients for equal powers of x:

$$\sum_{\ell=0}^{+\infty} a_\ell(\ell(\ell-1))x^\ell + a_\ell\,\ell x^\ell + a_\ell x^{\ell+2} = \sum_{\ell=0}^{+\infty}(a_\ell[\ell(\ell-1)] + a_\ell \ell + a_{\ell-2})x^\ell = 0 \tag{B6.4}$$

For this expression to evaluate to zero, each power of x must individually evaluate to zero. The result is a recursion relation for a_ℓ:

$$a_\ell[\ell(\ell-1)] + a_\ell\,\ell + a_{\ell-2} = 0 \Longrightarrow a_\ell = -\frac{a_{\ell-2}}{[\ell(\ell-1)]+\ell} = -\frac{a_{\ell-2}}{\ell^2} \tag{B6.5}$$

Note that this expression relates only the coefficients that differ in power by a factor of 2, and that the algebraic sign is the opposite of the next coefficient in the series. The values of the coefficients are determined by the boundary conditions used in the problem. The zeroth-order coefficient is the amplitude of the function at the origin, $a_0 = J_0[0]$, which is assumed to be unity. The subsequent coefficients for the even powers are:

$$a_2 = -\frac{1}{2^2} = -\frac{1}{4} \tag{B6.6a}$$

$$a_4 = -\frac{a_2}{4^2} = -\frac{(-\frac{1}{4})}{16} = +\frac{1}{64} \tag{B6.6b}$$

$$a_6 = -\frac{a_4}{6^2} = -\frac{(+\frac{1}{64})}{36} = -\frac{1}{2304} \tag{B6.6c}$$

$$a_8 = -\frac{a_6}{8^2} = -\frac{(-\frac{1}{2304})}{64} = +\frac{1}{147\,456} \tag{B6.6d}$$

$$\vdots$$

$$\begin{aligned} a_{2\ell} &= (-1)^\ell \frac{1}{((2\ell)^2\,[2(\ell-1)]^2(2[\ell-2])^2 \ldots 2^2)} \\ &= (-1)^\ell \frac{1}{(2^2)^\ell(\ell!)^2} = \frac{(-1)^\ell}{2^{2\ell}(\ell!)^2} \end{aligned} \tag{B6.6e}$$

The coefficients of the odd powers are assumed to be zero, which determines that $J_0[x]$ is an even function. The power series for the zero-order Bessel function of the first kind is:

$$\begin{aligned} J_0[x] &= \sum_{\ell=0}^{\infty} (-1)^\ell \frac{1}{(\ell!)^2}\left(\frac{x}{2}\right)^{2\ell} \\ &= 1 - \frac{x^2}{4} + \frac{x^4}{64} - \frac{x^6}{2304} + \frac{x^8}{147\,456} - \cdots \end{aligned} \tag{B6.7}$$

Note that the magnitudes of the coefficients decrease very rapidly with order, which means that the extrema of $J_0[x]$ decrease much more slowly with increasing x than the *SINC* function.

The series solution for $J_1[x]$ must satisfy Equation (6.73) for $v = 1$:

$$x^2\,\frac{d^2}{dx^2}(J_1[x]) + x\,\frac{d}{dx}(J_1[x]) + (x^2-1)J_1[x] = 0 \tag{B6.8}$$

The power-series solution for $J_1[x]$ has the form:

$$J_1[x] = \sum_{\ell=0}^{+\infty} b_\ell\, x^\ell \tag{B6.9}$$

$$\begin{aligned} &x^2\frac{d^2}{dx^2}\left(\sum_{\ell=0}^{+\infty} b_\ell\, x^\ell\right) + x\frac{d}{dx}\left(\sum_{\ell=0}^{+\infty} b_\ell\, x^\ell\right) + (x^2-1)\sum_{\ell=0}^{+\infty} b_\ell\, x^\ell \\ &= x^2\sum_{\ell=0}^{+\infty} b_\ell(\ell(\ell-1))x^{\ell-2} + \left(\sum_{\ell=0}^{+\infty} b_\ell\,\ell\, x^\ell\right) + \left(\sum_{\ell=0}^{+\infty} b_\ell\, x^{\ell+2} - \sum_{\ell=0}^{+\infty} b_\ell\, x^\ell\right) \end{aligned}$$

$$= \sum_{\ell=0}^{+\infty} (b_\ell(\ell(\ell-1)) + b_\ell \ell + b_{\ell-2} - b_\ell) x^\ell = 0$$

$$= \sum_{\ell=0}^{+\infty} (b_\ell((\ell(\ell-1)) + \ell - 1) + b_{\ell-2}) x^\ell = 0 \tag{B6.10}$$

The recursion relation that must be satisfied by the coefficients is:

$$b_\ell = -\frac{b_{\ell-2}}{\ell^2 - \ell + \ell - 1} = -\frac{b_{\ell-2}}{\ell^2 - 1} \tag{B6.11}$$

Again, the boundary conditions of the Bessel function determine the first two coefficients. Because $J_1[0] = 0$, all of the even-order coefficients vanish. The first-order coefficient is equal to the value of the slope at the origin, and is set to $\frac{1}{2}$. The rest of the odd-order coefficients are determined by the recursion relation in Equation (B6.11):

$$b_1 = \frac{1}{2} \tag{B6.12a}$$

$$b_3 = -\frac{b_1}{(3^2-1)} = -\frac{(\frac{1}{2})}{2 \cdot 4} = -\frac{1}{16} \tag{B6.12b}$$

$$b_5 = -\frac{b_3}{(5^2-1)} = \frac{(-1)^2}{(5^2-1)\cdot(3^2-1)\cdot 2} = \frac{(-1)^2}{(2^2 \cdot 4^2 \cdot 6)} = +\frac{1}{384} \tag{B6.12c}$$

$$b_7 = -\frac{b_5}{(7^2-1)} = \frac{(-1)^3}{(7^2-1)\cdot(5^2-1)\cdot(3^2-1)\cdot 2}$$

$$= -\frac{(\frac{1}{384})}{48} = -\frac{1}{18\,432} \tag{B6.12d}$$

$$\vdots$$

$$b_{2\ell+1} = (-1)^\ell \frac{1}{((2\ell+1)^2-1)\cdot((2\ell-1)^2-1)\cdot((2\ell-3)^2-1)\cdot\dots\cdot 2}$$

$$= \frac{(-1)^\ell}{(2^2 \cdot 4^2 \cdot \dots \cdot (2\ell)^2 \cdot (2\ell+2))} \tag{B6.12e}$$

Therefore, the power series for the first-order Bessel function of the first kind is:

$$J_1[x] = \frac{x}{2} - \frac{x^3}{(2^2 \cdot 4)} + \frac{x^5}{(2^2 \cdot 4^2 \cdot 6)} - \frac{x^7}{(2^2 \cdot 4^2 \cdot 6^2 \cdot 8)} + \cdots$$

$$= \frac{x}{2} - \frac{x^3}{16} + \frac{x^5}{384} - \frac{x^7}{18\,432} + \cdots \tag{B6.13}$$

Again note that the magnitudes of the coefficients decrease rapidly as the power of x increases.

After differentiating the series solution for $J_0[x]$ in Equation (B6.7) and comparing to Equation (B6.13), we obtain a useful relation that is valid for nonnegative values of x:

$$\frac{d}{dx}(J_0[x]) = 0 - \frac{x}{2} + \frac{x^3}{16} - \frac{x^5}{384} + \frac{x^7}{18\,432} - \cdots = -J_1[x] \tag{B6.14}$$

A more general discussion of the series solution for the Bessel function that is valid for all positive integer values of n (Stephenson and Radmore, 1990) demonstrates that:

$$J_n[x] = \sum_{\ell=0}^{+\infty} \frac{(-1)^\ell}{\ell!(n+\ell)!} \left(\frac{x}{2}\right)^{n+2\ell} \tag{B6.15}$$

PROBLEMS

6.1 For the 1-D function of x

$$f[x] = RECT\left[\frac{x+a}{b}\right] + RECT\left[\frac{x-a}{b}\right]$$

sketch $f[x]$ for the following parameters:

(a) $a = 2b$
(b) $a = b$
(c) $a = b/2$ (also write down a different and more concise expression for this function).

6.2 Sketch the following:

(a) $f_a[x] = 2\, TRI[(x-2)/2] \cdot SGN[2-x]$
$g_a[x] = SGN[f_a[x]]$
$h_a[x] = STEP[f_a[x]]$
(b) $f_b[x] = 2\, e^{-(4-x)} \cdot STEP[4-x] - 1$
$g_b[x] = SGN[f_b[x]]$
$h_b[x] = STEP[f_b[x]]$

6.3 Derive and sketch the even and odd parts of the following functions, where x_0 is a positive real number:

(a) $\delta[x - x_0]$
(b) $e^{-(x+x_0)}\, STEP[x + x_0]$

6.4 Evaluate $\delta[SINC[x]] = \delta[(\sin[\pi x])/\pi x]$.

6.5 Evaluate $\delta[\delta[x]]$.

6.6 Write down mathematical expressions using the special functions for these "objects":

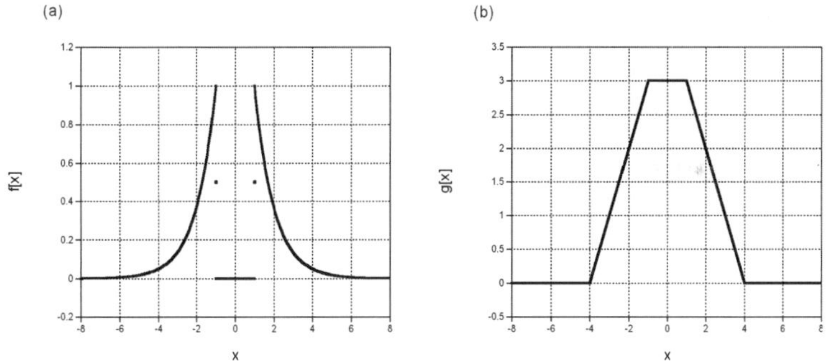

6.7 Use the definition of the Dirac delta function to evaluate the following in terms of one or more of the defined special functions:

(a) $\delta[2x + 2]$
(b) $\delta[-x^2 - 4]$
(c) $\delta[\cos[2\pi\xi_0 x]]$
(d) $\delta[x^3 - x]$

6.8 Evaluate the integral

$$\int_{-\infty}^{+\infty} \left(A_0 + A_1 \cos\left[2\pi\xi_0 \frac{x}{\alpha}\right]\right) \cdot RECT[x - \alpha]\, d\alpha$$

6.9 Rewrite the Gaussian function $f[x] = e^{-\pi x^2}$ in the form of a "chirp" function $e^{\pm i\pi(x/b_\pm)^2}$, where $b_\pm$ is the appropriate complex number.

6.10 Use the definitions of the Gaussian function $f[x] = e^{-\pi x^2}$ and of the gamma function $\Gamma[x] = \int_0^{+\infty} \alpha^x \, e^{-\alpha} \, d\alpha$ to construct a proof that the area of the Gaussian is unity. (HINT: change the variable of integration.)

6.11 Use the fact that $\int_{-\infty}^{+\infty} e^{-\pi\alpha^2} \, d\alpha = 1$ to determine the area of the "chirp" functions $e^{\pm i\pi(x/b)^2}$, where b is a real-valued parameter.

6.12 Derive an expression for and sketch (or plot) the even and odd parts of

$$e^{-(x+x_0)} \, STEP[x + x_0]$$

where x_0 is a positive real number.

6.13 Do the following:

(a) Sketch the function $f[x] = RECT[x + \frac{1}{2}] + STEP[x]TRI[x]$.

(b) Determine the "support" and the "area" of $f[x]$.

(c) Sketch $f[(-x + 1)/2]$ and $f[-2x + 4]$.

6.14 Calculate the area of $e^{\pm i\pi x^2}$ by computing the volume of $e^{\pm i\pi(x^2+y^2)}$ in polar coordinates. In other words, construct an analogous process to that used to compute $\Gamma[\frac{1}{2}]$.

6.15 Calculate the "full width" Δx of the positive central lobe of $\cos[+\pi(x/\alpha)^2]$, i.e., the distance along the x-axis between the two zeros closest to $x = 0$.

7

2-D Special Functions

A function of two independent spatial variables specifies the amplitude ("brightness") at each spatial coordinate in a plane. The particular importance of these functions arises because they fulfill the usual definition of an "image". For convenience, we will classify 2-D functions into three main categories: the Cartesian separable functions, the circularly symmetric functions, and general functions that are neither separable nor circularly symmetric. The first category includes the simplest cases; these may be represented as products of 1-D special functions along orthogonal Cartesian axes and are considered in Section 7.1. The circularly symmetric functions may be interpreted as the product of 1-D functions in the radial direction and the unit constant in the orthogonal azimuthal direction, as considered in Section 7.2. The arbitrary 2-D functions, such as normal pictorial scenes and 2-D stochastic functions, are also considered briefly.

7.1 2-D Separable Functions

The 2-D separable functions are defined as those generated by "orthogonal multiplication" of two 1-D functions $f_x[x]$ and $f_y[y]$ where the amplitude of the 2-D function at a coordinate is just the product of the 1-D functions evaluated at the individual coordinates. Though there might be some advantage to using a specific notation for the orthogonal product, we will not do so. Thus the separable function may be written:

$$f[x,y] = f_x[x]f_y[y] \tag{7.1}$$

The general expression for a separable function in terms of scaled and translated 1-D functions is:

$$f[x,y] = f_x\left[\frac{x-x_0}{a}\right]f_y\left[\frac{y-y_0}{b}\right] \tag{7.2}$$

where a and b are real-valued scale factors and x_0 and y_0 are real-valued translation parameters. Obviously, a large number of possible 2-D functions may be generated by constructing all possible permutations of the 1-D special functions defined in Chapter 6. We will briefly consider only those which have general application to imaging problems. Note that the volume of a 2-D separable function is equal to the product of the areas of the component 1-D functions:

$$\begin{aligned}\iint_{-\infty}^{+\infty} f[x,y]\,dx\,dy &= \iint_{-\infty}^{+\infty} f_x\left[\frac{x-x_0}{a}\right]f_y\left[\frac{y-y_0}{b}\right]dx\,dy\\ &= \left(\int_{-\infty}^{+\infty} f_x\left[\frac{x-x_0}{a}\right]dx\right)\left(\int_{-\infty}^{+\infty} f_y\left[\frac{y-y_0}{b}\right]dy\right)\end{aligned} \tag{7.3}$$

Fourier Methods in Imaging Roger L. Easton, Jr.

7.1.1 Rotations of 2-D Separable Functions

It is useful to generalize the concept of separable functions to include examples that are separable along orthogonal axes other than x and y, as by rotating a Cartesian-separable function about the origin of coordinates. For example, a square centered at the origin with sides parallel to the x- and y-axes may be rotated by $\pm\pi/4$ to generate a "baseball diamond" with vertices on the x- and y-axes.

The process of rotating a function about the origin may itself be viewed as an imaging system with 2-D input function $f[x, y]$ and 2-D output $g[x, y]$. The amplitude g of the rotated function at $[x, y]$ is identical to the original amplitude f at a different location $[x', y']$:

$$\mathcal{O}\{f[x, y]\} = g[x, y] = f[x', y'] \tag{7.4}$$

so the rotation operation is specified completely by a mapping that relates the coordinates $[x, y]$ and $[x', y']$. Consider a particular location with Cartesian coordinates $[x_0, y_0]$ and polar coordinates:

$$r_0 = \sqrt{x_0^2 + y_0^2}, \quad \theta_0 = \tan^{-1}\left[\frac{y_0}{x_0}\right]$$

The location after rotating the coordinates by ϕ radians is shown in Figure 7.1. Clearly, the radial coordinate r_0 is unchanged by the rotation, while the azimuthal angle of the same location expressed in the new system is $\theta' = \theta_0 - \phi$. These observations may be used to express the Cartesian coordinates of the original location in the rotated coordinates $[x_0', y_0']$ in terms of the original coordinates $[x_0, y_0]$:

$$\begin{aligned} x_0' &= |\underline{\mathbf{r}}_0| \cos[\theta_0 - \phi] = |\underline{\mathbf{r}}_0|(\cos[\theta_0]\cos[\phi] + \sin[\theta_0]\sin[\phi]) \\ &= (r_0 \cos[\theta_0])\cos[\phi] + (r_0 \sin[\theta_0])\sin[\phi] \\ &= x_0 \cos[\phi] + y_0 \sin[\phi] \end{aligned} \tag{7.5}$$

$$\begin{aligned} y_0' &= r_0 \sin[\theta_0 - \phi] = r_0(\sin[\theta_0]\cos[\phi] - \cos[\theta_0]\sin[\phi]) \\ &= (r_0 \sin[\theta_0])\cos[\phi] - (r_0 \cos[\theta_0])\sin[\phi] \\ &= -x_0 \sin[\phi] + y_0 \cos[\phi] \end{aligned} \tag{7.6}$$

7.1.2 Rotated Coordinates as Scalar Products

Yet another, and sometimes very convenient, notation for the function expressed in the rotated coordinate system is based upon the scalar products of particular vectors. The rotated x-coordinate in Equation (7.5) may be written as the scalar product of the position vector $\underline{\mathbf{r}} \equiv [x, y]$ and the unit vector directed along the azimuth angle ϕ, which has Cartesian coordinates $[\cos[\phi], \sin[\phi]]$. We denote this unit vector by $\underline{\hat{\mathbf{p}}}$:

$$x' = x\cos[\phi] + y\sin[\phi] = \begin{bmatrix} x \\ y \end{bmatrix} \bullet \begin{bmatrix} \cos[\phi] \\ \sin[\phi] \end{bmatrix} \equiv \underline{\mathbf{r}} \bullet \underline{\hat{\mathbf{p}}} \tag{7.7}$$

Though $\underline{\hat{\mathbf{p}}}$ may reside in any of the four quadrants, we will soon constrain it to the first or fourth. This notation for the rotated 1-D argument x' may seem a little "weird" because it is a function of both x and y through the scalar product of two vectors. However, the notation makes good sense; the rotated argument x' defines a set of points $\underline{\mathbf{r}} = [x, y]$ that fulfill the same conditions as the coordinate x in the original function. The rotated coordinate x' evaluates to 0 for all vectors $\underline{\mathbf{r}}$ such that $\underline{\mathbf{r}} \perp \underline{\hat{\mathbf{p}}}$, which is equivalent to the statement that $\underline{\mathbf{r}} \bullet \underline{\hat{\mathbf{p}}} = 0$. Since the vectors $\underline{\mathbf{r}}$ specified by this condition $\underline{\mathbf{r}} \bullet \underline{\hat{\mathbf{p}}} = 0$ must include coordinates on "both" sides of the origin, the complete set of possible azimuth angles may be specified by the angles ϕ in any azimuthal interval spanning π radians, e.g., $-\pi/2 \leq \phi < +\pi/2$. This

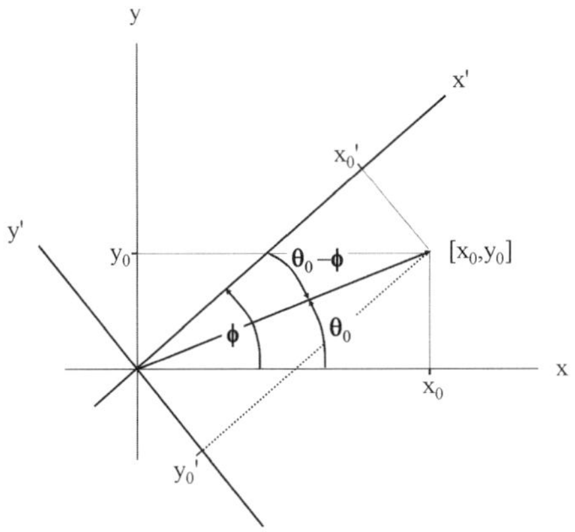

Figure 7.1 The Cartesian coordinates of a particular location $[x_0, y_0]$ are evaluated in the rotated coordinate system $[x', y']$ from the polar coordinates $(|\underline{\mathbf{r}}_0|, \theta_0 - \phi)$: $x'_0 = |\underline{\mathbf{r}}_0| \cos[\theta_0 - \phi] = x_0 \cos[\phi] + y_0 \sin[\phi]$, $y'_0 = |\underline{\mathbf{r}}_0| \sin[\theta_0 - \phi] = -x_0 \sin[\phi] + y_0 \cos[\phi]$.

means that the complete set of possible rotations may be specified by unit vectors $\underline{\hat{\mathbf{p}}}$ that are constrained to the first or fourth quadrants.

The polar form of the position vector $\underline{\mathbf{r}} = (|\underline{\mathbf{r}}|, \theta) \equiv (r, \theta)$ may be substituted into the expression for the scalar product in terms of the magnitudes of the vectors and the included angle in Equation (3.14):

$$\underline{\mathbf{r}} \bullet \underline{\hat{\mathbf{p}}} = |\underline{\mathbf{r}}|\,|\underline{\hat{\mathbf{p}}}| \cos[\theta - \phi] = r \cos[\theta - \phi] \tag{7.8}$$

where the fact that $|\underline{\hat{\mathbf{p}}}| = 1$ has been used.

The y-coordinate of the rotated function may be written in the same way as the scalar product of the position vector $\underline{\mathbf{r}}$ and the unit vector directed along azimuth angle $\phi + \pi/2$; we assign the notation $\underline{\hat{\mathbf{p}}}^{\perp}$ to this unit vector to indicate that it is perpendicular to $\underline{\hat{\mathbf{p}}}$:

$$y' = x \cos\left[\phi + \frac{\pi}{2}\right] + y \sin\left[\phi + \frac{\pi}{2}\right]$$

or

$$\begin{aligned} y' &= x \cos\left[\phi - \frac{\pi}{2}\right] + y \sin\left[\phi - \frac{\pi}{2}\right] \\ &= -x \sin[\phi] + y \cos[\phi] = \begin{bmatrix} x \\ y \end{bmatrix} \bullet \begin{bmatrix} -\sin[\phi] \\ \cos[\phi] \end{bmatrix} \equiv \underline{\mathbf{r}} \bullet \underline{\hat{\mathbf{p}}}^{\perp} \end{aligned} \tag{7.9}$$

The corresponding polar form for the rotated y-axis is:

$$\begin{aligned} \underline{\mathbf{r}} \bullet \underline{\hat{\mathbf{p}}}^{\perp} &= |\underline{\mathbf{r}}|\,|\underline{\hat{\mathbf{p}}}^{\perp}| \cos\left[\theta - \left(\phi \pm \frac{\pi}{2}\right)\right] \\ &= r \cos\left[(\theta - \phi) \mp \frac{\pi}{2}\right] = \pm r \sin[\theta - \phi] \end{aligned} \tag{7.10}$$

We select the sign of $\pi/2$ to place $\underline{\hat{\mathbf{p}}}^{\perp}$ in the first or fourth quadrants.

Equations (7.8) and (7.10) are substituted for x' and y' in Equation (7.1) to construct the expression for the separable function that has been rotated by ϕ radians about the origin:

$$g[x, y] = f_x[x'] f_y[y'] = f_x[\underline{\mathbf{r}} \bullet \underline{\hat{\mathbf{p}}}] f_y[\underline{\mathbf{r}} \bullet \underline{\hat{\mathbf{p}}}^{\perp}] \tag{7.11}$$

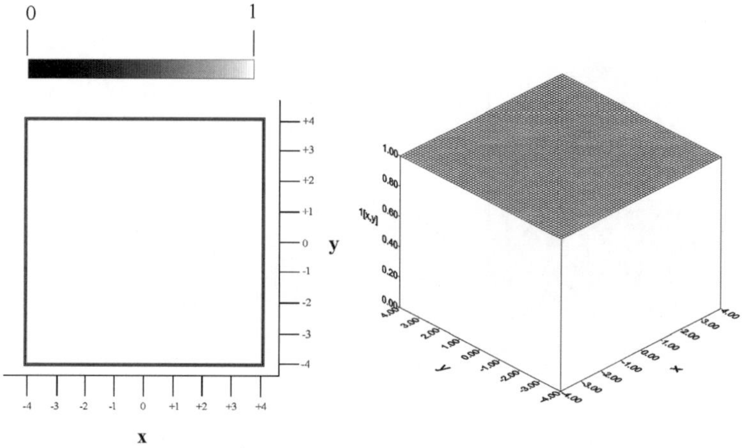

Figure 7.2 Representations of the 2-D unit constant function as a gray-scale image and as a surface, where the amplitude of the function determines the height of the 3-D surface.

This transformation may be applied to any of the 2-D separable special functions introduced in the next section.

7.2 Definitions of 2-D Special Functions

7.2.1 2-D Constant Function

The 2-D unit constant function obviously may be constructed via the orthogonal product of two 1-D unit constant functions.

$$1[x, y] = 1[x]1[y] \tag{7.12}$$

as shown in Figure 7.2. Just as obviously, the 2-D null constant can be expressed as the orthogonal product of a 1-D null constant function and any other 1-D function, e.g.:

$$0[x, y] = 0[x] f_y[y] \tag{7.13}$$

where $f_y[y]$ is any 1-D function with finite amplitude. The volume of a 2-D constant is its integral over all space, and is infinite if the amplitude is nonzero:

$$\iint_{-\infty}^{+\infty} 1[x, y]\, dx\, dy = \infty \tag{7.14}$$

Clearly, translations and rotations do not affect the amplitude of a 2-D constant function at any coordinate.

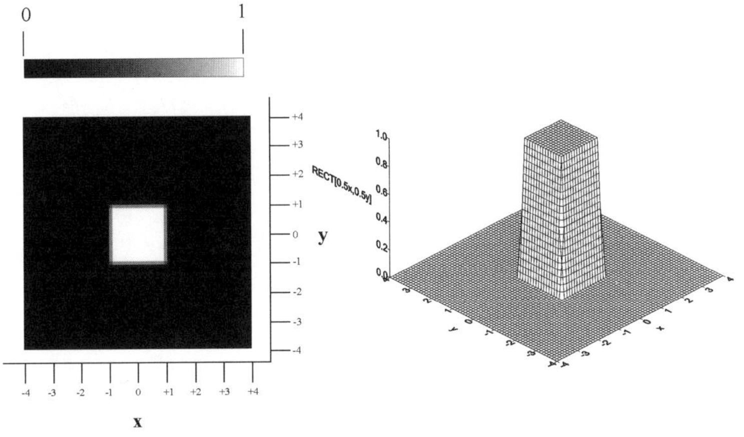

Figure 7.3 The 2-D rectangle function $f[x, y] = RECT[x/2, y/2]$.

7.2.2 Rectangle Function

The defining expression for the 2-D separable rectangle function evidently is the orthogonal product of two replicas of the 1-D rectangle defined in Equation (6.3):

$$RECT\left[\frac{x}{a}, \frac{y}{b}\right] = RECT\left[\frac{x}{a}\right] RECT\left[\frac{y}{b}\right] \tag{7.15a}$$

$$RECT\left[\frac{x - x_0}{a}, \frac{y - y_0}{b}\right] = RECT\left[\frac{x - x_0}{a}\right] RECT\left[\frac{y - y_0}{b}\right] \tag{7.15b}$$

This function is shown in Figure 7.3. Its volume obviously is finite if both a and b are finite:

$$\iint_{-\infty}^{+\infty} RECT\left[\frac{x - x_0}{a}, \frac{y - y_0}{b}\right] dx\, dy$$
$$= \int_{-\infty}^{+\infty} RECT\left[\frac{x - x_0}{a}\right] dx \int_{-\infty}^{+\infty} RECT\left[\frac{y - y_0}{b}\right] dy = |ab| \tag{7.16}$$

The 2-D *RECT* is often used to modulate other functions to constrain the domain of support.

A rotated rectangle is obtained by substituting the rotated coordinates of Equations (7.8) and (7.10) into the definition of the 2-D rectangle in Equation (7.15a):

$$RECT[x', y'] = RECT[\underline{\mathbf{r}} \bullet \hat{\underline{\mathbf{p}}}, \underline{\mathbf{r}} \bullet \hat{\underline{\mathbf{p}}}^{\perp}] = RECT[\underline{\mathbf{r}} \bullet \hat{\underline{\mathbf{p}}}]\, RECT[\underline{\mathbf{r}} \bullet \hat{\underline{\mathbf{p}}}^{\perp}]$$
$$= RECT[x \cos[\phi] + y \sin[\phi]]\, RECT[-x \sin[\phi] + y \cos[\phi]] \tag{7.17}$$

By substituting the coordinates of the 1-D *RECT* functions into the defining statement in Equation (6.3), we obtain the equation for the "transition" coordinates $[x', y']$ where $RECT[x', y'] = \frac{1}{2}$. These are the locations $[x, y]$ that satisfy two conditions:

$$x' = x \cos[\phi] + y \sin[\phi] = \pm\frac{1}{2} \Longrightarrow y = (-\cot[\phi])x \pm \frac{1}{2\sin[\phi]} \tag{7.18a}$$

$$y' = -x \sin[\phi] + y \cos[\phi] = \pm\frac{1}{2} \Longrightarrow y = (\tan[\phi])x \pm \frac{1}{2\cos[\phi]} \tag{7.18b}$$

These define the slopes and y-intercepts of two pairs of parallel lines in the 2-D plane with slopes $+\tan[\phi]$ and $-\cot[\phi] = \tan[\phi + \pi/2]$, as shown in Figure 7.4. The dotted lines enclose the support of the region where the amplitude of the rectangle is unity.

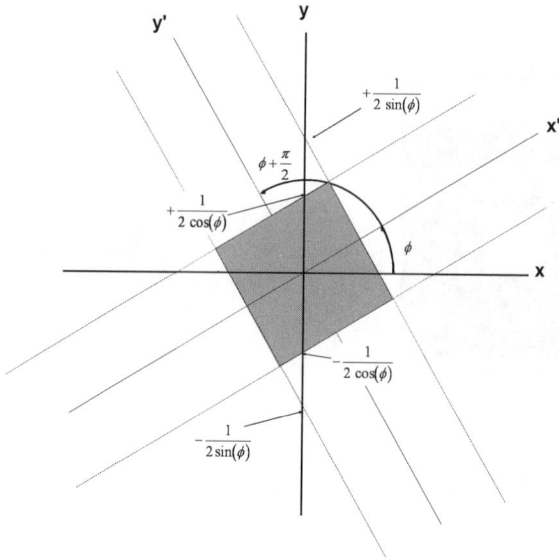

Figure 7.4 The edges of a rotated 2-D rectangle are defined by the loci of points that satisfy the conditions $y = x(-\cot[\phi]) \pm 1/(2\sin[\phi])$ and $y = x\tan[\phi] \pm 1/(2\cos[\phi])$.

7.2.3 Triangle Function

Obviously, the 2-D separable triangle is based on the 1-D triangle in Equation (6.5):

$$\begin{aligned} TRI\left[\frac{x}{a}, \frac{y}{b}\right] &= TRI\left[\frac{x}{a}\right] TRI\left[\frac{y}{b}\right] \\ &= \left(1 - \frac{|x|}{a}\right)\left(1 - \frac{|y|}{b}\right) RECT\left[\frac{x}{2a}, \frac{y}{2b}\right] \\ &= \left(1 - \frac{|x|}{a} - \frac{|y|}{b} + \frac{|xy|}{ab}\right) RECT\left[\frac{x}{2a}, \frac{y}{2b}\right] \end{aligned} \tag{7.19}$$

as shown in Figure 7.5. Note that profiles of the 2-D *TRI* function are composed of straight lines only along lines parallel to the x- or y-axes. The shape of an edge profile along any other radial line includes a quadratic function of the radial distance and thus is parabolic.

The volume of the 2-D *TRI* is obtained easily by direct integration of the separable parts:

$$\iint_{-\infty}^{+\infty} TRI\left[\frac{x}{a}, \frac{y}{b}\right] dx\, dy = \left(\int_{-\infty}^{+\infty} TRI\left[\frac{x}{a}\right] dx\right)\left(\int_{-\infty}^{+\infty} TRI\left[\frac{y}{b}\right] dy\right) = |ab| \tag{7.20}$$

7.2.4 2-D Signum and STEP Functions

Conditions requiring use of a 2-D separable version of the SIGNUM function are encountered relatively rarely and different definitions are possible. One possible choice is based on the "convention" that has been established in this section:

$$SGN[x, y] = SGN[x]\, SGN[y] \tag{7.21}$$

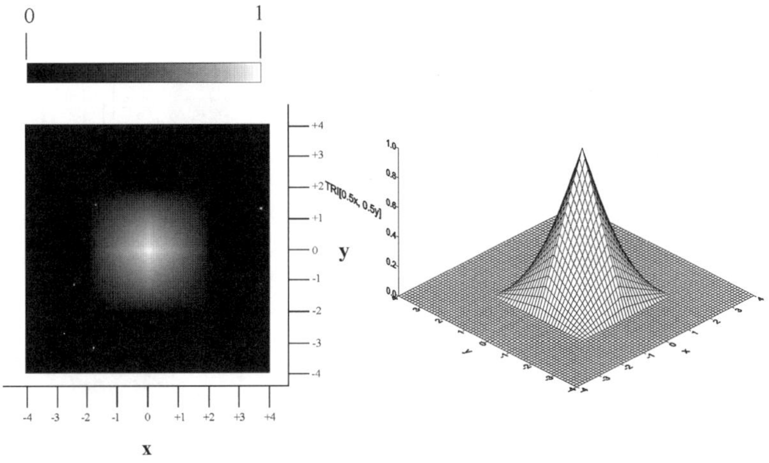

Figure 7.5 The 2-D triangle function $TRI[x/2, y/2]$, showing the parabolic character along profiles that do not coincide with the coordinate axes.

As shown in Figure 7.6a, the amplitude evidently is $+1$ in the first and third quadrants, -1 in the second and fourth quadrants, and 0 along the coordinate axes. Its volume is zero due to the cancellation of the equal-sized positive and negative regions.

It is useful on occasion to define a 2-D *STEP* that is constant in one direction, such as:

$$f[x, y] = STEP[x]1[y] \tag{7.22a}$$

or a rotated version:

$$f[x, y] = STEP[\underline{\mathbf{r}} \bullet \hat{\underline{\mathbf{p}}}]1[\underline{\mathbf{r}} \bullet \hat{\underline{\mathbf{p}}}^{\perp}] \tag{7.22b}$$

The notation for the unit constant also may seem weird, since we think of the 1-D unit constant as having the same unit amplitude everywhere in the 1-D domain. In this context, the 1-D unit constant assigns the amplitude of $STEP[\underline{\mathbf{r}} \bullet \hat{\underline{\mathbf{p}}}]$ at the value of $\underline{\mathbf{r}}_0$ that lies along the direction of $\hat{\underline{\mathbf{p}}}$ to all $\underline{\mathbf{r}}$ along the line perpendicular to $\hat{\underline{\mathbf{p}}}$, i.e., in the direction of $\hat{\underline{\mathbf{p}}}^{\perp}$. In other words, the action of the unit constant is to "spread" or "smear" the amplitude of the 1-D *STEP* evaluated at points in the direction of $\hat{\underline{\mathbf{p}}}$ along the perpendicular direction, thus generating a 2-D function. We will use this "rotated" form for the 1-D unit constant to construct 2-D functions that are constant along one direction. This expression will prove very useful for 2-D Fourier transforms of the rotated functions in Chapter 12.

As shown Figure 7.6b, the volume of Equation (7.22a) obviously is infinite. This definition reasonably could be assigned the name $STEP[x, y]$. However, to remain consistent with the notation for the 2-D separable functions already defined as the product of the identical 1-D function along each axis, this name will be assigned to the product of two 1-D *STEP* functions:

$$STEP[x, y] \equiv STEP[x]\, STEP[y] \tag{7.23}$$

as shown in Figure 7.6c. This function has unit amplitude in the first quadrant, half-unit amplitude along the positive x- or y-axis, and quarter-unit amplitude at the origin. This function is used as a modulator to restrict functions to one quadrant of the 2-D plane. Again, the volume obviously is infinite:

$$\iint_{-\infty}^{+\infty} STEP[x, y]\, dx\, dy = \infty \tag{7.24}$$

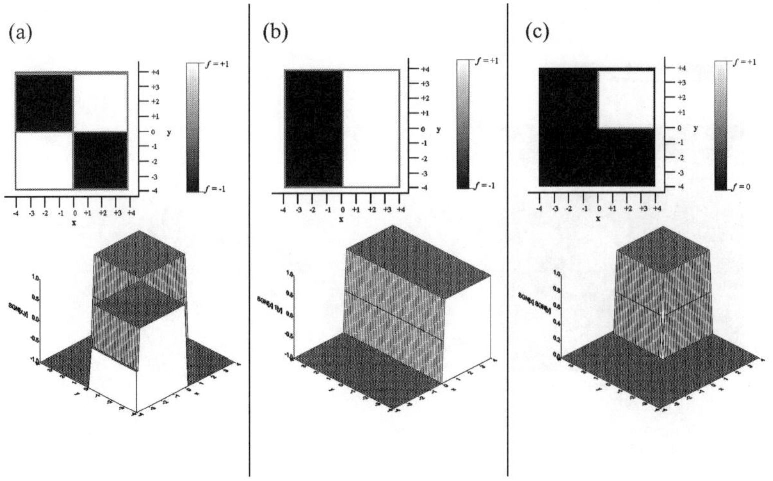

Figure 7.6 Different flavors of 2-D *SGN* and *STEP* functions: (a) $SGN[x] \cdot SGN[y]$; (b) $SGN[x] \cdot 1[y]$; (c) $STEP[x] \cdot STEP[y]$.

7.2.5 2-D *SINC* Function

The general expression for the "centered" 2-D *SINC* function evidently is:

$$SINC\left[\frac{x}{a}, \frac{y}{b}\right] = SINC\left[\frac{x}{a}\right] SINC\left[\frac{y}{b}\right] = \frac{\sin[\pi x/a]\sin[\pi y/b]}{(\pi x/a)(\pi y/b)} \tag{7.25}$$

The function is illustrated in Figure 7.7. Note that the amplitude is positive in regions where both of the constituent 1-D functions are positive or both are negative, and negative where either is negative. The transitions result in a "checkerboard-like" pattern of positive and negative regions, as shown in Figure 7.7. The volume of the 2-D *SINC* is the product of the areas of the individual functions:

$$\begin{aligned}\iint_{-\infty}^{+\infty} SINC\left[\frac{x}{a}, \frac{y}{b}\right] dx\, dy &= \left(\int_{-\infty}^{+\infty} SINC\left[\frac{x}{a}\right] dx\right)\left(\int_{-\infty}^{+\infty} SINC\left[\frac{y}{b}\right] dy\right) \\ &= |a| \cdot |b| = |ab| \end{aligned} \tag{7.26}$$

The 2-D *SINC* function appears in many contexts in optical imaging and diffraction problems.

7.2.6 $SINC^2$ Function

Following the discussion in the previous section, the general expression for the separable 2-D $SINC^2$ function centered on the origin is:

$$\begin{aligned}SINC^2\left[\frac{x}{a}, \frac{y}{b}\right] &= SINC^2\left[\frac{x}{a}\right] SINC^2\left[\frac{y}{b}\right] \\ &= \frac{\sin^2[\pi x/a]\sin^2[\pi y/b]}{(\pi x/a)^2(\pi y/b)^2}\end{aligned} \tag{7.27}$$

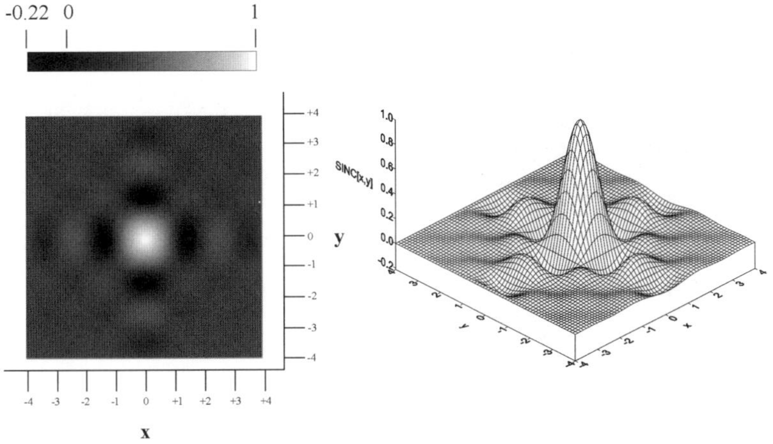

Figure 7.7 $SINC[x, y] = SINC[x] \cdot SINC[y]$, showing the regions of negative and positive amplitude.

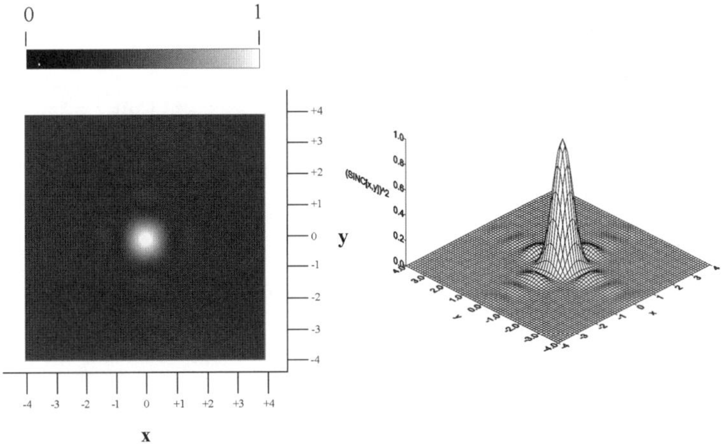

Figure 7.8 $SINC^2[x, y] = SINC^2[x] \cdot SINC^2[y]$, showing that it is nonnegative everywhere.

as shown in Figure 7.8. The $SINC^2$ function is everywhere nonnegative, and its volume matches that of $SINC[x/a, y/b]$:

$$\iint_{-\infty}^{+\infty} SINC^2\left[\frac{x}{a}, \frac{y}{b}\right] dx\, dy = \left(\int_{-\infty}^{+\infty} SINC^2\left[\frac{x}{a}\right] dx\right)\left(\int_{-\infty}^{+\infty} SINC^2\left[\frac{y}{b}\right] dy\right) \qquad (7.28)$$

$$= |a| \cdot |b| = |ab|$$

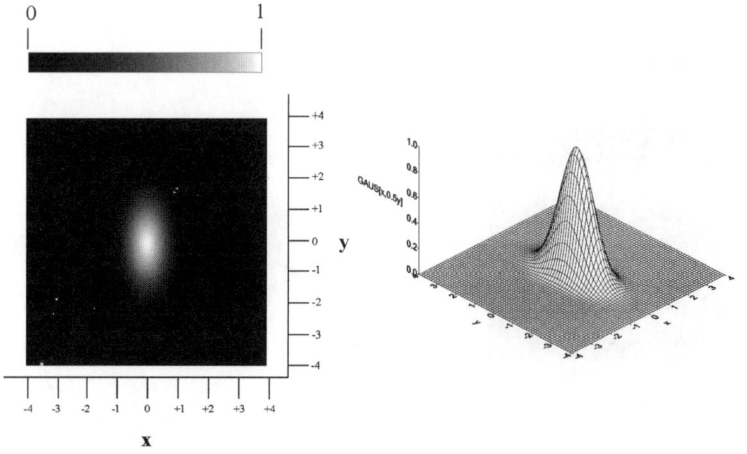

Figure 7.9 Example of 2-D separable Gaussian function: $GAUS[x, y/2] = GAUS[x] \cdot GAUS[y/2]$.

7.2.7 2-D Gaussian Function

The contours of constant amplitude of the 2-D separable Gaussian with different width parameters have elliptical profiles (Figure 7.9):

$$GAUS\left[\frac{x}{a}, \frac{y}{b}\right] = GAUS\left[\frac{x}{a}\right]GAUS\left[\frac{y}{b}\right] = \exp\left[-\frac{\pi x^2}{a^2}\right]\exp\left[-\frac{\pi y^2}{b^2}\right]$$
$$= \exp\left[-\pi\left(\frac{x^2}{a^2}+\frac{y^2}{b^2}\right)\right] \tag{7.29}$$

Again, the volume is the product of the areas of the separable component functions:

$$\iint_{-\infty}^{+\infty} GAUS\left[\frac{x}{a}, \frac{y}{b}\right] dx\, dy = \left(\int_{-\infty}^{+\infty} GAUS\left[\frac{x}{a}\right] dx\right)\left(\int_{-\infty}^{+\infty} GAUS\left[\frac{x}{a}\right] dy\right)$$
$$= |a| \cdot |b| = |ab| \tag{7.30}$$

The rotated version of the Gaussian function is:

$$GAUS\left[\frac{x'}{a}, \frac{y'}{b}\right] = GAUS\left[\frac{\mathbf{r} \bullet \hat{\mathbf{p}}}{a}, \frac{\mathbf{r} \bullet \hat{\mathbf{p}}^{\perp}}{b}\right] = \exp\left[-\pi\left(\left[\frac{\mathbf{r} \bullet \hat{\mathbf{p}}}{a}\right]^2 + \left[\frac{\mathbf{r} \bullet \hat{\mathbf{p}}^{\perp}}{b}\right]^2\right)\right] \tag{7.31}$$

A circularly symmetric function is generated when the two width parameters are equal:

$$GAUS\left[\frac{x}{d}, \frac{y}{d}\right] = \exp\left[-\pi\frac{(x^2+y^2)}{d^2}\right] = \exp\left[-\pi\left(\frac{r}{d}\right)^2\right] \tag{7.32}$$

The Gaussian function is considered in more detail during the discussion of circularly symmetric functions in Section 7.3.

7.2.8 2-D Sinusoid

Sinusoidal functions of two spatial variables may be defined in several ways, but the "useful" definition varies along one direction and is constant in the orthogonal direction. Consider the specific separable

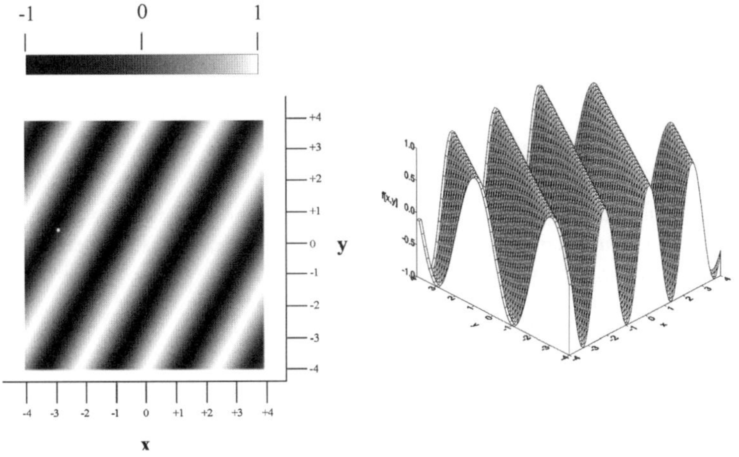

Figure 7.10 The 2-D sinusoid function rotated in azimuth about the origin by $\theta = -\pi/6$ radians.

form:

$$f[x, y] = A_0 \cos[2\pi\xi_0 x + \phi_0]1[y] = A_0 \cos[2\pi\xi_0 x + \phi_0] \cos[2\pi \cdot 0 \cdot y] \tag{7.33}$$

The parameters are identical to those in the 1-D case of Equation (6.19): A_0 is the amplitude, ξ_0 is the spatial frequency along the x-direction, and ϕ_0 is the initial phase. The function is shown in Figure 7.10. As a visual picture of the 2-D sinusoid, Gaskill (1978) suggests an image of this function of a field plowed with sinusoidal furrows. In these days when fewer students are familiar with traditional American farming practices, a more accessible (though perhaps less healthy) image is a "ridged" potato chip; the small-scale variation in the profile is approximately sinusoidal. Another appropriate image is of a snapshot of a water surface with sinusoidal waves that have been generated by a line source.

It is useful to define 2-D sinusoids that are constant along an arbitrary azimuth angle; this merely requires substitution of the rotated coordinates in Equation (7.11) into Equation (7.33):

$$\begin{aligned} f[x', y'] &= A_0 \cos(2\pi\xi_0 x' + \phi_0)1[y'] = A_0 \cos[2\pi\xi_0(\underline{\mathbf{r}} \bullet \hat{\underline{\mathbf{p}}}) + \phi_0]1[\underline{\mathbf{r}} \bullet \hat{\underline{\mathbf{p}}}^{\perp}] \\ &= A_0 \cos[2\pi(\xi_0 \cos[\theta]x + (\xi_0 \sin[\theta])y) + \phi_0]1[-x \sin[\theta] + y \cos[\theta]] \\ &\equiv A_0 \cos[2\pi(\xi_1 x + \eta_1 y) + \phi_0]1[-x \sin[\theta] + y \cos[\theta]] \end{aligned} \tag{7.34a}$$

The spatial frequencies of the rotated function along the "new" ξ- and η-axes are ξ_1 and η_1:

$$\xi_1 \equiv +\xi_0 \cos[\theta] \tag{7.34b}$$

$$\eta_1 \equiv +\xi_0 \sin[\theta] \tag{7.34c}$$

which specify the rates of sinusoidal variation along the x- and y-axes measured in cycles per unit length. By analogy with the 1-D expression in Equation (6.19), these also specify the periods of the 2-D sinusoid along the x- and y-directions, respectively:

$$f[x, y] = A_0 \cos[2\pi(\xi_1 x + \eta_1 y) + \phi_0] = A_0 \cos\left[2\pi\left(\frac{x}{X_1} + \frac{y}{Y_1}\right) + \phi_0\right] \tag{7.35a}$$

where the periods along the two Cartesian axes are:

$$X_1 = \frac{1}{\xi_0 \cos[\theta]} \tag{7.35b}$$

$$Y_1 = \frac{1}{\xi_0 \sin[\theta]} \tag{7.35c}$$

The expression in Equation (7.34a) for the 2-D sinusoid that oscillates along an arbitrary direction may be recast into a form where the 2-D spatial coordinate is the 2-D radius vector $\mathbf{\underline{r}} \equiv [x, y]$. The phase $[\xi_0 x + \eta_0 y]$ of the sinusoid is the scalar product of $\mathbf{\underline{r}}$ with a "polar spatial frequency vector" $\underline{\rho}_0$ with Cartesian coordinates $\underline{\rho}_0 \equiv [\xi_0, \eta_0]$. The polar representation of this vector is $\underline{\rho}_0 = (\rho_0, \psi_0)$, where the radial component has "length" $|\underline{\rho}_0|$:

$$|\underline{\boldsymbol{\rho}}_0| = \sqrt{\xi_0^2 + \eta_0^2} \tag{7.36}$$

and its azimuth angle ψ_0 is:

$$\psi_0 = \tan^{-1}\left[\frac{\eta_0}{\xi_0}\right] \tag{7.37}$$

The vector scalar product in Equation (3.15) is applied to construct a new expression for the 2-D sinusoid:

$$\begin{aligned} f(\mathbf{\underline{r}}) &= A_0 \cos[2\pi(\mathbf{\underline{r}} \bullet \underline{\rho}_0) + \phi_0] \\ &= A_0 \cos[2\pi(|\mathbf{\underline{r}}|\, |\underline{\rho}_0| \cos[\theta - \psi_0]) + \phi_0] \end{aligned} \tag{7.38}$$

By analogy with the expressions for the components of a 2-D vector in the space domain in Equations (3.13) and (3.14), the Cartesian components of the spatial frequency vector may be derived from the polar form:

$$\xi_0 = |\underline{\boldsymbol{\rho}}_0| \cos[\psi_0] \tag{7.39a}$$

$$\eta_0 = |\underline{\boldsymbol{\rho}}_0| \sin[\psi_0] \tag{7.39b}$$

Again, by analogy with the 1-D sinusoid of Equation (6.19), the spatial period of the 2-D sinusoid in units of length is the reciprocal of the magnitude of the spatial frequency vector:

$$R_0 = \frac{1}{|\underline{\rho}_0|} = \frac{1}{\sqrt{\xi_0^2 + \eta_0^2}} \tag{7.40}$$

In words, the variation of the 2-D sinusoid oscillates with period R_0 along the angle ψ_0 measured relative to the x-axis.

7.3 2-D Dirac Delta Function and its Relatives

In spaces of dimension two or higher, several flavors of the Dirac delta function are possible. For example, different separable versions of the 2-D Dirac delta function may be defined in Cartesian and polar coordinates. In this section, we consider forms of the 2-D Dirac delta function. The details and differences among the representations are significant and often may be confusing. Obviously, the 2-D Dirac delta function in Cartesian coordinates is a function of $[x, y]$ and must satisfy properties analogous to those of the 1-D function $\delta[x]$ in Equation (6.88). These properties are that the amplitude must be zero except where both arguments are zero, and the integral of the function over all space (the

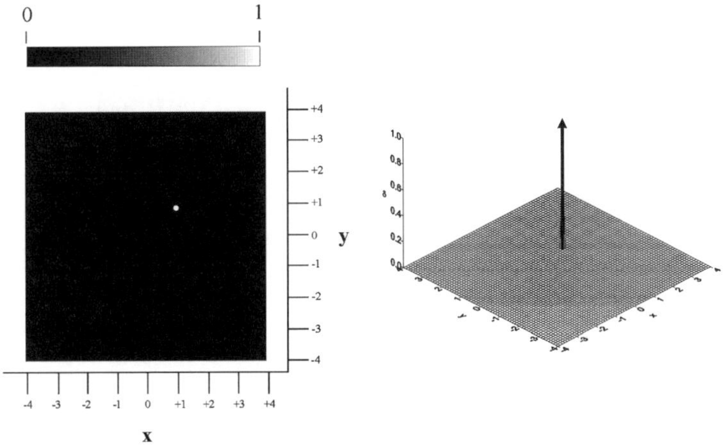

Figure 7.11 The 2-D separable Dirac delta function $\delta[x-1, y-1] = \delta[x-1] \cdot \delta[y-1]$.

volume, in this case) must be unity:

$$\delta[x - x_0, y - y_0] = 0 \quad for\ x - x_0 \neq 0\ or\ y - y_0 \neq 0 \tag{7.41a}$$

$$\iint_{-\infty}^{+\infty} \delta[x - x_0, y - y_0]\, dx\, dy = 1 \tag{7.41b}$$

Just as in the 1-D case, the 2-D Dirac delta function may be expressed as the limit of a sequence of 2-D functions, each having unit volume but with support that occupies a progressively smaller area of the 2-D domain. Appropriate representations for $\delta[x, y]$ may be generated from the 2-D *RECT*, *TRI*, *GAUS*, *SINC*, and *SINC*^{2} in Cartesian coordinates; the separability of these functions ensures that the corresponding representation of $\delta[x, y]$ also will be separable. Probably the most common representation of $\delta[x, y]$ is based upon the 2-D *RECT* function of unit volume in the limit of infinitesimal area:

$$\begin{aligned}\delta[x, y] &= \lim_{a\to 0} \lim_{b\to 0} \left\{ \frac{1}{ab}\, RECT\left[\frac{x}{a}, \frac{y}{b}\right] \right\} \\ &= \lim_{a\to 0} \left\{ \frac{1}{a} RECT\left[\frac{x}{a}\right] \right\} \cdot \lim_{b\to 0} \left\{ \frac{1}{b} RECT\left[\frac{y}{b}\right] \right\}\end{aligned} \tag{7.42}$$

The properties of the separable Cartesian forms of $\delta[x, y]$ are considered in the next section, and polar representations are described in Section 7.3.2.

7.3.1 2-D Dirac Delta Function in Cartesian Coordinates

Based on the discussion in Section 7.2.5, it is evident that the separable product of two orthogonal and symmetric 1-D Dirac delta functions satisfies the requirements in Equation (7.41):

$$\delta[x, y] = \delta[x]\, \delta[y] \tag{7.43}$$

where the specific 1-D representations may be selected from the choices in Section 6.2. Just as in the 1-D case, the particular functional form used is often unimportant. The graphical representation of the 2-D Dirac delta function is shown in Figure 7.11.

The expression in Equation (6.106) for the 1-D Dirac delta function may be used to obtain another representation for the 2-D Dirac delta function:

$$\begin{aligned}\delta[x]\,\delta[y] &= \left(\int_{-\infty}^{+\infty} e^{+2\pi i\xi x}\,d\xi\right)\left(\int_{-\infty}^{+\infty} e^{+2\pi i\eta y}\,d\eta\right)\\ &= \iint_{-\infty}^{+\infty} e^{+2\pi i\xi x}\,e^{+2\pi i\eta y}\,d\xi\,d\eta\\ &= \iint_{-\infty}^{+\infty} e^{+2\pi i(\xi x+\eta y)}\,d\xi\,d\eta = \delta[x,y]\end{aligned} \tag{7.44}$$

In words, the 2-D Dirac delta function may be synthesized by summing unit-amplitude 2-D complex linear-phase exponentials with all spatial frequencies. Because the areas of the odd imaginary parts of the complex exponentials are zero, a valid equivalent expression is:

$$\delta[x,y] = \iint_{-\infty}^{+\infty} \cos[2\pi(\xi x+\eta y)]\,d\xi\,d\eta \tag{7.45}$$

The scaling property of $\delta[x]$ in Equation (6.107) yields a simple expression for the 2-D Dirac delta function that has been scaled in "width":

$$\delta\left[\frac{x}{b},\frac{y}{d}\right] = \delta\left[\frac{x}{b}\right]\delta\left[\frac{y}{d}\right] = |b|\,\delta[x]\,|d|\,\delta[y] = |bd|\,\delta[x]\,\delta[y] = |bd|\,\delta[x,y] \tag{7.46}$$

where b and d are real valued. Of course, this separable representation of the 2-D Dirac delta function may be translated to any desired location $[x_0, y_0]$ by shifting the separable components appropriately:

$$\delta\left[\frac{x-x_0}{b},\frac{y-y_0}{d}\right] = |bd|\,\delta[x-x_0]\,\delta[y-y_0] \tag{7.47}$$

The sifting property of the 2-D Dirac delta function allows the amplitude of a function of two variables to be evaluated at a specific location. In the separable case, the expression is trivial to derive from Equation (6.110):

$$\iint_{-\infty}^{+\infty} f[x,y]\delta[x-x_0,y-y_0]\,dx\,dy = f[x_0,y_0] \tag{7.48}$$

7.3.2 2-D Dirac Delta Function in Polar Coordinates

The 2-D Dirac delta function also may be written as a separable function in polar coordinates, i.e., as the product of 1-D Dirac delta functions with radial and azimuthal coordinates, respectively. Consider first the 2-D Dirac delta function located on the x-axis at a distance $\alpha > 0$ from the origin, so that the polar coordinates are $r_0 = \alpha$ and $\theta_0 = 0$. The Cartesian form obviously is:

$$\delta[x-\alpha, y] = \delta[x-\alpha]\delta[y] \tag{7.49}$$

The polar representation of this function is easy to derive because the radial coordinate is directed along the x-axis and the azimuthal displacement due to the angle is parallel to the y-axis. We can identify the radial variable $r = x$, the radial parameter $r_0 = \alpha$, and that $r_0\theta = y$. Substitution into Equation (7.49) produces a polar representation for $\delta[x-\alpha, y]$:

$$\begin{aligned}\delta[x-\alpha]\delta[y] &= \delta[r-r_0]\delta[r_0\theta]\\ &= \delta[r-r_0]\left(\frac{1}{|r_0|}\delta[\theta]\right) = \left(\frac{\delta[r-r_0]}{r_0}\right)\delta[\theta]\end{aligned} \tag{7.50}$$

where the scaling property of the 1-D Dirac delta function from Equation (6.107) and the fact that the polar radial coordinate $r_0 = |r_0| \geq 0$ have been used. The domain of the azimuthal coordinate θ must span a continuous interval of 2π radians. Two common choices for this domain are the "single-sided" domain $0 \leq \theta < 2\pi$ and the "symmetric" domain $-\pi \leq \theta < +\pi$ (our choice).

This expression may be generalized for a 2-D Dirac delta function located at the same radial distance r_0 from the origin, but at a different azimuth θ_0. This location has Cartesian coordinates $x_0 = r_0 \cos[\theta_0]$ and $y_0 = r_0 \sin[\theta_0]$. Because this is identical to Equation (7.49) except for a rotation about the origin, the radial part of the polar form is unaffected, so the Dirac delta function evidently is:

$$\begin{aligned}\delta[x - x_0, y - y_0] &= \delta[x - r_0 \cos[\theta_0]]\, \delta[y - r_0 \sin[\theta_0]] \\ &= \left(\frac{\delta[r - r_0]}{r_0}\right) \delta[\theta - \theta_0] \equiv \delta(\underline{\mathbf{r}} - \underline{\mathbf{r}}_0)\end{aligned} \tag{7.51}$$

Note the interesting feature that the polar form of the 2-D Dirac delta function is the product of 1-D Dirac delta functions in the radial and azimuthal directions, but that the amplitude is scaled by the reciprocal of the radial distance. In other words, the area of the radial part of the Dirac delta function far from the origin is attenuated.

We should confirm that the expression in Equation (7.51) satisfies the criteria for the 2-D Dirac delta function in Equation (7.41). It is easy to show that the support is infinitesimal by substituting any of the limiting forms of the 1-D Dirac delta function from Equation (6.93), (6.95), or (6.96). The volume of Equation (7.51) is easily evaluated:

$$\begin{aligned}\iint_{-\infty}^{+\infty} \delta[\underline{\mathbf{r}} - \underline{\mathbf{r}}_0]\, d\underline{\mathbf{r}} &= \int_{-\pi}^{+\pi} d\theta \int_0^{+\infty} \left(\frac{\delta[r - r_0]}{r_0} \delta[\theta - \theta_0]\right) r\, dr \\ &= \left(\int_{-\pi}^{+\pi} \delta[\theta - \theta_0]\, d\theta\right) \cdot \left(\int_0^{+\infty} \frac{\delta[r - r_0]}{r_0} r\, dr\right) \quad \textit{where } r_0 > 0 \\ &= 1 \cdot \int_0^{+\infty} \frac{\delta[r - r_0]}{r_0} r_0\, dr = 1 \cdot \int_0^{+\infty} \delta[r - r_0]\, dr = 1\end{aligned} \tag{7.52}$$

where the integrals have been evaluated by applying the sifting property of the 1-D Dirac delta function in Equation (6.113) after appropriate modifications of the domains. The pictorial representation of the polar form of the "uncentered" Dirac delta function is shown in Figure 7.12.

We need to extend this derivation of the polar form of the 2-D Dirac delta function to the case of $r_0 = 0$ to express the function $\delta[\underline{\mathbf{r}} - \underline{\mathbf{0}}]$. The condition of Equation (6.88a) requires that a 1-D Dirac delta function be symmetric with respect to the origin. However, the domain of the argument r of Equation (7.51) is the semi-closed single-sided interval $[0, +\infty)$, which means that the 1-D radial part $\delta(r - r_0)$ of the 2-D Dirac delta function at the origin cannot be symmetric. This also creates a problem for the 1-D azimuthal part $\delta[\theta - \theta_0]$, because θ_0 is indeterminate at the origin. In other words, Equation (7.52) is not valid for $\underline{\mathbf{r}}_0 = \underline{\mathbf{0}}$. Note that the 2-D Dirac delta function at the origin must be circularly symmetric and therefore must be a function of r only, with no dependence on the azimuth angle θ. A valid form must be:

$$\delta(\underline{\mathbf{r}} - \underline{\mathbf{0}}) = \alpha \delta[|\underline{\mathbf{r}}|] \cdot 1[\theta] = \alpha \delta[r] \cdot 1[\theta] \tag{7.53}$$

where α is a scaling parameter that is set to ensure that $\delta(\underline{\mathbf{r}})$ has unit volume. The domains of the polar arguments were just stated to be $0 \leq r < +\infty$ and $-\pi \leq \theta < +\pi$. To simplify evaluation of the volume, we modify the domains of the polar variables to equivalent forms with a bipolar radial variable over the domain $(-\infty, +\infty)$ and an azimuthal domain that is constrained to an interval of π radians, such as $[0, +\pi)$ or $[-\pi/2, +\pi/2)$. Assuming the latter symmetric azimuthal domain, the volume of

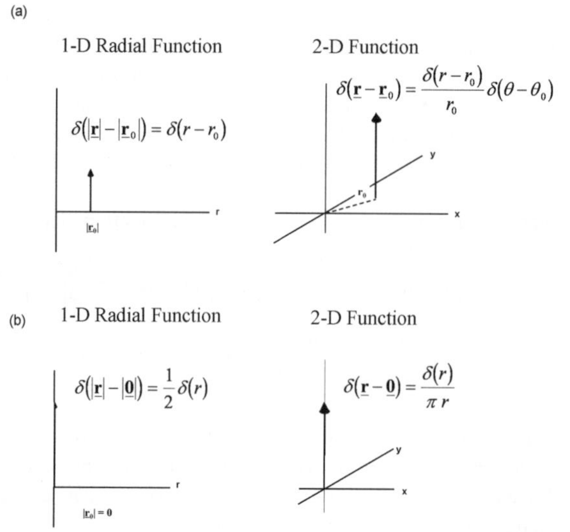

Figure 7.12 Radial representation of 2-D Dirac delta function $\delta[\underline{\mathbf{r}} - \underline{\mathbf{r}}_0]$. (a) If $\underline{\mathbf{r}}_0 \neq \underline{\mathbf{0}}$, the Dirac delta function may be decomposed into two 1-D Dirac delta functions in the radial and azimuthal directions. The azimuthal function is scaled by $|\underline{\mathbf{r}}_0|^{-1}$. (b) The 2-D Dirac delta function for $\underline{\mathbf{r}}_0 = \underline{\mathbf{0}}$ is circularly symmetric and thus may be decomposed into a radial part and a constant azimuthal part. The 1-D radial part is scaled by $\frac{1}{2}$ because of the symmetry.

the expression in Equation (7.53) is:

$$\begin{aligned} 1 &= \int_{-\pi}^{+\pi} \int_{0}^{+\infty} \alpha\delta(r)1(\theta) r\,dr\,d\theta = \int_{-\pi/2}^{+\pi/2} \int_{-\infty}^{+\infty} \alpha\delta(r)1(\theta) r\,dr\,d\theta \\ &= \int_{-\pi/2}^{+\pi/2} 1(\theta)\,d\theta \cdot \int_{-\infty}^{+\infty} \alpha\delta(r) r\,dr = \int_{-\infty}^{+\infty} (\alpha\pi r)\delta(r)\,dr \end{aligned} \tag{7.54}$$

The definition of the 1-D Dirac delta function in Equation (6.88b) ensures that the area of $\delta(r)$ over the complete domain $(-\infty, +\infty)$ is unity, which requires that $\alpha = (\pi r)^{-1}$ for the equation to be satisfied. The expression for the circularly symmetric Dirac delta function is:

$$\delta(\underline{\mathbf{r}} - \underline{\mathbf{0}}) = \delta(\underline{\mathbf{r}}) = \left(\frac{1}{\pi r}\right)\delta(r)1(\theta) = \frac{\delta(r)}{\pi r} \tag{7.55}$$

as shown in Figure 7.12. The two expressions in Equation (7.51) and Equation (7.55) may be combined to define the 2-D Dirac delta function in polar coordinates located at any point in the 2-D domain.

7.3.3 2-D Separable *COMB* Function

The separable 2-D *COMB* function evidently is defined as:

$$\begin{aligned} COMB[x, y] &= COMB[x]\,COMB[y] \\ &= \left(\sum_{n=-\infty}^{+\infty} \delta[x - n]\right)\left(\sum_{\ell=-\infty}^{+\infty} \delta[y - \ell]\right) \end{aligned} \tag{7.56}$$

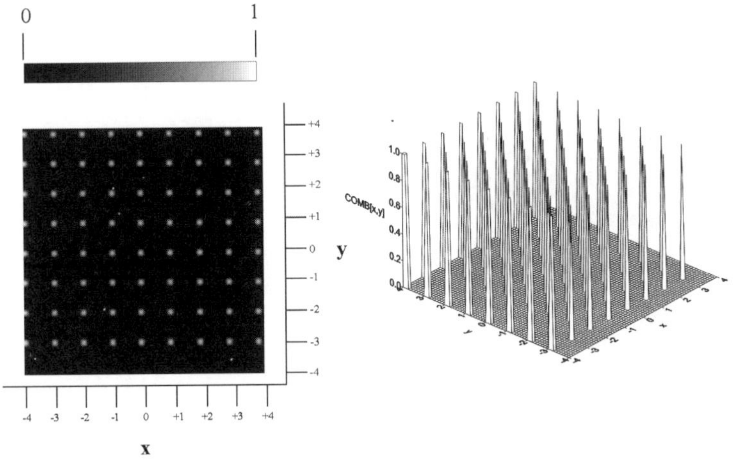

Figure 7.13 Image of the weights applied to the Dirac delta functions in the 2-D separable *COMB* function: $COMB[x, y] = COMB[x] \cdot COMB[y]$.

which, in Gaskill's words, resembles a "bed of nails". The graphical representation is shown in Figure 7.13. The volume of the 2-D *COMB* obviously is infinite. The most important application of the 2-D *COMB* is to model 2-D sampled functions. The product of $COMB[x, y]$ and a function with a continuous domain generates an array of equispaced sampled amplitudes. The process of sampling will be considered in detail in Chapter 14.

7.3.4 2-D Line Delta Function

Another 2-D separable variety of the Dirac delta function may be constructed by orthogonal multiplication of a 1-D Dirac delta function and a 1-D unit constant. These will be useful when describing 2-D coordinate transformations in Chapter 12 and are analogous to the 2-D sinusoid in Equation (7.33). One possible realization is:

$$m_1[x, y] = \delta[x]1[y] \tag{7.57}$$

which is a "line" or "wall" of 1-D Dirac delta functions directed along the y-axis, as shown in Figure 7.14. This type of function is often called a *line delta function*, *line mass*, or *straight-line impulse*. Most authors delete the explicit unit constant $1[y]$ from Equation (7.57), and therefore do not distinguish expressions for the 1-D Dirac delta function and the 2-D line delta function along the y-axis. We will retain the (seemingly cumbersome) notation for two reasons. First, it explicitly specifies that the "weighting" of the 1-D Dirac delta function is constant for all values of y. In other words, each section of unit width along the y-axis contributes the same unit contribution to the volume of $m_1[x, y]$. The second reason for using this notation will become evident in Chapter 12; it is very helpful when evaluating and interpreting the 2-D Fourier transform of the general line delta function.

It is easy to demonstrate that the volume of $f_1[x, y]$ is infinite:

$$\iint_{-\infty}^{+\infty} \delta[x]1[y]\,dx\,dy = \int_{-\infty}^{+\infty} \delta[x]\,dx \int_{-\infty}^{+\infty} 1[y]\,dy = 1 \cdot \infty \tag{7.58}$$

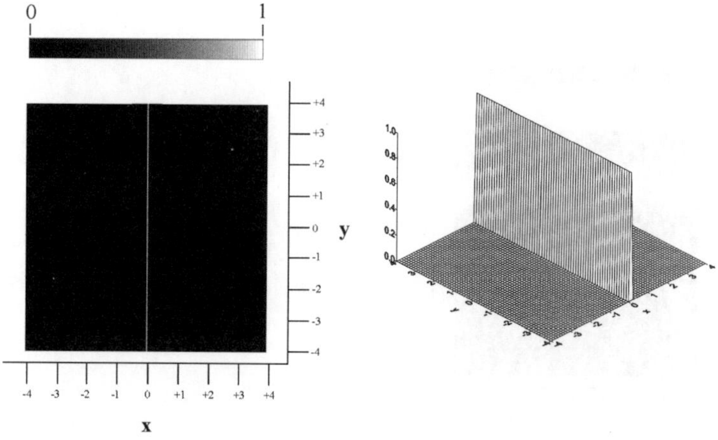

Figure 7.14 The 2-D line Dirac delta function $\delta[x] \cdot 1[y]$ produces a "wall" of Dirac delta functions along the y-axis.

Consider the product of an arbitrary function $f[x, y]$ with the line delta function in Equation (7.57). We can apply the sifting property in Equation (6.113) to obtain:

$$\begin{aligned} f[x, y](\delta[x]1[y]) &= (f[x, y]\delta[x])1[y] \\ &= (f[0, y]\delta[x])1[y] \\ &= f[0, y](\delta[x]1[y]) \end{aligned} \tag{7.59}$$

which is the product of the 1-D profile of the $f[x, y]$ along the y-axis and the line delta function. The volume of the product of these functions is easy to evaluate:

$$\begin{aligned} \iint_{-\infty}^{+\infty} f[x, y](\delta[x]1[y])\,dx\,dy &= \iint_{-\infty}^{+\infty} f[0, y]\delta[x]1[y]\,dx\,dy \\ &= \int_{-\infty}^{+\infty} f[0, y]1[y]\,dy \int_{-\infty}^{+\infty} \delta[x]\,dx \\ &= \int_{-\infty}^{+\infty} f[0, y]\,dy \end{aligned} \tag{7.60}$$

In words, the line delta function $\delta[x]1[y]$ "sifts out" the area of the profile of $f[x, y]$ along the y-axis.

Line delta functions oriented along arbitrary azimuth angles are constructed by determining the "new" locations in the 2-D domain where the argument of the Dirac delta function evaluates to zero. These are obtained by rotating the coordinates by $+\phi$ as in Equation (7.11):

$$\begin{aligned} m_2[x, y] &= \delta[\underline{\mathbf{r}} \bullet \underline{\hat{\mathbf{p}}}]1[\underline{\mathbf{r}} \bullet \underline{\hat{\mathbf{p}}}^{\perp}] \\ &= \delta[x\cos[\phi] + y\sin[\phi]]1[-x\sin[\phi] + y\cos[\phi]] \end{aligned} \tag{7.61}$$

where the symmetry of the unit constant has been used. In this notation, the line delta function "oriented" (i.e., nonzero) along the y-axis in Equation (7.57) is obtained by setting $\phi = 0$. Note that the complete set of line delta functions at all possible azimuth angles may be specified by selecting ϕ from the interval $-\pi/2 \leq \phi < +\pi/2$, which means that the defining unit vector $\underline{\hat{\mathbf{p}}}$ lies in the first or fourth quadrants.

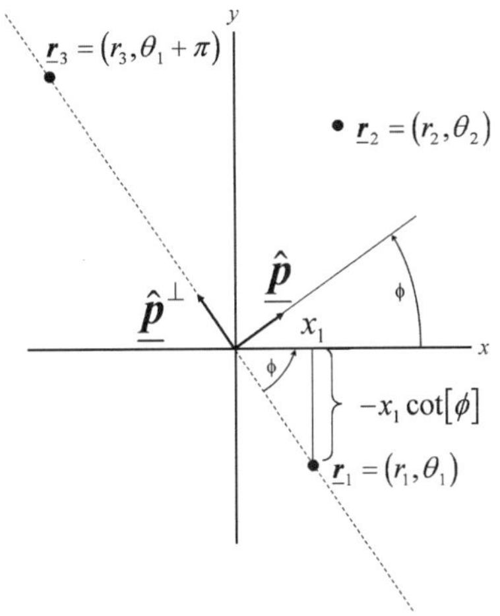

Figure 7.15 The rotated form of the line delta function $m_2[x, y] = \delta[\underline{\mathbf{r}} \bullet \widehat{\underline{\mathbf{p}}}] \cdot 1[\underline{\mathbf{r}} \bullet \widehat{\underline{\mathbf{p}}}^{\perp}]$, where $\widehat{\underline{\mathbf{p}}} = (\cos[\phi], \sin[\phi])$. Three examples of $\underline{\mathbf{r}}$ are shown: $\underline{\mathbf{r}}_1$ and $\underline{\mathbf{r}}_3$ are on the line perpendicular to $\widehat{\underline{\mathbf{p}}}$ and thus lie upon the line delta function. The location $\underline{\mathbf{r}}_2$ is such that $\underline{\mathbf{r}} \bullet \widehat{\underline{\mathbf{p}}} \neq 0$ and thus does not lie on the line, i.e., $\delta[\underline{\mathbf{r}}_2 \bullet \widehat{\underline{\mathbf{p}}}] = 0$.

Again, the action of the unit constant in Equation (7.61) is to "spread" or "smear" the amplitude of the 1-D Dirac delta function evaluated at points in the direction of $\hat{\underline{\mathbf{p}}}$ along the perpendicular direction, as shown in Figure 7.15. We will use this "rotated" form for the 1-D unit constant only to produce 2-D functions by this kind of orthogonal product to construct a 2-D function that is constant along one direction. This expression will prove very useful when considering the 2-D Fourier transform of the rotated line delta function and the Radon transform in Chapter 12.

If the polar forms of the vectors $\underline{\mathbf{r}}$, $\hat{\underline{\mathbf{p}}}$, and $\hat{\underline{\mathbf{p}}}^{\perp}$ (with angles in the domain $-\pi \leq \phi < +\pi$) are used in the alternative form of the scalar product in Equation (3.14), we obtain another useful expression for the line delta function through the origin:

$$\begin{aligned} m_2[x, y] &= \delta[\underline{\mathbf{r}} \bullet \hat{\underline{\mathbf{p}}}]1[\underline{\mathbf{r}} \bullet \hat{\underline{\mathbf{p}}}^{\perp}] = \delta[|\underline{\mathbf{r}}||\hat{\underline{\mathbf{p}}}| \cos[\theta - \phi]]1[\underline{\mathbf{r}} \bullet \hat{\underline{\mathbf{p}}}^{\perp}] \\ &= \delta[|\underline{\mathbf{r}}| \cos[\theta - \phi]]1[\underline{\mathbf{r}} \bullet \hat{\underline{\mathbf{p}}}^{\perp}] = \delta[r \cos[\phi - \theta]]1[\underline{\mathbf{r}} \bullet \hat{\underline{\mathbf{p}}}^{\perp}] \end{aligned} \tag{7.62}$$

where the fact that $|\hat{\underline{\mathbf{p}}}| = 1$ and the symmetry of the cosine function were used in the last step. Note that this expression determines the set of values of (r, θ) that lie on the radial line through the origin perpendicular to the azimuthal angle ϕ. The line delta function may be written as a function of the azimuthal angle ϕ by recasting the last expression into a (possibly) more convenient form by applying the expression for the Dirac delta function with a functional argument in Equation (6.138):

$$\begin{aligned} \delta[\underline{\mathbf{r}} \bullet \hat{\underline{\mathbf{p}}}]1[\underline{\mathbf{r}} \bullet \hat{\underline{\mathbf{p}}}^{\perp}] &= \delta[r \cos[\theta - \phi]]1[\underline{\mathbf{r}} \bullet \hat{\underline{\mathbf{p}}}^{\perp}] \\ &= \delta[g[\phi]]1[\underline{\mathbf{r}} \bullet \hat{\underline{\mathbf{p}}}^{\perp}] \\ &= \frac{1}{|g'(\phi_0)|} \delta[\phi - \phi_0]1[\underline{\mathbf{r}} \bullet \hat{\underline{\mathbf{p}}}^{\perp}] \\ &= \frac{1}{r|\sin[\theta - \phi_0]|} \delta[\phi - \phi_0]1[\underline{\mathbf{r}} \bullet \hat{\underline{\mathbf{p}}}^{\perp}] \end{aligned} \tag{7.63}$$

where ϕ_0 is the angle that satisfies the condition $\cos[\theta - \phi_0] = 0$, i.e., for $\phi_0 = \theta \pm \pi/2$ where the sign is selected to ensure that $-\pi/2 \leq \phi_0 < +\pi/2$ in the current convention. By applying the property of

Dirac delta functions in products from Equation (6.115), we obtain a concise form for the rotated line delta:

$$\begin{aligned}\delta[\underline{\mathbf{r}} \bullet \hat{\underline{\mathbf{p}}}]1[\underline{\mathbf{r}} \bullet \hat{\underline{\mathbf{p}}}^{\perp}] &= \frac{1}{r|\sin[\mp\pi/2]|}\delta\left[\phi - \left(\theta \pm \frac{\pi}{2}\right)\right] \\ &= \frac{1}{r}\delta\left[\phi - \left(\theta \pm \frac{\pi}{2}\right)\right] \\ &= \left(\frac{1}{r}\delta\left[\phi - \left(\theta \pm \frac{\pi}{2}\right)\right]\right)1[\underline{\mathbf{r}} \bullet \hat{\underline{\mathbf{p}}}^{\perp}]\end{aligned} \tag{7.64}$$

The same process may be used to derive an expression for the line delta function that is orthogonal to that in Equation (7.63). It is nonzero along the line defined by the unit vector $\hat{\underline{\mathbf{p}}}$:

$$\delta[\underline{\mathbf{r}} \bullet \hat{\underline{\mathbf{p}}}^{\perp}]1[\underline{\mathbf{r}} \bullet \hat{\underline{\mathbf{p}}}] = \left(\frac{\delta[\theta - \phi]}{r}\right)1[\underline{\mathbf{r}} \bullet \hat{\underline{\mathbf{p}}}] \tag{7.65}$$

In words, Equation (7.64) shows that the line delta function through the origin that lies along the radial line perpendicular to the azimuthal angle ϕ is equivalent to an amplitude-weighted 1-D Dirac delta function of the angle ϕ. The amplitudes are nonzero only where $\phi = \theta \pm \pi/2$; the sign is selected to ensure that ϕ lies within our usual domain of polar coordinates: $-\pi \leq \phi < +\pi$. However, the discussion of the "unrotated" line delta function $\delta[x]1[y]$ in Equation (7.57) noted that each unit increment of radial distance along the y-axis contributed the same unit volume. This expression seems to indicate otherwise for the rotated function; the contribution to the volume is affected by the factor r^{-1}. To resolve this apparent conundrum, we can substitute the expression for the Dirac delta function as the limit of a rectangle function from Equation (6.93).

$$\delta\left[\phi - \left(\theta \pm \frac{\pi}{2}\right)\right] = \lim_{b\to 0}\left\{\frac{1}{b}RECT\left[\frac{\phi - (\theta \pm \pi/2)}{b}\right]\right\} \tag{7.66}$$

where b is measured in radians; an example is shown in Figure 7.16 for $b = \pi/6$ radians and $\theta = 0$. The function consists of two symmetric "wedges" of fixed amplitude b^{-1}. The second "wedge" about the angle $\phi = -\pi/2$ in this example results from our convention that the domain of the azimuth angle is $-\pi \leq \phi < +\pi$. In the limit $b \to 0$, the angular "spread" of each wedge decreases while the amplitude increases. The contribution of segments of each wedge with unit radial extent clearly increases in proportion to the radial distance r from the origin. The factor of r^{-1} in Equation (7.65) compensates for the increase in volume to ensure that the contributions to the volume of segments with equal radial extent remain constant.

Though the expressions for the rotated line delta function in Equations (7.63) and (7.64) are arguably the most useful, still other forms may be derived by manipulating the argument to put it in the familiar slope–intercept form of a line in the 2-D plane. Two equivalent forms are:

$$\begin{aligned}\delta[\underline{\mathbf{r}} \bullet \hat{\underline{\mathbf{p}}}]1[\underline{\mathbf{r}} \bullet \hat{\underline{\mathbf{p}}}^{\perp}] &= \delta\left[+\sin[\phi]\left(y + \frac{x\cos[\phi]}{\sin[\phi]}\right)\right]1[x\sin[\phi] - y\cos[\phi]] \\ &= \left|\frac{1}{\sin[\phi]}\right|\delta[(y - (\cot[-\phi]x + 0))]1[x\sin[\phi] - y\cos[\phi]]\end{aligned} \tag{7.67a}$$

$$\begin{aligned}\delta[\underline{\mathbf{r}} \bullet \hat{\underline{\mathbf{p}}}]1[\underline{\mathbf{r}} \bullet \hat{\underline{\mathbf{p}}}^{\perp}] &= \delta\left[+\cos[\phi]\left(\frac{y\sin[\phi]}{\cos[\phi]} + x\right)\right]1[x\sin[\phi] - y\cos[\phi]] \\ &= \left|\frac{1}{\cos[\phi]}\right|\delta[x - (\tan[-\phi]\,y + 0)]1[x\sin[\phi] - y\cos[\phi]]\end{aligned} \tag{7.67b}$$

The Dirac delta functions in both expressions evaluate to zero except at those coordinates $[x, y]$ that satisfy the slope–intercept form of a straight line: $y - (sx + b) = 0$ where the y-intercept $b = 0$. The

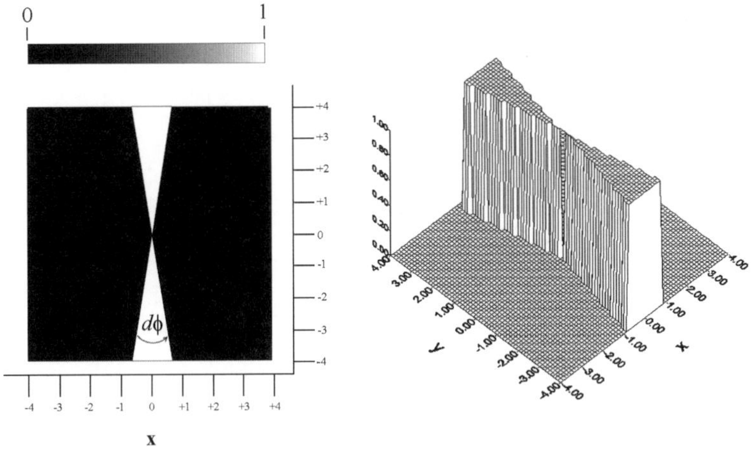

Figure 7.16 Angular delta function as limit of rectangle: $\lim_{d\phi \to 0} \left\{ \frac{1}{d\phi} RECT\left[\frac{\phi - \pi/2}{d\phi} \right] \right\}$.

slope in Equation (7.67a) is $s = \cot[-\phi] = -\cot[\phi]$, while that in Equation (7.67b) is $s = \tan[-\phi] = -\tan[\phi]$. It is again instructive to substitute the expression for the 1-D Dirac delta function as the limit of a rectangle of decreasing width and increasing amplitude:

$$\delta[\underline{\mathbf{r}} \bullet \hat{\underline{\mathbf{p}}}]1[\underline{\mathbf{r}} \bullet \hat{\underline{\mathbf{p}}}^{\perp}] = \left| \frac{1}{\cos[\phi]} \right| \delta[x - (y \tan[-\phi] + 0)]1[x \sin[\phi] - y \cos[\phi]]$$

$$= \left| \frac{1}{\cos[\phi]} \right| \lim_{b \to 0} \left\{ \frac{1}{b} RECT\left[\frac{x - (y \tan[-\phi] + 0)}{b} \right] \right\} 1[x \sin[\phi] - y \cos[\phi]] \quad (7.68a)$$

$$\delta[\underline{\mathbf{r}} \bullet \hat{\underline{\mathbf{p}}}]1[\underline{\mathbf{r}} \bullet \hat{\underline{\mathbf{p}}}^{\perp}] = \left| \frac{1}{\sin[\phi]} \right| \delta[y - (x \cot[-\phi] + 0)]1[x \sin[\phi] - y \cos[\phi]]$$

$$= \left| \frac{1}{\sin[\phi]} \right| \lim_{b \to 0} \left\{ \frac{1}{b} RECT\left[\frac{y - (x \cot[-\phi] + 0)}{b} \right] \right\} 1[x \sin[\phi] - y \cos[\phi]] \quad (7.68b)$$

The rectangle in Equation (7.68a) is a function of x with "edges" located at $x = -y \tan[\phi] \pm b/2$. As shown in Figure 7.17, the "perpendicular width" of the rectangle is $b \cos[\phi] < b$, which means that the volume of the segment of the 1-D rectangle with unit length in the "long" dimension is $\cos[\phi]$, where $|\phi| \geq \pi/2$ so that $\cos[\phi] \geq 0$. Similarly, the width of the rectangle in Equation (7.68b) is $b|\sin[\phi]|$. Therefore the leading weight factors of $|\sin[\phi]|^{-1}$ and $|\cos[\phi]|^{-1}$ are necessary to ensure that the area of the 1-D Dirac delta function is unity. The two expressions are respectively valid for $\cos[\phi] \neq 0$ (i.e., $\phi \neq (n + \frac{1}{2})\pi$, $n = 0, \pm 1, \ldots$) and for $\sin[\phi] \neq 0$ ($\phi \neq n\pi$). If the rotation angle is $\phi = \pm\pi/2$, the expressions in Equations (7.63) and (7.64) produce the same result:

$$\left| \frac{1}{\sin(\pm\pi/2)} \right| \delta\left[y - \left(x \cot\left[\mp\frac{\pi}{2} \right] + 0 \right) \right] 1\left[x \sin\left[\pm\frac{\pi}{2} \right] + y \cos\left[\pm\frac{\pi}{2} \right] \right] = 1[x]\delta[y] \quad (7.69a)$$

$$\delta\left[x \cos\left[\pm\frac{\pi}{2} \right] + y \sin\left[\pm\frac{\pi}{2} \right] \right] 1\left[x \sin\left[\pm\frac{\pi}{2} \right] - y \cos\left[\pm\frac{\pi}{2} \right] \right] = 1[x]\delta[y] \quad (7.69b)$$

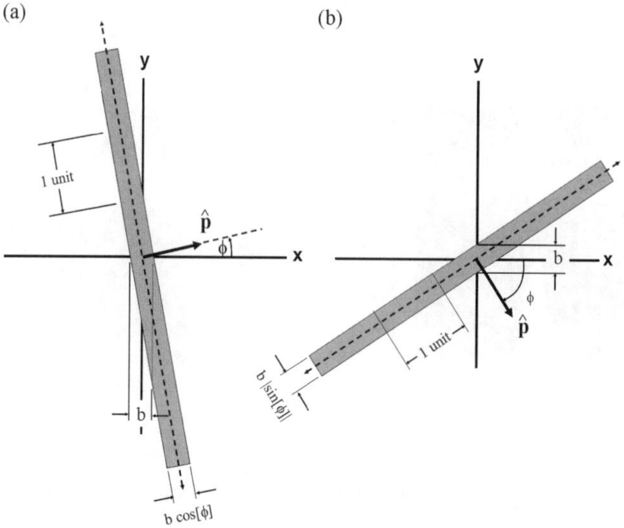

Figure 7.17 Two examples of limiting approximations to rotated line delta functions as 1-D rectangle functions of width b. The amplitude of the gray area in both cases is b^{-1}. (a) The rectangle is a function of x with width b measured on the x-axis and "perpendicular width" $b\cos(\phi)$. (b) The rectangle is a function of y with "perpendicular width" $b\sin[\phi]$.

Equation (7.68a) also may be used to evaluate the product of a 2-D function $f[x, y]$ and a rotated line delta function via the sifting property of the 1-D Dirac delta function:

$$\begin{aligned} f[x, y]\cdot(\delta[\underline{\mathbf{r}}\bullet\hat{\underline{\mathbf{p}}}]1[\underline{\mathbf{r}}\bullet\hat{\underline{\mathbf{p}}}^{\perp}]) &= f[x, y]\left|\frac{1}{\sin[\phi]}\right|\delta[y - x\cot[-\phi] + 0]1[x\sin[\phi] - y\cos[\phi]] \\ &= f[x, -x\cot[\phi]]\delta[\underline{\mathbf{r}}\bullet\hat{\underline{\mathbf{p}}}]1[\underline{\mathbf{r}}\bullet\hat{\underline{\mathbf{p}}}^{\perp}] \end{aligned} \tag{7.70}$$

In words, this line delta sifts out the values of $f[x, y]$ at these coordinates that lie along the line through the origin with slope $\cot[-\phi] = \tan[\pi/2 - \phi]$.

As a second example, consider the line delta function through the origin that is perpendicular to that just considered, i.e., at angle $\phi \pm \pi/2$. The mathematical expression is:

$$\begin{aligned} &\delta\left[x\cos\left[\phi \pm \frac{\pi}{2}\right] + y\sin\left[\phi \pm \frac{\pi}{2}\right]\right]1\left[x\sin\left[\phi \pm \frac{\pi}{2}\right] - y\cos\left[\phi \pm \frac{\pi}{2}\right]\right] \\ &= \delta[\mp x\sin[\phi] \pm y\cos[\phi]]1[\pm x\cos[\phi] \pm y\sin[\phi]] \\ &= \delta[x\sin[\phi] - y\cos[\phi]]1[x\cos[\phi] + y\sin[\phi]] \\ &= \delta[\underline{\mathbf{r}}\bullet\hat{\underline{\mathbf{p}}}^{\perp}]1[\underline{\mathbf{r}}\bullet\hat{\underline{\mathbf{p}}}] \\ &= \left|\frac{1}{\cos[\phi]}\right|\delta[x\tan[\phi] - y]1[x\cos[\phi] + y\sin[\phi]] \end{aligned} \tag{7.71}$$

Now consider the form of the product of this line delta function with an arbitrary 2-D function $f[x, y]$. The form may be recast via the sifting property of the Dirac delta function:

$$\begin{aligned} f[x, y]\left|\frac{1}{\cos[\phi]}\right|\delta[x\tan[\phi] - y]1[x\cos[\phi] + y\sin[\phi]] \\ = f[x, x\tan[\phi]]\left|\frac{1}{\cos[\phi]}\right|\delta[x\tan[\phi] - y]1[x\cos[\phi] + y\sin[\phi]] \\ = f[x, x\tan[\phi]](\delta[x\sin[\phi] - y\cos[\phi]]1[x\cos[\phi] + y\sin[\phi]]) \end{aligned} \tag{7.72}$$

This is the radial line delta function weighted by the amplitude of the 2-D function $f[x, y]$ along the radial line at azimuth angle ϕ. It will be convenient to substitute polar expressions for the Cartesian coordinates based on a bipolar radial variable $p = \pm\sqrt{x^2 + y^2} = \pm r$, so that $x = p\cos[\phi]$ and $y = p\sin[\phi]$, where the domain of the azimuthal angle now spans only π radians; we adopt the "symmetric" domain where $-\pi/2 \leq \phi < +\pi/2$. The product of $f[x, y]$ and the line delta function may be expressed as:

$$\begin{aligned} f[x, x\tan[\phi]] \cdot (\delta[x\sin[\phi] - y\cos[\phi]]1[x\cos[\phi] + y\sin[\phi]]) \\ = f[p\cos[\phi], p\sin[\phi]](\delta[x\sin[\phi] - y\cos[\phi]]1[x\cos[\phi] + y\sin[\phi]]) \\ = f[p\cos[\phi], p\sin[\phi]](\delta[\underline{\mathbf{r}} \bullet \hat{\underline{\mathbf{p}}}^{\perp}]1[\underline{\mathbf{r}} \bullet \hat{\underline{\mathbf{p}}}]) \\ = f[p\hat{\underline{\mathbf{p}}}](\delta[\underline{\mathbf{r}} \bullet \hat{\underline{\mathbf{p}}}^{\perp}]1[\underline{\mathbf{r}} \bullet \hat{\underline{\mathbf{p}}}]) \end{aligned} \tag{7.73}$$

In subsequent discussions, it will also be useful to consider rotated "walls" of Dirac delta functions that do not pass through the origin. For example, consider the line delta function that is parallel to the y-axis and that intersects the x-axis at x_0. Obviously, its form must be:

$$m_3[x, y] = \delta[x - x_0]1[y] = \delta[x_0 - x]1[y] \tag{7.74}$$

where the parameter x_0 is the "distance of closest approach" of the wall of delta functions to the origin. This "displaced" line delta function may be rotated by $+\phi$ radians. Since the distance of closest approach is along the direction of $\hat{\underline{\mathbf{p}}}$, it is convenient to substitute the "radial" distance p_0 for x_0 in the rotated version of Equation (7.74):

$$\begin{aligned} m_4[x, y] = \delta[p_0 - \underline{\mathbf{r}} \bullet \hat{\underline{\mathbf{p}}}]1[\underline{\mathbf{r}} \bullet \hat{\underline{\mathbf{p}}}^{\perp}] \\ = \delta[p_0 - (x\cos[\phi] + y\sin[\phi])]\,1[x\sin[\phi] - y\cos[\phi]] \end{aligned} \tag{7.75}$$

The line delta function at an arbitrary distance from the origin and along an arbitrary azimuthal direction may be written by analogy with Equation (7.64):

$$\begin{aligned} m_4[x, y] = \delta[(x\cos[-\phi] + y\sin[-\phi]) - p_0]\,1[x\sin[-\phi] - y\cos[-\phi]] \\ = \left|\frac{1}{\sin[\phi]}\right|\delta\left[y - \left(x\cot[-\phi] + \frac{p_0}{\sin[\phi]}\right)\right]1[x\sin[\phi] + y\cos[\phi]] \end{aligned} \tag{7.76}$$

where the argument of the Dirac delta function is a straight line with slope $\cot(-\phi)$ and y-intercept $x_0/\sin[\phi]$, as shown in Figure 7.18. To check this expression, substitute $\phi = \pm\pi/2$ to obtain the expected results:

$$\begin{aligned} m_4[x, y] = \left|\frac{1}{\sin[\mp\pi/2]}\right|\delta\left[y - \left(x\cot\left[\pm\frac{\pi}{2}\right] + \frac{p_0}{\sin[\mp\pi/2]}\right)\right]1\left[x\sin\left[\mp\frac{\pi}{2}\right] + y\cos\left[\mp\frac{\pi}{2}\right]\right] \\ = \delta\left[y - \left(0x + \frac{p_0}{(\mp 1)}\right)\right]1[(\mp 1)x + 0\,y] = \delta[y \mp x_0]\,1[x] \end{aligned} \tag{7.77}$$

Three examples of line delta functions are compared in Figure 7.18.

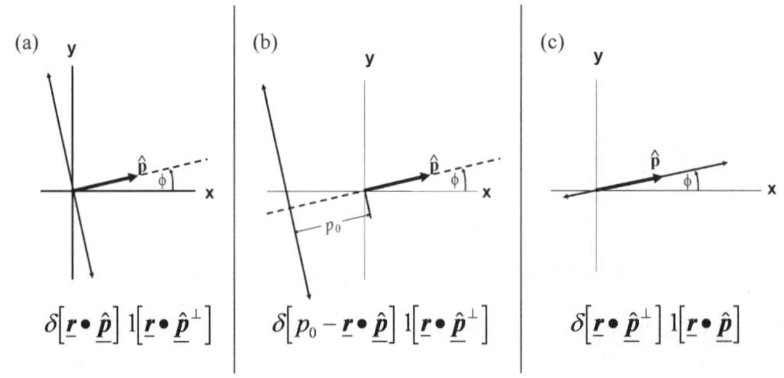

Figure 7.18 Different flavors of line delta functions: (a) $\delta[\underline{\mathbf{r}}\cdot\widehat{\underline{\mathbf{p}}}]1[\underline{\mathbf{r}}\cdot\widehat{\underline{\mathbf{p}}}^{\perp}]$ through the origin perpendicular to $\widehat{\underline{\mathbf{p}}}$; (b) $\delta[p_0-\underline{\mathbf{r}}\cdot\widehat{\underline{\mathbf{p}}}]1[\underline{\mathbf{r}}\cdot\widehat{\underline{\mathbf{p}}}^{\perp}]$ perpendicular to $\widehat{\underline{\mathbf{p}}}$ at a distance $p_0<0$ from the origin; (c) $\delta[\underline{\mathbf{r}}\cdot\widehat{\underline{\mathbf{p}}}^{\perp}]1[\underline{\mathbf{r}}\cdot\widehat{\underline{\mathbf{p}}}]$ through the origin in the direction of $\widehat{\underline{\mathbf{p}}}$.

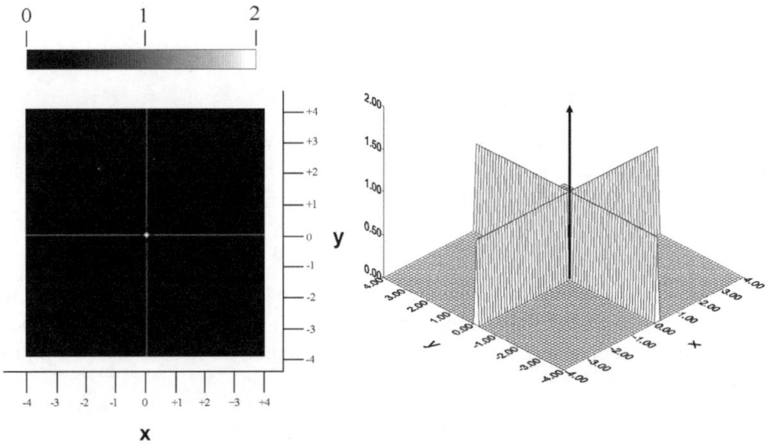

Figure 7.19 The "cross" function is composed of two orthogonal line delta functions: $CROSS[x,y]=\delta[x]1[y]+1[x]\delta[y]$, so that $CROSS[0,0]=2\cdot\delta[x,y]$.

7.3.5 2-D "Cross" Function

The sum of two orthogonal line delta functions also is a useful function. The simplest version is defined with the "lines" coincident with the Cartesian axes:

$$CROSS[x,y]\equiv\delta[x]1[y]+1[x]\delta[y] \tag{7.78}$$

as shown in Figure 7.19. Note that the Cross function at the origin is a 2-D Dirac delta function with a volume of two. The equation for the cross function with "arms" oriented along the vectors $\hat{\underline{\mathbf{p}}}$ and $\hat{\underline{\mathbf{p}}}^{\perp}$ is obtained by substituting $\underline{\mathbf{r}}\bullet\hat{\underline{\mathbf{p}}}$ for x and $\underline{\mathbf{r}}\bullet\hat{\underline{\mathbf{p}}}^{\perp}$ for y:

$$CROSS[\underline{\mathbf{r}}\bullet\hat{\underline{\mathbf{p}}},\,\underline{\mathbf{r}}\bullet\hat{\underline{\mathbf{p}}}^{\perp}]=\delta[\underline{\mathbf{r}}\bullet\hat{\underline{\mathbf{p}}}]1[\underline{\mathbf{r}}\bullet\hat{\underline{\mathbf{p}}}^{\perp}]+1[\underline{\mathbf{r}}\bullet\hat{\underline{\mathbf{p}}}]\delta[\underline{\mathbf{r}}\bullet\hat{\underline{\mathbf{p}}}^{\perp}] \tag{7.79}$$

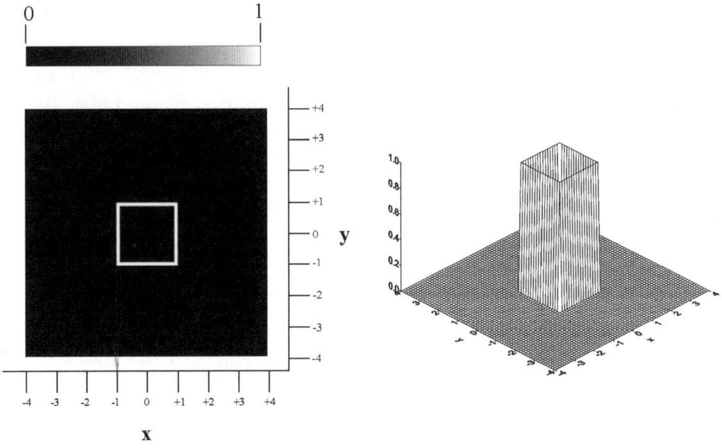

Figure 7.20 Weights applied to Dirac delta functions in 2-D "corral" function $COR[x/2, y/2]$.

7.3.6 "Corral" Function

We can superpose truncated line delta functions to construct a rectangular "stockade" of Dirac delta functions. If we follow Gaskill's theme of using names for special functions derived from the fabled American west, an appropriate name for this function is the "corral". One appropriate formula is obtained by summing four segments of line delta functions:

$$\begin{aligned} COR\left[\frac{x}{b}, \frac{y}{d}\right] &\equiv \delta\left[x + \frac{b}{2}\right] RECT\left[\frac{y}{d}\right] + RECT\left[\frac{x}{b}\right]\delta\left[y + \frac{d}{2}\right] \\ &\quad + \delta\left[x - \frac{b}{2}\right] RECT\left[\frac{y}{d}\right] + RECT\left[\frac{x}{b}\right]\delta\left[y - \frac{d}{2}\right] \\ &= \left(\delta\left[x + \frac{b}{2}\right] + \delta\left[x - \frac{b}{2}\right]\right) RECT\left[\frac{y}{d}\right] \\ &\quad + RECT\left[\frac{x}{b}\right]\left(\delta\left[y + \frac{d}{2}\right] + \delta\left[y - \frac{d}{2}\right]\right) \end{aligned} \tag{7.80}$$

The volume of $COR[x/b, y/d]$ is easy to compute from its separable parts:

$$\begin{aligned} \iint_{-\infty}^{+\infty} COR\left[\frac{x}{b}, \frac{y}{d}\right] dx\, dy &= \int_{-\infty}^{+\infty} \left(\delta\left[x + \frac{b}{2}\right] + \delta\left[x - \frac{b}{2}\right]\right) dx \int_{-\infty}^{+\infty} RECT\left[\frac{y}{d}\right] dy \\ &\quad + \int_{-\infty}^{+\infty} RECT\left[\frac{x}{b}\right] dx \int_{-\infty}^{+\infty} \left(\delta\left[y + \frac{d}{2}\right] + \delta\left[y - \frac{d}{2}\right]\right) dy \\ &= 2d + 2b = 2(b + d) \end{aligned} \tag{7.81}$$

Therefore the volume of $COR[x, y]$ is four units. The corral function will be used during the discussion of 2-D Fourier transforms in Chapters 10 and 12. It is illustrated in Figure 7.20.

7.4 2-D Functions with Circular Symmetry

Circularly symmetric functions vary along the radial direction and are constant in the azimuthal direction. This means that all profiles along radial lines are identical, including (obviously) those along the x- and y-axes. Such functions are particularly important in optics because many optical systems are constructed from lenses that have circular cross-sections.

Clearly, a circularly symmetric function must have the same amplitude at all points on a circle of radius r_0 centered at the origin. In other words, the amplitude of the radial function $f_r(r_0)$ is replicated at all locations $[x, y]$ that satisfy the relation $x^2 + y^2 = r_0^2$. Therefore, a circularly symmetric function may be expressed as the orthogonal product of its 1-D radial profile and the unit constant in the azimuthal direction:

$$f[x, y] \Longrightarrow f(\underline{\mathbf{r}}) = f_r(r)1[\theta], \quad 0 \le r < +\infty, \quad -\pi \le \theta < +\pi \tag{7.82}$$

Also, clearly, a rotation about the origin has no effect on the amplitude of such a function.

Because a circularly symmetric function is, by definition, symmetric with respect to the origin, then the domains of the radial and azimuthal variables may be recast so the radial interval is the symmetric infinite domain $-\infty < r < +\infty$, while the azimuthal domain now extends over only π radians. Again, we choose the symmetric interval $-\pi/2 \le \theta < +\pi/2$.

The volume of a circularly symmetric function may be calculated in polar coordinates by expressing it as the product of orthogonal Cartesian functions and then changing the integration variable to polar coordinates. The area element $dx\,dy$ must be replaced by the area element in polar coordinates $r\,dr\,d\theta$:

$$\iint_{-\infty}^{+\infty} f[x, y]\,dx\,dy = \int_{\theta=-\pi}^{\theta=+\pi} \int_{r=0}^{r=+\infty} f(\underline{\mathbf{r}})r\,dr = 2\pi \int_0^{+\infty} f_r(r)r\,dr \tag{7.83}$$

The center of symmetry of these circularly symmetric functions may be relocated to any point $[x_0, y_0]$ by adding the appropriate 2-D vector to the argument:

$$f(\underline{\mathbf{r}} - \underline{\mathbf{r}}_0) = f[x, y] = f[|\underline{\mathbf{r}}|\cos[\theta] - |\underline{\mathbf{r}}_0|\cos[\theta_0], |\underline{\mathbf{r}}|\sin[\theta] - |\underline{\mathbf{r}}_0|\sin[\theta_0]] \tag{7.84}$$

Several 2-D circularly symmetric functions are particularly helpful when describing imaging systems constructed from elements with circular cross-sections. Many of these functions are "analogous" to particular 2-D Cartesian-separable special functions.

7.4.1 Cylinder (Circle) Function

The cylinder function has unit amplitude inside its radius r_0 and null amplitude outside, and may be interpreted as the circularly symmetric version of the 2-D rectangle function. By reapplying the identical argument made for the 1-D rectangle in Equation (6.3), the amplitude of *CYL*(r) at its radius is defined to be half of the amplitude at the origin. The definition follows that of Gaskill (1978):

$$CYL\left(\frac{r}{d_0}\right) = \begin{cases} 1 & \textit{if } r < d_0/2 \\ \dfrac{1}{2} & \textit{if } r = d_0/2 \\ 0 & \textit{if } r > d_0/2 \end{cases} \tag{7.85}$$

The cylinder function for $d_0 = 2$ is pictured in Figure 7.21. Note that the scale parameter in the definition is the diameter of the circle $d_0 = 2r_0$ rather than the radius r_0. The area of the enclosed circle of unit diameter *CYL*(r) is $\pi/4 \cong 0.7854$, which obviously is smaller than the unit area of the rectangle with unit side *RECT*$[x, y]$. This also means that the volume of *CYL*(r/d_0) must be:

$$\int_{-\pi}^{+\pi} d\theta \int_0^{+\infty} CYL\left(\frac{r}{d_0}\right) r\,dr = \frac{\pi d_0^2}{4} \tag{7.86}$$

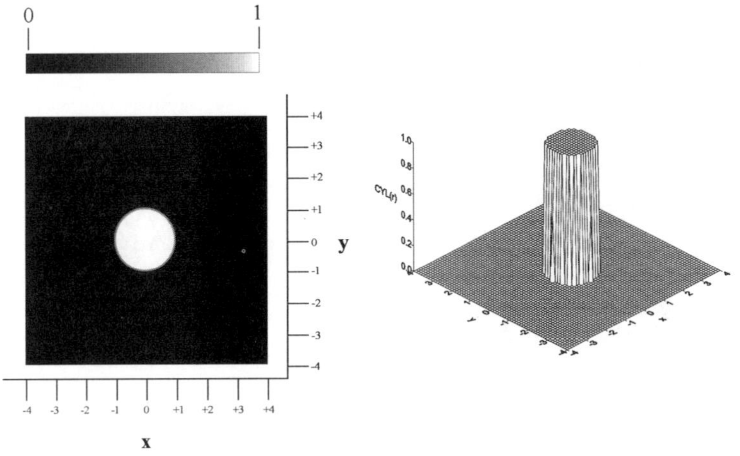

Figure 7.21 The 2-D cylinder function $CYL(r/2)$; the amplitudes of $\frac{1}{2}$ at the edge are not evident.

7.4.2 Circularly Symmetric Gaussian Function

We have already mentioned that the circularly symmetric Gaussian function is identical to the separable Gaussian function of Equation (7.29) if the scale parameters a and b along the x- and y-directions, respectively, are equal. For example:

$$\exp\left[-\frac{\pi x^2}{d^2}\right]\exp\left[-\frac{\pi y^2}{d^2}\right] = \exp\left[-\pi\frac{(x^2+y^2)}{d^2}\right] = \exp\left[-\pi\left(\frac{r}{d}\right)^2\right] \equiv GAUS\left(\frac{r}{d}\right) \tag{7.87}$$

This function is shown in Figure 7.22. The volume of the circularly symmetric Gaussian is found easily by integrating over polar coordinates and making the appropriate change of variable:

$$\iint_{-\infty}^{+\infty} GAUS\left(\frac{r}{d}\right)dx\,dy = \int_{-\pi}^{+\pi}\int_{0}^{+\infty} GAUS\left(\frac{r}{d}\right)r\,dr\,d\theta$$

$$= 2\pi\int_0^{+\infty} r\exp\left[-\pi\left(\frac{r}{d}\right)^2\right]dr \tag{7.88a}$$

By substituting a new integration variable $\alpha \equiv \pi(r/d)^2$, we obtain the desired volume:

$$\iint_{-\infty}^{+\infty} GAUS\left(\frac{\sqrt{x^2+y^2}}{d}\right)dx\,dy = 2\pi\int_0^{+\infty}\frac{d^2}{2\pi}e^{-\alpha}\,d\alpha = d^2\int_0^{+\infty}e^{-\alpha}\,d\alpha = d^2 \tag{7.88b}$$

7.4.3 Circularly Symmetric Bessel Function of Zero Order

The general form of a 2-D circularly symmetric form of the zero-order Bessel function may be generated by applying the recipe in Equation (7.82):

$$J_0[2\pi\rho_0 r]1[\theta] = J_0\left(2\pi\rho_0\cdot\sqrt{x^2+y^2}\right) \tag{7.89}$$

The argument of this Bessel function has been written in a form reminiscent of that used for sinusoids. The selectable parameter ρ_0 is analogous to the spatial frequency of a sinusoid: the larger the

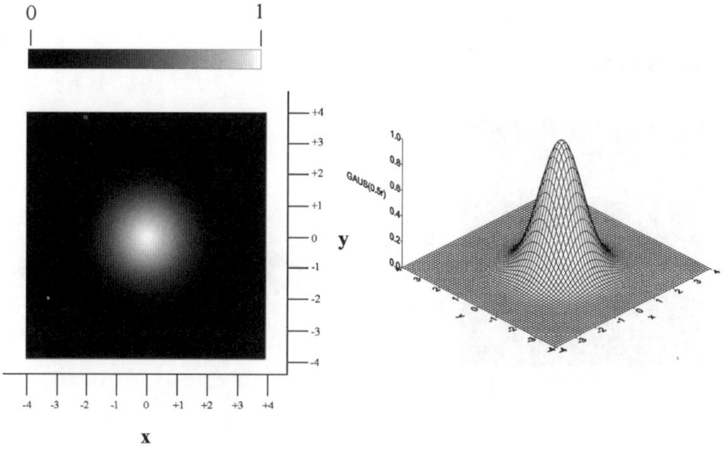

Figure 7.22 Example of 2-D circularly symmetric Gaussian $f(r) = e^{-\pi(r/2)^2}$.

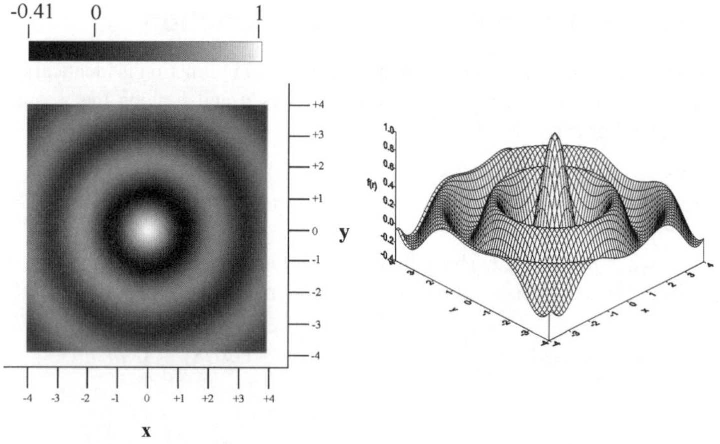

Figure 7.23 Circularly symmetric function $J_0(2\pi\rho_0 r)$ for $\rho_0 = \frac{1}{2}$.

value of ρ_0, the shorter the interval between successive maxima of the Bessel function. An example of the 2-D function $J_0(\pi r)$ is shown in Figure 7.23.

The circularly symmetric Bessel function of Equation (7.89) appears in several imaging applications. The demonstration of most of its useful properties will be postponed until Chapter 11, after the mathematical tools of the Fourier transform have been developed. However, one property is introduced here without proof because it leads to the useful means to construct the 1-D function $J_0[x]$ that was introduced in Chapter 6. The relationship will be demonstrated in Chapter 11 and may be used to define the 1-D and 2-D functions $J_0[x]$ and $J_0(r)$. As it happens, the 2-D circularly symmetric function in Equation (7.89) may be generated by summing 2-D cosine functions with the same period that are "directed" along all azimuthal directions. The constituent functions have the form $\cos[2\pi(\xi_0 x + \eta_0 y)]$ where the spatial frequency vectors are constrained to satisfy $(\xi_0^2 + \eta_0^2)^{\frac{1}{2}} = \rho_0^2$. Since the individual 2-D cosines are symmetric with respect to the origin of coordinates, rotation of the azimuth through π

radians yields an identical 2-D cosine. Therefore the azimuthal integral need only be evaluated over the "nonredundant" domain of azimuth angles that span a domain of π radians:

$$J_0[2\pi r\rho_0]1[\theta] = \int_{-\pi/2}^{+\pi/2} \cos[2\pi(\xi_0 x + \eta_0 y)]\, d\phi \quad \textit{where } \xi_0^2 + \eta_0^2 = \rho_0^2 \tag{7.90a}$$

The symmetry of the integrand may be used to evaluate the integral over the full domain of 2π radians by scaling the result by half:

$$\begin{aligned} J_0[2\pi r\rho_0] &= \frac{1}{2}\int_{-\pi}^{+\pi} \cos[2\pi(\xi_0 x + \eta_0 y)]\, d\phi \\ &= \frac{1}{2}\int_{-\pi}^{+\pi} \cos[2\pi(r\xi_0 \cos[\phi] + r\eta_0 \sin[\phi])]\, d\phi \end{aligned} \tag{7.90b}$$

We may rewrite the integrals over both domains by substituting the corresponding linear-phase complex exponential for the cosine function; the integral of the odd imaginary part vanishes, as will be demonstrated in Chapter 11:

$$\begin{aligned} J_0[2\pi r\rho_0]1[\theta] &= \frac{1}{\pi}\int_{-\pi/2}^{+\pi/2} e^{\pm 2\pi i(\xi_0 x + \eta_0 y)}\, d\phi = \frac{1}{2\pi}\int_{-\pi}^{+\pi} e^{\pm 2\pi i(\xi_0 x + \eta_0 y)}\, d\phi \\ &= \frac{1}{2\pi}\int_{-\pi}^{+\pi} \cos[2\pi r(\xi_0 \cos[\phi] + \eta_0 \sin[\phi])]\, d\phi \quad \textit{where } \xi_0^2 + \eta_0^2 = \rho_0^2 \end{aligned} \tag{7.91}$$

The profile of the 2-D J_0 Bessel function along the x-axis may be extracted from these equivalent expressions by setting $\eta_0 = 0$. The spatial frequency ξ along the x-axis becomes $\xi = \rho_0$:

$$\begin{aligned} J_0[2\pi r\rho_0]1[\theta]|_{\eta_0=0} = J_0[2\pi x\rho_0] &= \frac{1}{\pi}\int_{-\pi/2}^{+\pi/2} \cos[2\pi r\rho_0 \cos[\phi]]\, d\phi \\ &= \frac{1}{2\pi}\int_{-\pi}^{+\pi} \cos[2\pi r\rho_0 \cos[\phi]]\, d\phi \\ &= J_0[2\pi x\xi]|_{r=x,\xi=\rho_0} = J_0[2\pi r\rho_0] \end{aligned} \tag{7.92}$$

The term $\rho_0 \cos[\phi]$ in the argument of the integrand may be identified as the spatial frequency of the constituent 1-D cosines of the Bessel function. This suggests an alternate interpretation that the 1-D J_0 Bessel function is the summation of 1-D cosines with spatial frequencies in the interval $-\rho_0 \leq \xi \leq +\rho_0$ but weighted in "density" by $\cos[\phi]$. The largest spatial frequency exists when $\phi = 0$, while the cosine is the unit constant when $\phi = \pm\pi/2$.

Equivalent expressions for the 1-D Bessel function are obtained by instead setting $\xi_0 = 0$ in Equation (7.91), so that $\eta_0 = \rho_0$; this is equivalent to projecting the argument onto the y-axis:

$$\begin{aligned} J_0[2\pi y\eta]1[\theta]|_{\xi_0=0} &= \frac{1}{\pi}\int_{-\pi/2}^{+\pi/2} \cos[2\pi r(\rho_0 \sin[\phi])]\, d\phi \\ &= \frac{1}{2\pi}\int_{0}^{+\pi/2} \cos[2\pi r(\rho_0 \sin[\phi])]\, d\phi \\ &= J_0[2\pi y\eta]|_{r=y,\eta=\rho_0} = J_0[2\pi r\rho_0] \end{aligned} \tag{7.93}$$

The projection of the complex-valued formulations of the integral in Equation (7.91) onto the x-axis is:

$$(J_0[2\pi y\rho_0]1[\theta])|_{\xi=0} = \frac{1}{\pi}\int_{-\pi/2}^{+\pi/2} e^{\pm 2\pi i r\rho_0 \cos[\phi]}\, d\phi \quad \textit{where } \xi^2 + \eta^2 = \rho_0^2 \tag{7.94}$$

which may be used as an equivalent definition of the Bessel function.

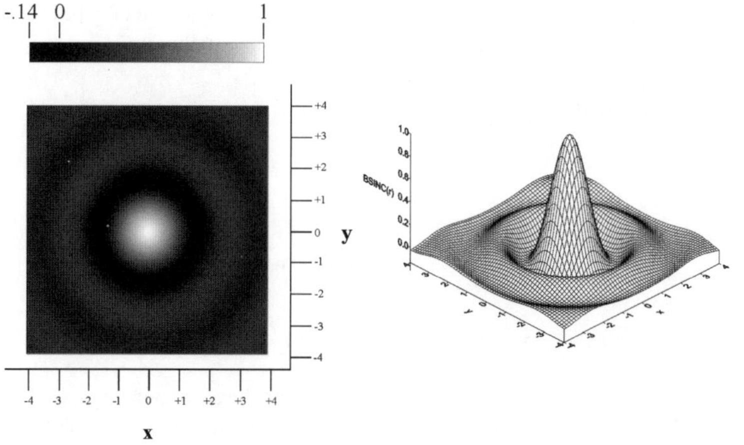

Figure 7.24 $SOMB(r) = (2/\pi r) J_1(\pi r)$.

7.4.4 Besinc or Sombrero Function

Just as the cylinder may be considered to be a circularly symmetric version of the *RECT* function, we also may define the circularly symmetric analogue to the $SINC$ function. This function is the ratio of two functions that each have null amplitude at the origin: the numerator is the Bessel function of the first kind of order unity $J_1(\pi r)$ and the denominator is a factor proportional to r:

$$SOMB(r) = \frac{2J_1(\pi r)}{\pi r} \tag{7.95}$$

The amplitude of $SOMB(r)$ at the origin is obtained again via L'Hôpital's rule by evaluating the ratio of the derivatives at the origin. The power-series expansion for $J_1[x]$ in Equation (6.78) provides a trivial path to the desired result:

$$\begin{aligned} SOMB(0) &= 2 \cdot \lim_{r\to 0} \left\{ \frac{(d/dr)[J_1(\pi r)]}{(d/dr)[\pi r]} \right\} \\ &= 2 \cdot \frac{(d/dr)\big[\pi r/2 - (\pi r)^3/16 + (\pi r)^5/384 - (\pi r)^7/18\,432 + \cdots\big]\big|_{r=0}}{(d/dr)[\pi r]|_{r=0}} \\ &= 2 \cdot \frac{(\pi/2)}{\pi} = 1 \end{aligned} \tag{7.96}$$

The 2-D plot of $SOMB(r)$ is shown in Figure 7.24. In tribute to the southwestern heritage of his university and because of the resemblance of $SOMB(r)$ to the fabled Mexican headgear, Gaskill named this function the "sombrero". Bracewell uses the name "*JINC*" for a function that differs in scaling constants; this draws upon analogies with the notation "J" for the Bessel function and the name "*SINC*".

The asymptotic form of $J_1[x]$ for large values of x that was presented in Equation (6.80) may be used to derive the expression for $SOMB(r)$ in the limit of large r:

$$\lim_{r\to+\infty} \{SOMB(r)\} = \frac{\sqrt{(2/\pi r)}\sin[\pi r - \pi/4]}{\pi r} \propto r^{-\frac{3}{2}} \sin\left[\pi r - \frac{\pi}{4}\right] \tag{7.97}$$

A comparison of the x-axis profiles $J_0[\pi x]$, $SINC[x]$, and $SOMB[x]$ is shown in Figure 7.25. Note that all have unit amplitude at the origin and appear to be the product of a periodic or pseudoperiodic oscillation and a decaying function of x. It is evident from these graphs that the peak amplitudes of

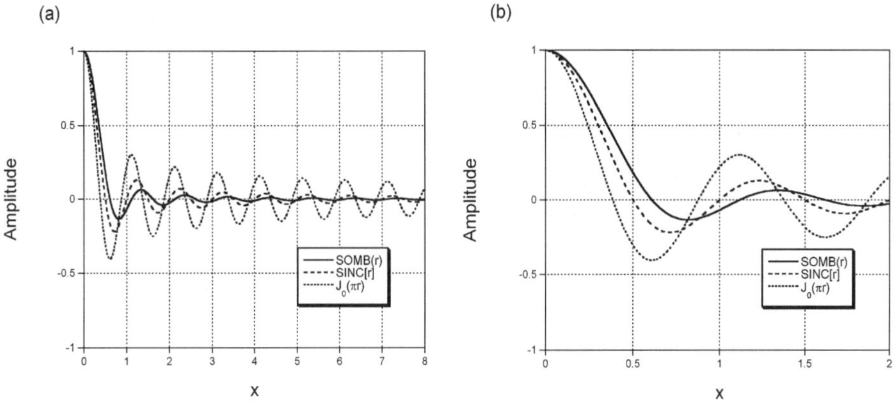

Figure 7.25 Comparison of profiles of *SOMB*[x], *SINC*[x], and $J_0[\pi x]$ (a) for $0 \leq x \leq 8$ and (b) magnified view for $0 \leq x \leq 2$. The amplitudes of *SOMB*[x] and $J_0[\pi x]$ decrease at the fastest and slowest rates, respectively, with increasing $|x|$.

J_0 decay most slowly with increasing x ($|J_0[\pi x]|$ decreases as $x^{-\frac{1}{2}}$, *SINC*[x] as x^{-1}, and *SOMB*[x] as $x^{-\frac{3}{2}}$). Also note the differences in locations of the zeros: those of *SINC*[x] are located at integer values of x. The first two zeros of $J_0[\pi x]$ are located at $x_1 \cong 2.4048/\pi \cong 0.7654$ and $x_2 \cong 5.5201/\pi \cong 1.7571$, and so the interval between them is just slightly less than unity. The interval between successive pairs of zeros of $J_0[\pi x]$ decreases and asymptotically approaches unity as $x \to \infty$. The first two zeros of $J_1[\pi x]$ (and therefore of *SOMB*(r)) are located at $x \cong 1.2197$ and $x \cong 2.2331$. The interval between zeros of *SOMB*(r) decreases with increasing r and asymptotically approaches unity as $r \to \infty$.

The volume of *SOMB*(r) is not as easily computed by direct methods as that of the 2-D *SINC* function, though again this computation is trivial after proving a few theorems of the Fourier transform. At this point, we again state the value for the volume without proof:

$$\int_{-\pi}^{+\pi} d\theta \int_{0}^{+\infty} SOMB\left(\frac{r}{d}\right) r\, dr = \frac{4d^2}{\pi} \tag{7.98}$$

7.4.5 Circular Triangle Function

This function is the circularly symmetric analogue of the 2-D triangle function and is most easily constructed from the definition of 2-D autocorrelation, though that is not considered until Chapter 10. Its primary use is to describe the spatial response of optical imaging systems constructed from elements with circular cross-sections. Some authors treat this as a special function. Though Gaskill is one of these, he does not assign it a special name; Bracewell (1995) used the same naming "convention" as Gaskill's "sombrero" by assigning the name of headwear: the "Chinese hat" function. We choose the shorthand name of "*CTRI*" for this function, which is introduced here only by presenting the mathematical expression; the derivation and a description of its properties are postponed until a discussion of the relationship between the Fourier transform and the correlation operation. Goodman (2005) derives the mathematical expression for the circularly symmetric analogue of the triangle:

$$CTRI(r) \equiv \frac{2}{\pi}\left(\cos^{-1}[r] - r\sqrt{1-r^2}\right) CYL\left(\frac{r}{2}\right) \tag{7.99}$$

The 2-D function is shown in Figure 7.26 and its profile in Figure 7.27. The amplitude has its maximum of unity at the origin and decreases monotonically to zero at unit radius. Note that the decrease in

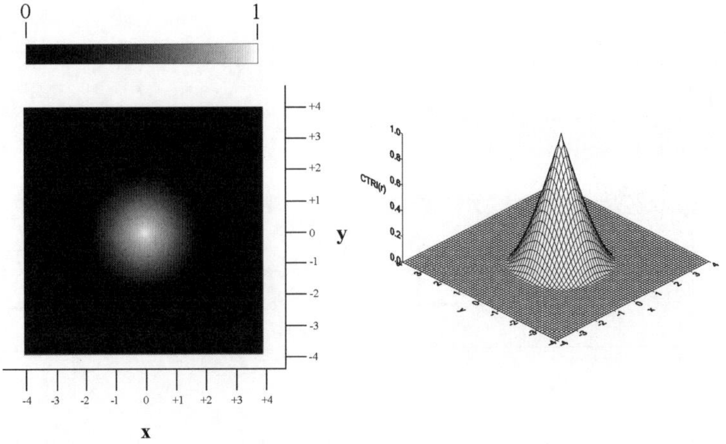

Figure 7.26 Circularly symmetric triangle function $CTRI(r) = (2/\pi)\left(\cos^{-1}[r] - r\sqrt{1-r^2}\right) \cdot CYL(r/2)$.

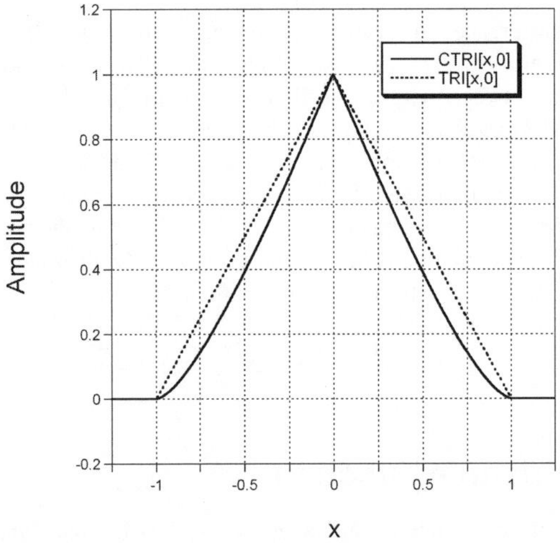

Figure 7.27 Comparison of the x-axis profiles of $CTRI(r)$ and $TRI[x, y]$, showing the deviation of the former profile from straightness, particularly for $|x| \lessapprox 1$.

amplitude is approximately linear until $r \cong 0.8$ where the slope "flattens out" slightly. The profiles of both the 2-D *CTRI* and 2-D *TRI* (except along the Cartesian axes) are *not* straight lines.

7.4.6 Ring Delta Function

The circularly symmetric analogue of the rectangular "corral" has the form:

$$f(\underline{\mathbf{r}}) = \delta(|\underline{\mathbf{r}}| - |\underline{\mathbf{r}}_0|) \equiv \delta(r - r_0) = \delta(r - r_0)1(\theta) \tag{7.100}$$

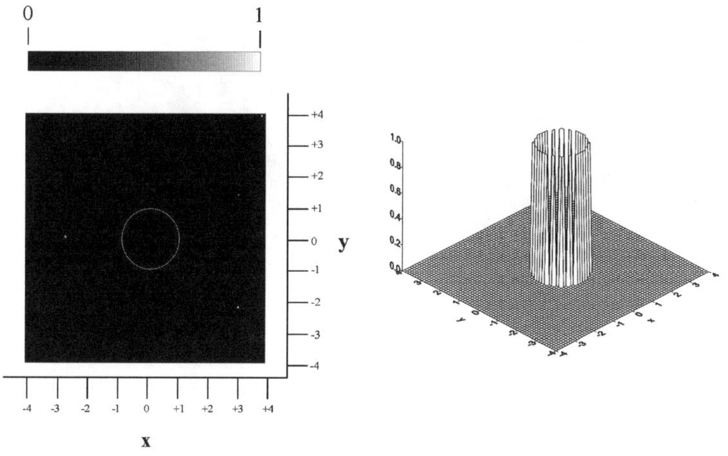

Figure 7.28 The 2-D ring Dirac delta function (which could also be called the "circular corral") $f(r) = \delta(r-1)1(\theta)$.

as shown in Figure 7.28. Though this resembles the polar representation of the 2-D Dirac delta function in Equation (7.55) at first glance, note that the radial arguments r and r_0 in Equation (7.100) are scalars, whereas the arguments in Equation (7.55) are 2-D vectors. The function in Equation (7.100) is a "circular wall" of Dirac delta functions about the origin at a distance r_0, as shown in Figure 7.27. The volume is evaluated easily in polar coordinates by applying the sifting property of the 1-D Dirac delta function:

$$\int_{-\pi}^{+\pi}\int_{0}^{+\infty}\delta(r-r_0)1[\theta]r\,dr\,d\theta = \int_{-\pi/2}^{+\pi/2}\int_{-\infty}^{+\infty}\delta(r-r_0)1(\theta)r\,dr\,d\theta$$
$$= \pi\int_{-\infty}^{+\infty}[\delta(r+r_0)+\delta(r-r_0)]r\,dr = 2\pi r_0 \qquad (7.101)$$

Just as the 1-D Dirac delta function may be written as the limit of a unit-area rectangle function of width b, the 2-D function is the scaled difference of two cylinder functions with diameters $2r_0+\Delta$ and $2r_0-\Delta$:

$$f(r) = \lim_{\Delta\to 0}\left\{\frac{1}{\Delta}\left[CYL\left(\frac{r}{2r_0+\Delta}\right) - CYL\left(\frac{r}{2r_0-\Delta}\right)\right]\right\} \qquad (7.102)$$

Its volume is:

$$\int_{-\pi}^{+\pi}\int_{0}^{+\infty} f(r)r\,dr\,d\theta = \lim_{\Delta\to 0}\left\{\frac{1}{\Delta}\left[\frac{\pi(2r_0+\Delta)^2}{4} - \frac{\pi(2r_0-\Delta)^2}{4}\right]\right\}$$
$$= \lim_{\Delta\to 0}\left\{\frac{\pi}{4\Delta}[(2r_0+\Delta)^2-(2r_0-\Delta)^2]\right\}$$
$$= \lim_{\Delta\to 0}\left\{\frac{\pi}{4\Delta}[8r_0\Delta]\right\} = \lim_{\Delta\to 0}\{2\pi\, r_0\} = 2\pi r_0 \qquad (7.103)$$

The expression for a ring delta function with unit volume therefore is $f(r) = (2\pi r_0)^{-1}\cdot\delta(r-r_0)$. The limiting expression in Equation (7.102) will be useful in Chapter 12 to compare the ring delta and cylinder functions with unit diameter to the 2-D rectangle and rectangular "corral" with unit 2-D support.

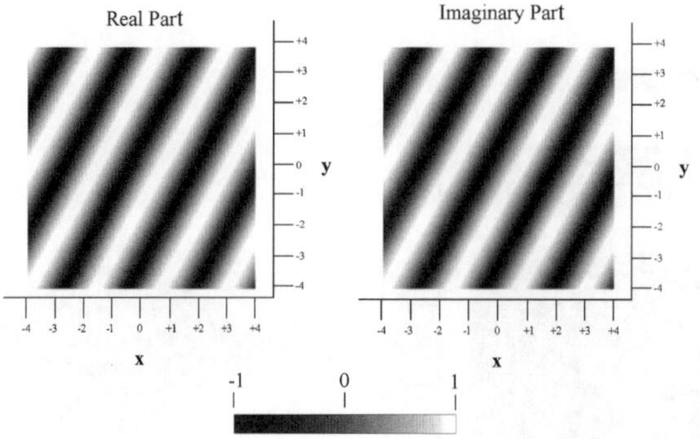

Figure 7.29 Real and imaginary parts of 2-D sinusoid that has been rotated so that it is constant along the azimuth $\phi \simeq \pi/3$, and thus it varies along the azimuth $\phi \simeq -\pi/6$.

7.5 Complex-Valued 2-D Functions

In a fashion identical to the 1-D case in Equation (6.143), a 2-D complex function is obtained by adding an arbitrary real-valued function to another real-valued function that has been scaled by $i = \sqrt{-1}$. As was the case in one dimension, several 2-D functions are conveniently defined and handled in complex form, and the analysis of the complex-valued function often leads to useful observations about the individual real and imaginary parts.

7.5.1 Complex 2-D Sinusoid

The extension of Equation (6.145) for 1-D sinusoids allows the real-valued 2-D symmetric cosine from Equation (7.33) to be combined with an imaginary-valued sinusoid of the same spatial frequency in quadrature. The result is a complex-valued 2-D function whose parts vary sinusoidally along one azimuthal direction and are constant in the orthogonal direction. The appropriate expression for such a function that oscillates along the x-direction is:

$$\begin{aligned} f[x, y] &= A_0(\cos[2\pi\xi_0 x + \phi_0] + i \sin[2\pi\xi_0 x + \phi_0]) \cdot 1[y] \\ &= A_0\left(\cos[2\pi\xi_0 x + \phi_0] + i \cos\left[2\pi\xi_0 x + \phi_0 - \frac{\pi}{2}\right]\right) \cdot 1[y] \end{aligned} \tag{7.104}$$

Of course, this also may be expressed as magnitude and phase by applying the Euler relation from Equation (4.19):

$$f[x, y] = A_0\, e^{+i(2\pi\xi_0 x+\phi_0)}\, 1[y] \tag{7.105}$$

The azimuth of the sinusoidal variation of the complex sinusoid may be oriented in an arbitrary direction:

$$\begin{aligned} f[x, y] &= \cos[2\pi\xi_0(\underline{\mathbf{r}} \bullet \hat{\underline{\mathbf{p}}})] + i \cos\left[2\pi\xi_0(\underline{\mathbf{r}} \bullet \hat{\underline{\mathbf{p}}}) - \frac{\pi}{2}\right] \\ &= \cos[2\pi(\xi_0 x + \eta_0 y)] + i \sin[2\pi(\xi_0 x + \eta_0 y)] = e^{+2\pi i(\xi_0 x+\eta_0 y)} \end{aligned} \tag{7.106}$$

as shown in Figure 7.29. The volumes of these three functions are zero because of the cancellation of the positive and negative amplitudes of both the real and imaginary parts.

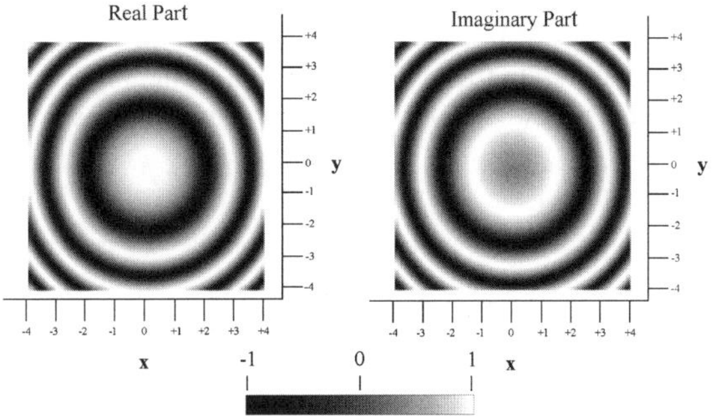

Figure 7.30 Real and imaginary parts of 2-D circularly symmetric chirp function: $f(r) = \exp[+i\pi(r/2)^2] = \cos[\pi(r/2)^2] + i\,\sin[\pi(r/2)^2]$.

7.5.2 Complex Quadratic-Phase Sinusoid

Several types of 2-D quadratic-phase sinusoidal function may be defined. One is obtained by direct analogy with the 2-D linear-phase sinusoid; it varies in one direction and is constant in the orthogonal direction to create a 1-D complex "chirp" function:

$$f[x,y] = A\,e^{\pm i\pi(x/a)^2}\,1[y] = A_0\left(\cos\left[\frac{\pi x^2}{a^2}\right] \pm i\,\sin\left[\frac{\pi x^2}{a^2}\right]\right)\cdot 1[y] \tag{7.107}$$

A 2-D separable complex chirp is obtained by orthogonal multiplication of 1-D complex-valued chirp functions with different chirp rates:

$$f[x,y] = A_0\exp\left[\pm i\pi\left(\frac{x}{a}\right)^2\right]B_0\exp\left[\pm i\pi\left(\frac{y}{b}\right)^2\right] = A_0B_0\exp\left[\pm i\pi\left(\left(\frac{x}{\alpha}\right)^2 \pm \left(\frac{y}{b}\right)^2\right)\right] \tag{7.108}$$

The magnitude of this function is not constant; the contours of constant magnitude are elliptical in the same sense of the 2-D separable Gaussian function in Equation (7.29). By setting axial chirp rates a and b to the identical value d, we obtain a circularly symmetric complex chirp function:

$$f[x,y] = A_0\exp\left[\pm i\pi\left(\frac{x}{d}\right)^2\right]B_0\exp\left[\pm i\pi\left(\frac{x}{d}\right)^2\right] = A_0B_0\exp\left[\pm i\pi\left(\frac{x}{d}\right)^2\right] \tag{7.109}$$

Clearly the chirp rate d is the radial distance from the origin where the phase changes by π radians compared to the phase at the origin. Just as for the 1-D case, a larger chirp rate d means that the amplitude oscillates more "slowly".

The 2-D chirp is a very important function in imaging, and particularly in optics, as will be demonstrated in Chapters 17 and 21. An example of the 2-D complex chirp function is shown in Figure 7.30.

7.6 Special Functions of Three (or More) Variables

Special functions obviously may be defined in spaces of dimension larger than two. Though such functions are not considered further in this book, we will say a few words about them. The easiest forms are generated in separable form from 1-D special functions, e.g., the 3-D Cartesian rectangle function is defined:

$$RECT[x,y,z] = RECT[x]\,RECT[y]\,RECT[z] \tag{7.110}$$

which obviously has unit 3-D volume. Separable functions in "spherical" form (r,θ,ϕ) also are feasible, though the interpretation is more complicated in 3-D than even in 2-D, particularly for the spherical version of the Dirac delta function.

PROBLEMS

7.1 Extend the discussion of the 2-D Dirac delta function as the product of orthogonal line delta functions to evaluate the product of two nonorthogonal line delta functions, e.g.:

$$f[x,y] = (1[x]\,\delta[y]) \cdot (\delta[\underline{\mathbf{r}} \bullet \underline{\boldsymbol{\rho}}]1[\underline{\mathbf{r}} \bullet \underline{\boldsymbol{\rho}}^{\perp}])$$

7.2 For each of the following functions, graph the "axial profiles", e.g., $f[x, 0]$ and $f[0, y]$, and the "diagonal profile", e.g., $f[x, y = x]$, as real and imaginary parts and as magnitude and phase. Graph the first by hand; you may plot the others with a software "toolbox" (e.g., Mathematica), but are not required to do so.

(a) $f[x,y] = RECT[x/4, y/4] \cdot e^{+2\pi i(x+y)}$

(b) $g[x,y] = RECT[x/4, y/4] \cdot e^{+2\pi i(\sqrt{x^2+y^2})}$

(c) $q[x,y] = e^{-2\pi i[x/5+y/12]}$

(d) $p[x,y] = (RECT[x+1, y] + RECT[x-1, y]) \cdot e^{+2\pi ix}$

7.3 Generate "images" of the following functions:

(a) $e[x,y] = SGN[RECT[x/4, y/8] + RECT[x/2, y/2] - RECT[(x-\frac{1}{2})/3, y/6]]$

(b) $f[r] = SGN[CYL(r/2) - CYL(2r)]$

7.4 Derive an expression for and sketch (or plot) the even and odd parts of the 2-D function $f[x, y] = CYL(\sqrt{(x-1)^2 + y^2}) + RECT[x+1, y]$.

7.5 Sketch $f[x, y] = \cos[2\pi(x/5 + y/12) + \pi/6]$ and find its spatial period.

7.6 For the 2-D circularly symmetric functions:

$$f_1(r) = \frac{\delta(r)}{\pi r}$$

$$f_2(r) = \delta(r-2)$$

(a) Sketch the functions.

(b) Calculate their volumes.

7.7 Derive a "new" expression for the 2-D separable Dirac delta function $\delta[\underline{\mathbf{r}} - \underline{\mathbf{r}}_0]$, where the position vector $\underline{\mathbf{r}}_0$ has Cartesian coordinates $[x_0, y_0]$ and polar coordinates (r_0, θ_0). The unit vector $\hat{\underline{\mathbf{p}}}$ that describes the rotation has coordinates $(1, \theta_0)$. The 2-D Dirac delta function may be separated into orthogonal 1-D Dirac delta functions whose rotated coordinates x' and y' are the scalar products of $\underline{\mathbf{r}} - \underline{\mathbf{r}}_0$ with $\hat{\underline{\mathbf{p}}}$ and $\hat{\underline{\mathbf{p}}}^{\perp}$, respectively.

(a) A 2-D Dirac delta function at an arbitrary location $[x_0, y_0] = (p_0, \phi_0)$ may be constructed as the product of a line delta function through the origin at the appropriate angle by a perpendicular line delta function that intersects at the appropriate radial distance. Consider first a 2-D Dirac delta function located at the origin, which has the form $\delta[x]1[y] \cdot 1[x]\delta[y] = \delta[x, y]$ in Cartesian coordinates. The same function may be expressed as the product of a line delta function through the origin "oriented" at azimuth angles ϕ_0 and $\phi_0 + \pi/2$. Demonstrate that:

$$(\delta[\underline{\mathbf{r}} \bullet \hat{\underline{\mathbf{p}}}_0]1[\underline{\mathbf{r}} \bullet \hat{\underline{\mathbf{p}}}_0^{\perp}])(1[\underline{\mathbf{r}} \bullet \hat{\underline{\mathbf{p}}}_0]\delta[\underline{\mathbf{r}} \bullet \hat{\underline{\mathbf{p}}}_0^{\perp}]) = \delta[x,y]$$

(b) A Dirac delta function located at the arbitrary location with Cartesian coordinates $[x_0, y_0]$ and polar coordinates (p_0, ϕ_0) may be written as the product of a line delta function through the origin at the angle ϕ_0 of the form $1[\underline{\mathbf{r}} \bullet \hat{\underline{\mathbf{p}}}_0]\,\delta[\underline{\mathbf{r}} \bullet \hat{\underline{\mathbf{p}}}_0^{\perp}]$ and a perpendicular line delta function at radial distance p_0, which is expressed as $\delta[p_0 - \underline{\mathbf{r}} \bullet \hat{\underline{\mathbf{p}}}_0]1[\underline{\mathbf{r}} \bullet \hat{\underline{\mathbf{p}}}_0^{\perp}]$. Demonstrate that:

$$(1[\underline{\mathbf{r}} \bullet \hat{\underline{\mathbf{p}}}_0]\delta[\underline{\mathbf{r}} \bullet \hat{\underline{\mathbf{p}}}_0^{\perp}])(\delta[p_0 - \underline{\mathbf{r}} \bullet \hat{\underline{\mathbf{p}}}_0]1[\underline{\mathbf{r}} \bullet \hat{\underline{\mathbf{p}}}_0^{\perp}]) = \delta[x-x_0, y-y_0]$$

8

Linear Operators

In Chapter 1, we introduced the concept of a mathematical operator $\mathcal{O}$ that specifies the action of a system on inputs to generate the corresponding outputs:

$$\mathcal{O}\{f[\underline{\mathbf{x}}]\} = g[\underline{\mathbf{x}}'] \tag{8.1}$$

where $f[\underline{\mathbf{x}}]$ is the input amplitude at location $\underline{\mathbf{x}}$ within the N-D continuous input domain and $g[\underline{\mathbf{x}}']$ is the output amplitude at location $\underline{\mathbf{x}}'$ within the M-D continuous output domain. In most imaging applications, the dimensionality N is larger than M, so that the input domain has more dimensions than the output domain. The most common example is photographic imaging of an input scene with three spatial dimensions and one spectral dimension: $[x, y, z, \lambda]$. A "black-and-white" photograph of the scene is described by two spatial dimensions with coordinates $[x', y']$.

The operator $\mathcal{O}$ symbolizes the mathematical recipe for the action of the system on the input function. Presumably, the output function differs from the input in some way. In fact, the "units" of the respective input and output amplitudes f and g need not be (and usually are not) identical. In other words, most imaging systems convert one type of signal amplitude or energy to another form; the systems act as energy "transducers". A familiar example is a photographic-film camera, which converts the electric field amplitude of the input signal (measured in volts per meter) to the dimensionless quantity of "film density".

The completely general system operator $\mathcal{O}$ may change the input amplitude f and/or affect its "location" $\underline{\mathbf{x}}$ in any arbitrary way. As an example of a simple imaging operator, consider a system that amplifies the input amplitude f at each location $\underline{\mathbf{x}}$ by a deterministic numerical factor without "moving" or "spreading" the amplitudes to other neighboring locations. A more complicated system operator might amplify the input amplitude at each location by numerical factors that vary randomly with position. Note that the spatial domains of the input and output signals of these two operators are identical. Yet another type of system operator affects only the position of the signal amplitude in a deterministic way (e.g., translation or rotation about the origin) or in a random fashion. The completely arbitrary system operator may mix any combination of such operations on the amplitude and location. The effect of the operator on the input image may be conceptually simple to describe but difficult or impossible to cast into a mathematical expression. For example, consider the second system just described whose only action is to "move" the amplitude from the 2-D input location $\underline{\mathbf{x}}$ to a new and unique 2-D output location $\underline{\mathbf{x}}'$ selected at random; a schematic of the process is shown in Figure 8.1. The resulting output image is a "scrambled" version of the input function, which suggests that such an imaging system might be used for image cryptography. The corresponding operator $\mathcal{O}$ is merely a transformation of the coordinates and may be expressed as a function whose domain and range are

Fourier Methods in Imaging Roger L. Easton, Jr.

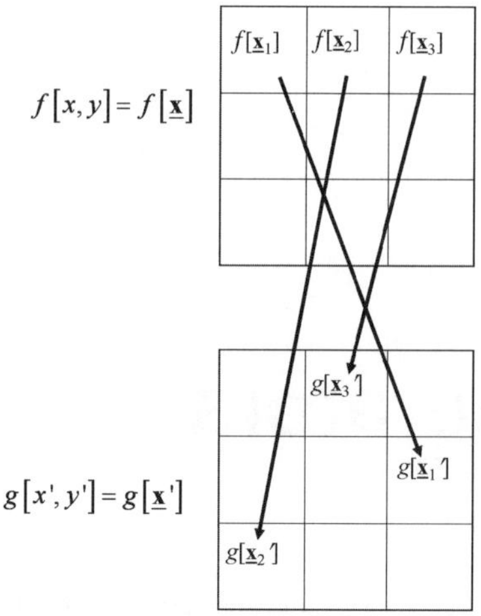

$$\mathcal{O}\{f[\underline{\mathbf{x}}]\} = g[\underline{\mathbf{x}}'] \Rightarrow \mathcal{O}\{f[\underline{\mathbf{x}}]\} = \underline{\mathbf{x}}'[\underline{\mathbf{x}}]$$

Figure 8.1 Schematic of the image "scrambling" operator $\underline{\mathbf{x}}'[\underline{\mathbf{x}}]$, which may be inverted if every location in the input maps to a unique location in the output.

both 2-D vectors, i.e., $\underline{\mathbf{x}}'[\underline{\mathbf{x}}]$. In the most general imaging case, both the domain $\underline{\mathbf{x}}$ and range $\underline{\mathbf{x}}'$ of this vector function are identically the infinite 2-D real plane. The infinity of points to be specified and the randomness of the mapping determine that no feasible expression for the vector function exists. In the discrete case where the input and output images are sampled at N picture elements specified by lexicographically ordered vectors $\underline{\mathbf{x}}$ and $\underline{\mathbf{x}}'$, then the vector function that describes the transformation operator may be specified (though it will be large).

It is useful to consider conditions that allow the "inverse" imaging task to be solved for this "sampled image scrambler". This would require that each sampled input pixel $\underline{\mathbf{x}}$ in the domain of the system function be mapped to a unique pixel $\underline{\mathbf{x}}'$ in the range. The corresponding inverse system operator exists in concept as the vector function $\underline{\mathbf{x}}[\underline{\mathbf{x}}']$.

This example suggests an imaging axiom that mathematically tractable operators exist only for systems whose actions are constrained from complete arbitrariness. The two most important such restrictions of linearity and shift invariance were introduced in Chapter 2. When these are invoked, mathematical operators may be constructed that are useful for solving particular imaging tasks. Recall that the quality of "linearity" may be loosely described as restricting the possible effect of the operator on the amplitude f of the input signal, while "shift invariance" constrains the action with respect to the coordinate x. These two restrictions are described individually in the next two sections, followed by derivation of the mathematical description of the imaging system operator $\mathcal{O}$ when both restrictions are applied simultaneously.

8.1 Linear Operators

Consider a particularly simple and yet still physically useful system operator whose action depends solely upon the input amplitude f. Put another way, the specific location of an amplitude within the domain has no influence upon the output. In this case, the domain of the output image must include the entire domain of the input to ensure that a unique output location exists for each input coordinate.

In such a case, the "primed" notation used for the output coordinate $\underline{\mathbf{x}}'$ in Equation (8.1) is superfluous. The action of such a system upon all amplitudes in the input signal with the same numerical value (say, f_0) must generate the same output amplitude (e.g., g_0) at the corresponding coordinate. Such an operator $\mathcal{O}$ may be described fully by specifying the single-valued functional relationship between the amplitudes of the input and output at the same location:

$$\mathcal{O}\{f[\underline{\mathbf{x}}]\} = g\{\underline{\mathbf{x}}\} \Longrightarrow \mathcal{O}\{f\} = g \Longrightarrow \mathcal{O}\{f\} = g[f] \tag{8.2}$$

The specification of the system operator $g[f]$ defines what might be called the "transfer characteristic" or "tone-transfer curve" of the imaging system. The action of this particularly simple operator is identical in form to the common "lookup table" in digital image processing where the output gray level is specified for each input gray level; the process is illustrated in Figure 8.2. Because the output amplitude at a single location in the output domain is determined by the input amplitude at a single (identical) location, imaging systems described by the form of Equation (8.2) may be called "point operations". In turn, these point operators may be classified based upon the shape of the transfer characteristic function $g[f]$. The analogy with a 1-D mathematical function in Section 1.3.1 suggests that a point operator $\mathcal{O}\{f[x]\} \Longrightarrow g[f]$ is linear when the output amplitude g is proportional to the input amplitude f:

$$\mathcal{O}\{f\} = g[f] \textit{ is linear if and only if } g[f] = k \cdot f \propto f \tag{8.3}$$

where k is a numerical proportionality constant with appropriate dimensions. This strict definition of a linear operator requires that the transfer characteristic $g[f]$ be a straight line that intersects the origin. The action of this operator upon a sum of multiple inputs is guaranteed to be identical to the sum of outputs obtained by applying the operator to the individual unweighted inputs. In mathematical jargon, this criterion of linearity is expressed as:

$$\textit{if } g[f_1] = g_1 = k \cdot f_1 \textit{ and } g[f_2] = g_2 = k \cdot f_2, \textit{ then } g[f_1 + f_2] = k \cdot (f_1 + f_2) = g_1 + g_2 \tag{8.4}$$

The formal definition of a linear operator is only slightly more general than this one that was based upon the shape of the graph of the lookup table. The action of the linear operator upon a superposition (weighted sum or linear combination) of inputs must be identical to the identically weighted sum of the individual outputs:

$$\textit{if } \mathcal{O}\{f_n[x]\} = g_n[x'], \textit{ then } \mathcal{O}\left\{\sum_n \alpha_n f_n[x]\right\} = \sum_n \alpha_n \mathcal{O}\{f_n[x]\} = \sum_n \alpha_n g_n[x'] \tag{8.5}$$

where the α_n are (generally complex-valued) weighting constants. In words, Equation (8.5) states that if the input/output relationship is defined for one input amplitude, then the relationship for *any* input amplitude is a proportional scaling. One corollary to the condition in Equation (8.5) is that the action of a linear system on a null input results in a null output, which was also required in the criterion of Equation (8.4).

Obviously, the linearity condition is a descriptor of the system operator $\mathcal{O}$ rather than of the input and/or output signals. To be "linear" in a strict sense, the operator $\mathcal{O}$ must satisfy Equation (8.5) over the infinite range of possible input amplitudes; any operator that does not fulfill this criterion obviously is "nonlinear". It should be noted that realistic systems cannot possibly satisfy the linearity condition over an infinite range of amplitudes. The response of any realistic detector must eventually "saturate", which means that addition of incremental input amplitude has no effect upon the output amplitude. A familiar example of a saturable (and therefore nonlinear) imaging system is a photographic emulsion, which converts the intensity of incident electromagnetic radiation (often represented by the symbol "E") to a spatial variation in the recorded optical density "D". The functional relationship between input and output of a photographic emulsion usually is shown as the "*H&D*" curve (for its originators Hurter and Driffield) or the *characteristic curve*, which is a graph of the image density versus the logarithm of the exposure. The sigmoidal shape of the *H&D* curve for most photographic films is depicted in Figure 8.3. The "flat" response at large exposures is evidence of detector saturation.

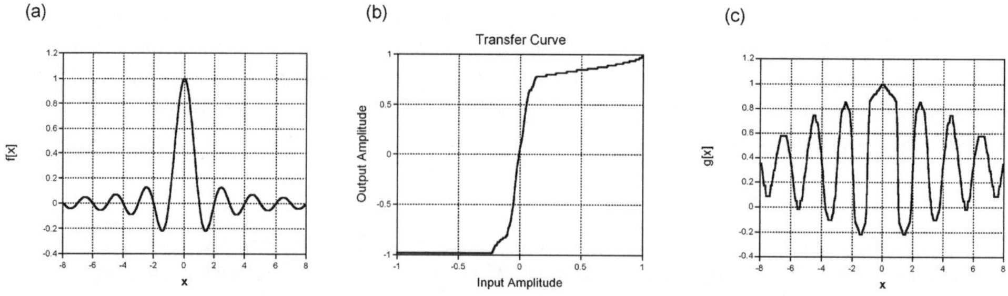

Figure 8.2 Effect of a nonlinear point operator $g[f]$ on the 1-D input function $f[x]$ to construct the output function $g[x]$: (a) input $f[x] = SINC[x]$; (b) nonlinear tone-transfer operator $g[f]$; (c) output image $g[x]$.

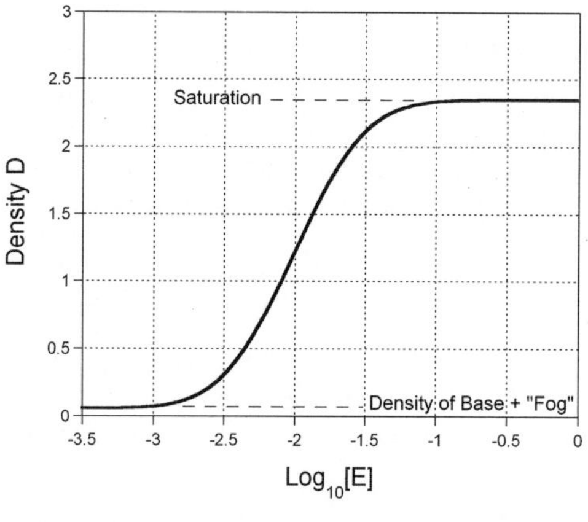

Figure 8.3 *H&D curve* (characteristic curve) of typical photographic emulsion. The "exposure" of light is plotted logarithmically on the x-axis; the developed density is plotted on the y-axis. The nonlinear performance of the photographic sensor is evident.

Though all realistic systems are nonlinear in the strict sense, some may obey the linearity condition to a good approximation over some finite range of input amplitudes. If the amplitude of the input scene is constrained to stay within this "linear region" of the operator, then the assumption of linearity may be valid.

Other types of common nonlinear imaging operations include "thresholders" (also called "hard limiters") and their more general siblings known as "quantizers". In thresholding, the measured output saturates "instantly" if the input exceeds some specific value. A quantization operator maps the continuously valued input to a finite set of discrete output amplitudes through a system that may be viewed as a collection of thresholders with different saturation levels. One example of a quantizer is the "greatest integer" operator introduced in Equation (2.6); this operator also is known as "truncation" and assigns an integer output to each real-valued input by lopping off any decimal part. For example, the output of the greatest-integer operator acting on the numerical value for "e" would yield:

$$\mathcal{T}\{e\} = \mathcal{T}\{2.718\,28\ldots\} = 2 \tag{8.6}$$

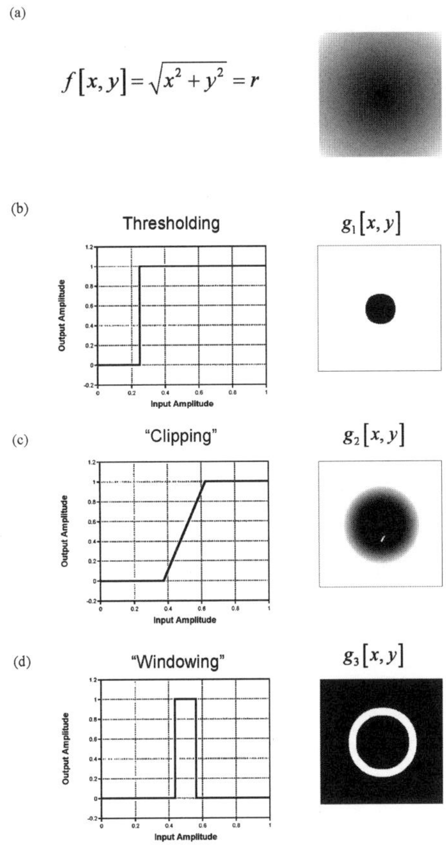

Figure 8.4 Examples of point operations common in digital imaging: (a) input object, the 2-D circularly symmetric function $f[x, y] = (x^2 + y^2)^{1/2} = r$; (b) image obtained via thresholding lookup table is a cylinder function; (c) image resulting from "clipping" lookup table is a "fuzzy" cylinder; and (d) image from window threshold is an annulus.

That this operation does not satisfy the criterion for linearity is easily demonstrated by applying $\mathcal{T}$ to $2e$:

$$\begin{aligned}\mathcal{T}\{e+e\} &= \mathcal{T}\{2 \cdot 2.718\,28\} = \mathcal{T}\{5.436\ldots\} = 5 \\ &\neq \mathcal{T}\{e\} + \mathcal{T}\{e\} = 4\end{aligned} \tag{8.7}$$

It is probably obvious that the shape of the transfer characteristic of the truncation operator establishes immediately that this operator is nonlinear. The transfer characteristics of several nonlinear operators are shown in Figure 8.4.

The linear regime of a realistic imaging system is determined by the qualities of the imaging sensor. Sensors can be fabricated that respond to the complex-valued amplitude f (magnitude and phase) of electromagnetic waves that have long wavelengths and thus small temporal frequencies (such as radio frequencies, called RF waves). Unfortunately, sensors for electromagnetic waves with larger frequencies, including wavelengths in the visible spectrum or shorter, cannot respond rapidly enough to record the phase. They respond to the incident "power" of the wave, which is the time-averaged squared magnitude of the amplitude and information about the phase of the light at different points in the output image is lost. It probably is obvious that a system that is linear in "amplitude" f cannot simultaneously be linear in "squared magnitude" $|f|^2$.

Several examples of linear operators will be used for illustration. The first has already been considered; it generates the output by applying a numerical scale factor to the input:

$$\mathcal{O}_1\{f[x]\} \equiv g[x] = k \cdot f[x] \tag{8.8a}$$

where k is a numerical (generally complex-valued) constant. The condition that must be satisfied for the operator $\mathcal{O}_1$ to be linear is:

$$\mathcal{O}_1\{\alpha_1 f_1[x] + \alpha_2 f_2[x]\} \stackrel{?}{=} \alpha_1 \mathcal{O}_1\{f_1[x]\} + \alpha_2 \mathcal{O}_1\{f_2[x]\} \tag{8.8b}$$

After inserting the form of the operator in Equation (8.8a) into both the left- and right-hand sides of this expression, we obtain expressions for both:

$$k \cdot \{\alpha_1 f_1[x] + \alpha_2 f_2[x]\} \stackrel{?}{=} \alpha_1 \cdot k \cdot f_1[x] + \alpha_2 \cdot k \cdot f_2[x] \tag{8.8c}$$

These obviously are equal, and thus the operation consisting of multiplication of the input amplitude by a constant is linear. This should be no surprise, as this operation defines the property of linearity.

The second operator to be tested for linearity is differentiation of the input function with respect to x. The system operator for the nth-order derivative is:

$$\mathcal{O}_2\{f[x]\} = g[x] = \frac{\partial^n}{\partial x^n} f[x] \tag{8.9a}$$

and the condition to be satisfied is:

$$\begin{aligned}\mathcal{O}_2\{\alpha_1 f_1[x] + \alpha_2 f_2[x]\} &= \frac{\partial^n}{\partial x^n}(\alpha_1 f_1[x] + \alpha_2 f_2[x]) \\ &= \alpha_1 \frac{\partial^n}{\partial x^n}(f_1[x]) + \alpha_2 \frac{\partial^n}{\partial x^n}(f_2[x]) \\ &= \alpha_1 \mathcal{O}_2\{f_1[x]\} + \alpha_2 \mathcal{O}_2\{f_2[x]\} \\ &= \alpha_1 g_1[x] + \alpha_2 g_2[x]\end{aligned} \tag{8.9b}$$

This establishes that differentiation is a linear operation.

The last operator to be tested for linearity is integration of the 1-D input function over some fixed interval (a, b). The expression for the operator is:

$$\mathcal{O}_3\{f[x]\} = g_3[x] = \int_a^b f[u]\, du \tag{8.10a}$$

The confirmation is quite easy:

$$\begin{aligned}\mathcal{O}_3\{\alpha_1 f_1[x] + \alpha_2 f_2[x]\} &= \int_a^b (\alpha_1 f_1[u] + \alpha_2 f_2[u]) du \\ &= \alpha_1 \int_a^b f_1[u]\, du + \beta \int_a^b f_2[u]\, du \\ &= \alpha_1 \mathcal{O}_3\{f_1[x]\} + \alpha_2 \mathcal{O}_3\{f_2[x]\}\end{aligned} \tag{8.10b}$$

Therefore integration of an input function over fixed limits is a linear operation. Note that this result is valid for all values of a and b, including $a = -\infty$ and $b = +\infty$.

The corollary to Equation (8.5) demonstrated that a null input must generate a null output for the operator to be linear in a strict sense: $g[x] = 0[x]$ if and only if $f[x] = 0[x]$. It is clear from this result that any operator that adds a constant "bias" to the input must be a nonlinear operation in the strict sense. An example of such an operator is:

$$\mathcal{O}\{f[x]\} = f[x] + b \equiv g[x] \text{ where } b \neq 0 \tag{8.11a}$$

Following Equation (8.8), the operator is applied to a weighted sum of inputs to generate an "image" consisting of the same input plus the bias term:

$$\mathcal{O}\{\alpha_1 f_1[x] + \alpha_2 f_2[x]\} = (\alpha_1 f_1[x] + \alpha_2 f_2[x]) + b \tag{8.11b}$$

On the other hand, the weighted sum of the individual outputs includes an "extra" bias term:

$$\begin{aligned}\alpha_1\mathcal{O}\{f_1[x]\} + \alpha_2\mathcal{O}\{f_2[x]\} &= \alpha_1(f_1[x] + b) + \alpha_2(f_2[x] + b)\\ &= \alpha_1 f_1[x] + \alpha_2 f_2[x] + (\alpha_1 + \alpha_2)\cdot b\end{aligned} \tag{8.11c}$$

Since Equations (8.11b) and (8.11c) are not equal, addition of a bias does not satisfy the criterion for a linear operator.

Often a subsequent operation may be performed on the output of a nonlinear process to create a cascaded process that is linear. For example, the nonlinear process of adding a bias to the input may be "undone" by applying a second operator that subtracts the bias, so that the cascade is a linear operator. As a second and more complicated example of a linear cascade of processes, consider the operation that computes the natural logarithm of the amplitude f. The process may be "made" linear by performing a subsequent point operation that "exponentiates" the amplitude:

$$\begin{aligned}g_2[x] &= \mathcal{O}_2\{g_1[x]\} = \mathcal{O}_2\{\mathcal{O}_1\{f[x]\}\}\\ &= \exp\{\log_e[f[x]]\} = \exp\{g_1[x]\} = f[x]\end{aligned} \tag{8.12}$$

The cascaded process $\mathcal{O} \equiv \mathcal{O}_2\mathcal{O}_1$ is identical to the identity operator, which is obviously linear.

8.2 Shift-Invariant Operators

The second constraint applied to imaging operators is "shift invariance", which describes the action of the system upon the location of amplitude within the input signal. Recall that the general system operator in Equation (8.1) may relocate the amplitude of the input distribution in any arbitrary way. For example, consider the (admittedly unphysical) system operator defined by the following conditions:

$$g[x] = \mathcal{O}\{f[x]\} = \begin{Bmatrix} f[x+2] & \text{if } INT[x] \text{ is even} \\ f[x-2] & \text{if } INT[x] \text{ is odd} \end{Bmatrix} \tag{8.13}$$

This (definitely weird) operator translates the amplitude leftwards or rightwards by two units depending on the evenness or oddness of the integer part of the coordinate x. The "response" of this system depends upon only the location within the input signal $f[x]$ and not at all on the amplitude. A system whose response varies with and/or is determined by the location in the input domain is called *space variant* or *shift variant*. A system whose action does not depend on the specific position in the input signal is easier to describe mathematically; these operators are *space invariant* or *shift invariant*. The mathematical criterion to be met by a shift-invariant operator $\mathcal{O}$ is expressed in terms of its action on a translated copy of the signal. For 1-D functions, the statement of the requirement that is satisfied by a shift-invariant operator $\mathcal{O}$ is:

$$\text{If } \mathcal{O}\{f[x]\} = g[x], \text{ then } \mathcal{O}\{f[x - x_0]\} = g[x - x_0] \tag{8.14a}$$

In words, the quality of "shift invariance" ensures that the only effect of translating the 1-D input by some distance x_0 is an identical translation of the original output. The corresponding statement for 2-D signals is:

$$\text{If } \mathcal{O}\{f[x, y]\} = g[x, y], \text{ then } \mathcal{O}\{f[x - x_0, y - y_0]\} = g[x - x_0, y - y_0] \tag{8.14b}$$

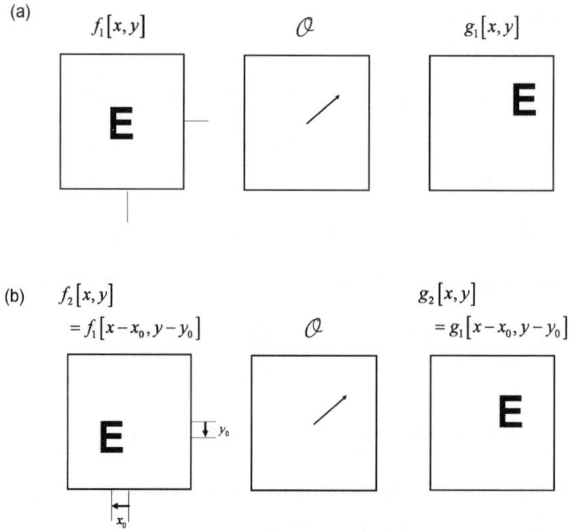

Figure 8.5 Test of shift properties of the translation operator: (a) $\mathcal{O}\{f_1[x,y]\} = g_1[x,y] = f_1[x - x_1, y - y_1]$; (b) $\mathcal{O}\{f_1[x - x_0, y - y_0]\} = g_1[x - x_0, y - y_0]$. Since the same result is obtained for the translated input, the translation operator is shift invariant.

Obviously, the quality of shift variance may be tested by measuring the system output before and after translating the input signal. Consider first a simple operator that translates the input image through a distance x_1:

$$\mathcal{O}_4\{f[x,y]\} = g_4[x,y] = f[x - x_1, y - y_1] \tag{8.15a}$$

The condition that must be satisfied for the operator $\mathcal{O}_4$ to be shift invariant is:

$$\begin{aligned}\mathcal{O}_4\{f[x - x_0, y - y_0]\} &= f[(x - x_0) - x_1, (y - y_0) - y_1]\\ &= f[(x - x_1) - x_0, (y - y_1) - y_0]\\ &= g_4[x - x_1, y - y_1]\end{aligned} \tag{8.15b}$$

Therefore, the criterion of Equation (8.14a) is satisfied by $\mathcal{O}_4$ and the operation of translating the input function is shift invariant. Again, this should be no surprise, as shift invariance is defined in terms of a translation operation such as $\mathcal{O}_4$. The test of shift invariance in this case is shown in Figure 8.5.

Now consider a system operator that scales the input image by a factor of b:

$$\mathcal{O}_5\{f[x]\} = g_5[x] = f\left[\frac{x}{b}\right] \tag{8.16a}$$

If $b > 1$, the output image is "larger" than the input (magnified image). The test of the shift-invariant condition yields two distinct expressions:

$$\mathcal{O}_5\{f[x - x_0]\} = f\left[\frac{x - x_0}{b}\right] \neq f\left[\frac{x}{b} - x_0\right] \tag{8.16b}$$

Because the results do not match, image magnification is not a shift-invariant operation.

The last operation to be tested for shift invariance is rotation of a 2-D function $f[x,y]$ about the origin by θ radians, which was introduced in Section 7.1:

$$\begin{aligned}\mathcal{O}_6\{f[x,y]\} &= g_6[x,y] = f[x', y']\\ &= f[x\cos[\theta] + y\sin[\theta], -x\sin[\theta] + y\cos[\theta]] = f[\underline{\mathbf{r}} \bullet \hat{\underline{\mathbf{p}}}, \underline{\mathbf{r}} \bullet \hat{\underline{\mathbf{p}}}^{\perp}]\end{aligned} \tag{8.17}$$

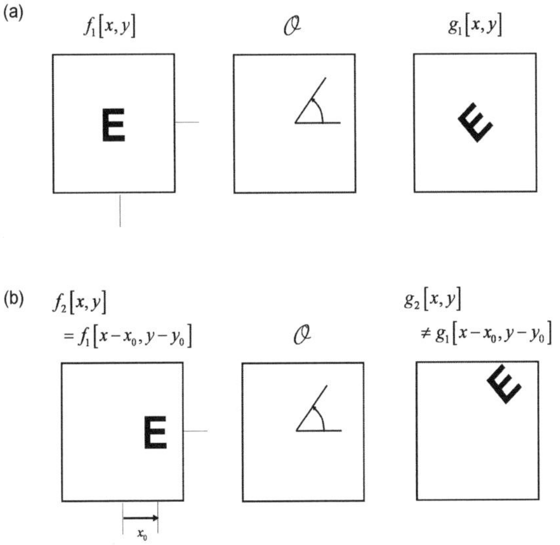

Figure 8.6 Demonstration that "rotation" about the origin of coordinates is a shift-variant operator: (a) $f[x, y]$ is rotated and then translated; (b) $f[x, y]$ is translated and then rotated. Because the output images differ, the rotation operation is shift variant.

To test the rotation operator for shift invariance, first compute the result from a cascade of translation followed by rotation:

$$\mathcal{O}_6\{f[x - x_0, y - y_0]\} = f[x'', y''] \tag{8.18}$$

where the coordinates $[x'', y'']$ are obtained by substituting $x - x_0$ and $y - y_0$ for x and y in Equation (8.17):

$$\begin{aligned} x'' &= (x - x_0)\cos[\theta] + (y - y_0)\sin[\theta] \\ &= (x\cos[\theta] + y\sin[\theta]) - (x_0\cos[\theta] + y_0\sin[\theta]) \\ &= \underline{\mathbf{r}} \bullet \underline{\hat{\mathbf{p}}} - \underline{\mathbf{r}}_0 \bullet \underline{\hat{\mathbf{p}}} = (\underline{\mathbf{r}} - \underline{\mathbf{r}}_0) \bullet \underline{\hat{\mathbf{p}}} \end{aligned} \tag{8.19a}$$

$$\begin{aligned} y'' &= -(x - x_0)\sin[\theta] + (y - y_0)\cos[\theta] \\ &= \underline{\mathbf{r}} \bullet \underline{\hat{\mathbf{p}}}^{\perp} - \underline{\mathbf{r}}_0 \bullet \underline{\hat{\mathbf{p}}}^{\perp} = (\underline{\mathbf{r}} - \underline{\mathbf{r}}_0) \bullet \underline{\hat{\mathbf{p}}}^{\perp} \end{aligned} \tag{8.19b}$$

If $\mathcal{O}_5$ is shift invariant, then these new coordinates must be identical to those obtained by translating the rotated coordinates x' and y' in Equation (8.18):

$$x' - x_0 = x\cos[\theta] - y\sin[\theta] - x_0 \neq x'' \tag{8.20a}$$

$$y' - y_0 = x\sin[\theta] + y\cos[\theta] - y_0 \neq y'' \tag{8.20b}$$

Because the coordinates generated by the two sequences of operations are not equal, then the process of coordinate rotation cannot be shift invariant.

That magnification and rotation are shift variant may be recognized at once by noting that the amplitude at the origin of coordinates is unchanged in both cases, though amplitudes at other locations are "moved" by amounts that vary with the distance from the fixed point. The test of shift invariance for object rotation is shown in Figure 8.6.

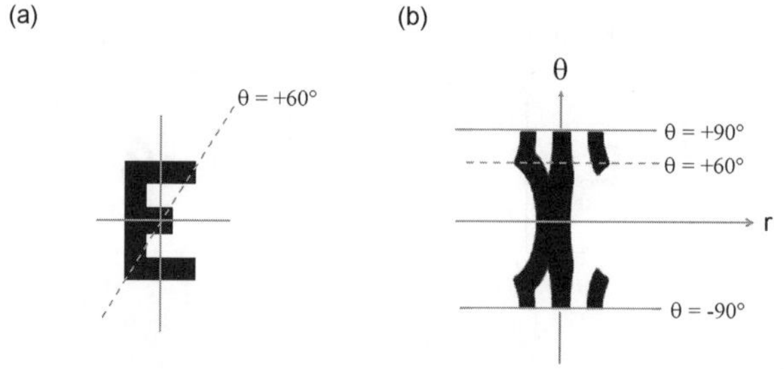

Figure 8.7 Transformation of input $f[x, y]$ from Cartesian to polar coordinates via Equation (8.21). The operation evidently is shift variant.

Just as nonlinear operations can be sometimes "made" linear by appropriate transformations of the amplitude, some shift-variant operations can be "made" shift invariant by appropriate change of coordinates. For example, a shift-variant rotation about the origin may be recast as a shift-invariant translation by transforming the input function to a polar coordinate representation. The amplitude at polar coordinates (r, θ) is plotted at Cartesian coordinates via the definition:

$$f[x, y] = f_1(r, \theta) \Longrightarrow f_1[x = r, y = \theta] \tag{8.21}$$

This process is shown in Figure 8.7. Note that the original operation of translation is not shift invariant after this transformation of coordinates.

8.3 Linear Shift-Invariant (LSI) Operators

We now consider the action of systems that are simultaneously linear and shift invariant. The combination of these two constraints ensures that the mathematical description of the system has a specific form that is useful for imaging systems and that facilitates solution of the inverse problem.

From the definition of linearity, we can calculate the output amplitude by scaling the output from an identical input function that has been normalized to unity. The property of shift invariance ensures that output function could be translated to any location in the domain. The output of a system that satisfies both constraints may be determined by translating, scaling, and summing the output generated from a "convenient" input function centered at the origin. In other words, the combined constraints of linearity and shift invariance allow the calculation of the output function to be simplified by decomposing the input function into the superposition of the outputs from a set of "basis" functions. The most common convenient decomposition of the input function uses the sifting property in Equation (6.113). The output from each of the scaled and translated Dirac delta functions is a scaled and translated replica of the output due to a unit-area Dirac delta function located at the origin:

$$\mathcal{O}\{\delta[x - 0]\} = \mathcal{O}\{\delta[x]\} \equiv h[x] \tag{8.22}$$

where $h[x]$ is the usual notation for this 1-D *impulse response*, so called for the obvious reason that it is the response of the system to an impulsive input. Since the system operator $\mathcal{O}$ is shift invariant, translation of the input results in an identical translation of the impulse response:

$$\mathcal{O}\{\delta[x - x_0]\} = h[x - x_0] \tag{8.23}$$

Because $\mathcal{O}$ also is linear, multiplicative scaling of the input results in the identical weighting applied to the output:

$$\begin{aligned}\mathcal{O}\{\alpha\,\delta[x-x_0]\} &= \alpha\,\mathcal{O}\{\delta[x-x_0]\}\\ &= \alpha\,h[x-x_0]\end{aligned} \tag{8.24}$$

As an aside, note that the response of the LSI system due to a pair of shifted and scaled impulses is the superposition of a pair of shifted and scaled impulse responses:

$$\mathcal{O}\{\alpha_0\,\delta[x-x_0]+\alpha_1\,\delta[x-x_1]\}=\alpha_0\,h[x-x_0]+\alpha_1\,h[x-x_1] \tag{8.25}$$

The general expression for the output of an LSI operator is obtained by decomposing the input function into its constituent set of weighted Dirac delta functions via the sifting property of the Dirac delta function in Equation (6.113):

$$f[x]=\int_{-\infty}^{+\infty} f[\alpha]\,\delta[x-\alpha]\,d\alpha \tag{8.26}$$

The response of a system $\mathcal{O}$ to this input is obtained by applying the operator:

$$g[x]=\mathcal{O}\{f[x]\}=\mathcal{O}\left\{\int_{-\infty}^{+\infty} f[\alpha]\,\delta[x-\alpha]\,d\alpha\right\} \tag{8.27}$$

The linearity property of the operator $\mathcal{O}$ allows it to be applied either after summing the inputs (i.e., outside the integral) or to the constituent Dirac delta functions before integrating:

$$\mathcal{O}\left\{\int_{-\infty}^{+\infty} f[\alpha]\,\delta[x-\alpha]\,d\alpha\right\}=\int_{-\infty}^{+\infty}\mathcal{O}\{f[\alpha]\,\delta[x-\alpha]\}\,d\alpha \tag{8.28}$$

Because the operator $\mathcal{O}$ acts on functions of x, the weighting term $f[\alpha]$ appears to $\mathcal{O}$ as a multiplicative constant. The linearity criterion allows the weighting to be applied after the action of the operator on $\delta[x-\alpha]$:

$$\int_{-\infty}^{+\infty}\mathcal{O}\{f[\alpha]\,\delta[x-\alpha]\}\,d\alpha=\int_{-\infty}^{+\infty} f[\alpha]\mathcal{O}\{\delta[x-\alpha]\}\,d\alpha \tag{8.29}$$

Finally, the shift invariance of $\mathcal{O}$ in Equation (8.23) is utilized to substitute the shifted replicas of the impulse response and the final result is obtained:

$$\int_{-\infty}^{+\infty} f[\alpha]\,\mathcal{O}\{\delta[x-\alpha]\}\,d\alpha=\int_{-\infty}^{+\infty} f[\alpha]h[x-\alpha]\,d\alpha \tag{8.30}$$

This expression for the output $g[x]$ of an LSI system arguably is the most important single equation in this book; it specifies the output of an LSI system as a specific integral – the *convolution* integral – of the input $f[x]$ and the impulse response $h[x]$. A common shorthand notation for the convolution of two 1-D functions $f[x]$ and $h[x]$ is an asterisk placed between the names of the two constituent functions:

$$g[x]=\int_{-\infty}^{+\infty} f[\alpha]h[x-\alpha]\,d\alpha\equiv f[x]*h[x] \tag{8.31}$$

Comparison of Equation (8.31) to Equation (8.26) demonstrates that the sifting property of the Dirac delta function is just the convolution of $f[x]$ with $\delta[x]$.

This output $h[x]$ of the system resulting from an impulsive input at the origin ($f[x]=\delta[x-0]$) is a very important descriptor of the system, so much so that it has been assigned different names by different scientific disciplines. Mathematicians may call $h[x]$ the *kernel* of the convolution, electrical engineers refer to it as the *impulse response* of the system, and optical scientists have named it the *point-spread function* (or *psf*).

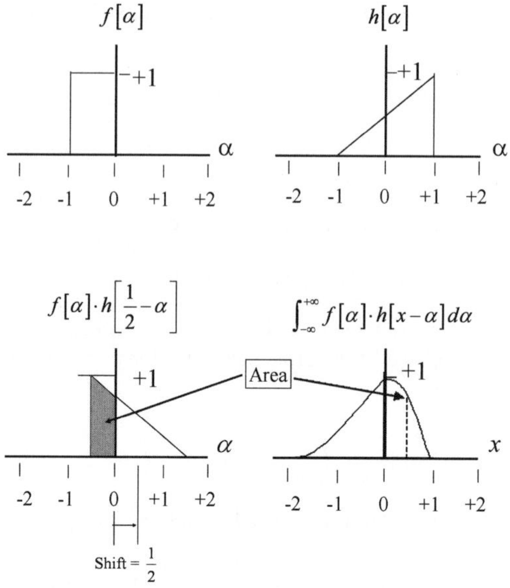

Figure 8.8 Demonstration of convolution: the convolution $f[x] * h[x]$ is evaluated at $x = \frac{1}{2}$.

The process of convolution is often called *linear filtering* of the input signal by the system. Note that the output of a convolution computed at a single coordinate (such as $g[x_0]$) is obtained by summing the products obtained by point-by-point multiplication of the input $f[\alpha]$ and a copy of $h[\alpha]$ that has been "reversed" (or "flipped" or "folded" about the origin) and then translated to place its central ordinate at $\alpha = x_0$. If both $f[x]$ and $h[x]$ have infinite support, then the amplitude g of the output at that single coordinate x_0 depends on the amplitudes of the constituent functions at all coordinates. Mathematicians refer to the output $g[x]$ as a "functional" of $f[x]$ and $h[x]$; this name distinguishes the convolution from the simple "function" of a single coordinate.

Note that the converse of this mathematical relationship also is true: any system that may be described by a convolution integral must be both linear and shift invariant.

The individual steps of the computation of the convolution integral are:

1. Copy $f[x]$ and $h[x]$ to the α-domain to create $f[\alpha]$ and $h[\alpha]$.
2. "Fold", "reflect", or "reverse" the impulse response $h[\alpha]$ about the origin to obtain $h[-\alpha]$.
3. Translate $h[-\alpha]$ by the distance x_0 to generate $h[-\alpha + x_0] = h[x_0 - \alpha]$.
4. Multiply $h[x_0 - \alpha]$ by $f[\alpha]$ to generate the product function.
5. Compute the area by integrating over the α-domain.
6. Assign this area to the output amplitude g at coordinate x_0.
7. Loop to step 3 for a new output coordinate x_1 and repeat for all values of x.

The graph of $g[x]$ is obtained after evaluating and plotting the area for all locations in the domain of the output $g[x]$, where $-\infty < x < +\infty$ (Figure 8.8).

The "reversal" (or "folding" or "flipping") operation in step 2 is the source of the German name of *Faltung* (folding) integral for the convolution. Other less commonly used names for convolution are the *composition product* and *running mean*.

The product of the two component functions (step 5) is often misconstrued by students because of special properties of the constituent functions in common examples used to illustrate the convolution process. In these examples, either the input $f[x]$ or the impulse response $h[x]$ (or both) is a unit-amplitude rectangle function. Consider the specific case where the impulse response $h[x] = RECT[x/b]$. The sole effect of $h[x]$ in the multiplication of step 4 is to truncate $f[\alpha]$ to the region between $x_0 - b/2$ and $x_0 + b/2$. The convolution evaluated at x_0 is merely the area of $f[\alpha]$ within this "window". The potential for confusion in the minds of students may be reduced by using a more general case to illustrate the action of an LSI system.

The results of this section may be summarized in a single sentence: "the output of any linear shift-invariant system is the convolution of the input with the impulse response of the system". This statement encapsulates the reason why the simplification of an arbitrary system operator to the LSI case is so desirable; the action of any N-D LSI system may be specified completely by a single N-D function – the impulse response. In fact, the "character" (shape) of the impulse response function determines much about the action of the imaging system, and we will expend significant effort to classify the action of systems based on this shape in the subsequent chapters on linear filtering (Chapters 16–19). However, a few words on the subject will be useful at this point to describe the general properties of "differencing" and "averaging" filters.

As mentioned in the discussion of Equation (8.31), the convolution integral reduces to the sifting property when $h[x]$ is a Dirac delta function:

$$g[x] = f[x] * \delta[x] = \int_{-\infty}^{+\infty} f[\alpha]\,\delta[x-\alpha]\,d\alpha = f[x] \tag{8.32}$$

In turn, this result may be used to prove several identities of the Dirac delta function:

$$\delta[x] * \delta[x] = \delta[x] \tag{8.33a}$$

$$\delta[x - x_0] * \delta[x] = \delta[x] * \delta[x - x_0] = \delta[x - x_0] \tag{8.33b}$$

$$\delta[x - x_0] * \delta[x - x_1] = \delta[x - x_0 - x_1] = \delta[x - (x_0 + x_1)] \tag{8.33c}$$

It also is easy to use the convolution integral to derive the sifting property for the derivative of the Dirac delta function by setting $h[x] = d\delta[x]/dx \equiv \delta'[x]$ in the convolution integral in Equation (8.31):

$$\begin{aligned} g[x] = f[x] * \delta'[x] &= \int_{-\infty}^{+\infty} f[\alpha]\frac{d}{dx}(\delta[x-\alpha])\,d\alpha \\ &= \int_{-\infty}^{+\infty} \frac{d}{dx}(f[\alpha]\,\delta[x-\alpha])\,d\alpha \quad \textit{by linearity} \\ &= \frac{d}{dx}\int_{-\infty}^{+\infty} (f[\alpha]\,\delta[x-\alpha])\,d\alpha \quad \textit{by linearity} \\ &= \frac{d}{dx} f[x] \equiv f'[x] \quad \textit{by the sifting property} \end{aligned} \tag{8.34}$$

In words, the derivative of the input function $f[x]$ is obtained by convolving with $\delta'[x]$. This expression may be generalized to compute the nth derivative of $f[x]$ via convolution of $f[x]$ with the nth derivative of $\delta[x]$:

$$\frac{d^n}{dx^n}(f[x]) = \int_{-\infty}^{+\infty} f[\alpha]\,\delta^{(n)}[x-\alpha]\,d\alpha = f[x] * \delta^{(n)}[x] \tag{8.35}$$

which was already surmised in Equation (6.132) for the derivative of a Dirac delta function consisting a "doublet" of Dirac delta functions with opposite sign and infinitesimal separation. The convolution of $f[x]$ with $\delta'[x]$ computes differences in the amplitude of $f[x]$ at coordinates separated by an infinitesimal distance. In other words, convolution of $f[x]$ with $\delta'[x]$ is a differencing operation over a very small local neighborhood.

As another example of convolution, consider the impulse response to be a unit-area rectangle function with support b:

$$h[x] = \frac{1}{|b|} RECT\left[\frac{x}{b}\right] \tag{8.36}$$

The resulting convolution $g[x]$ is:

$$g[x] = f[x] * h[x] = \int_{-\infty}^{+\infty} \frac{1}{|b|} RECT\left[\frac{x-\alpha}{b}\right] f[\alpha]\, d\alpha$$
$$= \frac{1}{|b|} \int_{x-b/2}^{x+b/2} f[\alpha]\, d\alpha \tag{8.37}$$

which is the uniformly weighted average of the input function $f[x]$ over a domain of width b. This describes a "local averaging operation" that tends to "smooth out" variations in amplitude of the input over distances smaller than b. Any impulse response with unit area, finite support, and with all nonnull amplitudes being positive (or negative) acts as a local averager.

For illustration, consider the convolution of this unit-area finite-support rectangle with an input function composed of a sinusoid with amplitude A_1 and an additive bias A_0. The bias satisfies the condition $A_0 \geq A_1$ to ensure that the input function is nonnegative. From Equation (6.25), the modulation of the input sinusoid is $m_f = A_1/A_0$. The convolution is straightforward to evaluate:

$$g[x] = (A_0 + A_1 \cos[2\pi\xi_0\alpha + \phi_0]) * \frac{1}{|b|} RECT\left[\frac{x}{b}\right]$$
$$= \frac{A_0}{|b|} \int_{x-b/2}^{x+b/2} 1[\alpha]\, d\alpha + \frac{A_1}{|b|} \int_{x-b/2}^{x+b/2} \cos[2\pi\xi_0\alpha + \phi_0]\, d\alpha$$
$$= \frac{A_0}{|b|} |b| + \frac{A_1}{|b|} \frac{\sin[2\pi\xi_0\alpha + \phi_0]}{2\pi\xi_0}\Bigg|_{\alpha=x-b/2}^{\alpha=x+b/2}$$
$$= A_0 + A_1\, SINC[\xi_0 \cdot b] \cos[2\pi\xi_0 x + \phi_0] \tag{8.38}$$

The process is illustrated in Figure 8.9 for different values of b. Because convolution is linear, the same result may be obtained by separately convolving the constant and oscillating parts of the input function and adding the results. The constant part of $f[x]$ (its DC component) is unchanged by the convolution. This should be no surprise because the average value of a constant over any interval is still the constant. The convolution of the oscillating part differs from the input by the scale factor $SINC[\xi_0 \cdot b]$. Of course, since the constant A_0 may be interpreted as a symmetric sinusoid with amplitude A_0 and spatial frequency $\xi_0 = 0$, then the expression for the scale factor of the oscillating part of $f[x]$ is valid also for the constant part, i.e., $SINC[0 \cdot b] = 1$.

The uniform average of the two parts of the biased sinusoid may be interpreted in the light of the discussion of eigenfunctions in Chapter 4. Since the uniform average of the individual parts results in the same function multiplied by appropriate scale factors, then these sinusoids satisfy the condition for eigenfunctions that was specified in Equation (5.103). In fact, any infinite-support sinusoid is an eigenfunction of any convolution, regardless of the form of the impulse response. This will be proven in Chapter 16.

The modulation of $g[x]$ in Equation (8.38) is easy to evaluate by applying Equation (6.27):

$$m_g = \left(\frac{g_{\max} - g_{\min}}{g_{\max} + g_{\min}}\right) = \frac{A_1\, SINC[\xi_0 \cdot b]}{A_0} = m_f \cdot SINC[\xi_0 \cdot b] \tag{8.39}$$

In words, it is the product of the modulation of the input sinusoid and the amplitude of the appropriate *SINC* function. Simple rearrangement of this result leads to a description of the action of this linear and shift-invariant "local averaging" system upon the input sinusoid:

$$\frac{m_g}{m_f} = SINC[\xi_0 \cdot b] \tag{8.40}$$

The ratio of output and input modulations at a particular spatial frequency ξ_0 measures "how well" the specific input sinusoid is "transferred" to the output by this LSI system; the output modulation varies

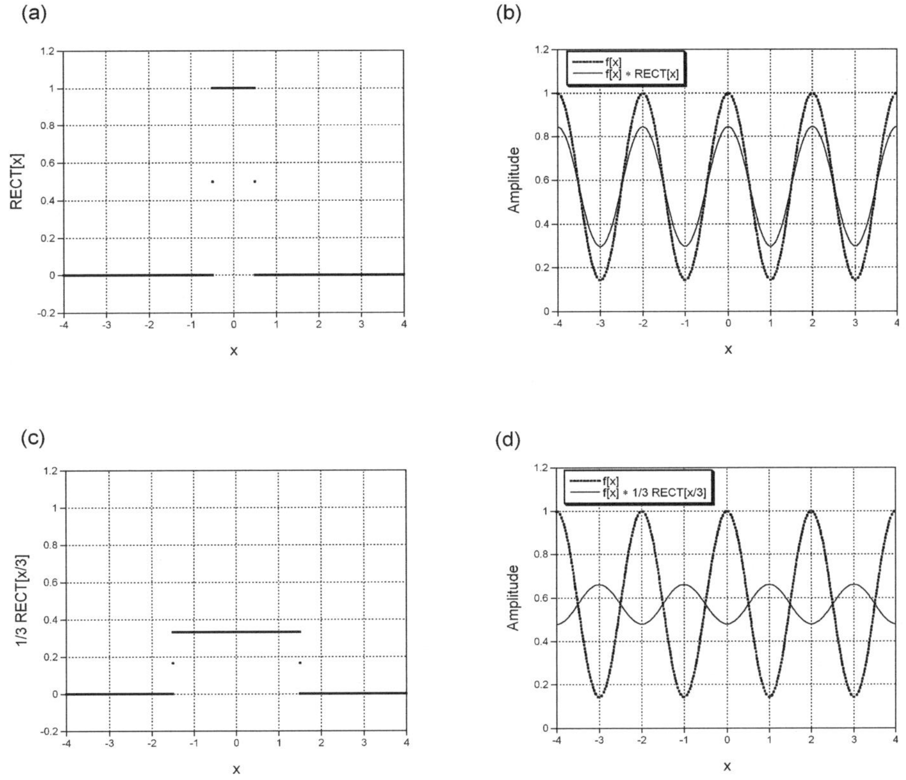

Figure 8.9 Convolution of a biased sinusoid with different unit-area rectangle functions. The sinusoidal function is $f[x] = \frac{4}{7} + \frac{3}{7}\cos[2\pi x/2]$, which has modulation $m_f = \frac{3}{4}$: (a) $h_1[x] = RECT[x]$; (b) $f[x]$ and $g_1[x] = f[x] * h_1[x]$, showing the reduction in modulation due to the averaging; (c) $h_2[x] = \frac{1}{3}RECT[x/3]$; (d) $g_2[x] = f[x] * h_2[x]$, showing the reduction in modulation and the change of phase by $-\pi$ radians.

with spatial frequency ξ_0 and width b. We can generate a complete descriptor of this LSI system by evaluating this "modulation transfer" for biased sinusoidal input functions with all spatial frequencies ξ, but we can see that it is determined by the value of ξ_0 in Equation (8.40). The graph of the modulation transfer of the system versus the spatial frequency of the sinusoids is the *modulation transfer function* (or *MTF*) of the imaging system, and will be considered in much more detail in following chapters.

Since $SINC[\xi_0 \cdot b] \leq 1$, $m_g < m_f$ in Equation (8.40) unless $\xi_0 = 0$ and/or $b = 0$. The former condition implies that the original function $f[x]$ is a constant of amplitude A_0, and therefore exhibits no modulation at all. The constraint $b = 0$ implies that the uniform averager has infinitesimal support and finite area, i.e., that $h[x] = \delta[x]$. In this case, the integral in Equation (8.38) is merely a statement of the sifting property of the Dirac delta function, and the output is identical to the input ($m_g = m_f$). The character of the convolution operations that are generated by different types of kernels will be discussed in detail in Chapter 16.

8.3.1 Linear Shift-Variant Operators

The definition of convolution in Equation (8.31) may be generalized to construct the general mathematical expression for linear and shift-variant (LSV) operations. The linearity ensures that the mathematical operation is a superposition that may be expressed as an integral. Shift variance may be modeled by assigning different impulse response functions to different locations in the output domain. In other words, each location in the 1-D output domain specified by x has an associated specific 1-D impulse response $h[\alpha]$, and therefore the space-variant analogue of the 1-D impulse response is the 2-D function $h[\alpha; x]$, which we just call $h[x, \alpha]$. The expression for the operation is:

$$\mathcal{O}\{f[x]\} \equiv g[x] = \int_{-\infty}^{+\infty} f[\alpha]h[x, \alpha]\, d\alpha \tag{8.41}$$

This is the 1-D *superposition integral*; note its similarity to the transformation between representations of a function based on different sets of basis functions that was presented in Equation (5.94). Also note that the independent variables x and α may have different dimensions ("units"), so that the coordinates of the input function $f[\alpha]$ and $g[x]$ may be different in a shift-variant operation. The similarity between Equation (5.94) and Equation (8.41) suggests an interpretation of space-variant superposition that is very similar to the inner product of $f[x]$ with a set of different 1-D functions, except for the explicit complex conjugation in Equation (5.94). The analogous process to Equation (8.41) for discrete vectors without the complex conjugation is the general form of matrix–vector multiplication already defined in Equation (3.23). This interpretation may be extended to obtain the discrete vector analogy of LSI convolution in Equation (8.31) as the product of a vector with a circulant matrix.

Note the difference between LSI convolution of two 1-D functions in Equation (8.31) and LSV superposition of a 1-D function with a 2-D function in Equation (8.41). The latter is analogous to matrix–vector multiplication in Equation (3.23), while the former is multiplication of a vector by a circulant matrix whose rows are replicas of $h[x]$ after shifting and "folding". The analogous process for discrete vectors would be the computation of the scalar product in Equation (3.9) of the input vector $\underline{\mathbf{x}}$ and a translated copy $\underline{\mathbf{x}}'$.

8.4 Calculating Convolutions

The complicated form of the convolution integral in Equation (8.30) ensures that analytic expressions for convolutions exist for only a few pairs of special functions. However, we often can evaluate some parts of the output. For example, if the two constituent functions have finite support, their convolution is null when evaluated at all translations that exceed the sum of their widths, as in the simple case already mentioned where $f[x] = h[x] = RECT[x]$; if the translation parameter x is less than -1 or greater than $+1$, the overlap of the component functions is zero and thus the product function has null amplitude. The particular convolution does have an analytic form, as seen by decomposing the integral into two cases:

$$\begin{aligned}
g[x] = RECT[x] * RECT[x] &= \int_{-\infty}^{+\infty} RECT[\alpha]\, RECT[x-\alpha]\, d\alpha \\
&= \begin{cases} \displaystyle\int_{-\frac{1}{2}}^{x+\frac{1}{2}} 1\, d\alpha = \left(x + \tfrac{1}{2}\right) + \tfrac{1}{2} = 1 + x & \textit{if } -1 < x < 0 \\ \displaystyle\int_{x-\frac{1}{2}}^{\frac{1}{2}} 1\, d\alpha = \tfrac{1}{2} - (x - \tfrac{1}{2}) = 1 - x & \textit{if } 0 < x < +1 \\ 0 & \textit{otherwise} \end{cases} \\
&= (1 - |x|)\, RECT\left[\frac{x}{2}\right] = TRI[x]
\end{aligned} \tag{8.42}$$

In words, the convolution of $RECT[x]$ with itself (what we might call the *autoconvolution*) yields the triangle function $TRI[x]$. Though straightforward, this example hints at the most tedious and difficult part of computing convolutions by direct integration; it is essential to keep track of the proper limits of the integrals that are valid for different regions of the domain of $g[x]$.

tions

of convolution, we will consider several examples with
nvolution of the decaying exponential in Equation (6.16):

$$x]) * (e^{-x} \cdot STEP[x])$$

$$\begin{aligned}
&= \int_{-\infty}^{+\infty} (e^{-\alpha}\, STEP[\alpha])(e^{-(x-\alpha)}\, STEP[x-\alpha])\, d\alpha \\
&= \int_{-\infty}^{+\infty} STEP[\alpha]\, STEP[x-\alpha]\, e^{-x}(e^{+\alpha}\, e^{-\alpha})\, d\alpha \\
&= e^{-x} \int_{-\infty}^{+\infty} STEP[\alpha]\, STEP[x-\alpha]\, d\alpha \\
&= e^{-x} \int_{0}^{+\infty} STEP[x-\alpha]\, d\alpha
\end{aligned} \tag{8.43}$$

where $STEP[\alpha]$ constrains the lower limit of the integral to $\alpha = 0$. Because the integral is over α, Equation (6.11) demonstrates that the second *STEP* function is centered at $\alpha = x$ and "reversed" so that it has unit amplitude for $\alpha < x$. Thus the integral in Equation (8.43) may be evaluated easily; the output is null for $x < 0$ because the two functions do not overlap:

$$\begin{aligned}
g_1[x] &= (e^{-x} \cdot STEP[x]) * (e^{-x} \cdot STEP[x]) \\
&= e^{-x} \int_{0}^{x} d\alpha = \begin{cases} x \cdot e^{-x} & \textit{if } x > 0 \\ 0 & \textit{if } x \leq 0 \end{cases}
\end{aligned} \tag{8.44}$$

which is shown in Figure 8.10. The output may be expressed in concise form by reapplying the definition of $STEP[x]$ to restrict $x \cdot e^{-x}$ to positive values of x:

$$g_1[x] = (e^{-x} \cdot STEP[x]) * (e^{-x} \cdot STEP[x]) = x \cdot (e^{-x} \cdot STEP[x]) \tag{8.45}$$

Another, perhaps nonintuitive, example of convolution is the case of two quadratic-phase functions with opposite signs, which may be solved directly by expanding the exponent:

$$\begin{aligned}
g_2[x] = e^{-i\pi x^2} * e^{+i\pi x^2} &= \int_{-\infty}^{+\infty} e^{-i\pi\alpha^2}\, e^{+i\pi(x-\alpha)^2}\, d\alpha \\
&= \int_{-\infty}^{+\infty} e^{-i\pi\alpha^2}\, e^{+i\pi x^2}\, e^{+i\pi\alpha^2}\, e^{-2\pi i\alpha x}\, d\alpha = e^{+i\pi x^2} \int_{-\infty}^{+\infty} e^{-2\pi i\alpha x}\, d\alpha \\
&= e^{+i\pi x^2}\, \delta[x] = e^{+i\pi x^2}\, \delta[x-0] = e^{+i\pi 0^2}\, \delta[x-0] = \delta[x]
\end{aligned} \tag{8.46}$$

where the integral formulation for the Dirac delta function in Equation (6.106) and the property of Dirac delta functions in products from Equation (6.115) have been used. In words, this particular convolution has infinitesimal support even though both constituent functions have infinite support. As we will see, this observation has great value when applied to many imaging systems, including radar and optical imaging.

8.5 Properties of Convolutions

We will now derive some theorems and analyze some properties of the convolution of 1-D continuous functions. The theorems are easily extended to discrete 1-D and to continuous and discrete 2-D functions.

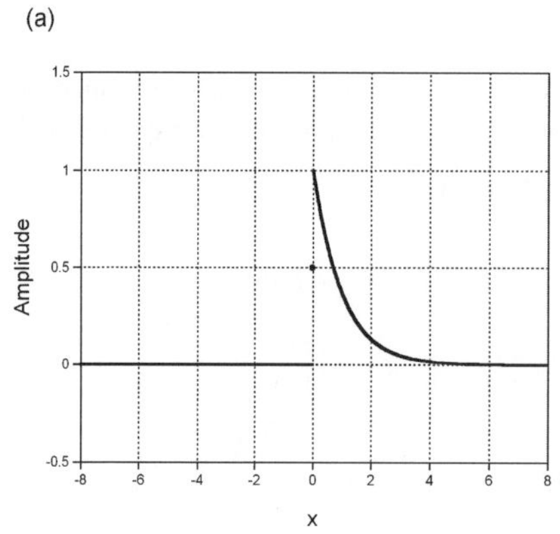

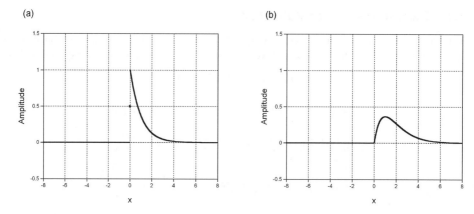

Figure 8.10 Autoconvolution of the decaying exponential: (a) $e^{-x} \cdot STEP[x]$; (b) $(e^{-x} \cdot STEP[x]) * (e^{-x} \cdot STEP[x]) = x(e^{-x} \cdot STEP[x])$.

The defining equation for the convolution of two continuous functions $f[x]$ and $h[x]$ in Equation (8.31) may be rewritten by changing the integration variable from α to $u = x - \alpha$, so that $\alpha = x - u$ and $du = -d\alpha$:

$$\begin{aligned} f[x] * h[x] &= \int_{u=+\infty}^{u=-\infty} f[x-u]h[u](-du) \\ &= \int_{u=-\infty}^{u=+\infty} f[x-u]\,h[u](+du) \\ &= \int_{-\infty}^{+\infty} h[\alpha]\,f[x-\alpha](+d\alpha) = h[x] * f[x] \end{aligned} \tag{8.47}$$

This demonstration that convolution is commutative means that the roles of the input and the impulse response are interchangeable in LSI systems; the same result is obtained by computing the response of the "system" $f[x]$ to the "stimulus" $h[x]$. This fact leads to an important observation about the imaging problems introduced in Chapter 1 for LSI systems. Both the inverse and system analysis tasks assume that the output $g[x]$ is known. In the former problem, the system function $h[x]$ is known, while the latter assumes that the input $f[x]$ is available. Because the roles of $f[x]$ and $h[x]$ are interchangeable, the LSI inverse and system analysis tasks are equivalent problems and may be solved by the same means. The solution to these for LSI systems will be discussed in detail in Chapter 19.

Because integration over fixed limits is a linear operation, then convolution with a specific impulse response $h[x]$ also is linear. One result of this observation is that amplification of the input amplitude by a complex-valued constant α results in equal amplification of the output:

$$\begin{aligned} &f[x] * h[x] = g[x] \\ &\Longrightarrow \alpha \cdot f[x] * h[x] = f[x] * \alpha \cdot h[x] = \alpha \cdot g[x] \end{aligned} \tag{8.48}$$

The linearity of convolution also means that it is distributive with respect to addition:

$$\begin{aligned} &(\alpha_1 f_1[x] + \alpha_2 f_2[x]) * (\beta_1 h_1[x] + \beta_2 h_2[x]) \\ &\quad = \alpha_1\beta_1(f_1[x] * h_1[x]) + \alpha_2\beta_1(f_2[x] * h_1[x]) + \alpha_1\beta_2(f_1[x] * h_2[x]) + \alpha_2\beta_2(f_2[x] * h_2[x]) \end{aligned} \tag{8.49}$$

Because the value of an integral with infinite limits is unchanged if the integrand is translated by a finite distance, the process of convolution also is shift invariant:

$$\begin{aligned}\mathcal{O}\{f[x]\} &\equiv f[x] * h[x] = g[x] \\ &\Longrightarrow \mathcal{O}\{f[x-x_0]\} = f[x-x_0] * h[x] = g[x-x_0]\end{aligned} \tag{8.50}$$

Convolutions also are associative:

$$(f_1[x] * f_2[x]) * f_3[x] = f_1[x] * (f_2[x] * f_3[x]) \tag{8.51}$$

This result may be combined with the commutativity of convolutions to show that cascades of convolutions may be computed in any order:

$$(f_1[x] * f_2[x]) * f_3[x] = f_3[x] * (f_2[x] * f_1[x]) \tag{8.52}$$

8.5.1 Region of Support of Convolutions

By examining the integral formulation for the convolution of $f[x]$ and $h[x]$, it can be shown that the support of the convolution is the sum of the supports of the constituent functions *if both have compact support*. This property has been confirmed by the autoconvolution of $RECT[x]$ (each with unit support) to generate $TRI[x]$ (whose support is two units). In general, if $f[x]$ and $h[x]$ have support b_f and b_h respectively, then the support of $g[x]$ is $b_g = b_f + b_h$. However, this simple expression for the support of a convolution fails if the support of either (or both) of the constituent functions is infinite or infinitesimal. We can apply Equation (8.34) to solve for the convolution of $f[x] = 1[x]$ (with $b_f = \infty$) and $h[x] = \delta'[x]$ (with $b_h \to 0$). The result is the derivative of $1[x]$:

$$g[x] = 1[x] * \delta'[x] = \frac{d}{dx}(1[x]) = 0[x] \tag{8.53}$$

Therefore, the support $b_g = 0$, and the support of the convolution clearly is not the sum of the supports of the input functions. We also have demonstrated that the convolution of two infinite-support complex quadratic-phase functions in Equation (8.46) is:

$$f[x] * h[x] = \exp[+i\pi x^2] * \exp[-i\pi x^2] = \delta[x] \tag{8.54}$$

In words, the convolution of these two particular functions with infinite support yields an output with finite area and infinitesimal (zero) support. In short, the region of a convolution must be considered with care.

8.5.2 Area of a Convolution

The area of the convolution of two 1-D functions $f[x]$ and $h[x]$ can be found by direct integration of Equation (8.27):

$$\begin{aligned}\int_{-\infty}^{+\infty} g[x]\,dx &= \int_{-\infty}^{+\infty} (f[x] * h[x])\,dx \\ &= \int_{-\infty}^{+\infty}\left(\int_{-\infty}^{+\infty} f[\alpha]h[x-\alpha]\,d\alpha\right)dx \\ &= \int_{-\infty}^{+\infty} f[\alpha]\left(\int_{-\infty}^{+\infty} h[x-\alpha]\,dx\right)d\alpha \\ &= \left(\int_{-\infty}^{+\infty} f[x]\,dx\right)\left(\int_{-\infty}^{+\infty} h[x]\,dx\right)\end{aligned} \tag{8.55}$$

where the invariance of an integral over infinite limits to translation of the integrand has been used and the dummy variable of integration has been renamed in the last step. In words, the area of a convolution is the product of the areas of the component functions.

8.5.3 Convolution of Scaled Functions

A not-so-intuitive property of convolutions is demonstrated by changing the scale of both $f[x]$ and $h[x]$ by the same factor b. Given that $f[x] * h[x] = g[x]$, then:

$$\begin{aligned} f\left[\frac{x}{b}\right] * h\left[\frac{x}{b}\right] &= \int_{-\infty}^{+\infty} f\left[\frac{\alpha}{b}\right] h\left[\frac{x-\alpha}{b}\right] d\alpha \\ &\equiv \int_{-\infty}^{+\infty} f[\beta] h\left[\frac{x}{b} - \beta\right] (|b|\, d\beta) = |b| g\left[\frac{x}{b}\right] \end{aligned} \tag{8.56a}$$

where the integration variable was changed to $\beta \equiv \alpha/b$ in the second step. Thus changes of scale applied to both the input and kernel change both the scale and the amplitude of the output function by the same scale factor; if the input and kernel are made "wider" by the same factor, the output is both "wider" and "taller" by that factor. The result is perhaps more clearly stated as:

$$f\left[\frac{x}{b}\right] * h\left[\frac{x}{b}\right] = |b| (f[u] * h[u])|_{u=x/b} \tag{8.56b}$$

This result may be extended to compute cascaded convolutions of functions scaled in width by identical factors. If $g[x]$ is obtained by the cascaded convolution of N functions $f_n[x]$, then the cascade of N replicas of $f[x/b]$ yields:

$$f_1\left[\frac{x}{b}\right] * f_2\left[\frac{x}{b}\right] * \cdots * f_N\left[\frac{x}{b}\right] = |b|^{N-1} g\left[\frac{x}{b}\right] \tag{8.57}$$

8.6 Autocorrelation

At this point we define a 1-D operator that resembles convolution, but that differs in a few significant details that may cause confusion. This operator is very useful in certain signal processing applications, such as finding the position of a known signal in an input object. This operator will be introduced as a special case and then generalized.

Recall that the inner product of an N-D discrete vector with complex components with itself was defined in Equation (5.2) for vectors with complex-valued components:

$$|\underline{\mathbf{x}}|^2 = \underline{\mathbf{x}}^* \bullet \underline{\mathbf{x}} = \sum_{n=1}^{N} |x_n|^2 \tag{8.58}$$

This inner product of $\underline{\mathbf{x}}$ with itself is the square of the length of the vector and is guaranteed to be real valued. The analogous construction for a complex-valued continuous function $f[x]$ is the integral of the squared magnitude:

$$\int_{-\infty}^{+\infty} (f[\alpha])^* f[\alpha]\, d\alpha = \int_{-\infty}^{+\infty} f[\alpha](f[\alpha])^*\, d\alpha = \int_{-\infty}^{+\infty} |f[\alpha]|^2\, d\alpha \tag{8.59}$$

which also must be real valued. In the terminology used for vectors, this operation yields a numerical value that is proportional to the *projection* of $f[x]$ onto itself, and as such is the analogue for the square of the "length" of a function.

Equation (8.59) may be generalized by translating the reference function by some distance x. The integral then computes the projection of the input function $f[x]$ onto a "shifted replica" of $f[x]$:

$$\int_{-\infty}^{+\infty} f[\alpha](f[\alpha - x_0])^*\, d\alpha = \int_{-\infty}^{+\infty} f[\alpha] f^*[\alpha - x_0]\, d\alpha \tag{8.60a}$$

The projection may be computed for all possible values of the translation parameter x, and the operation may be recast into the form of the convolution integral in Equation (8.31) by simple algebraic manipulation:

$$\int_{-\infty}^{+\infty} f[\alpha] f^*[\alpha - x]\, d\alpha = \int_{-\infty}^{+\infty} f[\alpha] f^*[-(x-\alpha)]\, d\alpha = f[x] * f^*[-x] \tag{8.60b}$$

From the interpretation of Equation (8.59) as a projection, Equation (8.60b) provides a measure of the "similarity" of $f[x]$ to its translated replica. Obviously, any function is most similar to its untranslated replica, so the amplitude of Equation (8.60b) must be largest for $x = 0$.

The operation defined in Equation (8.60) is the *autocorrelation* of $f[x]$ and is often signified by the "pentagram" (five-pointed star) "★":

$$\boxed{f[x] * f^*[-x] \equiv f[x] \bigstar f[x]} \tag{8.61}$$

This definition explicitly includes the complex conjugation of the "reference" function, unlike that of some authors (e.g., Gaskill). We use this definition to maintain the analogy with the length of complex-valued vectors.

Note that the autocorrelation integral in Equation (8.60) is not defined in many cases where $f[x]$ has infinite support or its integrated power is infinite. The trivial example of such a case is $f[x] = 1[x]$, but it also is true that the autocorrelation does not exist for some finite-amplitude periodic functions. If $f[x]$ has infinite support, the autocorrelation may be redefined to compute the "average" area of the product function over a finite but indefinite region of support and then extrapolated to infinite limits:

$$(f[x] \bigstar f[x])_{\text{infinite support}} = \lim_{B\to\infty} \left\{ \frac{1}{2B} \int_{-B}^{+B} f[\alpha] f^*[\alpha - x]\, d\alpha \right\} \tag{8.62}$$

This assumes that the expression converges in the limit $B \to \infty$.

In cases where $f[x]$ is periodic, a variant of the autocorrelation may be defined that is computed over a single period only:

$$(f[x] \bigstar f[x])_{\text{periodic}} \equiv \frac{1}{X} \int_{-\infty}^{+\infty} \left(f[\alpha] RECT\left[\frac{\alpha}{X}\right] \right) \left(f^*[\alpha - x] RECT\left[\frac{\alpha - x}{X}\right] \right) d\alpha \tag{8.63}$$

where X is the period of $f[x]$.

Because the maximum amplitude of the autocorrelation always occurs at the origin (no translation of the input function $f[x]$ relative to the "reference"), comparisons of the autocorrelations of different functions are facilitated by normalizing by the integrated power to yield unit amplitude at the origin. A common notation for the normalized autocorrelation is γ_{ff}, where the doubled subscript signifies that the same function is used as the "input" and as the "reference":

$$\gamma_{ff}[x] \equiv \frac{\displaystyle\int_{-\infty}^{+\infty} f[\alpha] f^*[\alpha - x]\, d\alpha}{\displaystyle\int_{-\infty}^{+\infty} |f[\alpha]|^2\, d\alpha} = \frac{f[x] \bigstar f[x]}{(f[x] \bigstar f[x])|_{x=0}} \tag{8.64}$$

Both the numerator and denominator in Equation (8.64) may be computed in the limit of Equation (8.62) if $f[x]$ has infinite support.

Any qualities of symmetry of the autocorrelation are determined by the character of $f[x]$. If $f[x]$ is either real valued or complex valued, then the operation of complex conjugation applied to the "reference" ensures that $\gamma_{ff}[x]$ will be real valued and symmetric with respect to the origin (i.e., $\gamma_{ff}[x] = \gamma_{ff}^*[x] = \gamma_{ff}[-x]$). Because the projection of a complex function onto itself at zero shift yields a real-valued "length", then $\gamma_{ff}[0]$ is guaranteed to be real. For the case of a complex-valued $f[x]$, then "cross-terms" arising from the product of the real part of $f[x]$ and the imaginary part of $f^*[x]$ generate an imaginary part in $\gamma_{ff}[x]$ that must be zero at $x = 0$ and form an odd function. In short, the

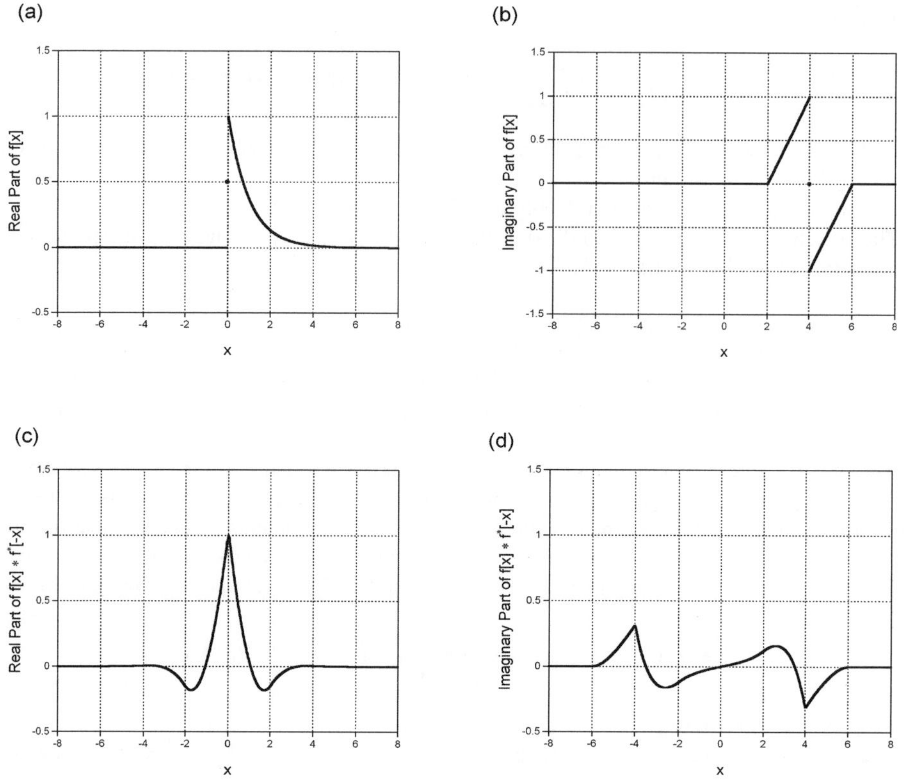

Figure 8.11 The autocorrelation of a complex-valued function $f[x]$ is Hermitian: (a) $\Re\{f[x]\} = e^{-(x+2)} \cdot STEP[x+2]$; (b) $\Im\{f[x]\} = -TRI[x-2] \cdot SGN[x-2]$; (c) $\Re\{f[x] \bigstar f[x]\}$ is even; (d) $\Im\{f[x] \bigstar f[x]\}$ is odd.

autocorrelation of a complex-valued $f[x]$ is Hermitian (even real part, odd imaginary part). An example is shown in Figure 8.11.

The interpretation of the autocorrelation demonstrates that the product of a function that is periodic over an interval X_0 with an unshifted replica of itself must be identical to the product with a replica that has been translated by a distance nX_0, where n is any integer. Therefore the generalized autocorrelation of a periodic function must be periodic over the same interval X_0.

The shape of the autocorrelation is a measure of the spatial "variability" of the function. Consider an input function $f[x]$ that varies "slowly" with x. In this case, the area of the product function also will vary "slowly" with translation parameter, so the autocorrelation "peak" is broad. Conversely, a function whose amplitude consistently varies over short distances will have a narrow autocorrelation peak. The autocorrelation and Fourier transform of a function $f[x]$ are intimately related via the Wiener–Khintchin theorem, which is considered in Chapter 9.

8.6.1 Autocorrelation of Stochastic Functions

One of the more useful applications of the autocorrelation is to describe the characteristics of stochastic functions. For example, consider a realization of Gaussian noise $f_1[x]$ with mean $\mu = 0$ and variance $\sigma^2 = 1$. If the noise function is truly random, then no systematic relationship between amplitudes at "adjacent" coordinates is expected; the central peak of the resulting autocorrelation function must be "narrow". However, if the amplitudes of the noise function at adjacent coordinates are "correlated", then the autocorrelation of the noise function will have a "wider" peak because the products of the amplitude of the function and a translated replica will be more likely to have the same sign.

In many cases of interest in imaging, the stochastic noise function is nonnegative everywhere. For example, photon-counting applications produce arrays of nonnegative integers that are selected from one of the discrete laws (binomial or Poisson) which must be nonnegative. In all such cases, the average amplitude of any nonnegative noise is itself guaranteed to be nonnegative. The autocorrelation of such a noise function will exhibit a large peak at zero shift due to the overlap of these positive values. For this reason, a variation of the autocorrelation called the "autocovariance" is often convenient.

8.6.2 Autocovariance of Stochastic Functions

The *autocovariance* of a signal is the autocorrelation of the bipolar function created by subtracting the mean amplitude $\langle f \rangle$. The expression is simplified by applying the generalized form of the autocorrelation integral in Equation (8.63):

$$\begin{aligned}
C_{ff}[x] &\equiv (f[x]-\langle f\rangle) \bigstar (f[x]-\langle f\rangle) = \int_{-\infty}^{+\infty} (f[\alpha]-\langle f\rangle)(f[\alpha-x]-\langle f\rangle)^* \, d\alpha \\
&= \lim_{X\to\infty}\left\{\frac{1}{2X}\int_{-X}^{+X} (f[\alpha]f^*[\alpha-x] - f[\alpha]\langle f\rangle^* - \langle f\rangle f^*[\alpha-x] + |\langle f\rangle|^2)\, d\alpha\right\} \\
&= \lim_{X\to\infty}\left\{\frac{1}{2X}\int_{-X}^{+X} (f[\alpha]f^*[\alpha-x])\, d\alpha\right\} - \langle f\rangle^* \lim_{X\to\infty}\left\{\frac{1}{2X}\int_{-X}^{+X} f[\alpha]\, d\alpha\right\} \\
&\quad - \langle f\rangle \lim_{X\to\infty}\left\{\frac{1}{2X}\int_{-X}^{+X} f^*[\alpha-x]\, d\alpha\right\} + |\langle f\rangle|^2 \\
&= f[x]\bigstar f[x] - (\langle f\rangle^*\langle f\rangle) - (\langle f\rangle\langle f\rangle^*) + |\langle f\rangle|^2 \\
&= f[x]\bigstar f[x] - |\langle f\rangle|^2
\end{aligned} \tag{8.65}$$

This has the advantage of removing any "bias" from the autocorrelation that may make it difficult to see subtle variations in its structure.

8.7 Crosscorrelation

The concept of autocorrelation may be generalized to compute a measure of "similarity" between $f[x]$ and an arbitrary "reference" function $m[x]$. At zero shift, the projection of $f[x]$ onto $m[x]$ is the (generally complex) number g_1:

$$g_1 = \int_{-\infty}^{+\infty} f[\alpha]m^*[\alpha]\, d\alpha \tag{8.66}$$

The numerical value of g_1 is large when $f[x]$ and $m[x]$ have similar "shapes". Just as for the autocorrelation, the central ordinate of the reference function $m^*[\alpha]$ is translated by adding the factor x_0 to the argument to measure the projection of $f[x]$ onto $m[x-x_0]$:

$$\int_{-\infty}^{+\infty} f[\alpha]m^*[\alpha-x_0]\, d\alpha = g_1[x_0] \tag{8.67}$$

This is a measure of the similarity between $f[x]$ and a translated replica of the reference. By evaluating this inner product at all possible shift parameters over the common domains of the two input functions, we construct a function of x that measures the similarity between $f[x]$ and the complete set of translated replicas of $m[x]$. This is the "crosscorrelation" of the input function $f[x]$ and the "reference" function $m[x]$:

$$g_1[x] = \int_{-\infty}^{+\infty} f[\alpha]m^*[\alpha - x]\, d\alpha \equiv f[x] \bigstar m[x] \tag{8.68}$$

Obviously this area must be finite at all translations x for the crosscorrelation integral to exist. The largest values of the inner product occur at the abscissa(s) x_n where the input function $f[x]$ is "most similar" to the translated replica of the reference function $m[\alpha - x]$, as measured by the inner product.

Note both the similarity and the two significant differences between the definition of crosscorrelation in Equation (8.68) and that of convolution in Equation (8.31). Both measure the area of the product of two functions after one has been translated by some distance. In the crosscorrelation operation, the input function is multiplied by shifted replicas of the complex conjugate of the reference function $m[x]$. In convolution, the analogue of the "reference function" is the "impulse response" $h[x]$, which is "reversed" or "flipped" prior to translation and multiplication, but without the complex conjugation operation. The argument of the reference function m in crosscorrelation is the negative of the argument of the impulse response h in convolution. The complex conjugation is applied to the reference function m in crosscorrelation for the same reason that it appears in the definition of the inner product of two vectors in Equation (5.4) to ensure that the crosscorrelation of a function $f[x]$ with itself (the "autocorrelation") is real valued.

Some authors define crosscorrelation with the "negative" of the shift parameter in Equation (8.60), so that the argument of m^* has a positive sign (Champeney, 1973):

$$g_2[x] \equiv \int_{-\infty}^{+\infty} f[\alpha]m^*[\alpha + x]\, d\alpha \tag{8.69}$$

The two definitions of crosscorrelation in Equations (8.68) and (8.69) produce functions that are "reversed" but otherwise identical, i.e., $g_1[x] = g_2[-x]$. We will use the former definition to maintain a larger degree of similarity to the definition of convolution. As already mentioned, Gaskill defines yet another variant of the crosscorrelation that does not include the complex conjugate of the "reference" function $m[x]$:

$$g_3[x] \equiv \int_{-\infty}^{+\infty} f[\alpha]m[\alpha - x]\, d\alpha \tag{8.70}$$

We might assign the name of "real-valued crosscorrelation" to this definition, which is arguably more flexible because it requires explicit inclusion of the complex conjugate to construct the expression in Equation (8.68). We choose to ignore this potential "advantage" and retain the similarity to the inner product of two complex-valued vectors. In short, the explicit definition of crosscorrelation that we will use is:

$$g_1[x] = f[x] \bigstar m[x] \equiv \int_{-\infty}^{+\infty} f[\alpha]m^*[\alpha - x]\, d\alpha \tag{8.71}$$

In this notation, Gaskill's "real-valued crosscorrelation" in Equation (8.70) is easily shown to be:

$$g_3[x] = f[x] \bigstar m^*[x] = \int_{-\infty}^{+\infty} f[\alpha]m[\alpha - x]\, d\alpha \tag{8.72}$$

where the fact that $(m^*[x])^* = m[x]$ has been used. Clearly the results obtained from the operations in Equations (8.70) and (8.72) are identical in any case where the reference function $m[x]$ is real valued. The definition of autocorrelation in Equation (8.61) may be generalized to construct the relation between crosscorrelation and convolution:

$$f[x] \bigstar m^*[x] = f[x] * m[-x] \tag{8.73a}$$

$$f[x] \bigstar m[x] = f[x] * m^*[-x] \tag{8.73b}$$

These relationships between crosscorrelation and convolution will be often used in subsequent discussions.

To soothe our fixation with the classification of mathematical operators, we now consider the question of linearity and shift invariance of the crosscorrelation. Just like convolution, crosscorrelation must be a linear operation because it is defined by an integral over fixed limits. The behavior of the operation under translation may be demonstrated easily:

$$\begin{aligned}\mathcal{O}\{f[x-x_0]\} = f[x-x_0] \bigstar m[x] &= \int_{-\infty}^{+\infty} f[\alpha - x_0] m^*[\alpha - x]\, d\alpha \\ &= \int_{-\infty}^{+\infty} f[\beta] m^*[\beta + x_0 - x]\, d\beta, \quad \textit{where } \beta \equiv \alpha - x_0 \\ &= \int_{-\infty}^{+\infty} f[\beta] m^*[\beta - (x - x_0)]\, d\beta \\ &= (f[x] \bigstar m[x])|_{x=x-x_0} \end{aligned} \tag{8.74}$$

Thus, crosscorrelation with $m[x]$ is a shift-invariant operation. Of course, Equation (8.73b) already showed us that crosscorrelation may be expressed as a convolution and therefore must be LSI. That would seem to suggest that the crosscorrelation operation commutes; we check this by changing the variable of integration to $\beta = \alpha - x$ in Equation (8.71). It is easy to show that the correlation of the "input function" $m[x]$ and the "reference function" $f[x]$ is:

$$\begin{aligned} m[x] \bigstar f[x] &= \int_{-\infty}^{+\infty} m[\beta + x] f^*[\beta]\, d\beta \\ &\neq f[x] \bigstar m[x] \end{aligned} \tag{8.75}$$

In words, the crosscorrelation operation does not commute; the roles of input and reference function are not so easily interchanged.

The amplitude of a crosscorrelation may be negative for some values of the shift parameter; at such locations the functions are said to be "anticorrelated" because the input function "resembles" the negative of the reference function.

It is useful to continue the analogy between the correlation and the dot (or scalar or inner) product of two vectors in Chapter 2 or of two continuous functions in Chapter 5. Recall in Equation (3.20) that the unambiguous measure of the projection of one real-valued vector onto a reference vector required that the latter be normalized to unit length. The analogous condition for the projection of an input function $f[x]$ onto a reference function $m[x]$ is that the latter be normalized to unit integrated power, as in Equation (5.98). The resulting operation evaluated at a particular coordinate is an unambiguous measure of the "similarity" of the two constituent functions. Following the notation of Gaskill, we assign the specific symbol $\gamma_{fm}[x]$ to this descriptor:

$$\gamma_{fm}[x] \equiv \frac{\displaystyle\int_{-\infty}^{+\infty} f[\alpha] m^*[\alpha - x]\, d\alpha}{\displaystyle\int_{-\infty}^{+\infty} |m[\alpha]|^2\, d\alpha} = \frac{f[x] \bigstar m[x]}{m[x] \bigstar m[x]|_{x=0}} \tag{8.76}$$

Note the similarity between this result and the Schwarz inequality in Equation (5.89) for the case $x = 0$. This implies that crosscorrelation is a generalized inner product evaluated for different translations of the reference function.

8.8 2-D LSI Operations

The sifting property of the 2-D Dirac delta function in Equation (7.48) may be applied in exactly the same way to evaluate the 2-D convolution integral. The criteria for linearity and for shift invariance of 2-D operators in Equations (8.2) and (8.14b), respectively, lead to the expression for the convolution of two 2-D functions $f[x, y]$ and $h[x, y]$:

$$\mathcal{O}\{f[x, y]\} = \iint_{-\infty}^{+\infty} f[\alpha, \beta] h[x - \alpha, y - \beta] \, d\alpha \, d\beta \equiv f[x, y] * h[x, y] \tag{8.77}$$

The 2-D kernel $h[x, y]$ determines the action of the system and is still called the impulse response or point-spread function of the system. In words, this expression for the convolution of two 2-D functions indicates that the kernel $h[x, y]$ is "reflected" through the origin (or equivalently, rotated about the origin by π radians), then translated over the input function $f[x, y]$, and the "volume" of the product function is determined by integrating over the 2-D domain. Again, the form of $h[x, y]$ determines the character of the result.

If the input and kernel functions are both separable, the 2-D convolution integral may be simplified to the product of two 1-D convolutions:

$$\begin{aligned} f[x, y] * h[x, y] &= (f_1[x] \cdot f_2[y]) * (h_1[x] \cdot h_2[y]) \\ &= \iint_{-\infty}^{+\infty} f_1[\alpha] f_2[\beta] \, h_1[x - \alpha] h_2[y - \beta] \, d\alpha \, d\beta \\ &= \int_{-\infty}^{+\infty} f_1[\alpha] h_1[x - \alpha] \, d\alpha \int_{-\infty}^{+\infty} f_2[\beta] h_2[y - \beta] \, d\beta \\ &= (f_1[x] * h_1[x])(f_2[y] * h_2[y]) \end{aligned} \tag{8.78}$$

As an example, consider the 2-D convolution of $f[x, y] = RECT[x, y]$ with itself, which evidently is:

$$\begin{aligned} RECT[x, y] * RECT[x, y] &= (RECT[x] * RECT[x])(RECT[y] * RECT[y]) \\ &= TRI[x] \cdot TRI[y] = TRI[x, y] \end{aligned} \tag{8.79}$$

We note that the triangle function is generated by (and may be *defined* by) the autoconvolution of the rectangle function in both the 1-D and 2-D cases. This observation may be extended to functions with dimensionality larger than two.

Several authors (e.g., Gaskill, 1978; Bracewell, 1986a) use "paired" asterisks to denote the convolution of 2-D functions that was defined by Equation (8.77). One rationale for this notation is that it distinguishes between 1-D and 2-D convolution operations when they are performed in the same context (or even within the same system). In other words, this notation may be used to describe the convolution of a 2-D function with either 1-D or 2-D impulse responses. Another arguable advantage of the paired asterisk notation is demonstrated when two 2-D separable functions are convolved, as in Equation (8.78). The two asterisks may be "divided" among the two 1-D convolutions to ensure "conservation of asterisks". Consider the example used by Gaskill and Bracewell to illustrate both "1-D convolution" and "2-D convolution" of a general 2-D function $f[x, y]$ with systems whose impulse response is the 1-D Dirac delta function $\delta[x - x_0]$. The two definitions of convolution give distinct outputs when applied to the same functions. For example, we may write the "1-D convolution" in Gaskill's notation as:

$$\begin{aligned} f[x, y] * \delta[x - x_0] &\equiv \int_{-\infty}^{+\infty} f[\alpha, y] \, \delta[(x - x_0) - \alpha] \, \delta[y - \beta] \, d\alpha \\ &= f[x - x_0, y] \end{aligned} \tag{8.80a}$$

Gaskill's notation for "2-D convolution" with a 1-D Dirac delta function is:

$$f[x, y] ** \delta[x - x_0] \equiv \iint_{-\infty}^{+\infty} f[\alpha, \beta]\, \delta[x - x_0 - \alpha] 1[y - \beta]\, d\alpha\, d\beta$$

$$= \iint_{-\infty}^{+\infty} f[\alpha, \beta]\, \delta[x - x_0 - \alpha] 1[\beta]\, d\alpha\, d\beta$$

$$= \int_{-\infty}^{+\infty} f[x - x_0, \beta]\, d\beta \tag{8.80b}$$

where the properties of the unit function have been used. In words, Equation (8.80a) is a generalization of the sifting property of the 1-D Dirac delta function and generates a translated replica of the 2-D function, while Equation (8.80b) is a 2-D convolution that yields a 2-D function that is constant along the y-direction.

Rather than have to remember the definition of 1-D convolution of a 2-D function, we will simplify the notation by using a single asterisk for all convolutions. This will require both component functions to have the same dimensions, and the two cases in Equation (8.80) will be distinguished by the impulse response. The equivalent operation to "1-D convolution" with the shifted Dirac delta function in Equation (8.80a) is denoted by:

$$f[x, y] * \delta[x - x_0] \Longrightarrow f[x, y] * (\delta[x - x_0]\, \delta[y])$$

$$= f[x, y] * \delta[x - x_0, y] = f[x - x_0, y] \tag{8.81a}$$

while the equivalent of "2-D convolution" is convolution with $\delta[x - x_0]\, 1[y]$:

$$f[x, y] ** \delta[x - x_0] \Longrightarrow f[x, y] * (\delta[x - x_0]\, 1[y]) = \int_{-\infty}^{+\infty} f[x - x_0, \beta]\, d\beta \tag{8.81b}$$

Experience has shown that the "single-asterisk" notation minimizes confusion by the readers because all functional forms are explicit.

It is easy to generalize the derivation of Equation (8.47) for 1-D convolutions to prove that 2-D convolutions commute:

$$f[x, y] * h[x, y] = h[x, y] * f[x, y] \tag{8.82}$$

Other theorems for convolution of 2-D functions that are analogous to the 1-D cases also are easily demonstrated.

8.8.1 Line-Spread and Edge-Spread Functions

The examples just used in the discussion of the notation for 2-D convolution in Equation (8.80) may be used to construct useful metrics other than the system impulse response $h[x, y]$. Though introduced here, these metrics will be considered in more detail in the discussion of image filtering in Chapter 16, after the description of the Fourier transform in Chapter 9. We know that the impulse response $h[x, y]$ of a 2-D system is defined as the output of the system when the input is the 2-D Dirac delta function:

$$\mathcal{O}\{\delta[x, y]\} = \iint_{-\infty}^{+\infty} \delta[x - \alpha, y - \beta] h[\alpha, \beta]\, d\alpha\, d\beta = h[x, y] \tag{8.83}$$

where the sifting property of the 2-D Dirac delta function in Equation (7.48) has been used.

Alternatively, consider the convolution of an input composed of the 2-D "line delta function" $\delta[x]\, 1[y]$ with the impulse response $h[x, y]$. The resulting 2-D function varies in the x-direction and is constant along lines parallel to the y-axis:

$$\begin{aligned}\mathcal{O}\{\delta[x]\,1[y]\} &= \iint_{-\infty}^{+\infty} (\delta[x-\alpha]\,1[y-\beta])h[\alpha,\beta]\,d\alpha\,d\beta \\ &= \int_{-\infty}^{+\infty} h[x,\beta]\,d\beta \equiv \ell[x] \end{aligned} \tag{8.84}$$

Since $\ell[x]$ is the output of a system when the input is a line delta function input, it is often called the *line-spread function* and may be visualized as the "line-integral projection" of the impulse response onto the x-axis, though the use of the term *projection* differs from that considered in this chapter up to this point. It is perhaps evident that the description of the impulse response that can be visualized from $\ell[x]$ is "incomplete"; no information about the variation of $h[x, y]$ along the y-axis is available from $\ell[x]$, and therefore $h[x, y]$ cannot be determined from $\ell[x]$ alone. However, we will see in Chapter 12 that the "complete set" of line-integral projections obtained at all azimuth angles does provide a complete representation of $h[x, y]$. If the imaging system is rotationally symmetric, the line-spread function is a complete description of the system and is useful when analyzing imaging systems.

Another useful descriptor of an imaging system is the output generated in response to an "edge" input, which is in turn the orthogonal product of a step function and the unit constant. For example, the output for an "edge" function that varies along x and is constant along y is:

$$\begin{aligned}\mathcal{O}\{STEP[x]\,1[y]\} &= \iint_{-\infty}^{+\infty} STEP[x-\alpha]\,1[y-\beta]h[\alpha,\beta]\,d\alpha\,d\beta \\ &= \int_{-\infty}^{+\infty}\left(\int_{-\infty}^{x} h[\alpha,\beta]\,d\alpha\right)1[y-\beta]\,d\beta \equiv e[x] \end{aligned} \tag{8.85}$$

The resulting output again varies along the x-axis and is constant along lines parallel to the y-axis. Again, the edge-spread function is a complete description of a rotationally symmetric imaging system.

8.9 Crosscorrelations of 2-D Functions

The extension of the definition of the crosscorrelation of two 1-D functions in Equation (8.60) evidently requires translation of the complex conjugate of the 2-D reference function $m[x, y]$ relative to the input function $f[x, y]$ and then evaluating the overlap volume. The result obtained for all possible translations produces a 2-D function. By analogy with the notation for 2-D convolution, we use a "single-pentagram" notation:

$$\begin{aligned} f[x,y] \bigstar m[x,y] &\equiv \iint_{-\infty}^{+\infty} f[\alpha,\beta]m^*[\alpha-x,y-\beta]\,d\alpha\,d\beta \\ &= f[x,y] * m^*[-x,-y] \end{aligned} \tag{8.86}$$

where again both component functions $f[x, y]$ and $m[x, y]$ must have the same dimensionality.

As an example, consider the crosscorrelation of an input function consisting of nine alphabetic characters with a reference function that is one of those characters, as shown in Figure 8.12. Note that the largest amplitudes in the crosscorrelation occur at the locations of replicas of $m[x, y]$. This property is the basis for the *matched filter* that may be used to identify the locations of replicas of some "reference" function, as discussed in Chapter 19.

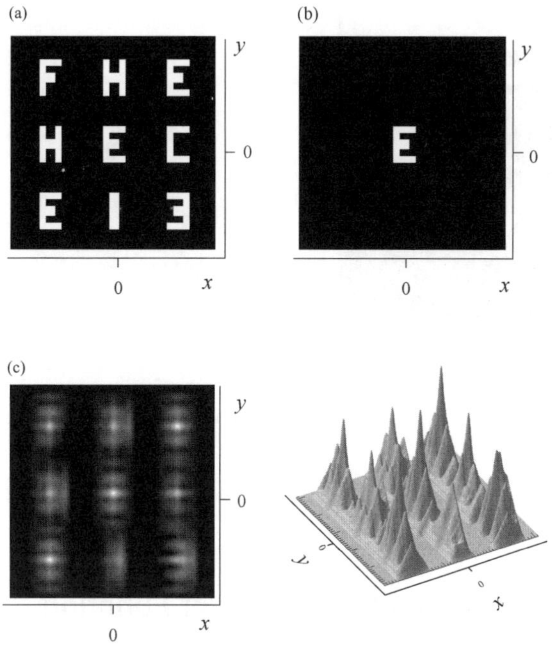

Figure 8.12 Crosscorrelation of the input object $f[x, y]$ with the "reference" function $m[x, y]$: (a) $f[x, y]$; (b) $m[x, y]$; (c) normalized crosscorrelation defined in Equation (8.76) displayed as image and as 3-D surface.

8.10 Autocorrelations of 2-D Functions

By analogy with Equation (8.86), the autocorrelation of a complex-valued 2-D function is:

$$f[x, y] \bigstar f[x, y] \equiv \iint_{-\infty}^{+\infty} f[\alpha, \beta] f^*[\alpha - x, y - \beta] \, d\alpha \, d\beta \tag{8.87}$$

which is guaranteed to be Hermitian (even real part, odd imaginary part). The amplitude of the autocorrelation at the origin is the area of the squared magnitude of the function:

$$\begin{aligned}(f[x, y] \bigstar f[x, y])|_{[x,y]=[0,0]} &= \iint_{-\infty}^{+\infty} f[\alpha, \beta] f^*[\alpha - 0, y - \beta] \, d\alpha \, d\beta \\ &= \iint_{-\infty}^{+\infty} |f[\alpha, \beta]|^2 \, d\alpha \, d\beta \end{aligned} \tag{8.88}$$

which is guaranteed to be real valued. The 2-D "normalized autocorrelation" is a straightforward generalization of Equation (8.76):

$$\begin{aligned}\gamma_{ff}[x, y] &\equiv \frac{\displaystyle\iint_{-\infty}^{+\infty} f[\alpha, \beta] f^*[\alpha - x, \beta - y] \, d\alpha \, d\beta}{\displaystyle\iint_{-\infty}^{+\infty} |f[\alpha, \beta]|^2 \, d\alpha \, d\beta} \\ &= \frac{f[x, y] \bigstar f[x, y]}{.[x, y] \bigstar f[x, y]|_{x=0, y=0}} \end{aligned} \tag{8.89}$$

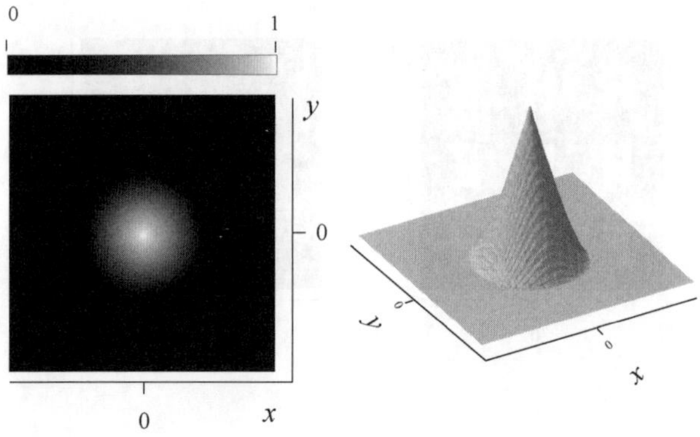

Figure 8.13 The normalized autocorrelation of 2-D cylinder function produces the circularly symmetric "circular triangle" function.

8.10.1 Autocorrelation of the Cylinder Function

As we will see in Chapter 22, the autocorrelation of a cylinder function will be useful when describing the action of optical systems with rotational symmetry. If the 2-D cylinder is interpreted to be the circularly symmetric analogue of the 2-D rectangle function, then the autocorrelation of $CYL(r)$ is the circularly symmetric analogue to the autocorrelation of $RECT[x, y]$. Because $RECT[x, y]$ is a real-valued even function, then its autocorrelation is identical to its autoconvolution which was shown in Equation (8.79) to be $TRI[x, y]$. The functional form of the "circular triangle" function was specified in Equation (7.99):

$$CYL(r) \bigstar CYL(r) = \frac{2}{\pi}\left(\cos^{-1}[r] - r\sqrt{1-r^2}\right) CYL\left(\frac{r}{2}\right) \tag{8.90}$$

The normalized autocorrelation is shown in Figure 8.13, which exhibits rotational symmetry; its sides are not "straight" lines.

PROBLEMS

8.1 Use "direct integration" to evaluate

$$\begin{aligned} f[x] &= TRI[x] * TRI[x] \\ &= \int_{-\infty}^{+\infty} \left((1-|\alpha|)\, RECT\left[\frac{\alpha}{2}\right]\right)\left((1-|x-\alpha|)\, RECT\left[\frac{x-\alpha}{2}\right]\right) d\alpha \end{aligned}$$

Plot this function on the same graph with:

$$g[x] = RECT[x]$$

$$h[x] = TRI[x] = RECT[x] * RECT[x] = (1-|x|)RECT\left[\frac{x}{2}\right]$$

$$r[x] = RECT[x] * TRI[x] = \begin{cases} 0 & \text{if } x < -\frac{3}{2} \\ \frac{x^2}{2} + \frac{3}{2}x + \frac{9}{8} & \text{if } -\frac{3}{2} \le x \le -\frac{1}{2} \\ \frac{3}{4} - x^2 & \text{if } -\frac{1}{2} \le x \le +\frac{1}{2} \\ \frac{x^2}{2} - \frac{3}{2}x + \frac{9}{8} & \text{if } +\frac{1}{2} \le x \le +\frac{3}{2} \\ 0 & \text{if } x > +\frac{3}{2} \end{cases}$$

8.2 Express the following product of two Gaussian functions as a Gaussian function, i.e., given x_0, evaluate A and d in the following expression:

$$GAUS\left[\frac{x + x_0}{b}\right] \cdot GAUS\left[\frac{x - x_0}{b}\right] = A\, GAUS\left[\frac{x}{d}\right]$$

8.3 Evaluate $\cos[2\pi\xi_0 x + \phi_0] * RECT[x/d]$.

8.4 Show that

$$RECT\left[\frac{x}{d_0}\right] * RECT\left[\frac{x}{b_0}\right] = \left(\frac{d_0 + b_0}{2}\right) \cdot TRI\left[\frac{x}{(d_0 + b_0)/2}\right] - \left(\frac{|d_0 - b_0|}{2}\right) \cdot TRI\left[\frac{x}{(|d_0 - b_0|)/2}\right]$$

8.5 You may substitute any function $f[x]$ and any impulse response $h[x]$ in the following expression. State the conditions on $f[x]$ and $h[x]$ that ensure that the result is correct. Explain your reasoning.

$$f[x] * h[x] = f[-x]$$

8.6 If $f[x] * h[x] = g[x]$, find an expression for $f^*[x] * h^*[x]$ in terms of $g[x]$.

8.7 Evaluate and sketch the 1-D convolutions:

(a) $f[x] = \delta[x+1] + \delta[x-1]$
(b) $g[x] = f[x] * f[x]$
(c) $h[x] = g[x] * f[x]$
(d) $r[x] = h[x] * f[x]$
(e) Explain the pattern present in the results of these convolutions.

8.8 Evaluate and sketch the 1-D convolutions:

(a) $STEP[x] * STEP[x]$
(b) $(x \cdot STEP[x]) * STEP[x]$
(c) $(x \cdot STEP[x]) * (x \cdot STEP[x])$
(d) $STEP[x] * (e^{-x} \cdot STEP[x])$
(e) $(e^{-x} \cdot STEP[x]) * (e^{-x} \cdot STEP[x])$
(f) $(e^{-x} \cdot STEP[x]) * RECT[x]$

8.9 Evaluate and sketch the 1-D convolutions:

(a) $RECT[2x] * (\delta[x+1] + \delta[x-1]) = RECT[2x] * \delta\delta[x]$
(b) $RECT[2x] * (\delta[x+1] - \delta[x-1]) \equiv RECT[2x] * \delta[x]$
(c) $(\delta[x+1] + \delta[x-1]) * (\delta[x+1] + \delta[x-1]) = \delta\delta[x] * \delta\delta[x]$
(d) $(\delta[x+1] - \delta[x-1]) * (\delta[x+1] - \delta[x-1]) = \delta[x] * \delta[x]$

(e) $(\delta[x+1]+\delta[x-1]) * TRI[x] = \delta\delta[x] * TRI[x]$
(f) $(\delta[x+1]-\delta[x-1]) * TRI[x] = \delta[x] * TRI[x]$
(g) $\delta'[x] * TRI[x]$
(h) $(\delta[x+1]+\delta[x-1]) * GAUS[x]$
(i) $(\delta[x+1]-\delta[x-1]) * GAUS[x]$
(j) $\delta'[x] * GAUS[x]$

8.10 Evaluate and sketch the 1-D convolutions and correlations:
(a) $RECT[x] * RECT[x-1]$
(b) $RECT[x] \bigstar RECT[x-1]$
(c) $RECT[x-1] * RECT[x+1]$
(d) $RECT[x-1] \bigstar RECT[x+1]$
(e) $RECT[x-1] \bigstar RECT[x-1]$
(f) $(\delta[x+2]+\delta[x]+\delta[x-1]) * (\delta[x+2]+\delta[x]+\delta[x-1])$
(g) $(\delta[x+2]+\delta[x]+\delta[x-1]) \bigstar (\delta[x+2]+\delta[x]+\delta[x-1])$
(h) $(e^{-x} \cdot STEP[x]) * (e^{-x} \cdot STEP[x])$
(i) $(e^{-x} \cdot STEP[x]) \bigstar (e^{-x} \cdot STEP[x])$

8.11 Evaluate and sketch the 1-D convolutions:
(a) $TRI[x, y] * (\frac{1}{2}\delta\delta[x/2]\,\delta[y])$
(b) $TRI[x, y] * (\frac{1}{2}\delta[x]\,\delta[y])$
(c) $TRI[x, y] * (STEP[x]\,1[y])$
(d) $RECT[x, y] * (\delta[x]\,1[y])$

8.12 Evaluate and sketch the 1-D convolutions:
(a) $RECT[x, y] * (\delta[x-1]\,1[y-1])$
(b) $RECT[x, y] * CROSS[x, y]$
(c) $COR[x, y] * COR[x, y]$

8.13 Prove that:

$$r[x-x_0, y-y_0] \bigstar r[x-x_0, y-y_0] = r[x, y] \bigstar r[x, y]$$

9

Fourier Transforms of 1-D Functions

Of the three imaging tasks introduced in Chapter 1 for imaging systems, we thus far have developed the necessary computational tools to solve only the first (the "direct" problem), and that only for linear systems. This task requires evaluation of the superposition integral in Equation (8.41) for shift-variant systems, or of the convolution integral in Equation (8.31) for less general shift-invariant systems. The remaining two imaging tasks, of finding the input function from knowledge of the output and deriving the imaging system operator from the input and output signals, are formidable problems even in the constrained case of shift-invariant convolution. Fortunately, the study of matrices and vectors in Chapters 3–5 hints that equivalent representations of the constituent functions of an imaging system may exist and that the inverse and system analysis problems may be more readily solved in these representations. This chapter considers the mathematical details of this alternative representation that provides a path to the solution of the inverse and system analysis imaging tasks.

9.1 Transforms of Continuous-Domain Functions

The analogy with the inner product of two vectors with complex-valued components defined in Equation (5.4) was used in Equation (5.93) to construct the inner product of a complex-valued 1-D function $f[x]$ and a 1-D "reference function" $m[x]$:

$$g = \int_{-\infty}^{+\infty} m^*[\alpha] f[\alpha] \, d\alpha \tag{9.1}$$

The conceptual parallel with vectors led to the interpretation that the (generally complex-valued) amplitude g measures the "projection" of the input function $f[x]$ onto the "reference" function $m[x]$. In other words, g is a measure of the "amount" of $m[x]$ that "exists" in $f[x]$, or, equivalently, the weighting that must be applied to $m[x]$ when synthesizing $f[x]$ as a weighted sum of "reference" functions. It is useful to interpret $m^*[x]$ as a multiplicative "mask" function that is applied to $f[x]$ before the integration in Equation (9.1). Recall also that the concept of "projection" was used in Chapter 8 to interpret the crosscorrelation operation that computed the projections of $f[x]$ onto translated replicas of the reference functions via the linear and shift-invariant operation:

$$f[x] \bigstar m[x] \equiv \int_{-\infty}^{+\infty} m^*[\alpha - x] f[\alpha] \, d\alpha = f[x] * m^*[-x] \tag{9.2}$$

Fourier Methods in Imaging Roger L. Easton, Jr.

We also may consider the concept of "projection" of the input function in the more general shift-variant case by evaluating the set of inner products of $f[x]$ with each of a set of *distinct* reference functions, instead of translated replicas of the same function. The operation may be specified in the manner of Equation (5.94):

$$g[u] \equiv \int_{-\infty}^{+\infty} m^*[\alpha; u] f[\alpha]\, d\alpha \tag{9.3}$$

The different 1-D reference functions are distinguished by the parameter u. Note how these two last equations are both similar and different. The argument of the reference function m in the crosscorrelation in Equation (9.2) is the 1-D linear combination of coordinates $[\alpha - x]$, which requires that the dimensions ("units") of α and x are identical. Contrast this with the reference function $m^*[\alpha; u]$ in the shift-variant case of Equation (9.3), which may be interpreted as a 2-D function of coordinates that may have different dimensions. In such a case, the domains of the "input" function $f[x]$ and the "output" function $g[u]$ may be different, which means that the dimensions of α and u may differ, e.g., α may have units of length and u of reciprocal length. We will often take advantage of functions $f[x]$ and $g[u]$ that are equivalent "representations" of the same entity if the reference function m satisfies certain criteria.

The (possible) difference in the coordinates of the "mask" or "reference" functions in Equations (9.2) and (9.3) is another example of the fundamental distinction between shift-invariant and shift-variant systems. We have seen the manifestation of this distinction in Chapter 5 for the discrete case where the system is represented by a matrix–vector multiplication; the discrete linear shift-variant system is modeled by a general matrix, while the linear shift-invariant matrix operator is constrained to be circulant or Toeplitz. The discussion in Chapter 5 showed that $g[u]$ in Equation (9.3) may be interpreted as a different (but equivalent) representation of $f[x]$ if the set of reference functions $m[x; u]$ satisfies certain requirements. To illustrate, consider the set of reference functions used in the sifting property of the Dirac delta function in Equation (6.113), so that $m^*[\alpha; u] = m^*[\alpha; x_0] = \delta^*[x_0 - \alpha]$. The resulting output function $g[x]$ is identical to the input function $f[x]$:

$$\begin{aligned} g[x] &= \int_{-\infty}^{+\infty} \delta^*[x_0 - \alpha] f[\alpha]\, d\alpha = f[x_0] \\ &= \int_{-\infty}^{+\infty} \delta[\alpha - x_0]\, f[\alpha]\, d\alpha \quad \textit{because } \delta[x] = \delta^*[-x] \end{aligned} \tag{9.4}$$

The output of the simple "imaging" system described by the sifting property is the amplitude of the input function at x_0. This shift-invariant operation may be compared to Equation (9.3) to see that the sifting property projects $f[x]$ onto the Dirac delta function located at the specific coordinate x_0. The complete representation of any 1-D continuous function $f[x]$ is constructed from its projections onto the set of Dirac delta functions at all coordinates x:

$$g[x] = \int_{-\infty}^{+\infty} \delta[x - \alpha] f[\alpha]\, d\alpha = f[x] \tag{9.5}$$

The domains of the input function $f[x]$ and the output function $g[x]$ must be identical; the "new" representation $g[x]$ described by this trivial "transformation" is identical to the original representation $f[x]$.

Though it merely restates the sifting property, Equation (9.5) may help develop some intuition about one condition that must be satisfied by a set of "reference" functions $m^*[\alpha; u]$. The set of all Dirac delta functions is "complete" because any $f[x]$ may be represented uniquely by the set of its projections onto each of the members of the set. This is exactly analogous to the concept of a complete set of basis vectors that was introduced in Chapter 3, and thus $m^*[\alpha; u]$ is often called a set of "basis" functions.

Another important property of the basis set of Dirac delta functions is demonstrated by projecting any one such function onto any other. In other words, assume that the input function and reference functions are distinct Dirac delta functions, say $f[x] = \delta[x - x_0]$ and $m[x; x_1] = \delta[x - x_1]$:

$$\int_{-\infty}^{+\infty} \delta[\alpha - x_0]\delta[\alpha - x_1]\, d\alpha = \int_{-\infty}^{+\infty} \delta[\alpha]\delta[\alpha - (x_1 - x_0)]\, d\alpha$$
$$= \delta[x_1 - x_0] = \delta[x_0 - x_1] = 0 \quad \textit{because } x_1 \neq x_0 \tag{9.6}$$

where the first property of the Dirac delta function in Equation (6.88a) has been used. In words, the projection of $\delta[x - x_0]$ onto $\delta[x - x_1]$ is zero if $x_0 \neq x_1$. If $x_0 = x_1$, then the projection is the inner product of $\delta[x - x_0]$ with itself, which yields the undefined (though positive) amplitude $\delta[0]$. By analogy with the properties of vectors, this demonstrates that members of the set of Dirac delta functions are "orthogonal" because inner products of distinct members produce null outputs.

The representation of some function $f[x]$ as the set of its projections onto the complete set of Dirac delta functions is such a natural concept that we are rarely conscious of it. Another and perhaps more appropriate name for the set of projections of $f[x]$ onto the Dirac delta functions is the "space-domain representation" because the particular reference functions $m[x; x_0] = \delta[x - x_0]$ are distinguished by coordinates that have dimensions of length. The primary goal of this chapter is to develop a specific equivalent representation of $f[x]$ obtained by projection onto a different complete set of basis functions whose members are distinguished by coordinates with dimensions of "reciprocal length", e.g., cycles per unit length, which we called *spatial frequencies* in Chapter 2. The explicit description of the space-domain representation just considered provides a useful model for the upcoming discussion of this alternative representation.

Because each of the "reference" functions $\delta[x; x_0]$ in the basis set for the space-domain representation has infinitesimal support, the calculation of each inner product in Equation (9.5) is determined by the amplitude of the input function at only a single coordinate. If the individual basis functions $m[x, u]$ have finite support, the inner product considers a larger domain of $f[x]$. As a simple example, consider $m[x, u]$ to be the set of real-valued rectangle functions with fixed support b and unit area centered at all locations in the domain. The projections of $f[x]$ onto this set of basis functions are evaluated by substitution into Equation (9.3):

$$g[u] = \int_{-\infty}^{+\infty} \left(\frac{1}{|b|} RECT\left[\frac{u - \alpha}{b}\right]\right) f[\alpha]\, d\alpha$$
$$= \frac{1}{|b|} \int_{u-b/2}^{u+b/2} f[\alpha]\, d\alpha \tag{9.7}$$

The output g at each u is the average amplitude of $f[x]$ evaluated over the finite support of the translated rectangle; in other words, the "output" $g[u]$ is the result of a "local" operation applied to $f[x]$. Comparison to Equation (8.31) demonstrates that Equation (9.7) evaluates the convolution of $f[x]$ with a rectangular impulse response and therefore is a linear and shift-invariant operation. We will see later that $g[u]$ generated by Equation (9.7) is not equivalent to an arbitrary $f[x]$ because the same output is produced by distinct input functions. This means that a unique $f[x]$ cannot be constructed from complete knowledge of $g[u]$, even though the rectangular impulse response in Equation (9.7) is known. In other words, the inverse transformation of Equation (9.7) from $g[u]$ to $f[x]$ does not exist.

The output $g[u]$ of the inner product in Equation (9.3) with an infinite-support mask function $m[x; u]$ is affected by the amplitude of $f[x]$ over its entire domain. By analogy with the terminology just introduced for Equation (9.7), this is an example of a "global operation" applied to $f[x]$. A simple example of a global mask function is the 2-D unit constant $m[x, u] = 1[x, u]$. The projection of $f[x]$ onto $1[x]$ is the area of $f[x]$ regardless of the value of u:

$$g[u] = \int_{-\infty}^{+\infty} 1^*[\alpha, u] f[\alpha]\, d\alpha = \int_{-\infty}^{+\infty} f[\alpha]\, d\alpha$$
$$= \left(\int_{-\infty}^{+\infty} f[\alpha]\, d\alpha\right) 1[u] \tag{9.8}$$

In words, the output amplitude g is unaffected by any translation of the input function $f[x]$, which means that the projections of both $f[x]$ and $f[x - x_0]$ onto the unit constant produce the same result, which means that these two functions cannot be distinguished from $g[u]$. Other examples of such global operators will be considered shortly.

In another variant, inner products of $f[x]$ may be calculated for the members of a discrete set of N distinct infinite-support reference masks. Specific reference functions in this set may be distinguished by an integer subscript n instead of the continuous variable u, e.g., $m[x, u] \rightarrow m_n[x]$. The resulting N discrete outputs g_n are:

$$g_n = \int_{-\infty}^{+\infty} m_n^*[\alpha] f[\alpha]\, d\alpha \tag{9.9}$$

and may be interpreted as the components of a vector. Equation (9.9) may be considered a hybrid operation because the domains of the input and output functions are different. An example of a discrete set of 1-D functions that defines a useful global operation is the set of integer powers x^n for $n = 0, 1, 2, \ldots, +\infty$. The resulting representation is a discrete vector with an infinite number of dimensions. This particular set of reference functions was introduced in Equation (6.165) to construct the complete set of *moments* of the continuous function $f[x]$. Note that the "zero-order" output g_0 is the inner product of $f[x]$ with $x^0 = 1$, which yields the area of $f[x]$ in the same fashion as Equation (9.8). The mask functions associated with $n > 1$ apply less weight to $f[x]$ near the origin and more for large values of $|x|$, which means that variation of $f[x]$ in the vicinity of the origin affects only the low-order moments.

Hausdorff showed that the countably infinite set of moments may be used to "reconstruct" a function $f[x]$ that has finite support and no discontinuities (Körner, 1988). The properties and applications of the moments of a function are considered in somewhat more detail in Chapter 13.

The concept of a discrete set of inner products of $f[x]$ may be generalized to construct continuous sets of reference functions for use in Equation (9.3). Specific reference functions are distinguished by the continuous parameter u whose dimensions depend on the specific set of reference functions $m[x, u]$. The amplitude g at a specific u is the projection of $f[x]$ onto that specific reference function. Many such sets $g[x, u]$ are possible, but we will use intuitive and visual arguments to derive the outputs that result from even and odd sinusoidal input functions for basis functions (weighting mask functions) $m[u, x]$ that are also even and odd sinusoids. This choice of basis functions is motivated by the observation in Chapter 5 that an N-sample vector is transformed to the basis of the diagonal form of a circulant matrix via the discrete Fourier transform. The continuous analogue to this operation was inferred in Equation (5.102).

9.1.1 Example 1: Input and Reference Functions are Even Sinusoids

Consider a 1-D real-valued cosine function with fixed period $X_1 = \xi_1^{-1}$ and arbitrary amplitude A_1, as shown in Figure 9.1:

$$f_1[x] = A_1 \cos[2\pi\xi_1 x] \tag{9.10}$$

We will compute projections of $f_1[x]$ onto the individual cosine functions for all spatial frequencies ξ over the interval $(-\infty, +\infty)$. For convenience, we substitute the usual notation ξ for spatial frequency for the parameter u in Equation (9.3):

$$m_1[x, u] = \cos[2\pi ux] \rightarrow m_1[x, \xi] = \cos[2\pi\xi x] \tag{9.11}$$

Because the reference functions are real valued, the complex-conjugation operation within the definition of the inner product has no effect in this case, though we need to keep it in mind. The two variables x and ξ that appear as a product in the argument of the cosine must have "reciprocal" dimensions of (length) and (length)$^{-1}$, respectively. The ensemble of reference functions may be visualized as a 2-D function $m_1[x, \xi]$ with a "mixed" domain in Figure 9.2.

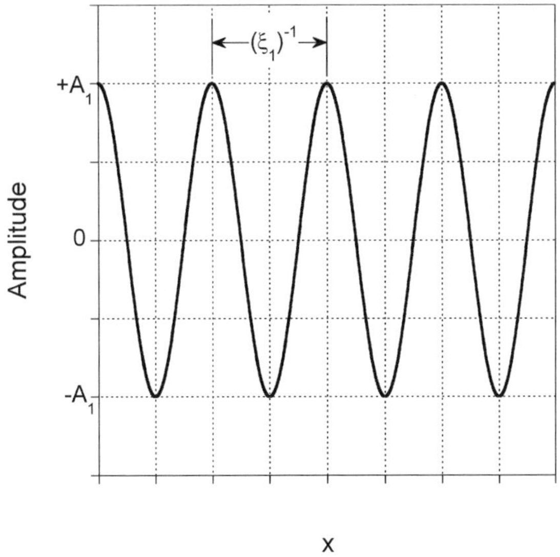

Figure 9.1 $f_1[x] = A_1 \cos[2\pi\xi_1 x]$.

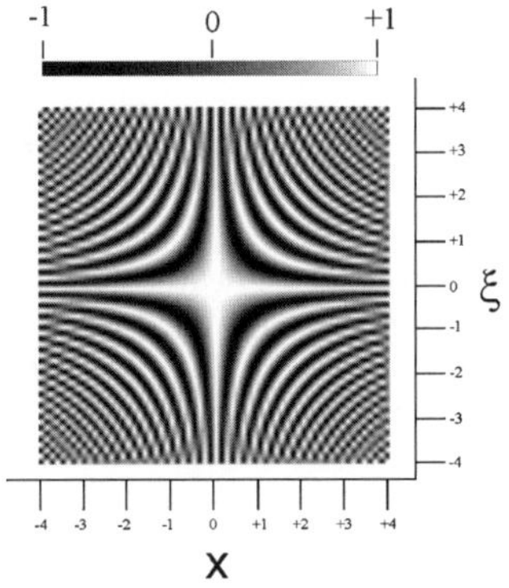

Figure 9.2 $m[x, \xi] = \cos[2\pi\xi x]$ displayed as an "image". The x-axis is horizontal and the ξ-axis is vertical. Note that the oscillation rate of the amplitude in the horizontal direction increases away from the origin and that the 2-D function is symmetric with respect to the origin of coordinates.

The inner product of $f_1[x]$ and $m_1[x, \xi]$ is evaluated for each spatial frequency ξ by computing the area of the product of $f_1[x]$ and the specific infinite-support 1-D reference "mask":

$$g_{11}[\xi] = \int_{-\infty}^{+\infty} m_1^*[x, \xi] f_1[x]\, dx$$

$$= \int_{-\infty}^{+\infty} \cos[2\pi\xi x]\, (A_1 \cos[2\pi\xi_1 x])\, dx \tag{9.12}$$

where the subscript "11" indicates the subscripts of the reference and input functions, respectively. Note that the "reference function" in this operation is *not* translated as it had been in convolution or crosscorrelation. Instead, this operation projects $f_1[x]$ onto each of the set of reference functions that are distinguished by the parameter ξ, so the output amplitude g generally varies with ξ. The integral in Equation (9.12) may be simplified by applying the trigonometric identity of Equation (4.42a):

$$\cos[\phi]\cdot\cos[\theta] = \frac{1}{2}(\cos[\phi-\theta]+\cos[\phi+\theta]) \tag{9.13}$$

This produces the sum of two integrals that may be easier to visualize:

$$g_{11}[\xi] = \frac{A_1}{2}\left[\int_{-\infty}^{+\infty}\cos[2\pi(\xi-\xi_1)x]\,dx + \int_{-\infty}^{+\infty}\cos[2\pi(\xi+\xi_1)x]\,dx\right] \tag{9.14}$$

The first integrand is the cosine whose spatial frequency is the difference of those in the "reference" and "input" functions: $[\xi-\xi_1]$. For all $\xi \neq \xi_1$, the areas of adjacent positive and negative lobes of the integrand cancel, so that the first integral evaluates to zero; this is the same result as in Equation (6.22). If $\xi = \xi_1$, the integrand of the first integral is the unit constant, so the area is indeterminate. An expression for this first integral that is valid for all spatial frequencies ξ may be derived by a procedure identical to that used to obtain Equation (6.105):

$$\begin{aligned}\frac{A_1}{2}\int_{-\infty}^{+\infty}\cos[2\pi(\xi-\xi_1)x]\,dx &= \frac{A_1}{2}\lim_{B\to\infty}\left\{\int_{-B}^{+B}\cos[2\pi(\xi-\xi_1)x]\,dx\right\}\\ &= \frac{A_1}{2}\lim_{B\to\infty}\left\{\frac{1}{2\pi[\xi-\xi_1]}\sin[2\pi(\xi-\xi_1)x]\Big|_{x=-B}^{x=B}\right\}\\ &= \frac{A_1}{2}\lim_{B\to\infty}\{2B\cdot SINC[2B[\xi-\xi_1]]\}\\ &= \frac{A_1}{2}\delta[\xi-\xi_1]\end{aligned} \tag{9.15}$$

where the area of the *SINC* function remains constant and the support decreases in the limit. In words, the first integral in Equation (9.14) yields a Dirac delta function located at $\xi = -\xi_1$ whose area is scaled by $A_1/2$ (Figure 9.3).

The second integrand in Equation (9.14) is a cosine that oscillates at the *sum* frequency $\xi + \xi_1$, and the integral produces a Dirac delta function with area $A_1/2$ located at $\xi = -\xi_1$ (Figure 9.4):

$$\begin{aligned}\frac{A_1}{2}\int_{-\infty}^{+\infty}\cos[2\pi(\xi+\xi_1)x]\,dx &= \frac{A_1}{2}\lim_{u\to 0}\left\{\frac{1}{u}SINC\left[\frac{\xi+\xi_1}{u}\right]\right\}\\ &= \frac{A_1}{2}\delta[\xi+\xi_1]\end{aligned} \tag{9.16}$$

Therefore the inner product of $f_1[x] = A_1\cos[2\pi\xi_1 x]$ with each of the global mask functions $m_1[x,\xi] = \cos[2\pi\xi x]$ yields a pair of symmetric Dirac delta functions (Figure 9.5). The integral evaluates to:

$$\begin{aligned}g_{11}[\xi] &= \frac{A_1}{2}[\delta[\xi-\xi_1]+\delta[\xi+\xi_1]]\\ &= \frac{A_1}{2}\left(\delta\left[\xi_1\left(\frac{\xi}{\xi_1}-1\right)\right]+\delta\left[\xi_1\left(\frac{\xi}{\xi_1}+1\right)\right]\right)\\ &= \frac{A_1}{2|\xi_1|}\delta\delta\left[\frac{\xi}{\xi_1}\right]\end{aligned} \tag{9.17}$$

where the definition of the even pair of Dirac delta functions in Equation (6.119) has been used. The output $g_{11}[\xi]$ may be interpreted as the "image" of $f_1[x]$ created by an "imaging" system that computes the ensemble of scalar products with $m_1[x,\xi]$ in Equation (9.12). In words, a symmetric sinusoidal input function with amplitude A_1 and spatial frequency ξ_1 generates an output "image" with zero amplitude at all output coordinates except $\pm\xi_1$; the amplitude ("brightness") of the "image" at those frequencies is indeterminate, but the "area" of each of the two nonzero image components is finite and equal to $A_1/2$.

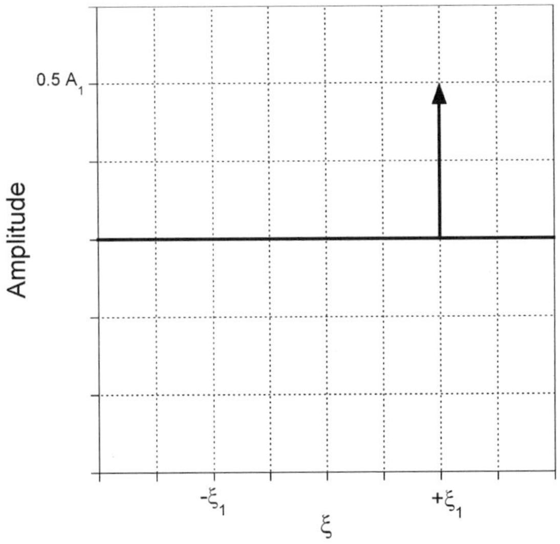

Figure 9.3 $(A_1/2)\int_{-\infty}^{+\infty} \cos[2\pi(\xi - \xi_1)x]\, dx = (A_1/2)\delta[\xi - \xi_1]$.

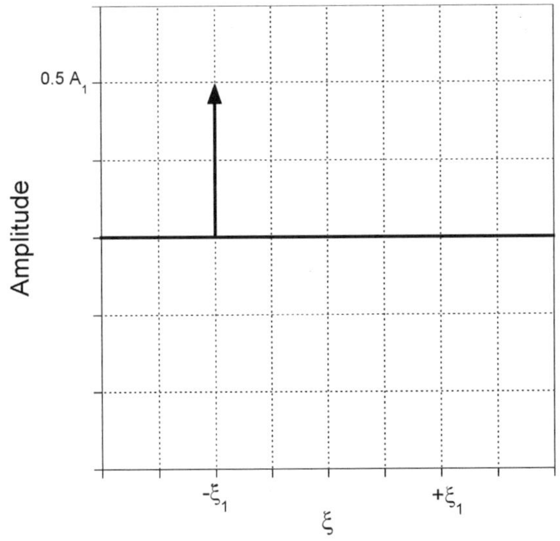

Figure 9.4 $(A_1/2)\int_{-\infty}^{+\infty} \cos[2\pi(\xi + \xi_1)x]\, dx = (A_1/2)\delta[\xi + \xi_1]$.

9.1.2 Example 2: Even Sinusoid Input, Odd Sinusoid Reference

The same symmetric input $f_1[x]$ in Equation (9.10) may be applied to a system that computes the inner product with each of a set of odd (antisymmetric) sinusoidal masks:

$$m_2[x, \xi] = \cos\left[2\pi\xi x - \frac{\pi}{2}\right] = \sin[2\pi\xi x] \tag{9.18}$$

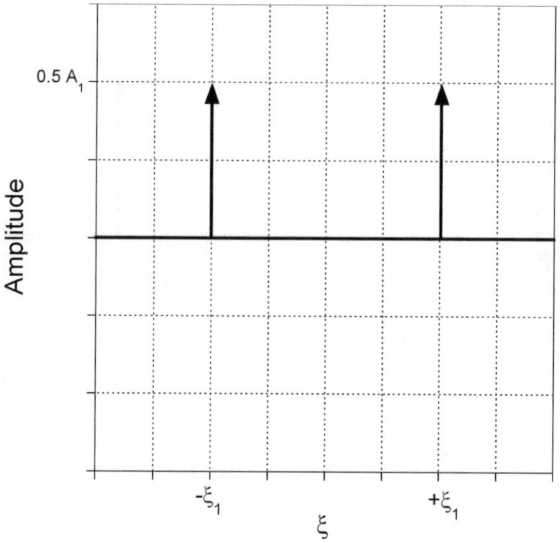

Figure 9.5 $g_{11}[\xi] = (A/2)(\delta[\xi + \xi_1] + \delta[\xi - \xi_1]) = (A_1/2|\xi_1|)\delta\delta[(\xi/\xi_1)]$.

The derivation is based on that used in Chapter 5 to demonstrate the orthogonality of the cosine and sine functions in Equation (5.91) and also is very similar to that used in the preceding example; a different trigonometric identity must be used to recast the form of the integral:

$$\sin[\phi]\cos[\theta] = \frac{1}{2}(\sin[\phi + \theta] + \sin[\phi - \theta]) \tag{9.19}$$

The specific result of this inner product is denoted by the subscript "21", which indicates that the second set of mask functions is applied to the first input function:

$$\begin{aligned} g_{21}[\xi] &= \int_{-\infty}^{+\infty} m_2^*[x,\xi] f_1[x]\,dx = A_1 \int_{-\infty}^{+\infty} \sin[2\pi\xi x]\cos[2\pi\xi_1 x]\,dx \\ &= \frac{A_1}{2}\int_{-\infty}^{+\infty} (\sin[2\pi(\xi+\xi_1)x] + \sin[2\pi(\xi-\xi_1)x])\,dx \\ &= \frac{A_1}{2}\int_{-\infty}^{+\infty} \sin[2\pi(\xi+\xi_1)x]\,dx + \frac{A_1}{2}\int_{-\infty}^{+\infty} \sin[2\pi(\xi-\xi_1)x]\,dx \end{aligned} \tag{9.20}$$

Both terms in Equation (9.20) are odd functions integrated over symmetric limits, and thus evaluate to zero for all ξ_1. The resulting output "image" is the null constant, as shown in Figure 9.6:

$$g_{21}[\xi] = A_1 \int_{-\infty}^{+\infty} \sin[2\pi\xi x]\cos[2\pi\xi_1 x]\,dx = 0[\xi] \tag{9.21}$$

The fact that the projection of $\cos[2\pi\xi_1 x]$ onto any odd sinusoidal function is null demonstrates that all members of the set of cosine functions are orthogonal to any of the sine functions.

9.1.3 Example 3: Odd Sinusoid Input, Even Sinusoid Reference

In the third example, the input function is the odd sinusoid with fixed spatial frequency ξ and amplitude A_1:

$$f_2[x] = A_1 \sin[2\pi\xi_1 x] \tag{9.22}$$

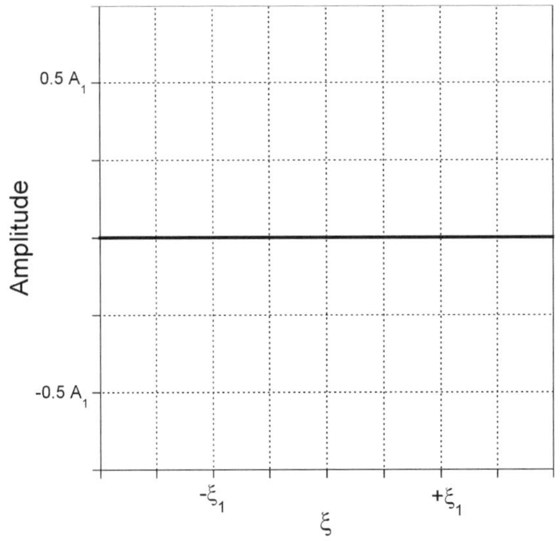

Figure 9.6 $g_{21}[\xi] = A_1 \int_{-\infty}^{+\infty} \sin[2\pi\xi x] \cos[2\pi\xi_1 x]\, dx = 0[\xi]$.

while the reference functions are the members of the original set of symmetric masks $m_1[x, \xi] = \cos[2\pi\xi x]$ from Equation (9.11). The analysis is identical to that of Example 2 where odd functions were integrated over symmetric limits. The output again is the null constant because the sine and cosine functions are orthogonal regardless of the spatial frequency ξ (Figure 9.7):

$$g_{12}[\xi] = A_1 \int_{-\infty}^{+\infty} \cos[2\pi\xi x] \sin[2\pi\xi_1 x]\, dx = 0[\xi] \tag{9.23}$$

9.1.4 Example 4: Odd Sinusoid Input and Reference

In this final example, both the input and reference functions are assumed to be odd sinusoids; the input function is $f_2[x]$ from Equation (9.22) while $m_2[x, \xi]$ is specified in Equation (9.18):

$$\begin{aligned} g_{22}[\xi] &= \int_{-\infty}^{+\infty} f_2[x]\, m_2[x, \xi]\, dx \\ &= A_1 \int_{-\infty}^{+\infty} \sin[2\pi\xi x] \sin[2\pi\xi_1 x]\, dx \end{aligned} \tag{9.24}$$

The analysis is identical to that of Example 1 after recasting the integral via the identity:

$$\sin[\phi]\sin[\theta] = \frac{1}{2}(\cos[\phi - \theta] - \cos[\phi + \theta]) \tag{9.25}$$

to obtain the difference of two integrals of even cosine functions over infinite limits:

$$g_{22}[\xi] = \frac{A_1}{2}\left(\int_{-\infty}^{+\infty} \cos[2\pi(\xi - \xi_1)\, x]\, dx - \int_{-\infty}^{+\infty} \cos[2\pi(\xi + \xi_1)\, x]\, dx\right) \tag{9.26}$$

These component integrals are identical to those in Equations (9.15) and (9.16). The output is the sum of two Dirac delta functions: one centered at $\xi = +\xi_1$ with area $A_1/2$ and one at $\xi = -\xi_1$ with area

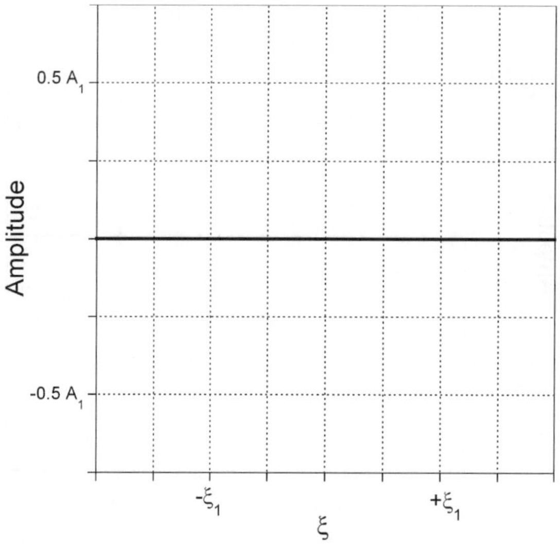

Figure 9.7 $g_{12}[\xi] = A_1 \int_{-\infty}^{+\infty} \cos[2\pi\xi x]\sin[2\pi\xi_1 x]\,dx = 0[\xi]$.

$-A_1/2$. The expression for $g_{22}[\xi]$ may be recast in terms of the odd pair of Dirac delta functions that was defined in Equation (6.121):

$$\begin{aligned} g_{22}[\xi] &= \frac{A_1}{2}(\delta[\xi-\xi_1] - \delta[\xi+\xi_1]) \\ &= \frac{A_1}{2|\xi_1|}\left(-\delta_\delta\left[\frac{\xi}{\xi_1}\right]\right) \end{aligned} \tag{9.27}$$

The "output" $g_{22}[\xi]$ is illustrated in Figure 9.8. The odd pair of Dirac delta functions resulting from the sine-wave input is the negative of the definition of the odd pair: the Dirac delta function located at the negative frequency $-\xi_1$ has negative area, while that at $+\xi_1$ has positive area.

Based on the discussion of $g_{11}[x]$, this result may be interpreted as a demonstration that the members of the set of odd sinusoids with positive frequencies are orthogonal.

To summarize the results for these four combinations of even and odd input and reference (or mask) functions, the output image $g_{ij}[\xi]$ is the null function $0[\xi]$ except when the reference function $m_i[x]$ and the input function $f_j[x]$ are *both* even or *both* odd. The output amplitude in these two cases is nonzero only at positive and negative values of the spatial frequency: $\xi = \pm\xi_1$.

The next step in this development is to consider the result when the input function $f[x]$ is a sinusoid with arbitrary initial phase, rather than being strictly even or odd. Because the inner product is an integral over infinite limits, it satisfies the condition for linearity:

$$\begin{aligned} &\int_a^b (\alpha f_1[x] + \beta f_2[x]) m[x;\xi]\,dx \\ &\quad = \alpha \int_a^b f_1[x] m[x;\xi]\,dx + \beta \int_a^b f_2[x] m[x;\xi]\,dx \end{aligned} \tag{9.28}$$

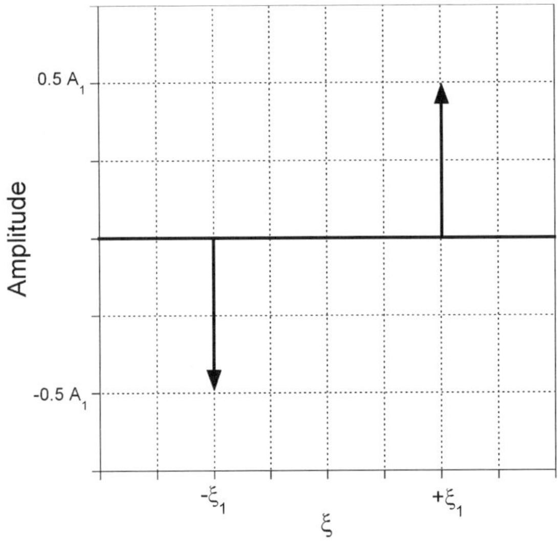

Figure 9.8 $g_{22}[\xi] = (A_1/2|\xi_1|)(-^{\delta}\delta[\xi/\xi_1])$.

This observation may be applied to decompose the new representation of an arbitrary sinusoidal input $f[x]$ into its even and odd parts by applying Equation (2.22):

$$\begin{aligned} f[x] &= A_1 \cos[2\pi\xi_1 x + \phi_1] \\ &= A_1 \cos[\phi_1] \cdot \cos[2\pi\xi_1 x] - A_1 \sin[\phi_1] \cdot \sin[2\pi\xi_1 x] \end{aligned} \tag{9.29}$$

The projection of the sinusoid with arbitrary phase onto the members of the two sets of reference functions may be decomposed into a weighted sum of the individual outputs from Equations (9.17) and (9.27). The even part of the sinusoid yields a symmetric pair of Dirac delta functions with areas $(A_1/2)\cos[\phi_1]$ from the inner products with the set of cosine masks $m_1[x, \xi]$, and a null output from the set of odd masks. The odd part of $f[x]$ and the odd masks m_2 generate an antisymmetric pair of Dirac delta functions with areas $\pm(A_1/2)\sin[\phi_1]$; a null output results from the symmetric masks. Thus the end result is:

$$g[\xi] = \frac{A_1}{2}|\xi_1|\left[\cos[\phi_1]\left(\delta\delta\left[\frac{\xi}{\xi_1}\right]\right) + \sin[\phi_1]\left(-^{\delta}\delta\left[\frac{\xi}{\xi_1}\right]\right)\right] \tag{9.30}$$

The output amplitude is zero at frequencies $\xi \neq \xi_1$ regardless of the amplitude and phase of $f[x]$. By itself, this result demonstrates that a sinusoidal function $f[x]$ is orthogonal to any sinusoid with a different spatial frequency. This is a generalization of the result in Equation (5.91) that the sinusoid $f[x]$ is orthogonal to any function with the same frequency and with the initial phase in quadrature. The relative areas of the Dirac delta functions in the even and odd parts of Equation (9.30) determine the initial phase ϕ_1 of the sinusoid via the arctangent of the ratio:

$$\tan^{-1}\left[\frac{(A_1 \sin[\phi_1])/2}{(A_1 \cos[\phi_1])/2}\right] = \tan^{-1}\left[\frac{\sin[\phi_1]}{\cos[\phi_1]}\right] = \phi_1 \tag{9.31}$$

Note that if $f[x]$ were composed of a sum of arbitrary sinusoids with different frequencies ξ_n, the independent application of the two families of mask operators $m_1[x, \xi] = \cos[2\pi\xi x]$ and $m_2(x, \xi) = \sin[2\pi\xi x]$ would generate pairs of output images $g_{11}[\xi]$ and $g_{22}[\xi]$ that are respectively composed of even and odd pairs of Dirac delta functions located at $\pm|\xi_n|$. The areas of these Dirac delta functions are proportional to the amplitudes of the even and odd parts of those sinusoids. Put another way, the output images $g_{11}[\xi]$ and $g_{22}[\xi]$ specify the frequencies, amplitudes, and initial phases of the sinusoids which, when summed, generate the image $f[x]$.

As final observations about this pair of sinusoidal mask functions, note that each of the input sinusoids has infinite support; they are "global" functions. However, the outputs have finite support because each is null outside of the interval with endpoints $\pm\xi_1$. In words, the operator constructed a "local" function with compact support from a "global" function. The location of the Dirac delta functions in the "images" $g[\xi]$ are determined by the spatial frequency ξ of the input sinusoid $f[x]$.

Summary of this global operator $g_{ij}[\xi] = \int_{-\infty}^{+\infty} m_i[x, \xi] f_j[x]\, dx$:

	Input	
"Reference"	$f_1[x] = A_1 \cos[2\pi\xi_1 x]$	$f_2[x] = A_1 \sin[2\pi\xi_1 x]$
$m_1[x, \xi] = \cos[2\pi\xi x]$	$g_{11}[\xi] = (A_1/2\vert\xi_1\vert)\delta\delta[\xi/\xi_1]$	$g_{12}[\xi] = 0[\xi]$
$m_2[\xi; x] = \sin[2\pi\xi x]$	$g_{21}[\xi] = 0[\xi]$	$g_{22}[\xi] = -(A_1/2)^{\delta}{}_{\delta}[\xi/\xi_1]$

1. $g_{ij}[\xi] = 0[\xi]$ unless *both* $m_i[x]$ *and* $f_j[x]$ have the same symmetry (even or odd).
2. If $g_{ij}[\xi] \neq 0$, the image is a pair of "spikes" of *magnitude* $|A_1|$, where A_1 is the amplitude of the sinusoid.
3. The "spikes" are located a distance ξ_i from the origin, where ξ_i is the spatial frequency of the sinusoid.
4. "Global" information is transformed to "local" information, e.g., sinusoids with infinite support map to Dirac delta functions with infinitesimal support.

This example illustrates the action of the transformation where "global" information in $f[x]$ becomes "local" information in $g[\xi]$. The converse is also true: local information in $g[\xi]$ becomes global information in $f[x]$.

9.2 Linear Combinations of Reference Functions

The linearity of the operation of integration over fixed limits ensures that the scalar (inner) product of the input function with the ensemble of sinusoidal mask operators is a linear operation. This allowed easy analysis of linear combinations of input functions in Equation (9.28), but also means that a new reference function $m[x, \xi]$ may be constructed from linear combinations of $m_1[x, \xi]$ and $m_2[x, \xi]$:

$$\begin{aligned} m[x, \xi] &= \alpha m_1[x, \xi] + \beta m_2[x, \xi] \\ &= \alpha \cos[2\pi\xi x] + \beta \sin[2\pi\xi x] \end{aligned} \tag{9.32}$$

where α and β are weights applied to the even and odd parts. The operator $\mathcal{O}$ defined by this new mask "responds" simultaneously to both even and odd sinusoidal inputs:

$$\begin{aligned} \mathcal{O}\{f[x]\} &= \int_{-\infty}^{+\infty} m^*[x, \xi] f[x]\, dx \\ &\equiv \alpha \int_{-\infty}^{+\infty} \cos[2\pi\xi x] f[x]\, dx + \beta \int_{-\infty}^{+\infty} \sin[2\pi\xi x] f[x]\, dx \end{aligned} \tag{9.33}$$

The output of the first integral is a response to the even part of $f[x]$, while that of the second arises from the odd part. Two examples of this general operator will be introduced that differ only in the values of the "weights" α and β.

9.2.1 Hartley Transform

We first set $\alpha = \beta = 1$, which produces a reference or "mask" function that may be written to produce a single sinusoid with magnitude $\sqrt{2}$ and initial phase $\phi_0 = -\pi/4$:

$$\begin{aligned} m[x, \xi] &= 1 \cdot m_1[x, \xi] + 1 \cdot m_2[x, \xi] \\ &= \cos[2\pi\xi x] + \sin[2\pi\xi x] \\ &= \cos[2\pi\xi x] + \cos\left(2\pi\xi x - \frac{\pi}{2}\right) \\ &= \sqrt{2}\cos\left(2\pi\xi x - \frac{\pi}{4}\right) \end{aligned} \tag{9.34}$$

which may be demonstrated easily by applying Equation (9.29). Each of the reference functions $m[x, \xi]$ is real valued, so that the complex conjugation present in the inner product of Equation (9.9) has no effect on the result. Some authors (notably Bracewell) assign the name $\cos[2\pi\xi x] + \sin[2\pi\xi x] \equiv CAS[x, \xi]$, where the notation is an abbreviation for Cosine And Sine. The integral operator in Equation (9.33) using Equation (9.34) is often called the "Hartley transform" to honor R.V.L. Hartley, who studied it in the 1940s (Bracewell, 1986b). We denote this operator by **H**:

$$\begin{aligned} \mathbf{H}\{f[x]\} &= \int_{-\infty}^{+\infty} m^*[x, \xi] f[x]\, dx \\ &= \int_{-\infty}^{+\infty} CAS[x, \xi] f[x]\, dx = g_h[\xi] \end{aligned} \tag{9.35}$$

The Hartley transform has received some attention in image processing circles because it has the arguable advantage that its character (real, imaginary, or complex) is the same as the input function $f[x]$:

$$\begin{aligned} g_H[\xi] &= \int_{-\infty}^{+\infty} CAS[x, \xi](\Re\{f[x]\} + i\,\Im\{f[x]\})\, dx \\ &= \Re\left\{\int_{-\infty}^{+\infty} CAS[x, \xi] f[x]\, dx\right\} + i\Im\left\{\int_{-\infty}^{+\infty} CAS[x, \xi] f[x]\, dx\right\} \end{aligned} \tag{9.36}$$

9.2.2 Examples of the Hartley Transform

As an obvious first example of the Hartley transform, consider the output generated from a real-valued sinusoidal input with arbitrary initial phase. We may apply Equation (9.29) to decompose it into its even and odd parts:

$$\begin{aligned} f[x] &= A_1 \cos[2\pi\xi_1 x + \phi_1] \\ &= (A_1 \cos[\phi_1]) \cos[2\pi\xi_1 x] + (-A_1 \sin[\phi_1]) \sin[2\pi\xi_1 x] \end{aligned} \tag{9.37}$$

The new representation is obtained by direct substitution of Equations (9.17) and (9.27):

$$\begin{aligned} \mathbf{H}\{f[x]\} &= \frac{A_1}{2}\cos[\phi_1](\delta[\xi - \xi_1] + \delta[\xi + \xi_1]) - \frac{A_1}{2}\sin[\phi_1](\delta[\xi - \xi_1] - \delta[\xi + \xi_1]) \\ &= \left(\frac{A_1}{2}(\cos[\phi_1] + \sin[\phi_1])\right)\delta[\xi + \xi_1] + \left(\frac{A_1}{2}(\cos[\phi_1] - \sin[\phi_1])\right)\delta[\xi - \xi_1] \end{aligned} \tag{9.38}$$

In words, the Hartley transform of a sinusoid is a pair of Dirac delta functions with areas proportional to the sum and difference of the cosine and sine of the initial phase, as shown in Figure 9.9. When the initial

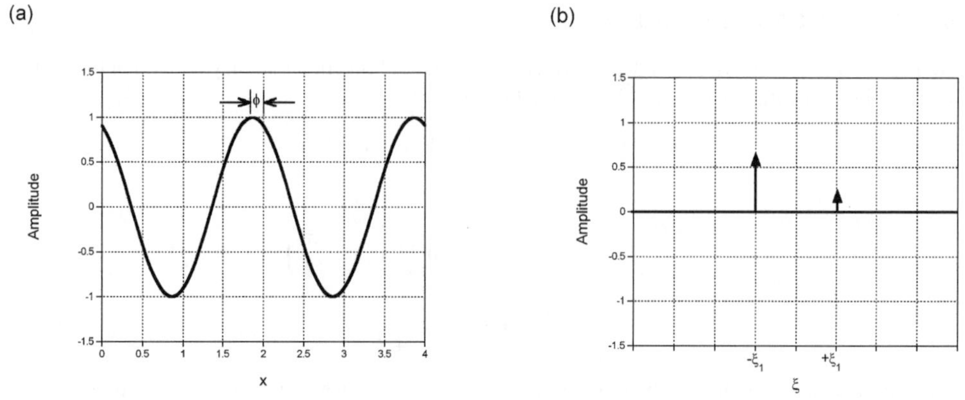

Figure 9.9 (a) $f[x] = \cos[2\pi\xi_0 x + \phi_0]$ and (b) Hartley transform $\mathbf{H}\{f[x]\} = [(A_1/2)(\cos[\phi_1] + \sin[\phi_1])] \cdot \delta[\xi + \xi_1] + ((A_1/2)(COS(\phi_1) - SIN(\phi_1)))\delta[\xi - \xi_1]$.

phase $\phi_1 = 0$, the input function $f[x]$ is a symmetric cosine function and the resulting transform is a symmetric pair of Dirac delta functions. If the initial phase $\phi_1 = \pm\pi/2$ radians, the Hartley transform is an odd pair of Dirac delta functions. The phase ϕ_1 of an input sinusoid may be evaluated by comparing the relative areas of the resulting Dirac delta functions.

Now consider the Hartley transform of the slightly more general case of real- and complex-valued sinusoids with arbitrary amplitude, initial phase, and frequency:

$$f[x] = A_1 \cos[2\pi\xi_1 x + \phi_1] + i\, A_2 \cos[2\pi\xi_2 x + \phi_2] \tag{9.39}$$

The Hartley transform is the complex sum of the Hartley transforms of the individual components, and thus consists of complex-valued pairs of Dirac delta functions (Figure 9.10):

$$\begin{aligned}\mathbf{H}\{f[x]\} &= \frac{A_1}{2}(\delta[\xi + \xi_1](\cos[\phi_1] + \sin[\phi_1]) + \delta[\xi - \xi_1](\cos[\phi_1] - \sin[\phi_1]))\\ &\quad + i\frac{A_2}{2}(\delta[\xi + \xi_2](\cos[\phi_2] + \sin[\phi_2]) + \delta[\xi - \xi_2](\cos[\phi_2] - \sin[\phi_2]))\end{aligned} \tag{9.40}$$

The Hartley transforms of some other functions will be considered after describing the Fourier transform.

9.2.3 Inverse of the Hartley Transform

We now seek to derive a means to determine $f[x]$ from its Hartley transform $\mathbf{H}\{f[x]\}$. If this inverse operation exists, then the Hartley transform satisfies the requirements for an *equivalent representation* of the coordinate-space representation of $f[x]$, meaning that complete knowledge of $f[x]$ may be obtained from $\mathbf{H}\{f[x]\}$, and vice versa. To demonstrate the inverse operation, first apply the sifting property of the Dirac delta function from Equation (6.113) to express $f[x]$ as the convolution of $f[x]$ and $\delta[x]$:

$$f[x] = f[x] * \delta[x] = \int_{-\infty}^{+\infty} f[\alpha]\delta[x - \alpha]\, d\alpha \tag{9.41}$$

We substitute an integral form for $\delta[x - \alpha]$ that is a straightforward generalization of Equation (6.106):

$$\delta[x - \alpha] = \int_{-\infty}^{+\infty} \cos[2\pi\xi(x - \alpha)]\, d\xi \tag{9.42}$$

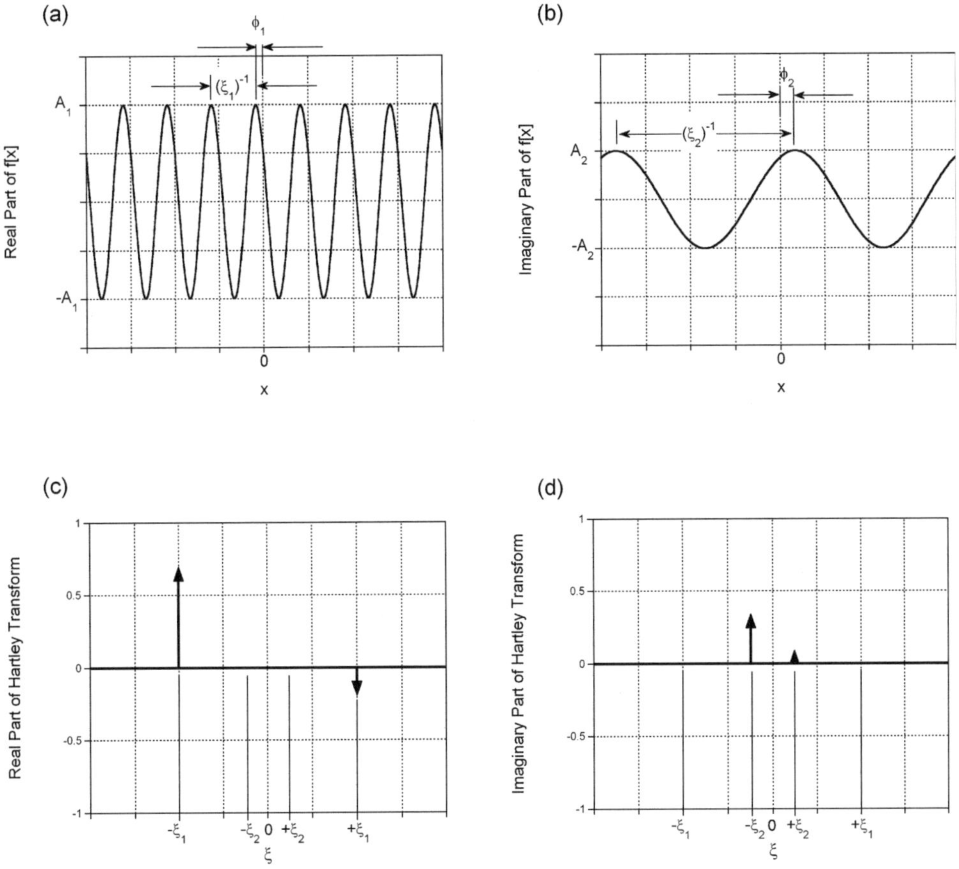

Figure 9.10 Hartley transform of complex function: (a) $\Re\{f[x]\} = A_1 \cos[2\pi\xi_1 x + \phi_1]$; (b) $\Im\{f[x]\} = A_2 \cos[2\pi\xi_2 x + \phi_2]$; (c) $\Re\{\mathbf{H}\{f[x]\}\}$, which shows two 1-D Dirac delta functions at larger values of ξ; (d) $\Im\{\mathbf{H}\{f[x]\}\}$, which consists of 1-D Dirac delta functions at smaller spatial frequencies.

For reasons that will be evident very soon, we will add the integral of an odd function of ξ over infinite limits, which we know to be zero:

$$\int_{-\infty}^{+\infty} \sin[2\pi\xi(x+\alpha)]\, d\xi = 0 \tag{9.43}$$

The equivalent expression for the Dirac delta function is obtained by substituting the trigonometric identities from Equation (4.42):

$$\begin{aligned}\delta[x-\alpha] &= \int_{-\infty}^{+\infty} (\cos[2\pi\xi(x-\alpha)] + \sin[2\pi\xi(x+\alpha)])\, d\xi \\ &= \int_{-\infty}^{+\infty} (\cos[2\pi\xi x] + \sin[2\pi\xi x])(\cos[2\pi\xi\alpha] + \sin[2\pi\xi\alpha])\, d\xi \end{aligned} \tag{9.44}$$

After substituting this expression for $\delta[x-\alpha]$ into the sifting property in Equation (6.106), we obtain an identity for $f[x]$ in the form of a double integral:

$$f[x] = \int_{-\infty}^{+\infty} f[\alpha]\left(\int_{-\infty}^{+\infty} \cos[2\pi\xi x] + \sin[2\pi\xi x])(\cos[2\pi\xi\alpha] + \sin[2\pi\xi\alpha])\, d\xi\right) d\alpha \tag{9.45}$$

After exchanging the order of integration and substituting the definition of the Hartley transform in Equation (9.35), an equivalent expression for $f[x]$ is obtained:

$$\begin{aligned} f[x] &= \int_{-\infty}^{+\infty}\left(\int_{-\infty}^{+\infty} f[\alpha](\cos[2\pi\xi\alpha] + \sin[2\pi\xi\alpha])\, d\alpha\right)(\cos[2\pi\xi x] + \sin[2\pi\xi x])\, d\xi \\ &= \int_{-\infty}^{+\infty} \mathbf{H}\{f[x]\}(\cos[2\pi\xi x] + \sin[2\pi\xi x])\, d\xi \\ &= \int_{-\infty}^{+\infty} \mathbf{H}\{f[x]\} CAS[2\pi\xi x]\, d\xi \equiv \mathbf{H}^{-1}\{\mathbf{H}\{f[x]\}\} \end{aligned} \tag{9.46}$$

In words, this defines the mathematical recipe for regenerating the space-domain representation $f[x]$ from its Hartley transform $\mathbf{H}\{f[x]\}$. Note that the expressions for the forward and inverse Hartley transforms in Equations (9.35) and (9.46) differ only in the variable of integration: x in the forward Hartley transform and ξ in the inverse transform.

9.3 Complex-Valued Reference Functions

Another variant of reference function (mask) that responds simultaneously to even and odd sinusoids is obtained by setting $\alpha = 1$ and $\beta = +i$ in Equation (9.32). It may be expressed in compact form by applying the Euler relation from Equation (4.19):

$$\begin{aligned} m[x,\xi] &= 1\cdot m_1[x,\xi] + i\cdot m_2[x,\xi] \\ &= \cos[2\pi\xi x] + i\,\sin[2\pi\xi x] = e^{+2\pi i\xi x} \end{aligned} \tag{9.47}$$

Because the argument of the exponential must be dimensionless, again we see that the coordinates must have "reciprocal units", i.e., a "space coordinate" x with dimensions of (length), and a "spatial frequency" coordinate ξ with dimensions of $(\text{length})^{-1}$. This yields a concise expression for a transformation operator that acts on functions in the space domain:

$$\begin{aligned} \int_{-\infty}^{+\infty} m^*[x,\xi] f[x]\, dx &= \int_{-\infty}^{+\infty} (\cos[2\pi\xi x] + i\,\sin[2\pi\xi x])^* f[x]\, dx \\ &= \int_{-\infty}^{+\infty} f[x]\, e^{-2\pi i\xi x}\, dx \equiv \mathcal{F}_1\{f[x]\} \equiv F[\xi] \\ \Longrightarrow \mathcal{F}_1\{f[x]\} &\equiv \int_{-\infty}^{+\infty} f[x]\, e^{-2\pi i\xi x}\, dx \equiv F[\xi] \end{aligned} \tag{9.48}$$

where the subscript on the operator $\mathcal{F}_1$ denotes that it acts on 1-D functions. Because the coordinates in the domain of the function $f[x]$ have dimensions of length, we call it the representation in the "space domain", whereas $F[\xi]$ is represented in the "frequency domain". Though it might be logical to say that the two representations are expressed in "reciprocal coordinates", they are more commonly considered to be "conjugate coordinates". We adopt a convention where the amplitude of the space-domain function is denoted by a lower-case letter (e.g., $f[x]$, $g[x]$), while the function expressed in the frequency domain is signified by the corresponding upper-case letter ($F[\xi]$, $G[\xi]$). The frequency-domain representation has some very nice properties when analyzing linear imaging systems and is the basis for many important signal and image processing operations.

Comparison of the operations in Equations (9.9) and (9.48) demonstrates that the latter may be interpreted as the ensemble of inner products (projections) of the input function $f[x]$ onto each of the orthogonal linear-phase complex exponentials $e^{+2\pi i \xi x}$. The amplitude F of the spectrum at frequency ξ_1 in Equation (9.48) measures "how much" of the infinite-support reference function $e^{+2\pi i \xi_1 x}$ is present in $f[x]$. For example, the amplitude of the spectrum evaluated at the origin is $F[\xi = 0]$, which is the projection of $f[x]$ onto $e^{+2\pi i \, 0 \, x} = 1$. Equation (9.8) established this to be the area of $f[x]$:

$$F[\xi = 0] = \int_{-\infty}^{+\infty} f[x] \; e^{-2\pi i \cdot 0 \cdot x} \; dx = \int_{-\infty}^{+\infty} f[x] \, dx \tag{9.49}$$

In words, the area of $f[x]$ is identical to the central ordinate of $F[\xi]$. This observation can be applied to easily evaluate the area of many functions and will be considered in more detail later in this chapter. The amplitude $F[0]$ is known as the "DC component" of $f[x]$ by analogy with the concepts of electric current. The amplitude of "direct current" (DC) is constant over time; it does not oscillate and so has temporal frequency $\nu = 0$. The central ordinate $F[0]$ is the amplitude of the "nonoscillating" (or constant) part of $f[x]$. The analogy may be generalized to say that the amplitudes of the oscillating components of $f[x]$ may be lumped into the category of "AC" terms, for "alternating current".

For any real-valued $f[x]$, the operation of Equation (9.48) projects $f[x]$ onto each of the even sinusoids (cosines) and places the result in the real part of $F[\xi]$; the projections of $f[x]$ onto the odd sinusoids form $\Im\{F[\xi]\}$. The extension to the general case where $f[x]$ is complex valued will be discussed shortly.

Because the operation in Equation (9.48) is an integral over fixed limits, the discussion of Chapter 8 demonstrates that it must be linear. The discussion surrounding Equation (9.3) establishes that the operation is shift *variant* because the output is represented in a different coordinate system specified by ξ. The "new" representation $F[\xi]$ is often called the *spectrum* of $f[x]$ because it specifies the amplitude of the sinusoidal functions which, when summed, generate $f[x]$. This name also suggests that $F[\xi]$ is analogous to the spectrum of white light created by a disperser such as a prism or diffraction grating; the "brightness" of a color in the spectrum is a measure of the amplitude of the sinusoidal component at that temporal frequency. The operation in Equation (9.48) may be interpreted as the *decomposition* or *analysis* of $f[x]$ into its sinusoidal constituents.

The analysis of a function or waveform into its component sinusoids in Equation (9.48) was derived and first applied by Baron Jean-Baptiste Joseph de Fourier in the early 1800s. The process has been named *Fourier analysis* and the operator is denoted by $\mathcal{F}_1$ to honor his achievement, which was one of the most significant developments in mathematical physics. Fourier developed the process as a tool for investigating heat conduction in solid bodies while designing cannon for Napoleon's army. He was able to demonstrate that a function with arbitrary amplitude over a finite domain could be represented as the summation of a possibly infinite number of distinct sinusoids, each with appropriate amplitude, spatial frequency, and initial phase. When presented to the French Academy in 1807, Fourier's claim was ridiculed by many members, who believed that such a superposition could yield only an analytic function (a function that may be differentiated an infinite number of times and therefore has no singularities). It has been reported that Fourier's paper proving the claim was rejected by the Academy for a decade, until the death of the original referee. Of course, Fourier's proof was accepted eventually, and the mathematical transformation now is an essential tool in most, if not all, scientific fields. In passing, we should also note that Fourier made other significant contributions in diverse areas of science; for example, he hypothesized that the average atmospheric temperature would rise due to increased concentration of atmospheric carbon dioxide from burning the fossil fuels that powered the Industrial Revolution. In other words, Fourier may have been the first to predict global warming, which also is accepted as scientific fact.

We have already demonstrated the existence of the inverse Hartley transform in Equation (9.46), which ensured that $f[x]$ and $\mathbf{H}\{f[x]\}$ are equivalent representations. Since the Fourier transform also is a linear combination of the two sinusoidal operations that were defined in Section 9.1, it should not be surprising that $F[\xi]$ also is a satisfactory representation of $f[x]$. The reasons for calling $f[x]$ the "space-domain representation" were described earlier in this chapter; the input function is projected onto a complete set of basis functions that are distinguished by a parameter x that has units of length in Equation (9.4).

The Fourier transforms of $\cos[2\pi\xi_1 x]$ (Figure 9.11) and $\sin[2\pi\xi_1 x]$ (Figure 9.12) may be derived simply by applying the results of the earlier discussion of the component transformations in Equations (9.15) and (9.27). The Fourier transform of the real-valued and symmetric cosine function is a symmetric pair of real-valued Dirac delta functions:

$$\begin{aligned}\mathcal{F}_1\{\cos[2\pi\xi_1 x]\} &= \frac{1}{2}\delta[\xi+\xi_1] + \frac{1}{2}\delta[\xi-\xi_1] \\ &= \frac{1}{2}\left(\delta\left[\xi_1\left(\frac{\xi}{\xi_1}+1\right)\right] + \delta\left[\xi_1\left(\frac{\xi}{\xi_1}-1\right)\right]\right) \\ &= \frac{1}{2|\xi_1|}\left(\delta\left[\frac{\xi}{\xi_1}+1\right] + \delta\left[\frac{\xi}{\xi_1}-1\right]\right) \equiv \frac{1}{2|\xi_1|}\delta\delta\left[\frac{\xi}{\xi_1}\right]\end{aligned} \tag{9.50}$$

The real-valued and even character of $\cos[2\pi\xi_1 x]$ ensure that its Fourier and Hartley transforms must be identical.

The Fourier transform of the real-valued odd sine is an odd pair of Dirac delta functions with imaginary weights:

$$\begin{aligned}\mathcal{F}_1\{\sin[2\pi\xi_1 x]\} &= -i\left(\frac{1}{2}\delta[\xi-\xi_1] - \frac{1}{2}\delta[\xi+\xi_1]\right) \\ &\equiv -i\frac{1}{2|\xi_1|}\left(-{}_\delta\delta\left[\frac{\xi}{\xi_1}\right]\right) \\ &= \frac{i}{2|\xi_1|}{}_\delta\delta\left[\frac{\xi}{\xi_1}\right]\end{aligned} \tag{9.51}$$

The negative coefficient applied to the odd imaginary part of the Fourier reference function in Equation (9.48) "inverts" the amplitude of the odd pair of Dirac delta functions. This ensures that the Fourier transform of the odd sinusoid with positive frequency ξ_1 is proportional to the odd pair of Dirac delta functions that was defined in Equation (6.121).

For obvious reasons, the individual global operators in Equation (9.48) are called the *Fourier cosine* and *Fourier sine* transforms by some authors. Note, however, that the so-called *discrete cosine transform* (DCT) commonly used in image data compression is *not* a discrete version of the Fourier cosine transform, as might be guessed from the name. The form and application of the DCT will be considered in Chapter 15.

Other frequency-domain coordinates also are used to define variants of the Fourier transform. For example, we defined the wavenumber in Equation (2.13) as the spatial frequency scaled by 2π radians per cycle, so that it has dimensions of radians per unit length. The corresponding definition of the Fourier transform is:

$$\int_{-\infty}^{+\infty} f[x]\, e^{-ikx}\, dx \equiv F[k] \tag{9.52}$$

9.4 Transforms of Complex-Valued Functions

The example in Section 9.1.1 demonstrated that the real part of the Fourier transform of a real-valued function $f[x]$ must be an even function of ξ, while the example in Section 9.1.4 established that the imaginary part of the Fourier transform of such a real-valued function must be odd. For a more general complex-valued function of the form $f[x] = f_R[x] + i f_I[x]$, its Fourier transform may be decomposed into real and imaginary parts by applying the linearity property of the Fourier transform as discussed

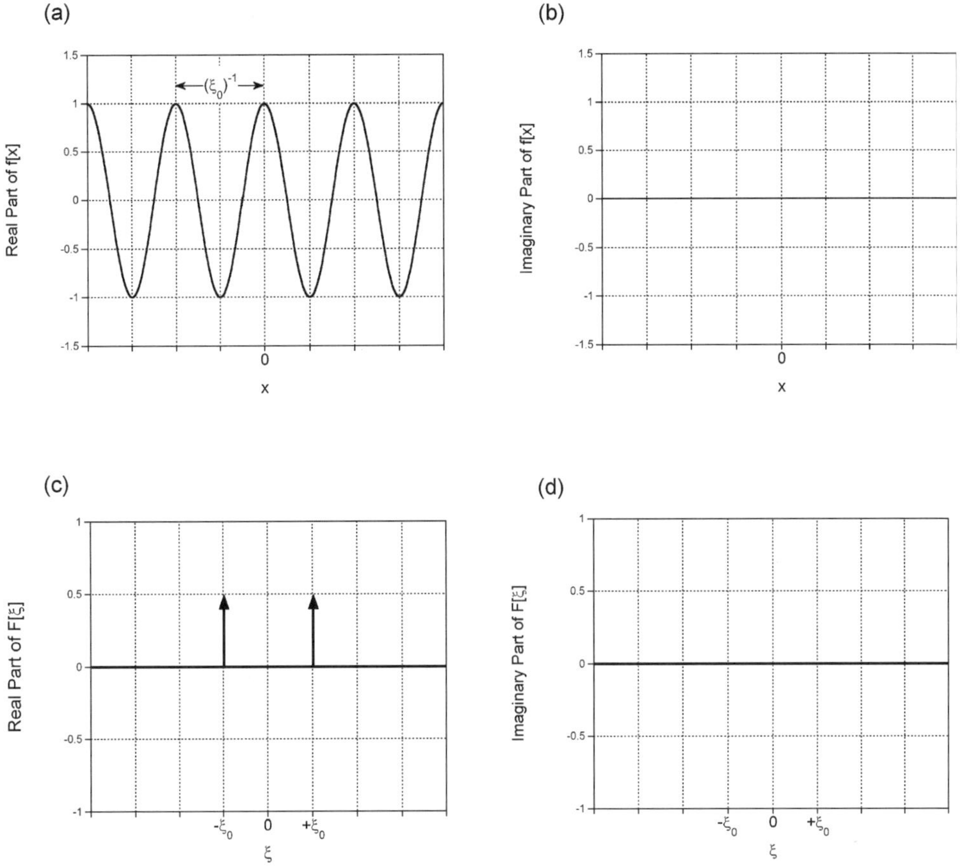

Figure 9.11 Fourier transform of the real-valued and even cosine function with spatial frequency ξ_0 is a real-valued and even pair of Dirac delta functions located at frequencies $\xi = \pm\xi_0$: (a) $\Re\{f[x]\} = \cos[2\pi\xi_0 x]$; (b) $\Im\{f[x]\} = 0[x]$; (c) $\Re\{F[\xi]\} = 0.5(\delta[\xi+\xi_1]+\delta[\xi-\xi_1])$; $\Im\{F[\xi]\} = 0[\xi]$.

following Equation (9.48):

$$
\begin{aligned}
F[\xi] &= F_R[x] + i\, F_I[\xi] \\
&= \int_{-\infty}^{+\infty} (\cos[2\pi\xi x] - i\,\sin[2\pi\xi x])(f_R[x] + i f_I[x])\, dx \\
&= \left(\int_{-\infty}^{+\infty} f_R[x]\cos[2\pi\xi x] + f_I[x]\sin[2\pi\xi x]\, dx\right) \\
&\quad + i\left(\int_{-\infty}^{+\infty} f_I[x]\cos[2\pi\xi x] - f_R[x]\sin[2\pi\xi x]\, dx\right) \\
&= \left(\int_{-\infty}^{+\infty} f_{R,e}[x]\cos[2\pi\xi x] + f_{I,o}[x]\sin[2\pi\xi x]\, dx\right) \\
&\quad + i\left(\int_{-\infty}^{+\infty} [f_{I,e}[x]\cos[2\pi\xi x] - f_{R,o}[x]\sin[2\pi\xi x]]\, dx\right)
\end{aligned}
\tag{9.53}
$$

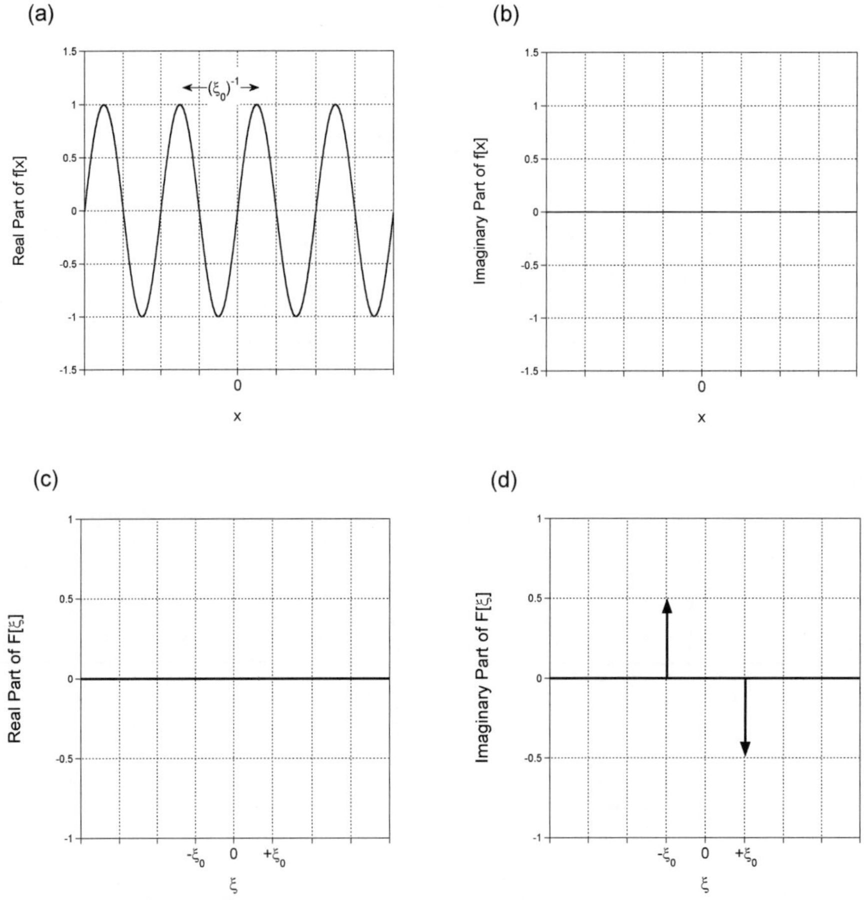

Figure 9.12 Fourier transform of the real-valued and odd sine function with spatial frequency ξ_0 is a complex-valued and odd pair of Dirac delta functions located at frequencies $\xi = \pm\xi_0$: (a) $\Re\{f[x]\} = \sin[2\pi\xi_0 x]$; (b) $\Im\{f[x]\} = 0[x]$; (c) $\Re\{F[\xi]\} = 0[\xi]$; (d) $\Im\{F[\xi]\} = 0.5(\delta[\xi + \xi_1] - \delta[\xi - \xi_1])$.

where the notations $f_{R,e}[x]$, $f_{I,o}[x]$, ..., denote respectively the even part of the real part of $f[x]$, the odd part of the imaginary part, etc. The last step follows because the area of any odd function, such as $(f_{R,e}[x] \sin[2\pi\xi x])$, is zero. Note that $\Re\{F[\xi]\} \equiv F_R[\xi]$ includes pieces due to the even real part and odd imaginary part of $f[x]$, while $\Im\{F[\xi]\} \equiv F_I[\xi]$ is derived from the even imaginary part and odd real part. This observation establishes that the Fourier transform of a real-valued function $f[x]$ must be *Hermitian*; the real part of the transform is generated from the even part of $f[x]$ and also must be even, while the imaginary part of the transform is due to the odd part of $f[x]$ and must also be odd.

Based on the discussion of general complex functions of real variables that led to Equation (4.26), we may construct the magnitude of the frequency-space representation $F[\xi]$ by evaluating the square root of the sum of the squares of the real and imaginary parts:

$$|F[\xi]| = \sqrt{F_R^2[\xi] + F_I^2[\xi]} \tag{9.54}$$

The phase of the Fourier transform at frequency ξ is the arctangent of the ratio of the imaginary and real parts:

$$\Phi\{F[\xi]\} = \tan^{-1}\left[\frac{F_I[\xi]}{F_R[\xi]}\right] \tag{9.55}$$

These two quantities evaluated at a particular spatial frequency ξ_1 completely describe the sinusoidal component of $f[x]$ that oscillates with that frequency. The magnitude $|F[\xi_1]|$ is identical to the maximum amplitude of the sinusoid, while $\Phi\{F[\xi_1]\}$ is the phase angle (in radians) of the sinusoidal component of $f[x]$ evaluated at the origin. In other words, $\Phi\{F[\xi]\}$ is more appropriately described as the "initial phase" rather than just the "phase". Later we will discuss the relative importance of the magnitude and initial phase of the sinusoidal components of $f[x]$.

Recall that the squared magnitude of the space-domain representation ($|f[x]|^2$) was called the *power* of the function in Equation (4.29). By analogy, the square of the magnitude of the Fourier transform in Equation (9.54) commonly is called the *power spectrum* of $f[x]$:

$$|F[\xi]|^2 = (F_R[\xi])^2 + (F_I[\xi])^2 \tag{9.56}$$

The power spectrum is a particularly useful construct for characterizing properties of stochastic functions (noise). The shapes of $|F[\xi]|$ and $|F[\xi]|^2$ are measures of the "variability" of $f[x]$; if the largest magnitudes of $|F[\xi]|$ are concentrated primarily at small spatial frequencies, then $f[x]$ is composed mainly of low-frequency sinusoids and thus must vary slowly with x. If the largest magnitudes of the spectrum occur at $|\xi| \gg 0$, then $f[x]$ must be a "busier" function that varies more rapidly with x. If the magnitude $|F[\xi]|$ is "flat" so that all sinusoidal frequencies are present with approximately equal amplitudes, then the behavior of the space-domain representation $f[x]$ is more difficult to characterize. Probably the most common example of a function with a "flat" spectrum is so-called *white noise*, but other examples will be considered later in this chapter.

9.5 Fourier Analysis of Dirac Delta Functions

The Fourier transform of the Dirac delta function $\delta[x]$ is simple to evaluate via the sifting property in Equation (6.113). This result will be used to derive the inverse Fourier transform via a scheme similar to that used to construct the inverse Hartley transform in Section 9.2.2. The spectrum of the Dirac delta function is:

$$\mathcal{F}_1\{\delta[x]\} = \int_{-\infty}^{+\infty} \delta[x]\, e^{-2\pi i\xi x}\, dx = e^{-2\pi i\xi \cdot 0} = 1[\xi] + i\, 0[\xi] \tag{9.57a}$$

The magnitude is the unit constant:

$$|\mathcal{F}_1\{\delta[x]\}| = \sqrt{1^2 + 0^2} = 1[\xi] \tag{9.57b}$$

and the initial phase is zero radians at all spatial frequencies:

$$\Phi\{\mathcal{F}_1\{\delta[x]\}\} = \tan^{-1}\left[\frac{0}{1}\right] = 0[\xi] \tag{9.57c}$$

These observations lead to the interpretation of the spectrum $F[\xi]$; an impulse located at the origin may be generated by summing complex linear-phase sinusoids with equal amplitudes and zero initial phase for each spatial frequency $-\infty < \xi < +\infty$, as shown in Figure 9.13. In fact, we already surmised the spectrum in Equation (6.106) based upon the approximation of $\delta[x]$ as $f[x]$ composed of the sum of four cosines with periods $X_1 = \infty$, $X_2 = 8$, $X_3 = 6$, and $X_4 = 4$. This approximation was shown in Figure 6.25:

$$\delta[x] = \int_{-\infty}^{+\infty} 1[\xi] \cdot e^{-2\pi i\xi x}\, d\xi = \int_{-\infty}^{+\infty} 1[\xi] \cdot e^{+2\pi i\xi x}\, d\xi \tag{9.58}$$

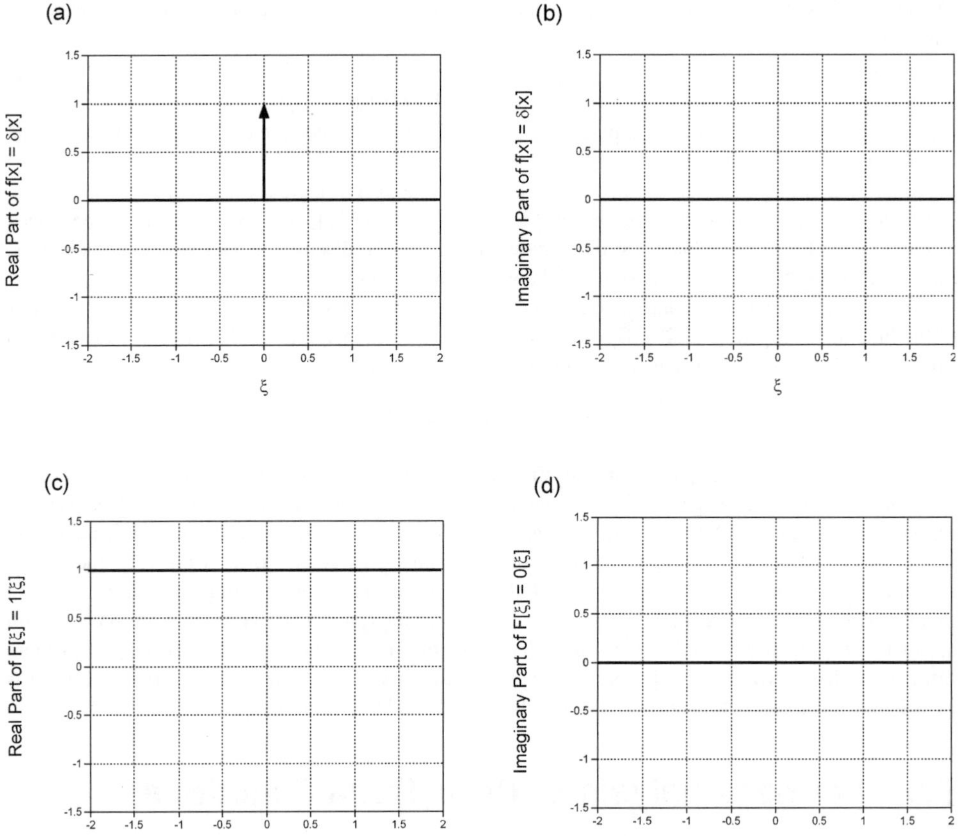

Figure 9.13 Fourier transform of the symmetric Dirac delta function $f[x] = \delta[x] + i \cdot 0[x]$ is $F[\xi] = 1[\xi] + i \cdot 0[\xi]$, indicating that the Dirac delta function is composed of the summation of the infinite set of cosine functions with unit amplitude at all spatial frequencies: (a) $\Re\{f[x]\} = \delta[x]$; (b) $\Im\{f[x]\} = 0[\xi]$; (c) $\Re\{F[\xi]\} = 1[\xi]$; (d) $\Im\{F[\xi]\} = 0[\xi]$.

The sifting property also may be used to evaluate the Fourier transform of a Dirac delta function that has been translated from the origin by some distance x_0:

$$\begin{aligned}\mathcal{F}_1\{\delta[x-x_0]\} &= \int_{-\infty}^{+\infty} \delta[x-x_0]\, e^{-2\pi i\xi x}\, dx = e^{-2\pi i\xi x_0}\\ &= \cos[2\pi\xi x_0] - i\,\sin[2\pi\xi x_0]\end{aligned} \tag{9.59a}$$

(this is the simplest example of the soon-to-be-proven shift theorem of the Fourier transform). The magnitude of the spectrum is unity at all frequencies:

$$|\mathcal{F}_1\{\delta[x-x_0]\}| = \sqrt{\cos^2[2\pi\xi x_0] + \sin^2[2\pi\xi x_0]} = 1[\xi] \tag{9.59b}$$

while the initial phase is a linear function of ξ:

$$\boldsymbol{\Phi}\{\mathcal{F}_1\{\delta[x-x_0]\}\} = \tan^{-1}\left[\frac{-\sin[2\pi\xi x_0]}{\cos[2\pi\xi x_0]}\right] = -2\pi\xi x_0 \tag{9.59c}$$

These results establish that a Dirac delta function centered at a coordinate other than the origin is the sum of unit-amplitude sinusoids with all spatial frequencies and whose initial phases increase (or decrease) in proportion to the spatial frequency ξ. The reason for these initial phases may be visualized by taking the same approximation $f[x]$ for the Dirac delta function in the discussion leading to Equation (6.106) and translating it to a new location, say $x_0 = 2$, so that $f[x] \cong \delta[x-2]$. Clearly, each of the four component cosines also must be translated by the same physical distance:

$$\cos\left[\frac{2\pi(x-x_0)}{X_n}\right] = \cos\left[\frac{2\pi x}{X_n} - \frac{2\pi x_0}{X_n}\right] = \cos[2\pi\xi_n x - (2\pi\xi_n x_0)]$$
$$= \cos[2\pi\xi_n x + \Delta\phi_n] \tag{9.60}$$

where x_0 is the required translation and X_n is the period of the particular sinusoidal component. If $x_0 = 2$ units, then the initial phase of each sinusoid must be incremented by the angle:

$$\Delta\phi_n = -\frac{2\pi \cdot x_0}{X_n} = -4\pi\xi_n \textit{ radians} \propto \xi_n \tag{9.61}$$

which confirms that the required increment of the initial phase necessary to translate the function is proportional to the spatial frequency. Note that the initial phase of the spectrum of $\delta[x-x_0]$ at $\xi = 0$ (DC) is zero, which means that the sinusoidal component with $X_1 = \infty$ (and $\xi_1 = 0$) need not be translated at all. This is because the DC component is the unit constant, which has the same amplitude at all x whether translated or not. The cosine with a period of eight units must be translated by one-fourth of a period to be centered at $x_0 = 2$. This converts the original cosine into a sine function, which requires that the initial phase be decremented by $\pi/2$ radians. In similar fashion, we may determine that $\Delta\phi_3$ and $\Delta\phi_4$ must be $-2\pi/3$ and $-\pi$ radians, respectively, to translate the corresponding sinusoids by two units. These sinusoids are shown in Figure 9.14.

9.6 Inverse Fourier Transform

We have shown that $F[\xi]$ specifies the amplitude and initial phase of each complex sinusoid in the set that generates $f[x]$ when superposed. It is probably obvious that the space-domain function $f[x]$ may be *synthesized* by generating and summing the appropriately weighted sinusoidal components specified by $F[\xi]$. The mathematical expression for this operation of "Fourier synthesis" is just the superposition integral of linear-phase exponentials weighted by the complex-valued scale factors $F[\xi]$. This logic leads us to surmise that the form of the inverse Fourier transform evidently is:

$$f[x] = \int_{-\infty}^{+\infty} F[\xi]\, e^{+2\pi i\xi x}\, d\xi$$
$$= \int_{-\infty}^{+\infty} \left(|F[\xi]|\, e^{i\Phi\{F[\xi]\}}\right) e^{+2\pi i\xi x}\, d\xi \tag{9.62}$$

A perhaps more convincing derivation of the inverse Fourier transform may be constructed that parallels that for the inverse Hartley transform in Section 9.2.2. The integral expression for the Dirac delta function in Equation (6.106) is applied to the specific realization of the sifting property in Equation (6.113):

$$f[x] = f[x] * \delta[x]$$
$$= \int_{-\infty}^{+\infty} f[\alpha]\, \delta[x-\alpha]\, d\alpha$$
$$= \int_{-\infty}^{+\infty} f[\alpha]\left(\int_{-\infty}^{+\infty} e^{+2\pi i\xi(x-\alpha)}\, d\xi\right) d\alpha$$

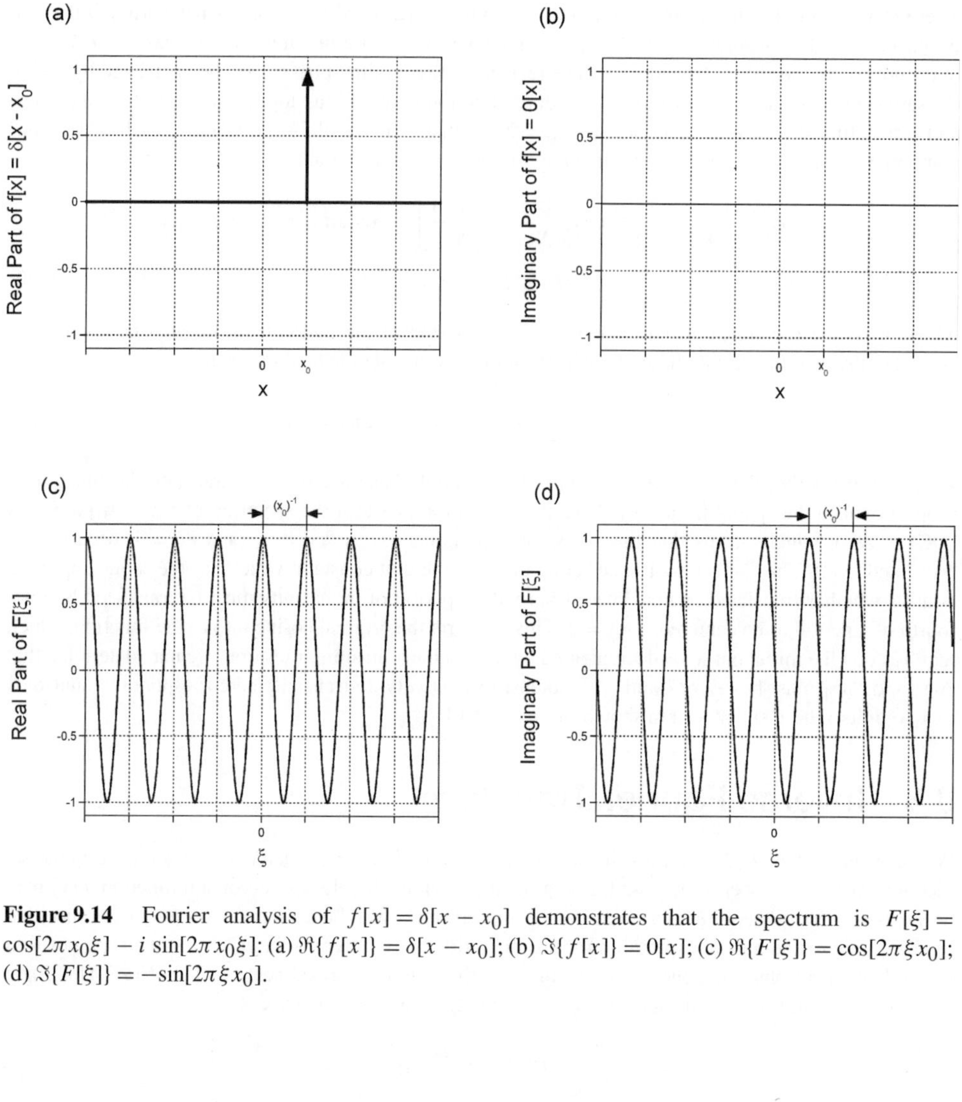

Figure 9.14 Fourier analysis of $f[x] = \delta[x - x_0]$ demonstrates that the spectrum is $F[\xi] = \cos[2\pi x_0\xi] - i\,\sin[2\pi x_0\xi]$: (a) $\Re\{f[x]\} = \delta[x - x_0]$; (b) $\Im\{f[x]\} = 0[x]$; (c) $\Re\{F[\xi]\} = \cos[2\pi\xi x_0]$; (d) $\Im\{F[\xi]\} = -\sin[2\pi\xi x_0]$.

$$= \int_{-\infty}^{+\infty} e^{+2\pi i\xi x}\left(\int_{-\infty}^{+\infty} f[\alpha]\,e^{-2\pi i\xi\alpha}\,d\alpha\right)d\xi$$

$$= \int_{-\infty}^{+\infty} e^{+2\pi i\xi x}\;F[\xi]\,d\xi = f[x] \equiv \mathcal{F}_1^{-1}\{F[\xi]\}$$

$$\Longrightarrow f[x] = \mathcal{F}_1^{-1}\{F[\xi]\} \equiv \int_{-\infty}^{+\infty} F[\xi]\,e^{+2\pi i\xi x}\,d\xi \tag{9.63}$$

The process of Fourier synthesis also may be interpreted as an inner product as defined in Equation (9.3), though this picture is perhaps not as intuitive. The inverse Fourier transform evaluates the set of projections of $F[\xi]$ onto each of the linear-phase complex exponential functions in the frequency domain with negative coordinates:

$$f[x] = \int_{-\infty}^{+\infty} (e^{+2\pi i \xi x})\, F[\xi]\, d\xi$$

$$= \int_{-\infty}^{+\infty} (e^{-2\pi i \xi x})^* F[\xi]\, d\xi = \int_{-\infty}^{+\infty} (e^{+2\pi i \xi(-x)})^* F[\xi]\, d\xi \tag{9.64}$$

Particularly note that the integrals for Fourier analysis in Equation (9.48) and Fourier synthesis in Equation (9.62) are performed over the different variables. The spectrum $F[\xi]$ is the area of $f[x]$ weighted by linear-phase exponentials in the space domain, while $f[x]$ is the area of $F[\xi]$ weighted by linear-phase exponentials in the frequency domain.

We may apply the "superposition" interpretation that was used to infer Fourier synthesis in Equation (9.62) to the forward Fourier transform in Equation (9.48). In words, the spectrum $F[\xi]$ is the superposition of the complex exponentials $e^{-2\pi i \xi x}$ over all space after weighting by the (generally complex-valued) factors $f[x]$.

The existence of the inverse Fourier transform establishes that the two functions $f[x]$ and $F[\xi]$ also must be equivalent representations of the *same* "information". The difference is due only to the choice of basis set. By analogy with the representations of vectors that were considered in Chapter 5, it is quite possible to interpret the transformations $\mathcal{F}_1$ and $\mathcal{F}_1^{-1}$ as rigid rotations of the basis functions to generate the alternate representations. Details of this interpretation are considered by Parker (1990). The two representations $f[x]$ and $F[\xi]$ are often called a *Fourier-transform pair.*

It is clear that properties of functions expressed in the space domain must also be true for the same functions expressed in the frequency domain. For example, the validity of the frequency-domain version of the sifting property of the Dirac delta function in Equation (6.113) is easy to demonstrate:

$$\int_{-\infty}^{+\infty} F[\xi]\delta[\xi - \xi_0]\, d\xi = \int_{-\infty}^{+\infty} F[\xi]\delta[\xi_0 - \xi]\, d\xi = F[\xi_0] \tag{9.65}$$

which leads directly to the space-domain representation of $\delta[\xi]$ via the inverse Fourier transform in Equation (9.63):

$$\mathcal{F}_1^{-1}\{\delta[\xi]\} = \int_{-\infty}^{+\infty} \delta[\xi - 0]\, e^{+2\pi i \xi x}\, d\xi$$

$$= e^{+2\pi i \cdot 0 \cdot x} = 1[x] \tag{9.66}$$

In other words, the spectrum of the unit-constant signal $1[x]$ is a real-valued Dirac delta function located at the origin in the frequency domain:

$$\mathcal{F}_1\{1[x]\} = \mathcal{F}_1\{\mathcal{F}_1^{-1}\{\delta[\xi]\}\}$$

$$= \delta[\xi] + i \cdot 0[\xi] \tag{9.67}$$

The linearity of the Fourier transform ensures that the spectrum of a complex-valued constant c is a Dirac delta function at the origin with area scaled by c. In other words, the constant function $c \cdot 1[x]$ is composed of the single sinusoid with zero frequency and amplitude c:

$$\mathcal{F}_1\{c\} = \mathcal{F}_1\{c \cdot 1[x]\} = c \cdot \mathcal{F}_1\{1[x]\} = c \cdot \delta[\xi] \tag{9.68}$$

9.7 Fourier Transforms of 1-D Special Functions

The spectra of the 1-D special functions defined in Chapter 6 are derived in this section; the Fourier transforms of functions of two or more independent variables will be considered in Chapter 10. As we will see, the spectra of only a few special functions may be derived easily by direct integration. These include *RECT*[x], *SGN*[x], *STEP*[x], $e^{\pm 2\pi i\xi_0 x}$, and *STEP*[x] $\cdot e^{-x}$. The Fourier transforms of some other functions (e.g., *GAUS*[x] and the quadratic-phase function $\exp[\pm i\pi\alpha\xi_0 x^2]$) may be derived rigorously via integration along an appropriate path in the complex domain. The spectra of some periodic functions such as *COMB*[x] may be determined from the spectrum of a single period.

Our strategy in this section will be to derive these easier transforms first and then prove several theorems satisfied by all Fourier transforms. Once these steps are completed, we will be able to derive the Fourier spectra of many functions constructed from scaled and shifted replicas of the special functions. The proofs of the theorems also should help the readers develop some intuition about the Fourier transform.

9.7.1 Fourier Transform of $\delta[x]$

The Fourier transform of a Dirac delta function located at the origin of coordinates already has been derived in Equation (9.58) and is included here for completeness:

$$\mathcal{F}_1\{\delta[x]\} = 1[\xi] + i \cdot 0[\xi] \tag{9.69}$$

The transform is shown in Figure 9.13. In words, the Dirac delta function located at the origin of the space domain is composed of the superposition of unit-amplitude complex linear-phase exponentials with all spatial frequencies. We have already mentioned that this result should be no surprise; it was inferred in the discussion of Equation (6.106). An equivalent expression results by inserting this spectrum into the inverse Fourier transform and combining the components with positive and negative spatial frequencies to form even and real-valued cosine functions with amplitude 2:

$$\begin{aligned}
\mathcal{F}_1^{-1}\{1[\xi]\} &= \int_{-\infty}^{+\infty} 1[\xi]\, e^{+2\pi i\xi x}\, d\xi = \int_{-\infty}^{+\infty} e^{+2\pi i\xi x}\, d\xi \\
&= \left(\int_{-\infty}^{0_-} e^{+2\pi i\xi x}\, d\xi\right) + (e^{+2\pi i\cdot 0\cdot x}) + \left(\int_{0_+}^{+\infty} e^{+2\pi i\xi x}\, d\xi\right) \\
&= 1 + \int_{+\infty}^{0_+} e^{+2\pi i(-\xi)x}\, d(-\xi) + \int_{0_+}^{+\infty} e^{+2\pi i\xi x}\, d\xi \\
&= 1 + \int_{0_+}^{+\infty} (e^{-2\pi i\xi x} + e^{+2\pi i\xi x})\, d\xi \\
&= 1[x] + 2\int_{0_+}^{+\infty} \cos[2\pi\xi x]\, d\xi = \delta[x]
\end{aligned} \tag{9.70}$$

where the notations 0_+ and 0_- are used to indicate that the integrals include all spatial frequencies up to, but not including, $\xi = 0$. In words, this result indicates that the Dirac delta function may be generated by summing the unit constant and cosines with amplitude 2 at all positive spatial frequencies.

9.7.2 Fourier Transform of Rectangle

Direct integration of the expression for $\mathcal{F}_1\{RECT[x]\}$ is easy because the rectangle restricts the limits of integration. The result in Equation (9.53) established that the spectrum of the real-valued and symmetric

RECT function must be real valued and symmetric (even):

$$
\begin{aligned}
\mathcal{F}_1\{RECT[x]\} &= \int_{-\infty}^{+\infty} RECT[x]\, e^{-2\pi i\xi x}\, dx \\
&= \int_{-\frac{1}{2}}^{+\frac{1}{2}} e^{-2\pi i\xi x}\, dx = \left(\frac{e^{-2\pi i\xi x}}{-2\pi i\xi}\right)\Bigg|_{x=-\frac{1}{2}}^{x=+\frac{1}{2}} \\
&= \frac{e^{-i\pi\xi} - e^{+i\pi\xi}}{-2\pi i\xi} \\
&= \frac{1}{\pi\xi}\,\frac{e^{+i\pi\xi} - e^{-i\pi\xi}}{2i} \\
&= \frac{\sin[\pi\xi]}{\pi\xi} + i\,0[\xi] = SINC[\xi] + i\,0[\xi]
\end{aligned}
\tag{9.71a}
$$

The magnitude spectrum is the nonnegative magnitude of a *SINC* function:

$$
|\mathcal{F}_1\{RECT[x]\}| = |SINC[\xi]| \tag{9.71b}
$$

which exhibits "cusps" at the locations of each sign change of the real-valued *SINC*. The initial phase may take on only two possible values in our convention: 0 radians at the frequencies where $SINC[\xi] \geq 0$ and $-\pi$ radians where $SINC[\xi] < 0$:

$$
\Phi\{\mathcal{F}_1\{RECT[x]\}\} = \begin{cases} 0 & \textit{for } 2n \leq |\xi| < 2n+1 \\ -\pi & \textit{for } 2n+1 \leq |\xi| < 2n+2 \end{cases} \tag{9.71c}
$$

As predicted, the Fourier transform of the real-valued and symmetric *RECT* function also is real and even. The imaginary part is zero at all frequencies ξ_0 because the product of $RECT[x]$ and the odd function $-\sin[2\pi\xi_0 x]$ is an odd function that integrates to zero over symmetric limits. The amplitude of the real part of the Fourier transform at frequency ξ_0 is the area of the product of $RECT[x]$ and $\cos[2\pi\xi_0 x]$. At $\xi_0 = 0$, the integral produces the unit area of $RECT[x]$. At nonzero frequencies, the area of the product function decreases from unity due to the decrease in the amplitude of the cosine function. At $\xi_0 = \pm 1$, the area is zero because the unit period of $\cos[2\pi \cdot (\pm 1) \cdot x]$ exactly matches the unit support of the *RECT*. The product of the two functions is a single period of the cosine, which has zero area. Note that the even character of the cosine guarantees that the reference functions at $\xi_0 = \pm 1$ are identical. At spatial frequencies in the range $1 < |\xi_0| < 2$ cycles per unit length, more than one and less than two periods of the reference cosine fit within the unit width of $RECT[x]$. The product function is a truncated cosine with more area in the two negative lobes than in the single positive lobe; the amplitude of the spectrum at these frequencies is this real-valued and negative area, which means that the initial phase of the constituent cosine functions is $\phi = -\pi$ radians. When summing the constituent sinusoids to generate $RECT[x]$, cosines with spatial frequencies in this range are *subtracted* in the superposition, as shown in Figure 9.15.

The spectrum of a rectangle function that has been scaled in width by a positive-valued scale factor b may be derived quite easily by generalizing Equation (9.71a):

$$
\begin{aligned}
\mathcal{F}_1\left\{RECT\left[\frac{x}{b}\right]\right\} &= \int_{-|b|/2}^{+|b|/2} e^{-2\pi i\xi x}\, dx \\
&= \frac{e^{-i\pi\xi|b|} - e^{i\pi\xi|b|}}{-2\pi i\xi} = \left(\frac{e^{+i\pi\xi|b|} - e^{-i\pi\xi|b|}}{2i}\right)\frac{|b|}{\pi\xi|b|} \\
&= \frac{|b|\,\sin[\pi|b|\xi]}{\pi|b|\xi} = |b|\; SINC[|b|\xi] = |b|\; SINC[b\xi]
\end{aligned}
\tag{9.72}
$$

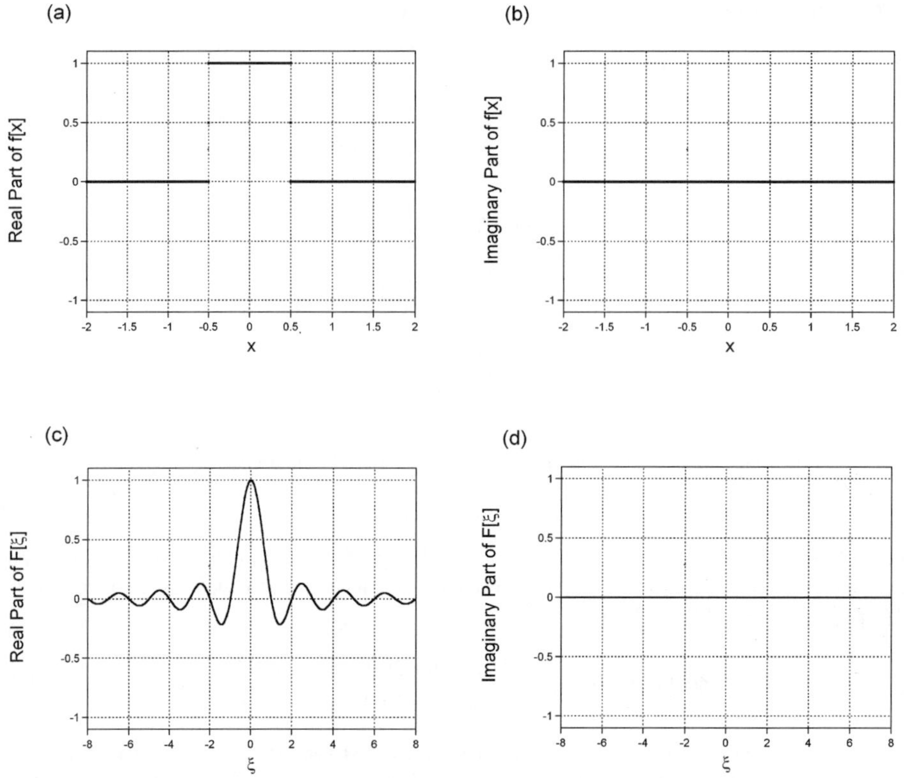

Figure 9.15 The real even function $f[x] = RECT[x] + i\ 0[x]$ is shown with real part in (a) and imaginary part in (b). Its Fourier transform is the real even function $F[\xi] = SINC[\xi] + i0[\xi]$, shown in (c) and (d).

where the symmetry of the *SINC* function has been used in the last step. In the limit $b \to +\infty$, the rectangle approaches the unit constant $1[x]$. The expression for the spectrum in this limit is:

$$\begin{aligned}\mathcal{F}_1\{1[x]\} &= \lim_{b\to\infty}\left\{\mathcal{F}_1\left\{RECT\left[\frac{x}{b}\right]\right\}\right\}\\ &= \lim_{b\to\infty}\{|b|\ SINC[b\xi]\} = \lim_{u\to 0}\left\{\frac{1}{|u|}SINC\left[\frac{\xi}{u}\right]\right\} = \delta[\xi]\end{aligned} \tag{9.73}$$

where $u \equiv 1/b$ and the limiting expression for the Dirac delta function in Equation (6.96a) has been used. This result is further confirmation of Equation (9.67). The amplitude of $1[x]$ is finite and the area is infinite, as demonstrated by applying Equation (9.49).

9.7.3 Fourier Transforms of Sinusoids

The Fourier transforms of the even and odd sinusoids have been derived already and are included here both for completeness and as steps in the derivation of the spectrum of the complex linear-phase exponential. The Fourier transform of the even real-valued cosine with spatial frequency $\pm\xi_0$ is the even pair of Dirac delta functions with scale parameter $\pm\xi_0$ cycles per unit length, as was shown in Equation

(9.50):

$$\mathcal{F}_1\{\cos[2\pi\xi_0 x]\} = \frac{1}{2}\delta[\xi+\xi_0] + \frac{1}{2}\delta[\xi-\xi_0] \equiv \frac{1}{2|\xi_0|}\delta\delta\left[\frac{\xi}{\xi_0}\right] \tag{9.74}$$

The spectrum of the real-valued and odd sine function is an odd pair of imaginary-valued Dirac delta functions:

$$\begin{aligned}\mathcal{F}_1\{\sin[2\pi\xi_0 x]\} &= -i\left(-\frac{1}{2}\delta[\xi+\xi_0] + \frac{1}{2}\delta[\xi-\xi_0]\right)\\ &= +i\left(\frac{1}{2}\delta[\xi+\xi_0] - \frac{1}{2}\delta[\xi-\xi_0]\right)\\ &= -i\frac{1}{2|\xi_0|}\left(-\delta_\delta\left[\frac{\xi}{\xi_0}\right]\right) = +\frac{i}{2|\xi_0|}\delta_\delta\left[\frac{\xi}{\xi_0}\right]\end{aligned} \tag{9.75}$$

The linearity property of the Fourier transform may be applied to show that the spectrum of the imaginary-valued and odd sinusoid $i \cdot \sin[2\pi\xi_0 x]$ is an odd pair of real-valued Dirac delta functions with scale parameter ξ_0:

$$\begin{aligned}\mathcal{F}_1\{i\ \sin[2\pi\xi_0 x]\} &= i(-i)\left(-\frac{1}{2}\delta[\xi+\xi_0] - \frac{1}{2}\delta[\xi-\xi_0]\right)\\ &= \frac{1}{2}(-\delta[\xi+\xi_0] + \delta[\xi-\xi_0])\\ &= \frac{1}{2|\xi_0|}\left(-\delta_\delta\left[\frac{\xi}{\xi_0}\right]\right)\\ &= -\frac{1}{2|\xi_0|}\delta_\delta\left[\frac{\xi}{\xi_0}\right]\end{aligned} \tag{9.76}$$

These two expressions may be interpreted in the light of Equation (9.49) to demonstrate that the area of an infinite-support sinusoidal function must be zero.

The linearity property of the Fourier transform ensures that the spectrum of the complex linear-phase exponential $e^{+2\pi i\xi_0 x}$ is the sum of the spectra of the cosine in Equation (9.74) and of the sine in Equation (9.76) weighted by a factor of i:

$$\begin{aligned}\mathcal{F}_1\{e^{+2\pi i\xi_0 x}\} &= \mathcal{F}_1\{\cos[2\pi\xi_0 x] + i\ \sin[2\pi\xi_0 x]\}\\ &= \frac{1}{2|\xi_0|}\delta\delta\left[\frac{\xi}{\xi_0}\right] - \frac{1}{2|\xi_0|}\delta_\delta\left[\frac{\xi}{\xi_0}\right]\\ &= \frac{1}{2}(\delta[\xi+\xi_0] + \delta[\xi-\xi_0]) + \frac{1}{2}(-\delta[\xi+\xi_0] + \delta[\xi-\xi_0])\\ &= \delta[\xi-\xi_0]\end{aligned} \tag{9.77}$$

In words, the spectrum of the unit-magnitude complex linear-phase exponential with spatial frequency $+\xi_0$ is a single Dirac delta function located at $+\xi_0$. By an identical process, the spectrum of the unit-magnitude complex linear-phase negative exponential is a single Dirac delta function located at $-\xi_0$. The two cases may be combined into a single equation:

$$\begin{aligned}\mathcal{F}_1\{e^{\pm 2\pi i\xi_0 x}\} &= \mathcal{F}_1\{\cos[2\pi\xi_0 x] \pm i\ \sin[2\pi\xi_0 x]\}\\ &= \delta[\xi \mp \xi_0]\end{aligned} \tag{9.78}$$

An example is shown in Figure 9.16.

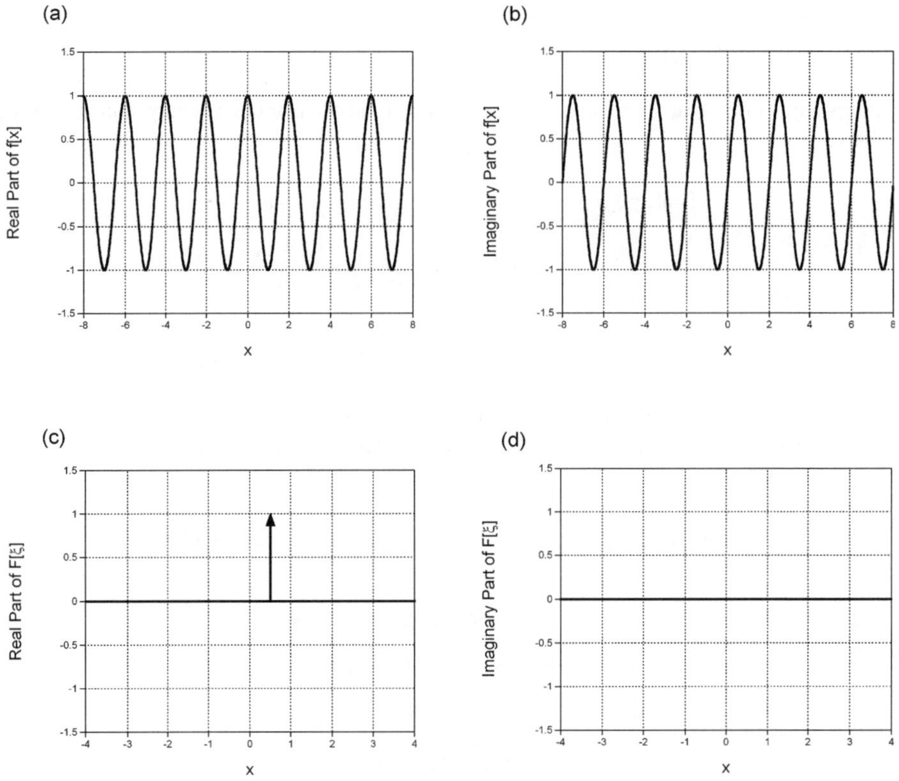

Figure 9.16 Fourier transform of the Hermitian complex linear-phase exponential function $f[x] = \exp[-2\pi i x/2]$ is the real-valued function $F[\xi] = \delta[\xi + \frac{1}{2}] + i0[\xi]$: (a) real part of $f[x]$; (b) imaginary part of $f[x]$; (c) real part of $F[\xi]$; (d) imaginary part of $F[\xi]$.

9.7.4 Fourier Transform of Signum and Step

Because the signum function in Equation (6.7) is real and odd, the discussion of Section 9.4 indicates that its spectrum must be imaginary and odd. The Fourier integral of *SGN*[x] may be recast into a different form by breaking the integral into segments with positive and negative unit amplitude for the respective positive and negative x:

$$\begin{aligned}\int_{-\infty}^{+\infty} e^{-2\pi i\xi x}\, SGN[x]\, dx &= \int_{-\infty}^{0} -e^{-2\pi i\xi x}\, dx + \int_{0}^{+\infty} +e^{-2\pi i\xi x}\, dx \\ &= \int_{-\infty}^{0_-} (-\cos[2\pi\xi x])\, dx + \int_{0_+}^{+\infty} \cos[2\pi\xi x]\, dx \\ &\quad + i\left(\int_{-\infty}^{0_-} (-\sin[2\pi\xi x])\, dx + \int_{0_+}^{+\infty} \sin[2\pi\xi x]\, dx\right)\end{aligned}$$

$$= \int_{0_+}^{+\infty} (-\cos[2\pi\xi x])\, dx + \int_{0_+}^{+\infty} \cos[2\pi\xi x]\, dx$$
$$+ i\left(\int_{0_+}^{+\infty} \sin[2\pi\xi x]\, dx + \int_{0_+}^{+\infty} \sin[2\pi\xi x]\, dx\right)$$
$$= 0[\xi] + i\,2\int_{0_+}^{+\infty} \sin[2\pi\xi x]\, dx = \frac{i}{\pi\xi} \cos[2\pi\xi x]\Big|_{x=0_+}^{x=+\infty} \tag{9.79}$$

The real parts of the two integrals have cancelled to leave an odd imaginary part of indeterminate area, which means that the spectrum of $SGN[x]$ is indeterminate in a strict sense. However, it is possible to develop a generalized solution to the spectrum of $SGN[x]$ by constructing a sequence of functions that converges to $SGN[x]$ and such that the Fourier spectrum of each member is well defined. One such function is the product of $SGN[x]$ and the symmetric decaying exponential $e^{-\tau|x|}$, where τ is a selectable positive real-valued parameter. If τ is small, the decaying function has approximately unit amplitude, and so the generalized signum function may be written in this limit:

$$SGN[x] = \lim_{\tau\to 0}\{e^{-\tau|x|}\, SGN[x]\} \tag{9.80}$$

The generalized Fourier transform of the functions in this sequence converges to $SGN[x]$ in the limit $\tau \to 0$:

$$\mathcal{F}_1\{SGN[x]\} = \lim_{\tau\to 0}\{\mathcal{F}_1\{e^{-\tau|x|}\, SGN[x]\}\} \tag{9.81}$$

The spectrum of each member in this sequence may be evaluated quite readily from well-known integrals:

$$\lim_{\tau\to 0}\left\{\int_{-\infty}^{+\infty} e^{-\tau|x|}\, SGN(x)\, e^{-2\pi i\xi x}\, dx\right\}$$
$$= \lim_{\tau\to 0}\left\{\int_{-\infty}^{0} -\, e^{+(\tau-2\pi i\xi)x}\, dx + \int_{0}^{+\infty} e^{-(\tau+2\pi i\xi)x}\, dx\right\}$$
$$= \lim_{\tau\to 0}\left\{\left[-\left(\frac{1}{\tau-2\pi i\xi}\right) e^{+(\tau-2\pi i\xi)x}\right]\Big|_{x=-\infty}^{x=0} + \left[-\left(\frac{1}{\tau+2\pi i\xi}\right) e^{-(\tau+2\pi i\xi)x}\right]\Big|_{x=0}^{x=+\infty}\right\}$$
$$= \lim_{\tau\to 0}\left\{-\left[\frac{1}{\tau-2\pi i\xi}\right][1-0] - \left[\frac{1}{\tau+2\pi i\xi}\right][0-1]\right\}$$
$$= \lim_{\tau\to 0}\left\{-\left[\frac{1}{\tau-2\pi i\xi}\right] + \left[\frac{1}{\tau+2\pi i\xi}\right]\right\}$$
$$= -\frac{1}{-2\pi i\xi} + \frac{1}{+2\pi i\xi} = \frac{1}{i\pi\xi} = -\frac{i}{\pi\xi} = 0[\xi] + i\left(-\frac{1}{\pi\xi}\right) \tag{9.82}$$

The spectrum is shown in Figure 9.17. Note that $(i\pi\xi)^{-1}$ is discontinuous at the origin, and we will soon show that the amplitude of the spectrum at the origin is zero. The magnitude of this generalized Fourier transform is indeterminate at DC and decays symmetrically as $|\xi| \to \infty$:

$$|\mathcal{F}_1\{SGN[x]\}| = \frac{1}{\pi|\xi|} \tag{9.83}$$

This result may be interpreted to provide an intuitive concept of the signum function. The low-frequency sinusoidal components of $SGN[x]$ are larger because they generate the physical "bulk" of $SGN[x]$. The peak amplitude of the sine function is located farther from the origin as the spatial frequency decreases. The high-frequency sinusoidal components primarily contribute to the "sharpness" of $SGN[x]$ at the transition.

The phase of the spectrum is proportional to a signum function:

$$\Phi\{\mathcal{F}_1\{SGN[x]\}\} = \begin{Bmatrix} +\dfrac{\pi}{2} & \textit{if } \xi < 0 \\ -\dfrac{\pi}{2} & \textit{if } \xi > 0 \end{Bmatrix} = -\frac{\pi}{2} SGN[\xi] \tag{9.84}$$

Again, the phase spectrum specifies the angle that is added to the phase of the sinusoidal component at each frequency ξ. For example, consider the components at $\xi = \pm|\xi_0|$; the magnitudes of these components are identically $(\pi|\xi_0|)^{-1}$ and the phases are respectively $\mp\pi/2$ radians. These particular sinusoidal components sum to:

$$\begin{aligned}
&\frac{1}{\pi|\xi_0|} \exp\left[+2\pi i \cdot (+|\xi_0|) \cdot x - i\frac{\pi}{2}\right] + \frac{1}{\pi|\xi_0|} \exp\left[+2\pi i \cdot (-|\xi_0|) \cdot x + i\frac{\pi}{2}\right] \\
&= \frac{1}{\pi|\xi_0|}\left(\exp i\left[+2\pi|\xi_0|x - \frac{\pi}{2}\right] + \exp\left[-2\pi|\xi_0|x + \frac{\pi}{2}\right]\right) \\
&= \frac{1}{\pi|\xi_0|}\left(\exp\left[+i\left(2\pi|\xi_0|x - \frac{\pi}{2}\right)\right] + \exp\left[-i\left(2\pi|\xi_0|x - \frac{\pi}{2}\right)\right]\right) \\
&= \frac{1}{\pi|\xi_0|} \cdot 2\cos\left[2\pi|\xi_0|x - \frac{\pi}{2}\right] \\
&= \frac{2}{\pi|\xi_0|} \cdot \sin[2\pi|\xi_0|x]
\end{aligned} \tag{9.85}$$

In words, the initial phases applied to the components with frequencies $\pm|\xi|$ translate these two complex exponential functions to create a real-valued sine wave whose amplitude is proportional to the reciprocal of the frequency. The signum function is therefore the superposition of real-valued sine functions only. If the spatial frequency is very small, the period of the sine wave is very large and the first lobes create the positive and negative "mass" of the signum function at large positive and negative distances from the origin. At larger spatial frequencies, the first lobes have smaller amplitude and provide the positive and negative "mass" at smaller positive and negative distances from the origin. The high-frequency components provide small-amplitude lobes that are "steep" in the vicinity of the origin, thus creating the "sharp corners" of the signum.

The fact that the Fourier transform is linear may be applied to derive the spectrum of *STEP*[x] via its decomposition into its even and odd parts in Equation (6.12):

$$STEP[x] = \frac{1}{2}(1[x] + SGN[x]) \tag{9.86}$$

The known spectrum of $1[x]$ in Equation (9.67) and of *SGN*[x] in Equation (9.82) merely are scaled and summed to produce the desired result:

$$\begin{aligned}
\mathcal{F}_1\{STEP[x]\} &= \mathcal{F}_1\left\{\frac{1}{2}(1 + SGN[x])\right\} \\
&= \frac{1}{2}\left[\delta[\xi] + \left(-\frac{i}{\pi\xi}\right)\right] = \frac{1}{2}\delta[\xi] + i\left(\frac{-1}{2\pi\xi}\right)
\end{aligned} \tag{9.87}$$

In words, the spectrum of the constant even part transforms to the real-valued and even Dirac delta function with area $\frac{1}{2}$, while the transform of the odd part is an odd imaginary function, as shown in Figure 9.18. Note that the phases of the spectra of the step and signum functions are identical.

9.7.5 Fourier Transform of Exponential

The unit constant $1[x]$ is finite everywhere and has infinite area. We have seen that its spectrum is the legitimate (though improper) function $\delta[\xi]$. The decaying exponential function e^{-x} in Equation (6.14)

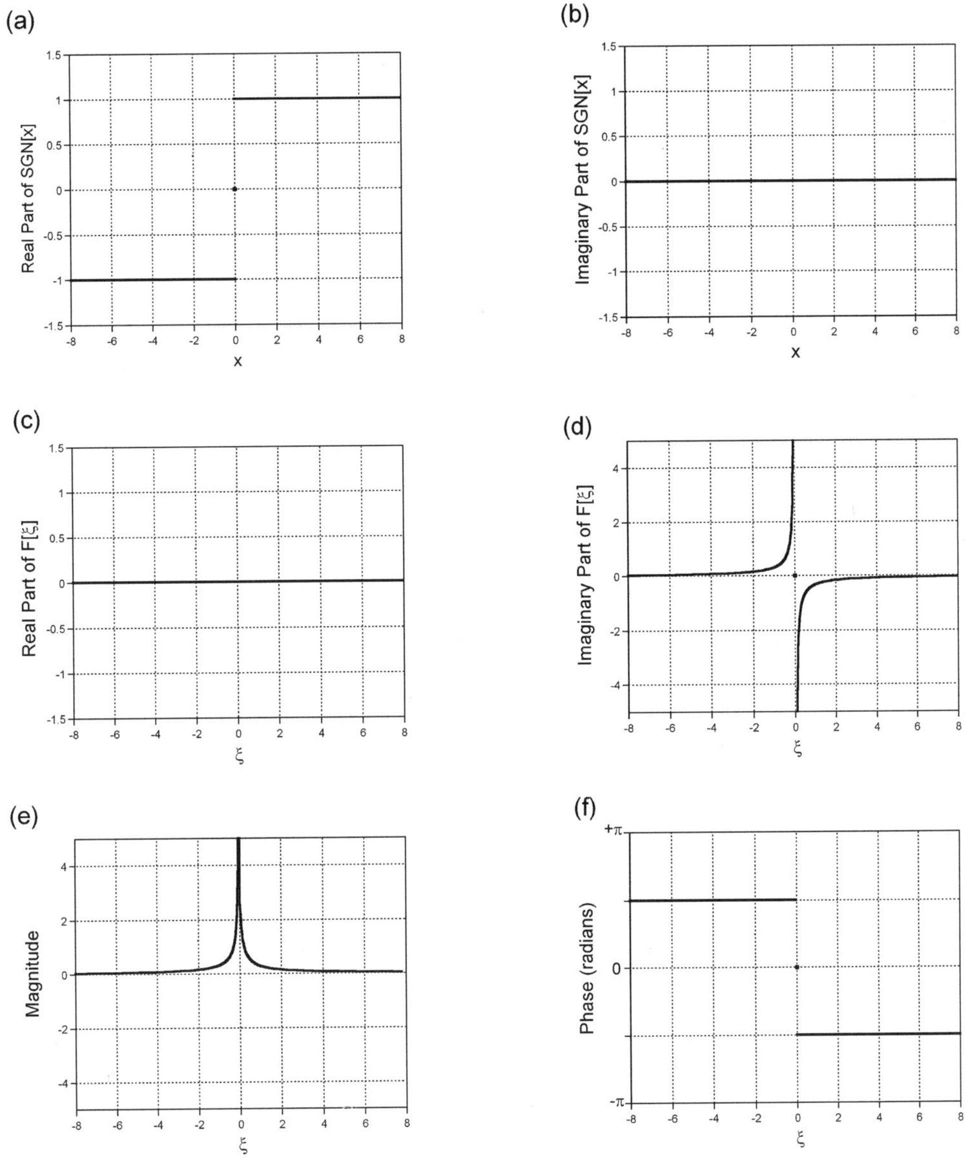

Figure 9.17 The Fourier transform of real-valued and odd function $SGN[x] + i\ 0[x]$ is the imaginary-valued and odd function $-i/(\pi\xi)$. The magnitude is $(\pi|\xi|)^{-1}$ and the phase is $-(\pi/2)\ SGN[\xi]$: (a) real part of $f[x]$; (b) imaginary part of $f[x]$; (c) real part of $F[\xi]$; (d) imaginary part of $F[\xi]$; (e) $|F[\xi]|$; (f) phase of $F[\xi]$.

also has infinite area, but its spectrum is not defined because its amplitude becomes indeterminate as $x \to -\infty$, and therefore the Fourier integral cannot be evaluated. However, the region of infinite amplitude may be eliminated by applying a multiplicative "window" function to e^{-x}; the obvious choice is $STEP[x]$. We will see shortly that the spectrum of the modulated function does exist.

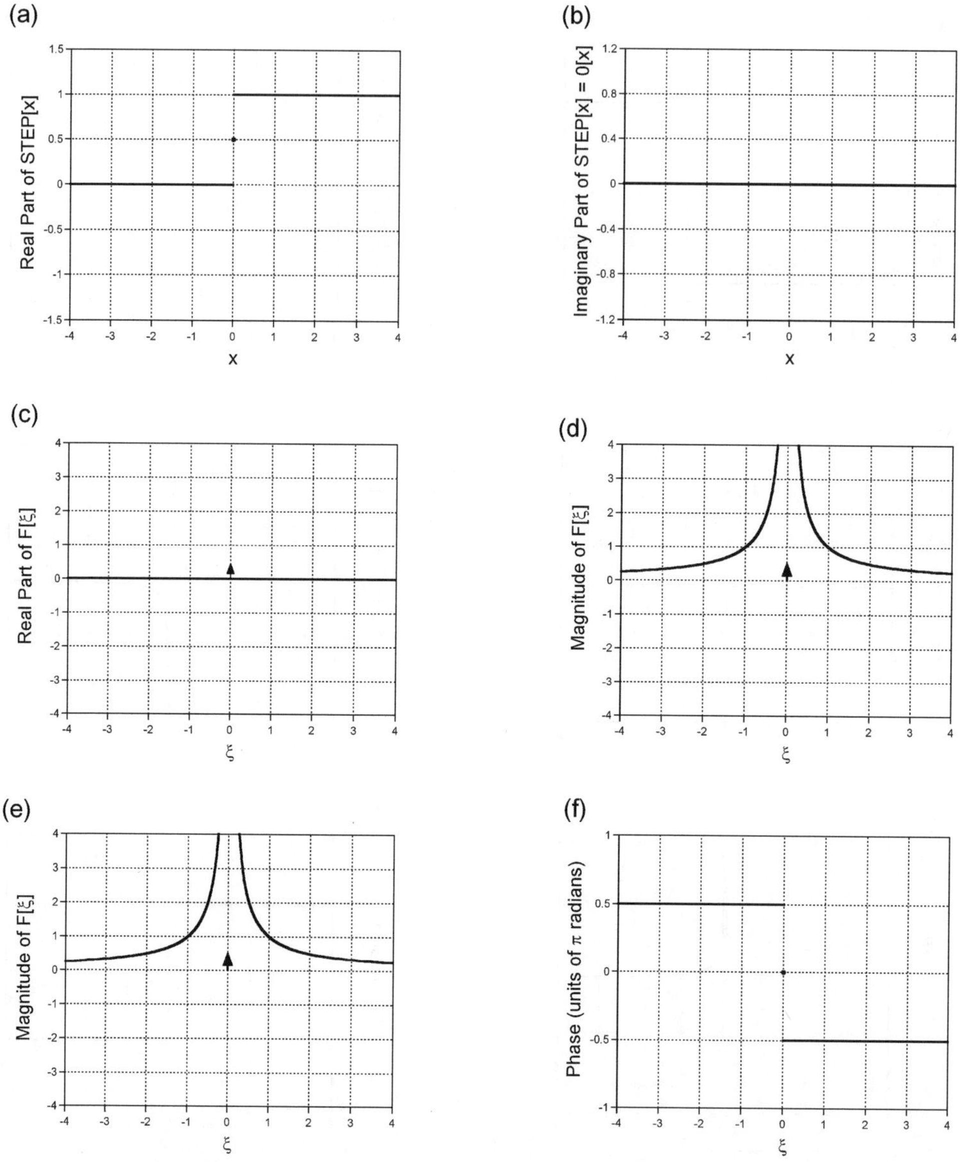

Figure 9.18 The real-valued function $STEP[x] + i \cdot 0[x]$ ((a) and (b)) and its Hermitian Fourier transform: (c) real part; (d) imaginary part; (e) magnitude; and (f) phase. The imaginary part of $\mathcal{F}\{STEP[x]\}$ is a scaled replica of the imaginary part of $\mathcal{F}\{SGN[x]\}$.

The truncated decaying exponential is featured in many physical problems. For example, the temporal version $e^{-t} \cdot STEP[t]$ acts as a causal damping term when describing the harmonic oscillator in classical mechanics. It also is very important in electromagnetic theory when modeling the damped motion of electric charges in a dielectric material when they respond to the influence of an oscillating electromagnetic wave. The Fourier transform of the motion of the charges leads to the variation of fractive index of the material with wavelength known as dispersion. Because this function is so on in physics, its spectrum is particularly useful to know.

e spectrum of the truncated decaying exponential has already been evaluated by direct integration of the Fourier transform of $SGN[x]$:

$$\begin{aligned}\mathcal{F}_1\{STEP[x]\cdot e^{-x}\} &= \int_{-\infty}^{+\infty} e^{-2\pi i\xi x}(STEP[x]\cdot e^{-x})\,dx\\ &= \int_{0}^{+\infty} e^{-2\pi i\xi x}\cdot e^{-x}\;dx = \int_0^{+\infty} e^{-(1+2\pi i\xi)x}\;dx\\ &= \left.\frac{e^{-(1+2\pi i\xi)x}}{-(1+2\pi i\xi)}\right|_0^{+\infty} = 0-\left(\frac{1}{-(1+2\pi i\xi)}\right)\\ &= \frac{1}{1+2\pi i\xi} = \frac{1}{1+2\pi i\xi}\cdot\left(\frac{1-2\pi i\xi}{1-2\pi i\xi}\right) = \frac{1-2\pi i\xi}{1+4\pi^2\xi^2}\\ &= LOR[\xi]\left(\frac{1}{2}+i(-\pi\xi)\right) = \frac{1}{2}CLOR[\xi] \end{aligned} \tag{9.88a}$$

where the definitions of the real- and complex-valued Lorentzian functions in Equations (6.83) and (6.159) have been used. Because $f[x]$ is real valued, its Fourier transform is Hermitian, as is evident from the graph in Figure 9.19. The real and imaginary parts of this spectrum are Lorentzian functions modulated by the unit constant and by a factor proportional to $-\xi$, respectively:

$$\Re\{\mathcal{F}_1\{STEP[x]\cdot e^{-x}\}\} = \frac{1}{1+4\pi^2\xi^2} = \frac{1}{2}LOR[\xi] \tag{9.88b}$$

$$\Im\{\mathcal{F}_1\{STEP[x]\cdot e^{-x}\}\} = -\frac{2\pi\xi}{1+4\pi^2\xi^2} = -(\pi\xi)LOR[\xi] \tag{9.88c}$$

Recall from Equation (6.160) that the magnitude of $CLOR[\xi]$ is the square root of its real part, and so it decays more slowly with increasing ξ than $LOR[\xi]$:

$$|\mathcal{F}_1\{STEP[x]\cdot e^{-x}\}| = \sqrt{\frac{1}{1+4\pi^2\xi^2}} \tag{9.88d}$$

The magnitude of the spectrum of the truncated decaying exponential is large at low spatial frequencies and decreases away from the origin. The initial phase of the spectrum of the decaying exponential has the functional form of the negative of the inverse tangent function, and so the range of possible initial phases is:

$$\Phi\{\mathcal{F}_1\{STEP[x]\cdot e^{-x}\}\} = \tan^{-1}[-2\pi\xi] = -\tan^{-1}[2\pi\xi] \tag{9.88e}$$

Interesting properties of this spectrum become apparent by computing the magnitude and phase after subtracting a "half-unit" constant:

$$\begin{aligned}\mathcal{F}_1\{STEP[x]\cdot e^{-x}\} - 0.5 &= \left(\frac{1}{1+(2\pi\xi)^2} + i\left(-\frac{2\pi\xi}{1+(2\pi\xi)^2}\right)\right) - \frac{1}{2}\\ &= \frac{1-(2\pi\xi)^2}{2(1+(2\pi\xi)^2)} + i\frac{-4\pi\xi}{2(1+(2\pi\xi)^2)}\end{aligned} \tag{9.89a}$$

The magnitude of the spectrum is the half-unit constant:

$$|\mathcal{F}_1\{STEP[x]\cdot e^{-x}\} - 0.5| = \sqrt{\frac{(1+(2\pi\xi)^2)^2}{4(1+(2\pi\xi)^2)^2}} = \sqrt{\frac{1}{4}} = \frac{1}{2}\cdot 1[\xi] \tag{9.89b}$$

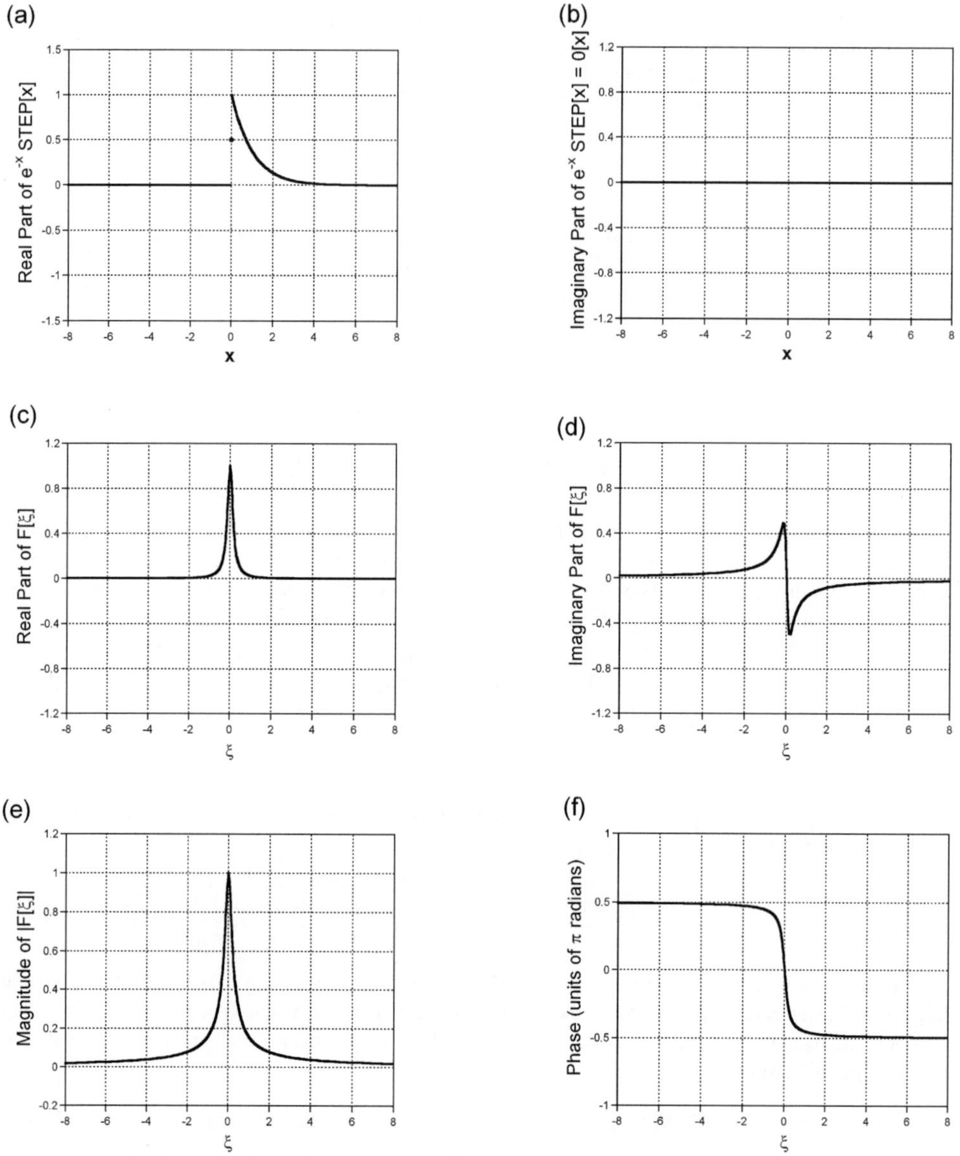

Figure 9.19 Decaying exponential and its Hermitian spectrum: (a) $\Re\{f[x]\} = e^{-x} \cdot STEP[x]$; (b) $\Im\{f[x]\} = 0[x]$; (c) $\Re\{F[\xi]\} = 1/[1 + (2\pi\xi)^2]$; (d) $\Im\{F[\xi]\} = -2\pi\xi/[1 + (2\pi\xi)^2]$; (e) $|F[\xi]| = \sqrt{1/[1 + (2\pi\xi)^2]}$; $\mathbf{\Phi}\{F[\xi]\} = -\tan^{-1}[2\pi\xi]$.

which means that the Argand diagram in the spatial frequency domain is the "half-unit circle", as shown in Figure 9.20. Now, we examine the phase of the spectrum of this shifted function on the Argand diagram. We saw in Chapter 2 that the rate of change of the phase determines the "speed" at which the phasor "travels" around the Argand diagram. The phase of the spectrum in Equation (9.89a) is very nonlinear, as may be seen by determining the "oscillation rate" of the spectrum at equally spaced phase

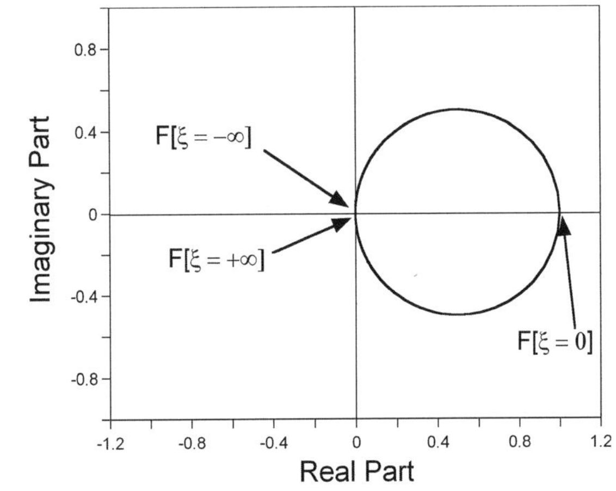

Figure 9.20 Argand diagram of $\mathcal{F}_1\{e^{-x} \cdot STEP[x]\} = (1 + 2\pi i\xi)^{-1}$, which is a circle of half-unit radius centered at $[\frac{1}{2}, 0]$. The amplitude of F at the coordinates $\xi = -\infty$, 0, and $+\infty$ are labeled.

angles. For example, the phase $\phi \cong +\pi$ radians where the real part is negative and the imaginary part approaches zero from the positive side. From Equation (9.89a), it is apparent that this situation occurs as $\xi \to -\infty$. The phase $\phi = +\pi/2$ where the real part is zero and the imaginary part is positive, which occurs for $\xi = -1/2\pi$. The phase $\phi = 0$ radians where the imaginary part is zero and the real part is positive, i.e., at $\xi = 0$. It is apparent that the phase changes by $-\pi/2$ radians as ξ moves from $-\infty$ to $-1/2\pi$, and again by $-\pi/2$ radians as ξ moves from $-1/2\pi$ to 0. The fact that the phase changes very rapidly for $\xi \cong 0$ and slowly for $\xi \to \pm\infty$ indicates the very nonlinear character of this spectrum.

The results just demonstrated may be applied to the "unshifted" spectrum to demonstrate that the Argand diagram of the spectrum of the decaying exponential is a circle of half-unit radius centered at $[\xi = 0.5, \eta = 0]$, as shown in Figure 9.20.

The behavior and applications of the truncated decaying exponential will be considered in more detail in Chapter 18.

9.7.6 Fourier Transform of Gaussian

The spectrum of the Gaussian may be evaluated in part by completing the square of the integrand:

$$\begin{aligned}\int_{-\infty}^{+\infty} e^{-2\pi i\xi x}\, e^{-\pi x^2}\, dx &= \int_{-\infty}^{+\infty} e^{-\pi(x^2+2i\xi x+(i\xi)^2-(i\xi)^2)}\, dx \\ &= e^{+\pi(i\xi)^2} \int_{-\infty}^{+\infty} e^{-\pi(x+i\xi)^2}\, dx \\ &= e^{-\pi\xi^2} \int_{-\infty+i\xi}^{+\infty+i\xi} e^{-\pi u^2}\, du \end{aligned} \tag{9.90}$$

This is a useful result even before solving the definite integral, which, after all, evaluates to a (possibly complex-valued) multiplicative constant. In words, this establishes that the spectrum of a space-domain Gaussian must be proportional to a frequency-domain Gaussian.

The definite integral in Equation (9.90) may be evaluated by establishing the appropriate contour on the complex plane and applying the Cauchy integral formula and the definition of the gamma function in Equation (6.37). The process is demonstrated in Appendix 9A and shows that the integral evaluates

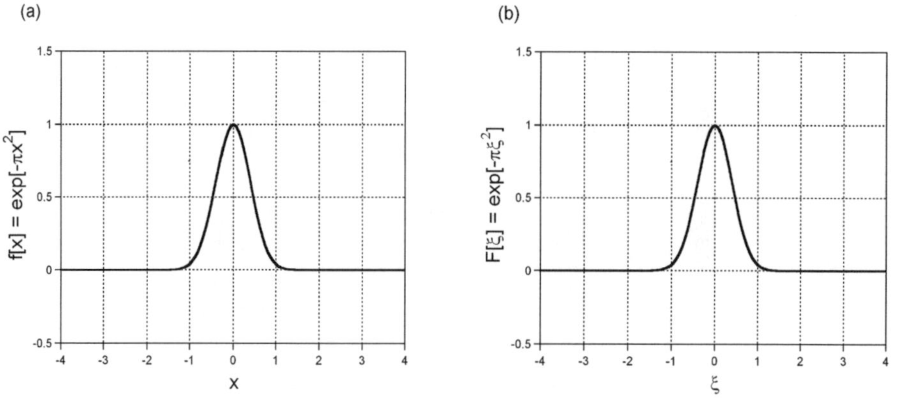

Figure 9.21 The Fourier transform of the real-valued and even function *GAUS*$[x]$ is the real-valued and even function *GAUS*$[\xi]$: (a) $\Re\{f[x]\} = GAUS[x]$ (imaginary part is $0[x]$); (b) $\Re\{F[\xi]\} = GAUS[\xi]$ (imaginary part is $0[\xi]$). This is the first example of a Fourier "self-transform".

to unity (Stephenson and Radmore, 1990):

$$\mathcal{F}_1\{GAUS[x]\} = e^{-\pi\xi^2} = GAUS[\xi] \tag{9.91}$$

The representations of the Gaussian in both domains are identical nonnegative and symmetric functions, as shown in Figure 9.21. This indicates that *GAUS*$[x]$ is composed of symmetric cosine functions whose amplitude decreases as $|\xi|$ increases. *GAUS*$[x]$ has the unique property thus far that the representations in both domains are identical. This "self-transform" character truly makes the Gaussian a "special" special function. The concept of the Fourier self-transform was explored by Lohmann (1994, pp. 155–157). Because all derivatives of the space-domain Gaussian are finite, the Gaussian truly is a "smooth" function.

9.7.7 Fourier Transforms of Chirp Functions

We defined the real-valued quadratic-phase sinusoid, or *chirp* function, in Equation (6.53) to be $f[x] = \cos[2\pi\alpha\xi_0 x^2 + \phi_0]$, where the constant α has units of inverse length to ensure that the entire argument is dimensionless. The complex quadratic-phase sinusoid was defined in Equation (6.146) by combining real- and imaginary-valued chirps via the Euler relation. As will be demonstrated in later chapters, the quadratic-phase signal is essential for describing the behavior of optical imaging systems, and its properties often defy naive intuition.

To derive the spectrum, consider the general complex-valued quadratic-phase sinusoid with arbitrary chirp rate α and initial phase ϕ_0:

$$f[x] = \exp\left[\pm i\left(\pi\left[\frac{x}{\alpha}\right]^2 + \phi_0\right)\right] = \cos\left[\frac{\pi x^2}{\alpha^2} + \phi_0\right] \pm i\,\sin\left[\frac{\pi x^2}{\alpha^2} + \phi_0\right] \tag{9.92}$$

The fact that the coordinate x appears only as its square ensures that both the real and imaginary parts of $f[x]$ are symmetric regardless of the value of ϕ_0. The discussion of Section 9.4 demonstrated that the Fourier transform of the even real part of $f[x]$ must generate an even real part of $F[\xi]$, while the transform of the even imaginary part of $f[x]$ produces an even imaginary part of $F[\xi]$. This observation will be applied to the quadratic-phase exponential to derive the spectra of the individual real-valued chirp functions. As was shown in Equation (6.55), the spatial frequency of the complex quadratic-phase function is proportional to the coordinate, and increases or decreases depending on the algebraic sign of the argument:

$$\xi[x] = \frac{1}{2\pi}\frac{\partial \Phi[x]}{\partial x} = \pm\frac{2x}{\alpha^2} \propto \pm x \tag{9.93}$$

which demonstrates that all spatial frequencies ξ are present in the complex chirp function.

We will derive the spectrum of a complex quadratic-phase sinusoid first for the specific case with $\alpha = 1, \phi_0 = 0$, and the negative sign in the argument, so that $f[x] = \exp[-i\pi x^2]$. The Fourier transform for arbitrary values of α and ϕ_0 may be obtained subsequently by applying the yet-to-be-derived scaling and shift theorems. The Fourier transform may be obtained by completing the square of the integrand and changing the variable of integration to obtain the form of the Gaussian function:

$$\begin{aligned}
\mathcal{F}_1\{e^{-i\pi x^2}\} &= \int_{-\infty}^{+\infty} e^{-i\pi x^2}\, e^{-2\pi i\xi x}\, dx \\
&= \int_{-\infty}^{+\infty} e^{-i\pi(x^2+2\xi x+\xi^2-\xi^2)}\, dx = \int_{-\infty}^{+\infty} e^{-i\pi(x^2+2\xi x+\xi^2)}\, e^{+i\pi\xi^2}\, dx \\
&= e^{+i\pi\xi^2}\int_{-\infty}^{+\infty} e^{-i\pi(x+\xi)^2}\, dx; \quad u \equiv (\sqrt{+i})(x+\xi), \quad du = \sqrt{+i}\, dx \\
&= e^{+i\pi\xi^2}\frac{1}{\sqrt{+i}}\int_{-\infty}^{+\infty} e^{-\pi u^2}\, du = 2\, e^{+i\pi\xi^2}(\sqrt{-i})\left(\frac{1}{2\sqrt{\pi}}\right)\Gamma\left[\frac{1}{2}\right] \\
&= e^{+i\pi\xi^2}\sqrt{-i}\frac{1}{\sqrt{\pi}}\sqrt{\pi} = e^{+i\pi\xi^2}\sqrt{e^{-i\pi/2}} \\
&= e^{-i\pi/4}\cdot e^{+i\pi\xi^2} = \frac{(1-i)}{\sqrt{2}}\, e^{+i\pi\xi^2}
\end{aligned} \tag{9.94a}$$

The same derivation may be applied to the Fourier transform of the chirp with the positive exponent to show that:

$$\mathcal{F}_1\{e^{+i\pi x^2}\} = e^{-i\pi\xi^2}\, e^{+i\pi/4} \tag{9.94b}$$

Both results can be combined into a single expression:

$$\mathcal{F}_1\{e^{\pm i\pi x^2}\} = e^{\pm i\pi/4}\, e^{\mp i\pi\xi^2} = e^{\mp i\pi(\xi^2 \mp \frac{1}{4})} \tag{9.95}$$

In words, the complex chirp function almost is a "self-transform"; its spectrum is the complex quadratic-phase sinusoid with the opposite sign multiplied by a unit-magnitude complex constant with phase of $\pm\pi/4$ radians. The magnitude of the spectrum of the complex-valued chirp with either algebraic sign is the unit constant:

$$|\mathcal{F}_1\{e^{\pm i\pi x^2}\}| = 1[\xi] \tag{9.96}$$

The magnitude spectra of the complex chirp is identical to that of any single unit-area Dirac delta function, which means that these space-domain functions may be synthesized by summing complex sinusoids with unit magnitude at each spatial frequency. The very significant differences between the complex chirp and the Dirac delta function are due only to the differences in phase spectra. Equation (9.57c) shows that all sinusoidal components of $\delta[x]$ have the same initial phase of zero radians, while the initial phases of the sinusoidal components of the quadratic-phase functions were specified to be quadratic functions of ξ in Equation (9.95):

$$\Phi\{\mathcal{F}_1\{e^{\pm i\pi x^2}\}\} = \mp\left(\pi\xi^2 \mp \frac{\pi}{4}\right) = \mp\pi\xi^2 \pm \frac{\pi}{4} \tag{9.97}$$

as shown in Figure 9.22, where the phase has been "unwrapped" to eliminate the "ratcheting" due to the computational ambiguity of the inverse tangent function, i.e., because $\tan[\pi] = \tan[-\pi]$. The initial phase at $\xi = 0$ is $-\pi/4$ radians, as predicted.

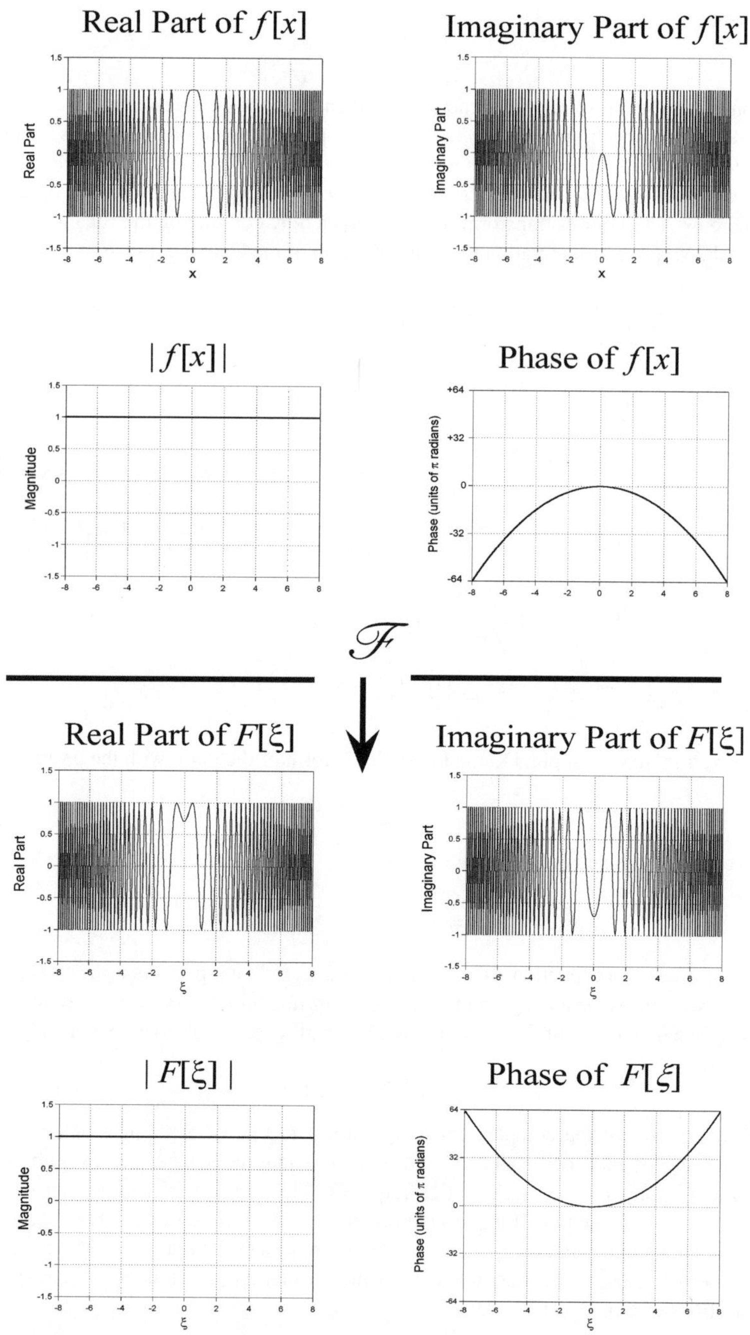

Figure 9.22 The complex and even "downchirp" function $\exp[-i\pi x^2]$ as real part, imaginary part, unit magnitude, and quadratic phase. Its complex and even Fourier transform is $\exp[+i\pi\xi^2]\exp[-i\pi/4]$, which is displayed as real part, imaginary part, unit magnitude, and quadratic "upchirp" phase (the phase at the origin is $-\pi/4$ radians). The unit-magnitude spectrum indicates that all sinusoidal frequencies are equally present in $\exp[-i\pi x^2]$.

The evenness of the real and the imaginary parts of both $f[x]$ and $F[\xi]$ in Equation (9.95) allows us to identify the Fourier transforms of the individual real-valued functions that comprise the respective real and imaginary parts of $f[x]$. The real part of Equation (9.95) is the spectrum of the "cosine chirp":

$$\Re\{\mathcal{F}_1\{e^{\pm(i\pi x^2)}\}\} = \mathcal{F}_1\{\cos[\pi x^2]\} = \Re\{e^{\mp i\pi(\xi^2-\frac{1}{4})}\}$$
$$= \cos\left[\pi\left(\xi^2 - \frac{1}{4}\right)\right] \tag{9.98a}$$

while the imaginary part yields the spectrum of the "sine chirp":

$$\Im\{\mathcal{F}_1\{e^{\pm(i\pi x^2)}\}\} = \mathcal{F}_1\{\pm\sin[\pi x^2]\} = \Im\{e^{\mp i\pi(\xi^2-\frac{1}{4})}\}$$
$$= \mp\sin\left[\pi\left(\xi^2 - \frac{1}{4}\right)\right] \tag{9.98b}$$

The uses of the quadratic-phase exponential functions are many and varied. In fact, it is easy to make the case that it is the most interesting function in this book. Their properties will be explored in greater detail in Chapters 17 and 21.

9.7.8 Fourier Transform of *COMB* Function

The set of equispaced Dirac delta functions that Gaskill named *COMB*[x] was introduced in Equation (6.124):

$$COMB[x] \equiv \sum_{n=-\infty}^{+\infty} \delta[x-n] \tag{9.99}$$

and will be particularly important in the discussion of sampling in Chapter 12. The linearity of the Fourier transform ensures that the spectrum of *COMB*[x] may be derived by summing the spectra of the individual Dirac delta functions. The individual spectra were derived in Equation (9.59):

$$\mathcal{F}_1\left\{\sum_{n=-\infty}^{+\infty} \delta[x-n]\right\} = \sum_{n=-\infty}^{+\infty} \mathcal{F}_1\{\delta[x-n]\} = \sum_{n=-\infty}^{+\infty} e^{-2\pi i\xi n}$$
$$= \sum_{n=-\infty}^{+\infty} \cos[2\pi\xi n] - i\sum_{n=-\infty}^{+\infty} \sin[2\pi\xi n]$$
$$= \sum_{n=-\infty}^{+\infty} \cos[2\pi\xi n] + i\cdot 0$$
$$= 1[\xi] + 2\sum_{n=1}^{+\infty} \cos[2\pi\xi n] \tag{9.100}$$

where the last step follows because $\sin[2\pi\xi n]$ is an odd function of n, which means that the pairs of sine waves indexed by $\pm n$ sum to zero. The maxima of the cosine indexed by $n = 1$ are located at the integer values of ξ. Each cosine with $n > 1$ has $n - 1$ additional equispaced maxima located between the integer values of ξ. In the infinite summation, the maxima of each cosine located at integer ξ will "reinforce" and generate infinite amplitude. At noninteger spatial frequencies, the amplitudes of the component cosine functions are distributed over the interval $[-1, +1]$ so that the sum is zero.

Now consider the contributions of the individual cosine functions to the area of $\mathcal{F}_1\{COMB[x]\}$. Because each individual cosine function with nonzero frequency has null area, the area of the infinite sum is entirely due to the infinite area of $1[\xi]$. In fact, because each cosine function oscillates through an integer number of periods within any interval of the frequency domain with unit "width" and centered

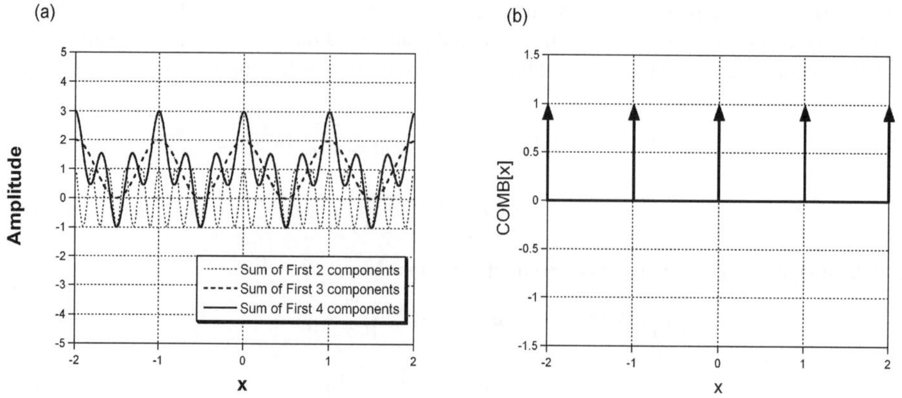

Figure 9.23 The first few constituent sinusoids of the real and even function $COMB[x]$: (a) the sums of the first two, four, and six cosine component functions; (b) the sum of all component sinusoids produces $COMB[x]$. Note that the amplitudes add at integer coordinates and tend to cancel at others.

about an integer frequency (e.g., $-\frac{3}{2} \leq \xi < -\frac{1}{2}$, $-\frac{1}{2} \leq \xi < +\frac{1}{2}$, $+\frac{1}{2} \leq \xi < +\frac{3}{2}$, etc.), each cosine contributes nothing to the area of the function within these regions. In short, the areas of $\mathcal{F}_1\{COMB[x]\}$ and of $1[\xi]$ within any unit-width interval are identically unity.

We have made three observations about the summation in Equation (9.100): that the amplitudes at noninteger frequencies tend to cancel, that the amplitudes at integer frequencies add, and that the area of the summation within each unit-width interval of the frequency domain centered at an integer frequency is unity. In other words, the summation within each unit-width interval of the frequency domain satisfies the criteria for the Dirac delta function, so the result of Equation (9.100) is a sequence of unit-area Dirac delta functions located at the integer values of ξ:

$$\mathcal{F}_1\{COMB[x]\} = COMB[\xi] \tag{9.101}$$

This is our second example of a Fourier "self-transform". The result of a partial sum of the spectral components using the first five cosine functions and the complete sum are shown in Figure 9.23.

9.8 Theorems of the Fourier Transform

Some easily proven theorems of the Fourier transform may be used to derive the spectra of many more functions from those few that were just derived. The most important theorems are proven in this section. Most proofs are quite simple and are based on one of two strategies: either by a change of integration variable or by substituting the inverse Fourier transform of space-domain representation(s). The notations of functions used in the derivations follow our usual convention of lower- and upper-case characters for the representations in the space and frequency domains, respectively. The initial phases of two representations are denoted by $\Phi\{f[x]\}$ and $\Phi\{F[\xi]\}$, respectively. The real-valued parameters b and x_0 represent the width parameter and shift of the function, respectively. Greek letters (e.g., α) are used to represent complex-valued constants.

9.8.1 Multiplication by Constant

The Fourier transform of a function $f[x]$ whose amplitude is scaled at all coordinates by a complex constant α is the spectrum $F[\xi]$ of the unscaled function multiplied by the same constant:

$$\mathcal{F}_1\{\alpha \cdot f[x]\} = \alpha \cdot F[\xi] \tag{9.102}$$

9.8.2 Addition Theorem (Linearity)

The linearity of integration over fixed limits may be combined with Equation (9.102) to establish that the Fourier transform of a weighted sum of functions is the weighted sum of the transforms:

$$\begin{aligned}\mathcal{F}_1\left\{\sum_{n=0}^{N} \alpha_n f_n[x]\right\} &= \int_{-\infty}^{+\infty} e^{-2\pi i\xi x}\left\{\sum_{n=0}^{N} \alpha_n f_n[x]\right\} dx \\ &= \sum_{n=0}^{N} \alpha_n \left\{\int_{-\infty}^{+\infty} f_n[x]\, e^{-2\pi i\xi x}\, dx\right\} \\ &= \sum_{n=0}^{N} \alpha_n\, F_n[\xi]\end{aligned} \tag{9.103}$$

9.8.3 Fourier Transform of a Fourier Transform

The very name of this theorem may be confusing at first because it appears to be based on representations of functions in the "wrong" domain. However, the concept actually is quite straightforward and very powerful: it doubles the number of known Fourier transform pairs. In words, the "transform-of-a-transform" theorem (or "transform^2 theorem", for short) states that if a function $f[x]$ has the associated Fourier transform $F[\xi]$, then a function in the space domain with the same form as $F[\xi]$ has an associated spectrum with the same form as $f[x]$ expressed in the frequency domain but "reversed" relative to the origin (whew!). The statement of the theorem in equation form is much more concise:

$$\mathcal{F}_1\{f[x]\} \equiv F[\xi] \Longrightarrow \mathcal{F}_1\{F[x]\} = f[-\xi] \tag{9.104}$$

The proof is easy, requiring only the definitions of the forward and inverse Fourier transforms:

$$\begin{aligned}\mathcal{F}_1\{F[x]\} &= \int_{-\infty}^{+\infty} F[x]\, e^{-2\pi i\xi x}\, dx \\ &= \int_{-\infty}^{+\infty} F[x]\, e^{+2\pi i(-\xi)x}\, dx = f[-\xi]\end{aligned} \tag{9.105}$$

Illustrative examples of this theorem are very useful during the first few encounters because the potential to confuse the notation is large. The first example is based on the Dirac delta function. Recall that the sifting property of $\delta[x - x_0]$ was used in Equation (9.59) to derive its spectrum:

$$\begin{aligned}\mathcal{F}_1\{f[x]\} &= \mathcal{F}_1\{\delta[x - x_0]\} \\ &= \int_{-\infty}^{+\infty} \delta[x - x_0]\, e^{-2\pi i\xi x}\, dx \\ &= e^{-2\pi i\xi x_0} = F[\xi]\end{aligned} \tag{9.106}$$

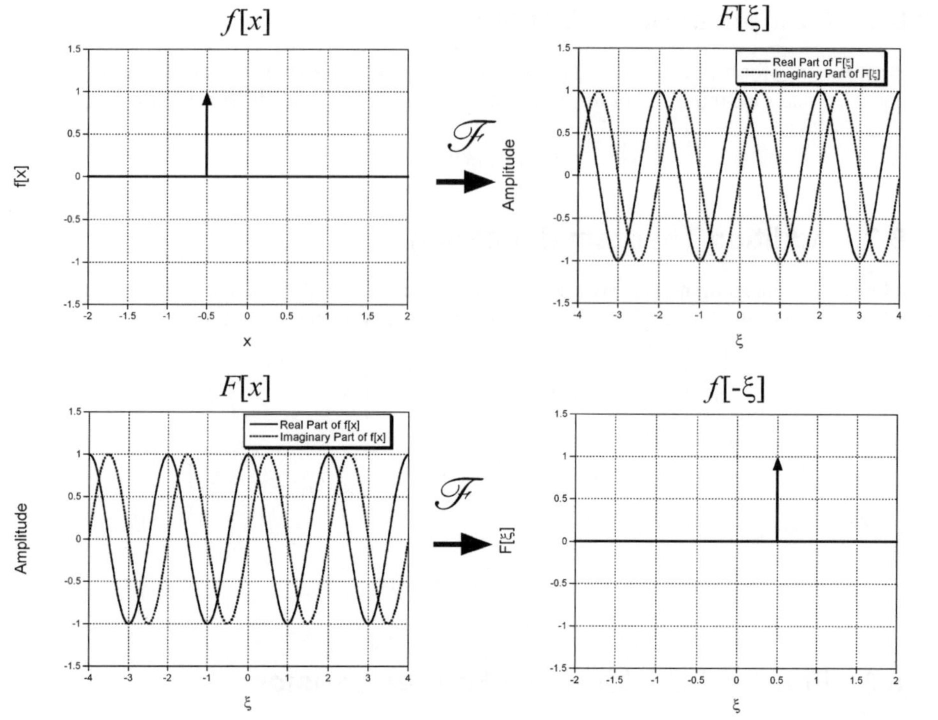

Figure 9.24 The "transform-of-a-transform" theorem: $\mathcal{F}_1\{\delta[x-\frac{1}{2}]\} = e^{-i\pi\xi}$ means that $\mathcal{F}_1\{e^{-i\pi x}\} = \delta[-\xi-\frac{1}{2}] = \delta[\xi+\frac{1}{2}]$. The general form is $\mathcal{F}_1\{F[x]\} = f[-\xi]$.

Equation (9.105) predicts that the spectrum of $F[x]$ is:

$$\begin{aligned}\mathcal{F}_1\{F[x]\} &= \mathcal{F}_1\{e^{-2\pi i\xi_0 x}\} = f[-\xi]\\ &= \delta[(-\xi)-\xi_0]\\ &= \delta[-(\xi+\xi_0)] = \delta[\xi+\xi_0]\end{aligned} \tag{9.107}$$

where the space-domain function $F[x]$ is obtained by substituting the coordinate ξ for the space-domain coordinate x and the parameter x_0 for the original parameter ξ_0. The symmetry of $\delta[\xi]$ has been used in the last step. The sifting property may be applied within the inverse Fourier transform to confirm this result, as shown in Figure 9.24:

$$\begin{aligned}\mathcal{F}_1^{-1}\{\delta[\xi+\xi_0]\} &= \int_{-\infty}^{+\infty} \delta[\xi-(-\xi_0)]\, e^{+2\pi i\xi x}\, d\xi\\ &= e^{+2\pi i(-\xi_0)x} = e^{-2\pi i\xi_0 x}\end{aligned} \tag{9.108}$$

As a second example, recall from Equation (9.71) that $\mathcal{F}_1\{RECT[x]\} = SINC[\xi]$. The theorem provides a direct avenue to the spectrum of $SINC[x]$:

$$\mathcal{F}_1\{SINC[x]\} = RECT[-\xi] = RECT[+\xi] \tag{9.109}$$

where the symmetry of $RECT[\xi]$ has been used. In words, this statement indicates that the *SINC* function is the superposition of complex linear-phase exponentials with equal amplitude and zero initial

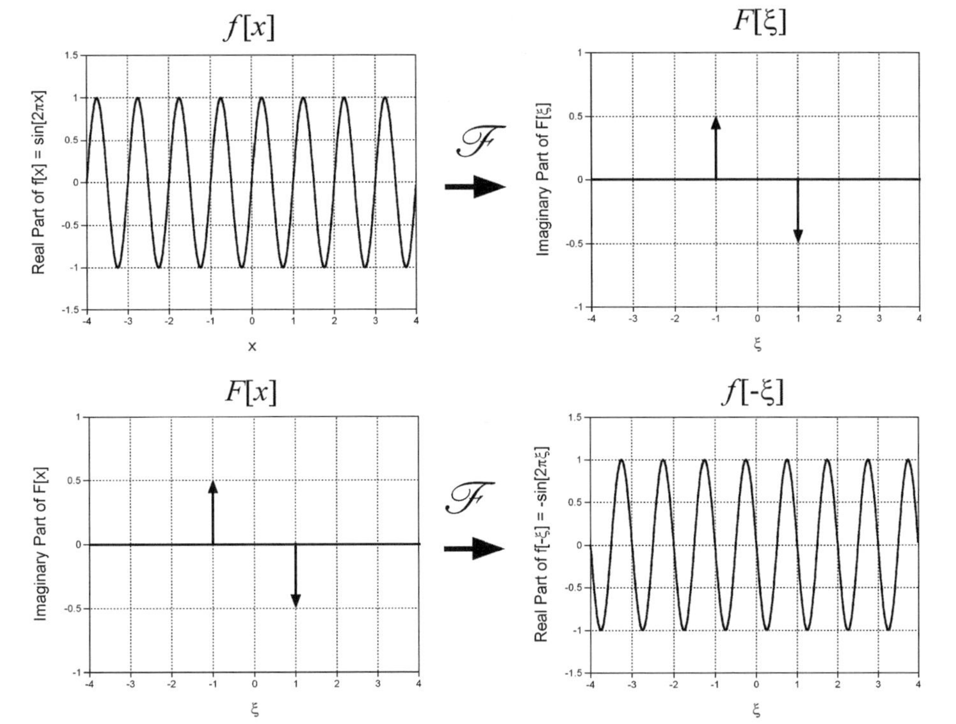

Figure 9.25 The "transform-of-a-transform" theorem for $f[x] = \sin[2\pi x]$. The transform is the odd imaginary function $F[\xi] = i(\delta[\xi+1] - \delta[\xi-1])$, and therefore $\mathcal{F}_2\{i(\delta[x+1] - \delta[x-1])\} = \sin[2\pi(-\xi)] = -\sin[2\pi\xi]$.

phase out to some maximum frequency. In the notation of Equation (9.70), we may write:

$$SINC[x] = 1[x] + \int_{0_+}^{+\frac{1}{2}} 2\cos[2\pi\xi x]\, d\xi \tag{9.110}$$

In other words, *SINC*[x] is a "lowpass" signal, since no sinusoidal components exist above some cutoff frequency.

As a third example of the "transform-of-a-transform" theorem, consider the spectrum of $\sin[2\pi x]$, which is obtained by substituting $\xi_0 = 1$ into Equation (9.75). This expression is recast by multiplying both sides by the complex-valued constant $2 \cdot i^{-1} = -2i$:

$$\begin{aligned} {}^{\delta}{}_{\delta}[\xi] &= \frac{2}{i}\mathcal{F}_1\{\sin[2\pi x]\} \\ &= (-2i)\,\mathcal{F}_1\{\sin[2\pi x]\} \end{aligned} \tag{9.111}$$

This result may be substituted directly into the theorem to derive the spectrum of an odd pair of Dirac delta functions in the space domain:

$$\mathcal{F}_1\{{}^{\delta}{}_{\delta}[x]\} = (-2i)\sin[2\pi(-\xi)] = +2i\sin[2\pi\xi] \tag{9.112}$$

as shown in Figure 9.25.

9.8.4 Central-Ordinate Theorem

Equation (9.49) shows that the amplitude of the spectrum at the origin (i.e., its central ordinate) is the area of the space-domain representation (which may be real or complex). For completeness, the result is included here along with the other theorems:

$$F[0] = \int_{-\infty}^{+\infty} f[x]\, e^{-2\pi i \cdot 0 \cdot x}\, dx$$
$$= \int_{-\infty}^{+\infty} f[x]\, dx \tag{9.113}$$

The equivalent property in the frequency domain is demonstrated in identical fashion by evaluating the expression for the inverse Fourier transform in Equation (9.63) at the origin $x = 0$:

$$f[0] = \int_{-\infty}^{+\infty} (F[\xi]\, e^{+2\pi i \xi \cdot 0})\, d\xi$$
$$= \int_{-\infty}^{+\infty} (F[\xi] \cdot 1)\, d\xi = \int_{-\infty}^{+\infty} F[\xi]\, d\xi \tag{9.114}$$

In words, the area of the spectrum is equal to the amplitude of the function at the origin. The two "parallel" results of Equation (9.113) and Equation (9.114) make it easy to evaluate the areas of many functions that are difficult to derive directly, such as $SINC[x]$ and $\exp[\pm i\pi x^2]$. The respective Fourier transforms of these functions are $RECT[\xi]$ and $\exp[\mp i\pi\xi^2]\, \exp[\pm i\pi/4]$. The respective central ordinates are $SINC[0] = 1$ and $\exp[\pm i\pi/4] = (1 \pm i)/\sqrt{2}$, which are the areas of the functions.

Note that the central ordinate of the spectrum of any sinusoid with nonzero spatial frequency will have null amplitude, which confirms the validity of the statement in Equation (6.22) that the area of any sinusoid with a finite period is zero.

9.8.5 Scaling Theorem

Recall from Chapter 6 that the independent variable of most special functions may be scaled by a "width parameter" b to construct $f[x/b]$. The scaling theorem determines the effect of this parameter b on the spectrum of the function. For the many special functions defined by rules that determine the amplitude in different segments of the domain, only real-valued scale factors are meaningful to ensure that x/b is real valued; examples include $RECT[x]$ and $TRI[x]$. The scaling theorem considered by most authors applies only to such cases. However, complex-valued scaling factors may be legitimately applied to some special functions. The spectra of the scaled functions are described by a variant of the scaling theorem.

9.8.5.1 Real-Valued Scale Factor

The amplitudes of most special functions in different regions of the domain are defined by "rules", such as the definition of $RECT[x]$ in Equation (6.3) that specifies the amplitude in five different regions: $x < -\frac{1}{2}$, $x = -\frac{1}{2}$, $-\frac{1}{2} < x < \frac{1}{2}$, $x = +\frac{1}{2}$, and $x > +\frac{1}{2}$. Only real-valued scale factors are meaningful for functions of this type because the scale factor applied to the coordinate x must produce a quantity with the same dimensions as x. The resulting scaled function is "wider" or "narrower" than $f[x]$ if $|b| > 1$ or $|b| < 1$, respectively, and thus the area is scaled by the same factor. Negative values of b "reverse" the function along the x-axis, which has no effect on the area. The effect of the scale factor upon the spectrum is easy to evaluate via an obvious substitution for the integration variable:

$$\mathcal{F}_1\left\{f\left[\frac{x}{b}\right]\right\} = \int_{-\infty}^{+\infty} f\left[\frac{x}{b}\right] e^{-2\pi i\xi x}\, dx = \int_{\alpha=-\infty/b}^{\alpha=+\infty/b} f[\alpha]\, e^{-2\pi i\xi(b\alpha)}\, b\, d\alpha$$

$$= b\int_{\alpha=-\infty/b}^{\alpha=+\infty/b} f[\alpha]\, e^{-2\pi i\xi(b\alpha)}\, d\alpha, \quad \textit{for } \alpha \equiv \frac{x}{b} \tag{9.115}$$

Consider the two conditions $b > 0$ and $b < 0$ (the case with $b = 0$ produces a function that is not well defined). The latter case of negative b "reverses" the function as well as scaling the width. These two cases may be subdivided further: into $0 < |b| < 1$ where the width is "compressed"; and $|b| > 1$, which makes the function "wider". Consider first the case $b > 0$; after changing the integration variable, the algebraic signs of the limits of the integral remain unchanged:

$$\mathcal{F}_1\left\{f\left[\frac{x}{b}\right]\right\} = b\int_{\alpha=-\infty}^{\alpha=+\infty} f[\alpha]\, e^{-2\pi i\xi(b\alpha)}\, d\alpha, \quad \textit{for } b > 0$$

$$= b\int_{\alpha=-\infty}^{\alpha=+\infty} f[\alpha]\, e^{-2\pi i(\xi b)\alpha}\, d\alpha = b\cdot F[b\xi] = +|b|\cdot F[b\xi] \tag{9.116a}$$

A negative scale factor may be written as $b = -|b|$, so that the sign of the integration variable α becomes negative:

$$\mathcal{F}_1\left\{f\left[\frac{x}{b}\right]\right\} = (-|b|)\int_{\alpha=+\infty}^{\alpha=-\infty} f[\alpha]\, e^{-2\pi i(\xi(-|b|))\alpha}\, d\alpha, \quad \textit{for } b < 0$$

$$= (-|b|)\left(-\int_{\alpha=-\infty}^{\alpha=+\infty} f[\alpha]\, e^{-2\pi i(\xi(-|b|))\alpha}\, d\alpha\right)$$

$$= +|b|\cdot F[-|b|\xi] = +|b|\cdot F[b\xi] \tag{9.116b}$$

The two cases of positive and negative b may be combined into a single expression:

$$\mathcal{F}_1\left\{f\left[\frac{x}{b}\right]\right\} = |b|\cdot F(b\xi) = |b|\cdot F\left(\frac{\xi}{b^{-1}}\right) \quad \textit{if } |b| \neq 0 \tag{9.117}$$

It is evident that scaling of the distance between "features" in $f[x]$ by b also scales the periods of its sinusoidal constituents by the same factor, and thus scales the spatial frequencies by the reciprocal factor b^{-1}. As an example, consider the effect of scaling the "width" of a cosine function:

$$f[x] = \cos[2\pi\xi_0 x]$$

$$\Longrightarrow f\left[\frac{x}{b}\right] = \cos\left[2\pi\xi_0\left(\frac{x}{b}\right)\right] = \cos\left[2\pi\left(\frac{\xi_0}{b}\right)x\right] \tag{9.118}$$

which produces a cosine function whose spatial frequency has been scaled by b^{-1} and thus whose period is scaled by b.

The scaling theorem for $b > 0$ is applied to a rectangle function in Figure 9.26 to show the change in amplitude and width parameter of the spectrum.

9.8.5.2 Fourier Transform of the "Reversed" Function

One useful corollary to the scaling theorem is obtained by setting the scale factor $b = -1$, which "reverses" $f[x]$ to generate $f[-x]$. By applying Equation (9.113), we obtain:

$$\mathcal{F}_1\{f[-x]\} = |-1|\, F[-\xi] = F[-\xi] \tag{9.119}$$

An example based on the decaying exponential function is shown in Figure 9.27.

This theorem provides the easiest means for deriving the spectrum of the crosscorrelation operation that was defined in Equation (8.68), as will be demonstrated shortly.

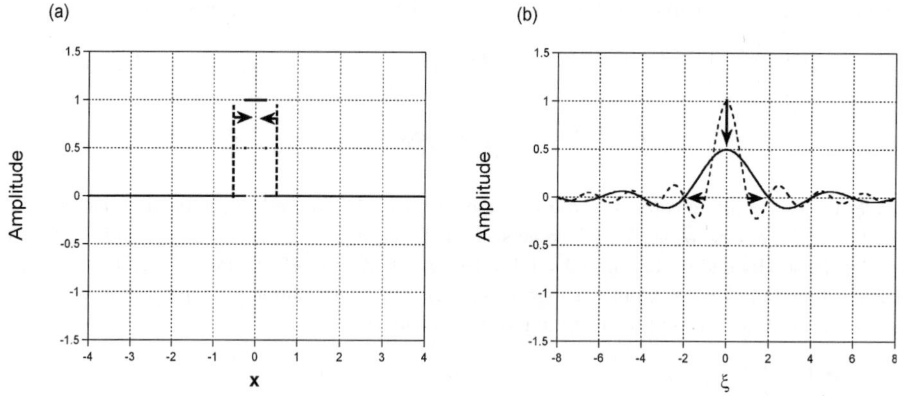

Figure 9.26 The scaling theorem for $f[x] = RECT[x]$ and $F[\xi] = SINC[\xi]$, with $b = 0.5$. The Fourier transform of $f[x/b] = f[2x]$ is $\frac{1}{2}SINC[\xi/2]$. As the space-domain function gets "skinnier", the amplitude decreases and separations between horizontal features (e.g., zero crossings) increase.

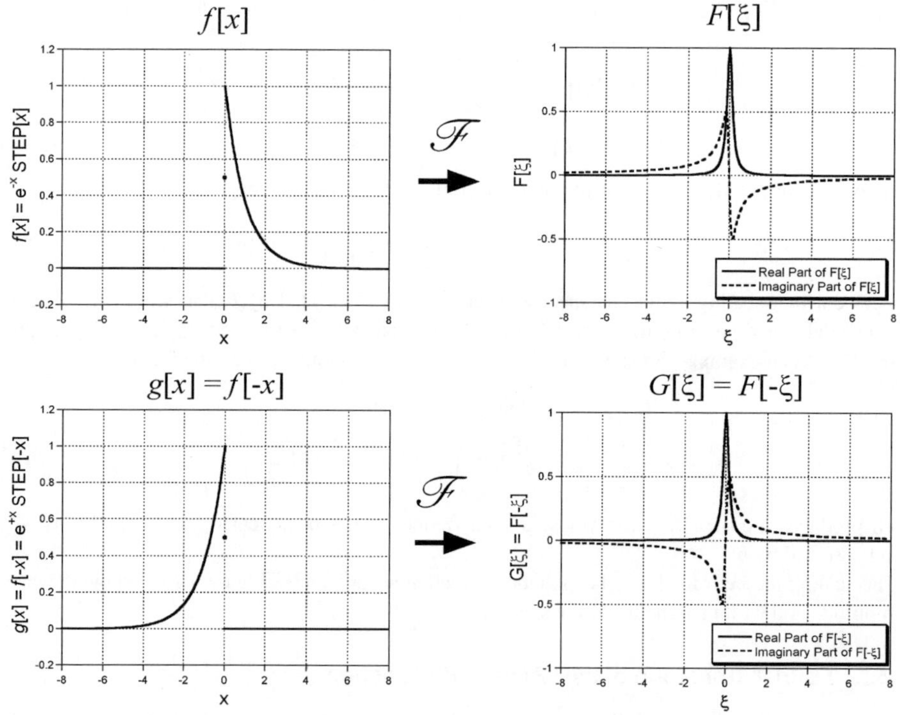

Figure 9.27 Fourier transform of a "reversed" function. Given that $f[x] = e^{-x} \cdot STEP[x]$ and $F[\xi] = [1 + 2\pi i\xi]^{-1}$, the "reversed" function is $f[-x] = e^{+x} \cdot STEP[-x]$ and its Fourier transform is $F[-\xi] = [1 + 2\pi i(-\xi)]^{-1}$.

9.8.5.3 Complex-Valued Scale Factor

Some functions, such as the Gaussian, may be legitimately scaled by complex-valued factors. The scaling completely changes the character of the function. The scaling theorem may be applied to such functions if complex factors are applied to coordinates raised to the second power. In such a case for a complex-valued scale factor $z = a + ib$, the scaling theorem is (Campbell and Foster, 1948):

$$\mathcal{F}_1\{e^{-\pi(x^2/z)}\} = \mathcal{F}_1\{e^{-\pi(x/\sqrt{z})^2}\} = (z)^{\frac{1}{2}}\, e^{-\pi(\xi\sqrt{z})^2} = (z)^{\frac{1}{2}}\, e^{-\pi z\xi^2} \tag{9.120}$$

This result is valid if two conditions are satisfied: the real part of z must be nonnegative (i.e., $a \geq 0$) and its magnitude must be finite ($|z| < \infty$). Note that these constraints are satisfied in the case $z = 0 \pm i$, which converts between Gaussian and chirp functions via Equation (6.149):

$$\begin{aligned} GAUS\left[\frac{x}{\sqrt{\pm i}}\right] &= e^{-\pi(x\sqrt{\mp i})^2} \\ &= e^{-\pi x^2(\mp i)} \\ &= e^{\pm i\pi x^2} \end{aligned} \tag{9.121}$$

Both space-domain functions $GAUS[x]$ and $e^{\pm i\pi x^2}$ have bounded amplitude and finite area, and so Fourier spectra exist for both. A valid scaling theorem for complex-valued scale factors applies in this case:

$$\begin{aligned} \mathcal{F}_1\{e^{\pm i\pi x^2}\} &= \mathcal{F}_1\left\{GAUS\left[\frac{x}{(\sqrt{(\pm i)})}\right]\right\} \\ &= (\sqrt{(\pm i)})\; GAUS[\xi(\sqrt{(\pm i)})] \\ &= e^{\pm i(\pi/4)}\; e^{-\pi(\xi^2(\pm i))} = e^{\pm i(\pi/4)}\; e^{\mp i\pi\xi^2} \end{aligned} \tag{9.122}$$

which agrees with the result of Equation (9.95). This result will be useful in some situations, as will be demonstrated shortly, but we emphasize again that the scaling theorem with complex factors is not valid in the general case and must be invoked with care.

9.8.6 Shift Theorem

The Fourier transform of a function that has been translated along the axis by a distance $\pm x_0$ may be derived from the Fourier transform by a simple change of variable:

$$\begin{aligned} \mathcal{F}_1\{f[x \pm x_0]\} &= \int_{-\infty}^{+\infty} f[x \pm x_0]\, e^{-2\pi i\xi x}\, dx, \quad u \equiv x \pm x_0 \\ &= \int_{-\infty}^{+\infty} f[u]\, e^{-2\pi i\xi(u \mp x_0)}\, du \\ &= e^{\pm 2\pi i\xi x_0}\; F[\xi] \\ &= |F[\xi]|\; e^{i(\Phi\{F[\xi]\} \pm 2\pi\xi x_0)} \end{aligned} \tag{9.123}$$

In words, the spectrum of a shifted function is the transform of the original function multiplied by a complex sinusoid whose phase is proportional to the translation. This is equivalent to saying that a *linear-phase factor* is applied to the original spectrum.

The spectrum of the translated Dirac delta function $f[x] = \delta[x - x_0]$ was derived in Equation (9.59), but we can now consider this result in the light of the shift theorem. Recall that the "centered" Dirac delta function $\delta[x]$ was approximated in Figure 6.25 by the sum of a few sinusoidal functions. Four

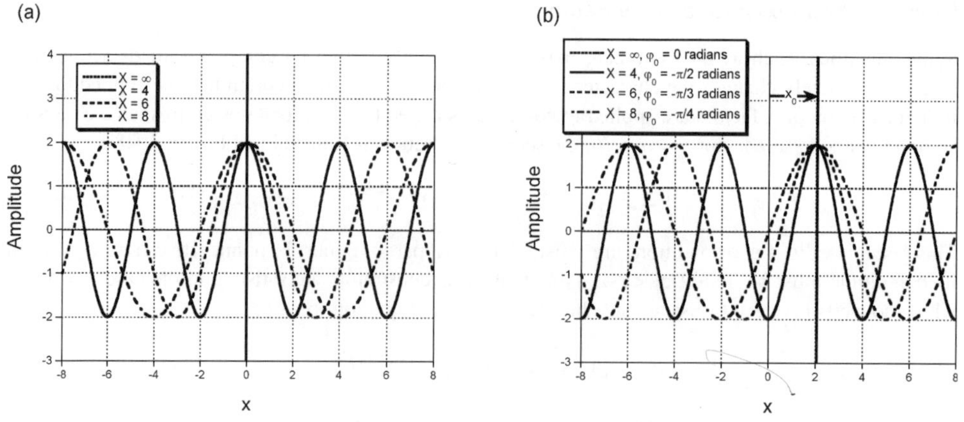

Figure 9.28 The principle of the shift theorem: (a) four cosine components of $\delta[x]$ with periods ∞, eight, six, and four units and initial phase $\phi_0 = 0$ radians so that each cosine has a maximum at $x = 0$; (b) the same four component sinusoids are translated to the right by $x_0 = 2$ units in $\delta[x-2]$. The initial phase of the cosine with period ∞ is still zero radians, but the initial phases of the terms with periods eight, six, and four units are respectively $-\pi/2$, $-\pi/4$, and $-\pi/8$ radians.

of these sinusoids are shown in Figure 9.28a: periods $X = \infty$, 8, 6, and 4 units with corresponding spatial frequencies $\xi = 0$, $\frac{1}{8}$, $\frac{1}{6}$, and $\frac{1}{4}$ cycle per unit. Note that the initial phase of each of these four sinusoids is zero so that their unit amplitudes all add at $x = 0$. The same four sinusoidal components of the Dirac delta function centered at $x = 2$ are shown in Figure 9.28b. All four were translated by the same physical distance $x_0 = 2$, but the initial phases of the four sinusoidal components now differ. The shorter the period, the larger the initial phase required to generate the same physical displacement, thus confirming the observation that the initial phase of the transform of the translated function is proportional to the spatial frequency ξ.

Another view of the shift theorem is obtained by considering the cosine expressed in complex notation via the Euler relation:

$$\cos[2\pi\xi_0 x] = \frac{1}{2}(e^{+2\pi i\xi_0 x} + e^{-2\pi i\xi_0 x}) = \frac{1}{2}(e^{+2\pi i(+\xi_0)x} + e^{+2\pi i(-\xi_0)x}) \tag{9.124}$$

which just demonstrates that the real-valued cosine is the average of two Hermitian complex sinusoids: one each with positive and negative spatial frequency. The Fourier transform of the cosine is obtained by applying the fact that the Fourier transform is linear and Equation (9.78):

$$\begin{aligned} \mathcal{F}_1\{\cos[2\pi\xi_0 x]\} &= \frac{1}{2}\mathcal{F}_1\{e^{+2\pi i(+\xi_0)x}\} + \frac{1}{2}\mathcal{F}_1\{e^{+2\pi i(-\xi_0)x}\} \\ &= \frac{1}{2}\delta[\xi - \xi_0]\, e^{i\cdot 0} + \frac{1}{2}\delta[\xi - (-\xi_0)]\, e^{i\cdot 0} \end{aligned} \tag{9.125}$$

which is a real-valued even pair of Dirac delta functions.

If the cosine is translated by the distance x_0, its complex expression is obtained by a simple substitution into Equation (9.124):

$$\begin{aligned}\cos[2\pi\xi_0(x-x_0)] &= \frac{1}{2}(e^{+2\pi i\xi_0(x-x_0)} + e^{-2\pi i\xi_0(x-x_0)})\\ &= \frac{1}{2}(e^{+2\pi i[(+\xi_0)x+(\xi_0 x_0)]} + e^{+2\pi i[(-\xi_0)x-(\xi_0 x_0)]})\\ &= \frac{1}{2}(e^{+2\pi i(+\xi_0)x+i\phi_0} + e^{+2\pi i(-\xi_0)x+i(-\phi_0)})\end{aligned} \tag{9.126}$$

where $\phi_0 \equiv 2\pi\xi_0 x_0 > 0$, which is (of course) measured in radians. The spectrum of the translated cosine is again an easy application of Equation (9.78):

$$\begin{aligned}\mathcal{F}_1\{\cos[2\pi\xi_0(x-x_0)]\} &= \frac{1}{2}\mathcal{F}_1\{e^{+2\pi i[(+\xi_0)x+\phi_0]}\} + \frac{1}{2}\mathcal{F}_1\{e^{+2\pi i[(+\xi_0)x+\phi_0]}\}\\ &= \frac{1}{2}\delta[\xi-\xi_0]\,e^{+i\phi_0} + \frac{1}{2}\delta[\xi-(-\xi_0)]\,e^{+i(-\phi_0)}\end{aligned} \tag{9.127}$$

In words, the cosine is translated in the positive direction ($x_0 > 0$) by adding the angle $\phi_0 = 2\pi\xi_0 x_0 > 0$ to the positive frequency component and the angle $-\phi_0 < 0$ to the negative frequency component. The steps in this process are shown in Figure 9.29.

The last example of the shift theorem compares the spectra of centered and translated *SINC* functions. The Fourier transform of *SINC*$[x]$ already was shown to be *RECT*$[\xi]$ in Equation (9.109). The spectrum of the translated *SINC* function may be evaluated via the shift theorem:

$$\begin{aligned}\mathcal{F}_1\{SINC[x-x_0]\} &= \mathcal{F}_1\{SINC[x]\}\,e^{-2\pi i\xi x_0}\\ &= RECT[\xi]\,e^{-2\pi i\xi x_0}\\ &= RECT[\xi](\cos[2\pi\xi x_0] - i\,\sin[2\pi\xi x_0])\end{aligned} \tag{9.128a}$$

The magnitude of the spectrum is identical to that in Equation (9.109):

$$|\mathcal{F}_1\{SINC[x-x_0]\}| = RECT[\xi] \tag{9.128b}$$

while the phase is linear within the region of support of the spectrum:

$$\Phi\{\mathcal{F}_1\{SINC[x-x_0]\}\} = -2\pi\xi x_0\; RECT[\xi] \tag{9.128c}$$

These results are illustrated in Figures 9.30 and 9.31.

The shift theorem also demonstrates that the Fourier transform is shift variant because the spectrum of a shifted function is *not* a replica that has been translated by the same distance. The mathematical expression for this property is $\mathcal{F}_1\{f[x-x_0]\} \neq F[\xi - x_0]$. The shift-variant character means that the Fourier transform operation cannot be implemented via a convolution. In retrospect, this result should have been (and perhaps was) obvious from the definition of shift invariance in Equation (8.14). Because the two parameters ξ and x_0 have different dimensions, the difference of the two coordinates $\xi - x_0$ is meaningless.

9.8.7 Filter Theorem

Without a doubt, the most important property of the Fourier transform in imaging applications is the relationship between the spectrum of a convolution and the spectra of the constituent functions. This "filter theorem" is so significant because many imaging systems (particularly optical systems) may be modeled as linear (at least approximately) and shift invariant. This theorem provides an extremely powerful tool for analyzing these imaging systems.

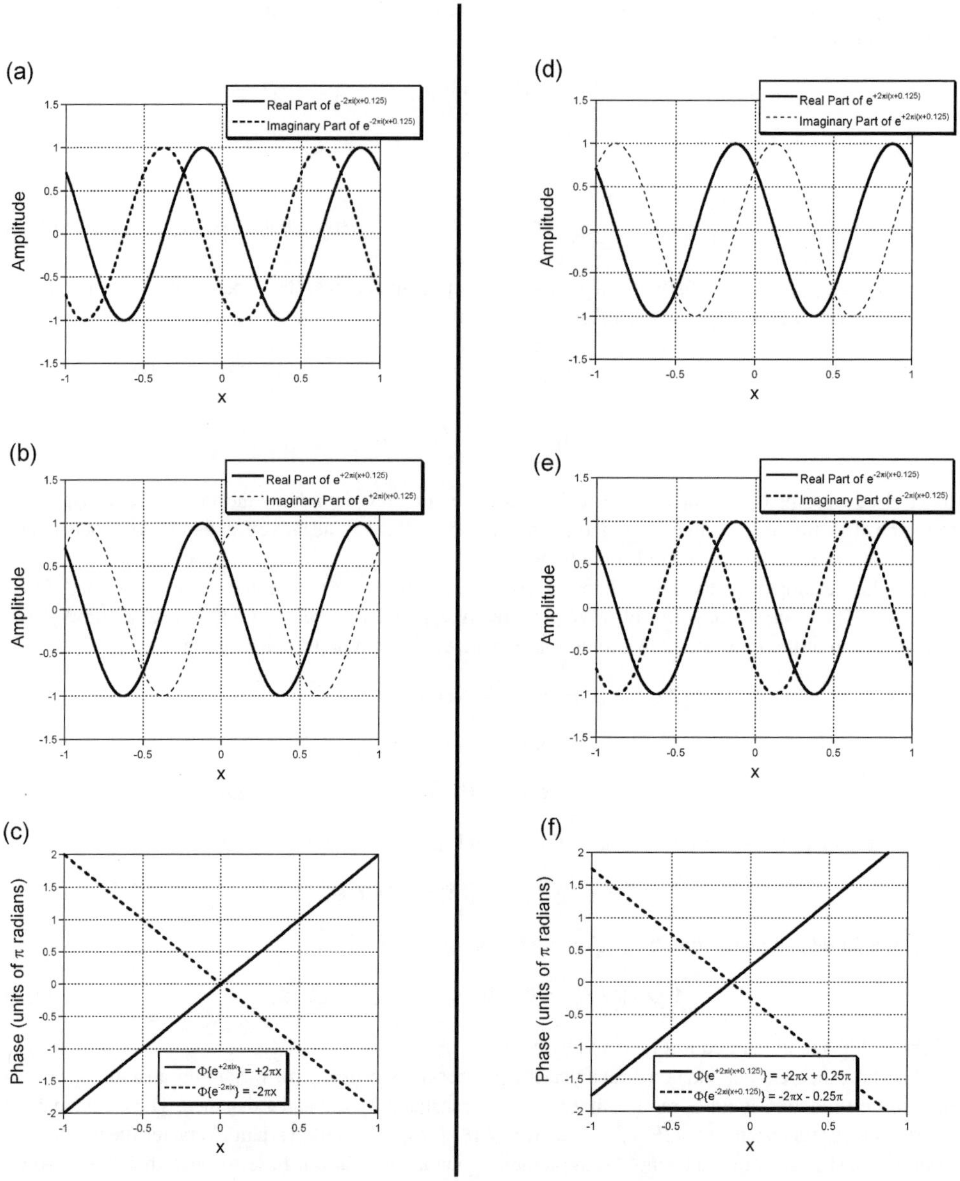

Figure 9.29 The phase factor in the shift theorem: (a), (b) the real-valued function $\cos[2\pi x]$ is the scaled sum of the two complex sinusoids $e^{\pm 2\pi i x}$; (c) the phases of these sinusoids are $\Phi[x] = \pm 2\pi x$; (d), (e) the component functions $\exp[\pm 2\pi i(x + \frac{1}{8})]$ of the cosine after translation by $x_0 = -\frac{1}{8}$; (f) the phases of the translated complex sinusoids. The phases of the positive frequency and negative frequency terms are respectively incremented by and decremented by $\pi/4$ radians.

The filter theorem is derived by changing the integration variable. Given a 1-D LSI system with input $f[x]$ and impulse response $h[x]$ such that the output is $g[x]$, and using the conventional upper-case notation for the individual spectra (e.g., that $\mathcal{F}_1\{f[x]\} = F[\xi]$ etc.), then the spectrum of $g[x]$ is

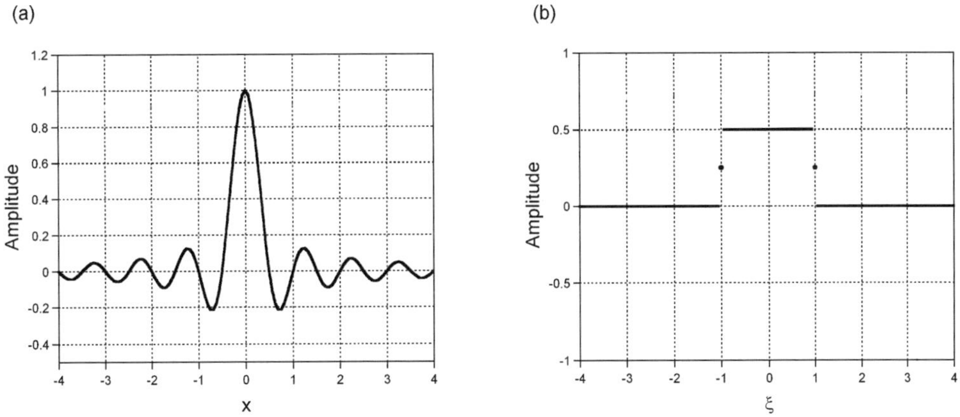

Figure 9.30 (a) The real and even function $f[x] = SINC[x/0.5]$; (b) its Fourier transform is the real and even function $F[\xi] = \frac{1}{2}RECT[\xi/2]$.

easily expressed in terms of $F[\xi]$ and $H[\xi]$:

$$\begin{aligned}
\mathcal{F}_1\{g[x]\} &= \mathcal{F}_1\{f[x] * h[x]\} \\
&= \int_{-\infty}^{+\infty} (f[x] * h[x])\, e^{-2\pi i\xi x}\; dx \\
&= \int_{-\infty}^{+\infty} \left(\int_{-\infty}^{+\infty} f[\alpha]h[x-\alpha]\, d\alpha\right) e^{-2\pi i\xi x}\; dx \\
&= \int_{-\infty}^{+\infty} f[\alpha]\left(\int_{-\infty}^{+\infty} h[x-\alpha]\, e^{-2\pi i\xi x}\; dx\right) d\alpha \\
&= \int_{-\infty}^{+\infty} f[\alpha](H[\xi]\, e^{-2\pi i\xi\alpha})\, d\alpha \\
&= H[\xi]\left(\int_{-\infty}^{+\infty} f[\alpha]\, e^{-2\pi i\xi\alpha}\; d\alpha\right) = H[\xi]\cdot F[\xi] = G[\xi]
\end{aligned} \tag{9.129}$$

where the shift theorem and the definition of convolution have been used. The transfer function $H[\xi]$ may be extracted from the integral over α because it is not a function of the integration variable α.

The filter theorem is so pervasive in imaging that conventional names have been established for the two representations of the system $h[x]$ and $H[\xi]$. We already have seen that the system function $h[x]$ is commonly known as the *impulse response* because it specifies the output function that is produced by an impulsive input $\delta[x]$. We also mentioned that $h[x]$ is commonly called the *point-spread function* (*psf*) in optics. The spectrum $H[\xi]$ of the psf generally is known as the system *transfer function*, because it specifies the effect upon the constituent sinusoids of the input function as they are transferred through the system to the output.

The filter theorem offers an alternative and frequently used route for computing the output of an LSI system:

$$\begin{aligned}
g[x] &= \mathcal{F}_1^{-1}\{F[\xi]\cdot H[\xi]\} \\
&= \mathcal{F}_1^{-1}\{\mathcal{F}_1\{f[x]\}\cdot \mathcal{F}_1\{h[x]\}\}
\end{aligned} \tag{9.130}$$

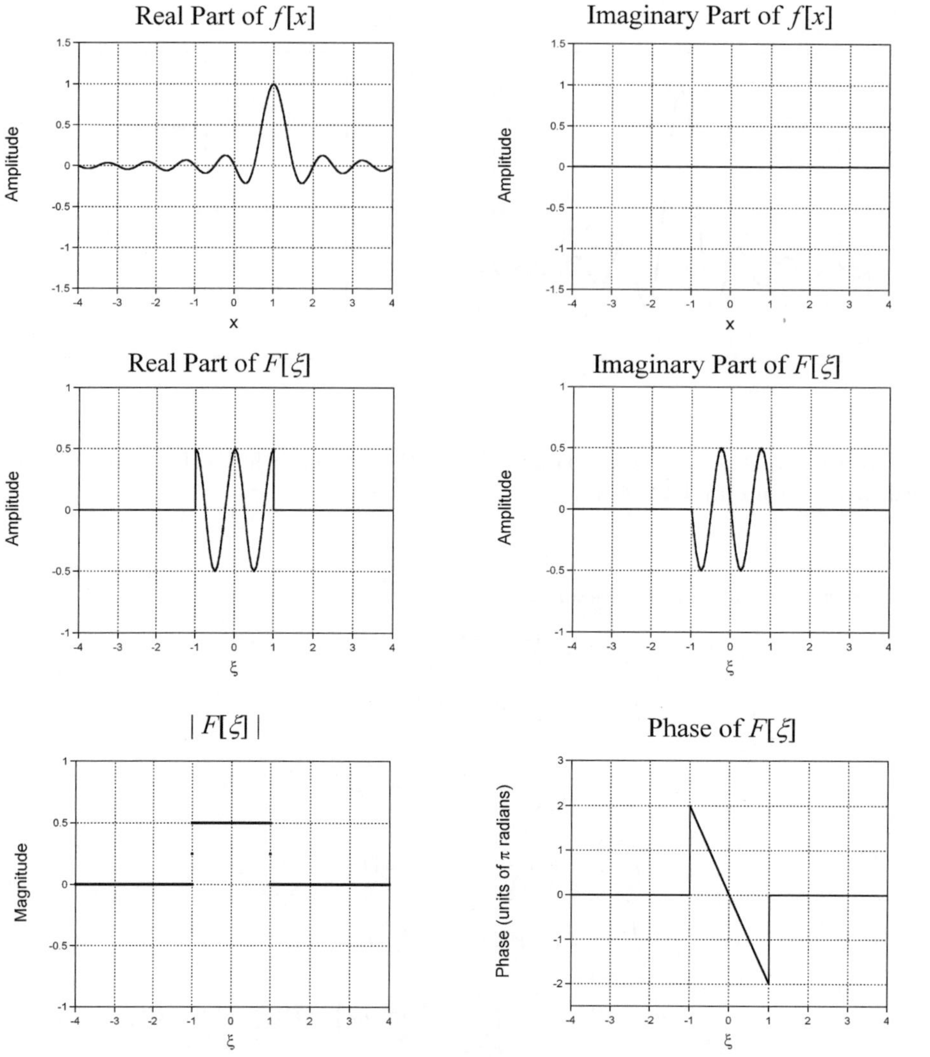

Figure 9.31 Fourier transform of the *SINC* function of Figure 9.30 after translation to create $SINC[x-1]$. The real part of the transform is $\frac{1}{2}RECT[\xi/2]\cdot\cos[2\pi\xi]$ and the imaginary part is $\frac{1}{2}RECT[\xi/2]\cdot(-\sin[2\pi\xi])$. The magnitude of the transform is preserved after translation of the function, while the new phase is $\Phi\{F[\xi]\}=-2\pi\xi$ where the magnitude is nonzero and arbitrarily set to zero where the magnitude is zero.

The schematic in Figure 9.32 demonstrates the reason why this route may be referred to as computing the convolution "around the block". Three Fourier transform operations are required: two in the "forward" direction followed by one "inverse" transform. The efficiency of this process must be compared to the space-domain convolution that requires computation of an infinite number of integrals over the infinite domain.

As a first example (of many to come), we can easily derive the spectrum of $TRI[x]$ using the filter theorem. The space-domain representation of $RECT[x]*RECT[x]$ has already been shown to

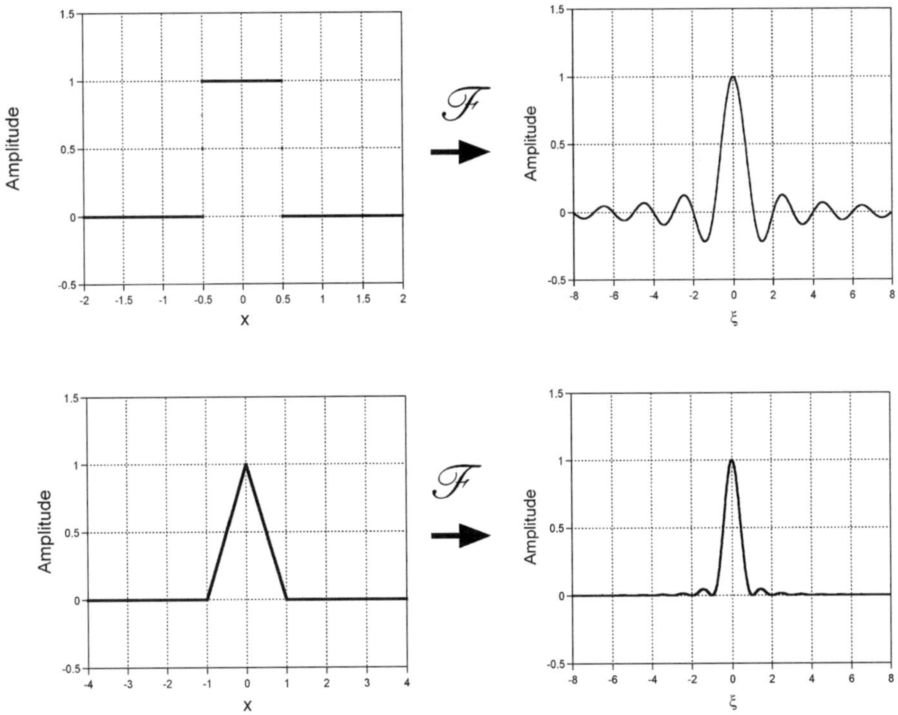

Figure 9.32 The filter theorem of the Fourier transform for $f[x] = h[x] = RECT[x]$. The Fourier transforms are $F[\xi] = H[\xi] = SINC[\xi]$. The convolution is $f[x] * h[x] = TRI[x]$, and its Fourier transform is $F[\xi] \cdot H[\xi] = SINC^2[\xi]$.

be $TRI[x]$ in Equation (8.42). The filter theorem leads directly to another Fourier transform pair, as shown in Figure 9.32:

$$\begin{aligned}\mathcal{F}_1\{TRI[x]\} &= \mathcal{F}_1\{RECT[x] * RECT[x]\} \\ &= \mathcal{F}_1\{RECT[x]\} \cdot \mathcal{F}_1\{RECT[x]\} \\ &= SINC[\xi] \cdot SINC[\xi] = SINC^2[\xi] \end{aligned} \tag{9.131}$$

This result may be inserted into the "transform-of-a-transform" theorem to obtain yet another transform pair:

$$\mathcal{F}_1\{SINC^2[x]\} = TRI[-\xi] = TRI[+\xi] \tag{9.132}$$

where the symmetry of the triangle function has been used in the last step. $TRI[x]$ is the second symmetric nonnegative function we have considered that has a symmetric nonnegative spectrum; the other is $GAUS[x]$. In other words, $TRI[x]$, $SINC^2[x]$, and $GAUS[x]$ are synthesized by summing cosine functions with nonnegative weights.

9.8.7.1 Classes of Transfer Functions

Theoretically, the transfer function of a linear filter may have any form; its action may amplify, attenuate, or reject ("block") any or all sinusoidal components and/or it may add any increment to the initial phase at any frequency. The transfer function cannot generate sinusoidal frequencies out of "thin air"; the input amplitude at the spatial frequency ξ_0 must be nonzero for the output amplitude at ξ_0 to be finite. Note that this property is not true for nonlinear systems, as will be shown later in this chapter.

We will find it convenient to group linear filters into categories that are distinguished by the shapes of their transfer functions. These classes will be considered in detail in Chapters 16–19, but they are introduced here to help describe sampled functions in Chapter 14. One possible classification scheme groups those filters that affect only the magnitude of the input spectrum, those that affect only the phase, and those that modify both. Subclasses also may be used within these groups. For example, magnitude filters are classified as *lowpass*, *highpass*, *bandpass*, or *bandstop* based upon the "shape" of the magnitude of the transfer function. *Lowpass* filters reject sinusoidal components with large spatial frequencies and pass those with small frequencies. *Highpass* filters are complementary; they reject small spatial frequencies. Pure *phase* filters do not affect the magnitude at all and thus may be classified as *allpass*.

A lowpass filter attenuates or rejects any rapid variations present in the amplitude of the input signal. This has the effect of reducing the variability of the input amplitude by "pushing" it toward the local mean amplitude of the input signal. The amplitude of the transfer function of the lowpass filter $H[\xi = 0] = 1$ and falls to zero at some *cutoff* frequency ξ_{cutoff}. A simple and common example of a lowpass filter transfer function is the symmetric rectangle function:

$$H_{LP}[\xi] = RECT[\xi] + i \cdot 0[\xi] \tag{9.133}$$

as shown in Figure 9.33. Clearly the MTF and the transfer function of this filter are identical because $H[\xi]$ is real valued and nonnegative. Since the imaginary part of is null and the real part is nonnegative, the phase transfer function of this filter is $0[\xi]$. Because $H_{LP}[\xi]$ passes all frequencies with magnitudes below the cutoff frequency $|\xi_{\text{cutoff}}| = 0.5$ without change, while rejecting all frequencies with $\xi > |\xi_{\text{cutoff}}|$, the transfer function in Equation (9.133) is an ideal lowpass filter.

From the spectrum of $RECT[x]$ in Equation (9.71) and the transform-of-a-transform theorem in Equation (9.105), the impulse response corresponding to $H_{LP}[\xi]$ obviously is the symmetric *SINC* function:

$$h_{LP}[x] = SINC[x] \tag{9.134}$$

The amplitude of the impulse response of the ideal lowpass filter is bipolar, but it is positive at coordinates near the origin. This means that positive weights are applied in the vicinity of the translated coordinate during convolution, so that the output amplitude is largely a "weighted average" of the amplitudes of the input function; a lowpass filter is a "local averager". The characteristics of lowpass filters will be discussed in more detail in Chapter 16.

Conversely, a highpass filter rejects components with small spatial frequencies while passing high-frequency terms with little or no attenuation. The transfer function of a "true" highpass filter completely rejects DC input signals, which means that the magnitude of the transfer function is zero at $\xi = 0$. The mean amplitude of the output $g[x]$ of such a filter must be zero, and therefore $g[x]$ must be bipolar. Many authors generalize the definition of highpass filters to include transfer functions that pass some small part of the DC component of the input signal.

The transfer function of the ideal highpass filter is the complement of Equation (9.133); it will reject all frequencies below the cutoff frequency while passing all frequencies above without change in magnitude. The transfer function of this ideal filter is:

$$H_{HP}[\xi] = 1[\xi] - H_{LP}[\xi] \tag{9.135}$$

while the linearity property of the inverse Fourier transform determines that the impulse response is:

$$h_{HP}[x] = \delta[x] - h_{LP}[x] \tag{9.136}$$

as shown in Figure 9.33. In this example, the algebraic sign of $h_{HP}[x = 0]$ is positive, while the signs at other coordinates in the vicinity of the origin are negative. In other words, this filter applies weights with

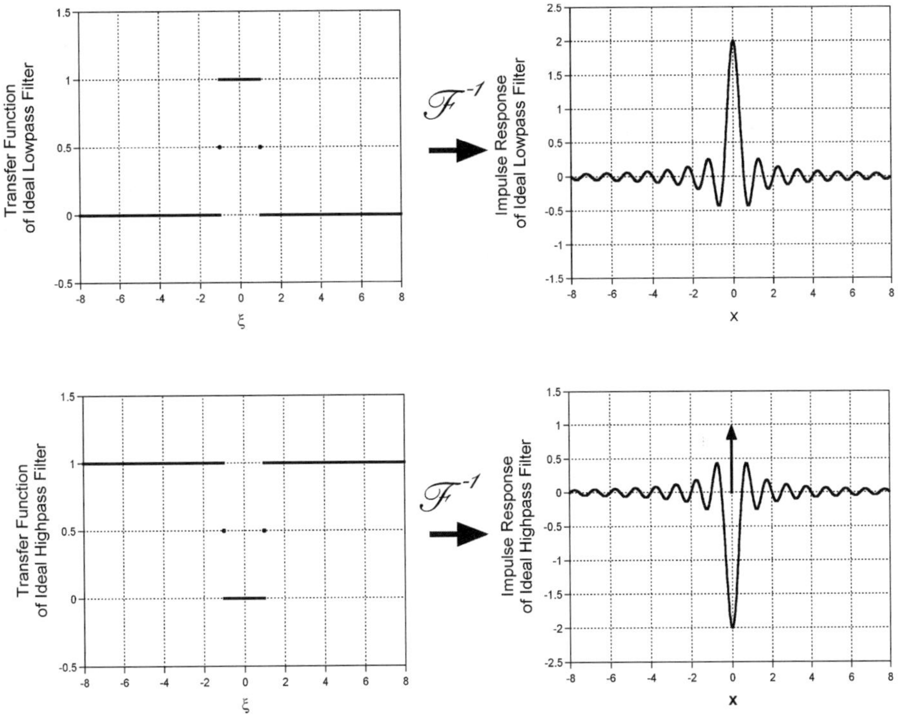

Figure 9.33 Transfer functions and corresponding impulse responses for ideal lowpass and highpass filters. The transfer function of the lowpass filter is $H_{LP}[\xi] = RECT[\xi/2]$ and the impulse response is $\mathcal{F}^{-1}\{RECT[x/2]\} = 2\,SINC[2\xi]$. The transfer function of the highpass filter is the complement $H_{HP}[\xi] = 1 - RECT[\xi/2]$ and the impulse response is $\delta[x] - 2\,SINC[2\xi]$.

opposite sign to nearby coordinates of the input function $f[x]$, and thus it computes "local differences" of amplitudes of $f[x]$. The differencing increases the local variability of the output of a highpass filter when compared to the input signal $f[x]$. From these two descriptions, it is clear that a bandpass filter blocks all sinusoidal components outside of some range of frequencies, while a bandstop filter rejects frequencies that fall within some range. Again, these types of filters will be described in more detail (perhaps better described as "tedious" detail) in Chapters 16–19.

9.8.8 Modulation Theorem

Now consider a system whose output $g[x]$ is the product of the input function $f[x]$ and a modulation $m[x]$:

$$g[x] = \mathcal{O}\{f[x]\} = f[x] \cdot m[x] \tag{9.137}$$

where we use "modulation" in the sense of "multiplication" rather than the sinusoidal analogue to "contrast". The spectrum of $g[x]$ is:

$$\mathcal{F}_1\{g[x]\} = \mathcal{F}_1\{f[x] \cdot m[x]\}$$

$$= \int_{-\infty}^{+\infty} (f[\alpha]m[\alpha])\, e^{-2\pi i\xi\alpha}\, d\alpha \tag{9.138}$$

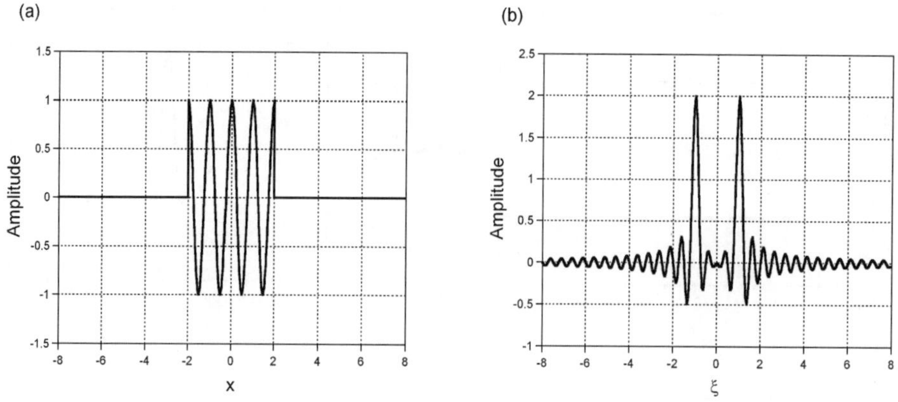

Figure 9.34 The modulation theorem: (a) input function $f[x] \cdot m[x] = RECT[x/4] \cdot \cos[2\pi x]$; (b) its spectrum $\mathcal{F}_1\{f[x] \cdot m[x]\} = F[\xi] * M[\xi]$, which in this case evaluates to: $2\,SINC[4\xi] * (\delta[\xi+1] + \delta[\xi-1])$.

The spectrum is most easily derived by substituting expressions for the inverse Fourier transforms for $f[x]$ and $m[x]$:

$$f[x] = \mathcal{F}_1^{-1}\{F[\xi]\} = \int_{-\infty}^{+\infty} F[v]\, e^{+2\pi i v x}\, dv \tag{9.139a}$$

$$m[x] = \mathcal{F}_1^{-1}\{M[\xi]\} = \int_{-\infty}^{+\infty} M[u]\, e^{+2\pi i u x}\, du \tag{9.139b}$$

Note that different integration variables u and v are used to allow their substitution into Equation (9.138):

$$\begin{aligned}
\mathcal{F}_1\{f[x] \cdot m[x]\} &= \int_{-\infty}^{+\infty} \left(\int_{-\infty}^{+\infty} F[v]\, e^{+2\pi i v x}\, dv\right)\left(\int_{-\infty}^{+\infty} M[u]\, e^{+2\pi i u x}\, du\right) e^{-2\pi i \xi x}\, dx \\
&= \int_{-\infty}^{+\infty} F[v]\left(\int_{-\infty}^{+\infty} M[u]\left(\int_{-\infty}^{+\infty} (e^{+2\pi i v x}\, e^{+2\pi i u x}\, e^{-2\pi i \xi x})\, dx\right) du\right) dv \\
&= \int_{-\infty}^{+\infty} F[v]\left(\int_{-\infty}^{+\infty} M[u]\left(\int_{-\infty}^{+\infty} e^{+2\pi i x(u+v-\xi)}\, dx\right) du\right) dv \\
&= \int_{-\infty}^{+\infty} F[v]\left(\int_{-\infty}^{+\infty} (M[u]\, \delta[u-(\xi-v)])\, du\right) dv \\
&= \int_{-\infty}^{+\infty} F[v]\, M[\xi - v]\, dv = F[\xi] * M[\xi]
\end{aligned} \tag{9.140}$$

where the integral over x is the known Fourier transform of the Dirac delta function from Equation (9.106).

In words, the spectrum of a modulation is the convolution of the constituent spectra. Of course, this result could have been deduced from the filter and "transform-of-a-transform" theorems. The results of the last two sections indicate that the processes of convolution and modulation are "dual operations" in the two representations; modulation in the space domain corresponds to convolution in the frequency domain, and vice versa. An example of the modulation theorem is shown in Figure 9.34.

9.8.9 Derivative Theorem

This theorem relates the spectra of $f[x]$ and of its derivatives, and again is proven most easily by substituting the expression for the inverse transform of $F[\xi]$ into the derivative. Given that $g[x] = d^n f/dx^n$, then:

$$\begin{aligned} g[x] &= \frac{d^n}{dx^n}(f[x]) = \frac{d^n}{dx^n}\left(\int_{-\infty}^{+\infty} F[\xi]\, e^{+2\pi i\xi x}\, d\xi\right) \\ &= \int_{-\infty}^{+\infty} \frac{d^n}{dx^n}(F[\xi]\, e^{+2\pi i\xi x})\, d\xi \quad \textit{(by linearity)} \\ &= \int_{-\infty}^{+\infty} F[\xi]\left[\frac{d^n}{dx^n}(e^{+2\pi i\xi x})\right] d\xi \quad \textit{(by linearity)} \\ &= \int_{-\infty}^{+\infty} [F[\xi](+2\pi i\xi)^n]\, e^{+2\pi i\xi x}\, d\xi = \mathcal{F}_1^{-1}\{(+2\pi i\xi)^n F[\xi]\} \\ &= \mathcal{F}_1^{-1}\{G[\xi]\} \end{aligned} \tag{9.141}$$

By evaluating the Fourier transform of both sides of this equation, the desired result is obtained:

$$\mathcal{F}_1\{g[x]\} = \mathcal{F}_1\left\{\frac{d^n}{dx^n} f[x]\right\} = (+2\pi i\xi)^n F[\xi] = G[\xi] \tag{9.142}$$

Comparison of this result to Equation (9.129) identifies the factor $(+2\pi i\xi)^n$ to be the transfer function of the process of nth-order differentiation. Since differentiation may be described by a transfer function, it must be an LSI operation and thus may be expressed as a convolution. Comparison to Equation (8.35) demonstrates that the impulse response of nth-order differentiation is the nth derivative of the Dirac delta function $\delta[x]$. These two observations provide an obvious path to the Fourier transform of the nth derivative of $\delta[x]$:

$$\mathcal{F}_1\left\{\frac{d^n}{dx^n}\delta[x]\right\} = \mathcal{F}_1\{\delta^{(n)}[x]\} = (2\pi i\xi)^n \tag{9.143a}$$

The transfer function of differentiation may be recast into magnitude and phase components in the manner of Equation (9.128). It becomes evident that the action of differentiation on $f[x]$ is an amplification of the magnitude of its sinusoidal components by a factor proportional to the nth power of the spatial frequency:

$$\left|\mathcal{F}_1\left\{\frac{d^n}{dx^n} f[x]\right\}\right| = |(+2\pi\xi)^n|\,|F[\xi]| \tag{9.143b}$$

The form of the phase transfer function of differentiation depends on the order n of the derivative; the phase is incremented or decremented by π radians if ξ^n is positive or negative, respectively:

$$\boldsymbol{\Phi}\left\{\mathcal{F}_1\left\{\frac{d^n}{dx^n} f[x]\right\}\right\} = \boldsymbol{\Phi}\{F[\xi]\} + \pi \cdot SGN[\xi^n] \tag{9.143c}$$

The effects of differentiation on the spectrum of $f[x]$ are shown in Figure 9.35. The derivative theorem may be combined with the filter, transform of the "reversed" function, and "transform-of-a-transform" theorems to evaluate the spectra of the powers of x. These are useful when evaluating the Fourier transform of a function expressed in the form of a Taylor series in powers of x. Equations (9.104) and (9.141) may be combined to show that:

$$\mathcal{F}_1\{(+2\pi i x)^n \cdot 1[x]\} = \left.\left(\frac{d^n}{du^n}\delta[u]\right)\right|_{u=-\xi} \tag{9.144}$$

Since $\delta[x]$ is an even function, its odd-order and even-order derivatives are respectively odd and even functions. "Reversal" of the frequency coordinate $(u = -\xi)$ in Equation (9.140) therefore multiplies

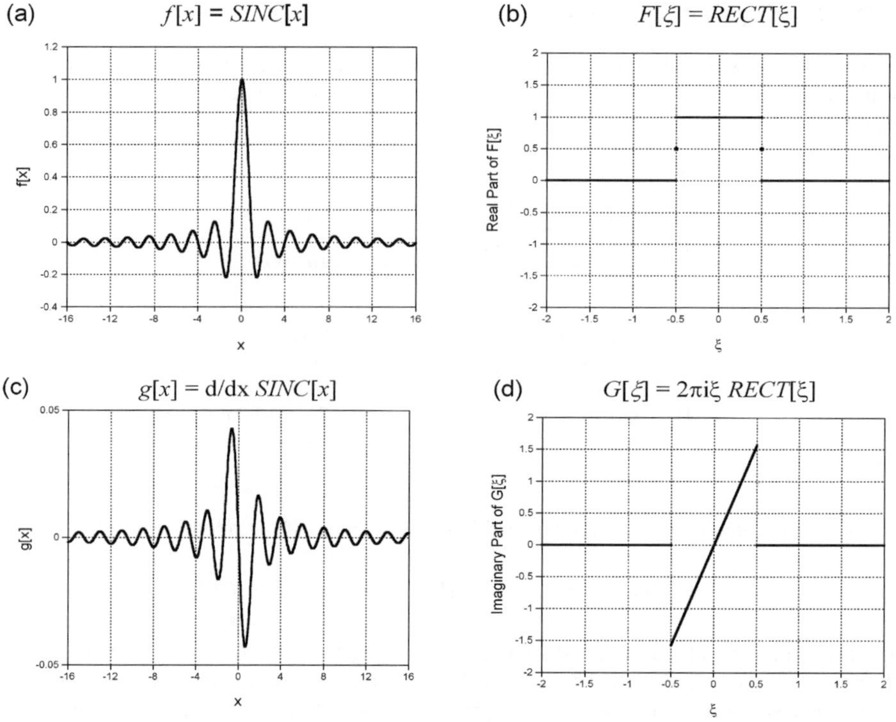

Figure 9.35 The derivative theorem: (a) $f[x] = SINC[x]$; (b) $F[\xi] = RECT[\xi] + 0i$ (imaginary part not shown). The Fourier transform of (c) $g[x] = (d/dx)(SINC[x])$ is (d) $0 + i(2\pi\xi)\,RECT[\xi]$ (real part not shown).

the spectrum by -1. This observation may be combined with the linearity of the Fourier transform to evaluate the "generalized" spectrum of $f[x] = x^n$:

$$\mathcal{F}_1\{x^n\} = \left(-\frac{1}{2\pi i}\right)^n \delta^{(n)}[\xi] = \left(+\frac{i}{2\pi}\right)^n \delta^{(n)}[\xi] \tag{9.145}$$

This expression may test our still-developing intuition of the Fourier transform, so we consider a few of its properties. First note that the symmetry of x^n is even or odd as n is even or odd. The resulting spectrum is respectively real or imaginary due to the leading factor of $(+i)^n$. The spectrum of x^n is proportional to the nth derivative of $\delta[\xi]$, which has infinitesimal support regardless of the value of n.

9.8.10 Fourier Transform of Complex Conjugate

If $f[x]$ is a complex-valued function with spectrum $F[\xi]$, the transform of the complex conjugate $f^*[x]$ also may be derived easily by making the appropriate change of variable:

$$\begin{aligned}\mathcal{F}_1\{f^*[x]\} &= \int_{-\infty}^{+\infty} f^*[x]\, e^{-2\pi i\xi x}\, dx \\ &= \left(\int_{-\infty}^{+\infty} f[x]\, e^{+2\pi i\xi x}\, dx\right)^*\end{aligned}$$

$$= \left(\int_{-\infty}^{+\infty} f[x]\, e^{-2\pi i(-\xi)x}\, dx \right)^*$$

$$= (F[-\xi])^* = F^*[-\xi] \tag{9.146}$$

where the fact that the complex conjugate of the sum is the sum of the complex conjugates has been used. As an example, consider the effect on the spectrum of a translated complex linear-phase exponential:

$$\mathcal{F}_1\{e^{+2\pi i\xi_0(x+x_0)}\} = \mathcal{F}_1\{e^{+2\pi i\xi_0 x} \cdot e^{+2\pi i\xi_0 x_0}\}$$

$$= e^{+2\pi i\xi_0 x_0}\ \mathcal{F}_1\{e^{+2\pi i\xi_0 x}\}$$

$$= e^{+2\pi i\xi_0 x_0}\ \delta[\xi - \xi_0] \equiv F[\xi] \tag{9.147}$$

The spectrum of the complex conjugate obeys the prediction of Equation (9.142):

$$\mathcal{F}_1\{(e^{+2\pi i\xi_0(x+x_0)})^*\} = \mathcal{F}_1\{e^{-2\pi i\xi_0 x}\ e^{-2\pi i\xi_0 x_0}\}$$

$$= e^{-2\pi i\xi_0 x_0}\ \mathcal{F}_1\{e^{-2\pi i\xi_0 x}\}$$

$$= e^{-2\pi i\xi_0 x_0}\ \delta[\xi + \xi_0] = F^*[-\xi] \tag{9.148}$$

The function $f[x]$ and its spectrum $F[\xi]$ are plotted in Figure 9.36 for $\xi_0 = \frac{1}{2}$ and $x_0 = +\frac{4}{5}$, so that $\exp[\pm 2\pi i\xi_0 x_0] \cong -0.809\,02 \pm i \cdot 0.587\,79$ and $F[\xi] = \exp[+i4\pi/5]\,\delta[\xi - \frac{1}{2}]$. The corresponding complex conjugate $f^*[x]$ and its spectrum $F^*[-\xi] = \exp[-i4\pi/5]\delta[\xi + \frac{1}{2}]$ are plotted in Figure 9.37.

The results of Equations (9.119) and (9.146) may be combined to derive the Fourier transform of the "reversed" complex conjugate of $f[x]$:

$$\mathcal{F}_1\{f^*[-x]\} = F^*[\xi] \tag{9.149}$$

To emphasize the similarities and differences between the spectra of the "reversed" function in Equation (9.119), of the complex conjugate in Equation (9.146), and of the "reversed" complex conjugate in Equation (9.149), they are compared in one place here:

$$\mathcal{F}_1\{f[-x]\} = F[-\xi] \tag{9.150a}$$

$$\mathcal{F}_1\{f^*[x]\} = F^*[-\xi] \tag{9.150b}$$

$$\mathcal{F}_1\{f^*[-x]\} = F^*[\xi] \tag{9.150c}$$

9.8.11 Fourier Transform of Crosscorrelation

The crosscorrelation at a coordinate was defined in Equation (8.71) as the integral of the product of an input function $f[x]$ and the translated and complex-conjugated replica of the "reference" function $m[x]$:

$$f[x] \bigstar m[x] = \int_{-\infty}^{+\infty} f[\alpha] m^*[\alpha - x]\, d\alpha$$

$$= f[x] * m^*[-x] \tag{9.151}$$

As discussed in Section 8.6, the crosscorrelation provides a measure of the "similarity" between the input and reference functions at different translations.

The Fourier transform of the crosscorrelation in terms of $F[\xi]$ and $M[\xi]$ is obtained most easily by inserting the transform of the reversed complex conjugate of $f[x]$ from Equation (9.149) into the filter

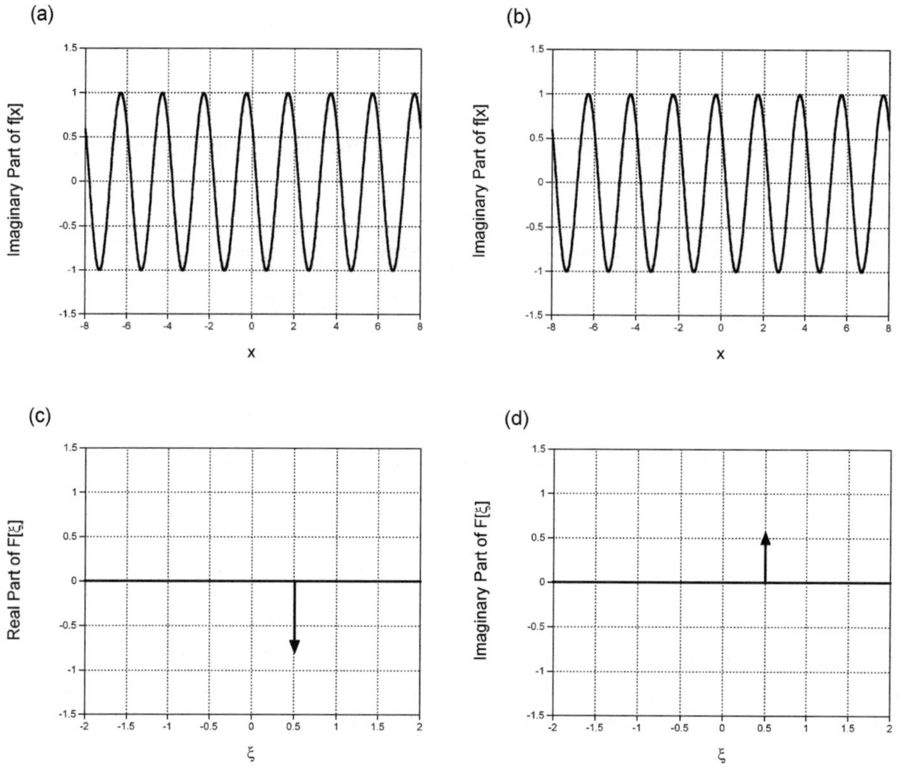

Figure 9.36 Complex function $f[x] = \exp[+2\pi i x/2] \exp[+i4\pi/5]$ and its Fourier transform, which is a complex-valued Dirac delta function $F[\xi] = \exp[+i4\pi/5] \cdot \delta[\xi - \frac{1}{2}]$.

theorem of Equation (9.129):

$$\begin{aligned}\mathcal{F}_1\{f[x] \bigstar m[x]\} &= \mathcal{F}_1\{f[x] * m^*[-x]\} \\ &= \mathcal{F}_1\{f[x]\} \cdot \mathcal{F}_1\{m^*[-x]\} \\ &= F[\xi] \cdot M^*[\xi] \end{aligned} \tag{9.152}$$

In words, the spectrum of the crosscorrelation is the product of the spectra of the input function and the complex conjugate of the "reference" function.

The variant of the crosscorrelation introduced in Equation (8.72) generates the ensemble of inner products of $f[x]$ and the complex conjugate of the "reference" function (this is what Gaskill defines as the "crosscorrelation"). The spectrum of this variant is easy to derive by applying Equation (9.140):

$$\begin{aligned}\mathcal{F}_1\{f[x] \bigstar m^*[x]\} &= \mathcal{F}_1\{f[x] * m[-x]\} = \mathcal{F}_1\{f[x]\} \cdot \mathcal{F}_1\{m[-x]\} \\ &= F[\xi] \cdot M[-\xi] \end{aligned} \tag{9.153}$$

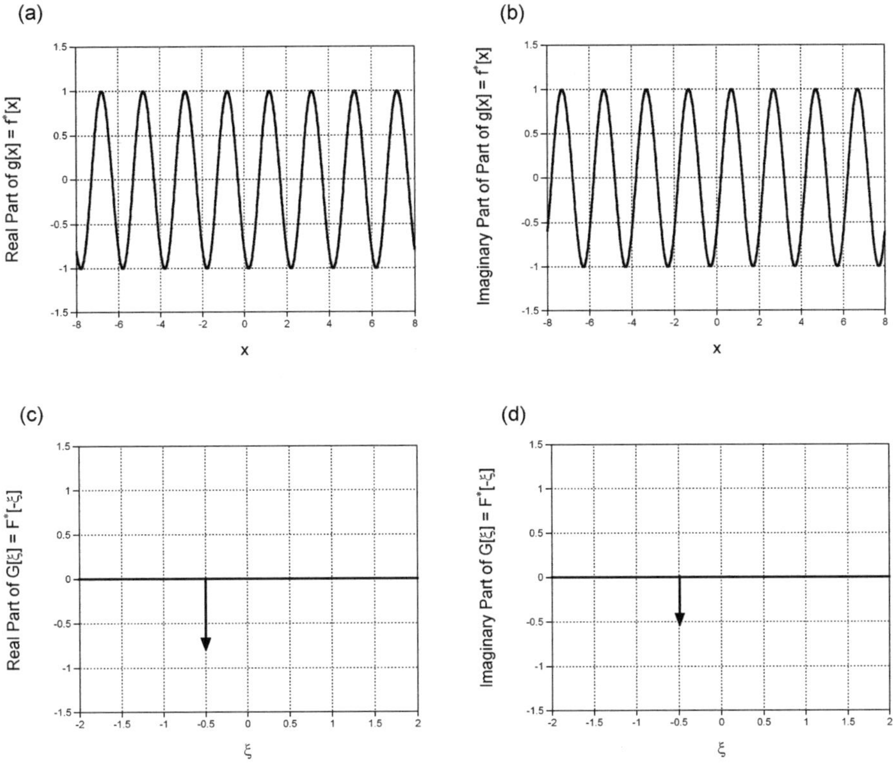

Figure 9.37 Complex function $f^*[x] = \exp[-2\pi i x/2]\exp[-i4\pi/5]$ and its Fourier transform, which is $F^*[-\xi] = \exp[-i4\pi/5]\cdot\delta[\xi+\frac{1}{2}]$.

The Fourier transforms of the two flavors of crosscorrelation are at once similar to and different from the filter theorem. The several results are compiled for reference:

$$\mathcal{F}_1\{f[x]*m[x]\} = F[\xi]\cdot M[\xi] \tag{9.154a}$$

$$\mathcal{F}_1\{f[x]\bigstar m[x]\} = F[\xi]\cdot M^*[\xi] \tag{9.154b}$$

$$\mathcal{F}_1\{f[x]\bigstar m^*[x]\} = F[\xi]\cdot M[-\xi] \tag{9.154c}$$

As an example, consider the correlation of the complex-valued input function and its spectrum:

$$f[x] = (1-i)RECT[x] \tag{9.155a}$$

$$F[\xi] = (1-i)SINC[\xi] \tag{9.155b}$$

and the "reference" function and its spectrum:

$$m[x] = (1+i)\ RECT[x+1] \tag{9.156a}$$

$$M[\xi] = (1+i)\ SINC[\xi]\cdot e^{+2\pi i\xi} \tag{9.156b}$$

The crosscorrelation of these functions is obtained by easy substitution into Equation (9.151):

$$\begin{aligned}g[x] &= \int_{-\infty}^{+\infty} f[\alpha]m^*[\alpha - x]\, d\alpha \\ &= \int_{-\infty}^{+\infty} ((1-i)\, RECT[\alpha])((1+i)\, RECT[(\alpha - x) + 1])^*\, d\alpha \\ &= (1-i)^2 \int_{-\infty}^{+\infty} RECT[\alpha]\, RECT[(x-1) - \alpha]\, d\alpha \\ &= -2i \cdot TRI[x-1] \end{aligned} \tag{9.157}$$

where $u \equiv \alpha - 1$ and the fact that $RECT[x]$ is real valued and even and Equation (8.42) have been used. The process is illustrated in Figure 9.38. The result indicates that $m[x]$ most resembles $f[x]$ if shifted by +1 unit and that the projection of $f[x]$ onto $m[x]$ is negative and imaginary at that location. That the corresponding spectra obey the relationship of Equation (9.152) may be confirmed by direct substitution:

$$\begin{aligned}G[\xi] &= F[\xi] \cdot M^*[\xi] \\ &= \left((1-i)\, SINC[\xi]\, e^{-2\pi i \xi (1)}\right)((1-i)\, SINC[\xi])^* \\ &= -2i\, SINC^2[\xi]\, e^{-2\pi i \xi} \\ &= 2\, SINC^2[\xi]\, e^{-2\pi i \xi}\, e^{-i\pi/2} \end{aligned} \tag{9.158}$$

9.8.12 Fourier Transform of Autocorrelation

As already described in Chapter 8, autocorrelation is the special case of crosscorrelation with identical input and "reference" functions: $h[x] = f[x]$. The spectra are related by:

$$\begin{aligned}\mathcal{F}_1\{f[x] \bigstar f[x]\} &= \mathcal{F}_1\{f[x] * f^*[-x]\} \\ &= F[\xi] \cdot F^*[\xi] = |F[\xi]|^2 \end{aligned} \tag{9.159}$$

This is the *Wiener–Khintchin theorem*, which says in words that the autocorrelation of $f[x]$ and $|F[\xi]|^2$ forms a Fourier transform pair. The spectrum of an autocorrelation is real, which ensures that the autocorrelation *must* be Hermitian (real part is even, imaginary part is odd). The area of the autocorrelation must be nonnegative because $|F[\xi]|^2 \geq 0$.

Figure 9.39 shows the process for $f[x] = e^{-(x+1)} \cdot STEP[x+1] + i\, e^{+(x+1)} \cdot STEP[-(x+1)]$ and demonstrates that the autocorrelation is Hermitian and the maximum value of the real part is located at the origin.

We noted in Chapter 6 that the squared magnitude of the space-domain function, $|f[x]|^2$, is called the "power" of the signal at x. We mentioned earlier that $|F[\xi]|^2$ is called the "power spectrum" of $f[x]$. The power spectrum is a useful metric for characterizing the correlation properties of stochastic signals. This nomenclature may be generalized still further to the output spectrum of the crosscorrelation in Equation (9.152). This entity $F[\xi] \cdot M^*[\xi]$ sometimes is called the "cross-energy spectrum" or "cross-spectral power" of the component functions.

9.8.13 Rayleigh's Theorem

Rayleigh's theorem relates the inner product projection of $f[x]$ onto a reference function $h[x]$ to the corresponding spectra. Equation (9.1) may be used to derive the relationship by substituting the inverse

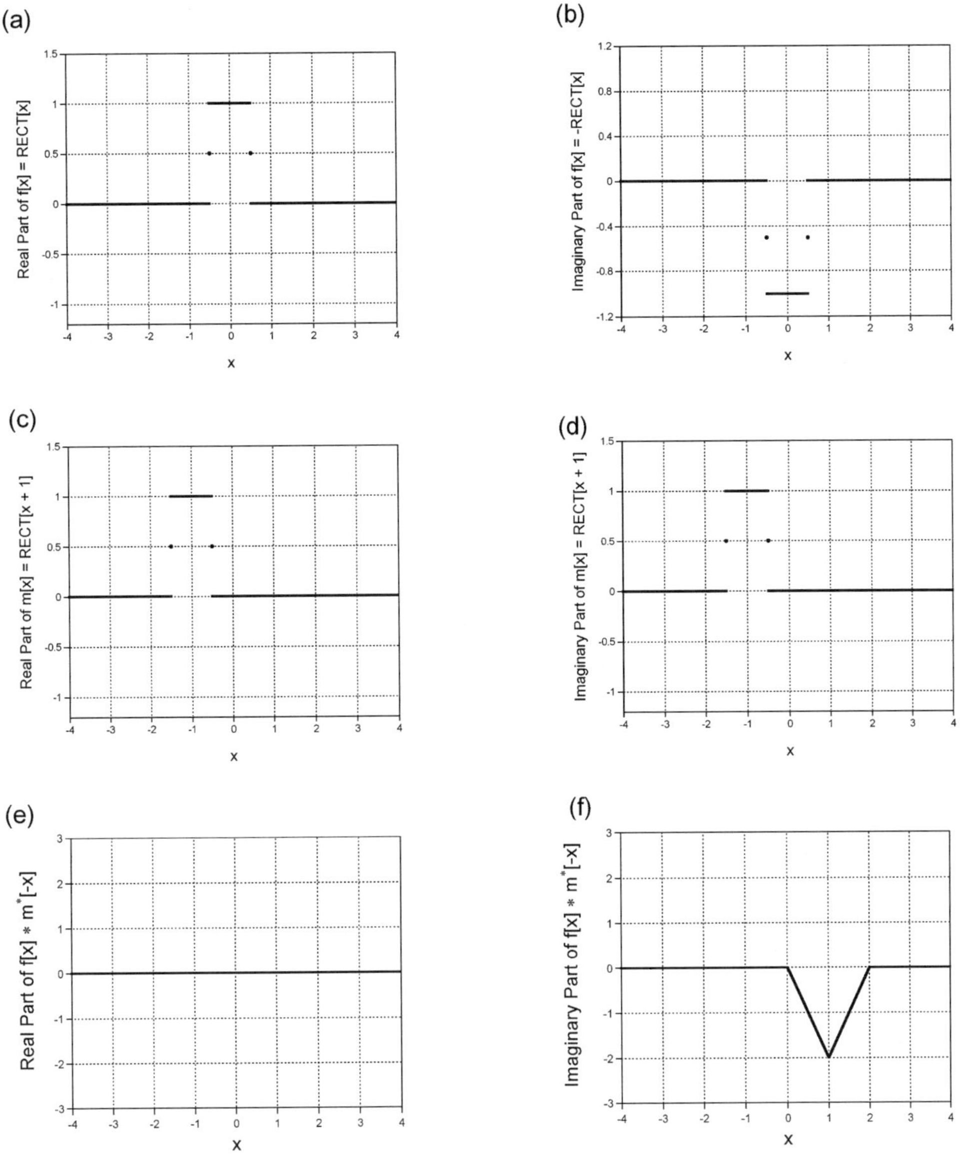

Figure 9.38 Illustration of the Fourier transform of a crosscorrelation. The input ("test object") is $f[x] = (1 - i)\, RECT[x]$ and the "reference object" is $m[x] = (1 + i)\, RECT[x + 1]$. The resulting output is $g[x] = f[x] \bigstar m[x] = f[x] * m^*[-x] = -2i \cdot TRI[x - 1]$.

Fourier transforms for the component functions from Equation (9.63) and applying Equation (9.146):

$$\int_{-\infty}^{+\infty} f[x]h^*[x]\, dx = \int_{-\infty}^{+\infty} \left(\int_{-\infty}^{+\infty} F[\xi]\, e^{+2\pi i \xi x}\; d\xi \right) \left(\int_{-\infty}^{+\infty} H[v]\, e^{+2\pi i v x}\; dv \right)^* dx$$

$$= \int_{-\infty}^{+\infty} \left(\int_{-\infty}^{+\infty} F[\xi]\, e^{+2\pi i \xi x}\; d\xi \right) \left(\int_{-\infty}^{+\infty} H^*[v]\, e^{-2\pi i v x}\; dv \right) dx$$

$$= \int_{-\infty}^{+\infty} F[\xi]\left(\int_{-\infty}^{+\infty} H^*[v]\left(\int_{-\infty}^{+\infty} e^{+2\pi i x(\xi-v)}\,dx\right)dv\right)d\xi$$

$$= \int_{-\infty}^{+\infty} F[\xi]\left(\int_{-\infty}^{+\infty} H^*[v]\delta[\xi-v]\,dv\right)d\xi$$

$$= \int_{-\infty}^{+\infty} F[\xi]H^*[\xi]\,d\xi \tag{9.160}$$

In words, Rayleigh's theorem says that the projection of $f[x]$ onto $h[x]$ is identical to the projection of $F[\xi]$ onto $H[\xi]$. By analogy with the nomenclature of the "power" for $f[x]\cdot f^*[x] = |f[x]|^2$ and with the "cross-spectral power" for $F[\xi]\cdot H^*[\xi]$, the quantity $f[x]\cdot h^*[x]$ is called the "cross-power" of the two space-domain functions. By setting $h[x] = e^{+2\pi i\xi x}$, Rayleigh's theorem yields the Fourier integral.

Rayleigh's theorem may be used to evaluate the areas of some possibly useful functions. Consider the problem of calculating the area of $SINC^3[x]$. The integrand may be decomposed into the product of two real-valued functions, $f[x] = SINC[x]$ and $h[x] = SINC^2[x]$:

$$\int_{-\infty}^{+\infty} SINC^3[x]\,dx = \int_{-\infty}^{+\infty} SINC[x]\cdot SINC^2[x]\,dx$$
$$= \int_{-\infty}^{+\infty} RECT[\xi]\cdot TRI[\xi]\,d\xi$$
$$= \int_{-\frac{1}{2}}^{+\frac{1}{2}} TRI[\xi]\,d\xi = \frac{3}{4} \tag{9.161}$$

which is easy to confirm graphically.

Note that Rayleigh's theorem may not be extended to calculate the area of the product of three functions:

$$\int_{-\infty}^{+\infty} f[x]h^*[x]g[x]\,dx \neq \int_{-\infty}^{+\infty} F[\xi]H^*[\xi]G[\xi]\,d\xi \tag{9.162}$$

9.8.14 Parseval's Theorem

This special case of Rayleigh's theorem relates the frequency-domain representation $F[\xi]$ to the projection of $f[x]$ onto itself. The proof is obtained by the trivial substitution of $f[x]$ for $h[x]$ in Equation (9.160):

$$\int_{-\infty}^{+\infty} |f[x]|^2\,dx = \int_{-\infty}^{+\infty} f[x]f^*[x]\,dx$$
$$= \int_{-\infty}^{+\infty} F[u]\,F^*[u]\,du = \int_{-\infty}^{+\infty} |F[\xi]|^2\,d\xi \tag{9.163}$$

In words, Parseval's theorem demonstrates that, if the area of $|f[x]|^2$ is finite, then $F[\xi]$ exists and the area of $|F[\xi]|^2$ is finite and equal to the area of $|f[x]|^2$. Equation (9.163) may be interpreted as a statement that the "power" is conserved in both representations of the function.

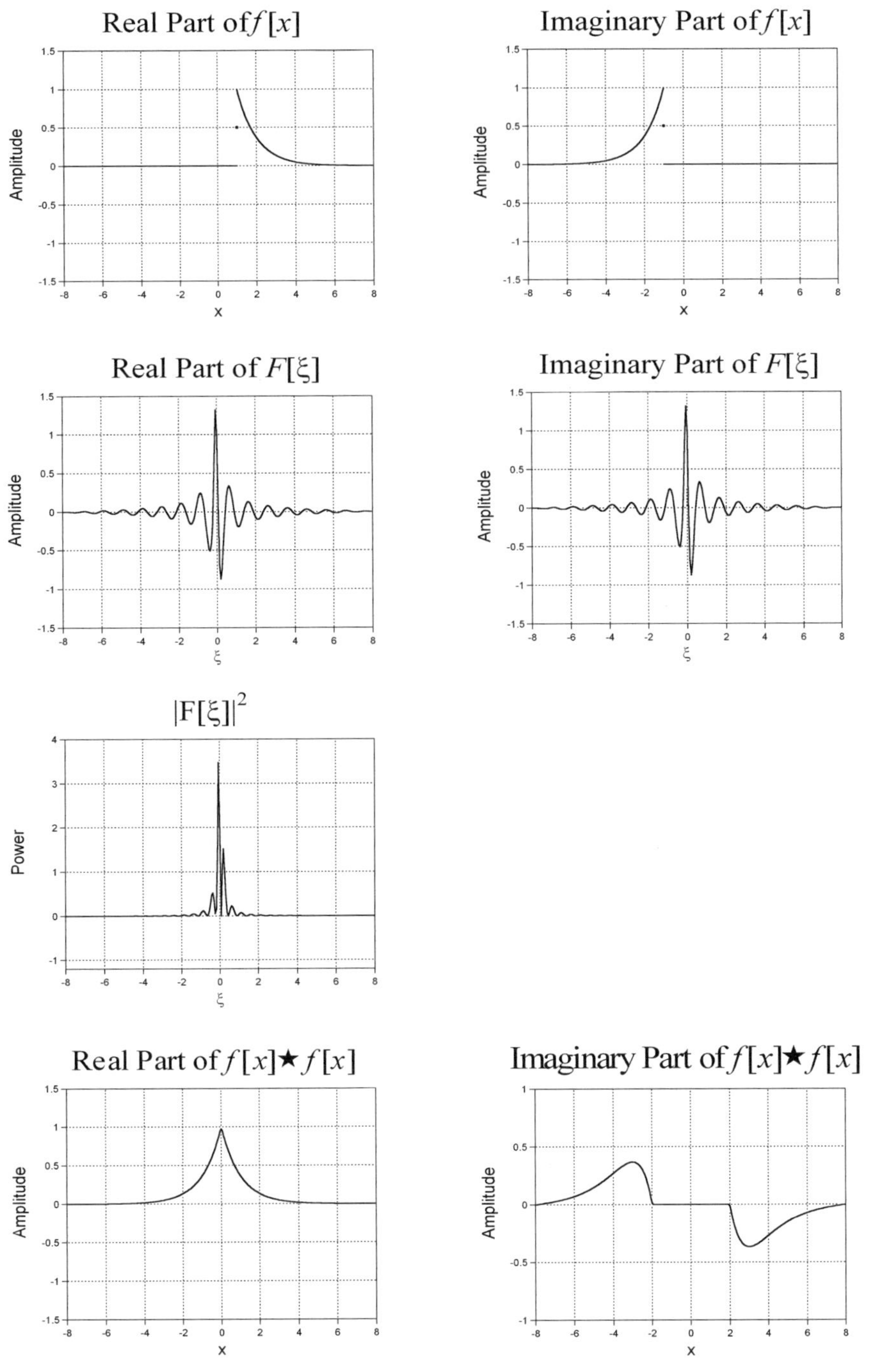

Figure 9.39 Evaluation of $f[x] \bigstar f[x]$ for $f[x] = \exp[-(x-1)] \cdot STEP[x-1] + i\ \exp[x+1] \cdot STEP[-(x+1)]$. The result is Hermitian.

Consider an example of Parseval's theorem with $f[x] = RECT[x/b]$, for which we know the spectrum to be $F[\xi] = |b| \cdot SINC[b\xi]$:

$$\int_{-\infty}^{+\infty} |f[x]|^2 \, dx = \int_{-\infty}^{+\infty} \left| RECT\left[\frac{x}{b}\right] \right|^2 dx = \int_{-\infty}^{+\infty} RECT\left[\frac{x}{b}\right] dx = |b| \tag{9.164a}$$

$$\int_{-\infty}^{+\infty} |F[\xi]|^2 \, d\xi = \int_{-\infty}^{+\infty} ||b| \cdot SINC[b\xi]|^2 \, d\xi = |b|^2 \int_{-\infty}^{+\infty} |SINC[b\xi]|^2 \, d\xi$$

$$= |b|^2 \int_{-\infty}^{+\infty} SINC^2[b\xi] \, d\xi = |b|^2 \cdot \frac{1}{|b|} = |b| \tag{9.164b}$$

The equality of the results of these two expressions confirms the theorem.

9.8.15 Fourier Transform of Periodic Function

The next few theorems have a different "flavor" from those already considered; they refer to the spectra of discrete and/or periodic signals. These results will be essential when considering the discrete Fourier transform in Chapter 15.

A periodic function $f[x]$ with period X_0 may be written as the convolution of a single period of the function with a set of unit-area Dirac delta functions separated by X_0. A single period is the product of $f[x]$ and a *RECT* function with "width" X_0:

$$f[x] \, RECT\left[\frac{x}{X_0}\right] \equiv w[x] \tag{9.165}$$

By applying the modulation theorem in Equation (9.140), the spectrum $W[\xi]$ is found to be:

$$W[\xi] = F[\xi] * |X_0| \, SINC[X_0\xi] = |X_0|(F[\xi] * SINC[X_0\xi]) \tag{9.166}$$

Because $SINC[X_0\xi]$ has infinite support, so must $W[\xi]$.

The periodic function $f[x]$ may be recast into the convolution of $w[x]$ with a *COMB* function:

$$\begin{aligned} f[x] &= w[x] * \sum_{n=-\infty}^{+\infty} \delta[x - nX_0] \\ &= w[x] * \frac{1}{|X_0|} COMB\left[\frac{x}{X_0}\right] \end{aligned} \tag{9.167}$$

The spectrum of the periodic function $f[x]$ is the Fourier transform of this expression, which may be evaluated via the filter theorem of Equation (9.129), the spectrum of the *COMB* function from Equation (9.101), and the scaling theorem of Equation (9.117):

$$\begin{aligned} \mathcal{F}_1\{f[x]\} &= \frac{1}{|X_0|} \mathcal{F}_1\left\{ w[x] * COMB\left[\frac{x}{X_0}\right]\right\} \\ &= \frac{1}{|X_0|} W[\xi] \cdot |X_0| \cdot COMB[\xi X_0] \\ &= W[\xi] \, COMB[\xi X_0] \\ &= (F[\xi] * |X_0| \, SINC[X_0\xi]) \, COMB[\xi X_0] \end{aligned} \tag{9.168}$$

This multiplication by the *COMB* function ensures that the spectrum is "sampled" or "discrete". One obvious example of a periodic function is the sinusoid, whose spectrum is the discrete pair of Dirac delta functions. Another example of a periodic function is the "biased" square wave, which may be

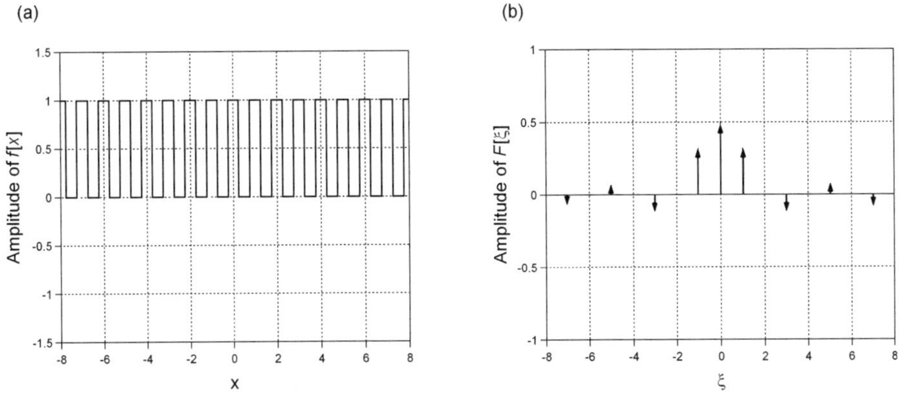

Figure 9.40 The Fourier transform of the real, even, and periodic function $f[x] = RECT[2x] * COMB[x]$ is the real, even, and discrete function $F[\xi] = (\frac{1}{2}SINC[\xi/2]) \cdot COMB[\xi]$. Note that the Dirac delta functions in the spectrum for even ξ (other than $\xi = 0$) coincide with the zeros of $SINC[\xi/2]$.

written in the form:

$$f[x] = RECT[2x] * COMB[x] \tag{9.169}$$

as shown in Figure 9.40. The Fourier transform is easy to evaluate:

$$F[\xi] = \frac{1}{2}SINC\left[\frac{\xi}{2}\right]COMB[\xi] \tag{9.170}$$

Note that the zeros of the *SINC* function are co-located with the Dirac delta functions at even values of ξ.

9.8.16 Spectrum of Sampled Function

The "transform-of-a-transform" theorem in Equation (9.105) may be immediately applied to the last result to demonstrate that the spectrum of a discrete function is periodic. This also may be demonstrated by expressing the sampled function $f[x]$ as the modulation of a continuous function (say, $w[x]$) with a *COMB* function whose elements have unit area and uniform spacing at the sampling interval Δx:

$$f[x] = w[x] \cdot \frac{1}{|\Delta x|}COMB\left[\frac{x}{\Delta x}\right] \tag{9.171}$$

The spectrum is obtained via the modulation theorem in Equation (9.140), the spectrum of the *COMB* function in Equation (9.101), and the scaling theorem of Equation (9.116):

$$\begin{aligned}
\mathcal{F}_1\{f[x]\} &= \frac{1}{|\Delta x|}\mathcal{F}_1\left\{w[x] \cdot COMB\left[\frac{x}{\Delta x}\right]\right\} \\
&= \frac{1}{|\Delta x|}W[\xi] * |\Delta x|\, COMB[\xi \cdot \Delta x] \\
&= W[\xi] * COMB[\xi \cdot \Delta x] = W[\xi] * COMB\left[\frac{\xi}{(\Delta x)^{-1}}\right] \\
&= W[\xi] * \frac{1}{\Delta x}\sum_{k=-\infty}^{+\infty}\delta\left[\xi - \frac{k}{\Delta x}\right] = \frac{1}{\Delta x}\sum_{k=-\infty}^{+\infty} W\left[\xi - \frac{k}{\Delta x}\right]
\end{aligned} \tag{9.172}$$

Because $w[x]$ has infinite support, its spectrum $W[\xi]$ has compact support. Therefore the spectrum of a discrete function is the convolution of this "local" function $W[\xi]$ with $COMB[\xi/(\Delta x)^{-1}]$. The resulting spectrum is guaranteed to be periodic. Note that the individual "periods" of the spectrum $W[\xi]$ are scaled in amplitude by $1/\Delta x$, so that the amplitude is decreased if $\Delta x > 1$ unit. An example is shown in Figure 9.41 for $f[x] = (e^{-x} \cdot STEP[x]) \cdot (2x\ COMB[2x])$.

9.8.17 Spectrum of Discrete Periodic Function

The last two theorems may be combined in an obvious way to see that the spectrum of a discrete periodic function is discrete and periodic. The period X_0 in the space domain determines the interval Δx between samples in the frequency domain via $(\Delta\xi)^{-1} = X_0$, while the sample interval Δx in the space domain determines the periodicity in the frequency domain via $\Xi_0 = (\Delta x)^{-1}$. The importance of these relations will become evident in the discussion of the discrete and fast Fourier transforms (DFTs and FFTs) in Chapter 15.

9.8.18 Spectra of Stochastic Signals

Several 1-D stochastic functions $n[x]$ were introduced in Section 6.4, where the amplitude n at each coordinate x is selected at random from some probability law. The noise signals generated by many real systems are "uncorrelated", which means that the amplitude at an arbitrary coordinate is unrelated to adjacent amplitudes; all values within the limits prescribed by the probability law are allowed. For example, consider random numbers selected from uniform distribution on the interval [0,1) that was introduced in Equation (6.184). If the amplitude n at coordinate x_0 is by chance $n[x_0] = 0.01$, then that at coordinate $x_0 + \epsilon$ (where ϵ is an arbitrarily small increment) may be any number in the allowed interval. If $n[x_0 + \epsilon] \cong 0.99$, then a high-frequency sinusoid must be present within the spectrum $N[\xi]$. If $n[x_0 + \epsilon] \cong 0.01$, then the spectrum must include slowly varying sinusoids. Of course, $N[\xi]$ also will "consider" the relative values of $n[x_0]$ and the amplitudes at all other x.

The Wiener–Khintchin theorem in Equation (9.159) demonstrated that the autocorrelation $n[x] \bigstar n[x]$ and the power spectrum $|N[\xi]|^2$ are Fourier transform pairs. Therefore, either representation may be used to characterize the noise. The power spectrum of a stochastic signal is often called its *Wiener spectrum*. Examples of power spectra of noise samples with different "colors" are shown in Figure 9.42.

The power of the sinusoidal components at different frequencies in $|N[\xi]|^2$ is a useful descriptor of the "character" of the noise and is often characterized as a "color". A "white" noise spectrum includes (at least approximately) equal power over the domain of spatial frequencies. "Pink" and "blue" noise spectra have larger powers at low and high frequencies, respectively. Because "pink" noise signals are dominated by low-frequency sinusoidal components, the amplitude $n[x]$ tends to vary "slowly" from coordinate to coordinate. The spectral power of "green" noise is largest at "middle" frequencies, while "blue" noise signals include more power at large frequencies. Examples of white, pink, green, and blue noise that were derived from uniformly distributed random noise with a mean value of 0 are shown in Figure 9.43. Note the variation in "oscillation rate" of the different samples of noise.

The autocorrelations of the samples of colored noise are shown in Figure 9.44. These were evaluated as a discrete summation rather than as a continuous integral. The amplitudes are normalized to the maximum amplitude for "white" noise. The autocorrelation of "white" noise is approximately a Dirac delta function, which indicates that the sample of white noise "resembles itself exactly" for zero translation, but the area of the product of the original samples and a translated replica is approximately zero for all other translations. In other words, "white" noise is perfectly correlated for zero translation and uncorrelated for other shifts. The autocorrelation of "blue" noise tends to change sign from one translation to the next. A negative amplitude of the autocorrelation for a particular translation means that the function at that shift resembles the negative of the original function; the amplitudes of the samples are "anticorrelated" at these translations. The autocorrelation of "pink" noise resembles a

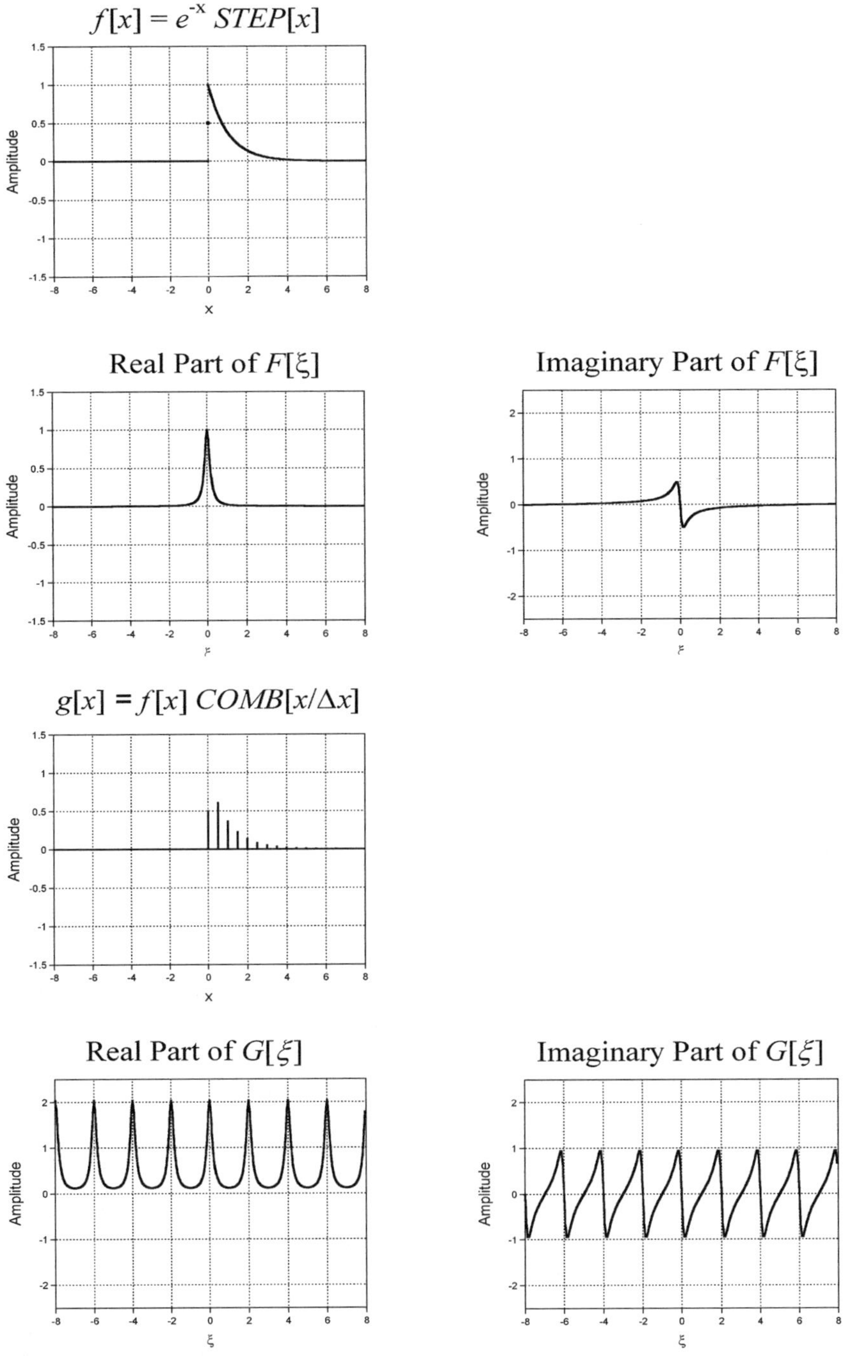

Figure 9.41 The Fourier transform of a sampled function. The continuous function is $w[x] = e^{-x} \cdot STEP[x]$ with $W[\xi] = (1 + 2\pi i \xi)^{-1}$. The function is sampled at half-integer intervals ($\Delta x = \frac{1}{2}$) by multiplying by $2COMB[2x]$. The spectrum of the sampled function is periodic at intervals $\Delta \Xi = (\Delta x)^{-1}$ and its amplitude is scaled by the same factor.

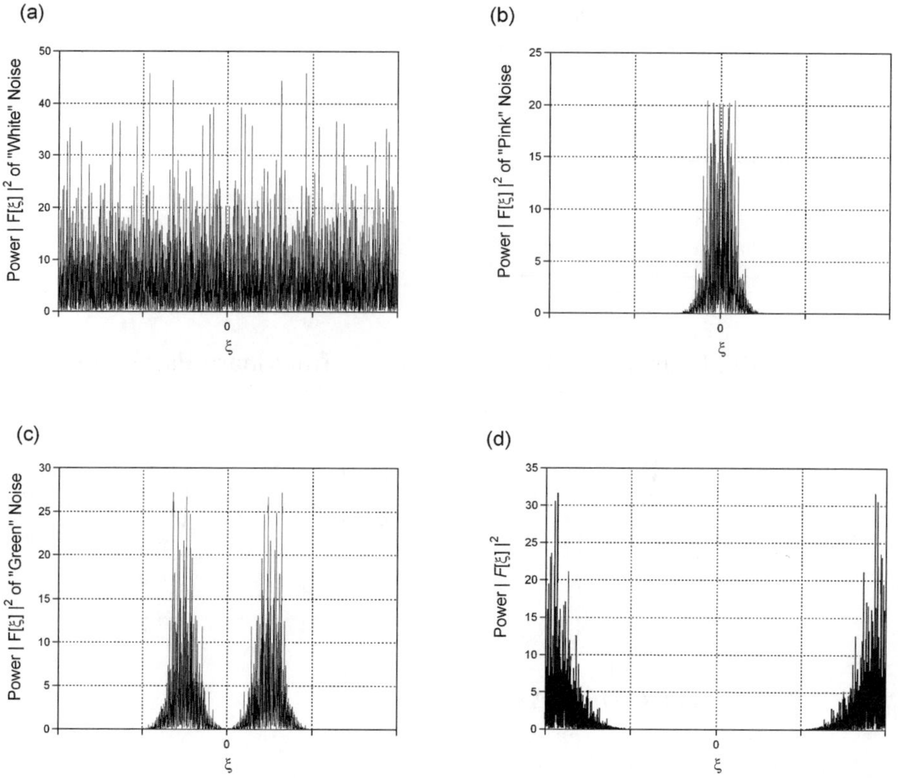

Figure 9.42 Power spectra of "colored" noise samples: (a) "white" noise contains approximately the same power at each spatial frequency, while the powers of (b) "pink" noise, (c) "green" noise, and (d) "blue" noise are respectively concentrated at low, medium, and high spatial frequencies.

Gaussian function; the amplitude decays from its maximum at zero shift. This indicates that "pink" noise resembles a translated replica over a range of translations because the amplitude in $n[x]$ tends to vary slowly with x. This is due to the dominance of sinusoidal components with small spatial frequencies (and thus that oscillate slowly).

9.8.19 Effect of Nonlinear Operations of Spectra

We have already shown that convolution is an LSI operation that is intimately related to the Fourier transform. We have also stated that no existing imaging system truly satisfies the necessary constraints that allow its action to be accurately described as a convolution. Specifically, no imaging system is truly linear. For that reason, this section considers the effect of nonlinear operations on the spectrum of a function.

The amplitude mapping introduced in Chapter 2 affects only the amplitude and has no effect on the coordinate x:

$$g[x] = \mathcal{O}\{f[x]\} \Longrightarrow g = \mathcal{O}\{f\} \tag{9.173}$$

Any functional operator (e.g., sin or log) may be substituted for $\mathcal{O}$, though only a proportional relation satisfies the strict requirements of linearity. We also described how the concept of linearity may be

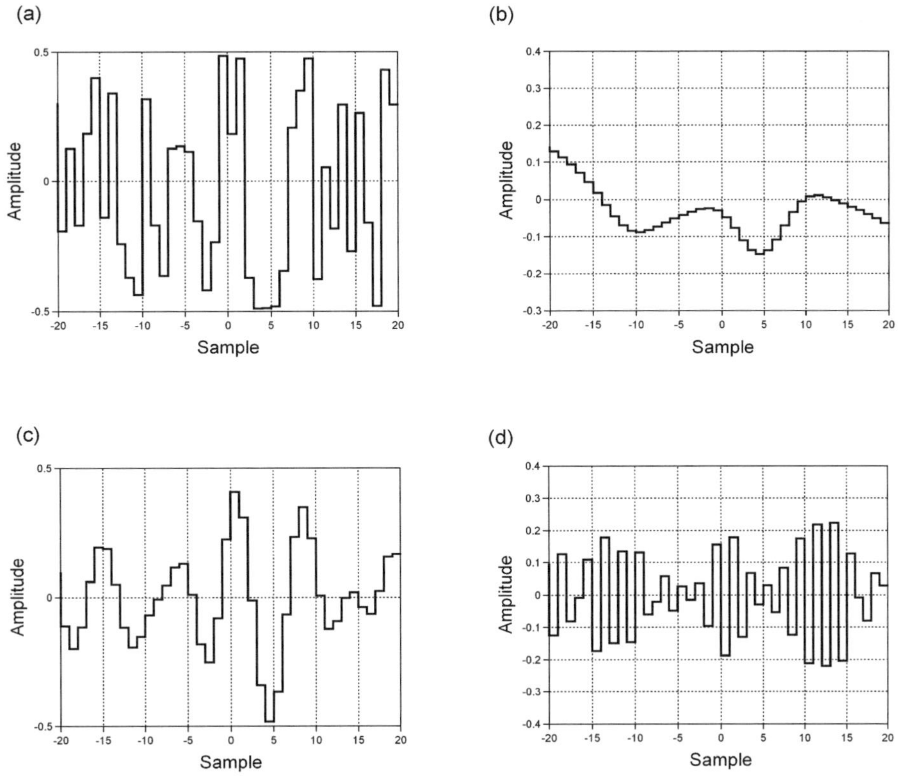

Figure 9.43 Samples of zero-mean stochastic signals with different "colors": (a) "white"; (b) "pink"; (c) "green"; and (d) "blue". Note that the algebraic sign of the amplitude at adjacent samples of "blue" noise tends to invert, because of the dominance of large frequency components. The algebraic sign of adjacent samples of "pink" noise tends to be preserved.

generalized to allow addition of a bias:

$$g[x] = a \cdot f[x] + b \tag{9.174}$$

where a and b are real-valued constants.

Any mapping that cannot be expressed in the form of Equation (9.174) must be nonlinear. The action of the nonlinearity on the spectrum of the input function may be quantified if the input $f[x]$ and output $g[x]$ of the nonlinear operation have known spectra or if $g[x]$ may be decomposed into functions with known spectra. Both classes include operations that are important in imaging applications.

9.8.19.1 Spectrum of a Thresholded Function

For example, consider the specific nonlinear operation:

$$\mathcal{O}\{f[x]\} = SGN[a \cdot f[x] + b] = g[x] \tag{9.175}$$

which generates a function with amplitude ± 1 or 0 depending on whether $a \cdot f[x] + b$ is positive, negative, or zero. This is a variation of *thresholding*, which we have already mentioned to be a basic part of the conversion from continuous to discrete amplitudes (*quantization*). The scaling factor a and

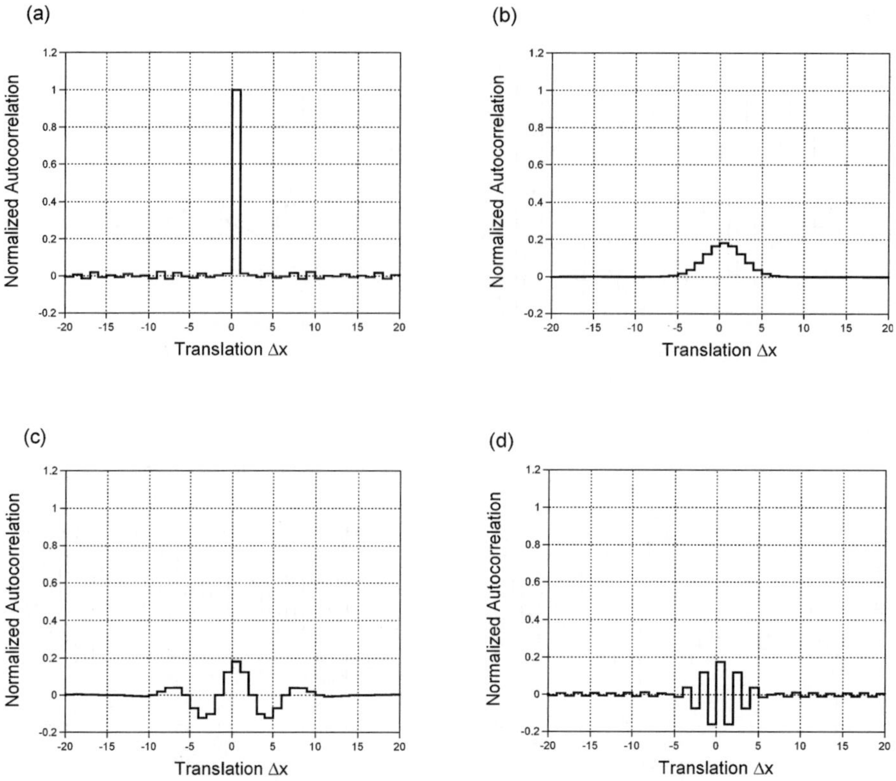

Figure 9.44 Autocorrelations of colored noises computed from the discrete samples and normalized to the maximum value for white noise: (a) the autocorrelation of white noise is approximately a Dirac delta function; (b) samples of the autocorrelation of pink noise have the same algebraic sign near $x = 0$; (c) autocorrelation of green noise; (d) autocorrelation of blue noise, whose sign oscillates from one sample to the next.

bias b modify the amplitude f in a generalized linear manner to vary the threshold relative to $f[x]$. Clearly the amplitude $g[x]$ resulting from Equation (9.175) can take on the three levels: 0, $+\frac{1}{2}$, and $+1$.

We now consider the action of the thresholding operation on a specific unbiased sinusoidal function:

$$f[x] = \cos[2\pi\xi_0 x] \tag{9.176}$$

Its spectrum has compact support with $|\xi|_{\max} = |\xi_0|$:

$$F[\xi] = \frac{1}{2}(\delta[\xi + \xi_0] + \delta[\xi - \xi_0]) \tag{9.177}$$

After applying the nonlinearity with $a = 1$ and $b = 0$, the output $g[x]$ is a square wave that "switches" between ± 1 with period ξ_0^{-1}. It may be written as:

$$g[x] = (2 \cdot RECT[2\xi_0 x] * \xi_0\, COMB[\xi_0 x]) - 1[x] \tag{9.178}$$

The output spectrum $G[\xi]$ is a *COMB* function modulated by a *SINC* with scale parameter $2\xi_0$:

$$\begin{aligned}
G[\xi] &= SINC\left[\frac{\xi}{2\xi_0}\right] \cdot \frac{1}{\xi_0} COMB\left[\frac{\xi}{\xi_0}\right] - \delta[\xi] \\
&= \left(\sum_{k=-\infty}^{+\infty} SINC\left[\frac{k}{2}\right]\delta[\xi - k\xi_0]\right) - \delta[\xi] \\
&= \sum_{k=1}^{+\infty} \frac{(-1)^k}{(2k-1)\pi}(\delta[\xi + k\xi_0] + \delta[\xi - k\xi_0])
\end{aligned} \tag{9.179}$$

This infinite-support discrete spectrum includes odd-integer multiples of the original sinusoidal frequency, which are the so-called *harmonics* of the original spectrum. The action of the thresholding operation on the bandlimited sinusoid produced harmonics to the spectrum with spatial frequencies out to $\pm\infty$, though with decreasing amplitude.

9.8.19.2 Spectrum of the Magnitude of a Function

The magnitude of a function is often evaluated in imaging problems and clearly is the result of a nonlinear operation:

$$\begin{aligned}
g[x] &= |f_1[x] + f_2[x]| \\
&= \sqrt{(\Re\{f_1[x] + f_2[x]\})^2 + (\Im\{f_1[x] + f_2[x]\})^2} \\
&\neq |f_1[x]| + |f_2[x]|
\end{aligned} \tag{9.180}$$

though it is occasionally approximated as a linear operation, as in the evaluation of the magnitude of the discrete gradient operation in Chapter 20. The magnitude of an arbitrary function $f[x]$ may be rewritten in terms of its phase function $\boldsymbol{\Phi}\{f[x]\}$:

$$\begin{aligned}
f[x] &= |f[x]| \cdot \exp[+i \cdot \boldsymbol{\Phi}\{f[x]\}] \\
&\Longrightarrow g[x] = |f[x]| = f[x] \cdot \exp[-i \cdot \boldsymbol{\Phi}\{f[x]\}]
\end{aligned} \tag{9.181}$$

If $f[x]$ is real valued, then $\boldsymbol{\Phi}\{f[x]\} = 0$ or $-\pi$; if $f[x]$ is imaginary, $\boldsymbol{\Phi}\{f[x]\} = \pm\pi/2$ or 0; if $f[x]$ is complex, then the phase may be any real number, though values obtained from the inverse tangent relationship in Equation (4.14) must be "unwrapped" for the support of phase to exceed 2π.

We can apply Parseval's theorem from Equation (9.163) to both $f[x]$ and $g[x] = |f[x]|$ in Equation (9.181):

$$\begin{aligned}
\int_{-\infty}^{+\infty} |f[x]|^2\, dx &= \int_{-\infty}^{+\infty} |(|f[x]|)|^2\, dx = \int_{-\infty}^{+\infty} |F[\xi]|^2\, d\xi = \int_{-\infty}^{+\infty} |G[\xi]|^2\, d\xi \\
&= \int_{-\infty}^{+\infty} |g[x]|^2\, dx
\end{aligned} \tag{9.182}$$

which shows that the areas of the squared magnitudes of the spectra of $f[x]$ and of $|f[x]|$ must be equal. This result also means that the spectrum of the magnitude of a function exists if the spectrum of the function does.

The spectrum of $g[x] = |f[x]|$ is obtained by applying the modulation theorem to Equation (9.181):

$$\begin{aligned}
G[\xi] &= \mathcal{F}_1\{|f[x]|\} = F[\xi] * \mathcal{F}_1\{\exp[-i \cdot \boldsymbol{\Phi}\{f[x]\}]\} \\
&= F[\xi] * \mathcal{F}_1\{(\exp[+i \cdot \boldsymbol{\Phi}\{f[x]\}])^*\}
\end{aligned} \tag{9.183}$$

which can be evaluated if the spectrum of the unit-magnitude phase function is known or can be expressed in terms of special functions with known spectra.

The discussion of convolution in Chapter 8 and the fact that the spectrum of the magnitude is the convolution of the spectra of the original function and of the unit-magnitude phase function seem to indicate that the spectrum of the magnitude will have wider support than that of $F[\xi]$. However, recall that this statement is guaranteed only if the functions have compact support, e.g., the spectrum of the chirp function $\exp[+i\pi(x/\alpha)^2]$ has infinite support, but the spectrum of its magnitude $1[x]$ has infinitesimal support.

In the case where $f[x]$ is real valued and bipolar, the magnitude and absolute value are identical and the phase function $\boldsymbol{\Phi}\{f[x]\}$ evaluates to 0 or $-\pi$ (in our convention). Therefore, $\exp[-i\cdot\Phi\{f[x]\}] = \exp[+i\cdot\boldsymbol{\Phi}\{f[x]\}]$ in Equation (9.183). In this case, the area of $|f[x]|$ must be larger than that of $f[x]$, and it is quite possible for the area of $|f[x]|$ to be infinite even if the area of $f[x]$ is zero, e.g., for $f[x] = \cos[2\pi\xi_0 x]$ (Problem 9.15). Any zero crossings of $f[x]$ produce "cusps" in $|f[x]|$ that require high-frequency sinusoids to generate, leading to the expectation that the support of $|F[\xi]|$ is wider than that of $F[\xi]$, and may be infinitely wide even in some cases where the support of $F[\xi]$ is finite.

An interesting and useful example of the spectrum of the magnitude of a real-valued function is $f[x] = SINC[x]$, which we well know (by now) to have unit area and a rectangular spectrum that satisfies Parseval's theorem:

$$f[x] = SINC[x] \tag{9.184a}$$

$$F[\xi] = RECT[\xi] \tag{9.184b}$$

$$\int_{-\infty}^{+\infty} |f[x]|^2\,dx = \int_{-\infty}^{+\infty} |SINC[x]|^2\,dx = \int_{-\infty}^{+\infty} SINC^2[x]\,dx = 1 \tag{9.184c}$$

$$\int_{-\infty}^{+\infty} |F[\xi]|^2\,d\xi = \int_{-\infty}^{+\infty} |RECT[\xi]|^2\,d\xi = \int_{-\infty}^{+\infty} RECT[\xi]\,d\xi = 1 \tag{9.184d}$$

Because the negative lobes of $SINC[x]$ are positive in $|SINC[x]|$, its area must exceed unity. The discontinuous "cusps" of $|SINC[x]|$ at the nonzero integer values of x ensure that its spectrum has infinite support. It is straightforward, though somewhat complicated, to evaluate the spectrum via Equation (9.183):

$$\begin{aligned} G[\xi] &= \mathcal{F}_1\{|SINC[x]|\} \\ &= RECT[\xi] * \mathcal{F}_1\{\exp[+i\cdot\boldsymbol{\Phi}\{SINC[x]\}]\} \\ &\equiv RECT[\xi] * P[\xi] \end{aligned} \tag{9.185}$$

where $P[\xi]$ is the spectrum of the unit-magnitude phase function of $SINC[x]$. We can substitute these terms into Parseval's theorem to see that the spectrum $G[\xi] = \mathcal{F}_1\{|SINC[x]|\}$ exists:

$$g[x] = |SINC[x]| \tag{9.186a}$$

$$G[\xi] = RECT[\xi] * P[\xi] \tag{9.186b}$$

$$\int_{-\infty}^{+\infty} |g[x]|^2\,dx = \int_{-\infty}^{+\infty} |(|SINC[x]|)|^2\,dx = \int_{-\infty}^{+\infty} SINC^2[x]\,dx = 1 \tag{9.186c}$$

$$\int_{-\infty}^{+\infty} |G[\xi]|^2\,d\xi = \int_{-\infty}^{+\infty} |RECT[\xi] * P[\xi]|^2\,d\xi = \int_{-\infty}^{+\infty} |g[x]|^2\,dx = 1 \tag{9.186d}$$

The central ordinate theorem already tells us that the area of $G[\xi]$ is unity. This spectrum may be expressed in terms of known spectra of special functions (Problem 9.16):

$$P[\xi] = \frac{1}{i\pi\xi} * \left(SINC[\xi]\cdot COMB[2\xi]\cdot\exp[-i\pi\xi] - \frac{1}{2}\delta[\xi]\right) \tag{9.187}$$

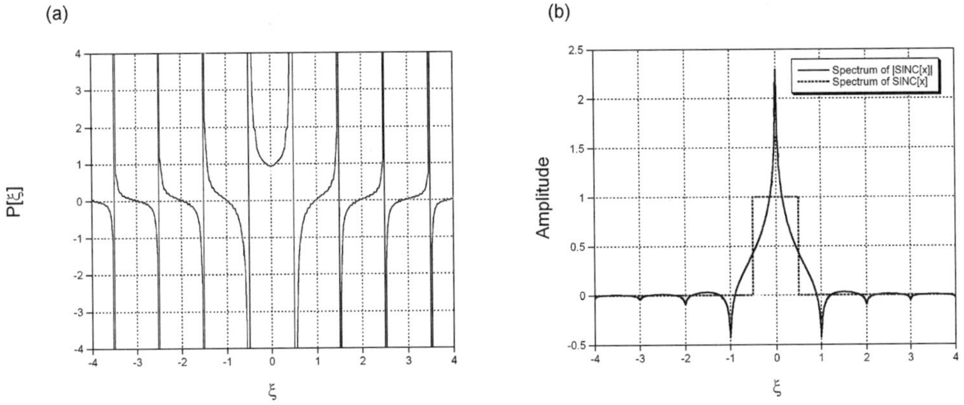

Figure 9.45 (a) Graphical representation of $P[\xi]$ from Equation (9.189), showing the discontinuities at half-integer values of ξ; (b) $\mathcal{F}_1\{|SINC[x]|\}$ compared to $\mathcal{F}_1\{SINC[x]\}$, showing the infinite amplitude at the origin, though it is not a Dirac delta function there.

which is composed of scaled replicas of ξ^{-1} located at half-integer values of ξ, as shown in Figure 9.45a. As a side comment for mathematicians, this spectrum also may be represented in terms of the so-called "digamma" or "psi" function $\psi[\xi] \equiv (d/d\xi)\log_e[\Gamma[\xi]]$:

$$P[\xi] = \frac{1}{\pi^2\xi}\left(\psi\left[\xi+\frac{1}{2}\right] - \psi\left[-\xi+\frac{1}{2}\right]\right) \tag{9.188a}$$

$$= \frac{1}{\pi^2\xi}\left[\frac{d}{d\xi}\left(\log_e\left[\Gamma\left[\xi+\frac{1}{2}\right]\right]\right) - \frac{d}{d\xi}\left(\log_e\left[\Gamma\left[-\xi+\frac{1}{2}\right]\right]\right)\right] \tag{9.188b}$$

The amplitude of $P[\xi]$ is not defined at these half-integer frequencies. The convolution with the rectangle displaces the discontinuities to the nonzero integer values of ξ:

$$\begin{aligned}\mathcal{F}_1\{|SINC[x]|\} &= RECT[\xi] * P[\xi] \\ &= RECT[\xi] * \left[\frac{1}{i\pi\xi} * \left(SINC[\xi]\cdot COMB[\xi]\cdot\exp[-i\pi\xi] - \frac{1}{2}\delta[\xi]\right)\right] \\ &= RECT[\xi] * \left(\frac{1}{\pi^2\xi}\left[\frac{d}{d\xi}\left(\log_e\left[\Gamma\left[\xi+\frac{1}{2}\right]\right]\right) - \frac{d}{d\xi}\left(\log_e\left[\Gamma\left[-\xi+\frac{1}{2}\right]\right]\right)\right]\right)\end{aligned} \tag{9.189}$$

For example, the area of the convolution evaluated at the origin (zero shift) is equal to the area of $|SINC[x]|$ and thus is infinite. However, the observation in Equation (9.186) that the area of $|G[\xi]|^2$ is finite ensures that $G[\xi]$ cannot include a Dirac delta function at the origin.

The spectra of $|SINC[x]|$ and $SINC[x]$ are compared in Figure 9.45b. Though the former has infinite support, the region of its largest amplitude is concentrated near the origin. This observation will be useful during the discussion of inverse and matched filters in Chapter 19.

9.8.19.3 Power-Law Nonlinearity

Another example of a nonlinear operation that is relevant to imaging is the "power-law" mapping of amplitude:

$$g[x] = (f[x])^\gamma \tag{9.190}$$

The real-valued exponent γ may be positive or negative, and (of course) a linear mapping results if $\gamma = 1$. Consider first the simple case where the input is composed of a single sinusoidal component with frequency ξ_0. The function $f[x]$ is (obviously) complex valued and bipolar, while the spectrum (also obviously) is a single Dirac delta function at ξ_0:

$$f[x] = e^{+2\pi i\xi_0 x} \Longrightarrow F[\xi] = \delta[\xi - \xi_0] \tag{9.191}$$

where ξ_0 may be positive or negative. The output of the nonlinear process is trivial to derive:

$$g[x] = (e^{+2\pi i\xi_0 x})^\gamma = e^{+2\pi i(\gamma\xi_0)x} \tag{9.192}$$

and the output spectrum is still a single Dirac delta function, but now located at $\xi_1 \equiv \gamma\xi_0$:

$$G[\xi] = \delta[\xi - (\gamma\xi_0)] = \delta[\xi - \xi_1] \tag{9.193}$$

The output spatial frequency ξ_1 is closer to the origin than ξ_0 if $|\gamma| < 1$ and farther away if $|\gamma| > 1$, and it has the opposite sign to ξ_0 if $\gamma < 0$. Also note that $\xi_1 = 0$ and $g[x]$ is a constant if $\gamma = 0$.

In the more general case of a complex-valued input composed of two sinusoidal components with different spatial frequencies ξ_0 and ξ_1, the input may be written:

$$f[x] = \alpha_0\, e^{+2\pi i\xi_0 x} + \alpha_1\, e^{+2\pi i\xi_1 x} \tag{9.194}$$

where α_0 and α_1 are real-valued weights. In the derivation, we assume that one weight is larger, e.g., $|\alpha_0| \geq |\alpha_1|$, but the algebraic signs and numerical values of ξ_0 and ξ_1 are arbitrary. The function $f[x]$ is complex and generally both the real and imaginary parts are bipolar. The output of the nonlinearity may be rewritten as the product of two terms:

$$g[x] = (\alpha_0\, e^{+2\pi i\xi_0 x})^\gamma \left(1 + \frac{\alpha_1}{\alpha_0}\, e^{+2\pi i(\xi_1-\xi_0)x}\right)^\gamma \tag{9.195}$$

If $|\alpha_0| \geq |\alpha_1|$, the second term may be expanded into a Taylor series:

$$\begin{aligned} g[x] &= \alpha_0^\gamma (e^{+2\pi i(\gamma\xi_0)x}) \\ &\quad\cdot \left(1 + \gamma\left(\frac{\alpha_1}{\alpha_0}\, e^{+2\pi i(\xi_1-\xi_0)x}\right) + \frac{\gamma(\gamma-1)}{2}\left(\frac{\alpha_1}{\alpha_0}\, e^{+2\pi i(\xi_1-\xi_0)x}\right)^2 + \cdots\right) \\ &= \alpha_0^\gamma\, e^{+2\pi i(\gamma\xi_0)x} + \alpha_0^\gamma \gamma\left(\frac{\alpha_1}{\alpha_0}\right) e^{+2\pi i(\xi_1-\xi_0+\gamma\xi_0)x} \\ &\quad + \alpha_0^\gamma \frac{\gamma(\gamma-1)}{2}\left(\frac{\alpha_1}{\alpha_0}\right)^2 e^{+2\pi i(2\xi_1-2\xi_0+\gamma\xi_0)x} + \cdots \end{aligned} \tag{9.196}$$

and the spectrum is easy to evaluate as an infinite series of single-frequency terms:

$$\begin{aligned} G[\xi] &= \alpha_0^\gamma \delta[\xi - \gamma\xi_0] + \alpha_0^\gamma\left(\gamma\frac{\alpha_1}{\alpha_0}\right) \cdot \delta[(\xi - \gamma\xi_0) - (\xi_1 - \xi_0)] \\ &\quad + \alpha_0^\gamma\left(\frac{\gamma(\gamma-1)}{2}\left(\frac{\alpha_1}{\alpha_0}\right)^2\right) \cdot \delta[(\xi - \gamma\xi_0) - 2(\xi_1 - \xi_0)] + \cdots \end{aligned} \tag{9.197}$$

The first Dirac delta function in this sequence is located at $\gamma\xi_0$ and is closest to the origin; we can consider this to be the "fundamental" frequency of $g[x]$. The action of the nonlinearity has generated additional frequency components at integer-multiple increments of $\xi_1 - \xi_0$, which may be positive or negative. The additional components are located on "one side" of the "fundamental" frequency, either larger than or smaller than $\gamma\xi_0$.

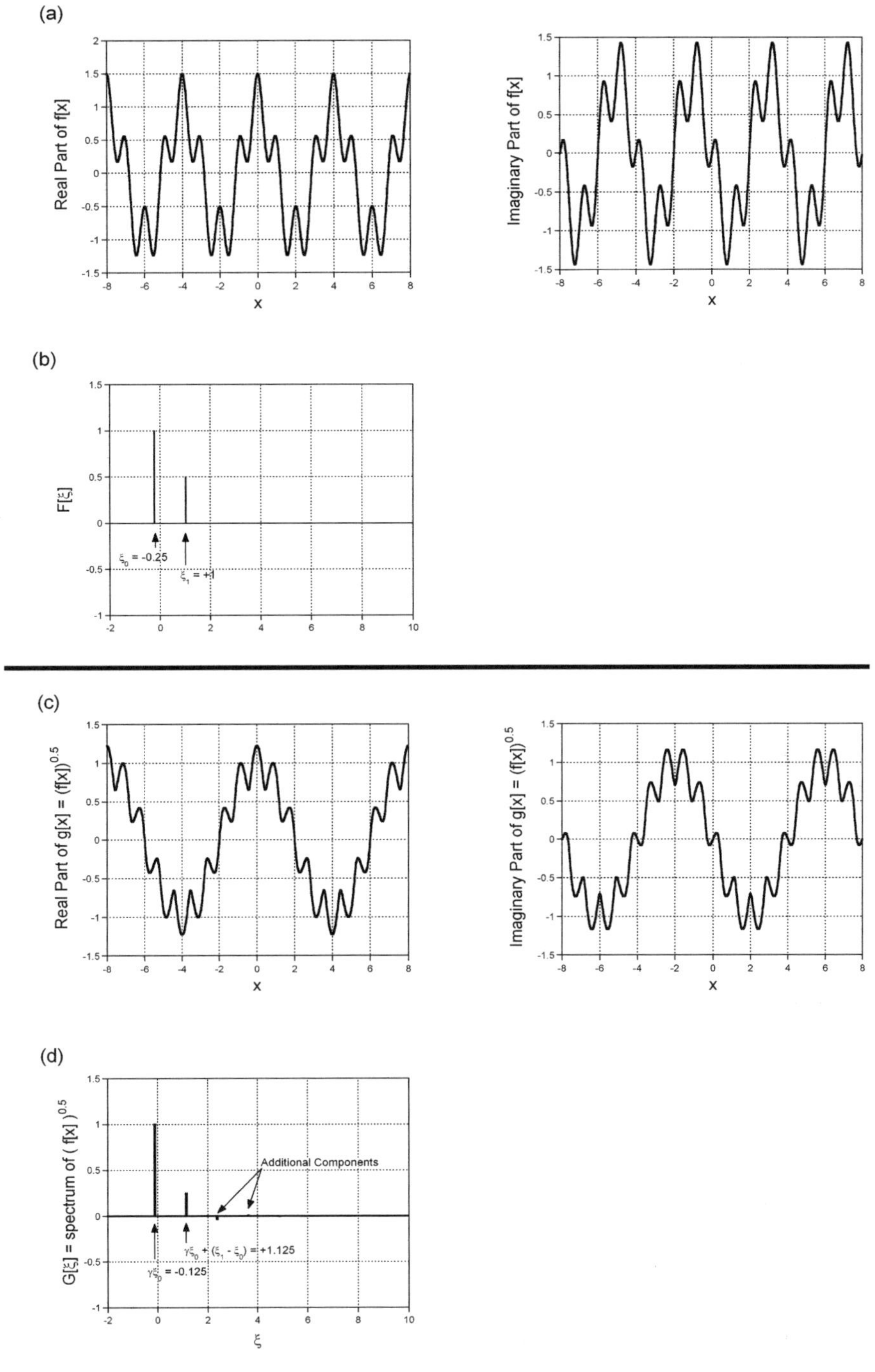

Figure 9.46 Action of a power-law nonlinearity for $f[x] = \exp[-2\pi i x/4] + \frac{1}{2}\exp[+2\pi i x]$ and $\gamma = +\frac{1}{2}$: (a) Hermitian function $f[x]$ as real and imaginary parts; (b) $F[\xi] = \delta[\xi + \frac{1}{4}] + \frac{1}{2}\delta[\xi - 1]$; (c) $g[x] = \sqrt{f[x]}$ as real and imaginary parts; (d) $\mathcal{F}_1\{\sqrt{f[x]}\}$, showing new "fundamental" frequency and components at increments of $\xi_1 - \xi_0$.

For illustration, consider $f[x] = \exp[-2\pi ix/4] + \frac{1}{2}\exp[+2\pi ix]$ with $\gamma = +0.5$, which means that $a_0 = 1$, $\xi_0 = -0.25$, $a_1 = +0.5$, and $\xi_1 = +1$. By direct substitution, the new "fundamental" frequency is $\gamma\xi_0 = -0.125$ and the frequency increment is $\xi_1 - \xi_0 = +1.25$. The functions before and after the nonlinearity are shown as are real and imaginary parts, along with the corresponding real-valued spectra in Figure 9.46.

Finally, consider the effect of the power-law nonlinearity in the common imaging case where $f[x]$ is real valued. The only such function composed of a single sinusoidal component is $f[x] = a \cdot 1[x]$, and the only possible action of a nonlinearity is a scaling by a constant:

$$f[x] = a \cdot 1[x] = e^{\pm 2\pi i(0)x} \Longrightarrow F[\xi] = \delta[\xi \mp 0] = \delta[\xi] \tag{9.198a}$$

$$g[x] = (a \cdot 1[x])^\gamma = a^\gamma \cdot 1[x] \Longrightarrow G[\xi] = a^\gamma \cdot \delta[\xi] \tag{9.198b}$$

The next simplest real-valued function is the cosine, which is the sum of equal-amplitude positive and negative frequencies:

$$f[x] = \cos[2\pi\xi_0 x] = \frac{1}{2}(e^{+2\pi i\xi_0 x} + e^{-2\pi i\xi_0 x})$$

$$\Longrightarrow F[\xi] = \frac{1}{2}(\delta[\xi + \xi_0] + \delta[\xi - \xi_0]) \tag{9.199}$$

The support of the spectrum is $2 \cdot \xi_0$. The spectrum of the integer powers of the cosine is easy to derive by applying the modulation theorem. The resulting spectrum is the n-fold convolution of the even pair of Dirac delta functions. For $n = 2$:

$$g[x] = (\cos[2\pi\xi_0 x])^2 = \frac{1}{2}(1 + \cos[2\pi(2\xi_0)x])$$

$$G[\xi] = \frac{1}{4}\delta[\xi + 2\xi_0] + \frac{1}{2}\delta[\xi] + \frac{1}{4}\delta[\xi - 2\xi_0] \tag{9.200}$$

which has support of $4 \cdot \xi_0$. The spectrum is also easy to evaluate if $n = 3$, producing a function with support $6 \cdot \xi_0$:

$$\mathcal{F}_1\{\cos^3[2\pi\xi_0 x]\} = \frac{1}{8}\delta[\xi + 3\xi_0] + \frac{3}{8}\delta[\xi + \xi_0] + \frac{3}{8}\delta[\xi - \xi_0] + \frac{1}{8}\delta[\xi - 3\xi_0] \tag{9.201}$$

If a general power-law nonlinearity with noninteger γ is applied to the bipolar cosine, the resulting image typically includes imaginary amplitudes. We will stick to the case where $g[x]$ is real, as it must be in many imaging applications. To ensure this property, a constant bias is added that is sufficiently large to ensure that $f[x] \geq 0$:

$$f[x] = \frac{1}{2}(1 + m\cos[2\pi\xi_0 x]) \tag{9.202}$$

where we assume that the modulation $m \leq 1$. The output of the nonlinear operation is:

$$g[x] = \left(\frac{1}{2}\right)^\gamma (1 + m\cos[2\pi\xi_0 x])^\gamma \tag{9.203}$$

The expression also may be written as a Taylor series:

$$g[x] = \left(\frac{1}{2}\right)^\gamma \left(1 + \gamma(m\cos[2\pi\xi_0 x]) + \frac{\gamma(\gamma - 1)}{2}(m^2\cos^2[2\pi\xi_0 x]) + \cdots\right) \tag{9.204}$$

If the power law is approximately linear ($\gamma \cong 1$), then the series may be truncated at a small power without introducing much error. The second-order approximation to the nonlinear output is:

$$\begin{aligned} g[x] &\cong \left(\frac{1}{2}\right)^\gamma \left(1 + \gamma m\cos[2\pi\xi_0 x] + \left(\frac{\gamma(\gamma - 1)}{4}m^2\right)(1 + \cos[2\pi(2\xi_0)x])\right) \\ &= \left(\frac{1}{2}\right)^\gamma \left[\left(1 + \frac{\gamma(\gamma - 1)}{4}m^2\right) + \gamma m\cos[2\pi\xi_0 x] + \left(\frac{\gamma(\gamma - 1)}{4}m^2\right)\cos[2\pi(2\xi_0)x]\right] \end{aligned} \tag{9.205}$$

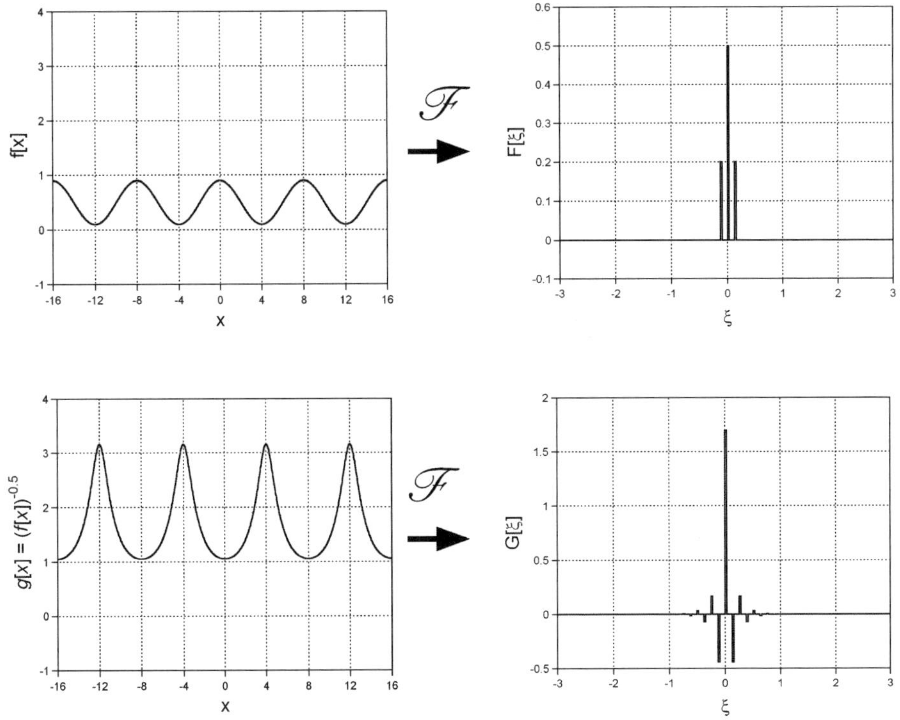

Figure 9.47 Effect of power-law nonlinearity on real-valued nonnegative function: $f[x] = \frac{1}{2} + \frac{2}{5}\cos[2\pi x/8]$ and its spectrum is $F[\xi] = \frac{1}{2}\delta[\xi] + \frac{1}{5}(\delta[\xi + \frac{1}{8}] + \delta[\xi - \frac{1}{8}])$; $g[x] = (f[x])^{-1/2}$; its spectrum $G[\xi]$ shows additional sinusoidal components at larger and smaller spatial frequencies.

The spectrum is easily evaluated by applying the known transforms and theorems:

$$G[\xi] \cong \left[\left(1 + \frac{\gamma(\gamma-1)}{4}m^2\right)\left(\frac{1}{2}\right)^{\gamma}\right]\delta[\xi] + \left[\frac{\gamma m}{2}\left(\frac{1}{2}\right)^{\gamma}\right](\delta[\xi + \xi_0] + \delta[\xi - \xi_0])$$
$$+ \left[\left(\frac{\gamma(\gamma-1)}{8}m^2\right)\left(\frac{1}{2}\right)^{\gamma}\right](\delta[\xi + 2\xi_0] + \delta[\xi - 2\xi_0]) \quad (9.206)$$

The fundamental frequencies of the result remain $\pm\xi_0$, but the nonlinear operation has introduced additional sinusoidal components at $\xi = \pm 2\xi_0$. In words, the nonlinearity applied to a real-valued signal added components to the spectrum on *both* sides of the fundamental frequency. If included in the series, the third- (and higher-)order terms will generate yet larger frequencies, though their amplitudes decrease with increasing order. An example is shown in Figure 9.47.

The result may be tested by substituting $m = 1$ and $\gamma = 1.05$ to produce:

$$F[\xi] = \frac{1}{2}\delta[\xi] + \frac{1}{4}(\delta[\xi + \xi_0] + \delta[\xi - \xi_0]) \quad (9.207a)$$

$$G[\xi] \cong 0.489\delta[\xi] + 0.253(\delta[\xi + \xi_0] + \delta[\xi - \xi_0])$$
$$+ 0.013\,75(\delta[\xi + 2\xi_0] + \delta[\xi - 2\xi_0]) \quad (9.207b)$$

Note that the amplitude of the DC component has *decreased*, while those at the fundamental frequencies have *increased* and the second-harmonic terms were created "out of thin air". In other words, the action of the nonlinearity redistributed energy among the existing spatial frequencies as well as creating new sinusoidal components at frequencies that did not exist before. This is fundamentally different from the action of a linear operator, which can remove energy from some frequency components but cannot "create" new ones.

Now consider the case $\gamma = +0.98$, so that the exponent is less than unity. The output spectrum for $m = 1$ is:

$$G[\xi] = 0.505\delta[\xi] + 0.248(\delta[\xi + \xi_0] + \delta[\xi - \xi_0]) - 0.0012(\delta[\xi + 2\xi_0] + \delta[\xi - 2\xi_0]) \quad (9.208)$$

The action of the nonlinearity *increased* the amplitude of the DC term while *decreasing* those at the fundamental frequencies.

In summary, a nonlinear operator applied to any function composed of more than one sinusoidal component will increase the support of the spectrum. These operators will appear during the discussion of optical holography in Chapter 23.

9.9 Appendix: Spectrum of Gaussian via Path Integral

The Fourier transform of $GAUS[x] = e^{-\pi x^2}$ may be rigorously evaluated via integration along the appropriate path in the complex plane. The desired 1-D integral is:

$$\int_{x=-\infty}^{+\infty} e^{-\pi x^2} e^{-2\pi i \xi x} \, dx = \lim_{x_1 \to \infty} \int_{x=-x_1}^{+x_1} e^{-\pi x^2} e^{-2\pi i \xi x} \, dx \quad (A9.1)$$

To evaluate this integral, consider the complex-valued integrand

$$f[z] = e^{-\pi z^2} = e^{-\pi (x+iy)^2} = e^{-\pi x^2} e^{-2\pi i x y} e^{+\pi y^2}$$

which has no singularities in the complex plane. This function is integrated around a path formed from four straight segments shown in Figure A9.1. The desired Fourier integral is along the line segment from $[-x_1, y_1]$ to $[+x_1, y_1]$ in the limit $x_1 \to +\infty$ for $y_1 \to \xi$. The integral around the entire closed path is:

$$\begin{aligned}
\oint_C e^{-\pi z^2} \, dz &= \int_{x=-x_1}^{+x_1} e^{-\pi (x+0i)^2} \, dx + \int_{y=0}^{y_1} e^{-\pi (x_1+iy)^2} (i \, dy) \\
&\quad + \int_{x=+x_1}^{-x_1} e^{-\pi (x+iy_1)^2} \, dx + \int_{y=y_1}^{0} e^{-\pi (-x_1+iy)^2} (i \, dy) \\
&= \int_{x=-x_1}^{+x_1} e^{-\pi x^2} \, dx + \int_{x=+x_1}^{-x_1} e^{-\pi (x^2+2ixy_1-y_1^2)} \, dx \\
&\quad + i\left(\int_{y=0}^{y_1} e^{-\pi (x_1^2+2ixy-y^2)} \, dy + \int_{y=y_1}^{0} e^{-\pi (x_1^2-2ix_1y-y^2)} \, dy \right)
\end{aligned} \quad (A9.2)$$

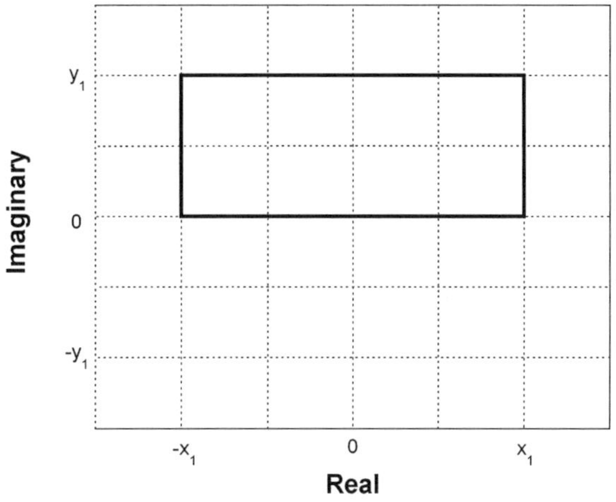

Figure A9.1 Integration path in the complex plane used to evaluate the Fourier transform of *GAUS*[x].

Cauchy's theorem determines that the integral of $e^{-\pi z^2}$ around any closed path must evaluate to zero because the integrand has no singularities. The integrals are rearranged by simple algebra to obtain:

$$\begin{aligned}0 &= \int_{x=-x_1}^{+x_1} e^{-\pi x^2}\,dx + e^{+\pi y_1^2}\int_{x=+x_1}^{-x_1} e^{-\pi(x^2+2ixy_1)}\,dx \\ &\quad + i\,e^{-\pi x_1^2}\left(\int_{y=0}^{y_1} e^{+\pi y^2}(e^{+2\pi i x_1 y} - e^{-2\pi i x_1 y})\,dy\right) \\ &= \int_{x=-x_1}^{+x_1} e^{-\pi x^2}\,dx + e^{+\pi y_1^2}\int_{x=+x_1}^{-x_1} e^{-\pi(x^2+2ixy_1)}\,dx \\ &\quad - 2\,e^{-\pi x_1^2}\left(\int_{y=0}^{y_1} e^{+\pi y^2}\sin[2\pi x_1 y]\,dy\right)\end{aligned} \tag{A9.3}$$

The integrals may be evaluated in the limit $x_1 \to \infty$, so that $e^{-\pi x_1^2} \to 0$, which eliminates the third integral:

$$0 = \int_{x=-\infty}^{+\infty} e^{-\pi x^2}\,dx - e^{+\pi y_1^2}\int_{x=-\infty}^{+\infty} e^{-\pi x^2}\,e^{-2\pi i x y_1}\,dx \tag{A9.4}$$

After the obvious rearrangement, choice of the parameter $y_1 = \xi$, and substitution of the known unit area of the Gaussian, the desired Fourier transform is obtained:

$$\int_{x=-\infty}^{+\infty} e^{-\pi x^2}\,e^{-2\pi i x \xi}\,dx = e^{-\pi \xi^2} \tag{A9.5}$$

PROBLEMS

9.1 Derive the formula for the phase of the input sinusoid from the areas of the Dirac delta functions in the Hartley transform in Equation (9.38).

9.2 Specify the common names for projections of $f[x]$ onto the basis functions defined by:

(a) $m[x, x'] = 1$ for $x = x'$

(b) $m[x, x'] = 0$ for $x \neq x'$

9.3 Use the general form of the derivative theorem in Equation (9.142) to derive an expression for the "half-order" derivative of $f[x]$, i.e., $(d^{1/2}/dx^{1/2})f[x]$.

9.4 Evaluate the areas of:

(a) $SINC[x] \cdot e^{-\pi x^2}$

(b) $SINC^2[x]$

(c) $SINC^3[x]$ (ANSWER: $\frac{3}{4}$)

(d) $SINC^4[x]$ (ANSWER: $\frac{2}{3}$)

(e) $SINC^5[x]$ (ANSWER: $\frac{115}{192}$)

9.5 Derive two equivalent expressions that are analogous to Rayleigh's theorem for:

$$\int_{-\infty}^{+\infty} f[x]h[x]\,dx$$

9.6 Find the Fourier transform of $f[x] = STEP[x + \frac{1}{2}] + STEP[x - \frac{1}{2}] - 1$ and of df/dx.

(a) Evaluate and sketch $f[x] = SINC[x] * 1/x$.

(b) Given that $f[x] = RECT[x] + i\,RECT[x-2]$ and $m[x] = f[x+2]$, find expressions for and sketch $f[x] \bigstar m[x]$.

9.7 Evaluate the following:

(a) $\mathcal{F}_1\{\cos(2\pi x/2) + \sin(2\pi x/2)\}$

(b) $\mathcal{F}_1\{\sin[(x/2)^2]\}$

(c) $\mathcal{F}_1\{SINC[x/3 - 1/2]\}$

9.8 Evaluate and sketch the following as real and imaginary parts and as magnitude and phase:

(a) $\mathcal{F}_1\{e^{-|x|}\}$

(b) $\mathcal{F}_1\{d\delta/dx\}$

(c) $\mathcal{F}_1\{x\}$

(d) $\mathcal{F}_1\{|x|\}$

(e) $\mathcal{F}_1\{1/x\}$

(f) $\mathcal{F}_1\{1/|x|\}$

(g) $\mathcal{F}_1\left\{\frac{1}{2}RECT[x] + \int_{x-\frac{1}{2}}^{x+\frac{1}{2}} \frac{1}{2\pi i\alpha}\,d\alpha\right\}$ (HINT: filter theorem.)

9.9 Evaluate the following Fourier transforms and sketch them as real and imaginary parts and as magnitude and phase:

(a) $RECT[x] * RECT[x]$

(b) $RECT[x-1] * RECT[x]$

(c) $RECT[x-1] * RECT[x+1]$

(d) $RECT[x-1] \bigstar RECT[x+1]$

(e) $GAUS[x/5] * GAUS[x/12]$

(f) $SINC[x/a] * SINC[x/b]$, where a, b are real valued and positive.

9.10 Use the theorems of the Fourier transform to prove that:

$$r[x-x_0, y-y_0] \bigstar r[x-x_0, y-y_0] = r[x, y] \bigstar r[x, y]$$

9.11 Evaluate the following Fourier transforms and sketch them as real and imaginary parts and as magnitude and phase:

(a) $SINC[x] \cdot SINC[x/2]$

(b) $\cos[\pi x] \cdot GAUS[x]$
(c) $\cos[2\pi \xi_0 x] \cdot RECT[x/b]$
(d) $\sin[2\pi \xi_0 x] \cdot RECT[x/b]$

9.12 Find general expressions for the spectra of $(\cos[2\pi \xi_0 x])^n$ and $(\sin[2\pi \xi_0 x])^n$, where n is a positive integer.

9.13 Derive the expression for the spectrum of the phase $SINC[x]$.

9.14 Evaluate the spectrum of $g[x] = |\cos[2\pi \xi_0 x]|$.
(ANSWER:

$$\frac{2}{\pi}\left(\sum_{k=-\infty}^{+\infty}\left[\frac{(-1)^k}{1-4k^2}\right]\delta[\xi - 2k\xi_0]\right)$$

The area of the Dirac delta function at the origin is $2/\pi < 1$.)

9.15 The expression

$$P[\xi] = \frac{1}{i\pi\xi} * \left(SINC[\xi] \cdot COMB[2\xi] \cdot \exp[-i\pi\xi] - \frac{1}{2}\delta[\xi]\right)$$

appeared during the derivation of the spectrum of $|SINC[x]|$. Evaluate $P[0]$.

10

Multidimensional Fourier Transforms

The next three chapters extend the development of equivalent representations of 1-D functions to 2-D domains, with a brief mention of the n-dimensional case with $n \geq 3$.

10.1 2-D Fourier Transforms

The coordinates of a 2-D spectrum may be specified by 2-D vectors, and so may be expressed either as Cartesian coordinates $[\xi, \eta]$ or polar coordinates (ρ, ϕ). The 2-D Fourier transformation operator $\mathcal{F}_2$ derives the complex amplitudes (magnitude and phase) of the spectrum at each spatial frequency from the mathematical expression for $f[x, y]$. By analogy with our interpretation of the 1-D spectrum, we can infer that the 2-D Fourier transform projects $f[x, y]$ onto each of the complete set of 2-D sinusoids defined in Section 7.2.8. We therefore surmise that the form of the 2-D Fourier transform is:

$$\mathcal{F}_2\{f[x, y]\} \equiv F[\xi, \eta] \equiv \iint_{-\infty}^{+\infty} (e^{+2\pi i(\xi x+\eta y)})^* f[x, y]\, dx\, dy$$
$$= \iint_{-\infty}^{+\infty} e^{-2\pi i(\xi x+\eta y)}\, f[x, y]\, dx\, dy \tag{10.1}$$

where the inner product of complex functions in Equation (5.87) and the Euler relation in Equation (4.19) have been used. The amplitude F of the 2-D spectrum at a particular spatial frequency $[\xi_0, \eta_0]$ is generally complex valued and thus may be expressed either as real and imaginary parts or as magnitude $|F[\xi_0, \eta_0]|$ and phase $\Phi_F[\xi_0, \eta_0]$. As in the 1-D case, the magnitude $|F[\xi_0, \eta_0]|$ specifies the maximum amplitude of that particular 2-D sinusoid "present" in $f[x, y]$, while the initial phase $\Phi_F[\xi_0, \eta_0]$ of the spectrum at that spatial frequency specifies the number of radians of phase of that sinusoid at the origin of coordinates. The initial phase at each spatial frequency varies in such a way as to force the sinusoidal components to "line up" properly to create features (such as "edges") in the space-domain representation $f[x, y]$.

The process of 2-D Fourier transformation at one spatial frequency is shown in Figure 10.1. In this case, the real-valued input function $f[x, y]$ is multiplied by $e^{-2\pi i(\xi_0 x+\eta_0 y)}$ to produce the complex function with real part $f[x, y] \cdot \cos[2\pi(\xi_0 x + \eta_0 y)]$ and imaginary part $f[x, y] \cdot \sin[-2\pi(\xi_0 x + \eta_0 y)]$. The volumes of the 2-D real and imaginary functions are respectively the real and imaginary

Fourier Methods in Imaging Roger L. Easton, Jr.

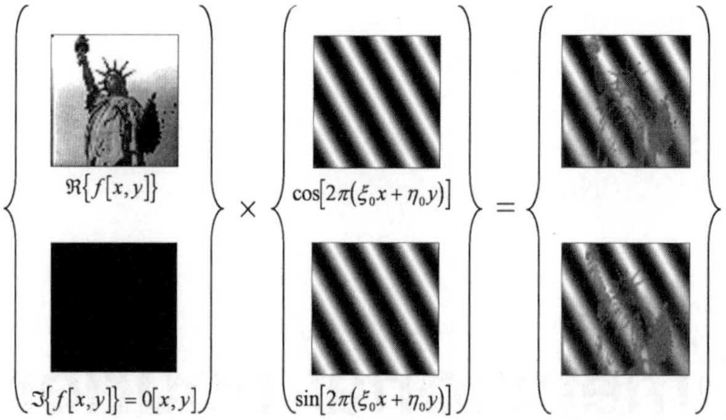

Figure 10.1 Product of $f[x, y]$ and $e^{-2\pi i[\xi_0 x+\eta_0 y]}$. The 2-D analogue of the central-ordinate theorem indicates that the volume of this product is $F[\xi_0, \eta_0]$.

parts of $F[\xi_0, \eta_0]$. The process is repeated for all 2-D spatial frequencies to evaluate the complete complex-valued 2-D function $F[\xi, \eta]$.

Because the hypothesized expression for the 2-D Fourier transform in Equation (10.1) is an integral over specific limits, it satisfies the necessary conditions for linearity. In words, the spectrum of the weighted sum (linear combination) of any number of 2-D functions is the weighted sum of the individual Fourier transforms:

$$\mathcal{F}_2\left\{\sum_{n=1}^{N} \alpha_n f_n[x, y]\right\} = \sum_{n=1}^{N} \alpha_n \mathcal{F}_2\{f_n[x, y]\} = \sum_{n=1}^{N} \alpha_n F_n[\xi, \eta] \tag{10.2}$$

We note from Equation (10.1) that the specific complex-valued linear-phase "reference" function (or "mask") of the Fourier transform is itself separable into orthogonal 1-D linear-phase functions:

$$e^{-2\pi i(\xi x+\eta y)} = e^{-2\pi i\xi x}\, e^{-2\pi i\eta y} \tag{10.3}$$

and so the 2-D Fourier transform may be performed as a cascade of 1-D transforms in orthogonal directions:

$$F[\xi, \eta] = \int_{-\infty}^{+\infty} \left(\int_{-\infty}^{+\infty} f[x, y]\, e^{-2\pi i\xi x}\, dx\right) e^{-2\pi i\eta y}\, dy \tag{10.4}$$

In words, this recipe for the 2-D Fourier transform requires evaluation of 1-D Fourier transforms for each of the infinite number of "rows" of the 2-D array $f[x, y]$, followed by 1-D Fourier transforms of each of the infinite number of resulting "columns". Clearly the order of the operations may be exchanged to compute the 1-D Fourier transforms in the y-direction first.

10.1.1 2-D Fourier Synthesis

The expression for the 1-D inverse Fourier transform operation was derived in Section 9.6 via the 1-D sifting property and the expression for the 1-D Dirac delta function as a superposition of complex linear-phase exponential functions with unit magnitude at all frequencies. An analogous procedure may be used to derive the corresponding expression for the 2-D inverse Fourier transform via the 2-D sifting property in Equation (7.48), the expression for the 1-D Dirac delta function as a superposition of complex linear-phase exponentials in Equation (6.106), and the separable Cartesian form of the 2-D

Dirac delta function in Equation (7.43):

$$
\begin{aligned}
f[x, y] = f[x, y] * \delta[x, y] &= \iint_{-\infty}^{+\infty} f[\alpha, \beta]\, \delta[x-\alpha, y-\beta]\, d\alpha\, d\beta \\
&= \iint_{-\infty}^{+\infty} f[\alpha, \beta](\delta[x-\alpha] \cdot \delta[y-\beta])\, d\alpha\, d\beta \\
&= \iint_{-\infty}^{+\infty} f[\alpha, \beta]\left(\int_{-\infty}^{+\infty} e^{+2\pi i\xi(x-\alpha)}\, d\xi \cdot \int_{-\infty}^{+\infty} e^{+2\pi i\eta(y-\beta)}\, d\eta\right) d\alpha\, d\beta \\
&= \iint_{-\infty}^{+\infty} f[\alpha, \beta]\left(\int_{-\infty}^{+\infty} e^{+2\pi i\xi x}\, e^{-2\pi i\xi\alpha}\, d\xi \cdot \int_{-\infty}^{+\infty} e^{+2\pi i\eta y}\, e^{-2\pi i\xi\beta}\, d\eta\right) d\alpha\, d\beta \\
&= \iint_{-\infty}^{+\infty} \left(\iint_{-\infty}^{+\infty} f[\alpha, \beta]\, e^{-2\pi i(\xi\alpha+\eta\beta)}\, d\alpha\, d\beta\right) e^{+2\pi i(\xi x+\eta y)}\, d\xi\, d\eta \\
&= \iint_{-\infty}^{+\infty} F[\xi, \eta]\, e^{+2\pi i(\xi x+\eta y)}\, d\xi\, d\eta \equiv \mathcal{F}_2^{-1}\{F[\xi, \eta]\}
\end{aligned}
\tag{10.5}
$$

In words, the 2-D space-domain representation $f[x, y]$ is generated by superposing complex linear-phase exponential functions with magnitude and phase specified by the complex amplitude of $F[\xi, \eta]$. As was true for the 1-D case, the only differences between the forward and inverse transforms are again the algebraic sign of the exponent in the reference function ("−" and "+" for the forward and inverse transforms, respectively) and the domain of integration determined by the differential element of area ($dx\, dy$ versus $d\xi\, d\eta$).

Equations (10.4) and (10.5) indicate that both the 2-D forward or inverse Fourier transform may be computed as sequences of 1-D forward or inverse Fourier transforms, respectively. As we will see, the dominant theme of any quest to evaluate a multidimensional Fourier transform is finding a method based upon 1-D computations. This is particularly true when evaluating spectra on digital computers that perform sequences of serial operations. However, we should note that systems capable of performing computations in parallel do exist; a single lens in coherent (laser) light is a notable example. This "parallel" computer may be "programmed" to efficiently evaluate the multidimensional Fourier transform by placing the 2-D object and image sensor at the proper positions, as will be considered in Chapter 21.

These chapters describe means for evaluating spectra of various 2-D functions based on the already considered expressions for the 1-D Fourier transform. This chapter discusses 2-D transforms of orthogonal functions based on the sequential implementation of 1-D transforms in Equation (10.4), including consideration of the relevant theorems and properties. In the next chapter, the forms of 2-D Fourier transforms of circularly symmetric functions are derived; these also reduce to 1-D integrals, though the operator is more complicated than the 1-D Fourier transform. Chapter 12 describes a recipe for evaluating spectra of arbitrary 2-D functions from Fourier transforms of each of a set of 1-D functions that are obtained by integrating $f[x, y]$ along parallel lines. The ensemble of these parallel line integrals for each azimuth angle constructs the invertible *Radon* transform of $f[x, y]$, and the relationship of the Radon and Fourier transforms may provide significant intuitive understanding of the nature of 2-D functions. The Radon transform also provides the mathematical basis for several important imaging techniques, including medical computed tomography and magnetic resonance imaging.

10.2 Spectra of Separable 2-D Functions

The discussion of Section 7.1 shows that a separable 2-D function may be expressed as the product of two 1-D functions along orthogonal axes:

$$f[x, y] = f_X[x] f_Y[y] \tag{10.6}$$

The 2-D Fourier transform of a separable function is obtained easily by decomposing the 2-D "mask" function into its own separable parts as in Equation (10.3):

$$\begin{aligned}\mathcal{F}_2\{f_x[x]f_y[y]\} &= \int_{-\infty}^{+\infty}\left(\int_{-\infty}^{+\infty}(f_x[x]f_y[y])\,e^{-2\pi i\xi x}\,dx\right)e^{-2\pi i\eta y}\,dy\\ &= \left(\int_{-\infty}^{+\infty} f_x[x]\,e^{-2\pi i\xi x}\,dx\right)\cdot\left(\int_{-\infty}^{+\infty} f_y[y]\,e^{-2\pi i\eta y}\,dy\right)\\ &= F_x[\xi]\cdot F_y[\eta]\end{aligned} \tag{10.7}$$

In words, the Fourier transform of a 2-D separable function evaluated at a particular frequency $[\xi_0, \eta_0]$ is the product of the amplitudes of the 1-D Fourier transforms of the two component parts evaluated at ξ_0 and at η_0, respectively. This result immediately allows the transforms of many useful 2-D functions to be derived directly from 1-D Fourier transforms that were evaluated in Chapter 9. A few of the more useful transforms of separable functions are presented next.

10.2.1 Fourier Transforms of Separable Functions

The Fourier transform of a 2-D function that varies along the x-axis and is constant in the y-axis is easy to evaluate by applying Equation (10.7) and the known transform of $1[y]$:

$$\mathcal{F}_2\{f_x[x]\,1[y]\} = F_x[\xi]\,\delta[\eta] \tag{10.8}$$

In words, the spectrum of a 2-D function $f[x, y]$ that is constant along y is zero for all 2-D spatial frequencies with $\eta \neq 0$; nonzero amplitudes appear only on the ξ-axis.

Perhaps the most useful example of such a function is the 2-D complex linear-phase exponential that varies in the x-direction:

$$f[x, y] = e^{+2\pi i\xi_0 x}\,1[y] = \cos[2\pi\xi_0 x]\,1[y] + i\,\sin[2\pi\xi_0 x]\,1[y] \tag{10.9}$$

Substitution of the 1-D spectra from Equations (9.50), (9.51), and (9.57) yields the 2-D Dirac delta function:

$$F[\xi, \eta] = \delta[\xi - \xi_0]\,\delta[\eta] = \delta[\xi - \xi_0, \eta] \tag{10.10}$$

In words, this probably obvious result indicates that the spectrum of a complex linear-phase exponential that varies along the x-direction only is a single Dirac delta function located at the spatial frequency ξ_0. The 2-D function and its spectrum are shown in Figure 10.2.

The unit step that varies along the x-direction and is constant for all y is another important separable function (Figure 10.3). Its 2-D spectrum evidently is:

$$\begin{aligned}\mathcal{F}_2\{STEP[x]\,1[y]\} &= \mathcal{F}_1\{STEP[x]\}\mathcal{F}_1\{1[y]\}\\ &= \frac{1}{2}\left(\delta[\xi] - \frac{i}{2\pi\xi}\right)\delta[\eta] = \left(\frac{1}{2}\delta[\xi, \eta]\right) + i\left(-\frac{1}{2\pi\xi}\,\delta[\eta]\right)\end{aligned} \tag{10.11}$$

10.2.2 Fourier Transform of $\delta[x, y]$

Though we have already surmised the spectrum of the 2-D Dirac delta function in Equation (7.44) and used it to derive the 2-D inverse Fourier transform in Equation (10.5), we include it here for completeness. The spectrum is obtained very easily by applying Equation (10.7) and the spectrum of the 1-D Dirac delta function from Equation (9.57):

$$\begin{aligned}\mathcal{F}_2\{\delta[x, y]\} &= \mathcal{F}_2\{\delta[x]\,\delta[y]\}\\ &= \mathcal{F}_1\{\delta[x]\}\mathcal{F}_1\{\delta[y]\}\\ &= 1[\xi]\,1[\eta] \equiv 1[\xi, \eta]\end{aligned} \tag{10.12}$$

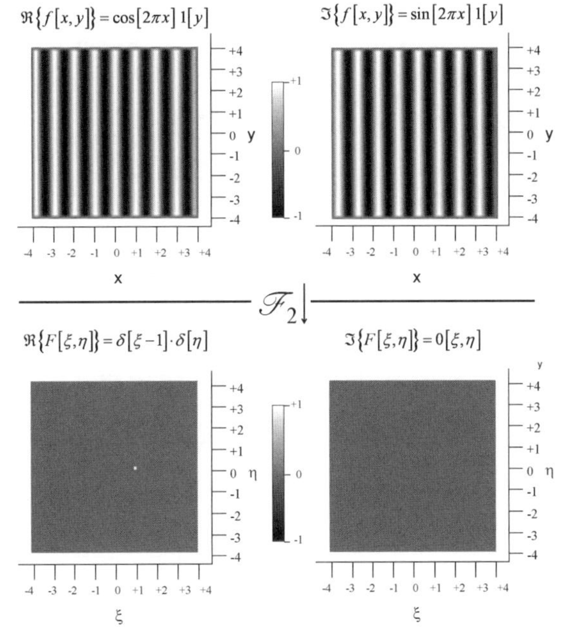

Figure 10.2 Fourier transform of complex exponential. Top: real and imaginary parts of $f[x] = e^{+2\pi ix}$. Bottom: real and imaginary parts of $\mathcal{F}\{f[x]\} = \delta[\xi - 1, \eta]$.

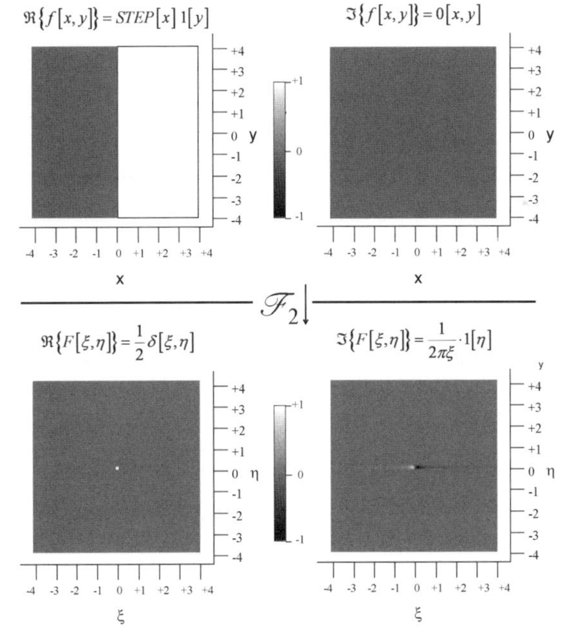

Figure 10.3 The real-valued function $f[x, y] = STEP[x]\ 1[y]$ and its spectrum $F[\xi.\eta] = \frac{1}{2}\delta[\xi, \eta] + (1/2\pi i\xi)\delta[\eta]$.

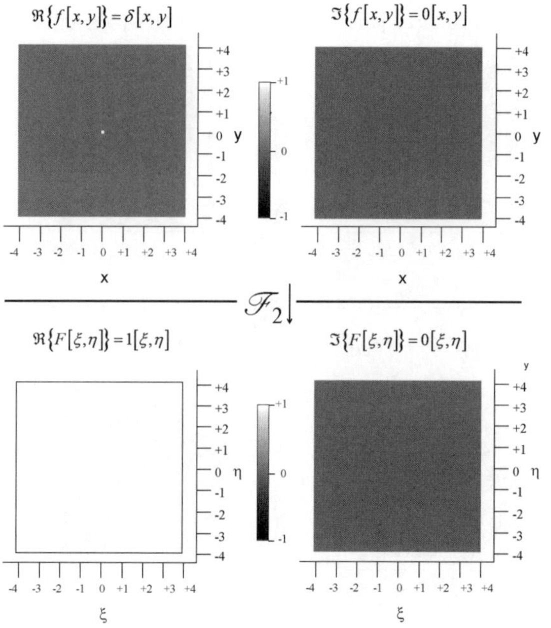

Figure 10.4 The 2-D real-valued Dirac delta function $f[x, y] = \delta[x, y] + i\,0[x, y]$ and its spectrum $F[\xi, \eta] = 1[\xi, \eta] + i0[\xi, \eta]$. The spectrum indicates that the 2-D Dirac delta function may be generated by summing unit-amplitude 2-D cosine functions with all spatial frequencies.

Because the 2-D Dirac delta function $\delta[x, y]$ centered at the origin is real valued and even, the spectrum must include only real-valued and even cosine terms. The spectrum of $\delta[x, y]$ is the 2-D real-valued unit constant, as shown in Figure 10.4. The phase of the spectrum is the null constant $0[\xi, \eta]$. The interpretation of this result is identical to the 1-D case in Equation (6.106) and Equation (9.58); the 2-D Dirac delta function located at the origin is generated by superposing 2-D unit-amplitude cosine functions at all 2-D spatial frequencies. In other words, this is the sum of unit-amplitude sinusoids that have all possible oscillation rates oriented at all azimuth angles.

10.2.3 Fourier Transform of $\delta[x - x_0, y - y_0]$

The Fourier transform of the 2-D Dirac delta function located at $[x_0, y_0]$ is the trivial generalization of Equation (10.12) obtained by applying the 1-D shift theorem from Equation (9.123) to each of the orthogonal component functions:

$$\begin{aligned}\mathcal{F}_2\{\delta[x - x_0, y - y_0]\} &= \mathcal{F}_1\{\delta[x - x_0]\}\mathcal{F}_1\{\delta[y - y_0]\}\\ &= (1[\xi]\, e^{-2\pi i\xi x_0})(1[\eta]\, e^{-2\pi i\eta y_0})\\ &\equiv 1[\xi, \eta]\, e^{-2\pi i(\xi x_0+\eta y_0)}\\ &= \cos[2\pi(\xi x_0 + \eta y_0)] - i\,\sin[2\pi(\xi x_0 + \eta y_0)] \qquad (10.13)\end{aligned}$$

The spectrum is a 2-D separable complex linear-phase exponential in the frequency domain with oscillation periods along the ξ- and η-axes equal to x_0^{-1} and y_0^{-1}, respectively. Because the function is real valued but with both even and odd parts, the spectrum is Hermitian, as shown in Figure 10.5.

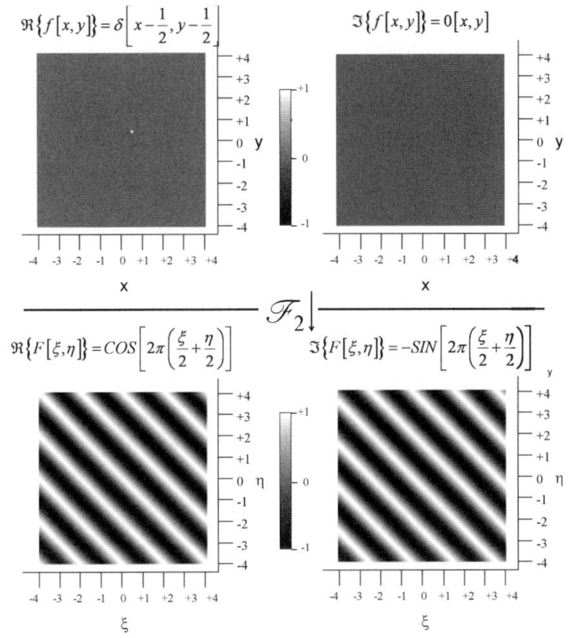

Figure 10.5 An "off-center" 2-D Dirac delta function $f[x, y] = \delta[x - \frac{1}{2}, y - \frac{1}{2}]$ and its spectrum $F[\xi, \eta] = \exp[-2\pi i(\xi/2 + \eta/2)] = \cos[\pi\xi + \pi\eta] - i\ \sin[\pi\xi + \pi\eta]$.

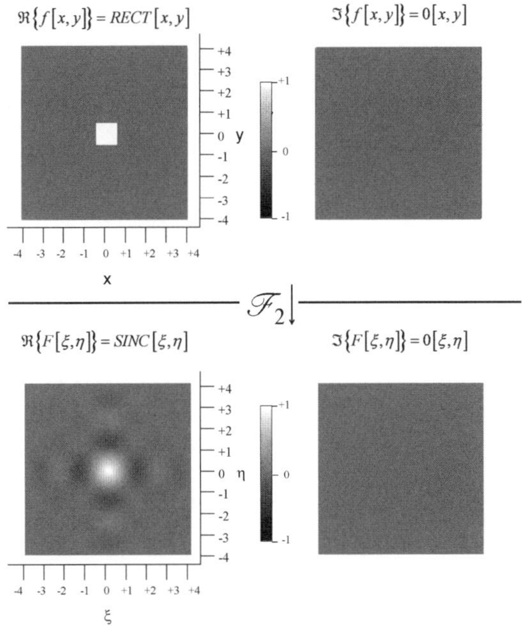

Figure 10.6 The real-valued and even 2-D rectangle function $f[x, y] = RECT[x, y] + i0[x, y]$ and its real-valued and even spectrum $F[\xi, \eta] = SINC[\xi, \eta] + i0[\xi, \eta]$.

10.2.4 Fourier Transform of *RECT*[*x*, *y*]

The spectrum of the centered 2-D *RECT* with unit width along both axes (and thus unit volume) is the orthogonal product of the two component 1-D *SINC* functions, as in Equation (7.25):

$$\begin{aligned}\mathcal{F}_2\{RECT[x, y]\} &= \mathcal{F}_2\{RECT[x]\,RECT[y]\}\\ &= \mathcal{F}_1\{RECT[x]\}\mathcal{F}_1\{RECT[y]\}\\ &= SINC[\xi]\,SINC[\eta] \equiv SINC[\xi, \eta]\end{aligned} \tag{10.14}$$

The spectrum is shown in Figure 10.6. This indicates that the *RECT* function is generated from the superposition of 2-D sinusoids whose amplitudes tend to decrease with increasing spatial frequency. The initial phase of each sinusoid is either 0 or $-\pi$ radians, which means that it is either a positive or negative cosine. Sinusoidal components with null amplitude in this spectrum are those that are orthogonal to the *RECT* function, and are therefore not necessary to generate it. These include sinusoids oriented along the x- or y-axis with periods that are integer multiples of the width of the *RECT*.

10.2.5 Fourier Transform of *TRI*[*x*, *y*]

By following the same steps that led to Equation (10.14), the spectrum of the centered 2-D triangle function is the orthogonal product of two 1-D $SINC^2$ functions, thus creating the 2-D $SINC^2$ introduced in Equation (7.27):

$$\begin{aligned}\mathcal{F}_2\{TRI[x, y]\} &= \mathcal{F}_2\{TRI[x]\,TRI[y]\}\\ &= \mathcal{F}_1\{TRI[x]\}\mathcal{F}_1\{TRI[y]\}\\ &= SINC^2[\xi]\,SINC^2[\eta] \equiv SINC^2[\xi, \eta]\end{aligned} \tag{10.15}$$

This also is real and even (Figure 10.7). Like the rectangle, the symmetric triangle function is composed only of real and even cosine functions, but the initial phases at all spatial frequencies are identically zero for the triangle.

10.2.6 Fourier Transform of *GAUS*[*x*, *y*]

The application of Equation (9.91) leads to the observation that the 2-D Gaussian with unit-width parameter is a "self-Fourier transform":

$$\begin{aligned}\mathcal{F}_2\{GAUS[x, y]\} &= \mathcal{F}_2\{GAUS[x]\,GAUS[y]\}\\ &= GAUS[\xi]\,GAUS[\eta] = GAUS[\xi, \eta]\end{aligned} \tag{10.16}$$

and the transform of the scaled 2-D Gaussian function is almost as easy to derive:

$$\begin{aligned}\mathcal{F}_2\left\{GAUS\left[\frac{x}{a}, \frac{y}{b}\right]\right\} &= \mathcal{F}_2\left\{GAUS\left[\frac{x}{a}\right]GAUS\left[\frac{y}{b}\right]\right\}\\ &= |ab|\;GAUS[a\xi]\,GAUS[b\eta]\\ &= |ab|\;GAUS\left[\frac{\xi}{a^{-1}}, \frac{\eta}{b^{-1}}\right]\end{aligned} \tag{10.17}$$

Like the triangle, the symmetric Gaussian function is composed of real-valued symmetric cosine functions because the initial phase hardcover is zero at all spatial frequencies.

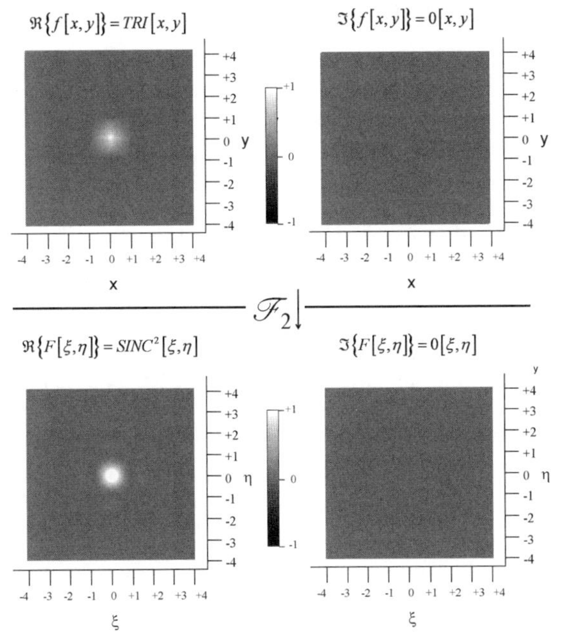

Figure 10.7 The real-valued and even 2-D triangle function $f[x, y] = TRI[x, y] + i0[x, y]$ and its real-valued and even 2-D spectrum $F[\xi, \eta] = SINC^2[\xi, \eta] + i0[\xi, \eta]$.

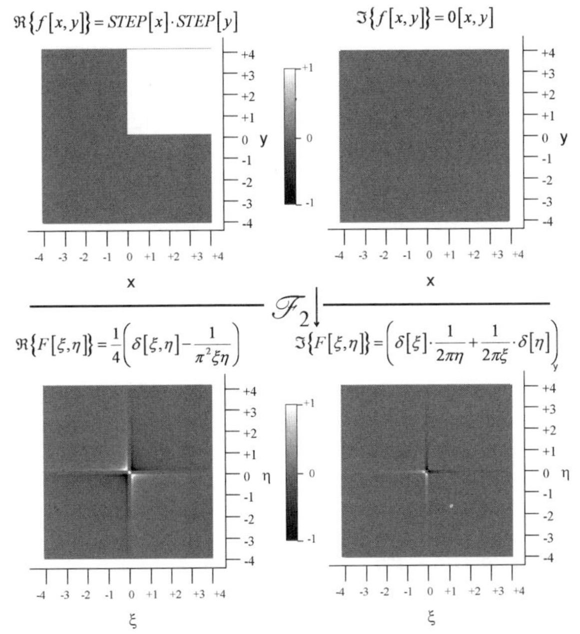

Figure 10.8 The real valued 2-D function $STEP[x, y] = STEP[x] \cdot STEP[y]$ and its 2-D spectrum.

10.2.7 Fourier Transform of $STEP[x] \cdot STEP[y]$

In Section 7.2.4, the separable 2-D *STEP* function was defined as the orthogonal product of two 1-D *STEP* functions in the x- and y-directions. The resulting function is nonzero only in the first quadrant. It may be decomposed into nonzero real-valued even and odd parts, and so its spectrum must have a real-valued even part and an imaginary-valued odd part. By applying Equation (10.7), the 2-D transform is easily shown to be:

$$\begin{aligned}\mathcal{F}_2\{STEP[x]\cdot STEP[y]\} &= \mathcal{F}_1\{STEP[x]\}\cdot\mathcal{F}_1\{STEP[y]\}\\ &= \left(\frac{1}{2}\delta[\xi]-\frac{i}{2\pi\xi}\right)\cdot\left(\frac{1}{2}\delta[\eta]-\frac{i}{2\pi\eta}\right)\\ &= \frac{1}{4}\left(\delta[\xi,\eta]-\frac{1}{\pi^2\xi\eta}\right)-i\left(\delta[\xi]\frac{1}{2\pi\eta}+\frac{1}{2\pi\xi}\delta[\eta]\right)\end{aligned} \tag{10.18}$$

which is shown in Figure 10.8.

10.2.8 Theorems of Spectra of Separable Functions

The theorems of the Fourier transforms of separable functions are trivial to derive by applying the appropriate 1-D theorems to the orthogonal component functions.

10.2.8.1 Scaling Theorem

$$\begin{aligned}\mathcal{F}_2\left\{f\left[\frac{x}{b},\frac{y}{d}\right]\right\} &= \mathcal{F}_2\left\{f_1\left[\frac{x}{b}\right]\cdot f_2\left[\frac{y}{d}\right]\right\}\\ &= \mathcal{F}_1\left\{f_1\left[\frac{x}{b}\right]\right\}\cdot\mathcal{F}_1\left\{f_2\left[\frac{y}{d}\right]\right\}\\ &= (|b|F_1[b\xi])\cdot(|d|F_2[d\eta])\\ &= |bd|F\left[\frac{\xi}{b^{-1}},\frac{\eta}{d^{-1}}\right]\end{aligned} \tag{10.19}$$

10.2.8.2 Shift Theorem

$$\begin{aligned}\mathcal{F}_2\{f[x-x_0,y-y_0]\} &= \mathcal{F}_2\{f_1[x-x_0]\cdot f_2[y-y_0]\}\\ &= (\mathcal{F}_1\{f_1[x-x_0]\})\cdot(\mathcal{F}_1\{f_2[y-y_0]\})\\ &= (F_1[\xi]\,e^{-2\pi i\xi x_0})\cdot(F_2[\eta]\,e^{-2\pi i\eta y_0})\\ &= F[\xi,\eta]\,e^{-2\pi i(\xi x_0+\eta y_0)}\end{aligned} \tag{10.20}$$

10.2.8.3 Central-Ordinate Theorem

$$\begin{aligned}\mathcal{F}_2\{f[x,y]\}|_{\xi=0,\eta=0} &= (\mathcal{F}_1\{f_1[x]\}|_{\xi=0})\cdot(\mathcal{F}_1\{f_2[y]\}|_{\eta=0})\\ &= \left(\int_{-\infty}^{+\infty} f_1[x]\,dx\right)\cdot\left(\int_{-\infty}^{+\infty} f_2[y]\,dy\right)\end{aligned}$$

$$= \iint_{-\infty}^{+\infty} f_1[x] f_2[y] \, dx \, dy$$

$$= \iint_{-\infty}^{+\infty} f[x, y] \, dx \, dy = F[0, 0] \tag{10.21}$$

In words, Equation (10.21) demonstrates that the amplitude of the central ordinate of the spectrum of a 2-D separable function is the *volume* of that function.

10.2.9 Superpositions of 2-D Separable Functions

The linearity of the 2-D Fourier transform ensures that the spectrum of the sum of distinct functions is the sum of the spectra of the component functions. This observation may be used to evaluate the spectra of complicated 2-D scenes. For example, a 2-D upper-case block letter "E" may be constructed by summing three rectangle functions with appropriate scale factors and translations, as shown in Figure 10.9:

$$e[x, y] = RECT\left[\frac{x}{7}, \frac{y}{11}\right] + RECT\left[\frac{x}{3}, \frac{y}{3}\right] - RECT\left[\frac{x-1}{5}, \frac{y}{7}\right] \tag{10.22}$$

The notation is a letter "*e*" to follow the convention that space-domain representations are indicated by lower-case symbols. Because the volume of each unit-amplitude 2-D *RECT* is the product of the length, width, and unit amplitude, the volume of the superposition is $77 + 9 - 35 = 51$. The superposition of the individual spectra is easy to derive by applying the scaling and shift theorems for 2-D separable functions from Equation (10.19) and Equation (10.20):

$$\mathcal{F}_2\{e[x, y]\} = |77| \, SINC[7\xi, 11\eta] + |9| \, SINC[3\xi, 3\eta] - |35| \, SINC[5\xi, 7\eta] \, e^{-2\pi i \xi} \tag{10.23a}$$

The spectrum $E[\xi, \eta]$ may be separated into its real and imaginary parts. The real part clearly is even:

$$\Re\{E[\xi, \eta]\} = |77| \, SINC[7\xi, 11\eta] + |9| \, SINC[3\xi, 3\eta] - |35| \, SINC[5\xi, 7\eta] \cos[2\pi\xi] \tag{10.23b}$$

Note that the imaginary part is due only to the "off-center" rectangle:

$$\Im\{E[\xi, \eta]\} = |35| \, SINC[5\xi, 7\eta] \cdot \sin[2\pi\xi] \tag{10.23c}$$

The central ordinate of $E[\xi, \eta]$ is easy to evaluate:

$$E[0, 0] = (|77| + |9| - |35|) + i \, |35| \, SINC[0, 0] \cdot (\sin[0] \cdot 1[0])$$

$$= 51 + 0 \cdot i \tag{10.24}$$

which is the sum of the volumes of the individual rectangles and thus confirms the 2-D central ordinate theorem of Equation (10.21).

10.3 Theorems of 2-D Fourier Transforms

The derivations of many properties of 1-D Fourier transforms may be extended easily to the 2-D operation. For example, the Fourier transform of a real-valued function $f[x, y]$ is Hermitian (even real part, odd imaginary part). Most theorems of the 2-D Fourier transform may be proven by parallel developments of the derivations in Section 9.8. The transforms of the sum of two 2-D functions or the product of a 2-D function by a constant have already been demonstrated in Equation (10.2).

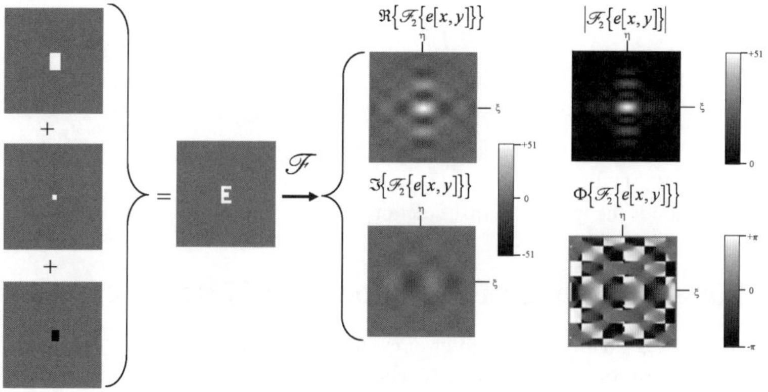

Figure 10.9 A unit-amplitude sans-serif character "E" represented as the sum of three rectangles, and its Fourier transform displayed as real/imaginary parts and as magnitude/phase. The imaginary part of the spectrum is due to the "off-center" rectangle with negative amplitude.

10.3.1 2-D "Transform-of-a-Transform" Theorem

Given that $\mathcal{F}_2\{f[x, y]\} = F[\xi, \eta]$, the "transform-of-a-transform" theorem for 2-D functions is derived easily and has the expected form:

$$\mathcal{F}_2\{F[x, y]\} = f[-\xi, -\eta] \tag{10.25}$$

In words, the spectrum is the original space-domain function expressed in the frequency domain and "reversed" by rotating about the origin in the frequency domain by π radians. An example is shown in Figure 10.10.

10.3.2 2-D Scaling Theorem

The general scaling and shift theorems also are easy to extend from separable to arbitrary 2-D functions. The amplitude of the spectrum of the scaled function is multiplied by the absolute value of the product of the scale factors:

$$\mathcal{F}_2\left\{f\left[\frac{x}{b}, \frac{y}{d}\right]\right\} = |bd| F[b\xi, d\eta] \tag{10.26}$$

In words, if the width of the 2-D function is increased along a particular axis, then the width of the 2-D spectrum along the corresponding axis is decreased and the amplitude is increased. An example is shown in Figure 10.11.

10.3.3 2-D Shift Theorem

The spectrum of the translated function is scaled by a unit-magnitude phase factor:

$$\mathcal{F}_2\{f[x - x_0, y - y_0]\} = e^{-2\pi i(\xi x_0 + \eta y_0)}\, F[\xi, \eta] \tag{10.27}$$

In words, the spectrum of a translated function is the Fourier transform of the original function modulated by a complex 2-D sinusoid which varies along one azimuth in the frequency domain and is constant in the orthogonal direction.

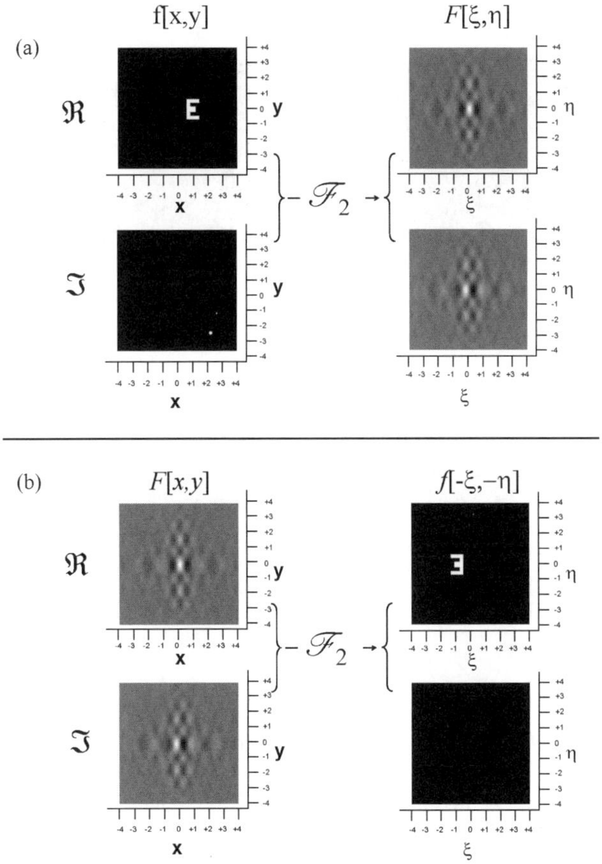

Figure 10.10 The 2-D "transform-of-a-transform" theorem. (a) $f[x, y]$ is an off-axis letter. Its 2-D spectrum $F[\xi, \eta]$ is Hermitian. (b) $F[x, y]$ and its 2-D spectrum $f[-\xi, -\eta]$.

10.3.4 2-D Filter Theorem

The proof of the 2-D filter theorem also is a straightforward generalization of the 1-D case:

$$\begin{aligned}
\mathcal{F}_2\{f[x, y] * h[x, y]\} &= \mathcal{F}_2\left\{\iint_{-\infty}^{+\infty} f[\alpha, \beta] h[x-\alpha, y-\beta]\, d\alpha\, d\beta\right\} \\
&= \iint_{-\infty}^{+\infty} \left(\iint_{-\infty}^{+\infty} f[\alpha, \beta] h[x-\alpha, y-\beta]\, d\alpha\, d\beta\right) e^{-2\pi i(\xi x+\eta y)}\, dx\, dy \\
&= \iint_{-\infty}^{+\infty} f[\alpha, \beta] \left(\iint_{-\infty}^{+\infty} h[x-\alpha, y-\beta]\, e^{-2\pi i(\xi x+\eta y)}\, dx\, dy\right) d\alpha\, d\beta \\
&= \iint_{-\infty}^{+\infty} f[\alpha, \beta] (H[\xi, \eta]\, e^{-2\pi i(\xi\alpha+\eta\beta)})\, d\alpha\, d\beta
\end{aligned}$$

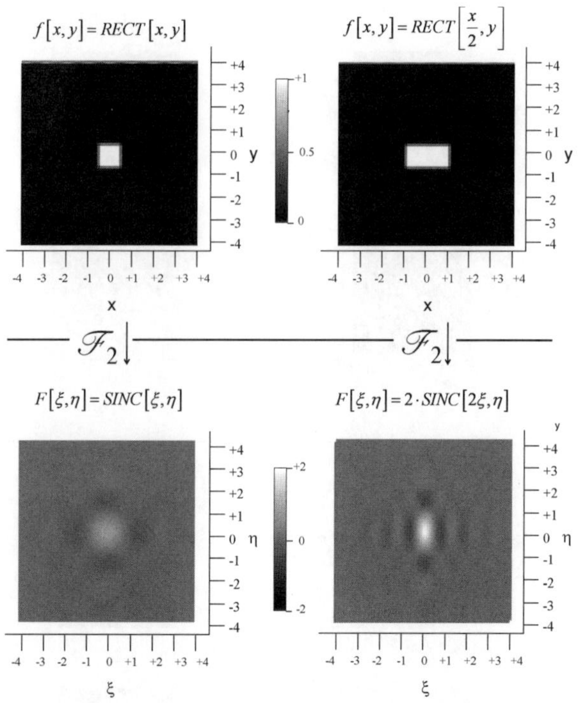

Figure 10.11 The 2-D scaling theorem. The original function is the 2-D separable even and real-valued rectangle $f_1[x, y] = RECT[x, y]$. Its spectrum is the even and real-valued $SINC[\xi, \eta]$. The width of the function along the x-axis is increased by a factor of 2 to produce $f_2[x, y] = RECT[x/2, y]$, with resulting spectrum $F_2[\xi, \eta] = 2\, SINC[2\xi, \eta]$.

$$
\begin{aligned}
&= H[\xi, \eta] \iint_{-\infty}^{+\infty} f[\alpha, \beta]\, e^{-2\pi i(\xi\alpha + \eta\beta)}\, d\alpha\, d\beta \\
&= H[\xi, \eta] F[\xi, \eta] = G[\xi, \eta] \\
&= \mathcal{F}_2\{f[x, y]\}\mathcal{F}_2\{h[x, y]\}
\end{aligned}
\tag{10.28}
$$

The filter theorem is used to demonstrate the shift theorem in Figure 10.12.

10.3.5 2-D Derivative Theorem

Several varieties of 2-D spatial derivative may be defined, but those of interest in most physical applications are partial derivatives with respect to the spatial variables x and y. Since the orthogonal spatial coordinates are independent, partial differentiations with respect to these variables are separable operations, which means that the operations may be cascaded in any order. The spectrum of the general 2-D partial derivative is easy to evaluate by applying the method of the 1-D derivative theorem in Equation (9.142):

$$
\begin{aligned}
\mathcal{F}_2\left\{\left(\frac{\partial}{\partial x}\right)^n \left(\frac{\partial}{\partial y}\right)^m f[x, y]\right\} &= (2\pi i\xi)^n (2\pi i\eta)^m F[\xi, \eta] \\
&= (2\pi i)^{n+m} \xi^n \eta^m F[\xi, \eta]
\end{aligned}
\tag{10.29}
$$

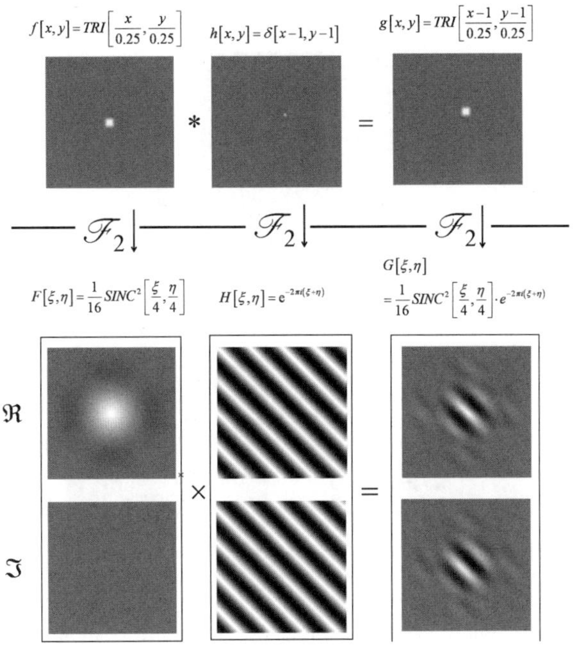

Figure 10.12 The 2-D shift theorem: $f[x, y]$ represented as the convolution of a symmetric triangle and an off-axis Dirac delta function. The spectrum of the shifted function is the product of the symmetric $SINC^2$ function and the complex linear-phase factor.

Many physical applications involve the evaluation of sums of partial derivatives. The linearity property and separability of the partial derivatives determine that the spectrum of such sums must be:

$$\mathcal{F}_2\left\{\left(\alpha\frac{\partial^n}{\partial x^n}+\beta\frac{\partial^m}{\partial y^m}\right)f[x,y]\right\}=(\alpha(2\pi i\xi)^n+\beta(2\pi i\eta)^m)F[\xi,\eta] \tag{10.30}$$

where α and β are generally complex-valued weights. The specific case with $n = m = 2$ and $\alpha = \beta = 1$ is particularly common in physical applications; this *Laplacian* operation computes the sum of the second partial derivatives along the two orthogonal directions. The operator is commonly denoted by the symbol ∇^2, which is pronounced "del squared". The spectrum of $\nabla^2 f[x, y]$ is:

$$\begin{aligned}\mathcal{F}_2\{\nabla^2 f[x,y]\}&\equiv\mathcal{F}_2\left\{\left(\frac{\partial^n}{\partial x^n}+\frac{\partial^m}{\partial y^m}\right)f[x,y]\right\}\\&=[(2\pi i\xi)^2+(2\pi i\eta)^2]F[\xi,\eta]\\&=(2\pi i)^2(\xi^2+\eta^2)F[\xi,\eta]=-4\pi\rho^2\cdot F[\xi,\eta]\end{aligned} \tag{10.31}$$

where the Pythagorean relationship between the Cartesian frequency coordinates $[\xi, \eta]$ and the polar radial frequency ρ has been used. In words, the transfer function of the linear and shift invariant Laplacian operator is the circularly symmetric function $-4\pi\rho^2 = +4\pi\rho^2 \cdot e^{-i\pi}$, which scales the magnitudes of the sinusoidal components by a factor proportional to ρ^2 and decrements the initial phase of each sinusoidal component by π radians. Note that the DC component of the input spectrum is completely "blocked" by the action of the Laplacian, which should be no surprise since this component is the constant part of $f[x, y]$ and any derivative of any constant function is null. All "low-frequency"

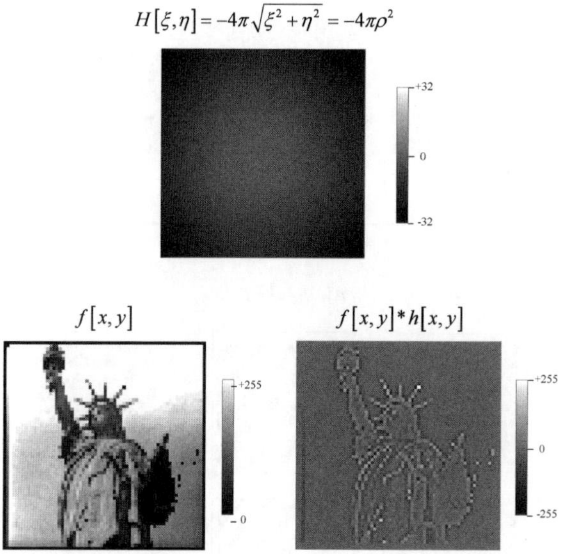

Figure 10.13 The action of the Laplacian operator $\nabla^2 = \partial^2/\partial x^2 + \partial^2/\partial y^2$: the transfer function is $H[\xi, \eta] = -4\pi^2\rho^2$. The output of the filter exhibits extrema at the locations of edges of objects in the scene.

terms in the input spectrum (i.e., those with spatial frequency $\xi^2 + \eta^2 < 1/4\pi$) are attenuated, while higher-frequency terms are amplified. The transfer function of the Laplacian operator and its effect on some 2-D functions are shown in Figure 10.13.

10.3.6 Spectra of Rotated 2-D Functions

One property of the 2-D Fourier transform that has no 1-D analogue is the effect of rotation about some point in the 2-D space on the spectrum of the function. Consider first the simple example of a 2-D Dirac delta function located at $[x_0, 0]$, which we will denote by $f_0[x, y]$:

$$f_0[x, y] = \delta[x - x_0, y] = \delta[x - x_0]\,\delta[y - 0] \tag{10.32}$$

Its 2-D Fourier transform is easy to derive by separability:

$$F_0[\xi, \eta] = e^{-2\pi i \xi x_0}\, e^{-2\pi i \eta \cdot 0} = e^{-2\pi i \xi x_0}\; 1[\eta] \tag{10.33}$$

The space-domain function $f_0[x, y]$ may be rotated by θ radians by applying the mapping from Equation (7.11) to produce $f_1[x, y]$:

$$f_1[x, y] = f_0[x', y'] = f_0[\underline{\mathbf{r}} \bullet \hat{\underline{\mathbf{p}}}_1, \underline{\mathbf{r}} \bullet \hat{\underline{\mathbf{p}}}_1^{\perp}] = \delta[\underline{\mathbf{r}} \bullet \hat{\underline{\mathbf{p}}}_1 - x_0]\,\delta[\underline{\mathbf{r}} \bullet \hat{\underline{\mathbf{p}}}_1^{\perp}] \tag{10.34}$$

where the unit vectors have Cartesian coordinates $\hat{\underline{\mathbf{p}}}_1 = [\cos[\theta], \sin[\theta]]$ and $\hat{\underline{\mathbf{p}}}_1^{\perp} = [-\sin[\theta], +\cos[\theta]]$. After some straightforward manipulation, this expression may be recast into the still-separable form:

$$f_1[x, y] = \delta[x - x_0 \cos[\theta], y - x_0 \sin[\theta]] \tag{10.35}$$

The 2-D spectrum is easy to obtain by applying the relevant 1-D transforms:

$$\begin{aligned}F_1[\xi,\eta] &= \mathcal{F}_2\{\delta[x - x_0\cos[\theta],\ y - x_0\sin[\theta]]\}\\ &= \mathcal{F}_{1\atop x\to\xi}\{\delta[x - x_0\cos[\theta]]\}\ \mathcal{F}_{1\atop y\to\eta}\{\delta[y - x_0\sin[\theta]]\}\\ &= \exp[-2\pi i\xi x_0\cos[\theta]]\exp[-2\pi i\eta x_0\sin[\theta]] = \exp[-2\pi i x_0(\xi\cos[\theta] + \eta\sin[\theta])]\\ &= \exp[-2\pi i x_0(\underline{\boldsymbol{\rho}}\bullet\hat{\underline{\mathbf{p}}}_1)]\ 1[\underline{\boldsymbol{\rho}}\bullet\hat{\underline{\mathbf{p}}}_1^{\perp}] = F_0[\underline{\boldsymbol{\rho}}\bullet\hat{\underline{\mathbf{p}}}_1,\ \underline{\boldsymbol{\rho}}\bullet\hat{\underline{\mathbf{p}}}_1^{\perp}]\end{aligned} \tag{10.36}$$

where the conjugate coordinates of the 1-D Fourier transforms are specified beneath the 1-D Fourier transform operators to avoid confusion due to the several coordinates in the expression. The spatial frequency vector is $\underline{\boldsymbol{\rho}} = [\xi,\eta]$, as shown in Figure 10.14. In words, the spectrum of the off-axis Dirac delta function rotated about the origin of the space domain by θ radians is a replica of the original linear-phase spectrum that has been rotated about the origin of the frequency domain by the same angle.

The 2-D sifting property in Equation (7.48) represents an arbitrary 2-D function as a weighted sum of 2-D Dirac delta functions. This may be combined with the linearity property of the 2-D Fourier transform to demonstrate that the 2-D Fourier transform of the general rotated function $f[x,y]$ is:

$$\mathcal{F}_2\{f[x,y]\} = F[\xi,\eta] \Longrightarrow \mathcal{F}_2\{f[\underline{\mathbf{r}}\bullet\hat{\underline{\mathbf{p}}},\ \underline{\mathbf{r}}\bullet\hat{\underline{\mathbf{p}}}^{\perp}]\} = F[\underline{\boldsymbol{\rho}}\bullet\hat{\underline{\mathbf{p}}},\ \underline{\boldsymbol{\rho}}\bullet\hat{\underline{\mathbf{p}}}^{\perp}] \tag{10.37}$$

The 2-D rotation theorem states that the 2-D transform of a rotated function with coordinates $x' = \underline{\mathbf{r}}\bullet\hat{\underline{\mathbf{p}}}$ and $y' = \underline{\mathbf{r}}\bullet\hat{\underline{\mathbf{p}}}^{\perp}$ is obtained by rotating the spectrum of the original function by substituting $\xi = \underline{\boldsymbol{\rho}}\bullet\hat{\underline{\mathbf{p}}}$ and $\eta = \underline{\boldsymbol{\rho}}\bullet\hat{\underline{\mathbf{p}}}^{\perp}$.

The 2-D "rotation" theorem may be combined with the linearity of the rotation operation and of Fourier transformation to demonstrate that 2-D rotations and 2-D Fourier transformations are commuting operations; their order may be exchanged without affecting the final result.

10.3.7 Transforms of 2-D Line Delta and Cross Functions

The theorems just derived may be used to evaluate some Fourier transforms that will be useful later in this chapter. Recall that the separable 2-D line delta function was defined in Equation (7.57) as the orthogonal product of a 1-D Dirac delta function and a unit constant:

$$f_1[x,y] = \delta[x]\ 1[y] \tag{10.38}$$

Because this function is separable in Cartesian coordinates, its Fourier transform is easy to evaluate:

$$\mathcal{F}_2\{f_1[x,y]\} = F_1[\xi,\eta] = \mathcal{F}_1\{\delta[x]\}\cdot\mathcal{F}_1\{1[y]\} = 1[\xi]\ \delta[\eta] \tag{10.39}$$

In words, the spectrum is obtained by "exchanging" the functional forms (the Dirac delta function and the unit constant are swapped: $\delta \leftrightarrows 1$) and changing the coordinates to the other domain ($x \to \xi$, $y \to \eta$). These two representations are shown in Figure 10.15.

This result may be combined with the rotation theorem of the 2-D Fourier transform that was just derived to obtain a result that will be useful in Chapter 12. Recall the expression for the line delta function that intersects the origin but is rotated by the angle θ from Equation (7.71):

$$\begin{aligned}f_2[x,y] &= \delta[\underline{\mathbf{r}}\bullet\hat{\underline{\mathbf{p}}}]\ 1[\underline{\mathbf{r}}\bullet\hat{\underline{\mathbf{p}}}^{\perp}]\\ &= \delta[x\cos[\theta] + y\sin[\theta]]\ 1[x\sin[\theta] - y\cos[\theta]]\end{aligned} \tag{10.40}$$

The coordinates $[x,y]$ with nonzero amplitude that lie on the "line mass" have position vectors $\underline{\mathbf{r}} = [x,y]$ that are perpendicular to the unit vector $\hat{\underline{\mathbf{p}}} = [\cos[\theta],\sin[\theta]]$. The expression for the rotated

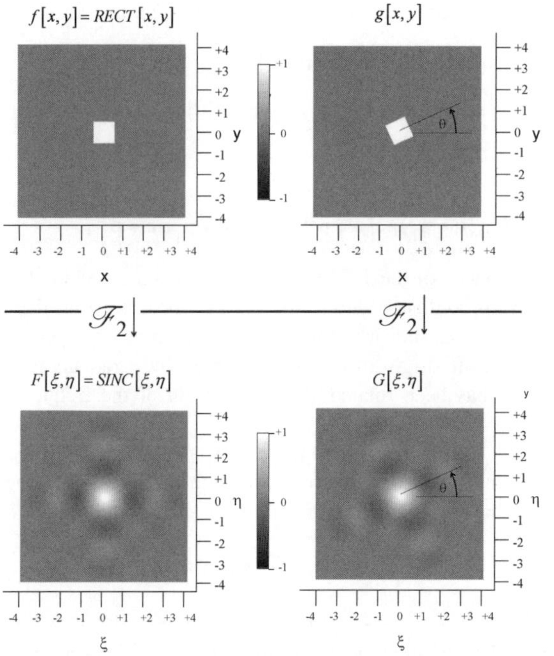

Figure 10.14 Demonstration that the Fourier transform of a rotated function is the original Fourier transform rotated in the same fashion. The original 2-D function is $f[x, y] = RECT[x, y]$, with spectrum $F[\xi, \eta] = SINC[\xi, \eta]$. The function is rotated by θ to produce the function $g[x, y]$ with rotated spectrum $G[\xi, \eta]$.

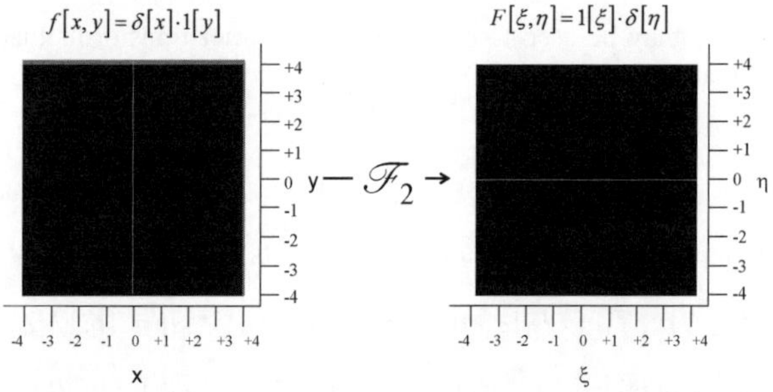

Figure 10.15 Line delta function $f_1[x, y] = \delta[x]\, 1[y]$ and its Fourier transform $F_1[\xi, \eta] = 1[\xi]\, \delta[\eta]$.

spectrum may be constructed by applying Equation (10.37); the spatial coordinate vector $\underline{\mathbf{r}}$ is replaced by the frequency vector $\underline{\boldsymbol{\rho}}$ and conjugate functions are again exchanged, $\delta \rightleftarrows 1$:

$$\begin{aligned}\mathcal{F}_2\{\delta[\underline{\mathbf{r}} \bullet \underline{\hat{\mathbf{p}}}]\; 1[\underline{\mathbf{r}} \bullet \underline{\hat{\mathbf{p}}}^{\perp}]\} &= \iint_{-\infty}^{+\infty} (\delta[\underline{\mathbf{r}} \bullet \underline{\hat{\mathbf{p}}}]\; 1[\underline{\mathbf{r}} \bullet \underline{\hat{\mathbf{p}}}^{\perp}])\, e^{-2\pi i(\xi x+\eta y)}\; dx\, dy \\ &= \iint_{-\infty}^{+\infty} (\delta[\underline{\mathbf{r}} \bullet \underline{\hat{\mathbf{p}}}]\; 1[\underline{\mathbf{r}} \bullet \underline{\hat{\mathbf{p}}}^{\perp}])\, e^{-2\pi i(\underline{\mathbf{r}} \bullet \underline{\boldsymbol{\rho}})}\; d^2 r\end{aligned}$$

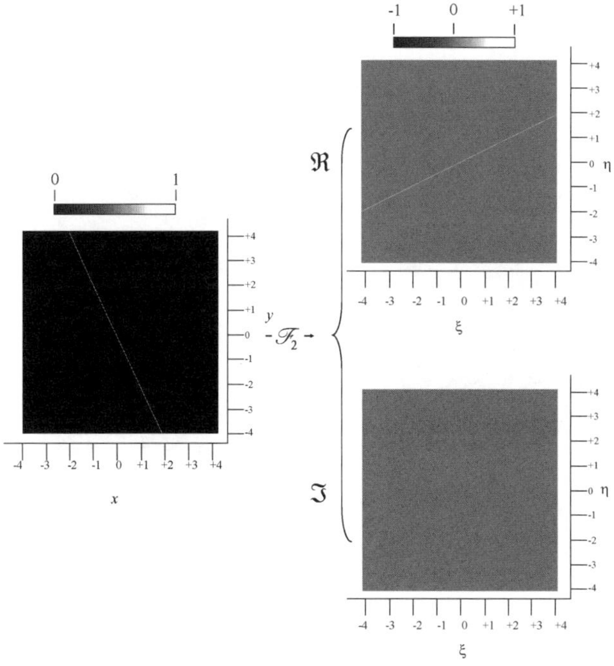

Figure 10.16 Symmetric line delta function after rotation by θ, so that $f_2[x, y] = \delta[\underline{\mathbf{r}} \cdot \widehat{\underline{\mathbf{p}}}]\ 1[\underline{\mathbf{r}} \cdot \widehat{\underline{\mathbf{p}}}^{\perp}]$ where $\widehat{\underline{\mathbf{p}}} = [\cos[\theta], \sin[\theta]]$, and its Fourier transform $F_2[\xi, \eta] = 1[\underline{\boldsymbol{\rho}} \cdot \widehat{\underline{\mathbf{p}}}]\ \delta[\underline{\boldsymbol{\rho}} \cdot \widehat{\underline{\mathbf{p}}}^{\perp}]$.

$$
\begin{aligned}
&= 1[\underline{\boldsymbol{\rho}} \bullet \hat{\underline{\mathbf{p}}}]\ \delta[\underline{\boldsymbol{\rho}} \bullet \hat{\underline{\mathbf{p}}}^{\perp}] \\
&= 1[\xi \cos[\theta] + \eta \sin[\theta]]\ \delta[\xi \sin[\theta] - \eta \cos[\theta]] \qquad (10.41)
\end{aligned}
$$

where the symmetry of the 1-D Dirac delta function has been used. This result confirms the rotation theorem: the 2-D Fourier transform in Equation (10.41) is identical to the spectrum in Equation (10.39) rotated by θ radians, so that the spectrum of the rotated line delta function is a line delta function in the frequency domain directed along an azimuth angle that is perpendicular to the space-domain azimuth (Figure 10.16).

The mathematical expression for a line delta function located at an arbitrary distance p_0 from the origin and oriented at an arbitrary angle θ was derived in Equation (7.77):

$$
\begin{aligned}
f_3[x, y] &= \delta[\underline{\mathbf{r}} \bullet \hat{\underline{\mathbf{p}}} - p_0]\ 1[\underline{\mathbf{r}} \bullet \hat{\underline{\mathbf{p}}}^{\perp}] = \delta[p_0 - \underline{\mathbf{r}} \bullet \hat{\underline{\mathbf{p}}}]\ 1[\underline{\mathbf{r}} \bullet \hat{\underline{\mathbf{p}}}^{\perp}] \\
&= \delta[p_0 - (x \cos[\theta] + y \sin[\theta])]\ 1[x \sin[\theta] - y \cos[\theta]] \qquad (10.42)
\end{aligned}
$$

where the symmetry of the Dirac delta function has been used. Its spectrum is evaluated by applying the shift theorem of the 2-D Fourier transform in Equation (10.27) to the result of Equation (10.41):

$$
\begin{aligned}
F_3[\xi, \eta] &= \mathcal{F}_2\{\delta[p_0 - \underline{\mathbf{r}} \bullet \hat{\underline{\mathbf{p}}}]\ 1[\underline{\mathbf{r}} \bullet \hat{\underline{\mathbf{p}}}^{\perp}]\} \\
&= \mathcal{F}_1\{\delta[p_0 - \underline{\mathbf{r}} \bullet \hat{\underline{\mathbf{p}}}]\}\mathcal{F}_1\{1[\underline{\mathbf{r}} \bullet \hat{\underline{\mathbf{p}}}^{\perp}]\} \\
&= (1[\underline{\boldsymbol{\rho}} \bullet \hat{\underline{\mathbf{p}}}]\ e^{-2\pi i p_0 (\underline{\boldsymbol{\rho}} \bullet \hat{\underline{\mathbf{p}}})})\ \delta[\underline{\boldsymbol{\rho}} \bullet \hat{\underline{\mathbf{p}}}^{\perp}] \\
&= (1[\underline{\boldsymbol{\rho}} \bullet \hat{\underline{\mathbf{p}}}]\ \delta[\underline{\boldsymbol{\rho}} \bullet \hat{\underline{\mathbf{p}}}^{\perp}])(\cos[2\pi i p_0 (\underline{\boldsymbol{\rho}} \bullet \hat{\underline{\mathbf{p}}})] - i \sin[2\pi i p_0 (\underline{\boldsymbol{\rho}} \bullet \hat{\underline{\mathbf{p}}})]) \qquad (10.43)
\end{aligned}
$$

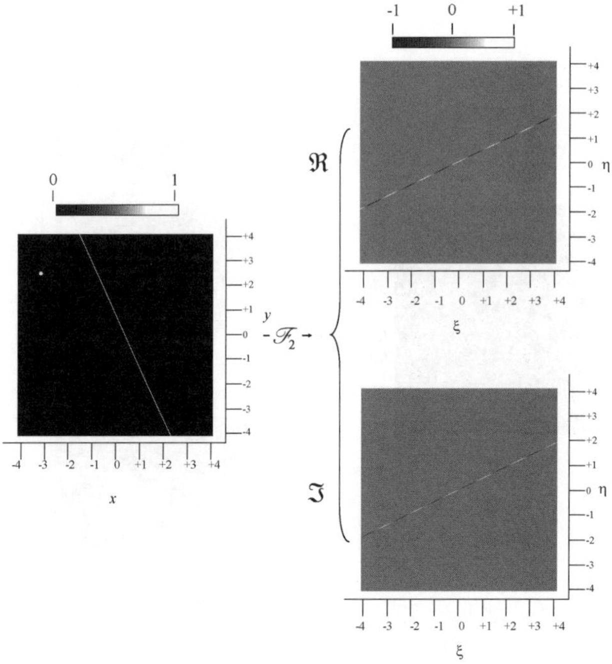

Figure 10.17 Rotated and translated 2-D line delta function: $f_3[x, y] = \delta[p_0 - \underline{\mathbf{r}} \cdot \widehat{\underline{\mathbf{p}}}]\, 1[\underline{\mathbf{r}} \cdot \widehat{\underline{\mathbf{p}}}^{\perp}]$ and its 2-D Fourier transform: $F_3[\xi, \eta] = 1[\underline{\boldsymbol{\rho}} \cdot \widehat{\underline{\mathbf{p}}}] \exp[-2\pi i p_0(\underline{\boldsymbol{\rho}} \cdot \widehat{\underline{\mathbf{p}}})]\delta[\underline{\boldsymbol{\rho}} \cdot \widehat{\underline{\mathbf{p}}}^{\perp}]$. Note that the amplitudes of the real and imaginary parts of $F_3[\xi, \eta]$ are bipolar with zero mean.

The spectrum is a line delta function through the origin of the frequency domain that is oriented in the direction orthogonal to the azimuth of the space-domain line delta function and that is modulated by the linear phase factor due to the translation of the space-domain function. An example is shown in Figure 10.17.

Since the Fourier transform of a line delta function through the origin is an orthogonal line delta function, the two line delta components of the cross function defined in Equation (7.78) transform create another cross; it is another example of a "self-Fourier transform":

$$\begin{aligned}\mathcal{F}_2\{CROSS[x, y]\} &= \mathcal{F}_2\{\delta[x]\, 1[y] + 1[x]\, \delta[y]\} = \delta[\xi]\, 1[\eta] + 1[\xi]\, \delta[\eta]\\ &= CROSS[\xi, \eta]\end{aligned} \tag{10.44}$$

The rotation theorem shows that the rotated "cross" function also is a self-Fourier transform:

$$\begin{aligned}\mathcal{F}_2\{CROSS[\underline{\mathbf{r}} \bullet \hat{\underline{\mathbf{p}}}_0, \underline{\mathbf{r}} \bullet \hat{\underline{\mathbf{p}}}_0^{\perp}]\} &= \mathcal{F}_2\{(\delta[\underline{\mathbf{r}} \bullet \hat{\underline{\mathbf{p}}}_0]\, 1[\underline{\mathbf{r}} \bullet \hat{\underline{\mathbf{p}}}_0^{\perp}]) + (1[\underline{\mathbf{r}} \bullet \hat{\underline{\mathbf{p}}}_0]\, \delta[\underline{\mathbf{r}} \bullet \hat{\underline{\mathbf{p}}}_0^{\perp}])\}\\ &= (1[\underline{\boldsymbol{\rho}} \bullet \hat{\underline{\mathbf{p}}}_0]\, \delta[\underline{\boldsymbol{\rho}} \bullet \hat{\underline{\mathbf{p}}}_0^{\perp}]) + (\delta[\underline{\boldsymbol{\rho}} \bullet \hat{\underline{\mathbf{p}}}_0]\, 1[\underline{\boldsymbol{\rho}} \bullet \hat{\underline{\mathbf{p}}}_0^{\perp}])\\ &= CROSS[\underline{\boldsymbol{\rho}} \bullet \hat{\underline{\mathbf{p}}}_0, \underline{\boldsymbol{\rho}} \bullet \hat{\underline{\mathbf{p}}}_0^{\perp}]\end{aligned} \tag{10.45}$$

Note that this means that the spectrum of the sum of any number of rotated *CROSS* functions is the same set of rotated *CROSS* functions in the frequency domain. In other words, any superposition of *CROSS* functions has a spectrum with the same functional form.

Recall that we demonstrated in Section 7.3.6 that individual 2-D Dirac delta functions may be expressed as the product of orthogonal line delta functions. For example, the 2-D Dirac delta function located at the origin my be written in the form:

$$\delta[x,y] = (\delta[\underline{\mathbf{r}} \bullet \hat{\underline{\mathbf{p}}}_0]\, 1[\underline{\mathbf{r}} \bullet \hat{\underline{\mathbf{p}}}_0^{\perp}])(1[\underline{\mathbf{r}} \bullet \hat{\underline{\mathbf{p}}}_0]\, \delta[\underline{\mathbf{r}} \bullet \hat{\underline{\mathbf{p}}}_0^{\perp}]) = \delta[\underline{\mathbf{r}} \bullet \hat{\underline{\mathbf{p}}}_0]\, \delta[\underline{\mathbf{r}} \bullet \hat{\underline{\mathbf{p}}}_0^{\perp}] \tag{10.46}$$

The 2-D Dirac delta function located at the arbitrary Cartesian coordinates $[x_0, y_0]$ is $\delta[x - x_0, y - y_0]$, which also may be written in polar coordinates as (p_0, θ_0), where $p_0 = (x_0^2 + y_0^2)^{1/2}$ and $\theta_0 = \tan^{-1}[y_0/x_0]$. The expression as the product of orthogonal line delta functions is:

$$(1[\underline{\mathbf{r}} \bullet \hat{\underline{\mathbf{p}}}_0]\, \delta[\underline{\mathbf{r}} \bullet \hat{\underline{\mathbf{p}}}_0^{\perp}])(\delta[p_0 - \underline{\mathbf{r}} \bullet \hat{\underline{\mathbf{p}}}_0]\, 1[\underline{\mathbf{r}} \bullet \hat{\underline{\mathbf{p}}}_0^{\perp}]) = \delta[p_0 - \underline{\mathbf{r}} \bullet \hat{\underline{\mathbf{p}}}_0]\, \delta[\underline{\mathbf{r}} \bullet \hat{\underline{\mathbf{p}}}_0^{\perp}] \tag{10.47}$$

The spectra of the "centered" and "off-axis" Dirac delta functions are easily evaluated by applying the "rule" in Equation (10.37):

$$\mathcal{F}_2\{\delta[\underline{\mathbf{r}} \bullet \hat{\underline{\mathbf{p}}}_0]\, \delta[\underline{\mathbf{r}} \bullet \hat{\underline{\mathbf{p}}}_0^{\perp}]\} = 1[\underline{\boldsymbol{\rho}} \bullet \hat{\underline{\mathbf{p}}}_0]\, 1[\underline{\boldsymbol{\rho}} \bullet \hat{\underline{\mathbf{p}}}_0^{\perp}] \tag{10.48}$$

$$\begin{aligned}
\mathcal{F}_2\{\delta[p_0 - \underline{\mathbf{r}} \bullet \hat{\underline{\mathbf{p}}}_0]\, \delta[\underline{\mathbf{r}} \bullet \hat{\underline{\mathbf{p}}}_0^{\perp}]\} &= (1[\underline{\boldsymbol{\rho}} \bullet \hat{\underline{\mathbf{p}}}]\, e^{-2\pi i p_0 (\underline{\boldsymbol{\rho}} \bullet \hat{\underline{\mathbf{p}}}_0)})\, 1[\underline{\boldsymbol{\rho}} \bullet \hat{\underline{\mathbf{p}}}_0^{\perp}] \\
&= e^{-2\pi i p_0 (\underline{\boldsymbol{\rho}} \bullet \hat{\underline{\mathbf{p}}}_0)} = e^{-2\pi i p_0 (\xi \cos[\theta_0] + \eta \sin[\theta_0])} \\
&= e^{-2\pi i (\xi p_0 \cos[\theta_0] + \eta p_0 \sin[\theta_0])}
\end{aligned} \tag{10.49}$$

The theorems just derived may be combined with the expression for the Fourier transform of separable functions to evaluate some Fourier transforms that will be useful later.

PROBLEMS

10.1 The function $f_0[x,y] = \delta[x - x_0]\, \delta[y]$ may be rotated by θ radians by substituting $\underline{\mathbf{r}} \bullet \hat{\underline{\mathbf{p}}}_1$ for x and $\underline{\mathbf{r}} \bullet \hat{\underline{\mathbf{p}}}_1^{\perp}$ for y to obtain:

$$f_1[x,y] = \delta[\underline{\mathbf{r}} \bullet \hat{\underline{\mathbf{p}}}_1 - x_0]\, \delta[\underline{\mathbf{r}} \bullet \hat{\underline{\mathbf{p}}}_1^{\perp}]$$

Show that this expression is equivalent to:

$$f_1[x,y] = \delta[x - x_0 \cos[\theta], \quad y - x_0 \sin[\theta]]$$

10.2 Consider the 2-D separable function and evaluate the functions listed:

$$f[x,y] = SINC[x-1]\, SINC^2\left[\frac{y}{2}\right]$$

(a) $F[\xi, \eta]$
(b) $f[x,y] * (1[x] \times \delta[y])$
(c) $f[x,y] * (\delta[x] \times 1[y])$

10.3 A 2-D "Fourier self-reciprocal transform" is a function $f[x,y]$ that satisfies the criterion:

$$\mathcal{F}_2\{f[x,y]\} = f[\xi, \eta]$$

where $\mathcal{F}_2\{\ \}$ is the 2-D Fourier transform operator. Describe how to construct a Fourier self-reciprocal transform from an arbitrary real-valued 2-D function $g[x,y]$ (i.e., $g[x,y]$ may exhibit even, odd, or no symmetry).

10.4 Find the Fourier transforms of the following 2-D separable functions and sketch them as profiles or as "images":

(a) $TRI[x]\, \delta[y]$
(b) $TRI[x]\, 1[y]$

(c) $TRI[x]\, RECT[y]$

(d) $TRI[x, y] \cdot \cos[2\pi x]\, 1[y]$

10.5 Find the Fourier transforms of the following 2-D separable functions and sketch them as profiles or as "images":

(a) $TRI[x, y] * (\frac{1}{2}\delta\delta[x/2]\, \delta[y])$

(b) $TRI[x, y] * (\frac{1}{2}\delta[x]\, \delta[y])$

(c) $TRI[x, y] * STEP[x]\, 1[y]$

10.6 Find the Fourier transforms of the following 2-D separable functions and sketch them as profiles or as "images":

(a) $RECT[x, y] * \delta[x]\, 1[y]$

(b) $RECT[x, y] * \delta[x-1]\, 1[y-1]$

(c) $RECT[x, y] * CROSS[x, y]$

(d) $COR[x, y] * COR[x, y]$

11

Spectra of Circular Functions

11.1 The Hankel Transform

One of the stated goals of this book is to develop a mathematical description of the properties of imaging systems. For this reason, we will take special note of the Fourier transforms of circularly symmetric 2-D functions because of their importance in describing optical systems composed of elements with circular symmetry. Images produced by these systems are not affected by rotations of the optical elements about the axis of symmetry. This observation indicates that the psf and OTF are circularly symmetric if the optical elements are.

We will see in a subsequent chapter that optical imaging systems are sufficiently linear and shift invariant to make useful predictions about the "quality" of the image from the optical layout. Briefly put, the spatial effect of the optical system may be derived in the space domain as the "point-spread function" (psf) or in the frequency domain as the "optical transfer function" (OTF). This observation confirms Equation (10.37), which showed that the only effect of rotation of the Fourier spectrum is an identical rotation of the function. This section is devoted to methods for determining 2-D spectra of circularly symmetric 2-D functions. The results will be used later to model optical systems and their performance. A mathematical tool for computing the Fourier spectra of circularly symmetric functions is the *Hankel transform*, also called the *Fourier–Bessel transform*.

A separable form for circularly symmetric functions was specified in Equation (7.84):

$$f[x,y] = f_r[|\underline{\mathbf{r}}|]1(\theta) = f_r\left(\sqrt{x^2+y^2}\right)1(\theta) \tag{11.1}$$

where the subscript r is used to distinguish the radial amplitude. This is substituted into the expression for the 2-D Fourier transform in Equation (10.1):

$$\mathcal{F}_2\{f[x,y]\} = \iint_{-\infty}^{+\infty} f_r\left(\sqrt{x^2+y^2}\right)1(\theta)\, e^{-2\pi i(\xi x+\eta y)}\,dx\,dy \tag{11.2}$$

where the 2-D Fourier "mask" function $e^{-2\pi i(\xi x+\eta y)}$ is not circularly symmetric, but may be recast into polar coordinates by recognizing that $\xi x + \eta y$ is the Cartesian representation of the dimensionless 2-D scalar product of two 2-D real-valued vectors, one each in the space and frequency domains:

$$\xi x + \eta y = \begin{bmatrix} x \\ y \end{bmatrix} \bullet \begin{bmatrix} \xi \\ \eta \end{bmatrix} = \underline{\mathbf{r}} \bullet \underline{\rho} \tag{11.3}$$

Fourier Methods in Imaging Roger L. Easton, Jr.

The space-domain vector $\underline{\mathbf{r}}$ may be recast into polar coordinates via the transformations that evaluate the radial and azimuthal coordinates r and θ in terms of the Cartesian coordinates x and y:

$$\underline{\mathbf{r}} = (|\underline{\mathbf{r}}|, \theta) = (r, \theta), \quad where \ |\underline{\mathbf{r}}| \equiv r = \sqrt{x^2 + y^2} \tag{11.4a}$$

$$\theta = \tan^{-1}\left[\frac{y}{x}\right] \tag{11.4b}$$

$$x = r\cos[\theta] \tag{11.4c}$$

$$y = r\sin[\theta] \tag{11.4d}$$

These same relations may be used to represent the spatial frequency vector $\underline{\rho}$ in terms of its magnitude $|\underline{\rho}| \equiv \rho$ and azimuth (phase) angle ϕ:

$$\underline{\rho} = [\xi, \eta] = (\rho, \phi), \quad where \ |\underline{\rho}| \equiv \rho = \sqrt{\xi^2 + \eta^2} \tag{11.5a}$$

$$\phi = \tan^{-1}\left[\frac{\eta}{\xi}\right] \tag{11.5b}$$

$$\xi = \rho\cos[\phi] \tag{11.5c}$$

$$\eta = \rho\sin[\phi] \tag{11.5d}$$

The last pair of relations in each group may be inserted into Equation (11.2) to generate an expression for the 2-D Fourier "mask" in polar coordinates:

$$\begin{aligned} e^{-2\pi i(\xi x + \eta y)} &= \exp[-2\pi i(r\cos[\theta]\cdot\rho\cos[\phi] + r\sin[\theta]\cdot\rho\sin[\phi])] \\ &= \exp[-2\pi i(r\rho(\cos[\theta]\cos[\phi] + \sin[\theta]\sin[\phi]))] \\ &= \exp[-2\pi i(r\rho\cos[\theta - \phi])] = \exp[-2\pi i\underline{\mathbf{r}} \bullet \underline{\rho}] = e^{-2\pi i\underline{\mathbf{r}}\bullet\underline{\rho}} \end{aligned} \tag{11.6}$$

where the trigonometric identity in Equation (2.21) has been used.

To solve the Fourier integral in polar coordinates, we also must convert the differential element of area and the limits of the integrals to appropriate polar representations. Differential geometry demonstrates that the new area element is $dx\,dy = r\,dr\,d\theta$. The domain of integration of the radial variable in polar coordinates is $0 \le r < +\infty$, while the azimuthal variable must cover a full circle of 2π radians. We choose the "symmetric" interval $[-\pi, +\pi)$ for the angles ϕ and θ, though the interval $[0, 2\pi)$ is used by many authors.

Substitution of these results yields the polar form of the 2-D Fourier transform:

$$\mathcal{F}_2\{f[|\underline{\mathbf{r}}|]\} = \int_{-\pi}^{+\pi}\left(\int_0^{+\infty} f[x, y]\, e^{-2\pi i(r\rho\cos[\theta-\phi])}\, r\,dr\right) d\theta \tag{11.7}$$

which is valid for any 2-D function $f[x, y]$ but reasonable to evaluate only if the function exhibits circular symmetry. In these cases, the azimuthal integral may be evaluated immediately by noting that the azimuthal angles θ and ϕ appear in the integrand only as a linear combination within the complex exponential. Therefore, the order of integration may be usefully swapped to evaluate the azimuthal integral first:

$$\mathcal{F}_2\{f[x, y]\} = \int_0^{+\infty} f_r\left(\sqrt{x^2 + y^2}\right)\left(\int_{-\pi}^{+\pi} e^{-2\pi i(r\rho\cos[\theta-\phi])}\, d\theta\right) r\,dr \tag{11.8}$$

By changing the variable of integration to $\psi \equiv \theta - \phi$ and using the fact that the cosine is periodic over 2π radians, the azimuthal integral may be rewritten as a complex-valued linear combination of two real-valued 1-D integrals:

$$\begin{aligned}\int_{-\pi}^{+\pi} e^{-2\pi i(r\rho\cos[\theta-\phi])}\,d\theta &= \int_{-\pi+\phi}^{+\pi+\phi} e^{-2\pi i(r\rho\cos[\psi])}\,d\psi \\ &= \int_{-\pi}^{+\pi} e^{-2\pi i(r\rho\cos[\psi])}\,d\psi \\ &= \int_{-\pi}^{+\pi} \cos[2\pi r\rho\cos[\psi]]\,d\psi - i\int_{-\pi}^{+\pi} \sin[2\pi r\rho\cos[\psi]]\,d\psi \qquad (11.9)\end{aligned}$$

where the periodicity of $\cos[\psi]$ over 2π radians has been used to remove the additive factor of ϕ from the limits of the integral. In words, the initial phase in the argument of the cosine has no effect on the area of a unit-magnitude complex exponential, even though the phase angle is a nonlinear function of the integration variable ψ. The real and imaginary parts of the integrand are respectively the cosine and the sine of an angle that is itself a cosine function that oscillates from $-2\pi r\rho$ at $\psi = -\pi$ through $+2\pi r\rho$ at $\psi = 0$ and back to $-2\pi r\rho$ at $\psi = +\pi$. The integral also clearly depends upon the particular radial spatial frequency ρ through the factor of $2\pi r\rho$ within the argument of the cosine.

Because the phase angles of the sinusoidal integrands in Equation (11.9) are nonlinear functions of ψ, this integral is more difficult to visualize and evaluate than if the phase had been linear. Fortunately, the power-series expansions for the sine and cosine come to our rescue and allow the expression to be simplified. To demonstrate, first consider the imaginary part of the integral. By substituting the Taylor series for $\sin[\theta]$ from Equation (4.18b), we obtain the summation of an infinite number of integrals:

$$\begin{aligned}&\int_{-\pi}^{+\pi} \sin[2\pi r\rho\cos[\psi]]\,d\psi \\ &= \int_{-\pi}^{+\pi}\left(2\pi r\rho\cos[\psi] - \frac{(2\pi r\rho\cos[\psi])^3}{3!} + \frac{(2\pi r\rho\cos[\psi])^5}{5!} - \cdots\right)d\psi \\ &= \frac{2\pi r\rho}{1!}\left(\int_{-\pi}^{+\pi}\cos[\psi]\,d\psi\right) - \frac{(2\pi r\rho)^3}{3!}\left(\int_{-\pi}^{+\pi}\cos^3[\psi]\,d\psi\right) + \cdots \\ &= \sum_{n=0}^{+\infty}(-1)^n\frac{(2\pi r\rho)^{2n+1}}{(2n+1)!}\left(\int_{-\pi}^{+\pi}\cos^{2n+1}[\psi]\,d\psi\right) \qquad (11.10)\end{aligned}$$

The integrands are odd powers of $\cos[\psi]$ and therefore have bipolar amplitudes over the domain of integration. When integrated over a single cycle of $\cos[\psi]$, each integral term in the sum evaluates to zero. The validity of this observation is demonstrated for several powers of $\cos[\psi]$ in Figure 11.1a. From these examples, we conclude that the imaginary part of the azimuthal integral in Equation (11.9) is zero for all values of ρ. This property could have been surmised earlier because we know that real-valued 2-D circularly symmetric functions are symmetric and therefore must have real-valued symmetric spectra, so the imaginary part must vanish.

Evaluation of the real part of the integral in Equation (11.9) is not quite so easy. We begin by inserting the series expansion for the cosine from Equation (4.18a):

$$\begin{aligned}&\int_{-\pi}^{+\pi}\cos[2\pi r\rho\cos[\psi]]\,d\psi = \sum_{\ell=0}^{+\infty}(-1)^\ell\frac{(2\pi r\rho)^{2\ell}}{2\ell!}\int_{-\pi}^{+\pi}(\cos[\psi])^{2\ell}\,d\psi \\ &= \left(\int_{-\pi}^{+\pi}d\psi\right) - \frac{(2\pi r\rho)^2}{2!}\left(\int_{-\pi}^{+\pi}\cos^2[\psi]\,d\psi\right) + \frac{(2\pi r\rho)^4}{4!}\left(\int_{-\pi}^{+\pi}\cos^4[\psi]\,d\psi\right) - \cdots \\ &\qquad (11.11)\end{aligned}$$

The individual integrands in the series are proportional to even powers of the cosine and therefore are symmetric functions of ψ. Their amplitudes are nonnegative at all azimuth angles ψ, and thus

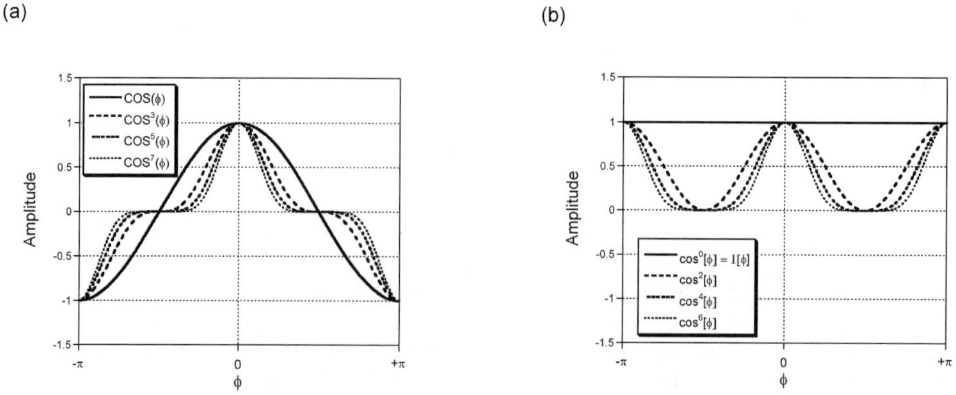

Figure 11.1 Graphs of $\cos^n[\phi]$ evaluated on the interval $[-\pi, +\pi]$, which are used in the series solution for the Bessel function $J_0[x]$. (a) Odd powers $n = 1, 3, 5, 7$, demonstrating that the area of each is zero. (b) Even powers $n = 0, 2, 4, 6$, demonstrating that the areas are positive and decrease with increasing order n.

the individual terms in the series evaluate to nonnegative real numbers. The "peaks" of all integrands are located at $\psi = 0, \mp\pi$, but those for larger values of ℓ become narrower, which ensures that the contribution of higher-order terms is smaller (Figure 11.1b). The individual integrals may be evaluated either by repeated integration by parts or by representation in terms of the gamma function. For the latter, the derivation in the appendix to this chapter demonstrates several equivalent forms for the individual integrals:

$$\begin{aligned}\int_{-\pi}^{+\pi} \cos^{2\ell}[\psi]\,d\psi &= \frac{2\Gamma[\frac{1}{2}]\Gamma[\ell+\frac{1}{2}]}{\Gamma[\ell+1]}\\ &= \frac{2\Gamma[\frac{1}{2}]\Gamma[\ell+\frac{1}{2}]}{\ell!}\\ &= 2\pi\,\frac{(2\ell-1)!!}{(2\ell)!!}\end{aligned} \tag{11.12}$$

where $(2\ell)!!$ and $(2\ell-1)!!$ are the common shorthand notations for the "even" and "odd" factorial products:

$$(2\ell)!! \equiv 2\ell\cdot(2\ell-2)\cdot(2\ell-4)\cdot\cdots\cdot 4\cdot 2 \tag{11.13a}$$

$$(2\ell-1)!! \equiv (2\ell-1)\cdot(2\ell-3)\cdot(2\ell-5)\cdot\cdots\cdot 3\cdot 1 \tag{11.13b}$$

Equation (11.12) is evaluated for the different orders and combined with Equation (11.11) to produce a series solution for Equation (11.8):

$$\begin{aligned}&\int_{-\pi}^{+\pi} e^{-2\pi i(r\rho\cos[\psi])}\,d\psi\\ &\quad= 2\pi\left(1 - \frac{(2\pi r\rho)^2}{4} + \frac{(2\pi r\rho)^4}{64} - \frac{(2\pi r\rho)^6}{2304} + \frac{(2\pi r\rho)^8}{147\,456} - \cdots\right)\end{aligned} \tag{11.14}$$

Readers with good memories may recognize the progression of numerical coefficients in the parentheses; this is the power series for $J_0[x]$ from Equation (A6.7), which means that the integral in Equation

(11.14) is:

$$\int_{-\pi}^{+\pi} e^{-2\pi i r\rho \cos[\psi]}\, d\psi = 2\pi\, J_0[2\pi r\rho] \tag{11.15}$$

In fact, the integral expression is often used to define the 1-D Bessel function $J_0[x]$.

Equation (11.15) may be substituted into Equation (11.7) to see that the 2-D Fourier transform of the circularly symmetric function $f_r(r)$ simplifies to a 1-D integral transformation of the radial profile:

$$\int_0^{+\infty} f_r(r)(2\pi r\, J_0[2\pi r\rho])\, dr \equiv F_r(\rho) \tag{11.16}$$

which defines the zero-order Hankel transform $\mathcal{H}_0$ of the 2-D circularly symmetric function $f_r(r)$. The form of the "mask function" $2\pi r\, J_0[2\pi r\rho]$ leads to the other common name for Equation (11.16) as the Fourier–Bessel transform.

In passing, we note that Equation (11.16) may be generalized by replacing $J_0[2\pi r\rho]$ with the Bessel function of the first kind with a different integer order ν to define the Hankel transform of order ν:

$$\mathcal{H}_\nu\{f_r(r)\} \equiv 2\pi \int_0^{+\infty} f_r(r)(r\, J_\nu[2\pi r\rho])\, dr \tag{11.17}$$

Being an integral over fixed limits, the Hankel transform of any order satisfies the linearity condition in Equation (2.1). Because the general Hankel transform of order ν is not applicable to most imaging problems, our discussion will be restricted to the zero-order operation in Equation (11.16). Readers who seek more details about the general Hankel transform should consult references such as Gaskill (1978) or Arfken (2000).

In words, the zero-order Hankel transform evaluates the area of the 1-D radial profile of $f_r(r)$ over the interval $0 \le r < +\infty$ after modulation by the 1-D weighting (or "mask") function $2\pi r\, J_0[2\pi r\rho]$. The 1-D radial spectrum generated by the transform is identical to the radial profile of the 2-D Fourier transform of the circularly symmetric function in the interval $0 \le \rho < +\infty$, as illustrated in Figure 11.2. The 2-D Fourier transform pair of a circularly symmetric function may be written as equivalent expressions:

$$\mathcal{F}_2\{f_r(r)1(\theta)\} = \mathcal{H}_0\{f_r(r)\}1(\theta) \equiv F_r(\rho)1(\theta) \tag{11.18}$$

The weighting function $2\pi r \cdot J_0[2\pi r\rho]$ is real valued and its amplitude at the origin is zero. Recall from the discussion of Chapter 6 that $J_0[x]$ vaguely resembles a *SINC* function except that its zeros are different and nonuniformly spaced zeros and its amplitude decays more slowly with increasing $|x|$. We showed that the amplitudes of the extrema of $J_0[x]$ decrease in proportion to $x^{-1/2}$ for large x, while those of *SINC*$[x]$ fall off as x^{-1}. The scaling of $J_0[r]$ by r in Equation (11.16) ensures that the amplitudes of the extrema of the mask function *increase* with increasing r as $r^{+1/2}$ (Figure 11.3). Compare this behavior to the complex-valued linear-phase exponential weight function for the Fourier transform: the real and imaginary parts of the mask function are bounded at ± 1 so that the maximum magnitude does not vary with distance from the origin. The variation in amplitude of the Hankel weighting function with radius r is one reason why its computation is less straightforward than the Fourier transform. We will show that the periodicity of the Fourier mask allows the spectrum of a discrete function to be computed "recursively", as will be considered in Chapter 12. Because the Hankel mask is only "pseudoperiodic", a recursive computational algorithm is not obvious.

As a final note, the real-valued character of the Hankel "mask" ensures that the character of the spectrum (real, imaginary, or complex valued) is the same as that of the input function.

11.1.1 Hankel Transform of Dirac Delta Function

The Dirac delta function located at the origin and expressed in polar coordinates was shown in Equation (7.55) to have the form:

$$f_r(|\underline{\mathbf{r}}|) = f_r(r) = \frac{\delta(r)}{\pi r} \tag{11.19}$$

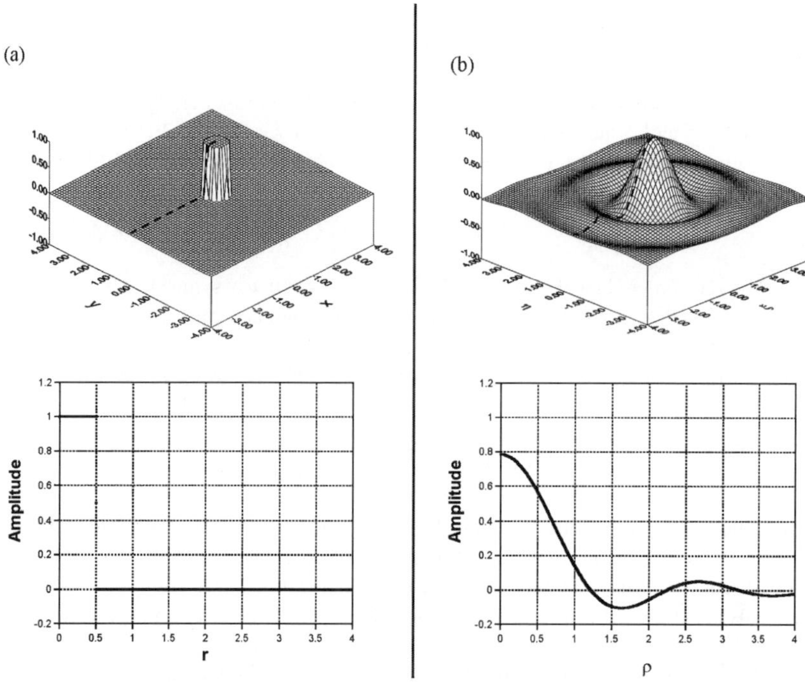

Figure 11.2 Comparison of the 2-D Fourier transform and the 1-D Hankel transform of the profile of $CYL(r)$: (a) $f(r) = CYL(r)$ and its radial profile for $r \leq 4$; (b) $F_r(\rho) = (\pi/4)\ SOMB(\rho)$ and its profile for $\rho \leq 4$.

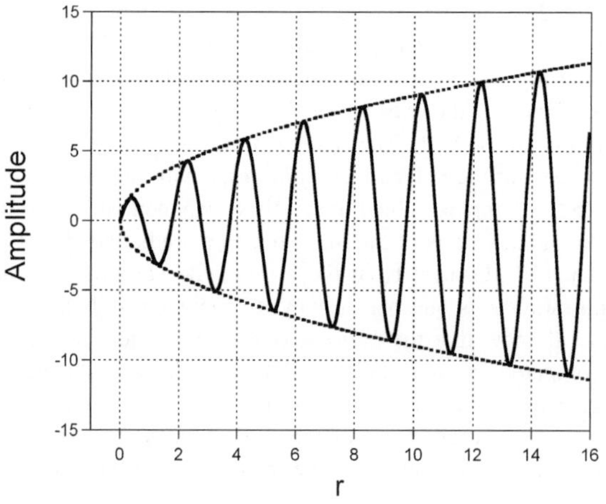

Figure 11.3 Amplitude of the kernel $2\pi r \cdot J_0[2\pi r]$ of the Hankel transform increases in proportion to $\pm\sqrt{r}$ (dashed lines).

This "one-sided" 1-D Dirac delta function is defined only for $r \geq 0$ and the normalization factor of $(\pi r)^{-1}$ ensures that the volume of the 2-D circularly symmetric function is unity. By substituting this expression into Equation (11.15) and using the fact that the 1-D profiles of a circularly symmetric function are symmetric, the spectrum of the 2-D circularly symmetric Dirac delta function is easily evaluated:

$$\begin{aligned}
\mathcal{H}_0\left\{\frac{\delta(r)}{\pi r}\right\} &= 2\pi \int_0^{+\infty} \frac{\delta(r)}{\pi r}(r J_0[2\pi r\rho])\, dr \\
&= 2\int_0^{+\infty} \delta(r) J_0[2\pi r\rho]\, dr \\
&= 2 \cdot \frac{1}{2}\int_{-\infty}^{+\infty} \delta(u) J_0[2\pi u\rho]\, du = J_0[0] = 1(\rho)
\end{aligned} \tag{11.20}$$

where the symmetry of the two functions and the sifting property of the 1-D Dirac delta function in Equation (6.113) were used in the last step. This result was anticipated because the circularly symmetric Dirac delta function in Equation (11.19) is the polar representation of the separable Dirac delta function in Equation (10.12). If the expression for the Hankel transform is valid, it must generate the polar form of the unit-constant spectrum. The equivalent expressions are:

$$\begin{aligned}
\mathcal{H}_0\left\{\frac{\delta(r)}{\pi r}\right\} &= \mathcal{F}_2\left\{\frac{\delta(r)}{\pi r}\, 1[\theta]\right\} = \mathcal{F}_2\{\delta[x, y]\} \\
&= 1[\xi, \eta] = 1(\rho) 1[\theta]
\end{aligned} \tag{11.21}$$

In fact, a derivation of the polar form of the Dirac delta function based on the Hankel transform may be substituted for the more complicated treatment that led to Equation (7.55).

11.2 Inverse Hankel Transform

It is useful at this point to summarize some observations about the 2-D Fourier transform and relate them to the Hankel transform:

1. The 2-D Fourier transform is a 2-D integral over x and y and generates a 2-D function of the spatial frequency coordinates $[\xi, \eta]$.
2. The Hankel transform is a 1-D integral over the radial distance r that generates a 1-D function of the radial spatial frequency ρ.
3. The 2-D inverse Fourier transform of Equation (10.5) is a 2-D integral over spatial frequency $[\xi, \eta]$ that yields a 2-D function of $[x, y]$.

It is logical to assume a fourth statement based on these three: that the inverse Hankel transform is a 1-D integral over radial spatial frequency ρ that constructs a 1-D function of the radial distance r. We now will surmise the form of the inverse Hankel transform by combining the result of Equation (11.20) with known properties of the 2-D Fourier transform. We know from Equation (10.12) that the spectrum of the 2-D Dirac delta function $\delta[x, y]$ centered at the origin is the unit constant:

$$\mathcal{F}_2\{\delta[x, y]\} = 1[\xi, \eta] \tag{11.22}$$

We also know from Equation (11.20) that the corresponding statement for the Hankel transform is:

$$\mathcal{H}_0\left\{\frac{\delta(r)}{\pi r}\right\} = 1(\rho) \tag{11.23}$$

The 2-D "transform-of-a-transform" theorem in Equation (10.25) demonstrates that the spectrum of the 2-D unit constant is a 2-D Dirac delta function located at the origin of coordinates:

$$\mathcal{F}_2\{1[x, y]\} = \delta[\xi, \eta] \Longrightarrow \mathcal{F}_2^{-1}\{\delta[\xi, \eta]\} = 1[-x, -y] = 1[x, y] \tag{11.24}$$

where the symmetry of the 2-D unit constant has been used. This indicates that the corresponding requirement on the Hankel transform must be:

$$\mathcal{H}_0\{1(r)\} = \frac{\delta(\rho)}{\pi\rho} \Longrightarrow \mathcal{H}_0^{-1}\left\{\frac{\delta(\rho)}{\pi\rho}\right\} = 1(r) \tag{11.25}$$

Since both the forward and inverse Hankel transforms must yield the same functional form when applied to radial Dirac delta functions in the appropriate domain, the actions of the "mask" functions in the two operations also must be identical. Equation (11.16) already demonstrated that the "mask" function that yields the unit constant when applied to the radial Dirac delta function is $r J_0[2\pi r\rho]$. Therefore, the inverse Hankel transform of a general circularly symmetric spectrum $F(\rho)$ differs from the forward Hankel transform operator only in the variable of integration:

$$\mathcal{H}_0^{-1}\{F_r(\rho)\} \equiv 2\pi \int_0^{+\infty} F_r(\rho)(\rho\, J_0[2\pi r\rho])\, d\rho \tag{11.26}$$

Note that the forward and inverse Hankel transforms are even more similar than the forward and inverse Fourier transforms, whose "mask functions" differ in the algebraic sign of the imaginary part and in the variable of integration. Because the weighting masks of the forward and inverse Hankel transforms are both real valued, the *character* (real, imaginary, or complex) of the spectrum is identical to that of the input function. Again, this should be no surprise, as the Hankel transform is valid only for 2-D functions that are symmetric with respect to the origin of coordinates.

11.3 Theorems of Hankel Transforms

The theorems of the 2-D Fourier transform may be applied to the Hankel transform after appropriate adjustments.

11.3.1 Scaling Theorem

The scaling theorem of the 2-D spectrum was specified in Equation (10.26). A circularly symmetric function must be scaled by identical factors along the x- and y-axes to maintain its symmetry. Given that $\mathcal{H}_0\{f(r)\} = F(\rho)$, the analogue of the 2-D scaling theorem for the Hankel transform must be:

$$\mathcal{H}_0\left\{f_r\left(\frac{r}{d}\right)\right\} = d^2 F_r(d\rho) \tag{11.27}$$

11.3.2 Shift Theorem

Because a circularly symmetric function may not be translated without destroying its symmetry, there is no analogue of the shift theorem for the Hankel transform.

11.3.3 Central-Ordinate Theorem

The central-ordinate theorem of the Hankel transform may be obtained directly from the operator, and gives the expected result that the Hankel transform evaluated at zero frequency is the volume of the circularly symmetric function:

$$\left(\int_{-\pi}^{+\pi} d\theta\right)\int_0^{+\infty} f_r(r) r\,dr = 2\pi \int_0^{+\infty} f_r(r) r\,dr$$
$$= \mathcal{H}_0\{f_r(r)\}|_{\rho=0} = F_r(\rho = 0) \tag{11.28}$$

11.3.4 Filter and Crosscorrelation Theorems

Because a function that is circularly symmetric must be symmetric with respect to the origin, the Hankel transform analogies to the filter theorem and the spectrum of the real-valued correlation of two functions are identical. The result is determined by applying the transform of the reversed function from Equation (9.119) and of the complex conjugate in Equation (9.146):

$$\begin{aligned}\mathcal{H}_0\{f_r(r) * h_r(r)\} &= \mathcal{H}_0\{f_r(r) \bigstar h_r^*(-r)\}\\ &= \mathcal{H}_0\{f_r(r) \bigstar h_r^*(+r)\}\\ &= F_r(\rho) \cdot H_r(\rho)\end{aligned} \tag{11.29}$$

where the symmetry of $h_r(r)$ has been used. The analogue of the 2-D crosscorrelation theorem in Equation (9.151) for the Hankel transform is:

$$\mathcal{H}_0\{f_r(r) \bigstar h_r(r)\} = F_r(\rho) \cdot H_r^*(\rho) \tag{11.30}$$

while the Hankel transform of the autocorrelation of a circularly symmetric function is:

$$\mathcal{H}_0\{f_r(r) \bigstar f_r(r)\} = F_r(\rho) \cdot F_r^*(\rho) = |F_r(\rho)|^2 \tag{11.31}$$

This is the circularly symmetric analogue of the Wiener–Khintchin theorem in Equation (9.159).

11.3.5 "Transform-of-a-Transform" Theorem

We used the 2-D Fourier "transform-of-a-transform" theorem to derive the form of the inverse Hankel transform. The corresponding theorem for the Hankel transform is:

$$\mathcal{H}_0\{f_r(r)\} = F_r(\rho) \Longrightarrow \mathcal{H}_0\{F_r(r)\} = f_r(-\rho) = f_r(+\rho) \tag{11.32}$$

where the symmetry of circular functions was used in the last step. This result could be inferred from the fact that the "mask" functions of the forward and inverse Hankel transforms have the same functional form.

11.3.6 Derivative Theorem

The symmetry of the 2-D functions ensures that all odd-order partial derivatives evaluate to zero. Therefore cascaded spatial derivatives of the form in Equation (10.29) evaluate to zero for circularly symmetric functions if either n or m is odd. When the orders of both derivatives are even, the corresponding Hankel transform is:

$$\mathcal{H}_0\left\{\left(\frac{\partial^n}{\partial x^n}\frac{\partial^m}{\partial y^m}\right) f_r(r)\right\} = \begin{cases}[(+2\pi i)^{n+m}(\xi^n \eta^m)] \cdot F(\rho) & \textit{for } n \textit{ and } m \textit{ even}\\ 0[\xi, \eta] & \textit{for } n \textit{ or } m \textit{ even}\end{cases} \tag{11.33}$$

11.3.7 Laplacian of Circularly Symmetric Function

Obviously, circularly symmetric functions may be expressed in polar coordinates. The expressions for differential operators often may be simplified by applying the principles of differentiation in curvilinear coordinates. For example, the Laplacian of a circularly symmetric function may be simplified to (Arfken, 2000):

$$\begin{aligned}\nabla^2\{f_r(r)\} &= \left(\frac{\partial^2}{\partial x^2}+\frac{\partial^2}{\partial y^2}\right)\{f_r(r)\} = \left(\frac{\partial^2}{\partial x^2}+\frac{\partial^2}{\partial y^2}\right)\left\{f_r\left(\sqrt{x^2+y^2}\right)\right\} \\ &= \frac{1}{r}\frac{d}{dr}\left[r\frac{df_r(r)}{dr}\right] = \frac{d^2f_r(r)}{dr^2}+\frac{1}{r}\frac{df_r(r)}{dr}\end{aligned} \tag{11.34}$$

The spectrum is obtained from Equation (11.33):

$$\begin{aligned}\mathcal{F}_2\{\nabla^2 f_r(r)\} &= \mathcal{F}_2\left\{\frac{1}{r}\frac{d}{dr}\left[r\frac{df_r(r)}{dr}\right]\right\} = \mathcal{H}_0\left\{\frac{1}{r}\frac{d}{dr}\left[r\frac{df_r(r)}{dr}\right]\right\} \\ &= -4\pi\rho^2 F_r(\rho)\end{aligned} \tag{11.35}$$

This result will be useful when considering mathematical descriptions of optical imaging systems in Chapter 21.

11.4 Hankel Transforms of Special Functions

11.4.1 Hankel Transform of $J_0(2\pi r\rho_0)$

The Hankel transform of a circularly symmetric zero-order Bessel function of the first kind is found easily by using Equation (11.26) to solve for the function $F_r(\rho)$ whose inverse Hankel transform is $J_0(2\pi r\rho)$:

$$J_0(2\pi r\rho_0) = \mathcal{H}_0^{-1}\{F_r(\rho)\} = 2\pi\int_0^{+\infty} F_r(\rho)(\rho\, J_0(2\pi r\rho))\,d\rho \tag{11.36}$$

Clearly, $F_r(\rho)$ must be proportional to the 1-D impulse $\delta(\rho-\rho_0)$, which will "sift" the factor of $J_0(2\pi r\rho_0)$ from the kernel of the Hankel transform. If we substitute $F_r(\rho)=\delta(\rho-\rho_0)$ and apply the sifting property, we obtain:

$$\begin{aligned}\mathcal{H}_0^{-1}\{\delta(\rho-\rho_0)\} &= \int_0^{+\infty}\delta(\rho-\rho_0)2\pi\rho J_0(2\pi r\rho)\,d\rho \\ &= 2\pi\rho_0 J_0(2\pi r\rho_0)\end{aligned} \tag{11.37}$$

The linearity of the Hankel transform allows both sides to be divided by the constant $2\pi\rho_0$ if $\rho_0\neq 0$:

$$\begin{aligned}\frac{1}{2\pi\rho_0}\mathcal{H}_0^{-1}\{\delta(\rho-\rho_0)\} = \mathcal{H}_0^{-1}&\left\{\frac{1}{2\pi\rho_0}\,\delta(\rho-\rho_0)\right\} = J_0(2\pi r\rho_0) \\ &\Longrightarrow \mathcal{H}_0\{J_0(2\pi r\rho_0)\} = \frac{1}{2\pi\rho_0}\,\delta(\rho-\rho_0)\end{aligned} \tag{11.38}$$

The 2-D circularly symmetric spectrum $F_r(\rho)=\delta(\rho-\rho_0)1(\theta)$ was shown in Chapter 9 to be a ring of 1-D Dirac delta functions centered at the origin with radius ρ_0. In words, the Hankel transform of the zero-order Bessel function is a ring delta function, as shown in Figure 11.4. Because the area of the 1-D radial part of the spectrum is finite and positive, so must be the volume of the 2-D function $J_0[2\pi r\rho_0]$. The central ordinate theorem for the Hankel transform of Equation (11.28) demonstrates

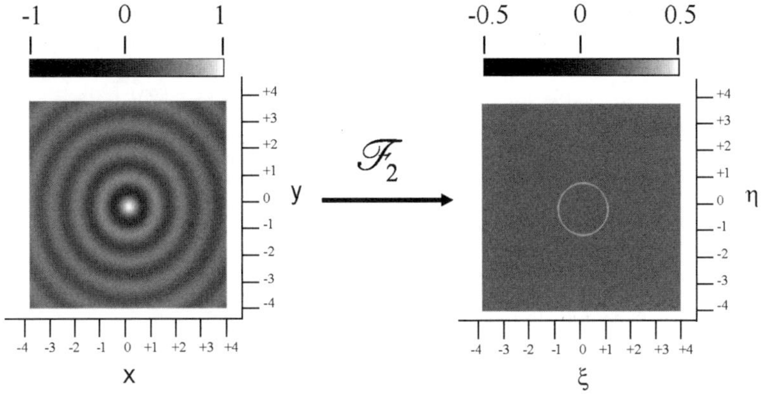

Figure 11.4 The 2-D Fourier transform of the circularly symmetric 2-D zero-order Bessel function of the first kind $J_0(2\pi r \cdot 1)$, which evaluates to the ring delta function: $\mathcal{F}_2\{J_0(2\pi r\rho_0)\} = (1/2\pi\rho_0)\delta(\rho - \rho_0)$.

that the volume of the 2-D circularly symmetric Bessel function $f[\underline{\mathbf{r}}] = J_0[2\pi r\rho_0]$ is zero:

$$\begin{aligned}\int_0^{+\infty}\int_{-\pi}^{+\pi} J_0[2\pi r\rho_0]r\,dr &= 2\pi\int_0^{+\infty} J_0(2\pi r\rho_0)r\,dr \\ &= \mathcal{H}_0\{J_0(2\pi r\rho_0)\}|_{\rho=0} \\ &= \frac{1}{2\pi\rho_0}\delta(0-\rho_0) = 0 \quad \textit{for } \rho_0 \neq 0\end{aligned} \tag{11.39}$$

except in the trivial case $\rho_0 = 0$, for which $f[\underline{\mathbf{r}}] = 1[\underline{\mathbf{r}}] = 1[x, y]$. The casual observer might generalize this result to assume that the area of the 1-D function $J_0[x]$ also is zero, but we will soon demonstrate that this is not so!

Equation (11.39) shows that the spectra of the 2-D Bessel functions $J_0(2\pi\rho_0 r)$ and $J_0(2\pi\rho_1 r)$ do not overlap if $\rho_0 \neq \rho_1$, which means that their product must evaluate to zero. This observation may be combined with the crosscorrelation theorem to show that the projection of the 2-D function $J_0(2\pi\rho_0 r)$ onto $J_0(2\pi\rho_1 r)$ must be zero if $\rho_0 \neq \rho_1$. In other words, distinct 2-D J_0 Bessel functions are orthogonal.

Equation (11.38) may be interpreted in another way by considering the properties of "radial profiles" of the spectrum constructed by "slicing" the ring delta function through the origin over the bipolar domain $(-\infty, +\infty)$. Mathematically, this is accomplished by multiplying by the appropriate line delta function. For example, the profile or "slice" of the 2-D spectrum evaluated along the ξ-axis is constructed by multiplying the ring delta function by $1[\xi]\delta[\eta]$ and recasting the expression into Cartesian coordinates:

$$\begin{aligned}\frac{1}{2\pi\rho_0}\delta(\rho-\rho_0)\cdot(1[\xi]\delta[\eta]) &= \frac{1}{2\pi\rho_0}\delta\left[\sqrt{\xi^2+\eta^2}-\rho_0\right]\delta[\eta] \\ &= \frac{1}{2\pi\rho_0}\delta\left[\sqrt{\xi^2+0^2}-\rho_0\right]\delta[\eta]\end{aligned} \tag{11.40}$$

where the sifting property Dirac delta function has been used. The 1-D Dirac delta function may be evaluated by applying the expression for a functional argument in Equation (6.137) with $g[\xi] = |\xi| - \rho_0$:

$$\begin{aligned}\frac{1}{2\pi\rho_0}\delta(|\xi| - \rho_0)\delta[\eta] &= \frac{1}{2\pi\rho_0}\left[\frac{\delta(-\xi - \rho_0)}{|-1|} + \frac{\delta(+\xi - \rho_0)}{|+1|}\right]\delta[\eta] \\ &= \frac{1}{2\pi\rho_0}[\delta(-\xi - \rho_0) + \delta(+\xi - \rho_0)]\delta[\eta] \\ &= \frac{1}{2\pi\rho_0}[\delta(\xi + \rho_0) + \delta(\xi - \rho_0)]\delta[\eta] \end{aligned} \tag{11.41}$$

We already know the space-domain representation of this 1-D "slice" of the 2-D spectrum; it is a 2-D cosine that oscillates along the x-axis:

$$\mathcal{F}_2^{-1}\left\{\frac{1}{2\pi\rho_0}(\delta[\xi + \rho_0] + \delta[\xi - \rho_0])\,\delta[\eta]\right\} = \frac{1}{\pi\rho_0}\cos[2\pi\rho_0 x]1[y] \tag{11.42}$$

The 1-D radial slice of the 2-D spectrum $\delta(\rho - \rho_0)$ along a different azimuth angle (say, θ_0) may be evaluated by multiplying by the appropriate 1-D Dirac delta function in azimuth to produce a pair of Dirac delta functions. This shows that the space-domain representation of each slice is a cosine that oscillates with spatial frequency $|\rho_0|$ along the azimuth direction θ_0. Because the Fourier transform is linear, the 2-D space-domain representation of $\delta(\rho - \rho_0)$ is the sum over azimuth angle θ of the complete set of "rotated" cosines with the same spatial frequency. Since the domain of the radial coordinate ρ is here assumed to be bipolar, the domain of the azimuthal integral includes only π radians. In short, the 2-D circularly symmetric Bessel function in the space domain is the azimuthal integral of the set of 2-D cosines with identical periods:

$$J_0[2\pi r\rho_0]1(\theta) = \frac{1}{\pi\rho_0}\int_{-\pi/2}^{+\pi/2}\cos[2\pi(\xi_0 x + \eta_0 y)]\,d\theta, \ \textit{where}\ \xi_0^2 + \eta_0^2 = \rho_0^2 \tag{11.43}$$

The limits of this integral may be changed by applying the symmetry of the cosine functions:

$$\begin{aligned}J_0[2\pi r\rho_0]1(\theta) &= \frac{1}{2}\cdot\frac{1}{\pi\rho_0}\int_{-\pi}^{+\pi}\cos[2\pi(\xi_0 x + \eta_0 y)]\,d\theta, \ \textit{where}\ \xi_0^2 + \eta_0^2 = \rho_0^2 \\ &= \frac{1}{2\pi\rho_0}\int_{-\pi}^{+\pi}\cos[2\pi(\underline{\rho}_0 \bullet \underline{\mathbf{r}})]\,d\theta \\ &= \frac{1}{2\pi\rho_0}\int_{-\pi}^{+\pi}\cos[2\pi|\underline{\rho}_0|\,|\underline{\mathbf{r}}|\cos[\theta]]\,d\theta \end{aligned} \tag{11.44}$$

In words, the 2-D J_0 Bessel function may be constructed by summing 2-D cosine functions with the same period over all azimuthal directions. A conceptual sketch of this result is shown in Figure 11.5. This result is intimately related to the "central-slice theorem" of the Radon transform that is considered in the next chapter.

11.4.2 Hankel Transform of *CYL*(*r*)

The cylinder function was defined in Equation (7.87):

$$CYL\left(\frac{r}{d}\right) = \begin{cases} 1 & \textit{if}\ r < d/2 \\ \frac{1}{2} & \textit{if}\ r = d/2 \\ 0 & \textit{if}\ r > d/2 \end{cases} \tag{11.45}$$

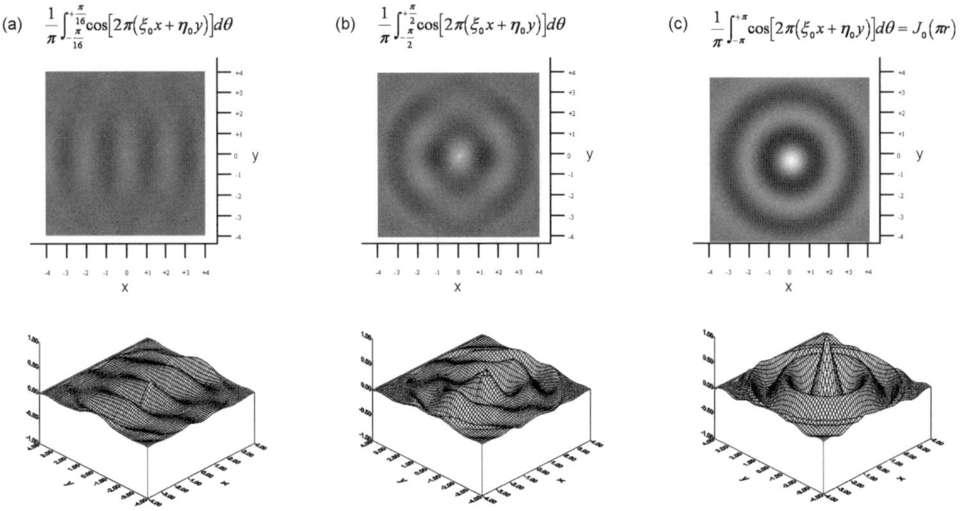

Figure 11.5 Generation of $J_0(\pi r)$ by summing rotated replicas of the 2-D sinusoid $(1/\pi)\cos[2\pi x/2]\cdot 1[y]$, shown as summations over different ranges of azimuth angle as images and as 3-D graphs: (a) summation over $-\pi/16 \le \theta < +\pi/16$ radians; (b) summation over $-\pi/2 \le \theta < +\pi/2$ radians; (c) summation over full domain of azimuths, $-\pi \le \theta < +\pi$ radians, which produces $J_0(\pi r)$.

and was interpreted to be a circularly symmetric "version" of the rectangle function. The Hankel transform of $CYL(r)$ is obtained by evaluating the integral:

$$\mathcal{H}_0\left\{CYL\left(\frac{r}{d}\right)\right\} = \int_0^{+\infty} CYL\left(\frac{r}{d}\right)(2\pi r\, J_0[2\pi r\rho])\, dr$$
$$= \int_0^{d/2} 2\pi r\, J_0[2\pi r\rho]\, dr \tag{11.46}$$

The integrand was graphed in Figure 11.3 for $\rho = 1$. The resulting areas are evaluated for different values of the diameter d in Figure 11.6; the amplitude increases from 0 at the origin and the amplitude maxima increase at a rate proportional to $\sqrt{d}$. After scaling by d^{-1}, the amplitude maxima of the integral decrease as $(\sqrt{d})^{-1}$, which is the same rate of decrease exhibited by all of the Bessel functions of the first kind. In fact, the integral is proportional to $J_1[2\pi r\rho]$, thus confirming an established identity for Bessel functions (Stephenson and Radmore, 1990):

$$\int_0^{x} u\, J_0[u]\, du = x\, J_1[x] \tag{11.47}$$

This result determines that the Hankel transform of the unit-diameter cylinder function is:

$$\mathcal{H}_0\left\{CYL\left(\frac{r}{1}\right)\right\} = \int_0^{\frac{1}{2}} 2\pi r\, J_0[2\pi r\rho]\, dr \Longrightarrow u \equiv 2\pi r\rho$$
$$= \frac{1}{2\pi\rho^2}\int_0^{\pi\rho} u\, J_0[u]\, du$$
$$= \frac{1}{2\pi\rho^2}(\pi\rho\ J_1[\pi\rho])$$

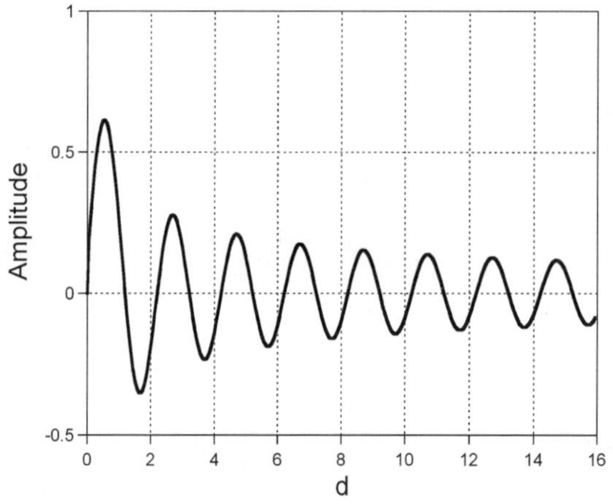

Figure 11.6 The function $\frac{1}{d}\int_0^{d/2} 2\pi r\, J_0[2\pi r]\, dr$ evaluated for different values of d. The resulting amplitude is proportional to $J_1[\pi d]$.

$$= \frac{1}{2\rho} J_1[\pi\rho]$$
$$= \frac{\pi}{4}\frac{2J_1[\pi\rho]}{\pi\rho} \equiv \frac{\pi}{4} SOMB(\rho) \tag{11.48}$$

where the definition from Equation (7.95) has been used. This pair of functions were used to illustrate the meaning of the Hankel transform in Figure 11.2. In words, the Hankel transform of the 1-D profile of the cylinder function is the 1-D profile of the sombrero function, which means that the 2-D Fourier transform of the cylinder function is a 2-D circularly symmetric function whose profile is a 1-D sombrero function. Recall from Chapter 7 that the magnitudes of the extrema of $SOMB(\rho)$ decrease as $\rho^{-3/2}$, which is a faster rate of "decay" with increasing ρ than the spectrum of either the ring delta function ($\max[|F(\rho)|] \propto \rho^{-1/2}$) or the profile of the spectrum of the rectangle along the ξ-axis ($\max[|F[\xi]|] \propto \xi^{-1}$). It also is clear from the "transform-of-a-transform" theorem in Equation (11.32) that the inverse Hankel transform of $CYL(\rho)$ must be the scaled Besinc function, which yields the transform of $SOMB(r)$:

$$\mathcal{H}_0^{-1}\{CYL(\rho)\} = \frac{\pi}{4} SOMB(r)$$
$$\Longrightarrow \mathcal{H}_0\{SOMB(r)\} = \frac{4}{\pi} CYL(\rho) \tag{11.49}$$

These spectra of cylinder functions with arbitrary diameters may be evaluated by applying the scaling property of the Hankel transform in Equation (11.27):

$$\mathcal{H}_0\left\{CYL\left(\frac{r}{d}\right)\right\} = \frac{\pi d^2}{4} SOMB(d\rho) \tag{11.50}$$

11.4.3 Hankel Transform of r^{-1}

The 2-D circularly symmetric function

$$f[x,y] = \frac{1}{\sqrt{x^2+y^2}} \Longrightarrow f_r(r) = r^{-1} \tag{11.51}$$

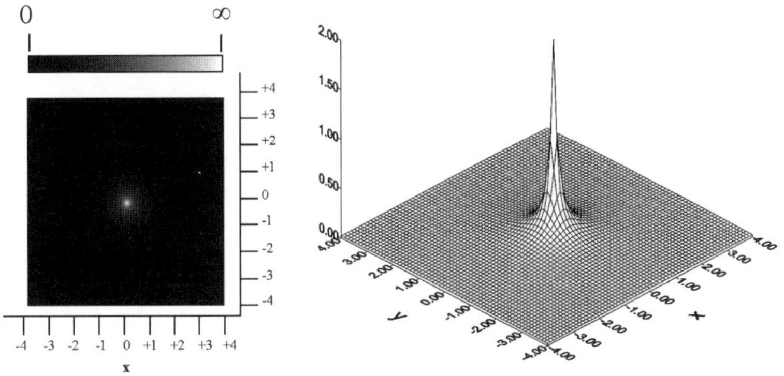

Figure 11.7 The real circularly symmetric function $f(r) = (x^2 + y^2)^{-1/2} = r^{-1}$ displayed in gray-scale and 3-D formats. Its Hankel transform is $F(\rho) = \rho^{-1} = (\xi^2 + \eta^2)^{-1/2}$, thus demonstrating that this is a 2-D Fourier "self-transform".

has a positive singularity at the origin and amplitude that decreases with increasing r (Figure 11.7), which shows that its volume must be positive. The coupling of the x- and y-dependencies and the square root make the 2-D Cartesian Fourier transform difficult to derive analytically, but the form of the corresponding Hankel transform is straightforward:

$$\begin{aligned}\mathcal{H}_0\left\{\frac{1}{r}\right\} &= \int_0^{+\infty} \left(\frac{1}{r}\right) 2\pi r\, J_0[2\pi r\rho]\, dr \\ &= 2\pi \int_{r=0}^{r=+\infty} J_0[2\pi r\rho]\, dr \Longrightarrow u \equiv 2\pi r\rho \\ &= 2\pi \int_{u=0}^{u=+\infty} J_0[u] \frac{du}{2\pi\rho} \\ &= \frac{1}{\rho} \int_0^{+\infty} J_0[u]\, du \end{aligned} \tag{11.52}$$

The definite integral is the real-valued "half area" of $J_0[u]$, which is a real-valued constant. This demonstrates immediately that the 2-D transform of r^{-1} is a scaled replica of ρ^{-1}. In fact, the "half area" of the 1-D Bessel function $J_0(u)$ is unity, as will be shown in the next chapter. This means that $\mathcal{H}_0\{r^{-1}\} = \rho^{-1}$; this is a "Hankel self-transform pair". The corresponding 2-D functions also must be "Fourier self-transforms":

$$\mathcal{F}_2\{|\underline{\mathbf{r}}|^{-1}\} = \mathcal{F}_2\left\{\frac{1}{r}\right\} = \mathcal{F}_2\left\{\frac{1}{\sqrt{x^2 + y^2}}\right\} = \frac{1}{\sqrt{\xi^2 + \eta^2}} = \frac{1}{|\underline{\rho}|} = \rho^{-1} \tag{11.53}$$

This result will be helpful in the discussion of the Radon transform in Chapter 12.

11.4.4 Hankel Transforms from 2-D Fourier Transforms

The known 2-D spectrum of any function that is simultaneously circularly symmetric and separable and that has known spectra may be used to derive the corresponding Hankel transform pair. For example, consider the 2-D circularly symmetric Gaussian and its 1-D radial profile:

$$f[x, y] = e^{-\pi(x^2+y^2)} \Longrightarrow f_r(r) = e^{-\pi r^2} \tag{11.54}$$

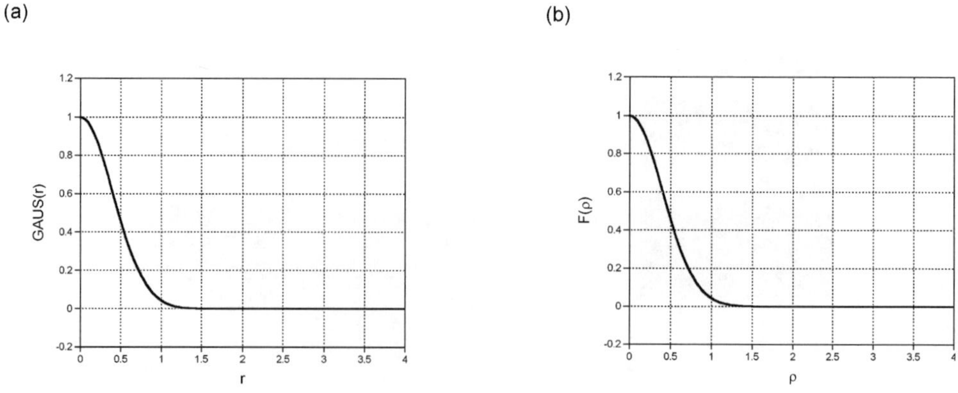

Figure 11.8 Hankel transform of Gaussian function: (a) profile of $GAUS(r)$; (b) 1-D profile of $\mathcal{H}_0\{\exp[-\pi r^2]\} = \exp[-\pi\rho^2]$, confirming that $GAUS[x, y]$ is a 2-D Fourier "self-transform".

The 2-D Fourier transform may be derived by using the separability of the function and the spectrum of the 1-D Gaussian from Equation (9.91):

$$F[\xi, \eta] = e^{-\pi(\xi^2+\eta^2)} \Longrightarrow F_r(\rho) = e^{-\pi\rho^2} \tag{11.55}$$

This demonstrates that the radial Gaussian also is a Hankel "self-transform", as shown in Figure 11.8:

$$\mathcal{H}_0\{e^{-\pi r^2}\} = 2\pi \int_0^{+\infty} (r\, e^{-\pi r^2}) J_0[2\pi r\rho]\, dr = e^{-\pi\rho^2} \tag{11.56}$$

The same avenue may be followed to derive the Hankel transform of the circularly symmetric chirp function:

$$f[x, y] = e^{\pm i\pi(x^2+y^2)} \Longrightarrow f_r(r) = e^{\pm i\pi r^2} \tag{11.57}$$

We can derive the 2-D Fourier transform of this separable function by applying the 1-D Fourier transform from Equation (9.95):

$$\begin{aligned} F[\xi, \eta] &= \mathcal{F}_1\{e^{\pm i\pi x^2}\}\mathcal{F}_1\{e^{\pm i\pi y^2}\} = (e^{\pm i\pi/4}\, e^{\mp i\pi\xi^2})(e^{\pm i\pi/4}\, e^{\mp i\pi\eta^2}) \\ &= e^{\pm i\pi/2}\, e^{\mp i\pi(\xi^2+\eta^2)} \\ &= \pm i\, e^{\mp i\pi(\xi^2+\eta^2)} = \pm i\, e^{\mp i\pi\rho^2} \end{aligned} \tag{11.58}$$

This result shows that the corresponding Hankel transform pair for the circularly symmetric chirp function must be:

$$\begin{aligned} \mathcal{H}_0\{e^{\pm i\pi r^2}\} &= \mathcal{H}_0\{\cos(\pi r^2) \pm i\, \sin(\pi r^2)\} \\ &= 2\pi \int_0^{+\infty} r\, e^{\pm i\pi r^2}\, J_0(2\pi r\rho)\, dr \\ &= \pm i\, e^{\mp i\pi\rho^2} \\ &= \pm i(\cos(\pi\rho^2) \mp i\, \sin(\pi\rho^2)) \\ &= \sin(\pi\rho^2) \pm i\, \cos(\pi\rho^2) \end{aligned} \tag{11.59}$$

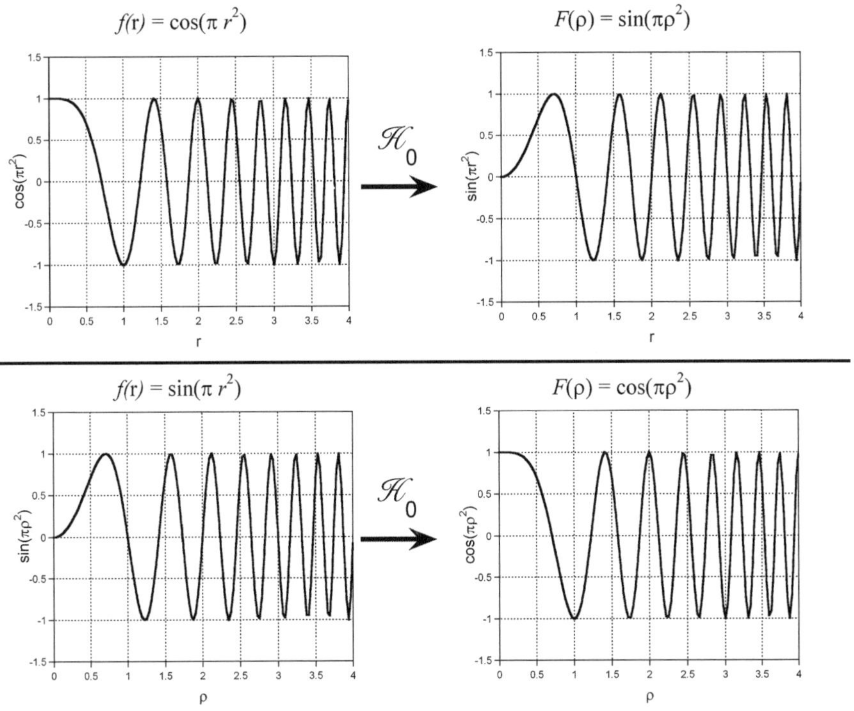

Figure 11.9 Hankel transform of the chirp functions derived from the known 2-D Fourier transform of the separable function $\exp[\pm i\pi(x^2+y^2)]$: $\mathcal{H}_0\{\exp[\pm i\pi r^2]\} = \pm i\ \exp[\mp i\pi\rho^2] \Longrightarrow \mathcal{H}_0\{\pm\cos(\pi r^2)\} = \pm\sin(\pi\rho^2)$ and $\mathcal{H}_0\{\pm\sin(\pi r^2)\} = \pm\cos(\pi\rho^2)$.

Because the real and imaginary parts of $e^{\pm i\pi r^2}$ are both even, the transform of $\cos(\pi r^2)$ and of $\pm\sin(\pi r^2)$ must both be real valued and even. This allows the Hankel transforms of the individual parts in Equation (11.59) to be equated to derive the two Hankel transform pairs of the real-valued circularly symmetric chirp functions:

$$\mathcal{H}_0\{\pm\cos(\pi r^2)\} = \pm\sin(\pi\rho^2) \tag{11.60a}$$

$$\mathcal{H}_0\{\pm\sin(\pi r^2)\} = \pm\cos(\pi\rho^2) \tag{11.60b}$$

as are shown in Figure 11.9.

11.4.5 Hankel Transform of $r^2\,GAUS(r)$

The discrete version of this circularly symmetric function is useful in digital image processing. It may be decomposed into the product of 2-D Cartesian functions:

$$\begin{aligned} r^2 GAUS(r) &= r^2\, e^{-\pi r^2} \\ &= (x^2+y^2)\, e^{-\pi(x^2+y^2)} \\ &= (x^2+y^2)\, e^{-\pi x^2}\, e^{-\pi y^2} \end{aligned} \tag{11.61}$$

The spectrum of this Cartesian representation is easily evaluated by applying the modulation and derivative theorems and the linearity of the 1-D Fourier transform:

$$\begin{aligned}\mathcal{F}_2\{r^2 GAUS(r)\} &= \mathcal{F}_2\{x^2+y^2\} * \mathcal{F}_2\{GAUS(x^2+y^2)\}\\ &= -\frac{1}{4\pi^2}(\delta''[\xi]+\delta''[\eta]) * e^{-\pi(\xi^2+\eta^2)}\\ &= -\frac{1}{4\pi^2}\left[\left(\frac{\partial^2}{\partial\xi^2}+\frac{\partial^2}{\partial\eta^2}\right)e^{-\pi(\xi^2+\eta^2)}\right]\\ &= \left(\frac{1}{\pi}-(\xi^2+\eta^2)\right)e^{-\pi(\xi^2+\eta^2)}\end{aligned} \tag{11.62}$$

The 2-D Fourier transform pair is illustrated in Figure 11.10. The corresponding "1-D" Hankel transform pair is:

$$\mathcal{H}_0\{r^2 GAUS(r)\} = \left(\frac{1}{\pi}-\rho^2\right)GAUS(\rho) \tag{11.63}$$

Another Hankel transform pair follows directly by applying the "transform-of-a-transform" theorem to Equation (11.32):

$$\mathcal{H}_0\left\{\left(\frac{1}{\pi}-r^2\right)GAUS(r)\right\} = \rho^2 GAUS(\rho) \tag{11.64}$$

The discrete forms of the space-domain representation of Equation (11.61) are called *Marr* functions (Niblack, 1986).

11.4.6 Hankel Transform of $CTRI(r)$

The circularly symmetric analogue of the triangle function was introduced without much motivation in Equation (7.99) and its defining equation was presented without proof:

$$CTRI(r) \equiv \frac{2}{\pi}\left(\cos^{-1}[r] - r\sqrt{1-r^2}\right) CYL\left(\frac{r}{2}\right) \tag{11.65}$$

Just as the 1-D triangle function is the autocorrelation of the 1-D rectangle and the 2-D separable triangle is the autocorrelation of the 2-D separable rectangle, $CTRI(r)$ is the autocorrelation of the circularly symmetric analogue of the 2-D rectangle, $CYL(r)$. The autocorrelation of this real and even function is identical to its autoconvolution:

$$k\cdot[CTRI(r)1(\theta)] = [CYL(r)1(\theta)] * [CYL(r)1(\theta)] \tag{11.66}$$

The scale factor k for the amplitude of the $CTRI$ is the volume of the autoconvolution evaluated at the origin, which clearly must be equal to the volume of the squared magnitude of the cylinder function. This is identical to the volume of the cylinder itself, because the squared magnitude of the cylinder function differs from the cylinder only on the boundary. These isolated finite values do not affect the calculation of the volume integral. The filter theorem in Equation (10.28) demonstrates that the Hankel transform of $CTRI(r)$ must be proportional to the squared magnitude of the Hankel transform of $CYL(r)$:

$$\begin{aligned}\mathcal{H}_0\left\{\frac{\pi}{4}CTRI(r)\right\} &= \mathcal{H}_0\{CYL(r) \star CYL(r)\}\\ &= |\mathcal{H}_0\{CYL(r)\}|^2 = \frac{\pi^2}{16}SOMB^2(\rho) \cong 0.617\cdot SOMB(\rho)\end{aligned} \tag{11.67}$$

The Hankel transform pair is shown in Figure 11.11 as topographic plots and as radial line scans.

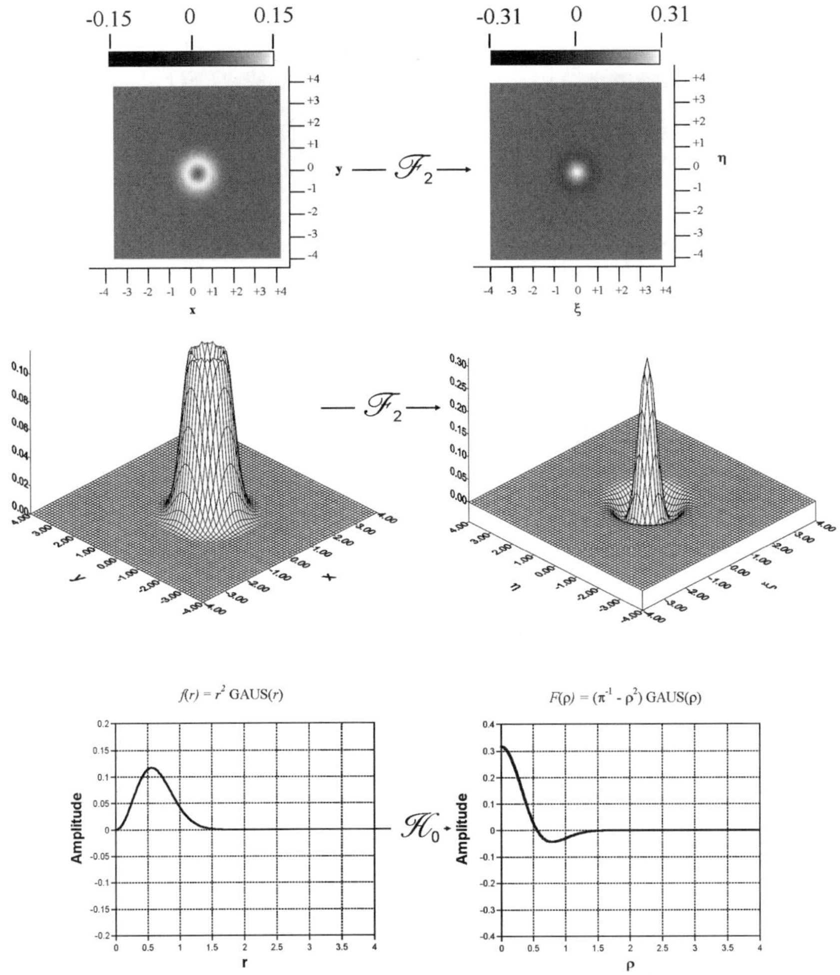

Figure 11.10 The 2-D real-valued and circularly symmetric function $f(r) = r^2 \exp[-\pi r^2]$ and its 2-D Fourier transform $F(\rho) = (\pi^{-1} - \rho^2) \exp[-\pi\rho^2]$ displayed in image and 3-D formats, and the corresponding Hankel transform relationship: $\mathcal{H}_0\{r^2 \exp[-\pi r^2]\} = (\pi^{-1} - \rho^2) \exp[-\pi\rho^2]$.

11.5 Appendix: Derivations of Equations (11.12) and (11.14)

We wish to evaluate an expression for the integral of the even powers of the cosine over one cycle of oscillation. We start with the definition of the gamma function from Equation (6.37):

$$\begin{aligned}\Gamma[n] &= \int_{-\infty}^{+\infty} STEP[\alpha]\, e^{-\alpha}\, \alpha^{n-1}\, d\alpha \\ &= \int_{0}^{+\infty} e^{-\alpha}\, \alpha^{n-1}\, d\alpha \qquad \text{(A11.1)}\end{aligned}$$

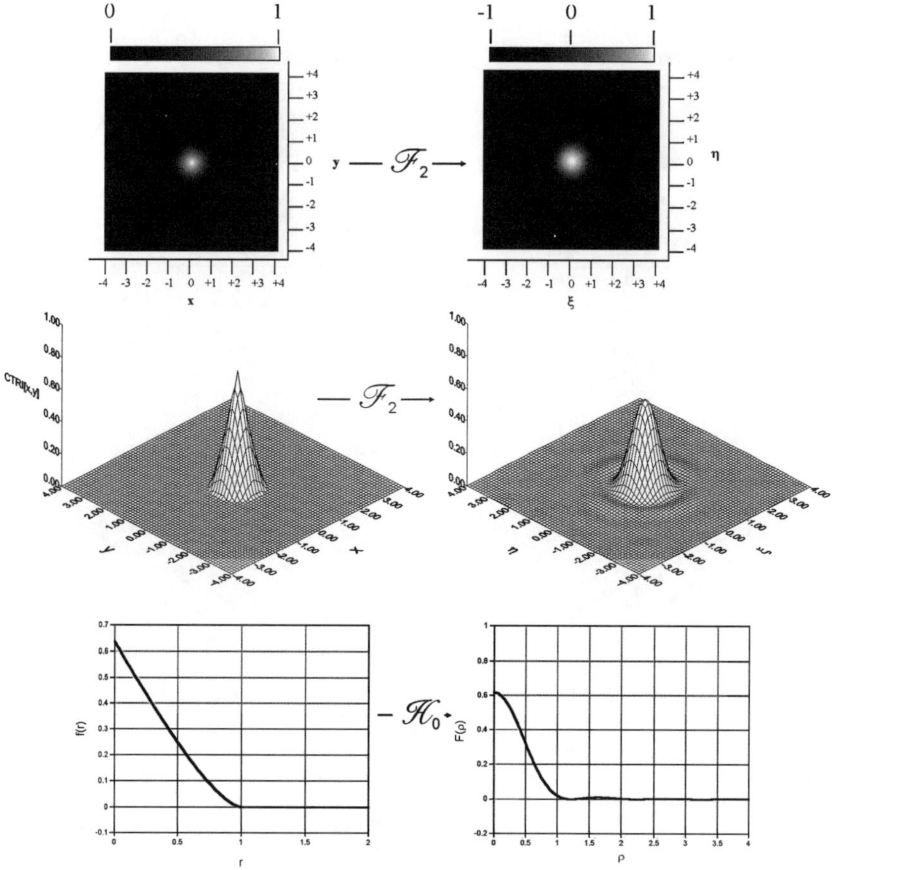

Figure 11.11 Circularly symmetric real-valued circular triangle $f(r) = (\pi/4)CTRI(r)$ and its 2-D Fourier transform $F(\rho) = (\pi/4)^2 SOMB^2(\rho)$ displayed in image and 3-D formats, and the relationship between the 1-D profiles through the Hankel transform.

and substitute $x^2 = \alpha$, so that $d\alpha = 2x \cdot dx$. The limits of integration are unchanged:

$$\begin{aligned}\Gamma[n] &= \int_0^{+\infty} e^{-x^2}\, x^{2(n-1)} 2x\, dx \\ &= 2\int_0^{+\infty} e^{-x^2}\, x^{2n-1}\, dx \qquad \text{(A11.2)}\end{aligned}$$

Then we multiply by an equivalent form for $\Gamma[m]$ with a different integration variable y:

$$\begin{aligned}\Gamma[n]\cdot\Gamma[m] &= \left(2\int_0^{+\infty} e^{-x^2}\, x^{2n-1}\, dx\right)\cdot\left(2\int_0^{+\infty} e^{-y^2}\, y^{2m-1}\, dy\right) \\ &= 4\int_{x=0}^{+\infty}\int_{y=0}^{+\infty} e^{-(x^2+y^2)}\, x^{2n-1} y^{2m-1}\, dx\, dy \qquad \text{(A11.3)}\end{aligned}$$

This double integral may be recast into polar coordinates by substituting $x = r\cos[\phi]$ and $y = r\sin[\phi]$, so that $r^2 = x^2 + y^2$ and $dx\, dy = r\, dr\, d\phi$. Because the domains of x and y are restricted to the first quadrant, the domain of the azimuth integral is $0 \leq \phi < +\pi/2$:

$$\Gamma[n] \cdot \Gamma[m] = 4 \int_{r=0}^{+\infty} \int_{\phi=0}^{+\pi/2} e^{-r^2}\, r^{2n-1} (\cos[\phi])^{2n-1} r^{2m-1} (\sin[\phi])^{2m-1} r\, dr\, d\phi$$

$$= 4 \int_{r=0}^{+\infty} e^{-r^2}\, r^{2(n+m)-1}\, dr \int_{\phi=0}^{+\pi/2} (\cos[\phi])^{2n-1} (\sin[\phi])^{2m-1}\, d\phi \qquad \text{(A11.4)}$$

where the integral over r may be identified as $\frac{1}{2}\Gamma[n+m]$ by substituting $u = r^2$, so that $dr = (1/2\sqrt{u})\, du$:

$$\int_{r=0}^{+\infty} e^{-r^2}\, r^{2(n+m)-1}\, dr \Longrightarrow \int_{u=0}^{+\infty} e^{-u}\, u^{(n+m)-\frac{1}{2}} \frac{1}{2u^{\frac{1}{2}}}\, du$$

$$= \frac{1}{2} \int_{u=0}^{+\infty} e^{-u}\, u^{(n+m)-1}\, du$$

$$= \frac{1}{2}\Gamma[n+m] \qquad \text{(A11.5)}$$

Thus Equation (A11.4) may be rewritten as:

$$\Gamma[n] \cdot \Gamma[m] = 2\Gamma[n+m] \int_{\phi=0}^{+\pi/2} (\cos[\phi])^{2n-1} (\sin[\phi])^{2m-1}\, d\phi$$

$$\Longrightarrow \int_{\phi=0}^{+\pi/2} (\cos[\phi])^{2n-1} (\sin[\phi])^{2m-1}\, d\phi = \frac{1}{2} \frac{\Gamma[n] \cdot \Gamma[m]}{\Gamma[n+m]} \qquad \text{(A11.6)}$$

To obtain the required integral, we substitute $m = \frac{1}{2}$ and $n = \ell + \frac{1}{2}$, where ℓ is a nonnegative integer. The symmetry of the cosine and the fact that $(\cos[\phi])^{2\ell} \geq 0$ ensure that:

$$\int_{\phi=0}^{+\pi/2} (\cos[\phi])^{2\ell} d\phi = \frac{1}{4} \int_{-\pi}^{+\pi} (\cos[\phi])^{2\ell} d\phi = \frac{1}{2} \frac{\Gamma[(2\ell+1)/2] \cdot \Gamma[\frac{1}{2}]}{\Gamma[\ell+1]}$$

$$\Longrightarrow \int_{-\pi}^{+\pi} (\cos[\phi])^{2\ell} d\phi = 2 \frac{\Gamma[\ell+\frac{1}{2}]\Gamma[\frac{1}{2}]}{\Gamma[\ell+1]} \qquad \text{(A11.7)}$$

It may be useful to further simplify this relation by evaluating the gamma functions. We know from Equation (6.43) and the fact that ℓ is an integer that $\Gamma[\ell+1] = \ell!$ in the denominator. Equation (6.49) shows that $\Gamma[\frac{1}{2}] = \sqrt{\pi}$. The recursion relation for the gamma function in Equation (6.42) may be used to show that:

$$\Gamma\left[\ell + \frac{1}{2}\right] = \left(\ell - \frac{1}{2}\right)\Gamma\left[\ell - \frac{1}{2}\right]$$

$$= \left(\ell - \frac{1}{2}\right)\left(\ell - \frac{3}{2}\right)\left(\ell - \frac{5}{2}\right)\cdots\left(\frac{1}{2}\right)\cdot\Gamma\left(\frac{1}{2}\right)$$

$$= \left(\left(\ell - \frac{1}{2}\right)\cdot\dots\cdot\frac{5}{2}\cdot\frac{3}{2}\cdot\frac{1}{2}\right)\cdot\sqrt{\pi}$$

$$= \frac{(2\ell-1)\cdot(2\ell-3)\cdot\dots\cdot 5\cdot 3\cdot 1}{2^{\ell}}\cdot\sqrt{\pi}$$

$$\equiv \frac{(2\ell-1)!!}{2^{\ell}}\cdot\sqrt{\pi} \qquad \text{(A11.8)}$$

where the notation $(2\ell - 1)!!$ for the "odd factorial" (the product of all odd integers less than or equal to $2\ell - 1$) is as defined in Equation (11.13b). Substitution of Equation (A11.9) into Equation (A11.7) yields:

$$\int_{\phi=-\pi}^{+\pi} (\cos[\phi])^{2\ell}\, d\phi = 2\frac{\Gamma[\ell+\frac{1}{2}]\Gamma[\frac{1}{2}]}{\Gamma[\ell+1]}$$
$$= 2\cdot\frac{([(2\ell-1)!!/2^{\ell}]\sqrt{\pi})(\sqrt{\pi})}{\ell!}$$
$$= 2\pi\frac{(2\ell-1)!!}{(2^{\ell})\cdot(\ell\cdot(\ell-1)\cdot(\ell-2)\cdot\cdots\cdot 1)}$$
$$= 2\pi\left(\frac{(2\ell-1)!!}{2\ell\cdot(2\ell-2)\cdot(2\ell-4)\cdot\cdots\cdot 2\cdot 1}\right)$$
$$= 2\pi\frac{(2\ell-1)!!}{(2\ell)!!} \tag{A11.9}$$

where $(2\ell)!!$ denotes the "even factorial" product that also was defined in Equation (11.13a).

The results of Equations (A11.7) and (A11.9) are identical, so either may be substituted into Equation (A11.7) to obtain the terms in the series. The first five terms are:

$$\ell=0 \Longrightarrow \int_{\phi=-\pi}^{+\pi} (\cos[\phi])^{0}\, d\phi = 2\pi\cdot 1 = 2\pi \tag{A11.10a}$$

$$\ell=1 \Longrightarrow \int_{\phi=-\pi}^{+\pi} (\cos[\phi])^{2}\, d\phi = 2\pi\cdot\frac{1}{2} = \pi \tag{A11.10b}$$

$$\ell=2 \Longrightarrow \int_{\phi=-\pi}^{+\pi} (\cos[\phi])^{4}\, d\phi = 2\pi\cdot\frac{3}{8} = \frac{3\pi}{4} \tag{A11.10c}$$

$$\ell=3 \Longrightarrow \int_{\phi=-\pi}^{+\pi} (\cos[\phi])^{6}\, d\phi = 2\pi\cdot\frac{15}{48} = \frac{5\pi}{8} \tag{A11.10d}$$

$$\ell=4 \Longrightarrow \int_{\phi=-\pi}^{+\pi} (\cos[\phi])^{8}\, d\phi = 2\pi\cdot\frac{35}{128} = \frac{35\pi}{64} \tag{A11.10e}$$

These terms are substituted into Equation (11.11) to obtain the desired power series, which has the same coefficients as the expansion for $2\pi\, J_0[2\pi r\rho]$ that was derived in Equation (B6.7):

$$\int_{-\pi}^{+\pi} e^{-2\pi i(r\rho\cos[\psi])}\, d\psi$$
$$= \int_{-\pi}^{+\pi} \cos[2\pi r\rho\cdot\cos[\phi]]\, d\phi$$
$$= \int_{-\pi}^{+\pi}\left(1-\frac{(2\pi r\rho)^2}{2!}\cos^2[\phi]+\frac{(2\pi r\rho)^4}{4!}\cos^4[\phi]-\frac{(2\pi r\rho)^6}{6!}\cos^6[\phi]+\cdots\right)d\phi$$
$$= 2\pi-\left(\frac{(2\pi r\rho)^2}{2}\cdot\pi\right)+\left(\frac{(2\pi r\rho)^4}{24}\cdot\frac{3\pi}{4}\right)-\left(\frac{(2\pi r\rho)^6}{720}\cdot\frac{5\pi}{8}\right)+\cdots$$
$$= 2\pi\cdot\left(1-\frac{(2\pi r\rho)^2}{4}+\frac{(2\pi r\rho)^4}{64}-\frac{(2\pi r\rho)^6}{2304}+\cdots\right)$$
$$= 2\pi\, J_0[2\pi r\rho] \tag{A11.11}$$

PROBLEMS

11.1 Find the results of the convolutions and sketch them:

(a) $CYL(r) * (\delta[x, y + 2] + \delta[x, y - 2])$

(b) $[CYL(r/2) - CYL(r)] * [CYL(r/2) - CYL(r)]$

11.2 Evaluate the Fourier transforms of the following functions and sketch them:

(a) $CYL(r) * (\delta[x](\delta[y + 2] + \delta[y - 2]))$

(b) $[CYL(r/2) - CYL(r)] * [CYL(r/2) - CYL(r)]$

(c) $GAUS(r) * \delta[x - 1, y]$

(d) $J_0(2\pi r) + J_0(4\pi r)$

(e) $\exp[-i\pi r^2/4] + \exp[+i\pi r^2/4]$

(f) $\exp[-i(\pi r^2 + \pi/2)]$

(g) $\exp[-i\pi r^2 - \pi/2]$

11.3 Find the transfer function of the imaging systems with the following impulse responses:

(a) $h_a(r) = J_0(2\pi r) + J_0(\pi r)$

(b) $h_b(r) = SOMB(r/10)$

(c) $h_c(r) = -r^2\, GAUS(r)$

11.4 Complete the proof that $\mathcal{F}_2\{r^2\, GAUS(r)\} = (\pi^{-1} - \rho^2)\, GAUS(\rho)$.

12

The Radon Transform

This chapter introduces another recipe for evaluating spectra of 2-D functions via 1-D operations that is interesting in its own right but also is important in several real-world imaging modalities, particularly in the medical imaging applications of X-ray computed tomography (commonly called *CT scans*) and magnetic resonance imaging (MRI). The algorithm is also used in other disciplines as diverse as geology, astronomy, and probability theory. This "Radon transform" is an invertible operation that produces a new 2-D function by integrating $f[x, y]$ along parallel lines at different orientations. The results of this discussion will have the side benefit of enabling us to calculate some additional 1-D Fourier transform pairs not previously available.

12.1 Line-Integral Projections onto Radial Axes

As an introduction to the Radon transform, consider methods for evaluating the spectrum of an arbitrary 2-D function $f[x, y]$ along the ξ-axis. The integral to be solved is obtained by substitution into the 2-D Fourier transform in Equation (10.1):

$$F[\xi, 0] = \iint_{-\infty}^{+\infty} f[x, y]\, e^{-2\pi i(\xi x+0y)}\, dx\, dy = \iint_{-\infty}^{+\infty} (f[x, y]\, e^{-2\pi i\xi x})\, dx\, dy \tag{12.1a}$$

$$= \int_{-\infty}^{+\infty} \left(\int_{-\infty}^{+\infty} f[x, y]\, e^{-2\pi i\xi x}\, dx \right) dy = \int_{-\infty}^{+\infty} \left(\mathcal{F}_{\substack{1 \\ x\to\xi}} \{f[x, y]\} \right) dy \tag{12.1b}$$

$$= \int_{-\infty}^{+\infty} \left(\int_{-\infty}^{+\infty} f[x, y]\, dy \right) e^{-2\pi i\xi x}\, dx = \mathcal{F}_{\substack{1 \\ x\to\xi}} \left\{ \int_{-\infty}^{+\infty} f[x, y]\, dy \right\} \tag{12.1c}$$

where the conjugate coordinates of the 1-D Fourier transform are specified beneath the operator to avoid confusion that may result from the several coordinates in the expression. These three equivalent expressions provide different paths for evaluating the ξ-axis profile of the 2-D spectrum:

1. By evaluating the volume of $f[x, y]\, e^{-2\pi i\xi x}$ directly in Equation (12.1a).
2. By evaluating 1-D Fourier transforms of $f[x, y]$ over the x-direction and integrating for each value of y in Equation (12.1b).
3. By computing the 1-D Fourier transform of a "new" 1-D function of x that is generated from $f[x, y]$ via 1-D integrals along paths parallel to the y-axis in Equation (12.1c).

Fourier Methods in Imaging Roger L. Easton, Jr.

The 1-D function produced by the line integrals in Equation (12.1c) is assigned the name ℓ_f:

$$\int_{-\infty}^{+\infty} f[x, y]\, dy \equiv \ell_f[x] \tag{12.2}$$

where the subscript specifies the original 2-D function. This is the "line-spread function" that was introduced in Section 8.8 and that is used as a metric for characterizing optical imaging systems. The integral operation is often visualized as "line-integral projection" of $f[x, y]$ onto the x-axis. This usage of the term *projection* is, at the same time, similar to and different from the concept of projecting a function onto the members of a set of basis functions that was introduced in Chapter 2. Insert Equation (12.2) into Equation (12.1) to obtain a concise relation for the spectrum evaluated only along the ξ-axis:

$$F[\xi, 0] = \underset{x\to\xi}{\mathcal{F}_1}\{\ell_f[x]\} \tag{12.3}$$

A 2-D function that is intimately related to the 1-D line-integral projection in Equation (12.2) is obtained by 2-D convolution of $f[x, y]$ with the 2-D line delta function parallel to the y-axis that was defined in Equation (7.57):

$$\begin{aligned} f[x, y] * (\delta[x] \cdot 1[y]) &= \iint_{-\infty}^{+\infty} f[\alpha, \beta](\delta[x-\alpha] \cdot 1[y-\beta])\, d\alpha\, d\beta \\ &= \left(\int_{-\infty}^{+\infty} \delta[x-\alpha]\left(\int_{-\infty}^{+\infty} f[\alpha, \beta]\, d\beta\right) d\alpha\right) \cdot 1[y] \\ &= \left(\int_{-\infty}^{+\infty} \delta[x-\alpha]\, \ell_f[\alpha]\, d\alpha\right) \cdot 1[y] = \ell_f[x] \cdot 1[y] \end{aligned} \tag{12.4}$$

where the fact has been used that the 1-D unit constant $1[y]$ is invariant under translation along the y-direction. In words, the operation in Equation (12.4) produces a 2-D function that is separable into the orthogonal product of a 1-D function of x and the unit constant along y. The amplitude of this 2-D function at all coordinates $[x_0, y]$ is the line-integral projection evaluated at x_0; an example is shown in Figure 12.1. In other words, the convolution of $f[x, y]$ with the line delta function produces a 2-D function by "smearing" the 1-D function $\ell_f[x]$ along the y-direction via the product with $1[y]$. The discussion of Chapter 5 also may be applied to interpret Equation (12.4) as the inner product of the 2-D "input function" $f[x, y]$ with the 2-D line Dirac delta functions that are constant along the y-axis.

The 2-D filter theorem in Equation (10.28) may be used to evaluate the 2-D spectrum of the specific 2-D "smeared line-integral projection" in Equation (12.4):

$$\begin{aligned} \mathcal{F}_2\{\ell_f[x] \cdot 1[y]\} &= \mathcal{F}_2\{f[x, y] * (\delta[x] \cdot 1[y])\} \\ &= F[\xi, \eta] \cdot (1[\xi] \cdot \delta[\eta]) = F[\xi, 0] \cdot (1[\xi] \cdot \delta[\eta]) \\ &= \underset{x\to\xi}{\mathcal{F}_1}\{\ell_f[x]\} \cdot (1[\xi] \cdot \delta[\eta]) \end{aligned} \tag{12.5}$$

where the property of Dirac delta functions in products in Equation (6.115) has been used. The frequency-domain representation of the line-integral projection of $f[x, y]$ obtained by 2-D convolution is a 2-D function that evaluates to zero everywhere except (possibly) on the ξ-axis, where the amplitude is identical to the spectrum of $F[\xi, \eta = 0]$. In short, the 2-D Fourier transform of the line-integral projection is a "slice" or "profile" of the 2-D spectrum through the origin of the frequency domain; it is often called a *central slice*.

We now generalize the concept to line-integral projections onto arbitrary "axes" by substituting the appropriately rotated line delta function from Equation (7.63) into the 2-D convolution in Equation (12.4). Recall that the line delta function through the origin is defined as the set of 1-D Dirac delta functions evaluated along a line perpendicular to a particular unit vector, say $\hat{\mathbf{p}}_0$, which is constrained

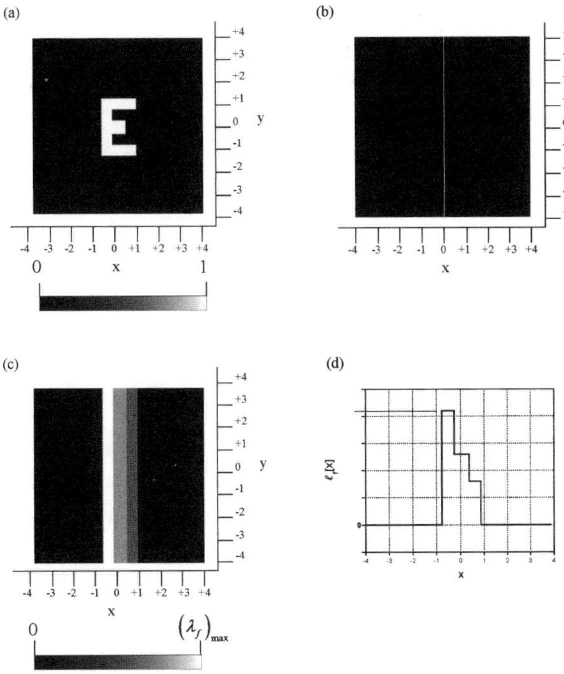

Figure 12.1 Line-integral projection of $f[x, y]$ onto the x-axis by computing its convolution with $\delta[x]1[y]$: (a) $f[x, y]$; (b) $\delta[x]1[y]$; (c) $f[x, y]* \delta[x]1[y] = g[x, y]$ is a 2-D function that is constant along the y-direction; (d) $g[x, 0]$, the profile of the 2-D convolution evaluated along the x-axis, which is equivalent to $g[x, y] \cdot (1[x]\delta[y])$.

in our convention to lie in the first or fourth quadrants. The unit vector may be expressed in equivalent Cartesian or polar coordinates:

$$\hat{\underline{\mathbf{p}}}_0 = [\cos[\phi_0], \sin[\phi_0]] = (1, \phi_0), \quad -\frac{\pi}{2} \leq \phi_0 < +\frac{\pi}{2} \tag{12.6}$$

The functional form of the line delta function is $\delta[\underline{\mathbf{r}} \bullet \hat{\underline{\mathbf{p}}}_0] \cdot 1[\underline{\mathbf{r}} \bullet \hat{\underline{\mathbf{p}}}_0^{\perp}]$, where the orthogonal unit vector $\hat{\underline{\mathbf{p}}}_0^{\perp}$ may be specified by the Cartesian or polar coordinates:

$$\hat{\underline{\mathbf{p}}}_0^{\perp} = \left[\cos\left[\phi_0 \pm \frac{\pi}{2}\right], \sin\left[\phi_0 \pm \frac{\pi}{2}\right]\right] = [\mp\sin[\phi_0], \pm\cos[\phi_0]] = \left(1, \phi_0 \pm \frac{\pi}{2}\right) \tag{12.7}$$

Since the vectors $\hat{\underline{\mathbf{p}}}_0$ and $\hat{\underline{\mathbf{p}}}_0^{\perp}$ appear within scalar products in the argument of symmetric 1-D Dirac delta functions, the algebraic sign within the scalar product does not matter. We may use this observation to constrain the orthonormal vectors $\hat{\underline{\mathbf{p}}}_0$ and $\hat{\underline{\mathbf{p}}}_0^{\perp}$ to two quadrants (say, the first and fourth). The geometric relationship that determines which locations in the 2-D space domain lie upon the line of integration is illustrated in Figure 12.2. The convolution of the input function with a specific line delta function is:

$$\begin{aligned} &f[x, y] * (\delta[\underline{\mathbf{r}} \bullet \hat{\underline{\mathbf{p}}}_0] \cdot 1[\underline{\mathbf{r}} \bullet \hat{\underline{\mathbf{p}}}_0^{\perp}]) \\ &\quad = f[x, y] * (\delta[x\cos[\phi_0] + y\sin[\phi_0]] \cdot 1[x\sin[\phi_0] - y\cos[\phi_0]]) \end{aligned}$$

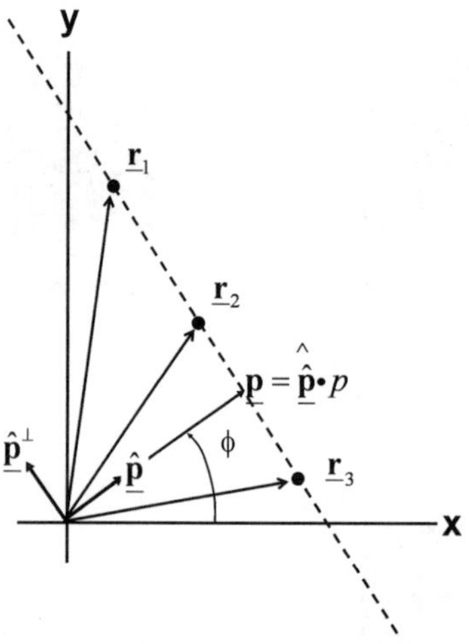

Figure 12.2 Relationship of the "pedal vector" $\underline{\mathbf{p}}$, the unit vectors $\hat{\underline{\mathbf{p}}}$ and $\hat{\underline{\mathbf{p}}}^\perp$, and the line of integration for that pedal vector (shown as dashes). The scalar products of $\hat{\underline{\mathbf{p}}}$ with each of the three vectors $\underline{\mathbf{r}}_n$ are identically $|\underline{\mathbf{p}}| = p$, and so all of these points lie on the line integration that is composed of the complete set of vectors $\underline{\mathbf{r}}$ that satisfy the relationship $\underline{\mathbf{r}} \bullet \hat{\underline{\mathbf{p}}} = p$, which is equivalent to $p - \underline{\mathbf{r}} \bullet \hat{\underline{\mathbf{p}}} = 0$. The points on the line of integration satisfy the conditions for the 1-D Dirac delta function $\delta[p - \underline{\mathbf{r}} \bullet \hat{\underline{\mathbf{p}}}]$.

$$= \iint_{-\infty}^{+\infty} f[\alpha,\beta]\,\delta[(x-\alpha)\cos[\phi_0] + (y-\beta)\sin[\phi_0]]$$

$$\cdot\, 1[(x-\alpha)\sin[\phi_0] - (y-\beta)\cos[\phi_0]]\,d\alpha\,d\beta$$

$$= \left(\iint_{-\infty}^{+\infty} f[\alpha,\beta]\delta[(\alpha-x)\cos[\phi_0] + (\beta-y)\sin[\phi_0]]\,d\alpha\,d\beta\right)\cdot 1[x\sin[\phi_0] - y\cos[\phi_0]]$$

$$\equiv \left(\iint_{-\infty}^{+\infty} f[\alpha,\beta]\delta[p - \underline{\mathbf{r}} \bullet \hat{\underline{\mathbf{p}}}_0]\,d\alpha\,d\beta\right)\cdot 1[\underline{\mathbf{r}} \bullet \hat{\underline{\mathbf{p}}}_0^\perp] \tag{12.8}$$

where $p \equiv \alpha\cos[\phi_0] + \beta\sin[\phi_0]$ and the symmetry of the 1-D Dirac delta function and invariance of the unit-constant function under translation have been used. By analogy with Equation (12.4), we can also write the convolution as:

$$f[x,y] * (\delta[\underline{\mathbf{r}} \bullet \hat{\underline{\mathbf{p}}}_0]\cdot 1[\underline{\mathbf{r}} \bullet \hat{\underline{\mathbf{p}}}_0^\perp]) = \left(\iint_{-\infty}^{+\infty} f[\alpha,\beta]\,\delta[p - \underline{\mathbf{r}} \bullet \hat{\underline{\mathbf{p}}}_0]\,d\alpha\,d\beta\right)\cdot 1[\underline{\mathbf{r}} \bullet \hat{\underline{\mathbf{p}}}_0^\perp]$$

$$\equiv \lambda_f(p,\phi_0)\cdot 1[\underline{\mathbf{r}} \bullet \hat{\underline{\mathbf{p}}}_0^\perp] \tag{12.9}$$

where the notation $\lambda_f(p,\phi_0)$ refers to the 1-D line-integral projection λ of the function $f[x,y]$ evaluated at the radial distance p for the specific azimuth angle parameter ϕ_0. In words, the convolution is the separable product of two 1-D functions: the 1-D function $\lambda_f(p,\phi_0)$ along the direction defined by $\hat{\underline{\mathbf{p}}}_0$ and the 1-D unit constant along the set of points $\underline{\mathbf{r}}$ that are perpendicular to $\hat{\mathbf{p}}_0$.

The line of integration is specified by the radial vector $\underline{\mathbf{p}}$ that has length $|\underline{\mathbf{p}}| \equiv p$ in the direction of $\hat{\underline{\mathbf{p}}}$, so that $\underline{\mathbf{p}} = \hat{\underline{\mathbf{p}}} \cdot p$. An archaic name for $\underline{\mathbf{p}}$ is the "pedal vector", so called probably because the set of vectors with a fixed value of p and increasing ϕ "revolves" about the origin at a fixed distance in the same manner as a bicycle pedal.

The line-integral projections evaluated for every azimuth angle produce a 2-D "polar-like" function λ_f of both the radial distance p and the azimuth angle ϕ. Equation (12.9) might tempt us to write a general expression for line-integral projection as the 2-D convolution of $f[x, y]$ and a rotated line delta function that produces a 2-D function $\lambda_f(p, \phi)$:

$$\lambda_f(p, \phi) \stackrel{?}{=} f[x, y] * (\delta[\underline{\mathbf{r}} \bullet \underline{\hat{\mathbf{p}}}] \cdot 1[\underline{\mathbf{r}} \bullet \underline{\hat{\mathbf{p}}}^{\perp}]) \tag{12.10}$$

This expression has a fatal flaw because it implies that the complete set of line-integral projections is a 2-D function $\lambda_f(p, \phi)$ with "polar" coordinates that is obtained by convolving $f[x, y]$ with the 2-D function $\delta[\underline{\mathbf{r}} \bullet \underline{\hat{\mathbf{p}}}] \cdot 1[\underline{\mathbf{r}} \bullet \underline{\hat{\mathbf{p}}}^{\perp}]$. If true, this would indicate that the Radon transform is shift invariant, which is *absolutely* not the case. Rather, a specific line-integral projection in the 2-D form $\lambda_f(p, \phi_0) \cdot 1[\underline{\mathbf{r}} \bullet \underline{\hat{\mathbf{p}}}_0^{\perp}]$ from Equation (12.9) is obtained by the 2-D convolution of $f[x, y]$ with a specific 2-D line delta function that has been *rotated* about the origin by ϕ_0. We then factor out the orthogonal constant $1[\underline{\mathbf{r}} \bullet \underline{\hat{\mathbf{p}}}_0^{\perp}]$ to generate the ensemble of line-integral projections $\lambda_f(p, \phi)$ for each azimuth angle ϕ. In other words, the Radon transform may be evaluated from an infinite number of 2-D convolutions, one for each azimuth angle, but the ensemble of operations is shift variant. The convolution in Equation (12.10) may create confusion because the specific azimuth angle is not specified in the expression for the line delta function. For this reason, it always is useful to apply subscripts to the vectors $\underline{\hat{\mathbf{p}}}$ and $\underline{\hat{\mathbf{p}}}^{\perp}$, as we will subsequently do.

The result of Equation (12.9) may be written in several equivalent ways:

$$f[x, y] * (\delta[\underline{\mathbf{r}} \bullet \underline{\hat{\mathbf{p}}}_0] \cdot 1[\underline{\mathbf{r}} \bullet \underline{\hat{\mathbf{p}}}_0^{\perp}]) = \lambda_f(\underline{\mathbf{p}}_0) \cdot 1[\underline{\mathbf{r}} \bullet \underline{\hat{\mathbf{p}}}_0^{\perp}] \tag{12.11a}$$

$$= \lambda_f(p\underline{\hat{\mathbf{p}}}_0) \cdot 1[\underline{\mathbf{r}} \bullet \underline{\hat{\mathbf{p}}}_0^{\perp}] \tag{12.11b}$$

$$= \lambda_f(p, \phi_0) \cdot 1[\underline{\mathbf{r}} \bullet \underline{\hat{\mathbf{p}}}_0^{\perp}] \tag{12.11c}$$

where $\underline{\mathbf{p}}_0$ is the radial vector constrained to the first or fourth quadrant that defines the azimuthal direction of the line of integration and p is a bipolar number that specifies the radial distance of a particular line of integration from the origin. That the first term on the right is in fact a 1-D function is apparent in (b) and (c) by the 1-D coordinate p. These notations make it clearer that each specific line-integral projection is a 1-D spatial function. The Radon transform is completely described by the set of 1-D functions $\lambda_f(p)$ for all angles ϕ, hence the notation $\lambda_f(p, \phi)$. Again, each 1-D function of the space variable p is the "varying part" of the 2-D convolution with each of the set of line delta functions at the azimuth angle ϕ. For example, the line-integral projection onto the x-axis is equivalent to the line-spread function in Equation (12.2):

$$\int_{-\infty}^{+\infty} f[x, y]\, dy = \ell_f[x] = \lambda_f(p, \phi = 0) = \lambda_f(p; 0) \tag{12.12}$$

An example of line-integral projection is shown in Figure 12.3 where $f[x, y]$ is a centered uppercase block letter "E"; the azimuth angle is $\phi_0 = +\pi/6$, so that $\underline{\hat{\mathbf{p}}}_0 = [\sqrt{3}/2, \frac{1}{2}]$ in the first quadrant and $\underline{\hat{\mathbf{p}}}_0^{\perp} = [\frac{1}{2}, -\sqrt{3}/2]$ in the fourth. Again note that the 2-D convolution produces a 2-D function that varies along the direction of $\underline{\hat{\mathbf{p}}}$ and is constant along the direction of $\underline{\hat{\mathbf{p}}}_0^{\perp}$, where the specific projection $\lambda_f(p, \phi = +\pi/6)$ is the 1-D varying part. By analogy with Equation (12.4), the 1-D projection $\lambda_f(p, \phi_0)$ is "smeared" along the direction of $\underline{\hat{\mathbf{p}}}_0^{\perp}$ by multiplying by the orthogonal constant function $1[\underline{\mathbf{r}} \bullet \underline{\hat{\mathbf{p}}}_0^{\perp}]$. The "smearing" operation constructs a 2-D function from a particular 1-D projection $\lambda_f(p, \phi_0)$ and is often called *back projection*.

It is often useful to visualize the ensemble of line-integral projections as a 2-D function of the radial coordinate p and the azimuthal coordinate ϕ, but represented in a Cartesian space, i.e., where the coordinates are plotted as $[p, \phi]$, with p on the "x"-axis and ϕ along the "y"-axis. This concept will be considered in detail shortly.

The Radon transform is a linear operation because it is defined by a set of integrals of $f[x, y]$ over fixed limits. It may be interpreted as a transformation of the 2-D function $f[x, y]$ to new "polar-like" coordinates (p, ϕ). Because λ_f is defined over a polar-like domain, with radial coordinate p and

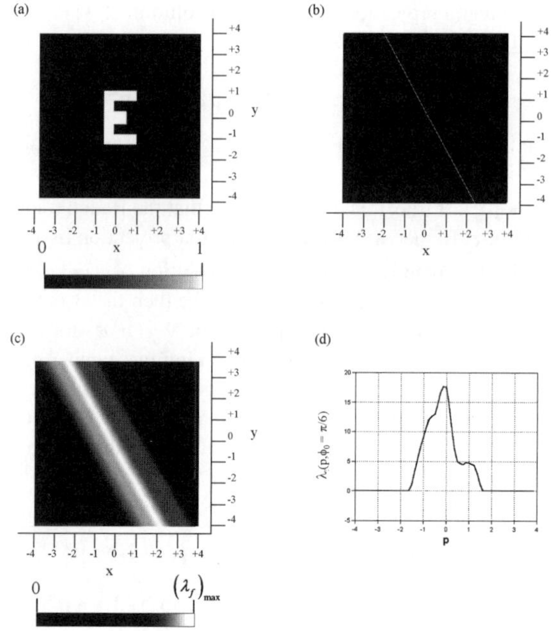

Figure 12.3 Line-integral projection of $f[x, y]$ onto the axis at $\phi_0 = +\pi/6$: (a) $f[x, y]$; (b) $\delta[\mathbf{r} \cdot \hat{\mathbf{p}}_0] \cdot 1[\mathbf{r} \cdot \hat{\mathbf{p}}_0^{\perp}]$; (c) 2-D convolution $f[x, y] * (\delta[\mathbf{r} \cdot \hat{\mathbf{p}}_0] \cdot 1[\mathbf{r} \cdot \hat{\mathbf{p}}_0^{\perp}]) = \lambda_f(p, +\pi/6) \cdot 1[\mathbf{r} \cdot \hat{\mathbf{p}}_0]$; (d) 1-D function $\lambda_f(p, \phi = +\pi/6)$.

azimuth ϕ, its coordinates are enclosed in parentheses to follow the convention set forth in Chapter 1. Much of the following discussion is devoted to the search for the inverse transformation from $\lambda_f(p, \phi)$ back to $f[x, y]$.

Some authors *define* the Radon transform by Equation (12.11c). The multiplication of $f[x, y]$ by $\delta[p - \mathbf{r} \bullet \hat{\mathbf{p}}]$ "assigns" each Cartesian location $[x_0, y_0]$ in the input function to a specific radial distance p_0 that varies with azimuth angle ϕ. The integral then sums the amplitudes for all input coordinates that are assigned to each specific value of p and thus generate the amplitude of the projection at that p for that azimuth ϕ.

The linear and shift-variant process defined in Equation (12.12) will be denoted by the operator $\mathcal{R}_2$ that acts on the 2-D space-domain function $f[x, y]$ to generate the "new" 2-D function $\lambda_f(p, \phi)$:

$$\mathcal{R}_2\{f[x, y]\} = \mathcal{R}_2\{f(\mathbf{r})\} = \lambda_f(\mathbf{p}) = \lambda_f(p\hat{\mathbf{p}}) = \lambda_f(p, \phi) \tag{12.13}$$

The subscript "2" specifies the number of dimensions in the domain of the input function; Radon transforms of functions with more dimensions may be defined, but are not considered here. We will demonstrate eventually that $\lambda_f(p, \phi)$ is equivalent to $f[x, y]$, which shows that the projection of $f[x, y]$ onto the set of line delta functions is unique and thus that the complete set of line delta functions may serve as a basis in the 2-D space domain.

Two useful choices are available for the domain of the polar-like coordinates (p, ϕ) of the line-integral projections, but we have already mentioned that our convention constrains the azimuth angle to the first and fourth quadrants $(-\pi/2 \leq \phi < +\pi/2)$, which requires that the domain of the radial "distance" p be bipolar: $-\infty < p < +\infty$. This choice allows convenient evaluation of the 1-D Fourier transform of the individual projections. The same convention will apply to the polar coordinates in the 2-D frequency domain; we will specify a particular spatial frequency by coordinates (ν, ψ), where ν is the bipolar "magnitude" ($-\infty < \nu < +\infty$) and ψ is the phase angle constrained to the first or fourth quadrants: $-\pi/2 \leq \psi < +\pi/2$.

The transformation from $f[x, y]$ to the "polar" line-integral-projection representation $\lambda_f(p, \phi)$ in Equation (12.12) was studied in detail by Johan Radon early in the twentieth century (Deans, 1983), and its common name honors his contribution. We will soon show that the Radon transform is invertible, thus proving the equivalence of the three representations $f[x, y]$, $F[\xi, \eta]$, and $\lambda_f(p, \phi)$. The Radon transform is important in imaging because of the number of situations where the original object $f[x, y]$ is not known but its line-integral projections may be measured. The most familiar examples are the medical imaging systems of computed tomography (CT or CAT) and magnetic resonance imaging (MRI), but similar situations exist in geophysical and astronomical applications (Iizuka, 1987).

We now consider the Radon transforms of the symmetric 2-D Dirac delta function and its "decentered" relatives.

12.1.1 Radon Transform of Dirac Delta Function

We already know that the "centered" 2-D Dirac delta function $\delta[x, y]$ is both Cartesian separable and circularly symmetric:

$$\delta[x, y] = \delta[x] \cdot \delta[y] = \frac{\delta(r)}{\pi r} \tag{12.14}$$

The circular symmetry ensures that projections at all azimuth angles ϕ are identical functions of the radial coordinate p. After substituting into Equation (12.11), we obtain the simple result:

$$\begin{aligned}
\lambda_{(\delta[x,y])}(p, \phi) &= \iint_{-\infty}^{+\infty} \delta[x, y] \cdot \delta[p - (x \cos[\phi] + y \sin[\phi])]\, dx\, dy \\
&= \iint_{-\infty}^{+\infty} \delta[x, y] \cdot \delta[p - (0 \cos[\phi] + 0 \sin[\phi])]\, dx\, dy \\
&= \delta[p - 0] \iint_{-\infty}^{+\infty} \delta[x, y]\, dx\, dy = \delta[p] \cdot 1 = \delta[p] \cdot 1[\phi]
\end{aligned} \tag{12.15}$$

where the sifting property of the 2-D Dirac delta function in Equation (7.48) and its volume in Equation (7.52) have been used to evaluate the integral. In words, this result shows that all projections of the centered 2-D Dirac delta function with unit volume are 1-D Dirac delta functions with unit area located at the origin. The projections $\lambda_{(\delta[x,y])}(p, \phi) = \delta[p] \cdot 1[\phi]$. The representations are shown in Figure 12.4.

The projections of a 2-D unit-volume Dirac delta function at an arbitrary location $\underline{\mathbf{r}}_0 \equiv [x_0, y_0] = (|\underline{\mathbf{r}}_0|, \theta_0)$ also are easy to evaluate by direct substitution into Equation (12.11):

$$\begin{aligned}
\lambda_{(\delta[\underline{\mathbf{r}}-\underline{\mathbf{r}}_0])}(p, \phi) \cdot 1[\underline{\mathbf{r}} \bullet \hat{\underline{\mathbf{p}}}^{\perp}] &= \delta[\underline{\mathbf{r}} - \underline{\mathbf{r}}_0] * (\delta[\underline{\mathbf{r}} \bullet \hat{\underline{\mathbf{p}}}] \cdot 1[\underline{\mathbf{r}} \bullet \hat{\underline{\mathbf{p}}}^{\perp}]) \\
&= \lambda_{(\delta[\underline{\mathbf{r}}-\underline{\mathbf{r}}_0])}(p, \phi) \cdot 1[\underline{\mathbf{r}} \bullet \hat{\underline{\mathbf{p}}}^{\perp}]
\end{aligned} \tag{12.16a}$$

$$\begin{aligned}
\lambda_{(\delta[\underline{\mathbf{r}}-\underline{\mathbf{r}}_0])}(p, \phi) &= \iint_{-\infty}^{+\infty} \delta[x - x_0, y - y_0] \cdot \delta[p - (x \cos[\phi] + y \sin[\phi])]\, dx\, dy \\
&= \iint_{-\infty}^{+\infty} \delta[x - x_0, y - y_0] \cdot \delta[p - (x_0 \cos[\phi] + y_0 \sin[\phi])]\, dx\, dy \\
&= \delta[p - (x_0 \cos[\phi] + y_0 \sin[\phi])] \cdot \left(\iint_{-\infty}^{+\infty} \delta[x - x_0, y - y_0]\, dx\, dy\right) \\
&= \delta[p - \underline{\mathbf{r}}_0 \bullet \hat{\underline{\mathbf{p}}}] \cdot 1 = \delta[p - |\underline{\mathbf{r}}_0| \cos[\theta_0 - \phi]]
\end{aligned} \tag{12.16b}$$

where $\hat{\underline{\mathbf{p}}} \equiv [\cos[\phi], \sin[\phi]] = (1, \phi)$ and the sifting property has been used. In words, the projection of the "decentered" 2-D Dirac delta function onto the axis defined by $\hat{\underline{\mathbf{p}}}$ is a 1-D Dirac delta function

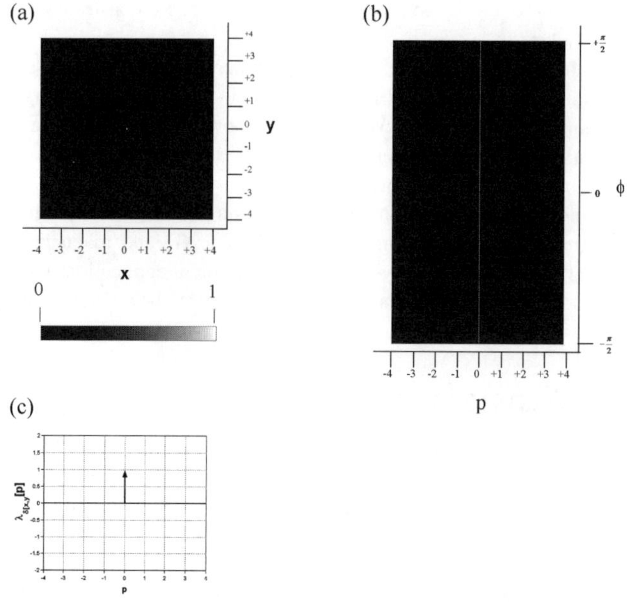

Figure 12.4 Radon transform of centered 2-D Dirac delta function: (a) $\delta[x, y]$; (b) $\lambda_{\delta[x,y]}(p, \phi) = \delta[p]1[\phi]$; (c) $\lambda_{\delta[x,y]}(p; \phi) = \delta[p]$.

whose argument is a sinusoidal function of azimuth angle ϕ. An example with $\mathbf{\underline{r}}_0 = [1, 1]$ is shown in Figure 12.5. For illustration, consider the projection evaluated for the angle $\phi = \theta_0$, i.e., the azimuth of the projection is the same as that of the 2-D Dirac delta function. In this case, $\mathbf{\underline{r}}_0 \bullet \hat{\mathbf{\underline{p}}} = |\mathbf{\underline{r}}_0| \cos[0] = |\mathbf{\underline{r}}_0|$, and the line-integral projection is $\delta[p - |\mathbf{\underline{r}}_0|]$. The projection evaluated along the azimuth angle $\phi = \theta_0 \pm \pi/2$ is the "centered" 1-D Dirac delta function $\delta[p]$. At other azimuth angles, the translation parameter in the argument of the 1-D Dirac delta function is a sinusoidal function of $[\phi - \theta_0]$. The nonzero amplitudes in the plot of $\lambda_{(\delta[\mathbf{\underline{r}}-\mathbf{\underline{r}}_0])}(p, \phi)$ form a sinusoidal wave, thus demonstrating the reason for the common name of "sinogram" often ascribed to this representation. It may be useful to determine the specific azimuth angle of the positive-valued extremum of p, which is analogous to the "initial phase" of a sinusoid in the 1-D space domain. The derivative of the radial distance p from the origin with respect to the azimuth angle ϕ evaluates to zero at this angle $\phi_{\max}$:

$$\left.\frac{\partial p}{\partial \phi}\right|_{\phi=\phi_{\max}} = x_0(-\sin[\phi_{\max}]) + y_0 \cos[\phi_{\max}] = 0$$

$$\Longrightarrow \phi_{\max} = \tan^{-1}\left[\frac{y_0}{x_0}\right] = \theta_0, \quad -\frac{\pi}{2} \leq \phi_{\max} < +\frac{\pi}{2} \tag{12.17}$$

where θ_0 is the azimuth angle of $\delta[x - x_0, y - y_0]$. The extremum of p in the set of projections of the 1-D Dirac delta function $\delta[x - x_0, y - y_0]$ is the hypotenuse of the right triangle with sides x_0 and y_0 and included angle θ_0:

$$p_{\max} = \sqrt{x_0^2 + y_0^2} = r_0 \tag{12.18}$$

This is, of course, the radial distance of the 2-D Dirac delta function from the origin on the $[x, y]$ plane.

The projection evaluated at a single point (p_0, ϕ_0) contains "information" from all points in the original function $f[x, y]$ along the line of projection. Note the difference from the 2-D Fourier transform, where the value at each point in the spectrum is a weighted sum of amplitudes at all points in the input function.

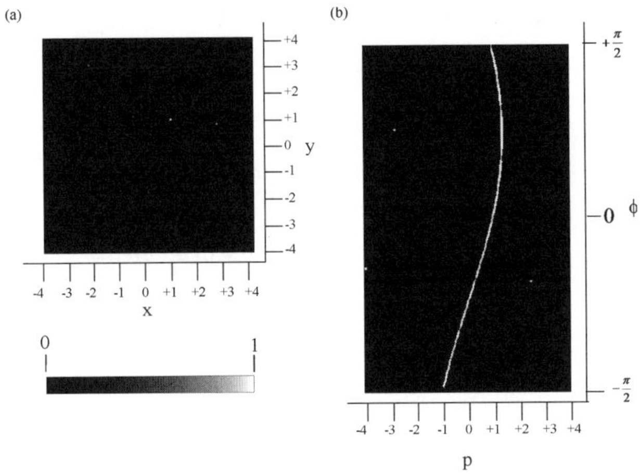

Figure 12.5 Radon transform of off-axis 2-D Dirac delta function: (a) $\delta[x-1, y-1]$; (b) $\lambda_{(\delta[x-1,y-1])}(p, \phi) = \sqrt{2} \cdot \cos(\phi - \pi/4)$. The representation $[p, \phi]$ is a section of a sinusoid, hence the name of "sinogram".

The sinogram representation of the projection $\delta[x - x_0, y - y_0]$ is:

$$\begin{aligned} \lambda_{(\delta[\underline{\mathbf{r}}-\underline{\mathbf{r}}_0])}(p, \phi) &= \delta[p - \underline{\mathbf{r}}_0 \bullet \hat{\underline{\mathbf{p}}}] \\ &= \delta\left[p - |\underline{\mathbf{r}}_0| \cos\left[\phi - \tan^{-1}\left[\frac{y_0}{x_0}\right]\right]\right] \\ &= \delta[p - |\underline{\mathbf{r}}_0| \cos[\phi - \theta_0]] \end{aligned} \tag{12.19}$$

which agrees with the result of Equation (12.16). The Radon transform may be interpreted as a one-to-one mapping between the input amplitude at a specific location $[x_0, y_0]$ and a specific sinusoid in sinogram space. The linearity of $\mathcal{R}_2$ ensures that scaling of the volume of $\delta[x - x_0, y - y_0]$ by a factor of α produces an identical scaling of the areas of the 1-D Dirac delta functions in the sinogram:

$$\begin{aligned} \mathcal{R}_2\{\alpha\delta[x - x_0, y - y_0]\} &= \alpha\delta[p - (x_0 \cos[\phi] + y_0 \sin[\phi])] \\ &= \alpha\delta[p - |\underline{\mathbf{r}}_0| \cos[\phi - \theta_0]] \end{aligned} \tag{12.20}$$

In other words, if the input 2-D Dirac delta function is made "brighter" by applying a multiplicative scale factor by $\alpha > 1$, then the associated sinusoid in the sinogram also will "brighten" by the same factor.

12.1.2 Radon Transform of Arbitrary Function

The linearity of the Radon transform may be combined with the projections of the Dirac delta functions just derived to formulate the Radon transform of an arbitrary function $f[x, y]$. We know that the function may be decomposed into its Dirac delta function components via the convolution:

$$f[x, y] = \iint_{-\infty}^{+\infty} f[\alpha, \beta] \cdot \delta[x - \alpha, y - \beta] \, d\alpha \, d\beta \tag{12.21}$$

Substitution of this expression into the Radon transform yields:

$$\mathcal{R}_2\{f[x,y]\} = \mathcal{R}_2\left\{\iint_{-\infty}^{+\infty} f[\alpha,\beta]\cdot\delta[x-\alpha,y-\beta]\,d\alpha\,d\beta\right\}$$
$$= \iint_{-\infty}^{+\infty} f[\alpha,\beta]\mathcal{R}_2\{\delta[x-\alpha,y-\beta]\}\,d\alpha\,d\beta$$
$$= \iint_{-\infty}^{+\infty} f[\alpha,\beta]\delta[p-\underline{\alpha}\bullet\hat{\underline{\mathbf{p}}}]\,d\alpha\,d\beta = \lambda_f(p,\phi) \tag{12.22a}$$

where $\underline{\alpha} \equiv [\alpha,\beta]$. This may be rewritten in a more intuitive form by substituting the space-domain coordinates $[x,y] = \underline{\mathbf{r}}$ for the dummy variables of integration to obtain:

$$\mathcal{R}_2\{f[x,y]\} = \iint_{-\infty}^{+\infty} f[x,y]\delta[p-\underline{\mathbf{r}}\bullet\hat{\underline{\mathbf{p}}}]\,dx\,dy = \lambda_f(p,\phi) \tag{12.22b}$$

which is (of course) identical to Equation (12.8) and the sinogram representation of $f[x,y]$ that was derived in Equation (12.11). It is the most useful form of the Radon transform for our purposes.

12.2 Radon Transforms of Special Functions

To assist with the development of some intuition, the Radon transforms of some of the special functions are now derived and discussed.

12.2.1 Cylinder Function *CYL(r)*

The circular symmetry of *CYL*(r) means that its line-integral projections are identical for all azimuth angles. Because the amplitude is unity within the support of the cylinder, the projection onto the x-axis is the length of the integration line within the support, which is easy to evaluate by recasting the limits of the integration in terms of x:

$$\lambda_{CYL(r)}(p,0) = \int_{-\infty}^{+\infty} CYL\left(\sqrt{x^2+y^2}\right)dy$$
$$= RECT[x]\cdot\int_{-\frac{1}{2}\sqrt{1-4x^2}}^{+\frac{1}{2}\sqrt{1-4x^2}} 1\,dy$$
$$= RECT[x]\cdot\sqrt{1-4x^2} \tag{12.23}$$

which is shown in Figure 12.6. The circular symmetry enables us to obtain the line-integral projection onto an arbitrary axis specified by $\underline{\mathbf{p}} = (p,\phi)$ by simple substitution of p for x in Equation (12.23):

$$\lambda_{CYL(r)}(p,\phi) = \iint_{-\infty}^{+\infty} CYL(r)\,\delta(p-(x\cos[\phi]+y\sin[\phi]))\,dx\,dy$$
$$= \left(RECT[p]\cdot\sqrt{1-4p^2}\right)1[\phi] \tag{12.24}$$

Figure 12.7 shows the Radon transform of *CYL*(r) plotted as an "image" (with amplitude displayed as gray value) and as a 3-D plot.

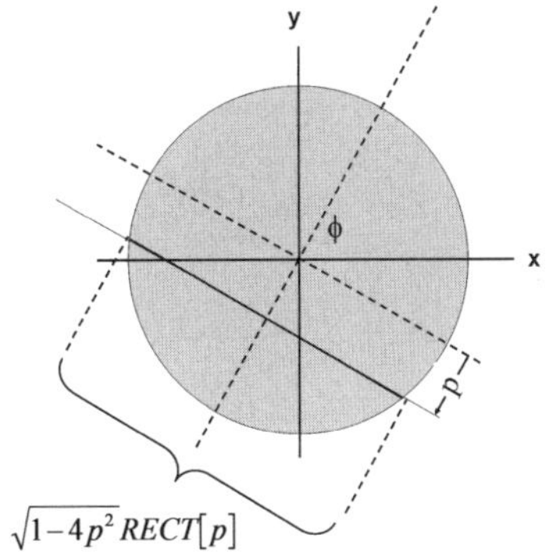

Figure 12.6 Evaluating the Radon transform of $CYL(r)$ at radial distance p is the length of the line within the support of the cylinder. The functional form is $\lambda_{CYL(r)}(p) = \sqrt{1-4p^2} \cdot RECT[p]$.

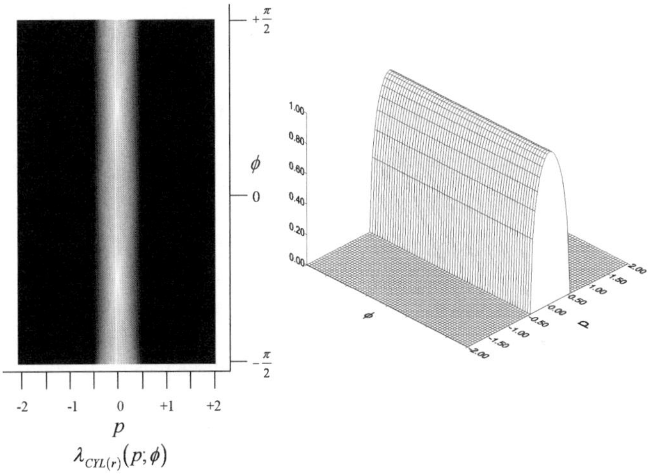

Figure 12.7 The Radon transform of $CYL(r)$ displayed in sinogram format as a Cartesian function of $[p;\,\phi]$, shown in gray-scale and surface formats.

The limits of integration for a cylinder with unit amplitude and diameter d are obviously scaled by the same factor, which means that *both* the amplitude and width of the projection are scaled by d:

$$\begin{aligned}\lambda_{CYL(r/d)}(p,\phi) &= \left(d\sqrt{1-4\left(\frac{p}{d}\right)^2}\right) RECT\left[\frac{p}{d}\right] \\ &= \left(\sqrt{d^2-4p^2}\right) RECT\left[\frac{p}{d}\right] \qquad (12.25)\end{aligned}$$

This result hints at the analogue of the scaling theorem for the Radon transform that will be considered shortly.

12.2.2 Ring Delta Function $\delta(r - r_0)$

The line-integral projections of the circularly symmetric ring delta function $\delta(r - r_0)$ also must be identical for all azimuth angles. The amplitude function of a projection may be evaluated from the limiting expression of $\delta(r - r_0)$ as the difference of two cylinder functions that was introduced in Equation (7.102):

$$\begin{aligned}\delta(r - r_0) &= \lim_{\Delta\to 0}\left\{\frac{1}{\Delta}\left[CYL\left(\frac{r}{2r_0 + \Delta}\right) - CYL\left(\frac{r}{2r_0 - \Delta}\right)\right]\right\}\\ &= \lim_{\Delta\to 0}\left\{\frac{1}{\Delta}\left[CYL\left(\frac{r}{d_0 + \Delta}\right) - CYL\left(\frac{r}{d_0 - \Delta}\right)\right]\right\}\end{aligned} \tag{12.26}$$

where $d_0 = 2r_0$. This limiting expression is shown in Figure 12.8, and the expression for the projection onto the x-axis is:

$$\begin{aligned}&\lambda_{[\delta(r-d_0/2)]}(p, 0)\\ &= \iint_{-\infty}^{+\infty} \lim_{\Delta\to 0}\left\{\frac{1}{\Delta}\left[CYL\left(\frac{\sqrt{x^2+y^2}}{d_0 + \Delta}\right) - CYL\left(\frac{\sqrt{x^2+y^2}}{d_0 - \Delta}\right)\right]\right\}\delta(p - x)\,dx\,dy\\ &= \lim_{\Delta\to 0}\left\{\int_{-\infty}^{+\infty}\left(\frac{1}{\Delta}\left[CYL\left(\frac{\sqrt{p^2+y^2}}{d_0 + \Delta}\right) - CYL\left(\frac{\sqrt{p^2+y^2}}{d_0 - \Delta}\right)\right]\right)dy\right\}\\ &= \lim_{\Delta\to 0}\left\{\frac{2}{\Delta}\left(\frac{\Delta}{\sin[\phi]}\right)\right\}\cdot RECT\left[\frac{p}{d_0}\right] = \frac{2}{\sin[\phi]}\cdot RECT\left[\frac{p}{d_0}\right]\\ &= \frac{2}{\sqrt{1 - (p/(d_0/2))^2}}\cdot RECT\left[\frac{p}{d_0}\right] = \frac{2}{\sqrt{1 - 4(p/d_0)^2}}\cdot RECT\left[\frac{p}{d_0}\right]\end{aligned} \tag{12.27}$$

The amplitude of the projection increases from its minimum value of 2 at the origin to $+\infty$ at $x = \pm 2r_0$. The image and surface plot of the projections are shown in Figure 12.9.

Note that the line-integral projections of the cylinder function and of the ring delta function have the same regions of support, but the amplitude of the projections of $CYL(r/d)$ decreases with increasing $|p|$, while that of the projections of $\delta(r - d/2)$ increases with increasing $|p|$ up to the limits set by $RECT[p/d]$.

The projections of the ring delta function may be usefully applied to an imaging-related application in probability theory. For example, consider the probability density function of the amplitude f of a 1-D sinusoidal function evaluated at a coordinate x selected at random:

$$f[x] = \cos[2\pi\xi_0 x + \phi_0] = \cos[\Phi[x]] \tag{12.28}$$

Are we more likely to select an amplitude near an extremum (i.e., $f \cong \pm 1$) or near the average value ($f \cong 0$)? If we assume that the domain of x is large enough to include many cycles of the sinusoid, then the probability density of f will be independent of ϕ_0, so we set $\phi_0 = 0$ in the remainder of this discussion. Since the phase angle Φ is proportional to x, the "wrapped" phase angle will be uniformly distributed over 2π radians:

$$p[\Phi[x]] = \frac{1}{2\pi}\,RECT\left[\frac{\Phi[x]}{2\pi}\right] \tag{12.29}$$

Of course, the real-valued cosine is the real part of the complex linear-phase function:

$$f[x] = \cos[2\pi\xi_0 x] + i\,\sin[2\pi\xi_0 x] = e^{+i\Phi[x]} \tag{12.30}$$

whose Argand diagram is the ring delta function with unit radius:

$$\delta\left(\sqrt{x^2 + y^2} - 1\right) = \delta(r - 1) \tag{12.31}$$

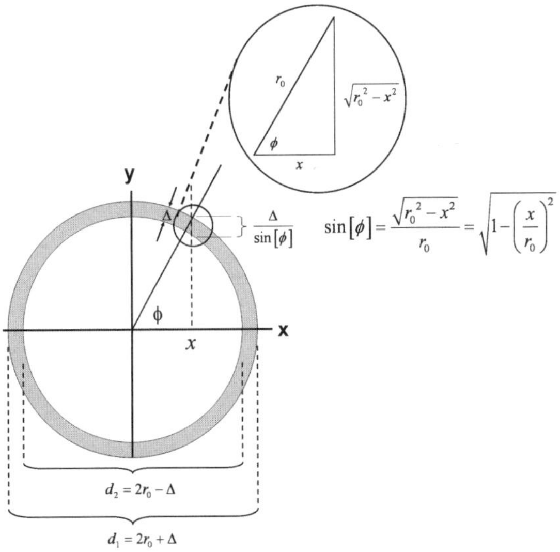

Figure 12.8 Radon transform of the ring delta function $\delta(r - r_0)$ calculated from the limiting expression. All projections have the same form: $\lambda_{\delta(r-r_0)}(p;\, \phi) = \frac{2}{\sqrt{1-4(p/2r_0)^2}}\, RECT[p/2r_0]$.

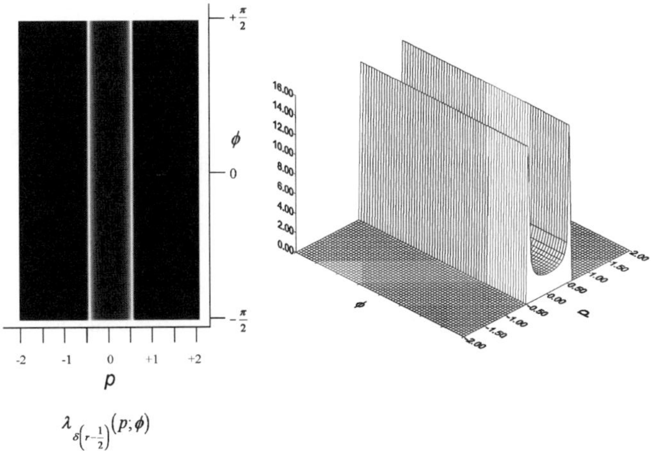

Figure 12.9 The Radon transform of $\delta(r - \frac{1}{2})$ in sinogram format.

The linear phase ensures that the phase angles are uniformly distributed around the ring, so that the probability density of the complex sinusoid is the ring delta normalized to unit volume:

$$p[f[x]] = \frac{1}{2\pi}\delta(r - 1) \tag{12.32}$$

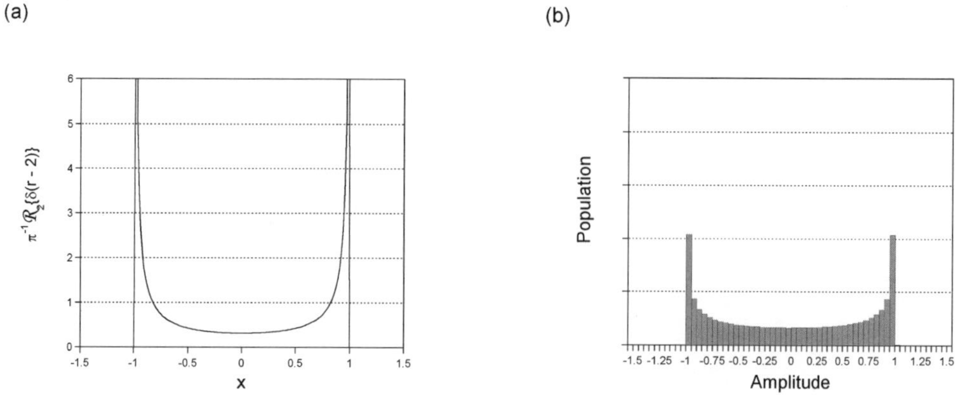

Figure 12.10 The normalized line-integral projection of $\delta(r-1)$ compared to the computed histogram of a 1-D sinusoid with unit amplitude. The discrete histogram approaches the continuous case as the width of the histogram bins is decreased toward zero.

The probability density of the real-valued cosine is the projection of the Argand diagram onto the real axis, after normalizing to unit area:

$$
\begin{aligned}
p[\cos[2\pi\xi_0 x]] &= \frac{1}{2\pi}\left(\frac{2}{\sqrt{1-\cos^2[2\pi\xi_0 x]}}\,RECT\left[\frac{\cos[2\pi\xi_0 x]}{2}\right]\right)\\
&= \frac{1}{\pi}\frac{1}{\sqrt{1-\cos^2[2\pi\xi_0 x]}}\,RECT\left[\frac{\cos[2\pi\xi_0 x]}{2}\right] \qquad (12.33)
\end{aligned}
$$

In words, this shows that the amplitude of a sinusoid evaluated at a randomly selected argument is more likely near the extrema than near the mean value. A histogram computed from a 1-D sinusoid and the line-integral projection of the ring delta function are compared in Figure 12.10.

12.2.3 Rectangle Function $RECT[x, y]$

As was the case for $CYL(r)$, evaluation of the Radon transform of $RECT[x, y]$ is eased by the fact that its amplitude is constant within the region of support. The projection is completely determined by the length ℓ of the line of integration within the support of $RECT[x, y]$. This length obviously varies with azimuth angle ϕ, as may be seen from Figure 12.11. Note that ℓ is identical for all values of p within the interval $0 \le p \le p_0$ and that ℓ decreases linearly with p for $p_0 \le p \le p_{\max}$. It is easy to see that $p_0 = (\cos[\phi] - \sin[\phi])/2$ and $p_{\max} = (\cos[\phi] + \sin[\phi])/2$.

It is straightforward, though somewhat tedious, to evaluate the Radon transform of $RECT[x, y]$ for all ϕ:

$$
\begin{aligned}
\lambda_{RECT[x,y]}\left(p, 0 \le |\phi| \le \frac{\pi}{2}\right) &= \left(\frac{\sqrt{2}\cos[\phi-\pi/4] - 2|p|}{\sin[2\phi]}\right) RECT\left[\frac{p}{\cos[\phi]+\sin[\phi]}\right]\\
&\quad - \left(\frac{\sqrt{2}\cos[\phi+\pi/4] - 2|p|}{\sin[2\phi]}\right) RECT\left[\frac{p}{|\cos[\phi]-\sin[\phi]|}\right] \qquad (12.34)
\end{aligned}
$$

which is shown in Figure 12.12 as both an "image" and as a surface plot. We can (and should) test this equation by evaluating it at angles where the projections are easily evaluated. At $\phi = 0$, we obtain the

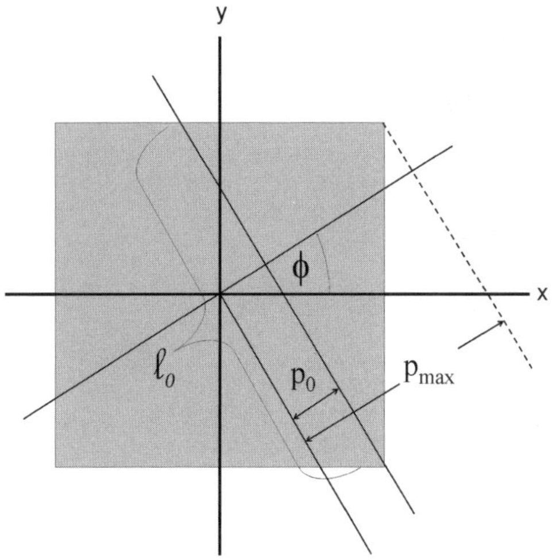

Figure 12.11 The parameters of the computation of $\mathcal{R}_2\{RECT[x, y]\}$. The projection is constant for $0 \le |p| \le p_0$ and decreases linearly for $p_0 \le |p| \le p_{\max}$.

expected rectangle function:

$$\begin{aligned}
&\lambda_{RECT[x,y]}(p, \phi = 0) \\
&= \lim_{\phi \to 0}\left\{\frac{(\cos[\phi] + \sin[\phi]) - 2|p|}{2\cos[\phi]\sin[\phi]}\right\} RECT\left[\frac{p}{1+0}\right] \\
&+ \lim_{\phi \to 0}\left\{\left(1 - \frac{(\cos[\phi] + \sin[\phi]) - 2|p|}{2\cos[\phi]\sin[\phi]}\right)\right\} RECT\left[\frac{p}{1-0}\right] \\
&= \lim_{\phi \to 0}\left\{\frac{(\cos[\phi] + \sin[\phi]) - 2|p|}{2\cos[\phi]\sin[\phi]} - \frac{(\cos[\phi] + \sin[\phi]) - 2|p|}{2\cos[\phi]\sin[\phi]} + 1\right\} RECT\left[\frac{p}{1}\right] \\
&= RECT[p]
\end{aligned} \tag{12.35}$$

while the projection at $\phi = +\pi/4$ is a "wider" and "taller" triangle:

$$\begin{aligned}
\lambda_{RECT[x,y]}\left(p, \frac{\pi}{4}\right) &= (\sqrt{2} - 2|p|)\, RECT\left[\frac{p}{\sqrt{2}}\right] + 2p\, RECT\left[\frac{p}{0}\right] \\
&= \sqrt{2}\, TRI\left[\frac{p}{(\sqrt{2})^{-1}}\right] = \sqrt{2}\, TRI[\sqrt{2}p]
\end{aligned} \tag{12.36}$$

The width of the triangle is determined from the support of p at this angle, which is $|p| \le \sqrt{2}/2$.

12.2.4 Corral Function $COR[x, y]$

The "unit corral" function was defined in Section 7.3.6 as the set of line delta functions along the boundary of $RECT[x, y]$:

$$COR[x, y] \equiv \left(\delta\left[x + \frac{1}{2}\right] + \delta\left[x - \frac{1}{2}\right]\right) RECT[y] + RECT[x]\left(\delta\left[y + \frac{1}{2}\right] + \delta\left[y - \frac{1}{2}\right]\right) \tag{12.37}$$

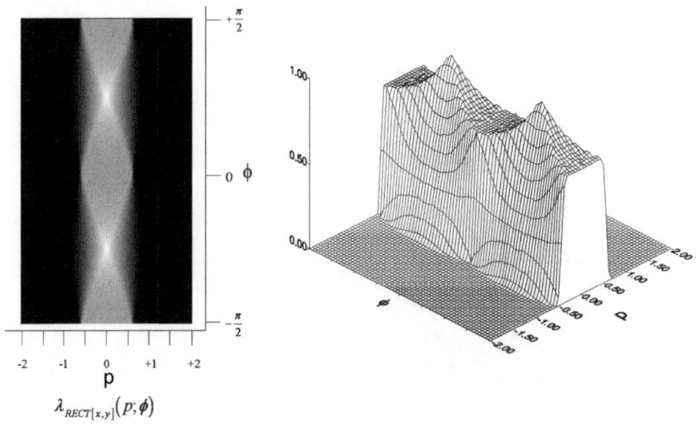

Figure 12.12 The sinogram of $RECT[x, y]$.

Its volume is four units. The line-integral projections of the corral onto the x- and y-axes are identical and easy to evaluate by direct integration:

$$\begin{aligned}\lambda_{COR[x,y]}(p,0) &= \lambda_{COR[x,y]}\left(p, \pm\frac{\pi}{2}\right)\\ &= \int_{-\infty}^{+\infty}\left(\delta\left[x+\frac{1}{2}\right]+\delta\left[x-\frac{1}{2}\right]\right) RECT[y]\,dy\\ &\quad+\int_{-\infty}^{+\infty} RECT[x]\left(\delta\left[y+\frac{1}{2}\right]+\delta\left[y-\frac{1}{2}\right]\right)dy\\ &= \delta\left[p+\frac{1}{2}\right]+\delta\left[p-\frac{1}{2}\right]+2\,RECT[p]\end{aligned} \tag{12.38}$$

The Radon transform of the corral onto an arbitrary axis is more difficult, but may be evaluated by following the analogy with the ring delta function by expressing $COR[x, y]$ as the difference of two rectangle functions in the limit where both widths approach unity. The limiting expression is:

$$COR[x,y] = \lim_{\Delta\to 0}\left\{\frac{1}{\Delta}\left(RECT\left[\frac{x}{1+\Delta},\frac{y}{1+\Delta}\right] - RECT\left[\frac{x}{1-\Delta},\frac{y}{1-\Delta}\right]\right)\right\} \tag{12.39}$$

It is easy to show that the volume of this function in the limit is the correct value of 4:

$$\begin{aligned}\iint_{-\infty}^{+\infty} COR[x,y]\,dx\,dy &= \lim_{\Delta\to 0}\left\{\frac{1}{\Delta}[(1+\Delta)^2-(1-\Delta)^2]\right\}\\ &= \lim_{\Delta\to 0}\left\{\frac{1}{\Delta}[2\Delta-(-2\Delta)]\right\} = 4\end{aligned} \tag{12.40}$$

The projections are evaluated over the same regions $0 \le |p| \le p_0$ and $p_0 \le |p| \le p_{\max}$ that were inferred from Figure 12.11 for $RECT[x, y]$. The line of integration passes through sections with constant, but different, width for these values p. Figure 12.13 shows that $\cos[\phi] = \Delta/\ell$ along the "top" of the corral, which means that $\ell = \Delta/\cos[\phi]$ there. Along the vertical side, $\sin[\phi] = \Delta/\ell \Longrightarrow \ell = \Delta/\sin[\phi]$. These cases are combined to produce the expression for the line-integral projection:

$$\lambda_{COR[x,y]}(p,\phi) = \begin{cases} 2/\cos[\phi] & \textit{if } 0 \le p \le p_0 = (\cos[\phi]-\sin[\phi])/2 \\ 1/\cos[\phi]+1/\sin[\phi] & \textit{if } p_0 < p < p_{\max} = (\cos[\phi]+\sin[\phi])/2 \end{cases} \tag{12.41}$$

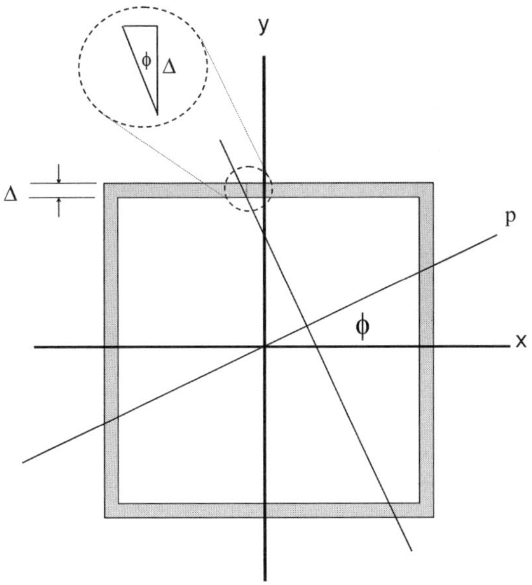

Figure 12.13 Graphical representation of the expression for $COR[x, y]$ as the difference of two rectangle functions.

which may be combined into a single expression:

$$\begin{aligned}\lambda_{COR[x,y]}(p, \phi) &= \left[\frac{1}{\cos[\phi]} + \frac{1}{\sin[\phi]}\right] RECT\left[\frac{p}{\cos[\phi] + \sin[\phi]}\right] \\ &\quad + \left[\frac{2}{\cos[\phi]} - \left(\frac{1}{\cos[\phi]} + \frac{1}{\sin[\phi]}\right)\right] RECT\left[\frac{p}{\cos[\phi] - \sin[\phi]}\right] \\ &= 2\frac{\sin[\phi]}{\sin[2\phi]}\left(RECT\left[\frac{p}{\cos[\phi] + \sin[\phi]}\right] + RECT\left[\frac{p}{\cos[\phi] - \sin[\phi]}\right]\right) \\ &\quad + 2\frac{\cos[\phi]}{\sin[2\phi]}\left(RECT\left[\frac{p}{\cos[\phi] + \sin[\phi]}\right] - RECT\left[\frac{p}{\cos[\phi] - \sin[\phi]}\right]\right) \end{aligned} \tag{12.42}$$

An "image" of the Radon transform of $COR[x, y]$ is shown in Figure 12.14.

12.3 Theorems of the Radon Transform

Just as was the case for the Fourier transform, several relationships satisfied by the Radon transform may be applied to evaluate Radon transforms of other functions. We now consider a few of these, leading up to the most important result, which is the so-called "central-slice theorem" that actually led us to this discussion in the first place. In each of the following sections, the notation $\mathcal{R}_2\{f_n[x, y]\} = \lambda_{(f_n)}(p, \phi)$ is used to represent the Radon transform of the 2-D function f_n.

12.3.1 Radon Transform of a Superposition

We have already considered the action of the Radon transform on a function whose amplitude is scaled by a constant factor. The fact that $\mathcal{R}_2$ is defined by an integral over fixed limits ensures that it satisfies

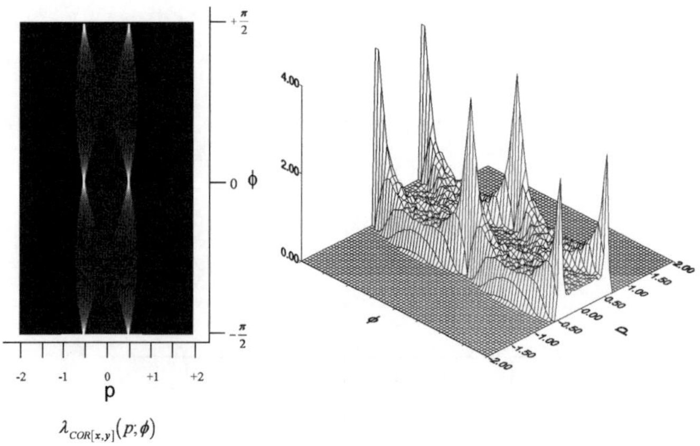

Figure 12.14 Radon transform of $COR[x, y]$.

the constraints for a linear operator:

$$\mathcal{R}_2\left\{\sum_{n=0}^{\infty} \alpha_n f_n[x, y]\right\}$$
$$= \sum_{n=0}^{\infty} \alpha_n \iint_{-\infty}^{+\infty} f_n[x, y]\delta[p - (x\cos[\phi] + y\sin[\phi])] \cdot 1[x\sin[\phi] - y\cos[\phi]]\,dx\,dy$$
$$= \sum_{n=0}^{\infty} \alpha_n \lambda_{(f_n)}(p, \phi) \tag{12.43}$$

12.3.2 Radon Transform of Scaled Function

In the earlier discussion of the Radon transform of a cylinder function, we noted how the functional form of the Radon transform is affected if the diameter is scaled by a factor d. At this point, we derive the corresponding result for an arbitrary function $f[x, y]$ if scaled by identical factors along both axes. Given that $\mathcal{R}_2\{f[x, y]\} = \lambda_f(p, \phi)$, then the Radon transform of $f[x/b, y/b]$ is easy to evaluate by changing the integration variables to $u = x/b$ and $v = y/b$:

$$\mathcal{R}_2\left\{f\left[\frac{x}{b}, \frac{y}{b}\right]\right\} = \iint_{-\infty}^{+\infty} f\left[\frac{x}{b}, \frac{y}{b}\right] \cdot (\delta[p - (x\cos[\phi] + y\sin[\phi])] \cdot 1[x\sin[\phi] - y\cos[\phi]])\,dx\,dy$$
$$= |b|^2 \iint_{-\infty}^{+\infty} f[u, v]\delta\left[b\left(\frac{p}{b} - (u\cos[\phi] + v\sin[\phi])\right)\right]$$
$$\cdot 1[bu\sin[\phi] - bv\cos[\phi]]\,du\,dv$$
$$= |b|\lambda_f\left(\frac{p}{b}, \phi\right) \tag{12.44}$$

where the invariance of the constant function under scaling has been used. In words, both the support and the amplitude of the projections of a scaled function are scaled by b. By substituting $b = -1$, we see that "reversing" $f[x, y]$ produces a corresponding reversal of the line-integral projections, i.e.,

$$\mathcal{R}_2\{f[x, y]\} = \lambda_f(p, \phi) \Longrightarrow \mathcal{R}_2\{f[-x, -y]\} = \lambda_f(-p, \phi) = \lambda_f(p, \phi + \pi) \tag{12.45}$$

12.3.3 Radon Transform of Translated Function

We have already concluded that the Radon transform is a shift-variant operation. The actual form of the Radon transform obtained from the translated function is easy to derive by direct substitution into Equation (12.22):

$$\begin{aligned}
&\mathcal{R}_2\{f[x-x_0, y-y_0]\} \\
&= \iint_{-\infty}^{+\infty} f[x-x_0, y-y_0]\delta[p-(x\cos[\phi]+y\sin[\phi])]\,dx\,dy \\
&= \iint_{-\infty}^{+\infty} f[u, v]\delta[p-((u+x_0)\cos[\phi]+(v+y_0)\sin[\phi])]\,du\,dv \\
&= \iint_{-\infty}^{+\infty} f[u, v]\delta[p-(x_0\cos[\phi]+y_0\sin[\phi])-(u\cos[\phi]+v\sin[\phi])]\,dx\,dy \\
&= \lambda_f(p-\underline{\mathbf{r}}_0 \bullet \underline{\mathbf{p}};\phi) \\
&= \lambda_f(p-(x_0\cos[\phi]+y_0\sin[\phi]),\phi)
\end{aligned} \tag{12.46}$$

This demonstrates that the spatial coordinate of the Radon transform of a translated function is itself translated by a parameter that varies with the azimuthal angle ϕ. Note that this result also may be inferred from the Radon transform of a translated 2-D Dirac delta function in Equation (12.20) and the linearity of the transform.

12.3.4 Central-Slice Theorem

We introduced the Radon transform by evaluating $F[\xi, 0]$ from $f[x, y]$ in Equation (12.3). The result of a similar process applied to the general formulation for the Radon transform in Equation (12.13) produces what is arguably the most important feature of the Radon transform.

Consider first the 1-D Fourier transform of a specific line-integral projection $\lambda_f(p, \phi_0)$:

$$\begin{aligned}
\mathop{\mathcal{F}_1}_{p\to\nu}\{\lambda_f(p,\phi_0)\} &= \int_{-\infty}^{+\infty} \lambda_f(p,\phi_0)\,e^{-2\pi i\nu p}\,dp \\
&= \int_{-\infty}^{+\infty}\left(\iint_{-\infty}^{+\infty} f[x,y]\,\delta[p-\underline{\mathbf{r}}\bullet\hat{\underline{\mathbf{p}}}_0]\,dx\,dy\right)e^{-2\pi i\nu p}\,dp \\
&= \iint_{-\infty}^{+\infty} f[x,y]\left(\int_{-\infty}^{+\infty}\delta[p-\underline{\mathbf{r}}\bullet\hat{\underline{\mathbf{p}}}_0]\,dp\right)e^{-2\pi i\nu(\underline{\mathbf{r}}\bullet\hat{\underline{\mathbf{p}}}_0)}\,dx\,dy \\
&= \iint_{-\infty}^{+\infty} f[x,y]\,e^{-2\pi i\nu(\underline{\mathbf{r}}\bullet\hat{\underline{\mathbf{p}}}_0)}\,dx\,dy \\
&= \iint_{-\infty}^{+\infty} f[x,y]\exp[-2\pi i\nu(x\cos[\phi_0]+y\sin[\phi_0])]\,dx\,dy \\
&= \iint_{-\infty}^{+\infty} f[x,y]\exp[-2\pi i(x(\nu\cos[\phi_0])+y(\nu\sin[\phi_0]))]\,dx\,dy \\
&= F[\nu\cos[\phi_0],\nu\sin[\phi_0]] = F[\nu\hat{\underline{\mathbf{p}}}_0]
\end{aligned} \tag{12.47}$$

where $|\nu| = \sqrt{\xi^2+\eta^2}$. Our previous convention for naming spectra suggests the notation of the upper-case Greek letter "lambda" for the amplitude of the 1-D spectrum of the line-integral projection evaluated at azimuth angle ϕ_0:

$$\mathcal{F}_{1\,p\to\nu}\{\lambda_f(p,\phi_0)\} = F[\nu\underline{\hat{\mathbf{p}}}_0] = F[\nu\cos[\phi_0],\ \nu\sin[\phi_0]] = \Lambda_f(\nu,\phi_0) \tag{12.48}$$

There is another, and arguably simpler, derivation of this feature that is based on Equation (12.8), the 2-D filter theorem, and the property of Dirac delta functions in products:

$$\begin{aligned}\mathcal{F}_2\{f[x,y] * (\delta[\underline{\mathbf{r}}\bullet\underline{\hat{\mathbf{p}}}_0]\cdot 1[\underline{\mathbf{r}}\bullet\underline{\hat{\mathbf{p}}}_0^{\perp}])\} &= F[\xi,\eta]\cdot(1[\underline{\boldsymbol{\rho}}\bullet\underline{\hat{\mathbf{p}}}_0]\cdot\delta[\underline{\boldsymbol{\rho}}\bullet\underline{\hat{\mathbf{p}}}_0^{\perp}])\\ &= F[\rho\cos[\phi_0],\ \rho\sin[\phi_0]]\cdot(1[\underline{\boldsymbol{\rho}}\bullet\underline{\hat{\mathbf{p}}}_0]\cdot\delta[\underline{\boldsymbol{\rho}}\bullet\underline{\hat{\mathbf{p}}}_0^{\perp}])\\ &= F[\nu\underline{\hat{\mathbf{p}}}_0]\cdot(1[\underline{\boldsymbol{\rho}}\bullet\underline{\hat{\mathbf{p}}}_0]\cdot\delta[\underline{\boldsymbol{\rho}}\bullet\underline{\hat{\mathbf{p}}}_0^{\perp}])\\ &= \Lambda_f(\nu,\phi_0)\cdot\delta[\underline{\boldsymbol{\rho}}\bullet\underline{\hat{\mathbf{p}}}_0^{\perp}]\end{aligned} \tag{12.49}$$

where the last step follows from the property of Dirac delta functions in products. In words, the 2-D spectrum of the 2-D convolution of $f[x, y]$ with any particular line delta function through the origin perpendicular to $\underline{\hat{\mathbf{p}}}_0$ is identical to the product of $F[\xi, \eta]$ and the line delta function through the origin along the direction of $\underline{\hat{\mathbf{p}}}_0$ (i.e., perpendicular to $\underline{\hat{\mathbf{p}}}_0^{\perp}$, where we recall that both $\underline{\hat{\mathbf{p}}}_0$ and $\underline{\hat{\mathbf{p}}}_0^{\perp}$ are constrained to the first or fourth quadrants); the result is a "slice" of the 2-D Fourier transform evaluated along this specific radial line. Note that the 2-D transform of the 2-D convolution results in the 2-D separable function $\Lambda_f(\nu, \phi_0)\delta[\underline{\boldsymbol{\rho}}\bullet\underline{\hat{\mathbf{p}}}_0^{\perp}]$, though the two components are not "Cartesian separable" because they are directed along orthogonal axes that do not coincide (in general) with the x- and y-axes. This observation may be used to recognize that $F[\nu\underline{\hat{\mathbf{p}}}_0]$ from Equation (12.47) is identical to the 1-D spatial Fourier transform of a specific line-integral projection $\Lambda_f(p, \phi_0)$:

$$\mathcal{F}_{1\,p\to\nu}\{\lambda_f(p,\phi_0)\} \equiv \Lambda_f(\nu,\phi_0) = F(\underline{\boldsymbol{\rho}})|_{\underline{\boldsymbol{\rho}}=\underline{\hat{\mathbf{p}}}_0\nu} = F(\nu\underline{\hat{\mathbf{p}}}_0) \tag{12.50}$$

By computing the 1-D Fourier transforms of each line-integral projection in sequence, we can evaluate each radial "slice" of the 2-D spectrum. This result is the essential intellectual step that demonstrates the equivalence of the set of line-integral projections $\lambda_f(p, \phi)$ and $F[\xi, \eta]$, and hence the equivalence to $f[x, y]$. These equivalent expressions are called the "central-slice" or "projection-slice" theorem. An example is shown in Figure 12.15.

Note that the amplitudes of all 1-D spectrum slices evaluated at $\nu = 0$ are identical to the "DC" component of the 2-D spectrum:

$$\Lambda_f(\nu = 0;\phi) = F[0, 0] \tag{12.51}$$

This observation will be useful when considering the inverse Radon transform.

12.3.5 Filter Theorem of the Radon Transform

The central-slice theorem leads naturally to the filter theorem of the Radon transform, which specifies the relationship between the Radon transforms of two functions and the Radon transform of their convolution.

$$\mathcal{F}_2\{g[x,y] * \delta[\underline{\mathbf{r}}\bullet\underline{\hat{\mathbf{p}}}]\cdot 1[\underline{\mathbf{r}}\bullet\underline{\hat{\mathbf{p}}}^{\perp}]\} = \Lambda_g(\nu,\phi)\cdot(1[\underline{\boldsymbol{\rho}}\bullet\underline{\hat{\mathbf{p}}}]\cdot\delta[\underline{\boldsymbol{\rho}}\bullet\underline{\hat{\mathbf{p}}}^{\perp}])$$

$$\begin{aligned}\mathcal{F}_2\{g[x,y] * \delta[\underline{\mathbf{r}}\bullet\underline{\hat{\mathbf{p}}}]\cdot 1[\underline{\mathbf{r}}\bullet\underline{\hat{\mathbf{p}}}^{\perp}]\} &= G[\xi,\eta]\cdot(1[\underline{\boldsymbol{\rho}}\bullet\underline{\hat{\mathbf{p}}}]\cdot\delta[\underline{\boldsymbol{\rho}}\bullet\underline{\hat{\mathbf{p}}}^{\perp}])\\ &= (F[\xi,\eta]\cdot H[\xi,\eta])\cdot(1[\underline{\boldsymbol{\rho}}\bullet\underline{\hat{\mathbf{p}}}]\cdot\delta[\underline{\boldsymbol{\rho}}\bullet\underline{\hat{\mathbf{p}}}^{\perp}])\\ &= (F[\xi,\eta]\cdot(1[\underline{\boldsymbol{\rho}}\bullet\underline{\hat{\mathbf{p}}}]\cdot\delta[\underline{\boldsymbol{\rho}}\bullet\underline{\hat{\mathbf{p}}}^{\perp}]))\cdot(H[\xi,\eta](1[\underline{\boldsymbol{\rho}}\bullet\underline{\hat{\mathbf{p}}}]\cdot\delta[\underline{\boldsymbol{\rho}}\bullet\underline{\hat{\mathbf{p}}}^{\perp}]))\end{aligned} \tag{12.52a}$$

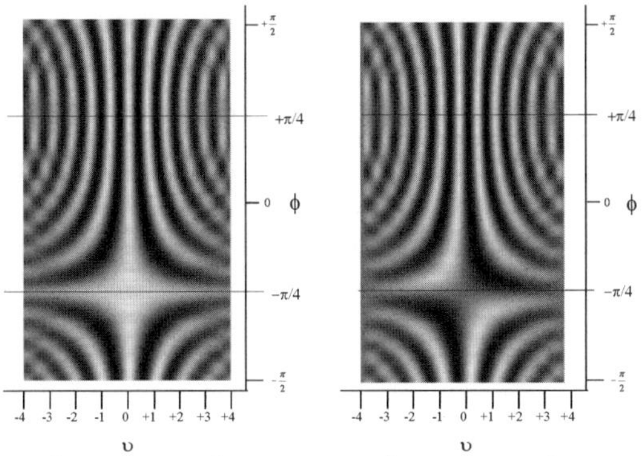

$\Re\{\Lambda_{\delta[x-1,y-1]}[\upsilon;\phi]\}$ $\quad$ $\Im\{\Lambda_{\delta[x-1,y-1]}[\upsilon;\phi]\}$

Figure 12.15 The central-slice theorem: the 1-D Fourier transforms of the line-integral projections in the sinogram of $\delta[x-1, y-1]$, shown in Figure 12.5. The projection at $\phi = -\pi/4$ is $\delta[p]$, and so its 1-D Fourier transform is $1[\upsilon]$. The projection at $\phi = +\pi/4$ is $\delta[p-\sqrt{2}]$; its 1-D spectrum is $\exp[-2\pi i(\sqrt{2}\upsilon)]$. At other angles, the spectra of the projections are complex sinusoids with larger spatial frequencies.

which is equal to the product of projections along the same azimuth:

$$\begin{aligned}\Lambda_g(\nu,\phi)\cdot(1[\underline{\rho}\bullet\hat{\underline{\mathbf{p}}}]\delta[\underline{\rho}\bullet\hat{\underline{\mathbf{p}}}^{\perp}]) &= (\Lambda_f(\nu,\phi)\cdot(1[\underline{\rho}\bullet\hat{\underline{\mathbf{p}}}]\cdot\delta[\underline{\rho}\bullet\hat{\underline{\mathbf{p}}}^{\perp}]))\\ &\quad\cdot(\Lambda_h(\nu,\phi)\cdot(1[\underline{\rho}\bullet\hat{\underline{\mathbf{p}}}]\cdot\delta[\underline{\rho}\bullet\hat{\underline{\mathbf{p}}}^{\perp}]))\\ &= \Lambda_f(\nu,\phi)\cdot\Lambda_h(\nu,\phi)\cdot(1[\underline{\rho}\bullet\hat{\underline{\mathbf{p}}}]\cdot\delta[\underline{\rho}\bullet\hat{\underline{\mathbf{p}}}^{\perp}])\\ &\Longrightarrow \Lambda_g(\nu,\phi) = \Lambda_f(\nu,\phi)\cdot\Lambda_h(\nu,\phi) \qquad (12.52b)\end{aligned}$$

$$\begin{aligned}\underset{\nu\to p}{\mathcal{F}_1^{-1}}\{\Lambda_g(\nu,\phi)\} &= \lambda_g(p,\phi)\\ &= \underset{\nu\to p}{\mathcal{F}_1^{-1}}\{\Lambda_f(\nu,\phi)\cdot\Lambda_h(\nu,\phi)\} = \lambda_f(p,\phi) * \lambda_h(p,\phi) \qquad (12.52c)\end{aligned}$$

where the property of Dirac delta functions in products has been used. Note that the convolution $\lambda_f(p,\phi) * \lambda_h(p,\phi)$ is, of course, one dimensional. In words, the line-integral projection evaluated at ϕ_0 of the 2-D convolution of $f[x, y]$ and $h[x, y]$ is identical to the 1-D convolution of the line-integral projections of the two functions evaluated at the same azimuth.

This very useful result demonstrates that the Radon transform of the 2-D convolution of two functions is identical to the 1-D convolution of the projections of the individual functions. If we can derive a method for "reconstructing" a 2-D function $f[x, y]$ from the complete set of its line-integral projections $\lambda_f(p,\phi)$, then this result may be used to compute the 2-D convolution. The central-slice and filter theorems of the Radon transform have been applied to process 2-D images via 1-D signal processors (Easton and Barrett, 1986).

12.4 Inverse Radon Transform

We now have sufficient information to determine whether $f[x, y]$ and $\lambda_f(p,\phi)$ are equivalent representations. If they are, then an operator $\mathcal{R}_2^{-1}$ must exist that "reconstructs" $f[x, y]$ from the complete set of projections $\lambda_f(p, \phi)$. This is, in fact, true and several equivalent recipes for $\mathcal{R}_2^{-1}$ exist; we will outline the developments of only a few. For a deeper discussion, consult the references, notably Barrett (1984) and Barrett and Myers (2004).

We will consider two routes to reconstructing $f[x, y]$. Perhaps the more easily understood path travels through the frequency domain via the central-slice theorem in Equation (12.50). Since any 1-D profile of the 2-D spectrum $F[\xi, \eta]$ may be evaluated from the appropriate line-integral projection, it seems reasonable that the 2-D spectrum $F[\xi, \eta]$ can be synthesized by summing the complete set of its 1-D "slices" over azimuth angle. A perhaps less circuitous and less easily comprehended recipe, commonly known as "filtered back projection", operates within the space domain only. We will derive the two sequences of operations for the projections of $\delta[x, y]$. Fortunately, the recipes for these two problems are based on some very similar mathematical operations. The algorithms will then be generalized to the recovery of $\delta[\underline{\mathbf{r}} - \underline{\mathbf{r}}_0]$ from its projections. We will see that the same recipes applying to both the on-axis and off-axis Dirac delta functions are identical, and thus will also be appropriate for recovering arbitrary functions from their projections.

12.4.1 Recovery of Dirac Delta Function from Projections

12.4.1.1 Synthesis of $F[\xi, \eta] = 1[\xi, \eta]$ from $\delta[p] \cdot 1[\phi]$

We first investigate the reconstruction of $\delta[x, y]$ from the set of its line-integral projections that were derived in Equation (12.15):

$$\mathcal{R}_2\{\delta[x, y]\} = \lambda_{(\delta[x,y])}(p, \phi) = \delta[p] \cdot 1[\phi] \tag{12.53}$$

The corresponding 1-D Fourier transforms are identically unit-constant functions of the bipolar radial frequency ν. For a specific azimuth ϕ_0 (in the first or fourth quadrants, so that $-\pi/2 \leq \phi_0 < +\pi/2$), the relation between the 2-D line-integral projection and the 2-D frequency-domain slice is:

$$\begin{aligned}\mathcal{F}_2\{\delta[x, y] * \delta[\underline{\mathbf{r}} \bullet \hat{\underline{\mathbf{p}}}_0] \cdot 1[\underline{\mathbf{r}} \bullet \hat{\underline{\mathbf{p}}}_0^{\perp}]\} &= 1[\xi, \eta] \cdot (1[\underline{\boldsymbol{\rho}} \bullet \hat{\underline{\mathbf{p}}}_0] \cdot \delta[\underline{\boldsymbol{\rho}} \bullet \hat{\underline{\mathbf{p}}}_0^{\perp}]) \\ &= 1[\nu \hat{\underline{\mathbf{p}}}_0] \cdot (1[\underline{\boldsymbol{\rho}} \bullet \hat{\underline{\mathbf{p}}}_0] \cdot \delta[\underline{\boldsymbol{\rho}} \bullet \hat{\underline{\mathbf{p}}}_0^{\perp}])\end{aligned} \tag{12.54}$$

where again the radial line is defined by $\hat{\underline{\mathbf{p}}}_0 = [\cos[\phi_0], \sin[\phi_0]]$. The goal is to construct the known 2-D unit-constant spectrum $\mathcal{F}_2\{\delta[x, y]\} = 1[\xi, \eta]$ from these 1-D unit-constant "central slices", which requires a summation of the contributions at the different azimuthal angles. Consider first the perhaps naive idea that the 2-D unit-constant spectrum is obtained by the azimuthal integral of the unit-amplitude 1-D "slices":

$$\int_{-\pi/2}^{+\pi/2} 1[\nu \hat{\underline{\mathbf{p}}}](1[\underline{\boldsymbol{\rho}} \bullet \hat{\underline{\mathbf{p}}}] \cdot \delta[\underline{\boldsymbol{\rho}} \bullet \hat{\underline{\mathbf{p}}}^{\perp}]) \, d\phi = \int_{-\pi/2}^{+\pi/2} \delta[\underline{\boldsymbol{\rho}} \bullet \hat{\underline{\mathbf{p}}}^{\perp}] \, d\phi \stackrel{?}{=} 1[\xi, \eta] \tag{12.55}$$

where the spatial frequency vector $\underline{\boldsymbol{\rho}}$ may be specified by its Cartesian coordinates $[\xi, \eta]$ or by its radial polar coordinates $(|\underline{\boldsymbol{\rho}}|, \psi)$ with $-\pi \leq \psi < +\pi$, or by its radial bipolar coordinates (ν, ψ) with $-\pi/2 \leq \psi < +\pi/2$. The constant functions have been deleted for simplicity. In fact, the two sides of Equation (12.55) are *not* equal because the amplitude of the integrated spectrum at a particular spatial frequency $[\xi_0, \eta_0]$ is determined in part by the amplitude of the spectral "slices" and in part by the "density" of radial spectral slices in the neighborhood of $[\xi_0, \eta_0]$. We use the expression for the line delta function in Equation (7.64) to rewrite the integrand:

$$\begin{aligned}\delta[\underline{\rho} \bullet \hat{\underline{\mathbf{p}}}^{\perp}] &= \delta\left[|\underline{\rho}|\,|\hat{\underline{\mathbf{p}}}^{\perp}|\cos\left[\psi - \left(\phi \pm \frac{\pi}{2}\right)\right]\right] \\ &= \delta\left[|\nu|\cos\left[\psi - \left(\phi \pm \frac{\pi}{2}\right)\right]\right] \\ &= \delta\left[|\nu|\cos\left[(\phi - \psi) \mp \frac{\pi}{2}\right]\right]\end{aligned} \tag{12.56a}$$

where the radial bipolar frequency coordinates are used so that the domains of ψ and ϕ are identical. The integration variable is ϕ, which means that $|\nu|$ is a constant in the integral. It may be extracted from the 1-D Dirac delta function via the scaling property in Equation (6.107):

$$\delta[\underline{\rho} \bullet \hat{\underline{\mathbf{p}}}^{\perp}] = \delta\left[|\nu|\,|\hat{\underline{\mathbf{p}}}^{\perp}|\cos\left[(\phi - \psi) \mp \frac{\pi}{2}\right]\right] = \frac{1}{|\nu|}\delta[\pm\sin[\phi - \psi]] \tag{12.56b}$$

where the unit length of $\hat{\underline{\mathbf{p}}}^{\perp}$ has also been used. The remaining expression is a 1-D Dirac delta function with a functional argument that was considered in Equation (6.138) and is very similar to Equation (7.66). We recognize that $\sin[\phi - \psi] = 0$ when $\phi = \psi$ to obtain:

$$\begin{aligned}\delta[\underline{\rho} \bullet \hat{\underline{\mathbf{p}}}^{\perp}] &= \frac{1}{|\nu|}\left(\frac{1}{\left|\pm d\sin[\phi - \psi]/d\phi|_{\phi=\psi}\right|}\right)\delta[\phi - \psi] \\ &= \frac{1}{|\nu|}\left(\frac{1}{|\cos[\phi - \psi]|_{\phi=\psi}|}\right)\delta[\phi - \psi] \\ &= \frac{1}{|\nu|}\left(\frac{1}{|\cos[0]|}\right)\delta[\phi - \psi] = \frac{1}{|\nu|}\delta[\phi - \psi]\end{aligned} \tag{12.56c}$$

The azimuthal integral is now easy to evaluate:

$$\begin{aligned}\int_{-\pi/2}^{+\pi/2} 1[\nu\hat{\underline{\mathbf{p}}}] \cdot (1[\underline{\rho} \bullet \hat{\underline{\mathbf{p}}}] \cdot \delta[\underline{\rho} \bullet \hat{\underline{\mathbf{p}}}^{\perp}])\, d\phi &= \frac{1}{|\nu|}\int_{-\pi/2}^{+\pi/2} \delta[\phi - \psi]\, d\phi \\ &= \frac{1}{|\nu|} \cdot 1 = \frac{1}{\sqrt{\xi^2 + \eta^2}} = \frac{1}{\rho}\end{aligned} \tag{12.57}$$

In words, the summation of the 1-D unit-constant "slices" produces the 2-D circularly symmetric "radially decaying" function $|\nu|^{-1} = \rho^{-1}$, instead of the required 2-D unit-constant spectrum, as shown in Figure 12.16. In other words, the summation of the constant slices over azimuth angles produces a spectrum with amplified amplitudes of sinusoidal components with "small" radial spatial frequencies ($\sqrt{\xi^2 + \eta^2} < 1$), while those components with large spatial frequencies $\sqrt{\xi^2 + \eta^2} > 1$ are attenuated. The reconstructed image in the space domain is the inverse 2-D Fourier transform of the spectrum in Equation (12.57), which was derived as a Hankel transform in Equation (11.53):

$$\mathcal{F}_2^{-1}\left\{\frac{1}{\rho}\right\} = \frac{1}{r} = \frac{1}{\sqrt{x^2 + y^2}} \neq \delta[x, y] \tag{12.58}$$

The 2-D function that is "recovered" by integrating the slices of the spectrum and the subsequent 2-D inverse Fourier transform is not the original 2-D Dirac delta function, but rather a "blurry" version r^{-1}. This demonstrates that our simple-minded guess of the formulation of the inverse Radon transform in Equation (12.55) was not correct. A modification that compensates for the attenuation of components with larger spatial frequencies will be discussed shortly.

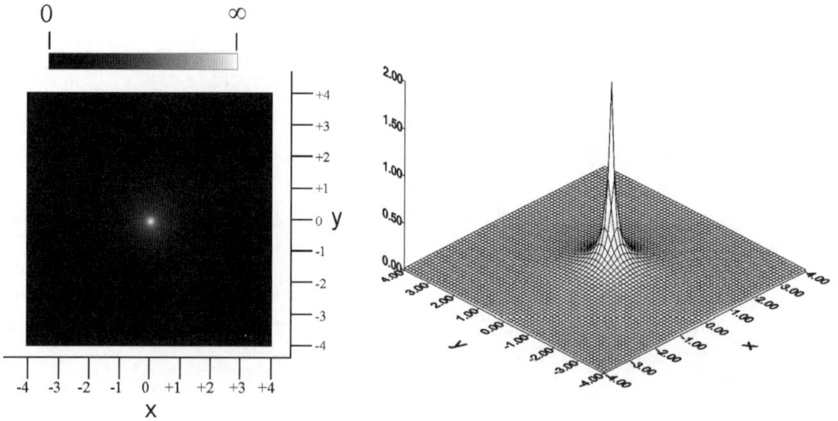

Figure 12.16 Summation of unit-amplitude central slices of the spectrum of $\delta[x, y]$ over azimuth angle produces the 2-D circularly symmetric function ρ^{-1}, which may be considered an approximation of the ideal spectrum $\mathcal{F}_2\{\delta[x, y]\} = 1[\xi, \eta]$.

12.4.1.2 Summation of Projections of $\delta[x, y]$ in Space Domain

The space-domain representation of an individual slice of the spectrum is:

$$\mathcal{F}_{1\atop \nu\to p}^{-1}\{1[\underline{\rho} \bullet \hat{\underline{\mathbf{p}}}_0] \cdot \delta[\underline{\rho} \bullet \hat{\underline{\mathbf{p}}}_0^{\perp}]\} = \delta[\underline{\mathbf{r}} \bullet \hat{\underline{\mathbf{p}}}_0] \cdot 1[\underline{\mathbf{r}} \bullet \hat{\underline{\mathbf{p}}}_0^{\perp}] \tag{12.59}$$

where the space-domain vector $\underline{\mathbf{r}} = [x, y] = (|\underline{\mathbf{r}}|, \theta)$. This separable product of the projection and the orthogonal unit constant is also a line delta function, and may be viewed as the 1-D projection $\lambda_{(\delta[x,y])}(p, \phi_0)$ "smeared" back along the orthogonal direction. The common term for such a smeared projection is a "back projection". The integral of these back projections over ϕ was used to obtain Equation (7.66) and is very similar to that used in Equation (12.57):

$$\begin{aligned}\int_{\phi=-\pi/2}^{+\pi/2} \delta[\underline{\mathbf{r}} \bullet \hat{\underline{\mathbf{p}}}] \cdot 1[\underline{\mathbf{r}} \bullet \hat{\underline{\mathbf{p}}}^{\perp}]\, d\phi &= \int_{\phi=-\pi/2}^{+\pi/2} \delta[|\underline{\mathbf{r}}||\hat{\underline{\mathbf{p}}}| \cos[\theta - \phi]]\, d\phi \\ &= \frac{1}{r} \int_{\phi=-\pi/2}^{+\pi/2} \delta\left[\phi - \left(\theta \pm \frac{\pi}{2}\right)\right] d\phi = \frac{1}{r}\end{aligned} \tag{12.60}$$

Each back projection contributes positive amplitude along the appropriate azimuth angle, so the summation of all of the back projections is positive everywhere, as shown in Figure 12.17. This is an important observation: the imaging system that cascades line-integral projection and integration over azimuth attenuates sinusoidal components with large spatial frequencies. This summation of back projections in the space domain is sometimes called a *layergram*.

The summation of back projections that produces r^{-1} instead of the $\delta[x, y]$ may be viewed as the effect of a filter that "blurs" the object as a result of the linear integration in the Radon transform. This linear summation of amplitudes is proportional to a linear "averaging" operator that was briefly discussed in Chapter 8. Because it is an "integrator", the Radon transform at each azimuth acts like a lowpass filter and thus attenuates some high-frequency components. This means that some kind of amplification of the attenuated components is required to compensate for the integration and recover the correct result.

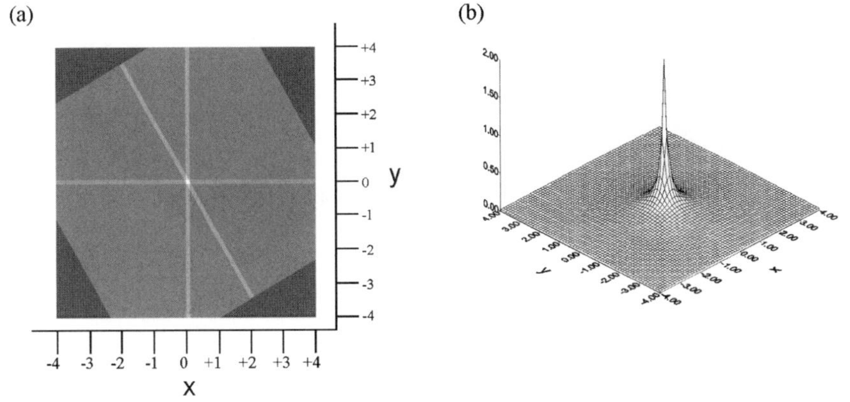

Figure 12.17 The concept of back projection. (a) Simulation of the summation of three 2-D line delta back projections of $\delta[x, y]$, showing that each back projection contributes positive amplitude to the summation image. The integral of a complete set of back projections yields $1/r$, shown in (b).

12.4.1.3 "High-Boost" Filtering of Back Projections

The remaining question is how to compensate the azimuthal integrals of both the frequency-domain "slices" and the space-domain "back projections" to produce the correct reconstructed images. The original summations in the two domains were calculated in Equations (12.57) and (12.60), respectively:

$$\int_{-\pi/2}^{+\pi/2} 1[\nu\hat{\underline{\mathbf{p}}}]\,(1[\underline{\rho}\bullet\hat{\underline{\mathbf{p}}}]\cdot\delta[\underline{\rho}\bullet\hat{\underline{\mathbf{p}}}^{\perp}])\,d\phi = \frac{1}{\rho} = \frac{1}{\sqrt{\xi^2+\eta^2}} \neq 1[\xi,\eta] \tag{12.61a}$$

$$\int_{\phi=-\pi/2}^{+\pi/2} \delta[\underline{\mathbf{r}}\bullet\hat{\underline{\mathbf{p}}}]\cdot 1[\underline{\mathbf{r}}\bullet\hat{\underline{\mathbf{p}}}^{\perp}]\,d\phi = \frac{1}{r} = \frac{1}{\sqrt{x^2+y^2}} \neq \delta[x,y] \tag{12.61b}$$

We need to filter these functions to obtain the correct functions in each domain: the unit constant in the frequency domain and the centered 2-D Dirac delta function in the space domain. The filtering process is easier to visualize in the frequency domain. Evidently, the correct spectrum is obtained by applying a transfer function $H[\xi, \eta]$ where:

$$\frac{1}{\sqrt{\xi^2+\eta^2}}\cdot H[\xi,\eta] = 1[\xi,\eta] \Longrightarrow H[\xi,\eta] = \sqrt{\xi^2+\eta^2} = \rho \tag{12.62}$$

as shown in Figure 12.18. This illustrates the reason for the common name of "rho filter" that is applied to the compensation. Note the obvious difficulty with the singularity of the spectrum located at the origin: it is not possible to uniquely scale the infinite amplitude there to unity.

The filter theorem of the Radon transform in Equation (12.52) may be applied to multiply the 1-D central slices of the spectrum by the (identical) 1-D slices of the rho filter before the azimuthal summation:

$$\left(\int_{-\pi/2}^{+\pi/2} 1[\nu\hat{\underline{\mathbf{p}}}]\,(1[\underline{\rho}\bullet\hat{\underline{\mathbf{p}}}]\cdot\delta[\underline{\rho}\bullet\hat{\underline{\mathbf{p}}}^{\perp}])\,d\phi\right)\cdot\rho = \int_{-\pi/2}^{+\pi/2} (1[\nu\hat{\underline{\mathbf{p}}}]\cdot|\nu|)(1[\underline{\rho}\bullet\hat{\underline{\mathbf{p}}}]\cdot\delta[\underline{\rho}\bullet\hat{\underline{\mathbf{p}}}^{\perp}])\,d\phi \tag{12.63}$$

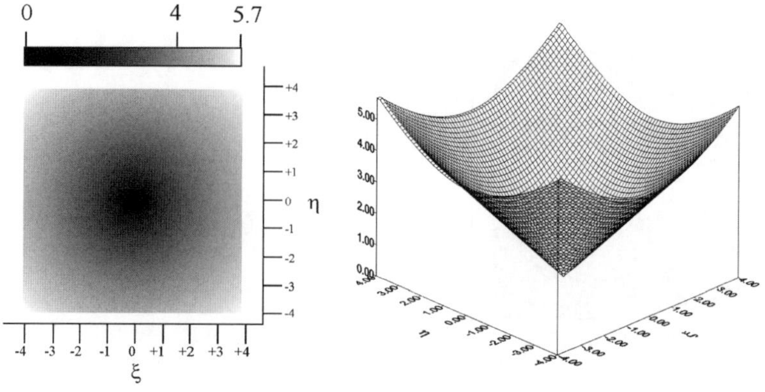

Figure 12.18 Graphical representation of the 2-D rho filter $H[\xi.\eta] = \rho = |\upsilon| = \sqrt{\xi^2 + \eta^2}$ as both gray-scale and surface plots.

The filter theorems produce two equivalent recipes for the appropriate impulse response in the space domain:

$$\left(\int_{\phi=-\pi/2}^{+\pi/2} \delta[\underline{\mathbf{r}} \bullet \underline{\hat{\mathbf{p}}}] \cdot 1[\underline{\mathbf{r}} \bullet \underline{\hat{\mathbf{p}}}^{\perp}]\, d\phi\right) * \mathcal{F}_2^{-1}\{\sqrt{\xi^2 + \eta^2}\}$$

$$= \int_{\phi=-\pi/2}^{+\pi/2} [(\delta[\underline{\mathbf{r}} \bullet \underline{\hat{\mathbf{p}}}] \cdot 1[\underline{\mathbf{r}} \bullet \underline{\hat{\mathbf{p}}}^{\perp}]) * \mathcal{F}_{1\,\nu\to p}^{-1}\{|\nu|\}]\, d\phi \tag{12.64}$$

Consider the 1-D case first; we need to find the 1-D function of p whose 1-D spectrum is $|\nu|$. The 1-D transfer function $H[\nu]$ is real valued and symmetric, and has infinite area and null amplitude at the origin. The 1-D theorems tell us that the impulse response $h[p]$ also must be real valued and even, and must have infinite amplitude at the origin and zero area. The inverse Fourier transform of $|\nu|$ is not defined in a conventional sense, but we may discern some of its features rather easily. For example, the filter theorem may be applied to $|\nu|$ by expressing it as product $\nu \cdot SGN[\nu]$:

$$\begin{aligned}\mathcal{F}_{1\,\nu\to p}^{-1}\{|\nu|\} &= \mathcal{F}_{1\,\nu\to p}^{-1}\{\nu \cdot SGN[\nu]\}\\ &= \mathcal{F}_{1\,\nu\to p}^{-1}\{\nu\} * \mathcal{F}_{1\,\nu\to p}^{-1}\{SGN[\nu]\}\\ &= \left(\frac{1}{2\pi i}\delta'[p]\right) * \left(\frac{1}{-i\pi p}\right)\\ &= \left(\frac{1}{2\pi^2}\right)\frac{d}{dp}(p^{-1})\end{aligned} \tag{12.65}$$

which is defined everywhere except at the origin:

$$\left(\frac{1}{2\pi^2}\right)\frac{d}{dp}(p^{-1}) = -\frac{1}{2\pi^2 p^2} < 0 \quad \mathit{if}\ p \neq 0 \tag{12.66}$$

This is real valued, symmetric, and negative, and its area is easily seen to be an indeterminate negative value. This means that the impulse response evaluated at $p = 0$ must have infinite positive area to cancel the infinite negative areas from the "wings". In other (loosely spoken) words, the impulse response at the origin is a 1-D Dirac delta function that has not only infinite amplitude, but infinite area as well.

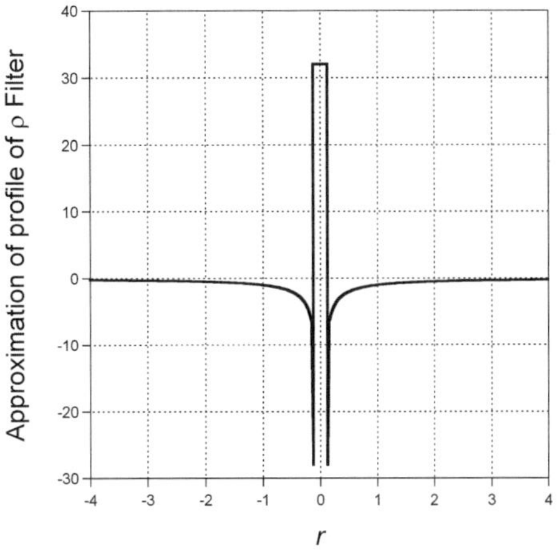

Figure 12.19 Profile of approximation of rho filter in the space domain.

Proponents of mathematical rigor are excused from the next equation, which specifies the impulse response of the 1-D rho filter:

$$h[p] = -\frac{1}{2\pi^2 p^2} + (A\ \delta[p]) \quad where\ A = 2\int_0^{+\infty} \frac{1}{2\pi p^2}\, dp \tag{12.67}$$

This may be approximated as the function $\hat{h}[p]$ by applying the scaled *RECT* function form of the Dirac delta function from Equation (6.93) and is shown in Figure 12.19.

Rather than concentrate on tying up the (not very formidable) mathematical loose ends here, we consider the ramifications of this 1-D impulse response. Readers who require a stricter mathematical framework should consult the references, especially Barrett and Swindell (1996). The filtering operation is now the summation of 2-D separable functions that are constant along the direction defined by $\underline{\hat{\mathbf{p}}}^{\perp}$ and whose profiles in the orthogonal direction $\underline{\hat{\mathbf{p}}}$ are identical to the 1-D impulse response $h[p]$ specified in Equation (12.67):

$$\begin{aligned}
&\int_{\phi=-\pi/2}^{+\pi/2} [(\delta[\underline{\mathbf{r}} \bullet \underline{\hat{\mathbf{p}}}]1[\underline{\mathbf{r}} \bullet \underline{\hat{\mathbf{p}}}^{\perp}]) * \mathcal{F}_{1\ \nu\to p}^{-1}\{|\nu|\}]\, d\phi \\
&= \int_{\phi=-\pi/2}^{+\pi/2} [\delta[\underline{\mathbf{r}} \bullet \underline{\hat{\mathbf{p}}}] * \mathcal{F}_{1\ \nu\to p}^{-1}\{|\nu|\}] \cdot 1[\underline{\mathbf{r}} \bullet \underline{\hat{\mathbf{p}}}^{\perp}]\, d\phi \\
&= \int_{\phi=-\pi/2}^{+\pi/2} h[p] \cdot 1[\underline{\mathbf{r}} \bullet \underline{\hat{\mathbf{p}}}^{\perp}]\, d\phi
\end{aligned} \tag{12.68}$$

Each of the 1-D component functions is a "filtered back projection" of the object. Because the volume of each such 2-D separable function is zero, the volume of the integrated function must be zero as well. During the process of azimuthal integration, the negative "wings" of the back projection cancel the off-axis positive amplitudes that were produced in the original layergram. A simulation of the summation for three filtered back projections is shown in Figure 12.20.

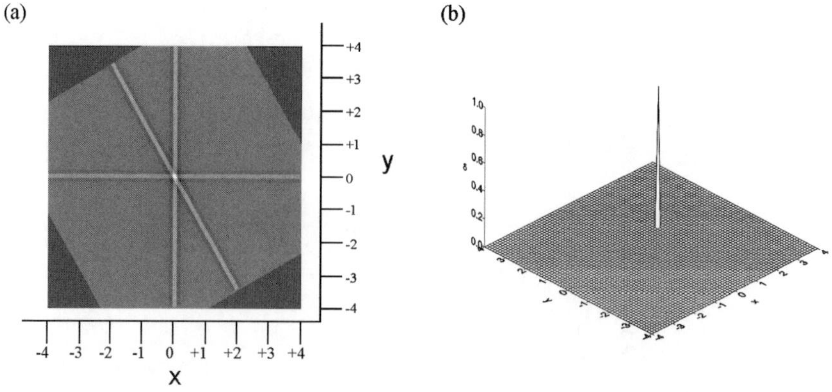

Figure 12.20 Summation of filtered back projections. (a) Three bipolar filtered back projections of the on-axis 2-D Dirac delta function. The summation of the negative "wings" of the filter cancels the extraneous positive amplitudes present in the original back-projected image. (b) Summation of the infinite set of filtered back projections reconstructs the ideal 2-D Dirac delta function.

12.4.2 Summation of Projections over Azimuths

Consider a 2-D Dirac delta function located at $\underline{\mathbf{r}}_0 = [x_0, y_0] = (r_0, \theta_0)$. The 2-D Fourier transform may be evaluated via the shift theorem and may be written in several equivalent ways:

$$\mathcal{F}_2\{\delta[\underline{\mathbf{r}} - \underline{\mathbf{r}}_0]\} = \exp[-2\pi i(\xi x_0 + \eta y_0)] \tag{12.69a}$$

$$= \exp[-2\pi i(\underline{\boldsymbol{\rho}} \bullet \underline{\mathbf{r}}_0)] \tag{12.69b}$$

$$= \exp[-2\pi i(|\underline{\boldsymbol{\rho}}||\underline{\mathbf{r}}_0| \cos[\psi - \theta_0])] \tag{12.69c}$$

The line-integral projections of the translated Dirac delta function were derived in Equation (12.20) or may be obtained from the shift theorem in Equation (12.46)

$$\lambda_{(\delta[\underline{\mathbf{r}}-\underline{\mathbf{r}}_0])}(p, \phi) = \delta[p - (x_0 \cos[\theta_0 - \phi] + y_0 \sin[\theta_0 - \phi])] \tag{12.70a}$$

$$= \delta[p - \underline{\mathbf{r}}_0 \bullet \hat{\underline{\mathbf{p}}}] \tag{12.70b}$$

$$= \delta[p - |\underline{\mathbf{r}}_0| \cos[\theta_0 - \phi]] \tag{12.70c}$$

The 1-D spectrum of a projection is obtained by applying the 1-D shift theorem to evaluate the "slice" $\Lambda_{(\delta[x-x_0, y-y_0])}(\nu, \phi)$ of the 2-D spectrum along the radial line defined by $\hat{\underline{\mathbf{p}}}$. The corresponding expressions to those in Equation (12.70) are:

$$\mathcal{F}_{\substack{1 \\ p\to\nu}}\{\delta[p - (x_0 \cos[\theta_0 - \phi] + y_0 \sin[\theta_0 - \phi])]\}$$

$$= 1[\nu] \exp[-2\pi i\nu(x_0 \cos[\theta_0 - \phi] + y_0 \sin[\theta_0 - \phi])] \tag{12.71a}$$

$$\mathcal{F}_{\substack{1 \\ p\to\nu}}\{\delta[p - (\underline{\mathbf{r}}_0 \bullet \hat{\underline{\mathbf{p}}})]\} = 1[\nu] \exp[-2\pi i\nu(\underline{\mathbf{r}}_0 \bullet \hat{\underline{\mathbf{p}}})]$$

$$= 1[\nu] \exp[-2\pi i(\underline{\mathbf{r}}_0 \bullet (\hat{\underline{\mathbf{p}}}\nu))] \tag{12.71b}$$

$$\mathcal{F}_{\substack{1 \\ p\to\nu}}\{\delta[p - |\underline{\mathbf{r}}_0| \cos[\theta_0 - \phi]]\} = 1[\nu] \exp[-2\pi i\nu(|\underline{\mathbf{r}}_0| \cos[\theta_0 - \phi])] \tag{12.71c}$$

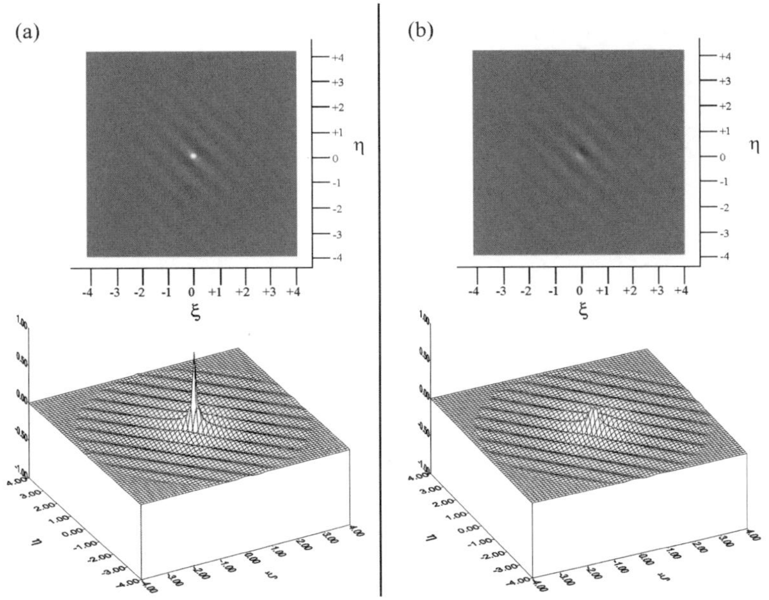

Figure 12.21 Spectrum of $\delta[x-1, y-1]$ reconstructed by the azimuthal sum of the central slices of the spectrum: (a) $\Re\{\hat{F}[\xi,\eta]\} = |\nu|^{-1}\cos[2\pi(\xi+\eta)]$; $\Im\{\hat{F}[\xi,\eta]\} = |\nu|^{-1}\sin[2\pi(\xi+\eta)]$.

The azimuthal integral of the slices is evaluated by expressing the 1-D Dirac delta function in Equation (12.71b) in the form of Equation (12.56):

$$\int_{-\pi/2}^{+\pi/2} \exp[-2\pi i(\underline{\mathbf{r}}_0 \bullet (\hat{\mathbf{p}}\nu))]\,(1[\underline{\boldsymbol{\rho}} \bullet \hat{\mathbf{p}}] \cdot \delta[\underline{\boldsymbol{\rho}} \bullet \hat{\mathbf{p}}^{\perp}])\,d\phi$$

$$= \frac{1}{|\underline{\boldsymbol{\rho}}|}\int_{-\pi/2}^{+\pi/2} \exp[-2\pi i(|\underline{\mathbf{r}}_0|\cdot|\boldsymbol{\nu}|\cos[\theta_0-\phi])]\cdot\frac{\delta[\psi-\phi]}{|\mp\sin[-\pi/2]|}\,d\phi$$

$$= \frac{1}{\rho}\int_{-\pi/2}^{+\pi/2} \exp[-2\pi i(|\underline{\mathbf{r}}_0|\cdot|\boldsymbol{\nu}|\cos[\theta_0-\phi])]\cdot\delta[\psi-\phi]\,d\phi$$

$$= \frac{1}{\rho}\exp[-2\pi i(|\underline{\mathbf{r}}_0|\cdot|\boldsymbol{\nu}|\cos[\theta_0-\psi])] = \frac{1}{|\nu|}\exp[-2\pi i(\underline{\mathbf{r}}_0 \bullet \underline{\boldsymbol{\rho}})] \tag{12.72}$$

where we have also applied the definition $\underline{\boldsymbol{\rho}} \equiv (\nu, \psi)$ for $-\pi/2 \le \psi < +\pi/2$. The reconstructed spectrum for the specific case $\underline{\mathbf{r}}_0 = [1, 1]$ is shown in Figure 12.21. Comparison to Equation (12.57) demonstrates that the azimuthal summations in both cases scale the correct spectra by $|\nu|^{-1}$. The inverse 2-D Fourier transform of Equation (12.72) is easy to evaluate by using the 2-D filter theorem of Equation (10.28) and the result of Equation (11.53):

$$\mathcal{F}_2^{-1}\left\{\frac{1}{|\nu|}\exp[-2\pi i(\underline{\mathbf{r}}_0 \bullet \underline{\boldsymbol{\rho}})]\right\} = \frac{1}{r} * \delta[\underline{\mathbf{r}} - \underline{\mathbf{r}}_0] = \frac{1}{|\underline{\mathbf{r}} - \underline{\mathbf{r}}_0|} \tag{12.73}$$

as shown in Figure 12.22.

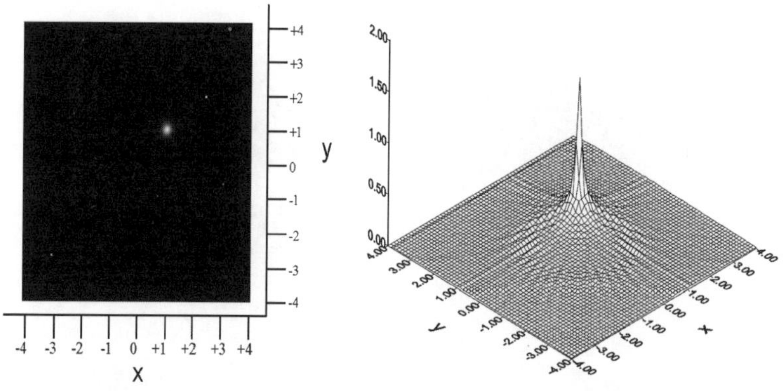

Figure 12.22 Space-domain representation of the reconstructed spectrum of $f[x, y] = \delta[x-1, y-1]$ via inverse Fourier transform without rho filtering produces the blurry reconstruction $\delta[x-1, y-1] * r^{-1}$.

We have demonstrated that integrating the central slices of spectra over angle for both "on-axis" and "off-axis" 2-D Dirac delta functions yields "blurry" reconstructions that differ only by the translation:

$$\mathcal{F}_2^{-1}\left\{\int_{-\pi/2}^{+\pi/2} \underset{p\to\nu}{\mathcal{F}_1}\{\mathcal{R}_2\{\delta[x, y]\}\}(1[\underline{\rho}\bullet\hat{\underline{\mathbf{p}}}]\cdot\delta[\underline{\rho}\bullet\hat{\underline{\mathbf{p}}}^{\perp}])\,d\phi\right\} = \delta[x, y] * \frac{1}{r} \tag{12.74a}$$

$$\mathcal{F}_2^{-1}\left\{\int_{-\pi/2}^{+\pi/2} \underset{p\to\nu}{\mathcal{F}_1}\{\mathcal{R}_2\{\delta[\underline{\mathbf{r}}-\underline{\mathbf{r}}_0]\}\}(1[\underline{\rho}\bullet\hat{\underline{\mathbf{p}}}]\cdot\delta[\underline{\rho}\bullet\hat{\underline{\mathbf{p}}}^{\perp}])\,d\phi\right\} = \delta[\underline{\mathbf{r}}-\underline{\mathbf{r}}_0] * \frac{1}{r} \tag{12.74b}$$

This means that the image of any 2-D Dirac delta function has the same form, and therefore that the imaging system composed of the cascade of line-integral projection and back projection is a linear and shift-invariant system with impulse response $h[x, y] = (x^2+y^2)^{-1/2} = r^{-1}$. Thus the cascade of the Radon transform, the 1-D Fourier transform of each projection, azimuthal summation, and the inverse 2-D Fourier transform produces a "blurry" replica of the original function due to the filtering by r^{-1}:

$$\mathcal{F}_2^{-1}\left\{\int_{-\pi/2}^{+\pi/2} \underset{p\to\nu}{\mathcal{F}_1}\{\mathcal{R}_2\{f[x, y]\}\}(1[\underline{\rho}\bullet\hat{\underline{\mathbf{p}}}]\cdot\delta[\underline{\rho}\bullet\hat{\underline{\mathbf{p}}}^{\perp}])\,d\phi\right\} = f[x, y] * \frac{1}{\sqrt{x^2+y^2}} \tag{12.74c}$$

An example of the reconstruction of the letter "E" by this system is shown in Figure 12.23. The "fuzziness" of the reconstruction due to the convolution of the object with the impulse response in Equation (12.74c) is evident:

Equation (12.74c) indicates that the rho filter provides the appropriate high-frequency compensation for the projections of any 2-D Dirac delta function, as shown in the filtered reconstruction of the letter "E" in Figure 12.24.

Because integration over azimuth is a linear operation, the same result is obtained by applying 1-D "rho" filters (i.e., the central slices of ρ) to each of the 1-D spectra before performing the azimuthal integral:

$$\begin{aligned}\mathcal{F}_2^{-1}\left\{|\nu|\cdot\int_{-\pi/2}^{+\pi/2} \underset{p\to\nu}{\mathcal{F}_1}\{\mathcal{R}_2\{f[x, y]\}\}\,d\phi\right\} &= \mathcal{F}_2^{-1}\left\{|\nu|\cdot\left(\int_{-\pi/2}^{+\pi/2}\Lambda_f[\nu,\phi]\,d\phi\right)\right\}\\ &= \mathcal{F}_2^{-1}\left\{|\nu|\cdot\left(\frac{1}{|\nu|}\cdot F[\xi,\eta]\right)\right\}\\ &= \mathcal{F}_2^{-1}\{F[\xi,\eta]\} = f[x, y]\end{aligned} \tag{12.75}$$

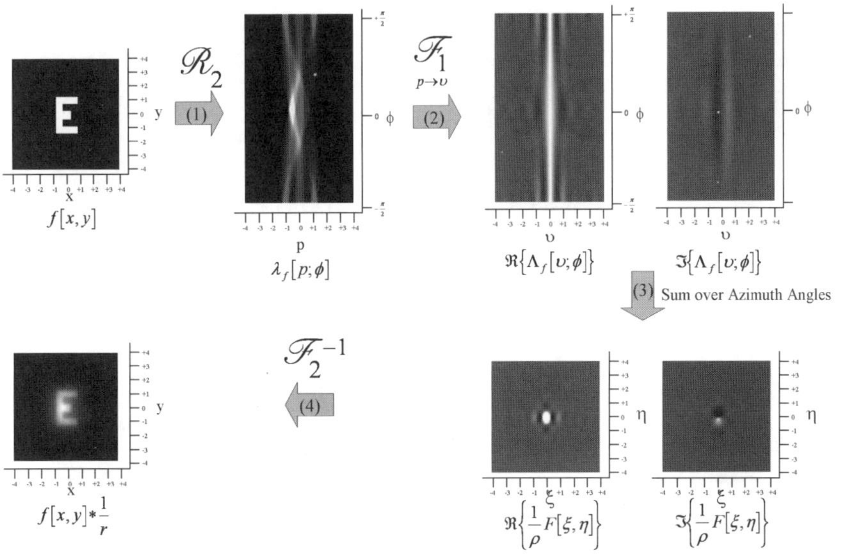

Figure 12.23 Reconstruction of "E" by cascade of operations: (1) $\mathcal{R}_2\{f[x,y]\} = \lambda_f(p;\phi)$; (2) $\mathcal{F}_{1\,p\to\upsilon}\{\lambda_f(p;\phi)\} = \Lambda_f(\upsilon;\phi)$; (3) summation over azimuth angles that results in $(1/\rho)F[\xi,\eta]$; and (4) inverse 2-D Fourier transform, which reconstructs the blurry image $f[x,y] * r^{-1}$.

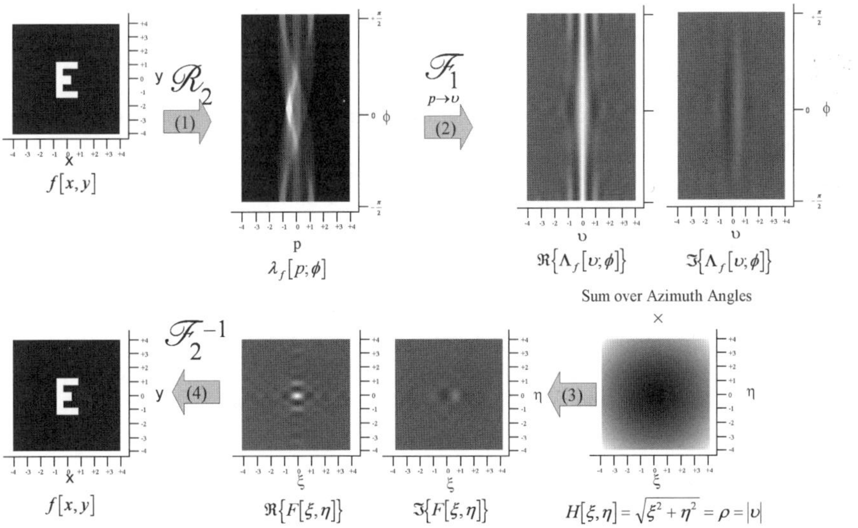

Figure 12.24 Reconstruction of "E" as cascade of operations with 2-D rho filter applied to the azimuthal integral of the central slices: (1) Radon transform; (2) 1-D Fourier transform of each projection; (3) sum over azimuths and rho filter; (4) inverse 2-D Fourier transform.

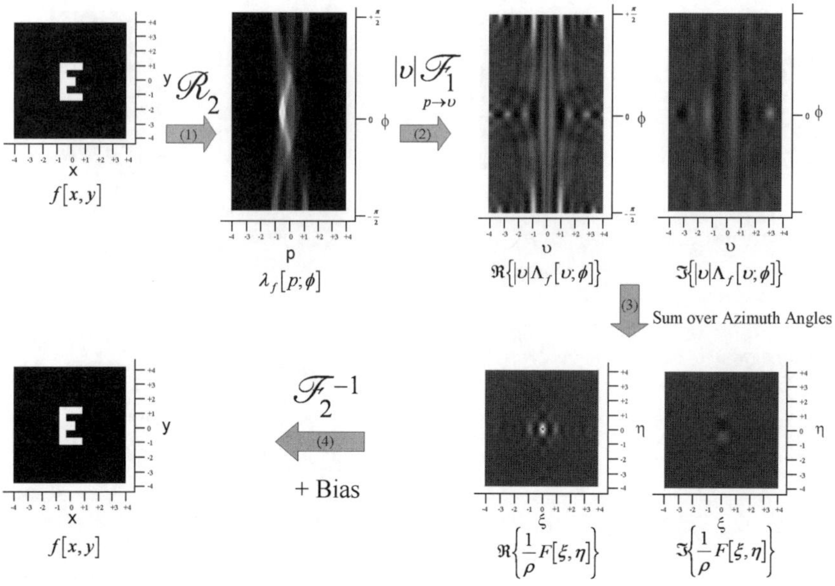

Figure 12.25 Reconstruction of "E" from its line-integral projections by applying 1-D rho filter to each projection prior to the azimuthal integral: (1) Radon transform; (2) filtered 1-D Fourier transform of each projection; (3) sum over azimuths; (4) inverse 2-D Fourier transform. Note that the amplitude of the reconstructed spectrum is zero at the origin, which means that the "bias" must be added to reconstruct the image.

This gives a 1-D space-domain recipe for the inverse Radon transform of the Dirac delta function, as shown in Figure 12.25.

12.5 Central-Slice Transform

The dual character of the operations of convolution and multiplication in the space and frequency domains also establishes a dual relationship between the radial profiles of $f[x, y]$ and the line-integral projections of $F[\xi, \eta]$. This observation suggests yet another invertible transformation of coordinates, which is really just the Radon transform applied to the frequency-domain representation.

12.5.1 Radial "Slices" of $f[x, y]$

The central-slice theorem proved that the 1-D Fourier transform of a line-integral projection of a 2-D function is identical to the corresponding central slice of the 2-D spectrum. The dual character of the space- and frequency-domain representations established by the "transform-of-a-transform" theorem in Section 9.8.3 demonstrates that a radial profile of the space-domain function and the line-integral projection of its frequency-domain representation also are Fourier transform pairs. We can "slice" the function in the space domain by modulating with a line delta function oriented along the azimuth ϕ_0 in the same manner as Equation (7.66) (and therefore perpendicular to the azimuth $\phi_0 \pm \pi/2$) to generate the new function that we denote by the name $\mu_f(p, \phi_0)$:

$$f[x, y](\delta[\underline{\mathbf{r}} \bullet \hat{\underline{\mathbf{p}}}_0^\perp] \cdot 1[\underline{\mathbf{r}} \bullet \hat{\underline{\mathbf{p}}}_0]) = f[p\hat{\underline{\mathbf{p}}}_0](\delta[\underline{\mathbf{r}} \bullet \hat{\underline{\mathbf{p}}}_0^\perp] \cdot 1[\underline{\mathbf{r}} \bullet \hat{\underline{\mathbf{p}}}_0])$$
$$\equiv \mu_f(p, \phi_0)(1[\underline{\mathbf{r}} \bullet \hat{\underline{\mathbf{p}}}_0] \cdot \delta[\underline{\mathbf{r}} \bullet \hat{\underline{\mathbf{p}}}_0^\perp]) \tag{12.76}$$

By analogy with the sinogram representation of the Radon transform in Figure 12.4, the complete set of "slices" $\{\mu_f(p, \phi)\}$ may be displayed as a Cartesian function in sinogram format with p and ϕ on the horizontal and vertical axes, respectively. However, note the difference: a specific location $[x_0, y_0]$ of the input function affects only the value of $\mu_f(p_0, \phi_0)$ that satisfies the conditions $p_0 = (x_0^2 + y_0^2)^{1/2}$ and $\phi_0 = \tan^{-1}[y_0/x_0]$.

The process of Equation (12.76) may be denoted by a "central-slice operator" $\mathcal{C}_2$, where the subscript again denotes that the operator acts on 2-D functions:

$$\mathcal{C}_2\{f[x, y]\} \equiv f[x, y](\delta[\underline{\mathbf{r}} \bullet \hat{\underline{\mathbf{p}}}^\perp] \cdot 1[\underline{\mathbf{r}} \bullet \hat{\underline{\mathbf{p}}}]) = \mu_f(p, \phi) \cdot 1[\underline{\mathbf{r}} \bullet \hat{\underline{\mathbf{p}}}] \tag{12.77}$$

The modulation theorem ensures that the 2-D Fourier transform of a single slice of $f[x, y]$ is equivalent to the line-integral projection of the 2-D spectrum $F[\xi.\eta]$ evaluated along the azimuth specified by $\hat{\underline{\mathbf{p}}}$:

$$\mathcal{F}_2\{f[x, y] \cdot (\delta[\underline{\mathbf{r}} \bullet \hat{\underline{\mathbf{p}}}^\perp]1[\underline{\mathbf{r}} \bullet \hat{\underline{\mathbf{p}}}])\} = F[\xi, \eta] * (1[\underline{\boldsymbol{\rho}} \bullet \hat{\underline{\mathbf{p}}}^\perp] \cdot \delta[\underline{\boldsymbol{\rho}} \bullet \hat{\underline{\mathbf{p}}}])$$
$$= \int_{-\infty}^{+\infty} F[\xi, \eta]\delta[\nu - \underline{\boldsymbol{\rho}} \bullet \hat{\underline{\mathbf{p}}}]\, d\xi\, d\eta$$
$$= \widetilde{\mathcal{R}}_2\{F[\xi, \eta]\} \tag{12.78}$$

where the tilde "~" over the operator is used to denote that this Radon transform operates on a function in the frequency domain.

12.5.2 Central-Slice Transforms of Special Functions

We now evaluate the central-slice transforms of the same special functions investigated in the discussion of the Radon transform.

12.5.2.1 Central-Slice Transform of $\delta[x, y]$

The 2-D Dirac delta function located at the origin may be written as the separable product of 1-D Dirac delta functions with rotated arguments:

$$\delta[x, y] = \delta[\underline{\mathbf{r}} \bullet \hat{\underline{\mathbf{p}}}] \cdot \delta[\underline{\mathbf{r}} \bullet \hat{\underline{\mathbf{p}}}^\perp] = \delta[\underline{\mathbf{r}} \bullet \hat{\underline{\mathbf{p}}}] \cdot (1[\underline{\mathbf{r}} \bullet \hat{\underline{\mathbf{p}}}] \cdot \delta[\underline{\mathbf{r}} \bullet \hat{\underline{\mathbf{p}}}^\perp]) \tag{12.79}$$

which already has the desired separable form of the central slices. The first component function $\delta[\underline{\mathbf{r}} \bullet \hat{\underline{\mathbf{p}}}]$ is a 1-D Dirac delta along the axis specified by $\hat{\underline{\mathbf{p}}}$. By again defining the bipolar radial coordinate p along this axis, the central slices are seen to be:

$$\mu_{(\delta[x,y])}(p, \phi) = \delta[p] \cdot 1[\phi] \tag{12.80}$$

The 1-D spectrum of each slice is the 1-D unit constant as a function of ν:

$$\mathcal{F}_{1}_{p \to \nu} \{\delta[p]\} = 1[\nu] \tag{12.81}$$

The central-slice transform and the 1-D Fourier transform of the slices are shown in Figure 12.26.

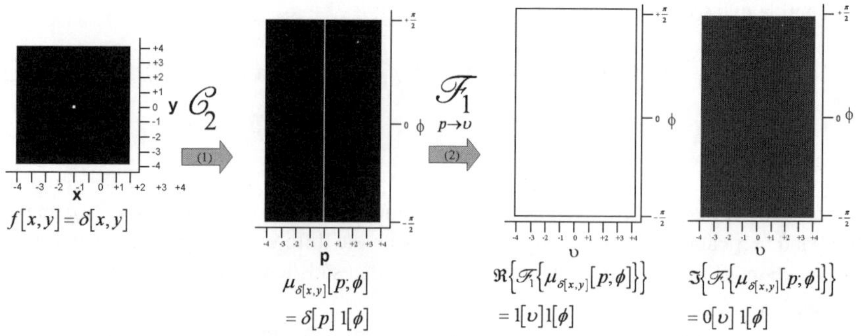

Figure 12.26 Central slice transform of $\delta[x, y]$: (1) $\mathcal{C}_2\{\delta[x, y]\} = \delta[p] \cdot 1[\phi]$; (2) the 1-D Fourier transforms of the individual central slices are identically $1[\upsilon]$.

12.5.2.2 Central-Slice Transform of $\delta[x - x_0, y - y_0] = \delta[\underline{\mathbf{r}} - \underline{\mathbf{r}}_0]$

All slices of the off-axis 2-D Dirac delta function evaluate to the 1-D null function, except along the azimuth angle $\phi_0 = \tan^{-1}[y_0/x_0]$. This nonnull slice is the 1-D Dirac delta function located at the bipolar radial distance $p_0 = |\underline{\mathbf{r}}_0|SGN[x_0]$:

$$\begin{aligned}
\delta[\underline{\mathbf{r}} - \underline{\mathbf{r}}_0] &= \frac{\delta[r - r_0]}{r_0}\delta[\theta - \theta_0] = \delta[r - r_0]\frac{\delta[\theta - \theta_0]}{r} \\
&= \delta[p - |\underline{\mathbf{r}}_0|SGN[x_0]]\frac{\delta[\theta - \theta_0]}{r} \\
&= \delta[p - p_0]\frac{\delta[\theta - \theta_0]}{r} = \delta[p - p_0] \cdot \delta[\underline{\mathbf{r}} \bullet \hat{\underline{\mathbf{p}}}_0^{\perp}] \\
&= (\delta[p - p_0])(\delta[\underline{\mathbf{r}} \bullet \hat{\underline{\mathbf{p}}}^{\perp}] \cdot \delta[\phi - \theta_0]) \\
&= (\delta[p - p_0]\,\delta[\phi - \theta_0]) \cdot \delta[\underline{\mathbf{r}} \bullet \hat{\underline{\mathbf{p}}}^{\perp}] \\
&\Longrightarrow \mu_{(\delta[\underline{\mathbf{r}} - \underline{\mathbf{r}}_0])}(p, \phi) = \delta[p - p_0] \cdot \delta[\phi - \theta_0]
\end{aligned} \tag{12.82}$$

where the property of Dirac delta functions in products from Equation (6.115) and the result of Equation (12.56) has been adapted to the space domain. The 1-D spectra of the slices are all zero except for that at the specific azimuth $\phi = \phi_0$, where the spectrum is a 1-D linear-phase exponential:

$$\begin{aligned}
\mathop{\mathcal{F}_1}_{p\to\nu}\{\delta[p - r_0\,SGN[x_0]] \cdot \delta[\phi - \phi_0]\} &= \delta[\phi - \phi_0]\left(\mathop{\mathcal{F}_1}_{p\to\nu}\{\delta[p - r_0\,SGN[x_0]]\}\right) \\
&= \exp[-2\pi i \nu r_0\,SGN[x_0]] \cdot \delta[\phi - \phi_0] \\
&= \widetilde{\mathcal{R}}_2\{\mathcal{F}_2\{\delta[x - x_0, y - y_0]\}\} \cdot \delta[\phi - \phi_0] \\
&= \widetilde{\mathcal{R}}_2\{\exp[-2\pi i(\xi x_0 + \eta y_0)]\} \cdot \delta[\phi - \phi_0]
\end{aligned} \tag{12.83}$$

The array of central slices and 1-D spectra are shown in Figure 12.27.

12.5.2.3 Central-Slice Transform of $CYL(r)$

The slices of a centered cylinder function are identical symmetric 1-D rectangle functions whose widths are equal to the diameter of the cylinder, as shown in Figure 12.28:

$$\mu_{[CYL(r)]}(p, \phi) = RECT[p] \cdot 1[\phi] \tag{12.84}$$

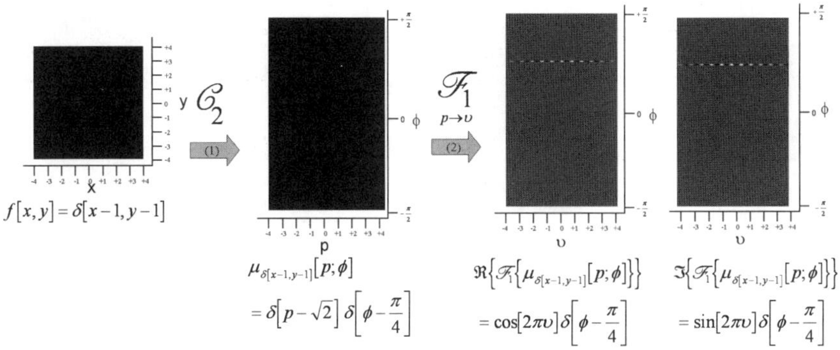

Figure 12.27 Central slice transform of off-axis 2-D Dirac delta function: (1) $\mathcal{C}_2\{\delta[x-1, y-1]\} = \delta[p-\sqrt{2}]\delta[\phi-\pi/4]$; (2) The 1-D Fourier transforms of the slices produces a single nonzero slice $\exp[-2\pi i \upsilon\sqrt{2}]\delta[\phi-\pi/4]$.

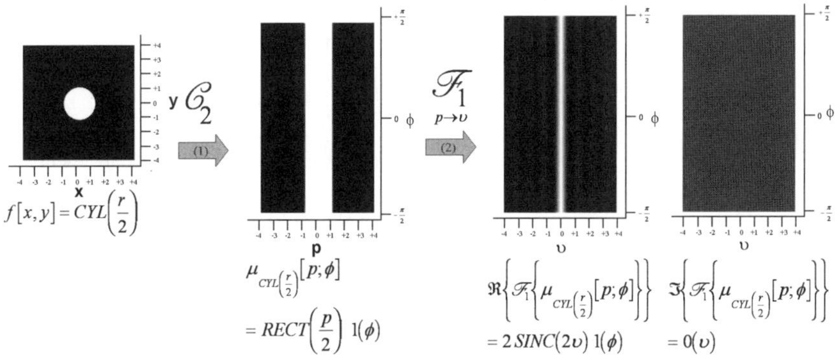

Figure 12.28 (1) The central slices of $CYL(r/2)$ are identically $RECT[p/2]$; (2) each 1-D Fourier transform is $2\,SINC[2\upsilon]$.

The 1-D spectra of the slices are identical *SINC* functions with unit "width":

$$\mathcal{F}_{1\,p\to\nu}\{RECT[p]\} = SINC[\nu] \tag{12.85}$$

Some consequences of this relation will be considered in Section 12.6.

12.5.2.4 Central-Slice Transform of $\delta(r-r_0)$

Recall that the ring delta function is the symmetric circular "corral" of delta functions. The transform "slices" the ring in the perpendicular direction along all azimuth angles, so there is no need to consider any angular variation in amplitude for the different slices, as was required when evaluating the Radon transform in Section 12.2.2. It is easy to see that the 1-D central slices are identical symmetric pairs of

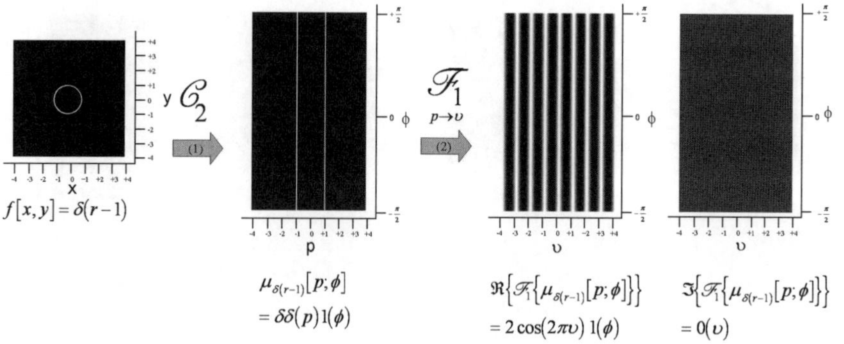

Figure 12.29 (1) Central slices of $\delta(r-1)$ are identically $\delta\delta(p)$; (2) the 1-D Fourier transforms are identically $2\cos[2\pi\upsilon]$.

1-D Dirac delta functions:

$$\begin{aligned}\mu_{[\delta(r-r_0)]}(p,\phi) &= \delta(r-r_0)\,(\delta[\mathbf{\underline{r}}\bullet\hat{\mathbf{\underline{p}}}^{\perp}]\cdot 1[\mathbf{\underline{r}}\bullet\hat{\mathbf{\underline{p}}}])\\ &= (\delta[p+r_0]+\delta[p-r_0])\cdot 1[\phi]\\ &= \left(\frac{1}{r_0}\,\delta\delta\left[\frac{p}{r_0}\right]\right)\cdot 1[\phi]\end{aligned}\tag{12.86}$$

where $r_0 \geq 0$ and $|\phi| \leq \pi/2$. The 1-D Fourier transforms of the individual slices are identical cosine functions:

$$\mathcal{F}_{\substack{1\\p\to\nu}}\{\mu_{\delta(r-r_0)}(p,\phi)\} = 2\cos[2\pi r_0\nu]\cdot 1[\phi]\tag{12.87}$$

as shown in Figure 12.29. This relation will be explored more fully in Section 12.6.

12.5.2.5 Central-Slice Transform of *RECT*[*x*, *y*]

All 1-D central slices of a symmetric rectangle function are 1-D symmetric rectangle functions:

$$\mu_{RECT[x,y]}(p,\phi) = RECT[x,y]\cdot\delta[\mathbf{\underline{r}}\bullet\hat{\mathbf{\underline{p}}}^{\perp}]\tag{12.88}$$

Our only task is to derive the width of the rectangle as a function of azimuth angle ϕ, which is rather easily shown to be the reciprocal of the larger of $\cos[\phi]$ or $\sin[|\phi|]$ for $|\phi| \leq \pi/2$:

$$\mu_{RECT[x,y]}(p,\phi) = \begin{cases} RECT\left[\dfrac{p}{1/\cos[\phi]}\right] = RECT[p\cos[\phi]] & \textit{if } |\phi| \leq \dfrac{\pi}{4}\\ RECT\left[\dfrac{p}{1/\sin[\phi]}\right] = RECT[p\sin[\phi]] & \textit{if } +\dfrac{\pi}{4} \leq |\phi| \leq +\dfrac{\pi}{2}\end{cases}\tag{12.89}$$

The 1-D spectra of these slices are obviously *SINC* functions with varying amplitude and width:

$$\begin{aligned}\mathcal{F}_{\substack{1\\p\to\nu}}\{\mu_{RECT[x,y]}(p,\phi)\} &= M_f(\nu,\phi)\\ &= \begin{cases}\left|\dfrac{1}{\cos[\phi]}\right| SINC\left[\dfrac{\nu}{\cos[\phi]}\right] & \textit{if } |\phi| \leq \dfrac{\pi}{4}\\ \left|\dfrac{1}{\sin[\phi]}\right| SINC\left[\dfrac{\nu}{\sin[\phi]}\right] & \textit{if } +\dfrac{\pi}{4} \leq |\phi| \leq +\dfrac{\pi}{2}\end{cases}\end{aligned}\tag{12.90}$$

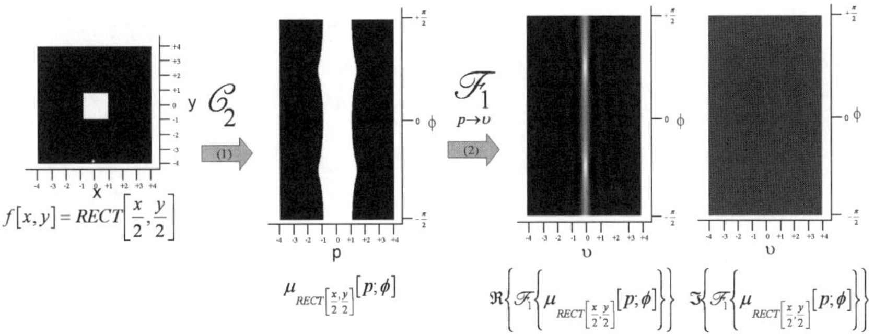

Figure 12.30 (1) Central slices of $RECT[x/2, y/2]$ are rectangle functions with widths that vary with ϕ; (2) the 1-D Fourier transforms of the slices are *SINC* functions with varying amplitudes and widths.

which we know must be equal to the line-integral projections of the 2-D spectra, e.g.:

$$\begin{aligned}
&\iint_{-\infty}^{+\infty} SINC[\xi,\eta]\delta(\nu - \underline{\rho} \bullet \hat{\underline{\nu}})\, d\xi\, d\eta \\
&= \iint_{-\infty}^{+\infty} SINC[\xi,\eta]\delta(\nu - (\xi\cos[\phi] + \eta\sin[\phi]))\, d\xi\, d\eta \\
&= \begin{cases} \left|\dfrac{1}{\cos[\phi]}\right| SINC\left[\dfrac{\nu}{\cos[\phi]}\right] & if\ |\phi| \le \dfrac{\pi}{4} \\ \left|\dfrac{1}{\sin[\phi]}\right| SINC\left[\dfrac{\nu}{\sin[\phi]}\right] & if\ +\dfrac{\pi}{4} \le |\phi| \le +\dfrac{\pi}{2} \end{cases}
\end{aligned} \tag{12.91}$$

The central slices of $RECT[x/2, y/2]$ and the spectra are shown in "sinogram" format in Figure 12.30.

12.5.2.6 Central-Slice Transform of $COR[x, y]$

Like the ring, the central slices of the "square corral" $COR[x, y]$ are symmetric pairs of Dirac delta functions, but now weighted by a function of the angle at which the line delta function of the "slicer" intercepts the line delta function of the corral. The expression for $COR[x, y]$ as the weighted difference of two 2-D rectangles leads to related expressions for the slices in two regions of the azimuthal domain:

$$\begin{aligned}
&\mu_{COR[x,y]}(p,\phi) \\
&= \begin{cases} \lim\limits_{\Delta\to 0}\left\{\dfrac{1}{\Delta}RECT\left[\dfrac{p+1/2\cos[\phi]}{(\Delta/\cos[\phi])}\right] + \dfrac{1}{\Delta}RECT\left[\dfrac{p-1/2\cos[\phi]}{(\Delta/\cos[\phi])}\right]\right\} & if\ |\phi| \le \dfrac{\pi}{4} \\ \lim\limits_{\Delta\to 0}\left\{\dfrac{1}{\Delta}RECT\left[\dfrac{p+1/2\sin[\phi]}{(\Delta/\sin[\phi])}\right] + \dfrac{1}{\Delta}RECT\left[\dfrac{p-1/2\sin[\phi]}{(\Delta/\sin[\phi])}\right]\right\} & if\ \dfrac{\pi}{4} \le |\phi| \le \dfrac{\pi}{2} \end{cases} \\
&= \begin{cases} \delta\left[p\cos[\phi] + \dfrac{1}{2}\right] + \delta\left[p\cos[\phi] - \dfrac{1}{2}\right] & if\ |\phi| \le \dfrac{\pi}{4} \\ \delta\left[p\sin[\phi] + \dfrac{1}{2}\right] + \delta\left[p\sin[\phi] - \dfrac{1}{2}\right] & if\ \dfrac{\pi}{4} \le |\phi| \le \dfrac{\pi}{2} \end{cases} \\
&= \begin{cases} \left|\dfrac{1}{\cos[\phi]}\right|\left(\delta\left[p + \dfrac{1}{2\cos[\phi]}\right] + \delta\left[p - \dfrac{1}{2\cos[\phi]}\right]\right) & if\ |\phi| \le \dfrac{\pi}{4} \\ \left|\dfrac{1}{\sin[\phi]}\right|\left(\delta\left[p + \dfrac{1}{2\sin[\phi]}\right] + \delta\left[p - \dfrac{1}{2\sin[\phi]}\right]\right) & if\ \dfrac{\pi}{4} \le |\phi| \le \dfrac{\pi}{2} \end{cases}
\end{aligned} \tag{12.92}$$

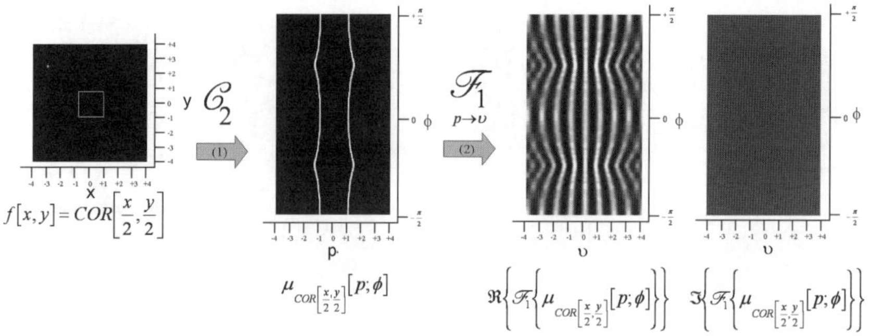

Figure 12.31 (1) Central slices of $COR[x/2, y/2]$ are symmetric pairs of 1-D Dirac delta functions; (2) the 1-D Fourier transforms are cosine functions with varying amplitudes and widths.

The 1-D Fourier transforms of the slices are amplitude-weighted cosines:

$$\underset{p\to\nu}{\mathcal{F}_1}\{\mu_{COR[x,y]}(p,\phi)\}$$

$$= \begin{cases} \left|\dfrac{1}{\cos[\phi]}\right|\left(2\cos\left[\dfrac{2\pi\nu}{2\cos[\phi]}\right]\right) = \left|\dfrac{2}{\cos[\phi]}\right|\cos\left[\dfrac{\pi\nu}{\cos[\phi]}\right] & \text{if } |\phi| \le \dfrac{\pi}{4} \\ \left|\dfrac{1}{\sin[\phi]}\right|\left(2\cos\left[\dfrac{2\pi\nu}{2\sin[\phi]}\right]\right) = \left|\dfrac{2}{\sin[\phi]}\right|\cos\left[\dfrac{\pi\nu}{\sin[\phi]}\right] & \text{if } \dfrac{\pi}{4} \le |\phi| \le \dfrac{\pi}{2} \end{cases} \tag{12.93}$$

which, again, are equal to the line-integral projections of the 2-D spectrum:

$$\underset{p\to\nu}{\mathcal{F}_1}\{\mu_{COR[x,y]}(p,\phi)\}$$

$$= \iint_{-\infty}^{+\infty} (2\cos[\pi\xi]\,SINC[\eta] + 2\,SINC[\xi]\cos[\pi\eta])\delta(\nu - \underline{\rho}\bullet\hat{\underline{\nu}})\,d\xi\,d\eta$$

$$= \iint_{-\infty}^{+\infty} (2\cos[\pi\xi]\,SINC[\eta] + 2\,SINC[\xi]\cos[\pi\eta])$$

$$\times\,\delta(\nu - (\xi\cos[\phi] + \eta\sin[\phi]))\,d\xi\,d\eta$$

$$= \begin{cases} \left|\dfrac{1}{\cos[\phi]}\right|\left(2\cos\left[\dfrac{2\pi\nu}{2\cos[\phi]}\right]\right) = \left|\dfrac{2}{\cos[\phi]}\right|\cos\left[\dfrac{\pi\nu}{\cos[\phi]}\right] & \text{if } |\phi| \le \dfrac{\pi}{4} \\ \left|\dfrac{1}{\sin[\phi]}\right|\left(2\cos\left[\dfrac{2\pi\nu}{2\sin[\phi]}\right]\right) = \left|\dfrac{2}{\sin[\phi]}\right|\cos\left[\dfrac{\pi\nu}{\sin[\phi]}\right] & \text{if } \dfrac{\pi}{4} \le |\phi| \le \dfrac{\pi}{2} \end{cases} \tag{12.94}$$

In words, all line-integral projections of the spectrum of the corral function are cosines with periods that vary between 2 cycles per unit length at $\phi = 0, \pm\pi/2$ and $\sqrt{2}$ cycles per unit length at $\phi = \pm\pi/4$. Note the similarity to the behavior of the central slices of the ring delta function, which are even pairs of Dirac delta functions with identical separations. The 2-D spectrum is $J_0[\pi\rho]$ and line-integral projections are identically $\pi^{-1}\cos[\pi\nu]$.

An example of the central slices of the corral function is shown in Figure 12.31.

12.5.2.7 Central-Slice Transform of Arbitrary $f[x, y]$

In exactly the same manner as was used to evaluate the Radon transform of an arbitrary function, the linearity of the central-slice transform enables it to operate on the decomposition of $f[x, y]$ into its delta function components:

$$\begin{aligned}\mathcal{C}_2\{f[x,y]\} &= f[x,y](\delta[\underline{\mathbf{r}}\bullet\hat{\underline{\mathbf{p}}}^{\perp}]\cdot 1[\underline{\mathbf{r}}\bullet\hat{\underline{\mathbf{p}}}])\\ &= (f[p\hat{\underline{\mathbf{p}}}]\cdot\delta[\underline{\mathbf{r}}\bullet\hat{\underline{\mathbf{p}}}^{\perp}])\cdot 1[\underline{\mathbf{r}}\bullet\hat{\underline{\mathbf{p}}}]\\ &\Longrightarrow \mu_f(p,\phi) = f[p\hat{\underline{\mathbf{p}}}]\end{aligned} \tag{12.95}$$

which is just the set of radial slices of $f[x, y]$. The example of the letter "E" is shown in Figure 12.32, which also demonstrates the equivalence of the 1-D Fourier transforms of the slices and the line-integral projections of the 2-D Fourier transforms.

12.5.3 Inverse Central-Slice Transform

We have already considered the steps necessary to reconstruct a 2-D function from its 1-D central slices in Section 12.4, where we found that the azimuthal integral of the central slices of the 2-D spectrum yields:

$$\int_{-\pi/2}^{+\pi/2} \Lambda_f(\nu,\phi)\,d\phi = \frac{1}{|\nu|}F[\xi,\eta] \tag{12.96}$$

The correct spectrum is reconstructed radially weighting the slices before or after the azimuthal integral. We can distinguish this operator implemented in the frequency domain as $\widetilde{\mathcal{C}}_2^{-1}$:

$$\begin{aligned}\widetilde{\mathcal{C}}_2^{-1}\{\Lambda_f(\nu,\phi)\} &= F[\xi,\eta]\\ &= \int_{-\pi/2}^{+\pi/2} |\nu|\cdot\Lambda_f(\nu,\phi)\,d\phi\\ &= |\nu|\cdot\int_{-\pi/2}^{+\pi/2} \Lambda_f(\nu,\phi)\,d\phi\end{aligned} \tag{12.97}$$

The first of these equivalent recipes for the inverse central-slice transform applies a 1-D rho filter to the slices of the spectrum which scales the amplitude by the radial spatial frequency. The second recipe applies the 2-D rho filter to the result of the azimuthal summation.

We can directly transfer these sequences of operations to the space domain. If we integrate the central slices $\mu_f(p, \phi)$ over azimuth angle, the resulting image is multiplied by the radial distance r^{-1}:

$$\int_{-\pi/2}^{+\pi/2} \mu_f(p,\phi)(\delta[\underline{\mathbf{r}}\bullet\hat{\underline{\mathbf{p}}}^{\perp}]\cdot 1[\underline{\mathbf{r}}\bullet\hat{\underline{\mathbf{p}}}])\,d\phi = \frac{1}{r}f[x,y] = \frac{1}{|p|}f[x,y] \tag{12.98}$$

The correct reconstruction is obtained by scaling the summation by $r^{-1} = |p|^{-1}$, again either before or after the azimuthal integral:

$$\begin{aligned}\int_{-\pi/2}^{+\pi/2} |p|\mu_f(p,\phi)(\delta[\underline{\mathbf{r}}\bullet\hat{\underline{\mathbf{p}}}^{\perp}]\cdot 1[\underline{\mathbf{r}}\bullet\hat{\underline{\mathbf{p}}}])\,d\phi &= |p|\int_{-\pi/2}^{+\pi/2} \mu_f(p,\phi)(\delta[\underline{\mathbf{r}}\bullet\hat{\underline{\mathbf{p}}}^{\perp}]\cdot 1[\underline{\mathbf{r}}\bullet\hat{\underline{\mathbf{p}}}])\,d\phi\\ &= \mathcal{C}_2^{-1}\{\mu_f(p,\phi)\}\end{aligned} \tag{12.99}$$

As a final comment, it is perhaps useful to explicitly note the difference between the space-domain azimuthal integrals of central slices and of line-integral projections: the integral of the central slices produces the original function scaled by the factor r^{-1}, while that of the line-integral projections produces the original function convolved with r^{-1}.

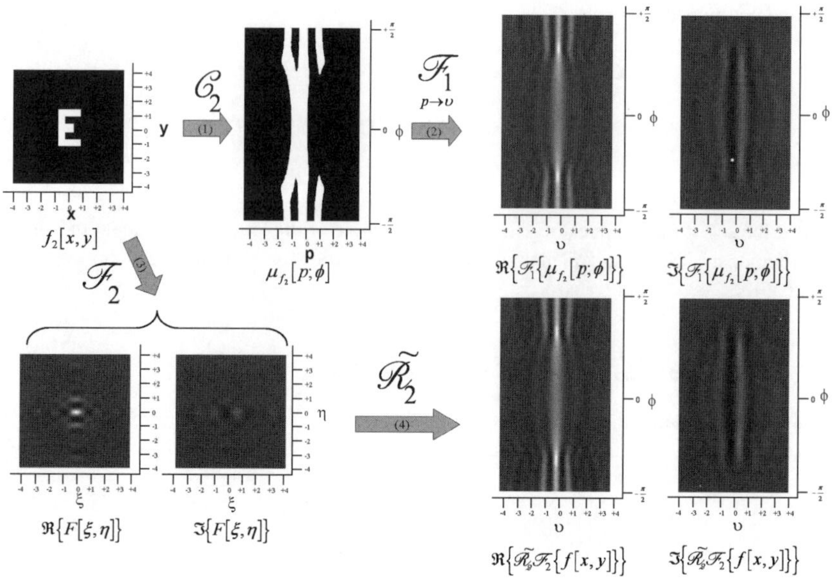

Figure 12.32 Comparison of two sequences of operations: (1) central slices of $f[x, y] =$ "E"; (2) 1-D Fourier transforms of central slices; (3) 2-D Fourier transform of $f[x, y] =$ "E"; (4) line-integral projections of the 2-D transform. Both sequences produce the 1-D Fourier transforms of the central slice.

12.6 Three Transforms of Four Functions

We will now use the relationships among the Fourier, Hankel, Radon, and central-slice transforms to compare the space-domain representations, frequency-domain representations, radial profiles (slices), and line-integral projections of four 2-D functions: $RECT[x, y]$, $CYL(r)$, $COR[x, y]$, and $\delta(r - \frac{1}{2})$. These observations will demonstrate some useful relationships among the functions that are perhaps not obvious and also may help develop additional intuition about the 2-D Fourier transform and its application to practical imaging problems.

Consider first the comparison of the 2-D unit-amplitude, unit-support rectangle and the unit-amplitude, unit-diameter cylinder, as shown in Figure 12.33. Their 2-D spectra were derived in Equation (10.14) and Equation (11.48), respectively:

$$f_1[x, y] = RECT[x, y] \Longrightarrow F_1[\xi, \eta] = SINC[\xi, \eta] \tag{12.100a}$$

$$f_2(r, \theta) = CYL(r)1(\theta) \Longrightarrow F_2(\rho, \phi) = \frac{\pi}{4}SOMB(\rho)1[\phi] \tag{12.100b}$$

The profiles of the 2-D spectra along the ξ-axis are different, though they have some "similar" characteristics:

$$F_1[\xi, 0] = SINC[\xi] \tag{12.101a}$$

$$F_2[\xi, 0] = \frac{\pi}{4}SOMB[\xi] = \frac{\pi}{4}\frac{2J_1[\pi\xi]}{\pi\xi} = \frac{J_1[\pi\xi]}{2\xi} \tag{12.101b}$$

Both 1-D profiles of the 2-D spectra along the ξ-axis have maximum amplitude at the origin and decay to zero for large values of $|\xi|$. The amplitude of the 2-D profiles may be applied to the 2-D central-ordinate theorem to show that the volumes of the 2-D space-domain functions are respectively unity and $\pi/4 \cong 0.785$.

Now consider the projections of the two functions onto the x-axis. That of $RECT[x, y]$ is trivial to evaluate because of its separability:

$$\begin{aligned}\lambda_{RECT[x,y]}(p, \phi = 0) &= \int_{-\infty}^{+\infty} RECT[p, y]\, dy \\ &= \int_{-\infty}^{+\infty} RECT[p] \cdot RECT[y]\, dy \\ &= RECT[p] \cdot \int_{-\infty}^{+\infty} RECT[y]\, dy \\ &= RECT[p] \end{aligned} \tag{12.102}$$

The projection of the cylinder onto the x-axis is almost as easy to determine, as was shown in Equation (12.23):

$$\begin{aligned}\lambda_{CYL(r)}(p; 0) &= \int_{-\infty}^{+\infty} CYL\left(\sqrt{p^2 + y^2}\right) dy \\ &= RECT[p] \cdot \int_{-\frac{1}{2}\sqrt{1-4p^2}}^{+\frac{1}{2}\sqrt{1-4p^2}} dy \\ &= RECT[p] \cdot \sqrt{1 - 4p^2} \end{aligned} \tag{12.103}$$

It is no surprise that the 1-D region of support is the same for both, and that the amplitude of the projection of the rectangle is constant, while that of the cylinder function decays smoothly to zero. The graphs are compared in Figure 12.34.

The central-slice theorem determines that the 1-D Fourier spectrum of a line-integral projection is the radial central slice of the 2-D spectrum, which leads to two 1-D Fourier transform pairs:

$$\mathcal{F}_{1\ p \to \nu} \{RECT[p]\} = SINC[\nu] \tag{12.104a}$$

$$\mathcal{F}_{1\ p \to \nu} \left\{ RECT[p]\sqrt{1 - 4p^2} \right\} = \frac{\pi}{4} SOMB[\nu] = \frac{J_1[\pi\nu]}{2\nu} \tag{12.104b}$$

Since $\phi = 0$, we can consider that the spectra are functions of ξ instead of the more general variable ν. The first of these Fourier transform pairs is now quite familiar, but the second is not. The discussions of Chapter 6 demonstrated that the envelopes of magnitudes of these 1-D spectra decrease as $|\xi|^{-1}$ and $|\xi|^{-3/2}$, respectively, which means that the magnitude of a sinusoidal component of the rectangle function of unit width decreases more slowly with $|\xi|$ than those of the cylinder function of unit diameter. From our interpretation of the Fourier transform, this observation implies that the high-frequency sinusoidal components necessary to generate edges in the function are present in the cylinder function in smaller amounts than in the rectangle function.

The profiles of the rectangle and cylinder evaluated along the x-axis are identical rectangle functions:

$$RECT[x, 0] = CYL\left(\sqrt{x^2 + 0^2}\right) = RECT[x] \tag{12.105}$$

as shown in Figure 12.35. The discussion of the central-slice transform demonstrates that the spectrum of the 1-D x-axis profiles are the line-integral projections of the 2-D spectra onto the ξ-axis:

$$\mathcal{F}_1\{f[x, 0]\} = \mathcal{F}_1\{\mu_f(p, \phi = 0)\} = \int_{-\infty}^{+\infty} F[\xi, \eta]\, d\eta \tag{12.106}$$

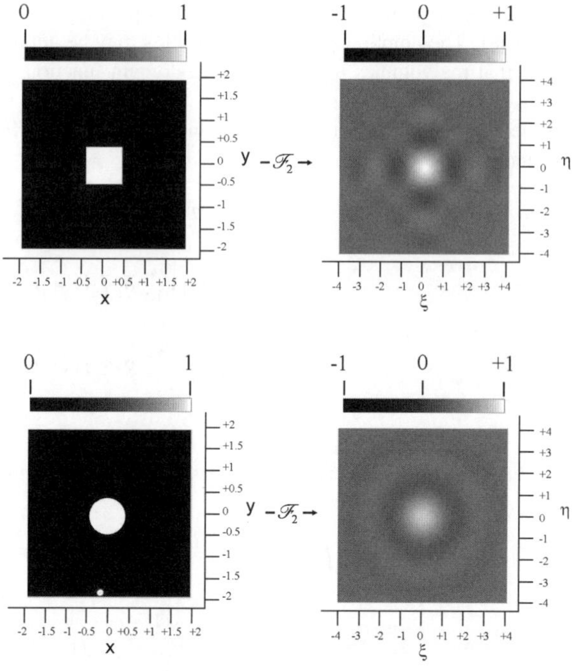

Figure 12.33 $RECT[x, y]$ and $CYL(r)$ and their respective 2-D spectra $SINC[\xi, \eta]$ and $(\pi/4)SOMB(\rho)$.

Because the x-axis profiles of $RECT[x, y]$ and $CYL(r)$ are identically $RECT[x]$, then the projections of their spectra onto the ξ-axis must be identical as well. This observation leads to a previously underived relation between the *SINC* and *SOMB* functions:

$$SINC[\xi] = \frac{\pi}{4} \int_{-\infty}^{+\infty} SOMB\left(\sqrt{\xi^2 + \eta^2}\right) d\eta \tag{12.107}$$

The circular symmetry guarantees that the line-integral projections of the sombrero function are identical *SINC* functions. This leads to another observation that the sombrero function may be generated by summing the infinite set of central slices $SINC[p] \cdot (\delta[\mathbf{\underline{r}} \bullet \hat{\mathbf{\underline{p}}}^{\perp}] 1[\mathbf{\underline{r}} \bullet \hat{\mathbf{\underline{p}}}])$ scaled by the necessary factor of $|p|$:

$$|p| \int_{-\pi/2}^{+\pi/2} SINC[p] \cdot (\delta[\mathbf{\underline{r}} \bullet \hat{\mathbf{\underline{p}}}^{\perp}] \cdot 1[\mathbf{\underline{r}} \bullet \hat{\mathbf{\underline{p}}}]) \, d\phi = \frac{\pi}{4} SOMB(r) \tag{12.108}$$

Now consider the circularly symmetric radial Dirac delta function with unit diameter, which has the same region of support in the space domain as $CYL(r)$ and may be visualized as a "circular stockade":

$$f_3(r, \theta) = \delta\left(r - \frac{1}{2}\right) \tag{12.109}$$

Its volume is π, which is significantly larger than the unit volume of $RECT[x, y]$ or the volume of $\pi/4$ for $CYL(r)$. The circularly symmetric spectrum of the ring delta function is derived easily by applying the "transform-of-a-transform" theorem to Equation (11.38):

$$\begin{aligned} \mathcal{H}_0\{f_3(r, \theta)\} &= \mathcal{H}_0\left\{\delta\left(r - \frac{1}{2}\right) \cdot 1(\theta)\right\} \\ &= F_3(\rho, \phi) = \pi \, J_0[\pi\rho] \cdot 1[\phi] \end{aligned} \tag{12.110}$$

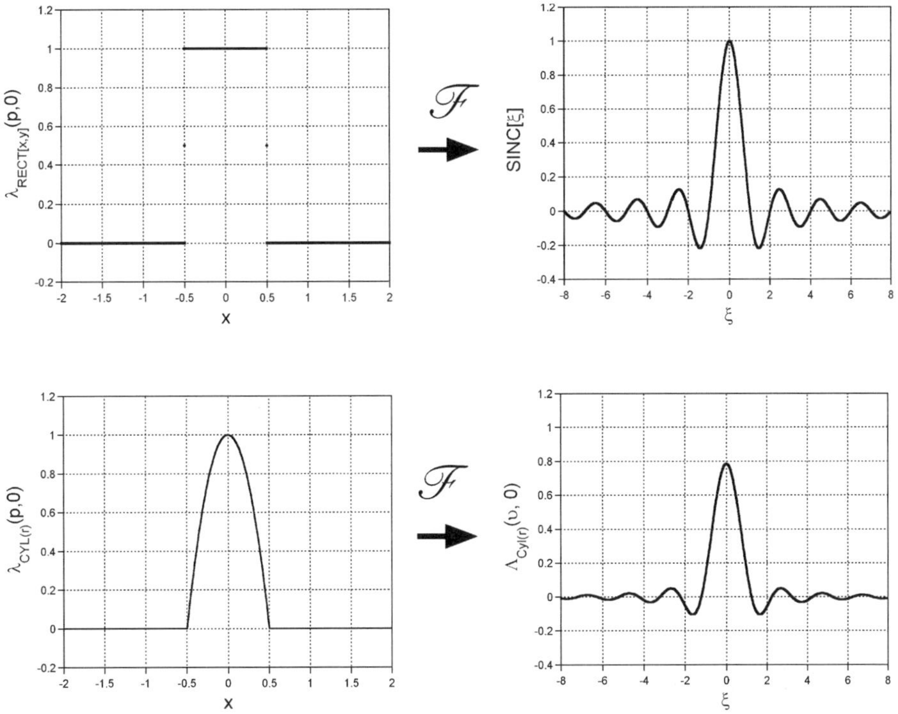

Figure 12.34 Line-integral projections of $RECT[x, y]$ and $CYL(r)$ onto the x-axis are respectively $RECT[x]$ and $(1 - 4x^2)^{1/2} \cdot RECT[x]$. Their 1-D Fourier transforms are respectively $SINC[\xi]$ and $(\pi/4)\ SOMB[\xi]$.

The "transform-of-a-transform" and central-ordinate theorems may be applied to show that the volume of the 2-D function $J_0[\pi r/d]$ is zero for $d \neq 0$, as already inferred in Equation (11.39).

Since the spectrum of the ring delta function is circularly symmetric, its ξ-axis profile is obtained by substituting $\rho = \xi$ in Equation (12.110):

$$F_3[\xi, 0] = \pi\ J_0[\pi\xi] \tag{12.111}$$

The discussion in Section 6.1.13 demonstrated that the envelope of magnitudes of the 1-D zero-order Bessel function decrease as $|\xi|^{-1/2}$, which is significantly "slower" than the decay of the envelopes of the spectra of $RECT[x]$ and of $CYL(r)$. In words, the 1-D profile of the spectrum of the ring delta function exhibits the largest amplitudes at high frequencies of the three functions. This quality of the ring delta function should be no surprise because its edge transitions are the "sharpest" of the three functions and thus require more amplitude at large spatial frequencies.

As an aside, compare the locations of the "low-order" zeros of the three spectra on the ξ-axis. The first zero of the *SINC* function is located at $\xi_1 = 1$, that of the cylinder is located at $\xi_1 \cong 1.22$, and that of the spectrum of the ring delta at $\xi_1 \cong 0.7655$. The first zero of the spectrum of the ring delta function is closest to the origin.

Now consider the relationships among the 1-D profiles and 1-D projections of the 2-D representations $\delta(r - \frac{1}{2})$ and $\pi\, J_0[\pi\rho]$. First, the 1-D profile of $\delta(r - \frac{1}{2})$ obviously is a symmetric pair of 1-D

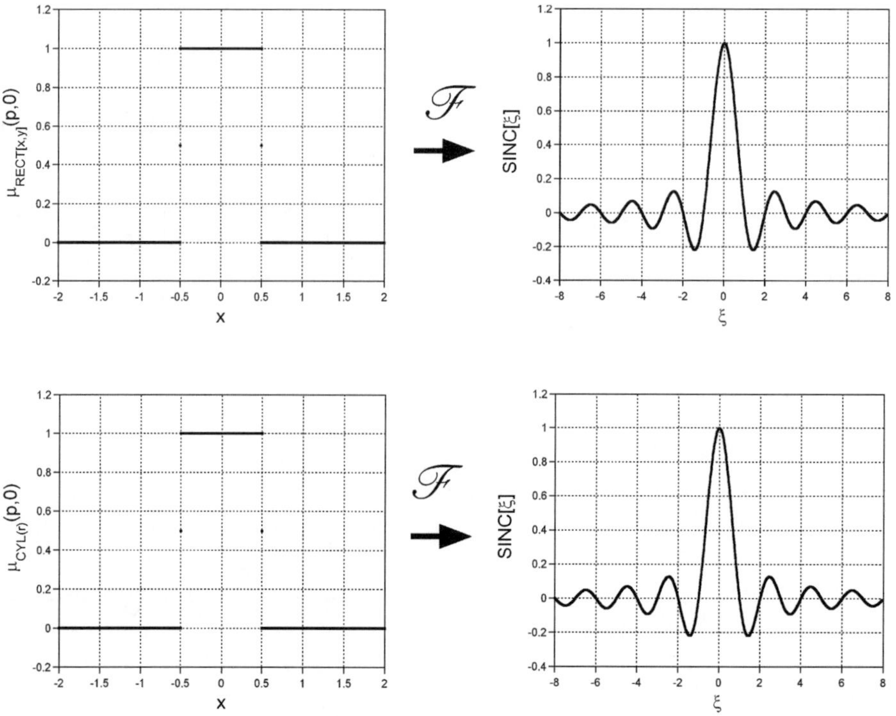

Figure 12.35 The central slices of $RECT[x, y]$ and $CYL(r)$ on the x-axis are identically $\mu_f[x] = RECT[x]$ and the two spectra are identically $SINC[\xi]$, which are the line-integral projections of the 2-D spectra $SINC[\xi, \eta]$ and $(\pi/4)\ SOMB(\sqrt{\xi^2 + \eta^2})$.

Dirac delta functions located at $x = \pm\frac{1}{2}$:

$$f_3[x, 0] = \delta\left(x + \frac{1}{2}\right) + \delta\left(x - \frac{1}{2}\right) = 2\ \delta\delta[2x] \tag{12.112}$$

The 1-D spectrum of this central slice is a cosine with period 2, and must be identical to the line-integral projection of the 2-D spectrum onto the ξ-axis:

$$\begin{aligned}\mathcal{F}_1\left\{\delta\left(x + \frac{1}{2}\right) + \delta\left(x - \frac{1}{2}\right)\right\} &= 2\cos\left[2\pi\xi \cdot \frac{1}{2}\right] = 2\cos[\pi\xi] \\ &= \pi \int_{-\infty}^{+\infty} J_0\left[\pi\sqrt{\xi^2 + \eta^2}\right] d\eta\end{aligned} \tag{12.113}$$

This result may be recast into the form:

$$\int_{-\infty}^{+\infty} J_0\left[2\pi\,\frac{\sqrt{\xi^2 + \eta^2}}{2}\right] d\eta = \frac{2}{\pi}\cos\left[2\pi \cdot \frac{1}{2} \cdot \xi\right] \tag{12.114}$$

In words, the projection of $J_0[\pi\rho]$ onto the ξ-axis is a cosine with period 2, and the circular symmetry of $J_0[\pi\rho]$ ensures that all projections have this same form. We can turn this observation around to say that $J_0[\pi\rho]$ may be synthesized by summing 2-D cosine functions with the same period along each of the complete set of azimuths:

$$\frac{2}{\pi}\int_{-\pi/2}^{+\pi/2}|p|\cos\left[2\pi\frac{p}{2}\right]\cdot\left(\delta[\underline{\mathbf{r}}\bullet\hat{\underline{\mathbf{p}}}^{\perp}]\cdot 1[\underline{\mathbf{r}}\bullet\hat{\underline{\mathbf{p}}}]\right)d\phi = J_0\left[2\pi\frac{r}{2}\right] = J_0[\pi r] \tag{12.115}$$

Note that the relative radial density of the slices is compensated by the scaling factor of $|p|$ in the integral.

We also know that the inverse Fourier transform of a 1-D central slice of $\pi J_0(\pi\rho)$ must be identical to the line-integral projection of $\delta(r-\frac{1}{2})$:

$$\mathcal{F}_1^{-1}\{\pi J_0[\pi\xi]\} = \int_{-\infty}^{+\infty}\delta\left(\sqrt{x^2+y^2}-\frac{1}{2}\right)dy \tag{12.116}$$

The functional form of the line-integral projection of the ring delta function was derived in Equation (7.102):

$$\delta\left(r-\frac{1}{2}\right) = \lim_{\Delta\to 0}\left\{\frac{1}{\Delta}\left(CYL\left(\frac{r}{1+\Delta}\right) - CYL\left(\frac{r}{1-\Delta}\right)\right)\right\} \tag{12.117}$$

The scale factor $1/\Delta$ ensures that the volume of the limiting expression is π. The line-integral projection of this function at azimuth angle ϕ was shown to be $2/\sin[\phi]$ in Figure 12.8. The angular dependence $\sin[\phi]$ may be rewritten as a function of x to establish another 1-D Fourier transform pair:

$$\mathcal{F}_1\left\{\frac{2}{\pi\sqrt{1-4x^2}}\cdot RECT[x]\right\} = \pi J_0[\pi\xi] \tag{12.118}$$

as shown in Figure 12.37. The linearity of the Fourier transform and the transform-of-a-transform theorem may be applied to obtain two relations:

$$\mathcal{F}_1\left\{\frac{2}{\pi\sqrt{1-4x^2}}\cdot RECT[x]\right\} = J_0[\pi\xi]$$

$$\mathcal{F}_1\{J_0[\pi x]\} = \frac{2}{\pi\sqrt{1-4(-\xi)^2}}\cdot RECT[-\xi] = \frac{2}{\pi\sqrt{1-4\xi^2}}\cdot RECT[\xi] \tag{12.119}$$

This last result may be combined with the central-ordinate theorem of the Hankel transform to evaluate the area of the 1-D J_0 Bessel function over infinite limits, which we have already used:

$$\int_{-\infty}^{+\infty}J_0[\pi x]\,dx = \frac{2}{\pi} \Longrightarrow \int_{-\infty}^{+\infty}J_0[x]\,dx = 2 \tag{12.120}$$

The symmetry of the Bessel function ensures that its half-area is unity:

$$\int_0^{+\infty}J_0[x]\,dx = 1 \tag{12.121}$$

We noted in Equation (11.39) that the volume of the 2-D function $J_0(|\underline{\mathbf{r}}|)$ is zero, but here we find that the area of the 1-D function $J_0[x]$ is not.

Finally, consider the corral function with the same support as $RECT[x, y]$:

$$COR[x,y] = \left(\delta\left[x+\frac{1}{2}\right]+\delta\left[x-\frac{1}{2}\right]\right)\cdot RECT[y] + RECT[x]\cdot\left(\delta\left[y+\frac{1}{2}\right]+\delta\left[y-\frac{1}{2}\right]\right) \tag{12.122}$$

The 2-D spectrum of $COR[x, y]$ also is easy to derive by applying known 1-D Fourier transforms:

$$\mathcal{F}_2\{COR[x,y]\} = 2\cos[\pi\xi]\cdot SINC[\eta] + 2\,SINC[\xi]\cdot\cos[\pi\eta] \tag{12.123}$$

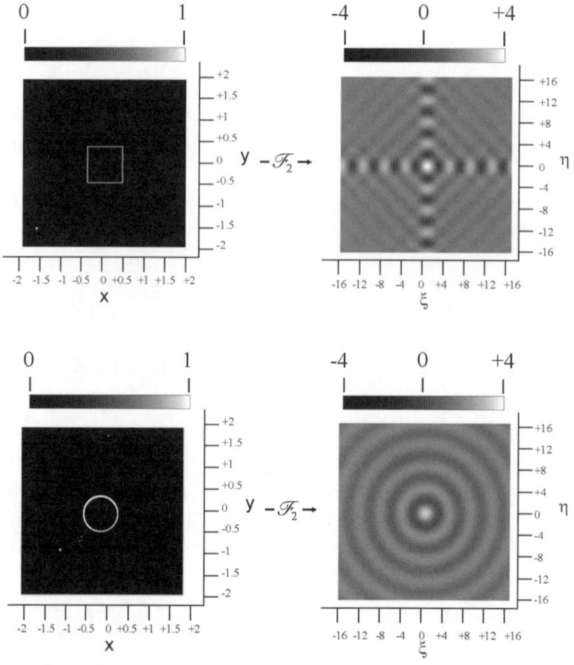

Figure 12.36 The square corral $COR[x, y]$, the ring delta $\delta(r - \frac{1}{2})$, and their 2-D spectra $2 \cdot SINC[\xi] \cos[\pi\eta] + 2 \cdot \cos[\pi\xi]\, SINC[\eta]$ and $\pi J_0(\pi\rho)$, respectively. Note the difference in scale of the coordinate axes in the two domains.

The central ordinate of the spectrum is 4, which confirms the volume from Equation (7.81). $COR[x, y]$ has the largest volume of the four functions considered in this section. $COR[x, y]$ and its spectrum are shown in Figure 12.36.

Now consider the line-integral projection and central slice of the corral evaluated on the x-axis. The projection is easy to compute from the properties of the Dirac delta and $SINC$ functions:

$$\begin{aligned}\lambda_{(COR[x,y])}(p, \phi = 0) &= \int_{-\infty}^{+\infty} COR[x, y]\, dy \\ &= 2\, RECT[x] + \left(\delta\left[x + \frac{1}{2}\right] + \delta\left[x - \frac{1}{2}\right]\right) \\ &= 2\, RECT[x] + 2\, \delta\delta[2x] \end{aligned} \tag{12.124}$$

The central slice of $COR[x, y]$ onto the x-axis is the same pair of Dirac delta functions in the central slice of $RECT[x, y]$:

$$\mu_{(COR[x,y])}(p, 0) = \delta\left[x + \frac{1}{2}\right] + \delta\left[x - \frac{1}{2}\right] \tag{12.125}$$

By the central-slice theorem, the 1-D Fourier transform of $\lambda_{(COR[x,y])}(p, \phi = 0)$ is identical to $F[\xi, 0]$:

$$\mathcal{F}_1\{2\, RECT[x] + 2\, \delta\delta[2x]\} = 2\, SINC[\xi] + 2 \cos\left[2\pi \frac{\xi}{2}\right] \tag{12.126}$$

The $SINC$ function is zero at all nonzero integer values of ξ, while the zeros of the cosine are located at half-integer values of ξ, i.e., at $\xi = (2m + 1)/2$ where m is an integer. Thus the amplitude of the slice

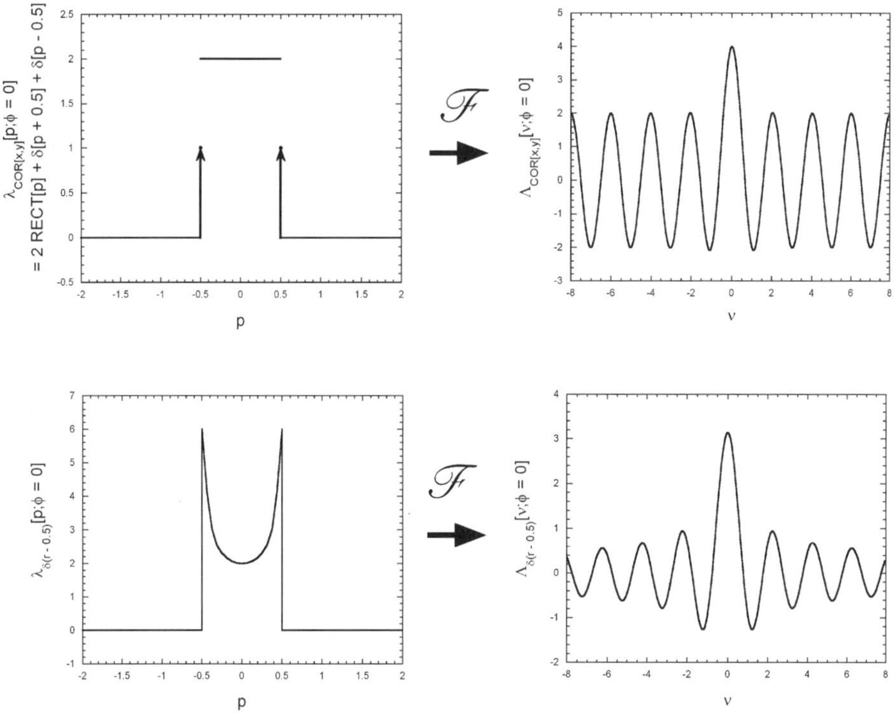

Figure 12.37 Line-integral projections of $COR[x, y]$ and $\delta(r - \frac{1}{2})$ onto the x-axis, and their respective spectra. The projection of the latter is a 1-D Fourier transform pair not previously derived: $\mathcal{F}_1\left\{\frac{2}{\pi\sqrt{1-4x^2}} \cdot RECT[x]\right\} = \pi\, J_0[\pi\xi]$.

of the spectrum of the corral function is ± 1 at integer ξ. The spectrum of the central slice obviously is a cosine, which must also be the line-integral projection of the spectrum of $COR[x, y]$ onto the ξ-axis:

$$\begin{aligned}\mathcal{F}_1\{\mu_f(p, 0)\} &= \int_{-\infty}^{+\infty} COR[\xi, \eta]\, d\eta \\ &= 2\int_{-\infty}^{+\infty} (\cos[\pi\xi] \cdot SINC[\eta] + SINC[\xi] \cdot \cos[\pi\eta])\, d\eta \\ &= 2\cos[\pi\xi] \cdot 1 + SINC[\xi] \cdot 0 \end{aligned} \tag{12.127}$$

The spectra of the ring delta and corral functions evaluated along the ξ-axis exhibit some similarities in form, as shown in Figure 12.37. Both functions have their maximum at the origin and both oscillate about zero away from the origin with similar periods, though the spectrum of the projection of the ring is only pseudoperiodic. The amplitudes of the extrema of J_0 decrease with $|\xi|$, while those of its rectangular analogue do not. Therefore we can state a loose analogue:

$$\pi\, J_0[\pi\xi] \sim 2\,SINC[\xi] + 2\cos[\pi\xi] \tag{12.128}$$

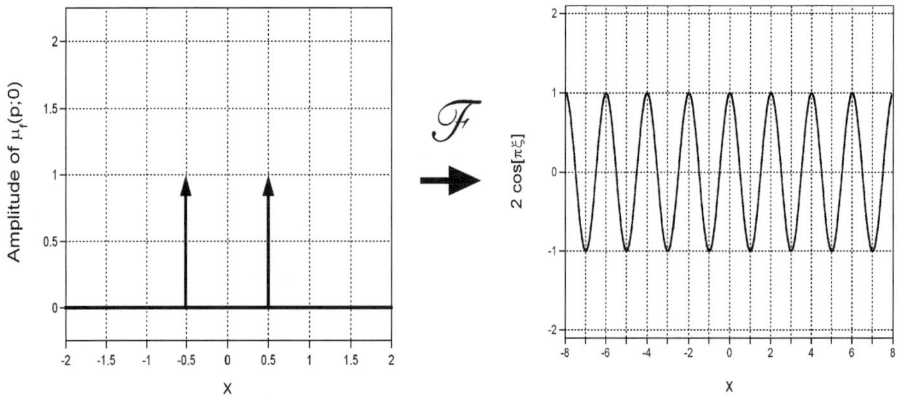

Figure 12.38 The central slices of $COR[x, y]$ and $\delta(r - \frac{1}{2})$ on the x-axis are identically $\delta[x + \frac{1}{2}] + \delta[x - \frac{1}{2}] = 2\delta[2x]$. The 1-D spectra are identically $2\cos[\pi x]$.

We can extend the analogy between $\delta(r - \frac{1}{2})$ and $COR[x, y]$ a bit further by recalling the series form of the Bessel function that was derived in Appendix 6B:

$$\begin{aligned} J_0[\pi\xi] &= 1 - \frac{(\pi\xi)^2}{2^2} + \frac{(\pi\xi)^4}{2^2 \cdot 4^2} - \frac{(\pi\xi)^6}{2^2 \cdot 4^2 \cdot 6^2} + \cdots \\ &= 1 - \frac{1}{4}(\pi\xi)^2 + \frac{1}{64}(\pi\xi)^4 - \frac{1}{2304}(\pi\xi)^6 + \frac{1}{147\,456}(\pi\xi)^8 - \cdots \end{aligned} \tag{12.129}$$

A similar series solution for the rectangular Bessel function may be constructed by substituting the series solutions for sine and cosine into Equation (12.126):

$$\begin{aligned} &2\,SINC[\xi] + 2\cos[\pi\xi] = 2\left(\frac{SINC[\pi\xi]}{\pi\xi} + \cos[\pi\xi]\right) \\ &= 2\left(1 - \frac{(\pi\xi)^2}{3!} + \frac{(\pi\xi)^4}{5!} - \frac{(\pi\xi)^6}{7!} + \cdots\right) + 2\left(1 - \frac{(\pi\xi)^2}{2!} + \frac{(\pi\xi)^4}{4!} - \frac{(\pi\xi)^6}{6!} + \cdots\right) \\ &= 2\left(\frac{(\pi\xi)^0}{(1/0! + 1/1!)^{-1}} - \frac{(\pi\xi)^2}{(1/2! + 1/3!)^{-1}} + \frac{(\pi\xi)^4}{(1/4! + 1/5!)^{-1}} - \frac{(\pi\xi)^6}{(1/6! + 1/7!)^{-1}} + \cdots\right) \\ &= 2\sum_{n=0}^{\infty}(-1)^n \frac{(\pi\xi)^{2n}}{(1/(2n)! + 1/(2n+1)!)^{-1}} \\ &= 4\left(1 - \frac{1}{3}(\pi\xi)^2 + \frac{1}{40}(\pi\xi)^4 - \frac{1}{1260}(\pi\xi)^6 + \frac{1}{72\,576}(\pi\xi)^8 - \cdots\right) \end{aligned} \tag{12.130}$$

The coefficients of the 1-D slice of the spectrum of $\delta(r - \frac{1}{2})$ fall off more rapidly than those of $COR[x, y]$, but both are generated from the even powers of ξ.

The central slices of the ring and corral are identically $2 \cdot \delta\delta[2x]$, which means that the projections of the 2-D spectra are identically $2 \cdot \cos[\pi x]$, as shown in Figure 12.38.

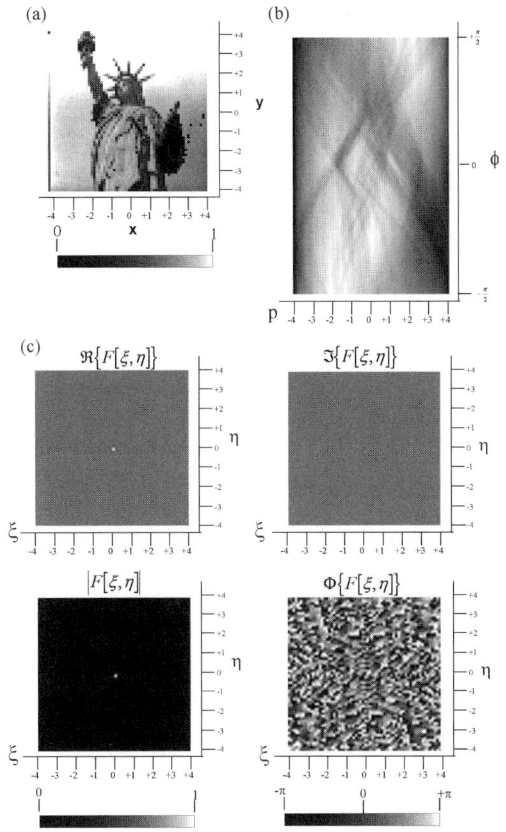

Figure 12.39 Comparison of the Radon and Fourier transforms for a pictorial image $f[x, y]$. Both transforms generate functions with wider dynamic ranges, but the amplitude of the Fourier transform is dominated by the large amplitude at the origin $[\xi, \eta] = [0, 0]$. The Radon transform of the image exhibits the sinusoidal character that led to the name of "sonogram", and the amplitude tends to be large at all (p, ϕ).

12.7 Fourier and Radon Transforms of Images

The amplitude of a "pictorial" image (e.g., a photograph) is generally nonnegative and real valued at each spatial coordinate. In such a case, the 2-D analogue of the central-ordinate theorem presented in Equation (10.21) demonstrates that the DC component of the Fourier transform is the "volume" of the image. Because the amplitude of the pictorial image is nonnegative everywhere, the DC component of the 2-D spectrum is often quite large compared to any "AC" term. The DC component often overwhelms any others when displayed with a linear "brightness" scale on a device and limited dynamic range (such as a CRT). An example is shown in Figure 12.39.

This problem may be expressed in other terms that are more familiar to readers with backgrounds in digital image processing. The histogram of the magnitude of the Fourier transform of realistic images is concentrated at small values. This is part of the explanation of why images may be compressed by requantizing and coding the Fourier transform of the image rather than the image itself. For these reasons, the dynamic range of Fourier transforms of pictorial images may be compressed by computing and displaying the logarithm or square root of the magnitude.

PROBLEMS

12.1 Find the Radon transform of a ring delta with a missing section (e.g., $\delta(r - r_0) \cdot STEP[x]\, 1[y]$ or $\delta(r - r_0) \cdot RECT[(x - r_0)/2r_0, (y - r_0)/2r_0]$.

12.2 The function $f_0[x, y] = \delta[x - x_0]\delta[y]$ may be rotated by θ radians by substituting $\underline{\mathbf{r}} \bullet \hat{\underline{\mathbf{p}}}_1$ for x and $\underline{\mathbf{r}} \bullet \hat{\underline{\mathbf{p}}}_1^\perp$ for y to obtain:

$$f_1[x, y] = \delta[\underline{\mathbf{r}} \bullet \hat{\underline{\mathbf{p}}}_1 - x_0]\, \delta[\underline{\mathbf{r}} \bullet \hat{\underline{\mathbf{p}}}_1^\perp]$$

Show that this expression is equivalent to:

$$f_1[x, y] = \delta[x - x_0 \cos[\theta], y - x_0 \sin[\theta]]$$

(HINT: $\delta[ax] = (1/|a|)\delta[x]$.)

12.3 Evaluate the results of the following operations, where the symbols "$*$" and "$\bigstar$" denote 2-D convolution and correlation, respectively:

- **(a)** $(\cos[\pi\xi] \cdot SINC[\eta] + SINC[\xi] \cdot \cos[\pi\eta])$
  $\bigstar(\cos[2\pi\xi]\, SINC[\eta/2] + SINC[\xi/2] \cos[2\pi\eta])$
- **(b)** $J_0(2\pi\rho_0 r) * J_0(2\pi\rho_1 r)$, where $\rho_0 \neq \rho_1$
- **(c)** $CYL(r) * (\delta[x] \cdot 1[y])$
- **(d)** $CYL(r) * (\delta[\underline{\mathbf{r}} \bullet \hat{\underline{\mathbf{p}}}] \cdot 1[\underline{\mathbf{r}} \bullet \hat{\underline{\mathbf{p}}}^\perp +])$ where $\hat{\underline{\mathbf{p}}} = \begin{bmatrix} \frac{\sqrt{3}}{2} \\ -\frac{1}{2} \end{bmatrix}$
- **(e)** $\delta(r - r_0) * (\delta[x] \cdot 1[y])$

12.4 Evaluate the following 2-D convolutions and make appropriate sketches of the results:

- **(a)** $(\cos[2\pi x/4]1[y]) * (\delta[x] \cdot 1[y])$
- **(b)** $(\cos[2\pi x/4]1[y]) * CROSS[x, y]$ (where $CROSS[x, y] \equiv \delta[x] \cdot 1[y] + 1[x] \cdot \delta[y]$)
- **(c)** $(\cos[2\pi[\underline{\mathbf{r}} \bullet \hat{\underline{\mathbf{p}}}]/4]1[\underline{\mathbf{r}} \bullet \hat{\underline{\mathbf{p}}}^\perp]) * (\delta[x] \cdot 1[y])$, where $\hat{\underline{\mathbf{p}}} = \left[\begin{smallmatrix} \cos[\theta] \\ \sin[\theta] \end{smallmatrix}\right]$ and $\theta \gtrsim 0$ (e.g., 10^{-4} radians)
- **(d)** $e^{+i\pi r^2} * \delta[x]1[y]$, where $r = \sqrt{x^2 + y^2}$
- **(e)** $e^{+i\pi r^2} * CROSS[x, y]$

12.5 Evaluate the 2-D Fourier transforms of these functions and make appropriate sketches:

- **(a)** $(\cos[2\pi x/4] \cdot 1[y]) * (\delta[x] \cdot 1[y])$
- **(b)** $(\cos[2\pi x/4] \cdot 1[y]) * CROSS[x, y]$
- **(c)** $(\cos[2\pi[\underline{\mathbf{r}} \bullet \hat{\underline{\mathbf{p}}}]/4] \cdot 1[\underline{\mathbf{r}} \bullet \hat{\underline{\mathbf{p}}}^\perp]) * (\delta[x] \cdot 1[y])$, where $\hat{\underline{\mathbf{p}}} = \left[\begin{smallmatrix} \cos[\theta] \\ \sin[\theta] \end{smallmatrix}\right]$ and $\theta \gtrsim 0$ (e.g., 10^{-4} radians)
- **(d)** $e^{+i\pi r^2} * \delta[x] \cdot 1[y]$, where $r = \sqrt{x^2 + y^2}$
- **(e)** $e^{+i\pi r^2} * CROSS[x, y]$

12.6 Evaluate the following 2-D operations and make appropriate sketches of the results:

- **(a)** $\mathcal{F}_2\{\cos[\pi r^2]\}$
- **(b)** $\mathcal{F}_2\{\cos[\pi r^2] \cdot CYL(r/d)\}$
- **(c)** $\mathcal{F}_2\{\cos[\pi r^2] * CYL(r/d)\}$

13

Approximations to Fourier Transforms

This chapter considers two methods for approximating spectra if the Fourier integral has no known closed-form solution. The approximations also may be useful in cases where the known functional form is complicated. Though the expressions are valid over limited regions of the frequency domain, they are useful in certain situations and also may help the reader develop intuition about the behavior of the Fourier transform of some other more complicated functions.

The first approximation is a series expansion for $F[\xi]$ in powers of ξ. The coefficients of the expansion are proportional to the *moments* of $f[x]$, which were introduced during the discussion of 1-D stochastic functions in Chapter 6. The moment expansion to be derived produces approximations of the spectrum of many functions that are most accurate for $|\xi| \cong 0$. For this reason, the series expansion may be considered to be an "asymptotic" formulation of the Fourier spectrum that is valid at small spatial frequencies.

The second approximate form of the Fourier transform obtained by the "method of stationary phase" may be evaluated only for functions that satisfy some specific criteria and is valid only for large values of $|\xi|$. However, the resulting expression is very useful in discussions of optical imaging.

The developments of these two approximations in this chapter are followed by the introduction of the so-called "central-limit theorem", which may provide an approximate expression for the action of a multistage imaging system.

13.1 Moment Theorem

The *moments* of a 1-D function were introduced in Section 6.4 as measures of the distribution of probability of a stochastic function about the mean value. The moments also are useful descriptors of deterministic functions and provide the path to a useful approximation of the Fourier transform. Recall from Equation (6.165) that the ℓth moment of the 1-D function $f[x]$ is its projection onto the 1-D "reference function" x^ℓ:

$$\mathrm{m}_\ell\{f[x]\} \equiv \int_{-\infty}^{+\infty} x^\ell f[x]\, dx \equiv \langle x^\ell \rangle_f \tag{13.1}$$

where the "bra–ket" notation with the subscript "$\langle\,\rangle_f$" denotes the *expectation value* of the enclosed quantity "averaged" over the amplitude density $f[x]$. Note that the weighting function x^ℓ in the integrand is even or odd as ℓ is even or odd and that moments of negative order are generally not defined because $x^{-|\ell|}$ is indeterminate at the origin.

As an aside, it is perhaps useful to compare the definition of moments in Equation (13.1) to that of the gamma function in Equation (6.37) to see that $\Gamma[n]$ for integer values produces the $(n-1)$st moment of the function $STEP[x] \cdot e^{-x}$:

$$\Gamma[n] = \int_{-\infty}^{+\infty} (STEP[\alpha] \cdot e^{-\alpha})\alpha^{n-1}\, d\alpha$$
$$= \mathrm{m}_{(n-1)}\{STEP[x] \cdot e^{-x}\} = (n-1)! \tag{13.2}$$

This result is useful in some physical problems involving the exponential probability law.

Obviously, the relative sizes of the moments of $f[x]$ reflect the "shape" of $f[x]$. For example, if $f[x]$ is even, then the product function $x^\ell f[x]$ is even for even ℓ and odd for odd ℓ. If $f[x]$ is odd, then the opposite situation applies: $x^\ell f[x]$ is odd for even ℓ and even for odd ℓ. Because the area of any odd function is zero, it is easy to see that all odd-order moments of even functions and even-order moments of odd functions evaluate to zero. In the general case of a function composed of both even and odd parts, the values of the even- and odd-order moments are determined by the even and odd parts of $f[x]$, respectively. Further, it is apparent that the moments of a complex-valued function are complex valued, in general.

The moment of order 0 is the projection of $f[x]$ onto $x^0 = 1[x]$, which applies unit weight to the amplitude of $f[x]$ at all coordinates before integrating to produce the area of $f[x]$:

$$\mathrm{m}_0\{f[x]\} = \int_{-\infty}^{+\infty} x^0 f[x]\, dx = \int_{-\infty}^{+\infty} f[x]\, dx = \langle x^0 \rangle_f = \langle 1 \rangle_f \tag{13.3}$$

In the common application of moments in the study of probability, $f[x]$ is a probability density function with unit area so that $\mathrm{m}_0 = 1$. The weighting function x^0 ensures that all the coordinates have equal "influence" upon the value of m_0. Note that x^0 is the weighting function with a nonzero amplitude at the origin, and thus m_0 is the only moment that includes the amplitude of $f[0]$. The central-ordinate theorem in Equation (9.113) may be applied directly to Equation (13.3) to demonstrate that:

$$\mathrm{m}_0\{f[x]\} = F[0] \tag{13.4}$$

The first moment of $f[x]$ is its projection onto the linear weighting function x^1:

$$\mathrm{m}_1\{f[x]\} = \int_{-\infty}^{+\infty} x^1 f[x]\, dx = \langle x^1 \rangle_f \tag{13.5}$$

The weighting is zero at the origin and its magnitude increases with $|x|$, which means that small amplitudes of $f[x]$ located at large distances from the origin may have more impact upon the value of m_1 than large amplitudes located near the origin.

The higher-order moments apply small weights to the amplitudes of $f[x]$ over increasingly wider regions near the origin and weights of quickly increasing magnitude distant from the origin, as shown in Figure 13.1. For example, the general expression for the moments of $RECT[x/b]$ of any order is easily derived:

$$\mathrm{m}_\ell\left\{RECT\left[\frac{x}{b}\right]\right\} = \int_{-b/2}^{+b/2} x^\ell\, dx = \left(\frac{x^{\ell+1}}{\ell+1}\right)\Bigg|_{-b/2}^{+b/2} = \frac{(b/2)^{\ell+1}}{\ell+1}(1-(-1)^{\ell+1})$$
$$= \begin{cases} \dfrac{b^{\ell+1}}{(\ell+1)\cdot 2^\ell} & \textit{for even } \ell \\ 0 & \textit{for odd } \ell \end{cases} \tag{13.6}$$

The amplitudes of the even moments grow without limit with increasing order ℓ if $b > 2$ and decrease with increasing ℓ if $b \leq 2$.

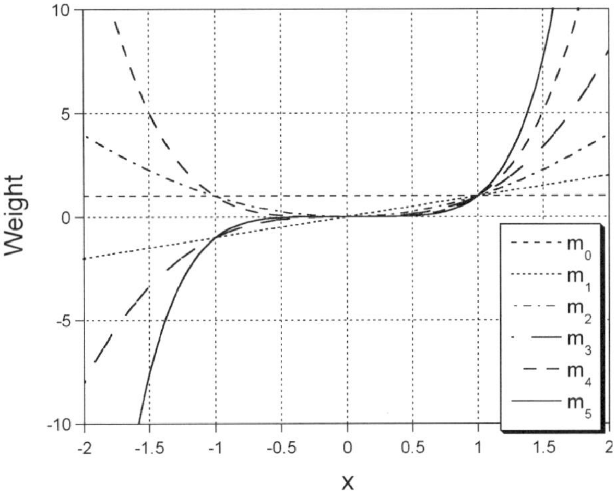

Figure 13.1 Weighting functions x^ℓ for moments of order 0–5. Note that only the zero-order moment applies a nonzero weighting at $x = 0$.

Equation (13.4) relates the zero-order moment and the central ordinate of the spectrum. The result may be generalized to construct a very useful relation between the moments of $f[x]$ and the spectrum $F[\xi]$. The 1-D "transform-of-a-transform" theorem in Equation (9.104) is used to construct a frequency-domain analogy of the derivative theorem in Equation (9.142):

$$\begin{aligned}\frac{d^\ell F[\xi]}{d\xi^\ell} &= \frac{d^\ell}{d\xi^\ell}\int_{-\infty}^{+\infty} f[x]\, e^{-2\pi i\xi x}\, dx = \int_{-\infty}^{+\infty} f[x]\cdot\left(\frac{d^\ell}{d\xi^\ell}\, e^{-2\pi i\xi x}\right) dx \\ &= \int_{-\infty}^{+\infty} f[x]\cdot(-2\pi i x)^\ell\cdot e^{-2\pi i\xi x}\, dx \\ &= (-2\pi i)^\ell \int_{-\infty}^{+\infty} (f[x]\cdot x^\ell)\cdot e^{-2\pi i\xi x}\, dx \end{aligned} \tag{13.7}$$

By combining this result and the linearity property of the Fourier transform, the ℓth moment is seen to be proportional to the ℓth-order derivative of the spectrum evaluated at the origin:

$$\mathrm{m}_\ell\{f[x]\} \equiv \int_{-\infty}^{+\infty} x^\ell f[x]\, dx = \langle x^\ell\rangle_f = \left(\frac{1}{-2\pi i}\right)^\ell \left(\frac{d^\ell F[\xi]}{d\xi^\ell}\right)\bigg|_{\xi=0} \tag{13.8}$$

This is the *moment theorem* of the Fourier transform, which immediately confirms the central-ordinate theorem and the observations that all odd-order moments of a symmetric function and all even moments of odd functions evaluate to zero because the odd-order derivatives of an even function and even-order derivatives of an odd function vanish.

Equation (13.8) is useful in statistics for computing the moments of probability laws, which are required to have unit area. The frequency-space representation of a probability law $p[x]$ is called the *characteristic function*. Other metrics derived from the moments that are important in statistics and mechanics may be evaluated from the spectrum of the function via the moment theorem. These may be easily defined in terms of "normalized" moments, defined as the ratio of the ℓth moment to the zero-order moment:

$$\frac{\displaystyle\int_{-\infty}^{+\infty} f[x] \cdot x^{\ell}\, dx}{\displaystyle\int_{-\infty}^{+\infty} f[x]\, dx} = \frac{m_{\ell}\{f\}}{m_0\{f\}} = \frac{\langle x^{\ell}\rangle_f}{\langle x^0\rangle_f} \equiv \overline{x_f^{\ell}} \tag{13.9}$$

where the "overbar" notation $\overline{x_f^{\ell}}$ distinguishes the normalized moment from its "unnormalized" sibling $\mathrm{m}_{\ell}\{f[x]\} = \langle x^{\ell}\rangle_f$. If $\mathrm{m}_0 = 1$, as in probability applications, then $\overline{x_f^{\ell}} = m_{\ell}\{f\}$. Note that all normalized moments remain unchanged if the amplitude is scaled, e.g., $\overline{x_g^{\ell}} = \overline{x_f^{\ell}}$ if $g[x] \propto f[x]$.

13.1.1 First Moment – Centroid

The abscissa of the *centroid* (or *center of mass*) of $f[x]$ defines the coordinate on the x-axis that divides the area of $f[x]$ into equal halves, hence its alternate name of *balance-point abscissa*. The centroid of an arbitrary 1-D function $f[x]$ is the normalized first moment:

$$\overline{x_f^1} \equiv \frac{\mathrm{m}_1}{\mathrm{m}_0} = \frac{\langle x^1\rangle_f}{\langle x^0\rangle_f} = \frac{\displaystyle\int_{-\infty}^{+\infty} x \cdot f[x]\, dx}{\displaystyle\int_{-\infty}^{+\infty} f[x]\, dx} = -\frac{1}{2\pi i}\frac{F'[0]}{F[0]} \tag{13.10}$$

In words, the centroid of $f[x]$ is proportional to the ratio of the slope and amplitude of the spectrum evaluated at the origin. This means that the centroid of any function whose spectrum has zero slope at the origin ($F'[0] = 0$) must be located at $x = 0$. The usefulness of the expression in Equation (13.10) for calculating the centroid depends on the form of $f[x]$; it may well be easier to evaluate the center of mass by direct integration in the space domain rather than via the derivative of $F[\xi]$ at the origin.

13.1.2 Second Moment – Moment of Inertia

The second moment measures the "spread" of amplitude of $f[x]$ about the centroid. This is a particularly important concept in statistics where $f[x]$ is the probability distribution of some random process so that the second moment may be estimated from the histogram of outcomes. The second moment also is a useful measure in physics, where it is called the *moment of inertia* of $f[x]$. Equation (13.8) establishes the relationship between m_2 and the spectrum $F[\xi]$:

$$\mathrm{m}_2\{f[x]\} = \langle x^2\rangle_f = \int_{-\infty}^{+\infty} x^2 \cdot f[x]\, dx = -\frac{1}{4\pi^2}F''[0] \tag{13.11}$$

i.e., the second moment is proportional to the *curvature* of the spectrum evaluated at the origin. In words, the second moment is positive if the second derivative of the spectrum at the origin is negative, which means that the spectrum is "concave down". The larger the curvature of the spectrum at the origin, the more concentrated the distribution of amplitude of the spectrum near the origin. The scaling theorem may then be used to demonstrate that the amplitude of the corresponding space-domain representation $f[x]$ is spread over a larger domain.

The normalized second moment is sometimes called the *mean-square abscissa* and has the dimensions of x^2:

$$\overline{x_f^2} \equiv \frac{\mathrm{m}_2}{\mathrm{m}_0} = \frac{\langle x^2\rangle_f}{\langle x^0\rangle_f} = \frac{\displaystyle\int_{-\infty}^{+\infty} x^2 \cdot f[x]\, dx}{\displaystyle\int_{-\infty}^{+\infty} f[x]\, dx} = -\frac{1}{4\pi^2}\frac{F''[0]}{F[0]} \tag{13.12}$$

Note that $\sqrt{\overline{x_f^2}}$ has the same dimensions as x and may be used as a measure of the "width" of $f[x]$.

13.1.3 Central Moments – Variance

It is often convenient to use moments measured about the centroid of the function $f[x]$ rather than about the origin of coordinates; these will be denoted by $\mu_\ell\{f[x]\}$ and are called the *central moments*:

$$\mu_\ell\{f[x]\} \equiv \int_{-\infty}^{+\infty} (x - \overline{x_f^1})^\ell f[x]\,dx$$

$$= \int_{-\infty}^{+\infty} \left(x - \frac{\mathrm{m}_1}{\mathrm{m}_0}\right)^\ell f[x]\,dx \tag{13.13}$$

The zero-order central moment μ_0 is identical to m_0 because translation has no effect on the area of $f[x]$:

$$\mu_0\{f[x]\} \equiv \int_{-\infty}^{+\infty} \left(x - \frac{\mathrm{m}_1}{\mathrm{m}_0}\right)^0 f[x]\,dx$$

$$= \int_{-\infty}^{+\infty} f[x]\,dx = \mathrm{m}_0 \tag{13.14}$$

The first-order central moment may be interpreted as the *centroid of the centroid* and always evaluates to zero:

$$\mu_1\{f[x]\} \equiv \int_{-\infty}^{+\infty} \left(x - \frac{\mathrm{m}_1}{\mathrm{m}_0}\right)^1 f[x]\,dx$$

$$= \left(\int_{-\infty}^{+\infty} xf[x]\,dx\right) - \frac{\mathrm{m}_1}{\mathrm{m}_0} \cdot \left(\int_{-\infty}^{+\infty} f[x]\,dx\right)$$

$$= \mathrm{m}_1 - \left(\frac{\mathrm{m}_1}{\mathrm{m}_0}\right) \cdot \mathrm{m}_0 = 0 \tag{13.15}$$

The second central moment also is easy to evaluate in terms of m_0, m_1, and m_2 and may be rewritten in terms of $\overline{x_f^2}$ and $\overline{x_f^1}$ by applying Equations (13.10) and (13.12):

$$\mu_2\{f[x]\} = \int_{-\infty}^{+\infty} \left(x - \frac{\mathrm{m}_1}{\mathrm{m}_0}\right)^2 f[x]\,dx$$

$$= \int_{-\infty}^{+\infty} \left(x^2 - 2x\frac{\mathrm{m}_1}{\mathrm{m}_0} + \left(\frac{\mathrm{m}_1}{\mathrm{m}_0}\right)^2\right) f[x]\,dx$$

$$= \left(\int_{-\infty}^{+\infty} x^2 f[x]\,dx\right) - 2\frac{\mathrm{m}_1}{\mathrm{m}_0}\left(\int_{-\infty}^{+\infty} xf[x]\,dx\right) + \left(\frac{\mathrm{m}_1}{\mathrm{m}_0}\right)^2\left(\int_{-\infty}^{+\infty} f[x]\,dx\right)$$

$$= \mathrm{m}_2 - 2\left(\frac{\mathrm{m}_1}{\mathrm{m}_0}\right)\mathrm{m}_1 + \left(\frac{\mathrm{m}_1}{\mathrm{m}_0}\right)^2 \mathrm{m}_0$$

$$= \mathrm{m}_2 - \mathrm{m}_0\left(\frac{\mathrm{m}_1}{\mathrm{m}_0}\right)^2$$

$$\mu_2\{f[x]\} = \mathrm{m}_0 \cdot (\overline{x_f^2} - (\overline{x_f^1})^2) \tag{13.16}$$

The second central moment normalized by the area of $f[x]$ is called the *variance* of $f[x]$ and is denoted by σ_f^2:

$$\frac{\mu_2\{f[x]\}}{\mathrm{m}_0} = \overline{x_f^2} - (\overline{x_f^1})^2 \equiv \sigma_f^2 \tag{13.17}$$

In words, the variance is the mean of the square of the deviation of $f[x]$ measured from its centroid. This expression may be recast in terms of the spectrum of $f[x]$ by substituting Equations (13.10) and (13.12):

$$\sigma_f^2 = \frac{\mathrm{m}_2}{\mathrm{m}_0} - \left(\frac{\mathrm{m}_1}{\mathrm{m}_0}\right)^2 = \frac{1}{4\pi^2}\left(-\frac{F''[0]}{F[0]} + \left(\frac{F'[0]}{F[0]}\right)^2\right) \tag{13.18}$$

For example, this expression may be used to confirm Equation (6.61b) for the variance of $GAUS[x/b]$:

$$\sigma^2_{(GAUS[x/b])} = \frac{b^2}{2\pi} \tag{13.19}$$

Note that it may be easier to compute the variance by direct integration in the space domain rather than by applying Equation (13.18). For example, the variance of $RECT[x/b]$ may be computed directly by evaluating the trivial integral:

$$\sigma^2_{(RECT[x/b])} = \frac{\int_{-\infty}^{+\infty} \left(x - \frac{\mathrm{m}_1}{\mathrm{m}_0}\right)^2 RECT\left[\frac{x}{b}\right] dx}{\int_{-\infty}^{+\infty} RECT\left[\frac{x}{b}\right] dx} = \frac{1}{|b|}\int_{-b/2}^{+b/2} x^2\, dx = \frac{b^2}{12} \tag{13.20}$$

whereas the computations via the spectrum and its derivatives are much more tedious.

Equation (13.12) may be applied to derive a convenient relationship between the mean-square abscissas of two functions $f[x]$ and $h[x]$ and the mean-square abscissa of their convolution $g[x] = f[x] * h[x]$:

$$\begin{aligned}\overline{x_g^2} &= -\frac{1}{4\pi^2}\frac{G''[0]}{G[0]} = \left(\frac{1}{2\pi i}\right)^2\left(\frac{1}{F[0]\cdot H[0]}\right)\cdot\left(\left(\frac{d^2}{d\xi^2}(F[\xi]\cdot H[\xi])\right)\bigg|_{\xi=0}\right)\\ &= \left(\frac{1}{2\pi i}\right)^2\left(\frac{F''[0]\cdot H[0] + 2(F'[0]\cdot H'[0]) + F[0]\cdot H''[0]}{F[0]\cdot H[0]}\right)\\ &= \overline{x_f^2} + 2\left(\frac{F'[0]}{2\pi i\, F[0]}\cdot\frac{H'[0]}{2\pi i\, H[0]}\right) + \overline{x_h^2}\\ &= \overline{x_f^2} + \overline{x_h^2} + 2\cdot\overline{x_f^1}\cdot\overline{x_h^1}\end{aligned} \tag{13.21}$$

where the common shorthand notation for derivatives via primed functions has been used for clarity. Recall that $F'[0] = 0$ if $F[\xi]$ is even, which also means that $f[x]$ is even. This leads to a particularly simple relationship for the mean-square abscissa if the input $f[x]$ and/or the impulse response $h[x]$ is even:

$$\overline{x^2_{(f*h)}} = \overline{x_f^2} + \overline{x_h^2} \quad \textit{if } f[x] \textit{ and/or } h[x] \textit{ is even} \tag{13.22}$$

In words, the mean-square abscissa of a convolution is the sum of the mean-square abscissas of the component functions in those cases where at least one of the component functions is symmetric.

The relationship analogous to Equation (13.21) for the variances of the component functions and of the resulting convolution does not require the symmetry constraint for $f[x]$ or $h[x]$ because the variance is measured relative to the centroid. In short, the variance of the output of a convolution is always the sum of the variances of the component functions:

$$\sigma^2_{(f*h)} = \sigma^2_f + \sigma^2_h \quad \textit{for arbitrary } f[x] \textit{ and } h[x] \tag{13.23}$$

which means that the variance of the autoconvolution of an arbitrary function $f[x]$ must be $2 \cdot \sigma^2_f$. For example, the variance of $TRI[x/b]$ may be evaluated by combining the results of Equations (13.20) and (13.23):

$$\begin{aligned}\sigma^2_{TRI[x/b]} &= 2\sigma^2_{(RECT[x/b])} \\ &= \frac{b^2}{6}\end{aligned} \tag{13.24}$$

Equation (13.23) may be easily generalized further to see that the variance of the convolution of N replicas of $f[x]$ is $N \cdot \sigma^2_f$. This result will be used later in this chapter in the discussion of the central-limit theorem.

13.1.4 Evaluation of 1-D Spectra from Moments

The moment theorem of Equation (13.8) may be applied to derive a useful relationship between $F[\xi]$ and the infinite discrete set of moments of $f[x]$. This relationship is valid for any infinitely differentiable ("analytic") spectrum $F[\xi]$, which is a stringent requirement because $F[\xi]$ must exhibit no discontinuities or singularities, such as Dirac delta functions or derivatives thereof. Spectra that satisfy this criterion may be expressed as a Taylor series with coefficients determined by its amplitude and its derivatives evaluated at some "base" frequency ξ_0:

$$F[\xi] = \sum_{\ell=0}^{+\infty} \left(\frac{1}{\ell!} \frac{d^\ell F}{d\xi^\ell}\bigg|_{\xi=\xi_0} \right) (\xi - \xi_0)^\ell \tag{13.25}$$

Often the base frequency ξ_0 is selected to be 0 cycles per unit length ("DC"). Because the ℓth coefficient of any such series is proportional to the ℓth derivative of $F[\xi]$ evaluated at the origin, the moments m_ℓ of the function may be substituted from Equation (13.8):

$$F[\xi] = \sum_{\ell=0}^{+\infty} \left(\frac{1}{\ell!} \frac{d^\ell F}{d\xi^\ell}\bigg|_{\xi=0} \right) (\xi - 0)^\ell = \sum_{\ell=0}^{+\infty} \left(\frac{(-2\pi i)^\ell \, \mathrm{m}_\ell}{\ell!} \right) \xi^\ell \tag{13.26}$$

Simply put, any "well-behaved" (analytic) spectrum $F[\xi]$ may be constructed from a power series based on the complete set of moments of its space-domain representation $f[x]$. Phrased yet another way, the discrete set of moments of $f[x]$ conveys all information necessary to generate an analytic spectrum $F[\xi]$, which is (of course) an equivalent representation of $f[x]$. In short, the discrete (though infinite) set of moments $\{\mathrm{m}_\ell\}$ is equivalent to $f[x]$ and to $F[\xi]$ if the spectrum is infinitely differentiable at the origin. The relationship of the three representations is shown schematically in Figure 13.2, which also lists the operations that convert between them. Also note that the dual character of the two representations allows $f[x]$ to be expressed in terms of the moments of $F[\xi]$.

In those cases where the coefficients of the power series decrease with increasing ℓ, it may be useful to truncate the series expansion for $F[\xi]$ in Equation (13.26) at some finite index to approximate $F[\xi]$. Mathematicians call such a truncated Taylor series a *Taylor polynomial*, which is discussed in many calculus textbooks. Any such approximation will be most accurate for small values of $|\xi|$ because the amplification of the amplitude by the factor ξ^ℓ produces a large error at frequencies far from the origin, even if the moment m_ℓ is small. This approximation is particularly useful when estimating the spectrum of a function $f[x]$ that had been measured in the presence of additive noise because the "averaging" effect of the integral that evaluates the moments reduces the effect of high-frequency noise. In other

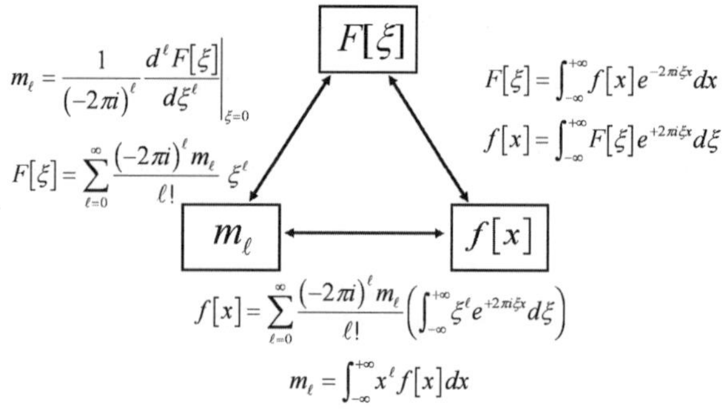

Figure 13.2 Relations between representations $f[x]$ in the space domain, $F[\xi]$ in the spatial frequency domain, and the infinite set of moments $\{\mathrm{m}_\ell\}$ in cases where the spectrum is infinitely differentiable at the origin.

words, the estimate of $F[\xi]$ obtained from moments is an approximation asymptotic solution that is valid for small ξ.

To illustrate the approximation of the spectrum obtained from the series, we evaluate the ℓth moment of the symmetric superGaussian $e^{-\pi|x|^n}$ from Equation (6.62):

$$\mathrm{m}_\ell\{e^{-\pi|x|^n}\} = \int_{-\infty}^{+\infty} e^{-\pi|x|^n}\, x^\ell\, dx = 2\cdot(-1^\ell)\cdot\int_0^{+\infty} e^{-\pi(x^n)}\, x^\ell\, dx \tag{13.27}$$

where the symmetry of the function has been used. This integral resembles the definition of the gamma function in Equation (6.37), and may be recast into that form by changing the variable of integration to $\alpha = \pi x^n$. The resulting expression for the ℓth moment is:

$$\mathrm{m}_\ell\{e^{-\pi|x|^n}\} = \begin{cases} \dfrac{2}{n}\cdot\pi^{-(\ell+1)/n}\cdot\Gamma\left[\dfrac{\ell+1}{n}\right] & \textit{for even } \ell \\ 0 & \textit{for odd } \ell \end{cases} \tag{13.28}$$

The odd-order moments vanish because $e^{-\pi|x|^n}$ is even. Substitution into Equation (13.26) yields the power series for the spectrum:

$$\begin{aligned}\mathcal{F}_1\{e^{-\pi|x|^n}\} &= \sum_{\ell=0}^{+\infty}\left(\frac{(-2\pi i)^\ell}{\ell!}\mathrm{m}_\ell\right)\xi^\ell \\ &= \sum_{p=0}^{+\infty}\left(\frac{(-2\pi i)^{2p}}{2p!}\cdot\frac{2}{n}(\pi^{-(2p+1)/n})\Gamma\left[\frac{2p+1}{n}\right]\right)\xi^{2p} \\ &= \sum_{p=0}^{+\infty}(-1)^p\left(\frac{2^{2p+1}}{n}\cdot\pi^{(2p(1-1/n)-1/n)}\cdot\frac{\Gamma[(2p+1)/n]}{\Gamma[2p+1]}\right)\xi^{2p}\end{aligned} \tag{13.29}$$

where the summation index has been changed to $p = 2\ell$ to reflect the fact that all odd moments are zero and the fact that $(2p)! = \Gamma[2p+1]$ has been used.

This expression for the spectrum of the nth-order symmetric superGaussian may be checked by evaluating the results for orders n for which the spectra are known, such as the first-order superGaussian:

$$e^{-\pi|x|^1} = \mathit{STEP}[x]\cdot e^{-\pi x} + \mathit{STEP}[-x]\cdot e^{+\pi x} \tag{13.30}$$

The spectrum may be evaluated by applying the linearity property of the Fourier transform from Equation (9.103), the scaling theorem from Equation (9.117), the reversal property from Equation (9.119), and the known spectrum of $STEP[x] \cdot e^{-x}$ from Equation (9.88):

$$\begin{aligned}\mathcal{F}_1\{e^{-\pi|x|}\} &= \frac{1}{\pi}\left(\frac{1}{1+2\pi i(\xi/\pi)}\right) + \frac{1}{\pi}\left(\frac{1}{1+2\pi i(-\xi/\pi)}\right)\\ &= \frac{2}{\pi}\left(\frac{1}{1+4\xi^2}\right) = \frac{2}{\pi}\left(\frac{1}{1-(-4\xi^2)}\right)\\ &= \frac{2}{\pi}(1-4\xi^2+(4\xi^2)^2-(4\xi^2)^3+\cdots) \quad \textit{if } |\xi| < \frac{1}{2}\\ &= \frac{2}{\pi} - \frac{8}{\pi}\xi^2 + \frac{32}{\pi}\xi^4 - \frac{128}{\pi}\xi^6 + \cdots \end{aligned} \tag{13.31}$$

where the well-known series for $(1-t)^{-1}$ has been used. The identical expression may be obtained directly from the series in Equation (13.29):

$$\begin{aligned}\mathcal{F}_1\{e^{-\pi|x|^1}\} &= \sum_{\ell=0}^{+\infty}((-1)^\ell \cdot (2^{2\ell+1}) \cdot \pi^{-1})\xi^{2\ell}\\ &= \frac{2}{\pi} - \frac{2^3}{\pi}\xi^2 + \frac{2^5}{\pi}\xi^4 - \frac{2^7}{\pi}\xi^6 + \cdots\\ &\cong 0.637 - 2.546\xi^2 + 10.186\xi^4 - 40.744\xi^6 + \cdots \end{aligned} \tag{13.32}$$

The zero-order term in the series is $2/\pi$, which is the area of the symmetric decaying exponential. Note that the magnitudes of the coefficients of the spectrum of $e^{-|x|}$ increase by a factor of 4 for each increment of ℓ, which limits the region of convergence of any truncated approximation of the series to small values of $|\xi|$. Also note that the sign changes of alternate coefficients tend to compensate for errors in the contributions from the smaller powers in the series. In other words, much of the amplitude generated by the ℓth moment at large values of ξ is cancelled by that from the $(\ell+1)$st moment, which in turn contributes error in the opposite direction. Therefore, a partial sum in the expansion for $F[\xi]$ evaluated at large ξ will oscillate between large positive and negative amplitudes as each additional term is added. In other words, the partial sum for $F[\xi]$ is unstable for large values of ξ, as shown in Figure 13.3a.

For a second confirmation, substitute $n = 2$ to evaluate the moments of the customary Gaussian function. After a bit of algebra, Equation (13.29) yields the correct power series for the spectrum:

$$\begin{aligned}\mathcal{F}_1\{e^{-\pi x^2}\} &= 1 - \pi\xi^2 + \frac{(\pi\xi^2)^2}{2} - \frac{(\pi\xi^2)^3}{6} + \cdots = \sum_{\ell=0}^{+\infty}(-1)^\ell\frac{(\pi\xi^2)^\ell}{\ell!}\\ &\cong 1 - 3.142\xi^2 + 4.935\xi^4 - 5.168\xi^6 + 4.059\xi^8 - 2.550\xi^{10} + 1.335\xi^{12} - \cdots \end{aligned} \tag{13.33}$$

The central-ordinate theorem ensures that the zero-order term is the unit area of the Gaussian. The magnitudes of the coefficients increase until the sixth-power term and then decrease toward zero for larger ℓ, which means that an approximation obtained by truncating this series for larger values of $|\xi|$ will be better than a corresponding approximation for the first-order superGaussian in Equation (13.32). This series is identical to the Taylor series for the known spectrum $\exp[-\pi\xi^2]$, as may be demonstrated by evaluating Equation (4.18c). Results obtained from these approximations are shown in Figure 13.3b.

We also can apply the power-series expression for the spectrum obtained via moments to functions for which the closed-form solution has not been derived. Consider first the spectrum of the third-order

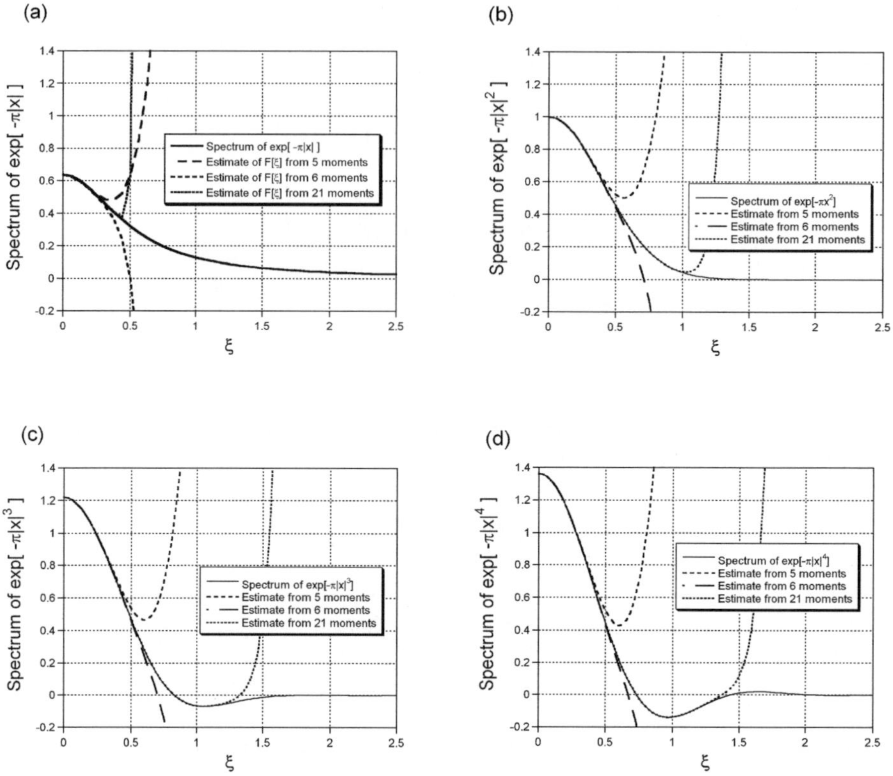

Figure 13.3 Estimates of the spectra of real-valued symmetric superGaussian functions via moments. Each example compares the spectrum to estimates obtained from 5, 6, and 21 moments: (a) $\exp[-\pi|x|^1]$; (b) normal Gaussian function $\exp[-\pi x^2]$; (c) $\exp[-\pi|x|^3]$; and (d) $\exp[-\pi x^4]$. Note the accuracy of the approximation for $|\xi| \simeq 0$ and the divergence from the true spectrum for larger $|\xi|$.

symmetric superGaussian, which may be evaluated by direct substitution of $n = 3$ into Equation (13.29):

$$\begin{aligned}\mathcal{F}_1\{e^{-\pi|x|^3}\} &= \sum_{\ell=0}^{+\infty}\left(\frac{2^{2\ell+1}}{3}(-1)^\ell \pi^{(2\ell(\frac{2}{3})-\frac{1}{3})}\frac{\Gamma[(2\ell+1)/3]}{\Gamma[2\ell+1]}\right)\xi^{2\ell} \\ &= \left(\frac{2}{3}\pi^{-\frac{1}{3}}\Gamma\left[\frac{1}{3}\right]\right)\xi^0 - \left(\frac{8}{3}\pi^1\frac{\Gamma[1]}{\Gamma[3]}\right)\xi^2 + \left(\frac{32}{3}\pi^{\frac{7}{3}}\frac{\Gamma[\frac{5}{3}]}{\Gamma[5]}\right)\xi^4 - \cdots \\ &\cong 1.219\cdot(1 - 3.436\xi^2 + 4.239\xi^4 - 3.190\xi^6 + 1.564\xi^8 - \cdots)\end{aligned} \tag{13.34}$$

The zero-order term in the series is larger than the unit area of the Gaussian and agrees with Equation (6.71d). Also note that the signs of the coefficients of adjacent terms in this series alternate in the same fashion as the first- and second-order superGaussians and that they decrease in magnitude after the fourth-power term. Finally note that the second-order coefficient of the third-order superGaussian is a larger negative number than the corresponding coefficient of the Gaussian in Equation (13.33). This additional negative weight is sufficient to "drive" the amplitude of the spectrum to negative values as the frequency increases. The positive fourth-order term compensates for this negative amplitude and forces the spectrum to zero as ξ is increased further. The spectrum exhibits a positive central lobe surrounded by a region of negative amplitude. Note the difference in behavior compared to the second-order superGaussian (the usual Gaussian function $e^{-\pi\xi^2}$), which is everywhere nonnegative and decays smoothly to zero. The estimated spectrum is compared to a discrete calculation in Figure 13.3c.

As a final example, the spectrum of the fourth-order superGaussian may be evaluated by substituting $n = 4$ into Equation (13.29):

$$\mathcal{F}_1\{e^{-\pi x^4}\} = \sum_{\ell=0}^{+\infty}\left(\frac{2^{2\ell}}{2}(-1)^\ell\,\pi^{(6\ell-1)/4}\,\frac{\Gamma[(2\ell+1)/4]}{\Gamma[2\ell+1]}\right)\xi^{2\ell}$$

$$= \left(\frac{1}{2}\pi^{-\frac{1}{4}}\Gamma\left[\frac{1}{4}\right]\right)\xi^0 - \left(\pi^{\frac{5}{4}}\Gamma\left[\frac{3}{4}\right]\right)\xi^2 + \left(\frac{1}{3}\pi^{\frac{11}{4}}\Gamma\left[\frac{5}{4}\right]\right)\xi^4 - \cdots$$

$$\cong 1.362\cdot(1 - 3.763\xi^2 + 5.167\xi^4 - 3.889\xi^6 + \cdots) \tag{13.35}$$

This spectrum has a positive central lobe and negative amplitudes for larger values of $|\xi|$, as shown in Figure 13.3d.

We have noted that the spectra of the first- and second-order superGaussians are nonnegative, while that of the third-order function is bipolar. Recall from Equation (6.65) that the superGaussian asymptotically approaches a rectangle function as $n \to \infty$:

$$\lim_{n\to\infty}\{e^{-\pi|x|^n}\} = RECT\left[\frac{x}{2}\right] \tag{13.36}$$

The corresponding spectrum is found by applying Equation (9.71) and the scaling theorem from Equation (9.117):

$$\mathcal{F}_1\left\{\lim_{n\to\infty}\{e^{-\pi|x|^n}\}\right\} = 2\cdot SINC[2\xi] \tag{13.37}$$

which has an infinite number of zero crossings between $\xi = 0$ and $\xi = +\infty$. This result will be considered in more detail in the next section and in the discussion of discrete spectra in Chapter 14.

13.1.5 Spectra of 1-D Superchirps via Moments

As discussed in Section 6.3.3, symmetric and Hermitian forms of the "superchirp" may be defined. The former version may be defined for any integer value n:

$$\exp[\pm i\pi|x|^n] = \cos[\pi x^n] \pm i\,\sin[\pi|x|^n] \tag{13.38}$$

while the Hermitian form in Equation (6.151) only exists for odd n:

$$f[x] = \cos[\pi|x|^n] \pm i(SGN[x]\,\sin[\pi|x|^n]) = f^*[-x] \tag{13.39a}$$

$$= \cos[\pi x^n] \pm i\,\sin[\pi x^n] = f^*[-x] \quad \textit{for odd } n \tag{13.39b}$$

The moments of the symmetric form may be evaluated by a development parallel to that for the superGaussian in the last section. After a bit of algebra:

$$\mathcal{F}_1\{\exp[\pm i\pi|x|^n]\}$$

$$= e^{\pm i\pi/2n}\sum_{\ell=0}^{+\infty}\left((-1)^\ell\left(\frac{2^{2\ell+1}}{n}\right)\pi^{(2\ell(1-1/n)-1/n)}\frac{\Gamma[(2\ell+1/2]}{\Gamma[2\ell+1]}\,e^{\pm i\pi\ell/n}\right)\xi^{2\ell} \tag{13.40}$$

where the factorial expression for the gamma function has been used and the summation index has been changed to $p = 2\ell$ to reflect the fact that all odd moments are zero. Note that the initial phase of the spectrum is the leading constant phase factor $\pm\pi/2n$ radians. For $n = 1$, 2, and 4, the corresponding initial phases are $\pm\pi/2$, $\pm\pi/4$, and $\pm\pi/6$ radians.

The validity of Equation (13.40) may be tested by substituting $n = 2$, applying the known power series for the exponential, and comparing the result to the spectrum of the chirp in Equation (9.95):

$$\begin{aligned}\mathcal{F}_1\{e^{\pm i\pi x^2}\} &= e^{\pm i\pi/4}\, e^{\mp i\pi\xi^2} \\ &= e^{\pm i\pi/4} \sum_{\ell=0}^{+\infty} \frac{(\mp i\pi\xi^2)^\ell}{\ell!} \\ &= e^{\pm i\pi/4} \cdot \left(1 \mp i\pi\xi^2 \pm \frac{\pi^2\xi^4}{2} \pm i\frac{\pi^3\xi^3}{6} \mp \cdots\right) \end{aligned} \tag{13.41}$$

where the expansion for the exponential function from Equation (4.18c) has been used. The expansion in Equation (13.41) may be expressed as:

$$\mathcal{F}_1\{e^{\pm i\pi x^2}\} = e^{\pm i\pi/4} \cdot \sum_{\ell=0}^{+\infty} \left((-1)^\ell (2^{2\ell})\pi^{(\ell-\frac{1}{2})} \frac{\Gamma[(2\ell+1)/2]}{\Gamma[2\ell+1]}\, e^{\pm i\pi\ell/2}\right)\xi^{2\ell} \tag{13.42}$$

These two expressions may be shown to be equivalent after a bit of algebraic manipulation and application of known properties of the gamma function. The approximations of the spectrum of the chirp are graphed in Figure 13.4.

The spectrum of the symmetric chirp for $n = 4$ has not been derived in closed form. The moment expansion in Equation (13.40) evaluates to:

$$\begin{aligned}\mathcal{F}_1\{e^{\pm i\pi|x|^4}\} &= e^{\pm i\pi/8} \sum_{\ell=0}^{+\infty} \left(\frac{2^{2\ell+1}}{4}(-1)^\ell \pi^{(3\ell/2-1/4)} \frac{\Gamma[(2\ell+1)/4]}{\Gamma[2\ell+1]}\, e^{\pm i\pi\ell/4}\right)\xi^{2\ell} \\ &\cong e^{\pm i\pi/8} \cdot (1.362 + 5.125\, e^{\mp i\pi 3/4}\, \xi^2 + 7.037\, e^{\pm i\pi/2}\, \xi^4 + 5.297\, e^{\mp i\pi/4}\, \xi^6 + \cdots)\end{aligned} \tag{13.43}$$

The approximations of the spectrum evaluated for different numbers of moments are plotted as real/imaginary parts and as magnitude/phase in Figure 13.5. Of particular note is the magnitude spectrum, which exhibits a maximum at the origin and decreases for larger values of $|\xi|$. This behavior contrasts with the constant-magnitude spectrum of the quadratic-phase signal in Equation (13.41).

Finally, consider the estimates of the spectrum of the Hermitian superchirp in Equation (13.39). Because the real and imaginary parts of $e^{\pm i\pi x^n}$ are even and odd, respectively, all odd moments of the real part and even moments of the imaginary part evaluate to zero. The moments of the Hermitian superchirps $e^{\pm i\pi x^n}$ (odd n) are:

$$\mathrm{m}_\ell = \begin{cases} \dfrac{2}{n}(\pi^{-(\ell+1)/n})\Gamma\left(\dfrac{\ell+1}{n}\right)\cos\left[\dfrac{\pi(\ell+1)}{2n}\right] & \textit{for odd } n, \textit{ even } \ell \\ \pm i\dfrac{2}{n}(\pi^{-(\ell+1)/n})\Gamma\left(\dfrac{\ell+1}{n}\right)\sin\left[\dfrac{\pi(\ell+1)}{2n}\right] & \textit{for odd } n, \textit{ odd } \ell \end{cases} \tag{13.44}$$

Pairs of terms of even order $\ell = 2p$ and of odd order $\ell = 2p + 1$ may be combined to construct an expression for the odd-order Hermitian superchirp:

$$\begin{aligned}&\mathcal{F}_1\{e^{\pm i\pi x^n}\} \quad \textit{(for odd } n\textit{)} \\ &= \frac{2}{n}\sum_{p=0}^{+\infty}\left((-1)^p(\pi^{-(2p+1)/n})\frac{(2\pi)^{2p}}{(2p)!}\right)\left(\Gamma\left[\frac{2p+1}{n}\right]\cos\left[\frac{\pi(2p+1)}{2n}\right]\xi^{2p}\right) \\ &\pm \frac{2}{n}\sum_{p=0}^{+\infty}\left((-1)^p(\pi^{-(2p+1)/n})\frac{(2\pi)^{2p}}{(2p)!}\right) \\ &\cdot\left(\frac{2\pi}{(2p+1)}(\pi^{-1/n})\Gamma\left[\frac{2p+2}{n}\right]\sin\left[\frac{\pi(2p+2)}{2n}\right]\xi^{2p+1}\right)\end{aligned} \tag{13.45}$$

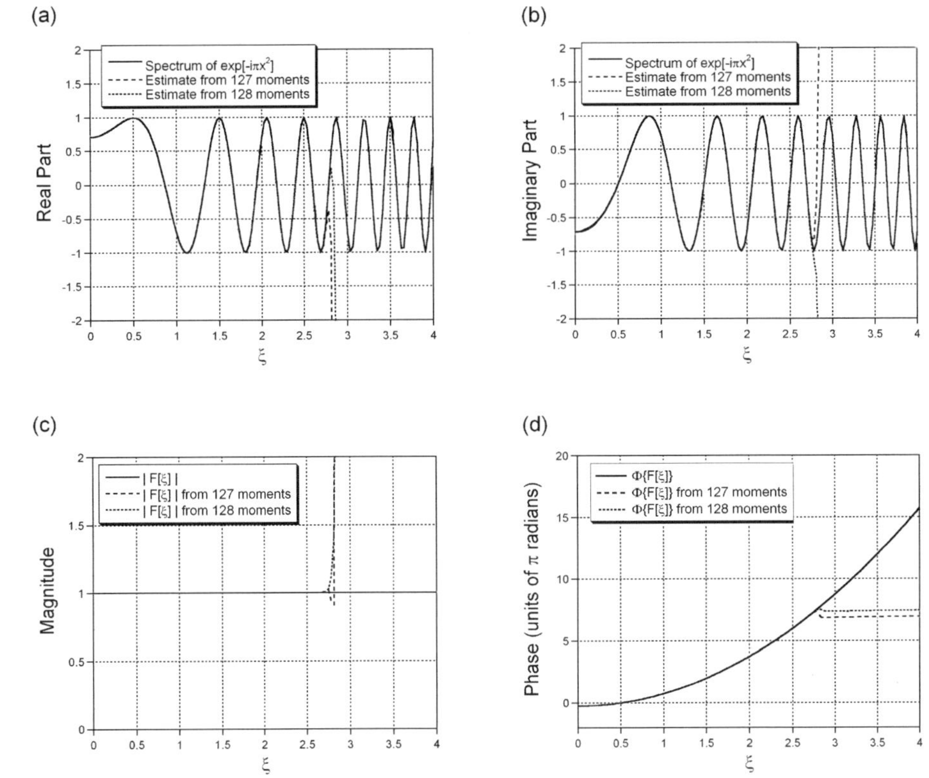

Figure 13.4 Approximations of spectrum of chirp function $f[x] = \exp[-i\pi x^2]$ from 127 and 128 moments: (a) real part; (b) imaginary part; (c) magnitude; (d) phase. The approximations are quite accurate for $|\xi| \lesssim 3$ cycles per unit length.

For $n = 3$, the expansion for $e^{+i\pi x^3}$ is:

$$\mathcal{F}_1\{e^{+i\pi x^3}\} \cong 1.056\,05 + 2.290\xi + 0\xi^2 - 4.632\xi^3 - 5.022\xi^4 + 0\xi^5$$
$$+ 4.064\xi^6 + 3.147\xi^7 + 0\xi^8 - 1.486\xi^9 - 0.920\,40\xi^{10} + 0\xi^{11} + \cdots \tag{13.46}$$

The central ordinate agrees with the area of the Hermitian chirp of order 3 that was computed in Equation (6.157b). The magnitudes of the coefficients decrease after the fourth power and some coefficients are zero. The coefficients appear in a pattern: two are positive, followed by a zero, followed by two that are negative, followed by a zero, etc. The spectrum obtained for the first 170 coefficients is shown in Figure 13.6. As expected, the spectrum is real valued, but also note the interesting fact that the amplitude of the spectrum is not zero for $\xi \lessapprox 0$, but rather decays rapidly to zero for negative frequencies.

13.1.6 2-D Moment Theorem

The definition of 1-D moments in Equation (13.1) is easily generalized to 2-D functions:

$$\mathrm{m}_{\ell k}\{f[x, y]\} \equiv \iint_{-\infty}^{+\infty} x^\ell y^k f[x, y]\, dx\, dy \tag{13.47}$$

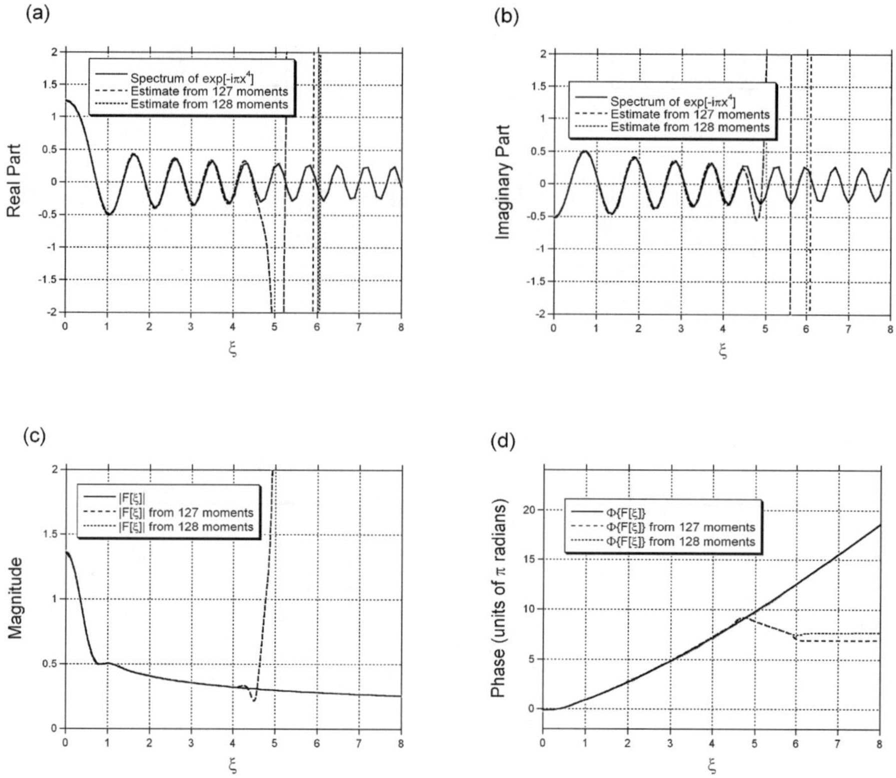

Figure 13.5 Approximations of spectrum of chirp function $f[x] = \exp[-i\pi x^4]$ from 127 and 128 moments: (a) real part; (b) imaginary part; (c) magnitude; (d) phase. The approximations are quite accurate for $|\xi| \lesssim 5$ cycles per unit length.

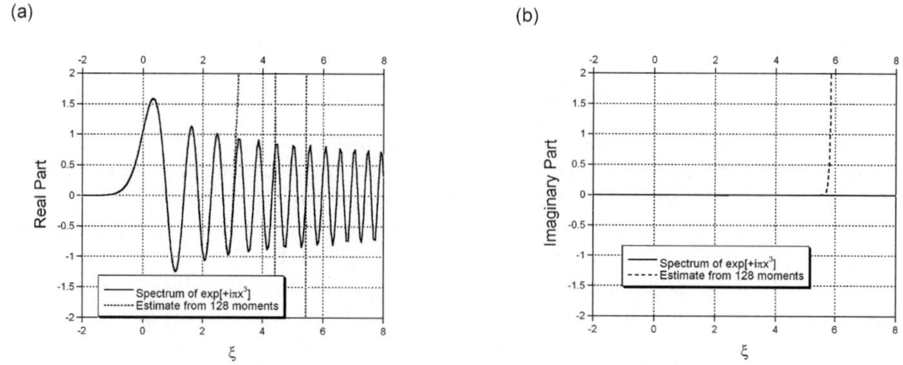

Figure 13.6 Real and imaginary parts of the spectrum of the Hermitian function $\exp[+i\pi x^3]$ and the estimate obtained from the first 128 moments. The spectrum is real and the estimate is accurate for $|\xi| \lesssim 3$ cycles per unit length.

The spectra of the 2-D moments may be evaluated directly:

$$\begin{aligned}\mathrm{m}_{\ell k}\{f[x,y]\} &\equiv \iint_{-\infty}^{+\infty} x^{\ell} y^{k}\, f[x,y]\,dx\,dy \\ &= \left(\frac{1}{-2\pi i}\right)^{\ell+k}\left(\frac{\partial^{\ell}\partial^{k}}{\partial\xi^{\ell}\partial\eta^{k}}F[\xi,\eta]\right)\Bigg|_{\xi=0,\eta=0}\end{aligned} \tag{13.48}$$

If all moments of the 2-D function $f[x,y]$ are finite, then an expression analogous to Equation (13.26) may be used to construct the 2-D spectrum $F[\xi,\eta]$ from the moments of $f[x,y]$:

$$F[\xi,\eta] = \sum_{\ell=0}^{+\infty}\sum_{k=0}^{+\infty}\frac{(-2\pi i)^{\ell+k}}{\ell!\cdot k!}\,\mathrm{m}_{\ell k}\,\xi^{\ell}\,\eta^{k} \tag{13.49}$$

13.1.7 Moments of Circularly Symmetric Functions

The 2-D moments may be recast into a form applicable to circularly symmetric functions by a simple substitution of the coordinate variables:

$$\begin{aligned}\mathrm{m}_{\ell k}\{f[x,y]\} \to \mathrm{m}_{\ell k}\{f(r)\} &= \iint_{-\infty}^{+\infty} x^{\ell} y^{k} f\left(\sqrt{x^2+y^2}\right)\,dx\,dy \\ &= \int_{-\pi}^{+\pi}\int_{0}^{\infty}(r\cos[\theta])^{\ell}(r\sin[\theta])^{k} f(r) r\,dr\,d\theta \\ &= \left[\int_{-\pi}^{+\pi}(\cos^{\ell}[\theta]\sin^{k}[\theta])\,d\theta\right]\left[\int_{0}^{\infty} r^{\ell+k+1} f(r)\,dr\right]\end{aligned} \tag{13.50}$$

The zero-order moment of the circularly symmetric function evaluates to the volume of the circularly symmetric function, as expected:

$$\begin{aligned}\mathrm{m}_{00}\{f(r)\} &= \left[\int_{-\pi}^{+\pi} 1(\theta)\,d\theta\right]\left[\int_{0}^{\infty} r\,f(r)\,dr\right] \\ &= 2\pi\int_{0}^{\infty} r f(r)\,dr = \mathcal{H}_0\{f(r)\}|_{\rho=0} = F_r(0) = F[0,0]\end{aligned} \tag{13.51}$$

The symmetry of $f(r)$ ensures that all "first-order" moments (with k and/or $\ell = 1$) are zero:

$$\begin{aligned}\mathrm{m}_{10}\{f(r)\} &= \left(\int_{-\pi}^{+\pi}\cos[\theta]\,d\theta\right)\left(\int_{0}^{\infty} r^2 f(r)\,dr\right) \\ &= 0\cdot\int_{0}^{\infty} r^2 f(r)\,dr = 0\end{aligned} \tag{13.52a}$$

$$\begin{aligned}\mathrm{m}_{01}\{f(r)\} &= \left(\int_{-\pi}^{+\pi}\sin[\theta]\,d\theta\right)\left(\int_{0}^{\infty} r^2 f(r)\,dr\right) \\ &= 0\cdot\int_{0}^{\infty} r^2 f(r)\,dr = 0\end{aligned} \tag{13.52b}$$

$$\mathrm{m}_{11}\{f(r)\} = \left(\int_{-\pi}^{+\pi}\cos[\theta]\cdot\sin[\theta]\,d\theta\right)\left(\int_{0}^{\infty} r^3 f(r)\,dr\right) = 0 \tag{13.52c}$$

These relationships also guarantee the self-evident property that the centroid of the 2-D circularly symmetric function is located at the origin.

The symmetry of circularly symmetric functions ensures that $\mathrm{m}_{20} = \mathrm{m}_{02}$ and they are not generally zero.

$$\begin{aligned}\mathrm{m}_{20}\{f(r)\} &= \int_{-\pi}^{+\pi} \cos^2[\theta]\, d\theta \int_0^{\infty} r^3 f(r)\, dr \\ &= \frac{1}{4\pi}\left(\int_0^{\infty} 2\pi r (r^2 f(r))\, dr\right) \\ &= \frac{1}{4\pi}\mathcal{H}_0\{r^2 f(r)\}|_{\rho=0} \qquad (13.53a)\end{aligned}$$

$$\begin{aligned}\mathrm{m}_{02}\{f(r)\} &= \left(\int_{-\pi}^{+\pi} \sin^2[\theta]\, d\theta\right)\left(\int_0^{\infty} r^3 f(r)\, dr\right) \\ &= \frac{1}{2}\int_0^{\infty} r^3 f(r)\, dr = \mathrm{m}_{20}\{f(r)\} \qquad (13.53b)\end{aligned}$$

13.2 1-D Spectra via Method of Stationary Phase

Now we consider a different approximation of the Fourier transform that is valid for certain 1-D functions if $|\xi| \gg 0$. The process is an application of the *method of stationary phase*, which was developed by Lord Kelvin in the 1800s to evaluate integrals encountered in the study of hydrodynamics. It is a variation of the *method of steepest descents* used to evaluate path integrals of complex functions. The method of stationary phase provides useful estimates of integrals of oscillating functions, and thus of integrands with imaginary-valued exponents. This method is particularly applicable to superchirp functions $\exp[\pm i\pi x^n]$ with $n > 2$, for which no closed form of the spectrum has been derived. The results obtained for these functions will be applied in several contexts later in the book. More detailed descriptions are available in Erdelyi (1956), Papoulis (1986), and Copson (2004).

The governing principle behind the method of stationary phase will be introduced by example. Consider an integral of the general form:

$$I[k] = \int_{-\infty}^{+\infty} r[x]\, e^{ik\cdot\mu[x]}\, dx \qquad (13.54)$$

where $r[x]$ and $\mu[x]$ are real-valued functions and k is a selectable real-valued parameter. The exponential function $e^{ik\cdot\mu[x]}$ oscillates at a rate that depends on both k and the functional form of $\mu[x]$. If k is large, the rate of oscillation of the exponential term must also be large in all regions of the domain where $\mu[x] \neq 0$. In such cases, the contribution to the oscillating function to the area will be small in any region where the exponential term oscillates more rapidly than the variation of $r[x]$, because the areas of the adjacent positive and negative lobes will approximately cancel. Conversely, in those regions of the domain where $k \cdot \mu[x]$ is small, the amplitude of the exponential term will approximate the unit constant. The area of those regions will be determined by the width of this region of the domain and the amplitude of the real-valued function $r[x]$. The real and imaginary parts of a sample integrand in these two regions are shown in Figure 13.7c,d. This shows that the integral in Equation (13.54) may be estimated by evaluating the integrand *only* in those regions where the exponential term oscillates slowly, i.e., wherever the derivative of the phase function is approximately 0, $d\mu/dx \cong 0$; these are the *stationary* points of $\mu[x]$. The "semi-infinite" integrals of the real and imaginary parts over the domain $-\infty < x \leq a$ are shown in Figure 13.7e,f. These illustrate that the primary contribution to the area in both cases arises from the integrand in the region of the stationary point.

The example in Figure 13.7 also shows that the requirement that $r[x]$ be real valued in Equation (13.54) creates no problem when evaluating the integral of a complex-valued function, because the linearity of integration allows the integrals of the individual parts to be performed separately and

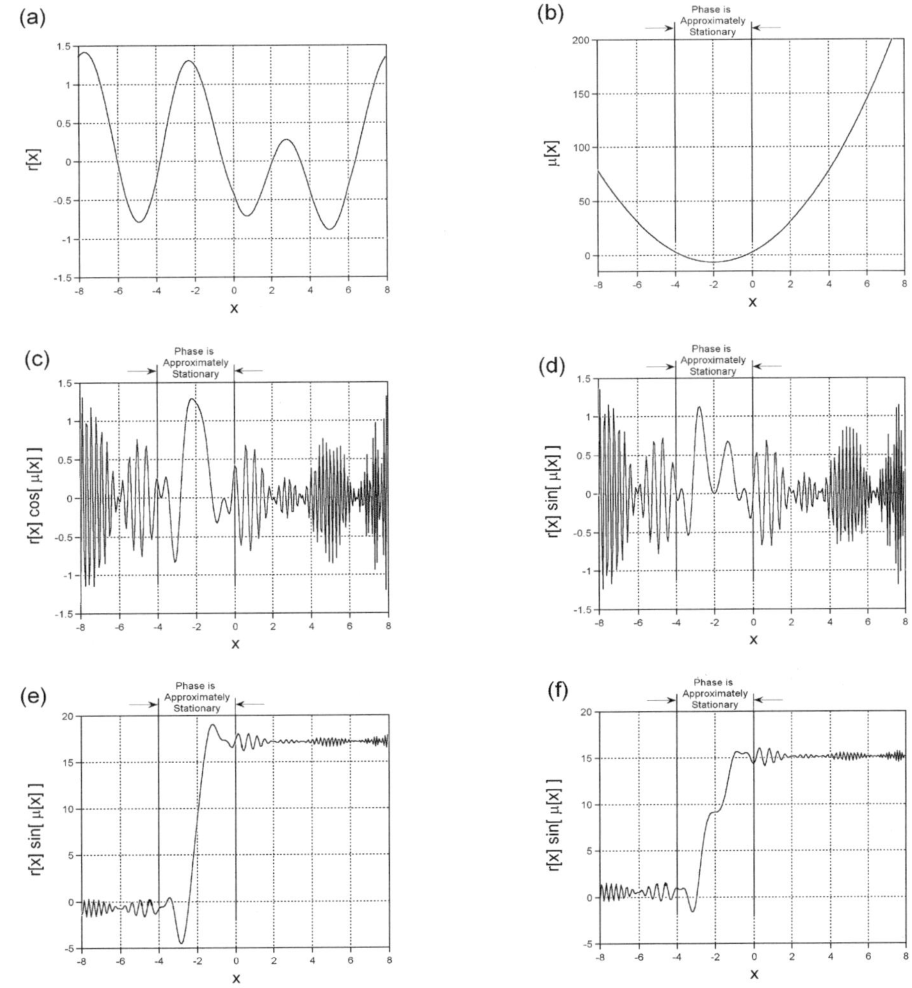

Figure 13.7 Principle of the method of stationary phase: (a) real-valued modulation $r[x]$; (b) phase function $\mu[x]$, which is approximately stationary in vicinity of $x = -2$; (c), (d) real and imaginary parts of $r[x]\, e^{i\mu[x]}$, showing rapid oscillations away from stationary point; (e), (f) real and imaginary parts of $\int_{-\infty}^{x} r[\alpha]\, e^{i\mu[\alpha]}\, d\alpha$, showing that the primary contribution to the area is from $r[x]$ in the vicinity of the stationary point.

summed:

$$
\begin{aligned}
I[k] &= \int_{-\infty}^{+\infty} (\Re\{r[x]\} + i\, \Im\{r[x]\})\, e^{ik\cdot\mu[x]}\, dx \\
&= \left(\int_{-\infty}^{+\infty} \Re\{r[x]\}\, e^{ik\cdot\mu[x]}\, dx\right) + i\left(\int_{-\infty}^{+\infty} \Im\{r[x]\}\, e^{ik\cdot\mu[x]}\, dx\right)
\end{aligned}
\tag{13.55}
$$

Under some circumstances, other and more subtle aspects of the method of stationary phase require special treatment. Since these occur rarely in the cases of greatest interest in imaging, they will be mentioned only in passing and not considered in detail. Interested readers should consult sources that concentrate on this subject, especially the work of Erdelyi (1956) and of Friedman (1969).

To illustrate use of the method of stationary phase in Fourier analysis, consider the Fourier transform of a 1-D complex-valued function $f[x]$ expressed in terms of its magnitude $|f[x]|$ and phase $\Phi\{f[x]\}$:

$$\begin{aligned}\mathcal{F}_1\{f[x]\} &= \int_{-\infty}^{+\infty} f[x]\, e^{-2\pi i\xi x}\, dx \\ &= \int_{-\infty}^{+\infty} (|f[x]|\, e^{i\boldsymbol{\Phi}\{f[x]\}})\, e^{-2\pi i\xi x}\, dx \\ &= \int_{-\infty}^{+\infty} |f[x]|\, e^{i\cdot(\boldsymbol{\Phi}\{f[x]\}-2\pi\xi x)}\, dx \end{aligned} \tag{13.56}$$

Note that the form of this integral is somewhat less general than that in Equation (13.54) because the modulation function $|f[x]|$ is not only real valued, but also nonnegative, whereas $r[x]$ in Equation (13.54) may be negative.

The Fourier integral in Equation (13.56) may be rewritten in the form of Equation (13.54) by defining:

$$\mu[x] \equiv \frac{1}{\xi}\boldsymbol{\Phi}\{f[x]\} - 2\pi x \tag{13.57}$$

and substituting the spatial frequency ξ for the parameter k and $|f[x]|$ for $r[x]$:

$$F[\xi] = \int_{-\infty}^{+\infty} |f[x]|\, e^{i\xi\cdot\mu[x]}\, dx \tag{13.58}$$

Note that the phase function $\mu[x]$ includes a factor of ξ^{-1}, which is reasonable because ξ is a parameter rather than a variable; the result is a function of this parameter.

If the phase function $\mu[x]$ has no singularities (i.e., if all of its derivatives are finite), then $\mu[x]$ may be expanded into a Taylor series about any arbitrary location x_0:

$$\begin{aligned}\mu[x] &= \mu[x_0] + \left((x-x_0)\frac{d\mu}{dx}\bigg|_{x=x_0}\right) + \left(\frac{(x-x_0)^2}{2}\frac{d^2\mu}{dx^2}\bigg|_{x=x_0}\right) + \cdots \\ &= \mu[x_0] + (x-x_0)\ \mu'[x_0] + \frac{(x-x_0)^2}{2}\ \mu''[x_0] + \cdots + \frac{(x-x_0)^n}{n!}\ \mu^{(n)}[x_0] + \cdots \end{aligned} \tag{13.59}$$

where again the concise "multiple-prime" notation for derivatives has been used. We now select x_0 to be a stationary point of the phase function so that $\mu'[x_0] = 0$. For now, assume that $\mu[x]$ has only one such stationary point; extension to cases with multiple stationary points is straightforward and will be considered later. The first-order term in the Taylor series vanishes:

$$\mu[x] = \mu[x_0] + 0 + \frac{(x-x_0)^2}{2!}\mu''[x_0] + \frac{(x-x_0)^3}{3!}\mu'''[x_0] + \cdots \tag{13.60}$$

and the Fourier integral in Equation (13.56) may be rewritten in terms of this series:

$$\begin{aligned}F[\xi] &= \int_{-\infty}^{+\infty} |f[x]|\exp\left[+i\xi\left(\mu[x_0] + \mu''[x_0]\frac{(x-x_0)^2}{2} + \cdots\right)\right] dx \\ &= \int_{-\infty}^{+\infty} |f[x]|\exp[+i\xi\cdot\mu[x_0]]\prod_{n=2}^{+\infty}\left(\exp\left[+i\xi\left(\mu^{(n)}[x_0]\frac{(x-x_0)^2}{2}\right)\right]\right) dx \end{aligned} \tag{13.61}$$

Of course, the zero-order term in the Taylor series is a constant with respect to x and may be extracted from the integral:

$$F[\xi] = \exp[+i\xi\cdot\mu[x_0]]\int_{-\infty}^{+\infty} |f[x]|\prod_{n=2}^{+\infty}\left(\exp\left[+i\xi\left(\mu^{(n)}[x_0]\frac{(x-x_0)^2}{2}\right)\right]\right) dx \tag{13.62}$$

If the spatial frequency ξ is assumed to be sufficiently large that the exponential term oscillates many times over the scale of variation of $|f[x]|$, then the integral may be approximated to evaluate the spectrum. Under this condition, the magnitude $|f[x]|$ contributes significant area to the Fourier integral only in the vicinity of the stationary point x_0, thus allowing the varying magnitude $|f[x]|$ to be approximated by the constant $|f[x_0]|$. In addition, this allows us to change the infinite limits of the integral to finite limits in the vicinity of the single stationary point. Finally, only the first nonzero term in the Taylor series of order 2 or larger is significant because $(x - x_0)^n \ll (x - x_0)^2$ for $n \geq 3$ when x is in the neighborhood of x_0. The resulting asymptotic form of the Fourier integral is:

$$\hat{F}[|\xi| \gg 0] \cong |f[x_0]| \exp[+i\xi \cdot \mu[x_0]] \int_{x_0-\epsilon}^{x_0+\epsilon} \exp\left[+i\xi \cdot \mu''[x_0] \frac{(x - x_0)^2}{2}\right] dx \tag{13.63}$$

where ϵ is a small positive number. Note that this assumes that $\mu''[x_0] \neq 0$. If the derivatives of second or larger order of $\mu[x]$ also are zero at x_0, then the derivative with the smallest order (other than 1) that does not vanish is used in the approximation. The remainder of the derivation must be appropriately modified; details are presented by Friedman.

In words, Equation (13.63) demonstrates that the Fourier integral of an oscillating function that includes a single stationary point in the infinite domain may be evaluated as the product of some easily evaluated constants and the finite integral of a quadratic-phase exponential. Since the area of the quadratic-phase factor also is concentrated in the vicinity of the stationary point, little additional error is incurred by returning to infinite limits on the integral and thus substituting the area of the quadratic-phase sinusoid over the infinite domain:

$$\int_{x_0-\epsilon}^{x_0+\epsilon} \exp\left[+i\xi \cdot \mu''[x_0] \frac{(x - x_0)^2}{2}\right] dx \cong \int_{-\infty}^{+\infty} \exp\left[+i\xi \left(\mu''[x_0] \frac{(x - x_0)^2}{2}\right)\right] dx \tag{13.64}$$

In short, the finite integral is approximated by the total area of the quadratic-phase exponential, which is easy to evaluate by changing the integration variable to $u \equiv \frac{1}{2}\mu''[x_0] \cdot (x - x_0)^2$ and applying the central-ordinate theorem from Equation (9.113). The area of this quadratic-phase term is:

$$\begin{aligned} \int_{x=-\infty}^{x=+\infty} \exp\left[+i\xi \left(\mu''[x_0] \frac{(x - x_0)^2}{2}\right)\right] dx &= \left(\sqrt{\frac{2\pi}{\xi\, \mu''[x_0]}}\right) \int_{u=-\infty}^{u=+\infty} \exp[+i\pi u^2]\, du \\ &= \left(\sqrt{\frac{2\pi}{\xi \cdot \mu''[x_0]}}\right) \exp\left[+i\frac{\pi}{4}\right] \end{aligned} \tag{13.65}$$

This result is substituted into Equation (13.63) to obtain the approximation for the spectrum that is valid in those cases where the phase of the integrand of the Fourier transform is stationary at a single coordinate:

$$\hat{F}[|\xi| \gg 0] \cong |f[x_0]| \left(\sqrt{\frac{2\pi}{\xi \cdot \mu''[x_0]}}\right) \exp\left[+i\frac{\pi}{4}\right] \exp[+i\xi \cdot \mu[x_0]] \tag{13.66a}$$

This complex amplitude may be expressed as magnitude and phase:

$$|\hat{F}[|\xi| \gg 0]| \cong |f[x_0]| \sqrt{\frac{2\pi}{\mu''[x_0]}}\, |\xi^{-\frac{1}{2}}| \tag{13.66b}$$

$$\mathbf{\Phi}\{\hat{F}[|\xi| \gg 0]\} \cong +\xi \cdot \mu[x_0] + \frac{\pi}{4} \tag{13.66c}$$

The approximation for the more general form of the Fourier integral with a real-valued bipolar modulation is obtained by a direct substitution of the real-valued modulation $r[x]$ for the nonnegative magnitude $|f[x]|$:

$$f[x] = r[x]\, e^{i\Phi\{f[x]\}}$$

$$\Longrightarrow \hat{F}[|\xi| \gg 0] \cong r[x_0]\sqrt{\frac{2\pi}{\xi \cdot \mu''[x_0]}}\exp\left[+i\frac{\pi}{4}\right]\exp[+i\xi \cdot \mu[x_0]] \tag{13.67}$$

Obviously, an integral in the frequency domain similar to that in Equation (13.56) may be constructed for the inverse Fourier transform, which will allow the asymptotic evaluation of $f[x]$ from a spectrum with an oscillating exponential. Such a development will be used in Chapter 16.

13.2.1 Examples of Spectra via Stationary Phase

The formulation in Equation (13.67) will be used to evaluate the asymptotic form of the Fourier integral for a few phase functions.

13.2.1.1 Unit-Magnitude Linear-Phase Exponential

Consider first the spectrum of the linear-phase exponential $f[x] = e^{+2\pi i\xi_0 x}$:

$$\begin{aligned} F_1[\xi] &= \int_{-\infty}^{+\infty} (1[x]\, e^{+2\pi i\xi_0 x})\, e^{-2\pi i\xi x}\, dx \\ &= \int_{-\infty}^{+\infty} e^{-2\pi i(\xi-\xi_0)x}\, dx \end{aligned} \tag{13.68}$$

which may be recast into the form of Equation (13.58) by changing the integration variable ξ to $\zeta = -(\xi - \xi_0)$ and identifying the phase function $\mu[x]$ to be $2\pi x$:

$$\begin{aligned} F_1[\zeta] &= \int_{-\infty}^{+\infty} e^{+i\zeta\,\mu[x]}\, dx \\ &= \int_{-\infty}^{+\infty} e^{+i\zeta(2\pi x)}\, dx \end{aligned} \tag{13.69}$$

The derivative of the phase function is the positive constant $\mu'[x] = 2\pi$, which confirms the observation that the integrand oscillates at the same rate over the entire domain; in other words, there is no point of stationary phase. Since the criterion for stationary phase is not fulfilled, the asymptotic solution for $F[\xi]$ does not exist. This result demonstrates that the method of stationary phase may be applied only if the phase of $f[x]$ includes terms of order 2 or higher.

13.2.1.2 Modulated Quadratic-Phase Exponential

Perhaps the most useful application of the method of stationary phase in Fourier analysis is to find an asymptotic expression for the spectrum of a function $f_2[x]$ that has quadratic phase:

$$f_2[x] = |f_2[x]|\, e^{+i\pi(x/\alpha)^2} \tag{13.70}$$

The scale factor α has units of length to ensure that the exponent is dimensionless. The Fourier integral is written in the form of Equation (13.58):

$$F_2[\xi] = \int_{-\infty}^{+\infty} |f_2[x]|\, e^{+i\pi[(x^2/\alpha^2)-2\xi x]}\, dx = \int_{-\infty}^{+\infty} |f_2[x]|\, e^{+i\xi[(\pi x^2/\alpha^2\xi)-2\pi x]}\, dx \tag{13.71}$$

The phase function and its first derivative are easy to evaluate, which leads to the expression for the stationary point x_0:

$$\mu[x] = \frac{\pi x^2}{\alpha^2 \xi} - 2\pi x \tag{13.72a}$$

$$\mu'[x] = 2\pi\left(\frac{x}{\alpha^2 \xi} - 1\right) \tag{13.72b}$$

$$\Longrightarrow \mu'[x_0] = 0 = 2\pi\left(\frac{x_0}{\alpha^2 \xi} - 1\right) \tag{13.72c}$$

$$\Longrightarrow x_0 = +\alpha^2 \xi \tag{13.72d}$$

Note that the stationary point $x_0 = +\alpha^2\xi$ in Equation (13.72d) has the required dimensions of length. The phase function and its derivatives evaluated at this stationary point are:

$$\mu[x_0] = \frac{\pi\alpha^4\xi^2}{\alpha^2\xi} - 2\pi\alpha^2\xi = -\pi\alpha^2\xi \tag{13.73a}$$

$$\mu'[x_0] = 0 \tag{13.73b}$$

$$\mu''[x] = \frac{2\pi}{\alpha^2\xi} \Longrightarrow \mu''[x_0] = \frac{2\pi}{\alpha^2\xi} \tag{13.73c}$$

$$\mu^{(n)}[x_0] = 0 \quad for\ n \geq 3 \tag{13.73d}$$

These results are substituted into the stationary-phase solution in Equation (13.66) to estimate the amplitude of the spectrum at spatial frequencies distant from the origin:

$$\begin{aligned}
\hat{F}[|\xi| \gg 0] &= |f_2[x_0]|\sqrt{\frac{2\pi}{\xi \cdot \mu''[x_0]}}\, e^{+i\pi/4}\, e^{+i\xi\ \mu[x_0]} \\
&= |f_2[\alpha^2\xi]|\sqrt{\frac{2\pi}{\xi \cdot (2\pi/\alpha^2\xi)}}\, e^{+i\pi/4}\, e^{+i\xi(-\pi\alpha^2\xi)} \\
&= |f_2[\alpha^2\xi]|\,\sqrt{\alpha^2}\, e^{+i\pi/4}\, e^{-i\pi\alpha^2\xi^2} \\
&= (|\alpha|\,|f_2[\alpha^2\xi]|)\, e^{+i\pi/4}\, e^{-i\pi\alpha^2\xi^2}
\end{aligned} \tag{13.74}$$

The validity of this expression may be confirmed for the case $|f_2[x]| = 1[x]$ and $\alpha = 1$ for an unmodulated chirp. The result may be compared to the known spectrum of the quadratic-phase exponential in Equation (9.95):

$$\mathcal{F}_1\{\exp[+i\pi x^2]\} \cong \hat{F}[|\xi| \gg 0] = 1[\xi]\exp\left[+i\frac{\pi}{4}\right]\exp[-i\pi\xi^2] = F[\xi] \tag{13.75}$$

In the case of a unit-magnitude quadratic-phase function, the asymptotic and exact forms of the spectrum are identical. This is because $\mu^{(n)}[x] = 0$ for $n > 3$, and thus no error in the phase function is incurred by truncating the Taylor series at the second-order term.

A more general quadratic-phase function is obtained by replacing the nonnegative modulation $|f[x]|$ with a real-valued bipolar function $r[x]$ after scaling by b and translating by x_1:

$$f_3[x] = r\left[\frac{x - x_1}{b}\right] e^{\pm i\pi(x/\alpha)^2} \tag{13.76}$$

Equation (13.67) may be applied directly to evaluate the asymptotic form of the spectrum:

$$\hat{F}_3[|\xi| \gg 0] \cong |\alpha| r\left[\frac{\alpha^2\xi \mp x_1}{b}\right] e^{\pm i\pi/4}\, e^{\mp i\pi\alpha^2\xi^2}$$

$$= |\alpha| r\left[\frac{\xi \mp (x_1/\alpha^2)}{(b/\alpha^2)}\right] e^{\pm i\pi/4}\, e^{\mp i\pi\alpha^2\xi^2} \tag{13.77a}$$

The estimate of the phase transfer function of the modulated quadratic-phase filter is identical to that of the unmodulated signal:

$$\Phi\{\hat{F}_3[\xi]\} = -\pi\left(\alpha^2\xi^2 - \frac{1}{4}\right) \tag{13.77b}$$

Equation (13.77a) exhibits some very interesting (and perhaps unexpected) features. Note that the magnitude of the Fourier spectrum is a scaled and translated replica of the *same* real-valued modulation r of the space-domain function. The frequency-domain modulation is translated and scaled by the respective factors x_1/α^2 and b/α^2, which have the required dimensions of "reciprocal length". In words, increasing the width parameter of the modulation of $f_3[x]$ *increases* the width parameter of the modulation of the spectrum estimate $\hat{F}_3[\xi]$ by a proportional factor. Similarly, translation of the modulation of the quadratic phase produces a proportional translation of the modulation of $\hat{F}_3[\xi]$. These features of the spectrum of a modulated quadratic-phase function may seem to violate the scaling and shifting theorems of the Fourier transform, but in fact are artifacts of the quadratic-phase function that will be discussed in more detail in Chapter 17 and will be useful in the discussion of optical systems in Chapter 20.

Because decreasing the scale factor α of the quadratic phase has the effect of increasing the oscillation rate of $f_3[x]$, this condition improves the accuracy of stationary-phase solution at a particular spatial frequency.

The prediction of Equation (13.77) can be tested; consider the specific case of a scaled and translated *SINC* function modulated by a quadratic-phase function with $\alpha = 2$:

$$f[x] = SINC\left[\frac{x-2}{4}\right] e^{+i\pi(x/2)^2} \tag{13.78}$$

which is graphed both as real/imaginary parts and as magnitude/phase in Figure 13.8. This function has a real-valued *bipolar* modulation, which is more general than the real-valued and nonnegative modulation by the magnitude in Equation (13.70). The stationary-phase solution for the spectrum is obtained by direct substitution into Equation (13.77):

$$\hat{F}[|\xi| \gg 0] \cong 2\, SINC\left[\frac{\xi - \frac{1}{2}}{1}\right] e^{+i\pi/4}\, e^{-i\pi(2\xi)^2} \tag{13.79}$$

where the scale factor $\alpha = 1$ in the denominator of the argument of the *SINC* function corrects the dimensions of both the translation and scale factor. The spectrum is compared to a discrete calculation in Figure 13.9. The differences between the approximations and the computed spectra are more apparent in the magnified views for $1 \leq \xi \leq 2$ cycles per unit length. Note that the approximate magnitude spectrum is zero at $\xi = 1$, while the "exact" computed spectrum is not.

13.2.1.3 Stationary-Phase Approximation for Symmetric Superchirps

We now briefly consider asymptotic forms of the spectra of unmodulated superchirp functions that were introduced in Equation (6.150). We defined two flavors of superchirps that differ in behavior for odd values of the order n: the intrinsically symmetric form $\cos[\pi|x|^n] \pm i \sin[\pi|x|^n]$ and the Hermitian variety $\cos[\pi|x|^n] \pm i\,SGN[x]\sin[\pi|x|^n]$. We consider the symmetric form first. Based upon the symmetry arguments developed in Section 9.1, we expect the spectra of these complex-valued and

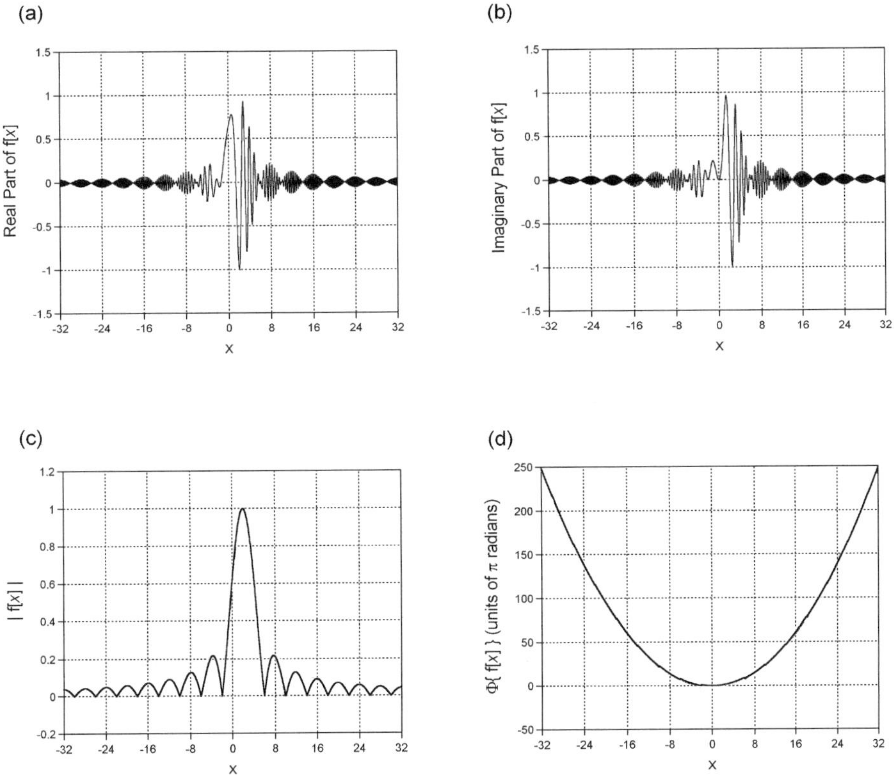

Figure 13.8 $f[x] = SINC[(x-2)/4]\exp[+i\pi(x/2)^2]$ as: (a) real part; (b) imaginary part; (c) magnitude; (d) phase. This function will be used to demonstrate the approximation of the Fourier transform via the method of stationary phase.

symmetric functions to be complex and symmetric. The Fourier integral is:

$$\mathcal{F}_1\{e^{+i\pi|x/\alpha|^n}\} = \int_{-\infty}^{+\infty} (e^{+i\pi|x/\alpha|^n})\, e^{-2\pi i\xi x}\, dx \tag{13.80}$$

The phase function and its first derivative are:

$$\mu[x] = +\frac{\pi}{\xi}\left(\frac{x}{\alpha}\right)^n - 2\pi x \tag{13.81}$$

$$\mu'[x] = +\frac{\pi}{\alpha\xi} n\left(\frac{x}{\alpha}\right)^{n-1} - 2\pi \tag{13.82}$$

It is convenient to consider the cases of even and odd values of n separately. When n is even, the exponent $n-1$ in $\mu'[x]$ is odd. The point(s) of stationary phase (if any) are the solutions to $\mu'[x_0] = 0$, which are determined by the spatial frequency ξ in the Fourier integral:

$$x_0 = \alpha\left(\frac{2\alpha\xi}{n}\right)^{1/(n-1)} \tag{13.83}$$

Note that x_0 has the required units of length. The stationary points of an even-order symmetric superchirp are the real-valued solutions of Equation (13.83), and thus are proportional to odd-order

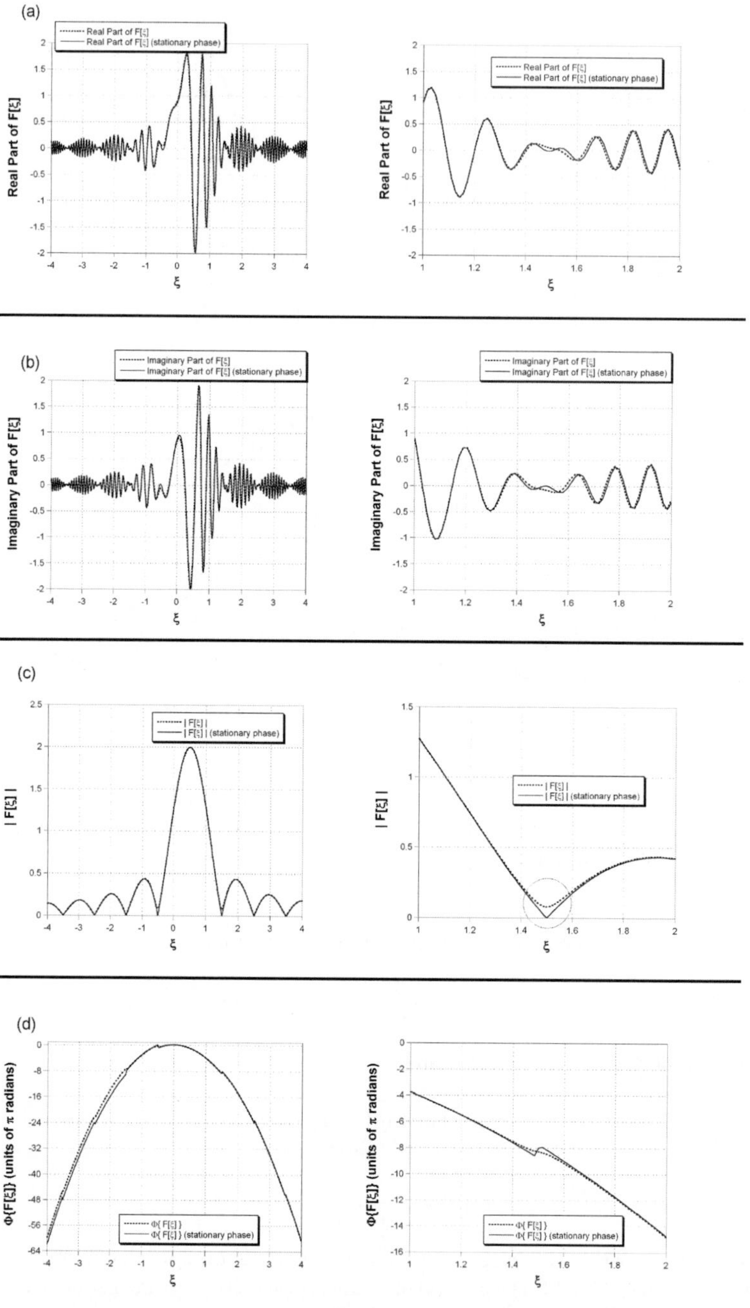

Figure 13.9 The stationary phase approximation to the Fourier transform of $f[x] = SINC[(x-2)/4] \exp[+i\pi(x/2)^2]$. The approximation is $\hat{F}[\xi] = 2\,SINC[\xi + \frac{1}{2}] \exp[-i\pi(2\xi)^2] \exp[+i\pi/4]$: (a) real part; (b) imaginary part; (c) magnitude; and (d) phase. The differences between the approximation and the computed spectra are more visible in the magnified views for $1 \leq \xi \leq 2$ cycles per unit length.

roots of the selected spatial frequency ξ. For example, if $n = 4$, the stationary points are proportional to the real-valued solutions of $\xi^{1/3}$. For $\xi > 0$, a single real-valued solution for $\xi^{1/3} \propto x_0$ exists and it is positive. When $\xi < 0$, the single real-valued solution for x_0 is negative.

When n is odd, the phase function of the symmetric superchirp may be decomposed into two functions, one each for positive and negative x:

$$\mu[x \geq 0] = +\frac{\pi}{\xi}\left(+\frac{x}{\alpha}\right)^n - 2\pi x \tag{13.84a}$$

$$\mu[x \leq 0] = +\frac{\pi}{\xi}\left(-\frac{x}{\alpha}\right)^n - 2\pi x \tag{13.84b}$$

The first derivatives of the phase in these two regions are:

$$\mu'[x \geq 0] = +\frac{\pi}{\alpha\xi}n\left(+\frac{x}{\alpha}\right)^{n-1} - 2\pi \tag{13.85a}$$

$$\mu'[x \leq 0] = +\frac{\pi}{\alpha\xi}n\left(-\frac{x}{\alpha}\right)^{n-1} - 2\pi \tag{13.85b}$$

Because n is odd, $n - 1$ is even. For positive values of x, the stationary point(s) must satisfy:

$$(x_0)_+ = +\alpha\left(\frac{2\alpha\xi}{n}\right)^{1/(n-1)} \tag{13.86a}$$

Because it is proportional to an even-order root of ξ, this expression is real valued only for $\xi > 0$. In other words, the stationary-phase estimate of the spectrum of an odd-order superchirp is nonzero only for positive frequencies.

Similarly, the stationary point for negative x must satisfy the condition:

$$(x_0)_- = -\alpha\left(\frac{2\alpha\xi}{n}\right)^{1/(n-1)} \tag{13.86b}$$

The real-valued even-order root of ξ again exists only for $\xi > 0$, and thus $(x_0)_- < 0$. In short, the integrand has a single stationary point if n is symmetric. An example is shown in Figure 13.10 for $n = 4$, $\alpha = 1$ unit, and $\xi = +4$ cycles per unit length. The stationary point is located at $x_0 = +(2 \cdot 1 \cdot 4/4)^{1/3} \cong 1.26$.

It may be convenient to recast the integral in Equation (13.80) into the form of Equation (13.58) by changing variables:

$$\int_{-\infty}^{+\infty} (e^{+i\pi|x|^n/\alpha})\, e^{-2\pi i\xi x}\, dx = \int_{-\infty}^{+\infty} e^{+iv(u^n-u)}\left(\frac{v}{\pi}\right)^{1/n} \alpha\, du \tag{13.87a}$$

where:

$$x = \alpha u\left(\frac{v}{\pi}\right)^{1/n} \tag{13.87b}$$

$$\xi = \frac{1}{2\alpha}\left(\frac{v}{\pi}\right)^{(n-1)/n}\frac{1}{2\alpha} \tag{13.87c}$$

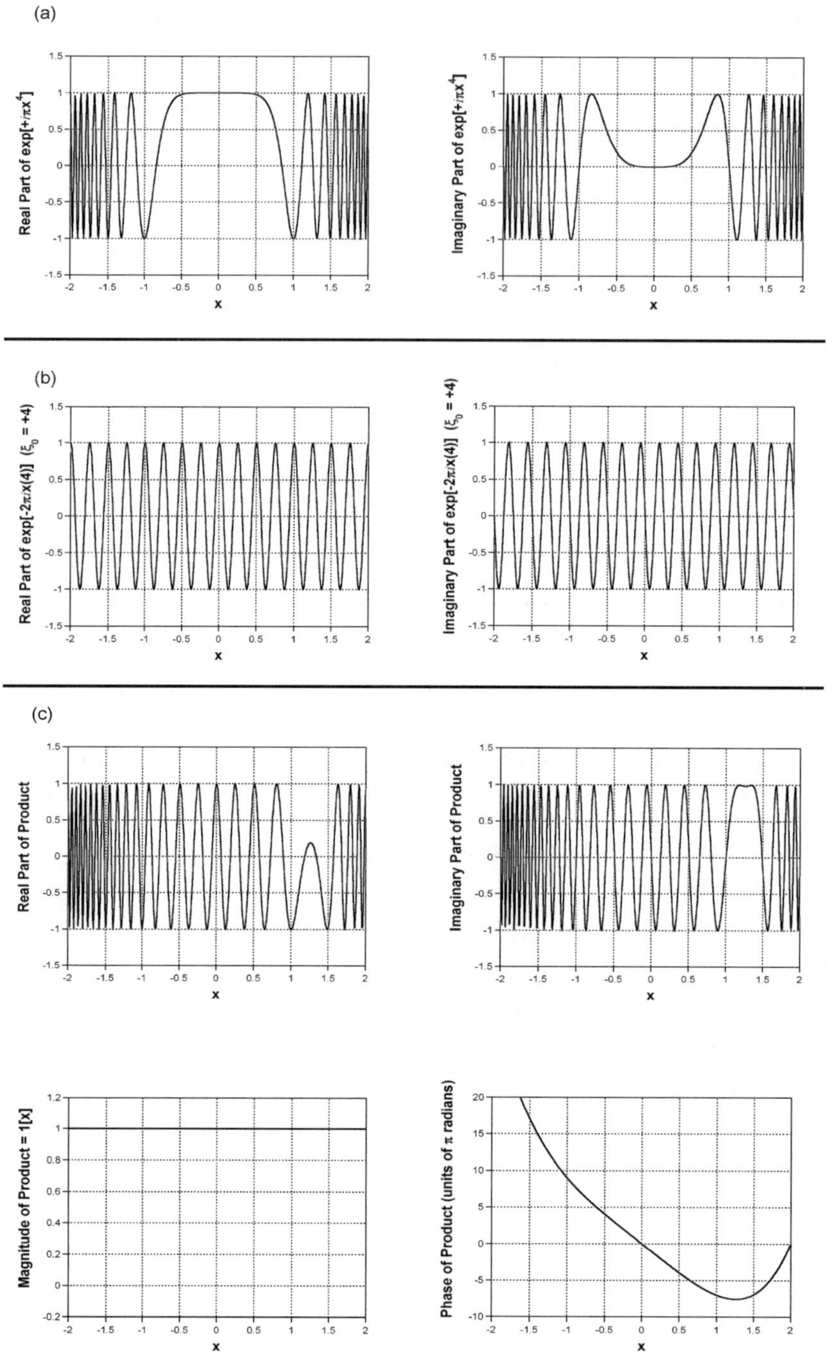

Figure 13.10 Stationary point of $\exp[+i\pi x^4] \cdot \exp[-2\pi i x \cdot 4]$: (a) real and imaginary parts of $\exp[+i\pi x^4]$; (b) real and imaginary parts of $\exp[-2\pi i \cdot x \cdot 4]$; (c) real part, imaginary part, unit magnitude, and phase of $\exp[+i\pi x^4] \cdot \exp[-2\pi i \cdot x \cdot 4]$, showing the stationary point at $x_0 = \sqrt[3]{2} \cong 1.26$.

The phase function and its derivatives are easily evaluated in terms of u and v:

$$\mu[u] = u^n - u \tag{13.88a}$$

$$\mu'[u] = nu^{n-1} - 1 \Longrightarrow u_0 = n^{1/(n-1)} \tag{13.88b}$$

$$\mu''[u] = n \cdot (n-1)\, u^{n-2} \tag{13.88c}$$

$$\mu[u_0] = (1-n) \cdot (n^{n/(1-n)}) \tag{13.88d}$$

$$\mu''[u_0] = n \cdot (n-1) \cdot (n^{-(n-2)/(n-1)}) \tag{13.88e}$$

Substitution of these terms into the stationary-phase solution of Equation (13.66) yields the approximate solution for the spectrum that is valid for large $|v|$:

$$\hat{F}[|v| \gg 0] = \left(\sqrt{\frac{2\pi}{v\,\mu''[u_0]}}\right) \exp\left[+i\frac{\pi}{4}\right] \exp[+iv \cdot \mu[u_0]]\alpha\left(\frac{v}{\pi}\right)^{1/n} \tag{13.89}$$

which may be written in the desired form by substituting the form of v from Equation (13.87b). The stationary-phase solution for the spectrum of the unmodulated symmetric superchirp function is the complex-valued symmetric function:

$$\hat{F}[\xi] = \alpha\left(\frac{2}{n(n-1)}\right)^{\frac{1}{2}} \left(\frac{2\alpha|\xi|}{n}\right)^{-(n-2)/(2n-2)} \cdot \exp\left[+i\frac{\pi}{4}\right]\left(\exp\left[+i\pi(1-n)\left(\frac{2\alpha|\xi|}{n}\right)^{n/(n-1)}\right]\right) \tag{13.90}$$

To check (if not confirm) the validity of this expression, substitute $n = 2$ and compare to the known spectrum of the quadratic-phase function:

$$\mathcal{F}_1\left\{\exp\left[+i\pi\left(\frac{x}{\alpha}\right)^2\right]\right\} \Longrightarrow \hat{F}[\xi] = \alpha^1\sqrt{1}(\alpha|\xi|)^0 \exp\left[+i\frac{\pi}{4}\right] \exp[-i\pi(\alpha|\xi|)^2]$$
$$= \alpha \exp\left[+i\frac{\pi}{4}\right] \exp[-i\pi(\alpha|\xi|)^2] \tag{13.91}$$

which we know to be the correct spectrum for all values of ξ via the scaling theorem applied to the known spectrum of the chirp.

It is perhaps instructive to examine the functional forms of the magnitude and phase of the stationary-phase solution to the superchirp spectrum. The magnitude spectrum is the even function:

$$|\hat{F}[\xi]| = \left|\alpha\sqrt{\frac{2}{n(n-1)}}\left(\frac{2\alpha|\xi|}{n}\right)^{-(n-2)/(2n-2)}\right| \tag{13.92a}$$

The fact that the magnitude spectrum is constant for $n = 2$ confirms our observation in Equation (9.96) that chirp functions are sums of sinusoids with all spatial frequencies with identical magnitudes. However, the magnitude of the spectrum is *not* constant for $n > 2$, but rather *decreases* with increasing frequency. The rate of decline in the magnitude spectrum is $|\xi|^{-1/4}$, $|\xi|^{-1/3}$, and $|\xi|^{-3/8}$ for $n = +3$, $+4$, and $+5$, respectively. In words, the magnitude falls off more quickly with increasing order n. This behavior is consistent with the observation that the amplitude of $f[x] = \exp[+i\pi(x/\alpha)^n]$ near the origin is approximately unity over wider regions and that the spatial frequency changes more quickly with x as the order n is increased. The sinusoidal components of $f[x]$ with larger frequencies therefore have smaller amplitudes.

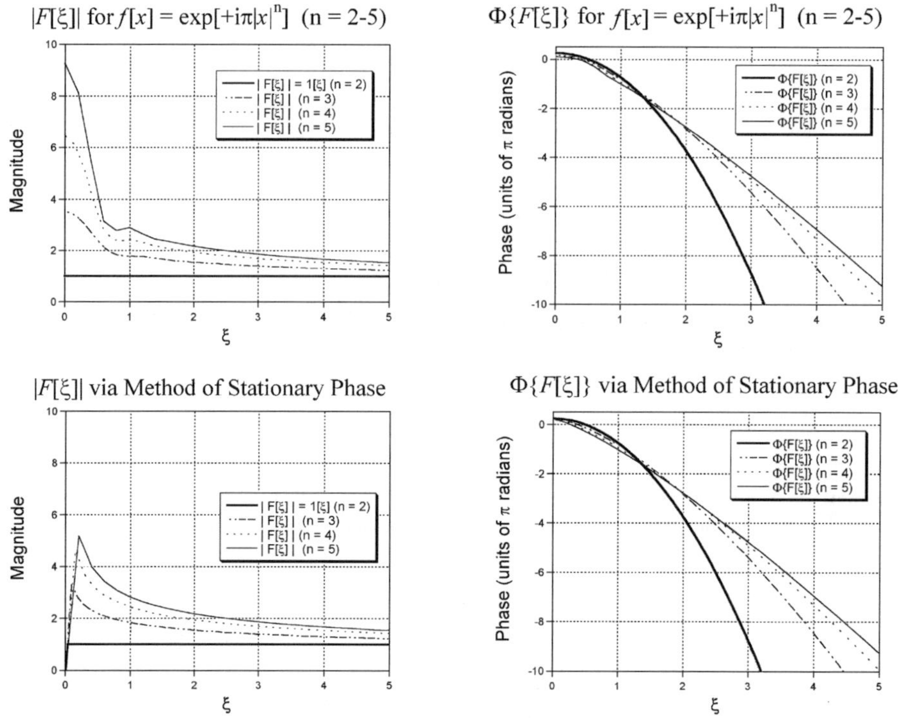

Figure 13.11 Magnitude and phase of spectra of symmetric superchirps $f[x] = \exp[+i\pi|x|^n]$ for $n = 2, 3, 4, 5$ by discrete computation and the approximation by the method of stationary phase from Equation (13.92). Note that the magnitude falls off more rapidly and the phase less rapidly with ξ as n increases.

The approximate phase spectrum of the symmetric superchirp is:

$$\Phi\{\mathcal{F}_1\{e^{+i\pi(x/\alpha)^n}\}\} \cong \pi\left(\frac{1}{4} + (1-n)\cdot\left(\frac{2\alpha|\xi|}{n}\right)^{n/(n-1)}\right) \tag{13.92b}$$

which varies as $-|\xi|^2$, $-|\xi|^{3/2}$, $-|\xi|^{4/3}$, and $-|\xi|^{5/4}$ for n = 2–5, respectively. In words, the phase spectrum varies *more slowly* with ξ as the order of the superchirp is *increased.* Examples are shown in Figure 13.11.

The initial phase of the stationary-phase approximation of the spectrum of all superchirps of the form $e^{+i\pi|x|^n}$ is $+\pi/4$ radians, though we know from the moment calculation in Equation (13.40) that the initial phase actually is $+\pi/2n$ radians. This again reminds us of the approximate form of the stationary-phase calculation and that it is strictly valid only for $|\xi| \gg 0$.

13.2.1.4 Spectra of Hermitian Superchirp Functions via Stationary Phase

The symmetry arguments of Section 9.1 and the expansion in terms of moments in Equation (13.49) demonstrate that the spectra of all odd-order Hermitian superchirps are real valued. When deriving the approximate forms of these spectra, we must consider multiple stationary points. The phase function of the Fourier integral of the Hermitian function is:

$$\mu[x] = \left(+\frac{\pi}{\xi}\left(\frac{x}{\alpha}\right)^n - 2\pi x\right) \tag{13.93a}$$

and its first and second derivatives are respectively:

$$\mu'[x] = \left(+\frac{\pi}{\alpha\xi}\right) n \left(\frac{x}{\alpha}\right)^{n-1} - 2\pi \tag{13.93b}$$

$$\mu''[x] = \left(+\frac{\pi}{\alpha^2\xi}\right) n(n-1)\left(\frac{x}{\alpha}\right)^{n-2} \tag{13.93c}$$

The stationary point(s) (if any) are the solutions to $\mu'[x_0] = 0$:

$$x_0 = \alpha\left(\frac{2\alpha\xi}{n}\right)^{1/(n-1)} \tag{13.94}$$

Since n is odd for all Hermitian superchirps, then $(n-1)$ is even. In words, the coordinates x_0 of the stationary points are proportional to the real-valued even-order roots of the spatial frequency ξ where the Fourier integral is evaluated. When ξ is large and positive ($\xi \gg 0$), there are two real-valued solutions for x_0 with identical magnitudes and opposite sign, i.e., at $x = \pm|x_0|$. The contributions from the two stationary points must be added to evaluate the asymptotic form of the Fourier integral for $\xi \gg 0$.

The situation is qualitatively different for $\xi < 0$, where the coordinates of the stationary points are proportional to even-order roots of negative numbers, which have *no* real-valued solutions. Therefore there are no stationary points if $\xi < 0$, and thus the stationary-phase approximation of the Fourier integral is zero for all negative ξ. Examples of the integrand of the Fourier integral of the cubic Hermitian superchirp are shown in Figure 13.12.

It remains to evaluate the phase function and its second derivative at the two stationary points of the odd-order Hermitian superchirp for $\xi > 0$. These may be inserted separately into Equation (13.66) and summed to evaluate the approximation. The positive and negative stationary points are labeled x_+ and x_-, respectively:

$$0 < x_+ = +\alpha\left(\frac{2\alpha\xi}{n}\right)^{1/(n-1)} \tag{13.95a}$$

$$0 > x_- = -\alpha\left(\frac{2\alpha\xi}{n}\right)^{1/(n-1)} \tag{13.95b}$$

where ξ, α, and n are all real-valued positive quantities. The corresponding phase functions are obtained by substitution of these derivatives into Equation (13.93a). The phase function evaluated for x_+ is:

$$\begin{aligned}\mu[x_+] &= (+1)^n\left(\frac{\pi(2\alpha\xi/n)^{n/(n-1)}}{\xi} - 2\pi\left(\frac{2\alpha\xi}{n}\right)^{1/(n-1)}\right)\\ &= \pi\left(\frac{2\alpha}{n}\right)^{n/(n-1)}\xi - 2\pi\left(\frac{2\alpha}{n}\right)^{1/(n-1)}\xi^{1/(n-1)}\end{aligned} \tag{13.96a}$$

Since ξ is assumed to be large and n is positive, the first term must be larger than the second, which means that $\mu[x_+]$ must be positive. The phase function evaluated at the negative stationary point is:

$$\begin{aligned}\mu[x_-] &= (-1)^n\left(\frac{\pi(2\alpha\xi/n)^{n/(n-1)}}{\xi}\right) - 2\pi\left(-\left(\frac{2\alpha\xi}{n}\right)^{1/(n-1)}\right)\\ &= -\left(\pi\left(\frac{2\alpha}{n}\right)^{n/(n-1)}\xi - 2\pi\left(\frac{2\alpha}{n}\right)^{1/(n-1)}\xi^{1/(n-1)}\right) = -\mu[x_+]\end{aligned} \tag{13.96b}$$

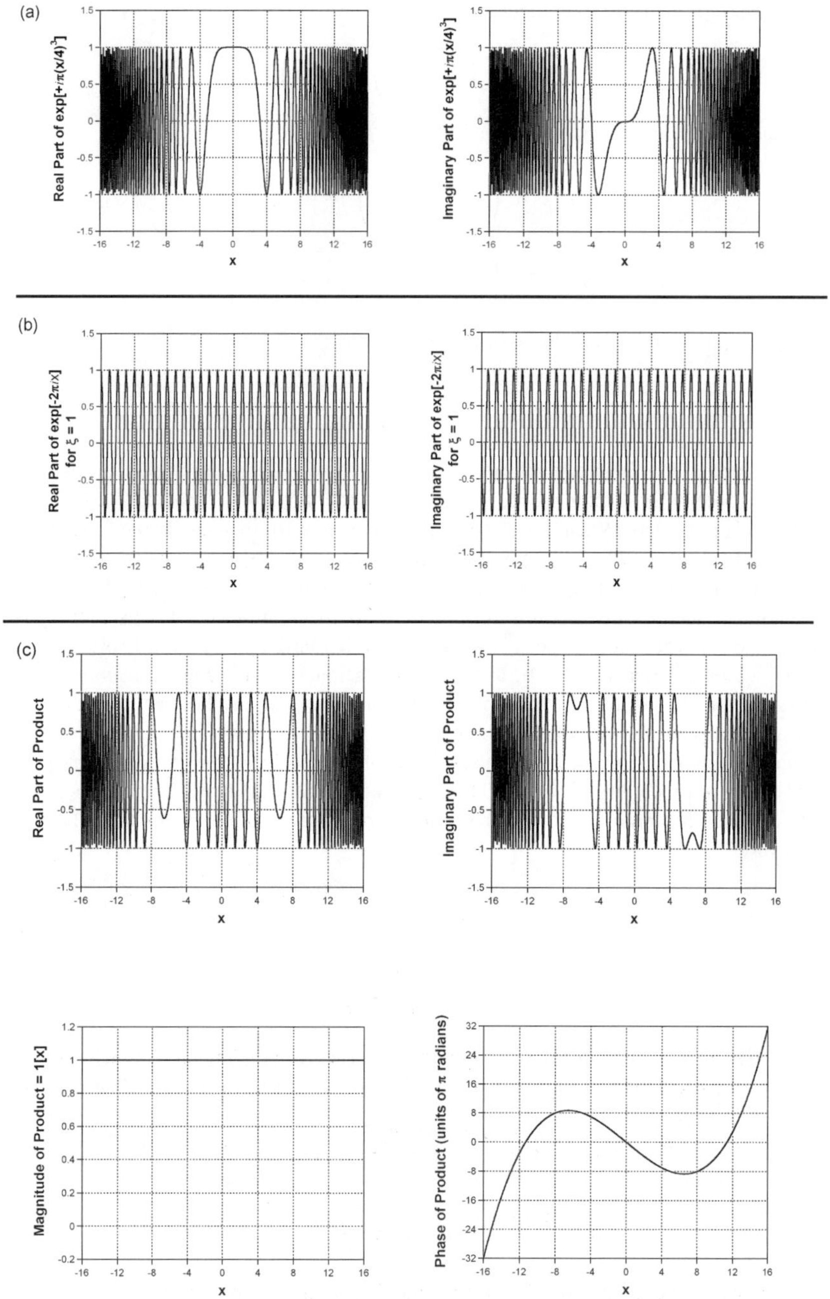

Figure 13.12 Stationary points of the product $\exp +i\pi(x/4)^3 \cdot \exp[-2\pi i x]$ ($n = 3$, $\alpha = 4$, $\xi = +1$). There are two stationary points located at $x_0 = \pm 4(2 \cdot 4 \cdot 1/3)^{1/2} \cong \pm 6.53$: (a) real and imaginary parts of $\exp[+i\pi(x/4)^3]$; (b) real and imaginary parts of $\exp[-2\pi i x]$; (c) real part, imaginary part, magnitude, and phase of the product, showing the two stationary points. The contributions from the two stationary points to the Fourier integral are complex conjugates, and so the imaginary parts cancel.

Note that the factors $e^{+i\xi\mu[x_0]}$ evaluated at these two stationary points are complex conjugates.

The corresponding solutions for the second derivative are:

$$\mu''[x_+] = -\frac{\pi}{\alpha^2\xi} n(n-1)\left(\frac{2\alpha\xi}{n}\right)^{(n-2)/(n-1)} \tag{13.97a}$$

$$\mu''[x_-] = +\frac{\pi}{\alpha^2\xi} n(n-1)\left(\frac{2\alpha\xi}{n}\right)^{(n-2)/(n-1)} = -\mu''[x_+] \tag{13.97b}$$

which also have the same magnitude and opposite sign; the second derivative evaluated at the negative stationary point x_- is positive, while that at x_+ is negative . Substitution of these results into Equation (13.66) yields the asymptotic form of the Fourier integral:

$$\begin{aligned}
\hat{F}[\xi \gg 0] &= e^{+i\pi/4}\left(e^{i\xi\mu[x_+]}\sqrt{\frac{2\pi}{\xi\mu''[x_+]}} + e^{i\xi\mu[x_-]}\sqrt{\frac{2\pi}{\xi\mu''[x_-]}}\right)\\
&= \sqrt{\frac{2\pi}{\xi}}\, e^{+i\pi/4}\left(\frac{e^{-i\xi\mu[x_-]}}{\sqrt{-\mu''[x_-]}} + \frac{e^{+i\xi\mu[x_-]}}{\sqrt{+\mu''[x_-]}}\right)\\
&= \sqrt{\frac{2\pi}{\xi\mu''[x_-]}}\, e^{+i\pi/4}\left(\left[\frac{e^{-i\xi\mu[x_-]}}{\sqrt{-1}}\right] + \left[\frac{e^{+i\xi\mu[x_-]}}{\sqrt{+1}}\right]\right)\\
&= \sqrt{\frac{2\pi}{\xi\mu''[x_-]}}\, e^{+i\pi/4}\left(\left[\frac{e^{-i\xi\mu[x_-]}}{+i}\right] + \left[\frac{e^{+i\xi\mu[x_-]}}{+1}\right]\right)\\
&= \sqrt{\frac{2\pi}{\xi\mu''[x_-]}}\, \frac{e^{+i\pi/4}}{e^{+i\pi/4}}\left(\left[\frac{e^{-i\xi\mu[x_-]}}{e^{+i\pi/4}}\right] + \left[\frac{e^{+i\xi\mu[x_-]}}{e^{-i\pi/4}}\right]\right)\\
&= \sqrt{\frac{2\pi}{\xi\mu''[x_-]}}\left(e^{-i(\xi\mu[x_-]+\pi/4)} + e^{+i(\xi\mu[x_-]+\pi/4)}\right)\\
&= \sqrt{\frac{2\pi}{\xi\mu''[x_-]}}\left(2\cos\left[\xi\,\mu[x_-] + \frac{\pi}{4}\right]\right)
\end{aligned} \tag{13.98}$$

After substituting the expressions for μ and μ'' from Equations (13.96) and (13.97), the asymptotic solutions for the spectra of odd-order Hermitian superchirps are obtained:

$$\mathcal{F}_1\{e^{+i\pi(x/\alpha)^n}\} \quad \textit{for odd } n$$

$$\cong \begin{cases} 2\alpha\sqrt{\dfrac{2}{n(n-1)}}\left(\dfrac{2\alpha\xi}{n}\right)^{-(n-2)/(2n-2)} \cdot \cos\left[\pi(1-n)\left(\dfrac{2\alpha\xi}{n}\right)^{n/(n-1)} - \dfrac{\pi}{4}\right] & \textit{if } \xi \gg 0 \\ 0 & \textit{if } \xi \ll 0 \end{cases} \tag{13.99}$$

In words, the asymptotic spectrum of the odd-order Hermitian superchirp function is real valued and also is zero for negative frequencies. The computed forms for $n = 3$ and $n = 5$ are shown in Figure 13.13. The graph of Equation (13.53) in Figure 13.6 shows that the moment expansion of the spectrum for $n = 3$ has nonzero amplitude for small negative frequencies. This behavior is not modeled by Equation (13.99) because it is valid only for large $|\xi|$. Note that the rate of oscillation of the spectrum with ξ decreases with the order n.

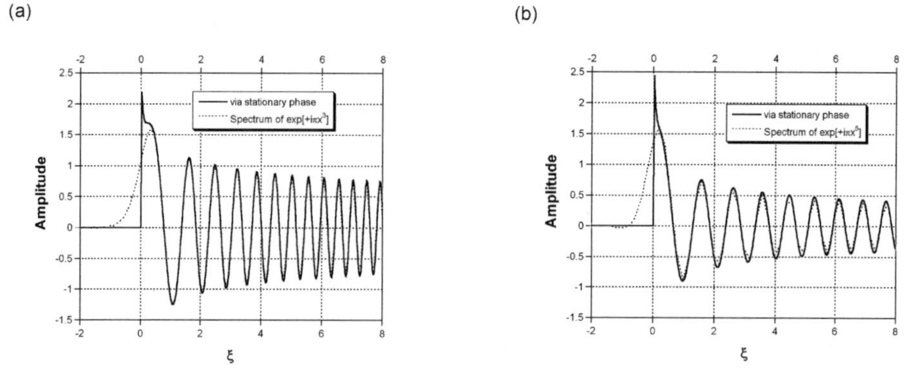

Figure 13.13 Fourier transforms of Hermitian superchirp functions compared to approximations from the method of stationary phase: (a) $\mathcal{F}\{\exp[+i\pi x^3]\}$; (b) $\mathcal{F}\{\exp[+i\pi x^4]\}$.

13.3 Central-Limit Theorem

We now divert our direction a bit to consider an expression that is neither an "approximation" of the Fourier transform nor valid for every function (or combination of functions), but it is applicable so often that it has become a very useful estimation tool in many disciplines of physical science. This, the "central-limit theorem", predicts that the convolution of multiple replicas of some function $f[x]$ often approximates a Gaussian:

$$(f[x])_1 * (f[x])_2 * (f[x])_3 * \cdots * (f[x])_N = g[x] \cong A\, e^{-\pi(x/d)^2} \tag{13.100}$$

where the subscripts on $(f[x])$ serve merely to count the replicas and the amplitude A and width parameter d are to be determined. The filter theorem and the self-transform character of the Gaussian establish the corresponding relationship in the frequency domain:

$$\hat{G}[\xi] = (F[\xi])_1 \cdot (F[\xi])_2 \cdot (F[\xi])_3 \cdot \cdots \cdot (F[\xi])_N = (F[\xi])^N \cong (A\,|d|)\, e^{-\pi(d\xi)^2} \tag{13.101}$$

This implies a constraint that must be satisfied by the magnitude spectrum of $f[x]$. $\hat{G}[\xi]$ may be approximated at small spatial frequencies by the power series based on its first few moments in Equation (13.40):

$$\hat{G}[\xi] = (A|d|) \cdot \sum_{n=0}^{\infty} \frac{(\pi d^2 \xi^2)^n}{n!} \cong (A|d|)[1 - (\pi d^2)\xi^2] \quad \text{if } |\xi| \gtrapprox 0 \tag{13.102}$$

As an example, consider the repeated convolution of the unit-width rectangle:

$$g[x] = (RECT[x])_1 * (RECT[x])_2 * \cdots * (RECT[x])_N \tag{13.103}$$

The spectrum is the Nth power of the *SINC* function:

$$G[\xi] = (F[\xi])^N = (SINC[\xi])^N \tag{13.104}$$

Obviously, the central ordinate $G[0]$ is the Nth power of $SINC[0] = 1$, which demonstrates that the area of the N-fold convolution of $RECT[x]$ is unity. The amplitude of $G[\xi]$ decreases with increasing $|\xi| \gtrapprox 0$ more rapidly than the spectrum $F[\xi]$ of an individual component, but $F[1] = G[1] = 0$. For spatial frequencies $|\xi| > 1$, the magnitude of $G[\xi]$ increases very much more slowly than $|F[\xi]|$ even

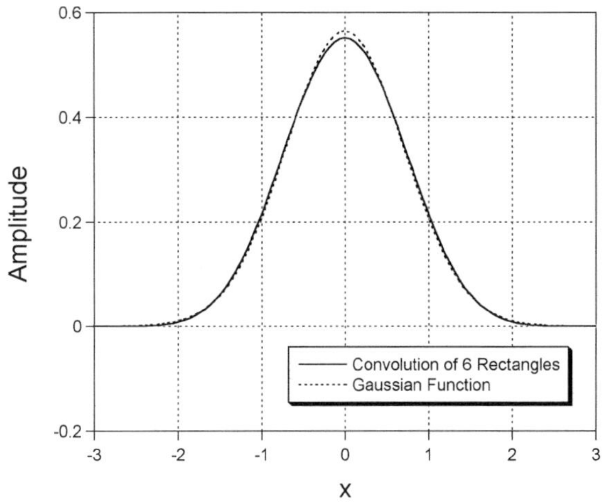

Figure 13.14 The convolution of six replicas of *RECT*$[x]$ compared to the Gaussian function $(1/\sqrt{\pi})GAUS[x/\sqrt{\pi}]$ from the central-limit theorem.

for medium-sized N because the amplitude of the *SINC* function is small. Therefore, $G[\xi] \cong 0$ for $|\xi| \geq 1$.

The series expansion for the *SINC* function in Equation (6.32) may be used to approximate Equation (13.104) for small values of $|\xi|$:

$$(SINC[\xi])^N = \left[\sum_{n=0}^{+\infty}(-1)^n \frac{(\pi\xi)^{2n}}{(2n+1)!}\right]^N$$
$$\cong \left[1 - \frac{(\pi\xi)^2}{3!}\right]^N \cong 1 - N\left(\frac{\pi^2}{6}\right)\xi^2 \quad \textit{if } |\xi| \gtrapprox 0 \tag{13.105}$$

This may be compared to Equation (13.102) to determine the width parameter d of the Gaussian in terms of N:

$$(\pi d^2)\xi^2 = N\left(\frac{\pi^2}{6}\right)\xi^2 \Longrightarrow d = \sqrt{\frac{N\pi}{6}} \tag{13.106}$$

This result may then be used to evaluate the amplitude A of $G[\xi]$ in terms of N:

$$A \cdot |d| = 1 \Longrightarrow A = |d|^{-1} = \frac{1}{\sqrt{N\pi/6}} \tag{13.107}$$

Thus the N-fold convolution of the unit-amplitude and unit-area rectangle function is approximated by:

$$(RECT[x])_1 * (RECT[x])_2 * \cdots * (RECT[x])_N \cong \frac{1}{\sqrt{N\pi/6}} GAUS\left[\frac{x}{\sqrt{N\pi/6}}\right] \tag{13.108}$$

The width and amplitude of this Gaussian approximation for the convolution are proportional to $\sqrt{N}$ and $\sqrt{N^{-1}}$, respectively. This is a fairly good approximation for a Gaussian for N as small as 3, though a common "rule of thumb" assumes that $N \gtrapprox 6$. The discrete calculation of the six-fold convolution of *RECT*$[x]$ is compared to the approximation in Equation (13.108) in Figure 13.14.

Equation (13.108) may be easily extended to the convolution of rectangle functions with arbitrary amplitude a and width parameter b:

$$\left(aRECT\left[\frac{x}{b}\right]\right)_1 * \left(aRECT\left[\frac{x}{b}\right]\right)_2 * \cdots * \left(aRECT\left[\frac{x}{b}\right]\right)_N$$

$$\cong \frac{|a|^N |b|^N}{|b|\sqrt{N\pi/6}} GAUS\left[\frac{x}{|b|\sqrt{N\pi/6}}\right] = \frac{|a|^N |b|^{N-1}}{\sqrt{N\pi/6}} GAUS\left[\frac{x}{|b|\sqrt{N\pi/6}}\right] \tag{13.109}$$

If convolving replicas of functions other than rectangles, we can surmise the conditions that must be satisfied by $f[x]$ for the central-limit theorem to be valid. Equation (13.102) may be used to determine the approximate form of the spectrum of $f[x]$:

$$F[\xi] \cong (G[\xi])^{1/N} = (A|d|)^{1/N} e^{-\pi(d\xi)^2/N}$$

$$= (A|d|)^{1/N}\left(1 - \pi\frac{(d\xi)^2}{N} + \pi^2\frac{(d\xi)^4}{2N} + \cdots\right) \tag{13.110}$$

In words, this shows that the second derivative of the spectrum evaluated near the origin must be negative to be satisfied; $|F[\xi]|$ is "concave down" near the origin. Since the magnitude of the sinusoidal components of $f[x]$ must *decrease* with increasing frequency, the "rapidity" of any excursions of $f[x]$ from its mean value are constrained. In other words, $f[x]$ must be "sufficiently smooth" for Equation (13.100) or Equation (13.101) to be valid. Also note that the coefficient of ξ^2 is proportional to N^{-1}, which loosens the "smoothness" constraint if the number of replicas convolved is large. Note that some functions $f[x]$ do not satisfy the conditions implied by Equation (13.102), including $1[x]$ and $\delta[x]$.

The multiple convolution of nonidentical functions often yields an approximate Gaussian as well. The discussion just concluded may be adapted to demonstrate that all constituent functions must have finite and positive area and that their second derivatives evaluated at the origin must be zero or negative to ensure that the product of their spectra satisfies Equation (13.102).

13.4 Width Metrics and Uncertainty Relations

The scaling theorem in Section 9.8.5 determined the relationship between width parameters of equivalent representations. For example, the product of the width parameters of the space- and frequency-domain representations of $RECT[x/b]$ is unity:

$$|b| \cdot \frac{1}{|b|} = 1 \tag{13.111}$$

This is one of the so-called *uncertainty relations* that appear frequently in modern physics, particularly in quantum mechanics. The specific value of the product of the two width metrics depends on the measure of width used. Several such measures are considered in this section.

13.4.1 Equivalent Width

The *equivalent width* of $f[x]$ will be denoted by Δx_f and is defined to be the width of the rectangle that has the same area and amplitude at the origin as $f[x]$:

$$\Delta x_f \equiv \frac{\int_{-\infty}^{+\infty} f[x]\,dx}{f[0]} \tag{13.112}$$

This measure has been used for many years by spectroscopists to estimate the widths of spectral absorption lines. It is easy to see that the equivalent width of a Dirac delta function centered at the origin is zero. All of the unscaled finite-support special functions that are defined with unit area and unit amplitude at the origin have unit equivalent width, including $RECT[x]$, $SINC[x]$, $SINC^2[x]$, $TRI[x]$, and $GAUS[x]$. If scaled by a width parameter b, the equivalent width of each of these functions is identical to b. The equivalent widths of the other special functions are scaled by some factor particular to the function; for example, the equivalent width of the real-valued chirp function $\cos[\pi x^2]$ is $(\sqrt{2})^{-1}$.

A corresponding metric evaluated for the spectrum $F[\xi]$ may be called the *equivalent spatial bandwidth*:

$$\Delta\xi_f \equiv \frac{\int_{-\infty}^{+\infty} F[\xi]\, dx}{F[0]} \tag{13.113}$$

Similar expressions for temporal functions $f[t]$ define the equivalent duration and equivalent temporal bandwidth. Clearly the equivalent width in the space domain is indeterminate for any function with $f[0] = 0$, while the equivalent bandwidth is not defined when $F[0] = 0$. An example of a function with an undefined equivalent width is $f[x] = \delta\delta[x]$.

13.4.2 Uncertainty Relation for Equivalent Width

If the equivalent widths of both representations of a function exist, then the central ordinate theorems in the two domains in Equations (9.113) and (9.114) may be combined to generate a simple relationship between these two metrics:

$$\Delta x_f \equiv \frac{\int_{-\infty}^{+\infty} f[x]\, dx}{f[0]} = \frac{F[0]}{f[0]} = \frac{F[0]}{\int_{-\infty}^{+\infty} F[\xi]\, d\xi} = (\Delta\xi_f)^{-1} \tag{13.114}$$

This relation may be recast into an invariant:

$$\Delta x_f \cdot \Delta\xi_f = 1 \tag{13.115}$$

This implies that scaling of the equivalent width of $f[x]$ by a factor of b must be compensated by the reciprocal scaling of $\Delta\xi_f$; if $f[x]$ gets wider, than the spectrum $F[\xi]$ must get narrower. Another interpretation of Equation (13.115) is that a function may not be localized to arbitrary precision in both domains simultaneously.

To illustrate the uncertainty relation for equivalent widths, consider the Fourier transform pair for the quadratic-phase function:

$$f[x] = e^{\pm i\pi(x/b)^2} \tag{13.116a}$$

which has unit amplitude at the origin and area equal to $|b|(e^{\pm i\pi/4})$, as may be shown by applying the central-ordinate theorem from Section 9.8.4 to the spectrum of the chirp that was derived in Section 9.7.7. The spectrum is:

$$F[\xi] = |b|\, e^{\pm i\pi/4}\, e^{\mp i\pi(b\xi)^2} \tag{13.116b}$$

which has unit area and central ordinate equal to $|b|(e^{\pm i\frac{\pi}{4}})$. The corresponding equivalent widths are:

$$\Delta x_f = \frac{|b|\, e^{\pm i\pi/4}}{1} \tag{13.117a}$$

$$\Delta\xi_f = \frac{1}{|b|\, e^{\pm i\pi/4}} \tag{13.117b}$$

which clearly satisfy Equation (13.115).

13.4.3 Variance as a Measure of Width

Many cases exist where the equivalent width is not defined in either or both domains. In these circumstances, other measures of the localization of the function sometimes may be constructed that exist and are finite. The corresponding uncertainty relation will have a different form than Equation

(13.115). One such useful alternative width metric is the second central moment (the variance) of $f[x]$ that was defined in Equation (13.18):

$$\sigma_f^2 = \frac{\mathrm{m}_2}{\mathrm{m}_0} - \left(\frac{\mathrm{m}_1}{\mathrm{m}_0}\right)^2 = \frac{1}{4\pi^2}\left(-\frac{F''[0]}{F[0]} + \left(\frac{F'[0]}{F[0]}\right)^2\right) \tag{13.118}$$

The "transform-of-a-transform" theorem in Equation (9.104) demonstrates that the corresponding width metric in the frequency domain is:

$$\sigma_F^2 = \frac{1}{4\pi^2}\left(-\frac{f''[0]}{f[0]} + \left(\frac{f'[0]}{f[0]}\right)^2\right) \tag{13.119}$$

The appropriate uncertainty relation is the product of these two quantities. Note that these metrics are defined only when $f[0]$ and $F[0]$ are both nonzero and $f'[0]$, $F'[0]$, $f''[0]$, and $F''[0]$ are finite. Since these constraints are not satisfied for any function that is discontinuous at the origin, the variance is not a useful width metric for many functions.

The use of the variance as a width metric will be demonstrated for the rectangle and Gaussian functions. It is easy to evaluate $RECT[x/b]$ and its first two derivatives at the origin:

$$\begin{aligned} f[x] &= RECT\left[\frac{x}{b}\right] \\ f[0] &= 1 \\ f'[0] &= 0 \\ f''[0] &= 0 \end{aligned} \tag{13.120a}$$

The first two derivatives of the Fourier transform evaluated at the origin are:

$$\begin{aligned} F[\xi] &= |b|\, SINC[b\xi] \\ F[0] &= |b| \\ F'[0] &= 0 \\ F''[0] &= -\frac{\pi^2|b|^3}{3} \end{aligned} \tag{13.120b}$$

Direct substitution of Equation (13.120b) into Equation (13.118) yields the variance of the rectangle function in the space domain:

$$\sigma_f^2 = \frac{b^2}{12} \tag{13.121a}$$

as was already demonstrated in Equation (13.20). However, substitution of the values in Equation (13.120) into Equation (13.119) demonstrates that the variance of the *SINC* function is zero:

$$\sigma_F^2 = 0 \tag{13.121b}$$

and therefore cannot be used to create a useful uncertainty relation for $RECT[x/b]$.

Now consider the variance of the Gaussian function and its Gaussian spectrum:

$$f[x] = e^{-\pi(x/b)^2} \tag{13.122a}$$

$$F[\xi] = |b|\, e^{-\pi(b\xi)^2} \tag{13.122b}$$

The central ordinates and first two derivatives of these functions are easy to evaluate at the origin:

$$\begin{aligned} f[0] &= 1 & F[0] &= |b| \\ f'[0] &= 0 & F'[0] &= 0 \\ f''[0] &= -2\pi/b^2 & F''[0] &= -2\pi|b|^3 \end{aligned} \tag{13.123}$$

The variances are evaluated by substitution into Equations (13.118) and (13.119):

$$\sigma_f^2 = \frac{b^2}{2\pi} \tag{13.124a}$$

$$\sigma_F^2 = \frac{1}{2\pi b^2} \tag{13.124b}$$

Note that the first of these agrees with the relationship between the width parameter b and the variance σ_f^2 that was determined in Equation (6.61). The product of these two width metrics is the numerical factor in the uncertainty relation:

$$\sigma_f^2 \cdot \sigma_F^2 = \left(\frac{1}{2\pi}\right)^2 \tag{13.125}$$

which is an important statement of the uncertainty relation in quantum mechanics.

PROBLEMS

13.1 Find an expression for the moments of $f[x] = \delta[x - x_0]$ and show that they produce the correct spectrum for $x_0 = 0$ and $x_0 = 1$.

13.2 Evaluate the moments and plot estimates of the spectrum of the following functions using the first five nonzero moments:

(a) $RECT[x/2]$

(b) $RECT[x - \frac{1}{2}]$

13.3 Determine the centroid of $RECT[x - 1]$ via direct integration and by applying the moment theorem.

13.4 Derive expressions for the centroid of a 2-D function $f[x, y]$.

13.5 Consider the real-valued, bipolar, and symmetric function $f[x]$ and its magnitude $g[x] = |f[x]|$. The respective spectra are $F[\xi]$ and $G[\xi]$.

(a) Show that the variances of $F[\xi]$ and $G[\xi]$ are equal.

(b) Find the condition that must be satisfied for the mean values of $F[\xi]$ and $G[\xi]$ to be equal. (ANSWER: $f[0] = |f[0]| \Longrightarrow f[0] > 0$)

13.6 Consider the 1-D function:

$$f[x] = RECT\left[\frac{x - 2.5}{5}\right] \exp[+i\pi x^2]$$

(a) Try to evaluate $f[x] * f[x]$. (I dare you!)

(b) Evaluate the stationary-phase approximation $\hat{F}[\xi]$ of the Fourier transform of $f[x]$.

(c) Use the result of part (b) to find an approximation for the convolution of $f[x]$ with itself.

(d) Use this result to approximately evaluate $f[x] * f^*[-x] = f[x] \bigstar f[x]$.

(e) Evaluate the Fourier transform of $\Re\{f[x]\}$.

13.7 Find expressions for the moments of the following functions and use them to evaluate the areas, mean values, and variances:

(a) $f[x] = SINC[x]$

(b) $g[x] = SINC^2[x]$

(c) $h[x] = STEP[x] \cdot \frac{1}{b} \exp\left[-\frac{x}{b}\right]$

(d) $s[x] = \frac{1}{\pi} \frac{a}{a^2 + x^2}$

(e) $t[x] = \cos\left[2\pi \frac{x}{2}\right] RECT[x]$

13.8 Equation (13.35) states that the initial phase of the superchirp of order n is $\pm\pi/2n$. Prove or disprove this result for $n = 1$.

13.9 Demonstrate the validity of Equation (13.110):

$$F[\xi] \cong (G[\xi])^{1/N} = (A\,|d|)^{1/N}\, e^{-\pi(d\xi)^2/N}$$
$$= (A|d|)^{1/N}\left(1 - \pi\frac{(d\xi)^2}{N} + \pi^2\frac{(d\xi)^4}{2N} + \cdots\right)$$

13.10 Use the expression for the variance of the rectangle and Gaussian functions to show that the N-fold convolution of $RECT[x/b]$ has the same variance as the approximation from the central-limit theorem.

13.11 Derive the moments of the 1-D superGaussian function in Equation (13.28):

$$\mathrm{m}_\ell\{e^{-\pi|x|^n}\} = \begin{cases} \frac{2}{n}\pi^{-(\ell+1)/n}\Gamma\left[\frac{\ell+1}{n}\right] & \text{for even } \ell \\ 0 & \text{for odd } \ell \end{cases}$$

14

Discrete Systems, Sampling, and Quantization

Up to this point in the discussion, both the space-domain function and its spectrum have been defined over continuous infinite domains. We have been able to derive closed-form solutions of 1-D Fourier transforms for only a few such functions, even after applying the theorems and/or considering superpositions of functions with known spectra. For this reason, it is rare that the spectrum of an arbitrary function in a real physical problem may be evaluated directly. This limitation on the applicability of frequency-domain methods in real-world problems provides the motivation for the next two chapters. The development of the discrete case will utilize notation based on the continuous case.

The practical application of Fourier methods is usually based on machine computation of spectra of sampled approximations of the original continuous signals via the so-called *discrete Fourier transform* (DFT). Discrete computation of Fourier spectra is now so common that many users forget (or never knew) that it was considered a very formidable task until the "fast Fourier transform" (FFT) was "invented" (or "rediscovered", in the minds of some observers) in the mid-1960s. Yet another two decades or so passed before the necessary computing capability became widely available to evaluate discrete spectra of "usefully large" arrays, but it now is possible to compute useful spectra on inexpensive personal computers. The very fact that the DFT is now common and easy to generate makes it even more essential to understand the differences between the continuous and discrete cases. Some of the distinctions are apparent at first glance, while others have more subtle impact.

The first step in the study of discrete linear systems is to thoroughly understand the process by which input signals (or images) defined over continuous domains are converted to approximations defined over discrete domains that have finite support. The conversion from continuous to discrete domains is called *sampling* and will be shown to be linear and shift variant. The equally important (though often less emphasized) conversion from discrete samples back to a continuous function is called *interpolation* because it "fills in" the continuous amplitudes between the samples of the function.

The amplitudes of a sampled function are still defined over a continuous range, but may be "remapped" to a discrete set of amplitudes by a nonlinear and shift-invariant process known as *quantization*. The effects of quantization on the signal will be discussed briefly at the end of the chapter. The particular amplitude of a quantized signal is specified by its integer index, which is sometimes called the *quantum number* or *gray value* of that amplitude.

An input signal that has been subjected to shift-variant sampling and nonlinear quantization results in an array of numbers that may be represented in a digital computer. Hence, the cascade of these two processes is often called *digitizing*, and the result is a *digital signal*.

Fourier Methods in Imaging Roger L. Easton, Jr.

This discussion of digital Fourier analysis will continue in Chapter 15, where the discrete transformations are considered, including the DFT and its now very common relative, the *discrete cosine transform* (DCT). Descriptions of the discrete versions of the theorems of the Fourier transform, efficient algorithms for computation, and practical considerations for applying the analysis are described in Chapter 15.

14.1 Ideal Sampling

Given a 1-D input function of continuous coordinates $f[x]$, the process of *sampling* specifies a discrete set of data at uniformly spaced intervals of the domain (sampling at nonuniform spacings may also be defined, but is not considered here). If we are lucky, the discrete function resembles the original in some sense. The mathematical expression for the operation of uniform sampling is the product of the continuous ("input") function $f[x]$ and a function that "measures" the input amplitude at discrete locations separated by a fixed interval, say Δx. The periodic sampling function will be denoted by $s[x;\ \Delta x]$, where the semicolon sets off the parameter Δx from the coordinate x (the corresponding spectrum is a function of spatial frequency whose notation also includes the parameter Δx, e.g., $S[\xi;\ \Delta x]$). The product $f[x] \cdot s[x;\ \Delta x]$ generates a third function containing the set of amplitudes at the sample locations $x = n \cdot \Delta x$, where n is an integer. The sampled function may be viewed as an "approximation" of the continuous function and is denoted by the same letter, but with a subscript "s" to emphasize that the amplitude of the sample may not be identical to the amplitude f of the original continuous function at the same coordinate x. Because sampling produces a new function, it is reasonable to denote the process by an operator $\boldsymbol{\mathcal{S}}$ acting on $f[x]$:

$$\boldsymbol{\mathcal{S}}\{f[x]\} = f[x] \cdot s[x;\ \Delta x] \equiv f_s[x;\ \Delta x] \tag{14.1}$$

The domain of the sampled function $f_s[x;\ \Delta x]$ may still be interpreted as being continuous because an amplitude f_s is specified at each coordinate in the continuous domain; the null amplitude $f_s = 0$ is assigned to all values of x that are noninteger multiples of Δx. Obviously, the amplitude of the sampling function $s[x;\ \Delta x]$ at the integer multiples of Δx determines the character of $f_s[x,\ \Delta x]$.

To illustrate these concepts, first consider an apparently logical definition for the sampling function that contains a set of isolated unit amplitudes located at integer multiples of Δx:

$$s_1[x;\ \Delta x] \equiv \begin{cases} 1 & \text{if } x = n \cdot \Delta x \\ 0 & \text{if } x \neq n \cdot \Delta x \end{cases} \quad (\text{for } n = 0, \pm 1, \pm 2, \ldots) \tag{14.2}$$

Obviously, the resulting sampled function is composed of the amplitudes of $f[x]$ at the specified coordinates:

$$(f_s[x;\ \Delta x])_1 = \begin{cases} f[n \cdot \Delta x] & \text{if } x = n \cdot \Delta x \\ 0 & \text{if } x \neq n \cdot \Delta x \end{cases} \tag{14.3}$$

The samples so defined have the nice property that their amplitudes are identical to those of $f[x]$ at those locations. The discussion surrounding Equation (6.89) demonstrates that this ensemble of samples has zero area, and thus the sampled function in Equation (14.3) differs from the original function $f[x]$ in a very important sense. A 1-D function with null area cannot interact with the physical world, so expression in Equation (14.3) is not very useful.

An alternative sampling strategy generates samples with finite areas that are "approximately equal" to the "local area" of the original continuous function $f[x]$. Consider a sampling function $s[x;\ \Delta x]$ composed of unit-area Dirac delta functions spaced at intervals Δx; that is, a *COMB* function defined in Equation (6.124):

$$s_{\text{ideal}}[x;\,\Delta x] = \sum_{n=-\infty}^{+\infty} \delta[x - n\cdot\Delta x] \equiv \frac{1}{\Delta x} COMB\left[\frac{x}{\Delta x}\right]$$
$$= \begin{cases} \delta[x - n\cdot\Delta x] & \text{if } x = n\cdot\Delta x \\ 0 & \text{if } x \neq n\cdot\Delta x \end{cases} \qquad (n = 0, \pm 1, \pm 2, \ldots) \tag{14.4}$$

The scaling factor $(\Delta x)^{-1}$ ensures that each Dirac delta function in the *COMB* function has unit area. The notation s_{ideal} is used to emphasize that Equation (14.4) defines the *ideal sampling function* for reasons that will be clear shortly. The product of the continuous input function $f[x]$ and the ideal sampling function $s_{\text{ideal}}[x;\,\Delta x]$ may be recast into a different and convenient form by applying the properties of the Dirac delta function that were described in Section 6.2:

$$(f_s[x;\,\Delta x])_2 = f[x]\cdot\frac{1}{\Delta x} COMB\left[\frac{x}{\Delta x}\right]$$
$$= \sum_{n=-\infty}^{+\infty} f[x]\cdot\delta[x - n\cdot\Delta x]$$
$$= \sum_{n=-\infty}^{+\infty} f[n\cdot\Delta x]\cdot\delta[x - n\cdot\Delta x]$$
$$= f[n\cdot\Delta x]\cdot\frac{1}{\Delta x} COMB\left[\frac{x}{\Delta x}\right] \tag{14.5}$$

where the property of Dirac delta functions in products in Equation (6.115) has been applied. The set of amplitudes $\{f[n\cdot\Delta x]\}$ that scale the elements of the *COMB* function are identical to the continuous function $f[x]$ evaluated at the integer multiples of Δx. This set $\{f[n\cdot\Delta x]\}$ is identical to the set of amplitudes that was generated by the naive "sampling function" $s_1[x;\,\Delta x]$ in Equation (14.2).

We will use the definition of Equation (14.5) for ideal sampling in this discussion from this point forward, so the subscript notation "ideal" that distinguished the recipe from Equation (14.3) will be dropped.

Though the ensemble of amplitudes $f[n\cdot\Delta x]$ is often called the set of *samples* of $f[x]$, readers should be careful to recognize that it only specifies the set of weights applied to the *COMB* function; it is not the ideally sampled function by itself.

14.1.1 Ideal Sampling of 2-D Functions

The sampling process is easily extended to 2-D functions by using the 2-D *COMB* function to generate a 2-D array of modulated Dirac delta functions:

$$f_s[x, y;\,\Delta x, \Delta y] = f[x, y]\cdot s[x, y;\,\Delta x, \Delta y]$$
$$= f[x, y]\cdot\left(\frac{1}{\Delta x} COMB\left[\frac{x}{\Delta x}\right]\right)\cdot\left(\frac{1}{\Delta y} COMB\left[\frac{y}{\Delta y}\right]\right)$$
$$= f[n\cdot\Delta x, m\cdot\Delta y]\cdot\left(\frac{1}{\Delta x\cdot\Delta y} COMB\left[\frac{x}{\Delta x}, \frac{y}{\Delta y}\right]\right) \tag{14.6}$$

The 2-D array $f[n\cdot\Delta x, m\cdot\Delta y]$ again specifies the weights applied to the unit volumes of the 2-D Dirac delta functions located at each *picture element*, or *pixel*. Often the sample spacings in the two directions are equal: $\Delta x = \Delta y$.

14.1.2 Is Sampling a Linear Operation?

At this point, it is useful to consider the characteristics of sampling, including the area and region of support of the sampled function, and whether sampling is a linear and shift-invariant operation. By analogy with the definition of the support of a continuous function in Chapter 2, the region of support of a sampled function may be defined as the interval between the samples with nonzero amplitude at the "ends" of the function, so that $f[n] \neq 0$ for $n = n_{\max}$ and for $n = n_{\min}$, and $f[n] = 0$ for $n > n_{\max}$ and for $n < n_{\min}$. The support of this sampled function may be defined as $(n_{\max} - n_{\min}) \cdot \Delta x$. It is easy to see that the unit constant $1[x]$ and its sampled version both have infinite support.

The area of the sampled function is easily shown to be the sum of the sample amplitudes:

$$\begin{aligned}\int_{-\infty}^{+\infty} \left(f[x] \cdot \frac{1}{\Delta x} COMB\left[\frac{x}{\Delta x}\right]\right) dx &= \sum_{n=-\infty}^{+\infty} \int_{-\infty}^{+\infty} f[x] \cdot \delta[x - n \cdot \Delta x]\, dx \\ &= \sum_{n=-\infty}^{+\infty} f[n \cdot \Delta x] \int_{-\infty}^{+\infty} \delta[x - n \cdot \Delta x]\, dx \\ &= \sum_{n=-\infty}^{+\infty} f[n \cdot \Delta x] \end{aligned} \tag{14.7}$$

To determine whether sampling is a linear operation, we calculate the amplitudes of the samples before and after scaling by an arbitrary complex number α. It is evident from Equation (14.7) that the weights of all Dirac delta functions in the sampled function must be scaled by this same factor:

$$\begin{aligned}\mathcal{S}\{\alpha f[x]\} &= (\alpha f[x]) \cdot \left(\frac{1}{\Delta x} COMB\left[\frac{x}{\Delta x}\right]\right) \\ &= \alpha \sum_{n=-\infty}^{+\infty} f[n \cdot \Delta x] \delta[x - n \cdot \Delta x] \\ &= \alpha f[n \cdot \Delta x] \cdot \left(\frac{1}{\Delta x} COMB\left[\frac{x}{\Delta x}\right]\right)\end{aligned} \tag{14.8}$$

This confirms that sampling is a linear operation in this simple example, as shown in Figure 14.1.

14.1.3 Is the Sampling Operation Shift Invariant?

To determine the character of the sampling operation under translation, we apply it to two functions: the unit constant and rectangle functions.

14.1.3.1 Sampling of $f[x] = 1[x]$

The set of ideal samples obtained from the 1-D unit constant function is obtained by inserting $f_1[x] = 1[x]$ into Equation (14.5), which generates the sampling *COMB* function:

$$\begin{aligned}\mathcal{S}\{1[x]\} &= 1[x] \cdot \frac{1}{\Delta x} COMB\left[\frac{x}{\Delta x}\right] \\ &= \sum_{n=-\infty}^{+\infty} 1[n \cdot \Delta x]\, \delta[x - n \cdot \Delta x] \\ &= 1[n \cdot \Delta x] \cdot \frac{1}{\Delta x} COMB\left[\frac{x}{\Delta x}\right] = \frac{1}{\Delta x} COMB\left[\frac{x}{\Delta x}\right]\end{aligned} \tag{14.9}$$

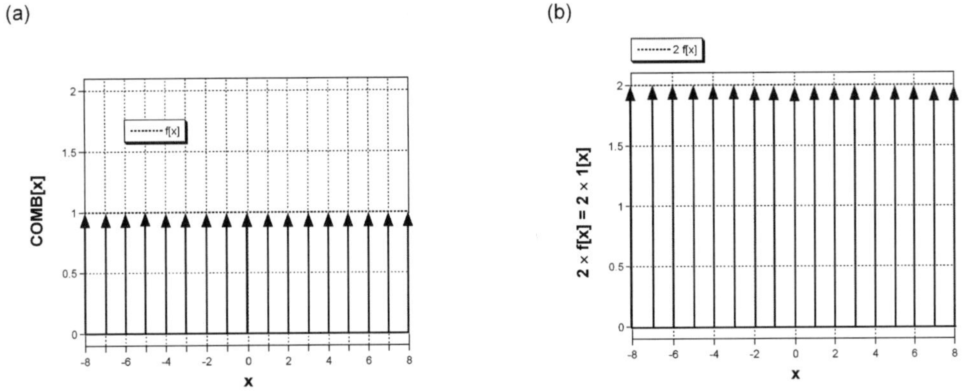

Figure 14.1 Illustration that sampling operation is linear. The continuous function $f[x] = 1[x]$. If sampled at intervals $\Delta x = 1$ unit, the result is $COMB[x]$. If the continuous function is scaled by a factor of 2, so is the sampled function, thus confirming that sampling is a linear operation.

Both the original unit constant and its samples obviously have infinite area, which may be shown to be identical by recognizing that each has the same area within subregions of width Δx. The areas of both the continuous and sampled functions within each subregion are identically unity, and so we can surmise that the unit constant and its discrete version sampled at unit intervals have the same area.

The discussion of operators in Chapter 8 demonstrated that the samples obtained from a translated copy of the input function must be identical to the translated samples of the input if the process is shift invariant. Since the areas of all Dirac delta functions generated by sampling the unit constant are identically unity, the requirement for shift invariance is met for the unit constant.

14.1.3.2 Sampling of $f[x] = RECT[x/b]$

In the second example, the input function is a unit-amplitude rectangle of width $4 \cdot \Delta x$ centered at the origin. Both the area and support of the continuous function are $4 \cdot \Delta x$. After sampling with the ideal *COMB* function, we obtain:

$$\begin{aligned}\mathcal{S}\left\{RECT\left[\frac{x}{4\cdot\Delta x}\right]\right\} &= RECT\left[\frac{x}{4\cdot\Delta x}\right]\cdot\frac{1}{\Delta x}COMB\left[\frac{x}{\Delta x}\right]\\ &= \sum_{n=-\infty}^{+\infty} RECT\left[\frac{x}{4\cdot\Delta x}\right]\delta[x-n\cdot\Delta x]\\ &= RECT\left[\frac{n\cdot\Delta x}{4\cdot\Delta x}\right]\cdot\frac{1}{\Delta x}COMB\left[\frac{x}{\Delta x}\right]\\ &= RECT\left[\frac{n}{4}\right]\cdot\frac{1}{\Delta x}COMB\left[\frac{x}{\Delta x}\right]\\ &= \frac{1}{2}\delta[x+2\cdot\Delta x]+\delta[x+\Delta x]+\delta[x]+\delta[x-\Delta x]+\frac{1}{2}\delta[x-2\cdot\Delta x]\end{aligned} \tag{14.10}$$

as shown in Figure 14.2a. Note that the weights of the "endpoint" samples of the discrete approximation to the rectangle are $\frac{1}{2}$ at $n = \pm 2$. The width of the region of support of this function is four units. The area of the sampled function is the sum of the weights applied to the unit-area Dirac delta functions:

$$\begin{aligned}
&\int_{-\infty}^{+\infty} \mathcal{S}\left\{RECT\left[\frac{x}{4\cdot\Delta x}\right]\right\} dx \\
&= \int_{-\infty}^{+\infty} \left(\frac{1}{2}\delta[x+2\cdot\Delta x] + \delta[x+\Delta x] + \delta[x] + \delta[x-\Delta x] + \frac{1}{2}\delta[x-2\cdot\Delta x]\right) dx \\
&= \frac{1}{2} + 1 + 1 + 1 + \frac{1}{2} = 4
\end{aligned} \tag{14.11}$$

To test the sampling operation for shift invariance, the rectangle function is translated by some arbitrary distance x_1 before sampling. The general expression for the sampled function is:

$$\begin{aligned}
\mathcal{S}\left\{RECT\left[\frac{x-x_1}{4\cdot\Delta x}\right]\right\} &= \sum_{n=-\infty}^{+\infty} RECT\left[\frac{x-x_1}{4\cdot\Delta x}\right]\delta[x-n\cdot\Delta x] \\
&= \sum_{n=-\infty}^{+\infty} RECT\left[\frac{(n\cdot\Delta x)-x_1}{4\cdot\Delta x}\right]\delta[x-n\cdot\Delta x] \\
&= RECT\left[\frac{(n\cdot\Delta x)-x_1}{4\cdot\Delta x}\right]\left(\sum_{n=-\infty}^{+\infty} \delta[x-n\cdot\Delta x]\right) \\
&= RECT\left[\frac{n}{4} - \frac{x_1}{4\cdot\Delta x}\right]\cdot\frac{1}{\Delta x}COMB\left[\frac{x}{\Delta x}\right]
\end{aligned} \tag{14.12}$$

The sampled function obtained for $x_1 = -0.25\cdot\Delta x$ is shown in Figure 14.2b. The coordinates of the endpoints of the translated continuous rectangle become $x_1 \pm 2\cdot\Delta x = -2.25\cdot\Delta x$ and $+1.75\cdot\Delta x$. Because these "endpoint" amplitudes are not located at integer multiples of Δx, no half-unit sample weights appear in Equation (14.12); the sampled function consists of four unit-area Dirac delta functions located at $x = -2\cdot\Delta x, -\Delta x, 0$, and $+\Delta x$. The aggregate area of the Dirac delta functions is 4, which means that the areas of the translated continuous function and its sampled "replica" still are equal. However, the support of the sampled translated function in Equation (14.12) is only $3\cdot\Delta x$, which is less than that of the continuous function. Obviously the sample amplitudes derived from the translated function differ from those that would be obtained by translating the original sample amplitudes:

$$\mathcal{S}\{f[x]\} = f_s[x;\Delta x] \Longrightarrow \mathcal{S}\{f[x-x_1]\} \neq f_s[x-x_1;\Delta x] \tag{14.13}$$

which means that sampling of an arbitrary function is a shift-variant process, with the consequence that sampling may not be implemented as a convolution; this may be obvious because $\mathcal{S}$ is defined as a product.

We have demonstrated that the region of support of the sampled function may be less than or equal to that of the continuous function. It should be evident that the region of support of the sampled function can never be larger than that of the continuous function.

In the examples just considered, the areas of the continuous and sampled functions have been equal. However, other examples may be constructed to demonstrate that this is not always the case. Consider the rectangle function of width 4 sampled at intervals of width $\Delta x = b$. The area of the sampled

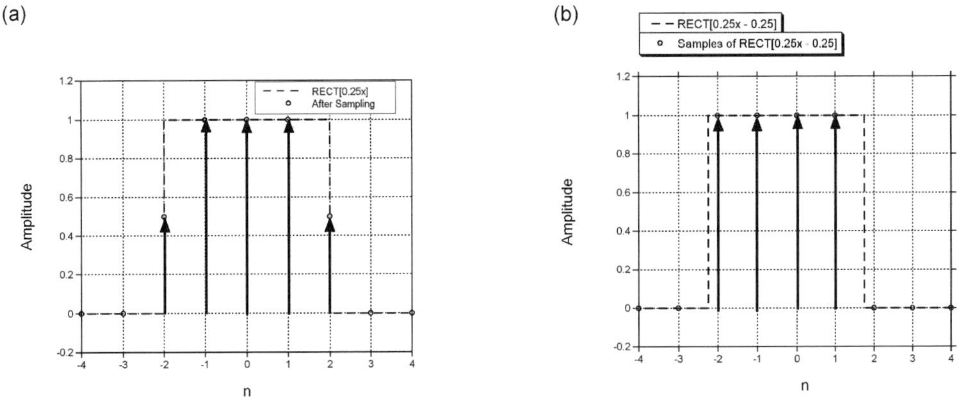

Figure 14.2 Illustration that sampling is shift variant for $f[x] = RECT[x/4]$. The samples of the original rectangle before and after translation by 0.25 units are shown. Since the two functions are not translated replicas, the process of sampling is shift variant.

function is:

$$\int_{-\infty}^{+\infty} \mathcal{S}\left\{RECT\left[\frac{x}{4}\right]\right\} dx = \int_{-\infty}^{+\infty} \left(\sum_{n=-\infty}^{+\infty} RECT\left[\frac{x}{4}\right] \delta[x - nb] \right) dx$$
$$= \sum_{n=-\infty}^{+\infty} RECT\left[\frac{nb}{4}\right] \int_{-\infty}^{+\infty} \delta[x - nb]\, dx$$
$$= \sum_{n=-\infty}^{+\infty} RECT\left[\frac{nb}{4}\right] \cdot 1 = \frac{4}{b} \tag{14.14}$$

This area is larger or smaller than that of $RECT[x/4]$ when b is respectively smaller than or larger than unity.

14.1.4 Aliasing Artifacts

This section considers the most important artifact of ideal sampling. We note first that this discussion is entirely academic; ideal sampling of "real" signals cannot be implemented in actual practice because it would require measurement of the finite amplitude of $f[x]$ over the infinitesimal support of the Dirac delta function. In the context of an imaging application, a finite "amount" of light must be measured over an infinitesimal area of the scene, which is not possible. Even so, a discussion of ideal sampling introduces some essential concepts and is useful for developing some intuition about realistic sampling.

Consider a cosine function with period X whose amplitude has been "biased up" and scaled to fit in the interval [0, 1]. The function is sampled at intervals Δx as shown in Figure 14.3:

$$f[x] = \frac{1}{2}\left(1 + \cos\left[\frac{2\pi x}{X} + \phi_0\right]\right)$$
$$\Longrightarrow f_s[x;\, \Delta x] = \frac{1}{2}\left(1 + \cos\left[\frac{2\pi x}{X} + \phi_0\right]\right) \cdot \left(\frac{1}{\Delta x} COMB\left[\frac{x}{\Delta x}\right]\right) \tag{14.15}$$

The action of sampling evaluates the amplitude factor applied to the unit-area Dirac delta functions, which may be evaluated easily at a particular coordinate $x = n_0 \cdot \Delta x$:

$$f_s[x = n_0 \cdot \Delta x; \Delta x] = \frac{1}{2}\left(1 + \cos\left[\frac{2\pi x}{X} + \phi_0\right]\right)\delta[x - n_0 \cdot \Delta x]$$
$$= \frac{1}{2}\left(1 + \cos\left[2\pi n_0 \frac{\Delta x}{X} + \phi_0\right]\right)\delta[x - n_0 \cdot \Delta x] \quad (14.16)$$

The dimensionless parameter $\Delta x/X$ is the ratio of the length of the sampling interval to the spatial period ("wavelength") of the continuous sinusoid. This parameter may be used to establish a metric of "fidelity" of the sampled image, as may be demonstrated by example. Sets of samples obtained from a sinusoid for different values of $\Delta x/X$ are shown in Figure 14.3. The variation of the sinusoid is most apparent when $\Delta x/X$ is small, and the sample amplitudes obtained for $\Delta x/X = \frac{1}{2}$ depend upon the initial phase of the sinusoid. This effect may be demonstrated by considering the "dynamic range" of the sample amplitudes, which may be defined as the difference between the maximum and minimum amplitudes within the set of sample weights. When the initial phase $\phi_0 = 0$ radians, the continuous function is a symmetric cosine, and the maximum and minimum weights are unity and zero. When the initial phase $\phi_0 = \pm\pi/2$ radians, the continuous function is an odd sinusoid and the weights of all samples are identically $\frac{1}{2}$. Note that the range may take on any intermediate value by appropriate selection of the initial phase ϕ_0.

Note the cases in Figure 14.3 where the sample spacing Δx is less than $X/2$, so that the dimensionless sampling parameter exceeds $\frac{1}{2}$. The resulting sample weights applied to the Dirac delta functions are identical to those that would have been obtained by sampling sinusoids with periods $X' > X$ at the same sample spacing Δx. In the example presented, the same set of weights is obtained for a sinusoid with period X and sampling parameter $\Delta x/X = \frac{3}{4}$ and for a sinusoid with period $X' = 3X$ and sampling parameter $\Delta x/X' = \frac{1}{4}$. Because the sampled data are ambiguous, the different input functions cannot be distinguished from the identical samples. However, we will see that if the sampling parameter $\Delta x/X$ is guaranteed to be less than the limiting value of one-half cycle per sample interval Δx for a sinusoidal input, then the set of sample weights is unique and thus may be used to uniquely specify the input function. Later, we will show how the continuous input function may be reconstructed from the complete set of unambiguous sample weights by "filling in the spaces", i.e., by interpolating amplitudes between the samples. If the function reconstructed from the samples does not match the original continuous function, then the sampled data have been *aliased*.

The limiting value for the sampling parameter defines the size of the *Nyquist sampling interval* for that particular sinusoid, which we will denote by $(\Delta x)_{\text{Nyquist}}$:

$$\frac{(\Delta x)_{\text{Nyquist}}}{X_0} = \frac{1}{2}$$
$$\Longrightarrow (\Delta x)_{\text{Nyquist}} = \frac{X_0}{2}\ \textit{units} = \frac{1}{2\xi_0} \quad (14.17)$$

where X_0 and ξ_0 are respectively the period and spatial frequency of the continuous sinusoid. This expression may be recast into a definition of the *Nyquist sampling frequency* ξ_{Nyquist}:

$$\xi_{\text{Nyquist}} = \frac{1}{(\Delta x)_{\text{Nyquist}}} = 2\xi_0 \frac{\textit{cycles}}{\textit{unit length}} \quad (14.18)$$

In words, the Nyquist sampling frequency is twice the spatial frequency of the continuous sinusoid. For emphasis, we repeat that the Nyquist sampling interval (or frequency) is the *limiting value*; if samples are generated at exactly this rate, the input function can only be reconstructed without error in special cases.

The Nyquist sampling rate also may be specified in terms of the phase of the spatially oscillating sinusoid that elapses within each sampling interval Δx. Since each period of a sinusoid spans 2π radians of phase, a spatial frequency of one-half cycle per sample corresponds to π radians of phase per sample:

$$\xi_{\text{Nyquist}} = \left(\frac{1}{2}\,\frac{\textit{cycle}}{\textit{sample}}\right) \cdot \left(2\pi\,\frac{\textit{radians}}{\textit{cycle}}\right) = \pi\,\frac{\textit{radians}}{\textit{sample}} \tag{14.19}$$

From the examples in Figure 14.3, it may seem that the correct value of an aliased frequency can be determined from knowledge of the original spatial frequency ξ_0 and the sampling parameter $\Delta x/X$. However, this is not true as will be demonstrated shortly.

Consider the case where $f[x]$ is a superposition of sinusoidal components where the most rapidly oscillating sinusoid has spatial frequency $\xi_{\max}$. The corresponding minimum spatial period is $X_{\min} = (\xi_{\max})^{-1}$. Clearly, if this minimum period sinusoid is sampled without aliasing, then all sinusoidal components of $f[x]$ will be. Any function that has a maximum spatial frequency is said to be *bandlimited*, which means that it satisfies the condition:

$$\textit{bandlimited function} \Longrightarrow F[\xi] = 0 \;\; \textit{for}\; |\xi| > |\xi_{\max}| \tag{14.20}$$

Any bandlimited function may be reconstructed without aliasing from the complete set of sample weights obtained at intervals $\Delta x < X_{\min}/2 = 1/2\xi_{\max}$. This result may be generalized to consider nonbandlimited functions, i.e., those for which $\xi_{\max} = +\infty$. Obvious examples are the Dirac delta function $\delta[x]$ with spectrum $1[\xi]$ and the rectangle function $RECT[x]$ with spectrum $SINC[\xi]$. The amplitudes of both spectra are nonzero out to $\xi = +\infty$, which means that the largest possible value of Δx that may be used to sample the functions without aliasing is $\Delta x = 0$. Evidently, any set of samples of a function whose spectrum has infinite support will be aliased.

14.1.5 Operations Similar to Ideal Sampling

An analogue of the ambiguity that occurs when a sinusoid is sampled with fewer than two samples per period occurs whenever periodic functions are multiplied or added. In other disciplines, these effects have been given names such as *beats* and *Moiré fringes*.

As one example, consider the product of two sinusoidal functions with distinct periods X_1 and X_2 (and thus spatial frequencies $\xi_1 = X_1^{-1}$, $\xi_2 = X_2^{-1}$). By using Equation (4.42), the product can be recast as the sum of two sinusoids that oscillate at the sum and difference frequencies:

$$\cos[2\pi\xi_1 x] \cdot \cos[2\pi\xi_2 x] = \frac{1}{2}\cos[2\pi(\xi_1 + \xi_2)x] + \frac{1}{2}\cos[2\pi(|\xi_1 - \xi_2|)x] \tag{14.21}$$

where the symmetry of the cosine has been used. The second cosine with spatial frequency $|\xi_1 - \xi_2|$ oscillates “slowly” and is analogous to the aliased frequency that results from sampling.

14.2 Ideal Sampling of Special Functions

The samples from each of the special functions are obtained by straightforward application of Equation (14.5), though the details of the resulting arrays sometimes may not be intuitive and should be noted. One common example results from ideal sampling of the *RECT* function, which was used to demonstrate that sampling is a shift-variant operation in Figure 14.2. Consider the symmetric function $f_b[x] = RECT[x/(b \cdot \Delta x)]$ where b is an integer. The area and the support of $f_b[x]$ are both $|b| \cdot \Delta x$. Consider the case where b is even, e.g., $b = 4$, as shown in Figure 14.4a. The amplitudes of the samples with $n = 0, \pm 1$ are unity, and those for $n = \pm 2$ are $\frac{1}{2}$. The discrete analogue to the area is the product of the sample interval Δx and the sum of the sample amplitudes, which evaluates to $\Delta x \cdot (3 \cdot 1 + 2 \cdot \frac{1}{2}) = 4 \cdot \Delta x$. If we define the support of the sampled function to be the product of Δx and the difference in index of the samples with nonzero amplitudes, then the support is $+2 \cdot \Delta x - (-2 \cdot \Delta x) = 4 \cdot \Delta x$ for the case $b = 4$. In the case $b = 5$, as shown in Figure 14.4b, the area and support of $f_b[x]$ are both $5 \cdot \Delta x$. The amplitudes of the sampled function are unity for $n = 0, \pm 1$, and ± 2 and zero elsewhere,

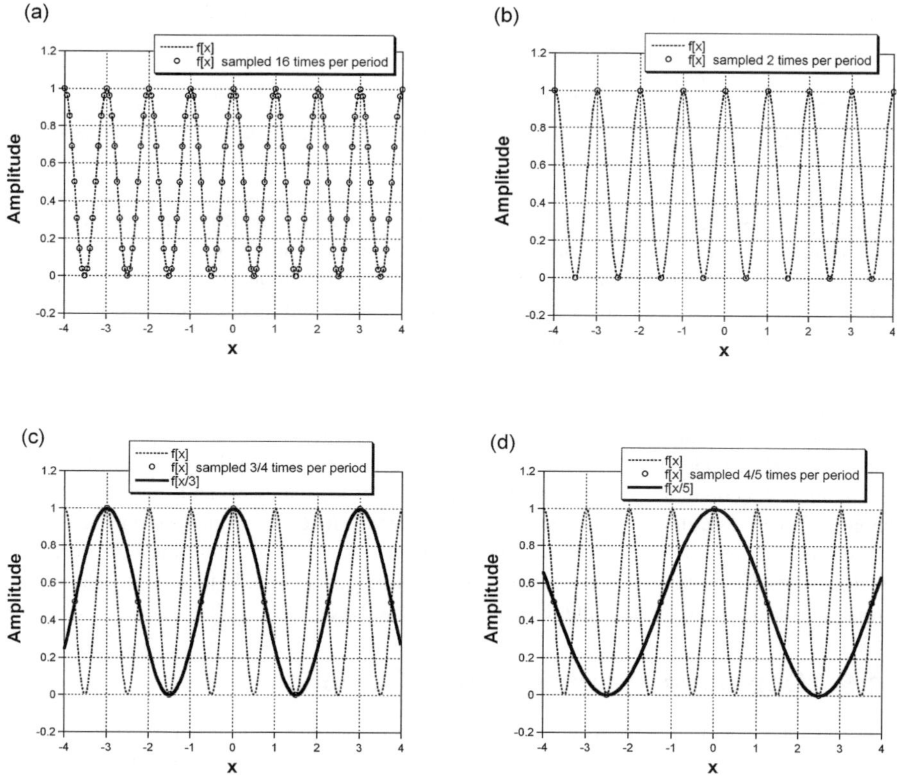

Figure 14.3 Samples of unit-modulation sinusoids with different Δx: (a) $f[x] = \frac{1}{2}(1 + \cos[2\pi x])$ sampled with $\Delta x = 1/16$, produces samples that are easily seen to be from the original sinusoid; (b) $f[x]$ with $\Delta x = 1/2$ produces samples at the extrema of amplitude; (c) sampled with $\Delta x = 4/3$, which produces the same set of samples that would be obtained from the lower-frequency function $\frac{1}{2}(1 + \cos[2\pi x/3])$; (d) $f_2[x] = \frac{1}{2}(1 + \sin[2\pi x])$ with $\Delta x = 1/2$, produces all samples with amplitude $f = \frac{1}{2}$.

which means that the endpoints of the continuous rectangle are not sampled. Since the five nonzero samples have unit amplitudes, the analogue to area for the sampled function is $5 \cdot \Delta x$. However, the same five samples have nonzero amplitudes as in the case $b = 4$, so the support again is $4 \cdot \Delta x$. In short, sampled rectangles with odd values of the scaling parameter b exhibit the same "area" as the continuous counterpart, but not the same support.

Now, consider ideal sampling of rectangle, triangle, and *SINC* functions that have identical width parameters $b = 1$. The resulting samples are easily obtained from Equation (14.5):

$$\mathcal{S}\left\{RECT\left[\frac{x}{1 \cdot \Delta x}\right]\right\} = RECT[n] \cdot \frac{1}{\Delta x} COMB\left[\frac{x}{\Delta x}\right] = 1 \cdot \delta[x] \tag{14.22a}$$

$$\mathcal{S}\left\{TRI\left[\frac{x}{1 \cdot \Delta x}\right]\right\} = TRI[n] \cdot \frac{1}{\Delta x} COMB\left[\frac{x}{\Delta x}\right] = 1 \cdot \delta[x] \tag{14.22b}$$

$$\mathcal{S}\left\{SINC\left[\frac{x}{1 \cdot \Delta x}\right]\right\} = SINC[n] \cdot \frac{1}{\Delta x} COMB\left[\frac{x}{\Delta x}\right] = 1 \cdot \delta[x] \tag{14.22c}$$

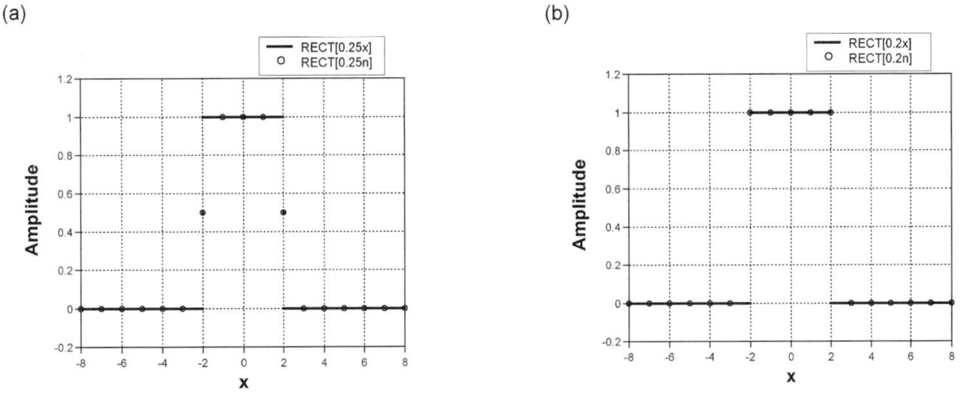

Figure 14.4 Rectangle functions sampled with $\Delta x = 1$ unit. (a) The amplitudes of the samples of $RECT[x/4]$ are $\frac{1}{2}$ at $n = \pm 2$ and unity at $n = 0, \pm 1$. The discrete analogues of the area and support of the sampled function are identically $4 \cdot \Delta x$. (b) Samples of $RECT[x/5]$ are unity for $n = 0, \pm 1, \pm 2$, and zero elsewhere. The discrete analogue to area is $5 \cdot \Delta x$ and the discrete support is $4 \cdot \Delta x$.

so the sample weights from all three functions are unity for $n = 0$ and zero elsewhere, as shown in Figure 14.5. The areas of the three continuous functions and of their discrete counterparts are identically Δx. The supports of the continuous functions are respectively Δx, $2 \cdot \Delta x$, and infinity, but the supports of the three discrete versions are all $0 \cdot \Delta x$, since all three have only one nonzero sample.

Because the same sets of sample weights result from all three functions, they may not be distinguished from their sample amplitudes alone. In other words, at least two of these functions must have been aliased by the sampling operation. The spectra of the continuous rectangle and triangle functions have infinite support, while that of the *SINC* function has support $(\Delta x)^{-1}$, which means that the first two must be aliased. The spectrum of the continuous function $SINC[x/\Delta x]$ is $\Delta x \cdot RECT[\Delta x \cdot \xi]$, which has finite support $(\Delta x)^{-1}$. The maximum frequency in its spectrum is $|\xi|_{max} = 1/(2 \cdot \Delta x)$, which means that $SINC[x/\Delta x]$ has been sampled *exactly* at its Nyquist frequency. This result provides a clue to the construction of the appropriate interpolation function used to regenerate the original function (or an "approximation") from the set of discrete samples.

While we are at it, the same sample amplitudes are obtained from the rectangle and triangle for the case $b = 2$ if both are sampled at the same interval Δx:

$$\mathcal{S}\left\{RECT\left[\frac{x}{2 \cdot \Delta x}\right]\right\} = RECT\left[\frac{n}{2}\right] \cdot \left(\frac{1}{\Delta x} COMB\left[\frac{n \cdot \Delta x}{\Delta x}\right]\right)$$
$$= \frac{1}{2}\delta[x + \Delta x] + \delta[x] + \frac{1}{2}\delta[x - \Delta x] \tag{14.23a}$$

$$\mathcal{S}\left\{TRI\left[\frac{x}{2 \cdot \Delta x}\right]\right\} = TRI\left[\frac{n}{2}\right] \cdot \left(\frac{1}{\Delta x} COMB\left[\frac{n \cdot \Delta x}{\Delta x}\right]\right)$$
$$= \frac{1}{2}\delta[x + \Delta x] + \delta[x] + \frac{1}{2}\delta[x - \Delta x] \tag{14.23b}$$

as shown in Figure 14.6. This observation means that these two functions also cannot be distinguished from their samples alone, which means that at least one must have been aliased. In fact, we already know that both functions must be aliased because the spectra of both have infinite support.

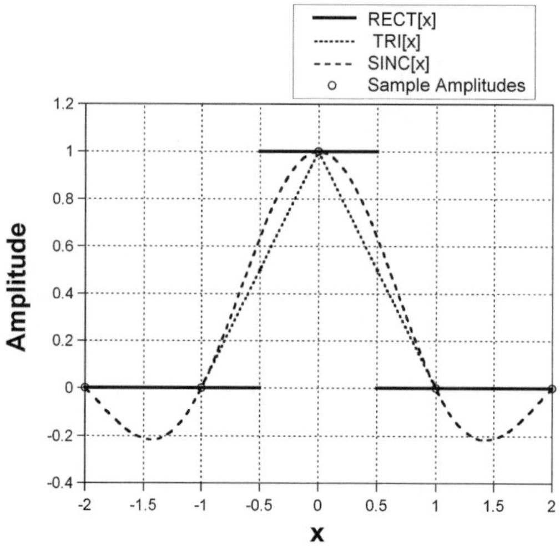

Figure 14.5 The special functions *RECT*$[x]$, *TRI*$[x]$, and *SINC*$[x]$ sampled with $\Delta x = 1$ unit all produce the same set of samples with unit amplitude at $n = 0$ and null amplitude for $n \neq 0$, indicating that at least two of these functions have been aliased by the sampling process.

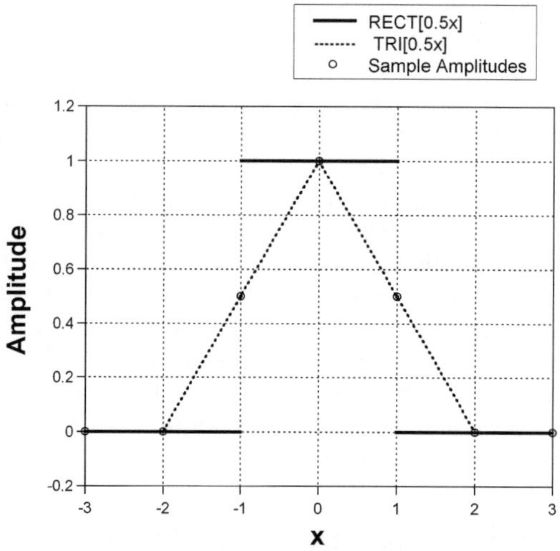

Figure 14.6 *RECT*$[x/2]$ and *TRI*$[x/2]$ sampled with $\Delta x = 1$ unit produce the same sets of sample amplitudes.

14.2.1 Ideal Sampling of $\delta[x]$ and $COMB[x]$

The continuous Dirac delta function was defined in Equation (6.88) to be any expression that satisfies two criteria:

$$\delta[x] = 0 \;\; \textit{if} \; x \neq 0 \tag{14.24a}$$

$$\int_{-\infty}^{+\infty} \delta[x] \, dx = 1 \tag{14.24b}$$

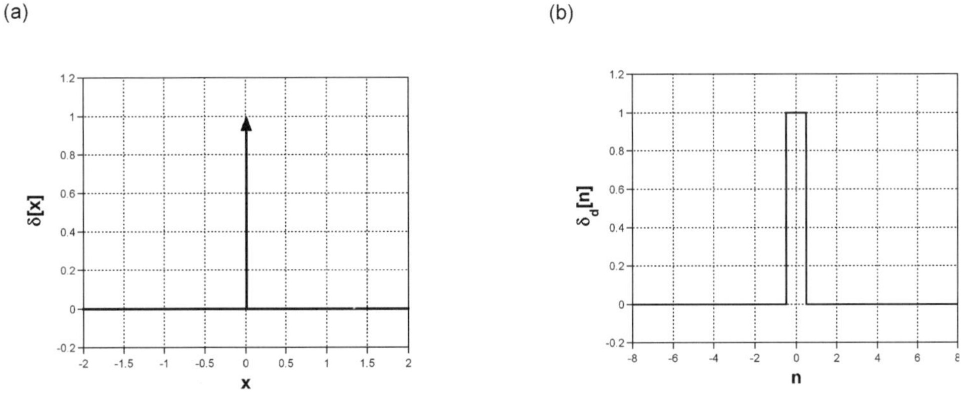

Figure 14.7 The continuous and discrete Dirac delta functions: (a) $\delta[x]$; (b) discrete delta function $\delta_d[n]$. The discrete delta function represents the weighting applied to *COMB*$[x]$ to reconstruct $\delta[x]$ at the sample coordinate.

Because its spectrum has infinite support, the Dirac delta function cannot be sampled over intervals $\Delta x > 0$ without aliasing. However, it is easy to see that a single Dirac delta function at the origin results when the sampling operation is applied to any continuous function $f[x]$ that satisfies the conditions:

$$f[x] = \begin{cases} 1 & \textit{if } n \cdot \Delta x = 0 \\ 0 & \textit{if } n \cdot \Delta x \neq 0 \end{cases} \tag{14.25}$$

where n is any integer. The sampled function is a 1-D Dirac delta function:

$$f[x] \cdot \frac{1}{\Delta x} COMB\left[\frac{x}{\Delta x}\right] = \delta[x] \tag{14.26}$$

The amplitude of $f[x]$ at any noninteger multiple of Δx is irrelevant, as the sampling process multiplies the amplitudes at these coordinates by zero. The three functions in Equation (14.22) satisfy the required conditions, as do many others. We can define a "discrete Dirac delta function" to be the sample weights:

$$\delta_d[n \cdot \Delta x - x_0] = \begin{cases} 1 & \textit{if } n \cdot \Delta x = x_0 \\ 0 & \textit{if } n \cdot \Delta x \neq x_0 \end{cases} \tag{14.27}$$

The difference between $\delta[x]$ and $\delta_d[n]$ is shown in Figure 14.7. From this point forward, graphs of sampled functions will depict only the *amplitudes* at each sample or coordinate, i.e., the set of weights applied to the Dirac delta function components of the sampled function. The Dirac delta function components of the implicit *COMB* function will not be included.

The product of this discrete function and the sampling *COMB* is a continuous Dirac delta function at the proper coordinate in all cases where x_0 is an integer multiple of Δx:

$$\delta_d[n \cdot \Delta x - x_0] \cdot \frac{1}{\Delta x} COMB\left[\frac{x}{\Delta x}\right] = \begin{cases} \delta[x - x_0] & \textit{if } n \cdot \Delta x = x_0 \\ 0 & \textit{if } n \cdot \Delta x \neq x_0 \end{cases} \tag{14.28}$$

The areas of both $\delta[x]$ and the discrete Dirac delta function that results from sampling by the *COMB* function are unity.

Just as was the case for the continuous Dirac delta function in Equation (6.88), the details of the form of the continuous function that satisfies Equation (14.25) are not important. Appropriate examples

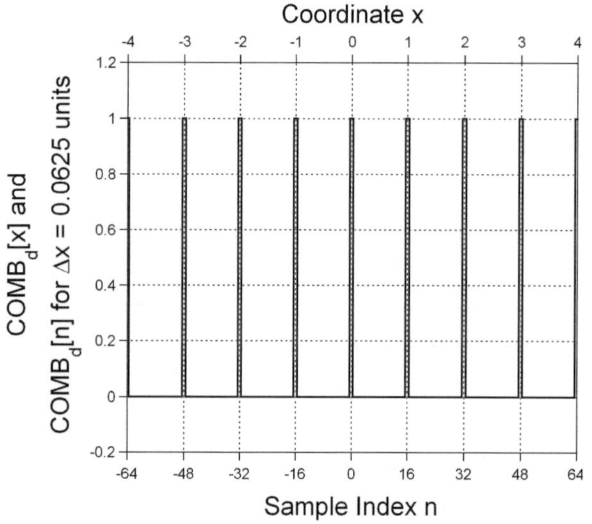

Figure 14.8 The discrete *COMB* function is the set of weights that are applied to $(1/\Delta x)COMB[x/\Delta x]$. The graph applies to $COMB_d[x]$ for $\Delta x = 1/16$ units.

include:

$$\delta_d\left[\frac{n}{1}\right] = RECT\left[\frac{n}{1}\right] = TRI\left[\frac{n}{1}\right] = SINC\left[\frac{n}{1}\right] \tag{14.29}$$

Note that any *SINC* function whose width parameter is $\Delta x/m$, where m is an integer, has zeros located at all integer multiples of Δx, and so satisfies the requirements of Equation (14.27).

The definition of the discrete *COMB* function is an obvious combination of the definitions for the continuous *COMB* and discrete Dirac delta functions:

$$COMB_d[n \cdot \Delta x] = \sum_{n=-\infty}^{+\infty} \delta_d[x - n \cdot \Delta x] \tag{14.30}$$

which obviously consists of unit-amplitude samples located at integer multiples of Δx as shown in Figure 14.8.

14.3 Interpolation of Sampled Functions

We now consider the question of whether a continuous function $f[x]$ can be reconstructed from a discrete set of samples. Any such process would have to “fill in” the amplitudes between the samples and, if it exists, acts as the *inverse* of the sampling operation.

Though most of the description of the Fourier transform of discrete functions is postponed to Chapter 15, the discussion of interpolation is facilitated by considering the spectrum of a bandlimited signal after sampling, which is easy to derive because we know the spectra of the constituent functions. We merely apply the modulation theorem of Equation (9.140) to the definition of sampling in Equation (14.5) to obtain:

$$\begin{aligned}
\mathcal{F}_1\{f_s[x;\ \Delta x]\} &\equiv F_s[\xi,\ \Delta x] \\
&= \mathcal{F}_1\left\{f[x]\cdot\frac{1}{\Delta x}COMB\left[\frac{x}{\Delta x}\right]\right\} \\
&= F[\xi] * COMB[\xi\cdot\Delta x] = F[\xi] * COMB\left[\frac{\xi}{(1/\Delta x)}\right] \\
&= F[\xi] * \frac{1}{\Delta x}\sum_{k=-\infty}^{+\infty}\delta\left[\xi-\frac{k}{\Delta x}\right] \\
\Longrightarrow F_s[\xi,\ \Delta x] &= \frac{1}{\Delta x}\sum_{k=-\infty}^{+\infty}F\left[\xi-\frac{k}{\Delta x}\right]
\end{aligned} \tag{14.31}$$

where the sifting property of the Dirac delta function has been used in the last step. In words, the spectrum $F_s[\xi;\ \Delta x]$ of an ideally sampled function is the convolution of the continuous spectrum $F[\xi]$ with the *COMB* function formed from identical Dirac delta functions with area $(\Delta x)^{-1}$ that are spaced by frequency intervals $(\Delta x)^{-1}$. The spectrum of the sampled function is the sum of the scaled replicas of $F[\xi]$ centered at integer multiples of $(\Delta x)^{-1}$, as shown in Figure 14.9. In other words, *the spectrum of a function sampled at uniformly spaced intervals is periodic*, as was already inferred in Equation (9.173). In the case of a compact continuous spectrum, the fact that the period of $F_s[\xi;\ \Delta x]$ in Equation (14.31) is $(\Delta x)^{-1}$ probably is obvious, but remains true even for all $F[\xi]$, even those with infinite support. This periodicity ensures that the spectrum of the discrete function is described completely by the amplitude within a single period of width $(\Delta x)^{-1}$.

Two specific semi-closed finite domains of "width" $(\Delta x)^{-1}$ that include all nonredundant spatial frequencies are commonly used to display the spectrum of the sampled function, and each has its own advantages and adherents. One choice is a semi-closed "symmetric" domain centered at $\xi = 0$. The endpoints of this spectrum "window" are $\pm 1/2\Delta x = \pm\xi_{\text{Nyquist}}$:

$$-\xi_{\text{Nyquist}} = -\frac{1}{2\cdot\Delta x} \le \xi < \frac{1}{2\cdot\Delta x} = +\xi_{\text{Nyquist}} \tag{14.32}$$

Note that either limit could be used as the "closed" endpoint; this choice follows the earlier convention. An obvious name for this domain is the *Nyquist window*.

The second choice of domain is the semi-closed nonnegative interval with zero as the included endpoint:

$$0 \le \xi < \frac{1}{\Delta x} \tag{14.33}$$

which is commonly used in applications involving temporal signals, where the analogue of the sampling interval is Δt and the temporal frequency is ν. The two window choices are compared in Figure 14.10. The second choice is used in the DFT, which is considered in Chapter 15. Simple transformations between the two domains are easy to derive, as will also be shown. The development in this chapter will concentrate on the "symmetric" window in Equation (14.32).

If the spectrum $F[\xi]$ of the original continuous function is bandlimited and if the sampling interval Δx satisfies the Nyquist condition of Equation (14.20), then the spectrum of the sampled function $F_s[\xi;\ \Delta x]$ consists of nonoverlapping replicas of the continuous spectrum $F[\xi]$ centered at integer multiples of $(\Delta x)^{-1}$. One replica of $F[\xi]$ may be segmented from this periodic spectrum by multiplying by an appropriate rectangular "window" whose amplitude is scaled by Δx to compensate for the fact that the area of each constituent Dirac delta functions in $COMB[\Delta x\cdot\xi]$ in Equation (14.31) is $(\Delta x)^{-1}$. If we isolate the central replica of $F[\xi]$, the mathematical expression for this "windowing" operation

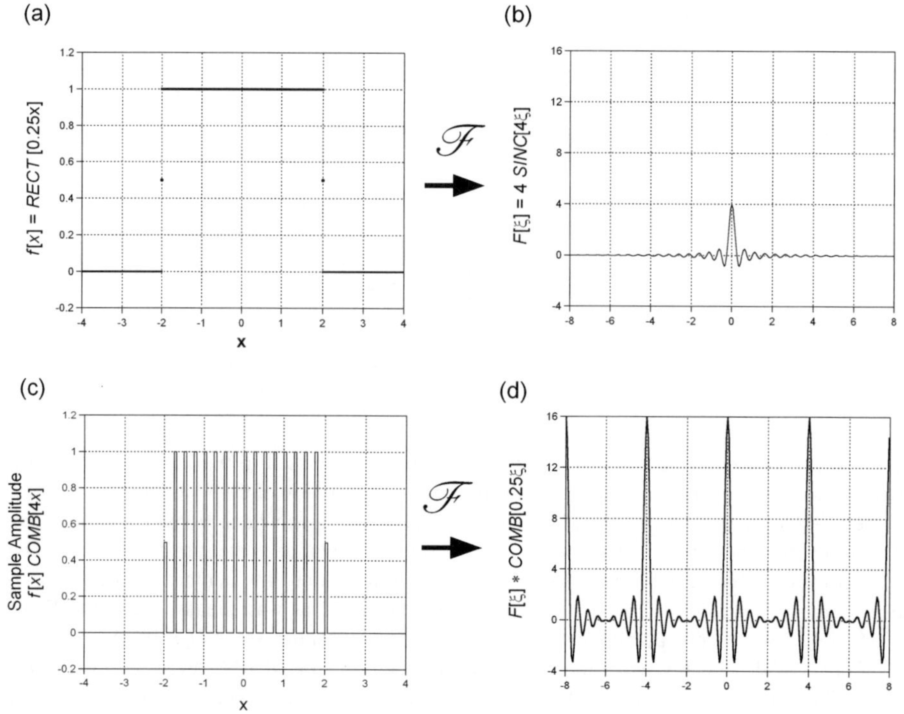

Figure 14.9 Effect of sampling on the spectrum: (a) continuous function $f[x] = RECT[x/4]$; (b) its spectrum $F[\xi] = 4\,SINC[4\xi]$. The sampling interval is $\Delta x = 1/4$ units, so that $s[x;\,\Delta x] = 4\,COMB[4x]$ to ensure that the individual Dirac delta functions have unit area. (c) Sampled function $f_s[x;\,\Delta x = 0.25]$; (d) its spectrum $F_s[\xi;\,\Delta x] = 4\,SINC[4\xi] * COMB[\xi/4]$. The period is $\Xi = 1.\Delta x = 4$ cycles per unit length. Also note that the areas of the continuous and sampled functions are different, as reflected in the central ordinates of the respective spectra.

becomes:

$$F_s[\xi;\,\Delta x]\cdot(\Delta x\cdot RECT[\xi\cdot\Delta x]) = \frac{1}{\Delta x}\sum_{k=-\infty}^{+\infty} F\left[\xi-\frac{k}{\Delta x}\right]\cdot\left(\Delta x\cdot RECT\left[\frac{\xi}{(1/\Delta x)}\right]\right)$$
$$\equiv \hat{F}[\xi;\,\Delta x] \tag{14.34}$$

where the circumflex indicates that the result is an "estimate" of the spectrum of the continuous function and is identical to $F[\xi]$ only if the original continuous function $f[x]$ is properly bandlimited:

$$\hat{F}[\xi;\,\Delta x] = (F[\xi] * COMB[\xi\cdot\Delta x])\cdot(\Delta x\cdot RECT[\xi\cdot\Delta x])$$
$$= F[\xi]\;\; if\;\; |\xi|_{\max} < \frac{1}{2\cdot\Delta x} \tag{14.35}$$

Of course, if $f[x]$ is not properly bandlimited (as is usually the case), then the amplitude of the spectrum at spatial frequencies within the rectangular window includes contributions from the adjacent periods of the spectrum; we speak of the resulting spectrum of the sampled function as "aliased". Figure 14.11 compares the aliased spectrum from the sampled function to that of the continuous function using the example of the rectangle from Figure 14.9.

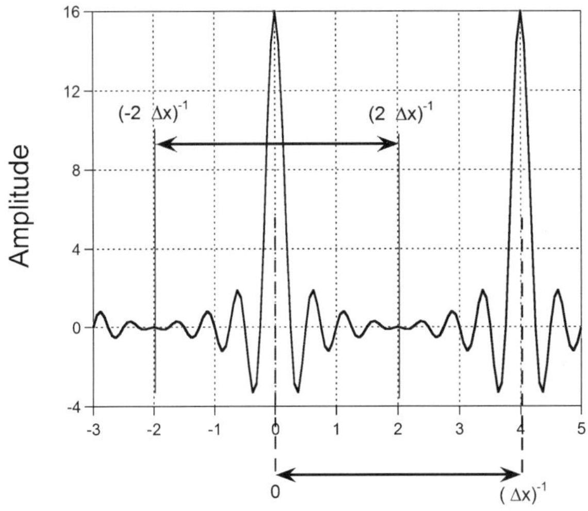

Figure 14.10 Two common choices of nonredundant domains for the spectrum of a sampled signal: $-1/2\ \Delta x \leq \xi < +1/2\ \Delta x$ and $0 \leq \xi < +1/\Delta x$. The scaled continuous spectrum (shown in dashes) is that of the sampled rectangle function in Figure 14.9, which exhibits aliasing.

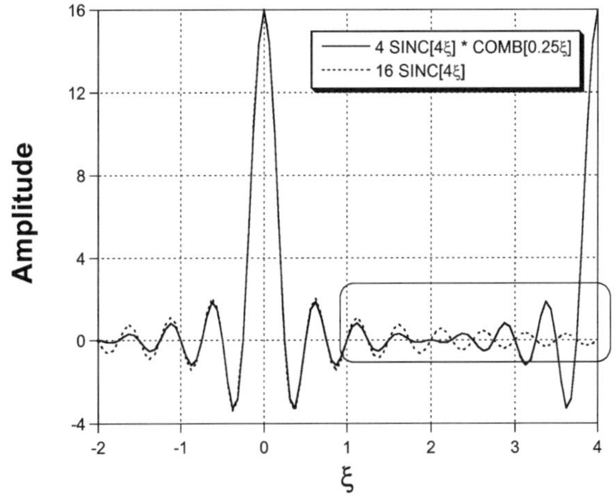

Figure 14.11 Aliasing in the spectrum of the sampled rectangle function (solid line). The amplitude of the continuous *SINC*-function spectrum (dashed line) has been scaled to match that of the sampled function. Note the significant differences in amplitude within the box where $|\xi| \gtrsim 1$ cycle per unit length.

The sampled spectrum modulated by the rectangle in Equation (14.34) may be expressed in the space domain by applying the inverse continuous Fourier transform and the filter theorem:

$$\begin{aligned}\hat{f}[x;\Delta x] &= \mathcal{F}_1^{-1}\left\{\frac{1}{\Delta x}\sum_{k=-\infty}^{+\infty} F\left[\xi - \frac{k}{\Delta x}\right]\cdot(\Delta x \cdot RECT[\xi\cdot\Delta x])\right\}\\ &= \mathcal{F}_1^{-1}\{(F[\xi] * COMB[\xi\cdot\Delta x])\cdot(\Delta x\cdot RECT[\xi\cdot\Delta x])\}\end{aligned}$$

$$
\begin{aligned}
&= \left(f[x] \cdot \frac{1}{\Delta x} COMB\left[\frac{x}{\Delta x}\right]\right) * SINC\left[\frac{x}{\Delta x}\right] \\
&= f_s[x;\ \Delta x] * SINC\left[\frac{x}{\Delta x}\right] \qquad (14.36)
\end{aligned}
$$

This defines ideal interpolation in the space domain that estimates the original continuous function from the samples. In words, each sample is replaced by a scaled *SINC* function and summed to generate $\hat{f}[x;\ \Delta x]$. Equation (8.55) demonstrated that the area of a convolution is the product of the areas of the constituent functions. This may be applied to Equation (14.36) to see that the area of the interpolated function $\hat{f}[x;\ \Delta x]$ is the product of Δx with the sum of the sample weights:

$$
\begin{aligned}
\int_{-\infty}^{+\infty} \hat{f}[x;\ \Delta x]\, dx &= \left(\int_{-\infty}^{+\infty} f_s[\alpha;\ \Delta x]\, d\alpha\right) \cdot \left(\int_{-\infty}^{+\infty} SINC\left[\frac{\beta}{\Delta x}\right] d\beta\right) \\
&= \left(\sum_{n=-\infty}^{+\infty} f[n \cdot \Delta x]\right) \cdot \Delta x \qquad (14.37)
\end{aligned}
$$

This leads to an interesting condition that the area of the continuous function and the sum of the sample amplitudes must be equal if the interpolated function is to be identical to the original continuous function:

$$
f[x] = \hat{f}[x;\ \Delta x] \Longrightarrow \int_{-\infty}^{+\infty} f[x]\, dx = \Delta x \sum_{n=-\infty}^{+\infty} f[n \cdot \Delta x] \qquad (14.38)
$$

The validity of this result may be checked by considering ideal interpolation of samples obtained from $f[x] = SINC[x/(b \cdot \Delta x)]$ for $b = 1$ and $b = 2$, whose areas are Δx and $2 \cdot \Delta x$, respectively. The sampled function for the first case was evaluated in Equation (14.22c) and found to be:

$$
\mathcal{S}\left\{SINC\left[\frac{x}{\Delta x}\right]\right\} = \delta[x] = \delta_d[x] \cdot \left(\frac{1}{\Delta x} COMB\left[\frac{x}{\Delta x}\right]\right) \qquad (14.39)
$$

while the sampled function obtained from the "wider" *SINC* function also is easy to evaluate:

$$
\begin{aligned}
\mathcal{S}\left\{SINC\left[\frac{x}{2 \cdot \Delta x}\right]\right\} &= SINC\left[\frac{x}{2 \cdot \Delta x}\right] \cdot \left(\frac{1}{\Delta x} COMB\left[\frac{x}{\Delta x}\right]\right) \\
&= \sum_{n=-\infty}^{+\infty} SINC\left[\frac{n}{2}\right] \cdot \delta[x - n \cdot \Delta x] \qquad (14.40)
\end{aligned}
$$

The areas of the respective sampled functions are:

$$
\int_{-\infty}^{+\infty} \mathcal{S}\left\{SINC\left[\frac{x}{\Delta x}\right]\right\} dx = 1 \qquad (14.41)
$$

and:

$$
\begin{aligned}
\int_{-\infty}^{+\infty} \mathcal{S}\left\{SINC\left[\frac{x}{2 \cdot \Delta x}\right]\right\} dx &= \sum_{n=-\infty}^{+\infty} SINC\left[\frac{n}{2}\right] \\
&= 1 + 2\left(\frac{2}{\pi}\right) + 2\left(-\frac{2}{3\pi}\right) + 2\left(+\frac{2}{7\pi}\right) + \cdots \\
&= 1 + \frac{4}{\pi}\left(1 - \frac{1}{3} + \frac{1}{5} - \frac{1}{7} + \frac{1}{9} - \cdots\right) \\
&= 1 + \frac{4}{\pi}\left(\frac{\pi}{4}\right) = 2 \qquad (14.42)
\end{aligned}
$$

where the well-known Taylor series for $\pi/4$ has been used. Therefore, the areas of the continuous functions obtained by convolving the samples with $SINC[x/\Delta x]$ are respectively Δx and $2 \cdot \Delta x$, as required.

The *SINC* function interpolator in Equation (14.36) is the impulse response of the system that reconstructs an estimate of the original function from the samples:

$$h[x;\, \Delta x] = \mathcal{F}_1^{-1}\{\Delta x \cdot RECT[\Delta x \cdot \xi]\} = SINC\left[\frac{x}{\Delta x}\right] \tag{14.43}$$

It is an example of the ideal lowpass filter introduced in Section 9.8. Also note that the estimate $\hat{f}[x;\, \Delta x]$ generated by Equation (14.36) is identical to the original continuous function $f[x]$ only if the sampling process satisfies the condition of Equation (14.34):

$$\begin{aligned} \hat{f}[x;\, \Delta x] &= f_s[x;\, \Delta x] * SINC\left[\frac{x}{\Delta x}\right] \\ &= f[x] \;\; if\, |\xi|_{\max} < \frac{1}{2 \cdot \Delta x} \end{aligned} \tag{14.44}$$

It is useful to examine the characteristics of the ideal *SINC* function interpolator in Equation (14.36). It evidently has unit amplitude at the origin, area Δx, infinite support, and regions of negative amplitude. The infinite support means that the amplitudes of the reconstructed continuous function between the sample coordinates are weighted sums of the amplitudes of all samples. The decay in the amplitude of the *SINC* interpolator ensures that the weighting applied to the amplitude tends to decrease with increasing distance from the sample. Also note that the reconstructed amplitude at a sample is guaranteed to be the sampled amplitude because all other sample amplitudes are scaled by the zeros of the *SINC* function. For an interpolated sample located at $x = n \cdot \Delta x$, negative amplitudes of the *SINC* interpolator occur at distances from the sample in the intervals:

$$SINC\left[\frac{x}{\Delta x}\right] < 0 \;\; if \;\; \begin{array}{c} \vdots \\ (n+3) \cdot \Delta x < x < (n+4) \cdot \Delta x \\ (n+1) \cdot \Delta x < x < (n+2) \cdot \Delta x \\ (-n-2) \cdot \Delta x < x < (-n-1) \cdot \Delta x \\ (-n-4) \cdot \Delta x < x < (-n-3) \cdot \Delta x \\ \vdots \end{array} \tag{14.45}$$

A portion of the amplitude of a sample is *subtracted* from the amplitude of the reconstructed continuous function at these locations. These allow the interpolated amplitude of the continuous function to vary more rapidly than would be possible if the weightings were nonnegative.

The process of ideal sampling in Equation (14.5) maps the uncountable infinite set of continuous amplitudes in $f[x]$ to a countable infinite set of discrete amplitudes $f[n \cdot \Delta x]$, while the interpolation process in Equation (14.36) performs the inverse mapping. That $f[x]$ may be reconstructed from $f_s[x;\, \Delta x]$ by convolution with $SINC[x/\Delta x]$ means that the amplitude of the original continuous function $f[x]$ at a coordinate x must be "related" to the amplitudes at adjacent samples; this correlation is described mathematically by the ideal *SINC* function interpolator. This observation means, at the very least, that arbitrary changes in the amplitude of the original continuous function cannot occur over infinitesimal increments of x; $f[x]$ must be a "smooth" curve if there is to be any hope of its exact reconstruction from the samples. This interpretation may give an intuitive understanding of the Nyquist sampling criterion in Equation (14.17); the amplitude of a bandlimited function may vary no more rapidly than its sinusoidal component with the largest spatial frequency.

In imaging systems, the display device interpolates the sampled data to derive an approximation of the original continuous signal. Unfortunately, negative weightings cannot be implemented in most (if not all) realistic image display systems, and therefore the true function cannot be reconstructed in fact.

The most common imaging display is the cathode-ray tube (CRT), which generates a blurred "spot" of light at the locations of the samples. The superposition of the "spots" is the approximation of the continuous image. Of course, an ideal display device would have to generate bipolar *SINC* functions at the location of each sample and thus "subtract" amplitude due to nearby bright samples. The limitations inherent in its capabilities often constrain the fidelity of the reconstructed image.

14.3.1 Examples of Interpolation

To illustrate that the correct amplitude of the bandlimited continuous function is obtained by the process of Equation (14.36), consider the infinite set of samples obtained from the unit constant $f[x] = 1[x]$:

$$\begin{aligned} f_s[x;\ \Delta x] &= 1[x] \cdot \frac{1}{\Delta x} COMB\left[\frac{x}{\Delta x}\right] \\ &= \sum_{n=-\infty}^{+\infty} \delta[x - n \cdot \Delta x] \end{aligned} \tag{14.46}$$

Equation (9.67) demonstrated that the spectrum of the continuous unit constant is the Dirac delta function:

$$F[\xi] = \mathcal{F}_1\{1[x]\} = \delta[\xi] \tag{14.47}$$

while the spectrum of the sampled unit constant is the spectrum of the *COMB* function in Equation (14.46), which is composed of Dirac delta functions with area $(\Delta x)^{-1}$ spaced at intervals $(\Delta x)^{-1}$:

$$F_s[\xi;\ \Delta x] = COMB[\xi \cdot \Delta x] = \frac{1}{\Delta x} \sum_{k=-\infty}^{+\infty} \delta\left[\xi - \left(\frac{k}{\Delta x}\right)\right] \tag{14.48}$$

After multiplying Equation (14.48) by the *RECT* transfer function, we obtain the estimate of the spectrum of the continuous function. In this case, the estimate and the original spectra are identical:

$$\begin{aligned} \hat{F}[\xi;\ \Delta x] &= COMB[\xi \cdot \Delta x] \cdot (\Delta x \cdot RECT[\xi \cdot \Delta x]) \\ &= \frac{1}{\Delta x} \sum_{k=-\infty}^{+\infty} \delta\left[\xi - \left(\frac{k}{\Delta x}\right)\right] \cdot (\Delta x \cdot RECT[\xi \cdot \Delta x]) \\ &= \sum_{k=-\infty}^{+\infty} \delta\left[\xi - \frac{k}{\Delta x}\right] \cdot RECT[\xi \cdot \Delta x] = \delta[\xi] \\ &\Longrightarrow \hat{f}[x;\ \Delta x] = \mathcal{F}_1^{-1}\{\delta[\xi]\} \end{aligned} \tag{14.49}$$

Therefore, the original continuous function and its estimate obtained by interpolating its infinite set of samples are identical:

$$\begin{aligned} \hat{f}[x;\ \Delta x] &= \mathcal{F}_1^{-1}\{COMB[\xi \cdot \Delta x] \cdot (\Delta x\ RECT[\xi \cdot \Delta x])\} \\ &= \left(\frac{1}{\Delta x} COMB\left[\frac{x}{\Delta x}\right]\right) * SINC\left[\frac{x}{\Delta x}\right] \\ &= \sum_{n=-\infty}^{+\infty} SINC\left[\frac{x - n \cdot \Delta x}{\Delta x}\right] = 1[x] \end{aligned} \tag{14.50}$$

In this derivation of the sampled spectrum, a single period of the continuous spectrum was extracted from the discrete spectrum by windowing with $RECT[\Delta x \cdot \xi]$. This recipe actually may be too restrictive in some cases. The width of the *RECT* function window in the frequency domain must only satisfy the constraint that it pass a single period of the continuous spectrum while rejecting all nonzero amplitudes due to other periods. In other words, the width of the *RECT* is not necessarily fixed at $\Delta\xi = (\Delta x)^{-1}$, but rather $\Delta\xi$ must just be larger than $2\xi_{max}$ (the width of the Nyquist window) and less than $2 \cdot ((\Delta x)^{-1} - \xi_{max})$. The corresponding impulse response of the ideal reconstruction filter must still have area Δx to preserve the amplitude of the reconstructed function, which means that its amplitude must vary in inverse proportion to its width. For example, a continuous sinusoid with period $16 \cdot \Delta x$ may be reconstructed from its samples by convolution with $SINC[x/\Delta x]$, which has unit amplitude and the width parameter equal to the sample spacing. The frequency-domain representation of this function is the transfer function $\Delta x\ RECT[\Delta x \cdot \xi]$. It is apparent from Equation (14.50) that $f[x]$ also may be reconstructed by using *SINC* functions with width parameters b such that $1/2\xi_{max} > b > \Delta x$, e.g., any width up to $8 \cdot \Delta x$ in this example. The general expression for a valid transfer function of interpolation is:

$$H[\xi] = \Delta x\ RECT[W \cdot \xi] \quad where\ \Delta x < W < \frac{1}{2\xi_{max}} \tag{14.51}$$

The expression for the corresponding impulse response is:

$$h[x] = \frac{\Delta x}{W} SINC\left[\frac{x}{W}\right] \tag{14.52}$$

At this point, perhaps it is easier than before to see that the frequency of a sinusoid must be strictly less than the Nyquist frequency to avoid aliasing. Amplitudes sampled from a cosine at exactly the Nyquist frequency will repeat every two samples. The actual amplitudes at samples at the Nyquist frequency are determined by the initial phase of the cosine. The amplitudes are the extrema of the sinusoid if the initial phase is $\phi_0 = 0$ radians, and therefore the recovered amplitude will be correct. In the case where the initial phase $\phi_0 = \pm\pi/2$, the amplitudes of both the sampled and reconstructed signal will be zero. If the spatial frequency of the cosine is infinitesimally smaller than the Nyquist sampling frequency, then the amplitudes of alternate samples will not be identical; the amplitudes will "march" along the sinusoid, eventually reaching the maximum amplitude at some sample index n_{max} and the minimum amplitude at a different (and possibly very distant) sample index n_{min}. The reconstructed function will have the correct extreme values at these samples, and the correct amplitude will be interpolated to the other coordinates by convolution with the ideal interpolator $SINC[x/\Delta x]$.

An example of ideal interpolation of a sinusoidal function is shown in Figure 14.12.

14.4 Whittaker–Shannon Sampling Theorem

The fact that a bandlimited function may be reconstructed from its samples without error is the outcome of the "Whittaker–Shannon sampling theorem". The theorem specifies the operation of sampling, the meaning of aliasing, and the reconstruction of the original function by interpolation. The cascade of processes defines an imaging system:

1. Samples of $f[x]$ are generated by multiplying by a *COMB* function composed of unit-area Dirac delta functions separated by intervals of width Δx.

2. If $f[x]$ is bandlimited such that its maximum frequency $\xi_{max} < 1/(2 \cdot \Delta x)$, then $f[x]$ can be recovered perfectly from the complete (infinite) set of samples obtained at intervals $n \cdot \Delta x$.

3. $f[x]$ is recovered from $f[n \cdot \Delta x]$ by convolving with the ideal lowpass filter that "cuts off" at ξ_{max}: $h[x] = SINC[x/\Delta x]$.

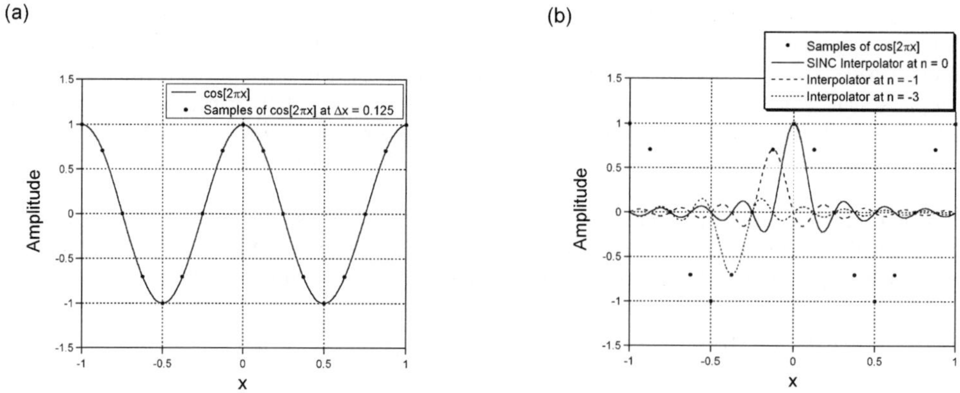

Figure 14.12 Ideal interpolation of the function $f[x] = \cos[2\pi x]$ sampled with $\Delta x = 1/16$ units. The weighted Dirac delta functions at each sample are replaced by weighted *SINC* functions (three shown, for $n = 0, -1, -3$), which are summed to reconstruct the original cosine function.

Sampling at rates greater than or less than the Nyquist limit is called *oversampling* or *undersampling*, respectively:

$$\Delta x > \frac{1}{2\xi_{\max}} \Longrightarrow \textit{undersampling} \tag{14.53a}$$

$$\Delta x < \frac{1}{2\xi_{\max}} \Longrightarrow \textit{oversampling} \tag{14.53b}$$

From this discussion, we know that undersampling produces a reconstructed function with "uncorrectable" errors. The downside of oversampling is the increased size of the data set.

14.5 Aliasing and Interpolation

14.5.1 Frequency Recovered from Aliased Samples

We know that the continuous signal reconstructed from an undersampled function is aliased. It may be useful on occasion to predict the aliased frequency of the interpolated signal. The process is perhaps best understood by considering a simple example. Assume that the input to an ideal sampler is a continuous linear-phase complex exponential with spatial frequency ξ_0, and thus period $X_0 = \xi_0^{-1}$:

$$f[x] = e^{+2\pi i\xi_0 x} = \cos[2\pi\xi_0 x] + i\,\sin[2\pi\xi_0 x] \tag{14.54}$$

The continuous spectrum consists of a single Dirac delta function at ξ_0:

$$F[\xi] = \delta[\xi - \xi_0] \tag{14.55}$$

as shown in Figure 14.13. The spectrum of the signal after sampling at intervals Δx is the convolution of $F[\xi]$ and $COMB[\xi \cdot \Delta x]$:

$$\begin{aligned} F_s[\xi;\,\Delta x] &= COMB[\xi \cdot \Delta x] * \delta[\xi - \xi_0] \\ &= COMB[(\xi - \xi_0) \cdot \Delta x] \\ &= \frac{1}{\Delta x} \sum_{k=-\infty}^{+\infty} \delta\left[\xi - \left(\xi_0 + \frac{k}{\Delta x}\right)\right] \end{aligned} \tag{14.56}$$

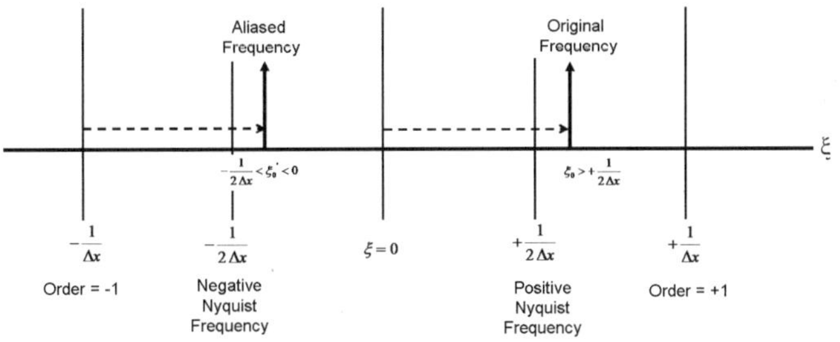

Figure 14.13 Apparent frequency of aliased signal: $f[x]=e^{+2\pi i\xi_0 x} \Longrightarrow F[\xi]=\delta[\xi-\xi_0]$. If $1/2\,\Delta x<\xi_0<1/\Delta x$, then the apparent frequency is $-1/2\,\Delta x<\xi_0'<0$.

In words, the Dirac delta functions in the periodic spectrum of the sampled function are located at all frequencies ξ' that satisfy the condition:

$$\xi'=\xi_0+\frac{k}{\Delta x} \tag{14.57}$$

where k is an integer. One of these Dirac delta functions is passed by the transfer function $RECT[\xi\cdot\Delta x]$ of the ideal interpolator; the specific frequency that is passed satisfies the constraint:

$$-\frac{1}{2\cdot\Delta x}\leq\left(\xi_0+\frac{k}{\Delta x}\right)<+\frac{1}{2\cdot\Delta x}\Longrightarrow-\frac{1}{2}\leq(\xi_0\cdot\Delta x+k)<+\frac{1}{2}\frac{cycles}{sample} \tag{14.58}$$

If the continuous frequency ξ_0 is positive and larger than the Nyquist frequency, then the component of the periodic spectrum that lies within the Nyquist window and thus is passed by the rectangle window arises from a replica of the continuous spectrum with order $k<0$. Conversely, if ξ_0 is negative and less than $-1/(2\cdot\Delta x)$, then the replica of the continuous spectrum that is passed by the rectangular interpolation filter has positive order, as shown in Figure 14.13. If ξ_0 and Δx are known, then the order of the "transmitted" spectrum replica may be calculated by solving for the integer that yields a spatial frequency that falls within the Nyquist window. It is easy to show that the order is:

$$k_0=-INT\left[\xi_0\cdot\Delta x+\frac{1}{2}\right] \tag{14.59}$$

where the greatest-integer function $INT[\,]$ determines the closest integer smaller than the argument, e.g., $INT[1.5]=1$ and $INT[-0.8]=-1$. Substitution of the order k into Equation (14.56) yields the "apparent" spatial frequency of the reconstructed complex exponential:

$$\begin{aligned}\xi'&=\xi_0-\left(\frac{1}{\Delta x}INT\left[\xi_0\cdot\Delta x+\frac{1}{2}\right]\right)\\&=\xi_0-\left(\frac{1}{\Delta x}INT\left[\frac{\Delta x}{X_0}+\frac{1}{2}\right]\right)\end{aligned} \tag{14.60}$$

Note that this expression is a function of the dimensionless sampling parameter $\Delta x/X_0$.

Consider some examples. We will use a sampling interval $\Delta x=1$ mm with a corresponding Nyquist frequency of $\xi_{\text{Nyquist}}=0.5$ cycles per mm. Therefore the period of the largest-frequency sinusoid that may be reconstructed without aliasing must be longer than $(\xi_{\text{Nyquist}})^{-1}=2$ mm. The period of a sinusoid at the Nyquist limit is 2 mm. If the spatial frequency of the continuous input function is $\xi_0=0.8$ cycles per mm, with a corresponding period of $X_0=1.25$ mm, then the dimensionless

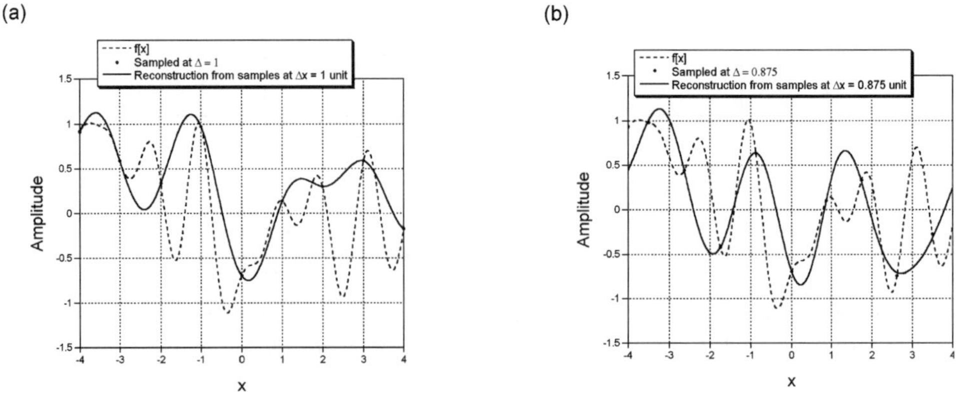

Figure 14.14 Ideal interpolation of the same signal if sampled with (a) $\Delta x = 1$ unit and (b) $\Delta x = 7/8$ units. The difference in the reconstructed signals demonstrates that the sampled signals are aliased.

sampling parameter is $\xi_0 \cdot \Delta x = 0.8$. The value of k_0 in Equation (14.59) is:

$$\begin{aligned} k_0 &= -INT[0.8 + 0.5] \\ &= -INT[1.3] = -1 \end{aligned} \tag{14.61}$$

and the reconstructed spatial frequency is obtained from Equation (14.57):

$$\begin{aligned} \xi' = \xi_0 + \frac{k}{\Delta x} &= 0.8 \, \frac{cycles}{\text{mm}} + \left(\frac{-1}{1} \, \frac{cycle}{\text{mm}} \right) = -0.2 \, \frac{cycles}{\text{mm}} \\ &\Longrightarrow X' = |\xi'|^{-1} = 5 \text{ mm} \end{aligned} \tag{14.62}$$

In this case, the period of the reconstructed sinusoid is very much longer than that of the original continuous function.

Now consider a second example where the sampling interval Δx is still 1 mm, but the continuous sinusoid oscillates at a larger rate of $\xi_0 = -2$ cycles per mm. The corresponding period of the sinusoid is $X_0 = 0.5$ mm. It is evident that alternate maxima of the continuous function are sampled and all samples have the same unit amplitude. This means that the reconstructed signal is predicted to be the unit constant. The dimensionless sampling parameter is $(\Delta x \cdot \xi) = -2$, which is substituted into Equation (14.59) to derive the order $k_0 = -2$. Substitution into Equation (14.60) demonstrates that the frequency of the reconstructed sinusoid is $\xi' = +0$ cycles per mm, which confirms our prediction.

The reconstructed frequency ξ' of an aliased signal depends upon both the original frequency ξ and the sampling interval Δx, while the reconstructed frequency of an unaliased signal is fixed at ξ. Therefore, the reconstructed frequencies of sampled signals that are computed using slightly different sampling intervals in an optical imaging system with a sampling detector may be used to determine if aliasing has occurred. This is illustrated in Figure 14.14.

14.5.2 "Unwrapping" the Phase of Sampled Functions

The ambiguity of the phase of a complex number was mentioned in the discussion of complex functions in Chapter 4, meaning that a complex number can be assigned phases that differ by increments of 2π radians. However, the direct calculation of the phase of a complex function in Equation (4.27b) is constrained to the interval $-\pi \leq \Phi < +\pi$, and thus the calculated phase function generally exhibits discontinuities as shown in Figure 4.3. However, it is useful (or even essential) in several imaging computations to "unwrap" the phase discontinuities of a sampled function by incrementing the phase by $\pm 2\pi$ radians at the transitions. The sign of the phase increment is determined by the phase derivative measured for already-unwrapped neighboring samples because the phase change between adjacent samples of an unaliased function must be smaller than π radians per sample. The trend in the phase derivative of an unaliased function is not ambiguous. However, the constraint in the phase derivative no longer applies if aliasing is present, so that the choice of the phase increment is no longer evident.

Note that the unwrapped phase only measures the "relative" phase of the samples; some "baseline" of the calculation where the phase angle is zero must be chosen. Generally, this is the sample indexed by $n = 0$.

The effect of aliasing on phase unwrapping is most easily seen in a function with known and smoothly varying phase, such as a complex chirp function. For example, consider the function with continuous domain:

$$f[x] = \exp\left[+i\frac{3}{2}\pi\left(\frac{x}{32\cdot\Delta x}\right)^2\right] \cong \exp\left[+i\pi\left(\frac{x}{26.128\cdot\Delta x}\right)^2\right] \tag{14.63}$$

In other words, this is a quadratic-phase function with chirp rate of just over $26\cdot\Delta x$ samples. If this function is sampled in an array with $N = 1024$ pixels over the domain $-512 \leq n \leq +511$ (Figure 14.15), the coordinate n_0 where the spatial frequency of the sampled chirp equals or just exceeds the Nyquist rate is determined by Equation (14.17):

$$\xi(n\cdot\Delta x) = \frac{1}{2\pi}\left(\frac{\partial}{\partial x}\Phi[x]\right) = \frac{1}{2\pi}\left(3\pi\frac{n\cdot\Delta x}{(32\cdot\Delta x)^2}\right) = \frac{3}{2}\cdot\frac{n}{1024\cdot\Delta x}$$

$$\xi(n_0\cdot\Delta x) \geq \frac{1}{2\cdot\Delta x} \Longrightarrow \pm n_0 \geq \frac{1024}{3} = 341\frac{1}{3} < 512 \tag{14.64}$$

This means that the derivative of the phase is larger than the maximum unaliased value of π radians per sample at the sample indexed by $n_0 = 342$, which is smaller than the index $n = 511$ at the edge of the array. The unwrapped phase increases quadratically with n until reaching the sample indexed by n_0, where the spatial frequency of the interpolated function starts to decrease. These artifacts are difficult to detect in sampled functions that are less obviously aliased. This observation will be an issue in the discussion of computer-generated (sampled) optical holograms in Chapter 23.

Phase unwrapping of 2-D functions is significantly more complicated than the 1-D case and the process is beyond the scope of this text. Interested readers should consult Ghiglia and Pritt (1998) and associated references.

14.6 "Prefiltering" to Prevent Aliasing

We now know that the function resulting from ideal interpolation of a sampled function will be aliased if the original continuous signal contains sinusoids with spatial frequencies that exceed the Nyquist limit. Aliasing may be prevented simply by removing the "offending" high-frequency signals that would be aliased before sampling. The ideal "antialiasing prefilter" is an ideal lowpass filter with cutoff frequency at (or less than) the Nyquist sampling frequency $\xi_{\text{Nyquist}} = (2\cdot\Delta x)^{-1}$. Note that the transfer functions of the ideal prefilter and ideal interpolation filter are both rectangle functions with "width" $(\Delta x)^{-1}$. Since the prefilter is applied first, its representations in the space or frequency domains will be denoted by a subscript "1":

$$H_1[\xi] = RECT[\xi\cdot\Delta x]$$

$$\Longrightarrow h_1[x] = \frac{1}{\Delta x}SINC\left[\frac{x}{\Delta x}\right] \tag{14.65}$$

This suggests the name of *postfilter* for the ideal interpolator. A complete "imaging" system consisting of the continuous ideal prefilter, ideal sampler, and ideal interpolator may be expressed as the cascade

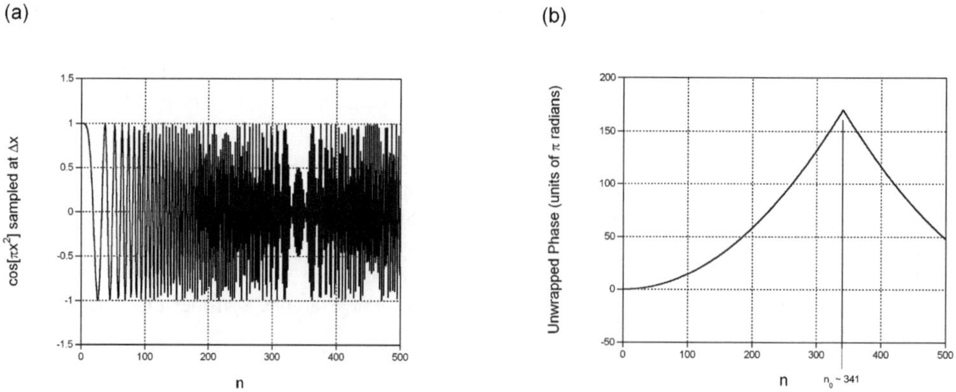

Figure 14.15 Effect of aliasing on phase of chirp function $f[x] = \exp[i\pi(x/b)^2]$: (a) the real part of $f[x]$ after sampling is the aliased function $\cos[\frac{3}{2}\pi(n/32)^2]$; (b) unwrapped phase calculated from aliased complex function, showing change in sign of derivative at $n_0 \gtrapprox 341$.

of three operations:

$$\hat{f}[x] = \left[\left(f[x] * \frac{1}{\Delta x}SINC\left[\frac{x}{\Delta x}\right]\right) \cdot \frac{1}{\Delta x}COMB\left[\frac{x}{\Delta x}\right]\right] * SINC\left[\frac{x}{\Delta x}\right]$$
$$= f[x] * \frac{1}{\Delta x}SINC\left[\frac{x}{\Delta x}\right] \tag{14.66}$$

Note that the two filtering operations in this sequence are linear and shift invariant, while the intervening sampling operation is linear and shift variant. Equation (14.66) is the complete mathematical statement of the Whittaker–Shannon sampling theorem. Of course, some already-mentioned caveats apply to this statement in real systems because the ideal infinite-support lowpass prefilter and postfilter do not exist and the required infinite number of samples cannot be collected in the real world. However, approximations of these steps may be implemented. For example, the effect of the ideal prefilter may be approximated in optical imaging systems with sampling detectors by defocusing the lens slightly. The blur attenuates the large spatial frequencies that might have been aliased.

Now consider the meaning of Equation (14.66). It indicates that the prefiltered signal may be recovered "exactly" by ideal interpolation of its samples. This result may be extended to make a more general statement that any continuous signal obtained from ideal lowpass filtering with cutoff frequency ξ_{cutoff} may be sampled with no loss of information. The sampling interval is identical to the width parameter of the ideal *SINC* function interpolator and must satisfy the constraint $\Delta x < (2\xi_{\text{cutoff}})^{-1}$. This more general statement will be useful in the discussion of the DFT in the next chapter.

14.6.1 Prefiltered Images Recovered from Samples

The set of samples obtained after prefiltering with an arbitrary impulse response $h_1[x]$ is:

$$f_{1s}[x; \Delta x] = (f[x] * h_1[x]) \cdot \left(\frac{1}{\Delta x}COMB\left[\frac{x}{\Delta x}\right]\right) \tag{14.67}$$

The ideal *SINC* interpolator acts on this infinite set of samples to recover the prefiltered continuous signal $f[x] * h_1[x]$ rather than the original continuous signal $f[x]$:

$$\hat{f}[x] = f_1[x;\, \Delta x] * SINC\left[\frac{x}{\Delta x}\right] \\ = f[x] * h_1[x] \tag{14.68}$$

If we wish to recover the frequency components of the original continuous image $f[x]$ out to the Nyquist sampling frequency, then the effect of the lowpass prefilter must be compensated by amplifying any sinusoidal components that were attenuated. The required compensating filter is often called an "inverse" filter because it is implemented by multiplying the recovered signal spectrum by the reciprocal of the prefilter transfer function (inverse filters are described in detail in Chapter 19). This will amplify frequency components of the recovered signal, including any unwanted "noise" that exists in any realistic situation. This amplification of "noise" ensures that the compensation process is fraught with peril. If the magnitude of the transfer function of the prefilter has small amplitudes within the Nyquist window, then the compensation filter will have to amplify those components by large factors.

The interpolation and compensation may be combined into a single operation, as shown in the frequency-domain representation:

$$(\hat{F}[\xi])_{\text{compensated}} = [(F[\xi] \cdot H_1[\xi]) * COMB[\Delta x \cdot \xi]] \cdot \Delta x \left(\frac{RECT[\Delta x \cdot \xi]}{H_1[\xi]} \right) \tag{14.69}$$

The *RECT* function interpolator also limits the support of the transfer function of the inverse filter to the Nyquist window in the frequency domain. The expression for the combined interpolator and compensator in the space domain is:

$$(\hat{f}[x])_{\text{compensated}} = f[n \cdot \Delta x] * SINC\left[\frac{x}{\Delta x}\right] * \mathcal{F}_1^{-1}\left\{\frac{1}{H_1[\xi]}\right\} \tag{14.70}$$

14.6.2 Sampling and Reconstruction of Audio Signals

Probably the most familiar example of sampling and reconstruction is the digital recording and playback of audio signals for entertainment. The appropriate sampling rate is determined by the maximum audible frequency of the human ear, which is approximately 20 kHz. The sampling frequency of digital audio recorders is fixed at 44 000 samples per second, which means that the sampling period of $1/44\,000\ \text{s} \cong 22.7\ \mu\text{s}$. At this sampling rate, sounds with periods greater than $2 \times 22.7\ \mu\text{s} = 45.4\ \mu\text{s}$, and thus frequencies less than $(45.4\ \mu\text{s})^{-1} = 22$ kHz may be reconstructed without error (at least in theory). This statement assumes (of course) that the original continuous signal $f[t]$ was sampled with an ideal *COMB* function and reconstructed with the ideal *SINC* function interpolator in the time domain.

If the original continuous audio signal includes sinusoidal components with frequency exceeding the Nyquist frequency of 22 kHz, then the samples of these formerly inaudible high-frequency signals will be aliased and the reconstructed continuous signal will appear at a lower frequency that may be in the audible range. For example, if a high-pitched "dog whistle" is sampled without a prefilter, its reconstruction will appear at an incorrect, yet audible, frequency. This is prevented by applying the lowpass prefilter to remove the signals with frequencies above the Nyquist limit before sampling.

The rectangular transfer function of the interpolation filter has a sharp "cutoff" in the transition region. These are "responsible for" the negative lobes in the ideal *SINC* function interpolator. Playback units are often configured to oversample the stored data. For example, units that oversample on playback by a factor of 4 will read the amplitude of each sample four times, at a rate of 176×10^3 samples per second. The spectrum of this oversampled array is still periodic, but the orders are spaced four times farther apart than those of the "raw" samples, and the support of each order is not affected. Because the orders are farther apart in the frequency domain, the higher-order spectral replicas may be rejected by a filter that does not have a perfectly rectangular passband.

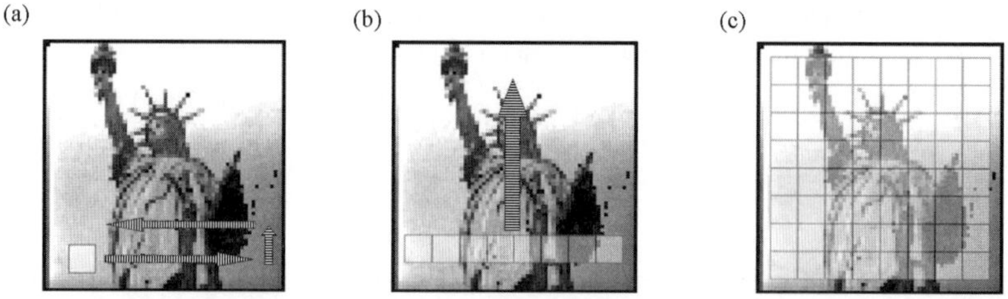

Figure 14.16 Sampling with different types of image scanners: (a) "whiskbroom"; (b) "pushbroom"; (c) "focal-plane array". In (a), a single-pixel detector is scanned in two dimensions across the scene. In (b), a line of detectors is scanned in one dimension. In (c), the entire scene is imaged at the same time.

14.7 Realistic Sampling

We have already seen that the reality of implementing ideal sampling is less rosy than might be assumed from the theory because all three operations required by the Whittaker–Shannon sampling theorem in Equation (14.66) are impossible to implement in realistic situations. First, real detector elements must have finite support because the amplitude of the measured signal is proportional to the area of the detector element. Thus it is not possible to evaluate the amplitude of $f[x]$ exactly at the sample locations. Second, it is not possible to measure the signal amplitude at the infinite number of samples required for any bandlimited signal to satisfy the Whittaker–Shannon sampling theorem. Finally, reconstruction of the continuous function requires filtering by the infinite-support bipolar *SINC* function interpolator of Equation (14.36). The impulse response of realistic filters must have finite support (the same problem obviously exists for the ideal lowpass prefilter). In addition, real display devices cannot deal with the necessary negative amplitudes of the ideal interpolator.

This section describes the methods used to implement approximations of ideal sampling, including a discussion of the effect of the approximation upon the interpolated signal. In imaging applications, the continuous input signal is often sampled by a 2-D discrete array of identical detectors; the common charge-coupled device (CCD) sensor is an example, even though the detectors can never really be "identical". Some digital imaging devices, particularly those from the dark ages of imaging in the 1970s and earlier, used 1-D arrays of "identical" detectors that were "pushed" across the scene in a direction (perpendicular to length of the array) rather like a broom, hence the common jargon of "pushbroom scanner" for these imagers. Other scanners used single-element detectors that had to be scanned across the scene in both directions and are sometimes called "whiskbroom scanners". Because the same detector is used at each pixel, any differences in response at different locations in the scene must be due to the optics of the scanning process.

The three types of discrete scanners are illustrated in Figure 14.16. The "pushbroom" and "whiskbroom" scanners have a long history, particularly in environmental remote sensing in the thermal infrared region of the spectrum ($8\ \mu\text{m} \leq \lambda \leq 14\ \mu\text{m}$) because 2-D arrays of detectors were quite expensive to build.

Regardless of the means used to scan across the input scene, the finite size of the detector element(s) ensures that the output of the detector(s) is an area integral of the input signal over the finite size of the detector. Clearly an analogous situation exists for temporal signals; the detected signal is the integral over some temporal period, called (obviously enough) the *integration time*. Our task now is to formulate a concise and useful mathematical description of the realistic sampling process. If the detector elements are assumed to be "identical", the process does not require evaluating individual integrals of the signal over each detector. Rather, realistic sampling is more easily modeled by a cascade of the two ideal operations of convolution and ideal sampling.

To illustrate the process of realistic sampling, consider first a simple model of a 1-D detector array constructed from identical rectangular elements of width d centered at integer multiples of the sampling interval Δx, as shown in Figure 14.17. The measured output signal at a sample is proportional to the integral of the input signal over the support of the individual detector. The signal evaluated at the element of the sample array indexed by $n = 0$ is:

$$\alpha \int_{-d/2}^{+d/2} f[x]\,dx = \int_{-\infty}^{+\infty} f[x]\left(\alpha\, RECT\left[\frac{x}{d}\right]\right) dx \tag{14.71}$$

where α is the amplification parameter of the process, which may be chosen to normalize the area of the rectangle function to unity, thus allowing the amplitudes of signals from different detectors to be compared directly. If $\alpha = d^{-1}$, then:

$$\frac{1}{d}\int_{-d/2}^{+d/2} f[x]\,dx = \int_{-\infty}^{+\infty} f[x]\cdot\left(\frac{1}{d} RECT\left[\frac{x}{d}\right]\right) dx \tag{14.72}$$

This integral evaluates the average value of $f[x]$ over the support d of the detector. The amplitude of the signal measured by the detector located at $x = n_0 \cdot \Delta x$ is obtained by merely translating the coordinates to that position:

$$\frac{1}{d}\int_{n_0\cdot\Delta x - d/2}^{n_0\cdot\Delta x + d/2} f[x]\,dx = \int_{-\infty}^{+\infty} f[x]\cdot\left(\frac{1}{d} RECT\left[\frac{x - n_0\cdot\Delta x}{d}\right]\right) dx \tag{14.73}$$

Because $RECT[x]$ is an even function, this expression may be recast into a familiar form:

$$\begin{aligned}\int_{-\infty}^{+\infty} f[x]\cdot\left(\frac{1}{d} RECT\left[\frac{x - n_0\cdot\Delta x}{d}\right]\right) dx &= \int_{-\infty}^{+\infty} f[x]\cdot\left(\frac{1}{d} RECT\left[\frac{n_0\cdot\Delta x - x}{d}\right]\right) dx\\ &= \left(f[x] * \frac{1}{d} RECT\left[\frac{x}{d}\right]\right)\Bigg|_{x=n_0\cdot\Delta x}\end{aligned} \tag{14.74}$$

In words, the signal averaged over the specific detector indexed by n_0 is the convolution of $f[x]$ with the normalized detector function evaluated at $n_0 \cdot \Delta x$. Note that this integral is *not* the sampled signal at that location, but rather the weighting applied to the Dirac delta function at that coordinate:

$$\begin{aligned}f_s[x = n_0\cdot\Delta x;\ \Delta x] &= \delta[x - n_0\cdot\Delta x]\cdot\left(f[x] * \frac{1}{d} RECT\left[\frac{x}{d}\right]\right)\Bigg|_{x=n_0\cdot\Delta x}\\ &= \delta[x - n_0\cdot\Delta x]\cdot\left(f[x] * \frac{1}{d} RECT\left[\frac{x}{d}\right]\right)\end{aligned} \tag{14.75}$$

where the behavior of the Dirac delta function in products from Equation (6.115) has been used in the last step. For emphasis, we repeat that Equation (14.75) describes a single sample of the input function as measured by the realistic rectangular detector. The general expression for the sampled function evaluated at any x is obtained by summing the Dirac delta functions located at all integer values of n and weighted by the convolution with the "detector response function" $h_1[x]$:

$$\begin{aligned}f_s[x;\ \Delta x] * h_1[x] &= \sum_{n=-\infty}^{+\infty} (\delta[x - n\cdot\Delta x](f[x] * h_1[x]))\\ &= \sum_{n=-\infty}^{+\infty}\left[\delta[x - n\cdot\Delta x]\left(f[x] * \frac{1}{d} RECT\left[\frac{x}{d}\right]\right)\right]\\ &= \left(f[x] * \frac{1}{d} RECT\left[\frac{x}{d}\right]\right)\cdot\left(\sum_{n=-\infty}^{+\infty}\delta[x - n\cdot\Delta x]\right)\\ &= \left(f[x] * \frac{1}{d} RECT\left[\frac{x}{d}\right]\right)\cdot\frac{1}{\Delta x} COMB\left[\frac{x}{\Delta x}\right]\end{aligned} \tag{14.76}$$

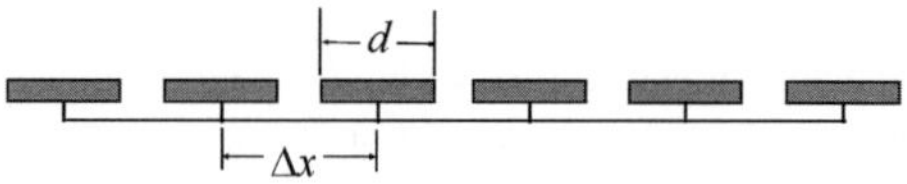

Figure 14.17 A 1-D detector composed of rectangular elements of width d separated by intervals Δx.

This is the mathematical expression for the sampled function obtained by integrating the continuous function over an array of identical rectangular detectors of width d. In words, the realistically sampled function is obtained by convolving the continuous input by the rectangular detector, followed by ideal sampling at the centers of each detector. This is a convenient expression because it is written in terms of operations and functions that are now familiar.

The expression for realistic sampling of $f[x]$ in Equation (14.76) was derived for the specific and simple case of a uniform "grid" of rectangular detectors. The process must be modified in cases where the response function of the detector is not symmetric. For convenience in comparisons, we assume that the area of this function is unity. If the detector response is $h_1[x]$, then the sampled signal evaluated at the detector indexed by n is:

$$f_s[x;\,\Delta x,\,h_1[x]] = \sum_{n=-\infty}^{+\infty}\left(\delta[x-n\cdot\Delta x]\cdot\int_{-\infty}^{+\infty} f[x]\cdot h_1[x-n\cdot\Delta x]\,dx\right)$$

$$= ((f[x] \bigstar h_1^*[x])|_{x=n\cdot\Delta x})\cdot\frac{1}{\Delta x}COMB\left[\frac{x}{\Delta x}\right] \quad (14.77a)$$

$$= ((f[x] * h_1[-x])|_{x=n\cdot\Delta x})\cdot\frac{1}{\Delta x}COMB\left[\frac{x}{\Delta x}\right] \quad (14.77b)$$

where the definition of crosscorrelation in Equation (8.68) and its relationship to convolution in Equation (8.73b) have been used. The complex-conjugation operation in Equation (14.77a) negates the explicit conjugation operation in the definition of crosscorrelation. Similarly, the reversion operation applied to $h_1[x]$ in Equation (14.77b) "undoes" that present in the definition of convolution. This rather subtle point usually is unimportant because most real imaging detectors have symmetric response functions.

From the discussion in Section 14.6.1, we can see that the effect of the finite-support detector function acts like a prefilter. In the usual cases where $h_1[x]$ is a nonnegative function with compact support, the prefilter is a local averager (lowpass filter) that attenuates the amplitudes of those sinusoidal components of $f[x]$ with large spatial frequencies more than those with small ones.

Since the realistically sampled function in Equation (14.77b) is modeled as a cascade of simple operations, it is easy to express it in the frequency domain:

$$F_s[\xi;\,\Delta x,\,h_1[x]] = \mathcal{F}_1\left\{(f[x]*h_1[-x])|_{x=n\cdot\Delta x}\cdot\frac{1}{\Delta x}COMB\left[\frac{x}{\Delta x}\right]\right\}$$

$$= (F[\xi]\cdot H_1[-\xi]) * COMB[\Delta x\cdot\xi] \quad (14.78)$$

where the Fourier transform of the reversed function in Equation (9.119) has been used. Again, note that $H[-\xi] = H[+\xi]$ in the usual case where the detector response function is real and symmetric. Many users describe $H_1[-\xi]$ as the transfer function of the sampling system. However, it would be more appropriate to call $H_1[-\xi]$ the transfer function of the sampling *detector*, since the shift-variant sampling process cannot be implemented as a convolution and thus does not have an associated transfer function.

To demonstrate the realistic sampling process and its effects, consider the measurements made by an array of uniform detectors of width d of a nonnegative sinusoidal input signal with spatial frequency

ξ_0. The expression for the signal is:

$$f[x] = \frac{1}{2}(1 + \cos[2\pi\xi_0 x + \phi_0]) = \frac{1}{2}\left(1 + \cos\left[\frac{2\pi x}{X_0} + \phi_0\right]\right) \tag{14.79}$$

where $X_0 = \xi_0^{-1}$. Substitution into Equation (14.76) yields the expression for the sampled signal:

$$f_s[x;\,\Delta x] = \frac{1}{2}\cdot\left(1[x] * \frac{1}{d}RECT\left[\frac{x}{d}\right]\right) + \frac{1}{2}\cdot\left(\cos[2\pi\xi_0 x + \phi_0] * \frac{1}{d}RECT\left[\frac{x}{d}\right]\cdot\frac{1}{\Delta x}COMB\left[\frac{x}{\Delta x}\right]\right) \tag{14.80}$$

The two convolutions were evaluated by direct integration in Equation (8.38), but the filter theorem of Equation (9.129) also may be used:

$$1[x] * \frac{1}{d}RECT\left[\frac{x}{d}\right] = \mathcal{F}_1^{-1}\{\delta[\xi]\cdot SINC[d\xi]\} = \mathcal{F}_1^{-1}\{\delta[\xi]\cdot SINC[d\cdot 0]\} = \mathcal{F}_1^{-1}\{\delta[\xi]\} = 1[x] \tag{14.81}$$

$$\cos[2\pi\xi_0 x + \phi_0] * \left(\frac{1}{d}RECT\left[\frac{x}{d}\right]\right) = \mathcal{F}_1^{-1}\left\{\frac{1}{2}\cdot(\delta[\xi+\xi_0]\cdot e^{+i\phi_0} + \delta[\xi-\xi_0]\cdot e^{-i\phi_0})\cdot SINC[d\xi_0]\right\} = SINC[d\xi_0]\cdot\cos[2\pi\xi_0 x + \phi_0] \tag{14.82}$$

where the property of the Dirac delta function in products from Equation (6.115) has been used to evaluate both terms. The realistically sampled function has the form:

$$f_s[x;\,\Delta x] = \frac{1}{2}(1 + SINC[d\xi_0])\cdot\cos[2\pi\xi_0 x + \phi_0]\cdot\frac{1}{\Delta x}COMB\left[\frac{x}{\Delta x}\right] \tag{14.83}$$

After applying the ideal interpolator, the reconstructed continuous signal is:

$$\hat{f}[x] = f_s[x;\,\Delta x] * SINC\left[\frac{x}{\Delta x}\right] = \frac{1}{2}(1 + SINC[d\xi_0]\cdot\cos[2\pi\xi_0 x + \phi_0]) \tag{14.84}$$

In words, this result indicates that the effect of a finite-support detector on a sinusoidal signal is a reduction in amplitude by the scale factor $SINC[d\xi_0]$, which obviously depends on both the spatial frequency of the sinusoid and the width of the detector. Since $|SINC[d\xi_0]| < 1$, the continuous output is guaranteed to be nonnegative. It is useful to establish the effect of the uniform averager on the modulation of the nonnegative sinusoid that was defined in Equation (6.25). The maximum and minimum amplitudes of the sinusoid and the resulting modulation are easily determined:

$$\hat{f}_{\max} = \frac{1}{2} + \frac{1}{2}SINC[d\xi_0] \tag{14.85a}$$

$$\hat{f}_{\min} = \frac{1}{2} - \frac{1}{2}SINC[d\xi_0] \tag{14.85b}$$

and the modulation of the continuous output signal is:

$$m_{\hat{f}} = \frac{\hat{f}_{\max} - \hat{f}_{\min}}{\hat{f}_{\max} + \hat{f}_{\min}} = \frac{SINC[d\xi_0]}{1} = SINC[d\xi_0] \tag{14.86}$$

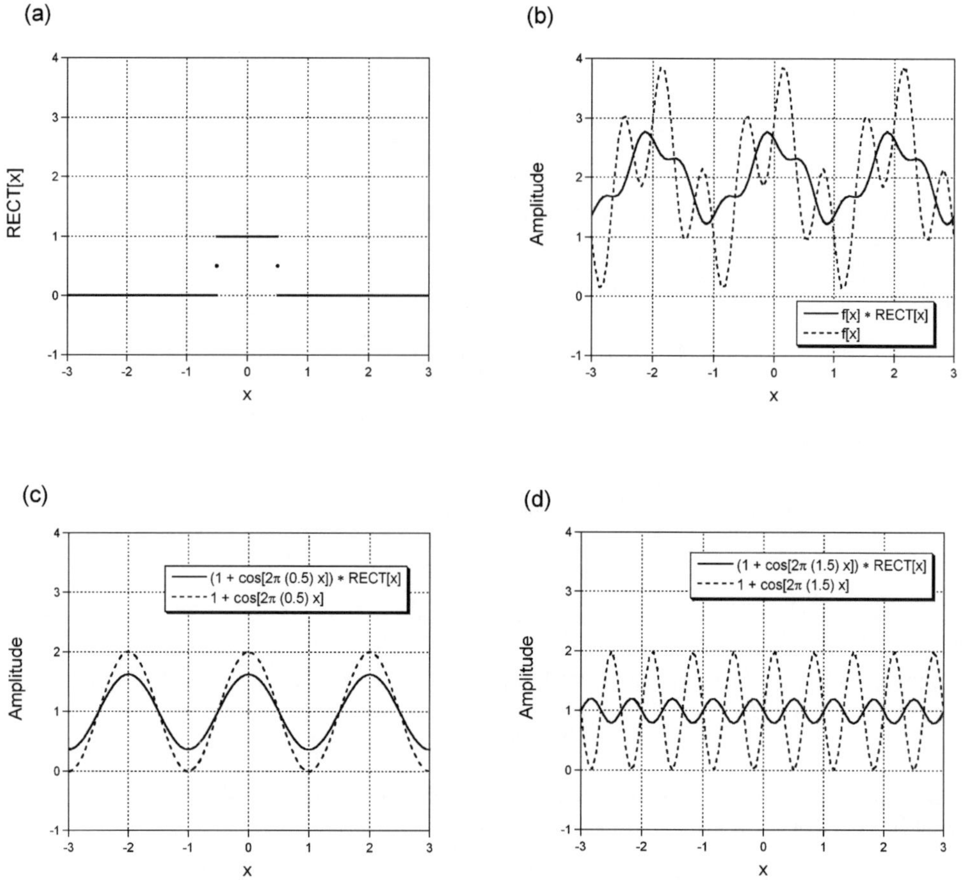

Figure 14.18 Demonstration of reduction in modulation of reconstructed sinusoidal signals as the spatial frequency is increased. The detector size $d = 1$ unit and the input signal is the sum of two sinusoids that have been biased to ensure nonnegative amplitude: $f[x] = f_1[x] + f_2[x]$, where $f_1[x] = 1 + \cos[2\pi x/2]$ and $f_2[x] = (1 + \cos[2\pi(x/(\frac{2}{3})]))$. (a) The detector function $h[x] = RECT[x]$; (b) $f[x] = f_1[x] + f_2[x]$ before and after convolution with $h[x]$; (c) $f_1[x]$ and $f_1[x] * h[x]$; (d) $f_2[x]$ and $f_2[x] * h[x]$, which shows the phase reversal of the output at this spatial frequency.

which confirms the result obtained in Equation (8.40). The modulation of the biased sinusoid is reduced by the factor $SINC[d\xi_0]$, which is positive for $d\xi_0$ in the interval $[2n, 2n+1]$, where n is an integer, and negative otherwise. The negative weight factors invert the reconstructed sinusoid, which is equivalent to changing the initial phase by $\pm\pi$ radians. This effect is illustrated by the examples in Figure 14.18.

The imaging system that consists of realistic sampling followed by interpolation decreased the modulation of the biased input sinusoid from unity to $SINC[d\xi_0]$. The ratio of the modulation of the output to that of the input is a metric of the "quality" of the imaging system; it specifies how well the modulation is "transferred" to the output by the action of the system. In this example, the "modulation transfer" metric is:

$$\textit{modulation transfer at } \xi_0 = \left(\frac{m_{\hat{f}}}{m_f}\right)\bigg|_{\xi_0} = \frac{SINC[d\xi_0]}{1} = SINC[d\xi_0] \tag{14.87}$$

Note that the value of the modulation transfer metric depends on the spatial frequency of the input sinusoid. Because this is generally true for imaging systems, the modulation transfer is usually represented as a function of ξ, and is called the *modulation transfer function.*

14.8 Realistic Interpolation

Just as for ideal sampling, the process of ideal interpolation cannot be implemented in the real world. The infinite support of the interpolation function $SINC[x/\Delta x]$ makes it impossible to apply realistically, while the negative weight factors of the *SINC* function are not feasible in many real display devices.

Some insight into the limitations of real devices may be illustrated by considering two hypothetical, but still unrealistic, situations: (1) the number of samples in the sampled function is finite, but the interpolator may act over the infinite domain; and (2) an infinite number of samples exist, but the interpolation function has finite support.

14.8.1 Ideal Interpolator for Compact Functions

A function composed of a finite number of nonzero samples is equivalent to the samples from a continuous function $f[x]$ that has been multiplied by a rectangle function in the space domain. The spectrum of the truncated continuous function must have infinite support. Even if ideal interpolation were possible, the original truncated continuous signal could not be recovered without aliasing error. The artifacts in the reconstructed continuous signal will be most visible near the limits of the support of the function, and are called *edge effects* for obvious reasons. This phenomenon is illustrated in Figure 14.19 for an input function consisting of a sinusoid modulated by a *RECT*. The windowed sinusoid is sampled eight times per period, which far exceeds the Nyquist sampling rate for a bandlimited sine wave. The continuous function was reconstructed by (approximately) ideal interpolation; the edge effects are quite evident.

14.8.2 Finite-Support Interpolators in Space Domain

If an infinite set of unaliased samples is interpolated with some function other than the ideal *SINC*, the reconstructed function obviously cannot exactly match the input (except perhaps for a few pathological cases). However, it is possible to construct interpolation functions that give reasonable approximations of the original function. To illustrate the concept, we will "derive" a sequence of realistic interpolation functions that generate reconstructions that are weighted sums of the amplitude of one sample only, of two samples, and of more than two. In the first and simplest case, the amplitude of the reconstructed continuous function at any coordinate is set to the amplitude of the nearest sample. In the ambiguous cases of coordinates located exactly midway between two samples, the interpolated amplitude becomes the average of the two neighboring sample amplitudes. This *nearest-neighbor* interpolation is equivalent to convolution of the sampled function $f_s[x;\, \Delta x]$ with a unit-amplitude *RECT* function whose width is the sample spacing Δx:

$$\begin{aligned}\hat{f}[x] &= f_s[x;\, \Delta x] * RECT\left[\frac{x}{\Delta x}\right] \\ &= \left(f[x] \cdot \frac{1}{\Delta x} COMB\left[\frac{x}{\Delta x}\right]\right) * RECT\left[\frac{x}{\Delta x}\right] \end{aligned} \tag{14.88}$$

This rectangle interpolator is the first of a sequence of interpolators to be derived. Because it computes the output amplitude based on one sample weight, an argument could be made to call this a "first-order" interpolator. However, we will call it the "zeroth-order" interpolator because the amplitude of the interpolator within its region of support is $1 = x^0$. This interpolator will be denoted by a roman (i.e.,

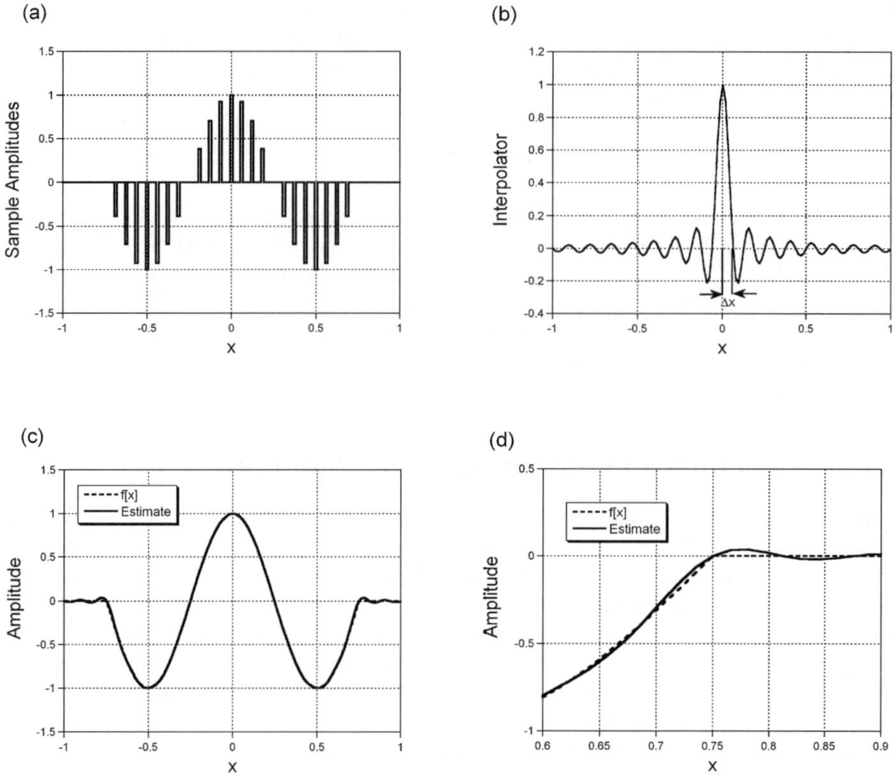

Figure 14.19 Ideal sampling and interpolation of $f[x] = \cos[2\pi x] \cdot RECT[2x/3]$ with $\Delta x = \frac{1}{16}$: (a) $f[x]$ after sampling; (b) ideal interpolator $SINC[x/\Delta x]$; (c) interpolated function compared to $f[x]$; (d) magnified view of interpolated function in vicinity of the second zero, showing "ringing" beyond the edge.

not italicized) character with a subscript indicating the order of the approximation:

$$\text{1-D } \hat{\mathrm{h}}_0[x; \Delta x] = RECT\left[\frac{x}{\Delta x}\right] \tag{14.89}$$

The circumflex again emphasizes that the interpolator is an approximation of the ideal impulse response. The transfer function of this filtering operation is trivial to derive by applying the scaling theorem to the known spectrum of $RECT[x]$, and is denoted by the upper-case roman character with the same subscript:

$$\begin{aligned}\hat{\mathrm{H}}_0[\xi; \Delta x] &= \Delta x \cdot SINC[\Delta x \cdot \xi] \\ &= \Delta x \cdot SINC\left[\frac{\xi}{\Delta x^{-1}}\right]\end{aligned} \tag{14.90}$$

Clearly, the transfer function of zeroth-order interpolation differs markedly from the ideal rectangular form in Equation (14.34), not least in the fact that it has infinite support. These two transfer functions do share the same amplitude Δx at the origin and the same unit area.

The zeroth-order interpolator in two dimensions obviously is the separable rectangle obtained by computing the orthogonal product of *RECT* functions along the x- and y-directions:

$$\text{2-D } \hat{h}_0[x, y; \Delta x, \Delta y] = RECT\left[\frac{x}{\Delta x}, \frac{y}{\Delta y}\right] \tag{14.91}$$

The volume of this 2-D interpolator is $\Delta x \cdot \Delta y$. A useful approximation of the 2-D interpolator is a circularly symmetric version of the zeroth-order interpolator in the form of a cylinder function, which in turn is an approximation of the circularly symmetric Gaussian:

$$\hat{h}[r; \Delta x = \Delta y] \cong CYL\left(\frac{r}{b}\right) \cong GAUS\left(\frac{r}{b}\right) \tag{14.92}$$

These two impulse responses approximate the action of many real display devices, such as CRTs which construct an image over a screen by superposing samples convolved with a blur spot determined by the amplitude profile of the electron beam and the response of the display phosphors.

The 1-D first-order interpolator computes a linearly weighted average of the two nearest sample amplitudes; the closer the coordinate is to a sample point, the larger the weight applied to the amplitude of that sample. In one dimension, the interpolator connects adjacent sample amplitudes with straight lines in a manner analogous to "connect-the-dots" pictures used as amusements for young children. The weight applied to a sample amplitude is the complement of the distance from that sample scaled to the sample spacing. The expression for the impulse response of the interpolator is:

$$\begin{aligned} \hat{h}_1[x; \Delta x] &= \left(1 - \frac{|x|}{\Delta x}\right) \cdot RECT\left[\frac{x}{2 \cdot \Delta x}\right] \\ &\equiv TRI\left[\frac{x}{\Delta x}\right] \end{aligned} \tag{14.93}$$

Within its region of support, the amplitude of the first-order interpolator varies as $\pm x^1$, hence the nomenclature, and thus may be called a "linear" interpolator. The impulse response of the linear interpolator has unit amplitude at the origin, support $2 \cdot \Delta x$, and area Δx. When exactly centered over a sample, the amplitude of the impulse response $TRI[x/\Delta x]$ is unity at the sample and zero at both adjacent samples, so the amplitudes of both the sampled and interpolated functions are evaluated at the sample locations. The impulse response of the first-order interpolator may be generated as an appropriately scaled self-convolution of the zeroth-order interpolator:

$$\begin{aligned} \hat{h}_1[x; \Delta x] = TRI\left[\frac{x}{\Delta x}\right] &= \frac{1}{\Delta x}\left(RECT\left[\frac{x}{\Delta x}\right] * RECT\left[\frac{x}{\Delta x}\right]\right) \\ &= RECT\left[\frac{x}{\Delta x}\right] * \left(\frac{1}{\Delta x} RECT\left[\frac{x}{\Delta x}\right]\right) \end{aligned} \tag{14.94}$$

as shown in Figure 14.20. The transfer function of the first-order interpolator is proportional to the square of the zeroth-order transfer function:

$$\begin{aligned} \hat{H}_1[\xi; \Delta x] &= \Delta x \cdot SINC^2[\xi \cdot \Delta x] \\ &= \Delta x \cdot SINC^2\left[\frac{\xi}{(\Delta x)^{-1}}\right] \end{aligned} \tag{14.95}$$

which may be interpreted as a first-order approximation to the ideal rectangular transfer function of interpolation. Note that the areas of $\hat{H}_1[\xi; \Delta x]$ and $\hat{H}_0[\xi; \Delta x]$ are both unity. The result of first-order interpolation for a sinusoidal function is shown in Figure 14.21.

The first-order transfer function for interpolation in Equation (14.94) may be applied in two dimensions to generate the so-called 2-D "bilinear" interpolator with impulse response:

$$\hat{h}_1[x, y; \Delta x, \Delta y] = RECT\left[\frac{x}{\Delta x}, \frac{y}{\Delta y}\right] * \left(\frac{1}{\Delta x \cdot \Delta y}\right) RECT\left[\frac{x}{\Delta x}, \frac{y}{\Delta y}\right]$$
$$= TRI\left[\frac{x}{\Delta x}, \frac{y}{\Delta y}\right] \tag{14.96}$$

The transfer function is just the 2-D Fourier transform of this 2-D triangle:

$$\hat{H}_1[\xi, \eta; \Delta x, \Delta y] = \Delta x \cdot \Delta y \cdot SINC^2[\xi \cdot \Delta x, \eta \cdot \Delta y] \tag{14.97}$$

The process in Equation (14.94) may be extended to generate higher-order interpolators. The impulse response of the second-order interpolator is proportional to the three-fold convolution of $RECT[x/\Delta x]$:

$$\hat{h}_2[x; \Delta x] = RECT\left[\frac{x}{\Delta x}\right] * \left(\frac{1}{\Delta x} RECT\left[\frac{x}{\Delta x}\right]\right) * \left(\frac{1}{\Delta x} RECT\left[\frac{x}{\Delta x}\right]\right) \tag{14.98}$$

as shown in Figure 14.20a. Where it is not zero, the amplitude of $\hat{h}_2[x; \Delta x]$ varies as x^2, hence the common name of "quadratic interpolator". The area of $\hat{h}_2$ is Δx, which is identical to that of $\hat{h}_0$ and $\hat{h}_1$, and its support is $3 \cdot \Delta x$, which ensures that the amplitude of the interpolated function at any coordinate is a weighted sum of the amplitudes of three samples. Like $\hat{h}_0$ and $\hat{h}_1$, the quadratic interpolator is nonnegative everywhere and thus no scaled sample amplitudes are subtracted during interpolation. The amplitude of the quadratic interpolator at the origin is $\frac{3}{4}$ rather than unity and that at the adjacent samples is $\frac{1}{8}$ rather than zero, which means that the amplitude of the interpolated function at the location of a sample is a weighted sum of the amplitudes of that sample and its two nearest neighbors. The interpolated amplitude that is computed at the sample indexed by n is:

$$\hat{f}[x = n \cdot \Delta x] = \frac{1}{8} f[(n-1) \cdot \Delta x] + \frac{3}{4} f[n \cdot \Delta x] + \frac{1}{8} f[(n+1) \cdot \Delta x] \tag{14.99}$$

The quadratic interpolator computes a weighted average of the sample amplitudes in the neighborhood of the coordinate, and thus is a form of the lowpass filter introduced in Section 9.8. Its action as an "averager" of the samples tends to "blur" or "smear" the small-scale variations in the sample amplitudes, and thus reduces the modulation of high-frequency sinusoidal components of the original continuous function. Another useful interpretation of second-order interpolation is the fitting of a semi-flexible "spline" to the data points. The constraint on the flexibility of the spline may prevent the interpolated amplitude from exactly matching the amplitude of the original function at some of the sample locations, and thus the interpolated function may vary more smoothly than the original continuous function.

The transfer function of the second-order interpolator in Equation (14.98) is proportional to the cube of the *SINC* function:

$$\hat{H}_2[\xi; \Delta x] = \Delta x \cdot SINC^3[\xi \cdot \Delta x] \tag{14.100}$$

and has the same unit area as $\hat{H}_0$ and $\hat{H}_1$. Its graph is shown in Figure 14.20b. The extension of the transfer function to two dimensions is straightforward:

$$\hat{H}_2[\xi, \eta; \Delta x, \Delta y] = \Delta x \cdot \Delta y \cdot SINC^3[\xi \cdot \Delta x, \eta \Delta y] \tag{14.101}$$

and the impulse response is the inverse 2-D Fourier transform of this expression.

The third-order (cubic) interpolator is a convolution of four rectangle functions:

$$\hat{h}_3[x; \Delta x] = RECT\left[\frac{x}{\Delta x}\right] * \left(\frac{1}{\Delta x} RECT\left[\frac{x}{\Delta x}\right]\right) * \left(\frac{1}{\Delta x} RECT\left[\frac{x}{\Delta x}\right]\right) * \left(\frac{1}{\Delta x} RECT\left[\frac{x}{\Delta x}\right]\right) \tag{14.102}$$

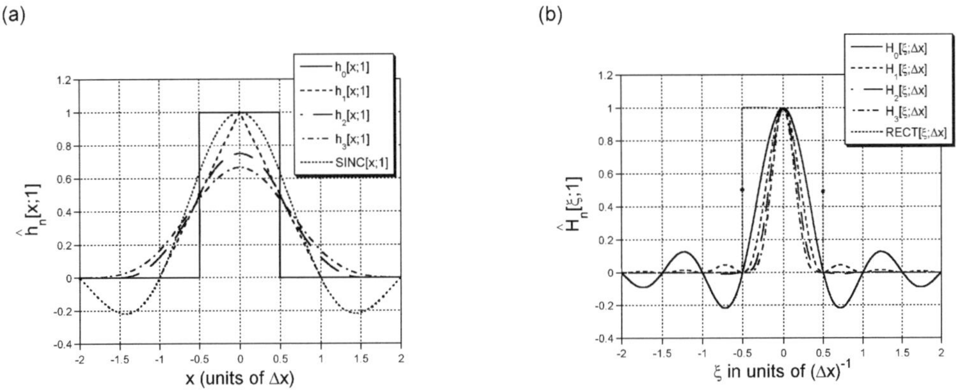

Figure 14.20 Realistic interpolators derived in the space domain for $n = 0$–3 compared to the ideal interpolator $SINC[x/\Delta x]$: (a) impulse responses of the realistic interpolators are everywhere nonnegative; (b) transfer functions of the realistic interpolators are bipolar, unlike the ideal rectangle.

which has area Δx, a central ordinate of $\frac{2}{3}$, and support $4 \cdot \Delta x$. This last ensures that the interpolated amplitude generally is a weighted sum of four sample amplitudes, thus "smoothing" out variations in the sample amplitudes even more than $\hat{\text{h}}_2[x;\, \Delta x]$. The corresponding transfer function is:

$$\hat{\text{H}}_3[\xi;\, \Delta x] = \Delta x\; SINC^4[\xi \cdot \Delta x] \tag{14.103}$$

which has unit area.

From the sequence of operations that generate these interpolators, it is evident that the impulse response of the Nth-order interpolator is the convolution of $N+1$ $RECT$ functions; the result has area Δx and support $n \cdot \Delta x$. The central-limit theorem derived in Section 13.3 may be applied to approximate the normalized N-fold self-convolution of $RECT[x/\Delta x]$ as a Gaussian with area Δx and width parameter $\Delta x \cdot \sqrt{2\pi N/12}$:

$$\hat{\text{h}}_N[x;\, \Delta x] \cong \frac{1}{\sqrt{N\pi/6}}\, GAUS\left[\frac{x}{\sqrt{N\pi/6}}\right] \tag{14.104}$$

The Fourier transform of this expression is approximately proportional to the Nth power of $SINC[\xi/(\Delta x)^{-1}]$ and also is approximately Gaussian:

$$\begin{aligned} \hat{\text{H}}_N[\xi;\, \Delta x] &= \Delta x\left(SINC\left[\frac{\xi}{(\Delta x)^{-1}}\right]\right)^N \\ &\cong \Delta x\; GAUS\left[\frac{\xi}{(\sqrt{N\pi/6}\Delta x)^{-1}}\right] \end{aligned} \tag{14.105}$$

The interpolator applies small weights to the spectrum of the sampled function at large spatial frequencies, which is just another way of saying that the Nth-order interpolator is a lowpass filter.

14.8.3 Realistic Frequency-Domain Interpolators

In the previous section, we constructed impulse responses of interpolators as convolutions of identical rectangle functions. These approximations have the desirable characteristic of finite support, but their nonnegative amplitude at all x ensures that they are poor approximations of the ideal $SINC$ interpolator. This section explores the alternative approach of approximating the ideal rectangular transfer function,

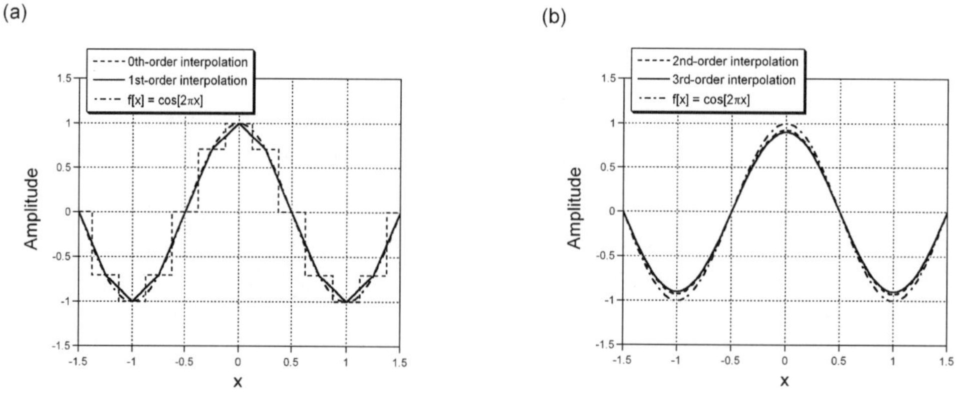

Figure 14.21 Performance of the approximate interpolators for the input function $f[x] = \cos[2\pi x]$ sampled with $\Delta x = 1/16$ units: (a) outputs obtained from zeroth- and first-order interpolators; (b) outputs from second and third orders. For this function, the fidelity improves with increasing order n.

with the goal of producing an impulse response that has bipolar amplitude but finite support. Though these bipolar interpolators may not be feasible to implement in real imaging display hardware (e.g., CRTs), they are useful for graphical regression (curve fitting).

One convenient sequence of functions that may be used to approximate the ideal rectangular transfer function is the family of superGaussian functions introduced in Equation (6.62). The superGaussian approaches the ideal rectangle as the index $n \to +\infty$, and therefore the nth-order function may be interpreted as an nth-order approximation to the ideal frequency-domain rectangle. We have derived closed-form solutions for the Fourier transform for only the first few functions in the sequence, but were able to construct approximations for the spectra via the moment expansion in Chapter 13. In this section, we discuss the use of this sequence of functions to derive possible approximate interpolators, not with the expectation of producing a practical result, but rather because the derivation unites many of the concepts of linear systems that have been considered thus far. In short, the process is a useful academic exercise.

To model the transfer function of ideal interpolation, we must scale the width parameter of the infinite-order frequency-domain superGaussian function in Equation (6.62) to produce the desired rectangle with amplitude Δx and width $(\Delta x)^{-1}$:

$$H[\xi;\, \Delta x] = \Delta x\, RECT[\Delta x \cdot \xi]$$

$$= \lim_{n \to \infty} \{\Delta x \exp[-\pi(2 \cdot \Delta x \cdot |\xi|)^n]\} \tag{14.106}$$

At the Nyquist frequencies $\xi = \pm(2 \cdot \Delta x)^{-1}$, the amplitude of the superGaussian is $e^{-\pi} \cong 0.043\,21$ regardless of the order n. As n is decreased from $+\infty$, the frequency-domain superGaussian is "smoothed" or "blurred" to provide successively poorer approximations to the ideal rectangular form. Though the area of the transfer function in Equation (14.106) increases with n, the central-ordinate theorem guarantees that the area of the corresponding impulse response has the correct value of Δx. The nth-order approximation to the ideal transfer function $H[\xi;\, \Delta x]$ generated from the superGaussian will be denoted by italic upper-case "H" to distinguish it from the nth-order approximation derived in the space domain used in Section 14.8.2. The corresponding nth-order approximation to the impulse response is the inverse Fourier transform of $\hat{H}_n[\xi;\, \Delta x]$ and will be specified by the italic notation $\hat{h}_n[x;\, \Delta x]$. Though we have not derived closed-form solutions for the inverse Fourier transforms for $n > 2$, the asymptotic forms derived in Chapter 13 provide useful approximations.

Several examples of these approximations are compared in Figure 14.21 and will be used to illustrate the process. Each reconstructs the constant function $1[x]$ from the infinite set of its samples without error because all have unit central ordinates. However, the decrease in amplitude of the approximations with increasing ξ means that higher-frequency sinusoids will be reconstructed with less modulation than exhibited by the continuous input function.

The expression for the first-order superGaussian approximation to the ideal transfer function is:

$$\begin{aligned}\hat{H}_1[\xi;\ \Delta x] &= \Delta x\ GAUS[2\cdot\Delta x\cdot\xi;\ 1]\\ &= \Delta x \exp\left[-\frac{|\xi|}{(2\pi\cdot\Delta x)^{-1}}\right]\end{aligned} \tag{14.107}$$

This symmetric decaying-exponential function may be created by summing an appropriately scaled version of $STEP[x]\cdot e^{-x}$ from Equation (6.16) and a "reversed" replica. The corresponding impulse response is the sum of the appropriately scaled complex-valued Lorentzian and its reversed replica:

$$\hat{h}_1[x;\ \Delta x] = \frac{1}{\pi}\left(\frac{1}{1+(x/\Delta x)^2}\right) \tag{14.108}$$

Because this impulse response is nonnegative, it computes a weighted average of the sampled values like the approximate impulse responses derived in the previous section. Because its amplitude at the origin is only $\pi^{-1}\cong 0.3183$ (and thus much smaller than the unit amplitude of the ideal *SINC* function interpolator), the amplitude of the interpolated function may differ significantly from $f[x]$ at the location of a sample. The amplitude of $\hat{h}_1[x;\ \Delta x]$ decays more slowly than the ideal *SINC* function, dropping to $(2\pi)^{-1}\cong 0.1592$ at $x=\Delta x$. The existence of the sharp "cusp" in $\hat{H}_1[\xi;\ \Delta x]$ at the origin ensures that $\hat{h}_1[x;\ \Delta x]$ has infinite support, which means that the amplitude of the interpolated function is a weighted sum of the amplitudes of all samples.

The second-order approximation to $\hat{H}[\xi,\ \Delta x]$ in Equation (14.106) obviously is the standard frequency-domain Gaussian with width parameter $(2\cdot\Delta x)^{-1}$ and central ordinate Δx:

$$\begin{aligned}\hat{H}_2[\xi;\ \Delta x] &= \Delta x\cdot GAUS[2\cdot\Delta x\cdot\xi;\ 2] = \Delta x\exp[-\pi(2\cdot\Delta x\cdot\xi)^2]\\ &= \Delta x\cdot GAUS\left(\frac{\xi}{(2\cdot\Delta x)^{-1}}\right)\end{aligned} \tag{14.109}$$

Therefore, the corresponding impulse response also is a nonnegative Gaussian, with central ordinate and width parameter equal to $\frac{1}{2}$ and $(2\cdot\Delta x)$:

$$\begin{aligned}\hat{h}_2[x;\ \Delta x] &= \Delta x\cdot\frac{1}{2\cdot\Delta x}\cdot GAUS\left(\frac{x}{2\cdot\Delta x}\right)\\ &= \frac{1}{2}GAUS\left(\frac{x}{2\cdot\Delta x}\right)\end{aligned} \tag{14.110}$$

The central-ordinate theorem demonstrates that the area of $\hat{h}_2[x;\ \Delta x]$ is Δx, as expected. The two representations of this interpolator are compared to those for the ideal interpolator in Figure 14.22. The Gaussian impulse response is half as "tall" and about twice as "wide" as the central lobe of the ideal *SINC* function. Note that the Gaussian impulse response decays more quickly with increasing x than the first-order Lorentzian interpolator $\hat{h}_1[x;\ \Delta x]$, which means that the weightings applied to distant samples during the interpolation are smaller for the Gaussian.

For $n>2$, the impulse responses of the higher-order superGaussian transfer functions are not known in closed form, but they may be evaluated from the moment expansion in Equation (13.29) by applying the "transform-of-a-transform" theorem. The corresponding form of the impulse response is an adaptation of Equation (13.29):

$$\mathcal{F}_1^{-1}\{\Delta x\, GAUS[2\cdot\Delta x\cdot\xi;\, n]\} = h_n[x;\,\Delta x] = \sum_{\ell=0}^{+\infty} a_\ell |x|^\ell$$
$$= \sum_{\ell=0}^{+\infty}\left((-1)^\ell \frac{2^{2\ell+1}}{n}\pi^{(2\ell(1-1/n)-1/n)}\frac{\Gamma[(2\ell+1)/n]}{\Gamma[2\ell+1]}\right)\left(\frac{x}{\Delta x}\right)^{2\ell} \quad (14.111)$$

For $n \geq 2$, the magnitude of the coefficients in this summation increases with ℓ until reaching a maximum and then decreases. The magnitude of the coefficients of the first-order Lorentzian impulse response for $n = 1$ is $|a_\ell| = 2^{2\ell+1}/\pi$, which increases monotonically and rapidly with increasing ℓ. Examples of the nth-order impulse responses are shown in Figure 14.22. The area of each is Δx, but the "form" of the interpolators follows a distinct trend with increasing n. The central ordinate of the nth-order interpolator is $\pi^{-1/n}\cdot\Gamma[1+1/n]$, which increases monotonically from $\cong 0.3183$ for $n = 1$ to unity as $n \to +\infty$; this fact merely reflects the increase in area of the transfer functions. The amplitude of both the first- and second-order interpolators is positive everywhere and the lobes are symmetric and "wider" (i.e., decay more slowly with $|x|$) than the central lobe of the ideal *SINC* function interpolator.

As the order n is increased from 2, all of the resulting impulse responses are bipolar and the height and width of the central lobe respectively increase and decrease with increasing n. The properties of these impulse responses may be measured from cases that are calculated via discrete transforms. The zero crossings of the amplitude of the third-order impulse response are located at $x \cong \pm 1.67\cdot\Delta x$ and $x \cong \pm 3.67\cdot\Delta x$.

The impulse response for $n = 4$ has a central lobe that is taller and narrower yet, with the zero crossings at $x \cong \pm 1.46\cdot\Delta x$. There are two smaller positive lobes beyond zero crossings located at $x \cong \pm 2.88\cdot\Delta x$. For $n = 5$, the zero crossings are located at $x \cong \pm 1.36\cdot\Delta x$, $\pm 2.68\cdot\Delta x$, $\pm 3.88\cdot\Delta x$, and $\pm 5\cdot\Delta x$. As n is increased, the zero crossings decrease toward $x = k\cdot\Delta x$. Roughly speaking, a pair of lobes with "significant" amplitude is added for each increment of the order n. The region of support of the "significant" amplitudes increases by approximately Δx on each side of the origin, so that the support of the nth-order impulse response is approximately $n\cdot\Delta x$ on each side of the origin, and so that the total region of support is approximately $2n\cdot\Delta x$. As $n \to \infty$, the impulse responses more closely approximate the ideal *SINC* function interpolator.

The performances of superGaussian interpolators of order 1–4 are compared in Figure 14.23.

14.8.3.1 Realistic Interpolators for Specific Conditions

Other approximations of the ideal *SINC* function interpolator may be constructed by fitting curves that satisfy specific requirements to the amplitudes of the sampled function. For example, the interpolator may be required to have finite support and null amplitude at all integer multiples of Δx other than $x = 0$. Consider the following set of conditions for the space-domain representation of an approximate interpolation function $\hat{h}[x;\,\Delta x]$:

$$\hat{h}[x=0;\,\Delta x] = 1 \quad (14.112a)$$

$$\hat{h}[x=\pm\Delta x;\,\Delta x] = \hat{h}[x=\pm 2\cdot\Delta x;\,\Delta x] = 0 \quad (14.112b)$$

$$\hat{h}[x;\,\Delta x] = 0 \quad \textit{for}\ |x| > 2\cdot\Delta x \quad (14.112c)$$

$$\left.\frac{d\hat{h}[x;\,\Delta x]}{dx}\right|_{x=0} = 0 \quad (14.112d)$$

$$\left.\frac{d\hat{h}[x;\,\Delta x]}{dx}\right|_{x=\pm 2\cdot\Delta x} = 0 \quad (14.112e)$$

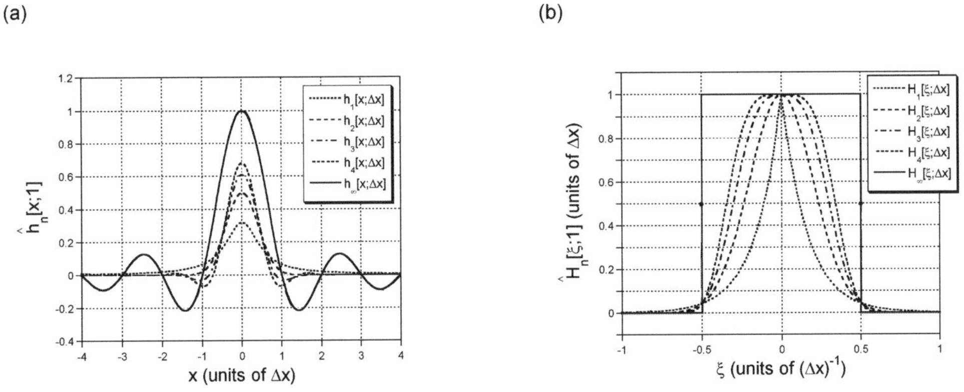

Figure 14.22 The impulse responses (a) and transfer functions (b) of the approximate interpolators derived as superGaussian approximations of $RECT[\Delta x\ \xi]$ in the frequency domain for orders $n = 1$–4. Note that the impulse responses for $n > 2$ have negative regions and thus resemble the ideal interpolator $SINC[x/\Delta x]$.

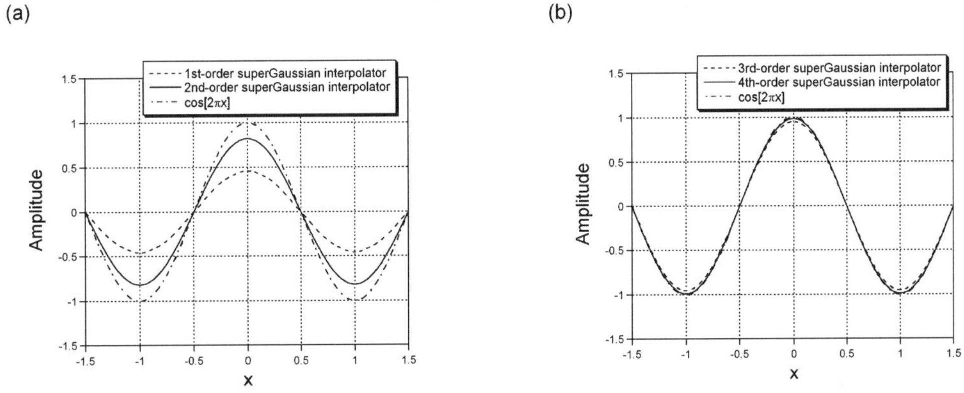

Figure 14.23 Comparison of the performance of the superGaussian interpolators of order $n = 1$–4 with ideal interpolation. The input function is $f[x] = \cos[2\pi x]$ sampled with $\Delta x = 1/16$.

$$\left.\frac{d\hat{h}[x;\ \Delta x]}{dx}\right|_{x=\pm|\Delta x_+|} = \left.\frac{d\hat{h}[x;\ \Delta x]}{dx}\right|_{x=\pm|\Delta x_-|} \tag{14.112f}$$

$$\int_{-\infty}^{+\infty} \hat{h}[x;\ \Delta x]\,dx = \int_{-\infty}^{+\infty} h[x;\ \Delta x]\,dx = \int_{-2\cdot\Delta x}^{+2\cdot\Delta x} \hat{h}[x;\ \Delta x]\,dx = \Delta x \tag{14.112g}$$

In words, the first two conditions determine that this interpolator has unit amplitude at the origin and null amplitude at all other distances from the origin that are integer multiples of Δx. This second requirement ensures that the interpolated function will have the same amplitude as the sample weight at the sample location. The third requirement constrains the support of the interpolator. The fourth requirement is met if $\hat{h}[x;\ \Delta x]$ is even. The derivative condition of Equation (14.112e) ensures a "smooth" transition to zero amplitude at the edges of the interpolator. The derivative condition of Equation (14.112f) ensures a smooth transition of the interpolator through the zero at $\pm\Delta x$. Equation

(14.112g) ensures that the area of the interpolated function must be identical to that generated by the ideal interpolator.

The impulse response $\hat{h}[x,\ \Delta x]$ may be generated by assuming power-series solutions with unknown coefficients for $|x| \leq \Delta x$ and for $\Delta x < |x| \leq 2 \cdot \Delta x$. The series may be terminated at any power, but the number of conditions constrains the number of coefficients that may be determined. The general form for the interpolator is:

$$\hat{h}[x;\ \Delta x] = \begin{cases} a_3|x/\Delta x|^3 + a_2(x/\Delta x)^2 + a_1|x/\Delta x| + a_0 & \textit{if } 0 \leq |x/\Delta x| \leq 1 \\ b_3|x/\Delta x|^3 + b_2(x/\Delta x)^2 + b_1|x/\Delta x| + b_0 & \textit{if } 1 < |x/\Delta x| \leq 2 \\ 0 & \textit{otherwise} \end{cases} \tag{14.113}$$

The seven conditions of Equation (14.112) determine seven of the coefficients a_n and b_n, leaving one undetermined parameter (call it α). The solution to the differential equation with these constraints is:

$$\hat{h}[x;\ \Delta x,\ \alpha] = \begin{cases} (\alpha+2)|x/\Delta x|^3 - (\alpha+3)(x/\Delta x)^2 + 1 & \textit{if } 0 \leq |x/\Delta x| \leq 1 \\ \alpha|x/\Delta x|^3 - 5\alpha(x/\Delta x)^2 + 8\alpha|x/\Delta x| - 4\alpha & \textit{if } 1 < |x/\Delta x| \leq 2 \\ 0 & \textit{otherwise} \end{cases} \tag{14.114}$$

Interpolators that satisfy these conditions may be constructed for all possible real-valued α, but the desired negative amplitude in the region $1 < |x/\Delta x| \leq 2$ is assured only for $\alpha < 0$. For example, setting $\alpha = -1$ yields the expressions:

$$\hat{h}[x;\ \Delta x,\ \alpha = -1] = \begin{cases} 1 - 2(x/\Delta x)^2 + |x/\Delta x|^3 & \textit{if } -\Delta x \leq x \leq +\Delta x \\ 4 - 8|x/\Delta x| + 5(x/\Delta x)^2 - |x/\Delta x|^3 & \textit{if } \Delta x < |x| \leq 2 \cdot \Delta x \\ 0 & \textit{otherwise} \end{cases} \tag{14.115}$$

which has the same slope at $x = \pm\Delta x$ as $\mathit{SINC}[x/\Delta x]$. The interpolator for $\alpha = -\frac{1}{2}$ is:

$$\hat{h}[x;\ \Delta x,\ \alpha = -\tfrac{1}{2}] = \begin{cases} 1 - \frac{5}{2}(x/\Delta x)^2 + \frac{3}{2}|x/\Delta x|^3 & \textit{if } -\Delta x \leq x \leq +\Delta x \\ 2 - 4|x/\Delta x| + \frac{5}{2}(x/\Delta x)^2 - \frac{1}{2}|x/\Delta x|^3 & \textit{if } \Delta x < |x| \leq 2 \cdot \Delta x \\ 0 & \textit{otherwise} \end{cases} \tag{14.116}$$

which minimizes the squared error between the original continuous function and the reconstructed function. These interpolators and their spectra are shown in Figure 14.24, and reconstruction of a sinusoidal wave by the interpolator with $\alpha = -\frac{1}{2}$ and $\alpha = -1$ are shown in Figure 14.25. The transfer function for $\alpha = -1$ exhibits a narrower positive lobe and larger negative lobes which generate "less smooth" reconstructions than those obtained from $\hat{h}[x;\ \Delta x,\ \alpha = -0.5]$. Higher-order interpolators also may be derived that have wider regions of support, and thus which compute weighted sums of more samples.

14.9 Quantization

A signal or image to be read and processed by a digital computer must have both a finite and discrete domain (spatial extent or support) and a finite and discrete range (possible amplitudes or "brightnesses"). The subject of this section is the effect of restricting the range to a finite set of possible amplitudes, which will be distinguished by an integer index in the same manner as the samples. In a

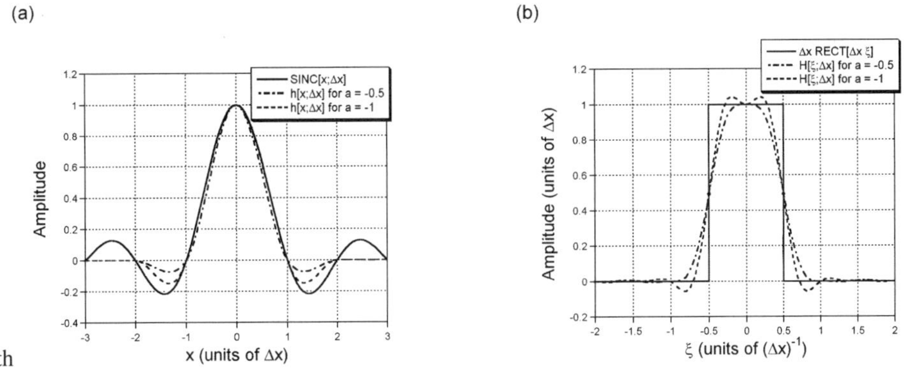

width

Figure 14.24 Impulse responses (a) and transfer functions (b) of realistic interpolators derived from the differential equation for $\alpha = -\frac{1}{2}$ and -1, compared to the ideal interpolator in both domains. Note that the support of both transfer functions is larger than the ideal rectangle and the amplification for $|\xi| \gtrsim 0$ in the transfer function for $\alpha = -1$.

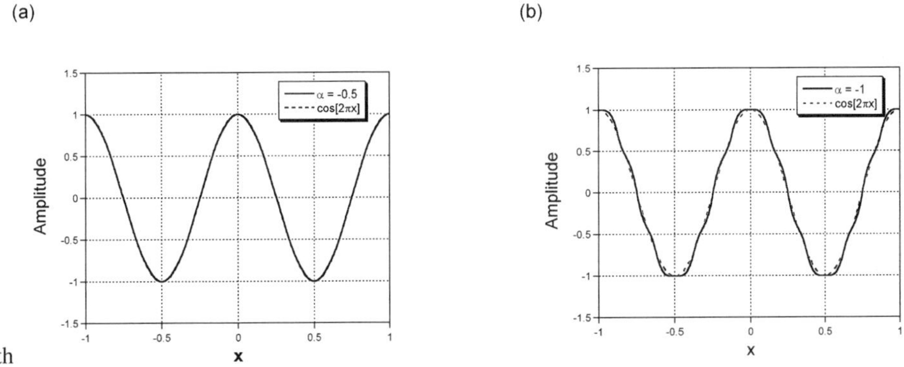

width

Figure 14.25 Performance of the interpolators derived from the differential equation with (a) $\alpha = -\frac{1}{2}$ and (b) $\alpha = -1$ for $f[x] = \cos[2\pi x]$ sampled with $\Delta x = 1/8$ units. Note the deviation in the interpolated function for $\alpha = -1$.

way, we are going to "sample" the amplitude of the signal via the process of *quantization*. We have seen that the data "between" the samples may be interpolated, so that the continuous signal may be recovered from the samples without error if a certain condition can be met. Unfortunately, the same cannot be said of quantization; the data "between" the samples are lost.

To introduce quantization, we assume that the measured amplitude of the sample evaluated at the origin of a some continuous function is a real-valued number, say $f[0,0] = 1.234\,567\,890\ldots\ \mathrm{W/mm^2}$. This amplitude is to be converted to an integer between zero and some maximum value (say, 255) by an *analog-to-digital converter* (A/D or ADC); the specific integer is often called the *quantum number* of the sample. The number of possible quantization levels is determined by the number of binary digits ("bits") available in the ADC. A quantizer with m bits specifies $M = 2^m$ possible amplitudes ("gray levels" or "digital counts"). Currently, the most common image quantizers have $m = 8$ bits (1 *byte*) per sample and thus each sample may take on one of 256 possible levels, usually numbered from 0 to 255, where $0 \equiv$ black and $255 \equiv$ white. The 1-D signals, such as in audio recording, often are digitized to 16 or even 18 bits, yielding 65 536 and 262 144 levels, respectively. As we will see at the end of this section, the number of "effective" bits available in image digitizers depends on the noise characteristics of the detector. For example, it does little good to digitize a signal to 4096 levels (12 bits) if the maximum signal amplitude is only a factor of 100 larger than the uncertainty in the amplitude due to noise. With the continuous improvement in image detectors, images with 12 bits of data have become much more common.

Following our informal operator convention, the quantization operation is denoted by $\mathcal{Q}$ and acts on the amplitude f of the sampled input signal to yield a quantized amplitude f_q:

$$\mathcal{Q}\{f[n \cdot \Delta x]\} = f_q[n \cdot \Delta x] = q \cdot \Delta f \tag{14.117}$$

if Δf is the interval between levels of the quantizer and q is the quantum number of the state. All signal amplitudes within a range of "width" Δf in the neighborhood of the level $f = q \cdot \Delta f$ will be assigned the same quantum number, which means that the conversion to a discrete range entails *thresholding* by truncating or rounding the real-valued amplitudes. For example, all measured irradiances between 0.76 and 0.77 $\mathrm{W/mm^2}$ might be assigned to quantum number ("gray level") 59. The threshold conversion clearly is a *nonlinear* operation because:

$$\mathcal{Q}\{f_1 + f_2\} \neq \mathcal{Q}\{f_1\} + \mathcal{Q}\{f_2\} \tag{14.118}$$

in general. The *resolution* of the quantizer is the interval Δf of amplitude between adjacent quantum levels. The intervals between quantum numbers in most quantizers is fixed. For example, it may be chosen to be the ratio of the range of amplitudes to the number of intervals between levels:

$$\Delta f = \frac{f_{\max} - f_{\min}}{2^m - 1} \tag{14.119}$$

where $f_{\max}$ and $f_{\min}$ are the extrema of the measured irradiances of the image samples and m is the number of bits of the quantizer.

The nonlinear character of quantization makes it inappropriate to analyze a complete analog-to-digital-to-analog system by linear methods. This problem is often ignored, particularly in cases where the number of quantum levels is sufficiently large and b is small compared to the noise of the detector; under these conditions, the range of the digitized image "appears" to be continuous. Recall that the discussion in Section 9.8.19 shows that the support of the spectrum of the quantized function is wider than the support of the original function (sampled or not).

If the darkest and brightest samples of a continuous-tone image have measured irradiances $f_{\min}$ and $f_{\max}$ respectively, and if the image is to be quantized to m bits (2^m gray levels), then the quantized signal may be defined via:

$$\begin{aligned} f_q[n \cdot \Delta x] &\equiv q \cdot \Delta f \\ &= \mathcal{Q}\left\{\frac{f[n \cdot \Delta x] - f_{\min}}{\Delta f}\right\} \\ &= \mathcal{Q}\left\{\left(\frac{f[n \cdot \Delta x] - f_{\min}}{f_{\max} - f_{\min}}\right) \cdot (2^m - 1)\right\} \end{aligned} \tag{14.120}$$

The type of nonlinearity in the quantizer $\mathcal{Q}$ determines the locations of the *decision levels*, where the quantizer "jumps" from one level to the next. The image irradiances are "reconstructed" by assigning all pixels with a particular gray level f_q to the same irradiance value $\hat{f}$, which might be considered as the (approximate) "inverse" of quantization. The reconstruction level is often placed between the decision levels by adding a factor of half the step size:

$$\hat{f}[n \cdot \Delta x] = f_q[n \cdot \Delta x] \cdot \left(\frac{f_{\max} - f_{\min}}{2^m - 1}\right) + f_{\min} + \frac{\Delta f}{2} \tag{14.121}$$

Of course, the estimated amplitude $\hat{f}$ generally differs from the original continuous amplitude f due to the quantization. The difference $f[n \cdot \Delta x] - \hat{f}[n \cdot \Delta x]$ is the *quantization error* at the sample, which will be discussed more completely in the next section. The goal of so-called *optimum* quantization is to adjust the quantization scheme (i.e., the locations of the decision and reconstruction levels) to reconstruct the set of image irradiances which "most closely approximates" the ensemble of original values. The criterion which defines the "goodness" of the approximation, combined with the statistics of the original sample irradiances, determine the parameters of the optimum quantizer.

To this point, the quantizer determines the quantum number of the amplitude based only on the amplitude of the specific sample being considered. Such a quantizer is *memoryless*, i.e., the quantization level at each sample is computed independently of that at all other pixels. It also is possible to design and implement quantizers *with memory*; these base the choice of quantum level on the amplitudes of other samples as well as that under consideration. Quantizers with memory have significant advantages in some applications, as will be discussed shortly.

14.9.1 Quantization "Noise"

The difference between the true input irradiance (or "brightness") and the quantized image irradiance is the quantization error at that sample:

$$\epsilon[n \cdot \Delta x] \equiv f[n \cdot \Delta x] - \hat{f}[n \cdot \Delta x] \tag{14.122}$$

Note that the quantization error ϵ is generally a real number and has the same units as the amplitude f (e.g., irradiance in $\mathrm{W/m^2}$). If the rounding nonlinearity is used, the resulting quantization error is bipolar. If the nonlinearity is "truncation" by discarding the noninteger part, then the continuous amplitude f is at least as large as the quantized amplitude f_q. This means that the quantization error cannot be negative in this case. If the quantizer performs rounding, the error is bipolar.

It is often useful to describe the statistical properties of the quantization error, which depends on both the type of quantizer (rounding or truncation) and on the statistics of the continuous amplitudes. Though we give no proof, it is often true that the probability distribution of quantization error of a rounding quantizer with fixed step size Δf applied to a realistic signal approximates a zero-mean uniform distribution. The variance of the error is at least approximately $\langle(\epsilon_1[n])^2\rangle = (\Delta f)^2/12$, as stated in Section 6.4. That this error distribution is usually valid will be demonstrated by examining the statistics of quantization error in two 1-D examples. The first is a section of a scaled cosine whose amplitude lies in the range [0, 1] which has been quantized to eight levels with the maximum at 1 and minimum at 0, so that $\Delta f = \frac{1}{7}$. The input and quantized images are shown in Figure 14.26a. The calculated mean of the error for a large number of such samples is $\langle\epsilon_1[n]\rangle = 9.5 \times 10^{-4} \cong 0$ and the variance for this case is $\langle\epsilon_1^2[n]\rangle = 1.66 \times 10^{-3}$, whereas the calculated value is $\sigma^2 \cong (\Delta f)^2/12 \cong 1.70 \times 10^{-3}$. That the distribution of error is approximately uniform is shown by examining the histogram, as shown in Figure 14.26b.

The second image used to test for the statistics of the quantization error is composed of 256 samples of Gaussian distributed random noise in the interval [0,63] and quantized to 64 levels so that $\Delta f =$ 1; examples of the samples are shown in Figure 14.27. The mean and variance of the error in this case were $\mu_f \cong 3.06 \times 10^{-3} \cong 0$ and $\sigma_f^2 \cong 8.52 \times 10^{-2}$, which is close to the predicted value $\sigma_f^2 \cong (\Delta f)^2/12 = 8.33 \times 10^{-2}$. Again, the histogram of the quantization error is approximately rectangular, which indicates that the error is approximately uniformly distributed.

The *total quantization error* of a quantized signal is the sum of the individual errors over the entire image:

$$\epsilon = \sum_n \epsilon[n \cdot \Delta x] \tag{14.123}$$

It is easy to see that the total quantization error may be small for an image due to cancellations of large positive and negative excursions. The *mean-squared error* (the average of the squared error, also called

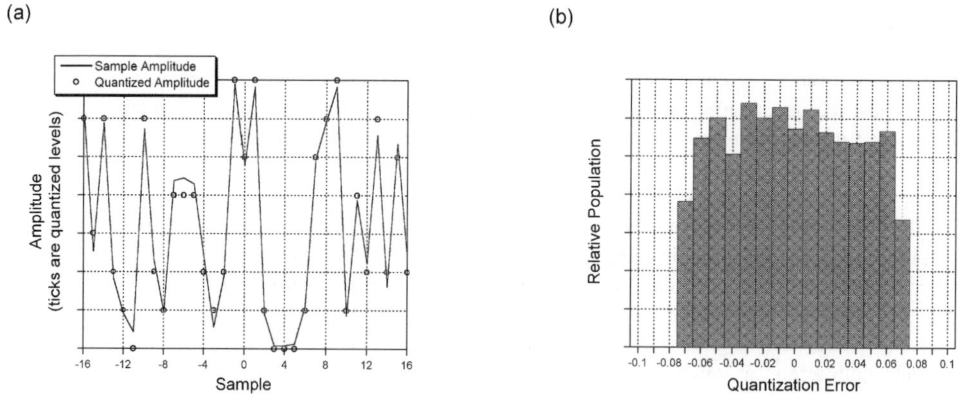

Figure 14.26 Quantization of uniformly distributed noise. (a) Samples of noise $f[n]$ from a uniform distribution over the range [0, 1] before and after quantizing to eight levels. The maximum and minimum quantized levels are 1 and 0, respectively, so that $\Delta f = \frac{1}{7}$. (b) Histogram of the quantization error $f[n] - \mathcal{Q}\{f[n]\}$, which is uniformly distributed over the interval $[-\Delta f/2, +\Delta f/2]$.

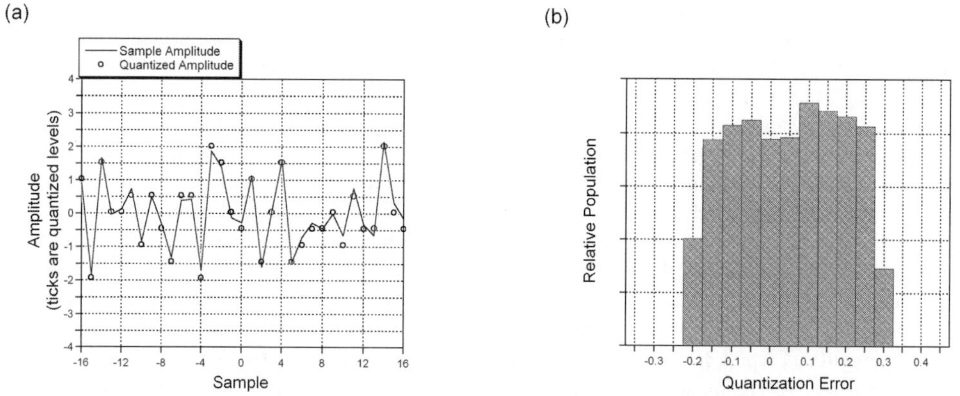

Figure 14.27 Quantization of Gaussian noise. (a) Samples of noise from a Gaussian distribution with $\mu = 0$ and $\sigma^2 = 1$ before and after quantizing to 17 levels over the interval $[-4, +4]$, so that $\Delta f = \frac{1}{2}$. (b) Histogram of the quantization error $f[n] - \mathcal{Q}\{f[n]\}$, which is uniformly distributed over the interval $[-\Delta f/2, +\Delta f/2]$.

the "MSE") is a better descriptor of the fidelity of the quantization:

$$\epsilon^2 = \frac{1}{N} \sum_n \epsilon^2[n \cdot \Delta x] \tag{14.124}$$

where N is the number of samples in the image. The dimension of the MSE is the square of the dimensions of the signal amplitude. For example, if the image irradiance is measured in W/mm^2, the units of the MSE ϵ^2 are $(W/mm^2)^2$. The dimensional disparity is removed by using the *root-mean-squared error* (RMSE), which is the square root of Equation (14.123) and has the same dimensions as the error:

$$RMSE = \sqrt{\overline{\epsilon^2}} = \sqrt{\frac{1}{N}\sum_n \epsilon^2[n \cdot \Delta x]} \tag{14.125}$$

Several properties should be obvious: that it is desirable to minimize the RMSE due to quantization, that the RMSE for a particular image is a function of the quantizer used, and that the RMSE resulting from a single quantizer will be different for different input images. The brute-force method for minimizing quantization error is to add more bits in the ADC, at least to the point where the quantization step size is equal to the RMS noise of the detector. This tactic will increase both the cost of the quantizer and the memory required to store the image.

A second method to reduce quantization error is to use "nonuniform" (also called "tapered") quantization, where the interval between quantization steps is varied to take advantage of the particular statistics of the signal amplitude. This may be accomplished either by directly changing the step size for different amplitudes or by performing a nonlinear mapping of the amplitude f to obtain a "new" amplitude f', which is then quantized uniformly. For example, if many pixels in an image are "dark" (small amplitudes) and relatively few are "bright", then uniform quantization will result in a significant RMSE. If the image is passed through a logarithmic transformation, so that $f' = \log_e[f]$, then the difference between small amplitudes will be amplified, while large amplitudes will be compressed together. After uniform quantization, the nonlinear scaling is compensated before display by passing through a nonlinear mapping of the form $\hat{f} = \exp[f']$ to regenerate the approximately "correct" output levels. This scheme is the operating principle of the *compander* (compressor–expander), which is often used to maintain fidelity of signals that must be passed through transmission lines with limited dynamic range. Clearly, the reduction in error is achieved at the cost of extra computation.

14.9.2 *SNR* of Quantization

Since quantization introduces errors ("noise") that may be characterized statistically, it may be useful to express these errors in terms of the *SNR* that was introduced in Chapter 6. Though the input signal and the type of quantizer must be known to evaluate the probability density function of the quantization error in a strict sense, the quantization errors in the two examples just considered suggest that we can approximate the distribution of quantization error as approximately uniform with a rectangular probability. In the case of an m-bit uniform quantizer (2^m gray levels) with levels spaced by intervals of width Δf over the full analog dynamic range of the signal, the error due to quantization will be (approximately) uniformly distributed over this interval Δf. If the nonlinearity of the quantizer is *rounding*, the mean value of the error should be approximately 0; the error should as often be negative as positive. If the quantizer *truncates* to the next lower integer, the mean value of the quantization error should be approximately $-\Delta f/2$. In either case, we showed in Chapter 6 that the variance of uniformly distributed noise is:

$$\sigma_n^2 = \frac{(\Delta f)^2}{12} \tag{14.126}$$

The "width" Δf of a quantization level for an m-bit quantizer and signal f with maximum and minimum amplitudes f_{max} and f_{min} is:

$$\Delta f = \frac{f_{\text{max}} - f_{\text{min}}}{2^m} \tag{14.127}$$

If the quantization noise is (assumed to be) uniformly distributed, the variance of the quantization noise is:

$$\sigma_n^2 = \frac{(\Delta f)^2}{12} = \frac{(f_{\text{max}} - f_{\text{min}})^2}{12 \cdot (2^m)^2} = \frac{(f_{\text{max}} - f_{\text{min}})^2}{12 \cdot 2^{2m}} \tag{14.128}$$

The resulting *SNR* is the ratio of the variance of the signal to that of the quantization noise:

$$SNR \equiv \frac{\sigma_f^2}{\sigma_n^2} = \sigma_f^2 \cdot \frac{12 \cdot 2^{2m}}{(f_{\text{max}} - f_{\text{min}})^2} \tag{14.129}$$

which may be expressed on a logarithmic scale as dB:

$$\begin{aligned}
SNR &= 10\ \log_{10}[\sigma_f^2 \cdot 12 \cdot 2^{2m}] - 10\ \log_{10}[(f_{\max} - f_{\min})^2] \\
&= 10\ \log_{10}[\sigma_f^2] + 10\ \log_{10}[12] + 20m\ \log_{10}[2] - 10\log_{10}[(f_{\max} - f_{\min})^2] \\
&\cong 10\log_{10}[\sigma_f^2] + 10 \cdot 1.079 + 20m \cdot 0.301 - 10\log_{10}[(f_{\max} - f_{\min})^2] \\
&\cong 6.02\ m + 10.8 + 10\ \log_{10}\left[\frac{\sigma_f^2}{(f_{\max} - f_{\min})^2}\right] \quad [\text{dB}] \qquad (14.130)
\end{aligned}$$

The last term obviously depends on both the signal and the quantizer, but the first terms do not. This result demonstrates that the *SNR* of quantization increases by $\gtrapprox 6\,\text{dB}$ for every additional bit added to the quantizer, so that the step size b decreases. If using the (poor) estimate that $\sigma_f^2 = (f_{\max} - f_{\min})^2$, then the third term evaluates to zero and the approximate *SNR* is:

$$\begin{aligned}
\textit{SNR for quantization to m bits} &\cong 6.02m + 10.8 + 10\log_{10}[1] \\
&= 6.02m + 10.8 \quad [\text{dB}] \qquad (14.131)
\end{aligned}$$

The statistics of the signal (and thus its variance σ_f^2) may be approximated for many types of signals (e.g., music, speech, realistic images) as outcomes of random processes. The histograms of these signals usually are peaked at or near the mean value μ_f and the probability of a gray level decreases for values away from the mean; the signal approximately is the output of a Gaussian random process with variance σ_f^2. By selecting the dynamic range of the quantizer $f_{\max} - f_{\min}$ to be sufficiently larger than σ_f, few (if any) levels should be saturated, and therefore clipped by the quantizer. We assume that virtually no values are clipped if the maximum and minimum levels of the quantizer are four standard deviations from the mean level:

$$\mu_f - f_{\min} = f_{\max} - \mu_f = \frac{f_{\max} - f_{\min}}{2} = 4 \cdot \sigma_f \qquad (14.132)$$

In other words, we may choose the step size between levels of the quantizer to satisfy the criterion:

$$f_{\max} - f_{\min} = 8 \cdot \sigma_f \Longrightarrow \frac{\sigma_f^2}{(f_{\max} - f_{\min})^2} = \frac{1}{64} \qquad (14.133)$$

Under this condition, the *SNR* of the quantization process becomes:

$$\begin{aligned}
SNR &= 6.02m + 10.8 + 10\log_{10}\left[\frac{1}{64}\right] \\
&= 6.02m + 10.8 + 10 \cdot (-1.806) \\
&= 6.02m - 7.26 \quad [\text{dB}] \qquad (14.134)
\end{aligned}$$

which is 18 dB less than the estimate obtained by assuming that $\sigma_f^2 \cong (f_{\max} - f_{\min})^2$. This again demonstrates that the original estimate of *SNR* was optimistic.

We can use this result to determine the "useful" number of bits of quantization for different imaging systems. The best *SNR* that can be obtained from analog recording (such as on magnetic tape) is approximately 65 dB, which is equivalent to that from a signal digitized to 12 bits per sample or 4096 gray levels. To effectively digitize at 16 bits per sample (as is common in audio recording for CD players), the *SNR* of the recording medium must be approximately 89 dB.

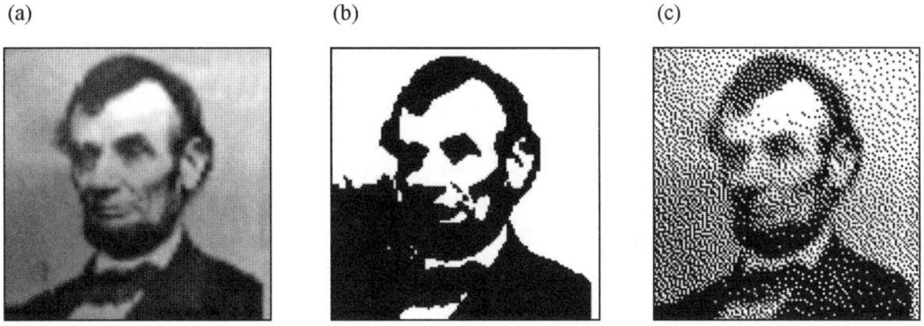

Figure 14.28 Error-diffused quantization: (a) original gray-scale image; (b) after thresholding at the 50% level; (c) after error-diffused quantization using the original algorithm of Floyd and Steinberg. When viewed at a sufficient distance, the blurring action of the human visual system produces the illusion of gray scale.

14.9.3 Quantizers with Memory – "Error Diffusion"

A computationally simpler means than nonlinear companding has been developed to reduce quantization error. It acts by adding "memory" to the quantizer, so that the quantized value at a sample is determined in part by the quantization error at nearby samples. A simple method which generally results in reduced total error without a priori knowledge of the statistics of the input image and without much additional computational complexity was introduced in imaging by Floyd and Steinberg (1975). The method may be used to simulate gray-level images on binary image displays by sacrificing spatial resolution to obtain more quantization levels. The method compensates the amplitude of the sample to be quantized based on the quantization error of neighboring samples. For this reason, the method is known as *error diffusion*. In the simplest form of error-diffused quantization, the error at a sample is added to the amplitude of the next sample before quantization. In the 1-D case, the quantization level at sample location x is the gray level of the sample plus the error $\epsilon[x-1]$ at the preceding sample. The process is implemented by two alternating processes. The quantization performed at each sample is:

$$f_q[x] = Q\{f[x] + \epsilon[x-1]\} \tag{14.135a}$$

and the error at that sample is:

$$\epsilon[x] = f[x] - f_q[x] \tag{14.135b}$$

Note that the second calculation is linear and thus may be investigated by applying linear methods (Knox and Eschbach, 1993)

In the 2-D case, the error may be weighted and propagated in different directions. A discussion of the use of error diffusion in AD conversion was presented by Anastassiou (1989).

Examples of 2-D error diffusion in Figure 14.28 demonstrate the process. Note that the MSEs of the two quantized images may be very similar, but more "fine detail" is preserved in binary images from the quantizer with memory.

It is easy to generalize the error-diffused quantization to more than two levels, and it may be used to perform useful processing of images.

14.10 Discrete Convolution

We now relate the continuous convolution of two bandlimited functions $f[x]$ and $h[x]$ and their sampled relatives $f_s[x;\,\Delta x]$ and $h_s[x;\,\Delta x]$. The goal of this discussion is to construct an equivalent form of the continuous convolution as a summation. Both input functions have infinite support and thus so will the output $g[x]$. The discrete arrays are assumed to be sampled from $f[x]$ and $h[x]$ without aliasing, which is ensured if this criterion is satisfied:

$$\Delta x < \frac{1}{2 \cdot \xi_{\max}} \quad \textit{where } \xi_{\max} = \max(|\xi_f|, |\xi_h|) \tag{14.136}$$

where the maximum frequencies of the two functions are respectively $|\xi_f|$ and $|\xi_h|$. Both sampled functions may be written in the form of Equation (14.5):

$$\begin{aligned} f_s[x;\ \Delta x] &= f[x] \cdot \left(\frac{1}{\Delta x} COMB\left[\frac{x}{\Delta x}\right]\right) \\ &= \sum_{p=-\infty}^{+\infty} f[x] \cdot \delta[x - p \cdot \Delta x] = \sum_{p=-\infty}^{+\infty} f[p \cdot \Delta x] \cdot \delta[x - p \cdot \Delta x] \end{aligned} \tag{14.137}$$

Since both $f[x]$ and $h[x]$ are sampled with the same Δx, each may be reconstructed from its infinite set of ideal samples by convolution with the same ideal interpolator using Equation (14.35):

$$\begin{aligned} f[x] &= f_s[x;\ \Delta x] * SINC\left[\frac{x}{\Delta x}\right] \\ &= \left(\sum_{p=-\infty}^{+\infty} f[p \cdot \Delta x] \cdot \delta[x - p \cdot \Delta x]\right) * SINC\left[\frac{x}{\Delta x}\right] \end{aligned} \tag{14.138a}$$

$$\begin{aligned} h[x] &= h_s[x;\ \Delta x] * SINC\left[\frac{x}{\Delta x}\right] \\ &= \left(\sum_{q=-\infty}^{+\infty} h[q \cdot \Delta x] \cdot \delta[x - q \cdot \Delta x]\right) * SINC\left[\frac{x}{\Delta x}\right] \end{aligned} \tag{14.138b}$$

The continuous convolution produces an output $g[x]$ that is guaranteed by the filter theorem to be bandlimited at the same spatial frequency $\xi_{\max}$ and thus may be sampled at the same Δx without aliasing:

$$g[x] = \left(\sum_{n=-\infty}^{+\infty} g[n \cdot \Delta x] \cdot \delta[x - n \cdot \Delta x]\right) * SINC\left[\frac{x}{\Delta x}\right] \tag{14.139}$$

We now want to evaluate the continuous convolution of the sampled functions to look for the relationship of sample weights of g with f and h. Direct substitution of Equation (14.139) into the definition of convolution in Equation (8.31) produces:

$$\begin{aligned} f[x] * h[x] &= \left(f_s[x;\ \Delta x] * SINC\left[\frac{x}{\Delta x}\right]\right) * \left(h_s[x;\ \Delta x] * SINC\left[\frac{x}{\Delta x}\right]\right) \\ &= (f_s[x;\ \Delta x] * h_s[x;\ \Delta x]) * \left(SINC\left[\frac{x}{\Delta x}\right] * SINC\left[\frac{x}{\Delta x}\right]\right) \\ &= \left(\sum_{p=-\infty}^{+\infty} f[p \cdot \Delta x] \cdot \delta[x - p \cdot \Delta x]\right) \\ &\quad * \left(\sum_{q=-\infty}^{+\infty} h[q \cdot \Delta x] \cdot \delta[x - q \cdot \Delta x]\right) * \Delta x\ SINC\left[\frac{x}{\Delta x}\right] \end{aligned} \tag{14.140}$$

where the convolution of the two *SINC* functions was evaluated via the filter theorem. Now we convolve the sampled functions:

$$\left(\sum_{p=-\infty}^{+\infty} f[p\cdot\Delta x]\cdot\delta[x-p\cdot\Delta x]\right)*\left(\sum_{q=-\infty}^{+\infty} h[q\cdot\Delta x]\cdot\delta[x-q\cdot\Delta x]\right)$$

$$=\sum_{p=-\infty}^{+\infty} f[p\cdot\Delta x]\sum_{q=-\infty}^{+\infty} h[q\cdot\Delta x]\cdot(\delta[x-p\cdot\Delta x]*\delta[x-q\cdot\Delta x])$$

$$=\sum_{p=-\infty}^{+\infty} f[p\cdot\Delta x]\sum_{q=-\infty}^{+\infty} h[q\cdot\Delta x]\cdot\delta[x-(p+q)\cdot\Delta x]$$

$$=\sum_{p=-\infty}^{+\infty}(f[p\cdot\Delta x]\cdot h[x-p\cdot\Delta x])\cdot\delta[x-p\cdot\Delta x] \tag{14.141}$$

After substituting into Equation (14.140) and comparing with Equation (14.139), we see that:

$$g[x]=\left(\sum_{p=-\infty}^{+\infty}(f[p\cdot\Delta x]\cdot h[x-p\cdot\Delta x])\cdot\delta[x-p\cdot\Delta x]\right)*\Delta x\ SINC\left[\frac{x}{\Delta x}\right]$$

$$=\left(\Delta x\left(\sum_{p=-\infty}^{+\infty} f[p\cdot\Delta x]\cdot h[x-p\cdot\Delta x]\right)\cdot\frac{1}{\Delta x}\,COMB\left[\frac{x}{\Delta x}\right]\right)*SINC\left[\frac{x}{\Delta x}\right] \tag{14.142}$$

The sampled output may be written as a discrete summation of the sampled input functions:

$$g[n\cdot\Delta x]=\Delta x\cdot\sum_{p=-\infty}^{+\infty}(f[p\cdot\Delta x]\cdot h[(n-p)\cdot\Delta x])$$

$$\Longrightarrow g[n]=\Delta x\cdot\sum_{p=-\infty}^{+\infty} f[p]\cdot h[n-p] \tag{14.143}$$

It is important to recognize that Equation (14.143) relates dimensionless arrays and that the factor of Δx is required to obtain the "correct" amplitudes of the discrete convolution of data arrays composed of amplitudes that are not pure numbers.

The limits of the summation of the discrete convolution are $\pm\infty$ because the support of the bandlimited continuous functions is so. The complete description of the output array $g[n]$ requires that Equation (14.143) be evaluated at an infinite number of samples, where each computation is an infinite sum. Thus the discrete convolution also cannot be calculated in real life. If either or both of the bandlimited constituent functions $f[x]$ and $h[x]$ are periodic, then so will be $g[x]$. In these cases, only a single period of $g[n]$ need be evaluated.

In the more realistic cases where either or both of the input and system functions have finite support, and therefore infinite bandwidth, the samples of $f_s[x;\,\Delta x]$ and/or $h_s[x;\,\Delta x]$ will be aliased. The discussion just concluded indicates that the samples of $g_s[x;\,\Delta x]$ also must be aliased, and thus the computed amplitudes of the discrete convolution will be incorrect. That being said, the results obtained in such cases may still be (and usually are) useful approximations to the true function.

PROBLEMS

14.1 Find expressions for the following:

(a) $COMB[x/\Delta x]*RECT[x/\Delta x]$

(b) $COMB[x/\Delta x]*TRI[x/\Delta x]$

(c) $COMB[x/\Delta x] * SINC[x/\Delta x]$

(d) $COMB[x/\Delta x] * SINC^2[x/\Delta x]$

14.2 Derive the expression for the moments of the transfer function of the first-order superGaussian interpolator. Discuss any problems that occur if these are used to derive the form of the impulse response.

14.3 Analyze the sampling functions shown and specified below to determine their sampling characteristics, e.g., the associated Nyquist sampling frequency (if any), whether a function sampled with $s[x;\ \Delta x]$ may be recovered from the samples, the corresponding interpolation function, etc.

(a) $s_1[x;\ \Delta x] = ((1/\Delta x)\ COMB[x/\Delta x]) \cdot (TRI[x/b] * (1/2b)\ COMB[x/2b])$ where $b > \Delta x$. Consider any interplay of the factors Δx and b.

(b) $s_2[x;\ \Delta x]$ as shown in the figure.

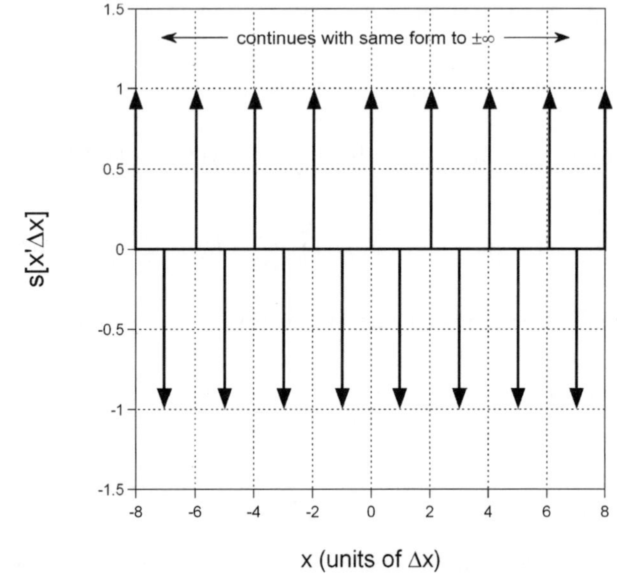

14.4 What are the amplitudes of the samples of $SINC[x]$ and $SINC^2[x]$ when sampled exactly at the Nyquist frequency?

14.5 Describe how to sample a function most efficiently without aliasing (i.e., fewest number of samples) if the spectrum is asymmetric.

14.6 Compute the discrete convolution of the following functions and plot the results:

(a) $f[n] = RECT[n/16],\ h[n] = RECT[(n-10)/16]$

(b) $f[n] = h[n] = SINC[n/16]$

(c) $f[n] = \sin[2\pi n/32] \cdot RECT[n/128],\ h[n] = RECT[n/4]$

(d) $f[n] = \sin[2\pi n/32] \cdot RECT[n/128],\ h[n] = RECT[n/8]$

(e) $f[n] = \sin[2\pi n/32] \cdot RECT[n/128],\ h[n] = RECT[n/16]$

(f) $f[n] = \sin[2\pi n/32] \cdot RECT[n/128],\ h[n] = RECT[n/32]$

(g) $f[n] = \sin[2\pi n/32] \cdot RECT[n/128],\ h[n] = RECT[n/48]$

15

Discrete Fourier Transforms

This chapter extends the discussion of discrete functions that began in Chapter 14 to include explicit consideration of the discrete Fourier transform (DFT). Some of the relevant properties of these spectra were previewed in Chapter 9 and used in Chapter 14; in particular, Equation (9.171) demonstrated that the Fourier transform of any infinite-support discrete function is periodic. This chapter includes descriptions of the modifications to the DFT that are necessary if the discrete space-domain function has finite support.

As a brief review of the relevant results of Chapter 14, consider an arbitrary continuous 1-D function $f[x]$ that has been sampled by the ideal *COMB* function at uniform intervals $\Delta x > 0$. Equation (14.5) presented the form of the sampled function:

$$f_s[x;\,\Delta x] = f[x] \cdot \frac{1}{\Delta x} COMB\left[\frac{x}{\Delta x}\right] = \sum_{n=-\infty}^{+\infty} f[n \cdot \Delta x] \cdot \delta[x - n \cdot \Delta x] \tag{15.1}$$

where the property of Dirac delta functions in products from Equation (6.115) has been used. The modulation theorem of Equation (9.140) is used to construct a simple expression for the spectrum of this discrete function in terms of the two continuous component functions:

$$\begin{aligned}\mathcal{F}_1\left\{f[x] \cdot \frac{1}{\Delta x} COMB\left[\frac{x}{\Delta x}\right]\right\} &= F[\xi] * COMB[\xi \cdot \Delta x] \\ &= F[\xi] * \left(\sum_{k=-\infty}^{+\infty} \delta\left[\Delta x \cdot \left(\xi - \frac{k}{\Delta x}\right)\right]\right) \\ &= \frac{1}{\Delta x} \sum_{k=-\infty}^{+\infty} F\left[\xi - \frac{k}{\Delta x}\right]\end{aligned} \tag{15.2}$$

where the scaling property of the Dirac delta function and the fact that $\Delta x > 0$ have been used. This result confirms the discussion of Equation (9.172) that the spectrum of a sampled function is periodic at intervals $(\Delta x)^{-1}$, regardless of the character of the continuous spectrum $F[\xi]$.

Fourier Methods in Imaging Roger L. Easton, Jr.

Alternatively, the spectrum of the sampled function $f_s[x;\,\Delta x]$ may be computed by direct integration of Equation (15.1) via the sifting property of the Dirac delta function from Equation (6.113):

$$\begin{aligned}\mathcal{F}_1\{f_s[x;\,\Delta x]\} \equiv F_s[\xi;\,\Delta x] &= \int_{-\infty}^{+\infty}\left(f[x]\cdot\frac{1}{\Delta x}COMB\left[\frac{x}{\Delta x}\right]\right)\cdot e^{-2\pi i\xi x}\,dx\\ &= \int_{-\infty}^{+\infty} f[x]\cdot\left(\sum_{n=-\infty}^{+\infty}\delta[x-n\cdot\Delta x]\right)\cdot e^{-2\pi i\xi x}\,dx\\ &= \sum_{n=-\infty}^{+\infty}\left(\int_{-\infty}^{+\infty}(f[x]\,e^{-2\pi i\xi x})\cdot\delta[x-n\cdot\Delta x]\,dx\right)\\ F_s[\xi;\,\Delta x] &= \sum_{n=-\infty}^{+\infty} f[n\cdot\Delta x]\,e^{-2\pi i\xi\cdot n\cdot\Delta x}\end{aligned}\tag{15.3}$$

Recall from Chapter 14 that the array $f[n\cdot\Delta x]$ is the set of discrete amplitudes that modulate the components of the *COMB* function to create the sampled function $f_s[x;\,\Delta x]$. Equation (15.3) demonstrates that the spectrum of a discrete function at the specific frequency ξ_0 is the sum of the infinite set of products of $f[n\cdot\Delta x]$ and of the corresponding samples of the "mask" function $e^{-2\pi i\xi_0\cdot n\cdot\Delta x}$. Though the spectrum $F_s[\xi;\,\Delta x]$ is generated from a sampled array, it is a function of the continuous variable ξ; in other words, Equation (15.3) may be evaluated at any value of ξ in the infinite frequency domain. The continuous periodic and infinite-support spectrum $F_s[\xi;\,\Delta x]$ is the DFT of the sampled function $f_s[x;\,\Delta x]$.

As an aside, recognize that this periodic spectrum $F_s[\xi;\,\Delta x]$ generally is complex valued and may be represented either as its real/imaginary parts or as magnitude/phase:

$$|F_s[\xi;\,\Delta x]| \equiv \sqrt{(\Re\{F_s[\xi;\,\Delta x]\})^2+(\Im\{F_s[\xi;\,\Delta x]\})^2}\tag{15.4a}$$

$$\Phi[\xi] \equiv \tan^{-1}\left[\frac{\Im\{F_s[\xi;\,\Delta x]\}}{\Re\{F_s[\xi;\,\Delta x]\}}\right]\tag{15.4b}$$

We already know from Equation (9.172) that the period of the discrete spectrum $F_s[\xi;\,\Delta x]$ is identical to the period $(\Delta x)^{-1}$ of the spectrum of the ideal sampling function (the *COMB*). This is confirmed by noting that the "mask" function $e^{-2\pi i\xi\cdot n\cdot\Delta x}$ in Equation (15.3) evaluates to unity at all frequencies ξ that satisfy the condition:

$$\xi\cdot\Delta x = k \Longrightarrow e^{-2\pi i\xi\cdot(n\cdot\Delta x)} = 1\tag{15.5}$$

where k is any integer. This means that the amplitudes of the spectrum evaluated at any integer multiple of $(\Delta x)^{-1}$ are identically:

$$\begin{aligned}F_s\left[\xi=\frac{k}{\Delta x};\,\Delta x\right] &= \sum_{n=-\infty}^{+\infty} f[n\cdot\Delta x](e^{-2\pi i(k/\Delta x)})^{(n\cdot\Delta x)}\\ &= \sum_{n=-\infty}^{+\infty} f[n\cdot\Delta x](e^{-2\pi i})^{nk}\\ &= \sum_{n=-\infty}^{+\infty} f[n\cdot\Delta x]\end{aligned}\tag{15.6}$$

where the fact that $e^{-2\pi i}=1$ was used in the last step. In words, the discrete spectrum evaluated at these integer multiples is identically the sum of the sample amplitudes, which is the discrete analogue of the area of the function. Equation (15.6) evaluated at $k=0$ ($\Longrightarrow \xi=0$) bears an obvious similarity to the central-ordinate theorem for continuous functions in Equation (9.113). More generally, it also is easy to show that the spectrum evaluated at ξ_0 is equal to that at $\xi_0 + k/\Delta x$, where k is any integer.

As discussed in Section 14.3, the periodicity of $F_s[\xi;\ \Delta x]$ ensures that all nonredundant information in the spectrum of the discrete function is contained in any single region of the frequency domain with support equal to the period $(\Delta x)^{-1}$. Equations (14.32) and (14.33) specified the two most common choices for the frequency interval; these are respectively centered at $\xi = 0$ and at $\xi = +1/(2 \cdot \Delta x) = \xi_{\text{Nyquist}}$. Electrical engineers who process temporal signals often use the second "single-sided" interval in Equation (14.33) (after substituting Δt for Δx and ν for ξ). All temporal frequencies within this window are nonnegative, which is considered (at least by some) to be advantageous. However, this spectrum includes frequencies that apparently exceed the Nyquist limit; this quandary is remedied by "folding" all such frequencies back into the "allowed" domain. This is the reason for the term *folding frequency* that sometimes is applied to $\xi = (2 \cdot \Delta x)^{-1}$. The periodicity of the spectrum ensures that this interpretation is "correct", but the concept of the folding frequency is occasionally confusing. In imaging applications, the concept of a "negative" spatial frequency is not considered to be unusual, and so the symmetric spectrum interval with negative frequencies is generally preferred. We will use the symmetric period when describing the DFT, and the "single-sided" period when constructing the efficient computational algorithm called the fast Fourier transform (FFT). Simple methods for converting between the conventions also will be described.

15.1 Inverse of the Infinite-Support DFT

The DFT in Equation (15.3) maps a discrete coordinate-space representation $f_s[x;\ \Delta x]$ to a continuous and periodic frequency-space representation $F_s[\xi;\ \Delta x]$ that has infinite support. The continuous inverse Fourier transform of this periodic spectrum $F_s[\xi;\ \Delta x]$ therefore must regenerate the sampled function $f_s[x;\ \Delta x]$. This statement may be proven by substituting the expression for the forward DFT within the continuous inverse Fourier transform:

$$\begin{aligned}
\mathcal{F}_1^{-1}\{F_s[\xi;\ \Delta x]\} &= \int_{-\infty}^{+\infty} F_s[\xi;\ \Delta x]\, e^{+2\pi i\xi x}\, d\xi \\
&= \int_{-\infty}^{+\infty} (F[\xi] * COMB[\Delta x \cdot \xi])\, e^{+2\pi i\xi x}\, d\xi \\
&= \int_{-\infty}^{+\infty} \left(\sum_{n=-\infty}^{+\infty} f[n \cdot \Delta x]\, e^{-2\pi i\xi(n\cdot\Delta x)} \right) e^{+2\pi i\xi x}\, d\xi \\
&= \sum_{n=-\infty}^{+\infty} f[n \cdot \Delta x] \cdot \int_{-\infty}^{+\infty} e^{-2\pi i\xi(n\cdot\Delta x - x)}\, d\xi \\
&= \sum_{n=-\infty}^{+\infty} f[n \cdot \Delta x] \cdot \delta[n \cdot \Delta x - x] \\
&= f[x] \cdot \left(\frac{1}{\Delta x} COMB\left[\frac{x}{\Delta x}\right] \right)
\end{aligned} \tag{15.7}$$

where the spectrum of the Dirac delta function from Equation (6.106) has been used.

We have already seen in Equation (14.36) that an estimate $\hat{f}[x]$ of the original continuous function $f[x]$ is obtained from the inverse continuous Fourier transform of a single period of the spectrum of the discrete function:

$$\begin{aligned}\hat{f}[x] &= \mathcal{F}_1^{-1}\{F_s[\xi;\,\Delta x]\cdot(\Delta x\, RECT[\Delta x\cdot\xi])\}\\ &= \Delta x\cdot\int_{-1/(2\cdot\Delta x)}^{+1/(2\cdot\Delta x)} F_s[\xi;\,\Delta x]\,e^{+2\pi i\xi x}\,d\xi\\ &= \mathcal{F}_1^{-1}\{F_s[\xi;\,\Delta x]\} * SINC\left[\frac{x}{\Delta x}\right]\\ &= f_s[x;\,\Delta x] * SINC\left[\frac{x}{\Delta x}\right] \end{aligned} \tag{15.8}$$

In words, the inverse continuous Fourier transform of the periodic spectrum generates the convolution of the sampled function with the ideal interpolator. Of course, the estimate $\hat{f}[x]$ is identical to the original continuous function $f[x]$ only if the sampling interval $\Delta x < (2\xi_{\max})^{-1}$.

To summarize our surmises so far, the forward DFT of an infinite-support bandlimited sampled signal is an infinite sum, while the inverse transform is a continuous integral over the finite limits of a single period of the spectrum. Neither process is susceptible to computation for arbitrary functions because both require infinite numbers of operations.

15.2 DFT over Finite Interval

In practice, data arrays must have finite support whether the continuous function is bandlimited or not. We often encounter 1-D arrays that have been sampled at some set of N coordinates separated by Δx and indexed by n. This is conceptually equivalent to sampling the product of a bandlimited infinite-support function and a rectangular "window" of width $N\cdot\Delta x$. Though N may be any integer, it is often constrained to be an even number, and often to an integer power of two, for reasons that will become evident shortly. The DFT of the finite array may be expressed as the infinite-support DFT of an array that has been modulated by a finite-support rectangle function:.

$$\begin{aligned}F_{\text{finite}}[\xi;\,N\cdot\Delta x,\,x_0] &= \mathcal{F}_1\left\{f_s[x;\,\Delta x]\cdot RECT\left[\frac{x-x_0}{N\cdot\Delta x}\right]\right\}\\ &= \mathcal{F}_1\left\{\left(f[x]\cdot\frac{1}{\Delta x}COMB\left[\frac{x}{\Delta x}\right]\right)\cdot RECT\left[\frac{x-x_0}{N\cdot\Delta x}\right]\right\}\end{aligned} \tag{15.9}$$

where the "shift parameter" x_0 translates the rectangular window function relative to the origin of coordinates, as shown in Figure 15.1a. For the moment, $N\cdot\Delta x$ and x_0 are included as parameters in the notation $F_{\text{finite}}[\xi;\,N\cdot\Delta x,\,x_0]$ for the spectrum, though the shift parameter x_0 will be deleted in short order.

The spectrum of Equation (15.9) is easy to derive from known Fourier transform pairs via the modulation theorem of Equation (9.140):

$$\begin{aligned}&\mathcal{F}_1\left\{\left(f[x]\cdot\frac{1}{\Delta x}COMB\left[\frac{x}{\Delta x}\right]\right)\cdot RECT\left[\frac{x-x_0}{N\cdot\Delta x}\right]\right\}\\ &= (F[\xi] * COMB[\Delta x\cdot\xi]) * (N\cdot\Delta x)\,SINC[N\cdot\Delta x\cdot\xi]\,e^{-2\pi i\xi x_0}\\ &= (N\cdot\Delta x)\left(F[\xi] * COMB\left[\frac{\xi}{(\Delta x)^{-1}}\right] * SINC\left[\frac{\xi}{(N\cdot\Delta x)^{-1}}\right]\right)e^{-2\pi i\xi x_0}\end{aligned} \tag{15.10}$$

If N is an even integer and x_0 is an integer multiple of the sampling interval Δx (e.g., $x_0 = 0$), then the amplitudes of the pair of samples of $f[n\cdot\Delta x]$ at the endpoints of the rectangle are scaled by $\frac{1}{2}$ and the array contains $N+1$ samples (an odd number). This possibly undesirable situation may be avoided by

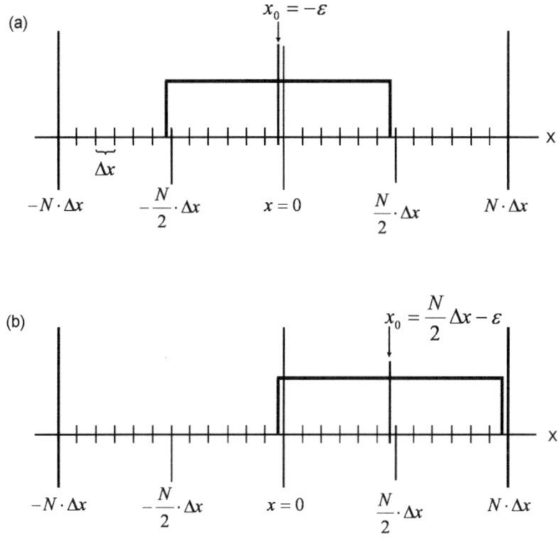

Figure 15.1 Rectangular windows of the form $RECT[(x - x_0)/(N \cdot \Delta x)]$ used to truncate DFT to N samples: (a) symmetric window with $x_0 = -\epsilon$, with samples indexed over the interval $-N/2 \leq n \leq (N/2) - 1$ (b) one-sided window with $x_0 = ((N/2) \cdot \Delta x) - \epsilon$, with sample indices $0 \leq n \leq N - 1$, generally used for functions of time.

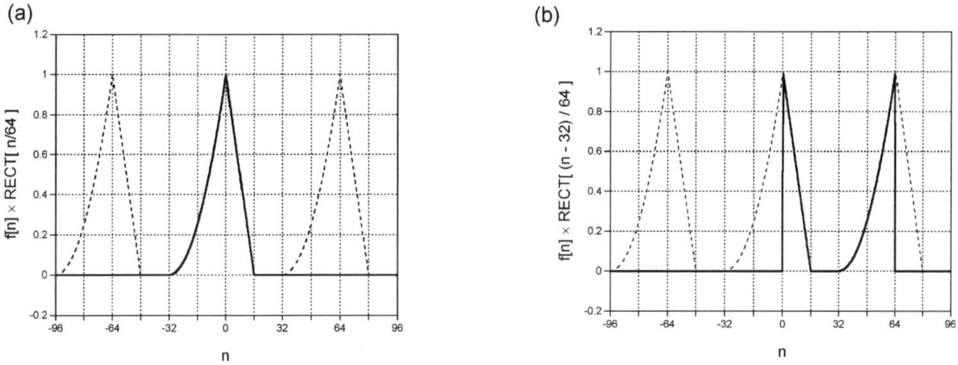

Figure 15.2 Sampled functions resulting from the two choices of windows: (a) symmetric window with $-N/2 \leq n \leq N/2 - 1$; (b) one-sided window with $0 \leq n \leq N - 1$. The dashed line is the periodic function $f[x]$ and the solid line is the domain of the finite-support sampled function in both cases.

requiring x_0 to be a noninteger multiple of Δx so that the endpoints of the rectangle do not coincide with any samples.

Two choices of the translation parameter x_0 are common and the one used is often based upon the type of signal. For 1-D temporal signals, the first sample usually is assumed to have been "grabbed" at time $t = 0$ and the index of this sample is $n_0 = 0$. The corresponding coordinate of the Nth sample is $(N - 1) \cdot \Delta t$. For this set of samples to be generated by Equation (15.10), x_0 must satisfy the condition $(N/2 - 1) < x_0/\Delta x < N/2$. On the other hand, the origin of coordinates of a space-domain image usually is assumed to coincide with the axis of symmetry of the collecting optical system. If it coincides with a specific sample, then the remaining samples cannot be arranged symmetrically if N is even;

if $N/2$ samples are placed on one side of $n = 0$, then $N/2 - 1$ will be placed on the other. With no obvious motivation (yet), we choose the domain $-N/2 \leq n \leq N/2 - 1$, so that the sample indexed by $n = -N/2$ has no partner indexed by $n = +N/2$ in the finite array. This choice requires that x_0 satisfy the condition $-\Delta x < x_0 < 0$, which thus "nudges" the originally centered rectangular window to the left and displaces its endpoints from the coordinates of the endpoint samples, as shown in Figure 15.1a. Since an infinitesimal shift is sufficient, we choose $x_0 = -\epsilon$ where ϵ is an arbitrarily small positive number.

Though this "pseudosymmetric" group of samples indexed in the interval $-N/2 \leq n < N/2 - 1$ will be assumed in the derivation of the finite-support DFT, conversion to the single-sided array with $0 \leq n \leq N - 1$ is easy to perform, as will be demonstrated later.

With the translation parameter $x_0 = -\epsilon$, the spectrum of the finite-support function is the cascade of continuous operations:

$$\begin{aligned} F_{\text{finite}}[\xi; N \cdot \Delta x, \epsilon] &= \int_{-\infty}^{+\infty} \left(f[x] \cdot \frac{1}{\Delta x} COMB\left[\frac{x}{\Delta x}\right]\right) \cdot RECT\left[\frac{x+\epsilon}{N \cdot \Delta x}\right] \cdot e^{-2\pi i \xi x}\, dx \\ &= \sum_{n=-\infty}^{+\infty} f[n \cdot \Delta x]\, RECT\left[\frac{n \cdot \Delta x + \epsilon}{N \cdot \Delta x}\right] e^{-2\pi i \xi n \cdot \Delta x} \\ &= \sum_{n=-N/2}^{N/2-1} f[n \cdot \Delta x]\, e^{-2\pi i \xi n \cdot \Delta x} \end{aligned} \tag{15.11}$$

where the sifting property of the Dirac delta function from Equation (6.113) has been used. It is important to note that the frequency variable ξ of the finite DFT still is continuous, so that the spectrum may be evaluated at any spatial frequency ξ in the infinite domain $(-\infty, +\infty)$. The recipe for evaluating the DFT requires that the specific spatial frequency ξ_0 be selected first, followed by evaluation of the sampled cosine and sine functions with frequency ξ_0 at the N sample coordinates $n \cdot \Delta x$. The sample weights $f[n \cdot \Delta x]$ are then multiplied independently by the sampled cosine and sine functions, and both sets of N products are summed independently to produce the real and imaginary parts of the spectrum evaluated at spatial frequency ξ_0. If the sample weights $f[n \cdot \Delta x]$ are complex valued, the real part $f_R[n \cdot \Delta x]$ and imaginary part $f_I[n \cdot \Delta x]$ may be processed separately to obtain the complex spectrum:

$$\begin{aligned} &F_{\text{finite}}[\xi; N \cdot \Delta x] \\ &\quad = \sum_{n=-N/2}^{N/2-1} (f_R[n \cdot \Delta x] + i\, f_I[n \cdot \Delta x])(\cos[2\pi\xi(n \cdot \Delta x)] - i \sin[2\pi\xi(n \cdot \Delta x)]) \\ &\quad = \sum_{n=-N/2}^{N/2-1} (f_R[n \cdot \Delta x] \cdot \cos[2\pi\xi(n \cdot \Delta x)] + f_I[n \cdot \Delta x] \cdot \sin[2\pi\xi(n \cdot \Delta x)]) \\ &\qquad + i \sum_{n=-N/2}^{N/2-1} (-f_R[n \cdot \Delta x] \cdot \sin[2\pi\xi(n \cdot \Delta x)] + f_I[n \cdot \Delta x] \cdot \cos[2\pi\xi(n \cdot \Delta x)]) \end{aligned} \tag{15.12}$$

The modulation theorem of the continuous Fourier transform in Equation (9.140) may be applied to relate the functional expression for the finite DFT in Equation (15.9) to the infinite DFT of

Equation (15.3):

$$\begin{aligned} F_{\text{finite}}[\xi;\, N\cdot\Delta x,\, \epsilon] &= \mathcal{F}_1\left\{\left(f[x]\cdot\frac{1}{\Delta x}COMB\left[\frac{x}{\Delta x}\right]\right)RECT\left[\frac{x+\epsilon}{N\cdot\Delta x}\right]\right\} \\ &= F_s[\xi;\,\Delta x] * (N\cdot\Delta x)\,SINC[N\cdot\Delta x\cdot\xi]\,e^{+2\pi i\xi\epsilon} \\ &\cong F_s[\xi;\,\Delta x] * (N\cdot\Delta x)\,SINC\left[\frac{\xi}{(N\cdot\Delta x)^{-1}}\right] \end{aligned} \tag{15.13}$$

where the last step assumes that ϵ may be arbitrarily small, thus making the linear-phase term $e^{+2\pi i\xi\epsilon}$ arbitrarily close to $1+0i$. In parallel with the definition of the width parameter of the *COMB* function in the space domain, the width parameter of the frequency-domain *SINC* function is:

$$\Delta\xi \equiv \frac{1}{N\cdot\Delta x} \tag{15.14}$$

which allows Equation (15.13) to be rewritten as:

$$F_{\text{finite}}[\xi;\, N\cdot\Delta x] = F_s[\xi;\,\Delta x] * \left(\frac{1}{\Delta\xi}SINC\left[\frac{\xi}{\Delta\xi}\right]\right) \tag{15.15}$$

The convolution "averages" the amplitudes of the infinite-support and periodic DFT $F_s[\xi;\,\Delta x]$ over the infinite support of the *SINC* function, as illustrated in Figure 15.3. The amplitude of the spectrum of the sampled function at a specific frequency is "spread" over the entire frequency domain, though most remains concentrated within the main lobe of the *SINC* function. Because the convolution evaluated at a particular frequency is a "mixture" of spectrum amplitudes from the entire domain of frequencies, its effect is a "correlation" of amplitudes. For example, the amplitude of the spectrum of the infinite-domain sampled function evaluated at a large spatial frequency is "spread" over the infinite domain by the action of the convolution when the space domain is restricted to N samples, though most of the "influence" of the convolution appears in nearby samples of the spectrum. Therefore, some portion of this high-frequency amplitude will contribute to the spectrum of the sampled function at small spatial frequencies. At first glance, the effect of "mixing" these spatial frequencies may resemble aliasing, because some of the amplitude from high-frequency sinusoidal components "appears" at small spatial frequencies. However, note that the current phenomenon acts in "both directions"; it can also transfer amplitude from low-frequency components to high frequencies, which aliasing cannot do. This artifact in the finite DFT due to the convolution with the *SINC* function is called *leakage* of the spectrum amplitudes and will be discussed in more detail shortly.

The width parameter of the frequency-domain *SINC* function in Equation (15.14) is $(N\cdot\Delta x)^{-1}$, which decreases with increasing N. In words, this means that if the domain of the sampled function is widened, the scale of the *SINC* narrows and the region over which "significant" averaging occurs in the DFT spectrum decreases. In the limit $N\to\infty$, the *SINC* function approaches the unit-area Dirac delta function:

$$\lim_{N\to\infty}\left\{\frac{1}{(N\cdot\Delta x)^{-1}}SINC\left[\frac{\xi}{(N\cdot\Delta x)^{-1}}\right]\right\} = \delta[\xi] \tag{15.16}$$

The discrete spectrum in Equation (15.15) approaches the infinite-support DFT in Equation (15.3) in this limit because any "averaging" of amplitudes due to convolution with $SINC[N\cdot\Delta x\cdot\xi]$ disappears.

The finite-support DFT of $f_s[x;\, N\cdot\Delta x]$ may be expressed in terms of the continuous spectrum $F[\xi]$ in several equivalent ways:

$$\begin{aligned} F_{\text{finite}}[\xi;\, N\cdot\Delta x] &= \sum_{n=-N/2}^{N/2-1} f[n\cdot\Delta x]\,e^{-2\pi i\xi n\cdot\Delta x} \\ &= F[\xi] * (COMB[\Delta x\cdot\xi] * (N\cdot\Delta x)\,SINC[N\cdot\Delta x\cdot\xi]) \\ &= (N\cdot\Delta x)((F[\xi] * COMB[\Delta x\cdot\xi]) * SINC[N\cdot\Delta x\cdot\xi]) \end{aligned} \tag{15.17}$$

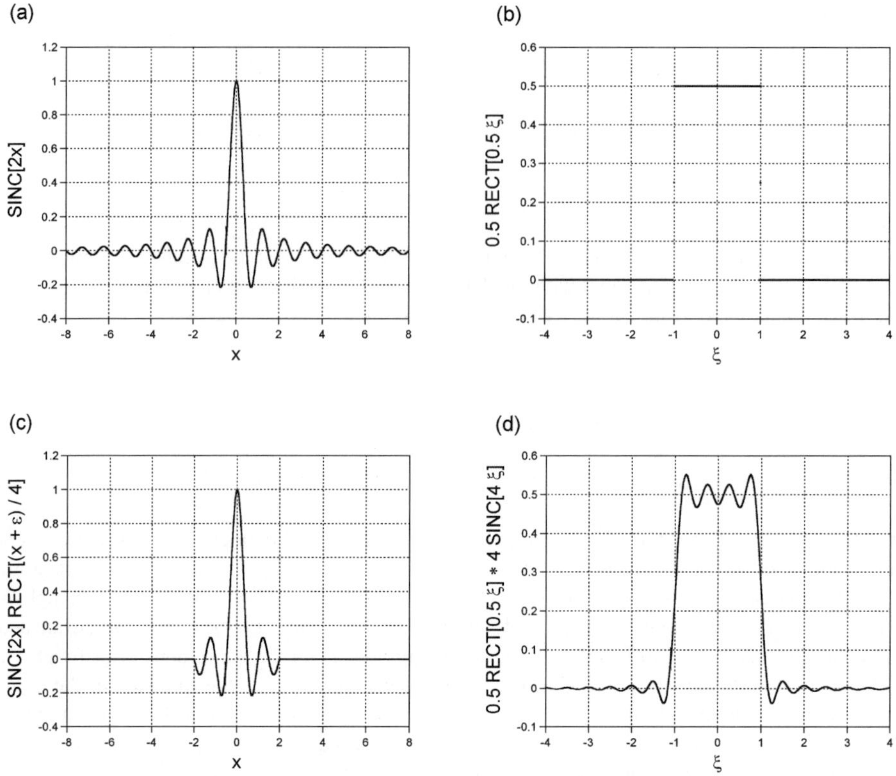

Figure 15.3 Blurring of the spectrum due to truncation of space-domain function: (a) 1-D continuous function $f[x] = SINC[2x]$, which has infinite support; (b) its spectrum $F[\xi] = \frac{1}{2} RECT[\xi/2]$; (c) $f[x]$ modulated by $RECT\,[x/4]$; (d) spectrum of windowed function is a scaled replica of the convolution of the continuous spectrum with $4\,SINC[4\xi]$, which has blurred edges.

where ϵ is assumed to be arbitrarily close to zero to cancel any residual phase factor, thus removing the parameter $x_0 = \epsilon \cong 0$ from the notation. The commutativity of convolution in Equation (8.47) ensures that the finite DFT in Equation (15.15) may be interpreted in two ways: as the result of averaging (lowpass filtering) of the periodic infinite DFT $F_s[\xi;\, \Delta x]$; or as the convolution of the continuous (and generally nonperiodic) spectrum $F[\xi]$ first with the *SINC* function filter, followed by convolution with a *COMB* function with width parameter $(\Delta x)^{-1}$. In this second interpretation, the convolution with the *SINC* function "averages" the continuous spectrum, and the subsequent convolution with the *COMB* function replicates and sums the averaged spectrum at intervals $(\Delta x)^{-1}$. In either case, the resulting continuous spectrum is periodic over the interval $(\Delta x)^{-1}$ and the *SINC* function filter "spreads" the amplitude of the continuous spectrum evaluated at any frequency throughout the infinite support of the frequency domain.

We may now apply the observation noted after Equation (14.66) that the output from any ideal lowpass filter may be sampled in a way that allows the continuous function to be recovered exactly. In the current context, $F_{\text{finite}}[\xi;\, N \cdot \Delta x]$ may be sampled without any loss of information if the sampling interval $\Delta\xi$ is sufficiently small. The lowpass-filtered continuous signal is reconstructed from the samples by applying the ideal *SINC* function interpolator:

$$\left(F_{\text{finite}}[\xi;\, N\cdot\Delta x]\cdot\frac{1}{\Delta\xi}COMB\left[\frac{\xi}{\Delta\xi}\right]\right)*SINC\left[\frac{\xi}{\Delta\xi}\right]=F_{\text{finite}}[\xi;\, N\cdot\Delta x] \tag{15.18}$$

The sampled spectrum may be written in several equivalent ways:

$$\begin{aligned}F_{\text{finite}}[\xi;\, N\cdot\Delta x]\cdot\frac{1}{\Delta\xi}COMB\left[\frac{\xi}{\Delta\xi}\right]&=F_{\text{finite}}[\xi;\, N\cdot\Delta x]\sum_{k=-\infty}^{+\infty}\delta[\xi-k\cdot\Delta\xi]\\&=\sum_{k=-\infty}^{+\infty}F_{\text{finite}}[\xi;\, N\cdot\Delta x]\cdot\delta[\xi-k\cdot\Delta\xi]\\&=\sum_{k=-\infty}^{+\infty}F_{\text{finite}}[k\cdot\Delta\xi;\, N\cdot\Delta x]\cdot\delta[\xi-k\cdot\Delta\xi]\end{aligned} \tag{15.19}$$

Because the spectrum has been sampled, its representation in the space domain is periodic and may be evaluated via the inverse continuous Fourier transform of Equation (15.9):

$$\begin{aligned}&\mathcal{F}_1^{-1}\left\{F_{\text{finite}}[\xi;\, N\cdot\Delta x]\cdot\frac{1}{\Delta\xi}COMB\left[\frac{\xi}{\Delta\xi}\right]\right\}\\&=\mathcal{F}_1^{-1}\{F_{\text{finite}}[\xi;\, N\cdot\Delta x]\}*\mathcal{F}_1^{-1}\left\{\frac{1}{\Delta\xi}COMB\left[\frac{\xi}{\Delta\xi}\right]\right\}\\&=\left(\left[f[x]\cdot\frac{1}{\Delta x}COMB\left[\frac{x}{\Delta x}\right]\right]RECT\left[\frac{x-x_0}{N\cdot\Delta x}\right]\right)*COMB\left[\frac{x}{(1/\Delta\xi)}\right]\\&=\left(f_s[x;\,\Delta x]\,RECT\left[\frac{x-x_0}{N\cdot\Delta x}\right]\right)*COMB\left[\frac{x}{N\cdot\Delta x}\right]\\&=(N\cdot\Delta x)\sum_{\ell=-\infty}^{+\infty}\left(f_s[x;\,\Delta x]\,RECT\left[\frac{x-x_0}{N\cdot\Delta x}\right]\right)*\delta\left[x-\frac{\ell}{N\cdot\Delta x}\right]\end{aligned} \tag{15.20}$$

where $f_s[x;\,\Delta x]$ is the original signal after ideal sampling over the infinite domain. In words, the space-domain representation of the sampled function is a scaled and periodic "version" of the finite-support sampled function, where the period is equal to the region of support $N\cdot\Delta x$. Therefore, the effect of truncating the DFT summation to a finite domain containing N samples is to "blur" the frequency spectrum, which thus may be sampled without loss of information. This sampling operation makes the space-domain function periodic over an interval equal to the region of support $N\cdot\Delta x$. Therefore, *both representations of a finite-support sampled function are sampled and periodic.*

Equation (15.14) relates the number of samples N with the sampling intervals Δx and $\Delta\xi$ in the two domains. Simple rearrangement yields two equivalent forms:

$$N\cdot\Delta x\cdot\Delta\xi=1 \tag{15.21a}$$

$$\Delta x\cdot\Delta\xi=\frac{1}{N} \tag{15.21b}$$

The second of these resembles the *uncertainty relation* for continuous functions that was introduced in Section 13.4 and may be considered to be its discrete form. It shows that the two representations $f[n\cdot\Delta x;\, N\cdot\Delta x]$ and $F_{\text{finite}}[\xi;\, N\cdot\Delta x]$ each contain N independent sample amplitudes. This is an intellectually satisfying statement because it confirms the principle of linear algebra that N independent equations are required to determine N unknown quantities. In other words, weighted sums of the N-element input vector $f[n\cdot\Delta x]$ produce at most N linearly independent numbers; the amplitude of the $(N+1)$st spatial frequency may be expressed as a linear combination (weighted sum) of the amplitudes of the N independent elements of $F[k \cdot \Delta\xi]$. The weights that produce the $(N+1)$st frequency from the N independent frequencies are the amplitudes of the interpolator $SINC[N \cdot \Delta x \cdot \xi]$ evaluated at the appropriate distances from the N known sample amplitudes.

The sampled frequencies may be entered into the complex exponential in Equation (15.11) as integer multiples of $\Delta\xi$, e.g., $\xi = k \cdot \Delta\xi$. The discrete uncertainty relation in Equation (15.21) also may be applied:

$$\begin{aligned} F[k \cdot \Delta\xi; N \cdot \Delta x] &= \sum_{n=-N/2}^{N/2-1} f[n \cdot \Delta x; N \cdot \Delta x]\, e^{-2\pi i (n \cdot \Delta x)(k \cdot \Delta\xi)} \\ &= \sum_{n=-N/2}^{N/2-1} f[n \cdot \Delta x; N \cdot \Delta x]\, e^{-2\pi i n k (\Delta x \cdot \Delta\xi)} \\ &= \sum_{n=-N/2}^{N/2-1} f[n \cdot \Delta x; N \cdot \Delta x]\, e^{-2\pi i n k / N} \end{aligned} \tag{15.22}$$

where the descriptive subscript "finite" that was formerly applied to the spectrum is now implied by use of the discrete index k and is no longer included explicitly. Note that the intervals Δx and $\Delta\xi$ between independent samples in the space and frequency domains appear only within the arguments of the equivalent space- and frequency-domain representations in Equation (15.22); they have no impact on the amplitudes in the summation and thus may be "ignored" to define the finite DFT of vectors with dimensionless integer arguments:

$$F[k] = \sum_{n=-N/2}^{N/2-1} f[n]\, e^{-2\pi i n k / N} \quad \textit{for} \quad -\frac{N}{2} \leq k \leq \frac{N}{2} - 1 \tag{15.23}$$

This expression strongly resembles that of the DFT transform that was derived in the discussion of matrix operators in Equation (5.72); the only difference is the domain of the indices. The conversion between these two choices of domain will be considered shortly.

Because a particular value of the complex exponential "mask" function $e^{-2\pi i n k/N}$ is specified by two values $[n, k]$, the mask function may be displayed as an $N \times N$ matrix where n and k specify the column and row, respectively. Except for the different (but easily handled) limits of the summation, Equation (15.23) has the form of the matrix–vector product in Equation (3.24). A particular sample $F[k_0]$ of the output spectrum is the inner product of a row of the matrix $[\exp[-2\pi i n k_0/N]]$ and the vector $f[n]$. The interpretation of matrix–vector multiplication in Chapter 3 may be used to recognize that $F[k_0]$ is the projection of the N-D vector $f[n]$ onto the N-D vector with components $(\exp[-2\pi i n k_0/N])^* = \exp[+2\pi i n k_0/N]$.

A disadvantage of removing the explicit notation of Δx and $\Delta\xi$ from Equation (15.23) is that they remind us of the scale factors applied to the spatial frequency. It may be useful to emphasize that the spatial frequency of the spectrum in Equation (15.23) is not *equal* to k but rather is *indexed* by k:

$$\xi = k \cdot \Delta\xi = \frac{k}{N \cdot \Delta x} = \left(\frac{k}{N}\right)\left(\frac{1}{\Delta x}\right) \tag{15.24}$$

The quantity k/N is the number of cycles per sample of the space-domain sinusoid with spatial frequency ξ. If the spatial interval Δx between samples is known, then ξ may be specified in cycles per unit length. For example, if $\Delta x = 10\ \mu$m, then $\xi = 100 \cdot k$ cycles per mm. The number of cycles of a sinusoid in the N-sample domain is $(k/N) \cdot N = k$. This provides a convenient mnemonic: the components of the spectrum indexed by $\pm k$ refer to sinusoids in the space domain that oscillate through k cycles over the domain of N samples. Thus the sample $F[k = 0]$ is the amplitude of the sinusoid that

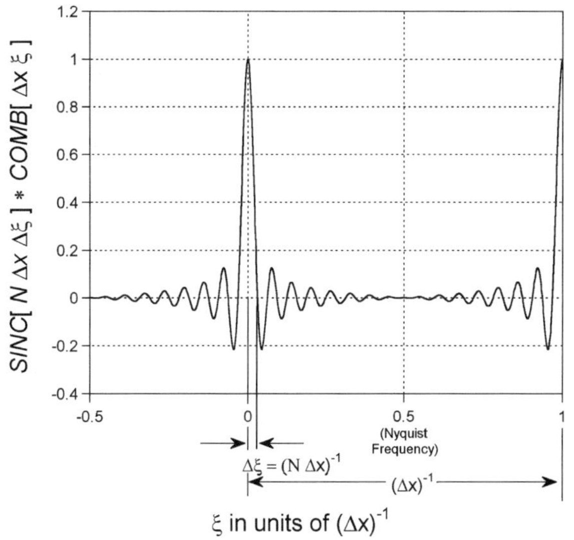

Figure 15.4 The discrete uncertainty relation. The convolution of the spectrum with $COMB[\Delta x\xi] * SINC[N \cdot \Delta x \cdot \xi]$ produces a function whose amplitude is correlated over the interval $(N \cdot \Delta x)^{-1} \equiv \Delta\xi$.

oscillates through 0 cycles in N samples; $F[0]$ is the part of $f[x]$ with constant amplitude. The sample $F[k = 1]$ is that of the sinusoid that oscillates through one cycle in N samples, and so forth.

The discrete uncertainty relations in Equation (15.21) may be inferred in another way by comparing two intervals in the frequency domain. We know that the distance between redundant frequency components is the period of the sampled spectrum $(\Delta x)^{-1}$, while the distance between independent frequencies is equal to the width parameter $\Delta\xi$ of the *SINC* function that correlates the spectrum amplitudes, where $\Delta\xi = (N \cdot \Delta x)^{-1}$. The ratio of these two numbers is the number of independent samples in the spectrum:

$$\frac{(\Delta x)^{-1}}{\Delta\xi} = \frac{1}{\Delta x \cdot \Delta\xi} = N \tag{15.25}$$

This calculation is shown schematically in Figure 15.4.

If N is small, then the width parameter $\Delta\xi$ of the correlating *SINC* function in the frequency domain is large. In this case, the weights applied to the amplitudes of the continuous spectrum are approximately equal over a "long" distance, which means that the amplitudes of the continuous spectrum are averaged together with approximately equal weights so that the spectrum is severely "blurred". If N is large, the interpolating *SINC* function is "skinny" and large weights are applied to amplitudes of the continuous spectrum $F[\xi]$ over only a short distance from the frequency under consideration. From Equation (15.16):

$$\lim_{N\to\infty} \{(N \cdot \Delta x)\, SINC[(N \cdot \Delta x) \cdot \xi]\} = \delta[\xi] \tag{15.26}$$

which means that the spectrum approaches the DFT of a bandlimited signal computed from an infinite number of samples.

15.2.1 Finite DFT of $f[x] = 1[x]$

The constant function is equivalent to a cosine function with infinite period ($\xi = 0$) and arbitrary initial phase:

$$1[x] = \cos\left[2\pi \cdot \frac{x}{\infty}\right] \tag{15.27}$$

Because it is composed of the single finite sinusoidal frequency, $1[x]$ clearly is bandlimited and therefore may be sampled without aliasing. Because its continuous spectrum is $\delta[\xi]$, any interval with finite width Δx must satisfy the Nyquist sampling condition in Equation (14.17). The form of the sampled function is:

$$1_s[x, \Delta x] = \sum_{n=-\infty}^{+\infty} 1[n \cdot \Delta x]\delta[x - n \cdot \Delta x] \tag{15.28}$$

If limited to 16 samples in the interval $-8 \leq n \leq +7$, the sample weights are:

$$1_s[n \cdot \Delta x] = \begin{cases} 1 & \textit{if} \ -8 \leq n \leq +7 \\ 0 & \textit{otherwise} \end{cases} \tag{15.29}$$

The spectrum of this finite sampled function is completely described by 16 equispaced independent frequency samples in the Nyquist window such that $\Delta\xi = 1/(16 \cdot \Delta x)$. Though an infinite number of possible sets of 16 independent frequencies exist in this domain, only one includes a sample at $\xi = 0$ (DC):

$$\xi = \frac{k}{16 \cdot \Delta x} \quad \textit{for } k = 0, \pm 1, \ldots, \pm 7, -8 \tag{15.30}$$

The spatial frequency indexed by $k = -8$ is $\xi = -1/(2 \cdot \Delta x)$, which is the "negative" Nyquist frequency. Now it is easy to see why there is no need for a "matching" sample of the spectrum located at $\xi = +1/(2 \cdot \Delta x)$; the periodicity of the spectrum over intervals of $1/\Delta x$ ensures that the amplitudes at $\xi = \pm 1/(2 \cdot \Delta x)$ are identical. The corresponding statement in the space domain also is true; because the space-domain function is assumed to be periodic over the distance $N \cdot \Delta x = 16 \cdot \Delta x$, the samples located at $x = \pm 8 \cdot \Delta x$ also must have the same amplitudes (as they obviously do for the unit constant). The amplitude of the DFT at zero frequency is found by substituting the sample weights into Equation (15.11) and evaluating for $k = 0$:

$$F[k = 0] = \sum_{n=-8}^{+7} f[n]\, e^{-2\pi i n \cdot \frac{0}{16}} = \sum_{n=-8}^{+7} 1[n] = 16 = N \tag{15.31}$$

which is the sum of the 16 sample amplitudes, as predicted by Equation (15.6).

The amplitudes of the spectrum at the spatial frequencies indexed by $k = \pm 1$ are:

$$F[k = \pm 1] = \sum_{n=-8}^{+7} 1[n]\, e^{\mp 2\pi i n \cdot \frac{1}{16}} \tag{15.32}$$

The real parts evaluated at ± 1 are identically the sum of the amplitudes of the samples of a cosine whose spatial frequency is $\frac{1}{16}$ cycles per sample. Since one cycle fills the 16 samples of the array, the sum of amplitudes is zero. The imaginary parts of the two sums in Equation (15.32) also are zero, as they are sums of the amplitudes of the samples of one cycle of a sine wave.

Similarly, the real and imaginary parts of the summation for $k = \pm 2$ are respectively the sums of the samples of cosine and sine waves that oscillate with spatial frequencies $\mp \frac{1}{8}$ cycles per sample, corresponding to 8 samples per period or to two cycles within the 16 samples of the array. These amplitudes also sum to zero. It is easy to see that substitution of the other nonzero integer values of k in Equation (15.32) compute sums of amplitudes of integer numbers of periods of sampled sinusoids and evaluate to zero for all $k \neq 0$. The discrete spectrum evaluated for a frequency index k is the sum of the input array modulated by a complex linear-phase exponential that oscillates k times over the N samples of the array. The result of the 16 sums is the spectrum of the unit constant evaluated at 16 samples:

$$F[k \cdot \Delta\xi] = 16 \cdot \delta_d[k \cdot \Delta\xi] \tag{15.33}$$

where the definition of the discrete delta function in Equation (14.27) has been used. This result also may be derived directly by substituting $f[x] = 1[x]$, $N = 16$, $\Delta x = 1$, and $\Delta\xi = (16)^{-1}$ into Equation (15.22):

$$\begin{aligned} F[k \cdot \Delta\xi;\ N \cdot \Delta x] &= ((\delta[\xi] * COMB[\xi]) * 16\, SINC[16\xi])\, COMB[16\xi]\, RECT[\xi] \\ &= (COMB[\xi] * 16\, SINC[16\xi])\, COMB[16\xi]\, RECT[\xi] \\ &= 16\left(\sum_{k=-\infty}^{+\infty} \delta[\xi - k] * SINC[16\xi]\right) COMB[16\xi]\, RECT[\xi] \\ &= 16\left(\sum_{k=-\infty}^{+\infty} SINC[(16\xi - k)]\right) COMB[16\xi]\, RECT[\xi] \\ &= 16\, \delta_d[k \cdot \Delta\xi]\, COMB[16\xi]\, RECT[\xi] = 16\, \delta_d[k \cdot \Delta\xi] \end{aligned} \tag{15.34}$$

The delta function components of $COMB[\xi \cdot \Delta x]$ are separated by the interval $(\Delta x)^{-1}$ and each has area $(\Delta x)^{-1}$. Replicas of $SINC[\xi/(16 \cdot \Delta x)^{-1}]$ are spaced by $(\Delta x)^{-1}$ and summed to create the convolution. Each of the *SINC* functions evaluates to zero at integer multiples of $(16 \cdot \Delta x)^{-1}$, except at their central ordinates where they evaluate to unity. The superposition of these *SINC* functions generates the unit constant, which then is modulated by $COMB[\xi/(16 \cdot \Delta x)^{-1}]$. The resulting function consists of Dirac delta functions located at integer multiples of $(16 \cdot \Delta x)^{-1}$ and that are scaled by $(16 \cdot \Delta x)^{-1}$.

We have said that only 16 frequencies of the DFT of a 16-sample function can be independent. It is useful to consider the result when the DFT of the 16-sample unit constant is evaluated at more than 16 independent spatial frequencies within the Nyquist window. For example, consider the values obtained when evaluating the spectrum at 32 equispaced spatial frequencies in the interval $-1/(2 \cdot \Delta x) \le \xi < +1/(2 \cdot \Delta x)$. This may be done by evaluating the spectrum at the midpoints of the frequency intervals, such as at $\xi = \frac{1}{32}$ cycles per sample. This is equivalent to evaluating the spectrum at a frequency-domain sample indexed by $k = +\frac{1}{2}$. The amplitude of the spectrum at this frequency is:

$$\begin{aligned} F\left[\xi = \frac{1}{32}\right] &= \sum_{n=-8}^{+7} 1[n]\, e^{-2\pi i n/32} = \sum_{n=-8}^{+7} e^{-i\pi n/16} \\ &= e^{-2\pi i \frac{(-8)}{32}} + e^{-2\pi i \frac{(-7)}{32}} + e^{-2\pi i \frac{(-6)}{32}} + \cdots + e^{-2\pi i \frac{0}{32}} + \cdots + e^{-2\pi i \frac{(+7)}{32}} \\ &= e^{-2\pi i \frac{(-8)}{32}} + (e^{-2\pi i \frac{(-7)}{32}} + e^{-2\pi i \frac{(+7)}{32}}) + \cdots \\ &\quad + (e^{-2\pi i \frac{(-1)}{32}} + e^{-2\pi i \frac{(+1)}{32}}) + e^{-2\pi i \frac{0}{32}} \\ &= e^{+i\pi/2} + 2 \cdot \cos\left[2\pi \frac{7}{32}\right] + 2 \cdot \cos\left[2\pi \frac{6}{32}\right] + \cdots + 2 \cdot \cos\left[\frac{2\pi}{32}\right] + 1 \\ &= i + 2\left(\cos\left[2\pi \frac{7}{32}\right] + \cos\left[2\pi \frac{6}{32}\right] + \cdots + \cos\left[\frac{2\pi}{32}\right]\right) + 1 \\ &\cong 10.153 + i \end{aligned} \tag{15.35}$$

In words, the amplitude of the finite DFT of the sampled unit constant evaluated at this nonzero spatial frequency is *not* zero, as someone versed only in the continuous Fourier transform might expect, but

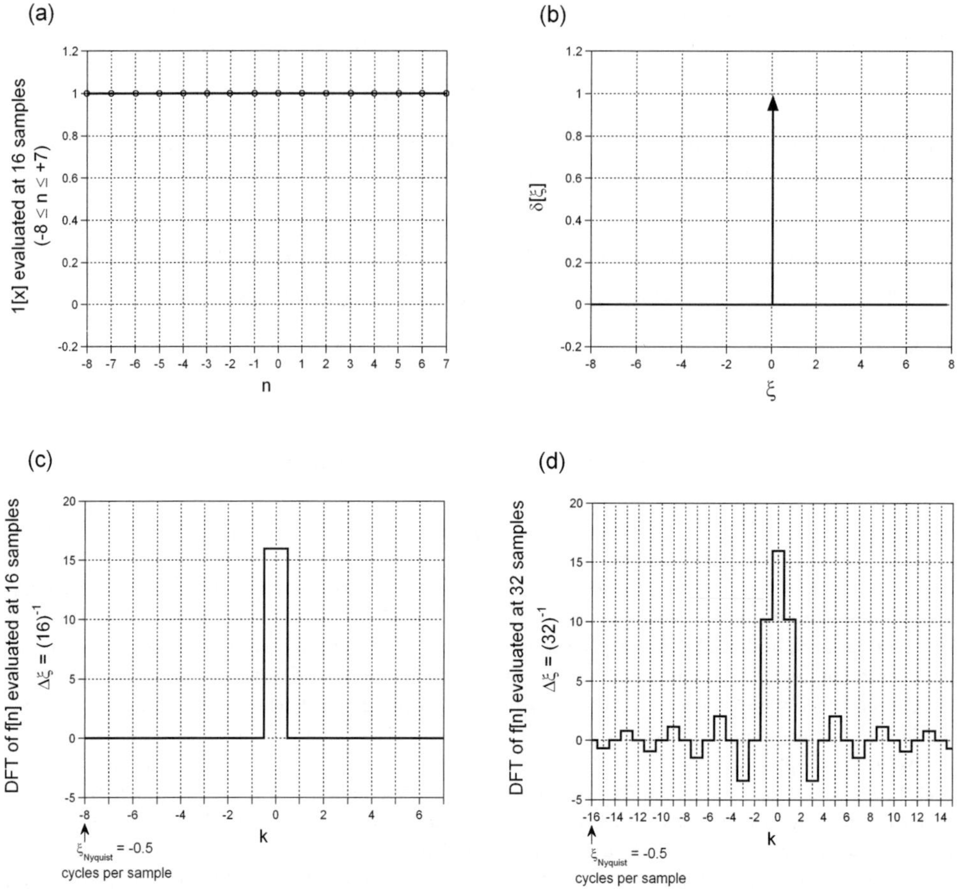

Figure 15.5 Evaluation of an N-point DFT at $M > N$ points: (a) $f[x] = 1[x]$ sampled at 16 locations separated by Δx yields $f[n] = 1[n]$; (b) spectrum of continuous function $\mathcal{F}_1\{1[x]\} = \delta[\xi]$; (c) DFT of $1[n]$ evaluated at $N = 16$ frequencies separated by $\Delta\xi = 1/16$ cycles per sample; (d) DFT of $1[n]$ evaluated at $M = 32$ frequencies separated by $\Delta\xi = 1/32$ cycles per sample. The continuous spectrum has been convolved with $SINC[32 \cdot \Delta x \cdot \xi]$.

rather a superposition of amplitudes of sinusoidal functions. This is due to the convolution with the *SINC* function in Equation (15.15), as is shown by the graph in Figure 15.5d, which resembles a *SINC* function.

15.2.2 Scale Factor in DFT

Equation (15.31) demonstrated that the finite DFT of $f[n] = 1$ is not equal to the continuous transform $\mathcal{F}_1\{1[x]\} = \delta[\xi]$, but rather to a discrete Dirac delta function located at $k = 0$ with amplitude N. This does not agree with an intuitive analogue of the central-ordinate theorem where the amplitude of the discrete Dirac delta function “should” be unity. However, the linearity of the DFT allows us to include any multiplicative constants c_1 and c_2 into the definitions of the forward and inverse DFT:

$$F[k \cdot \Delta\xi] = c_1 \sum_{n=-N/2}^{N/2-1} f[n \cdot \Delta x]\, e^{-2\pi i n k/N} \tag{15.36a}$$

$$f[n \cdot \Delta x] = c_2 \sum_{k=-N/2}^{N/2-1} F[k \cdot \Delta\xi]\, e^{+2\pi i n k/N} \tag{15.36b}$$

The values of c_1 and c_2 must ensure that the finite inverse DFT of $F[k \cdot \Delta\xi]$ be $f[n \cdot \Delta x]$. We can substitute Equation (15.36a) into Equation (15.36b) after changing the dummy summation variable:

$$\begin{aligned}
f[n \cdot \Delta x] &= c_2 \sum_{k=-N/2}^{N/2-1} \left(c_1 \sum_{m=-N/2}^{N/2-1} f[m \cdot \Delta x]\, e^{-2\pi i m k/N} \right) e^{+2\pi i n k/N} \\
&= c_1 c_2 \sum_{k=-N/2}^{N/2-1} \sum_{m=-N/2}^{N/2-1} f[m \cdot \Delta x]\, e^{-2\pi i k(m-n)/N} \\
&= c_1 c_2 \sum_{m=-N/2}^{N/2-1} f[m \cdot \Delta x] \sum_{k=-N/2}^{N/2-1} e^{-2\pi i k(m-n)/N} \\
&= c_1 c_2 \sum_{m=-N/2}^{N/2-1} f[m \cdot \Delta x] \cdot N \cdot \delta_d[n-m] = c_1 \cdot c_2 \cdot N \cdot f[n \cdot \Delta x] \\
&\Longrightarrow c_1 \cdot c_2 = \frac{1}{N}
\end{aligned} \tag{15.37}$$

We may select any "convenient" values for c_1 and c_2 as long as this constraint is satisfied, though there are "preferred" sets of constants for different applications, such as discrete convolution (considered later in this chapter). For example, the choice of $c_1 = 1$ ensures that the DC component of the discrete spectrum is the sum of the amplitudes of the samples and thus obeys the analogous property of the continuous central-ordinate theorem:

$$F[k \cdot \Delta\xi] = \sum_{n=-N/2}^{N/2-1} f[n \cdot \Delta x]\, e^{-2\pi i n k/N} \quad \textit{for } c_1 = 1 \tag{15.38a}$$

$$f[n \cdot \Delta x] = \frac{1}{N} \sum_{k=-N/2}^{N/2-1} F[k \cdot \Delta\xi]\, e^{+2\pi i n k/N} \tag{15.38b}$$

Other authors (including Bracewell, 1986a) apply a factor of N^{-1} in the forward DFT:

$$F[k \cdot \Delta\xi] = F\left[\frac{k}{N \cdot \Delta x}\right] = \frac{1}{N} \sum_{n=-N/2}^{N/2-1} f[n \cdot \Delta x]\, e^{-2\pi i n k/N} \quad \textit{for } c_1 = \frac{1}{N} \tag{15.39a}$$

$$f[n \cdot \Delta x] = \sum_{k=-N/2}^{N/2-1} F[k \cdot \Delta\xi]\, e^{+2\pi i n k/N} \tag{15.39b}$$

The amplitudes of the resulting Fourier components are "spatial averages" of the original signal, i.e., the amplitude of the DC component in this convention is the *average* value of the input samples. In this convention, the finite DFT of the unit constant is a discrete delta function with unit weight. The DFT of a unit-amplitude cosine is a pair of delta functions with amplitude $\frac{1}{2}$. The disadvantage of this convention will become evident shortly.

The linearity of the DFT also allows a third convention that applies factors of $(\sqrt{N})^{-1}$ to both the forward and inverse transform. Though the consequential symmetry of the expressions for the DFT and inverse DFT is pleasing, this choice is less common because $\sqrt{N}$ is often not an integer.

15.2.3 Finite DFT of Discrete Dirac Delta Function

The discrete Dirac delta function was defined in Equation (14.27) to be:

$$\delta_d[(n-n_0)\cdot\Delta x] = \begin{cases} 1 & \textit{if } n = n_0 \\ 0 & \textit{otherwise} \end{cases} \tag{15.40}$$

Its DFT is very easy to evaluate:

$$\begin{aligned} \frac{1}{N}\sum_{n=-N/2}^{N/2-1} \delta_d[(n-n_0)\cdot\Delta x]\cdot e^{-2\pi ink/N} &= \frac{1}{N}\,e^{-2\pi in_0k/N} \\ &= \frac{1}{N}\left(\cos\left[2\pi\cdot\frac{k}{(N/n_0)}\right] - i\,\sin\left[2\pi\cdot\frac{k}{(N/n_0)}\right]\right) \end{aligned} \tag{15.41}$$

The discrete spectrum is a complex-valued function with real and imaginary parts in the forms of a cosine and a sine, respectively, that oscillate with a period of N/n_0 samples. For example, if the discrete Dirac delta function is located one pixel from the origin ($n_0 = +1$), the real part of the spectrum is a cosine with a period of N samples (one cycle over the domain). As n_0 is incremented or decremented from 0, the number of cycles of the sinusoid in the N-sample transform increases by one until reaching the limit at $n_0 = -N/2$:

$$\begin{aligned} F\left[k;\, n_0 = -\frac{N}{2}\right] &= \frac{1}{N}\left(\cos\left[2\pi\left(-\frac{N}{2}\right)\frac{k}{N}\right] - i\,\sin\left[2\pi\left(-\frac{N}{2}\right)\frac{k}{N}\right]\right) \\ &= \frac{1}{N}(\cos[-\pi k] - i\,\sin[-\pi k]) \\ &= \frac{1}{N}[(-1)^k - 0\cdot i] = \frac{1}{N}(-1)^k \end{aligned} \tag{15.42}$$

In words, the transform of a discrete Dirac delta function located at the edge of the array is a cosine at the Nyquist frequency, so that the sample amplitude oscillates between ± 1.

15.2.4 Summary of Finite DFT

$$\textit{Forward transform:}\ F[k\cdot\Delta\xi] = F\left[\frac{k}{N\cdot\Delta x}\right] = \sum_{n=-N/2}^{N/2-1} f[n\cdot\Delta x]\,e^{-2\pi ink/N} \tag{15.43}$$

$$\textit{Inverse transform:}\ f[n\cdot\Delta x] = \frac{1}{N}\sum_{k=-N/2}^{N/2-1} F[k\cdot\Delta\xi]\,e^{+2\pi ink/N} \tag{15.44}$$

$$\textit{No. of independent samples in both domains:}\ N = \frac{2\cdot X_{\max}}{\Delta x} = \frac{2\cdot\xi_{\text{Nyquist}}}{\Delta\xi} \tag{15.45}$$

$$\textit{Nyquist frequency:}\ \xi_{\text{Nyquist}} = \pm\frac{N}{2}\cdot\Delta\xi \tag{15.46}$$

$$\textit{Largest space-domain coordinate: } X_{\max} = +\frac{(N-1)}{2} \cdot \Delta x \tag{15.47}$$

$$\textit{Sample spacing: } \Delta x = \frac{2 \cdot X_{\max}}{N} = \frac{1}{2 \cdot \xi_{\text{Nyquist}}} \tag{15.48}$$

$$\textit{Period in space domain: } X_0 = \frac{1}{\Delta \xi} \tag{15.49}$$

$$\textit{Sample interval: } \left[-\frac{N \cdot \Delta x}{2} \le \Delta x \le +\frac{(N-1) \cdot \Delta x}{2}\right] \tag{15.50}$$

$$\textit{Frequency spacing: } \Delta \xi = \frac{2 \cdot \xi_{\text{Nyquist}}}{N} = \frac{1}{2 \cdot X_{\max}} \tag{15.51}$$

$$\textit{Period in frequency domain: } \Xi = \frac{1}{\Delta x} \tag{15.52}$$

$$\textit{Frequency interval: } -\frac{N}{2} \cdot \Delta \xi \le \xi \le +\frac{(N-1)}{2} \cdot \Delta \xi \tag{15.53}$$

$$\textit{Uncertainty relation: } \Delta x \cdot \Delta \xi = \frac{1}{N} \tag{15.54}$$

15.3 Fourier Series Derived from Fourier Transform

In Section 14.7, we demonstrated that the amplitudes of the sinusoidal components of a periodic function are evaluated at a discrete set of sinusoidal frequencies. This limiting case of the general Fourier transform integral is the *Fourier series*. Many authors derive the Fourier series first and consider the Fourier transform to be a limiting case. For better or worse, our discussion takes the complementary point of view that the Fourier series is a special case of the continuous Fourier transform.

Recall that a periodic function $f[x]$ (with infinite support and period X_0) may be represented as the convolution of a *COMB* function and a function $w[x]$ that has support equal to a single period of $f[x]$. This case was considered in Equation (9.167):

$$\begin{aligned} f[x] &= \left(f[x]\,RECT\left[\frac{x}{X_0}\right]\right) * \left(\frac{1}{X_0} COMB\left[\frac{x}{X_0}\right]\right) \\ &\equiv w[x] * \left(\frac{1}{X_0} COMB\left[\frac{x}{X_0}\right]\right) \end{aligned} \tag{15.55}$$

The local function $w[x]$ vanishes outside the semi-closed interval $[-X_0/2, +X_0/2)$. The amplitudes of the sinusoidal components are easily found via the forward continuous Fourier transform:

$$\begin{aligned} F[\xi] &= W[\xi]\,COMB[\xi \cdot X_0] = COMB\left[\frac{\xi}{(X_0)^{-1}}\right] W[\xi] \\ &= \left(\frac{1}{X_0} \sum_{n=-\infty}^{+\infty} \delta\left[\xi - \frac{n}{X_0}\right]\right)\left(\int_{-\infty}^{+\infty} w[\alpha]\, e^{-2\pi i \xi \alpha}\, d\alpha\right) \\ &= \sum_{n=-\infty}^{+\infty} \frac{1}{X_0}\, \delta\left[\xi - \frac{n}{X_0}\right] \int_{-X_0/2}^{+X_0/2} f[\alpha]\, e^{-2\pi i \xi \alpha}\, d\alpha \end{aligned} \tag{15.56}$$

The dummy variable of integration is changed to α to simplify future steps in the analysis. The Fourier-series form of the space-domain representation may be found via the continuous inverse Fourier transform of $F[\xi]$:

$$\begin{aligned}
f[x] &= \int_{-\infty}^{+\infty} F[\xi]\, e^{+2\pi i\xi x}\, d\xi \\
&= \int_{-\infty}^{+\infty} \left(\sum_{n=-\infty}^{+\infty} \frac{1}{X_0} \delta\left[\xi - \frac{n}{X_0}\right] \int_{-X_0/2}^{+X_0/2} f[\alpha]\, e^{-2\pi i\xi\alpha}\, d\alpha \right) e^{+2\pi i\xi x}\, d\xi \\
&= \sum_{n=-\infty}^{+\infty} \frac{1}{X_0} \left(\int_{-X_0/2}^{+X_0/2} f[\alpha] \left(\int_{-\infty}^{+\infty} \delta\left[\xi - \frac{n}{X_0}\right] e^{+2\pi i\xi(x-\alpha)}\, d\xi \right) d\alpha \right) \\
&= \sum_{n=-\infty}^{+\infty} \frac{1}{X_0} \int_{-X_0/2}^{+X_0/2} f[\alpha] \exp\left[+2\pi i \frac{n(x-\alpha)}{X_0}\right] d\alpha \\
&= \sum_{n=-\infty}^{+\infty} \frac{1}{X_0} \int_{-X_0/2}^{+X_0/2} f[\alpha] \left(\cos\left[2\pi \frac{n(x-\alpha)}{X_0}\right] + i \sin\left[2\pi \frac{n(x-\alpha)}{X_0}\right] \right) d\alpha \\
&= \sum_{n=-\infty}^{+\infty} \frac{1}{X_0} \int_{-X_0/2}^{+X_0/2} f[\alpha] \left(\cos\left[2\pi \frac{nx}{X_0}\right] \cos\left[2\pi \frac{n\alpha}{X_0}\right] + \sin\left[2\pi \frac{nx}{X_0}\right] \sin\left[2\pi \frac{n\alpha}{X_0}\right] \right) d\alpha \\
&\quad + i \sum_{n=-\infty}^{+\infty} \frac{1}{X_0} \int_{-X_0/2}^{+X_0/2} f[\alpha] \left(\sin\left[2\pi \frac{nx}{X_0}\right] \cos\left[2\pi \frac{n\alpha}{X_0}\right] + \cos\left[2\pi \frac{nx}{X_0}\right] \sin\left[2\pi \frac{n\alpha}{X_0}\right] \right) d\alpha
\end{aligned} \tag{15.57}$$

If we restrict our attention to real-valued functions (that are guaranteed to have Hermitian spectra), the imaginary part of the summation vanishes, and the Fourier series reduces to the sum of two integrals that evaluate to real numbers for each value of the index n. We denote the numerical value of the first integral by α_n and the second by β_n:

$$\alpha_n \equiv \frac{1}{X_0} \int_{-X_0/2}^{+X_0/2} f[\alpha] \cdot \cos\left[2\pi \frac{n\alpha}{X_0}\right] d\alpha \tag{15.58a}$$

$$\beta_n \equiv \frac{1}{X_0} \int_{-X_0/2}^{+X_0/2} f[\alpha] \cdot \sin\left[2\pi \frac{n\alpha}{X_0}\right] d\alpha \tag{15.58b}$$

These expressions for the coefficients may be simplified further by recognizing that integrals of odd functions over symmetric limits evaluate to zero. Therefore, only the even part of $f[x]$ contributes to α_n and only the odd part to β_n. This also demonstrates that the cosine coefficients are even ($\alpha_n = \alpha_{-n}$) and the sine coefficients are odd ($\beta_n = -\beta_{-n}$). Therefore we may define new coefficients $a_n \equiv \alpha_n + \alpha_n$ (for $n > 0$) and $b_n \equiv b_n - b_{-n}$ that take advantage of this symmetry:

$$f[x] = \frac{a_0}{2} + \sum_{n=1}^{+\infty} a_n \cos\left[2\pi \frac{nx}{X_0}\right] + \sum_{n=1}^{+\infty} b_n \sin\left[2\pi \frac{nx}{X_0}\right] \tag{15.59a}$$

$$\textit{where } a_n \equiv \frac{2}{X_0} \int_{-X_0/2}^{+X_0/2} f[\alpha] \cdot \cos\left[2\pi \frac{n\alpha}{X_0}\right] d\alpha \tag{15.59b}$$

$$\textit{and } b_n \equiv \frac{2}{X_0} \int_{-X_0/2}^{+X_0/2} f[\alpha] \cdot \sin\left[2\pi \frac{n\alpha}{X_0}\right] d\alpha \tag{15.59c}$$

We compute the coefficients of the Fourier series by multiplying by the cosine and sine masks and integrating over the period of $f[x]$. The discrete spectrum contains only those sinusoidal components whose frequencies are integral multiples of the fundamental frequency $\xi_0 = 1/X_0$.

15.4 Efficient Evaluation of the Finite DFT

This section considers algorithms for computing the DFT. In particular, an efficient computation will be demonstrated that takes advantage of the periodicity of the complex "mask" function in the Fourier transform. The efficiency is obtained by restricting the choice of the number of samples in the input array.

As already discussed, the DFT of an N-pixel array may be evaluated at any frequency ξ in the infinite domain $(-\infty, +\infty)$, though only N equispaced samples within the Nyquist interval are independent. The amplitude at any other frequency within the Nyquist interval may be evaluated by interpolating these N values, and the periodicity of the spectrum ensures that a spectrum amplitude at a frequency outside the Nyquist interval is a replica of the amplitude of a frequency within. We also saw that the discrete finite representations in both domains ($f[n \cdot \Delta x]$ and $F[k \cdot \Delta\xi]$) are effectively periodic over N samples.

Computation of each sample of the DFT of an N-pixel real function requires computing N values each of the cosine and sine, followed by $2N$ multiplications and $2 \cdot (N - 1)$ sums. The total number of calculations is of order N^2, and so rises very quickly as the number of samples is increased. This intensity of computation made calculation of the DFT a tedious and rarely performed task before digital computers. For example, Fourier deconvolution of seismic traces has been a tool for petroleum exploration for many years. Enders Robinson (1982) reports that he spent the entire summer of 1951 deriving spectra of 32 seismic records by hand. This task can now be performed even by a hopelessly obsolete personal computer in a small fraction of a second. Discrete Fourier analysis was unusual even with mainframe digital computers well into the 1960s, because of memory requirements and computation times. This all changed in 1965 when Cooley and Tukey developed the fast Fourier transform (FFT) algorithm, based on a suggestion by Richard Garwin. Though the DFT may be evaluated at any number of frequencies in the infinite domain, the set of amplitudes evaluated at N equispaced frequencies includes all of the available data about the spectrum. The FFT is a recursive algorithm that eliminates repetitive computations in DFTs evaluated for arrays of specific sizes ($N = 2^m$, where m is a positive integer). The recursive calculation means that the FFT calculations are not independent; the spectrum must be computed at all frequencies to obtain the amplitude of any one component.

To derive the FFT algorithm, recall that the periodicity of the sequences $f[n]$ and $F[k]$ over N samples allows us to use any period of the sampled data; for reasons that will become evident, we adopt the "one-sided" convention where the indices n and k of the respective space- and frequency-domain representations reside in the interval $0 \leq n, k \leq N - 1$. The independent discrete frequencies in the sequence $F[k \cdot \Delta\xi]$ are $k \cdot \Delta\xi = k/(N \cdot \Delta x)$, which are equally spaced within the domain:

$$0 \leq \xi \leq \left((N-1) \cdot \Delta\xi = \frac{N-1}{N \cdot \Delta x} = \frac{1}{\Delta x} - \frac{1}{N \cdot \Delta x} \right) < \frac{1}{2 \cdot \Delta x} \tag{15.60}$$

Though the "upper half" of this range exceeds the Nyquist limit, the periodicity of the spectrum guarantees that these frequencies have the same amplitudes as those at the corresponding negative frequencies within the Nyquist interval; this "folding" of the spectrum leads to the oft-used term *folding frequency* as a synonym for the Nyquist frequency.

Most readily available ("canned") FFT programs are computed over the "one-sided" or "decentered" interval in Equation (15.60) instead of the "centered" interval in Equation (15.11). However, the algorithms for computing the spectra of "uncentered" data are modified easily to derive DFTs of centered data. For convenience, we will ignore the scaling factors Δx and $\Delta\xi$ in the expressions for the signal and spectrum sequences. Consider again the expression for the DFT of a "decentered" 1-D array from Equation (15.23), which we will denote by the operator $\mathcal{F}_{\text{DFT}}$:

$$\mathcal{F}_{\text{DFT}}\{f[n]\} = F[k] = \sum_{n=0}^{N-1} f[n]\, e^{-2\pi i n k/N} \tag{15.61}$$

As we saw in Chapter 5, this may be written as the product of an $N \times N$ matrix $A[n, k]$ with the N-dimensional vector $f[n]$ in the form of Equation (5.67), except for the normalizing constant $(\sqrt{N})^{-1}$:

$$F[k] = \sum_{n=0}^{N-1} A[n, k] f[n] \tag{15.62}$$

The elements in $A[n, k]$ may be written as the complex constant $\exp[-2\pi i/N]$ raised to the $n \cdot k$ power; the complex factor will be denoted by the shorthand W_N:

$$A[n, k] = \left(\exp\left[-\frac{2\pi i}{N}\right]\right)^{nk} \equiv (W_N)^{nk} \tag{15.63}$$

This definition recasts the DFT in Equation (15.62) into the form:

$$F[k] = \sum_{n=0}^{N-1} (W_N)^{nk} f[n] \tag{15.64}$$

15.4.1 DFT of Two Samples – The "Butterfly"

We now evaluate the DFT for small N to demonstrate its effect on computational efficiency. Consider first the simplest case with $N = 2$, so that the possible values of the indices n and k in Equation (15.64) are 0 and 1. The complex constant W_2 used to evaluate the matrix operator is $e^{-i\pi} = -1$, which means that the two possible numerical values for W_2^{nk} are ± 1 and that $W_2^0 = -W_2^1$. The 2-pixel DFT may be represented as the product of a 2×2 matrix and a two-element vector $f[n]$ to generate a two-element vector $F[k]$:

$$\begin{aligned}
\begin{bmatrix} F[0] \\ F[1] \end{bmatrix} &= \begin{bmatrix} A[0, 0] & A[0, 1] \\ A[1, 0] & A[1, 1] \end{bmatrix} \begin{bmatrix} f[0] \\ f[1] \end{bmatrix} = \begin{bmatrix} W_2^0 & W_2^0 \\ W_2^0 & W_2^1 \end{bmatrix} \begin{bmatrix} f[0] \\ f[1] \end{bmatrix} \\
&= \begin{bmatrix} +1 & +1 \\ +1 & -1 \end{bmatrix} \begin{bmatrix} f[0] \\ f[1] \end{bmatrix} = \begin{bmatrix} f[0] + f[1] \\ f[0] - f[1] \end{bmatrix}
\end{aligned} \tag{15.65}$$

In words, the elements of the 2-pixel DFT are just the sum and difference of the amplitudes at the two samples:

$$F[0] = f[0] + f[1] \tag{15.66a}$$

$$F[1] = f[0] - f[1] \tag{15.66b}$$

Evaluation of the 2-pixel FFT may be represented by the flow chart in Figure 15.6. The paths form a pattern that led to the common name of *butterfly* for the calculations in Equation (15.66). A butterfly is composed of a single complex-valued multiplication (by -1) followed by two complex additions. The computational power of the FFT arises because the DFT of large arrays may be computed via recursive application of the butterfly.

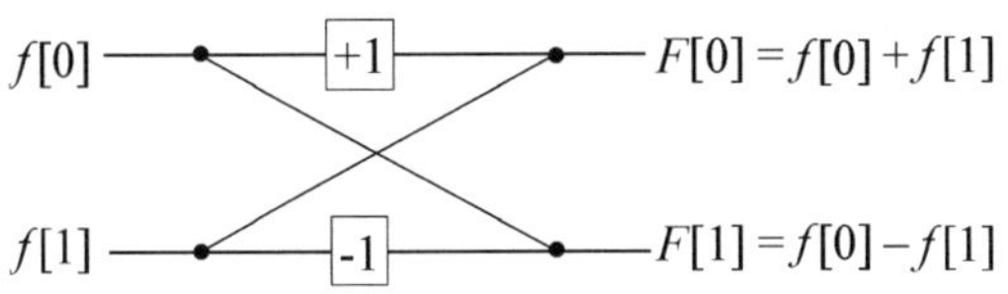

Figure 15.6 Two-point DFT evaluated via a "butterfly": the indices of the two-point array are $n = 0, +1$ and the elements are combined via sum and difference to produce $F[0] = f[0] + f[1]$ and $F[1] = f[0] - f[1]$.

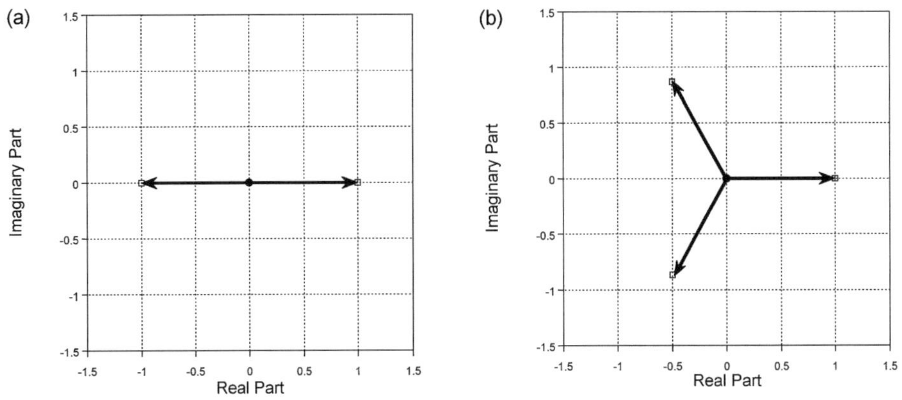

Figure 15.7 Argand diagrams of the complex weights applied in the DFT for (a) $N = 2$ and (b) $N = 3$. The weights in the latter case cannot be implemented as sums and differences.

15.4.2 DFT of Three Samples

Now consider an input sequence that consists of three samples indexed by $n = 0$, 1, and 2. The possible values of W_3^{nk} are 0, $e^{-2\pi i/3}$, and $e^{-4\pi i/3} = e^{+2\pi i/3}$. The expansion for the spectrum is:

$$
\begin{aligned}
F[k = 0, 1, 2] &= \sum_{n=0}^{3} f[n]\, e^{-2\pi i nk/N} \\
&= f[0]\, e^{-2\pi i \cdot 0 \cdot k/3} + f[1]\, e^{-2\pi i \cdot 1 \cdot k/3} + f[2]\, e^{-2\pi i \cdot 2 \cdot k/3} \\
&= f[0] + f[1]\, e^{-2\pi i k/3} + f[2]\, e^{-2\pi i \cdot 2 \cdot k/3} \\
&= f[0] + f[1]\left(-\frac{1}{2} - i\frac{\sqrt{3}}{2}\right)^k + f[2]\left(-\frac{1}{2} + i\frac{\sqrt{3}}{2}\right)^k
\end{aligned} \tag{15.67}
$$

Note that the output sequence cannot be decomposed into simple sums and differences of component terms because the complex exponential terms W_3^{nk} do not include pairs of terms that differ only in their algebraic sign. This can be easily seen on the Argand diagram (Figure 15.7b).

This example should make it clear that the DFT of sequences with odd numbers of samples cannot be decomposed into butterflies because the elements of the matrix operator cannot be decomposed into pairs of positive and negative terms with identical magnitudes.

15.4.3 DFT of Four Samples

The DFT of a four-sample sequence may be expressed as the product of a 4×4 matrix and a four-element vector. The elements of the matrix are $W_4^0 = 1$, $W_4^1 = -i$, $W_4^2 = -1$, and $W_4^3 = i$, so that $W_4^0 = -W_4^2$ and $W_4^1 = -W_1^3$. The form of the matrix operator is:

$$\begin{bmatrix} W_4^0 & W_4^0 & W_4^0 & W_4^0 \\ W_4^0 & W_4^1 & W_4^2 & W_4^3 \\ W_4^0 & W_4^2 & W_4^4 & W_4^6 \\ W_4^0 & W_4^3 & W_4^6 & W_4^9 \end{bmatrix} = \begin{bmatrix} +1 & +1 & +1 & +1 \\ +1 & -i & -1 & +i \\ +1 & -1 & +1 & -1 \\ +1 & +i & -1 & -i \end{bmatrix} \tag{15.68}$$

and the elements of the 4-pixel sequence $F[k]$ are:

$$F[0] = f[0] + f[1] + f[2] + f[3] = (f[0] + f[2]) + (f[1] + f[3]) \tag{15.69a}$$

$$F[1] = f[0] - if[1] - f[2] + if[3] = (f[0] - f[2]) - i(f[1] - f[3]) \tag{15.69b}$$

$$F[2] = f[0] - f[1] + f[2] - f[3] = (f[0] + f[2]) - (f[1] + f[3]) \tag{15.69c}$$

$$F[3] = f[0] + if[1] - f[2] - if[3] = (f[0] - f[2]) + i(f[1] - f[3]) \tag{15.69d}$$

Note that the first pair of terms in each element is the sum or difference of $f[0]$ and $f[2]$ and the second pair is the sum or difference of $f[1]$ and $f[3]$, sometimes weighted by $\pm i$. The elements of the DFT are the sums and differences of the sums and differences of these two pairs. The four possible coefficients are shown in Figure 15.8.

The 4-pixel DFT is evaluated by constructing two stages with two butterflies per stage, as shown in the flow chart in Figure 15.8b. Again, each butterfly consists of a complex multiplication and two complex additions. The correct spectrum amplitudes are returned, albeit in scrambled order; the first and last terms $F[0]$ and $F[3]$ are positioned correctly, but $F[2]$ and $F[1]$ are exchanged. The proper sequence may be obtained easily by expressing the numerical value of k as a binary number, which will be indicated by the "binary point" notated with the symbol "$_\wedge$", so that $0. \Longrightarrow 00_\wedge$, $1. \Longrightarrow 01_\wedge$, $2. \Longrightarrow 10_\wedge$, and $3. \Longrightarrow 11_\wedge$. The proper sequence of amplitudes is obtained by reversing the order of the bits of k. This process of *bit reversal* yields the respective values $00_\wedge \Longrightarrow 0.$, $10_\wedge \Longrightarrow 2.$, $01_\wedge \Longrightarrow 1.$, and $11_\wedge \Longrightarrow 3.$

15.4.4 DFT of Six Samples

The powers of $W_6 = e^{-i\pi/3}$ that occur in the matrix operator are $W_0 = 1$, $W_6^1 = e^{-i\pi/3}$, $W_6^2 = e^{-2\pi i/3}$, $W_6^3 = e^{-i\pi} = -1$, $W_6^4 = -W_6^1$, and $W_6^5 = -W_6^2$, as shown in Figure 15.9. All integer powers of W_6 evaluate to one of these six values, as seen in the Argand diagram in Figure 15.9:

$$\begin{bmatrix} F[0] \\ F[1] \\ F[2] \\ F[3] \\ F[4] \\ F[5] \end{bmatrix} = \begin{bmatrix} W_6^0 & W_6^0 & W_6^0 & W_6^0 & W_6^0 & W_6^0 \\ W_6^0 & W_6^1 & W_6^2 & W_6^3 & W_6^4 & W_6^5 \\ W_6^0 & W_6^2 & W_6^4 & W_6^6 & W_6^8 & W_6^{10} \\ W_6^0 & W_6^3 & W_6^6 & W_6^9 & W_6^{12} & W_6^{15} \\ W_6^0 & W_6^4 & W_6^8 & W_6^{12} & W_6^{16} & W_6^{20} \\ W_6^0 & W_6^5 & W_6^{10} & W_6^{14} & W_6^{20} & W_6^{25} \end{bmatrix} \begin{bmatrix} f[0] \\ f[1] \\ f[2] \\ f[3] \\ f[4] \\ f[5] \end{bmatrix}$$

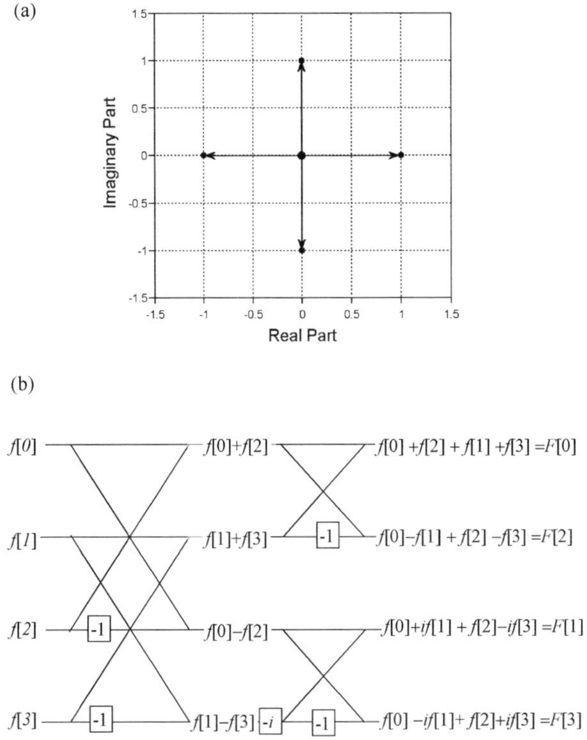

Figure 15.8 DFT for $N = 4$: (a) the Argand diagram of the four complex-valued weights, which are negatives of each other; (b) the flow chart of the four-sample DFT, which consists of two sets of two butterflies, though the output array is in "bit-reversed" order (note that the factors of -1 are applied in the horizontal paths only).

$$= \begin{bmatrix} +1 & +1 & +1 & +1 & +1 & +1 \\ +1 & e^{-i\pi/3} & e^{-i2\pi/3} & -1 & e^{+i2\pi/3} & e^{+i\pi/3} \\ +1 & e^{-i2\pi/3} & e^{+i2\pi/3} & +1 & e^{-i2\pi/3} & e^{+i2\pi/3} \\ +1 & -1 & +1 & -1 & +1 & -1 \\ +1 & e^{+i2\pi/3} & e^{-i2\pi/3} & +1 & e^{+i2\pi/3} & e^{-i2\pi/3} \\ +1 & e^{+i\pi/3} & e^{+i2\pi/3} & -1 & e^{-i2\pi/3} & e^{-i\pi/3} \end{bmatrix} \begin{bmatrix} f[0] \\ f[1] \\ f[2] \\ f[3] \\ f[4] \\ f[5] \end{bmatrix} \tag{15.70}$$

Within each of the spectrum components, the sample amplitudes separated by $N/2$ appear in pairs as sums or differences. In $F[0]$, $f[0]$ and $f[3]$, $f[1]$ and $f[4]$, and $f[2]$ and $f[5]$ are paired as sums. The same combinations occur within $F[2]$ and $F[4]$. The same pairs appear as differences within $F[1]$, $F[3]$, and $F[5]$. The six amplitudes of $F[k]$ may be combined via butterflies without repetition via three banks of three butterflies summed within $F[0]$, $F[2]$, and $F[4]$ and as a difference within each $F[1]$, $F[3]$, and $F[5]$. The 6-pixel DFT may be decomposed into some butterflies, but some operations must be performed independently, and thus the 6-pixel DFT is partially recursive.

15.4.5 DFT of Eight Samples

Like the 2- and 4-pixel cases, the 8-pixel DFT may be performed in a fully recursive manner using only butterflies and complex multiplications. The flow diagram for the FFT of a sequence of eight samples

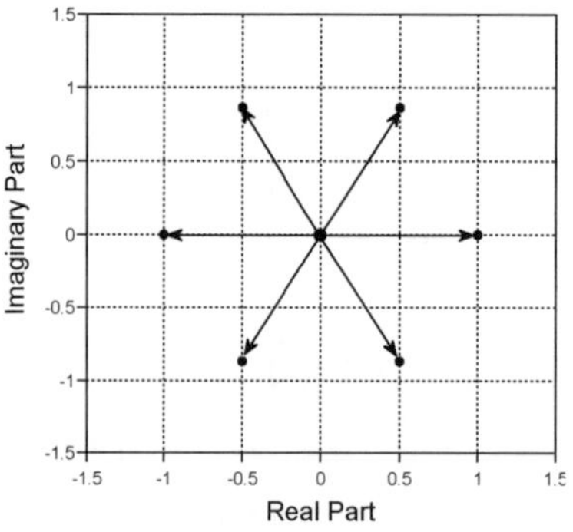

Figure 15.9 The six complex-valued weights used to evaluate the six-point DFT may be decomposed into three sets of weights whose complex amplitudes are negatives of each other.

in Figure 15.10 shows that $N/2$ pairs of samples are added and subtracted in three groups ("banks"), for a total of 12 butterflies. Again, the output samples appear in bit-reversed order. This format of FFT combines pairs of samples that are separated by $\Delta n = N/2$ at the input end and that are adjacent at the output end. It is called a *decimation-in-frequency* FFT. Because the inverse transform could be computed by reversing the procedure, and because the forward and inverse transforms differ only in the sign of the W_N weights and a constant scale factor, it is possible to compute the forward transform by a procedure similar to the mirror image of this process, where adjacent samples are combined by addition and subtraction on the input end, and terms separated by $N/2$ are combined on the output. Because it was most commonly used for temporal inputs, such an algorithm is a *decimation-in-time* FFT.

15.4.6 Complex Matrix for Computing 1-D DFT

The matrices for computing the DFT may be represented as "images" by converting the values to gray levels. The matrices for a 1-D DFT of a 64-sample sequence evaluated at 64 frequencies are shown in Figure 15.11 as real/imaginary parts and as magnitude/phase. Note that the magnitudes of all elements in the array are identical.

15.5 Practical Considerations for DFT and FFT

The differences among the continuous, discrete, and fast Fourier transforms require some explanation to ensure that the results are properly interpreted. In this discussion, the number of samples N in the space-domain array is assumed to be a power of two.

15.5.1 Computational Intensity

We already have demonstrated that each of the N frequency samples of the N-pixel DFT requires of the order of N operations, so the complete process requires of the order of N^2 operations. We have just seen that the FFT with $N = 2$, 4, and 8 requires respectively one bank of one butterfly, two banks of two butterflies, and three banks of four. This sequence may be extended to larger powers of two to show that the FFT requires $\log_2 N$ banks of $N/2$ butterflies, each with one complex multiplication and two complex additions, for a total of $N \cdot \log_2 N$ operations. When N is not a power of two, hybrid algorithms have been written that utilize any available recursive sequences of operations to decrease the number of calculations to something between the DFT and the power-of-two FFT. In these cases,

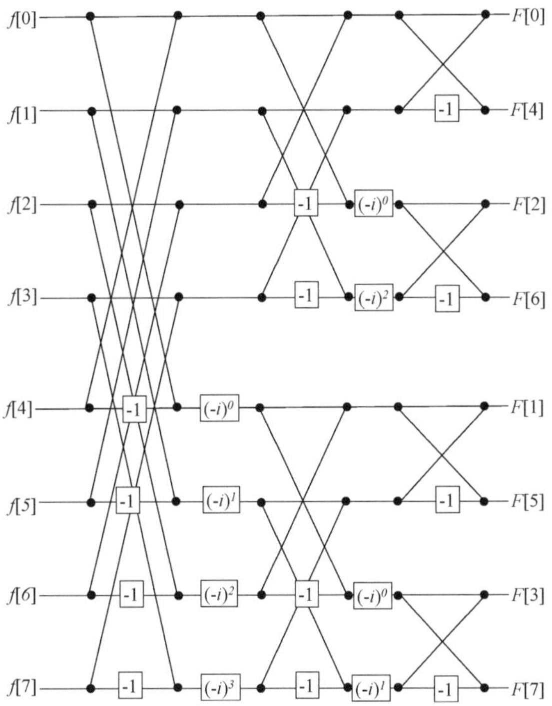

Figure 15.10 The flow chart of the eight-point DFT, which includes three banks of four butterflies. Again note that the output array is in "bit-reversed" order, i.e., the second output term indexed by $1. = 001_\wedge$ contains the term with the reversed binary index $100_\wedge = 4$.

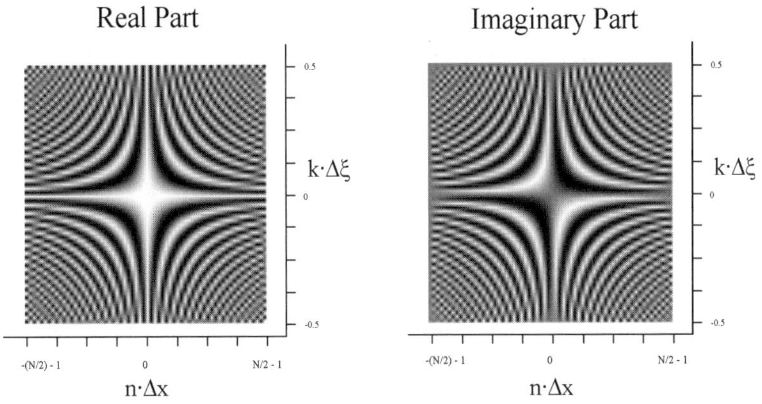

Figure 15.11 Real and imaginary parts of complex matrix used to evaluate the N-point DFT of a "symmetric" sequence ($-N/2 \leq n \leq +N/2 - 1$). The gray value at a location represents the amplitude (white $= 1$, black $= -1$).

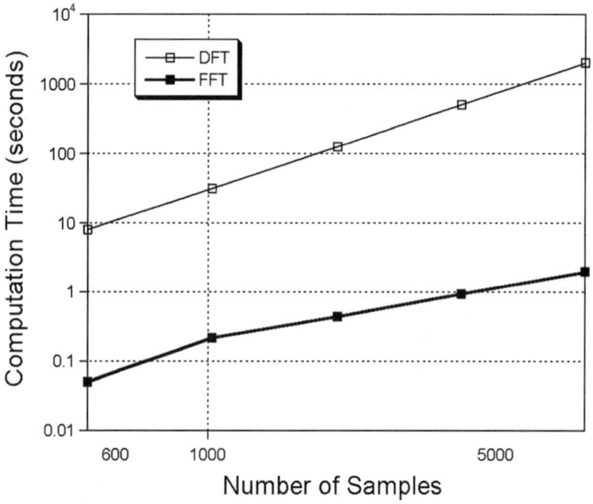

Figure 15.12 Comparison of computation times of 1-D DFTs and FFTs for $N = 2^9$–2^{13} on a log–log scale. The slope of the DFT graph is 2, indicating that the DFT requires of the order of N^2 computations.

the number of operations is of the order of N^2 only when N is prime. Plots of measured computation times versus number of samples for the DFT and FFT are shown in Figure 15.12. The algorithms were coded in BASIC and run on a slow computer. Note that the slope of the DFT time in the log–log plot is approximately twice that of the FFT time, as expected because of the relative number of computations (N^2 versus $N \cdot \log_2 N$).

15.5.2 "Centered" versus "Uncentered" Arrays

The coordinate system generally used in imaging applications places the origin on the optical axis, thus requiring negative indices of the coordinates. When N is an even number, the appropriate domain of sample indices is $-N/2 \leq n \leq N/2 - 1$. However, most FFT algorithms have been written for "uncentered" arrays that are indexed in the interval $0 \leq n \leq N - 1$. Fortunately, the assumed periodicities of both the input array $f[n]$ and the discrete spectrum $F[k]$ make it easy to convert from "centered" to "uncentered" data. In the "forward" direction, N is added to negative indices to construct a data set with $N/2 \leq n \leq N - 1$. The subsequent FFT generates an "uncentered" spectrum indexed by k, where $0 \leq k \leq N - 1$. The indices are converted to "centered" data by subtracting N from those between $N/2$ and $N - 1$. Schemes to convert data arrays from "centered" to "uncentered" format require extra operations and computer memory to perform, and would thus slow down the process. Fortunately, a mathematically equivalent method exists that is based on the discrete shift theorem and operates on data "in place", and so does not require additional memory. Consider the FFT of a "one-sided" set of N samples of data indexed from $n = 0$ to $n = N - 1$:

$$F[k] = \sum_{n=0}^{N-1} f[n]\, e^{-2\pi i n k/N} \quad \text{if } 0 \leq k \leq N - 1 \tag{15.71}$$

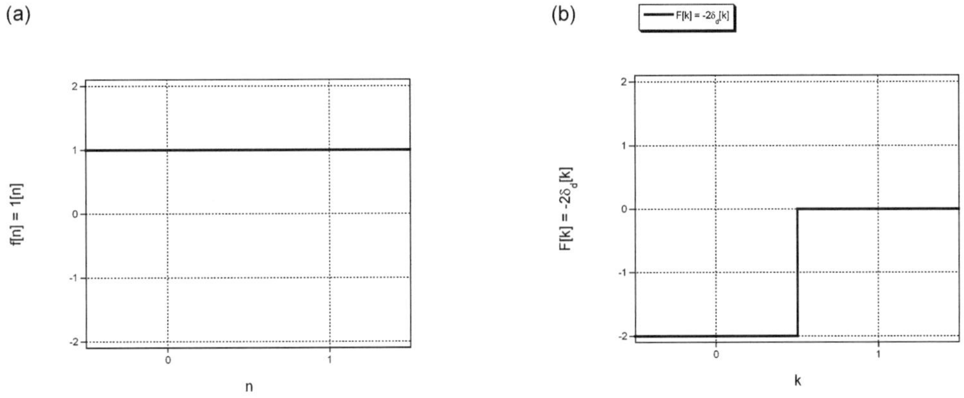

Figure 15.13 FFT of $f[n] = 1[n]$ for $N = 2$. The output array $F[k]$ is the negative of the correct result due to the presence of the constant-phase factor $e^{-i\pi N/2} = e^{-i\pi} = -1$ in Equation (15.73).

To obtain the same result for data indexed over $[0, N-1]$ but with the center of symmetry located at $n = N/2$, the sample and frequency indices must be decremented by $N/2$:

$$F[\mathrm{v}] = \sum_{\mathrm{u}=-N/2}^{N/2-1} f[\mathrm{u}]\, e^{-2\pi i \mathrm{uv}/N} \quad \mathit{if}\ -\frac{N}{2} \le \mathrm{v} \le \frac{N}{2} - 1 \tag{15.72}$$

By changing the quantities in this expression to $n \equiv \mathrm{u} + N/2$, and $k = v + N/2$, we obtain the equivalent operation over the interval $[0, N-1]$:

$$\begin{aligned} F\left[k - \frac{N}{2}\right] &= \sum_{n=0}^{N-1} f\left[n - \frac{N}{2}\right] e^{-2\pi i((1/N)(n-N/2)(k-N/2))} \\ &= \sum_{n=0}^{N-1} f\left[n - \frac{N}{2}\right] (e^{-2\pi ink/N})(e^{+i\pi n})(e^{+i\pi k})(e^{-i\pi N/2}) \\ &= e^{-i\pi N/2}\left(e^{+i\pi k}\left(\sum_{n=0}^{N-1}\left(f\left[n - \frac{N}{2}\right] e^{+i\pi n}\right) e^{-2\pi nk/N}\right)\right) \end{aligned} \tag{15.73}$$

Because the samples of $e^{+i\pi n}$ and $e^{+i\pi k}$ alternate between $+1$ and -1, they resemble 1-D "checkerboards" and multiplication by these factors is called "checkerboarding". For all even values of $N > 2$, the leading constant-phase factor $e^{-i\pi N/2}$ evaluates to $+1$ and thus this term may be ignored except in the (extremely unusual) case of a 2-pixel FFT. In this rare situation, the output amplitudes must be multiplied by $e^{-i\pi} = -1$ to generate the correct values of $F[k]$. To illustrate this situation, consider two samples with identical amplitudes of unity. The discrete spectrum should be a discrete delta function with unit amplitude at the origin. As shown in Figure 15.13, the output sequence obtained by the FFT yields the amplitudes $F[-1] = 0$ and $F[0] = -1$.

The checkerboarding FFT algorithm in Equation (15.73) for space-domain data indexed from 0 to $N-1$ but with the "center" of the array at $n = +N/2$ is:

1. Checkerboard the input data (multiply the amplitude of odd-indexed samples of $f[n]$ by -1).
2. Perform the standard FFT on the interval $0 \le n \le N-1$.
3. Checkerboard the output spectrum data (multiply the amplitude of odd-indexed samples of $F[k]$ by -1).

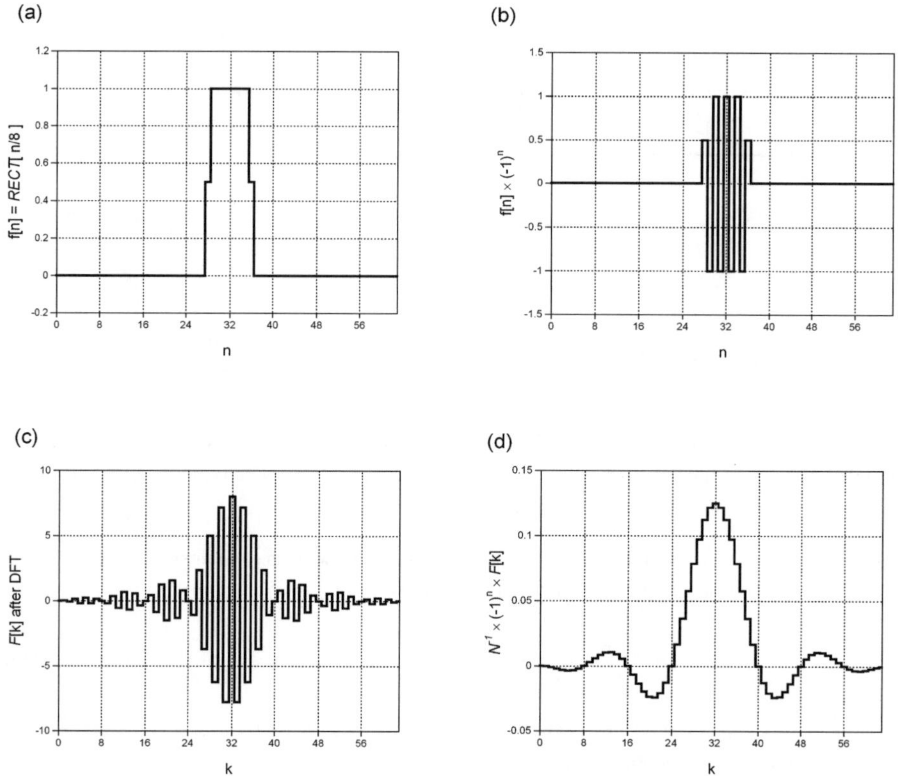

Figure 15.14 FFT with checkerboarding of array with $N = 64$ indexed from $0 \leq n \leq 63$: (a) original function $f[n] = RECT[(n - N/2)/8]$; (b) $f[n] \cdot (-1)^n$ (after first "checkerboard"); (c) $F[k] = (-1)^n \sum_{n=0}^{N-1} f[n]\, e^{-2\pi i n k/N}$; (d) $(N)^{-1}(-1)^k F[k]$, which yields the DFT with $\xi = 0$ translated to the sample $k = N/2$.

4. Multiply each amplitude of $F[k]$ by $1/N$ (if desired).
5. In the (ridiculous) case of the FFT of a two-sample array, multiply the amplitudes of $F[k]$ by -1.

A sample of the results after each step of this algorithm is shown in Figure 15.14.

15.5.3 Units of Measure in the Two Domains

Sampled data collected from a real process have an associated sampling interval Δx or Δt. The summary of the finite DFT in Section 15.2.3 lists the conversions of the dimensions in the two domains. We mention one additional observation that may be helpful when processing real data. Consider the selection of the sampling interval:

$$\Delta x = \frac{1}{\sqrt{N}} \text{ units of length (e.g., mm)} \tag{15.74}$$

The corresponding spacing between independent spatial frequencies is obtained from the discrete uncertainty relation:

$$\frac{1}{N} = \Delta x \cdot \Delta\xi = \frac{1}{\sqrt{N}} \cdot \Delta\xi$$
$$\Longrightarrow \Delta\xi = \frac{1}{\sqrt{N}} \frac{cycles}{unit\ length}$$
$$\Longrightarrow \sqrt{N} \cdot \Delta\xi = 1 \frac{cycle}{unit\ length} \tag{15.75}$$

This result is most convenient if N is an *even* power of two, e.g., $N = 2^4 = 16$, $2^6 = 64$, or $2^8 = 256$. For example, if $N = 2^8 = 256$ and the sample spacing is $\Delta x = \frac{1}{16}$ mm, then spacing between samples in the frequency domain is $\Delta\xi = \frac{1}{16}$ cycles per mm and the respective supports in the two domains are $N \cdot \Delta x = \sqrt{N} = 16$ mm and $N \cdot \Delta\xi = \sqrt{N} = 16$ cycles per mm. This "symmetry" of numerical values is often convenient.

15.5.4 Ensuring Periodicity of Arrays – Data "Windows"

The assumptions inherent in the FFT algorithm can produce artifacts that may be difficult for the unsuspecting user to interpret. Consider the two sets of unaliased samples shown in Figure 15.15 with $N = 256$. The first array is a cosine with period $X_1 = 16$ samples. The resulting FFT is the expected pair of Dirac delta functions located at $k = \pm 16$ samples, which correspond to spatial frequencies of $\xi = \pm\frac{1}{16}$ cycles per unit length. The second array is a cosine with period $X_2 = 17$ samples. The FFT is not a pair of Dirac delta functions, but rather a pair of functions that have *SINC*-like character and exhibit "ringing" or "ripple" (rapid oscillations in amplitude between samples). We now consider the source of this effect.

Recall that the FFT assumes that both the sampled function $f[n \cdot \Delta x]$ and sampled spectrum $F[k \cdot \Delta\xi]$ are periodic over N samples. The space-domain representation of the continuous form of the sampled function in Equation (15.19) is:

$$f_s[x; N \cdot \Delta x] = \left[f[x] \cdot \frac{1}{\Delta x} COMB\left[\frac{x}{\Delta x}\right] \cdot RECT\left[\frac{x}{N \cdot \Delta x}\right]\right] * COMB\left[\frac{x}{N \cdot \Delta x}\right] \tag{15.76}$$

and therefore the associated spectrum is:

$$\begin{aligned}
&F_s[\xi; N \cdot \Delta x] \\
&= \left(F[\xi] * COMB[\Delta x \cdot \xi] * \left((N \cdot \Delta x) \cdot SINC\left[\frac{\xi}{N \cdot \Delta x}\right]\right)\right) \cdot (N \cdot \Delta x \cdot COMB[N \cdot \Delta x \cdot \xi]) \\
&= \left(F[\xi] * \left((N \cdot \Delta x) \cdot SINC\left[\frac{\xi}{\Delta\xi}\right]\right) * \sum_{\ell=-\infty}^{+\infty} \frac{1}{\Delta x} \cdot \delta\left[\xi - \frac{\ell}{\Delta x}\right]\right) \cdot \sum_{k=-\infty}^{+\infty} \delta[\xi - k \cdot \Delta\xi] \\
&= N \cdot \left(\sum_{\ell=-\infty}^{+\infty} F\left[\xi - \frac{\ell}{\Delta x}\right] * SINC\left[\frac{\xi}{\Delta\xi}\right]\right) \cdot \sum_{k=-\infty}^{+\infty} \delta[\xi - k \cdot \Delta\xi] \\
&= N \cdot \sum_{\ell=-\infty}^{+\infty} \sum_{k=-\infty}^{+\infty} \left(F\left[\xi - \frac{\ell}{\Delta x}\right] * SINC\left[\frac{\xi}{\Delta\xi}\right]\right) \cdot \delta[\xi - k \cdot \Delta\xi]
\end{aligned} \tag{15.77}$$

The replicas of the continuous spectrum centered about the integer multiples of $(\Delta x)^{-1}$ are convolved with the infinite-support *SINC* function with width parameter $\Delta\xi = (N \cdot \Delta x)^{-1}$, followed by frequency-domain sampling at intervals $\Delta\xi$. In the (fortunate) case where the continuous function $f[x]$

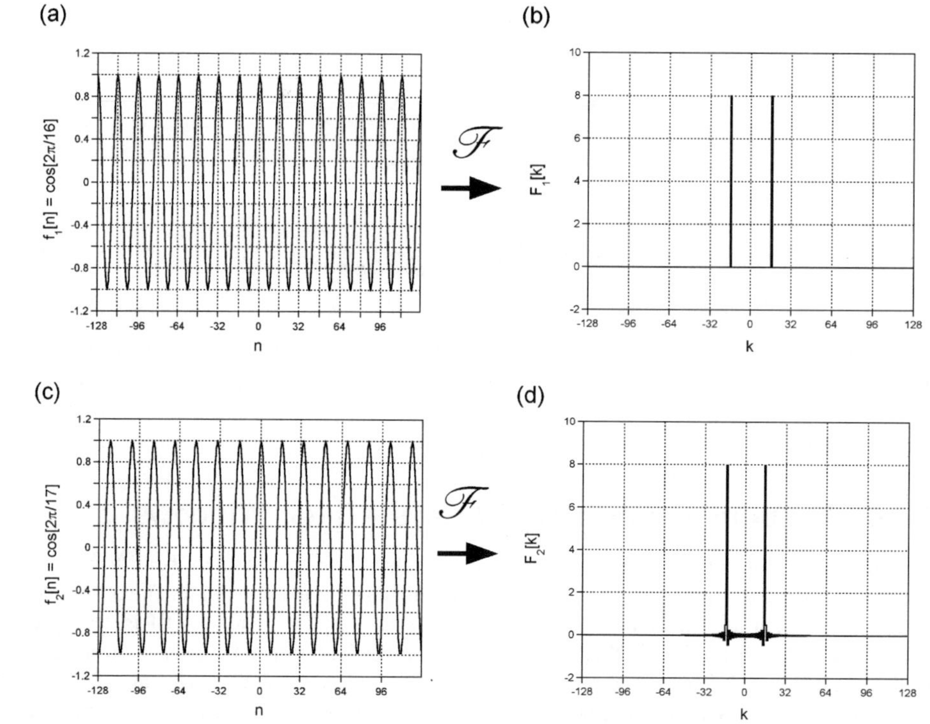

Figure 15.15 DFT of cosines for $N = 256$: (a) the DFT of $f_1[n] = \cos[2\pi n/16]$ is $F_1[k] = 8\,\delta[k + 16] + 8\,\delta[k - 16]$; (b) DFT of $f_2[n] = \cos[2\pi n/17]$ is $F_2[k]$, which shows leakage because $f_2[n]$ is not periodic over N samples.

is "truly" periodic over N samples, then the discrete components of its spectrum $F[\xi]$ are located at integer multiples of the frequency-domain sample interval $\Delta\xi$, and the convolution with the *SINC* function has no effect on the sampled spectrum. If the continuous function $f[x]$ is not periodic, or is periodic over a region of size other than $N \cdot \Delta x$, then the convolution of $F[\xi - \ell/\Delta x]$ with the *SINC* function generates "leakage" artifacts in the spectrum as was described in Section 15.2. To illustrate, the two input functions $f_1[n] = \cos[2\pi n/32]$ and $f_2[n] = \cos[2\pi n/24]$ are sampled at unit intervals $(\Delta x = 1)$ in an array with $N = 32$ indexed over the interval $0 \le n \le 31$. Two adjacent periods of the sampled array for each example are shown in Figure 15.16. One period of f_1 exactly fills the 32-pixel array, so the transition in amplitude between the adjacent replicas of the "synthesized" array is "smooth". However, this is not true for f_2: the "synthesized" periodic function assumed by the FFT exhibits a discontinuity in amplitude at each edge of the original 32-pixel array. These sharp transitions of the synthesized function f_2 produce "incorrect" information that appears in the single period of the computed discrete spectrum. This also may be visualized in the frequency domain. The continuous spectrum of a cosine with period X_n is composed of two Dirac delta functions located at $\xi = \pm(N/X_n) \cdot \Delta\xi$. These frequencies are $\pm 2 \cdot \Delta\xi$ for f_1 and $\cong \pm 1.8823 \cdot \Delta\xi$ for f_2. Because the continuous Dirac delta functions are located at integer multiples of $\Delta\xi$ in the first case, the amplitudes at all other samples due to the convolution with the *SINC* function are zero. In the second case, the convolution with the *SINC* function produces nonzero amplitudes that decay as $(\Delta\xi)^{-1}$ at the other samples in the discrete frequency domain.

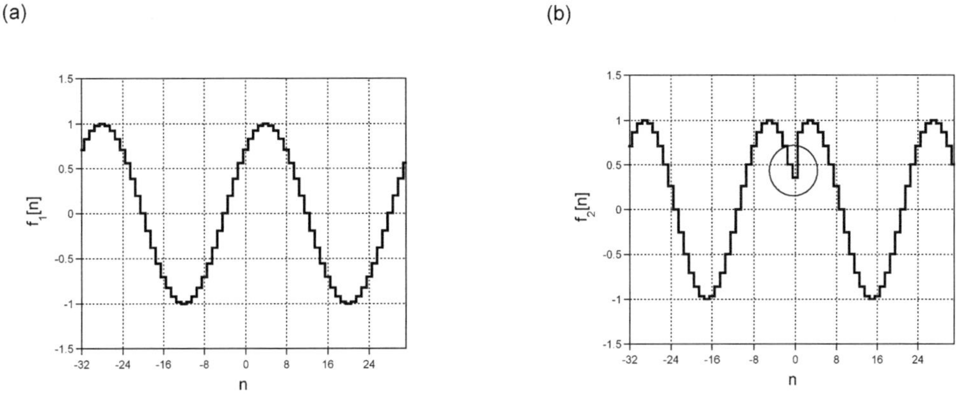

Figure 15.16 The source of leakage illustrated by two functions originally sampled over "single-sided" arrays with $0 \leq n \leq 31$. The functions are $f_1[n] = \cos[2\pi n/32]$ and $f_2[n] = \cos[2\pi n/24]$. The FFT assumes that the arrays are periodic over 32 samples. In (a), the array really is periodic over 32 samples, so the transition between the replicas at $-32 \leq n \leq -1$ and at $0 \leq n \leq 31$ is "smooth". In (b), the transition between the two replicas of $f_2[n]$ exhibits a discontinuity (circled), which generates artifacts at other spatial frequencies above and below the actual frequency.

Leakage may be reduced or even eliminated by replacing the rectangular window in Equation (15.76) with a different function $w[x]$ that (usually) decays smoothly to zero at the array "endpoints" $x = \pm N \cdot \Delta x/2$. The window function may be specified either by its continuous or sampled forms $w[x]$ or $w[n \cdot \Delta x]$; the former choice is advantageous for this analysis. The modulation theorem establishes the effect of the window function upon the spectrum of the continuous functions:

$$\mathcal{F}_1\left\{(f[x] \cdot w[x]) * \left(\frac{1}{\Delta x} COMB\left[\frac{x}{\Delta x}\right]\right)\right\} = (F[\xi] * W[\xi]) \cdot COMB[\Delta x \cdot \xi] \tag{15.78}$$

Since $w[x]$ is a smoothly varying nonnegative function that is conceptually "narrower" than the rectangle, the spectrum $W[\xi]$ is smoothly varying and "wider" than the spectrum $SINC[\xi/\Delta\xi]$ of the rectangular "window" in Equation (15.77). The convolution of the continuous input spectrum $F[\xi]$ with $W[\xi]$ averages (or "blurs") the oscillating amplitudes of the "unwindowed" spectrum, thus reducing the visible leakage. In other words, the process may improve the visibility of small spectral components by sacrificing some ability to distinguish neighboring components in the spectrum.

Note that leakage is not a problem if $f[x]$ consists of a "transient" that is completely contained within the support $N \cdot \Delta x$ samples. Windowing is only necessary if the original signal is longer than N samples, as for a periodic $f[n]$.

Many different windowing functions $w[x]$ may be constructed that differ in the rate of decay of the amplitude to (or near) zero at the edges of the array. The scaling theorem ensures that decreasing the "effective width" of $w[x]$ increases the width of $W[\xi]$ and thus will average the spectrum to a greater extent and further reduce the resolution. Because the amplitude of "narrow" windows can decay slowly and smoothly to zero, their spectra may be smooth and also nonnegative. Wider windows that decay "rapidly" at the edges will have narrower spectra, thus averaging less of the original spectrum $F[\xi]$, but will have oscillating "sidelobes" (like the *SINC* function) that may hide the presence of small signals near other and larger components.

The action of the window is often characterized by the width of its central lobe and the relative amplitude of its first sidelobe. The width of the lobe is the interval between frequencies on either side of the origin with amplitude at or closest to zero. For the rectangle window, this interval is

$$2 \cdot \Delta\xi \frac{cycles}{unit\ length} = \frac{2}{N \cdot \Delta x} \frac{cycles}{unit\ length} = \frac{2}{N} \frac{cycles}{sample} = \frac{4\pi}{N} \frac{radians}{sample} \tag{15.79}$$

The widths of the other windows will, of course, be larger.

The relative amplitude of the first sidelobe is usually described by the power ratio of the central ordinate A_0 of the continuous spectrum to the extremum of the first lobe A_1, which is often expressed in logarithmic units of power from Equation (6.192):

$$Power\ ratio = -20\ \log_{10}\left[\left|\frac{A_1}{A_0}\right|\right]\ \text{dB} \tag{15.80}$$

For the "unwindowed" case (the "rectangular" window), the first sidelobe is located at $\xi \cong 1.5 \cdot \Delta\xi$ and the power ratio of its amplitude is:

$$-20\ \log_{10}\left[\left|\frac{-0.217\,23}{1}\right|\right] \cong 13.26\ \text{dB} \tag{15.81}$$

where the amplitude of the first lobe was retrieved from Section 6.1.8. We say that the power of the largest sidelobe is about "13 dB down". This is a small reduction in power; the spectrum of a sampled function that is not periodic over N samples will be noticeably affected by the convolution with the *SINC* function. If the first sidelobe is attenuated to a greater extent (larger values of the power ratio), then the windowed spectrum will exhibit less "ringing".

15.5.4.1 Hann (or Hanning) Window

Perhaps the most common window used with the FFT is a biased cosine with a period of N samples. The function was introduced by Julius Von Hann, an Austrian meteorologist. It also is often called the *Hanning* window, though there is some danger of aural confusion with the *Hamming* window that is discussed next.

The coordinate-space representation of the continuous Hann window is:

$$\begin{aligned} w[x] &= \frac{1}{2}\left(1 + \cos\left[\frac{2\pi x}{N \cdot \Delta x}\right]\right) RECT\left[\frac{x}{N \cdot \Delta x}\right] \\ &= \frac{1}{2}(1 + \cos[2\pi \cdot \Delta\xi \cdot x])\, RECT[\Delta\xi \cdot x] \\ &= \cos^2[\pi \cdot \Delta\xi \cdot x]\, RECT[\Delta\xi \cdot x] \end{aligned} \tag{15.82}$$

where the discrete "uncertainty" relation in Equation (15.21) has been used. More flexible forms of the window that include free parameters for the power of the cosine and for the relative amplitudes of the constant and sinusoidal term will be considered shortly.

The continuous expression for the spectrum is easy to obtain by substituting known transforms:

$$\begin{aligned} W[\xi] &= \frac{1}{2}\left(\delta[\xi] + \frac{1}{2 \cdot \Delta\xi}\delta\delta\left[\frac{\xi}{\Delta\xi}\right]\right) * \frac{1}{\Delta\xi} SINC\left[\frac{\xi}{\Delta\xi}\right] \\ &= (\Delta\xi)^{-1}\left(\frac{1}{2}\delta[\xi] + \frac{1}{4}\delta[\xi + \Delta\xi] + \frac{1}{4}\delta[\xi - \Delta\xi]\right) * SINC\left[\frac{\xi}{\Delta\xi}\right] \\ &= (\Delta\xi)^{-1}\left(\frac{1}{2} SINC\left[\frac{\xi}{\Delta\xi}\right] + \frac{1}{4} SINC\left[\frac{\xi + \Delta\xi}{\Delta\xi}\right] + \frac{1}{4} SINC\left[\frac{\xi - \Delta\xi}{\Delta\xi}\right]\right) \end{aligned} \tag{15.83}$$

It is instructive to compare this expression to the transfer function for the rectangle window in Equation (15.77). The spectrum of the Hann window produces a wider "*SINC*-like" function with smaller

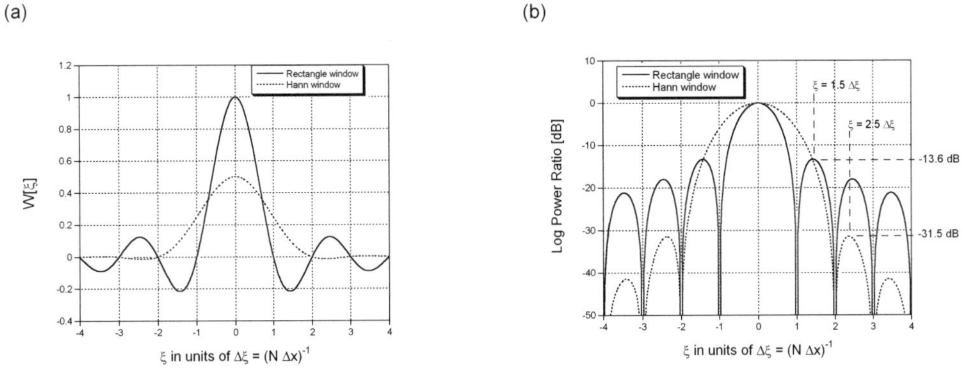

Figure 15.17 Comparison of the spectrum of the unwindowed rectangle and the Hann window: (a) with linear amplitude scaling, showing the reduced amplitude at the origin and larger width of the spectrum of the Hann window; (b) displayed as the logarithm of the power ratio, showing the different locations and reduced power in the sidelobes of the Hann window.

sidelobes than $SINC[\xi/\Delta\xi]$. The width between the first zeros is $4 \cdot \Delta\xi$ cycles per unit length, which is twice as wide as the corresponding measurement for the spectrum of the rectangle. The first sidelobe of the spectrum of the Hann window is located at $\xi \cong 2.5 \cdot \Delta\xi$ instead of $\xi \cong 1.5 \cdot \Delta\xi$ and the power ratio of the first lobe to the central ordinate is about -31.47 dB, which is significantly smaller than the ratio of -13.26 dB for the "unwindowed" rectangle. The two cases are compared in Figure 15.17.

The discrete representation of the window spectrum is obtained by sampling $W[\xi]$ at integer multiples of $\Delta\xi$ via multiplication by the appropriate *COMB* function:

$$\begin{aligned}
W[k \cdot \Delta\xi] &= W[\xi] \cdot COMB\left(\frac{\xi}{\Delta\xi}\right) \\
&= W[\xi] \sum_{k=-N/2}^{N/2} \delta\left[\frac{\xi - k \cdot \Delta\xi}{\Delta\xi}\right] \equiv \sum_{k=-N/2}^{N/2} \delta\left[\frac{\xi - k \cdot \Delta\xi}{\Delta\xi}\right] W[k] \\
&= (\Delta\xi)^{-1}\left(\frac{1}{2}SINC\left[\frac{\xi}{\Delta\xi}\right] + \frac{1}{4}SINC\left[\frac{\xi + \Delta\xi}{\Delta\xi}\right] + \frac{1}{4}SINC\left[\frac{\xi - \Delta\xi}{\Delta\xi}\right]\right) \\
&\quad \cdot (\Delta\xi) \sum_{k=-N/2}^{N/2} \delta[\xi - k \cdot \Delta\xi] \\
&= \sum_{k=-N/2}^{N/2} \delta[\xi - k \cdot \Delta\xi] \\
&\quad \cdot \left(\frac{1}{2}SINC\left[\frac{k \cdot \Delta\xi}{\Delta\xi}\right] + \frac{1}{4}SINC\left[\frac{k \cdot \Delta\xi + \Delta\xi}{\Delta\xi}\right] + \frac{1}{4}SINC\left[\frac{k \cdot \Delta\xi - \Delta\xi}{\Delta\xi}\right]\right) \qquad (15.84a)
\end{aligned}$$

where the property of Dirac delta functions in products from Equation (6.115) has been used. Since the *SINC* functions are centered at frequencies that are integer multiples of $\Delta\xi$, their values at all other

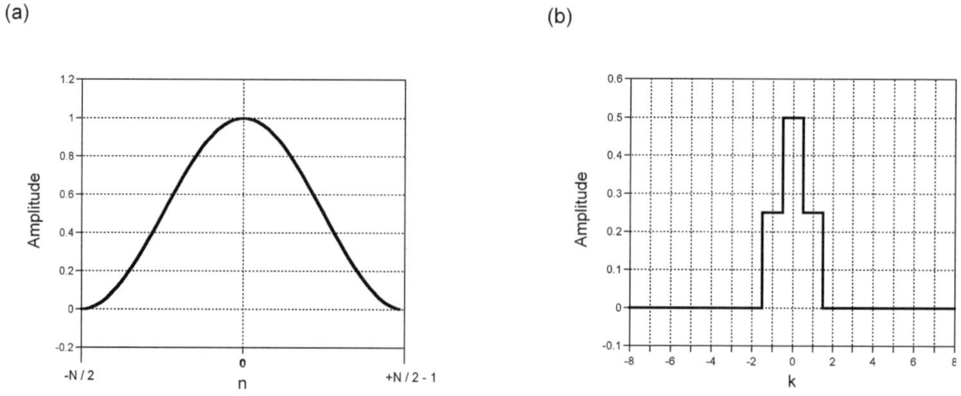

Figure 15.18 Hann window: (a) space-domain function $w[n] = \frac{1}{2}(1 + \cos[2\pi n/N])$; (b) discrete spectrum $W[k]$, which consists of three discrete Dirac delta functions located at $k = 0$ with weight $\frac{1}{2}$ and at $k = \pm 1$ with weight $\frac{1}{4}$.

samples are zero. Therefore the discrete spectrum of the Hann window may be simplified to:

$$\begin{aligned} W[k] &= \left(\frac{1}{4}SINC[k+1] + \frac{1}{2}SINC[k] + \frac{1}{4}SINC[k-1]\right) \\ &= \left(\frac{1}{4}\delta_d[k+1] + \frac{1}{2}\delta_d[k] + \frac{1}{4}\delta_d[k-1]\right) \end{aligned} \tag{15.84b}$$

The space-domain Hann window and the corresponding sampled $W[k]$ are shown in Figure 15.18. The action of the window function on the spectrum is to convolve the "unwindowed" spectrum with this set of three discrete Dirac delta functions. This "spreads" the amplitude of each discrete frequency component over the two adjacent frequency samples. Note that the sum of the amplitudes of $W[k]$ is unity, which means that the sum of the spectrum amplitudes before and after convolution with $W[k]$ will be identical.

15.5.4.2 Example of Hann Window

The effect of applying a Hann window to the function before computing the FFT is shown for two cases. In Figure 15.19a, the input function is $\cos[(2\pi x/(16 \cdot \Delta x))]$ sampled 256 times. The array includes 16 complete periods of the sampled cosine, so that the infinite array created from the N samples is truly periodic. Therefore, the discrete spectrum evaluated at $N = 256$ frequencies consists of a pair of Dirac delta functions located at $k = \pm N/16 = \pm 16$. In Figure 15.19b, the same cosine is multiplied by the Hann window, which decreases the amplitudes and increases the width of the components of the windowed spectrum because the amplitudes of neighboring components have been "blurred together".

15.5.4.3 Windowing to Improve Detectability of Small Signals

A practical application of windowing the FFT is to improve the ability to detect small sinusoidal signals in the presence of sinusoidal "noise". Consider 256 samples of the signal shown in Figure 15.20a:

$$f[x] = \cos\left[\frac{2\pi x}{24}\right] + 0.01 \cdot \cos\left[\frac{2\pi x}{8}\right] \tag{15.85}$$

The spatial frequencies of the two components are $\frac{1}{24}$ and $\frac{1}{8}$ cycles per sample, respectively, which would be located at samples $256/24 = 10\frac{2}{3}$ and $256/8 = 32$. The spectrum of the first term "rings"

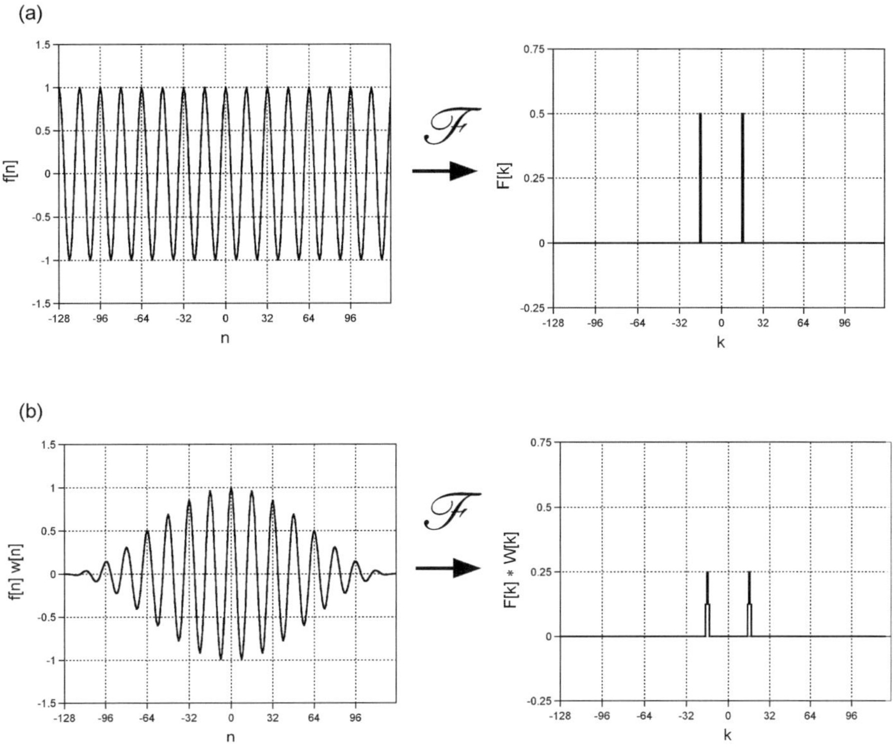

Figure 15.19 Effect of Hann window on spectrum computed via FFT: (a) $f[n] = \cos[2\pi x/16]$ for $N = 256$ and the FFT $F[k]$ exhibits a pair of discrete Dirac delta functions located at $k = \pm 16/256$ with amplitude of $\frac{1}{2}$ after appropriate normalization; (b) $f[n]$ after multiplication by a Hanning window $w[n]$ and the FFT of $f[n] \cdot w[n]$, showing that the discrete Dirac delta functions have been "blurred" into replicas of $W[k]$ scaled by the half-unit amplitude of the original components.

because of the noninteger number of periods in the array. The smaller-amplitude component in the spectrum is difficult to see due to this "noise", as shown in the spectrum in Figure 15.20a. The Hann window (Figure 15.20b) convolves the original discrete spectrum with the 3-pixel averager of Equation (15.84b), which averages the amplitude of the oscillatory "ringing" and thus improves the detectability of the smaller component, as shown in the spectrum of Figure 15.20b.

15.5.5 A Garden of 1-D FFT Windows

15.5.5.1 Windows Derived from Hann Window

Relatives of the Hann window may be constructed by changing the power applied to the cosine function in Equation (15.82):

$$w[x] = \cos^n[\pi \cdot \Delta\xi \cdot x]\, RECT[\Delta\xi \cdot x], \quad n \geq 1 \tag{15.86}$$

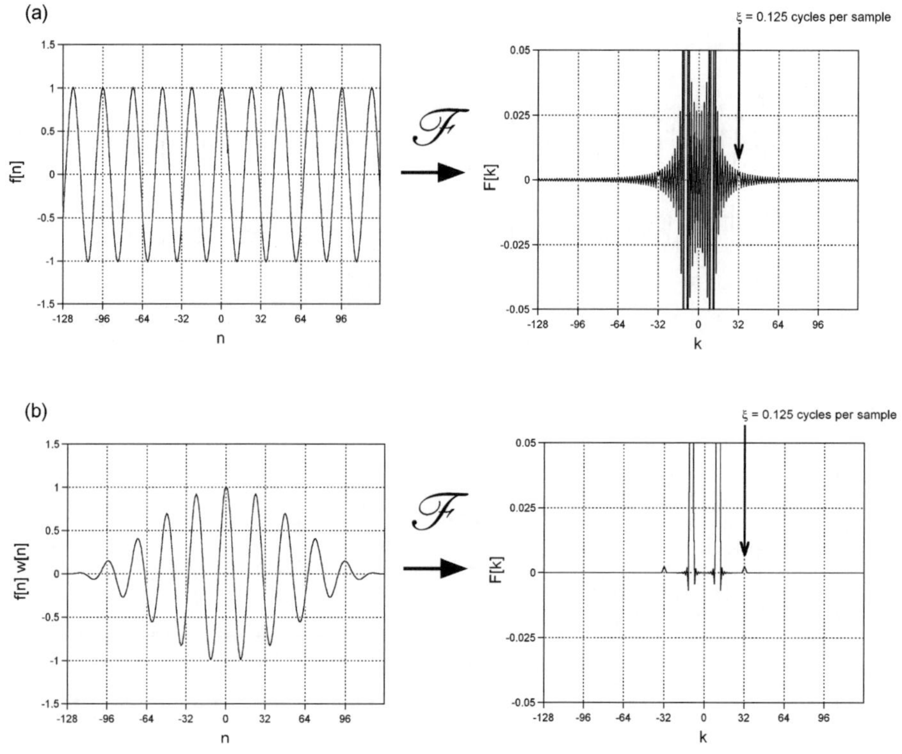

Figure 15.20 Use of Hann window to improve visibility of small frequency components: (a) $f[n] = \cos[2\pi x/24] + 0.01 \cos[2\pi x/8]$ and the FFT of $f[n]$ produces $F[k]$ that exhibits severe "ringing" due to leakage from the discontinuity at the edge of the array. The amplitude is magnified to show the effect in the vicinity of $\xi_0 = \frac{1}{8}$ cycles per sample; (b) $f[n]$ after multiplication by a Hann window $w[n]$ and the FFT of $f[n] \cdot w[n]$, showing the more visible component at $\xi_0 = \frac{1}{8}$.

Examples are shown in Figure 15.21. Consider first the case $n = 1$:

$$\begin{aligned} w[x] &= \cos[\pi \cdot \Delta\xi \cdot x]\, RECT[\Delta\xi \cdot x] \\ &= \cos\left[\frac{2\pi x}{2 \cdot N \cdot \Delta x}\right] RECT\left[\frac{x}{N \cdot \Delta x}\right] \end{aligned} \tag{15.87}$$

This is the central lobe of a cosine whose amplitude decays to zero at the edges of the array. The period of the cosine is $2 \cdot N \cdot \Delta x$, but the truncation by the rectangle of width $N \cdot \Delta x$ leads to the appropriate name of *half-cycle cosine* window (if used with a single-sided array with $0 \leq n \leq N - 1$, it is commonly called the *half-cycle sine* window). The sampled version is obtained by setting $x = n \cdot \Delta x$ in Equation (15.87):

$$w[n] = \cos\left[\frac{\pi n}{N}\right] RECT\left[\frac{n}{N}\right] \tag{15.88}$$

The continuous spectrum of this window is easy to evaluate:

$$\begin{aligned}W[\xi] &= \frac{1}{2(2N \cdot \Delta x)^{-1}}\delta\delta\left[\frac{\xi}{(2N \cdot \Delta x)^{-1}}\right] * (N \cdot \Delta x)\, SINC[N \cdot \Delta x \cdot \xi] \\ &= \left(\frac{1}{2}\delta\left[\xi + \frac{\Delta\xi}{2}\right] + \frac{1}{2}\delta\left[\xi - \frac{\Delta\xi}{2}\right]\right) * (\Delta\xi)^{-1}\, SINC\left[\frac{\xi}{\Delta\xi}\right] \\ &= \frac{1}{2 \cdot \Delta\xi}\left(SINC\left[\xi + \frac{\Delta\xi}{2}\right] + SINC\left[\xi - \frac{\Delta\xi}{2}\right]\right) \qquad (15.89)\end{aligned}$$

It is the sum of two *SINC* functions in the frequency domain, each displaced from the origin by half of the sampling interval $\Delta\xi$. The maximum amplitude (at $\xi = 0$) is $2\, SINC[\frac{1}{2}] = 4/\pi \cong 1.27$. The width parameter of this spectrum is $6 \cdot \Delta\xi$ cycles per unit length. The amplitude extremum of the first sidelobe is approximately -0.09, so the sidelobe ratio from Equation (15.80) is approximately -23 dB, which is an improvement over that of the unwindowed "rectangle" function at -13 dB, but this is significantly larger than the ratio of $\cong -31.5$ dB of the Hann window with $n = 2$. The differences can be explained by noting that the cusp at the edges of the window for the half-cycle cosine is less "sharp" than for the rectangle but "sharper" than for the Hann window.

The discrete spectrum of the half-cycle cosine is obtained by multiplying by the *COMB* function and evaluating at the sample frequencies $\xi = k \cdot \Delta\xi$:

$$W[k] = \frac{1}{2}\left(SINC\left[k + \frac{1}{2}\right] + SINC\left[k - \frac{1}{2}\right]\right) \qquad (15.90)$$

as shown in Figure 15.21a. This spectrum is fundamentally different from those already considered because the *SINC* functions are centered between samples. The amplitude of the spectrum of the window oscillates about zero due to the "sharp" transitions at the edges of the window which generate the negative lobes of the *SINC* functions. Evaluation of the amplitudes of the *SINC* function for "half-integer" arguments allows Equation (15.90) to be simplified to:

$$W[k] = (-1)^k\left(\frac{2}{\pi}\right)\frac{1}{1 - 4k^2} \qquad (15.91)$$

The square of the Hann window is obtained by setting $n = 4$ in Equation (15.86). The resulting function decays more "quickly" away from the edges of the array and more "smoothly" close to them:

$$w[x] = \cos^4\left[\frac{\pi x}{N \cdot \Delta x}\right] = \left(\frac{1}{2} + \frac{1}{2}\cos\left[\frac{2\pi x}{N \cdot \Delta x}\right]\right)^2 RECT\left[\frac{x}{N \cdot \Delta x}\right] \qquad (15.92)$$

as shown in Figure 15.21b. The discrete version in the space domain is obviously:

$$w[n] = \cos^4\left[\frac{\pi n}{N}\right] \cdot RECT\left[\frac{n}{N}\right] \qquad (15.93)$$

The spectrum of the continuous function in Equation (15.92) is the autoconvolution of the spectrum of the Hann window from Equation (15.82):

$$\begin{aligned}W[\xi] &= \left(\frac{1}{4}\delta[\xi + \Delta\xi] + \frac{1}{2}\delta[\xi] + \frac{1}{4}\delta[\xi - \Delta\xi]\right) \\ &\quad * \left(\frac{1}{4}\delta[\xi + \Delta\xi] + \frac{1}{2}\delta[\xi] + \frac{1}{4}\delta[\xi - \Delta\xi]\right) * (\Delta\xi)^{-1}\, SINC\left[\frac{\xi}{\Delta\xi}\right] \\ &= \left(\frac{1}{16}\delta[\xi + 2 \cdot \Delta\xi] + \frac{1}{4}\delta[\xi + \Delta\xi] + \frac{3}{8}\delta[\xi] + \frac{1}{4}\delta[\xi - \Delta\xi] + \frac{1}{16}\delta[\xi - 2 \cdot \Delta\xi]\right) \\ &\quad * (\Delta\xi)^{-1}\, SINC\left[\frac{\xi}{\Delta\xi}\right] \qquad (15.94)\end{aligned}$$

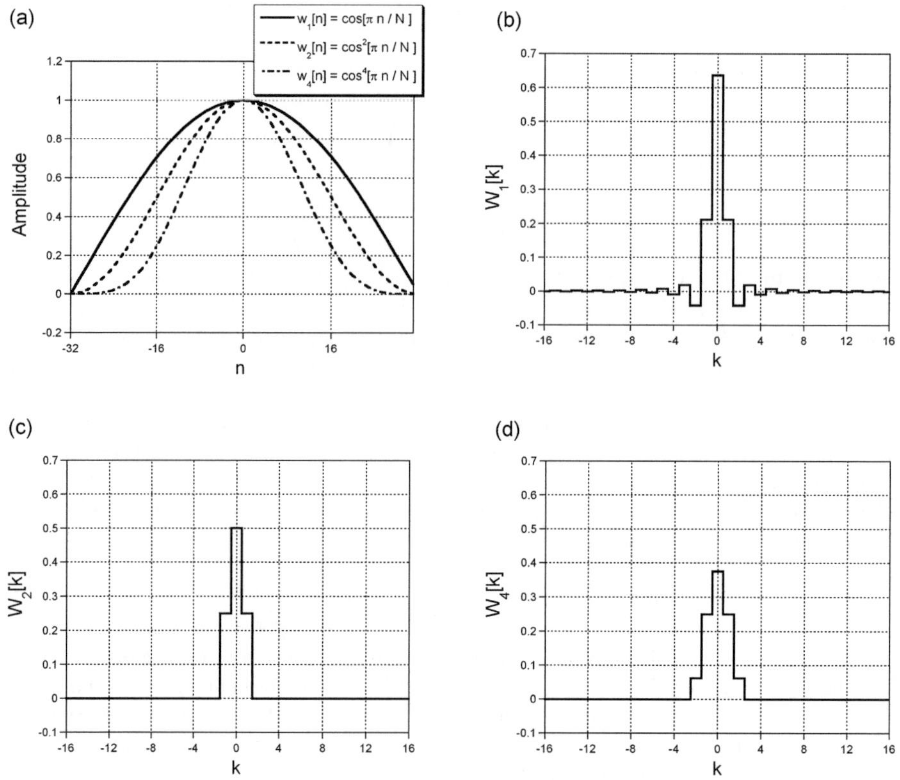

Figure 15.21 Variations of the Hann window: (a) $w_\ell[n] = \cos^\ell[\pi n/N]$ for $\ell = 1, 2$ (normal Hann window), and 3; (b) $W_1[k]$, which is the FFT of $w_1[n]$, which exhibits leakage from the "hard edges" of the window; (c) $W_2[k]$, which is the set of three discrete Dirac delta functions seen for the Hann window in Equation (15.84b); (d) $W_4[k]$, which is a set of more discrete Dirac delta functions. $W_4[k]$ is approaching a Gaussian function.

The sidelobe amplitude ratio is −46 dB.

The amplitudes of the sampled spectrum after windowing with the $\cos^4$ function are obtained by sampling $f[x] \cdot w_4[x]$ with $COMB[\xi/\Delta\xi]$ to obtain:

$$W[k] = \frac{1}{16}\delta_d[k+2] + \frac{1}{4}\delta_d[k+1] + \frac{3}{8}\delta_d[k] + \frac{1}{4}\delta_d[k-1] + \frac{1}{16}\delta_d[k-2] \tag{15.95}$$

as shown in Figure 15.21d. Note that increasing the power n of the cosine function in Equation (15.86) produces "smoother" space-domain functions that approach Gaussian functions; the corresponding spectra exhibit smaller sidelobe levels.

15.5.5.2 Hamming Window

The Hamming window is a common variation of the Hann window whose amplitude does not decay to zero at the endpoints. A more general form of the Hann function may be defined with a selectable real-valued scaling parameter α applied to the amplitude:

$$w[x] = \frac{1}{2}\left(\alpha + (1-\alpha)\cos\left[2\pi\frac{x}{N\cdot\Delta x}\right]\right) RECT\left[\frac{x}{N\cdot\Delta x}\right] \tag{15.96}$$

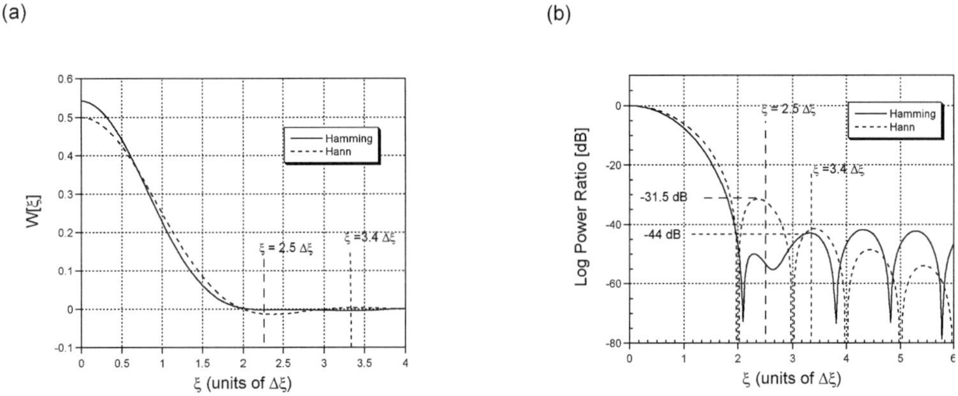

Figure 15.22 Comparison of the spectra of the Hamming and Hann windows: (a) linear amplitude scaling; (b) logarithm of the power ratios, showing the approximate cancellation of the Hamming window at the sidelobe frequency of the Hann window. The "new" first sidelobe of the Hamming window is at a larger frequency and smaller level (approximately -44 dB). Also note that subsequent lobes of the Hamming window decay more slowly than those of the Hann.

The continuous spectrum is a generalization of Equation (15.82):

$$\begin{aligned} W[\xi] &= \frac{1}{2}\left(\alpha\cdot\delta[\xi] + \frac{(1-\alpha)}{2\cdot\Delta\xi}\delta\delta\left[\frac{\xi}{\Delta\xi}\right]\right) * (\Delta\xi)^{-1}\,SINC\left[\frac{\xi}{\Delta\xi}\right] \\ &= (\Delta\xi)^{-1}\left(\alpha\cdot\frac{1}{2}\delta[\xi] + \frac{(1-\alpha)}{4}(\delta[\xi+\Delta\xi]+\delta[\xi-\Delta\xi])\right) * SINC\left[\frac{\xi}{\Delta\xi}\right] \\ &= (\Delta\xi)^{-1}\left(\frac{\alpha}{2}SINC\left[\frac{\xi}{\Delta\xi}\right] + \frac{(1-\alpha)}{4}\left(SINC\left[\frac{\xi+\Delta\xi}{\Delta\xi}\right] + SINC\left[\frac{\xi-\Delta\xi}{\Delta\xi}\right]\right)\right) \end{aligned} \tag{15.97}$$

The free parameter α is chosen to make the amplitudes of the three *SINC* functions cancel at the first sidelobe of the Hann window located at $\xi \cong \pm 2.5\cdot\Delta\xi$. This cancellation "moves" the sidelobe peak to a frequency farther from DC and its amplitude is reduced. The solution is obtained by setting Equation (15.97) to zero, which occurs for $\alpha = 25/46 \cong 0.54$. The formula for the Hamming window is obtained by using this value for α:

$$w[x] = \frac{1}{2}\left(0.54 + 0.46\cdot\cos\left[2\pi\frac{x}{N\cdot\Delta x}\right]\right) RECT\left[\frac{x}{N\cdot\Delta x}\right] \tag{15.98}$$

The spectrum of this window is:

$$W[\xi] = \frac{1}{2\cdot\Delta\xi}\left(0.54\,SINC\left[\frac{\xi}{\Delta\xi}\right] + 0.23\left(SINC\left[\frac{\xi+\Delta\xi}{\Delta\xi}\right] + SINC\left[\frac{\xi-\Delta\xi}{\Delta\xi}\right]\right)\right) \tag{15.99}$$

which is compared to the Hann window on both linear and logarithmic scales in Figure 15.22. The approximate cancellation at the Hann sidelobe frequency of approximately $\pm 2.5\cdot\Delta\xi$ means that the "new first sidelobe" is located farther from the origin, at approximately $\pm 3.4\cdot\Delta\xi$, and the sidelobe power ratio is reduced to approximately -44 dB.

The continuous space-domain Hamming window is:

$$w[x] = \left(0.54 + 0.46\cdot\cos\left[\frac{2\pi x}{N\cdot\Delta x}\right]\right) RECT\left[\frac{x}{N\cdot\Delta x}\right] \tag{15.100}$$

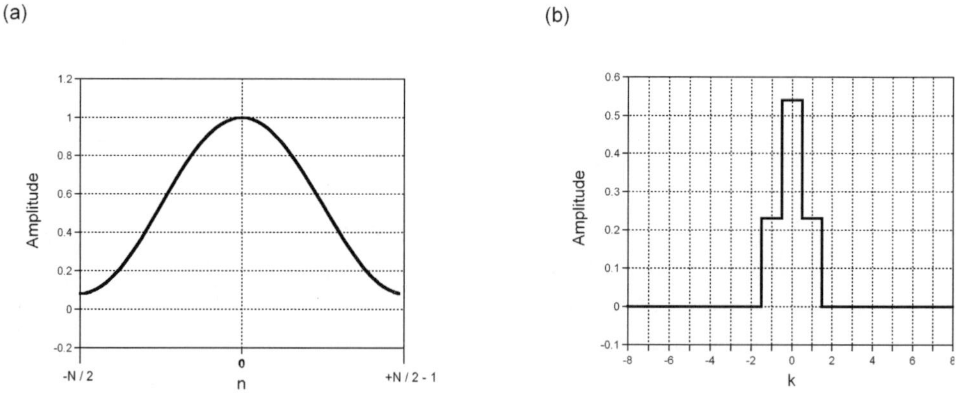

Figure 15.23 Hamming window in both domains: (a) $w[n] = 0.54 + 0.46\cos[2\pi n/N]$; (b) discrete spectrum of $w[n]$, which consists of three discrete Dirac delta functions: one located at $k = 0$ with weight 0.54 and two at $k = \pm 1$ with weight 0.23.

and so the discrete form is:

$$w[n] = 0.54 + 0.46 \cdot \cos\left[\frac{2\pi n}{N}\right] \tag{15.101}$$

as shown in Figure 15.23. Since the "component" functions of this window are scaled versions of those in the Hann window, the continuous and discrete spectra may be evaluated easily. The continuous spectrum of Equation (15.100) is:

$$\begin{aligned} W[\xi] &= \left(0.54\delta[\xi] + 0.46\left[\frac{N \cdot \Delta x}{2}\right]\delta\delta[N \cdot \Delta x \cdot \xi]\right) * (N \cdot \Delta x)\, SINC[N \cdot \Delta x \cdot \xi] \\ &= (\Delta\xi)^{-1}\left(0.23\, SINC\left[\frac{\xi + \Delta\xi}{\Delta\xi}\right] + 0.54\, SINC\left[\frac{\xi}{\Delta\xi}\right] + 0.23\, SINC\left[\frac{\xi - \Delta\xi}{\Delta\xi}\right]\right) \end{aligned} \tag{15.102}$$

The corresponding discrete form of the spectrum of the Hamming window is:

$$W[k] = 0.23\delta_d[k+1] + 0.54\delta_d[k] + 0.23\delta_d[k-1] \tag{15.103}$$

In words, the unwindowed discrete spectrum is convolved with this three-element kernel. The resulting averaging of the spectrum is slightly less effective than that from the Hann window.

15.5.5.3 Bartlett (Triangle) Window

The Bartlett window is a triangle function that decays to zero at the endpoints of the centered array. The continuous form of the window is:

$$w[x] = TRI\left[\frac{x}{(N \cdot \Delta x/2)}\right] = TRI\left[\frac{2x}{N \cdot \Delta x}\right] \tag{15.104}$$

The "cusp" at the origin and the transitions at the edges of this function lead to the suspicion that the sidelobe suppression will be less noticeable for the Bartlett window compared to the Hann.

The discrete form of the Bartlett window is obtained by substituting $x = n \cdot \Delta x$:

$$w[n] = TRI\left[\frac{2 \cdot n \cdot \Delta x}{N \cdot \Delta x}\right] = TRI\left[\frac{2n}{N}\right] \tag{15.105}$$

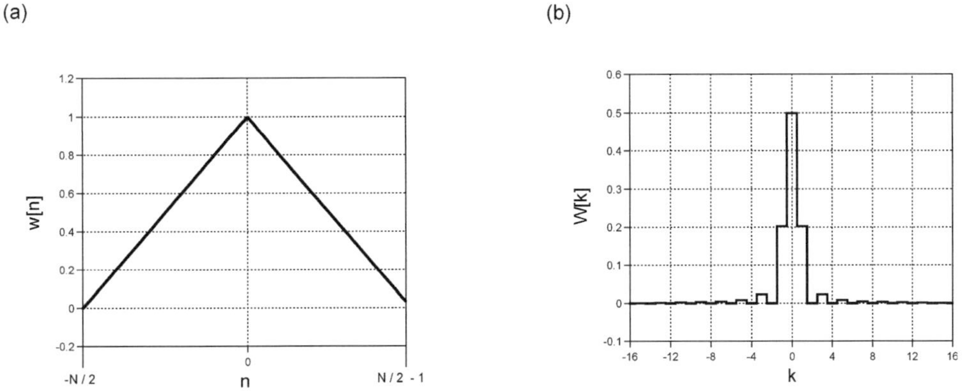

Figure 15.24 Bartlett (triangle) window in both domains: (a) $w[n] = TRI[n/N]$; (b) discrete spectrum $W[k] = \frac{1}{2} SINC^2[k/2]$.

Since the amplitude of the triangle goes to zero at the edges of the N-pixel array, there is no need to additionally apply the truncating *RECT* function.

The continuous spectrum is a $SINC^2$ function with the same width of $2 \cdot \Delta\xi$ cycles per unit length as the *SINC* function for the "unwindowed" spectrum. The amplitude is nonnegative at all spatial frequencies.

$$W[\xi] = \left(\frac{N \cdot \Delta x}{2}\right) SINC^2\left[\frac{N \cdot \Delta x \cdot \xi}{2}\right] \tag{15.106}$$

The amplitude of the first lobe of $SINC^2[\xi]$ is approximately $(-0.217)^2 \cong 0.047$ and the corresponding sidelobe power ratio is approximately -26.5 dB.

By substituting $\xi = k \cdot \Delta\xi$, we obtain the discrete form of the spectrum:

$$W[k] = \frac{1}{2} SINC^2\left[\frac{(N \cdot \Delta x)(k \cdot \Delta\xi)}{2}\right] = \frac{1}{2} SINC^2\left[\frac{k}{2}\right] \tag{15.107}$$

where the discrete "uncertainty relation" from Equation (15.21) has been used. The sampled amplitudes in the spectrum of the Bartlett window are nonzero at $k = 0$ and all odd values of k. Each spectrum component is "spread" over the entire spectrum, and so some leakage still occurs. The Bartlett window and its discrete frequency spectrum are shown in Figure 15.24.

Many other window functions have been studied. Interested readers should consult Harris (1978) for further details.

15.5.6 Undersampling and Aliasing

In practical use, aliasing of the input signal may be prevented either by resampling at a rate exceeding the Nyquist criterion (by reducing Δx) or by eliminating aliased frequencies *before sampling* by applying a lowpass filter with cutoff at or below the Nyquist frequency. As an example of the DFT of an aliased function, consider the transform pairs generated from data sampled from two symmetric square waves, as shown in Figure 15.25. The general form of the symmetric square wave is:

$$f[x] = SGN\left[\cos\left[\frac{2\pi x}{X_0}\right]\right] = \frac{2}{X_0} COMB\left[\frac{x}{X_0}\right] * RECT\left[\frac{2x}{X_0}\right] - 1[x] \tag{15.108}$$

and the corresponding continuous spectrum has infinite support, thus demonstrating that $f[x]$ cannot be sampled without aliasing:

$$F[\xi] = (2\,COMB[X_0 \cdot \xi]) \cdot \left(\frac{X_0}{2} SINC\left[X_0 \cdot \frac{\xi}{2}\right]\right) - \delta[\xi]$$

$$= \left(X_0\,COMB[X_0 \xi] \cdot SINC\left[X_0 \cdot \frac{\xi}{2}\right]\right) - \delta[\xi] \tag{15.109a}$$

$$= \left(\sum_{k=-\infty}^{+\infty} \delta\left[\xi - \frac{k}{X_0}\right] \cdot SINC\left[\xi \cdot \frac{X_0}{2}\right]\right) - \delta[\xi]$$

$$= \sum_{k=+1}^{+\infty} \left(\delta\left[\xi + \frac{k}{X_0}\right] + \delta\left[\xi - \frac{k}{X_0}\right]\right) SINC\left[\xi \cdot \frac{X_0}{2}\right]$$

$$= \sum_{k=+1}^{+\infty} \left(\delta\left[\xi + \frac{k}{X_0}\right] + \delta\left[\xi - \frac{k}{X_0}\right]\right) SINC\left[\frac{k}{2}\right] \tag{15.109b}$$

The *SINC* function evaluates to zero for all even values of k, and so the only frequencies with nonzero amplitude in the spectrum are the odd multiples of X_0^{-1}.

The two continuous square waves in this example are sampled at $N = 256$ data points separated by the same interval Δx. The periods were chosen to ensure that there is an integer number within the 256 samples to prevent leakage effects due to "artificial" transitions at the edges of the arrays:

$$X_1 = \frac{256 \cdot \Delta x}{16} = 16\ samples \tag{15.110a}$$

$$X_2 = \frac{256 \cdot \Delta x}{15} \cong 17.067\ samples \tag{15.110b}$$

The fundamental frequencies of the continuous square waves are therefore:

$$\xi_1 = \frac{1}{X_1} = \frac{1}{16} \frac{cycles}{sample} \tag{15.111a}$$

$$\xi_2 = \frac{1}{X_2} = \frac{15}{256} \cong \frac{1}{17.067} \frac{cycles}{sample} \tag{15.111b}$$

The FFTs evaluate the spectra at 256 sampled frequencies indexed by k, where $-128 \leq k \leq +127$, and consequently the spatial frequencies are $\xi = k \cdot \Delta\xi = k/(N \cdot \Delta x)$. The fundamental frequency components in the two cases are located respectively at indices $\pm k_1 = \pm 16$ and $\pm k_2 = \pm 15$.

The frequency components of these continuous square waves should appear only at odd harmonics of the fundamental frequency:

$$\xi = \pm\frac{1}{X}, \pm\frac{3}{X}, \pm\frac{5}{X}, \ldots, \pm\frac{2k+1}{X}, \ldots. \tag{15.112}$$

The amplitudes of the continuous spectra at these harmonics should be $F[k] = SINC[(2k+1)/2] = (-1)^k 2/k\pi$, for $k = 0, 1, 2, \ldots$, giving $F[k=1] \cong +0.637$, $F[2] \cong -0.212$, $F[3] \cong +0.127$, etc. The spectral orders of $F_1[k]$ are located at the "correct" spatial frequencies, but the amplitudes are "incorrect". In this case, the aliased components (due to the higher-order periods of the spectrum centered about multiples of $(\Delta x)^{-1}$) happen to "line up" at the same samples because the number of periods in the continuous square wave happened to be a power of two. The fact that the amplitudes are incorrect provides evidence of the aliasing.

The second sampled square wave also is aliased, as shown by the fact that components appear at incorrect frequencies. The amplitude at a sample of the DFT is the superposition of the "proper"

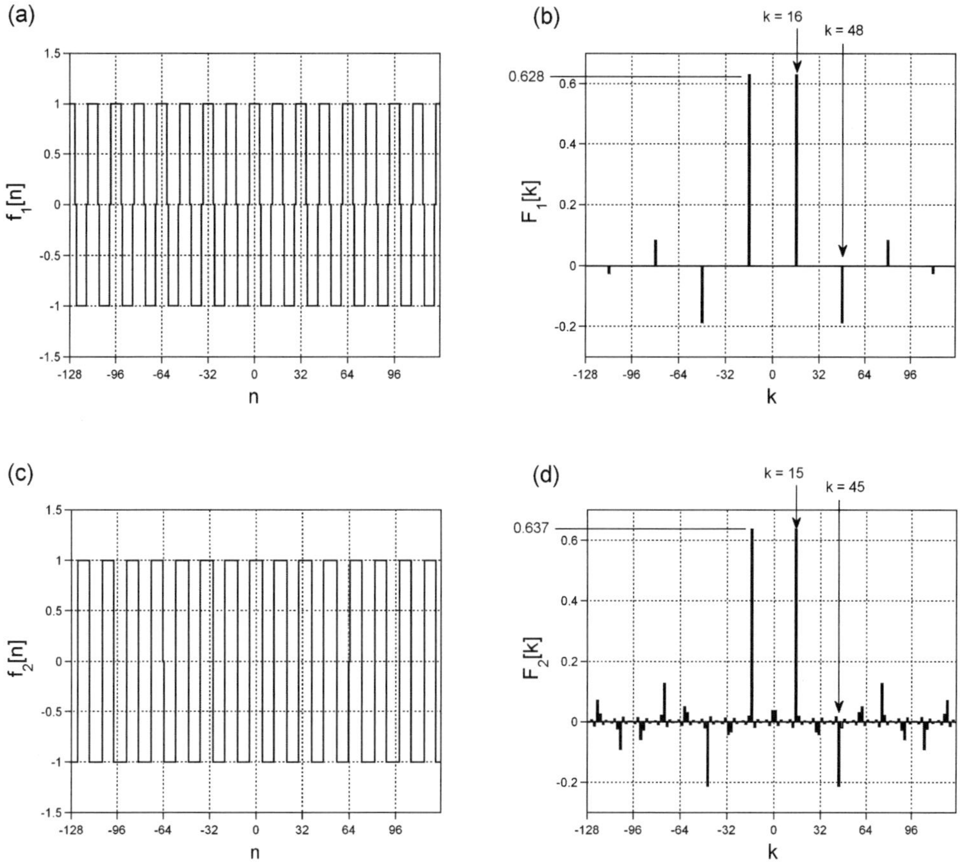

Figure 15.25 The two manifestations of aliasing for square waves: (a) $f_1[n] = SGN[\cos[2\pi n/16]]$ for $N = 256$, 16 periods within the array; (b) $F_1[k]$, which consists of discrete Dirac delta functions at the correct locations but with amplitudes that are too small; (c) $f_2[n] = SGN[\cos[2\pi n/X]]$, where $X = 256/15$, 15 periods within the array; (d) $F_2[k]$, which consists of many discrete Dirac delta functions at many samples. The amplitude of the fundamental component is very nearly correct.

amplitude and components at distant aliased frequencies that have very small amplitudes. The amplitudes at the proper frequencies are approximately correct, but aliased components appear at other samples where the amplitude of the continuous spectrum is zero.

The aliased frequencies may be removed before sampling by a lowpass filter with cutoff at the Nyquist rate. The prefiltered square wave of period 16 pixels and its FFT are shown in Figure 15.26. Note the overshoot at the edges (the Gibbs phenomenon) due to the filter and that the spectrum amplitudes are correct.

These spectra of the aliased square waves could be interpreted in another way if the two sets of 256 samples were obtained from the same square wave by using sample spacings such that $\Delta x_1 = 15/(16 \cdot \Delta x_2)$. The difference in sample spacing ensures that the Nyquist frequencies and hence the apparent aliased frequencies will differ between the two cases. From knowledge of Δx_1, Δx_2, and the two spectra, it may be possible to infer the true frequencies of aliased information.

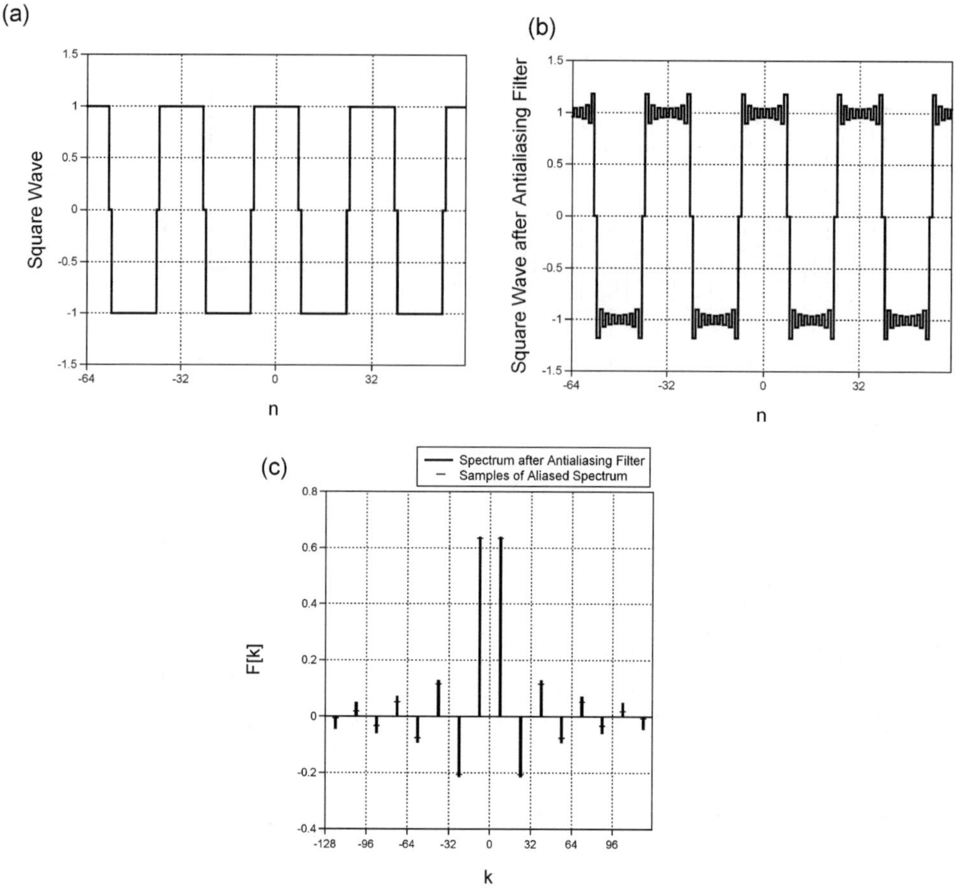

Figure 15.26 Effect of antialiasing filter on 1-D square wave with $N = 256$: (a) central region of sampled square wave, which is aliased; (b) square wave with antialiasing filter applied before sampling; (c) comparison of the spectra showing evidence of aliasing near the edges of the array.

15.5.7 Phase

When collecting real data, the position of a sequence of samples will be unrelated to the origin of coordinates of the continuous function unless precautions are taken. The absolute phase computed by the DFT will have no significance in such a situation. Information about relative phase angle is often still useful.

15.5.8 Zero Padding

The spectral resolution of the DFT is the interval $\Delta\xi$ between independent samples in the frequency domain, and is determined by N and Δx via the uncertainty relation $\Delta\xi = (N \cdot \Delta x)^{-1}$. The frequency resolution may be increased (so that $\Delta\xi$ is decreased) by increasing either Δx or N. The former strategy decreases the Nyquist frequency with the same number of samples, thus shrinking $\Delta\xi$. If Δx is fixed, increasing the number of samples N results in the same ξ_{Nyquist}, but with more samples in the interval and thus a smaller $\Delta\xi$. Thus the apparent spectral resolution is increased simply by padding the sampled

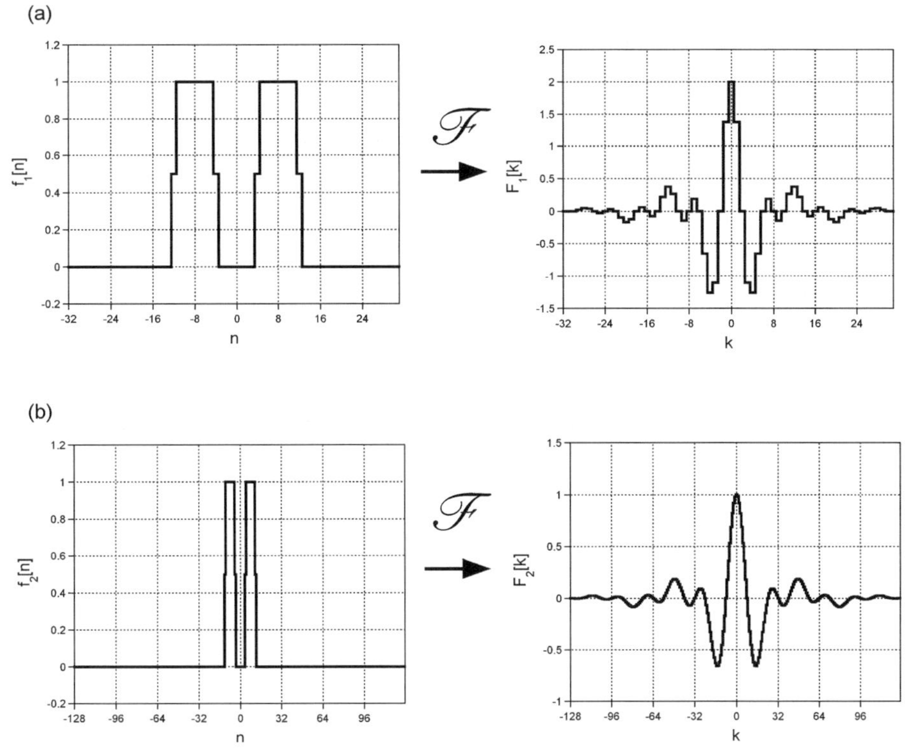

Figure 15.27 Effect of zero padding on the FFT: (a) $f_1[n] = RECT[(n+8)/8] + RECT[(n-8)/8]$ for $N_1 = 64$ and its FFT $F_1[k]$, which is approximately a *SINC* function of width 8 modulating a cosine with period 8 samples; (b) $f_2[n] = RECT[(n+8)/8] + RECT[(n-8)/8]$ in an array with $N_2 = 256$, which is identical to $f_1[n]$ padded with 192 zeros and its FFT $F_2[k]$, which is a "smoothed" version of that in (a), but with no added "information".

function with zeros. However, this statement is misleading, as shown by the examples in Figure 15.27 for a pair of sampled rectangle functions. In the first case, the number of samples is $N = 64$, so the frequency interval in the spectrum is $\Delta\xi = \frac{1}{64}$ cycles per sample. In the second example, the same function has been sampled with the same interval Δx in an array with $N = 256$; the number of samples has been increased by a factor of 4 by adding samples with null amplitudes. The frequency interval $\Delta\xi$ in the spectrum is one-fourth as large in the second case, and so the *apparent* spectral resolution is four times as good. However, no additional "information" has been added by padding the array with zeros, so it may be apparent that the "new" samples of $F[k]$ are not independent and thus could have been generated by interpolation with the appropriate *SINC* function. In other words, zero padding produced a "smoother" spectrum that contains precisely the same information.

Padding of arrays is indisputably helpful when convolving two discrete arrays using the DFT via the filter theorem. This is considered in the next section.

15.5.9 Discrete Convolution and the Filter Theorem

The filter theorem for continuous functions states that the Fourier transform of a convolution is the product of the transforms of the component functions. A corresponding result for the discrete case may be obtained, but the periodicity of the functions that is assumed by the FFT requires some modifications. The discrete analogue of continuous convolution was derived in Equation (14.143) for bandlimited input and system functions:

$$g[n \cdot \Delta x] = \Delta x \sum_{p=-\infty}^{+\infty} f[p \cdot \Delta x] h[(n-p) \cdot \Delta x] \tag{15.113}$$

If the continuous input function and impulse response are bandlimited, then they have infinite support and the amplitude of the discrete convolution may well be infinite at some samples. If the input and/or impulse response have compact support, then the resulting aliasing ensures that the output of the convolution is not the sampled continuous convolution.

In the case of finite arrays, we must define what happens to the samples of the shifted kernel that are "off the edge" of the unshifted array $f[n]$. If values off the edge are assumed to be zero, the number of products computed decreases by one for each increment of the shift and so the average amplitude of the convolution will decrease. This process is sometimes called *linear convolution*. The support of the resulting function is one sample less than the sum of the supports of $f[n]$ and $h[n]$. This is qualitatively different from the result obtained by performing discrete convolution in the frequency domain via the FFT, which assumes that the representations in both the space and frequency are periodic over N samples. This means that samples of the space-domain function that are shifted "off the edge" of the other array reappear on the opposite side. The result is called *circular convolution*:

$$f[n] * h[n] = g[n] \equiv \sum_{m=-N/2}^{N/2-1} f[m] h[n-m] \tag{15.114}$$

where $f[m + \ell N] = f[m]$ and $h[m + \ell N] = h[m]$ for all integer values of ℓ. The process may be visualized by displaying the N-pixel functions on strips of paper with the ends taped together to form circularly periodic functions, as shown in Figure 15.28. The outputs of linear and circular convolutions of the same pair of functions are compared in Figure 15.29.

If we perform discrete convolution via the FFT, the result is scaled by some factor C that depends on the normalization used in the FFT:

$$f[n] * h[n] = g[n] = C \cdot \mathcal{F}_{\mathrm{DFT}}^{-1}\{F[k] \cdot H[k]\} \tag{15.115}$$

If we apply arbitrary normalization factors c_1 and c_2 used in Equation (15.36) to the forward and inverse transforms, respectively:

$$F[k] \equiv c_1 \sum_{n=0}^{N-1} f[n]\, e^{-2\pi i n k/N} \tag{15.116a}$$

$$f[n] \equiv c_2 \sum_{k=0}^{N-1} F[k]\, e^{+2\pi i n k/N} \tag{15.116b}$$

and compute the convolution via the filter theorem, we obtain:

$$g[n] = f[n] * h[n] = \mathcal{F}_{\mathrm{DFT}}^{-1}\{F[k] \cdot H[k]\}$$

$$g[n] = c_2 \sum_{k=0}^{N-1} \left[\left(c_1 \sum_{\ell=0}^{N-1} f[\ell]\, e^{-2\pi i \ell k/N} \right) \cdot \left(c_1 \sum_{m=0}^{N-1} h[m]\, e^{-2\pi i n k/N} \right) e^{+2\pi i n k/N} \right]$$

$$= c_1^2 \cdot c_2 \sum_{k=0}^{N-1} \sum_{\ell=0}^{N-1} \sum_{m=0}^{N-1} f[\ell] h[m]\, e^{-2\pi i (\ell+m-n)k/N}$$

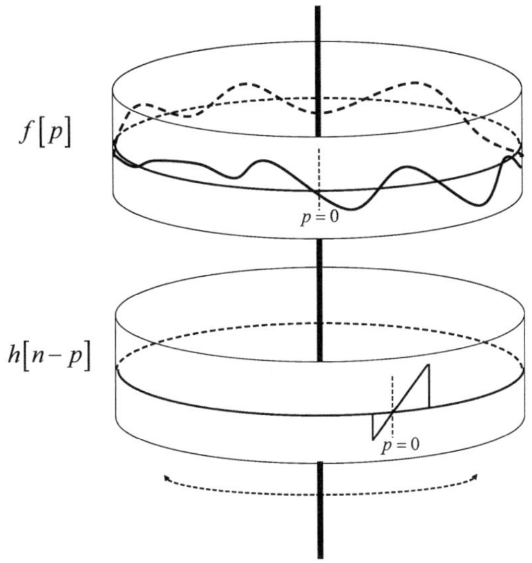

Figure 15.28 Schematic of circular convolution: the "reversed" impulse response $h[p]$ is rotated about the axis relative to the origin of $f[p]$. The circular convolution is the N-sample sum of the product of the two functions.

$$= c_1^2 \cdot c_2 \sum_{\ell=0}^{N-1} \sum_{m=0}^{N-1} f[\ell]h[m] \sum_{k=0}^{N-1} e^{-2\pi i(\ell+m-n)k/N}$$

$$= c_1^2 \cdot c_2 \sum_{\ell=0}^{N-1} \sum_{m=0}^{N-1} f[\ell]h[m][N \cdot \delta[m-(n-\ell)]]$$

$$= (N \cdot c_1^2 \cdot c_2) \sum_{\ell=0}^{N-1} f[\ell]h[n-\ell] \tag{15.117}$$

Recall that Equation (15.37) required that $c_1 \cdot c_2 = N^{-1}$ to ensure that the cascade of the forward and inverse transforms produce the proper amplitudes for $f[n]$. If this criterion is satisfied, then the general expression for the discrete convolution is:

$$f[n] * h[n] = g[n] = c_1 \sum_{\ell=0}^{N-1} f[\ell]h[n-\ell] \tag{15.118}$$

The properly normalized convolution of two functions results from the "discrete" filter theorem if $c_1 = 1$ and $c_2 = N^{-1}$. This is the preferred normalization for evaluating the DFT of functions with compact support. The normalization with $c_1 = N^{-1}$ and $c_2 = 1$ results for the spectra of periodic functions such as sinusoids and square waves that are more "continuous-like", but the amplitudes of the convolution via DFT would be too small by a factor of N^{-1}. The scaling problem may be fixed while retaining this normalization by including a factor of N:

$$f[n] * h[n] = \mathcal{F}_{\text{DFT}}^{-1}\{N\mathcal{F}_{\text{DFT}}\{f[n]\} \cdot \mathcal{F}_{\text{DFT}}\{h[n]\}\} = g[n] \tag{15.119}$$

An alternative method with this normalization applies inverse DFTs to the two "space-domain" arrays $f[n]$ and $h[n]$ followed by a forward transform (with one factor of N^{-1}):

$$f[n] * h[n] = \mathcal{F}_{\text{DFT}}^{-1}\{N\mathcal{F}_{\text{DFT}}\{f[n]\} \cdot \mathcal{F}_{\text{DFT}}\{h[n]\}\} = g[n] \tag{15.120}$$

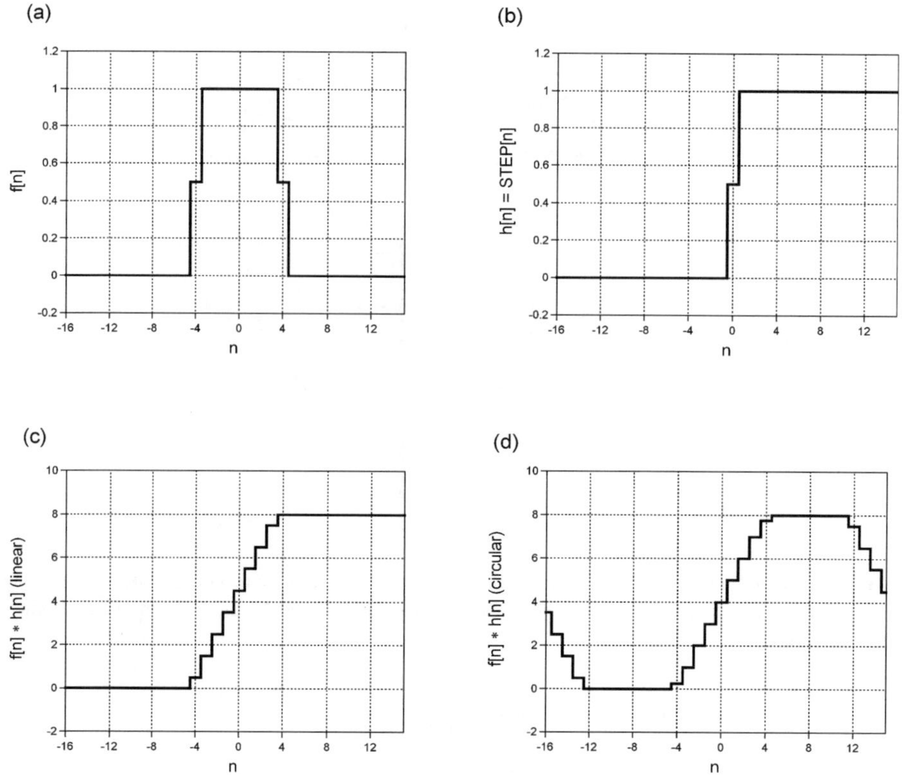

Figure 15.29 Difference between 1-D linear and circular convolution, $N = 32$: (a) $f[n] = RECT[n/8]$; (b) $h[n] = STEP[n]$; (c) linear convolution $f[n] * h[n]$, where the translated step function is assumed to maintain the value of unity; (d) circular convolution $f[n] * h[n]$, where the amplitudes of the translated impulse response "wrap around" the array.

This is perfectly feasible for the arrays $f[n]$ and $h[n]$ with purely numerical indices and no units in the coordinates.

Because the discrete convolution computed via the FFT is inherently circular, other artifacts may result are best understood by example. Consider the convolution of two functions $f[n]$ and $h[n]$ that have compact supports $2M_f$ and $2M_h$, respectively. In other words, $f[n] = 0$ for $|n| > M_f$ and $h[n] = 0$ for $|n| > M_h$, so that f and h have respectively no more than $2M_f + 1$ and $2M_h + 1$ samples with nonzero amplitudes. The resulting convolution $g[n]$ will also have compact support, but over $2(M_f + M_h) + 1$ pixels. In other words, the width of g is the sum of the widths of the input f and the kernel h (minus 1 pixel). This same result holds for convolution via the FFT, but it is important to recognize that the array size of the computation is fixed at N pixels; if $N < 2(M_f + M_h) + 1$, then (for example) nonzero amplitudes will "bleed" into pixels with coordinates greater than $n = N/2 - 1$, i.e., into the next period of the FFT. These amplitudes will "appear" at negative frequencies and will be added to the true amplitude there. For this reason, the convolution via the FFT is *circular*. To ensure that there is no "wrap-around" when convolving via FFT, it is necessary to pad the input and kernel sequences with zeros to fill an array of size $N = 2^m$, where m is an integer and $N \geq 2(M_f + M_h) + 1$.

15.5.9.1 Example of Discrete Convolution via FFT

Consider the autoconvolution of $f[n] = RECT[n/b]$ via the FFT in an array with $N = 256$ (Figure 15.30); the analogous result for a continuous function is obviously a triangle. If $b = 120 < N/2$, the circular convolution obtained via the FFT yields the "expected" result. The width of the resulting triangle is $2 \cdot b - 1 = 239$ and the amplitude at the origin is:

$$F[k=0] = \sum_{n=-128}^{+127} \left(RECT\left[\frac{n}{120}\right]\right)^2$$
$$= (1 \cdot 1 \cdot 119 + 0.5 \cdot 0.5 \cdot 2)\ samples = 119.5 \tag{15.121}$$

If the width of the rectangle is increased to $b = 140 > N/2$, the width of the linear convolution would be $2 \cdot b - 1 = 279 > N$. We might expect to obtain a similar result for the autoconvolution of $RECT[n/70]$. However, the width of the function generated by linear convolution is 141 pixels, which is larger than the array size of 128. Amplitudes are correct near the center of the array. Those near the edge are the sum of amplitudes of 2 pixels: one correctly placed and one that has "wrapped around" from the other side. To ensure that this problem does not occur, the original array size should be larger than the sum of the widths of the two functions to be convolved.

15.5.9.2 Computation Time for Discrete Convolution

Convolution of a 1-D array of length N with a 1-D kernel of length M by direct computation requires of the order of M complex multiplications and additions at each of N samples, for a total number of NM computations. Circular convolution via the FFT requires that the M-element kernel be embedded in an N-element array, and that two forward transforms, a complex multiplication, and the inverse transform be computed. The number of computations is of the order of:

$$N \log_2 N + N \log_2 N + N + N \log_2 N = 3N \log_2 N + N \tag{15.122}$$

so the FFT algorithm will be faster if:

$$N \cdot M > 3N \log_2 N + N \Longrightarrow M > 3 \log_2 N + 1 \tag{15.123}$$

If $N = 512$, then circular convolution via the FFT will be faster if $M \geq 27$.

If the linear convolution is required (rather than circular), the array of length N and the kernel of length M must be padded so that each contains $2N$ samples. After convolution, N samples of the output array are retrieved. The FFT method will be faster if:

$$N \cdot M > 3 \cdot 2N \log_2[2N] + 2N \Longrightarrow M > 6 \log_2[2N] + 2$$
$$\Longrightarrow M > 62 \text{ for } N = 512 \tag{15.124}$$

15.5.10 Discrete Transforms of Quantized Functions

The discussion of Section 14.10 demonstrated that the nonlinear process of quantizing a sampled signal restricts the range of possible amplitudes to a discrete set of levels separated by (usually uniform) intervals. This resulting error in the recorded amplitude is uniformly distributed in most cases, and the standard deviation of the quantization noise is $\sigma = b/\sqrt{12}$ for a quantizer with step size b.

The qualitative effect of quantization on the DFT is easy to see by recognizing that the same digital signal is produced regardless of the order of the sampling and quantization operations. If quantization is performed first, then the quantized signal may be decomposed into the sampled unquantized function

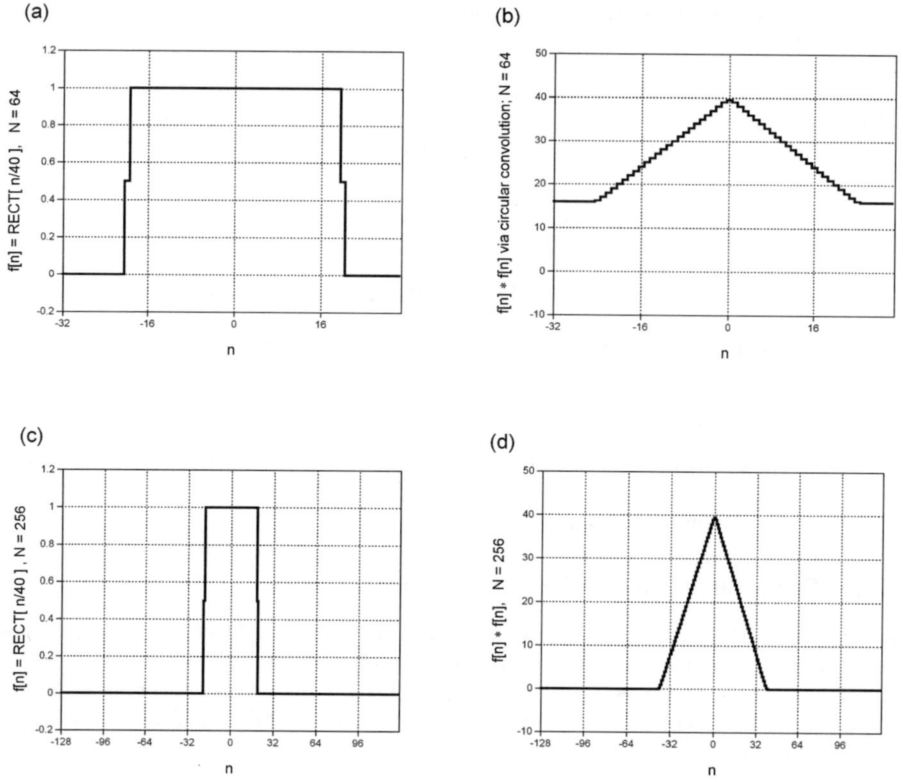

Figure 15.30 Effect of circular convolution: (a) $f[n] = RECT[n/40]$ in array with $N = 64$; (b) circular convolution $f[n] * f[n]$ evaluated via DFT, which is the triangle "sitting" on a bias due to the wrapped-around amplitudes; (c) the same rectangle $f[n] = RECT[n/40]$ in an array with $N = 256$; (d) circular convolution, which produces the "correct" linear convolution.

and the sampled error from the nonlinear quantization. Even if the unquantized sampled signal is bandlimited, the discussion of Section 9.8.19 shows that the support of the continuous spectrum of the quantization error has infinite support and thus will be aliased. An example of a comparison for sampled and quantized sinusoids is shown in Figure 15.31.

15.5.11 Parseval's Theorem for DFT

Parseval's theorem for continuous linear systems in Equation (9.163) demonstrates that the integrated powers in both domains are identical:

$$\int_{-\infty}^{+\infty} |f[x]|^2 \, dx = \int_{-\infty}^{+\infty} |F[\xi]|^2 \, d\xi \tag{15.125}$$

The generalization of this theorem to the discrete case must account for the normalization. If we again use the arbitrary normalization of c_1 for the forward DFT and c_2 for the inverse DFT, such that $c_1 \cdot c_2 = N^{-1}$, we can derive the general form of the discrete version of Parseval's theorem. The power at a

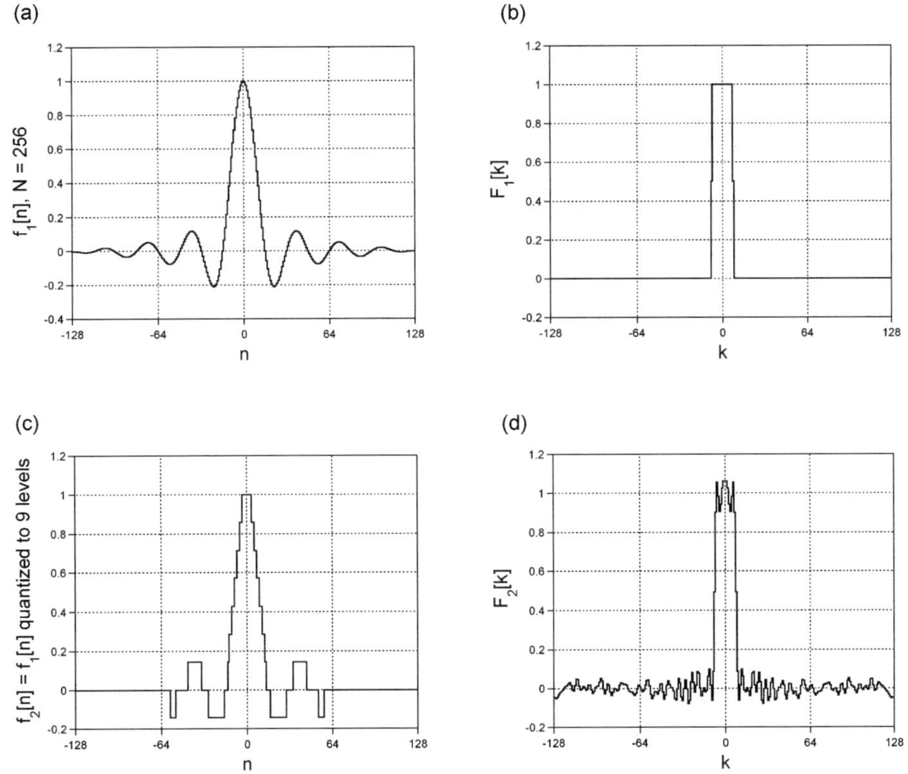

Figure 15.31 Effect of quantization on the DFT for $N = 256$: (a) input function $f_1[n]$; (b) DFT $F_1[k] = RECT[k/16]$; (c) $f_2[n]$, which is $f_1[n]$ quantized to nine equally spaced levels; (d) $F_2[k]$, which shows quantization noise at all spatial frequencies and showing that amplitude quantization increases the support of the spectrum.

sample in the frequency domain is:

$$
\begin{aligned}
|F[k]|^2 &= \left(c_1 \sum_{n=0}^{N-1} f[n]\, e^{-2\pi i n k/N} \right) \left(c_1 \sum_{m=0}^{N-1} f[m]\, e^{-2\pi i m k/N} \right)^* \\
&= |c_1|^2 \sum_{n=0}^{N-1} \sum_{m=0}^{N-1} f[n] f^*[m]\, e^{-2\pi i (n-m)k/N}
\end{aligned}
\tag{15.126}
$$

The analogue to the integrated power is the summation over k:

$$
\begin{aligned}
\sum_{k=0}^{N-1} |F[k]|^2 &= |c_1|^2 \sum_{k=0}^{N-1} \sum_{n=0}^{N-1} \sum_{m=0}^{N-1} f[n] f^*[m]\, e^{-2\pi i (n-m)k/N} \\
&= |c_1|^2 \sum_{n=0}^{N-1} \sum_{m=0}^{N-1} f[n] f^*[m] \sum_{k=0}^{N-1} e^{-2\pi i (n-m)k/N}
\end{aligned}
$$

$$= |c_1|^2 \sum_{n=0}^{N-1} \sum_{m=0}^{N-1} f[n] f^*[m] (N\delta[n-m])$$

$$= N|c_1|^2 \sum_{n=0}^{N-1} |f[n]|^2 \qquad (15.127)$$

When using the convention with the normalization $c_1 = N^{-1}$ in the forward DFT, which maximizes the similarity between the continuous and discrete transforms of periodic functions, the discrete form of Parseval's theorem is:

$$\sum_{k=0}^{N-1} |F[k]|^2 = \frac{1}{N} \sum_{n=0}^{N-1} |f[n]|^2 \quad \textit{for}\ c_1 = \frac{1}{N},\ c_2 = 1 \qquad (15.128)$$

If using the normalization $c_1 = 1,\ c_2 = N^{-1}$, which produces the "correct" analogue to the discrete central-ordinate theorem for finite-support functions, the discrete expression of Parseval's theorem is:

$$\frac{1}{N} \sum_{k=0}^{N-1} |F[k]|^2 = \sum_{n=0}^{N-1} |f[n]|^2 \quad \textit{for}\ c_1 = 1,\ c_2 = \frac{1}{N} \qquad (15.129)$$

And in the last scaling convention with $c_1 = c_2 = (\sqrt{N})^{-1}$, the expression for Parseval's theorem is produced that best resembles the continuous case:

$$\sum_{k=0}^{N-1} |F[k]|^2 = \sum_{n=0}^{N-1} |f[n]|^2 \quad \textit{for}\ c_1 = c_2 = \frac{1}{\sqrt{N}} \qquad (15.130)$$

To demonstrate the discrete Parseval's theorem, consider the first normalization in Equation (15.128) with $c_1 = N^{-1}$ and $c_2 = 1$ for $f[n] = \cos[2\pi n/32]$ and $N = 256$. The power of the space-domain function at sample n is:

$$|f[n]|^2 = \cos^2\left[\frac{2\pi n}{32}\right] = \frac{1}{2}\left(1 + \cos\left[\frac{2\pi n}{16}\right]\right) \qquad (15.131)$$

The sum of the power over $N = 256$ samples is $N/2 = 128$; if scaled by N^{-1}, the resulting integrated power is 0.5.

The discrete spectrum in this normalization is:

$$F[k] = \mathcal{F}_{\text{DFT}}\left\{\cos\left[\frac{2\pi n}{32}\right]\right\} = \frac{1}{2}\delta_d[k-4] + \frac{1}{2}\delta_d[k+4] \qquad (15.132)$$

and the resulting discrete power spectrum is:

$$|F[k]|^2 = \frac{1}{4}(\delta_d[k-4] + \delta_d[k+4]) \qquad (15.133)$$

with integrated power equal to $0.25 + 0.25 = 0.5$, thus confirming the theorem.

15.5.12 Scaling Theorem for Sampled Functions

The scale factor of the space-domain representation in the continuous case is inversely related to that in the frequency domain, as demonstrated by Equation (9.117):

$$\mathcal{F}_1\{f[x]\} = F[\xi] \Longrightarrow \mathcal{F}_1\left\{f\left[\frac{x}{b}\right]\right\} = |b| \cdot F[b\xi] \qquad (15.134)$$

The spacing Δx between samples is a fixed length that must be considered when scaling a discrete function. Consider an example of "minifying" an N-sample discrete function by the factor $b = \frac{1}{4}$. This may be implemented in two ways, First, the function may be sampled at the new spacing $(\Delta x)' = \Delta x/4$, which retains the same sample amplitudes and reduces the physical domain of the sampled function by $\frac{1}{4}$. The sampling interval in the frequency domain is also changed: $(\Delta\xi)' = 4 \cdot \Delta\xi$. If the original sampled function had compact support, the physical domain could be enlarged by padding with zeros, for which we have already analyzed the result in the last section.

If the discrete function is "stretched" by the factor $b = 4$ while retaining the original sample spacing Δx, three "undefined" amplitudes are created between the original samples. The samples of the function are "spread apart" so that samples with undefined amplitudes are created in between. After assigning the value "0" to these samples, the DFT of the stretched function is computed over $b \cdot N$ samples. The function has effectively been "sampled" by a new *COMB* function with a larger separation between samples. The transform of the stretched function exhibits b periods of the spectrum in the Nyquist window.

15.6 FFTs of 2-D Arrays

Two-dimensional arrays are transformed using a 1-D FFT algorithm because of the separability of the transform, as shown in Chapter 10:

$$F\left[\frac{k}{N \cdot \Delta x} \frac{\ell}{N \cdot \Delta y}\right] = \sum_m \sum_n f[n \cdot \Delta x, m \cdot \Delta y]\, e^{-2\pi i(nk+m\ell)/N}$$

$$= \sum_m \left(\sum_n f[n \cdot \Delta x, m \cdot \Delta y]\, e^{-2\pi i nk/N} \right) e^{-2\pi i m\ell/N} \qquad (15.135)$$

Thus the 2-D FFT consists of a cascade of 1-D FFTs of the rows and 1-D FFTs of the resulting columns. The format of the data arrays depends on the computer language, e.g., the default for 2-D BASIC arrays is "column-major" format, where array elements with the same value of the second index (which specifies the column, or x-coordinate) are stored in adjacent memory locations. The elements of a 3×3 array named "X" would be stored thus:

$$X[0, 0],\ X[1, 0],\ X[2, 0],\ X[0, 1],\ X[1, 1],\ X[2, 1],\ X[0, 2],\ X[1, 2],\ X[2, 2]$$

In computer languages that use "row-major" format (e.g., C), the same array would be stored as follows:

$$X[0, 0],\ X[0, 1],\ X[0, 2],\ X[1, 0],\ X[1, 1],\ X[1, 2],\ X[2, 0],\ X[2, 1],\ X[2, 2]$$

To optimize the speed of the FFT, the data should be retrieved from adjacent memory locations; in BASIC, the columns of data should be transformed first. To retrieve those data for the second set of transforms from nonadjacent locations, often the array is transposed to reorder the data into adjacent locations. The 2-D procedure for the FFT in a column-major language such as BASIC would be:

1. FFT of each column (the 1-D vectors with the same x-coordinate).
2. Transpose the matrix (exchange the columns and rows).
3. FFT of each "new" column.
4. Retranspose the matrix to the proper format.

If using the standard 1-D transform indexed over the range $[0, N - 1]$ on data in image format (origin of coordinates at the center of the 2-D array), the data must be recast to obtain a centered transform. The 2-D analogy of the translation process divides the $N \times N$ array into four quadrants of size $N/2 \times N/2$ and exchanges the diagonal quadrants. This must be performed both before and after the FFT sequence just discussed. This torusing procedure requires substantial processing time and many

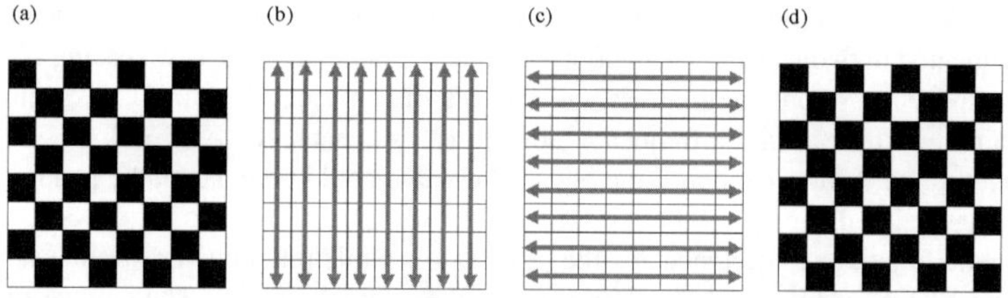

Figure 15.32 Schematic of 2-D FFT of array with $-N/2 \leq n, m \leq +N/2 - 1$ using a 1-D FFT for arrays defined over $0 \leq n \leq N - 1$ via "checkerboarding": (a) multiply $f[n, m]$ by 2-D checkerboard (black squares $\Longrightarrow -1$, white squares $\Longrightarrow +1$); (b) evaluate 1-D FFT of each column; (c) evaluate 1-D FFT of each row that is produced by step (b); (d) multiply by 2-D checkerboard.

temporary memory locations. Two-dimensional "checkerboarding" (multiplication of alternate samples by -1) is much more efficient. The entire procedure for transforming an $N \times N$ image array $f[n, m]$ indexed over the domain $0 \leq n, m \leq N - 1$ but with centered data and to obtain a transform with the zero-frequency term in the center is:

1. Checkerboard by multiplying $f[n, m]$ by $(-1)^{n+m}$.
2. FFT of each column (the 1-D vectors with the same x-coordinate).
3. FFT of each row (the 1-D vectors with the same y-coordinate).
4. Checkerboard the output by multiplying by $(-1)^{n+m}$.

This process is shown in schematic form in Figure 15.32.

A 2-D FFT of an $N \times N$ image requires that $2N$ 1-D FFTs be computed, for a total of $2N \cdot (N \log_2[N])$ operations. In the 2-D case of an $N \times N$ image and an $M \times M$ kernel, the number of operations required for direct circular convolution is N^2M^2, while the FFT method requires:

$$3 \cdot 2N \cdot N \log_2 N + N^2 = 6N_2^2 \log_2 N + N^2 \tag{15.136}$$

This will be faster if this condition is satisfied:

$$M^2 \cdot N^2 > 6N^2 \log_2 N + N^2 \Longrightarrow M^2 > 6 \log_2 N + 1 \tag{15.137}$$

If $N = 512^2$, computation via the FFT is quicker for $M > 7$.

If the arrays are padded to compute the linear convolution, the FFT will be faster if:

$$M^2 \cdot N^2 > (3 \cdot 2N \cdot 2N) \log_2[2N] + (2N)^2 = 12N^2 \log[2N] + 4N^2 \tag{15.138a}$$

$$M^2 > 12 \log_2[2N] + 4 \Longrightarrow M > 11 \text{ for } N = 512 \tag{15.138b}$$

Note that the accuracy of these estimates will depend on other factors in a particular situation, e.g., on time-shared systems.

15.6.1 Interpretation of 2-D FFTs

The spatial frequency in the ξ- or η-directions of the 2-D FFT array $F[k, \ell]$ is determined from the indices of the sample via Equation (15.30):

$$\xi = k \cdot \Delta\xi = \frac{k}{N \cdot \Delta x} \tag{15.139a}$$

$$\eta = \ell \cdot \Delta\eta = \frac{\ell}{N \cdot \Delta y} \tag{15.139b}$$

where $-N/2 \leq k, \ell \leq N/2 - 1$. These Cartesian coordinates may be converted to a radial spatial frequency $\boldsymbol{\rho} = (\rho, \psi)$ via:

$$|\underline{\rho}| = \sqrt{\xi^2 + \eta^2} = \frac{1}{N}\sqrt{\left(\frac{k}{\Delta x}\right)^2 + \left(\frac{\ell}{\Delta y}\right)^2} \tag{15.140a}$$

$$\psi = \tan^{-1}\left[\frac{(\ell/\Delta y)}{(k/\Delta x)}\right] = \tan^{-1}\left[\frac{\ell \cdot \Delta x}{k \cdot \Delta y}\right] \tag{15.140b}$$

In the usual case where the sampling intervals $\Delta x = \Delta y$, these simplify to:

$$\rho = \frac{1}{N \cdot \Delta x}\sqrt{k^2 + \ell^2} \tag{15.141a}$$

$$\psi = \tan^{-1}\left[\frac{\ell}{k}\right] \tag{15.141b}$$

The spatial frequency vector with the maximum magnitude has indices $k = \ell = -N/2$, and corresponds to the sinusoid with the shortest period that may be represented in the 2-D space domain. A spectrum composed of this single frequency may be expressed as separable 1-D functions:

$$\begin{aligned} F[k, \ell] &= \delta_d[k - (-32), \ell - (-32)] \\ &= \delta_d[k + 32] \cdot \delta_d[\ell + 32] \end{aligned} \tag{15.142}$$

The sampled space-domain function is easily evaluated:

$$\begin{aligned} f[n, m] &= \cos\left[\frac{2\pi n}{2}\right] \cdot \cos\left[\frac{2\pi k}{2}\right] \\ &= \cos[\pi n] \cdot \cos[\pi k] = (-1)^n(-1)^k = (-1)^{n+k} \end{aligned} \tag{15.143}$$

which is our (now-familiar) checkerboard pattern. The continuous function from which these samples were obtained is a cosine oscillating along $\psi = \pi/4$ with period $\sqrt{2} \cdot \Delta x$. By applying Equation (15.140), the magnitude and azimuth of the radial spatial frequency vector are:

$$\left|\underline{\rho}\left(k = -\frac{N}{2}, \ell = -\frac{N}{2}\right)\right| = \frac{1}{N \cdot \Delta x}\left(\sqrt{2} \cdot \frac{N}{2}\right) = \frac{1}{\sqrt{2} \cdot \Delta x} \tag{15.144a}$$

$$\psi = +\frac{\pi}{4} \tag{15.144b}$$

The corresponding period is $R = |\underline{\rho}|^{-1} = \sqrt{2} \cdot \Delta x$. This may seem incorrect because R is smaller than the shortest unaliased period ($X_{\min} = 2 \cdot \Delta x$) for 1-D signals sampled along each axis at the same interval $\Delta x = \Delta y$ (Figure 15.33). This result indicates that the concept of aliasing for multidimensional functions must be interpreted with more care. If the 2-D function is sampled by a *COMB* function that is separable into orthogonal components, the separable information in the spectrum allows 1-D functions to be adequately sampled at a rate less than required in one dimension by "dividing" the information between the two orthogonal sampling directions. Put another way, the maximum value of the *projection* of the continuous spectrum onto a frequency axis determines the Nyquist sampling rate in that direction. If the frequency projected to the axis is located at an azimuth of ϕ, then the necessary sampling interval is reduced by a factor of $\cos[\phi]$ along the x-axis and by $\sin[\phi]$ along the y-axis. For $\phi = \pi/4$, the factor along each axis is $(\sqrt{2})^{-1}$. A multidimensional function sampled by a multidimensional *COMB* function is aliased only when one (or both) of the orthogonal spatial directions is undersampled.

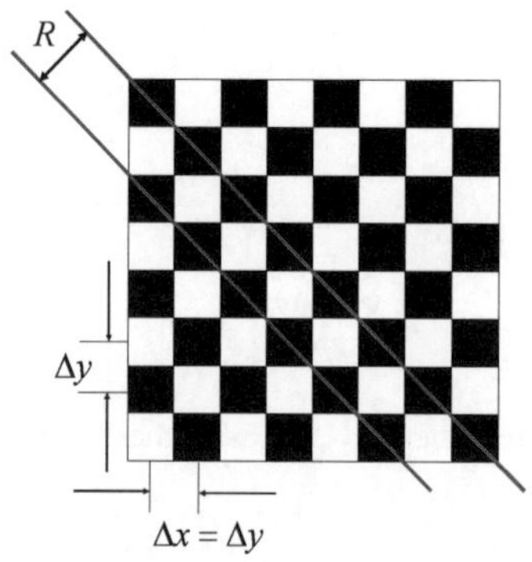

Figure 15.33 Spatial frequency of 2-D FFT illustrated by the checkerboard pattern $f[n, m] = (-1)^{n+m}$. The period of the array along the diagonal direction ($\theta = \pi/4$ radians) is $R = \sqrt{2} \cdot \Delta x$, and thus the associated spatial frequency is $\rho = R^{-1} = (\sqrt{2} \cdot \Delta x)^{-1}$.

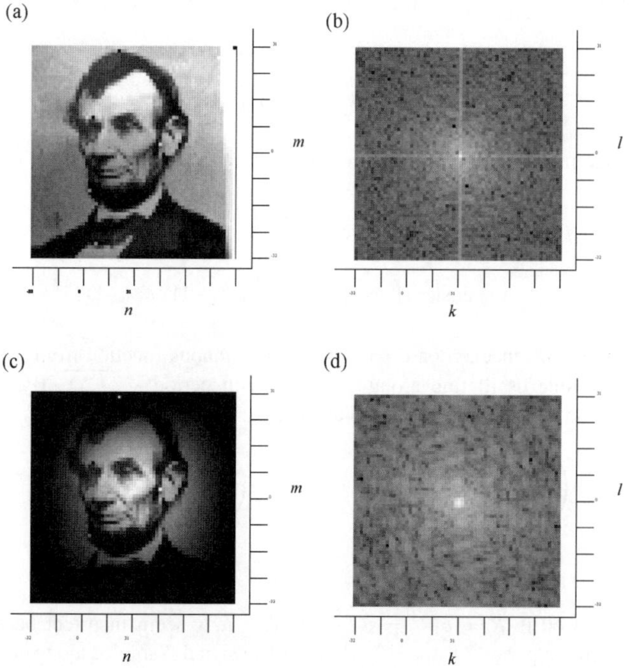

Figure 15.34 Leakage exhibited by 2-D arrays and the effect of the 2-D Hann window: (a) the 64 × 64 array $f[n, m]$, which is darker at the left and bottom than at the right and top; (b) logarithm of the squared magnitude of the FFT $F[k, l]$, showing a "cross" pattern due to leakage in the horizontal and vertical directions; (c) $f[n, m]$ after multiplying by the Hann window $w[n.m]$; (d) logarithm of the squared magnitude of $f[n, m] \cdot w[n, m]$, showing the "blurring" of the frequency components and the reduction in leakage.

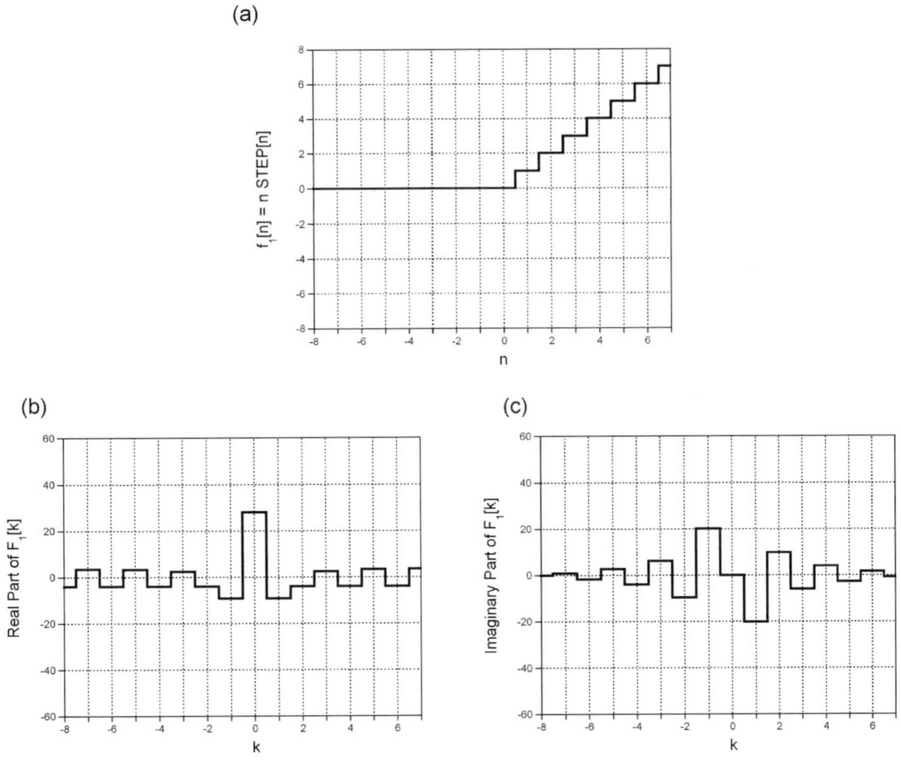

Figure 15.35 (a) $f_1[n] = n \cdot STEP[n]$ for $2N = 16$; (b) $\Re\{F_1[k]\}$; (c) $\Im\{F_1[k]\}$, showing the redundancy of the complex spectrum due to its Hermitian character.

15.6.2 2-D Hann Window

A 2-D Hann window may be defined as the separable product of two 1-D Hann windows:

$$
\begin{aligned}
w[n, m] &= \frac{1}{2}\left(1 + \cos\left[2\pi \frac{n}{N}\right]\right) \cdot \frac{1}{2}\left(1 + \cos\left[2\pi \frac{m}{M}\right]\right) \\
&= \cos^2\left[\frac{\pi n}{N}\right] \cos^2\left[\frac{\pi m}{M}\right] \rightarrow \cos^2\left[\frac{\pi n}{N}\right] \cos^2\left[\frac{\pi m}{N}\right] \quad \textit{for } N = M \qquad (15.145)
\end{aligned}
$$

Consider the 2-D 64 × 64 example shown in Figure 15.34. The gray values of the sampled image of Lincoln are dark at the bottom and on the left and bright at the top and on the right, thus causing leakage in the 2-D FFT, which is evident in the magnitude of the spectrum. The leakage was averaged by the blurring evident in the spectrum after applying the Hann window.

15.7 Discrete Cosine Transform

A variant of the finite-support DFT plays an important role in many algorithms for *compressing* digital images, i.e., processes for reducing the number of bits necessary to store data arrays or images so the original array/image may be retrieved without significant error. Image compression is essential in any application involving storage in systems with limited capacity and/or transmission through

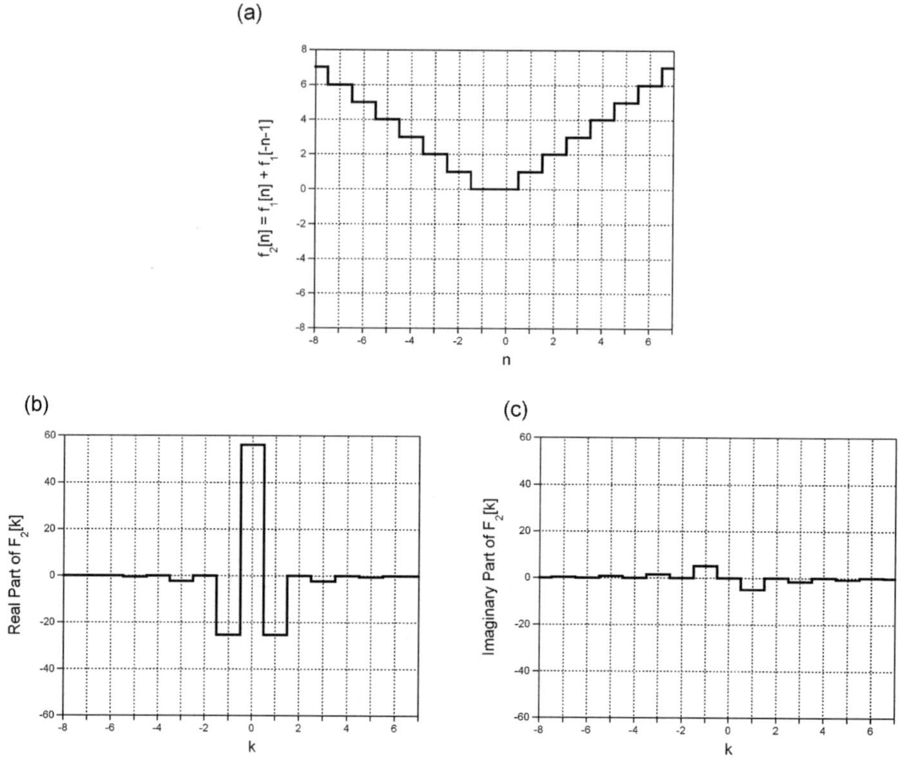

Figure 15.36 (a) $f_2[n] = f_1[n] + f_1[-(n+1)]$ for $2N = 16$; (b) $\Re\{F_2[k]\}$; (c) $\Im\{F_2[k]\}$, showing the reduction in the relative amplitude of the imaginary part compared to $F_1[k]$.

channels with a limited data rate. The importance of data compression is amply demonstrated by the pervasiveness of JPEG image compression and MP3 audio compression. Though an exceedingly important topic in imaging, data compression is far too broad to be covered here. We only introduce the basic principles of this transformation.

The *discrete cosine transform* (DCT) decomposes a space-domain representation into the amplitudes of a specific set of constituent cosine functions. It has the useful property that the amplitudes of the transform are real valued for real-valued input arrays. The 1-D DCT of an N-pixel array can be expressed as a modification of the DFT of a $2N$-pixel array, where the modification ensures that the amplitudes are real valued. The DCT algorithm will be developed by first deriving an analogous process for continuous functions with semi-infinite support and relating this process to finite discrete functions. The specific discrete function with semi-infinite support that will be used for illustration is the sampled version of x defined over the domain $0 \leq x \leq (N-1) \cdot \Delta x \cdot (x \cdot STEP[x])$, which we can call $RAMP[x]$:

$$f_0[x] = x \cdot STEP[x] \equiv RAMP[x] \qquad (15.146)$$

This function will be useful again in the discussion of filtering in Chapter 18. If sampled at N points, the function is:

$$f_0[n] = n \quad \text{if } 0 \leq n \leq N-1 \qquad (15.147)$$

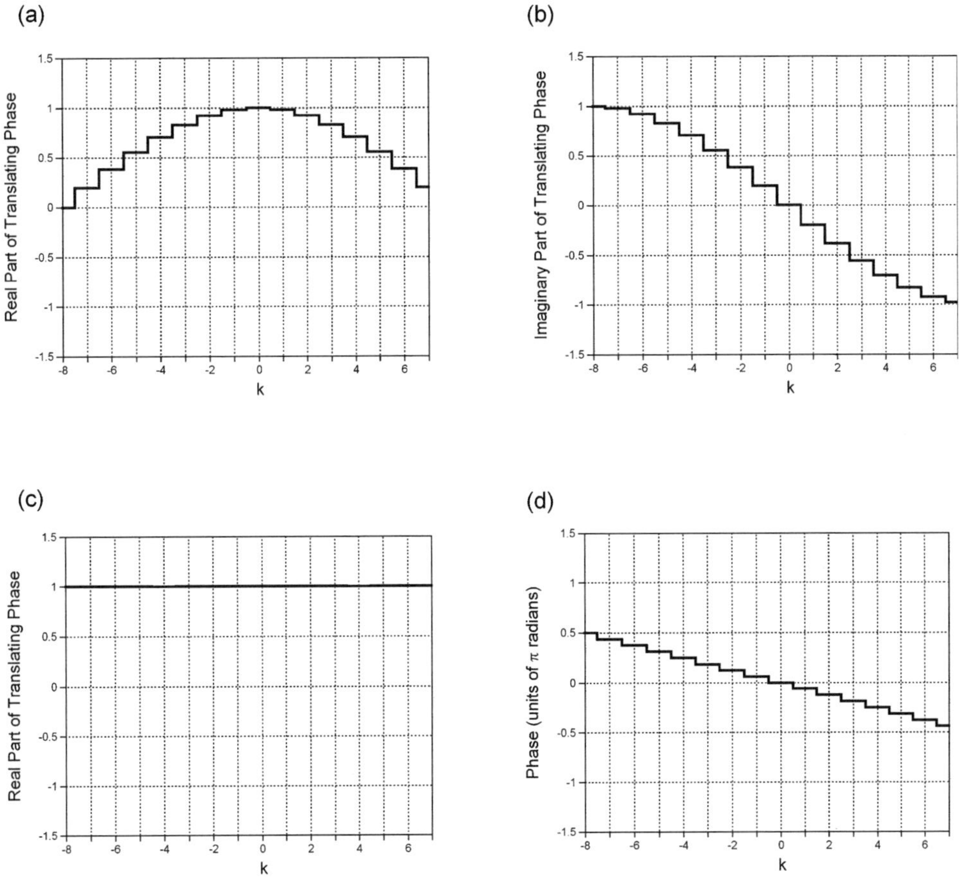

Figure 15.37 Spectrum of the translation term in the derivation of the DCT for $N = 16$ (a) $\Re\{e^{-i\pi k/8}\} = \cos[\pi k/8]$; (b) $\Im\{e^{-i\pi k/8}\} = -\sin[\pi k/8]$; (c) $| e^{-i\pi k/8} | = 1[k]$; (d) $\Phi\{e^{-i\pi k/8}\} = -\pi k/8$.

The function may be inserted into an array with $2N$ samples in the domain $-N \leq n \leq N-1$:

$$f_1[n] = \begin{cases} n & \text{if } 0 \leq n \leq N-1 \\ 0 & \text{if } -N \leq n < 0 \end{cases} \tag{15.148}$$

Note that the discrete function $f_1[n]$ is real valued and asymmetric, and so its DFT is Hermitian. The Hermitian spectrum over $2N$ samples is redundant and may be specified completely by the complex amplitudes of half of its samples. The discrete function and its DFT are shown in an array with $2N = 16$ in Figure 15.35, which shows the "ringing" due to the wrap-around of $f[n]$.

It is possible to reduce the ringing by constructing a unique discrete function $f_2[n]$ defined over $2N$ points from $f_1[n]$ that is "almost symmetric" relative to the origin of coordinates:

$$f_2[n] = \begin{cases} f_1[n] & \text{if } n \geq 0 \\ f_1[-(n+1)] & \text{if } n < 0 \end{cases} \tag{15.149}$$

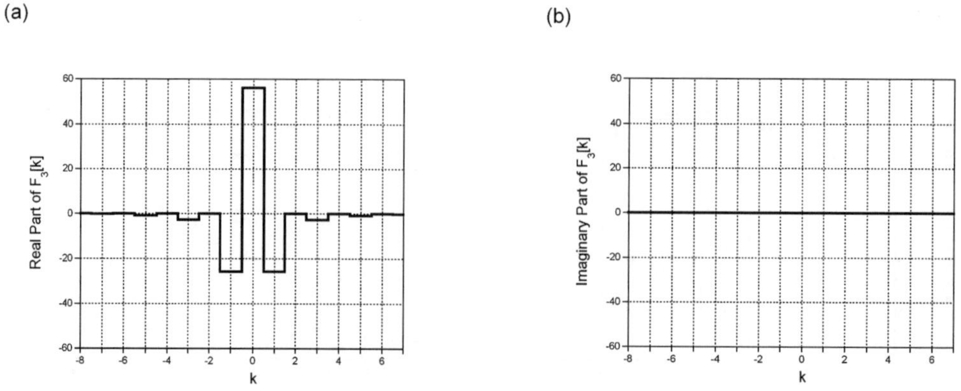

Figure 15.38 Spectrum of $F_2[k] \cdot e^{-i\pi k/N}$: (a) real part; (b) imaginary part. The translation due to the linear phase produces a real-valued and symmetric function over $2N$ points, which is equivalent to the original N-point function $f[n]$.

In this example, $f_2[-1] = f_1[0] = 0$, $f_2[-2] = f_1[+1] = +1$, etc., as shown in Figure 15.36. Evidently, a continuous function that yields $f_2[n]$ when sampled at integer coordinates is:

$$f_2[x] = RAMP[x] + RAMP[-x - 1] \tag{15.150}$$

which also is not symmetric with respect to the origin of coordinates, and therefore which has a complex-valued Fourier transform. The sampled function and its spectrum are shown in Figure 15.36, which clearly reduced ringing.

A symmetric continuous version of $f_1[n]$ may be constructed by translating the continuous function $f_2[x]$ by half of the sampling interval:

$$\begin{aligned} f_3[x] &= f_2[x] * \delta\left[x - \left(-\frac{\Delta x}{2}\right)\right] \\ &= RAMP\left[x - \frac{\Delta x}{2}\right] + RAMP\left[-\left(x - \frac{\Delta x}{2}\right)\right] \end{aligned} \tag{15.151}$$

The real-valued and even continuous Fourier transform of this symmetric function is the product of the complex-valued transform of $f_2[x]$ and the complex linear-phase factor that accounts for the translation by $\Delta x/2$, as determined by the shift theorem of Equation (9.123):

$$F_3[\xi] = \mathcal{F}_1\left\{f_2[x] * \delta\left[x - \frac{1}{2}\right]\right\} = F_2[\xi]\, e^{-2\pi i\xi\cdot\frac{1}{2}} = F_2[\xi]\, e^{-i\pi\xi} \tag{15.152}$$

However, it is not possible to translate the discrete function by $\Delta x/2$, but we can apply a sampled linear phase factor in the frequency domain from Equation (15.152) to the $2N$-pixel DFT of $f_2[n]$. This produces a real-valued and symmetric discrete transform:

$$\begin{aligned} &F_3[\xi] = F_2[\xi]\, e^{-i\pi\xi} \\ \Longrightarrow\ &F_3[k\cdot\Delta\xi] = F_2[k\cdot\Delta\xi]\, e^{-i\pi k\cdot\Delta\xi} = F_2[k\cdot\Delta\xi]\, e^{-i\pi k/(N\cdot\Delta x)} \end{aligned} \tag{15.153}$$

The linear-phase term for the $2N$-pixel DFT is shown in Figure 15.37 and its product with the spectrum in Figure 15.36 is shown in Figure 15.38. The spectrum of the translated function is an even function defined at $2N$ samples, and thus contains N real-valued pieces of data. In other words, one-half of $F_3[k]$ in Figure 15.38 evaluated at the N samples indexed by $0 \le k \le N - 1$ contains the same information as the original N-pixel function $f_0[n]$.

PROBLEMS

15.1 Sketch the functions:

$$f_1[x] = 2\,RECT\left[\frac{x}{b}\right] - 1$$

$$f_2[x] = 1 - \frac{2}{b}\left(RECT\left[\frac{2x}{b}\right] * \delta\delta\left[\frac{x}{b}\right]\right)$$

15.2 A 1-D continuous function $f[x]$ is sampled at N pixels. The continuous function may be translated by the distance x_0 to produce $f[x - x_0]$ by convolution with $h[x] = \delta[x - x_0]$. We want to find an expression for the discrete translated function if the sample locations do not change. In the simple case where the function is translated by an integer multiple of the sampling interval Δx, we can consider the impulse response of the corresponding discrete operation is $\delta_d[n - n_0]$, where $x_0 = n_0 \cdot \Delta x$ and the "discrete delta function" is defined:

$$\delta_d[n - n_0] = \begin{cases} 1 & \text{if } n = n_0 \\ 0 & \text{if } n \neq n_0 \end{cases}$$

However, this formulation for discrete translation "works" only when n_0 is an integer. It is often necessary to translate a 1-D discrete function by a noninteger number of samples, as when registering images.

- **(a)** Derive a rigorous formulation for translating $f[n]$ by a noninteger number of samples if the discrete function $f[n]$ has infinite support. Include sketches of the discrete transfer function and impulse response. (HINT: filter theorem.)
- **(b)** Sketch the result you would expect if $f[x] = SINC[x/(4 \cdot \Delta x)]$.
- **(c)** Adapt the process derived in part (a) for the case where $f[n]$ is defined over a finite array of size N samples. Again, include sketches of the transfer function and impulse response, and also discuss the limitations of the process compared to that in part (a).

15.3 A four-element 1-D array contains the elements $[A, B, C, D]$, where the individual components are complex numbers.

- **(a)** Evaluate the DFT of this four-sample sequence if the array is assumed to be indexed over the domain $n = [0, 1, 2, 3]$.
- **(b)** Evaluate the DFT of this four-sample sequence if the array is assumed to have both "symmetric" and "single-sided" array indices.
- **(c)** Determine the condition that must be satisfied for the DFTs to be real valued.
- **(d)** If the DFT of $[A, B, C, D]$ is $[P, Q, R, S]$, evaluate the DFT of $[C, D, A, B]$.
- **(e)** If the DFT of $[A, B, C, D]$ is $[P, Q, R, S]$ and the DFT of $[1, -1, 0, 0]$ is $[0, 1 + i, 2, 1 - i]$, evaluate the DFT of $[A - B, B - C, C - D, D - A]$.

15.4 Calculate the DFTs of the following functions by analysis, i.e., not by computer, though you need not do the multiplications and additions by hand:

- **(a)** $f[n] = \cos[2\pi x/8]$, $N = 32$
- **(b)** $f[n] = \cos[2\pi x/8]$, $N = 128$
- **(c)** $f[n] = \cos[2\pi x/16]$, $N = 64$
- **(d)** $f[n] = \delta[n + 32]$, $N = 64$

15.5 Given that:

$$F[k] = \mathcal{F}_{\mathrm{DFT}}\{f[n]\} \equiv \frac{1}{N} \sum_{n=-N/2}^{N/2-1} f[n]\, e^{-2\pi i n k/N}$$

$$f[n] = \mathcal{F}_{\mathrm{DFT}}^{-1}\{F[k]\} = \sum_{k=-N/2}^{N/2-1} F[k]\, e^{+2\pi i n k/N}$$

Derive the discrete analogues of the following theorems:

(a) Fourier transform of a Fourier transform

(b) Shift theorem

(c) Transform of crosscorrelation

(d) Central-ordinate theorem.

15.6 Explain the similarities and differences between the discrete Fourier transforms obtained from $f_1[n] = RECT[n/16]$ and $f_2[n] = RECT[n/15]$ if $N = 256$ samples.

16

Magnitude Filtering

This is the first of a sequence of chapters that consider the properties of linear filters and their application to imaging. The filter classifications are based on their characteristics in the frequency domain. This chapter begins the discussion of filtering and concentrates on the 1-D case because the 2-D case usually is a simple generalization.

Without doubt, the most significant aspect of linear systems in the context of imaging is the filter theorem of Equation (9.129), which demonstrated that the spectrum of the convolution of two functions is the product of the individual spectra:

$$\begin{aligned}\mathcal{F}_1\{f[x] * h[x]\} &= \mathcal{F}_1\{g[x]\} \\ &= \mathcal{F}_1\{f[x]\} \cdot \mathcal{F}_1\{h[x]\} \\ &= F[\xi] \cdot H[\xi]\end{aligned} \tag{16.1}$$

In the context of optical imaging, the impulse response $h[x]$ is often called the *point-spread function* (abbreviated to *psf*) and the transfer function $H[\xi]$ is called the *optical transfer function* (or *OTF*).

It is no doubt obvious by now that the generally complex-valued output spectrum resulting from frequency-by-frequency multiplication of the input spectrum $F[\xi]$ and the transfer function $H[\xi]$ may be expressed either as real/imaginary parts:

$$\Re\{G[\xi]\} = \Re\{F[\xi]\} \cdot \Re\{H[\xi]\} - \Im\{F[\xi]\} \cdot \Im\{H[\xi]\} \tag{16.2a}$$

$$\Im\{G[\xi]\} = \Re\{F[\xi]\} \cdot \Im\{H[\xi]\} + \Im\{F[\xi]\} \cdot \Re\{H[\xi]\} \tag{16.2b}$$

or as magnitude and phase (with shorthand notation $\boldsymbol{\Phi}\{F[\xi]\} \equiv \Phi_F[\xi]$):

$$|G[\xi]| = |F[\xi] \cdot H[\xi]| = |F[\xi]| \cdot |H[\xi]| \tag{16.3a}$$

$$\Phi_G[\xi] = \Phi_F[\xi] + \Phi_H[\xi] \tag{16.3b}$$

The description of $H[\xi]$ as magnitude and phase often provides a more convenient and intuitive description of the action of the system in imaging problems. For this reason, the magnitude and phase of the transfer function have been assigned their own names, though with some variation in definition by different authors. The phase $\Phi_H[\xi]$ of the transfer function specifies the number of radians "added" by the system to the initial phase of the input spectrum at each spatial frequency. It is sometimes called the *phase transfer function* (or **Φ**TF), though this definition is not universal (e.g., Gaskill (1978) extracts a multiplicative factor of -1). Later, we will extract a factor of π and expand the **Φ**TF into a power series, thus decomposing it into a unique set of terms based on the shape of $\Phi_H[\xi]$. As will become evident in that discussion, a system with a nonlinear **Φ**TF has a very significant (and usually detrimental) impact on the quality of the output image $g[x]$.

The magnitude $|H[\xi]|$ of the transfer function specifies the scale factor applied to the amplitude of each sinusoidal component of $f[x]$ and is often called the *modulation transfer function* (*MTF*) of the system, though many authors (particularly those from backgrounds in optical imaging) reserve this name for the magnitude of the transfer function that has been normalized by its central ordinate: $MTF[\xi] \equiv |H[\xi]|/|H[0]|$. The reason for this definition arises from the use of the MTF in optics, which will be described in some detail in Chapter 21. The normalized definition is appropriate for optical imaging systems used in natural (incoherent) light, because the maximum of $|H[\xi]|$ always occurs at $\xi = 0$ (DC). This MTF describes the effectiveness of transfer of modulation of a biased (nonnegative) sinusoid from the input to output and ensures that the modulation cannot increase. Clearly this normalized definition of "MTF" is inappropriate for any system with $|H[0]| = 0$. We adopt the less stringent definition that $MTF[\xi] = |H[\xi]|$, even though this convention does not accurately predict the output modulation of a sinusoid.

Obviously, a filter amplifies sinusoids with spatial frequency ξ if $|H[\xi]| > 1$. At frequencies where $|H[\xi]| = 1$, the sinusoidal components of the input are passed with no change in magnitude (though of course the phase may be affected). At frequencies where the MTF is less than unity or zero, the sinusoidal components are attenuated and rejected ("blocked"), respectively.

The entire point of this introductory discussion is to demonstrate how $H[\xi]$ modifies both the magnitude and phase of the spectrum of the input function. It is convenient to group filters into more restrictive classes based on similar performance. For example, we will consider the group of filters that affect only the "magnitude" of the input spectrum, the group that affects only the phase, and those that affect both.

16.1 Classes of Filters

16.1.1 Magnitude Filters

Under a strict definition, a pure magnitude filter would have no effect upon the phase spectrum, thus requiring that the phase transfer function be zero:

$$H[\xi] = |H[\xi]|\, e^{i\cdot 0[\xi]} = |H[\xi]| \quad (\textit{strict definition of magnitude filter}) \tag{16.4}$$

Filters in this class may only scale the magnitude by amounts specified by the functional form of the MTF. Under this strict definition, the transfer function is nonnegative and real-valued at all frequencies ξ:

$$\Re\{H[\xi]\} = |H[\xi]| \cdot \cos[0] = |H[\xi]| \geq 0 \tag{16.5a}$$

$$\Im\{H[\xi]\} = |H[\xi]| \cdot \sin[0] = 0[\xi] \tag{16.5b}$$

Very few (if any) filters that are significant in imaging satisfy this constraint. We therefore relax the criterion to include filters where $H[\xi]$ is real valued and (thus perhaps) bipolar. The initial phase of the resulting transfer function may take on the values of 0 and $-\pi$ radians in this convention. We already know that the inverse Fourier transform of any real-valued spectrum must be Hermitian, which means that the real and imaginary parts of the impulse response are even and odd, respectively. One useful example of a useful magnitude filter in this relaxed definition is $H[\xi] = SINC[\xi]$.

In the introduction to the filter theorem in Chapter 9, we mentioned that filters in this larger category of magnitude filters often are classified further into subgroups based on the "shape" of the MTF. Common categories include the *lowpass*, *highpass*, *bandpass*, and *bandstop* filters, which attenuate or remove some of the sinusoidal components. The definitions of these classes also are "fuzzy"; authors

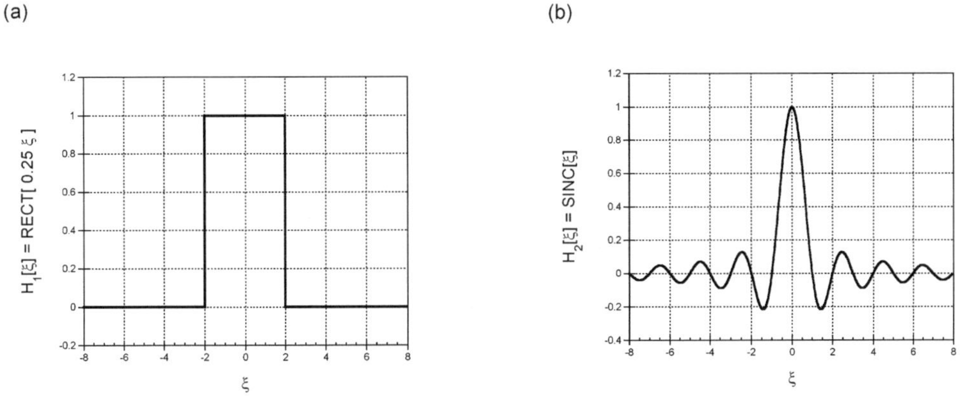

Figure 16.1 Transfer functions of lowpass filters under the "strict" and "relaxed" definitions: (a) $H_1[\xi] = RECT[\xi/4]$ satisfies the strict requirements that $H[0] = 1$ and $H[\pm\infty] = 0$; (b) $H_2[\xi] = SINC[\xi]$ satisfies the relaxed condition. It has no definite cutoff frequency but decays to zero at $\xi = \pm\infty$.

who take the trouble to define the categories often do not agree on the details of the definitions. For example, one strict definition of a lowpass filter requires that all sinusoidal components be completely blocked for all spatial frequencies whose absolute value exceeds some "cutoff". In other words, the transfer function of all strict lowpass filters would have compact support. However, it is common to characterize some filters as "lowpass" even if they do not have definitive cutoff frequencies. A more relaxed, yet still appropriate, definition of a lowpass filter requires only that the MTF at zero frequency be finite while its amplitude at $\xi = \pm\infty$ would be zero. Sample MTFs under these two definitions are compared in Figure 16.1.

The MTF of a highpass filter is complementary to that of a lowpass filter. We apply the condition that the MTF at $\pm\infty$ only be nonzero (rather than unity), which allows for amplification at infinite frequency and so includes differentiation operations, though they might be more rigorously called "high-boost".

The transfer functions of bandpass and bandstop filters also are complementary: the former remove sinusoids with spatial frequencies outside some band, while the latter remove those within the specified band. Components with frequencies outside of the specified band are passed without change. Again, looser complementary definitions may be applied; the MTF of a bandpass filter is often merely required to be nonzero at some spatial frequency ξ_0 and zero at $\xi = 0$ and $\pm\infty$, as shown in Figure 16.1.

16.1.2 Phase ("Allpass") Filters

A filter with unit MTF ($|H[\xi]| = 1$) at all frequencies passes all sinusoidal components with no change in amplitude, though possibly with alteration of the initial phases. These filters are appropriately called *allpass* or *phase-only* filters and are considered in Chapter 18. Their only possible action is translation of the component sinusoids by distances specified by the initial phase angles. We can classify these filters based on the "shape" of the **Φ**TF. The simplest pure phase filter has $\boldsymbol{\Phi}\{H[\xi]\} = 0[\xi]$, which is (of course) the identity operator with impulse response $h[x] = \delta[x]$. The next simplest allpass filter increments the initial phase of each component by an angle proportional to the spatial frequency; we know from the shift theorem that these *linear-phase* filters translate $f[x]$ by some fixed distance. We also have shown that the spectrum of a *quadratic-phase* signal has quadratic phase and infinite support. We will demonstrate that the output of an allpass filter differs significantly from the input when the **Φ**TF is a nonlinear function of ξ. Such filters significantly affect the fidelity (or *image quality*) of $g[x]$

measured relative to the input $f[x]$, even though all sinusoidal components are passed to the output without attenuation or amplification.

16.2 Eigenfunctions of Convolution

We now digress for a moment to demonstrate that all linear and shift-invariant filters have the same eigenfunctions, i.e., the action of the filter on the eigenfunction is merely a change of scale by the constant eigenvalue. This relates to the observation about the eigenvectors of circulant matrices in Chapter 5 and the observation in Chapter 8 that sinusoids are the eigenfunctions of the filter that convolves the input function with a unit-area rectangle. We now may apply the tools developed in subsequent chapters to show that these are in fact eigenfunctions of any convolution.

The eigenvalue condition for a 1-D operation $\mathcal{O}$ is:

$$\mathcal{O}\{f[x;\lambda]\} = \lambda \cdot f[x;\lambda] \tag{16.6}$$

where λ is the corresponding eigenvalue of the function $f[x]$ and is included as the parameter in the notation. The interpretation of this condition for vectors in Chapter 5 was that both the input and output vectors "point" in the same "direction" in the hyperspace of vectors. Now consider the space-domain operator to describe the 1-D convolution with the impulse response $h[x]$:

$$\begin{aligned}\mathcal{O}\{f[x;\lambda]\} &= g[x] = f[x;\lambda] * h[x] \\ &= \lambda \cdot f[x;\lambda]\end{aligned} \tag{16.7}$$

The equivalent operation in the frequency domain is easy to derive because the Fourier transform is linear:

$$\begin{aligned}\mathcal{F}_1\{\mathcal{O}\{f[x;\lambda]\}\} &= \mathcal{F}_1\{\lambda \cdot f[x;\lambda]\} \\ &= \lambda \cdot \mathcal{F}_1\{f[x;\lambda]\} = \lambda \cdot F[\xi;\lambda]\end{aligned} \tag{16.8}$$

The combination of these two results gives a relationship that must be satisfied by the frequency-domain representation of the eigenfunction:

$$F[\xi;\lambda] \cdot H[\xi] = \lambda \cdot F[\xi;\lambda] \tag{16.9}$$

The property of the Dirac delta function in Equation (6.115) suggests that one function $F[\xi;\lambda]$ that satisfies this condition is $\delta[\xi - \xi_0]$:

$$\begin{aligned}\delta[\xi - \xi_0] \cdot H[\xi] &= \delta[\xi - \xi_0] \cdot H[\xi_0] \\ &= \delta[\xi - \xi_0] \cdot \lambda\end{aligned} \tag{16.10}$$

where the eigenvalue λ is identified as $H[\xi_0]$. The corresponding relationship in the space domain is obtained via the inverse Fourier transform:

$$\begin{aligned}\lambda \cdot f[x;\lambda] &= \mathcal{F}_1^{-1}\{\delta[\xi - \xi_0] \cdot H[\xi]\} \\ &= \mathcal{F}_1^{-1}\{\delta[\xi - \xi_0] \cdot H[\xi_0]\} \\ &= H[\xi_0] \cdot e^{+2\pi i \xi_0 x}\end{aligned} \tag{16.11}$$

where the linearity of the Fourier transform has been used. This result demonstrates that $\exp[+2\pi i \xi_0 x]$ is an eigenfunction of any convolution. That this is the appropriate result is confirmed by direct substitution into the convolution integral:

$$\begin{aligned}\mathcal{O}\{e^{+2\pi i\xi_0 x}\} = \lambda \cdot e^{+2\pi i\xi_0 x} &= e^{+2\pi i\xi_0 x} * h[x] \\ &= \int_{-\infty}^{+\infty} e^{+2\pi i\xi_0(x-\alpha)}\, h[\alpha]\, d\alpha \\ &= e^{+2\pi i\xi_0 x} \int_{-\infty}^{+\infty} e^{-2\pi i\xi_0\alpha}\, h[\alpha]\, d\alpha \\ &= e^{+2\pi i\xi_0 x} \cdot \mathcal{F}_1\{h[x]\}|_{\xi=+\xi_0} \\ &= e^{+2\pi i\xi_0 x} \cdot H[+\xi_0] \end{aligned} \tag{16.12}$$

In words, this demonstrates that if a linear-phase complex exponential with frequency $+\xi_0$ is applied to an LSI system, then the output must be the linear-phase complex exponential with the same frequency, though possibly with different magnitude and phase. The changes in magnitude and phase are specified by the complex-valued eigenvalue $\lambda = H[+\xi_0]$.

That the frequency-domain representation of the general eigenfunction of the LSI operator is a Dirac delta function indicates immediately that the complete set of eigenfunctions is orthogonal. The inner product of two distinct eigenfunctions with spatial frequencies ξ_0 and ξ_1 is:

$$\begin{aligned}\int_{-\infty}^{+\infty} e^{+2\pi i\xi_0 x} (e^{+2\pi i\xi_1 x})^* \, dx &= \int_{-\infty}^{+\infty} e^{+2\pi i(\xi_0-\xi_1)x}\, dx \\ &= \delta[\xi_0 - \xi_1] \end{aligned} \tag{16.13}$$

where the representation of the Dirac delta function from Equation (6.106) has been used.

16.3 Power Transmission of Filters

Parseval's theorem in Equation (9.163) demonstrates that the integrated powers of the space- and frequency-domain representations are equal:

$$\int_{-\infty}^{+\infty} |f[x]|^2\, dx = \int_{-\infty}^{+\infty} |F[\xi]|^2\, d\xi \tag{16.14}$$

If applied to the output of a convolution, the theorem establishes a relation between the integrated spectra of the input and output functions:

$$\begin{aligned}\int_{-\infty}^{+\infty} |g[x]|^2\, dx = \int_{-\infty}^{+\infty} |G[\xi]|^2\, d\xi &= \int_{-\infty}^{+\infty} |f[x] * h[x]|^2\, dx \\ = \int_{-\infty}^{+\infty} |F[\xi] \cdot H[\xi]|^2\, d\xi &\leq \int_{-\infty}^{+\infty} |F[\xi]|^2 \cdot |H[\xi]|^2\, d\xi \end{aligned} \tag{16.15}$$

where the Schwarz inequality in Equation (5.89) was used in the last step. Most realistic LSI systems used in imaging contexts are *passive* processes, meaning that the system does not inject power into the input spectrum from external sources. In such a case, the magnitude of the output of a passive filter at any frequency may be no larger than the input magnitude, so that:

$$|H[\xi]|^2 \leq 1 \Longrightarrow |H[\xi]| \leq 1 \quad \textit{for "passive" filters} \tag{16.16}$$

One example of passive filtering in an imaging context is a glass lens used for imaging. An ideal lens transmits all incident radiation to the output image, thus preserving the integral of the incident power, but some incident radiation is lost in a realistic lens due to Fresnel reflections at the air–glass interfaces and perhaps absorption within the glass. This fact may be expressed in terms of Parseval's relation:

$$\begin{aligned}\int_{-\infty}^{+\infty} |g[x]|^2\,dx &= \int_{-\infty}^{+\infty} |G[\xi]|^2\,d\xi \\ &\leq \int_{-\infty}^{+\infty} (|F[\xi]|^2 \cdot |H[\xi]|^2)\,d\xi \\ &\leq \int_{-\infty}^{+\infty} |F[\xi]|^2\,d\xi = \int_{-\infty}^{+\infty} |f[x]|^2\,dx \end{aligned} \tag{16.17}$$

where Equation (16.16) has been used. This shows that the integrated power (total intensity) of the output of a passive filter must be less than or equal to the integrated power of the input:

$$\int_{-\infty}^{+\infty} |g[x]|^2\,dx \leq \int_{-\infty}^{+\infty} |f[x]|^2\,dx \quad \textit{for “passive” filters} \tag{16.18}$$

In an allpass (phase-only) filter, $|H[\xi]| = 1[\xi]$ and the integrated intensity of the input and output must match:

$$\begin{aligned}\int_{-\infty}^{+\infty} |g[x]|^2\,dx &= \int_{-\infty}^{+\infty} |F[\xi] \cdot H[\xi]|^2\,d\xi \\ &= \int_{-\infty}^{+\infty} |(|F[\xi]|\, e^{i\Phi\{F[\xi]\}}) \cdot 1 \cdot e^{i\Phi\{H[\xi]\}}|^2\,d\xi \\ &= \int_{-\infty}^{+\infty} |F[\xi]|^2 |e^{i(\Phi\{F[\xi]\}+\Phi\{H[\xi]\})}|^2\,d\xi \\ &= \int_{-\infty}^{+\infty} |F[\xi]|^2\,d\xi = \int_{-\infty}^{+\infty} |f[x]|^2\,dx \end{aligned}$$

$$\int_{-\infty}^{+\infty} |g[x]|^2\,dx = \int_{-\infty}^{+\infty} |f[x]|^2\,dx \quad \textit{for allpass filters} \tag{16.19}$$

In words, the integrated power is preserved by an allpass filter, though this does *not* mean that the powers evaluated at a particular coordinate are equal:

$$\int_{-\infty}^{+\infty} |f[x]|^2\,dx = \int_{-\infty}^{+\infty} |g[x]|^2\,dx \quad \textit{if} \;\; |H[\xi]| = 1[\xi] \tag{16.20a}$$

$$|f[x]|^2 \neq |g[x]|^2 \quad \textit{in general} \tag{16.20b}$$

An *active* filter can inject power from an external source into the spectrum of the input function, and so may amplify any or all of its frequency components. If $|H[\xi]|$ exceeds unity at *all* frequencies, then the integrated intensity of the output image must exceed that of the input; the integrated output image is "brighter" than the integrated input:

$$\int_{-\infty}^{+\infty} |g[x]|^2\,dx \geq \int_{-\infty}^{+\infty} |f[x]|^2\,dx \quad \textit{for “active” filter with } |H[\xi]| \geq 1 \tag{16.21}$$

The general active filter may amplify some frequency components and attenuate others, so that $|H[\xi]|$ may be greater than unity at some frequencies, less than unity at others, and perhaps equal to unity at still others. In this general case, the relation between $|f[x]|$ and $|h[x]|$ cannot be determined without knowledge of both $f[x]$ and $|H[\xi]|$.

16.4 Lowpass Filters

In our generalized definition of a magnitude filter that includes all real-valued transfer functions, the action of the filter on a particular sinusoidal component is a scaling of the magnitude at each frequency ξ by a real-valued and possibly bipolar weight factor. If the weighting applied at a particular spatial frequency is negative, then the initial phase of that sinusoidal component of the input is decremented by π radians. The transfer functions of the lowpass filters considered in this section are real valued and symmetric, though it is easy to construct lowpass filters with asymmetric MTFs.

The ideal lowpass filter has already been introduced and was applied in the discussion of ideal interpolation in Chapter 14. Its MTF is a symmetric rectangle:

$$H_1[\xi] = RECT[\xi b] = RECT\left[\frac{\xi}{b^{-1}}\right] \tag{16.22}$$

Obviously, this definition satisfies the strict requirements for a lowpass filter with cutoff frequency $\xi_{\text{cutoff}} = (2b)^{-1}$. The area of the transfer function is $b^{-1} = 2\xi_{\text{cutoff}}$, which also is the central ordinate of the impulse response:

$$h_1[x] = \frac{1}{|b|} SINC\left[\frac{x}{b}\right] = (2 \cdot |\xi_{\text{cutoff}}|)\, SINC[\xi_{\text{cutoff}} \cdot x] \tag{16.23}$$

which has unit area and infinite support. Convolution with this impulse response produces a bipolar weighted average of $f[x]$ over the infinite domain. The convolution evaluated at x_0 includes large positive weights applied to amplitudes in the vicinity of x_0 (within a distance b), which means that the dominant action of the filter is a local averaging of the input amplitude. Smaller negative weights are applied to signal amplitudes at distances between b and $2b$ from x_0, so that some weighted amplitude is subtracted from the summation. The bipolar character of the impulse response means that the output of the ideal lowpass filter may be negative at some coordinates even if $f[x] > 0$. Also, the output of an ideal lowpass filter often oscillates between positive and negative amplitudes in the vicinity of edges of the input function. This oscillation of the output signal is called "ringing" and is characteristic of ideal lowpass filters.

As an example of a lowpass filter with different character, consider the unit-area rectangular impulse response:

$$h_2[x] = \frac{1}{|b|} RECT\left[\frac{x}{b}\right] \tag{16.24}$$

This computes a uniformly weighted local average of $f[x]$ over domains of support b, and thus is proportional to the integral of the input signal over a local region. The uniform weighting in the average ensures that the output signal resulting from a nonnegative input must also be nonnegative; in such a case there can be no bipolar oscillatory "ringing" of the output. Local averaging "attracts" the output toward the mean amplitude of the input, which means that the effect of the filter decreases the amplitude in the vicinity of maxima and increases it near minima. This reduction in the amplitude of oscillating sinusoidal components is shown in Figure 16.2. In other words, the averaging filter reduces local variability in amplitude of the input. The amplitude of a sinusoid whose period is small compared to b is attenuated more than slowly oscillating components. The high-frequency components that are necessary to generate "sharp" edges in a signal are attenuated more than low-frequency terms, thus producing an output $g[x]$ that looks "fuzzy" or "blurred" when compared to the input $f[x]$.

The transfer function of the uniform averager in Equation (16.24) clearly is:

$$H_2[\xi] = SINC[b\xi] = SINC\left[\frac{\xi}{b^{-1}}\right] \tag{16.25}$$

which has unit amplitude at the origin and asymptotically approaches zero at $\xi = \pm\infty$, so it satisfies the relaxed definition of a lowpass filter. Note that any oscillating sinusoidal component (for which $\xi \neq 0$)

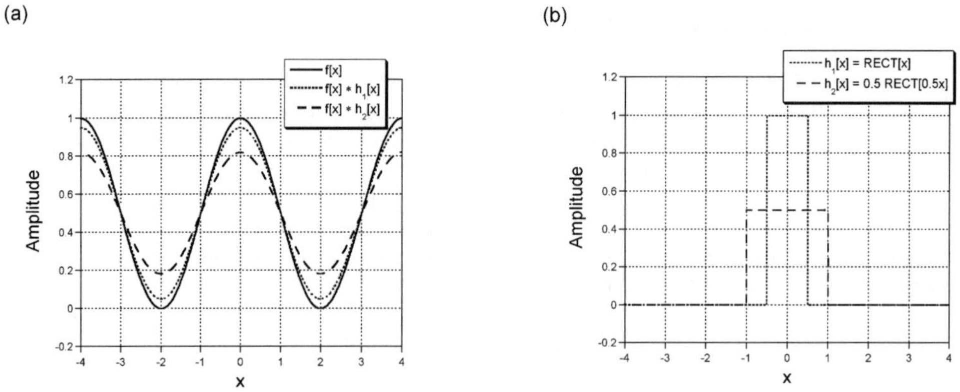

Figure 16.2 Action of "lowpass filter" on the unit-modulation sinusoid $f[x] = \frac{1}{2}(1 + \cos[2\pi x/4])$: (a) $h_1[x] = RECT[x]$ and $h_2[x] = \frac{1}{2}RECT[x/2]$; (b) $f[x]$, $f[x] * h_1[x]$, and $f[x] * h_2[x]$. The modulation of the output sinusoids has decreased from unity to 0.900 and 0.637, respectively.

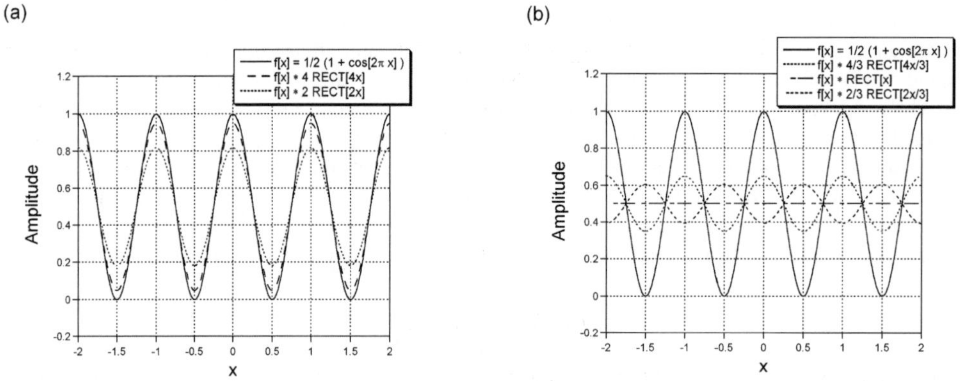

Figure 16.3 The effect of the uniform averager $h[x] = b^{-1}RECT[x/b]$ on the modulation of $f[x] = \frac{1}{2}(1 + \cos[2\pi x])$: (a) $f[x]$ and the output after convolution with $h[x]$ with $b = \frac{1}{4}$ and $\frac{1}{2}$, showing the reduction in modulation with increasing width b; (b) $f[x]$ and the output after convolution with $b = \frac{3}{4}$, 1, and $\frac{4}{3}$, which shows the "contrast reversal" of the sinusoid for $1 < b < 2$, when the width of the rectangle exceeds the period of the sinusoid.

is attenuated to some degree by the action of the filter. The bipolar and real-valued character of this transfer function ensure that the initial phases of sinusoids within particular bands of spatial frequency are decremented by π radians. The cutoff frequency is not well defined: frequencies that are nonzero integer multiples of b^{-1} are completely blocked, but attenuated spatial frequencies out to $\xi = \pm\infty$ do reach the output. These particular sinusoids are averaged to zero by the rectangular impulse response because an integer number of periods fits exactly within its support b, as shown in Figure 16.3.

Parseval's theorem in Equation (9.163) may be used to relate the integrated powers in both domains for this filter:

$$\int_{-\infty}^{+\infty} |SINC[b\xi]|^2 \, d\xi = \int_{-\infty}^{+\infty} \frac{1}{b^2} \left| RECT\left[\frac{x}{b}\right] \right|^2 dx$$
$$= \frac{1}{b^2} \int_{-\infty}^{+\infty} RECT\left[\frac{x}{b}\right] dx$$
$$= \left(\frac{1}{b^2}\right) \cdot |b| = \frac{1}{|b|} \tag{16.26}$$

where the scaling theorem of Equation (9.117) also has been used. Because $|H_2[\xi]| \leq 1$ at all frequencies, the system need not "inject" any power during the filtering process.

Many varieties of lowpass filter may be defined that satisfy either the strict or the relaxed definition. A common and useful example is the Gaussian impulse response with unit area:

$$h_3[x] = \frac{1}{|b|} GAUS\left[\frac{x}{b}\right] \tag{16.27}$$

which is a nonuniformly weighted averager with no negative amplitudes. Derivation of the corresponding transfer function is straightforward:

$$H_3[\xi] = GAUS[\xi b] = GAUS\left[\frac{\xi}{b^{-1}}\right] \tag{16.28}$$

which has no cutoff frequency under the strict definition.

16.4.1 1-D Test Object

To illustrate the action of various filters, consider the specific input function $f[x]$ composed of three sets of five biased sinusoidal "bars" shown in Figure 16.4a. The period of the coarsest bars is two units, so its "dominant" spatial frequency is $\xi_1 = \frac{1}{2}$ cycles per unit length. The sinusoidal pattern is truncated by a rectangular window, thus ensuring that other spatial frequencies also exist in the spectrum of the coarse bars. The dominant spatial frequencies in the other sets of bars are respectively 1 and 2 cycle(s) per unit length, but they are also truncated by rectangular windows that pass five "bars", and thus are $\frac{1}{2}$ and $\frac{1}{4}$ as wide as the window for the coarse bars. Figure 16.4b–d shows the spectrum of $f[x]$ as real part, imaginary part, magnitude, and unwrapped phase. The windowing of higher-frequency bar patterns by narrower rectangles appears in the magnitude spectrum of Figure 16.4d as the convolution of the Dirac delta functions at larger frequencies with wider *SINC* functions.

16.4.2 Ideal 1-D Lowpass Filter

In Figure 16.5, the bars have been filtered by the ideal lowpass transfer function $H_1[\xi] = RECT[\xi/1.5]$, which passes average value ("DC") and the fundamental frequency of the coarse bars. It removes all sinusoidal components with spatial frequencies $|\xi| > \frac{3}{4}$ cycles per unit length, and thus "blocks" the dominant spatial frequency of the other two sets of bars. The only evidence of their presence in $g_1[x]$ is the existence of regions of "incorrectly" modulated positive amplitude. Though $f[x] \geq 0$ everywhere, the filtered output image exhibits bipolar oscillations, particularly near the locations of edges in the input. This "ringing" would have been cancelled by the "missing" high-frequency components of $f[x]$.

16.4.3 1-D Uniform Averager

In the second example of lowpass filtering in Figure 16.6, the impulse response is the uniform averager $h_2[x] = \frac{4}{3} RECT[\frac{4}{3}x]$. Because both $f[x]$ and $h_2[x]$ are nonnegative everywhere, so is the output

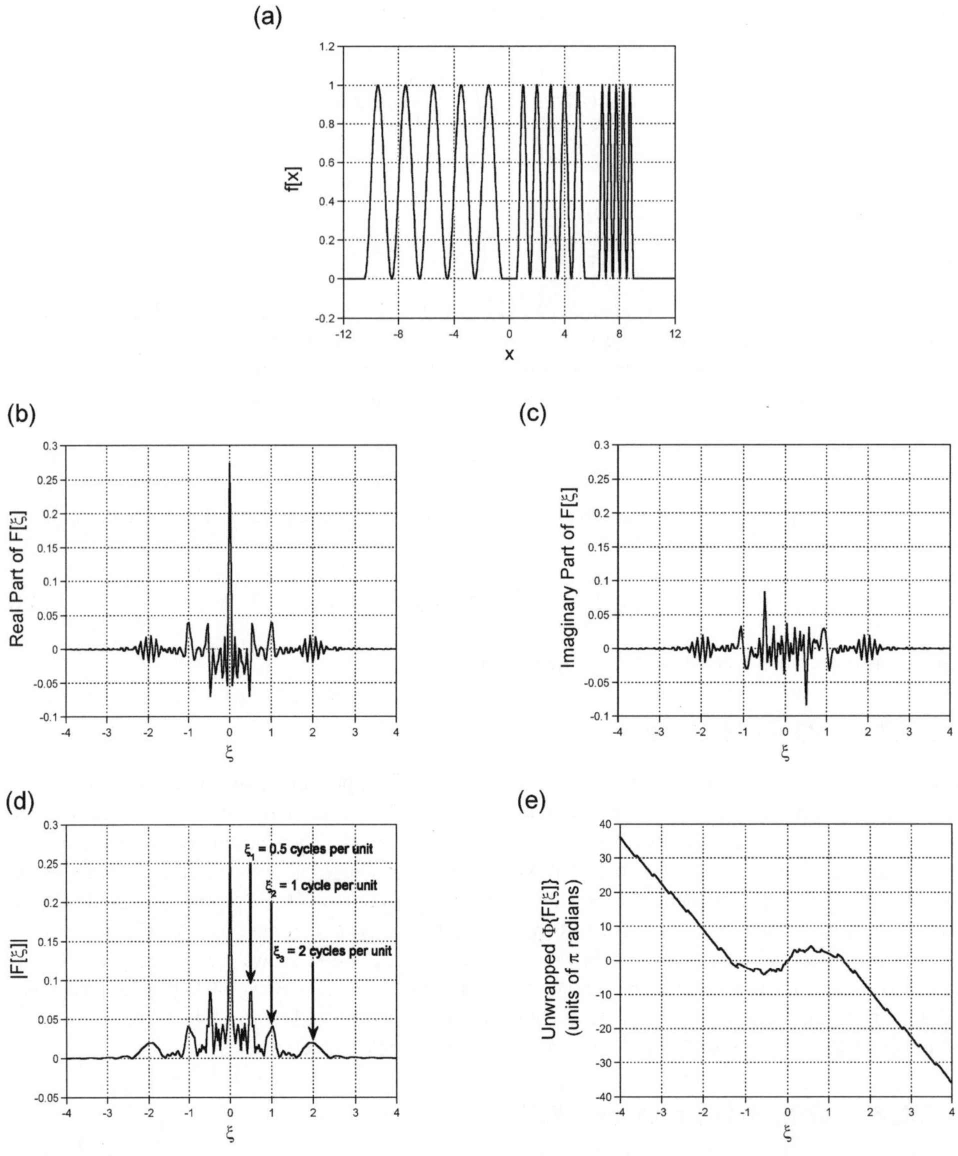

Figure 16.4 (a) Test object $f[x]$ composed of three sets of five nonnegative sinusoidal "bars" with different spatial frequencies ($\frac{1}{2}$, 1, and 2 cycles per unit length); (b) $\Re\{F[\xi]\}$; (c) $\Im\{F[\xi]\}$; (d) $|F[\xi]|$, showing local maxima centered on the spatial frequencies of the "bars" as well as convolution with *SINC* functions from the rectangular windows; (e) the unwrapped phase $\Phi\{F[\xi]\}$.

amplitude. The corresponding transfer function is $H_2[\xi] = SINC[\frac{3}{4}\xi]$, which attenuates all spatial frequencies other than $\xi = 0$, has zeros at the other multiples of $\frac{4}{3}$, and "reverses" the phase within alternate intervals between the zeros of spatial frequency. The phase applied by the filter to the dominant

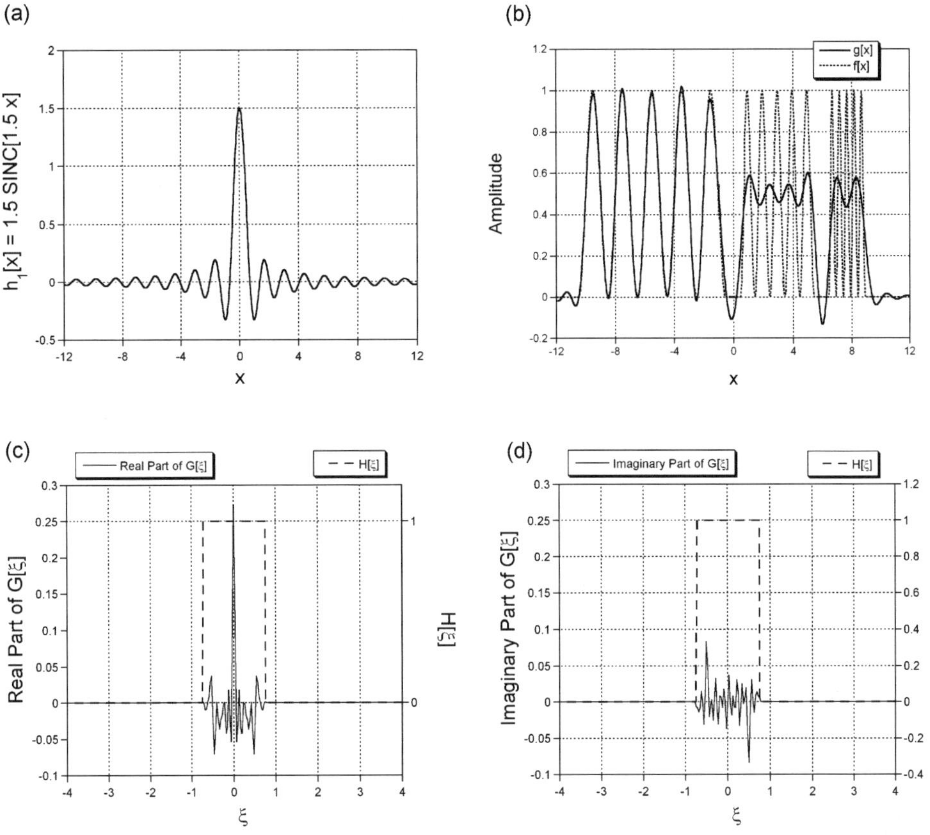

Figure 16.5 Ideal lowpass filtering: (a) $h_1[x] = \frac{3}{2}\, SINC[\frac{3}{2}x]$, with transfer function $H_1[\xi] = RECT[\frac{2}{3}\xi]$; (b) filtered image $f[x] * h_1[x] = g_1[x]$; (c) $\Re\{G_1[\xi]\}$, showing cutoff frequency; (d) $\Im\{G_1[\xi]\}$. The two largest spatial frequencies are missing from $g_1[x]$, having been "averaged out".

frequencies of the two coarser bars is 0 radians, while the phase applied to the fine bars with $\xi_3 = 2$ cycles per unit is $-\pi$ radians.

16.4.4 2-D Lowpass Filters

The discussion of 2-D Fourier transforms in Chapters 10–12 leads directly to the description and analysis of 2-D filters. For this reason, we only give a couple of illustrative examples and comments. The separable transfer function and impulse response for the ideal 2-D lowpass with cutoff frequencies $\xi_{\text{cutoff}} = (2b)^{-1}$ and $\eta_{\text{cutoff}} = (2d)^{-1}$ are obtained from Equation (10.14) and the "transform-of-a-transform" theorem in Equation (10.25):

$$H_4[\xi, \eta] = RECT\left[\frac{\xi}{b^{-1}}, \frac{\eta}{d^{-1}}\right] \tag{16.29a}$$

$$h_4[x, y] = \frac{1}{|bd|} SINC\left[\frac{x}{b}, \frac{y}{d}\right] \tag{16.29b}$$

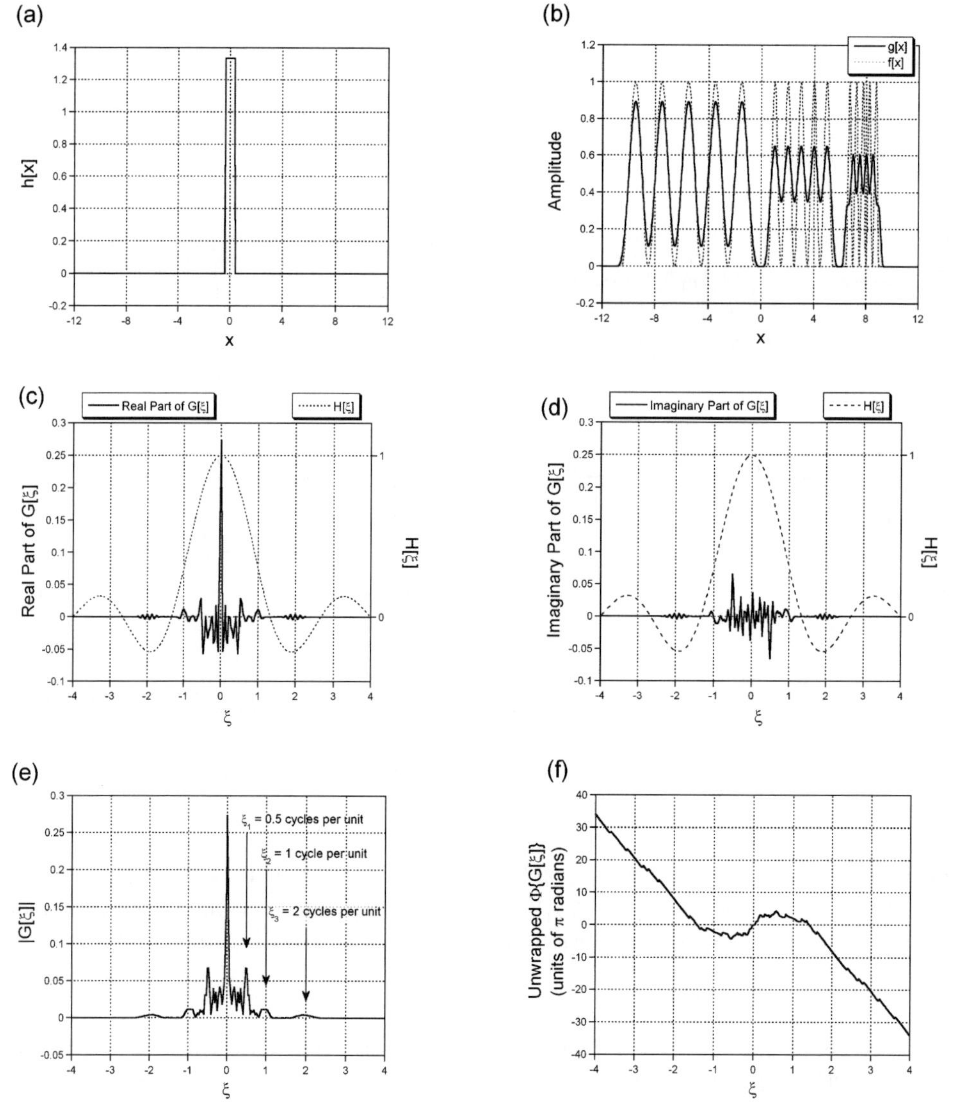

Figure 16.6 Effect of uniform averaging: (a) $h_1[x] = \frac{4}{3}RECT[\frac{4}{3}x]$, so that $H_1[\xi] = SINC[\frac{3}{4}\xi]$; (b) $g_1[x]$, showing that the phase of the highest-frequency "bars" has been reversed; (c) $\Re\{G_1[\xi]\}$, showing reduction in amplitude with increasing frequency and reversed phase for $\frac{4}{3} < |\xi| < \frac{8}{3}$; (d) $\Im\{G_1[\xi]\}$; (e) $|G_1[\xi]|$; (f) unwrapped $\Phi\{G_1[\xi]\}$, which shows subtle differences from $\Phi\{F[\xi]\}$. Note that the transfer function applies a phase of $-\pi$ at the fundamental frequency of $\xi = 2$ cycles per unit for the fine bars.

The transfer function and impulse response of the analogous circularly symmetric filter are:

$$H_5(\rho) = CYL\left(\frac{\rho}{d^{-1}}\right) \tag{16.30a}$$

$$h_5(r) = \frac{\pi}{4d^2} SOMB\left(\frac{r}{d}\right) \tag{16.30b}$$

Other variations of Equation (16.29) may be constructed by rotating the transfer function via Equation (7.11), with the resulting impulse response following from the rotation theorem of Equation (10.37).

16.5 Highpass Filters

Filters that transmit no information about the sinusoidal component with $\xi = 0$ (DC term) to the output but apply nonzero weighting at $\xi = \pm\infty$ satisfy the relaxed criteria for a highpass filter. An easy way to construct a highpass filter is to construct the complement of the transfer function of a valid lowpass filter:

$$H_{\text{highpass}}[\xi] = (H_{\text{lowpass}}[\xi])_{\max} - H_{\text{lowpass}}[\xi] \rightarrow 1[\xi] - H_{\text{lowpass}}[\xi] \tag{16.31}$$

where the last expression is valid if the lowpass transfer function has been normalized so that its maximum amplitude is unity. The corresponding impulse response is:

$$h_{\text{highpass}}[x] = \delta[x] - h_{\text{lowpass}}[x] \tag{16.32}$$

The output of such a highpass filter may be expressed as the difference between the original input and a lowpass-filtered replica:

$$\begin{aligned} g_{\text{highpass}}[x] &= f[x] * h_{\text{highpass}}[x] = f[x] * (\delta[x] - h_{\text{lowpass}}[x]) \\ &= f[x] * \delta[x] - f[x] * h_{\text{lowpass}}[x] \\ &= f[x] - (f[x] * h_{\text{lowpass}}[x]) \\ &= f[x] - g_{\text{lowpass}}[x] \end{aligned} \tag{16.33}$$

We surmised in Section 14.3 that lowpass filters compute some kind of local average of the input signal. In other words, the highpass filter is a *local differencer* that increases the local variation of amplitude of the input signal. This behavior is complementary to that of the lowpass filter, which is a *local averager* or *integrator* that reduces local variability in the signal amplitude. A particular highpass filter computes the derivative of the input function, and thus is a *differentiator*.

16.5.1 Ideal 1-D Highpass Filter

The ideal highpass filter removes all frequency components from the signal with $|\xi| \leq \xi_{\text{cutoff}}$; its transfer function is the complement of that of the ideal lowpass filter in Equation (16.22):

$$H_6[\xi] = 1[\xi] - RECT\left[\frac{\xi}{2\xi_{\text{cutoff}}}\right] \tag{16.34}$$

Because the amplitude of the transfer function is zero at $\xi = 0$, the DC component is removed from the output spectrum $G_6[\xi]$. The central-ordinate theorem demonstrates that the areas of both the ideal highpass impulse response $h_6[x]$ and the output signal $g_6[x]$ must be zero. In other words, the output image $g_6[x]$ generated from a highpass filter either is bipolar or has null amplitude; it cannot be nonnegative. An example is shown in Figure 16.7, where the transfer function is the complement of that used in Figure 16.6.

The linearity of the Fourier transform ensures that the impulse response $h_6[x]$ of the ideal highpass filter is the difference between the unit-area Dirac delta function located at the origin and the unit-area *SINC* function with width parameter b. The cancellation of the areas of the two functions ensures that the impulse response $h_6[x]$ must have null area:

$$h_6[x] = \delta[x] - \frac{1}{b}SINC\left[\frac{x}{b}\right] \tag{16.35}$$

The amplitude of $h_4[x]$ is positive and undefined at $x = 0$ and negative in the immediate vicinity of the origin, which means that it computes a particular local difference of signal amplitude.

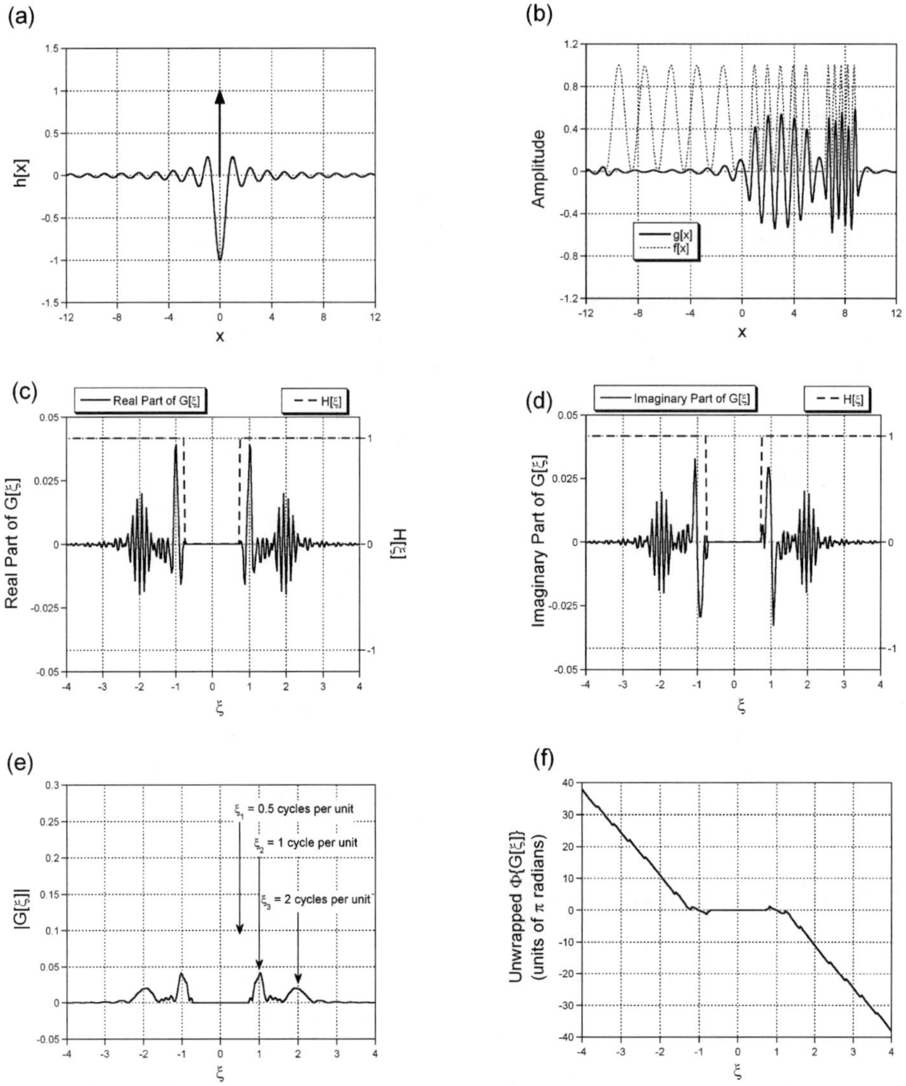

Figure 16.7 Action of ideal high-pass filter with $H_6[\xi] = 1[\xi] - RECT[\xi/1.5]$: (a) $h_6[x] = \delta[x] - \frac{3}{2}SINC[\frac{3}{2}x]$; (b) $g_6[x]$, showing bipolar amplitude due to the removal of $F[0]$ and missing low-frequency "bars"; (c) $\Re\{G_6[\xi]\}$ (note change of amplitude scale); (d) $\Im\{G_6[\xi]\}$; (e) $|G_6[\xi]|$, showing the complete removal of the DC and dominant low-frequency components; and (f) $\Phi\{G_6[\xi]\}$.

16.5.2 1-D Differentiators

One filter that satisfies the relaxed condition for a highpass filter is the derivative operator. The impulse response and transfer function were derived by applying the filter theorem in Equation (9.129):

$$g_7[x] = \frac{d^n f}{dx^n} \Longrightarrow h_7[x] = \left(\frac{d^n}{dx^n}\delta[x]\right) \tag{16.36a}$$

$$\Longrightarrow H_7[\xi] = (2\pi i\xi)^n \tag{16.36b}$$

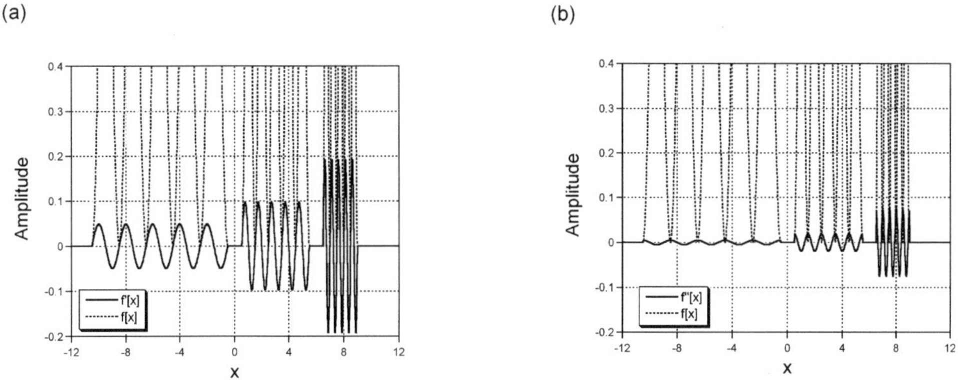

Figure 16.8 First and second derivatives of $f[x]$: (a) $g[x] = (d/dx) f[x]$, showing the linear increase in amplitude of the derivative with increasing spatial frequency and the phase shift; (b) the second derivative $(d^2/dx^2) f[x]$, which amplifies in proportion to ξ^2.

The transfer functions of the odd-order derivatives are imaginary, and so the phase of the transfer function at any frequency is 0 and $\pm\pi/2$ radians. Therefore, these odd-order derivatives do not satisfy even our relaxed criterion for inclusion into the category of magnitude filters. However, we will just sweep this problem under the rug. Readers who object to this lack of rigor may be assuaged somewhat because we will concentrate most often upon the second-order derivative for which both $h[x]$ and $H[\xi]$ are real valued and symmetric, and so do satisfy the most relaxed criterion. The outputs of the first- and second-order derivatives applied to the three-bar input signal are shown in Figure 16.8. These give bipolar "maps" of the edge transitions of the input signal.

16.5.3 2-D Differentiators

The partial derivative of a 2-D function may be evaluated via convolution. For example, the 2-D impulse response of differentiation along the x-axis will be denoted ∂_x:

$$g[x] = f[x, y] * \partial_x[x, y] = \frac{\partial f[x, y]}{\partial x} = f[x, y] * (\delta'[x] \cdot \delta[y]) \tag{16.37a}$$

The corresponding transfer function is:

$$H_{\partial_x}[\xi, \eta] = (+2\pi i \xi) \cdot 1[\eta] \tag{16.37b}$$

which is imaginary and odd, as expected. The corresponding impulse response and transfer function of partial differentiation along the y-axis obviously are:

$$\partial_y[x, y] = \delta[x] \cdot \delta'[y] \tag{16.38a}$$

$$H_{\partial_y}[\xi, \eta] = 1[\xi] \cdot (+2\pi i \eta) \tag{16.38b}$$

The extra degree of freedom of the 2-D case allows construction of 2-D differentiators along any azimuthal direction by applying the rotation operator in Equation (7.11) and the rotation theorem in Equation (10.37). It also is useful to construct combinations of derivative operators, such as the *gradient*, which will be briefly considered during the discussion of discrete operators in Chapter 20.

16.5.4 High-Frequency Boost Filters – Image Sharpeners

Filters that preserve low-frequency content while amplifying high-frequency structure are "high-frequency boosters". In imaging contexts, these filters preserve the large-scale structure in the image that is generated by low-frequency sinusoids while enhancing the high-frequency components that are responsible for edge detail. In other words, a "high-boost" filter is an image "sharpening" operator. The transfer function of a high-boost filter may be generated by adding the unit constant to any highpass transfer function so that the magnitude of the transfer function exceeds unity at all spatial frequencies:

$$H_{\text{sharpen}}[\xi] = 1[\xi] + H_{\text{highpass}}[\xi]$$
$$\Longrightarrow |H_{\text{sharpen}}[\xi]| \geq 1[\xi] \tag{16.39a}$$

The corresponding impulse response is:

$$h_{\text{sharpen}}[x] = \delta[x] + h_{\text{highpass}}[x] \tag{16.39b}$$
$$= \delta[x] + (\delta[x] - h_{\text{lowpass}}[x]) = 2 \cdot \delta[x] - h_{\text{lowpass}}[x] \tag{16.39c}$$

where the relationship between a highpass filter and the complementary lowpass filter in Equation (16.32) has been used. In other words, the sum of the input and the output of some flavor of local differencing operator produces a sharpened version of the input. A physical rationale for a specific sharpening operator is considered next.

Images are blurred if the incident irradiance diffuses into some neighborhood about the ideal image point on the sensor. The process is similar conceptually (but not identical mathematically) to the process of heat diffusion from a source into a 3-D volume, which is described by the differential heat diffusion equation that is derived in thermodynamics:

$$\frac{\partial}{\partial t} u[x, y, z, t] = k \cdot \left(\frac{\partial^2 u}{\partial x^2} + \frac{\partial^2 u}{\partial y^2} + \frac{\partial^2 u}{\partial z^2} \right) \tag{16.40}$$

where $u = u[x, y, z, t]$ is the distribution of temperature within the body as a function of position and time and k is a constant proportional to the thermal conductivity of the material. The 2-D version of this equation describes the diffusion of heat over a surface, where the temperature $u[x, y, t]$ is a function of two space coordinates and time. The corresponding diffusion equation is:

$$\frac{\partial u}{\partial t} = k \cdot \left(\frac{\partial^2 u}{\partial x^2} + \frac{\partial^2 u}{\partial y^2} \right) \equiv k \cdot \nabla^2 u \tag{16.41}$$

In words, the temporal variation in temperature at a particular location is proportional to the Laplacian of the spatial temperature distribution. A long time after the source is removed ($t \to \infty$), the heat will have diffused throughout the object so that $\partial u/\partial t = k \cdot \nabla^2 u = 0$, which means that the temperature in the body has become uniform.

The subject of interest in imaging is the distribution of irradiance $f[x, y]$, which is analogous to the 2-D temperature distribution $u[x, y]$. Consider a system that produces a sharp image at $z = 0$ (the *focal plane*), as shown schematically in Figure 16.9. If the focus is adjusted without moving the sensor, so that the image distance is $z \neq 0$, the resulting image is blurred due to *defocus*. The blur worsens as $|z|$ is increased. Recall from the heat equation that a uniform temperature was obtained for $t \to +\infty$. The analogous situation in imaging occurs if the image plane is so badly defocused that the image becomes uniform, $f[x, y; z \to \infty] \propto 1[x, y]$. The continuous diffusion equation for imaging is:

$$\frac{\partial f[x, y; z]}{\partial z} = k \cdot \left(\frac{\partial^2}{\partial x^2} + \frac{\partial^2}{\partial y^2} \right) f[x, y; z] \quad \textit{where } k > 0 \tag{16.42}$$

The available measured image is the defocused function $f[x, y; z]$, and we wish to solve this differential equation to evaluate the "sharply focused" image distribution $f[x, y; 0]$ from this blurry image. This differential equation may be solved by expressing the unknown deblurred image $f[x, y; 0]$ as a Taylor series based on the known blurry image $f[x, y; z_0]$:

$$f[x, y; 0] = f[x, y; z_0] + \frac{(0 - z_0)}{1!}\left(\frac{\partial f}{\partial z}\bigg|_{z=z_0}\right) + \frac{(0 - z_0)^2}{2!}\left(\frac{\partial^2 f}{\partial z^2}\bigg|_{z=z_0}\right) + \cdots \tag{16.43}$$

If there is little image blur because the observation plane is close to the "in-focus" plane, then $|z| \gtrsim 0$ and $z \gg z^2 \gg z^3$ etc. Therefore the Taylor series may be truncated at the linear term without incurring much error in the estimate of the original image:

$$f[x, y; 0] \cong f[x, y; z_0] - z_0\left(\frac{\partial f}{\partial z}\bigg|_{z=z_0}\right) + 0 + \cdots \tag{16.44}$$

We can use the diffusion equation to substitute for $\partial f/\partial z$:

$$\begin{aligned} f[x, y; 0] &\cong f[x, y; z_0] - z_0 \cdot (k \cdot \nabla^2 f[x, y; z_0]) \\ &\equiv f[x, y; z_0] - \alpha \cdot \nabla^2 f[x, y; z_0] \end{aligned} \tag{16.45}$$

where $z_0 \cdot k \equiv \alpha$. Note that α increases as z_0 increases, i.e., the weighting factor must be increased to recover imagery from systems with more blur. In words, the first approximation of the sharply focused image $f[x, y; 0]$ is obtained by computing the difference between the blurred image $f[x, y; z_0]$ and a scaled replica of its Laplacian. This may be expressed as the weighted sum of the convolution of $f[x, y; z]$ with $\delta[x, y]$ and with the Laplacian operator:

$$f[x, y; 0] \cong f[x, y; z_0] * (\delta[x, y] - \alpha \cdot (\delta''[x] \cdot \delta[y] + \delta[x] \cdot \delta''[y])) \tag{16.46}$$

The corresponding transfer function is:

$$\begin{aligned} H[\xi, \eta] &= 1[\xi, \eta] - \alpha \cdot (+2\pi i)^2 \cdot (\xi^2 + \eta^2) \\ &= 1 + \alpha \cdot 4\pi^2 \rho^2 \end{aligned} \tag{16.47}$$

The transfer function is circularly symmetric and positive at all spatial frequencies.

This analysis demonstrates that a convolution operator based on the Laplacian may be used to sharpen "blurry" digital images. The ad hoc process is not "tuned" to the details of the blurring process and thus does not reconstruct the original sharp image. Rather it "steepens" the slope of pixel-to-pixel changes in gray level, thus making the edges appear sharper. The discrete version of the Laplacian filter is considered in Chapter 20.

A 1-D analogue of the Laplacian sharpener is constructed by setting $\eta = 0$ in Equation (16.47). The transfer function and impulse response are:

$$H_8[\xi] = 1[\xi] + \alpha(4\pi^2\xi^2) \tag{16.48a}$$

$$h_8[x] = \delta[x] - \alpha\left(\frac{d^2}{dx^2}\right) \tag{16.48b}$$

An example of the action of the 1-D Laplacian sharpener is shown in Figure 16.10.

16.6 Bandpass Filters

The MTF of a strictly defined bandpass filter reaches its maximum at one (or more) nonzero frequencies and drops to zero at $\xi = 0$ and $\xi = \pm\infty$. If the transfer function has a single maximum, say at $\xi = +|\xi_0|$, the transfer function has a single *passband*. Before considering this case, we examine the simpler case of symmetric passbands described by a symmetric function $R[\xi/\Xi]$. The scale factor denoted by Ξ

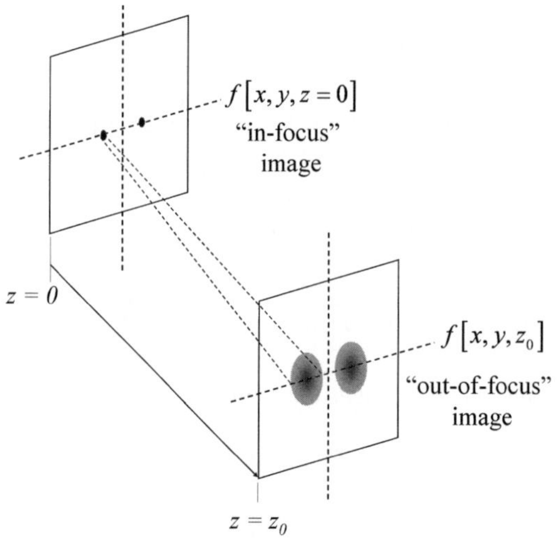

Figure 16.9 Schematic view of the diffusion model of defocus. The focused image is located at $z = 0$. The image in the plane located at $z \neq 0$ is "out of focus" because the irradiance has diffused into the neighborhood of the correct location. The Laplacian sharpening model assumes that the irradiance diffuses under the same model as heat.

(upper-case ξ) is the *bandwidth* of the passband, which is used instead of "$\Delta\xi$" to avoid confusion with the sample spacing in the frequency domain. Replicas of $R[\xi/\Xi]$ are positioned at the *center* frequencies $\pm\xi_0$, and thus the transfer function has the equivalent forms:

$$
\begin{aligned}
H_{\text{bandpass}}[\xi] &= R\left[\frac{\xi+\xi_0}{\Xi}\right] + R\left[\frac{\xi-\xi_0}{\Xi}\right] \\
&= R\left[\frac{\xi}{\Xi}\right] * (\delta[\xi+\xi_0] + \delta[\xi-\xi_0]) \\
&= R\left[\frac{\xi}{\Xi}\right] * \frac{1}{|\xi_0|}\delta\delta\left[\frac{\xi}{\xi_0}\right]
\end{aligned}
\tag{16.49a}
$$

The impulse response of this filter is the inverse Fourier transform of $H_{\text{bandpass}}[\xi]$:

$$h_{\text{bandpass}}[x] = \mathcal{F}_1^{-1}\{H_{\text{bandpass}}[\xi]\} = (|\Xi|\, r[\Xi \cdot x]) \cdot (2\cos[2\pi\xi_0 x]) \tag{16.49b}$$

In words, the impulse response of a bandpass filter is the product of two terms: one whose width is inversely proportional to the passband width Ξ; and a symmetric sinusoid with spatial frequency equal to the center frequency ξ_0. If the center frequency ξ_0 is much larger than the width of the passband, so that $1/\xi_0 \ll 1/\Xi$, then the impulse response will exhibit several oscillation periods of length ξ_0^{-1} within the distance $(\Xi)^{-1}$. In the example of Figure 16.11, the passband is the ideal rectangle with width Ξ centered at $\pm\xi_0$:

$$H_9[\xi] = RECT\left[\frac{\xi}{\Xi}\right] * |\xi_0|\delta\delta\left[\frac{\xi}{\xi_0}\right] \quad \textit{where } \xi_0 \gg \Xi \tag{16.50a}$$

and the corresponding impulse response is:

$$h_9[x] = (|\Xi|\, SINC[\Xi \cdot x]) \cdot (2\cos[2\pi\xi_0 x]) \tag{16.50b}$$

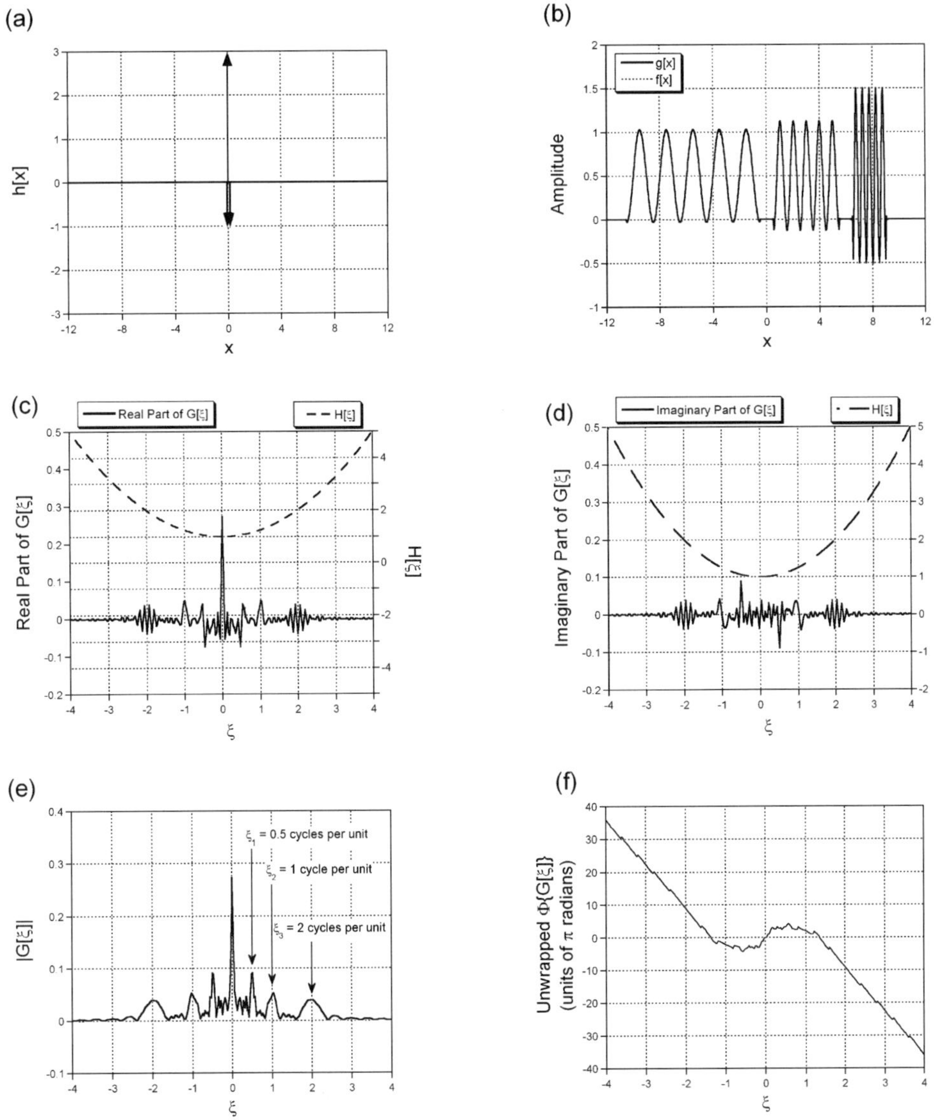

Figure 16.10 Image sharpening using 1-D analogue of Laplacian: (a) impulse response $h[x] = \lim_{\varepsilon \to 0}(-\delta[x+\varepsilon] + 3\delta[x] - \delta[x-\varepsilon])$; (b) $g[x]$, which shows increased modulation at larger spatial frequencies; (c) $\Re\{G[\xi]\}$ shown with $H[\xi]$; (d) $\Im\{G[\xi]\}$; (e) $|G[\xi]|$; (f) $\Phi\{G[\xi]\}$.

The impulse response of the bandpass filter is a cosine function whose amplitude is modulated by a *SINC* function. The impulse response has infinite support. In the example in Figure 16.11, $\xi_0 = 1$ cycle per unit and $\Xi = 0.5$ cycles per unit.

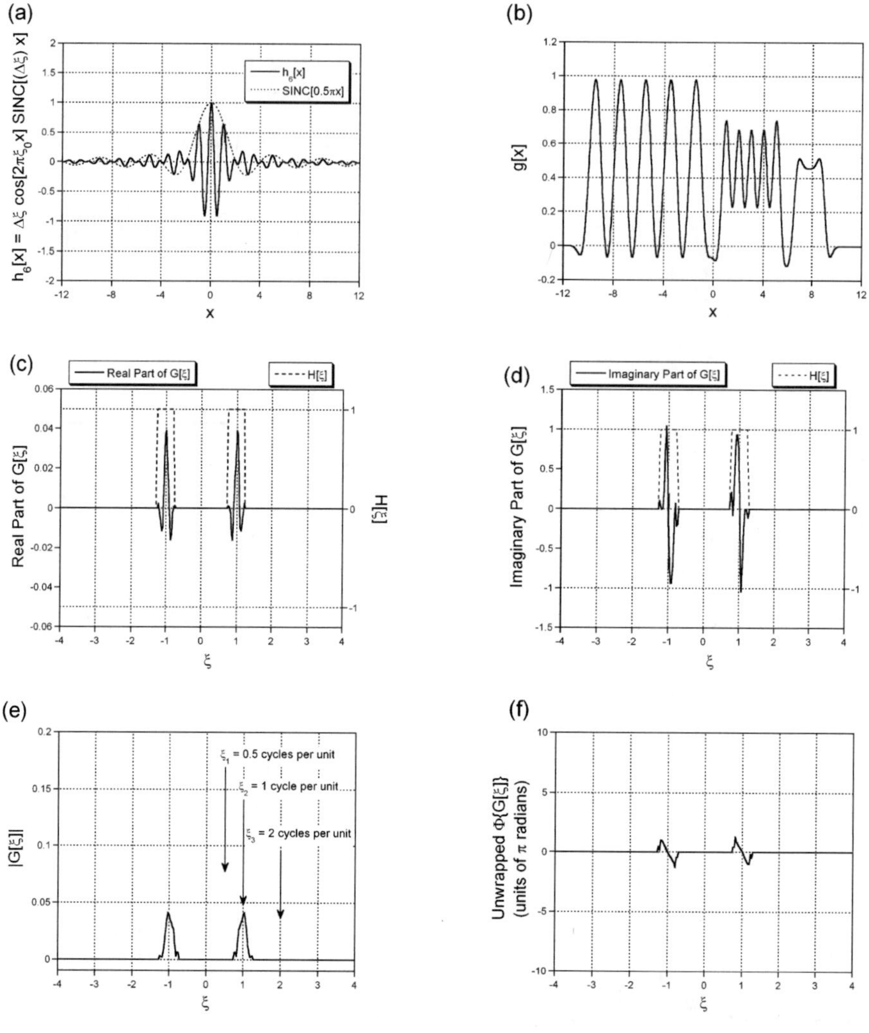

Figure 16.11 Bandpass filtering with $H[\xi] = RECT[(x+1)/0.5] + RECT[(x-1)/0.5]$: (a) $h[x] = \cos[2\pi x] \cdot SINC[x/2]$, with *SINC* modulation shown as dashed line; (b) $g[x]$, showing the bipolar character and the visibility of the dominant spatial frequency with $\xi_2 = 1$ cycle per unit; (c) $\Re\{G[\xi]\}$; (d) $\Im\{G[\xi]\}$; (e) $|G[\xi]|$; and (f) $\Phi\{G[\xi]\}$.

If the passband function $R[\xi/\Xi]$ is a Gaussian with width parameter Ξ, then the transfer function and impulse response are:

$$H_{10}[\xi] = GAUS\left[\frac{\xi}{\Xi}\right] * \left(|\xi_0|\,\delta\delta\left[\frac{\xi}{\xi_0}\right]\right) \quad \xi_0 < \Xi \tag{16.51a}$$

$$h_{10}[x] = (|\Xi| GAUS[\Xi \cdot x]) \cdot (2\cos[2\pi\xi_0 x]) = 2|\Xi|\, GAUS\left[\frac{x}{\Xi^{-1}}\right] \cdot \cos[2\pi\xi_0 x] \tag{16.51b}$$

The DC component of the transfer function will be positive due to the slow decay of the Gaussian passbands. A particular example:

$$H_{11}[\xi] = GAUS\left[\xi - \frac{1}{2}\right] + GAUS\left[\xi + \frac{1}{2}\right] \tag{16.52a}$$

$$h_{11}[x] = 2\cos\left[2\pi\frac{x}{2}\right] \cdot GAUS[x] \tag{16.52b}$$

is illustrated in Figure 16.12.

This transfer function may be used as an approximate model of the response of the human visual system (HVS), though such a model is not strictly valid because the HVS actually is not linear. That said, this model has some use if the eye has adapted to a small range of "brightnesses" so that the HVS is approximately linear. A measurement of the eye response under these conditions leads to some valid conclusions. A simple experiment demonstrates that the response (i.e., the output modulation) of the HVS is larger than zero at DC, increases to its peak response at a frequency of about six cycles per angular degree ($\cong$ 343 cycles per radian), and falls off at larger spatial frequencies. The corresponding impulse response resembles that shown in Figure 16.12. In this model, the impulse response of the HVS is positive at the origin, but becomes *negative* within a short distance. In words, this indicates that the response of an eye receptor to a bright source *inhibits* the response of nearby receptors. Therefore, the other receptors must be stimulated more strongly to elicit the same response as the first receptor. The HVS "processor" that generates this response is the network of neural connections behind the retina, the so-called *visual neural net*. We can now use this model for the HVS to calculate the visual response to a specific input image. If the stimulus is a simple "stairstep" input irradiance, with regular steps of increasing brightness, the output illustrates the phenomenon of *Mach bands*, which result in a nonuniform visual appearance of uniform areas in the vicinity of a transition in gray level. The eye perceives an "overshoot" of the actual brightness distribution, and so the action of the HVS enhances edges, as shown in Figure 16.13.

A single-sideband bandpass filter has the form:

$$H[\xi] = R\left[\frac{\xi - \xi_0}{\Xi}\right] \tag{16.53a}$$

where $\xi_0 \neq 0$. The corresponding impulse response is:

$$h[x] = \left(\Xi \cdot r\left[\frac{x}{\Xi^{-1}}\right]\right) \cdot e^{+2\pi i \xi_0 x} = (\Xi \cdot r[\Xi \cdot x]) \cdot (\cos[2\pi\xi_0 x] + i\,\sin[2\pi\xi_0 x]) \tag{16.53b}$$

If the bandpass function $R[\xi/\Xi]$ is real and even, so must be the impulse response $r[\Xi \cdot x]$. The real and imaginary parts of the transfer function are copies of $r[\Xi \cdot x]$ modulated by sinusoids in quadrature.

A more general form of a double-passband filter may be constructed from an arbitrary (and even complex-valued) $R[\xi/\Xi]$. A symmetric transfer function may be constructed by modifying Equation (16.50) to ensure symmetry:

$$H[\xi] = R\left[\frac{\xi - \xi_0}{\Xi}\right] + R\left[\frac{-(\xi - \xi_0)}{\Xi}\right] = R\left[\frac{\xi - \xi_0}{\Xi}\right] + R\left[\frac{-\xi + \xi_0}{\Xi}\right] \tag{16.54a}$$

The corresponding impulse response is:

$$h[x] = \Xi \cdot \left(r\left[\frac{x}{\Xi^{-1}}\right] \cdot e^{-2\pi i \xi_0 x} + r\left[\frac{-x}{\Xi^{-1}}\right] \cdot e^{+2\pi i \xi_0 x}\right) \tag{16.54b}$$

which is still real valued and even but cannot be expressed as a cosine modulation of a real-valued function.

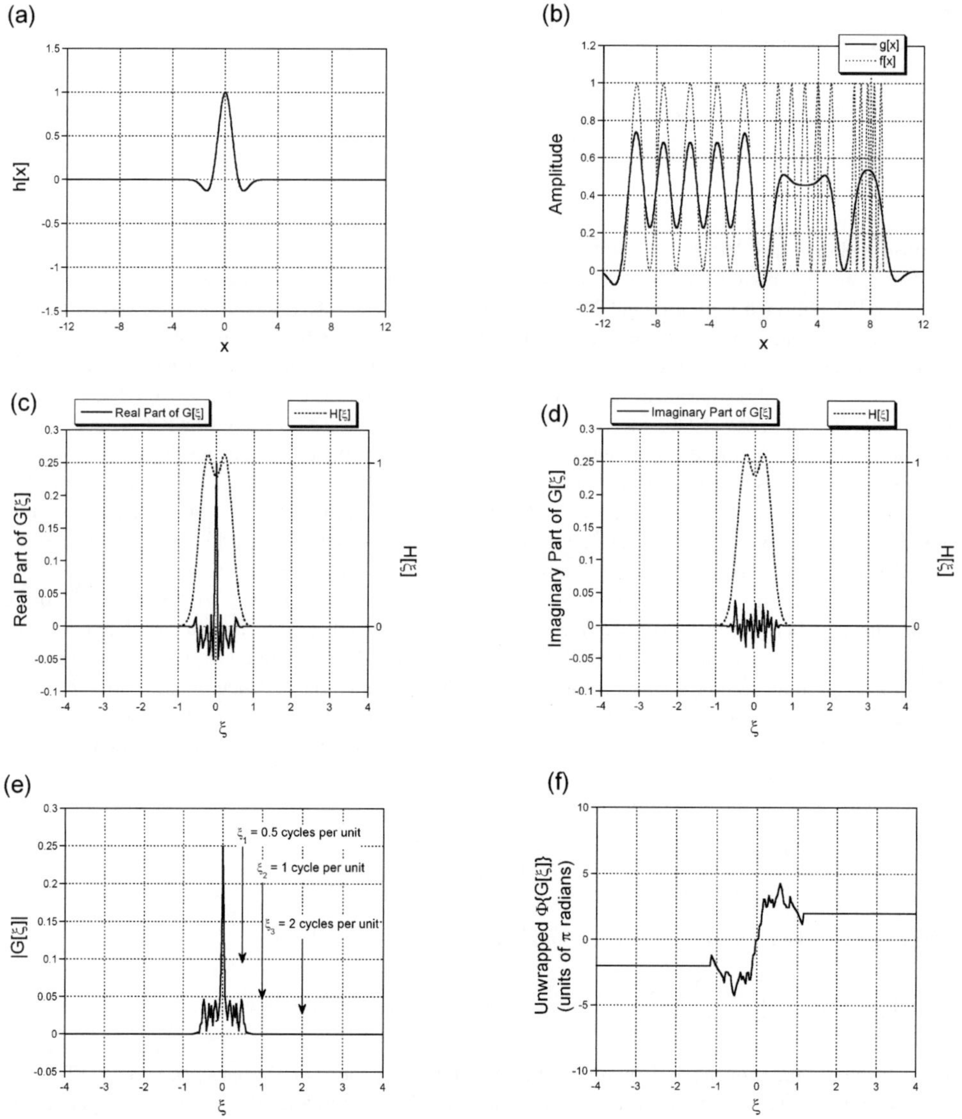

Figure 16.12 Gaussian "lowpass–bandpass" filter: (a) $h[x] = 2\cos[2\pi x] \cdot GAUS[x]$, which passes the DC component and slightly amplifies the lowest-frequency AC component; (b) $g[x]$; (c) $\Re\{G[\xi]\}$; (d) $\Im\{G[\xi]\}$; (e) $|G[\xi]|$; (f) $\Phi\{G[\xi]\}$.

16.7 Fourier Transform as a Bandpass Filter

Consider a bandpass filter whose passband function is a single Dirac delta function:

$$H_{12}[\xi] = \delta[\xi - \xi_0] \tag{16.55a}$$

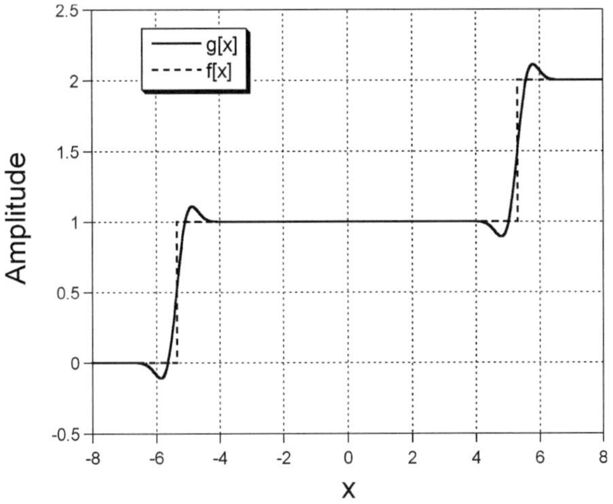

Figure 16.13 Effect of lowpass–bandpass filter on *STEP* function input $f[x]$ (shown as dashed line). The output $g[x]$ "overshoots" $f[x]$ at both sides of an edge transition, illustrating the edge "enhancement" of the human visual system.

which means that only a single spatial frequency component is passed to the image. The corresponding impulse response is:

$$h_{12}[x] = e^{+2\pi i\xi_0 x} \tag{16.55b}$$

We have already seen that this is an eigenfunction of the convolution operation. The fact that convolution commutes means that the output of the filter may be inferred directly from Equation (16.12). The space-domain representation is:

$$g_{12}[x] = f[x] * h_{12}[x] = \int_{-\infty}^{+\infty} f[\alpha]\, e^{+2\pi i\xi_0(x-\alpha)}\, d\alpha \tag{16.56a}$$

while the corresponding spectrum is:

$$\begin{aligned} G_{12}[\xi] &= F[\xi_0]\, e^{+2\pi i\xi_0 X} \\ &= (\Re\{F[\xi_0]\} + i\,\Im\{F[\xi_0]\}) \cdot (\cos[2\pi\xi_0 x] + i\,\sin[2\pi\xi_0 x]) \end{aligned} \tag{16.56b}$$

The real part of the output $g[x]$ includes contributions from both the real and imaginary parts of $F[\xi_0]$, which modulate a sinusoid of frequency ξ_0. The two sinusoids $\cos[2\pi\xi_0 x]$ and $\sin[2\pi\xi_0 x]$ are *in quadrature*.

In many imaging applications, particularly those involving optical systems, the measurable quantity is the squared magnitude of the signal. The power of the output of the filtering operation is the squared magnitude of $g[x]$, followed by a spatial average over a large area X:

$$\begin{aligned} \langle |g[x]|^2 \rangle &= \lim_{X\to\infty} \frac{1}{X} \int_{-X/2}^{+X/2} |g[x]|^2\, dx \\ &= \frac{1}{X} \int_{-X/2}^{+X/2} |F[\xi_0]|^2 (|\cos[2\pi\xi_0 x] + i\,\sin[2\pi\xi_0 x]|^2)\, dx \\ &= |F[\xi_0]|^2 \cdot \frac{1}{X} \cdot \int_{-X/2}^{+X/2} dx = |F[\xi_0]|^2 \end{aligned} \tag{16.57}$$

In words, the output of this single-sided bandpass filter is proportional to the spectral power in the original signal at ξ_0. This sequence of operations is the basis for the *filter-bank spectrum analyzer* that computes the power spectrum of an input signals by measuring the amplitude of each component frequency by frequency.

To evaluate the complex-valued Fourier component at each frequency, the output of the bandpass filter in Equation (16.56b) must be *demodulated* by multiplying by the complex conjugate of the complex sinusoidal modulation:

$$g_{12}[x] \cdot e^{-2\pi i\xi_0 x} = (e^{+2\pi i\xi_0 x}\ F[\xi_0]) \cdot e^{-2\pi i\xi_0 x} = \Re\{F[\xi_0]\} + i\,\Im\{F[\xi_0]\} \tag{16.58}$$

The desired result can be obtained by multiplying the real part of $g[x]$ separately by $\cos[2\pi\xi_0 x]$ and $\sin[2\pi\xi_0 x]$ and lowpass filtering:

$$\begin{aligned}
&\langle g_{12}[x]\cos[2\pi\xi_0 x]\rangle \\
&= \frac{1}{X}\int_{-X/2}^{+X/2} (\Re\{F[\xi_0]\}\cos[2\pi\xi_0 x] - \Im\{F[\xi_0]\}\sin[2\pi\xi_0 x])\cos[2\pi\xi_0 x]\,dx \\
&= \Re\{F[\xi_0]\}\frac{1}{X}\int_{-X/2}^{+X/2} \cos^2[2\pi\xi_0 x]\,dx - \Im\{F[\xi_0]\}\frac{1}{X}\int_{-X/2}^{+X/2} \sin[2\pi\xi_0 x]\cos[2\pi\xi_0 x]\,dx \\
&= \Re\{F[\xi_0]\}\cdot\frac{1}{2} - \Im\{F[\xi_0]\}\cdot 0
\end{aligned} \tag{16.59a}$$

$$\begin{aligned}
&\langle g_{12}[x]\sin[2\pi\xi_0 x]\rangle \\
&= \frac{1}{X}\int_{-X/2}^{+X/2} (\Re\{F[\xi_0]\}\cos[2\pi\xi_0 x] - \Im\{F[\xi_0]\}\sin[2\pi\xi_0 x])\sin[2\pi\xi_0 x]\,dx \\
&= \Re\{F[\xi_0]\}\frac{1}{X}\int_{-X/2}^{+X/2} \cos[2\pi\xi_0 x]\sin[2\pi\xi_0 x]\,dx - \Im\{F[\xi_0]\}\frac{1}{X}\int_{-X/2}^{+X/2} \sin^2[2\pi\xi_0 x]\,dx \\
&= \Re\{F[\xi_0]\}\cdot 0 - \Im\{F[\xi_0]\}\cdot\frac{1}{2}
\end{aligned} \tag{16.59b}$$

16.8 Bandboost and Bandstop Filters

The *bandboost* filter is analogous to the "high-frequency boost" filter described in Section 16.5.2. Its impulse response may be written as the sum of the original function and the output of a bandpass filter $h_{\text{bandpass}}[x]$:

$$\begin{aligned}
g_{\text{bandboost}}[x] &= f[x] + f[x] * h_{\text{bandpass}}[x] \\
&= f[x] * (\delta[x] + h_{\text{bandpass}}[x]) \\
&\equiv f[x] * h_{\text{bandboost}}[x]
\end{aligned} \tag{16.60}$$

The corresponding transfer function $H_{16}[\xi]$ of the bandboost filter is the sum of the unit constant and the transfer function of the bandpass filter:

$$H_{\text{bandboost}}[\xi] = \mathcal{F}_1\{\delta[x] + h_{\text{bandpass}}[x]\} = 1[\xi] + H_{\text{bandpass}}[\xi] \tag{16.61}$$

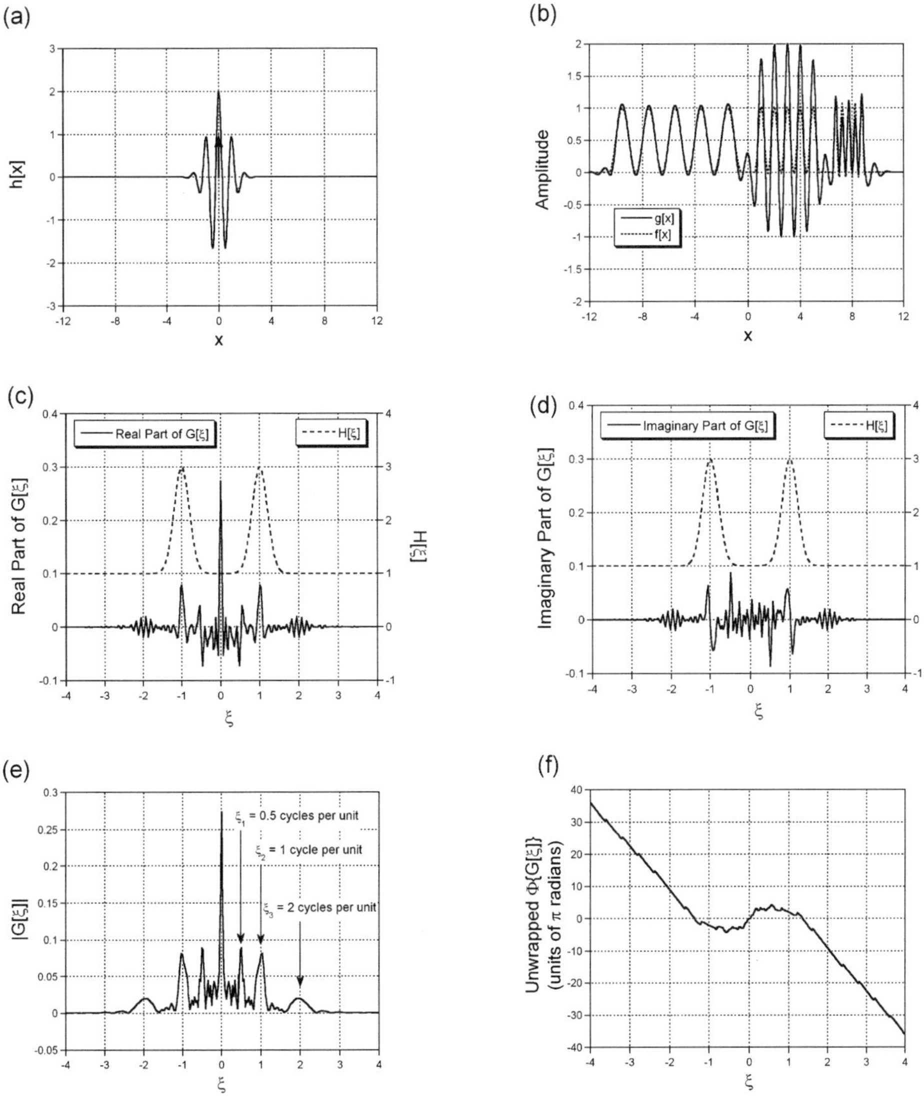

Figure 16.14 Bandboost filtering: (a) $h[x] = \delta[x] + 2\cos[2\pi x]GAUS[x/2]$, the corresponding transfer function is $H[\xi] = 1 + 2(GAUS[(\xi + 1)/0.5] + GAUS[(\xi - 1)/0.5])$; (b) $g[x]$, showing amplified signal with $\xi_1 = 1$ cycle per unit length; (c) $\Re\{G[\xi]\}$ with $H[\xi]$; (d) $\Im\{G[\xi]\}$ with $H[\xi]$; (e) $|G[\xi]|$; (f) $\Phi\{G[\xi]\}$.

The output of a 1-D bandboost filter for the mid-frequency bars applied to our 1-D sample function is shown in Figure 16.14. Such a filter can be used in image processing to accentuate structures with fixed spatial frequency content. For example, an aerial photograph of agricultural crops may be processed to emphasize crops that are planted with a specific separation (spatial period).

The complement of a bandpass filter is a *bandstop* or *notch* filter that prevents certain ranges of spatial frequencies from reaching the output. Such filters are useful for removing sinusoidal interference, e.g., 60 Hz "hum" from audio signals. The transfer function of a filter $H[\xi]$ that blocks the

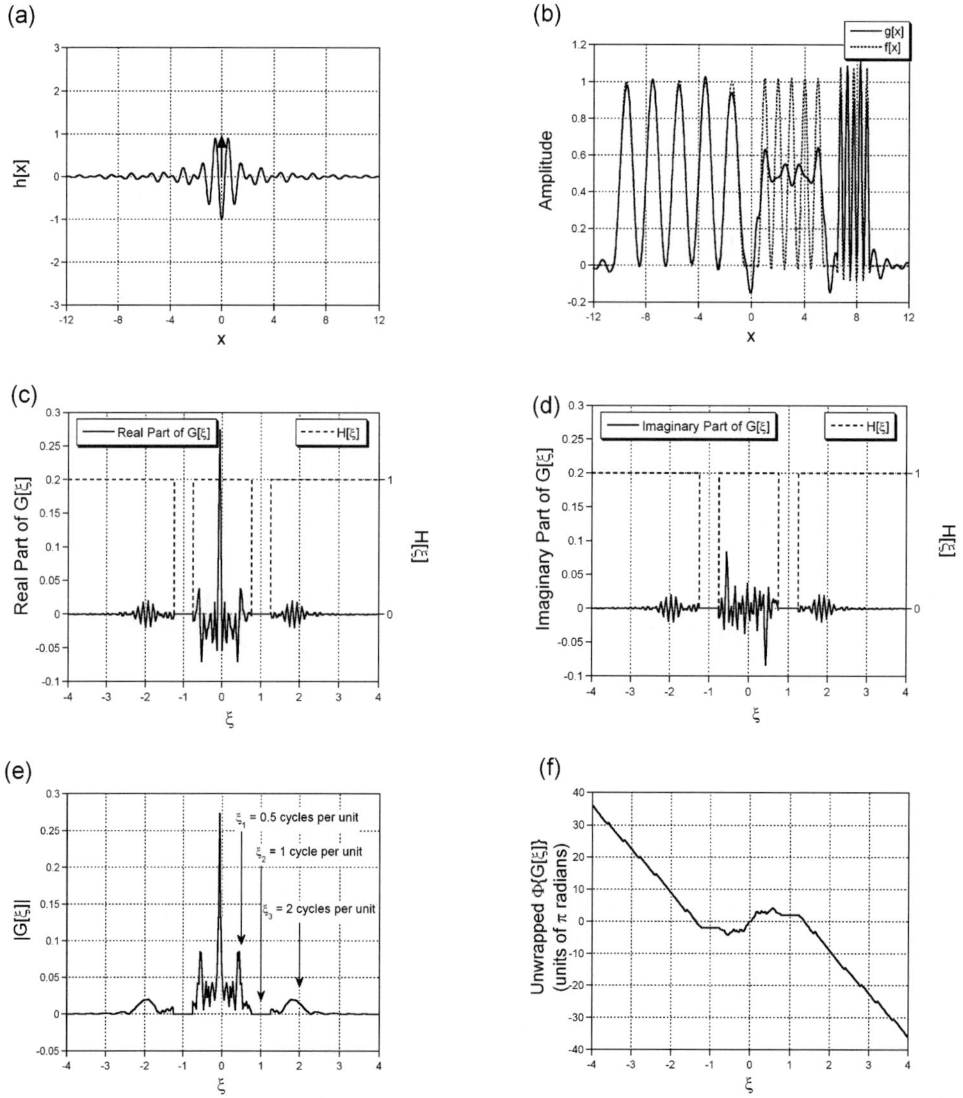

Figure 16.15 Bandstop filtering: (a) $h[x] = \delta[x] - 2\cos[2\pi x]\,SINC[x/2]$, the corresponding transfer function is $H[\xi] = 1 - 2(RECT[(\xi+1)/0.5] + RECT[(\xi-1)/0.5])$; (b) $g[x]$, showing attenuation of signal with $\xi_1 = 1$ cycle per unit length; (c) $\Re\{G[\xi]\}$ with $H[\xi]$; (d) $\Im\{G[\xi]\}$ with $H[\xi]$; (e) $|G[\xi]|$; (f) $\Phi\{G[\xi]\}$.

bands that were passed by $H[\xi]$ is:

$$H_{\text{bandstop}}[\xi] = 1 - H_{\text{bandpass}}[\xi] \tag{16.62a}$$

$$\Longrightarrow h_{\text{bandstop}}[x] = \delta[x] - h_{\text{bandpass}}[x] \tag{16.62b}$$

A bandstop filter for the mid-frequency bars in our sample is shown in Figure 16.15.

The filter has blocked the frequency corresponding to the fundamental of the center bars; note the reduced contrast of those structures. Also note that any "real" information at those frequencies is also removed at other locations in the scene, and thus the resulting image may not be a good representation of the original function.

16.9 Wavelet Transform

The wavelet transform is a straightforward generalization of bandpass filtering that has found wide application in recent years, especially to digital image compression. The filtered images obtained from a specific set of bandpass filters define the transformation. We will briefly discuss the wavelet transform in the 1-D case. Consider the transfer function of an ideal bandpass filter with center frequency ξ_1 and bandwidth Ξ_1 that has already been described:

$$\begin{aligned} H_1[\xi] &= RECT\left[\frac{\xi+\xi_1}{\Xi_1}\right] + RECT\left[\frac{\xi-\xi_1}{\Xi_1}\right] \\ &= RECT\left[\frac{\xi}{\Xi_1}\right] * (\delta[\xi+\xi_1] + \delta[\xi-\xi_1]) \\ &= RECT\left[\frac{\xi}{\Xi_1}\right] * \frac{1}{|\xi_1|}\delta\delta\left[\frac{\xi}{\xi_1}\right] \end{aligned} \tag{16.63}$$

The impulse response of this filter is a *SINC* function with width parameter Ξ_1 modulated by a cosine with frequency ξ_1:

$$\begin{aligned} h_1[x] &= \mathcal{F}_1^{-1}\{H[\xi]\} = |\Xi_1|\, SINC[\Xi_1 \cdot x] \cdot 2\cos[2\pi\xi_1 x] \\ &= 2|\Xi_1|(SINC[\Xi_1 \cdot x] \cdot \cos[2\pi\xi_1 x]) \end{aligned} \tag{16.64}$$

If the transfer function is scaled so that both the bandwidth and center frequency are increased by the same multiplicative factor b, the resulting impulse response is determined by the scaling theorem:

$$\begin{aligned} h_2[x] &= \mathcal{F}_1^{-1}\left\{RECT\left[\frac{\xi+b\cdot\xi_1}{b\cdot\Xi_1}\right] + RECT\left[\frac{\xi-b\cdot\xi_1}{b\cdot\Xi_1}\right]\right\} \\ &= (|b\ \Xi_1|\, SINC[b\cdot\Xi_1\cdot x]) \cdot (2\cos[2\pi(b\cdot\xi_1)x]) \\ &= 2b|\Xi_1|(SINC[\Xi_1\cdot(bx)] \cdot \cos[2\pi\xi_1\cdot(bx)]) \\ &= b\cdot h_1[bx] = b\cdot h_1\left[\frac{x}{b^{-1}}\right] \end{aligned} \tag{16.65}$$

In words, the impulse response of the bandpass filter whose center frequency and bandwidth are scaled by a factor b is a replica of the original impulse response with amplitude amplified by b and width scaled by b^{-1}; the "shape" of the impulse response is not changed. The factor Ξ_1/ξ_1 is identical in concept to the *quality factor* q used to describe the damped harmonic oscillator in classical mechanics. In this context, q is defined to be the dimensionless ratio of the bandwidth to the center frequency:

$$\frac{bandwidth}{center\ frequency} = \frac{\Xi_1}{\xi_1} \equiv q \tag{16.66}$$

It is possible to subdivide the entire frequency domain into a discrete set of nonoverlapping rectangular passbands with identical q factors. The filters will be indexed by k. In the jargon of computer graphics, the frequency axis is *tiled* by the transfer functions of the filters. The impulse responses $h_k[x]$ of the filters thus obtained are identical (but for scale). They also are *orthogonal*, which means that two additional criteria are satisfied:

$$h_k[x] * h_m[x] = 0 \quad for\ k \neq m \tag{16.67a}$$

$$h_k[x] * h_k[x] = h_k[x] \tag{16.67b}$$

Therefore, the set of bandpass-filtered output images $g_k[x]$ obtained from an input image $f[x]$ are themselves orthogonal:

$$g_k[x] * g_m[x] = 0 \quad for\ k \neq m \tag{16.68}$$

The individual bandpass impulse responses $h_k[x]$ are often called *wavelets*, and the set of images $g_k[x]$ is the decomposition of $g[x]$ into the orthogonal wavelets $h_k[x]$. The set of orthogonal images $g_k[x]$ defines the *wavelet transform* of $f[x]$. If the entire frequency axis is tiled by $h_k[x]$, the sum of the set of output images $g_k[x]$ generates the original input image $f[x]$:

$$f[k] = \sum_k g_k[x] = \sum_k (f[x] * h_k[x]) \tag{16.69}$$

Such a wavelet decomposition is *complete* because all spatial frequencies are included; any input image $f[x]$ is reconstructed without error from the set $\{g_k[x]\}$. The bandpass filtered images contain information about the location of the spatial frequencies in the passband, i.e., the images $g_k[x]$ for large k will emphasize the edges of $f[x]$.

It is also possible to construct wavelets $h_k[x]$ which are complete and not orthogonal by using bandpass filters with overlapping transfer functions. In this case, the output images are not orthogonal, in general. In other words, the crosscorrelation between output images $g_k[x]$ may be nonzero.

16.9.1 Tiling of Frequency Domain with Orthogonal Wavelets

The conditions that must be satisfied by complete sets of orthogonal scaled wavelets are easy to derive. The task requires that the frequency axis be tiled with nonoverlapping ideal bandpass filters with constant q. As just discussed, the discrete set of filters will be indexed by the integer k, so that the center, minimum, and maximum spatial (positively valued) frequencies within the passband are denoted by $(\xi_k)_C$, $(\xi_k)_{\min}$, and $(\xi_k)_{\max}$, respectively. To tile the entire plane, the maximum and minimum frequencies must satisfy:

$$(\xi_N)_{\max} = (\xi_{N+1})_{\min} \tag{16.70}$$

Also, to ensure that q is constant, the maximum and minimum frequencies of a particular band must satisfy the conditions:

$$\frac{(\xi_N)_{\max} + (\xi_N)_{\min}}{2} = (\xi_N)_{\max} - (\xi_N)_{\min} = (\xi_N)_0 \tag{16.71}$$

These conditions lead to the expression for the center frequency of the kth filter in terms of the center frequency of the filter with index $k = 0$. The derivation is straightforward, and yields the value for the kth-order filter:

$$(\xi_k)_C = \left(\frac{2+q}{2-q}\right)(\xi_{k-1})_C = \left(\frac{2+q}{2-q}\right)^k (\xi_0)_C \tag{16.72}$$

The center frequency of the zeroth-order filter is now a variable parameter. Note that the center frequency is indeterminate when $q = 2$, i.e., when the bandwidth is twice the center frequency. However, we can tile the 1-D frequency plane for other positive values of q. Consider the case where the quality factor $q = 1$, so that the bandwidth is equal to the center frequency, and assume that $(\xi_0)_C = 1$. The center frequency of the second-order filter is:

$$(\xi_1)_C = \left(\frac{2+1}{2-1}\right)(\xi_0)_C = 3 = (\Xi)_1 \tag{16.73}$$

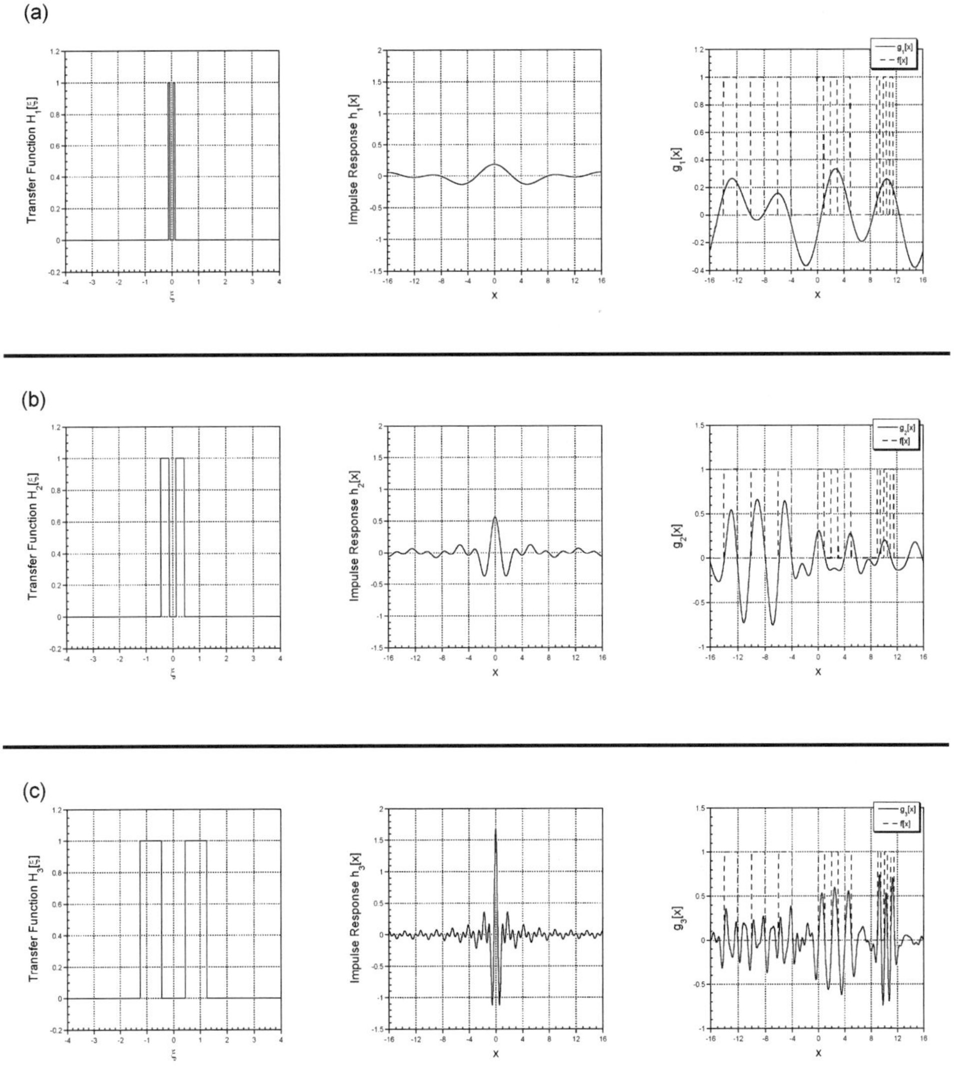

Figure 16.16 Wavelet transform with $q = 1$ for three-bar chart, evaluated for three scaled orthogonal wavelets. The transfer function, impulse response, and a comparison of the input and output signals are shown for each case: (a) the output generated by the lowest-frequency wavelet shows little information about $f[x]$; (b) the output of the "mid-frequency" wavelet shows the position of the coarsest bars; (c) the output of the next wavelet includes some modulation from the other two sets of bars.

and its maximum and minimum frequencies are

$$(\xi_1)_{\min} = (\xi_1)_C - \frac{(\Xi)_1}{2} = (\xi_1)_C - \frac{(\xi_1)_C}{2} = \frac{1}{2}(\xi_1)_C \tag{16.74a}$$

$$(\xi_1)_{\max} = (\xi_1)_C + \frac{(\Xi)_1}{2} = \frac{3}{2}(\xi_1)_C \tag{16.74b}$$

The transfer function and impulse response of the kth filter are:

$$H_k[\xi] = RECT\left[\frac{\xi + 3^k}{3^k}\right] + RECT\left[\frac{\xi - 3^k}{3^k}\right] \tag{16.75}$$

$$\begin{aligned} h_k[x] &= (3^k\ SINC[x \cdot 3^k]) \cdot (2\cos[2\pi \cdot 3^k \cdot x]) \\ &= (2 \cdot 3^k)\left(SINC\left[\frac{x}{3^{-k}}\right] \cdot \cos\left[2\pi\left(\frac{x}{3^{-k}}\right)\right]\right) \end{aligned} \tag{16.76}$$

These expressions are defined for all integer values of k, and thus define an infinite but countable set of bandpass filters. For negative k, the center frequency and bandwidth of the filters are quite small and thus little energy is transmitted from the input function $f[x]$ to the output functions $g_k[x]$.

16.9.2 Example of Wavelet Decomposition

Consider the wavelet transform of a three-bar chart for $q = 1$; the space-domain representations of the object, filters, and outputs are shown in Figure 16.16. Note that the x-axis is scaled differently for the impulse responses and for the input and outputs. The amplitudes of the filters are normalized to the same scale, and illustrate the increase in amplitude with index k because the area of the transfer function increases as 3^k.

PROBLEMS

16.1 Find the outputs of the 1- D systems with the listed system functions when applied to the input function:

$$f[x] = COMB[x] * RECT[2x]$$

You may use appropriate approximations if necessary.

- **(a)** $H[\xi] = RECT[\xi]$
- **(b)** $h[x] = RECT[x]$
- **(c)** $H[\xi] = -RECT[4\xi + 2] + RECT[4\xi - 2]$
- **(d)** $H[\xi] = 1 + \xi^2$

16.2 Find the transfer function and impulse response of a bandpass filter that passes only the first-order spatial frequencies of:

$$f[x] = \frac{1}{2} COMB\left[\frac{x}{2}\right] * RECT[x]$$

Sketch the output of the filter.

16.3 Find the output (exact or approximate) of the operation:

$$\cos\left[\pi\left(\frac{x}{2}\right)^2\right] * RECT[2x]$$

16.4 Derive the sequence of center frequencies and passbands for the 1-D wavelets for $q = 3$.

17

Allpass (Phase) Filters

According to the classification scheme for filters adopted in Chapter 16, a filter that affects only the phase of the input spectrum is an *allpass* or *phase* filter whose MTF is a unit constant. Such a filter may translate the individual sinusoidal components of $f[x]$ by the number of radians specified by the phase transfer function (ΦTF) at that frequency, but does not affect the magnitude of these components. The shift theorem of the Fourier transform in Equation (9.123) is an example of one flavor of allpass filter. In words, translation adds some number of radians to the initial phase that is proportional to the spatial frequency ξ. This ensures that each sinusoidal component of $f[x]$ is translated by the same physical distance relative to the origin, and consequently preserves the relative positions of the constituent sinusoids. From this interpretation, it is not difficult to envision the effect of a phase transfer function that is a nonlinear function of ξ, which will translate the component sinusoids of $f[x]$ by different physical distances and not preserve their relative positions. After a nonlinear phase filter, the sinusoidal constituents of $f[x]$ generally do not "line up" correctly, thus distorting high-frequency structures in the image, such as edges. In short, allpass filters with nonlinear phase transfer functions may have significant detrimental effects on the "quality" of the resulting output image $g[x]$.

In this chapter, the actions of specific allpass filters are described and the corresponding impulse responses are derived. From there, it is easy to use the "transform-of-a-transform" theorem derived in Equation 9.105 to derive spectra of "allpass signals" with space-domain representations of the form $f[x] = 1[x]\, e^{i\Phi[x]}$. The results of these discussions will be applied to several imaging problems, including the development of a useful algorithm for computing the Fourier transform of input images based on quadratic-phase functions, a method for creating "intensity" images of phase objects, and methods for extracting information from images made through transparent distorting media.

The action of the general filter on the magnitude and phase of the input spectrum is apparent from Equation (16.3):

$$\begin{aligned} G[\xi] = F[\xi] \cdot H[\xi] &= |F[\xi]| \cdot |H[\xi]| \exp[+i \cdot (\Phi\{F[\xi]\} + \Phi\{H[\xi]\})] \\ &= (|F[\xi]| \cdot |H[\xi]|) \exp[+i \cdot (\Phi\{F[\xi]\} + \pi W[\xi])] \end{aligned} \tag{17.1}$$

where $\Phi\{H[\xi]\} \equiv \pi W[\xi] = \tan^{-1}[\Im\{H[\xi]\}/\Re\{H[\xi]\}]$. The factor of π has been extracted from the phase transfer function for convenience in the forthcoming analysis. This scaled phase transfer function $W[\xi]$ specifies the number of "half cycles" that the sinusoidal component is translated by the action of the filter at spatial frequency ξ. The range of $W[\xi]$ is $-\infty < W < +\infty$ for "unwrapped" phase, but $-1 \leq W < +1$ in our convention if the phase is not unwrapped.

Because the MTF of an allpass filter is unity at all frequencies, the magnitude of the input spectrum is preserved by the action of the filter:

$$|G[\xi]| = |F[\xi]| \cdot |H[\xi]| = |F[\xi]| \cdot 1[\xi] = |F[\xi]| \tag{17.2}$$

and therefore the power spectra of the input and output functions also must be identical:

$$|G[\xi]|^2 = |F[\xi]|^2 \tag{17.3}$$

Application of the Wiener–Khintchin theorem of Equation (9.159) produces a simple relationship between the space-domain representations of the input and output on all allpass filters:

$$\begin{aligned}&\mathcal{F}_1^{-1}\{|G[\xi]|^2\} = \mathcal{F}_1^{-1}\{|F[\xi]|^2\} \quad \textit{for an allpass filter}\\ &\Longrightarrow g[x] \bigstar g[x] = g[x] * g^*[-x] = f[x] * f^*[-x] = f[x] \bigstar f[x]\end{aligned} \tag{17.4}$$

In words, allpass filters have no effect on the autocorrelation of the signal, which means in turn that the autocorrelation of the impulse response of any allpass filter must be a Dirac delta function:

$$\begin{aligned}&\mathcal{F}_1^{-1}\{|H[\xi]|^2\} = \mathcal{F}_1^{-1}\{1[\xi]\}\\ &\qquad\Longrightarrow h[x] \bigstar h[x] = h[x] * h^*[-x] = \delta[x]\end{aligned} \tag{17.5}$$

Because the power spectrum is preserved by the action of an allpass filter, Parseval's theorem for Equation (9.163) may be invoked to show that the integrated power in the space domain is so also:

$$\textit{if } |F[\xi]|^2 = |G[\xi]|^2 \tag{17.6a}$$

$$\Longrightarrow \int_{-\infty}^{+\infty} |F[\xi]|^2 \, d\xi = \int_{-\infty}^{+\infty} |G[\xi]|^2 \, d\xi \tag{17.6b}$$

$$\Longrightarrow \int_{-\infty}^{+\infty} |f[x]|^2 \, dx = \int_{-\infty}^{+\infty} |g[x]|^2 \, dx \tag{17.6c}$$

It is important to note that the power at a specific coordinate x_0 is generally not preserved, but rather is redistributed over the space domain.

17.1 Power-Series Expansion for Allpass Filters

In general, the scaled phase transfer function $W[\xi]$ of the general allpass filter may have any shape, including discontinuities and singularities at any number of frequencies. However, the analysis is simplified significantly and still produces useful results if $W[\xi]$ is assumed to be a "smoothly varying" function of ξ, so that all of the derivatives exist and are finite. Under those conditions, $W[\xi]$ may be expanded into a Taylor series in positive powers of ξ to obtain an expression in terms of the phase transfer function and its derivatives evaluated at the origin:

$$\begin{aligned}\frac{1}{\pi} \cdot \Phi\{H[\xi]\} = W[\xi] &= a_0 + a_1\xi + a_2\xi^2 + \cdots\\ &= \sum_{n=0}^{+\infty} a_n \xi^n\end{aligned} \tag{17.7}$$

where $a_n = (1/n!)(d^n W[\xi]/d\xi^n)|_{\xi=0}$. The coefficients a_n in the expansion are real valued and generally bipolar. The expansion may be recast into a more convenient form by redefining the expansion coefficients to be $\alpha_n \equiv (a_n)^{1/n}$:

$$W[\xi] = a_0 + (a_1\xi)^1 + ((a_2)^{\frac{1}{2}}\xi)^2 + ((a_3)^{\frac{1}{3}}\xi)^3 + \cdots$$
$$\equiv \alpha_0 + (\alpha_1\xi)^1 + (\alpha_2\xi)^2 + (\alpha_3\xi)^3 + \cdots \tag{17.8}$$

Since the a_n may be negative, the corresponding α_n may be imaginary. Note that α_0 is dimensionless and that α_n has dimensions of length for $n \geq 1$. This expansion for $W[\xi]$ may be substituted into the general expression for the transfer function of an allpass phase filter:

$$H[\xi] = 1[\xi]\exp[+i\pi W[\xi]] = \exp\left[+i\pi \sum_{n=0}^{+\infty}(a_n\xi^n)\right]$$
$$= e^{+i\pi\alpha_0} \cdot e^{+i\pi\alpha_1\xi} \cdot e^{+i\pi(\alpha_2\xi)^2} \cdot e^{+i\pi(\alpha_3\xi)^3} \cdot \cdots$$
$$= \prod_{n=0}^{+\infty} \exp[+i\pi(\alpha_n\xi)^n] \tag{17.9}$$

In words, an allpass transfer function whose phase $W[\xi]$ is a "smoothly varying" continuous function may be expressed as the product of component transfer functions for each integer power of the spatial frequency. The filter theorem may be applied to this result to construct the impulse response as the convolution of the individual allpass impulse responses for each order in the expansion:

$$h[x] = \mathcal{F}_1^{-1}\{e^{+i\pi W[\xi]}\}$$
$$= \mathcal{F}_1^{-1}\{e^{+i\pi\alpha_0}\} * \mathcal{F}_1^{-1}\{e^{+i\pi\alpha_1\xi}\} * \mathcal{F}_1^{-1}\{e^{+i\pi(\alpha_2\xi)^2}\} * \mathcal{F}_1^{-1}\{e^{+i\pi(\alpha_3\xi)^3}\} * \cdots \tag{17.10}$$

In this chapter, the corresponding impulse responses for the first few powers of ξ will be derived. The phases of the transfer functions with $n = 0$, 1, and 2 are respectively constant, linear, and quadratic functions of ξ and the Fourier transform pairs needed to derive these impulse responses were considered in Chapter 9. The impulse responses of higher-order filters will be derived by applying the asymptotic forms based upon the moment theorem and the stationary-phase approximations from Chapter 13. This will be followed by a discussion of allpass filters with stochastic (random) phase.

17.2 Constant-Phase Allpass Filter

The simplest allpass filter has zero phase at all frequencies, with transfer function:

$$H[\xi] = 1[\xi]\, e^{i \cdot 0[\xi]} \tag{17.11a}$$

and the corresponding impulse response clearly is a Dirac delta function located at the origin of coordinates:

$$h[x] = \delta[x] \tag{17.11b}$$

In words, this transform pair demonstrates that a system whose impulse response is an "on-axis" impulse passes the full amplitude at all spatial frequencies with no change of phase, and therefore the output is identical to the input:

$$g[x] = f[x] * h[x]$$
$$= f[x] * \delta[x] = f[x] \tag{17.12}$$

No surprise here; if all spatial frequencies are passed to the output without change, the output must be identical to the input. However, if the phase of the input spectrum is altered even without changing the magnitudes, the input object and output image must be different. Trivially, the phase transfer function of the zero-phase allpass filter in Equation (17.11) may be cast in the form of Equation (17.9) simply by setting $\alpha_n = 0$ for all n.

A slightly more general version of allpass filter increments the phase of all frequency components in the spectrum by some constant number of radians. In Equation (17.9), this means that $\alpha_0 \neq 0$ and $\alpha_n = 0$ for $n \geq 1$. The corresponding impulse response is computed trivially via the multiplication-by-constant theorem in Equation (9.102). Since $e^{+i\pi\alpha_0}$ generally is a complex-valued constant, we have:

$$H_0[\xi] = 1[\xi]\, e^{+i\pi\alpha_0} \tag{17.13a}$$

$$\Longrightarrow h_0[x] = \delta[x]\, e^{+i\pi\alpha_0} = (\cos[\pi\alpha_0] \cdot \delta[x]) + i \cdot (\sin[\pi\alpha_0] \cdot \delta[x]) \tag{17.13b}$$

The impulse response is a Dirac delta function with a complex-valued area located at the origin of coordinates. In words, the action of the constant-phase allpass filter is to redistribute the complex amplitude of the input object (or equivalently, of the input spectrum) between the real and imaginary parts without changing the magnitude, so that the regions of support of the input and output functions are equal. The equality of the input and output magnitudes is a stricter condition than that in Equation (17.6), which indicates that only the integrated squared magnitude (power) of the space-domain signal is preserved.

Figure 17.1 shows the impulse response and transfer function of a constant-phase allpass filter with $\alpha_0 = +0.794$. The transfer function, impulse response, and output for this filter are obtained by substitution:

$$H_0[\xi] = 1[\xi]\, e^{+i\pi(0.794)} \cong -0.798 + 0.643i \tag{17.14a}$$

$$\Longrightarrow h_0[x] \cong (-0.798 + 0.643i) \cdot \delta[x] \tag{17.14b}$$

The space-domain representation of the output signal is:

$$\begin{aligned} g[x] &= f[x] * h_0[x] \cong (\Re\{f[x]\} + i\Im\{f[x]\}) * (-0.798 + 0.643\, i)\, \delta[x] \\ &= (-0.798\, \Re\{f[x]\} - 0.643\Im\{f[x]\}) + i(0.643\Re\{f[x]\} - 0.798\Im\{f[x]\}) \end{aligned} \tag{17.15}$$

17.3 Linear-Phase Allpass Filter

The phase transfer function of the allpass filter in Equation (17.9) with $\alpha_1 \neq 0$ is proportional to ξ. The shift theorem determines that the impulse response is a Dirac delta function translated along the x-axis by a distance proportional to α_1:

$$H_1[\xi] = e^{+i\pi\alpha_1\xi} = e^{+2\pi i(\alpha_1/2)\xi} \tag{17.16a}$$

$$\Longrightarrow h_1[x] = \delta\left[x + \frac{\alpha_1}{2}\right] \tag{17.16b}$$

Consider the action on a sinusoid with frequency $|\xi|$ in the particular case of $\alpha_1 > 0$. The sinusoid is composed of positive and negative frequency components. The phase of the component with $\xi > 0$ is incremented by $\pi\alpha_1\xi$ radians while that with $\xi < 0$ is decremented by the same number.

In words, the magnitudes of all frequency components are transmitted to the output without change, but the linear dependence of the initial phase of the output spectrum on ξ effectively has translated the

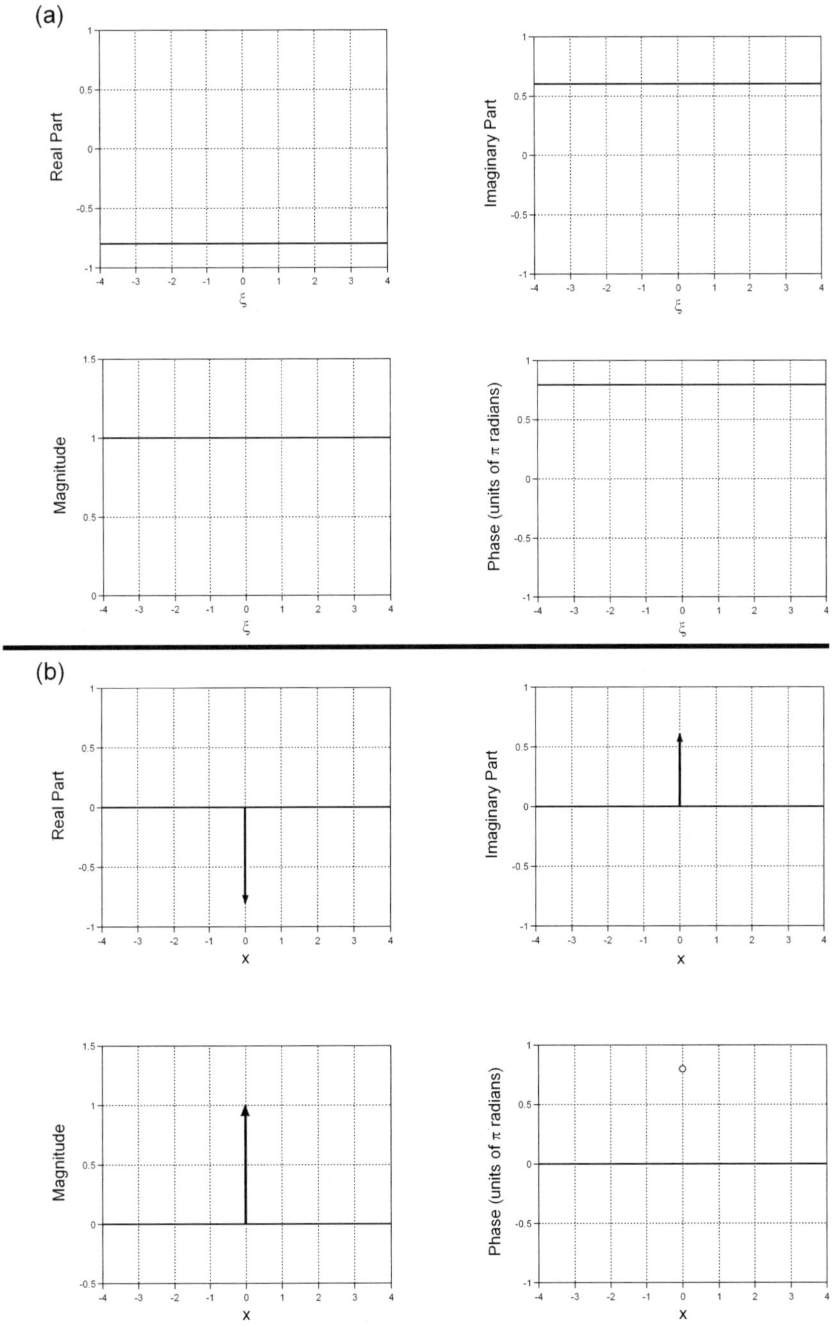

Figure 17.1 The constant-phase allpass filter: (a) transfer function $H_0[\xi] = \exp[+i\pi\alpha_0]$ as real part, imaginary part, magnitude, and phase; (b) impulse response as real part etc. The impulse response $h_0[x] = \exp[+i\pi\alpha_0] \cdot \delta[x]$ is a complex-valued Dirac delta function located at the origin with area $\exp[+i\pi\alpha_0]$.

origin of coordinates of the space-domain signal to $x_0 = -\alpha_1/2$. Therefore the output of the linear-phase allpass filter is an identical but translated replica of the input:

$$f[x] * h_1[x] = g[x] = f[x] * \delta\left[x + \frac{\alpha_1}{2}\right]$$
$$= f\left[x + \frac{\alpha_1}{2}\right] * \delta[x] = f\left[x + \frac{\alpha_1}{2}\right] \tag{17.17}$$

Obviously the size of the region of support is unchanged by the action of the linear-phase filter, though the limits of the support are changed. The transfer function and impulse response of a linear-phase allpass filter are shown in Figure 17.2.

From Equation (17.9), it is evident that the impulse response of an allpass phase filter with both constant and linear phase is the convolution of the individual impulse responses:

$$h_1[x] = \mathcal{F}_1^{-1}\{e^{+i\pi W[\xi]}\} = \mathcal{F}_1^{-1}\{e^{+i\pi\alpha_0}\} * \mathcal{F}_1^{-1}\{e^{+i\pi\alpha_1\xi}\}$$
$$= (e^{+i\pi\alpha_0}\,\delta[x]) * \delta\left[x + \frac{\alpha_1}{2}\right]$$
$$= e^{+i\pi\alpha_0}\,\delta\left[x + \frac{\alpha_1}{2}\right] \tag{17.18}$$

where the last step is an obvious application of the sifting property of the Dirac delta function from Equation (6.113). Thus the impulse response of this low-order allpass filter reduces to a single Dirac delta function that is scaled by a complex factor proportional to α_0 and translated by a distance proportional to α_1. The impulse response of this filter has infinitesimal support, and the autocorrelation obviously is preserved.

17.4 Quadratic-Phase Filter

17.4.1 Impulse Response and Transfer Function

The phase transfer function of what is arguably the most interesting allpass filter is a *quadratic* function of ξ. The interest in this filter is more than purely academic, as its applications in imaging range from optical imaging (Chapters 21–22), through holography, to synthetic-aperture radar (Chapter 23).

For a quadratic-phase filter, the only nonzero coefficient of the expansion in Equation (17.9) is α_2:

$$|H_2[\xi]| = 1[\xi]$$
$$\boldsymbol{\Phi}\{H_2[\xi]\} = \pm\pi(\alpha_2\xi)^2$$
$$\Longrightarrow H_2[\xi] = 1[\xi]\cdot\exp[\pi\alpha_2^2\xi^2] = \cos[\pi\alpha_2^2\xi^2] \pm i\,\sin[\pi\alpha_2^2\xi^2] \tag{17.19}$$

The quadratic-phase sinusoids were introduced in Section 6.1.11, where they were given the name of "chirp" because of their linear variation of the frequency. It was shown in Equation (6.55) that the local rate of oscillation of its real and imaginary parts are both linear functions of ξ. The representations of the transfer function are shown in Figure 17.3a.

The corresponding impulse response of the quadratic-phase allpass filter is derived by applying the complex scaling theorem of Equation (9.120) to Equation (9.95):

$$h_2[x] = \frac{1}{|\alpha_2|}\exp\left[\pm i\frac{\pi}{4}\right]\exp\left[\mp i\pi\left(\frac{x}{\alpha_2}\right)^2\right] = \frac{1}{|\alpha_2|}\exp\left[\mp i\pi\left(\left(\frac{x}{\alpha_2}\right)^2 - \frac{\pi}{4}\right)\right]$$
$$= \frac{1}{|\alpha_2|}\cos\left[\frac{\pi x^2}{\alpha_2^2} - \frac{\pi}{4}\right] \mp i\,\sin\left[\frac{\pi x^2}{\alpha_2^2} - \frac{\pi}{4}\right] \tag{17.20}$$

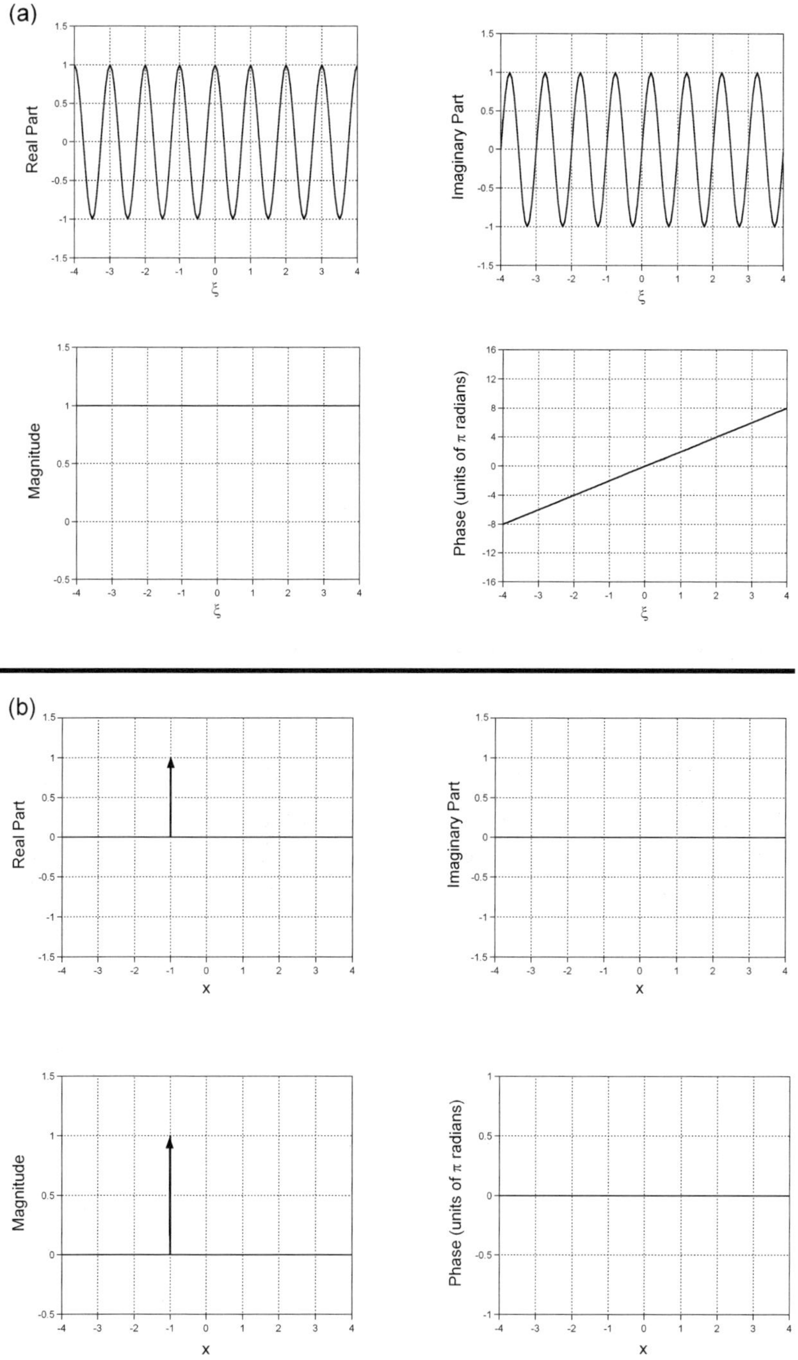

Figure 17.2 The linear-phase allpass filter: (a) transfer function $H_1[\xi] = e^{+i\pi\alpha_1\xi}$ as real part, imaginary part, magnitude, and phase, showing the linear variation in phase with ξ; (b) impulse response $h_1[x] = \delta[x + \alpha_1/2]$ as real part, imaginary part, magnitude, and phase. The impulse response is a single Dirac delta function located at $x_0 = -\alpha_2/2$.

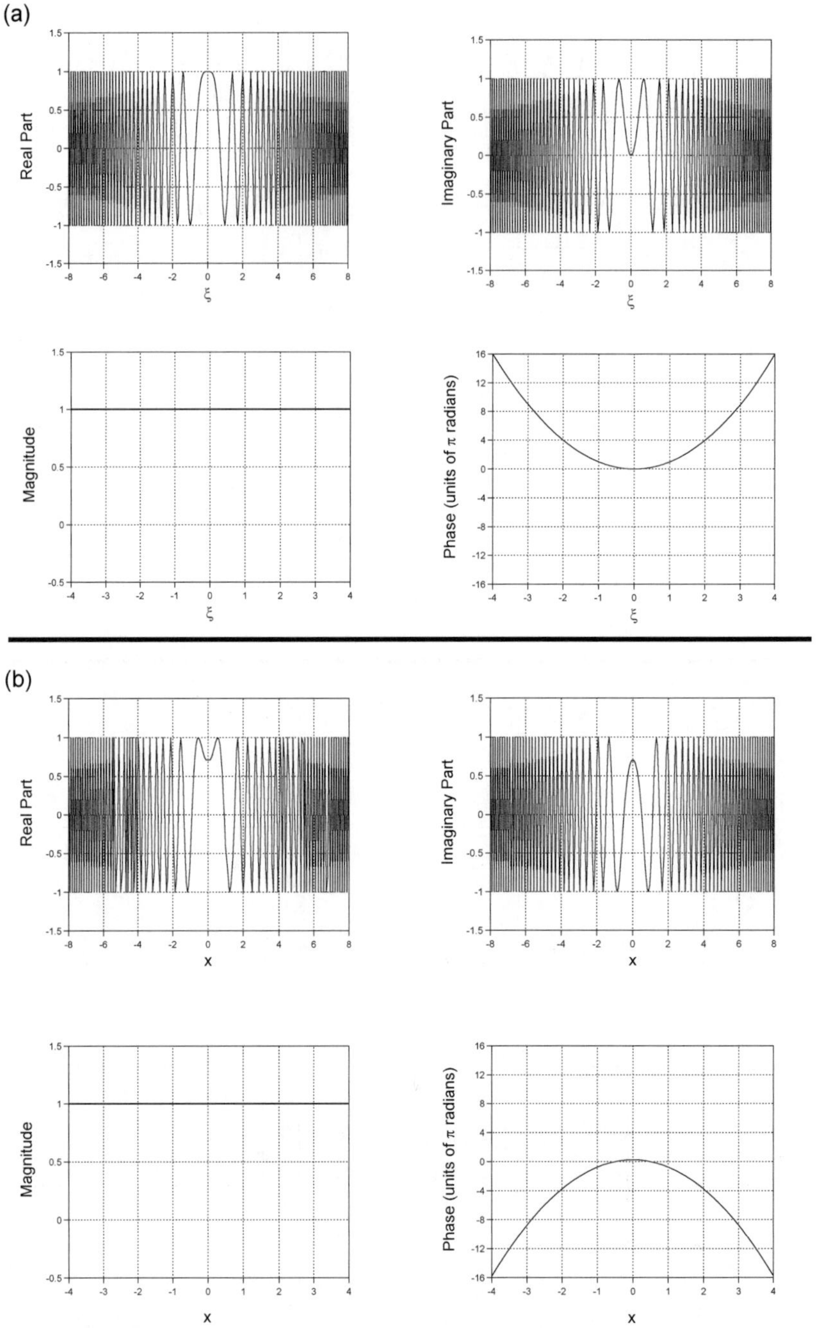

Figure 17.3 The quadratic-phase allpass filter: (a) transfer function $H_2[\xi] = \exp[+i\pi(\alpha_2\xi)^2]$, where $\alpha_2 = 1$, as real part, imaginary part, unit magnitude, and phase; (b) the impulse response $h_2[x] = (1/|\alpha_2|)\exp[+i\pi/4]\cdot\exp[-i\pi(x/\alpha_2)^2]$ as real part etc. Note that the magnitude is constant in both domains.

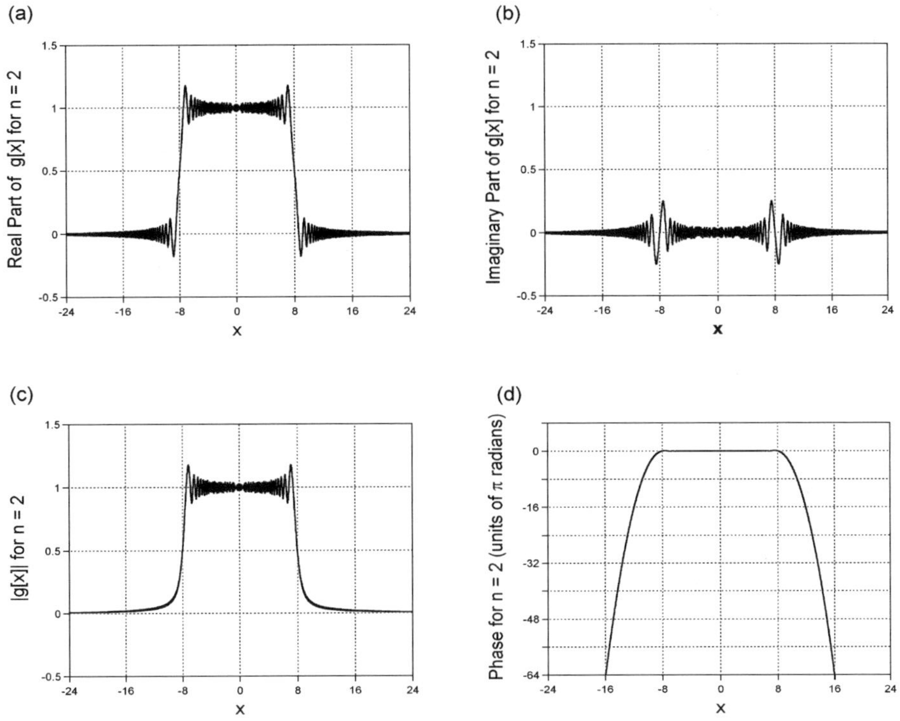

Figure 17.4 The image $g[x]$ of $f[x] = RECT[x/16]$ through an allpass filter with $\Phi\{H[\xi]\} = \pi\xi^2$: (a) real part; (b) imaginary part; (c) magnitude; (d) phase. Note that the magnitude of the "image" has infinite support, but is concentrated between $x = \pm 8$.

Both real and imaginary parts are symmetric complex-valued functions with unit magnitude and infinite support, as shown in Figure 17.3b.

We have noted that the phase of the transfer function $H_2[\xi]$ is approximately zero at the origin and increases slowly as $|\xi|$ grows. Therefore, the initial phase of low-frequency components is approximately unchanged, while larger increments that change rapidly with ξ are applied to higher-frequency components. If the phase increment ϕ_1 is added to a sinusoidal component with large spatial frequency ξ_1 ($|\xi_1| \gg 0$), then the phase added to the component with frequency $\xi_2 = \xi_1 + \Delta\xi \cong \xi_1$ will be very different from ϕ_1. In short, low-frequency components that produce the "bulk" character of objects in the scene appear at approximately the correct locations in $g[x]$, but high-frequency components that must "line up" to produce edges and other fine detail will not. For this reason, the low-frequency "structure" of the input object is generally more discernible in the filtered image $g[x]$ than the high-frequency structure, even though no frequency components have been attenuated. This observation illustrates why the action of the allpass quadratic-phase filter can resemble that of a lowpass filter (Figure 17.4). The common experience of defocusing an optical system approximates the action of the allpass quadratic-phase filter; the image looks "blurry" because the phases of large spatial frequencies are translated by larger distances than the low frequencies (it should be noted that optical systems actually must be lowpass filters because their aperture sizes must be finite, so this is not a perfect analogy). The characterization of optical systems in terms of the transfer function will be considered in Chapters 21–22.

The infinite-support quadratic-phase filter may be used to perform useful signal processing operations, including computation of the Fourier transform of a space-domain signal. This application will be considered in Section 17.8.

17.4.2 Scaling of Quadratic-Phase Transfer Function

The fact that the representations in both the space and frequency domains have infinite support may seem to violate the scaling theorem in Equation (9.117), but this is merely a shortcoming of naive intuition rather than fact. We are most familiar with functions that have compact support in one domain and infinite support in the other, rather than infinite support in both. In fact, the two representations of chirp functions do obey the scaling theorem based on the applicable metric of "width", which is the "chirp rate" α_2. As already mentioned in Section 6.1.11, α_2 has dimensions of length and specifies the coordinate x where the phase difference is $\pm\pi$ radians measured relative to the phase at $x = 0$. The reciprocal α_2^{-1} has dimensions of inverse length and specifies the smallest coordinate in the frequency domain for which the phase difference is $\pm\pi$ radians, again measured relative to the phase at $\xi = 0$. For large values of α_2, the space-domain chirp oscillates "slowly" near the origin, meaning that the rate of change of its oscillation frequency in the space domain is small. Conversely, the frequency-domain representation oscillates as α_2^{-1}, which is small if α_2 is large. This means that the oscillation period of the frequency-domain chirp will change by relatively large amounts over small changes in ξ. For illustration, consider the specific case $\alpha_2 = 2$. The impulse response and corresponding transfer function are:

$$h[x] = \exp\left[+i\left(\frac{\pi x^2}{2^2}\right)\right] \tag{17.21a}$$

$$\begin{aligned} H[\xi] &= |2| \exp\left[+i\frac{\pi}{4}\right] \exp[-i\pi \cdot (2\xi)^2] \\ &= 2 \exp\left[+i\frac{\pi}{4}\right] \exp\left[-i\pi \cdot \frac{\xi^2}{0.25}\right] \end{aligned} \tag{17.21b}$$

The real and imaginary parts of $H[\xi]$ and $h[x]$ are shown in Figure 17.5. Note the difference in the "oscillation rate" of the chirp functions in the two domains. A phase difference of π radians is measured in the transfer function between $\xi = 0$ and $\xi = 0.5$, while the phase difference of π radians occurs at $x = 0$ and $x = 2$ in the space domain.

Before analyzing the action of the allpass quadratic-phase filter, it is useful to derive some properties of its transfer function and impulse response. For example, the area of the transfer function is determined by the central-ordinate theorem in Equation (9.113):

$$\begin{aligned} \int_{-\infty}^{+\infty} H_2[\xi]\, d\xi &= \int_{-\infty}^{+\infty} e^{+i\pi(\alpha_2\xi)^2}\, d\xi = \frac{1}{|\alpha_2|}\, e^{+i\pi/4} \exp\left[-i\pi\left(\frac{x}{\alpha_2}\right)^2\right]\Bigg|_{x=0} \\ &= \frac{1}{|\alpha_2|}\, e^{+i\pi/4} = \frac{1}{|\alpha_2|}\left(\frac{1+i}{\sqrt{2}}\right) \end{aligned} \tag{17.22}$$

The symmetry of the chirp functions may be used to derive the corresponding semi-infinite integrals, and the results also may be decomposed into real and imaginary parts:

$$\begin{aligned} \int_{-\infty}^{+\infty} e^{+i\pi\alpha_2^2\xi^2}\, d\xi &= \int_{-\infty}^{+\infty} \cos[\pi\alpha_2^2\xi^2]\, d\xi + i\int_{-\infty}^{+\infty} \sin[\pi\alpha_2^2\xi^2]\, d\xi \\ &= 2\left(\int_0^{+\infty} \cos[\pi\alpha_2^2\xi^2]\, d\xi + i\int_0^{+\infty} \sin[\pi\alpha_2^2\xi^2]\, d\xi\right) \end{aligned} \tag{17.23}$$

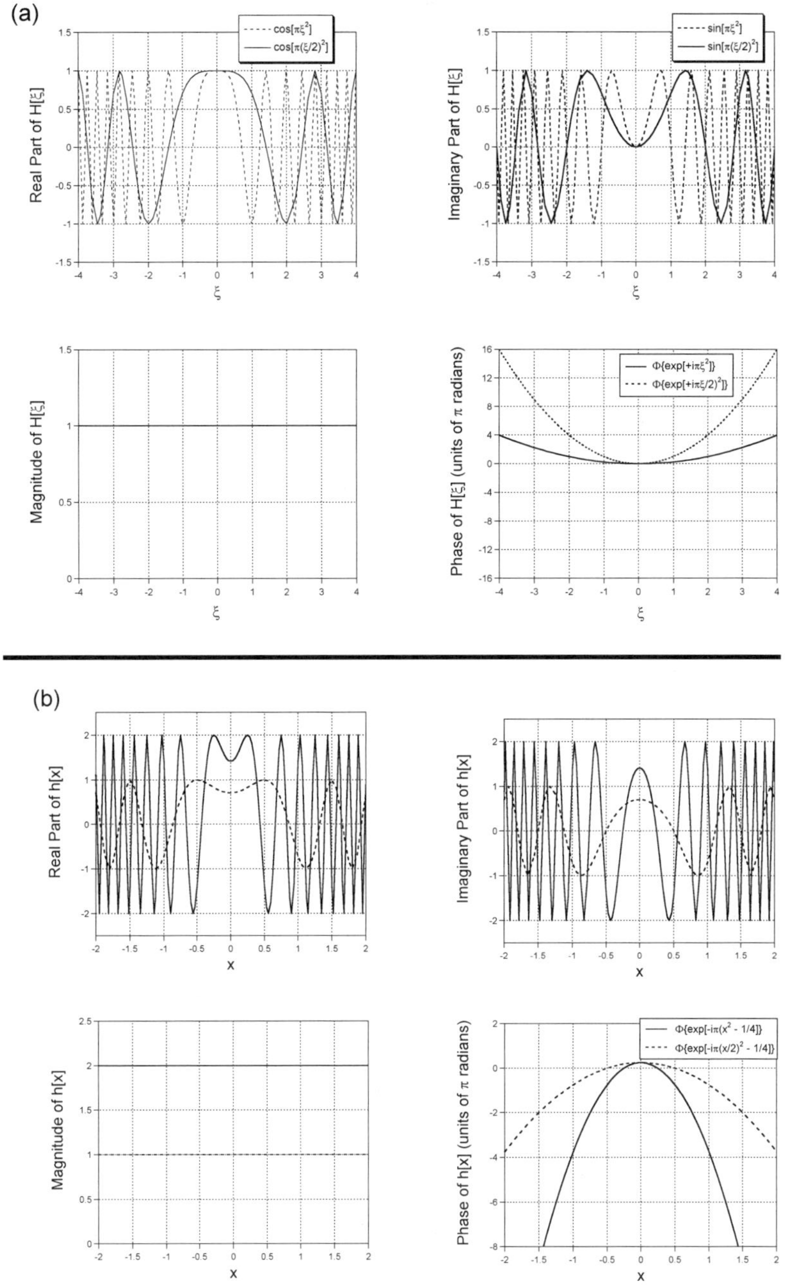

Figure 17.5 The scaling theorem applied to quadratic-phase allpass filter: (a) comparison of the transfer functions $H_2[\xi] = \exp[+i\pi(\alpha_2\xi)^2]$ for $\alpha_2 = 1$ (dashed line) $\alpha_2 = \frac{1}{2}$ (solid line) as real part, imaginary part, magnitude, and phase; (b) the corresponding impulse responses $h_2[x] = |1/\alpha_2|$ $\exp[+i\pi/4] \cdot \exp[-i\pi(x/\alpha_2)^2]$ for $\alpha_2 = 1$ (dashed line) and $\alpha_2 = \frac{1}{2}$ (solid line). The scaled impulse response exhibits larger magnitude and more rapidly changing phase.

The combination of Equation (17.22) and Equation (17.23) implies that:

$$\int_0^{+\infty} \cos[\pi(\alpha_2\xi)^2]\,d\xi = \int_0^{+\infty} \sin[\pi(\alpha_2\xi)^2]\,d\xi$$
$$= \frac{1}{2\sqrt{2}|\alpha_2|} \tag{17.24}$$

These two definite integrals were evaluated by Fresnel in his study of optical diffraction. The area of the transfer function is therefore complex valued and finite, even though both the transfer function and impulse response have infinite support. It may be instructive to compare this result to the area of the linear-phase sinusoid, which was shown to be zero by applying the central-ordinate theorem in Section 9.8.14.

Because the quadratic-phase filter is allpass, the autocorrelation of its impulse response must be a Dirac delta function. The autocorrelation may be written:

$$h_2[x] \bigstar h_2[x] = h_2[x] * h_2^*[-x]$$
$$= \frac{1}{|\alpha_2|}\,e^{+i\pi/4}\exp\left[-i\pi\left(\frac{x}{\alpha_2}\right)^2\right] * \frac{1}{|\alpha_2|}\,e^{-i\pi/4}\exp\left[+i\pi\left(\frac{x}{\alpha_2}\right)^2\right]$$
$$= \frac{1}{|\alpha_2|^2}\left(\exp\left[-i\pi\left(\frac{x}{\alpha_2}\right)^2\right] * \exp\left[+i\pi\left(\frac{x}{\alpha_2}\right)^2\right]\right) \tag{17.25}$$

This may be demonstrated very easily in the frequency domain by applying the filter theorem to show that the resulting spectrum is the unit constant:

$$\mathcal{F}_1\{h_2[x] \bigstar h_2[x]\} = H_2[\xi]\cdot H_2^*[\xi]$$
$$= \exp[+i\pi(\alpha_2\xi)^2]\cdot\exp[-i\pi(\alpha_2\xi)^2]$$
$$= 1[\xi] \tag{17.26}$$

where the transform of the complex conjugate from Equation (9.152) has been used. The space-domain representation of this function obviously is the Dirac delta function:

$$h_2[x] \bigstar h_2[x] = \mathcal{F}_1^{-1}\{1[\xi]\} = \delta[x] \tag{17.27}$$

The autocorrelations of both the infinitesimal-support Dirac delta function and the infinite-support quadratic-phase function are identically Dirac delta functions. The latter result may violate an already developed intuition that the support of the autocorrelation should be twice the support of the input, but we have already seen that this applies only if $h[x]$ has finite support. The more generally applicable statement of the width of the autocorrelation is made obvious by the Wiener–Khintchin theorem in Equation (9.159). The spectrum of the autocorrelation of the chirp is $|H[\xi]|^2 = 1[\xi]$ because $|H[\xi]| = 1[\xi]$. The region of support of the magnitude spectrum is the bandwidth of $f[x]$, and the support of the autocorrelation of the chirp is inversely proportional to its bandwidth. The result of Equation (17.27) is extended easily to derive the convolution of two chirp functions in the space domain with the same nonunit chirp rate α_2 but opposite sign:

$$\exp\left[-i\pi\left(\frac{x}{\alpha_2}\right)^2\right] * \exp\left[+i\pi\left(\frac{x}{\alpha_2}\right)^2\right] = |\alpha_2|^2\,\delta[x] \tag{17.28}$$

This result also applies in the frequency domain:

$$\exp[-i\pi(\alpha_2\xi)^2] * \exp[+i\pi(\alpha_2\xi)^2] = \frac{1}{|\alpha_2|^2}\,\delta[\xi] \tag{17.29}$$

17.4.3 Limiting Behavior of the Quadratic-Phase Allpass Filter

Further insight into the quadratic-phase filter may be gained by examining its behavior as the chirp rate $\alpha_2 \to 0$:

$$H_2[\xi] = \lim_{\alpha_2 \to 0} \{\exp[+i\pi\alpha_2^2\xi^2]\} = \lim_{\alpha_2 \to 0} \left\{ \exp\left[+\pi \left(\frac{\xi}{\alpha_2^{-1}} \right)^2 \right] \right\} \tag{17.30}$$

In words, the scale factor of the transfer function gets "wider", so that its phase varies more slowly with ξ. The corresponding impulse response is:

$$h_2[x] = \lim_{\alpha_2 \to 0} \left\{ \frac{1}{|\alpha_2|} \exp\left[-i\pi \left(\left(\frac{x}{\alpha_2} \right)^2 - \frac{1}{4} \right) \right] \right\} \tag{17.31}$$

As α_2 decreases and α_2^{-1} increases, the amplitude of $h_2[x]$ increases and its scale factor decreases. As the limit is approached, the frequency-domain chirp function approximates the unit constant $1 + 0i$ and $h_2[0]$ approaches infinity. However, the area of the space-domain chirp remains fixed at unity because $H_2[0] = 1$ regardless of the value of α_2. Thus the limiting expression for the impulse response $h_2[x]$ satisfies the requirements for the Dirac delta function in the limit, as expected.

If the relationship between the space- and frequency-domain representations of the ideal allpass quadratic-phase filter is counterintuitive, that of the more realistic finite-support quadratic-phase filter is even more so. These filters will be described in some detail in Chapter 18, and are very important because they emulate the action of an out-of-focus optical imaging system.

17.4.4 Impulse Response of Allpass Filters of Order 0, 1, 2

The impulse response of the allpass filter whose phase may be described by a superposition of constant, linear, and quadratic components is the convolution of the three component impulse responses. The result is an appropriately scaled and translated quadratic-phase exponential, as shown in Figure 17.6. The filter changes the distribution of amplitude among the real and imaginary parts, translates the function, and seems to "blur" it because the high-frequency components are displaced more than the low-frequency components:

$$\begin{aligned} h[x] &= \mathcal{F}_1^{-1}\{e^{+i\pi W[\xi]}\} \\ &= \mathcal{F}_1^{-1}\{e^{+i\pi\alpha_0}\} * \mathcal{F}_1^{-1}\{e^{+i\pi\alpha_1\xi}\} * \mathcal{F}_1^{-1}\{e^{+i\pi(\alpha_2\xi)^2}\} \\ &= e^{+i\pi\alpha_0}\,\delta[x] * \delta\left[x + \frac{\alpha_1}{2}\right] * \frac{1}{|\alpha_2|}\, e^{+i\pi/4}\, e^{-i\pi[x/\alpha_2]^2} \\ &= \frac{1}{|\alpha_2|} \exp\left(+i\pi \left[\alpha_0 - \left(\frac{x + \alpha_1/2}{\alpha_2} \right)^2 + \frac{1}{4} \right] \right) \end{aligned} \tag{17.32}$$

17.5 Allpass Filters with Higher-Order Phase

The transfer functions of allpass filters with phase of order 3 or larger are frequency-domain "superchirps", which were introduced in Equation (6.150):

$$H_n[\xi] = e^{+i\pi(\alpha_n\xi)^n} = \cos[\pi(\alpha_n\xi)^n] + i\,\sin[\pi(\alpha_n\xi)^n], \quad n > 2 \tag{17.33}$$

We observed in Chapter 6 that these functions are even when n is an even number and Hermitian (even real, odd imaginary) when n is odd. In all cases, the amplitude of the even real part near $\xi = 0$ is the cosine of a small argument and thus approximately unity. The width of this "constant" region increases

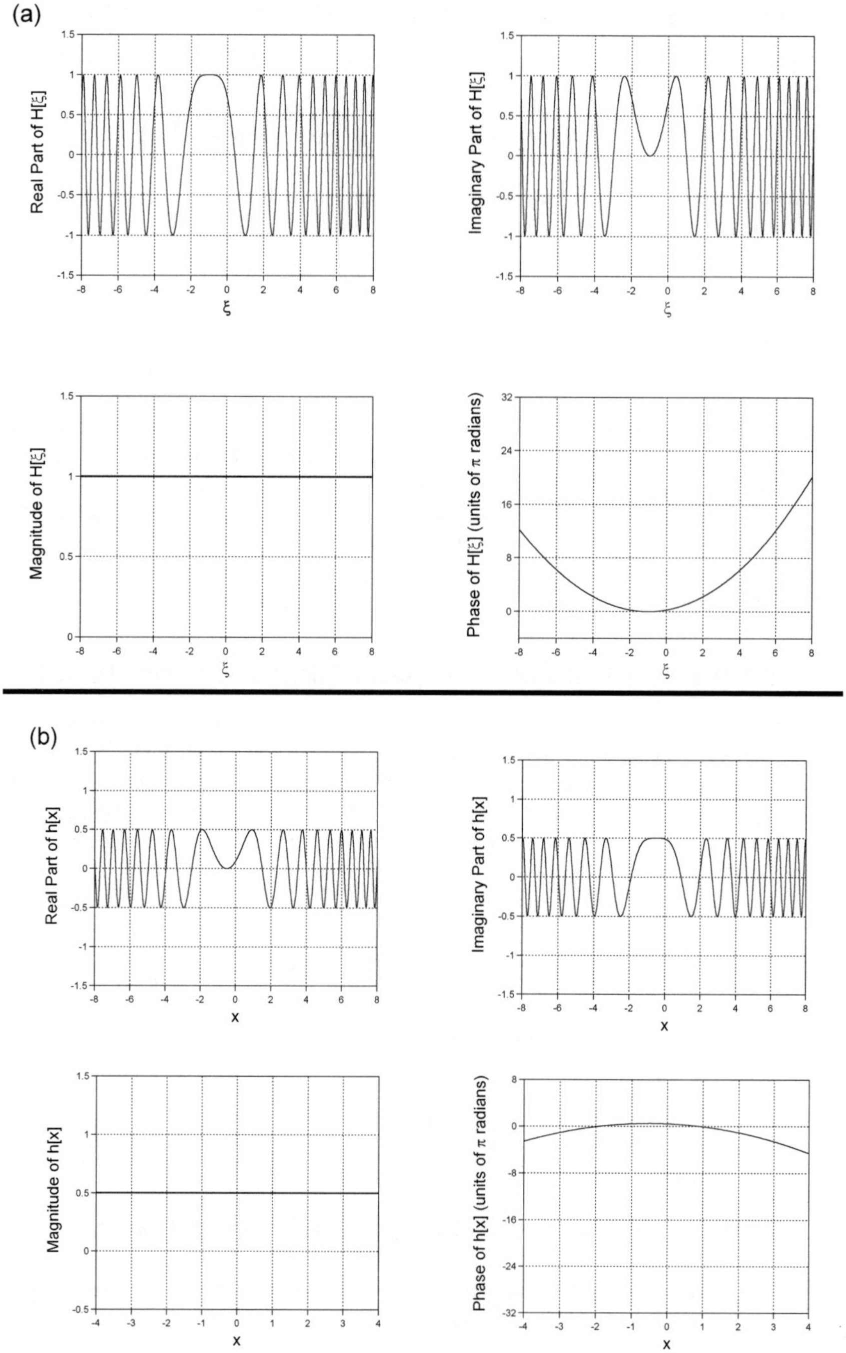

Figure 17.6 Example of allpass filter with constant, linear, and quadratic phase, with $\alpha_0 = 1/4, \alpha_1 = 1$, and $\alpha_2 = 2$: (a) transfer function; (b) impulse response. The magnitude functions in both domains are constant and the chirp rates have opposite signs.

with the order n. Similarly, the amplitude of the imaginary part is approximately zero over the same region, and the width of this region also increases with n. The phase of $H_n[\xi]$ is $\pm\pi$ radians at $\xi = \pm\alpha_n^{-1}$ regardless of the value of n. In short, the real part of the complex amplitude of all allpass transfer functions is unity at the origin and -1 at $\xi = \pm\alpha_n^{-1}$. The locations of the adjacent maxima of the real part occur at the spatial frequencies for which the phase is $\pm 2\pi$ radians, which are easily evaluated to be $\xi = \pm\alpha_n^{-1}(2^{1/n})$.

The impulse responses of the higher-order allpass filters are obtained via the inverse Fourier transform:

$$h_n[x] = \mathcal{F}_1^{-1}\{e^{+i\pi(\alpha_n\xi)^n}\} = \int_{-\infty}^{+\infty} 1[\xi]\, e^{+i\pi[(\alpha_n\xi)^n+2\xi x]}\, d\xi \tag{17.34}$$

Though analytic expressions have not been derived, the asymptotic solutions that were developed in Chapter 13 may be applied to obtain useful results. The moment expansion gives an approximate impulse response that is valid for $x \cong 0$, while the method of stationary phase is valid for large values of $|x|$. The moment expansion for the 1-D even-order superchirp was derived in Equation (13.40):

$$\begin{aligned}&\mathcal{F}_1\{e^{\pm i\pi x^n}\} \quad (\textit{even } n)\\ &= e^{\pm i\pi/2n}\sum_{\ell=0}^{+\infty}\left((-1)^\ell\left(\frac{2^{2\ell+1}}{n}\right)(\pi^{(2\ell(1-1/n)-1/n)})\frac{\Gamma[(2\ell+1)/n]}{\Gamma[2\ell+1]}\,e^{\pm i\pi\ell/n}\right)\xi^{2\ell} \quad \textit{if } |\xi| \gg 0\end{aligned} \tag{17.35}$$

while that for the odd-order sibling is:

$$\begin{aligned}&\mathcal{F}_1\{e^{\pm i\pi x^n}\} \quad (\textit{odd } n)\\ &= \frac{2}{n}\sum_{p=0}^{+\infty}\left\{(-1)^p(\pi^{-(2p+1)/n})\frac{(2\pi)^{2p}}{(2p)!}\xi^{2p}\right\}\cdot\left\{\Gamma\left[\frac{2p+1}{n}\right]\cos\left[\frac{\pi(2p+1)}{2n}\right]\right.\\ &\quad\left.+\frac{2\pi}{(2p+1)}(\pi^{-1/n})\cdot\Gamma\left[\frac{2p+2}{n}\right]\sin\left[\frac{\pi(2p+2)}{2n}\right]\xi\right\}\end{aligned} \tag{17.36}$$

The "transform-of-a-transform" theorem in Equation (9.104) and the scaling theorem may be applied to derive the impulse responses as an infinite series of moments. The even-order impulse response is:

$$\begin{aligned}&h_n[x] \quad (\textit{even } n)\\ &= \mathcal{F}_1^{-1}\{e^{\pm i\pi(\alpha_n\xi)^n}\}\\ &= \frac{1}{|\alpha_n|}e^{\pm i\pi/2n}\sum_{\ell=0}^{+\infty}\left((-1)^\ell\left(\frac{2^{2\ell+1}}{n}\right)(\pi^{(2\ell(1-1/n)-1/n)})\frac{\Gamma[(2\ell+1)/n]}{\Gamma[2\ell+1]}\,e^{\pm i\pi\ell/n}\right)\left(-\frac{x}{\alpha_n}\right)^{2\ell}\\ &= \frac{2}{n|\alpha_n|}\sum_{\ell=0}^{+\infty}\left((-1)^\ell(\pi^{(2\ell(1-1/n)-1/n)})\frac{\Gamma[(2\ell+1)/n]}{\Gamma[2\ell+1]}\,e^{\pm i\pi\ell/n}\right)\left(\frac{2x}{\alpha_n}\right)^{2\ell}\end{aligned} \tag{17.37}$$

The presence of the term $e^{\pm i\pi\ell/n}$ ensures that the impulse response is complex valued. The corresponding expression for the odd-order impulse response may be written as the sum of adjacent pairs of real-valued terms, i.e., each term in the summation includes an odd-order term that is a function of $(2p+1)$ and an even-order term that is a function of $(2p+2)$:

$$h_n[x] \quad (odd\ n)$$
$$= \mathcal{F}_1^{-1}\{\exp[\pm i\pi(\alpha_n\xi)^n]\}$$
$$= \frac{2}{n|\alpha_n|}\sum_{p=0}^{+\infty}\left\{(-1)^p(\pi^{-(2p+1)/n})\frac{(2\pi)^{2p}}{(2p)!}\left(\frac{x}{\alpha_n}\right)^{2p}\right\}$$
$$\cdot\left\{\Gamma\left[\frac{2p+1}{n}\right]\cos\left[\frac{\pi(2p+1)}{2n}\right]\mp\frac{2\pi}{(2p+1)}(\pi^{-1/n})\Gamma\left[\frac{2p+2}{n}\right]\sin\left[\frac{\pi(2p+2)}{2n}\right]\left(\frac{+x}{\alpha_n}\right)\right\} \tag{17.38}$$

Thus the impulse response of the odd-order Hermitian superchirp is seen to be real valued.

17.5.1 Odd-Order Allpass Filters with $n \geq 3$

Because the phase transfer functions of all odd-order Hermitian filters are odd, the phase increment at $\xi = -\xi_1$ is the negative of that at $\xi = +\xi_1$. As considered during the discussion of the shift theorem in Chapter 9, the complex components of a real-valued sinusoid at ξ_1 will be translated by the phase filter in the same direction (i.e., toward or away from $x = +\infty$) just as in the case of the linear-phase filter. To illustrate, consider the action of an odd-order filter on the real-valued sinusoid $f_0[x] = \cos[2\pi\xi_0 x]$. Its spectrum $F_0[\xi]$ is a pair of (generally complex-valued) Dirac delta functions located at $\pm\xi_0$. If the coefficient $\alpha_n > 0$, then the phase transfer function is positive for $\xi > 0$ and negative for $\xi < 0$. The filter adds the constant phase $\pi\alpha_n^n\xi_0^n > 0$ to the complex sinusoidal component with positive frequency and decrements the phase of the negative-frequency component by the same amount (because $\alpha_n > 0$ and $\xi_0^n < 0$ for $\xi_0 < 0$). The equation for the action of the odd-order filter in such a case is:

$$\cos[2\pi\xi_0 x] * h_n[x] = \mathcal{F}_1^{-1}\left\{\left(\frac{1}{2}\delta[\xi+\xi_0] + \frac{1}{2}\delta[\xi-\xi_0]\right)\cdot e^{+i\pi\alpha_n^n\xi^n}\right\}$$
$$= \frac{1}{2}\mathcal{F}_1^{-1}\{\delta[\xi+\xi_0]\cdot e^{+i\pi\alpha_n^n(-\xi_0)^n} + \delta[\xi-\xi_0]\cdot e^{+i\pi\alpha_n^n(+\xi_0)^n}\}$$
$$= \frac{1}{2}\mathcal{F}_1^{-1}\{\delta[\xi+\xi_0]\}\cdot e^{-i\pi\alpha_n^n\xi_0^n} + \mathcal{F}_1^{-1}\{\delta[\xi-\xi_0]\}\cdot e^{+i\pi\alpha_n^n\xi_0^n}$$
$$= \frac{1}{2}\,e^{-2\pi i\xi_0 x}\cdot e^{-i\pi\alpha_n^n\xi_0^n} + \frac{1}{2}\,e^{+2\pi i\xi_0 x}\cdot e^{+i\pi\alpha_n^n\xi_0^n}$$
$$= \frac{1}{2}\exp\left[-2\pi i\xi_0\left(x+\frac{1}{2}\alpha_n^n\xi_0^{n-1}\right)\right] + \frac{1}{2}\exp\left[+2\pi i\xi_0\left(x+\frac{1}{2}\alpha_n^n\xi_0^{n-1}\right)\right]$$
$$= \cos\left[2\pi\xi_0\left(x+\frac{1}{2}(\alpha_n)^n(\xi_0)^{n-1}\right)\right] \quad for\ odd\ n \tag{17.39}$$

In words, this result indicates that real-valued sinusoids with period $(\xi_0)^{-1}$ are translated by the distance $x_0 = \frac{1}{2}(\alpha_n)^n(\xi_0)^{n-1}$; the translation is toward the negative direction if $\alpha_n > 0$. The translation distance is positive if $\alpha_n > 0$. In short, all real-valued sinusoidal inputs will be translated in one direction by distances proportional to ξ_0^{n-1}, which means that the impulse response is real valued and apparently "one sided", i.e., it appears that $h_n[x] = 0$ for $x > 0$ if $\alpha_n > 0$. This result conforms to the prediction from the method of stationary phase in Equation (13.99).

The transfer function and an estimate of the impulse response calculated via the inverse discrete Fourier transform are shown in Figure 17.7 for the cubic Hermitian allpass filter with $\alpha_3 = 1$. The amplitude at the origin is the correct value of approximately 1.056 that is predicted by the central-ordinate theorem. Note that the amplitude of $h_3[x]$ for $x > 0$ is not strictly zero, as was inferred from

Equation (13.100), but rather decays rapidly to zero for $x \gtrsim 0$. This behavior conforms to the prediction of the moment expansion of the impulse response, as may be seen by applying the transform-of-a-transform theorem to Equation (13.41). In other words, if a finite-support input function $f[x]$ is filtered by $h_3[x]$, some of the amplitude of $f[x]$ "bleeds" to the right of the edge of support because $h_3[x]$ has some nonzero amplitude for $x > 0$. All of the odd-order allpass filters exhibit similar behavior. This apparent paradox is explained by recognizing that the zero-frequency term of the input spectrum is unaffected by the action of the phase filter.

As a side comment, we know that the autocorrelation of the real-valued and almost "single-sided" impulse response $h_3[x]$ in Figure 17.7 is a Dirac delta function at the origin:

$$h_3[x] \bigstar h_3[x] = \delta[x] \tag{17.40}$$

as predicted by Equation (17.5). This is perhaps surprising during the first glance at the impulse response. Of course, the reason is that the product of $h_3[x]$ with translated replicas yields functions with zero area except for zero shift, where the perfect match creates a function with unit area.

The image of $f[x] = RECT[x/16]$ through the cubic-phase allpass filter is shown in Figure 17.8.

17.5.2 Even-Order Allpass Filters with $n \geq 4$

We will finish this discussion with a brief mention of the fourth-order case for comparison to the quadratic-phase filter. The moment expansion for the impulse response is obtained from Equation (13.40):

$$\begin{aligned} h_4[x] &= \mathcal{F}_1^{-1}\{e^{\pm i\pi(\alpha_4\xi)^4}\} \\ &= e^{\pm i\pi/8} \sum_{\ell=0}^{+\infty} \left(\frac{2^{2\ell+1}}{4} (-1)^\ell \, \pi^{\left(\frac{3}{2}\ell - \frac{1}{4}\right)} \frac{\Gamma[(2\ell+1)/4]}{\Gamma[2\ell+1]} \, e^{\pm i\pi\ell/4} \right) x^{2\ell} \\ &\cong \frac{1}{|\alpha_4|} \, e^{\pm i\pi/8} (1.362 + 5.125 \, e^{\mp i\pi\frac{3}{4}} (\alpha_4 x)^2 + 7.037 \, e^{\pm i\pi/2} (\alpha_4 x)^4 + \cdots) \end{aligned} \tag{17.41}$$

but this expression does not give much insight into the structure of the impulse response. The stationary-phase solution is obtained by applying the "transform-of-a-transform" theorem to Equation (13.90):

$$\hat{h}_4[x] = (2^{-\frac{1}{6}})(3^{-\frac{1}{2}})(|\alpha_4|^{-\frac{4}{3}})(|x|^{-\frac{1}{3}}) \, e^{+i\pi/4} \exp\left[+3\pi i \left(-\frac{x}{2\alpha}\right)^{\frac{4}{3}}\right] \tag{17.42}$$

The transfer function and impulse response are shown in Figure 17.9 as real parts, imaginary parts, magnitude, and phase. The image of the rectangle function is shown in Figure 17.10.

The impulse response has infinite support, though its magnitude is concentrated near the origin.

17.6 Allpass Random-Phase Filter

As a final example of the allpass filter, we consider some aspects of a filter that makes random changes to the initial phase at each frequency without absorbing any energy from the incident signal. This system is a useful model for several optical imaging phenomena, including imaging through both a turbulent atmosphere and a "ground glass" screen. The MTF of the system is assumed to be unity at all spatial frequencies, thus satisfying the requirement for an allpass filter. In the simplest example, the phase transfer function is an ensemble of real numbers selected from a uniform statistical distribution $N[\xi]$ which has unit range. The range of possible initial phases spans the cycle of 2π radians and they are uniformly distributed, so that the probability density function is an appropriately scaled rectangle:

$$p[\Phi] = \frac{1}{2\pi} RECT\left[\frac{\Phi}{2\pi}\right] \tag{17.43}$$

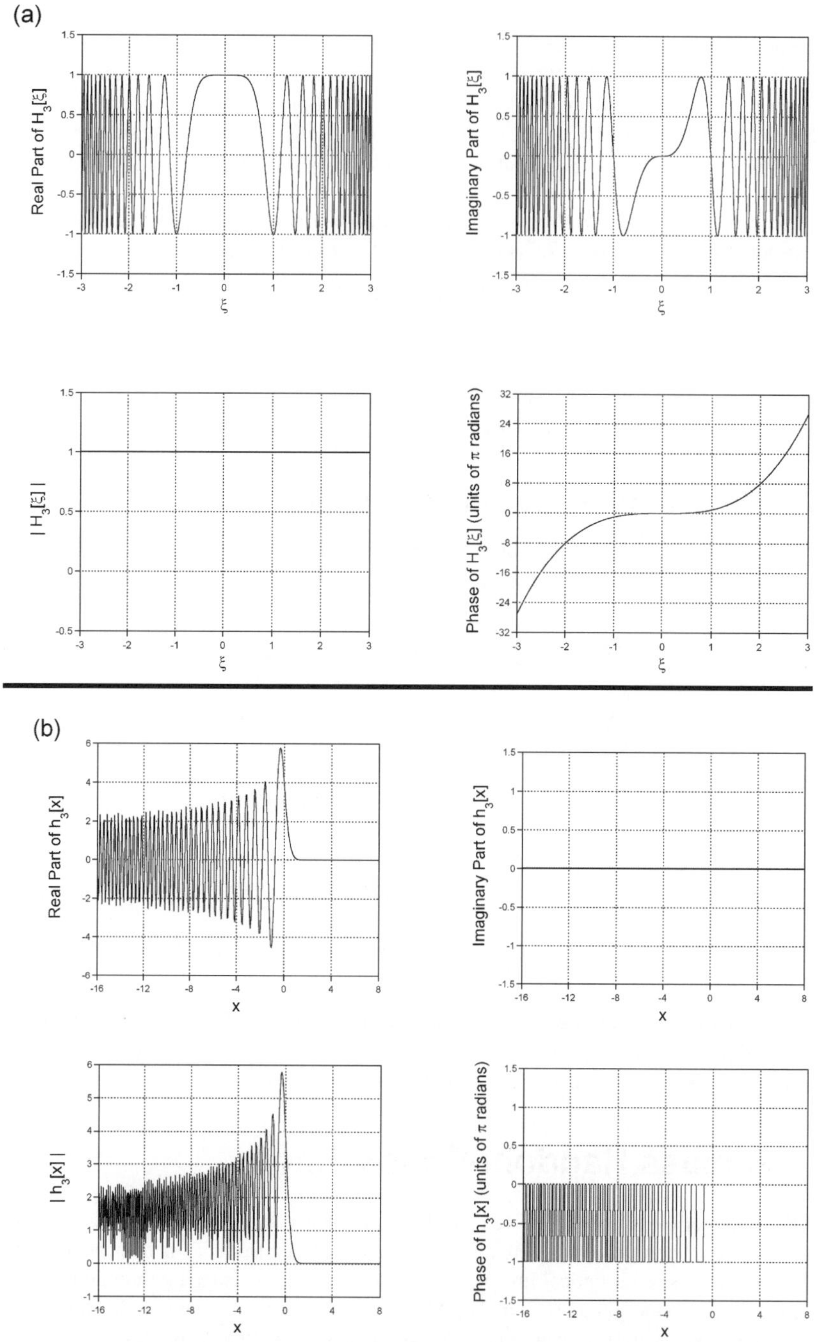

Figure 17.7 The allpass filter of order $n = 3$: (a) transfer function $H_3[\xi]$ with $\alpha_3 = +1$ as (even) real part, (odd) imaginary part, unit magnitude, and (odd) phase; (b) the impulse response $h_3[x]$ evaluated via the discrete Fourier transform as real part, (null) imaginary part, magnitude, and phase. Note that $h_3[0] \simeq 1.056$, as predicted in Equation (13.46), and that $h_3[x]$ decays to zero for $x > 0$.

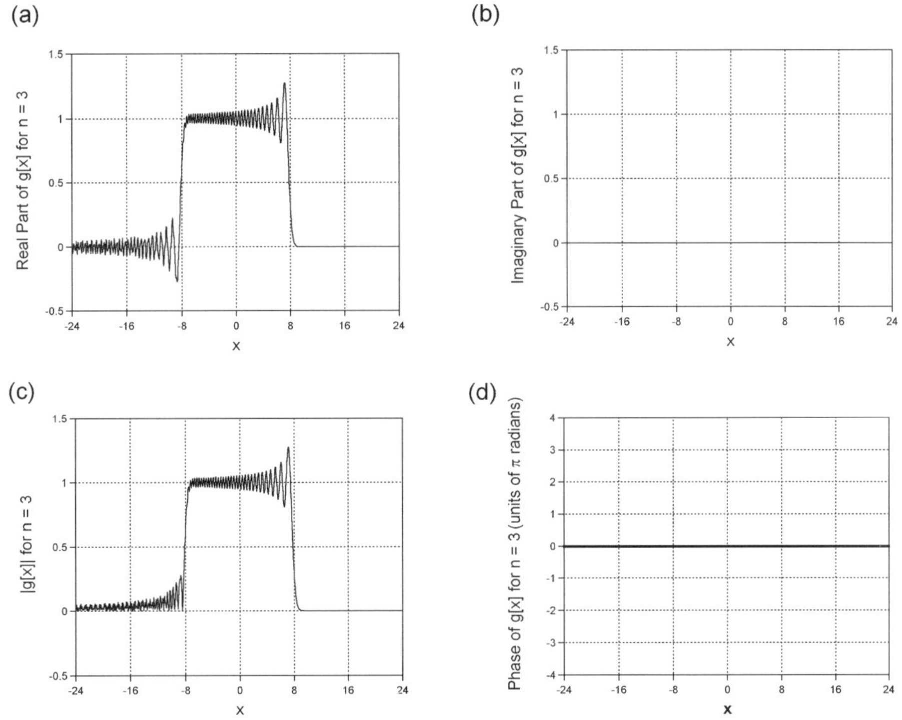

Figure 17.8 Image $g[x]$ of $f[x] = RECT[x/16]$ through allpass filter with $\Phi\{H[\xi]\} = +\pi\xi^3$: (a) real part; (b) imaginary part; (c) magnitude; (d) phase. Note that the image is real valued and exhibits "ringing" on the left-hand side of the edges of the rectangle.

The real and imaginary parts of the transfer function will be respectively the cosine and sine of these random phases, with identical ranges in the closed interval $[-1, +1]$. Since the magnitude $|H[\xi]| = 1$, its histogram is a Dirac delta function. The exact form of the transfer function cannot be predicted, but the probability distribution of the real and imaginary parts of the transfer function must be identical to the probability distribution of amplitudes of a pure sinusoid. The conditions are shown in Figure 17.11. The phase at frequency ξ may be assigned the value $2\pi \cdot N[\xi]$. The form of the transfer function of the filter in Equation (17.9) therefore is:

$$\begin{aligned} H_{\text{random}}[\xi] &= 1[\xi] \exp[+i\pi W[\xi]] 1[\xi] \exp[+2\pi i\, N[\xi]] \\ &= \cos[2\pi N[\xi]] + i \sin[2\pi N[\xi]] \end{aligned} \tag{17.44}$$

The transfer function is displayed as real and imaginary parts in Figure 17.11(a) and (b), along with the histograms of the amplitudes. The functional forms of these histograms were derived during the discussion of the Radon transform of the ring delta function in Section 12.2.2. The probability density function of the real or imaginary parts is:

$$p[\Re\{H_{\text{random}}[\xi]\}] = p[\Im\{H_{\text{random}}[\xi]\}] = p[n] = \frac{1}{\pi\sqrt{1-n^2}} RECT\left[\frac{n}{2}\right] \tag{17.45}$$

The random number added to the initial phase at each frequency "scrambles" the sinusoidal frequency components that otherwise would have "lined up" to create edges or other fine detail in

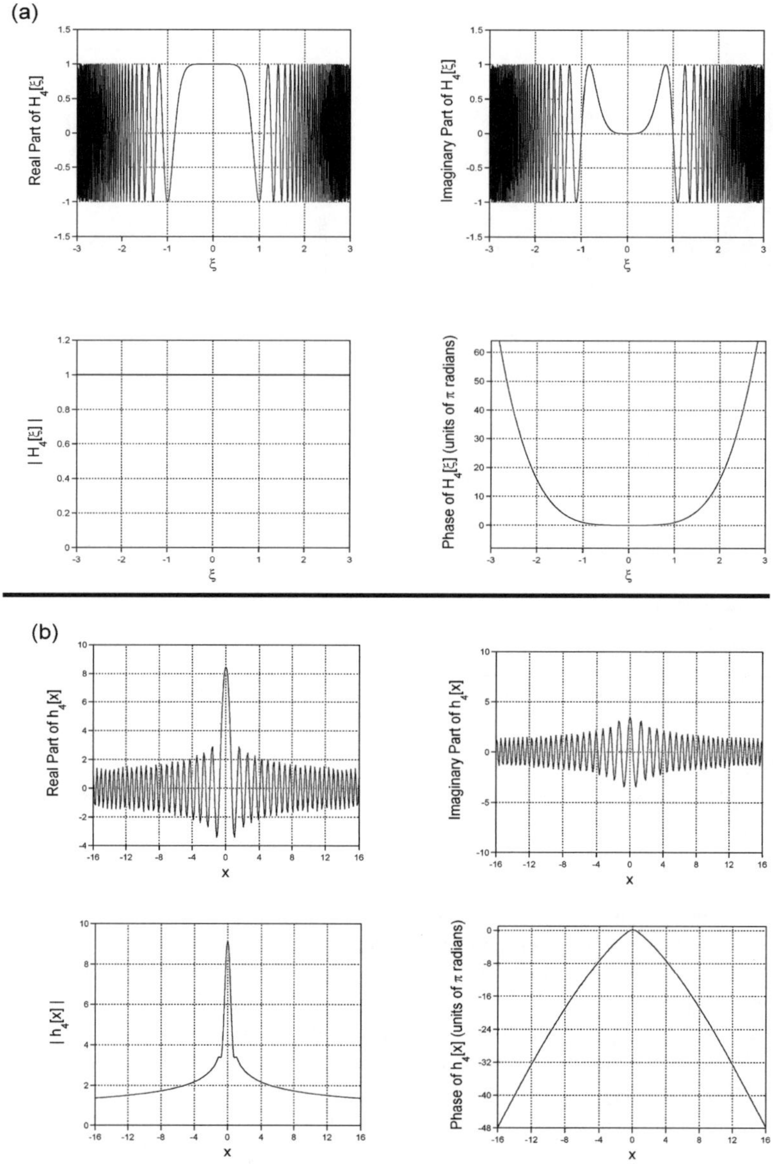

Figure 17.9 Allpass filter of order $n = 4$ with $\alpha_4 = 1$: (a) transfer function $H_4[\xi] = \exp[+i\pi\xi^4]$ as (even) real part, (even) imaginary part, unit magnitude, and phase; (b) impulse response $h_4[x]$ evaluated via the inverse discrete Fourier transform as (even) real part, (even) imaginary part, magnitude, and phase. Note that $|h_4[x]|$ "peaks" at the origin, unlike the constant magnitude of $h_2[x]$.

the original image, without affecting the magnitudes of those components. The quality of the output image will be degraded without changing its integrated energy (total "brightness"). The randomness of the phase transfer function ensures that the exact form of the impulse response is not predictable, but it will have infinite support.

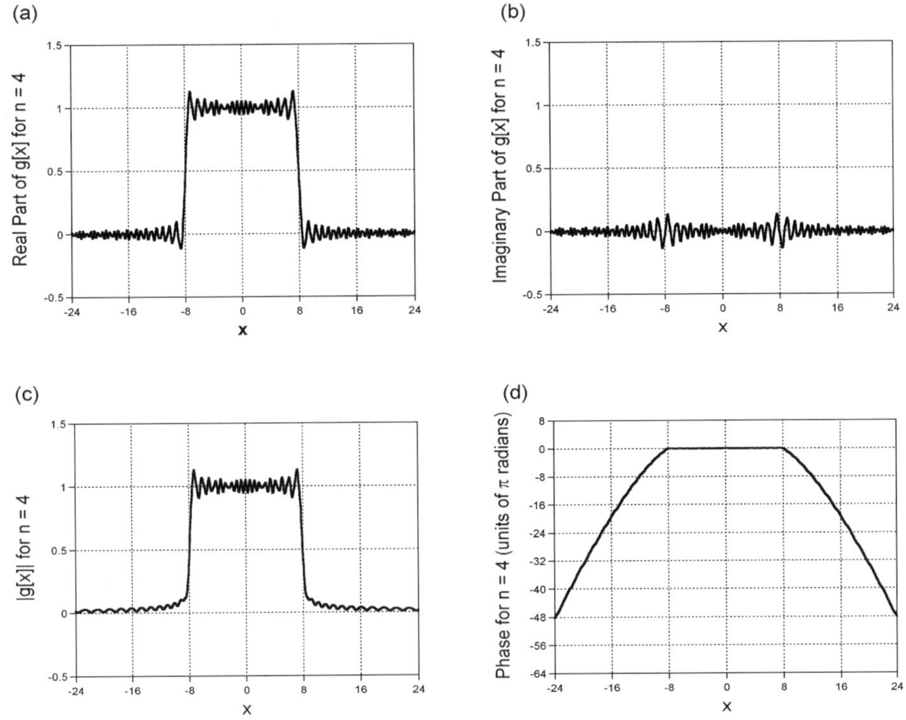

Figure 17.10 The image of $f[x] = RECT[x/16]$ through an allpass filter with $\Phi\{H[\xi]\} = \pi\xi^4$, displayed as real part, imaginary part, magnitude, and phase. The output is complex valued and the magnitude has infinite support, though the image is concentrated within the region $x = \pm 8$.

The impulse response of the filter in Figure 17.11 was evaluated via a numerical inverse Fourier transform and is shown in Figure 17.12 as real part, imaginary part, magnitude, and phase, along with their histograms. The randomness within the transfer function precludes the prediction of the form of the impulse response of the random-phase filter from the initial conditions, but it is still possible to make some valid statements about the character of $h[x]$. Since a particular sample of the real or imaginary parts is a weighted sum of a large number of samples from the real and imaginary parts of $H[\xi]$, it is reasonable to assume that the central-limit theorem applies, which is confirmed by the Gaussian distribution of the individual real and imaginary parts of $h[x]$. This result also may be demonstrated rigorously (Goodman,1985; Frieden, 2002). The histogram of the phase of the impulse response exhibits a uniform distribution over the interval $[-\pi, +\pi)$. The magnitude of the impulse response is random noise from a Rayleigh distribution with probability density function:

$$p[|h_{\text{random}}|] = \frac{h_{\text{random}}}{\sigma^2} \cdot \exp\left[-\frac{(h_{\text{random}})^2}{2\sigma^2}\right] \cdot STEP[h_{\text{random}}]$$

$$= \frac{h_{\text{random}}}{\sigma^2} \cdot GAUS\left[\frac{h_{\text{random}}}{\sqrt{2\pi\sigma^2}}\right] \cdot STEP[h_{\text{random}}] \tag{17.46}$$

where σ^2 is the variance of the Gaussian noise in the real or imaginary part. If this random-phase filter is applied to an arbitrary input function, the randomness will destroy any meaningful structure even though all energy from the input is transmitted to the final image.

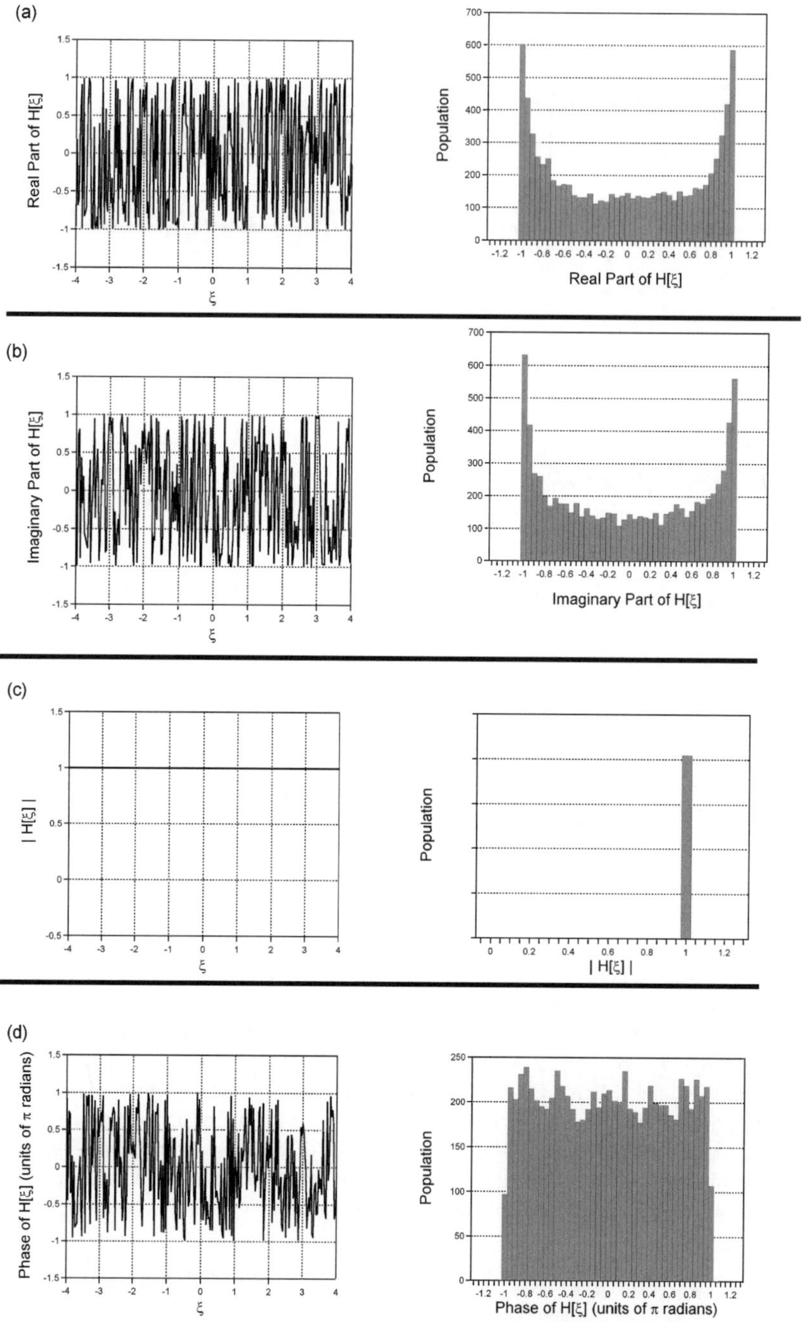

Figure 17.11 Transfer function $H_{\text{random}}[\xi]$ of a discrete realization of a random-phase allpass filter: (a) real part and its histogram, which is identical to the histogram of sinusoidal function; (b) imaginary part and histogram; (c) unit magnitude and its "Dirac delta function" histogram; (d) random phase and its histogram, demonstrating that the phase is uniformly distributed over 2π radians.

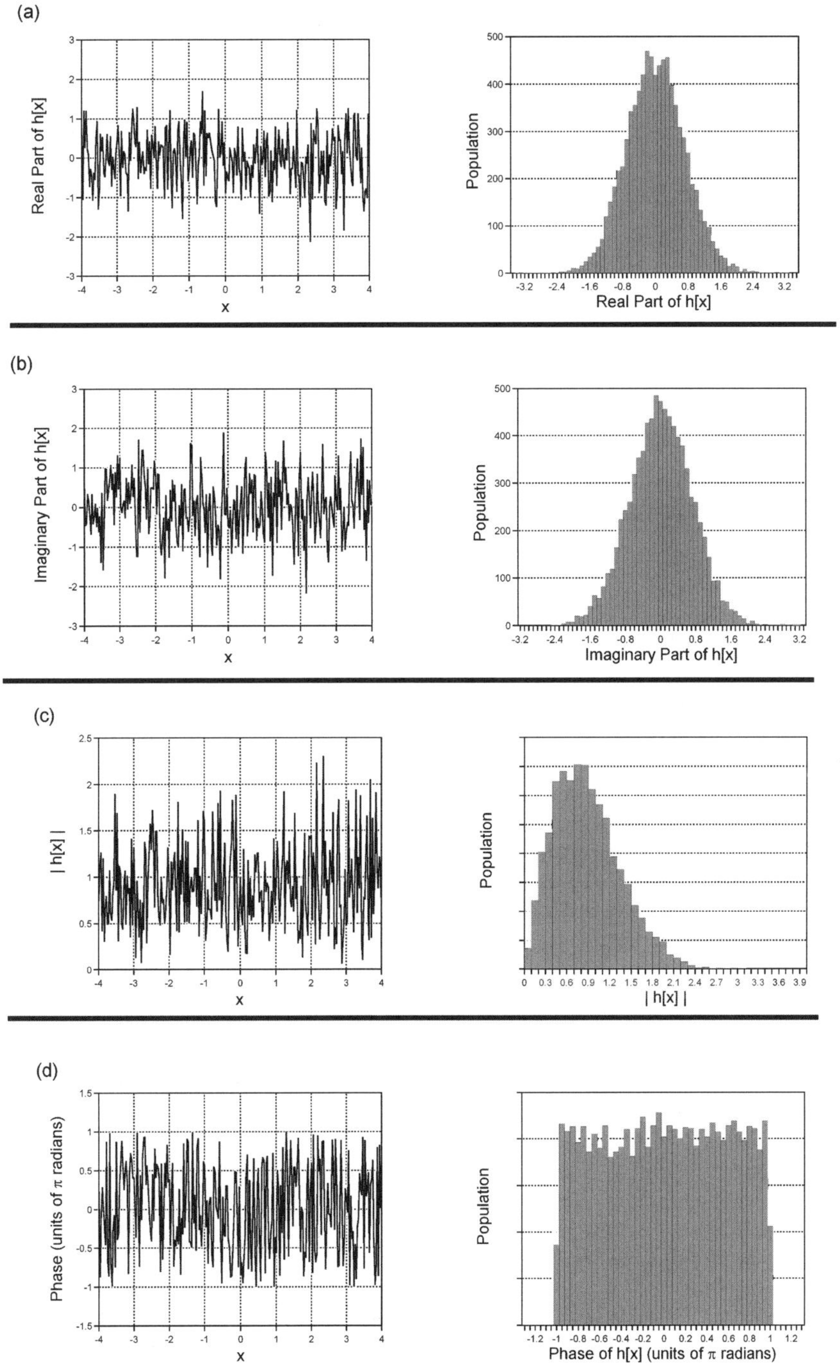

Figure 17.12 Impulse response $h[x]$ of the random phase filter in Figure 17.11, including histograms of the parts. The real and imaginary parts (a), (b) are Gaussian distributed with $\sigma \simeq 0.71$; (c) the magnitude exhibits a Rayleigh distribution with $\sigma \simeq 0.46$; (d) the phase is uniformly distributed over 2π radians.

An optical analogue of the random-phase filter is a glass screen where the thickness varies randomly with position and which scatters incident light throughout the image. The image is unidentifiable, even though all light from the original distribution is present. However, it is possible to glean useful information from the scrambled image because the magnitude spectrum has not been disturbed:

$$g[x] = f[x] * h[x] \tag{17.47a}$$

$$G[\xi] = F[\xi]H[\xi] = |F[\xi]|\, e^{i(\Phi_F[\xi]+\Phi_H[\xi])} \tag{17.47b}$$

We may discard the scrambled phase information by computing the power spectrum of the output signal. The inverse Fourier transform yields the autocorrelation of the input function, which still carries useful information about the structure of the signal:

$$\mathcal{F}_1^{-1}\{|G[\xi]|^2\} = g[x] \bigstar g[x] = \mathcal{F}_1^{-1}\{|F[\xi]|^2\} = f[x] \bigstar f[x] \tag{17.48}$$

In words, even though the information about the true spatial distribution of energy has been lost due to the disruption of the phase, the autocorrelation of the input signal $f[x]$ is preserved.

17.6.1 Information Recovery after Random-Phase Filtering

We now consider the information that may be recovered from the images obtained through allpass random-phase filters, such as that shown in Figure 17.11. The input object is the "three-bar" chart consisting of three patterns with different periods. The object is imaged using an allpass filter whose phase is selected from a uniform distribution over the interval $-\pi \leq \phi < +\pi$. The action of the filter is shown in Figure 17.13.

The broad support of the corresponding impulse response produces an image that is so random as to render the structure completely unrecognizable, even though the full magnitude of the sinusoidal components is transmitted out to the maximum frequency. Though the image bears no resemblance to the input object, we know that the Wiener–Khintchin theorem ensures that the inverse Fourier transform of the power spectrum is identical to the autocorrelation of the input signal. In this example where $f[x]$ is composed of two impulses, the autocorrelation is sufficient to measure the separation and relative brightnesses of the components. This is the basic principle behind "stellar speckle interferometry", which was introduced by Labeyrie in 1970. The autocorrelation of the star image(s) may be recovered even though the images are scrambled severely.

17.7 Relative Importance of Magnitude and Phase

Since we have spent so much effort to examine the effect of a filter that changes either the magnitude spectrum or the phase spectrum, it is natural to ask the question of which type of information is more important to the image quality. In other words, we wish to determine which has the greater impact on the fidelity of the output image. The answer to this question is of more than academic interest because imaging systems often transmit magnitude information well while distorting or discarding the initial phases (Hayes, 1987).

Consider a system that transmits the magnitude information to the output but discards the phase. The lost information specifies where the "peak" of each sinusoidal frequency component is positioned relative to the origin. Since there is no information available to distinguish the differences in phase at each frequency, we can only assign the same phase to each. For example, we might guess that the initial phase at each frequency is 0 radians. The output image would be composed of sinusoids with the correct amplitude but all would be assumed to be symmetric, thus guaranteeing the output to be an even function of x. The resulting image usually is unrecognizable, as shown in the example in Figure 17.14b.

In the case where initial phases are known but the magnitude spectrum is not, one possible strategy is to set the magnitude spectrum to unity at all frequencies: $|F[\xi]| = 1[\xi]$. This means that the real and

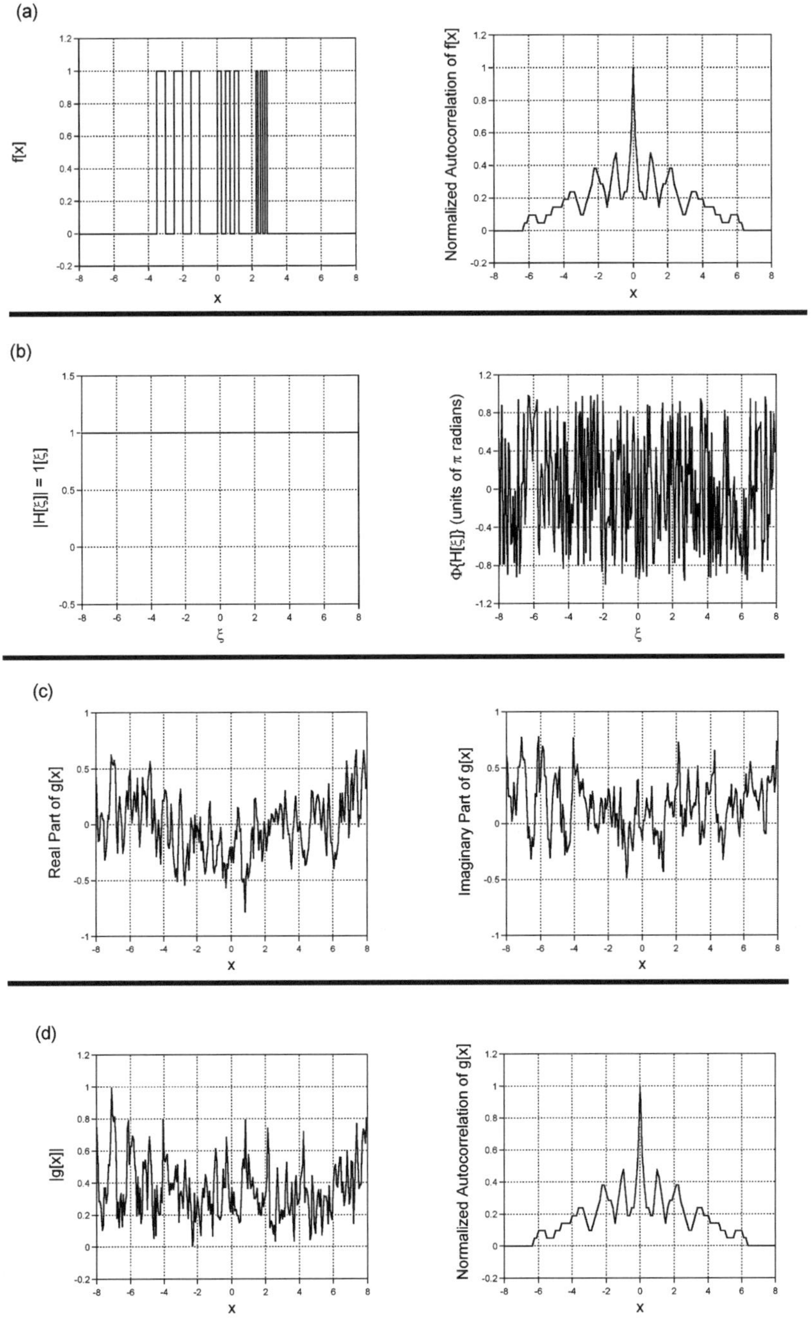

Figure 17.13 Information recovery from random-phase filter: (a) $f[x]$ and its normalized autocorrelation; (b) $|H[\xi]| = 1[\xi]$ and $\Phi\{H[\xi]\}$, showing that all spatial frequencies are passed to the output without attenuation; (c) $\Re\{g[x]\}$ and $\Im\{g[x]\}$, showing the apparent randomness of the output; (d) $|g[x]|$ and normalized autocorrelation of $g[x]$, which is identical to that of $f[x]$.

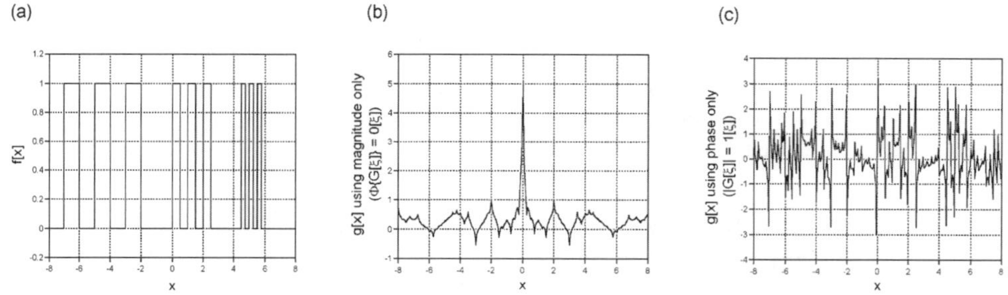

Figure 17.14 Relative importance of magnitude and phase information: (a) $f[x]$; (b) $\hat{g}[x]$ computed from $\hat{G}[\xi] = |F[\xi]| \exp[i \cdot 0]$, showing no resemblance to $f[x]$; (c) $\hat{g}[x]$ computed from $\hat{G}[\xi] = 1[\xi] \exp[i\Phi\{F[\xi]\}]$, which has larger amplitudes at larger values of $|\xi|$ than $F[\xi]$, so the edges of the output image are enhanced.

imaginary parts of the estimated spectrum are respectively the cosine and sine of the initial phase. Because the magnitude usually diminishes with increasing $|\xi|$ for real images, the high-frequency amplitude of the output image obtained from this estimate will be too large. This high-frequency amplification enhances the edges of features in the image.

From this analysis, it is apparent that the phase information in an image is "more important" than the magnitude. In fact, it is often possible to obtain adequate reconstruction of an image by using the correct phase and the magnitude after filtering of the image, or by using the magnitude from another image. Examples are shown in Figure 17.15. However, it is much more common to have better knowledge of the magnitude of the spectrum than of the phase, which will limit the success of the reconstruction by simple linear means.

17.8 Imaging of Phase Objects

We now consider a variation on the idea of an allpass filter based on the "transform-of-a-transform" theorem and the fact that convolution is a commuting operation. In this situation, the space-domain representation $f[x]$ is a pure phase object, i.e., a complex function with constant magnitude. If imaged by an optical system with a square-law detector, the output is a constant:

$$|f[x]|^2 \propto 1[x] \quad \textit{for a pure phase object} \tag{17.49}$$

In other words, all phase objects produce identical outputs. An example of a phase object is turbulent air, which transmits the incident radiation with virtually no attenuation and may change the direction of propagation, but the effect may not be evident if the refractive effects are subtle. However, we can use an analogy with phase filters to construct an impulse response $h[x]$ that renders the object phase "visible" when viewed with a square-law detector. Methods were developed by August Toepler in the 1860s and Fritz Zernike in the 1930s that rely on frequency-domain magnitude-only filtering to convert the pure phase modulation of the input function to an intensity modulation at the output. Toepler's system is now called *Schlieren* imaging.

We have seen that the impulse response of a phase object generally does not have constant magnitude. The exception is the quadratic-phase allpass object, whose magnitude is constant in both domains. The magnitude of the spectrum of allpass signals with higher-order phase is not constant:

$$f[x] = 1[x]\, e^{i\Phi_f[x]} \Longrightarrow |F[\xi]| \neq 1[\xi] \quad \textit{unless}\ \Phi_f[x] \propto x^2 \tag{17.50}$$

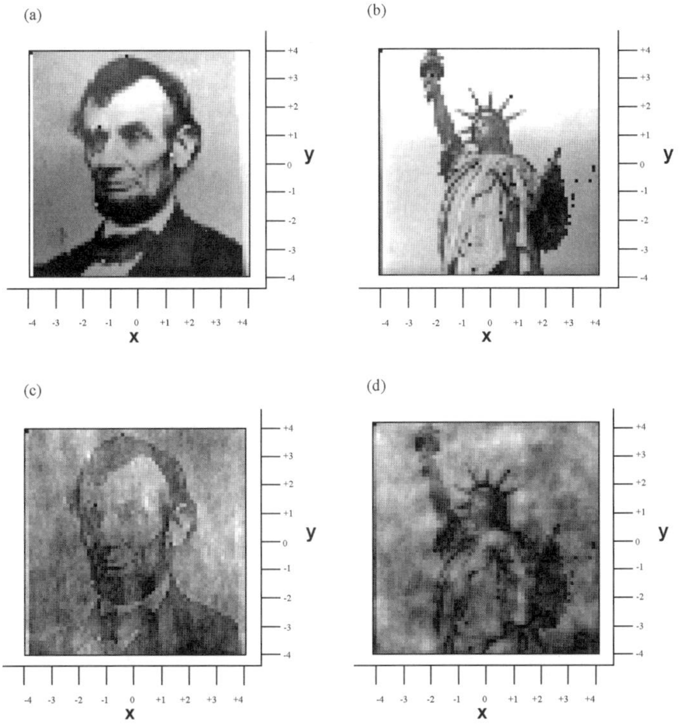

Figure 17.15 Effect of exchanging magnitude and phase of two input functions: (a) $f_1[x, y]$; (b) $f_2[x, y]$; (c) $g_1[x, y] = |\mathcal{F}_2{}^{-1}\{|F_2[\xi, \eta]| \exp[+i\Phi\{F_1[\xi, \eta]\}]\}|$, which more closely resembles $f_1[x, y]$; (d) $|\mathcal{F}_2{}^{-1}[|F_1[\xi, \eta]| \exp[+i\Phi\{F_2[\xi, \eta]\}]]|$, which more closely resembles $f_2[x, y]$.

If the phase modulation $\Phi_f[x]$ is "small", so that the phase angles are changed by small increments, then the phase function may be approximated by its Taylor series truncated at the first order:

$$f[x] = 1[x]\, e^{i\Phi_f[x]} \cong 1[x] \cdot (1 + i \cdot \Phi_f[x]) = 1[x] + i \cdot \Phi_f[x] \quad \textit{if } |\Phi_f[x]| \gtrapprox 0 \tag{17.51}$$

The corresponding Fourier spectrum is the sum of a Dirac delta function and the spectrum of the phase function:

$$F[\xi] \cong \delta[\xi] + i \cdot \mathcal{F}_1\{\Phi_f[x]\} \tag{17.52}$$

The imaging system consists of a highpass transfer function that removes the DC component and passes the AC terms. Of course, this filter also removes the DC component of the spectrum of the phase term, i.e., the "area" of the phase function. However, given the assumption used to approximate $\Phi_f[x]$ by its truncated Taylor series and the fact that the phase function may be considered to be distributed over the interval $-\pi \leq \phi < +\pi$, this area will likely be small. The spectrum of the output signal is approximately:

$$\begin{aligned} G[\xi] &\cong F[\xi] - F[0] \\ &= i \cdot \mathcal{F}_1\{\Phi_f[x]\} - i \cdot \int_{-\infty}^{+\infty} \Phi_f[x]\, dx \\ &\cong i \cdot \mathcal{F}_1\{\Phi_f[x]\} \end{aligned} \tag{17.53}$$

where the assumption that the phase function is bipolar and approximately uniformly distributed has been used. In this case, the complex amplitude of the filtered output is proportional to the phase function:

$$g[x] \cong \mathcal{F}_1^{-1}\{i \cdot \mathcal{F}_1\{\Phi_f[x]\}\} \cong i \cdot \Phi_f[x] \tag{17.54}$$

When measured by a square-law detector, the image is approximately the squared magnitude of the phase variation:

$$|g[x]|^2 \cong |\Phi_f[x]|^2 \tag{17.55}$$

and thus the phase modulation has become visible.

To illustrate, consider the 1-D sinusoidal phase "grating" shown as magnitude/phase and real/imaginary parts in Figure 17.16:

$$f[x] = 1[x] \exp\left[+i \cdot \pi\left(0.15 + 0.1 \cos\left[2\pi \frac{x}{4}\right]\right)\right] \tag{17.56}$$

Note that the squared magnitude is constant; there is no visible modulation of intensity if measured by a square-law detector.

The input may be decomposed into its real and imaginary parts:

$$\begin{aligned} f[x] &= \cos\left[2\pi \cdot \left(0.075 + 0.05 \cos\left[2\pi \frac{x}{4}\right]\right)\right] + i \sin\left[2\pi \cdot \left(0.075 + 0.05 \cos\left[2\pi \frac{x}{4}\right]\right)\right] \\ &\cong 0.891 \cdot \cos\left[\frac{\pi}{10} \cdot \cos\left[2\pi \frac{x}{4}\right]\right] - 0.454\, i \cdot \sin\left[\frac{\pi}{10} \cdot \cos\left[2\pi \frac{x}{4}\right]\right] \\ &\cong 0.891 \cdot \left(1 - \frac{\pi}{10} \cdot \cos\left[2\pi \frac{x}{4}\right]\right) - 0.454\, i \cdot \left(\frac{\pi}{10} \cdot \cos\left[2\pi \frac{x}{4}\right]\right) \end{aligned} \tag{17.57}$$

where the small-angle approximation for the sine and cosine have been used. This expression for $f[x]$ is even because the argument of the sinusoids is real and even. Evidence of the higher-order terms in the power series is visible in the graphs of the real and imaginary parts in Figure 17.16, but the approximate spectrum $F[\xi]$ is easy to derive from this result:

$$F[\xi] \cong 0.891 \cdot \delta[\xi] - (0.891 + 0.454\, i) \cdot \left(\frac{\pi}{20} \cdot \delta\left[\xi + \frac{1}{4}\right] + \delta\left[\xi - \frac{1}{4}\right]\right) \tag{17.58}$$

A highpass filter that blocks the component at $\xi = 0$ is applied to the spectrum of the input, thus producing an approximate output spectrum:

$$G[\xi] \cong -(0.891 + 0.454\, i) \cdot \left(\frac{\pi}{20} \cdot \delta\left[\xi + \frac{1}{4}\right] + \delta\left[\xi - \frac{1}{4}\right]\right) \tag{17.59}$$

In an optical system, this is accomplished by placing a small opaque "spot" at the focal plane of an imaging lens so that only the light diffracted around the spot reaches the image plane. The approximate filtered image $g[x]$ (real and imaginary parts) is:

$$g[x] \cong -(0.891 + 0.454\, i) \cdot \left(\frac{\pi}{10} \cdot \cos\left[2\pi \frac{x}{4}\right]\right) \tag{17.60}$$

and the output of a square-law detector is:

$$|g[x]|^2 \cong \left(\frac{\pi}{10}\right)^2 \cdot \cos^2\left[2\pi \frac{x}{4}\right] \tag{17.61}$$

As shown in Figure 17.16d, an approximation of the phase grating (including higher-order frequency components) appears in the image of the squared magnitude. This principle is often used in microscopy to create so-called "dark field" images.

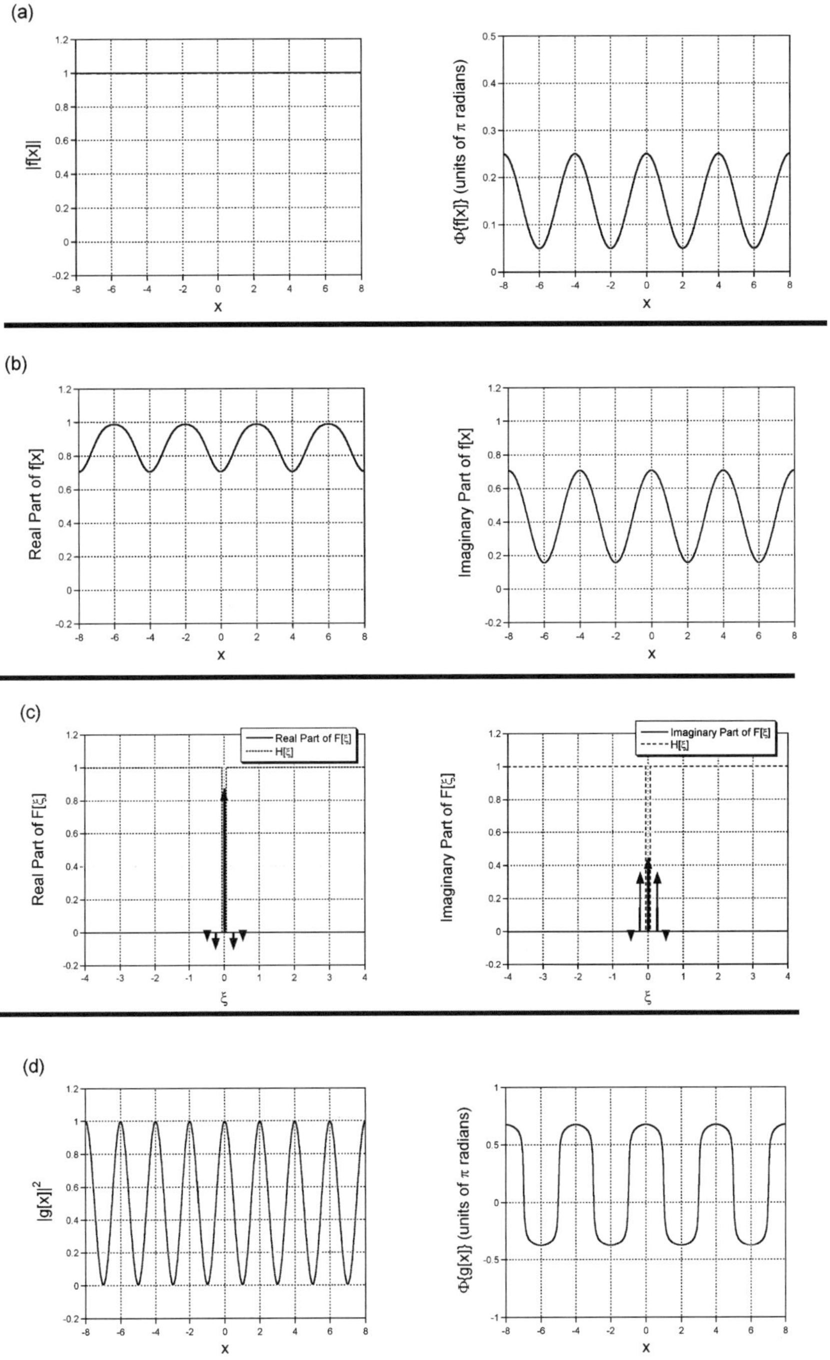

Figure 17.16 Schlieren Imaging: (a) unit magnitude and varying phase of input object $f[x]$, the squared magnitude $|f[x]|^2 = 1[x]$; (b) real and imaginary parts of input $f[x]$; (c) real and imaginary parts of $F[\xi]$, with transfer function $H[\xi]$ of highpass filter; (d) squared magnitude $|g[x]|^2$ and phase $\Phi\{g[x]\}$ of output, showing a distorted rendering of the phase modulation.

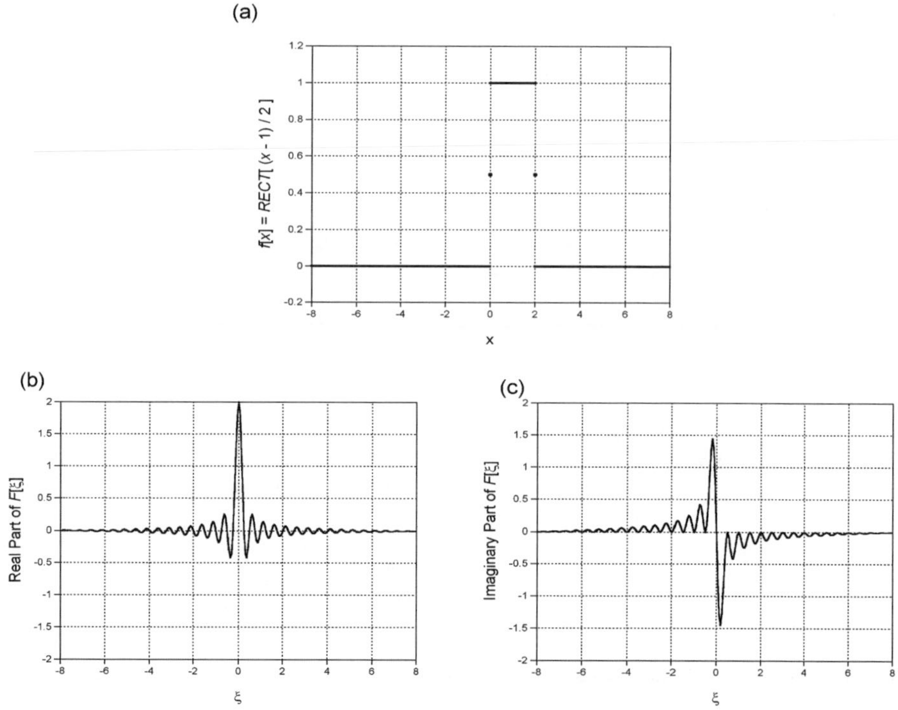

Figure 17.17 Function used in demonstration of chirp Fourier transform: (a) $f[x] = RECT[(x-1)/2]$; (b) $\Re\{F[\xi]\} = 2\, SINC[2\xi] \cdot \cos[-2\pi\xi]$; (c) $\Im\{F[\xi]\} = 2\, SINC[2\xi] \cdot \sin[-2\pi\xi]$.

17.9 Chirp Fourier Transform

This section considers the application of allpass quadratic-phase signals and filters to perform useful signal processing on arbitrary inputs. The most important examples are two particular sets of operations that evaluate the so-called "chirp Fourier transform". Partial implementations of these sequences yield useful expressions for operations that relate a function and a quadratic-phase signal, as will be demonstrated during the discussion of the finite-support quadratic-phase filter in Chapter 18 and mathematical models of optical imaging systems in Chapter 21. Consult Mertz (1965) and Gaskill (1978) for other discussions. The input function used to test the algorithms and its spectrum are shown in Figure 17.17.

17.9.1 1-D "M–C–M" Chirp Fourier Transform

As we have now seen many times, the 1-D Fourier transform of $f[x]$ is the integral of its product with the complex linear-phase exponential:

$$\mathcal{F}_1\{f[x]\} \equiv F[\xi] = \int_{-\infty}^{+\infty} f[x]\, e^{-2\pi i\xi x}\, dx \tag{17.62}$$

This expression may be recast into a different form by recognizing that the exponent is the sum of three terms:

$$(\xi - x)^2 = x^2 + \xi^2 - 2x\xi$$
$$\Longrightarrow -2x\xi = (\xi - x)^2 - x^2 - \xi^2 \tag{17.63}$$

Though this expression is valid for functions of pure numbers, the two coordinates x and ξ that are actually used have dimensions that do not match (length versus length^{-1}), which means that the subtractions in Equation (17.63) are not legitimate. We may create a valid expression that relates these terms by normalizing by the factor α_2 that we defined to be the "chirp rate" with dimensions of length in Equation (17.19). The subscript on the chirp rate will be dropped from this point forward to simplify the notation. The two terms x/α and $\alpha\xi$ are dimensionless and may be subtracted without metaphysical difficulties to produce the expression:

$$-2\xi x = \left(\alpha\xi - \frac{x}{\alpha}\right)^2 - \left(\frac{x}{\alpha}\right)^2 - (\alpha\xi)^2 \tag{17.64}$$

This result is substituted into Equation (17.62) and the "constant" term $e^{-i\pi\alpha^2\xi^2}$ is extracted from the integral:

$$F[\xi] = \int_{-\infty}^{+\infty} f[x] \cdot (e^{+i\pi(\alpha\xi - x/\alpha)^2} \cdot e^{-i\pi(x/\alpha)^2} \cdot e^{-i\pi(\alpha\xi)^2})\, dx$$
$$= e^{-i\pi\alpha^2\xi^2} \cdot \int_{-\infty}^{+\infty} (f[x] \cdot e^{-i\pi(x/\alpha)^2}) \cdot e^{+i\pi(\alpha\xi - x/\alpha)^2}\, dx \tag{17.65}$$

By changing the frequency parameter to $\mathrm{v} \equiv \alpha^2\xi$ (which has dimensions of length), we obtain the expression:

$$F[\xi] = e^{-i\pi(\mathrm{v}/\alpha)^2} \cdot \int_{-\infty}^{+\infty} (f[x] \cdot e^{-i\pi(x/\alpha)^2}) \cdot e^{+i\pi((\mathrm{v}-x)/\alpha)^2}\, dx \Bigg|_{\mathrm{v}\to\alpha^2\xi} \tag{17.66}$$

which is a convolution in the space domain that produces a function of v:

$$\int_{-\infty}^{+\infty} (f[x] \cdot e^{-i\pi(x/\alpha)^2}) \cdot e^{+i\pi((\mathrm{v}-x)/\alpha)^2}\, dx$$
$$= \int_{-\infty}^{+\infty} (r[x] \cdot h[\mathrm{v} - x])\, dx = r[x] * h[x]\Bigg|_{x\to\mathrm{v}} \tag{17.67a}$$

where

$$h[x] = \exp\left[+i\pi\left(\frac{x}{\alpha}\right)^2\right] \tag{17.67b}$$

This demonstrates that the Fourier transform may be written as a cascade of three operations: multiplication by a "downchirp" quadratic-phase function with chirp rate α and a negative sign, convolution with an "upchirp" quadratic phase with the same chirp rate and evaluated at v, and finally a second multiplication by a "downchirp". The coordinate of the result of the convolution is remapped as $\mathrm{v} \to \alpha^2\xi$ to obtain an equivalent expression for the Fourier transform:

$$F[\xi] = \left(\left[f[x] \cdot \exp\left[-i\pi\left(\frac{x}{\alpha}\right)^2\right]\right] * \exp\left[+i\pi\left(\frac{x}{\alpha}\right)^2\right]\Bigg|_{x\to\mathrm{v}} \cdot \exp\left[-i\pi\left(\frac{\mathrm{v}}{\alpha}\right)^2\right]\right)\Bigg|_{\mathrm{v}\to\alpha^2\xi} \tag{17.68}$$

Alternatively, the coordinates may be remapped from space to frequency after the second multiplication, thus performing all operations with quadratic-phase signals in the space domain:

$$F[\xi] = \left(\left[f[x] \cdot \exp\left[-i\pi\left(\frac{x}{\alpha}\right)^2\right]\right] * \exp\left[+i\pi\left(\frac{x}{\alpha}\right)^2\right]\right) \cdot \exp\left[-i\pi\left(\frac{x}{\alpha}\right)^2\right]\Bigg|_{x\to\alpha^2\xi} \tag{17.69}$$

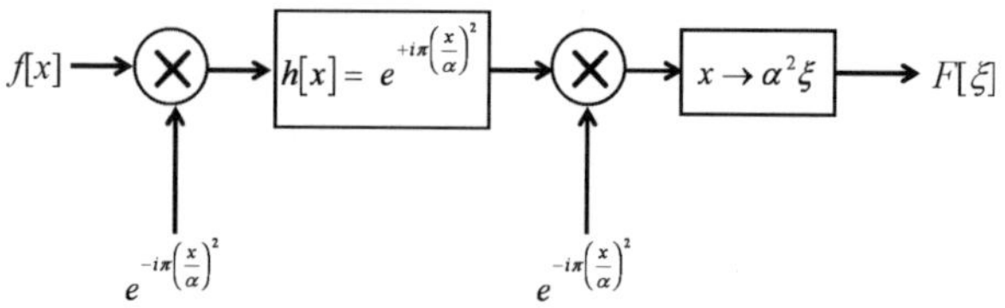

Figure 17.18 Schematic of 1-D M–C–M chirp Fourier transformer.

The sequence of operations suggests the obvious name of "M–C–M" chirp Fourier transformer" for this process. In words, the recipe for the M–C–M chirp Fourier transform is:

1. Multiply $f[x]$ by the 1-D "downchirp" $\exp[-i\pi(x/\alpha)^2]$.
2. Filter the result by convolving with the "upchirp" impulse response $\exp[+i\pi(x/\alpha)^2]$.
3. Multiply the result by a "downchirp" $\exp[-i\pi(x/\alpha)^2]$.
4. Redefine the coordinates by replacing x with $\alpha^2\xi$.

The schematic diagram of the M–C–M transformer is shown in Figure 17.18, and the outputs at the various stages in the process are shown in Figure 17.19. Note that only the phase of the output is affected by the third step in the process, which therefore can be eliminated if only the magnitude (or power) spectrum is required.

The coordinate "remapping" to the frequency domain may be deleted to obtain a relationship between two functions in the space domain. The "output" has the same functional form as the Fourier transform of the input:

$$([f[x]\cdot e^{-i\pi(x/\alpha)^2}] * e^{+i\pi(x/\alpha)^2})\cdot e^{-i\pi(x/\alpha)^2} = \mathcal{F}_1\{f[x]\}|_{\xi=x/\alpha^2} = F[x/\alpha^2] \tag{17.70}$$

In words, this cascade of the three space-domain operations applied to $f[x]$ yields a function *in the space domain* whose amplitude at coordinate x is equal to the amplitude of the Fourier transform of $f[x]$ evaluated at frequency $\xi = x/\alpha^2$. This is an interesting and very useful result in itself, as we will see later, but it also provides the starting point for deriving yet another Fourier relationship involving chirp functions in the next section.

An equivalent expression to Equation (17.70) in the frequency domain is obtained by remapping the coordinates of the input function and the chirps to the frequency domain first, via $x = \alpha^2\xi$, to obtain a relation between two frequency-domain functions $f[\alpha^2\xi]$ and $F[\xi]$:

$$\begin{aligned} F[\xi] &= e^{-i\pi\alpha^2\xi^2}\cdot\int_{-\infty}^{+\infty}(f[\alpha^2\xi]\cdot e^{-i\pi\alpha^2\xi^2})\cdot e^{+i\pi\alpha^2(\xi-x/\alpha^2)^2}\,d(\alpha^2\xi)\\ &= |\alpha|^2((f[\alpha^2\xi]\cdot e^{-i\pi\alpha^2\xi^2}) * e^{+i\pi\alpha^2\xi^2})\cdot e^{-i\pi\alpha^2\xi^2} \end{aligned} \tag{17.71}$$

17.9.2 1-D "C–M–C" Chirp Fourier Transform

The fact that convolution and multiplication in the spatial domain transform to multiplication and convolution in the frequency domain means that there is an alternate algorithm for the chirp transform. We can derive it by applying the "transform-of-a-transform" theorem in Equation (9.104) and the reversal corollary of Equation (9.119) to the "space-domain" chirp Fourier transform in Equation (17.70). There are some subtle features of the derivation that are easy to miss, so we proceed through the steps with care. Consider the M–C–M transform of the specific input function $F[-x/\alpha^2]$; the scaling

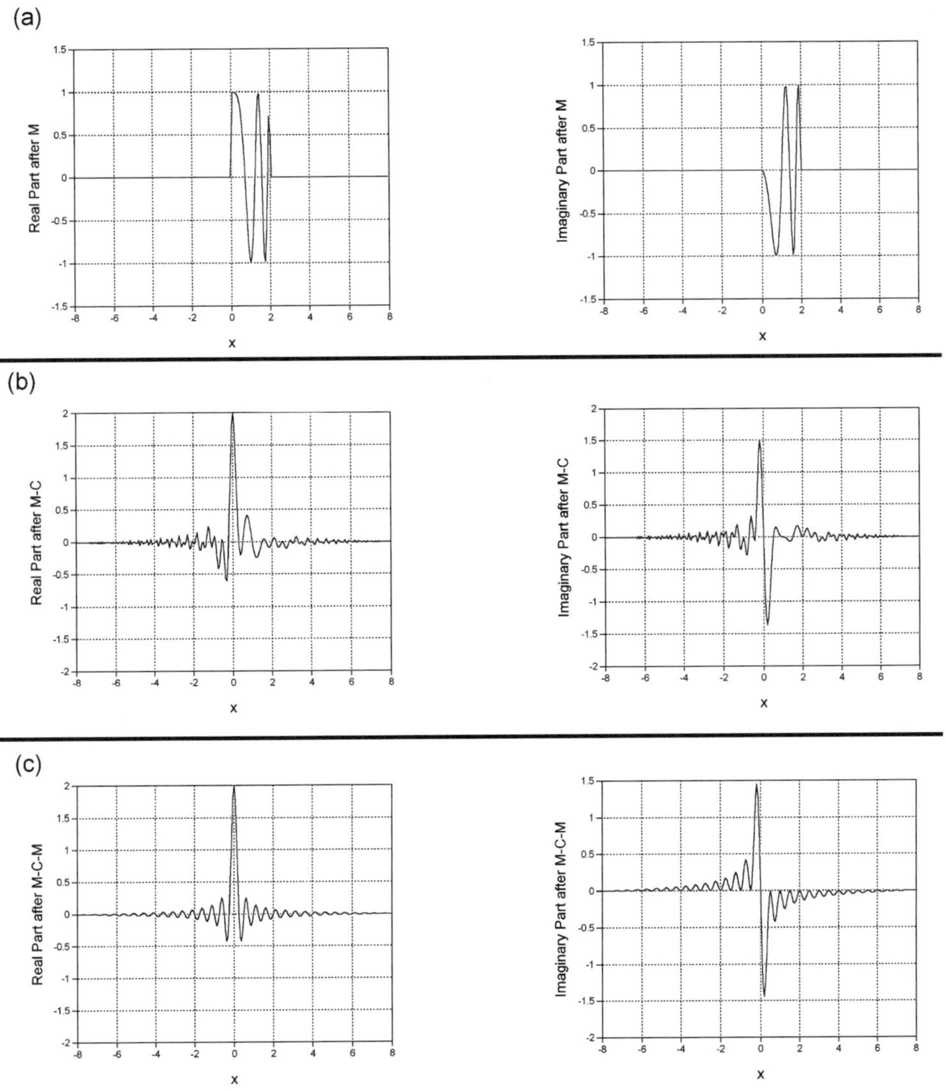

Figure 17.19 Steps in evaluation of M–C–M chirp Fourier transform using the chirp functions $\exp[\pm i\pi x^2]$, so that $\alpha = 1$: (a) real and imaginary parts after M, i.e., of $f[x] \cdot \exp[-i\pi x^2]$; (b) real and imaginary parts after M–C; (c) real and imaginary parts after M–C–M, which produces the amplitude of the spectrum of $f[x]$, but in the space domain, i.e., $F[x]$.

and "transform-of-transform" theorems may be used to find the 1-D spectrum:

$$\left(\left[F\left[-\frac{x}{\alpha^2}\right] \cdot e^{-i\pi(x/\alpha)^2}\right] * e^{+i\pi(x/\alpha)^2}\right) \cdot e^{-i\pi(v/\alpha)^2} = \mathcal{F}_1\left\{F\left[-\frac{x}{\alpha^2}\right]\right\}\Big|_{\xi=+x/\alpha^2}$$
$$= |\alpha|^2 \mathcal{F}_1\{f[\alpha^2\xi]\}|_{\xi=+x/\alpha^2} = \alpha^2 f[x] \tag{17.72}$$

After performing the 1-D Fourier transform on both sides:

$$\mathcal{F}_1\left\{\left(\left[F\left[-\frac{x}{\alpha^2}\right]\cdot e^{-i\pi(x/\alpha)^2}\right]*e^{+i\pi(x/\alpha)^2}\right)\cdot e^{-i\pi(x/\alpha)^2}\right\}=\mathcal{F}_1\{\alpha^2\, f[x]\}$$

$$=\alpha^2\, F[\xi] \qquad (17.73a)$$

This expression may be evaluated via the known transforms of the chirp functions and the filter and modulation theorems:

$$\mathcal{F}_1\left\{\left(\left[F\left[-\frac{x}{\alpha^2}\right]\cdot e^{-i\pi(x/\alpha)^2}\right]*e^{+i\pi(x/\alpha)^2}\right)\cdot e^{-i\pi(x/\alpha)^2}\right\}$$

$$=((|\alpha|^2 f[+\alpha^2\xi]*\mathcal{F}_1\{e^{-i\pi(x/\alpha)^2}\})\cdot\mathcal{F}_1\{e^{+i\pi(x/\alpha)^2}\})*\mathcal{F}_1\{e^{-i\pi(x/\alpha)^2}\}$$

$$=\alpha^2((f[\alpha^2\xi]*[|\alpha|\,e^{-i\pi/4}\,e^{+i\pi\alpha^2\xi^2}])\cdot[|\alpha|\,e^{+i\pi/4}\,e^{-i\pi\alpha^2\xi^2}])*[|\alpha|\,e^{-i\pi/4}\,e^{+i\pi\alpha^2\xi^2}] \qquad (17.73b)$$

Equating the right-hand sides of the expressions in Equations (17.73a) and (17.73b):

$$\Longrightarrow \alpha^2 F[\xi]=|\alpha|^5\,e^{-i\pi/4}[([f[\alpha^2\xi]*e^{+i\pi\alpha^2\xi^2}]\cdot e^{-i\pi\alpha^2\xi^2})*e^{+i\pi\alpha^2\xi^2}]$$

$$\Longrightarrow F[\xi]=|\alpha|^3\,e^{-i\pi/4}[([f[\alpha^2\xi]*e^{+i\pi(\alpha^2\xi)^2/\alpha^2}]\cdot e^{-i\pi\alpha^2\xi^2})*e^{+i\pi\alpha^2\xi^2}] \qquad (17.73c)$$

Because the arguments of the component functions in the first convolution on the right are scaled by α^2, Equation (8.56) must be applied to evaluate the innermost convolution:

$$f[\alpha^2\xi]*e^{+i\pi(\alpha^2\xi)^2/\alpha^2}=\frac{1}{\alpha^2}[f[u]*e^{+i\pi(u/\alpha)^2}]|_{u\to\alpha^2\xi} \qquad (17.74)$$

where u has dimensions of length. The Fourier transform of the M–C–M sequence of operations in Equation (17.73c) becomes:

$$F[\xi]=|\alpha|^3\,e^{-i\pi/4}\left[\left(\frac{1}{\alpha^2}[f[u]*e^{+i\pi(u/\alpha)^2}]|_{u\to\alpha^2\xi}\cdot e^{-i\pi\alpha^2\xi^2}\right)*e^{+i\pi\alpha^2\xi^2}\right]$$

$$=|\alpha|\,e^{-i\pi/4}[([f[u]*e^{+i\pi(u/\alpha)^2}]|_{u\to\alpha^2\xi}\cdot e^{-i\pi\alpha^2\xi^2})*e^{+i\pi\alpha^2\xi^2}] \qquad (17.75)$$

As in the derivation of the M–C–M transform in Equation (17.69), the output coordinates of the first convolution may be "remapped" *after* multiplication by the downchirp:

$$F[\xi]=|\alpha|\,e^{-i\pi/4}([([f[u]*e^{+i\pi(u/\alpha)^2}]\cdot e^{-i\pi(u/\alpha)^2})]|_{u\to\alpha^2\xi}*e^{+i\pi\alpha^2\xi^2})$$

$$=|\alpha|\,e^{-i\pi/4}([([f[u]*e^{+i\pi(u/\alpha)^2}]\cdot e^{-i\pi(u/\alpha)^2})]|_{u\to\alpha^2\xi}*e^{+i\pi(\alpha^2\xi)^2/\alpha^2}) \qquad (17.76)$$

The second scaled convolution may be recast into the form of Equation (8.56):

$$F[\xi]=|\alpha|\,e^{-i\pi/4}\left(\frac{1}{\alpha^2}[([f[u]*e^{+i\pi(u/\alpha)^2}]\cdot e^{-i\pi(u/\alpha)^2})*e^{+i\pi(u/\alpha)^2}]|_{u\to\alpha^2\xi}\right)$$

$$=\frac{1}{|\alpha|}\,e^{-i\pi/4}([([f[u]*e^{+i\pi(u/\alpha)^2}]\cdot e^{-i\pi(u/\alpha)^2})*e^{+i\pi(u/\alpha)^2}]|_{u\to\alpha^2\xi}) \qquad (17.77)$$

We can use any desired symbol for the dummy space-domain variable of integration u. The obvious choice is x, since we usually think of the function f as being in the space domain. This produces an expression for the Fourier transform of $f[x]$ as the cascade of a convolution, a multiplication, and a convolution with appropriate chirp functions:

$$F[\xi]=\frac{1}{|\alpha|}\,e^{-i\pi/4}([([f[x]*e^{+i\pi(x/\alpha)^2}]\cdot e^{-i\pi(x/\alpha)^2})*e^{+i\pi(x/\alpha)^2}]|_{x\to\alpha^2\xi}) \qquad (17.78)$$

In words, this second route to the Fourier transform via quadratic-phase operations is the sequence of operations:

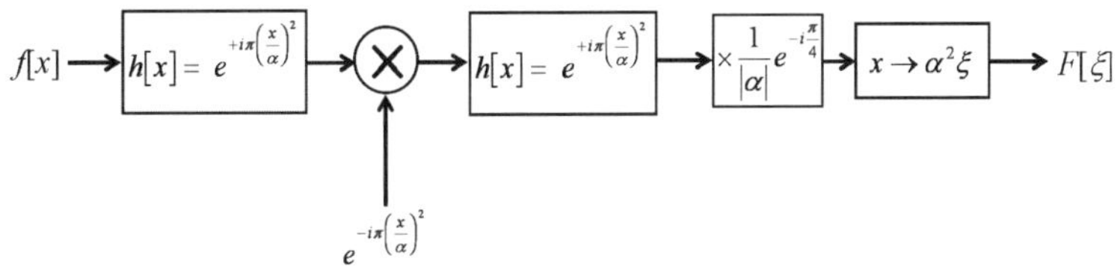

Figure 17.20 Schematic of 1-D C–M–C chirp Fourier transformer.

1. Filter the input $f[x]$ with the "upchirp" impulse response $\exp[+i\pi(x/\alpha)^2]$.
2. Multiply the filtered signal by the "downchirp" $\exp[-i\pi(x/\alpha)^2]$.
3. Filter the result with the "upchirp" $\exp[+i\pi(x/\alpha)^2]$.
4. Multiply the result by the constant factor $|\alpha|^{-1} \cdot e^{-i\pi/4}$.
5. Finally, redefine the coordinates by replacing x with $\alpha^2\xi$.

The obvious shorthand name for this sequence of operations is the "C–M–C" chirp Fourier transform; this interpretation will prove very useful for describing the action of imaging systems in monochromatic (coherent) light and will be examined in detail in Chapter 21.

As before, we may skip the coordinate remapping to produce an operation that is wholly in the space domain:

$$F\left[\frac{x}{\alpha^2}\right] = \left(\frac{1}{|\alpha|} \cdot e^{-i\pi/4}\right) \cdot [([f[x] * e^{+i\pi(x/\alpha)^2}] \cdot e^{-i\pi(x/\alpha)^2}) * e^{+i\pi(x/\alpha)^2}] \tag{17.79}$$

Note that all multiplications are by upchirps and all convolution filters have downchirp impulse responses in both the M–C–M and C–M–C algorithms. In any system where a filter with a quadratic-phase impulse response may be constructed, either algorithm may be implemented in a system with three such filters; the multiplicative upchirps may be generated by applying impulses to filters with upchirp responses, and a similar device with a downchirp response may be used as the convolution filter.

The block diagram schematic of the C–M–C chirp Fourier transformer is shown in Figure 17.20, and the steps in the illustrative example in Figure 17.21.

17.9.3 M–C–M and C–M–C with Opposite-Sign Chirps

The two flavors of chirp were defined with specific signs on the chirp functions: multiplicative chirps always have negative signs and convolution chirps have positive signs. This suggests the question of the effect of reversing the signs of these chirps. The result may be obvious – it is another manifestation of the "transform-of-a-transform" theorem. Consider the M–C–M case:

$$([f[x] \cdot e^{+i\pi(x/\alpha)^2}] * e^{-i\pi(x/\alpha)^2}) \cdot e^{+i\pi(x/\alpha)^2}$$

$$= e^{+i\pi(x/\alpha)^2} \cdot \int_{-\infty}^{+\infty} f[u] \cdot e^{+i\pi(u/\alpha)^2} \cdot e^{-i\pi((x-u)/\alpha)^2}\, du$$

$$= e^{+i\pi(x/\alpha)^2} \cdot \int_{-\infty}^{+\infty} f[u] \cdot e^{+i\pi(u/\alpha)^2} \cdot e^{-i\pi(x^2-2xu+u^2)/\alpha^2}\, du$$

$$= e^{+i\pi(x/\alpha)^2} \cdot \int_{-\infty}^{+\infty} f[u] \cdot e^{-i\pi((x^2-2xu)/\alpha^2)}\, du$$

$$= (e^{+i\pi(x/\alpha)^2} \cdot e^{-i\pi(x/\alpha)^2}) \cdot \int_{-\infty}^{+\infty} f[u]\, e^{-2\pi i(-x/\alpha^2)u}\, du$$

$$= F\left[\xi = -\frac{x}{\alpha^2}\right] \tag{17.80}$$

In words, swapping the algebraic signs of the chirps "reverses" the transform. A similar result applies to the C–M–C case, though the constant phase factor also is conjugated:

$$\frac{1}{|\alpha|}\, e^{+i\pi/4}[([f[x] * e^{-i\pi(x/\alpha)^2}] \cdot e^{+i\pi(x/\alpha)^2}) * e^{-i\pi(x/\alpha)^2}] = F\left[-\frac{x}{\alpha^2}\right] \tag{17.81}$$

17.9.4 2-D Chirp Fourier Transform

The separability of the chirp functions allows the analysis for the M–C–M chirp Fourier transformer to be easily extended to the 2-D case:

$$\mathcal{F}_2\{f[x, y]\} = \iint_{-\infty}^{+\infty} f[x, y]\, e^{-2\pi i(\xi x+\eta y)}\, dx\, dy$$

$$= \iint_{-\infty}^{+\infty} f[x, y]\, e^{i\pi((\alpha\xi - x/\alpha)^2 - (\alpha\xi)^2 - (x/\alpha)^2)}\, e^{i\pi((\alpha\eta - y/\alpha)^2 - (\alpha\eta)^2 - (y/\alpha)^2)}\, dx\, dy$$

$$= \iint_{-\infty}^{+\infty} f[x, y]\, e^{-i\pi((x^2+y^2)/\alpha^2)}\, e^{-i\pi\alpha^2(\xi^2+\eta^2)}\, e^{i\pi(\alpha\xi - x/\alpha)^2}\, e^{i\pi(\alpha\eta - y/\alpha)^2}\, dx\, dy$$

$$= e^{-i\pi\alpha^2\rho^2} \cdot \iint_{-\infty}^{+\infty} (f[x, y]\, e^{-i\pi r^2/\alpha^2}) \cdot e^{i\pi((\alpha\xi - x/\alpha)^2 + (\alpha\eta - y/\alpha)^2)}\, dx\, dy$$

$$= e^{-i\pi\alpha^2\rho^2} \cdot [(f[x, y] \cdot e^{-i\pi r^2/\alpha^2}) * e^{i\pi((x/\alpha)^2 + (y/\alpha)^2)}]|_{x\to\alpha^2\xi, y\to\alpha^2\eta}$$

$$F[\xi, \eta] = ([(f[x, y] \cdot e^{-i\pi(r/\alpha)^2}) * e^{+i\pi(r/\alpha)^2}] \cdot e^{-i\pi(r/\alpha)^2})|_{x\to\alpha^2\xi, y\to\alpha^2\eta} \tag{17.82}$$

The corresponding relation between functions in the space domain is:

$$([(f[x, y] \cdot e^{-i\pi(r/\alpha)^2}) * e^{+i\pi(r/\alpha)^2}] \cdot e^{-i\pi(r/\alpha)^2}) = F\left[\frac{x}{\alpha^2}, \frac{y}{\alpha^2}\right] \tag{17.83}$$

It is easy to show that the 2-D C–M–C transform includes two multiplicative factors of $(|\alpha|\, e^{+i\pi/4})^{-1}$, and so evaluates to:

$$\frac{1}{i \cdot |\alpha|^2}([f[x, y] * e^{+i\pi(r/\alpha)^2}] \cdot e^{-i\pi(r/\alpha)^2}) * e^{+i\pi(r/\alpha)^2} = F\left[\frac{x}{\alpha^2}, \frac{x}{\alpha^2}\right] \tag{17.84}$$

17.9.5 Optical Correlator

As a prelude to the discussion of an optical implementation of the M–C–M Fourier transform, we digress to consider an optical system that evaluates the correlation of two real-valued functions. The system depends on the model of straight-line propagation of light that is the basis for *ray optics* (also called *geometrical optics*). The correlation of the 2-D input $f[x]$ and the 1-D "reference" function $m[x]$ was defined in Equation (8.86):

$$g[x_0, y_0] = f[x, y] \bigstar m[x, y]|_{x=x_0, y=y_0} = \iint_{-\infty}^{+\infty} f[\alpha, \beta]\, m^*[\alpha - x_0, \beta - y_0]\, d\alpha\, d\beta$$

$$\to \iint_{-\infty}^{+\infty} f[\alpha, \beta] m[\alpha - x_0, \beta - y_0]\, d\beta \tag{17.85}$$

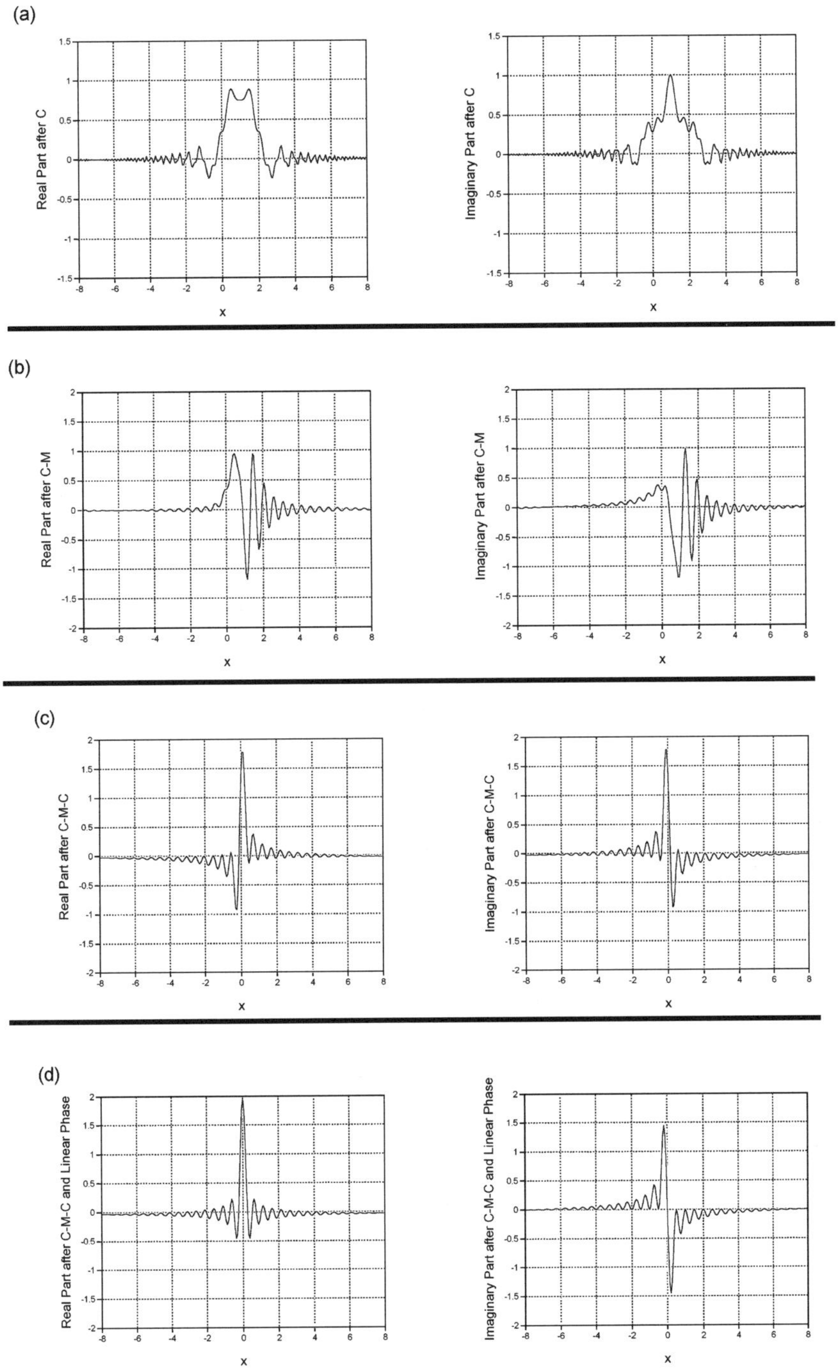

Figure 17.21 Steps in C–M–C chirp Fourier transform algorithm: (a) real and imaginary after C, i.e., $f[x] * \exp[+i\pi x^2]$; (b) real and imaginary parts after C–M; (c) real and imaginary parts after C–M–C; (d) real and imaginary parts after multiplication by constant phase $\exp[-i\pi/4]$ to produce $F[x]$.

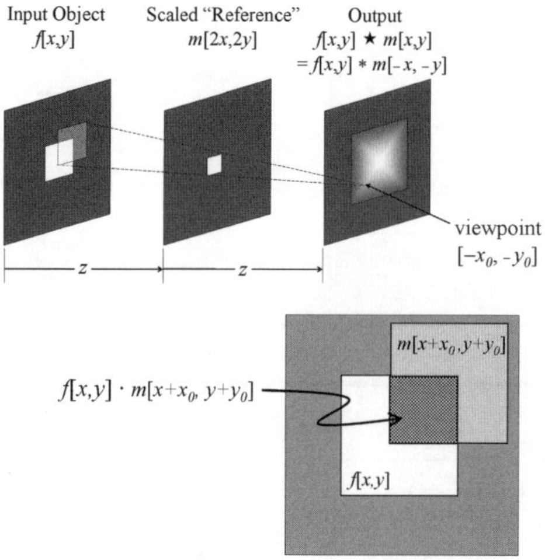

Figure 17.22 Schematic of optical correlator based on ray propagation. The object transparency $f[x, y]$ is placed at the "input" plane on the left and a scaled replica of the reference function $m[2x, 2y]$ at the midplane; the apparent projection of the reference plane onto the input matches their scales. A specific output location (labeled as "viewpoint" in the figure) is the sum of light from all points on the input through the projected reference function. The integrated light observed at the viewpoint is the crosscorrelation of the two signals evaluated at $[-x_0, -y_0]$.

where the last step is valid if the reference function is real valued. In words, the amplitude of the correlation at coordinate $[x_0, y_0]$ is obtained by translating $m[\alpha, \beta]$ by $[x_0, y_0]$, multiplying by the unshifted function $f[x, y]$, and summing the point-by-point products. The corresponding 2-D process is implemented in the *optical correlator* shown in Figure 17.22. The input function $f[x, y]$ is placed in the first plane and a transparency of the scaled replica of the reference function $m[x, y]$ at the midplane so that the scales of the input and projected replica match at the input plane. Each location at the output plane views the intensity at a coordinate of the input $f[x, y]$ in the distant aperture through a specific location in the projected reference function $m[x, y]$ near the aperture, thus evaluating the point-by-point products of the intensity of the distant (input-plane) aperture and the transmittance of the midplane aperture. Because of parallax, the projected image of the midplane reference function is shifted relative to the input, thus providing the translation parameters $[x_0, y_0]$ in Equation (17.85).

As an example, we place a small aperture (approximating a Dirac delta function) at the midplane plane as $m[x, y]$ and some arbitrary $f[x, y]$ at the middle plane. The light distribution at the "detector" plane will be a scaled and reversed replica of $f[x, y]$, meaning that the output evaluated at $[-x_0, -y_0]$ is:

$$f[x, y] \bigstar m[x, y]|_{x=-x_0, y=-y_0} = \iint_{-\infty}^{+\infty} f[\alpha + x_0, \beta + y_0]\, \delta[\alpha, \beta]\, d\alpha\, d\beta$$
$$= f[x_0, y_0] \tag{17.86}$$

This is the principle of the pinhole camera.

If the apertures are exchanged so that $f[2x, 2y]$ is at the midplane, the system projects the image onto the output plane:

$$\delta[x, y] \bigstar f[x, y] = \iint_{-\infty}^{+\infty} f[\alpha + x_0, \beta + y_0]\delta[\alpha, \beta]\, d\alpha\, d\beta = f[+x_0, +y_0] \tag{17.87}$$

Equation (8.37) demonstrated that a crosscorrelator can be used to compute the convolution of f and h by "flipping" the complex conjugate of $h[x]$ before evaluating the crosscorrelation:

$$f[x, y] \bigstar m^*[-x, -y] = f[x, y] * m[x, y] \tag{17.88}$$

17.9.6 Optical Chirp Fourier Transformer

We can use the ray optics correlator to evaluate an approximation to the real part of the 2-D Fourier transform. Though practical systems based on monochromatic wave optics that use lenses accurately generate the 2-D Fourier transform (as we will see in Chapter 21), the system described here may be useful in specialized situations, such as to evaluate the 2-D Fourier transform of the distribution of gamma rays for which no lenses or mirrors exist and for which ray optics is a valid approximation.

We know from Chapter 10 that the 2-D Fourier transform requires multiplication of the input function by 2-D sinusoids with each spatial frequency followed by summing of the point-by-point products. Because of the limitation of the optical correlator, we constrain the input $f[x, y]$ to be real valued and will evaluate only the real part of the transform:

$$\Re\{F[\xi, \eta]\} = \iint_{-\infty}^{+\infty} f[x, y] \cos[2\pi(\xi x + \eta y)]\, dx\, dy \tag{17.89}$$

As we will show, the complete transform may be evaluated by inserting the imaginary part and the odd sinusoids in sequence.

As noted by Mertz (1965), appropriately scaled chirp functions may be inserted at the two planes in the correlator to produce an optical M–C–M chirp Fourier transform. In fact, it is more usual to delete the second multiplication, so that the output is modulated by the last quadratic phase. The mathematical expression for the 1-D M–C–M chirp transform is obtained by multiplying both sides of Equation (17.70) by $e^{+i\pi(x/\alpha)^2}$, thus cancelling the last term on the left-hand side:

$$\left(f[x] \cdot \exp\left[-i\pi\left(\frac{x}{\alpha}\right)^2\right]\right) * \exp\left[+i\pi\left(\frac{x}{\alpha}\right)^2\right] = \exp\left[+i\pi\left(\frac{x}{\alpha}\right)^2\right] \cdot F[\xi]|_{\xi=x/\alpha^2} \tag{17.90}$$

Because convolution is commutative, the left-hand side may be rewritten:

$$e^{+i\pi(x/\alpha)^2} * \left(f[x] \cdot \exp\left[-i\pi\left(\frac{x}{\alpha}\right)^2\right]\right) = F\left(\frac{x}{\alpha^2}\right) \cdot \exp\left[+i\pi\left(\frac{x}{\alpha}\right)^2\right] \tag{17.91}$$

and application of Equation (17.88) yields an output $g[x]$ of the form:

$$\begin{aligned} g[x] &\equiv F\left[\frac{x}{\alpha^2}\right] \cdot \exp\left[+i\pi\left(\frac{x}{\alpha}\right)^2\right] \\ &= \exp\left[+i\pi\left(\frac{x}{\alpha}\right)^2\right] \bigstar \left(f[-x] \cdot \exp\left[-i\pi\left(\frac{x}{\alpha}\right)^2\right]\right) \end{aligned} \tag{17.92}$$

Since it is only possible to use real-valued functions in the optical correlator, we consider the effect of using biased cosine chirps of the form:

$$\cos\left[\frac{\pi x^2}{\alpha^2}\right] = \frac{1}{2}\left(\exp\left[+i\pi\left(\frac{x}{\alpha}\right)^2\right] + \exp\left[-i\pi\left(\frac{x}{\alpha}\right)^2\right]\right) \tag{17.93}$$

so that the output is:

$$g[x] = \frac{1}{2}\left(1 + \cos\left[\frac{\pi x^2}{\alpha^2}\right]\right) \bigstar \left[f[x] \cdot \frac{1}{2}\left(1 + \cos\left[\frac{\pi x^2}{\alpha^2}\right]\right)\right] \tag{17.94}$$

The correlation may be expanded into the sum of nine terms, of which two are proportional to the desired Fourier transform in Equation (17.92):

$$\begin{aligned}
g[x] &= \left[\frac{1}{2}\left(1+\frac{1}{2}[e^{+i\pi(x/\alpha)^2}+e^{-i\pi(x/\alpha)^2}]\right)\right] \bigstar \left[f[x]\cdot\frac{1}{2}\left(1+\frac{1}{2}[e^{+i\pi(x/\alpha)^2}+e^{-i\pi(x/\alpha)^2}]\right)\right] \\
&= \frac{1}{2}(1[x] \bigstar f[x]) \\
&\quad + \frac{1}{8}\left(\exp\left[+i\pi\left(\frac{x}{\alpha}\right)^2\right] \bigstar f[x]\right) \\
&\quad + \frac{1}{8}\left(\exp\left[-i\pi\left(\frac{x}{\alpha}\right)^2\right] \bigstar f[x]\right) \\
&\quad + \frac{1}{8}\left(1[x] \bigstar \left(f[x]\exp\left[+i\pi\left(\frac{x}{\alpha}\right)^2\right]\right)\right) \\
&\quad + \frac{1}{16}\left(\exp\left[+i\pi\left(\frac{x}{\alpha}\right)^2\right] \bigstar \left(f[x]\cdot\exp\left[+i\frac{\pi}{4}\left(\frac{x}{\alpha}\right)^2\right]\right)\right) \\
&\quad + \boxed{\frac{1}{16}\left(\exp\left[-i\pi\left(\frac{x}{\alpha}\right)^2\right] \bigstar \left(f[x]\cdot\exp\left[+i\pi\left(\frac{x}{\alpha}\right)^2\right]\right)\right)} \\
&\quad + \frac{1}{8}\left(1[x] \bigstar \left(f[x]\cdot\exp\left[-i\pi\left(\frac{x}{\alpha}\right)^2\right]\right)\right) \\
&\quad + \boxed{\frac{1}{16}\left(\exp\left[+i\pi\left(\frac{x}{\alpha}\right)^2\right] \bigstar \left(f[x]\cdot\exp\left[-i\pi\left(\frac{x}{\alpha}\right)^2\right]\right)\right)} \\
&\quad + \frac{1}{16}\left(\exp\left[-i\pi\left(\frac{x}{\alpha}\right)^2\right] \bigstar \left(f[x]\cdot\exp\left[-i\pi\left(\frac{x}{\alpha}\right)^2\right]\right)\right)
\end{aligned} \tag{17.95}$$

The terms in boxes are those we seek, while the others are correlations with unit-magnitude functions that contribute "stray light" to the output, thus reducing the contrast of the desired Fourier transform. The output is the real part of the Fourier transform modulated by the real part of a thresholded quadratic phase factor $\cos[r^2/\alpha^2]$. This modulation allows the inference of the positive and negative regions of the transform.

It is helpful to look at the schematic of the M–C optical correlator with "no" input, or, rather, $f[x, y] = 1[x, y]$ in Figure 17.23; this will show how the 2-D cosine functions in Equation (17.89) are generated. The midplane chirp projects onto the input chirp with a displacement proportional to the off-axis distance of the observation point. The zone-plate patterns match exactly if viewed on axis, so this location will be brightly illuminated. When viewed at any other location in the output plane, the apparent displacement results in a function that includes a linear-phase term whose spatial frequency is proportional to the off-axis distance of the observation point $[x_0, y_0]$. In other words, any input function viewed at the output location $[x_0, y_0]$ has been multiplied by a function that includes the cosine with spatial frequency proportional to $(x_0^2 + y_0^2)^{1/2}$; the light measured at that location is the integral of the product of $f[x, y]$ and that cosine, hence computing the 2-D Fourier transform. In Figure 17.23, the displacement is "small" at point A and large at point B, so the input signal is multiplied by linear-phase terms whose frequency is proportional to the off-axis location of the observation point.

The "resolution" of the system (i.e., its ability to distinguish closely spaced spatial frequencies in the Fourier transform) is determined by the bandwidth of the zone plates, i.e., the difference between the maximum and minimum frequencies in the zone plate. The bandwidth of the zone plates may be increased by including higher-frequency terms, but the limit is the assumption of rectilinear propagation

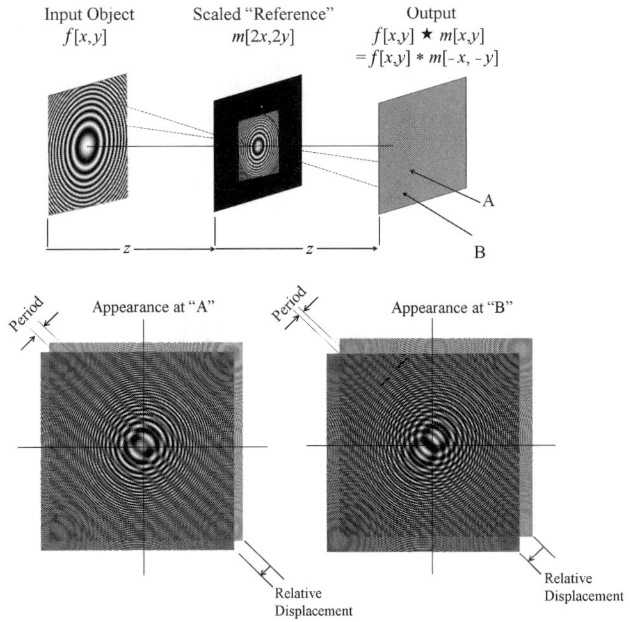

Figure 17.23 Optical correlator implementation of 2-D chirp Fourier transform with $f[x, y] = 1[x, y]$; the appearance of the input chirp seen through the midplane chirp at output locations A and B is shown, demonstrating that the product includes a linear-phase term proportional to the apparent displacement. The integral of the product of these two chirps and the input includes the Fourier transform at the corresponding spatial frequency.

of light from the input to the observation point. Large spatial frequencies imply small structures which in turn amplify any diffraction.

Since we can represent only real functions in the optical correlator, the real and imaginary parts must be computed in separate channels or in sequence (which is analogous to computing the full complex discrete Fourier transform in a computer language that does not support complex arrays). If the input function $f[x]$ is real, then two signal channels are required; if $f[x]$ is complex, a total of four channels are necessary and appropriate outputs must be summed to obtain the final real/imaginary parts of the transform:

$$\begin{aligned} F[\xi] &= \Re\{F[\xi]\} + i\Im\{F[\xi]\} \\ &= \int_{-\infty}^{+\infty} \Re\{f[x]\}\cos[2\pi\xi x]\,dx - \int_{-\infty}^{+\infty} \Im\{f[x]\}\sin[2\pi\xi x]\,dx \\ &\quad + i\int_{-\infty}^{+\infty} \Im\{f[x]\}\cos[2\pi\xi x]\,dx + i\int_{-\infty}^{+\infty} \Re\{f[x]\}\sin[2\pi\xi x]\,dx \end{aligned} \tag{17.96}$$

Mertz's description of the system used biased thresholded quadratic-phase functions, which are called *Fresnel zone plates*:

$$c(r) = \frac{1}{2}\left(1 + SGN\left[\cos\left[\frac{r^2}{\alpha^2}\right]\right] + \phi_0\right) \tag{17.97}$$

For this reason, the system is often called a *Fresnel sandwich*. In actual use, it is common to use "sideband" zone plates whose spatial frequency $\xi_0 \neq 0$ at the center. Such zone plates are not symmetric, and the modulated input function at the source plane must be "flipped" to compute the convolution.

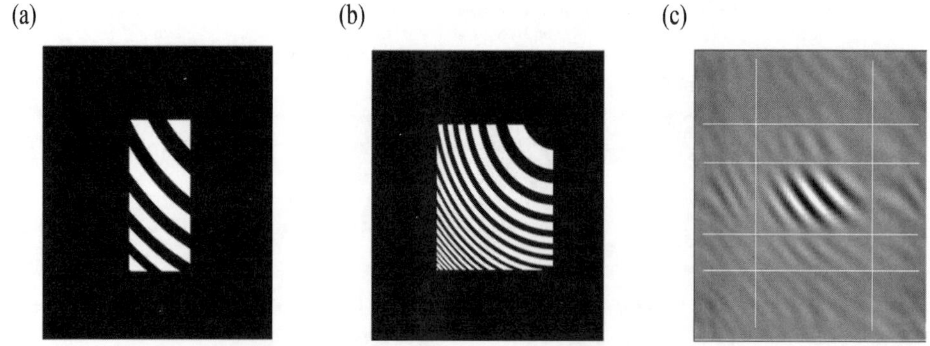

Figure 17.24 Output of optical correlator for $f[x, y] = RECT[x, y/2]$: (a) input plane, showing $f[x, y] \cdot \frac{1}{2}(1 + SGN[\cos[\pi r^2/\alpha^2]])$; (b) midplane function $m[x, y] = \frac{1}{2}(1 + SGN[\cos[\pi r^2/\alpha^2]])$; (c) magnified view of output function $g[x, y] = f[x, y] \bigstar m[x, y]$, showing bipolar Fourier modulated by grating function. The white lines are located at the zeros of the *SINC* function.

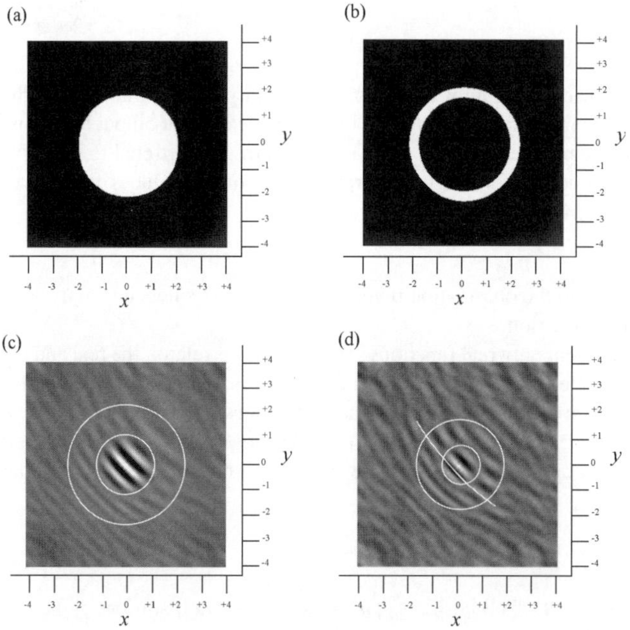

Figure 17.25 Comparison of outputs of optical Fourier transformer for (a) $f(r) = CYL(r/d)$ and (b) $f(r) = CYL(r/d_1) - CYL(r/d_2)$, which approximates the ring delta function. The outputs of the correlator are shown in (c) and (d), respectively, and resemble the sombrero function and the 2-D J_0 Bessel function. The diameters of the central lobe and first ring are outlined and one fringe is tracked in (d) to show the phase change of the grating at the zero crossing.

The right-hand side of Equation (17.91) indicates that the Fourier transform is modulated by a quadratic-phase signal. To visualize the effect of this modulating chirp, consider the case where the input is a centered Dirac delta function $f[x] = \delta[x]$. The impulse source casts a shadow on the output plane of the midplane zone plate, which may be interpreted as a constant modulated by a quadratic phase. Note that the quadratic phase modulation of the constant is a scaled copy of the midplane chirp with the same orientation, as predicted by the right-hand side of the expression for the processor output:

$$\exp\left[+i\pi\left(\frac{x}{\alpha}\right)^2\right] \bigstar \left(f[-x] \cdot \exp\left[-i\pi\frac{(-x)^2}{\alpha^2}\right]\right) = F\left[\frac{x}{\alpha^2}\right] \cdot \exp\left[+i\pi\left(\frac{x}{\alpha}\right)^2\right] \tag{17.98}$$

The signals at the three planes of a Fresnel-sandwich correlator are shown in Figure 17.24 for $f[x, y] = RECT[x, y/2]$, and the outputs for cylinder and ring functions in Figure 17.25. The modulation of the outputs is clearly visible.

As a second example, consider $f[x] = 1[x]$, i.e., the source is a uniformly bright field. The output will "see" the source-plane zone plate through the midplane zone plate. Since the zone-plate patterns match exactly when seen by an observer at the center of the output plane, this location will be brightly illuminated. When viewed at any other location in the output plane, the source-plane zone plate will be partially obscured. In short, the output pattern is a bright spot at the center surrounded by a uniform background. The central bright spot is the DC component, and its spatial extent (e.g., full width at half maximum) is determined by the bandwidth of the zone plates, i.e., the difference between the maximum and minimum frequencies in the zone plate. The bandwidth of the zone plates may be increased by including higher-frequency terms. The distance over which the zone-plate patterns approximately "match" will be shorter, and the size of the DC spot will decrease. In this way, the bandwidth of the zone plates determines the frequency resolution of the chirp Fourier transformer. The performance of this system is limited by the requirement that functions be positive, by the bandwidth of the zone plates, and by light diffracted by the zone plates.

PROBLEMS

17.1 Show that the autocorrelations of the impulse response of the constant-phase and linear-phase allpass filters are on-axis Dirac delta functions.

17.2 Evaluate Equation (17.28) using both the filter theorem and the result for the convolution of two scaled functions:

$$\exp\left[-i\pi\left(\frac{x}{\alpha_2}\right)^2\right] * \exp\left[+i\pi\left(\frac{x}{\alpha_2}\right)^2\right]$$

17.3 Prove Equation (17.29):

$$\exp[-i\pi(\alpha_2\xi)^2] * \exp[+i\pi(\alpha_2\xi)^2] = \frac{1}{|\alpha_2|^2}\delta[\xi]$$

17.4 Show that the magnitude of the function at a coordinate is not preserved under the action of the allpass quadratic-phase filter, i.e., $|g[x]| \neq |f[x]|$, as was true for the constant-phase filter.

17.5 Show that the autocorrelation is preserved for an allpass filter that is the combination of constant and linear phases.

17.6 Show that the magnitude of the function at a coordinate is not preserved under the action of the allpass quadratic-phase filter, i.e., $|g[x]| \neq |f[x]|$, as was true for the constant-phase filter.

17.7 Derive Equation (17.38):

$$\mathcal{F}_1^{-1}\{\exp[\pm i\pi(\alpha_n\xi)^n]\}$$

$$= \frac{2}{n|\alpha_n|}\sum_{p=0}^{+\infty}\left\{(-1)^p(\pi^{-(2p+1)/n})\frac{(2\pi)^{2p}}{(2p)!}\left(\frac{x}{\alpha_n}\right)^{2p}\right\}$$

$$\cdot\left\{\Gamma\left[\frac{2p+1}{n}\right]\cos\left[\frac{\pi(2p+1)}{2n}\right]\mp\frac{2\pi}{(2p+1)}(\pi^{-1/n})\Gamma\left[\frac{2p+2}{n}\right]\sin\left[\frac{\pi(2p+2)}{2n}\right]\left(\frac{+x}{\alpha_n}\right)\right\}$$

17.8 Evaluate the M–C–M or C–M–C chirp Fourier transforms

$$F\left[\frac{x}{\alpha^2}\right] = ((f[x]\cdot e^{-i\pi(x/\alpha)^2}) * e^{+i\pi(x/\alpha)^2})\cdot e^{-i\pi(x/\alpha)^2}$$

$$= \frac{1}{|\alpha|}e^{-i\pi/4}\cdot(\{(f[x] * e^{+i\pi(x/\alpha)^2})\cdot e^{-i\pi(x/\alpha)^2}\} * e^{+i\pi(x/\alpha)^2})$$

for the following functions:

(a) $f[x] = \delta[x]$
(b) $f[x] = \delta[x - x_0]$
(c) $f[x] = 1[x]$
(d) $f[x] = \exp[+2\pi i\xi_0 x]$
(e) $f[x] = \exp[+i\pi x^2]$

17.9 Use the C–M–C chirp Fourier transform to find an expression for $f[x] * \exp[+i\pi(/x\alpha)^2]$ in terms of $F[x/\alpha^2]$.

17.10 Find expressions for "incomplete" chirp Fourier transforms where the last operations in both the M–C–M and C–M–C cases have been deleted.

17.11 Show that the product of biased zone plates placed at the input plane and midplane of the Fresnel sandwich includes a linear-phase term whose spatial frequency is proportional to the off-axis observation location $[x_0, y_0]$.

18

Magnitude–Phase Filters

In the discussion of linear filters in the previous two chapters, the transfer function could modify either the magnitude spectrum or the phase spectrum of the input signal, but not both. It may be apparent that all realistic LSI systems affect both attributes of the input spectrum simultaneously, if only because real materials are used to make the filters. For example, electronic filters cannot have infinite bandwidth due to the physical behavior of the materials. Consider a temporal electronic signal of the form $f[t]$ (with spectrum $F[\nu]$), which usually is a current of electrons carried by conductive wires or a voltage measured across a device. The current or voltage may be constant with time ($\nu = 0$) or may oscillate at a nonzero frequency. One class of electronic device produces an output current proportional to the voltage across the device; the input and output signals (voltages or currents) of these *resistors* always have the same phase, which means that the output signal is proportional to the input at all frequencies ν. Real resistors do not behave in the ideal manner due to the physical limitations of the constituent materials. For example, electron forces in real resistors prevent the output voltage from exactly following a high-frequency oscillating input voltage. It also is not possible to construct realistic time-domain filters with infinitesimal bandwidth, because the length of the corresponding impulse response would approach infinity. These two constraints (among others) mean that the magnitude of the filtering action will vary with frequency. These bounding constraints on realistic filters provide some motivation for considering filters that affect both magnitude and phase simultaneously.

The first examples of "mixed" filters discussed in this chapter are basic operations in mathematics that we have already considered in other contexts: the identity operator, the derivative, and the semi-infinite integral. The identity operator is obviously an allpass filter with zero phase, and so neither the magnitude nor the phase can be affected. The other two operators affect both the magnitude and phase in well-defined ways, and the investigation of the character of the corresponding transfer functions may enhance understanding of these common operations. Next, a common problem from classical mechanics will be considered in terms of linear filters to introduce the *causal* filters. Though the concept of causality rarely arises in imaging applications, the study of this problem is a useful exercise. Magnitude filters with linear, random, and quadratic phases are discussed next, with some emphasis placed on the ideal lowpass filter with quadratic phase because of its importance in optical imaging, where it models the action of an aberration-free lens in the Fresnel diffraction region.

Equation (17.4) showed that the action of an allpass filter preserves the autocorrelation of the input function. The corresponding relationship between the input and output signals for the more general linear filter also is easy to derive because the magnitude of the output spectrum is the product of the magnitude of the input spectrum and the MTF:

$$|\mathcal{F}_1\{g[x]\}| = |G[\xi]| = |\mathcal{F}_1\{f[x] * h[x]\}| = |F[\xi] \cdot H[\xi]| = |F[\xi]| \cdot |H[\xi]| \tag{18.1}$$

Fourier Methods in Imaging Roger L. Easton, Jr.

which clearly shows that the output spectral power is the product of the squared magnitudes of the input spectrum and transfer function:

$$|G[\xi]|^2 = |F[\xi]|^2 \cdot |H[\xi]|^2 \tag{18.2}$$

The autocorrelations of the input, output, and impulse response are related through the Wiener–Khintchin theorem of Equation (9.159):

$$\begin{aligned} g[x] \bigstar g[x] &= \mathcal{F}_1^{-1}\{|G[\xi]|^2\} \\ &= \mathcal{F}_1^{-1}\{|F[\xi]|^2 \cdot |H[\xi]|^2\} \\ &= (f[x] \bigstar f^*[x]) * (h[x] \bigstar h^*[x]) \\ &= (f[x] * f^*[-x]) * (h[x] * h^*[-x]) \end{aligned} \tag{18.3}$$

In words, the autocorrelation of the output of the general linear filter is the convolution of the autocorrelations of the input and the impulse response.

18.1 Transfer Functions of Three Operations

Consider three specific LSI operators that may be implemented as convolutions:

$$\textit{Identity operator:} \quad g_1[x] = \mathcal{O}_1\{f[x]\} = f[x] \tag{18.4}$$

$$\textit{Differentiator:} \quad g_2[x] = \mathcal{O}_2\{f[x]\} = \frac{df}{dx} \tag{18.5}$$

$$\textit{Integrator:} \quad g_3[x] = \mathcal{O}_3\{f[x]\} = \int_{-\infty}^{x} f[\alpha]\, d\alpha \tag{18.6}$$

We have already considered the associated transfer functions in Chapter 9, but list them here for convenience.

18.1.1 Identity Operator

The impulse response of the identity operator is derived trivially from the sifting theorem:

$$\begin{aligned} \mathcal{F}_1\{f[x]\} = \mathcal{F}_1\{f[x] * \delta[x]\} &= F[\xi] \cdot 1[\xi] \\ &= F[\xi] \cdot H_1[\xi] \end{aligned} \tag{18.7}$$

The expressions for the transfer function as real/imaginary parts and as magnitude/phase also are trivial to derive, as was shown in Equation (9.57):

$$H_1[\xi] = \mathcal{F}_1\{\delta[x]\} = 1[\xi] + i \cdot 0[\xi] = |1|\, e^{i \cdot 0} \tag{18.8}$$

As we have already seen, the identity operator passes all spectrum information to the output without change. The impulse response and transfer function of the identity operator are shown in Figure 18.1a,b.

18.1.2 Differentiation

The derivative of $f[x]$ was expressed as a convolution in Equation (6.132):

$$g_2[x] = \frac{df}{dx} = f[x] * \delta'[x] = \int_{-\infty}^{+\infty} f[\alpha]\, \delta'[x - \alpha]\, d\alpha \tag{18.9}$$

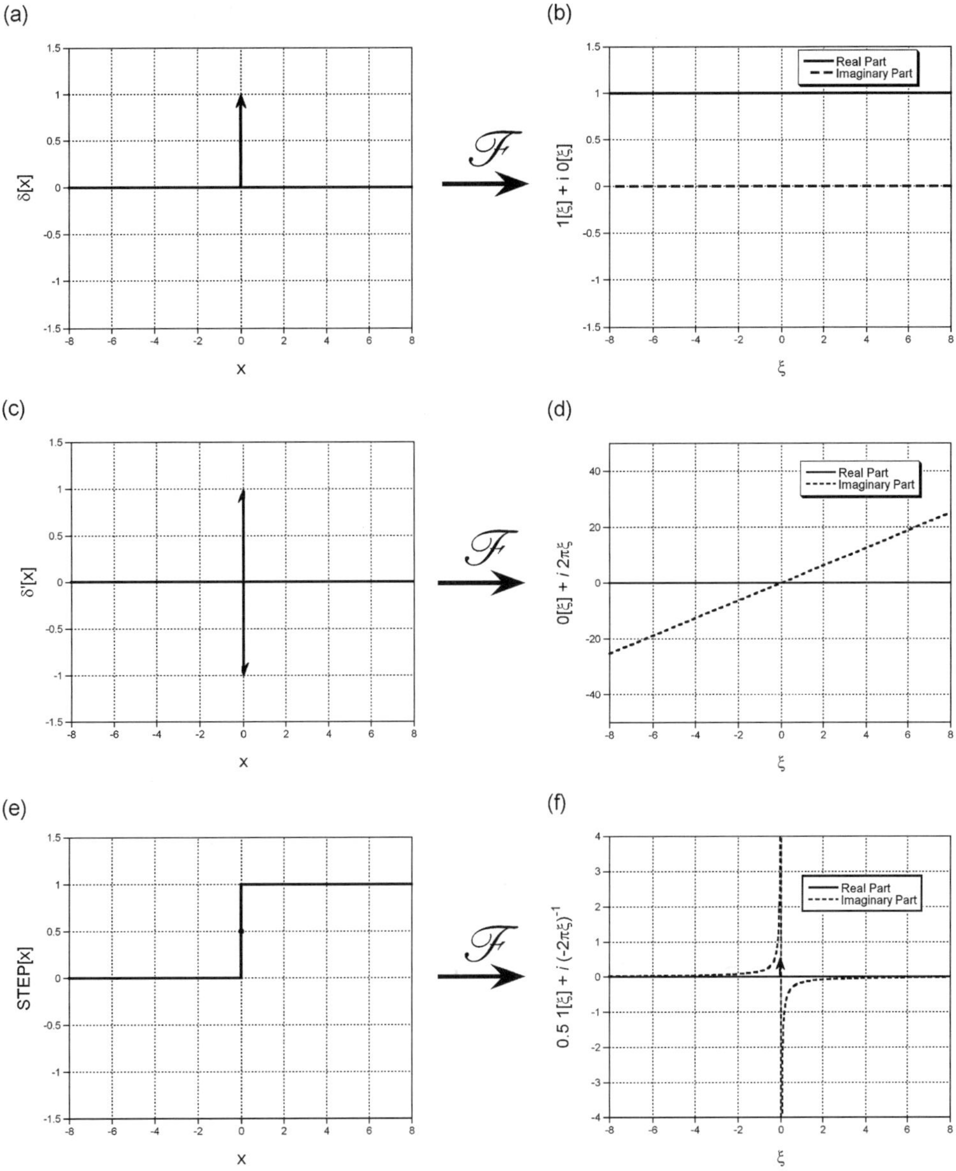

Figure 18.1 Three mathematical operators expressed as impulse responses and transfer functions: (a) impulse response of the identity operator $h[x] = \delta[x] + i \cdot 0[x]$; (b) transfer function $H[\xi] = 1[\xi] + i \cdot 0[\xi]$; (c) impulse response of differentiation; (d) transfer function of differentiation $H[\xi] = 0[\xi] + i(2\pi\xi)$; (e) impulse response of integration $h[x] = STEP[\xi]$; (f) transfer function of integration $H[\xi] = \frac{1}{2}\delta[\xi] + i(-1/(2\pi\xi))$.

which means that the impulse response of the derivative filter is the derivative of the Dirac delta function:

$$h_2[x] = \delta'[x] \tag{18.10}$$

The transfer function is easy to evaluate by applying the derivative theorem in Equation (9.142):

$$\mathcal{F}_1\left\{\frac{df}{dx}\right\} = 2\pi i\xi \cdot F[\xi] \equiv F[\xi] \cdot H_2[\xi]$$

$$\Longrightarrow H_2[\xi] = 0[\xi] + i \cdot (2\pi\xi) \tag{18.11}$$

Therefore, the MTF and **Φ**TF of the derivative filter are:

$$|H_2[\xi]| = 2\pi|\xi| \tag{18.12a}$$

$$\Phi\{H_2[\xi]\} = +\frac{\pi}{2} \cdot SGN[\xi] \tag{18.12b}$$

The impulse response and transfer function of the differentiator are shown in Figure 18.1c,d. This demonstrates the effect of differentiation on the spectrum of the input function. First, the magnitude of the input spectrum is amplified by a factor proportional to $|\xi|$; the constant of proportionality is chosen so that the magnitudes of sinusoidal components with spatial frequency $\xi = \pm 1/2\pi$ are not changed. The phase of a harmonic component with spatial frequency greater than zero is incremented by $\pi/2$, while the phase of negative frequency components is decremented by the same amount. The phase at $\xi = 0$ is not relevant, since $|H_2[\xi = 0]| = 0$ so that the constant part of the output must be zero.

The transfer function of differentiation could have been inferred from observations in the study of calculus. Recall the expression for the derivative of a cosine of positive spatial frequency ξ_0:

$$\frac{d}{dx}\cos[2\pi|\xi_0|x + \phi_0] = 2\pi|\xi_0|(-\sin[2\pi|\xi_0|x + \phi_0])$$

$$= 2\pi|\xi_0|\cos\left[(2\pi|\xi_0|x + \phi_0) + \frac{\pi}{2}\right] \tag{18.13a}$$

where the identity $\sin[\theta] = -\cos[\theta + \pi/2]$ has been used. This demonstrates that the amplitude of this sinusoid is amplified by the factor ξ_0 and its phase is incremented by $\pi/2$ radians.

The derivative of a negative frequency sinusoid is:

$$\frac{d}{dx}(\cos[2\pi(-|\xi_0|)x + \phi_0]) = \frac{d}{dx}\cos[2\pi|\xi_0|x - \phi_0]$$

$$= 2\pi|\xi_0|[-\sin[2\pi|\xi_0|x - \phi_0]]$$

$$= 2\pi|\xi_0|[\sin[-(2\pi|\xi_0|x + \phi_0)]]$$

$$= 2\pi|\xi_0|\cos\left[(2\pi(-|\xi_0|)x + \phi_0) - \frac{\pi}{2}\right] \tag{18.13b}$$

where the symmetry of the cosine has been used. This result establishes that the MTF for $\xi_0 < 0$ is the same as for $\xi_0 > 0$ except that the phase is decremented by $\pi/2$ radians instead of incremented. The MTF ensures that the amplitude of the derivative of a sinusoid will be large if the frequency is large. This is because the differences of "nearby" amplitudes are larger for rapidly oscillating signals.

Since differentiation is linear, the derivative of the sum of multiple sinusoids is identical to the sum of the derivatives. The differentiator will amplify each component in proportion to its frequency, while changing the phase of positive frequencies by $+\pi/2$ and of negative frequencies by $-\pi/2$.

18.1.3 Integration

The last basic mathematical operator to be considered is the integral:

$$\int_{-\infty}^{x} f[\alpha]\,d\alpha = \int_{-\infty}^{+\infty} f[\alpha]\,STEP[x-\alpha]\,d\alpha$$

$$= f[x] * STEP[x] \tag{18.14}$$

The corresponding transfer function has already been derived in Equation (9.87):

$$
\begin{aligned}
H_3[\xi] &= \mathcal{F}_1\{STEP[x]\} \\
&= \mathcal{F}_1\left\{\frac{1}{2}(1 + SGN[x])\right\} \\
&= \frac{1}{2}\delta[\xi] + \frac{1}{2\pi i\xi} \\
&= \frac{1}{2}\delta[\xi] + i\cdot\left(-\frac{1}{2\pi\xi}\right)
\end{aligned}
\tag{18.15}
$$

The MTF and **Φ**TF also were shown to be:

$$
|H_3[\xi]| = \frac{1}{2}\delta[\xi] + \frac{1}{2\pi|\xi|} \tag{18.15a}
$$

$$
\boldsymbol{\Phi}\{H_3[\xi]\} = \begin{Bmatrix} -\dfrac{\pi}{2} & & \xi > 0 \\ 0 & \text{if} & \xi = 0 \\ +\dfrac{\pi}{2} & & \xi < 0 \end{Bmatrix} = -\frac{\pi}{2}SGN[\xi] \tag{18.15b}
$$

The impulse response and transfer function are shown in Figure 18.1e,f.

Consider the action of this transfer function on the input spectrum $F[\xi]$:

$$
\begin{aligned}
G_3[\xi] = F[\xi]\cdot H_3[\xi] &= \frac{1}{2}(F[\xi]\cdot\delta[\xi]) - \frac{i}{2\pi}\left(F[\xi]\cdot\frac{1}{\xi}\right) \\
&= \frac{1}{2}(\delta[\xi]\cdot F[0]) - \frac{i}{2\pi}\left(F[\xi]\cdot\frac{1}{\xi}\right) \\
&= \frac{1}{2}\left(\delta[\xi]\cdot\int_{-\infty}^{+\infty} f[x]\,dx\right) - \frac{i}{2\pi}\left(F[\xi]\cdot\frac{1}{\xi}\right)
\end{aligned}
\tag{18.16}
$$

The DC term $G_3[0]$ is the sum of a Dirac delta function weighted by the area of $f[x]$. The central-ordinate theorem shows that the area of $g_3[x]$ is infinite if the area of $f[x]$ is nonzero. The algebraic sign of the area of $g_3[x]$ is positive or negative depending on whether the area of $f[x]$ is positive or negative.

The transfer function indicates that the integrator amplifies sinusoidal components with $|\xi| < 1/2\pi$ and attenuates those with larger frequencies. Also, integration respectively decrements and increments the phase of positive and negative frequency sinusoids by $\pi/2$ radians. Again, this results in the form of integrals of sinusoids that we already know from integral calculus:

$$
\begin{aligned}
\int_{-\infty}^{+\infty} \cos[2\pi|\xi_0|x + \phi_0]\,dx &= \frac{1}{2\pi}|\xi_0|\sin[2\pi|\xi_0|x + \phi_0] \\
&= \frac{1}{2\pi}|\xi_0|\cos\left[2\pi|\xi_0|x + \phi_0 - \frac{\pi}{2}\right]
\end{aligned}
\tag{18.17a}
$$

$$\int_{-\infty}^{+\infty} \cos[2\pi(-|\xi_0|)x + \phi_0]\,dx = \int_{-\infty}^{+\infty} \cos[(2\pi|\xi_0|x) - \phi_0]\,dx$$
$$= \frac{1}{2\pi}|\xi_0| \sin[2\pi|\xi_0|x - \phi_0]$$
$$= \frac{1}{2\pi}|\xi_0| \cos\left[(2\pi|\xi_0|x - \phi_0) - \frac{\pi}{2}\right]$$
$$= \frac{1}{2\pi}|\xi_0| \cos\left[2\pi(-|\xi_0|)x + \phi_0 + \frac{\pi}{2}\right] \tag{18.17b}$$

which shows that the phase transfer function of the integrator is the negative of that for the differentiator.

The effect of integration on the harmonic components of the input function is well illustrated by applying the process to a Dirac delta located at the origin, so that $f[x] = \delta[x]$. The sifting property of the Dirac delta function is applied to show trivially that the output is a *STEP* function:

$$\int_{-\infty}^{x} \delta[\alpha]\,d\alpha = g[x] = f[x] * h[x]$$
$$= \delta[x] * STEP[x] = STEP[x] \tag{18.18}$$

and the output spectrum was derived in Equation (9.87):

$$G[\xi] = \frac{1}{2}\left(\delta[\xi] + \frac{1}{i\pi\xi}\right) \cdot F[\xi] \tag{18.19}$$

We have demonstrated several times that the Dirac delta function at the origin is composed of the sum of cosines with equal amplitudes at all frequencies. Equation (6.102) demonstrates that the semi-infinite integral of the Dirac delta function is *STEP*$[x]$, which is itself composed of the weighted sum of a scaled copy of *SGN*$[x]$ and the half-unit constant. Note that the amplitude of the signum is positive for positive x and negative for negative x, which is behavior shared by $\sin[2\pi\xi x]$ near the origin. Thus the sum of sine waves with different spatial frequencies will be expected to have the same character as *SGN*$[x]$ near the origin. The sine waves are created by decrementing the phase of the original cosine components by $\pi/2$ radians to make them "line up" with zero amplitude at the origin. The initial phase of the constant component ($\xi = 0$) always is zero because there is no need to translate a constant. Unlike the components of $\delta[x]$, the amplitudes of the sine-wave components of *SGN*$[x]$ vary with spatial frequency. Low-frequency components have the large amplitude necessary to create the "mass" of the signum away from the origin; the small-amplitude high-frequency components are responsible for creating the "sharp corners" of *SGN*$[x]$ at the origin.

If the operations of integration and differentiation are cascaded, the spectrum of the output can be expressed as the product of the input spectrum and the two transfer functions:

$$G[\xi] = F[\xi] \cdot (H_2[\xi] \cdot H_3[\xi]) \tag{18.20}$$

$$H_2[\xi] \cdot H_3[\xi] = (i \cdot 2\pi\xi) \cdot \left(\frac{1}{2}\delta[\xi] - \frac{i}{2\pi\xi}\right)$$
$$= i\pi\xi \cdot \delta[\xi] + \frac{2\pi\xi}{2\pi\xi}$$
$$= \begin{cases} 1 & \textit{if } \xi \neq 0 \\ +\infty & \textit{if } \xi = 0 \end{cases} \tag{18.21}$$

It probably is not surprising that the transfer functions of the integration and differentiation filters are mutually reciprocal at all frequencies except $\xi = 0$. This ensures that the transfer function of the cascaded operators is unity for $\xi \neq 0$, but not that its value is not well defined at $\xi = 0$. In words,

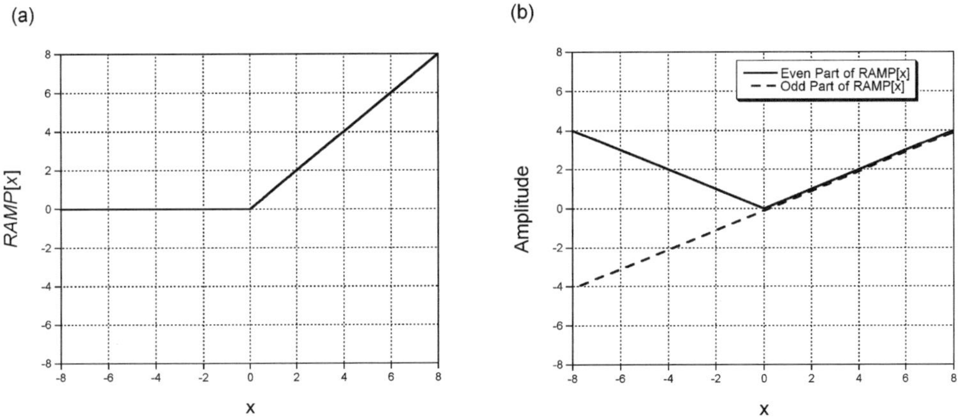

Figure 18.2 (a) $RAMP[x] = x \cdot STEP[x]$; (b) even and odd parts: the even part is $|x|/2$ and the odd part is $x/2$.

the DC component of the input function $f[x]$ is not passed by a system that cascades differentiation and integration; it is "blocked" by the derivative operator. Therefore, the original function cannot be recovered perfectly from its derivative by integration alone; we also need a priori information about the DC component that is rejected by the derivative. In other words, the integration operator is the *pseudoinverse* of differentiation, and vice versa. In the study of differential equations, this necessary information about the constant value is introduced as a boundary condition.

18.2 Fourier Transform of Ramp Function

At this point, we make another digression to derive the spectrum of $x \cdot STEP[x]$, which may be assigned its own name, $RAMP[x]$:

$$RAMP[x] \equiv x \cdot STEP[x] = \begin{cases} x & \text{if } x \geq 0 \\ 0 & \text{if } x < 0 \end{cases} \tag{18.22}$$

This function may be decomposed into its even and odd parts, which respectively are the sources of the real and imaginary parts of the spectrum:

$$RAMP[x] = \frac{|x|}{2} + \frac{x}{2} \tag{18.23}$$

shown in Figure 18.2. The even part is a "V"-shaped function centered at the origin and the odd part is a scaled replica of x.

The spectrum of $RAMP[x]$ is useful when deriving the properties of certain physical systems involving causal filters and resonance. The spectrum may be derived by applying the known spectra of $STEP[x]$ from Equation (9.87) and of x^n from Equation (9.145) to the modulation theorem of Equation (9.140). We have already derived a similar result during the discussion of the inverse Radon transform in Chapter 12. The resulting spectrum is:

$$\begin{aligned}\mathcal{F}_1\{RAMP[x]\} &= \mathcal{F}_1\{x \cdot STEP[x]\} \\ &= \mathcal{F}_1\{x\} * \mathcal{F}_1\{STEP[x]\} \\ &= -\frac{1}{2\pi i}\delta'[\xi] * \frac{1}{2}\left(\delta[\xi] + \frac{1}{i\pi\xi}\right)\end{aligned}$$

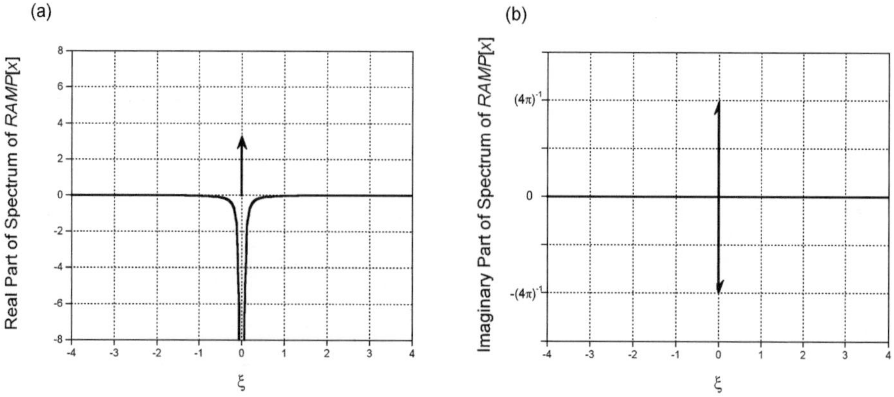

Figure 18.3 Spectrum of *RAMP*[x]: (a) Real part, which is proportional to $|\xi|^{-1}$ for $|\xi| \neq 0$ and an indeterminate-area Dirac delta function; (b) imaginary part, which is a scaled "doublet" function.

$$= \left(-\frac{1}{2\pi i}\delta'[\xi] * \frac{1}{2}\delta[\xi]\right) - \left(\frac{1}{2\pi i}\delta'[\xi] * \frac{1}{2\pi i}\xi^{-1}\right)$$

$$= +i\left(\frac{1}{4\pi}\delta'[\xi]\right) + \frac{1}{4\pi^2}\frac{d(\xi^{-1})}{d\xi}$$

$$= \begin{cases} -\dfrac{1}{4\pi^2\xi^2} + i\left(\dfrac{1}{4\pi}\delta'[\xi]\right) & \textit{if } \xi \neq 0 \\ \displaystyle\int_0^{+\infty} x\,dx + i \cdot 0 & \textit{if } \xi = 0 \end{cases} \tag{18.24}$$

where the central-ordinate theorem is the source of the second condition. The spectrum at the origin is a Dirac delta function and is Hermitian, as required. By equating the transform of the real part of *RAMP*[x] to the even part of Equation (18.23), we obtain the spectrum of $|x|$:

$$\mathcal{F}_1\{|x|\} = \begin{cases} -\dfrac{1}{2\pi^2\xi^2} & \textit{if } \xi \neq 0 \\ 2 \cdot \displaystyle\int_0^{+\infty} x\,dx & \textit{if } \xi = 0 \end{cases} \tag{18.25}$$

which is a real-valued even function, as expected. Application of the "transform-of-a-transform" theorem produces the corresponding expression:

$$\mathcal{F}_1\{x^{-2}\} = -2\pi^2|-\xi| = -2\pi^2|\xi| \quad \textit{if } \xi \neq 0 \tag{18.26}$$

The spectrum of *RAMP*[x] is illustrated in Figure 18.3.

18.3 Causal Filters

Though the action of the simple harmonic oscillator in classical mechanics may seemingly bear no relation to linear filtering, the interpretation of the former in terms of the latter may provide insight into both. This investigation requires that the temporal impulse response of the filter be constrained by the concept of *causality*, which is essential for modeling many physical phenomena that are relevant to imaging, such as the frequency dependence of the optical index of refraction. This discussion may cast the phase behavior of the driven ("forced") harmonic oscillator into a more intuitive light.

From basic classical mechanics, we know that harmonic motion results from the balanced interaction of kinetic and potential energy in a system where the force applied to a body is proportional to its displacement from a point of equilibrium. The situation is often illustrated by the action of a mass m attached to a spring that applies a force proportional to a restoring constant k. The restoring force is given by Hooke's law:

$$\underline{\mathbf{F}} = -k \cdot (\underline{\mathbf{x}}[t] - \underline{\mathbf{x}}_0) = m\frac{d^2\underline{\mathbf{x}}}{dt^2} \tag{18.27}$$

where $\underline{\mathbf{x}}[t]$ is the position of the mass at time t and $\underline{\mathbf{x}}_0$ is the equilibrium position. The constant k has dimensions of force per unit distance, e.g., grams per second^2 in cgs units. This expression may be recast in terms of an equation of motion:

$$m\frac{d^2\underline{\mathbf{x}}}{dt^2} + k \cdot (\underline{\mathbf{x}}[t] - \underline{\mathbf{x}}_0) = 0 \tag{18.28}$$

The origin of coordinates may be selected so that $\underline{\mathbf{x}}_0 = 0$. For simplicity in this illustration, we assume that the motion is in one spatial dimension only; the system is a 1-D oscillator and the vector notation may be dropped so that $\underline{\mathbf{x}} \to x$. The standard techniques for solving differential equations usually are applied to Equation (18.28). For example, a trial solution of the form:

$$x[t] = A\, e^{qt} \tag{18.29}$$

may be assumed, which leads to a condition on the parameter q:

$$q = \pm i\sqrt{\frac{k}{m}} \equiv \pm i\omega_0 \tag{18.30}$$

The parameter ω_0 is an angular temporal frequency measured in radians per second. The general form of the equation of motion is:

$$x[t] = A_1\, e^{+i\omega_0 t} + A_2\, e^{-i\omega_0 t} = A_0 \cos[\omega_0 t + \phi_0] \tag{18.31}$$

The amplitudes A_1 and A_2, or the equivalent "polar" quantities (the magnitude of oscillation $A_0 \geq 0$ and the initial phase ϕ_0), may be evaluated from the boundary conditions of the problem. The response of the system that satisfies the differential equation in Equation (18.28) apparently is a sinusoid that oscillates at the angular temporal frequency ω_0 that is determined by the physical conditions of the system. Equation (2.14) may be used to recast the expression in terms of the temporal frequency $\nu_0 = \omega_0/2\pi$. This is the natural oscillation frequency of the oscillator, also called its *resonant* frequency. In the case where the maximum amplitude of the system occurs at $t = 0$, then $\phi_0 = 0$ and Equation (18.31) reduces to:

$$x_e[t] = A_0 \cos[\omega_0 t] = A_0 \cos[2\pi\nu_0 t] \tag{18.32}$$

where the subscript "e" indicates that the motion is an even (symmetric) function of time. A more common initial condition in physical problems requires that the amplitude of the oscillator at $t = 0$ be $x[0] = 0$, which means that $\phi_0 = \pm\pi/2$ and that the subsequent motion is specified by an odd sinusoid. We select $\phi_0 = -\pi/2$, which merely means that the amplitude increases in the positive direction after $t = 0$. The equation of motion of the oscillator is:

$$x_o[t] = A_0 \sin[\omega_0 t] = A_0 \sin[2\pi\nu_0 t] \tag{18.33}$$

where, again, the subscript "o" is applied to x to indicate that this is an odd sinusoid.

In words, the action of the harmonic oscillator may be described as the output of a linear system with a particular impulse response. If an oscillator is initially at rest and is then disturbed by the action of an impulsive input, the system "responds" by oscillating about the equilibrium point at the resonant frequency ν_0. The output in Equation (18.31) may be interpreted in a loose sense as the convolution of the impulsive input and an impulse response. For simplicity, assume that the impulse occurs at $t = 0$ so that the time-domain representation of the system is an infinite-support cosine function:

$$f[t] * h[t] \Longrightarrow \delta[t] * h_e[t] = x_e[t]$$
$$\Longrightarrow h_e[t] = x_e[t] = A_0 \cos[2\pi \nu_0 t] \tag{18.34}$$

where the subscript "e" again indicates that the impulse response is an even function of time. At this point, we emphasize that $x_e[t]$ in Equation (18.32) would have the form of the impulse response under the stated conditions of the problem, but this not a realistic situation, as will be seen very shortly. For the moment, we continue to ignore reality and find the form of the corresponding transfer function by applying Equation (9.74):

$$H_e[\nu] = \frac{A_0}{2|\nu_0|}\delta\delta\left[\frac{\nu}{\nu_0}\right] = \frac{A_0}{2}\delta[\nu + \nu_0] + \frac{A_0}{2}\delta[\nu - \nu_0] \tag{18.35}$$

Now consider the behavior of $x_o[t]$ if the system is assumed to be at rest at the equilibrium point before being disturbed by the action of an impulsive input. Again, we might interpret that the impulse response is a sine wave with infinite support:

$$h_o[t] = x_o[t] = A_0 \sin[2\pi \nu_0 t] = A_0 \cos\left[2\pi \nu_0 t - \frac{\pi}{2}\right] \tag{18.36}$$

The form of the assumed system transfer function may be found by the straightforward application of the spectrum of the odd sinusoid in Equation (9.75):

$$H_o[\nu] = i\frac{A_0}{2|\nu_0|}\delta_\delta\left[\frac{\nu}{\nu_0}\right] = i \cdot \frac{A_0}{2}(\delta[\nu + \nu_0] - \delta[\nu - \nu_0]) \tag{18.37}$$

Once the impulse response or transfer function of the system has been derived, the motion of the system may be determined for any input "driving force" as the convolution with the impulse response. From this point forward, we return to the customary notation $g[t]$ instead of $x[t]$ to represent the output, so that:

$$F[\nu] \cdot H[\nu] = G[\nu] \tag{18.38}$$

The input $f[t]$ to the system is the *driving* motion applied to the oscillator, and may be impulsive, oscillatory, or some other continuous function. For illustration, consider that the input to the system is a cosinusoidal function of time that oscillates at temporal frequency ν_1 with amplitude A_1. The spectrum of the driving force has the same form as the assumed transfer function in Equation (18.35) except for the difference in the temporal frequencies:

$$f_e[t] = A_1 \cos[2\pi \nu_1 t]$$
$$\Longrightarrow F_e[\nu] = \frac{A_1}{2|\nu_1|}\delta\delta\left[\frac{\nu}{\nu_1}\right] = \frac{A_1}{2}\delta[\nu + \nu_1] + \frac{A_1}{2}\delta[\nu - \nu_1] \tag{18.39}$$

If instead the input were an odd sinusoid, the spectrum would have a form identical to Equation (18.36) except at a different frequency:

$$f_o[t] = A_1 \cos\left[2\pi \nu_1 t - \frac{\pi}{2}\right] = A_1 \sin[2\pi \nu_1 t]$$
$$\Longrightarrow F_o[\nu] = i \cdot \frac{A_1}{2}(\delta[\nu + \nu_1] - \delta[\nu - \nu_1]) \tag{18.40}$$

If either the even input in Equation (18.39) or the odd input in Equation (18.40) is applied to a system whose impulse response has infinite support, as in Equation (18.34) or Equation (18.36), and if

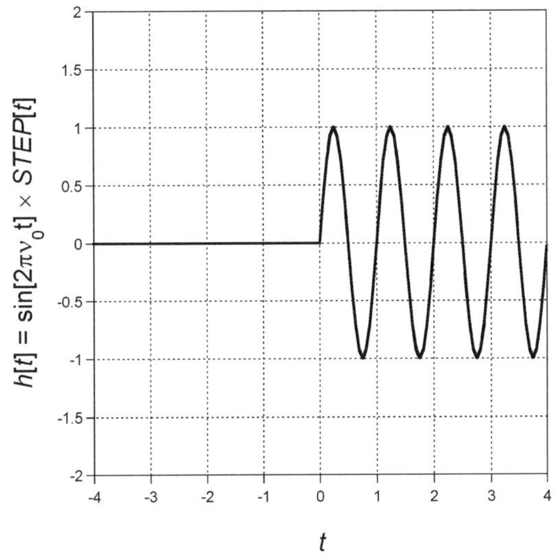

Figure 18.4 Impulse response of casual system that remains at rest until disturbed. The amplitude has a sinusoidal form for $t > 0$, so its functional form is $h[t] = \sin[2\pi\nu_0 t] \cdot STEP[t]$.

the resonant frequency ν_0 of the system is different from the driving frequency ν_1, then the frequency-domain picture demonstrates that the amplitude of the output oscillation will be zero because the two frequencies are different. In other words, no output is produced from a system with a purely sinusoidal impulse response that is driven "off resonance". However, any number of laboratory experiments in freshman physics indicate that this is not the true behavior of realistic systems, which implies that something about the analysis must be incorrect.

The error in the analysis actually becomes rather obvious if we reconsider the assumed impulse responses in Equations (18.34) and (18.36). Both assume that application of an impulsive input to the temporal system produces a sinusoidal function of time with infinite support, which implies that the output oscillation would exist for all times t, including $t < 0$ (before the impulsive driving force had been applied). This obviously is an unphysical assumption. The error was in our assumed output in Equation (18.31), which specified no initial conditions. In realistic systems, the system impulse response must satisfy the constraint of *causality*, so that no output is produced before application of an input. The appropriate causal impulse response is obtained by modulating the appropriate harmonic impulse response by a *STEP* function, as shown in Figure 18.4. In the realistic case where the system is at rest at its equilibrium position until disturbed by the input, the impulse response is:

$$h[t] = A_0 \sin[2\pi\nu_0 t] \cdot STEP[t] \tag{18.41}$$

An impulsive input applied at $t = t_0$ has the form $f[t] = \delta[t - t_0]$. After the impulse is applied, the "mass" (or its equivalent) begins to move and the restoring force increases in opposition to the inertia until the former is sufficiently strong to reverse the direction of the net force, thus reversing the direction of motion. The amplitude returns to zero where the inertial force is maximized and the restoring force is zero. The amplitude then becomes negative until again the restoring force overcomes the inertia and again reverses the direction of motion.

The associated transfer function of this "causal" impulse response is obtained by applying the modulation theorem in Equation (9.140):

$$\begin{aligned} H(\nu) &= \mathcal{F}_1\{A_0 \sin[2\pi\nu_0 t] \cdot STEP[t]\} \\ &= A_0 \cdot \mathcal{F}_1\{\sin[2\pi\nu_0 t]\} * \mathcal{F}_1\{STEP[t]\} \end{aligned}$$

$$= A_0 \cdot \left(\frac{i}{2}(\delta[\nu+\nu_0] - \delta[\nu-\nu_0])\right) * \frac{1}{2}\left(\delta[\nu] + i\left(-\frac{1}{\pi\nu}\right)\right)$$

$$= \frac{A_0}{4}\left[\frac{1}{\pi}\left(\frac{1}{\nu+\nu_0} - \frac{1}{\nu-\nu_0}\right) + i \cdot (\delta[\nu+\nu_0] - \delta[\nu-\nu_0])\right]$$

$$= \frac{A_0}{2\pi}\left(\frac{\nu_0}{\nu_0^2 - \nu^2}\right) + i \cdot \frac{A_0}{4}(\delta[\nu+\nu_0] - \delta[\nu-\nu_0]) \tag{18.42}$$

as shown in Figure 18.5. The impulse response is real valued and therefore the associated transfer function is Hermitian. The scale factor applied to the odd pair of Dirac delta functions in the imaginary part is $\frac{1}{4}$ because the impulse response is a sinusoid over only half of the infinite domain.

Note that the amplitude of the transfer function is not defined at $\pm\nu_0$ in both the "ideal" case and the causal case, but the real part is nonzero at all other finite frequencies for the "causal" oscillator. The "broad" response of the "causal" harmonic oscillator means that input functions that oscillate at frequencies other than the resonant frequencies $\pm\nu_0$ can produce nonzero outputs. If the causal oscillator is driven with a true infinite-support harmonic oscillation "on resonance" ($\nu_1 = \nu_0$), then the amplitude of the output spectrum at the resonant frequency is undefined due to the singularities in the transfer function, which means that the amplitude of the time-domain output function is infinitely large.

The output of the oscillator when driven by an infinite-support harmonic wave off resonance is illustrated in Figure 18.6 for the two cases of the driving frequency below resonance ($\nu_0 > \nu_1$) and above resonance ($\nu_0 < \nu_1$). The output for ν_1 below resonance is "in phase" with the input, so that both the input and output reach a positive maximum at the same time. The output for $\nu_1 > \nu_0$ is "out of phase" by π radians so that when the input is a positive maximum, the output is a negative maximum. These results may also be obtained by classical (and more tedious) methods described in many texts on college physics or mechanics (e.g., Marion and Thornton, 1995).

However, this analysis also is not realistic because it assumes that the input $f[t]$ existed at all times t. This is (of course) also not possible; the driving sinusoid must start at a specific time, e.g., $f[t] = A_1 \cos[2\pi\nu_1 t + \phi_1] \cdot STEP[t]$. By following the same logic that led to Equation (18.41), so that the oscillation "starts" from its equilibrium point and increases, we set $\phi_1 = -\pi/2$ radians:

$$f[t] = A_1 \sin[2\pi\nu_1 t] \cdot STEP[t] \tag{18.43}$$

Note that we could have also chosen $\phi_1 = +\pi/2$.

Now consider the output that would result if the frequency of the input oscillation were to match the resonant frequency of the oscillator, so that $\nu_1 = \nu_0$. The output $g[t]$ is easily evaluated in the time domain:

$$g[t] = (A_0 \sin[2\pi\nu_0 t] \cdot STEP[t]) * (A_1 \sin[2\pi\nu_0 t] \cdot STEP[t])$$

$$= \frac{A_0 A_1}{2} \cdot \left(RAMP[t] \cdot \cos[2\pi\nu_0 t - \pi] + \frac{STEP[t]}{2\pi\nu_0} \cdot \sin[2\pi\nu_0 t]\right) \text{ (“on” resonance)} \tag{18.44}$$

In words, the output of the causal harmonic oscillator on resonance includes two terms. The more noticeable is in quadrature with the driving force and increases in amplitude over time without limit. The second term is proportional to the driving sinusoid and its amplitude decreases with increasing frequency. The output of a harmonic oscillator driven on resonance is a classic example in classical mechanics and has natural manifestations as lake seiches. It is also famous as the physical mechanism that tore apart the Tacoma Narrows suspension bridge in 1940.

18.4 Damped Harmonic Oscillator

In yet-more-realistic problems, the amplitude of the impulse response of the causal oscillator decreases with time due to additional "damping" forces, which are proportional to the velocity of the mass and

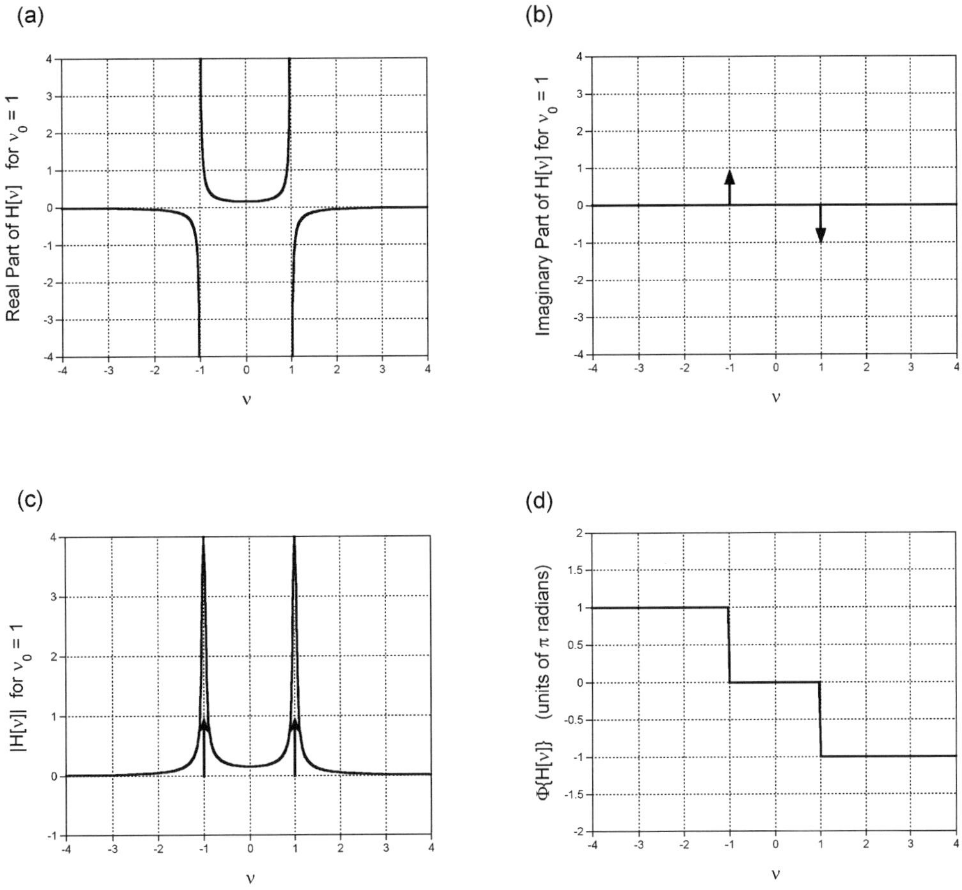

Figure 18.5 Transfer function of causal filter with $h[t] = \sin[2\pi t] \cdot STEP[t]$, showing response at frequencies other than $\nu_0 = 1$: (a) real part $(1/2\pi)[\nu_0/(1-\nu^2)]$; (b) imaginary part $\frac{1}{4}(\delta[\nu+\nu_0] - \delta[\nu-\nu_0])$; (c) Magnitude; (d) Phase.

which add a third term to the differential equation:

$$m\frac{d^2x}{dt^2} - b\frac{dx}{dt} + kx[t] = 0 \tag{18.45a}$$

This may be rewritten by setting $\kappa \equiv \sqrt{k/m}$ and $\gamma \equiv b/2m > 0$:

$$\frac{d^2x}{dt^2} - 2\gamma\frac{dx}{dt} + \kappa^2 x[t] = 0 \tag{18.45b}$$

The form of the equation of motion depends on the size of the damping term $\gamma > 0$, which has dimensions of reciprocal time; γ is the reciprocal of the *decay time* of the oscillator, which is the time required for the amplitude to drop by a factor of $e^{-1} \cong 0.367$.

The causal impulse response of the system is:

$$\begin{aligned} h[t] &= A_0(e^{-\gamma t} \cdot \sin[\omega' t]) \cdot STEP[t] \\ &= A_0(e^{-\gamma t} \cdot STEP[\gamma t]) \cdot \sin[\omega' t] \end{aligned} \tag{18.46}$$

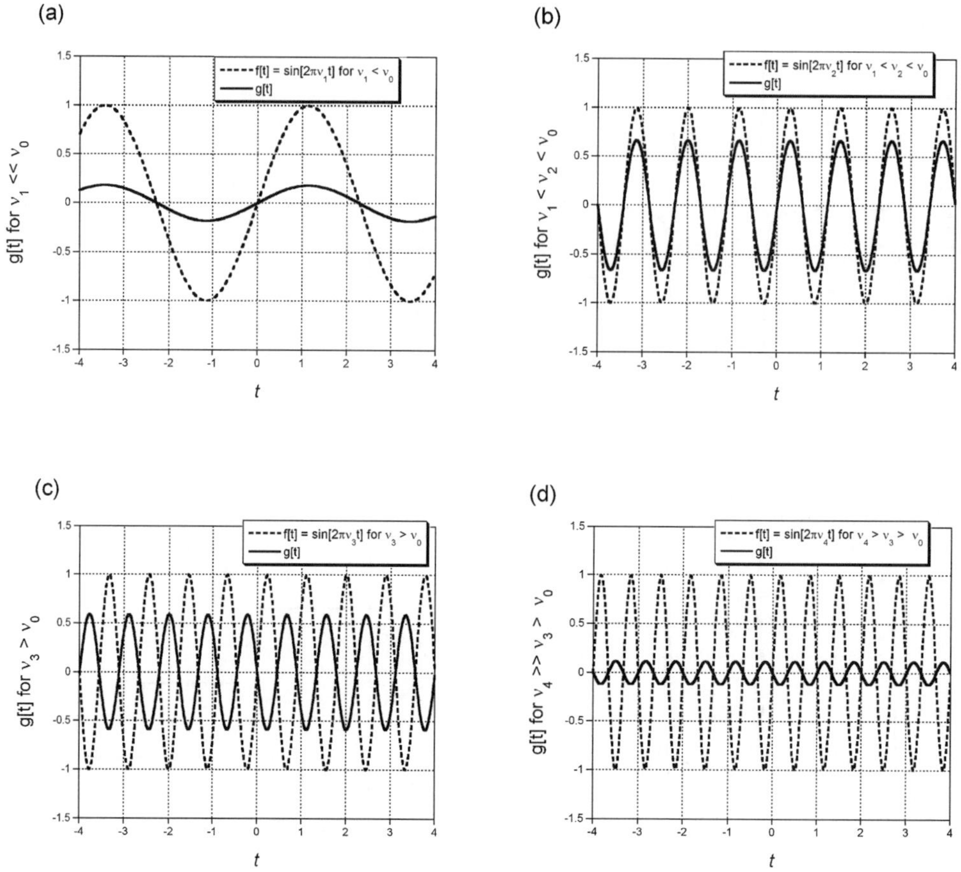

Figure 18.6 Steady-state output of causal filter with $h[t] = \sin[2\pi\nu_0 t] \cdot STEP[t]$ for input frequencies that are smaller and larger than the resonant frequency: (a) $f[t] = \sin[2\pi\nu_1 t]$ for $\nu_1 \ll \nu_0$; (b) $f[t] = \sin[2\pi\nu_2 t]$ for $\nu_1 < \nu_2 < \nu_0$; (c) $f[t] = \sin[2\pi\nu_3 t]$ for $\nu_3 > \nu_0$; (d) $f[t] = \sin[2\pi\nu_4 t]$ for $\nu_4 \gg \nu_0$. The outputs of (a) and (b) with $\nu < \nu_0$ are "in phase", while those with $\nu > \nu_0$ are "out of phase".

The parameter $\omega' = 2\pi\nu'$ is the *natural angular frequency* of the oscillator, which includes contributions from both the *resonant frequency* ν_0 of the undamped oscillator and the damping term γ:

$$(\omega')^2 \equiv \frac{k}{m} - \gamma^2 \Longrightarrow \nu' = \sqrt{\nu_0^2 - \left(\frac{\gamma}{2\pi}\right)^2} \tag{18.47}$$

Clearly $\nu' < \nu_0$, so the natural oscillations with damping are slower than those without damping. The difference between ν_0 and ν' is more noticeable for larger values of γ, which means that the damping time becomes shorter.

The associated transfer function is the convolution of the odd pair of Dirac delta functions with a complex Lorentzian:

$$\begin{aligned} H(\nu) &= A_0\mathcal{F}_1\{\sin[2\pi\nu' t]\cdot(e^{-\gamma t}\cdot STEP[\gamma t])\} \\ &= i\frac{A_0}{2\gamma}(\delta[\nu+\nu'] - \delta[\nu-\nu']) * \left(\frac{1}{1+2\pi i(\nu/\gamma)}\right) \\ &= \frac{2\pi A_0}{\gamma^2}\frac{\nu'(1+(4\pi^2/\gamma^2)(\nu'^2-\nu^2)) - i(4\pi\nu/\gamma)}{(1+(4\pi^2/\gamma^2)(\nu'^2-\nu^2))^2 + (4\pi\nu/\gamma)^2} \end{aligned} \tag{18.48}$$

as shown in Figure 18.7. By comparing to the transfer function for the undamped case in Equation (18.42), we see that the effect of damping "broadens" the real part of the transfer function from a pair of Dirac delta functions to the sum of a pair of Lorentzians; this means that the transfer function passes a broader range of input spatial frequencies to the output. Consequently, if a damped harmonic oscillator is driven by an off-resonance input oscillation, part of the output will be in phase with the input. The effect of damping on the imaginary part of the transfer function averages the two singularities at $\pm\nu_0$ to produce a smoother curve without singularities. The output representation in the time domain is obtained by the Fourier transform of the filtered spectrum and is definitely not harmonic.

18.5 Mixed Filters with Linear or Random Phase

The simplest mixed (magnitude and phase) filters are described easily in both the coordinate and frequency representations. For example, consider a filter with linear **Φ**TF and MTF $M[\xi]$:

$$H[\xi] = M[\xi]\cdot e^{+2\pi i x_0\xi} \tag{18.49}$$

The impulse response is easy to derive via the filter theorem and is a translated replica of the space-domain representation of the MTF:

$$h[x] = m[x] * \delta[x+x_0] = m[x+x_0] \tag{18.50}$$

The situation is not so simple if the **Φ**TF is random, but we can still draw some useful conclusions (Frieden, 2002):

$$H[\xi] = M[\xi]\cdot e^{+2\pi i\cdot N[\xi]} \tag{18.51}$$

where $N[\xi]$ consists of amplitudes selected randomly from some statistical distribution. The **Φ**TF is identical to that of the allpass random-phase filter considered in Chapter 17. The impulse response is the convolution of the two component functions:

$$h[x] = m[x] * \mathcal{F}_1^{-1}\{e^{+2\pi i\cdot N[\xi]}\} \tag{18.52}$$

We know from Chapter 9 that the impulse response of the phase function is complex-valued uncorrelated Gaussian random noise with infinite support. If $M[\xi]$ is a lowpass MTF, then $m[x]$ acts as a local averager that reduces the variance of the noise in $h[x]$. If $M[\xi]$ is highpass, then the variance of the space-domain Gaussian noise will increase. An example for a lowpass MTF is shown in Figure 18.8.

18.6 Mixed Filter with Quadratic Phase

Consider a filter with MTF $M[\xi]$ and quadratic phase with chirp rate α (dimensions of length). The transfer function is:

$$H[\xi] = M[\xi]\cdot e^{+i\pi\alpha^2\xi^2} \tag{18.53}$$

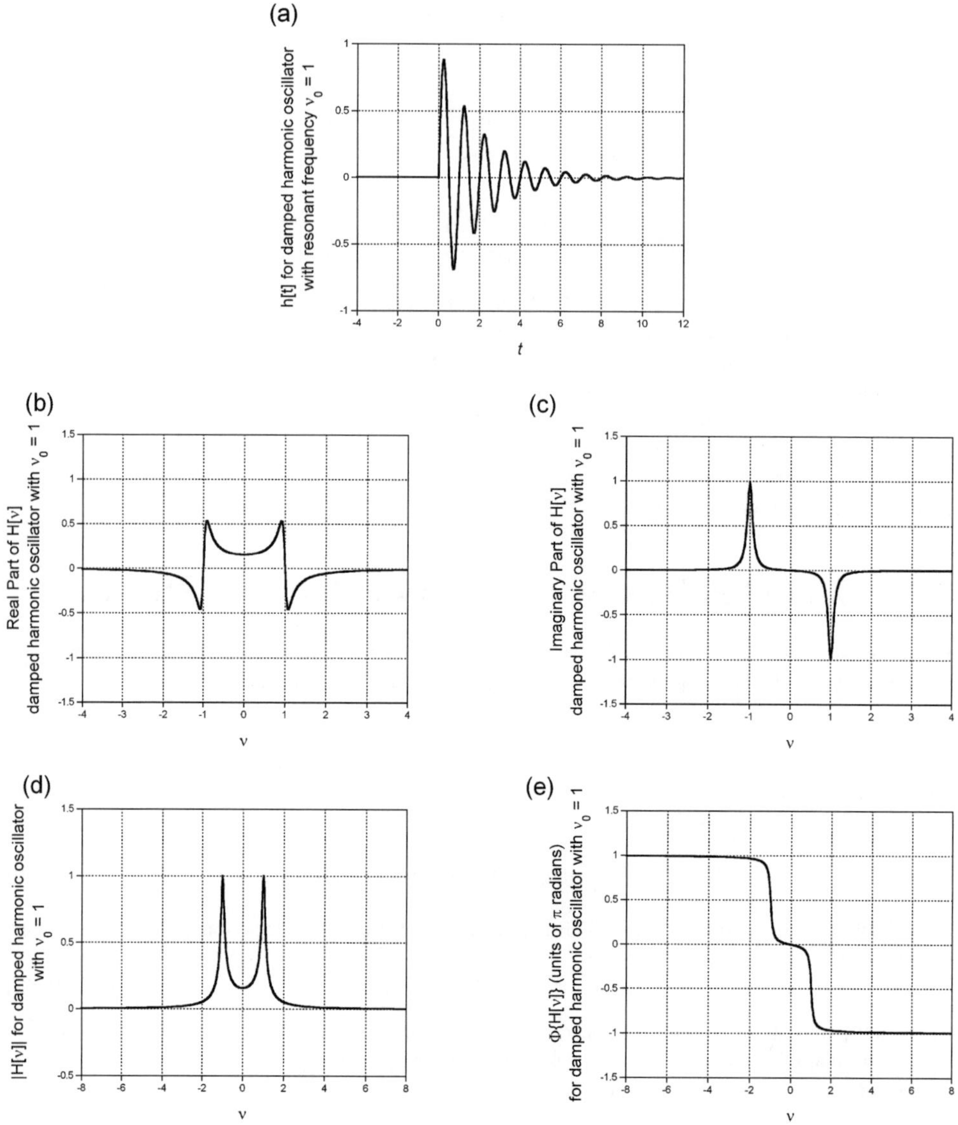

Figure 18.7 Impulse response and transfer function of damped harmonic oscillator: (a) impulse response $h[t] = \sin[2\pi\nu_0 t] \cdot \exp[-\gamma t] \cdot STEP[t]$; (b) $\Re\{H[\nu]\}$; (c) $\Im\{H[\nu]\}$; (d) $|H[\nu]|$; (e) $\Phi\{H[\nu]\}$, which is approximately in phase for driving frequencies $|\nu| < \nu'$ and out of phase for $|\nu| > \nu'$. Note the "broadened" response of the transfer function due to the damping $\exp[-\gamma t]$.

If α is large, then the frequency-domain chirp oscillates "rapidly" and the space-domain chirp oscillates "slowly", i.e., the coordinate where the phase difference is π radians measured relative to the origin is small in the frequency domain and large in the space domain. We can write the corresponding impulse response in several ways, as both exact and approximate expressions, via the filter theorem of Equation (9.129), the chirp Fourier transforms in Equations (17.69) and (17.78), and the stationary-phase approximation in Equation (13.64). The first is the most obvious, though perhaps least effective,

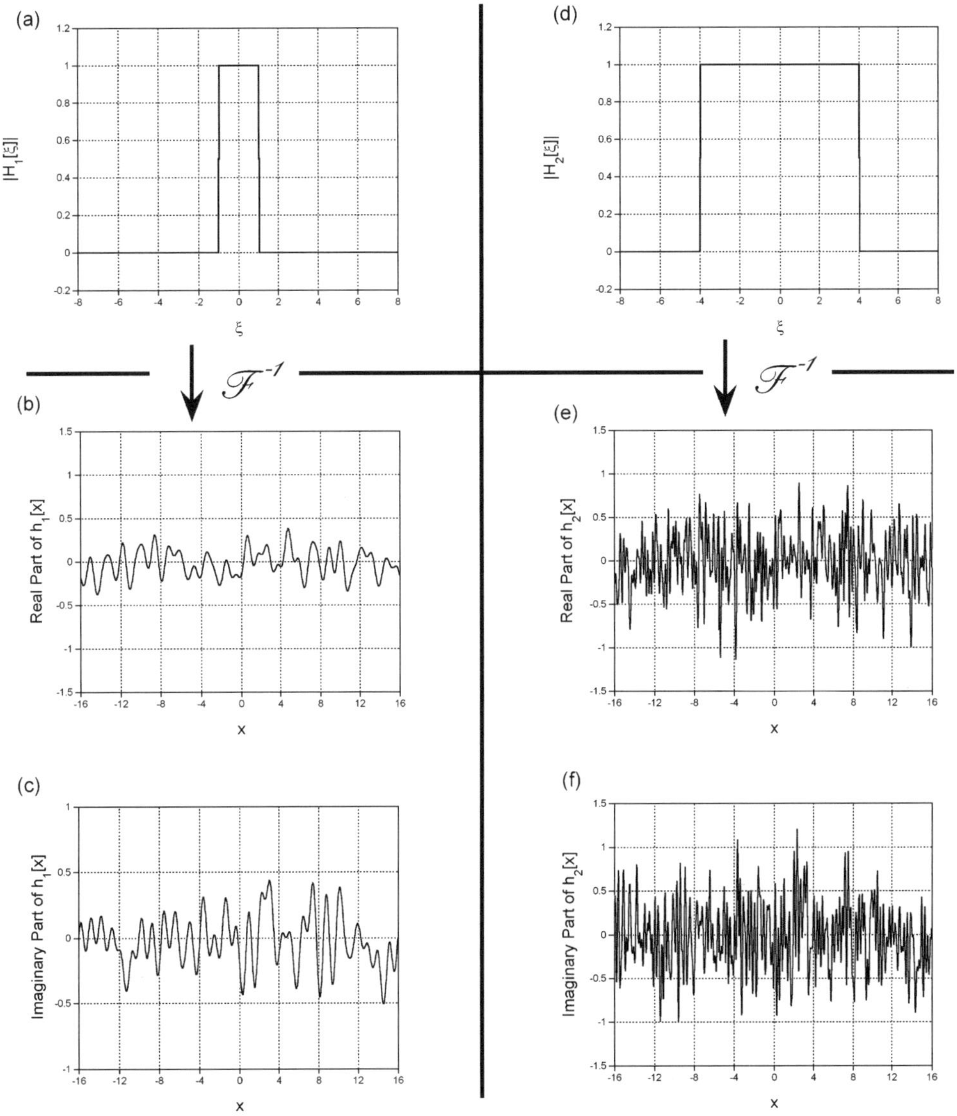

Figure 18.8 Impulse responses of two lowpass filters with random phase that differ in cutoff frequency: (a) MTF $|H_1[\xi]| = RECT[\xi/2]$; (b) $\Re\{h_1[x]\}$, showing low-frequency oscillations; (c) $\Im\{h_1[x]\}$; (d) MTF $|H_2[\xi]| = RECT[\xi/8]$; (e) $\Re\{h_2[x]\}$, showing high-frequency oscillations; (f) $\Im\{h_2[x]\}$. Note that the variance of the impulse response increases with increasing ξ_0.

strategy, and yields:

$$\begin{aligned} h[x] &= \boldsymbol{\mathcal{F}}^{-1}\{M[\xi] \cdot e^{+i\pi(\alpha\xi)^2}\} \\ &= \frac{1}{|\alpha|}\, e^{+i\pi/4} \cdot (m[x] * e^{-i\pi(x/\alpha)^2}) \end{aligned} \tag{18.54}$$

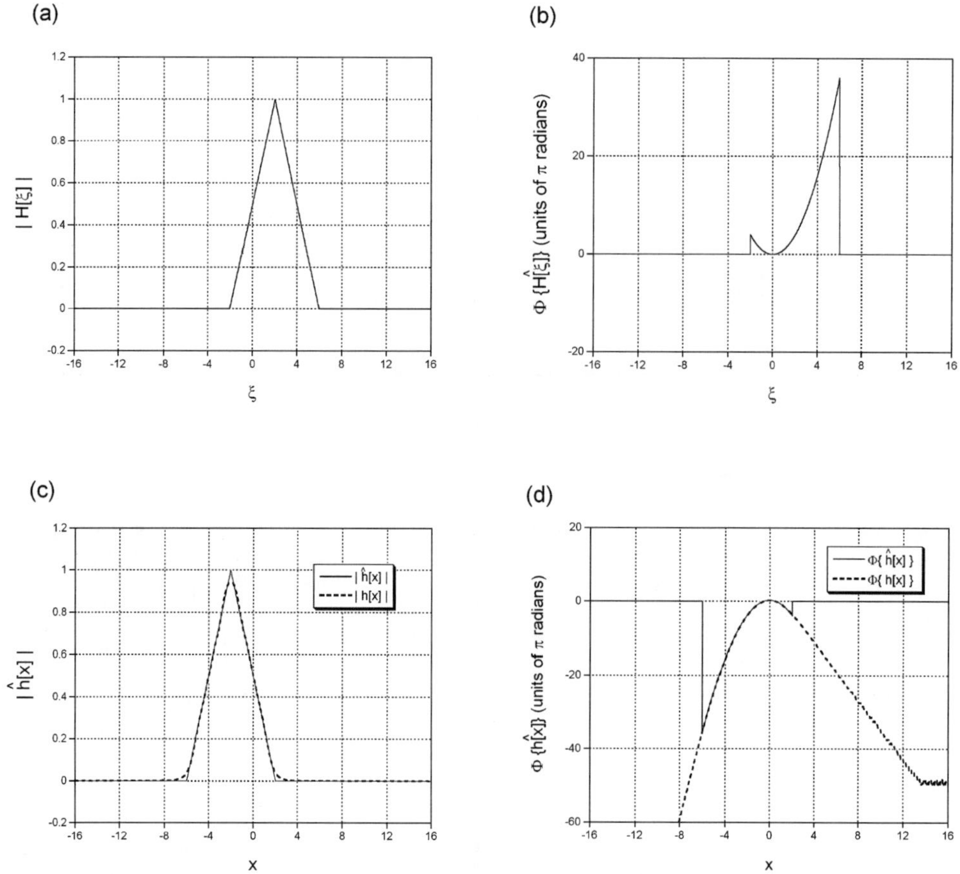

Figure 18.9 Approximation of the impulse response of a quadratic-phase filter modulated by an off-axis triangular MTF: (a) $|H[\xi]| = TRI[(\xi - 2)/4]$; (b) $\Phi\{H[\xi]\} = +\pi\xi^2$; (c) $|\hat{h}[x]|$ from method of stationary phase compared to calculation of $h[x]$ via DFT; (d) $\Phi\{\hat{h}[x]\}$ compared to $\Phi\{h[x]\}$ from DFT.

The method of stationary phase in Section 13.2 may be used to derive an approximation for the impulse response of the general modulated chirp function:

$$h[x] \cong \frac{1}{|\alpha|} e^{+i\pi/4} \cdot \left(M\left[-\frac{x}{\alpha^2}\right] \cdot e^{-i\pi(x/\alpha)^2} \right) \tag{18.55}$$

An example is shown in Figure 18.9.

Note that an equivalent expression for the impulse response also may be obtained by multiplying both sides of the frequency-domain expression for the chirp Fourier transform algorithm in Equation (17.71) by $e^{+i\pi\alpha^2\xi^2}$ to obtain:

$$M[\xi] \cdot e^{+i\pi\alpha^2\xi^2} = |\alpha|^2((m[\alpha^2\xi] \cdot e^{-i\pi\alpha^2\xi^2}) * e^{+i\pi\alpha^2\xi^2}) \tag{18.56}$$

The left-hand side is the transfer function of the filter in Equation (18.51) and the right-hand side is an equivalent expression in terms of $m[x]$, but this is a function in the *frequency* domain rather than the space domain. The inverse Fourier transforms of the left- and right-hand sides yield equivalent expressions for the impulse response. The result for the left-hand side is identical to that obtained by applying the filter theorem to Equation (18.53):

$$h[x] = \mathcal{F}_1^{-1}\{M[\xi]\, e^{+i\pi\alpha^2\xi^2}\}$$
$$= \frac{1}{|\alpha|}\, e^{+i\pi/4} \cdot (m[x] * e^{-i\pi(x/\alpha)^2}) \tag{18.57}$$

The inverse transform of the right-hand side of Equation (18.55) gives an expression in terms of a scaled replica of the MTF, but expressed in the space domain:

$$h[x] = \mathcal{F}_1^{-1}\{|\alpha|^2 (m[\alpha^2\xi] \cdot e^{-i\pi\alpha^2\xi^2}) * e^{+i\pi\alpha^2\xi^2}\}$$
$$= \frac{1}{|\alpha|^2}\left(M\left[\frac{x}{\alpha^2}\right] * e^{+i\pi(x/\alpha)^2}\right) \cdot e^{-i\pi(x/\alpha)^2} \tag{18.58}$$

This expression for the impulse response includes a convolution of a scaled replica of the MTF in the space domain with a chirp function.

We can also apply the method of stationary phase to generate an approximate form for the impulse response:

$$h[x] = \int_{-\infty}^{+\infty} (M[\xi] \cdot e^{+i\pi\alpha^2\xi^2})\, e^{+2\pi i\xi x}\, d\xi$$
$$= \int_{-\infty}^{+\infty} M[\xi] \cdot e^{+ix((\pi\alpha^2/x)\xi^2 + 2\pi\xi)}\, d\xi$$
$$= \int_{-\infty}^{+\infty} M[\xi] \cdot e^{+ix\,\mu[\xi]}\, d\xi \tag{18.59}$$

Equation (13.67) may be applied to obtain the approximate form of the impulse response:

$$h[x] \cong M[\xi_0]\sqrt{\frac{2\pi}{x \cdot \mu''[\xi_0]}}\, e^{+i\pi/4}\, e^{+ix\mu[\xi_0]} \tag{18.60}$$

Because the stationary point of $\mu[\xi]$ is located at $\xi_0 = -x/\alpha^2$, we obtain:

$$h[x] \cong \left(\frac{1}{|\alpha|}\, e^{+i\pi/4}\right) M\left[-\frac{x}{\alpha^2}\right] \cdot e^{-i\pi(x/\alpha)^2} \quad \textit{for } |x| \gg 0 \tag{18.61}$$

which shows that the impulse response of the mixed quadratic-phase filter is a quadratic-phase function modulated by a scaled and reversed replica of the MTF. If applied to an ideal lowpass filter with $M[\xi] = RECT[\xi/2\xi_0]$, the corresponding impulse response is:

$$h[x] \cong \left(\frac{1}{|\alpha|}\, e^{+i\pi/4}\right) RECT\left[\frac{x}{\alpha^2\xi_0}\right] \cdot e^{-i\pi(x/\alpha)^2} \quad \textit{for } |x| \gg 0 \tag{18.62}$$

so that most of the amplitude is concentrated within a region with support $\alpha^2\xi_0$, so that the support of the impulse response increases with increasing cutoff frequency ξ_0 and/or increasing chirp rate α.

PROBLEMS

18.1 Confirm Equation (18.44):

$$\begin{aligned} g[t] &= f[t] * g[t] \\ &= (A_0 \sin[2\pi\nu_0 t] \cdot STEP[t]) * (A_1 \sin[2\pi\nu_0 t] \cdot STEP[t]) \\ &= -\frac{A_0 A_1}{2}(RAMP[t] \cdot \cos[2\pi\nu_0 t]) \\ &= \frac{A_0 A_1}{2}(RAMP[t] \cdot \cos[2\pi\nu_0 t + \pi]) \end{aligned}$$

18.2 Evaluate the convolution of the causal oscillator when driven on resonance but "out of phase" by ϕ radians, i.e., evaluate:

$$(STEP[t] \cdot \sin[2\pi\nu_0 t + \phi]) * (STEP[t] \cdot \sin[2\pi\nu_0 t])$$

18.3 Use Equation (18.55) with $\alpha = 1$ to find a relation between the support of $M[\xi]$ and $m[\xi]$.

18.4 Find an expression (exact or approximate) for the autocorrelation of $f[x]$ where $F[\xi] = RECT[\xi/\Delta\xi] \cdot e^{+i\pi(x/\alpha)^2}$.

19

Applications of Linear Filters

19.1 Linear Filters for the Imaging Tasks

The goal of this chapter is to apply the concepts of linear filtering, which were developed in previous chapters, to construct linear shift-invariant systems that perform the three imaging tasks mentioned in Chapter 1. We will demonstrate the processes for the 1-D case, but the results are easily adapted to 2-D images. In these examples, the input "scene" is a deterministic (and often unknown) input function $f[x]$ centered about the (also often unknown) location x_0. In general, this signal is contaminated by an additional random "noise" function $n_1[x]$ that is added to $f[x - x_0]$ before encountering the system. A second additive noise function $n_2[x]$ may be added after the convolution to produce the final result:

$$(f[x - x_0] + n_1[x]) * h[x] + n_2[x] = g[x] \tag{19.1}$$

The filter $h[x]$ tends to act upon the spectra of both f and n_1 in the same manner, thus correlating the two signals; for example, a lowpass filter will "push" the amplitudes of both $n_1[x]$ and $f[x]$ toward their respective local means. The second additive noise term $n_2[x]$ is not filtered and therefore likely is not correlated to the input f unless generated by a similar mechanism. The action of the filter on the signal and the first noise ensures that $f[x]$ is more difficult to distinguish from $n_1[x]$ than from $n_2[x]$. For this reason, most models of imaging systems assume that no "preloaded" noise function so that $n_1[x] = 0$. We accept this assumption and delete $n_1[x]$, thus allowing us to drop the subscript from the second noise term from this point forward; the general mathematical model of imaging systems considered in this chapter is:

$$g[x] = f[x - x_0] * h[x] + n[x] \tag{19.2}$$

The corresponding expression for the spectrum of $g[x]$ is easy to derive by applying the shift theorem from Equation (9.123) and the linearity property:

$$G[\xi] = (F[\xi] \cdot e^{-2\pi i \xi x_0}) \cdot H[\xi] + N[\xi] \tag{19.3}$$

The goal of this chapter is to design transfer functions of linear shift-invariant filters that will be applied to $G[\xi]$ in Equation (19.3) to generate the "best" solution for two types of imaging problems that may be classified as "estimation" or "detection" tasks:

1A. Estimate the unknown input signal $f[x]$ from complete knowledge of the output $g[x]$ and of the impulse response $h[x]$, assuming no additive noise ($n[x] = 0[x]$). This is the "inverse" task introduced in Chapter 1; the solution is the "inverse filter".

Fourier Methods in Imaging Roger L. Easton, Jr.

1B. Estimate the original system impulse response $h[x]$ from complete knowledge of $f[x]$ and $g[x]$, again assuming no noise. This "system analysis" task also was introduced in Chapter 1. We will see that the commutativity of convolution ensures that the form of the proper filter for this task is identical to that of the inverse filter.

2A. Estimate the unknown input function $f[x]$ in the presence of additive uncorrelated noise $n[x]$. The system is assumed to pass all spatial frequencies, so that $h[x] = \delta[x]$. As we will see, the autocorrelations of the input and noise signals (or equivalently, their power spectra) also must be available. The system constructed from this knowledge is the "Wiener filter".

2B. Estimate the unknown input $f[x]$ from an output that was filtered with a known impulse response $h[x]$ prior to corruption by unknown additive noise $n[x]$. In the common situation, the autocorrelations of the input and noise signals (or, equivalently, their power spectra) also must be available. This "Wiener–Helstrom filter" is a generalization of task 2A.

3. Detect the presence of the known signal $f[x]$ and estimate its unknown position (the translation parameter x_0) for the case where $n[x] \neq 0$ and $h[x] = \delta[x]$. The impulse response that solves this problem is the "matched filter".

As we will see, the filters for estimation tasks 1 and 2 and for detection in 3 actually are quite similar in principle, even though the tasks are seemingly quite different.

As one example, if the appropriate impulse response in task 2A is $w[x]$, then the output of the system for this task is an estimate of $f[x]$:

$$g[x] * w[x] = (f[x] + n[x]) * w[x] = \hat{f}[x] \tag{19.4}$$

Situations where the desired parameter or function for a task may be recovered without error, so that $\hat{f}[x] = f[x]$, are rare. The success of the estimate is assessed by some "quality metric" or "cost function" derived from $\hat{f}[x]$ and $f[x]$. Many metrics are available, but the most common is the total squared error of the recovered signal $\hat{f}[x]$ compared to the correct signal $f[x]$.

Obviously, the error at coordinate x is:

$$\epsilon[x] = f[x] - \hat{f}[x] = f[x] - (g[x] * w[x]) \tag{19.5}$$

and the total error ϵ_1 is the area of $\epsilon[x]$:

$$\epsilon_1 = \int_{-\infty}^{+\infty} \epsilon[x]\, dx \tag{19.6}$$

(the integral obviously becomes a sum in the discrete case). For real-valued functions $f[x]$, the error $\epsilon[x]$ is usually (but not always) bipolar, so that ϵ may be quite small even if the "local" errors $\epsilon[x]$ are large. For this reason alone, the total error generally is a poor measure of fidelity, but the situation is even worse in the more general case of complex-valued signals because complex numbers do not exhibit the property of ordered size as considered in the discussion leading to Equation (4.12).

A more reasonable fidelity metric is the area of the real-valued (and nonnegative) squared magnitude of the local error because the squared-magnitude operation eliminates any problems with ordering and bipolar error:

$$\epsilon \equiv \int_{-\infty}^{+\infty} |f[x] - \hat{f}[x]|^2\, dx = \int_{-\infty}^{+\infty} |\epsilon[x]|^2\, dx \tag{19.7}$$

If ϵ is minimized, we say that the estimate is "optimal" in a squared-error sense. We must note that squared error is a rather notorious criterion in imaging applications because the value can be very misleading. It is quite possible to produce two estimates $\hat{f}_1[x]$ and $\hat{f}_2[x]$ such that the image with larger squared error appears to be "better" subjectively (we already have an example of such a case during the discussion of error-diffused quantization in Chapter 14). Despite its flaws, the squared-error metric is easy to calculate and so well entrenched in the imaging community that it is unlikely to be displaced soon.

The strict derivations of the forms of the optimum filters requiring statistical analyses are somewhat tedious though not terribly difficult. We will consider the cases on an intuitive level first, but also relate them to the results obtained in a more rigorous derivation later. Derivations are available in many sources, including Papoulis (1962), who demonstrated that the optimum filter is that for which the error is orthogonal to the available signal $g[x]$. Other useful references include Barrett and Swindell (1986), Blackledge (1989), and Castleman (1996).

19.2 Deconvolution – "Inverse Filtering"

We have seen that a linear shift-invariant system can alter the amplitude and/or phase of some or all of the sinusoidal components of the input function. Knowledge of the system impulse response often provides the mathematical means for recovering information about the input spectrum and thus about the input function. The problem may be expressed in a sentence: *Given a measured image* $g[x]$ *that is the convolution of an unknown object* $f[x]$ *and a known system impulse response* $h[x]$*, derive a "better" estimate* $\hat{f}[x]$ *from* $g[x]$ and $h[x]$ *via a linear shift-invariant filter with impulse response* $w[x]$.

In the nomenclature of Equation (19.2), we assume that the translation parameter $x_0 = 0$ and that there is no added noise ($n[x] = 0[x]$). In other words, the task is to *deconvolve* the system impulse response from the image to obtain a better estimate of $f[x]$. This is an example of the pervasive *inverse problem* in imaging, which is the second in the list of three tasks introduced in Chapter 1. Solution requires derivation of the *inverse filter* from knowledge of the output $g[x]$ and the system $h[x]$. As an example, consider a 2-D image $g[x, y]$ that has been "blurred" by some mechanism, such as defocus, object motion, or camera motion. If the "blur" is described by a known impulse response $h[x, y]$, then it may be possible to construct an operator that acts on $g[x, y]$ to construct a "better" estimate of $f[x, y]$. Solutions of inverse problems are important in many areas of imaging, and particularly in medical imaging where internal pathology must be inferred from image data obtained via X-ray, acoustic, or magnetic measurements. Realistic inverse problems often are *ill posed*, meaning that the available information is insufficient to determine the specific input function from a set of possibilities that each satisfies the given constraints.

Because the original system is assumed to be linear and shift invariant, the operation to be inverted is a convolution:

$$f[x] * h[x] = g[x] = \int_{-\infty}^{+\infty} f[\alpha]h[x-\alpha]\, d\alpha \tag{19.8}$$

If it exists, the ideal impulse response $w[x]$ of the inverse filter would satisfy the relation:

$$g[x] * w[x] = (f[x] * h[x]) * w[x] = f[x] * (h[x] * w[x]) = f[x] \tag{19.9}$$

Clearly the convolution of $h[x]$ and $w[x]$ must yield a Dirac delta function for Equation (19.9) to be satisfied:

$$h[x] * w[x] = \delta[x] \tag{19.10}$$

The filter theorem ensures that the corresponding condition in the frequency domain is that the product of the two transfer functions be the unit constant:

$$H[\xi] \cdot W[\xi] = 1[\xi] \tag{19.11}$$

which implies that the transfer function of the inverse filter must be the reciprocal of the transfer function of the original system:

$$W[\xi] = \frac{1}{H[\xi]} \tag{19.12}$$

In the (very rare) case where $|H[\xi]| \neq 0$ at all frequencies, then a potentially valid expression for the impulse response of the inverse filter is:

$$w[x] = \mathcal{F}_1^{-1}\left\{\frac{1}{H[\xi]}\right\} \tag{19.13}$$

If this inverse Fourier transform exists, then the original signal $f[x]$ may be recovered via:

$$f[x] = g[x] * \mathcal{F}_1^{-1}\left\{\frac{1}{H[\xi]}\right\} \tag{19.14}$$

It is much more common that $|H[\xi]| = 0$ at some (or many) frequencies. In these cases, we can construct the *pseudoinverse*, by analogy with the Moore–Penrose pseudoinverse matrix in Equation (5.79):

$$\hat{W}[\xi] \equiv \begin{cases} (H[\xi])^{-1} & \textit{if } H[\xi] \neq 0 \\ 0 & \textit{if } H[\xi] = 0 \end{cases} \tag{19.15}$$

The corresponding pseudoinverse impulse response is:

$$\hat{w}[x] = \mathcal{F}_1^{-1}\{\hat{W}[\xi]\} \tag{19.16}$$

In many, if not most, cases, the impulse response of the ideal inverse filter or the pseudoinverse does not exist. We can decompose the transfer function of the ideal inverse filter $W[\xi]$ in several ways that may lead to useful approximations. If it exists, the transfer function of the inverse filter in Equation (19.12) may be rewritten in these equivalent forms:

$$W[\xi] = |H[\xi]|^{-1} \cdot e^{-i\Phi\{H[\xi]\}} \tag{19.17a}$$

$$= H^*[\xi] \cdot \frac{1}{|H[\xi]|^2} \tag{19.17b}$$

$$= (|H[\xi]| \cdot e^{-i\Phi\{H[\xi]\}}) \cdot \frac{1}{|H[\xi]|^2} \tag{19.17c}$$

The phase contribution is explicit in the first and third expressions and buried in $H^*[\xi]$ in the second. The factors of $|H[\xi]|^{-2}$ in the second and third equations are real valued (and symmetric if $h[x]$ is real), and also are large at those frequencies where $|H[\xi]| \gtrapprox 0$. Corresponding expressions for the impulse response may be evaluated by applying the filter theorem in Equation (9.129) and the spectrum of the complex conjugate in Equation (9.146):

$$w[x] = \mathcal{F}_1^{-1}\left\{\frac{1}{|H[\xi]|}\right\} * \mathcal{F}_1^{-1}\{1[\xi]\, e^{-i\Phi\{H[\xi]\}}\} \tag{19.18a}$$

$$= \mathcal{F}_1^{-1}\{H^*[\xi]\} * \mathcal{F}_1^{-1}\left\{\frac{1}{|H[\xi]|^2}\right\} = h^*[-x] * \mathcal{F}_1^{-1}\left\{\frac{1}{|H[\xi]|^2}\right\} \tag{19.18b}$$

$$= \mathcal{F}_1^{-1}\{|H[\xi]|\} * \mathcal{F}_1^{-1}\{1[\xi]\, e^{-i\Phi\{H[\xi]\}}\} * \mathcal{F}_1^{-1}\left\{\frac{1}{|H[\xi]|^2}\right\} \tag{19.18c}$$

Each form for the inverse filter expresses the impulse response as the convolution of component functions. In Equation (19.18a), the "reciprocal magnitude" $|H[\xi]|^{-1}$ compensates for amplification or attenuation of frequency components by $h[x]$, while any phase contribution due to the original filter $h[x]$ is negated by the "allpass" filter with transfer function $e^{-i\Phi\{H[\xi]\}}$. The action of Equation (19.18b) is obtained by substitution into Equation (19.9):

$$\begin{aligned} \hat{f}[x] &= f[x] * (h[x] * w[x]) \\ &= f[x] * \left(h[x] * h^*[-x] * \mathcal{F}_1^{-1}\left\{\frac{1}{|H[\xi]|^2}\right\}\right) \\ &= f[x] * (h[x] \bigstar h[x]) * \mathcal{F}_1^{-1}\left\{\frac{1}{|H[\xi]|^2}\right\} \end{aligned} \tag{19.19}$$

where the definition of autocorrelation in Equation (8.61) has been used. In words, the output of this inverse filter produces the convolution of $f[x]$ with the Hermitian autocorrelation of $h[x]$ and then with $\mathcal{F}^{-1}\{|H[\xi]|^{-2}\}$. The autocorrelation $h[x] \bigstar h[x]$ acts as magnitude filter under our "generalized" definition (with phase increments of 0 and $-\pi$ radians). It has twice the impact on the amplitudes of the sinusoidal components of $f[x]$ than would be true of $h[x]$ acting alone; if $h[x]$ is, say, a local averaging (lowpass) filter, then the filter with impulse response $h[x] \bigstar h[x]$ performs the local averaging of sinusoids twice and thus doubles the attenuation. The filter $\mathcal{F}^{-1}\{|H[\xi]|^{-2}\}$ compensates the magnitude of the sinusoidal components for this "double" impulse response to ensure that the impulse response of the entire sequence has the form of the desired Dirac delta function. As we will see, it is useful in some situations to delete the last "amplification" term to produce an approximate inverse filter:

$$w[x] = h^*[-x] * \mathcal{F}_1^{-1}\left\{\frac{1}{|H[\xi]|^2}\right\} \to \hat{w}[x] = h^*[-x] \tag{19.20}$$

We might call $\hat{w}[x]$ the "unamplified" inverse filter. This approximation will be considered in more detail shortly.

As a further illustration of the action of the inverse filter, consider the case where the system impulse response is translated by a distance x_0, thus adding a linear phase factor to sinusoidal components. The reciprocal of the transfer function includes the complex conjugate of the linear phase factor, which negates the linear phase due to the translation:

$$\begin{aligned} h_1[x] &= h[x - x_0] = h[x] * \delta[x - x_0] \\ &\Longrightarrow w_1[x] = w[x] * \mathcal{F}_1\{\exp[+2\pi i\xi x_0]\} = w[x + x_0] \end{aligned} \tag{19.21}$$

The corresponding impulse response of the inverse filter clearly includes a translation in the opposite direction.

19.2.1 Conditions for Exact Recovery via Inverse Filtering

Based on the discussion of the previous section, we can see that the original object $f[x]$ may be recovered exactly from the output by inverse filtering if two conditions are satisfied simultaneously:

1. $|H[\xi]| \neq 0$ at all spatial frequencies.
2. There is no uncertainty in the knowledge of the image $g[x]$.

The second condition is equivalent to requiring that there be no "noise" in the measured image; the output must be deterministic. In practice, the existence of any zeros in the transfer function (any frequencies ξ_n where $H[\xi_n] = 0$) and/or the existence of noise in the image $g[x]$ constrain the possible success of the process. Specifically, the transfer function of any physically "interesting" system will have finite support and thus must have a cutoff frequency $\xi_{\max}$. In other words, the measured image $g[x]$ will not include any sinusoidal components with $|\xi| > \xi_{\max}$, so that perfect recovery of the input is not possible.

We emphasize one more time that inverse filtering may be applied only to the effects of shift-invariant convolutions; any degradation must be applied identically at each location in the scene. Such situations certainly exist; for example, an image may be blurred by uniform motion relative to the sensor. For example, consider the image of a planar object (such as a billboard) as taken with a camera at a fixed distance z_0. The camera travels at constant speed v_0 and exposes the sensor for t_0 seconds. In that time interval the camera travels a distance $b_0 = v_0 t_0$. The normalized impulse response $h[x]$ of the entire object is:

$$h[x] = \frac{1}{|b_0|} RECT\left[\frac{x}{b_0}\right] \tag{19.22a}$$

As shown in Figure 19.1, the 2-D impulse response for linear-motion blur along the x-axis is:

$$h[x, y] = \frac{1}{|b_0|} RECT\left[\frac{x}{b_0}\right] \cdot \delta[y] \tag{19.22b}$$

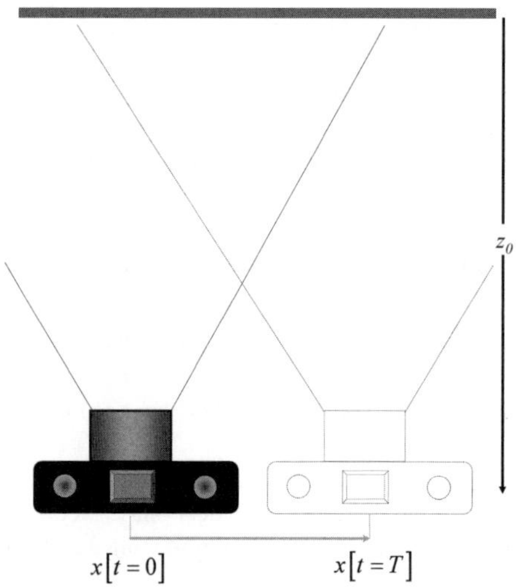

Figure 19.1 Shift-invariant blur due to camera motion. The 2-D object $f[x, y]$ is located at a fixed distance z_0 from the camera. The blur of the system is characterized by the impulse response $h[x, y]$ and the image $g[x, y]$ is the convolution $f[x, y] * h[x, y]$.

In most real-life situations, the action of the imaging system is more likely to be shift variant. For example, the angular subtense of motion will vary for objects at different distances from the camera and therefore the "blur length" also varies, as depicted in Figure 19.2. In other words, planes within the scene at different distances from the camera will be blurred by distinct 2-D impulse responses. The inverse filter that is appropriate to correct blur of objects at one distance from the camera will distort objects located at other distances. Images taken under such conditions cannot be corrected by basic inverse filtering.

A second space-variant degradation is related to that in the simplified view of Figure 19.2. In the optical imaging equation in Equation (1.4) for a fixed sensor location (i.e., a fixed image distance z_2), only the plane of the object at distance z_1 will appear "in focus"; planes at other object distances will appear to be "out of focus". In other words, the images at these other planes will have been filtered by different impulse responses. Optical imaging is considered in more detail in Chapter 21.

Some images degraded by shift-variant impulse responses may still be processed via convolutions with the inverse filter by first remapping the image coordinates appropriately. For example, consider an image taken of an object in the plane at a fixed distance that rotates beneath the camera (Figure 19.3). We briefly described the process during the discussion of system operators in Chapter 8 and showed how the image representation could be transformed to polar coordinates to convert shift-variant rotation to shift-invariant translation. We will touch on this again briefly at the end of this chapter to implement inverse filtering of rotational blur.

Three examples of the 1-D inverse filter are considered in the next sections. These were chosen to illustrate both the useful characteristics as well as common problems with the method.

19.2.2 Inverse Filter for Uniform Averager

Consider again the 1-D imaging system that computes a uniformly weighted average of the input $f[x]$ over some real-valued width interval b; the impulse response is:

$$h_1[x] = \frac{1}{|b_0|} RECT\left[\frac{x}{b_0}\right] \tag{19.23}$$

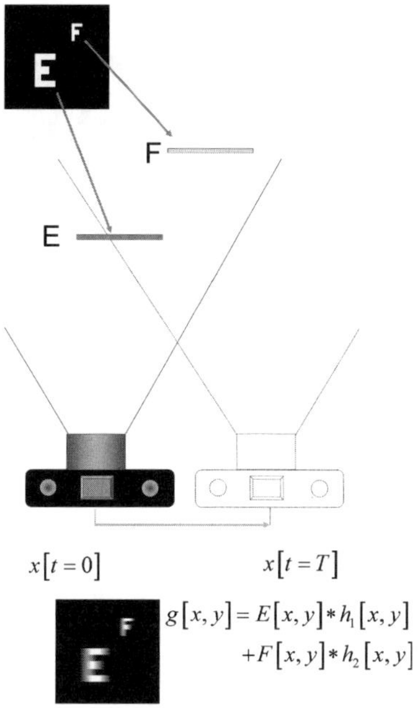

Figure 19.2 Shift-variant blur: two planar objects are located at different distances from the same moving camera shown in Figure 19.1. Different blurs are imposed on the two objects, thus producing a shift-variant operation.

where the normalizing factor $|b_0|^{-1}$ ensures that $H[0] = 1$, which means that the constant part of the input signal is passed to the output without change. The frequency-domain representation of the imaging equation is:

$$G_1[\xi] = F[\xi] \cdot H_1[\xi] = F[\xi] \cdot SINC[b_0\xi] \tag{19.24}$$

If there is no noise, the output spectrum must be zero at frequencies $\xi_n = \pm n/b_0$ $(n = 1, 2, 3, \ldots)$, and thus no estimate of the amplitudes $F[\xi_n]$ can be obtained from the available information. For this reason, we adopt the analogy to the pseudoinverse matrix in Equation (5.79) to evaluate the transfer function:

$$\hat{W}_1[\xi] \equiv \begin{cases} (H_1[\xi])^{-1} & \textit{if } H_1[\xi] \neq 0 \\ 0 & \textit{if } H_1[\xi] = 0 \end{cases} \tag{19.25a}$$

$$\hat{w}_1[x] = \mathcal{F}_1^{-1}\{\hat{W}_1[\xi]\} \tag{19.25b}$$

This function is analogous to the Moore–Penrose pseudoinverse matrix that was considered in Equation (5.79), and so we adopt the same nomenclature. The steps in the inverse filtering process for this system impulse response $h_1[x]$ are illustrated in Figure 19.4, with the recovered image in Figure 19.4d.

In the example of Figure 19.4, $\hat{F}$ is assumed to be zero at the set of frequencies $\xi_n = \pm n \cdot b^{-1}$; the difference of these null amplitudes from the actual amplitude F distorts the recovered image $f[x]$. The error $\epsilon[x]$ of the estimate is composed primarily of the larger-amplitude sinusoids that have spatial frequencies $\xi_1 = b_0^{-1}$ and $\xi_2 = 2 \cdot b_0^{-1}$. Clearly the inverse filter must amplify any frequency components that were attenuated because $|H_1[\xi]| < 1$.

The significant limitation of inverse filtering becomes apparent from estimates obtained from an output image $g[x]$ that has been corrupted by additive noise or if the assumed system impulse response

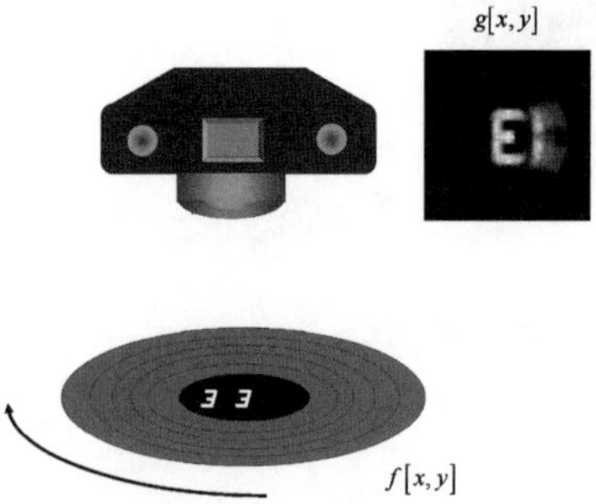

Figure 19.3 Schematic of rotational blurring. The object rotates on a turntable during the exposure. The resulting image $g[x, y]$ suffers from shift-variant blur.

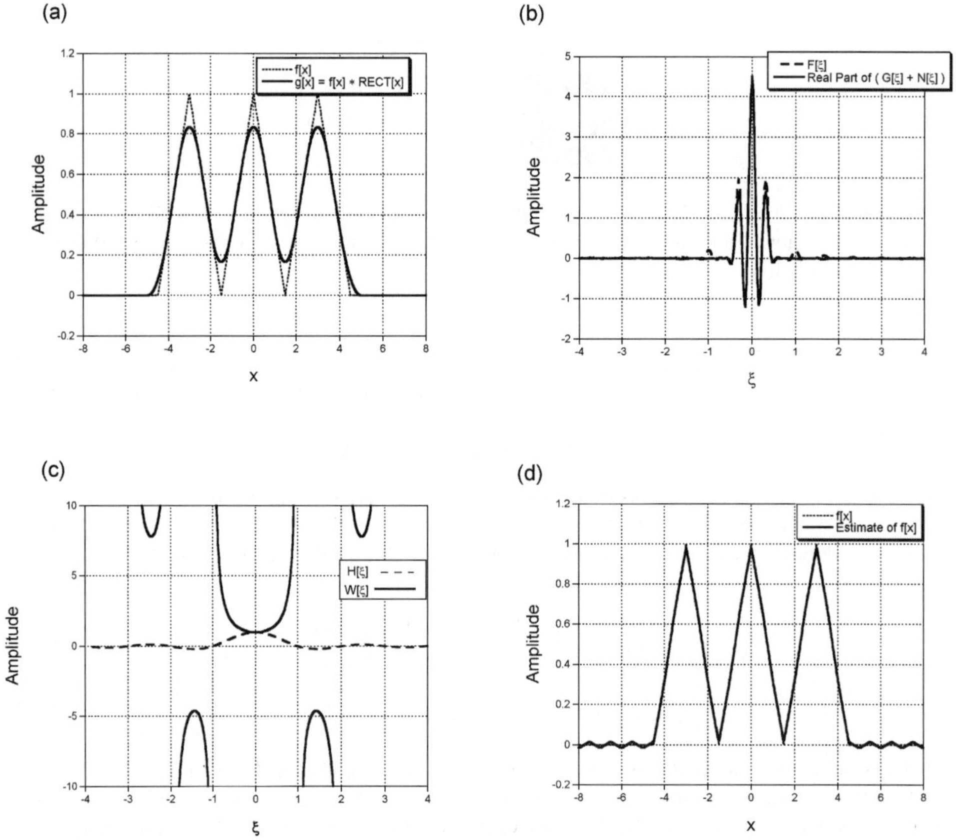

Figure 19.4 Noise-free implementation of pseudoinverse filter for $h[x] = RECT[x]$: (a) $f[x] = TRI[2x/3] * [\delta[x+3] + \delta[x] + \delta[x-3]]$ and $g[x] = f[x] * h[x]$; (b) $F[\xi]$ and $G[\xi]$; (c) $H[\xi]$ and $W[\xi]$, which is the "pseudoinverse" filter for $H[\xi]$. In this case, (d) is the reconstructed signal $\hat{f}[x]$ compared to $f[x]$. The oscillating amplitude of $\hat{f}[x]$ outside the support of $f[x]$ is due to the sinusoidal components with frequencies $\xi = \pm 1, \pm 2, \ldots$ that are missing from $f[x]$.

$h_1[x]$ is incorrect. The latter case may be modeled as an impulse response that contains noise. The former case is illustrated in Figure 19.5. The substantial amplification necessary to recover the input spectrum at frequencies where $|H_1[\xi]|$ is small (such as in the vicinity of its zeros) also amplifies any noise components at those frequencies. The amplified noise in the frequency domain is "spread" over the entire space domain by the inverse Fourier transform during the evaluation of $\hat{f}[x]$, thus producing a noisy recovered image. In the presence of noise, the measured $g_1[x]$ may be considered to be actually an estimate $\hat{g}_1[x]$ such that:

$$\hat{g}_1[x] = g_1[x] + n[x] \tag{19.26}$$

Any local error in $\hat{g}_1[x]$ transforms to global errors in the estimated object spectrum $\hat{G}[\xi]$:

$$\hat{G}_1[\xi] = G_1[\xi] + N[\xi] \tag{19.27}$$

so the estimate of the input spectrum is:

$$\hat{F}_1[\xi] = \frac{G_1[\xi]}{H[\xi]} + \frac{N[\xi]}{H[\xi]} \tag{19.28}$$

The recipe for the pseudoinverse in Equation (19.25) takes care of the first term by setting $\hat{F}[\xi] = 0$ at frequencies where $H[\xi] = 0$. However, the amplitude of the noise spectrum $N[\xi]$ likely differs from zero at these frequencies; the estimated object spectrum $\hat{F}[\xi]$ will "blow up" in the vicinity of these frequencies and the resulting estimate $f[x]$ of the input object likely will be worthless. In commonly used language, the inverse filter is *ill conditioned* in the presence of noise.

$$\hat{F}_1[\xi] = \frac{G_1[\xi]}{SINC[b_0\xi]} \tag{19.29}$$

The zeros of the *SINC* function lead to singularities of the inverse filter at all frequencies that are nonzero integer multiples of b^{-1}.

Swindell (1970) implemented an approximation of the inverse filter for uniform-motion blur that was analyzed by Honda and Tsujiuchi (1975). In effect, this method approximately corrected for the amplitude only near the zero crossings of the impulse response.

19.2.3 Inverse Filter for Ideal Lowpass Filter

The second example of the inverse filter applies to the ideal lowpass filter:

$$h_2[x] = \frac{1}{|b_0|} SINC\left[\frac{x}{b_0}\right] \tag{19.30}$$

where the normalization factor $|b_0|^{-1}$ again ensures that the DC component of $f[x]$ is passed to the output without change. The corresponding transfer function is:

$$H_2[\xi] = RECT[b_0\xi] \tag{19.31}$$

Sinusoidal components with frequencies smaller than $\xi_{\text{cutoff}} = |2b_0|^{-1}$ pass to the output image $g_2[x]$ without change, but all other frequency components are blocked from the output. *The input spectrum at frequencies outside the passband may not be recovered by any subsequent filter*. The expression for the pseudoinverse filter is:

$$\hat{W}_2[\xi] = \begin{cases} (RECT[b_0\xi])^{-1} = 1 & \text{if } \xi < \dfrac{1}{2b_0} \\ \left(RECT\left[\dfrac{1}{2}\right]\right)^{-1} = 2 & \text{if } \xi = \dfrac{1}{2b_0} \\ 0 & \text{if } \xi > \dfrac{1}{2b_0} \end{cases} \tag{19.32}$$

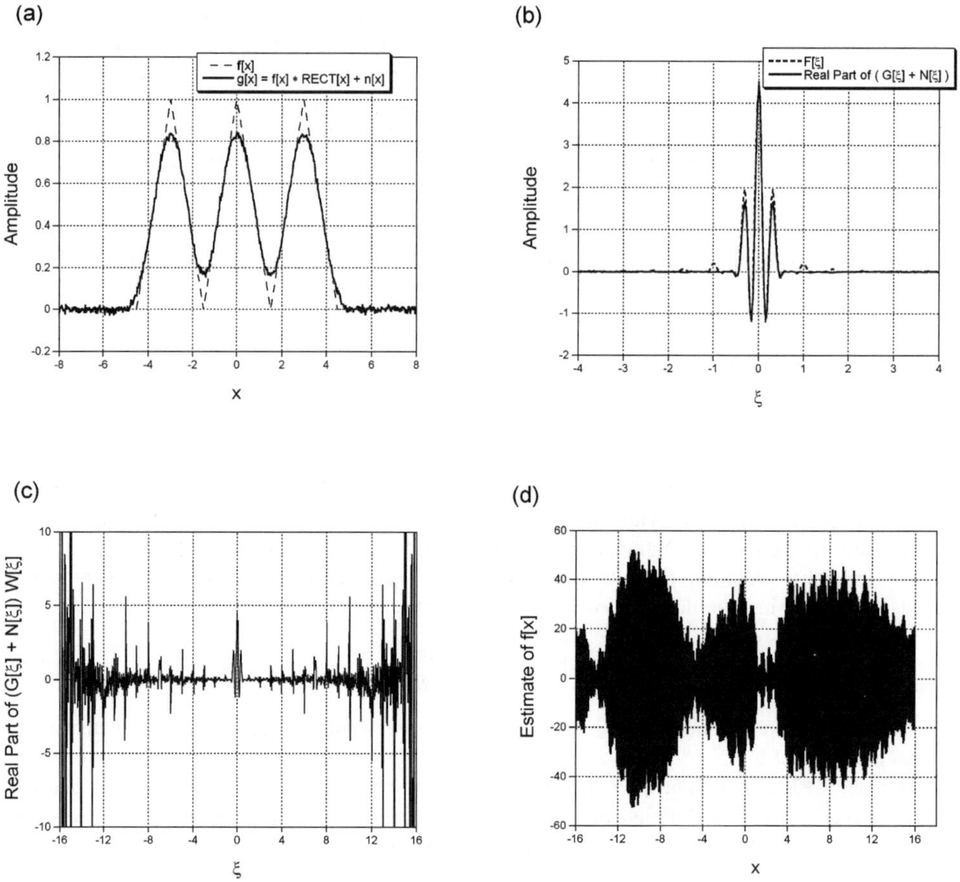

Figure 19.5 Implementation of inverse filter in the presence of noise: (a) $f[x]$ and $g[x] = f[x] * RECT[x] + n[x]$, where the noise is Gaussian distributed with $\mu = 0$, and $\sigma = 0.01$ – the amplitude of the noise is quite small compared to the signal; (b) $F[\xi]$; (c) $G[\xi]$ and $G[\xi] \cdot W[\xi]$, showing the large amplification of high-frequency noise components (note the presence of $F[\xi]$ near $\xi = 0$); (d) $f[x]$ and $\hat{f}[x]$, showing the domination of the amplified high-frequency noise over the entire space domain.

as shown in Figure 19.6. The isolated frequencies at the edges of the *RECT* function amplify sinusoidal components with those frequencies by factors of 2, but these isolated amplitudes have no impact on the areas calculated by the inverse Fourier transform (because they are not weightings applied to Dirac delta functions). Therefore the impulse response $\hat{w}_2[x]$ of the inverse filter for the ideal lowpass filter is functionally equivalent to the system impulse response $h_2[x]$. In effect, the cascade of the transfer functions of the system and pseudoinverse filter is equivalent to the transfer function of the system alone; the estimate of the input spectrum is identical to the output spectrum:

$$\hat{F}_2[\xi] = F[\xi] \cdot (H_2[\xi] \cdot \hat{W}_2[\xi])$$

$$\Longrightarrow F[\xi] \cdot (H_2[\xi] \cdot H_2[\xi]) \cong F[\xi] \cdot H_2[\xi] = G_2[\xi] \tag{19.33a}$$

and the corresponding relationship in the space domain is:

$$\hat{f}_2[x] = g_2[x] \tag{19.33b}$$

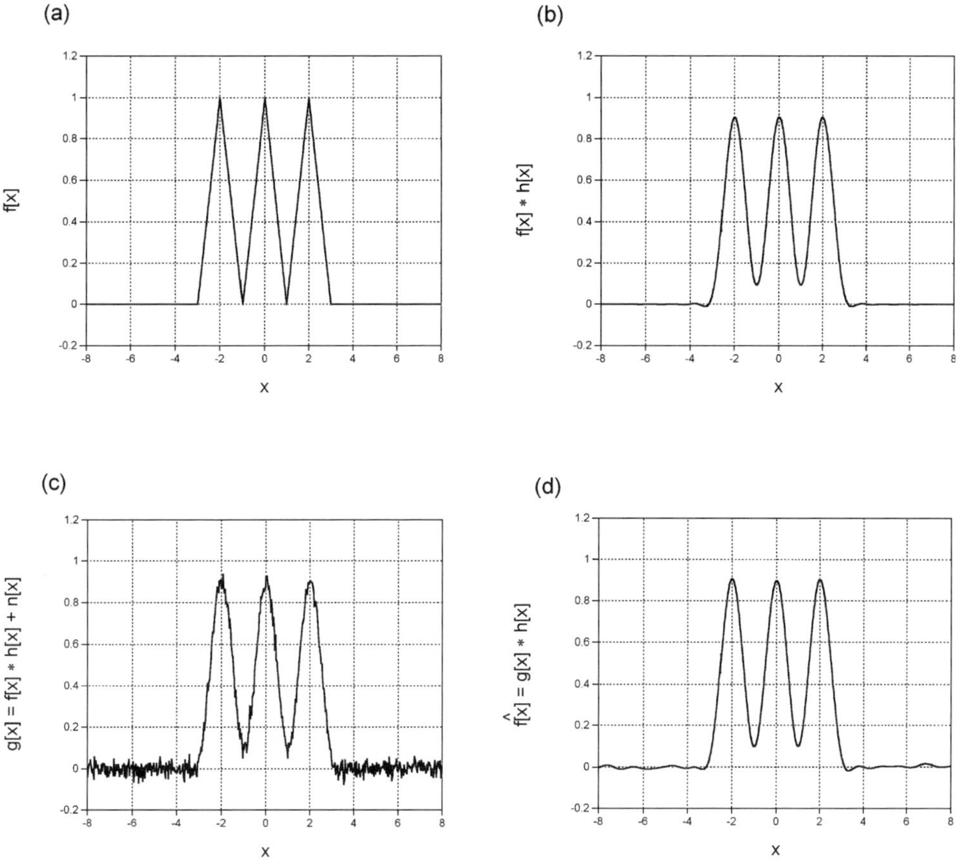

Figure 19.6 Inverse filter for ideal LPF: (a) $f[x]$; (b) $f[x] * 2 \cdot SINC[2x]$; (c) $g[x] = f[x] * h[x] + n[x]$, where $n[x]$ is Gaussian distributed random noise with $\sigma = 0.025$; (d) $g[x] * w[x] = g[x] * h[x]$, showing reduced visibility of the additive random noise, though residual effects are apparent outside the support of $f[x]$.

In words, the annihilation of the sinusoidal components at spatial frequencies outside $|\xi|_{\text{cutoff}}$ determines that no improvement in the estimate of the original signal is possible; the best estimate of the input function in the case of the ideal lowpass filter is the original output $g_2[x]$.

If the output of the ideal lowpass filter included broadband noise:

$$g_2[x] = f[x] * \frac{1}{|b_0|} SINC\left[\frac{x}{b_0}\right] + n[x] \tag{19.34}$$

the action of the pseudoinverse filter $\hat{W}_2[\xi] = RECT[b_0 \xi]$ from Equation (19.25) would average (and thus remove) noise at any frequencies larger than the cutoff frequency, thereby improving the appearance of the image:

$$\hat{f}_2[x] = \left(f[x] * \frac{1}{|b_0|} SINC\left[\frac{x}{b_0}\right] \right) + \left(n[x] * \frac{1}{|b_0|} SINC\left[\frac{x}{b_0}\right] \right) \tag{19.35}$$

19.2.4 Inverse Filter for Decaying Exponential

As a third example of the inverse filter, consider the impulse response to be the unit-area decaying exponential function with scaled width in Equation (6.16):

$$h_3[x] = STEP[x] \cdot \frac{1}{|b_0|} \exp\left[-\frac{x}{b_0}\right] = STEP\left[\frac{x}{b_0}\right] \cdot \frac{1}{|b_0|} \exp\left[-\frac{x}{b_0}\right] \tag{19.36}$$

where the amplitude scaling factor of $|b_0|^{-1}$ constrains the area of the impulse response to unity. The transfer function is easily obtained by combining Equation (9.88) with the scaling theorem in Equation (9.117):

$$H_3[\xi] = \frac{1}{1 + 2\pi i \xi b_0} = \sqrt{\frac{1}{1 + (2\pi \xi b_0)^2}} \cdot \exp[-i \cdot \tan^{-1}[2\pi \xi b_0]] \tag{19.37}$$

The ideal inverse filter $W_3[\xi]$ exists because $|H_3[\xi]|$ has no zeros at finite values of $|\xi|$:

$$W_3[\xi] = (H_3[\xi])^{-1} = 1 + 2\pi i \xi b_0 = \sqrt{1 + (2\pi \xi b_0)^2} \cdot \exp[+i \cdot \tan^{-1}[2\pi \xi b_0]] \tag{19.38}$$

as shown in Figure 19.7.

The impulse response of the inverse filter for continuous functions is easy to derive from known Fourier transforms and the linearity property:

$$\begin{aligned} w_3[x] &= \mathcal{F}_1^{-1}\{1[\xi] + b_0 \cdot 2\pi i \xi\} \\ &= \mathcal{F}_1^{-1}\{1[\xi]\} + b_0 \cdot \mathcal{F}_1^{-1}\{2\pi i \xi\} \\ &= \delta[x] + b_0 \cdot \delta'[x] \end{aligned} \tag{19.39}$$

where $b_0 > 0$ and Equation (9.143a) has been used. The validity of this result may not be obvious at first glance, but is straightforward to confirm by applying the basic principles of differential calculus:

$$\begin{aligned} h_3[x] * w_3[x] &= \left(STEP\left[\frac{x}{b_0}\right] \cdot \frac{1}{|b_0|} \exp\left[-\frac{x}{b_0}\right]\right) * (\delta[x] + b_0 \cdot \delta'[x]) \\ &= \left(STEP\left[\frac{x}{b_0}\right] \cdot \frac{1}{|b_0|} \exp\left[-\frac{x}{b_0}\right]\right) * \delta[x] \\ &\quad + \frac{b_0}{|b_0|} \cdot \left(STEP\left[\frac{x}{b_0}\right] \cdot \frac{d}{dx}\left(\exp\left[-\frac{x}{b_0}\right]\right) + \exp\left[-\frac{x}{b_0}\right] \cdot \frac{d}{dx}\left(STEP\left[\frac{x}{b_0}\right]\right)\right) \\ &= \left(STEP\left[\frac{x}{b_0}\right] \cdot \frac{1}{|b_0|} \exp\left[-\frac{x}{b_0}\right]\right) \\ &\quad + SGN[b_0] \cdot \left(STEP\left[\frac{x}{b_0}\right] \cdot \left(-\frac{1}{b_0} \exp\left[-\frac{x}{b_0}\right]\right) + \exp\left[-\frac{x}{b_0}\right] \cdot (SGN[b_0] \cdot \delta[x])\right) \\ &= \left(STEP\left[\frac{x}{b_0}\right] \cdot \frac{1}{|b_0|} \exp\left[-\frac{x}{b_0}\right]\right) \\ &\quad - \left(STEP\left[\frac{x}{b_0}\right] \cdot \frac{1}{|b_0|} \exp\left[-\frac{x}{b_0}\right]\right) + (SGN[b_0])^2 \cdot \exp[0] \cdot \delta[x] \\ &= 1 \cdot 1 \cdot \delta[x] = \delta[x] \end{aligned} \tag{19.40}$$

where Equation (6.103) and the fact that $b_0/|b_0| = SGN[b_0] = 1$ for $b_0 > 0$ were used. In words, the original object may be recovered from an image "blurred" by convolution with a decaying exponential by adding the blurry image to an amplitude-scaled derivative. A longer blur "length" b_0 requires a proportional increase in the amplitude scaling of the derivative.

We now consider a more rigorous derivation of the linear filter for estimating a filtered signal in the presence of noise.

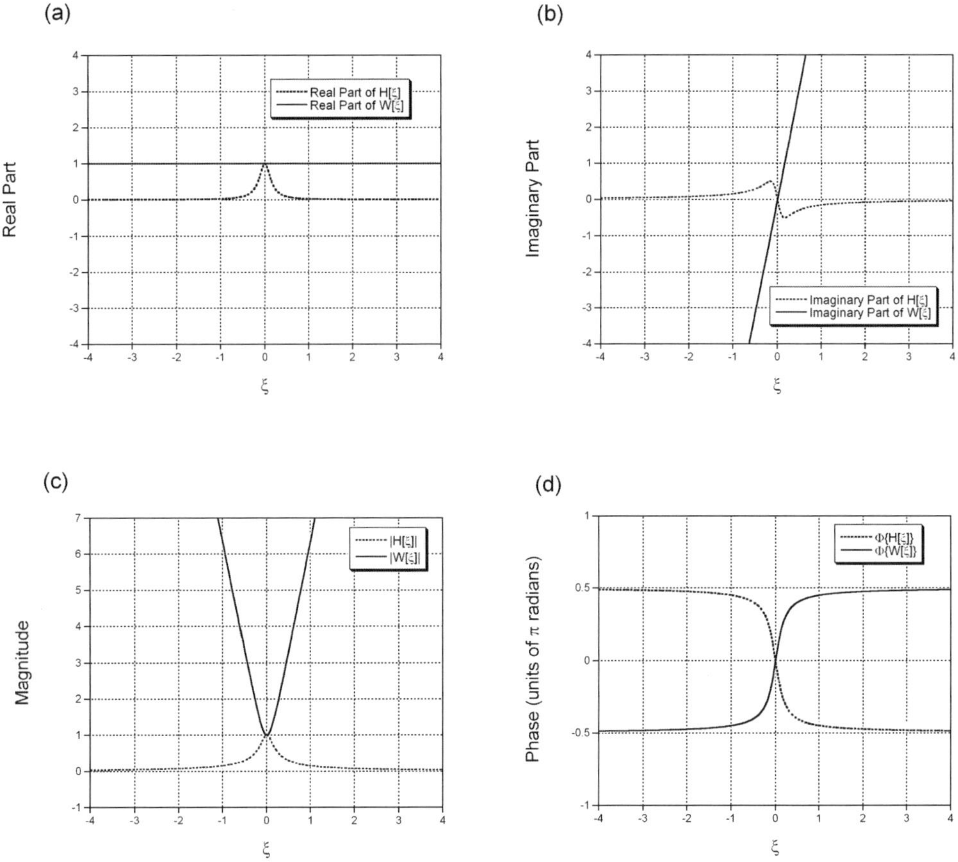

Figure 19.7 Transfer function of filter and inverse filter for $h[x] = STEP[x] \cdot (1/|b|)\exp[-x/b]$: (a) $\Re\{W[\xi]\} = 1[\xi]$; (b) $\Im\{W[\xi]\} = +2\pi b\xi$; (c) $|W[\xi]| = \sqrt{1 + (2\pi\xi b)^2}$; (d) $\Phi\{W[\xi]\} = \tan^{-1}[2\pi\xi b]$.

19.3 Optimum Estimators for Signals in Noise

We have seen that the inverse filter attempts to estimate the input from knowledge of the system but performs poorly in the presence of noise. However, a linear and shift-invariant filter may be designed that produces a better estimate of the input signal from a noisy (and possibly blurred) image $g[x]$, assuming that we have additional knowledge of the input signal and the noise. The appropriate variant of Equation (19.2) and its spectrum from Equation (19.3) for this task are:

$$g[x] = f[x] * h[x] + n[x] \tag{19.41a}$$

$$G[\xi] = F[\xi] \cdot H[\xi] + N[\xi] \tag{19.41b}$$

As with the inverse filter, we wish to derive the linear filter $w[x]$ that may be applied to $g[x]$ to produce a "better" estimate of $f[x]$ than exists in $g[x]$. The equations for the cascaded systems are:

$$g[x] * w[x] = (f[x] * h[x] * w[x]) + (n[x] * w[x]) = \hat{f}[x] \tag{19.42a}$$

$$G[\xi] \cdot W[\xi] = (F[\xi] \cdot H[\xi] \cdot W[\xi]) + (N[\xi] \cdot W[\xi]) = \hat{F}[\xi] \tag{19.42b}$$

Some criterion, i.e., a "quality metric" or "cost function", is applied to judge the success of this filter. As already mentioned, the most common metric is the total squared error of the recovered signal $\hat{f}[x]$ compared to the correct signal $f[x]$ that was defined in Equation (19.7):

$$\begin{aligned}\epsilon &= \int_{-\infty}^{+\infty} |f[x] - \hat{f}[x]|^2 \, dx \\ &= \int_{-\infty}^{+\infty} |f[x] - (g[x] * w[x])|^2 \, dx\end{aligned} \tag{19.43}$$

The optimum filter $w[x]$ is constructed to minimize ϵ, which implies that appropriate a priori knowledge must exist about $f[x]$ and $n[x]$. It is convenient to consider two cases: $h[x] = \delta[x]$ where noise is added to the unfiltered signal, and $h[x] \neq \delta[x]$, which typically means that the signal is "blurred" (or perhaps "sharpened") before adding the noise. In the first case with:

$$g[x] = f[x] + n[x] \tag{19.44}$$

the formula for $w[x]$ that minimizes the squared error was derived by Norbert Wiener and is called the *Wiener filter* in his honor. Helstrom (1967) extended Wiener's derivation to the second case where $h[x] \neq \delta[x]$ via the *Wiener–Helstrom filter*. We will derive the Wiener filter rigorously (more or less) and then adapt it to the more general case.

19.3.1 Wiener Filter

If $h[x] = \delta[x]$, the system equations in the two domains in Equation (19.42) simplify to:

$$g[x] * w[x] = (f[x] + n[x]) * w[x] = \hat{f}[x] \tag{19.45a}$$

$$G[\xi] \cdot W[\xi] = (F[\xi] + N[\xi]) \cdot W[\xi] = \hat{F}[\xi] \tag{19.45b}$$

Right away we can apply the observations of the previous section to see that the ideal transfer function would have the form:

$$(F[\xi] + N[\xi]) \cdot W[\xi] = F[\xi] \Longrightarrow W[\xi] = \frac{F[\xi]}{F[\xi] + N[\xi]} \tag{19.46}$$

This expression is not very practical in actual use, but for now we leave it to the readers to ponder the reasons as we derive this expression. The intermediate steps in the derivation may be adapted to realistic situations.

From the frequency-space representation for $\hat{F}[\xi]$ in Equation (19.45b), we can easily see that $W[\xi]$ must simultaneously satisfy two conditions to evaluate the original input spectrum:

$$N[\xi] \cdot W[\xi] = 0[\xi] \Longrightarrow W[\xi] = 0[\xi] \tag{19.47a}$$

$$F[\xi] \cdot W[\xi] = F[\xi] \Longrightarrow W[\xi] = 1[\xi] \tag{19.47b}$$

If the two spectra $F[\xi]$ and $N[\xi]$ are *disjoint*, i.e., if $N[\xi] = 0$ at all frequencies where $F[\xi] \neq 0$ and vice versa, then these two conditions are satisfied simultaneously by setting $W[\xi] = 1$ at all frequencies where $N[\xi] = 0$ and/or $F[\xi] \neq 0$, and set $W[\xi] = 0$ where $N[\xi] \neq 0$. Of course, this implies that both $F[\xi]$ and $N[\xi]$ are known, but the same recipe results for the bitonal transfer function $W[\xi]$ if we substitute the (more easily estimated) magnitudes (or even the squared magnitudes) for the complex-valued spectra $F[\xi]$ and $N[\xi]$. The corresponding impulse response $w[x]$ of the Wiener filter is (of course) obtained via the inverse Fourier transform of the bitonal transfer function $W[\xi]$.

Note the ambiguity in the definition of $W[\xi]$ implied by the use of "and/or" in the criterion for $W[\xi]$ for disjoint spectra. In other words, there is a question whether to set $W[\xi] = 0$ or $W[\xi] = 1$ at frequencies where $F[\xi] = N[\xi] = 0$. Clearly either choice produces the same estimate $\hat{F}[\xi]$ if the noise spectrum in the data is identical to that used to evaluate $W[\xi]$. We can make the case that it is more appropriate to set $W[\xi] = 0$ where $F[\xi] = N[\xi] = 0$, because this choice would block any additional noise outside the spectrum of the original sample.

The desirable situation of disjoint signal and noise spectra rarely (if ever) exists in real applications. In fact, $N[\xi]$ is often nonzero at all ξ and thus will overlap the entire signal spectrum (e.g., noise generated by quantization error considered in Chapter 14). The realistic task in such a case is to derive a transfer function that accounts for both the ideal "disjoint" case and the realistic "overlapping" case to minimize the total squared error in Equation (19.43). From the limiting cases, it is evident that the desired transfer function will transmit very little signal amplitude at spatial frequencies where the noise power spectrum is much larger than the signal power spectrum:

$$|N[\xi]|^2 \gg |F[\xi]|^2 \Longrightarrow W[\xi] \cong 0 \tag{19.48}$$

This equation may be rearranged to construct an expression from the two known quantities that is small in the desired region:

$$1 \gg \frac{|F[\xi]|^2}{|N[\xi]|^2} \equiv \Gamma[\xi] \cong 0[\xi] \quad \textit{if } |N[\xi]|^2 \gg |F[\xi]|^2 \tag{19.49}$$

where $\Gamma[\xi]$ is the *noise-to-signal power ratio*, which (obviously) is the reciprocal of the *signal-to-noise power ratio*. At frequencies where the signal power is much larger than the noise power, it is logical to set the transfer function approximately to unity so that it transmits most of this "useful" amplitude:

$$|N[\xi]|^2 \ll |F[\xi]|^2 \Longrightarrow W[\xi] \cong 1 \tag{19.50}$$

Our task is to find a relation between the signal and noise power spectra that "smoothly transitions" between the limiting cases of small and large signal-to-noise power ratios in Equations (19.48) and (19.50) that minimizes the squared error of the final estimate of the entire function. Wiener rigorously proved the appropriate relation to minimize the squared error. The derivation is based on a statistical approach and requires consideration of the properties of random variables and their spectra from Chapter 9. We outline the steps here. The squared error in Equation (19.43) may be expanded:

$$\begin{aligned}
\epsilon &= \int_{-\infty}^{+\infty} |f[x] - \hat{f}[x]|^2 \, dx \\
&= \int_{-\infty}^{+\infty} (f[x] - \hat{f}[x]) \cdot (f[x] - \hat{f}[x])^* \, dx \\
&= \int_{-\infty}^{+\infty} |f[x]|^2 + |\hat{f}[x]|^2 - (f[x] \cdot \hat{f}^*[x]) - (f^*[x] \cdot \hat{f}[x]) \, dx \\
&= \int_{-\infty}^{+\infty} |f[x]|^2 \, dx + \int_{-\infty}^{+\infty} |\hat{f}[x]|^2 \, dx - 2 \cdot \Re\left\{\int_{-\infty}^{+\infty} (f[x] \cdot \hat{f}^*[x]) \, dx\right\}
\end{aligned} \tag{19.51}$$

where a simple generalization of Equation (4.4) has been used:

$$z_1 \cdot z_2^* + z_1^* \cdot z_2 = 2 \cdot \Re\{z_1 \cdot z_2^*\} = 2 \cdot \Re\{z_1^* \cdot z_2\} \tag{19.52}$$

The first term may be rewritten in a different convenient form based on the autocorrelation of the "unknown" signal $f[x]$:

$$\int_{-\infty}^{+\infty} |f[x]|^2 \, dx = f[x] \bigstar f[x]|_{x=0} \tag{19.53}$$

The second term in Equation (19.51) may be expanded and rearranged:

$$\begin{aligned}\int_{-\infty}^{+\infty} |\hat{f}[x]|^2\,dx &= \int_{x=-\infty}^{+\infty} (g[x]*w[x])\cdot(g[x]*w[x])^*\,dx\\&= \int_{x=-\infty}^{+\infty}\left(\int_{\beta=-\infty}^{+\infty} g[x-\beta]w[\beta]\,d\alpha\right)\cdot\left(\int_{\gamma=-\infty}^{+\infty} g[x-\gamma]w[\gamma]\,d\gamma\right)^*dx\\&= \int_{x=-\infty}^{+\infty}\left(\int_{\beta=-\infty}^{+\infty} g[x-\beta]w[\beta]\,d\beta\right)\cdot\left(\int_{\gamma=-\infty}^{+\infty} g^*[x-\gamma]w^*[\gamma]\,d\gamma\right)dx\\&= \int_{\beta=-\infty}^{+\infty} w[\beta]\left(\int_{\gamma=-\infty}^{+\infty} w^*[\gamma]\left(\int_{x=-\infty}^{+\infty} g[x-\beta]g^*[x-\gamma]\,dx\right)d\gamma\right)d\beta\end{aligned} \tag{19.54}$$

We substitute the translated variable $u \equiv x-\gamma$ into the integral over x:

$$\begin{aligned}\int_{x=-\infty}^{+\infty} g[x-\beta]g^*[x-\gamma]\,dx &= \int_{u=-\infty}^{+\infty} g[u-(\beta-\gamma)]g^*[u]\,du\\&= g[x]\bigstar g[x]|_{x=\beta-\gamma}\end{aligned} \tag{19.55}$$

where Equation (8.60) has been used. Substituting Equation (19.55) into Equation (19.54) produces a new expression for the second term in Equation (19.51):

$$\int_{x=-\infty}^{+\infty} |\hat{f}[x]|^2\,dx = \int_{\beta=-\infty}^{+\infty} w[\beta]\left(\int_{\gamma=-\infty}^{+\infty} w^*[\gamma]\cdot(g[x]\bigstar g[x]|_{x=\beta-\gamma})\,d\gamma\right)d\beta \tag{19.56}$$

The third term in Equation (19.51) may be recast by expanding the convolution, rearranging, and rewriting as a crosscorrelation:

$$\begin{aligned}-2\cdot\Re\left\{\int_{-\infty}^{+\infty}(f[x]\cdot\hat{f}^*[x])\,dx\right\} &= -2\cdot\Re\left\{\int_{-\infty}^{+\infty} f[x](g[x]*w[x])^*\,dx\right\}\\&= -2\cdot\Re\left\{\int_{-\infty}^{+\infty} f[x]\left(\int_{-\infty}^{+\infty} g^*[x-\alpha]w^*[\alpha]\,d\alpha\right)dx\right\}\\&= -2\cdot\Re\left\{\int_{-\infty}^{+\infty}\left(\int_{-\infty}^{+\infty} f[x]g^*[x-\alpha]\,dx\right)\cdot w^*[\alpha]\,d\alpha\right\}\\&= -2\cdot\Re\left\{\int_{-\infty}^{+\infty}(f[x]\bigstar g[x])|_{x=\alpha}\cdot w^*[\alpha]\,d\alpha\right\}\end{aligned} \tag{19.57}$$

The squared error due to the filter is the sum of the three terms in Equations (19.53), (19.56), and (19.57):

$$\begin{aligned}\epsilon = (f[x]\bigstar f[x]|_{x=0}) &+ \int_{\beta=-\infty}^{+\infty} w[\beta]\left(\int_{\gamma=-\infty}^{+\infty} w^*[\gamma]\cdot(g[x]\bigstar g[x]|_{x=\beta-\gamma})\,d\gamma\right)d\beta\\&- 2\cdot\Re\left\{\int_{-\infty}^{+\infty} w^*[\alpha]\cdot(f[x]\bigstar g[x])|_{x=\alpha}\,d\alpha\right\}\end{aligned} \tag{19.58}$$

which is now expressed in terms of the autocorrelation of the (unknown) input function $f[x]$, the autocorrelation of the measured output $g[x]$, and the crosscorrelation of the unknown input and the measured output. Note that ϵ is a nonnegative constant and is not a function of the variables x, α, β, or γ.

Our task now is to find the expression for the impulse response $w[x]$ that may be applied to $g[x]$ to produce the estimate $\hat{f}[x]$ that minimizes Equation (19.58). We expect that only one specific case of the set of all possible filters is optimal, which means that an arbitrary filter selected by the user includes the "ideal" optimum filter and an additional nonoptimum piece:

$$w[x] \equiv w_o[x] + w_n[x] \tag{19.59}$$

where $w_o[x]$ is the desired optimum filter and $w_n[x]$ is the additional nonoptimum part. We substitute this into the expression for the squared error in Equation (19.58) and perform some straightforward (though tedious) expansion and rearrangement:

$$\begin{aligned}
\epsilon &= (f[x] \bigstar f[x]|_{x=0}) \\
&\quad + \int_{\beta=-\infty}^{+\infty} (w_o[\beta] + w_n[\beta]) \left(\int_{\gamma=-\infty}^{+\infty} (w_o^*[\gamma] + w_n^*[\gamma]) \cdot (g[x] \bigstar g[x]|_{x=\beta-\gamma})\, d\gamma \right) d\beta \\
&\quad - 2 \cdot \Re\left\{ \int_{-\infty}^{+\infty} (w_o^*[\alpha] + w_n^*[\alpha]) \cdot (f[x] \bigstar g[x])|_{x=\alpha}\, d\alpha \right\} \\
&= (f[x] \bigstar f[x]|_{x=0}) \\
&\quad + \int_{\beta=-\infty}^{+\infty} \int_{\gamma=-\infty}^{+\infty} (w_o[\beta]w_o^*[\gamma] + w_n[\beta]w_n^*[\gamma] + w_o[\beta]w_n^*[\gamma] + w_n[\beta]w_o^*[\gamma]) \\
&\quad \cdot (g[x] \bigstar g[x]|_{x=\beta-\gamma})\, d\gamma\, d\beta \\
&\quad - 2 \cdot \Re\left\{ \int_{\alpha=-\infty}^{+\infty} w_o^*[\alpha] \cdot (f[x] \bigstar g[x])|_{x=\alpha}\, d\alpha \right\} \\
&\quad - 2 \cdot \Re\left\{ \int_{\alpha=-\infty}^{+\infty} w_n^*[\alpha] \cdot (f[x] \bigstar g[x])|_{x=\alpha}\, d\alpha \right\} \\
&= (f[x] \bigstar f[x]|_{x=0}) + \int_{\beta=-\infty}^{+\infty} \int_{\gamma=-\infty}^{+\infty} (w_o[\beta]w_o^*[\gamma])(g[x] \bigstar g[x]|_{x=\beta-\gamma})\, d\gamma\, d\beta \\
&\quad - 2 \cdot \Re\left\{ \int_{\alpha=-\infty}^{+\infty} w_o^*[\alpha] \cdot (f[x] \bigstar g[x])|_{x=\alpha}\, d\alpha \right\} \\
&\quad + \int_{\beta=-\infty}^{+\infty} \int_{\gamma=-\infty}^{+\infty} (w_n[\beta]w_n^*[\gamma] + w_o[\beta]w_n^*[\gamma] + w_n[\beta]w_o^*[\gamma])(g[x] \bigstar g[x]|_{x=\beta-\gamma})\, d\gamma\, d\beta \\
&\quad - 2 \cdot \Re\left\{ \int_{\alpha=-\infty}^{+\infty} w_n^*[\alpha] \cdot (f[x] \bigstar g[x])|_{x=\alpha}\, d\alpha \right\}
\end{aligned} \tag{19.60}$$

We identify the first three terms to be the (constant) optimum squared error and replace them by the symbol $\epsilon_{\text{optimum}}$:

$$\begin{aligned}
\epsilon_{\text{optimum}} &\equiv (f[x] \bigstar f[x]|_{x=0}) + \int_{\beta=-\infty}^{+\infty} \int_{\gamma=-\infty}^{+\infty} (w_o[\beta]w_o^*[\gamma])(g[x] \bigstar g[x]|_{x=\beta-\gamma})\, d\gamma\, d\beta \\
&\quad - 2 \cdot \Re\left\{ \int_{\alpha=-\infty}^{+\infty} w_o^*[\alpha] \cdot (f[x] \bigstar g[x])|_{x=\alpha}\, d\alpha \right\}
\end{aligned} \tag{19.61}$$

The resulting shorter expression for Equation (19.60) is:

$$\begin{aligned}\epsilon = \epsilon_{\text{optimum}} &+ \int_{\beta=-\infty}^{+\infty}\int_{\gamma=-\infty}^{+\infty} (w_o[\beta]w_n^*[\gamma] + w_n[\beta]w_o^*[\gamma])(g[x] \bigstar g[x]|_{x=\beta-\gamma})\,d\gamma\,d\beta \\ &+ \int_{\beta=-\infty}^{+\infty}\int_{\gamma=-\infty}^{+\infty} w_n[\beta]w_n^*[\gamma](g[x] \bigstar g[x]|_{x=\beta-\gamma})\,d\gamma\,d\beta \\ &- 2\cdot\Re\left\{\int_{\alpha=-\infty}^{+\infty} ((f[x] \bigstar g[x])|_{x=\alpha}\cdot w_n^*[\alpha])\,d\alpha\right\}\end{aligned} \tag{19.62}$$

For the moment, consider the second term in Equation (19.62). We can split it into the sum of two integrals, expand the autocorrelations via Equation (19.55), and rearrange both to perform the integrals over x last:

$$\begin{aligned}&\int_{\beta=-\infty}^{+\infty}\int_{\gamma=-\infty}^{+\infty} (w_o[\beta]w_n^*[\gamma] + w_n[\beta]w_o^*[\gamma])(g[x] \bigstar g[x]|_{x=\beta-\gamma})\,d\gamma\,d\beta \\ &= \int_{\beta=-\infty}^{+\infty}\int_{\gamma=-\infty}^{+\infty} w_o[\beta]w_n^*[\gamma]\int_{x=-\infty}^{+\infty} g[x-\beta]g^*[x-\gamma]\,dx\,d\gamma\,d\beta \\ &\quad+ \int_{\beta=-\infty}^{+\infty}\int_{\gamma=-\infty}^{+\infty} w_n[\beta]w_o^*[\gamma]\int_{x=-\infty}^{+\infty} g[x-\beta]g^*[x-\gamma]\,dx\,d\gamma\,d\beta \\ &= \int_{x=-\infty}^{+\infty}\left(\int_{\beta=-\infty}^{+\infty} w_o[\beta]g[x-\beta]\,d\beta\right)\left(\int_{\gamma=-\infty}^{+\infty} w_n^*[\gamma]g^*[x-\gamma]\,d\gamma\right)dx \\ &\quad+ \int_{x=-\infty}^{+\infty}\left(\int_{\beta=-\infty}^{+\infty} w_n[\beta]g[x-\beta]\,d\beta\right)\left(\int_{\gamma=-\infty}^{+\infty} w_o^*[\gamma]g^*[x-\gamma]\,d\gamma\right)dx \\ &= \int_{x=-\infty}^{+\infty}\left(\int_{\beta=-\infty}^{+\infty} w_o[\beta]g[x-\beta]\,d\beta\right)\left(\int_{\gamma=-\infty}^{+\infty} w_n[\gamma]g[x-\gamma]\,d\gamma\right)^* dx \\ &\quad+ \left(\int_{x=-\infty}^{+\infty}\left(\int_{\gamma=-\infty}^{+\infty} w_o[\gamma]g[x-\gamma]\,d\gamma\right)\left(\int_{\beta=-\infty}^{+\infty} w_n[\beta]g[x-\beta]\,d\beta\right)^* dx\right)^* \\ &= 2\cdot\Re\int_{x=-\infty}^{+\infty}\left(\int_{\beta=-\infty}^{+\infty} w_o[\beta]g[x-\beta]\,d\beta\right)\left(\int_{\gamma=-\infty}^{+\infty} w_n[\gamma]g[x-\gamma]\,d\gamma\right)^* dx\end{aligned} \tag{19.63}$$

We rewrite this term in the form of Equation (19.56) with the autocorrelation of $g[x]$:

$$\begin{aligned}&2\cdot\Re\int_{x=-\infty}^{+\infty}\left(\int_{\beta=-\infty}^{+\infty} w_o[\beta]g[x-\beta]\,d\beta\right)\left(\int_{\gamma=-\infty}^{+\infty} w_n[\gamma]g[x-\gamma]\,d\gamma\right)^* dx \\ &= 2\cdot\Re\left\{\int_{\beta=-\infty}^{+\infty} w_o[\beta]\int_{\gamma=-\infty}^{+\infty} w_n^*[\gamma]\left(\int_{x=-\infty}^{+\infty} g[x-\beta]g^*[x-\gamma]\,dx\right)d\gamma\,d\beta\right\} \\ &= 2\cdot\Re\left\{\int_{\beta=-\infty}^{+\infty}\int_{\gamma=-\infty}^{+\infty} w_o[\beta]w_n^*[\gamma](g[x] \bigstar g[x]|_{x=\beta-\gamma})\,d\gamma\,d\beta\right\}\end{aligned} \tag{19.64}$$

This is resubstituted in Equation (19.62), which is then rearranged:

$$\epsilon = \epsilon_{\text{optimum}} + 2 \cdot \Re\left\{\int_{\beta=-\infty}^{+\infty} \int_{\gamma=-\infty}^{+\infty} w_o[\beta] w_n^*[\gamma] (g[x] \bigstar g[x]|_{x=\beta-\gamma})\, d\gamma\, d\beta\right\}$$
$$- 2 \cdot \Re\left\{\int_{\alpha=-\infty}^{+\infty} (f[x] \bigstar g[x])|_{x=\alpha} \cdot w_n^*[\alpha]\, d\alpha\right\}$$
$$+ \int_{\beta=-\infty}^{+\infty} \int_{\gamma=-\infty}^{+\infty} w_n[\beta] w_n^*[\gamma] (g[x] \bigstar g[x]|_{x=\beta-\gamma})\, d\gamma\, d\beta \tag{19.65}$$

Now we remind ourselves of the goal: to find the filter function $w[x]$ that makes last two terms vanish. Toward this end, we rearrange the last term so that the integral over x is evaluated last:

$$\int_{\beta=-\infty}^{+\infty} \int_{\gamma=-\infty}^{+\infty} w_n[\beta] w_n^*[\gamma] (g[x] \bigstar g[x]|_{x=\beta-\gamma})\, d\gamma\, d\beta$$
$$= \int_{\beta=-\infty}^{+\infty} \int_{\gamma=-\infty}^{+\infty} w_n[\beta] w_n^*[\gamma] \left(\int_{x=-\infty}^{+\infty} g[x-\beta] g^*[x-\gamma]\, dx\right) d\gamma\, d\beta$$
$$= \int_{x=-\infty}^{+\infty} \left(\int_{\alpha=-\infty}^{+\infty} w_n[\beta] g[x-\beta]\, d\beta\right) \left(\int_{\beta=-\infty}^{+\infty} w_n^*[\gamma] g^*[x-\gamma]\, d\gamma\right) dx$$
$$= \int_{x=-\infty}^{+\infty} \left(\int_{\alpha=-\infty}^{+\infty} w_n[\beta] g[x-\beta]\, d\beta\right) \left(\int_{\beta=-\infty}^{+\infty} w_n[\gamma] g[x-\gamma]\, d\gamma\right)^* dx$$
$$= \int_{x=-\infty}^{+\infty} (w_n[x] * g[x])(w_n[x] * g[x])^*\, dx$$
$$= \int_{x=-\infty}^{+\infty} |w_n[x] * g[x]|^2\, dx \tag{19.66}$$

Since this term is the area of the squared magnitude of a convolution, it must be nonnegative; we assign it the new name ϵ_1:

$$\epsilon_1 \equiv \int_{x=-\infty}^{+\infty} |w_n[x] * g[x]|^2\, dx \geq 0 \tag{19.67}$$

The calculation of the minimum squared error is now reduced to minimizing the sum of the nonnegative optimum squared error, the nonnegative value ϵ_1, and two remaining terms:

$$\epsilon = (\epsilon_{\text{optimum}} + \epsilon_1)$$
$$+ 2 \cdot \Re\left\{\int_{\gamma=-\infty}^{+\infty} w_n^*[\gamma] \left(\int_{\beta=-\infty}^{+\infty} w_o[\beta] \cdot (g[x] \bigstar g[x]|_{x=\beta-\gamma})\, d\beta\right)\right\} d\gamma$$
$$- 2 \cdot \Re\left\{\int_{\alpha=-\infty}^{+\infty} w_n^*[\alpha] \cdot (f[x] \bigstar g[x])|_{x=\alpha}\, d\alpha\right\} \tag{19.68}$$

This is a minimum if the difference of the last two terms is minimized. We combine them by redefining the dummy variable of integration in the last term:

$$\begin{aligned}
&\Re\left\{\int_{\gamma=-\infty}^{+\infty} w_n^*[\gamma]\left(\int_{\beta=-\infty}^{+\infty} w_o[\beta]\cdot(g[x]\bigstar g[x]|_{x=\beta-\gamma})\,d\beta\right)d\gamma\right\}\\
&\quad-\Re\left\{\int_{\alpha=-\infty}^{+\infty} w_n^*[\alpha]\cdot(f[x]\bigstar g[x])|_{x=\alpha}\,d\alpha\right\}\\
&=\Re\left\{\int_{\gamma=-\infty}^{+\infty} w_n^*[\gamma]\left(\int_{\beta=-\infty}^{+\infty} w_o[\beta]\cdot(g[x]\bigstar g[x]|_{x=\beta-\gamma})\,d\beta\right)d\gamma\right\}\\
&\quad-\Re\left\{\int_{\gamma=-\infty}^{+\infty} w_n^*[\gamma]\cdot(f[x]\bigstar g[x])|_{x=\gamma}\,d\gamma\right\}\\
&=\Re\left\{\int_{\gamma=-\infty}^{+\infty} w_n^*[\gamma]\left(\int_{\beta=-\infty}^{+\infty} w_o[\beta]\cdot(g[x]\bigstar g[x]|_{x=\beta-\gamma})\,d\beta-(f[x]\bigstar g[x])|_{x=\gamma}\right)d\gamma\right\}
\end{aligned}\tag{19.69}$$

This vanishes if the inner integral is zero, i.e., if:

$$\int_{\beta=-\infty}^{+\infty} w_o[\beta]\cdot(g[x]\bigstar g[x]|_{x=\beta-\gamma})\,d\beta=(f[x]\bigstar g[x])|_{x=\gamma}\tag{19.70}$$

The autocorrelation in the integral on the left-hand side is a function of $\beta-\gamma$, which means that the integral is the convolution between the impulse response of the optimum filter w_o and the autocorrelation of the measured signal g:

$$\int_{\beta=-\infty}^{+\infty} w_o[\beta]\cdot(g[x]\bigstar g[x]|_{x=\beta-\gamma})\,d\beta=w_o[\gamma]*(g[\gamma]\bigstar g[\gamma])\tag{19.71}$$

This may be substituted into Equation (19.69) to find the condition for minimizing the squared error:

$$\begin{aligned}
&w_o[\gamma]*(g[\gamma]\bigstar g[\gamma])=(f[x]\bigstar g[x])|_{x=\gamma}\\
&\Longrightarrow w[x]*(g[x]\bigstar g[x])=f[x]\bigstar g[x]
\end{aligned}\tag{19.72}$$

where we have renamed the dummy variable γ and dropped the subscript "o" that denoted optimality from the notation. In words, the optimum filter impulse response $w[x]$ is that whose convolution with the autocorrelation of the measured signal is equal to the crosscorrelation of the unknown input with the measured signal.

We can evaluate the Fourier transforms of both sides of Equation (19.72) via Equations (9.152) and (9.159):

$$W[\xi]\cdot|G[\xi]|^2=F[\xi]\cdot G^*[\xi]\tag{19.73}$$

This is rearranged to obtain an expression for the transfer function of the optimum filter:

$$\begin{aligned}
W[\xi]&=\frac{F[\xi]\cdot G^*[\xi]}{|G[\xi]|^2}=\frac{F[\xi]\cdot G^*[\xi]}{G[\xi]\cdot G^*[\xi]}=\frac{F[\xi]}{G[\xi]}\\
&=\frac{F[\xi]}{F[\xi]+N[\xi]}
\end{aligned}\tag{19.74}$$

which matches our naive expression in Equation (19.46); we can now revisit the exercise left to the readers at that point. Clearly Equation (19.74) is not helpful in practical use because the specification of $W[\xi]$ requires full knowledge of $F[\xi]$ and $N[\xi]$ (and therefore of $f[x]$ and $n[x]$). In other words, we need to know that which we seek to determine the filter needed to find it! Even so, the expression may be useful in those cases where samples of $G[\xi]$ may be measured or estimated with and without added noise.

The expression in Equation (19.74) does suggest the question of the reason for constructing a filter. If the noise spectrum is known, then why not just subtract it from $G[\xi]$ to find $F[\xi]$? One answer to this question is that the goal of the derivation was to construct the "best" LSI filter; this did not allow use of subtraction, which may not be implemented as convolution.

19.3.1.1 Wiener Filter for Uncorrelated Signal and Noise

Though not very useful in the general case, the transfer function in Equation (19.74) is the solution to the problem in Equation (19.45) for any possible signal $f[x]$ and noise $n[x]$, including those cases where the two functions are correlated. We now consider the special case where the signal and noise are not correlated, i.e., such that:

$$f[x] \bigstar n[x] = 0[x] \tag{19.75}$$

The derivation leads to a different and more practical solution. Equation (19.75) may be inserted into the crosscorrelation on the right-hand side of Equation (19.72), which expands to the sum of two terms, of which one vanishes:

$$\begin{aligned} f[x] \bigstar g[x] &= \int_{\alpha=-\infty}^{+\infty} f[\alpha] g^*[\alpha - x]\, d\alpha \\ &= \int_{\alpha=-\infty}^{+\infty} f[\alpha](f^*[\alpha - x] + n^*[\alpha - x])\, d\alpha \\ &= \int_{\alpha=-\infty}^{+\infty} f[\alpha] f^*[\alpha - x]\, d\alpha + \int_{\alpha=-\infty}^{+\infty} f[\alpha] n^*[\alpha - x]\, d\alpha \\ &= f[x] \bigstar f[x] + f[x] \bigstar n[x] \\ &= f[x] \bigstar f[x] + 0 \end{aligned} \tag{19.76}$$

The autocorrelation of the measured signal in Equation (19.72) is the sum of four correlations, two of which vanish via Equation (19.75):

$$\begin{aligned} g[x] \bigstar g[x] &= ((f[x] + n[x]) \bigstar (f[x] + n[x])) \\ &= f[x] \bigstar f[x] + f[x] \bigstar n[x] + n[x] \bigstar f[x] + n[x] \bigstar n[x] \\ &= f[x] \bigstar f[x] + 0 + 0 + n[x] \bigstar n[x] \\ &= f[x] \bigstar f[x] + n[x] \bigstar n[x] \end{aligned} \tag{19.77}$$

After substituting Equations (19.76) and (19.77) into the condition of Equation (19.72) that optimizes the filter, the impulse response of the optimum filter in the case of uncorrelated noise must satisfy the relation:

$$w[x] * (f[x] \bigstar f[x] + n[x] \bigstar n[x]) = f[x] \bigstar f[x] \tag{19.78}$$

Now we evaluate the Fourier transforms of both sides and rearrange to find the recipe for the transfer function of the Wiener filter in the case of uncorrelated noise:

$$W[\xi] \cdot (|F[\xi]|^2 + |N[\xi]|^2) = |F[\xi]|^2 \Longrightarrow \boxed{W[\xi] = \frac{|F[\xi]|^2}{|F[\xi]|^2 + |N[\xi]|^2}} \tag{19.79}$$

Compare Equation (19.79) to the more general expression in Equation (19.74) to see that the expression for the Wiener filter with uncorrelated noise requires knowledge only of the *power spectra* of the signal and the noise and no knowledge of the phase. We have a fighting chance of estimating or modeling the power spectra, but the phase spectra necessary to evaluate Equation (19.74) are much more difficult to estimate (as was considered in Chapter 17).

The common expression for the Wiener filter in Equation (19.79) may be rewritten in the equivalent form:

$$W[\xi] = \frac{1}{1 + |N[\xi]|^2/|F[\xi]|^2} = \frac{1}{1 + \Gamma[\xi]} \tag{19.80}$$

where we have used the expression for the noise-to-signal power ratio from Equation (19.49). Now we may apply the model of the condition on the power spectra in Equation (19.50) for the factor of unity in the denominator of Equation (19.80) as the ratio of the known signal power spectrum to itself:

$$1[\xi] = \frac{|F[\xi]|^2}{|F[\xi]|^2} \tag{19.81}$$

Note that this condition is satisfied only at those frequencies where the signal power spectrum $|F[\xi]|^2$ is nonzero; the model of the pseudoinverse is followed if $|F[\xi]|^2 = 0$, so that the ratio is assigned the value of 0. We insert Equation (19.81) into the denominator in Equation (19.79) to see that the Wiener filter may be interpreted as "blending" the expressions at the extrema as the reciprocal of the sum of the reciprocals:

$$W[\xi] = \left[\left(\frac{|F[\xi]|^2}{|F[\xi]|^2} \right)^{-1} + \left(\frac{|F[\xi]|^2}{|N[\xi]|^2} \right)^{-1} \right]^{-1} \tag{19.82}$$

Readers may well (and understandably) be scratching their temples at this expression; what is the possible advantage of writing the simple expression in Equation (19.79) in this manner? The answer is that the same blending scheme may be applied in the more general case with $h[x] \neq \delta[x]$ that we will consider in the next section. In fact, the combination of terms as the reciprocal of the sum of reciprocals appears in other disciplines: the object and image distances in the imaging equation in Equation (1.4) and when evaluating impedances of electronic resistors acting in parallel.

The transfer function of the Wiener filter in Equation (19.82) requires knowledge of the power spectra (or, equivalently, of the magnitude spectra) of both the noise $n[x]$ *and* the desired signal $f[x]$. For the former, the power spectrum of the additive noise often may be estimated from measurements made when the signal is known to be absent. The assumption of a priori knowledge of the magnitude spectrum of the signal is more difficult to swallow. If $|F[\xi]|$ is known, then the Wiener filter only need estimate the phase spectrum $\Phi_F[\xi]$ to recover $\hat{f}[x]$, so it seems as though we are assuming half of the required knowledge up front. However, the discussion of the relative importance of magnitude and phase spectra in Chapter 17 demonstrated that the phase information is arguably more important than the magnitude spectrum.

19.3.2 Wiener Filter Example

The two expressions for the Wiener filter outside of the noise region are ambiguous about the behavior outside of this range: Equation (19.47a) implies that $W[\xi] = 0$ where $|F[\xi]|^2 = |W[\xi]|^2 = 0$, while Equation (19.47b) implies that $W[\xi] = 1$ under those conditions. Either choice has some logic, but we choose the former to ensure that any additional noise at different frequencies is not passed to the output. The latter choice was used in these examples.

In the first case, the signal and noise power spectra are assumed to be symmetric functions based on Gaussians. The signal and noise spectra are dominant at small spatial frequencies and the noise at large $|\xi|$:

$$|F[\xi]|^2 = 2 \cdot GAUS[\xi] \tag{19.83a}$$

$$|N[\xi]|^2 = 3 \cdot GAUS[\xi + 1] + 3 \cdot GAUS[\xi - 1] \tag{19.83b}$$

A simple analytic expression for $W[\xi]$ in terms of the special functions does not exist, but we can easily evaluate and graph it, as shown in Figure 19.8. The Wiener filter transfer function $W[\xi]$ blocks frequency components centered about $\xi = \pm 1$.

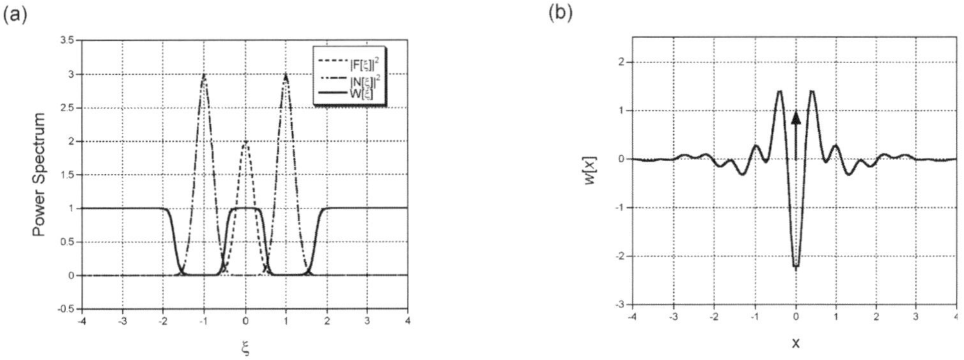

Figure 19.8 Example of the Wiener filter: (a) $|F[\xi]|^2 = 2 \cdot GAUS[2\xi]$, $|N[\xi]|^2 = 3 \cdot GAUS[\xi + 1] + 3 \cdot GAUS[\xi - 1]$, and $W[\xi]$, which is a bandstop filter; (b) impulse response $w[x]$.

The second example uses the same signal power spectrum and an asymmetric noise power spectrum shown in Figure 19.9:

$$|F[\xi]|^2 = 2 \cdot GAUS[\xi] \tag{19.84a}$$

$$|N[\xi]|^2 = 3 \cdot GAUS[\xi + 1] \tag{19.84b}$$

Again, a simple analytic expression for the transfer function escapes us, but we can graph the Wiener filter to see that it blocks the single sideband noise spectrum. The impulse response is complex valued and its phase is dominated by the linear phase that compensates for the asymmetry of the noise spectrum.

Now consider examples of the signal retrieved by the Wiener filter. The object $f[x]$ is a complex-valued chirp signal that has been modulated by a lowpass Gaussian, as shown in Figure 19.10. The signal spectrum $F[\xi]$ is dominated by low-frequency components and the noise spectrum $N[\xi]$ by high frequencies. The output of the Wiener filter is shown in Figure 19.11. Note that the transfer function of the Wiener filter resembles a superGaussian function, and the associated impulse response resembles an approximation to a *SINC* function. The output of the Wiener filter is certainly a better rendition of $f[x]$ than exists in $g[x] = f[x] + n[x]$.

19.3.3 Wiener–Helstrom Filter

In the more general case of signal estimation, the input to the filter is composed of the input function $f[x]$ filtered by some impulse response $h[x]$ before adding noise with a known power spectrum. The transfer function derived under these conditions is a modification to the Wiener filter by Carl Helstrom in 1967. Some authors continue to lump this more general filter under the original name, but others honor Helstrom's contribution by calling this the *Wiener–Helstrom filter*.

The filtered images in the two domains are:

$$g[x] * w[x] = (f[x] * h[x] * w[x]) + (n[x] * w[x]) = \hat{f}[x] \tag{19.85a}$$

$$G[\xi] \cdot W[\xi] = (F[\xi] \cdot H[\xi] \cdot W[\xi]) + (N[\xi] \cdot W[\xi]) = \hat{F}[\xi] \tag{19.85b}$$

The Wiener–Helstrom filter also assumes the same knowledge as the Wiener and inverse filters: complete knowledge of the power (or magnitude) spectra of the input and noise of the system transfer function $H[\xi]$. The noise power spectrum again is often estimated by measuring the noise without any signal present. The power spectrum of the signal is often not available directly, but is either modeled or assumed.

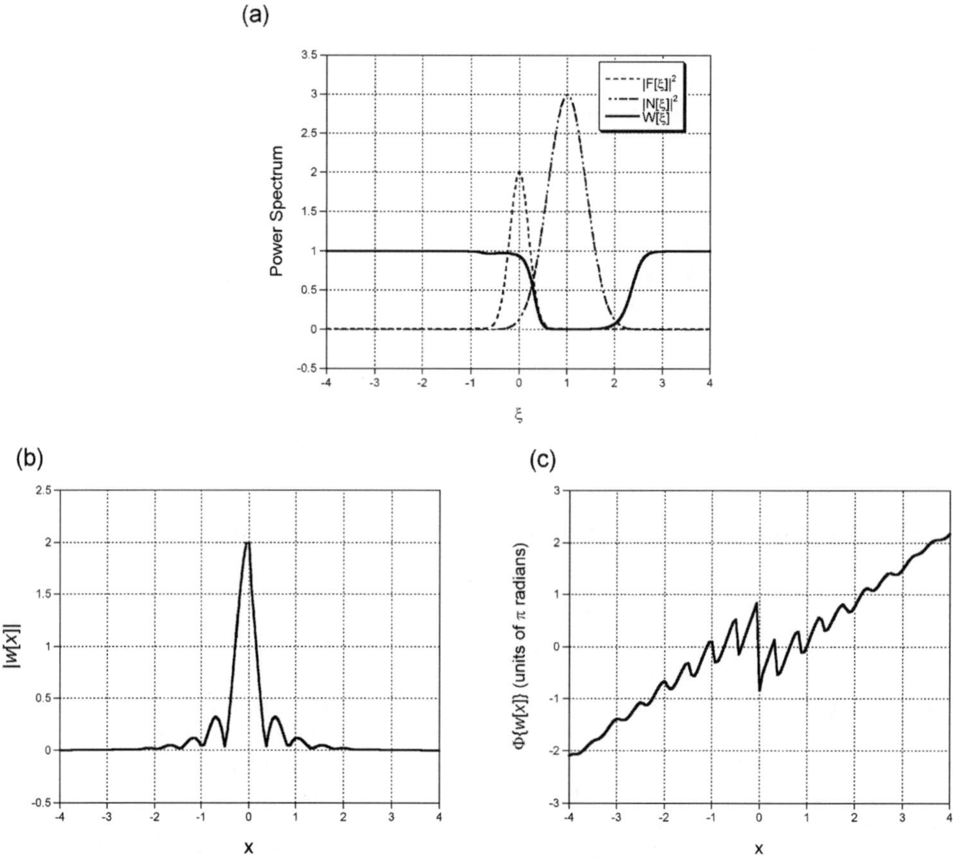

Figure 19.9 Wiener filter for an asymmetric noise power spectrum: (a) $|F[\xi]|^2 = 2 \cdot GAUS[2\xi]$, $|N[\xi]|^2 = 3 \cdot GAUS[\xi - 1]$, and $W[\xi]$, which is a "sideband" bandstop filter; (b) $|w[x]|$; (c) phase of the impulse response, showing the linear-phase term necessary to compensate for the asymmetry.

Limiting cases analogous to those of the Wiener filter in Equation (19.47) may be established after accounting for the effect of the known system transfer function $H[\xi]$ and the power spectra of the noise and of the filtered "noise-free" signal. Clearly this filter must be closely related to the inverse filter. We can use the interpretation of the Wiener filter as the "reciprocal of the sum of the reciprocals" of the limiting expressions. The signal power spectrum includes the contribution of the system transfer function $H[\xi]$; otherwise, the spectrum would be incorrectly compensated at any frequency where $|F[\xi]|^2$ is large compared to $|N[\xi]|^2$ even though $|F[\xi] \cdot H[\xi]|^2$ may be small.

At frequencies where the noise is small, so that $|F[\xi] \cdot H[\xi]|^2 \gg |N[\xi]|^2$, the transfer function of the Wiener filter must satisfy the criterion:

$$F[\xi] \cdot H[\xi] \cdot W[\xi] = F[\xi] \text{ if } |F[\xi] \cdot H[\xi]|^2 \gg |N[\xi]|^2 \tag{19.86a}$$

so the recipe for the transfer function at frequencies with large filtered signal power is:

$$W[\xi] \cong \frac{1}{H[\xi]} = \frac{1}{H[\xi]} \cdot 1[\xi] = \frac{1}{H[\xi]} \cdot \left(\frac{|F[\xi] \cdot H[\xi]|^2}{|F[\xi] \cdot H[\xi]|^2} \right) \tag{19.86b}$$

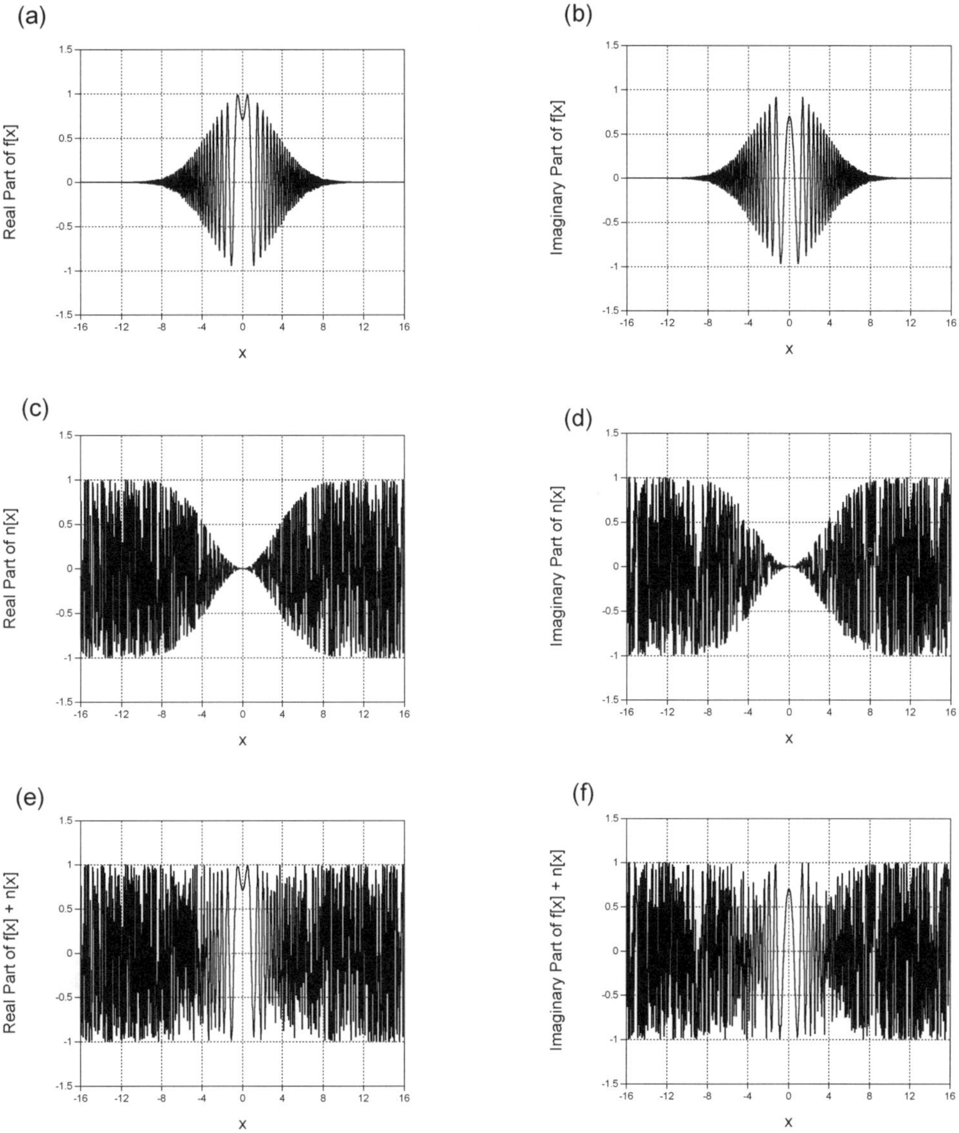

Figure 19.10 Signal and noise used to test the Wiener filter. The signal $f[x]$ is a chirp function modulated by a wide Gaussian. The noise is a complex random-phase signal modulated by a complementary Gaussian: (a) $\Re\{f[x]\}$; (b) $\Im\{f[x]\}$; (c) $\Re\{n[x]\}$; (d) $\Im\{n[x]\}$; (e) $\Re\{g[x]\} = \Re\{f[x] + n[x]\}$; (f) $\Im\{g[x]\} = \Im\{f[x] + n[x]\}$. The presence of the signal is arguably just detectable in $g[x]$.

At frequencies where the noise power dominates the filtered signal power, the transfer function is proportional to the unamplified approximation of the inverse filter $H^*[\xi]$ in Equation (19.20) that corrects for the phase but that "doubly attenuates" any sinusoidal components for which $|H[\xi]| < 1$:

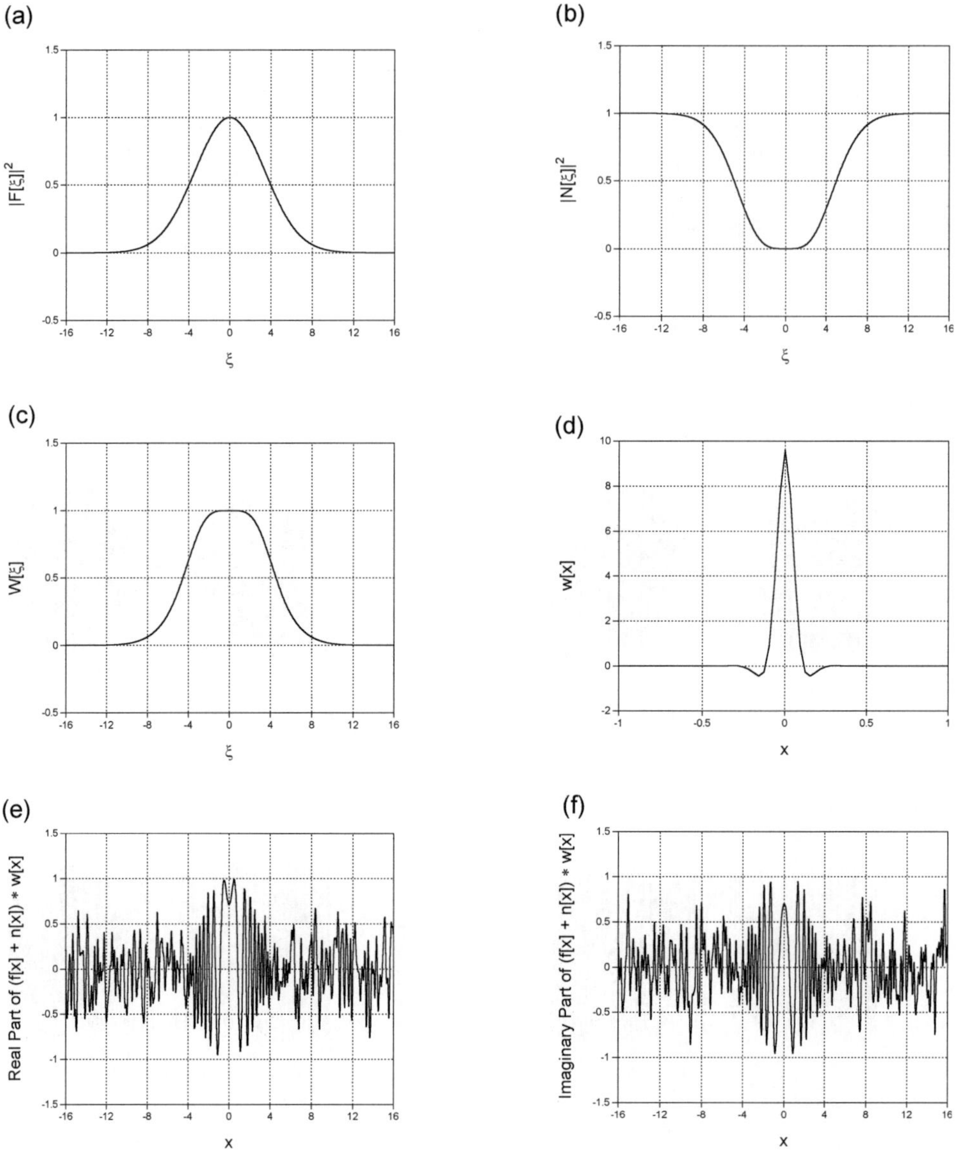

Figure 19.11 The components of the transfer function of the Wiener filter, the resulting impulse response, and the output: (a) $|F[\xi]|^2$; (b) $|N[\xi]|^2$; (c) the transfer function $W[\xi]$, which is lowpass; (d) impulse response $w[x]$. The output is shown in (e) and (f) as $\Re\{\hat{f}[x]\}$ and $\Im\{\hat{f}[x]\}$.

$$N[\xi] \cdot W[\xi] \cong 0 \cong \frac{1}{H[\xi]} \cdot \left(\frac{|F[\xi] \cdot H[\xi]|^2}{|N[\xi]|^2}\right) \quad \textit{if } |F[\xi] \cdot H[\xi]|^2 \ll |N[\xi]|^2 \tag{19.87a}$$

The transfer function of the filter has the form:

$$W[\xi] \cong H^*[\xi] \cdot \left(\frac{|F[\xi]|^2}{|N[\xi]|^2}\right) = H^*[\xi] \cdot (\Gamma[\xi])^{-1} \tag{19.87b}$$

where the noise-to-signal power ratio $\Gamma[\xi]$ has been substituted from Equation (19.49).

These two bounding representations of the transfer function in Equations (19.85b) and (19.86b) may be blended in the same manner as in the Wiener filter in Equation (19.82) as the reciprocal of the sum of reciprocals, which may be expressed in several equivalent ways:

$$W[\xi] = \left[\left(\frac{|F[\xi]|^2}{|F[\xi]|^2 \cdot H[\xi]}\right)^{-1} + \left(\frac{H^*[\xi] \cdot |F[\xi]|^2}{|N[\xi]|^2}\right)^{-1}\right]^{-1}$$

$$= \frac{H^*[\xi]}{|H[\xi]|^2 + |N[\xi]|^2/|F[\xi]|^2} = \frac{H^*[\xi]}{|H[\xi]|^2 + \Gamma[\xi]} = \frac{H^*[\xi] \cdot |F[\xi]|^2}{|H[\xi]|^2 \cdot |F[\xi]|^2 + |N[\xi]|^2} \tag{19.88}$$

We should check the limiting cases. In the limit $|H[\xi]|^2 \cdot |F[\xi]|^2 \gg |N[\xi]|^2$ (good *SNR*), then:

$$W[\xi] \cong \frac{H^*[\xi]}{|H[\xi]|^2 + 0} = \frac{H^*[\xi]}{|H[\xi]|^2} = \frac{1}{H[\xi]} \tag{19.89}$$

so the Wiener–Helstrom filter has the form of the inverse filter at frequencies where the signal dominates the noise.

In the "bad" limit where $|H[\xi]|^2 \cdot |F[\xi]|^2 \ll |N[\xi]|^2$, the expression reduces to:

$$W[\xi] = \frac{H^*[\xi]}{|H[\xi]|^2 + |N[\xi]|^2/|F[\xi]|^2} \cong \frac{H^*[\xi]}{(|N[\xi]|^2/|F[\xi]|^2)}$$

$$= \frac{|F[\xi]|^2}{|N[\xi]|^2} \cdot H^*[\xi] \equiv k \cdot H^*[\xi] \tag{19.90}$$

where k is a small positive value. In words, the transfer function of the Wiener–Helstrom filter at frequencies where noise dominates signal is proportional to the inverse filter in Equation (19.20) without the amplification term $\mathcal{F}_1^{-1}\{|H[\xi]|^{-2}\}$.

As an aside, note that this expression may be expanded into a Taylor series at frequencies where the noise-to-signal power spectrum is smaller than the squared magnitude of the transfer function:

$$W[\xi] = \frac{H^*[\xi]}{|H[\xi]|^2} \cdot \left(\frac{1}{1 + \Gamma[\xi]/|H[\xi]|^2}\right) = \frac{1}{H[\xi]} \cdot \left(\frac{1}{1 - (-\Gamma[\xi]/|H[\xi]|^2)}\right)$$

$$= \frac{1}{H[\xi]} \cdot \sum_{n=0}^{\infty}\left(\frac{-\Gamma[\xi]}{|H[\xi]|^2}\right)^n \quad \textit{if } |N[\xi]|^2 < |F[\xi]|^2 \cdot |H[\xi]|^2$$

$$= \frac{1}{H[\xi]} \cdot \left(1 - \frac{\Gamma[\xi]}{|H[\xi]|^2} + \left(\frac{\Gamma[\xi]}{|H[\xi]|^2}\right)^2 - \cdots\right) \tag{19.91}$$

If the noise-to-signal power ratio is small (*SNR* is large), the series may be truncated after the first term to produce an inverse filter. For larger values of $\Gamma[\xi]$ (worse *SNR*), the series will "roll off" the amplification of the inverse filter.

The expressions for the Wiener and Wiener–Helstrom filters as the reciprocal of the sum of reciprocals suggest that different limiting cases may be substituted for the limiting cases. In other words, instead of using the limits where the noise-to-signal power ratios are zero and infinite, we could use series solutions for different values of N as the limiting expressions.

The Wiener–Helstrom filter is optimum for *estimating* a signal in noise in that it maximizes the signal-to-noise power. However, "optimum" estimation often does not mean "good" estimation, as demonstrated in the following examples.

19.3.4 Wiener–Helstrom Filter Example

The object is the "three-bar" chart that has been filtered by the asymmetric function $h[x] = e^{-2x} \cdot STEP[x]$, as shown in Figure 19.12. The mean and standard deviation of the noise are $\mu = 0$ and

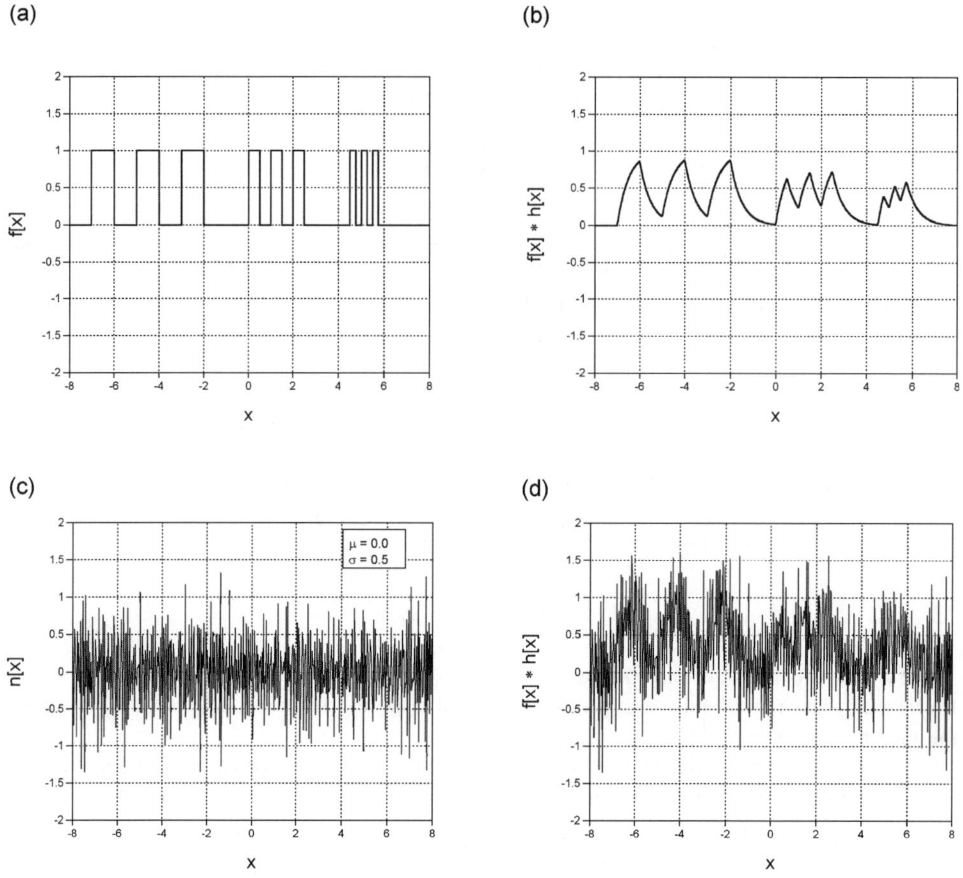

Figure 19.12 Functions used to illustrate the Wiener–Helstrom filter: (a) $f[x]$; (b) $f[x] * h[x]$, where the impulse response is the asymmetric blur $h[x] = 2\,e^{-2x} \cdot STEP[x]$; (c) $n[x]$, which is real-valued "blue" noise with $\mu = 0$ and $\sigma = \frac{1}{2}$; (d) $f[x] * h[x] + n[x]$.

$\sigma = 0.5$. The noise is the real part of a signal generated from the power spectrum $|N[\xi]|^2 = (1 - GAUS[\xi/12])^2$ with phase randomly distributed over 2π radians. The attenuation of the low-frequency components means that the noise is very "blue". The signal is very well hidden by the noise.

The transfer function of the original blur and of the Wiener–Helstrom filter are shown in Figure 19.13. Because the noise dominates at large spatial frequencies ($|\xi| \gg 0$), the transfer function $W[\xi]$ "rolls off" to attenuate these components. For small values of $|\xi|$, the compensating action on the attenuated amplitudes of $|H[\xi]|$ is apparent in the high-boost character of $|W[\xi]|$, while the cancellation of the phase of the impulse response is seen by comparing $\mathbf{\Phi}\{H[\xi]\}$ and $\mathbf{\Phi}\{W[\xi]\}$.

The impulse response $w[x]$ of the Wiener–Helstrom filter and the resulting output are shown in Figure 19.14. The impulse response is real valued and asymmetric in the "opposite" orientation relative to $h[x]$, i.e., the largest amplitudes in $w[x]$ appear for $x < 0$, which compensates for the asymmetry. The output is a much better rendition of the original function than the blurred signal with the noise, and even when compared to the original blurred signal alone. Of course, the noise was "designed" to ensure a good result by severely attenuating the components with small spatial frequencies. In real life, the results tend to be much less dramatic. It is much easier to make a matched filter to locate an exactly

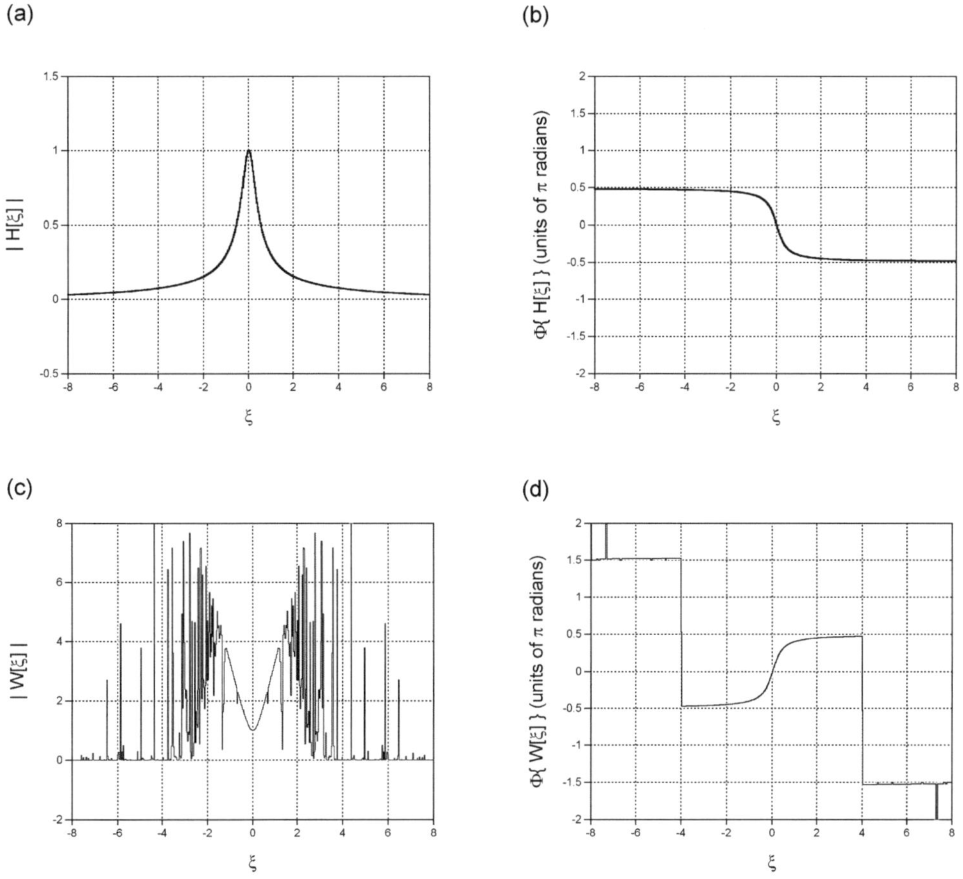

Figure 19.13 The transfer function of the original blur $h[x] = 2\exp[-2x] \cdot STEP[x]$ and the Wiener–Helstrom filter: (a) $|H[\xi]|$, which is a Lorentzian function showing the lowpass character of the blur; (b) $\Phi\{H[\xi]\}$, showing that the phase is nonlinear but dominated by a linear term due to the asymmetry of $h[x]$; (c) $|W[\xi]|$, showing the amplification at small spatial frequencies and the rolloff to block the dominant noise frequencies at large values of $|\xi|$; (d) $\Phi\{W[\xi]\}$, showing the compensation for the phase of $H[\xi]$ at small spatial frequencies where the signal is dominant.

known signal in noise than to estimate an unknown signal in additive noise, even if the signal power spectrum is known.

19.3.5 Constrained Least-Squares Filter

An approximation to the Wiener–Helstrom filter may be constructed when the exact forms of the power spectra of the signal and noise are not known, but rather are assumed to have a constant ratio. The resulting transfer function is the constrained least-squares filter:

$$W[\xi] = \frac{H^*[\xi]}{|H[\xi]|^2 + \Gamma[\xi]} \to \frac{H^*[\xi]}{|H[\xi]|^2 + \Gamma} \tag{19.92}$$

where the parameter Γ is the (assumed constant) noise-to-signal power ratio.

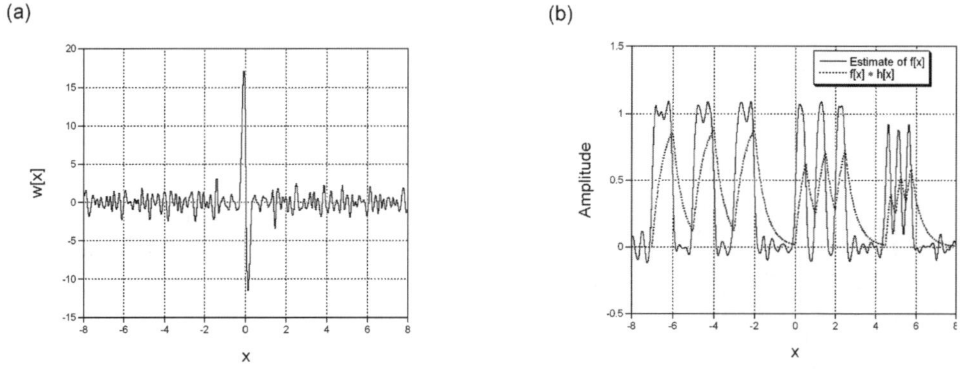

Figure 19.14 The impulse response of the Wiener–Helstrom filter and the resulting output: (a) $w[x]$, which is real valued and asymmetric; (b) $\hat{f}[x]$ shown along with $f[x] * h[x]$ to demonstrate the enhanced symmetry of the bars and the reduction of the noise.

19.4 Detection of Known Signals – Matched Filter

Some imaging tasks require detecting the presence and/or location of a known signal in a typically unknown background that may be considered "noise". Obvious examples are radar and sonar systems that locate "echoes" of the known transmitted signal in a corrupting background of natural or deliberate noise ("clutter"). Another task for such a filter would be the identification of occurrences of a particular character or word on a page of text. In this case, the noise includes any markings on the page other than the chosen pattern, obviously including other valid words. In all of these examples, the filter is "matched" to the desired signal, hence its name and the symbol $m[x]$ we will use for the impulse response.

As an introductory example, consider a measured signal $g[x]$ that is the sum of a single copy of the known input $f[x]$ scaled by the parameter α_1 and centered about some unknown coordinate x_1 in the presence of additive (and usually uncorrelated) noise $n[x]$:

$$g[x] = \alpha_1 \cdot f[x - x_1] + n[x] \tag{19.93}$$

The task is to construct a filter with impulse response $m[x]$ to be applied to $g[x]$ to produce an output that leads to the "best estimate" of x_1 (call it $\hat{x}_1$). The appropriate $m[x]$ clearly must be a function of $f[x]$; its output is:

$$g[x] * m[x] = \alpha_1 \cdot f[x - x_1] * m[x] + n[x] * m[x] \tag{19.94}$$

The next step after applying the filter is to identify the value of x_1. A seemingly reasonable choice would be the coordinate location ("argument") of the maximum output from the filter:

$$\hat{x}_1 = \arg[\max[g[x] * m[x]]] \tag{19.95}$$

However, our experience with highpass and phase filters demonstrates that $g[x] * m[x]$ may well be complex and bipolar, and thus it is possible for the output to have an imaginary or negative extremum at x_1. For this reason, we consider using the criterion of the argument of the maximum *magnitude* for locating x_1:

$$\begin{aligned}\hat{x}_1 &= \arg[\max[|g[x] * m[x]|]] \\ &= \arg[\max[|\alpha_1 \cdot f[x - x_1] * m[x] + n[x] * m[x]|]]\end{aligned} \tag{19.96}$$

Of course, the argument of the maximum of the magnitude is identical to the argument of that of the squared magnitude of the output. For reasons to be considered shortly, this criterion is usually expressed in this form:

$$\hat{x}_0 = \arg[\max[|g[x] * m[x]|^2]] \tag{19.97}$$

In words, the location of the known ("reference") signal is assumed to be the coordinate of the matched-filtered output with maximum "signal-to-noise power".

We will consider the derivation of the filter on two levels. First, we infer some properties of the impulse response $m[x]$ by a simple thought experiment that considers the "ideal" output of such a filter based on an intuitive criterion. The filter resembles the inverse filter discussed in the previous section, and the similarity will lead to some other useful variants of both filters. After considering useful approximate forms for both the inverse and matched filter, we will then derive the form for the matched filter in a more rigorous fashion.

If we use the criterion of Equation (19.42), the output of the ideal matched filter with impulse response $m_{\text{ideal}}[x]$ would produce its maximum (and ideally infinite) amplitude at x_1 and small (ideally zero) amplitude everywhere else. Clearly, the best possible output would be a Dirac delta function located at x_1:

$$\begin{aligned} g[x] * m_{\text{ideal}}[x] &= (\delta[x - x_1] * \alpha_1 \cdot f[x] * m_{\text{ideal}}[x]) + (n[x] * m_{\text{ideal}}[x]) \\ &= \alpha_1 \cdot \delta[x - x_1] + 0[x] \end{aligned} \tag{19.98}$$

The corresponding expression in the frequency domain is:

$$\begin{aligned} G[\xi] \cdot M_{\text{ideal}}[\xi] &= e^{-2\pi i \xi x_1} \cdot \alpha_1 \cdot F[\xi] \cdot M_{\text{ideal}}[\xi] + N[\xi] \cdot M_{\text{ideal}}[\xi] \\ &= \alpha_1 \cdot e^{-2\pi i \xi x_1} + 0[\xi] \end{aligned} \tag{19.99}$$

We can see that two conditions must be satisfied simultaneously in both domains:

$$\alpha_1 \cdot f[x] * m_{\text{ideal}}[x] = \delta[x] \Longrightarrow \alpha_1 \cdot F[\xi] \cdot M_{\text{ideal}}[\xi] = 1[\xi] \tag{19.100a}$$

$$n[x] * m_{\text{ideal}}[x] = 0[x] \Longrightarrow N[\xi] \cdot M_{\text{ideal}}[\xi] = 0[\xi] \tag{19.100b}$$

The first criterion is identical to that for the ideal inverse filter in Equation (19.12), and can only be satisfied if $F[\xi] \neq 0$ at all ξ. In most (if not all) realistic situations, the noise spectrum effectively has infinite support, and thus the second criterion is only satisfied if either $N[\xi] = 0$ or $M[\xi] = 0$. The former precludes the existence of noise in $g[x]$, while the latter produces a null output; both are uninteresting cases. Nevertheless, it still is useful to consider the form of $M_{\text{ideal}}[\xi]$ that satisfies the condition in Equation (19.100a), in part because this solution will provide clues to the form of the filter that is appropriate for more realistic cases. This condition requires that the transfer function of the ideal matched filter be the reciprocal of the signal spectrum, which is the identical form of the transfer function of the inverse filter, except that it is the reciprocal of the spectrum of a different function. For this reason, we may consider the inverse and matched filters to be examples of the general class of "reciprocal filters", where the term *reciprocal* is understood to apply to the transfer function. Awwal et al. (1990), Carnicer et al. (1993), and Iftekharuddin et al. (1996) noted the value of the amplification by the reciprocal filter.

As for the inverse filter in Equation (19.17), the transfer function of the matched filter may be written in several forms:

$$M_{\text{ideal}}[\xi] = \frac{1}{\alpha_1 \cdot F[\xi]} = \frac{1}{\alpha_1} \frac{F^*[\xi]}{|F[\xi]|^2} = \frac{1}{\alpha_1} \frac{e^{-i\Phi\{F[\xi]\}}}{|F[\xi]|} \tag{19.101}$$

The three cases in Equation (19.18) may be substituted to construct three expressions for the matched-filter impulse response $m_{\text{ideal}}[x]$. The second of these divides the ideal impulse response into two terms and provides some insight into the required behavior of the matched filter:

$$\begin{aligned} m_{\text{ideal}}[x] &= \mathcal{F}_1^{-1}\{F^*[\xi]\} * \mathcal{F}_1^{-1}\left\{\frac{1}{\alpha_1 \cdot |F[\xi]|^2}\right\} \\ &= \frac{1}{\alpha_1} \cdot f^*[-x] * \mathcal{F}_1^{-1}\left\{\frac{1}{|F[\xi]|^2}\right\} \end{aligned} \tag{19.102}$$

The factor of α_1^{-1} compensates for the amplitude of the input function and would also amplify any noise by the same amount. It may (and will) be eliminated without affecting the relative output values. The term $f^*[-x]$ cancels the intrinsic phase of the reference signal $f[x]$, so that any residual phase in the output will be a linear term due to the translation of $f[x]$ by (the unknown distance) x_1. However, the convolution with $f^*[-x]$ also compounds the effect of the magnitude spectrum of $f[x]$; if $f[x]$ is dominated by low-frequency sinusoids (as is usually true in imaging systems), then the convolution with $f^*[-x]$ produces a signal that is even more "lowpass" in character. This is compensated by convolution with the highpass filter with impulse response $\mathcal{F}_1^{-1}\{|F[\xi]|^{-2}\}$.

We will discuss the analogy between the inverse and matched filters indicated by Equations (19.19) and (19.102) in more detail shortly, as well as relate them to the Wiener and Wiener–Helstrom filters, but we can see already that any technique developed to improve the performance of the inverse filter may be applied or adapted to the matched filter and vice versa. For example, if the magnitude spectrum $|F[\xi]|$ of the input signal has zeros at some frequencies, then these spectral components of $G[\xi]$ cannot be amplified to generate a finite amplitude by the ideal matched filter. However, we can construct a "pseudomatched" filter by analogy with Equation (19.25):

$$\hat{M}[\xi] \equiv \begin{cases} \dfrac{1}{F[\xi]} & \textit{if } F[\xi] \neq 0 \\ 0 & \textit{if } F[\xi] = 0 \end{cases} \tag{19.103a}$$

$$\hat{m}[x] = \mathcal{F}_1^{-1}\{\hat{M}[\xi]\} \tag{19.103b}$$

where the constant factor α_1^{-1} has been ignored, as promised. If the zeros of $|F[\xi]|$ are isolated, then the sinusoidal components at these frequencies will be "missing" from the output spectrum of the pseudomatched filter:

$$F[\xi] \cdot \hat{M}[\xi] = \begin{cases} 1 & \textit{if } F[\xi] \neq 0 \\ 0 & \textit{if } F[\xi] = 0 \end{cases} \tag{19.104}$$

The output of $f[x] * \hat{m}[x]$ approximates the ideal Dirac delta function. This approximation can often be characterized further based on the signal spectrum $F[\xi]$. For example, if $F[\xi]$ is "lowpass" and compact, so that $F[\xi] = 0$ for $|\xi| \geq \xi_{\max}$, then the best possible output of the pseudomatched filter is the *SINC* function that does not contain the sinusoidal components at those frequencies.

If wideband random noise is present in $g[x]$, the pseudomatched filter suffers from the same problem as the pseudoinverse filter in Equation (19.25); it will amplify any noise components at spatial frequencies where $|F[\xi]| < 1$. The subsequent inverse Fourier transform "spreads" this noise throughout the space domain and possibly obscures the desired output centered at $x = x_1$. One strategy for avoiding this problem in the realistic case is to delete the amplification term $\mathcal{F}_1^{-1}\{|F[\xi]|^{-2}\}$ in Equation (19.102) when applying the matched filter in the presence of wideband noise. The approximation to the matched filter becomes:

$$\hat{M}[\xi] = F^*[\xi] \Longrightarrow \hat{m}[x] = f^*[-x] \tag{19.105}$$

This might be called the *classical matched filter* based upon the treatment in standard textbooks. The output of the classical approximation to the ideal matched filter in the space domain is:

$$\begin{aligned}(\alpha_1 \cdot f[x-x_1]+n[x]) * \hat{m}[x] &= (f[x-x_1]+n[x]) * f^*[-x] \\ &= (\alpha_1 \cdot \delta[x-x_1] * f[x] * f^*[-x]) + (n[x] * f^*[-x]) \\ &= \alpha_1 \cdot \delta[x-x_1] * (f[x] \bigstar f[x]) + n[x] \bigstar f[x] \end{aligned} \tag{19.106}$$

Instead of the ideal Dirac delta function, this "realistic" matched filter produces a scaled replica of the autocorrelation of $f[x]$ centered at the unknown location x_1 amid the crosscorrelation of $n[x]$ with $f[x]$. We have already seen that the autocorrelation of a compact signal is a Hermitian function with finite amplitude and finite support. The maximum amplitude of the output is the area of $|f[x]|^2$ scaled by the amplitude α_1 of the replica and its support is twice that of $f[x]$. In the common case where $|F[\xi]|$ is a "lowpass" spectrum, then the convolution with $f^*[-x]$ averages the noise signal and decreases its variance, thus helping to "pull" the autocorrelation signal out of the noise.

Because the filtering process is linear and shift invariant, a matched filter will locate any number of replicas of the "reference" function $f[x]$ with different amplitudes in $g[x]$:

$$\begin{aligned} g[x] &= \left(\sum_{\ell=1}^{N} \alpha_\ell \cdot f[x-x_\ell]\right) + n[x] \\ &= \left(\sum_{\ell=1}^{N} \alpha_\ell \cdot (\delta[x-x_\ell] * f[x])\right) + n[x] \end{aligned} \tag{19.107}$$

If using the "unamplified" approximation that we called the "classical matched filter" from Equation (19.105):

$$\begin{aligned} g[x] * \hat{m}[x] &= \left(\sum_{\ell=1}^{N} \alpha_\ell \cdot (\delta[x-x_\ell] * f[x])\right) * f^*[-x] + (n[x] * f^*[-x]) \\ &= \left(\sum_{\ell=1}^{N} \alpha_\ell \cdot \delta[x-x_\ell]\right) * f[x] \bigstar f[x] + n[x] \bigstar f[x] \end{aligned} \tag{19.108}$$

In words, we obtain scaled replicas of the autocorrelation of $f[x]$ on top of the crosscorrelation of the noise and signal.

Now consider some examples of the matched filter based on the same illustrations of the inverse filter in the previous section. In the first, the input $f[x]$ is the unit-area *RECT* function:

$$f_1[x] = \frac{1}{|b_0|} RECT\left[\frac{x}{b_0}\right] \Longrightarrow F_1[\xi] = SINC[b_0 \cdot \xi] \tag{19.109}$$

The corresponding pseudomatched filter is analogous to the pseudoinverse filter for the uniform averager in Equation (19.25), which recovers all sinusoidal frequencies except the nonzero integer multiples of b_0^{-1}. If $f_1[x]$ is centered upon an (assumed unknown) coordinate x_0, the output spectrum includes the linear phase:

$$\begin{aligned} F_1[\xi] \cdot \hat{M}_1[\xi] &= (SINC[b_0 \cdot \xi]\, e^{-2\pi i \xi x_0}) \cdot (SINC[b_0 \cdot \xi])^{-1} \\ &= \hat{1}[\xi]\, e^{-2\pi i \xi x_0} \end{aligned} \tag{19.110}$$

where the notation $\hat{1}[\xi]$ signifies the approximation to the ideal unit constant produced by the cascade of the input spectrum and the matched-filter transfer function. Since the spectrum misses only "occasional" sinusoidal components, this expression produces a "reasonable" approximation to the ideal unit-magnitude spectrum. The pseudomatched filter also amplifies all nonzero spatial frequencies, whether belonging to signal or noise, and thus is not likely to be useful in realistic cases.

In the second case, the matched filter searches for replicas of the *SINC* function, whose spectrum is compact:

$$f_2[x] = \frac{1}{|b_0|} SINC\left[\frac{x}{b_0}\right] \tag{19.111a}$$

$$\Longrightarrow F_2[\xi] = |F_2[\xi]| = RECT[b_0 \cdot \xi] \tag{19.111b}$$

The corresponding matched filter is analogous to the inverse filter for the ideal lowpass filter in Equation (19.34). The transfer function of the "pseudomatched" filter is identical to the magnitude spectrum of the input except for the two isolated amplitudes at the edges of the rectangle:

$$\begin{aligned}(F_2[\xi]\, e^{-2\pi i\xi x_0}) \cdot \hat{M}_2[\xi] &= (RECT[b_0 \cdot \xi]\, e^{-2\pi i\xi x_0}) \cdot RECT[b_0 \cdot \xi] \\ &= RECT[b_0 \cdot \xi]\, e^{-2\pi i\xi x_0} \\ &= \hat{1}[\xi]\, e^{-2\pi i\xi x_0}\end{aligned} \tag{19.112}$$

This pseudomatched filter is not an amplifier. The output of the "ideal" matched filter in the space domain is identical to the original signal:

$$\begin{aligned}f_2[x - x_0] * \hat{m}_2[x] &= \mathcal{F}_1^{-1}\{RECT[b_0 \cdot \xi]\, e^{-2\pi i\xi x_0}\} \\ &= \frac{1}{|b_0|} SINC\left[\frac{x}{b_0}\right] * \delta[x - x_0] \\ &= \frac{1}{|b_0|} SINC\left[\frac{x - x_0}{b_0}\right] = \hat{\delta}[x - x_0]\end{aligned} \tag{19.113}$$

The lowpass character of this matched filter can reduce the variance of wideband noise, and thus possibly enhance the visibility of the correlation peak relative to the noisy background.

As a third illustration of the ideal matched filter, assume that the reference function is the decaying exponential in Equation (19.37):

$$f_3[x] = STEP\left[\frac{x}{b_0}\right] \cdot \frac{1}{|b_0|} \exp\left[-\frac{x}{b_0}\right] \tag{19.114}$$

The spectrum of the reference function has no zeros, but its amplitude is smaller than unity at all nonzero spatial frequencies. These frequencies are amplified by the ideal matched filter, which has the same form as the inverse filter derived in Equation (19.39):

$$m_3[x] = \delta[x] + b_0 \cdot \delta'[x] \tag{19.115}$$

The amplification of the derivative increases with spatial frequency ξ, thus again limiting the utility of the ideal matched filter for the decaying exponential.

An example based on the decaying exponential function is shown in Figure 19.15. The "reference function" $f[x]$ is a single-sided decaying exponential and the "position function" $s[x]$ is composed of a set of impulses located at random positions and with random (and bipolar) amplitudes. The measured noise-free image is $g[x] = f[x] * s[x]$. If the classical matched filter from Equation (19.105) is applied to $g[x]$, the replicas of $f[x]$ are replaced by scaled copies of the autocorrelation functions $f[x] * f^*[-x] = f[x] \bigstar f[x]$. The lowpass action of the unamplified matched filter produces rather broad (and thus perhaps indistinct) correlation peaks at the location of replicas of the reference function. Any replicas of $f[x]$ scaled by smaller amplitudes may be masked by the sidelobes of replicas with larger amplitudes.

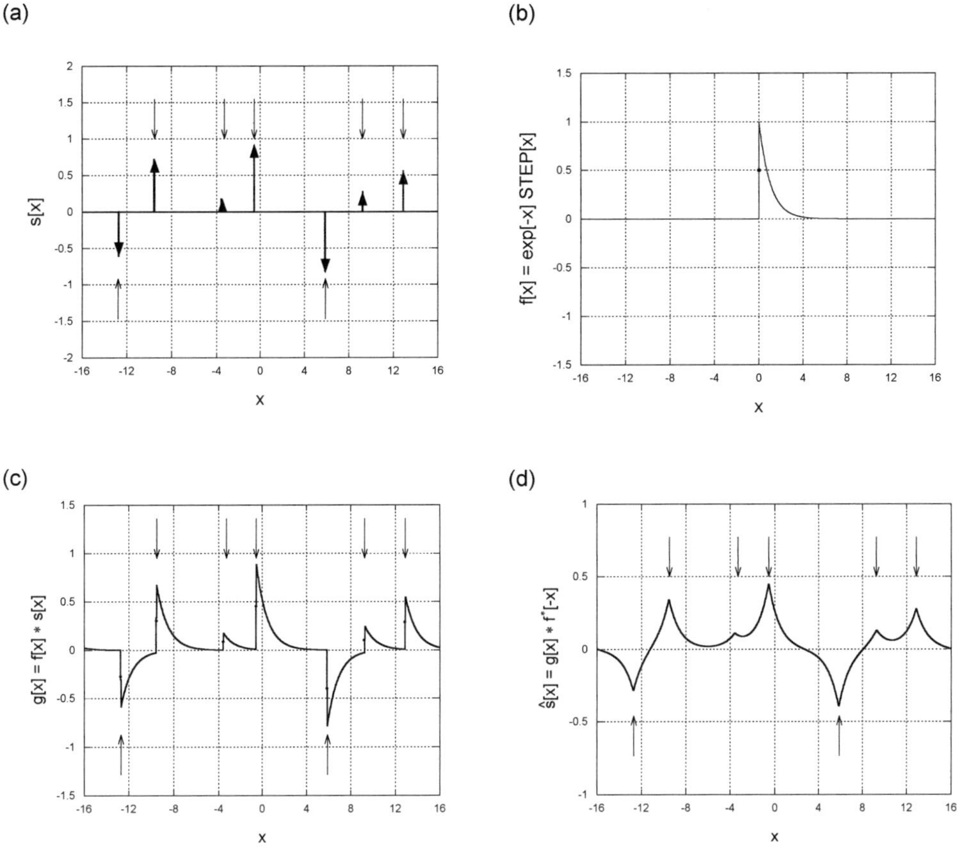

Figure 19.15 Action of unamplified matched filter with no noise: (a) $s[x]$, composed of a set of Dirac delta functions at random positions and with areas selected at random from the interval $-1 \leq f < +1$; (b) $f[x] = e^{-x} \cdot STEP[x]$; (c) $g[x] = f[x] * s[x]$; (d) output of "unamplified" matched filter, $g[x] * f^*[-x]$, showing the symmetric maxima due to the autocorrelations $f[x - x_n] * f^*[-x]$.

The utility of (or even "necessity for") the unamplified matched filter becomes apparent if $g[x]$ is corrupted with additive Gaussian noise. The results for several sample noise functions are shown in Figure 19.16. The lowpass character of the "unamplified" matched filter $m[x] = f^*[-x]$ reduces the variance of the noise sufficiently that the output is very similar to that obtained for the noise-free case.

19.4.1 Inputs for Matched Filters

We digress for a moment to consider how matched filters are used in "active" imaging systems, such as radar, sonar, and medical ultrasound. The known reference signal is generated and transmitted into a specific azimuthal direction in object space. The system monitors the interaction of a transmitted signal with the "scene" (usually by reflection) and attempts to detect replicas of the transmitted signal along that azimuthal direction. The positions of the replicas of the transmitted reference signal are determined by matched filtering and then mapped.

The "reference" function to be located by matched filters is usually selectable by the user, and its desirable characteristics are easy to visualize. In the previous section, we observed that the matched

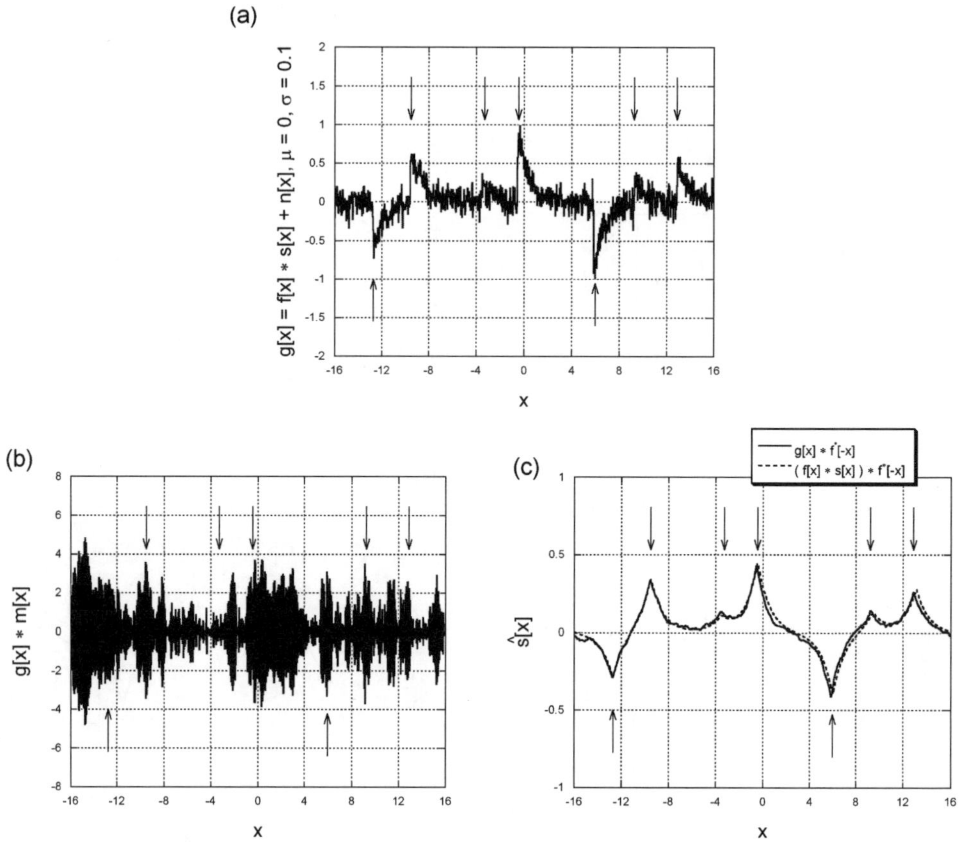

Figure 19.16 Performance of matched filters in the presence of noise: (a) $g[x] = f[x] * s[x] + n[x]$, where $n[x]$ is Gaussian noise with $\mu = 0$ and $\sigma = 0.1$, the arrows showing the locations of the replicas of $f[x]$; (b) output of amplified matched filter $g[x] * \mathcal{F}_1^{-1}\{F[\xi]^{-1}\}$, showing the detrimental effect of the amplified noise; (c) output of unamplified matched filter with and without noise, showing the effect of averaging the random noise.

filter can produce an ideal Dirac delta function at the location of the reference object $f[x]$ if $F[\xi]$ has no zeros. In this case, the ideal matched filter in Equation (19.102) may be used:

$$M[\xi] = \frac{F^*[\xi]}{|F[\xi]|^2} \tag{19.116}$$

where again the constant scale factor has been ignored. The denominator of this transfer function may be interpreted as compensating for the attenuation or amplification of all frequency components that have nonunit amplitude, thus ensuring that the output is the desired Dirac delta function. Obviously, any such compensation (usually amplification) is also applied to any noise at those frequencies, which may make estimation of x_0 more difficult. The requirement for amplifying the spectrum may be eliminated by selecting a signal whose power spectrum has unit magnitude at all frequencies, including the "allpass" signals considered in Chapter 17. These signals are distinguished by their phase spectra:

$$F[\xi] = 1[\xi]\, e^{i\Phi\{F[\xi]\}}$$
$$\Longrightarrow |F[\xi]|^2 = 1[\xi] \tag{19.117}$$

We now briefly consider matched filtering with allpass signals with constant phase, linear phase, quadratic phase, and random phase. A unit-magnitude signal with a constant- or linear-phase spectrum has the form of a Dirac delta function:

$$f_1[x] = (\alpha + i\beta) \cdot \delta[x - x_0] \quad (\textit{for constant- and linear-phase functions}) \tag{19.118}$$

These signals are only of academic interest because they cannot be generated in real systems; it is not possible to create a signal with infinitesimal width, finite area, and infinite bandwidth. Even if they could be generated, the matched filter would also be a Dirac delta function and the filtered signal would be identical to the measured signal $g[x]$.

A more useful signal in matched-filter systems is based upon the quadratic-phase "chirp" introduced in Equation (6.146). The ideal chirp has constant magnitude and infinite bandwidth, and its space- (or time-) domain representation has infinite support, as was shown in Equation (9.94):

$$f_2[x] = \frac{1}{|\alpha|} \exp\left[\pm\frac{i\pi}{4}\right] \exp\left[\mp i\pi\left(\frac{x}{\alpha}\right)^2\right] \quad (\textit{for quadratic-phase functions}) \tag{19.119}$$

The ideal infinite-support temporal chirp is not a practical signal either, as it would require an infinite time to transmit. The support of the function (and thus its effective bandwidth) must be truncated, which limits the "sharpness" of the correlation peak and thus the ultimate resolution of the system. We saw in Chapter 17 that the bandwidth of the chirp signal is determined by the width of its region of support, so that matched filtering with a wider chirp signal produces a narrower correlation peak.

Quadratic-phase signals are commonly used in radar and sonar systems, and sometimes are called *spread-spectrum signals*. Note that the transmitted signal must be complex valued to produce the ideal Dirac delta function autocorrelation peak because of the zeros in the spectrum of a real-valued chirp signal. A noise-free example that models a chirp function with infinite support is shown in Figure 19.17, while filtering of a chirp function in the presence of additive noise is shown in Figure 19.18.

In realistic situations, the support of the chirp must be finite, thus constraining the bandwidth of the reference signal. We can apply the discussion of the stationary-phase solution to the spectrum of the modulated quadratic-phase signal in Chapter 13 to determine the approximate autocorrelation of the finite-support chirp function. When used in radar and sonar applications, matched filtering of a chirp function is sometimes called *pulse compression*.

Any constant-magnitude signal, including the higher-order "superchirp" functions and complex-valued white noise, also are suitable inputs for matched filtering, but the results are more difficult to analyze without closed-form expressions for the impulse responses.

19.5 Analogies of Inverse and Matched Filters

We noted that the transfer functions of the ideal matched filter for a known input $f[x]$ and of the ideal inverse filter for the system impulse response $h[x]$ are identically reciprocals of the corresponding spectra, so from this point forward we apply the name "reciprocal filter" to both. We have already used the observed analogy to construct the "pseudomatched" filter in Equation (19.103). We can also apply the analogy between the filters in the opposite direction by constructing an approximation to the inverse filter in Equation (19.20) based on the classical matched filter that assumes $|H[\xi]|^2 = 1$:

$$\hat{W}[\xi] = H^*[\xi] = |H[\xi]| \exp[-i\Phi\{H[\xi]\}]$$
$$\Longrightarrow \hat{w}[x] = \mathcal{F}_1^{-1}\{H^*[\xi]\} = h^*[-x] \tag{19.120}$$

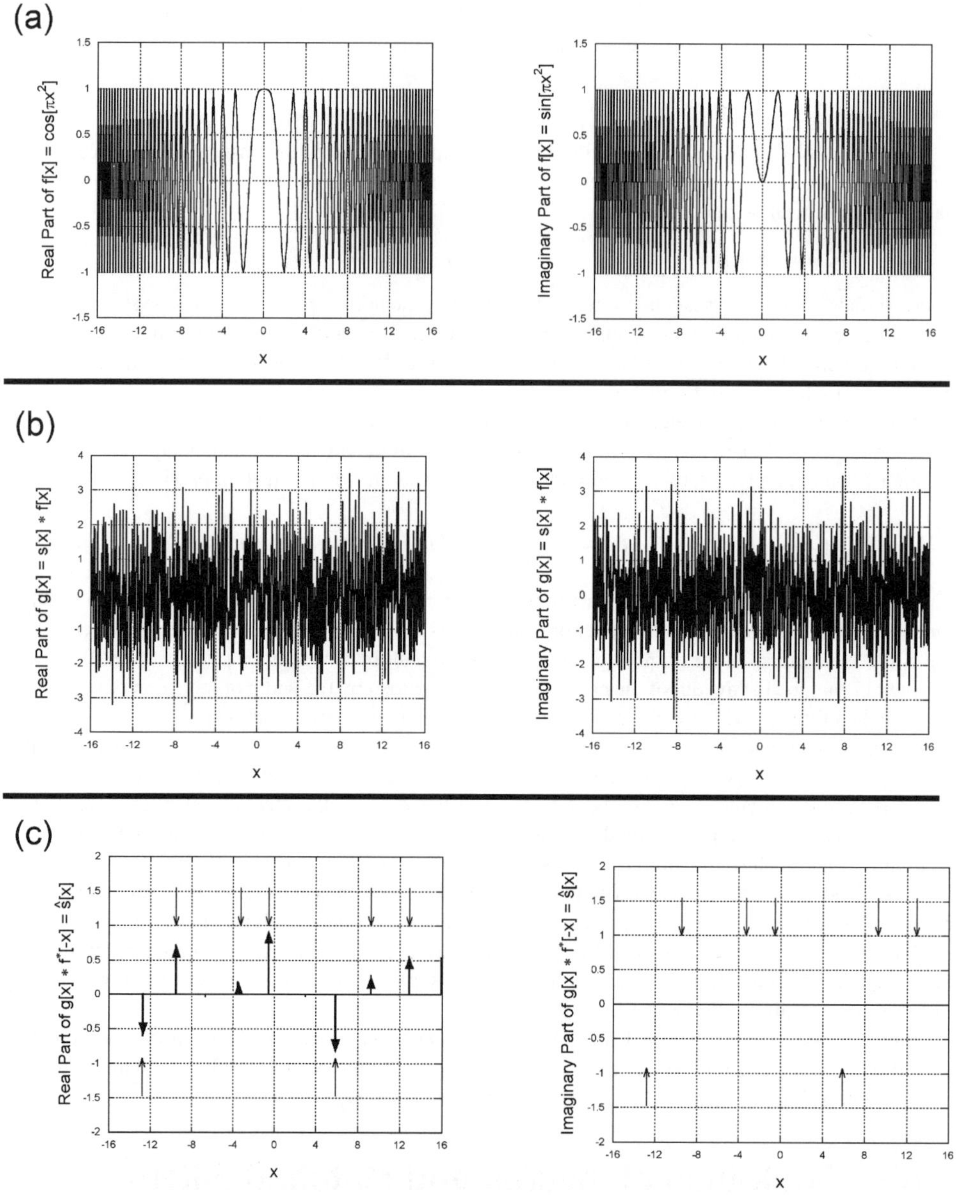

Figure 19.17 Matched filtering of a noise-free chirp signal placed at locations specified by $s[x]$ in Figure 19.8: (a) real and imaginary parts of $f[x] = \exp[+i\pi x^2]$; (b) real and imaginary parts of $g[x] = f[x] * s[x]$, showing little indication of $s[x]$; (c) real and imaginary parts of $g[x] * f^*[-x]$, showing the locations of the Dirac delta functions in $s[x]$. Since there is no noise, the performance is ideal.

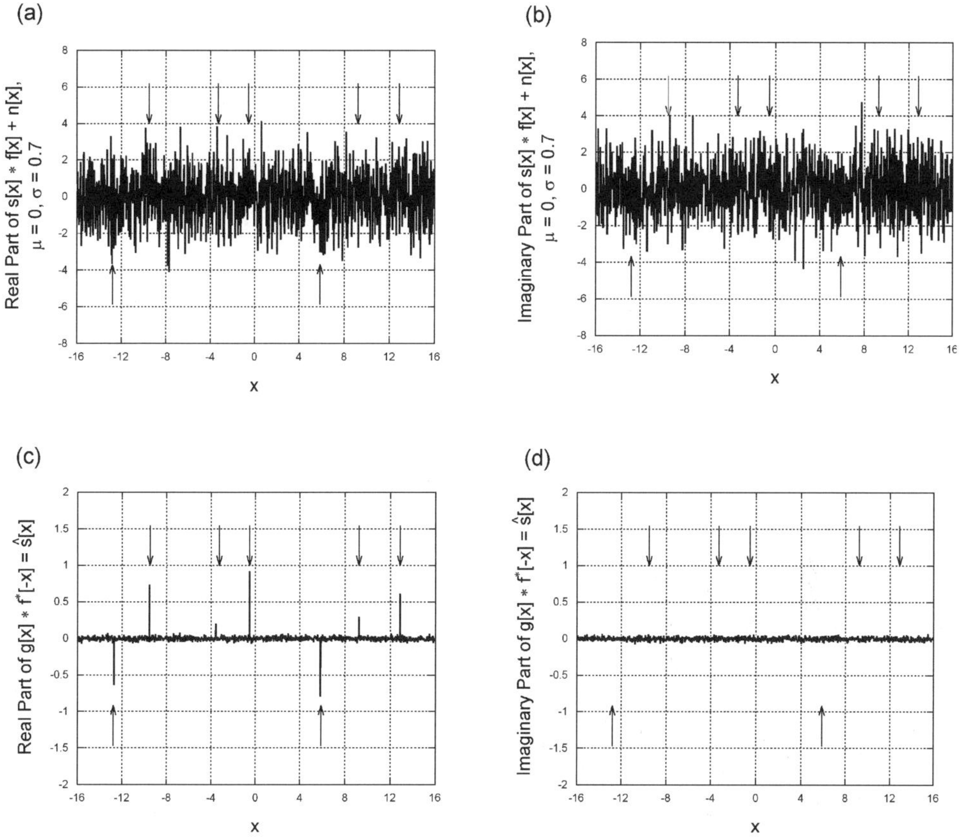

Figure 19.18 Matched filtering of a chirp function in the presence of noise: (a) real and imaginary parts of $g[x] = f[x] * s[x] + n[x]$, where $n[x]$ is Gaussian noise with $\mu = 0, \sigma = 0.7$; (b) real and imaginary parts of $g[x] * f^*[-x]\} \cong s[x]$. The wide bandwidth and constant amplitude of $F[\xi]$ preclude noise amplification and ensure excellent recovery of $s[x]$.

This corrects for the intrinsic phase of the signal to produce a symmetric output but exacerbates any decrease in magnitude of the signal spectrum.

The spectrum of the resulting output is:

$$\begin{aligned}\hat{F}[\xi] &= G[\xi] \cdot H^*[\xi] \\ &= (F[\xi] \cdot H[\xi]) \cdot H^*[\xi] = F[\xi] \cdot |H[\xi]|^2 \end{aligned} \tag{19.121}$$

and the corresponding estimate of $f[x]$ is its convolution with the autocorrelation of the impulse response:

$$\begin{aligned}\hat{f}[x] &= g[x] * \hat{w}[x] = (f[x] * h[x]) * h^*[-x] \\ &= f[x] * (h[x] \bigstar h^*[x]) \end{aligned} \tag{19.122}$$

The character (highpass, lowpass, etc.) of the "unamplified" inverse filter $\hat{w}[x]$ is the same as that of the original filter $h[x]$. If $h[x]$ is lowpass, then the variance of any noise present in the signal will be attenuated by the averaging action of $\hat{w}[x]$. The wider the bandwidth of $H[\xi]$, the narrower the width of the autocorrelation peak of $h[x]$, and the better the estimate of $f[x]$ from this modified inverse

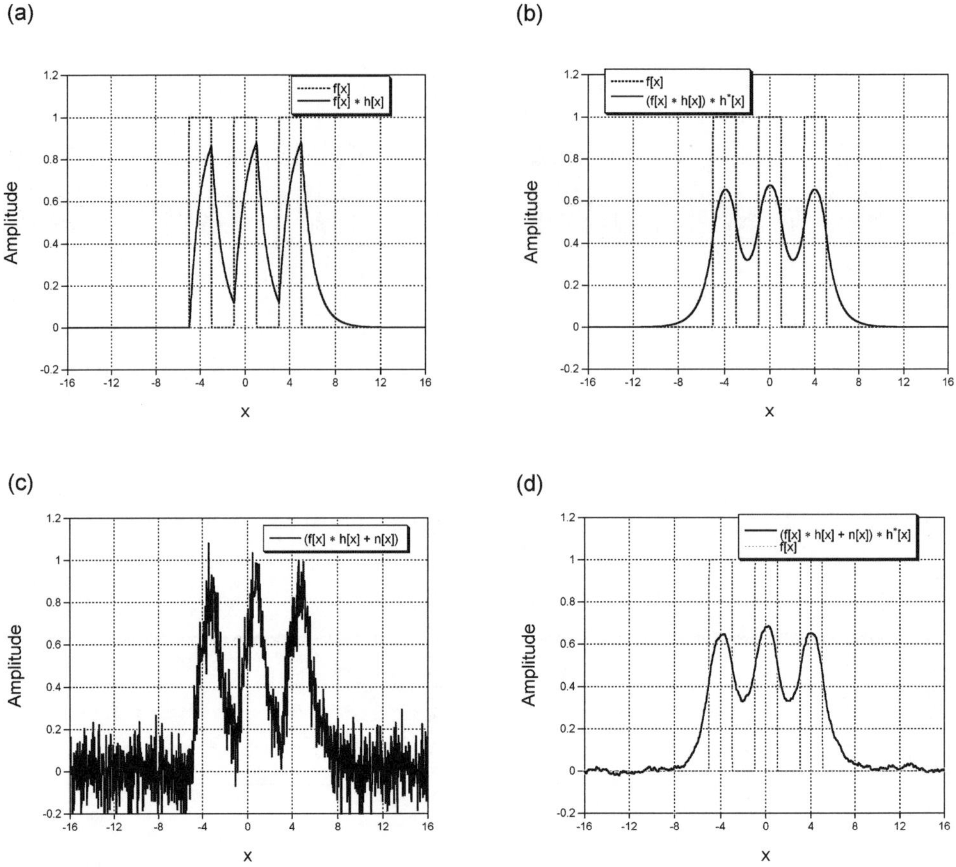

Figure 19.19 Analogy of "unamplified" matched filter for inverse filtering: (a) input object $f[x]$ is a set of three rectangular bars, along with $f[x] * h[x]$, where $h[x] = \exp[-x] \cdot STEP[x]$, producing a "skewed" output; (b) $(f[x] * h[x]) * h^*[-x]$, showing the restoration of symmetry by this approximate inverse filter; (c) $f[x] * h[x] + n[x]$, where $n[x]$ is Gaussian random noise with $\sigma = 0.1$; (d) $(f[x] * h[x] + n[x]) * h^*[-x]$, showing the reduction in noise and the restoration of symmetry.

filter. We have already seen how this alternative inverse filter is useful when interpreting the Wiener–Helstrom filter for estimating the input function $f[x]$. An example of the "unamplified" inverse filter for an asymmetric impulse response is shown in Figure 19.19; note that the unamplified filter recovers the symmetry of the original object and also attenuates the white Gaussian noise.

19.5.1 Wiener and Wiener–Helstrom "Matched" Filter

We can extend the analogy between the inverse and matched filters to construct an example of the latter that is analogous to the Wiener–Helstrom filter. We might think of the amplitudes and positions of the replicas of $f[x]$ as a "position function" $s[x]$ that is the sum of Dirac delta functions with the appropriate translations and amplitudes:

$$s[x] \equiv \sum_{\ell=1}^{N} \alpha_\ell \cdot f[x - x_\ell] \Longrightarrow g[x] = s[x] * f[x] + n[x] \tag{19.123}$$

which is functionally equivalent to Equation (19.41a) for the Wiener–Helstrom filter. In those terms, we can think of the matched filter task as constructing a linear filter $m[x]$ to apply to $g[x]$ to determine or estimate the form of $s[x]$ (and therefore of the values x_ℓ) from complete knowledge of the input function $f[x]$ and some knowledge of $n[x]$ and $s[x]$. Clearly we can apply the lessons of the previous section to derive related filter functions.

In the case of a single instance of $f[x]$ in $g[x]$, so that $s[x] = \alpha_1 \cdot \delta[x - x_1]$, the task is equivalent to estimating the coordinate x_1, so that the ideal output would be identical to $s[x]$. The spectrum has unit magnitude:

$$S[\xi] = \alpha_1 \cdot 1[\xi] \cdot \exp[-2\pi i \xi x_1] \implies |S[\xi]|^2 = \alpha_1^2 \cdot 1[\xi] \tag{19.124}$$

Therefore the two conditions implied by Equation (19.100) are:

$$S[\xi] \cdot F[\xi] \cdot M_{\text{ideal}}[\xi] = S[\xi] \implies M_{\text{ideal}}[\xi] = \frac{1}{F[\xi]} = \frac{F^*[\xi]}{|F[\xi]|^2} \tag{19.125a}$$

$$\mathcal{F}_1^{-1}\{M_{\text{ideal}}[\xi]\} = m_{\text{ideal}}[x] = f^*[-x] * \mathcal{F}_1^{-1}\left\{\frac{1}{|F[\xi]|^2}\right\} \tag{19.125b}$$

The output of the matched filter in the space domain is:

$$s[x] * f[x] * f^*[-x] * \mathcal{F}_1^{-1}\left\{\frac{1}{|F[\xi]|^2}\right\} + n[x] * f^*[-x] * \mathcal{F}_1^{-1}\left\{\frac{1}{|F[\xi]|^2}\right\}$$
$$= s[x] * (f[x] \bigstar f[x]) * \mathcal{F}_1^{-1}\left\{\frac{1}{|F[\xi]|^2}\right\} + (n[x] \bigstar f[x]) * \mathcal{F}_1^{-1}\left\{\frac{1}{|F[\xi]|^2}\right\} \tag{19.126a}$$

The analogue to the expression for the Wiener–Helstrom filter in Equation (19.88) for the transfer function of the matched filter in the case of a single replica of the input is:

$$M[\xi] = \frac{F^*[\xi]}{|F[\xi]|^2 + |N[\xi]|^2/|S[\xi]|^2} \to \frac{F^*[\xi]}{|F[\xi]|^2 + |N[\xi]|^2/\alpha_1^2} = \alpha_1^2 \cdot \frac{F^*[\xi]}{\alpha_1^2 \cdot |F[\xi]|^2 + |N[\xi]|^2} \tag{19.127}$$

where Equation (19.124) has been used. The leading constant factor α_1^2 will be retained until testing the limiting cases. At frequencies where the noise power is smaller than the signal power, the transfer function for the matched filter becomes:

$$M[\xi] \cong \alpha_1^2 \cdot \frac{F^*[\xi]}{\alpha_1^2 \cdot |F[\xi]|^2} = \frac{1}{F[\xi]} \tag{19.128}$$

Again we see that the transfer function of the matched filter is the "inverse filter" for the known signal. At frequencies where the noise power is large, the transfer function has the form:

$$M[\xi] \cong \left(\frac{\alpha_1^2}{|N[\xi]|^2}\right) F^*[\xi] = k \cdot F^*[\xi] \tag{19.129}$$

where k is a weighting factor that will be small in the case where $|N[\xi]|^2 \gg \alpha_1^2$. In words, the transfer function of the matched filter at noisy frequencies is analogous to the "unamplified" inverse filter for $F[\xi]$ from Equation (19.90).

Finally, reconsider the more general case of multiple replicas of the known input at different locations and with different amplitudes:

$$g[x] = \left(\sum_{\ell=1}^{N} \alpha_\ell \cdot f[x - x_\ell]\right) + n[x] \implies s[x] = \sum_{\ell=1}^{N} \alpha_\ell \cdot \delta[x - x_\ell] \tag{19.130}$$

where α_ℓ is the (assumed real-valued) amplitude weighting factor for the ℓth replica. The power spectrum of the position function may be expanded and approximated:

$$
\begin{aligned}
|S[\xi]|^2 &= \left(\sum_{\ell=1}^{N} \alpha_\ell \cdot \exp[-2\pi i x_\ell \xi]\right)\left(\sum_{m=1}^{N} \alpha_m \cdot \exp[-2\pi i x_m \xi]\right)^* \\
&= \sum_{\ell=1}^{N}\sum_{m=1}^{N} \alpha_\ell \alpha_m \cdot \exp[+2\pi i (x_m - x_\ell)\xi] \\
&= \sum_{\ell=1}^{N} \alpha_\ell^2 + \sum_{\ell>m}^{N} 2 \cdot \alpha_\ell \cdot \alpha_m \cdot (\exp[+2\pi i (x_m - x_\ell)\xi] - \exp[+2\pi i (x_m - x_\ell)\xi]) \\
&= \sum_{\ell=1}^{N} \alpha_\ell^2 + \sum_{\ell>m}^{N} 2 \cdot \alpha_\ell \cdot \alpha_m \cdot \cos[+2\pi (x_m - x_\ell)\xi] \\
&\cong \sum_{\ell=1}^{N} \alpha_\ell^2
\end{aligned} \tag{19.131}
$$

where the last step is valid if the phases are uniformly distributed, so that the sum of the phase differences for the different replicas should approximately cancel.

If Equation (19.131) is true, the transfer function of the matched filter is:

$$
M[\xi] \cong \left(\sum_{\ell=1}^{N} \alpha_\ell^2\right) \cdot \frac{F^*[\xi]}{(\sum_{\ell=1}^{N} \alpha_\ell^2)|F[\xi]|^2 + |N[\xi]|^2} \tag{19.132}
$$

Note the similarity to the matched filter for a single replica of $f[x]$ in Equation (19.127).

An example of the Wiener–Helstrom matched filter is shown in Figure 19.20 where $|S[\xi]|^2$ and $|N[\xi]|^2$ are assumed to be known exactly, which again is unlikely unless we have control over $s[x]$. The noise is "blue" and the position function $s[x]$ has a fairly flat power spectrum. Compare the real part of the result in Figure 19.20e to the result using the classical matched filter in Figure 19.15d, showing the "sharper" peaks resulting from the Wiener–Helstrom "matched" filter.

19.6 Approximations to Reciprocal Filters

The ideal form of either reciprocal filter (the inverse or matched filter) corrects for any amplification and/or attenuation of the magnitude spectrum, which contrasts with the approximate forms in Equations (19.122) and (19.105) that exacerbate the same amplification and attenuation. This difference in behaviors suggests that these might be limiting cases of a more general series expression for the transfer functions. We will show this to be true by deriving such a series for a reciprocal filter. The series may be truncated to produce a "suboptimum" controllable approximation of the filter (optimum methods are considered next) that will constrain the amplification of the magnitude. This procedure may be useful if noise is present or if the filtering device is not capable of amplifying the signal to the levels required of the inverse filter (as in a passive imaging system), but its primary benefit may be its impact on intuition about the filters.

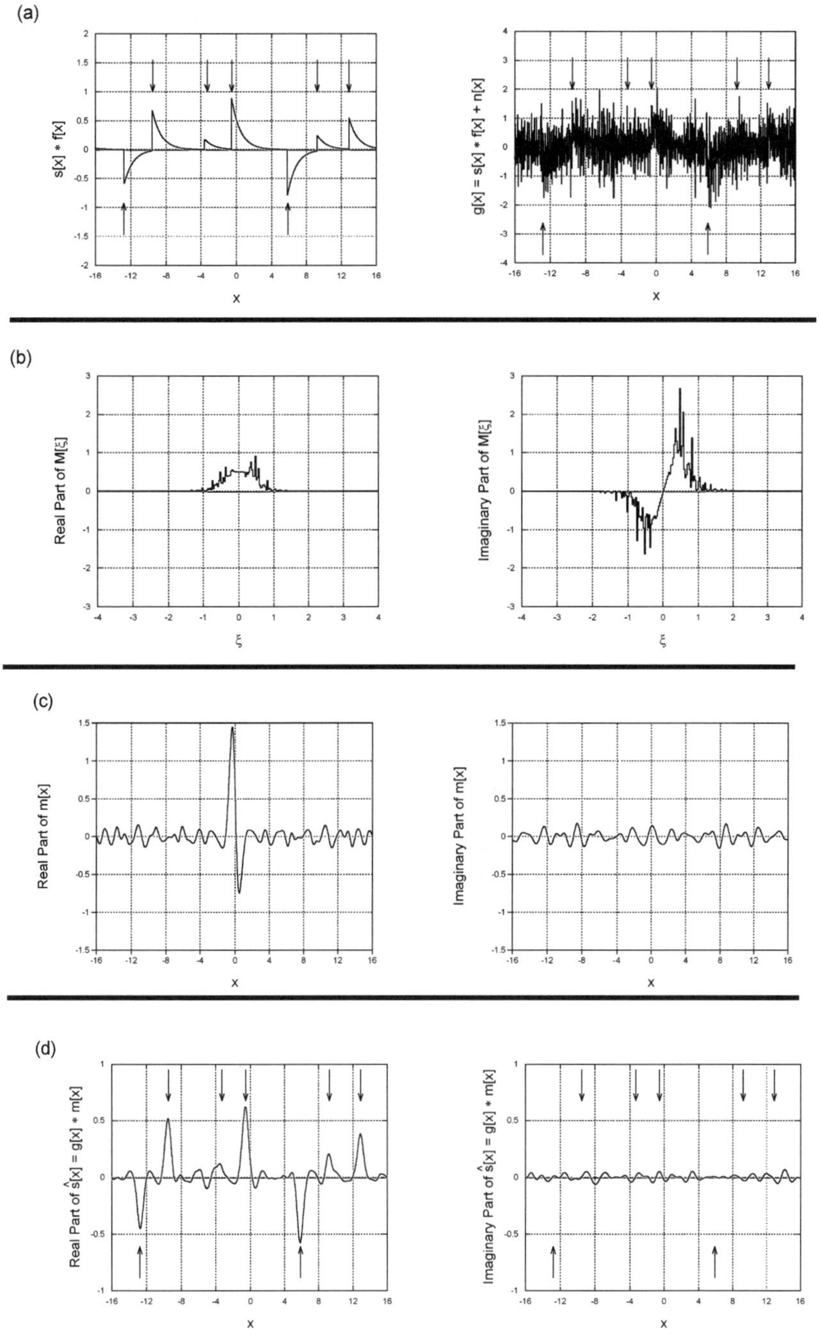

Figure 19.20 Wiener–Helstrom matched filter: (a) $s[x] * f[x]$ and $s[x] * f[x] + n[x]$, where $s[x]$ is the same set of Dirac delta functions used in Figure 19.15, $f[x] = STEP[x] \cdot \exp[-x]$, and $n[x]$ is that used in the Wiener matched filter; (b) real and imaginary parts of $M[\xi]$, showing the lowpass character; (c) real and imaginary parts of $m[x]$; (d) real and imaginary parts of $\hat{s}[x] = g[x] * m[x]$, showing good recovery of $s[x]$ and attenuation of the added noise.

Consider the magnitude–phase expression for a transfer function $H[\xi]$; the magnitude may be split into the product of two terms and the phase $\exp[+i\Phi\{H[\xi]\}]$ is carried along "for the ride":

$$\begin{aligned} H[\xi] &= |H[\xi]|\, e^{+i\Phi\{H[\xi]\}} \\ &= [|H|_{\max} + (|H[\xi]| - |H|_{\max})]\, e^{+i\Phi\{H[\xi]\}} \\ &= (|H|_{\max} \cdot e^{+i\Phi\{H[\xi]\}}) \cdot \left(1 - \frac{|H|_{\max} - |H[\xi]|}{|H|_{\max}}\right) \end{aligned} \tag{19.133}$$

where $|H|_{\max}$ is the maximum of the magnitude of the transfer function, which must be finite for the derivation to be valid. The corresponding inverse filter is the reciprocal of this expression:

$$W[\xi] = \frac{1}{H[\xi]} = \left(\frac{1}{|H|_{\max}}\, e^{-i\Phi\{H[\xi]\}}\right) \cdot \left(1 - \frac{|H|_{\max} - |H[\xi]|}{|H|_{\max}}\right)^{-1} \tag{19.134}$$

Any frequency-dependent amplification and/or attenuation of the magnitude spectrum by the inverse filter is encapsulated in the last term of the form $(1-\alpha)^{-1}$ where $|\alpha| \leq 1$. This term may be expanded into the well-known power series:

$$\frac{1}{1-\alpha} = \sum_{n=0}^{+\infty} \alpha^n \text{ if } |\alpha| < 1 \tag{19.135}$$

The corresponding series solution for the reciprocal filter is:

$$\begin{aligned} W[\xi] &= \left(\frac{1}{|H|_{\max}}\, e^{-i\Phi\{H[\xi]\}}\right) \sum_{n=0}^{+\infty} \left(\frac{|H|_{\max} - |H[\xi]|}{|H|_{\max}}\right)^n \\ &= \left(\frac{1}{|H|_{\max}}\, e^{-i\Phi\{H[\xi]\}}\right)\left(1 + \left[\frac{|H|_{\max} - |H[\xi]|}{|H|_{\max}}\right] + \left[\frac{|H|_{\max} - |H[\xi]|}{|H|_{\max}}\right]^2 + \cdots\right) \end{aligned} \tag{19.136}$$

which is valid for $|H[\xi]| > 0$ and $|H|_{\max} < \infty$. At those frequencies where $|H[\xi]| = 0$, the series expansion does not converge because the corresponding magnitude of the inverse filter "should be" infinite. This expression also cannot be evaluated in any case with $|H|_{\max} = \infty$. One such example is the derivative operator acting on continuous functions, though the approximation does exist for discrete differentiation because $|H|$ is bounded at the Nyquist frequency. This case will be considered briefly in Chapter 20.

An Nth-order approximation to the inverse filter is generated by truncating the series after N terms:

$$\hat{W}_N[\xi] = \left(\frac{1}{|H|_{\max}}\, e^{-i\Phi\{H[\xi]\}}\right) \sum_{n=0}^{N} \left(\frac{|H|_{\max} - |H[\xi]|}{|H|_{\max}}\right)^n \tag{19.137}$$

As mentioned in Chapter 13, such a truncated Taylor series yields a "Taylor polynomial" that is generally not an optimum approximation to the transfer function in a squared-error sense. Even so, the series may contribute to the user's intuition about reciprocal filters (and to the Wiener and Wiener–Helstrom filters).

We now consider the approximations for small N and illustrate them for the symmetric uniform averager in Equation (19.22), the ideal lowpass filter in Equation (19.30), the decaying-exponential filter in Equation (19.36), and the first derivative.

19.6.1 Small-Order Approximations of Reciprocal Filters

19.6.1.1 Zero-Order Approximation: Phase-Only Filter

We substitute $N = 0$ into Equation (19.137) to obtain the zero-order approximation to the reciprocal filter:

$$\hat{W}_0[\xi] = \frac{1}{|H|_{\max}} e^{-i\Phi\{H[\xi]\}} \tag{19.138}$$

which produces a constant-magnitude allpass filter that compensates for the phase introduced by $h[x]$. This transfer function is perhaps better described as a phase-only approximation to the filter. Because the transfer function has constant magnitude, it is easy to see that its only impact on the mean and standard deviation of any noise in $g[x]$ is a scaling by $|H|_{\max}^{-1}$. In the common case where the impulse response has unit area, so that $|H|_{\max} = H[0] = 1$, the mean and standard deviation of any noise in $f[x]$ are unchanged by this filter, though the noise itself is "scrambled" by any nonlinear phase term.

The cascade of the original transfer function and that of the ideal reciprocal filter should yield the unit constant spectrum:

$$H[\xi] \cdot W[\xi] = 1[\xi] \tag{19.139}$$

We can characterize the performance of the approximate reciprocal filter by comparing the resulting spectrum to this ideal form:

$$\begin{aligned} H[\xi] \cdot \hat{W}_0[\xi] &= (|H[\xi]|\, e^{+i\Phi\{H[\xi]\}}) \cdot \left(\frac{1}{|H|_{\max}} e^{-i\Phi\{H[\xi]\}}\right) \\ &= \frac{|H[\xi]|}{|H|_{\max}} \leq 1[\xi] \end{aligned} \tag{19.140}$$

The corresponding impulse response of the phase-only approximate inverse filter is:

$$\hat{w}_0[x] = \frac{1}{|H|_{\max}} \cdot \mathcal{F}_1^{-1}\{1[\xi]\, e^{-i\Phi\{H[\xi]\}}\} \tag{19.141}$$

The cascade of $h[x]$ and the ideal inverse filter $w[x]$ yields the ideal Dirac delta function $\delta[x]$, so the success of the approximate filter may be characterized by how closely its convolution with $h[x]$ resembles the ideal Dirac delta function:

$$h[x] * \hat{w}_0[x] = \frac{1}{|H|_{\max}} \mathcal{F}_1^{-1}\{|H[\xi]|\} \cong \delta[x] \tag{19.142}$$

We briefly investigated the effect of the magnitude on the Fourier transform in Section 9.8.19. From the example considered in that discussion, we saw that the support of $h[x] * \hat{w}_0[x]$ generally is "wider" than that of $h[x]$, which seems to contradict the goal of the filtering process. However, the examples considered below show that the amplitude of $h[x] * \hat{w}_0[x]$ is more "concentrated" than $h[x]$ alone.

19.6.1.2 First-Order Approximation

Substitution of $N = 1$ into Equation (19.137) produces the first-order approximation to the reciprocal filter:

$$\begin{aligned} \hat{W}_1[\xi] &= \left(\frac{1}{|H|_{\max}} e^{-i\Phi\{H[\xi]\}}\right)\left(1 + \frac{|H|_{\max} - |H[\xi]|}{|H|_{\max}}\right) \\ &= \hat{W}_0[\xi] \cdot \left(1 + \frac{|H|_{\max} - |H[\xi]|}{|H|_{\max}}\right) \end{aligned} \tag{19.143}$$

The amplification of $\hat{W}_1[\xi]$ is constrained to be no larger than $2 \cdot (|H|_{\max})^{-1}$, which proves to be a useful guide to the performance of the filter if noise exists in the input signal. In the common case where the area of the impulse response $h[x]$ has been normalized to unity (so that $|H|_{\max} = 1$), the first-order estimate of the reciprocal filter may be expressed in several simple and equivalent ways:

$$\hat{W}_1[\xi] = e^{-i\Phi\{H[\xi]\}}(1 + (1 - |H[\xi]|)) \text{ if } |H|_{\max} = 1 \tag{19.144a}$$

$$= (1[\xi]\, e^{-i\Phi\{H[\xi]\}}) + (1 - |H[\xi]|)\; e^{-i\Phi\{H[\xi]\}} \tag{19.144b}$$

$$= \hat{W}_0[\xi] + (1 - |H[\xi]|)\, e^{-i\Phi\{H[\xi]\}} \tag{19.144c}$$

$$= 2 \cdot \hat{W}_0[\xi] - H^*[\xi] \tag{19.144d}$$

Equation (19.144b) shows that the first-order approximation is the sum of the phase-only filter and a filter with the same phase and complementary magnitude $1 - |H[\xi]|$. The effect of this filter on any noise in $g[x]$ depends on the character of $H[\xi]$ and $N[\xi]$, but the constraint on the amplification ensures that the standard deviation σ of the noise is amplified by no more than a factor of 2. If the signal-to-noise power ratio of the signal spectrum is known, this observation suggests a useful rule of thumb for the maximum order of the filter that may be used. The maximum amplification may be used to "tune" the transfer function if the gain of the filtering device is limited.

To characterize the performance of the first-order approximation to the reciprocal filter, we can again consider how well the cascade of the transfer functions approximates the ideal unit constant:

$$H[\xi] \cdot \hat{W}_1[\xi] = 2 \cdot |H[\xi]| - |H[\xi]|^2 \cong 1[\xi] \tag{19.145}$$

or, equivalently, how closely the cascade of impulse responses resembles a Dirac delta function:

$$2 \cdot (h[x] * \hat{w}_0[x]) - (h[x] \bigstar h[x]) \cong \delta[x] \tag{19.146}$$

19.6.1.3 Unamplified First-Order Filter: the "Complement" Filter

The approximations to the transfer function of order larger than zero amplify the magnitude spectrum and thus cannot be implemented in passive filtering systems (e.g., in optical correlators), where the magnitude of the transfer function is constrained: $|\hat{W}[\xi]| \leq 1$. However, the form of Equation (19.144c) suggests the form of a filter that exhibits some of the frequency "character" of the reciprocal filter without amplification. This filter has the form:

$$\hat{W}_{1C}[\xi] \equiv |H|_{\max}\left(1 - \frac{|H[\xi]|}{|H|_{\max}}\right) e^{-i\Phi\{H[\xi]\}} = \hat{W}_0[\xi] - H^*[\xi] \tag{19.147}$$

where the second subscript "C" is used because this filter is based on the complement of the normalized magnitude of the original transfer function. The cascade of the original transfer function and this "complementary" approximation to the reciprocal filter a function that may expressed in several equivalent ways:

$$H[\xi] \cdot \hat{W}_{1C}[\xi] = (|H[\xi]|\;\, e^{+i\Phi\{H[\xi]\}}) \cdot |H|_{\max}\left(1 - \frac{|H[\xi]|}{|H|_{\max}}\right) e^{-i\Phi\{H[\xi]\}} \tag{19.148a}$$

$$= (|H[\xi]|\;\, e^{+i\Phi\{H[\xi]\}}) \cdot (|H|_{\max} - |H[\xi]|)\, e^{-i\Phi\{H[\xi]\}} \tag{19.148b}$$

$$= (|H|_{\max} \cdot |H[\xi]|) - |H[\xi]|^2 \tag{19.148c}$$

In the space domain, the output of the "complement" reciprocal inverse filter yields an approximation of the original input function:

$$\hat{f}_1[x] = f[x] * (h[x] * \hat{w}_0[x]) - f[x] * (h[x] \bigstar h[x]) \tag{19.149}$$

19.6.1.4 Second-Order Approximation

The second-order approximation to the filter with $|H|_{max} = 1$ has the form:

$$\hat{W}_2[\xi] = (1 + [1 - |H[\xi]|] + [1 - |H[\xi]|]^2)\, e^{-i\Phi\{H[\xi]\}}$$
$$= (3 \cdot [1 - |H[\xi]|] + |H[\xi]|^2)\, e^{-i\Phi\{H[\xi]\}} \quad (19.150)$$

The magnitude evaluates to unity at all frequencies where $|H[\xi]| = |H|_{max} = 1$ and to 3 where $|H[\xi]| = 0$, which again illustrates the constrained gain of the approximate reciprocal filter.

19.6.2 Examples of Approximate Reciprocal Filters

19.6.2.1 Uniform Averager

The phase-only approximation to the reciprocal filter for uniform averaging over finite support b_0 in Equation (19.22) has transfer function:

$$(\hat{W}_0[\xi])_1 = \begin{Bmatrix} +1 & \text{if } SINC[b_0\xi] > 0 \\ 0 & \text{if } SINC[b_0\xi] = 0 \\ -1 & \text{if } SINC[b_0\xi] < 0 \end{Bmatrix} = SGN[SINC[b_0\xi]] \quad (19.151)$$

as shown in Figure 19.21a. Because $SINC[\xi]$ is real valued, the phase transfer function is either 0 or $-\pi$ radians in our convention. The phase transfer function of the corresponding reciprocal filter must be 0 and $-\pi$ radians, respectively, though (of course) phase shifts of $\pm\pi$ radians are equivalent. The product of the transfer function and the approximation to the reciprocal filter is:

$$H_1[\xi] \cdot (\hat{W}_0[\xi])_1 = SINC[b_0\xi] \cdot SGN[SINC[b_0\xi]]$$
$$= |SINC[b_0\xi]| \neq 1[\xi] \quad (19.152)$$

The corresponding impulse response of the cascade of the system and its phase-only reciprocal filter is the inverse Fourier transform of the magnitude of the *SINC* function, which was evaluated in Equation (9.189):

$$h_1[x] * \hat{w}_0[x] = \mathcal{F}_1^{-1}\{|SINC[b_0\xi]|\}$$
$$= \frac{1}{|b_0|} RECT\left[\frac{x}{b_0}\right] * \left(\frac{|b_0|}{i\pi x} * \left(SINC[b_0\xi] \cdot COMB[2b_0\xi] \cdot \exp[+i\pi b_0\xi] - \frac{|b_0|}{2}\delta[\xi]\right)\right) \quad (19.153)$$

which has infinite support but is more "concentrated" near the origin than the impulse response $h_1[x]$ alone. This increased "concentration" is a useful feature for matched filtering.

The first-order approximation of the reciprocal filter for uniform averaging is:

$$(\hat{W}_1[\xi])_1 = (2 - |SINC[b_0\xi]|) \cdot SGN[SINC[b_0\xi]] \quad (19.154)$$

The transfer function is shown in Figure 19.21b. The corresponding impulse response does not have an obvious analytical form, but the transfer function indicates that it must be a highpass filter and thus computes a local difference in amplitude. The cascade of the transfer function for uniform averaging and the first-order approximate inverse filter is:

$$H_1[\xi] \cdot (\hat{W}_1[\xi])_1 = SINC[b_0\xi] \cdot (2 - |SINC[b_0\xi]|) \cdot SGN[SINC[b_0\xi]]$$
$$= |SINC[b_0\xi]| \cdot (2 - |SINC[b_0\xi]|)$$
$$= 2|SINC[b_0\xi]| - SINC^2[b_0\xi] \quad (19.155)$$

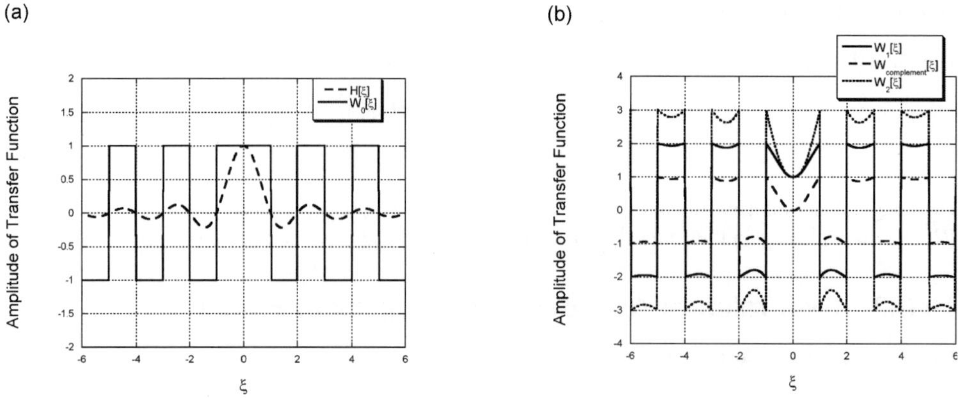

Figure 19.21 Transfer functions of several approximations to the inverse filter for uniformly weighted blur: (a) zero-order (phase-only) approximation, which is constrained to $|\hat{W}_0[\xi]| \leq 1$, shown along with $H[\xi]$; (b) $\hat{W}_1[\xi]$, $\hat{W}_{\text{complement}}[\xi]$, and $\hat{W}_2[\xi]$, showing the increase in amplification with increasing order (note change of vertical scale).

The unamplified "complement" approximation to the reciprocal filter for uniform averaging is:

$$(\hat{W}_1[\xi])_1 = (1 - |SINC[b_0\xi]|) \cdot SGN[SINC[b_0\xi]] \tag{19.156}$$

which evaluates to zero at the origin and its magnitude is constrained to be no larger than unity. The cascade of the original transfer function and the first-order approximate reciprocal filter is:

$$H_1[\xi] \cdot \hat{W}_1[\xi] = |H[\xi]| - |H[\xi]|^2 = |SINC[b_0\xi]| - SINC^2[b_0\xi] \tag{19.157}$$

which differs from Equation (19.142) only in the factor applied to $|SINC[b_0\xi]|$.

The second-order approximation of the reciprocal filter for uniform averaging is:

$$(\hat{W}_2[\xi])_1 = (3 - 3|SINC[b_0\xi]| + SINC^2[b_0\xi]) \cdot SGN[SINC[b_0\xi]] \tag{19.158}$$

The three examples of $H_1[\xi] \cdot (\hat{W}_n[\xi])_1$ for $n = 0, 1, 2$ are shown in Figure 19.21b. Note the increased amplitude of the high-frequency components with increasing order, which means that the filtered spectrum more closely approximates the ideal unit constant.

The actions of the approximate reciprocal filters are shown in Figure 19.22 acting on the same object function used in Figure 19.5. Note the increasing "sharpness" of the reconstruction and increasing noise with increasing order.

19.6.2.2 Ideal Lowpass Filter

The zero-order approximation to the transfer function of the reciprocal filter for an ideal lowpass signal also is easy to derive. Since the phase of the transfer function is zero at all frequencies, then the phase-only approximation to the filter is:

$$(\hat{W}_0[\xi])_2 = \frac{1}{|H|_{\max}} e^{-i\Phi\{H[\xi]\}} = 1[\xi] \tag{19.159}$$

The corresponding transfer function is obviously a Dirac delta function:

$$(\hat{w}_0[x])_2 = \delta[x] \tag{19.160}$$

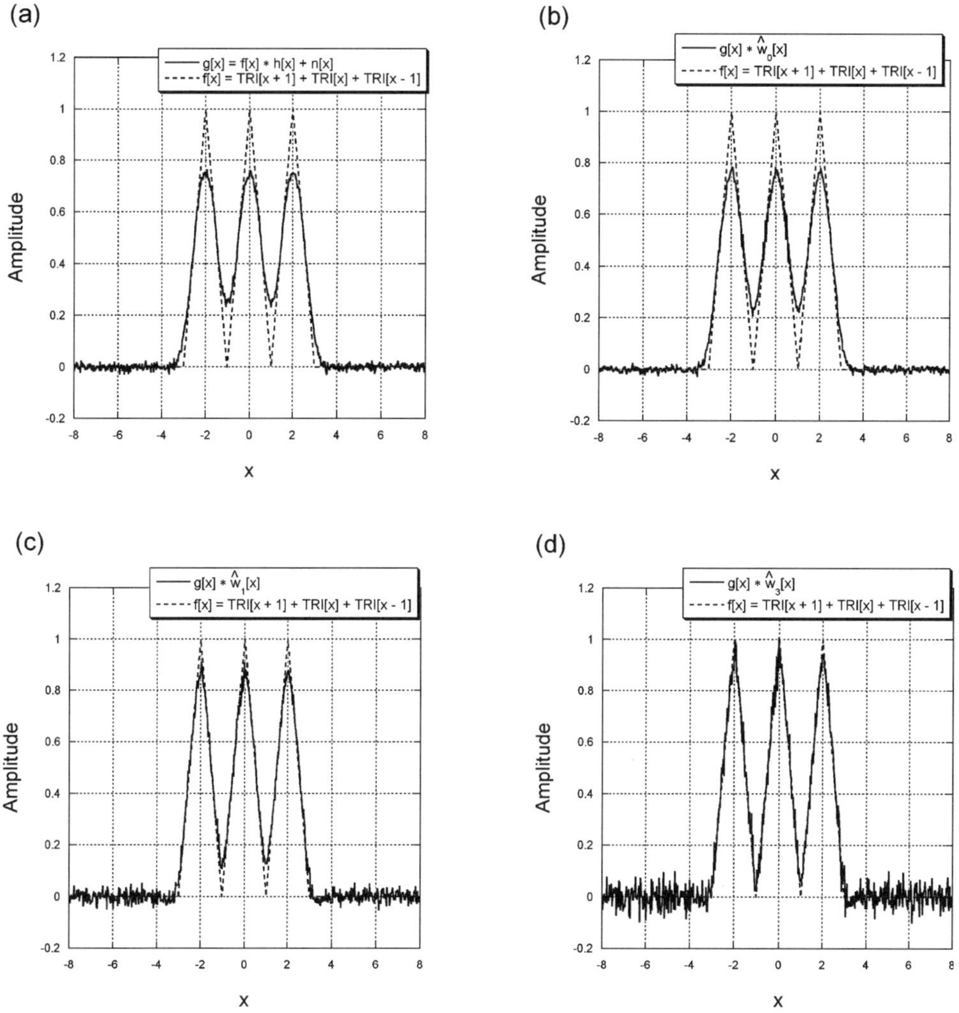

Figure 19.22 Action of the approximations to the inverse filter for $h[x] = RECT[x]$: (a) $f[x]$ and $g[x] = f[x] * h[x] + n[x]$, where $n[x]$ is Gaussian-distributed random noise with mean $\mu = 0$ and standard deviation $\sigma = 0.01$; (b) $g[x] * \hat{w}_0[x]$, the standard deviation of the filtered noise is $\sigma_0 = 0.01$; (c) $g[x] * \hat{w}_1[x]$, with resulting $\sigma_1 = 0.02$; (d) $g[x] * \hat{w}_3[x]$, with $\sigma_3 = 0.04$.

Thus the phase-only approximation to the ideal inverse filter has no effect; the output of the phase-only approximation to the reciprocal filter is identical to the blurred signal.

The transfer function of the first-order approximation of the inverse filter for the ideal lowpass filter may be expressed in the form of Equation (19.38):

$$(\hat{W}_1[\xi])_2 = 2 - |RECT[b_0\xi]| \tag{19.161}$$

which evaluates to unity within the passband and to two outside, where $G[\xi] = 0$. Thus we see that:

$$(F[\xi] \cdot H_2[\xi]) \cdot (\hat{W}_1[\xi])_2 = F[\xi] \cdot H_2[\xi] \tag{19.162}$$

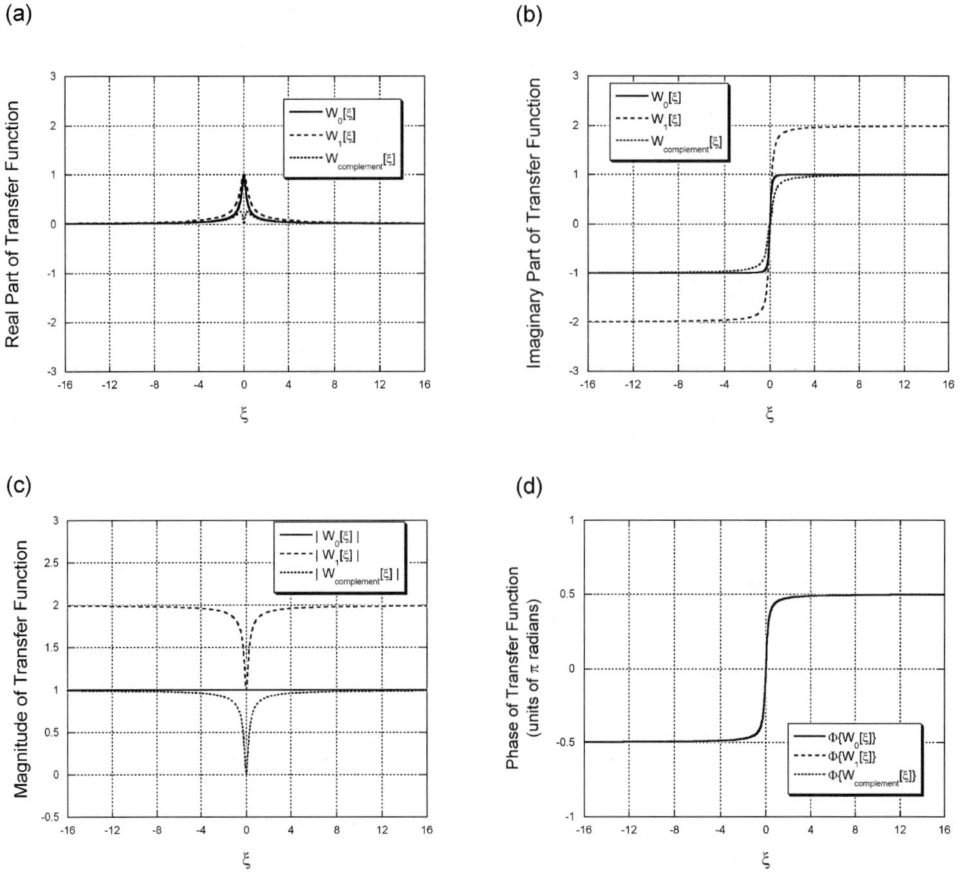

Figure 19.23 Approximate transfer functions for the inverse filter for the decaying exponential $h[x] = STEP[x]\exp[-x]$ for zero order, complement filter, and first order: (a) real part; (b) imaginary part; (c) magnitude; and (d) phase. Note the increase in magnitude with order.

which means that the first-order approximation to the ideal reciprocal filter also has no effect on the recovered signal. This trend persists for higher-order approximations. These results make intuitive sense, because there is no information to be recovered from the spectrum after ideal lowpass filtering.

19.6.2.3 Decaying Exponential Averager

The impulse response of the decaying exponential filter was listed in Equation (19.36):

$$h_3[x] = \frac{1}{|b_0|}\exp\left[-\frac{x}{b_0}\right] \cdot STEP\left[\frac{x}{b_0}\right] \tag{19.163}$$

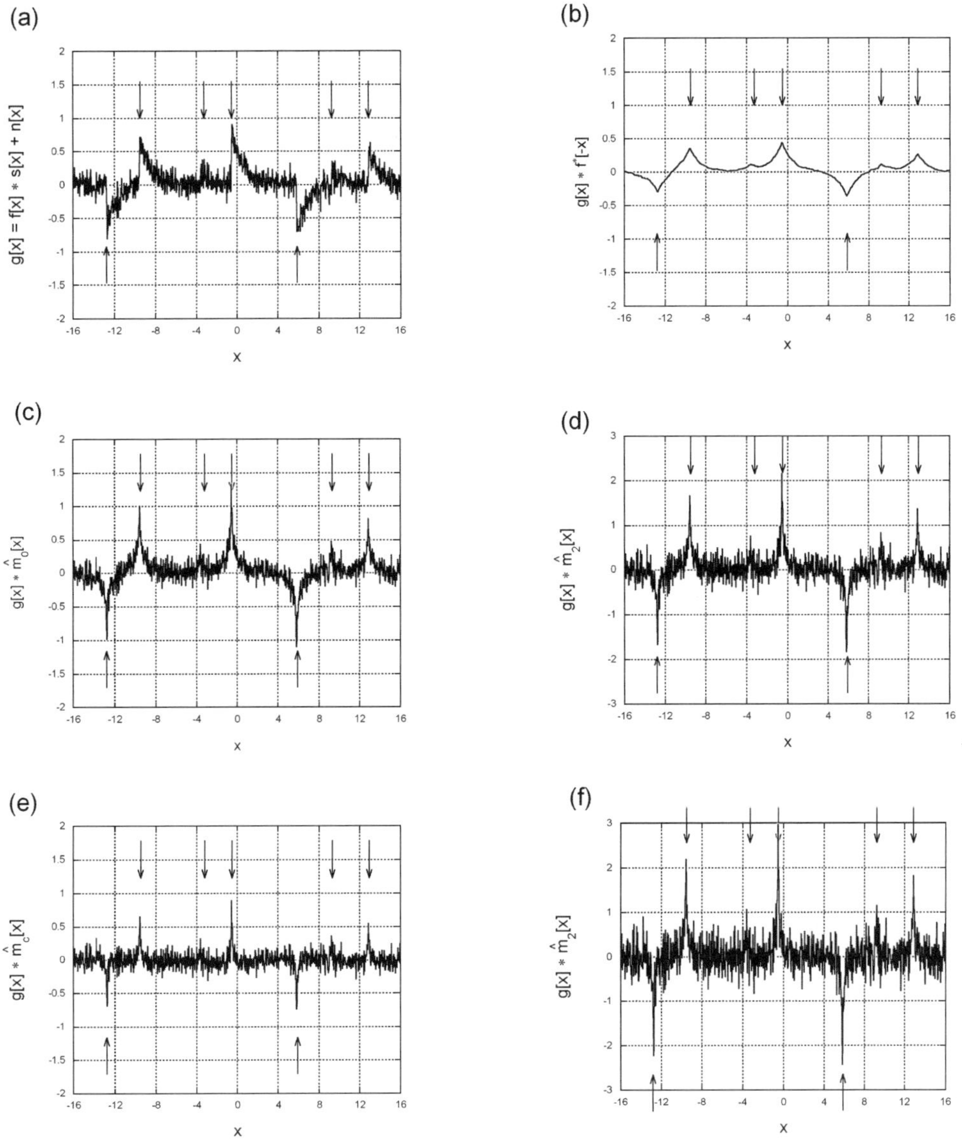

Figure 19.24 Action of the approximate matched filter for the decaying exponential: (a) noisy input $g[x] = f[x] * s[x] + n[x]$, where $n[x]$ is Gaussian distributed with zero mean and $\sigma = 0.1$ (arrows indicate locations of replicas of $f[x]$); (b) $g[x] * f^*[-x]$, showing the lowpass action of the classical matched filter; (c) $g[x] * \hat{m}_0[x]$; (d) $g[x] * \hat{m}_1[x]$; (e) $g[x] * \hat{m}_c[x]$; (f) $g[x] * \hat{m}_2[x]$. The amplitude and sharpness of the correlation peaks both increase with increasing order. The output of the complement filter is also noticeably sharper than that of the classic matched filter.

The associated transfer function is easy to derive by combining the spectrum derived in Equation (9.88) with the scaling theorem:

$$H_3[\xi] = \frac{1}{1 + 2\pi i b_0 \xi} = \frac{1 + i(-2\pi b_0 \xi)}{1 + (2\pi b_0 \xi)^2} \tag{19.164a}$$

$$|H_3[\xi]| = \sqrt{\frac{1}{1 + (2\pi b_0 \xi)^2}} \Longrightarrow |H|_{\max} = 1 \tag{19.164b}$$

$$\mathbf{\Phi}\{H_3[\xi]\} = \tan^{-1}[2\pi b_0 \xi] \tag{19.164c}$$

The phase-only approximation to the reciprocal filter for the decaying exponential is

$$(\hat{W}_0[\xi])_3 = \frac{1}{|H|_{\max}} \exp[-i\mathbf{\Phi}\{H_3[\xi]\}] = 1[\xi] \cdot \exp[-i \cdot \tan^{-1}[2\pi \xi b_0]] \tag{19.165}$$

The cascade of the transfer function and the zero-order approximation to the filter cancels the phase and leaves the magnitude of the original transfer function:

$$H_3[\xi] \cdot (\hat{W}_0[\xi])_3 = |H_3[\xi]| = \sqrt{\frac{1}{1 + (2\pi b_0 \xi)^2}} \tag{19.166}$$

We have not derived the space-domain form of this spectrum, but the discussion of the impact of nonlinear operators on the Fourier transform in Section 9.8.19 indicates that its space-domain form should be "wider" than the Lorentzian spectrum for which we derived the inverse transform:

$$\mathcal{F}_1^{-1}\left\{\frac{1}{1 + (2\pi b_0 \xi)^2}\right\} = \frac{1}{2} \cdot \mathcal{F}_1^{-1}\left\{\frac{2}{1 + (2\pi b_0 \xi)^2}\right\} = \frac{1}{2} \exp\left[-\left|\frac{x}{b_0}\right|\right] \tag{19.167}$$

We thus expect that the cascade of the decaying-exponential impulse response and the zero-order approximation to the inverse filter will be "narrower" than this function:

$$h[x] * \hat{w}_0[x] = \mathcal{F}_1^{-1}\left\{\sqrt{\frac{1}{1 + (2\pi b_0 \xi)^2}}\right\} \tag{19.168}$$

The transfer function of the first-order approximation to the reciprocal filter for the decaying exponential is:

$$(\hat{W}_1[\xi])_3 = \left(2 - \sqrt{\frac{1}{1 + (2\pi b_0 \xi)^2}}\right) \cdot \exp[+i \cdot \tan^{-1}[+2\pi b_0 \xi]] \tag{19.169}$$

The same argument is used to construct a "complement" matched filter based on Equation (19.148). From the experience obtained with the reciprocal filter, we would expect low-order complement matched filters to produce "sharper" peaks of the amplitude at the locations of the replicas of $f[x]$, while being less sensitive to noise. Examples of the approximate transfer functions of the matched filter are shown in Figure 19.23, and the performance of the filters is compared in Figure 19.24.

The discussion suggests that the approximations to the reciprocal filters may be applied as limiting cases in the interpretation of the Wiener–Helstrom filter. In other words, high- and low-order approximations in Equation (19.134) may be substituted for the low-noise and high-noise limits in Equation (19.86) to "tune" the response of the Wiener–Helstrom filter (Walvoord and Easton, 2004).

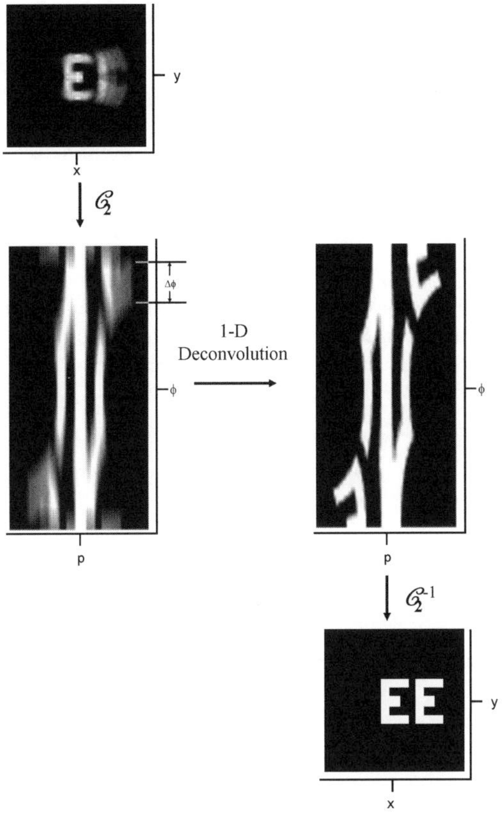

Figure 19.25 Deblurring a rotated image via the central-slice transform: the object rotated through $\pi 4$ radians during the exposure. The blur is obviously shift variant. The central slices through the center of rotation are calculated, producing the "image" $C_f(r, \phi)$ that has been blurred in the azimuthal direction only. The 2-D array is deblurred by the inverse filter $w(r, \phi)$, and then remapped to Cartesian coordinates by the inverse central-slice transform of Equation (12.99).

19.7 Inverse Filtering of Shift-Variant Blur

Rotational image blur shown in Figure 19.2 is one example of a shift-variant process that may be converted to shift-invariant convolution by a geometrical transformation of the image coordinates. The central-slice transform of Chapter 12 provides the required conveyance from shift variance to shift invariance. Consider the example of Figure 19.2, where the object function $f[x, y]$ rotated through θ_0 radians during the exposure time. The slices of the scene through the axis of rotation are calculated to evaluate the central-slice transform of Equation (12.77), as shown in Figure 19.20. This expresses $f[x, y]$ in polar coordinates of the form $\mu_f(p, \phi)$, where the blur is a 1-D function along the azimuthal direction only, i.e., it may be expressed in the form:

$$h(p, \phi) = \delta(p) \cdot \frac{1}{\theta_0} RECT\left[\frac{\phi}{\theta_0}\right] \tag{19.170}$$

where the coordinates are now considered to be Cartesian, and thus the 2-D Fourier transform is separable. The associated transfer function is:

$$\underset{p\to\nu}{\mathcal{F}_1}\ \underset{\phi\to\chi}{\mathcal{F}_1}\ \{h(p,\phi)\} = H(\nu,\chi) = 1[\nu]\cdot SINC[\theta_0\cdot\chi] \tag{19.171}$$

The corresponding pseudoinverse filter is:

$$w(p,\phi) = \delta(p)\cdot \underset{\chi\to\phi}{\mathcal{F}_1^{-1}}\{(SINC[\theta_0\cdot\chi])^{-1}\} \tag{19.172}$$

The 2-D array is deblurred by convolving $\mu_f(p,\phi)$ with $w(p,\phi)$ in Cartesian coordinates. The 2-D array of central slices is then transformed back to the usual Cartesian domain via the inverse central-slice transform of Equation (12.99). An example is shown in Figure 19.25.

PROBLEMS

19.1 The transfer functions listed below describe the action of different LSI systems. The goal of this problem is to find the corresponding "inverse filter" in both domains, i.e.:

$$W[\xi] = (H[\xi])^{-1}$$

$$w[x] = \mathcal{F}^{-1}\{W[\xi]\}$$

In the situations where the inverse filter does not exist, we will instead evaluate the "pseudoinverse" filter:

$$\hat{W}[\xi] = \begin{cases} (H[\xi])^{-1} & \mathit{for}\ H[\xi]\neq 0 \\ 0 & \mathit{for}\ H[\xi]=0 \end{cases}$$

$$\hat{w}[x] = \mathcal{F}^{-1}\{\hat{W}[\xi]\}$$

In each case, determine which of the inverse filters ($W_n[\xi]$ or $\hat{W}_n[\xi]$) is appropriate and sketch its transfer function. Classify the action of the appropriate filter as lowpass, highpass, etc. *Also*, in those cases where it is possible, evaluate and sketch the appropriate impulse response ($w_n[x]$ or $\hat{w}_n[x]$) of the appropriate inverse filter. You may use reasonable approximations where appropriate – the sketches may be helpful here.

(a) $H[\xi] = GAUS[\xi]$

(b) $H[\xi] = e^{+i\pi\xi}$

(c) $H[\xi] = e^{+i\pi(1-RECT[\xi])}$

19.2 A 1-D image $g[x]$ has been created by a double exposure of the original object $f[x]$. The original scene was translated between the exposures by the known distance $+b_0$. The object was stationary during the time that each image was collected, and the exposure time was the same in both cases.

(a) Design the inverse filter for this system in the frequency domain. Comment about the potential of success of the deblurring process, particularly if noise is present.

(b) Find an exact or approximate expression for the inverse filter in the space domain.

19.3 Find the matched filter for $f[x] = GAUS[x]$ such that the recorded signal is:

$$\begin{aligned} g[x] &= GAUS[x-x_0] + n[x] \\ &= GAUS[x] * \delta[x-x_0] + n[x] \end{aligned}$$

19.4 Design the Wiener or Wiener–Helstrom filter for the following input signals, impulse responses, and noise power spectra:

(a) $f[x] = 2\,GAUS[x]$, $h[x] = \delta[x]$, $|N[\xi]|^2 = GAUS[\xi + \xi_0] + GAUS[\xi - \xi_0]$

(b) $f[x] = GAUS[x/b_0] \cdot \exp[+i\pi x^2]$, $h[x] = RECT[x]$, $|N[\xi]|^2 = GAUS[\xi + \xi_0] + GAUS[\xi - \xi_0]$

(c) $f[x] = GAUS[x/2] \cdot \cos[10\pi x]$, $h[x] = \delta[x]$, $|N[\xi]|^2 = GAUS[x]$

19.5 Equation (19.70) showed that the phase-only approximation to the inverse filter for uniform averaging is:

$$\hat{W}_0[\xi] = SGN[SINC[b_0\xi]]$$

(a) Write this in the form of a square wave to derive an analytical expression for the impulse response $\hat{w}_0[x]$.

(b) Find the transfer function and impulse response of the approximation of the inverse filter for differentiation based on the first-order complement filter.

20

Filtering in Discrete Systems

This chapter considers the practical application of linear imaging systems to modeling impulse responses and transfer functions of some useful 1-D and 2-D discrete linear filters. One goal is to more completely understand the differences between the continuous and discrete cases. The 1-D examples assume that the functions have been sampled and quantized to create an N-pixel b-bit array $f_q[n]$, where N typically is a power of two with domain $-N/2 \leq n \leq +N/2 - 1$, though the results may be generalized to other values of N rather easily. In both 1-D and 2-D cases, the domain of allowed "gray values" (also called "digital counts" or "quantum numbers") in the digital input images is $0 \leq f_q \leq 2^b - 1$.

The expression for discrete convolution of two N-D sampled functions was noted in Equation (15.114):

$$g[n] \equiv \sum_{m=-N/2}^{N/2-1} f[m]h[n-m] \tag{20.1}$$

which shows that the convolution kernel h is "reversed" and translated before applying the weights to the input function. Most often we will assume that convolution is performed via the DFT (or FFT), and thus computes circular convolution where the samples of the impulse response that are translated "off the edge" of the 1-D finite array "reappear" on the other side. Equation (15.117) demonstrated that the amplitude of the output is scaled by $N \cdot c_1^2 \cdot c_2$, where c_1 and c_2 are the respective normalization constants applied in the forward and inverse DFTs. This scale factor is unity for $c_1 = 1$ and $c_2 = N^{-1}$, which is the preferred normalization for evaluating the DFT of functions with compact support. We choose the normalization $c_1 = N^{-1}$ and $c_2 = 1$ to generate results that resemble continuous convolution if the constituent functions are periodic.

The impulse response of a 1-D convolution is represented by the array $h[n]$, which may be considered to be a set of weights applied to the sample amplitudes; the weights are specified before the "reversal" in the convolution. The kernel $h[n]$ is often represented as a smaller array that has been truncated to the smallest possible odd number of elements centered on the sample indexed by $n = 0$, e.g., the three-element vector

a	b	c

with indices $n = -1$ for the sample with amplitude a, $n = 0$ for that with b, and $n = +1$ for c. When convolving via the DFT, this three-element vector is padded with zeros to create an N-pixel array, where again N typically is a power of two.

The 2-D discrete kernels and the associated transfer functions are easy to derive by applying the separability property. Consider a 2-D image that has been sampled and quantized to create an $N \times N$

Fourier Methods in Imaging Roger L. Easton, Jr.

b-bit digital image $f_q[n, m]$. The choice of the domain of indices is not significant, but we will continue to use the convention $-N/2 \leq n, m \leq +N/2 - 1$. In the examples considered here, the origin of coordinates is placed at the center of the image in both the space and frequency domains. The positive x- (or ξ-) and y- (or η-) directions are toward the right and top, respectively. We specify these directions explicitly because some graphics applications define the positive y-direction toward the bottom of the page since the top of the page is printed first. This difference may be confusing when interpreting printed data.

By analogy with the 1-D case, 2-D kernels are represented by the smallest square table of $M \times M$ weights that includes all of the nonzero elements, where M is an odd number and the index $[n = 0, m = 0]$ is located at the center of the table.

Because discrete impulse responses may be constructed by scaling, translating, and summing discrete delta functions, it will be convenient to consider the translation operation as a convenient starting point for introducing general discrete impulse responses. The derivation closely parallels the discussion of "leakage" in Chapter 15, as they really describe the same phenomenon.

20.1 Translation, Leakage, and Interpolation

20.1.1 1-D Translation

Recall that the impulse response and transfer function for translation of a 1-D continuous input function by the distance x_0 are:

$$h[x] = \delta[x - x_0] \tag{20.2a}$$

$$H[\xi] = \exp[-2\pi i \xi x_0] \tag{20.2b}$$

The Fourier shift theorem relates the continuous impulse response and transfer function of the 1-D translation operator. A discrete form of the impulse response that has been sampled at integer multiples of Δx evidently is:

$$h[n \cdot \Delta x] = \delta_d[n \cdot \Delta x - n_0 \cdot \Delta x] = \delta_d[(n - n_0) \cdot \Delta x] \equiv \begin{cases} 1 & \textit{if } n = n_0 \\ 0 & \textit{otherwise} \end{cases} \tag{20.3}$$

where we have identified $x = n \cdot \Delta x$ and $x_0 = n_0 \cdot \Delta x$. Obviously the impulse response of the identity operator is the discrete delta function with $n_0 = 0$. Because of the "reversal" operation inherent in the convolution, the kernel that translates an N-pixel input function by one sample toward positive n (to the right) is:

$$h[x] = \delta[x - \Delta x] \implies h[n] = \delta_d[n - 1] = \boxed{0 \mid 0 \mid +1} \tag{20.4}$$

The discussion in Chapter 14 may be applied to evaluate the corresponding transfer function:

$$H[k] = c_1 \exp\left[-2\pi i \frac{k}{N}\right] \cdot RECT\left[\frac{k + \varepsilon}{N}\right] \tag{20.5}$$

where again c_1 is the normalization factor of the DFT and ε is a small positive parameter that ensures that the full amplitudes of samples indexed by $k = -N/2$ and $+N/2 - 1$ are included. The real and imaginary parts of the transfer function are respectively one cycle of a cosine and of a sine with identical amplitudes determined by the DFT normalization (Figure 20.1). We can easily extend this analysis to show that the real and imaginary parts of the transfer function of a kernel that translates by a distance equal to an integer number n_0 of samples, so that $x_0 = n_0 \cdot \Delta x$. is composed of n_0 cycles of the cosine and of the sine, respectively.

This simple discrete translation operator exists only for integer values of n_0, which means that the translation is an integer multiple of the sampling interval Δx. However, it is easy to construct the transfer

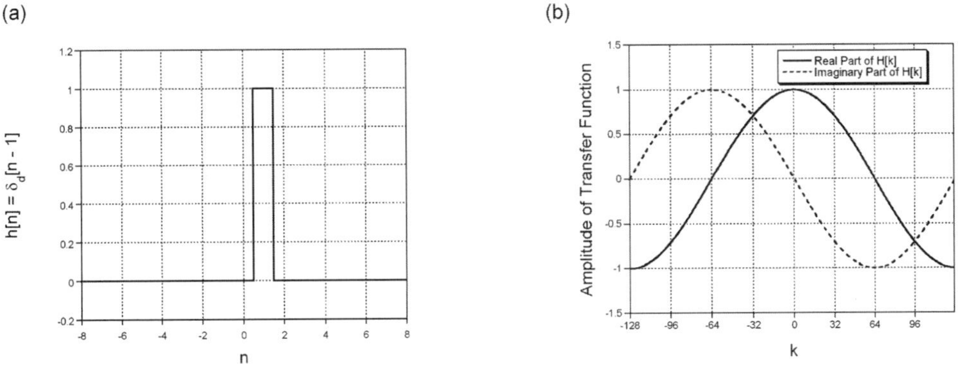

Figure 20.1 The 1-D translation by one sample: (a) impulse response $h[n] = \delta_d[n-1]$ for $N = 256$, where only the central region $-8 \le n \le +8$ is shown; (b) real and imaginary parts of the transfer function $H[k] = \exp[-2\pi k/N]$, which are respectively one cycle each of the cosine and sine over $N = 256$ samples.

function of a more general discrete translation operator by sampling the continuous transfer function at frequencies that are integer multiples of $\Delta\xi = (N \cdot \Delta x)^{-1}$. From the discussion of Chapter 14, we know that the corresponding discrete spectrum is sampled and periodic:

$$H_s[\xi] = \left(e^{-2\pi i\xi x_0} \cdot \frac{1}{\Delta\xi} COMB\left[\frac{\xi}{\Delta\xi}\right] \cdot RECT\left[\frac{\xi}{N \cdot \Delta\xi}\right]\right) * COMB\left[\frac{\xi}{N \cdot \Delta\xi}\right] \tag{20.6}$$

The corresponding impulse response is obtained from $H_s[\xi]$ via the inverse continuous Fourier transform and is also sampled and periodic:

$$\begin{aligned} h_s[x] &= \left(\delta[x - x_0] * COMB\left[\frac{x}{N \cdot \Delta x}\right] * SINC[N \cdot \Delta\xi \cdot x]\right) \\ &\quad \cdot (N \cdot \Delta\xi) \cdot COMB[N \cdot \Delta\xi \cdot x] \\ &= \left(\left(\delta[x - x_0] * SINC\left[\frac{x}{\Delta x}\right]\right) * COMB\left[\frac{x}{N \cdot \Delta x}\right]\right) \cdot \frac{1}{\Delta x} COMB\left[\frac{x}{\Delta x}\right] \\ &= \left(SINC\left[\frac{x - x_0}{\Delta x}\right] * COMB\left[\frac{x}{N \cdot \Delta x}\right]\right) \cdot \frac{1}{\Delta x} COMB\left[\frac{x}{\Delta x}\right] \end{aligned} \tag{20.7}$$

The width parameter of the *SINC* function is identical to the space-domain sampling interval Δx, and it has been translated by the continuous distance x_0. The convolution with the first *COMB* function produces replicas of the *SINC* at intervals of $N \cdot \Delta x$ that are sampled at integer multiples of Δx by the second *COMB* function.

If the shift parameter x_0 is an integer multiple of Δx (say $x_0 = n_0 \cdot \Delta x$), the *SINC* function is sampled at its central peak and at its zeros so that the discrete impulse response $h[n]$ is the discrete Dirac delta function located at $x = n_0 \cdot \Delta x$:

$$\mathcal{O}\{f[n]\} = f[n - n_0] \Longrightarrow h[n] = \delta_d[n - n_0] * COMB\left[\frac{n}{N}\right] \tag{20.8}$$

In the (more interesting) case where x_0 is not an integer multiple of Δx, the impulse response is the *SINC* function sampled at points other than its maximum and zeros. This sampled *SINC* function has "infinite" support (though decreasing amplitude with distance), and the convolution with the *COMB*

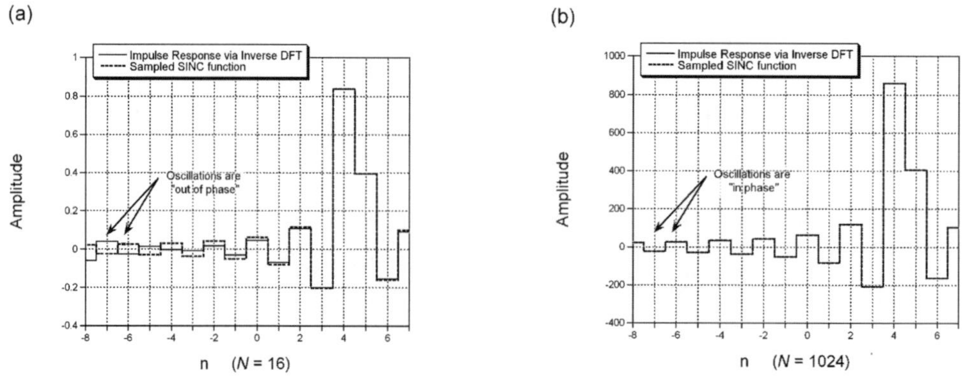

Figure 20.2 Comparison of impulse responses for translation by $x_0 = 4.32 \cdot \Delta x$ for two values of N: (a) $N = 16$ and (b) a magnified view of the central region for $N = 1024$ samples. In both cases, the impulse response of translation obtained via the inverse DFT is compared to the sampled ideal *SINC* function interpolator. The ideal and computed responses in (a) are noticeably different due to the significant leakage for $N = 16$, but they are difficult to distinguish in the vicinity of the maximum for $N = 1024$ in (b).

function again ensures that the discrete impulse response $h[n]$ is periodic. When viewed in a "window" of support N, the periodic *SINC* function impulse response "wraps around" the array:

$$h[n] = \left(\left(SINC\left[\frac{n \cdot \Delta x - x_0}{\Delta x}\right] * COMB\left[\frac{x}{N \cdot \Delta x}\right]\right) \cdot \frac{1}{\Delta x} COMB\left[\frac{x}{\Delta x}\right]\right) \cdot RECT\left[\frac{x + \varepsilon}{N \cdot \Delta x}\right] \tag{20.9}$$

If the number of samples N is large, the amplitudes of the "wrapped-around" samples of the *SINC* function are small and the impulse response is approximately equal to the N samples of the translated *SINC*:

$$h[n] \cong SINC\left[\frac{n \cdot \Delta x - x_0}{\Delta x}\right] = SINC\left[n - \frac{x_0}{\Delta x}\right] \quad \text{if } N \text{ is large} \tag{20.10}$$

Discrete impulse responses for translation by 4.32 samples are compared to the sampled *SINC* function for $N = 16$ and $N = 1024$ in Figure 20.2. The impulse responses are quite different in the former case of small N, but nearly indistinguishable in the latter case.

As an example of the action of the discrete translation operator, consider the translation of $f[n] = STEP[n]$ by both an integer and noninteger number of samples in Figure 20.3. In the former case, the shift is $x_0 = 8 \cdot \Delta x$ and the output is a replica of the input except for the samples that were "wrapped around" due to the circular convolution. In the second example, the translation distance is $x_1 = 8.32 \cdot \Delta x$ and the impulse response is the periodic *SINC* function that has been sampled. The resulting output is interpolated by the *SINC* function, thus producing "ringing" artifacts at both the "explicit" and "implied" edges of the *STEP* function.

20.1.2 2-D Translation

The 2-D kernel that translates the image by 1 pixel to the right may be written as a 3×3 kernel constructed from the orthogonal product of the 1-D translation kernel in the x-direction and the 1-D

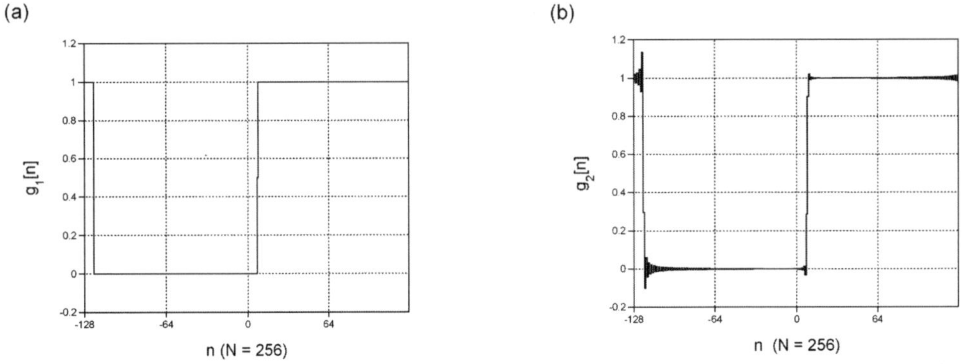

Figure 20.3 The discrete translation of $f[n] = STEP[n]$ for $N = 256$: (a) $x_0 = 8 \cdot \Delta x$, the impulse response is translated discrete delta function and the output $g_1[n]$ is a "wrapped-around" replica of $f[n]$; (b) $x_0 = 8.32 \cdot \Delta x$, the impulse response is the sampled and periodic *SINC* function and the output $g_2[n]$ exhibits "ringing" at the edges.

identity kernel in the y-direction:

$$h_1[n, m] = \delta_d[n-1, m] = \delta_d[n-1] \cdot \delta_d[m] = \begin{array}{|c|c|c|} \hline 0 & 0 & 0 \\ \hline 0 & 0 & +1 \\ \hline 0 & 0 & 0 \\ \hline \end{array} \tag{20.11}$$

The associated transfer function is:

$$H[k, \ell] = \exp\left[-2\pi \frac{k}{N}\right] \cdot 1[\ell] \cdot RECT\left[\frac{k+\varepsilon}{N}, \frac{\ell+\varepsilon}{N}\right] \tag{20.12}$$

which includes one cycle each of the sinusoid (cosine in the real part, sine in the imaginary part) along the k-direction and is constant along the ℓ-direction.

Kernels that translate in other directions are obvious extensions, though we make one comment about translation along a diagonal direction, e.g.:

$$h_2[n, m] = \begin{array}{|c|c|c|} \hline 0 & 0 & +1 \\ \hline 0 & 0 & 0 \\ \hline 0 & 0 & 0 \\ \hline \end{array} = \delta_d[n-1, m-1] \tag{20.13}$$

In the usual situation where $\Delta x = \Delta y$, the corresponding continuous impulse response translates the image by $\sqrt{2} \cdot \Delta x$ along the diagonal. From the scaling theorem, we can see that the period of the sinusoidal transfer function is smaller than for translation along a Cartesian direction. The associated transfer function is:

$$H_2[k, \ell] = \lim_{\varepsilon \to 0} \left\{ \exp\left[-2\pi \frac{k+\ell}{N}\right] \cdot RECT\left[\frac{k+\varepsilon}{N}, \frac{\ell+\varepsilon}{N}\right] \right\} \tag{20.14}$$

20.2 Averaging Operators – Lowpass Filters

20.2.1 1-D Averagers

The impulse response of the linear operator that averages uniformly over a finite region is a unit-area rectangle:

$$h[x] = \frac{1}{|b|} RECT\left[\frac{x}{b}\right] \tag{20.15}$$

where b is the width of the averaging region. The corresponding continuous transfer function is:

$$H[\xi] = SINC\left[\frac{\xi}{(1/b)}\right] \tag{20.16}$$

In the discrete case, the rectangular impulse response is sampled and the width b is measured in units of Δx. If $b/\Delta x$ is even, the amplitudes of the endpoint samples are $b/2$. We consider the cases where $b = 2 \cdot \Delta x$, $3 \cdot \Delta x$, and $4 \cdot \Delta x$. The discrete impulse responses of uniform averagers that are 2 and 3 pixels wide have three nonzero samples:

$$b = 2 \cdot \Delta x : h_2[n] = \boxed{+\tfrac{1}{4}}\boxed{+\tfrac{1}{2}}\boxed{+\tfrac{1}{4}} = \frac{1}{4}(\delta_d[n+1] + 2 \cdot \delta_d[n] + \delta_d[n-1]) \tag{20.17a}$$

$$b = 3 \cdot \Delta x : h_3[n] = \boxed{+\tfrac{1}{3}}\boxed{+\tfrac{1}{3}}\boxed{+\tfrac{1}{3}} = \frac{1}{3}(\delta_d[n+1] + \delta_d[n] + \delta_d[n-1]) \tag{20.17b}$$

The linearity of the DFT ensures that the corresponding transfer function may be constructed by summing transfer functions for the identity operator and for translations by one sample each to the left and right. The resulting transfer functions may be viewed as discrete approximations to corresponding continuous *SINC* functions:

$$\begin{aligned} H_2[k] &= \frac{1}{4} e^{-2\pi ik\cdot\Delta\xi} + \frac{1}{2} + \frac{1}{4} e^{+2\pi ik\cdot\Delta\xi}, \quad -\frac{N}{2} \leq k \leq \frac{N}{2} - 1 \\ &= \frac{1}{2}(1 + \cos[2\pi(k \cdot \Delta\xi)]) \equiv SINC_d\left[k; \frac{N}{2}\right] \end{aligned} \tag{20.18a}$$

$$\begin{aligned} H_3[k] &= \frac{1}{3} e^{-2\pi ik\cdot\Delta\xi} + \frac{1}{3} + \frac{1}{3} e^{+2\pi ik\cdot\Delta\xi} \\ &= \frac{1}{3}(1 + 2\cos[2\pi(k \cdot \Delta\xi)]) \equiv SINC_d\left[k; \frac{N}{3}\right] \end{aligned} \tag{20.18b}$$

Note that both $H_2[k]$ and $H_3[k]$ have unit amplitude at the origin because of the discrete central-ordinate theorem for the assumed normalization.

The "zero crossings" of the 2-pixel averager are located at the spatial frequencies $\xi = \pm\frac{1}{2}$ cycles per sample (the positive and negative Nyquist frequencies), but the index convention admits only the negative value $k = -N/2$. The sinusoid with this frequency oscillates with a period of two samples and is thus averaged to zero by the 2-pixel averager. From this result, it seems reasonable to assume that the 3-pixel averager should also "block" any sinusoid with a period $3 \cdot \Delta x$, which certainly can be constructed without aliasing because the Nyquist condition would be satisfied. The zero crossings of the continuous transfer function are located at $\xi = \pm\frac{1}{3}$ cycles per sample, so the discrete zeros should occur at indices $k = \pm N/3$. However, $\pm N/3$ are not integers for even N, so the zero crossings of the discrete spectrum occur *between* samples. In other words, the discrete transfer function of the 3-pixel averager has no zeros and will "pass" all sampled and unaliased sinusoidal components with some attenuated amplitude. This seems to be a paradox – a sinusoid with a period of 3 pixels is passed by a 3-pixel averager, but the 2-pixel sinusoid is blocked by a 2-pixel averager. The reason is because there are a noninteger number of periods of length $3 \cdot \Delta x$ in an array where N is even. Thus there must be "leakage" in the spectrum of the sampled function. The resulting "spurious" frequency components

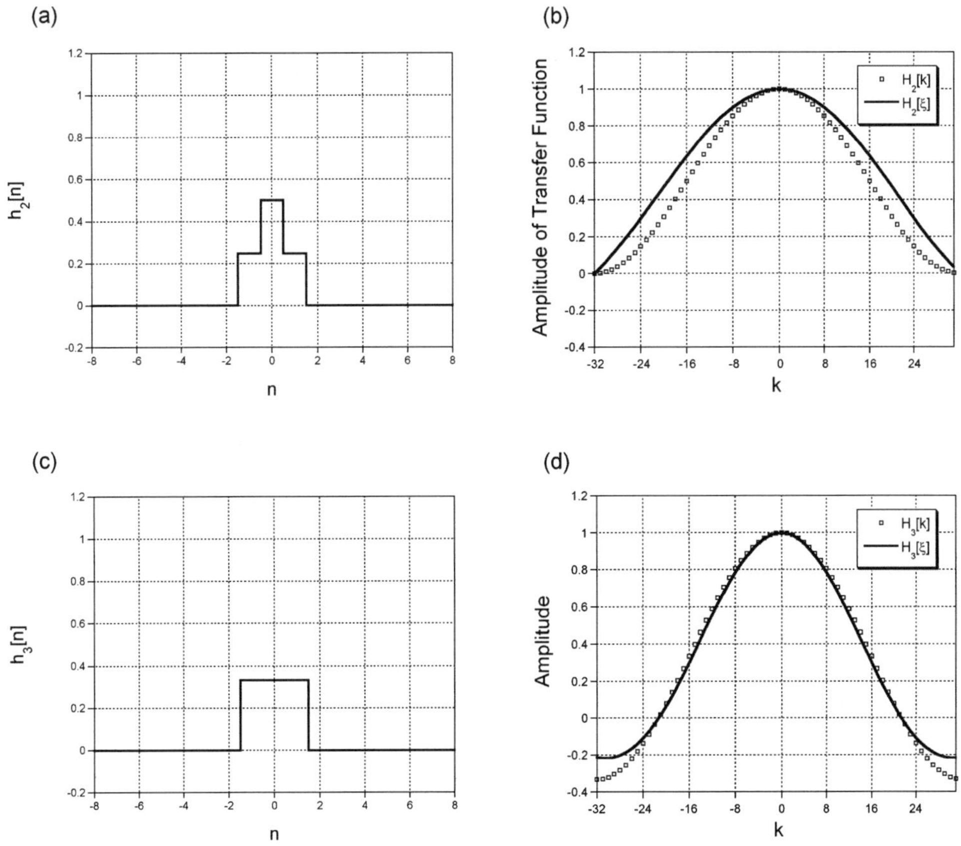

Figure 20.4 Comparison of 2- and 3-pixel uniform averagers for $N = 64$: (a) central region of the impulse response $h_2[n] = \frac{1}{2}RECT[n/2]$; (b) discrete transfer function $H_2[k]$ compared to the continuous analogue $H_2[\xi] = SINC[2\xi]$ out to the Nyquist frequency $\xi = -\frac{1}{2}$ cycles per sample; (c) $h_3[n] = \frac{1}{3}RECT[n/3]$, which has the same support as $h_2[n]$; (d) $H_3[k]$ compared to $H_3[\xi] = SINC[3\xi]$, showing the smooth transition at the edge of the array.

reach the output. As a final observation, note also that the discrete transfer functions approach the edges of the array "smoothly" (without "cusps") in both cases, as shown in Figure 20.4.

The discrete impulse response of the 4-pixel averager has five nonzero samples:

$$b = 4 \cdot \Delta x : h_4[n] = \begin{array}{|c|c|c|c|c|} \hline +\frac{1}{8} & +\frac{1}{4} & +\frac{1}{4} & +\frac{1}{4} & +\frac{1}{8} \\ \hline \end{array} \tag{20.19}$$

The linearity of the DFT ensures that the corresponding transfer function may be constructed by summing the transfer function of the 3-pixel averager scaled by $\frac{3}{4}$ with the transfer functions for translation by two samples each to the left and right:

$$\begin{aligned} H_4[k] &= \frac{1}{4}\left(\frac{1}{2}\, e^{-2\pi ik\cdot 2\Delta\xi} + e^{-2\pi ik\cdot\Delta\xi} + 1 + e^{+2\pi ik\cdot\Delta\xi} + \frac{1}{2}\, e^{+2\pi ik\cdot 2\Delta\xi}\right) \\ &= \frac{1}{4}(1 + 2\cos[2\pi(k\cdot\Delta\xi)] + \cos[2\pi(k\cdot 2\Delta\xi)]) \end{aligned} \tag{20.20}$$

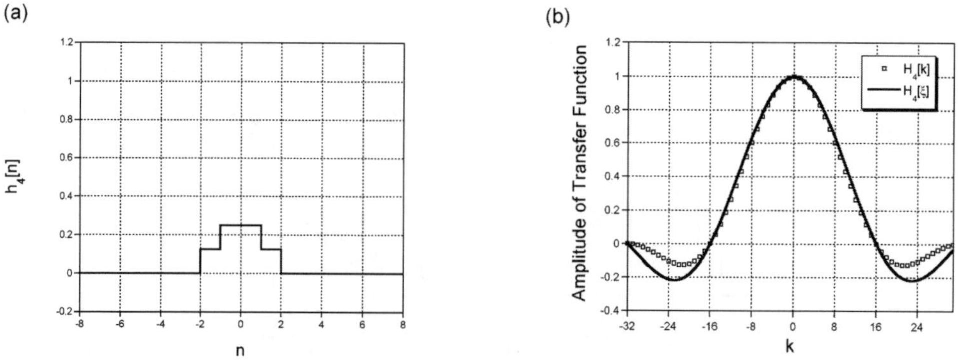

Figure 20.5 The 4-pixel averager for $N = 64$: (a) central region of impulse response $h_4[n] = \frac{1}{4}RECT[n/4]$; (b) Discrete transfer function $H_4[k]$ compared to the continuous transfer function $SINC[4\xi]$, showing the smooth decay of the discrete case at the edges of the array.

which also may be thought of as a discrete approximation of a *SINC* function: $SINC_d[k; N/4]$. This discrete transfer function has zeros located at $\xi = \pm\frac{1}{4}$ cycles per sample, which correspond to $k = \pm N/4$, so it removes any sampled sinusoid with period $4 \cdot \Delta x$ from the output. Again the transfer function has "smooth" transitions of amplitude at the edges of the array, thus preventing "cusps" in the periodic spectrum (Figure 20.5).

The general expression for the discrete *SINC* function in the frequency domain suggested by these results for $-N/2 \leq k \leq N/2 - 1$ is:

$$SINC_d\left[k; \frac{N}{w}\right] = \begin{cases} \dfrac{1}{w}\left(1 + 2\displaystyle\sum_{\ell=1}^{(w-1)/2} \cos[2\pi(k \cdot \ell \cdot \Delta\xi)]\right) & \textit{if } w \textit{ is odd} \\ \dfrac{1}{w}\left(1 + \cos[2\pi(k \cdot w/2 \cdot \Delta\xi)] + 2\displaystyle\sum_{\ell=1}^{w/2-1} \cos[2\pi(k \cdot \ell \cdot \Delta\xi)]\right) & \textit{if } w \textit{ is even} \end{cases} \tag{20.21}$$

20.2.2 2-D Averagers

The 2-D kernel that averages three adjacent samples along the x-direction is constructed by padding the 1-D kernel in Equation (20.17b) with zeros in the vertical direction:

$$h[n, m] = \begin{array}{|c|c|c|} \hline 0 & 0 & 0 \\ \hline +\frac{1}{3} & +\frac{1}{3} & +\frac{1}{3} \\ \hline 0 & 0 & 0 \\ \hline \end{array} \tag{20.22}$$

The associated transfer function has the same variation along the k-direction as Equation (20.18b) and is constant along ℓ:

$$H[k, \ell] = \left(\frac{1}{3}(1 + 2\cos[2\pi(k \cdot \Delta\xi)]) \cdot 1[\ell]\right) \cdot RECT\left[\frac{k+\varepsilon}{N}, \frac{\ell+\varepsilon}{N}\right] \tag{20.23}$$

Other 2-D discrete averaging kernels may be constructed by orthogonal multiplication, e.g., we can construct the common 3×3 uniform averager via the outer product of two orthogonal 1-D uniform averagers:

$$\begin{array}{|c|c|c|}\hline +\frac{1}{3} & +\frac{1}{3} & +\frac{1}{3} \\ \hline \end{array} \cdot \begin{array}{|c|}\hline +\frac{1}{3} \\ \hline +\frac{1}{3} \\ \hline +\frac{1}{3} \\ \hline \end{array} = \begin{array}{|c|c|c|}\hline +\frac{1}{9} & +\frac{1}{9} & +\frac{1}{9} \\ \hline +\frac{1}{9} & +\frac{1}{9} & +\frac{1}{9} \\ \hline +\frac{1}{9} & +\frac{1}{9} & +\frac{1}{9} \\ \hline \end{array} \tag{20.24}$$

The associated transfer function is the orthogonal product of the individual 1-D transfer functions. Note that the 1-D kernels need not be identical, e.g., a 3-pixel uniform averager along the n-direction and a 2-pixel uniform averager along m:

$$\begin{array}{|c|c|c|}\hline +\frac{1}{3} & +\frac{1}{3} & +\frac{1}{3} \\ \hline \end{array} \cdot \begin{array}{|c|}\hline +\frac{1}{4} \\ \hline +\frac{1}{2} \\ \hline +\frac{1}{4} \\ \hline \end{array} = \begin{array}{|c|c|c|}\hline +\frac{1}{12} & +\frac{1}{12} & +\frac{1}{12} \\ \hline +\frac{1}{6} & +\frac{1}{6} & +\frac{1}{6} \\ \hline +\frac{1}{12} & +\frac{1}{12} & +\frac{1}{12} \\ \hline \end{array} \tag{20.25}$$

20.3 Differencing Operators – Highpass Filters

20.3.1 1-D Derivative

The discrete analogue of differentiation may be derived from the definition of the continuous derivative:

$$\frac{df}{dx} \equiv \lim_{\tau \to 0} \left(\frac{f[x+\tau] - f[x]}{\tau} \right) \tag{20.26}$$

In the discrete case, the smallest nonzero value of τ is the sampling interval Δx, and thus the corresponding expression is:

$$\frac{1}{\Delta x}(f[(n+1)\cdot \Delta x] - f[n \cdot \Delta x]) = \frac{1}{\Delta x} f[n \cdot \Delta x] * (\delta[n+1] - \delta[n]) \tag{20.27}$$

In words, the discrete derivative is the scaled difference of the value at the sample indexed by $n+1$ and that indexed by n. By assuming that $\Delta x = 1$ sample, the leading scale factor may be ignored. The 1-D derivative operator may be implemented by discrete convolution with a 1-D kernel that has two nonzero elements; we will write it with three elements to clearly denote the sample indexed by $n = 0$:

$$f[n] * (\delta[n+1] - \delta[n]) = f[n] * \begin{array}{|c|c|c|}\hline +1 & -1 & 0 \\ \hline \end{array} \equiv f[n] * \partial[n] \tag{20.28}$$

where

$$\partial[n] \equiv \begin{array}{|c|c|c|}\hline +1 & -1 & 0 \\ \hline \end{array}$$

is defined to be the discrete impulse response of differentiation, which is perhaps better called a *differencing operator*. Note that $\partial[n]$ may be decomposed into even and odd parts:

$$\begin{aligned} \partial_{\text{even}}[n] &= \begin{array}{|c|c|c|}\hline +\frac{1}{2} & -1 & +\frac{1}{2} \\ \hline \end{array} \\ &= \left(+\frac{1}{2}\right) \begin{array}{|c|c|c|}\hline +1 & +1 & +1 \\ \hline \end{array} + \left(-\frac{3}{2}\right) \begin{array}{|c|c|c|}\hline 0 & +1 & 0 \\ \hline \end{array} \end{aligned} \tag{20.29a}$$

$$\partial_{\text{odd}}[n] = \begin{array}{|c|c|c|}\hline +\frac{1}{2} & 0 & -\frac{1}{2} \\ \hline \end{array} = \left(+\frac{1}{2}\right) \begin{array}{|c|c|c|}\hline +1 & 0 & -1 \\ \hline \end{array} \tag{20.29b}$$

The even part is a weighted sum of the identity operator and the 3-pixel averager, while the odd part computes differences of pixels separated by two sample intervals.

The corresponding 1-D transfer function may be derived by the appropriately weighted combination of transfer functions for the translation operator:

$$\mathcal{F}_1\left\{\boxed{+1}\boxed{-1}\boxed{0}\right\} = H_\partial[k] = (-1)\cdot 1[k] + (+1)\cdot e^{+2\pi ik/N} = -1 + e^{+2\pi ik/N}$$

$$= \left(-1+\cos\left[2\pi\frac{k}{N}\right]\right) + i\left(\sin\left[2\pi\frac{k}{N}\right]\right) \tag{20.30a}$$

$$|H_\partial[k]| = \sqrt{2-2\cos\left[2\pi\frac{k}{N}\right]} = 2\cdot\left|\sin\left[2\pi\frac{k}{2N}\right]\right| \tag{20.30b}$$

$$\Phi\{H_\partial[k]\} = \tan^{-1}\left[\frac{\sin[2\pi k/N]}{-1+\cos[2\pi k/N]}\right] \tag{20.30c}$$

This discrete result is compared to the transfer function of differentiation in the continuous case in Figure 20.6. The real and imaginary parts of $H_\partial[k\cdot\Delta\xi]$ again "decay" smoothly to zero at the edge of the array, thus ensuring periodicity in the frequency domain, which means that the magnitude of the discrete transfer function amplifies components with nonzero frequencies less than the continuous derivative.

20.3.2 2-D Derivative Operators

We can construct a 2-D kernel for the difference along the x-direction by analogy with the translation operator. The kernel is a 3×3 array that contains the 1-D derivative operator from Equation (20.28) padded with zeros. We will denote this discrete operator by ∂_x, where the subscript denotes the direction in which the difference is evaluated:

$\partial_x =$ (20.31)

0	0	0
+1	−1	0
0	0	0

The associated discrete transfer function is the straightforward extension of Equation (20.30):

$$\mathcal{F}_2\{\partial_x\} = H_{(\partial_x)}[k,\ell] = (-1+e^{+2\pi ik/N})\cdot 1[\ell]$$

$$= \left(-1[k,\ell]+\cos\left[2\pi\frac{k}{N}\right]\cdot 1[\ell]\right) + i\left(\sin\left[2\pi\frac{k}{N}\right]\cdot 1[\ell]\right) \tag{20.32}$$

The differencing operator in the y-direction and its associated transfer function are obtained by rotating the expressions just derived by $+\pi/2$ radians. The kernel is:

$\partial_y =$ (20.33)

0	0	0
0	−1	0
0	+1	0

The corresponding transfer function is obtained by applying the rotation theorem in Equation (10.37):

$$\mathcal{F}_2\{\partial_y\} = H_{(\partial_y)}[k,\ell] = 1[k]\cdot(-1+e^{+2\pi i\ell/N}) \tag{20.34}$$

We can also define 2-D derivatives along the diagonal directions:

$\partial_{(\theta=-\pi/4)} =$ (20.35a)

0	0	0
0	−1	0
+1	0	0

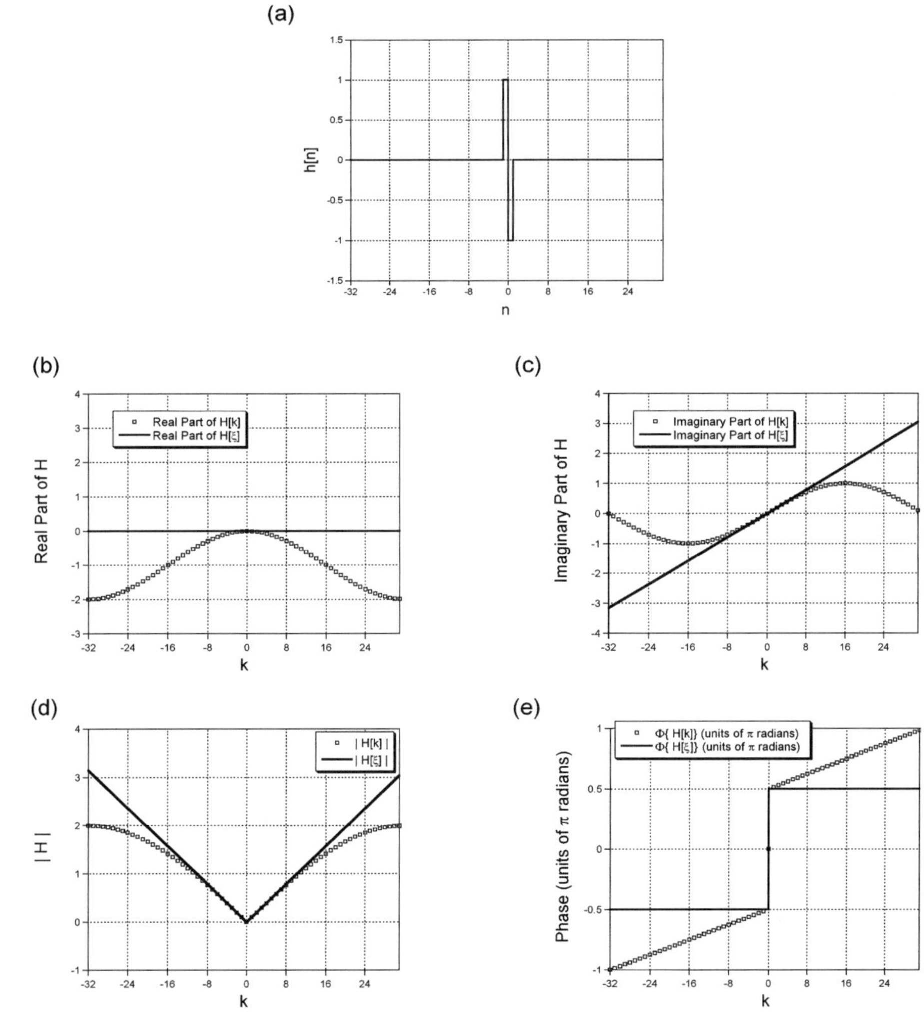

Figure 20.6 The 1-D discrete derivative operator and the corresponding transfer function: (a) 1-D impulse response $h[n]$. The samples of the discrete transfer function are shown as (b) real part, (c) imaginary part, (d) magnitude, and (e) phase, along with the corresponding continuous transfer function $H[\xi] = i2\pi\xi$, which is plotted out to the Nyquist frequency $|\xi| = 1/(2 \cdot \Delta x)$. Note that the real part of the transfer function is not the ideal null constant, that the imaginary part decays to zero at the Nyquist frequency, that the magnitude is attenuated near the Nyquist frequency, and that the deviates phase from the ideal signum function.

$$\partial_{(\theta=-3\pi/4)} = \begin{array}{|c|c|c|} \hline 0 & 0 & 0 \\ \hline 0 & -1 & 0 \\ \hline 0 & 0 & +1 \\ \hline \end{array} \tag{20.35b}$$

where the angle in radians measured from the x-axis has been substituted for the subscript, which means that $\partial_y = \partial_{(\theta=-\pi/2)}$. For diagonal derivatives the continuous distance between the elements must be scaled by $\sqrt{2}$, as mentioned in the discussion of Equation (20.13).

20.3.3 1-D Antisymmetric Differentiation Kernel

The odd part of the discrete differencing kernel in Equation (20.29b) suggests that we can construct a discrete differentiator with odd symmetry by placing the components of the discrete "doublet" at samples $n = \pm 1$:

$$(\partial_x)_2 = \boxed{+1} \boxed{0} \boxed{-1} \tag{20.36}$$

This impulse response is proportional to the odd part of the original 1-D differentiator. The corresponding transfer function is again easy to evaluate via the appropriate combination of translation operators. Because $(\partial_x)_2$ is odd, the discrete transfer function is imaginary (Figure 20.7):

$$H[k] = e^{+2\pi i(k/N)} - e^{+2\pi i(-k/N)} = 2i \sin\left[2\pi \frac{k}{N}\right] \tag{20.37a}$$

$$|H[k]| = 2\left|\sin\left[2\pi \frac{k}{N}\right]\right| \tag{20.37b}$$

$$\Phi\{H[k]\} = +\frac{\pi}{2}\left(SGN[k] - \delta_d\left[k + \frac{N}{2}\right]\right) \tag{20.37c}$$

Note that this transfer function evaluated at the Nyquist frequency is:

$$\left|H\left[k = -\frac{N}{2}\right]\right| = 2|\sin[-\pi]| = 0 \tag{20.38}$$

which means that this differentiator also "blocks" the Nyquist frequency. This may be seen by convolving $(\partial_x)_2$ with a sinusoid function that oscillates with a period of two samples. Adjacent positive extrema are multiplied by ± 1 in the kernel and thus cancel. Also note that the transfer function amplifies lower frequencies more and larger frequencies less than the continuous transfer function.

20.3.4 Second Derivative

20.3.4.1 1-D Second Derivative

The impulse response and transfer function of the continuous second derivative are easily obtained from the derivative theorem:

$$h[x] = \delta''[x] \tag{20.39a}$$

$$H[\xi] = (2\pi i \xi)^2 = -4\pi^2\xi^2 \tag{20.39b}$$

Again, different forms of the discrete second derivative may be defined. One form is obtained by differentiating the first derivative operator via discrete convolution of two replicas of ∂_x. The result is a 5-pixel kernel including two null weights:

$$\partial_x * \partial_x = \boxed{+1} \boxed{-1} \boxed{0} * \boxed{+1} \boxed{-1} \boxed{0} = \boxed{+1} \boxed{-2} \boxed{+1} \boxed{0} \boxed{0} \tag{20.40}$$

The corresponding discrete transfer function is obtained by substituting results from the translation operator:

$$\begin{aligned} H[k] &= e^{+2\pi i 2k/N} - 2\, e^{+2\pi i k/N} + 1[k] \\ &= e^{+2\pi i k/N}(e^{+2\pi i k/N} - 2 + e^{-2\pi i k/N}) \\ &= 2\, e^{+2\pi i k/N}\left(\cos\left[2\pi \frac{k}{N}\right] - 1\right) \end{aligned} \tag{20.41}$$

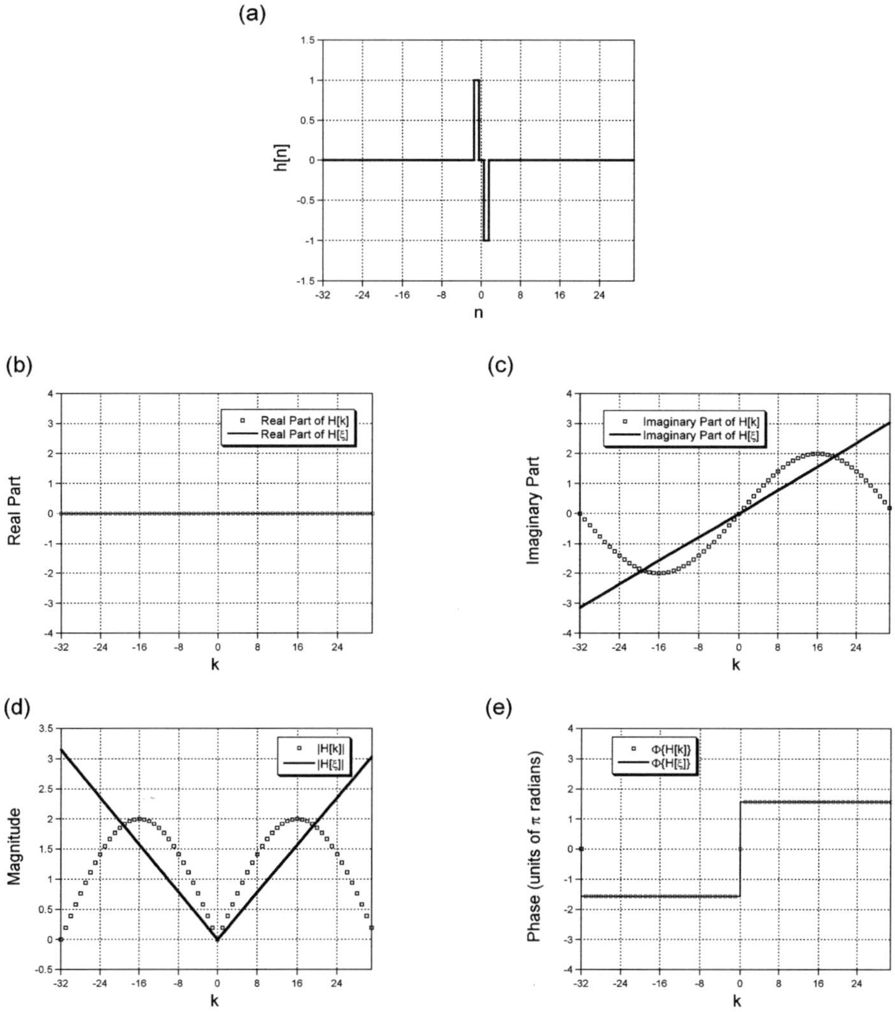

Figure 20.7 The 1-D discrete antisymmetric derivative operator and its transfer function: (a) 1-D impulse response $h[n]$. Because $h[n]$ is odd, $H[k]$ is imaginary and odd. The samples of the discrete transfer function are shown as (b) real part, (c) imaginary part, (d) magnitude, and (e) phase, along with the corresponding continuous transfer function $H[\xi] = i2\pi\xi$. Note that the magnitude of the discrete transfer function is attenuated near the Nyquist frequency and that its phase is identical to that of the transfer function of the continuous derivative.

The leading linear-phase factor usually is discarded to produce the real-valued and symmetric discrete transfer function:

$$H[k] = 2\left(\cos\left[2\pi\frac{k}{N}\right] - 1\right) \tag{20.42}$$

Deletion of the linear phase is the same as translation of the original discrete second derivative kernel by 1 pixel to the right. The discrete impulse response for this symmetric discrete kernel is also real valued

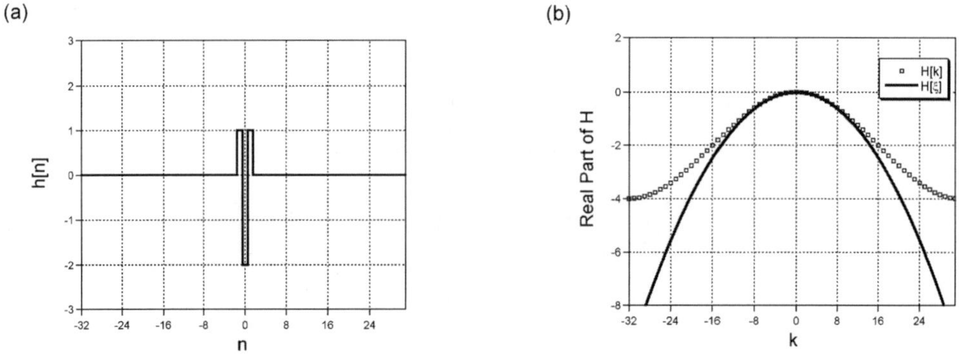

Figure 20.8 The 1-D discrete second derivative: (a) impulse response ∂_x^2; (b) comparison of discrete and continuous transfer functions.

and symmetric:

$$h[n] = \partial_x^2 \equiv \boxed{0 \mid +1 \mid -2 \mid +1 \mid 0} = \boxed{+1 \mid -2 \mid +1}$$

$$= \delta_d[n+1] - 2\delta_d[n] + \delta_d[n-1] \tag{20.43}$$

so that the magnitude and phase of the transfer function are:

$$|H[k]| = 2\left(1 - \cos\left[2\pi \frac{k}{N}\right]\right) \tag{20.44a}$$

$$\mathbf{\Phi}\{H[k]\} = \begin{cases} -\pi & \mathit{for}\ k \neq 0 \\ 0 & \mathit{for}\ k = 0 \end{cases} = \pi(-1 + \delta_d[k]) \tag{20.44b}$$

as shown in Figure 20.8. The amplitude of the discrete transfer function at the Nyquist frequency is:

$$H\left[k = -\frac{N}{2}\right] = 2 \cdot (\cos[-\pi] - 1) = -4 \tag{20.45}$$

while that of the continuous transfer function is $-4\pi^2(-\frac{1}{2})^2 = -\pi^2 \cong -9.87$, so the discrete second derivative does not amplify large spatial frequencies as much as the continuous second derivative. The transfer function is a discrete approximation of the parabola with values at the edges of the array that ensure smooth periodicity.

Higher-order discrete derivatives may be derived by repeated discrete convolution of ∂_x and discarding any common linear-phase factors.

20.3.5 2-D Second Derivative

The 2-D derivative in the x-direction may be defined by convolving the 1-D operator in Equation (20.31) with itself. Just as in the 1-D case, the discrete kernel is "off center", but the linear-phase factor in the transfer function may be deleted to produce a symmetric 3 × 3 kernel that may be denoted by ∂_x^2:

$\partial_x^2 \equiv$ (20.46a)

0	0	0
+1	−2	+1
0	0	0

The corresponding second derivatives in the y- and diagonal directions obtained by rotating this kernel:

$$\partial_y^2 \equiv \begin{array}{|c|c|c|}\hline 0 & +1 & 0 \\ \hline 0 & -2 & 0 \\ \hline 0 & +1 & 0 \\ \hline\end{array} \tag{20.46b}$$

$$\partial^2_{(+\pi/4)} = \begin{array}{|c|c|c|}\hline 0 & 0 & +1 \\ \hline 0 & -2 & 0 \\ \hline +1 & 0 & 0 \\ \hline\end{array} \tag{20.46c}$$

$$\partial^2_{(+3\pi/4)} = \partial^2_{(-\pi/4)} = \begin{array}{|c|c|c|}\hline +1 & 0 & 0 \\ \hline 0 & -2 & 0 \\ \hline 0 & 0 & +1 \\ \hline\end{array} \tag{20.46d}$$

where the analysis of the rotations to the diagonal directions are again complicated by the implicit scaling mentioned in the discussion of Equation (20.13).

20.3.6 Laplacian

The continuous Laplacian operator was introduced in Equation (10.31) as the sum of the orthogonal second derivatives:

$$\nabla^2 f[x, y] = \left(\frac{\partial^2}{\partial x^2} + \frac{\partial^2}{\partial y^2}\right) f[x, y] \tag{20.47}$$

and its associated transfer function is the negative quadratic that evaluates to zero at DC, which again demonstrates that constant terms are blocked by differentiation:

$$H[\xi, \eta] = -4\pi^2(\xi^2 + \eta^2) \tag{20.48}$$

The discrete Laplacian operator is the sum of the 2-D kernels in Equations (20.46a) and (20.46b):

$$\partial_x^2 + \partial_y^2 \equiv \nabla_d^2 = \begin{array}{|c|c|c|}\hline 0 & 0 & 0 \\ \hline +1 & -2 & +1 \\ \hline 0 & 0 & 0 \\ \hline\end{array} + \begin{array}{|c|c|c|}\hline 0 & +1 & 0 \\ \hline 0 & -2 & 0 \\ \hline 0 & +1 & 0 \\ \hline\end{array} = \begin{array}{|c|c|c|}\hline 0 & +1 & 0 \\ \hline +1 & -4 & +1 \\ \hline 0 & +1 & 0 \\ \hline\end{array} \tag{20.49}$$

The discrete transfer function of this "standard" discrete Laplacian kernel is:

$$\begin{aligned} H[k, \ell] &= 2\left(\cos\left[2\pi\frac{k}{N}\right] - 1\right) + 2\left(\cos\left[2\pi\frac{\ell}{N}\right] - 1\right) \\ &= 2\left(\cos\left[2\pi\frac{k}{N}\right] + \cos\left[2\pi\frac{\ell}{N}\right] - 2\right) \end{aligned} \tag{20.50}$$

The amplitude at the origin is $H[k = 0, \ell = 0] = 0$ and decays in the horizontal or vertical directions to -6 at the "edge" of the array and to -8 at the corners.

20.3.6.1 Rotated Laplacian

The second derivatives along the diagonals in Equations (20.46c) and (20.46d) may be added to create a version of the Laplacian that is "nearly" equivalent to rotating the operator of Equation (20.49) by

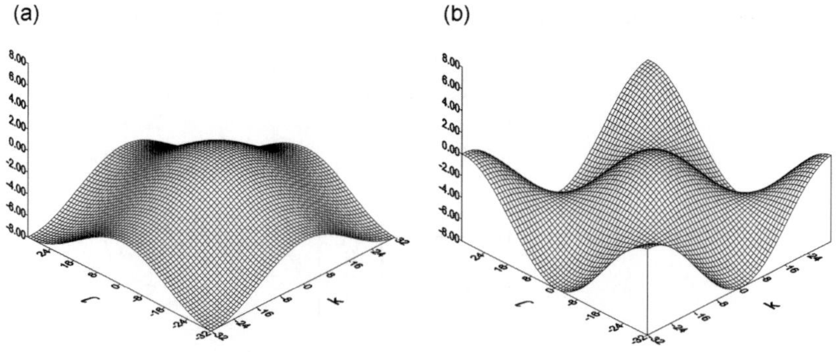

0	+1	0
+1	−4	+1
0	+1	0

+1	0	+1
0	−4	0
+1	+1	+1

Figure 20.9 The real and symmetric transfer functions of Laplacian operators: (a) normal Laplacian from Equation (20.50); (b) rotated Laplacian from Equation (20.51), showing that the amplitude rises back to zero at the corners of the array.

$\theta = +\pi/4$ radians, except for the factor of $\sqrt{2}$ in the distance:

$$(\partial^2_{(+\pi/4)} + \partial^2_{(+3\pi/4)}) = \begin{array}{|c|c|c|}\hline 0 & 0 & +1 \\ \hline 0 & -2 & 0 \\ \hline +1 & 0 & 0 \\ \hline \end{array} + \begin{array}{|c|c|c|}\hline +1 & 0 & 0 \\ \hline 0 & -2 & 0 \\ \hline 0 & 0 & +1 \\ \hline \end{array}$$

$$= \begin{array}{|c|c|c|}\hline +1 & 0 & +1 \\ \hline 0 & -4 & 0 \\ \hline +1 & 0 & +1 \\ \hline \end{array} \tag{20.51}$$

Derivation of the transfer function is left to readers as one of the problems. The magnitude of the transfer function is zero at the origin and its maximum negative values are located at the horizontal and vertical edges, but the transfer function is zero at the corners (Figure 20.9). In words, the rotated Laplacian "blocks" the largest frequency components along the diagonal. A 2-D example is shown in Figure 20.10, where the input function is nonnegative and the bipolar output $g[n, m]$ is displayed as amplitude and as magnitude $|g[n, m]|$, which shows that the response is largest at the edges and corners.

20.3.6.2 Isotropic Laplacian

A common generalization of the Laplacian is obtained by adding the original and rotated Laplacian kernels:

$$(\partial_x^2 + \partial_y^2) + (\partial^2_{(+\pi/4)} + \partial^2_{(+3\pi/4)}) = \begin{array}{|c|c|c|}\hline 0 & +1 & 0 \\ \hline +1 & -4 & +1 \\ \hline 0 & +1 & 0 \\ \hline \end{array} + \begin{array}{|c|c|c|}\hline +1 & 0 & +1 \\ \hline 0 & -4 & 0 \\ \hline +1 & 0 & +1 \\ \hline \end{array}$$

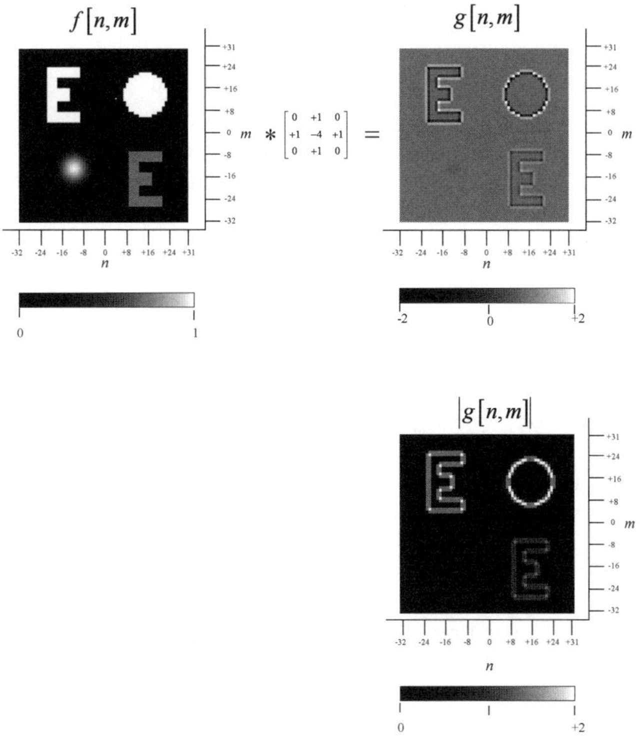

Figure 20.10 Action of the 2-D discrete Laplacian. The input amplitudes are nonnegative in the interval $0 \leq f \leq 1$, and the output amplitude is bipolar in the interval $-2 \leq g \leq +2$ in this example. As shown in the magnitude image, the extrema of output amplitude occur at the edges and corners of the input.

$$= \begin{array}{|c|c|c|} \hline +1 & +1 & +1 \\ \hline +1 & -8 & +1 \\ \hline +1 & +1 & +1 \\ \hline \end{array} \tag{20.52}$$

The linearity of the DFT ensures that the transfer function of the isotropic Laplacian is the real-valued and symmetric sum of the "normal" and rotated Laplacians.

20.3.6.3 Generalized Laplacian

The isotropic Laplacian may be written as the difference of a 3 × 3 average and a scaled discrete delta function:

$$\begin{array}{|c|c|c|} \hline +1 & +1 & +1 \\ \hline +1 & -8 & +1 \\ \hline +1 & +1 & +1 \\ \hline \end{array} = \begin{array}{|c|c|c|} \hline +1 & +1 & +1 \\ \hline +1 & +1 & +1 \\ \hline +1 & +1 & +1 \\ \hline \end{array} - 9 \cdot \begin{array}{|c|c|c|} \hline 0 & 0 & 0 \\ \hline 0 & +1 & 0 \\ \hline 0 & 0 & 0 \\ \hline \end{array} \tag{20.53}$$

which suggests that the Laplacian operator may be generalized to include all operators that compute differences between a weighted replica of the original image and a copy blurred by some averaging kernel. One example of a generalized Laplacian may be constructed from the 2-D circularly symmetric continuous Gaussian impulse response:

$$h[x, y] = A \exp\left[-\pi\left(\frac{x^2+y^2}{\alpha^2}\right)\right] \tag{20.54}$$

where the decay parameter α determines the rate of attenuation of kernel values with radial distance from the origin. The amplitude parameter A is often selected to normalize the sum of the elements of the kernel to unity, thus ensuring that the process computes a weighted average. The amplitudes are often set to zero outside of some window. For example, a normalized discrete approximation of the Gaussian kernel with $\alpha = 2 \cdot \Delta x$ is:

$$h_1[n, m] = \frac{1}{2047}\begin{array}{|c|c|c|c|c|}\hline 1 & 11 & 23 & 11 & 1 \\ \hline 11 & 111 & 244 & 111 & 11 \\ \hline 23 & 244 & 535 & 244 & 23 \\ \hline 11 & 111 & 244 & 111 & 11 \\ \hline 1 & 11 & 23 & 11 & 1 \\ \hline\end{array} \cong \frac{1}{21}\begin{array}{|c|c|c|c|c|}\hline 0 & 0 & 1 & 0 & 0 \\ \hline 0 & 1 & 2 & 1 & 0 \\ \hline 1 & 2 & 5 & 2 & 1 \\ \hline 0 & 1 & 2 & 1 & 0 \\ \hline 0 & 0 & 1 & 0 & 0 \\ \hline\end{array} \tag{20.55}$$

A corresponding generalized Laplacian operator is constructed by subtracting the 5×5 identity kernel from this discrete Gaussian:

$$h_2[n, m] = \frac{1}{21}\begin{array}{|c|c|c|c|c|}\hline 0 & 0 & +1 & 0 & 0 \\ \hline 0 & +1 & +2 & +1 & 0 \\ \hline +1 & +2 & -16 & +2 & +1 \\ \hline 0 & +1 & +2 & +1 & 0 \\ \hline 0 & 0 & +1 & 0 & 0 \\ \hline\end{array} \tag{20.56}$$

Note that the sum of the elements of this generalized Laplacian kernel is zero because of the normalization of the Gaussian kernel, which means the output will be a null image if the input is a uniform gray field.

20.4 Discrete Sharpening Operators

20.4.1 1-D Sharpeners

As mentioned in Chapter 16, the transfer function of a "sharpener" passes all sinusoidal components without change in phase while amplifying those with large spatial frequencies. This action will tend to correct the spectra of images that have been lowpass filtered. One example of a 1-D continuous sharpener constructed from the second derivative has transfer function:

$$H[\xi] = 1 + 4\pi^2\xi^2 \tag{20.57}$$

The corresponding continuous 1-D impulse response is:

$$h[x] = \delta[x] - \delta''[x] \tag{20.58}$$

A discrete version of the impulse response may be generated by substituting the discrete Dirac delta function and the "centered" discrete second derivative operator:

$$\begin{aligned} h[n] = \delta_d - \partial_x^2 &= \begin{array}{|c|c|c|}\hline 0 & +1 & 0 \\ \hline\end{array} - \begin{array}{|c|c|c|}\hline +1 & -2 & +1 \\ \hline\end{array} \\ &= \begin{array}{|c|c|c|}\hline -1 & +3 & -1 \\ \hline\end{array} = \begin{array}{|c|c|c|}\hline 0 & +4 & 0 \\ \hline\end{array} - \begin{array}{|c|c|c|}\hline +1 & +1 & +1 \\ \hline\end{array} \\ &= 4 \cdot \delta_d[n] - RECT\left[\frac{n}{3}\right] \end{aligned} \tag{20.59}$$

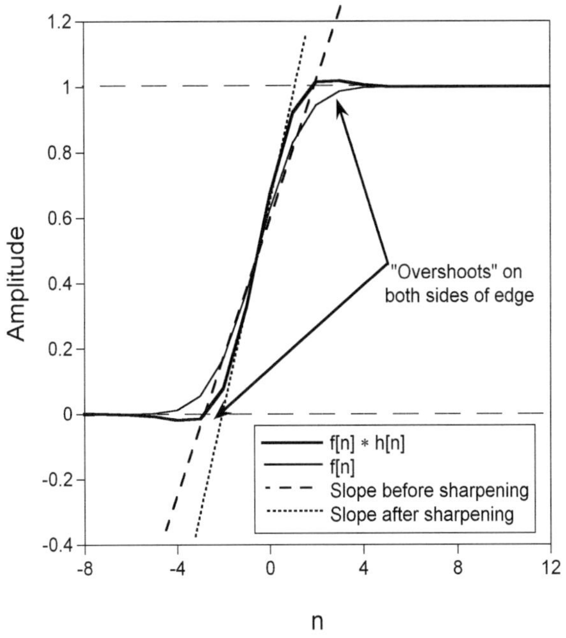

Figure 20.11 Action of 1-D second-derivative sharpening operator on a blurred edge. The angle of the slope of the sharpened edge is "steeper", but the amplitude overshoots the correct value on both sides of the edge. This operator only approximates the ideal inverse filter, so the output is not the ideal sharp edge.

The derivation demonstrates that the action of the 1-D sharpener may be interpreted as computing the difference between a scaled replica of the "blurred" input image and a 3-pixel uniformly weighted sum of gray values of the neighboring samples, which is of course a lowpass filter. This interpretation may be generalized to construct sharpeners based on other lowpass filters.

The transfer function of the discrete 1-D sharpener is:

$$H[k] = 4 \cdot 1[k] - \left(1 + 2\cos\left[2\pi\frac{k}{N}\right]\right)$$
$$= 3 - 2\cos\left[2\pi\frac{k}{N}\right] \tag{20.60}$$

The amplitudes of the transfer function at DC and at the Nyquist frequency are:

$$H[k=0] = +1 \tag{20.61a}$$

$$H\left[k = -\frac{N}{2}\right] = +5 \tag{20.61b}$$

In words, the "second-derivative sharpener" amplifies the amplitude of the sinusoidal component that oscillates at the Nyquist frequency by a factor of 5.

The action of this sharpening operator on a "blurry" edge is shown in Figure 20.11. The slope of the edge is "steeper" after sharpening, but the edge also "overshoots" the correct amplitude at both sides. In other words, an image sharpener is not an ideal inverse filter.

20.4.2 2-D Sharpening Operators

A sharpening operator based on the Laplacian was introduced in Chapter 16. The operator implemented sharpening of images based on the heat-diffusion model of image blur. The discrete version of this operator is often used to sharpen "blurry" digital images. The process is ad hoc in the sense that it is not "tuned" to the details of the blurring process and so is not an "inverse" filter. It does not reconstruct the original sharp image in general, but it "steepens" the slope of pixel-to-pixel changes in gray level, thus making the edges appear "sharper". In Equation (16.46), the impulse response of the Laplacian sharpener was shown to be:

$$f[x, y, 0] \cong f[x, y, z] - \alpha \cdot \nabla^2 f[x, y, z] \tag{20.62}$$

where α is a real-valued free parameter that allows the sharpener to be "tuned" to the amount of blur. Obviously, the corresponding discrete solution is:

$$g[n, m] = f[n, m] - \alpha \cdot \nabla_d^2 f[n, m] = (\delta_d[n, m] - \alpha \cdot \nabla_d^2) * f[n, m] \tag{20.63}$$

where $\nabla_d^2[n, m]$ is one of the variants of the Laplacian kernel already considered. A single discrete sharpening kernel $h[n, m]$ may be constructed from the simplest form for the Laplacian in Equation (20.49):

$$h_1[n, m; \alpha] = \begin{array}{|c|c|c|}\hline 0 & 0 & 0 \\ \hline 0 & +1 & 0 \\ \hline 0 & 0 & 0 \\ \hline\end{array} - \alpha \cdot \begin{array}{|c|c|c|}\hline 0 & +1 & 0 \\ \hline +1 & -4 & +1 \\ \hline 0 & +1 & 0 \\ \hline\end{array} = \begin{array}{|c|c|c|}\hline 0 & -\alpha & 0 \\ \hline -\alpha & 1-4\alpha & -\alpha \\ \hline 0 & -\alpha & 0 \\ \hline\end{array} \tag{20.64}$$

The parameter α may be increased to enhance the sharpening. Selection of $\alpha = +1$ produces a sharpening kernel:

$$h_1[n, m; \alpha = +1] = \begin{array}{|c|c|c|}\hline 0 & -1 & 0 \\ \hline -1 & +5 & -1 \\ \hline 0 & -1 & 0 \\ \hline\end{array} \tag{20.65}$$

The weights in the kernel sum to unity, which means that the average gray value of the image is preserved by the sharpening operation. This process amplifies differences in gray level of adjacent pixels while preserving the mean gray value.

The corresponding discrete transfer function for the parameter α is:

$$H_1[k, \ell; \alpha] = (1 + 4\alpha) - 2\alpha\left(\cos\left[2\pi \frac{k}{N}\right] + \cos\left[2\pi \frac{\ell}{N}\right]\right) \tag{20.66}$$

In the case $\alpha = +1$, the resulting transfer function is:

$$H_1[k, \ell; \alpha = 1] = 5 - 2\left(\cos\left[2\pi \frac{k}{N}\right] + \cos\left[2\pi \frac{\ell}{N}\right]\right) \tag{20.67}$$

which has its maximum amplitude of $(H_1)_{\max} = 9$ at the corners of the array.

A sharpening operator also may be derived from the isotropic Laplacian:

$$h_2[n, m; \alpha] = \begin{array}{|c|c|c|}\hline -\alpha & -\alpha & -\alpha \\ \hline -\alpha & 1+8\alpha & -\alpha \\ \hline -\alpha & -\alpha & -\alpha \\ \hline\end{array} \tag{20.68}$$

The sum of the elements in the kernel is unity, which ensures that the average value of the image is preserved by the action of the sharpener. If the weighting factor is again selected to be unity, the

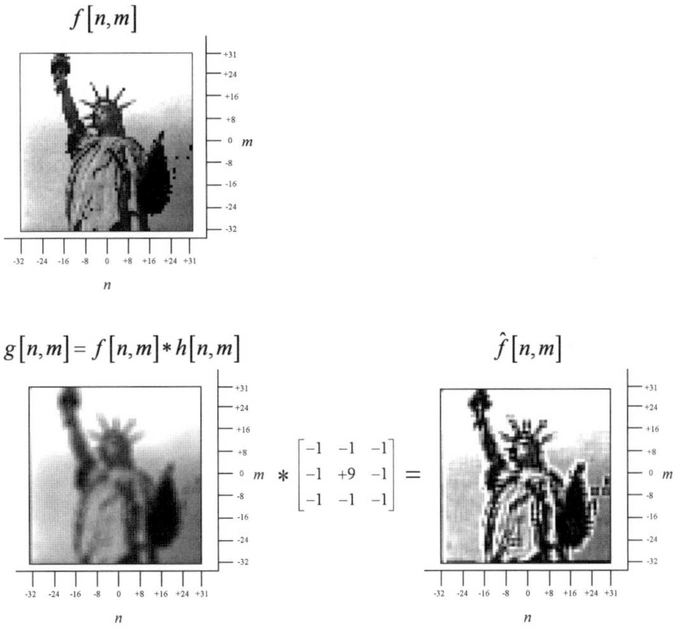

Figure 20.12 Action of the 2-D sharpening operator based on the Laplacian. The original image $f[n, m]$ has been blurred by a 3×3 uniform averager to produce $g[n, m]$. The action of the 3×3 Laplacian sharpener on $g[n, m]$ produced the bipolar image $\hat{f}[n, m]$, which was clipped at the original dynamic range. The overshoots at the edges give the impression of a sharper image.

kernel is:

$$h_2[n, m; 1] = \begin{array}{|c|c|c|} \hline -1 & -1 & -1 \\ \hline -1 & +9 & -1 \\ \hline -1 & -1 & -1 \\ \hline \end{array} \tag{20.69}$$

This type of process has been called *unsharp masking* by photographers. By printing the sandwich of a transparency of the original image and a blurred negative, the result is a sharpened image of the original. This difference of the blurred image and the original is easily implemented in a digital system as a single convolution.

An example of 2-D sharpening is shown in Figure 20.12.

20.5 2-D Gradient

The *gradient* of a 2-D continuous function $f[x, y]$ constructs a 2-D vector at each coordinate whose components are the x- and y-derivatives:

$$\underline{\mathbf{g}}[x, y] = \nabla f[x, y] = \left[\frac{\partial f}{\partial x}, \frac{\partial f}{\partial y}\right] \tag{20.70}$$

This is a discrete version of a common operation in physics (particularly in electromagnetism). The image $f[n, m]$ is a *scalar* function which assigns a numerical gray value f to each coordinate $[n, m]$. The gray value f is analogous to terrain "elevation" in a map. In physics, the gradient of a scalar "field" $f[x, y]$ is the product of a vector operator ∇ (pronounced *del*) and the scalar "image" f, yielding

$\nabla f[x, y]$. This process calculates a *vector* for each coordinate $[x, y]$ whose Cartesian components are $\partial f/\partial x$ and $\partial f/\partial y$. Note that the 2-D vector ∇f may be represented in polar form as magnitude $|\nabla f|$ and direction $\mathbf{\Phi}\{\nabla f\}$:

$$|\nabla f[x, y]| = \sqrt{\left(\frac{\partial f}{\partial x}\right)^2 + \left(\frac{\partial f}{\partial y}\right)^2} \tag{20.71a}$$

$$\mathbf{\Phi}\{\nabla f[n, m]\} = \tan^{-1}\left[\frac{(\partial f/\partial y)}{(\partial f/\partial x)}\right] \tag{20.71b}$$

The vector points "uphill" in the direction of the maximum "slope" in gray level:

$$\underline{\mathbf{g}}[n, m] = \nabla f[n, m] = \begin{bmatrix} (\partial_x * f[n, m]) \\ (\partial_y * f[n, m]) \end{bmatrix} \tag{20.72}$$

In image processing, the magnitude of the gradient is often approximated as the sum of the magnitudes of the components:

$$\begin{aligned} |\underline{\mathbf{g}}[n, m]| = |\nabla f[n, m]| &= \sqrt{(\partial_x * f[n, m])^2 + (\partial_y * f[n, m])^2} \\ &\cong |\partial_x * f[n, m]| + |\partial_y * f[n, m]| \end{aligned} \tag{20.73}$$

The magnitude $|\nabla f|$ is the "slope" of the 3-D surface f at pixel $[n, m]$. The azimuth $\mathbf{\Phi}\{\nabla f[n, m]\}$ defines the compass direction where this slope points "uphill". The gradient is not a linear operator, and thus can neither be evaluated as a convolution nor described by a transfer function. The largest values of the magnitude of the gradient correspond to the pixels where the gray value "jumps" by the largest amount, and thus the thresholded magnitude of the gradient may be used to identify such pixels. In this way the gradient may be used as an "edge detection operator". An example of the gradient operator is shown in Figure 20.13.

20.6 Pattern Matching

The principles of the matched filter in Chapter 19 may be applied to design kernels for locating specific gray-level patterns, such as edges at particular orientations, corners, isolated pixels, particular shapes – you name it. Particularly in the early days of digital image processing when computers were less capable than they are today, the computational intensity of the calculation often was an important issue. It was desirable to find the least intensive method for common tasks such as pattern detection, which generally meant that the task was performed in the space domain using a small convolution kernel rather than calculating a better approximation to the ideal result in the frequency domain. That said, the process of designing and applying a pattern-matching kernel illuminates some of the concepts and thus is worth some time and effort.

A common technique for pattern matching is based on the "unamplified" matched filter in Equation (19.105), where the input image is convolved with a kernel of the same size as the reference pattern. The process and its limitations will be illustrated by example. Consider an input image $f[n, m]$ that is composed of two replicas of a real-valued nonnegative pattern $p[n, m]$ centered at coordinates $[n_1, m_1]$ and $[n_2, m_2]$ with respective amplitudes A_1 and A_2. The image also includes a bias $b \cdot 1[n, m]$:

$$f[n, m] = A_1 \cdot p[n - n_1, m - m_1] + A_2 \cdot p[n - n_2, m - m_2] + b \cdot 1[n, m] \tag{20.74}$$

The continuous expression for the unamplified or realistic matched filter in Equation (19.105) suggests that the appropriate kernel of the discrete filter is:

$$\hat{m}[n, m] = p[-n, -m] \tag{20.75}$$

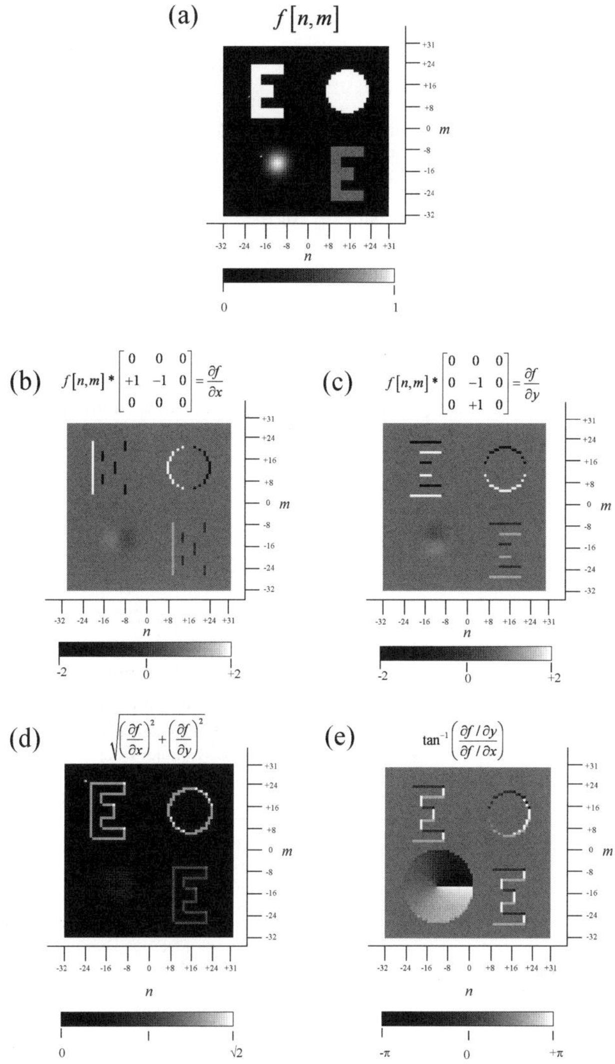

Figure 20.13 Example of the discrete gradient operator $\nabla f[n, m]$. The original object is the nonnegative function $f[n, m]$ shown in (a), which has amplitude in the interval $0 \leq f \leq +1$. The gradient at each pixel is the 2-D vector with components bipolar $[\partial f/\partial x, \partial f/\partial y]$. The two component images are shown in (b) and (c). These also may be displayed as magnitude $\sqrt{(\partial f/\partial x)^2 + (\partial f/\partial y)^2}$ in (c) and the angle $\phi = \tan^{-1}[(\partial f/\partial y)/(\partial f/\partial x)]$ in (d). The extrema of the magnitude are located at corners and edges in $f[n, m]$.

which also is real valued and nonnegative within its region of support. The output from this matched-filter autocorrelation of the pattern centered at those coordinates is:

$$\begin{aligned} g[n, m] &= f[n, m] * \hat{m}[n, m] \\ &= A_1 \cdot p[n, m] \bigstar p[n, m]|_{n=n_1, m=m_1} + A_2 \cdot p[n, m] \bigstar p[n, m]|_{n=n_2, m=m_2} \\ &\quad + b \cdot (1[n, m] * p[-n, -m]) \end{aligned}$$

$$= A_1 \cdot p[n,m] \bigstar p[n,m]|_{n=n_1,m=m_1} + A_2 \cdot p[n,m] \bigstar p[n,m]|_{n=n_2,m=m_2} + b \cdot \sum_{n,m} p[n,m] \tag{20.76}$$

The last term is the spatially invariant output due to the constant bias convolved with the matched filter, which produces the sum of the product of the bias and the weights at each sample. The spatially varying autocorrelation functions rest on a bias proportional to the sum of the gray values p in the pattern. If the output bias is large, it can reduce the "visibility" of the autocorrelations in exactly the same way as the modulation of a nonnegative sinusoidal function in Equation (6.25). Therefore it is convenient to construct a matched-filter kernel whose weights sum to zero by subtracting the average value:

$$\hat{m}[n,m] = p[-n,-m] - p_{\text{average}} \Longrightarrow \sum_{n,m} \hat{m}[-n,-m] = \sum_{n,m} \hat{m}[n,m] = 0 \tag{20.77}$$

This condition ensures that the constant bias in the third term in Equation (20.76) vanishes. This result determines the strategy for designing convolution kernels that produce outputs that have large magnitudes at pixels centered on neighborhoods that contain these patterns and small magnitudes in neighborhoods where the feature does not exist. For example, consider an image containing an "upper-right corner" of a brighter object on a darker background:

$$f[n,m] = \begin{array}{|c|c|c|c|c|c|c|} \hline \ddots & \vdots & \vdots & \vdots & \vdots & \vdots & \cdots \\ \hline \cdots & 50 & 50 & 50 & 50 & 50 & \cdots \\ \hline \cdots & 50 & 50 & 50 & 50 & 50 & \cdots \\ \hline \cdots & 100 & 100 & 100 & 50 & 50 & \cdots \\ \hline \cdots & 100 & 100 & 100 & 50 & 50 & \cdots \\ \hline \cdots & 100 & 100 & 100 & 50 & 50 & \cdots \\ \hline \vdots & \vdots & \vdots & \vdots & \vdots & \vdots & \ddots \\ \hline \end{array} \tag{20.78}$$

The task is to design a 3×3 kernel for locating this pattern in a scene:

$$p[n,m] = \begin{array}{|c|c|c|} \hline 50 & 50 & 50 \\ \hline 100 & 100 & 50 \\ \hline 100 & 100 & 50 \\ \hline \end{array} \tag{20.79}$$

The recipe tells us to rotate the pattern by π radians about its center to create $p[-n,-m]$:

$$p[-n,-m] = \begin{array}{|c|c|c|} \hline 50 & 100 & 100 \\ \hline 50 & 100 & 100 \\ \hline 50 & 50 & 50 \\ \hline \end{array} \tag{20.80}$$

The average weight in this 3×3 kernel is $650/9 \cong 72.222$, which is subtracted from each element:

$$\begin{array}{|c|c|c|} \hline -22.222 & +27.778 & +27.778 \\ \hline -22.222 & +27.778 & +27.778 \\ \hline -22.222 & -22.222 & -22.222 \\ \hline \end{array} = (+22.222) \begin{array}{|c|c|c|} \hline -1 & +1.25 & +1.25 \\ \hline -1 & +1.25 & +1.25 \\ \hline -1 & -1 & -1 \\ \hline \end{array} \tag{20.81}$$

The multiplicative factor may be ignored since it just scales the output of the convolution by this constant. Thus one realization of the unamplified 3×3 matched filter for upper-right corners is:

$$\hat{m}[n,m] \cong \begin{array}{|c|c|c|} \hline -1 & +1.25 & +1.25 \\ \hline -1 & +1.25 & +1.25 \\ \hline -1 & -1 & -1 \\ \hline \end{array} \tag{20.82}$$

Though not really an issue now with faster computers, it was once considered more convenient to restrict the weights in the kernel to integer values. This may be done by redistributing the weights slightly. In this example, the fraction of the positive weights is often concentrated in the center pixel to produce the *Prewitt corner detector*:

$$\hat{m}[n,m] \cong \begin{array}{|c|c|c|}\hline -1 & +1 & +1 \\ \hline -1 & +2 & +1 \\ \hline -1 & -1 & -1 \\ \hline \end{array} \tag{20.83}$$

Note that the upper-right corner detector contains a bipolar pattern that looks like a lower-left corner because of the rotation ("reversal") inherent in the convolution. Because $\hat{m}$ is bipolar, so generally is the output of the convolution with the input $f[n,m]$. The linearity of convolution ensures that the output amplitude at a pixel is proportional to the contrast of the feature. If the contrast of the upper-right corner is large and "positive", meaning that the corner is much brighter than the dark background, the output at the corner pixel will be a large and positive extremum. Conversely, a dark object on a very bright background will produce a large negative extremum. The magnitude of the image shows the locations of features with either contrast. The output image may be thresholded to specify the pixels located at the desired feature.

This method of feature detection is not ideal. The output of this unamplified filter at a corner is the autocorrelation of the feature rather than the ideal 2-D discrete Dirac delta function. If multiple copies of the pattern with different contrasts are present in the input, it will be difficult or impossible to segment the desired features by thresholding the convolution alone. Another consequence of the unamplified matched filter is that features other than the desired pattern produce nonnull outputs, as shown in the output of the corner detector applied to a test object consisting of "E" at two different amplitudes as shown in Figure 20.14a. The threshold properly locates the upper-right corners of the bright "E" and one point on the sampled circle, but misses the corners of the fainter "E". This shows that corners of some objects are missed (false negatives). If the threshold were set at a lower level to detect the corner of the fainter "E", other pixels would be incorrectly identified as corners (false positives). A simple method for reducing misidentified pixels is considered in the next section.

20.6.1 Normalization of Contrast of Detected Features

The recipe just developed allows the creation of kernels for detecting pixels in neighborhoods that are "similar" to some desired pattern. However, the sensitivity of the process to feature contrast can significantly limit its utility. A simple modification can improve the classification. A normalized correlation measure $R[n,m]$ was defined by Hall (1979):

$$R[n,m] = \frac{f[n,m] * h[n,m]}{\sqrt{\sum_{n,m}(f[n,m])^2}\sqrt{\sum_{n,m}(h[n,m])^2}} \tag{20.84}$$

where the sums in the denominator are over *only* the values of the input and kernel within the support of the latter. Note that the sum of the squares of the elements of the kernel $h[n,m]$ results in a constant scale factor k and may be ignored:

$$R[n,m] = k\left(\frac{f[n,m]}{\sqrt{\sum_{n,m}(f[n,m])^2}}\right) * h[n,m] \tag{20.85}$$

In words, this operation divides the convolution by the geometric sum of gray levels under the kernel. The modification to the filter makes the entire process shift variant and thus may not be performed by a simple convolution. The denominator may be computed by convolving $(f[n,m])^2$ with a uniform averaging kernel $s[n,m]$ of the same size as the original kernel $h[n,m]$ and then evaluating the square root:

$$R[n,m] = k\frac{f[n,m] * h[n,m]}{\sqrt{(f[n,m])^2 * s[n,m]}} \tag{20.86}$$

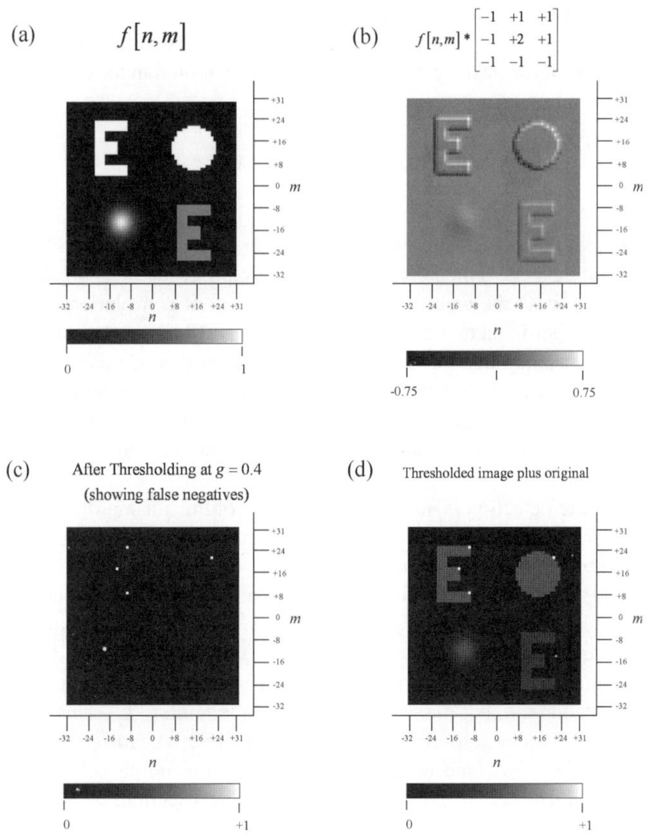

Figure 20.14 Thresholding to locate features in the image: (a) $f[n, m]$, which is the nonnegative function with $0 \leq f \leq 1$; (b) $f[n, m]$ convolved with the "upper-right corner detector", producing the bipolar output $g[n, m]$ where $-5 \leq g \leq 4$. The largest amplitudes occur at the upper-right corners, as shown in (c), output after thresholding at level 4; (d) thresholded output combined with "ghost" replica of original image. This demonstrates detection of the upper-right corners of the high-contrast "E" and circle, but the corners of the low-contrast "E" are missed by the thresholding.

The upper-right corner detector with normalization is shown in Figure 20.15, where the features of both "E"s are located with a single threshold.

20.6.2 Amplified Discrete Matched Filters

The unamplified pattern-matching kernel just derived has a small region of support (3×3) and thus may be implemented as fast space-domain operations. This property was desirable in the early days of digital image processing on slow computers. However, the resulting crosscorrelation output is only an approximation of the ideal discrete delta function at the locations of the features. We know how to modify the process to amplify the appropriate frequency components and thus produce closer approximations to the ideal output, but the resulting kernels would no longer be small. The reason is easy to understand because the amplification (or addition) of frequency components in the transfer function of the matched filter in Equation (19.101) implies the existence of additional full-support sinusoids in the formerly small-support kernel. Thus the amplification generally results in a "large" ($N \times N$) kernel.

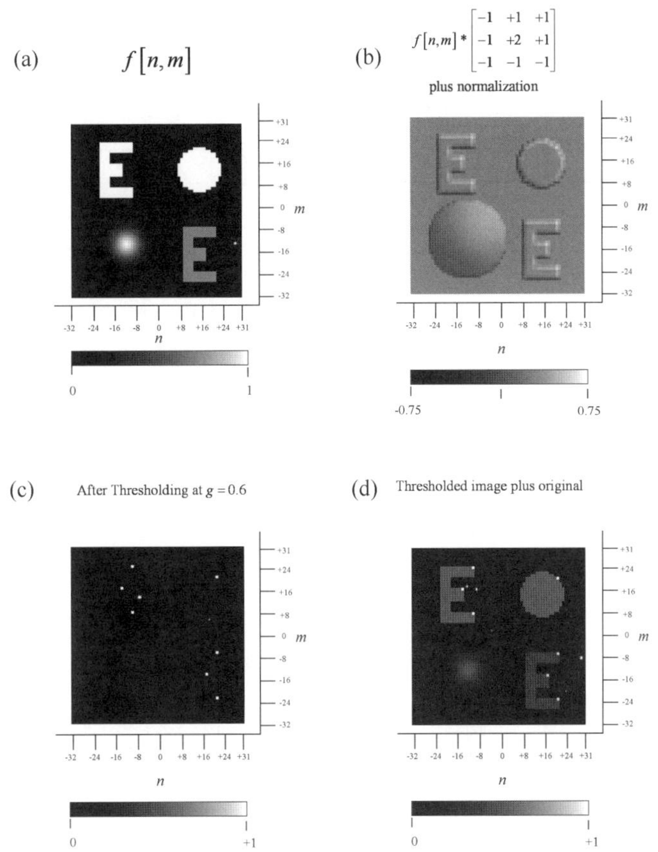

Figure 20.15 The action of the nonlinear normalization of detected features using the same object: (a) $f[n, m]$; (b) after convolution with upper-right corner detector using the normalization in Equation (20.86) to produce the bipolar output $g[n, m]$ where $-0.60 \leq g \leq +0.75$; (c) image after thresholding at $+0.6$; (d) thresholded image plus "ghost" of original image, showing detection of the upper-right corners of both "E"s despite their different contrasts.

Fortunately computational speed is happily less of an issue than formerly and it is very feasible in many (if not most) applications to operate in the discrete frequency domain. In situations where noise is a factor, the approximations to the transfer function obtained from the "complement" expansion in Equation (19.147) may be useful.

20.7 Approximate Discrete Reciprocal Filters

20.7.1 Derivative

As an example of the discrete "reciprocal" filter (either the inverse or ideal matched filter), consider the discrete reciprocal filter for the derivative, which was derived in Equation (20.27):

$$\frac{1}{\Delta x}(f[(n+1) \cdot \Delta x] - f[n \cdot \Delta x]) = \frac{1}{\Delta x} f[n \cdot \Delta x] * (\delta[n+1] - \delta[n]) \qquad (20.87)$$

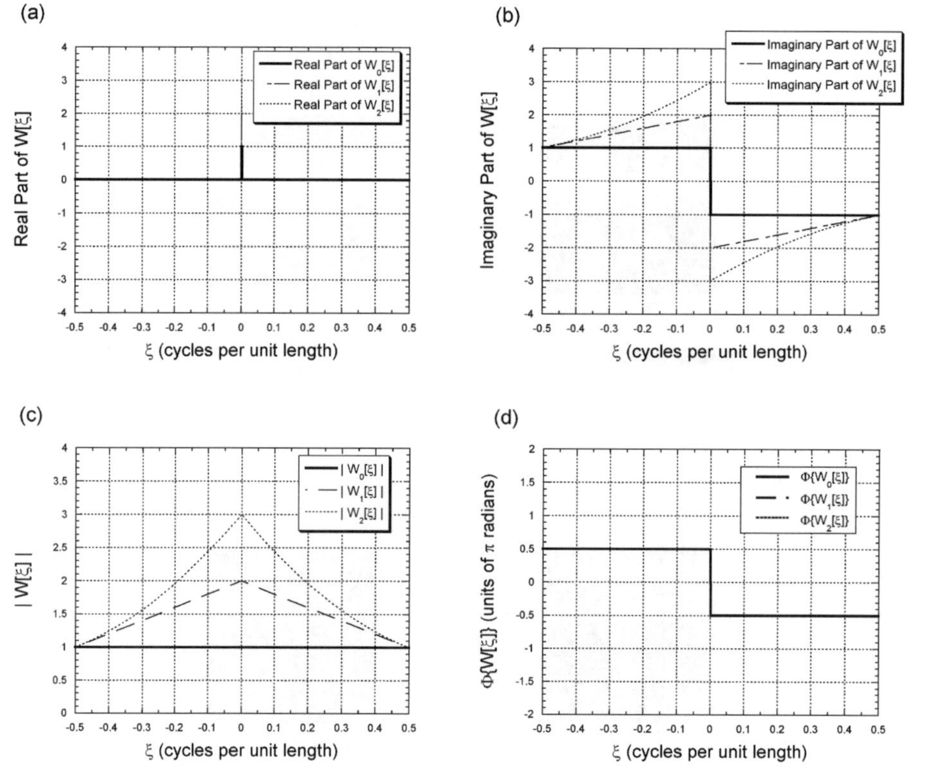

Figure 20.16 Approximate transfer functions for the inverse filter of discrete differentiation evaluated for the first three orders: (a) real part; (b) imaginary part; (c) magnitude; (d) phase. Note the increase in maximum magnitude with order and that the phase transfer functions are identical.

The corresponding discrete transfer function of differentiation is:

$$H_\partial[k] = -1 + e^{+2\pi ik/N} \tag{20.88a}$$

$$|H_\partial[k]| = \sqrt{2 - 2\cos\left[2\pi\frac{k}{N}\right]} = 2\cdot\left|\sin\left[2\pi\frac{k}{2N}\right]\right| \tag{20.88b}$$

$$\mathbf{\Phi}\{H_\partial[k]\} = \tan^{-1}\left[\frac{\sin[2\pi k/N]}{\cos[2\pi k/N]-1}\right] \tag{20.88c}$$

For the discrete derivative, the maximum amplitude occurs at the maximum frequency within the filter:

$$|H_\partial[k]|_{\max} = 2\cdot\left|\sin\left[\frac{\pi}{2}\right]\right| = 2 \tag{20.89}$$

The approximation to the transfer function is obtained by direct substitution into Equation (19.136):

$$\hat{W}_N[k] = \frac{1}{2}\exp\left[-i\cdot\tan^{-1}\left[\frac{\sin[2\pi k/N]}{\cos[2\pi k/N]-1}\right]\right]\cdot\sum_{n=0}^{N}\left(1-\left|\sin\left[2\pi\frac{k}{2N}\right]\right|\right)^n \tag{20.90}$$

The transfer function of the zero-order approximation is the leading term:

$$\hat{W}_0[k] = \frac{1}{2}\exp\left[-i \cdot \tan^{-1}\left[\frac{\sin[2\pi k/N]}{\cos[2\pi k/N] - 1}\right]\right] \tag{20.91}$$

which is shown in Figure 20.16. Our understanding of differentiation considered in Chapter 18 suggests that the corresponding pseudoinverse operation is semi-infinite integration, which can be implemented by convolution with a *STEP* function impulse response:

$$\int_{-\infty}^{x} f'[\alpha]\, d\alpha = f[x] * STEP[x] \tag{20.92}$$

so we can interpret $\hat{W}_0[k]$ as the zero-order approximation to the spectrum of $STEP[x]$.

The cascade of differentiation and its zero-order inverse filter yields:

$$H_\partial[k] \cdot \hat{W}_0[k] = 2 \cdot \left|\sin\left[2\pi \frac{k}{2N}\right]\right| \tag{20.93}$$

which has the same form as the magnitude spectrum of the discrete derivative, so that the zero-frequency term is blocked and the low-frequency terms are attenuated.

The first-order approximation to the inverse filter for discrete differentiation is easy to evaluate:

$$\hat{W}_1[k] = \frac{1}{2}\left(2 - \left|\sin\left[2\pi\frac{k}{2N}\right]\right|\right) \cdot \exp\left[-i \cdot \tan^{-1}\left[\frac{\sin[2\pi k/N]}{\cos[2\pi k/N] - 1}\right]\right] \tag{20.94}$$

The transfer function is shown in Figure 20.16 and exhibits amplification of less than a factor of 2 at low frequencies. The transfer function of the second-order approximation is also shown in the figure and amplifies by factors less than 3, as expected.

PROBLEMS

20.1 Find the output of a discrete system whose impulse response is a 3-pixel averager if the input is a sinusoid with a period of 3 pixels in an array where N is an even number.

20.2 Derive the discrete transfer function of the rotated discrete and isotropic Laplacian operators.

20.3 Design the 3×3 unamplified matched filter kernels for detecting the following "structures" in a 2-D digital image:

- **(a)** horizontal lines 3 pixels wide
- **(b)** diagonal lines 1 pixel wide along the direction from lower left to upper right, i.e., along the azimuthal direction "/".

20.4 Find the impulse response of the approximate inverse filter $\hat{w}_n[n]$ if $h[n] = STEP[n]$.

20.5 Find the approximate inverse filter for the second derivative operator

$h[n] =$

+1	−2	+1

21

Optical Imaging in Monochromatic Light

An *optical imaging system* is composed of lenses, mirrors, and (perhaps) other elements that collect electromagnetic radiation emitted by or reflected from an object and then redirects that radiation to a sensor that measures the spatial distribution of energy that defines the "image". This chapter mathematically models the optical radiation and the imaging system to enable solution of the imaging tasks set forth in Chapter 1. A few physical models for radiation and its interaction with matter are investigated that differ in computational difficulty and accuracy. In those cases where the mathematical model of light propagation is linear and shift invariant, the tools of linear systems may be applied to solve the imaging problems.

We begin this discussion by returning to the very simple model of light, its propagation, and its interaction with matter that was briefly considered in the introductory chapter. This geometrical (or ray) model is adapted from the first theories of light propagation proposed in pre-Christian times. It assumes that light propagates in straight lines ("rays") within "media" that are assumed to have uniform physical properties. One of these properties is the velocity of light v, which is assumed to be constant throughout a specific medium; this velocity is most commonly expressed in terms of the *refractive index* of the medium ($n = c/\text{v}$), where c is the velocity of light in vacuum. In this model, a ray of light within a medium described by n_1 travels in a fixed direction with no spread of energy until the ray is either absorbed within the medium or encounters an interface with a different medium described by the different refractive index n_2. At such an interface, the ray divides into *refracted* and *reflected* components that generally travel in directions distinct from that of the incident ray. These interactions will be exploited to construct optical systems that redirect light rays diverging from a source to make them *converge* to form an "image" of that source on the sensor. Such a system is denoted by the operator $\mathcal{O}$ that acts on light emitted by the (generally 3-D) distribution of point sources $f[x, y, z]$. The "image" is the 3-D spatial distribution of energy $g[x, y, z]$ at the sensor. The action of the imaging "system" is the mathematical mapping between these two functions:

$$\mathcal{O}\{f[x, y, z]\} = g[x, y, z] \tag{21.1}$$

as was shown schematically in Figure 1.1. In real life, image sensors generally are 2-D devices, which means that information about the "depth" of the source along the z-axis is not deliberately recorded, though it may create artifacts in the image (e.g., "defocus blur"). To eliminate defocused images, the source object is often constrained to a plane, so that its mathematical form is $f[x, y; z_0]$, where the parameter z_0 is set off from the variables by the semicolon. The resulting image is measured in a

Fourier Methods in Imaging Roger L. Easton, Jr.

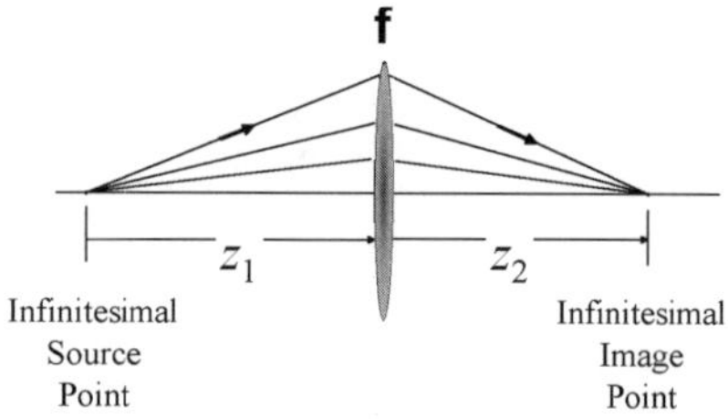

Figure 21.1 Optical imaging system with a real object and a real image. All three distances z_1, $\mathbf{f}$, and z_2 are positive.

different plane, say z_1, so the imaging equation has the form:

$$\mathcal{O}\{f[x, y; z = z_0 = 0]\} = g[x, y; z = z_1 \neq 0] \tag{21.2}$$

This 3-D spatial notation is cumbersome but useful because it explicitly denotes the axial positions of the planes. Otherwise, it would be necessary to denote the input and output coordinates in a specific plane by subscripts or other means, e.g.:

$$\mathcal{O}\{f[x_0, y_0]\} = g[x_1, y_1] \tag{21.3}$$

In the simple geometrical model of light propagation, rays diverge from a source point until they encounter a single optical element (often a "thin" lens) located at the distance z_1 from the source. The rays are refracted and redirected to converge to an image point located at a distance z_2 from the lens. For the rays to create an "image" on a light-sensitive detector, they must converge to a "real" image point in space, rather than create a "virtual" image point that is not accessible. The simple mathematical model of the imaging system relates the distances z_1 and z_2:

$$\frac{1}{z_1} + \frac{1}{z_2} = \frac{1}{\mathbf{f}} \tag{21.4}$$

where $\mathbf{f}$ is the "focal length" of the lens. The algebraic signs of the distances z_1 and z_2 determine whether the object and image are "real" or "virtual". All examples considered here assume that $z_1 > 0$, which means that diverging rays from a "real" object are collected by the system. If $z_2 > 0$, then the image rays from a point object converge to a "real" point image, where a sensor may be placed to record the spatial distribution of collected energy, as shown in Figure 21.1. Note that the rays diverge beyond this location (Figure 21.2). If $z_2 < 0$, then the image rays diverge from a point "behind" or "within" the imaging system; in this case it is not possible to place the sensor at the point of convergence. Such a virtual image cannot be recorded on a sensor without adding another optical element that creates a corresponding real image. The principles of geometric optics also demonstrate that the image is "magnified" by $M_T = -z_2/z_1$, which means that the rays emitted from one point on the object, say at location $[x_0, y_0]$, will "reconverge" at the point $[M_T x_0, M_T y_0]$ in the image plane.

21.1 Imaging Systems Based on Ray Optics Model

21.1.1 Seemingly "Plausible" Models of Light in Imaging

Before considering mathematical models for imaging systems based on both the reflection and the refraction of rays diverging from a point source, we first consider a simple optical imaging "system" model that is based on ray optics that includes no optical elements. The intent is to demonstrate how the tools of linear systems may be applied in theory to evaluate the "response" of the system and use

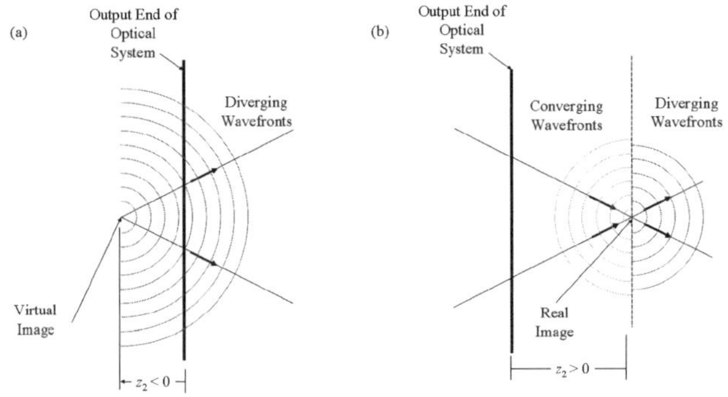

Figure 21.2 Wavefronts emerging from an optical imaging system: (a) diverging wavefronts emerge from the imaging system and appear to be emitted by a "virtual" image; (b) converging wavefronts emerge from the system to form a "real" image and then diverge.

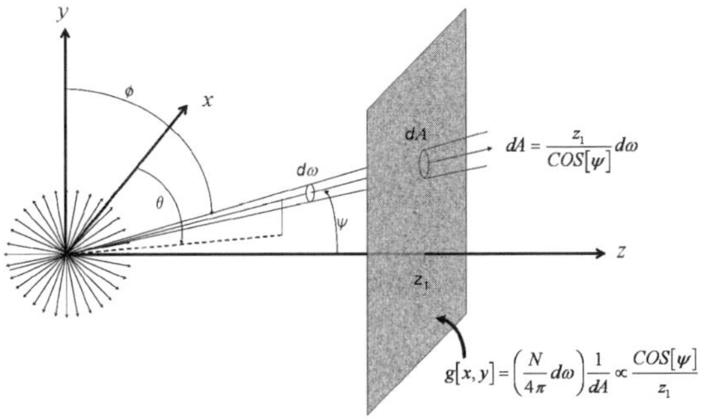

Figure 21.3 Ray optics model of point source. The flux of rays from the point source through an element of solid angle $d\omega$ does not vary with angle (θ or ϕ), but the flux of rays through an element of area dA on the detector is a function of the azimuth angle ψ.

it to deduce the distribution of light of the original object from measurements of the "image". The object is assumed to be a point source of light rays located at the origin of coordinates (Figure 21.3). In this simple model, the source emits a "family" of rays distributed "uniformly" in space, meaning that the same number of rays passes through every element of solid angle $d\omega$. In other words, this model assumes that the "angular density" of rays is independent of the azimuthal direction specified by (θ, ϕ) in spherical coordinates. The number of rays emitted into a unit of solid angle (the "flux") measures the "strength" or "intensity" of the source. The assumed isotropy of the source means that a source of "strength" N emits $N \cdot d\omega$ rays into any solid angle element of size $d\omega$.

The light rays are assumed to be measured by a detector that converts the number of rays per unit area of the sensor (the "area density" of incident rays) into a change in some physical parameter (such as electrical voltage or current, visible change in lightness or color of a material, or whatever). The change in an ideal detector is assumed to be proportional to the number of rays absorbed per unit area centered about the location with coordinates $[x_0, y_0]$. The sensor is assumed to be a planar device that absorbs and "counts" the light rays incident upon it at each location $[x_0, y_0]$ and this number is reported as the "image". Because all rays are "counted", this hypothetical detector does not "saturate" and truly obeys the requirements for linearity. The sensor is located in the plane specified by $z = z_1$, as shown in

Figure 21.3. At first, we assume that the sensor has infinite support and thus absorbs all photons emitted into the hemisphere with $z > 0$.

Because light rays emitted by a single point source travel different distances to reach different locations on the detector, our experience with the inverse square law would lead us to expect that the signal measured at a particular location $[x_0, y_0]$ is a function of the distance from the source. In this model, the mathematical relationship is derived by evaluating the density of rays absorbed by each area element of the sensor. The "projection" of the element of solid angle $d\omega$ onto an area element dA is called the "projected area". From Figure 21.3, it is evident that the area element dA on the sensor increases with magnitude of the azimuth angle $|\psi|$ that describes the position of the detector element. It is evident from the figure that $|\psi| \leq \pi/2$, and in fact it is easy to show that $dA \propto z_1/\cos[\psi]$. Since the same number of rays passes through each element of solid angle $d\omega$, and since the projected area due to $d\omega$ varies as $z_1/\cos[\psi]$, the signal measured per unit area by this ideal sensor is proportional to the *reciprocal* of the projected area. In other words, we assume that the largest-amplitude signal is measured at the point on the detector that is closest to the source and will decrease with increasing $|\psi|$. In this simple model of light and sensor, the "image" of the point source is a continuously varying and 2-D "smear" of signal that is circularly symmetric about the z-axis:

$$h[x, y; z_1] = k\frac{\cos[\psi]}{z_1} = k\frac{1}{\sqrt{x^2 + y^2 + z_1^2}}$$

$$= k\frac{1}{z_1\sqrt{1 + (x^2 + y^2)/z_1^2}} = \left(\frac{k}{z_1}\right)\left(\sqrt{1 + \frac{r^2}{z_1^2}}\right)^{-1} \tag{21.5}$$

where k is a constant of proportionality that relates the strength of the output signal to the input flux of rays, $r \equiv \sqrt{x^2 + y^2}$ is the radial distance in the plane of the sensor measured from the origin of coordinates, and the fact that

$$\cos[\psi] = \frac{z_1}{\sqrt{x^2 + y^2 + z_1^2}}$$

has been used. Note that $h[x, y; z]$ is the square root of a circularly symmetric 2-D Lorentzian function whose profile is specified in Equation (6.83). The square root ensures that $h[x, y; z_1]$ decays "more slowly" with increasing radial distance than the corresponding Lorentzian. Graphical examples of the impulse response of this simple optical imaging model are shown in Figure 21.4. In words, the contribution from a single-source location spreads out over the entire half plane of the sensor in the form of the square root of a Lorentzian. Therefore each point on the sensor measures a weighted sum of the contributions from all points on the object.

Now consider some of the consequences of using this model for optics and the sensor in an imaging system. In particular, it would be useful to know whether the imaging system is shift invariant. It is fairly easy to see that the shape of the "image" of a point source does not change if the source is constrained to the plane $[x, y, z = 0]$, i.e., if the source moves in the plane parallel to the sensor. Since the amplitude of the response to an "on-axis" source (located at $[x, y] = [0, 0]$) varies as the reciprocal of the distance z between the source and sensor, the maximum amplitude of the measured signal due to a particular point source "decreases" as the sensor is moved farther from the source (increasing z). Thus the shape of the "image" function *does* change if the source is moved along the z-axis (in "depth"). Therefore this imaging "system" is shift invariant only if all point sources are constrained to lie in the plane parallel to the detector; the system will be shift variant if the distribution of source points is three dimensional. Since the impulse response in Equation (21.5) has infinite support, the detector must "measure" the entire impulse response to ensure that the system is shift invariant, and thus the detector must be infinitely large.

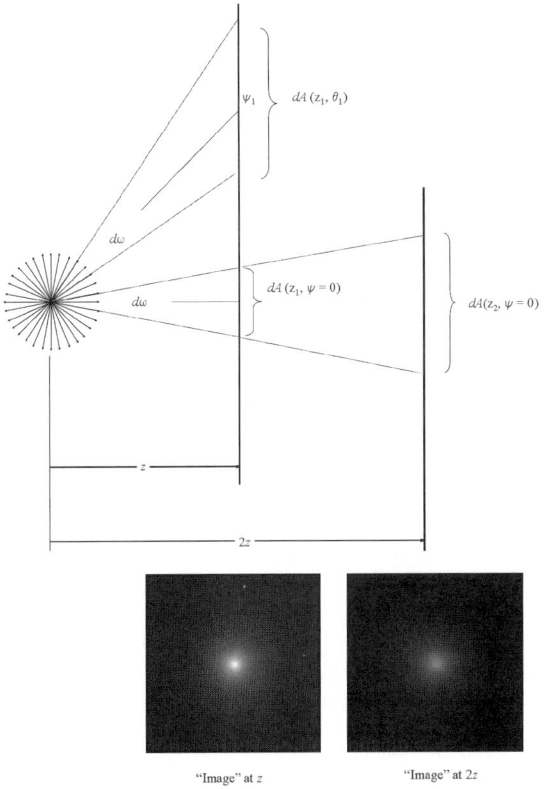

Figure 21.4 Impulse responses in this optical model evaluated for different distances z from the source. Since the flux from the point source is uniformly distributed over angle, the measured intensity of the flux is inversely proportional to the projected area, and thus to the distance z.

If the necessary constraints for shift invariance are assured, then the 2-D output "image" is the convolution of the arbitrary planar distribution of sources and of the 2-D "impulse response" $h[x, y; z_1]$:

$$\begin{aligned} g[x, y; z_1] &= f[x, y; 0] * h[x, y; z_1] \\ &= \left(\frac{c}{z_1}\right)\left(f[x, y; 0] * \left[1 + \frac{x^2 + y^2}{z_1^2}\right]^{-\frac{1}{2}}\right) \end{aligned} \tag{21.6}$$

Since the impulse response is everywhere positive, each point on the sensor sees a weighted sum of signals from the points on the object. This hypothetical imaging system "averages" the signal and thus is a lowpass filter.

This simple imaging system has (at least) one obvious drawback compared to the optical systems in our common experience: the measured amplitude distribution (the "image") $g[x, y; z_1]$ in no way resembles the original object $f[x, y]$. However, we may recover the original object to compensate for the blurring action of the system by evaluating the inverse filter $W[\xi, \eta; z_1] = (H[\xi, \eta; z_1])^{-1}$, if it exists.

If the object consists of multiple point sources located at different distances from the sensor and if the detection process is linear, then the "image" will be the sum of replicas of the impulse response that have been scaled both in amplitude and in "width" due to its dependence on z. In other words, the inverse filter for a particular 2-D object located at z_1 will not be correct for any other distance. Thus it will not be possible to reconstruct the exact 3-D distribution of object points by applying a single 2-D filter. However, we could construct a family of inverse filters, each "matched" to a specific distance z between the source and the sensor, and apply each filter individually to the recorded "image". The output of a particular filter would contain images of the Dirac delta functions located at that specific distance and "out-of-focus" images for points in other planes. However, the relatively slow variation of $h[x, y; z]$ with depth z means that the signal from nearby sources will be similarly amplified and these "unmatched" amplitudes will appear as noise artifacts in the reconstruction.

21.1.2 Imaging Systems Based on Ray "Selection" by Absorption

The process of imaging via inverse filtering that was just described is a "system" only under the broadest of definitions. Because all rays emitted by the object are collected by the sensor without "selection" or "redirection", the coordinates of the original sources may be deduced only via the post-detection inverse filter. The usual definition of an imaging "system" employs some physical interaction of light rays with matter that selects a set of rays that reaches the sensor, or that redirects rays from their original path to the sensor. Many such interaction mechanisms exist; among those that may be used in an imaging system are absorption, refraction, reflection, and diffraction.

The introduction in Chapter 1 described an imaging system in the ray optics model that is based on simple absorption to select rays that travel along specific paths. This useful model introduces some essential concepts that may be applied to image radiation that cannot be focused by lenses or mirrors, such as gamma rays in nuclear medicine (Barrett and Swindell, 1996). The pinhole camera system shown in Figure 1.8 "selects" gamma rays with a single aperture in an absorber. A schematic of the system used to image a single point source located at the plane specified by $z = 0$ is shown in Figure 21.5. The "system" consists of an absorptive medium located in the plane z_1 with a small pinhole drilled in the center. At this point we make an unrealistic simplifying assumption: that the pinhole has infinitesimal area but transmits some light, so that the transmission of the plate is the 2-D Dirac delta function $\delta[x, y; z_1]$. This clearly is unrealistic, but we will consider a more reasonable model shortly. A planar sensor is placed behind the absorber in the plane specified by z_2. The output distribution of rays has the form:

$$\begin{aligned} g[x, y; z_2] &= f\left[x \bigg/ \left(-\frac{z_2}{z_1}\right), y \bigg/ \left(-\frac{z_2}{z_1}\right)\right] \\ &= f[x, y] * \delta[x, y]|_{x \to -(z_2/z_1)x,\ y \to -(z_2/z_1)y} \end{aligned} \tag{21.7}$$

In words, the image is an inverted and scaled ("magnified") replica of the object, so the impulse response is a 2-D Dirac delta function. The distance of the image from the origin is scaled by the factor $-z_2/z_1$, which means that the system "magnifies" (or "minifies") the scale of the image if $z_2 \neq z_1$. The system magnification depends on the object distance z_1, and thus the system is shift variant if used to image nonplanar objects.

A fundamental shortcoming of the pinhole camera is its small efficiency; a very small percentage of emitted photons are selected for imaging and recorded. The statistical nature of the process is useful to consider. Emitted and detected photons exhibit Poisson statistics, which means that the standard deviation (the "noise") of N measured photons (per unit area) is $\sqrt{N}$. The "signal-to-noise ratio" is the ratio of the mean to the standard deviation:

$$SNR = \frac{N}{\sqrt{N}} = \sqrt{N} \tag{21.8}$$

In short, to ensure a large *SNR*, the number of detected photons must be large. There are several ways to increase N. Perhaps the most obvious is to increase the number of rays emitted by the source.

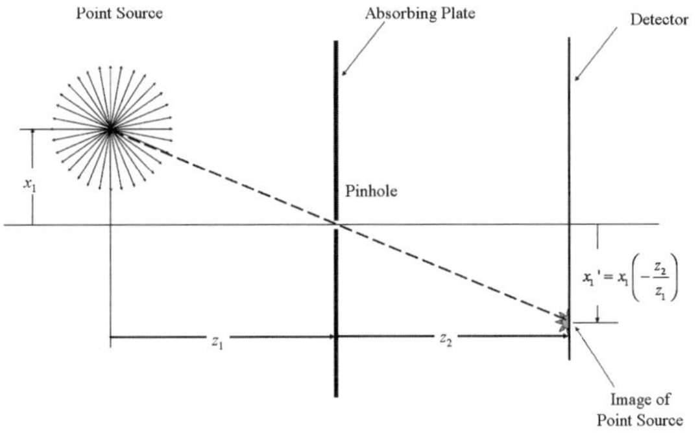

Figure 21.5 Schematic of pinhole camera: the object point is located at coordinates $[x_1, 0, 0]$. Photons are selected by a pinhole at $[0, 0, z_1]$ and imaged by a detector in the plane $z = z_1 + z_i$. The image is magnified by the factor $-z_2/z_1$, and thus is a function of the "depth" z.

In diagnostic nuclear medicine, this requires a larger radiation dose to the patient, which is hardly desirable. A second strategy to increase N is to lengthen the exposure time, which would require the entire system (including the patient) to remain stationary for a longer time, which also is less than desirable. A third alternative is to increase the diameter of the aperture, thus allowing more photons to reach the detector. However, this option changes the spatial impulse response as well as the statistics of the system response. It is easy to see that the larger pinhole is described by the finite-support binary function $h[x, y]$. As a simple example, consider that the aperture is a rectangle function (though a circle is actually more common). The system impulse response and the image are:

$$h[x, y] = RECT\left[\frac{x}{b_0}, \frac{y}{d_0}\right] \tag{21.9a}$$

$$g[x, y; z_2] = \left(f[x, y] * RECT\left[\frac{x}{b_0}, \frac{y}{d_0}\right]\right)\Bigg|_{x \to -(z_2/z_1)x,\ y \to -(z_2/z_1)y} \tag{21.9b}$$

The larger aperture allows photons from a neighborhood of the object point to reach the same location on the sensor, thus "blurring" the image and decreasing its "quality". However, we can apply some of the tools we have developed to improve the measured image. For example, the appropriate inverse filter for "sharpening" the image is:

$$\begin{aligned} W[\xi, \eta] &= (|b_0 d_0|\ SINC[b_0\xi, d_0\eta])^{-1} \\ &= \left(\frac{1}{|b_0|\ SINC[b_0\xi]}\right)\left(\frac{1}{|d_0|\ SINC[d_0\xi]}\right) \end{aligned} \tag{21.10}$$

However, the small magnitudes of H in the vicinity of its zeros and at large values of $\rho = \sqrt{\xi^2 + \eta^2}$ require large amplifications by W, which also will amplify any Poisson noise.

The number of imaged photons also may be increased by boring more small pinholes in the absorber. By applying the same approximation used in Equation (21.7), the resulting array of disjoint holes is modeled as an array of 2-D Dirac delta functions:

$$h[x, y] = \sum_{n=1}^{N} \delta[x - x_n, y - y_n] \tag{21.11}$$

The output of the system is a set of disjoint (though possibly overlapping) pinhole images. The transfer functions of the system and of the corresponding inverse filter are easy to derive:

$$H[\xi, \eta] = \sum_{n=1}^{N} \exp[-2\pi i(\xi x_n + \eta y_n)] \tag{21.12}$$

$$W[\xi, \eta] = \left(\sum_{n=1}^{N} \exp[-2\pi i(\xi x_n + \eta y_n)]\right)^{-1} \tag{21.13}$$

Note that both are generally complex valued (magnitude and phase). Again, any attenuated magnitudes must be amplified by the inverse filter, and thus any Poisson noise that exists in the measured image will also be boosted. The level of amplification could be controlled by truncating a power-series approximation for $W[\xi, \eta]$ based on Equation (19.136).

The number of pinholes may be increased still further to the point where they cover a large percentage of the area of the absorber and are perhaps no longer disjoint. The resulting "mask" function is often called a *coded aperture*. The required amplification of the inverse filter can be minimized by judicious design of the aperture pattern $h[x, y]$ based on the discussion of the matched filter in Chapter 19. There we saw that a pure phase filter did not amplify any frequency components. Of course, a pure phase filter is complex valued, whereas the transmission of the pinhole aperture is purely real. We proceed by replacing the pinhole with a real-valued and thresholded 2-D chirp function, the "Fresnel zone plate":

$$h[x, y] = \frac{1}{2}\left(1 + SGN\left[\cos\left[\pi\left(\frac{x^2 + y^2}{d^2}\right)\right]\right]\right) \tag{21.14}$$

An example of a Fresnel zone plate is shown in Figure 17.23. The impulse response of the inverse filter is approximately $h[x, y]$. The resulting image is quite sharp, though a significant uniform "background" bias results because $h[x, y]$ is not bipolar.

21.1.3 Imaging System that Selects and Reflects Rays

Imaging by "ray selection" using pinholes or coded apertures can be very useful for highly energetic photons, but the statistical limit on photon counts constrains the image fidelity that may be achieved. We need to further increase the number of detected photons without increasing either the number of emitted photons or the size of the selecting aperture. It would be very useful to use some physical interaction based on a property of matter that "redirects" photons diverging from a source point back to a single image point. Fortunately, systems to do exactly that can be designed using reflection, refraction, and diffraction to redirect the light.

Consider first the specific case of reflection; a mirror diverts incoming rays to different outgoing paths that may be controlled by changing the direction of the normal to the mirror surface. The ray angles obey Snell's law for reflection:

$$\theta' = -\theta \tag{21.15}$$

where θ and θ' are the angles of the entering and exiting rays measured from the normal to the mirror. By decreasing the area of the reflective surface, we can create a hypothetical mirror that selects *and* redirects specific rays. In the limit, the area of the mirror is reduced until it becomes infinitesimal, but we assume that it still has a "direction" defined by the normal to its surface. Since this is just a thought experiment, we ignore the metaphysical quandary suggested by such a mirror. If we construct a planar array of these "pinhole mirrors" such that all normal vectors lie in a plane, then each mirror selects and reflects a single ray emitted by the source. The reflected rays converge to the same location, thus creating a point image as shown schematically in Figure 21.6. The same set of pinhole mirrors creates images of multiple point sources that have the same geometrical relationship in three dimensions as the original distribution. Rays at different heights from the axis of symmetry are reflected at the appropriate angles to make the rays converge.

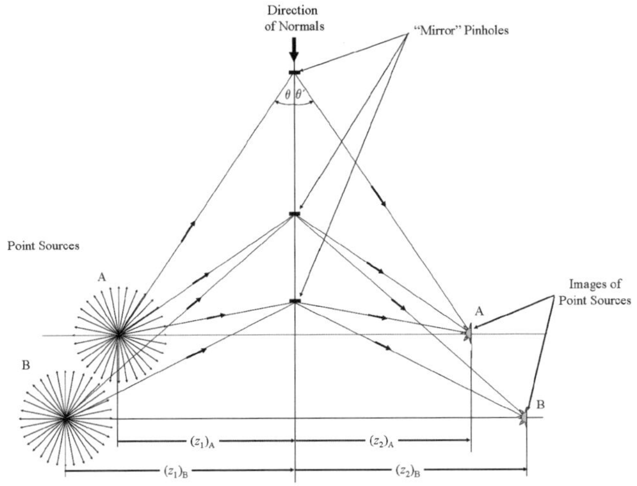

Figure 21.6 Principle of pinhole mirror imaging. An array of infinitesimal mirrors is placed such that the normals to the mirrors are in the same plane. The rays selected by the individual mirrors from the two point sources are redirected to converge and create images of the sources. The perpendicular distances from object to mirror and mirror to image are equal.

Because the number of recorded photons is proportional to the number of selected rays, the statistical noise in the image created by this system is reduced from that created by a single pinhole or pinhole mirror.

21.1.4 Imaging Systems Based on Refracting Rays

Perhaps the most familiar example of an optical imaging system is a single refracting lens. This system avoids the issue of ray "selection" because it redirects *all* rays incident on the aperture at angles that depend on their angle of incidence. The angles of the output rays are determined by Snell's law for refraction:

$$n_1 \sin[\theta_1] = n_2 \sin[\theta_2] \tag{21.16}$$

where n_1 and n_2 are the refractive indices of the two media and θ_1 and θ_2 are the angles of the ray within the two media measured from the surface normal. Snell's law is a direct consequence of Fermat's principle that light rays follow the path that requires the least time to traverse. Proofs are presented in many optics texts, e.g:, Hecht (2002). The shapes of the surfaces are chosen to ensure that rays from a point object converge to form a point image (within limits). We have seen that the parameters of the lens, including the shapes of the surface(s) and the refractive indices, can be combined into a single descriptor of the system: the focal length **f**. It is generally easy to find the location of the point image using the geometrical imaging equation that was introduced in Equation (1.4). It is not so easy to characterize the "quality" of the image in the simple model of light propagating as rays; a more sophisticated model for the propagation of light is required. For this reason, we now divert the discussion to consider the wave model of light.

21.1.5 Model of Imaging Systems

Though we have already described a system that can create images of 3-D emitting objects, most of the subsequent discussion will deal with 2-D sources, objects, images, and even systems. A simple

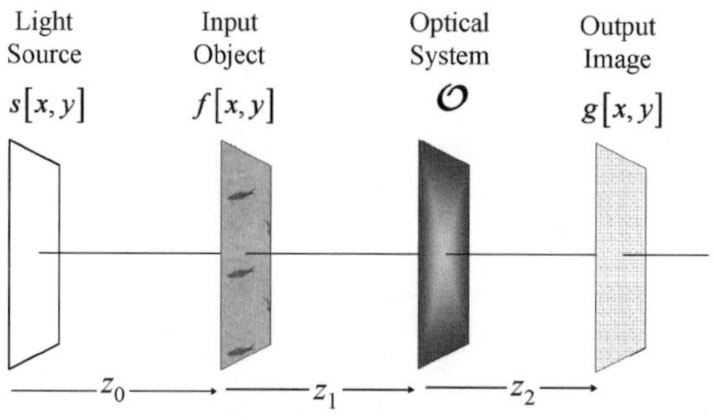

Figure 21.7 Schematic of general optical imaging system, consisting of planar light source $s[x, y]$, planar input object $f[x, y]$, optical system (which is often not planar but whose response is reduced to two dimensions), and the output image $g[x, y]$. Not all components will be considered in all imaging systems.

schematic model of the optical system is shown in Figure 21.7, though not all components are required in all systems to be considered. The model consists of 2-D planes specified by coordinates $[x, y]$ or (r, θ) that are separated by distances down the z-axis which is called the *optical axis*. The light interacts with the various components at the planes and then propagates to the next plane. Implicit in the mathematical model of propagation between planes is the concept of *diffraction*, which is often loosely characterized as a "spreading" of the light and which is considered in detail in the next section.

In the model, light generated by the source of light that is confined to the plane $z = 0$ propagates a distance z_0 to the 2-D object $f[x, y]$. The light interacts with the object by some physical mechanism (typically absorption) and then propagates a distance z_1 to the imaging system. In this schematic, the system is specified by an operator $\mathcal{O}$, which may require more than two spatial dimensions. In the useful shift-invariant case, the operator is a 2-D impulse response or the equivalent 2-D transfer function. Finally, the light emerging from the system propagates the distance z_2 to the observation plane, where the time-averaged incident power is measured by a sensor to produce the image.

Note that a more general system involving 3-D objects and images will be considered in the next chapter.

21.2 Mathematical Model of Light Propagation

21.2.1 Wave Description of Light

An *oscillation* is a periodic variation in some parameter of a physical system that results from the balance between the forces of *inertia* and some *restorative* force provided by an external medium. A *wave* is an oscillation that travels through space, so that the temporal variation in the parameter is also a function of position. Since the function describing a wave is periodic in both space and time, we might expect that its spatial and temporal derivatives are related in some way. Though we will not derive the relationship (as is done in any number of physics texts), it is specified (coincidentally enough) by the *wave equation*:

$$\begin{aligned}\nabla^2(f[x, y, z, t]) &\equiv \left(\frac{\partial^2}{\partial x^2} + \frac{\partial^2}{\partial y^2} + \frac{\partial^2}{\partial z^2}\right) f[x, y, z, t] \\ &= \frac{1}{c^2}\frac{\partial^2}{\partial t^2} f[x, y, z, t] \qquad (21.17)\end{aligned}$$

where c is the velocity of the wave. This differential equation may be solved rigorously by integration after applying Green's theorem, though the solution is often simplified by assuming certain conditions. We will take a different tack by inferring the impulse response from physical observations to demonstrate its important features. Readers requiring a deeper discussion of the strict solution to the wave equation should consult some of the standard texts in optics.

Image information is transferred from the source to the image by electromagnetic radiation that propagates through vacuum as well as through media. The character of the illuminating radiation is an important (if not *the* important) factor in the response of any imaging system. Therefore we must consider the mathematical model of radiation and its propagation in this discussion. James Clerk Maxwell established that light is an *electromagnetic* traveling wave; both electric ($\underline{\mathbf{E}}$) and magnetic ($\underline{\mathbf{B}}$) waves must be present simultaneously for the radiation to propagate, and thus Equation (21.17) applies to both individually. The oscillations travel in a particular direction and are mutually perpendicular and transverse to the direction of travel. The electric field is a vector whose direction defines the *polarization* of the field. This is not important in the imaging methods described in most of the following discussion (holography is the only exception), so the vector notation will be dropped.

An electromagnetic wave with temporal frequency ν (wavelength $\lambda = c/\nu$) moving along the straight-line path in the direction of the vector $\hat{\underline{\mathbf{k}}}$ in space may be described as a sinusoidal traveling wave whose phase angle is a function of both spatial and temporal coordinates. The spatial component of the phase evaluated at coordinate z_1 and at time t_1 specifies the number of radians of phase in the wave between $z = 0$ and $z = z_1$. In other words, for a "snapshot" of the wave taken at time t_1, it is the number of wavelengths of the spatial sinusoid between these two locations multiplied by 2π radians of phase per cycle:

$$\Phi_{\text{spatial}}[z_1, t_1] = 2\pi \frac{z_1}{\lambda} \equiv k \cdot z_1 \tag{21.18}$$

The factor $k \equiv 2\pi/\lambda$ is the *angular wavenumber* of the wave, i.e., the number of radians of phase per unit length of the spatial variation in the amplitude.

The temporal part of the phase of an electromagnetic wave measured at coordinate z_1 at time t_1 specifies the number of radians of phase that have passed that point in space between $t = 0$ and $t = t_1$, and therefore is the product of the time interval and 2π radians per temporal cycle:

$$\Phi_{\text{temporal}}[z_1, t_1] = 2\pi \frac{t_1}{T} = 2\pi \nu t_1 \equiv \omega t_1 \tag{21.19}$$

where ω is the *angular temporal frequency*, measured in radians per second.

The phase angle of the traveling wave is the sum or difference of the phases of the spatial and temporal components, where the choice of sign determines the direction of travel of the wave, whether toward positive or negative ∞:

$$\Phi[z_1, t_1] = \begin{cases} \Phi_{\text{spatial}}[z_1, t_1] \pm \Phi_{\text{temporal}}[z_1, t_1] \\ \textit{or} \\ \Phi_{\text{temporal}}[z_1, t_1] \pm \Phi_{\text{spatial}}[z_1, t_1] \end{cases} = \begin{cases} kz_1 \pm \omega t_1 \\ \textit{or} \\ \omega t_1 \pm kz_1 \end{cases} \tag{21.20}$$

The direction of propagation is the direction of motion of a particular point on the wave (often chosen to be one of the maxima) that is specified by its phase angle. Thus the direction of motion is determined by a "point of constant phase" that defines the direction of travel of the wave. For a wave traveling toward $z = +\infty$, the spatial coordinate of the point of constant phase increases with time, which means that the spatial and temporal parts of the phase must have opposite signs:

$$\Phi[z, t] = \begin{cases} kz - \omega t \\ \textit{or} \\ \omega t - kz \end{cases} \quad \textit{for wave traveling toward } z = +\infty \tag{21.21}$$

The selection of convention from these two options has no practical impact on the calculation or the physics; different authors make different choices. An informal census of optics books seems to indicate that authors of more recent books often apply the positive sign to the spatial part of the phase:

$$\begin{aligned}\textit{Older books}: E[x, y, z, t] &= E_0[x, y] \cos\left[2\pi\left(\nu t - \frac{z}{\lambda}\right)\right] \\ &= E_0[x, y] \cos[\omega t - kz] = \Re\{E_0[x, y]\, e^{i[\omega t - kz]}\}\end{aligned} \tag{21.22a}$$

$$\begin{aligned}\textit{Newer books}: E[x, y, z, t] &= E_0[x, y] \cos\left[2\pi\left(\frac{z}{\lambda} - \nu t\right)\right] \\ &= E_0[x, y] \cos[kz - \omega t] = \Re\{E_0[x, y]\, e^{i[kz - \omega t]}\}\end{aligned} \tag{21.22b}$$

We follow the current trend, which means that the phase of radiation of light at a specific point in space and traveling toward $z = +\infty$ *decreases* with increasing time. Also, the phase of the radiation observed at the same instant of time *increases* at observation points located at larger values of z. Similarly, if the phase at a location is increased by some physical mechanism (as by inserting a thickness of glass that "slows down" the light), this "delays" the arrival time of light at a location farther "downstream" (at a larger value of z). In other words, augmenting the phase is equivalent to a phase "delay".

In the 3-D case, the amplitude of the electric field must be specified for each coordinate $[x, y, z, t] \equiv [\underline{\mathbf{R}}, t]$. The phase of the field is a linear function of both spatial and temporal dimensions. A particular point on the wave located at 3-D spatial location $\underline{\mathbf{R}}$ at time t moves in the direction specified by the 3-D *wavevector* $\underline{\mathbf{k}}$, where $|\underline{\mathbf{k}}| = 2\pi/\lambda$. The spatial part of the phase is the scalar product of the wavevector with the spatial coordinate $\underline{\mathbf{R}}$:

$$\begin{aligned}\underline{\mathbf{k}} \bullet \underline{\mathbf{R}} &= \begin{pmatrix} k_x \\ k_y \\ k_z \end{pmatrix} \bullet \begin{pmatrix} x \\ y \\ z \end{pmatrix} \\ &= k_x x + k_y y + k_z z \textit{ where } \sqrt{k_x^2 + k_y^2 + k_z^2} = \frac{2\pi}{\lambda}\end{aligned} \tag{21.23}$$

For example, the wavevector of a traveling wave moving parallel to the z-axis with wavelength λ is:

$$\underline{\mathbf{k}} = \begin{pmatrix} 0 \\ 0 \\ 2\pi/\lambda \end{pmatrix} \tag{21.24}$$

and the phase of the light measured at $\underline{\mathbf{R}}_0$ is the spatiotemporal function:

$$\underline{\mathbf{k}} \bullet \underline{\mathbf{R}}_0 - \omega t + \phi_0 = \begin{pmatrix} 0 \\ 0 \\ 2\pi/\lambda \end{pmatrix} \bullet \begin{pmatrix} x_0 \\ y_0 \\ z_0 \end{pmatrix} - 2\pi\nu t + \phi_0 = 2\pi\left(\frac{z_0}{\lambda} - \nu t\right) + \phi_0 \tag{21.25}$$

where the initial phase ϕ_0 is evaluated at the origin of both space and time:

$$\phi_0 = \Phi[\underline{\mathbf{R}} = \underline{\mathbf{0}}, t = 0] \tag{21.26}$$

Several equivalent expressions for the electric field are:

$$\begin{aligned}E[x, y, z, t] = E[\underline{\mathbf{R}}, t] &= E_0 \cos[\boldsymbol{\Phi}[\underline{\mathbf{R}}, t]] = E_0 \cos[\underline{\mathbf{k}} \bullet \underline{\mathbf{R}} - \omega t + \phi_0] \\ &= \Re\left\{E_0 \exp\left[+2\pi i\left(\hat{\underline{\mathbf{k}}} \bullet \frac{\underline{\mathbf{R}}}{\lambda} - \nu t + \frac{\phi_0}{2\pi}\right)\right]\right\} \\ &= \Re\{E_0\, e^{i\boldsymbol{\Phi}[\underline{\mathbf{R}}, t]}\}\end{aligned} \tag{21.27}$$

From this point forward, we will assume that the initial phase $\phi_0 = 0$, though extension to the general case is easy.

21.2.2 Irradiance

The temporal oscillation frequency of visible light is very large ($\nu \cong 10^{14}$ Hz), which means that the phase changes by 2π radians in time intervals of the order of 10^{-14} s. Because visible light waves oscillate many times over the time required by optical sensors to respond to the radiation, detectors of visible light are insensitive to the optical phase. They instead respond to the time average of the squared magnitude of the incident electric field amplitude, which is called the *irradiance* and is specified by I:

$$I[x, y, z, T_0] = 1/T_0 \int_{-T_0/2}^{+T_0/2} |E[x, y, z, t]|^2 \, dt \tag{21.28}$$

where the time interval T_0 is characteristic of the sensor. For example, the temporal response of the human visual system is constrained by the photochemistry in the receptors, so that $T_0 \cong 30$ ms.

21.2.3 Propagation of Light

In the 1600s, Christiaan Huygens proposed a model for light propagation where each point on a wavefront is the source of a "secondary spherical wavelet". The electric field at a point farther "downstream" from the source is the sum of all of these wavelet fields. To conserve the energy in the spherical wavelets as they propagate over a larger surface area, the amplitude of the electric field decreases as the wave propagates. Using our convention for phase, the mathematical equation of a spherical wave with wavelength λ (angular temporal frequency $\omega = 2\pi\nu$) that expands from a point source with amplitude $\mathcal{E}_0$ located at the origin and observed at the plane specified by $z = z_1$ is:

$$\begin{aligned} f[x, y, z_1, t] &= \frac{\alpha}{\sqrt{x^2 + y^2 + z_1^2}} \exp\left(+2\pi i \left[\frac{\sqrt{x^2 + y^2 + z_1^2}}{\lambda} - \nu t\right]\right) \\ &= \frac{\alpha}{|\underline{\mathbf{R}}|} \exp[+i(\underline{\mathbf{k}} \bullet \underline{\mathbf{R}} - \omega t)] \end{aligned} \tag{21.29}$$

where the constant of proportionality α appropriate for electric fields must be determined. As depicted in Figure 21.8, the configuration for diffraction includes a subset of the system components in Figure 21.7. The amplitude and phase of a spherical traveling electrical wave emitted from a point source located in the x–y plane at $\underline{\mathbf{R}}_0 = [x_0, y_0, z_0 = 0]$ with frequency $\nu = \omega/2\pi$ (wavelength $\lambda = c/\nu$) and observed at location $\underline{\mathbf{R}}_1 = [x_1, y_1, z_1]$ and at time t_1 is mathematically described by:

$$f[x_1, y_1, z_1, t_1] = \frac{\alpha}{|\underline{\mathbf{R}}_1 - \underline{\mathbf{R}}_0|} \exp[+i(\underline{\mathbf{k}} \bullet (\underline{\mathbf{R}}_1 - \underline{\mathbf{R}}_0) - \omega(t_1 - t_0))] \tag{21.30}$$

The wave observed at time t_1 was emitted by the source at the earlier time t_0, to account for the travel time. The time dependence $\exp[-i\omega(t_1 - t_0)]$ of the wave merely describes the harmonic oscillation of the electric field amplitude due to the temporal frequency of the light. Since it has no effect on spatial calculations and integrations, it will be ignored in the following discussion. It is possible to sum the contributions from different source points $[x_0, y_0, z_0]$ to evaluate the amplitude at the observation point $[x_1, y_1, z_1, t_1]$, but since the distance $|\underline{\mathbf{R}}_1 - \underline{\mathbf{R}}_0|$ from each source point to the observation plane will be different, then the calculation is a space-variant superposition, rather than a space-invariant convolution.

The task now is to determine the constant of proportionality and make appropriate approximations that simplify the evaluation of Equation (21.30). The most common procedure (e.g., Goodman, 2005) is to rigorously solve Maxwell's equations by applying Green's theorem within appropriate volumes. We choose an arguably simpler method based on synthesis of the electric field from its sinusoidal components in such a manner that the wave equation is satisfied:

$$\nabla^2 \underline{\mathbf{E}} - \frac{1}{v^2} \frac{\partial^2 \underline{\mathbf{E}}}{\partial t^2} = 0 \tag{21.31}$$

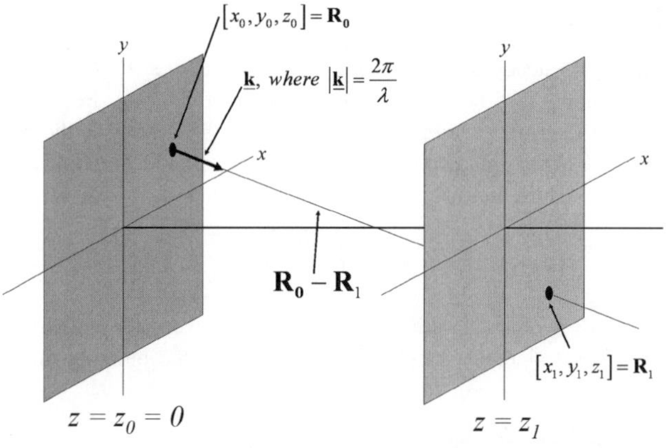

Figure 21.8 Coordinate system for optical diffraction: the source function is located at the plane $z = z_0 = 0$ with coordinates specified either by the 3-D vector $\mathbf{R}_0 = [x_0, y_0, 0]$ or by the 2-D vector $[x_0, y_0]$. The light from the source propagates to an observation plane located at $z = z_1$ with 3-D coordinates $\mathbf{R}_1 = [x_1, y_1, z_1]$ or the corresponding 2-D vector $[x_1, y_1]$. The distance from one source point $[x_0, y_0]$ to one observation point $[x_1, y_1]$ is $|\mathbf{R}_0 - \mathbf{R}_1|$.

The velocity v is the "phase velocity" of the wave, i.e., the velocity of a "point of constant phase" on the wave. The phase velocity is defined $v \equiv c/n$, where n is the refractive index of the medium. In the situation of interest, the light is propagating in vacuum, so that $n = 1$. The same equation must be satisfied by the individual vector components that determine the "polarization" of the electric field:

$$\nabla^2 E_j - \frac{n^2}{c^2}\frac{\partial^2 E_j}{\partial t^2} = 0 \tag{21.32}$$

where $j = x, y, z$ specifies the vector component under consideration. The complete solution for the electric field is the superposition of the three polarizations summed over all values of $\underline{\mathbf{k}}$ and ω after weighting by the (generally complex-valued) function $F[\underline{\mathbf{k}}, \omega] = F[k_x, k_y, k_z, \omega]$. The weighting function F is analogous to the frequency spectrum in the Fourier transform. In words, we are decomposing the electric field into a set of plane waves, each propagating in a direction specified by the wavevector $\underline{\mathbf{k}} = [k_x, k_y, k_z]$. This is the so-called *angular spectrum of plane waves*. The electric field E evaluated at the space–time coordinate $[x, y, z_1, t]$ is:

$$\underline{\mathbf{E}}[x, y, z_1, t] = \sum_{j=1}^{3} \hat{\mathbf{a}}_j \cdot \int_{-\infty}^{+\infty} d\omega \iiint_{-\infty}^{+\infty} dk_x\, dk_y\, dk_z (F[k_x, k_y, k_z, \omega]\, e^{+i(k_x x + k_y y + k_z z_1 - \omega t)}) \tag{21.33}$$

An electric field with an arbitrary polarization can be expressed by this relationship. From this point forward, we consider only one polarization and ignore the summation over j because the linearity of Fourier synthesis allows any additional contributions to be added at the end if needed. If we apply the wave equation to the well-known sinusoidal trial solution for the electric field:

$$\underline{\mathbf{E}}[x, y, z_1, t] = \underline{\mathbf{E}}_0 \exp[+i(\underline{\mathbf{k}} \bullet \underline{\mathbf{r}} - \omega t)] \tag{21.34}$$

the spatial derivative evaluates to:

$$\begin{aligned}\nabla^2\underline{\mathbf{E}} &= \nabla^2(\underline{\mathbf{E}}_0 \exp[+i(\underline{\mathbf{k}}\bullet\underline{\mathbf{r}} - \omega t)])\\ &= (\underline{\mathbf{E}}_0 \exp[-i\omega t])\left(\frac{\partial^2}{\partial x^2} + \frac{\partial^2}{\partial y^2} + \frac{\partial^2}{\partial z^2}\right) e^{+i(k_x x + k_y y + k_z z - \omega t)}\\ &= (\underline{\mathbf{E}}_0 \exp[-i\omega t]) \cdot ((ik_x)^2 + (ik_y)^2 + (ik_z)^2)\, e^{+i(k_x x + k_y y + k_z z - \omega t)}\\ &= i^2(k_x^2 + k_y^2 + k_z^2)\underline{\mathbf{E}} = -k^2\underline{\mathbf{E}}\end{aligned} \tag{21.35a}$$

while the temporal derivative yields:

$$\frac{\partial^2\underline{\mathbf{E}}}{\partial t^2} = (-i\omega)^2\underline{\mathbf{E}} = -\omega^2\underline{\mathbf{E}} \tag{21.35b}$$

The wave equation establishes a condition that must be satisfied by the length of the wavevector $|\underline{\mathbf{k}}| = k$ and the angular temporal frequency ω:

$$\nabla^2\underline{\mathbf{E}} - \frac{1}{\mathrm{v}^2}\frac{\partial^2\underline{\mathbf{E}}}{\partial t^2} = -k^2\underline{\mathbf{E}} - \frac{\omega^2}{c^2}\underline{\mathbf{E}} = 0 \Longrightarrow k^2 = \left(\frac{\omega}{c}\right)^2 \Longrightarrow k = \frac{\omega}{c} \tag{21.36}$$

In words, the *wavenumber* k (the number of radians in the oscillating electric field per unit length) is proportional to ω; the faster the oscillation, the more radians of phase per unit length. This equivalence relation introduces a constraint in the 4-D Fourier synthesis that may be inserted via the appropriate 1-D Dirac delta function $\delta[k - \omega/c]$:

$$\begin{aligned}\underline{\mathbf{E}}[x, y, z, t] = &\int_{-\infty}^{+\infty} d\omega\\ &\cdot \iiint_{-\infty}^{+\infty} dk_x\, dk_y\, dk_z (F[k_x, k_y, k_z, \omega]\, e^{i(k_x x + k_y y + k_z z - \omega t)}) \cdot \delta\left[k - \frac{\omega}{c}\right]\end{aligned} \tag{21.37}$$

21.2.3.1 Constraint: Monochromatic Light

At this point we constrain the light to be *monochromatic* with frequency ω_0 (wavelength $\lambda_0 = 2\pi c/\omega_0$), which will significantly simplify the subsequent derivation. We will maintain this assumption through the derivation of the impulse response and transfer function of the imaging system, and then will generalize the discussion to allow polychromatic light.

The temporal frequency dependence of the monochromatic field may be specified by introducing the appropriate Dirac delta function:

$$\begin{aligned}F[k_x, k_y, k_z, \omega] &= F[k_x, k_y, k_z, \omega] \cdot \delta[\omega - \omega_0]\\ &= F[k_x, k_y, k_z, \omega_0] \cdot \delta[\omega - \omega_0]\end{aligned} \tag{21.38}$$

where the property of Dirac delta functions in products from Equation (6.115) has been used. The resulting expression for the constrained electric field is:

$$\begin{aligned}\underline{\mathbf{E}}[x, y, z, t] = &\iiint_{-\infty}^{+\infty} dk_x\, dk_y\, dk_z\\ &\cdot \left(\int_{-\infty}^{+\infty} d\omega\, F[k_x, k_y, k_z, \omega_0] \cdot \delta[\omega - \omega_0] \cdot \delta\left[k - \frac{\omega}{c}\right] e^{+i(k_x x + k_y y + k_z z - \omega t)}\right)\\ = &\, e^{-i\omega_0 t} \iiint_{-\infty}^{+\infty} dk_x\, dk_y\, dk_z (F[k_x, k_y, k_z, \omega_0]\, e^{+i(k_x x + k_y y + k_z z)})\, \delta\left[k - \frac{\omega_0}{c}\right]\end{aligned}$$

$$= \exp[-i\omega_0 t] \iiint_{-\infty}^{+\infty} dk_x\, dk_y\, dk_z$$

$$\cdot \left(F[k_x, k_y, k_z, \omega_0]\, e^{+i(k_x x + k_y y + k_z z)}\right) \delta\left[\sqrt{k_x^2 + k_y^2 + k_z^2} - \frac{\omega_0}{c}\right] \tag{21.39}$$

where the sifting property of the Dirac delta function has been used to evaluate the integral over ω and the temporal phase term has been factored out of the 3-D integral over $\underline{\mathbf{k}}$. The remaining 1-D Dirac delta function of k may now be used to reduce the 3-D integral to 2-D. We need to integrate over any one of the three components of $\underline{\mathbf{k}}$ in the Dirac delta function. For reasons of convention, we will integrate over k_z, which is consistent with specifying a 2-D source distribution f in the x–y plane.

To integrate over k_z, we rewrite the 1-D Dirac delta function $\delta[\sqrt{k_x^2 + k_y^2 + k_z^2} - \omega_0/c]$ as $\delta[k_z - (k_z)_0]$, where $(k_z)_0$ is the positive value at the location where the argument of the Dirac delta function is zero, i.e., it satisfies the constraint $k = \omega_0/c$. In this case, we can write:

$$(k_z)_0 = \left[\frac{\omega_0^2}{c^2} - (k_x^2 + k_y^2)\right]^{\frac{1}{2}} \tag{21.40}$$

The argument of the 1-D Dirac delta function is now a *function* of k_z, which requires Equation (6.138) to evaluate; in the present context, we can write:

$$\delta[g[k_z]] = \frac{\delta[k_z - (k_z)_0]}{|dg/dk_z|_{k_z=(k_z)_0}} \tag{21.41}$$

This allows the expression for the 1-D Dirac delta function in terms of $(k_z)_0$ to be rewritten:

$$\delta\left[k - \frac{\omega_0}{c}\right] = \frac{1}{[1 - (c^2/\omega_0^2)(k_x^2 + k_y^2)]^{\frac{1}{2}}}\, \delta\left[k_z - \left(\frac{\omega_0^2}{c^2} - (k_x^2 + k_y^2)\right)^{\frac{1}{2}}\right] \tag{21.42}$$

where the positive root of the denominator is understood because of Equation (21.41).

We use the expression for the 1-D Dirac delta function in Equation (21.42) to evaluate the 1-D integral over k_z in Equation (21.39) by substituting $(k_z)_0 = ((\omega_0^2/c^2) - (k_x^2 + k_y^2))^{\frac{1}{2}}$ for every instance of k_z in the integrand:

$$\int_{-\infty}^{+\infty} dk_z (F[k_x, k_y, k_z, \omega_0] \exp[i(k_x x + k_y y + k_z z)])\, \delta\left[k - \frac{\omega_0}{c}\right]$$

$$= \int_{-\infty}^{+\infty} dk_z (F[k_x, k_y, k_z, \omega_0] \exp[i(k_x x + k_y y + k_z z)])$$

$$\cdot \frac{1}{[1 - (c^2/\omega_0^2)(k_x^2 + k_y^2)]^{\frac{1}{2}}} \cdot \delta\left[k_z - \left(\frac{\omega_0^2}{c^2} - (k_x^2 + k_y^2)\right)^{\frac{1}{2}}\right]$$

$$= \left(F[k_x, k_y, (k_z)_0, \omega_0] \frac{1}{[1 - (c^2/\omega_0^2)(k_x^2 + k_y^2)]^{\frac{1}{2}}}\right)$$

$$\cdot \exp\left[+i\left(k_x x + k_y y + \left(\frac{\omega_0^2}{c^2} - (k_x^2 + k_y^2)\right)^{\frac{1}{2}} \cdot z\right)\right]$$

$$= \left(F[k_x, k_y, (k_z)_0, \omega_0] \frac{1}{[1 - (c^2/\omega_0^2)(k_x^2 + k_y^2)]^{\frac{1}{2}}}\right)$$

$$\cdot \exp[+i(k_x x + k_y y)] \exp\left[+i\left(\frac{\omega_0^2}{c^2} - (k_x^2 + k_y^2)\right)^{\frac{1}{2}} \cdot z\right] \tag{21.43}$$

Note that the only dependence on the distance z "downstream" from the source plane specified by $z = 0$ appears in the last exponent. To simplify (and shorten!) this expression, we rename part of the integrand:

$$U[k_x, k_y; z] \equiv F[k_x, k_y, (k_z)_0, \omega_0] \cdot \left(\frac{1}{[1 - (c^2/\omega_0^2)(k_x^2 + k_y^2)]^{\frac{1}{2}}} \exp\left[+i \left(\frac{\omega_0^2}{c^2} - (k_x^2 + k_y^2) \right)^{\frac{1}{2}} \cdot z \right] \right) \tag{21.44}$$

If evaluated at the source plane $z = 0$, U is a 2-D function of $[k_x, k_y]$:

$$U[k_x, k_y; z = 0] = F\left[k_x, k_y, \left(\frac{\omega_0^2}{c^2} - (k_x^2 + k_y^2) \right)^{\frac{1}{2}} \right] \frac{1}{[1 - (c^2/\omega_0^2)(k_x^2 + k_y^2)]^{\frac{1}{2}}} \tag{21.45}$$

This is substituted into Equation (21.39) to obtain the more concise form:

$$\begin{aligned} \underline{\mathbf{E}}[x, y, z, t] &= e^{-i\omega_0 t} \cdot \left(\int_{-\infty}^{+\infty} dk_x \int_{-\infty}^{+\infty} dk_y \; U[k_x, k_y; z_1] \, e^{+i(k_x x + k_y y)} \right) \\ &= e^{-i\omega_0 t} \cdot \left(\iint_{-\infty}^{+\infty} dk_x \, dk_y \left(U[k_x, k_y; 0] \exp\left[+i \left(\frac{\omega_0^2}{c^2} - (k_x^2 + k_y^2) \right)^{\frac{1}{2}} z \right] \right) e^{+i(k_x x + k_y y)} \right) \end{aligned} \tag{21.46}$$

which may be identified as an inverse 2-D Fourier transform if we specify $\omega_0/c = k = |\underline{\mathbf{k}}| = 2\pi/\lambda_0$ and the frequency variables $k_x = 2\pi\xi$ and $k_y = 2\pi\eta$. The dependence on the propagation distance z is carried in the phase factor:

$$\begin{aligned} \exp[+i(k_z)_0 z] &= \exp\left[+i \left(\frac{\omega_0^2}{c^2} - (k_x^2 + k_y^2) \right)^{\frac{1}{2}} \cdot z \right] \\ &= \exp\left[+i \frac{2\pi}{\lambda_0} \left(\sqrt{1 - \lambda_0^2(\xi^2 + \eta^2)} \right) \cdot z \right] \end{aligned} \tag{21.47}$$

The other exponential factor becomes the kernel of the 2-D Fourier transform:

$$\exp[i(k_x x + k_y y)] = \exp[+2\pi i(\xi x + \eta y)] \tag{21.48}$$

The coordinates of the spectrum are scaled by constant factors of 2π, so for clarity we define the new spectrum:

$$A[\xi, \eta; z = 0] \equiv U[2\pi\xi, 2\pi\eta; 0] \tag{21.49}$$

With these substitutions, Equation (21.39) now is written as:

$$\underline{\mathbf{E}}[x, y, z, t] = (2\pi)^2 \cdot e^{-i\omega_0 t} \cdot \iint_{-\infty}^{+\infty} d\xi \, d\eta \left(A[\xi, \eta; 0] \cdot \exp\left[+i \frac{2\pi}{\lambda_0} \left(\sqrt{1 - \lambda_0^2(\xi^2 + \eta^2)} \right) z \right] \right) e^{+2\pi i(\xi x + \eta y)} \tag{21.50}$$

Since Equation (21.50) is now in the explicit form of a 2-D spatial inverse Fourier transform, we can evaluate the expression for the electric field at the plane z_1 in the frequency domain:

$$\mathcal{F}_2\{\underline{\mathbf{E}}[x, y, z]\} = (2\pi)^2 \cdot \left(A[\xi, \eta; 0] \cdot \exp\left[+2\pi i \left(\frac{z}{\lambda_0} \sqrt{1 - \lambda_0^2(\xi^2 + \eta^2)} - \nu_0 t \right) \right] \right) \tag{21.51}$$

In problems involving monochromatic light, we often can safely ignore the leading temporal part of the phase because it "disappears" in the calculation of the squared magnitude for the irradiance and thus contributes nothing to the spatial form of the field. For this reason, we will ignore the time dependence from this point forward for monochromatic systems, though we will revisit it in the next chapter when considering imaging systems in nonmonochromatic light.

The right-hand side of Equation (21.51) is the product of the 2-D spectrum of the electric field in the plane $z = 0$ with a term that depends on ξ, η, and z_1; then this second term has the form of a transfer function, which applies to light propagation over the distance between the plane $z = 0$ and the plane $z = z_1$:

$$H[\xi, \eta; z] = \exp\left[+i\frac{2\pi}{\lambda_0}\left(\sqrt{1 - \lambda_0^2(\xi^2 + \eta^2)}\right) z\right] \tag{21.52}$$

This is the result sought for the action of light propagation based on decomposing the light at the plane $z = 0$ into its constituent plane waves. However, the square root complicates the evaluation of the impulse response via the inverse Fourier transform since we have no analytic expression for $\mathcal{F}_2{}^{-1}\{\exp[\sqrt{1 - \xi^2}]\}$. Our task now is to make appropriate approximations that allow the impulse response to be evaluated. The obvious strategy is to remove the square root from the exponential that contains z. To do this, recall the binomial expansion:

$$(1 + \epsilon)^n = 1 + n\epsilon + \frac{n(n-1)}{2!}\epsilon^2 + \cdots \tag{21.53}$$

Again, we recognize that the series can be approximated by the first two terms if $|\epsilon| \ll 1$:

$$(1 + \epsilon)^n \cong 1 + n\epsilon \quad for\ |\epsilon| \ll 1 \tag{21.54}$$

In the current example, $n = \frac{1}{2}$ and $\epsilon = -\lambda_0^2(\xi^2 + \eta^2)$, so the approximation requires that $\xi^2 + \eta^2 \ll \lambda_0^{-2}$. If so, the exponential approximates to:

$$\begin{aligned}\exp\left[+\frac{2\pi i}{\lambda_0}\left(\sqrt{1 - \lambda_0^2(\xi^2 + \eta^2)}\right) \cdot z\right] &\cong \exp\left[+\frac{2\pi i}{\lambda_0}\left(1 - \frac{1}{2}\lambda_0^2(\xi^2 + \eta^2)\right) \cdot z\right] \\ &= \exp\left[+i\frac{2\pi z_1}{\lambda_0}\right] \cdot \exp[-i\pi\lambda_0 z \cdot (\xi^2 + \eta^2)]\end{aligned} \tag{21.55}$$

From this we see that the transfer function for propagation of the electric field over the distance z_1 from the source plane to the observation plane in the *Fresnel* approximation of Equation (21.55) is:

$$H[\xi, \eta; z] \cong \exp\left[+i\frac{2\pi z_1}{\lambda_0}\right] \cdot \exp[-i\pi(\lambda_0 z_1) \cdot (\xi^2 + \eta^2)]\ if\ \xi^2 + \eta^2 \ll \lambda_0^{-2} \tag{21.56}$$

This is a very significant result that specifies the transfer function of light propagation in the Fresnel region. It is sometimes called the *transfer function of free space*, though another (and perhaps clearer name) is the *transfer function of light propagation*. It is a quadratic-phase "allpass" filter, with which we have some experience. The variation in chirp rate with propagation distance z_1 means that the "image" produced by light propagation varies with the distance z to the observation plane. The allpass character of the filter makes intuitive sense because the total energy in the signal must be conserved by propagation in the Fresnel diffraction region.

21.2.3.2 Impulse Response of Monochromatic Propagation in Fresnel Approximation

We can now use the known Fourier transform of the quadratic-phase function in Equation (9.95) and the scaling theorem in Equation (9.117) to evaluate the corresponding impulse response in the Fresnel diffraction region:

$$
\begin{aligned}
h[x, y; z_1] &= \exp\left[+i\frac{2\pi z_1}{\lambda_0}\right] \cdot \mathcal{F}_2^{-1}\{\exp[-i\pi(\lambda_0 z_1)\cdot(\xi^2+\eta^2)]\} \\
&= \exp\left[+i\frac{2\pi z_1}{\lambda_0}\right] \cdot \mathcal{F}_1^{-1}\{\exp[-i\pi(\lambda_0 z_1)\cdot\xi^2]\} \cdot \mathcal{F}_1^{-1}\{\exp[-i\pi(\lambda_0 z_1)\cdot\eta^2]\} \\
&= \exp\left[+i\frac{2\pi z_1}{\lambda_0}\right] \cdot \left(\frac{1}{\sqrt{\lambda_0 z_1}}\, e^{-i\pi/4}\, e^{+i\pi x^2/\lambda_0 z_1}\right) \cdot \left(\frac{1}{\sqrt{\lambda_0 z_1}}\, e^{-i\pi/4}\, e^{+i\pi y^2/\lambda_0 z_1}\right) \\
&= \exp\left[+i\frac{2\pi z_1}{\lambda_0}\right] \cdot \left(\frac{1}{\sqrt{\lambda_0 z_1}}\right)^2 \left(\exp\left[-i\frac{\pi}{4}\right]\right)^2 \exp\left[+i\pi\frac{(x^2+y^2)}{\lambda_0 z_1}\right] \\
h[r; z_1] &= \left(\frac{1}{i\lambda_0 z_1} \cdot \exp\left[+2\pi i\frac{z_1}{\lambda_0}\right]\right) \cdot \exp\left[+i\pi\left(\frac{r}{\sqrt{\lambda_0 z_1}}\right)^2\right]
\end{aligned}
\tag{21.57}
$$

where the "off-axis" radial distance $r \equiv \sqrt{x^2+y^2}$. In words, the impulse response includes a constant phase due to propagation directly down the optical axis by the distance z_1, a multiplicative factor of z_1^{-1} (that represents the inverse square law in the Fresnel approximation), and a quadratic-phase factor with chirp rate $\alpha_0 = \sqrt{\lambda_0 z_1}$. Recall from Chapter 7 that the rate α_0 of a 2-D chirp is the radial distance from the origin to the point where the phase has changed by π radians. For larger distances z_1 from the source plane, the chirp rate is larger, which means that the chirp function varies "more slowly". As $z_1 \to +\infty$, the quadratic phase approaches a plane wave; this case satisfies the criteria Fraunhofer diffraction, which is considered below.

We have now seen that the electric field g at the plane $z = z_1$ may be derived from knowledge of the field $f[x, y]$ evaluated at the source plane $z = 0$ via a 2-D linear shift-invariant convolution:

$$
g[x, y; z = z_1] \cong \left(\frac{1}{i\lambda_0 z_1}\exp\left[+i\frac{2\pi z_1}{\lambda_0}\right]\right) \cdot \left(f[x, y; z=0] * \exp\left[+i\pi\frac{(x^2+y^2)}{\lambda_0 z_1}\right]\right) \tag{21.58}
$$

where the relation "$\cong$" indicates that this is the Fresnel approximation to the true amplitude. The impulse response in the Fresnel diffraction region is separable, which means that the convolution may be evaluated as the product of two 1-D convolutions in cases where the input function $f[x, y; 0]$ is separable. This also means that 1-D profiles of diffraction patterns calculated by 1-D convolution will have the correct functional form.

21.2.3.3 Interpretation of Fresnel Diffraction

The Fresnel approximation to diffraction has replaced the spherical wavefronts in Huygens' model with paraboloidal wavefronts. The amplitude and irradiance of the impulse response decreases by respective factors of z^{-1} and z^{-2}, as required by the inverse square law. The limitations of the Fresnel approximation are demonstrated by considering a single wavefront emitted from a point source located at the origin. The approximation assumes that the amplitudes observed on axis at $[r = 0; z_1]$ and off axis at $[r \neq 0; z_1]$ are identical, even though the propagation distance to the off-axis point is longer than to the on-axis point. In other words, the Fresnel approximation assumes that the effect of the inverse square law for light that travels at any angle relative to the z-axis is sufficiently small to be neglected.

The constant-phase factor $\exp[+i2\pi z_1/\lambda_0]$ represents the change in phase of light along the optical axis and the quadratic phase $\exp[+i\pi(r/\sqrt{\lambda_0 z_1})^2]$ is due to observation off the optical axis. Using the jargon of Chapter 17, the transfer function in Equation (21.56) demonstrates that light propagation in the Fresnel region has the form of an allpass quadratic-phase filter. The allpass character is due to the fact that all light emitted from the source plane reaches the observation plane; no energy is lost by the filter. The phase is "scrambled" by the increments added to the phase from the nonlinear function of spatial frequency. We will use the tools developed in Chapters 13 and 17 to interpret Fresnel diffraction.

The factor of i^{-1} in the impulse response is a constant phase due to the two factors of $\exp[-i\pi/4]$ in the inverse Fourier transform of the quadratic-phase transfer function. The corresponding physical interpretation is less obvious. This factor is often explained by using a model of the source function at the plane $z = 0$ as an aperture that obstructs light incident on this plane from a monochromatic point source located a very large distance away, i.e., the location of the point source is $[0, 0, z \ll 0]$. This light from this source at the aperture is a plane wave so that all points in the aperture experience the same optical phase at the same instant of time. However, the only actual "sources" of light in the aperture plane exist at the "edges" of the aperture where the light from the original point source is absorbed and reradiated as expanding spherical waves. The phase delay of the diffracted light may be interpreted as due to the delay in the absorption and reradiation.

As described in Section 17.5, the "image" of the diffracted light somewhat resembles that from a lowpass filter since components with small spatial frequencies are less affected by the filter. The effect is evident in the "blurry" images of sharp-edged objects, as will be demonstrated in the examples following.

When using the Fresnel diffraction model, it is important to check the validity of the approximations, which are confined to a relatively small region near the center of the observation plane, i.e., close to the optical axis of the system. The region where the Fresnel approximation for diffraction is valid is called the "near field".

21.2.3.4 Fresnel Transform

Equation (21.58) provides the formula for propagating the 2-D distribution of light $f[x, y; z = 0]$ to the plane $z = z_1$:

$$\begin{aligned} g[x, y; z_1] &\cong \left(\frac{1}{i\lambda_0 z_1} \exp\left[i\frac{2\pi z_1}{\lambda_0}\right]\right)\left(f[x, y; 0] * \exp\left[+i\pi\frac{(x^2+y^2)}{\lambda_0 z_1}\right]\right) \\ &= \left(\frac{1}{i\lambda_0 z_1} \exp\left[i\frac{2\pi z_1}{\lambda_0}\right]\right)\left(f[x, y; 0] * \exp\left[+i\pi\frac{r^2}{\lambda_0 z_1}\right]\right) \end{aligned} \tag{21.59}$$

where $r^2 = x^2 + y^2$. This operation is the *Fresnel transform* (Gori, 1994), which differs in concept from the Fourier transform: both the input and output are functions of spatial coordinates. Though we may not yet realize it, we have already developed some mathematical tools that recast the free-space propagation into other useful forms. The chirp Fourier transform that was described in Section 17.9 is particularly useful.

21.2.4 Examples of Fresnel Diffraction

We now evaluate the electric field in the Fresnel diffraction region for various source functions: a point source, a uniform planar source, a uniform planar source that is truncated by a "knife edge", and a rectangle.

21.2.4.1 Doubly Diffracted Light in Fresnel Region

The output field at the plane specified by $z = z_1 + z_2$ in the Fresnel region due to a point source at the origin may be derived in one step by inserting this distance into the impulse response in Equation (21.57):

$$\begin{aligned} \delta[x, y] * h[x, y; z_1 + z_2] &= h[x, y; z_1 + z_2] \\ &= \frac{1}{i\lambda_0(z_1+z_2)} \exp\left[+2\pi i\frac{z_1+z_2}{\lambda_0}\right] \exp\left[+i\pi\frac{x^2+y^2}{\lambda_0(z_1+z_2)}\right] \end{aligned} \tag{21.60}$$

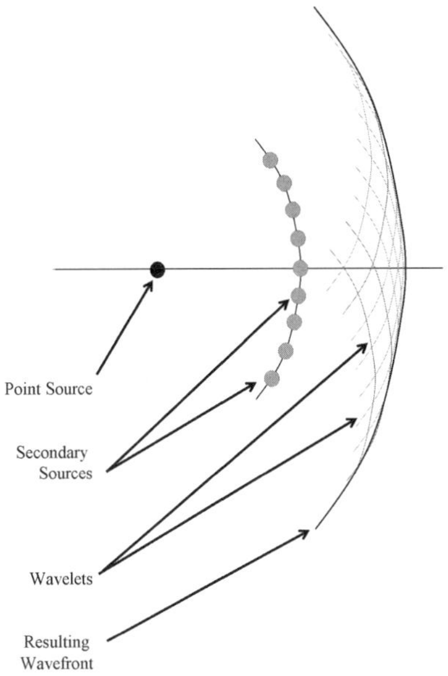

Figure 21.9 Huygens' principle and Fresnel diffraction from a point source. Each point on the spherical "wavelet" at the distance z_1 acts as a "secondary" source of spherical waves (shown in gray). The superposition evaluated at a distance z_2 "downstream" creates the next wavefront. The impulse response of diffraction evaluated for the distance $z_1 + z_2$ must be identical to the convolution of the impulse responses for propagation to z_1 and from z_1 to z_2.

In this case, we assume that both z_1 and z_2 are positive distances that satisfy the constraints for Fresnel diffraction, though it is useful for readers to consider the effect if $z_2 < 0$ (and particularly if $z_2 = -z_1$). If this description of propagation using the Fresnel transform is appropriate for Huygens' model of light propagation, then the same result should be obtained by propagating from the source plane to an intermediate plane at the distance z_1 followed by propagation over the distance z_2 to the final observation plane. The process is shown schematically in Figure 21.9:

$$\begin{aligned}\delta[x, y] * h[x, y; z_1 + z_2] &= (\delta[x, y] * h[x, y; z_1]) * h[x, y; z_2]\\ &= h[x, y; z_1] * h[x, y; z_2]\end{aligned} \tag{21.61}$$

After substituting the appropriate impulse responses from Equation (21.57) for the two propagation distances:

$$\begin{aligned}&\left(\frac{1}{i\lambda_0 z_1}\, e^{+2\pi i z_1/\lambda_0}\, e^{+i\pi(x^2+y^2)/\lambda_0 z_1}\right) * \left(\frac{1}{i\lambda_0 z_2}\, e^{+2\pi i z_2/\lambda_0}\, e^{+i\pi(x^2+y^2)/\lambda_0 z_2}\right)\\ &= \left(\frac{1}{i\lambda_0}\right)^2 \left(\frac{1}{z_1 z_2}\right) e^{+2\pi i(z_1+z_2)/\lambda_0} \left(e^{+i\pi x^2/\lambda_0 z_1} * e^{+i\pi x^2/\lambda_0 z_2}\right) \cdot \left(e^{+i\pi y^2/\lambda_0 z_1} * e^{+i\pi y^2/\lambda_0 z_2}\right)\end{aligned} \tag{21.62}$$

where the separability of the two functions has been used. The 1-D convolutions may be performed directly or in the frequency domain via the filter theorem; we choose the latter. The spectra of the 1-D chirp functions have the same form that can be obtained using the known transform in Equation (9.95) and the scaling theorem in Equation (9.117):

$$\mathcal{F}_1\left\{\exp\left[+i\pi\frac{x^2}{(\sqrt{\lambda_0 z_1})^2}\right]\right\} = (\sqrt{\lambda_0 z_1})\, e^{+i\pi/4} \exp[-i\pi(\sqrt{\lambda_0 z_1})^2\xi^2] \tag{21.63a}$$

$$\mathcal{F}_1\left\{\exp\left[+i\pi\frac{x^2}{(\sqrt{\lambda_0 z_2})^2}\right]\right\} = (\sqrt{\lambda_0 z_2})\, e^{+i\pi/4} \exp[-i\pi(\sqrt{\lambda_0 z_2})^2\xi^2] \tag{21.63b}$$

The spectrum of the 1-D convolution over x is easy to derive:

$$\begin{aligned}&(\sqrt{\lambda_0 z_1})(\sqrt{\lambda_0 z_2})(e^{+i\pi/4})^2(\exp[-i\pi\lambda_0 z_1\xi^2]\cdot\exp[-i\pi\lambda_0 z_2\xi^2])\\&\quad= (\lambda_0\sqrt{z_1 z_2})\, e^{+i\pi/2}\exp[-i\pi[\lambda_0(z_1+z_2)]\xi^2]\end{aligned} \tag{21.64}$$

which is a scaled "downchirp" function. The 1-D inverse Fourier transform yields the 1-D convolution over x:

$$\begin{aligned}\exp\left[+i\pi\frac{x^2}{\lambda_0 z_1}\right] * \exp\left[+i\pi\frac{x^2}{\lambda_0 z_2}\right] &= \left(\frac{\lambda_0\sqrt{z_1 z_2}}{\sqrt{\lambda_0(z_1+z_2)}}\right) e^{+i\pi/4}\exp\left[+i\pi\frac{x^2}{\lambda_0(z_1+z_2)}\right]\\&= \left(\sqrt{\lambda_0}\sqrt{\frac{z_1 z_2}{z_1+z_2}}\right) e^{+i\pi/4}\exp\left[+i\pi\left(\frac{x^2}{\lambda_0(z_1+z_2)}\right)\right]\end{aligned} \tag{21.65}$$

The product of this result with the leading multiplicative factors and the analogous convolution over y yields the desired result:

$$\begin{aligned}&(\delta[x,y] * h[x,y;z_1]) * h[x,y;z_2]\\&\quad= \left(\frac{1}{i\lambda_0}\right)^2\left(\frac{1}{z_1 z_2}\right) e^{+2\pi i/(z_1+z_2)}\left(\sqrt{\lambda_0}\sqrt{\frac{z_1 z_2}{z_1+z_2}}\, e^{+i\pi/4}\right)^2 \exp\left[+i\pi\frac{x^2+y^2}{\lambda_0(z_1+z_2)}\right]\\&\quad= \frac{1}{i\lambda_0(z_1+z_2)}\, e^{+2\pi i/(z_1+z_2)}\exp\left[+i\pi\frac{x^2+y^2}{\lambda_0(z_1+z_2)}\right]\end{aligned} \tag{21.66}$$

Because this expression is identical to that in Equation (21.60), the propagation of light over the distance $z_1 + z_2$ in the Fresnel diffraction region in one step is equivalent to propagation in two steps. This is (or should be) comforting because it confirms the consistency of the mathematical model of Fresnel diffraction. Note also that the correct leading amplitude term is obtained only by combining terms from the two orthogonal 1-D convolutions of quadratic-phase factors. In other words, the convolution for the Fresnel approximation produces two consistent expressions in two dimensions $[x, y]$, but not in one dimension; we need both spatial coordinates to produce the correct result.

21.2.4.2 Fresnel Diffraction from 2-D Planar Source

Now consider light diffraction from a constant (uniform) planar source of amplitude $\boldsymbol{\mathcal{E}}_0$ at the plane $z = 0$; the irradiance at this plane is proportional to the squared magnitude $|\boldsymbol{\mathcal{E}}_0|^2 \cdot 1[x, y]$. The radiation field obtained after propagating the distance $z_1 > 0$ in the Fresnel diffraction region is obtained by direct substitution into Equation (21.58):

$$g[x,y;z_1] = (\boldsymbol{\mathcal{E}}_0 \cdot 1[x,y]) * \left(\frac{1}{i\lambda_0 z_1}\exp\left[+2\pi i\frac{z_1}{\lambda_0}\right]\exp\left[+i\pi\frac{(x^2+y^2)}{\lambda_0 z_1}\right]\right) \tag{21.67}$$

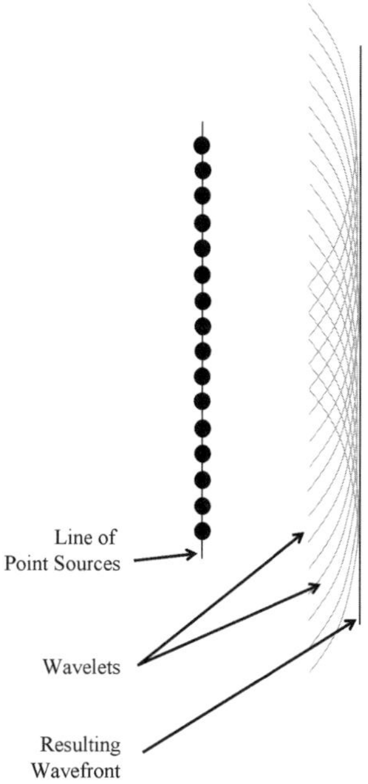

Figure 21.10 Profile of diffraction of light from a planar source. The planar wavefront begets another planar wavefront by superposition of Huygens' spherical wavelets.

The process is shown schematically in Figure 21.10. The convolution is easy to evaluate in the frequency domain by applying the property of Dirac delta functions in products from Equation (6.115):

$$\begin{aligned}
g[x, y; z_1] &= (\mathcal{E}_0 \cdot \delta[\xi, \eta]) \cdot \left(\exp\left[+2\pi i \frac{z_1}{\lambda_0}\right] \cdot \exp[-i\pi\lambda_0 z_1(\xi^2 + \eta^2)]\right) \\
&= \mathcal{E}_0 \cdot \delta[\xi, \eta] \cdot \exp\left[+2\pi i \frac{z_1}{\lambda_0}\right] \cdot \exp[-i\pi\lambda_0 z_1(0^2 + 0^2)] \\
&= \left(\mathcal{E}_0 \cdot \exp\left[+2\pi i \frac{z_1}{\lambda_0}\right]\right) \cdot \delta[\xi, \eta] \\
&\Longrightarrow g[x, y; z_1] = \left(\mathcal{E}_0 \cdot \exp\left[+2\pi i \frac{z_1}{\lambda_0}\right]\right) \cdot 1[x, y] \qquad (21.68)
\end{aligned}$$

In words, the amplitude obtained by convolving the uniform distribution with the impulse response for Fresnel diffraction is the same uniform distribution multiplied by the constant-phase factor $\exp[+2\pi i z_1/\lambda_0]$, which is the phase change due to propagation of light straight down the z-axis. The resulting irradiance at the observation plane is proportional to the squared magnitude:

$$I[x, y; z_1] = \left|\left(\mathcal{E}_0 \cdot \exp\left[+2\pi i \frac{z_1}{\lambda_0}\right]\right) \cdot 1[x, y]\right|^2 = |\mathcal{E}_0|^2 \cdot 1[x, y] \qquad (21.69)$$

In words, the irradiance of an infinite-support plane wave is unchanged by propagation; the inverse square law appears to have no impact. In fact, the inverse square law still applies to the individual wavelets, but the summation of the wavelets over the 2-D infinite support produces the constant result.

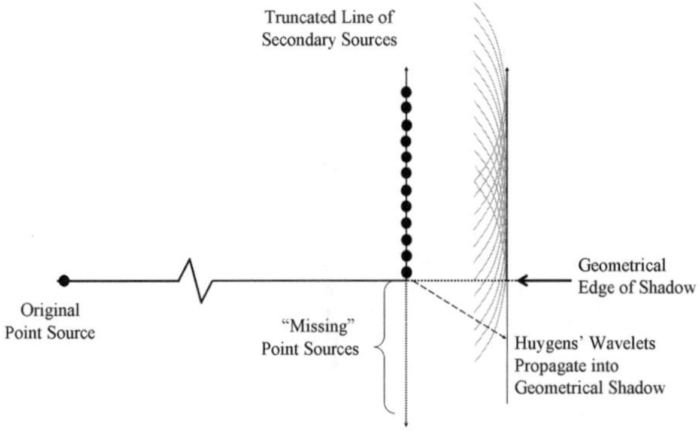

Figure 21.11 Fresnel diffraction of light into the geometrical shadow of a "knife edge". The line of secondary Huygens' wavelets from the original point source in the plane of the knife edge has been truncated. Light from spherical wavelets penetrates within the shadow of the knife edge.

21.2.4.3 Fresnel Diffraction from a Knife Edge

The functional form for a 2-D uniform "half-planar" source located at the plane $z = 0$ is $f[x, y; 0] = STEP[x] \cdot 1[y]$. The diagram in Figure 21.11 suggests that light from Huygens' wavelets generated by the sources in the plane of the knife edge will be visible "within" the geometrical shadow of the edge. The amplitude of the electric field evaluated at the observation plane $z = z_1$ is again the convolution of $f[x, y; 0]$ with the separable Fresnel impulse response $h[x, y; z_1]$:

$$\begin{aligned} g[x, y; z_1] &= \mathcal{E}_0 \cdot (STEP[x] \cdot 1[y]) * \left(\frac{1}{i\lambda_0 z_1} e^{+2\pi i z_1/\lambda_0} \exp\left[+i\pi\left(\frac{x^2+y^2}{\lambda_0 z_1}\right)\right]\right) \\ &= \frac{\mathcal{E}_0}{i\lambda_0 z_1} e^{+2\pi i z_1/\lambda_0}\left(STEP[x] * \exp\left[+i\pi\frac{x^2}{\lambda_0 z_1}\right]\right) \cdot \left(1[y] * \exp\left[+i\pi\frac{y^2}{\lambda_0 z_1}\right]\right) \\ &= \frac{\mathcal{E}_0}{i\lambda_0 z_1} e^{+2\pi i z_1/\lambda_0}\left(STEP[x] * \exp\left[+i\pi\frac{x^2}{\lambda_0 z_1}\right]\right) \cdot (e^{+i\pi/4}\sqrt{\lambda_0 z_1}\, 1[y]) \\ &= \left(\frac{\mathcal{E}_0}{\sqrt{\lambda_0 z_1}} e^{+2\pi i z_1/\lambda_0} e^{-i\pi/4}\right)\left(STEP[x] * \exp\left[+i\pi\frac{x^2}{\lambda_0 z_1}\right]\right) 1[y] \end{aligned} \tag{21.70}$$

where the central-ordinate theorem has been used. The 1-D convolution of $STEP[x]$ with the expanding quadratic phase may be evaluated by direct integration in the space domain:

$$\begin{aligned} STEP[x] * \exp\left[+i\pi\frac{x^2}{\lambda_0 z_1}\right] &= \int_{-\infty}^{+\infty} STEP[x-u] \exp\left[+i\pi\frac{u^2}{\lambda_0 z_1}\right] du \\ &= \int_{-\infty}^{x} \cos\left[+\pi\left(\frac{u}{\sqrt{\lambda_0 z_1}}\right)^2\right] du + i\int_{-\infty}^{x} \sin\left[+\pi\left(\frac{u}{\sqrt{\lambda_0 z_1}}\right)^2\right] du \end{aligned} \tag{21.71}$$

This is the "semi-infinite" integral of a complex chirp function $\exp[+\pi(x/\alpha)^2]$, where the chirp rate $\alpha = \sqrt{\lambda_0 z_1}$ is the distance measured along the x-axis from the center of symmetry (the origin of coordinates in this case) to the point where the phase differs by π radians. We already know that integrands are symmetric functions. Integrals of this form appear frequently in problems involving Fresnel diffraction

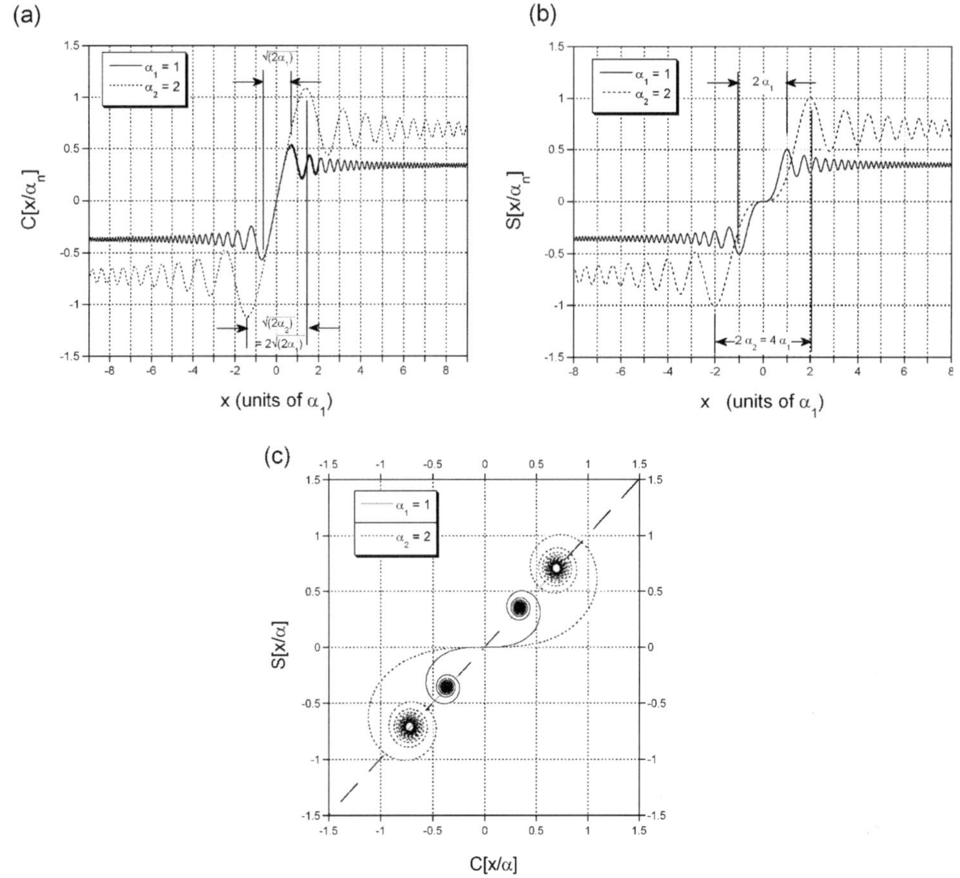

Figure 21.12 The Fresnel integrals: (a) $C[x/\alpha]$; (b) $S[x/\alpha]$ for two values of the chirp rate $\alpha_1 = \sqrt{\lambda_0 z_1}$ and $\alpha_2 = \sqrt{\lambda_0 z_2} = \sqrt{\lambda_0 \cdot (2z_1)}$, plotted on the same scale with x expressed in units of α_1. The width of the transition region increases with chirp rate α, and thus so do the values of the Fresnel integrals. (c) The Argand diagrams of $C[x/\alpha] + i\ S[x/\alpha]$ with axes plotted in units of α_1. These are the Cornu spirals.

and are often decomposed into a constant part and an integral from the origin of coordinates to x:

$$\begin{aligned}
&\int_{-\infty}^{x} \exp\left[+i\pi \frac{u^2}{\lambda_0 z_1}\right] du \\
&= \int_{-\infty}^{0} \exp\left[+i\pi \frac{u^2}{\lambda_0 z_1}\right] du + \int_{0}^{x} \exp\left[+i\pi \frac{u^2}{\lambda_0 z_1}\right] du \\
&= \frac{1}{2}\int_{-\infty}^{+\infty} \exp\left[+i\pi \frac{u^2}{\lambda_0 z_1}\right] du + \int_{0}^{x} \exp\left[+i\pi \frac{u^2}{\lambda_0 z_1}\right] du \\
&= \frac{1}{2}\sqrt{\lambda_0 z_1}\, e^{+i\pi/4} + \left(\int_0^x \cos\left[\pi \frac{u^2}{\lambda_0 z_1}\right] du + i \int_0^x \sin\left[\pi \frac{u^2}{\lambda_0 z_1}\right] du\right) \\
&= \left(\frac{1}{2}\sqrt{\frac{\lambda_0 z_1}{2}} + \int_0^x \cos\left[\pi \frac{u^2}{\lambda_0 z_1}\right] du\right) + i\left(\frac{1}{2}\sqrt{\frac{\lambda_0 z_1}{2}} + \int_0^x \sin\left[\pi \frac{u^2}{\lambda_0 z_1}\right] du\right) \qquad (21.72)
\end{aligned}$$

where the symmetry of the chirp function and the central-ordinate theorem have been used. Note that x may be positive or negative in the integral. We know from our study of the method of stationary phase in Chapter 13 that the largest contributions to the areas of both chirps occur in the vicinity of the origin. This observation gives a tool to understand the main features of the integral. For $x \ll 0$, the spatial frequency of both integrands is large, and the rapid oscillation of amplitude ensures that the areas of both the real and imaginary parts do not vary far from zero. As x increases, the spatial frequency decreases and thus the areas deviate from the null value by larger amounts before returning to zero and changing sign. At the coordinate $x < 0$ closest to the origin where $\cos[\pi(x/\sqrt{\lambda_0 z_1})] = 0$, the positive area of the central lobe of the cosine chirp "kicks in" and the area increases rapidly with x until the chirp reaches its first zero with $x > 0$. The two zeros of the cosine chirp occur at the points where the phase is $\pm\pi/2$, so that $x = \pm\sqrt{\lambda_0 z_1/2}$. For yet larger x, the spatial frequency of the integrand again is large and the area oscillates rapidly with x about a positive average value. The imaginary part of the integral exhibits similar behavior, but the positive lobe of the integrand is wider; it is located between the coordinates where the phase is $\pm\pi$, so that $x = \pm\sqrt{\lambda_0 z_1}$. The "widths" of the transition zones where the areas change from approximately zero to the approximate maxima are thus approximately the distances between the zeros. For the cosine chirp, the transition width is approximately $\sqrt{2\alpha} = \sqrt{2\lambda_0 z_1}$ and for the sine chirp it is approximately $2\alpha = 2 \cdot \sqrt{\lambda_0 z_1}$. The two indefinite integrals appear frequently enough that they are often assigned the names of the Fresnel cosine and sine integrals, respectively:

$$C[x] \equiv \int_0^x \cos\left[\pi \frac{\beta^2}{2}\right] d\beta \tag{21.73a}$$

$$S[x] \equiv \int_0^x \sin\left[\pi \frac{\beta^2}{2}\right] d\beta \tag{21.73b}$$

It should be noted that some authors define the Fresnel sine integral with a leading negative sign. The coordinate x can be any real number, positive or negative, which means that $C[x]$ and $S[x]$ are both odd and thus that $C[0] = S[0] = 0$. The computational tools of the Fourier transform may be applied to evaluate these integrals for some values of x:

$$C[+\infty] = \frac{1}{2}\int_{-\infty}^{+\infty} \cos\left[\pi \frac{u^2}{2}\right] du = \frac{1}{2}\Re\{\sqrt{2}\, e^{+i\pi/4}\} = +\frac{1}{2} = -C[-\infty] \tag{21.74a}$$

$$S[+\infty] = \frac{1}{2}\int_{-\infty}^{+\infty} \sin\left[\pi \frac{u^2}{2}\right] du = -\frac{1}{2}\Im\{\sqrt{2}\, e^{+i\pi/4}\} = +\frac{1}{2} = -S[-\infty] \tag{21.74b}$$

The amplitudes of both of these functions "oscillate" with the transverse coordinate x, as shown in Figure 21.12a and b for two values of $\sqrt{\lambda_0 z_1/2}$. As the distance z_1 to the observation plane increases, the leading amplitude constants increase and the rates of "oscillation" of the Fresnel integrals decrease. The Argand diagram of $C[x] + i\ S[x]$ is the *Cornu spiral*, which is useful to graphically estimate the Fresnel transform generated by a knife edge. The spiral "unwraps" from the value $-((1+i)/2)\sqrt{\lambda_0 z_1/2}$ at $x = -\infty$ and passes smoothly through the origin before "rewrapping" to the point $+((1+i)/2)\sqrt{\lambda_0 z_1/2}$, as shown in Figure 21.12c for the same two values of $\sqrt{\lambda_0 z_1/2}$ that differ by a factor of 2. Note that the real and imaginary parts of the amplitudes are larger for the larger value of α (and thus of z).

The Fresnel integrals may be inserted into Equation (21.70) to obtain a concise expression for the amplitude diffracted from the knife edge that is observed at the plane z_1:

$$\begin{aligned} g[x, y; z_1] = \boldsymbol{\mathcal{E}}_0 \exp\left[+\frac{2\pi i z_1}{\lambda_0}\right] \cdot \exp\left[-\frac{i\pi}{4}\right] \\ \cdot\left[\left(\sqrt{\frac{1}{8}} + \frac{1}{\sqrt{\lambda_0 z_1}} C\left[\frac{x}{\sqrt{\lambda_0 z_1/2}}\right]\right) + i\left(\sqrt{\frac{1}{8}} + \frac{1}{\sqrt{\lambda_0 z_1}} S\left[\frac{x}{\sqrt{\lambda_0 z_1/2}}\right]\right)\right] \end{aligned} \tag{21.75}$$

where the constant dependence on y is implicit. The observed "image" of the knife edge in the Fresnel diffraction region is proportional to the squared magnitude of this amplitude distribution:

$$\begin{aligned} I[x,y;z_1] &= |g[x,y;z_1]|^2 \\ &= |\mathcal{E}_0|^2 \cdot \left[\left(\sqrt{\frac{1}{8}} + \sqrt{2}\frac{1}{\sqrt{\lambda_0 z_1/2}} C\left[\frac{x}{\sqrt{\lambda_0 z_1/2}} \right] \right)^2 \right. \\ &\quad \left. + \left(\sqrt{\frac{1}{8}} + \sqrt{2}\frac{1}{\sqrt{\lambda_0 z_1/2}} S\left[\frac{x}{\sqrt{\lambda_0 z_1/2}} \right] \right)^2 \right] \\ &= \frac{|\mathcal{E}_0|^2}{4} + \frac{|\mathcal{E}_0|^2}{\lambda_0 z_1} \left[\left(C\left[\frac{x}{\sqrt{\lambda_0 z_1/2}} \right] \right)^2 + \left(S\left[\frac{x}{\sqrt{\lambda_0 z_1/2}} \right] \right)^2 \right] \\ &\quad + \sqrt{\frac{2}{\lambda_0 z_1}} |\mathcal{E}_0|^2 \left(C\left[\frac{x}{\sqrt{\lambda_0 z_1/2}} \right] + S\left[\frac{x}{\sqrt{\lambda_0 z_1/2}} \right] \right) \end{aligned} \tag{21.76}$$

If evaluated "on axis" (at $[0, 0; z_1]$) by substituting $C[0] = S[0] = 0$, the irradiance at the edge of the geometrical shadow is found to be the constant part:

$$|g[0,0;z_1]|^2 = \frac{|\mathcal{E}_0|^2}{4} \tag{21.77}$$

This is one-quarter of the irradiance that would be present at this location if the knife edge were not present, i.e., it is one-quarter of the value of the constant-irradiance plane wave that propagates down the z-axis to produce Equation (21.69). This value is independent of the propagation distance z_1, as long as the distance satisfies the criterion for Fresnel diffraction. Also note that the irradiance is not zero within the geometrical shadow (for $x < 0$), which agrees with the observation suggested by the schematic of Huygens' principle in Figure 21.11. Computed profiles of the irradiance patterns diffracted from a knife edge that is illuminated by a plane wave with $|\mathcal{E}_0|^2 = 1$ are shown in Figure 21.13 for two propagation distances, z_1 and $z_2 = 4z_1$. The x-axis of the graph is measured in the same units of $\sqrt{\lambda_0 z_1}$. The qualitative features of the irradiance patterns are identical; they rise slowly and smoothly from small values as the edge is approached. The pattern observed at the larger distance from the edge is larger in the shadow, showing that "more light" has been able to propagate into the shadow. The irradiances are equal at the edge of the geometrical shadow, as predicted by Equation (21.77). Away from the geometrical edge, both irradiance patterns exhibit "oscillations" that start slowly with a large amplitude and increase in rate and decrease in amplitude. This reflects the source of the oscillations as due to the integral of the complex chirp function in Equation (21.73). Also note that the "oscillation rate" of the pattern observed at the larger distance is slower; in fact, we can see that the scale of the first "oscillation" is of the order of the chirp rate evaluated for the particular propagation distance (i.e., $\sqrt{\lambda_0 z_1}$ and $\sqrt{4\lambda_0 z_1}$). Far from the edge, the amplitudes of both patterns converge to the irradiance ($|\mathcal{E}_0|^2 = 1$ in this case) that would have existed had the knife edge not been present.

21.2.4.4 Fresnel Diffraction from a Rectangular Aperture

As a final example, consider the "image" of a rectangle function observed at the same locations z_1 and $4z_1$ within the Fresnel diffraction region. Because the pattern from a knife edge is obtained by a linear shift-invariant convolution of a *STEP* function input and a rectangle may be written as a sum of displaced *STEP* functions, the amplitude diffracted from a rectangle function is the difference of copies of the diffracted amplitude from a knife edge at the appropriate distance z_1 and with appropriate translations along the x-axis. We know from Equation (6.13) that the rectangle function of width b may be written as:

$$RECT\left[\frac{x}{b}\right]1[y] = \left(STEP\left[x + \frac{b}{2}\right] - STEP\left[x - \frac{b}{2}\right] \right) 1[y] \tag{21.78}$$

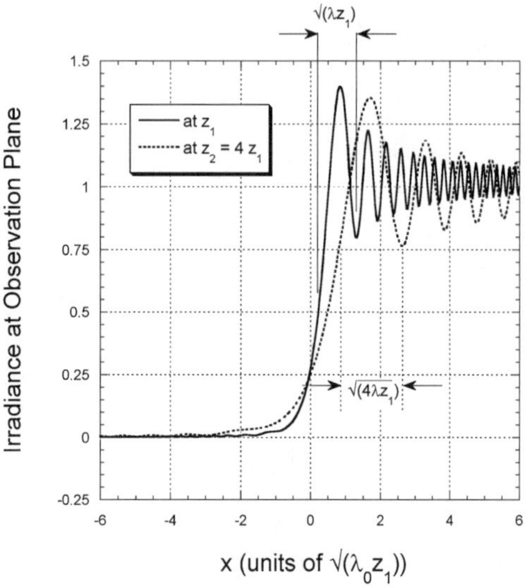

Figure 21.13 Intensity profiles of "images" of a knife edge in the Fresnel diffraction images at two distances z_1 and $4z_1$. The profiles have been normalized to the same scale. Note that the width of the first "oscillation" of the irradiance is of the order of $\sqrt{\lambda_0 z_1}$ and $\sqrt{\lambda_0 z_1}$, respectively. The "period" of the "ringing oscillation" of the irradiance decreases as the observation distance increases and the irradiances for both propagation distances converge to the value that would have been obtained had the edge not been present.

The diffracted amplitude at that distance is:

$$g[x, y; z_1] = \left(\frac{\mathcal{E}_0}{\sqrt{i\lambda_0 z_1}}\, e^{+2\pi i z_1/\lambda_0}\right) \cdot \left[\left(C\left[\frac{x+b/2}{\sqrt{\lambda_0 z_1/2}}\right] - C\left[\frac{x-b/2}{\sqrt{\lambda_0 z_1/2}}\right]\right) + i\left(S\left[\frac{x+b/2}{\sqrt{\lambda_0 z_1/2}}\right] - S\left[\frac{x-b/2}{\sqrt{\lambda_0 z_1/2}}\right]\right)\right] \tag{21.79a}$$

and the diffracted irradiance is easily obtained from Equation (21.75):

$$|g[x, y; z_1]|^2 = \frac{|\mathcal{E}_0|^2}{\lambda_0 z_1} \cdot \left[\left(C\left[\frac{x+b/2}{\sqrt{\lambda_0 z_1/2}}\right] - C\left[\frac{x-b/2}{\sqrt{\lambda_0 z_1/2}}\right]\right)^2 + \left(S\left[\frac{x-b/2}{\sqrt{\lambda_0 z_1/2}}\right] - S\left[\frac{x+b/2}{\sqrt{\lambda_0 z_1/2}}\right]\right)^2\right] \tag{21.79b}$$

In words, the two edges generate knife-edge patterns that vary with the propagation distance z_1 (and wavelength λ_0). The oscillating amplitudes generated by the two edges "combine" constructively and destructively to produce the diffracted irradiance. Just as was the case for the knife edge, the spatial frequency of the "ringing" from the edges decreases with increased distance from the source. If the rectangle is "wide", then the frequency of the oscillations in largest in the center of the patter. The examples in Figure 21.14 are evaluated at the same distances used for the knife edge in Figure 21.13.

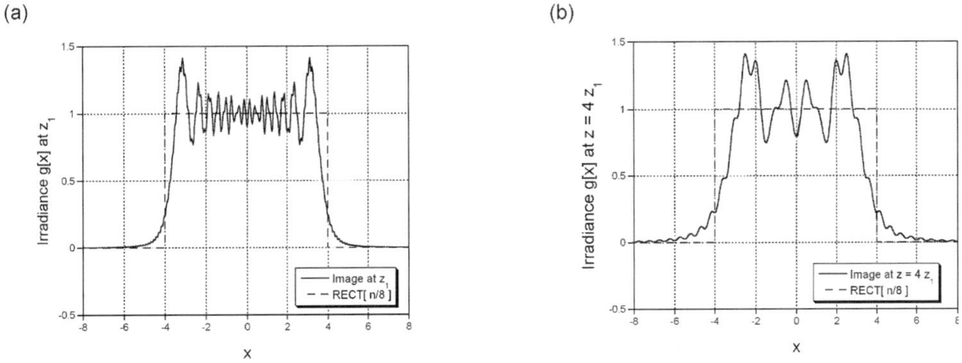

Figure 21.14 Computed profiles of the squared magnitude due to Fresnel diffraction from 1-D rectangular aperture at distances (a) z_1 and (b) $z = 4 \cdot z_1$. The oscillation period of the ringing increases with increasing propagation distance.

21.2.4.5 Fresnel Transform as a C–M–C Chirp Fourier Transform

The 2-D C–M–C chirp Fourier transform in Equation (17.84) of the function $f[x, y]$ has the form:

$$F\left[\frac{x}{\alpha^2}, \frac{y}{\alpha^2}\right] = \left(\frac{1}{\alpha^2}\, e^{-i\pi/2}\right)\left[\left(\left[f[x, y] * e^{+i\pi(r/\alpha)^2}\right] \cdot e^{-i\pi(r/\alpha)^2}\right) * e^{+i\pi(r/\alpha)^2}\right] \tag{21.80}$$

The first convolution on the right-hand side is the Fresnel transform of $f[x, y]$. We can eliminate the other terms on that side by a sequence of operations beginning with convolution from the right with the complex conjugate of the propagation function:

$$\begin{aligned} F\left[\frac{x}{\alpha^2}, \frac{y}{\alpha^2}\right] * e^{-i\pi(r/\alpha)^2} &= \left(\frac{-i}{\alpha^2}\right)[([f[x, y] * e^{+i\pi(r/\alpha)^2}] \cdot e^{-i\pi(r/\alpha)^2}) * e^{+i\pi(r/\alpha)^2}] * e^{-i\pi(r/\alpha)^2} \\ &= \left(\frac{-i}{\alpha^2}\right)([f[x, y] * e^{+i\pi(r/\alpha)^2}] \cdot e^{-i\pi(r/\alpha)^2}) * (e^{+i\pi(r/\alpha)^2} * e^{-i\pi(r/\alpha)^2}) \\ &= \left(\frac{-i}{\alpha^2}\right)([f[x, y] * e^{+i\pi(r/\alpha)^2}] \cdot e^{-i\pi(r/\alpha)^2}) * |\alpha|^2\, \delta[x, y] \\ &= -i[f[x, y] * e^{+i\pi(r/\alpha)^2}] \cdot e^{-i\pi(r/\alpha)^2} \end{aligned} \tag{21.81}$$

where Equation (17.28) has been used. Now we multiply both sides by $-i \;\, e^{+i\pi(r/\alpha)^2}$ and substitute $\alpha^2 = \lambda_0 z_1$ $(z_1 > 0)$:

$$f[x, y] * \exp\left[+i\pi\frac{r^2}{\lambda_0 z_1}\right] = +i\left(F\left[\frac{x}{\lambda_0 z_1}, \frac{y}{\lambda_0 z_1}\right] * \exp\left[-i\pi\frac{r^2}{\lambda_0 z_1}\right]\right) \cdot \exp\left[+i\pi\frac{r^2}{\lambda_0 z_1}\right] \tag{21.82}$$

After applying the amplitude scale factor from Equation (21.57), we obtain an equivalent expression for the Fresnel transform:

$$\begin{aligned} g[x, y; z_1] &= \mathcal{E}_0 \cdot \frac{1}{i\lambda_0 z_1}\, e^{+2\pi i z_1/\lambda_0}\left(f[x, y; 0] * e^{+i\pi r^2/\lambda_0 z_1}\right) \\ &= \mathcal{E}_0 \cdot \frac{1}{i\lambda_0 z_1}\, e^{+2\pi i z_1/\lambda_0} \cdot i\left(F\left[\frac{x}{\lambda_0 z_1}, \frac{y}{\lambda_0 z_1}\right] * e^{-i\pi r^2/\lambda_0 z_1}\right) \cdot e^{+i\pi r^2/\lambda_0 z_1} \\ &= \left(\mathcal{E}_0 \cdot \frac{1}{\lambda_0 z_1}\, e^{+2\pi i z_1/\lambda_0} \cdot e^{+i\pi r^2/\lambda_0 z_1}\right) \cdot \left(F\left[\frac{x}{\lambda_0 z_1}, \frac{y}{\lambda_0 z_1}\right] * e^{-i\pi r^2/\lambda_0 z_1}\right) \end{aligned} \tag{21.83}$$

In words, the Fresnel transform of $f[x, y]$ may be written as the convolution of a scaled replica of its spectrum with a downchirp, followed by multiplication by an upchirp. This gives a useful equivalent expression for the light diffracted by the aperture function $f[x, y]$ in the Fresnel region.

21.2.4.6 Stationary-Phase Approximation to Fresnel Transform

We can also evaluate the Fresnel diffraction integral in the frequency domain. The spectrum of the source function at the origin is $F[\xi, \eta; z=0]$ and the transfer function was given in Equation (21.56). The space-domain representation of the diffracted light is the inverse 2-D Fourier transform:

$$g[x, y; z_1] = e^{+2\pi i z_1/\lambda_0} \cdot \mathcal{F}_2^{-1}\{\mathcal{E}_0 \cdot F[\xi, \eta; z=0]\, e^{-i\pi\lambda_0 z_1(\xi^2+\eta^2)}\}$$
$$= \mathcal{E}_0 \cdot e^{+2\pi i z_1/\lambda_0} \iint_{-\infty}^{+\infty} (F[\xi, \eta; z=0]\, e^{-i\pi\lambda_0 z_1(\xi^2+\eta^2)})\, e^{+2\pi i(\xi x+\eta y)}\, d\xi \quad (21.84)$$

The integral may be approximated via the method of stationary phase that was described in Chapter 13. For simplicity, we first consider a 1-D complex-valued input spectrum $F[\xi]$, which can be decomposed into real and imaginary parts:

$$F[\xi] = \Re\{F[\xi]\} + i\Im\{F[\xi]\} \quad (21.85)$$

The 1-D amplitude at z_1 in the space domain is:

$$g[x; z_1] = \mathcal{E}_0 \cdot e^{+2\pi i z_1/\lambda_0}\, \mathcal{F}^{-1}\{(\Re\{F[\xi]\} + i\Im\{F[\xi]\})\, e^{-i\pi\lambda_0 z_1 \xi^2}\}$$
$$= \mathcal{E}_0 \cdot e^{+2\pi i z_1/\lambda_0} \cdot \left[\left(\int_{-\infty}^{+\infty} \Re\{F[\xi]\}\, e^{-i\pi\lambda_0 z_1 \xi^2}\, e^{+2\pi i\xi x}\, d\xi\right)\right.$$
$$\left. + i\left(\int_{-\infty}^{+\infty} \Im\{F[\xi]\}\, e^{-i\pi\lambda_0 z_1 \xi^2}\, e^{+2\pi i\xi x}\, d\xi\right)\right] \quad (21.86)$$

Since both $\Re\{F[\xi]\}$ and $\Im\{F[\xi]\}$ are real valued, the integrands satisfy the necessary constraints for the method of stationary phase. We can rewrite the exponents using the notation of Equation (13.55):

$$g[x; z_1] = \mathcal{E}_0 \cdot e^{+2\pi i z_1/\lambda_0} \cdot \left[\left(\int_{-\infty}^{+\infty} \Re\{F[\xi]\}\, e^{+ix(-\pi(\lambda_0 z_1/x)\xi^2+2\pi\xi)}\, d\xi\right)\right.$$
$$\left. + i\left(\int_{-\infty}^{+\infty} \Im\{F[\xi]\}\, e^{ix(-\pi(\lambda_0 z_1/x)\xi^2+2\pi\xi)}\, d\xi\right)\right]$$
$$= \mathcal{E}_0 \cdot e^{+2\pi i z_1/\lambda_0} \cdot \left[\left(\int_{-\infty}^{+\infty} \Re\{F[\xi]\}\, e^{+ix\mu[\xi]}\, d\xi\right) + i\left(\int_{-\infty}^{+\infty} \Im\{F[\xi]\}\, e^{+ix\mu[\xi]}\, d\xi\right)\right] \quad (21.87)$$

where $\mu[\xi] \equiv \pi(\lambda_0 z_1/x)\xi^2 + 2\pi\xi$. The approximate solution for the 1-D integral using the method of stationary phase in Equation (13.67) is:

$$g[x; z_1] \cong \mathcal{E}_0 \cdot e^{+2\pi i z_1/\lambda_0}\left(\Re\left\{F\left[\frac{x}{\lambda_0 z_1}\right]\right\} + i\,\Im\left\{F\left[\frac{x}{\lambda_0 z_1}\right]\right\}\right) e^{+i\pi/4}\, e^{+i\pi x^2/\lambda_0 z_1} \cdot \frac{1}{\sqrt{-\lambda_0 z_1}}$$
$$= \mathcal{E}_0 \cdot e^{+2\pi i z_1/\lambda_0}\, \frac{1}{\sqrt{+i\lambda_0 z_1}}\left(\Re\left\{F\left[\frac{x}{\lambda_0 z_1}\right]\right\} + i\,\Im\left\{F\left[\frac{x}{\lambda_0 z_1}\right]\right\}\right) e^{+i\pi x^2/\lambda_0 z_1}$$
$$= \mathcal{E}_0 \cdot \frac{1}{\sqrt{+i\lambda_0 z_1}} F\left[\frac{x}{\lambda_0 z_1}\right] e^{+2\pi i z_1/\lambda_0}\, e^{+i\pi x^2/\lambda_0 z_1} \quad (21.88)$$

which is proportional to the Fourier transform of the input amplitude and a quadratic-phase factor with chirp rate $\sqrt{\lambda_0 z_1}$.

The 2-D solution is a straightforward generalization of this result, including a second factor of $(+i\lambda_0 z_1)^{-1/2}$:

$$g[x, y; z_1 \to \infty] \cong \frac{\mathcal{E}_0}{i\lambda_0 z_1} \exp\left[+2\pi i \frac{z_1}{\lambda_0}\right] \exp\left[+i\pi \frac{(x^2+y^2)}{\lambda_0 z_1}\right] F\left[\frac{x}{\lambda_0 z_1}, \frac{y}{\lambda_0 z_1}\right] \tag{21.89}$$

The constraint of the Fresnel approximation, that the propagation distance z_1 must be large compared to the scale of the function $f[x, y]$, also applies here. In words, the stationary-phase approximation of the Fresnel transform produces a function in the space domain with amplitude proportional to a scaled replica of the Fourier transform of the object function. The space-domain coordinates are scaled by $\lambda_0 z_1$, thus ensuring that the coordinates of F have the proper dimensions of spatial frequency. Note the additional quadratic-phase factor due to off-axis propagation in the observation and that the constant phase due to the propagation down the z-axis and the amplitude factor $(\lambda_0 z_1)^{-1}$ are still present. This result will be helpful when interpreting some results in the next section.

21.3 Fraunhofer Diffraction

Now consider observation of diffracted light at a plane yet more distant from the source than in the Fresnel approximation so that the expanding spherical wavefronts may be accurately modeled as planar. The mathematical form is derived by writing out the convolution in Equation (21.58):

$$f[x, y; 0] * \exp\left[+i\pi \frac{(x^2+y^2)}{\lambda_0 z_1}\right] = \iint_{-\infty}^{+\infty} f[\alpha, \beta; 0] \cdot \exp\left[+i\pi \frac{(x-\alpha)^2 + (y-\beta)^2}{\lambda_0 z_1}\right] d\alpha\, d\beta \tag{21.90}$$

where the coordinates $[x, y]$ specify the location in the observation plane $z = z_1$ and the integral is over the coordinates $[\alpha, \beta]$ in the source (or "input") plane $z = 0$. The argument of the exponential can be expanded into three terms:

$$\frac{(x-\alpha)^2 + (y-\beta)^2}{\lambda_0 z_1} = \left(\frac{x^2+y^2}{\lambda_0 z_1}\right) - \left(\frac{2(x\alpha + y\beta)}{\lambda_0 z_1}\right) + \left(\frac{\alpha^2+\beta^2}{\lambda_0 z_1}\right) \tag{21.91}$$

The last is the quadratic phase due to off-axis locations at the source and may be ignored if z_1 is sufficiently large and the pupil sufficiently small to ensure that the quadratic-phase factor is approximately unity:

$$\frac{\alpha^2+\beta^2}{\lambda_0 z_1} \cong 0 \Longrightarrow z_1 \gg \frac{\alpha^2+\beta^2}{\lambda_0} \tag{21.92}$$

In a realistic situation where the aperture diameter is 10 mm (so that $\alpha^2 + \beta^2 = 5$ mm) and the illumination source is a He:Ne laser operating at $\lambda_0 = 623.8$ nm, then the phase difference of light emitted from the center and from the edge of the pupil is 1 radian if the propagation distance $z_1 \cong 40$ m. For the quadratic-phase factor to evaluate to unity, the phase difference must be much smaller yet. The region where this, the *Fraunhofer approximation*, is valid is called the *far field*, and is effectively an "infinite distance" from the source since the spherical waves are modeled as plane waves. This large physical distance may be reduced optically by inserting a suitable lens to "bring" the distant observation plane closer to the source.

If Equation (21.92) is satisfied, then the amplitude of the electric field observed in the plane specified by z_1 in Equation (21.58) simplifies to:

$$E\left[x, y; z_1 \gg \frac{\alpha^2+\beta^2}{\lambda_0}\right] \cong \frac{\mathcal{E}_0}{i\lambda_0 z_1} \cdot \exp\left[+2\pi i \frac{z_1}{\lambda_0}\right] \exp\left[+i\pi \frac{(x^2+y^2)}{\lambda_0 z_1}\right] \cdot \iint_{-\infty}^{+\infty} f[\alpha, \beta] \exp\left[-2\pi i \frac{(x\alpha + y\beta)}{\lambda_0 z_1}\right] d\alpha\, d\beta \tag{21.93}$$

At a specific plane, the multiplicative factors include a constant $\mathcal{E}_0/\lambda_0 z_1$ due to the source "strength" and the distance propagated along the optical axis, a linear-phase term $\exp[+2\pi i z_1/\lambda_0]$ due to the distance propagated along the optical axis, a quadratic-phase factor $\exp[+i\pi(x^2+y^2)/\lambda_0 z_1]$ that describes the additional phase due to the "off-axis" distance traveled from the source point to the observation point, and a superposition integral over the source coordinates $[\alpha, \beta]$.

To help visualize the superposition integral, consider the result for a single point source located at $[x_0, y_0]$ with "strength" $\mathcal{E}_0$:

$$f[x, y] = \mathcal{E}_0\, \delta[x - x_0, y - y_0] \tag{21.94}$$

The electric field evaluated at the location $[x_1, y_1]$ in the observation plane $z = z_1$ is:

$$\begin{aligned} E&\left[x, y;\ z_1 \gg \frac{\alpha^2+\beta^2}{\lambda_0}\right] \\ &\cong \left(\frac{\mathcal{E}_0}{i\lambda_0 z_1} e^{+2\pi i z_1/\lambda_0}\, e^{+i\pi(x^2+y^2)/\lambda_0 z_1}\right) \iint_{-\infty}^{+\infty} \delta[\alpha - x_0, \beta - y_0]\, e^{-2\pi i(x\alpha + y\beta)/\lambda_0 z_1}\, d\alpha\, d\beta \\ &= \left(\frac{\mathcal{E}_0}{i\lambda_0 z_1} e^{+2\pi i z_1/\lambda_0}\, e^{+i\pi(x^2+y^2)/\lambda_0 z_1}\right) e^{-2\pi i((x/\lambda_0 z_1)x_0 + (y/\lambda_0 z_1)y_0)} \end{aligned} \tag{21.95}$$

The superposition integral yields a function that has linear phase along the x- and y-axes. In short, the spherical waves in the superposition integral are approximated by plane waves that are "tilted" relative to the optical axis. Put yet another way, the additional approximation allowed by propagating large distances has changed the integrand to a function of the product of the positions in the source and observation coordinates ($x \cdot x_0$ and $y \cdot y_0$), rather than of their differences ($x - x_0$ and $y - y_0$), as in convolution. In short, the superposition integral is no longer a convolution, and the "system" is now linear and shift *variant*.

We can recast the last expression in Equation (21.93) by recognizing that the scale factors in the bilinear phase have the dimensions of reciprocal length and may be considered to be spatial frequencies. We make the substitutions $\xi = x/\lambda_0 z_1$ and $\eta = y/\lambda_0 z_1$:

$$\begin{aligned} E[x, y; z_1] &\cong \frac{\mathcal{E}_0}{i\lambda_0 z_1} e^{+2\pi i z_1/\lambda_0}\, e^{+i\pi(x^2+y^2)/\lambda_0 z_1} \iint_{-\infty}^{+\infty} f[\alpha, \beta]\, e^{-2\pi i(\xi\alpha + \eta\beta)}\, d\alpha\, d\beta \\ &= \frac{\mathcal{E}_0}{i\lambda_0 z_1} e^{+2\pi i z_1/\lambda_0}\, e^{+i\pi(x^2+y^2)/\lambda_0 z_1}\, F[\xi, \eta]|_{(\xi = x/\lambda_0 z_1, \eta = y/\lambda_0 z_1)} \\ &= \frac{\mathcal{E}_0}{i\lambda_0 z_1} e^{+2\pi i z_1/\lambda_0} \exp\left[+i\pi \frac{(x^2+y^2)}{\lambda_0 z_1}\right] F\left[\frac{x}{\lambda_0 z_1}, \frac{y}{\lambda_0 z_1}\right] \end{aligned} \tag{21.96}$$

This important expression demonstrates that diffraction of light in the far field generates an amplitude proportional to the Fourier transform of the spatial distribution of the source function. Note that the expression in Equation (21.96) is identical to that in Equation (21.89) obtained from the stationary-phase approximation to the Fresnel approximation.

Again recognize that sensors operating at visible wavelengths actually measure the time average of the squared magnitude of the electric field, which eliminates the leading constant- and quadratic-phase factors:

$$I[x, y; z_1] \propto |E[x, y; z_1]|^2 = \frac{|\mathcal{E}_0|^2}{\lambda_0^2 z_1^2} \left| F\left[\frac{x}{\lambda_0 z_1}, \frac{y}{\lambda_0 z_1}\right]\right|^2 \tag{21.97}$$

From our study of Fourier transforms, we can now infer some important (and possibly counterintuitive) properties of Fraunhofer diffraction:

1. Scaling theorem: if the scale factor of the aperture function $f[x, y]$ increases, then the resulting diffraction pattern becomes brighter and "smaller", i.e., its scale factor is decreased.

2. Shift theorem: translation of $f[x, y]$ adds a linear term to the phase of the diffraction pattern, which is not visible in the irradiance. Thus translation of the input has no visible effect on the diffraction pattern.
3. Modulation theorem: if an aperture can be expressed as the product of two functions, the amplitude of the diffraction pattern is their convolution.
4. Filter theorem: if an aperture pattern is the convolution of two patterns, the amplitude of the resulting diffraction pattern is the product of the amplitudes of the individual patterns.

21.3.1 Examples of Fraunhofer Diffraction

21.3.1.1 On-Axis Point Source

The 2-D function for a point source located at the origin including the time dependence is:

$$s[x, y, t; z = 0, \nu_0] = A_0 \exp[-2\pi i \nu_0 t] \cdot \delta[x, y] \tag{21.98}$$

If propagated a distance z_0 into the Fraunhofer region, we already know that the illumination at the plane $z = z_0$ is proportional to the Fourier transform of the source function. The time required for light to propagate the distance z_0 to the source plane means that the temporal part of the phase of the light at the observation plane is equal to the phase at the source plane evaluated at the earlier time $t - z_0/c$. The amplitude at the observation plane observed at the time t is therefore:

$$\begin{aligned} g[x, y, t; z = z_0, \nu_0] &= \frac{1}{i\lambda_0 z_0} \exp\left[+2\pi i\left(\frac{z_0}{\lambda_0} - \nu_0 t\right)\right] \cdot \mathcal{F}_2\{A_0 \cdot \delta[x, y]\}|_{\xi \to x/\lambda_0 z_0, \eta \to y/\lambda_0 z_0} \\ &= A_0 \cdot \frac{1}{i\lambda_0 z_0} \exp\left[+2\pi i\left(\frac{z_0}{\lambda_0} - \nu_0\left(t - \frac{z_0}{c}\right)\right)\right] \cdot 1\left[\frac{x}{\lambda_0 z_0}, \frac{y}{\lambda_0 z_0}\right] \\ &= A_0 \cdot \frac{1}{i\lambda_0 z_0} \exp\left[+2\pi i\left(\frac{z_0}{\lambda_0} - \nu_0 t + \left(\frac{\nu_0 z_0}{c}\right)\right)\right] \cdot 1[x, y] \\ &\equiv A_0 \cdot K_0 \cdot \exp\left[-i\left(2\pi \nu_0 t - \left(\frac{\nu_0 z_0}{c}\right)\right)\right] \cdot 1[x, y] \end{aligned} \tag{21.99}$$

where we see that the effect of the propagation time is to add a constant to the temporal part of the phase. The leading constants are collapsed into K_0:

$$K_0 \equiv \frac{1}{i\lambda_0 z_0} \exp\left[+2\pi i \frac{z_0}{\lambda_0}\right] \tag{21.100}$$

The measured signal is the time average of the squared magnitude, or the irradiance. Because the source is monochromatic, the squared magnitude eliminates the temporal term, so the time average in the calculation of the observed irradiance pattern just scales the squared magnitude by an ignorable constant, so we can say that:

$$\begin{aligned} I[x, y; z = z_0, \nu_0, T_0] &\propto |g[x, y, t; z = z_0, \nu_0]|^2 \\ &= A_0^2 \cdot |K_0|^2 \cdot \left|\exp\left[-i\left(2\pi \nu_0 t - \left(\frac{\nu_0 z_0}{c}\right)\right)\right]\right|^2 \cdot |1[x, y]|^2 \\ &= \left(\frac{A_0}{\lambda_0 z_0}\right)^2 \cdot 1[x, y] \end{aligned} \tag{21.101}$$

where the factor of z_0^{-2} represents the inverse square law in the Fraunhofer region. In words, the irradiance from a monochromatic on-axis point source observed in the Fraunhofer diffraction region is spatially constant. The 1-D example is shown in Figure 21.15.

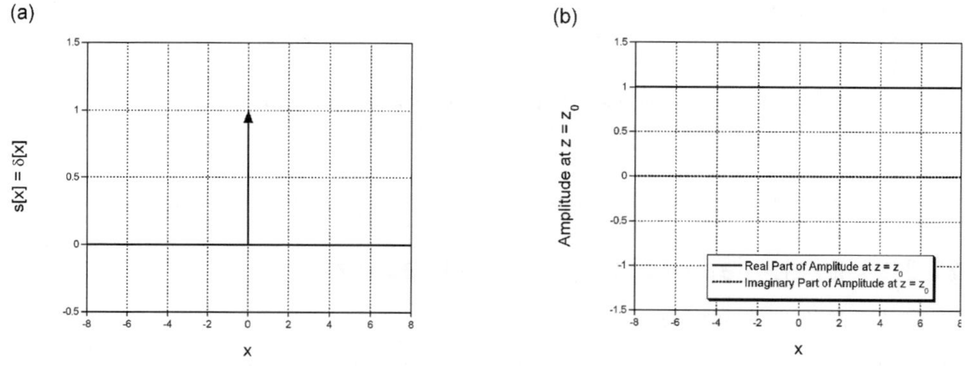

Figure 21.15 Fraunhofer diffraction of on-axis point source: (a) source function $s[x] = \delta[x]$; (b) amplitude at plane $z = z_0$ in Fraunhofer diffraction region is proportional to $S[x/\lambda_0 z_0] = 1[x] + i \cdot 0[x]$, so that the irradiance at $z = z_0$ is constant.

21.3.1.2 Example: Off-Axis Point Source

For a reason that will be clear shortly, we choose the off-axis distance to be $d_0/2$. The source function is:

$$s[x, y; z = 0] = A_0 \cdot \exp[-2\pi i \nu_0 t] \cdot \delta\left[x - \frac{d_0}{2}, y\right] \tag{21.102}$$

After repeating the steps just followed, the observed irradiance pattern is seen to be a replica of that from a source at the origin multiplied by a linear-phase factor:

$$\begin{aligned}
g[x, y; z = z_0] \\
&= \frac{1}{i\lambda_0 z_0} \exp\left[+2\pi i\left(\frac{z_0}{\lambda_0} - \nu_0 t\right)\right] \cdot \mathcal{F}_2\left\{A_0 \cdot \delta\left[x - \frac{d_0}{2}, y\right]\right\}\bigg|_{\xi \to x/\lambda_0 z_0, \eta \to y/\lambda_0 z_0} \\
&= A_0 \cdot \frac{1}{i\lambda_0 z_0} \exp\left[+2\pi i \frac{z_0}{\lambda_0}\right] \\
&\quad \cdot \exp\left[-i\left(2\pi \nu_0 t - \left(\frac{\nu_0 z_0}{c}\right)\right)\right] \cdot \exp\left[-2\pi i \cdot \frac{x}{\lambda_0 z_0} \cdot \frac{d_0}{2}\right] \cdot 1\left[\frac{y}{\lambda_0 z_0}\right] \\
&= A_0 \cdot K_0 \cdot \exp\left[-i\left(2\pi \nu_0 t - \left(\frac{\nu_0 z_0}{c}\right)\right)\right] \cdot \exp\left[-2\pi i \cdot \frac{x}{(2\lambda_0 z_0/d_0)}\right] \cdot 1[x, y]
\end{aligned} \tag{21.103}$$

The observed irradiance is again proportional to the spatially constant squared magnitude:

$$\begin{aligned}
I[x, y; z = z_0] &\propto \left|A_0 \cdot K_0 \cdot \exp[-2\pi i \nu_0 t] \cdot \exp\left[-2\pi i \cdot \frac{x}{(2\lambda_0 z_0/d_0)}\right] \cdot 1[x, y]\right|^2 \\
&= \left(\frac{A_0}{\lambda_0 z_0}\right)^2 \cdot 1[x, y]
\end{aligned} \tag{21.104}$$

where, again, the time dependence vanishes in the squared magnitude because the temporal frequencies are identical. In words, the two cases of on-axis and off-axis point sources produce identical results in the Fraunhofer diffraction region; we cannot discriminate between the two source functions based only on the observed irradiance (Figure 21.16).

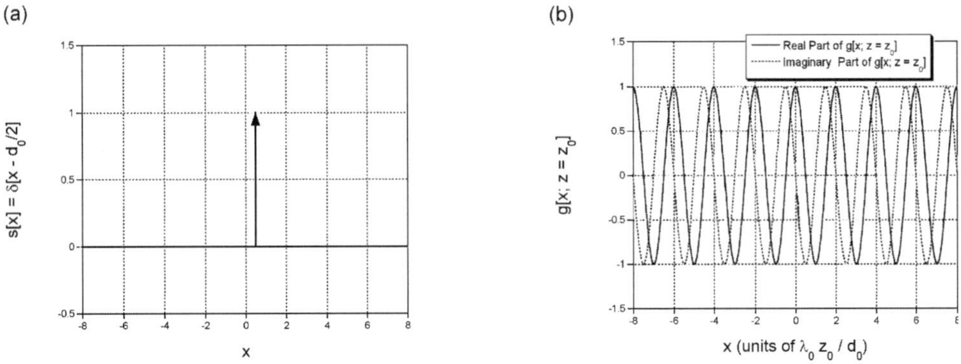

Figure 21.16 Fraunhofer diffraction of off-axis point source: (a) source function $s[x] = \delta[x - 1]$; (b) amplitude at plane $z = z_0$ in Fraunhofer diffraction region is proportional to $S[x/\lambda_0 z_0] = \cos[2\pi x/\lambda_0 z_0] + i \cdot \sin[2\pi x/\lambda_0 z_0]$, so that the irradiance at $z = z_0$ is constant.

21.3.1.3 Two Point Sources, Equal Amplitudes (Young's Experiment)

Now consider two identical monochromatic point sources displaced along the x-axis, as shown in Figure 21.17a. This is the classic two-aperture experiment performed by Thomas Young in the early 1800s that demonstrated the wave nature of light. The source function in the calculation using the Fourier interpretation is:

$$s[x, y; z = 0] = A_0 \cdot \exp[-2\pi i \nu_0 t] \cdot \left(\delta\left[x + \frac{d_0}{2}, y\right] + \delta\left[x - \frac{d_0}{2}, y\right]\right) \tag{21.105}$$

It is easy to see that the diffracted amplitude at the distance z_0 in the Fraunhofer diffraction region is a cosine pattern (Figure 21.17b):

$$\begin{aligned} g[x, y; \lambda_0, z_0] &= (K_0 \cdot A_0) \cdot \exp\left[-i\left(2\pi\nu_0 t - \left(\frac{\nu_0 z_0}{c}\right)\right)\right] \\ &\cdot \left(\exp\left[+2\pi i \frac{d_0}{2}\frac{x}{\lambda_0 z_0}\right] + \exp\left[-2\pi i \frac{d_0}{2}\frac{x}{\lambda_0 z_0}\right]\right) \cdot 1\left[\frac{y}{\lambda_0 z_0}\right] \\ &= (2 \cdot K_0 \cdot A_0) \cdot \exp\left[-i\left(2\pi\nu_0 t - \left(\frac{\nu_0 z_0}{c}\right)\right)\right] \cdot \cos\left[2\pi \frac{x}{(2\lambda_0 z_0/d_0)}\right] \cdot 1[y] \end{aligned} \tag{21.106}$$

The observed irradiance is proportional to the squared magnitude, which is a biased cosine (Figure 21.17b):

$$\begin{aligned} I[x, y; z_0, \lambda_0] &\propto |g[x, y; \lambda_0, z_0]|^2 \\ &= \frac{4A_0^2}{(\lambda_0 z_0)^2} \cdot \cos^2\left[2\pi \frac{x}{(2\lambda_0 z_0/d_0)}\right] \cdot 1[y] \\ &= \frac{2A_0^2}{(\lambda_0 z_0)^2} \cdot \left(1 + \cos\left[2\pi \frac{x}{(\lambda_0 z_0/d_0)}\right]\right) \cdot 1[y] \\ &= \frac{2A_0^2}{(\lambda_0 z_0)^2} \cdot \left(1 + \cos\left[2\pi \frac{x}{D_0}\right]\right) \cdot 1[y] \propto A_0^2\left(1 + \cos\left[2\pi \frac{x}{D_0}\right]\right) \end{aligned} \tag{21.107}$$

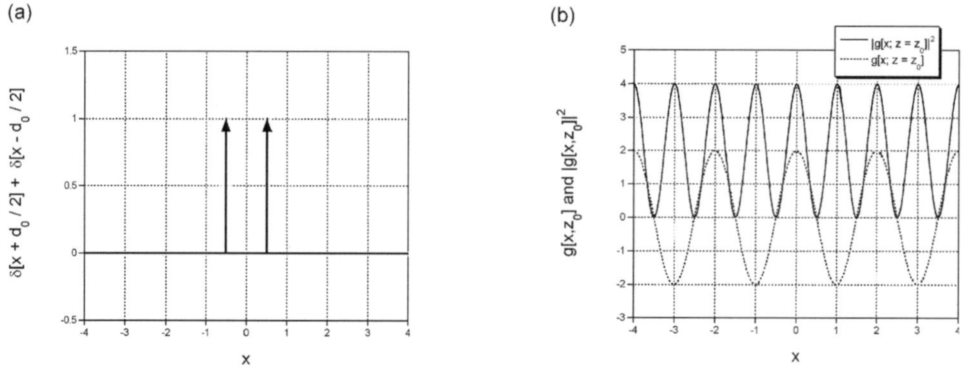

Figure 21.17 Fraunhofer diffraction of symmetric pair of point sources: (a) source function $s[x] = \delta[x + d_0] + \delta[x - d_0]$; (b) amplitude at plane $z = z_0$ in Fraunhofer diffraction region is proportional to $S[x/\lambda_0 z_0] = 2 \cdot \cos[2\pi x/(2\lambda_0 z_0/d_0)]$; the irradiance is proportional to $4 \cdot \cos^2[2\pi x/(2 \cdot \lambda_0 z_0)] = 2 + 2\cos[2\pi x/(\lambda_0 z_0/d_0)]$ constant.

where the identity in Equation (6.24) has been used. The scaling theorem of the Fourier transform from Equation (9.117) is apparent in this expression; the smaller the separation d_0 of the point sources, the wider the period D_0 of the fringe pattern. The modulation of the cosine pattern is unity because the two sources emit identical amplitudes:

$$m = \frac{I_{\max} - I_{\min}}{I_{\max} + I_{\min}} = 1 \tag{21.108}$$

We assign the symbol D_0 to the period of the irradiance fringe pattern:

$$D_0 \equiv \frac{\lambda_0 z_0}{d_0} \Longrightarrow \lambda_0 \cdot z_0 = D_0 \cdot d_0 \tag{21.109}$$

which is a convenient mnemonic for the distances involved; the product of the two "longitudinal" distances λ_0 and z_0 is equal to the product of the two "transverse" distances d_0 and D_0 for Fraunhofer diffraction from two monochromatic point sources.

The sinusoidal nature of the irradiance produced by two monochromatic point sources differs markedly from that produced by either alone and provides evidence of the form of the source distribution. We now consider how the sinusoidal pattern is altered by changes in the character of the two sources, which will lead us to the concept of coherence.

The steps in the superposition process, including the time average, are shown in Figure 21.18.

The effect of time averaging on the squared magnitude is shown in Figure 21.19.

21.3.1.4 Two Extended Sources with Equal Amplitudes

As a last example, consider the case where the two apertures have identical finite widths b_0 (that must be smaller than the separation d_0). The source function is:

$$\begin{aligned} s[x, y] &= A_0 \cdot \exp[-2\pi i \nu_0 t] \cdot \left(RECT\left[\frac{x + d_0/2}{b_0}\right] + RECT\left[\frac{x - d_0/2}{b_0}\right]\right) \cdot \delta[y] \\ &= A_0 \cdot \exp[-2\pi i \nu_0 t] \cdot \left(RECT\left[\frac{x + d_0/2}{b_0}\right] \cdot \delta[y]\right) \\ &\quad * \left(\delta\left[x + \frac{d_0}{2}, y\right] + \delta\left[x - \frac{d_0}{2}, y\right]\right) \end{aligned} \tag{21.110}$$

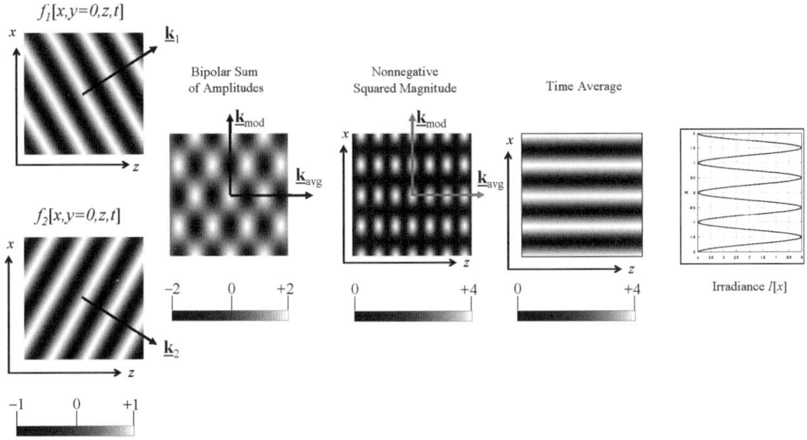

Figure 21.18 Interference of two "tilted plane waves" with the same wavelength: from the left, the two component traveling waves shown as "snapshots" at one instant of time (white = 1, black = −1); the sum of the amplitudes is shown (white = 2, black = −2), which moves to the left at the velocity of light. The squared magnitude on the right (white = 4, black = 0), showing the sinusoidal form of the interference fringes. The modulation in the vertical direction is constant, while that in the horizontal direction is a traveling wave and "averages" out to a constant value that forms the cosine fringes.

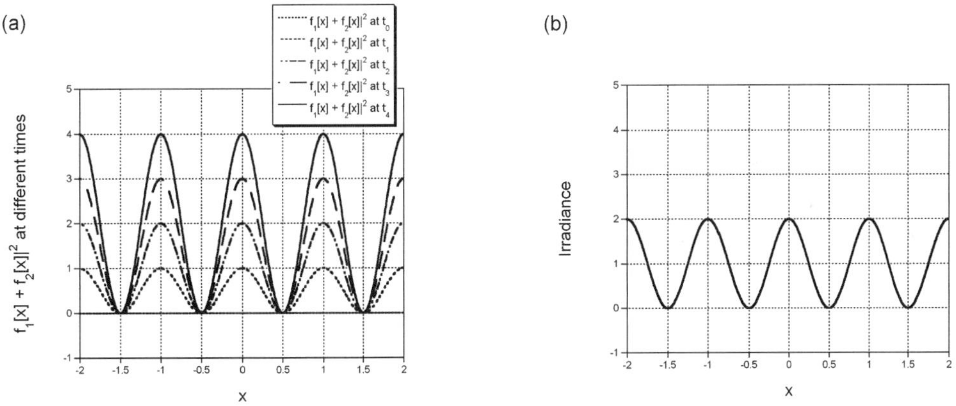

Figure 21.19 Effect of time averaging on the calculation of the irradiance: (a) the squared magnitude observed at the output plane at several instants of time, showing that the spatial variation of the irradiance is preserved; (b) the time average reduces the maximum by half.

The amplitude at the observation plane in the Fraunhofer diffraction region is:

$$g[x, y; \lambda_0, z_0] = \frac{1}{i\lambda_0 z_0} \exp\left[+2\pi i\left(\frac{z_0}{\lambda_0} - \nu_0 t\right)\right] \cdot \left(\exp\left[+2\pi i\frac{d_0}{2}\frac{x}{\lambda_0 z_0}\right] + \exp\left[-2\pi i\frac{d_0}{2}\frac{x}{\lambda_0 z_0}\right]\right)$$
$$\cdot\, b_0 \cdot SINC\left[b_0 \cdot \frac{x}{\lambda_0 z_0}\right]$$
$$= \frac{2b_0}{i\lambda_0 z_0} \exp\left[+2\pi i\left(\frac{z_0}{\lambda_0} - \nu_0 t\right)\right] \cdot \cos\left[2\pi \frac{x}{(2\lambda_0 z_0/d_0)}\right] \cdot SINC\left[\frac{x}{(\lambda_0 z_0/b_0)}\right] \tag{21.111}$$

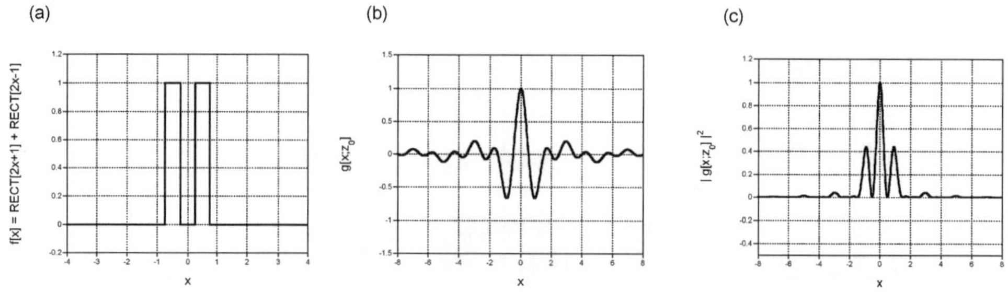

Figure 21.20 Profiles of the interference in the Fraunhofer region from rectangular slits: (a) object function $s[x, y] = RECT[(x + d_0/2)/(d_0/4)] + RECT[(x - d_0/2)/(d_0/4)]$; (b) amplitude at plane $z = z_0$ in the Fraunhofer diffraction region; (c) the irradiance pattern is a cosine fringe modulated by the scaled $SINC^2$ pattern from the rectangular sources.

where the width of the *SINC* function is wider than the period of the cosine. The time dependence again vanishes in the calculation of the squared magnitude, so the irradiance pattern is:

$$\begin{aligned} I[x, y; z_0, \lambda_0] &\propto \left(\frac{2b_0}{\lambda_0 z_0}\right)^2 \cos^2\left[2\pi \frac{x}{(2\lambda_0 z_0/d_0)}\right] \cdot SINC^2\left[\frac{x}{(\lambda_0 z_0/b_0)}\right] \\ &= \frac{2b_0^2}{(\lambda_0 z_0)^2} \cdot \left(1 + \cos\left[2\pi \frac{x}{(\lambda_0 z_0/d_0)}\right]\right) \cdot SINC^2\left[\frac{x}{(\lambda_0 z_0/b_0)}\right] \end{aligned} \tag{21.112}$$

Note that the width parameter $\lambda_0 z_0/b_0$ of the $SINC^2$ modulation is larger than the period $\lambda_0 z_0/d_0$ of the cosine fringes. In words, the diffraction pattern generated by apertures with finite widths (and therefore from waves that are not planar) is an apodized cosine function, as shown in Figure 21.20.

This result is easily generalized to different aperture functions, e.g., triangles instead of rectangles. In short, if the two apertures have finite widths or shapes specified by a spatial function, then the period interference pattern is determined by the separation of the centers of the apertures and the fringes are modulated by the Fourier transform of the aperture spatial function.

21.4 Imaging System based on Fraunhofer Diffraction

Now consider a particularly simple imaging "system" shown in Figure 21.21a. The system may be modeled in three stages:

1. Propagation from the input object $f[x, y]$ to the Fraunhofer diffraction region over the distance z_1,
2. Multiplication by the (possibly complex-valued) transmittance function

$$t[x, y] = |t[x, y]| \exp[+i\Phi_t[x, y]]$$

   that specifies the aperture (or *pupil*).

3. A second propagation over the distance z_2 into the Fraunhofer diffraction region, where the criterion on z_2 is determined by the size of the aperture.

To eliminate an awkward notation, we will substitute the notation $p[x, y]$ for the magnitude of the pupil function $|t[x, y]|$. In this example, we assume that the pupil has no phase component, so that $\Phi_t[x, y] = 0$, though solution of the more general case is straightforward. The 2-D input function $f[x, y; z = 0]$ is illuminated by a unit-amplitude monochromatic plane wave with wavelength λ_0. The light propagates into the Fraunhofer diffraction region at a distance z_1, where the resulting amplitude pattern is easily obtained from Equation (21.96):

$$E[x, y; z_1] = \frac{\mathcal{E}_0}{i\lambda_0 z_1} e^{+2\pi i z_1/\lambda_0} e^{+i\pi(x^2+y^2)/\lambda_0 z_1} F\left[\frac{x}{\lambda_0 z_1}, \frac{y}{\lambda_0 z_1}\right] \tag{21.113}$$

This pattern illuminates the 2-D pupil function $p[x, y]$ and then propagates the distance z_2 into the Fraunhofer diffraction region (determined by the support of p). A second application of Equation (21.96) produces the amplitude at the observation plane:

$$\begin{aligned}
&E[x, y; z_1 + z_2] \\
&= \mathcal{E}_0\left(\frac{1}{i\lambda_0 z_1} e^{+2\pi i z_1/\lambda_0} e^{+i\pi(x^2+y^2)/\lambda_0 z_1}\right)\left(\frac{1}{i\lambda_0 z_2} e^{+2\pi i z_2/\lambda_0} e^{+i\pi(x^2+y^2)/\lambda_0 z_2}\right) \\
&\quad \cdot \mathcal{F}_2\left\{F\left[\frac{x}{\lambda_0 z_1}, \frac{y}{\lambda_0 z_1}\right] \cdot p[x, y]\right\}\Bigg|_{\xi = x/\lambda_0 z_2,\, y/\lambda_0 z_2} \\
&= \mathcal{E}_0\left(-\frac{1}{\lambda_0^2 z_1 z_2}\right) e^{+2\pi i(z_1+z_2)/\lambda_0} e^{+i\pi[(x^2+y^2)/\lambda_0](1/z_1+1/z_2)} \\
&\quad \cdot (\lambda_0 z_1)^2 (f[-\lambda_0 z_1 \xi, -\lambda_0 z_1 \eta] * P[\xi, \eta])|_{\xi = x/\lambda_0 z_2,\, y/\lambda_0 z_2} \\
&= \mathcal{E}_0\left(-\frac{z_1}{z_2}\right) e^{+2\pi i(z_1+z_2)/\lambda_0} e^{+i\pi[(x^2+y^2)/\lambda_0](1/z_1+1/z_2)} \\
&\quad \times \left(f\left[\left(-\frac{z_1}{z_2}\right)x, \left(-\frac{z_1}{z_2}\right)y\right] * P\left[\frac{x}{\lambda_0 z_2}, \frac{y}{\lambda_0 z_2}\right]\right) \\
&= \frac{\mathcal{E}_0}{M_T} e^{+2\pi i(z_1+z_2)/\lambda_0} e^{+i\pi[(x^2+y^2)/\lambda_0](1/z_1+1/z_2)} \left(f\left[\frac{x}{M_T}, \frac{y}{M_T}\right] * P\left[\frac{x}{\lambda_0 z_2}, \frac{y}{\lambda_0 z_2}\right]\right)
\end{aligned} \tag{21.114}$$

where the theorems of the Fourier transform and the definition of the transverse magnification from geometrical optics, $M_T = -z_2/z_1$, have been used. Note that if the propagation distances z_1 and z_2 must both be positive in Fraunhofer diffraction, which requires that $M_T < 0$, so that the image is "reversed".

The irradiance of the image is proportional to the squared magnitude of the amplitude in Equation (21.114):

$$|E[x, y; z_1 + z_2]|^2 = \left|\frac{\mathcal{E}_0}{M_T}\right|^2 \left|f\left[\frac{x}{M_T}, \frac{y}{M_T}\right] * P\left[\frac{x}{\lambda_0 z_2}, \frac{y}{\lambda_0 z_2}\right]\right|^2 \tag{21.115}$$

In words, the output amplitude created by this imaging "system" is the product of some constants, a quadratic-phase function of $[x, y]$, and the convolution of the input amplitude scaled by the transverse magnification and the scaled replica of the spectrum of the aperture function, $P[x/\lambda_0 z_2, y/\lambda_0 z_2]$. Since the output is the result of a convolution, we identify the spectrum as the impulse response of a shift-invariant convolution that is composed of two shift-variant Fourier transforms and multiplication by a quadratic-phase factor of $[x, y]$. This system does not satisfy the strict conditions for shift invariance because of the leading quadratic-phase factor and the fact that the input to the convolution is a scaled and reversed replica of the input to the system. That said, these details are often ignored to allow the process to be considered shift invariant. We will revisit this conceptual imaging system after considering the mathematical models for optical elements. The action of the system is demonstrated for inputs consisting of Dirac delta functions in Figure 21.21.

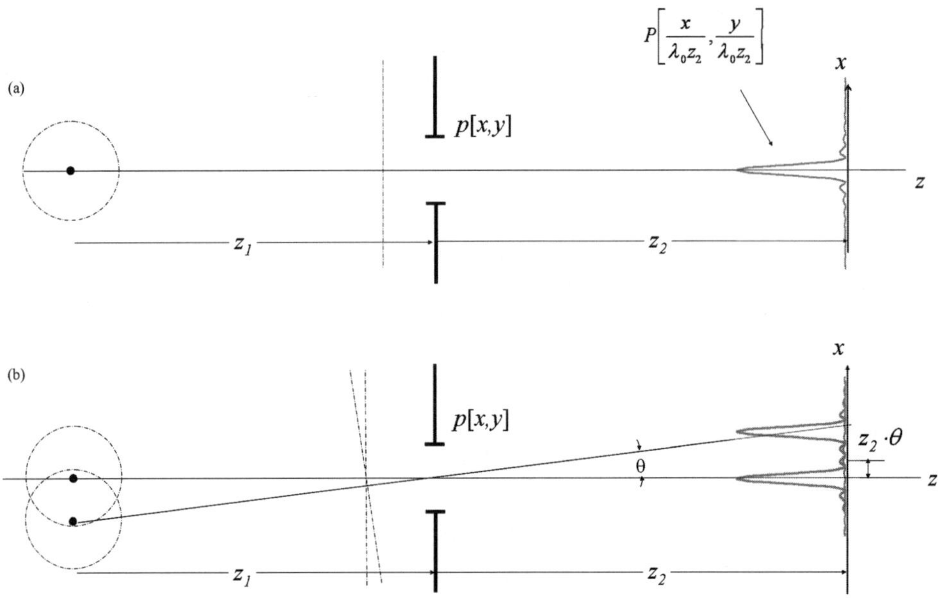

Figure 21.21 Imaging "system" based on Fraunhofer diffraction: (a) light with wavelength λ_0 propagates from a point source at the origin to the Fraunhofer region a distance z_1 away, where it encounters an aperture with transmittance function $p[x, y]$. The transmitted light propagates the distance z_2 in the Fraunhofer diffraction region, where it creates the image function that is proportional to $P[x/\lambda_0 z_2, y/\lambda_0 z_2]$. (b) Image of two point sources separated in angle by θ created by the same system is composed of two replicas of $P[x/\lambda_0 z_2, y/\lambda_0 z_2]$ separated by the same angle.

21.5 Transmissive Optical Elements

The imaging "systems" that we have considered so far are as simple as you can get; the light in the first cases merely propagates from one plane to another by Huygens' principle, and the amplitude is calculated using one of the two approximations: the Fresnel transform in the near field and the Fourier transform in the far field. The shift-variant expression in Equation (21.37) simplified to a convolution with a quadratic-phase factor in the Fresnel region, and to a scaled Fourier transform in the Fraunhofer region. In the second case, light propagated from the source to the aperture to the observation plane.

At this point, we consider methods for manipulating or modifying the propagating light waves by inserting optical elements at the pupil plane that are based on the physical interaction of *refraction*; elements may also be constructed based on the mechanisms of *reflection* and *diffraction*. The physical mechanism of refraction is the result of the fact that light waves (or rays) within a medium slow down in inverse proportion to a parameter known as the refractive index of the medium, i.e.:

$$\mathrm{v}_{\text{glass}} = \frac{c}{n} \textit{ where } c \cong 3 \cdot 10^8 \text{ m/s} \tag{21.116}$$

If the glass has thickness τ (measured in mm, say), then the "arrival time" of a fixed point (point of constant phase) on a light wave that travels through the glass will be delayed when compared to that for a wave traveling the same distance through vacuum. The time delay is:

$$\Delta t = t_{\text{glass}} - t_{\text{vacuum}} = \frac{\tau}{\mathrm{v}_{\text{glass}}} - \frac{\tau}{c} = \frac{\tau(n-1)}{c} \textit{seconds} \tag{21.117}$$

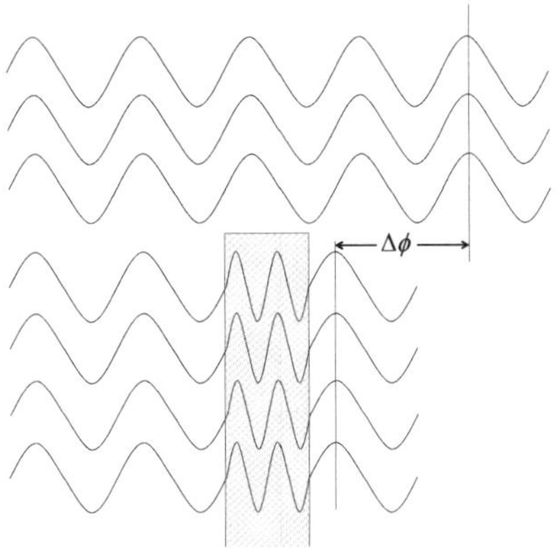

Figure 21.22 Phase shift induced by a flat piece of glass compared to the phase of light propagating in vacuum. The locus of points of constant phase on the entering wavefront is a vertical plane. The relative phase shift of light traveling through the glass is constant (not a function of x or y) and the locus of points of constant phase is now two discontinuous planes.

This time delay produces a difference in the phase of the light that passes through the glass when compared to that traveling through vacuum:

$$\Delta\phi = \phi_{\text{glass}} - \phi_{\text{vacuum}} = (c \cdot \Delta t)\frac{2\pi}{\lambda_0} = 2\pi(n-1)\frac{\tau}{\lambda_0} \text{ } radians \tag{21.118}$$

An ideal nonabsorptive transmissive optical element of refractive index n and circularly symmetric physical thickness $\tau(r)$ may be described by a complex-valued transmittance function:

$$t(r) = 1 \cdot e^{+2\pi i\, n \cdot \tau(r)/\lambda_0} = e^{+2\pi i(n-1)\tau(r)/\lambda_0}\, e^{+2\pi i \tau(r)/\lambda_0} \tag{21.119}$$

Our task now is to evaluate this transmittance function for different phase functions.

21.5.1 Optical Elements with Constant or Linear Phase

If the optical element is a slab of glass with uniform thickness τ, then the phase measured at all points after the glass exhibits an additional constant factor of $\exp[+2\pi i\, n \cdot \tau/\lambda_0]$. Since the phase change is constant, the effect of the glass plate on the observed irradiance is undetectable. An example is shown in Figure 21.22.

In a slightly more complicated case where the glass thickness varies with position $[x, y]$, so will the phase delay. For example, if the thickness is a linear function of x, then the transmittance function is:

$$\tau[x, y] = \alpha x \cdot 1[y] \to t[x, y] = e^{+2\pi i(n-1)\alpha x/\lambda_0}\, 1[y] \tag{21.120}$$

Such an optical element is called a *prism*, as shown in Figure 21.23. The term $(n-1)\alpha/\lambda_0$ has dimensions of reciprocal length and thus may be identified as a spatial frequency:

$$t[x, y] = e^{+2\pi i \xi_0 x}\, 1[y] \quad where\ \xi_0 = (n-1)\frac{\alpha}{\lambda_0} \tag{21.121}$$

Assume that the prism located at $z_1 \gg 0$ (in the Fraunhofer region) is illuminated by a point source at $z = 0$. The light emerging from the prism propagates to a plane at $z = z_1 + z_2$ that is in the Fraunhofer region measured from z_1. The "image" generated at the observation plane is proportional to the Fourier

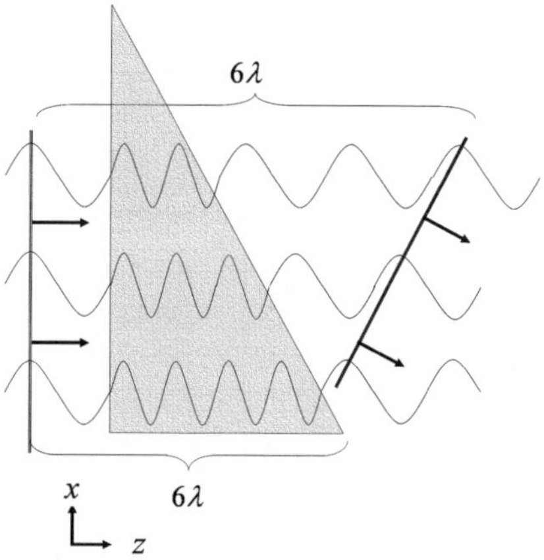

Figure 21.23 Schematic of the phase shift induced by a prism with thickness function $\tau[x, y] = \alpha x$. The locus of the points with constant phase on the entering wavefront is a vertical plane. Distances between several points where the optical path difference is 6λ (phase difference of 12π radians) are shown, which confirm that the wavefront was tilted "down" by the action of the prism.

transform of the unit-magnitude "aperture" function $t[x_1, y_1]$:

$$\begin{aligned}E[x, y; z_1 + z_2] &= \mathcal{F}_2\{t[x_1, y_1]\}|_{(\xi\to x/\lambda_0 z_2, \eta\to y/\lambda_0 z_2)}\\ &= \mathcal{F}_2\{e^{+2\pi i\xi_0 x}\, 1[y]\}\\ &= \delta[\xi - \xi_0]\,\delta[\eta]|_{(\xi\to x/\lambda_0 z_2, \eta\to y/\lambda_0 z_2)}\\ &= \delta\left[\frac{x}{\lambda_0 z_2} - (n-1)\frac{\alpha}{\lambda_0}\right]\delta\left[\frac{y}{\lambda_0 z_2}\right]\\ &= |\lambda_0 z_2|^2\,\delta[x - (n-1)\alpha z_2]\,\delta[y]\end{aligned} \tag{21.122}$$

The action of the prism is a change in position of the displayed "image" from the origin to the off-axis location $[x_2, y_2] = [(n-1)\alpha z_2, 0]$.

21.5.2 Lenses with Spherical Surfaces

Optical systems typically are used to form images of the source distribution by redirecting the radiation from point sources to point images. For example, consider a slab of glass with maximum thickness τ and with a convex spherical face with radius of curvature R (i.e., a plano-convex lens). The thickness function of the glass is found by using the Pythagorean formula in the following picture:

$$(R - s(r))^2 + r^2 = R^2 \Longrightarrow r^2 - 2R\cdot s(r) + s^2(r) = 0 \tag{21.123}$$

If both R and r are much larger than $s(r)$, then the factor $s^2(r)$ may be ignored. The result is a simple formula, the *sag formula*, for a spherical lens:

$$s(r) \cong \frac{r^2}{2R} \tag{21.124}$$

as shown in Figure 21.24. In words, the sag formula approximates the spherical surface by a paraboloid. This process is analogous to the approximation of a spherical wavefront as a paraboloid in the Fresnel diffraction region.

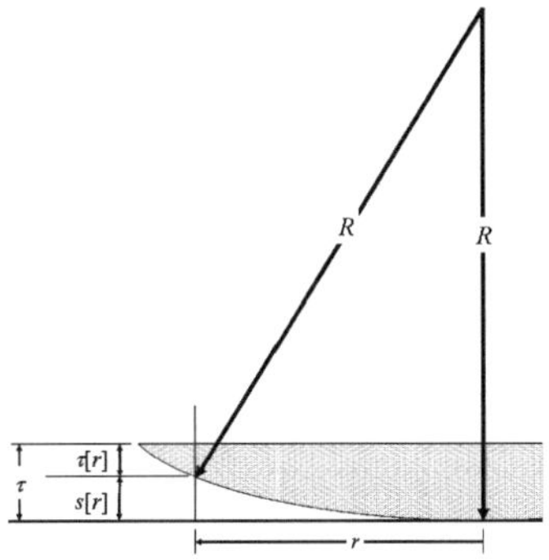

Figure 21.24 Parameters in the sag formula for a spherical lens, showing that $r^2 - 2R \cdot s(r) + s^2(r) = 0$, which implies that $s[r] \simeq r^2/2R$ for small values of $s(r)$.

The thickness of glass as a function of radial distance from the optical axis is the difference between the maximum thickness and the parabolic sag:

$$\tau(r) = \tau - s(r) \cong \tau - \frac{r^2}{2R} \tag{21.125}$$

which measures the thickness of glass as a function of radial distance r from the optical axis. The phase delay as a function of radial distance from the optical axis is the sum of the phase delay through the glass and the phase delay through air in the sag region. The delay due to the glass is the number of wavelengths multiplied by the refractive index:

$$\begin{aligned}\phi(r) &= \left(\frac{2\pi}{(\lambda_0/n)} \times \textit{distance traveled in glass}\right) + \left(\frac{2\pi}{\lambda_0} \times \textit{distance traveled in vacuum}\right)\\ &= \left(\frac{2\pi}{(\lambda_0/n)} \cdot \tau(r)\right) + \left(\frac{2\pi}{\lambda_0} \cdot s(r)\right)\\ &= 2\pi\frac{n}{\lambda_0}\left(\tau - \frac{r^2}{2R}\right) + \frac{2\pi}{\lambda_0}\left(\frac{r^2}{2R}\right)\\ &= \frac{2\pi}{\lambda_0} n\tau - \pi(n-1)\frac{r^2}{\lambda_0 R}\end{aligned} \tag{21.126}$$

The first (constant) term is the fixed phase delay of light rays traveling through the maximum thickness τ of glass, while the second term is a quadratic function that decreases as a function of the radial distance r measured from the optical axis due to the more rapid velocity of light in the sag region. The quadratic approximation to the spherical surface is the source of the quadratic phase contribution of a spherical lens.

The electric field exiting the lens as a function of radial position r from the optical axis is the product of the incident field and the phase delays, leading to constant- and quadratic-phase terms:

$$E[r,t] = E_0[r] \exp\left[+2\pi i \frac{n\tau}{\lambda_0}\right] \exp\left[-i\pi(n-1)\frac{r^2}{\lambda_0 R}\right] \tag{21.127}$$

In the common case of a lens having two convex spherical surfaces with radii R_1 and R_2, the radius of curvature of one surface must have the opposite sign because it is oriented in the opposite direction. By convention, a surface with center of curvature to the right of the surface has positive radius. The phase transmittance of the lens has contributions from each sag:

$$
\begin{aligned}
t[r] &= \exp\left[+2\pi i\frac{n\tau}{\lambda_0}\right]\exp\left[-i\pi(n-1)\frac{r^2}{\lambda_0 R_1}\right]\exp\left[\equiv i\pi(n-1)\frac{r^2}{\lambda_0(-R_2)}\right] \\
&= \exp\left[+2\pi i\frac{n\tau}{\lambda_0}\right]\exp\left[-i\pi(n-1)\frac{r^2}{\lambda_0}\left(\frac{1}{R_1}-\frac{1}{R_2}\right)\right]
\end{aligned}
\qquad (21.128)
$$

The expression may be simplified by combining terms to define the factor $\mathbf{f}$ that has dimensions of length:

$$
\frac{1}{\mathbf{f}} \equiv (n-1)\left(\frac{1}{R_1}-\frac{1}{R_2}\right) \qquad (21.129)
$$

This possibly is a familiar equation; it is the *lens-maker's formula* that relates the *focal length* $\mathbf{f}$ of the lens to the index of refraction n and the radii of curvature R_1 and R_2 of the two surfaces. The effect of the lens in Equation (21.127) is a change in the phase of the transmitted light due to the optical transmission function $t(r)$:

$$
\begin{aligned}
t(r) &= \exp\left[+2\pi i\frac{n\tau}{\lambda_0}\right]\exp\left[-i\pi\frac{r^2}{\lambda_0\mathbf{f}}\right] \\
&= \exp\left[+2\pi i\frac{n\tau}{\lambda_0}\right]\exp\left[-i\pi\frac{r^2}{\lambda_0|\mathbf{f}|}\right] \quad \textit{if } \mathbf{f}>0
\end{aligned}
\qquad (21.130)
$$

Note that if $\mathbf{f}$ is negative, the algebraic sign of the quadratic phase due to the lens is positive:

$$
t(r) = \exp\left[+2\pi i\frac{n\tau}{\lambda_0}\right]\exp\left[+i\pi\frac{r^2}{\lambda_0|\mathbf{f}|}\right] \quad \textit{if } \mathbf{f}<0 \qquad (21.131)
$$

which means that the phase of the transmitted light *increases* with increasing radial distance from the optical axis. At this point, the transmittance is a pure phase function and thus has infinite support. The finite diameter of a realistic lens may be introduced into Equation (21.130) as a real-valued function $p[x, y]$ (which may be circularly symmetric or not) that defines the *pupil* of the lens. The pupil function usually is binary ($p[x, y] = 0, 1$) but we can (and will) use pupil functions whose transmission varies with radial distance. A circularly symmetric spherical lens with positive focal length (converging lens) will be modeled by the transmittance function:

$$
t(r) = p(r)\exp\left[+2\pi i\frac{n\tau}{\lambda_0}\right]\exp\left[-i\pi\frac{r^2}{\lambda_0\mathbf{f}}\right] \qquad (21.132)
$$

as shown in Figure 21.25. In words, a lens changes the radius of curvature of the incident parabolic wave by adding or subtracting a phase factor that is a quadratic function of the distance from the optical axis. The variable part of the phase contribution from a spherical lens is quadratic with a negative sign, while the (approximate) impulse response of light propagation in the Fresnel diffraction region is a quadratic-phase factor with a positive sign.

21.6 Monochromatic Optical Systems

21.6.1 Single Positive Lens with $z_1 \gg 0$

We now have collected all of the tools necessary to model optical imaging systems. The first example we will consider is a variation of the Fraunhofer diffraction imaging "system" that was considered earlier. In this case, the transmittance function is modified to include the quadratic-phase factor for a positive lens. The three steps in the process are:

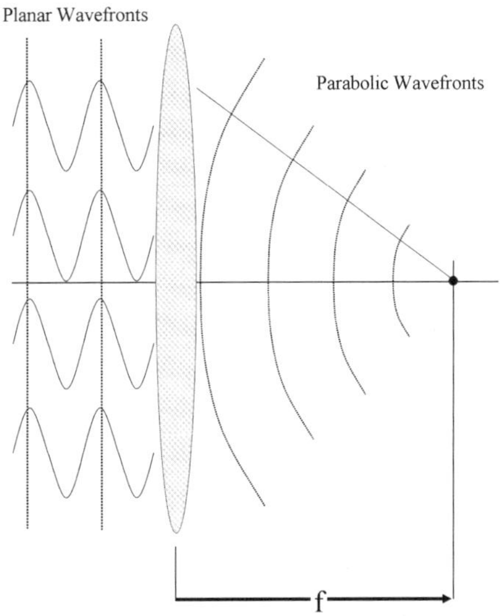

Figure 21.25 A negative quadratic phase induced by a positive lens converts plane waves to converging parabolic waves at the distance **f**.

1. Propagation from the source function $f[x, y]$ into the Fraunhofer diffraction region over the distance $z_1 \gg 0$.
2. Multiplication by the transmittance function $t[x, y]$ of a positive lens with focal length **f**.
3. Propagation to a plane in the Fresnel diffraction region at a distance z_2 measured from the lens.

The amplitude at the "front" of the lens is identical to that at the aperture in the Fraunhofer "system" that was derived in Equation (21.96):

$$E[x, y; z_1] = \frac{\mathcal{E}_0}{i\lambda_0 z_1} e^{+2\pi i z_1/\lambda_0} \exp\left[+i\pi \frac{(x^2+y^2)}{\lambda_0 z_1}\right] F\left[\frac{x}{\lambda_0 z_1}, \frac{y}{\lambda_0 z_1}\right] \tag{21.133a}$$

$$\cong \frac{\mathcal{E}_0}{i\lambda_0 z_1} e^{+2\pi i z_1/\lambda_0} F\left[\frac{x}{\lambda_0 z_1}, \frac{y}{\lambda_0 z_1}\right] \quad \textit{if } z_1 \gg 0 \tag{21.133b}$$

where the large value of z_1 ensures that the leading quadratic-phase term is approximately unity and may be neglected. This field is multiplied by the pupil function $t[x, y]$ of the lens, which includes a magnitude part $p[x, y]$ and a converging quadratic phase with focal length **f**:

$$E[x, y; z_1] \cdot t[x, y] = E[x, y; z_1] \cdot \left(p[x, y] \cdot \exp\left[-i\pi \frac{(x^2+y^2)}{(\sqrt{\lambda_0 \mathbf{f}})^2}\right]\right)$$

$$\cong \frac{\mathcal{E}_0}{i\lambda_0 z_1} e^{+2\pi i z_1/\lambda_0} F\left[\frac{x}{\lambda_0 z_1}, \frac{y}{\lambda_0 z_1}\right] \cdot \left(p[x, y] \exp\left[-i\pi \frac{(x^2+y^2)}{\lambda_0 \mathbf{f}}\right]\right) \tag{21.134}$$

Again because z_1 is large, the geometrical optics imaging relation in Equation (21.4) demonstrates that an image is formed at the distance $z_2 \cong \mathbf{f}$ from the lens. In other words, the image is formed at the image-space *focal point* of the lens. If the focal length is relatively "small" compared to z_1 then we must use Fresnel diffraction to calculate the field at the image plane:

$$\begin{aligned}
&E[x,y;z_1+z_2]\\
&=\frac{\mathcal{E}_0}{i\lambda_0 z_1}\,e^{+2\pi i z_1/\lambda_0}\left(F\left[\frac{x}{\lambda_0 z_1},\frac{y}{\lambda_0 z_1}\right]\cdot p[x,y]\,e^{-i\pi(x^2+y^2)/\lambda_0\mathbf{f}}\right)\\
&\quad *\left(\frac{1}{i\lambda_0\mathbf{f}}\,e^{+2\pi i\mathbf{f}/\lambda_0}\,e^{+i\pi(x^2+y^2)/\lambda_0\mathbf{f}}\right)\\
&=\mathcal{E}_0\left(-\frac{1}{\lambda_0^2 z_1\mathbf{f}}\right)e^{+2\pi i(z_1+\mathbf{f})/\lambda_0}\left(F\left[\frac{x}{\lambda_0 z_1},\frac{y}{\lambda_0 z_1}\right]\cdot p[x,y]\,e^{-i\pi(x^2+y^2)/\lambda_0\mathbf{f}}\right)*e^{+i\pi(x^2+y^2)/\lambda_0\mathbf{f}}
\end{aligned}\tag{21.135}$$

Though the expression in Equation (21.135) looks quite complicated, the M–C–M chirp Fourier transform comes to our assistance by supplying a useful equivalent expression. Recall from Equation (17.84) that the M–C–M transform of the 2-D function $r[x,y]$ processed by chirps with the scale factor $\alpha=\sqrt{\lambda_0\mathbf{f}}$ has the form:

$$(r[x,y]\cdot e^{-i\pi(x^2+y^2)/\lambda_0\mathbf{f}})*e^{+i\pi(x^2+y^2)/\lambda_0\mathbf{f}}\cdot e^{-i\pi(x^2+y^2)/\lambda_0\mathbf{f}}=R\left[\frac{x}{\lambda_0\mathbf{f}},\frac{y}{\lambda_0\mathbf{f}}\right]\tag{21.136}$$

This can be modified to obtain an expression in the same form as Equation (21.135) by multiplying both sides by $\exp[+i\pi(x^2+y^2)/\lambda_0\mathbf{f}]$ to obtain an expression in terms of the M–C–M chirp Fourier transform of the pupil function that was derived in Equation (17.70):

$$(r[x,y]\cdot e^{-i\pi(x^2+y^2)/\lambda_0\mathbf{f}})*e^{+i\pi(x^2+y^2)/\lambda_0\mathbf{f}}=R\left[\frac{x}{\lambda_0\mathbf{f}},\frac{y}{\lambda_0\mathbf{f}}\right]\cdot e^{+i\pi(x^2+y^2)/\lambda_0\mathbf{f}}\tag{21.137}$$

In our example, the "input" function to the M–C–M chirp transform and its Fourier transform are:

$$r[x,y]=F\left[\frac{x}{\lambda_0 z_1},\frac{y}{\lambda_0 z_1}\right]\cdot p[x,y]\tag{21.138}$$

$$R[\xi,\eta]=(\lambda_0 z_1)^2\cdot f[-\lambda_0 z_1\xi,-\lambda_0 z_1\eta]*P[\xi,\eta]\tag{21.139}$$

where the scaling and "transform-of-a-transform" theorems have been used. The scaled Fourier transform therefore is:

$$R\left[\frac{x}{\lambda_0\mathbf{f}},\frac{y}{\lambda_0\mathbf{f}}\right]=(\lambda_0 z_1)^2\left(f\left[\frac{x}{(-\mathbf{f}/z_1)},\frac{y}{(-\mathbf{f}/z_1)}\right]*P\left[\frac{x}{\lambda_0\mathbf{f}},\frac{y}{\lambda_0\mathbf{f}}\right]\right)\tag{21.140}$$

In words, this is the convolution of a scaled and reversed replica of the input and a scaled replica of the pupil function. The amplitude at the image plane therefore is:

$$\begin{aligned}
&E[x,y;z=z_1+\mathbf{f}]\\
&=-\frac{\mathcal{E}_0}{\lambda_0^2 z_1\mathbf{f}}\,e^{+2\pi i(z_1+\mathbf{f})/\lambda_0}(\lambda_0 z_1)^2\\
&\quad\times\left[\left(f\left[\frac{x}{(-\mathbf{f}/z_1)},\frac{y}{(-\mathbf{f}/z_1)}\right]*P\left[\frac{x}{\lambda_0\mathbf{f}},\frac{y}{\lambda_0\mathbf{f}}\right]\right)\cdot e^{+i\pi(x^2+y^2)/\lambda_0\mathbf{f}}\right]\\
&=\mathcal{E}_0\left(\left(-\frac{z_1}{\mathbf{f}}\right)e^{+2\pi i(z_1+\mathbf{f})/\lambda_0}\,e^{+i\pi(x^2+y^2)/\lambda_0 z_1}\cdot e^{+i\pi(x^2+y^2)/\lambda_0\mathbf{f}}\right)\\
&\quad\times\left(f\left[-\frac{z_1}{\mathbf{f}}\cdot x,-\frac{z_1}{\mathbf{f}}\cdot y\right]*P\left[\frac{x}{\lambda_0\mathbf{f}},\frac{y}{\lambda_0\mathbf{f}}\right]\right)
\end{aligned}$$

$$= \left(\left(-\frac{z_1}{\mathbf{f}} \right) \mathcal{E}_0 \, e^{+2\pi i (z_1 + \mathbf{f})/\lambda_0} \, e^{+i\pi (x^2+y^2)/\lambda_0 (1/z_1 + 1/\mathbf{f})} \right)$$
$$\times \left(f\left[\frac{x}{(-\mathbf{f}/z_1)}, \frac{y}{(-\mathbf{f}/z_1)} \right] * P\left[\frac{x}{\lambda_0 \mathbf{f}}, \frac{y}{\lambda_0 \mathbf{f}} \right] \right)$$
$$= \left(\frac{\mathcal{E}_0}{M_T} \, e^{+2\pi i (z_1 + \mathbf{f})/\lambda_0} \, e^{+i\pi [(x^2+y^2)/\lambda_0](1/z_1 + 1/\mathbf{f})} \right) \left(f\left[\frac{x}{M_T}, \frac{y}{M_T} \right] * P\left[\frac{x}{\lambda_0 \mathbf{f}}, \frac{y}{\lambda_0 \mathbf{f}} \right] \right) \tag{21.141}$$

where we have included the quadratic-phase term in Equation (21.133a) and the substitution of the geometrical transverse magnification has been applied. In words, the amplitude at the image is proportional to the convolution of a scaled and inverted replica of the input object and a scaled replica of the Fourier transform of the pupil function. If the pupil function $p[x, y]$ is "wide", then the impulse response $P[x/\lambda_0\mathbf{f}, y/\lambda_0\mathbf{f}]$ is "narrow"; if the focal length $\mathbf{f}$ is increased, then the impulse response P becomes "wider". The irradiance produced at the observation plane is easy to see:

$$|E[x, y; z = z_1 + \mathbf{f}]|^2 = \left| \frac{\mathcal{E}_0}{M_T} \right|^2 \left| f\left[\frac{x}{M_T}, \frac{y}{M_T} \right] * P\left[\frac{x}{\lambda_0 \mathbf{f}}, \frac{y}{\lambda_0 \mathbf{f}} \right] \right|^2 \tag{21.142}$$

Compare these results to those for the Fraunhofer–Fraunhofer imaging "system" that was derived in Equation (21.114):

$$E[x, y; z_1 + z_2] = \frac{\mathcal{E}_0}{M_T} \, e^{+2\pi i (z_1 + z_2)/\lambda_0} \, e^{+i\pi [(x^2+y^2)/\lambda_0](1/z_1 + 1/z_2)}$$
$$\cdot \left(f\left[\frac{x}{M_T}, \frac{y}{M_T} \right] * P\left[\frac{x}{\lambda_0 z_2}, \frac{y}{\lambda_0 z_2} \right] \right) \tag{21.143}$$

The outputs of these two systems are identical except for the change of the second propagation distance from z_2 to $\mathbf{f}$. The diagram of the Fraunhofer–Fresnel imaging process is shown in Figure 21.26 for two different inputs composed of impulse functions.

21.6.2 Single-Lens System, Fresnel Description of Both Propagations

Now consider the more interesting case where both propagation distances are sufficiently long to apply the Fresnel approximation, but not so long that we can use the Fraunhofer approximation. In our first case, the two distances and the focal length $\mathbf{f}$ are positive, though we will consider other cases shortly. In this example, the object distance $z_1 > 0$ so that the object is "real", the lens has a positive focal length $\mathbf{f}$, so that the lens makes parallel rays converge, and the image distance $z_2 > 0$ so that the image is "real".

The three stages of propagation are:

1. Convolution of the amplitude distribution in the source plane with the quadratic-phase impulse response in Equation (21.57) for propagation over the distance $z_1 > 0$ to the lens.
2. Multiplication by the complex-valued transmittance function of the lens of refractive index n and focal length $\mathbf{f}$, including the pupil function and the quadratic-phase factor of the lens.
3. Convolution with the quadratic-phase impulse response from Equation (21.57) for the distance z_2 from the lens to the observation plane.

This sequence of operations is reminiscent of the C–M–C chirp Fourier transformer in Chapter 17, and would in fact implement a Fourier transform if $z_1 = z_2 = \mathbf{f}$. Even though the output of this imaging system is not a Fourier transform, the model is nonetheless useful and will be used several times during this discussion.

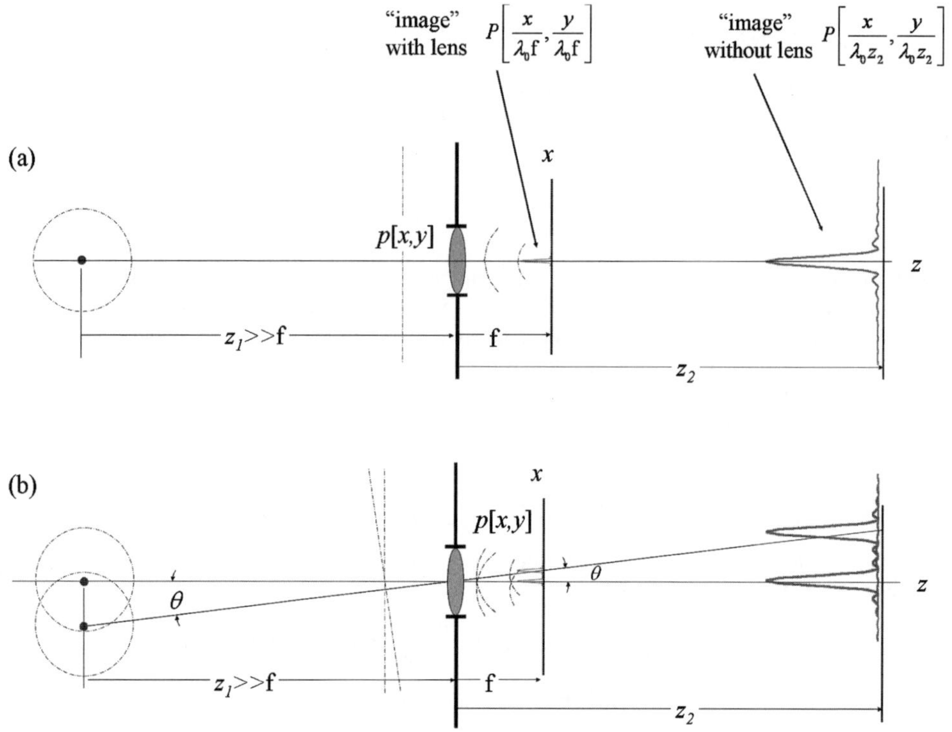

Figure 21.26 Imaging system with a lens of focal length **f** and aperture function $p[x, y]$: (a) light from point source at origin propagates to the Fraunhofer region at distance z_1. The image is located approximately at the focal point at the distance **f** from the lens. The image of the impulse is $P[x/\lambda_0\mathbf{f}, y/\lambda_0\mathbf{f}]$. If the light is propagated from the pupil without the lens the distance z_2, the pattern is $P[x/\lambda_0 z_2, y/\lambda_0 z_2]$. (b) Image of two point sources separated in angle by θ consists of two replicas of $P[x/\lambda_0\mathbf{f}, y/\lambda_0\mathbf{f}]$ separated by the same angle θ.

We anticipate that the imaging process is linear, meaning that a doubling of the input amplitude produces a doubling of the output amplitude. We now consider whether the process also is shift invariant by comparing images of on-axis and off-axis sources. If so, then the action of the lens system may be described by an impulse response or transfer function.

21.6.2.1 Image of Off-Axis Point Source

As an example, consider the "image" of a point source located off axis at $[x_0, y_0, z = 0]$. Direct application of the sifting property of the 2-D Dirac delta function yields a simple expression for the amplitude after propagating to the plane specified by z_1 at the "front" of the lens:

$$\begin{aligned} g[x, y; z_1] &= \delta[x - x_0, y - y_0; 0] * h[x, y; z_1] \\ &= \frac{1}{i\lambda_0 z_1} e^{+2\pi i z_1/\lambda_0} e^{+i\pi[(x-x_0)^2+(y-y_0)^2]/\lambda_0 z_1} \\ &= \frac{1}{i\lambda_0 z_1} e^{+2\pi i z_1/\lambda_0} e^{+i\pi(x^2+y^2)/\lambda_0 z_1} e^{+i\pi(x_0^2+y_0^2)/\lambda_0 z_1} e^{-2\pi i(x x_0+y y_0)/\lambda_0 z_1} \end{aligned} \quad (21.144)$$

The shift invariance of Fresnel diffraction ensures that the amplitude distribution at the front of the lens is an appropriately translated replica of the impulse response. The second quadratic-phase factor $\exp[+i\pi(x_0^2 + y_0^2)/\lambda_0 z_1]$ is due to the off-axis source location $[x_0, y_0]$.

This field is multiplied by the complex-valued transmittance function of the lens to produce $g[x, y; z_1, \mathbf{f}]$, where the two parameters z_1, $\mathbf{f}$ are set off from the coordinates by the semicolon. The quadratic-phase factors in the output coordinates $[x, y]$ may be combined immediately:

$$\begin{aligned} g[x, y; z_1, \mathbf{f}, \lambda_0] &= g[x, y; z_1, \lambda_0] \cdot t[x, y] \\ &= \left(\frac{1}{i\lambda_0 z_1} e^{+2\pi i z_1/\lambda_0} e^{+i\pi(x^2+y^2)/\lambda_0 z_1} e^{+i\pi(x_0^2+y_0^2)/\lambda_0 z_1} e^{-2\pi i(xx_0+yy_0)/\lambda_0 z_1}\right) \\ &\quad \cdot \left(p[x, y]\, e^{+2\pi i n\tau/\lambda_0} e^{-i\pi(x^2+y^2)/\lambda_0 \mathbf{f}}\right) \\ &= \frac{1}{i\lambda_0 z_1} e^{+2\pi i(z_1+n\tau)/\lambda_0} e^{+i\pi(x_0^2+y_0^2)/\lambda_0 z_1} \\ &\quad \cdot p[x, y]\, e^{+i\pi(x^2+y^2)/\lambda_0(1/z_1-1/\mathbf{f})} e^{-2\pi i(xx_0+yy_0)/\lambda_0 z_1} \end{aligned} \tag{21.145}$$

The light that emerges from the lens propagates the distance z_2 to the observation plane, i.e., it is convolved with another Fresnel impulse response. Before delving into the mathematical details of this last convolution, we briefly digress to consider another (and arguably more pleasing and intuitive) interpretation. For simplicity, consider the specific case where the original point source is on the line of symmetry, so that $[x_0, y_0] = [0, 0]$. We also assume that the pupil function has infinite support: $p[x, y] = 1[x, y]$, which ignores the criteria for Fresnel diffraction. The light from the on-axis point source that emerges from the back side of the lens is a simplified version of Equation (21.145):

$$g[x, y; z_1, \mathbf{f}, x_0 = y_0 = 0] = \left(\frac{1}{i\lambda_0 z_1} e^{+2\pi i(z_1+n\tau)/\lambda_0}\right) e^{+i\pi[(x^2+y^2)/\lambda_0](1/z_1-1/\mathbf{f})} \tag{21.146}$$

which is a positive quadratic-phase factor with chirp rate:

$$\alpha = \sqrt{\lambda_0 \cdot \left(\frac{1}{z_1} - \frac{1}{\mathbf{f}}\right)} \tag{21.147}$$

Equation (21.146) describes the action of the system consisting of the object distance z_1 and the lens focal length on an impulse, and so may be called the impulse response under those conditions; in other words, we could substitute the notation h for g in Equation (21.146):

$$h[x, y; z_1, \mathbf{f}, x_0 = y_0 = 0] = \left(\frac{1}{i\lambda_0 z_1} e^{+2\pi i(z_1+n\tau)/\lambda_0}\right) e^{+i\pi[(x^2+y^2)/\lambda_0](1/z_1-1/\mathbf{f})} \tag{21.148}$$

Since the goal is to reconstruct the original point source, we can think of the task as an example of the inverse filter described in Chapter 19 for the impulse response in Equation (21.148):

$$h[x, y; z_1, \mathbf{f}, x_0 = y_0 = 0] * w[x, y; z_2, \lambda_0] = \delta[x, y] \tag{21.149}$$

(we could also make a comparable interpretation as a matched filter for the input point object). The propagation over the distance z_2 has a positive quadratic-phase impulse response. Equation (17.25) shows that the inverse filter for the impulse response in Equation (21.149) has the same chirp rate with a negative sign:

$$\begin{aligned} w[x, y; z_2, \lambda_0] &= \left(\frac{1}{i\lambda_0 z_2} e^{+2\pi i z_2/\lambda_0}\right) e^{+i\pi(x^2+y^2)/\lambda_0 z_2} \\ &= \left(\frac{1}{i\lambda_0 z_2} e^{+2\pi i z_2/\lambda_0}\right) \left(e^{+i\pi[(x^2+y^2)/\lambda_0](1/z_1-1/\mathbf{f})}\right)^* \end{aligned} \tag{21.150}$$

We equate the chirp rates to find the relation between the object distance z_1, the focal length $\mathbf{f}$ of the lens, and the image distance z_2:

$$+\frac{1}{\lambda_0 z_2} = -\frac{1}{\lambda_0}\left(\frac{1}{z_1} - \frac{1}{\mathbf{f}}\right)$$
$$\Longrightarrow \frac{1}{z_1} + \frac{1}{z_2} = \frac{1}{\mathbf{f}} \tag{21.151}$$

which is the well-known equation that must be satisfied to form an image at the distance z_2 from the lens.

We now return to the more rigorous solution of the imaging equation by convolving the output at the back of the lens in Equation (21.145) with the impulse response for Fresnel propagation over the distance z_2. The amplitude of the resulting "image" is $g[x, y; z_1, \mathbf{f}, z_2, \lambda_0]$:

$$\begin{aligned}
g[x, y; z_1, \mathbf{f}, z_2, \lambda_0] \\
&= g[x, y; z_1, \mathbf{f}, \lambda_0] * h[x, y; z_2] \\
&= \left(\frac{1}{i\lambda_0 z_1} p[x, y]\, e^{+2\pi i(z_1+n\tau)/\lambda_0}\, e^{+i\pi[(x^2+y^2)/\lambda_0](1/z_1-1/\mathbf{f})}\right. \\
&\quad \left.\times e^{+i\pi(x_0^2+y_0^2)/\lambda_0 z_1}\, e^{-2\pi i(xx_0+yy_0)/\lambda_0 z_1}\right) * \left(\frac{1}{i\lambda_0 z_2}\, e^{+2\pi i z_2/\lambda_0}\, e^{+i\pi(x^2+y^2)/\lambda_0 z_2}\right) \\
&= \left(-\frac{1}{\lambda_0^2 z_1 z_2}\, e^{+2\pi i(z_1+n\tau+z_2)/\lambda_0}\, e^{+i\pi(x_0^2+y_0^2)/\lambda_0 z_1}\right) \\
&\quad \times [(p[x, y]\, e^{+i\pi[(x^2+y^2)/\lambda_0](1/z_1-1/\mathbf{f})}\, e^{-2\pi i(xx_0+yy_0)/\lambda_0 z_1}) * e^{+i\pi(x^2+y^2)/\lambda_0 z_2}]
\end{aligned} \tag{21.152}$$

To reduce notational clutter in the next several steps, the leading amplitude term and the constant phase are collapsed into the numerical constant K:

$$K \equiv -\frac{1}{\lambda_0^2 z_1 z_2}\, e^{+2\pi i(z_1+n\tau+z_2)/\lambda_0} \tag{21.153}$$

We can solve this convolution in either the space or frequency domains; we choose the latter because many of the resulting terms may be extracted from the integral. After substitution of the definition of 2-D convolution over the "dummy" integration variables [u, v], we obtain a rather prodigious expression for the amplitude distribution in the output plane:

$$\begin{aligned}
g[x, y; z_1, \mathbf{f}, z_2, \lambda_0] \\
&= K\, e^{+i\pi(x_0^2+y_0^2)/\lambda_0 z_1} \iint_{-\infty}^{+\infty} p[\mathrm{u}, \mathrm{v}]\, e^{+i\pi[(\mathrm{u}^2+\mathrm{v}^2)/\lambda_0](1/z_1-1/\mathbf{f})} \\
&\qquad \times e^{-2\pi i(\mathrm{u}x_0+\mathrm{v}y_0)/\lambda_0 z_1}\, e^{+i\pi[(x-\mathrm{u})^2+(y-\mathrm{v})^2]/\lambda_0 z_2}\, d\mathrm{u}\, d\mathrm{v} \\
&= K\, e^{+i\pi(x_0^2+y_0^2)/\lambda_0 z_1}\, e^{+i\pi(x^2+y^2)/\lambda_0 z_2} \\
&\quad \times \iint_{-\infty}^{+\infty} p[\mathrm{u}, \mathrm{v}]\, e^{+i\pi\{[(\mathrm{u}^2+\mathrm{v}^2)/\lambda_0](1/z_1-1/\mathbf{f}+1/z_2)\}} \\
&\qquad \times e^{-(2\pi i/\lambda_0)[\mathrm{u}(x_0/z_1+x/z_2)+\mathrm{v}(y_0/z_1+y/z_2)]}\, d\mathrm{u}\, d\mathrm{v}
\end{aligned} \tag{21.154}$$

Note that this expression is valid if both distances z_1 and z_2 individually satisfy the conditions required of Fresnel propagation. The leading quadratic-phase terms arise from the off-axis location of the source Dirac delta function with chirp rate $\alpha_1 = \sqrt{\lambda_0 z_1}$ and from the observation location with chirp rate

$\alpha_2 = \sqrt{\lambda_0 z_2}$. These represent what might be called a "residual" phase term that will be rewritten in simplified form shortly. The integrand includes a quadratic phase whose chirp rate is a function of z_1, z_2, and **f**, along with a "bilinear" phase that is a function of the source location $[x_0, y_0]$, the observation location $[x, y]$, and the propagation distances z_1 and z_2. If this expression is to represent a shift-invariant impulse response, there can be no explicit dependence of the integral on the source location $[x_0, y_0]$.

21.6.3 Amplitude Distribution at Image Point

Consider first the case where the quadratic-phase factor of $[u, v]$ in the integrand of Equation (21.154) evaluates to unity, which is assured by setting:

$$\frac{1}{z_1} - \frac{1}{\mathbf{f}} + \frac{1}{z_2} = 0 \Longrightarrow \frac{1}{z_1} + \frac{1}{z_2} = \frac{1}{\mathbf{f}} \Longrightarrow z_2 = \frac{z_1 \mathbf{f}}{z_1 - \mathbf{f}} \tag{21.155}$$

This is the same imaging equation from ray optics that resulted from the interpretation of the inverse filter in Equation (21.151) An infinite number of pairs z_1 and z_2 exist that satisfy Equation (21.155). A particular pair may be specified by defining their ratio:

$$-\frac{z_2}{z_1} \equiv M_T \tag{21.156}$$

which is also a well-known parameter in geometrical optics: the *transverse magnification* of the image. These last two equations may be combined to establish a condition on the transverse magnification and the focal length:

$$z_2 = (1 - M_T)\mathbf{f} \tag{21.157}$$

which is easily confirmed.

In geometrical optics, an object point located at position $[x_0, y_0]$ is imaged at $[M_T x_0, M_T y_0]$. If we select z_2 to satisfy Equation (21.155), then Equation (21.154) becomes:

$$g\left[x, y; z_1, \mathbf{f}, z_2 = \frac{z_1 \mathbf{f}}{z_1 - \mathbf{f}}\right] = K\left(e^{+i\pi(x_0^2+y_0^2)/\lambda_0 z_1}\, e^{+i\pi(x^2+y^2)/\lambda_0 z_2}\right)$$
$$\times \iint_{-\infty}^{+\infty} p[\mathrm{u}, \mathrm{v}]\, e^{-(2\pi i/\lambda_0)[\mathrm{u}(x_0/z_1 + x/z_2) + \mathrm{v}(y_0/z_1 + y/z_2)]}\, d\mathrm{u}\, d\mathrm{v} \tag{21.158}$$

We can now insert the transverse magnification to simplify several of the factors. First, the leading constant factor K may be rewritten in terms of M_T:

$$\begin{aligned} K &= -\frac{1}{\lambda_0^2 z_1 z_2} \exp\left[+2\pi i \frac{z_1 + n\tau + z_2}{\lambda_0}\right] \\ &= \frac{M_T}{(\lambda_0 z_2)^2} \exp\left[+2\pi i \frac{z_2(1 - 1/M_T) + n\tau}{\lambda_0}\right] \end{aligned} \tag{21.159}$$

Similarly, the source coordinates $[x_0, y_0]$ in the leading "residual" quadratic-phase factors may be remapped by applying the relation for the transverse magnification:

$$x = M_T x_0 \Longrightarrow x_0 = \frac{x}{M_T} \tag{21.160a}$$

$$y = M_T y_0 \Longrightarrow y_0 = \frac{y}{M_T} \tag{21.160b}$$

to obtain:

$$\begin{aligned}
&\exp\left[+i\pi\frac{x_0^2+y_0^2}{\lambda_0 z_1}\right]\cdot\exp\left[+i\pi\frac{x^2+y^2}{\lambda_0 z_2}\right]\\
&=\exp\left[-\frac{i\pi M_T}{\lambda_0 z_2}\frac{x^2+y^2}{M_T^2}\right]\cdot\exp\left[+i\pi\frac{x^2+y^2}{\lambda_0 z_2}\right]\\
&=\exp\left[-\frac{i\pi}{\lambda_0 z_2}\frac{x^2+y^2}{M_T}\right]\cdot\exp\left[+i\pi\frac{(x^2+y^2)}{\lambda_0 z_2}\right]\\
&=\exp\left[+\frac{i\pi}{\lambda_0 z_2}\left((x^2+y^2)\left(1-\frac{1}{M_T}\right)\right)\right] \qquad (21.161a)\\
&=\exp\left[+\frac{i\pi}{\lambda_0[(1-M_T)\mathbf{f}]}\left((x^2+y^2)\left(\frac{M_T-1}{M_T}\right)\right)\right]\\
&=\exp\left[-i\pi\frac{(x^2+y^2)}{M_T\lambda_0\mathbf{f}}\right] \qquad (21.161b)
\end{aligned}$$

where Equation (21.157) has been used. This shows that the residual quadratic phase at an image point is a function of both the location $[x, y]$ in the output image plane and the magnification M_T. The dependence on the input coordinates $[x_0, y_0]$ is now implicit in the magnification. Note that the effect of the quadratic phase is largest at a particular frequency if $|M_T| \cong 0$, which means that $z_1 \to +\infty$ and $z_2 \to \mathbf{f}$:

$$\begin{aligned}
\exp\left[+\frac{i\pi}{\lambda_0 z_2}\left((x^2+y^2)\left(1-\frac{1}{M_T}\right)\right)\right] &\to \exp\left[+\frac{i\pi}{\lambda_0\mathbf{f}}\left((x^2+y^2)\left(1+\frac{\mathbf{f}}{z_1}\right)\right)\right]\\
&\cong \exp\left[+i\pi\frac{x^2+y^2}{\lambda_0\mathbf{f}}\right]\exp\left[+i\pi\frac{x^2+y^2}{\lambda_0\cdot\infty}\right]\\
&\cong \exp\left[+i\pi\frac{x^2+y^2}{\lambda_0\mathbf{f}}\right]\cdot 1[x, y] \qquad (21.162)
\end{aligned}$$

This is the same quadratic-phase factor we saw in Equation (21.137).

The transverse magnification also may be substituted into the "bilinear" exponent in the integrand to remove the explicit dependence on the object distance z_1:

$$\frac{\mathrm{u}}{\lambda_0}\left(\frac{x_0}{z_1}+\frac{x}{z_2}\right)+\frac{\mathrm{v}}{\lambda_0}\left(\frac{y_0}{z_1}+\frac{y}{z_2}\right)=\mathrm{u}\left(\frac{x-M_T x_0}{z_2\lambda_0}\right)+\mathrm{v}\left(\frac{y-M_T y_0}{z_2\lambda_0}\right) \qquad (21.163)$$

The bilinear exponential is identical to the Fourier transform kernel evaluated at the particular spatial frequency $[\xi_0, \eta_0] = [(x - M_T x_0)/\lambda_0 z_2, (y - M_T y_0)/\lambda_0 z_2]$.

After combining all of these results and identifying that the bilinear term may be interpreted as the kernel of the Fourier integral, we obtain a new (and still prodigious!) expression for the output amplitude distribution produced at an "image plane" by the off-axis source point:

$$\begin{aligned}
&g\left[x, y; z_1, \mathbf{f}, z_2 = \frac{z_1\mathbf{f}}{z_1-\mathbf{f}}\right]\\
&= \mathcal{O}\{\delta[x-x_0, y-y_0]\}\\
&= K\, e^{-i\pi(x^2+y^2)/M_T\lambda_0\mathbf{f}} \iint_{-\infty}^{+\infty} p[\mathrm{u}, \mathrm{v}]\, e^{-2\pi i[\mathrm{u}(x-M_T x_0)/\lambda_0 z_2+\mathrm{v}(y-M_T y_0)/\lambda_0 z_2]}\, d\mathrm{u}\, d\mathrm{v}
\end{aligned}$$

$$= K\, e^{-i\pi(x^2+y^2)/M_T\lambda_0\mathbf{f}}(P[\xi,\eta]|_{\xi=(x-M_Tx_0)/\lambda_0z_2,\eta=(y-M_Ty_0)/\lambda_0z_2})$$

$$= \left(\frac{M_T}{(\lambda_0z_2)^2}\,e^{+2\pi i(z_1+n\tau+z_2)/\lambda_0}\right)e^{-i\pi(x^2+y^2)/M_T\lambda_0\mathbf{f}}\,P\left[\frac{x-M_Tx_0}{\lambda_0z_2},\frac{y-M_Ty_0}{\lambda_0z_2}\right] \quad (21.164)$$

where $P[\xi,\eta]$ is the 2-D Fourier transform of the pupil function $p[x,y]$. The coordinates of the spectrum of the pupil function have been scaled by factors such that the variables are space-domain coordinates $[x,y]$.

The amplitude of the "image" of an off-axis point source located at $[x_0,y_0]$ includes the scale factor $M_T/(\lambda_0z_2)^2$, the constant phase due to propagation "down" the z-axis, a quadratic-phase factor that is a function of the coordinates $[x,y]$ in the observation plane and the system magnification M_T, and a replica of the spectrum of the pupil function that is scaled by λ_0z_2 and whose translation relative to the origin is scaled by the system magnification M_T.

For a strictly shift-invariant system, the input and output coordinates $[x_0,y_0]$ and $[x,y]$ should appear only as differences $[x-x_0,y-y_0]$. This is clearly not true in this case. For example, the quadratic-phase factor is a function only of the output coordinates, while the spectrum of the aperture function includes differences of scaled input coordinates $[M_Tx_0,M_Ty_0]$ and the output coordinates $[x,y]$. To illustrate, consider a specific example with $M_T=-2$, which requires that $z_1=\frac{3}{2}\mathbf{f}$ and $z_2=-M_Tz_1=3\mathbf{f}$. A point source located on the axis at $[x_0,y_0]=[0,0]$ produces an image that is also "on axis", while the off-axis source located at $[x_0,y_0]=[+1,-2]$ is imaged at $[x,y]=[-2,+4]$. Note that either characteristic, "reversal" and/or magnification, would be sufficient to ensure that this optical system violates the strict definition of shift invariance.

With this analysis in mind, we consider practical applications of this imaging equation. The common situation of optical imaging at visible wavelengths is considered in detail in the next section, but we can introduce some of the important aspects now. The radiation at the output plane specified by z_2 is measured by a sensor that evaluates the *power* (squared magnitude) of the electric field amplitude averaged over many temporal cycles. In short, the sensor is insensitive to the phase of the radiation, so the leading shift-variant quadratic-phase function may be ignored. Therefore the shift variance evident in the leading quadratic-phase factor $(x^2+y^2)(1-1/M_T)$ is considered to be insignificant.

Just as was true for the Fraunhofer diffraction imaging "system", we can generalize the concept of shift invariance to deal with the magnification factor in the spectrum of the pupil function. We merely redefine the output coordinates to include "reversal" and magnification, thus ensuring that the translations of the input and output are "equivalent". In this way, it is possible to consider a system with distances z_1 and z_2 that satisfy the imaging equation to be shift invariant.

In a second, more general, imaging system, visible radiation continues to propagate through subsequent optical elements instead of being detected at the output plane specified by z_2. Such a system may be divided into C–M–C "stages" that each include the leading quadratic-phase term and thus do not individually satisfy the constraint of shift invariance. In other words, it is not possible to characterize the action of individual C–M–C stages by shift-invariant transfer functions, and thus the action of the ensemble of stages may not be expressed as the product of constituent transfer functions. That said, it is possible to instead characterize the action of the entire system as an "equivalent" single lens and pupil function that can be analyzed as a single C–M–C stage that is described using the shift-invariant picture of the elementary system. This is analogous to the situation in the Fraunhofer diffraction imaging "system", which produced shift-invariant imaging by cascading two shift-variant operations.

The third scenario deals with systems that use analogues of optical lenses to redirect and "image" electromagnetic radiation with smaller temporal frequencies (and thus longer wavelengths), such as radio waves. Because the waves oscillate more slowly in time, the detectors used in such systems often *are* sensitive to the "optical" phase. Such systems are shift variant due to the leading quadratic-phase term.

21.6.4 Shift-Invariant Description of Optical Imaging

If the radiation at the output plane is measured by a square-law detector (which is thus insensitive to the rapidly varying phase of the electric field at visible wavelengths), we may immediately discard all of the leading unit-magnitude phase factors in the impulse response since they will evaluate to unity in the final measured image:

$$\begin{aligned}\mathcal{O}\{\delta[x-x_0,y-y_0]\} &\equiv h[x-x_0,y-y_0;z_1,\mathbf{f},z_2]\\ &= \frac{M_T}{(\lambda_0 z_2)^2}\iint_{-\infty}^{+\infty} p[\alpha,\beta]\exp\left[-2\pi i\left(\alpha\left(\frac{x-M_T x_0}{\lambda_0 z_2}\right)+\beta\left(\frac{y-M_T y_0}{\lambda_0 z_2}\right)\right)\right]d\alpha\,d\beta \end{aligned}\tag{21.165}$$

To put this expression in the more familiar form, we substitute $\mathrm{u}\equiv\alpha/\lambda_0 z_2$, $\mathrm{v}\equiv\beta/\lambda_0 z_2$, $x_0'=M_T x_0$, $y_0'=M_T y_0$:

$$\begin{aligned}\mathcal{O}\{\delta[x-x_0,y-y_0]\}\\ &= \frac{M_T}{(\lambda_0 z_2)^2}\cdot|\lambda_0 z_2|^2\cdot\iint_{-\infty}^{+\infty} p[\lambda_0 z_2\mathrm{u},\lambda_0 z_2\mathrm{v}]\exp[-2\pi i(\mathrm{u}\cdot(x-x_0')+\mathrm{v}\cdot(y-y_0'))]\,d\mathrm{u}\,d\mathrm{v}\\ &= M_T P[\lambda_0 z_2(x-x_0'),\lambda_0 z_2(y-y_0')]\end{aligned}\tag{21.166}$$

The magnified image of an extended object $f[x,y]$ may be evaluated by first expressing it in the form of a convolution:

$$f\left[\frac{x}{M_T},\frac{y}{M_T}\right]*\delta[x,y]=\iint_{-\infty}^{+\infty} f\left[\frac{\mathrm{u}}{M_T},\frac{\mathrm{v}}{M_T}\right]\delta[x-\mathrm{u},y-\mathrm{v}]\,d\mathrm{u}\,d\mathrm{v}\tag{21.167}$$

We now apply the imaging operation to this expression, and obtain:

$$\begin{aligned}\mathcal{O}\left\{f\left[\frac{x}{M_T},\frac{y}{M_T}\right]\right\} &= \mathcal{O}\left\{\iint_{-\infty}^{+\infty} f\left[\frac{\mathrm{u}}{M_T},\frac{\mathrm{v}}{M_T}\right]\delta[x-\mathrm{u},y-\mathrm{v}]\,d\mathrm{u}\,d\mathrm{v}\right\}\\ &= \iint_{-\infty}^{+\infty} f\left[\frac{\mathrm{u}}{M_T},\frac{\mathrm{v}}{M_T}\right]\mathcal{O}\{\delta[x-\mathrm{u},y-\mathrm{v}]\}\,d\mathrm{u}\,d\mathrm{v}\\ &\propto M_T\iint_{-\infty}^{+\infty} f\left[\frac{x_0}{M_T},\frac{y_0}{M_T}\right]P\left[\frac{x-M_T x_0}{\lambda_0 z_2},\frac{y-M_T y_0}{\lambda_0 z_2}\right]dx_0\,dy_0\end{aligned}\tag{21.168}$$

The complete expression for the complex amplitude in the image plane includes the leading quadratic-phase factor:

$$\begin{aligned}\mathcal{O}\left\{f\left[\frac{x}{M_T},\frac{y}{M_T}\right]\right\} &= e^{+2\pi i(z_1+n\tau+z_2)/\lambda_0}\,e^{-i\pi(x^2+y^2)/M_T\lambda_0\mathbf{f}}\\ &\quad\times\iint_{-\infty}^{+\infty}\left(\frac{1}{M_T}f\left[\frac{x_0}{M_T},\frac{y_0}{M_T}\right]\right)\frac{(M_T)^2}{(\lambda_0 z_2)^2}P\left[\frac{x-M_T x_0}{\lambda_0 z_2},\frac{y-M_T y_0}{\lambda_0 z_2}\right]dx_0\,dy_0\end{aligned}\tag{21.169}$$

In words, the action of a single-lens imaging system in monochromatic light may be written as the convolution of the input function with an impulse response that is a scaled replica of the Fourier transform of the pupil function. The form of the impulse response is:

$$h\left[x,y;z_1,\mathbf{f},z_2=\left(\frac{1}{\mathbf{f}}-\frac{1}{z_1}\right)^{-1}\right]=\frac{M_T}{(\lambda_0 z_2)^2}\cdot P\left[\frac{x}{\lambda_0 z_2},\frac{y}{\lambda_0 z_2}\right]\tag{21.170}$$

By applying the scaling and "transform-of-a-transform" theorems, the corresponding transfer function is seen to be proportional to the reversed pupil function:

$$H\left[\xi,\eta;z_1,\mathbf{f},z_2=\left(\frac{1}{\mathbf{f}}-\frac{1}{z_1}\right)^{-1}\right]=M_T\cdot p[-\lambda_0 z_2\xi,-\lambda_0 z_2\eta]\tag{21.171}$$

Since the transfer function of the single-lens system in monochromatic light is a scaled replica of the pupil, it is apparent that the action of a realistic imaging system with a finite pupil must be a lowpass filter. Using the same analogy, an optical system whose pupil includes a central obscuration acts as a bandpass filter and thus rejects the constant part of the input function. Examples for a few cases will be considered shortly.

21.6.5 Examples of Single-Lens Imaging Systems

We will look at a few computed examples of imaging with single-lens systems by examining the amplitude step by step through the system to help develop intuition about the process. The first examples are the simplest: the object is an on-axis point and the object and image distances are equal. The subsequent examples will consider an extended object at different distances.

21.6.5.1 On-Axis Point Source, Infinite Aperture, Unit Magnification

Consider an on-axis point source imaged at "equal conjugates", where the object and image distances are equal. Equation (21.155) shows that $z_1 = z_2 = 2\mathbf{f}$. Profiles of the amplitude at different points in the process are shown in Figure 21.27. Though the concept of transverse magnification has no meaning for an on-axis point, it would be $M_T = -1$ for this system. The amplitude at the lens is the impulse response from Equation (21.57):

$$h[x, y; z_1 = 2\mathbf{f}] = \left(\frac{1}{i\lambda_0 \cdot 2\mathbf{f}}\, e^{+2\pi i z_1/\lambda_0}\right) e^{+i\pi(x^2+y^2)/\lambda_0(2\mathbf{f})} \tag{21.172}$$

We will not be concerned with the leading amplitude and phase factors, so we ignore them from this point forward in this discussion. If the aperture diameter is infinite, then all of the light from the source into the hemisphere with $z > 0$ is collected by the lens. The amplitude distribution immediately after the lens is:

$$h[x, y; z_1 = 2\mathbf{f}, \mathbf{f}] \propto e^{+i\pi(x^2+y^2)/\lambda_0(2\mathbf{f})}\, e^{-i\pi(x^2+y^2)/\lambda_0\mathbf{f}} = e^{-i\pi(x^2+y^2)/\lambda_0(2\mathbf{f})} \tag{21.173}$$

Thus the action of the lens changes the diverging parabolic wave with chirp rate $\sqrt{\lambda_0(2\mathbf{f})}$ to a converging wave with the same chirp rate. This wave is propagated by convolution with the same impulse response in Equation (21.170):

$$\begin{aligned} g[x, y; z_1 = 2\mathbf{f}, \mathbf{f}, \mathbf{z_2} = 2\mathbf{f}] &\propto e^{-i\pi(x^2+y^2)/\lambda_0(2\mathbf{f})} * e^{+i\pi(x^2+y^2)/\lambda_0(2\mathbf{f})} \\ &= e^{-i\pi(x^2+y^2)/\lambda_0(2\mathbf{f})} \bigstar e^{-i\pi(x^2+y^2)/\lambda_0(2\mathbf{f})} \end{aligned} \tag{21.174}$$

The output amplitude is proportional to the autocorrelation of the chirp function, which we showed to be proportional to a Dirac delta function in Equation (17.28). The image plane is located at the plane specified by $z = 4\mathbf{f}$:

$$g[x, y; z_1 = 2\mathbf{f}, \mathbf{f}, \mathbf{z_2} = 2\mathbf{f}] \propto \delta[x, y; z = 2\mathbf{f} + 2\mathbf{f} = 4\mathbf{f}] \tag{21.175}$$

In words, the image of the on-axis point source created by the infinite-aperture lens is a replica Dirac delta function (Figure 21.27). In a different way, this result also illustrates the reason for the geometrical imaging condition in Equation (21.155); the constraint on z_2 ensures that the chirp rate of propagation from the lens to the image plane "cancels" that of the wavefront that exits the lens. If this condition is not met, then the image is "out of focus".

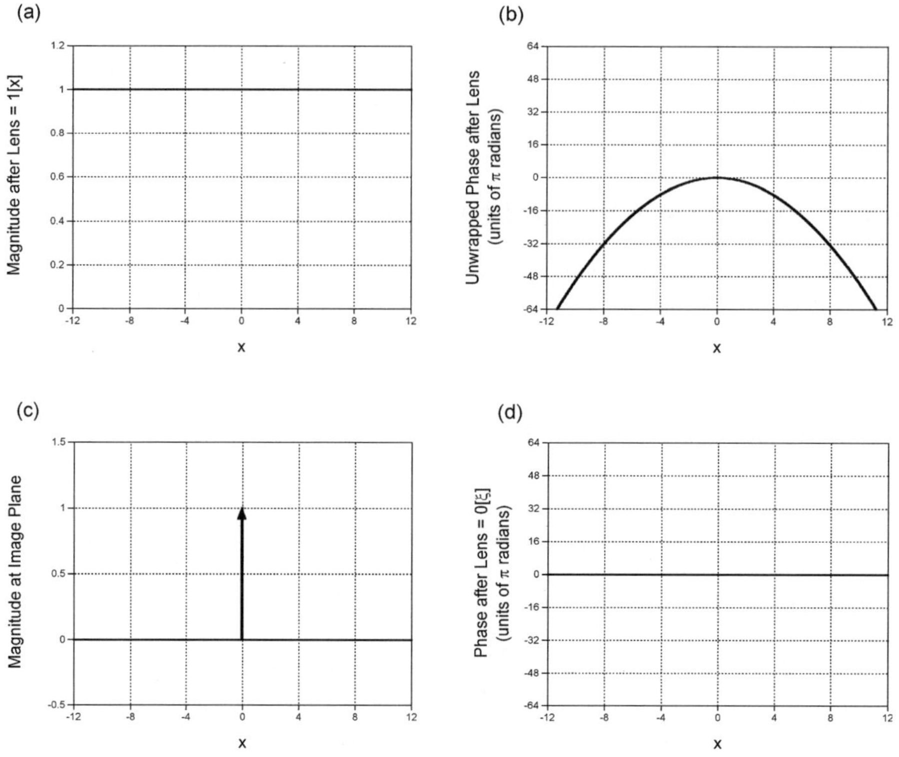

Figure 21.27 Profiles of amplitude due to on-axis point source with equal conjugates: (a) unit-constant magnitude after lens; (b) converging quadratic phase after lens; (c) magnitude at image plane is the ideal Dirac delta function; and (d) phase at image plane is zero.

21.6.5.2 On-Axis Point Source, Finite Aperture, Unit Magnification

A finite aperture diameter is described by the pupil function $p[x, y]$, which may be applied after the phase of the lens, thus modifying Equation (21.172):

$$h[x, y; 2\mathbf{f}, \mathbf{f}] \propto e^{+i\pi(x^2+y^2)/\lambda_0(2\mathbf{f})} \cdot (p[x, y]\, e^{-i\pi(x^2+y^2)/\lambda_0\mathbf{f}}) = p[x, y]\, e^{-i\pi(x^2+y^2)/\lambda_0(2\mathbf{f})} \tag{21.176}$$

The amplitude at the image plane is the convolution of this pattern with the same impulse response for propagation in Equation (21.170):

$$\begin{aligned} g[x, y; 2\mathbf{f}, \mathbf{f}, 2\mathbf{f}] &\propto (p[x, y]\, e^{-i\pi(x^2+y^2)/\lambda_0(2\mathbf{f})}) * e^{+i\pi(x^2+y^2)/\lambda_0(2\mathbf{f})} \\ &\propto P\left[\frac{x}{\lambda_0(2\mathbf{f})}, \frac{y}{\lambda_0(2\mathbf{f})}\right] \cdot e^{+i\pi(x^2+y^2)/\lambda_0(2\mathbf{f})} \end{aligned} \tag{21.177}$$

where the 2-D M–C–M chirp Fourier transform from Equation (17.83) has been used to rewrite the result in terms of the spectrum of the pupil function. The image of the on-axis point source is a scaled replica of the 2-D Fourier transform of the aperture function multiplied by a diverging chirp function. The image of the point source has been "blurred" by the finite aperture, as shown by the profiles of the amplitude in Figure 21.28.

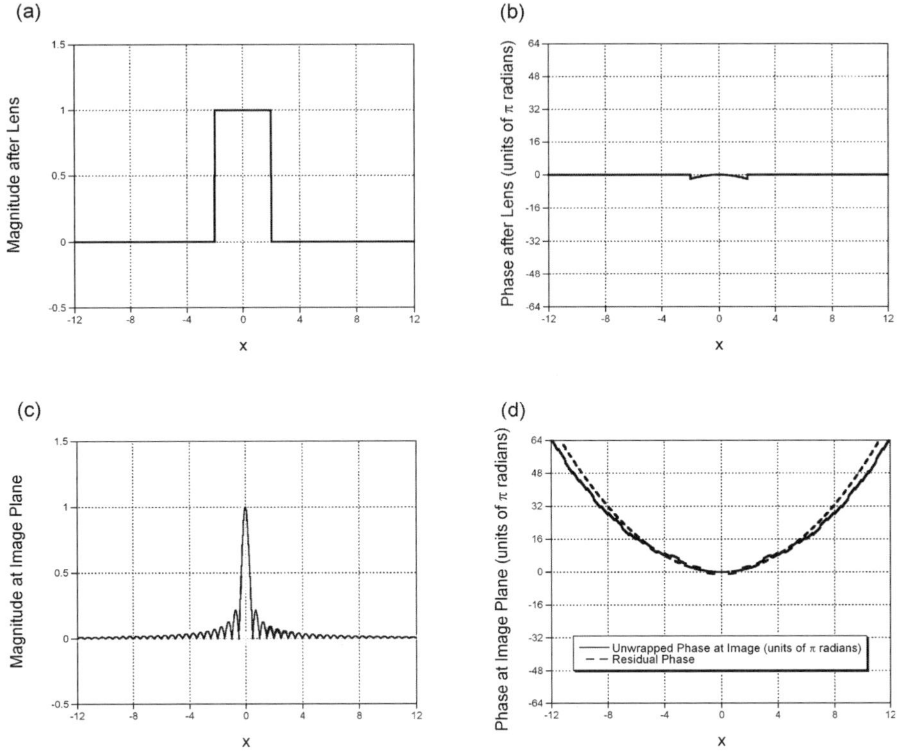

Figure 21.28 Profiles of amplitude due to on-axis point source with equal conjugates in system with finite pupil: (a) magnitude after lens, showing small pupil function; (b) converging quadratic phase after lens; (c) magnitude of image, showing blur due to finite pupil size; (d) phase at image compared to residual phase factor.

21.6.5.3 Imaging of Extended Object

Now consider imaging of an extended object, so that the concept of transverse magnification is more obvious. In these examples, the objects, systems, and images are displayed as 1-D profiles, but this is valid because the coordinates are separable. The test object is a real-valued function with intensity (squared magnitude) $|f[x]|^2$ as shown in Figure 21.29. The intensity of the input function may be written in several equivalent ways, including:

$$|f[x]|^2 = RECT\left[\frac{x-2}{2}\right] + \left(TRI\left[\frac{x-3}{3}\right] \cdot STEP[x-3]\right) \tag{21.178}$$

This "gray-scale" function is "off axis"; its region of support is $+1 \leq x \leq +6$. Note that the amplitude of the object is the square root of the intensity, and we assume that the phase of the light is assumed to be zero for all x. The mathematical equations are applied to the complex amplitude that includes the optical phase, rather than the "phase-free" intensity. The magnitude is the square root of the intensity. The phase is assumed to be constant across the object, which means that the object is illuminated by a plane wave.

Imaging at Equal Conjugates In the first example, assume that the object and image distances are equal ($z_1 = z_2 = 2\mathbf{f}$), so the transverse magnification is $M_T = -1$. Since the aperture is assumed to be

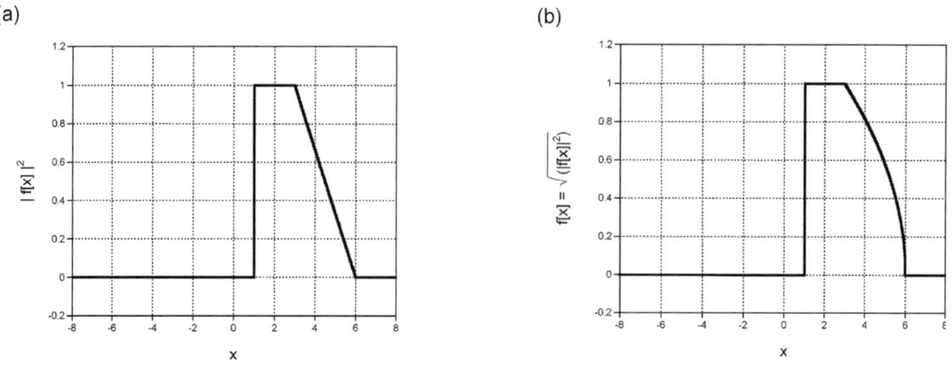

Figure 21.29 The 1-D extended object used to demonstrate optical imaging: (a) observed squared magnitude $|f[x]|^2$ of input function; (b) nonnegative amplitude $f[x] = \sqrt{|f[x, y]|^2}$, where the phase is assumed to be $0[x]$.

infinite, all light reflected from (or emitted by) the object into the hemisphere is collected and imaged. The impulse responses of propagation from the object to the lens and from the lens to the image plane therefore are identical "upchirps":

$$h[x, y; z_1 = 2\mathbf{f}, \mathbf{f}, z_2 = 2\mathbf{f}] = \frac{1}{i\lambda_0(2\mathbf{f})} \exp\left[+2\pi i \frac{2\mathbf{f}}{\lambda_0}\right] \exp\left[+i\pi \frac{(x^2 + y^2)}{\lambda_0(2\mathbf{f})}\right] \tag{21.179}$$

We first assume that the lens has infinite aperture, so its complex-valued amplitude transmittance is the unit-amplitude "downchirp":

$$t[x, y] = 1[x, y] \exp\left[-i\pi \frac{(x^2 + y^2)}{\lambda_0 \mathbf{f}}\right] \tag{21.180}$$

The residual phase factor in Equation (21.161b) becomes:

$$\exp\left[-i\pi \frac{(x^2 + y^2)}{M_T \lambda_0 \mathbf{f}}\right] = \exp\left[+i\pi \frac{(x^2 + y^2)}{\lambda_0 \mathbf{f}}\right] \tag{21.181}$$

which has a positive chirp rate corresponding to a diverging wave.

The steps in the imaging process are shown in Figures 21.30–21.32, and the final output irradiance is compared to the input in Figure 21.33.

Imaging of Extended Object with $|M_T| < 1$ Now consider a system that "minifies" the image; the object distance is $z_1 = 4\mathbf{f}$ and the image distance is $z_2 = \frac{4}{3}\mathbf{f} < z_1$. The transverse magnification is $M_T = -z_2/z_1 = -\frac{1}{3}$. Because the image is "minified", the full radiant power from an area element on the object converges into a smaller area, so that the image is "brighter" than the object, as shown in Figure 21.34.

Note that the residual phase from Equation (21.160) evaluates to:

$$\exp\left[+\frac{i\pi}{\lambda_0 z_2}(x^2 + y^2)(1 + 3)\right] = \exp\left[+i\pi \frac{(x^2 + y^2)}{(\lambda_0 \mathbf{f}/3)}\right] \tag{21.182}$$

which describes a diverging wave at the output image with "chirp rate" $\alpha = \sqrt{\lambda_0 \mathbf{f}/3}$.

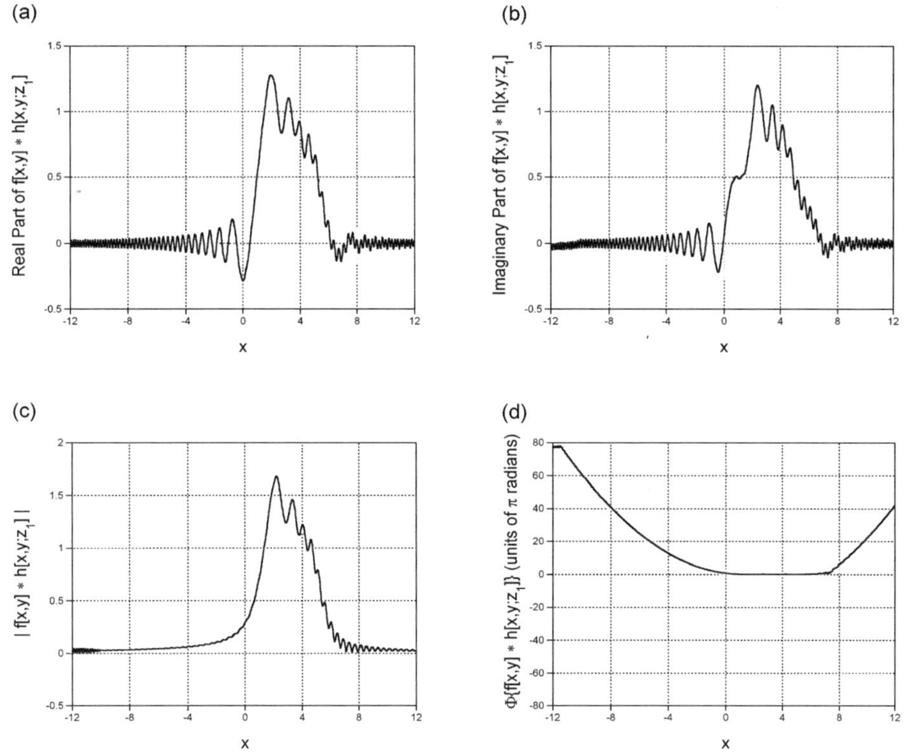

Figure 21.30 Amplitude distribution of light diffracted from $f[x]$ calculated at the front of the lens for equal conjugates, $z_1 = z_2 = 2\mathbf{f}$: (a) real part; (b) imaginary part; (c) magnitude; (d) phase, which is approximately zero where the magnitude is large.

Virtual Images Consider imaging where the object is located "inside" the focal length of the lens, e.g., $z_1 = \mathbf{f}/2$. Equation (21.155) indicates that the image distance $z_2 = -\mathbf{f}$, meaning that the image is "virtual" (located "behind" the lens). The transverse magnification in this case is $M_T = +2$, so that the image is "upright". The residual quadratic-phase factor in Equation (21.161b) evaluates to $\exp[-i\pi(x^2 + y^2)/2\lambda_0\mathbf{f}]$. Since the sign of this phase is negative, the quadratic-phase term describes a "converging" wave, as shown in Figure 21.35.

21.7 Shift-Variant Imaging Systems

21.7.1 Response of System at "Nonimage" Point

Now consider the output due to an off-axis point source if the image distance does not satisfy the geometrical optics imaging condition in Equation (21.155). For an off-axis source point, the phase measured at different locations in the observation plane is the vector sum of contributions from the same source point that have traveled different paths. If the point source moves around in the source plane, the different paths traveled from the same source point ensure that the amplitude at the output plane cannot be described by a simple translation, and therefore the "out-of-focus" imaging process is

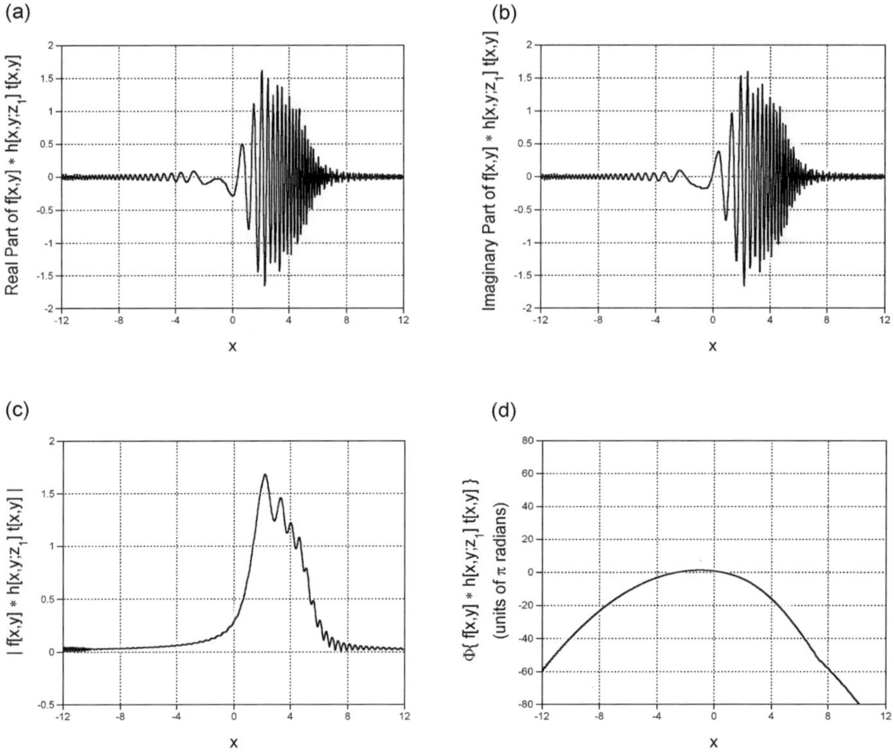

Figure 21.31 Distribution of amplitude of light at the output side of the lens for equal conjugates with infinite aperture: (a) real part of amplitude; (b) imaginary part; (c) magnitude; (d) phase, which is dominated by the downchirp of the lens.

not shift invariant. Even so, there still are useful concepts to be gleaned from further examination of this situation.

To evaluate the result, we must perform the sequence of steps in the imaging equation for an off-axis source in Equation (21.154):

$$
\begin{aligned}
g[x, y; z_1, \mathbf{f}, z_2]
&= K\, e^{+i\pi(x_0^2+y_0^2)/\lambda_0 z_1}\, e^{+i\pi(x^2+y^2)/\lambda_0 z_2} \\
&\quad \cdot \iint_{-\infty}^{+\infty} p[\alpha, \beta]\, e^{+i\pi\{[(\alpha^2+\beta^2)/\lambda_0](1/z_1-1/\mathbf{f}+1/z_2)]} \\
&\quad \cdot e^{-2\pi i[\alpha(x_0/\lambda_0 z_1+x/\lambda_0 z_2)+\beta(y_0/\lambda_0 z_1+y/\lambda_0 z_2)]}\, d\alpha\, d\beta \\
&= K\, e^{+i\pi(1/\lambda_0)[(x_0^2+y_0^2)/z_1+(x^2+y^2)/z_2]} \\
&\quad \cdot (\mathcal{F}_2\{p[x, y]\, e^{+i\pi[(x^2+y^2)/\lambda_0](1/z_1-1/\mathbf{f}+1/z_2)}\}|_{\xi\to x_0/\lambda_0 z_1+x/\lambda_0 z_2,\, \eta\to y_0/\lambda_0 z_1+y/\lambda_0 z_2})
\end{aligned}
\tag{21.183}
$$

Since no image is formed in the out-of-focus case, the concept of transverse magnification becomes "fuzzy" in this context (pun intended), but we retain the definition of $M_T = -z_2/z_1$.

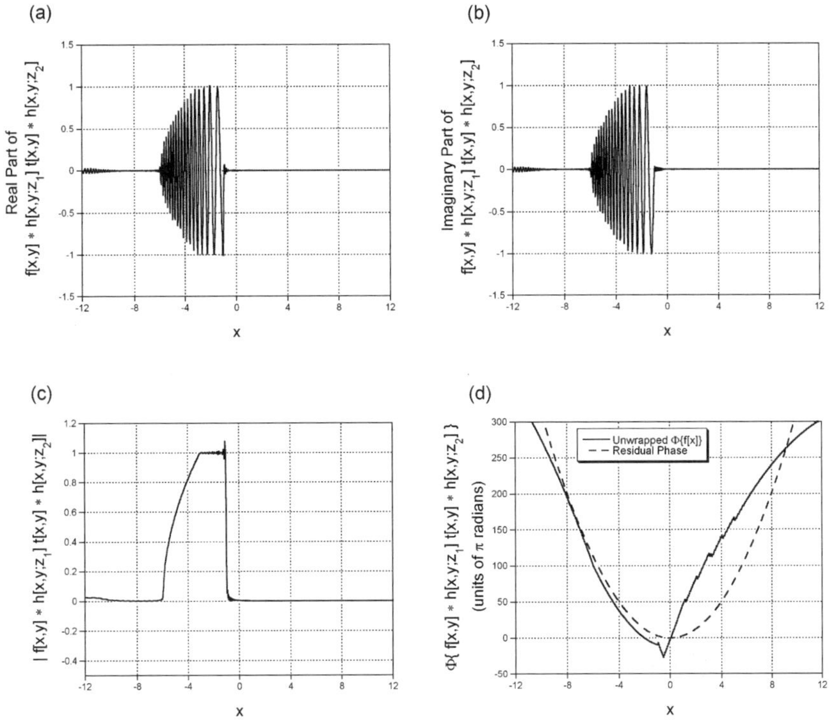

Figure 21.32 Distribution of amplitude of light at the "output" side of the lens for equal conjugates with infinite aperture: (a) real part of amplitude; (b) imaginary part; (c) magnitude; (d) phase, which is dominated by the downchirp of the lens.

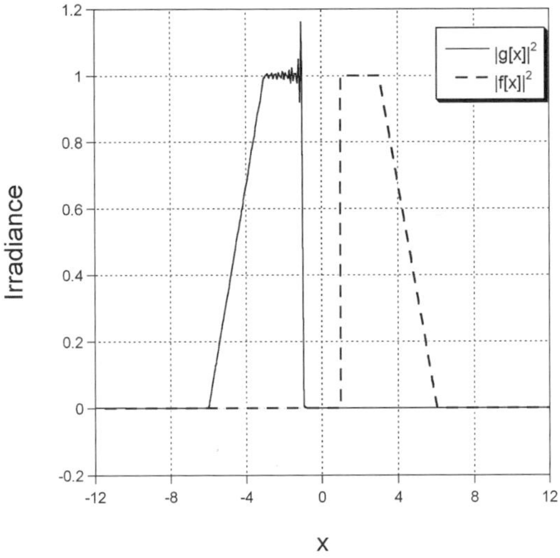

Figure 21.33 Comparison of the output irradiance $|g[x]|^2$ (calculated via FFTs) compared to $|f[x]|^2$; the transverse magnification $M_T = -1$. The ringing apparent in $|g[x]|^2$ is due to the constraints of the discrete computation.

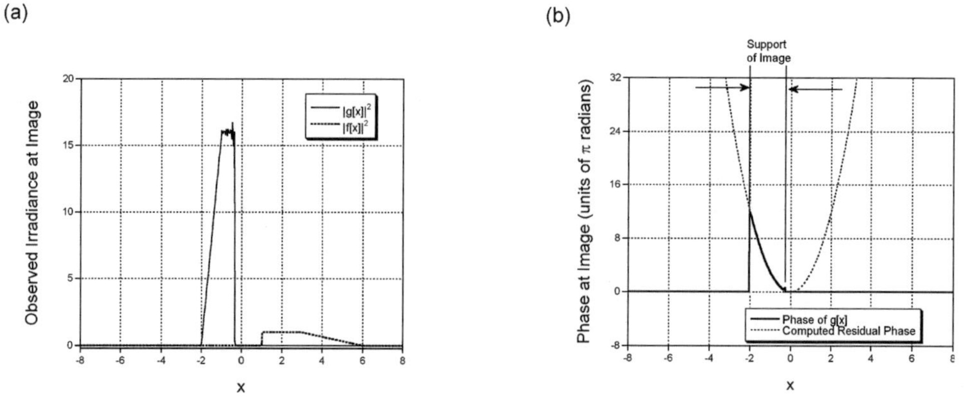

Figure 21.34 Imaging with $z_1 = 4\mathbf{f}$, $z_2 = \frac{4}{3}\mathbf{f}$, $M_T = -\frac{1}{3}$: (a) comparison of irradiance of image to object intensity, showing reduced support due to magnification. Also note the scaling of the irradiance necessary to conserve energy. (b) Phase of image compared to computed residual phase factor.

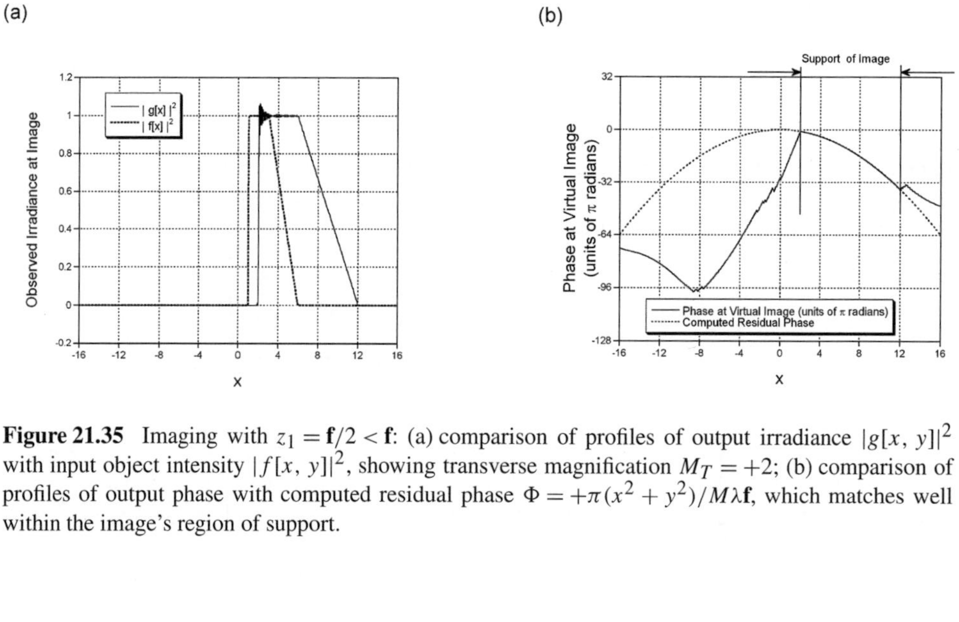

Figure 21.35 Imaging with $z_1 = \mathbf{f}/2 < \mathbf{f}$: (a) comparison of profiles of output irradiance $|g[x, y]|^2$ with input object intensity $|f[x, y]|^2$, showing transverse magnification $M_T = +2$; (b) comparison of profiles of output phase with computed residual phase $\Phi = +\pi(x^2 + y^2)/M\lambda\mathbf{f}$, which matches well within the image's region of support.

Consider first the Fourier transform of the quadratic-phase factor modulated by the pupil function of the lens. For simplicity, first consider that the pupil function is separable in Cartesian coordinates and the chirp rate of the quadratic-phase factor is α:

$$\sqrt{\frac{\lambda_0}{1/z_1 - 1/\mathbf{f} + 1/z_2}} \equiv \alpha$$

$$\Longrightarrow \exp\left[+i\pi \frac{(x^2 + y^2)}{\lambda_0}\left(\frac{1}{z_1} + \frac{1}{z_2} - \frac{1}{\mathbf{f}}\right)\right] = \exp\left[+i\pi\left(\frac{x^2 + y^2}{\alpha^2}\right)\right] \tag{21.184}$$

Note that $\alpha = \infty$ if z_2 satisfies the geometrical optics imaging equation, so that the quadratic-phase factor evaluates to the unit constant. As the observation plane is moved away from the image plane, then α decreases so that the chirp applied to the pupil function $p[x, y]$ oscillates "faster".

We have not found an analytic expression for the Fourier transform in Equation (21.183), but we can apply the method of stationary phase from Chapter 13 to approximate it. The resulting expression is:

$$\mathcal{F}_2\{p[x,y]\,e^{+i\pi(x^2+y^2)/\alpha^2}\}|_{\xi\to x_0/\lambda_0 z_1 + x/\lambda_0 z_2,\,\eta\to y_0/\lambda_0 z_1 + y/\lambda_0 z_2}$$
$$\cong i\lambda_0 z_2\left(\frac{1}{z_2(1/z_1+1/z_2-1/\mathbf{f})}\right)p\left[\frac{x+(z_2/z_1)x_0}{z_2(1/z_1+1/z_2-1/\mathbf{f})},\frac{y+(z_2/z_1)y_0}{z_2(1/z_1+1/z_2-1/\mathbf{f})}\right]$$
$$\cdot\exp\left[-i\pi\frac{(x_0/z_1+x/z_2)^2+(y_0/z_1+y/z_2)^2}{\lambda_0(1/z_1+1/z_2-1/\mathbf{f})}\right]\exp\left[+i\pi\frac{x_0^2+y_0^2}{\lambda_0 z_1}\right]\exp\left[+i\pi\frac{x^2+y^2}{\lambda_0 z_2}\right] \tag{21.185}$$

Note that the coordinates of the pupil function have been translated by a factor proportional to $-z_2/z_1$, which we again define to be the "transverse magnification" M_T, even though we could argue that the concept of magnification at a "nonimage" point is not well defined. The diameter of the pupil function p has been scaled by a factor that will be called the "pupil magnification" M_p and that may be written in several equivalent ways:

$$M_p \equiv z_2\left(\frac{1}{z_1}+\frac{1}{z_2}-\frac{1}{\mathbf{f}}\right)=1+z_2\left(\frac{1}{z_1}-\frac{1}{\mathbf{f}}\right)=1-M_T-\frac{z_2}{\mathbf{f}} \tag{21.186}$$

M_p evaluates to zero at the location of the image. The complete approximate expression for the defocused image of the off-axis point source is:

$$\mathcal{F}_2\{p[x,y]\,e^{+i\pi(x^2+y^2)/\alpha^2}\}|_{\xi\to x_0/\lambda_0 z_1 + x/\lambda_0 z_2,\,\eta\to y_0/\lambda_0 z_1 + y/\lambda_0 z_2}$$
$$\cong\frac{i\lambda_0 z_2}{M_p}p\left[\frac{x-M_T x_0}{M_p},\frac{y-M_T y_0}{M_p}\right]\exp\left[+i\pi\frac{x_0^2+y_0^2}{\lambda_0 z_1}\left(1+\frac{M_T}{M_p}\right)\right]$$
$$\times\exp\left[+i\pi\frac{x^2+y^2}{\lambda_0 z_2}\left(1-\frac{1}{M_p}\right)\right]\exp\left[-2\pi i\frac{x_0 x+y_0 y}{\lambda_0 M_p z_1}\right]$$
$$\propto p\left[\frac{x-M_T x_0}{1-M_T-z_2/\mathbf{f}},\frac{y-M_T y_0}{1-M_T-z_2/\mathbf{f}}\right]\exp\left[+i\pi\frac{x_0^2+y_0^2}{\lambda_0 z_1[M_p/(M_p+M_T)]}\right]$$
$$\times\exp\left[+i\pi\frac{x^2+y^2}{\lambda_0 z_2[M_p/(M_p-1)]}\right]\exp\left[-2\pi i\frac{x_0 x+y_0 y}{\lambda_0 M_p z_1}\right]$$
$$=p\left[\frac{x-M_T x_0}{M_p},\frac{y-M_T y_0}{M_p}\right]\exp\left[-i\pi\frac{x_0^2+y_0^2}{\lambda_0 z_2[M_p/M_T(M_p+M_T)]}\right]$$
$$\times\exp\left[+i\pi\frac{x^2+y^2}{\lambda_0 z_2[M_p/(M_p-1)]}\right]\exp\left[+2\pi i\frac{x_0 x+y_0 y}{\lambda_0(M_p/M_T)z_2}\right] \tag{21.187}$$

At "out-of-focus" locations, the support of this approximation of the amplitude pattern is constrained by the scaled and translated replica of the pupil, as shown in Figure 21.36. If the observation plane is closer to the lens than the image (the observation plane is "inside" focus), the scale factor for the pupil's width parameter is finite and positive; if observed farther from the lens than the image, the scale factor is negative so that the image of the pupil is "upside down". The two quadratic-phase factors are due respectively to the location of the source and observation points. The former evaluates to unity if the source is on axis, while the latter is the "residual" phase at the observation point.

The defocused image of an extended object is composed of appropriately scaled and translated replicas of the pupil for each point on the object.

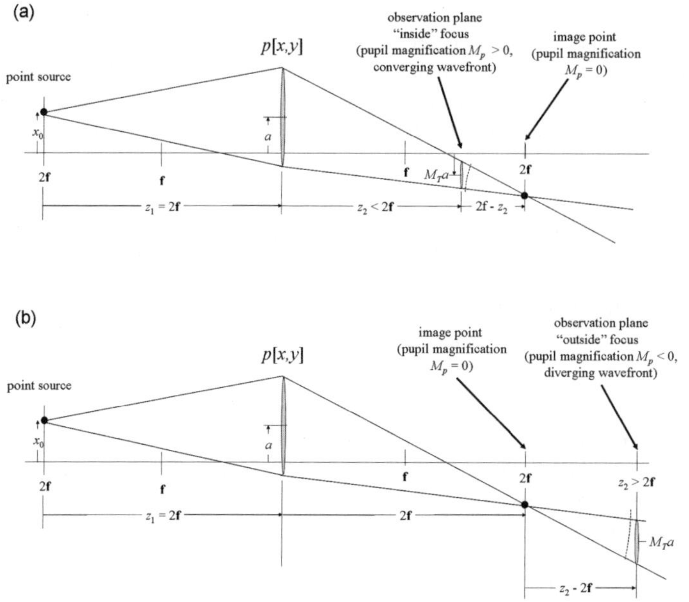

Figure 21.36 Approximation of amplitude due to "out-of-focus" off-axis point source. The source is located in the object plane at coordinates $[x_0, y_0]$. The distance to the lens is $2\mathbf{f}$, so the image plane is at the same distance from the lens. If viewed inside focus with $z_2 < 2\mathbf{f}$ in (a), the magnification of the pupil is positive and less than unity and it is translated off axis by the factor $M_T x_0$ and modulated by a converging quadratic-phase factor with a negative chirp rate. (b) If viewed outside focus with $z_2 > 2\mathbf{f}$, the pupil magnification is negative and it is modulated by a diverging quadratic phase.

Consider an example where the object distance $z_1 = 2\mathbf{f}$, so the image plane would be located at the same distance $z_2 = 2\mathbf{f}$. However, if the observation plane is "inside" the focus at $z_2 = \frac{3}{2}\mathbf{f}$, then the scale factors are:

$$M_T = -\frac{z_2}{z_1} = -\frac{3}{4} \tag{21.188a}$$

$$M_p = 1 - M_T - \frac{z_2}{\mathbf{f}} = 1 + \frac{3}{4} - \frac{3}{2} = +\frac{1}{4} \tag{21.188b}$$

so the "image" of the pupil is upright and scaled by a factor of $\frac{1}{4}$.

21.7.2 Chirp Fourier Transform and Fraunhofer Diffraction

We have derived the image $g[x, y]$ at the observation plane due to a coherent source distribution $f[x]$ located at the origin and generated by a lens of focal length $\mathbf{f}$ located at $z = z_1$. Implicit in the derivation was the assumption that light emitted from all points in the source had the same phase. The output is obtained by the sequence of convolve, multiply, and convolve operations:

$$g[x, y] \propto [(f[x, y] * e^{+i\pi(x^2+y^2)/\lambda_0 z_1}) \cdot (p[x, y]\, e^{-i\pi(x^2+y^2)/\lambda_0 \mathbf{f}})] * e^{+i\pi(x^2+y^2)/\lambda_0 z_2} \tag{21.189}$$

which described a system that could be considered to be linear and shift invariant if the geometrical imaging equation is satisfied and if the irradiance (squared magnitude) is the measured quantity. If the

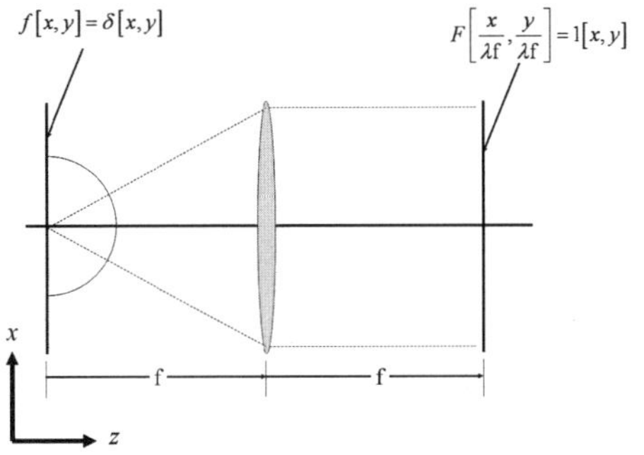

Figure 21.37 Optical chirp Fourier transform of on-axis point source $f[x, y] = \delta[x, y]$. The combination of (1) propagation over the distance **f**, (2) multiplication by a lens with focal length **f**, and (3) propagation over **f** produces the unit-constant spectrum.

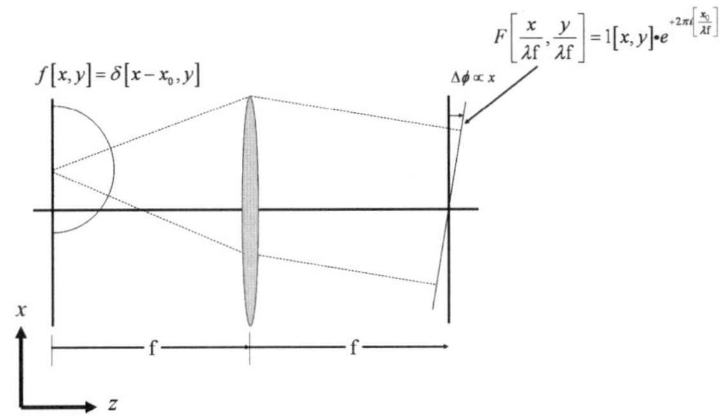

Figure 21.38 Optical chirp Fourier transform of on-axis point source $f[x, y] = \delta[x - x_0, y]$ produces a "tilted" plane wave whose phase is proportional to x_0.

geometrical equation is not satisfied, then the image is defocused and the system must be considered shift variant.

Consider first a few cases that assume the pupil is infinitely large ($p[x, y] = 1[x, y]$), as shown in Figure 21.37. If the source is located on the optical axis in the front focal plane of the lens, then $[x_0, y_0] = [0, 0]$ and $z_1 = \mathbf{f}$. From the geometrical imaging equation, the image distance is $z_2 = +\infty$, which implies that the transverse magnification $M_T = -z_2/z_1 = -\infty$ and the irradiance at the output plane is uniform. Note that the irradiance must also be uniform at any observation plane located at a finite distance from the lens. Moreover, the light waves arriving at all points in the observation plane will have traveled the same distance, and therefore required the same travel time; the sinusoidal waves measured at all points in the output plane have the same phase.

If the position of the source is moved off axis, then the observation plane will still be uniformly illuminated, but the light waves will now arrive at different times and thus have different phases. In fact, the phase is proportional to the distance off axis, as shown in Figure 21.38.

In both of these two examples, the same amplitude pattern is measured in any observation plane located at a finite distance from the source. What if we observe at $z_2 = \mathbf{f}$ (so that $z_2 = +z_1$)? Clearly the object and "image" distances do not satisfy the geometrical optics imaging equation. The expression for the observed amplitude is identical to the 2-D C–M–C chirp Fourier transform in Equation (17.84):

$$([f[x, y] * e^{+i\pi(x^2+y^2)/\lambda_0\mathbf{f}}] \cdot e^{-i\pi(x^2+y^2)/\lambda_0\mathbf{f}}) * e^{+i\pi(x^2+y^2)/\lambda_0\mathbf{f}} = (i\lambda_0\mathbf{f}) \cdot F\left[\frac{x}{\lambda_0\mathbf{f}}, \frac{x}{\lambda_0\mathbf{f}}\right] \tag{21.190}$$

which (of course) is a shift-variant system. The output is a function of $[x, y]$, and therefore is in the space domain, but the spatial frequency of the spectrum is obtained by applying the scale factor of $\lambda_0\mathbf{f}$ to the coordinates. In words, for an object located one focal length "in front" of the lens and illuminated with a monochromatic plane wave, so that the optical phase of the light is uniform across the input plane, the output amplitude observed at a plane one focal length "behind" the lens is a scaled replica of the 2-D Fourier transform of the input. It is useful to compare the output of this single-lens system to the Fraunhofer diffraction pattern in Equation (21.96):

$$\mathcal{O}\{\mathcal{E}_0 f[x, y]\}|_{x_1,y_1;z_1\to+\infty} = \frac{\mathcal{E}_0}{i\lambda_0 z_1} e^{+2\pi i z_1/\lambda_0} e^{+i\pi(x_1^2+y_1^2)/\lambda_0 z_1} F\left[\frac{x_1}{\lambda_0 z_1}, \frac{y_1}{\lambda_0 z_1}\right] \tag{21.191}$$

The single-lens system produces a similar result without propagating the large distances necessary in Fraunhofer diffraction. The lens "compresses" the propagation distance from the very large value of z_1 to the more reasonable and useful value of $\mathbf{f}$.

21.7.2.1 Optical Implementation of the "Transform of a Transform"

If the output of the first system in Equation (21.190) is applied to a second identical system, the final output is proportional to the "transform of a transform":

$$\begin{aligned}&(i\lambda_0\mathbf{f})\left[\left(F\left[\frac{x}{\lambda_0\mathbf{f}}, \frac{y}{\lambda_0\mathbf{f}}\right] * e^{+i\pi(x^2+y^2)/\lambda_0\mathbf{f}}\right) \cdot e^{-i\pi(x^2+y^2)/\lambda_0\mathbf{f}}\right] * e^{+i\pi(x^2+y^2)/\lambda_0\mathbf{f}}\\ &= (i\lambda_0\mathbf{f})^2\left(|\lambda_0\mathbf{f}|^2 f\left[\lambda_0\mathbf{f}\cdot\left(-\frac{x}{\lambda_0\mathbf{f}}\right), \lambda_0\mathbf{f}\cdot\left(-\frac{y}{\lambda_0\mathbf{f}}\right)\right]\right)\\ &= -(\lambda_0\mathbf{f})^4 \cdot f[-x, -y]\end{aligned} \tag{21.192}$$

In words, the output of this "4f" optical system is proportional to an inverted replica of the input, as shown in Figure 21.39.

The Fourier spectrum $F[x/\lambda_0\mathbf{f}, y/\lambda_0\mathbf{f}]$ in Equation (21.190) produced at the midplane of this 4**f** optical system is accessible and may be modified. For example, an optical transparency of the form $H[x/\lambda_0\mathbf{f}, y/\lambda_0\mathbf{f}]$ that represents a magnitude function may be placed at the Fourier plane to modify the input spectrum. The resulting amplitude observed at the output plane is $f[-x, -y] * h[-x, -y]$. It is theoretically possible to construct transparencies with variable optical thickness that models a complex-valued transfer function.

21.7.2.2 Effect of Finite Apertures

The pupil function of the lens in the realistic optical C–M–C chirp Fourier transformer must be compact, thus truncating the domain of spatial frequencies that it can pass. This significantly complicates the system analysis, though some simple cases illustrate the qualitative effect. If the source is an infinite-support on-axis plane wave, the amplitude at the front of the lens may be evaluated from Equation (21.67):

$$g[x, y; \mathbf{f}] = 1[x, y] \cdot e^{+2\pi i\mathbf{f}/\lambda_0} \tag{21.193}$$

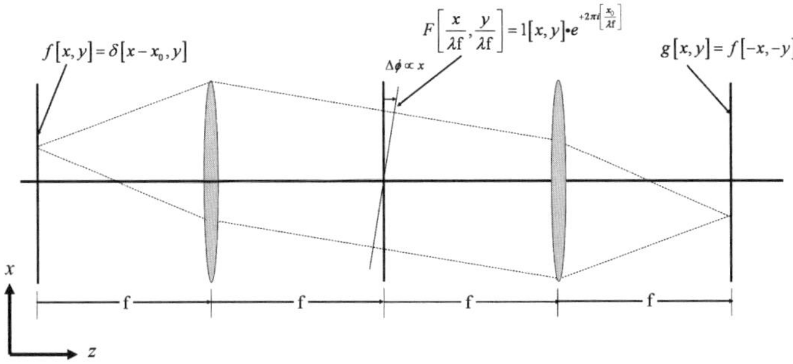

Figure 21.39 Optical implementation of the "transform-of-a-transform" theorem, showing that the output of the system is an inverted replica of the input.

If the lens pupil is infinitely large, the output is easy to evaluate by applying Equation (17.28):

$$
\begin{aligned}
g[x, y; z_1 = \mathbf{f}, \mathbf{f}, \mathbf{z}_2 = \mathbf{f}] &= ((1[x, y] \cdot e^{+2\pi i\mathbf{f}/\lambda_0}) \cdot e^{-i\pi(x^2+y^2)/\lambda_0 \mathbf{f}}) \\
&\quad * \left(\frac{1}{i\lambda_0 \mathbf{f}} e^{+2\pi i\mathbf{f}/\lambda_0} e^{+i\pi(x^2+y^2)/\lambda_0\mathbf{f}}\right) \\
&= \frac{1}{i\lambda_0\mathbf{f}} e^{+2\pi i 2\mathbf{f}/\lambda_0} \cdot (e^{-i\pi(x^2+y^2)/\lambda_0\mathbf{f}} * e^{+i\pi(x^2+y^2)/\lambda_0\mathbf{f}}) \\
&= ((-i\lambda_0\mathbf{f})\, e^{+2\pi i 2\mathbf{f}/\lambda_0}) \cdot \delta[x, y]
\end{aligned} \tag{21.194}
$$

Thus the output is the ideal 2-D Dirac delta function.

In the realistic case of a finite lens pupil $p[x, y]$, if the input is the unit constant, then the amplitude at the Fourier plane is:

$$
\begin{aligned}
g[x, y; \mathbf{f}, \mathbf{f}, \mathbf{f}] &= \mathcal{O}\{1[x, y]\} \\
&= e^{+2\pi i\mathbf{f}/\lambda_0}(p[x, y] \cdot e^{-i\pi(x^2+y^2)/\lambda_0\mathbf{f}}) * \left(\frac{1}{i\lambda_0\mathbf{f}} \cdot e^{+2\pi i\mathbf{f}/\lambda_0} \cdot e^{+i\pi(x^2+y^2)/\lambda_0\mathbf{f}}\right) \\
&= \left(\frac{1}{i\lambda_0\mathbf{f}} e^{+2\pi i 2\mathbf{f}/\lambda_0}\right) \cdot (p[x, y] \cdot e^{-i\pi(x^2+y^2)/\lambda_0\mathbf{f}}) * (e^{+i\pi(x^2+y^2)/\lambda_0\mathbf{f}})
\end{aligned} \tag{21.195}
$$

which may be evaluated by applying the M–C–M chirp Fourier transform from Equation (17.70):

$$
g[x, y; \mathbf{f}, \mathbf{f}, \mathbf{f}] = \left(\frac{1}{i\lambda_0\mathbf{f}} e^{+2\pi i 2\mathbf{f}/\lambda_0}\right) \cdot \left(P\left[\frac{x}{\lambda_0\mathbf{f}}, \frac{y}{\lambda_0\mathbf{f}}\right] \cdot e^{+i\pi(x^2+y^2)/\lambda_0\mathbf{f}}\right) \tag{21.196}
$$

This is the product of the 2-D Fourier transform of the pupil function modulated by a quadratic-phase factor. The scaling theorem determines that decreasing the size of the pupil increases the scale factor of P, thus "blurring" the spectrum from the ideal 2-D Dirac delta function. A realistic input object that is more complicated than $1[x, y]$ will be more complicated to analyze, but the qualitative effect is the same: the finite aperture blurs the output of the 2-D transform.

PROBLEMS

21.1 Derive the amplitude and the irradiance of the Fresnel diffraction patterns observed in light with wavelength λ_0 at the plane $z = z_1$ for the following source distributions:

(a) $f[x, y; 0] = (\delta[x + x_0] + \delta[x - x_0]) \cdot \delta[y]$

(b) $f[x, y; 0] = (\delta[x + x_0] + \delta[x - x_0]) \cdot 1[y]$

(c) $f[x, y; 0] = (RECT[(x + x_0)/b] + RECT[(x - x_0)/b]) \cdot 1[y]$ where $x_0 > b > 0$

21.2 Prove the relation in Equation (21.42) for $\delta[k - \omega_0/c]$.

21.3 Sketch an optical system that implements the M–C–M chirp transform algorithm.

21.4 For Fresnel diffraction from a rectangular aperture of width b propagated over the distance z_1, find the width of the aperture that produces the same relative pattern if z_1 is doubled.

21.5 Consider the "imaging system" based on two stages of Fraunhofer diffraction over the distance z_1 to a pupil plane and then over the distance z_2 to the observation plane. The description in Equation (21.122) indicates that the width of the impulse response observed at the plane $z = z_1 + z_2$ is proportional to the reciprocal of the width of the aperture function $t[x, y; z_1]$. In other words, if the width of the aperture function increases, the width of the impulse response decreases. Yet if the aperture function grows to infinite width ($t[x, y] \rightarrow 1[x, y]$), then it is clear that the "image" is just the Fraunhofer diffraction pattern of the source observed at the plane $z = z_1 + z_2$. Explain this apparent paradox. (SOLUTION: As the pupil size increases, eventually the condition for Fraunhofer diffraction in Equation (21.92) becomes invalid. For larger pupils, it is incorrect to assume that the diffraction pattern is proportional to the appropriately scaled Fourier transform of the pupil function.)

21.6 Plane waves of light with wavelength λ_0 travel down the z-axis to a circular aperture of diameter d_0. The light then propagates a further distance z_1 to the observation plane in the Fresnel diffraction region. Derive the formula for the on-axis irradiance and show that the expression confirms the inverse square law if $z_1 \rightarrow \infty$.

21.7 Virtually all of the examples of optical systems considered in this chapter consisted of a single optical element with the object distance z_1 and observation plane located at z_2; the only exception was the cascade of such systems to produce the so-called "4f" optical correlator. We found that we could evaluate an impulse response (and thus a transfer function) of the system at an image point. Consider the two-lens system shown in the figure, which is illuminated by monochromatic radiation. In words, the object $f[x, y]$ is illuminated with a converging wave from the first lens. The pupil functions of the two lenses are respectively $p_1[x, y]$ and $p_2[x, y]$:

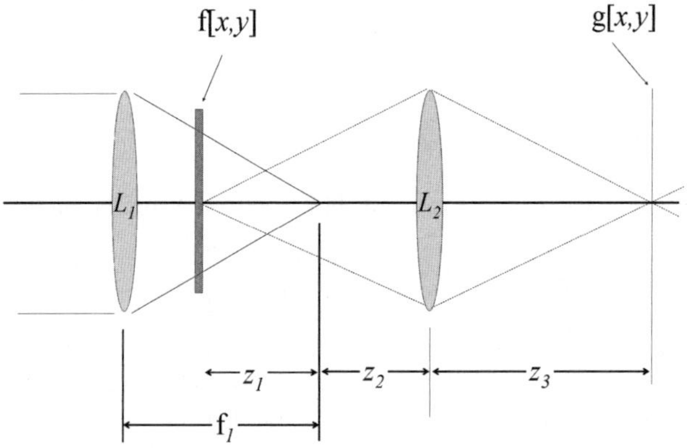

(a) Write down the mathematical equation for the amplitude located at the distance z_1 from the object.

(b) Evaluate the amplitude of the output image $g[x, y]$ in terms of $f[x, y]$ and the parameters of the system.

(c) Use the result of part (b) to find a condition on the distances z_1, z_2, and z_3 that must be satisfied for an image to be formed at the plane where $g[x, y]$ is observed.

(d) Locate the point in this system where the Fourier transform of the object exists and may be modified by introducing a multiplicative transfer function for the system.

21.8 Two telescopes used for imaging objects illuminated with coherent light have pupil functions:

$$p_1[x, y] = CYL\left(\frac{r}{d}\right) * \left(\delta\delta\left[\frac{x}{2d}\right] \cdot \delta\delta\left[\frac{y}{2d}\right]\right)$$

$$p_2[x, y] = CYL\left(\frac{r}{d}\right) * (\delta[x - 2d] \cdot \delta[y - 2d] + \delta[x + 2d] \cdot \delta[y - 2d]\, e^{i\pi/2})$$

$$+ CYL\left(\frac{r}{d}\right) * (\delta[x + 2d] \cdot \delta[y + 2d]\, e^{i\pi} + \delta[x - 2d] \cdot \delta[y + 2d]\, e^{i3\pi/2})$$

For both cases, *sketch* the pupil function and transfer function, find an expression for the impulse responses, and sketch them along the x- (or y-)axis and along one diagonal.

21.9 The Moon occasionally passes in front of background stars located along the path of the Moon's orbit on the celestial sphere. Consider the distance to the star to be infinite and the distance from the Earth to the Moon to be 400 000 km. Assume that the star emits monochromatic light with $\lambda_0 = 550$ nm and the dark edge (the dark "limb") passes in front of the star (assume that the dark limb of the Moon reflects no light):

(a) Assuming the Fresnel model of diffraction, sketch the irradiance pattern that would be measured as a function of time at a point on the Earth as the limb of the Moon passes in front of the star. Be as quantitative as you can, but be sure to show the qualitative features.

(b) What qualitative changes would you expect in this pattern in the more realistic case where the star emits white light over the visible spectrum?

22

Incoherent Optical Imaging Systems

The discussion of the previous chapter considered imaging systems where the object is illuminated by a point source of monochromatic light at a sufficient distance from the object that the waves are planar. In this way, the phase of the light at all points in space is deterministic. We found that the imaging system could be considered to be linear and shift invariant after an appropriate adjustment to account for the transverse magnification. The impulse response and transfer function of the imaging system under these conditions have simple, even elegant, forms. We extend that discussion in this chapter to consider optical systems acting in light for which the phase is not deterministic, because the source of illumination is either not monochromatic and/or not a point source. As a prelude to the analysis of the imaging system, we first need to consider the mathematical treatment of the phase effects, which are lumped into the topic of *optical coherence*. A source of light with deterministic phase is coherent, and a source producing light whose phase is random is incoherent.

22.1 Coherence

The discussions in the previous chapter assumed that light is monochromatic, i.e., it is described by a single temporal frequency ν_0 and a single wavelength λ_0. Another important, yet somewhat more subtle, consideration has been implicit in this discussion: that the phase of the light is deterministic at each point in the system. In other words, the phase at any point in the system and at any time is determined by (and may be computed from) the phase at any other point and at any other time; the light is *coherent*. Our goal now is to derive the impulse response and transfer function of light with a nondeterministic phase. The discussion requires a short digression to consider optical interference, specifically of monochromatic light, monochromatic light with random phase, and polychromatic light. These digressions will of necessity be brief and readers requiring more details will need to consult other sources, such as Goodman (1985) or Reynolds et al. (1989).

22.1.1 Optical Interference

As a prelude to the discussion of coherence, we first consider the simple, well-known, and widely understood phenomenon of optical *interference*, which is the effect of optical phase on observed irradiance. We have already done most of the work to calculate the relevant interference patterns in

Fourier Methods in Imaging Roger L. Easton, Jr.

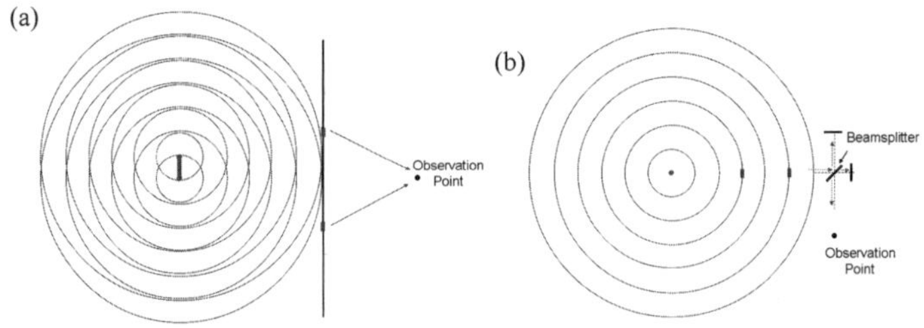

Figure 22.1 Two classes of interference used to measure statistical variation in the phase. (a) Division-of-wavefront interference, where light from two locations on the wavefront (from extended source in this example) is combined by some means (e.g., diffraction due to propagation) and summed; measured irradiance is a function of spatial coherence. (b) Division-of-amplitude interference, where light from two different wavefronts (often from a polychromatic source) is combined using a beamsplitter; the resulting pattern is function of temporal coherence.

this discussion because they may be visualized as diffraction patterns generated by a small number of sources. In this discussion, we need to consider only the Fraunhofer approximation, so the calculation is quite easy.

The game of interference is to find expressions for combinations of two (or more) sinusoidal traveling waves that differ in some attribute: propagation directions, path lengths, different locations on a source, etc. For example, the waves may have been produced by sampling different points on the same source wavefront by using an opaque screen with two (or more) apertures. The light through the apertures propagates and diffracts so that the new wavefronts overlap to produce "division-of-wavefront" interference (Figure 22.1a). Alternatively, light amplitude from different source wavefronts may be combined by dividing the incident amplitude with a beamsplitter and then allowing the pair of beams to propagate (generally different) distances and recombine; the result is "division-of-amplitude" interference (Figure 22.1b).

We will base our treatment of optical coherence on the picture of division-of-wavefront interference in Figure 22.1a under the assumption that the propagation distance from the apertures to the observation plane is sufficiently large so that the Fraunhofer approximation applies. The wavefronts from point emitters are approximated as plane waves and the observed amplitude pattern is proportional to the appropriately scaled Fourier transform of the source function. The results arising from monochromatic point and extended sources and from nonmonochromatic point sources will demonstrate the concepts of spatial coherence and temporal coherence, respectively. We start by extending the results of the previous chapter for Fraunhofer diffraction from a pair of point sources with equal amplitudes to consider the effects of varying the phase and the amplitude of the two sources on the irradiance pattern.

22.1.1.1 Two Point Sources, Different Phases

Two point sources separated by the distance ℓ_0 emit light with the same "strength" (amplitude) A_0 and the same temporal frequency ν_0. The *optical axis* is the line of symmetry that bisects the line between the two sources. The sources emit light with the same frequency but with different constant phases, as would happen if one (or both) sources were covered by pieces of glass with different thicknesses that "slow down" the light by different amounts. We can write the general expression for the light from the source at the plane $z = 0$:

$$s[x, y, t; z = 0, \nu_0] = (A_0 \cdot \exp[-2\pi i\nu_0 t]) \cdot \left(\delta\left[x + \frac{\ell_0}{2}, y\right] \exp[+i\phi_1] + \delta\left[x - \frac{\ell_0}{2}, y\right] \exp[+i\phi_2]\right) \tag{22.1}$$

where the temporal part of the phase is common to both sources because the light from both has the same frequency. The diffracted amplitude at the observation plane $z = z_0$ in the Fraunhofer diffraction region is proportional to the scaled Fourier transform:

$$g[x, y; \lambda_0, z_0] = \left(K_0 \cdot A_0 \exp\left[-i\left(2\pi\nu_0 t - \left(\frac{\nu_0 z_0}{c}\right)\right)\right]\right) \cdot \exp\left[-2\pi i \nu_0 \left(t - \frac{z_0}{c}\right)\right]$$
$$\cdot\left(\left(\exp\left[+2\pi i \frac{\ell_0}{2}\frac{x}{\lambda_0 z_0}\right]\exp[+i\phi_1]\right) + \exp\left[-2\pi i \frac{\ell_0}{2}\frac{x}{\lambda_0 z_0}\right]\exp[+i\phi_2]\right) \quad (22.2)$$

where the constant factors in Fraunhofer diffraction have been collapsed into the term K_0 that was originally defined in Equation (21.100):

$$K_0 \equiv \frac{1}{i\lambda_0 z_0} \cdot \exp\left[+2\pi i \frac{z_0}{\lambda_0}\right] \quad (22.3)$$

and the phase of the light is evaluated at the earlier time $t - z_0/c$ due to the propagation down the optical axis. The constant phase due to the propagation time may be factored out and thus is often ignored.

The phases may be decomposed into the common part and the difference term via the expressions:

$$\phi_1 = \left(\frac{\phi_1 + \phi_2}{2}\right) + \left(\frac{\phi_1 - \phi_2}{2}\right) \equiv \phi_{avg} + \phi_{mod} \quad (22.4a)$$

$$\phi_2 = \left(\frac{\phi_1 + \phi_2}{2}\right) - \left(\frac{\phi_1 - \phi_2}{2}\right) \equiv \phi_{avg} - \phi_{mod} \quad (22.4b)$$

where the constituent terms ϕ_{avg} and ϕ_{mod} are the *average* and *modulation* phases, respectively.

The resulting expression for the amplitude is:

$$g[x, y; \lambda_0, z_0] = (K_0 \cdot A_0) \cdot \exp[+i\phi_{avg}] \cdot \exp\left[-2\pi i \nu_0\left(t - \frac{z_0}{c}\right)\right]$$
$$\cdot \exp\left[+2\pi i \frac{\ell_0}{2}\frac{x}{\lambda_0 z_0}\right] \cdot \exp[+i\phi_{mod}] + \exp\left[-2\pi i \frac{\ell_0}{2}\frac{x}{\lambda_0 z_0}\right]\exp[-i\phi_{mod}]$$
$$\equiv (K_0 \cdot A_0) \cdot \exp\left[-2\pi i \nu_0\left(t - \frac{z_0}{c}\right)\right] \cdot \exp[+i\phi_{avg}] \cdot 2\cos\left[2\pi \frac{x}{(2\lambda_0 z_0/\ell_0)} + \phi_{mod}\right] \quad (22.5)$$

The irradiance is proportional to the squared magnitude, which eliminates the leading phase terms; the time-dependent term again vanishes because the temporal frequencies are identical:

$$I[x, y; z_0, \lambda_0] = \frac{1}{T_0}\int_{t-T_0/2}^{t+T_0/2} |g[x, y; \lambda_0, z_0]|^2\, dt$$
$$\propto |K_0|^2 \cdot A_0^2 \cdot \left(2 \cdot \cos\left[2\pi \frac{x}{(2\lambda_0 z_0/\ell_0)} + \phi_{mod}\right]\right)^2$$
$$= \frac{2A_0^2}{(\lambda_0 z_0)^2} \cdot \left(1 + \cos\left[2\pi \frac{x}{(\lambda_0 z_0/\ell_0)} + \Delta\phi\right]\right) \quad (22.6)$$

where Equation (6.24) has been used. To envision the effect of the phase on the irradiance pattern, it may be useful to collect the phase difference and scale factors into a term x_0 that has dimensions of length:

$$I[x, y; z_0, \lambda_0] \propto \frac{2A_0^2}{(\lambda_0 z_0)^2}\left(1 + \cos\left[2\pi \frac{x - (-\Delta\phi \cdot \lambda_0 z_0/\ell_0)}{(\lambda_0 z_0/\ell_0)}\right]\right)$$
$$\equiv \frac{2A_0^2}{(\lambda_0 z_0)^2}\left(1 + \cos\left[2\pi \frac{x - x_0}{(\lambda_0 z_0/\ell_0)}\right]\right) \quad (22.7)$$

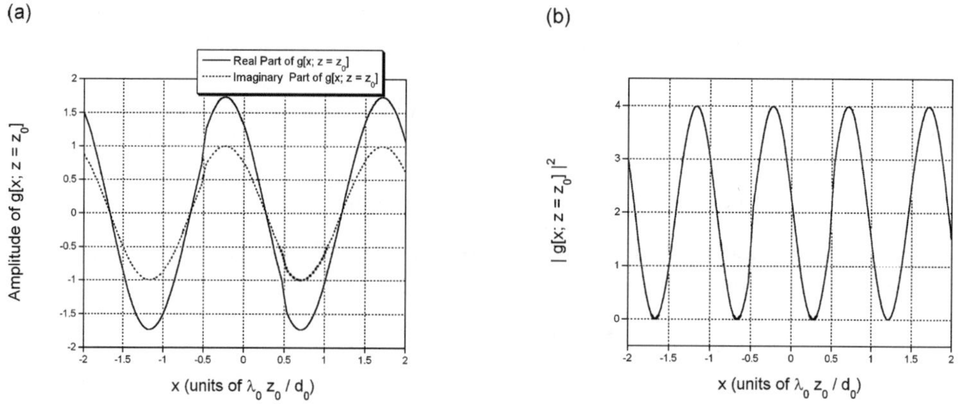

Figure 22.2 The effect of a phase difference between the apertures. The source function is $s[x] = \delta[x + d_0/2]\, e^{+i\pi/3} + \delta[x - d_0/2]$, so that the source to the left has larger phase: (a) amplitude at plane $z = z_0$ in the Fraunhofer diffraction region; (b) the sinusoidal irradiance pattern has been translated by $x_0 = -\frac{1}{6}\lambda_0 z_0/d_0$, i.e., toward the aperture with the larger phase.

An example is shown in Figure 22.2. In words, the irradiance generated by two sources with the same amplitude but different initial phases is a cosine fringe pattern with the same period D_0:

$$D_0 = \frac{\lambda_0 z_0}{\ell_0} \tag{22.8}$$

that has been translated from the origin of coordinates at the observation plane by a distance:

$$x_0 \equiv -\frac{\Delta\phi}{2\pi} \cdot \left(\frac{\lambda_0 z_0}{\ell_0}\right) \tag{22.9}$$

that is proportional to the phase difference at the two apertures: the larger the phase difference, the larger the translation of the center of symmetry of the irradiance pattern. The center of symmetry of the cosine pattern is translated along the x-axis "toward" the aperture with the larger phase. This makes sense because increasing the phase of one source increases the delay in its arrival time. Light through the aperture without the phase increment travels a larger distance to achieve the same phase delay, thus forming an irradiance maximum. Note that the translation of the irradiance fringe provides evidence of the relative values of the initial phases of the sources.

Note that phase differences that are odd multiples of π radians move the sinusoidal pattern by a half period:

$$x_0 \equiv \pm\frac{1}{2} \cdot \left(\frac{\lambda_0 z_0}{\ell_0}\right) \tag{22.10}$$

so that the on-axis irradiance is now a minimum instead of a maximum. In other words, the modulation of the resulting irradiance pattern is negative.

22.1.1.2 Point Objects, Different Amplitudes, Same Phase

Now consider the pattern produced by two point sources with different amplitudes; for example, identical sources could be covered by neutral-density filters with different attenuations but the same phase delay. The form of such a source function is:

$$s[x, y; \lambda_0] = \left(A_1 \cdot \delta\left[x + \frac{\ell_0}{2}, y\right] + A_2 \cdot \delta\left[x - \frac{\ell_0}{2}, y\right]\right) \cdot \exp[-2\pi i \nu_0 t] \tag{22.11}$$

Since the strengths $A_1 \neq A_2$, we can split one into the sum of a source with equal strength plus leftover amplitude:

$$\begin{aligned}
s[x, y] &= \left(A_1 \cdot \delta\left[x + \frac{\ell_0}{2}, y\right] + A_1 \cdot \delta\left[x - \frac{\ell_0}{2}, y\right]\right) \cdot \exp[-2\pi i \nu_0 t] \\
&\quad + (A_2 - A_1) \cdot \delta\left[x - \frac{\ell_0}{2}, y\right] \cdot \exp[-2\pi i \nu_0 t] \\
&= A_1 \cdot \left(\delta\left[x + \frac{\ell_0}{2}, y\right] + \delta\left[x - \frac{\ell_0}{2}, y\right]\right) \cdot \exp[-2\pi i \nu_0 t] \\
&\quad + (A_2 - A_1) \cdot \delta\left[x - \frac{\ell_0}{2}, y\right] \cdot \exp[-2\pi i \nu_0 t]
\end{aligned} \tag{22.12}$$

Note that the leftover term may be positive or negative depending on the relative sizes of A_1 and A_2.

The diffracted amplitude is proportional to the scaled Fourier transform, which is the sum of a cosine term and a linear-phase term:

$$\begin{aligned}
g[x, y; z_0] &= K_0 \cdot A_1 \cdot \exp\left[-2\pi i \nu_0\left(t - \frac{z_0}{c}\right)\right] \\
&\quad \cdot \left(\exp\left[+2\pi i \cdot \left(\frac{\ell_0}{2}\right) \cdot \left(\frac{x}{\lambda_0 z_0}\right)\right] + \exp\left[-2\pi i \cdot \left(\frac{\ell_0}{2}\right) \cdot \left(\frac{x}{\lambda_0 z_0}\right)\right]\right) \\
&\quad + K_0 \cdot (A_2 - A_1) \cdot \exp\left[-2\pi i \nu_0\left(t - \frac{z_0}{c}\right)\right] \cdot \exp\left[-2\pi i \cdot \left(\frac{\ell_0}{2}\right) \cdot \left(\frac{x}{\lambda_0 z_0}\right)\right] \\
&= K_0 \cdot \exp\left[-2\pi i \nu_0\left(t - \frac{z_0}{c}\right)\right] \\
&\quad \cdot \left(2A_1 \cos\left[2\pi \frac{x}{(2\lambda_0 z_0/\ell_0)}\right] + (A_2 - A_1) \cdot \exp\left[-2\pi i \frac{x}{(2\lambda_0 z_0/\ell_0)}\right]\right)
\end{aligned} \tag{22.13}$$

The squared magnitude is straightforward to calculate (though perhaps a bit tedious):

$$\begin{aligned}
I[x, y; z_1] &\propto g[x, y; z_0] \cdot g^*[x, y; z_0] \\
&= \frac{1}{(\lambda_0 z_0)^2}\left((A_1^2 + A_2^2) + 2A_1 A_2 \cdot \cos\left[2\pi \frac{x}{(\lambda_0 z_0/\ell_0)}\right]\right) \\
&= \frac{(A_1^2 + A_2^2)}{(\lambda_0 z_0)^2}\left(1 + \left(\frac{2A_1 A_2}{A_1^2 + A_2^2}\right) \cdot \cos\left[2\pi \frac{x}{(\lambda_0 z_0/\ell_0)}\right]\right)
\end{aligned} \tag{22.14}$$

In words, the observed light is a biased cosine fringe pattern with the same period D_0 as for the point sources with equal amplitudes in Equation (21.109):

$$D_0 = \frac{\lambda_0 z_0}{\ell_0} \tag{22.15}$$

The amplitude $2A_1 A_2$ and bias $A_1^2 + A_2^2$ are different in this case, so that the modulation ("visibility") of the sinusoidal fringe pattern in Equation (6.27) is:

$$m = \mathrm{V} = \frac{\textit{amplitude}}{\textit{bias}} = \frac{2A_1 A_2}{(A_1^2 + A_2^2)} \tag{22.16}$$

which is smaller than unity if $A_1 \neq A_2$. In words, for different amplitudes at the two apertures, the fringe period and center of symmetry are unchanged, but the fringe visibility (modulation) is decreased.

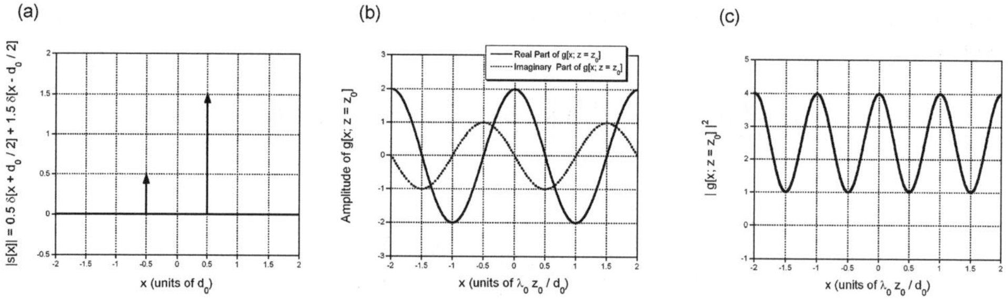

Figure 22.3 The effect of an amplitude difference between the apertures: (a) source function $s[x;\, z = 0] = \frac{1}{2}\delta[x + d_0/2] + \frac{3}{2}\delta[x - d_0/2]$; (b) amplitude at $z = z_0$ in the Fraunhofer diffraction region; (c) irradiance at $z = z_0$, which is a biased cosine with period $D_0 = \lambda_0 z_0/d_0$ and modulation $m = 0.6$.

In the 1-D example shown in Figure 22.3, the source function is:

$$\begin{aligned} s[x;\, z = 0] &= \frac{1}{2}\delta[x + 1] + \frac{3}{2}\delta[x - 1] \\ &= \frac{1}{2}(\delta[x + 1] + \delta[x - 1]) + \delta[x - 1] \end{aligned} \tag{22.17}$$

so the amplitude and irradiance at the plane $z = z_0$ in the Fraunhofer diffraction region are:

$$\begin{aligned} g[x;\, z = z_0] &= K_0 \cdot \left(\cos\left[2\pi \frac{x}{\lambda_0 z_0} \cdot 1\right] + \exp\left[-2\pi i \frac{x}{\lambda_0 z_0}\right]\right) \\ I[x;\, z = z_0] \propto |g[x;\, z = z_0]|^2 &= \left(\frac{1}{\lambda_0 z_0}\right)^2 \cdot \left(1 + 3\cos^2\left[2\pi \frac{x}{\lambda_0 z_0}\right]\right) \\ &= \left(\frac{1}{\lambda_0 z_0}\right)^2 \cdot \left(1 + \frac{3}{2}\left(1 + \cos\left[2\pi \frac{x}{(\lambda_0 z_0/2)}\right]\right)\right) \\ &= \left(\frac{1}{\lambda_0 z_0}\right)^2 \cdot \left(\frac{5}{2} + \frac{3}{2}\cos\left[2\pi \frac{x}{(\lambda_0 z_0/2)}\right]\right) \end{aligned} \tag{22.18}$$

The modulation of the irradiance is:

$$m = \frac{2 \cdot \frac{1}{2} \cdot \frac{3}{2}}{(\frac{1}{2})^2 + (\frac{3}{2})^2} = \frac{3}{5} \tag{22.19}$$

22.1.2 Spatial Coherence

To evaluate the ability of monochromatic light to create irradiance fringes, we now consider a variation of division-of-wavefront interference just considered; the system is shown in Figure 22.4. Two monochromatic point sources with possibly different phases, and separated by the distance ℓ_0, are located in the plane $z = 0$. The light propagates a distance z_0 to a plane in the Fraunhofer diffraction region where it encounters an absorbing screen with transmittance function $f[x, y]$ that represents two apertures of infinitesimal area and finite transmittance separated by the distance d_0:

$$f[x, y] = \delta\left[x + \frac{d_0}{2}\right] + \delta\left[x - \frac{d_0}{2}\right] \tag{22.20}$$

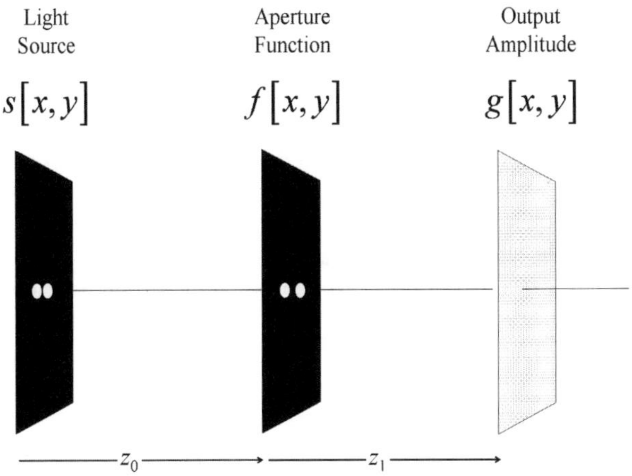

Figure 22.4 Schematic of optical system for measuring the spatial coherence of the light source $s[x, y]$. The light from the source propagates the distance z_0 in the Fraunhofer diffraction region to the aperture plane with transmittance $f[x, y]$. The transmitted light propagates the distance z_1 to the observation plane where the amplitude has the form $g[x, y]$.

We will vary the relative separations ℓ_0 and d_0 in the course of this discussion. The complex amplitude (magnitude and phase) through each aperture propagates the distance z_1 into the Fraunhofer diffraction region where the irradiance pattern is observed. Since the sources are assumed to be monochromatic with the same temporal frequency, we have seen already that the temporal part vanishes in the calculation of the squared magnitude, so there is no need to explicitly evaluate the time average. Our task here is to evaluate the effect of changing the separation ℓ_0 and phase difference between the two sources on the observed interference pattern.

We have already derived most of the expressions that are needed to analyze this system. The amplitudes through each aperture may add "in phase" or "out of phase" to produce an interference maximum or minimum at the observation plane; we speak of *constructive* or *destructive* interference, respectively.

22.1.2.1 Example: Off-Axis Monochromatic Point Source

As a first example, consider an off-axis monochromatic point source located at $x = -\ell$ with initial phase ϕ_1:

$$s[x, y; z = 0, \lambda_0] = \delta[x + \ell, y; z, \lambda_0] \cdot \exp[-i\phi_1] \tag{22.21}$$

The amplitude of the Fraunhofer diffraction pattern evaluated at the plane $z = z_0$ is:

$$\begin{aligned} g[x, y; z = z_0, \lambda_0] &= \frac{1}{i\lambda_0 z_0} \exp\left[+2\pi i \frac{z_0}{\lambda_0}\right] \exp\left[+i\left(2\pi \frac{x}{\lambda_0 z_0} \cdot \ell - \phi_1\right)\right] \\ &\equiv K_0 \cdot \exp[-i\phi_1] \cdot 1[x] \cdot \exp\left[+i\left(2\pi \frac{x}{\lambda_0 z_0} \cdot \ell\right)\right] \end{aligned} \tag{22.22}$$

where K_0 is defined as before. In words, the apertures at $x = \pm d_0/2$ are illuminated by a uniform magnitude with linear phase, and thus act as sources with a phase difference between them:

$$g[x, y; z \gtrapprox z_0, \lambda_0]$$
$$= K_0 \cdot \exp[-i\phi_1] \cdot \left(\exp\left[-2\pi i \frac{\ell d_0}{2\lambda_0 z_0}\right] \cdot \delta\left[x + \frac{d_0}{2}\right] + \exp\left[+2\pi i \frac{\ell d_0}{2\lambda_0 z_0}\right] \cdot \delta\left[x - \frac{d_0}{2}\right]\right) \tag{22.23}$$

The result of Equation (22.5) immediately shows that the amplitude at the observation plane is translated from the optical axis. We can evaluate the same expression here by propagating to the Fraunhofer diffraction region at the distance z_1 from the aperture plane:

$$g[x, y; z = z_0 + z_1, \lambda_0]$$
$$= K_0 \cdot K_1 \cdot \exp[-i\phi_1] \cdot \left(\exp\left[-2\pi i \frac{\ell d_0}{2\lambda_0 z_0}\right] \cdot \delta\left[x + \frac{d_0}{2}\right] + \exp\left[+2\pi i \frac{\ell d_0}{2\lambda_0 z_0}\right] \cdot \delta\left[x - \frac{d_0}{2}\right]\right)$$
$$= K_0 \cdot K_1 \cdot \exp[-i\phi_1]$$
$$\cdot \left(\exp\left[-2\pi i \frac{\ell d_0}{2\lambda_0 z_0}\right] \cdot \exp\left[-2\pi i \frac{x}{\lambda_0 z_1} \cdot \left(-\frac{d_0}{2}\right)\right] + \exp\left[+2\pi i \frac{\ell d_0}{2\lambda_0 z_0}\right]\right.$$
$$\left. \cdot \exp\left[-2\pi i \frac{x}{\lambda_0 z_1} \cdot \left(+\frac{d_0}{2}\right)\right]\right)$$
$$= K_0 \cdot K_1 \cdot \exp[-i\phi_1] \cdot \left(\exp\left[-2\pi i \frac{\ell d_0}{2\lambda_0 z_0}\right] \cdot \exp\left[+2\pi i \frac{x}{(2\lambda_0 z_1/d_0)}\right]\right.$$
$$\left. + \exp\left[+2\pi i \frac{\ell d_0}{2\lambda_0 z_0}\right] \cdot \exp\left[-2\pi i \frac{x}{(2\lambda_0 z_1/d_0)}\right]\right)$$
$$= 2 \cdot K_0 \cdot K_1 \cdot \exp[-i\phi_1] \cdot \cos\left[2\pi \left(\frac{x - \ell \cdot z_1/z_0}{(2\lambda_0 z_1/d_0)}\right)\right] \tag{22.24}$$

where K_1 is the corresponding constant factor from Equation (22.3). This result demonstrates that the amplitude pattern at the observation plane is a replica of the cosine fringe pattern in Equation (22.5) but translated so that the center of symmetry is now located at:

$$x_0 = -\ell \cdot \frac{z_1}{z_0} \tag{22.25}$$

The new center of symmetry of the irradiance pattern is located at the point where the path lengths through the two apertures are equal, and is located at the "projection" of the original source point through the center of the aperture plane to the observation plane (Figure 22.5). The translation is the direct result of the linear phase that illuminated the apertures, via Equation (22.9).

The initial phase ϕ_1 has no impact upon measured irradiance at the observation plane:

$$|g[x, y; z = z_0 + z_1, \lambda_0]|^2 = 4 \cdot |K_0|^2 \cdot |K_1|^2 \cdot \cos^2\left[2\pi \left(\frac{x - \ell \cdot z_1/z_0}{(2\lambda_0 z_1/d_0)}\right)\right]$$
$$= 2 \cdot |K_0|^2 \cdot |K_1|^2 \cdot \left(1 + \cos\left[\frac{2\pi}{(\lambda_0 z_1/d_0)}\left(x - \ell \cdot \frac{z_1}{z_0}\right)\right]\right)$$
$$= 2 \cdot |K_0|^2 \cdot |K_1|^2 \cdot \left(1 + \cos\left[\frac{2\pi}{D_1}\left(x - \ell \cdot \frac{z_1}{z_0}\right)\right]\right) \tag{22.26}$$

where the period of the irradiance fringe is:

$$D_1 \equiv \frac{\lambda_0 z_1}{d_0} \tag{22.27}$$

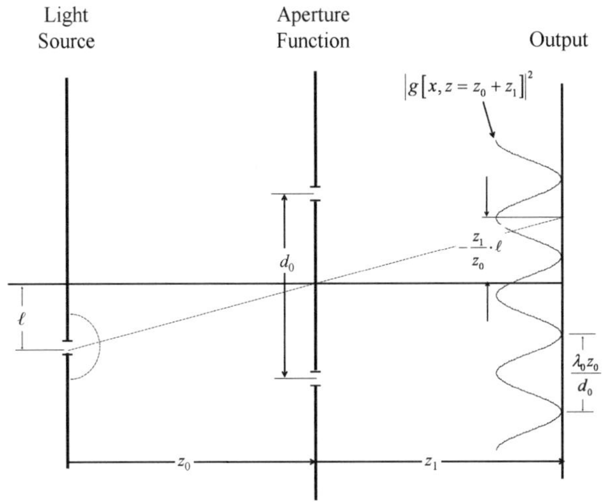

Figure 22.5 Interference of light from off-axis source located at $x = -\ell$ from two apertures separated by d_0. The interference pattern is translated by the distance $(-\ell) \cdot (-z_1/z_0)$.

In words, the irradiance is a unit-modulation sinusoidal fringe pattern that has been translated from the center of symmetry by the increment in Equation (22.25), as expected.

22.1.2.2 Example: Extended Monochromatic Source

Now we generalize the analysis to an "extended" source, which might be a pair of sources separated by ℓ_0, a "monolithic" rectangular source of width ℓ_0, etc. We specify the source as $s[\ell]$. We also assume that all points on the source emit light with the same phase. It is clear that each source point produces a sinusoidal amplitude pattern of the form in Equation (22.24) centered at the corresponding value of x_0 in Equation (22.25). The weighting of each cosine amplitude pattern is specified by the value of s. The amplitude at the observation plane is the superposition of the cosine terms, which has the form of an integral over the coordinate ℓ at the source plane:

$$\begin{aligned} g[x, y; z = z_0 + z_1, \lambda_0] &= 2 \cdot K_0 \cdot K_1 \cdot \int_{-\infty}^{+\infty} s[\ell] \cdot \cos\left[+2\pi \frac{d_0}{2\lambda_0 z_1}\left(x - \ell \cdot \frac{z_1}{z_0}\right)\right] d\ell \\ &= K_0 \cdot K_1 \\ &\quad \cdot \int_{-\infty}^{+\infty} s[\ell] \cdot \left(\exp\left[+2\pi i \cdot \frac{d_0}{2\lambda_0 z_0} \cdot \left(\ell - x \cdot \frac{z_0}{z_1}\right)\right]\right. \\ &\quad \left. + \exp\left[-2\pi i \cdot \frac{d_0}{2\lambda_0 z_0} \cdot \left(\ell - x \cdot \frac{z_0}{z_1}\right)\right]\right) d\ell \end{aligned} \tag{22.28}$$

The phase factors that are functions of x are extracted from the integrals:

$$\begin{aligned} &g[x, y; z = z_0 + z_1, \lambda_0] \\ &= K_0 \cdot K_1 \cdot \exp\left[-2\pi i\left(\frac{x}{2\lambda_0 z_1/d_0}\right)\right] \cdot \int_{-\infty}^{+\infty} s[\ell] \cdot \exp\left[-2\pi i \cdot \left(-\frac{d_0}{2\lambda_0 z_0}\right) \cdot \ell\right] d\ell \\ &\quad + K_0 \cdot K_1 \cdot \exp\left[+2\pi i\left(\frac{x}{2\lambda_0 z_1/d_0}\right)\right] \cdot \int_{-\infty}^{+\infty} s[\ell] \cdot \exp\left[-2\pi i\left(\frac{d_0}{2\lambda_0 z_0}\right) \cdot \ell\right] d\ell \end{aligned} \tag{22.29}$$

and the integrals evaluated via the Fourier transform:

$$\begin{aligned}g[x, y; z = z_0 + z_1, \lambda_0] &\\ = K_0 \cdot K_1 &\\ \cdot \left(\exp\left[-2\pi i \frac{x}{(2\lambda_0 z_1/d_0)}\right] \cdot S\left[\xi = -\frac{d_0}{2\lambda_0 z_0}\right] + \exp\left[+2\pi i \left(\frac{x}{2\lambda_0 z_1/d_0}\right)\right] \cdot S\left[\xi = +\frac{d_0}{2\lambda_0 z_0}\right]\right)\end{aligned} \tag{22.30}$$

Take note of this relation; it indicates that some characteristics of the output amplitude are determined by the spatial Fourier transform of the source function. We will return to this thought shortly.

If the spectrum $S[\xi]$ is even, which means that the source distribution function $s[\ell]$ is symmetric, then the amplitude simplifies to the product of the cosine amplitude fringe pattern and a scaling factor:

$$g[x, y; z = z_0 + z_1, \lambda_0] = (2 \cdot K_0 \cdot K_1) \cdot S\left[\frac{d_0}{2\lambda_0 z_0}\right] \cdot \cos\left[2\pi \frac{x}{(2\lambda_0 z_1/d_0)}\right] \tag{22.31}$$

The squared magnitude is:

$$\begin{aligned}|g[x, y; z = z_0 + z_1, \lambda_0]|^2 &= 2 \cdot |K_0|^2 \cdot |K_1|^2 \cdot \left|S\left[\frac{d_0}{2\lambda_0 z_0}\right]\right|^2 \cdot \left(1 + \cos\left[2\pi \frac{x}{(\lambda_0 z_1/d_0)}\right]\right)\\ &= 2 \cdot |K_0|^2 \cdot |K_1|^2 \cdot \left|S\left[\frac{d_0}{2\lambda_0 z_0}\right]\right|^2 \cdot \left(1 + \cos\left[2\pi \frac{x}{D_1}\right]\right)\end{aligned} \tag{22.32}$$

where Equation (22.15) has been used to define D_0. The effect of the source function is a scaling of the irradiance fringe pattern by the constant $|S[d_0/2\lambda_0 z_0]|^2$, which is guaranteed to be nonnegative, but may be zero.

If the source function consists of multiple point sources, then we can see that the amplitude pattern is the superposition of cosine fringe patterns with the same period but displaced by the distance $-\ell \cdot z_2/z_1$. To see the impact on the fringe pattern, consider two specific examples.

22.1.2.3 Example: Two Point Sources

Now consider the fringe pattern generated by two point sources separated by the distance ℓ_0; we can easily assign different initial phases to each. The source function may be written:

$$\begin{aligned}s[\ell] &= A_0 \cdot \left(\delta\left[\ell + \frac{\ell_0}{2}\right] \cdot \exp[+i\phi_1] + \delta\left[\ell - \frac{\ell_0}{2}\right] \cdot \exp[+i\phi_2]\right)\\ &= A_0 \cdot \exp[+i\phi_{\text{avg}}]\left(\delta\left[\ell + \frac{\ell_0}{2}\right] \exp[+i\phi_{\text{mod}}] + A_0 \cdot \delta\left[\ell - \frac{\ell_0}{2}\right] \exp[-i\phi_{\text{mod}}]\right)\end{aligned} \tag{22.33}$$

where the definitions of the average and modulation phases from Equation (22.4) have been used and the common phase term has been factored out. Since s is not real, the amplitude at the observation plane is obtained from the general expression in Equation (22.29). The spectrum of the source function is:

$$S\left[\frac{d_0}{2\lambda_0 z_0}\right] = 2A_0 \cdot \exp[+i\phi_{\text{avg}}] \cdot \cos\left[2\pi \cdot \frac{d_0 \ell_0}{4\lambda_0 z_0} + \phi_{\text{mod}}\right] \tag{22.34a}$$

$$S\left[-\frac{d_0}{2\lambda_0 z_0}\right] = 2A_0 \cdot \exp[+i\phi_{\text{avg}}] \cdot \cos\left[2\pi \cdot \frac{d_0 \ell_0}{4\lambda_0 z_0} - \phi_{\text{mod}}\right] \tag{22.34b}$$

The amplitude at the observation plane is found by applying Equation (22.30):

$$g[x, y; z = z_0 + z_1, \lambda_0]$$
$$= 2A_0 \cdot K_0 \cdot K_1 \cdot \exp[+i\phi_{avg}] \cdot \left(\cos\left[2\pi \cdot \frac{d_0 \ell_0}{4\lambda_0 z_0} - \phi_{mod}\right] \cdot e^{-2\pi i x/(2\lambda_0 z_1/d_0)} \right.$$
$$\left. + \cos\left[2\pi \cdot \frac{d_0 \ell_0}{4\lambda_0 z_0} + \phi_{mod}\right] \cdot e^{+2\pi i x/(2\lambda_0 z_1/d_0)} \right) \tag{22.35}$$

where K_1 is the corresponding constant factor from Equation (22.3). In words, sources with different initial phases produce an amplitude pattern at the apertures that has been translated from the original symmetric position. This means that the apertures separated by d_0 are illuminated by different amplitudes and therefore act as sources with different "strengths". We saw in Equation (22.13) that sources with different amplitudes produce an irradiance fringe pattern with the same period, but the modulation (visibility) of the fringe is reduced. We can see this by evaluating the squared magnitude:

$$|g[x; z = z_0 + z_1]|^2$$
$$= 4A_0^2 \cdot |K_1 \cdot K_0|^2 \cdot \left| \cos\left[\frac{\pi}{2} \frac{d_0 \ell_0}{\lambda_0 z_0} - \phi_{mod}\right] \cdot \exp\left[-2\pi i \frac{x}{2\lambda_0 z_1/d_0}\right] \right.$$
$$\left. + \cos\left[\frac{\pi}{2} \frac{d_0 \ell_0}{\lambda_0 z_0} + \phi_{mod}\right] \cdot \exp\left[+2\pi i \frac{x}{2\lambda_0 z_1/d_0}\right] \right|^2 \tag{22.36}$$

Trigonometric identities may be used to simplify this expression (Problem 22.7):

$$|g[x; z = z_0 + z_1]|^2$$
$$= \frac{4A_0^2}{\lambda_0^4 z_0^2 z_1^2}$$
$$\cdot \left(\left(1 + \cos\left[2\pi \frac{d_0 \ell_0}{2\lambda_0 z_0}\right] \cdot \cos[2 \cdot \phi_{mod}] \right) + \left(\cos\left[2\pi \frac{d_0 \ell_0}{2\lambda_0 z_0}\right] + \cos[2 \cdot \phi_{mod}] \right) \cos\left[2\pi \frac{x}{D_1}\right] \right) \tag{22.37}$$

where the period of the irradiance fringe is the same as that in Equation (22.27). It is convenient to resubstitute the definition of $\phi_{mod}[t]$:

$$2 \cdot \phi_{mod}[t] = \phi_2[t] - \phi_1[t] \equiv \Delta\phi[t] \tag{22.38}$$

so that the irradiance pattern generated by two sources with the same amplitude but different phases is seen to be a biased cosine with the period given in Equation (22.27), but with modulation that depends on the phase difference $\Delta\phi$:

$$|g[x; z = z_0 + z_1]|^2 = 4A_0^2 \cdot \left(\frac{1}{\lambda_0 z_0}\right)^2 \cdot \left(\frac{1}{\lambda_0 z_1}\right)^2 \cdot \left(\left(1 + \cos\left[\pi \frac{d_0 \ell_0}{\lambda_0 z_0}\right] \cdot \cos[\Delta\phi] \right) \right.$$
$$\left. + \left(\cos\left[\pi \frac{d_0 \ell_0}{\lambda_0 z_0}\right] + \cos[\Delta\phi] \right) \cos\left[2\pi \frac{x}{D_1}\right] \right) \tag{22.39}$$

Note that the argument of the cosine function of x in Equation (22.39) includes no additive constant, which means that the sinusoidal irradiance pattern is symmetric with respect to the origin of coordinates. In other words, the value of the irradiance at $x = 0$ must be an extremum; it is a maximum or minimum if the modulation is positive or negative, respectively. We know from Equation (6.27) that the modulation is the ratio of the amplitude and bias:

$$m = \frac{\cos[\pi d_0 \ell_0/\lambda_0 z_0] + \cos[\Delta\phi]}{1 + \cos[\pi d_0 \ell_0/\lambda_0 z_0] \cdot \cos[\Delta\phi]} \tag{22.40}$$

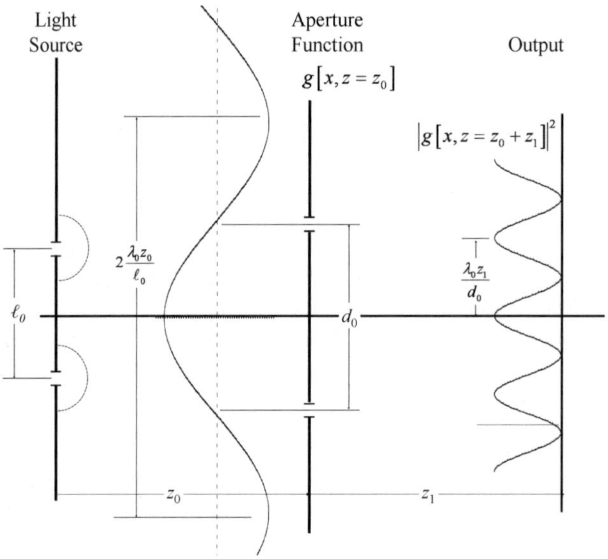

Figure 22.6 Schematic of diffraction of light from point sources with equal amplitudes and phases separated by distance ℓ_0. The light propagates the distance z_0 in the Fraunhofer diffraction region where the period of the amplitude fringe is $2\lambda_0 z_0/\ell_0$. The apertures separated by d_0 sample the amplitude, which then propagates the distance z_1 to the observation plane, where fringes are created with period $\lambda_0 z_1/d_0$.

We see that the effect of the phase difference of the sources is to change the modulation of the observed fringe pattern, but not change its center of symmetry. It is easy to see from Equation (22.39) that different values of $\Delta\phi$ produce modulations anywhere in the interval $-1 \leq m \leq +1$.

Consider the results in a few cases. If both sources have the same phase, so that $\Delta\phi = 0$, as in Figure 22.6, then the modulation is zero (no fringes visible) if:

$$m = 0 \Longrightarrow \cos\left[\pi \frac{d_0 \ell_0}{\lambda_0 z_0}\right] = -1 \Longrightarrow d_0 = (2n+1) \cdot \frac{\lambda_0 z_0}{\ell_0} \tag{22.41}$$

where n is an integer. The modulation will be positive and nonzero if d_0 is smaller than the distance to the first zero in Equation (22.41):

$$d_0 < \frac{\lambda_0 z_0}{\ell_0} \Longrightarrow m > 0 \tag{22.42}$$

The limiting value for d_0 is the *coherence width* of the two-point source; it is the width of possible aperture separations such that fringes are visible.

If the phase difference $\Delta\phi = \pi$, as in Figure 22.7, then a zero of the amplitude pattern generated by the sources appears between the apertures, so that the amplitudes sampled by the apertures are equal but with opposite signs. The modulation of the resulting irradiance pattern is:

$$m = \frac{\cos[\pi d_0 \ell_0/\lambda_0 z_0] + \cos[\pi]}{1 + \cos[\pi d_0 \ell_0/\lambda_0 z_0] \cdot \cos[\pi]} = \frac{\cos[\pi d_0 \ell_0/\lambda_0 z_0] - 1}{1 - \cos[\pi d_0 \ell_0/\lambda_0 z_0]} = -1 \tag{22.43}$$

This means that the observed irradiance oscillates with the same period D_1 in Equation (22.27), but the irradiance at the origin is zero instead of the maximum value. This important observation will assist interpretation of the case of a time-varying phase difference of light emitted by the two sources.

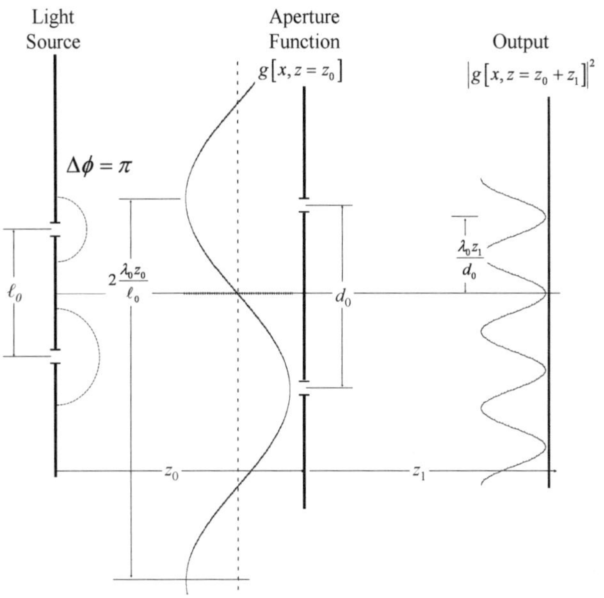

Figure 22.7 Schematic of diffraction of light from point sources with equal amplitudes but a phase difference of $\Delta\phi$. The amplitude fringe at the aperture plane has the same period of $2\lambda_0 z_0/\ell_0$, but has a zero at $x = 0$ so that the amplitudes sampled by the apertures have different signs. The resulting irradiance pattern has the same period $\lambda_0 z_1/d_0$ but has a zero instead of a maximum at $x = 0$, so that the modulation is negative.

22.1.2.4 Extended Source of Width ℓ_0

Now consider a monochromatic and monolithic source that emits light with the same phase at all locations and whose width is identical to the separation of the point sources in Equation (22.33). The source function is:

$$s[\ell] = A_0 \cdot RECT\left[\frac{\ell}{\ell_0}\right] \tag{22.44}$$

The scaling factor corresponding to that in Equation (22.34) is:

$$S\left[\xi = \frac{d_0}{2\lambda_0 z_0}\right] = A_0 \cdot \ell_0 \cdot SINC\left[\frac{\ell_0 d_0}{2\lambda_0 z_0}\right] \tag{22.45}$$

which cannot be nonnegative. This result may be interpreted by noting that the extended source consists of symmetric pairs of point sources with separations ranging from $\ell \gtrsim 0$ up to $\ell = \ell_0$. Each pair produces an amplitude of the form of Equation (22.35) with $\phi_{avg} = \phi_{mod} = 0$. The argument of the scale factor varies in proportion to the separation ℓ of each pair, and the summation of those cosine terms produces a *SINC* function, as shown in Figure 22.8.

By analogy with Equation (22.42), we can define the coherence width as the limiting value of the aperture separation d_0 such that the $SINC^2$ function in Equation (22.45) is nonzero. It is easy to see that:

$$d_0 < 2 \cdot \frac{\lambda_0 z_0}{\ell_0} \tag{22.46}$$

which is twice as large as that in the case of two point sources separated by ℓ_0.

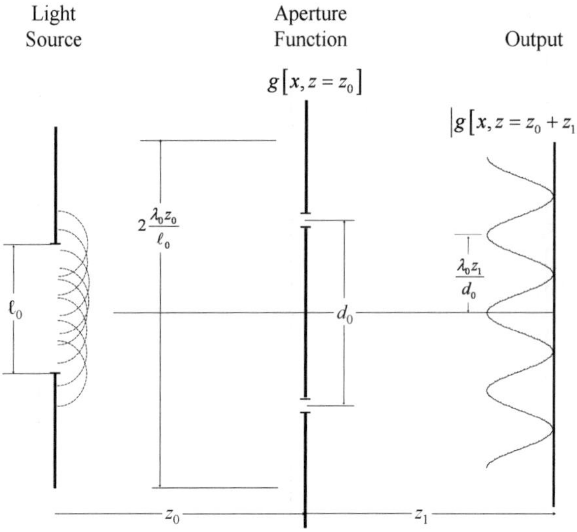

Figure 22.8 Schematic of interference from a monochromatic extended source of width ℓ_0. The amplitude at the aperture plane $z = z_0$ is a *SINC* function with width parameter $\lambda_0 z_0/\ell_0$. The irradiance pattern at the observation plane is a cosine fringe pattern scaled by the factor $SINC^2[\ell_0 d_0/2\lambda_0 z_0]$ due to the aperture illumination.

The irradiance of the resulting fringe pattern is proportional to the squared magnitude:

$$|g[x, y; z = z_0 + z_1, \lambda_0]|^2 = 4 \cdot A_0 \cdot |K_0|^2 \cdot |K_1|^2 \cdot \ell_0^2 \cdot SINC^2\left[\frac{\ell_0 d_0}{2\lambda_0 z_0}\right] \cdot \left(1 + \cos\left[2\pi \frac{x}{D_1}\right]\right) \tag{22.47}$$

which has the same spatial period D_1, which is determined by the spacing d_0 and wavelength λ_0, rather than the spatial characteristics of the source.

22.1.2.5 Two Point Sources, Time-Varying Phase Difference

In the examples considered up to this point, the light emitted by the sources was assumed to be monochromatic, possibly with a fixed phase difference. In other words, the optical phase from the sources was predictable at all times and locations in space. This ensured that the time average had no significant impact on the irradiance calculation. At this point we generalize the assumption a bit by allowing the phase of the monochromatic sources to vary in a random fashion. In the case of the two point sources with a phase difference that can vary in time, the time average in the irradiance calculation applies only to the varying phase difference:

$$\langle |g[x; \lambda_0, z = z_0 + z_1]|^2 \rangle = \frac{4}{\lambda_0^4 z_0^2 z_1^2} \cdot \left(1 + \cos\left[\pi \frac{d_0 \ell_0}{\lambda_0 z_0}\right] \cdot \langle \cos[\Delta\phi[t]] \rangle\right) \cdot \left(1 + \frac{\cos[\pi d_0 \ell_0/\lambda_0 z_0] + \langle \cos[\Delta\phi[t]] \rangle}{1 + \cos[\pi d_0 \ell_0/\lambda_0 z_0] \cdot \langle \cos[\Delta\phi[t]] \rangle}\right) \cdot \cos\left[2\pi \frac{x}{(\lambda_0 z_1/d_0)}\right] \tag{22.48}$$

The modulation (or visibility) of the fringe pattern now includes the explicit time average:

$$m = \frac{\cos[\pi d_0 \ell_0/\lambda_0 z_0] + \langle\cos[\Delta\phi[t]]\rangle}{1 + \cos[\pi d_0 \ell_0/\lambda_0 z_0] \cdot \langle\cos[\Delta\phi[t]]\rangle}$$
$$= \frac{\cos[\pi d_0 \ell_0/\lambda_0 z_0] + (1/T_0)\int_{-T_0/2}^{+T_0/2} \cos[\Delta\phi[t]]\,dt}{1 + (\cos[\pi d_0 \ell_0/\lambda_0 z_0] \cdot (1/T_0)\int_{-T_0/2}^{+T_0/2} \cos[\Delta\phi[t]]\,dt)} \tag{22.49}$$

Clearly the result depends on the range of $\Delta\phi$ and the measurement time T_0. Consider some limiting cases.

If T_0 is much smaller than the time over which the phase changes, then the modulation depends on the dominant value of $\Delta\phi$ in that time interval. For example, if the phase difference is zero at all times (purely monochromatic light), then the modulation is unity:

$$\langle\cos[\Delta\phi[t]]\rangle = \langle\cos[0]\rangle = 1 \Longrightarrow m = \frac{\cos[\pi d_0 \ell_0/\lambda_0 z_0] + 1}{1 + \cos[\pi d_0 \ell_0/\lambda_0 z_0] \cdot 1} = 1 \tag{22.50}$$

If the phase difference is a random number that spans the full range of 2π radians, then each forms a squared-magnitude pattern with a different modulation in the interval $-1 \le m \le +1$. If all possible values are equally likely within the measurement time T_0, then the time average of the cosine is zero and the modulation reduces to:

$$\langle\cos[\Delta\phi[t]]\rangle = 0 \Longrightarrow m = \cos\left[\pi \frac{d_0 \ell_0}{\lambda_0 z_0}\right] \tag{22.51}$$

so that the observed irradiance for two point sources with random phases is proportional to:

$$\langle |g[x; \lambda_0, z = z_0 + z_1]|^2 \rangle = \frac{4}{\lambda_0^4 z_0^2 z_1^2} \cdot \left(1 + \cos\left[\pi \frac{d_0 \ell_0}{\lambda_0 z_0}\right]\right) \cdot \cos\left[2\pi \frac{x}{D_1}\right] \tag{22.52}$$

As we saw in the purely monochromatic case, this function sets the limits on the separation d_0 of the apertures such that interference fringes are visible in the time-averaged irradiance. To ensure that fringes are readily visible, we should select d_0 to be much narrower than the central lobe of this cosine function:

$$\cos\left[\pi \frac{d_0 \ell_0}{\lambda_0 z_0}\right] > 0 \Longrightarrow \frac{d_0 \ell_0}{\lambda_0 z_0} < 1 \Longrightarrow d_0 < \frac{\lambda_0 z_0}{\ell_0} \tag{22.53}$$

This ensures that both apertures are illuminated by approximately equal amplitudes, producing fringes with unit modulation with zeros at the same widely separated locations. Also, if d_0 is small, only occasionally will a zero crossing of the amplitude appear between the apertures, so that they are illuminated by small amplitudes with opposite signs. The squared magnitude in this (rare) case is "faint" and with reversed modulation. If d_0 is small, then the time average of the squared magnitude is more likely to exhibit visible fringes.

We can also interpret this condition on d_0 in terms of the apparent source size:

$$\frac{\lambda_0 z_0}{\ell_0} = \frac{\lambda_0}{(\ell_0/z_0)} \equiv \frac{\lambda_0}{\theta_0} \tag{22.54}$$

where θ_0 is the approximate angular subtense of the source as "seen" from the aperture plane. In words, the irradiance fringe will be visible if the two apertures are separated by a small fraction of the distance to the first zero of the cosine, which ensures that both apertures "see" the same lobe of the sinusoidal pattern. This constraint on the separation d_0 of the apertures is called the *coherence width* of the monochromatic source; it is a measure of the largest possible separation of apertures such that the transmitted light can interfere to produce visible fringes.

22.2 Polychromatic Source – Temporal Coherence

We now consider the interference patterns produced by sources that include more than one wavelength. Because the sources emit light with different temporal frequencies, we can no longer assume that the temporal dependence vanishes when evaluating the squared magnitude. To introduce the process, we calculate the diffraction pattern generated by two point sources with different temporal frequencies in a fashion analogous to that used for sources with different amplitudes or phases in the last section.

Consider a point source of monochromatic light that is located on the optical axis. The source function at wavelength λ has the form:

$$s[x, y; z=0, \lambda] = \delta[x, y; z] \exp[-2\pi i \nu t] = \delta[x, y; z] \exp\left[-2\pi i \frac{ct}{\lambda}\right] \tag{22.55}$$

The resulting plane wave at the plane $z = z_0$ in the Fraunhofer diffraction region illuminates the two apertures separated by d_0. Based on the discussion leading to Equation (22.24), we can see that the amplitude pattern produced in the Fraunhofer diffraction region at the distance z_1 from the aperture plane is a cosine pattern whose period is proportional to the wavelength λ and may be written in terms of the temporal frequency $\nu = c \cdot \lambda^{-1}$:

$$\begin{aligned} g[x, y; z = z_0 + z_1, \lambda] &= 2 \cdot K_0 \cdot K_1 \cdot \cos\left[2\pi \left(\frac{x}{2\lambda z_1/d_0}\right)\right] \\ &= 2 \cdot K_0 \cdot K_1 \cdot \cos\left[2\pi \left(\frac{x d_0}{2 c z_1} \nu\right)\right] \end{aligned} \tag{22.56}$$

Note that the leading terms K_0 and K_1 are functions of wavelength in both magnitude and phase and may also be written in terms of the temporal frequency:

$$\begin{aligned} K_0 \cdot K_1 &= \frac{1}{i\lambda z_0} \exp\left[+2\pi i \frac{z_0}{\lambda}\right] \cdot \frac{1}{i\lambda z_1} \exp\left[+2\pi i \frac{z_1}{\lambda}\right] \\ &= -\left(\frac{\nu^2}{c^2}\right) \frac{1}{z_0 z_1} \exp\left[+2\pi i \frac{z_0 + z_1}{c} \cdot \nu\right] \end{aligned} \tag{22.57}$$

The advantage of the notations in Equations (22.56) and (22.57) is that the frequency dependence appears in the numerator, which allows easy factoring.

We now want to derive the amplitude and irradiance patterns at the observation plane if the source emits more than one wavelength, and thus different temporal frequencies. We can see right away that the spatial period of the amplitude fringe is proportional to the wavelength and therefore inversely proportional to the temporal frequency. The amplitude due to all wavelengths will be the summation of these cosine fringe patterns. At some values of x, the amplitudes will sum to large values (constructive interference), while at other coordinates the summation will produce a small value (destructive interference), as shown in Figure 22.9. Note that each amplitude fringe is scaled by the corresponding factors $K_0 \cdot K_1$, which includes contributions to both the magnitude and phase that are proportional to the frequency. We will dispose of these frequency-dependent terms by a reasonable approximation.

Consider the amplitude generated by a point source of light that emits a limited range of temporal frequencies over the interval $\nu_{\min} \leq \nu \leq \nu_{\max}$ (i.e., the spectrum has compact support). The amplitude at the observation plane is the summation of the sinusoidal amplitude patterns generated by the light

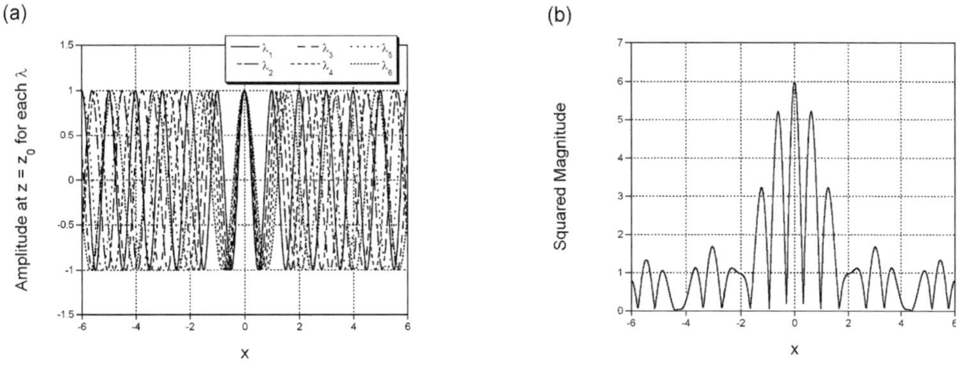

Figure 22.9 Two-aperture interference of light generated by point source that emits multiple wavelengths: (a) each wavelength produces a cosine amplitude pattern at the observation plane in the Fraunhofer diffraction region; (b) squared magnitude of the sum of amplitudes from six wavelengths, showing the support of the region where fringes are visible.

with each temporal frequency:

$$g[x, y; z = z_0 + z_1, \lambda]$$
$$= -\frac{2}{z_0 z_1 c^2} \cdot \int_{-\infty}^{+\infty} RECT\left[\frac{\nu - \nu_{avg}}{\nu_{max} - \nu_{min}}\right] \cdot \nu^2 \cdot \exp\left[+2\pi i \frac{z_0 + z_1}{c} \cdot \nu\right] \cdot \cos\left[2\pi\left(\frac{x d_0}{2 c z_1}\nu\right)\right] d\nu$$
$$= -\frac{2}{z_0 z_1 c^2} \cdot \int_{\nu_{min}}^{\nu_{max}} \nu^2 \cdot \exp\left[+2\pi i \frac{z_0 + z_1}{c} \cdot \nu\right] \cdot \cos\left[2\pi\left(\frac{x d_0}{2 c z_1}\nu\right)\right] d\nu \tag{22.58}$$

We now substitute the so-called *average* and *modulation* frequencies:

$$\nu_{avg} \equiv \frac{\nu_{max} + \nu_{min}}{2} \tag{22.59a}$$

$$\nu_{mod} \equiv \frac{\nu_{max} - \nu_{min}}{2} = \frac{\Delta\nu}{2} \tag{22.59b}$$

where $\Delta\nu$ is the frequency *bandwidth* of the source. We can use these terms to specify the limits of the frequency interval:

$$\nu_{min} = \nu_{avg} - \frac{\Delta\nu}{2} \leq \nu \leq \nu_{max} = \nu_{avg} + \frac{\Delta\nu}{2} \tag{22.60}$$

Clearly this expression may be generalized to cases where the source emits light with different strengths at each frequency by using a function other than the rectangle.

After substitution, Equation (22.58) becomes:

$$g[x, y; z = z_0 + z_1, \lambda]$$
$$= -\frac{2}{z_0 z_1 c^2} \cdot \int_{\nu_{avg} - \Delta\nu/2}^{\nu_{avg} + \Delta\nu/2} \nu^2 \cdot \exp\left[+2\pi i \frac{z_0 + z_1}{c} \cdot \nu\right] \cdot \cos\left[2\pi\left(\frac{x d_0}{2 c z_1}\nu\right)\right] d\nu \tag{22.61}$$

We now make a further assumption that $\Delta\nu$ is sufficiently small compared to ν_{avg} that the leading magnitude and the constant-phase terms may be approximated by inserting the average frequency ν_{avg}; light that satisfies this condition is called *quasimonochromatic*. As we will see, it is convenient to evaluate the wavelength corresponding to this average temporal frequency:

$$\frac{c}{\nu_{avg}} \equiv \lambda_\mu \tag{22.62}$$

where the notation λ_μ is used instead of λ_{avg} to avoid potential confusion since the latter is often defined by analogy with Equation (22.59a):

$$\lambda_{avg} \equiv \frac{\lambda_1 + \lambda_2}{2} \neq \lambda_\mu \equiv \frac{c}{\nu_{avg}} \tag{22.63}$$

Under the quasimonochromatic assumption, Equation (22.61) simplifies to:

$$\begin{aligned} g[x, y; z = z_0 + z_1, \lambda_\mu, \Delta\nu] &= \left(-\frac{2 \cdot \nu_{avg}^2}{z_0 z_1 c^2} \cdot \exp\left[+2\pi i \frac{z_0 + z_1}{c} \cdot \nu_{avg}\right]\right) \cdot \int_{\nu_{avg}-\Delta\nu/2}^{\nu_{avg}+\Delta\nu/2} \cos\left[\pi \frac{x d_0}{c z_1} \nu\right] d\nu \\ &= \left(-\frac{4}{z_0 z_1 \lambda_\mu^2} \cdot \exp\left[+2\pi i \frac{z_0 + z_1}{\lambda_\mu}\right]\right) \cdot \left(\pi \frac{x d_0}{c z_1}\right)^{-1} \sin\left[\pi \frac{x d_0}{c z_1} \frac{\Delta\nu}{2}\right] \cdot \cos\left[2\pi \frac{x d_0}{2 c z_1} \nu_{avg}\right] \\ &= \left(-\frac{2 \cdot \Delta\nu}{z_0 z_1 \lambda_\mu^2} \cdot \exp\left[+2\pi i \frac{z_0 + z_1}{\lambda_\mu}\right]\right) \cdot SINC\left[\frac{x d_0}{z_1} \cdot \frac{\Delta\nu}{2c}\right] \cdot \cos\left[2\pi \frac{x}{(2\lambda_\mu z_1 / d_0)}\right] \end{aligned} \tag{22.64}$$

In words, the amplitude at the observation plane in the quasimonochromatic case includes a cosine fringe pattern along the x-direction with period determined by the wavelength λ_μ and a leading *SINC* function whose argument is a function of the bandwidth $\Delta\nu$ and the location x in the observation plane. Parenthetically, note that the *SINC* function is a scaled Fourier transform of the function $RECT[(\nu - \nu_{avg})/(\nu_{max} - \nu_{min})]$ in Equation (22.58) that describes the source spectrum. We see right away that the *SINC* function weights the amplitude of the fringe pattern and is small for large values of x and/or $\Delta\nu$. We may also interpret the argument of the *SINC* function via a calculation based on Figure 22.10. The path lengths from the lower and upper apertures, respectively, to the observation point x are R_1 and R_2. It is easy to use the Pythagorean theorem and a Fresnel-like approximation to evaluate the difference in path length, labeled ΔR:

$$\begin{aligned} \Delta R[x] = R_1[x] - R_2[x] &= \sqrt{\left(z_1^2 + \left(x + \frac{d_0}{2}\right)^2\right)} - \sqrt{\left(z_1^2 + \left(x - \frac{d_0}{2}\right)^2\right)} \\ &\cong z_1\left(1 + \frac{1}{2}\frac{(x + d_0/2)^2}{z_1^2}\right) - z_1\left(1 + \frac{1}{2}\frac{(x - d_0/2)^2}{z_1^2}\right) = \frac{x d_0}{z_1} \end{aligned} \tag{22.65}$$

This may be inserted into the argument of the *SINC* function, leaving the ratio of the path length difference and the velocity of light, which is a time interval denoted by t, that is, in turn, a function of the observation point x:

$$\frac{x d_0}{z_1} \cdot \frac{\Delta\nu}{2c} = \frac{\Delta R[x]}{c} \cdot \frac{\Delta\nu}{2} \equiv \Delta t[x] \cdot \frac{\Delta\nu}{2} \quad \textit{where } \Delta t[x] \equiv \frac{x d_0}{c z_1} \tag{22.66}$$

In words, $\Delta t[x]$ is the time difference required for light to travel from each aperture to the observation point x. Though polychromatic light is sampled by two apertures in this example and the transverse interference pattern is formed in the Fraunhofer diffraction region, the same concept also applies if point objects are placed at different longitudinal distances from the source. In other words, the time differences for light that is scattered and then recombined at the observation plane will affect the ability of the light to interfere. For example, two objects placed sufficiently close together can create interference fringes, even in quasimonochromatic light.

We can now write the expression for the irradiance at the observation plane in terms of the source bandwidth and the time difference for light through the two apertures:

$$\begin{aligned} |g[x, y; z = z_0 + z_1, \lambda]|^2 &\cong \left(\frac{2 \cdot (\Delta\nu)^2}{z_0^2 z_1^2 \lambda_\mu^4}\right) \cdot SINC^2\left[\Delta t[x] \cdot \frac{\Delta\nu}{2}\right] \cdot \left(1 + \cos\left[2\pi \frac{x}{(\lambda_\mu z_1 / d_0)}\right]\right) \end{aligned} \tag{22.67}$$

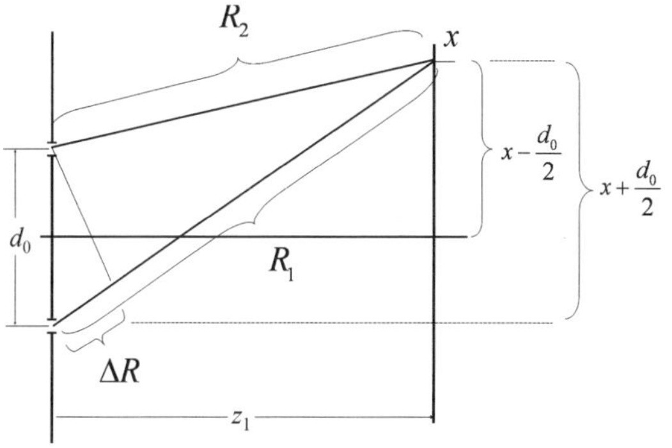

Figure 22.10 The difference in path lengths from the two apertures to the observation point is the difference between the hypotenuses R_1 and R_2 of the two triangles. Equation (22.65) shows that $\Delta R \cong x d_0 / z_1$.

We see that the $SINC^2$ function defines the region where fringes are visible. The first zero of the $SINC^2$ is located at x_0 such that:

$$\frac{x_0 d_0}{2cz_1} \cdot \Delta\nu = +1 \Longrightarrow x_0 = \left(\frac{2cz_1}{d_0}\right) \cdot \frac{1}{\Delta\nu} \tag{22.68}$$

The distance x_0 to the first zero where the fringes "disappear" is proportional to $(\Delta\nu)^{-1}$, which is the *coherence time* of the source:

$$\tau \equiv \frac{1}{\Delta\nu} \tag{22.69}$$

This is the limiting time difference from the two apertures where the phases of the light traveling the two paths are no longer correlated. The distance traveled by the light in this time is the limit to the separation of two object points from which interference may be produced; it is called the *coherence length* of the source:

$$L \equiv c \cdot \tau = \frac{c}{\Delta\nu} > \Delta R \tag{22.70}$$

Now consider a variation of Figure 22.10 shown in Figure 22.11a obtained by rotating the aperture plane so that one aperture is closer to the observation plane than the other and the irradiance is observed at the center of the plane at the distance z_1 measured from one aperture. If the path length difference ΔR exceeds the coherence length of the quasimonochromatic source, then the fringes will not be visible. The system in Figure 22.11a may be modified to the form of Figure 22.11b, where the two apertures are located close to the axis and still separated by ΔR. Again, if the path difference of light from the two apertures exceeds the coherence length L, then the phase of light arriving at the output plane will not be correlated and fringes will not be visible. Finally, we replace the two apertures by two objects that scatter light from the distant quasimonochromatic source. Again, the phase of the light will not be correlated if $\Delta R > L$ and fringes will not be visible.

For a nearly monochromatic source, such as a laser, the frequency bandwidth can be very narrow so that fringes may be created from objects separated by large distances. For a visibly white light source with $\lambda_{max} = 700$ nm and $\lambda_{min} = 400$ nm, the frequency bandwidth and coherence length are approximately 3×10^{14} Hz and 900 nm, respectively; the optical path difference must be extremely short to be able to see interference fringes. Light that is "perfectly polychromatic" has $\Delta\nu = \infty$, so that the phases are completely uncorrelated and no interference fringes may be generated.

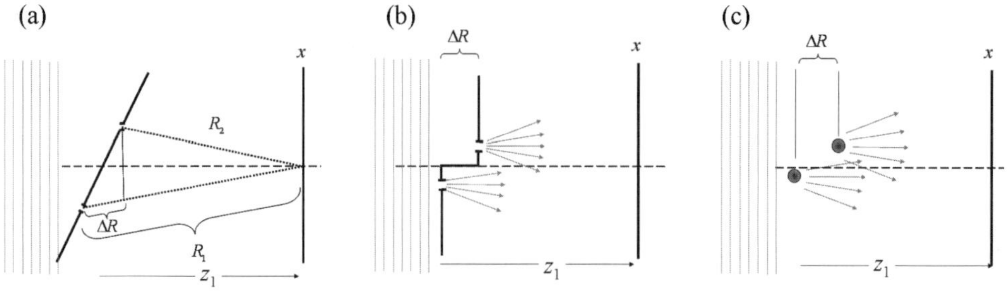

Figure 22.11 Interpretation of coherence length in terms of two-aperture interference: (a) the opaque planar object with the two apertures is rotated so that one aperture is closer to the output plane than the other. Interference effects are visible if the coherence length L of the source exceeds ΔR. (b) The rotated object is converted into two apertures at two locations on the z-axis. Again, interference effects are visible if $L > \Delta R$. (c) The apertures are replaced by two objects that scatter incident light. Again, interference effects are visible if $L > \Delta R$.

22.2.1 Coherence Volume

We can generalize the concept of coherence width for a monochromatic source in Equation (22.53) with quasimonochromatic light by substituting the expression for the wavelength corresponding to the average frequency in Equation (22.62) into Equation (22.53):

$$d_{\max} = \frac{\lambda_\mu z_0}{\ell_0} \tag{22.71}$$

We can consider this to be the diameter of the area where apertures (objects) may be placed such that interference fringes are visible in proportion to the square of the coherence width. The corresponding cross-sectional coherence area of a quasimonochromatic source with finite area is:

$$A_{\max} = \frac{\pi}{4} \cdot \left(\frac{\lambda_\mu z_0}{\ell_0}\right)^2 \tag{22.72}$$

We can add the concept of coherence length to define the depth of the region where objects may be placed such that interference fringes are visible, thus defining the *coherence volume*:

$$V_{\max} = \frac{\pi}{4} \cdot \left(\frac{\lambda_\mu z_0}{\ell_0}\right)^2 \cdot \frac{c}{\Delta\nu} \tag{22.73}$$

This will be a useful concept in the discussion of optical holography in the next chapter.

22.3 Imaging in Incoherent Light

The imaging examples in Chapter 21 were based on the assumption of illumination by monochromatic plane waves from a source observed at a distance that satisfies the criteria for Fraunhofer diffraction. In other words, the optical phases of light from that source measured at all locations in space are perfectly correlated, which, in turn, means that interference fringes may be formed at all points in space and at all times. Images created under this condition are affected by this interference, which led to the impulse response in Equation (21.170) and transfer function in Equation (21.171). At the other extreme, the phases of light from any pair of points on the source are completely uncorrelated, which means that interference is not observable. Though a (much) closer approximation to the behavior of imaging systems that act in "natural" light, this is still not completely realistic, since a realistic source is bandlimited and of finite size, which means that interference fringes may be visible if the conditions are right.

To consider the action of imaging systems in light with uncorrelated phases, we return to the definition of irradiance in Equation (21.28), which is a result of the fact that detectors of light measure the time average of the squared magnitude, and thus are insensitive to the optical phase. The irradiance of the image produced by a linear shift-invariant system with impulse response $h[x, y]$ is:

$$I[x, y; z_1, \mathbf{f}, z_2, T_0] = \frac{1}{T_0}\int_{-T_0/2}^{+T_0/2} (f[x, y, t] * h[x, y; z_1, \mathbf{f}, z_2]) \cdot (f[x, y, t] * h[x, y; z_1\mathbf{f}, z_2])^* \, dt, \tag{22.74}$$

where $f[x, y, t] = f[x, y] \cdot \exp[-2\pi i \nu_0 t]$ describes the spatial variation of the wave with temporal frequency ν and $h[x, y; z_1, \mathbf{f}, z_2]$ is the impulse response of the system in coherent light with object and image distances z_1 and z_2 and focal length $\mathbf{f}$; the pupil function is not listed explicitly. For simplicity, we delete these parameters from the notation in the following discussion. The integration time T_0 is determined by characteristics of the sensor; it may be the response time of the human visual system, the exposure time of the image, or the *RC* time constant of the electronics. For the human eye–brain system, $T_0 \lesssim \frac{1}{20}$ s. The integration ensures that the output is not a function of time.

The complex amplitude of the object distribution at $z = 0$ may be written as magnitude/phase:

$$f[x, y, t] = |f[x, y]| \; \exp[+i \cdot \Phi_f[x, y, t]] \tag{22.75}$$

where the temporal oscillation of the phase $\exp[-2\pi i \nu_0 t]$ is included in the phase $\Phi_f[x, y, t]$. The convolutions in the integrand may be rewritten as:

$$f[x, y, t] * h[x, y] = (|f[x, y]| \; \exp[+i \cdot \Phi_f[x, y, t]]) * h[x, y] = \int_{-\infty}^{+\infty} (|f[x', y']| \; \exp[+i \cdot \Phi_f[x, y, t]]) \, h[x - x', y - y'] \, dx' \, dy' \tag{22.76}$$

where the primed coordinates are the "dummy" variables of integration. The complex conjugate of the convolution is:

$$(f[x, y, t] * h[x, y])^* = (|f[x, y]| \exp[+i \cdot \Phi_f[x, y, t]] * h[x, y])^* = \int_{-\infty}^{+\infty} |f[x'', y'']| \; \exp[+i \cdot \Phi_f[x'', y'', t]] \, h^*[x - x'', y - y''] \, dx'' \, dy'' \tag{22.77}$$

using "double-primed" coordinates for the variables of integration. After combining these expressions and rearranging the orders of integration to perform the time integral first, we obtain the expression for the observed irradiance as a function of the spatial coordinates $[x, y]$:

$$I[x, y] = \left(\int_{-\infty}^{+\infty} |f[x', y']| \, h[x - x', y - y']\right) \cdot \int_{-\infty}^{+\infty} |f[x'', y'']| \, h^*[x - x'', y - y''] \cdot \frac{1}{T_0}\int_{-T_0/2}^{+T_0/2} e^{+i\cdot(\Phi_f[x',y',t]-\Phi_f[x'',y'',t])} \, dt \frac{x}{\lambda_0 \mathbf{f}} \, dx'' \, dy'' \, dx' \, dy' \tag{22.78}$$

The integrand of the time integral is proportional to the phase difference observed at the same time t at two points in space: $\underline{\mathbf{R}}' = [x', y']$ and $\underline{\mathbf{R}}'' = [x'', y'']$. The time integral is a function of the integration time T_0, which is retained as a parameter in the shorthand notation $\Gamma[\underline{\mathbf{R}}', \underline{\mathbf{R}}''; T_0]$:

$$\frac{1}{T_0}\int_{-T_0/2}^{+T_0/2} \exp[i\cdot(\Phi_f[x', y', t] - \Phi_f[x'', y'', t])]\, dt \equiv \Gamma[\underline{\mathbf{R}}', \underline{\mathbf{R}}''; T_0] \tag{22.79}$$

This is the *mutual coherence function* (or *MCF*), which measures the correlation between phases measured at two points at the object averaged over the measurement time T_0.

Clearly, the MCF of phases measured at the same point (with $\underline{\mathbf{R}}' = \underline{\mathbf{R}}''$) must be unity:

$$\begin{aligned}\Gamma[\underline{\mathbf{R}}', \underline{\mathbf{R}}'; T_0] &= \frac{1}{T_0}\int_{-T_0/2}^{+T_0/2} \exp[i\cdot(\Phi_f[\underline{\mathbf{R}}', t] - \Phi_f[\underline{\mathbf{R}}', t])]\, dt \\ &= \frac{1}{T_0}\int_{-T_0/2}^{+T_0/2} e^{i\,0}\, dt = \frac{1}{T_0}\int_{-T_0/2}^{T_0/2} 1\, dt = 1\end{aligned} \tag{22.80}$$

In words, the phase of light measured at the same point in space must be perfectly correlated at all times, so that the MCF evaluates to unity regardless of the measurement time T_0.

Now we consider the values of $\Gamma[\underline{\mathbf{R}}', \underline{\mathbf{R}}''; T_0]$ in the two limiting cases where the action of the system is easy to evaluate. The first such case requires that the phase difference measured at two points in space is a function *only* of $\underline{\mathbf{R}}'$ and $\underline{\mathbf{R}}''$. This means that phase difference measured at two points in space is deterministic (over the time period T_0 of the measurement). The resulting MCF evaluates to a complex constant that is independent of time interval T_0:

$$\begin{aligned}\Gamma[\underline{\mathbf{R}}', \underline{\mathbf{R}}''; T_0] &= \frac{1}{T_0}\int_{-T_0/2}^{+T_0/2} \exp[+i\cdot(\Phi_f[\underline{\mathbf{R}}', t] - \Phi_f[\underline{\mathbf{R}}'', t])]\, dt \\ &= \exp[+i\cdot\Delta\Phi[\underline{\mathbf{R}}' - \underline{\mathbf{R}}'']]\frac{1}{T_0}\int_{-T_0/2}^{T_0/2} dt \\ &= \exp[+i\cdot\Delta\Phi[\underline{\mathbf{R}}' - \underline{\mathbf{R}}'']]\end{aligned} \tag{22.81}$$

A light field with perfectly correlated phases is "coherent".

The other limiting case requires that the phase of the light measured at two distinct points $\underline{\mathbf{R}}'$ and $\underline{\mathbf{R}}''$ varies in a completely random fashion over the measurement interval T_0, even if very short. Strictly speaking, the phase is completely random only if the source emits all wavelengths with equal amplitudes, i.e., the spectrum of the light is truly "white". We assume that the phase difference is uniformly distributed over the wrapped interval $-\pi \leq \Delta\Phi[\underline{\mathbf{R}}' - \underline{\mathbf{R}}''] < +\pi$; then the MCF will evaluate approximately to zero regardless of the size of T_0:

$$\frac{1}{T_0}\int_{-T_0/2}^{+T_0/2} \exp[i\cdot(\Phi_f[\underline{\mathbf{R}}', t] - \Phi_f[\underline{\mathbf{R}}'', t])]\, dt = 0 \quad \textit{for } \underline{\mathbf{R}}' - \underline{\mathbf{R}}'' \neq 0 \textit{ in incoherent light} \tag{22.82}$$

Because the phase measured at the same location $\underline{\mathbf{R}}' = \underline{\mathbf{R}}''$ must be correlated perfectly, the MCF is proportional to a Dirac delta function of the spatial coordinates:

$$\lim_{T_0\to\infty}\left\{\frac{1}{T_0}\int_{-T_0/2}^{+T_0/2} \exp[i\cdot(\Phi_f[\underline{\mathbf{R}}', t] - \Phi_f[\underline{\mathbf{R}}'', t])]\, dt\right\} \propto \alpha\cdot\delta[\underline{\mathbf{R}}' - \underline{\mathbf{R}}''] \quad \textit{in incoherent light} \tag{22.83}$$

where the constant α is chosen to satisfy the volume constraint on the 2-D Dirac delta function.

Equation (22.83) is the only result required to derive the linear properties of an imaging system in incoherent light. Note that the MCF is not a function of the integration time T_0 in both the ideal coherent and ideal incoherent cases.

22.4 System Function in Incoherent Light

After our digression into the depths of the MCF, we may now return to the subject of interest, the properties of a linear shift-invariant imaging system in incoherent light. The parameters in the expressions for the impulse response and transfer function will be deleted to reduce notational clutter, but the observation plane always is that which satisfies the geometrical imaging condition. We will be assuming that the "incoherent" light may be considered to be quasimonochromatic, so that the impact of the wavelength on the magnitude and constant-phase terms can be approximated by the average temporal frequency, for which we derived an expression for the corresponding wavelength λ_μ in Equation (22.62). In words, the bandwidth is sufficiently broad that optical interference does not occur, but the light may still be described by a single wavelength.

The output irradiance is:

$$\begin{aligned} I[x, y; T_0] = \int_{-\infty}^{+\infty} & dx'\, dy' |f[x', y']| h[x - x', y - y'] \\ & \cdot \left(\int_{-\infty}^{+\infty} |f[x'', y'']| h^*[x - x'', y - y''] \, dx''\, dy'' \right. \\ & \times \left. \left[\frac{1}{T_0} \int_{-T_0/2}^{+T_0/2} \Gamma[x', y', x'', y'', T_0] \, dt \right] \right) \end{aligned} \tag{22.84}$$

In the coherent case already considered, the MCF is constant, so:

$$\begin{aligned} I[x, y] &\propto \left(\int_{-\infty}^{+\infty} |f[x', y']| h[x - x', y - y'] \, dx'\, dy' \right) \\ &\quad \cdot \left(\int_{-\infty}^{+\infty} |f[x'', y'']| \, h^*[x - x'', y - y''] \, dx''\, dy'' \right) \\ &= (|f[x, y]| * h[x, y]) \cdot (|f[x, y]| * h[x, y])^* = |g[x, y]|^2 \end{aligned} \tag{22.85}$$

Now we substitute the MCF in the incoherent case from Equation (22.83) to evaluate the output irradiance, which we denote by $I_g[x, y]$ to distinguish from $I_f[x, y]$ for the input:

$$\begin{aligned} I_g[x, y] &= \left(\int_{-\infty}^{+\infty} |f[x', y']| h[x - x', y - y'] \, dx'\, dy' \right) \\ &\quad \cdot \left(\int_{-\infty}^{+\infty} |f[x'', y'']| h^*[x - x'', y - y''] \delta[x' - x'', y' - y''] \, dx''\, dy'' \right) \\ &= \int_{-\infty}^{+\infty} |f[x', y']|^2 |h[x - x', y - y']|^2 \, dx'\, dy' \\ &= |f[x, y]|^2 * |h[x, y]|^2 = I_f[x, y] * |h[x, y]|^2 \end{aligned} \tag{22.86}$$

In words, this shows that incoherent imaging systems are linear *in the squared magnitude* of the complex amplitude. Contrast this with optical systems acting in coherent light, which are linear in complex amplitude. The input to the incoherent system is the nonnegative function $|f[x, y]|^2$, which is mapped to the nonnegative intensity $|g[x, y]|^2$ by convolution with an "incoherent impulse response" $\mathfrak{h}$:

$$\mathfrak{h}[x, y] \equiv |h[x, y]|^2 \tag{22.87}$$

For example, the incoherent impulse response for propagation of incoherent light in the Fresnel diffraction region is proportional to the unit constant:

$$\mathfrak{h}[x, y] = |h[x, y]|^2 = \left| \frac{1}{i\lambda_\mu z_1} e^{+2\pi i z_1/\lambda_\mu} e^{+i\pi(x^2+y^2)/\lambda_\mu z_1} \right|^2 = \frac{1}{\lambda_\mu^2 z_1^2} \cdot 1[x, y] \tag{22.88}$$

In words, the intensity at a plane in the Fresnel diffraction region due to an incoherent source is uniform across the plane.

Therefore the impulse response of the C–M–C single-lens imaging system with pupil function $p[x, y]$ in incoherent light is the squared magnitude of the coherent impulse response:

$$\mathfrak{h}[x, y; z_1, \mathbf{f}, z_2] = \left| \frac{1}{-\lambda_\mu^2 z_1 z_2} e^{+2\pi i(z_1+\tau+z_2)/\lambda_\mu} e^{+i\pi(x^2+y^2/\lambda_\mu z_2} P\left[\frac{x}{\lambda_\mu z_2}, \frac{y}{\lambda_\mu z_2}\right]\right|^2$$

$$= \left(\frac{1}{\lambda_\mu^2 z_1 z_2}\right)^2 \left| P\left[\frac{x}{\lambda_\mu z_2}, \frac{y}{\lambda_\mu z_2}\right]\right|^2 \quad \textit{at image point, incoherent} \tag{22.89}$$

If the object is located a large distance from the imaging lens, then $z_2 \cong \mathbf{f}$. The 2-D Fourier transform of $\mathfrak{h}[x, y; z_1, \mathbf{f}, z_2]$ leads to the observation that the transfer function is proportional to the autocorrelation of the pupil function:

$$\mathfrak{H}[\xi, \eta] = \mathcal{F}_2\{\mathfrak{h}[x, y]\} = \mathcal{F}_2\{|h[x, y]|^2\} = H[\xi, \eta] \bigstar H[\xi, \eta]$$

$$= \left(\frac{1}{\lambda_\mu z_1 z_2}\right)^2 \cdot (p[u, v] * p^*[-u, -v]|_{u\to -\lambda_\mu z_2 \xi, v \to -\lambda_\mu z_2 \eta}) \quad \textit{at image point} \tag{22.90}$$

where the 2-D generalization of Equation (8.56b) has been used. Because the maximum of an autocorrelation always occurs at the origin of coordinates, the sinusoidal component that is best imaged by an incoherent optical system is that with $[\xi, \eta] = [0, 0]$ (the "DC" component). This also means that incoherent imaging systems with finite pupils must act as lowpass filters.

22.4.1 Incoherent MTF

Because the maximum of the incoherent transfer function is located at the origin of the frequency domain, it is feasible to normalize the transfer function:

$$MT[\xi, \eta] \equiv \frac{\mathfrak{H}[\xi, \eta]}{\mathfrak{H}[0, 0]} = \frac{p[-\lambda_\mu z_2 \xi, -\lambda_\mu z_2 \eta] \bigstar p[-\lambda_\mu z_2 \xi, -\lambda_\mu z_2 \eta]}{p[-\lambda_\mu z_2 \xi, -\lambda_\mu z_2 \eta] \bigstar p[-\lambda_\mu z_2 \xi, -\lambda_\mu z_2 \eta]|_{\xi=0,\eta=0}} \tag{22.91}$$

The theoretical range of this function is:

$$-1 \le MT[\xi, \eta] \le +1 \tag{22.92}$$

The normalization ensures that $MT[0, 0] = 1$. If the input function $f[x, y]$ is nonnegative, then it must include an additive bias A_0 that is sufficiently large to ensure that each component sinusoid with spatial frequency $[\xi_0, \eta_0]$ and amplitude A_1 is nonnegative. For example, if the input consists of one nonzero spatial frequency, then:

$$f[x, y] = A_0 + A_1 \cos[2\pi(\xi_0 x + \eta_0 t) + \phi_0] \ge 0 \tag{22.93}$$

The modulation of this sinusoid is easy to evaluate from Equation (6.25):

$$m_f \equiv \frac{f_{\max} - f_{\min}}{f_{\max} + f_{\min}} = \frac{A_1}{A_0} \tag{22.94}$$

The action of the system on this sinusoidal component generates output sinusoids scaled by the incoherent transfer function evaluated at that spatial frequency:

$$g[x, y] = \mathfrak{H}[0, 0] \cdot A_0 + \mathfrak{H}[\xi_0, \eta_0] \cdot A_1 \cos[2\pi(\xi_0 x + \eta_0 t) + \phi_0]$$

$$= \mathfrak{H}[0, 0] \cdot \left(A_0 + \frac{\mathfrak{H}[\xi_0, \eta_0]}{\mathfrak{H}[0, 0]} \cdot A_1 \cos[2\pi(\xi_0 x + \eta_0 t) + \phi_0]\right)$$

$$= \mathfrak{H}[0, 0] \cdot (A_0 + MT[\xi_0, \eta_0] \cdot A_1 \cos[2\pi(\xi_0 x + \eta_0 t) + \phi_0]) \tag{22.95}$$

The modulation of $g[x, y]$ is easy to evaluate:

$$m_g = \frac{g_{\max} - g_{\min}}{g_{\max} + g_{\min}} = \frac{MT[\xi_0, \eta_0] \cdot A_1}{A_0} = MT[\xi_0, \eta_0] \cdot m_f \tag{22.96}$$

which implies that the ratio of the input and output modulations is the function MT evaluated at that spatial frequency $[\xi_0, \eta_0]$:

$$\frac{m_g}{m_f} = MT[\xi_0, \eta_0] \tag{22.97}$$

Thus $MT[\xi, \eta]$ describes the "transfer of modulation" from the input to the output in the incoherent case, hence the notation. It is (obviously) called the *modulation transfer function*, and (just as obviously) does not consider any effects of the system on the initial phases of the sinusoidal components.

22.4.2 Comparison of Coherent and Incoherent Imaging

22.4.2.1 Circular Pupil

Consider an imaging system with object and image distances z_1 and z_2, and a lens with focal length $\mathbf{f}$ with circular pupil of diameter d:

$$p[x, y] \to p_r(r) = CYL\left(\frac{r}{d_0}\right) \tag{22.98}$$

We have seen that the coherent transfer function is proportional to a scaled replica of the pupil:

$$H[\rho; z_1, \mathbf{f}, z_2, \lambda_0] \propto CYL\left(-\frac{\lambda_0 z_2 \rho}{d_0}\right) = CYL\left(\frac{\rho}{(d_0/\lambda_0 z_2)}\right) \tag{22.99}$$

where λ_0 is the wavelength of the light from the coherent source. This system transmits all spatial frequencies up to the cutoff frequency $\rho_{\max} = d/2\lambda_0 z_2$ without change; the system acts as an ideal lowpass filter that "blurs" the image. The impulse response of this coherent imaging system is a "sombrero" function:

$$h[r; z_1, \mathbf{f}, z_2, \lambda_0] \propto \mathcal{F}_2^{-1}\left\{CYL\left(\frac{\rho}{(d_0/\lambda_0 z_2)}\right)\right\} \propto SOMB\left(\frac{r}{(\lambda_0 z_2/d_0)}\right) \tag{22.100}$$

Incoherent light may be described by its mean wavelength λ_μ, and the impulse response of the same system is the squared magnitude of the coherent impulse response. The incoherent impulse response for a circular pupil is well known in optics:

$$\mathfrak{h}(r) \propto SOMB^2\left(\frac{r}{(\lambda_\mu z_2/d_0)}\right) \tag{22.101}$$

The central lobe of the pattern is called the *Airy disk*. The discussion of the sombrero function in Section 7.4.4 showed that the diameter of the Airy disk is approximately 2.44 $\lambda_\mu \mathbf{f}/d_0$, which decreases with increasing aperture diameter d_0. The diameter of the Airy disk is often used as a metric of image quality.

The transfer function of the incoherent system with the circular pupil in light centered upon this wavelength is a circular triangle:

$$\begin{aligned}\frac{\mathfrak{H}(\rho)}{\mathfrak{H}(\rho=0)} &\propto CYL\left(\frac{r}{d_0}\right) \bigstar CYL\left(\frac{r}{d_0}\right)\Bigg|_{r\to-\lambda_\mu z_2 \rho} \\ &= \frac{2}{\pi}\left(\cos^{-1}\left[\frac{\lambda_\mu z_2 \rho}{d_0}\right] - \frac{\lambda_\mu z_2 \rho}{d_0}\sqrt{1-\left(\frac{\lambda_\mu z_2 \rho}{d_0}\right)^2}\right) CYL\left(\frac{\lambda_\mu z_2 \rho}{2d_0}\right) \\ &= \frac{2}{\pi}\cdot\left(\cos^{-1}\left[\frac{\rho}{(d_0/\lambda_\mu z_2)}\right] - \frac{\rho}{(d_0/\lambda_\mu z_2)}\sqrt{1-\left(\frac{\rho}{(d_0/\lambda_\mu z_2)}\right)^2}\right) CYL\left(\frac{\rho}{(2d_0/\lambda_\mu z_2)}\right)\end{aligned} \tag{22.102}$$

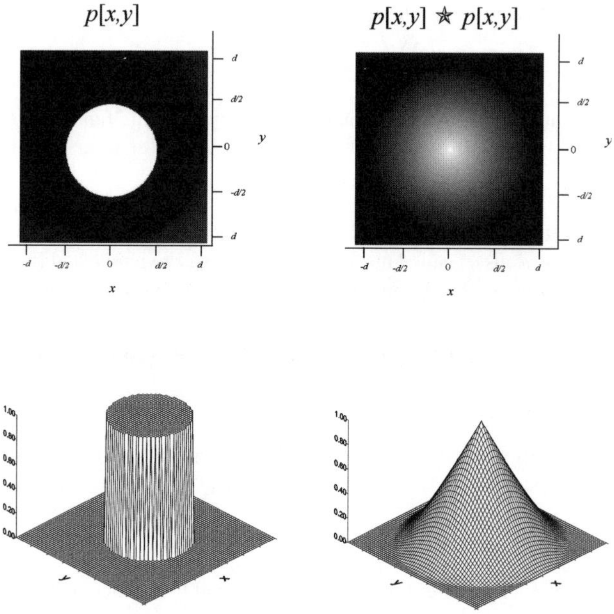

Figure 22.12 Pupil function of circular aperture $p[x, y] = CYL(r/d_0)$ and its autocorrelation, which are respectively proportional to the transfer function of the coherent and incoherent imaging system with this pupil if the coordinates are mapped to $[x, y] \to [-\lambda_0 z_2 \xi, -\lambda_0 z_2 \eta]$. Both systems act as lowpass filters. The incoherent system passes information with larger spatial frequencies, though with less modulation.

where the formulas for the scaled convolution from Equations (8.56b) and (7.99) have been used. The result is shown in Figure 22.12. The profiles of the point spread functions in coherent and incoherent light are shown in Figure 22.13. Note that since the autocorrelation of the circular pupil is twice as wide as the pupil itself, the cutoff frequency in incoherent light (i.e., the spatial frequency at which $H(\rho) = 0$) is twice as large as in coherent light. However, the incoherent OTF drops from unity for all spatial frequencies larger than zero. The incoherent imaging system with a finite aperture also acts as a lowpass filter and blurs the image.

22.4.2.2 Circularly Symmetric Systems with Obscured Apertures

A very common optical system used to gather imagery at a distance is the *Cassegrain* telescope, which cascades a large concave mirror with a smaller convex mirror to collect the radiation. The light path through a Cassegrain system is shown in Figure 22.14. The smaller mirror obstructs the primary mirror, and thus the aperture function of a Cassegrain is:

$$p(r) = CYL\left(\frac{r}{d_0}\right) - CYL\left(\frac{r}{d_1}\right), \quad d_0 > d_1 \tag{22.103}$$

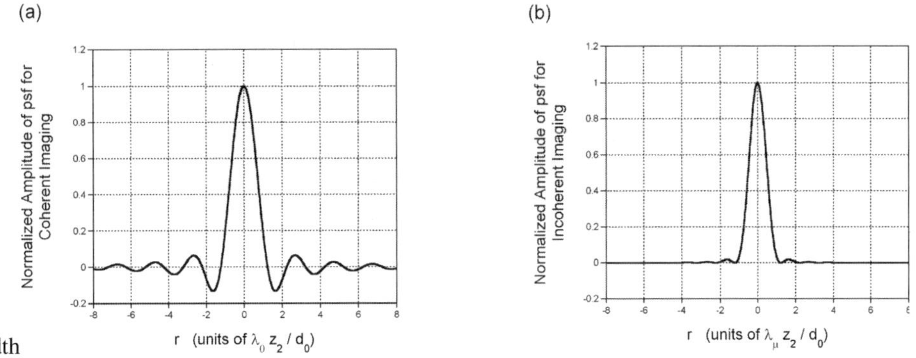

width

Figure 22.13 Profiles of the normalized point spread functions of an optical imaging system with a circular pupil of diameter d_0: (a) coherent imaging; (b) incoherent imaging.

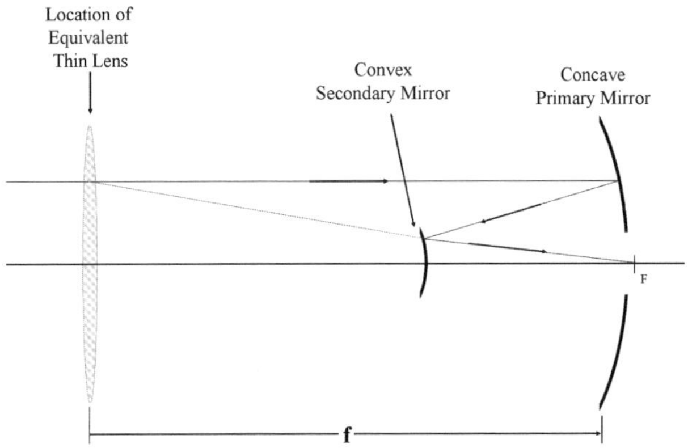

Figure 22.14 Optical layout of a Cassegrain imaging system consisting of circular primary and secondary mirrors of diameters d_0 and d_1, respectively. The focal length of the equivalent thin lens is much longer than the "tube length" between the primary and secondary. The pupil function is $p(r) = CYL(r/d_0) - CYL(r/d_1)$.

The coherent transfer function and impulse response are:

$$H(\rho) = CYL\left(\frac{\lambda_\mu z_2 \rho}{d_0}\right) - CYL\left(\frac{\lambda_\mu z_2 \rho}{d_1}\right) \tag{22.104}$$

$$h(r) \propto SOMB\left(\frac{r d_1}{\lambda_\mu z_2}\right) - \left(\frac{d_1}{d_0}\right)^2 SOMB\left(\frac{r d_1}{\lambda_\mu z_2}\right) \tag{22.105}$$

The centrally obscured system in coherent light acts as a *bandpass filter*; it passes those spatial frequencies in the interval:

$$\frac{d_0}{2\lambda_\mu z_2} > \rho > \frac{d_1}{2\lambda_\mu z_2} \tag{22.106}$$

to the output.

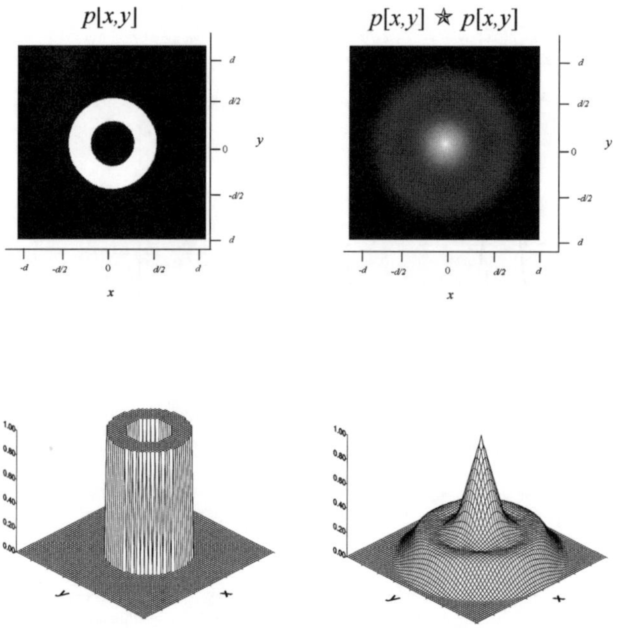

Figure 22.15 Pupil function of Cassegrain imaging system: $p[x, y] = CYL(r/d_0) - CYL(r/d_1)$. The incoherent transfer function does not decay smoothly to zero with increasing $\rho = \sqrt{\xi^2 + \eta^2}$, but exhibits a shoulder region where the modulation flattens out.

The incoherent psf of the Cassegrain is the squared magnitude of the difference of two sombreros, and the OTF is the autocorrelation of the aperture function, as shown in Figure 22.15. Because the subtraction of two terms causes the psf to drop faster for increasing r, the incoherent psf of an optical system with a central obscuration is "skinnier" than that of an equal-diameter clear-aperture system. However, the first lobe of the psf is "brighter" for a Cassegrain than for a clear-aperture system, as shown in the profiles in Figure 22.16:

$$\mathfrak{h}(r) \propto \left| SOMB\left(\frac{rd_0}{\lambda_\mu z_2}\right) - \left(\frac{d_1}{d_0}\right)^2 SOMB\left(\frac{rd_1}{\lambda_\mu z_2}\right)\right|^2 \tag{22.107}$$

$$\mathfrak{H}(\rho) = \left[CYL\left(\frac{\lambda_\mu z_2 \rho}{d_0}\right) - CYL\left(\frac{\lambda_\mu z_2 \rho}{d_1}\right)\right] \bigstar \left[CYL\left(\frac{\lambda_\mu z_2 \rho}{d_0}\right) - CYL\left(\frac{\lambda_\mu z_2 \rho}{d_1}\right)\right] \tag{22.108}$$

22.4.2.3 Multiaperture Systems

We can also construct optical systems whose pupil function contains multiple apertures that may be regularly placed or "scattered" about. Such systems are sometimes called "sparse" or "dilute" apertures. The "Very Large Array" (VLA) of radio telescopes in New Mexico and the former "Multiple-Mirror Telescope" (MMT) on Mount Hopkins near Tucson are perhaps the best-known examples (the multiple mirrors of the MMT have been replaced with a single large mirror). One interesting potential feature of a multiaperture system is the possibility of changing the phase of one or more of the component apertures.

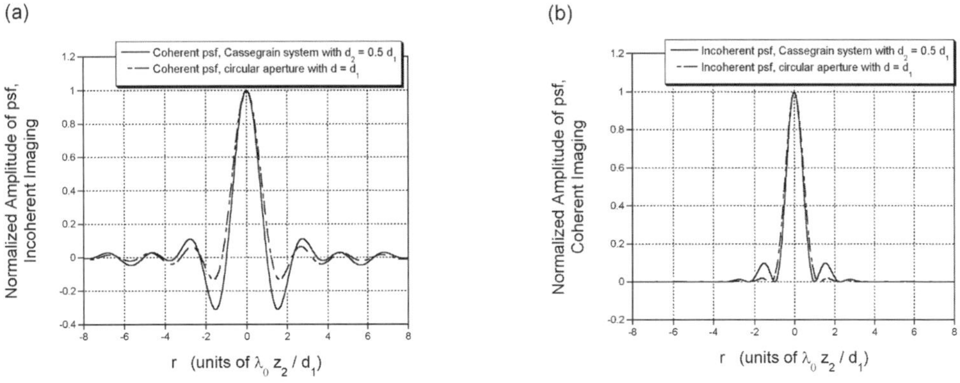

Figure 22.16 Profiles of normalized circularly symmetric psfs of a Cassegrain optical imaging system with primary diameter d_0 and secondary diameter $d_1 = d_0/2$, compared to the profiles for the unobscured pupil: (a) coherent imaging; (b incoherent imaging. Note the increase in amplitude of the first sidelobe for the Cassegrain compared to the full pupil in the incoherent case.

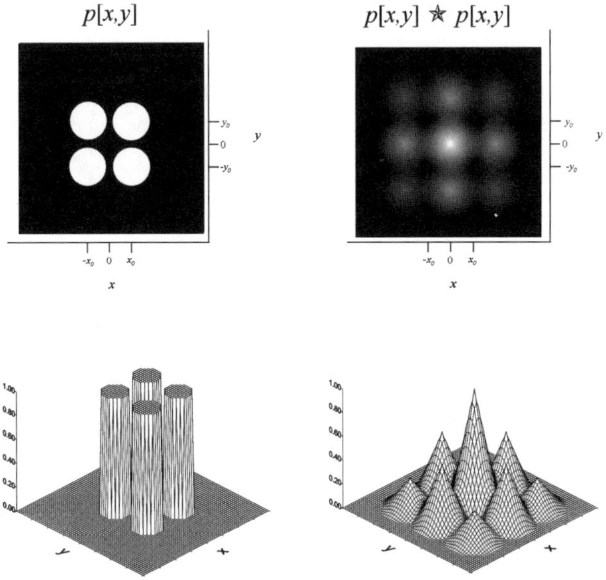

Figure 22.17 Pupil function of a multiaperture imaging system composed of four circular subpupils placed at the vertices of a square. The transfer function of the system (OTF) in coherent light is proportional to the suitably scaled pupil function, and thus the system acts as a bandpass filter. The OTF in incoherent (yet still quasimonochromatic) illumination is proportional to the scaled autocorrelation of $p[x, y]$, and thus the incoherent system is a lowpass filter.

If the pupil is created by an array of identical individual apertures, it may be written as a convolution of the individual aperture and an array of Dirac delta functions that describe their locations, e.g.:

$$p[x, y] = \sum_{n=1}^{N} a[x - x_n, y - y_n] = a[x, y] * \sum_{n=1}^{N} \delta[x - x_n, y - y_n] \tag{22.109}$$

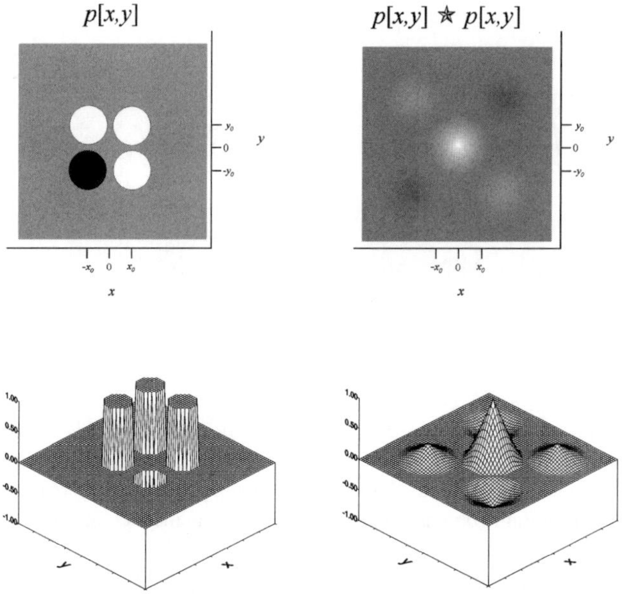

Figure 22.18 OTF of the multiaperture system if the phase of one subpupil is changed by π radians. The pupil function is real valued and bipolar (midgray $\Longrightarrow$ 0 amplitude), and thus so is the coherent OTF. The OTF of the corresponding incoherent system is bipolar and thus can "reverse" the phase of some frequency components.

So the coherent transfer function is:

$$H[\xi,\eta]=p[-\lambda_\mu z_2\xi,-\lambda_\mu z_2\eta]=\sum_{n=1}^{N}a[-\lambda_\mu z_2\xi-x_n,-\lambda_\mu z_2\eta-y_n]$$

$$=\sum_{n=1}^{N}a[-\lambda_\mu z_2\xi-x_n,-\lambda_\mu z_2\eta-y_n]$$

$$=a[-\lambda_\mu z_2\xi,-\lambda_\mu z_2\eta]*\sum_{n=1}^{N}\delta[-\lambda_\mu z_2\xi-x_n,-\lambda_\mu z_2\eta-y_n]$$

$$=\left(\frac{1}{\lambda_\mu z_2}\right)^2 a[-\lambda_\mu z_2\xi,-\lambda_\mu z_2\eta]*\sum_{n=1}^{N}\delta\left[\xi+\frac{x_n}{\lambda_\mu z_2},\eta+\frac{y_n}{\lambda_\mu z_2}\right] \tag{22.110}$$

An example is shown in Figure 22.17.

Multiaperture mirror systems exhibit an interesting effect if one of the subaperture mirrors "slips" in its mounting, say by a distance Δz. The light traverses the additional distance Δz twice and thus incurs a phase change of approximately:

$$\Delta\phi \cong 2\cdot\frac{\Delta z}{\lambda_\mu} \tag{22.111}$$

where λ_μ is the wavelength of the light. The phase term affects the transfer function of the system. An example is shown in Figure 22.18, where the phase delay of one subaperture is $-\pi$ radians. The incoherent transfer function exhibits regions of negative amplitude, which means that the phase of any sinusoidal component imaged by the system is changed by π radians, thus "reversing" the modulation of the imaged component.

PROBLEMS

22.1 The MTF is defined for nonnegative sinusoidal functions and specifies how well the modulation is transferred as a function of spatial frequency. The analogous metric for square-wave functions is the *contrast transfer function* (CTF), which is measured for square waves with 50% duty cycle (50% "on" and 50% "off"). Derive an expression for the 1-D CTF $C[\xi]$ in terms of 1-D MTF $M[\xi]$.

(a) Consider the following optical systems that operate in both monochromatic (coherent) and "quasimonochromatic" (incoherent) light centered at wavelength $\lambda_\mu = 500$ nm. The systems operate at "equal conjugates" so that the object and image distances are identically $2\mathbf{f}$, where $\mathbf{f} = +200$ mm. In other words, the complex amplitude transmittance of the lens is:

$$t[x, y] = p[x, y] \exp\left[-i\pi\left(\frac{x^2 + y^2}{\lambda_\mu \mathbf{f}}\right)\right]$$

where $p[x, y]$ may be complex valued. Evaluate and plot graphical profiles of the impulse responses of the following optical photographic systems along the x- and y-axes *and* of the transfer functions along the ξ- and η-axes.

(b) $p[x, y]$ is a circular aperture of diameter 50 mm. Also, find the spatial frequency where the MTF is 50% and the "cutoff" frequency where the MTF first reaches zero.

(c) $p[x, y]$ is a square aperture with sides of length 50 mm. Also, find the spatial frequency where the MTF is 50% and the "cutoff" frequency where the MTF first reaches zero.

(d) $p[x, y]$ is a square aperture with sides of length 50 mm where the left half (for $x < 0$) is covered with a sheet of glass of refractive index $n = 1.5$ and thickness such that the phase of the light is delayed by π radians.

(e) $p[x, y]$ is a square aperture as in part (c) but with an additional phase:

$$p[x, y] = RECT\left[\frac{x}{a}, \frac{y}{a}\right] \exp\left[+i\left(2\pi\left|\frac{x}{a}\right|\right)\right] \quad \textit{where } a = 50 \text{ mm}$$

(f) $p[x, y]$ consists of two square apertures with sides of 10 mm that are symmetrically placed about the optical axis along the x-axis with centers separated by 30 mm. Find the frequencies where the MTF first reaches 50% and 0%.

(g) The pupil function is identical to that in part (b), except that one of the two square apertures is covered with the same sheet of glass of refractive index $n = 1.5$ described above.

(h) The pupil function consists of four square apertures with sides of 10 mm whose centers are arranged to form a square with sides 30 mm that is symmetrically placed about the optical axis.

(i) The pupil function is identical to that in part (e) but one of the four apertures is overlaid with the same sheet of glass in part (c).

22.2 A 1-D linear shift-invariant optical system forms an image at a distance $z = 200$ mm and has the following pupil function:

$$p[x, y] = \left(RECT\left[\frac{x}{100 \text{ mm}}\right] SGN[x]\right) 1[y]$$

(a) Derive expressions for and sketch the OTF and psf if the illumination is coherent.

(b) Sketch the transfer function and impulse response of this system for incoherent illumination.

(c) What images will be produced by this optical system in the two cases if $f(x) = 1$?

(d) Describe how such an imaging system may be constructed.

22.3 We often hear claims that images taken by spy satellites have sufficient resolution to allow license numbers to be read (of course, relatively few cars have license plates on the roof!).

(a) What lens diameter is required if the altitude of the camera is 300 km ($\cong$ 185 miles), $\lambda_\mu =$ 700 nm, and resolution on the ground is required to be 20 mm?

(b) What other physical processes will degrade the image?

(c) From the results of (a) and (b), speculate on the image quality if coherent illumination is used.

22.4 Light with wavelength $\lambda_0 = 600$ nm encounters an opaque screen with two point apertures separated by the distance d_0. Light through one hole then encounters a variable neutral-density filter whose transmittance cycles from 1 to 0 and back to 1 in a sinusoidal fashion at the rate $\nu_0 = 1$ cycles per second, where 1 second is much longer than the measurement time T_0. At all transmittances, the neutral-density filter induces a constant phase of $\pi/4$ radians. Derive the expression for the modulation of the irradiance pattern as a function of time that is observed on the observation screen located a distance z_1 "downstream" from the aperture plane in the Fraunhofer diffraction region.

22.5 Derive Equation (22.36) from Equation (22.35):

$$I[x; z_0 + z_1] \propto |g[x; z = z_0 + z_1]|^2$$

$$= 4A_0^2 \cdot \left(\frac{1}{\lambda_0 z_0}\right)^2 \cdot \left(\frac{1}{\lambda_0 z_1}\right)^2$$

$$\cdot \left(\left(1 + \cos\left[\pi \frac{d_0 \ell_0}{\lambda_0 z_0}\right] \cdot \cos[\Delta\phi]\right) + \left(\cos\left[\pi \frac{d_0 \ell_0}{\lambda_0 z_0}\right] + \cos[\Delta\phi]\right) \cos\left[2\pi \frac{x}{(\lambda_0 z_1/d_0)}\right]\right)$$

22.6 Assume the following parameters for the light from the Sun:

$$\lambda_{max} = 700 \text{ nm}, \ \lambda_{min} = 400 \text{ nm}$$

$$\text{diameter of Sun} \cong 1.4 \times 10^6 \text{ km}$$

$$\text{distance from Earth to Sun} \cong 150 \times 10^6 \text{ km}$$

Assume that the amount of light at each wavelength that reaches the Earth is constant:

(a) Determine the coherence volume of light from the Sun observed at the Earth; if helpful, you may assume that the cross-section of the Sun is square with the given diameter.

(b) Describe the qualitative difference in the coherence volume of light from the Sun as seen from Saturn, which is about 8.8× as far from the Sun as the Earth.

22.7 Derive a function for the irradiance pattern produced by two point sources separated by the distance ℓ_0 where the light emitted by the two sources is monochromatic but with different wavelengths.

22.8 Consider an imaging system with a square pupil of width d_0 that has an added quadratic-phase term that moves the location of the image point:

$$p[x, y] = RECT\left[\frac{x}{d_0}, \frac{y}{d_0}\right] \cdot \exp\left[+2\pi i \frac{x^2 + y^2}{(d_0/2)^2} \cdot W_{max}\right]$$

where W_{max} is the maximum error of the wavefront measured in units of λ_0; in other words, the system suffers from defocus.

(a) Perform the autocorrelation to find the normalized incoherent transfer function.

(b) Graph the transfer function along the ξ-axis for $W_{max} = 0, \frac{1}{2}, \frac{3}{4}$, 1, and 2 waves of defocus.

(c) Evaluate the approximate transfer function in the case where the defocus error is very large.

(d) Find an expression for the approximate impulse response for severe defocus error and explain it.

23

Holography

In the previous two chapters, the equations for light propagation were used to design optical systems capable of producing replicas of the light distribution from planar objects. Because imaging sensors that act in visible light, including the eye, observe the irradiance pattern rather than the amplitude, the explicit phase of the light is lost. In other words, the imaging system reproduces the gray-scale pattern of the object. In this chapter, we revisit those same equations to describe methods for reproducing the wavefronts generated by the interaction of light with the object. If the reproduced wavefronts are identical to those that would emerge directly from the object, a viewer would not be able to distinguish between the two.

A process for recording and reproducing wavefronts of light was first described by Dennis Gabor in the 1940s as a result of his efforts to improve the spatial resolution of electron microscopy. He recorded the pattern created by light created from the sum of wavefronts that have interacted with an object (e.g., by reflection or refraction) and wavefronts emerging directly from the source (the "reference" wave). Gabor (1948) called the process *holography* (from the Greek, *holos* = whole + *gramma* = message). The best sources of light for this purpose available to Gabor were spectral emissions in low-pressure gas lamps. The narrow bandwidth of the light ensures that the coherence length is sufficiently long for interference fringes to be visible. Holography became much more practical after a source of light that could produce interference patterns over much longer distances became available in the 1960s; of course, this is the laser. In addition, the geometry of Gabor's process was modified by Emmett Leith and Juris Upatnieks (1962). The combination of these two advances made holography much more practical and useful.

Holography is a two-step process: (1) recording of the interference pattern; and (2) reconstruction of the image from the recording. In this discussion, the object will be denoted by the 3-D spatial function $f[x, y, z]$ and its irradiance observed at the sensor emulsion by $s[x, y]$. The hologram processed from the recording of $s[x, y]$ will be denoted by $t[x, y]$ and the amplitude produced by the reconstruction process by $g[x, y]$. For ease of analysis, the function $f[x, y, z]$ includes the light both from the "object" and from the source of the reference wave. The object may be assumed to consist of point sources that emit light, but in actual use the object more often reflects or transmits light from a single monochromatic point source that is generally assumed to be sufficiently distant such that planar wavefronts illuminate $f[x, y, z]$. In such a case, all points on a planar object $f[x, y, z_0]$ are illuminated by light with the same phase, which simplifies evaluation of the interference pattern. In a more realistic model of light, the original illuminating source is neither a point nor monochromatic, which means that different phases are seen by different object points. The discussion in the previous chapter indicates that the locations of points in a 3-D object function $f[x, y, z]$ that produce wavefronts capable of interfering are determined

Fourier Methods in Imaging   Roger L. Easton, Jr.
© 2010 John Wiley & Sons, Ltd

by the spatial and temporal coherence properties of the source. We have seen that the transverse "width" between interfering points is determined by the spatial coherence and the depth along the z-axis by the temporal coherence.

This discussion demonstrates the relationship of holography to the chirp Fourier transforms and will be cursory. We will consider in detail only the simplest cases consisting of two source points (a "reference" and an "object"). The extension of the discussion to realistic objects composed of multiple points will be brief, but is straightforward. Readers requiring a deeper discussion should consult a text devoted to the subject, e.g., Collier et al. (1971), or Reynolds et al. (1989).

We will first consider holography of planar objects in the Fraunhofer diffraction model and then extend this result to 3-D objects in the Fresnel model. We extend the discussion to a brief consideration of methods for computer-generated holograms (CGHs), which render calculated holograms so that they may be introduced into an optical system. Finally we will briefly consider synthetic-aperture radar, which is an active imaging system related to 3-D Fresnel holography that nicely ties together several of the concepts we have considered.

23.1 Fraunhofer Holography

23.1.1 Two Points: Object and Reference

We begin by considering the simplest hologram imaginable: the interference pattern of two point sources in the same plane viewed in the Fraunhofer diffraction region. The mathematical description of the entire process is very straightforward and will provide the framework for more complicated cases in the Fresnel region. This model is similar to the basis for the original holograms by Gabor (1948).

We saw in Equation (21.97) that the transmittance of a source function $f[x, y]$ that has compact support constrained to a plane perpendicular to the optical axis determines the resulting diffracted irradiance $|g[x, y]|^2$ in the Fraunhofer diffraction region via the appropriately scaled Fourier transform:

$$|g[x, y]|^2 = \left|\frac{\boldsymbol{\mathcal{E}}_0}{\lambda_0 z_1}\right|^2 \left|F\left[\frac{x_1}{\lambda_0 z_1}, \frac{y_1}{\lambda_0 z_1}\right]\right|^2 \tag{23.1}$$

where $\boldsymbol{\mathcal{E}}_0$ is the amplitude of the illuminating function. This expression is valid if light from all points on the planar object propagates the same distance down the optical axis to reach the output plane at $z = z_1$; the sources must be in the same plane. For example, consider the simple case of two point sources located in the plane $z = 0$ that emit the same monochromatic wavelength λ_0. For convenience, one source is located "on axis" (at $x = y = 0$) with unit amplitude, while the complex amplitude of the second ("off-axis") source is $\alpha_0\, e^{+i\phi_0}$. For convenience in the discussion, we choose $0 \leq \alpha_0 \leq 1$ so that the off-axis source is no "brighter" than the on-axis source, though this constraint is not essential. We can think of the on-axis source as the "reference" and the off-axis source as the "object", but this distinction is arbitrary. If the relative phase of the two sources is $\phi_0 = 0, \pm\pi$, or $\pm\pi/2$, then we speak of the sources as being "in phase", "out of phase", or "in quadrature", respectively.

Though the object function is assumed to be planar, it is convenient to include the third coordinate dimension so that the extension to 3-D objects becomes more straightforward. The third dimension for planar objects is specified by the appropriate 1-D Dirac delta functions of z:

$$\begin{aligned} f[x, y; z = 0] &= \delta[x, y, z] + (\alpha_0\, e^{+i\phi_0})\delta[x - x_0, y, z] \\ &= (\delta[x, y] + (\alpha_0\, e^{+i\phi_0})\delta[x - x_0, y]) \cdot \delta[z - 0] \end{aligned} \tag{23.2}$$

Since the source is assumed to be monochromatic, we ignore the time dependence of the wave.

As before, the radiation is observed at the plane $z = z_1$, which is assumed to be sufficiently distant so that the spherical waves are suitably approximated by plane waves in the Fraunhofer model. In such a case, Equation (21.96) demonstrates that the observed amplitude $g[x, y; z = z_1]$ is proportional to the

Fourier transform of the source function evaluated at suitably scaled coordinates:

$$s[x, y; z = z_1] \propto \mathcal{F}_2\{f[x, y; z]\}|_{\xi = x/\lambda_0 z_1, \eta = y/\lambda_0 z_1}$$
$$= (1 + (\alpha_0 \, e^{+i\phi_0}) \, e^{-2\pi i (x_0/\lambda_0 z_1) x}) 1\left[\frac{x}{\lambda_0 z_1}, \frac{y}{\lambda_0 z_1}\right] \tag{23.3}$$

The measurable quantity at the observation plane is the irradiance $|s[x, y; z_1]|^2$, which is proportional to a biased sinusoidal function of x:

$$|s[x, y; z_1]|^2 \propto (1 + (\alpha_0 \, e^{+i\phi_0}) \, e^{-2\pi i (x_0/\lambda_0 z_1) x})(1 + (\alpha_0 \, e^{+i\phi_0}) \, e^{-2\pi i (x_0/\lambda_0 z_1) x})^*$$
$$= (1 + \alpha_0^2) + 2\alpha_0 \cos\left[2\pi \left(\frac{x_0}{\lambda_0 z_1}\right) x - \phi_0\right] \cdot 1[y]$$
$$= (1 + \alpha_0^2) \cdot \left(1 + \frac{2\alpha_0}{1 + \alpha_0^2} \cos[2\pi \xi_0 x - \phi_0]\right) \cdot 1[y] \tag{23.4}$$

where $\xi_0 \equiv x_0/\lambda_0 z_1$ is the spatial frequency of the sinusoidal irradiance pattern observed at the recording plane $z = z_1$. The amplitude of this sinusoidal "grating" oscillates over the amplitude range $(1 + \alpha_0^2) \pm 2\alpha_0$ and its modulation from Equation (6.25) evaluates to:

$$m \equiv \frac{|s[x, y; z_1]|^2_{\max} - |s[x, y; z_1]|^2_{\min}}{|s[x, y; z_1]|^2_{\max} + |s[x, y; z_1]|^2_{\min}}$$
$$= \frac{[(1 + \alpha_0^2) + 2\alpha_0] - [(1 + \alpha_0^2) - 2\alpha_0]}{[(1 + \alpha_0^2) + 2\alpha_0] + [(1 + \alpha_0^2) - 2\alpha_0]} = \frac{2\alpha_0}{1 + \alpha_0^2} \tag{23.5}$$

This may be substituted into Equation (23.4) and the process is shown schematically in Figure 23.1a:

$$|s[x, y; z_1]|^2 = (1 + \alpha_0^2) \cdot (1 + m \cos[2\pi \xi_0 x - \phi_0] \cdot 1[y]) \tag{23.6}$$

If the sources have equal "strength" ($\Longrightarrow \alpha_0 = 1 \Longrightarrow m = 1$), the sinusoidal irradiance "grating" oscillates over the range of amplitudes $0 \leq |s|^2 \leq 2 \cdot (1 + \alpha_0^2)$. In the limit where the amplitude of the second source is "small" ($\alpha_0 \gtrapprox 0$), the modulation of the recorded irradiance approaches $m \cong 2\alpha_0$, which means that $|g|^2$ oscillates over the small range $\pm 2\alpha_0$ about the mean value $1 + \alpha_0^2$.

The sinusoidal variation of irradiance in the plane $z = z_1$ conveys evidence of the existence of the two point sources. The measurable parameters of the grating (orientation, period, phase at the origin, and modulation) are determined by corresponding physical properties of the point sources (orientation, separation, relative optical phase, and relative amplitude, respectively). In other words, a recording of the irradiance pattern preserves evidence of the physical properties of the object. This means that the original configuration of light sources may be inferred from the measurable properties of the grating, which suggests the possibility of finding a formula for the inverse filter of the original "imaging" process that will "reconstruct" the original sources from the recorded pattern.

23.1.1.1 Processing to create the transparency $t[x, y]$

The first task in this process is recording the irradiance pattern $|s[x, y; z_1]|^2$. Based on (formerly common) photographic experience (rapidly becoming less common in this age of digital imaging), the goal of photographic recording is to produce a transparency whose optical transmittance t, such that $0 \leq t \leq 1$, is a faithful rendition of incident irradiance in some sense. Of course, the processed emulsion is actually a "negative" of the irradiance, which is a rather unfortunate term since it implies the possibility of recorded values less than zero. In fact, the ideal processed photographic emulsion is a record of the nonnegative real-valued scaled "complement" of the irradiance. For a single-frequency

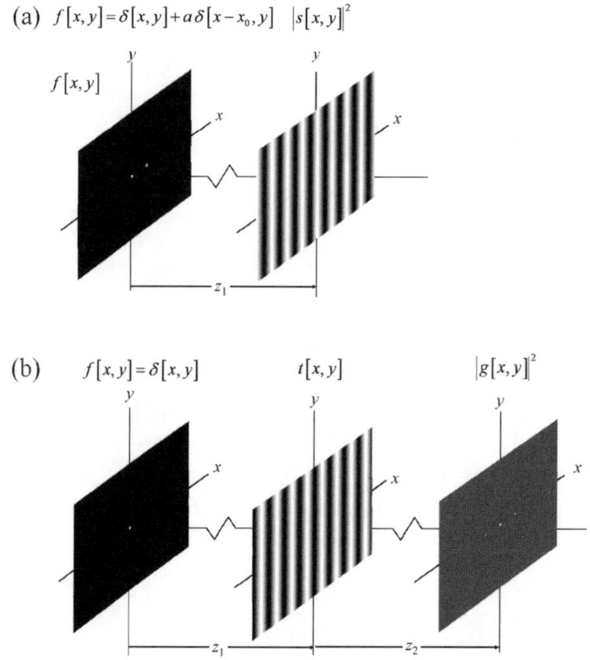

Figure 23.1 Sequence of operations in Fraunhofer holography: (a) recording of interference pattern from reference and object points on emulsion at a distance $z = z_1$; (b) reconstruction by illuminating processed hologram with original reference point to reconstruct two replicas of the object point at distance z_2 from hologram.

sinusoidal grating, we assume an "ideal" recording of the transmittance that has been processed to map the "average" irradiance (the "bias" of the sinusoidal grating) to the midpoint of the transmittance scale ($t = \frac{1}{2}$) and that the resulting grating exhibits the same modulation as the incident irradiance. The functional form of the transmittance is:

$$t[x, y; z_1] = \frac{1}{2}(1 - m\cos[2\pi\xi_0 x - \phi_0]) \cdot 1[y] \tag{23.7a}$$

$$= \begin{cases} \left(\frac{1}{2} - \frac{1}{2}\cos[2\pi\xi_0 x - \phi_0]\right) \cdot 1[y] & \textit{if } \alpha_0 = 1 \Longrightarrow m = 1 \\ \left(\frac{1}{2} - \alpha_0 \cdot \cos[2\pi\xi_0 x - \phi_0]\right) \cdot 1[y] & \textit{if } \alpha_0 \gtrapprox 0 \Longrightarrow m \cong 2\alpha_0 \end{cases} \tag{23.7b}$$

where the spatial frequency $\xi_0 \equiv x_0/\lambda_0 z_1$. If $\alpha_0 = 1$, so that the reference and object sources have equal "strength", the transmittance oscillates about the mean value of $\frac{1}{2}$ with an amplitude of $\frac{1}{2}$, so that the full range of transmittance is utilized. If $\alpha_0 \gtrapprox 0$, the transmittance of the developed hologram oscillates over the approximate range $\frac{1}{2} \mp \alpha_0$. Note that this model of the emulsion assumes that only the interference pattern in the plane specified by z_1 is recorded, which means that the emulsion is "thin" compared to a wavelength of light. Realistic "thick" emulsions record interference patterns in depth and thus require additional consideration.

A side comment about "reality" is useful here. We have blithely stated that the emulsion is processed so that the resulting transmittance is proportional to the scaled complement of the irradiance. This is easily said, but not so easily done. Processed photographic emulsions are inherently nonlinear in transmittance; even if carefully processed, deviations from the desired linear behavior can easily creep into the processing and complicate the analysis. We will consider the qualitative effects of nonlinearities after analyzing reconstruction of the ideal thin hologram.

23.1.1.2 Reconstruction

After processing, the transparency is replaced in its original position and reilluminated by light only from the on-axis "reference" source; the object "source" has been removed. The illumination from this reference source is (approximately) a plane wave traveling down the z-axis until it reaches the transparency, which modulates the uniform amplitude to produce a pattern of light proportional to $t[x, y, z_1]$. This amplitude then propagates a distance z_2 to an observation plane farther "downstream" in the Fraunhofer region, as shown in Figure 23.1b. The discussion of Fraunhofer diffraction is used again to see that the observed amplitude pattern is proportional to the Fourier transform of the plane reference wave modulated by the transmittance function $t[x, y; z_1]$, which evaluates a scaled replica of its spectrum:

$$\begin{aligned} g[x, y; z_1 + z_2] &\propto T[\xi, \eta]|_{\xi=x/\lambda_0 z_2, \eta=y/\lambda_0 z_2} \\ &= \frac{1}{2} \cdot \mathcal{F}_2\{(1 - m\cos[2\pi\xi_0 x - \phi_0])\, 1[y]\}|_{\xi=x/\lambda_0 z_2, \eta=y/\lambda_0 z_2} \\ &\propto \delta[x, y] - \frac{\alpha_0}{1+\alpha_0^2}\left(\delta\left[x + \frac{z_2}{z_1}x_0, y\right] e^{-i\phi_0} + \delta\left[x - \frac{z_2}{z_1}x_0, y\right] e^{+i\phi_0}\right) \end{aligned} \tag{23.8}$$

where the scaling property of the Dirac delta function in Equation (6.107) has been used and the leading constant factors have been dropped. Note that the amplitude pattern consists of three 2-D Dirac delta functions. The output irradiance at z_2 is proportional to the squared magnitude of this pattern, where the squaring affects only the weighting factors of the Dirac delta functions:

$$\begin{aligned} |g[x, y; z_1]|^2 &\propto \delta[x, y] + \left(-\frac{m}{2}\right)^2 \left(\delta\left[x + \frac{z_2}{z_1}x_0, y\right] + \delta\left[x - \frac{z_2}{z_1}x_0, y\right]\right) \\ &= \delta[x, y] + \left(\frac{\alpha_0}{1+\alpha_0^2}\right)^2 \cdot \left(\delta\left[x - (-1)\cdot\frac{z_2}{z_1}x_0, y\right] + \delta\left[x - (+1)\cdot\frac{z_2}{z_1}x_0, y\right]\right) \end{aligned} \tag{23.9}$$

where the cross-terms drop out because the products of Dirac delta functions with different arguments evaluate to zero and the fact that α_0 is real valued has been used. In words, the resulting irradiance "image" is a set of three Dirac delta functions: one "on axis" with unit weight and two separated from the origin by $x = \pm(z_2/z_1)x_0$ and weighted by $(m/2)^2$, which evaluates to $\frac{1}{4}$ if $\alpha_0 = 1$ and approximately to α_0^2 for small positive values of α_0. The individual reconstructed object points are labeled by their numerical "orders": the image of the object point "reconstructed" on the correct side of the origin has order $+1$, while the order of its symmetric replica is -1. The "reconstructed" image is a scaled replica of the autocorrelation of the source function that has been scaled in both position and "brightness".

Note that the measured irradiance in Equation (23.9) is proportional to the squared magnitude of the input amplitude. This means that the magnitude of the original input signal $s[x, y]$ must be the square root of the desired irradiance.

In this example, the holographic recording is assumed to be infinitely large, so that its "pupil" function has no effect on the reconstruction. Though this assumption seems to be unrealistic, it is no more so than the assumption of Fraunhofer diffraction itself, which is only valid if the object points are close to the optical axis.

23.1.1.3 Effect of Positive Processing

Consider the reconstruction resulting from an emulsion that produces a "positive" image, i.e., one for which the transmittance is directly proportional to the incident irradiance rather than to its "complement". In this case, the transmittance function that was evaluated in Equation (23.7) would include a positive sign:

$$t[x, y; z_1] \propto \left(1 + \frac{|s[x, y; z_1]|^2}{|s|^2_{\max}}\right)$$
$$= \frac{1}{2}(1 + m\cos[2\pi\xi_0 x - \phi_0]) \tag{23.10}$$

so that the amplitudes of the reconstructed Dirac delta functions are positive instead of negative:

$$g[x, y; z_1 + z_2]$$
$$\propto \delta[x, y] + \frac{\alpha_0}{1+\alpha_0^2}\left(\delta\left[x + \frac{z_2}{z_1}x_0, y\right]e^{-i\phi_0} + \delta\left[x - \frac{z_2}{z_1}x_0, y\right]e^{+i\phi_0}\right) \tag{23.11}$$

The reconstructed irradiance is proportional to the squared magnitude:

$$|g[x, y; z_1]|^2 \propto \delta[x, y] + \left(\frac{m}{2}\right)^2\left(\delta\left[x + \frac{z_2}{z_1}x_0, y\right] + \delta\left[x - \frac{z_2}{z_1}x_0, y\right]\right) \tag{23.12}$$

which is identical to that in Equation (23.9) for normal "negative" processing because the phase of the reconstructed amplitude had no effect on the measured irradiance. This is because the average transmittance is assumed to be 0.5 for both, so that the Dirac delta function at the origin has the same amplitude in both cases. For an extended object consisting of multiple points, the average value of the transmittance is generally different for positive and negative processing, which changes the relative amplitude of the reconstructed reference point source, but not that of the reconstructed object points.

23.1.1.4 Effect of Nonlinear Processing

We have just seen that the goal in photographic recording of holograms is to produce a transparency with transmittance proportional to the normalized complement of the recorded irradiance. We also mentioned that photographic processes are inherently nonlinear so that the gray values of the recorded image are not proportional to the incident exposure. In pictorial photography, this actually is a "feature" rather than a "bug" because the photograph simulates the logarithmic gray-scale response of the eye, and is related to the reason why photographic emulsions are generally characterized in terms of nonnegative numerical "density" D rather than "transmittance" t, where $D = -\log_{10}[t]$, which can theoretically range from $D = 0 \Longrightarrow t = 1$ to $D = +\infty \Longrightarrow t = 0$. As was just stated, processing nonlinearities disrupt the simple description of holographic reconstruction. Our goal in this section is to consider the effect of nonlinear processing, which is a difficult task for a general nonlinearity, but not so bad for the power-law nonlinearities that we considered in Chapter 9.

A power-law nonlinear mapping of irradiance may be expressed as an operator $\mathcal{O}$ such that:

$$\mathcal{O}\{|s[x, y; z_1]|^2\} = \left(1 - \frac{|s[x, y; z_1]|^2}{|s|^2_{\max}}\right)^\gamma \tag{23.13}$$

If $\gamma \cong 1$, then we can expand the nonlinear operation into a Taylor series of the form of Equation (9.196) that may be truncated without incurring much error. For example, the series resulting from truncation after the quadratic term is:

$$t[x, y; z_1] \cong \left[\frac{1}{2}\left(1 - m\cos\left[2\pi x\left(\frac{x_0}{\lambda_0 z_1}\right)\right]\right)\right]^\gamma$$
$$\cong \left(\frac{1}{2}\right)^\gamma\left(1 - \gamma m\cos\left[2\pi x\left(\frac{x_0}{\lambda_0 z_1}\right)\right] + \frac{\gamma^2-\gamma}{2}\cdot\frac{m^2}{2}\left(1 + \cos\left[2\pi x\left(\frac{2x_0}{\lambda_0 z_1}\right)\right]\right)\right) \tag{23.14}$$

where the trigonometric identity for $\cos^2[\theta]$ in Equation (6.24) has been used. The Fourier transform of this approximate transmittance is easy to evaluate as five 2-D Dirac delta functions with weights determined by m and γ:

$$T[\xi,\eta;z_1] \cong \frac{1}{2^\gamma}\cdot\delta[\xi,\eta] - \frac{1}{2^\gamma}\frac{m}{4}\cdot\gamma\cdot\left(\delta\left[\xi+\frac{x_0}{\lambda_0 z_1},\eta\right]+\delta\left[\xi-\frac{x_0}{\lambda_0 z_1},\eta\right]\right)$$
$$+\frac{1}{2^\gamma}\frac{m^2(\gamma^2-\gamma)}{4}\left(\delta[\xi,\eta]+\frac{1}{2}\delta\left[\xi+\frac{2x_0}{\lambda_0 z_1},\eta\right]+\frac{1}{2}\delta\left[\xi-\frac{2x_0}{\lambda_0 z_1},\eta\right]\right) \quad (23.15)$$

The reconstructed amplitude at the observation plane is obtained by substituting the scaled space-domain coordinates and may be expressed in terms of the analogue of transverse magnification from geometrical optics, $M_T = -z_2/z_1$:

$$T\left[\frac{x}{\lambda_0 z_2},\frac{y}{\lambda_0 z_2}\right] \cong \frac{1}{2^\gamma}\cdot\delta\left[\frac{x}{\lambda_0 z_2},\frac{y}{\lambda_0 z_2}\right]$$
$$-\frac{1}{2^\gamma}\frac{m}{4}\gamma\left(\delta\left[\frac{x}{\lambda_0 z_2}+\frac{x_0}{\lambda_0 z_1},\frac{y}{\lambda_0 z_2}\right]+\delta\left[\frac{x}{\lambda_0 z_2}-\frac{x_0}{\lambda_0 z_1},\frac{y}{\lambda_0 z_2}\right]\right)$$
$$+\frac{1}{2^\gamma}\left(\frac{m^2\gamma^2}{4}\right)\left(\delta\left[\frac{x}{\lambda_0 z_2},\frac{y}{\lambda_0 z_2}\right]+\frac{1}{2}\delta\left[\frac{x}{\lambda_0 z_2}+2\frac{x_0}{\lambda_0 z_1},\frac{y}{\lambda_0 z_2}\right]\right.$$
$$\left.+\frac{1}{2}\delta\left[\frac{x}{\lambda_0 z_2}-2\frac{x_0}{\lambda_0 z_1},\frac{y}{\lambda_0 z_2}\right]\right)$$
$$=\frac{(\lambda_0 z_2)^2}{2^\gamma}\left(\delta[x,y]-\left(\frac{m}{4}\gamma\right)\left(\delta\left[x+x_0\left(\frac{z_2}{z_1}\right),y\right]+\delta\left[x-x_0\left(\frac{z_2}{z_1}\right)\right]\right)\right)$$
$$+\frac{(\lambda_0 z_2)^2}{2^\gamma}\left(\frac{m^2(\gamma^2-\gamma)}{4}\right)\left(\delta[x,y]+\frac{1}{2}\delta\left[x+x_0\left(2\frac{z_2}{z_1}\right),y\right]\right.$$
$$\left.+\frac{1}{2}\delta\left[x-x_0\left(2\frac{z_2}{z_1}\right),y\right]\right)$$
$$\Longrightarrow T\left[\frac{x}{\lambda_0 z_2},\frac{y}{\lambda_0 z_2}\right] \propto \left(\frac{1}{2^\gamma}+m^2\frac{(\gamma^2-\gamma)}{4\cdot 2^\gamma}\right)\delta[x,y]$$
$$-\left(\frac{m}{4\cdot 2^\gamma}\gamma\right)(\delta[x-M_T x_0,y]+\delta[x+M_T x_0,y])$$
$$+\left(m^2\frac{(\gamma^2-\gamma)}{4\cdot 2^\gamma}\right)(\delta[x-2\cdot M_T x_0,y]+\delta[x+2\cdot M_T x_0,y]) \quad (23.16)$$

The resulting "image" irradiance is proportional to the squared magnitude of this amplitude, which evaluates the squares of the weights applied to each Dirac delta function:

$$\left|T\left[\frac{x}{\lambda_0 z_2},\frac{y}{\lambda_0 z_2}\right]\right|^2 \propto \left(\frac{1}{2}+m^2\frac{(\gamma^2-\gamma)}{4}\right)^2\delta[x,y]$$
$$-\left(\frac{m}{4}\gamma\right)^2(\delta[x-M_T x_0,y]+\delta[x+M_T x_0,y])$$
$$+\left(m^2\frac{(\gamma^2-\gamma)}{8}\right)^2(\delta[x-2M_T x_0,y]+\delta[x+2M_T x_0,y]) \quad (23.17)$$

In words, the approximate image reconstructed from this nonlinear recording is composed of five Dirac delta functions located at $x=0$, $x=\pm x_0/\lambda_0 z_2$, and $x=\pm 2x_0/\lambda_0 z_2$ instead of three located at $x=0$

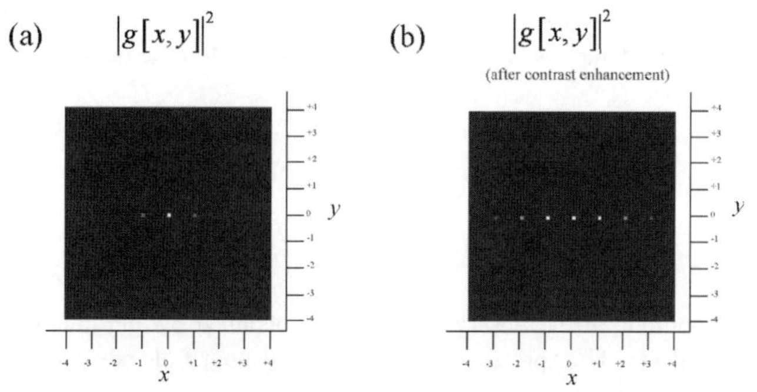

Figure 23.2 Effect of nonlinear processing on reconstruction of Fraunhofer hologram: (a) output of same two-point object used in Figure 23.1 after nonlinear processing with $\gamma = 0.95$ – the reconstructed images due to the nonlinearity are still rather faint; (b) image after contrast enhancement to emphasize reconstructions due to nonlinearity.

and $x = \pm x_0/\lambda_0 z_2$. If additional terms in the infinite series in Equation (23.14) are included in this analysis, each adds a pair of Dirac delta functions separated from the origin by an integer multiple of $M_T x_0$, as well as tweaking the scale factors applied to the lower-order terms.

For example, the reconstructed images for two sources at the same locations with $\alpha_0 = 0.8$ ($\Longrightarrow m = 0.976$) and $\gamma = 0.95$ are shown in Figure 23.2; the relative "brightnesses" of the reconstructed images are $I[0] \cong 1$ for the zero-order term, $I[\pm 1] \cong 0.188$ for the first-order terms, and $I[\pm 2] \cong 1.7 \times 10^{-3}$ for the second-order terms. Contrast these values with those for linear processing: $I[0] = 1$, $I[\pm 1] \cong 0.39$, and $I[\pm 2] = 0$, which leads to the interpretation that the nonlinearity "moved" light from the first order to the second.

We will revisit the effect of nonlinear processing on a hologram of an object consisting of multiple points.

23.1.2 Multiple Object Points

Successful reconstruction of images of single-point objects is hardly worth writing home about (though it may merit an instant message). To extend this discussion to interesting objects with multiple points, we start small by adding only one. At first, the two object sources are assumed to be located in the same plane as the reference point: at $[x_1, y_1]$ with amplitude α_1 and at $[x_2, y_2]$ with amplitude α_2. For simplicity, we assume that all sources are "in phase", so that $\phi_1 = \phi_2 = 0$, but extension to the more general case is straightforward. The source function in the plane $z = 0$ is:

$$f[x, y; 0] = \delta[x, y] + \alpha_1 \delta[x - x_1, y - y_1] + \alpha_2 \delta[x - x_2, y - y_2] \tag{23.18}$$

The computation of the reconstructed image in such a case is somewhat more tedious, though not difficult. The amplitude pattern of the Fraunhofer hologram is proportional to the Fourier transform evaluated at appropriately scaled positions:

$$\begin{aligned} s[x, y; z_1] &\propto 1\left[\frac{x}{\lambda_0 z_1}, \frac{y}{\lambda_0 z_1}\right] + \alpha_1\, e^{-2\pi i (x x_1/\lambda_0 z_1 + y y_1/\lambda_0 z_1)} + \alpha_2\, e^{-2\pi i (x x_2/\lambda_0 z_1 + y y_2/\lambda_0 z_1)} \\ &\equiv 1 + \alpha_1\, e^{-2\pi i (\xi_1 x + \eta_1 y)} + \alpha_2\, e^{-2\pi i (\xi_2 x + \eta_2 y)} \end{aligned} \tag{23.19}$$

where $\xi_n = x_n/\lambda_0 z_1$ and $\eta_n = y_n/\lambda_0 z_1$ for $n = 1, 2$. The irradiance includes a constant term and *three* sinusoidal gratings, one for each pair of the three sources:

$$\begin{aligned}|s[x, y; z_1]|^2 &= (1 + \alpha_1\, e^{-2\pi i(\xi_1 x+\eta_1 y)} + \alpha_2\, e^{-2\pi i(\xi_2 x+\eta_2 y)}) \\ &\quad \cdot (1 + \alpha_1\, e^{-2\pi i(\xi_1 x+\eta_1 y)} + \alpha_2\, e^{-2\pi i(\xi_2 x+\eta_2 y)})^* \\ &= (1 + \alpha_1^2 + \alpha_2^2) + 2\alpha_1 \cos[2\pi(\xi_1 x + \eta_1 y)] + 2\alpha_2 \cos[2\pi(\xi_2 x + \eta_2 y)] \\ &\quad + 2\alpha_1\, \alpha_2 \cos[2\pi([\xi_1 - \xi_2]x + [\eta_1 - \eta_2]y)] \end{aligned} \tag{23.20}$$

The transmittance of the ideally processed negative image includes three cosine functions with the same modulation but with phases of π radians, plus the additive constant bias of $\frac{1}{2}$:

$$\begin{aligned} t[x, y; z_1] &= \frac{1}{2} \cdot 1[x, y] - \left(\frac{\alpha_1}{1 + \alpha_1^2 + \alpha_2^2}\right) \cos[2\pi(\xi_1 x + \eta_1 y)] \\ &\quad - \left(\frac{\alpha_2}{1 + \alpha_1^2 + \alpha_2^2}\right) \cos[2\pi(\xi_2 x + \eta_2 y)] \\ &\quad - \left(\frac{\alpha_1\, \alpha_2}{1 + \alpha_1^2 + \alpha_2^2}\right) \cos[2\pi([\xi_1 - \xi_2]x + [\eta_1 - \eta_2]y)] \end{aligned} \tag{23.21}$$

The spectrum of this function is composed of seven Dirac delta functions: one at the origin and three weighted symmetric pairs:

$$\begin{aligned} &T[\xi, \eta; z_2] \\ &= \frac{1}{2}\delta[\xi, \eta] - \frac{1}{2}\left(\frac{\alpha_1}{1 + \alpha_1^2 + \alpha_2^2}\right)(\delta[\xi + \xi_1, \eta + \eta_1] + \delta[\xi - \xi_1, \eta - \eta_1]) \\ &\quad - \frac{1}{2}\left(\frac{\alpha_2}{1 + \alpha_1^2 + \alpha_2^2}\right)(\delta[\xi + \xi_1, \eta + \eta_1] + \delta[\xi - \xi_2, \eta - \eta_2]) \\ &\quad - \frac{1}{2}\left(\frac{\alpha_1 \alpha_2}{1 + \alpha_1^2 + \alpha_2^2}\right)(\delta[\xi + (\xi_1 - \xi_2), \eta + (\eta_1 - \eta_2)] + \delta[\xi - (\xi_1 - \xi_2), \eta - (\eta_1 - \eta_2)]) \end{aligned} \tag{23.22}$$

The reconstructed image is a replica of the squared magnitude of $T[\xi, \eta; z_2]$ after mapping the coordinates back to the space domain via $\xi = x/\lambda_0 z_2$ and $\eta = y/\lambda_0 z_2$:

$$\begin{aligned} &I[x, y] \propto |g[x, y; z_1, z_2]|^2 \\ &= \delta[x, y] + \left(\frac{\alpha_1}{1 + \alpha_1^2 + \alpha_2^2}\right)^2 (\delta[x - M_T x_1, y - M_T y_1] + \delta[x + M_T x_1, y + M_T y_1]) \\ &\quad + \left(\frac{\alpha_2}{1 + \alpha_1^2 + \alpha_2^2}\right)^2 (\delta[x - M_T x_2, y - M_T y_2] + \delta[x + M_T x_2, y + M_T y_2]) \\ &\quad + \left(\frac{\alpha_1 \cdot \alpha_2}{1 + \alpha_1^2 + \alpha_2^2}\right)^2 \cdot \delta[x - M_T(x_1 - x_2), y - M_T(y_1 - y_2)] \\ &\quad + \left(\frac{\alpha_1 \cdot \alpha_2}{1 + \alpha_1^2 + \alpha_2^2}\right)^2 \cdot \delta[x + M_T(x_1 - x_2), y + M_T(y_1 - y_2)] \end{aligned} \tag{23.23}$$

where the transverse magnification has been substituted, $M_T = -z_2/z_1$. Again, the squared-magnitude operation only affects the weighting of the disjoint Dirac delta functions. In words, the "ideal" reconstruction of three point sources (assuming linear processing) is a set of seven Dirac delta functions in the Fraunhofer domain: the reconstruction of the "reference" source at the origin, the symmetric pair of reconstructions of the first object source located at $[\pm(z_2/z_1)x_1, \pm(z_2/z_1)y_1]$ with amplitude $\alpha_1/(1+\alpha_1^2+\alpha_2^2)^2$, the corresponding symmetric pair of reconstructions of the second object source, and a symmetric pair of reconstructions located at $[\pm M_T(x_1-x_2), \pm M_T(y_1-y_2)]$ with amplitude $((\alpha_1\cdot\alpha_2)/(1+\alpha_1^2+\alpha_2^2))^2$.

It is instructive to consider the reason for the last pair of Dirac delta functions in Equation (23.23). They are not reconstructions of the original objects, but rather are generated by the cross-term in the measured irradiance at the hologram plane. They also may be interpreted as reconstructions created by using one object point as the reference source for the other object point; these are "self-reconstructions" that did not exist in the original object. Note that the weight factor applied to the last term is proportional to $(\alpha_1\alpha_2)^2$, whereas the amplitudes of the "true" images are proportional to α_1^2 and α_2^2, respectively. This means that the relative "brightness" of the false reconstructions is very small if both α_1 and α_2 are small compared to the unit weight of the reference source. In other words, the visibility of the cross-term is minimized if the reference source (located at the origin in our example) is much brighter than any of the object points in multiple-source holography. In actual use, the reference beam is often adjusted to be at least five times brighter than the light from the object.

An object $f[x, y]$ consisting of three sources, a unit-amplitude point at the origin and two additional sources with $\alpha_1=\alpha_2=1$ so that all points are equally bright, and the distribution $|g[x, y]|^2$ reconstructed from a hologram are shown in Figure 23.3. The additional pair of reconstructed Dirac delta functions due to the cross-terms are identified.

23.1.3 Fraunhofer Hologram of Extended Object

We can interpret the ideal reconstruction of the Fraunhofer hologram after linear processing as the cascade of four operations on the amplitude: (1) Fourier transform, (2) squared magnitude, (3) Fourier transform, and (4) squared magnitude:

$$\begin{aligned} I[x, y] &\propto |\mathcal{F}_2\{|\mathcal{F}_2\{f[x, y]\}|_{\xi\to x/\lambda_0 z_1, \eta\to x/\lambda_0 z_1}|^2\}|_{\xi\to x/\lambda_0 z_2, \eta\to x/\lambda_0 z_2}|^2 \\ &= \left|\mathcal{F}_2\left\{\left|F\left[\frac{x}{\lambda_0 z_1}, \frac{x}{\lambda_0 z_1}\right]\right|^2\right\}\right|_{\xi\to x/\lambda_0 z_2, \eta\to x/\lambda_0 z_2}\Bigg|^2 \end{aligned} \tag{23.24}$$

We can use the theorems of the Fourier transform from Chapter 9 to recast the last expression into a simpler form:

$$I[x, y] \propto |f[u, v] \bigstar f[u, v]|_{u\to -(z_1/z_2)x,\ v\to -(z_1/z_2)y}|^2 \tag{23.25}$$

In words, the hologram reconstructs the squared magnitude of a scaled replica of the autocorrelation of the source function, including the reference.

Now consider the result if the input function consists of a real-valued extended object $r[x, y]$ centered at $[x_0, y_0]$:

$$f[x, y] = \delta[x, y] + \alpha_0 \cdot r[x-x_0, y-y_0] \tag{23.26}$$

It may already be evident that the squared magnitude of $r[x, y]$ is reconstructed, so we choose r to be the square root of the desired output. The autocorrelation of the input function is:

$$\begin{aligned} f[x, y] \bigstar f[x, y] &= \delta[x, y] \bigstar \delta[x, y] + \alpha_0^2(r[x-x_0, y-y_0] \bigstar r[x-x_0, y-y_0]) \\ &\quad + \alpha_0 r[x-x_0, y-y_0] \bigstar \delta[x, y] + \delta[x, y] \bigstar \alpha_0 r[x-x_0, y-y_0] \\ &= (\delta[x, y] + \alpha_0^2(r[x, y] \bigstar r[x, y])) + \alpha_0 r[x-x_0, y-y_0] \\ &\quad + \alpha_0 r^*[-(x+x_0), -(y+y_0)] \end{aligned}$$

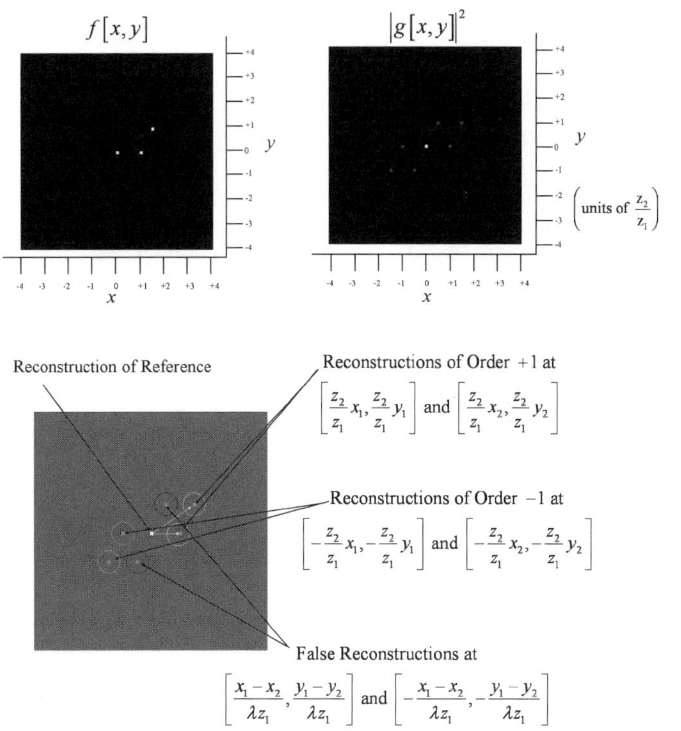

Figure 23.3 Reconstruction of $f[x, y]$ consisting of two source points and the reference. In this example, all three points have the same weighting. The reconstructed image $|g[x, y]|^2$ contains seven points: the reconstruction of the reference, pairs of reconstructions of the two object sources, and a pair of spurious reconstructions.

$$= (\delta[x, y] + \alpha_0^2(r[x, y] \bigstar r[x, y])) + \alpha_0 r[x, y] * \delta[x - x_0, y - y_0] + \alpha_0 r^*[-x, -y] * \delta[x + x_0, y + y_0] \qquad (23.27)$$

where the easily determined fact that:

$$r[x - x_0, y - y_0] \bigstar r[x - x_0, y - y_0] = r[x, y] \bigstar r[x, y] \qquad (23.28)$$

has been used.

The resulting irradiance is easy to calculate if there is no overlap of the autocorrelation and crosscorrelation terms. Because the autocorrelation has twice the support of the object, the criterion of no overlap is guaranteed if the distance between the origin and the offset location $[x_0, y_0]$ is larger than the support of the object scaled by $\frac{3}{2}$. The calculation of the squared magnitude also ensures that the complex conjugate applied to the object amplitude r in the fourth term in Equation (23.27) has no effect on the observed result:

$$I[x, y] \propto |f[x, y] \bigstar f[x, y]|^2$$
$$= \delta[x, y] + \alpha_0^4(r[x, y] \bigstar r[x, y])^2 + (\alpha_0^2|r[x, y]|^2 * \delta[x - x_0, y - y_0]) + (\alpha_0^2|r[-x, -y]|^2 * \delta[x + x_0, y + y_0]) \qquad (23.29)$$

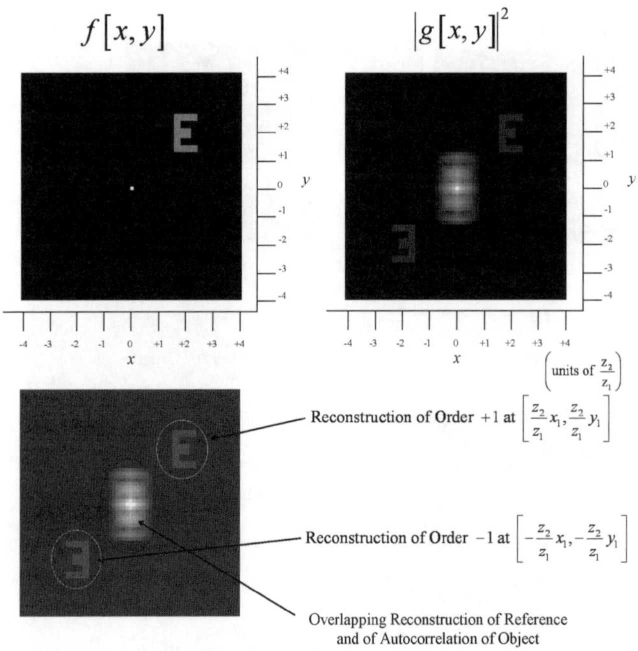

Figure 23.4 Fraunhofer hologram of extended object: the input function $f[x, y] = \delta[x, y] + \alpha_0 r[x - x_0, y - y_0]$, where $r[x, y]$ is an upper-case "E". The output irradiance $|g[x, y]|^2$ includes the autocorrelations of the reference and object plus two replicas of the squared magnitude of the object located at $\pm[x_0, y_0]$ and scaled by α_0^2. The parts of the reconstruction are labeled in the bottom image.

In words, the reconstructed irradiance is the sum of the undiffracted light from the reference (centered at the origin), a scaled replica of the autocorrelation of the object (also centered at the origin), a reversed replica of the squared magnitude of the object centered about $[-x_0, -y_0]$ (the "conjugate image"), and a replica of the object centered at its original location $[+x_0, +y_0]$. An example is shown in Figure 23.4.

23.1.4 Nonlinear Fraunhofer Hologram of Extended Object

We now briefly consider the effect of nonlinear processing on the reconstructed hologram. The result derived for the simple case of two point sources provides some guidance toward the result. From this we can surmise that the nonlinear processing will generate additional replicas of each object point at multiples of their distances from the reference. If we model the power-law nonlinearity in Equation (9.190), then the processed reconstruction result in Equation (23.24) becomes:

$$I[x, y] \propto \mathcal{F}_2\left\{\left(\left|F\left[\frac{x}{\lambda_0 z_1}, \frac{y}{\lambda_0 z_1}\right]\right|^2\right)^{\gamma}\right\} \tag{23.30}$$

By analogy with the discussion leading to Equation (23.14), we assume that $|R[x/\lambda_0 z_1, y/\lambda_0 z_1]|^2 < 1$ and expand the expression for the spectrum using the binomial theorem:

$$\begin{aligned} I[x, y] &\propto \mathcal{F}_2\{1[\xi, \eta]\} \\ &\quad + \gamma \cdot \mathcal{F}_2\{(|R[\xi, \eta]|^2 + R[\xi, \eta]\, e^{-2\pi i(\xi x_0 + \eta y_0)} + R^*[\xi, \eta]\, e^{+2\pi i(\xi x_0 + \eta y_0)})\} \end{aligned}$$

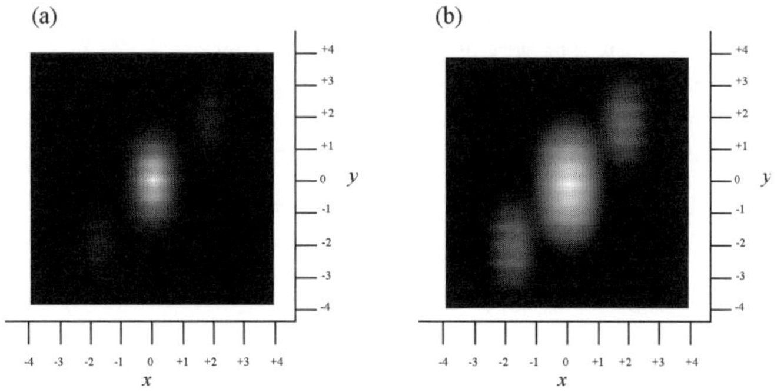

Figure 23.5 Effect of nonlinear processing on reconstruction of the same object in Figure 23.4: (a) reconstruction after processing with $\gamma = 2$; (b) after contrast enhancement, showing the distortion of the reconstructions of the object due to the additional terms centered about $\pm[x_0, y_0]$.

$$
\begin{aligned}
&+ \frac{\gamma(\gamma-1)}{2} \cdot \mathcal{F}_2\{(|R[\xi,\eta]|^2 + R[\xi,\eta]\, e^{-2\pi i(\xi x_0+\eta y_0)} + R^*[\xi,\eta]\, e^{+2\pi i(\xi x_0+\eta y_0)})^2\} \\
&+ \cdots \qquad (23.31)
\end{aligned}
$$

We now apply the linearity property of the Fourier transform and the filter theorem to see that the output of the nonlinear processing produces additional terms, some located at the origin and others at integer multiples of $\pm[x_0, y_0]$:

$$
\begin{aligned}
I[x,y] \propto\ & \delta[x,y] + \gamma^2\alpha_0^2 \cdot r[x,y] \bigstar r[x,y] \\
&+ \frac{\gamma(\gamma-1)}{2}\alpha_0^4 \cdot (r[x,y] \bigstar r[x,y] \bigstar r[x,y] \bigstar r[x,y]) \\
&+ (\gamma\alpha_0 \cdot r[x,y] + \gamma(\gamma-1)\alpha_0^3 \cdot (r[x,y] \bigstar r[x,y]) * r[x,y]) * \delta[x-x_0, y-y_0] \\
&+ (\gamma\alpha_0 \cdot r^*[-x,-y] + \gamma(\gamma-1)\alpha_0^3 \cdot (r[x,y] \bigstar r[x,y]) \\
&* r^*[-x,-y]) * \delta[x+x_0, y+y_0] \\
&+ \frac{\gamma(\gamma-1)}{2}\alpha_0^2 \cdot (r[x,y] * r[x,y]) * \delta[x-2x_0, y-2y_0] \\
&+ \frac{\gamma(\gamma-1)}{2}\alpha_0^2 \cdot (r^*[-x,-y] * r^*[-x,-y]) * \delta[x+2x_0, y+2y_0] + \cdots \qquad (23.32)
\end{aligned}
$$

If we just consider the first-order reconstruction centered about $[x_0, y_0]$, we see an additional term consisting of the convolution of $r[x, y]$ with its autocorrelation scaled by $\gamma(\gamma-1)\alpha_0^3$. The additional term due to the nonlinearity overlaps and "blurs" the ideal reconstruction, as shown in the example in Figure 23.5.

23.2 Holography in Fresnel Diffraction Region

The mathematical tools just developed may also be used to model holography in the Fresnel diffraction region. The analysis is (somewhat) more complicated, but also much more interesting as it allows 3-D objects to be recorded and reconstructed. In the Fraunhofer case just discussed, all source points were assumed to lie in the same plane so that the propagation distances z_1 were identical. In the Fresnel region, the different propagation distances along the z-axis affect the phases, and thus influence the recorded hologram and the resulting reconstructed image. The tools of the chirp Fourier transform that were developed in Chapter 17 will be very useful in this discussion.

Though we will certainly examine the effect of source displacements along z on Fresnel holograms, we first consider the same simple case of two sources located in the same plane that was the basis of the discussion of Fraunhofer holography.

23.2.1 Object and Reference Sources in Same Plane

The object shown in Figure 23.6 includes the same pair of Dirac delta functions used in the Fraunhofer case in Equation (23.2):

$$f[x, y, z] = \delta[x, y, z] + (\alpha_1 e^{+i\phi_1})\delta[x - x_0, y, z] \tag{23.33}$$

The propagation of light from monochromatic sources located in the same plane to an observation plane in the Fresnel diffraction region may be evaluated as a convolution with the impulse response for distance z_1 in Equation (21.57):

$$h[x, y; z_1] = \frac{1}{i\lambda_0 z_1} \exp\left[+2\pi i \frac{z_1}{\lambda_0}\right] \cdot \exp\left[+i\pi \frac{x^2 + y^2}{\lambda_0 z_1}\right] \tag{23.34}$$

The diffracted amplitude observed at the distance z_1 is:

$$\begin{aligned} s[x, y; z_1] &= (\delta[x, y, z] + \alpha_1 e^{+i\phi_1}\, \delta[x - x_0, y, z]) * h[x, y; z_1] \\ &= (\delta[x, y] + \alpha_1 e^{+i\phi_1}\, \delta[x - x_0, y]) * \left(\frac{1}{i\lambda_0 z_1}\, e^{+2\pi i z_1/\lambda_0}\, e^{+i\pi(x^2+y^2)/\lambda_0 z_1}\right) \\ &= K_0(e^{+i\pi x^2/\lambda_0 z_1} + \alpha_1\, e^{+i\phi_1}\, e^{+i\pi(x-x_0)^2/\lambda_0 z_1})\, e^{+i\pi y^2/\lambda_0 z_1} \end{aligned} \tag{23.35}$$

where the complex constant K_0 was defined in Equation (21.100) to include the constant scale factors for amplitude and phase. Because the light is assumed to be coherent, the irradiance measured at the photographic emulsion is proportional to the squared magnitude of this amplitude:

$$\begin{aligned} |s[x, y; z_1]|^2 &= \frac{1}{(\lambda_0 z_1)^2}\left((1 + \alpha_1^2) + 2\alpha_1 \cos\left[\frac{\pi}{\lambda_0 z_1}(x^2 - (x - x_0)^2) - \phi_1\right]\right) 1[y] \\ &\propto \left(1 + \frac{2\alpha_1}{1 + \alpha_1^2} \cos\left[2\pi\xi_0\left(x - \frac{x_0}{2}\right) - \phi_1\right]\right) 1[y] \end{aligned} \tag{23.36}$$

where $\xi_0 \equiv x_0/\lambda_0 z_1$ is the spatial frequency of the recorded sinusoidal grating. Note that the observed irradiances at the observation plane in the Fresnel case in Equation (23.36) and the Fraunhofer case in Equation (23.4) are very similar; both patterns vary sinusoidally with the same spatial frequency $\xi_0 = x_0/\lambda_0 z_1$ and the same modulation $m = 2\alpha_1/(1 + \alpha_1^2)$, but the Fresnel grating is translated from the origin by $+x_0/2$, which means that the irradiance pattern is symmetric about the point halfway between the two sources (Figure 23.7).

This irradiance pattern is again recorded photographically and chemically processed to produce a transparency whose transmittance function is proportional to the normalized complement of the incident irradiance. We assume at first that the photographic emulsion has infinite support, eliminating any "pupil function" $p[x, y]$ from the grating; the realistic restriction to compact support is easy to add and will be considered shortly. Though the real and imaginary parts of the Fresnel hologram differ greatly from that in the Fraunhofer case in Equation (23.7), the recorded transmittance is identical except for the translation of the center of symmetry in Equation (23.36), as shown in Figure 23.7:

$$t[x, y] = \frac{1}{2}\left(1 - \frac{2\alpha_1}{1 + \alpha_1^2} \cos\left[2\pi\xi_0\left(x - \frac{x_0}{2}\right) - \phi_1\right]\right) 1[y] \tag{23.37}$$

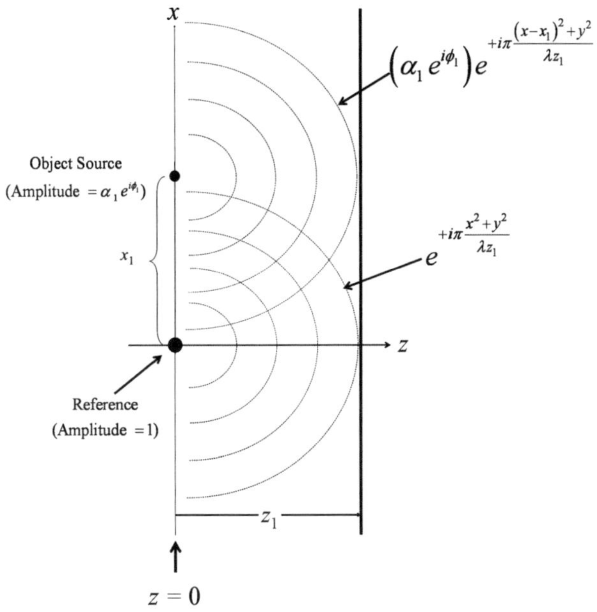

Figure 23.6 Schematic of Fresnel holography for object and reference points in the same plane $z = 0$. The irradiance in the plane $z = z_1$ is recorded.

The holographic image is again "reconstructed" by illuminating the transparency with light from the on-axis reference source and evaluating the squared magnitude. Because the reference source is at the origin, the amplitude of its illumination at the hologram plane is identical to the impulse response $h[x, y; z_1]$:

$$h[x, y; z_1] \cdot t[x, y] = \left(\frac{1}{i\lambda_0 z_1} e^{+2\pi i z_1/\lambda_0} e^{+i\pi(x^2+y^2)/\lambda_0 z_1}\right) \cdot \frac{1}{2} \cdot \left(1 - \frac{2\alpha_1}{1+\alpha_1^2} \cos\left[2\pi\xi_0\left(x - \frac{x_0}{2}\right) - \phi_1\right]\right) 1[y] \tag{23.38}$$

As before, we must propagate this modulated amplitude to the observation plane. In the Fraunhofer case, the light propagated "forward" to an observation plane located a large distance z_2 from the hologram. We have more freedom in the Fresnel case; the only constraint is that the propagation distance z_2 be sufficiently large to allow spherical waves to be approximated by quadratic-phase terms. For $z_2 > 0$, the propagation is "forward" from the hologram to the observation plane, creating a real image analogous to that created by Fraunhofer holography or by a lens. The second alternative is to propagate "backward" over the distance $z_2 < 0$ (i.e., *toward* the illuminating source), which produces a "virtual" reconstruction "behind" the hologram. We denote the option to propagate the light in either direction from the hologram to a suitable observation plane by writing the propagation distance $z_2 = |z_2| \cdot SGN[z_2]$ in the equation for the amplitude pattern:

$$g[x, y; z_1, z_2] \propto \left(\frac{1}{i\lambda_0 z_1} e^{+2\pi i z_1/\lambda_0} e^{+i\pi(x^2+y^2)/\lambda_0 z_1} \cdot t[x, y]\right) * \left(\frac{1}{i\lambda_0 z_2} e^{+2\pi i z_1/\lambda_0} e^{+i\pi(x^2+y^2)/\lambda_0 |z_2| SGN[z_2]}\right)$$

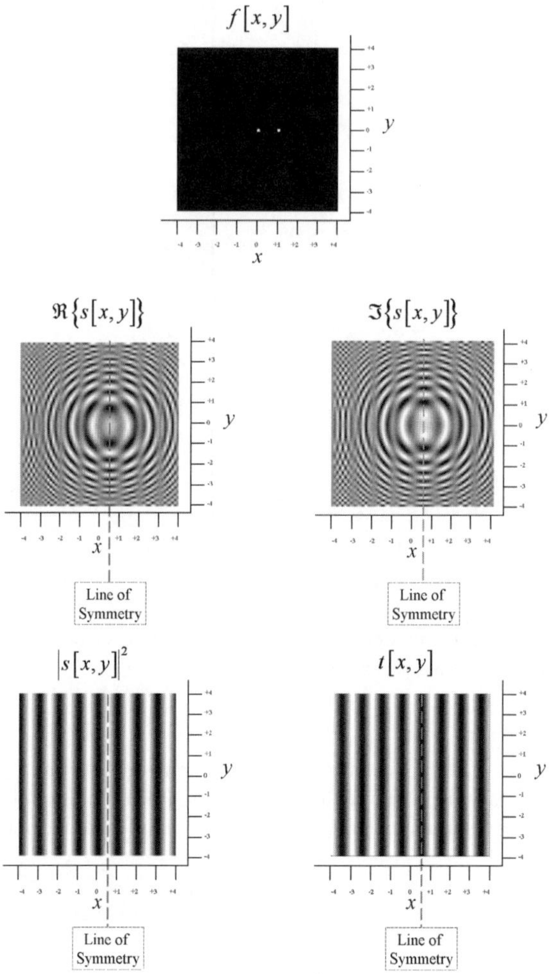

Figure 23.7 Schematic of reconstruction of virtual images. The hologram $t[x, y]$ is illuminated by the original on-axis "reference" source. Two images of the second "object" source are generated by diffraction in the hologram. The intensity of the reconstructions of the object sources is $\alpha^2/(1 + a^2)^2$, which evaluates to $\frac{1}{4}$ if $a = 1$.

$$= \left(-\frac{1}{\lambda_0^2 z_1 z_2}\, e^{+2\pi i(z_1+z_2)/\lambda_0}\right) \cdot \left([e^{+i\pi(x^2+y^2)/\lambda_0 z_1} \cdot t[x, y]] * e^{+i\pi[(x^2+y^2)/\lambda_0|z_2|]SGN[z_2]}\right) \tag{23.39}$$

where we have used the obvious relation $SGN[z_2] = (SGN[z_2])^{-1}$ for $z_2 \neq 0$. Because our primary interest is the spatial form of the amplitude at the observation plane, we ignore (yet again!) the leading constant factors in the expression for the output amplitude:

$$g[x, y; z_1, z_2] \propto [t[x, y] \cdot e^{+i\pi(x^2+y^2)/\lambda_0 z_1}] * e^{+i\pi[(x^2+y^2)/\lambda_0|z_2|]SGN[z_2]} \tag{23.40}$$

Of course, the measurable quantity is the irradiance $|g[x, y; z_1, z_2]|^2$. The apparently complicated form of $g[x, y]$ in Equation (23.40) suggests evaluation in the frequency domain, which is certainly feasible even if seemingly difficult and tedious. However, before plunging blindly down this path, instead take note of the resemblance of Equation (23.40) to the M–C–M chirp Fourier transform algorithm that was considered in detail in Chapter 17. The 2-D generalization of the M–C–M transform with inverted chirp rates in Equation (17.80) may be rewritten in the form of Equation (23.40) by simple cross-multiplication:

$$[(t[x, y] \cdot e^{+i\pi(x^2+y^2)/\lambda_0 z_1}) * e^{-i\pi(x^2+y^2)/\lambda_0 z_1}] \cdot e^{+i\pi(x^2+y^2)/\lambda_0 z_1} = T\left[-\frac{x}{\lambda_0 z_1}, -\frac{y}{\lambda_0 z_1}\right]$$

$$\Longrightarrow (t[x, y] \cdot e^{+i\pi(x^2+y^2)/\lambda_0 z_1}) * e^{-i\pi(x^2+y^2)/\lambda_0 z_1} = T\left[-\frac{x}{\lambda_0 z_1}, -\frac{y}{\lambda_0 z_1}\right] \cdot e^{-i\pi(x^2+y^2)/\lambda_0 z_1} \tag{23.41}$$

The quadratic-phase factor on the right vanishes when the squared magnitude is evaluated:

$$|(t[x, y] \cdot e^{+i\pi(x^2+y^2)/\lambda_0 z_1}) * e^{-i\pi(x^2+y^2)/\lambda_0 z_1}|^2 = \left|T\left[-\frac{x}{\lambda_0 z_1}, -\frac{y}{\lambda_0 z_1}\right]\right|^2 \tag{23.42}$$

Note that this simple expression applies *only* if both quadratic-phase factors have the same chirp rate $\sqrt{\lambda_0 z_1}$; if not, then the expression is much more complicated. In holography, equal chirp rates imply that $z_2 = -z_1$ so that $SGN[z_2] = -1$. In words, the observation plane where Equation (23.42) is valid is the original source plane; the reconstructed amplitude distribution appears as a virtual image "behind" the hologram. The right-hand side of Equation (23.42) is the squared magnitude of a scaled replica of the Fourier transform of the transmittance of the hologram processed as a negative image:

$$\begin{aligned} T[\xi, \eta] &\propto \delta[\xi, \eta] - \frac{\alpha_1}{1+\alpha_1^2}(\delta[\xi+\xi_0, \eta]\, e^{+i\phi_1} + \delta[\xi-\xi_0, \eta]\, e^{-i\phi_1})\, e^{-2\pi i \xi(x_0/2)} \\ &= \delta[\xi, \eta] - \frac{\alpha_1}{1+\alpha_1^2}(\delta[\xi+\xi_0, \eta]\, e^{+i\phi_1} + \delta[\xi-\xi_0, \eta]\, e^{-i\phi_1})\, e^{-i\pi\xi x_0} \\ &= \delta[\xi, \eta] - \frac{\alpha_1}{1+\alpha_1^2}(\delta[\xi+\xi_0, \eta]\, e^{-i(\pi\xi_0 x_0 - \phi_1)} + \delta[\xi-\xi_0, \eta]\, e^{-i(\pi\xi_0 x_0 + \phi_1)}) \end{aligned} \tag{23.43}$$

The resulting irradiance consists of three 2-D Dirac delta functions with scale factors equal to the squared magnitudes of the weights in Equation (23.43):

$$\begin{aligned} |g[x, y; z_1, -z_1]|^2 &= \left|T\left[-\frac{x}{\lambda_0 z_1}, -\frac{y}{\lambda_0 z_1}\right]\right|^2 \\ &\propto (1)^2 \cdot \delta[x, y] + \left(\frac{\alpha_1}{1+\alpha_1^2}\right)^2 (\delta[x+x_0, y] + \delta[x-x_0, y]) \end{aligned} \tag{23.44}$$

If the amplitudes of the two point sources are equal ($\Longrightarrow \alpha_1 = 1$), then the "reconstructed" irradiance of the virtual image is:

$$|g[x, y; z_1, -z_1]|^2 \propto \delta[x, y] + \frac{1}{4}(\delta[x+x_0, y] + \delta[x-x_0, y]) \tag{23.45}$$

The reconstruction process is illustrated in Figure 23.8.

In words, our analysis of the chirp Fourier transform led to a fairly simple expression for the "image" generated by the illuminated hologram observed back at the source plane (i.e., "behind" the hologram). This virtual image includes three point sources: one at the location of the original on-axis source and two with smaller amplitudes at $x = \pm x_0$.

We will locate the reconstructed real images after a brief detour to consider the effect of the size of the hologram on the image.

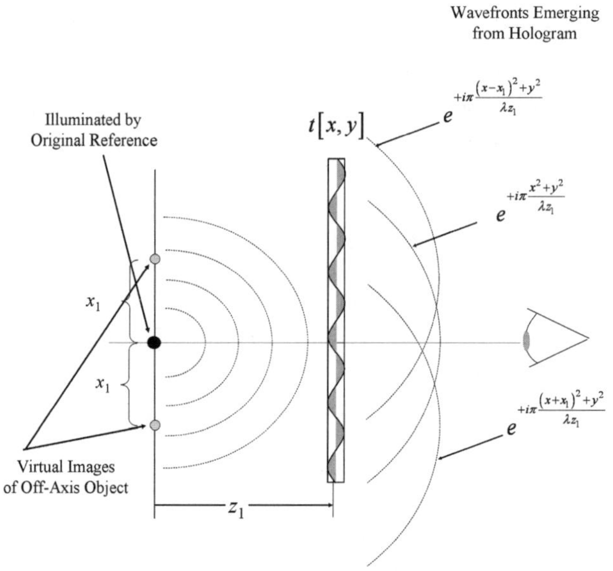

Figure 23.8 Schematic of reconstruction of virtual images. The hologram $t[x, y]$ is illuminated by the original on-axis "reference" source. Two images of the "object" point source are generated by diffraction in the sinusoidal hologram. The intensity of the reconstructions of the object sources is $\alpha^2/(1+a^2)^2$, which evaluates to $\frac{1}{4}$ for $\alpha = 1$.

23.2.2 Reconstruction of Virtual Image from Hologram with Compact Support

A realistic hologram cannot have infinite support, so we need to truncate the transparency function $t[x, y]$ by a "pupil function" $p[x, y]$ (typically a rectangle) that models the size and shape of the photographic emulsion. The effect on the reconstructed image is easily generalized:

$$\begin{aligned} g[x, y; z_1, -z_1] &= ((t[x, y] \cdot p[x, y]) \cdot e^{+i\pi(x^2+y^2)/\lambda_0 z_1}) * e^{-i\pi(x^2+y^2)/\lambda_0 z_1} \\ &= \left(T\left[-\frac{x}{\lambda_0 z_1}, -\frac{y}{\lambda_0 z_1}\right] * P\left[-\frac{x}{\lambda_0 z_1}, -\frac{y}{\lambda_0 z_1}\right]\right) \cdot e^{-i\pi(x^2+y^2)/\lambda_0 z_1} \end{aligned} \quad (23.46)$$

If $p[x, y] = RECT[x/b_0, y/d_0]$ and if $z_2 = -z_1$, then the Dirac delta functions are replaced by scaled replicas of $SINC[b_0 x/\lambda_0 z_1, d_0 y/\lambda_0 z_1]$. In a typical example, z_1 is of the order of 1 meter, b_0 and d_0 are of the order of centimeters, and (of course) λ_0 is of the order of 0.5 μm. In such a case, the *SINC* functions are quite narrow so that their approximation by Dirac delta functions is realistic. If b_0, d_0, and/or z_1 are smaller, then the effect of the support of the hologram would have to be more carefully considered.

23.2.3 Reconstruction of Real Image: $z_2 > 0$

A process very similar to that used to reconstruct the virtual image may be used to generate the corresponding real image by generalizing the "usual" 2-D M–C–M chirp transform in Equation (17.70):

$$(e^{-i\pi(r/\lambda_0 z_1)^2} \cdot t[x, y]) * e^{+i\pi(r/\lambda_0 z_1)^2} = T\left[\frac{x}{\lambda_0 z_1}, \frac{y}{\lambda_0 z_1}\right] \cdot e^{+i\pi(r/\lambda_0 z_1)^2} \quad (23.47)$$

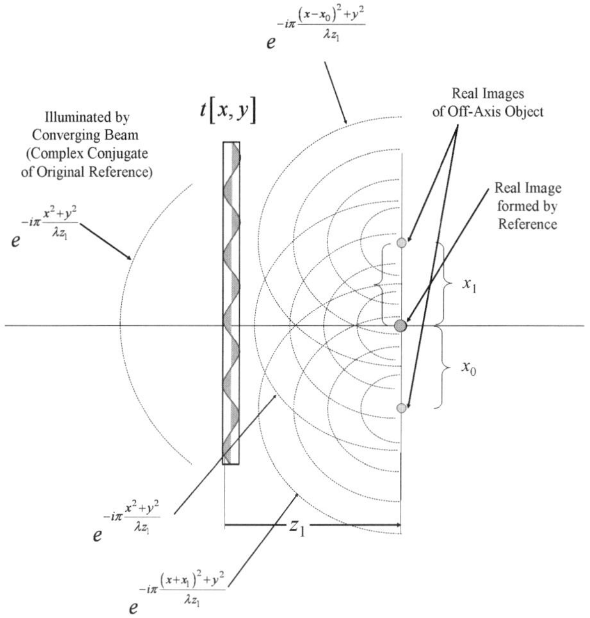

Figure 23.9 Reconstruction of real images from a Fresnel hologram. The transparency $t[x, y]$ is illuminated by the complex conjugate of the original reference. The diffraction by the hologram generates two converging object wavefronts that form real images.

The sign of the first quadratic phase (the "illuminating" beam) is negative, which physically means that the hologram is illuminated by the (so-called) "conjugate wave" to the original illumination, which converges to a point at the distance $z_2 = |z_1|$ "downstream".

Again, the irradiance is the squared magnitude of the appropriately scaled Fourier transform of the transmittance function:

$$|(e^{-i\pi(x^2+y^2)/\lambda_0 z_1} \cdot t[x, y]) * e^{+i\pi(x^2+y^2)/\lambda_0 z_1}|^2 = \left| T\left[\frac{x}{\lambda_0 z_1}, \frac{y}{\lambda_0 z_1}\right] \cdot e^{+i\pi(x^2+y^2)/\lambda_0 z_1} \right|^2$$

$$= \left| T\left[\frac{x}{\lambda_0 z_1}, \frac{y}{\lambda_0 z_1}\right] \right|^2 \tag{23.48}$$

In words, the light of the converging reference beam is diffracted by the hologram to form real images in space that may be viewed on a projection screen or recorded on a light-sensitive detector; this reconstruction is analogous to that generated by a Fraunhofer hologram when illuminated by the original reference. A schematic reconstruction of the real images is shown in Figure 23.9.

23.2.4 Object and Reference Sources in Different Planes

Now consider the much more interesting case of two monochromatic point sources located in different planes; the source distribution is now "three dimensional" as seen from the sensor (Figure 23.10). The recording emulsion is located in the plane $z = z_1$, the unit-amplitude "reference" source is again at the origin, but an object source with weight β_0 and initial phase ϕ_1 is placed at $[x_0, 0]$ at distance z_0 measured from the hologram:

$$f[x, y; z] = \delta[x, y]\delta[z] + \beta_1\, e^{+i\phi_1}\, \delta[x - x_0, y]\delta[z - (z_1 - z_0)] \tag{23.49}$$

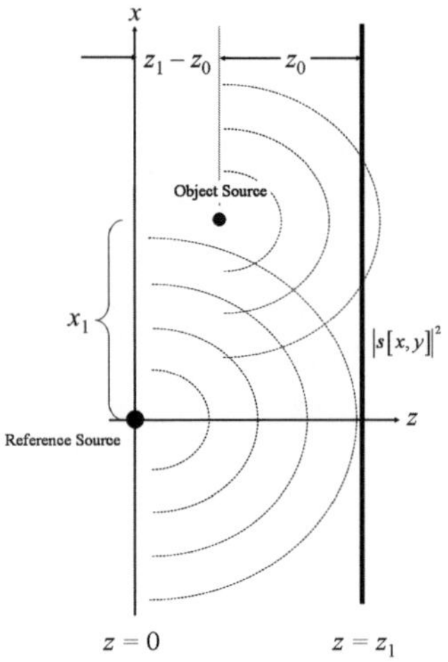

Figure 23.10 Schematic of Fresnel holography with object point in a different plane than the source point.

The distance between the object and reference source planes is $z_1 - z_0 \neq 0$. The amplitude evaluated at the plane $z_1 > 0$ is the superposition of the convolutions of the sources with different "impulse responses" due to the different propagation distances, which means that the optical propagation from the sources to the sensor is linear but not shift invariant. The amplitude at the sensor is:

$$s[x,\, y;\, z_1] = \left(\frac{1}{i\lambda_0 z_1}\; e^{+2\pi i z_1/\lambda_0}\; e^{+i\pi(x^2+y^2)/\lambda_0 z_1}\right) + \beta_1\, e^{+i\varphi_1}\left(\frac{1}{i\lambda_0 z_0}\; e^{+2\pi i z_0/\lambda_0}\; e^{+i\pi[(x-x_0)^2+y^2]/\lambda_0 z_0}\right) \tag{23.50}$$

For simplicity, we collect the scaling constants for the second source into a single amplitude factor α_1 and the constant phase factors into ϕ_1 to obtain an expression similar to that in the Fraunhofer case in Equation (23.3):

$$\beta_1 \frac{z_1}{z_0}\; e^{+i\varphi_1}\; e^{+2\pi i(z_0-z_1)/\lambda_0} \equiv \alpha_1\, e^{+i\phi_1} \tag{23.51a}$$

$$\Longrightarrow s[x,\, y;\, z_1] = \left(\frac{1}{i\lambda_0 z_1}\; e^{+2\pi i z_1/\lambda_0}\right)\left(e^{+i\pi(x^2+y^2)/\lambda_0 z_1} + (\alpha_1\, e^{+i\phi_1})\, e^{+i\pi[(x-x_0)^2+y^2]/\lambda_0 z_0}\right) \tag{23.51b}$$

Though both quadratic-phase factors in $s[x,\, y;\, z_1]$ are positive, their "chirp rates" differ.

The observed irradiance is the squared magnitude of s, which is a more complicated expression than that for the two sources in the same plane given in Equation (23.36):

$$\begin{aligned}|s[x,\, y;\, z_1]|^2 &\propto (e^{+i\pi(x^2+y^2)/\lambda_0 z_1} + \alpha_1\, e^{+i\phi_1}\, e^{+i\pi[(x-x_0)^2+y^2]/\lambda_0 z_0})\\ &\quad\cdot (e^{+i\pi(x^2+y^2)/\lambda_0 z_1} + \alpha_1\, e^{+i\phi_1}\, e^{+i\pi[(x-x_0)^2+y^2]/\lambda_0 z_0})^*\\ &= (1+\alpha_1^2)\left(1 + \frac{2\alpha_1}{1+\alpha_1^2}\cos\left[\frac{\pi}{\lambda_0 z_1}(x^2+y^2) - \frac{\pi}{\lambda_0 z_0}([x-x_0]^2+y^2) - \phi_1\right]\right)\end{aligned} \tag{23.52}$$

In words, the irradiance from these two point sources again has the form of a biased sinusoidal grating with modulation $2\alpha_1/(1+\alpha_1^2)$, but the argument of the sinusoid is now the sum of two quadratic-phase factors. If $z_0 = z_1$ so that the two sources are located in the same plane, this expression reduces to the now-familiar linear-phase function of x with spatial frequency $\xi_0 = x_0/\lambda_0 z_1$ centered at $x_0/2$ in Equation (23.36):

$$\lim_{z_0 \to z_1} \{|s[x, y; z_1]|^2\} \propto 1 + \frac{2\alpha_1}{1+\alpha_1^2} \cos\left[2\pi\left(\frac{x_0}{\lambda_0 z_1}\right)\left(x - \frac{x_0}{2}\right) - \phi_1\right] \tag{23.53}$$

In the more interesting case where the two sources are in different planes ($z_0 \neq z_1$), the argument of the sinusoid may be rewritten as the sum of quadratic-, linear-, and constant-phase functions:

$$\begin{aligned}
|s[x, y; z_1]|^2 &\propto 1 + \frac{2\alpha_1}{1+\alpha_1^2} \cos\left[\frac{\pi}{\lambda_0 z_1}(x^2+y^2) - \frac{\pi}{\lambda_0 z_0}[(x-x_0)^2 + y^2] - \phi_1\right] \\
&= 1 + \frac{2\alpha_1}{1+\alpha_1^2} \cos\left[\frac{\pi}{\lambda_0}(x^2+y^2)\left(\frac{1}{z_1} - \frac{1}{z_0}\right) - \frac{\pi}{\lambda_0 z_0}[-2xx_0 + x_0^2] - \phi_1\right] \\
&= 1 + \frac{2\alpha_1}{1+\alpha_1^2} \cos\left[\frac{\pi}{\lambda_0(z_1 z_0/(z_0 - z_1))}(x^2+y^2) - \frac{\pi}{\lambda_0 z_0}[-2xx_0 + x_0^2] - \phi_1\right] \\
&\equiv 1 + \frac{2\alpha_1}{1+\alpha_1^2} \cos\left[\frac{\pi}{\lambda_0 z_3}(x^2+y^2) + \frac{2\pi x_0}{\lambda_0 z_0}\left[x - \frac{x_0}{2}\right] - \phi_1\right] \\
&= 1 + \frac{2\alpha_1}{1+\alpha_1^2} \cos\left[\frac{\pi}{\lambda_0 z_3}(x^2+y^2) + 2\pi\xi_0\left[x - \frac{x_0}{2}\right] - \phi_1\right]
\end{aligned} \tag{23.54}$$

where $\xi_0 \equiv x_0/\lambda_0 z_0$ and $z_3 \equiv z_1 z_0/(z_0 - z_1)$. This expression for z_3 may be rewritten in a form with some familiar features:

$$\frac{1}{z_1} = \frac{1}{z_0} + \frac{1}{z_3} \tag{23.55}$$

Again, if the two sources are in the same plane, then $z_3 \to \infty$, which means that the quadratic part of the phase term vanishes from the irradiance function.

If the "object" source is located "on axis" (i.e., along the line from the reference source to the hologram plane), then $x_0 = 0$ and the expression for the irradiance simplifies to:

$$\begin{aligned}
|s[x, y; z_1, x_1 = 0]|^2 &\propto 1 + \frac{2\alpha_1}{1+\alpha_1^2} \cos\left[\frac{\pi}{\lambda_0}(x^2+y^2)\left(\frac{1}{z_1} - \frac{1}{z_0}\right) - \phi_1\right] \\
&= 1 + \frac{2\alpha_1}{1+\alpha_1^2} \cos\left[\frac{\pi}{\lambda_0 z_3}(x^2+y^2) - \phi_1\right]
\end{aligned} \tag{23.56}$$

which is a biased circularly symmetric quadratic-phase function with chirp rate $\sqrt{\lambda_0 |z_3|}$. The distance $z_1 - z_0$ determines the chirp rate of the irradiance pattern; the farther apart the two source points, the *smaller* the chirp rate of the hologram. This means that the quadratic phase oscillates more rapidly at the same radial distance from its center, as shown in Figure 23.11.

Alternatively, the argument of the cosine in Equation (23.46) may be recast as the sum of a translated quadratic-phase function and an additional constant phase that includes a quadratic function of the radial

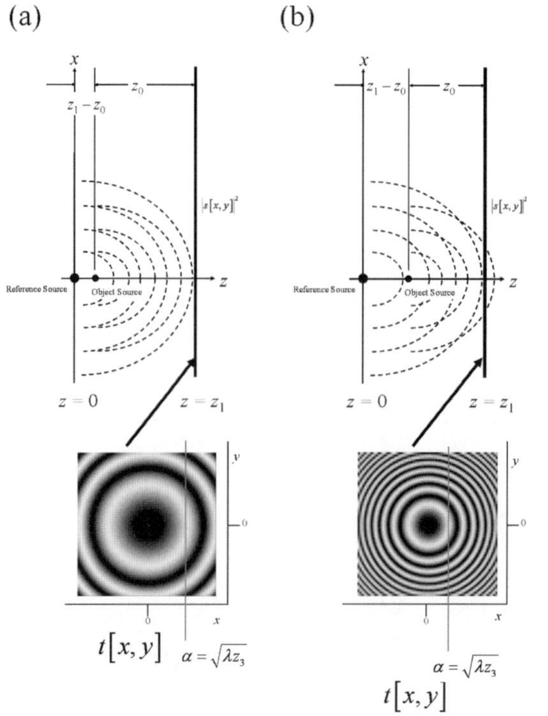

Figure 23.11 Fresnel hologram created with reference and object sources on the axis of symmetry. In (a), the reference and object sources are close together, and the chirp rate $\sqrt{\lambda z_3}$ of the quadratic-phase factor in the hologram is large. As the object source moves toward the hologram and away from the reference in (b), the chirp rate of the recorded hologram decreases.

displacement x_0 of the source:

$$
\begin{aligned}
|s[x, y; z_1]|^2 &\propto 1 + \frac{2\alpha_1}{1+\alpha_1^2}\cos\left[\frac{\pi}{\lambda_0 z_1}(x^2+y^2) - \frac{\pi}{\lambda_0 z_0}([x-x_0]^2+y^2) - \phi_1\right] \\
&= 1 + \frac{2\alpha_1}{1+\alpha_1^2}\cos\left[\frac{\pi}{\lambda_0 z_3}\left(\left[x+\frac{z_1}{z_0-z_1}x_0\right]^2 - \left[\frac{z_3}{z_0-z_1}\right]x_0^2\right)\right. \\
&\quad \left. + \frac{\pi}{\lambda_0}\left(\frac{1}{z_1}-\frac{1}{z_0}\right)y^2 - \phi_1\right] \\
&= 1 + \frac{2\alpha_1}{1+\alpha_1^2}\cos\left[\frac{\pi}{\lambda_0 z_3}\left[\left(x-\left[\frac{z_1}{z_1-z_0}\right]x_0\right)^2 + y^2\right] + \frac{\pi}{\lambda_0(z_1-z_0)}x_0^2 - \phi_1\right] \\
&= 1 + \frac{2\alpha_1}{1+\alpha_1^2}\cos\left[\frac{\pi}{\lambda_0 z_3}\left(\left[x+\frac{z_3}{z_0}x_0\right]^2 - z_3\left[\frac{1}{z_0-z_1}\right]x_0^2\right) - \phi_1\right] \qquad (23.57)
\end{aligned}
$$

where again z_3 was defined in Equation (23.55). The center of symmetry of the quadratic-phase factor is located at $[x_0 z_1/(z_1 - z_0), 0]$, which is the projection of the line connecting the two sources to the hologram plane, as shown in Figure 23.12. In other words, the chirp irradiance at the sensor plane is

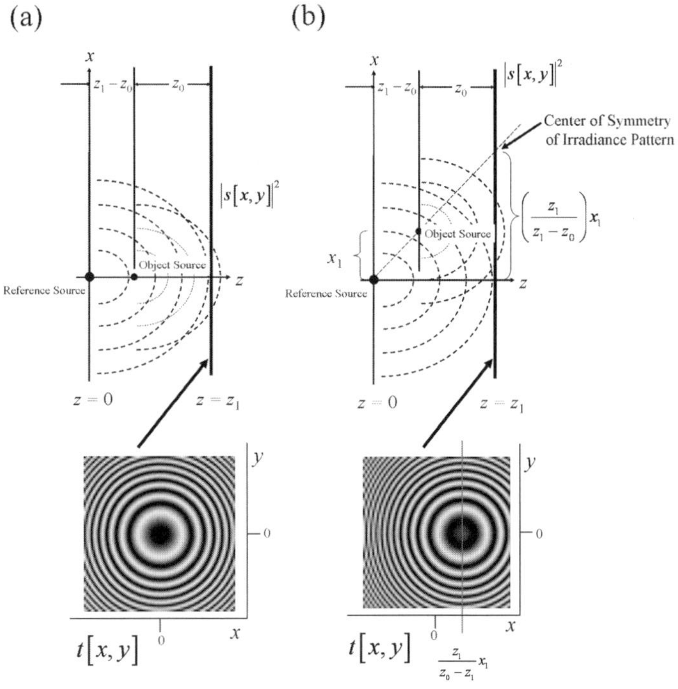

Figure 23.12 Fresnel hologram of two point sources as the object source moves off axis by the distance x_0. The center of symmetry of the hologram moves off axis by the scaled distance $(z_1/(z_1 - z_0))x_0$.

centered on the projection of the line connecting the sources. The grating frequency increases with radial distance away from this point with a "chirp rate" of $\sqrt{\lambda_0|z_3|} = \sqrt{\lambda_0 z_1 z_0/|z_0 - z_1|}$.

Again, the transmittance of the processed hologram ideally is proportional to the complement of the irradiance. Obviously any one of the three expressions for the irradiance in Equation (23.54), Equation (23.56), or Equation (23.57) may be used as the basis to express the transmittance, but the first is the simplest to analyze:

$$\begin{aligned} t[x,y] &\propto \frac{1}{2} - \frac{\alpha_1}{1+\alpha_1^2}\cos\left[\frac{\pi}{\lambda_0 z_1}(x^2+y^2) - \frac{\pi}{\lambda_0 z_0}([x-x_0]^2+y^2) - \phi_1\right] \\ &= \frac{1}{2} - \frac{1}{2}\left(\frac{\alpha_1}{1+\alpha_1^2}e^{-i\phi_1}\right)e^{+i\pi(x^2+y^2)/\lambda_0 z_1}\,e^{-i\pi[(x-x_0)^2+y^2]/\lambda_0 z_0} \\ &\quad - \frac{1}{2}\left(\frac{\alpha_1}{1+\alpha_1^2}e^{+i\phi_1}\right)e^{-i\pi(x^2+y^2)/\lambda_0 z_1}\,e^{+i\pi[(x-x_0)^2+y^2]/\lambda_0 z_0} \end{aligned} \tag{23.58}$$

In words, the hologram transparency includes three additive components: a constant term (chirp with infinite rate), the scaled product of an "upchirp" with rate appropriate for propagation by a distance $+z_1$ and a translated "downchirp" with propagation rate for the distance $-z_0$, and the complex conjugate with respective chirp rates corresponding to propagation by $-z_1$ and $+z_0$. We will see that the chirp functions in $t[x, y]$ act as "lenses" with different focal lengths.

23.2.4.1 Coherence Length and Coherence Volume

It is important to note an assumption implicit in Equation (23.58): that the interference pattern can be created in the first place. In other words, the phases of light arriving at the sensor from the object and reference points in Figure 23.10 must be sufficiently correlated to interfere. In common applications, the interference pattern is created by light from a single source that has passed through a small aperture ("pinhole") to ensure spatial coherence. The light reflects or scatters from the object and reference points and then recombines at the sensor. If the source is not monochromatic, Equation (22.70) indicates that interference occurs only if the difference in path lengths traveled from the source to the emulsion is shorter than the coherence length of the source:

$$L_{\text{reference}} - L_{\text{object}} \equiv \Delta L < \frac{c}{\Delta\nu} \tag{23.59}$$

We can account for both the spatial and temporal properties of the source by constructing its coherence volume in Equation (22.73). Light from the object and reference must originate within this volume to ensure that interference fringes are formed at the hologram.

23.2.5 Reconstruction of Point Object

To reconstruct the image of the object point, the hologram is illuminated by an appropriate reference beam and the diffracted light propagated to the reconstruction location. Our experience in the previous section suggests that the M–C–M chirp Fourier transform provides a suitable mathematical model for the reconstruction process, but several chirps with possibly different rates are involved. We consider the cases in turn.

23.2.5.1 Reconstruction of Virtual Image

Based on the results obtained in the previous section, we expect that illumination of the hologram by the original expanding "reference" beam (i.e., multiplication of the transparency function by the quadratic phase for propagation by the distance $+z_1$) will produce virtual images of the original source points. The amplitude of the illuminated hologram is:

$$\begin{aligned} t[x,y]\cdot e^{+i\pi r^2/\lambda_0 z_1} = e^{+i\pi r^2/\lambda_0 z_1} &- \left(\frac{\alpha_1}{1+\alpha_1^2}\,e^{-i\phi_1}\right)(e^{+i\pi(x^2+y^2)/\lambda_0 z_1})^2\, e^{-i\pi[(x-x_0)^2+y^2]/\lambda_0 z_0} \\ &- \left(\frac{\alpha_1}{1+\alpha_1^2}\,e^{+i\phi_1}\right) e^{+i\pi[(x-x_0)^2+y^2]/\lambda_0 z_0} \end{aligned} \tag{23.60}$$

Note that the phase of the expanding illumination chirp is cancelled by the quadratic-phase factor in the third term, leaving a quadratic-phase function with chirp rate $\sqrt{\lambda_0 z_{01}}$, but is not canceled in the other two terms.

The transmitted light is propagated over the distance z_2 (which may be positive or negative) to the observation plane by convolving with the appropriate impulse response for Fresnel diffraction at that distance:

$$\begin{aligned} (t[x,y]\cdot e^{+i\pi r^2/\lambda_0 z_1}) * e^{+i\pi(r^2/\lambda_0|z_2|)SGN[z_2]} &\propto (e^{+i\pi r^2/\lambda_0 z_1} * e^{+i\pi(r^2/\lambda_0|z_2|)SGN[z_2]}) \\ &- \left(\frac{\alpha_1}{1+\alpha_1^2}\,e^{-i\phi_1}\right)([e^{+i\pi(x^2+y^2)/\lambda_0 z_1}]^2\, e^{-i\pi[(x-x_0)^2+y^2]/\lambda_0 z_0} * e^{+i\pi(r^2/\lambda_0|z_2|)SGN[z_2]}) \\ &- \left(\frac{\alpha_1}{1+\alpha_1^2}\,e^{+i\phi_1}\right)(e^{+i\pi[(x-x_0)^2+y^2]/\lambda_0 z_0} * e^{+i\pi(r^2/\lambda_0|z_2|)SGN[z_2]}) \end{aligned} \tag{23.61}$$

where again the leading constant-magnitude and phase factors have been ignored. Our experience with convolutions of chirp functions has taught us that the first term is proportional to the object Dirac delta function $\delta[x, y]$ *if* we choose $z_2 = -z_1$, which means that we propagate "back" to the original reference plane. The result is a "virtual image" of the original "reference" point source at its original location. The second and third terms are messy combinations of quadratic-phase factors that may evaluated in a straightforward (though tedious) manner. The resulting amplitude at the reference plane is:

$$
\begin{aligned}
&(t[x, y] \cdot e^{+i\pi r^2/\lambda_0 z_1}) * e^{-i\pi r^2/\lambda_0 z_1} \\
&= (\lambda_0 z_1)^2 \delta[x, y] - \left(\frac{\alpha_1}{1+\alpha_1^2}\, e^{-i\phi_1}\right)([e^{+i\pi(x^2+y^2)/\lambda_0 z_1}]^2\, e^{-i\pi[(x-x_0)^2+y^2]/\lambda_0 z_0} * e^{-i\pi r^2/\lambda_0 z_1}) \\
&\quad - \left(\frac{\alpha_1}{1+\alpha_1^2}\, e^{+i\phi_1}\right)(e^{+i\pi[(x-x_0)^2+y^2]/\lambda_0 z_0} * e^{-i\pi r^2/\lambda_0 z_1}) \quad \textit{if } z_2 = -z_1
\end{aligned}
\tag{23.62}
$$

In the same way, if the propagation distance to the observation plane is $z_2 = -z_0$, which is the plane where the original "object" point was located, the amplitude includes a Dirac delta function that is the virtual replica of the object point:

$$
\begin{aligned}
&(t[x, y] \cdot e^{+i\pi r^2/\lambda_0 z_1}) * e^{-i\pi r^2/\lambda_0 z_0} \\
&= i\lambda_0 \left(\frac{z_0 z_1}{z_0 - z_1}\right) e^{-i\pi r^2/\lambda_0(z_0 - z_1)} \\
&\quad - \left(\frac{\alpha_1}{1+\alpha_1^2}\, e^{-i\phi_1}\right)([e^{+i\pi r^2/\lambda_0 z_1}]^2\, e^{-i\pi[(x-x_0)^2+y^2]/\lambda_0 z_0} * e^{-i\pi r^2/\lambda_0 z_0}) \\
&\quad - (\lambda_0 z_0)^2 \left(\frac{\alpha_1}{1+\alpha_1^2}\, e^{+i\phi_1}\right) \delta[x - x_0, y] \quad \textit{if } z_2 = -z_0
\end{aligned}
\tag{23.63}
$$

The first term is the quadratic-phase factor for the diffracted light from the reference point source, while the second is a quadratic-phase factor for diffracted light from the object point that creates the "real" image (considered next). The third term is proportional to $\delta[x - x_0, y]$, which is a reproduction of the original "object" point. The action of the hologram in this case diffracts the light from the reference source at the distance z_1 to create a virtual image at a distance $-z_0$. In other words, the hologram acts as a lens with negative focal length. The undiffracted wavefront can be viewed as creating an image of the reference point source at $z = \infty$, so the hologram also may be interpreted as acting as a lens with infinite focal length.

A schematic of the reconstruction of the virtual image point is shown in Figure 23.13. The results after the various steps in the process are shown in Figure 23.14, though the scale of features in realistic holograms is much smaller due to the very small value of λ_0.

23.2.5.2 Reconstruction of Real Image

We can use the analogy with Section 23.2.3 to reconstruct real images of the sources in the Fresnel region. A schematic of the process is shown in Figure 23.15. The illumination wave is the complex conjugate of the reference wave; that is, a spherical wavefront that converges to an on-axis point image at a distance z_1 from the hologram:

$$
\begin{aligned}
&(t[x, y] \cdot e^{-i\pi r^2/\lambda_0 z_2}) * e^{+i\pi(r^2/\lambda_0|z_2|)SGN[z_2]} \\
&= (e^{-i\pi r^2/\lambda_0 z_1} * e^{+i\pi(r^2/\lambda_0|z_2|)SGN[z_2]}) \\
&\quad - \left(\frac{\alpha_1}{1+\alpha_1^2}\, e^{-i\phi_1}\right)(e^{-i\pi[(x-x_0)^2+y^2]/\lambda_0 z_0} * e^{+i\pi(r^2/\lambda_0|z_2|)SGN[z_2]}) \\
&\quad - \left(\frac{\alpha_1}{1+\alpha_1^2}\, e^{+i\phi_1}\right)([e^{-i\pi r^2/\lambda_0 z_1}]^2\, e^{+i\pi[(x-x_0)^2+y^2]/\lambda_0 z_0} * e^{+i\pi(r^2/\lambda_0|z_2|)SGN[z_2]})
\end{aligned}
\tag{23.64}
$$

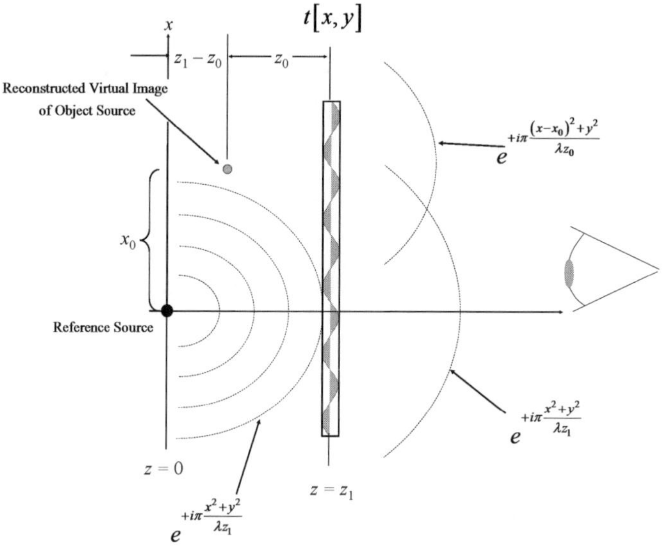

Figure 23.13 Schematic of reconstruction of virtual image of the object point. The hologram is illuminated by the original reference source. The hologram diffracts light to create three waves (only two shown). One appears to emerge from the original reference source and one from the original object. The wavefront generating the second term in Equation (23.61) is not shown.

The phase of the illuminating beam cancels the quadratic-phase factor in the second term. The first two terms are convolutions of pairs of chirps that will evaluate to Dirac delta functions if we choose the propagation distance z_2 correctly, while the third term is a "mess" of quadratic-phase factors. If we propagate a distance $z_2 = +z_1$ (the distance from the reference point source to the hologram), the resulting amplitude pattern is:

$$
\begin{aligned}
&(t[x, y] \cdot e^{-i\pi r^2/\lambda_0 z_2}) * e^{+i\pi r^2/\lambda_0 z_1} \\
&\quad = (e^{-i\pi r^2/\lambda_0 z_1} * e^{+i\pi r^2/\lambda_0 z_1}) - \left(\frac{\alpha_1}{1+\alpha_1^2}\, e^{-i\phi_1}\right)(e^{-i\pi[(x-x_0)^2+y^2]/\lambda_0 z_0} * e^{+i\pi r^2/\lambda_0 z_1}) \\
&\qquad - \left(\frac{\alpha_1}{1+\alpha_1^2}\, e^{+i\phi_1}\right)([e^{-i\pi r^2/\lambda_0 z_1}]^2\, e^{+i\pi[(x-x_0)^2+y^2]/\lambda_0 z_0} * e^{+i\pi r^2/\lambda_0 z_1}) \\
&\quad = |\lambda_0 z_1|\delta[x, y] + \textit{messy quadratic-phase “stuff”}, \textit{ for } z_2 = +z_1
\end{aligned}
\tag{23.65}
$$

This is the reconstruction of the original "reference" point source as a real image. Because the hologram diffracts light to create a real image, we can interpret its action as a lens with positive focal length, as well as having the infinite and negative focal lengths discussed in the previous section.

If the propagation distance from the hologram is $z_2 = +z_0$ (the distance from the object point to the hologram), the resulting amplitude pattern is:

$$
\begin{aligned}
&(t[x, y] \cdot e^{-i\pi r^2/\lambda_0 z_2}) * e^{+i\pi r^2/\lambda_0 z_0} \\
&\quad = (e^{-i\pi r^2/\lambda_0 z_1} * e^{+i\pi r^2/\lambda_0 z_0}) - \left(\frac{\alpha_1}{1+\alpha_1^2}\, e^{-i\phi_1}\right)(e^{-i\pi[(x-x_0)^2+y^2]/\lambda_0 z_0} * e^{+i\pi r^2/\lambda_0 z_0}) \\
&\qquad - \left(\frac{\alpha_1}{1+\alpha_1^2}\, e^{+i\phi_1}\right)([e^{-i\pi r^2/\lambda_0 z_1}]^2\, e^{+i\pi[(x-x_0)^2+y^2]/\lambda_0 z_0} * e^{+i\pi r^2/\lambda_0 z_0})
\end{aligned}
$$

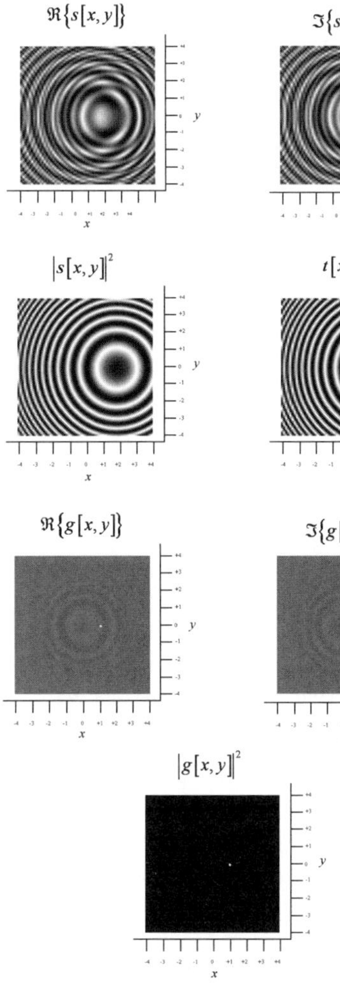

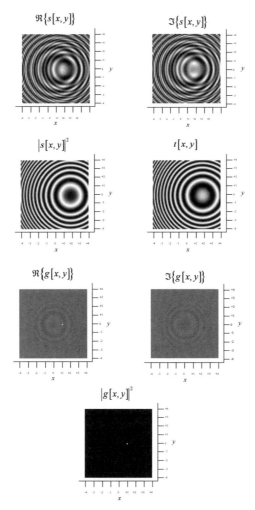

Figure 23.14 Steps in making and reconstructing a Fresnel hologram. The "reference" source is located at the origin of coordinates, the hologram is at the plane z_1, and the object source is located at $[1, 0, z_1 - z_0]$, where $z_0 = 0.8z_1$. The amplitudes at the hologram plane z_1 are shown as real and imaginary parts and as the irradiance (squared magnitude). The hologram transparency $t[x, y]$ is the complement of the transmittance. Note that the center of symmetry in the hologram is translated to the point $x = x_0 \cdot z_1/z_0 = +1.6$, in this example. The original reference is used to reconstruct the virtual image $g[x, y]$, which is shown as real and imaginary parts, where the chirp of the out-of-focus image of the reference source can be seen. The measurable image is the irradiance $|g[x, y]|^2$, where the object source is visible but light from the reference source is much fainter.

$$= \left(\frac{\alpha_1}{1+\alpha_1^2}\, e^{-i\phi_1}\right) |\lambda_0 z_0| \delta[x - x_0, y]$$

$$+ \textit{different messy quadratic-phase "stuff", for } z_2 = +z_0 \tag{23.66}$$

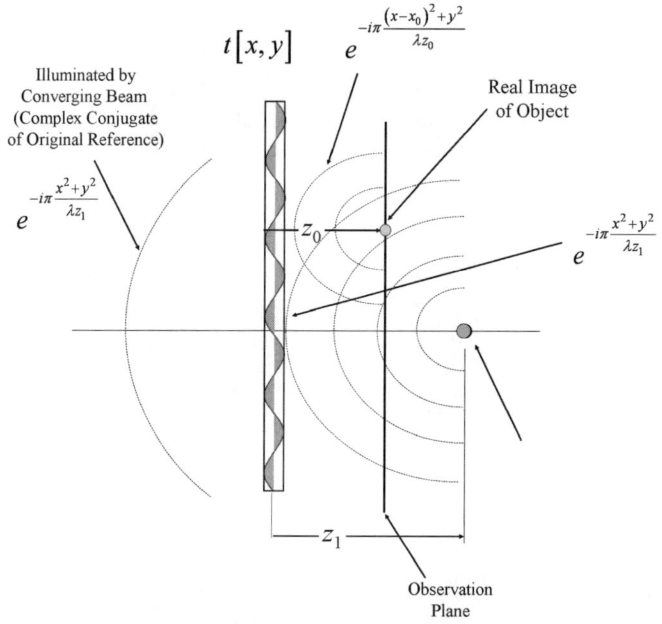

Figure 23.15 Schematic of reconstruction of real image of object point. The hologram $t[x, y]$ is illuminated by the converging complex conjugate of the reference wave. Diffraction by the hologram creates a wavefront converging at a distance z_0 from the hologram, which is reconstructed by convolving by the appropriate quadratic-phase factor. The other object wavefront from the hologram, listed as "messy quadratic-phase stuff" in Equation (23.65), is not shown.

The Fresnel hologram may be generalized to multiple object points by analogy with the derivation of the Fraunhofer hologram evaluated earlier. This is considered in one of the problems.

23.2.6 Extended Object and Planar Reference Wave

Now we generalize this analysis for the case of an extended planar object placed at the plane $z = z_1 - z_0$, so that the propagation distance from $f[x, y]$ is z_0. The reference wave is assumed to be generated by a point source placed at the coordinate $[-x_r, 0]$ in a plane at a sufficiently large distance z_r to be in the Fraunhofer diffraction region; then the reference is a plane wave incident on the recording medium at angle:

$$\theta_r = \tan^{-1}\left[\frac{x_r}{z_r}\right] \cong \sin^{-1}\left[\frac{x_r}{z_r}\right] \cong \frac{x_r}{z_r} \tag{23.67}$$

The object function including the reference is:

$$f[x, y; z = 0] = \exp\left[+2\pi i\left(\frac{\theta_r}{\lambda_0}x\right)\right] + r[x, y] \tag{23.68}$$

The plane wave at the hologram has the same form, so the amplitude is:

$$s[x, y; z = z_1] \propto \exp\left[+2\pi i\left(\frac{\theta_r}{\lambda_0}x\right)\right] + (r[x, y] * h[x, y; z = z_1]) \tag{23.69}$$

where $h[x, y; z = z_1]$ is the impulse response for light propagation in the Fresnel diffraction region. The measured irradiance at the recording medium is proportional to the squared magnitude, which has four terms:

$$|s[x,y;z=z_1]|^2 = (1+|r[x,y]*h[x,y;z=z_1]|^2) + \exp\left[+2\pi i\left(\frac{\theta_r}{\lambda_0}x\right)\right]\cdot(r[x,y]*h[x,y;z=z_1])^* + \exp\left[-2\pi i\left(\frac{\theta_r}{\lambda_0}x\right)\right]\cdot(r[x,y]*h[x,y;z=z_1]) \quad (23.70)$$

To reconstruct the virtual image, we illuminate with the reference wave at angle θ_r:

$$\exp\left[+2\pi i\left(\frac{\theta_r}{\lambda_0}x\right)\right]\cdot|s[x,y;z=z_1]|^2 = \exp\left[+2\pi i\left(\frac{\theta_r}{\lambda_0}x\right)\right]\cdot(1+|r[x,y]*h[x,y;z=z_1]|^2) + \exp\left[+2\pi i\left(\frac{2\theta_0}{\lambda_0}x\right)\right]\cdot(r[x,y]*h[x,y;z=z_1])^* + (r[x,y]*h[x,y;z=z_1]) \quad (23.71)$$

The first two terms are plane waves that continue to propagate at angle θ_r: undiffracted light through the hologram due to the average transmittance and modulated by the spatially varying transmittance. The third term is a plane wave that is diffracted at angle $2\theta_0$ modulated by the complex conjugate of the convolution of the object and impulse response of propagation. The last term is the convolution of the object and the impulse response of propagation that propagates at angle $\theta = 0$. To reconstruct the object, we simple propagate back to the object plane by convolving with $h[x,y;z=-z_1] = h^*[x,y;z=+z_1]$:

$$(r[x,y]*h[x,y;z=z_1])*h^*[x,y;z=+z_1] = r[x,y] \quad (23.72)$$

To reconstruct the real image, we illuminate with the conjugate of the reference wave, which is equivalent to changing the angle of incidence of the plane reference wave to $-\theta_r$:

$$\left(\exp\left[+2\pi i\left(\frac{\theta_r}{\lambda_0}x\right)\right]\right)^*\cdot|s[x,y;z=z_1]|^2 = \left(\exp\left[-2\pi i\left(\frac{\theta_r}{\lambda_0}x\right)\right]\right)\cdot|s[x,y;z=z_1]|^2 = \exp\left[+2\pi i\left(\frac{-\theta_r}{\lambda_0}x\right)\right]\cdot(1+|r[x,y]*h[x,y;z=z_1]|^2) + (r[x,y]*h[x,y;z=z_1])^* + \exp\left[+2\pi i\left(\frac{-2\theta_0}{\lambda_0}x\right)\right]\cdot(r[x,y]*h[x,y;z=z_1]) \quad (23.73)$$

Note that the same result is produced if the reference wave is unchanged but the recorded hologram is "reversed" so that the recorded transmittance is on the side opposite the incident reference wave. The undiffracted light propagates as plane waves at angle $-\theta_r$, the convolution of the object and the impulse response of propagation is impressed on the plane wave at angle $-2\theta_0$, and the complex conjugate of the convolution emerges parallel to the optical axis at angle $\theta = 0$.

23.2.7 Interpretation of Fresnel Hologram as Lens

The hologram of the single object point in the Fresnel region demonstrated that the chirp function in the hologram acts as lenses with three focal lengths, as shown schematically in Figure 23.16. The average transmittance of the hologram passes "undiffracted" light from the reference, and thus acts as a "pane

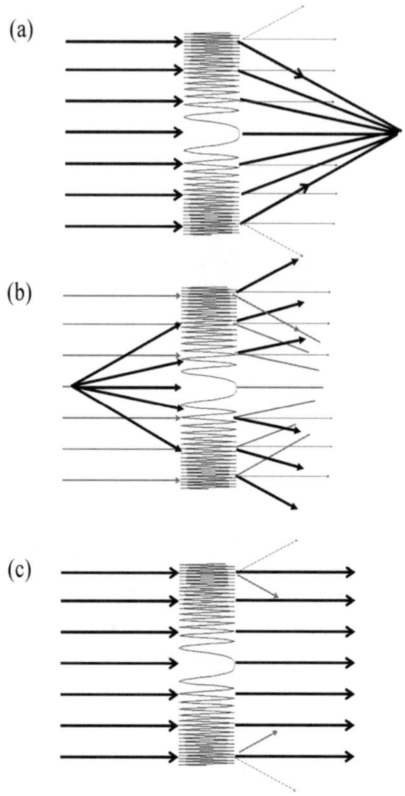

Figure 23.16 Diffraction by chirp function in hologram for plane-wave illumination: (a) diffracted light forms a real image in front of the hologram; (b) virtual image behind the hologram; (c) undiffracted light. The hologram acts as a lens with positive, negative, and infinite focal lengths. It is important to note that all three images are reconstructed simultaneously, so that the undiffracted light affects the visibility of the images of the point sources.

of glass", or a lens with infinite focal length. The light diffracted away from the optical axis creates a virtual image of the object "behind" the hologram, identical to a lens with negative focal length. The light diffracted toward the optical axis creates a real image of the object point "in front" of the hologram, as would a lens with positive focal length. The lesson from this observation and the transmittance pattern of the hologram in Equation (23.58) is that a hologram that can focus light, i.e., one with optical power, includes chirp functions with the appropriate rates proportional to the focal lengths. A hologram that does not include chirp functions can only divert light to the sides, e.g., a sinusoidal grating acting in the Fraunhofer region.

In the case of the "on-axis" Fresnel hologram shown, the light diffracted to create the three images all appears "on axis" and thus the viewer sees all three at the same time. At best, the mixing of the light from the three sources reduces the contrast of the image that the viewer desires to see, e.g., the light from the virtual image and the undiffracted light may overwhelm the real image. This was a fundamental shortcoming of the original method for holography by Gabor. In 1962, Leith and Upatnieks demonstrated holography with an off-axis reference source where the diffracted beams of light from the other reconstructions travel in different directions and thus do not overlap.

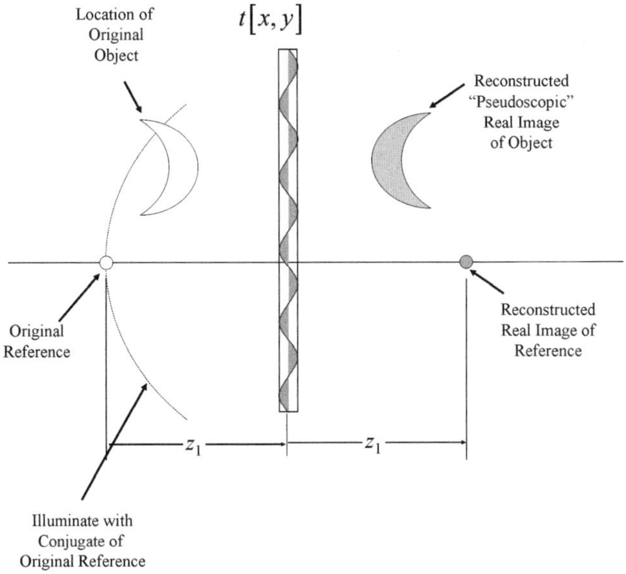

Figure 23.17 "Pseudoscopic" nature of the real-image reconstruction; the original object and reference sources are shown. The processed hologram is illuminated with the complex conjugate of the reference and forms a real image of the reference and a pseudoscopic real image of the object.

23.2.8 Reconstruction of Real Image of 3-D Extended Object

The analysis of Fraunhofer holograms of objects consisting of more than one point source that led to Equation (23.23) may be directly extended to the Fresnel case. Again, every point source in the object acts as a "reference" for reconstructing the other point sources, leading to "phony" reconstructions. As before, the visibility of these false reconstructions is minimized if the reference source is much brighter than the ensemble of object points.

Reconstructed real images of 3-D objects in the Fresnel region exhibit an interesting feature that may not be obvious at first glance. Object points that were positioned closer to the hologram plane also reconstruct closer to the hologram. In other words, the relative orientations of the reconstructed real image points are "reversed" with respect to the original configuration, so that the real-image reconstructions of points in the object that were closer to the sensor appear at locations closer to the hologram. This is the opposite of the images generated by a lens; object points closer to the lens create real images farther from the lens. This is the so-called "pseudoscopic" feature of a hologram. The image is often described as being "inside out", but a perhaps better description is "frontside back". A schematic of the pseudoscopic quality of real-image reconstructions is shown in Figure 23.17.

23.3 Computer-Generated Holography

The mathematical operations just considered in the 2-D Fraunhofer and 3-D Fresnel cases may be implemented digitally, which allows evaluation of the sampled and quantized complex amplitude of the hologram. The resulting patterns may be rendered as optical transparencies and used in optical systems either as source objects or filters. Just as is true for analogue holograms, these computer-generated holograms (CGHs) are hardly new, but advances in computing and rendering technologies have increased their capability in several applications, such as to generate wavefronts for interferometric assessment of the quality of large aspherical optics. In our context, the methods for creating CGHs nicely encapsulate many concepts of linear systems in imaging, which provides the motivation for this brief introduction to the relevant concepts.

Recall that the process of evaluating the hologram depended on the particular diffraction region to be used, and that it is significantly more difficult for 3-D objects in the Fresnel case than for 2-D objects in the Fraunhofer domain. In the former case, the interpretation of the previous section demonstrates that a hologram of an object may be viewed as an ensemble of chirp functions that act as negative or positive lenses to reconstruct the virtual or real image, respectively. The Fresnel CGH may be calculated by propagating light from the desired image pattern "backward" in space via the Fresnel transform with a negative propagation distance to the location of the hologram. This amplitude is combined with that from the assumed reference beam (usually a constant- or linear-phase amplitude from a planar reference) to calculate the interference pattern. The real-valued transmittance pattern resulting from the sum of these individual chirp patterns is then rendered to make the CGH. Of issue here is the observation made in Chapter 14 that any sampled chirp function will be aliased if sampled sufficiently far from the center of symmetry. The severity of the problem is reduced if the propagation distances z_n are "large" so that the chirp rate and the off-axis aliasing distance also are large, but this would ensure that only part of the chirp function can be rendered. Clearly, such a tactic would eliminate large spatial frequencies, which thus affect the ability to localize point objects in the reconstruction; put another way, the resolution of the reconstruction would suffer.

In the simpler 2-D Fraunhofer case that we will emphasize, Equation (21.96) showed that the Fraunhofer diffraction pattern of $f[x, y]$ is proportional to a scaled replica of its Fourier transform. In this case, we wish to compute the desired function that will generate $f[x, y]$ after Fraunhofer diffraction via the Fourier transform. In other words, we need to propagate the light "back" from the Fraunhofer diffraction region for the unknown "input". The "transform-of-a-transform" theorem of Equation (9.104) showed that the object whose Fraunhofer diffraction pattern is $f[x/\alpha^2, y/\alpha^2]$ must be proportional to a scaled replica of $F[-x, -y]$, where the scale factor α is determined by the wavelength of the coherent light and the propagation distance. In a CGH, the corresponding discrete calculation uses the tools developed in Chapters 14 and 15, where the desired output is $f[n, m]$ and the input is proportional to $\mathcal{F}_2{}^{-1}\{f[n, m]\} = F[-k, -\ell]$. We also must find a means to render the resulting complex-valued pattern as a transparency for use in a coherent optical system.

The calculational intensity required to evaluate Fresnel holograms is such that much of the early work on CGHs in the 1960s was focused (pun intended) on Fraunhofer holography using the (then new) FFT algorithm. This was a sufficiently significant constraint that one of the early references on CGH explicitly compared computation times with and without the FFT. Of course, modern computer technology is now capable of evaluating even large Fresnel CGHs quickly and easily. In addition, new types of devices capable of introducing computed holograms into optical systems have been introduced and remain in development. For these reasons, we can expect that the uses of CGHs will become even more widespread.

23.3.1 CGH in the Fraunhofer Diffraction Region

A flow chart of the processes in a 1-D model of a Fraunhofer CGH is shown in Figure 23.18. The desired function $f[x]$ is assumed to be an irradiance (i.e., proportional to the squared magnitude), so that its positive square root is the magnitude of the complex amplitude that serves as the input to the calculation. As we will see, it is often useful to apply a random phase to this array to produce a complex-valued amplitude $s[x]$. This amplitude at the reconstruction is projected back to the hologram plane over the distance $-z_1$ via the inverse Fourier transform (with a positive exponential term) to produce the desired complex amplitude at the hologram, which we will call $S[x/\lambda_0 z_1]$. This is quantized and rendered as the real-valued function $t[x]$ with transmittance t in the range $0 \leq t \leq 1$ by methods yet to be described. The transparency is illuminated with the reference beam and the light is propagated over some large distance z_2 that generally is equal to z_1.

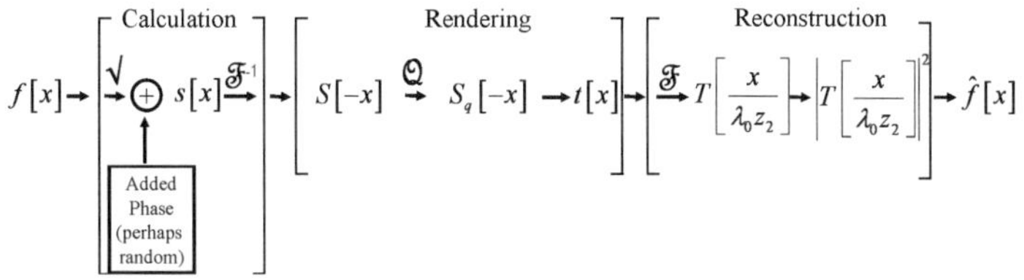

Figure 23.18 Processes in a Fraunhofer CGH where the squared magnitude of the input is $f[x]$. The calculation step includes the evaluation of the amplitude via the square root (and possibly a random phase) to generate $s[x]$. The scaled inverse Fourier transform leads to the second step, which includes quantization and rendering as a transparency with transmittance $t[x]$. The reconstruction step is composed of the Fourier transform and squared magnitude in the optical system to create the estimate $\hat{f}[x]$.

23.3.1.1 Rendering the CGH: Detour Phase

The complex-valued sampled hologram must be converted to a real-valued function that may be rendered as a transparency. In the early days of holography in the 1960s, the only economically feasible output devices were plotters that could print only black lines of fixed width and fixed sampled lengths at sampled locations on white paper. The problem of rendering the complex-valued hologram on such a bitonal device was solved in a simple and elegant way by Adolf Lohmann (Brown and Lohmann 1969). Subsequent variations of Lohmann's process were introduced by Lee (1970) and by Burckhardt (1970). All of these methods group arrays of bitonal pixels into 2-D "cells" within which the complex-valued samples of the CGH are rendered. Figure 23.19 illustrates the principle; a plane wave of light encounters a hologram (shown in cross-section) where groups of pixels in cells are opened to form "apertures". The incident wave is sampled by the apertures and the diffracted light from each propagates beyond the aperture plane. If the apertures are equally spaced, then the diffracted wavefronts recombine with interference maxima formed where the ensemble summation is in phase. The emerging wavefront is a plane wave possibly at a different angle that depends on the separation of the apertures and the wavelength of light. Alternatively, if the apertures are displaced by varying distances, then the wavefronts through the displaced apertures must travel longer paths to the observation point, thus changing their optical phases and the locations of the interference maxima, thus changing the shape of the emerging wavefront. In short, the shape of the emerging wavefront may be altered by displacing the apertures.

Lohmann Algorithm In the Fraunhofer case, the desired 2-D reconstruction is sampled to form $s[n, m]$ (we maintain the notation where $s[n, m]$ is the input array and $S[k, \ell]$ is its inverse FFT). This is inserted into an $N \times N$ array (often zero padded), where N traditionally is a power of two to facilitate evaluation of the spectrum via the inverse FFT. The inverse Fourier transform is calculated to construct $S[k, \ell]$, which is equivalent to propagating "back" in space. The magnitudes are normalized to the maximum value so that each complex-valued sample lies within a circle of unit radius in the complex plane and the phases are evaluated such that their range of values is $-\pi \leq \phi < +\pi$. Lohmann rendered the complex amplitude of each sample of $S[k, \ell]$ in the $N \times N$ array as an $M \times M$ "cell" of bitonal "subpixels". In other words, the hologram is rendered in an array of size $NM \times NM$. The magnitude is quantized to $M + 1$ magnitude levels via:

$$|S[k, \ell]|_q = CINT\left[M \cdot \frac{|S[k, \ell]|}{|S|_{\max}}\right] \Longrightarrow 0 \leq |S[k, \ell]|_q \leq M \tag{23.74a}$$

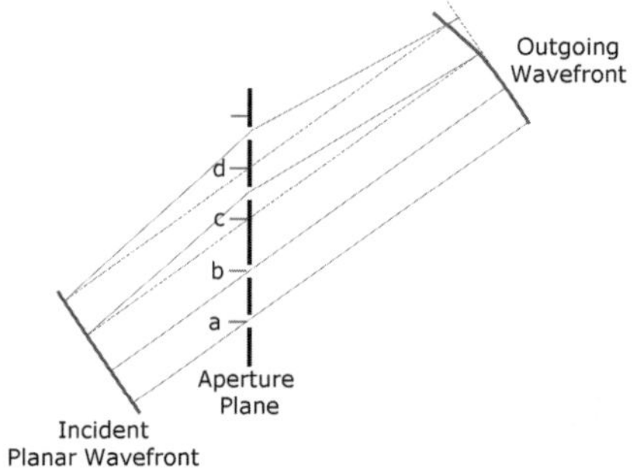

Figure 23.19 Principle of detour phase: the plane wave incident from the left encounters the opaque object with transparent apertures in cells labeled "a"–"d" (the tick marks in the hologram are equally spaced). Light passes through the apertures in cells a and b with no delay, but light passing through the displaced apertures in cells c and d is delayed, thus changing the curvature of the departing wavefront.

where *CINT* outputs the "closest integer" to the argument. The phases are independently quantized to eight levels in the range $-\pi \leq \Phi\{S[k, \ell]\} < +\pi$ via:

$$\Phi_q\{S[k, \ell]\} = INT\left[M \cdot \frac{\Phi\{S[k, \ell]\}}{2\pi}\right]$$
$$\Longrightarrow -\frac{M}{2} \leq \Phi_q\{S[k, \ell]\} < +\frac{M}{2} - 1 \qquad (23.74b)$$

where the integer function *INT* truncates the argument to its integer part. The quantized complex amplitudes are the samples of the Argand diagram shown in Figure 23.20b where $M = 8$. The quantized phases are equally spaced in azimuth angle around the complex plane and the nine magnitude levels (including 0) are equally spaced along radial lines. Obviously 65 normalized states are available in the complex plane if using an 8×8 cell of bitonal pixels, as shown for one specific complex amplitude in Figure 23.20a. Based on our experience in Section 9.8.19, we expect that the nonlinearity inherent in the quantization will generate additional spatial frequencies that will appear as artifacts in the reconstruction.

The quantized complex amplitude of a sample is rendered in a cell by "opening" a vertical line of bitonal pixels of length equal to $|S[k, \ell]|_q$ in Equation (23.74a) along the abscissa of the cell specified by the quantized phase Φ_q in Equation (23.74b). The result is a transparent "slit" 1 pixel wide for that cell (Figure 23.20b). The complex values are rendered for all samples to produce a bipolar array with semiperiodic structure along both axes due to M locations for the phase. The Fourier transform inherent in the Fraunhofer propagation ensures that the reconstruction also is semiperiodic.

Not surprisingly, this simple rendering scheme has some practical problems. For one, the largest possible fraction of light in the reconstruction beam that is transmitted to the output is M^{-1} for an $M \times M$ cell, which means that the reconstructed image will be "dark". Lohmann addressed this problem by opening lines of apertures with the same magnitude at phases to either side of the calculated value, thus increasing the transmitted light by a factor of 3 while retaining the same average value for the quantized phase in each cell (Figure 23.20c, d). In this modified scheme, the adjacent "slits" in cells with quantized phases of $-\pi$ or $+3\pi/4$ create apertures "split" in two (Figure 23.20e, f).

A second related problem of the simple rendering is a direct consequence of the magnitude quantization for realistic real-valued objects. The magnitude spectrum tends to be large at small spatial frequencies and small at large frequencies, which means that many of the latter terms are quantized to zero. In other words, the quantization acts as an effective lowpass filter and produces a "blurry" reconstruction. Lohmann addressed this issue by scaling the normalized magnitude spectrum by a

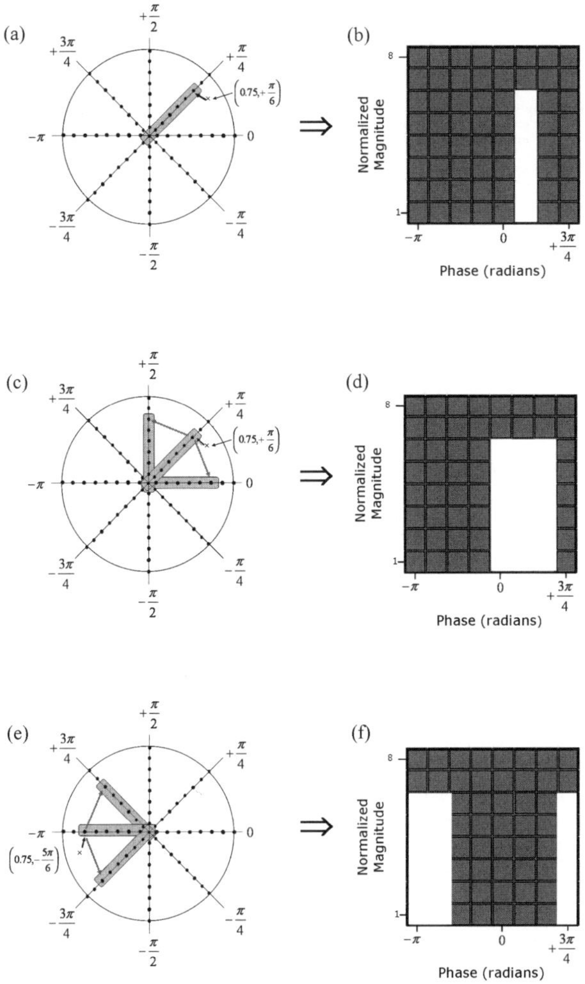

Figure 23.20 Lohmann CGH rendering algorithms: (a) the available states in one cell, showing the quantization for the complex amplitude $(0.75, \pi/6)$; (b) the corresponding rendered cell; (c) amplitude of cell with two adjacent apertures opened; (d) rendered cell; (e) amplitude $(0.75, -5\pi/6)$ and (f) rendered cell with two disjointed apertures.

constant (typically 3) before quantization, thus amplifying the high-frequency terms and "clipping" large magnitudes at the smallest spatial frequencies.

Lee Algorithm Lee (1970) modified the Lohmann algorithm to project the complex amplitude of each sample onto the real and imaginary axes in the complex plane before quantizing those projections with positive amplitude. In other words, he decomposed the complex amplitude into its positive components along the closest two phase angles of the possible values $\phi = 0$, $\pm\pi/2$, and $-\pi$. The individual apertures along these two phase angles generally have different lengths. For a fixed number of magnitude levels, the Lee algorithm effectively increases the density of available states in the complex plane, up to 197 for $M = 8$ (Figure 23.21a), compared to 65 available states in the Lohmann algorithm.

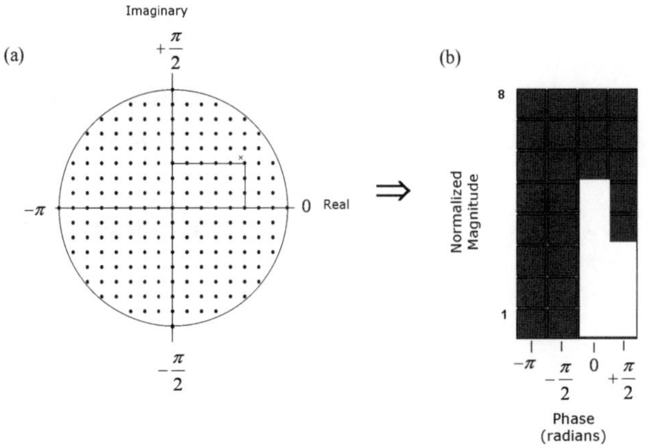

Figure 23.21 Lee variation of Lohmann CGH: (a) available states in one cell, showing a larger number of available states compared to the Lohmann algorithm for the same quantization – the normalized complex amplitude $(0.75, \pi/6)$ is quantized; (b) the corresponding rendered cell.

The increase in the number of states effectively reduces the quantization error. The cell corresponding to the quantized normalized magnitude is shown in Figure 23.21b.

Lee addressed the problem of many small quantized values by applying a random phase to the original real-valued object $s[n, m]$ before evaluating the Fourier transform; this is analogous to placing a ground glass screen over the object before evaluating the Fraunhofer diffraction pattern. The random phase has the effect of "flattening" the probability distribution (i.e., the histogram) of Fourier magnitudes (Figure 23.22), which effectively reduces the dynamic range of the Fourier transform and thus reduces the quantization error. Though helpful for creating recognizable reconstructions, the use of a random phase eliminates the possibility of using the CGH in phase-sensitive applications, e.g., matched filtering.

Burckhardt Algorithm Burckhardt (1970) further modified the Lee variation by projecting the complex amplitude onto the three equally spaced phase angles $\phi = 0, \pm 2\pi/3$ and quantizing the positive projections (Figure 23.23). By projecting the values parallel to the nonorthogonal axes, larger quantized values are obtained, which partially addresses the small-magnitude problem. The Burckhardt algorithm further increases the density of available normalized states in the complex plane over the Lohmann and Lee algorithms for the same number of magnitude levels (217 states for $M = 8$).

23.3.2 Examples of Cell CGHs

23.3.2.1 Effect of Quantization of Spectrum on 1-D Reconstruction

A few 1-D examples will be presented to illustrate the process of the cell CGH, the addition of a random phase, and the value of error diffusion in a CGH. For all three, the 1-D distribution of intensity $f[x]$ consists of rectangles and a half triangle, as shown in Figure 23.24a. Since the calculated hologram represents the amplitude of the object, the real-valued square root of the original intensity of a gray-scale object must be calculated first (Figure 23.24b).

As preparation for examples of the rendering, consider first the effect of the quantization on the reconstruction. The Fraunhofer diffraction pattern of the object amplitude without an added random phase is shown in Figure 23.25 as real part, imaginary part, and Argand diagram. The spectrum after quantizing to nine magnitude and eight phase levels is shown in Figure 23.26. Clearly the sinusoidal

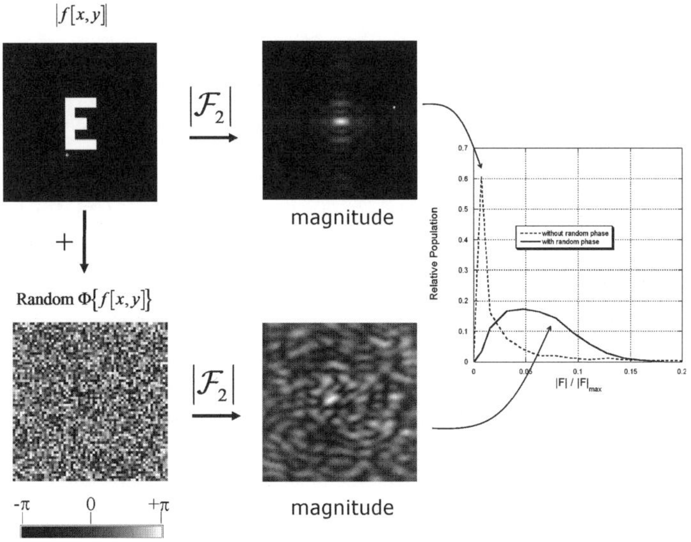

Figure 23.22 Effect of Lee's random phase on the histogram of the Fourier magnitude: magnitude spectrum of the "E" without a random phase is clustered at "darker" gray values, whereas the histogram of the magnitude with the additive random phase exhibits a Rayleigh distribution with many more "lighter" values, resulting in less clipping at large spatial frequencies during quantization.

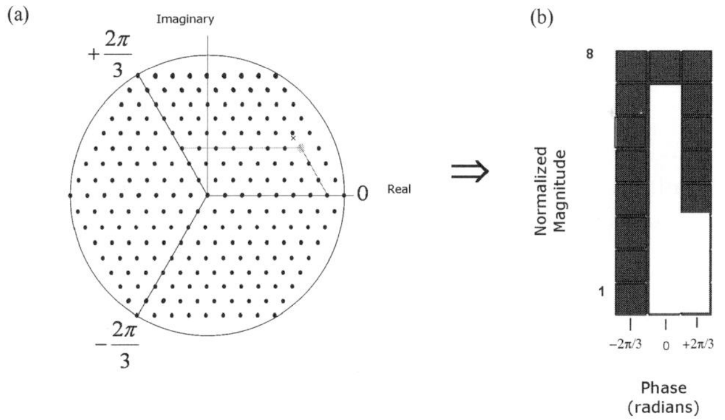

Figure 23.23 Burckhardt algorithm for a CGH: (a) the complex amplitude is projected onto the three phase angles $\phi = 0, \pm 2\pi/3$ radians, giving 217 available normalized states for eight magnitude levels; (b) the corresponding rendered cell.

components of the function with large spatial frequencies have been quantized to zero, which means that the action of quantization resembles that of a lowpass filter.

A magnified view of the reconstructed image is compared to the original object in Figure 23.27, showing that the "edges" of the reconstructed rectangles have been blurred due to the lowpass action of the quantization.

The random phase suggested by Lee (1970) to reduce the quantization error was applied to the same object before evaluating the spectrum shown as real part, imaginary part, and Argand diagram in

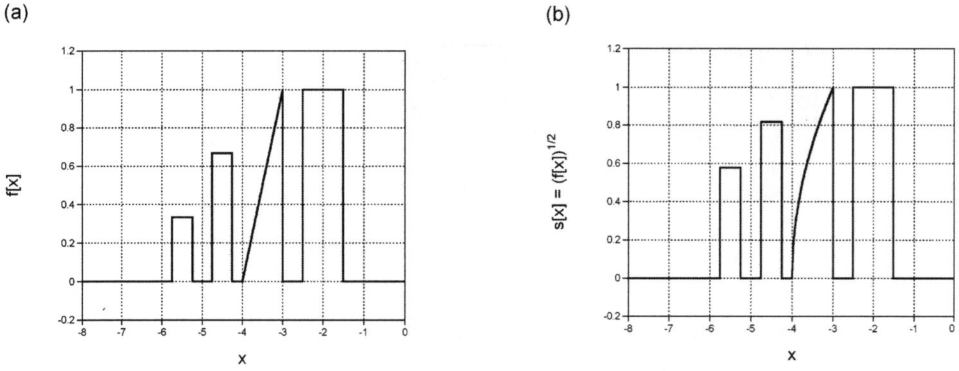

Figure 23.24 Object for 1-D CGH demonstrations: (a) $f[x]$ consists of three rectangles and a half triangle with different intensities; (b) magnitude of complex amplitude, which is the square root of the desired irradiance, $s[x] = \sqrt{f[x]}$; note that no random phase has been applied in this case.

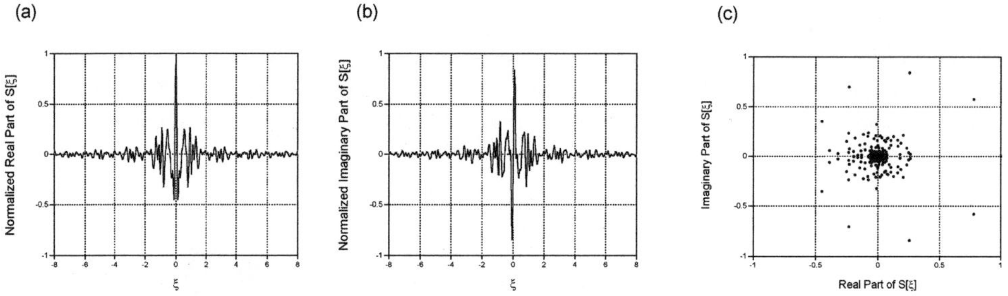

Figure 23.25 Amplitude of normalized Fraunhofer diffraction pattern of object $s[x] = \sqrt{f[x]}$: (a) real part; (b) imaginary part; (c) Argand diagram of amplitude before quantizing, showing that distribution of sample values of complex spectrum is concentrated at small spatial frequencies.

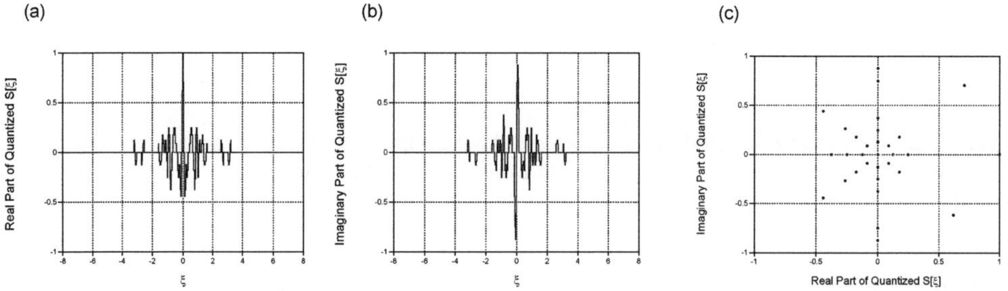

Figure 23.26 Amplitude of the diffraction pattern quantized to nine magnitude and eight phase levels, which should be compared to those in Figure 23.24: (a) real part; (b) imaginary part; (c) Argand diagram of the amplitude showing the small number of available phases.

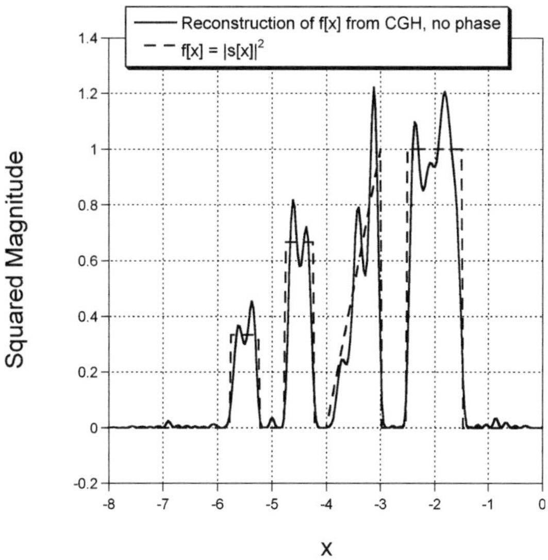

Figure 23.27 Squared magnitude of the reconstruction from the quantized spectrum compared to the object function.

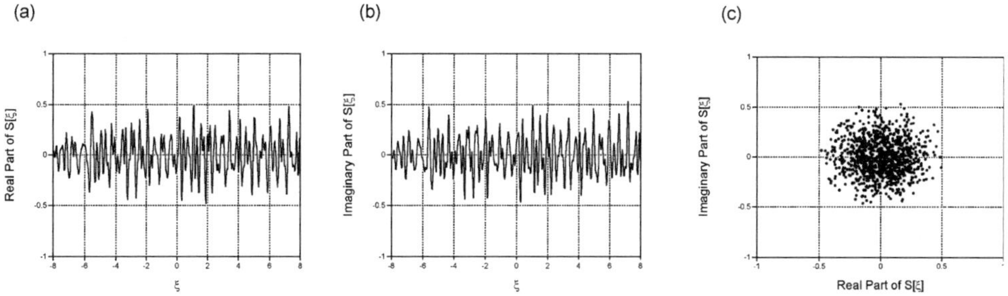

Figure 23.28 Spectrum of 1-D function $s[x] = \sqrt{f[x]}$ after addition of random phase: (a) real part; (b) imaginary part; (c) Argand diagram of the amplitude before quantizing, showing that the distribution of sample values of the complex spectrum is spread over a much larger domain than that in Figure 23.24.

Figure 23.28. The reconstructed image is compared to the original object in Figure 23.29, showing the sharper edges characteristic of larger amplitudes at large spatial frequencies.

23.3.2.2 1-D Lohmann Holograms

When applied to the Lohmann CGH rendering algorithm, the random phase increases the magnitudes of the sinusoidal components with large spatial frequencies. The bitonal renderings of the 1-D quantized Fourier transforms with and without the added random phase are shown in Figure 23.30. The holograms were produced by evaluating the inverse Fourier transform of the discrete array $f[n]$, which was then quantized and rendered as a Lohmann hologram with single-pixel aperture widths. Note that the addition of the random phase increases the light transmitted to the output because many more of the apertures have larger magnitudes. Figure 23.31a is the squared magnitude of the FFT of Figure 23.30a,

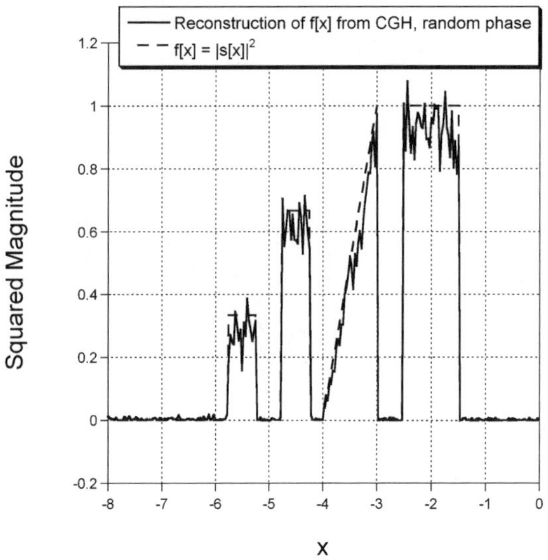

Figure 23.29 Comparison of squared magnitude of original object to reconstruction of quantized hologram with random phase, showing improvement in fidelity, and particularly in the response at edges, compared to result in Figure 23.27.

showing the periodicity of the reconstruction. The magnified view in Figure 23.31b is centered about the order -1 and shows the effects of the quantization and the lowpass character of the quantization. The magnified view of the reconstruction with the random phase in Figure 23.31d shows the improved high-frequency performance due to the random phase.

23.3.3 2-D Lohmann Holograms

The 2-D Lohmann hologram of an Archimedean spiral in a 32×32 array using an 8×8 cell and a random phase is shown in Figure 23.32, along with its optical and digital reconstructions. The cell size ensures that the digital reconstruction is nearly periodic with eight orders in the horizontal direction and eight replicas in the vertical direction. The optical reconstruction was created by placing a laser print of the hologram on transparency material in an expanded beam from an He:Ne laser and recording the irradiance on a CCD sensor placed at the focus of the beam in the Fraunhofer diffraction region. The angular separation between the orders is determined by the pixel pitch in the rendered hologram via the scaling theorem of the Fourier transform; the smaller the pixel pitch, the larger the angle between orders. The dots in the vertical line in the center are reconstructions of undiffracted light through the hologram (the "DC term"); these will be significant during our discussion of the optical matched filter in the next section. Note that these reconstructions are not exactly Dirac delta functions, but also include other nearby frequencies due to the nonlinear quantization. The primary reconstructions of the object are located at the orders ± 1. The digital reconstruction in Figure 23.32d is rendered as the logarithm of the squared magnitude to simulate the visual appearance due to the logarithmic response of the human visual system; the similarity between the optical and digital reconstructions is evident. The random variations in brightness of spiral and of background, called *speckle*, are primarily due to the additive random phase. The eight replicas of the reconstructions along the vertical direction in the discrete reconstruction are due to cells 8 pixels tall in the CGH. Note that an approximate replica of the squared magnitude of the object $|f[x, y]|^2$ is reconstructed at order -1, while the reconstruction at order $+1$

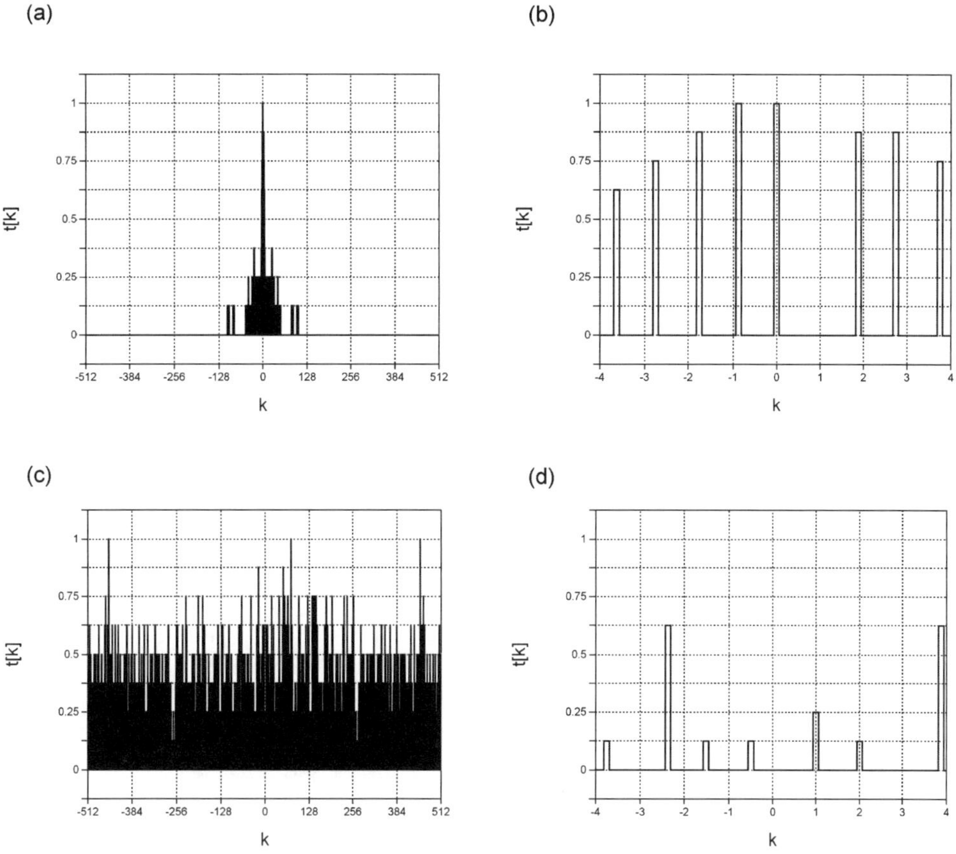

Figure 23.30 The 1-D Lohmann holograms of 1024-pixel input object with and without a random phase: (a) full hologram without random phase, showing lowpass character of quantization; (b) magnified view of central region with dashed lines between adjacent cells; (c) full hologram with random phase, showing significant magnitudes in most cells; (d) magnified view of central region.

has been rotated by π radians, i.e., it has the form $|f[-x, -y]|^2$. The distinction between these two reconstructions will prove important in the discussion of the holographic matched filter in the next section.

23.3.4 Error-Diffused Quantization

All quantizations are nonlinear and produce errors; quantization error in CGHs produces error in the reconstructions. In a Fraunhofer hologram, the reconstruction by optical Fourier transformation "spreads" this quantization error over the entire space domain, where it appears as "noise". The process of *error-diffused quantization* that was mentioned briefly in Chapter 14 usefully improves the appearance by subtracting the quantization error at one sample from as-yet-unquantized amplitude at one (or more) surrounding unquantized samples. In this way, the average value of the quantized quantity is preserved. Error diffusion was originally used to render gray-scale or color images on displays with limited capability, but is easily adapted to complex-valued quantities. The quantization process is shown schematically in Figure 23.33.

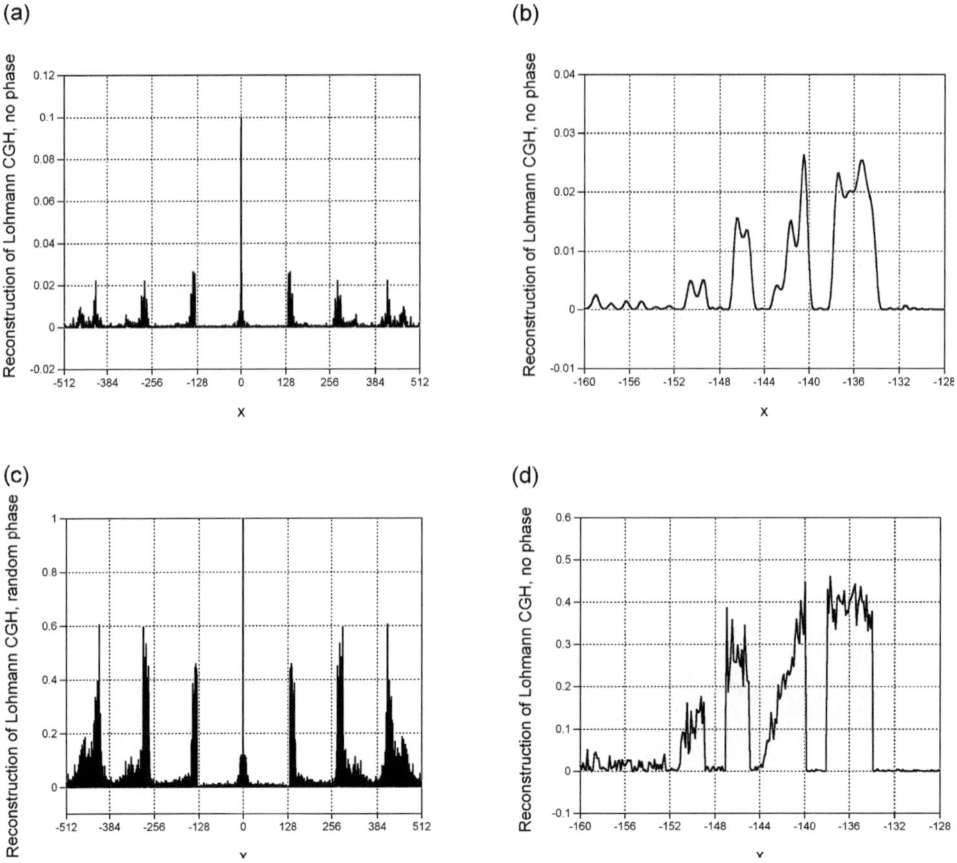

Figure 23.31 Simulated reconstructions of 1-D Lohmann CGHs: (a) without random phase, showing the periodicity, a large DC term, and other amplitudes in the vicinity of the origin; (b) magnified view of reconstruction with order -1, showing "blurry" reconstruction due to missing high-frequency components; (c) reconstruction of CGH with random phase, showing increase in output intensity; (d) magnified view, showing sharpened output due to additional components with large spatial frequencies.

The transfer of error from a quantized sample to its unquantized neighbor(s) ensures that the constant part of the Fourier transform is accurately rendered. The quantization error increases with the spatial frequency of the oscillation, which means that the noise in the reconstruction becomes more noticeable farther from the accurately rendered DC term at the center. In other words, error-diffused quantization produces "blue" noise in the nomenclature of Chapter 9. Even so, some rapidly oscillating information in the spectrum is preserved in the reconstruction, even without Lee's added random phase. This means that error-diffused quantization may be useful for phase-sensitive applications, such as inverse or matched filtering.

For comparison, the reconstruction of the object in Figure 23.24b without the added random phase is shown in Figure 23.34. No random phase was added and the Fourier transform was quantized to the same nine magnitude and eight phase levels. Comparison to the independently quantized holograms without the added random phase in Figure 23.27 and with the random phase in Figure 23.29 shows the improvement in fidelity made possible by this technique.

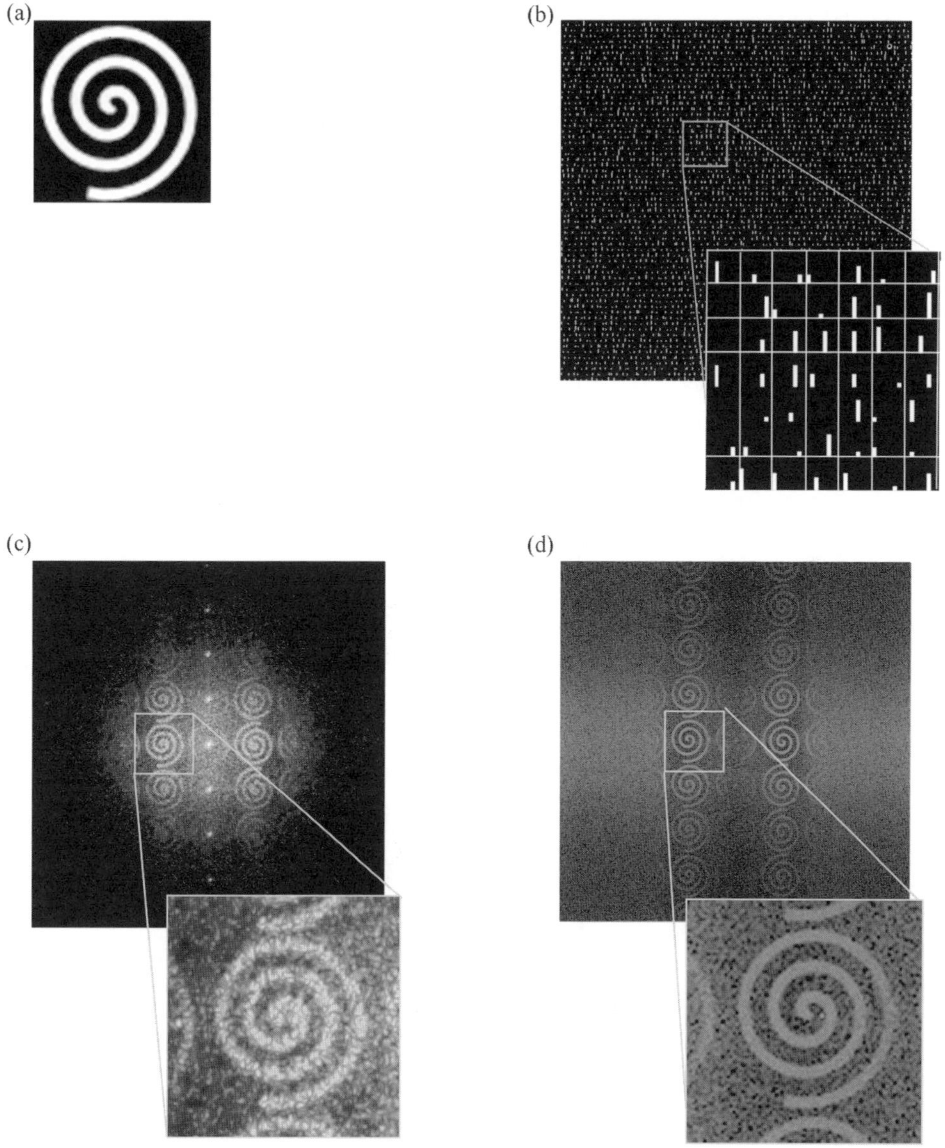

Figure 23.32 The 2-D Lohmann CGH with random phase: (a) input object is a rendering of an Archimedean spiral in a 32×32 array; (b) Fraunhofer hologram of inverse Fourier transform of object with 8×8 cell, with magnified section showing apertures and cell boundaries; (c) optical reconstruction in Fraunhofer diffraction region showing periodicity and reconstructions at orders ± 1 (the "speckle" in the reconstruction is primarily due to the random phase); (d) digital reconstruction rendered as logarithm of squared magnitude of computed DFT, thus simulating the visual appearance.

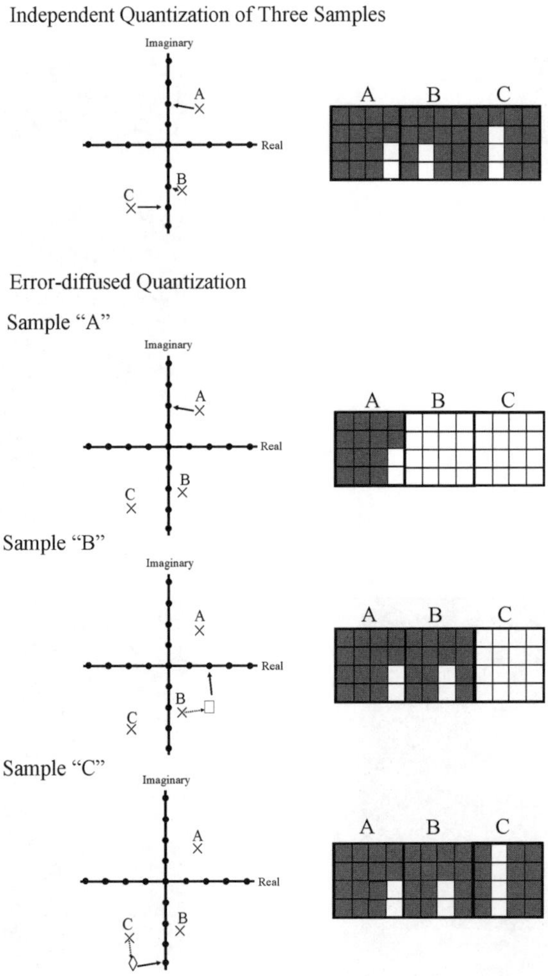

Figure 23.33 Error-diffused quantization in Lohmann CGH: normalized amplitude of sample "A" is quantized first (shown in row 1) to render the first cell. The resulting quantization error is subtracted from amplitude of sample "B" and then quantized (row 2) to render the second cell. The process is repeated to render the third amplitude ("C") etc. This preserves the average value of the quantized quantity.

Simulated results (FFT reconstruction) of holograms using independent and error-diffused quantization are shown in Figure 23.35 (Easton et al., 1996). The reconstruction from the error-diffused quantization shows less error though the number of available quantization states was reduced by a factor of 4.

23.4 Matched Filtering with Cell-Type CGH

The limitation on dynamic range due to quantization in CGHs limits the applications that require amplification (such as the reciprocal filters – inverse and matched – considered in Chapter 19). The

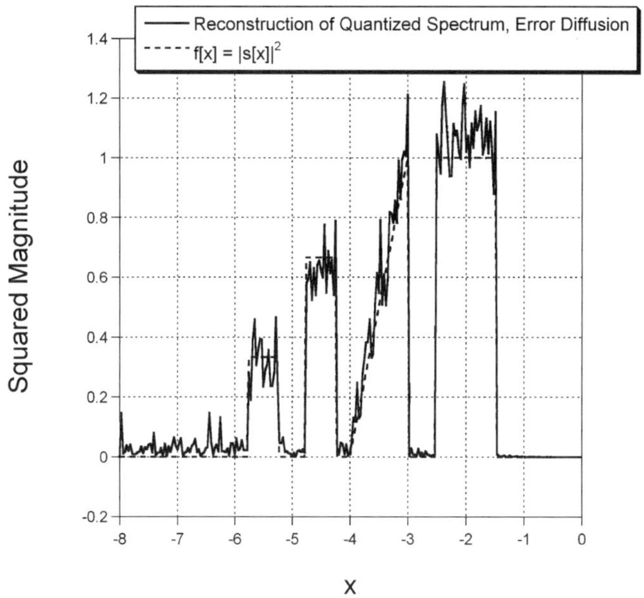

Figure 23.34 Reconstruction of 1-D hologram with random phase formed with error-diffused quantization compared to original object function.

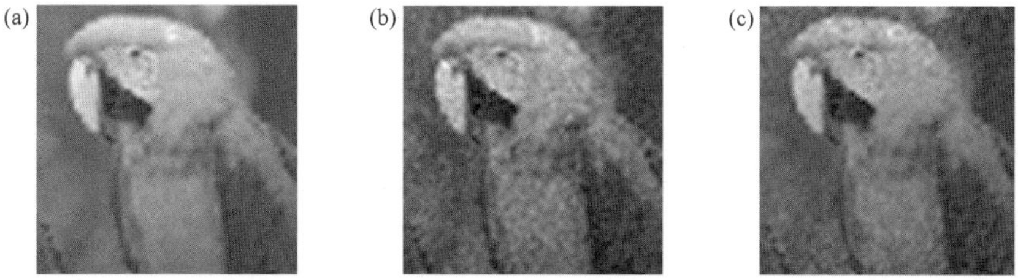

Figure 23.35 Simulated CGH reconstructions using independent and error-diffused quantization: (a) original object; (b) reconstruction via FFT of CGH with Lohmann single-phase rendering and independent quantization to nine magnitude and eight phase levels (RMS error = 0.282); (c) digital reconstruction using error-diffused quantization with two magnitude and eight phase levels (RMS error = 0.246).

addition of the random phase to compensate for the quantization prevents CGH use in phase-sensitive applications, such as the matched filter. We have derived some tools that may help in such applications, including error-diffused quantization and the power-series approximation for a reciprocal filter based on Equation (19.137).

The optical implementation of the matched filter in the basic 4**f** optical filtering system considered in Chapter 21 is shown in Figure 23.36. The input $f[x, y]$ is placed one focal length from a lens and another lens is placed two focal lengths from the first. The output is observed one focal length from the second lens. The amplitude at the "midplane" between the lenses is proportional to the scaled Fourier transform of the object, which may be multiplied by multiplicative masks of the form $M[x/\lambda_0\mathbf{f}, y/\lambda_0\mathbf{f}]$. The second lens evaluates the Fourier transform of this transform to produce a reversed replica of the filtered object, i.e., the output $g[x, y] = f[x, y] * m[-x, -y]$. In the case of the classical matched filter for the object $r[x, y]$, the impulse response is $m[x, y] = r^*[-x, -y]$ so that the transfer function is $M[\xi, \eta] = R^*[\xi, \eta]$.

We can use a cell-type CGH as the transfer function of the filter. For example, if the matched object is an upper-case "E", the example of Figure 23.31 shows that the reconstructed image (the CGH "impulse response") includes an approximate replica of the impulse response $r^*[-x, -y]$ centered on order -1, an approximate Dirac delta function at order 0 in the vicinity of the origin, and a rotated replica of the approximate impulse response $r^*[+x, +y]$ at order $+1$ (of course, higher-order terms exist due to the nonlinearities inherent in the quantization and the processing, which we will ignore in this simple analysis).

If the input is a Dirac delta function located at the origin, then the CGH is illuminated by a light with constant amplitude and phase and the second lens generates a reconstruction of the hologram centered on the output plane, as shown in the first row of Figure 23.36. The second row of the figure shows the reconstruction if the input is a Dirac delta function displaced from the origin (by $+y_0$ in the example), which reconstructs three orders that have been displaced by $-y_0$. The linearity of the system ensures that the output for a general function includes three terms:

$$|g[x, y; -1]|^2 \cong |f[-x, -y] * m^*[x, y]|^2 = |(f[x, y] * m^*[-x, -y])|_{x\to -x, y\to -y}|^2$$
$$= |(f[x, y] \bigstar m^*[x, y])|_{x\to -x, y\to -y}|^2 = |f[x, y] \bigstar m[x, y]|^2 \tag{23.75a}$$

$$|g[x, y; 0]|^2 \cong |f[-x, -y] * \delta[x, y]|^2 = |f[-x, -y]|^2 \tag{23.75b}$$

$$|g[x, y; +1]|^2 \cong |f[-x, -y] * m^*[-x, -y]|^2$$
$$= |(f[-x, -y] * m[x, y])^*|_{x\to -x, y\to -y}|^2 = |f[x, y] * m[x, y]|^2 \tag{23.75c}$$

In words, if the input is the object to which the filter is matched, the output at order -1 is the rotated autocorrelation of the object, the output at order 0 is a rotated replica of the object (due to the inversion by the optical system), while that at order $+1$ is the rotated convolution of the object with itself (the "autoconvolution" of the object).

In practical use, the input may include a number of replicas of the object at different locations, as shown in the example in Figure 23.37. The bitonal reference function is an upper-case "E", while the input object consists of several upper-case letters, including three examples of the reference object and one reversed replica. The quantization without an added random phase ensures that the CGH matched filter is dominated by low-frequency information. The corresponding output is shown in Figure 23.38, which shows the inverted reconstruction of the input object in the zero-order term, their correlation in the term of order -1, and their convolution at order $+1$. The magnified view of these three terms is shown as images and as surface plots, where the relative sizes of the correlation peaks are more easily seen.

The last column in Figure 23.27 shows the discrete version of the complement matched filter in Equation (19.147), where the enhanced edges of the reference function result in additional amplitude at larger spatial frequencies. The simulated output in Figure 23.39 exhibits the expected sharper correlation peaks due to the reduced attenuation of the sinusoidal components with large spatial frequencies.

23.5 Synthetic-Aperture Radar (SAR)

To conclude the discussion of the mathematical model of holography, we consider an imaging system based on the Fresnel diffraction model that maps object points to specific chirp functions distinguished either by location of the center of symmetry and/or by chirp rate. Because the same analysis applies to the chirp functions in the Fresnel hologram of a 3-D object, it is reasonable to surmise that the original

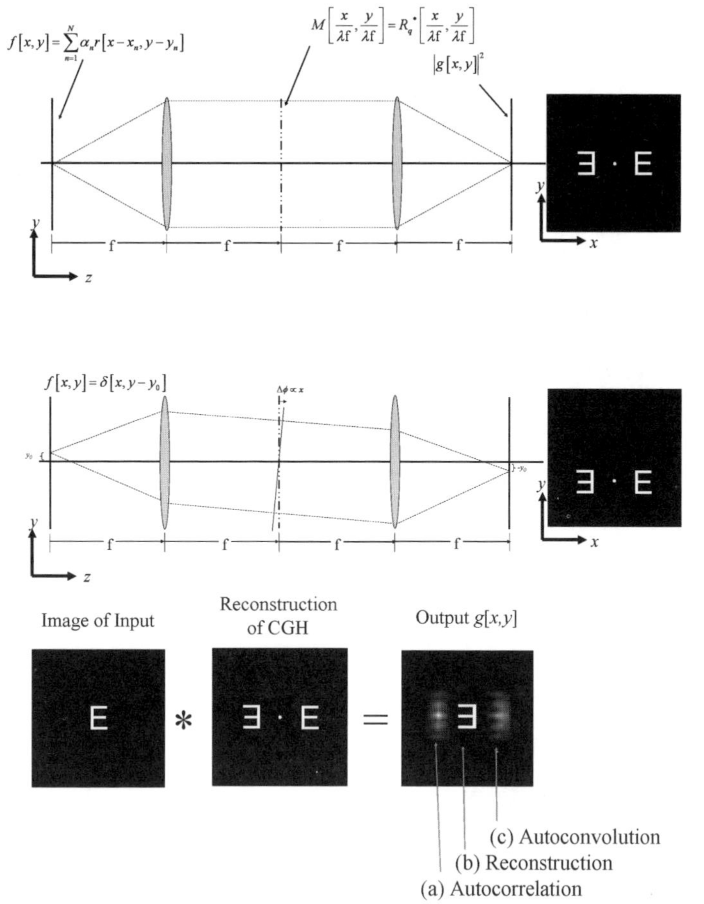

Figure 23.36 Schematic of matched filtering with a CGH in 4**f** optical correlator. The transfer function of the matched filter for the character "E" is placed at the midplane and the second lens computes the 2-D Fourier transform to generate the output. In the first row, the input is a Dirac delta function at the origin so that the matched filter is illuminated by the constant amplitude to produce a copy of the CGH reconstruction at the output. If the Dirac delta function is above the optical axis, the reconstruction is generated below the axis. If the input is the same "E", then the output is the convolution of the reconstruction of the CGH with the input, which includes the inverted convolution of the input with the centered Dirac delta, plus translated replicas of the autocorrelation and autoconvolution.

distribution of object points may be determined from the collected data via a scheme analogous to Fresnel reconstruction. In fact, the original version of this system reconstructed the imagery by optical methods (Leith and Upatnieks, 1962).

We have just seen that reconstruction of virtual and real images of point objects in Fresnel holograms is based on the observation that the convolution of two chirp functions with the same chirp rate and opposite signs produces a Dirac delta function, e.g., the 1-D case in Equation (17.29):

$$\exp\left[\pm i\pi\left(\frac{x}{\alpha}\right)^2\right]*\exp\left[\mp i\pi\left(\frac{x}{\alpha}\right)^2\right]=|\alpha|\cdot\delta\left[\frac{x}{\alpha}\right]=|\alpha|^2\cdot\delta[x] \tag{23.76}$$

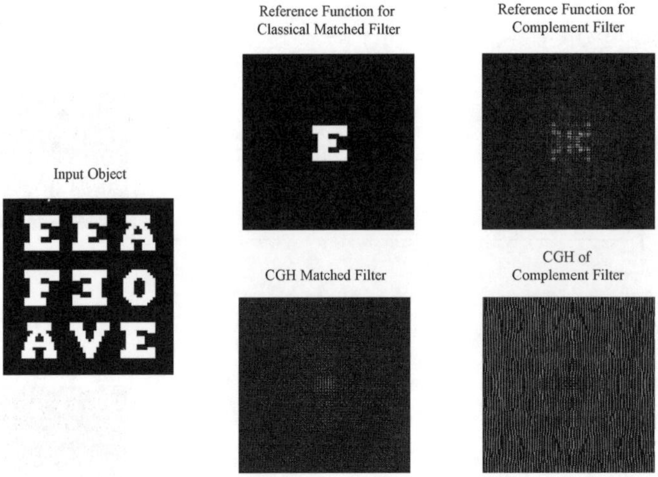

Figure 23.37 Input object, reference function, and CGH for the classical matched filter, and edge-enhanced reference function and CGH for the complementary matched filter. Note the additional high-frequency information transmitted by the latter CGH.

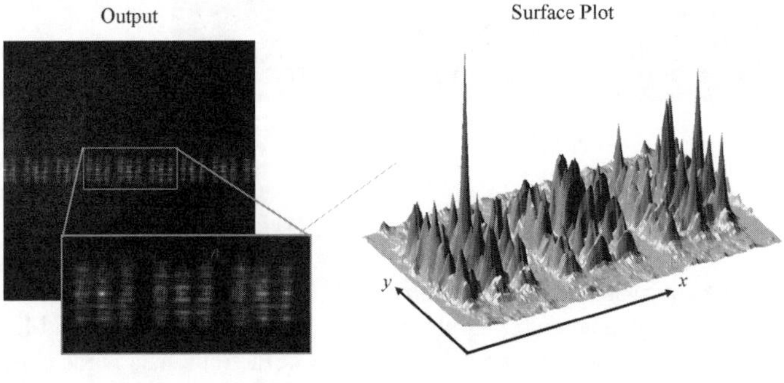

Figure 23.38 Simulated optical output of optical correlator with CGH for the classical matched filter: magnified view shows "E" convolved with the rotated object (producing the crosscorrelation of the object and reference) at order -1, the inverted replica of the object at order 0, and the convolution of object and reference at order $+1$. The surface plot shows the relative sizes of the correlation peaks.

In words, this means that the space-domain form of the inverse filter for any imaging system with a quadratic-phase impulse response $h[x]$ is proportional to the chirp function with the opposite sign. Given the output $g[x]$ from such an imaging system, the original object $f[x]$ may be reconstructed by convolving with the inverse filter:

$$
\begin{aligned}
g[x] * w[x] &= (f[x] * h[x]) * w[x] \\
&= \left(f[x] * \exp\left[\pm i\pi\left(\frac{x}{\alpha}\right)^2\right]\right) * \frac{1}{\alpha^2}\exp\left[\mp i\pi\left(\frac{x}{\alpha}\right)^2\right]
\end{aligned}
$$

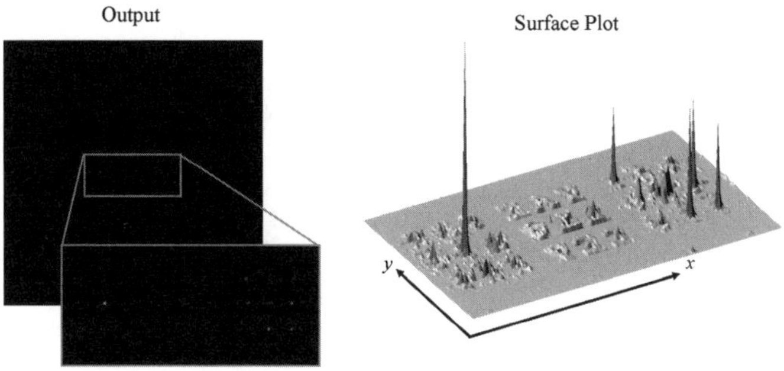

Figure 23.39 Simulated optical output of optical correlator with CGH for the complement matched filter as image and as surface, showing the "sharper" correlation peaks that are better approximations of Dirac delta functions due to reduced attenuation of sinusoidal components with large spatial frequencies.

$$= f[x] * \frac{1}{\alpha^2}\left(\exp\left[\pm i\pi\left(\frac{x}{\alpha}\right)^2\right] * \exp\left[\mp i\pi\left(\frac{x}{\alpha}\right)^2\right]\right)$$

$$= f[x] * \frac{1}{\alpha^2}(|\alpha|^2 \cdot \delta[x]) = f[x] \tag{23.77}$$

We showed in Chapter 21 that we could construct an inverse filter for Fresnel diffraction with a lens with pupil diameter d_0 and focal length $\mathbf{f}$ located at a distance z_1 from an object if the light observed at the distance z_2 satisfies the condition in Equation (21.134):

$$\frac{1}{z_1} - \frac{1}{\mathbf{f}} + \frac{1}{z_2} = 0 \Longrightarrow \frac{1}{z_1} + \frac{1}{z_2} = \frac{1}{\mathbf{f}} \tag{23.78}$$

We also showed that the chirp rate of the impulse response depends on the diameter d_0 of the lens; that the impulse response is "narrower" for aberration-free lenses with larger apertures. We also used this same construction earlier in this chapter to find the observation distance z_2 where the light diffracted by a Fresnel hologram will reconstruct the image of the original point object. We now apply these results to consider yet another imaging system that depends on the property of Equation (23.76) to reconstruct the original object. As for holography, this system depends on active illumination of the object and therefore is a type of *radar*, an acronym for "RAdio Detection And Ranging". The newer system is variously called "synthetic-aperture radar" (sometimes abbreviated to the second-order acronym "SAR") or "side-looking radar". The same principles may be applied to imaging systems with quadratic-phase impulse responses that act on other types of energy, such as acoustic waves.

We know from the discussion of the effect of varying pupil sizes on the image from Section 21.6.4 that a larger aperture produces images with improved spatial resolution. As the name implies, SAR creates a "synthetic" pupil that is much larger than the actual size of the antenna. We consider only a simplified view of SAR that is based on optical imaging and holography in the Fresnel diffraction region; readers needing more detail may consult references such as Elachi et al. (1982).

Just as in Fresnel holography, each point source in the scene generates a quadratic-phase function whose chirp rate varies with distance of the object point from the sensor; the chirp rate is larger for more distant sources. Because the impulse response is not identical for source points at different distances, the system is *shift variant* and the image may not be constructed from the data via a single filter. As we will see, we can theoretically apply different inverse filters for objects at different distances to reconstruct the image.

Before delving into the details of SAR, it may be helpful to first consider the basic principles of traditional radar. As many readers probably know already, radar is an active imaging system that

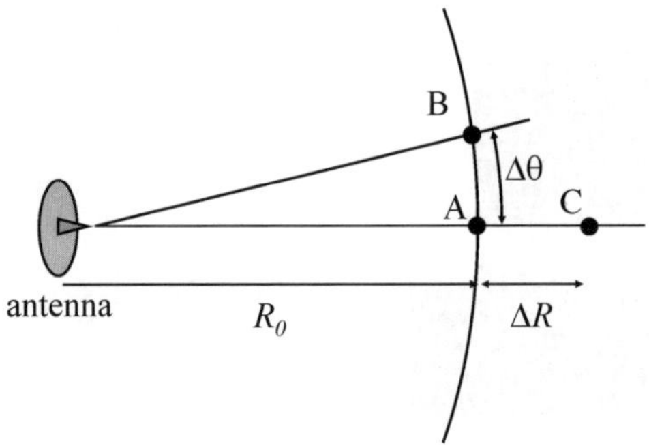

Figure 23.40 Configuration of azimuthal and range resolution in radar: the smallest angular separation $\Delta\theta$ between resolvable point reflectors "A" and "B" is the azimuthal resolution. The smallest separation ΔR between resolvable targets "A" and "C" is the range resolution.

transmits pulses of radio-frequency (RF) radiation (with temporal frequency of the order of 10^6–10^9 Hz) toward some compass azimuth and then "listens" for reflections from targets along that direction. The traditional radar system is characterized by two types of resolution: the *azimuthal resolution* is the ability to distinguish two targets at the same distance but along different compass directions, while the *range resolution* is the ability to distinguish targets at different distances along the path traveled by the radiation (Figure 23.40). We now consider the effects of the system parameters on the two measures of resolution in a traditional radar system.

23.5.1 Range Resolution

23.5.1.1 Pulsed Radar

In the simplest picture of radar, the range resolution is determined by the temporal "width" of the transmitted pulse. Consider that the system emits pulses in the form of narrow rectangles of width Δt separated by the interval t_0:

$$s[t] = \sum_n A_n \cdot RECT\left[\frac{t - n \cdot t_0}{\Delta t}\right] \tag{23.79}$$

For simplicity, we examine the result for a single emitted pulse of width Δt centered about $t = 0$:

$$s[t] = A_0 \cdot RECT\left[\frac{t}{\Delta t}\right] \tag{23.80}$$

This signal is reflected from a point target at distance R_0 so that the center of the pulse arrives back at the antenna after a time delay τ equal to twice the travel time:

$$\tau = 2 \cdot \frac{R_0}{c} \tag{23.81}$$

where c is (of course) the speed of light. The attenuation of the received signal due to the inverse square law and the reflectivity of the target is collected into a single "reflectivity" term σ_0:

$$g[t] = \sigma_0 \cdot A_0 \cdot RECT\left[\frac{t - 2 \cdot R_0/c}{\Delta t}\right] + n[t] \tag{23.82}$$

where $n[t]$ is the noise in the measurement.

Our goal is to determine the range R_0 from the measured signal $g[t]$ and knowledge of the transmitted signal $s[t]$; this is one of the most obvious examples of the matched-filter problem. If noise is present, the impulse response of the unamplified ("classical") matched filter in Equation (19.52) is appropriate:

$$m[t] = s^*[-t] = \left(A_0 \cdot RECT\left[\frac{-t}{\Delta t}\right]\right)^* = A_0 \cdot RECT\left[\frac{t}{\Delta t}\right] = s[t] \tag{23.83}$$

In words, the impulse response of the matched filter is identical to the transmitted signal and the output is the convolution:

$$\begin{aligned} g[t] * m[t] &= \sigma_0 \cdot A_0 \cdot \left(RECT\left[\frac{t - 2 \cdot R_0/c}{\Delta t}\right] * RECT\left[\frac{t}{\Delta t}\right]\right) + A_0 \cdot \left(n[t] * RECT\left[\frac{t}{\Delta t}\right]\right) \\ &= \sigma_0 \cdot A_0 \cdot \left(\Delta t \cdot TRI\left[\frac{t}{\Delta t}\right]\right) * \delta\left[t - 2 \cdot \frac{R_0}{c}\right] + n[t] * \left(A_0 \cdot RECT\left[\frac{t}{\Delta t}\right]\right) \\ &= A_0 \cdot \left(\sigma_0 \cdot \Delta t \cdot TRI\left[\frac{t - 2 \cdot R_0/c}{\Delta t}\right] + n[t] * RECT\left[\frac{t}{\Delta t}\right]\right) \end{aligned} \tag{23.84}$$

The deterministic first term is a triangle whose support is twice that of the emitted rectangle, which means that the range resolution ΔR is of the order $c \cdot \Delta t$. If the noise is sufficiently reduced by the averaging over the rectangular impulse response, then the maximum of the deterministic triangle may be easy to locate on the filtered noisy background and the position of its maximum amplitude provides the estimate of the time delay of the returned signal:

$$\begin{aligned} (g[t] * m[t])_{\max} &= \arg\left\{\max\left\{\sigma_0 \cdot \Delta t \cdot TRI\left[\frac{t - 2 \cdot R_0/c}{\Delta t}\right] + n[t] * RECT\left[\frac{t}{\Delta t}\right]\right\}\right\} \\ &\Longrightarrow t_{\max} \cong 2 \cdot \frac{R_0}{c} \Longrightarrow R_0 \cong \frac{c \cdot t_{\max}}{2} \end{aligned} \tag{23.85}$$

If two targets are located at ranges that differ by a distance of the order of $\Delta R < c \cdot \Delta t$, then the triangular filtered signals will overlap, possibly impairing the ability to distinguish them. By analogy with the resolution of an optical system, an approximate metric for the range resolution is the width parameter of the triangular filtered output. By shortening Δt, we can improve the ability of the system to distinguish point targets at different distances from the radar, and thus Δt provides a measure of the range resolution of the system; the smaller its value, the better the range resolution. That said, also note that the power in the pulse is proportional to the area of $|s[t]|^2$, so that the power in the transmitted pulse decreases as the range resolution increases. We could compensate the measured signal in Equation (23.82) by amplifying its amplitude in a fashion analogous to the limiting form of the Dirac delta function in Equation (6.93), but the amplitude of the transmitted signal will eventually overload the RF circuitry. These observations effectively limit the range resolution of pulsed radar.

23.5.1.2 Chirp Radar

Fortunately, we have already found the means to circumvent the pulse-width limit during the discussion of the best signals for matched filters in Chapter 19. There we recognized that the fundamental constraint is not the *width* of the pulse in $s[t]$, but rather the *bandwidth* of $s[t]$; the wider the bandwidth of $s[t]$, the narrower the support of the deterministic part of the filtered output. Based on the discussion surrounding Equation (19.119), we can transmit a modulated chirp with finite support:

$$s[t] = RECT\left[\frac{t}{\Delta t}\right] \cdot \exp\left[+i\pi\left(\frac{t}{\alpha_0}\right)^2\right] \tag{23.86}$$

where both parameters Δt and α_0 have dimensions of time. The maximum magnitude $|s[t]|_{\max} = 1$, so that the maximum amplitude is constrained. The discussion of stationary phase leading to Equation (13.77) shows that the spectrum of the transmitted signal in Equation (23.86) is approximately:

$$S[\nu] \cong RECT\left[\frac{\nu}{(\Delta t/\alpha_0^2)}\right] \cdot \exp\left[+i\frac{\pi}{4}\right] \cdot \exp\left[-i\pi\left(\frac{\nu}{(\Delta t/\alpha_0^2)}\right)^2\right] \tag{23.87}$$

The bandwidth of the transmitted signal is approximately $\Delta t/\alpha_0^2$, so that the width of the resulting correlation peak emerging from the matched filter is of order $\alpha_0^2/\Delta t$. This allows the range resolution to be "tuned" by varying the ratio of the pulse width Δt and the chirp rate α_0. A target at distance R_0 produces a noise-free return signal:

$$\begin{aligned} g[t] &= \sigma_0 \cdot RECT\left[\frac{t - 2 \cdot R_0/c}{\Delta t}\right] \cdot \exp\left[+i\pi\left(\frac{t - 2 \cdot R_0/c}{\alpha_0}\right)^2\right] \\ &= \sigma_0 \cdot s[t] * \delta\left[t - 2 \cdot \frac{R_0}{c}\right] \end{aligned} \tag{23.88}$$

Equation (19.102) shows that the matched filter for $s[t]$ is:

$$\begin{aligned} m[t] = (s[-t])^* &= RECT\left[\frac{-t}{\Delta t}\right] \cdot \exp\left[-i\pi\left(\frac{-t}{\alpha_0}\right)^2\right] \\ &= RECT\left[\frac{t}{\Delta t}\right] \cdot \exp\left[-i\pi\left(\frac{t}{\alpha_0}\right)^2\right] \end{aligned} \tag{23.89}$$

The filtered output may be approximately evaluated by applying the stationary-phase approximation of Equation (13.77) for the spectrum of the modulated quadratic-phase term:

$$\begin{aligned} g[t] * m[t] &= \sigma_0 \cdot \left(s[t] * \delta\left[t - 2 \cdot \frac{R_0}{c}\right]\right) * s^*[-t] \\ &= \sigma_0 \cdot \delta\left[t - 2 \cdot \frac{R_0}{c}\right] * \left(RECT\left[\frac{t}{\Delta t}\right] \cdot \exp\left[+i\pi\left(\frac{t}{\alpha_0}\right)^2\right]\right) \\ &\quad * \left(RECT\left[\frac{t}{\Delta t}\right] \cdot \exp\left[-i\pi\left(\frac{t}{\alpha_0}\right)^2\right]\right) \\ &= \sigma_0 \cdot \delta\left[t - 2 \cdot \frac{R_0}{c}\right] * \left(\mathcal{F}_1^{-1}\left\{\left|\mathcal{F}_1\left\{RECT\left[\frac{t}{\Delta t}\right] \cdot \exp\left[+i\pi\left(\frac{t}{\alpha_0}\right)^2\right]\right\}\right|^2\right\}\right) \\ &\cong \sigma_0 \cdot \delta\left[t - 2 \cdot \frac{R_0}{c}\right] * \left(\mathcal{F}_1^{-1}\left\{\left|RECT\left[\frac{\alpha_0^2 \cdot \xi}{\Delta t}\right] \cdot e^{+i\pi/4}\, e^{-i\pi\alpha_0^2\xi^2}\right|^2\right\}\right) \\ &\cong \sigma_0 \cdot \delta\left[t - 2 \cdot \frac{R_0}{c}\right] * \mathcal{F}_1^{-1}\left\{RECT\left[\frac{\xi}{(\Delta t/\alpha_0^2)}\right]\right\} \\ &= \sigma_0 \cdot \left(\frac{\Delta t}{\alpha_0^2}\right) \cdot \delta\left[t - 2 \cdot \frac{R_0}{c}\right] * SINC\left[\left(\frac{\Delta t}{\alpha_0^2}\right)t\right] \\ &= \sigma_0 \cdot \frac{1}{(\alpha_0^2/\Delta t)} SINC\left[\frac{t - 2 \cdot R_0/c}{(\alpha_0^2/\Delta t)}\right] \end{aligned} \tag{23.90}$$

In words, the filtered output is a *SINC* function in the time domain of width $\alpha_0^2/\Delta t$ and amplitude $(\alpha_0^2/\Delta t)^{-1}$ centered about the coordinate at the time delay $t_0 = 2 \cdot R_0/c$. This may be interpreted as an approximation to the Dirac delta function, and the support of the filtered output decreases as the width Δt of the chirp pulse increases. In other words, the range resolution *improves* as the pulse width is increased (for a fixed chirp rate), which is the opposite of the case for pulsed radar."

23.5.2 Azimuthal Resolution

The ability of a radar system to distinguish targets at the same range but along different radial lines from the antenna is the azimuthal resolution. In the discussion of Chapter 21, we saw that the angular resolution of an optical system is determined by the angular extent of the impulse response, which is proportional to the reciprocal of the pupil diameter. For a radar system acting at long wavelengths, the impulse response generated by a finite-sized antenna will be very wide. For example, the pupil function of a radar system with a square antenna with side d_0 is:

$$p[x, y] = RECT\left[\frac{x}{d_0}, \frac{y}{d_0}\right] \tag{23.91}$$

Equation (21.170) shows that the associated impulse response in the Fraunhofer diffraction region is proportional to the appropriately scaled 2-D *SINC* function:

$$h[x, y] \propto P\left[\frac{x}{\lambda_0 z_1}, \frac{y}{\lambda_0 z_1}\right] \propto SINC\left[\frac{x}{(\lambda_0 z_1/d_0)}, \frac{x}{(\lambda_0 z_1/d_0)}\right] \tag{23.92}$$

The combination of the small value of d_0 and the long wavelength λ_0 means that the width parameter of the central lobe of the *SINC* may be very wide. This means that the radar antenna will receive signals of similar strength from point reflectors spread over a wide angle. This means, in turn, that the direction from the antenna to a reflecting point object cannot be determined to good accuracy because the amplitude of the signals will have similar "strength". In other words, the radar "optical" system with a small antenna has poor *azimuthal resolution*.

SAR was designed to improve the azimuthal resolution by conveying a small system pupil (with a large impulse response) on a moving platform. The measured signal is recorded over time as the platform moves past the scene. The recorded signal is processed to produce an image of the objects as though obtained from a system with a large aperture that may be interpreted as "synthesized" from the moving small aperture.

23.5.3 SAR System Architecture

A SAR system consists of a stable RF oscillator and a small antenna mounted on a moving platform (often an aircraft or spacecraft). The antenna emits an RF signal generated by the oscillator as the platform moves and measures the reflected radiation; it acts as the pupil of the system. Because the linear dimension of the pupil is small, the emitted antenna pattern is very wide so that the system has little or no ability to resolve targets in azimuth. The relative motion of the pupil and target(s) as the platform moves ensures that the temporal frequency of signal(s) reflected from the object(s) changes relative to the transmitted signal, i.e., the measured signal is Doppler shifted. If the platform approaches the object, the frequency of the received signal is larger than that transmitted ("blue shift"); as the platform recedes, the received frequency is smaller ("red shift"). The time-varying Doppler-shifted frequency impresses a time-varying phase shift on the received signal relative to that of the transmitted signal. This recorded phase shift produces a chirp function that is characteristic of the object location and range.

23.5.3.1 Theoretical SAR System

To introduce the principles of SAR, we first consider a system that transmits a signal from a stable oscillator at frequency ν_0, with a value typically of the order of hundreds of megahertz:

$$s[t] = A_0 \exp[+2\pi i \nu_0 t] \tag{23.93}$$

The signal could be generalized to include a time-varying modulation and to include a time-dependent initial phase $\phi[t]$, but this simpler model is sufficient for this case. Because the system is assumed to be linear, we can set $A_0 = 1$ without loss of generality. At this point, we assume that this signal is emitted continuously, and so the reflected signals return continuously as well. Later, we will consider a more realistic system that emits short pulses.

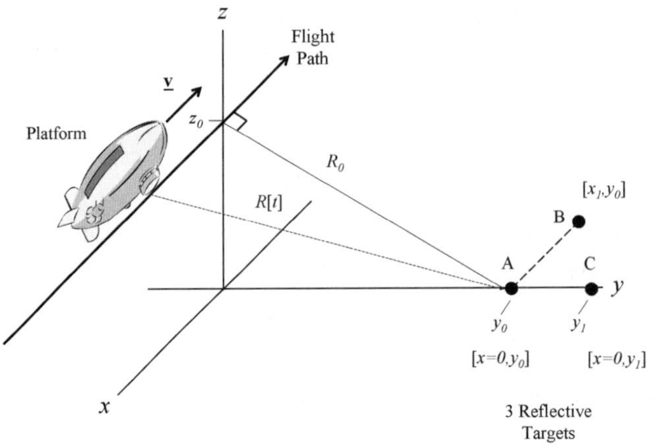

Figure 23.41 Configuration for SAR imaging system: platform (depicted realistically as a blimp) travels parallel to x-axis at altitude z_0. Three reflective targets (A, B, C) are shown in the plane $[x, y; z = 0]$. The range from the platform to target A is $R[t]$, which is a minimum at the slant range R_0.

Assume an object that consists of three point reflectors at different locations shown in Figure 23.41. The platform carrying the system moves parallel to the x-axis at velocity v_0 and at a fixed altitude z_0. The platform moves past point reflector "A" located on the y-axis at distance y_0 from the origin, so that its location is $[0, y_0, 0]$. The range at closest point of approach is the so-called "slant range", which is labeled R_0 in the figure. The object function for reflector "A" has the form:

$$f[x, y; z = 0] = \delta[x, y - y_0] \tag{23.94}$$

The slant range to this target is $(R_A)_0 = \sqrt{y_0^2 + z_0^2}$ and occurs when the platform is over the origin of coordinates. The general expression for the distance from the platform to target "A" as a function of the platform coordinate x is:

$$\begin{aligned} R_A[x; y = y_0, z = z_0] &= \sqrt{x^2 + y_0^2 + z_0^2} \\ &= \sqrt{x^2 + R_{A0}^2} = R_{A0} \cdot \sqrt{1 + \frac{x^2}{R_{A0}^2}} \end{aligned} \tag{23.95}$$

If $|x| < R_A$, we can expand the square root in a Taylor series that is truncated by analogy with the Fresnel approximation and the sag formula:

$$\begin{aligned} R_A[x] &= R_{A0}\left(1 + \frac{1}{2}\frac{x^2}{R_{A0}^2} - \frac{1}{8}\left(\frac{x^2}{R_{A0}^2}\right)^2 + \cdots\right) \\ &= R_{A0} + \frac{x^2}{2R_{A0}} - \frac{x^4}{8R_{A0}^3} + \cdots \\ &\cong R_{A0} + \frac{x^2}{2R_{A0}} \end{aligned} \tag{23.96}$$

In words, the range is approximated by the sum of the constant slant ("shortest") range and a varying quadratic part.

As the platform moves at speed v_0, the distance R_A to the point reflector varies with time. Because the platform moves with a (stable) speed v_0, the position x satisfies $x = v_0 t$ and the range $R_A[x]$ may be expressed as a function of time:

$$R_A[x] \cong R_{A0} + \frac{x^2}{2R_{A0}} \to R_A[t] \cong R_{A0} + \left(\frac{v_0^2}{2R_{A0}}\right) t^2 \tag{23.97}$$

The range coincides with the slant range for this object at $x = 0$ and therefore at $t = 0$. The travel time τ_A from the transmitter to point reflector "A" and back is twice as long as the one-way travel time:

$$\tau_A[t] = 2 \cdot \frac{R_A[x]}{c} \cong \frac{2}{c} \cdot \left(R_{A0} + \frac{x^2}{2R_{A0}}\right) = \frac{2R_{A0}}{c} + \frac{(v_0 t)^2}{R_{A0} c} \tag{23.98}$$

where c is the speed of light and the approximation in Equation (23.96) has been used.

The reflected and measured signal is a time-delayed version of the emitted signal with reduced amplitude. The combination of the amplitude reflectance of the target and the decrease in amplitude due to the inverse square law is encapsulated in the "reflectivity" factor σ_A:

$$\begin{aligned} g_A[t] &= \sigma_A \exp\left[+2\pi i \nu_0 \left(t - \frac{2R_A[t]}{c}\right)\right] \\ &= \sigma_A \exp\left[+2\pi i \nu_0 \left(t - \frac{2R_{A0}}{c} - \frac{v_0^2}{R_{A0} c} t^2\right)\right] \\ &= \sigma_0 \cdot \exp[+2\pi i \nu_0 t] \cdot \exp\left[-2\pi i \nu_0 \frac{2R_{A0}}{c}\right] \cdot \exp\left[-i\pi \left(\frac{t}{\alpha_A}\right)^2\right] \end{aligned} \tag{23.99}$$

where the chirp rate of the quadratic-phase term has been collected into the parameter:

$$\alpha_A = \sqrt{\frac{R_{A0} \cdot c}{2{v_0}^2 \nu_0}} \tag{23.100}$$

which has dimensions of time and which clearly increases with increasing slant range R_{A0}. This observation confirms that the signal recorded by the SAR system is shift variant because objects at different slant ranges map to quadratic-phase functions with different chirp rates. In words, the (approximate) measured temporal signal from target "A" includes a constant phase, a linear phase, and a quadratic phase. The signal reflected from the target also has infinite support, which means that the target has been and will be illuminated and the reflection has been measured "forever" (again, we will later consider a system based on realistic assumptions). The real part of the return from target "A" is shown in Figure 23.42a.

As an aside, note that Equation (23.99) may be rewritten in terms of the temporal frequency ν_0 of the transmitted signal or as an explicit Doppler shift due to the motion of the platform at velocity $\underline{\mathbf{v}} = \hat{\underline{\mathbf{x}}} v_0$:

$$g_A[t] = \sigma_A \exp\left[+2\pi i \left(\nu_0 - \frac{{v_0}^2}{R_{A0} c} t\right) t + i\left(4\pi \nu_0 \frac{R_{A0}}{c}\right)\right] \tag{23.101}$$

It may be helpful to rewrite this expression in a simpler form:

$$g_A[t] = \sigma_A \exp[+i(2\pi \cdot \nu[t] \cdot t + \phi_{A0})] \tag{23.102a}$$

where the (constant) initial phase is proportional to the slant range:

$$\phi_{A0} \equiv +4\pi \nu_0 \frac{R_{A0}}{c} \tag{23.102b}$$

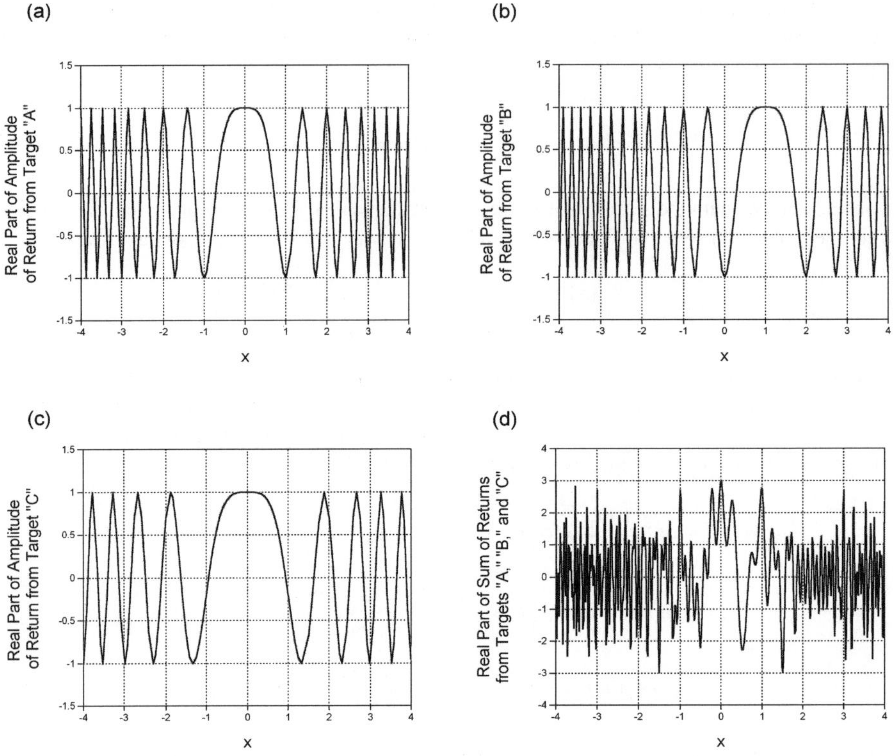

Figure 23.42 The real parts of complex-valued measured signals from the three targets in Figure 23.41: (a) from target "A" located at $[0,\, y_0]$ with slant range R_{A0}; (b) from target "B" located at $[x_0 = 1,\, y_0]$ with the same slant range; (c) from target "C" located at $[0,\, y_1 \cong (\frac{4}{3})^2 \cdot y_0]$, so that $\alpha_C = \frac{4}{3}\alpha_A$; (d) sum of the real parts of the three returns.

and the Doppler-shifted frequency $\nu[t]$ is:

$$\nu_A[t] \equiv \nu_0 - \left(\frac{\mathrm{v}_0{}^2}{R_{A0}c}\right)t \tag{23.102c}$$

Note that the frequency $\nu_A[t] > \nu_0$ for $t < 0$ (corresponding to a "blue shift" in frequency as the platform approaches the target), $\nu_A[t = 0] = \nu_0$ (when there is no relative motion of platform and target), and $\nu_A[t] < \nu_0$ for $t > 0$ (a "red shift" as the target recedes).

After the platform takes an infinitely long time to pass over the scene, the infinite-length recorded signal $g_A[t]$ is processed to recover "image(s)" of the target(s). Because the measured signal in Equation (23.98) includes a quadratic phase, our previous experience suggests that we should apply the corresponding matched filter from Equation (19.105) to "compress" the chirp signal back into a Dirac delta function (a process obviously called "pulse compression). The impulse response of the appropriate matched filter is proportional to the complex conjugate of the transmitted signal, i.e., a chirp with the same rate but opposite sign:

$$w_A[t] \propto g_A^*[t] \propto \exp\left[+i\pi\frac{t^2}{\alpha_A^2}\right] \tag{23.103}$$

If the signal has infinite support, the output of the matched filter is easy to obtain by applying Equation (17.29):

$$g_A[t] * w_A[t] = \left(\sigma_A \exp\left[-2\pi i \nu_0 \frac{2R_{A0}}{c}\right] \cdot \exp[+2\pi i \nu_0 t]\right) \cdot \alpha_A^2 \cdot \delta[t] \propto \sigma_A \cdot R_{A0} \cdot \delta[t] \quad (23.104)$$

The location of the Dirac delta function in the time domain gives the azimuth of the target (on the y-axis in this example) and the slant range R_{A0} may be derived from the measured initial phase via Equation (23.102b).

Now consider the effect of this filter on the signal reflected from reflector "B" in Figure 23.41 that has the same slant range but a different azimuth location. It is easy to see that the measured chirp signal $g_B[t]$ is identical to that for $g_A[t]$ in Equation (23.99) except its center is translated to the location $t = x_1/\mathrm{v}_0$:

$$\begin{aligned} g_B[t] &= \sigma_B \exp\left[+2\pi i \nu_0 \left(\left[t - \frac{x_1}{\mathrm{v}_0}\right] - \frac{2R_{A0}}{c} - \frac{\mathrm{v}_0{}^2}{R_{A0}c}\left[t - \frac{x_1}{\mathrm{v}_0}\right]^2\right)\right] \\ &= \sigma_B \exp\left[+2\pi i \nu_0 t - 2\pi i \nu_0 \left(\frac{x_1}{\mathrm{v}_0} + \frac{2R_{A0}}{c}\right) - 2\pi i \nu_0 \frac{\mathrm{v}_0{}^2}{R_{A0}c}\left(t^2 - 2\frac{x_1 t}{\mathrm{v}_0} + \frac{x_1^2}{\mathrm{v}_0{}^2}\right)\right] \\ &= \sigma_B \exp\left[-2\pi i \nu_0 \left(\frac{2R_{A0}}{c} + \frac{x_1}{\mathrm{v}_0} + \frac{x_1^2}{R_{A0}c}\right) + 2\pi i \nu_0 t \left(1 + \frac{2x_1 \mathrm{v}_0}{R_{A0}c}\right) - i\pi \frac{t^2}{\alpha_A^2}\right] \end{aligned} \quad (23.105)$$

The real part of the measured signal from this target, with the obvious translation, is shown in Figure 23.42b. Since the signal includes the same quadratic-phase term present in Equation (23.99) due to target "A" at the same slant range, the matched filter is identical to that in Equation (23.104) and the output will consist of a Dirac delta function at the correct location $t = x_1/\mathrm{v}_0$. This shows that the radar system is shift invariant for targets at the same slant range.

Now add a third target "C" at the same azimuth as target "A", but with a longer slant range R_{0C}. The measured signal still has infinite support, but its chirp rate due to the Doppler shift is "larger" ("slower" chirp). The return $g_C[t]$ from target "C" has the same form as Equation (23.83) after substituting R_{0C} for R_{0A}, which means that the chirp rate of the measured signal is larger than for targets "A" and "B" (Figure 23.42c). In other words, the impulse response for the matched filter $w_C[t]$ that recovers the "image" of target "C" is given in Equation (23.103) after the same substitution:

$$w_C[t] \propto g_C^*[t] \propto \exp\left[+i\pi \frac{t^2}{(\sqrt{R_{0C}/2\mathrm{v}_0{}^2 \nu_0})^2}\right] \equiv \exp\left[+i\pi \frac{t^2}{\alpha_C^2}\right] \quad (23.106)$$

Since the system is linear, we can sum the responses from the three signals to derive the signal $g_{ABC}[t]$ resulting from the three targets viewed simultaneously. The real part of the sum is shown in Figure 23.42d. The processed signals are shown in Figure 23.43. The system depends on the Doppler effect to create the chirp functions, but has an additional complication that source objects at different minimum distances create infinite-support quadratic-phase functions with different chirp rates. The reconstructions of targets at a specific slant range are obtained by applying the appropriate quadratic-phase matched filter. If targets at other slant ranges are present, the output will also include "messy quadratic-phase stuff" analogous to Equations (23.65) and (23.66). By analogy with the reconstruction of the Fresnel hologram, we can reconstruct the signal by applying the filter for each slant range in turn and saving the resulting line of filtered data. The lines of filtered data may be butted together to create an "image" of the scene. This process is analogous to measuring the reconstruction of the Fresnel hologram on observation planes at each distance z_2 from the hologram. Targets at other ranges produce "out-of-focus" light in each line of data.

Note that the point-source antenna assumed in this development would illuminate targets on both sides of the flight path, which means that point targets at the same slant range would produce returns

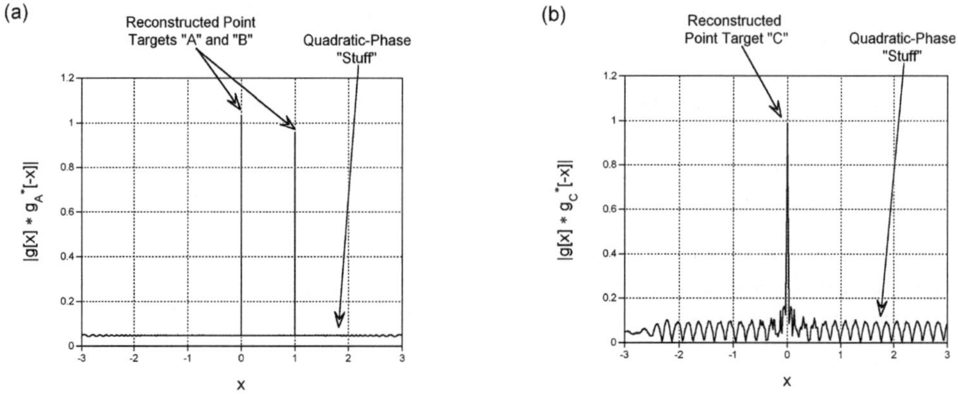

Figure 23.43 The output of the matched filters for the two slant ranges: (a) magnitude of the output of the filter matched to the slant range R_{A0} that applies to targets "A" and "B", showing the messy quadratic-phase "stuff" due to the "out-of-focus" target "C"; (b) magnitude of the output of the filter matched to R_{C0}, again showing the quadratic-phase "stuff" due to "A" and "B".

that are indistinguishable. For this reason, the antenna is designed and oriented to illuminate only one side of the flight path of the platform, which is the source of the common name for such a system as "side-looking radar".

23.5.3.2 Realistic SAR System

The system just considered is not realistic for several reasons, not least because the emitted and measured signals both have infinite support, which means that the target remains within the "field of view" of the system for an infinitely long time. A realistic SAR system emits "pulses" of RF radiation rather than the continuous tone in Equation (23.93). Also, the antenna must transmit *and* receive signals simultaneously, which is not easily done. The functional shape of the energy pulse could be a narrow (and tall) rectangle function (the usual vision of a pulse), a short burst of a sinusoid with linear phase, or a short burst of a sinusoidal signal with quadratic phase (a short "chirp"). The first case is perhaps the easiest to model, the second is perhaps easiest to implement in hardware, and the third combines useful aspects of both. A simplified view of a system for the second case is shown in Figure 23.44. We assume that all reflectors are relatively nearby so that the time from the RF pulse transmission to the measurement of the final return is relatively small. The antenna transmits the RF pulse and then switches to "listen" for the reflections. The received signal is amplified and multiplied by the original sinusoidal signal in a *mixer*. It is easy to see that the output of the ideal mixer for equal-amplitude transmitted and received sinusoids yields additive terms with the sum and difference frequencies; the sum-frequency term is removed by lowpass filtering:

$$\cos\left[2\pi\nu_0\left(t-\frac{2R[x]}{c}\right)\right]\cdot\cos[2\pi\nu_0 t]$$
$$=\frac{1}{2}\cos\left[2\pi\nu_0\left(2t-\frac{2R[x]}{c}\right)\right]+\frac{1}{2}\cos\left[2\pi\nu_0\left(\frac{2R[x]}{c}\right)\right]\rightarrow\frac{1}{2}\cos\left[2\pi\nu_0\left(\frac{2R[x]}{c}\right)\right] \quad (23.107)$$

The output of the mixer is a sinusoid whose phase is determined by the range to the target that varies with the platform position x. In the Fresnel approximation already considered in Equation (23.96), we

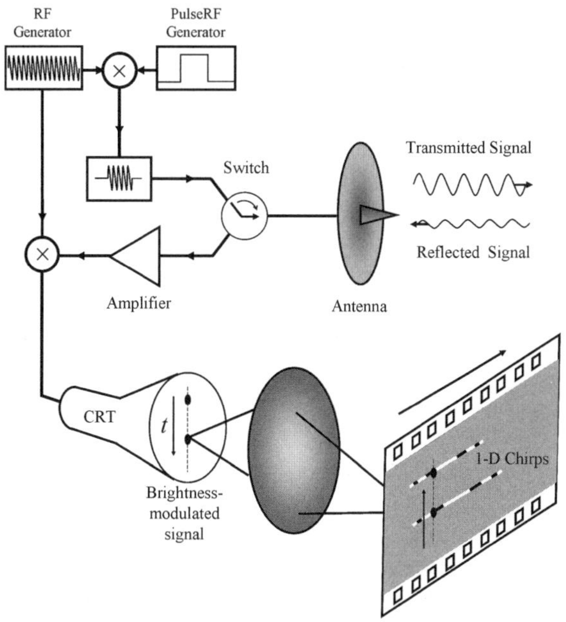

Figure 23.44 Schematic of realistic early SAR system: the measured reflection of the RF pulse is amplified and its phase used to modulate the brightness of the vertically scanned cathode-ray tube (CRT) display. The modulated line is imaged onto film. The phase changes of the reflected signal write 1-D chirp functions on the film, which focus light in a subsequent reconstruction.

see that the range includes constant and quadratic terms with values determined by the slant range R_0:

$$\cos\left[4\pi\frac{\nu_0}{c}\cdot\left(R_0+\frac{x^2}{2R_0}\right)\right]=\cos\left[\pi\cdot\left(\frac{x^2}{R_0\cdot c/2\nu_0}\right)+4\pi\frac{\nu_0}{c}R_0\right]$$
$$\equiv\cos\left[\pi\cdot\left(\frac{x^2}{R_0\cdot c/2\nu_0}\right)+\phi_0\right] \tag{23.108}$$

In words, the "brightness" of the return measured from the target for a specific platform position x is determined by the phase of the chirp function. As the coordinate x varies due to motion of the platform, the measured signal is a chirp with rate determined by the slant range. This output modulates the "brightness" of a CRT display that is scanned in proportion to the elapsed time, and thus to the instantaneous range. In this simple (and historically fairly accurate) system, the line scan is recorded on a piece of film. The next modulated pulse is transmitted and the process repeated. After the platform has moved past the scene, the phases of the returns from a point reflector generate a 1-D chirp function on the film of the form of Equation (23.108) transverse to the line scan; the chirp rate is proportional to the slant range. Point reflectors at different slant ranges and azimuths produce a set of 1-D chirp functions that are stacked on the film at "heights" and with chirp rates determined by the slant range, as shown schematically in Figure 23.44. The 1-D chirp functions act as lenses in a coherent optical system that focus light at different distances in a fashion analogous to the different chirp rates in the system with infinite-support chirps.

PROBLEMS

23.1 Derive the holographic reconstruction obtained in the Fraunhofer region if the hologram is illuminated by light from the off-axis point source.

23.2 Derive the final result in Equation (23.8).

23.3 Find the expression for the hologram of a pair of Dirac delta functions: one located at $x = 0$ with unit weight and the other at $x = +\frac{1}{2}$ with weight $\alpha_0 = \frac{1}{3}$.

23.4 Prove Equation (23.57):

$$|s[x, y; z_1]|^2 \propto 1 + \frac{2\alpha_0}{1+\alpha_0^2} \cos\left[\frac{\pi}{\lambda_0 z_3}\left(\left[x + \frac{z_3}{z_0}x_0\right]^2 - z_3\left[\frac{1}{z_0 - z_1}\right]x_0^2\right)\right]$$

23.5 Determine the functional form of a hologram of a single point source and an off-axis planar reference wave.

23.6 Consider two point sources of monochromatic light with the same wavelength λ_0 located *on* the z-axis: the first is at the origin with unit amplitude, while the second is located at $z = z_0 > 0$ (i.e., "closer" to the plane of the holographic recording). The second source has amplitude b where $0 \leq b \leq 1$ so that the second source may be "fainter" than the first. The light from the two sources propagates to the plane defined by $z = z_1 > z_0 \geq 0$ in the Fresnel diffraction region. Ignore any leading constant-phase factors in the derivation. You also may combine the amplitude factors for the second source into a single factor α_0. The radiance at the plane z_1 is recorded on a photographic plate and developed so that the "amplitude transmittance" is proportional to the complement of the incident irradiance, i.e., the result is a photographic "negative". The developed hologram is replaced at its original location and illuminated with the original "on-axis" reference source at $z = 0$. The diffracted light is observed at an arbitrary plane specified by z_2 in the Fresnel diffraction region, where $z_2 \neq z_1$.

- **(a)** Write down the mathematical formula for the 2-D amplitude pattern generated at this plane by the two sources. The formula is (obviously) a function of z_1, z_0, b, and λ_0.
- **(b)** Evaluate the 2-D radiance at the plane z_1 as a function of z_1, z_0, λ_0, and α_0.
- **(c)** Evaluate the 2-D radiance of the two cases in part (b): the point sources have equal amplitudes, and the second source is much fainter than the first.
- **(d)** Find an expression for this transmittance. Describe the features of this transmittance pattern in words.
- **(e)** Write down the expression for the amplitude and irradiance of the light diffracted by the hologram that is observed at two different planes: those specified by $z_2 = 0$ and $z_2 = z_0$.
- **(f)** Evaluate the reconstructed image for the two cases of part (c).

23.7 Consider the Fresnel hologram of the two point sources that emit the same wavelength λ_0. The hologram is processed, replaced in the original location, and illuminated by both sources. The "real-image" reconstruction of the fainter object source is measured by a detector placed at its location. The fainter source is then moved at a constant velocity in some direction while both continue to illuminate the hologram in any convenient direction. Describe the brightness measured by the detector as a function of time. Be as quantitative as possible, but a qualitative description is valuable too.

23.8 Consider the Fresnel hologram of three point sources: the reference point at the origin and two point sources in the plane at the distance $z_1 - z_0$ from the origin. The two points in the "object" are located at $[x_1, y_1]$ and $[x_2, y_2]$ with respective amplitudes β_1 and β_2.

- **(a)** Generalize Equation (23.20) based on the derivation leading to Equation (23.57) to find the irradiance recorded at the plane $z = z_1$ for these three points.
- **(b)** Find the reconstructed amplitudes at the plane of the object points and at the plane of the reference point.

23.9 Qualitatively describe the differences in reconstructions from Lohmann CGHs in an 8×8 cell with apertures of width 1 and 3 pixels.

23.10 Consider the hologram in Equation (23.58) of two point sources: the reference point at the origin and the object point at $[x_0, 0, z_1 - z_0]$ created in light of wavelength λ_0. Find the locations of the images if the hologram is illuminated by light with wavelength λ_1 from a point source at the origin.

23.11 Find the angle between orders if a CGH is printed with a sample spacing of 600 dots per inch.

References

Anastassiou, D., 1989. Error diffusion coding for A/D conversion. *IEEE Transactions on Circuits and Systems*, 36, pp. 1175–1186.

Andrews, L.C., 1998. *Special Functions of Mathematics for Engineers*, 2nd edn. Oxford: Oxford University Press.

Arfken, G.B., 2000. *Mathematical Methods for Physicists*, 5th edn. New York: Harcourt/Academic Press.

Awwal, A.A.S., Karim, M.A., and Jahan, S.R., 1990. Improved correlation discrimination using an amplitude-modulated phase-only filter. *Applied Optics*, 29, pp. 233–236.

Barrett, H.H., 1984. The Radon transform and its applications. In E. Wolf, ed., *Progress in Optics XXI*. Amsterdam: Elsevier, pp. 219–286.

Barrett, H.H. and Myers, K., 2004. *Foundations of Image Science*. Hoboken, NJ: Wiley-Interscience.

Barrett, H.H. and Swindell, W., 1996. *Radiological Imaging: The theory of image formation, detection, and processing*, rev. edn. New York: Academic Press.

Baxes, G.A., 1994. *Digital Image Processing: Principles and Applications*. New York: John Wiley & Sons, Inc.

Bender, C.M. and Orszag, S.A., 1978. *Advanced Mathematical Methods for Scientists and Engineers*. New York: McGraw-Hill.

Blackledge, J.M., 1989. *Quantitative Coherent Imaging: Theory, methods, and some applications*. New York: Academic Press.

Bracewell, R.N., 1995. *Two-dimensional Imaging*. Englewood Cliffs, NJ: Prentice Hall.

Bracewell, R.N., 1986a. *The Fourier Transform and Its Applications*, 2nd rev. edn. New York: McGraw-Hill.

Bracewell, R.N., 1986b. *The Hartley Transform*. New York: Oxford University Press.

Bracewell, R.N., 1989. The Fourier transform. *Scientific American*, 260(6), pp. 86–95.

Brigham, E.O., 1988. *The Fast Fourier Transform and its Applications*. Englewood Cliffs, NJ: Prentice Hall.

Brown, B.R. and Lohmann, A., 1969, Computer-generated binary holograms. *IBM Journal of Research and Development*, 13, p. 160.

Burckhardt, C.B., 1970. A simplification of Lee's method of generating holograms by computer. *Applied Optics*, 9, p. 1949.

Campbell, G.A. and Foster, R.A., 1948. *Fourier Integrals for Practical Applications*. Princeton, NJ: Van Nostrand.

Carnicer, A., Juvells, I., and Vallmitjana, S., 1993. Design of inverse filter for pattern recognition. *Journal of Modern Optics*, 40, pp. 391–400.

Cartwright, M., 1990. *Fourier Methods for Mathematicians, Scientists, and Engineers*. New York: Ellis Horwood.

Fourier Methods in Imaging Roger L. Easton, Jr.

Casasent, D.C., ed., 1978. *Optical Data Processing, Applications*. Berlin: Springer-Verlag.

Castleman, K.R., 1996. *Digital Image Processing*. Englewood Cliffs, NJ: Prentice Hall.

Champeney, D.C., 1973. *Fourier Transforms and Their Physical Applications*. New York: Academic Press.

Champeney, D.C., 1987. *A Handbook of Fourier Theorems*. London: Cambridge University Press.

Collier, R.J., Burckhardt, C.B., and Lin, L.H., 1971. *Optical Holography*. New York: Academic Press.

Cooley, J.W., 1988. The re-discovery of the fast Fourier transform algorithm. *Mikrochimica Acta*, III, pp. 33–45.

Cooley, J.W. and Tukey, J.W, 1965. An algorithm for the machine calculation of Fourier series. *Mathematics of Computation*, 19, pp. 297–301.

Copson, E.T., 2004. *Asymptotic Expansions*. Cambridge: Cambridge University Press.

Davis, H.F., 1963. *Fourier Series and Orthogonal Functions*. New York: Dover.

Deans, S.R., 1983. *The Radon Transform and Some of its Applications*. New York: John Wiley & Sons, Inc.

Deller, J.R., Jr., 1994. Tom, Dick, and Mary discover the DFT. *IEEE Signal Processing Magazine*, 11(4), pp. 36–50.

Dudgeon, D.E. and Mersereau, R.M., 1984. *Multidimensional Digital Signal Processing*. Englewood Cliffs, NJ: Prentice Hall.

Easton, R.L., Jr., 1993. SIGNALS: interactive software for one-dimensional signal processing. *Computer Applications in Engineering Education*, 1(6), pp. 489–501.

Easton, R.L., Jr. and Barrett, H.H., 1986. Tomographic transformations in optical signal processing. In J.L. Horner, ed., *Optical Signal Processing*. San Diego: Academic Press, pp. 335–386.

Easton, R.L., Jr., Wang, S.-G and Nagarajan, R., 1996. Error-diffused quantization in computer-generated holography incorporating a centering model for dot-overlap correction. *Proceedings of the SPIE*, 2652, pp. 345–356.

Elachi, C., Bicknell, T., Jordan, R.L., and Wu, C., 1982. Spaceborne synthetic-aperture imaging radars: applications, techniques, and technology, *Proceedings of the IEEE*, 70, p. 1174.

Erdelyi, A., 1956. *Asymptotic Expansions*. New York: Dover.

Floyd, R.W. and Steinberg, L., 1975. An adaptive algorithm for spatial greyscale. *Proceedings of the SID*, 17, pp. 75–77.

Fowles, G.R. and Cassiday, G.L., 2004. *Analytical Mechanics*, 7th edn. New York: Brooks/Cole.

Frieden, B.R., 2002. *Probability, Statistical Optics, and Data Testing*, 3rd edn. Berlin: Springer-Verlag.

Friedman, B., 1969. *Lectures on Applications-oriented Mathematics*. San Francisco: Holden-Day.

Gabor, D., 1948. A new microscopic principle. *Nature*, 161, pp. 777–778.

Gaskill, J.D., 1978. *Linear Systems, Fourier Transforms, and Optics*. New York: John Wiley & Sons, Inc.

Geckinli, N.C. and Yavuz, D., 1978. Some novel windows and a concise tutorial comparison of window families. *IEEE Transactions on Acoustics, Speech, and Signal Processing*, 24, pp. 501–507.

Ghiglia, D.C. and Pritt, M.D., 1998. *Two-Dimensional Phase Unwrapping: Theory, algorithms, and software*. New York: John Wiley & Sons, Inc.

Goldstein, H., Poole, C.P., and Safko, J.L., 2002. *Classical Mechanics*, 3rd edn. Englewood Cliffs, NJ: Prentice Hall.

Golub, G.H. and Van Loan, C.F., 1996. *Matrix Computations*, 3rd edn. Baltimore, MD: Johns Hopkins University Press.

Gonzalez, R.C. and Woods, R.E., 2002. *Digital Image Processing*, 2nd edn. Upper Saddle River, NJ: Prentice Hall.

Goodman, J.W., 1985. *Statistical Optics*. New York: Wiley-Interscience.

Goodman, J.W., 2005. *Introduction to Fourier Optics*, 3rd edn. Englewood, CO: Roberts.

Gori, F., 1994. Why is the Fresnel transform so little known? In C.J. Dainty, ed., *Current Trends in Optics*. New York: Academic Press, pp. 139–148.

Gray, R.M. and Goodman, J.W., 1995. *Fourier Transforms: An introduction for engineers*. Boston: Kluwer Academic.

Hall, E.L., 1979. *Computer Image Processing and Recognition*. San Diego: Academic Press.

Hariharan, P., 1984. *Optical Holography*. Cambridge: Cambridge University Press.

Harris, F., 1978. On the use of windows for harmonic analysis with the DFT. *Proceedings of the IEEE*, 66, pp. 51–82.

Hayes, M.H., 1987. The unique reconstruction of multidimensional sequences from Fourier transform magnitude or phase. In H. Stark, ed., *Image Recovery: Theory and Application*. New York: Academic Press, pp. 198–230.

Hecht, E., 2002. *Optics*, 4th edn. Reading, MA: Addison-Wesley.

Helstrom, C.W., 1967. Image restoration by the method of least squares. *Journal of the Optical Society of America*, 57, pp. 297–303.

Herman, G.T., ed., 1979. *Image Reconstruction from Projections*. Berlin: Springer-Verlag.

Herman, G.T., 1980. *Image Reconstruction from Projections*. New York: Academic Press.

Honda, T. and Tsujiuchi, J., 1975. Restoration of linear-motion blurred pictures by image scanning method. *Optica Acta*, 22, pp. 537–549.

Iftekharuddin, K.M., Karim, M.A., and Awwal, A.A.S., 1996. Optimization of amplitude-modulated inverse filter. *Mathematical and Computer Modelling*, 24, pp. 103–112.

Iizuka, K., 1987. *Engineering Optics*, 2nd edn. Berlin: Springer-Verlag.

Jain, A.K., 1988. *Fundamentals of Digital Image Processing*. Englewood Cliffs, NJ: Prentice Hall.

James, J.F., 1995. *A Student's Guide to Fourier Transforms*. Cambridge: Cambridge University Press.

Jennison, R.C., 1961. *Fourier Transforms and Convolutions for the Experimentalist*. New York: Pergamon.

Kak, A.C., 1984. Image reconstruction from projections. In M.P. Ekstrom, ed., *Digital Image Processing Techniques*. New York: Academic Press.

Klauder, J.R., Price, A.C., Darlington, S., and Albersheim, W.J., 1960. The theory and design of chirp radars. *Bell System Technical Journal*, 34, pp. 745–808.

Knox, K.T. and Eschbach, R., 1993. Threshold modulation in error diffusion. *Journal of Electronic Imaging*, 2, pp. 185–192.

Körner, T.W., 1988. *Fourier Analysis*. Cambridge: Cambridge University Press.

Kraniauskas, P., 1994. A plain man's (sic) guide to the FFT. *IEEE Signal Processing Magazine*, 11(2), pp. 24–35.

Kyrala, A., 1972. *Applied Functions of a Complex Variable*. New York: Wiley-Interscience.

Lee, S.H., ed., 1981. *Optical Information Processing: Fundamentals*. Berlin: Springer-Verlag.

Lee, W.H., 1970. Sampled Fourier transform hologram generated by computer. *Applied Optics*, 9, p. 639.

Leith, E., and Upatnieks, J., 1962. Reconstructed wavefronts and communications theory. *Journal of the Optical Society of America*, 52, p. 1123.

Lim, J.S., 1990. *Two-dimensional Signal and Image Processing*. Englewood Cliffs, NJ: Prentice Hall.

Lohmann, A., 1994. Fourier curios. In C.J. Dainty, ed., *Current Trends in Optics*. New York: Academic Press, pp. 149–161.

Marion, J. and Thornton, S.T., 1995. *Classical Dynamics of Particles and Systems*, 4th edn. New York: Brooks/Cole.

Marsden, J.E., 1973. *Basic Complex Analysis*. San Francisco: W.H. Freeman.

Mertz, L., 1965. *Transformations in Optics*. New York: John Wiley & Sons, Inc.

Merzbacher, E., 1997. *Quantum Mechanics*. New York, John Wiley & Sons, Inc.

Miles, J.W., 1971. *Integral Transforms in Applied Mathematics*. London: Cambridge University Press.

Nahin, P.J., 1998. *An Imaginary Tale*. Princeton, NJ: Princeton University Press.

Niblack, W., 1986. *An Introduction to Digital Image Processing*. Englewood Cliffs, NJ: Prentice Hall.

Olver, F.W.J., 1974. *Asymptotes and Special Functions*. New York: Academic Press.

Papoulis, A., 1962. *The Fourier Integral and its Applications*. New York: McGraw-Hill.

Papoulis, A., 1986. *Systems and Transforms with Applications in Optics*. Malabar, FL: Robert E. Krieger.

Parker, J.A., 1990. *Image Reconstruction in Radiology*. Boca Raton, FL: CRC Press.

Pratt, W.A., 1991. *Digital Image Processing*, 2nd edn. New York: John Wiley & Sons, Inc.

Ramirez, R.W., 1985. *The FFT: Fundamentals and concepts*. Englewood Cliffs, NJ: Prentice Hall.

Reynolds, G.O., DeVelis, J.B., Parrent, G.B., Jr., and Thompson, B.J., 1989. *The New Physical Optics Notebook*. Bellingham, WA: SPIE Optical Engineering Press.

Robinson, E.A., 1982. A historical perspective of spectrum estimation. *Proceedings of the IEEE*, 70, pp. 885–907.

Rosenfeld, A. and Kak, A.C., 1982. *Digital Picture Processing*, 2nd edn. San Diego: Academic Press.

Rudin, W., 1976. *Principles of Mathematical Analysis*. New York, McGraw-Hill.

Schneider, H. and Barker, G.P., 1989. *Matrices and Linear Algebra*. New York: Dover.

Scott, C., 1998. *Introduction to Optics and Optical Imaging*. New York: IEEE Press.

Silverman, R.A., 1984. *Complex Analysis with Applications*. New York: Dover.

Smith, J.M., 1975. *Scientific Analysis on the Pocket Calculator*. New York: John Wiley & Sons, Inc.

Stephenson, G. and Radmore, P.M., 1990. *Advanced Mathematical Methods for Engineering and Science Students*. Cambridge: Cambridge University Press.

Steward, E.G., 1987. *Fourier Optics: An introduction*, 2nd edn. New York: John Wiley & Sons, Inc.

Strang, G., 2005. *Linear Algebra and its Applications*, 4th edn. New York: Brooks/Cole.

Swindell, W.K., 1970. A noncoherent optical analog image processor. *Applied Optics*, 9, pp. 2459–2469.

Walker, J.S., 1996. *Fast Fourier Transforms*, 2nd edn. Boca Raton, FL: CRC Press.

Walvoord, D.J. and Easton, R.L., Jr., 2004. Variations of matched filtering for reduced noise amplification, *Proceedings of the SPIE*, 5298, pp. 59–69.

Watson, G.N., 1944. *A Treatise on the Theory of Bessel Functions*, 2nd edn. London: Cambridge University Press.

Williams, C.S. and Becklund, O.A., 1989. *Introduction to the Optical Transfer Function*. New York: John Wiley & Sons, Inc.

Wilson, R., 1995. *Fourier Series and Optical Transform Techniques in Contemporary Optics*. New York: John Wiley & Sons, Inc.

Index

abscissa
balance-point, 424
mean-square, 424
absolute value, 54, 119, 122, 139, 575
Airy disk, 847
aliasing, 465–467, 469, 474, 479–484, 491, 507, 508, 517, 522, 539, 552, 553, 556, 560, 565
2-D, 565
analog-to-digital converter (ADC), 501
analogy of inverse and matched filter, 703
analysis
Fourier, 255, 259, 263
analyzer spectrum, 596
angular temporal frequency, 24, 655, 763
angular wavenumber, 763
antialiasing prefilter, 483
anticorrelation, 231, 308
aperture, 758, 772, 779, 782, 784, 790, 792, 794, 797, 799, 805, 808, 819
circular, 783, 847
coded, 760
rectangular, 779
Argand
Jean-Robert, 54
Argand diagram, 54, 62, 77, 143, 274, 382, 384, 531, 532, 778, 888, 890
argument, 54
associativity, 41
atmosphere
turbulent, 619
autoconvolution, 222, 225, 232, 236, 364, 427, 547, 559
autocorrelation, 201, 226–230, 235, 308, 355, 364, 604, 608, 614, 619, 626, 647, 648, 668, 670, 671, 682, 684, 686, 687, 699, 700, 703, 705, 745–747, 807, 846, 848, 850, 859, 864–867, 900
Fourier transform of, 302
normalized, 227
of cylinder function, 236, 848
autocovariance, 229
averager
uniform, 220, 221, 489, 579, 581, 672, 699, 710, 713, 714, 721
azimuth, 9, 54, 181, 182, 188, 193, 204, 327, 343, 348, 358, 367, 374, 376
summation over, 392, 394, 395, 399, 400
azimuthal resolution, 906

back projection, 375, 394, 400
filtered, 392, 395, 397
bandwidth, 509, 590, 614, 642, 645
equivalent spatial, 455, 599, 600, 602, 614, 645, 647, 703, 705, 839–841, 845
equivalent temporal, 455
spatial, 455
temporal, 455
beats, 467
bel, 160, 542
bell curve, 117
Bernoulli trials, 153, 154, 156
Bessel function, 121, 122, 124, 166, 197
series solution, 166, 168
bias, 107, 108, 125, 161, 213, 220, 229, 311, 312, 318, 465
binomial coefficient, 154
binomial probability distribution, 153, 154
bit reversal (FFT), 532
Burckhardt CGH algorithm, 890
butterfly (FFT), 530, 532, 536

Cassegrain telescope, 848
causality, 657
center of mass, 424
central moment, 152
central-ordinate theorem, 166, 284, 776
centroid, 152, 424
CGH, 885
Burckhardt algorithm, 890
error-diffused quantization, 895
Lee algorithm, 889
Lohmann algorithm, 887, 893, 894
charge-coupled device (CCD), 486, 894
checkerboarding (FFT), 537, 564, 565
chirp, 25, 115, 143, 608, 632, 703
chirp Fourier transform, 632, 638, 641
chirp radar, 905
chirp rate, 612, 614, 615, 633, 661, 665, 770, 771, 776, 779, 782, 801–803, 807, 810, 814, 871, 874, 875, 877, 878, 886, 900, 901, 903, 906, 909, 911, 913
circular triangle function, 201, 236, 364, 847
circularly symmetric function, 195
classical matched filter, 699
closure, 41
coherence length, 841

coherence of light
spatial, 828, 856
temporal, 838, 856
coherence time, 841
coherence volume, 842
COMB function, 186
commutativity, 32, 38, 41, 224, 225, 518, 641, 668
complement filter, 712, 718, 900
completeness, 42
complex conjugate, 52, 298
complex Lorentzian function, 147, 273, 497, 498, 718
complex numbers
difference, 52
equality, 52
magnitude, 54
modulus, 54
product, 53
ratio, 53
reciprocal, 53
sum, 52
composition product, 218
compression
image, 567, 599
pulse, 703
computation time, 559
DFT, 536
computed tomography, 14, 371, 377
conjugate coordinates, 52
conjugate variables, 24
consistent data, 86
constrained least-squares filter, 695
contrast normalization, 747
contrast transfer function (CTF), 108
converter
analog to digital (ADC), 501
convolution, 69, 217–223, 477, 480, 486, 487, 517, 573, 576, 579
area of, 225, 476
calculating, 222
circular, 556, 558
discrete, 507, 509, 525, 555–559
computation time, 559
discrete linear, 556
of chirp functions, 223
of decaying exponential, 223
of rectangle functions, 220
of scaled functions, 226
properties, 223
region of support of, 225
convolution integral, 217
coordinate system, 16
Cornu spiral, 778
correlator
optical, 640, 899
COSTAR, 7
critically determined imaging task, 84
CROSS function, 194, 195
cross-spectral power, 302
crosscorrelation, 229–231, 234, 682, 686, 699, 865
Fourier transform of, 299
normalized, 231
cryptography, 207
CTF, 108
curve
H&D, 209
cutoff frequency, 294, 575
cylinder function, 196

damped harmonic oscillator, 107, 659
dark-field microscopy, 630
data
consistent, 86
inconsistent, 86
DC component, 220
DCT, 460
De Moivre's theorem, 61
decibel, 160, 542
decimation (FFT)
in frequency, 534
in time, 534
decision level, 502
decomposition, 255
deconvolution, 669
defocus, 588
degenerate eigenvalue, 73
demodulation, 596
density
photographic, 207
derivative
1-D Dirac delta function, 137, 139
operator, 586
partial, 338, 587
theorem, 297
detection of signal, 667
determinant, 38, 40, 47, 72, 75
detour phase, 887
DFT, 83, 93, 308, 459, 460, 473, 484, 511–514, 516, 517, 519, 521–526, 529–534, 538, 551, 552, 554, 557, 559, 560, 562, 563, 568–570, 723, 724, 728, 729, 739
diagonalization of circulant matrix, 75
diagonalization of matrix, 73, 86, 87
differentiation, 212, 648
as convolution, 219
differentiator, 585, 586
diffraction, 6, 792
Fraunhofer, 783
Fresnel, 647, 771
diffusion
heat, 588
digital count, 723
digitization, 21, 459
dimension, 16, 30
dimensionality
vector, 29
Dirac delta function, 126, 264, 606, 711
Dirac P.A.M., 127
discrete cosine transform (DCT), 460, 568
discrete Fourier transform (DFT), 83, 93, 308, 459, 460, 473, 484, 511–514, 516, 517, 519, 521–523, 525, 526, 529–534, 538, 551, 552, 554, 557, 559, 560, 562, 563, 568–570, 723, 724, 728, 729, 739
scale factor, 524

distribution
    delta, 127
domain, 17
    continuous, 18
    discrete, 18
dot product, 32, 66, 231

edge effects, 491
eigenfunction, 93, 220, 576, 595
eigenvalue, 70, 71, 75, 76, 80, 87
    circulant matrix, 80, 81
    degenerate, 73
eigenvector, 70–75, 80, 84, 87, 88
    circulant matrix, 80, 81, 87, 93
    normalized, 76, 77, 81
    ordering, 76
electromagnetic wave, 763
emulsion
    photographic, 209
equation
    secular, 72, 75
    wave, 105, 762
equivalent width, 454
error, 668
    local, 668
    mean squared (MSE), 504
    quantization, 503
    root mean squared (RMS), 505
    total, 668
estimation of signal, 667
Euler relation, 56
expectation value, 422
exponential function, 104, 716

factorial function, 113
fast Fourier transform (FFT), 308, 459, 513, 529, 534, 723, 886
FFT, 308, 459, 513, 529, 534, 723, 886
filter
    active, 578
    allpass, 294, 575, 578, 603, 711
        chirp, 608
        constant phase, 605
        linear phase, 606
        quadratic phase, 608, 611, 612, 614, 615, 619, 628, 632–634, 637
        random phase, 619
    bandboost, 596
    bandpass, 574, 589, 849
    bandstop, 574, 597
    causal, 647, 653, 655
    complement, 712, 718
    constrained least-squares, 695
    derivative
        1-D discrete, 731, 734
        1-D discrete second derivative, 734
        2-D discrete, 732
        2-D discrete second derivative, 736
    highboost, 588
    highpass, 294, 574, 585
    inverse, 667, 669, 672, 675, 678, 679, 703, 902
    Laplacian
        discrete, 737, 742
        discrete generalized, 740
        discrete rotated, 738
    linear, 218
    linear phase, 575
    lowpass, 294, 574, 579, 848
        1-D discrete, 728
        2-D discrete, 730
        ideal, 579, 714
    magnitude, 574
    matched, 234, 668, 696, 699, 701, 703, 706, 898
        discrete, 744
        ideal, 708
        unamplified, 699
    notch, 597
    passive, 577
    phase, 294, 575, 711
    phase-only, 603
    power transmission, 577
    pseudoinverse, 670, 673
    pseudomatched, 700, 703
    reciprocal, 703
        approximation, 708
        discrete approximation, 749
    rho, 395, 397
    sharpening
        1-D discrete, 740
        2-D discrete, 742
        unsharp masking, 743
    translation, 727
        discrete, 724
    Wiener, 680, 681, 688, 693
    Wiener–Helstrom, 689, 693, 694, 706
filter theorem, 289, 555, 573, 785
focal length, 754, 761, 796, 797, 799, 801–803, 811, 815, 816, 818, 877
focal plane, 588
Fourier series, 527
Fourier transform
    2-D, 325
    C–M–C chirp, 634
    chirp, 632, 638, 641
    M–C–M chirp, 632
    of autocorrelation, 302
    of complex conjugate, 298
    of CROSS function (2-D), 341
    of crosscorrelation, 299
    of discrete periodic function, 308
    of exponential, 270
    of Fourier transform, 281
    of Gaussian (1-D), 275
    of Gaussian (2-D), 332
    of line delta (2-D), 341
    of magnitude of function, 313
    of power-law nonlinearity, 315
    of rectangle (2-D), 332
    of rotated function, 340
    of sampled function, 307
    of separable function, 327, 328
    of step function (2-D), 334

of thresholded function, 311
of triangle (2-D), 332
Fourier transform pair, 263
Fourier transformer
optical, 641
Fourier, Baron Jean-Baptiste Joseph, 255
Fourier–Bessel transform, 351
Fraunhofer diffraction, 783, 886
frequency
angular spatial, 24, 105
angular temporal, 24, 655, 763, 767
cutoff, 294, 575
instantaneous, 62
negative, 62
Nyquist sampling, 466
resonant, 655, 660
spatial, 23, 241
temporal, 24, 763
Fresnel approximation, 770, 771
Fresnel diffraction, 771, 794
example
double diffraction, 772
knife edge, 776
planar source, 774
rectangular aperture, 779
Fresnel integrals, 778
Fresnel sandwich, 643
Fresnel transform, 772
Fresnel zone plate (FZP), 643, 760
function, 16
3-D, 205
antisymmetric, 26
bandlimited, 467
basis, 91, 92
Besinc (2-D), 200, 847
Bessel (1-D), 121, 122, 124, 413
Bessel (2-D), 197
characteristic, 424
chirp, 115, 608, 632
chirp (1-D), 276, 603
circular triangle (2-D), 201, 236, 364, 847
circularly symmetric (2-D), 195, 347
circularly symmetric Gaussian (2-D), 197
COMB (1-D), 135
COMB (1-D), 279, 461, 511
complete set, 91
complex Lorentzian (1-D), 147, 273, 497, 498, 718
complex sinusoid (1-D), 58, 143
complex valued (1-D), 142
complex-valued, 56
CORRAL (2-D), 195
CROSS (2-D), 194, 341, 344
cylinder (2-D), 196
digital, 21
Dirac delta (1-D), 126, 264, 606, 711
integral expression, 131
limiting expressions, 129
of functional argument, 139
property in products, 133
sifting property, 132
Dirac delta (2-D), 182, 328
edge-spread (esf), 233
even, 26
even pair of Dirac deltas (1-D), 133, 244, 249, 266, 267, 288, 318, 408
exponential (1-D), 104, 270, 678, 716
factorial $x!$, 113
gamma, 97, 112, 350, 365, 428, 617, 619
Gaussian (1-D), 117, 275, 320, 426, 581
Gaussian (2-D), 180, 332
harmonic, 21, 107
Hermitian, 58, 228, 235, 258, 302
line delta (2-D), 187, 341
line-spread (lsf), 233, 372
linear, 17
Lorentzian (1-D), 124, 147, 273, 497, 498, 661, 694, 718, 756
modulation transfer (MTF), 221, 491, 604, 647, 650, 651, 661, 846
moment, 151
mutual coherence (MCF), 844
nonlinear, 349
norm, 92
null constant (1-D), 99
odd, 26
odd pair of Dirac deltas (1-D), 133, 134, 248, 249, 252, 256, 267, 283, 658, 661
optical transfer (OTF), 573
orthogonal, 90
orthonormal, 92
periodic, 17, 21, 306
phase transfer, 573, 650, 651
point-spread (psf), 291, 573
polar Dirac delta (2-D), 184
polar separable (2-D), 195
position, 706
probability density (pdf), 151
projection, 90, 226
pupil, 799
quadratic (1-D), 17
quadratic phase, 25, 115, 143, 575, 603, 608, 632
quadratic phase (1-D), 703
quadratic phase (2-D), 205
quantized, 20
ramp (1-D), 568, 653
rectangle (1-D), 99, 119, 133, 156, 265, 289
rectangle (2-D), 175, 190, 191, 203, 332
reference, 239, 701
representation, 91, 93
ring delta (2-D), 202
rotation of, 172
sampled, 307
separable (2-D), 171, 327
separable Dirac delta (2-D), 183
signum (1-D), 101, 268
signum (2-D), 176
$SINC^2$ (2-D), 178
SINC (1-D), 265, 289
area, 162
$SINC^2$ (1-D), 111, 293
$SINC^2$ (2-D), 332
SINC (2-D), 178
SINC (1-D), 57, 109

SINC (2-D), 332
sinusoid (1-D), 105, 267
sinusoid (2-D), 180, 204
Sombrero (2-D), 200
sombrero (2-D), 847
STEP[x], 102
STEP (1-D), 102, 268
STEP (2-D), 176
stochastic, 149, 229, 308
superchirp (1-D), 145, 431
superGaussian (1-D), 119, 428
symmetric, 26
thresholded, 125
tone transfer (TTF), 209
transfer, 291, 294, 650, 697
triangle (1-D), 101, 232
triangle (2-D), 176, 332
unit constant (1-D), 99
unit constant (2-D), 174
zeros, 17
functional, 218

Gabor, Dennis, 856
Gaussian function, 117, 180, 426, 581
Gaussian noise, 623
Gaussian probability distribution, 154
gradient, 743
ground glass, 619

Hankel transform, 347, 351
from 2-D Fourier transform, 361
of r^{-1}, 360
of Bessel function, 356
of chirp function, 363
of circular triangle, 364
of cylinder function, 358
of Gaussian, 362, 363
of quadratic phase, 363
harmonic, 313
Hartley transform, 251, 254
examples, 251
heat diffusion, 588
Hermitian function, 58, 258
highpass filter
ideal, 585
histogram, 151, 154
hologram reconstruction, 859, 878
holography, 608, 855
computer generated, 885
Fraunhofer, 856, 886
Fresnel, 868
Hooke's law, 655
human visual system (HVS), 593
Hurter and Driffield (H&D) curve, 209
Huygens, Christiaan, 765, 771

ideal sampling, 486
identity matrix, 37, 38
identity operator, 648
ill-conditioned, 675
image
real, 869, 871, 879
virtual, 869, 872, 878
image compression, 567, 599
imaginary number, 51
imaginary part, 51, 52
imaging
gamma-ray, 9
magnetic resonance (MRI), 371, 377
medical computed tomography, 14, 371, 377
optical, 4, 608
radiography, 11
Schlieren, 628
synthetic-aperture radar (SAR), 223, 608
imaging chain, 1
imaging of phase object, 628
imaging system
pinhole, 758
imaging task, 667
direct, 3
inverse, 3, 84, 669, 671
system analysis, 3
impulse response, 216, 217, 291
incoherent light, 853
inconsistent data, 86
inequality
Schwarz, 33, 67, 90, 231
triangle, 33
inner product, 66, 88, 231
instantaneous frequency, 116
integral
convolution, 217–219, 222, 577
Faltung, 218
Fresnel, 778
superposition, 222
integration, 650
intensity, 107
interference, 823
constructive, 829
destructive, 829
division-of-amplitude, 824
division-of-wavefront, 824
extended monochromatic source, 831
two point sources, 824
interferometry
stellar speckle, 8, 626
interpolation
nearest-neighbor, 491
of sampled function, 459, 472
realistic, 491
zero-order, 491
interpolator
bilinear, 494
first-order, 493
frequency domain, 496
linear, 493
inverse filter, 703
inverse filtering, 667, 669, 672, 675, 678, 679
inverse Hartley transform, 261
inverse imaging task, 84
inverse matrix, 39
inverse problem, 3, 669, 671
irradiance, 765, 843, 861

Kelvin, Lord, 436
Kronecker delta, 43

Labeyrie, Antoine, 8, 626
Laplacian, 339, 340, 588, 589, 765
  of circularly symmetric function, 356
layergram, 394
leakage, 517, 728
Lee CGH algorithm, 889
length
  coherence, 841
  focal, 754, 761, 796, 797, 799, 801–803, 811, 815, 816, 818, 877
lens
  thin, 754, 792
lens-maker's formula, 796
lexicographic ordering, 31
light
  coherent, 816
  incoherent, 845, 848, 853
  quasimonochromatic, 839–842, 845, 853
line delta function (2-D), 187
linear dependence, 42
linear independence, 42
linear-phase factor, 287
linear-phase filter, 575
linearity, 15
Lissajous figure, 62
local differencing operator, 585
logarithm
  complex amplitude, 149
Lohmann, Adolf, 887
Lorentzian function, 124, 147, 273, 497, 498, 661, 694, 718, 756

Mach band, 593
magnetic resonance imaging (MRI), 371, 377
magnification
  pupil, 815
  transverse, 803
magnitude, 250
  squared, 31
matched filter, 696, 703, 898
  input function, 701
  Wiener, 706
matched filtering, 668
matrix, 35
  circulant, 69, 75, 222, 240
  determinant, 38, 40, 47, 72, 75
  diagonal, 37
  diagonalization, 73
  Hermitian, 79
  identity, 37, 38, 71
  inverse, 39
  orthogonal, 40
  orthonormal, 40
  product, 67
  symmetric, 77
  Toeplitz, 70, 240
  transpose, 38
matrix–matrix multiplication, 36
Maxwell's equations, 765
mean value, 118, 152, 160
  of binomial probability distribution, 154
  of normal distribution, 158
  of Poisson distribution, 156
  of Rayleigh distribution, 158
  of sinusoid, 152
  of uniform distribution, 156
method of stationary phase, 436, 617, 664, 782
method of steepest descent, 436
microscopy
  dark-field, 630
modulation, 54, 97, 107
  of sinusoid, 220
modulation theorem, 295
modulation transfer function (MTF), 108, 221, 491, 604, 647, 650, 651, 661, 846
Moiré fringes, 467
moment, 151, 242, 421
  central, 152, 425
  of inertia, 424
Moore–Penrose pseudoinverse, 86, 673
MTF, 108, 221, 491, 604, 647, 650, 651, 661
  incoherent, 846
Multiple-Mirror Telescope (MMT), 850
multiplication
  matrix–matrix, 36
  matrix–vector, 35, 36, 45
  matrix-vector, 35
multiplication theorem, 295
mutual coherence function (MCF), 844

neural net, 593
noise, 149
  blue, 308
  colored, 308
  Gaussian, 623
  green, 308
  pink, 310
  quantization, 503
  white, 308
noise-to-signal power ratio, 681
nonlinearity
  magnitude, 250, 315
  power-law, 315, 860
norm, 31
normal probability distribution, 152
notation
  Cartesian coordinates, 17
  domain intervals, 17
  polar coordinates, 17
null constant function, 99
null vector, 41
Nyquist sampling frequency, 481
Nyquist sampling interval, 466
Nyquist window, 473

object wave, 855
operator
  derivative, 586
  gradient, 587
  identity, 648
  integration, 212
  Laplacian, 339, 340, 356, 588, 589
  linear, 209, 281
  linear shift-invariant, 216
  linear shift-invariant (2-D), 232
  linear shift-variant, 222

nonlinear, 310, 860
rotation, 214
scaling, 214
sharpening, 588, 589
shift-invariant, 213
shift-variant, 255
system, 207
optical correlator, 640, 899
optical elements, 792
optical Fourier transformer, 641
optical imaging system, 753
based on Fraunhofer diffraction, 790
examples
equal conjugates, 810
extended object, 809
magnified, 810
on-axis source finite-aperture unit-magnification, 808
on-axis source infinite-aperture unit-magnification, 807
virtual images, 811
optical transfer function, 573
optics
geometrical, 753
ray, 4, 753, 755
wave, 5
orthogonal matrix, 40
orthogonal vectors, 34
orthonormal matrix, 40
orthonormal vectors, 34
oscillation, 762
oscillator
damped harmonic, 107, 659
overdetermined imaging task, 84, 85
oversampling, 480

Parseval's theorem, 304, 313, 314, 578, 581
partial derivative, 587
passband, 589
pentagram "★", 227
periodic function, 17
phase
angle, 22, 54
average, 825
detour, 887
initial, 22, 61, 261
modulation, 825
stationary, 436, 617, 664, 782
unwrapped, 60, 483
phase filter, 711
phase object
imaging of, 628
phase transfer function, 573, 650, 651
phasor diagram, 54
pinhole, 760
pinhole camera, 758
point mass, 127
point spread function (psf), 216, 217
point-spread function (psf), 291
Poisson probability distribution, 153, 154
polynomial
Taylor, 427, 710
postfilter, 483
power, 259, 805
power spectrum, 259, 302
prefiltering
antialiasing, 483
Prewitt corner detector, 747
principal component, 84
principal value, 55, 60
prism, 793
probability density function, 151
probability distribution
binomial, 153, 154
Gaussian, 154, 623
normal, 152, 157
Poisson, 153, 154, 156
Rayleigh, 152, 158, 623
uniform, 152, 156, 308, 619, 626
product
composition, 218
dot, 32
inner, 79, 88, 231
scalar, 29, 32, 35, 79
projection, 34
line-integral, 372, 374, 389
vector, 29, 35
property sifting, 512
pseudoinverse, 87, 670
Moore–Penrose, 86, 673
pseudomatched filter, 700, 703
pulse compression, 703
pupil, 790, 796
pupil function, 799

quadratic-phase function, 703
quadrature, 595
quality factor, 599
quantization, 20, 312, 459, 501
nonuniform, 505
tapered, 505
quantization error, 503
quantizer, 210
error diffusion, 507
rounding, 505
truncating, 505
quantum mechanics, 457
quantum number, 501, 723
quasimonochromatic light, 839–842, 845, 853

radar
chirp, 905
synthetic aperture (SAR), 608, 900, 903
radiography, 11
range
slant, 908
range of function, 16
discrete, 20
range resolution, 904
rate
chirp, 612, 614, 615, 633, 661, 665, 770, 771, 776, 779, 782, 801–803, 807, 810, 814, 871, 874, 875, 877, 878, 886, 900, 901, 903, 906, 909, 911, 913
Rayleigh probability distribution, 152
Rayleigh's theorem, 166, 302, 304

real part, 51, 52
reciprocal filter, 703
reconstruction
hologram, 859
real image, 879
virtual image, 878
rectangle function (1-D), 119, 133, 156
rectangle function (2-D), 175, 190, 191, 203
reference wave, 879
reflection, 792
refraction, 792
refractive index, 753
relation
uncertainty, 454, 519
representation, 91, 93
resolution
azimuthal, 906
range, 904
resonance, 653
resonant frequency, 655, 660
response
impulse, 216, 217
ring delta function, 202
ringing, 539, 542, 545, 569, 570, 579, 726
running mean, 218

sag formula, 794, 908
sampling, 307, 459
ideal, 460, 486
linearity property, 462
realistic, 486
shift variance of, 462
sampling interval, 460
sandwich
Fresnel, 643
SAR, 608, 900, 903
scalar product, 29, 32, 65, 172, 231
linearity, 85
scaling operator, 214
scaling theorem, 284, 581
scanner
pushbroom, 486
whiskbroom, 486
Schlieren imaging, 628
Schwarz inequality, 33, 67, 90, 231
secular equation, 72, 75
series
Fourier, 527
Taylor, 140, 316, 349, 427, 438, 477, 589, 604, 629, 693, 710, 860, 908
sharpener, 588
shift invariance, 16, 68, 80, 806
shift theorem, 284, 285, 287
side-looking radar, 903, 912
sideband, 643
sifting property, 512
signal
audio, 485
detection, 667
estimation, 667
spread spectrum, 703
signal-to-noise power ratio, 681
signal-to-noise ratio (*SNR*), 160, 505
approximations, 161
signum function $SGN[x]$, 101
signum function $SGN[x]$, 268
signum function (2-D), 176
sinogram, 378
sinusoid function (1-D), 105
sinusoid function (2-D), 180, 204
slant range, 908
slice
central, 372
Snell's law, 760, 761
spatial bandwidth, 455
spatial power, 107
spectrum
operator, 71
Wiener, 308
spectrum analyzer, 596
spline interpolator, 494
squared magnitude, 58
standard deviation, 118, 119
of normal distribution, 158
of Rayleigh distribution, 158
of uniform distribution, 156
stationary phase, 436
steepest descent, 436
stellar speckle interferometry, 8, 626
STEP function (1-D), 268
subspace, 42
left null, 46
superchirp function, 145, 431
superGaussian function, 119, 428
superposition, 15
superposition of 2-D functions, 335
synthesis
Fourier, 261, 263
synthetic-aperture radar (SAR), 608, 900, 903
system
linear, 6
optical imaging, 753
system analysis, 3

Tacoma Narrows bridge, 658
tapered quantization, 505
Taylor polynomial, 427, 710
Taylor series, 55, 710
theorem
1-D Fourier transform
central ordinate, 284
linearity, 281
multiplication by constant, 281
2-D Fourier transform
central ordinate, 334
derivative, 338
filter, 337
of Fourier transform, 336
scaling, 334, 336
shift, 334
central limit, 623
central ordinate, 166, 651, 776
central-limit, 452
central-slice of Radon transform, 389
filter of Radon transform, 390

Fourier transform
discrete scaling, 563
scaling, 581
Fourier transform of autocorrelation, 302
Fourier transform of complex conjugate, 298
Fourier transform of convolution, 289, 555, 573, 785
Fourier transform of crosscorrelation, 299
Fourier transform of derivative, 297
Fourier transform of Fourier transform, 281, 818
Fourier transform of modulated function, 295, 785
Fourier transform of periodic function, 306
Fourier transform of power-law nonlinearity, 315
Fourier transform of reversed function, 285
Fourier transform of rotated functon, 340
Fourier transform of sampled function, 307
Fourier transform of scaled function, 284, 612, 784
complex-valued scaling factor, 287
real-valued scaling factor, 284
Fourier transform of shifted function, 287, 785
Fourier transform of thresholded function, 311
Hankel transform
"transform-of-a-transform", 355
central ordinate, 354
crosscorrelation, 355
filter, 355
partial derivative, 355
scaling, 354
shift, 354
linearity of Radon transform, 388
moment, 423
2-D, 433
Parseval's, 304, 313, 314, 578, 581
discrete functions, 561
Radon transform of scaled function, 388
Radon transform of translated function, 389
Rayleigh's, 166, 302, 304
shift
2-D, 336
Whittaker–Shannon sampling, 479
Wiener–Khintchin, 302, 604, 614, 648
thin lens, 754
threshold, 125, 210, 312
thresholding, 502
Toepler, August, 628
tomography, 14, 371, 377
transducer, 207
transfer function, 650
transform
1-D Fourier
chirp, 276
COMB, 279
complex sinusoid, 266
Dirac delta, 264
even sinusoid, 266
exponential, 273
inverse, 261
odd sinusoid, 266
rectangle, 264
signum, 269
step, 269
2-D Fourier, 410
2-D Hankel, 410
2-D central-slice, 403, 404, 406, 408, 410
example, 407
inverse, 409
2-D fast Fourier, 563
chirp Fourier
C–M–C, 637, 638, 799, 805, 818, 819, 846
M–C–M, 634, 637, 781, 798, 808, 871, 872, 878
discrete cosine (DCT), 256, 460, 568
discrete Fourier (DFT), 83, 93, 459, 460, 473, 484, 511–514, 516, 517, 519, 521–523, 525, 526, 529–534, 538, 551, 552, 554, 557, 559, 560, 562, 563, 568–570, 723, 724, 728, 729, 739
scale factor, 524
discrete of quantized functions, 559
fast Fourier (FFT), 459, 513, 529, 534, 723
Fourier
2-D image, 419
of Dirac delta function, 259
of exponential, 270
of modulated quadratic-phase exponential, 440
Fourier–Bessel, 347, 351
Fresnel, 772
Hankel, 347, 351
from 2-D Fourier transform, 361
Hankel of r^{-1}, 360
Hankel of Bessel function, 356
Hankel of cylinder function, 358
Hankel of Gaussian, 362
Hartley, 254
inverse 1-D Hartley, 261
inverse Fourier
2-D, 327
inverse Hankel, 353
inverse Radon, 392
Radon of 2-D image, 419
Radon of corral, 385
Radon of cylinder function, 380
Radon of Dirac delta function, 377
Radon of rectangle function, 384
Radon of ring delta function, 382
wavelet, 599
transformation
similarity, 73
transpose, 29, 38
transverse magnification, 803, 861
triangle function, 101, 176, 232
triangle inequality, 33

uncertainty relation, 519
underdetermined imaging task, 84
undersampling, 480, 552
uniform averager, 220, 579, 581, 672, 699, 710, 713, 714, 721
uniform probability law, 152

unit constant function, 99
unsharp masking, 743

variance, 118, 119, 152, 160, 425, 456
 of binomial probability distribution, 154
 of normal distribution, 158
 of Poisson probability distribution, 156
 of Rayleigh distribution, 158
 of sinusoid, 161
 of square wave, 161
 of uniform distribution, 156
vector
 basis, 43
 column subspace, 46
 difference, 31
 inner product, 66
 length, 31, 65
 null, 41, 46
 null subspace, 45
 orthogonal, 34
 outer product, 66
 pedal, 374
 principal, 70
 representation, 44
 row subspace, 45
 scalar product, 32, 65
 sum, 30
 transpose, 29
 unit, 32, 71
vector space, 41
vectors
 orthonormal, 34
Very Large Array (VLA), 850

wave
 traveling, 23
wave equation, 105, 762
wavelet, 765
wavelet transform, 599
wavenumber, 24
 angular, 763
width
 equivalent, 454
width of function, 424
Wiener, 680, 681
Wiener filter, 680, 681, 688, 693
Wiener spectrum, 308
Wiener–Helstrom filter, 689, 693, 694, 706
Wiener–Khintchin theorem, 302, 604, 648
window
 Bartlett (1-D), 550
 data, 539, 544
 Hamming (1-D), 548
 Hann, 542, 545
 Hann (2-D), 567
 Hanning, 542, 545
 Hanning (2-D), 567
 Nyquist, 473
 triangle (1-D), 550

Zernike, Fritz, 628
zero padding, 555
zero-order interpolator, 491